ASAE Publication
PROC-275

# Managing Livestock Wastes

PROCEEDINGS

3rd INTERNATIONAL SYMPOSIUM ON LIVESTOCK WASTES — 1975

April 21-24, 1975

University of Illinois
Urbana-Champaign, Illinois

Published by

AMERICAN SOCIETY OF AGRICULTURAL ENGINEERS
BOX 410
ST. JOSEPH, MICHIGAN 49085
PHONE (616) 429-0300

Library of Congress Card Number: 75-37164
and
International Standard Book Number: 0-916150-00-3

The International Symposium on Livestock Wastes was financed in part with Federal funds from the Environmental Protection Agency under grant number R803699-01-0. The contents of these Proceedings do not necessarily reflect the views and policies of the Environmental Protection Agency or the American Society of Agricultural Engineers, nor does mention of trade names or commercial products constitute endorsement or recommendation for use.

# Managing Livestock Wastes

PROCEEDINGS

3rd INTERNATIONAL SYMPOSIUM
ON LIVESTOCK WASTES — 1975

## Table of Contents

## VIII. Systems for Dairy



## IX. Systems for Swine





## X. Utilization as Livestock Feed

ACKNOWLEDGMENTS

The success of this Symposium was the result of diligent work and close cooperation of many people and organizations. The following individuals and their organizations are acknowledged for contributions of time and resources which were so important in making this information exchange possible. Special acknowledgment is due the University of Illinois, the Cooperative State Research Service of the U.S. Department of Agriculture, the Deere and Company, and the U.S. Environmental Protection Agency for their major contributions.

EXECUTIVE PLANNING COMMITTEE

D. L. Day, University of Illinois — Chairman
J. C. Nye, Purdue University — Secretary
G. L. Pratt, North Dakota State University — Program
A. J. Muehling, University of Illinois — Local Arrangements
N. K. Newtson, Borkholder Buildings — Exhibits
E. A. Olson, University of Nebraska — Publicity
E. P. Taiganides, Ohio State University — International Cooperation
R. E. Larson, ARS—USDA, University of Minnesota — Finance
F. R. Hore, Engineering Research Service, Agriculture Canada — Publications
R. G. Yeck, ARS, USDA — Past Chairman
R. H. Hahn, ASAE — Coordinator

PROGRAM COMMITTEE

G. L. Pratt, North Dakota State University, Fargo, North Dakota — Chairman
F. G. Viets, ARS—USDA, Fort Collins, Colorado — representing the American Society of Agronomy
H. S. Teague, U. S. Meat Animal Research Center, Clay Center, Nebraska — representing the American Society of Animal Science
J. R. Ogilvie, McGill University, Quebec, Canada — representing the Canadian Society of Agricultural Engineering
M. R. Overcash, North Carolina State University, Raleigh, North Carolina — representing the American Institute of Chemical Engineers

F. J. Humenik, North Carolina State University, Raleigh, North Carolina — representing the American Society of Civil Engineers
G. R. Fisher, North Dakota State University, Fargo, North Dakota — representing the American Dairy Science Association
D. D. Badger, Oklahoma State University, Stillwater, Oklahoma — representing the American Agricultural Economics Association
V. Davis, ERS—USDA, Washington, D.C. — representing the Economic Research Service, USDA
T. J. Brevik, University of Wisconsin, Madison, Wisconsin — representing the American Society of Agricultural Engineers
L. R. Shuyler, Robert S. Kerr Environmental Research Laboratory, Ada, Oklahoma — representing the U.S. Environmental Protection Agency
H. G. Geyer, Washington, D.C. — representing the Extension Service, USDA
K. Porter, Starline, Inc., Harvard, Illinois — representing the Farm Equipment Manufacturers Association
C. J. Schacht, Clay Equipment Corp., Cedar Falls, Iowa — representing the Farm and Industrial Equipment Institute
E. P. Taiganides, Ohio State University, Columbus, Ohio — representing the International Commission of Agricultural Engineers
S. M. Morrison, Colorado State University, Fort Collins, Colorado — representing the American Society for Microbiology

W. R. Jenkins, Federal Extension Service, USDA, Washington,
D.C. — representing the Poultry Science Association
K. D. Quist, Department of Health, Education, and Welfare,
Atlanta, Georgia — representing the U.S. Public Health
Service
R. G. Yeck, ARS, USDA, Beltsville, Maryland — representing
the Agricultural Research Service, USDA
P. E. Schleusener, CSRS—USDA, Washington, D.C. —
representing the Cooperative State Research Service, USDA
C. E. Fogg, SCS—USDA, Washington, D.C. — representing the
Soil Conservation Service, USDA
B. A. Stewart, Southwestern Great Plains Research Center,
Bushland, Texas — representing the Soil Conservation
Society of America
F. G. Viets, ARS—USDA, Fort Collins, Colorado — representing
the Soil Science Society of America
R. C. Loehr, Cornell University, Ithaca, New York —
representing the Water Pollution Control Federation

## LOCAL ARRANGEMENTS COMMITTEE
University of Illinois, Urbana

A. J. Muehling, Agricultural Engineering Department —
Chairman
O. G. Bentley, Dean, College of Agriculture — Administrative
Advisor
D. G. Jedele, Agricultural Engineering Department —
Demonstrations
H. J. Hirning, Agricultural Engineering Department —
Exhibit Space
E. L. Hansen, Agricultural Engineering Department —
Hospitality
W. D. Lembke, Agricultural Engineering Department —
Housing and Transportation
J. K. Mitchell, Agricultural Engineering Department —
Housing and Transportation
C. J. W. Drablos, Agricultural Engineering Department —
Meals, Banquets, and Entertainment
D. H. Vanderholm, Agricultural Engineering Department —
Meeting Rooms and Visual Aids Equipment
P. N. Walker, Agricultural Engineering Department —
Publications Display
C. M. Scherer, Extension Editor, Cooperative Extension
Service — Publicity
J. O. Curtis, Agricultural Engineering Department —
Registration
S. R. Aldrich, Assistant Director, Agricultural Experiment
Station — Special Guests
Barbara Mathews, Convention Coordinator — Ramada Inn
Convention Center
Wayne Hecht, Assistant Director — Assembly Hall

## EXHIBITS COMMITTEE

N. K. Newtson, Borkholder Buildings, Nappanee, Indiana —
Chairman
R. C. Evans, Bacon Bin. Inc., Bluffton, Indiana — Farm
Equipment
R. K. White, Ohio State University, Columbus, Ohio —
Laboratory Equipment
B. C. Horsfield, University of California, Davis, California —
Publications
W. A. Fischer, Fischer Enterprises, Pittsburgh, Pennsylvania —
Displays

## PUBLICITY COMMITTEE

E. A. Olson, University of Nebraska, Lincoln, Nebraska —
Chairman
H. G. Geyer, Extension Service, USDA, Washington, D.C.
L. Jackson, Soil Conservation Service, Lincoln, Nebraska
D. Magdanz, National Livestock Feeders Association, Omaha,
Nebraska
M. G. Garner, National Pork Producers Council, Des Moines,
Iowa
J. B. Adams, National Milk Producers Association, Washington,
D.C.
W. H. Stone, Livestock Equipment Council, Chicago, Illinois
R. H. Hahn, American Society of Agricultural Engineers
E. P. Taiganides, Ohio State University, Columbus, Ohio
R. Berglund, American National Cattlemens Association,
Denver, Colorado
W. C. LaVeille, U.S. Environmental Protection Agency,
Washington, D.C.
C. M. Scherer, University of Illinois, Cooperative Extension
Service, Urbana, Illinois

## FINANCE COMMITTEE

R. E. Larson, ARS—USDA, University of Minnesota, St. Paul,
Minnesota — Chairman
J. A. Moore, University of Minnesota, St. Paul, Minnesota
C. A. Cramer, University of Wisconsin, Madison, Wisconsin

## PUBLICATIONS COMMITTEE

F. R. Hore, Agriculture Canada, Ottawa, Canada — Chairman

Abstracts Subcommittee

G. B. Willson, ARS—USDA, University of Maryland, College
Park, Maryland — Chairman
J. Pos, University of Guelph, Canada

Proceedings Subcommittee

J. R. Miner, Oregon State University, Corvallis, Oregon —
Chairman
C. D. Barth, Clemson University, Clemson, South Carolina
V. Bonecutter, Clay Equipment Corporation, Cedar Falls, Iowa
A. F. Butchbaker, Oklahoma State University, Stillwater,
Oklahoma
P. R. Goodrich, University of Minnesota, St. Paul, Minnesota
R. E. Graves, University of Wisconsin, Madison, Wisconsin
J. J. Kolega, University of Connecticut, Storrs, Connecticut
D. C. Ludington, Cornell University, Ithaca, New York
J. Lunin, ARS—USDA, Beltsville, Maryland
H. F. Mayes, ARS, USDA, University of Missouri, Columbia,
Missouri
J. A. Moore, University of Minnesota, St. Paul, Minnesota
R. A. Nordstedt, University of Florida, Gainesville, Florida
M. R. Overcash, North Carolina State University, Raleigh,
North Carolina
J. I. Sewell, University of Tennessee, Knoxville, Tennessee
J. C. Ward, Colorado State University, Fort Collins, Colorado
T. L. Willrich, Oregon State University, Corvallis, Oregon

# THE 3rd INTERNATIONAL SYMPOSIUM ON LIVESTOCK WASTES

## was sponsored by

The American Society of Agricultural Engineers

### in cooperation with the following organizations:

American Agricultural Economics Association
American Dairy Science Association
American Institute of Chemical Engineers
American Society of Agronomy
American Society of Animal Science
American Society of Civil Engineers
American Society for Microbiology
Canadian Society of Agricultural Engineering
Farm Equipment Manufacturers Association
Farm and Industrial Equipment Institute

International Commission of Agricultural
Engineering (CIGR)
The Poultry Science Association, Inc.
Soil Conservation Society of America
Soil Science Society of America
U.S. Department of Agriculture
(ARS, CSRS, ERS, ES, SCS)
U.S. Environmental Protection Agency
U.S. Department of Health, Education,
and Welfare — Public Health Service
Water Pollution Control Federation

# Opening Remarks


D. L. Day
MEMBER
ASAE


WELCOME to the Third International Symposium on Livestock Wastes. The beginning of this Conference culminates over three years of intense planning and preparation.

Waste management is not a very glamorous subject; however, acceptable solutions must be achieved so that our countries may continue to grow livestock to furnish meat protein for human diets. Since finding acceptable solutions is our work, it is our responsibility to do the best job possible. This Symposium is designed to help us do that, and an excellent program is available.

This Symposium is the third of its kind. The first was held in 1966 at Michigan State University where 55 papers were presented with about 300 people attending. The second was in 1971 at Ohio State University with 100 papers presented and about 600 people in attendance. About 600 people have registered for this Symposium, which is very good, considering the recent budgetary restrictions imposed on many organizations.

The technical program for this Symposium contains 180 papers that were carefully selected by representatives of over 20 cooperating societies and organizations from over 250 proposals submitted to the Program Committee. The papers cover three main areas: (a) reports of major research studies, (b) state-of-the-art and literature reviews, and (c) case studies of livestock farms with complete waste management systems. The papers will be presented at three concurrent sessions over the four days of the Symposium.

In addition to the technical papers, there is an exhibit area that includes field equipment, laboratory equipment, and field demonstrations. Over 50 companies are exhibiting. The exhibit also includes publications from non-commercial organizations.

Technical sessions will be dismissed Tuesday afternoon so that Symposium registrants may visit the exhibit show at the Assembly Hall. There will be a general session inside the Assembly Hall from 1:30 to 2:30 p.m., with an address given by Under-Secretary of Agriculture, J. Phil Campbell of Washington, D.C. Also, there will be a similar session on Wednesday afternoon for the general public, with the program consisting of a panel of specialists on utilizing livestock wastes, covering: spreading on land as a fertilizer, refeeding processed wastes, generating methane, and exporting chemically treated liquid manure. Orion Samuelson of Chicago WGN radio and television stations will moderate the panel discussion. He will also be the speaker at our banquet on Tuesday evening, where a historical, delicious full-pound pork chop will be served to each person.

This Symposium is truly international, with 30 of the technical papers coming from outside the U.S.A. In addition, there are Monday and Wednesday evening forums for discussing activities and trends of waste management in at least 13 other countries.

The 600 Symposium registrants include 111 from the following 18 countries, represented as follows: Australia, 2; Canada, 46; Czechoslovakia, 1; Denmark, 2; England, 12; Finland, 1; Germany: FRG, 4, GDR, 1; Ireland, 2; Japan, 14; Mexico, 3; Netherlands, 5; Northern Ireland, 2; Scotland, 5; South Africa, 3; Sweden, 4; Taiwan, 1; and the Soviet Union, 3. All 10 of the Canadian provinces are represented as well as 41 of the 50 United States of America and the District of Columbia.

The main commodity of this Conference is knowledge and information about livestock waste management. Please help yourself to your share. The Proceedings from this Symposium are a collection of the best papers in the world on our subject.

Because of the many people from numerous organizations who have helped with this Symposium, it is not possible to name them all and give proper and just credit to their contributions. They have made this a very successful meeting. Note, however, the list of persons with major responsibilities included in the Proceedings. Please try to get acquainted with these people and express your appreciation for what they have done for this meeting.

Lastly, the people attending the Symposium are, in the final analysis, the most important group for which this meeting has been planned. I am very glad you are here. Thanks for coming. May you be richly rewarded for your efforts to attend.

---


The author is: D. L. Day, Professor, Agricultural Engineering Dept., University of Illinois, Urbana, and Chairman, Executive Planning Committee, ISLW-75.

# The Livestock Waste Program in Agricultural Colleges

Orville G. Bentley

TODAY, the entire livestock industry is aware and concerned about the complexities associated with an age-old costly and demanding chore—the disposal of animal wastes. In the United States, concern about waste management accelerated after World War II. And the concerns of producers stimulated research by agricultural experiment stations, the USDA research agencies—particularly ARS—and private concerns such as those in the farm machinery and material-handling industries.

In the late sixties, livestock waste management became a subsidiary interest of agencies concerned with human waste disposal, especially the federal and state environmental protection agencies. And since that time, the waste management problem has continued to edge its way up the priority ladder. Ten years ago only the large producers near urban centers were really concerned about the problem. But today, waste management is of direct concern to all producers, and how wastes are managed has an effect on people everywhere. Here is why:

Population and income levels are increasing worldwide. That means an increased demand for meat and poultry. But, at the same time, competition for land and land prices increase. Costs go up, and efficiency becomes imperative. All signs point toward even more confinement: more livestock on fewer acres—and consequently, an even more intense waste-management problem.

The thrust of this Symposium is on education, research, and planning for future developments in the livestock waste-management field. In the United States, the land-grant universities' agricultural experiment stations have given leadership to research on waste management. In the mid-sixties for example, experiment stations in the North Central Region organized a regional research committee on "Farm Animal Waste Disposal" that carried the numerical designation "NC-69." Representatives from that early committee were prime movers in organizing an early, if not the first, major conference on waste management in the United States held at East Lansing, Mich. in May, 1966. Each of the follow-up conferences are designed to review the "state of the art," highlight new research developments, and pinpoint new research and educational needs. From this perspective, I will be discussing the role the land-grant universities have in this country.

It is abundantly clear that the problem of waste management is not simple. Progress depends on the combined efforts of scientists and engineers from a variety of disciplines. Biochemists, biologists, animal and poultry scientists, engineers, computer scientists, and others can all contribute to the solution of the problems being dealt with in this symposium.

From one point of view, animal wastes are simply a materials handling problem. But it is not that simple, because the problem generates a complex mix of interrelationships among the livestock and poultry industry, the Environmental Protection Agency, consumers and other groups. And ever more complicating is the overshadowing need to use resources efficiently.

The complexity and the interrelationships involved in the development of an effective system of animal waste management suggest the need for a research strategy that recognizes the whole system rather than its component parts.

That strategy must continuously be studied and altered as developments occur that directly affect it in other areas. For example, consider the potential impact of waste management practices on water quality standards. That example also points out the important role our research programs play in helping establish such standards. The water quality standards were established with the benefit of a unique national and world perspective, shaped and based largely on our research.

Another example affecting our research strategy is the world food situation. Although the world now uses only half the land area potentially suitable for crop production, most of the additional land lies outside the densely populated countries. Consequently, looking at the supply side of the equation, a good part of future food production gains will have to come from yield-increasing techniques—more fertilizer, improved varieties, and improved cultural practices, for example.

The demand side of the equation is greatly affected by population growth. World population is currently growing at the rate of 70 million people a year, with the developing countries accounting for 86 percent of the annual increase. Another factor affecting demand is income, because as incomes rise consumers buy more food.

Changes in either the supply of or demand for food greatly affect the prices of commodities and of the various necessary production inputs. Such changes have direct impact on our waste management research strategies.

Consider the effect of the energy crisis. Increased energy costs and shifts in petroleum supplies sent nitrogen costs soaring. The result was a renewed interest in reevaluating animal and human wastes as sources of nitrogen fertilizer. During the era of low crude petroleum prices it was more advantageous to buy nitrogen fertilizers than to put manure on the land. But will that relationship always be true? We do not know, and consequently, we must now give increased attention to efficiency of labor. That area has evolved as a part of the waste handling research strategy.

Perhaps all of this reflects a critical point that unfortunately has been overlooked by some and forgotten by others: Environmental quality costs. It is not available for free.

A "concerned citizen" attending our Second Allerton Conference on Environmental Qaulity and Agriculture in 1971 made the point so well when she said, "Part of the problem has been that everybody wants a cleaner environment, but nobody is willing to pay the price. Some of us are now realizing that we must pay back our debt to the environment and add these costs to the consumer's food bill."

The author is: ORVILLE G. BENTLEY, Dean, College of Agriculture, University of Illinois, Urbana.

It is fine to talk about developing a responsive research strategy that changes pace and direction as new needs become evident and as new problems emerge. But most often, it is difficult to document that such a responsiveness does in fact exist. I am pleased to report that in considering Land-Grant college research programs in animal waste management there is evidence that our strategy today reflects that essential ingredient—responsiveness.

I will be referring to data presented in the Report of the North Central Regional Planning Committee. The report deals with projected changes in science man years in the various research problem areas during the five-year period from 1973 to 1978. In the area of environmental quality, the report indicates that if the budget remains the same, the allocation of science-man years (SMY's) will increase by 32. But it must be noted that at that level of budgeting the report predicts a decrease of 18 science-man years in the regional total for all research areas.

For example, if the North Central Regional Experiment Stations' budgets increase by 10 percent, the report predicts an increase of 72 science-man years in environmental quality. And that would involve putting about one-fifth of all added science-man years in that particular research area.

Of course, not all research in environmental quality deals in waste management. But at the same time, shifts in other research areas suggest increased attention to the waste management problem. For example, the report predicts an increase in beef and swine research of 30 science-man years given the same budgeting level, and an increase of 145 science-man years if budgets increase 10 percent. Waste management is a critical problem to the industry, and no doubt that is reflected in research efforts relating to livestock management and building design. In addition, some part of the 94 science-man year increase in soil and water management might also be related to waste management. I hope, for example, that our Illinois research with sewage sludge might result in some findings on applicaiton rates and viruses that will provide insights to the animal waste problem.

Obviously, agricultural experiment stations recognize the need to find answers to pressing questions involved in waste management. Mounting a research program requires planning by the "bench scientists" and the administrators on an individual basis, as well as nationally, to develop a balanced program and to maximize the returns from the resources allocated to these programs. Avoiding undue duplication is one way to improve research productivity.

Agricultural experiment station research is approached on an interdisciplinary basis. One can cite the many successful efforts of the chemists, physiologists, biochemists, geneticists, and animal scientists working cooperatively on many multifaceted research projects. The application of an integrated approach to studies on waste disposal does not then require a new dimension to research planning. It does, however, require appreciation of scope, urgency and relevance of research on waste management to livestock production systems. It means, also, that the station director must place projects in this area of waste-management systems in competition for funds and other resources along with the more traditional programs in the areas of production livestock management, housing, soil fertility, and others.

The concept of regional research is one way to accomplish the research planning needed among the state stations. Experiment stations have traditionally exchanged information on program planning and research progress. About 30 years ago, however, an added formal arrangement was provided for planning cooperative studies by two or more states on problems of mutual interest. Planning is done in each of the four regional associations of directors—North Central, Northeastern, Southern and Western. National coordination is achieved through a single committee made up of nine representatives selected by the regional associations. The activities within this organizational structure are coordinated in part by the Cooperative State Research Service for the USDA. The regional program provides ample opportunity for cooperation and joint planning with the Agricultural Research Service and Economic Research Service.

So, a system for planning and conducting a responsive research program in waste management does exist. It is "at home" with the experiment stations in the Land-Grant Colleges of America. But in addition, other factors favor research in agricultural experiment stations. Universities are the educational and training grounds for future workers, with many scientists and trained personnel located there. Equipment and many laboratories also are available at agricultural experiment stations. Research is the reason agricultural experiment stations are in business.

Finally, the experiment stations have another advantage as facilitators of waste management research and that advantage is their close linkage with the Cooperative Extension Service (CES). The Cooperative Extension Service is based on the belief that human progress can be enhanced if the products of research can be translated to lay language and made available to help individuals make decisions.

Key to the success of the CES has been its unique structure as a partnership of federal, state and county governments with strong guidance in its priorities from those it serves. Local volunteers extend its range and influence. The system has survived and grown because of its objectivity and its ability to adapt short-range priorities to the longer-range needs of the public. Its strength lies in its ability to use research-based facts in logical relationship to national or state goals without being dominated by any of the levels of government which support it.

Today, the CES, with more than 3,000 offices throughout the nation, remains the only system with a statewide organization, national affiliation, and local support dedicated to the transfer and use of knowledge to solve problems. The system has great flexibility.

And so, as waste management research programs unfold, the information we learn will be passed on to those who can benefit from our findings. And in fact, it is this two-step flow that has made the Land-Grant college system so effective in our nation that it serves as a model for countries throughout the world.

I find our work in this area both challenging and satisfying. It is challenging because the work is complex and dependent upon so many variables. But it is satisfying because we know we are working on an important problem. And solutions to these problems may provide breakthroughs that will have a lasting impact on all mankind.

# Who Is Sensitive and Why?

Irving Kass, M.D.

THIS presentation is not going to focus on noxious gases generated by wastes, i.e., hydrogen sulfide, ammonia, carbon dioxide, methane. In a majority of instances the host responds to these just as one would to any chemical irritant. Some investigators, however, believe that before these agents initiate a tissue reaction involving the pulmonary tree (bronchiolitis) they must be adsorbed by a respirable particle (<5 microns) which lodges in the bronchial tree beyond the 16th generation of bronchi. (There are 65 500 such bronchi each <0.6 mm in internal diameter.) In Table 1 (Bouheys 1974) lists the gases, vapors, and fumes which affect the lower airways and lungs. Occasionally chemical fumes can cause asthma as is the case with amino-ethyl ethanolamine in aluminum soldering flux (Pepys and Pickering 1972).

In this discussion concerning who is sensitive and why one will be primarily concerned with other factors i.e., genetic, immunologic, etc. From the beginning we have always viewed our environment as essentially hostile, and that humans, with varying degrees of resistance and adaptive capacity, are in continuous struggle with it. Any factor in the environment which increases its hostility or anything which lowers the resistance of humans, decreases adaptability. Thus for those individuals with genetic defects, allergic problems, heart and lung disease, there can be no well defined threshhold or safe level. Some of the factors which decrease the host's resistance to environmental stressors have been listed in Table 2 (American Lung Association 1974).

The precise role of alpha-1 antitrypsin (Bell and Currell 1973) is unknown at the present time. It apparently has the ability to inhibit a variety of enzymes. The Z gene produces a much smaller amount of alpha-1 antitrypsin; individuals with two Z genes produce only 10 percent of the normal level of alpha-1 antitrypsin in the serum. A marked decrease in this glycoprotein (homozygote ZZ phenotype deficiency) has been associated with early development of chronic obstructive pulmonary diseases i.e., emphysema, chronic bronchitis, bronchial asthma, and in some instances with cirrhosis of the liver. The homozygote as well as subjects with an intermediate decrease (heterozygote SZ, MZ, etc. phenotypes) in alpha-1 antitrypsin serum levels may be more susceptible to pulmonary damage by cigarette smoke and noxious agents than an individual with a normal MM phenotype. Because this is an inherited deficiency, the members of a given kindred who are similarly affected can be identified by electrophoretic techniques (Fagerhol and Tenfjord 1968).

A child with cystic fibrosis or an individual with a deficiency in his T dependent lymphocytes (cellular defense) or B dependent lymphocytes (humoral defense) is certainly more prone to develop respiratory infections than one in whom these deficiencies don't exist (Wybran and Fudenberg 1974). Some observers believe that the onset of cancer can be associated with a decrease in the T dependent lymphocytes — and currently some of the therapies are directed toward raising the level of T dependent lymphocytes.

We have begun to achieve some control over the acquired defects. Because of antimycobacterial therapy, tuberculosis is no longer a major public health problem. Some of the other acquired public health problems i.e., hypertension, coronary artery disearse, will decline if the public will cooperate in efforts to screen for these problems as well as insist upon more sophisticated levels of care and adequate funding for basic research into the natural history of these diseases.

As a stressor, all forms of smoking not only have a harmful effect upon the smoker but also upon the nearby nonsmoker. Blood levels of carbon monoxide (CO), 3-4 benzo(a)pyrene and cadmium have also been found to be

The author is: IRVING KASS, Larson Professor of Medicine, Department of Internal Medicine and Regional Chest Center, College of Medicine, University of Nebraska, Omaha.

TABLE 1. GASES, VAPORS, AND FUMES WITH EFFECTS
ON LOWER AIRWAYS AND LUNGS*

| Compound | Principal effect on airways and lungs | Examples of usage and exposures |
|---|---|---|
| Ammonia | B, E† | — |
| Cadmium oxide fumes | P | Electroplating, alloys, welding |
| Chlorine | B, E | Bleaching |
| Hydrogen sulfide | E, P, RP | — |
| Mercury vapor | E, P | Power plants; laboratories |
| Nitrogen dioxide | P | Welding; component of photochemical smog |
| Ozone | E | Welding; component of photochemical smog |
| Phosgene | E, P | Organic synthesis; product of combustion of chlorinated solvents |
| Selenium compounds | A, B, P | Metal refining |
| Sulfur dioxide | B | Product of fossil fuel combustion; oil refining; paper industry |
| Toluene-2, 4-diisocyanate (TDI) | A | Polyurethane manufacture |

*Partial list
†A = asthma, B = chemical bronchitis, E = lung edema, P = chemical pneumonitis, RP = respiratory paralysis

**TABLE 2. FACTORS WHICH DECREASE THE HOST'S RESISTANCE
TO ENVIRONMENTAL STRESSORS.**

Genetic defects
    Hyperlipidemia i.e., hypercholesterolemia
    Alpha-1 antitrypsin deficiency
    Cystic fibrosis
    T and B lymphocyte deficiency
    Congenital heart defects i.e., tetralogy of Fallot
    Allergic diathesis
    Bleeding diathesis
Acquired defects
    Cor pulmonale (secondary to chronic obstructive pulmonary diseases)
    Coronary artery disease
    Rheumatic heart disease with valvular defects
    Hypertension with left ventricular hypertrophy
    Infectious disease i.e., tuberculosis
Personal habits
    Cigarette smoking
    Alcoholism
Physical condition
    Aging and/or debilitation
    Obesity
Living conditions
    Overcrowding
    Malnutrition
    Air pollution

elevated in the latter. One of the most impressive studies on smoking and mortality, involving 41 000 physicians, showed a strong association between smoking and death from lung cancer, chronic bronchitis, etc. Physiologically, smoking a cigarette induces an increase in airway resistance, a decrease in maximum expiratory flow rates, impaired exercise performance, and an increase in carboxy-hemoglobin (COHb) content in blood. Because of the increased affinity of CO for hemoglobin (Hb), COHb decreases the capacity of Hb to carry oxygen. Even among non-asthmatic children, a team of researchers found that respiratory illnesses happen twice as often in young children whose parents smoked at home compared to those with non-smoking parents.

Both alcohol and cigarettes lessen the clearance of particles from the airway, decrease the activity of the alveolar macrophages, and in general lower the host's defense mechanisms against bacterial infections.

Occupation-wise perhaps the clinical problem which may be of the greatest interest to this group is the allergic reaction to the inhalation of organic dusts. Ramazzini in 1713 (Ramazzini 1940) described the occurrence of respiratory infections in workers who were exposed to the dust of over-heated cereal grains in a paper entitled, "Diseases of Sifters and Measurers of Grain". A long interval followed until the report of Towey (Towey et al. 1932) on an unusual pneumonitis (extrinsic allergic alveolitis) occurring in maple bark strippers. During the past 15 years there has been a profusion of papers describing a wide spectrum of pulmonary diseases secondary to a variety of inhaled organic dusts. Recently Schleuter (Schleuter et al. 1972) described a respiratory disease resulting from prolonged exposure to logs contaminated with a mold called Alternaria.

Table 3 (Bouhuys 1974) lists these diseases and includes the type of exposure and the specific offending agent where it has been determined. This list has grown rapidly as physicians have become more aware of those diseases which previously were diagnosed as acute pneumonias. The antigenic material in a majority of cases has been identified as fungal spores but protein materials have been implicated in pigeon breeders and pituitary snuff takers. Because of the very small size of dust particles (four microns or less), extremely large quantities of antigenic material can be delivered during a brief exposure. It has been estimated that the number of spores retained in the lung of a farmer working with

**TABLE 3. OCCUPATIONAL LUNG DISEASE***
**ORGANIC DUSTS**

| Agent | Industries | Disease† |
|---|---|---|
| Castor bean | Castor bean manufacture | |
| Flour | Bakeries | |
| Guinea pig hair | Laboratories | A |
| Gum acacia | Printing | |
| Tamarind seed | Yarn finishing | |
| Moldy sugar cane | Sugar cane waste processing | Bagassosis |
| Cotton, flax, and hemp dust | Textile | Byssinosis |
| Enzyme detergents (Bacillus subtilis) | Laundry products | A, EAA |
| Moldy hay (Micropolyspora faeni) | Farming | A, EAA, F, G |
| Cheese mold (penicillium) | Cheese manufacturing | EAA |
| Maple bark strippers disease (Cryptostroma corticale) | Lumber | EAA, F |
| Moldy redwood sawdust (Graphium) | Lumber | EAA, F |
| Bird fanciers disease (Avian dust) | Pigeon breeders / Bird fanciers | EAA, F, G |
| Pituitary snuff takers disease | Treatment of diabetes insipidus | EAA |

*Partial list
†A = asthma, EAA = extrinsic allergic alveolitis, F = fibrosis, G = granulomatosis

TABLE 4. OCCUPATIONAL LUNG DISEASE
MINERAL DUSTS*.

| Agent | Industries | Disease† |
|---|---|---|
| Aluminum and aluminum oxide | Bauxite mining; corundum smelting; abrasives; explosives; paints; fireworks | Aluminosis; bauxite lung; F |
| Asbestos | Insulation; brake linings; lagging; construction; shipyards | Asbestosis; mesothelioma; cancer |
| Barium sulfate | Barite mining | Baritosis; D |
| Beryllium compounds | Aircraft manufacturing; metallurgy; rocket fuels | Beryllium disease; F; G; P |
| Silica ($SiO_2$) | Mining (gold, tin, copper, graphite; coal); pottery; sand blasting; foundaries; quarries (granite, slate, sandstone) | Silicosis; F |

*Partial List
†A = asthma, B = bronchitis, F = fibrosis, G = granulomas, P = pneumonitis

moderately moldy hay may reach 750 000/minute*. Needless to say as the definition of farmer's lung implies other moldy vegetable dusts may give rise to similar conditions and presumably to a similar morbid immunopathologic process. Bizzare associations elicited to date include thatched roofs in tropical climates, black fat tobacco, malting, coffee roasting, processing of cork, and the making of toy soldiers.

If the reaction (pulmonary or other-wise) to an organic dust allergen is immediate it is a Type I† immunological response (Gell and Coombs 1963). If the reaction (pulmonary or other-wise) to the organic dust occurs several hours after exposure it usually indicates an inflammation of the tissue involved. Such a reaction is known as a Type III‡ immunologic response (Gell and Coombs 1963). This is generally an acquired reaction to an allergen which has previously been encountered by the patient.

Inhalation of mineral dusts (Table 4, Bouhuys 1974) may involve a Type IV§ immunologic response (Gell and Coombs 1963). One of the first mineral dust responses to be recognized as an occupational hazard was silicosis. This is the fibrosis traceable to the inhalation of silica or quartz dust which is a hazard in mining for minerals or from handling materials containing quartz. Recently a decrease in the incidence of silico-tuberculosis has been documented. This can be attributed to dust control measures, decline in the incidence of tuberculosis and an improvement in the nutritional status of workers.

Asbestosis is another story. It is a disease of latency in which the worker may not develop symptoms until 20 years after his first exposure. More than 2000 of the 200 000 workers in asbestos in the United States die each year, and it is anticipated that 40 percent of those directly involved will die of a malignancy involving the lungs, stomach, throat, or peritoneum. Other inorganic dusts that present occupational hazards include beryllium, which is widely used with metals as alloys; iron dust, which can cause siderosis or welder's disease; tin dust which is responsible for stannosis; and barium which can cause baritosis.

Diseases caused by inorganic dusts are also diseases of synergism. Asbestos workers have an increased risk of lung cancer that is closely linked to their smoking habits. Nonsmoking asbestos workers rarely get lung cancer. Sulfur dioxide and ozone have been found to act synergistically, affecting the lungs more adversely when both are present than when they are present singly in equivalent amounts.

The response of individuals to environmental pollutants is quite variable. It depends upon the individual's immunological reactivity i.e., atopic or non-atopic, the nature of the dust, the size of the particles, and the intensity of the exposure particularly whether it is regular or intermittent. Because the presence of atopy, a constitutional or hereditary tendency to develop immediate hypersensitivity states is more frequently associated with some form of immune reaction, it might be well to screen individuals before they are allowed to work in a known sensitizing environment i.e., manufacturing of detergents, industries in which toluene 2,4 diisocyanate is used.

A positive reaction to allergens such as grass, trees, dust, molds, or the substance *per se* when applied by the prick or intradermal technique might serve as a warning that these individuals should not be employed in these environments.

Hyper-responsiveness of one's bronchial tree may be reflected by a decrease in pulmonary functions i.e., first-second forced expiratory volume ($FEV_{1.0}$), specific airway conductance ($S_{GA}$) (Bartell 1973). A reaction to methacholine‖ may serve to warn us that this individual can develop reversible bronchospasm (asthma) to a hostile ambient environment even though no demonstrable allergy has been recorded.

Whenever an organic dust has been incriminated the most specific diagnostic test available is controlled inhalation challenge (provocative test) using a specific antigen and monitoring the pulmonary changes with serial pulmonary function studies. In the sensitive individual, the signs and symptoms as well as laboratory abnormalities will be accurately reproduced. Cigarette smoking should be dis-

---

*This can be prevented by ensiling dried crops or to spray bales of hay with 2 percent proprionic acid.

†Type I immediate reaction. Antigen (allergen) and non-precipitating antibody (Reagin IgE) fix to mast cell with release of histamine, slow reacting substance anaphylaxis (SRS-A), etc. This can be prevented by the prophylactic administration of cromolyn sodium.

‡Type III intermediate reaction (6-24 hours). Antigen (allergen) and precipitating antibody (IgG or IgM) plus complement cause tissue damage. The complexes formed in moderate antigen excess must be small enough to be soluble and therefore able to circulate, large enough to be capable of reacting with complement and yet possess the ability to fix to tissues. Depending on the location the reaction can be prevented by corticosteroids administered as an aerosol or orally.

§Type IV delayed reaction. Allergen plus sensitized lymphocytes (cells which have already encountered the allergen) attract macrophages which cause tissue changes, i.e., induration, granulomas, fibrosis.

‖Methacholine is a synthetic choline ester which in the susceptible host causes the release of cyclic guanosine monophosphate (cyclic GMP) which causes broncho-constriction.

*(Continued on page 14)*

# Airborne Health Hazards Generated While Treating and Land Disposing Waste

Philip R. Goodrich, Stanley L. Diesch, Larry D. Jacobson

MEMBER
ASAE

ASSOC. MEMBER
ASAE

HEALTH and environmental concerns as well as economic considerations have encouraged recent changes in housing and manure management systems for animal production enterprises. Drastic changes in housing systems pose the potential for unforeseen results that may be detrimental to the economic well-being of the livestock enterprise. Transmission of disease by aerosols from animal to animal or from animal to man should be kept minimal or severe hardships may result.

The microbioaerosol production from housing units, treatment systems and disposal units has not been thoroughly studied. The public health impact from potential aerosol-producing units requires consideration. This paper presents the hazards in several situations associated with animal agriculture which produce aerosol contamination of the environment. The aerosols found in a field-scale operating oxidation ditch study quantify the results one could expect when utilizing this waste treatment device in a confinement beef facility. The aerosols generated in a floor litter system of raising turkeys are quite different, but demonstrate the levels of contamination that may occur on farms. Aerosol levels in a dairy barn and near manure spray disposal are also presented.

## AEROSOLS ABOVE A FIELD OXIDATION DITCH

A field oxidation ditch was studied which was located at the Agricultural Engineering Farmstead, University of Minnesota's Experiment Station, Rosemount. Two enclosed housing units over the oxidation ditch channel each house 18 beef cattle fed a concentrate ration. The wastes pass through the slotted floors, and a brush rotor at the west end of the channel aerates the waste slurry. Air is exhausted at each end of the ditch moving from wall inlets down through the slotted floor and out each end of the ditch. Details of the study were described by Goodrich (1974).

The all glass impinger (AGI) was used to obtain samples once every seven days at seven sites in and around the oxidation ditch. At each site, three samples were taken concurrently at 30 cm (ground level), 100 cm (cattle respiratory level), and 160 cm (human respiratory level). Determinations were made of total colonies (on plate count agar), coliform colonies (on Levine's-EMB), and fecal streptococci (on M-enterococcus). There was a definite pattern (Table 1) in both the east and west vents where the 100 cm samplers gave higher counts. This is probably because the

Miscellaneous Journal Paper No. 1578, University of Minnesota Agricultural Experiment Station, St. Paul.

Support for part of the investigation reported here was provided by Environmental Protection Agency Grant No. R802205.

The authors are: PHILIP R. GOODRICH, Associate Professor, Agricultural Engineering Dept., STANLEY L. DIESCH, Professor, Veterinary Clinical Sciences Dept., and LARRY D. JACOBSON, Extension Agricultural Engineer, Agricultural Engineering Dept., University of Minnesota, St. Paul.

**TABLE 1. TOTAL VIABLE BACTERIAL AEROSOLS AT DIFFERENT ELEVATIONS WITH CATTLE PRESENT AND ABSENT**

| Location | Height above ground, cm | Average colony forming units per liter on PCA | | | |
| --- | --- | --- | --- | --- | --- |
| | | Cattle present, CFU/1 | | Cattle absent, CFU/1 | |
| West vent | 30 | 17 | | 3.8 | |
| | 100 | 52 | (31) | 6.9 | (5.4) |
| | 160 | 22 | | 5.6 | |
| East vent | 30 | 11 | | 24 | |
| | 100 | 50 | (28) | 5.9 | (12) |
| | 160 | 19 | | 4.2 | |
| West barn | 30 | 132 | | 42 | |
| | 100 | 119 | (117) | 20 | (24) |
| | 160 | 102 | | 12 | |
| East barn | 30 | 242 | | 3.2 | |
| | 100 | 215 | (204) | 4.9 | (10) |
| | 160 | 160 | | 21 | |
| Upwind | 30 | 6.5 | | 0.75 | |
| | 100 | 9.5 | (7) | 2.4 | (1.6) |
| | 160 | 4.8 | | 1.6 | |
| Downwind | 30 | 4.5 | | 1.8 | |
| | 100 | 9.1 | (6) | 2 | (1.7) |
| | 160 | 3.6 | | 1.2 | |
| Rotor | | 55 | | 16 | |

Notes: Numbers in ( ) are the averages of 3 heights. Cattle present data taken weekly May 25, 1972 to June 13, 1973. Cattle absent data taken August 9 to September 13, 1972 and March 1 to March 5, 1973.

100 cm height was most directly in line with the vent. In the barns, the highest bioaerosol concentration was at the 30 cm height, and the lowest at the 160 cm height. Therefore, there was some air contamination that did not reach the elevation associated with a human in the housing unit. The ventilation system forced air down through the slats and probably restricted flow of these aerosols to the higher elevations. The air associated with the animal housing units contained a rich bioaerosol. Counts of 100 to 200 total colony-forming units per liter were between 1 to 2 orders of magnitude higher than those normally experienced by humans in nonagricultural occupations. These high counts can be attributed almost entirely to the active, crowded presence of the livestock and not to the manure disposal system. The tests run when the cattle were present and when the cattle were absent show that the bio-load decreased within less than one hour to levels approximating those of the outside environment even though the manure in the ditch was still being aerated by the rotor. Therefore, the oxidation ditch contributes an insignificant number of microbes to the air when compared to the contribution of the living animals. The indicator fecal organisms show also that the concentration of aerobic fecal contamination is high when the animals are in the barn and insignificant when the animals are not in the barn. Therefore, the contribution from the oxidation ditch itself is insignificant. The oxidation ditch was not deemed to be a source of the

**TABLE 2. AEROSOLS FROM SPRAY IRRIGATION
OF BEEF OXIDATION DITCH WASTE**

| | November 22, 1971 | | April 18, 1972 | |
|---|---|---|---|---|
| | Meters downwind* | CFU/l | Meters downwind | CFU/l |
| Before spray | N.D. | N.D. | 0 | 1 |
| During spray | 0 | 0.4 | 37 | 3 |
| During spray | 61 | 15 | 52 | 4 |
| During spray | 82 | 120 | 67 | 2 |
| During spray | 113 | 25 | 98 | 5 |
| After spray | 82 | 4 | N.D. | N.D. |
| During spray | N.D. | N.D. | 37 (upwind) | 0.4 |

*   From nozzle

| | | |
|---|---|---|
| Temperature: | -3 C | 8 C |
| Wind: | 8-16 km/hr | 16 km/hr |
| Sky: | overcast | overcast |
| Sampler: | AGI | AGI |
| Media: | plate count agar | plate count agar |

bacterial aerosols within the housing units. Tests taken during cleaning operations when a dry floor existed and persons were scraping within the environment resulted in a very hazardous environment. Unusually high, potentially hazardous levels of aerosols, (2000 CFU/l) existed within the barn during cleaning operations. Therefore, it would be strongly recommended that protective masks be worn by personnel exposed to the environment in the barn during cleaning operations.

## SPRAY-IRRIGATION OF ANIMAL WASTES

Spray-irrigation of waste material plays an important role in the management of animal wastes. Solids build up in the oxidation ditch or a holding pit as time progresses. When the pit is full, the material must be removed and disposed. Spray-irrigation pumping is one such method for disposing this material onto the land. The mechanical ejection of manure into the air by a sprinkler irrigating system provided the opportunity to examine the production of airborne bacteria. Aerosol monitoring indicated the relative degree of public health hazard associated with this method of handling waste. Samples of air were taken with the all glass impinger (AGI) and with agar-filled petri dishes laid on the ground.

Previous aerosol studies of operations similar to spray-irrigation have been reported (Albrecht 1958, Randall 1966, Adams 1970, Napolitano 1966). Previous studies sampled the visible spray from sewage trickling filters, aeration tanks and laboratory generated sprays.

## RESULTS AND DISCUSSION

The three spray-irrigation studies are presented in Tables 2 and 3. Each study was analyzed separately because daily and even hourly varying conditions prevented overall generalizations. In November (Table 2), the airborne bacterial counts were lower at the edge of the spray than at distances farther downwind. The high count at 82 m downwind may have been due to an unnoticed wind change or due to an unknown spreading pattern which had a high concentration of bacteria at this distance and height. The count of the sample taken after spraying was lower than all the counts except that taken at the edge of spray.

For April however, Table 2 shows no pattern of zonal concentration; but the total counts during spraying were only slightly higher than the counts before spraying and the counts upwind.

Table 3 shows that there was no significant difference in numbers of airborne bacteria between the 100 and 30 cm above-ground samples at any distance. There was no zonal

**TABLE 3. AEROSOLS FROM SPRAY IRRIGATION
OF BEEF OXIDATION DITCH WASTE AT ROSEMOUNT**

| | | Height of all glass impinger above ground | | | | | | Agar plates on ground CFU/5 min exposure | | |
|---|---|---|---|---|---|---|---|---|---|---|
| | | 30 cm | | | 100 cm | | | | | |
| | Downwind m | Total CFU/l | Coli CFU/l | F–Strep CFU/l | Total CFU/l | Coli CFU/l | F–Strep CFU/l | PCA | L-EMB | M-ent |
| Before spray | 0 | 1 | 0 | 0 | 4 | 0 | 0 | ovgr* | 0 | 0 |
| During spray | 15 | 2 | 0 | 0 | 6 | 0 | 0 | 348 | 28 | 27 |
| During spray | 15† | 23 | 0 | 0 | 30 | 0 | 0 | 2000 | 98 | 188 |
| During spray | 33 | 23 | 0 | 0 | 26 | 0 | 0 | 221 | 1 | 5 |
| After spray | 50 | 0 | 0 | 0 | 0 | 0 | 0 | ovgr* | 0 | 0 |
| | Upwind | | | | | | | | | |
| During spray | 33 | 2 | 0 | 0 | 0 | 0 | 0 | 77 | 0 | 1 |

* ovgr denotes overgrowth with a spreading organism.
† during this sampling the wind speed was greater than the other 15 meter downwind sample.

Bacterial counts of ditch liquid collected during irrigation:
$92 \times 10^6$ total aerobic bacteria per ml
$20 \times 10^4$ coliform bacteria per ml
$32 \times 10^4$ f. streptococci per ml

| | | |
|---|---|---|
| Temperature: | 11 C | |
| Wind: | 8 km/hr | |
| Sky: | Clear, 57 percent RH | |
| Sampler: | Samples with AGI were taken simultaneously at two heights, 30 and 100 cm, plus exposed plates on ground. | |

concentration apparent, but, comparing two, 15-m downwind sample results, the counts downwind increased when the wind increased. No indicator organisms (coliform or f. streptococci) were detected from AGI samples, but they were found on all exposed plates during irrigation— implying that particles from the ditch were airborne, but not detectable in sizes that would penetrate the lung (1-5 microns). The total count, however, increased from 0 before spraying to 30 CFU/l during spraying, showing that some small particles are airborne. The afterspray counts show that the small and large fallout particles are quickly removed from or diluted out of the air.

The ditch liquid samples (92 x 10$^6$ total aerobic bacteria, 20 x 10$^4$ coliform bacteria, 32 x 10$^4$ f. streptococci bacteria per ml ditch liquid) showed that the enteric indicator organisms were approximately 300 times less than the concentration of the total number of organisms present. This same concentration proportion was borne out in the AGI samples of the air, but the open plates showed proportionately higher concentrations of enteric organisms recovered. One explanation of this may be that because open plates collect falling particles (therefore large particles), each colony probably resulted from a group of bacteria, thus each colony on the petri dish represents more than one collected bacterium. The number of bacteria collected was much higher than enumerated by the technique. Also, since there are 300 times the number of total bacteria than enteric bacteria in the sprayed liquid, the chances of having many enteric organisms per falling particle are small. So, the number of enteric colonies on the plates approximates the actual number. Unlike other air samples taken near the animal housing unit where f. streptococci outnumbered coliform, the fallout plates showed similar numbers of coliform and f. streptococci present. A speculated reason for this is that in spite of the fact that coliform organisms do not survive long when airborne (Higgins 1964), it is possible that they are protected by a coating of ditch material when they are in the fallout droplets, thus accounting for the coliform counts approaching those of the fecal streptococci.

The assessment of the public health hazard of spray-irrigating liquid from an oxidation ditch would involve many factors, many of which are unknown.

In none of the bioaerosol studies with the AGI during irrigation were any enteric organisms present (less than 1 CFU/l) and total counts increased from 1 CFU/l before spraying to a maximum of 120 CFU/l and an average of 20 CFU/l during and downwind of spray. Other cited researchers present this same approximate number when sampling downwind in the vicinity of sewage spray. The counts of samples taken seconds after spraying stopped were the same as before spraying samples, indicating that there is no build-up of bacteria. The previously reported sampling of animal housing units showed that the units frequently have enteric f. strep. The average Resemount animal housing unit total count was 161. In addition, the samples of a dairy barn (Diesch 1975) showed averages of 196 total and 0.5 f. strep. Therefore, the relative hazard of spraying compared to confined housing is small as far as cattle health and people working in the housing areas are concerned. As far as the respiratory hazard of workers in the area, there is no denial of the increase in bioaerosol from the ditch during spraying in the immediate area, but this hazard is present for a short while. Therefore, it is recommended that, during spraying, workers in the area wear protective masks. A more possible danger is through contact with large droplets

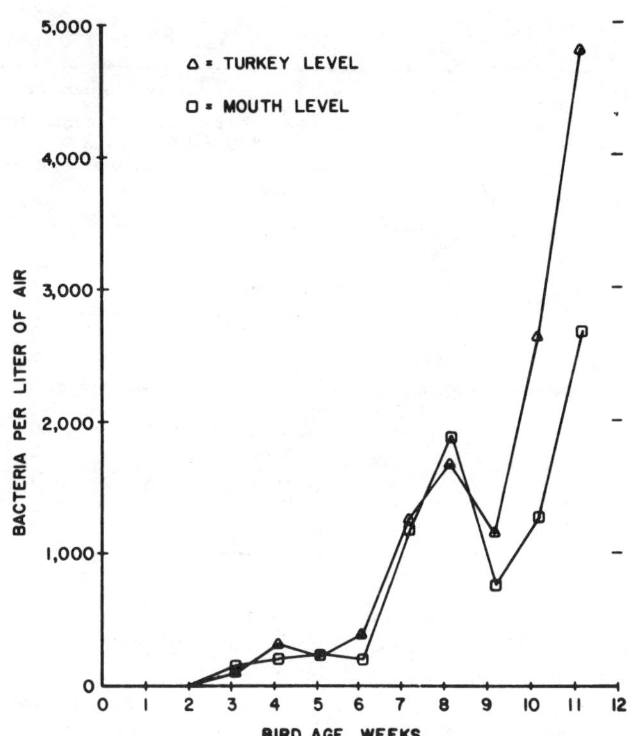



FIG. 1 Average standard plate count levels in the turkey house.

of spray which are visible during the operation. Diseases can arise in the nasal mucosa, tonsils, or respiratory mucosa of the upper respiratory tract through inhalation of large particles (National Biocentric, Inc. 1972). A mask in the areas during spraying is recommended.

## BACTERIAL AEROSOLS IN A TURKEY HOUSE

The environment of the Turkey Research Building at the Rosemount Experiment Station was sampled for bacterial aerosols. Concentrations of bacteria, at 4 separate locations in the building, were observed using the all glass impinger (AGI). Details were presented by Jacobson (1974).

The average bacterial concentration for turkey level (50 cm above the floor) and mouth level (160 cm above the floor) are shown in Fig. 1. The concentrations increased as the turkeys matured. Also, the bacterial concentrations at turkey level were generally higher than those taken at the human mouth level.

The results reveal the large concentration of bacteria in a turkey building when compared with other environments discussed previously. When the turkeys were about ten weeks old, bacteria concentrations of 2700 per liter of air were observed. Environments whose bacterial concentrations are nearest to those of a turkey building are beef cattle over an oxidation ditch, 100 to 200 colony-forming units per liter of air (Goodrich et al. 1974) and bed-making in a hospital ward, 70 bacterial-carrying particles per liter of air (Pelczar and Reid 1965). The bacterial concentrations are at least 10 times (or 1 order of magnitude) as great in a turkey building as in these environments.

However, *Escherichia coli (E. coli)* was not significantly detected in areosol form during the entire sampling period. *E. coli* bacteria were found on the turkeys and solid surfaces in the pens using a swab technique.

## BACTERIAL AEROSOLS IN A DAIRY BARN

A program of aerosol sampling at a St. Paul Campus

Aerosol data taken weekly February 21, 1973 - June 11, 1973 — When cattle present

Viable bacterial aerosols during different conditions (AGI counts) on PCA (total),
L-EMB (coliform), and M-ent (fecal streptococcus), 37 C, 24 hr
Counts given in colony-forming units/liter (CFU/l)

| Height above floor | Organisms | Quiet conditions | | | Sleeping and/or eating | | |
| | | Number of samples | Range CFU/l | Average CFU/l | Number of samples | Range CFU/l | Average CFU/l |
|---|---|---|---|---|---|---|---|
| 30 cm | Total | 14 | 38-266 | 112 | 12 | 75-826 | 294 |
| | Coliform | 12 | 0-0 | 0 | 12 | 0-3 | 0.25 |
| | F. Strep. | 14 | 0-3 | 0.43 | 12 | 0-5 | 0.67 |
| 100 cm | Total | 15 | 22-222 | 92 | 12 | 57-1168 | 370 |
| | Coliform | 13 | 0-0 | 0 | 12 | 0-0 | 0 |
| | F. Strep. | 15 | 0-2 | 0.2 | 12 | 0-3 | 0.67 |
| 160 cm | Total | 15 | 40-240 | 107 | 12 | 78-820 | 261 |
| | Coliform | 13 | 0-1 | 0.15 | 12 | 0-1 | 0.08 |
| | F. Strep | 15 | 0-2 | 0.4 | 12 | 0-3 | 0.67 |

dairy barn was conducted from February 21, 1973, through June 11, 1973, on a weekly basis. The dairy barn manure disposal is by gutter cleaner, shovel, washing when required, and sweeping. Continuously operating fans exhaust air from the barn. Samples were taken in the center of the alleyway between the two rows of producing cows. The samples were analyzed for total plate count, coliform, and fecal streptococci in the same manner as at Rosemount.

Samples in the dairy barn were taken consecutively starting at the lowest level (30 cm) and proceeding to the highest (160 cm) at a single location, once in the morning and once in the afternoon, in order to monitor variable conditions. In the morning, the animals were quiet; in the afternoon the animals were eating and the floors were being swept. Table 4 shows that the counts were considerably higher during the afternoon activity. The counts show that the animals and workers in the area were exposed to comparable numbers of total and fecal bioaerosols as at the Rosemount area. This is and has been a traditional system that has withstood any changes of its environment due to a public health necessity.

## CONCLUSIONS

Bacterial aerosols are generated in housing units, treatment systems and land disposal of wastes. The hazards are higher than would normally exist in the absence of these operations. Due caution should be exercised when animals and humans are exposed to the more extreme conditions as evidenced by elevated microbial counts shown in this study. Proper respiratory protection devices should be used in some cases for workers in the more hazardous situations.

## References

1   Adams, P. and J. C. Spendlove. 1970. Coliform aerosols emitted by sewage treatment plants. Science. 169:1218.

2   Albrecht, C. R. 1958. Bacterial air pollution associated with the sewage treatment process. M.S. thesis. University of Florida. Gainsville, Florida.

3   Diesch, S. L., P. R. Goodrich, B. S. Pomeroy and L. A. Will. 1975. Survival of pathogens in animal manure disposal. Final project report to U.S. Environmental Agency Project R802205, 136 pp.

4   Goodrich, P. R., S. L. Spier, S. L. Diesch and L. A. Will 1974. Microbial aerosol monitoring of a beef housing oxidation ditch. Proceeding of the International Livestock Environment Symposium. ASAE, p. 182-188. St. Joseph, Mich. 49085.

5   Higgins, F. B. 1964. Bacterial aerosols from bursting bubbles. Ph.D. dissertation, Georgia Tech University. Atlanta, Ga.

6   Jacobson, L. D. 1974. Bacterial and particulate concentration in a turkey house environment. Masters thesis, University of Minnesota. St. Paul, Minn.

7   Napolitano, P. J., and D. R. Rowe. 1966. Microbial content of air near sewage treatment plants. Water and Sewage Works. 113:480-483.

8   National Biocentric, Inc. 1972. Evaluation of potential environmental health problems associated with virus transmission from spray irrigation of waste water treatment plant effluent. Report submitted to Sanitary Sewer Board of Alexandria Lake Area Sanitary District, Alexandria, Minnesota. National Biocentric Inc. St. Paul, Minn.

9   Pelczar, M. J. Jr., and R. D. Reid. 1965. Microbiology, 2nd edition, McGraw-Hill Book Company, New York.

10   Randall, C. W. and J. O. Ledbetter. 1966. Bacterial air pollution from activated sludge units. American Industrial Hygiene Association Journal. 27:506.

# Survival of Salmonellae, Total Coliforms and Fecal Coliforms in Swine Waste Lagoon Effluents

D. J. Krieger, J. H. Bond, C. L. Barth

MEMBER
ASAE

MANAGEMENT of agricultural solid wastes is a problem which must be considered in any livestock production system. One method of managing swine wastes is by treatment in an anaerobic lagoon. Lagoons will contain any organism common to the intestinal tract of the animals they serve. Certain bacterial pathogens, particularly *Salmonellae*, are found in swine waste material (Coley 1972, Galton et al. 1954, and Vassiliadis et al. 1970). Due to an accelerated consumption of natural resources, the need to conserve vital water supplies is becoming increasingly important. Thus, it is worthwhile to investigate the feasibility of recycling lagoon supernatant to flush swine pens rather than using fresh water. Survival times of the pathogens in the lagoon will be important when this practice is used. Few studies have investigated the survival of pathogenic bacteria in animal waste treatment lagoons. Sidio et al. (1961) investigated pathogen survival in a domestic two-cell experimental oxidation pond which had a 99.1 percent reduction in coliform bacteria. They found a maximum survival time of 12 days for a heavy load of Salmonella in either influent or effluent samples. Neel et al. (1961) showed a coliform reduction of 99.9 percent in all cells of a domestic six-cell lagoon system. Coetzee and Fourie (1965) demonstrated a 99.9 percent reduction of *Escherichia coli* and a 99.5 percent reduction of *Salmonella typhi* after 35 days retention in a domestic sewage stabilization pond. Slanetz et al. (1972) determined that bacteriophage, Bdellovibrio, and toxic products produced by other organisms appeared to be significant factors in the death of certain indicator organisms in domestic sewage lagoons.

This report describes the quantitative determination of the survival of the common swine pathogens, S. *Choleraesuis* and S. *typhimurium*, in relation to standard indicator organisms under typical lagoon conditions and warm temperatures. A second aspect was to determine those factors, such as presence of bacteriophage, antagonistic products or nutrient depletion, which affected the death rate of Salmonella.

## MATERIALS AND METHODS

### Cultures

S. *choleraè-suis* (ATCC 5001) was obtained from the American Type Culture Collection. S. *typhimurium* was obtained from the Clemson University stock culture collection.

### Culture Media

All culture media employed in this investigation were obtained from Difco Laboratories, Inc.

Approved as Technical Contribution No. 1248 by the South Carolina Agricultural Experiment Station.

The authors are: D. J. KRIEGER, Director of Quality Control, Fisher Cheese Co., Wapakoneta, Ohio; J. H. BOND, Associate Professor, Microbiology Dept., and C. L. BARTH, Associate Professor, Agricultural Engineering Dept., Clemson University, Clemson, S. C.

### Source of Samples

Lagoon samples employed in this study were collected from a single cell swine waste lagoon in operation at the Clemson University Swine Research Center. The pH of each sample was taken within one hour using a Corning model 7 pH meter. The lagoon was anaerobic and had a depth of 1.5 m. Lagoon temperatures varied from 28.5-31.5 C when samples were collected.

### Experimental Columns

Nine hundred ninety milliliters of the lagoon material were added to sterile glass tubes of 1200 ml capacity. These columns, used to simulate the reaction of a column of water in the lagoon, were incubated at ambient room temperature of 23-28 C. Separate columns were individually loaded with S. *cholerae-suis* and S. *typhimurium* to moderate, high and low population levels. Actual numbers in each 10 ml inoculum (in 0.85 percent saline) were determined by viable plate count on Bacto-Tryptic Soy Agar. Columns were checked for the presence of Salmonella before loading and all columns were duplicated.

### Enumeration of Salmonella from Loaded Columns

A modification of the most probable number method (MPN) was used to monitor the survival of Salmonella in the columns (Cheng et al. 1971 and Spino 1966). A 1 ml sample was withdrawn from the column and diluted through tenfold serial dilutions in 0.85 percent saline. One ml portions of each dilution were placed into each of five tubes containing 10 ml of Bacto-Tetrathionate Broth made with 25 percent of the recommended amount of standard iodine solution. This concentration of iodine was found in earlier tests to be optimum for the recovery of S. *choleraesuis*. All tubes were incubated at 40 C for 24 hours. A sample from each tube was streaked onto a Bacto-Brilliant Green Agar plate. Colonies typical of Salmonella on BG Agar were transferred to Bacto-Triple Sugar Iron Agar slants. Cultures producing the proper reactions for Salmonella on TSI Agar were subjected to slide agglutination tests with Bacto-Salmonella 0 poly A-1 antiserum for final confirmation. Since S. *cholerae-suis* often gave weak or delayed reactions with polyvalent antiserum, it was necessary to use Bacto-Salmonella 0 group $C_1$ antiserum to obtain a proper agglutination (Gallagher and Spino 1968 and Galton et al. 1954). Therefore, only colonies on BG Agar which could be confirmed as Salmonella were counted as positive in the MPN series.

### Enumeration of Total Coliforms

Survival of natural populations of total coliforms in the columns was determined by periodic enumeration employing a five tube MPN series with lactose broth (APHA 1969). All tubes were incubated at 37 C for 48 hours. Gas production after this time constitued a positive test. Results were confirmed by plating on Bacto-Eosin Methylene Blue (EMB) Agar.

### Enumeration of Fecal Coliforms

Survival of the natural populations of fecal coliforms in the columns was determined by periodic enumeration employing a five tube MPN series utilizing Bacto EC Medium (Standard Methods, 1960). All tubes were incubated in a 44.5 C water bath for 24 hours. Gas production after this time constituted a positive test. Results were confirmed by plating on EMB Agar.

### Nutrient Depletion as a Factor in the Die-Off of Salmonella

Sixty days after the death of the original inoculum of *S. cholerae-suis*, the column was reloaded with this same organism. Similarly another column was reloaded with the organism three days after the death of the original inoculum also at a moderate level. The die-off rate was monitored by the MPN technique.

In another experiment, 200 ml of fresh lagoon material and 200 ml of fluid from a column three days after the complete die-off of the original inoculum of *S. cholerae-suis* were individually centrifuged (20 000 x g for one hr) and filter sterilized with a 0.65 $\mu$m Millipore filter. The filtrates were placed in 500 ml erlenmeyer flasks, inoculated with *S. cholerae-suis* to a moderate level and incubated at 27-28 C. Initial cell concentrations and die-off were monitored by plate count with Bacto Veal Infusion Agar incubated at 37 C for 24 hours.

RESULTS

### Determination of Die-Off Rates of Salmonella, Total and Fecal Coliforms

Initial concentration of *S. cholerae-suis* at the high population level was $4 \times 10^6$ organisms/ml. As shown in Fig. 1, *S. cholerae-suis* could not be recovered after 33 days. The organism died off at a rate of 5.1 days/log cycle. In the same columns, the natural population of total coliforms was found initially to be $5.2 \times 10^3$. As shown in Fig. 2, these organisms could not be recovered after 18 days, a death rate of 5.1 days/log cycle. For moderate population levels, *S. cholerae-suis* was loaded at $2.5 \times 10^3$ organisms/ml, and in another set of columns, *S. typhimurium* was loaded at $7 \times 10^4$ organisms/ml. As shown in Fig. 1, both organisms died off within 24 days, a death rate of 6.6 days/log cycle for *S. cholerae-suis* and 5.3 days/log cycle for *S. typhimurium*. In *S. cholerae-suis* columns, the initial natural population level of total coliforms was $3.3 \times 10^3$ organisms/ml. These organisms died off in 21 days, for a death rate of 6.6 days/log cycle. Fecal coliforms, with an initial density of 20 organisms/ml, showed complete die-off in 10 days (Fig. 2). Separate columns were loaded with low population levels of *S. cholerae-suis* and *S. typhimurium* to be more representative of Salmonella densities to be found in actual lagoons. Population of *S. cholerae-suis* was initially 72 organisms/ml and complete die-off as shown in Fig. 1 occurred within 8 days, a death rate of 6.1 days/log cycle. The test with *S. typhimurium*, loaded initially to a population level of 340 organisms/ml was terminated after 8 days due to the overgrowth of interfering organisms and yielded a death rate of 5.3 days/log cycle. As shown in Fig. 2, the natural population of total coliforms, $3.2 \times 10^3$ organisms/ml, died off within 20 days for a death rate of 5.5 days/log cycle. The natural population of fecal coliforms, only 4 organisms/ml, died off within 5 days, a death rate of 8.3 days/log cycle. These numbers were too small for an accurate death rate determination.

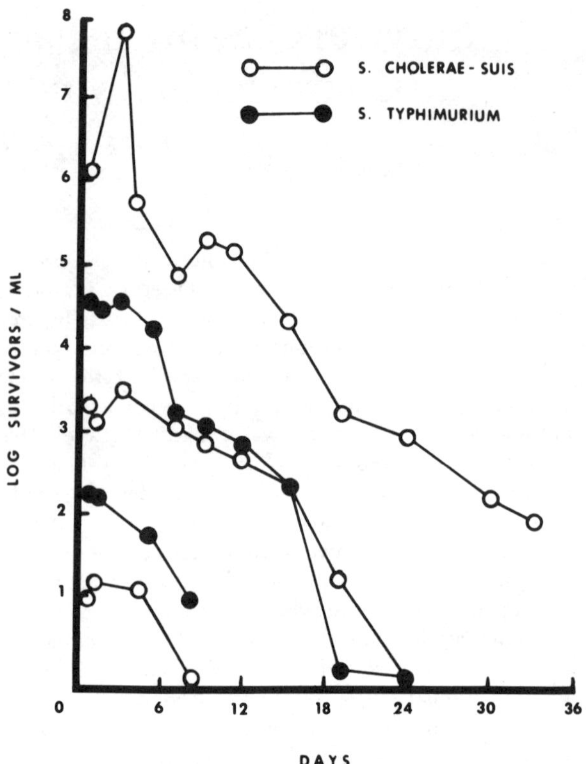


FIG. 1 $4 \times 10^6$, $25 \times 10^3$ and 72 organisms/ml *S. cholerae-suis*; $7 \times 10^4$ and $3.4 \times 10^2$ organisms/ml *S. typhimurium*. Loading rate determined by plating on veal infusion agar.

### Factors Influencing the Die-Off of Salmonella

Bacteriophages specific for *E. coli* and *S. cholerae-suis* could not be isolated from the column material after die-off, or from fresh lagoon material. A phage was found only

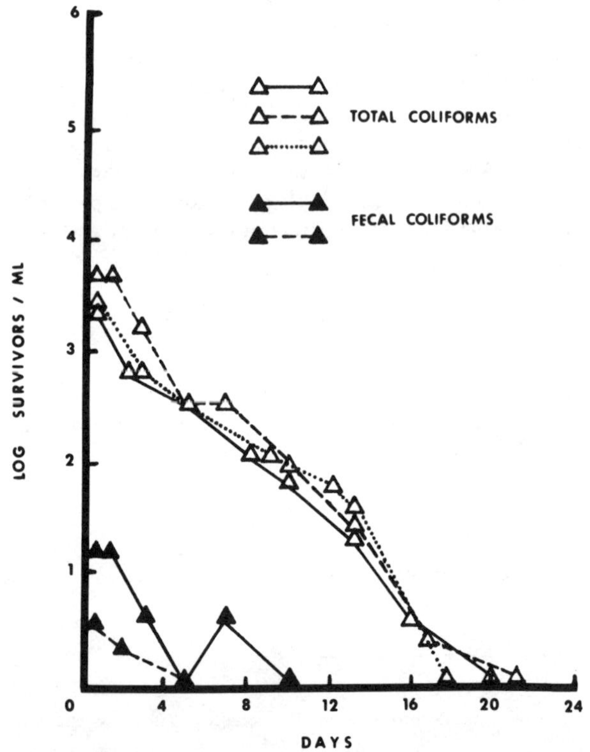


FIG. 2 Survival times of natural populations of total and fecal coliforms in columns with different initial loading rates of *Salmonella*.

for *Citrobacter freundii* previously isolated from the lagoon. The pH of all samples taken from the loaded columns, initially 6.95-7.2 and stabilizing at 7.7-7.8 within 5-8 days, was unlikely to cause death in any enteric organism. As shown in Fig. 3, in the column re-inoculated 3 days after die-off, the new inoculum of *S. cholerae-suis* died-off within 24 hours. The column re-inoculated 60 days after die-off maintained its new inoculum at near the initial population level for over 21 days before being terminated. Results of the second test of nutrient depletion are shown in Fig. 4. Contamination by chromagens occurred due to a faulty seal in the filtering process. With chromagens appearing in significant numbers, the Salmonella inoculum showed growth rather than death. Prior to the appearance of these contaminants, however, *S. cholerae-suis* was showing very similar death rates in both the lagoon filtrate and the filtrate from a column after the die-off of an *S. cholerae-suis* inoculum.

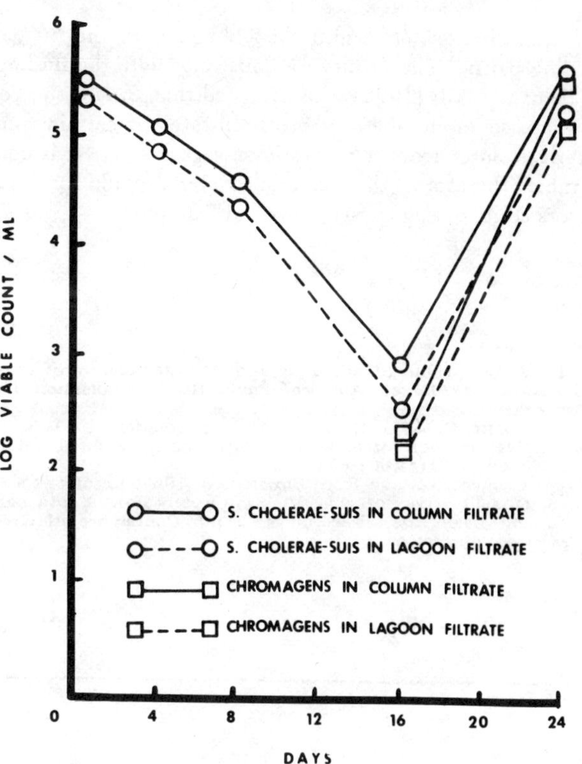



FIG. 4 Survival of *Salmonella cholerae-suis* and contaminant chromagens in filtrates of fresh lagoon material and from holding column 3 days after death of original inoculum of *Salmonella cholerae-suis*.

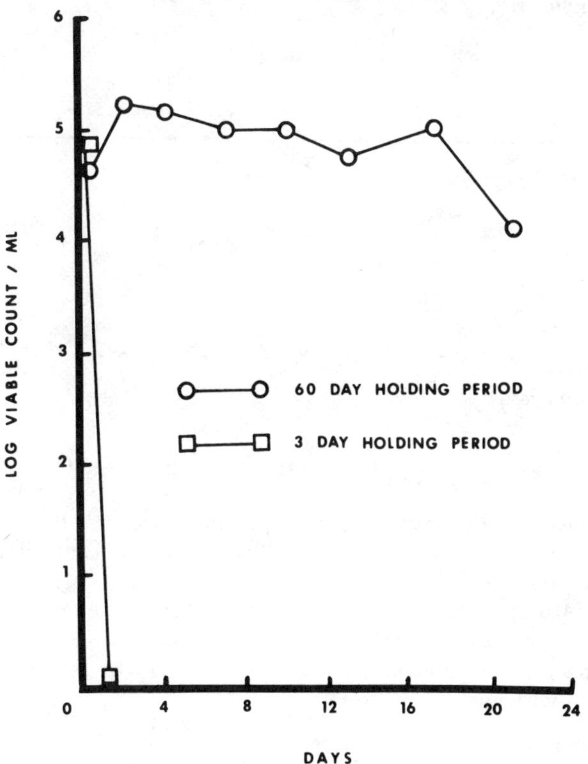



FIG. 3 Survival of *Salmonella cholerae-suis* in columns held for different times after the death of their original inocula of *S. cholerae-suis*.

## DISCUSSION

Numerous reports note the survival of Salmonella in rivers and domestic sewage lagoons (Cheng et al. 1971, Clemente and Christensen 1967, Coetzee and Fourie 1965 and Slanetz et al. 1972). This report is the first to examine the survival of Salmonella in an anaerobic swine waste lagoon. In all cases, the survival times of all organisms observed were determined by the initial numbers; the death rate of Salmonella, total and fecal coliforms were independent of numbers in the initial inoculum. This does not correlate with the results of Sidio et al. (1961) who reported die-off of a heavy inoculum of Salmonella in domestic sewage in 12 days. It was evident that total coliforms had very similar death rates regardless of the population of Salmonella inoculated into the columns. In addition, the death rates were logarithmic, like the death rates obtained from organsims growing in batch cultures. Die-off rates of fecal coliforms were not affected by the size of the initial Salmonella inoculum. Fig. 1 shows very similar die-off rates of 5.1, 6.6 and 6.1 days/log cycle for *S. cholerae-suis* in three different experiments. Both Salmonella organisms exhibited logarithmic death typical of batch cultures indicating that, under normal conditions, and low influent population level of Salmonella, a lagoon retention time of 28 days would be adequate at the temperatures tested. Further studies are needed to determine the effect of winter temperatures on pathogen survival.

Figs. 1 and 2 show that total coliforms are more reliable indicator organisms than fecal coliforms. In that experiment, total coliforms survived longer than a low load of Salmonella, whereas, the fecal coliforms did not survive as long. Even though total coliforms have a die-off rate roughly equivalent to Salmonella, they should survive longer than Salmonella by virtue of their greater influent density.

If nutrient depletion were a factor in die-off, fresh lagoon filtrate would support the inoculum of *S. cholerae-suis* longer than the column filtrate. The effective difference between the first (Fig. 3) and second experiments (Fig. 4) to test for nutrient depletion was the removal of competition in the latter. New inoculum of *S. cholerae-suis* placed in a column immediately after the death of the original inoculum showed almost immediate death, whereas, this same organism had a much slower, logarithmic die-off in the filtrate of a similar column. This indicates that the die-off rate is actually a matter of competition for nutrients. The sudden increase in the growth of Salmonella with the appearance of significant numbers of chromagens tends to verify this assumption. It appeared that these chromagens supplied some essential factor for the growth of Salmonella. This would help explain how *S. cholerae-suis* survived so well when reinoculated into the column with the 60 day holding period. The results are consistent with the findings of Slanetz et al. (1972) who reported that *E. coli* showed an increase in numbers in broth filtrates of certain non-enteric isolates from a domestic sewage lagoon. It is conceivable therefore, that certain bacteria could produce factors contributing to the growth of Salmonella.

### References


1   APHA. 1960. Standard methods for examination of water and wastewater. 11th ed. American Public Health Association, Inc. New York.
2   Cheng, C. M., W. C. Boyle and J. M. Goepfert. 1971. Rapid quantitative methods for Salmonella detection in polluted waters. Appl. Microbiol. 21(4):661-666.
3   Clemente, J. and R. G. Christensen. 1967. Results of a recent Salmonella survey of some Michigan waters flowing into Lake Huron and Lake Erie. Proceedings, Tenth Conference of Great Lakes Research. January 11.
4   Coetzee, O. J. and N. A. Fourie. 1965. The efficiency of conventional sewage purification works, stabilization ponds and maturation ponds with respect to the survival of pathogenic bacteria and indicator organisms. J. Inst. Sew. Purif. Part 3:210-215.
5   Coley, J. E. 1972. Salmonella in pigs in Papua New Guinea. Aust. Vet. J. 48:601-604.
6   Gallagher, T. P. and D. F. Spino. 1968. The significance of numbers of coliform bacteria as an indicator of enteric pathogens. Water Research. 2:169-175.
7   Galton, M. M., W. V. Smith, H. B. Elrath and A. B. Hardy. 1954. Salmonella in swine, cattle and the environment of abattoirs. J. Infect. Dis. 95(6):236-245.
8   Neel, J. K., J. H. McDermott and C. A. Monday, Jr. 1961. Experimental lagooning of raw sewage at Fayette, Missouri. J.W.P.C.F. 33(6):603-608.
9   Sidio, A. D., R. T. Hartman and P. Fugazzatto. 1961. First domestic waste stabilization pond in Pennsylvania. Pub. Health Repts. 76:201-208.
10   Slanetz, L. W., C. H. Bartley, T. G. Metcalf and R. Nesman. 1972. Survival of enteric bacteria and viruses in oxidation pond systems. Water Resource Research Center (New Hampshire) Research Report No. 6.
11   Spino, D. J. 1966. Elevated-temperature technique for the isolation of Salmonella from streams. Appl. Microbiol. 14(4):591-595.
12   Vassiliadis, P., D. Trichopoulos, J. Papadakis and G. Politi. 1970. Salmonella isolations in abattoirs in Greece. J. Hyg. 68:601-610.


---

## Who is Sensitive?

*(Continued from page 6)*

couraged or smokers not employed in those environments where synergism and/or diseases of latency have been observed.

Finally, the *in vitro* platelet factor 3 immunoinjury test (Garg and Sarker 1974) may prove to be of value where a generalized antigen — antibody reaction is suspected. Such an antiplatelet antibody was observed in a patient who developed thrombocytopenia purpura 12 hours after he had taken two tablets containing acetylsalicylic acid and caffeine.

In addition to a chest radiograph the following pulmonary function studies should be performed on an annual basis in environments where particularly the respiratory system is challenged.

1   First-second forced expiration volume ($FEV_{1.0}$)
2   Single breath diffusing capacity with CO ($DL_{CO_{SB}}$)
3   Volume of isoflow (V iso V) (Hutcheon et al. 1974).

### SUMMARY

We presently cannot completely characterize the individual who will be sensitive to a given environmental stressor. In our opinion anyone can be sensitive to any factor in his environment. The etiologic agent may be singular, a synergistic interaction, or a combination of factors in which the total body burden must be considered. Because the heart, lungs, liver, brain, etc. are not self-replicating and thus cannot reproduce when they are the recipients of bodily injury, one must continually look upon the body's atmosphere as a finite region with a limited capacity for self-cleansing that varies with different substances. If the introduction of pollutants exceeds the self-cleansing capacity potentially hazardous concentrations may remain in the body.

Although the problem is a formidable one, one should be more than comforted by the words of Frederick Nietzsche when he said, "That he who has a why to live can bear with almost any how".

### References


1   American Lung Association. Testimony in defense of the Clean Air Act. June 11, 1974.
2   Bartels, H., P. Dejours, R. K. Kellogg, J. Mead. 1973. Glossary on respiration and gas exchange. J Applied Physiol 34(4):549-558.
3   Bell, O. F. and R. W. Carrell. 1973. Basis of the defect of a-1 antitrypsin deficiency. Nature 243 (5407):410-411.
4   Bouhuys, A. 1974. Breathing physiology, environment and lung disease. Grune and Stratton. New York pp 342-384.
5   Fagerhol, M. K. and O. W. Tenfjord. 1968. Serum pi types in some European, American, Asian and African populations. Acta Path Microbiol Scand 72(4):G01-G08.
6   Garg, S. K. and C. R. Sarker. 1974. Case report aspirin induced thrombocytopenia on an immune basis. Amer J Med Sci 267(2):129-132.
7   Gell, P. H. G. and R. R. A. Coombs. 1963. Clinical aspects of immunology. Blackwell Scientific Publication. Oxford.
8   Hutcheon, M., P. Griffin, H. Levison and N. Zamel. 1974. Volume of isoflow. A new test in detection of mild abnormalities of lung mechanics. Amer Rev Resp Dis 110(4):458-464.
9   Pepys, J and C. A. C. Pickering. 1972. Asthma due to exhaled chemical funes — amino-ethyl ethanolamine in aluminum soldering flux. Clinical Allergy 2(2):197-204.
10   Ramazzini, B. 1940. De Morbus Artificum Diatribal 700. The Sutin Text of 1713 revised with translation and notes by Wright, W. C. University of Chicago Press. Chicago.
11   Schleuter, D. P., J. N. Fink and G. T. Hensley. 1972. Wood pulp workers' disease: A hypersensitivity pneumonitis caused by alternaria. Ann Intern Med 77(6):907-914.
12   Towey, J. W., H. C. Sweeny and W. H. Huron. 1932. Severe bronchial asthma apparently due to fungal spores found in maple bark. JAMA 99(6):453-359.
13   Wybran, J. and H. Fundenberg. 1974. How clinically useful is T and B cell quantitation. Ann Intern Med 20(6):765-767.

# Mosquito Production and Control in Animal Waste Lagoons

R. C. Axtell, D. A. Rutz, M. R. Overcash, F. J. Humenik

ASSOC. MEMBER
ASAE

MEMBER
ASAE

THE high organic content of animal waste lagoons provides a favorable environment for the production of those species of mosquitoes which prefer polluted waters. The most common species in lagoons in the southern United States is *Culex pipiens quinquefasciatus* Say which is known as the southern house mosquito (Steelman et al. 1967a,b; Weidhaas et al. 1973). Other closely related forms in the *Culex pipiens* mosquito species complex are common in polluted waters throughout the world (Barr 1957, Laven 1967, Mattingly 1951). Control of these mosquitoes in animal waste lagoons, as well as other polluted waters, is often necessary to reduce the potential for disease transmission and the level of annoyance from mosquito bites. Therefore, we are investigating the factors affecting mosquito production from animal waste lagoons and methods for mosquito control.

This report summarizes our data on (a) the relationship between swine and poultry lagoon manure-loading rates and the numbers of *C. p. quinquefasciatus* produced, and (b) the effectiveness of various insecticides and insect growth regulators for mosquito control in swine waste lagoons.

## MATERIALS AND METHODS

### Loading Rates Versus Mosquito Production

Mosquito numbers were determined for various lagoon loading rates of swine manure added weekly to large tanks (11.5 ft diameter X 6 ft water depth) and to 55-gal drums (2 ft diam X 3 ft water depth). Both the tanks and drums were recessed in the ground. The tanks were in operation 10 months and the drums 1 month prior to initiation of the mosquito counts. There was 1 tank per loading rate and 3 drums per loading rate. The poultry manure (white leghorn laying hens) lagoon study utilized the same size drums (3 per loading rate). Table 1 presents the loading rates used in these studies. Organic pollution levels were measured weekly using as parameter concentrations the chemical oxygen demand (COD) and the total organic carbon (TOC) as described in Standard Methods (American Public Health Assoc. 1971) and subsequent adaptations (Overcash et al. 1974).

The numbers of mosquito larvae and pupae were determined weekly by a standard dipping method using a 4-in. diameter water dipper (6 dips per tank, 3 dips per drum). The tanks were sampled for 32 weeks (May - Nov., 1973) and the swine drums for 20 weeks (June - Oct., 1974). The poultry drums were sampled for 17 weeks (July - Oct., 1974).

The authors are: R. C. AXTELL and D. A. RUTZ, Department of Entomology; M. R. OVERCASH and F. J. HUMENIK, Department of Biological and Agricultural Engineering, North Carolina Agricultural Experiment Station, North Carolina State University, Raleigh, N.C. 27607, USA.

To compare the results from the loading rates versus mosquito production studies in these tanks and drums to farm scale operating swine lagoons, a series of 5 lagoons in Johnston County, North Carolina were sampled for 19 weeks (June - Oct., 1974). Weekly, the abundance of mosquito larvae and the COD and TOC values were determined. Mosquito sampling sites included vegetated and nonvegetated or clear areas along the margin of each lagoon with a total of 6 sample areas per lagoon (9 dips per area).

**TABLE 1. ANAEROBIC REACTOR LOADING RATES USED IN MOSQUITO PRODUCTION STUDIES.**

| Reactor colume (ft$^3$) | Waste type | Loading rate | |
| --- | --- | --- | --- |
| | | ft$^3$ lagoon volume per 100 lb hog or per 4 lb bird | lb COD per ft$^3$ lagoon volume per week* |
| 625 | Swine | 80 | 58 |
| 625 | | 160 | 29 |
| 625 | | 320 | 14.5 |
| 625 | | 640 | 7.2 |
| 625 | | 1280 | 3.6 |
| 7.4 | Swine | 160 | 29 |
| 7.4 | | 320 | 14.5 |
| 7.4 | | 640 | 7.2 |
| 7.4 | | 1280 | 3.6 |
| 7.4 | Poultry | 1.27 | 400 |
| 7.4 | | 5.1 | 100 |
| 7.4 | | 20 | 25 |
| 7.4 | | 82 | 12.5 |

*Based on 0.71 lb COD per day per 100 lb hog and 0.08 lb COD per day per 4 lb bird.

### Mosquito Control

The use of several insecticides and insect growth regulators to control mosquitoes in swine waste lagoons was investigated with 55-gal drums and with farm scale lagoons. The drums used were separate from the study of loading rate described above. Each drum test included 3 treatment replicates and 3 untreated controls. Randomization of treatments and controls was used within the test site. Swine manure was added to the drums weekly at loading rates equivalent to 80 ft$^3$ lagoon volume per 100 lb hog in 1973 (3 tests); 160 ft$^3$ and 320 ft$^3$ lagoon volume per 100 lb hog in 1974 (1 test each). For the farm scale units, there was 1 lagoon per treatment and 2 untreated controls. The farm lagoons (all in Columbus County, North Carolina) varied in size from 0.05 to 0.17 acres and 8-10 ft deep with steeply sloping sides. The loading rates could not be accurately determined but the COD values for these farm lagoons ranged from 336 to 1407 mg/l and the TOC values from 70 to 395 mg/l. All of the farm lagoons as well as the drums were heavy producers of *C. p. quinquefasciatus*.

The insecticides and insect growth regulators were applied to the water surface in the drums by means of a pressurized spray applicator. Farm lagoons were treated by a 3-gal compressed air sprayer which allowed treatment of the water surface to 6-8 ft away from the margin. Most of the chemicals were diluted with water to provide a convenient volume for application (3 oz per drum, 2-3 gal per farm lagoon). The Flit MLO® is a refined petroleum oil and it was applied undiluted. Abate® was used only on the farm lagoons and it was in granular form which was distributed by hand.

The chemicals used, their identities and sources are as follows:

Flit MLO® 100 percent active ingredient: a refined petroleum product. Exxon Corp.

Altosid® SR-10, 10 percent active ingredient and Altosid® 20 percent F, 20 percent active ingredient: isopropyl 11-methoxy-3,7,11-trimethyldodeca-2,4-dienoate. Zoecon Corp.

malathion (Cythion®), 4 lb/gal emulsifiable concentrate: diethyl mercaptosuccinate, S-ester with 0,0-dimethyl phosphorodithioate. American Cyanamid Co.

chlorphrofos (Dursban®), 2 lb/gal emulsifiable concentrate: 0,0-diethyl 0-(3,5,6-trichloro-2-pyridyl) phosphorothioate. Dow Chemical Co.

Abate®, 2 percent granules: 0,0,0́,0́-tetramethyl 0, 0́ thiodi-p-phenylene phosphorothioate. American Cyanmid Co.

Dimilin® (TH6040), 25 percent wettable powder: 1-(4-chlorophenyl)-3-(2,6-difluorobenzoyl) -urea Thompson-Haywood Chemical Co.

The number of mosquito larvae and pupae in the treated and untreated drums and lagoons was determined by the standard dipping method (3 dips per drum, 10 dips per farm lagoon) before treatment and at regular intervals after treatment. The insect growth regulators, Altosid® and Dimilin®, at low concentrations, exhibit delayed effectiveness, i.e. they interfere with the normal developmental processes so that 4th instar larvae fail to molt into pupae or healthy adults fail to emerge from the pupal stage (Schaefer and Wilder 1972). Therefore, in addition to the dip counts, pupae and 4th-instar larvae (when present) from these treatments were held in the laboratory to determine the percent successful adult emergence which gives the best measure of the duration of effectiveness of these growth regulators.

Different combinations of chemicals and dosage rates were used in the different tests. There were 5 drum tests and 4 farm scale tests. Percent mosquito control was calcu-

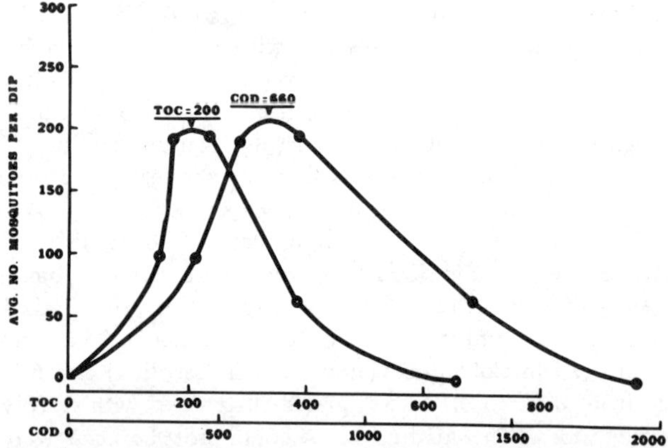

FIG. 1 Abundance (avg. no/dip/wk for 32 weeks) of larvae and pupae of the mosquito and pupae of the mosquito *Culex pipiens quinquefasciatus* in swine waste tanks (625 ft³) in relation to the degree of pollution expressed as supernatant COD (chemical oxygen demand) in mg/l and TOC (total organic carbon) in mg/l.

lated on the basis of the number of pupae present (or, in the case of Altosid and Dimilin, those successfully yielding healthy adults in the laboratory) in the treated drums or lagoons compared to the number in the untreated controls, at the various intervals of time posttreatment.

## RESULTS AND DISCUSSION

### Loading Rates Versus Mosquito Production

Fig. 1 illustrates the relationship between the number of mosquito larvae and pupae and the manure loading rate in the swine manure tanks over the entire 32 week period. The greatest mosquito production occurred at the loading rates of 320 ft³ of lagoon volume per 100 lb hog (supernatant COD = 761 mg/l, TOC = 176 mg/l) and 640 ft³ of lagoon volume per 100 lb hog (supernatant COD = 562 mg/l, TOC = 176 mg/l). The optimal pollution level for production of *C. p. quinquefasciatus* was estimated to be a a level resulting in the COD = 660 mg/l and TOC = 200 mg/l which corresponds to about 480 ft³ of lagoon volume per 100 lb hog. No mosquito production occurred in the unpolluted

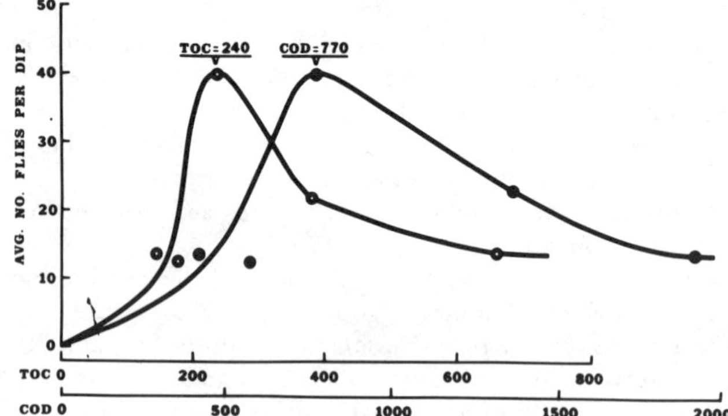

FIG. 2 Abundance (avg. no/dip/wk for 32 weeks) of larvae and pupae of the ephyridid fly *Brachydeutera argentata* in swine waste tanks (625 ft³) in relation to the degree of pollution expressed as supernatant COD and TOC in mg/l.

water in the control tank. Very low numbers of mosquitoes were found in the tank with the loading rate of 1280 ft³ per hog (COD = 428 mg/l, TOC = 151 mg/l). At the high pollution level from the loading rate of 80 ft³ per hog (COD = 1912 mg/l, TOC = 658 mg/l) there was no mosquito breeding. The tank with 4 times this loading rate had a dense surface scum and no mosquito breeding.

Along with the mosquitoes, the tanks produced different numbers of shore flies (Family Ephyridae) at different loading rates. These were identified by W. Wirth as *Brachydeutera argentata* (Walker), a species which is common in polluted waters (Sturtevant and Wheeler 1954) and, although nonbloodsucking may become extremely annoying when present in large numbers. The larvae and pupae of these flies were counted in the same dips obtained for the mosquito samples. As shown in Fig. 2, this fly was most abundant at the loading rate of 320 ft³ of lagoon per 100 lb hog (supernatant COD = 770 mg/l, TOC = 240 mg/l). It was relatively more tolerant of the higher pollution levels than were the mosquitoes.

The results obtained with the smaller volumes of swine waste in the drums were similar to those from the tanks (Fig. 3). The optimal pollution level for mosquito production over the 20-week period was at the loading rate of 320 ft³ of lagoon per 100 lb hog which yielded a COD = 593 mg/l and a TOC = 228 mg/l. Mosquito production was appreciable although lower at one-half this loading rate (COD = 839 mg/l, TOC = 300 mg/l). The data suggest that there



MANAGING LIVESTOCK WASTES

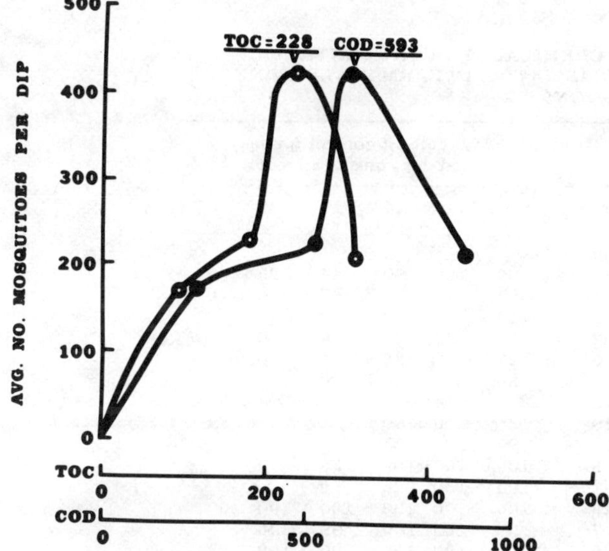

FIG. 3 Abundance (avg. no/dip/wk for 20 weeks) of larvae and pupae of the mosquito *Culex p. quinquefasciatus* in swine waste drums (7.4 ft$^3$) in relation to the degree of pollution expressed as supernatant COD and TOC in mg/l.

would be little or no mosquito production at a loading rate of 80 or less ft$^3$ per hog.

The poultry waste drums yielded similar results (Fig. 4). Optimal mosquito production occurred at the loading rate of 5.1 ft$^3$ lagoon per 4 lb bird which yielded pollution levels of supernatant COD = 669 mg/l and TOC = 243 mg/l.

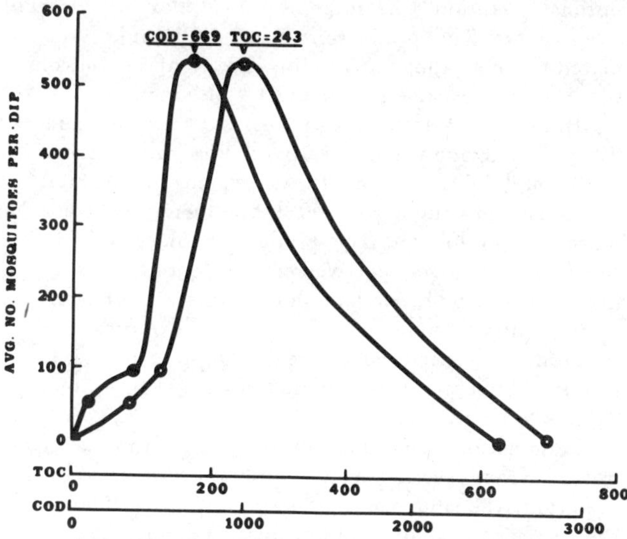

FIG. 4 Abundance (avg. no/dip/wk for 17 weeks) of larvae and pupae of the mosquito *Culex p. quinquefasciatus* in poultry waste drums (7.4 ft$^3$) in relation to the degree of pollution expressed as supernatant COD and TOC in mg/l.

These are very close to the COD and TOC values found for optimal mosquito production in swine waste. There was very little mosquito production at the loading rate of 82 ft$^3$ per bird (COD = 76 mg/l, TOC = 181 mg/l). At the loading rate of 20 ft$^3$ per bird (COD = 329 mg/l, TOC = 122 mg/l) there was appreciable mosquito production (100 per dip). At the high pollution level produced by the loading rate of 1.27 ft$^3$ of lagoon per bird (COD = 2529 mg/l, TOC = 669 mg/l) there was no mosquito production.

Examination of mosquito production in the farm scale swine lagoons revealed that the mosquito abundance is also affected by factors other than the degree of pollution. The average values (mg/l) for the season for the COD and TOC determinations were: Lagoon no. 1 — COD = 600, TOC = 221, Lagoon no. 2 — COD = 734, TOC = 254; Lagoon no. 3 — COD = 755, TOC = 257; Lagoon no. 4 — COD = 1500, TOC = 470; Lagoon no. 5 — COD = 1628, TOC = 547. The highest average number of mosquito larvae per dip (210.8) was found in Lagoon no. 3, while lagoons of similar pollution levels had fewer mosquitoes per dip (no. 1 = 49.9, no. 2 = 15.5). Further, the high pollution lagoons differed greatly in the average number of mosquito larvae per dip (no. 4 = 0.6, no. 5 = 195.5). The difference in degree of vegetation and/or floating debris was the major factor causing this lack of simple correlation of mosquito numbers with the degree of pollution. A certain level of pollution is necessary for mosquito production but the degree of protection afforded by emergent vegetation and debris determines to a great extent the amount of mosquito larvae that will be present. This relationship is being investigated further.

### Mosquito Control

A summary of the drum tests is presented in Table 2. Dimilin at 0.1 lb/A gave complete mosquito control for 12 days and was about 50 percent effective after an estimated 16 days. At 0.03 lb/A, Dimilin gave partial control (80 percent) for up to 5 days and was 50 percent effective for about 7 days. Altosid SR-10 at 0.1 lb/A was about equal in effectiveness to the Dimilin at 0.03 lb/A. The 20 percent slow release formulation of Altosid also gave partial control for up to 5 days but retained its effectiveness slightly longer (50 percent control after about 15 days) than the SR-10 formulation. Malathion at 1.0 lb/A gave complete control for 8 days and about 50 percent control after an estimated 10 days. Flit MLO gave essentially complete control for 8 days and then quickly lost any effectiveness. Chlorpyrifos at 0.1 lb/A gave the most long-lasting effectiveness with 100 percent control for 29 days; a few mosquito larvae were found at 35 days posttreatment.

A summary of the tests in farm swine lagoons is pre-

**TABLE 2. EFFECTIVENESS OF VARIOUS CHEMICALS FOR THE CONTROL OF MOSQUITOES (*CULEX PIPIENS QUINQUEFASCIATUS*) IN SWINE WASTE IN SIMULATED LAGOONS (55-GAL DRUMS).**

| Chemical | lb AI/ acre | No. tests | Avg. no. mosquitoes pretreatment | | Avg. percent control (pupa) and days posttreatment | | | | | | | |
|---|---|---|---|---|---|---|---|---|---|---|---|---|
| | | | Larvae | Pupae | 2—3 | 4—5 | 7—8 | 11—12 | 14—15 | 17—18 | 21—22 | 28—29 |
| Untreated | 0.0 | 5 | 208 | 26 | | | | | | | | |
| Dimilin | 0.1 | 2 | 299 | 55 | 100 | 100 | 100 | 99 | 66 | 25 | 22 | 2 |
| Dimilin | 0.03 | 2 | 179 | 47 | 80 | 81 | 46 | 3 | 18 | 16 | 12 | 1 |
| Altosid SR—10 | 0.1 | 4 | 220 | 27 | 85 | 82 | 68 | 27 | 6 | 0 | | |
| Altosid 20F | 0.1 | 1 | 198 | 32 | | 87 | 70 | 57 | 55 | 0 | | |
| malathion | 1.0 | 3 | 159 | 9 | 100 | 100 | 100 | 46 | 12 | 0 | | |
| chlorpyrifos | 0.1 | 2 | 218 | 52 | 100 | 100 | 100 | 100 | 100 | 100 | 100 | 100 |
| Flit MLO | 5 gal/A | 3 | 154 | 17 | 99 | 98 | 94 | 40 | 3 | 0 | | |

**TABLE 3. EFFECTIVENESS OF VARIOUS CHEMICALS FOR THE CONTROL OF MOSQUITOES (*CULEX PIPIENS QUINQUEFASCIATUS*) IN FARM SCALE SWINE WASTE LAGOONS.**

| Chemical | lb AI/ acre | No. tests | Avg. no. mosquitoes pretreatment | | Avg. percent control (pupae) and days posttreatment | | | | | |
|---|---|---|---|---|---|---|---|---|---|---|
| | | | Larvae | Pupae | 3-4 | 7 | 14 | 21 | 28 | 35 |
| Untreated | | 4 | 735 | 39 | | | | | | |
| Dimilin | 0.1 | 2 | 326 | 17 | 100 | 100 | 90 | 44 | 25 | 7 |
| | 0.06 | 2 | 619 | 44 | 93 | 74 | 75 | 37 | 6 | 0 |
| | 0.03 | 1 | 766 | 1 | 29 | 6 | 0 | | | |
| Altosid SR-10 | 0.4 | 1 | 825 | 52 | 81 | 69 | 35 | 4 | 0 | |
| | 0.2 | 1 | 271 | 6 | 27 | | 25 | 4 | 0 | |
| Altosid 20F | 0.5 | 1 | 112 | 10 | 80 | 84 | 66 | 0 | | |
| | 0.25 | 1 | 74 | 5 | | 4 | 0 | | | |
| | 0.1 | 1 | 435 | 19 | | 8 | 0 | | | |
| malathion | 1.0 | 2 | 109 | 13 | 94 | 100 | 55 | 0 | | |
| Abate | 0.5 | 1 | 127 | 8 | 100 | 100 | 100 | 38 | 0 | |
| | 0.3 | 1 | 158 | 1 | 100 | 100 | 28 | 0 | | |
| Chlorpyrifos | 1.0 | 1 | 841 | 128 | 100 | 100 | 100 | 100 | 100 | 100 |
| | 0.5 | 1 | 132 | 7 | 99 | 100 | 100 | 99 | 90 | 78 |
| | 0.4 | 1 | 1606 | 35 | 100 | 100 | 100 | 100 | 100 | 85 |
| | 0.2 | 1 | 317 | 6 | 100 | 100 | 98 | 54 | 0 | |
| | 0.1 | 1 | 1575 | 97 | 100 | 90 | 0 | | | |
| | 0.05 | 1 | 136 | 3 | 100 | 100 | 0 | | | |
| Flit MLO | 7 gal/A | 1 | 250 | 36 | 100 | 93 | 29 | 0 | | |
| | 5 gal/A | 1 | 203 | 13 | 100 | 97 | 37 | 0 | | |

sented in Table 3. Dimilin at 0.1 lb/A gave nearly complete mosquito control for 14 days and was 50 percent effective after an estimated 20 days; at 0.06 lb/A the control was less and at 0.03 lb/A very poor control was obtained. Altosid SR-10 gave partial control (81 percent) for 4 days and was 50 percent effective after an estimated 10 days; at 0.2 lb/A the control was poor. The Altosid 20F formulation at 0.5 lb/A gave partial control (80 percent) for 7 days and was 50 percent effective after an estimated 16 days; at 0.25 and 0.1 lb/A there was essentially no control at 7 days posttreatment. Malathion at 1.0 lb/A was nearly 100 percent effective for 7 days and gave 50 percent control for an estimated 15 days. Abate at 0.5 lb/A gave complete control for 14 days and 50 percent control for an estimated 18 days while at 0.3 lb/A it was totally effective for 7 days and 50 percent effective for an estimated 10 days. Flit MLO at 7 and 5 gal/A gave nearly complete control for 7 days and then rapidly lost its effectiveness. Chlorpyrifos was the most long-lasting control agent giving essentially 100 percent control for 35 days at the 1.0 lb/A rate and for 28 days at the 0.5 and 0.4 lb/A rates (which still gave high levels of control after 35 days). Control by chlorpyrifos was less at lower rates with 0.2 lb/A giving 100 percent control for 14 days and 54 percent control after 21 days while the 0.1 and 0.05 lb/A rates were effective for 7 days.

In these drum and on-farm tests, chlorpyrifos consistently gave the most long-lasting mosquito control and 0.4 lb/A is probably an average dose needed to give mosquito control for a month. Abate and malathion are effective but only for up to 2 weeks at reasonable dosage rates. Flit MLO is an effective temporary control chemical which breaks down rapidly in the environment. The insect growth regulators can be effective in these highly polluted waters. In our tests, Dimilin was more effective and longer lasting than the Altosid formulations.

Weekly, COD and TOC measurements were made in the drum tests and there was no evidence of interference with the normal swine waste degradation processes by the treatments with insecticides and insect growth regulators. No abnormal effects on the farm lagoons were observed in the course of these chemical treatments.

## SUMMARY AND CONCLUSIONS

Simulated swine and poultry waste lagoons were operated for several months at different weekly manure loading rates which resulted in different organic pollution levels as expressed by the supernatant COD and TOC values. The optimal pollution level for mosquito, *Culex pipiens quinquefasciatus* Say, production in 625 ft$^3$ tanks was estimated to be at a loading rate of 480 ft$^3$ of lagoon volume per 100 lb hog (supernatant COD = 660 mg/l, TOC = 200 mg/l); in 7.4 ft$^3$ drums the optimal was at a loading rate of 320 ft$^3$ of lagoon volume per 100 lb hog (supernatant COD = 593 mg/l, TOC = 228 mg/l). At a loading rate of 80 ft$^3$ or less of lagoon volume per 100 lb hog there was little or no mosquito production. The production of ephyridid flies, *Brachydeutera argentata* (Walker), was optimal at a slightly higher organic pollution level in the swine waste tanks.

In poultry manure in 7.4 ft$^3$ drums, the optimal mosquito production occurred at the loading rate of 5.1 ft$^3$ of lagoon volume per 4 lb bird (supernatant COD = 669 mg/l, TOC = 243 mg/l).

Examination of mosquito production in farm scale swine waste lagoons revealed that, in addition to level of organic pollution, the abundance of mosquitoes is influenced by the degree of vegetation and/or floating debris.

The effectiveness for mosquito control in swine waste in drums and in farm scale lagoons was determined for the insect growth regulators Altosid® and Dimilin® and for the insecticides malathion, chlorpyrifos, Abate® and Flit MLO®. No impaired lagoon performance was evident from the addition of these chemicals at the dosage rates used. Mosquito control was obtained for periods of a few days to over a month, depending upon the chemical and dosage rate, with chlorpyrifos being effective for the longest period.

### References


1 American Public Health Association. 1971. Standard methods for the examination of water and wastewater, 13th edition, Wash., D.C.
2 Barr, A. R. 1957. The distribution of *Culex p. pipiens* and *C. p. quinquefasciatus* in North America. Amer. J. Trop. Med. Hyg. 6:153-65.
3 Laven, H. 1967. Speciation and evolution in *Culex pipiens*,


*(Continued on page 21)*

# Pathogenic Microorganisms in the Environment

Glenn B. Van Ness

**P**ATHOGENIC microorganisms from livestock enter the environment by way of urine and fecal wastes as well as secretions from the mouth and respiratory tract. The Task Force on Environmental Quality (1968) reported "the survival and possible multiplication in soil and water of pathogenic microorganisms that may infect plants, animals or man is controlled to a significant extent by the presence or absence of other organisms in that same environment. Non-pathogenic organisms with a close genetic relationship to a pathogen may specifically eliminate the pathogen from an environment. Many soil microorganisms produce metabolites that are toxic to other microorganisms". The laboratory study of livestock pathogens provides some of the ecological requirements which together with epidemiological data, tells us what to expect of pathogenic microorganisms in the environment.

A soil organism, *Bacillus anthracis*, causes anthrax, which leads to sudden death in livestock, and malignant carbuncle in man. Anthrax is spread by contaminated feed, hides, hair, and by movement of infected livestock. In 1974, tourist items and horse blankets were found to be contaminated with the organism. Herbivorous animals become infected, and infected meat fed to omnivores and carnivores spreads the disease. The epidemiology and ecology was studied by Van Ness (1956, 1971), without finding evidence of spread by sewage or composted manure. Outbreaks are associated with "incubator areas", although some diffuse outbreaks may occur on favorable soils.

An anthrax incubator area is a ponded depression (Fig. 1), with a high water mark of killed brown vegetation, and a zone of silted vetetative trash just above the pond mud area. Water still in the depression may become anaerobic, favoring livestock losses due to *Clostridium* species, causing malignant edema and related anaerobic infections.

In the Northwest U. S., Van Ness (1964), found that some pastures, streams and irrigated areas have an alkaline wetland environment suitable for bacillary hemoglobinuria caused by *Clostridium hemolyticum*. The disease may be carried by infected cattle into other alkaline environments which may occur from the Platte river of Nebraska, northward through the glaciated states, and to some areas in Florida. The disease is already known on the Gulf Coast west of the Mississippi river. A suitable environment was described by Rastas (1974) in a Wisconsin outbreak. Cattle grazed on the suitable environment may then pick up the organism, and die later when the liver is damaged or stressed. The bacillary hemoglobinuria niche may also harbor anthrax, and both diseases may occur on the same pasture.

Some microorganisms, such as the enteric *Salmonella*, *Escherichia* and *Vibrio* are intimately associated with fecal wastes. The *Salmonella* tend to be host-adapted in causing infection, but the many serotypes and their environmental adaptability have so far made total Salmonella control an impossibility. The salmonella and related enteric organisms will grow in numbers wherever nutrients, moisture, temperature and other factors are favorable. The dead organisms release powerful endotoxins, which in turn may lead to tissue invasion and disease by the living organisms. An epidemic can follow the introduction of very few organisms into a susceptible herd or flock. Morse (1974) in a concise review of the problem, indicates environmental factors exert a profound impact on the occurrence, epizootiology, prevention and control of salmonellosis. In controlled studies of poultry litter, Turnbull (1972), suggested high pH due to ammonia in moist litter eliminated *Salmonella typhimurium* within 11 days. Reused litter taken to a clean house by Duff (1973), failed to transmit *Salmonella typhimurium* to disease-free chicks.

With our present knowledge and ability to isolate salmonella and related organisms, the coliform criteria for water is of limited value in water contaminated with organic wastes and nutrients. Direct isolation of suspect pathogens is essential to properly evaluate the role of the enteric organisms in waste management. Pathogenicity tests may also be necessary.

Tuberculosis in cattle, swine and poultry is an important disease with an environmental impact.

Tuberculosis reactor rates in cattle have dropped from 4.88 percent in 1918 to 0.03 percent in 1974 (Tuberculosis Statistical Tables), with reactors reported from 141 lots of cattle, and 1779 lots contained suspects not considered infected with *Mycobacterium bovis*. Many of these lots occurred in states with a long history of suspect herds, and isolation studies are either negative, or yield other mycobacteria.

FIG. 1 A dead grass "incubator area" is seen in a pasture where eight cattle died of anthrax.

The author is: GLENN B. VAN NESS, Senior Veterinarian, Veterinary Services, Animal and Plant Health Inspection Service, USDA, Beltsville, Md. 20705.

When tuberculous livestock are identified, the premises and vehicles are cleaned and disinfected. Disposition of manure is by composting, or spreading to ground not to be used soon by livestock. The threat of infection from sewage and wastes, as Greenburg (1957) in an extensive reveiw, points out, cannot be disregarded. The tuberculosis organisms survive well in the environment.

Despite the value of the tuberculin test in cattle, Roswurm (1973) pointed to the problem of cattle which do respond to the test but are not infected with *M. bovis*. These responses could be more closely related to *M. avium* tuberculin antigens, and the Comparative Cervical (CC) test using balanced bovine and avian purified protein derived tuberculins is now used to retest tuberculin responders. The tuberculin test was applied to 27 institutional herds having no recent history of bovine tuberculosis, with the percentage of response in each herd reported in Fig. 2. Background on this map are soils which Van Ness (1966) introduced in 1959 for ecological study of the occurrence of deviators. Under natural conditions, an alkaline wetland environment is found in the habitat of the suspect. Alkaline environments may also be created by new concrete structures, disposal of barn lime and spent hypochlorite solution, and by alkaline wash water. Animal manure itself, although alkaline, has not been incriminated.

The organism of swine erysipelas, *Erysipelothrix rhusiopathiae*, may persist in the soil, and Wood (1972), found the organism in the soil of 18 of 19 Iowa hog farms. Soil receiving drainage at the edge of concrete platforms was 48.6 percent positive, while open swine pasture was negative. Wood found a lower incidence of 27.1 percent and 20 percent in the pens. In my own experience, pools of water on turkey range was also linked with turkey erysipelas. Under laboratory controlled conditions, Wood (1973), was unable to confirm a soil survival phase, and particularly noted poor survival in soil with high organic content. Erysipelas has not been associated with general waste disposal methods, however, and limited but undefined soil eco-

logical factors may permit survival in the environment.

Leptospirosis in livestock is caused by an organism dependent on a water environment for survival. Shed in the urine of the infected animal, most infections are associated with ponds or slow moving streams. Direct infection could occur by urine splash, or contamination of feed. The host-adapted *Leptospira pomona* spreads in cattle and swine, but all others produce disease or serological evidence of disease without further spread in livestock. Control depends upon water management practices. There is not yet evidence that solid manure management practices favor the spread of leptospirosis.

Viruses may spread in the environment, although incapable of increasing in numbers outside the host cell. The virus particles with protein coats but without lipids are more resistant, and dry or cold environments offers the most protection. Generally, animal viruses are unable to resist the biological activity of the manure pile or lagoon. The ether-resistant enteroviruses may pass through sewage treatment and be found in the effluent. The arthropod-borne arbovirus has a growth phase in an arthropod, in addition to causing infection in a mammalian host. Ecological factors favorable for the arthropod host control the distribution of arboviruses. Pox virus and other viruses transferred mechanically by arthropods do not increase in the arthropod host, but Kligler (1929), showed fowl pox could be transmitted by mosquitoes after 16-19 days.

The appearance and spread of foot and mouth disease (FMD) virus in the United States would seriously affect the trade in livestock, and most of our international regulatory programs for livestock are directed at its control. The behavior of FMD virus in the environment is an example of ether-resistant viruses, which also include human polio, coxsackie viruses, and other picornaviruses. Shahan (1962) describes the virus with an outer protein coat, and a ribonucleic acid (RNA) core. Removal of the protein coat by phenol does not change infectivity of the RNA core, but environmental ribonucleases will quickly destroy the core.

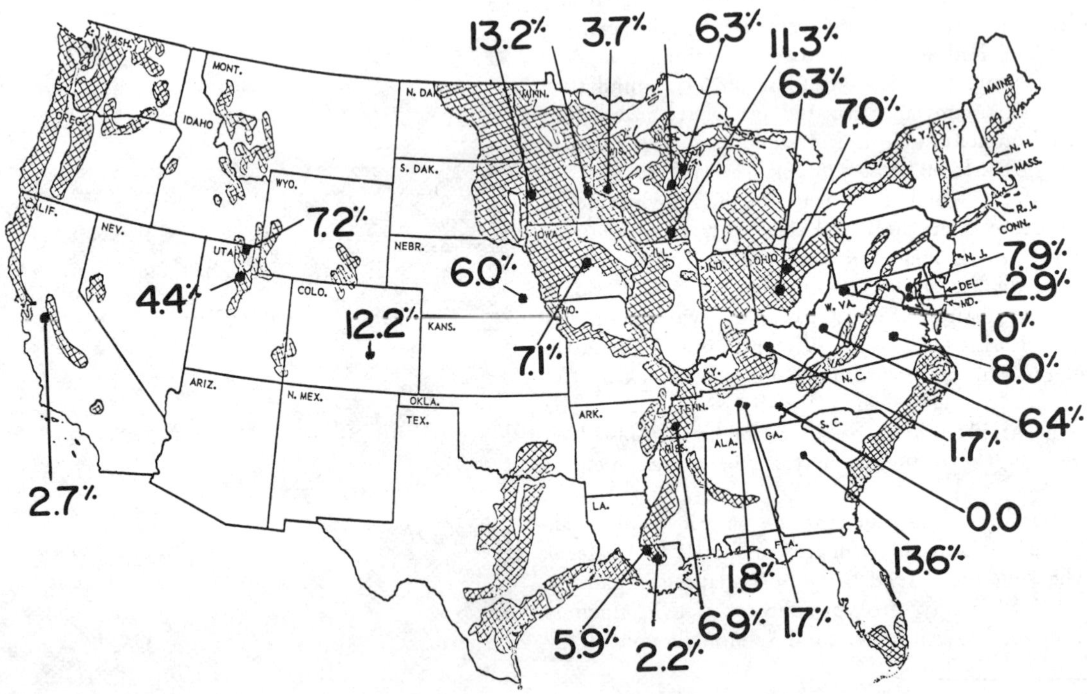


FIG. 2 **Percentage of tuberculin response in cattle in 27 institutional herds, located on alkaline soils map for ecological study of non-infected herds.**

Manure contaminated by FMD virus would be required by regulation to be buried or composted for a period, dependent upon ambient temperature and pile temperature. Composted manure should not be spread to pasture where there would be contact with livestock. Solid waste control practices for FMD should prevent the spread of all other livestock disease viruses.

Another ether-resistant enterovirus causing Swine Vesicular Disease (SVD) is not present in the United States, and regulations are directed at imported pork products which might introduce this disease. The virus resembles the Coxsackie group of human viruses, and Graves (1973) points to a strong evidence that this swine virus has much in common with human Coxsackie B-5 virus. It can be spread by swine manure or contaminated garbage and may be more resistant than F & M virus to cleaning and disinfection. The virus survived 138 days (Dawe 1974), stored in 50 kg plastic bags at an ambient temperature between 12 and 17 C.

For the enteric cytopathic porcine virus (ECPO-1), Meyer (1971), found experimentally survival was 5 days within the anaerobic digester.

Transmissible Gastroenteritis (TGE) may be the most important virus problem of modern swine production. Haelterman (1973), identifies the cause as a coronavirus, which is rapidly inactivated at warm temperatures, and by sunlight. Cold and cloudy weather provides some protection. It can be a problem where fecal wastes are recycled across the swine feeding floor. It is primarily a disease of nursing pigs on non-immune sows, or just after weaning, or by the mixing of susceptible feeder pigs at concentration points. Breaking a cycle of spread by not adding new contact swine for two weeks should bring the disease under control.

New emphasis on manure management, including pollution and degradation of the environment now includes the recognition of the energy value and nutrient potential of this resource. We should know how and why pathogenic microorganisms in this waste are able to survive and multiply in the natural environment. At the same time, knowledge of the physical makeup, metabolic requirements and environmental needs may give us practical control of pathogenic microorganisms in the environment.

## References


1   Dawe, P. S. 1974. Viability of swine vesicular disease in carcasses and feces. The Veterinary Record 94(19):430.

2   Duff, R. H., J. G. Ross and D. D. Brown. 1973. The influence of litter on Salmonella typhimurium infection in poultry. Avian Path. 2:163-268.

3   Graves, J. H. and P. D. McKercher. 1973. Swine vesicular disease. U. S. Animal Health Assn. 77th Meeting. 155-159.

4   Greenberg, A. E., E. Kupka. 1957. Tuberculosis transmission by waste waters—a Review. Sewage and Industrial Wastes 29(5):524-537.

5   Haelterman, E. O. 1973. On the control of transmissible gastroenteritis of swine. U. S. Animal Health Assn. 77th Meeting. 345-349.

6   Kligler, I. J., M. Asner. 1929. Transmission of fowl pox by mosquitoes. Brit. J. Expt. Path. 10:347-349.

7   Meyer, R. C., F. C. Hinds, H. R. Isaacson and T. D. Hinesly. 1971. Porcine entovirus survival and anaerobic sludge digestion. Proc. Int. Symp. Livestock Waste Management. 183-184. ASAE, St. Joseph, Mich. 49085.

8   Morse, E. V. and M. A. Duncan. 1974. Salmonellosis—an environmental health problem. J.A.V.M.A. 165:1015-1019.

9   Rastas, V. P., G. H. Myers, Stan Lesar. 1974. J.A.V.M.A. 165:1203-1204.

10   Roswurm, J. D. and L. D. Konyha. The comparative cervical tuberculin test as an aid to diagnosing bovine tuberculosis. Proc. U. S. Animal Health Assn. 77th Meeting: 368-389.

11   Shahan, M. S. 1962. The virus of foot-and-mouth disease. Anis. N. Y. Acad. Sci. 101:444-454.

12   Task Force on Environmental Quality—Pollution in relation to agriculture and forestry. 1968. USDA, Washington, D. C. 20250.

13   Tuberculosis statistical tables for 1974. VS, APHIS, USDA.

14   Turnbull, P. C. B. and G. H. Snoeyenbos. 1973. The roles of ammonia, water activity, and pH in the salmonellacidal effect of long-used poultry litter. Avian Dis. 17:72-86.

15   Van Ness, G. B., K. Erickson. 1964. Ecology of bacillary hemoglobinuria. J.A.V.M.A. 144:492-496.

16   Van Ness, G. B. 1966. The role of meat inspection in a disease reporting system. U. S. Livestock Sanitary Assn. 70th Meeting: 396-402.

17   Van Ness, Glenn B. 1971. Ecology of anthrax. Sci. 172:1303-1307.

18   Wood, R. L. and R. A. Packer. 1972. Isolation of *Erysipelothrix rhusiopathiae* from soil and manure of swine-raising premises. Am. J. Vet. Res. 33:1611-1620. Chicago, Ill. 60605.

19   Wood, R. L. 1973. Survival of *Erysipelothrix rhusiopathiae* in soil under various environmental conditions. Cornell Vet. 63:390-410. Ithaca, N. Y. 14851.


---

## Mosquito Control

*(Continued from page 18)*


p. 251-275. In: J. W. Wright and R. Pal (eds.) Genetics of insect vectors of disease. Elsevier, Amsterdam.

4   Mattingly, P. R. 1951. the *Culex pipiens* complex. Trans. Royal Entomol. Soc. (London) 102:331-342.

5   Overcash, M. R., D. L. Reddell and D. L. Day. 1974. Chemical analysis. ASAE Paper No. 74-4546, ASAE, St. Joseph, Michigan 49085.

6   Schaefer, C. H. and W. H. Wilder. 1972. Insect developmental inhibitors: A practical evaluation as mosquito control agents. J. Econ. Entomol. 65(4):1066-1071.

7   Steelman, C. D., J. M. Gassie and B. R. Craven. 1967a. Laboratory and field studies on mosquito control in waste disposal lagoons in Louisiana. Mosquito News. 27(1):57-59.

8   Steelman, C. D., A. R. Colmer, L. Cabes, H. T. Barr and B. A. Tower. 1967b. Relative toxicity of selected insecticides to bacterial populations in waste disposal lagoons. J. Econ. Entomol. 60(2):467-468.

9   Sturtevant, A. H. and M. R. Wheeler. 1954. Synopses of nearctic Ephydridae (Diptera). Amer. Entomol. Soc. Trans. 79:151-261.

10   Weidhaas, D. E., B. J. Smittle, R. S. Patterson, H. R. Ford and C. S. Lofgren. 1973. Survival, reproductive capacity, and migration of adult *Culex pipiens quinquefasciatus* Say. Mosquito News. 33(1):83-87.

# Engineering and Economic Overview of Alternative Livestock Waste Utilization Techniques

Judson M. Harper, David Seckler
MEMBER
ASAE

## INTRODUCTION

THERE are three possible used of manure: (a) as a fertilizer; (b) as a fuel; and (c) as livestock feed. Each use and all published technologies of which we are aware within each use are evaluated below. The methodology employed borrows from the mining industry: (a) What is the assay or potential value of manure as an "ore" *in situ* for each use? and (b) What is the cost and value of products obtained in each "benefaction" process? Each category is briefly discussed and an overall evaluation is provided at the end in Table 4 in terms of the following criteria: (a) degree of utilization; (b) economic value; (c) break even scale; (d) marketability of products; and (e) environmental impact.

## MANURE AS A FERTILIZER

Table 1 shows the fertilizer ingredients of typical beef feedlot manure and its value at current market prices of chemical fertilizers.

The value of the humus factor is dubious because of the use of crop residues to fill this need. Inclusion of this factor provides a safe upper limit.

Costs of benefaction (gathering, transporting and spreading) rise severely with concentration of manure supply in relation to land area. Small farms may "benefact" for $1.50-$2.00 per ton (as is basis). Estimated costs for a 15 000-head feedlot in Colorado vary between $4 and $6 per ton. Thus, net value varies between -$0.83 and $3.67 depending on scale.

**TABLE 1. ASSAY VALUE OF ONE TONE OF MANURE AS A FERTILIZER IN FEBRUARY 1975 PRICES**

| Element | Number lb per ton of manure 40 percent moisture basis (Graber 1974 p. 42) | Value per ton | Reference value of commercial fertilizer |
|---|---|---|---|
| N | 10 | $1.64 | Anhydrous ammonia @ $270 per ton (82 percent N) |
| $P_2O_5$ | 5 | $1.35 | Superphosphate @ $238 per ton (44 percent P) |
| $K_2O$ | 10 | $1.18 | Muriate of potash @ $122.50 per ton (52 percent K) |
| Trace, estimated Organic matter | | $0.50 | |
| "Humus value" | | $0.50 | |
| As is basis | | $5.17 | |
| Dry matter basis | | $8.62 | |

The authors are: JUDSON M. HARPER, Professor of Agricultural Engineering and DAVID SECKLER, Professor of Economics, at Colorado State University and Consultants to Ceres Ecology Corporation.

Some areas such as the Chino Valley Milk Shed are so concentrated that manure cannot be disposed of as fertilizer. Also episodic gathering creates air pollution through putrefying piles of manure.

## MANURE AS A FUEL

The B.T.U. rating of beef feedlot manure is as high as 6500 per pound or 13 x $10^6$ B.T.U. per ton (D.M.B.)* (Halligan and Sweazy 1972). Its value in terms of other energy sources is shown in Table 2. There are three means of utilizing manure as a fuel: (a) direct combustion; (b) as a substrate for methane production; and (c) for the production of synthetic gas.

**TABLE 2. VALUE OF ONE TON OF MANURE AS A FUEL.**

| Fuel source | Cost per unit | Btu per unit | Cost 1 x $10^6$ Btu | Fuel value of 1 ton of manure 13 x $10^6$ Btu |
|---|---|---|---|---|
| Coal* (subbituminous low-sulpher) | 0.8 cents/pound | 1 x $10^4$ | $0.80 | $10.40 |
| Natural gas | 90 cents/1000 ft$^3$ | $10^6$ | $0.90 | $11.70 |
| Diesel | 35 cents/gallon | 1.3 x $10^5$ | $2.63 | $34.21 |

*Coal is sold delivered to an industrial user in Craig, Colorado, at $5 per ton (long-term price—25 years). Transportation costs are about $11 per ton per thousand miles. Thus delivered price is about $16 per ton in most areas of the West.

### Direct Combustion

This alternative is possible only when the moisture content of manure is sufficiently low (<25 percent) to sustain combustion. "Trash" type furnaces and air poolution equipment are required. The ash will be valuable as a fertilizer since only N and humus is lost in combustion. The fertilizer value of the ash per ton of manure input is $2.50; or per ton of ash about $9. The total fuel and fertilizer value per ton of manure is, thus, about $5.50.

### Manure as a Substrate for Methane

On Farm: manure must be diluted to 5 percent solids and held at temperatures of 110-140 F for 15-30 day detentions. A 100-head herd would require 5000-6000 ft$^3$ of fermentation tanks. This system would produce about 4000 ft$^3$ of low (500 BTU/ft$^3$) gas per day worth about $1.80. The total system would cost $15 000. A lagoon is required to handle the sludge since only 50 percent of the solids would be converted to gas (Fairbank 1974).

*By way of comparison, bagasse is rated at 8000-9000 B.T.U.'s per pound and straw at about 6000 B.T.U.'s per pound.

Off Farm: proposals exist for large units handling the manure of over 100 000-head of beef cattle. Such a system would cost $6-$10 x $10^6$ (Varany 1974) and break even at $3 per 1000 ft$^3$ equivalent cost of natural gas. Such a system would produce some 85 x $10^6$ gallons of sludge at 20 percent moisture requiring 5000 acres of land at 10 tons per acre (D.M.B.) for disposal.

## Synthetic Fuels

Of the three technologies contemplated in this area, liquefaction, hydrogasification and pyrolysis for synthesis gas, the latter is generally agreed to be the most promising (Walawender et al. 1973). Only this alternative is considered here for reasons which shall later become apparent.

Pyrolysis requires high temperature heating in the absence of air to produce a variety of products including CO, $CO_2$, $CH_4$ and $H_2$. Economies of scale require plants in the 200 000-head range to break even at about $3 per 1000 ft$^3$ equivalent natural gas prices.

MANURE AS A LIVESTOCK FEED

Typical nutrient analyses of beef and poultry manure are shown in Table 3. The assay value of whole manure depends on a variety of factors the resolution of which can only be determined in a computer least cost ration formulation. Based on our results in computer evaluation at feed prices prevailing over the past six months the assay value of fresh beef manure is between $35-$45 per ton (D.M.B.) and poultry manure is $50-$70 per ton. It is this relatively large potential value that has spawned a variety of manure-as-feed techniques.

Use of manure as feed requires satisfaction of several subcriteria:

1  Manure contains residues of potentially harmful substances (heavy metals, antibiotics and pesticides) and residues of indigestible materials (lignin and mineral matter, amounting to 20-40 percent of the total) which would rapidly accumulate in recycling to prevent its use as a feed.

This accumulation cycle must be broken. There are three ways: (a) dispersion to animals other than those producing manure; (b) dilution through using only a fraction of the manure as feed and disposal of the balance, and (c) extraction and disposal of these residues as a continuous "blow down" feature of the process.

2  Manure can contain pathogens and safety against infection can only be assured through continuous thermal and/or chemical processing.

3  The feed products must be palatable to livestock and possess good "shelf life" in storage and in the feed bunk.

4  Since on-farm livestock will typically not consume all the manure derived feeds produced, it is necessary that at least a fraction be in the form of readily transportable and marketable products if all the manure is to be so utilized.

There are four basic technologies: (a) whole manure drying; (b) the wasteledge system; (c) fractionation with partial recovery; and (d) fractionation with full recovery.

## Whole Manure Drying

In arid areas of the world air-dried manure has been directly used in feed rations. Since there is no pathogen control, this is not a feasible technology.

Dried poultry waste (DPW) is dried with a rotary drier and used as feed. Costs of production depends largely on the moisture content of the manure. Using natural gas prices of $0.90 per 1000 ft$^3$ we estimate (Smith 1974) costs of between $17 per ton for 44 percent solids poultry manure to $60 per ton at 25 percent solids.

Particle temperatures in drying do not typically attain pasteurization levels. Certainly the very dry state of the feed inhibits biological acitvity and no cases of infection have been reported to date.

## The Wasteledge System

The wasteledge system (Anthony 1971) consists of blending 40 percent wet (70-80 percent moisture) manure with dry standard feed ingredients and then ensiling the mixture for over 10 days detention. Preferably fermentation takes place in a top loading bottom unloading airtight silo to insure uniform fermentation. The pH drops to near 4.0. This acidic state, while not theoretically sufficient to insure destruction of all possible pathogens, certainly destroys most and inhibits biological activity throughout the feed cycle. Excellent feed results with cattle have been obtained using wasteledge. This system utilizes only about 25 percent of the manure produced and does not present an opportunity for utilization for all the manure produced.

## Fractionation with Partial Recovery

In these systems manure is washed or settled and one or the other fraction is refed.

At Illinois University (Harmon et al. 1972) swine manure is treated in an oxidation ditch and the protein containing water is fed to swine as drinking water. While excellent feed results have been obtained, it is hard to visualize biological control in this wet, basic medium. Evaluation of this technique must wait further research results.

Coral Industries of Phoenix, Arizona (Gross 1975) manufacturers a system for screening dilute solutions of manure and then pressing and chemically sterilizing the fiberous fraction for cattle feed as a roughage replacer in the ration. The liquid fraction is then pumped to lagoons for eventual disposal on fields as a fertilizer.

Ceres Ecology Corporation (Seckler 1975) manufacturers a similar system—consisting of the CI line of Fig. 1—for use in small to medium size livestock operations.

The value of the roughage feed produced in these systems is about equal to that of low to medium grade corn silage $40-$50 per ton (D.M.B.). Costs of production are about $10 per ton (assuming break even on the liquid fertilizer fraction) at small to medium scale.

**TABLE 3. FEED ANALYSIS OF MANURE (DMB)**

| Fresh cattle manure average of ceres ecology tests | | Dehydrated poultry manure, rounded |
|---|---|---|
| Crude protein (N x 6.25) | 15 | 30 |
| Fat | 15 | n.a. |
| Crude fiber | 24 | 25 |
| Ash | 25 | 25 |
| NFE | 34.5 | 24 |
| P | 0.75 | 2.0 |

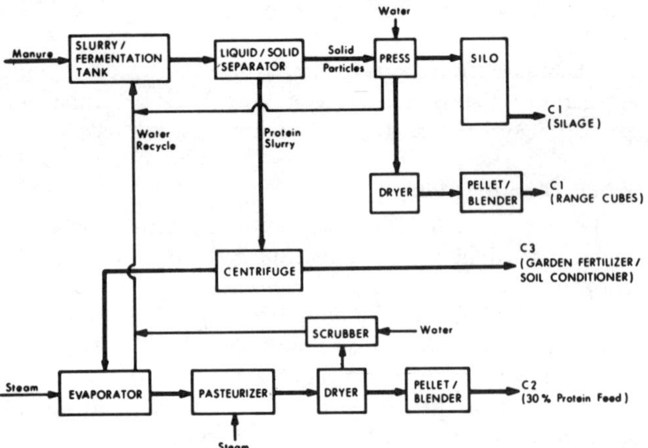

FIG. 1 Ceres Ecology Corporation manure fractionation process schematic.

## Fractionation with Full Recovery

Feed Cycling Company of Blyth, California (Senior 1974) has developed a system for extracting sand from beef feedlot manure and ending with a feed product (82 percent of the input) in a dry pelleted form containing about 20 percent crude protein, 39 percent cellulose and lignin and 14 percent ash. Senior estimates the value of this product in beef cattle rations at about $45 per ton with a total processing cost of about $27 per ton. The system entails a brine discharge into salt beds.

Ceres Ecology Corporation in cooperation with W. Brady Anthony and the Auburn Research Foundation have developed a system which is designed to produce different feeds for ruminate and monogastric animals (Fig. 1).† "CI" (20-40 percent of input D.M.B.) consists of the grain and fiber particles in manure. It is either fermented into a silage product for feeding feedlot cattle or dried, blended and pelleted for range cattle. It is worth $40-$50 per ton. "CII" (40 percent) is a dry pelleted product containing 27-30 percent crude protein, 4 percent fat and 25 percent ash. This product is fermented in the liquid phase to encourage production of "single cell protein." Its value varies from $70 per ton in cattle rations to over $100 in monogastric rations. "CIII" (20-40 percent) is a compost-like material suitable as a soil conditioner of about the same value as manure or $8.62 per ton (D.M.B.).

The value of products obtained from one ton of manure (D.M.B.) is thus between $45-65. Costs of production vary geographically but vary between $20-$35 per ton.

Hanilton Standard Corporation (Turk 1972) in conjunction with the Northern Regional Research Laboratory (U.S.D.A.) has proposed a joint methane-protein system. We do not have sufficient information to evaluate this proposal other than to observe that (as shown under Manure as a Fuel) methane production would increase costs of feed production over systems using other fuel sources and that these high costs of methane generation would not likely be overcome by cost saving and/or added product value in protein production.

General Electric Corporation is also working on a protein fermentation system about which we have little information (Anonymous 1972).

---

†Fig. 1 is included here because it has not been published elsewhere. The flow diagram has been simplified and will change according to various conditions.

## SUMMARY

Table 4 presents our overall assessment of the various techniques of manure utilization. The figures represent our best estimates from published sources and inferences based on standard engineering and economic estimators.

It is apparent that three variables strongly influence the viability of alternative waste utilization systems: (a) climatic conditions which determine the moisture content of manure and the stability of nutrients; (b) waste management systems (e.g., slatted floors vs. open lots) which largely determine ash content of manure; and (c) scale of enterprise which determines: (1) break even points; (2) environmental impact, and (3) the form feeds must be in for feed management requirements and marketability of products.

These considerations together with estimates surrounding Table 4 lead to the following strong conclusions: (a) fertilizer utilization of manure is attractive only at small scale; (b) the only attractive fuel utilization technique is direct combustion at medium to large scale where dry manure is available; (c) feed utilization techniques are generally superior to fertilizer and fuel uses; and (d) within feed utilization techniques the choice depends on scale.

We believe that these conclusions are not sensitive to future rises in the prices of petroleum (coal prices are not likely to rise dramatically over present levels): (a) the value of manure as a fertilizer will be directly affected by a rise in petroleum prices only with respect to N or 32 percent of total value; (b) direct combustion of manure as fuel will only compete with coal and other alternatives require very large increases in natural gas prices even to break even; (c) most of the energy used in manure for feed is for drying and coal-steam dryers can be used in most cases (DPW a possible exception); and (d) lastly, agricultural crops depend on petroleum energy—thus, as this cost rises so will the prices of feeds with which manure-derived feedstuffs compete.

Of course, this is a relatively new field in technology and one can confidently expect exciting new discoveries and applications over the years ahead. The strongest conclusion that emerges from this analysis is that the brief period of technological advance to date has definitely taken manure out of the waste category and into the category of an extremely valuable agricultural resource. Thus, any technology which depends on "disposal" subsidies is now obsolete.


### References

1 Anonymous. 1972. This plant will convert waste into protein. Feedlot Management 14:70-71. May.

2 Anonymous. 1974. Energy resources report. p. 419.

3 Anthony, W. B. 1971. Cattle manure as feed for cattle. Proceedings ISLW. P. 293-296.

4 Fairbank, W. C. 1974. Fuel from feces? The Dairyman. p. 8-11. May.

5 Graber, Richard. 1974. Agricultural animals and the environment feedlot waste management regional extension project. Oklahoma State University, Stillwater, Oklahoma.

6 Gross. D. 1975. Feed reclamation. Calf News. p. 36-37. February.

7 Halligan, J. E. and R. M. Sweazy. 1972. Themochemical evaluation of animal waste conversion processes. Paper presented to AICHE Annual Meeting.

8 Harmon, B. G., D. L. Day, A. H. Jensen and D. H. Baker. 1972. Nutritive value of aerobically sustained swine excrement, Journal of Animal Science 34:403-407.

| Utilization process | Percent manure used in non-fertilizer application | Process costs of primary product per ton, DMB | Manure gathering transportation and fertilizer residue disposal costs per ton, DMB | Total |
|---|---|---|---|---|
| I. Fertilizer | 0 | — | $3-$10 | $3-$10 |
| II. Fuel | | | | |
| A. Direct combustion | 70 | $6[†] | $5 | $11 |
| B. Methane | | | | |
| 1. on farm | 50 | $30 | $2 | $32 |
| 2. off farm | 58 | $5-$12.50[‖] | $13 | $18-$25.50 |
| C. Gas synthesis[#] | 50 | $12 | $2 | $14 |
| III. Feed | | | | |
| A. DPW | 100 | $42 | $3 | $45 |
| B. Wasteledge[**] | 100[††] | $5 | $2 | $7 |
| C. Fractionation with partial recovery | 40 | $5 | $2-$4 | $7-$9 |
| D. Fractionation with full recovery | 70-80 | $20-$35[‡‡] | $2-$4 | $22-$39 |

| Utilization process | Value of primary Products obtained from 1 ton manure, DMB | Value of fertilizer residue | Total | Net value per ton manure in use, DMB | Minimal size tons per year, (or head of beef cattle) | Market ability of primary product | Environmental Impact |
|---|---|---|---|---|---|---|---|
| I. Fertilizer | — | $9 | $9 | -$1 to $6 | — | poor-fair | poor-good |
| II. Fuel | | | | | | | |
| A. Direct combustion | $14 | $2.50 | $16.50 | $5.50 | 3 000[‡] | poor-fair | fair-good |
| B. Methane | | | | | | | |
| 1. on farm | $3[§] | $8 | $11 | -$21 | 100 | poor | fair |
| 2. off farm | $3 | $8 | $11 | -$9 to -$14.50 | 100 000 | good | poor |
| C. Gas synthesis[#] | $7 | $5 | $12 | -$2 | 200 000 | good | fair-good |
| III. Feed | | | | | | | |
| A. DPW | $50-$70 | — | $50-$70 | $5 to $25 | 2 200[§§] | fair-good | good |
| B. Wasteledge[**] | $40 | — | $40 | $33 | 720 | poor | good |
| C. Fractionation with partial recovery | $16 | $4 | $20 | $11 to $13 | 720 | poor | poor-good |
| D. Fractionation with full recovery | $40-$65 | $2 | $42-$67 | $3 to $45 | 10 000 | good | good |

[*]Poor at high concentrations—good at low.

[†]Gathering and transporting = $5. Boiler, scrubber $60 000 or $12 000 P.I. per year = $4/ton  $2/ton labor.

[‡]100 BHP. 24 hr/day = 80 x $10^6$ Btu or 10 tons manure per day. Economics of scale possible.

[§]Seven $ft^3$/lb solids fermented; 1000 lb/ton fermented or 700 $ft^3$ gas per ton at 500 Btu/$ft^3$ = 3.5 x $10^6$ Btu x $9/$10^6$ Btu = $3/ton plus fertilizer value $8 = $11/ton.

[‖]Published estimates state break even at between $1 (Anonymous 1974) and $2.50 (Varany 1974) per 1000 $ft^3$ of natural gas equivalent. This converts to $5 to $12.50 per ton of manure. These estimates do not appear to include either gathering manure and transportation to the plant nor disposal of sludge. Assuming $5 per ton for the former and $8 for the latter (20 percent solids) total costs per ton (DMB) amount to $18-$25.50 per ton.

[#]Only gas synthesis is included here as it appears to be the most promising of the synthetic fuel processes (Walawender et al. 1973).

[**]Assuming 120 ton airtight silo; 20 tons (DMB) manure, 10 day retention = 720 tons manure per year. Cost of silo about $18 000 = $3600/yr or $5/ton. Plus gathering and management (slatted floors) $2. Total = $7.

[††]However, only 25 percent of the manure produced is so utilized.

[‡‡]Senior (1974) and Seckler (1975).

[§§]50 000 layers.


9   Seckler, David. 1975. The cereco process. Ceres Ecology Corporation. Denver, Colorado.

10   Senior, F. C. 1974. The feed recycle process. Presented at Agricultural Waste Symposium, University of Saskatchewan, Regina, Canada. November.

11   Smith, L. W. 1974. Dehydrated poultry excreta as crude protein suppliment for ruminants. World Animal Review. No. 11. p. 6-11.

12   Turk, M. 1972. Production of power fuel by anaerobic digestion of feedlot waste. USDA CRIS No. 0022215. p. 1.

13   Varany, W. 1974. Personal correspondence on Zimpro process.

14   Walawender, W. P., L. T. Fan, C. R. Engler, L. E. Erickson. 1973. Feedlot manure and other agricultural wastes for future materials and energy resources. III. Economic Evaluations. Institute for Systems Design and Optimization, Kansas State University, Manhattan, Kansas. July.

# An Economic Analysis of Methane Generation Feasibility on Commercial Egg Farms

Thomas C. Slane, Robert L. Christensen, Cleve E. Willis, Robert G. Light
MEMBER
ASAE

THIS study involves an economic analysis of a poultry waste management system that includes a methane generation component. The system decomposes manure and forms methane gas which in turn serves as a supplement to the operation's present energy source.

A series of commercial egg producing operations were specified as benchmarks. Flock sizes considered in the analysis were 20,000, 40,000 and 80,000 layers each with a specific semi-automatic equipment system. The initial focus was placed on the effect of adoption of a waste management system upon the internal costs of the firm. The cost of final disposal of the liquid manure handling system was considered to be essentially the same regardless of the intermediate (methane generation) treatment received and hence was ignored in this analysis. A second phase involved the estimation of external benefits, resulting from innovations of this type and the determination of public subsidies.

Economic-engineering techniques were used to develop design parameters and the associated costs. The basic system satisfied the technical requirements for methane generation, although an actual generator does not exist. Several attempts to obtain data from operating systems were not productive and financial limitations precluded visiting actual sites. However, the authors feel that the system design and associated costs and output characteristics fairly represent the current state of the arts.

## DESCRIPTION OF AN INTERMEDIATE METHANE GENERATING WASTE TREATMENT STAGE

This section of the study involved determination of the costs associated with adopting a methane generating anaerobic digester. This system would decompose poultry wastes and form methane gas to be used as a fuel or power source supplement to the operation's present sources. (A more complete description of the hypothetical methane generating system is found in Slane et al. 1975.)

For this system, agitation within the digester was accomplished by means of a displacement effect. Further, heating the digester was accomplished by circulating heated water through pipes or coils placed within the digester. The engine-generator(s), budgeted as a standby power source(s) was assumed to run directly and continuously on the fuel gas produced. The electricity generated was then used to supply the electricity for the operation of the digester as well as the lighting and ventilating fan requirements of the production operation. Commercial electricity would be considered the secondary or standby power source. All other electric requirements of the production operation would be served by commercial electricity.

## COSTS OF THE INTERMEDIATE METHANE GENERATING WASTE TREATMENT STAGE

The fixed and variable costs associated with the intermediate methane generating waste treatment stage were estimated for the three benchmark operations by identifying the fixed and variable factors associated with the system, estimating input rates, and assigning prices to each unit of input. Annual total costs were derived by summing the various items of fixed and variable costs associated with each benchmark firm.

Fixed costs were defined as those remaining constant whether or not the system was in operation. These costs were further divided into digester and equipment costs. The parameters cited in Table 1 were used to estimate the digester volumes needed to handle the accumulation of slurry. Technical limitations influenced not only the design but the structural costs of the digester. Table 2 presents the total volume of slurry accumulated in the digester over a ten day detention time and the required digester dimensions for each benchmark operation. The annual fixed costs associated with the digester and equipment are also presented in Table 2 for each benchmark operation. (Slane et al. 1975, for further detail.)

Variable costs of the systems are those which vary with the volume or size of the digester. Due to the nature of this system and the fact that final disposal is exogenous in this study, the input categories of operating costs and tractor time required for handling and disposal of the manure need not be considered. Thus, the only variable input category considered was labor.

The labor input was further divided into three functions: (a) loading/unloading of the digester, which relies upon the capacity of the loading pump, (b) agitation of the digester contents, and (c) monitoring the system to insure the bio-

The authors are THOMAS C. SLANE, former Research Assistant (now a Graduate Assistant at Pennsylvania State University); CLEVE E. WILLIS, Assistant Professor; ROBERT L. CHRISTENSEN, Professor (on leave with Research Division, Economics Branch, Agriculture Canada) of Food and Resource Economics; and ROBERT G. LIGHT, Professor of Food and Agricultural Engineering, University of Massachusetts, Amherst.

Research supported by University of Massachusetts Agricultural Experiment Station.

TABLE 1.* GUIDELINES FOR MAXIMUM LOADING RATES OF HEATED MIXED ANAEROBIC DIGESTERS HELD AT 95 F (35 C) WHEN FED FRESH LAYER† MANURE

| Unit | Value |
|---|---|
| Weight of raw (wet) manure (lb day$^{-1}$ [1,000 lb animal]$^{-1}$) | 53 |
| Volume of raw (wet) manure (ft$^3$ day$^{-1}$ [1,00 lb animal]$^{-1}$) | 0.86 |
| Dilution as manure/(manure and water) | 1:8.3 |
| Estimated dilution water required (lb water [1,000 lb animal]$^{-1}$) | 387 |
| Hydraulic detention time (day) | 10 |
| Loading rate ‡ (lb VS day$^{-1}$ ft$^{-3}$) | 0.13 |
| Digester volume per animal unit (ft$^3$ [1,000 lb animal]$^{-1}$) | 72 |

*Criteria used in this table were avoidance of ammonia toxicity (total ammonia-N < 1,200 mg/liter), avoidance of excess production of volatile fatty acids (lb VS day$^{-1}$ ft$^{-3}$ < 0.37), and 7.9 percent solid material.
†Assumes a layer to be approximately 5 lb in weight.
‡Assumes 9.4 lb VS day$^{-1}$ (1,000 lb animal)$^{-1}$.



MANAGING LIVESTOCK WASTES

TABLE 2. TOTAL VOLUME OF SLURRY ACCUMULATION
AND REQUIRED DIGESTER VOLUME, WITH DIMENSIONAL
BREAKDOWN, TO HANDLE SLURRY ACCUMULATIONS, AND
ANNUAL FIXED COSTS ASSOCIATED WITH THE DIGESTER
AND EQUIPMENT FOR EACH BENCHMARK OPERATION

|  | 20,000 layers | 40,000 layers | 80,000 layers |
|---|---|---|---|
| Volume of slurry accumulated in a 10 day detention time (cubic foot or feet$^3$ | 7,195.104 | 14,390.208 | 28,780.416 |
| Required digester volume to handle slurry accumulations (feet$^3$) | 7,288.615 | 14,577.231 | 29,154.462 |
| Dimensional breakdown  L  W  D (ft  x ft x ft) | 23 x 20 x 12 (7,404.96) | 53 x 20 x 12 (14,604.96) | 114 x 20 x 12 (29,244.96) |
| Digester costs, dollars | 1,760 | 3,070 | 5,740 |
| Equipment costs, dollars | 965 | 1,460 | 2,430 |
| Total annual fixed cost, dollars | 2,725 | 4,530 | 8,170 |

logical and mechanical processes are functioning properly. The annual hours of labor and dollar costs associated with these functions for each benchmark operation are presented in Table 3.

TABLE 3. ANNUAL HOURS OF LABOR AND COSTS (IN
DOLLARS) ASSOCIATED WITH VARIOUS FUNCTIONS OF THE
INTERMEDIATE METHANE GENERATING WASTE
TREATMENT STAGE FOR EACH BENCHMARK OPERATION

| Labor function | 20,000 layers | 40,000 layers | 80,000 layers |
|---|---|---|---|
| Loading/unloading* of the digester | 260 (780) | 550 (1,650) | 1,095 (3,285) |
| Agitation of digester contents | 625 (1,875) | 915 (2,745) | 1,460 (4,380) |
| Monitoring the system | 365 (1,095) | 365 (1,095) | 365 (1,095) |
| Total annual hours of labor | 1,250 (3,750) | 1,830 (5,490) | 2,920 (8,760) |

*Assumes capacity of centrifugal pump to be 7,200 gallons per hour and labor costs at $3.00 per hour.

Annual total cost associated with the methane generating stage for each benchmark operation was then calculated by combining the total fixed and total variable costs of each operation as shown in Table 4. However, since the methane is used to generate electricity for the production operation's lighting and ventilating requirements and the system's internal use, a credit or cost reduction is created.

TABLE 4. ANNUAL TOTAL COST ASSOCIATED WITH AN
INTERMEDIATE METHANE GENERATING WASTE
TREATMENT STAGE FOR EACH BENCHMARK OPERATION
(IN DOLLARS)

|  | Costs | | |
|---|---|---|---|
|  | 20,000 layers | 40,000 layers | 80,000 layers |
| Total annual fixed cost | 2,725 | 4,530 | 8,170 |
| Total annual variable cost | 3,750 | 5,490 | 8,760 |
| Annual total cost | 6,475 | 10,020 | 16,930 |

TABLE 5. DAILY AND ANNUAL ELECTRICITY
REQUIREMENTS FOR EACH OF THE BENCHMARK EGG
PRODUCTION OPERATIONS

|  | Kilowatt-hours used | | |
|---|---|---|---|
|  | 20,000 layers | 40,000 layers | 80,000 layers |
| Lights | 68 | 136 | 272 |
| Fans | 120 | 240 | 480 |
| Automatic feeders and augers | 20 | 40 | 80 |
| Egg coolers | 20 | 40 | 80 |
| Egg collectors | 18 | 36 | 72 |
| Water pumps | 18 | 36 | 72 |
| Miscellaneous tools, augers, etc. | 4 | 8 | 16 |
| Annual kilowatt-hour usage | 97,820 | 195,640 | 391,280 |

Table 5 presents the daily and annual electricity requirements for each of the benchmark production operations. The credit or cost reduction, based on the commercial rate of 2.3 cents per kilowatt-hour is presented in Table 6 for each benchmark operation. (This was the rate prevailing in the Spring of 1974.) The cost reduction was then subtracted from the respective annual total cost associated with the methane generating stage to yield the annual net total cost (or return) that can be expected for each benchmark operation. These costs are presented in Table 7 for each benchmark operation.

TABLE 6. COST REDUCTIONS ACCRUED TO EACH
BENCHMARK EGG PRODUCTION OPERATION USING
METHANE-GENERATED ELECTRICITY TO FULFILL
LIGHTING AND VENTILATING ELECTRICITY
REQUIREMENTS

|  | Cost reduction, dollars | | |
|---|---|---|---|
|  | 20,000 layers | 40,000 layers | 80,000 layers |
| Lighting | 570 | 1,140 | 2,280 |
| Ventilating | 1,005 | 2,015 | 4,030 |
| Annual cost reduction | 1,575 | 3,155 | 6,310 |

TABLE 7. ANNUAL TOTAL COST, INCLUDING COST
REDUCTIONS ACCRUED, ASSOCIATED WITH AN
INTERMEDIATE METHANE GENERATING WASTE
TREATMENT STAGE FOR EACH BENCHMARK OPERATION
(IN DOLLARS)

|  | Costs | | |
|---|---|---|---|
|  | 20,000 layers | 40,000 layers | 80,000 layers |
| Gross annual total cost | 6,475 | 10,020 | 16,930 |
| Cost reduction accrued by each benchmark | 1,575 | 3,155 | 6,310 |
| Net annual total cost | 4,900 | 6,865 | 10,620 |
| Present value of cost stream* | 41,716 | 58,308 | 90,414 |

*Assumed 20 year life and 10 percent discount rate.

The data in Table 7 indicate net increases in annual total costs associated with adoption of a methane generating system. These cost increases range from about \$5,000 for the 20,000 layer flock to \$10,600 for the 80,000 layer flock size. The data do indicate some economies of size.

Clearly, the value of the annual cost reductions is directly related to the cost of commercial electricity. In this analysis a price of 2.3 cents per kilowatt-hour was assumed. However, with continuing energy shortages and rising rates for commercial electricity, higher rates may soon be in evidence. Breakeven prices for commercial electricity were computed for each size and are as follows: 20,000 layers — 9.44 cents/kW-hr; 40,000 layers — 7.30 cents/kW-hr; 80,000 layers — 6.17 cents/kW-hr.

Commercial rates currently approach the 6 cents per kilowatt-hour level. As these rates rise and the technology and yield associated with methane production from animal wastes become more efficient, it may shortly become an economically feasible alternative.

## LEARNING AND EXTERNAL BENEFITS

According to the foregoing analysis, costs of commercially constructed methane generating waste treatment systems exceed expectations of returns from the methane generated. That is, although the process is technically feasible, it is not economically feasible. However, if the accumulation of construction and operating experience produces dividends in the form of future cost reductions, the use of public funding to stimulate such experience may be justified, particularly if alternative power sources should exhibit substantial price increases. The provision of a public subsidy can be based on the value of the accrued learning. Then, if the experience (learning) is made available on a broad basis, subsequent external cost reductions become externalities which ought to be internalized from an economic efficiency standpoint. The magnitude of these external benefits depends on the slope of the relevant learning function (Slane et al. 1975, for a fuller discussion of the learning function phenomenon), amount of experience, and projected future production which can benefit from this cost reduction.

### External Benefits

Since learning derived from the experience of a methane production operation directly alters the production functions of succeeding operations elsewhere, an externality is involved (Mishan 1971). In evaluating the external learning benefits to be gained from the potential operation, the present value of projected cost reductions is considered as the measure of benefits. If this estimated value ($V_1$) exceeds the value of the public transfer necessary to induce the poultryman to adopt a methane generating system ($V_2$), then societal net benefits would justify a subsidy less than $V_1$ but greater than $V_2$.

External benefits (B) are, then:

$$B = 6{,}570 \sum_{t=p}^{p+T} \lambda^t \left[ a \left[\!\left[ K(1+g)^t \right]\!\right]^b - a \left[\!\left[ K(1+g)^t + \Delta \right]\!\right]^b \right] \left[ M(1+g)^t \right] \quad \dots \dots [1]$$

where multiplication by 6,570 presumes the system is on-line an average of 328.5 days per year (90 percent) and this magnified by the useful digester life of 20 years, produces a constant of 6,570, and

- a  is the intercept of the learning function,
- b  is the slope of the learning function,
- $\Delta$  is the capacity of the proposed digester in cubic feet per day,
- M  is initial (current) industry-wide daily capacity additions subject to learning,
- g  denotes expected growth rate in new capacity,
- p  denotes the number of years by which the realization of benefits from learning is presumed to lag behind the encouragement and provision of experience,
- T  is the number of years during which the incremental value of the learning is positive, and industry-wide capacity is projected by $K(1+g)^t$, where K is 1974 experience.

### Empirical Results

As mentioned previously, the proxy for experience is cumulative digester capacity. It is unlikely, however, that learning increases proportionally with digester size and there are indications that suggest that after some point further size increases provide no incremental learning. Accordingly, the case may also be examined in which experience is proportional to capacity up to a 2,000 cf/d size but no additional experience is provided by plants larger than this.

Estimations of B for selected values of the parameters in Equation [1] are presented in Table 8. The discount rates (r) assumed are 0.10 and 0.15. The expected additions to new capacity subject to learning (M) was 2,500 cf/d. A lower limit of 1,000 cf/d was also examined. Likewise, while the estimates of b for the cumulative capacity proxy and the modified cumulative capacity proxy (2,000 cf/d limit to learning from a single operation) were -0.4215 and -0.3883, respectively, their counterpart lower limits were presumed to be -0.3793 and -0.3494. (These estimations are based on very limited data. A more complete treatment of this process is contained in Slane et al. 1975).

Using the ten percent discount rate, construction and operation of the largest digester is estimated to generate a stream of external benefits of roughly 2.7 million dollars. This could be as low as approximately 600 thousand dollars under the least favorable assumptions regarding the parameters b and M. Assuming a fifteen percent discount rate, the range becomes 300 thousand to 1.4 million dollars. For the smaller ($\Delta = 1,000$ and 200) operations these benefits are, of course, somewhat less. For the most optimistic values of b and M, the estimated benefits are even greater.

## CONCLUSIONS

In the first section we suggested that the benchmark poultry operations would incur discounted net additional costs of roughly 42 to 90 thousand dollars and that it would likely require a public transfer in the range of this amount* to induce a poultryman to include methane generation in his operation. We further pointed out that such a transfer may be in order if external learning benefits were valued in excess of this range. The estimates of a measure of

---

*This presumes that experience by this time will have lowered costs sufficiently close to alternative energy prices so that poultrymen will choose methane generation without further subsidy. The authors are currently investigating a fuller set of alternative assumptions.

| Experience proxy | Size of unit ($\Delta$) | Learning parameter (b) | Discount rate (r) and additions to new capacity (M) | | | |
| | | | 0.10 | | 0.15 | |
| | | | 1,000 | 2,500 | 1,000 | 2,500 |
|---|---|---|---|---|---|---|
| Cumulative capacity | 2,000 | -0.3793 | 0.5823 | 1.4558 | 0.3069 | 0.7674 |
| | | -0.4215 | 1.0869 | 2.7173 | 0.5645 | 1.4114 |
| | 1,000 | -0.3793 | 0.5656 | 1.4141 | 0.3000 | 0.7501 |
| | | -0.4215 | 1.0503 | 2.6258 | 0.5494 | 1.3735 |
| | 200 | -0.3793 | 0.5011 | 1.2529 | 0.2728 | 0.6822 |
| | | -0.4215 | 0.9156 | 2.2891 | 0.4927 | 1.2317 |
| Modified cumulative capacity | 2,000 | -0.3494 | 0.6098 | 1.5247 | 0.3174 | 0.7935 |
| | | -0.3883 | 1.0802 | 2.7005 | 0.5545 | 1.3864 |
| | 1,000 | -0.3494 | 0.5897 | 1.4744 | 0.3090 | 0.7727 |
| | | -0.3883 | 1.0390 | 2.5975 | 0.5375 | 1.3438 |
| | 200 | -0.3494 | 0.5153 | 1.2882 | 0.2777 | 0.6943 |
| | | -0.3883 | 0.8933 | 2.2332 | 0.4762 | 1.1905 |

external benefits, even under the most unfavorable assumptions regarding the presumed values of the parameters in Equation [1], are extremely high in comparison with the required subsidy and indicate such transfers would seem to be well motivated. Note that since all benchmarks examined would produce more than 2,000 cf/d of methane, all comparisons should be made with the estimated benefits from the 2,000 cf/d size ($\Delta$) operation. The modified experience proxy, at least, presumes no additional learning (benefits) above this level.

To be sure, the analysis has been performed on the basis of currently available (and limited) data. Since the estimates of external benefits B obviously depend critically on the estimates of the learning parameter b (and other values), the operational use of this measure in such public decisions requires the estimates of b (and the other parameters) to be modified as additional information unfolds.

Finally, a factor leading to a possible overstatement of benefits is inherent in the learning function formulation. When cost reductions are related to a measure of experience, a part of those cost reductions must be attributed to non-methane generation research and development, such as improved pumps, new digester technology, microbiology technology, new displacement technology, etc. To some extent this may be offset by the omission of by-product benefits or new technologies made in the methane generating industry that are disseminated outside the industry.

### References

1    Mishan, E. J. 1971. The postwar literature on externalities: an interpretative essay. Journal of Economic Literature 9(1):1-28.
2    Slane, T. C. et al. 1975. An economic analysis of methane generation: internal costs and external benefits. Experiment Station Bulletin 618. University of Massachusetts.

# Economics of Substitution and the Demand for Beef Feedlot Wastes: One Alternative for Solving Environmental Quality Problems

Daniel D. Badger

FOUR years ago, many accusations were being made concerning the large accumulations of feedlot wastes by large beef feedlot operations in the southwestern United States. Manufactured fertilizer was abundant at that time, and prices were relatively low, particularly for nitrogen. Also at that time, national agricultural programs required millions of acres of land to be in set-aside programs. It became increasingly difficult to convince farmers to buy and use manure on their cropland. Many feedlot operators were unable to move the "mountains" of solid wastes being generated, although the selling price per ton and transportation costs were nominal. Many feedlot operators were giving the manure away (Howell 1970) (Badger and Cross 1971).

Probably the foremost problem with beef feedlot wastes in Texas and Oklahoma related to esthetics i.e., the visual and smell senses of highway travelers and other visitors to the area. Feedlot wastes caused very little actual pollution of streams; some fly and other insect problems were prevalent immediately after rainy periods.

## CHANGES IN FERTILIZER PRICES FROM RECENT PAST

Mathers and Stewart (1971) stated: "Manure was once a valued fertilizer. Today, many people consider it strictly as waste to be disposed of. This change is due to increased technology, which has lowered the cost of chemical fertilizers and rising labor costs that have made manure handling more expensive." At that time, the cost of a pound of Nitrogen applied to cropland was about $0.08. In the spring of 1975 the cost of a pound of Nitrogen in the form of Ammonium Nitrate is $0.25, a cost increase of 300 percent to the farmer top-dressing wheat (Table 1).

**TABLE 1. AVERAGE PRICES PER TON AND PER POUND OF NITROGEN PAID BY FARMERS FOR FERTILIZER MATERIALS FOR PERIOD 1970-74 (Prices on September 15)**

| | Oklahoma | | | | Texas | | | |
|---|---|---|---|---|---|---|---|---|
| | Ammonium Nitrate* | | Anhydrous Ammonia† | | Ammonium Nitrate* | | Anhydrous Ammonia† | |
| Year | Ton | Lb. N. | Ton | Lb. N. | Ton | Lb. N. | Ton | Lb. N. |
| 1970 | $ 55 | $0.08 | $ 72 | $0.04 | $ 65 | $0.10 | $ 76 | $0.05 |
| 1971 | 57 | 0.09 | 70 | 0.04 | 68 | 0.10 | 77 | 0.05 |
| 1972 | 59 | 0.09 | 70 | 0.04 | 67 | 0.10 | 77 | 0.05 |
| 1973 | 75 | 0.11 | 89 | 0.05 | 78 | 0.12 | 84 | 0.05 |
| 1974 | 170 | 0.25 | 220 | 0.13 | 175 | 0.26 | 235 | 0.14 |

*Ammonium Nitrate is 33.5 percent N.
†Anhydrous Ammonia is 82 percent N.
Source: <u>Agricultural Prices</u>, Crop Reporting Board, SRS, USDA, Washington, D.C. (September 15 issue for each year cited)

Article approved as Journal Article J-2983 of the Oklahoma Agricultural Experiment Station.
The author is: DANIEL D. BADGER, Agricultural Economics Dept., Oklahoma State University, Stillwater.

The chemical fertilizer-manure situation has changed dramatically since 1971. The government farm program has been changed and crop producers are being encouraged to plant all acres that previously were in set-aside programs. Many marginal acres have come back into production; these acres can be improved in productivity by generous applications of beef feedlot and other animal wastes.

Concurrently, due to the increased demand for all types of fertilizer to bring 40 million acres of land back into production, and due to price controls imposed on domestic fertilizer prices in 1972 and early 1973, fertilizer shortages began to occur. Overseas markets also bid higher prices, resulting in increased exports of fertilizer during this period. Farmers increasingly looked to alternative nutrient sources for their crop lands, and the large mountains of beef feedlot wastes that had been accumulating for several years began to disappear.

After the government price controls were lifted in 1973, fertilizer prices skyrocketed, increasing 150 to 200 percent, or more. As the manufactured fertilizer prices have increased and with supplies of fertilizer still limited, cropland farmers have been willing once again to pay for beef feedlot wastes, as well as to pay higher transportation costs to have these wastes spread on their land.

In the fall of 1974, a survey of beef cattle feedlot operators, farmers and commercial manure handlers was taken in the Oklahoma and Texas panhandle to determine the supply and demand situation for beef feedlot wastes and resulting environmental quality implications. Indications are that recent economic and other events have combined to eliminate much of the livestock waste solids problem in the southwestern beef feeding states, as related to both state and federal environmental quality requlations.

## ECONOMIC VALUE OF MANURE FROM TEXAS-OKLAHOMA FEEDLOTS

It is almost impossible to derive "average" nutrient values for N, $P_2O_5$ and $K_2O$ for beef feedlot wastes in the Texas-Oklahoma area. According to Tucker, Burton and Baker (1972), Oklahoma researchers used, as a rule of thumb, nutrient values for beef cattle wastes as follows: 0.5 percent N, 0.25 percent $P_2O_5$, and 0.5 percent $K_2O$; or in pounds per ton of waste, 10 pounds of N, 5 pounds of $P_2O_5$ and 10 pounds of $K_2O$. However, dramatic changes in beef ration formulations for confined feeding operations occurred in the early and mid-1960's. These changes have resulted in increased nutrient values in beef manure. While it is still difficult to derive an average, due to climatic conditions (rain and temperature), season of year and type of ration, typical nutrient values for beef solid wastes in the Texas-Oklahoma area appear to be: 1.5 percent for N; 1.0 percent for $P_2O_5$; and 1.5 percent for $K_2O$.

In addition to the big-three nutrients above, beef feedlot wastes also contain other valuable nutrients essential for crop and pasture production such as sulfur, boron, zinc, copper, manganese and iron. Based on early 1975 prices of purchased chemical inputs, the equivalent economic value of a ton of beef feedlot wastes is $14.80 (Table 2). Informed observers estimate that only 50 percent of major nutrients ($N$, $P_2O_5$, $K_2O$) are available in first year of application and one half of remainder is available in subsequent years. Thus 50 percent of $13.90 = $6.95 + $0.90 = $7.85 is a closer estimate of the immediate economic value of a ton of beef feedlot wastes. After adjusting for the loss of one-half of the remaining primary nutrients, the economic value of that same ton of beef feedlot wastes in succeeding years is an additional $3.48. Thus, total economic value of the wastes is $7.85 + $3.48 = $11.33 per ton.

TABLE 2. ECONOMIC VALUE OF NUTRIENTS PER TON OF BEEF FEEDLOT WASTES FOR A SELECTED SAMPLE, OKLAHOMA–TEXAS AREA, SPRING 1975

| Nutrient | Percent composition | Lbs. per ton | Value of nutrient per lb. | Total value of nutrient in ton* |
|---|---|---|---|---|
| Nitrogen | 1.500 | 30.0 | $0.25 | $ 7.50 |
| Phosphate | 1.000 | 20.0 | 0.20 | 4.00 |
| Potash | 1.500 | 30.0 | 0.08 | 2.40 |
| Zinc | 0.005 | 0.1 | 0.50 | 0.05 |
| Manganese | 0.020 | 0.4 | 0.50 | 0.20 |
| Iron | 0.005 | 0.1 | 0.50 | 0.05 |
| Sulfur | 0.250 | 5.0 | 0.10 | 0.50 |
| Boron | 0.002 | 0.04 | 0.50 | 0.02 |
| Copper | 0.004 | 0.08 | 1.00 | 0.08 |
| Total Nutrient Value Per Ton | | | | $14.80 |

*This implies all nutrients become available for plant growth immediately upon application. This point is explained in the text.

Cattle and calves on feed in Texas and Oklahoma have declined by 938 000 head or by 37 percent from the first quarter of 1974 to the first quarter of 1975. Oklahoma numbers declined from 292 000 to 232 000 and Texas numbers declined from 2 205 000 to 1 327 000 to 1 327 000 head. Essentially, this decline, which reversed a several year upward trend, results from low fed cattle prices and high feed grain prices. Despite the bleak outlook, an estimated 1.6 million head of cattle are in Texas and Oklahoma feedlots at the present time. It seems reasonable to assume that unless national economic conditions become even worse than at present, at least another 1.6 million head of cattle will be fattened in the Texas-Oklahoma area in calendar year 1975, for a total of 3.2 million.

The consensus of Oklahoma-Texas feedlot operators and commercial manure handlers in the fall 1974 survey is that a beef animal will generate one ton of solid wastes during a typical feeding period. Using the very low 1975 estimate of 3.2 million head fed in the Oklahoma-Texas area, approximately 3.2 million tons of beef feedlot wastes will need to be moved from these feedlots and applied on land. Based on the generally recommended rate of 10 tons of manure applied per acre, 320 000 acres of cropland and pasture land near these feedlots could be fertilized. The total economic value of this quantity of manure, based on current market prices for both primary and secondary nutrients is 3 200 000 tons x $11.33 = $36 256 000 for the Oklahoma-Texas area.

Some will argue that this value is over inflated since many of the soils in the Texas-Oklahoma cattle feeding areas do not need these quantities of N, $P_2O_5$ or K applied per acre, if the beef wastes application rate is 10 tons per acre. The author does not wish to debate the vagaries associated with beef wastes nutrient analysis, loss of nutrients by evapotranspiration, loss of nutrients from root zone, and unneeded applications of selected nutrients based on soil analyses. These soils do need, and can effectively utilize, increasing quantities of these nutrients, if crop and pasture yields are to be increased.

It is pertinent that the survey indicated prevailing cost for beef feedlot wastes of $2.50 per ton applied on the land within a 10 mile radius of the feedlot. For lands farther away than 10 miles, the rate varies from $0.05 to $0.07 per ton mile. Thus for those crop and pasture lands as far away as 35 miles, just the first year economic value of the N in a ton of beef feedlot wastes — one-half the value indicated in Table 2 — would justify the $3.75 per ton cost ($2.50 + 25 additional miles X $0.05 = $2.50 + $1.25).

Recently, more and more cattle being slaughtered are coming directly off pastures, rather than being fattened for extended periods in feedlots. One implication of this trend is the need to increase the fertility of both native and improved pastures, and small grain fields used for winter grazing. Thus depressed cattle prices, high feed grain prices, and high chemical fertilizer prices have opened up millions of "new" acres for sizeable applications of beef wastes from those lots continuing to fatten cattle in feedlots in the Texas-Oklahoma area. By again selling the wastes, beef feedlot operators still fattening cattle are able to recoup a higher proportion of their solid wastes handling costs, which provides at least a little relief on the cost side of the price-cost squeeze they are facing.

An interesting aspect of beef feedlot waste handling has become more prevalent in the last three or four years. This is the emergence of at least 20 commercial feedlot waste haulers in the Texas-Oklahoma confined beef feeding area. Three of these were interviewed to get their concept of waste handling costs and manure value. One method involves the feedlot operator scraping the pens and stockpiling the manure. The commercial hauler loads his own trucks from the stockpile. In this situation, the feedlot operator is selling the manure at $1.00 to $1.25 per ton to the hauler. Twelve of the 24 feedlots surveyed (ranging from 10 000 to 100 000 head capacity) sold all of their manure in this manner. A second method is where the feedlot operator contracts with a commercial manure hauler to rip and load the manure using the hauler's equipment completely. All types of pricing arrangements exist with this method; three feedlot operators charged nothing and paid nothing to have the manure moved, whereas two feedlot operators received $0.50 to $0.75 per ton from the commercial hauler to have their pens cleaned in this manner. This is a distinct change from 1971 when some feedlot operators paid $0.40 to $0.50 per ton to have their pens cleaned and the manure hauled away.

## SOME ADDITIONAL BENEFITS OF BEEF FEEDLOT WASTES

Most of the 24 feedlot operators and two of the farmers interviewed commented on advantages of using manure at the rate of 10 tons per acre on their fields in addition to its substitution role for commercial fertilizer. About one-half of the feedlot operators used some of the wastes on their own fields. In addition to increasing yields, consensus comments were: micro-nutrients or trace elements in manure improve appearance of crop and pastures (look healthier, greener, etc.); improves moisture intake and water holding capacity of sandy and loam type soils; improves texture or tilth of soil after a short period of use; improves appearance or "richness" of light colored tight soil typical in parts of Panhandle areas.

Most of the farmers commented that with the relatively inexpensive commercial fertilizer prices of the late 1960's and early 1970's, they had taken the easy way out and gone to the use of purchased chemicals. Many had tended to forget the side benefits of manure, and did not realize the relative ease of application and incorporation into the soil with new, larger equipment. In response to the question, "Will you continue to use manure no matter what happens to commercial fertilizer prices?", 8 of 10 farmers and all 12 of the feedlot operators with land adjacent to their feedlots responded yes.

### SOME PROBLEMS ASSOCIATED WITH THE USE OF MANURE

Some of the farmers indicated problems in using manure. However, none of them indicated the problems were major, or would prevent them from continuing to use manure. Problems listed were: compaction of soil due to large, heavy trucks which deliver and spread manure; slower rate of application compared to commercial fertilizer application; need to work the field 1 to 2 times more to incorporate manure into the soil; unevenness of application or "skip" problems; some imbalance in ratio of nutrients applied relative to soil needs; introduction of new weeds, and/or weed seeds; and, salts in manure possibly causing some soil damage. The salt problem is being slowly eliminated as more feedlot operators change to salt free rations.

### ENVIRONMENTAL QUALITY IMPACTS OF FEEDLOT WASTE APPLICATION

Indications are that in the future beef feedlot operators will have fewer problems with federal and state environmental agencies concerning odor, insect and health related issues, and fewer nuisance doctrine lawsuites are likely concerning the solid wastes aspects of beef feeding operations. This is a direct result of the proper disposal via land application of the large quantities of beef animal wastes in concentrated areas of beef feeding operations, such as in the Texas-Oklahoma panhandle area. State and Federal agencies have not slackened their efforts to implement new federal environmental guidelines which affect feeding operations of 1000 head capacity or larger (Environmental Protection Agency 1974). Significant capital outlays have been expended by beef feedlots in recent years to control runoff.

### SUMMARY

Fertilizer prices have increased 200 to 300 percent in the last three years. Consequently, farmers have turned increasingly to beef feedlot wastes as a valuable source of plant nutrients. Beef feedlot operators had a difficult time disposing of these wastes in the late 1960's and early 1970's. Some even reported giving away the manure just to reduce the environmental and esthetic problems. Most were losing money handling the manure.

Now farmers in the survey are convinced that the nutrient value of beef feedlot wastes is sufficiently high to pay the current rate of $2.50 to $3.50 per ton of beef feedlot waste applied to their land. The 1975 economic value of beef feedlot wastes produced in the Texas-Oklahoma feeding area is likely to exceed $36 million.

Increasing quantities of beef feedlot wastes in dryer climate zones, such as the Southwest will likely be applied to both irrigated and dryland crops and pastures. The price structure of chemical fertilizers and transportation costs will determine if these wastes will be transported farther than the current 15-20 mile radius of beef feeding feedlots.


**References**

1    Badger, D. D. and G. R. Cross. 1971. Economic implications of environmental quality legislation for confined animal feeding operations. Livestock Waste Management and Pollution Abatement. Proceedings 1971 ISLW. ASAE, St. Joseph, Mich. 49085. pp 204-206.

2    Environmental Protection Agency. 1974. Feedlots point source category: effluent guidelines and standards. Federal Register. Vol 39. No. 32 Part II. Washington, D.C. February 14.

3    Howell, D. 1970. Wastes increase at cattle yards. Lubbock-Avalanche Journal. February 7.

4    Mathers, A. C. and B. A. Stewart. 1971. Crop production and soil analyses as affected by applications of cattle feedlot waste. Livestock Waste Management and Pollution Abatement. Proceedings 1971 ISLW. ASAE, St. Joseph, Mich. 19085. pp 229-231.

5    Tucker, B. B., C. H. Burton and J. M. Baker. 1972. The use and value of animal waste as fertilizer for crop production. Extension Circular E-815, Oklahoma State University. Stillwater, OK. 6 p.

# Economic Research Pertaining to Problems of Livestock Waste Management and Pollution Control


L. J. Connor, J. B. Johnson


CONSIDERABLE economic research pertaining to livestock waste management and pollution control has been conducted. This paper describes the state of the art in the economic analyses of such problems. Major economic research findings are categorized and summarized. An appraisal of this economic research in terms of conclusions which can be drawn and research voids is presented. Suggestions for future economic research are put forth.

## REVIEW OF RESEARCH

Economic research on these problem issues is summarized in seven categories. No attempt was made to summarize all research conducted in each of the categories; emphasis is on selected studies indicative of problems addressed in each category.

### Industry Structure

A major problem which has confronted economists in researching problems of waste management and pollution control has been inadequate data on the structure of the livestock industries. Secondary data typically presents the distribution of livestock enterprises by herd size or output levels. To research the aggregate economic impacts of specified pollution control measures, other structural data were needed on the distribution of production technologies and waste systems by enterprise size.

State-level studies were conducted to determine the structure of specified livestock industries in Illinois and Michigan (Gamble 1972 and Hoglund et al. 1972). Data obtained on the distribution of livestock enterprise by size, production technology, and waste management practices provided the basis for related studies that assessed the economic impacts of selected pollution control rules.

On a national scale, several studies were conducted to determine the structure of the livestock industries. Estimates of the structure of the beef feeding industry were made from information obtained from university, Federal and State agency personnel knowledgeable about the cattle feeding industry in their respective States (Johnson and Davis 1974). A distribution of feedlots by size and housing technology was obtained along with the estimated number of feedlots with surface water control problems or the potential for runoff problems. The study indicated that the majority of the surface water control problems are concentrated on the smaller feedlots (less than 1000 head capacity).

Similar techniques were used to estimate industry struc-

ture characteristics for hog production (Van Arsdall and Smith 1974). Nearly 22 percent of the hog producers in fifteen major hog producing States were estimated to have runoff problems requiring additional control. In 1969, these producers produced nearly a third of the total output in these fifteen States.

In a related study by Buxton and Ziegler (1974), survey data taken by the National Milk Producers Federation were used to identify the percentage of diary farms with potential runoff problems. Although this sample was not representative of all United States milk producers, some 38 percent of the sample respondents were expected to have to construct runoff control facilities.

### Nonmarket Control Measures

The usual progression in methods of solution to environmental quality problems is from more to less reliance on market forces (Randall 1972). An array of alternative means have been considered to effectuate the control of pollution originating from livestock production. The most prevalent means are collective actions, generally far removed from market forces, which can be initiated by local, State and Federal governments. Such collective actions avoid in part the shortcomings associated with other types of action. Since no single collective action may be expected to minimize pollution from livestock wastes and maintain an economic climate in which livestock producers may effectively operate, it has been argued that a variety of collective actions might be pursued (Johnson and Connor 1971). Other alternative forms of action include: (a) voluntary actions initiated by industry groups, (b) individual legal actions initiated by those personally damaged, and (c) legal actions initiated by those not personally damaged (Johnson and Connor 1971).

A 1971 survey was made of major beef feeding and dairy States to determine the types of actions applicable to the management of livestock wastes (Johnson et al. 1972). A variety of collective actions were found to be in existence at the State level. Survey results indicated that all the states had statute provisions, or rules derived therefrom, that focused on the surface water control problems at livestock production facilities. One-third of the States had statutes or rules applicable to the control of groundwater pollution. Only a few States had specific statutes or rules applicable to problems of water pollution associated with the field application of livestock wastes.

In 1973, the United States Environmental Protection Agency (EPA) presented a report proposing effluent limitations guidelines for the feedlots point source industrial discharge category (United States Environmental Protection Agency 1973). This report provided the basis for EPA to promulgate performance standards and limitations which require "no discharge" of pollutants to navigable waters


Michigan Agricultural Experiment Station Journal Article No. 7167.



The authors are: L. J. CONNOR, Professor, Agricultural Economics Dept., Michigan State University, East Lansing; and J. B. JOHNSON, Agricultural Economist, Economic Research Service, USDA, East Lansing, MI.

except that in overflow caused by excessive chronic or catastrophic precipitation from control systems designed and
operated to control runoff from usual storms and runoff
from the 10-year, 24-hour or 25-year, 24-hour storms.

The question of segmenting the feedlots category on a
size criterion (the minimum capacity cut-off) has remained
an open question. EPA gave consideration to further seg-
mentation of the livestock industries in its 1974 announcement of final guidelines for feedlots with one-time capacities of 1000 animal units or more. A subcommittee of the
Committee on Government Operations of the House held
hearings "to attempt to ascertain whether Federal pollution
control laws are being administered efficiently, economically, and so implemented as to control point sources such
as animal feedlots" (Anonymous 1973). In part, these hearings dealt with the proposed EPA guidelines reissued in
May, 1973, which exempted all feedlots of less than 1000
animal units. The subcommittee asserted in the 1973 hearing that such an exemption was not only contrary to the
specific language of the Federal Water Pollution Control
Amendments of 1972, but also not rationally related to the
statutory objective (Anonymous 1973).

### Economic Impact Studies of Nonmarket Control Measures

A number of studies have been conducted to determine
the firm-level and aggregate economic impacts of imposing
selected nonmarket control measures upon livestock firms.
Economic impacts considered have been the increases in
capital requirements and changes in annual production
costs for individual livestock firms, and the resulting implications such as changes in livestock commodity supplies
and consumer prices. With one exception, these studies
have generally employed a static analysis (ignored the time
dimensions of adjustments).

One study analyzed the economic impacts of applying
control measures on Michigan dairy farms for the (a) mandatory control of surface runoff from the production sites,
(b) prohibition of winter spreading of dairy wastes, and (c)
mandatory subsurface disposal of dairy wastes (subsurface
injection of liquid dairy wastes or immediate plowdown of
solid dairy wastes). Study results indicated that all the control measures would necessitate additional capital and increase annual production costs, but none would generate
additional revenue (Good et al. 1973). A similar study was
conducted by Ashraf and Christensen (1974). A study of
wider geographic coverage analyzed the economic impacts
of controlling surface water runoff from United States
dairy farms (Buxton and Ziegler 1974). Their findings indicated that industry investment for runoff control could
reach $312 million if the estimated 40 percent of the dairy
producers with runoff problems adopted control systems.
Such adjustments would hasten the exit of small dairy
farms and stimulate the shift to fewer, but larger dairy
farms.

The economic impacts of imposing EPA effluent guidelines on the United States cattle feeding industry have also
been analyzed (Johnson and Davis 1974). Study results
show that per head runoff control costs would be less for
larger capacity operations than for smaller capacity operations. Runoff control costs were generally greatest within
each capacity level for land-extensive, open lot systems and
lowest for the land intensive, drylot paved housing systems.
Other economic impact studies of runoff control on the
cattle feeding industry were conducted by Badger and Cross
(1971), and Forster, et al. (1975).

The study by Forster et al. (1975) is a dynamic analysis
of adjustments by livestock producers. The economic performance of feedlots of less than 1,000 head capacity was
simulated under four alternative surface water pollution
control rules. Sample feedlots were simulated over the
1974-85 period. The average simulated feedlot would realize from $3,700 to $4,800 less accumulation of equity (in
present value terms) over the 1974-85 period under the
alternative rules compared with equity levels achieved by
the average feedlot if it had continued to operate in the
absence of rule implementation through this period. These
reductions in the absolute levels of equity accumulation
reflect slightly lower levels of production because of the
implementation of alternative rules. Implementation of the
rules would result in about 1 percent less production over
the period by the beef feeding industry than would have
otherwise been achieved. At the farm level, fed beef prices
would be expected to increase by less than 1 1/2 percent, as
a result of slightly smaller beef supplies.

Van Arsdall and Smith (1974) analyzed the economic
impacts of controlling surface water runoff from point
sources in United States hog production. Their results indicated that of the 112,000 hog farms estimated to have
uncontrolled runoff, 86,000 use open lot systems. Under
their assumptions, runoff control facilities for these open
lot hog production systems would require investments ranging from $254 to $290 million. Annual costs of runoff
control per head marketed would be over 10 times as large
for the smallest as for the largest hog enterprise class. Their
study suggests that the imposition of runoff control measures would likely result in a short-run decrease in pork
supplies. This, in turn, would result in a more than proportionate increase in pork prices during the adjustment
period.

### Least Cost Waste Management Systems

Economic research pertaining to waste management
systems has compared costs of different systems for specified output levels. In such analyses, the waste management
system has been viewed as a sub-system of the overall livestock production system. In a study conducted by Johnson
et al. (1973), major manure handling and housing systems
for typical herd sizes were identified for major dairy regions. Least cost manure handling systems were specified
for various herd size-technology combinations. Systems
considered were those which would limit on-site runoff
problems, and runoff and odor problems associated with
field application of wastes. The least cost system was dependent on the particular housing system in use.

An early study for similar concerns in beef feeding was
conducted to determine least cost waste management systems for Texas feedlots (Owens and Griffin 1968). Systems
compared were open field and playa lake modifications,
and evaporative discharge systems for 5,000, 10,000 and
25,000 head feedlot capacity levels. In an EPA funded
study (Butchbaker et al., 1971), various waste management
systems were evaluated for confined housing and open beef
feedlot systems. Of particular interest was the delineation
of United States climatic zones for beef feeding and waste
management system selection. Systems were ranked according to their potential for pollution control and least cost
performance. Systems which had the highest potential for
pollution control were not the least cost system.

A least cost analysis conducted by Morris (1971) differed from those discussed previously because it included



MANAGING LIVESTOCK WASTES

the valuation of manure based on its plant nutrient content. The valuation was imputed from the prices of nutrient elements obtained from alternative sources.

## Recycling Livestock Wastes

Several studies have attempted to analyze the economic considerations involved in recycling livestock wastes. A USDA study (Agricultural Research Service 1971) examined the nutritive value and potential problems from feed additives associated with animal waste reuse. This study was a literature review, and was nearly void of economic analyses.

The feed value of hog wastes was analyzed (Anonymous 1972). Results indicated that feeding nutrients from aerobic hog wastes offers little to most existing producers without a substantial change in their present facilities and size of operation. For example, only 8 percent of Illinois' hog producers had confinement facilities in 1972. It was concluded that additional research would be needed before the commercial feasibility of recycling could be adequately judged.

The use of ruminant manure as feed for fattening cattle was analyzed (Anonymous 1972). It was concluded that economic analyses of experimental trials do not provide evidence that wastage feeding at hgih rates is either economically or technically efficient. Land disposal of manure for recapture of plant nutrient value was considered a more attractive economic alternative.

In assessing the economic feasibility of recycling, the plant nutrient values of wastes, labor savings associated with recycling wastes, and health regulations were important considerations.

## Energy Requirements Associated with Waste Management Systems

A relatively new research area for agriculturalists is the conservation and utilization of energy for alternative livestock production systems. Studies have estimated energy requirements associated with livestock waste managment systems. Energy sources considered have been electricity, fuel, labor, and land (as a proxy for solar energy).

Hughes, Holtman and Connor (1974) estimated energy and monetary costs for beef feedlots using liquid and solid manure handling systems. Their study shows that economies of size were present for electricity and labor for feedlots employing these types of waste handling systems. Diseconomies of size were noted for fossil fuel consumption for both handling systems. That is, as a feedlot capacity increased, so did fossil energy requirements per hundredweight of gain. Adriano et al. (1974) provided estimates of energy coefficients by alternative beef housing systems, waste handling systems, rations, and animal type and sex.

## Other Related Studies

Nelson (1972) evaluated Oregon tax options on pollution control facilities availabe to livestock producers. His publications provides a format to assist livestock producers in comparing tax relief options.

A Minnesota study of Pherson (1974) used a linear programming model to determine optimal feedlot organization for Minnesota farmer-feeders with alternative waste management and housing systems for beef feedlots with a 500 acre cropland base. This study considered the optimal organization for the entire farm, with particular consideration

given to time period constraints of operator, family, and hired labor. Study results indicated that the implicit value of beef feedlot wastes for plant nutrients was highly dependent on existing labor constraints and rule requirements relative to acceptable times for field spreading.

Schaffer et al. (1974) conducted an economic analysis of policies to control nutrient and soil losses from a small watershed in New York. Agricultural activities in this study included livestock production. Nitrogen and soil losses and costs incurred by farmers in reducing such losses were estimated.

Connor (1973) summarized previous research efforts on livestock wastes and discussed difficulties in conducting such research. Selected conclusions drawn in this study are presented in the following section.

### APPRAISAL OF RESEARCH

Most economic research on livestock waste and pollution control issues was conducted in response to pollution control measures imposed or thought likely to be imposed on livestock farms. The wide range of economic research which has been conducted on these problem issues was summarized by category in the previous section. Some general conclusions can be drawn relative to this research, and what is implied for additional research efforts.

Some initial work on animal waste management research indicated that a primary need of economists was to conduct benefit/cost analyses. However, primary emphasis has been on abatement costs borne by producers and the resulting implications on livestock supply functions and consumer prices. Little attention has been given to the benefits of pollution control relative to livestock waste. Public costs associated with pollution control program development, implementation, monitoring, and enforcement have not been considered. Economic benefits (in a quantitative sense) derived from alleviating environmental degradation occurring from improper management of livestock wastes have not been developed because there is little knowledge as to how livestock wastes affect the environment. Various USDA studies have provided some indications of the potential number of livestock producers with surface water runoff problems. How these "problem" firms affect the environment in unknown. This fact, coupled with the need for information on possible economic impacts of control measures on livestock farms, has resulted in concentrating research on producer abatement costs and the likely industry supply and price effects.

Adjustments in waste handling systems and practices will have varying impacts on United States livestock farms. Economic impacts will vary according to firm size, technology employed, and location. Adjustments will require additional investments and will be cost increasing. None of these investments in pollution abatement facilities will be revenue generating. The new result will be one of effectively reducing net farm income, assuming the same output levels are retained. Economies of size will be realized in the adjustment processes; thus the distribution of economic impacts must be noted within each livestock industry. Smaller producers will generally be most adversely affected by control measures.

Conclusions on the economic impacts of selected control measures generally hold for individual firm adjustments; however, estimated aggregate industry effects have not been consistent nor known with a great deal of certainty. Most studies have been partial-budget type economic analyses of producer adjustments. Additionally, most studies have been comparative static analyses, assessing situations at the industry level before and after to control implementation. Economic studies based on comparative statics are not sufficient to empirically assess the aggregate industry impacts.

A dynamic analysis conducted by Forster et al. (1975) indicated that various pollution control measures for small beef feedlots (less than 1,000 head) would have little impact upon the industry supply and resulting consumer prices. Whether or not these conclusions would hold for dairying and for hog production is debatable. It is a difficult research question to answer, because the likely adjustments by farmers to pollution control measures need to be sorted out from the farm adjustments in response to recent inflationary impacts on farm input prices. Conclusions on the economic impacts of pollution control measures on industry supply and consumer prices should always be viewed in conjunction with current economic conditions.

A few studies have examined the possible economic impacts of imposing nonpoint pollution control measures (Forster et al. 1975, Good et al. 1973, Schaffer et al. 1974, Johnson et al. 1973). Measures such as prohibiting winter spreading, designating where waste may be spread, and requiring subsurface disposal of wastes could result in rather severe impacts on some livestock farms. Additional research is needed on the economic impacts of such measures for livestock farming. It would be useful to have meaningful economic research as an input to the development of nonpoint pollution control guidelines and recommendations. Public agencies have a definite obligation to engage in such economic analyses. In the past, pollution control guidelines for the livestock industries have in large part been based on studies conduted by private consulting firms. Task force reports of the Council of Agricultural Science and Technology on the empirical findings of economic impact studies conducted by private consulting firms, suggest dissatisfaction is the rule rather than the exception.

The USDA economic impact studies for beef feeding, dairying, and hog production have indicated that there may be some interregional impacts in the adjustment to pollution control rules. For example, the proportion of hog farms having a runoff problem was estimated to be more than twice as great in the Corn Belt-Lake States as in the Plain States where precipitation is relatively low (Van Arsdall and Smith 1974). Johnson and Davis (1974) found that a high proportion of the beef feedlots with runoff problems were in the Eastern States. Such data indicated that imposition of selected pollution control measures will more adversely affect some production regions than others. However, the magnitude of possible shifts in production between or among regions is still a question which has not been answered.

Surprisingly, limited economic research has been conducted on whole farm analyses for livestock farms that include waste management systems as one subsystem of the overall farm business. Instead of just identifying least cost systems, this type of research is directed to selecting the waste management system which would fit into the overall farm plan for maximizing profits. Such research would also consider capital restraints, labor availabilities, and timeliness constraints, as well as environmental concerns. Obviously, for Cooperative Extension purposes, additional research needs to be conducted in this area. Not to be overlooked in such analyses are some of the more "exotic" waste handling systems. These include composting, methane generation, recycling, and other advanced systems being developed.

Additional research needs to be devoted to the energy requirements associated with alternative agricultural production systems, including waste management systems for livestock production. To date, energy requirements associated with alternative waste management systems and livestock production systems have received only minimal attention. Fuel, electricity, labor, and land requirements associated with alternative livestock production and waste handling systems will necessarily be a focal point of analysis.

Lastly, economists need to work with researchers from other disciplines to identify "workable systems" for livestock waste management that are technically and economically feasible and acceptable with regard to ecological and energy considerations. Livestock production systems can no longer be evaluated in purely technical and economic realms. The identification and implementation of such "workable systems" on United States livestock farms will require much greater cooperation among the various disciplines in agriculture than has previously been the case.


## References

1  Adriano, D. C., et al. 1974. Ecosystem design and management: technology alternatives in beef production. Internal Rept. No. GI-20 of the National Science Foundation prepared by Michigan State University. 91 p.

2  Anonymous. 1971. Animal waste reuse—nutritive value and potential problems from feed additives: a review. USDA. Agric. Res. Serv. ARS 44-224. 56 p.

3  Anonymous. 1973. Control of pollution from animal feedlots. Hearings before a subcommittee on government operations. House of Representatives, Ninety-third Congress. U. S. Govt. Printing Office. 1,268 p.

4  Anonymous. 1973. Development document for proposed effluent limitations guidelines and new service performance standards for the feelot point source category. U. S. Environmental Protection Agency Rept. No. EPA 440/1-13/004. 302 p.

5  Anonymous. 1972. Economic feasibility of processing and handling animal wastes as feed. Summary report of Subgroup C of USDA task force on use of animal wastes in feed for animals. A compilation of reports.

6  Ashraf, M. and R. L. Christensen. 1974. An analysis of the impact of manure disposal regulations on dairy farms. AJAE. 56:331-337.

7  Badger, D. D. and G. R. Cross. 1971. Economic implications of environmental quality legislation for confined animal feeding operations. Proc. Int. Symp. Livestock Waste Mgt. ASAE. 199-204. St. Joseph, Mich. 49085.

8  Butchbaker, A. F., J. E. Garton, G. W. A. Maloney, and M. D. Paine. 1971. Evaluation of beef cattle feedlot waste management alternatives. Report 13040 FXG for U. S. Environmental Protection Agency. 322 p.

9  Buxton, B. M., S. J. Ziegler. 1974. Economic impact of controlling surface water runoff from U. S. dairy farms. USDA, ERS, Agric. Econ. Rept. No. 260. 40 p.

10  Connor, L. J. 1973. Environmental economics research on animal wastes. Empirircal research in environmental economics. Southern Land Econ. Res. Com. Bul. No. 11. p. 21-37.

11  Forster, D. L., L. J. Connor, and J. B. Johnson. 1975. Economic impacts of selected pollution control rules on Michigan beef feedlots of less than 1,000 head capacity. Michigan State University, farm sci, series, res. rept. No. 270. p.

12  Gamble, John Clark. 1972. Economic evaluation of the impact of alternative environmental regulations and waste management systems on Illinois hog producers. Unpublished Ph.D. thesis, University of Illinois. 317 p.

13  Good, D. L., C. R. Hoglund, L. J. Connor, and J. B. Johnson. 1973. Economic impacts of applying selected pollution control measures on Michigan dairy farms. Michigan State University, farm sci. series, res. rept. No. 225. 12 p.

14  Hoglund, C. R., J. S. Boyd, L. J. Connor, and J. B. Johnson. 1972. Waste management practices and systems on Michigan dairy farms. Michigan State University. Agric. Econ. Rept. No. 208. 15 p.

15  Hughes, H. A., J. B. Holtman, and L. J. Connor. 1974. Proc. and mgt. of agric. waste. Proceedings of the 1974 Cornell Agric. Waste Mgt. Conf. p. 271-282.

16  Johnson, J. B., and L. J. Connor. 1971. Origins and implications of environmental quality standards for animal production firms. Proc. Int. Symp. Livestock Waste Mgt. ASAE publ. p. 102-104. St. Joseph, Mich. 49085.

17  Johnson, J. B., L. J. Conner, and C. R. Hoglund. 1972. Summary of state air and water quality statutes applicable to the managment of livestock wastes. Michigan State University, Agric. Econ. Rept. No. 231. 71 p.

18  Johnson, J. B. and G. A. Davis. 1974. The economic im-


*(Continued on page 40)*

# Economics of Alternative Beef Waste Management Systems

## Maurice Baker

SIX waste management systems for beef feeding facilities are briefly described in this paper. Two systems for open feedlots and four for totally housed facilities were studied. The costs associated with each system are identified. Finally, implications of waste management for the beef feeding industry are discussed.

Investment figures for the six systems include excavation, materials, land for the collection facilities, and labor necessary to install and operate the facilities. The investment for disposal equipment includes tractors, spreaders, irrigation equipment, pumps and other necessary machinery.

Physical requirements for constructing and operating waste management systems were determined by interviews with operators of beef feeding facilities; however, economic data were obtained from contractors, dealers and suppliers. Facility capacity was based upon 400 and 18 square feet per head for open feedlots and totally housed facilities, respectively.

## WASTE MANAGEMENT SYSTEMS FOR OPEN DRYLOT FEEDING FACILITIES

Both systems were designed with a minimum storage capacity for debris basins and detention ponds based on a 10-year, 24-hour storm. Solids and liquids were separated by both of the systems. The major difference between the two systems was in where the settleable solids are stopped and how the liquids are transported to the detention ponds.

### Continuous Flow System

The settling basin was located outside of the feedlot and the liquids flowed overland to the detention pond (Fig. 1). The solids were removed from the settling basin with conventional manure handling equipment and the liquids were pumped from the detention pond to cropland with conventional irrigation equipment.

Investment per head capacity declined rapidly until the capacity of the feedlot reached approximately 1000 head (Table 1). The decline was because of more fully utilizing disposal equipment as well as economies in construction of the detention pond as the capacity increased. Investments per head capacity were $31.84 for the lot with a capacity of 72. Investments per head capacity were approximately $9.75, $5.75 and $4.60 for lots of 350-450, 750, and 7500 head capacity respectively.

Investment per head capacity was not large but the

Based in part upon: Daiss, Bill J. 1971. Economics of water pollution abatement from beef feedlots. Unpublished MS thesis, University of Nebraska, Lincoln; and Holtorf, Roger C. 1974. Economics of confinement feeding of beef cattle in eastern Nebraska. Unpublished MS thesis, University of Nebraska, Lincoln.

Published as Paper No. 3990 Journal Series, Nebraska Agricultural Experiment Station.

The author is: MAURICE BAKER, Professor, Agricultural Economics Dept., Institute of Agriculture and Natural Resources, University of Nebraska, Lincoln.

total investment for waste management was nearly $35,000 for the 7500 head capacity lot. Total investments of $3000-$4000 were required for lots of 350-750 head capacity.

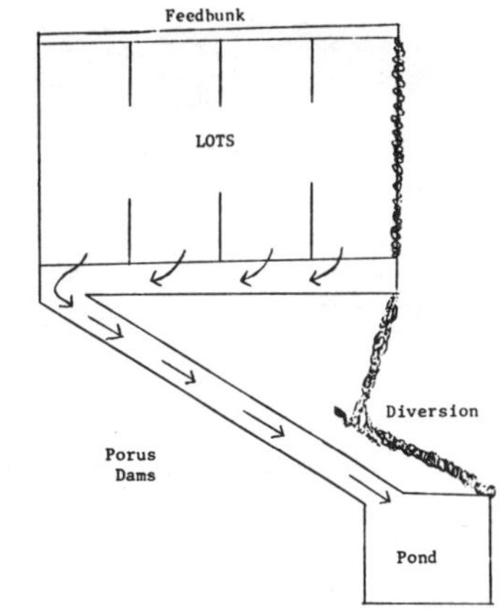


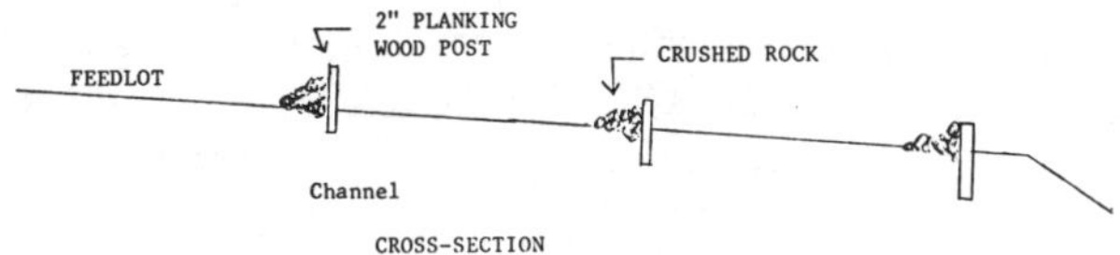


FIG. 1 Continuous flow runoff control system.
Courtesy of C. B. Gilbertson, Department of Agricultural Engineering, University of Nebraska-Lincoln.

**TABLE 1. INVESTMENT FOR CONTINUOUS FLOW RUNOFF CONTROL SYSTEMS FOR OPEN DRYLOT FEEDING FACILITIES**

| Lot capacity, head | Investment | | | | | |
| | Collection facilities | | Disposal equipment | | | |
| | Total | Per head capacity | Total | Per head capacity | Total | Per head capacity |
|---|---|---|---|---|---|---|
| 72 | $ 976 | $13.56 | $ 1 317 | $18.29 | $ 2 293 | $31.84 |
| 370 | 1 785 | 4.82 | 1 883 | 5.09 | 3 668 | 9.91 |
| 436 | 2 366 | 5.43 | 1 883 | 4.32 | 4 248 | 9.74 |
| 447 | 1 794 | 4.01 | 2 471 | 5.53 | 4 265 | 9.54 |
| 545 | 1 919 | 3.52 | 1 883 | 3.46 | 3 802 | 6.98 |
| 750 | 2 433 | 3.24 | 1 883 | 2.51 | 4 316 | 5.75 |
| 7500 | 23 047 | 3.07 | 11 546 | 1.54 | 34 593 | 4.61 |

Source: 1971 survey data.

### Terraced System

Debris basins for a terraced system were located within the feedlot (Fig. 2). They were broad basin terraces which trap the runoff. The solids settle in the basin and the liquids are drained off to a detention pond through a series of risers connected to underground plastic pipe. The solids are removed from the basin and mounded in the lots. Again,

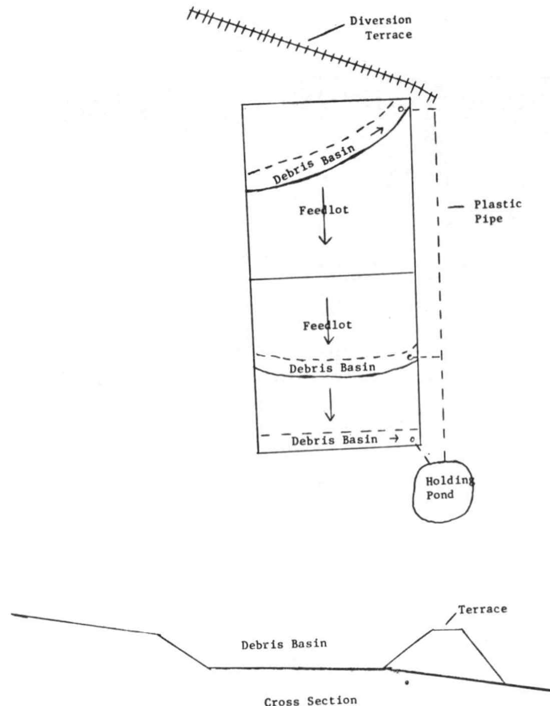



FIG. 2 Terraced runoff control system.

Basic Design by N. P. Swanson, L. N. Mielke and J. C. Lorimor, Department of Agricultural Engineering, University of Nebraska-Lincoln.

the liquids are pumped to cropland by irrigation equipment.

Investment for terraced systems was approximately $4000 and $28,000 for lots with nearly 450 and 6000 head capacity, respectively (Table 2). This results in per head capacity investment of $9.22 for the smaller lot and $4.81 per head capacity for the larger lot. These were the only lots using this system at the time of the study.

**TABLE 2. INVESTMENT FOR TERRACED RUNOFF CONTROL SYSTEMS FOR OPEN DRYLOT BEEF FEEDING FACILITIES**

| Lot capacity, head | Collection facilities | | Disposal equipment | | | |
|---|---|---|---|---|---|---|
| | Total | Per head capacity | Total | Per head capacity | Total | Per head capacity |
| 436 | $ 2 138 | $4.90 | $ 1 883 | $4.32 | $ 4 021 | $9.22 |
| 5777 | 17 224 | 2.98 | 10 596 | 1.83 | 27 821 | 4.81 |

Source: 1971 survey data.

## WASTE MANAGEMENT SYSTEMS FOR HOUSED FEEDING FACILITIES

The housed feeding facilities included in the study were of three sizes — 500, 1000, and 1667 head capacity. The buildings were 48 feet wide and 300, 600 or 1000 feet long. Three waste management systems were used with all three sizes of facilities; however, the oxidation ditch was used with only the 500 and 1000 head capacity facilities.

### Oxidation Ditch

The oxidation ditch was located under slotted floors of the feeding facilities. Oxidation ditches have not always functioned properly; therefore they are not widely used. When the systems functioned properly, removal of solids was seldom required. Liquids flowed into a detention pond and were pumped onto cropland with irrigation equipment.

Total investment in an oxidation ditch system for a 500 head capacity facility was over $40,000 (Table 3). The per head investment cost for a 500 head capacity facility was nearly $88 but declined approximately 10 percent to $78 for a 1000 head capacity facility.

### Deep Pit System

The wastes were stored in an 8-foot deep pit under slotted floors of the feeding facilities. The wastes were periodically pumped out of the pit and spread on cropland with liquid manure handling equipment.

The total investment for deep pit systems varied from approximately $54,000 for 500 head capacity facilities to nearly $150,000 for the 1667 head capacity ones. Per head investment varied from $109.58 to $88.93 for the smallest and largest facilities, respectively.

While the per head investment declined with increases in capacity of the feeding facilities, the deep pit was still the most expensive system. This was because of the need for additional excavation and materials for this system.

### Shallow Pit System

The shallow pit system differed from the deep pit one by being only half as deep; therefore, the shallow pit had to be emptied more frequently.

The total investment for the shallow pit system was over $43,000 for the 500 head capacity facilities. This was nearly $11,000 less than for the deep pit system for 500 head capacity facilities. The total investment for the 1667

**TABLE 3. TOTAL AND PER HEAD INVESTMENTS FOR WASTE MANAGEMENT SYSTEMS USED WITH TOTALLY HOUSED BEEF FEEDING FACILITIES**

| System | Size, head | Investment | | | | | |
|---|---|---|---|---|---|---|---|
| | | Collection facilities | | Disposal equipment | | | |
| | | Total | Per head capacity | Total | Per head capacity | Total | Per head capacity |
| Oxidation ditch | 500 | $ 38 810 | $77.62 | $ 5 000 | $10.00 | $ 43 810 | $ 87.62 |
| | 1000 | 73 690 | 73.69 | 5 000 | 5.00 | 78 690 | 78.69 |
| Deep pit | 500 | 40 790 | 81.58 | 14 000 | 28.00 | 54,790 | 109.58 |
| | 1000 | 81 222 | 81.22 | 14 000 | 14.00 | 95 222 | 95.22 |
| | 1667 | 134 250 | 80.53 | 14 000 | 8.40 | 148 250 | 88.93 |
| Shallow pit | 500 | 29 724 | 59.45 | 14 000 | 28.00 | 43 724 | 87.45 |
| | 1000 | 60 323 | 60.32 | 14 000 | 14.00 | 74 323 | 74.32 |
| | 1667 | 99 215 | 59.52 | 14 000 | 8.40 | 113 215 | 67.92 |
| Slot flush | 500 | 23 063 | 46.13 | 5 000 | 10.00 | 28 063 | 56.13 |
| | 1000 | 40 353 | 40.35 | 5 000 | 5.00 | 45 353 | 45.35 |
| | 1667 | 63 354 | 38.00 | 5 000 | 3.00 | 63 354 | 41.01 |

Source: 1973 survey data.

head capacity shallow pit facilities was over $113,000 but was still less than the investment for the deep pit system by approximately $45,000.

The per head investment varied from $87.45 for the 500 head capacity facilities to $67.92 for the 1667 head capacity facilities.

### Slot Flush System

The slot-flush system utilized 8 to 12-inch diameter flumes recessed into the concrete floors to carry the wastes out of the building. These flumes ran full length of the building and were covered by the floor except for 2-inch wide slots in he top fo each. The floor of the building slopes toward the flumes in a series of "hills and valleys". Recirculated water was used to flush the wastes into the lagoon.

Liquids were pumped from the lagoon onto cropland but no plans were made to remove the solids from the lagoon.

The slot flush system was the only one used with totally housed facilities where the cattle were not supported on slotted floors. Therefore, the investment in this system was considerably less than for others.

The per head investment varied from $56.13 for 500 head capacity facilities to $41.01 for the 1667 head capacity ones. This was approximately one-half as much as for the deep pit system and about three-fourths as much as for the other two systems.

## ANNUAL COSTS FOR ALTERNATIVE SYSTEMS

Depreciation, interest on ivestment, insurance, taxes and operating costs were included in annual costs. These costs were used to indicate the impact of waste management upon cost of production for the beef industry. No credit was given to the beef enterprise for irrigation or fertilizer value of the liquids or solids.

### Continuous Flow System

Annual costs per head capacity declined with increases in size (Table 4). Costs for the lot with less than 100 head capacity was $5.06 per head capacity. Those lots with 350 to 500 head capacity had annual costs per head capacity of approximately $1.80. The annual costs for the 7500 head capacity lots was $1.10 per head.

Differences in the cost per head marketed from the small vs. the larger lots may be even greater than the annual costs per head capacity. Small lots were usually farmer feeders and only one lot of cattle were fed each year. The large lots may turn cattle 2 to 3 times per year; thus, the cost per head marketed may be considerably less than per head capacity cost figures.

### Terraced System

The 436 and 5777 head capacity facilities had annual per head costs of $1.79 and $1.13, respectively (Table 5). As with the continuous flow systems, the costs per head marketed may be different from the costs per head capacity. The costs for the terraced system was similar to that for continuous flow systems of similar size.

### Oxidation Ditch

Per head annual costs were $21.78 for the 500 head capacity facilities and $14.12 for the 1000 head capacity

**TABLE 4. TOTAL AND PER HEAD ANNUAL COSTS FOR CONTINUOUS FLOW RUNOFF CONTROL SYSTEMS FOR OPEN DRYLOT BEEF FEEDING FACILITIES**

| Lot capacity, head | Annual costs | | | | | |
| | Collection facilities | | Disposal equipment | | | |
| | Total | Per head capacity | Total | Per head capacity | Total | Per head capacity |
|---|---|---|---|---|---|---|
| 72 | $ 150 | $2.08 | $ 215 | $2.98 | $ 364 | $5.06 |
| 370 | 227 | 0.61 | 427 | 1.16 | 654 | 1.77 |
| 436 | 365 | 0.84 | 445 | 1.02 | 810 | 1.86 |
| 447 | 359 | 0.81 | 448 | 1.00 | 807 | 1.81 |
| 545 | 283 | 0.52 | 476 | 0.87 | 758 | 1.39 |
| 750 | 376 | 0.50 | 534 | 0.71 | 910 | 1.21 |
| 7500 | 3514 | 0.56 | 4800 | 0.64 | 8233 | 1.10 |

Source: Derived from Table 1.

facilities. (Table 6) Costs for the systems were markedly higher than other systems because of the energy necessary to operate the paddle wheels. Energy costs alone were over $6000 per year for both size of systems since the number and size of paddles were not increased with the larger system. These costs will likely increase substantially in the future.

### Deep Pit System

Deep pit systems varied in annual costs from $16.63 per head for 500 head capacity facilities to $12.09 per head for 1667 head capacity ones. This decline was largely due to the disposal equipment being used more fully with the larger facility.

The added cost for this system over the shallow pit one was justified only when conditions did not permit spreading the wastes as frequently as was necessary with the smaller pits. This may be the case if weather and cropping patterns limit times for spreading.

**TABLE 5. TOTAL AND PER HEAD ANNUAL COSTS FOR TERRACED RUNOFF CONTROL FOR OPEN DRYLOT BEEF FEEDING FACILITIES**

| Lot capacity, head | Annual costs | | | | | |
| | Collection facilities | | Disposal equipment | | | |
| | Total | Per head capacity | Total | Per head capacity | Total | Per head capacity |
|---|---|---|---|---|---|---|
| 436 | $ 336 | $0.77 | $ 445 | $1.02 | $ 781 | $1.79 |
| 5777 | 2741 | 0.48 | 3760 | 0.65 | 6501 | 1.13 |

Source: Derived from Table 2.

### Shallow Pit System

Annual costs for shallow pit systems were approximately $2 per head less than for the deep pit systems. The difference was due mostly to the need for less excavating and fewer materials. The costs declined from $14.59 per head for the smallest capacity system to $10.15 for the largest one.

### Slot Flush System

The slot flush system was the least costly of the waste management systems used with totally housed feeding facilities. This was largely because the animals do not have to be supported above ground level. Thus, the costs of slotted floors and supports were avoided.

The per head annual costs are less than one-half as great for the slot flush system than for the next lower cost one. The 500 head capacity system costs $5.65 per head and the 1667 head capacity facilities cost only $3.95 per head.

| System | Size, head | Annual costs | | | | | |
| --- | --- | --- | --- | --- | --- | --- | --- |
| | | Collection facilities | | Disposal equipment | | | |
| | | Total | Per head capacity | Total | Per head capacity | Total | Per head capacity |
| Oxidation ditch | 500 | $ 3 775 | $ 7.55 | $7113 | $14.23 | $10 888 | $21.78 |
| | 1000 | 6 988 | 6.99 | 7136 | 7.14 | 14 125 | 14.12 |
| Deep pit | 500 | 5 048 | 10.10 | 3267 | 6.53 | 8 314 | 16.63 |
| | 1000 | 8 772 | 8.77 | 4636 | 4.64 | 13 409 | 13.41 |
| | 1667 | 13 658 | 8.19 | 6494 | 3.90 | 20 152 | 12.09 |
| Shallow pit | 500 | 4 028 | 8.06 | 3267 | 6.53 | 7 295 | 14.59 |
| | 1000 | 6 663 | 6.66 | 4636 | 4.64 | 11 299 | 11.30 |
| | 1667 | 10 430 | 6.26 | 6494 | 3.90 | 16 924 | 10.15 |
| Slot flush | 500 | 2 124 | 4.25 | 702 | 1.40 | 2 826 | 5.65 |
| | 1000 | 3 717 | 3.72 | 724 | 0.72 | 4 442 | 4.44 |
| | 1667 | 5 836 | 3.50 | 747 | 0.45 | 6 583 | 3.95 |

Source: Derived from Table 3.

## IMPLICATIONS OF WASTE MANAGEMENT FOR THE BEEF FEEDING INDUSTRY

A waste management system is one of the requirements of a beef feeding enterprise and there is a wide range in investment and annual costs of the various systems; therefore, it is necessary for the operator to know the costs associated with each system before deciding on which one he will install. Even the least expensive waste management systems for open dry lot and housed feeding facilities require a substantial investment.

Present investments of more than $50,000 and $4000 for waste management systems for totally housed and open dry lot feeding facilities, respectively, are not uncommon. Investments of this magnitude indicate the importance of carefully evaluating the various alternatives.

Annual per head capacity costs of waste management systems decline as size of feeding operation increases. Large capacity feeding facilities are more likely to be used all year long than are the smaller ones; therefore, the cost per head marketed will be less for the larger sized lots than for the smaller ones. Thus, the small feeder is at a disadvantage compared to the larger operator in meeting costs of waste management.

While annual waste management costs are not great when compared to total production costs, they are of concern to feeders because they do increase production costs with no associated increases in revenue. However, waste management costs are not likely to force many operators out of the feeding business.

## Economic Research
*(Continued from page 36)*

pacts of imposing EPA effluent guidelines on the U. S. fed-beef industry. Proc. and mgt. of agric. waste. Proceedings of the 1974 Cornell Agric. Waste Mgt. Conf. p. 59-71.

19    Johnson, J. B., C. R. Hoglund, and B. Buxton. 1973. An economic appraisal of alternative dairy waste management systems designed for pollution control. Journ. of Dairy Sci. 56:1354-1366.

20    Morris, W. H. M. 1971. Economics of waste disposal from confined livestock. Proc. Int. Symp. Livestock Waste Mgt. ASAE pub. p. 195-197. St. Joseph, Mich. 49085.

21    Nelson, A. G. 1972. Evaluating Oregon's tax options on pollution control facilities. Oregon STate University Ext. Serv. spec. rept. 361. 9 p.

22    Owens, T. R. and W. L. Griffin. 1968. The economics of water pollution control for cattle feedlot operations. Texas Tech. Univ., Dept. of Agric. Econ. special rept. No. 9. 62 p.

23    Pherson, C. L. 1974. Beef waste management economics for Minnesota farmer-feeders. Proc. and mgt. of agric. waste. Proceedings of the 1974 Cornell Agric. Waste Mgt. Conf. p. 250-271.

24    Randall, A. 1972. Market solutions to externality problems. AJAE. 54:175-184.

25    Schaffer, W. H., J. J. Jacobs, and G. L. Casler. 1974. Proc. and mgt. of agric. waste. Proceedings of the 1974 Cornell Agric. Waste Mgt. Conf. p. 200-210.

26    Van Arsdall, R. N. and R. B. Smith. 1974. Economic impact of controlling surface water runoff from point sources in U. S. hog production. USDA, ERS, Agric. Econ. Rept. No. 263. 56 p.

# Economic Impacts of Alternative Water Pollution Control Rules on Beef Feedlots of Less Than 1000 Head Capacity


D. L. Forster, L. J. Connor, J. B. Johnson


PASSAGE of the Federal Water Pollution Control Act Amendments of 1972 provided additional authority to the U.S. Environmental Protection Agency (EPA) for controlling the pollution of the nation's waters (United States Congress 1972). The Act authorized EPA to establish effluent limitations guidelines to be achieved by point sources of waste discharge. In addition, the Act authorized EPA to establish a set of suggested methods and practices for the control of pollution from nonpoint sources. Runoff from farmland to which livestock wastes have been applied has been included as one of these nonpoint sources.

The legalistic approach for controlling water pollution places increasing importance on research to analyze the economic impacts of alternative effluent limitations guidelines, suggested methods, and practices. Decision makers developing these guidelines, suggested methods, and practices need information concerning the costs of abatement and the expected societal benefits in order to appraise their desirability.

The objective of this research was to analyze the economic impacts of alternative water pollution control rules on the long-run behavior of beef feedlots of less than 1000 head capacity. A computerized simulation model was developed to examine the effects of a set of alternative guidelines and suggested methods and practices. The model simulated the behavior of feedlots over the 1974-85 period in order to observe the impact of the rules on their long-run performance. The model was designed to simulate feedlots found in Michigan and similar areas in the North Central and Eastern States.

The impacts of the alternative rules were measured by their effects on (a) the equity positions of simulated firms, (b) the capital structure of the simulated firms, and (c) the numbers of cattle produced in the simulated feedlots. Observing the effects of the rules on the equity position of the simulated firms allowed the cost of the rules to feedlot operators to be approximated. The capital structure of the simulated firms under alternative rules was compared in order to detect any changes in the proportion of durable assets (buildings, tractors, etc.). It has been hypothesized that increasing the proportion of durable to total assets would tend to impair operators' abilities to shift production patterns and would lead to some rigidity in production (Johnson 1972). The numbers of cattle fattened were observed to approximate the shifts in fed cattle supplies which would result from the imposition of alternative rules.

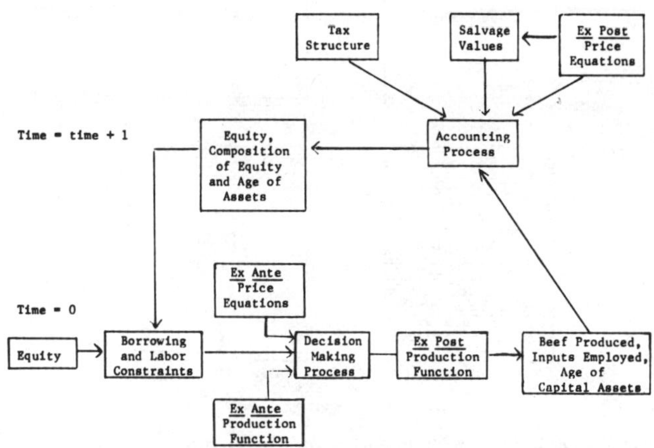



FIG. 1 Model of feedlot behavior over multiple year horizon.

## SIMULATION MODEL

The model was designed to represent the behavior of individual simulated feedlots over a multi-period horizon. Various financial and physical characteristics were assumed for individual feedlots at the beginning of the planning period. Then the simulated firm was allowed to form expectations concerning future price for its product (slaughter cattle) and for the prices of its inputs (labor, equipment, crop production inputs, etc.). The simulated firm also developed expectations of its production alternatives and the input-output relationship of each of these alternatives. Based on these expectations or *ex ante* relationships, the firm made a decision concerning the inputs to use in the farm-feedlot operation. The success of the firm was based on actual or *ex post* price and production relationships.* The alternative rules affected the performance of the simulated firm by changing the input requirements in the *ex ante* and *ex post* production functions. These changes affected profitability of various production activities and influenced the decisions concerning the farm-feedlot inputs and the level of slaughter cattle production.

The model was comprised of five components, illustrated in Fig. 1. The first component was a farm-feedlot production component (Hughes 1973 and Adriano et al. 1974). The component assumed a whole farm approach to feedlot production by simulating feedlot design, the production of crops necessary to feed the cattle, transportation of the crops from field to storage facilities, the design of feed storage facilities, the removal of wastes to the fields,

---


Michigan Agricultural Experiment Station Journal Article No. 7163.

The authors are: D. L. FORSTER, Assistant Professor, Agricultural Economics Dept., Ohio State University, Columbus; L. J. CONNOR, Professor, Agricultural Economics Dept., Michigan State University, East Lansing; and J. B. JOHNSON, Agricultural Economist, Economic Research Service, USDA, East Lansing, MI.


*Space considerations preclude a detailed explanation of these relationships. For a detailed explanation of ex ante and ex post relationships, see Forster ref. 3.

†The component was a modified product of a joint effort of personnel in the Electrical Engineering and Systems Science, Agricultural Engineering, Crop and Soil Science, and Agricultural Economics Depts. at Michigan State University.

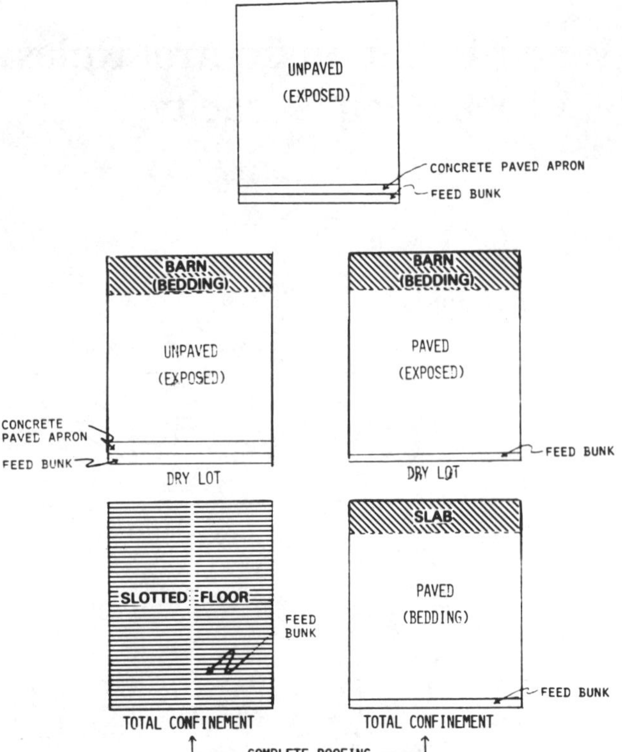

FIG. 2 Beef housing technologies used in simulation model.



and the production of the beef.† Beef housing technologies used in the simulation model are shown in Fig. 2.

The four other components of the model were:

1  An *ex ante* price relationship component was used to determine operators' expected input and output prices.

2  A decision making component was used to determine the inputs used in the production processes. The objective of this component was to select those inputs which maximized profits; a linear programming model was used to incorporate the *ex ante* price relationships and the *ex ante* production function to accomplish this objective.

3  Price realization relationships component was used to provide estimates of actual prices paid and prices received; these prices were based on historical price cycles and trends.

4  An accounting component was used to compute the annual net earnings, taxes, user cost of durable assets, and the financial position of the firm.

## ALTERNATIVE WATER POLLUTION CONTROL RULES ANALYZED

Alternative rules analyzed were those which might be employed by State or Federal agencies in the control of runoff from beef feedlots and the control of runoff from farmland to which wastes are applied (Table 1).

Rules A through D assumed that the runoff from a 10-year, 24-hour storm was equal to 5 inches of rainfall over the exposed feedlot surface; the runoff from a 25-year, 24-hour storm was equal to 6 inches of rainfall over the exposed feedlot surface; and the six-month storage requirements was equal to 16 inches of rainfall over the exposed feedlot surface. Rule D's requirement that no winter spreading occur assumed that solid waste storage, loading and spreading activities were to be equipped to handle waste accumulation over a 180-day interval. (It was assumed that solid waste accumulated for 90 days under the other rules.)

## TABLE 1. ALTERNATIVE WATER POLLUTION CONTROL RULES INCLUDED IN ANALYSIS.*

| Rule | Provisions |
|---|---|
| A | Current EPA effluent limitation guidelines would be expanded to all beef feedlots. Facilities must be constructed to control the runoff from a 10-year, 24-hour rainfall event by 1977 and the runoff from a 25-year, 24-hour rainfall event by 1983. |
| B | All feedlots must construct facilities by 1977 to control all runoff from a 25-year, 24-hour rainfall event. |
| C | All feedlots must construct facilities by 1977 to control all runoff from the rainfall occurring in any six-month interval. |
| D | The same feedlot runoff abatement provisions as in C. Also, the feedlot may not spread wastes in winter months. |
| E | The feedlot is not required to retain runoff nor is it prevented from spreading wastes. (no water pollution control rule assumed). |

*Rule A has been established for feedlots with capacities greater than 1,000 head. Rule B is what some producers may consider as a plausible technical and economic alternative to the sequential adjustment required in rule A. Rule C is what the authors have viewed as a suggested adjustment path by leading practitioners in humid States. Rule D incorporates both rule C and one of several acceptable methods and practices for nonpoint source runoff control recently announced by EPA.

## RESULTS

A random sample of twenty feedlots was drawn from the population of Michigan feedlots of less than 1000 head for the 1960 production year.‡ These twenty firms were simulated over the 1960-85 period. The 1960-71 period was used to assess the accuracy of the simulation model to represent actual feedlot performance, and the 1974-85 period was used to analyze the economic impacts of the alternative rules.

The mean net worth of the simulated firms was nearly $79 000 in 1960, feedlot capacity was 207 head, and the

‡The rationale for drawing a random sample was based on the distribution of Michigan feedlots by size and housing type. About 96 percent of the feedlots are less than 500 head capacity and about 93 percent are drylots.

## TABLE 2. PERFORMANCE OF THE AVERAGE SIMULATED MICHIGAN FEEDLOT FIRM OF LESS THAN 1,000 HEAD CAPACITY FOR THE 1960-73 PERIOD PRIOR TO IMPOSITION OF ALTERNATIVE WATER POLLUTION CONTROL RULES.

| Year | Average feedlot capacity | Weighted average return to equity* | Average annual equity per firm† | Mean annual equity change |
|---|---|---|---|---|
| | (Head) | (Ratio) | ($) | ($) |
| 1960 | 207 | 0.0847 | 78,541 | ---- |
| 1961 | 207 | 0.0766 | 82,900 | 4,359 |
| 1962 | 206 | 0.0702 | 87,354 | 4,454 |
| 1963 | 214 | 0.0680 | 92,800 | 5,446 |
| 1964 | 215 | 0.0550 | 97,412 | 4,612 |
| 1965 | 224 | 0.0643 | 104,225 | 6,813 |
| 1966 | 225 | 0.0656 | 111,724 | 7,499 |
| 1967 | 232 | 0.0564 | 119,121 | 7,397 |
| 1968 | 239 | 0.0533 | 127,416 | 8,295 |
| 1969 | 246 | 0.0597 | 137,773 | 10,357 |
| 1970 | 257 | 0.0570 | 149,321 | 11,548 |
| 1971 | 266 | 0.0609 | 163,301 | 13,980 |
| 1972 | 278 | 0.0692 | 180,488 | 17,187 |
| 1973 | 287 | 0.0690 | 199,825 | 19,337 |

*Returns include net income after taxes. Changes in the market value of assets are not included.
†Annual equity position is the market value of all assets (measured as salvage value) minus debts.

TABLE 3. MEASURES OF PERFORMANCE FOR THE AVERAGE SIMULATED MICHIGAN FEEDLOT FIRM AND THE AVERAGE FEEDLOT FIRM IN THE MICHIGAN TELFARM RECORD SYSTEM (MEDIUM-SIZED FEEDLOTS), SELECTED YEARS.*

| Year | Feedlot capacity, simulated | No. head bought, actual | Production cost/asset, simulated | Production cost/asset, actual | Average return to net worth simulated | Average return to operator's investment actual | Average equity simulated | Average operator's investment actual |
|---|---|---|---|---|---|---|---|---|
| | (Head) | (Head) | (Ratio) | (Ratio) | (Ratio) | (Ratio) | ($) | ($) |
| 1967 | 232 | 189 | 0.0936 | 0.106 | 0.0564 | 0.0 | 119,121 | 140,066 |
| 1969 | 246 | 222 | 0.1008 | 0.106 | 0.0597 | 0.063 | 137,773 | 168,745 |
| 1970 | 257 | 206 | 0.1048 | 0.093 | 0.057 | 0.05 | 149,321 | 181,561 |
| 1972 | 278 | 262 | 0.1130 | 0.116 | 0.0692 | 0.07 | 180,488 | 228,146 |

*Only selected years are shown because of space limitations. Results for other years in model validation period are comparable to results shown.

average return to equity was approximately 8.5 percent. The simulation results indicate that capacity expanded to 287 head by 1973 with the mean net worth increasing to nearly $200 000 (Table 2). Historical data indicate that actual feedlot performance during the 1960-73 period closely parallel the simulated performance. The correlation coefficient of simulated feedlot capacity and actual cattle on feed between 1960 and 1971 is 0.95. Furthermore, the simulated performance is consistent with data from Michigan State University's Telfarm record system (Table 3).

If feedlots would be allowed to continue operation without any mandatory adjustment to water pollution control rules (rule E, Table 1), the simulated capacity over the average sample feedlot in 1985 would be approximately 29 percent greater than in 1974 (Table 4). Most of this increase would occur in the 1974-79 period, reflecting the cyclical *ex post* beef prices incorporated in the simulation model. The average simulated feedlot operator would have an average net worth of nearly $750 000 by 1985 which reflects an average annual growth rate of approximately 11 percent.

When any of the four alternative water pollution control rules are imposed on the simulated firms, feedlot production, return to operator's net worth, and the annual net worth of the feedlot operators are slightly impaired. In the event rule A (extending current EPA guidelines to include feedlots with less than 1000 head capacity, Table 1) is imposed, production would be 7 head less for the average simulated firm over the entire 1974-85 period compared with simulated production when no water pollution control rules are imposed (rule E, Table 1). This slowdown in expansion is only 0.167 percent less than total production if no water pollution control rules were imposed. Impairment of the operator's return to equity and equity position was determined by calculating the "equity loss" caused by a particular water pollution control rule.§ Under rule A, the present value of "equity loss" would be $3724 over the 1974-85 period for the average simulated Michigan beef feedlot (Table 5).

Rule B, the requirement that runoff from a 25-year, 24-hour rainfall event be controlled by 1977, has a similar effect on production and net worth. When compared with the level of beef production when no rule is imposed, production would be 7.2 head (0.17 percent) less for the average Michigan feedlot under rule B over the entire 1974-85 period. The "equity loss" under rule B would be $3911 per firm compared with the continued operation of the average simulated firm in the absence of rule implementation.

Rules C and D are more severe in their effects than rules A and B. When compared with production when no rule is imposed, production for the average sample firm would be

---

§The "equity loss" is calculated by determining the present value of annual differences in the simulated firm's equity under a particular water pollution control rule and under no water pollution control rule. The discount rate used is 8 percent.

TABLE 4. PERFORMANCE OF THE AVERAGE SIMULATED MICHIGAN FEEDLOT FIRM OF LESS THAN 1,000 HEAD CAPACITY FOR 1974-85 PERIOD WITHOUT THE IMPOSITION OF ALTERNATIVE WATER POLLUTION CONTROL RULES.

| Year | Average feedlot capacity | Weighted average return to equity* | Average annual equity per firm† | Annual equity change per firm |
|---|---|---|---|---|
| | (Head) | (Ratio) | ($) | ($) |
| 1974 | 284 | 0.0669 | 219,889 | 20,064 |
| 1975 | 319 | 0.0724 | 245,904 | 26,015 |
| 1976 | 330 | 0.0753 | 276,037 | 30,133 |
| 1977 | 344 | 0.0730 | 309,790 | 33,753 |
| 1978 | 352 | 0.0764 | 348,605 | 38,815 |
| 1979 | 362 | 0.0730 | 391,224 | 42,619 |
| 1980 | 361 | 0.0747 | 439,280 | 48,056 |
| 1981 | 367 | 0.0728 | 492,332 | 53,052 |
| 1982 | 361 | 0.0726 | 550,829 | 58,497 |
| 1983 | 367 | 0.0713 | 615,093 | 64,264 |
| 1984 | 367 | 0.0690 | 683,935 | 68,842 |
| 1985 | 368 | 0.0679 | 758,687 | 74,752 |

*Returns include net income after taxes. Changes in market values of assets are not included.
†Annual equity position is market value of all assets (measured as salvage values) minus debts.

TABLE 5. PERFORMANCE OF THE AVERAGE SIMULATED MICHIGAN FEEDLOT FIRM UNDER THE ALTERNATIVE RULES EXPRESSED AS THE FIRST DIFFERENCE FROM THE EXPECTED PERFORMANCE, THROUGH CONTINUED OPERATION IN THE ABSENCE OF IMPLEMENTATION OF THESE PERFORMANCE RULES.

| Rule* | Lower production per firm over the 12-year period | Present value of average equity loss per firm† |
|---|---|---|
| | (Head) | ($) |
| A | 7.0 | 3,724 |
| B | 7.2 | 3,911 |
| C | 37.7 | 4,800 |
| D | 38.3 | 5,990 |

*Refer to Table 1 for explanation of rules.
†Present value of average equity loss per firm consists of the discounted values for the relevant years of the difference between the average firm's equity under a particular rule and the equity in the absence of any pollution control rule.

TABLE 6. COMPARISON OF EQUITY LOSSES OVER THE
1974-85 PERIOD THROUGH THE IMPOSITION OF
ALTERNATIVE WATER POLLUTION CONTROL
RULES FOR THE AVERAGE SIMULATED
FEEDLOT UNDER ALTERNATIVE
FIRM NET WORTH LEVELS.

| | Average Simulated firm equity in 1974 equals $105,000 | | Average Simulated firm equity in 1974 equals $220,000 | |
|---|---|---|---|---|
| Rule | Equity over 1974-85 period | Equity loss/ 1974 equity | Equity over 1974-85 period | Equity loss/ 1974 equity |
| | ($) | (Ratio) | ($) | (Ratio) |
| A | 3,281 | 0.0313 | 3,724 | 0.0169 |
| B | 3,479 | 0.0331 | 3,911 | 0.0178 |
| C | 4,983 | 0.0474 | 4,800 | 0.0218 |
| D | 5,746 | 0.0546 | 5,990 | 0.0272 |

37.7 head (0.90 percent) less under rule C (requiring the control of runoff from a six-month rainfall, would result in a production level for the average firm that is 38.3 head less (0.91 percent) than that achieved with no rule implementation and "equity losses" of $5990 for the average simulated firm.

None of the four rules have a noticeable impact on the asset structure of the simulated firms. The proportion of durable assets to total assets is nearly the same under all five water pollution control rules (Table 1). Thus, asset fixity will likely not be a problem through implementation of any of the water pollution control rules.

Public decision makers need to be concerned not only with the costs of alternative rules, but also with who will bear these costs. All four rules (rules A, B, C, and D) are regressive in nature, i.e., the present value of "equity losses" as a percent of the firm's original 1974 equity, becomes larger with lower equity levels. To identify the impacts of varying equity levels of average net worth. One level had an average net worth of $220 000 in 1974. The other had an average net worth level of $105 000 in 1974. Both levels were subjected to the five water pollution control rules, and a comparison was made between (a) the equity losses under the two levels of net worth and (b) the proportion that these "equity losses" were of 1974 average net worth. Generally, the present value of equity loss per dollar of 1974 equity was approximately twice as large for firms with $105 000 net worth as with firms with $220 000 net worth (Table 6). Thus, smaller firms are apparently disadvantaged in that they cannot take advantage of economies of size in waste handling facilities and the effects of reductions in capital for expansion purposes are more severe.

In making the above estimates, sample feedlots at the beginning of the simulation were randomly selected from a uniform distribution with feedlot ages ranging from one to fifteen years. Alternative rules may have a differential impact on feedlots of different ages in that older feedlots may have facilities which are depreciated out and the technologies that are obsolete. In order to test this hypothesis, a twenty-firm sample was simulated several times over the 1960-85 period with feedlot age in the initial period fixed at a specified level (e.g. all simulated feedlots were 5 years old in 1960, 10 years old in 1960, etc.). Although the various assumptions concerning feedlot age in the initial period did affect feedlot performance over the 1960-85 period, the implementation of alternative water pollution control rules had similar economic impacts regardless of the initial feedlot age.

## IMPLICATIONS

The focus of this study was to investigage some impacts of alternative water pollution control rules on Michigan beef feetlots. Estimates were made of the effects of these rules by comparing the performance of simulated beef feedlots operating in the presence of alternative rules with performance in the absence of the implementation of the rules.

The potential effect of the rules on the performance of the feedlots has implications for beef producers, beef consumers, and users of water supplies. This study further clarified the implications of water pollution control rules for beef producers and beef consumers.

The net worth positions of beef producers would be affected by each of the four rules investigated. Extending current EPA guidelines (rule A) to producers with less than 1000 head feedlot capacity would cost the average simulated producer approximately $3700 (in present value terms) over the 1974-85 period. This averages about $13 per head of 1973 capacity (Table 2). Other rules investigated had a more severe impact on the net worth position of operators than rule A. The most severe rule required runoff control from a six-month rainfall event plus no winter spreading of solid wastes. This rule would cost the average simulated firm $6000 (in present value terms) over the 1974-85 period.

The net worth distribution of beef feedlot operators will be changed by each of the rules. The impact of the rules was found to be more severe on feedlot operators with smaller equity positions than those operators with larger net worth positions.

The rules had little effect on the asset structure of simulated feedlots. This implies that the alternative rules would not lead to increased rigidity in the production patterns of Michigan beef feedlots.

If all beef feedlots of less than 1000 head capacity make adjustments similar to those simulated for Michigan feedlots, and feedlots of 1000 head or more capacity in all regions of the U.S. do not increase production in response to pollution control rules, beef production over the 1974-85 period can be expected to be only one percent less than what it would have been with no rules. The most severe rule (rule D) would retard growth in aggregate supplies by 0.91 percent over the 1974-85 period compared with production when no rules are imposed. With an estimated price flexibility for beef at the farm level of 1.7, consumers could expect price increases of approximately 1.5 percent as a result of the most severe rule (Trimble 1973 and Forster 1974). Rules A, B, and C would result in even smaller price increases for the consumer.

Lastly, it should be emphasized that this was a dynamic analysis of potential firm adjustments to pollution control rules. The price realization component of the model was based on historical price cycles and trends. Recent price movements point to the difficulties in making projections of future prices. A sampling procedure was used to evaluate the economic impacts because of the relative housing and size homogeneity of Michigan feedlots. However, it was impossible to model all feedlots of varying financial and technological characteristics. Marginal feedlots with small equity positions and/or low capacity levels can be forced out of production. These considerations should be weighed against the restrictive assumptions used in analyzing aggregate supply and price effects. Findings from this study

*(Continued on page 48)*

# Effects of Environmental Legislation on Cattle Feedlot Location


D. L. Byrkett, E. P. Taiganides, R. A. Miller

MEMBER
ASAE


ON October 18, 1972, the United States Congress (1972) passed the Federal Water Pollution Control Act Amendments of 1972. The legislation called for the establishment of effluent guidelines to be achieved by both new and existing "point sources" of water discharge into navigable waters. Animal feedlots were classified as one of these point sources.

On September 7, 1973, the Environmental Protection Agency published proposed effluent limitations for the animal feedlot industry (USEPA 1973). The proposed effluent limitation for existing cattle feedlots was that by July 1, 1977, there will be no discharge of process wastewater, except for precipitation events in excess of the ten year, twenty-four hour storm. By July 1, 1983, there was to be no discharge except for precipitation events in excess of the twenty-five year, twenty-four hour storm. New cattle feedlots were required to meet the 1983 effluent limitation.

During the following months, the Environmental Protection Agency received public comments and criticisms concerning these proposed limitations. When the final effluent guidelines and standards were published (USEPA 1974), on February 14, 1974, one major change had taken place; cattle feedlots with fewer than one thousand head capacity were excluded from the effluent limitations. One of the fears expressed was that the new restrictive environmental regulations would result in relocation of cattle feedlots from humid to arid regions. The objective of this research was to determine what effect this new legislation will have on the location of the cattle feeding industry.

To investigate the interactions of the many factors affecting the location of cattle feeding throughout the United States, a computer model has been developed (Byrkett 1974). The model contains features that are similar to other location models (King 1963, Dietrich 1971, Goodwin 1973) and considers both traditional factors that influence feedlot location, such as feeders and grain, as well as environmental factors.

Some of the more important factors considered by the computer model are delineated in this paper plus the final results on our study of the effect of the Federal Water Pollution Control Act Amendments of 1972 on the location of cattle feeding. Two cases are analyzed: one, in which the effluent limitations are directed only at commercial feedlots, as specified by the Environmental Protection Agency, and two, in which the effluent limitations are directed at all cattle feedlots, regardless of size.


This article was approved as Journal Article No. 30-75 of the Ohio Agricultural Research and Development Center.



The authors are: D. L. BYRKETT, Former Graduate Student, Industrial and Systems Engineering, Ohio State University, E. P. TAIGANIDES, Professor, Agricultural Engineering Dept., Ohio State University and Ohio Agricultural Research and Development Center, and R. A. MILLER, Assistant Professor, Industrial and Systems Engineering, Ohio State University, Columbus.

FIG. 1 Sixteen cattle feeding regions.

## MODELLING CONSIDERATIONS

Three of the major considerations that went into the design of the model to study the location of cattle feeding are:

1  The model must be spatially oriented in order to estimate varying effects in different parts of the country.

2  The model must consider all of the major factors affecting feedlot location so that tradeoffs between these factors and the new legislation can be observed.

3  The model must overcome the problem of a continually changing industry so that individual factors may be studied and cases may be consistently compared.

To model the spatial characteristics of the country, the United States was divided into 16 cattle feeding regions as outlined in Figure 1. These 16 cattle feeding regions were selected by combining states with similar cattle feeding characteristics. Individual states were not subdivided in order to facilitate data accumulation. In a like manner, the resources required for cattle feeding, such as feeders, grain, roughage, and slaughter facilities and the beef demand sites were grouped into regions, but not necessarily with the same boundaries as those of the 16 cattle feeding regions. The model uses 30 feeder cattle regions, 29 grain regions, 32 roughage regions, 30 slaughter regions, and 32 beef demand regions. This greater definition of regions increases the detail with which spatial characteristics may be input into the model, which has a significant effect on the results (Byrkett 1974).

The location of cattle feeding depends on a large number of factors, including regional nonfeed costs, transportation costs, beef demand, and the availability of feeders, grain, roughage, slaughter facilities, land, water, power, labor, and equipment. Our research was aimed at studying the location of cattle feeding from a very aggregated, or industry-wide, point of view. From this very aggregated point of view, it was determined that the most important factors affecting cattle feedlot locations are regional nonfeed costs, transportation costs of carcass beef, cattle, grain, and roughage, beef demand, and the availability of feeders, grain, roughage, slaughter facilities, and land. The other factors were judged to be more important from a micro, or individual operator, point of view. Nonfeed costs and transportation costs were estimated from regression equations. Feeder, grain, roughage and slaughter availability were computed from a variety of United States Department of Agriculture (1972, 1973a, 1973b) publications.

All data used in this study were estimated for the year 1972.

One factor in the model, which was not included in previous studies, is the effect that competition for alternate uses of the land has on the expansion of the cattle feeding industry. This factor is important for two reasons:

1 If the cost of land is extremely high due to the competition for alternate uses, it may be uneconomical to use the land for beef production.

2 In areas of high population density, the conflict between animals and people may limit production. This conflict may arise over such things as odor, noise, or the disposal of wastes.

This factor was reflected in the model by increasing nonfeed costs as a given region's animal and human populations approach higher densities (Brykett 1974).

In measuring the effect of a particular variable on feedlot location, one problem which arises is how to isolate this effect from other changes taking place in the industry. To circumvent such a problem, an optimization model was used to provide a consistent comparison between cases. That is, a cost minimization linear programming algorithm was used to determine the least cost equilibrium locations of cattle feeding. Then, appropriate parameters are adjusted to model the desired effect and new equilibrium locations are calculated. These equilibrium locations may then be compared to measure the desired effect.

## BASE CASE

Table 1 contains the actual production of fed cattle in 1972, the equilibrium production of fed cattle as predicted from the model, and the percentage deviation of the model results from the actual production. The reason that Regions 11 through 15 (Eastern Corn Belt and Eastern United States) show no deviation from actual is that it is advantageous to produce as much beef as possible in these regions, but that they are limited by the competition of alternate uses for the land to their current production level. In spite of some of the large deviations from actual, the results are considered acceptable because most of the large deviations occur in regions of very small production. The most disturbing result is the large deviation of Region 2 (California) from actual. Recent years have shown a decline in California feeding, but not to the extent indicated in Table 1. The results do support the conclusion of continued growth in the central and southern plains (Regions 6, 7, 9, and 10).

It was concluded from the results shown in Table 1 that the model fits reasonably well the current trends in cattle feeding location and that this case may be used as a consistent basis for comparison to study the effect of various changes in the cattle feeding industry. The equilibrium locations established in Table 1 will be referred to as the Base Case. A thorough analysis of this model, including verifica-

**TABLE 1. COMPARISON OF ACTUAL AND EQUILIBRIUM PRODUCTION LEVELS FOR THE CALENDAR YEAR 1972.**

| Cattle feeding region | Actual* production level, 1000's of head | Equilibrium† production level, 1000's of head | Percentage deviation from actual, percent |
|---|---|---|---|
| 1 | 518 | 718 | 39 |
| 2 | 2,062 | 872 | - 58 |
| 3 | 637 | 399 | - 37 |
| 4 | 899 | 569 | - 37 |
| 5 | 321 | 54 | - 83 |
| 6 | 2,291 | 2,605 | +14 |
| 7 | 5,310 | 5,939 | +12 |
| 8 | 1,581 | 1,414 | - 11 |
| 9 | 3,990 | 4,590 | +15 |
| 10 | 2,405 | 3,305 | +37 |
| 11 | 3,986 | 3,986 | 0 |
| 12 | 638 | 638 | 0 |
| 13 | 1,217 | 1,217 | 0 |
| 14 | 1,256 | 1,256 | 0 |
| 15 | 194 | 194 | 0 |
| 16 | 628 | 319 | - 49 |

*Estimated from USDA (1973b).
†Computed by model considering all factors except new environmental legislation.

tion, validation, and sensitivity analysis, is contained in Byrkett (1974).

## EFFECT OF ENVIRONMENTAL LEGISLATION

The major effect of the effluent limitations imposed by the 1972 Amendments of PL92-500 will be an increase in the nonfeed costs of commercial feedlots to provide facilities to meet the effluent standards. The amount of this increase will vary by region for several reasons:

1 The rainfall in each region is different, and thus, the investment in retention facilities, pumping equipment, and land requirements will vary.

2 Many feedlots are already meeting the new effluent standards, and thus, average regional costs will vary according to the number of feedlots with pollution problems.

3 Economics of scale are present so that regions with predominantly large feedlots will have a lower increment in nonfeed cost per head.

Johnson (1974) used these considerations to estimate the economic impact of imposing the 1977 effluent guidelines on the cattle feeding industry in the eastern and western United States. Supporting data, which Johnson made available, contains similar data for each of 17 major

**TABLE 2. INCREMENTAL CHANGES IN NONFEED COSTS* REQUIRED TO MEET THE 1977 EFFLUENT GUIDELINES.**

| Cattle feeding region | Average cost increment per head for guidelines applied only to problem commercial feedlots | Average cost increment per head for guidelines applied to all problem feedlots |
|---|---|---|
| 1 | $0.15 | $0.19 |
| 2 | 0.07 | 0.08 |
| 3 | 0.12 | 0.41 |
| 4 | 0.07 | 0.08 |
| 5 | 0.08 | 0.63 |
| 6 | 0.15 | 0.19 |
| 7 | 0.05 | 0.37 |
| 8 | 0.02 | 1.07 |
| 9 | 0.26 | 0.73 |
| 10 | 0.03 | 0.95 |
| 11 | 0.00 | 0.21 |
| 12 | 0.05 | 3.52 |
| 13 | 0.07 | 5.69 |
| 14 | 0.02 | 1.02 |
| 15 | 0.03 | 1.60 |
| 16 | 0.05 | 3.52 |

*Derived from supporting work in Johnson (1974).

TABLE 3. EQUILIBRIUM PRODUCTION LEVELS WITH EFFLUENT GUIDE-
LINES APPLIED ONLY TO PROBLEM COMMERCIAL FEEDLOTS.

| Cattle feeding region | Equilibrium production level before guidelines, 1000's of head | Equilibrium production level after guidelines, 1000's of head | Percentage deviation with change, percent |
|---|---|---|---|
| 1 | 718 | 718 | 0 |
| 2 | 872 | 872 | 0 |
| 3 | 399 | 399 | 0 |
| 4 | 569 | 569 | 0 |
| 5 | 54 | 54 | 0 |
| 6 | 2,605 | 2,605 | 0 |
| 7 | 5,939 | 5,939 | 0 |
| 8 | 1,414 | 1,414 | 0 |
| 9 | 4,590 | 4,590 | 0 |
| 10 | 3,305 | 3,305 | 0 |
| 11 | 3,986 | 3,986 | 0 |
| 12 | 638 | 638 | 0 |
| 13 | 1,217 | 1,217 | 0 |
| 14 | 1,256 | 1,256 | 0 |
| 15 | 194 | 194 | 0 |
| 16 | 319 | 319 | 0 |

feeding states. These data include the incremental change in production cost for each state by feedlot size range, as well as estimates of the number of head marketed from feedlot operations not meeting the new guidelines. Thus, the average change in production costs attributable to the new guidelines was calculated for these 17 major feeding states and adapted to the 16 cattle feeding regions used in the model. The estimated average cost changes for each cattle feeding region are listed in Table 2. The second column contains the average incremental change in cost per head marketed as a result of the effluent guidelines directed at commercial lots. The third column contains the incremental costs that would result if the legislation had been directed at all feedlots.

To evaluate the effect on feedlot location of the Federal Water Pollution Control Act amendments of 1972, the regional nonfeed cost parameters are adjusted by the incremental costs contained in the second column of Table 2. The equilibrium production levels that result from making these changes are contained in the third column of Table 3. The previous equilibrium production levels and the percentage deviations are also listed for comparison. Table 3 shows that the new legislation will have negligible, if any, effect on the aggregate distribution of fed cattle production.

For the second case, the equilibrium production levels were calculated assuming the legislation had been applied with equal force to all cattle feedlots, regardless of size. For this case, the incremental costs contained in the third column of Table 2 were used. The equilibrium production levels that result from making these changes are given in Table 4 along with the base case equilibrium production levels and the percentage deviations. This case does indicate significant changes in the equilibrium position of cattle feeding. The western regions, containing predominantly commercial feedlots, (Regions 2, 4, 6, 7, and 9) show substantial increases, while Wisconsin, Illinois, and Missouri (Regions 12 and 13) show substantial decreases. One of the major reasons for the large decline in this area is that a large proportion of the marketings are from small, problem feedlots. This is reflected in the unusually large increment to average nonfeed costs for Region 13. Thus, the decision to direct the effluent limiatations only at the commercial feedlots reduced the impact this legislation might have on the distribution of the cattle feeding industry.

## SUMMARY

A computer model was used to describe the operation of the economic system in allocating cattle production throughout the United States in such a way as to minimize costs. It is an aggregated model and is developed from the point of view of an average feedlot in each region. The effect of new environmental legislation on the equilibrium location of cattle feeding was modelled by adding average

TABLE 4. EQUILIBRIUM PRODUCTION LEVELS WITH EFFLUENT GUIDE-
LINES APPLIED TO ALL PROBLEM FEEDLOTS.

| Cattle feeding region | Equilibrium production level before guidelines, 1000's of head | Equilibrium production level after guidelines, 1000's of head | Percentage deviation with change, percent |
|---|---|---|---|
| 1 | 718 | 718 | 0 |
| 2 | 872 | 964 | 11 |
| 3 | 399 | 387 | -3 |
| 4 | 569 | 676 | 19 |
| 5 | 54 | 54 | 0 |
| 6 | 2,605 | 2,691 | 3 |
| 7 | 5,939 | 6,110 | 3 |
| 8 | 1,414 | 1,542 | 9 |
| 9 | 4,590 | 4,685 | 2 |
| 10 | 3,305 | 3,305 | 0 |
| 11 | 3,986 | 3,986 | 0 |
| 12 | 638 | 611 | -4 |
| 13 | 1,217 | 591 | -51 |
| 14 | 1,256 | 1,256 | 0 |
| 15 | 194 | 194 | 0 |
| 16 | 319 | 319 | 0 |

incremental costs, as a result of the new legislation, to the average nonfeed cost, prior to the legislation. The model was again used to describe the operation of the economic system, considering the newly calculated costs, in allocating production throughout the United States. On the basis of these two cases, it was concluded that, in the aggregate, the new environmental legislation will have little effect on the distribution of cattle production in the United States.

## References


1   Byrkett, D. L. 1974. Modelling the optimal location of the cattle feeding industry with particular emphasis on environmental consideration. Ph.D. diss., Ohio State University.

2   Dietrich, R. A. 1971. Interregional competition in the cattle feeding industry with special emphasis on economies of size. Tex. Agr. Exp. Sta. Bull. 1115.

3   Goodwin, J. W. and J. R. Crow. 1973. Optimal regional locations of beef production and processing enterprises. Okla. St. Univ. Agr. Exp. Sta. Bull. 707.

4   Johnson, J. B. and G. A. Davis. 1974. The economic impacts of imposing EPA effluent guidlines on the U.S. fed beef industry. Cornell Waste Mgmt. Conf.

5   King, G. A. and L. F. Schrader. 1963. Regional location of cattle feeding—a spatial equilibrium analysis. Hilgardia. 34:331-416.

6   U.S. Congress. House. 1972. Federal water pollution control act Amendments of 1972. Pub. law 92-500, 92nd Cong., 1st sess.

7   U.S. Department of Agriculture. 1972. National and state livestock-feed relationships. ERS, SRS, Stat. Bull. No. 446.

8   U.S. Department of Agriculture. 1973a. Agricultural statistics 1973.

9   U.S. Department of Agriculture. 1973b. Livestock and meat statistics. ERS, SRS, Stat. Bull. No. 522.

10   U.S. Environmental Protection Agency. 1973. Effluent limitations guidelines for existing sources and standards of performance and pre-treatment stnadards for new sources--notice of proposed rule-making. Federal Register, 38(173).

11   U.S. Environmental Protection Agency. 1974. Feedlots point source category--effluent guidelines and standards. Federal Register, 39(32).


---

## Water Pollution Control Rules

*(Continued from page 44)*

should therefore be viewed in relative, not absolute, terms.

## References


1   Adriano, D. C. et al. 1974. Ecosystem design and management: technology alternatives in beef production. Internal Rept. No. GI-20 of the National Science Foundation prepared by Michigan State University. 91 p.

2   Black, J. Roy and Harlan D. Ritchie. 1973. Average daily gain and daily dry matter intake of various kinds of cattle fed three different rations under several environmental situations. Staff paper 1973-1, Dept. of Agricultural Economics, Michigan State University. 7 p.

3   Forster, David Lynn. 1974. The effects of selected water pollution control rules on the simulated behavior of beef feedlots. Unpublished Ph.D. thesis, Michigan State University. 202 p.

4   Hughes, Harold A. 1973. Energy consumption in beef production systems are influenced by technology and size. Unpublished Ph.D. thesis, Michigan State University. 180 p.

5   Johnson, Glenn L. 1972. Alternatives to the neo-classical theory of the firm. American Journal of Agricultural Economics. 54(2):295-303.

6   Johnson, J. B. and G. A. Davis. 1974. The economic impacts of imposing EPA effluent guidelines on the U.S. fed-beef industry. Cornell University Waste Management Conference.

7   Kyle, Leonard R. Business analysis summary for cattle feeding farms. Agricultural Economics Report, Michigan State University. Selected annual issues.

8   Trimble, Richard L. 1973. An economic analysis of the effect of monetary policy on the beef industry. Unpublished Ph.D. thesis, Michigan State University. 174 p.

9   U.S. Congress. 1972. Public law 92-500. Ninety-second Congress. 89 p.

# Economic Impacts of Implementing EPA Water Pollution Control Rules on the United States Beef Feeding Industry

J. B. Johnson, G. A. Davis

THE Federal Water Pollution Control Act Amendments of 1972 (Act) provide a mandate for the U. S. Environmental Protection Agency (EPA) to work towards improving the quality of navigable water (United States Congress 1972). The Act provides specific reference to point source water pollution problems associated with feedlots. In February, 1974, EPA announced final effluent limitations guidelines for control of point source discharges from feedlots with one-time capacities of 1000 animal units or more (Anonymous 1974). This paper presents an analysis of certain economic impacts of implementing quidelines for the control of runoff from beef feedlot production facilities.

The objectives of this paper are to: (a) estimate the number of beef feedlots that have surface water control problems and (b) estimate the economic impacts of implementing effluent limitations guidelines on the beef feeding industry. Economic impacts explicitly considered are the aggregate industry capital outlays for runoff control, and the per head investment and production cost changes.

Estimates of the number of feedlots with problems and the economic impacts of implementing guidelines are presented for feedlots of all capacity levels to provide a preliminary assessment for the entire industry. Although guidelines announced in February, 1974, apply to only larger capacity feedlots, EPA has indicated it may develop guidelines for beef feedlots of lesser capacity. EPA may specify a feedlot size cutoff less than the current minimum capacity of 1000 head steers or heifers. If EPA takes such action, the cutoff level for feedlot capacity and the nature of other possible point source control requirements can affect the economic viability of individual feedlots and the economic impacts on the entire industry.

By July 1, 1977, feedlots with discharges will be required to have in use the "best practicable control technology currently available" (BPT). By July 1, 1983, guidelines require the use of "best available technology economically achievable" (BAT). Provisions of the Act suggest attainment of levels of substantial improvement in discharge quality are expected prior to these dates. Specifics of the final guidelines for feedlots are:* (Anonymous 1974).

After application of BPT there shall be no discharge of process waste water pollutants to navigable waters except that process waste waters pollutants in the overflow may be discharged to navigable waters whenever rainfall events, either chronic or catastrophic, cause an overflow of process waste water from a facility designed, constructed, and operated to contain all process generated waste waters plus the runoff from a 10-year, 24-hour rainfall for the location of the point source.

After application of BAT there shall be no discharge of process waste water pollutants to navigable waters except that process waste water pollutants in the overflow may be discharged to navigable waters whenever rainfall events, either chronic or catastrophic, cause an overflow of process waste water from a facility designed, constructed, and operated to contain all process generated waste waters plus the runoff from a 25-year, 24-hour rainfall for the location of the point source.

An understanding of the intent of these guidelines requires an understanding of the terminology used. Specialized definitions are presented in Appendix A.

The final feedlot guidelines are performance standards. Although specific references are made to the 10-year, 24-hour and 25-year, 24-hour rainfall events, the language in the announcement of final guidelines indicates that the expected level of performance is "no discharge" of process waste water pollutants to navigable waters except that overflow from control systems (designed and operated in accordance with guidelines) caused by excessive chronic or catastrophic precipitation may be discharged without regard to pollutants in, or volume of, the overflow.

Guidelines for feedlots had not been announced by EPA when the research on which this paper is based was initiated. Indicators of the nature of guidelines that could be expected were derived from available sources. Although the base information used for the analysis does not coincide exactly with the final guidelines, it is sufficiently close to allow meaningful judgements of industry economic impacts. As a prelude to identifying that segment of the beef feeding industry with surface water control problems, a description of the beef feeding industry is provided to place the adjustments to the guidelines in the context of the entire industry.

## THE BEEF FEEDING INDUSTRY IN 18 MAJOR FEEDING STATES

The beef feeding industry is widely dispersed over the United States. This study focuses on problems encountered by feedlots in 18 of the major feeding States (Fig. 1). In 1969, these 18 States accounted for nearly 98 percent of all beef feedlots and 95 percent of the beef cattle marketed from feedlots in the United States.

About 1 percent of the nearly 183 000 beef feedlots in these 18 States were in the 1000 head or more capacity class (Table 1). Feedlots of 1000 head or more accounted for nearly one-half of the marketings from all feedlots in the 18 States.

Although feedlots are often classified by capacity, more indicative of surface water problems is a classification by

---

Unless otherwise cited, information in this paper was adapted from an ERS manuscript entitled "Economic Impacts of Controlling Surface Water Runoff from Fed-beef Production Facilities," by J. B. Johnson, G. A. Davis, J. Rod Martin, and C. Kerry Gee.

The authors are: J. B. JOHNSON and G. A. DAVIS, Agricultural Economists, Economic Research Service, USDA, East Lansing, Mich.

*Both these guidelines apply to existing feedlots. The latter guideline is essentially the standard of performance for new entrants to the industry.

| Capacity class, by State groupings | Population of beef feedlots | | | | | Feedlots with runoff problems | | | | | Proportion of population of feedlots with runoff problems | | | | |
|---|---|---|---|---|---|---|---|---|---|---|---|---|---|---|---|
| | Type of housing | | | | Total feed-lots, by capacity class | Type of housing | | | | Total feed-lots, by capacity class | Type of housing | | | | Total feed-lots, by capacity class |
| | Total confinement | Open-lot | Dry-lot paved | Dry-lot unpaved | | Total confinement | Open-lot | Dry-lot paved | Dry-lot unpaved | | Total confinement | Open-lot | Dry-lot paved | Dry-lot unpaved | |
| | Number | | | | | Number | | | | | Percent | | | | |
| **Eastern States:** | | | | | | | | | | | | | | | |
| <100 | 3 511 | 54 886 | 22 663 | 67 094 | 148 154 | 0 | 13 779 | 7 444 | 16 906 | 38 129 | 0 | 25 | 33 | 25 | 26 |
| 100-199 | 216 | 6 200 | 2 884 | 7 251 | 16 551 | 0 | 1 299 | 1 201 | 1 748 | 4 248 | 0 | 21 | 42 | 24 | 26 |
| 200-499 | 235 | 4 619 | 1 651 | 3 968 | 10 473 | 0 | 1 116 | 761 | 1 019 | 2 896 | 0 | 24 | 46 | 26 | 28 |
| 500-999 | 61 | 1 348 | 312 | 735 | 2 456 | 0 | 362 | 142 | 202 | 706 | 0 | 27 | 46 | 28 | 29 |
| 1 000 and more | 16 | 862 | 53 | 75 | 1 006 | 0 | 435 | 10 | 26 | 471 | 0 | 50 | 19 | 35 | 47 |
| Sub-total | 4 039 | 67 915 | 27 563 | 79 123 | 178 640 | 0 | 16 991 | 9 558 | 19 901 | 46 450 | 0 | 25 | 35 | 25 | 26 |
| **Western States:** | | | | | | | | | | | | | | | |
| <1 000 | 0 | 3 473 | 0 | 0 | 3 473 | 0 | 2 244 | 0 | 0 | 2 244 | 0 | 65 | 0 | 0 | 65 |
| 1 000- 7 999 | 0 | 597 | 0 | 0 | 597 | 0 | 107 | 0 | 0 | 107 | 0 | 18 | 0 | 0 | 18 |
| 8 000-15 999 | 0 | 110 | 0 | 0 | 110 | 0 | 19 | 0 | 0 | 19 | 0 | 17 | 0 | 0 | 17 |
| 16 000 and more | 0 | 97 | 0 | 0 | 97 | 0 | 13 | 0 | 0 | 13 | 0 | 13 | 0 | 0 | 13 |
| Sub-total | 0 | 4 277 | 0 | 0 | 4 277 | 0 | 2 383 | 0 | 0 | 2 383 | 0 | 56 | 0 | 0 | 56 |
| 18 state total | 4 039 | 72 192 | 27 563 | 79 123 | 182 917 | 0 | 19 374 | 9 558 | 19 901 | 48 833 | 0 | 27 | 35 | 25 | 27 |

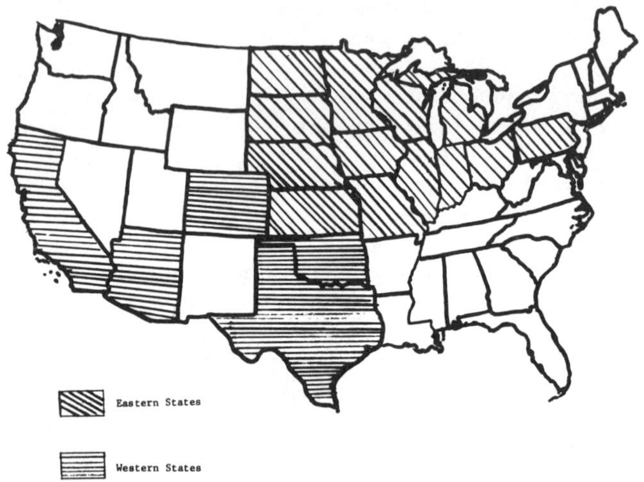


FIG. 1 Eighteen of the major beef feeding states by state groupings.

production technology for which housing type is taken as a major indicator. Ample identification of surface water problems or runoff potential was achieved by considering four housing types. These are:

1 Total confinement. Such systems consist of either cold-covered shelters enclosed on three sides with the fourth side fenced, or warm-enclosed shelters with all sides enclosed.

2 Dry-lot paved. Such systems combine shelter and exposed areas. Exposed areas are totally surfaced.

3 Dry-lot unpaved. Such systems combine shelter and exposed areas. Exposed areas are not surfaced except for surfaced feedlot aprons.

4 Open-lot. Such systems have no roofed shelters. The most prevalent open-lot system consists of fenceline bunk feeding with no surfacing except in front feed bunks.

Open-lot systems were used in operations of all capacity levels in the Western States (Table 1). For operations with marketings of 1000 head or more in the Eastern States, open-lot systems were very prevalent. Dry-lot unpaved housing systems were prevalent for operations in the less than 100 and 100-199 head capacity classes in the Eastern States. In these States, open-lot systems were most prevalent in the 200-499 head and 500-999 head capacity classes.

## BEEF FEEDLOTS WITH SURFACE WATER CONTROL PROBLEMS

University, Federal, and State agency personnel knowledgeable about waste management practices provided estimates of the numbers of feedlots that had surface water control problems. Problem situations considered were feedlots with existing problems of runoff or other discharges from production facilities, and feedlots with the potential for runoff in the event of the occurrence of a 10-year, 24-hour rainfall.

Some 48,833 beef feedlots were identified as having problem situations (Table 1). Nearly 27 percent of the feedlots in the 18 States had problem situations (hereafter referred to as runoff problems). Collectively, the feedlots with runoff problems marketed 25 percent of the beef produced in these 18 States in 1972. About a third of the 1810 feedlots with one-time capacities of 1000 head or more had runoff problems.

Only nominal runoff problems were associated with total confinement housing. Those few surface water problems associated with discharges from total confinement systems were thought to be rectifiable through waste management practice modifications.

Some 35 percent of those feedlots with dry-lot paved housing systems, because of their location in the higher rainfall Eastern States, had runoff problems. About 27 percent of those feedlots using open-lot systems had runoff problems. Of feedlots with dry-lot unpaved housing systems, some 25 percent had runoff problems (Table 1).



MANAGING LIVESTOCK WASTES

## THE ECONOMIC IMPACTS OF RUNOFF CONTROL

Estimates are provided of the capital outlays and changes in production costs that would result from the imposition of guidelines requiring the use of BPT and BAT technology levels by feedlots with runoff problems (measured in 1972 dollars). The means of runoff control selected by feedlot operators can be influenced by the technical compatibility of the control technology with existing resource situations, the economic viability of the feedlot firm preceding the adjustment, and the nature of the requirements specified in the guidelines. In this analysis a four-component runoff control system, judged as exemplifying available control technology, was evaluated. The components of this system are: diversion terraces to divert extraneous flows away from exposed feedlot surfaces; settling basins to prevent solids carried in runoff from entering retention ponds; retention ponds for containment of runoff; and pump-irrigation equipment for emptying retention ponds and distributing controlled runoff onto farmland.

### Capital Outlays

Guidelines announced by EPA are specific relative to the quantities of runoff to be controlled, and do not directly pertain to the organic and nutrient content of the runoff or the degradation inflicted on receiving waters. Runoff leaving feedlots is primarily determined by the land area exposed to storm events and storm magnitudes. Capital outlays for retention ponds (and associated lining and fencing) and pump-irrigation equipment are a function of the land area exposed and quantities of runoff arising from design storm events.† Capital outlays for diversion terraces and settling basins are primarily a function of feedlot layout, and are not directly dependent on the level of precipitation the system is designed to control.

Differences in capital outlay requirements for control systems by capacity class by region are apparent when expressed on a per head basis (Table 2). As feedlot capacity increases, economies of size are realized in capital outlays on a per head capacity basis for diversion terrace construction and lining and fencing of retention ponds.

If feedlot operators were required to own pump-irrigation equipment with the capabilities of meeting the management practice considered would require a minimum capital outlay in most cases of $2000 to $2500. Consequently, investments per head in pump-irrigation equipment are much higher for small capacity feedlots than for large capacity feedlots.

The range in per head capital outlays for any particular feedlot capacity class is primarily attributable to the feedlot location and housing type (Table 2). For feedlots of the same capacity class and housing type, capital outlays for retention ponds and pump-irrigation equipment are generally highest in the high rainfall States. For feedlots of the same capacity class and in similar locations, capital outlays for control systems are generally highest for land-extensive, open-lot systems and lowest for the land-intensive, dry-lot paved housing systems.

---

†Capital outlays are also a function of the price per unit charged for excavation and construction charges. These per unit charges will vary by location (among states). For example, excavation charges varied from less than $0.50 to more than $1.00 per cu yd. However, pump-irrigation equipment prices for equipment of equal size were assumed to be the same for producers in all states.

**TABLE 2. CAPITAL OUTLAYS PER HEAD OF FEEDLOT CAPACITY FOR RUNOFF CONTROL SYSTEMS, BY FEEDLOT CAPACITY CLASS, FOR FEEDLOTS WITH RUNOFF PROBLEMS, 1972.***

| Capacity class, by state groupings (head) | Per head capital outlay (weighted average)† | Range in per head capital outlay† |
|---|---|---|
| | Dollars | Dollars |
| Eastern States: | | ‡ |
| <100 | 145.20 | |
| 100-199 | 21.00 | 13.42 to 46.84 |
| 200-499 | 11.60 | 8.66 to 37.61 |
| 500-999 | 8.18 | 4.44 to 20.73 |
| 1 000 and more | 3.13 | 2.62 to 30.71 |
| Western States: | | |
| <1 000 | 21.65 | 11.85 to 29.75§ |
| 1 000- 7 999 | 2.92 | 2.15 to 4.26 |
| 8 000-15 999 | 1.61 | 1.08 to 2.59 |
| 16 000 and more | 1.38 | 0.88 to 2.41 |

*These estimates should not be extrapolated to the entire beef feeding industry, as weighting reflects the characteristics of feedlots with problem situations and not those of the total feedlot population.
†These estimates closely approximate those that could be expected by feedlot firms adopting the four-component system capable of meeting the guideline requiring use of the best practicable control technology currently available. Estimates for feedlot firms using the four-component runoff control system to be in compliance with the guideline requiring the use of the best available technology economically achievable would be expected to be slightly greater. Feedlots using the four-component system to be in compliance with July 1, 1983 requirements would incur larger expenditures for retention pond excavation, lining, and fencing to accommodate control of the larger extreme storm runoff. As the diversion terrace and settling basin components are designed independently of the storm event magnitude, capital outlays would be similar for these two components under both guidelines. Pump-irrigation equipment with capacity to handle runoff from the 10-year, 24-hour storm would generally accommodate runoff distribution for the 25-year, 24-hour storm.
‡The range in per head capital outlays reflects differences in housing type, location, and per unit charges for excavation and construction. In addition, as feedlot capacity for capacity classes of less than 1,000 head was determined by marketings, the actual capacity of feedlots within a particular capacity class may vary by State; therefore, the per head capital outlay range for operations of less than 100 head has been omitted.
§In the Western States, the range does not reflect differences in housing type, as all are open-lot systems.

### Production Cost Changes

The addition of runoff control systems will cause associated changes in beef production costs. Once the runoff control systems are in place, total costs of runoff control will remain unchanged for a feedlot of fixed physical dimensions, irrespective of the level of fed-beef production.‡ Net effects of beef production cost increases on net income levels will depend on the income and tax positions of the individual feedlot firms.

---

‡Amortization rates reflecting a 7.5 percent interest rate for a 20-year life on all system components, except pump-irrigation equipment, were applied to capital outlays to determine ownership costs. A one percent per-annum maintenance charge was included in ownership costs. Once the runoff system is in place, the total costs of runoff control are largely fixed; that is, they depend on the capital outlays in the system, and are invariant with the number of cattle being fed. For example, a feedlot with physical capacity for 150 head in an Eastern state could easily require an initial capital outlay of $3300 for a runoff control system. Annual total costs of operating each year would be about $480. Be feeding 150 head per year, production costs attributable to runoff control would be $3.20 per head; by feeding 100 head per year, per head costs would be $4.80.

Differences in production cost increases for feedlot firms that adopt control systems and continue to operate at historical production levels are primarily attributable to initial levels of capital outlay for the control systems (Table 3). Size economies are reflected in the expected changes in production costs. Larger capacity feedlots will realize lower cost increases per head marketed. The range in per head production cost increases within any capacity class is primarily a function of varying levels of initial capital outlays for the control systems which are influenced by feedlot location and housing type. Largest cost increases within each capacity class are expected for the land-extensive housing systems.

**TABLE 3. INCREASES IN FED-BEEF PRODUCTION COSTS ATTRIBUTABLE TO THE OWNERSHIP AND OPERATION OF RUNOFF CONTROL SYSTEMS, BY FEEDLOT CAPACITY CLASS, FOR FEEDLOTS WITH RUNOFF PROBLEMS, 1972***

| Capacity class, by state groupings (head) | Cost increase per head marketed (weighted average)[†] | Range in cost increases per head marketed[†] |
|---|---|---|
| | Dollars | Dollars |
| Eastern States: | | [‡] |
| <100 | 21.17 | |
| 100-199 | 3.19 | 2.57 to 6.55 |
| 200-499 | 1.84 | 0.64 to 5.04 |
| 500-999 | 1.28 | 0.68 to 2.49 |
| 1 000 and more | 0.69 | 0.40 to 4.03 |
| Western States: | | |
| <1 000 | 5.79 | 2.00 to 8.04 [§] |
| 1 000- 7 999 | 0.57 | 0.47 to 0.83 |
| 8 000-15 999 | 0.40 | 0.34 to 0.60 |
| 16 000 and more | 0.36 | 0.31 to 0.58 |

*These estimates should not be extrapolated to the entire beef feeding industry, as weighting reflects the characteristics of feedlots with problem situations and not those of the population of all feedlots.
†These estimates closely approximate those that could be expected by feedlot firms adopting the best practicable control technology currently available. Production cost increases would be expected to be slightly higher for feedlot firms adopting a control system that would represent the best available technology economically achievable. For a more complete explanation, see footnote †, Table 2.
‡As feedlot capacity was determined from actual marketings, the ranges in the estimates for feedlot firms of the capacity class are subject to misinterpretation and have been omitted.
§In the Western States, the range does not reflect differences in housing type as all are open-lot systems.

### Industry Capital Outlays

To provide some insight into the magnitude of the industry adjustment problem, aggregate estimates of total capital outlays were prepared. These estimates presuppose that all beef feedlots with runoff problems adopt the four-component control system and retain historical production levels (Table 4).

The total capital outlays for runoff control systems for feedlots with one-time capacities of 1000 head or more steers or heifers with runoff problems was estimated to be nearly $7.4 million in 1972 (Table 4). The additional capital outlay for runoff control for those feedlots currently subject to EPA guidelines is nominal compared with the existing level of investments in feedlot production facilities for this segment of the beef feeding industry.

Of major concern is the burden of adjustment if the announced EPA guidelines currently applicable to only larger capacity feedlots were applied to feedlots of lower capacity levels.

If these guidelines were applied to the more than 1800 beef feedlots of 500-999 head one-time capacities, an additional $7.4 million (split about evenly between the Eastern and Western States) would have to be invested by feedlots with runoff problems in this capacity class (Table 4). (To estimate the Western States' capital requirements, it was assumed that one-half the 2244 feedlots reported in the less than 1000 head capacity class in the Western States were in the 500-999 head capacity range). Industry capital outlays for runoff control systems would total nearly $15 million (in 1972 dollars) if EPA guidelines were applied to all feedlots of one-time capacities of 500 head or more with runoff problems (Table 4). Thus, dropping the feedlot capacity cutoff from 1000 to 500 head would more than double the required capital outlay.

Most striking is what would happen if EPA guidelines currently applicable to feedlots with one-time capacities of 1000 head or more were applied to all feedlots with runoff problems, irrespective of feedlot capacity. Approximately 95 percent of the investment would be required by feedlots with capacities of less than 1000 head. And over two-thirds of the industry's capital outlay for runoff control systems would be made by feedlots with one-time capacities of less than 100 head (Table 4). These estimates show vividly that the $133 million capital outlay required by the beef feeding industry to adopt the four-component runoff control system would not be uniformly distributed across feedlot capacity classes.

**TABLE 4. NUMBER OF FEEDLOTS WITH RUNOFF PROBLEMS, CAPITAL OUTLAYS REQUIRED FOR RUNOFF CONTROL SYSTEMS, AND PERCENT AND CUMULATIVE PERCENT OF CAPITAL OUTLAYS, BY FEEDLOT CAPACITY CLASS.**

| Capacity class, by state groupings (head) | Feedlots with problem situations | Capital outlays | Percent of total capital outlays | Cumulative percent of total capital outlays |
|---|---|---|---|---|
| | Number | Thousand dollars | Percent | Percent |
| Eastern States: | | | | |
| <100 | 38 129 | 91 789 | 69.2 | 69.2 |
| 100-199 | 4 248 | 12 435 | 9.3 | 78.5 |
| 200-499 | 2 896 | 10 053 | 7.6 | 86.1 |
| 500-999 | 706 | 3 736 | 2.8 | 88.9 |
| Western States: | | | | |
| <1 000 | 2 244 | 7 413 | 5.6 | 94.5 |
| Eastern States: | | | | |
| 1 000 and more | 471 | 5 212 | 3.9 | 98.4 |
| Western States: | | | | |
| 1 000- 7 999 | 107 | 771 | 0.6 | 99.0 |
| 8 000-15 999 | 19 | 434 | 0.3 | 99.3 |
| 16 000 and more | 13 | 943 | 0.7 | 100.0 |
| Total | 48 833 | 132 780 | 100.0 | |

## IMPLICATIONS

Additional costs will accrue to those feedlot firms that invest in control systems. These additional costs become fixed when systems are in place. For the individual feedlot, the total costs of runoff control cannot be altered with changes in the level of production.

In the Eastern States, the costs of runoff control can be expected to be substantial for small capacity feedlots. Within the same capacity level, per head costs are generally greatest for the land-extensive, open-lot systems and lowest

*(Continued on page 60)*

# Economic and Environmental Aspects of Daily and Annual Dairy Manure Spreading Systems in a Small Watershed

W. Harry Schaffer, George L. Casler, James J. Jacobs

CONCERN for water quality and lake eutrophication has brought livestock operations under critical review as a possible source of polluting nutrients. However, nutrient measurements taken from a water resource of a small watershed do not identify the amount of losses originating from a particular manure handling system. Stream measurements indicate only the total loading, i.e., the sum of many contributing factors. Although the nutrient losses from a manure handling system cannot be measured, they can be simulated.

This paper reports the simulated nitrogen, phosphorus, and soil loss from a small watershed where daily manure spreading is practiced. It also reports the simulated results when the system is changed to annual storage where manure is applied several days prior to crop planting and plowed down within 24 hours after application. The simulated nitrogen, phosphorus and soil losses from the two systems are incorporated into an analysis to determine the economic and environmental impact of controlling nutrient losses from the watershed.

The physical model was constructed from published laboratory and field data (Schaffer, Jacobs and Casler 1974) and had sub-components for soil moisture-temperature and for soil, nitrogen, and phosphorus movement. The nitrogen and phosphorus sub-models were superimposed in part upon the soil loss model and all three were superimposed on the first sub-model. Sections of the soil moisture model were computed, where needed, in one-tenth hour segments to more closely determine runoff amounts. Nitrogen, phosphorus and soil losses were calculated daily to account for specific climatic and hydrological conditions during the period of measurement. Three soil zones were maintained for hydrologic accounting and two soil layers for nutrient accounting. A system similar to double entry bookkeeping was used to keep track of the stocks and flows of the many nutrient forms and transformations. Nutrient and soil movement into and within a stream were computed using a proportionate delivery ratio for each combination of soil type, crop activity, fertilizer amount and manure handling system.

The economic and environmental analysis was made on the 6900 acre Mink Creek basin of the Canadarago Lake Watershed in east-central New York. The area was 64 percent in agriculture with 21 out of 22 farmers engaged in dairying. The watershed was selected because previous studies had developed some soil and land use data and because measurements of stream flow and nutrient loading were available for one hydrological year. Additional data on agricultural production practices was collected by survey during July 1973.

The basic economic model was structured to be representative of the kinds, amounts and intensities of agriculture found in the watershed. This required some consolidation of soil types, crop and livestock activities. The physical model, implemented by using the collected data, was used to determine the residual losses to water for these representative agricultural activities. The amounts of nitrogen and phosphorus lost from the watershed, as measured at the mouth of the stream, were divided by the sum of the computed losses of these nutrients from all watershed activities. This proportion was used as the delivery ratio to modify all preliminary cropping loss coefficients and was approximately one-third for nitrogen and eitht-tenths for phosphorus. Altered in this way the computed economic model representing agriculture in Mink Creek delivered the measured nutrient losses.

Table 1 presents the adjusted loss coefficients by crop activity, soil type and manure practice. Differences in loss

## TABLE 1. COMPUTED NITROGEN, PHOSPHORUS, AND SOIL LOSSES FOR MINK CREEK AGRICULTURAL ACTIVITIES

| Crop | Year | Soil | Daily manure spreading<br>Loss/Acre | | | Annual storage, direct manure plow down<br>Loss/Acre | |
|---|---|---|---|---|---|---|---|
| | | | N lb | P lb | Soil tons | N lb | P lb |
| Corn silage | 1 | Honeoye | 23.9 | 0.33 | 0.17 | 27.2 | 0.247 |
| | | Lima | 22.5 | 0.247 | 0.11 | 24.4 | 0.165 |
| | | Lansing | 23.6 | 0.33 | 0.16 | 30.8 | 0.247 |
| | | Conesus | 21.1 | 0.247 | 0.10 | 24.6 | 0.165 |
| | | Appleton | 20.2 | 0.165 | 0.06 | 22.9 | 0.165 |
| | | Farmington | 26.0 | 0.247 | 0.10 | 27.2 | 0.165 |
| Corn silage | 2 | Honeoye | 26.6 | 0.989 | 0.51 | 31.9 | 0.66 |
| | | Lansing | 29.5 | 0.907 | 0.45 | 33.3 | 0.66 |
| | | Farmington | 28.4 | 0.577 | 0.25 | 29.1 | 0.495 |
| Potato | | Lansing | 63.1 | 0.165 | 0.18 | 63.1 | 0.165 |
| Corn grain | | Conesus | 21.6 | 0.247 | 0.17 | 21.6 | 0.247 |
| Oats* | | Honeoye | 24.1 | 0.33 | 0.26 | 27.7 | 0.247 |
| | | Lima | 20.7 | 0.247 | 0.15 | 22.4 | 0.165 |
| | | Lansing | 22.8 | 0.33 | 0.21 | 29.8 | 0.247 |
| | | Conesus | 22.5 | 0.247 | 0.15 | 26.2 | 0.165 |
| | | Farmington | 16.8 | 0.165 | 0.11 | 17.5 | 0.165 |
| Alfalfa | 1-3 | Honeoye | 10.4 | 0.082 | 0.01 | 10.4 | 0.082 |
| | | Lima | 10.1 | 0.082 | 0.01 | 10.1 | 0.082 |
| | | Lansing | 11.7 | 0.082 | 0.01 | 11.7 | 0.082 |
| | | Conesus | 10.8 | 0.082 | 0.01 | 10.8 | 0.082 |
| | | Farmington | 7.3 | 0.082 | 0.01 | 7.3 | 0.082 |
| Alfalfa | 4-5 | Honeoye | 7.7 | 0.082 | 0.01 | 11.1 | 0.082 |
| | | Lima | 6.9 | 0.082 | 0.01 | 11.0 | 0.082 |
| | | Lansing | 7.4 | 0.082 | 0.01 | 11.8 | 0.082 |
| | | Conesus | 7.2 | 0.082 | 0.01 | 11.6 | 0.082 |
| | | Farmington | 5.7 | 0.082 | 0.01 | 8.0 | 0.082 |
| Birdsfoot trefoil | | Appleton | 9.1 | 0.082 | 0.01 | 9.1 | 0.082 |
| Improved pasture | | Honeoye | 4.4 | 0.082 | 0.01 | 4.4 | 0.082 |
| | | Lima | 3.4 | 0.082 | 0.01 | 3.4 | 0.082 |
| Permanent pasture | | | 1.6 | 0.124 | 0.03 | 1.6 | 0.124 |

*Nutrient losses from oats for annual manure storage and direct soil incorporation were computed from the ratio of daily to direct first year corn loss rather than via the physical model.

The authors are: W. HARRY SCHAFFER, Pennsylvania State University, Cooperative Extension Service, Agricultural Center, R.D. 1, Leesport, Pa. 19533; GEORGE L. CASLER, Department of Agricultural Economics, Cornell University, Ithaca, N.Y. and JAMES J. JACOBS, the Economic Research Service, stationed at Cornell University, respectively.

## TABLE 2. EFFECT OF NITROGEN LOSS RESTRICTIONS ON FARM ORGANIZATION AND INCOME, MINK CREEK WATERSHED

|  | | Daily manure spreading | | | | Annual manure storage with direct plow down | | | |
|  | | Restrictions on nitrogen loss, lbs | | | | Restrictions on nitrogen loss, lbs | | | |
|  | Unit | Initial | 34 536 | 25 773 | 17 010 | Initial | 34 536 | 25 773 | 17 010 |
|---|---|---|---|---|---|---|---|---|---|
| Net return | Dollars | 447 272 | 388 957 | 347 022 | 266 015 | 401 977 | 339 074 | 295 688 | 214 132 |
| Cows | No. | 935 | 935 | 838 | 733 | 935 | 838 | 831 | 646 |
| Heifers | No. | 234 | 53 | — | — | 234 | — | — | — |
| Buy heifers | No. | — | 181 | 209 | 183 | — | 210 | 208 | 162 |
| Beef | No. | 200 | — | — | — | 200 | — | — | — |
| Potatoes | Acres | 55 | 55 | — | — | 55 | 55 | — | — |
| Corn | Acres | 800 | 420 | 354 | 169 | 800 | 444 | 238 | 149 |
| Oats | Acres | 499 | 272 | 222 | 169 | 379 | 201 | 208 | 149 |
| Hay | Acres | 1 976 | 1 390 | 1 119 | 851 | 2 095 | 1 199 | 1 236 | 754 |
| Permanent pasture | Acres | 970 | 970 | 970 | 970 | 970 | — | — | — |
| Buy corn | Bu | 21 520 | 28 037 | 24 532 | 53 564 | 25 985 | 18 799 | 39 883 | 44 089 |
| N purchased | Lb | 32 736 | 20 569 | 9 801 | 2 529 | 20 800 | 15 458 | 3 566 | 2 239 |
| $P_2O_5$ purchased | Lb | 95 145 | 60 948 | 35 640 | 17 853 | 87 923 | 50 131 | 23 343 | 15 809 |
| N loss | Lb | 54 501 | 34 536 | 25 773 | 17 010 | 57 732 | 34 536 | 25 773 | 17 010 |
| P loss | Lb | 690 | 471 | 402 | 294 | 575 | 384 | 324 | 251 |
| Soil loss | Tons | 320 | 195 | 158 | 103 | 272 | 175 | 126 | 94 |
| Buy labor | Hr | 25 604 | 12 352 | 1 157 | — | 25 056 | 2 575 | — | — |
| Sell hay | Tons | 584 | — | — | — | 791 | — | — | — |
| Idle | Acres | — | 1 193 | 1 635 | 2 141 | — | 1 431 | 1 648 | 2 278 |

coefficients between soil types are caused by different pH levels, water handling characteristics, slope, and productivity. Loss differences between crops on the same soil depend on the yield and nutrient uptake, ground cover provided, manuring and amount of manure and/or fertilizer. Loss difference between manure practices on the same soil and for the same crop largely result from the timing of manure application.

The first column with numbers in Table 2 presents the representative watershed organization. The agricultural return over variable costs was computed to be $447 272 or approximately $20 300 per farm. Net farm income derived by subtracting fixed cost (such as depreciation, interest and taxes) was just under $10 000 per farm. Column 5, Table 2 lists the analogous computed description and return over variable costs with the changed manure practice of annual storage and direct plow down. Note that nitrogen losses are higher, phosphorus losses are lower, and return over variable costs to the watershed are lower than with daily manure spreading. Reduced income was anticipated because of added manure handling costs for annual storage. The reduction of $45 295 represents a 15 percent loss of watershed income in order to satisfy the annual manure storage with direct incorporation requirement. This income loss would be more palatable if all environmental losses affecting water quality were reduced. But this was not the case! Less manure phosphorus was lost through direct plow down, but nitrogen losses were greater with storage than with daily spreading. The reasons are clear from the simulation model, but as yet unsubstantiated from experimentation. Large ammonia volatilization losses occur with daily manure spreading. Only small amounts occur when manure is directly incorporated into the soil.

For example, on Honeoye soil with a 10 ton manure application, post-spreading, pre-plow ammonia volatilization was simulated to be 53.5 pounds of nitrogen equivalent per acre compared to 9.3 pounds for the manure system of annual storage combined with next day soil incorporation. Plow dates were identical. It should be cautioned that different climatic conditions from the measured year may have produced different results. After soil incorporation these larger amounts of nitrogen per ton of manure are converted to leachable nitrates. Higher initial nitrate inventories particularly before crop uptake allow greater leaching losses.

## NITROGEN AND PHOSPHORUS LOSS RESTRICTIONS

Restrictions on nitrogen losses are reported for the two manure handling methods in Table 2. Each column represents rational farmer response, given full knowledge, to a designated nitrogen loading level. Thus, the watershed is reorganized to produce maximum farm income.

Crop acres are decreased somewhat more at each nitrogen restriction level for annual manure storage than for daily manure spreading. This is an expected occurrence because nitrogen losses are greater for similar crop activities with the stored manure option (i.e., as predicted by Schaffer 1974). In general, for both manure handling systems, as nitrogen restrictions become more severe, farm income, crop acres, livestock numbers, phosphorus and soil losses are reduced. But the decline is greater for the system of annual manure storage combined with direct plow down.

Restricting the nitrogen loss from agricultural activities in the watershed to 2/3 the initial loss with daily manure spreading reduces farm income about $58 000 or nearly $2600 per farm. Approximately $2.90 of farm income was sacrificed per reduced pound of nitrogen in the stream (Table 3). Phosphorus and soil losses were also reduced. Most of the reductions in losses were accomplished by decreasing the amount of crop acres; in particular, corn acres were approximately halved. The same nitrogen restriction, 34 536 pounds, for annual storage and direct soil incorporation, reduces farm income about $63 900 or $2860 per dairy farm, assuming the farmers already have storage. This amounts to about $2.70 reduced farm income per decreased pound of nitrogen. However, the combination of annual storage and a 34 536 pound restriction reflects a cost of $4.65 per reduced pound of nitrogen in the stream when compared to the initial situation embracing the typical daily manure handling practice. Table 3 summarizes the marginal cost to the watershed of reducing nitrogen loss at different restriction levels for the two manure systems.



MANAGING LIVESTOCK WASTES

TABLE 3. REDUCTION IN NET FARM INCOME PER POUND
OF DECREASED NITROGEN LOSS

| Restriction on N loss, pounds | Daily manure spreading | Annual storage with direct plowdown | |
|---|---|---|---|
| | | Not including storage cost | Including storage cost |
| None to 34 536 | $2.90 | $2.70 | $4.65 |
| 34 536 to 25 773 | 4.80 | 5.00 | 6.35 |
| 25 773 to 17 010 | 9.20 | 9.30 | 10.40 |

Note that for both manure systems, the reduction in watershed income increases as the nitrogen restrictions become more severe.

Table 4 is a summary of the consequences of application of restrictions on phosphorus loss. As more stringent restrictions were placed on phosphorus losses, nitrogen and soil losses, farm income and crop acres were reduced for both manure handling systems. Generally, livestock numbers declined. At comparable restriction levels, fewer crop acres and less farm income was realized for annual manure storage and direct plow down. The major reason is that annual manure storage—direct soil incorporation is more costly than daily manure spreading. The rate of decline in farm income and crop acres was less for the annual manure storage—direct soil incorporation.

The marginal cost per pound of reduced phosphorus loss is compiled in Table 5. These costs are greater than for nitrogen because both the initial losses and the reductions in P loss are much smaller than for N. At each restriction

TABLE 5. REDUCTION IN NET FARM INCOME PER POUND
OF DECREASED PHOSPHORUS LOSS,
MINK CREEK WATERSHED

| Restriction on phosphorus loss, lb | Daily manure spreading | Annual storage with direct plowdown | |
|---|---|---|---|
| | | Not including storage cost | Including storage cost |
| None to 442 | $166 | $126 | $466 |
| 442 to 330 | 414 | 308 | 493 |
| 330 to 218 | 614 | 559 | 686 |

level, it is more costly to reduce phosphorus losses with daily manure spreading than with annual storage and soil incorporation within 24 hours, assuming the farmers already have the storage. If storage must be constructed, the restrictions are more costly for the storage system.

Agricultural activities appear to be an effective method for removing phosphorus from the watershed. In the initial solution with daily manure spreading, elementary phosphorus (P) purchased was 41 522 pounds. Manure production contributed 27 028 pounds of P. The estimated loss from farming was 690 pounds. Farm activities removed 99.0 percent of the phosphorus. By changing to annual manure storage with direct soil incorporation the percentage phosphorus removal was 99.1 percent. The removal rates for both systems surpassed that which is achieved by most tertiary sewage treatment plants.

## ENVIRONMENTAL TRADEOFFS

The above modeling effort indicates that farm income is lower for a given level of nitrogen or phosphorus loss with manure storage than with daily spreading. The rationale for storing manure and incurring this reduction in farm income is that it will reduce nutrient losses and thereby enhance water quality. If manure storage is adopted the inherent assumption is that the benefit from reduced nutrient losses (i.e., improved water quality) is at least as great as the reduction in farm income.

However, the relationship between water quality, nutrient losses and the method of handling manure is not quantified. Furthermore, the results of the model suggest that manure storage may reduce phosphorus losses, but at the same time tends to increase nitrogen losses to water when compared with daily spreading. If the model is correct, the relevance of manure storage as a means to improve water quality will depend on whether nitrogen or phosphorus is the main concern. In addition, the overall environmental effects of alternative dairy manure handling systems should be considered. Environmental characteristics, in addition to nutrient losses and water quality, that might be considered are odor, flies, appearance and noise. To obtain

TABLE 4. EFFECT OF PHOSPHORUS LOSS RESTRICTIONS ON FARM ORGANIZATION AND INCOME, MINK CREEK WATERSHED

| | Unit | Daily manure spreading | | | | Annual manure storage with direct plow down | | | |
|---|---|---|---|---|---|---|---|---|---|
| | | Restrictions on phosphorus loss, lbs | | | | Restrictions on phosphorus loss, lbs | | | |
| | | Initial | 442 | 330 | 218 | Initial | 442 | 330 | 218 |
| Net return | Dollars | 447 272 | 406 216 | 359 830 | 291 016 | 401 977 | 385 213 | 350 760 | 288 149 |
| Cows | No. | 935 | 935 | 850 | 835 | 935 | 935 | 935 | 825 |
| Heifers | No. | 234 | 192 | — | — | 234 | 234 | 96 | — |
| Buy heifers | No. | — | 42 | 212 | 209 | — | — | 137 | 206 |
| Beef | No. | 200 | — | — | — | 200 | 200 | — | — |
| Potatoes | Acres | 55 | 55 | 55 | 55 | 55 | — | — | — |
| Corn | Acres | 800 | 519 | 393 | 197 | 800 | 800 | 700 | 449 |
| Oats | Acres | 499 | 363 | 259 | 206 | 379 | 378 | 260 | 174 |
| Hay | Acres | 1 976 | 2 021 | 1 493 | 1 228 | 2 095 | 2 095 | 1 499 | 1 061 |
| Permanent pasture | Acres | 970 | — | — | — | 970 | 59 | — | — |
| Buy corn | Bu | 21 520 | 28 874 | 23 625 | 79 001 | 25 985 | 25 785 | 29 746 | 54 827 |
| N purchased | Lb | 32 736 | 25 534 | 20 996 | 11 765 | 20 800 | 20 800 | 19 300 | 15 532 |
| $P_2O_5$ purchased | Lb | 95 145 | 75 861 | 59 592 | 39 368 | 87 923 | 85 535 | 70 258 | 46 995 |
| N loss | Lb | 54 501 | 42 944 | 32 759 | 23 519 | 57 732 | 55 755 | 44 370 | 30 599 |
| P loss | Lb | 690 | 422 | 330 | 218 | 575 | 442 | 330 | 218 |
| Soil loss | Tons | 302 | 192 | 145 | 84 | 272 | 232 | 178 | 111 |
| Buy labor | Hr | 25 604 | 17 387 | 3 974 | 688 | 25 056 | 23 975 | 13 294 | — |
| Sell hay | Tons | 584 | — | — | — | 791 | 14 | — | — |
| Idle | Acres | — | 1 342 | 2 100 | 2 614 | — | 913 | 2 259 | 2 559 |

| Manure handling systems | Odor | Flies | Appear-ance | Noise | Loss of nutrients | | Risk to water quality | Estimated environ-mental impact | Cost/cow/year dollar |
| --- | --- | --- | --- | --- | --- | --- | --- | --- | --- |
| | | | | | During storage | After spreading | | | |
| Stanchion housing: | | | | | | | | | |
| 1. Gutter cleaner - spreader (daily) | 36 | 27 | 19 | 15 | 4† | 29 | 49 | 179 | 42 |
| 2. Gutter flush - liquid storage | 65 | 28 | 19 | 13 | 11 | 20 | 31 | 187 | 72 |
| 3. Gutter flush - liquid storage - soil injection | 48 | 24 | 16 | 13 | 10 | 14 | 20 | 145 | 74 |
| Free stall housing: | | | | | | | | | |
| 1. Tractor scraper - spreader (daily) | 37 | 32 | 22 | 15 | 4† | 28 | 46 | 184 | 38 |
| 2. Tractor - scraper - liquid storage | 64 | 28 | 18 | 15 | 8 | 20 | 29 | 182 | 81 |
| 3. Tractor scraper - liquid - soil injection | 51 | 25 | 13 | 15 | 10 | 17 | 23 | 154 | 83 |

*The data in this table are taken from Tables 7 and 8 of (Jacobs and Casler 1972).
†Since daily spreading involves no storage, four represents the minimum value a characteristic could receive.

a measure of the importance of these environmental characteristics for alternative manure handling systems, an opinion survey was conducted and reported by (Jacobs and Casler 1972). Scores representing the environmental impact were developed from this survey. Table 6 summarizes the results of this survey for daily spreading systems and liquid storage systems with and without soil injection.

Several observations can be made by comparing impact scores of individual environmental characteristics for the three manure handling systems. One of the most interesting comparisons is between the characteristics "odor" and "risk to water quality." Table 6 points out that the characteristic "risk to water quality" receives the highest score for daily spreading systems and "odor" receives the highest score for storage systems. Considering the tradeoff between these two characteristics, Table 6 indicates that the adoption of liquid manure storage without soil injection may improve water quality but its effect on odor is considerably higher than for daily spreading. These tradeoffs between characteristics for daily spreading and liquid storage without direct soil incorporation result in similar environmental impact scores for these two systems. However, when the liquid manure is immediately incorporated into the soil there is a reduction in the environmental impact score.

The purpose of the above discussion is to point out that when choices among alternative systems are made tradeoffs among environmental characteristics of these systems occur. Realizing these tradeoffs exist, choices among alternatives must consider several environmental factors that may be of concern. Not to do so may lead to an alternative system that improves the particular environmental characteristic being considered and at the same time creates another environmental impact that is unacceptable. The net result may well be a system which increases cost with no improvement in overall environmental quality.

## SUMMARY

The modeling suggests that farm costs of reducing nutrient losses to water are substantial. Reduced phosphorus losses appear particularly difficult and costly to achieve. More severe restrictions on nitrogen and phosphorus losses to water result in progressively declining farm income, reduction in crop acres and livestock numbers. Either nitrogen or phosphorus restrictions are accompanied by reduced losses of the other nutrient and by reduced soil losses, illustrating the interdependencies between these constituents. When the typical manure practice of daily spreading is changed to annual manure storage with soil incorporation within 24 hours, farm income, phosphorus and soil losses decreased. But nitrogen losses increased. Therefore, this research casts doubt on whether reducing some or all of daily manure spreading as practiced throughout the year would produce improved environmental conditions. In particular, it seems questionable to require farmers to invest in large manure storage facilities.


## References

1    Jacobs, J. J. and G. L. Casler. 1972. Economic and environmental considerations in dairy manure management systems. A.E. Res. 72-18, Department of Agricultural Economics, Cornell University, Ithaca, New York.

2    Schaffer, W. H. 1974. An economic analysis of nutrient and soil loss from a small New York watershed. PhD Thesis, Department of Agricultural Economics, Cornell University, Ithaca, New York.

3    Schaffer, W. H., J. J. Jacobs and G. L. Casler. 1974. An economic analysis of policies to control nutrient and soil losses from a small watershed in New York State. Proceeding, Cornell Agricultural Waste Management Conference.

# Implications of Selected Non-Point Source Pollution Regulations for U.S. Dairy Farms

Boyd M. Buxton, Stephen J. Ziegler

THE Federal Water Pollution Control Act Amendments of 1972 require of states "a description of the nature and extent of non-point sources of pollutants, and recommendations as to programs which must be undertaken to control each category of such sources". Non-point sources of pollution refer to those pollutants from non-specific or unidentified origins. One of these potential sources is the disposal of animal manure on land.

Several states currently have regulations either imposed or under consideration. Other states are expected to follow, and in time federal guidelines may be established. The pollution control regulations under consideration or already imposed in several states were examined in early 1974. Three general categories of guidelines emerged:

1 Guidelines that restrict application rates per acre. These restrictions usually limit the amount of nitrogen, manure tonnage, or number of animals per acre, and generally vary by type of crop and soil.

2 Guidelines that restrict manure disposal in certain locations. These restrictions limit manure disposal on land adjacent to water or residences, or on land exceeding a specified slope.

3 Guidelines that restrict manure disposal under certain climatic conditions. In most instances these restrictions limit or prohibit manure disposal on frozen or snow-covered ground. They were often combined with guidelines restricting manure disposal on land exceeding a specified slope, or with guidelines restricting application rates per acre.

Based on these state recommendations, hypothetical guidelines were selected. They included restrictions that could be required of U. S. dairy farmers who:

1 have more than 3, 5, or 10 dairy cows per acre;

2 have more than 3, 5, or 10 animal units per acre;*

3 apply more than 10, 20, 30, or 40 tons of manure per acre annually;†

4 apply more than 200, 250, 300, and 400 pounds of nitrogen per acre annually;†

5 spread manure on frozen land; and

6 have farm boundaries located within 300, 600, or 1100 feet of the nearest farm residence, non-farm residence, public area, or body of water used for recreation.

A survey to determine the animal waste handling characteristics and manure disposal practices on dairy farms was conducted in early 1973. The survey obtained information from 2 652 dairy farmers in ten regions covering the continental United States.‡ The proportion of farmers surveyed who would be affected by these hypothetical guidelines is reported in this paper.

## GUIDELINES RESTRICTING APPLICATION RATES PER ACRE

Farmers surveyed were asked how many acres they had available and how many acres they actually used for manure disposal. They were also asked what proportion of the acres actually used could be classified as flat or nearly flat.

The amount of land available on each farm would be equal to or greater than the amount of land actually used. For all farmers responding to the survey an average of 34 percent of land available was actually used for manure disposal in a typical year. Land actually used would be equal to or greater than the amount of flat land used on each farm. For all farmers about 49 percent of the land actually used was nearly flat.

The acreage available for manure disposal is probably the most realistic to use in determining whether or not a farmer could comply with a particular guideline. Most farmers, if required, probably could use most of the land available for manure disposal. However, this is an upper limit on the amount of land that could be used. Some of this land may be unavailable during certain times of the year. The acreage of land actually used probably would be the more appropriate basis for determining whether present application rates exceed a particular guideline. The acreage of nearly flat land actually used is of more limited interpretation. However, it does provide a general idea about whether or not slope restrictions for spreading would affect a substantially greater proportion of the U.S. dairy industry.

The following guidelines, based on rates per acre, are shown for each of the three acreages reported: (a) land available, (b) land actually used, and (c) nearly flat land actually used.

### Guidelines Limiting Milk Cows Per Acre

Two percent of the farmers surveyed would have exceeded three cows per acre if all land available for manure disposal had been utilized (Table 1). On the basis of land actually used, one percent exceeded ten dairy cows per acre, while ten percent exceeded three cows per acre. Based on nearly flat land actually used, a much larger proportion of farmers would have exceeded all possible restrictions. Thirty-six percent of the farmers would have exceeded ten cows per acre, and 50 percent would have exceeded three

Paper presented at the Third International Symposium on Livestock Wastes, University of Illinois, April 11, 1975.

The authors are: BOYD M. BUXTON, Agricultural Economist, Economic Research Service, stationed at the University of Minnesota, St. Paul, Minnesota; and STEPHEN J. ZIEGLER, Research Specialist, Department of Agricultural and Applied Economics, University of Minnesota.

*To compute animal units (AU) the following conversions were used: one milk or dry cow, 1.4 AU; one dairy replacement, 0.7 AU; one farrowing sow, 0.4 AU; one feeder pig, 0.1 AU; and one beef cow, 1 AU.

†Includes manure and nitrogen produced from all livestock enterprises including non-dairy.

‡About 5000 questionnaires were distributed by the National Milk Producers Federation to fieldmen of their affiliated member cooperatives. The degree of confidence of the sample estimates could not be determined, because (a) only 2652 farmers responded, (b) the Federation represented only about 60 percent of U.S. milk produced, and (c) possible non random selection of farmers may have been made by the fieldmen.

**TABLE 1. PROPORTION OF SURVEY RESPONDENTS WHO WOULD EXCEED ALTERNATIVE DAIRY COWS-TO-LAND RESTRICTIONS**

| | Dairy cows per acre of land available | | | Dairy cows per acre of land actually used | | | Dairy cows per acre of nearly flat land actually used | | |
|---|---|---|---|---|---|---|---|---|---|
| | 10 | 5 | 3 | 10 | 5 | 3 | 10 | 5 | 3 |
| | | | | | percent | | | | |
| U.S. total | 0 | 1 | 2 | 1 | 4 | 10 | 36 | 42 | 50 |
| U.S.—by herd size | | | | | | | | | |
| 1-49 cows | 0 | 0 | 0 | 0 | 1 | 7 | 37 | 40 | 48 |
| 50-99 cows | 0 | 0 | 0 | 2 | 3 | 10 | 37 | 45 | 53 |
| 100+ cows | 3 | 6 | 9 | 5 | 14 | 26 | 28 | 40 | 55 |

**TABLE 2. PROPORTION OF SURVEY RESPONDENTS WHO WOULD EXCEED ALTERNATIVE ANIMAL UNITS-TO-LAND RESTRICTIONS**

| | Animal units per acre of land available | | | Animal units per acre of land actually used | | | Animal units per acre of nearly flat land actually used | | |
|---|---|---|---|---|---|---|---|---|---|
| | 10 | 5 | 3 | 10 | 5 | 3 | 10 | 5 | 3 |
| | | | | | percent | | | | |
| U.S. total | 1 | 2 | 4 | 4 | 13 | 29 | 41 | 54 | 67 |
| U.S.—by herd size | | | | | | | | | |
| 1-49 cows | 0 | 0 | 1 | 2 | 9 | 24 | 41 | 51 | 64 |
| 50-99 cows | 0 | 1 | 2 | 3 | 13 | 28 | 41 | 57 | 70 |
| 100+ cows | 5 | 8 | 15 | 12 | 29 | 46 | 34 | 54 | 66 |

cows per acre. The proportion of farmers exceeding three cows per acre on both land available and land actually used was much higher for those with large herds than for those with small herds.§ However, based on nearly flat land actually used, approximately one-half of the herds in each herd size group would have exceeded three cows per acre.

### Guidelines Limiting Animal Units Per Acre

Non-point guidelines could also be based on the number of animal units per acre. Conversion ratios for the various types of farm animals were used to determine the total animal unit equivalents on the farm.

One percent of the farmers would have exceeded ten animal units per acre and four percent would have exceeded three animal units per acre if all land available for manure disposal had been utilized (Table 2). Based on land actually used, four percent of the farmers surveyed exceeded ten animal units per acre while 29 percent exceeded three animal units per acre. If spreading had been limited to flat land actually used, 41 percent would have exceeded ten animal units per acre and 67 percent would have exceeded three animal units per acre.

The proportion of farmers exceeding animal unit restrictions, except for the nearly flat land category, was greater for those with large herds than for those with small herds.

### Guidelines Limiting Manure Tonnage Per Acre

Measuring manure disposal by tonnage applied would more directly relate to waste concentrations and pollution potential. If all land available for manure disposal had been utilized, three percent of the farmers surveyed would have exceeded 40 tons of manure applied per acre. Thirty-six percent would have exceeded ten tons per acre (Table 3).

Based on land actually used, nearly one-fourth of the farmers surveyed exceeded the least restrictive 40 tons of manure per acre. Ninety-two percent exceeded ten tons per acre. If disposal had been limited to flat land actually used, 64 percent of the farmers would have exceeded 40 tons per acre and 97 percent would have exceeded ten tons per acre.

For most manure tonnage restrictions, the larger the herd size, the larger the proportion of farmers who would exceed the hypothetical guidelines.

### Guidelines Limiting Nitrogen Application Per Acre

Three percent of the farmers surveyed would have exceeded an application rate of 200 pounds of nitrogen per acre if all land available for manure disposal had been utilized (Table 4). Seven percent of these farmers applied more than 400 pounds of nitrogen per acre annually on land actually used for manure disposal. Twelve percent exceeded 300 pounds and 22 percent applied more than 200 pounds per acre of land actually used for manure disposal. If spreading had been limited to flat land actually used, 47 percent would have exceeded 400 pounds and 63 percent would have exceeded 200 pounds of nitrogen per acre.

The proportion of farmers exceeding all four levels of nitrogen application rate restrictions was markedly higher for those with large herds than for those with small herds, except for the category "nearly flat land actually used".

### MANURE DISPOSAL ON FROZEN OR SNOW-COVERED GROUND

Some states recommended that manure not be spread on frozen or snow-covered ground to avoid waste flow into streams and lakes during spring runoff. Disposal on rolling

---

§ For purposes of this analysis, it was assumed that small herds had less than 50 dairy cows, medium herds had 50 to 99 dairy cows, and large herds had 100 or more dairy cows.

**TABLE 3. PROPORTION OF SURVEY RESPONDENTS WHO WOULD EXCEED ALTERNATIVE MANURE TONNAGE-TO-LAND RESTRICTIONS**

| | Manure tonnage per acre of land available | | | | Manure tonnage per acre of land actually used | | | | Manure tonnage per acre of nearly flat land actually used | | | |
|---|---|---|---|---|---|---|---|---|---|---|---|---|
| | 40 | 30 | 20 | 10 | 40 | 30 | 20 | 10 | 40 | 30 | 20 | 10 |
| | | | | | | percent | | | | | | |
| U.S. total | 3 | 6 | 12 | 36 | 24 | 40 | 65 | 92 | 64 | 73 | 86 | 97 |
| U.S.—by herd size | | | | | | | | | | | | |
| 1-49 cows | 1 | 2 | 5 | 21 | 20 | 32 | 58 | 89 | 61 | 70 | 81 | 96 |
| 50-99 cows | 1 | 4 | 12 | 39 | 25 | 47 | 69 | 95 | 68 | 79 | 90 | 99 |
| 100+ cows | 13 | 20 | 37 | 72 | 43 | 57 | 83 | 98 | 70 | 77 | 95 | 99 |



MANAGING LIVESTOCK WASTES

**TABLE 4. PROPORTION OF FARMERS SURVEY RESPONDENTS WHO WOULD EXCEED ALTERNATIVE NITROGEN-TO-LAND RESTRICTIONS**

| | Nitrogen applied per acre of land available | | | | Nitrogen applied per acre of land actually used | | | | Nitrogen applied per acre of nearly flat land actually used | | | |
|---|---|---|---|---|---|---|---|---|---|---|---|---|
| | 400 | 300 | 250 | 200 | 400 | 300 | 250 | 200 | 400 | 300 | 250 | 200 |
| | | | | | | | | percent | | | | |
| U.S. total | 1 | 1 | 2 | 3 | 7 | 12 | 16 | 22 | 47 | 53 | 57 | 63 |
| U.S. by herd size | | | | | | | | | | | | |
| 1-49 cows | 0 | 0 | 0 | 0 | 5 | 8 | 11 | 18 | 46 | 51 | 54 | 60 |
| 50-99 cows | 0 | 1 | 1 | 1 | 4 | 12 | 16 | 22 | 51 | 56 | 60 | 66 |
| 100+ cows | 7 | 8 | 10 | 12 | 19 | 28 | 33 | 42 | 45 | 53 | 62 | 68 |

or steeply sloping ground would pose the greatest threat. Survey results indicated that 84 percent of the farmers in the northern United States did spread manure during winter months. Nearly 50 percent of northern farmland used for manure disposal was either rolling or steep. As a result, many farmers spread manure on frozen or snow-covered ground which, by nature of its slope, may lend itself to surface runoff.

To avoid winter manure disposal, farmers could construct a storage system for solid manure. This system is comprised of a manure stacker and a concrete storage structure. Total investment costs for this type of system would range from about $3550 for a 30 cow herd to over $7300 for a 150 cow herd (Table 5).‖

If farmers who spread manure in the winter were to install a manure storage system the total industry investment could exceed $844 million (Table 6). Exempting farmers with less than 20 cows would reduce the aggregate investment to about $600 million; exempting farmers with less than 100 cows would drop the total investment to about $40 million.

## ODOR AND ESTHETIC PROBLEMS

Esthetic concerns, such as offensive odors and unsightly mounds of manure, are potential pollution problems. Urban development and the emergence of large, concentrated livestock production facilities are bringing the non-agricultural sector into closer contact with dairy production.

---

‖For more detailed treatment of costs see: Buxton, Boyd M. and Stephen J. Ziegler, Economic Impact of Controlling Surface Water Runoff from U.S. Dairy Farms, Agricultural Economic Report No. 260. Economic Research Service, U.S. Department of Agriculture, June 1974.

The farm boundary of about 50 percent of the dairy farms surveyed was less than 1100 feet from the nearest farm residence; about five percent were within 300 feet (Fig. 1). About 35 percent of the farm boundaries were within 1100 feet of the nearest non-farm residence. The boundary of less than three percent of the farms was within 1100 feet of the nearest park, picnic area, lake or reservoir used for recreation.

## SUMMARY AND CONCLUSIONS

If all land available had been utilized for manure disposal, few farmers surveyed would have been affected by livestock-to-land, manure tonnage-to-land, or nitrogen-to-land restrictions. However, in the case of nearly flat land actually used, as many as 50 percent of the respondents would have been affected by a limitation of three dairy cows per acre. Sixty-s even percent of the farmers would have exceeded a limitation of three animal units per acre. Ninety-seven percent of the farmers would have exceeded ten tons per acre of nearly flat land.

In most cases, there was a direct relationship between herd size and the proportion of farmers affected by those guidelines selected.

Over 84 percent of the farmers surveyed in the northern United States indicated that they spread manure on frozen ground. If these northern dairy farmers were to install

**TABLE 6. TOTAL INVESTMENT FOR DAIRY FARMERS WITH WINTER MANURE DISPOSAL PROBLEMS**

| | Aggregate investment | |
|---|---|---|
| | All farmers who spread manure during winter months | Farmers who spread manure on rolling or steep ground during winter months |
| | million dollars | |
| Total farmers with problem | 844.5 | 5.79.3 |
| Exempting farmers with less than 20 cows (percent decrease) | 604.3 (28.5) | 421.3 (27.2) |
| Exempting farmers with less than 100 cows (percent decrease) | 38.8 (95.4) | 26.4 (95.4) |

* Northern regions only. Assumes the proportion of farmers surveyed reflects the true proportion of all farmers spreading manure on frozen ground.

**TABLE 5. CHANGES IN INVESTMENT, ANNUAL COST, AND COST OF PRODUCING MILK FOR SOLID MANURE STORAGE SYSTEMS ON TYPICAL NORTHERN DAIRY FARMS**

| | 15 cows | 30 cows | 80 cows | 150 cows |
|---|---|---|---|---|
| Total investment* | $4,538† | $3,557 | $4,884 | $7,304 |
| Investment per cow | 303 | 119 | 61 | 49 |
| Annual cost with 5-yr depreciation | | | | |
| Total cost | 1,333 | 1,133 | 1,703 | 2.659 |
| Cost per cow | 89 | 38 | 21 | 18 |
| Cost per cwt. of milk | 0.74 | 0.31 | 0.18 | 0.15 |
| Annual cost with useful life depreciation‡ | | | | |
| Total cost | 705 | 644 | 1,035 | 1.663 |
| Cost per cow | 47 | 21 | 13 | 11 |
| Cost per cwt of milk | 0.39 | 0.18 | 0.11 | 0.09 |

* Assumes a semicircular, poured concrete structure with 6 ft walls and a storage requirement of 1.5 cu ft per cow per day. Manure would be allowed to accumulate in a conical shape above the walls. Costs were adjusted to reflect 1974 prices.

† Investment reflects additional equipment and facilities needed to reduce runoff pollution for illustrative purposes only. Costs will vary depending on local topographical conditions as well as bedding content of the manure. Per cow investment is higher for operators of 15-cow farms because it was assumed that smaller farms would require a manure loader while the larger farms already have one.

‡ 15-yr depreciation schedule.

manure storage systems, ranging in cost from about $3500 to $7300 per farm, the aggregate industry investment could reach $844 million.

Boundaries of 50 percent of the farms were located within 1100 feet of the nearest farm residence. Less than three percent were within 1100 feet of the nearest park, picnic area, lake, or reservoir used for recreation.

Prudent animal waste management can help reduce or eliminate non-point sources of pollution. Most farmers apparently have adequate land for manure disposal, based on the guidelines evaluated in this paper. This gives them substantial flexibility to properly manage manure disposal to minimize pollution. Many of the management changes could be made without severe economic hardship; others would increase the farmers' production costs. Not spreading on land subject to runoff or near surface water could be a low-cost change. However, not spreading on frozen land could require a major expense. Immediate incorporation of manure into the soil to minimize runoff would be another cost to dairy farmers. Additional cost to farmers would be incurred if it was necessary to spread on a larger acreage because of increased travel distances. The timing of manure application with cropping patterns would also become more difficult.

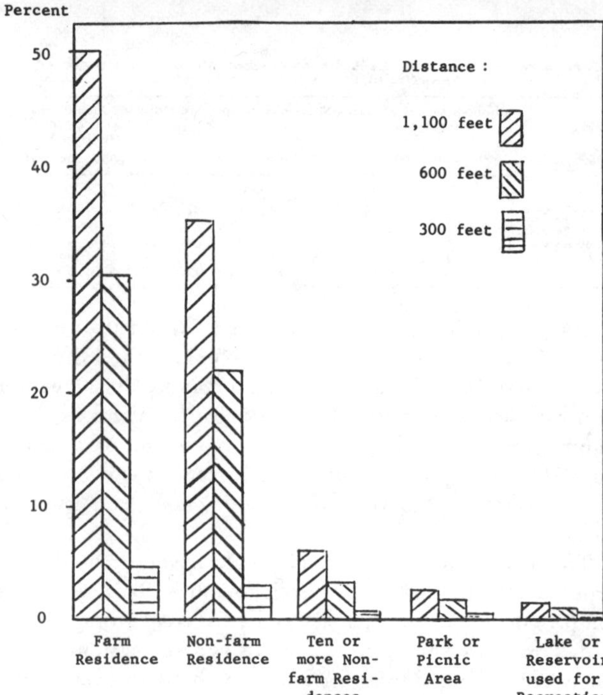


FIG. 1 Proportion of farms whose boundary was within 300, 600, and 1,100 feet of the nearest farm and non-farm residence, surface waters, and recreational areas.

---

## EPA Water Pollution Control Rules

(Continued from page 52)

for the land-intensive, dry-lot paved systems. Production cost increases could severely encumber the economic viability of smaller-sized feedlots with land-extensive housing systems. Some feedlots, particularly smaller feedlots, may cease production.

Final guidelines announced by EPA are performance standards. Feedlot operators can select from a variety of available control technologies. Allowing operators to select control systems that are most consistent with existing resource positions and specified discharge limits could result in a somewhat more economical industry solution than those estimated for the use of the four-component system.

Reactions of individual feedlot firms to whom guidelines apply cannot be fully determined from information considered in this research. Some may discontinue producing beef and turn to other farm or nonfarm activities. Others may make adjustments and continue to operate at historical production levels. Other paths of adjustment are also possible. Additional knowledge of production system-cost relationships, operator equity positions, expected beef and input price changes, and access to capital sources would be needed to make such determinations. These factors are considered in a related paper included in these proceedings entitled "Economic Impacts of Alternative Water Pollution Control Rules on Beef Feedlots" by D. L. Forster et al.

### References

1  Anonymous. 1974. Effluent limitations guidelines and standards, feedlots point source category, p. 5704-5708. In Federal Register, 39(32), subchapter N, Part 412. U. S. Government Printing Office, Washington, D. C. February 14.

2  United States Congress. 1972. Federal water control act amendments of 1972. Public Law 92-500, 92nd Congress, S 2770. 89 pages.

### Appendix A

Specialized Definitions for Terms Used in Final Effluent Limitations Guidelines

**Feedlot**

The term feedlot shall mean a concentrated, confined animal or poultry growing operation for meat, milk or egg production or stabling, in pens or houses where the animals or poultry are fed at the place of confinement and crop or forage production or growth is not sustained in the area of confinement.

**Process Waste Water**

The term process waste water shall mean any process generated waste water and any precipitation (rain or snow) which comes into contact with manure, litter, or bedding, or any other raw material or intermediate or final material or product used in or resulting from the production of animal or poultry or direct products (e.g. milk, eggs).

**Process Generated Waste Water**

The term process generated waste water shall mean any water directly or indirectly used in the operation of a feedlot for any or all of the following: Spillage or overflow from animal or poultry watering systems; washing, cleaning or flushing pens, barns, manure pits or other feedlot facilities; direct contact swimming, washing or spray cooling of animals; and dust control.

**Ten-year, 24-hour Rainfall**

The term 10-year, 24-hour rainfall event shall mean a rainfall event with a probable recurrence interval of once in ten years. (For analysis, a 10-year, 24-hour rainfall event is a rainfall of a specific magnitude or greater that has a 1-in-10 chance of occurring in a 24-hour period in any given year.)

**Twenty-five year, 24-hour Rainfall**

The term 25-year, 24-hour rainfall event shall mean a rainfall event with a probable recurrence interval of once in 25 years. (For analysis, a 25-year, 24-hour rainfall event is a rainfall of a specific magnitude or greater that has a 1-in-25 chance of occurring in a 24-hour period in any given year.)

# Feedlot Effluent Limitations Based Upon Exemplary Operations

Jeffery D. Denit
ASSOC. MEMBER
ASAE

PASSAGE of the Federal Water Pollution Control Act, as Amended, 1972 (the Act) served to focus Federal regulatory attention upon livestock operations in a manner which was substantially foreign to this segment of the nation's economy and population. In this regard, the Act defined concentrated livestock and poultry growing operations (feedlots) as "point" sources of "industrial" pollution and further required that permits be issued for these operations. The permits provide a mechanism for both administrative and enforcement control of water pollution discharges by the Environmental Protection Agency (EPA). Livestock producers were thus confronted by regulation of much of the land-oriented aspects of their enterprises. As the primary carrier of pollutants, precipitation runoff from the normally land-based feedlot now required control.

From the point of view of the feedlot operator the permit provides the vehicle for conveying the degree of pollution control expected of him; i.e., a guideline for effluent limitations on existing sources of pollution and standards of performance for new sources. The regulation which provides the basis for the permit, and which achieves the result indicated is codified at 40 CFR 412, 39 F.R. 5704. All the detailed considerations underlying the new regulation cannot be covered in this discussion and are presented elsewhere. It is possible, however, to summarize the essential thrust of regulatory standards and limitations which are based upon an analysis of what is generally termed, "exemplary" feedlot operations. This designation applies to those operations currently practicing "best practicable technology" (1977), best available technology (1983), and best available demonstrated technology (new sources) in accordance with the above cited statute and regulation.

For a useful descriptive definition, "exemplary" operations may be thought of as those feedlot operations which exemplified and utilized a high standard of management of solid and liquid feedlot waste control concepts. Compared with other facilities, the exemplary operations routinely reflected several outstanding traits of the feedlot operators:

1   perception of need for, and role of, pollution control in overall feedlot operation schemes;

2   concern for environmental protection;

3   initiative in resolving problems frequently before a firm legal requirement was imposed; and

4   imaginative use of existing site characteristics and available complementary technology.

As a practical matter, the major thrust of this analysis of exemplary operations was to document methods for minimizing water pollution from precipitation runoff (open lots for beef, dairy and swine) and hydraulic manure removal

(pen flushing in swine and dairy operations). As described below, the specified storm serves to establish the flexible end point based upon existing technology. Nevertheless, it is inherent that the control facilities once constructed, must function in a manner which precludes subsequent discharges each and every time waste water enters them. There is the expectation by EPA that a deliberate attempt be made to reasonably assure: (a) that the control facilities are available for recurring runoff; (b) that such actions as may be taken to assure availability do not result in a direct discharge to navigable waters; and (c) that established principles of sound agricultural or waste disposal practice are not compromised to an extent which fosters ancillary problems of soil contamination, malodorous conditions or the like (Denit 1974).

Fortunately, facilities which proximate the required levels of control at reasonable cost are already in place and in use at feedlots throughout the United States. The regulation reflects the EPA determination of the general nature of these facilities in keeping with the requirements of the Act. While the following discussion is necessarily restricted to open feedlots applicable to beef cattle, swine and sheep, the suggested principles are the same for all feedlots (Development Document 1974).

## THE FEEDLOTS POINT SOURCE CATEGORY

The feedlots regulation evolved from a highly intensive thirteen month undertaking involving a technical study of the feedlots industry and public review of tentative technical reports and regulations. As would be expected of such an expansive industry, feedlots have diverse practices and frequently include unique circumstances. Nevertheless, there are sufficient consistent similarilities, (although broad generalization must at times be applied) to allow reasonable assimilation of the diversity and uniqueness into a manage-

**TABLE 1. CATEGORIZATION OF THE FEEDLOTS POINT SOURCE CATEGORY.**

| | | |
|---|---|---|
| Beef Cattle | - | Open Lots |
| Beef Cattle | - | Housed Lots |
| Dairy Cattle | - | Stall Barn |
| Dairy Cattle | - | Free Stall Barn |
| Dairy Cattle | - | Cowyard |
| Swine | - | Open Dirt or "Pasture" Lots |
| Swine | - | Housed, Slotted Floor |
| Swine | - | Solid Concrete Floor, Open or Housed |
| Sheep | - | Open Lots |
| Sheep | - | Housed Lots |
| Horses | - | Stables (Racetracks) |
| Chickens | - | Broilers Housed |
| Chickens | - | Layers (Egg Production), Housed |
| Chickens | - | Layer Breeding or Replacement Stock, Housed |
| Turkeys | - | Open Lots |
| Turkeys | - | Housed Lots |
| Ducks | - | Wet Lots |
| Ducks | - | Dry Lots |

The author is: JEFFERY D. DENIT, Chief, Impact Analysis Section, Effluent Guidelines Div., Environmental Protection Agency, Washington, D. C.

| Animal type | Number/type visted* | General location |
|---|---|---|
| Beef cattle | 38 | Throughout major producing states |
| Dairy cattle | 27 | New England, Southeast Wisconsin, Southwest |
| Swine | 31 | Cornbelt, Southeast Great Plains |
| Sheep | 7 | Texas, Colorado |
| Chickens | 17 | Southeast, Texas Midwest |
| Turkeys | 6 | Texas, North Carolina Indiana |
| Ducks | 8 | Long Island, Indiana Wisconsin |
| Horses | 46† | New England, Maryland |

*Totals do not include sites visited or known to project consultants or general information visits by author.

†Includes trade association questionnaire data from 41 racetracks.

TABLE 3. SUMMARY OF RUNOFF CONTROL
SYSTEMS IN USE (1972-73).

| General basis for control capability | Production system where used | Example area being practiced |
|---|---|---|
| Rainfall Runoff: | | |
| 5-year, 48-hour storm | Open livestock operations | Oklahoma |
| 10-year, 24-hour storm | Open livestock operations | Kansas, California |
| 25-year, 24-hour storm | Open livestock operations | Texas, Minnesota |
| long term storage (60, 90, 120 days) | Open livestock operations | Illinois, Indiana Oregon |
| Washdown, Flushing: | | |
| 125-400 cubic feet per head | swine, open or housed lots | Midwest |
| 1500 cubic feet per head | cattle, open or housed | Midwest |
| hydraulic load | hydraulic flush poultry systems | widely dispersed |

able framework.

As this assessment proceeded, it became apparent that the raw wastes (both fresh manure and diluted runoff or hydraulically flushed wastes) from all types of operations were similar on the basis of production units (e.g., pounds of BOD per 1000 pounds of animals). At the same time, it was clear that failure to develop subcategories simply because of similarities in wastes, would have fostered considerable inequity by oversimplifying practical variations. Consequently, the feedlots industry was subcategorized as shown in Table 1 to account primarily for differences in production method, final product, and indirectly for variations in site and location.

For all of the animal types, on-site visits to exemplary operations were made and specific site reports were used to verify the principal substantive characteristics and generalized literature data applicable to feedlots as a whole. A summary of the specific site information thus developed is shown in Tables 2 and 3. Table 3 is particularly useful as a display of the variety and capability of control procedures that were already in use. As expected, runoff control overwhelmingly prevailed as the pollution control concepts of choice. Moreover, runoff control using classical water conservation and management practices is most important for the foreseeable future. As a consequence of considerable analyses of evidence on runoff control concepts, the requirements shown in Table 4 provided the basis for the feedlot effluent limitations and standards.

In terms of the fundamental technical requirements, the thrust of the present regulation was directed at larger feedlot operations; i.e., operations with a capacity of 1000 beef cattle, 700 dairy cattle, 2500 swine, and other sizes as specified in the regulation. For the present, those operations which are smaller than the stipulated sizes would have direct regulation only on the basis of water quality; that is, in-stream water quality standards are being violated by discharges from the (smaller) feedlot operation. The level of control and nature of the technology required may con-

form to those upon which the effluent limitations are based; on the other hand, both may vary toward either more stringent or less stringent as dictated by the water quality circumstances. Regardless of the legal basis by which feedlots require pollution control, success in implementing the controls (whatever their nature) is predicated upon judicious planning and design for the specific site conditions. Climate, soil type, slope, proximity to streams, and land availability are but a few of the many factors involved. The principles applied to account for many of these factors are applied by the exemplary facilities.

One way to minimize runoff of feedlot pollutants is to divert "outside" or uphill runoff away from the feedlot site. Fig. 1, shows an example where the natural topography (hill) was used to advantage with berms to divert outside runoff. This technique resulted in a reduction of 80 percent in the size of the impoundments (now being constructed) required to achieve "no discharge" from the swine and cattle lots. Problems encountered due to the close proximity of feedlots to streams are often rather easily overcome by simply constructing a dike which serves both to keep runoff from direct discharge and to preclude frequent flooding of the feedlot site (Fig. 2). Even if the specific site conditions are difficult, reasonably simple concepts are usually the most feasible. In Fig. 3, for example, outside runoff is controlled by natural topography and an upstream structure. The feedlot runoff is isolated and controlled by employing small settling basins within the feedlot itself and a large holding pond which prevents direct stream discharges. Ingenuity of site layout, even with constraints of available space, is frequently the key to successful production coincident with efficient pollution control. A good illustration of this is shown by the production/runoff control system shown in Fig. 4. In this instance, the feedlot operator has taken full advantage of site conditions, and in

TABLE 4. SUMMARY OF BASELINE CONTROL REQUIREMENTS.

| Technology Level | Date applicable | Control level |
|---|---|---|
| Best practicable technology for existing sources | July 1, 1977 | 10-year, 24-hour storm |
| Best available technology for existing sources | July 1, 1983 | 25-year, 24-hour storm |
| Best available demonstrated technology for new sources | September 7, 1973 | 25-year, 24-hour storm |



MANAGING LIVESTOCK WASTES

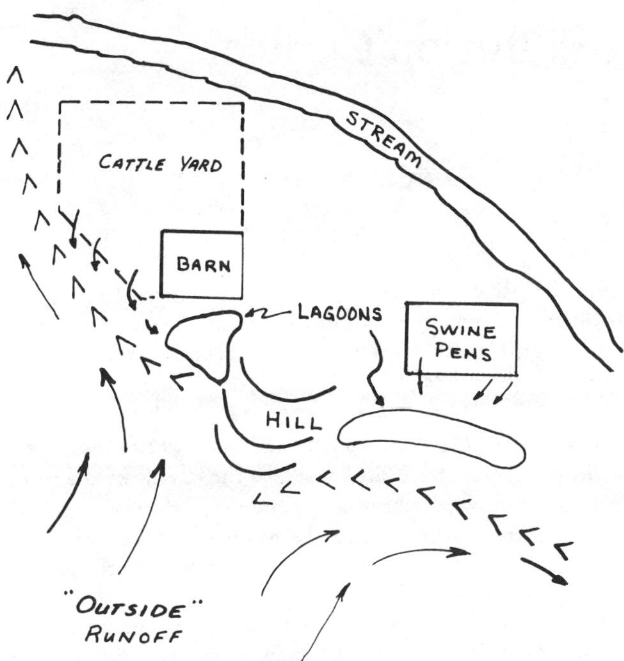

FIG. 1 Sketch depicting diversion of "outside" runoff from combined dairy/swine operation in Wisconsin.

a minimum space with little expense, developed a pollution control system which achieves required levels of "no discharge."

All of the above examples serve to emphasize two points most salient to feedlot pollution control. First, in keeping with requirements of the Act regarding technology based limitations, the industry is already achieving a practical level of no discharge of pollutants by intelligent implementation of fundamental soil and water management principals. Secondly, the exemplary plants manifest one primary difference from other operations in the industry: reasonably simple concepts have been applied to make the feedlot site work in the operator's favor. Even with rather stringent constraints, runoff control systems have been incorporated successfully.

### References


1   Denit, J. D. 1974. Effluent regulations for livestock and poultry feedlots, Proc. Sixth National Conference on Agricultural Waste Management, pp. 51-58. Cornell University, Ithaca, New York.
2   Development Document for Effluent Limitations Guidelines and Standards of Performance for New Sources for the Feedlots Point Source Category. U.S. Environmental Protection Agency, Washington D. C., February, 1974.


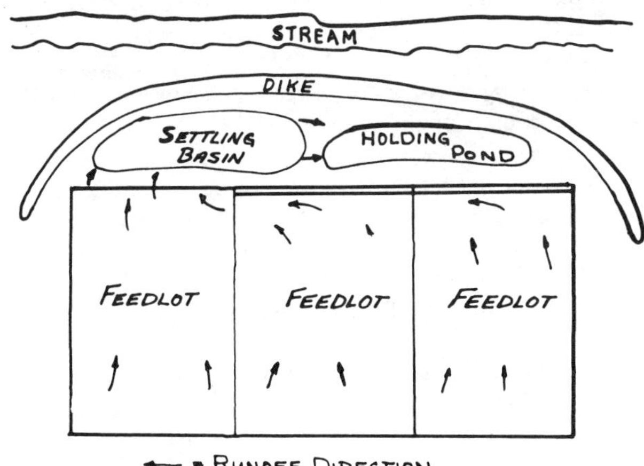

FIG. 2 Illustration of measures to control runoff from a cattle feedlot located near a stream, beef cattle facility in Nebraska.

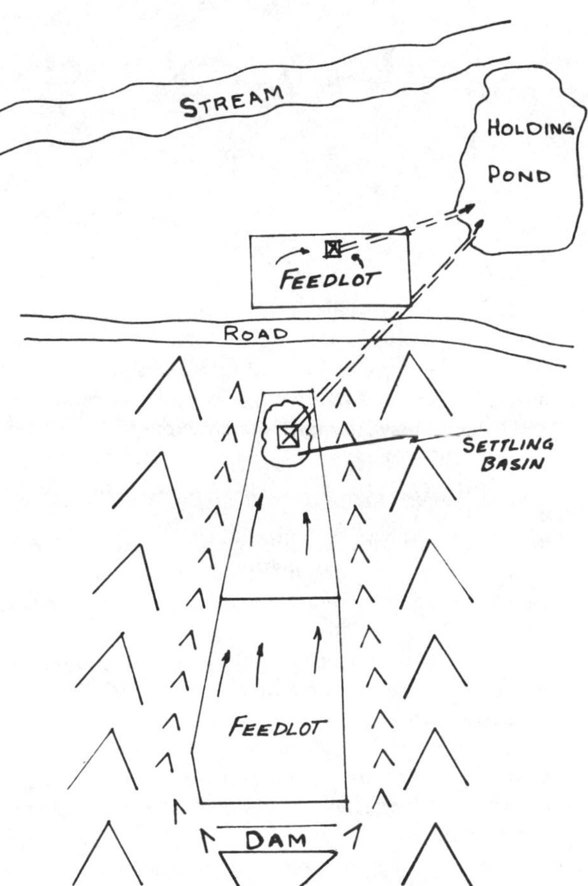

FIG. 3 Illustration of the use of several types of control structures and concepts at a beef cattle facility in Nebraska.

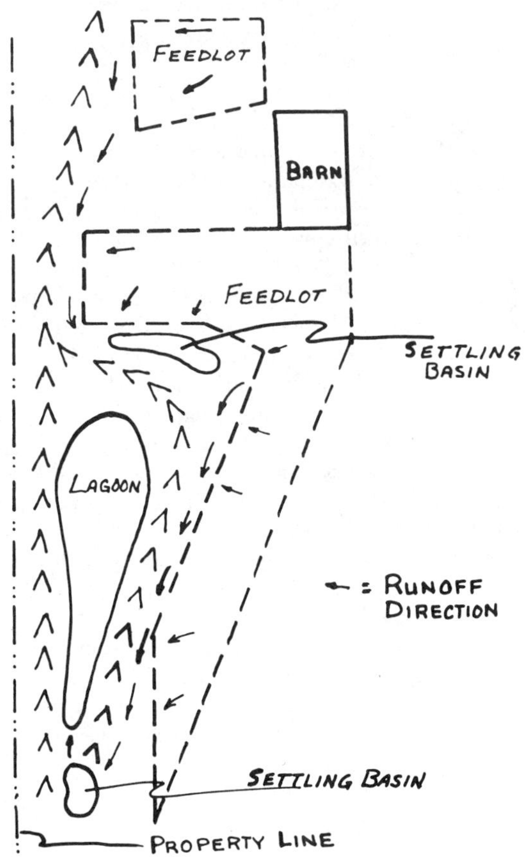

FIG. 4 Maximum beneficial use of site conditions for a beef cattle facility in Iowa.

# Legal Aspects of Odor Pollution Control

Pamela Giblin

POLLUTION was labeled a nuisance as early as 1611 when an English Court affirmed the granting of an injunction and damages to the plaintiff upon a showing that defendant had erected a "hogstye so near the house of the plaintiff that the air thereof was corrupted". At common law, it was not a personal right to breathe clean air, but rather a property right which was protected in a nuisance action.

Most clean air legislation still contains a general nuisance provision. In fact the definition of air pollution in most state statutes is set out in nuisance language similar to the following:

> "Air Pollution" means the presence in the ambient air of one or more air contaminants in quantities, of characteristics, and of such duration as are or may tend to be injurious to or adversely affect human health or welfare, animal life, vegetation or property, or as to interfere with the normal use and enjoyment of animal life, vegetation or property.

The essence of a definition of this type is that the emission, in order to be classified as pollution, must cause damage or harm. Under a nuisance theory, the mere presence in the atmosphere of air contaminants does not mean that air pollution is present. Some harmful effect must result. This principle is distinguishable from the emission standard concept. When prosecuting for violation of an emission standard, no harm must be proven because the legislature has, in effect, declared that any waste discharge exceeding the emission standard will produce harm.

While there has been some experimentation in the area of odor pollution standards, most states simply use a nuisance theory to control odor. There are some problems associated with this approach. One of the difficulties with nuisance is that it is ill-suited to guide engineering and business decisions. It is difficult for a person constructing a source or facility to know how much abatement equipment he needs to prevent his neighbors from being offended. Another problem associated with nuisance is the subjectiveness of the test. For example, what would be offensive to one person might be enjoyable to another. For this reason, the court in nuisance cases usually balances the equities. That is, the court looks at all facets of the problem rather than applying a strict standard.

The purpose of this paper is to analyze the approach taken by a typical air pollution control agency, the Texas Air Control Board, in handling odor problems. There have been a number of enforcement actions from which examples can be drawn to show the balance that has been struck between the rights of persons to live free from offensive odors and the rights of persons to conduct certain activities, which by their very nature emit a certain amount of odor.

Texas, like most states, does not have a specific odor regulation. In view of this fact, the staff of the Texas Air Control Board relies heavily on citizen complaints in order to determine whether an odor problem exists. It is important to distinguish between "grudge-type" complaints and those which are legitimate. Whenever a complaint is received, a representative of the Texas Air Control Board staff contacts the complainant and asks specific questions in order to determine the cause and the extent of the problem. Using a standard nuisance interview form, the investigator gathers information on the following questions:

1   What is the direction and distance of the source from the complainant?

2   How does this odor affect the complainant and his family, their health, or physical well-being or enjoyment of their property?

3   How long has this condition existed?

4   How often do these conditions occur?

5   Can the complainant identify the source of the odor?

6   How does the complainant isolate the source?

7   Are the odors associated with any particular weather condition?

8   Can the complainant describe conditions prior to the development of odors?

9   Has the complainant contracted anyone at the source of the odor?

10   Have visitors to the complainant's home commented on the odor during their presence?

11   Can the complainant describe the specific dates and times when the odor conditions were experienced?

12   Would the complainant be willing to testify in Court?

During the initial contact with the complainant, the investigator usually requests that the complainant keep a record of the dates and times he experiences the odor and, if possible, the wind direction. It is extremely important and often very difficult to identify the specific source of the odor. The investigator from the Texas Air Control Board staff surveys the area around the complainant's home to determine if there are any other sources of odor which could affect the person or to which the complaint could be attributed.

The investigator usually visits the facility suspected of being the source of the odor. If possible, he obtains a written statement from the person in charge of the facility regarding the nature and the source of the odor. While at the facility, the investigator examines potential sources of odors. Observations upwind and downwind of the facility are helpful in determining whether the facility is in fact the source. If the investigator experiences the same odor which precipitated the complaint, he describes the odor and its effect on him. It is possible that the odor may not be present at the time the agency representative conducts his investigation. This does not necessarily mean that the complaints are not justified. The investigator should try to be present at the times and places at which complainants in the community have perceived the odor.

It is often extremely useful to conduct a public hearing. This affords an opportunity for all complainants to testify and it allows the owner or operator of the source in question to present his side. There have been instances where the source owner was not aware of the nuisance conditions he was creating. The public hearing provides a forum for both sides to present their problems before a state agency having expertise in the area of air pollution.

It is unlike a trial in that strict rules of evidence are not followed. A hearing is an administrative fact-finding proceeding. The hearing examiner, usually an attorney with the staff of the Texas Air Control Board, receives testimony and evidence and on the basis of the record compiled at the hearing, the Agency can select a course of action.

A hearing may indicate that there are so many complainants and that the probelm is so severe that legal action must be taken. This was the situation in the case of State of Texas vs. Hollingsworth Feed Mill, Inc. A hearing was held on the feed mill in question and it became apparent that no voluntary solution could be reached.

Hollingsworth had expanded the sheep feeding operations bringing them in close proximity to homes. At the public hearing, numerous citizens testified that they were unable to enjoy their homes and yards. The conditions, particularly during wet weather, were such that windows had to be kept shut and activities of surrounding home owners were severely curtailed. Many of the persons who attended the hearing indicated that they would gladly serve as witnesses in any court action.

After the hearing, the Texas Air Control Board entered an order finding that the operation of Hollingsworth Feed Mill and Feed Lot resulted in emissions of odorous air contaminants, causing a condition of air pollution. The odors were caused by the very nature of operating a sheep feed lot with resulting concentrations of sheep manure in a relatively small area. The problem was aggravated by the close proximity of the feed mill and feed lot to a residential area and further by occasionally lax housekeeping procedures, including failure to dispose of dead sheep carcasses daily and permitting feed in the troughs to sour and ferment.

When Hollingsworth failed to comply with the remedial measures set out in the Board Order, the Texas Air Control Board, through the State Attorney General, filed a civil action pursuant to the Texas Clean Air Act. The Act provides for injunctive relief and penalties of $50.00 to $1,000.00 for each day of violation. This provision is very similar to those contained in air pollution control legislation throughout the country.

The trial was held in the community where the feed lot was located and as a result of the jury's findings, the Court ordered that the feed lot pay penalties totaling $1,250.00 plus all court costs. Furthermore, the defendant feed lot was permanently enjoined from causing or permitting emissions of dust, or odor in such concentration and duration as might cause a nuisance. The Court ordered Hollingsworth to comply with the provisions of the Board Order that had been entered following the public hearing.

This was the first jury verdict in an odor pollution suit brought under the Texas Clean Air Act. There was heavy reliance on the testimony of complainants. Approximately 200 residents of the community where the feed lot was located had sent a petition to the Texas Air Control Board. In addition to citizen testimony, agricultural engineers on the Texas Air Control Board staff were instrumental in setting out specific housekeeping conditions, which, if followed, could significantly reduce odors from the facility.

In some cases where a feed lot is extremely close to a residential area, relocation may be the only solution. However, as in all nuisance actions, the Judge weighs the equities in the case. For example, he will look at whether the complainants moved to the nuisance or whether the nuisance expanded close to the complainants.

Most recently, the Texas Air Control Board has received complaints concerning a hog feeding operation. A hearing was held by the Texas Air Control Board staff and it appears that the problem may be corrected by the use of better housekeeping measures and by restricting the placement of hogs to a portion of the lot downwind from the homes of the complainants. If the proper corrections can be achieved by an administrative Board Order, there will probably not be a need for court action.

# Partnership in Pollution Control

## Russell T. Odell

ILLINOIS, a major agricultural and industrial state, leads all otherstates in exports of both agricultural and manufactured products. Intensive industrial and agricultural production can cause significant deterioration of the environment unless care is taken to control pollution.

### ILLINOIS ENVIRONMENTAL PROTECTION ACT AND AGENCIES

In 1970, the Illinois General Assembly passed the Illinois Environmental Protection Act which consolidated and reoriented environmental programs in Illinois (reference 3). Under this Act, three organizations were established. The Pollution Control Board (Board) establishes regulations to protect the environment. It also sits as a quasi-judicial body to rule on cases of alleged violations of regulations and decides requests for variances from regulations. Regulations are now operative for air pollution, water pollution, mine wastes, public water supplies, solid wastes, and noise. The 5-man Board is unique in that it "is the only appointive anti-pollution board in the country that works full-time and has its own staff and funds" (Haskell and Price 1973). The second organization, the Illinois Environmental Protection Agency (IEPA), is the monitoring agency which, along with citizens, brings alleged polluters before the Board. The IEPA also makes recommendations to the Board concerning requests for variances from regulations. The third organization, the Illinois Institute for Environmental Quality (Institute), collates environmental information, especially for the IEPA and the Board.

### DEVELOPMENT OF LIVESTOCK WASTE REGULATIONS

Livestock produce large quantities of wastes, and these cause pollution if they are not managed properly. The current Illinois Livestock Waste Regulations were developed during a 2-yr period. Proposals were made by the Board in 1972, but testimony during hearings indicated that these proposals were not suitable for adoption unless they were altered significantly. Also, the final guidelines and regulations governing animal feedlots had not promulgated by the U.S. Environmental Protection Agency under the Federal Water Pollution Control Act (FWPCA) of 1972, as amended. Because of these problems, the Board agreed to delay final action until further study could be given to the problem.

Early in 1973 the Institute covened an Agricultural Advisory Committee to draft proposed livestock waste regulations that would (a) be in compliance with federal guidelines, (b) meet the requirements of the Illinois Environmental Protection Act, and (c) be generally acceptable to the agricultural community. This Agricultural Advisory Committee comprised 22 members and represented state and federal agencies, eight agricultural producer organizations, agricultural lending institutions, and three environmental groups. Their proposal was submitted to the Board on November 6, 1973, and after four hearings early in 1974 and subsequent refinements, it was adopted by the Illinois Pollution Control Board on August 29 and September 5, 1974 (reference 4).

These Livestock Waste Regulations are a part of the National Pollutant Discharge Elimination System (NPDES) program in Illinois. They have been submitted to the U.S. Environmental Protection Agency for review for the purpose of obtaining approval for the state of the NPDES program in Illinois. The effective data of these Regulations will be the date (mid 1975 anticipated) when the Board files with the Illinois Secretary of State a copy of a letter approving the Livestock Waste Regulations by the Administrator of the U.S. Environmental Protection Agency pursuant to Section 402(b) of the FWPCA of 1972, as amended.

Although these Livestock Waste Regulations are concerned primarily with preventing water pollution, livestock feedlots must also comply with provisions of the Illinois Environmental Protection Act (reference 3) and applicable Air Pollution Regulations which deal with odor and dust nuisance problems.

### FOUNDATION FOR THE REGULATIONS

These Livestock Waste Regulations provide for a balance between a wholesome environment and the efficient production of adequate livestock products. The Opinion (reference 5) which supports these Regulations explains in detail the need for abatement of water pollution from feedlots, the properties of livestock wastes, nitrogen transformations and movement in soils, and the estimated economic implications of the Regulations.

Costs for water pollution control will not fall equally on all farmers and service firms, nor on all regions of the state. Livestock producers with the smallest enterprises will be confronted with the highest unit costs for pollution control. Most of the fed cattle, hogs, and milk produced in Illinois come from relatively small enterprises. Generally, the economic impact of controlling pollution from livestock feedlots will be to accelerate adjustment trends that are already in progress. These include (a) a decrease in the number of small livestock enterprises and an increase in the size of business among continuing enterprises, (b) an accelerated shift to confined systems of production, and (c) a continuing shift to drier climates where water pollution control is simpler, especially in cattle feeding. The long-range impacts of pollution control on the supply of livestock products and costs to consumers are expected to be small. However, economic impacts during the adjustment period may be rather sever, especially for some smaller livestock enterprises and associated service businesses.

### EXPLANATION OF SELECTED RULES

The Opinion (reference 5) explains the basis for the various rules in the Livestock Waste Regulations (reference 4). The groups of rules concerning general provisions and permits deserve special emphasis.

The author RUSSELL T. ODELL is member, Illinois Pollution Control Board, 309 West Washington Street, Suite 300, Chicago, Illinois 60606; and Professor of Agronomy, Emeritus, University of Illinois, Urbana, Illinois 61801.

The most important general provisions provide for the handling, storage, and field application of livestock wastes; for existing and new livestock facilities to be constructed so as to prevent excessive outside surface waters from flowing through the feedlot and to direct feedlot runoff to an appropriate disposal or storage area; and the location of new livestock facilities with regard to surface waters, flood plains, soil conditions, and population centers. Livestock waste-handling facilities include a variety of constructions and devices, such as manure-holding pits and lagoons, as well as acceptable disposal areas in fields. The contents of manure-holding pits and lagoons are to be kept at levels such that there is adequate storage capacity so that overflow does not occur except in the case of precipitation in excess of a 25-yr 24-hr storm. Acceptable field disposal areas include pasture or cultivated land which can serve as an adequate filtering device to settle out and assimilate any pollutants before clarified water reaches a stream. Much attention was given to developing a procedure to enable inspectors to enter premises for investigation, but to prevent the introduction or transmission of disease.

A permit system was established which distinguishes three groups of livestock feedlots on the basis of size. Large livestock operators with more than 1000 animal units are required to have an NPDES Permit. This is identical to federal regulations and applies to both new and existing feedlots. For livestock operations in which there are 100 to 1000 animal units, an NPDES Permit is not required unless the IEPA determines that the livestock facility is causing or threatening to cause a violation of applicable regulations; in which case, the operator must make application for an NPDES Permit. An NPDES Permit is not required for livestock operations in which there is a total of less than 100 animal units unless the Board determines that the facility is causing pollution, pursuant to an enforcement action. Since the economic implications of these Regulations are greatest on small producers whose volume of waste is small, the Board believes that they should be spared costs beyond what is necessary to control pollution. NPDES Permits are issued for fixed terms not to exceed 5 yr, as required in federal regulations. However, almost all bankers, farm managers, and farmers who testified at the hearings indicated that, because of credit and amortization considerations, the NPDES Permits should be for periods longer than 5 yr.

## IMPLEMENTATION

The achievement of pollution abatement will involve action and understanding by many people and organizations. There are approximately 121 000 farms in Illinois. Two-thirds of these farms have some kind of livestock, of which 40 000 have enough livestock to be considered as feedlots. Farmers should become informed (through the Cooperative Extension Service and others) concerning situations where pollution is often a problem and various methods for controlling it. Each farmer should evaluate his livestock operations and, if additional measures for pollution control are needed, he can consult with representatives of the Extension Service, Soil Conservation Service, Environmental Protection Agency, agricultural businesses, and other sources to plan effective pollution control measures for his farm. Each livestock operation will require individualized attention. After making financial arrangements, the necessary physical alterations should be made to adequately control pollution from livestock operations.

Pollution abatement strategy in agriculture should be different from many other businesses. The numbers are large—at least 40 000 farms are subject to the Illinois Livestock Waste Regulations. They are scattered throughout the state and many of the livestock operations are small. Fortunately, agriculture has public research and educational facilities in the Illinois Agricultural Experiment Station, Cooperative Extension Service, and Soil Conservation Service which develop new techniques and transmit this information to farmers. During the past year, "expenditures in the Agricultural Experiment Station for environmental research programs approach \$667 000" (Bentley 1974). These established and widely respected public organizations should be utilized to the maximum in developing and implementing sound pollution abatement practices by voluntary action by farmers. Pollution control standards should be adhered to, but enforcement action should be restricted to recalcitrant livestock operators who do not show good faith in meeting standards.

Farmers and these public agricultural agencies have been deeply involved in the development of the Illinois Livestock Waste Regulations. This spirit of cooperation can and should be continued as the Regulations are implemented.

## SUMMARY

In 1970, the Illinois General Assembly passed the Environmental Protection Act which consolidated and reoriented environmental programs in Illinois. Under this Act, three organizations—Pollution Control Board, Illinois Environmental Protection Agency, and Illinois Institute for Environmental Quality—were established to develop and administer environmental controls.

The Illinois Livestock Waste Regulations were developed by an Agricultural Advisory Committee (composed of representatives of livestock producers, environmental groups, and public agencies) with subsequent hearings and action by the Pollution Control Board. The Regulations provide for the proper handling, storage, and field application of livestock wastes and the proper location of new livestock facilities. A procedure was developed to enable inspectors to enter premises for investigation, but to prevent the introduction or transmission of diseases. A permit system was established which requires a permit for (a) all livestock operations with a total of more than 1000 animal units, (b) operations with 1000 to 100 animal units which are believed by the IEPA to be causing pollution, and (c) livestock operations with less than 100 animal units which are found by the Board to be causing pollution.

The smooth implementation of this program in Illinois will require the cooperation of farmers, agricultural extension and research workers, Soil Conservation Service, and the organizations established by the Illinois Environmental Protection Act.


**References**

1    Bentley, O. G. November-December, 1974. The college of agriculture: Looking at the old; ringing in the new. Current Affairs No. 52. 4p.
2    Haskell, E. H. and V. S. Price. 1973. State environmental management— Case studies of nine states. Praeger Publishers. New York. 283p.
3    Illinois General Assembly. 1970. The environmental protection act. Illinois rev. stat., ch. 111 1/2, sec. 1001-1051 as amended.
4    Illinois Pollution Control Board. 1974. Livestock waste regulations. R72-9, 13 PCB 451-469.
5    Illinois Pollution Control Board. 1974. Livestock waste regulations opinion. R72-9, 14 PCB 429-457.

# The NPDES Discharge Permit Program for Agricultural Point Sources

John C. Nye
ASSOC. MEMBER
ASAE

ON October 18, 1972, the 92nd Congress of the United States passed the most comprehensive water pollution control legislation ever envisioned. Section 402 of Public Law 92-500, the Federal Water Pollution Control Act Amendments of 1972, established a permit program to regulate the discharge of pollutants to the nation's waters. Under this National Pollutant Discharge Elimination System, NPDES, permit program, all industries, municipalities, and agricultural point sources must obtain a permit before they can discharge any pollutants to any of the nation's waters. A point source was defined as any ". . . . . discernible, confined and discrete conveyance, including . . . concentrated animal feeding operations, . . . from which pollutants are or may be discharged." As a result of the Refuse Act of 1899 there was a good inventory of the industrial dischargers but no such inventory existed for agricultural point sources. The agricultural point sources had to be identified before the law could be implemented in the agricultural community. This paper describes the process by which Region V of the U.S. Environmental Protection Agency implemented this law.

## THE REGULATIONS

The U.S. Environmental Protection Agency was required to formulate the filing requirements for the agricultural point sources. The December 5, 1972 Federal Register announced the first proposal for the agricultural portion of the NPDES permit program. Almost all farms would have had to apply for a permit under the original proposal. Due to the tremendous negative response that overwhelmed the Agency, this proposal was radically changed. The May 3, 1973 Federal Register announced a revised program along with a proposed agricultural application form. This May 3 proposal was clarified slightly and in the July 5, 1973 Federal Register, the promulgated filing requirements for the agricultural portion of NPDES permit program were published. The July 5, 1973 regulations required that any agricultural operation had to apply for a permit, if: (a) there was a discharge or a potential discharge, and (b) there were more than one thousand animal units of livestock on hand for 30 days or more during the previous year. One thousand animal units of livestock was defined as 1000 slaughter or feeder cattle, 700 mature dairy cattle, 2500 swine weighing over 55 pounds, 10 000 sheep, 55 000 turkeys, 5000 ducks, 100 000 laying hens or broilers with a continuous overflow watering system, or 30 000 laying hens or broilers with a liquid manure handling system. A formula was provided for combining several different types of livestock. Any agricultural point source that was identified as a "significant contributor of pollutants" would also have to apply for a permit.

"The Efflunet Guidelines and Limitations for the Feedlot Point Source Category" were proposed in the September 7, 1973 Federal Register and finally promulgated in the February 14, 1974 Federal Register. These regulations stated that a feedlot with a NPDES permit would have to be able to contain all process generated waste water plus the runoff from the 10-year, 24-hour rainfall event by July 1, 1977. By July 1, 1983, these agricultural permittees must be able to contain all process generated waste water plus the runoff from the 25-year, 24-hour rainfall event.

In addition to the filing requirements and effluent limitations, there were general procedural regulations by which a NPDES permit would be issued. These were published in the May 22, 1973 Federal Register and later revised in the July 24, 1974 Federal Register. These regulations specify the minimum requirements for the NPDES permit program, and the procedure through which a permit would be issued.

## IMPLEMENTING THE LAWS IN REGION V

From these regulations and Public Law 92-500, it was the responsibility of the 10 Regions of the United States Environmental Protection Agency to administer the NPDES program to all industries, municipalities and agricultural point sources. This discussion relates how the agricultural program was implemented in Region V which includes the states of Minnesota, Wisconsin, Illinois, Michigan, Indiana and Ohio. Many approaches and techniques were used to implement the requirements of the law. The first step was the identification of the livestock producers who would be required to obtain a permit. There was no readily-available inventory of livestock producers who had facilities for over 1000 animal units of livestock. Many of the Land Grant Universities had been requested to estimate the industry structure and status of pollution controls in the states. These educated guesses were used as targets for the number of permits to be issued. The large majority of midwestern feedlots were not issued an NPDES permit. Many people stated that any livestock producer that had over 1000 animal units of livestock had a potential discharge. Some of these potential discharges are the result of structural failures or mismanagement which would never be authorized in a permit. The only authorizable discharge would be one caused by extremely rare rainfall or other precipitation events that overflowed a normally adequate animal waste management facility.

In addition to the basic problem of defining what an authorizable discharge was, and therefore, who would have to obtain a permit, there was a great deal of misunderstanding about the date on which all livestock producers should apply for a permit. The Law required that 180 days after a category had been identified as a point source, an application for a permit must be filed. The feedlot category was identified in the Act which was passed on October 18, 1972, and by the middle of April, 1973, all agricultural point sources should have applied for a permit. Because of administrative problems, the application forms were not

---

The author is: JOHN C. NYE, Extension Agricultural Engineer, Purdue University, West Lafayette, Ind.

available until the agriculture filing requirements were promulgated on July 5. Some people identified the December 31, 1974 deadline from Section 402(K) of Public Law 92-500 as the date when application should be made. The December 31, 1974 deadline was a deadline for the Environmental Protection Agency to have all permits issued.

Since the regulations were in litigation in Federal courts from the date they were announced, many people questioned the status of the regulations even when they were promulgated. The law suit against the EPA for arbitrarily excluding the smaller feedlots has still not been decided, as of February 26, 1975.

All of these conflicting problems created misunderstandings about the agricultural portion of the NPDES program. In order to help clarify these misunderstandings in Region V, the Soil Conservation Service, the Cooperative Extension Service, the livestock producer organizations and the farm press were contacted and given as much information as possible about the regulations. The Cooperative Extension Service was asked to assist by informing the agricultural community of the filing requirements under the NPDES program. The Soil Conservation Service was informed of the program and asked to make the information known to their cooperators who had over 1000 animal units of livestock. The livestock producer organizations were contacted and given information about the NPDES program. They were asked to inform their membership of the filing requirements. The agricultural news media was sent news stories. These informational methods were largely unsuccessful in securing applications.

The majority of the applications arrived through direct contact from the State water pollution control agency. Most of the states in Region V had some type of agricultural program and a list of livestock operations was available. This original list from the 6 State agencies was used as a guide in starting the NPDES program. Each operator was sent an application form and later contacted by telephone to determine the status of his livestock operation. In many cases, the livestock producer were not required to obtain a permit because they had no potential discharge that could be authorized under the NPDES permit program. This was a very difficult point to make clear to the agricultural community. Many people felt that if they had over 1000 animal units of livestock, they should get a permit. They really did not understand that the permit was a tool to control discharges to the nation's water.

After the applications were obtained and before the effluent limitations were promulgated, the Regions were responsible for developing the permit for the agricultural point sources. Even when the effluent guideline was available, it was difficult to explain to the agricultural community. The effluent guidelines specified that there would be no discharge except for one caused by a chronic or catastrophic rainfall that overflowed facilities that were capable of controlling all the process generated waste water and the 10-year, 24-hour rainfall event. Questions came up like "How much process generated waste water is all process generated waste water?" and "What is a chronic rainfall?" For the livestock producer that had an adequate waste management program it was difficult to explain the purpose of the permit. The most effective permits were those prepared for livestock operations with no control facilities and where an implementation schedule was established as a result of the permit.

The effluent guideline was vague enough that several interpretations were made in the Regions and used for developing permit conditions. The feedlots industry was the first industry where the standard was no discharge and an exception was then authorized. The exception stated that the facility could overflow if it was " . . . designed, constructed and operated to contain all process generated waste water plus the runoff from the 10-year, 24-hour rainfall". The overflow had to be caused by "a chronic or catastrophic rainfall event". In Region V and Region VII (Kansas, Iowa, Nebraska and Missouri) a specific operation requirement was stated based on the available storage in the runoff retention facility. The operational requirements stated a rate of dewatering. This was too specific according to the developers of the effluent guideline. As a result, a national form for agricultural permits was developed in Washington and sent to the Regions in September, 1974, after almost all the agricultural permits were prepared and the majority were issued.

One of the major problems in developing a uniform national permit was the difference in climatic conditions. This problem was recognized in the development document for the effluent guidelines. The climatic differences sizably influenced the requirements for a pollution control system.

## IMPROVING THE POLLUTION CONTROL PROGRAM

There were many problems associated with the agricultural portion of Public Law 92-500. The definition of point sources was vague in the Law. The burden of identification of the agricultural point sources was left to the Environmental Protection Agency. There are several ways of going about identifying point sources. One way would be to conduct an aerial survey of the agricultural communities. Another method would be to conduct exhaustive water quality surveys. Either approach would be far superior to asking either the Land Grant Universities or the Soil Conservation Service to guess at the status of pollution control on livestock farms.

The development of a national standard of performance was a difficult task for the Environmental Protection Agency. The contract for the development document for this effluent guideline was let to an aerospace firm that had been working on an anaerobic fermentation process to recover livestock feed from feces. The use of this firm was questioned by many livestock groups. One comment was made that "I sure wouldn't feel very safe it I was riding in an airplane that was designed by an Ag Engineer after he had visited a few airplane manufacturers." There certainly is more to livestock waste management than what this firm saw during a few visits to feedlots. There were Federal and State agencies that had much experience in the area of feedlot waste management systems. The Soil Conservation Service had been responsible for approving the cost sharing funds for pollution abatement systems under the Rural Environment Assistance Program administered by the Agricultural Stabilization and Conservation Service. Their experience throughout the country and the design standards that they had developed would surely have been as good a starting point for an effluent guideline as the final development document.

The difficulties in interlacing the State and Federal Programs was and is a serious problem. The Federal Regulations establish the minimum standard for all feedlots across the country. More stringent State requirements can be added to the Federal requirements. Several states were re luctant to use the term "discharge" to describe the over-flows caused by rare "catastrophic or chronic" rainfall events. It is important that these problems are 'ironed out' at the local level so that the livestock producers see a single pollution control program.

The massive number of livestock producing farms with varying degrees of pollution problems makes it difficult to implement any "all inclusive" pollution control program. It is difficult to identify all livestock farms that should develop a new effective waste management plan, but it must be accomplished if the goal of Public Law 92-500 "to restore and maintain the biological, chemical and physical integrity of our nation's water", is to be realized.

### References


1    The 92nd Congress of the United States. 1972. The Federal Water Pollution Control Act Amendments of 1972, Public Law 92-500. October 18, 1972.

2    Environmental Protection Agency. 1972. National pollutant discharge elimination system, proposed forms and guidelines for acquisition of information from owners and operators of point sources. 40CFR, Part 126. December 5, 1972. Federal Register 37:25898-25902.

3    Environmental Protection Agency. 1973. National pollutant discharge elimination system, notice of proposed forms and proposed rule making regarding agricultural and silvicultural activities. CFR Part 124, 125, May 3, 1973 Federal Register 38:10960-10968.

4    Environmental Protection Agency. 1973. National pollutant discharge elimination system, forms and guidelines regarding agricultural and silvicultural activities. 40CFR Part 124 and 125, July 5, 1973 Federal Register 38:18000-18004.

5    Environmental Protection Agency. 1973. Feedlot category, effluent limitation guidelines for existing sources and standards of performance and pretreatment standards for new sources. 40CFR Part 412, September 7, 1973 Federal Register 38:24466-24470.

6    Environmental Protection Agency. 1974. Development document for effluent limitations guidelines and new source performance standards. January 1974, EPA-440/1-74-004a, Superintendent of Documents, U.S. Government Printing Office, Washington, D.C.

7    Environmental Protection Agency. 1974. Feedlot point source category. Effluent guidelines and standards. 40CFR Part 124, February 14, 1974 Federal Register 39:5704-5710.

8    Environmental Protection Agency. 1974. National pollutant discharge elimination system miscellaneous amendments. 40CFR Part 125, July 24, 1974 Federal Register 39:27077-27084.

# Techniques That Are Solving Pollution Problems for Poultrymen

Charles E. Ostrander

**P**OULTRYMEN throughout the United States have made great progress in alleviating the pollution problems created by high density large volume poultry operations. This has not been by any one practice or procedure. What works best for one operation with a particular set of circumstances may not necessarily be best for another, with a different set of circumstances. This has been one of the biggest stumbling blocks encountered. Each situation is different.

One needs to consider — Location, Climate, Size of Operation, Availability of Land, Cropping Possibilities, Markets, and perhaps others before deciding which system might be best for their particular situation.

Systems that have been used successfully by poultrymen are:

| Dry Systems | Liquid Systems |
|---|---|
| | Aerobic |
| High Rise Houses | Oxidation Ditch |
| In-House Drying | Surface Aeration |
| Dehydration | Anaerobic |
| | Soil Injection |
| | Anaerobic Digestion |

## DESCRIPTION AND OPERATION OF SYSTEMS

### Dry Systems

**High Rise House**—This system is probably one of the least costly and most efficient in the use of labor. It is very dependent on good drainage, prevention of spilling of water on to the manure, and good air circulation over the manure mass.

Ideally, the high rise house is built above ground on a pad of gravel, cinders or other porous material. The ground is graded away from the house forming a diversion ditch to carry away any surface and roof waters. A concrete floor is advisable for easy cleaning but some do not use it because of the extra cost.

Drying underneath the cages can be enhanced by using circulating fans every 30 to 50 ft to circulate air over the "cones" formed (Fig. 1) or by using slats hung under the cages to increase the drying surface.

If circulating fans are used they should face one direction on one side of the house and the opposite direction on the other side, causing the air to flow around the house in a circular pattern.

Every effort should be made to prevent spilling of fountain water into the pit area. Any excess water requires more evaporation and may overload the capability of evaporation causing wet areas and odors to develop.

The manure is usually removed once a year or once in

FIG. 1 Cone development under flat-deck cages in a high-rise poultry house. Cones increase the surface area to enhance drying.

two to three years or longer. The moisture content should range from 40 percent to 60 percent with very little "house odor" or odor at cleaning time if everything is operating properly. A front end loader or lift on a tractor, or better yet a short-coupled four-wheel drive vehicle is ideal for cleaning. The material can be handled and spread with conventional manure handling equipment. It is usually done in the spring or fall when land is most accessible. Plowing soon after spreading is advised to conserve the nutrients, particularly the nitrogen, to prevent odor problems, and to prevent non-point source pollution.

The house can be cleaned with the birds in the cages if desired but most poultrymen clean between flocks (Sprague et al. 1972).

Summary Cornell Study
November 20, 1970 - October 13, 1971

| | |
|---|---|
| Average number of birds | 26 000 |
| Size of pit area, square feet | 12 096 |
| Number of fans, 1/2 hp | 6 |
| Air flow rate, cfm/fan | 10 000 |
| Total energy used, kW hours | 18 118 |
| Operating cost, cents per dozen | .07 |
| Operating cost, cents per bird | 1.4 |
| Loads of manure removed | 172 |
| Average weight per load, lbs | 4 700 |
| Average moisture content of manure, percent | 50 |
| Lbs manure removed, 50 percent moisture | 808 400 |
| Total man days for manure removal | 8.5 |
| Estimated water removed during drying, lbs | 895 800 |

We would not recommend this system in a highly populated area but it should be ideal for a rural area where land is available on which to place the manure.

The key is to keep the manure dry by preventing seep-

---

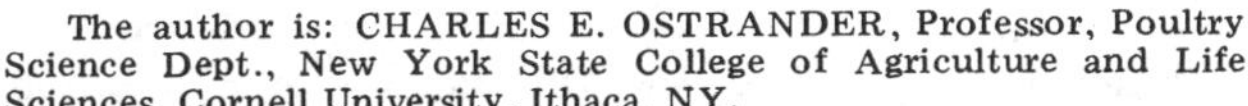

The author is: CHARLES E. OSTRANDER, Professor, Poultry Science Dept., New York State College of Agriculture and Life Sciences, Cornell University, Ithaca, NY.

age, no spilling of water into the pit area and having good air circulation over the manure.

**In-House Drying**—is very similar to the high rise house except that storage is not provided. Where frequent cleaning does not present a problem or if the manure is to be dehydrated this system is satisfactory. Fans and/or stirring are used to reduce the moisture content and consequently the odor.

Total labor and costs are greater than with the high rise house and is dependent upon frequency of removal. Many areas will not tolerate frequent spreading.

**Dehydration**—Considerable interest has been developed in dehydrating poultry manure for use as an organic fertilizer or to be recycled through livestock. Dehydrated manure probably cannot compete with inorganic fertilizer on a cost per unit of plant nutrient for commercial agriculture. There is a limited market, primarily to the organic enthusiast. If inorganic fertilizers continue to increase in cost dehydrated manure may become competitive.

With the increase in the cost of protein supplements this material may be very competitive particularly for feeding to ruminants providing it is approved by the Food and Drug Administration (FDA).

Many trials have shown very satisfactory and efficient gains when fed to ruminants, up to 30 percent of the ration (Blair and Knight 1973) (Bull and Reid 1971). Ruminants are capable of synthesizing the non-protein nitrogen very satisfactorily. Monogastric species are not capable of using this product as efficiently as they are not capable of converting the non-protein nitrogen to a usable nutrient.

New dehydrators have been developed and others have been improved considerably. One of the real problems has been to dry a product with 75 percent to 85 percent moisture, efficiently. The cost for fuel rises very rapidly above 50 percent moisture. With the increasing cost of energy this becomes increasingly important.

Pre-drying under the livestock or some form of dewatering shows a great deal of promise in reducing the overall cost of dehydrating. This may be accomplished by circulating air over the manure, agitation, or both.

Odors from dehydrators have been a problem but this can be controlled by using afterburners. However, these increase the cost considerable, about doubling it in most instances.

If profitable markets can be obtained, dehydration provides a good solution for disposal of poultry wastes, particularly for the large operation with very little land.

### Liquid Systems

An aerobic system might be the best system for a poultry operation in a highly populated area. When operated properly there will be little or no odor. Aerobic systems might logically be accomplished by using mechanical aeration equipment such as an oxidation ditch, surface aerators, aerob-A-jets or others.

Aeration equipment entrains oxygen (air) into the liquid mass which oxidizes it to the point where odors are minimal or non-existent. The resulting mass can be spread on the land any time and most any place without offending anyone. It is not suitable for discharge into streams or waterways because of the color, mineral salts and residual oxygen demand of the effluent.

**Oxidation Ditch**(Loehr and Ostrander 1972) (Jones et al. 1970)—The oxidation ditch is one form of aeration system developed originally in the Netherlands. Its use,

FIG. 2 Floating surface aerator. No need to adjust emersion depth. Aerator goes up or down with water depth.

structure, etc. has been modified and used in many other areas of the world. Originally oxidation ditches were outside the building but in some areas this reduces efficiency. In the United States the oxidation ditch is often incorporated as a part of the building housing the livestock. The ditches are placed directly under the livestock enabling the manure to drop directly into the aerated mass. They have a fairly long detention time and considerable reduction of solids occurs. Compared to other liquid systems much less volume has to be handled.

The oxidation ditch is particularly adapted to areas of dense population where odors could be a problem. Even though the capital costs might be somewhat higher than some systems this may be justified with such a situation to avoid sociological problems due to odors.

Key points in operating an oxidation ditch are: loading rate, solids contents of the mass, depth of immersion and velocity.

**Aerated Ponds** (Jones 1970). The principle is similar to that of the oxidation ditch in that oxygen is entrained into the waste mass to control odors and reduce solids. Usually a larger volume of area is used than with the oxidation ditch and the pond is outside the building. Because of this it is much more subject to environmental conditions. These are more adaptable in warm climates because of greater efficiency and less problems due to freezing, etc. Aerobic bacteria are much more active in temperatures above 4.5 C (40 F).

Usually floating aerators are used to aerate the pond (Fig. 2). There is no adjusting for immersion depth as it rises and falls with changes in depth of water. Aerated ponds are usually 12 to 16 feet deep.

Settling ponds in conjunction with the aerated pond can be used to remove solids or solids can be removed by pumping directly from the aerated pond. The settling pond system uses less makeup water because the effluent from the settling pond can be pumped back into the aeration pond after settling occurs.

Other aeration systems being used include the Aerob-A-Jet, flush and aerate, diffused aeration, and perhaps others. They all work on the principle of aerobic treatment to reduce odors and solids.

**Anaerobic Digestion**—Straight anaerobic digestion by use of anaerobic pond can be used to reduce solids but has the disadvantage of being odorous, therefore it is advised only in isolated areas. Today with the shortage of energy and



MANAGING LIVESTOCK WASTES

FIG. 3 Rolling coulter, soil opening shoe and coverers on soil injector.

FIG. 4 Soil injector from the rear showing good covering. No odors or fly breeding.

fertilizer it has little to offer unless one captures the methane for energy production.

Anaerobic digestion in closed containers enabling the collection of methane gas and an activated sludge which is minimal in odor and that retains its fertilizing value is getting considerable attention. Refinements are needed but if energy continues to be short this system may have a great deal to offer.

New developments are occurring regularly. One of the newest is to treat the effluent from anaerobic digestion to produce single cell bacteria which can be harvested as a high quality protein. This is still in the developmental stage but may have potential possibilities.

**Soil Injection**—Soil injection is not a treatment system to reduce odors of the material itself, but it can be very helpful where spreading odors are a problem (Ostrander 1972). Laboratory and field tests have shown that six inches of a moist soil proves to be a very good filter to prevent odors from permeating the area (Burnett and Dondero 1969).

In many situations odors are a problem only at spreading time. When this situation exists soil injection can aid materially in eliminating the problem.

Soil injectors have been improved greatly during the past few years. Most of these will inject the material from four to twelve inches into the soil. Many will not inject in untilled soil. If injected in sod a rolling coulter is recommended. This prevents tearing of the sod and provides for better covering (Fig. 3).

Most equipment inject slurries at approximately one gallon per lineal foot per injector. Few, if any, have flow controls, therefore, the rate of injection is determined by the speed of travel. Two to three miles per hour will usually provide the correct flow for good covering (Fig. 4). If operated too slowly the slurry will be forced to the surface where odors and fly breeding may create a problem.

Spring loaded soil openers are recommended. With these if an obstruction is hit the opener will trip and then go right back into the soil. This prevents a great deal of equipment breakage. Soil injection requires a tractor with a minimum of 60 horsepower. It also requires storage facilities, in most areas of the country, either under the livestock or in separate storage tanks. There are times when it is impractical to soil inject because of crops growing or because the ground is frozen or covered with snow. Facilities for storing three to four months manure production is suggested.

Soil injection has been very helpful to those who have populated areas near their crop land where they want to utilize the poultry manure but have had complaints because of odors at time of spreading.

More refinement is needed to develop more accurate flow control to be able to control proper rate of application for various soils and crops. There is interest in developing equipment capable of applying the soil injection principle for side dressing corn (maize). Some agronomists feel this should be very applicable, particularly for poultry manure, because of its continuous and gradual release of nitrogen.


### References

1   Blair, Robert and David W. Knight. 1973. Recycling animal wastes. Feedstuffs 45(10):32-34, 45(12):31-36.
2   Burnett, William E. and Norman C. Dondero. 1969. Soil filtration to remove odors. p. 71-86. In: Odors, gases and particulate matter from high density poultry management systems as they relate to air pollution. N.Y. State Coll. of Agr. and Life Sciences, Depts. of Agr. Engr., Food Sci. and Poultry Sci. final report to N.Y. State Dept. of Health, Contract No. C-1101, Cornell Univ., Ithaca, N.Y., 14853, 106 p.
3   Bull, L. S. and J. T. Reid. 1971. Nutritive value of chicken manure for cattle. Livestock waste management and pollution abatement. Proc. of the Intern. Symp. on Livestock Wastes, ASAE: 297-300.
4   Jones, D. D., D. L. Day and A. C. Dale. 1970. Aerobic treatment of livestock wastes. Univ. of Illinois Bull. 737, Univ. of Illinois, Urbana-Champaign, Illinois. 55 p.
5   Loehr, Raymond C. and Charles E. Ostrander. 1972. The oxidation ditch. N.Y. State Coll. of Agr. and Life Sciences Info. Bull. 30, Cornell Univ., Ithaca, N.Y., 14853, 4 p.
6   Ostrander, C. E. 1972. Soil injection of farm manures. N.Y. State Coll. of Agr. and Life Sciences Mimeo, Cornell Univ., Ithaca, N.Y., 14853. 3 p.
7   Sprague, D. C., A. T. Sobel, H. R. Davis and T. L. Todd. 1972. Performance observations of a high-rise system for collecting, conditioning, handling and utilizing caged layer manure. N.Y. State Coll. of Agr. and Life Sciences, Dept. of Agr. Engr. Bull. AWN 72-06, Cornell Univ., Ithaca, N.Y., 14853, 42 p.

# Modifications on the Michigan State Poultry In-House Drying System


C. C. Sheppard, C. J. Flegal, H. C. Zindel, T. S. Chang, J. B. Gerrish,  M. L. Esmay, F. Walton

EXEC. AFFILIATE      ASSOC. MEMBER   MEMBER
ASAE                ASAE        ASAE


THE caged layer house described by Esmay et al. (1975) was modified during November and December of 1974. Modifications were made to: (a) improve the in-house manure drying capability, (b) decrease or eliminate the pollution emissions from the house and (c) eliminate the need for the afterburner on the oil-fired manure dehydrator.

Four changes were made in the poultry house and equipment design. They were:

1   Modification of the original stirring device through a new design.

2   Addition of six fans and duct work, to allow the air in the house to be recirculated after being blown over the manure.

3   A filter chamber or scrubbing tower added to "wash" all of the air leaving the house, including that from the oil-fired manure dryer, with or without the afterburner in operation.

4   The watering system was designed, constructed, and installed to reduce water spillage in the house.

## STIRRING DEVICE-METHODS

The original stirring device (Fig. 1) did not produce the desired effect of reducing the moisture content of the stirred manure and did not operate as expected. This device consisted of three rows of one centimeter diameter steel rods mounted on a wheeled harrow. There were four spikes spaced approximately 15 centimeters apart in each row. The spikes in each row were staggered to plough a different furrow than the row of spikes in front of it. In January, 1975, a new stirring device was installed. It was made of 10-gauge galvanized rolling discs, 25 centimeters in diameter. There were six discs, 10.6 centimeters apart, on an axle that was set on an angle approximately 60 degrees to the length of the belt. The purpose of the disc stirring device was to turn, move and cut the excreta on the belt in order to expose more surface area to the air for drying. The data on stirring, in both cases, were only gathered from one-half (that nearest the exhaust fans) of the belt.

Each day when data were taken, manure was placed on the belt. The excreta was then stirred two times in the afternoon, at approximately 1:30 and 3:30 p.m. It was stirred again two times in the evening at approximately 7:00 and 9:00 p.m. The following morning the excreta was again stirred at 7:00 a.m. A complete round trip was made by the stirring device each time the excreta was stirred. Samples of manure, for moisture determination, were taken


Michigan Agricultural Experiment Station Journal Article No. 7182.

The authors are: C. C. SHEPPARD, C. J. FLEGAL, H. C. ZINDEL, T. S. CHANG, Poultry Science Dept., J. B. GERRISH, M. L. ESMAY and F. WALTON, Agricultural Engineering Dept., Michigan State University, East Lansing.


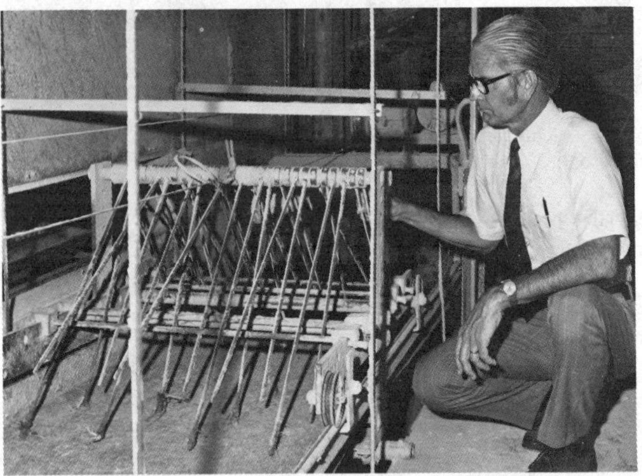

FIG. 1 Dr. Merle Esmay examining the original stirring device.

twice. The manure was first sampled when loaded on the belt (between 10:00 and 11:00 a.m.) and again from the same manure the following morning at approximately 9:30 a.m.

## MOISTURE CONTENT OF STIRRED VS. NON-STIRRED MANURE

Moisture content of the stirred and non-stirred manure was very similar when the first stirring device was used. An average moisture content of the manure samples of 16 observations in the cool months of March, October and November of 1974 was 67.4 for the stirred and 65.6 percent for the non-stirred manure. The purpose of the spikes was to lift and turn the manure.

The new stirring device produced tracks, valleys and ridges through all the manure on the belt as the belt was

**TABLE 1. PERCENT MOISTURE CONTENT (M.C.) OF EXCRETA ON THE BELT 24 HOURS.**

| | | | Non-stirred | | | |
|---|---|---|---|---|---|---|
| Sample locations | 1 | 2 | 3 | 4 | 5 | 6 |
| No. readings | 5 | 5 | 5 | 5 | 5 | 5 |
| Avg. percent moisture | 61.1 | 53.1 | 64.8 | 54.8 | 59.6 | 51.3 |
| | | | Stirred | | | |
| Sample locations | 7 | 8 | 9 | 10 | 11 | 12 |
| No. readings | 15 | 15 | 15 | 15 | 15 | 4 |
| Avg. percent moisture | 66.1 | 54.2 | 65.6 | 59.5 | 63.4 | 60.2 |

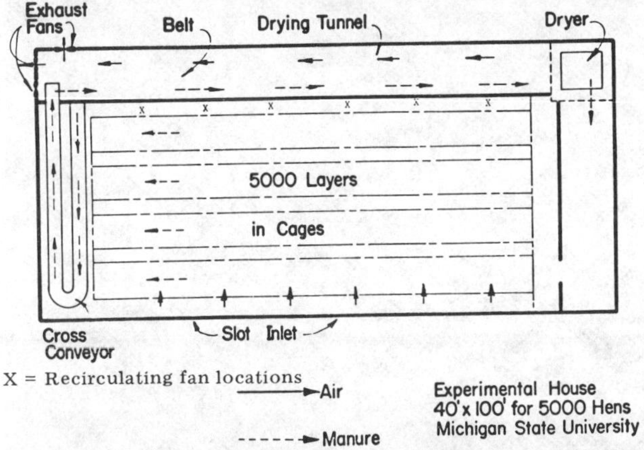


FIG. 2 An experimental in-house system for handling and drying of poultry manure.

being loaded, as well as when the manure was stirred. The new stirring device did not result in additional drying of the manure (Table 1) when compared with moisture of the manure from the non-stirred area. The average moisture content of the manure samples on the half of the belt farthest from the exhaust fans (not stirred) was 57.4 percent after 24 hours on the belt. The manure on half the belt nearest the exhaust fans (the stirred portion) contained 61.7 percent moisture content after 24 hours on the belt.

## RECIRCULATING AIR

Six, 35 centimeter, six-bladed fans with one-quarter horsepower motors were installed in the wall between the

FIG. 3 The scrubbing tower under construction.

FIG. 4 Scrubbing tower contained six of these trays.

birds and the drying tunnel (just below the ceiling), (see floor plan, Fig. 2). They were placed approximately 4 meters apart, with the first fan being placed about 4 meters from the end of the cage row. Each fan was placed in a 15 x 60 centimeter duct that was extended approximately 5 meters to the middle of the aisle in the center of the house. The fans were hooded on the tunnel side of the wall with a 42.5 x 42.5 centimeter hood extending down to within approximately 60 centimeters of the belt, directing the air onto the manure.

Table 1 presents the results of the use of recirculation fans to help dry the manure on the belt. The odd-numbered sample locations refer to manure samples taken between fan outlets. The even-numbered sample locations refer to manure samples taken under the fan outlets. The average moisture content of the manure between fan outlets on the half of the manure not stirred was 61.5 percent and the half of the manure which was stirred contained 65.0 percent moisture. The average moisture content of the manure not stirred under the fan outlets was 53.1 percent and of the manure stirred was 57.5 percent.

Data from three days, February 11, 12, 1975 and February 19, 1974 resulted in manure moisture contents of 65.0 - 68.5 percent after 24 hours on the belt when no recirculating fans were used. The 65.0 - 68.5 percent moisture content of the manure was reduced under the recirculation fans to 53.0 -58.0 percent. The 10.0 - 16.0 percent moisture reduction means that up to 50 percent of the water in the 67 percent moisture content manure, (typically present in manure on the belt in the winter), has been removed by the air from the recirculation fans.

## PARTICULATE EMISSION

A tower 2.4 meters x 2.4 meters x 7.2 meters (Fig. 3) was constructed in such a manner to receive all the exhaust air from a 5,000 bird caged layer house. The exhaust air consisted of ventilation air as well as the exhaust of a small, oil-fired dryer and afterburner that was used to dry poultry manure to anaphage.

At the base of the tower was a tank that was approximately 2.4 meters x 2.4 meters x 0.8 meters. The purpose of this tank was to hold the waste water from the drinking troughs and to serve as a sump for the scrubber. Tap water was added when necessary to maintain an adequate water level.

There was a space 2.4 meters x 2.4 meters x 2.4 meters, immediately above the tank into which the exhaust air was introduced. The upper 3.6 meters (also 2.4 meters x 2.4 meters) contained about 418 square meters of surface area. This surface area consisted of quart plastic bottles with part of the neck removed (to avoid some constriction). The bottoms of the bottles were also removed to allow air passage through the bottle.

The quart plastic bottles were placed in a honeycomb arrangement on trays (Fig. 4) approximately 1 x 2 meters. The bottles were stacked three tiers high (approximately 65 centimeters) in each tray. Two trays were placed on each level; one at 2.4 meters above the top of the tank; the second set of trays were placed approximately 1.25 meters above the first; and the third set, approximately 1.25 meters above the second set of trays. Between each tier of trays there was an open space of approximately 0.6 meters to permit some air turbulence to develop.

The water from the tank was pumped to the top of the tower. It was then dispersed over the 418 square meters of surface area to trap (or attach) particulate matter in the exhaust stream. Measurements were made on the particulates entering and leaving the filtration tower.

The original watering system consisted of a V-trough continuous flow waterer. After six months of operation, due to excessive water spillage, a new system was installed. The new waterer consisted of a 10 centimeter wide aluminum seamless trough. This was a standing water system with a 2.5 centimeter high overflow pipe at the end of each trough. Installation of this new system minimized water spillage onto the floor and manure.

When the tower was built, the waste water line was intercepted. When the watering troughs were cleaned, the discharged water was pumped into the tank on an almost daily basis to maintain the water level in the tank.

Emission measurements were made by using high volume air samplers. Samples were taken at two locations. The first was taken just ahead of the ventilation fans that forced the air up the tower. The second sample was taken in the air stream at the top of the tower. The samples were taken with the dryer off and with the dryer on.

The exhaust air at the top of the tower contained droplets of water. These droplets contained dissolved material and were measured as a part of the particulate emissions. The volume of air flowing through the tower was varied by changing the number of 283 cubic meters per minute fans that were operating.

In mid-January, 1975, the afterburner was detached from the dryer by turning the elbow going from the dryer to the horizontal afterburner, one quarter of a turn. This allowed the dryer to discharge its emissions directly into the drying tunnel.

FIG. 5 Some particulate emissions (debris) in the drying tunnel.

Presented in Table 2 are the data of airborne particulates for inlet and outlet concentration for 1 fan (5 m³/sec), 2 fans (10 m³/sec), and 3 fans (15 m³/sec) with the dryer operating and with the dryer not operating. In all cases, the tower reduced the particulate load by 67 to 75 percent.

A large amount of particulate matter (Fig. 5) exhausted by the dryer was too heavy to be carried the length of the drying tunnel by the air stream. This was not measured but was evident as debris in the drying tunnel.

## ODOR

Odors are one of the major problems of large commercial egg production farms. Several attempts were made to measure odors in the house and at the top of the tower. One method that was used was personal observation by three researchers.

Another method of odor detection used was described by Jacobs (1960). This method consisted of bubbling a measured amount of air through a dilute sulfuric acid solu-

**TABLE 2. EFFECT OF THE ADDITION OF A SCRUBBER TOWER ON PARTICULATE EMISSIONS.**

| Dryer | Flow m³/sec | Inlet particulate concentration µg/m³ | Outlet particulate concentration µg/m³ |
|---|---|---|---|
| Off | 5 | 4080 | 1350 |
| | 10 | 3870 | 1810 |
| | 15 | 3850 | 1320 |
| On | 5 | 5990 | 1400 |
| | 10 | 5310 | 1440 |
| | 15 | 7100* | 2780 |

*Smoke episode. The poultry house was clear, but the dryer exhaust was smokey.
The afterburner was bypassed during these tests. This test was run January 25, 1975.



MANAGING LIVESTOCK WASTES

|  | Start up | 20 days | 60 days | 80 days | Typical make-up water |
|---|---|---|---|---|---|
| Chemical oxygen demand (mg/l) | 3900 | 2300 | 4400 | 8500 | 9000 |
| Total solids (mg/l) | 4000 | 4700 | 4500 | 3500 | 5400 |
| Volatile solids (mg/l) | 2400 | 3500 | 3200 | 2000 | 3800 |
| $NH_4+$ (mg/l) | 150 | 75 | 220 | 170 | 75 |
| Total nitrogen (as N) (mg/l) | 580 | 280 | 500 | 560 | 270 |
| Electrical conductivity (micromho/cm) | 2600 | 2300 | 2600 | 3500 | 2000 |

tion for ammonia detection. Air samples were taken at three sites; one, at the center of the laying house, approximately three-tenths of a meter above the floor level; the second site was located in the air stream as it entered the ventilation fans; the third site was located at the top of the tower to test the air as it left the scrubber.

When the dryer was operating the odor of fuel oil could be detected at two sampling sites (at the ventilation fans and at top of tower). The odor of drying manure was also detectable at both sites when the dryer was operating. When the dryer was operating the odors at the top of the tower were markedly less than at the inlet site. None of these odors were really objectionable to the odor panel (Gerrish, Sheppard and Flegal). During drying, there were some episodes of heavy smoke emissions. These were only partially clarified by the scrubber unit.

The results of the ammonia determination indicated that there was less than 1 ppm at all sites tested (0.54 -0.72 ppm).

## WATER QUALITY MEASUREMENTS OF LIQUID MEDIUM IN TOWER

Several measurements were made on the liquid medium used in the scrubber tower. The tests included chemical oxygen demand, total solids, volatile solids, electrical conductivity, total nitrogen, ammonia and pH. Initially, the tank at the base of the tower was filled with tap water. Due to evaporation and droplet emissions, liquid was added to maintain a depth of 72.5 centimeters. Most of the additional water added was the discharge from the bird watering system. This water contained feed, feathers, fecal material, etc. (Table 2). Samples of the liquid in the tank and of the discharge water from trough cleaning, were taken periodically.

It can be seen in Table 3 that the chemical oxygen demand was 3900 mg/l at the start of operation of the scrubber unit. This dropped to 2300 mg/l after 20 days and then steadily increased to 8500 mg/l at 80 days. This same general pattern was observed in the electrical conductivity. Volatile solids in the liquid medium of the scrubber unit decreased from 3500 mg/l at 20 days to 2000 mg/l at 80 days. Total solids also dropped from 4700 mg/l at 20 days to 3500 mg/l at 80 days. This reduction of total solids would correlate with droplet emissions from the top of the tower. Total nitrogen and $NH_4+$ were variable during the time that measurements were taken.

When the project was ended a sludge blanket was found at the bottom of the liquid tank. The accumulated solids would account for the decrease in volatile solids observed toward the end of the operation. This sludge accumulated may eventually have necessitated a clean-out of the tank and a fresh start with new liquid.

## SUMMARY

Based on the evidence obtained in this series of experiments, the following observations were made:

1    The stirring devices used did not result in increased moisture removal from the manure.

2    The recirculation fans resulted in increased moisture removal from the manure.

3    The addition of the scrubber unit resulted in lower odor and particulate emissions from the laying house with and without the oil-fired dehydrator operating.

4    The trough (standing water) type watering system resulted in a minimum of water spillage.

**References**

1    Esmay, M. L., C. J. Flegal, J. B. Gerrish, J. E. Dixon, C. C. Sheppard, H. C. Zindel and T. S. Chang. 1975. In-house handling and dehydration of poultry manure from a caged layer operation; a project overview. Presented at the 3rd Int. Symp. on Livestock Waste, April 21-24, 1975, University of Illinois.

2    Flegal, C. J., M. L. Esmay, J. B. Gerrish, J. E. Dixon, C. C. Sheppard, H. C. Zindel and T. S. Chang. 1974. A complete system for collecting, handling, air drying and machine dehydration of poultry manure in a caged layer production unit. Proc. 1974 Cornell Ag. Waste Mgt. Conf.

3    Jacobs, M. B. 1960. The chemical analysis of air pollutants. Interscience Pub. Inc., New York, NY.

# Design of Poultry Manure Drying System for a 155 000 Layers Egg Factory

K. Koskuba
MEMBER
ASAE

ONE of the most advanced Czechoslovakian poultry farms started limited operation at the end of 1969 and achieved full scale operation of 15 poultry houses in June, 1970. The original project called for the scraping and conveying of the manure into the pits at each house with later field spreading. The first year of operation proved the original idea to be unreal and too costly for such a high concentration of poultry. In 1971 a British-made flash dryer with pulverizing effect was installed. The machine is now drying all manure produced on the farm.

The farm is owned by the Association of Co-operative and State Farms of the District Kutna Hora and located near the town Caslav. It is now supplying the equivalent of the former egg production of all agricultural enterprises of the District. Having a capacity of 155 000 layers, it has a yearly production of 31 million eggs, 240 metric tons of meat, and 1300 metric tons of dried poultry manure used for feeding cattle at the farms of the District.

## DRYING SYSTEM

The fresh manure from the cages is automatically scraped into a horizontal screw conveyor (Diam. 320 mm) at the end of the house and by means of another similar conveyor delivered into a trailer. A tractor operator serves all houses, periodically hauling the manure into a 25 m³ pit at the drying plant. The drying is performed in an ATRITOR* flash dryer-pulverizer. The following reasons were taken into account when choosing this type of dryer:

1 The machine can handle wet and sticky material using feed back of the required amount of the dry material taken from the end of the process.

---

The author is: K. KOSKUBA, CSc, Engineering Consultant, Prague, Czechoslovakia.

*Atritor machines are manufactured in different sizes by Herbert Machine Tools Ltd., Coventry, England, and they are widely used for drying and grinding of many materials in the poultry industry.

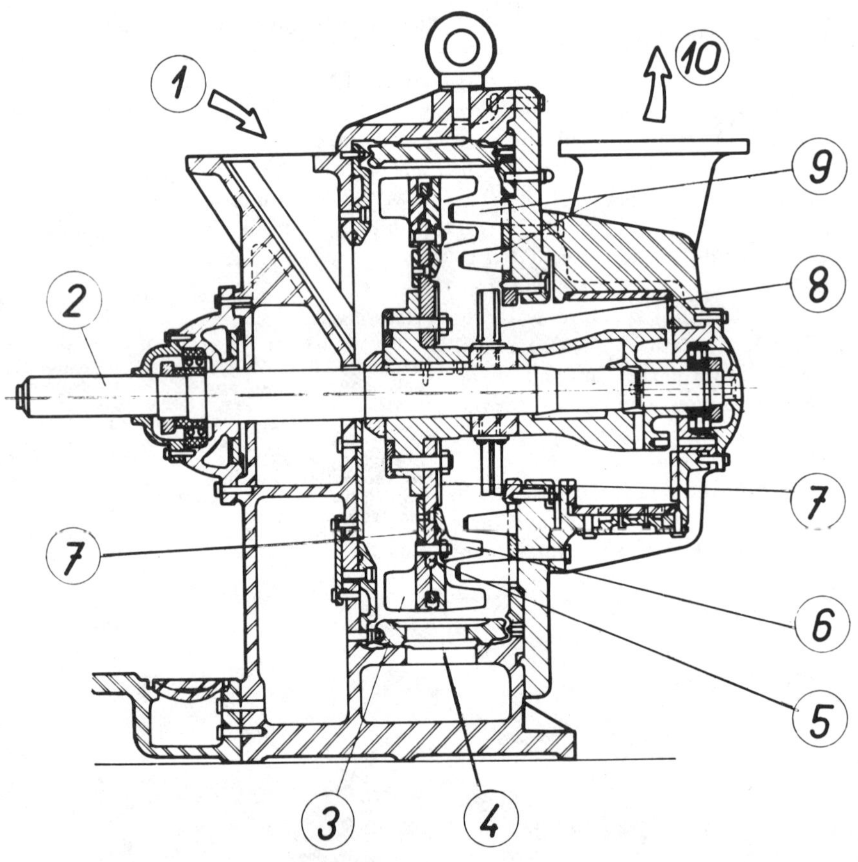

FIG. 1 Section drawing of the dryer-pulverizer — (1) wet material and hot air entry; (2) rotor shaft; (3) first effect hammers; (4) trampmetal hole; (5) rotor disc; (6) second effect peg segment; (7) wear plates; (8) rejector arm; (9) stator pegs; (10) outlet.

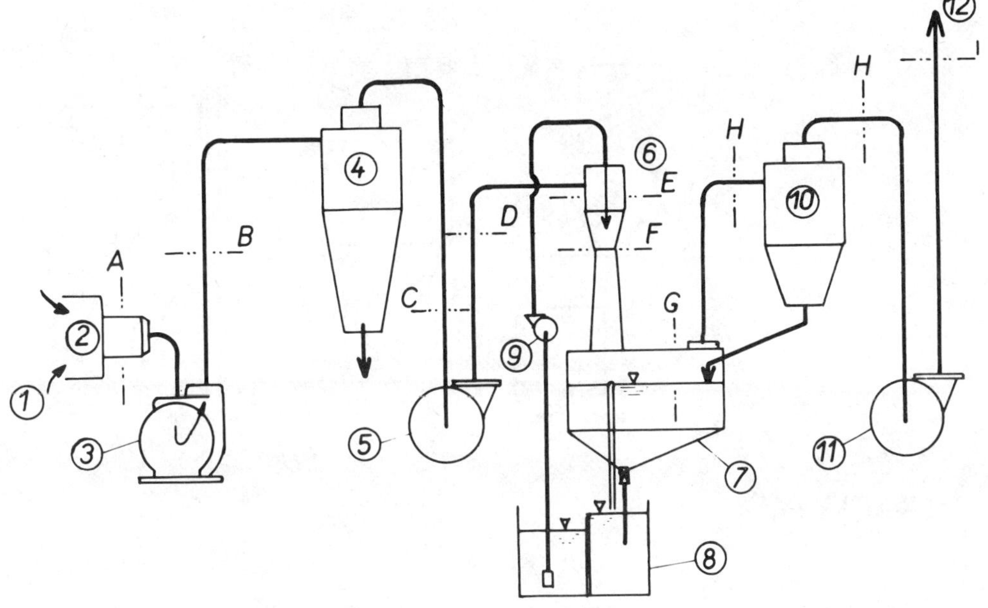

FIG. 2 The flow of gases — (1) atmospheric air entry; (2) oil fired air heater (capacity up to 160 l/h of oil); (3) dryer-pulverizer; (4) cyclone 2713 mm diameter, 8230 mm high; (5) radial blade centrifugal fan (rotor, 1270 mm diameter, speed 1454 rpm, motor 60 kW, capacity 5.2 $m^3$/sec at 56 C and total pressure of 6470 $N/m^2$); (6) jet scrubber; (7) water tank; (8) pit (10 $m^3$); (9) pumps; (10) cyclone droplet eliminator; (11) radial blade centrifugal fan (rotor diameter 1000 mm, speed 1465 rpm, motor 25 kW, capacity 5.5 $m^3$/sec at 100 C and total pressure of 3040 $N/m^2$).

2    The dried material is submitted to a disintegrating action and heavy turbulence. The surface exposed to the drying influence of the hot gas is high, the heat transfer being rapid and very uniform.

3    The machine is very compact with low wear, easy and quick replacement of wearing parts, and clean and quiet in operation. A section drawing of the machine is shown in Fig. 1.

The heat is supplied by an oil fired heater with toroidal combustion. The oil pump of the heater is controlled by the outlet air temperature (set at a range of 100-110 C) and by the inlet temperature (set on maximum of 510 C).

The backmixing of already dried material with the wet material has a disadvantage: the higher the moisture content of the fresh manure, the greater the quantity of recirculating dried material, and the higher the cyclone losses. This is due to low cyclone efficiency and the fact that higher circulation means an increased amount of fine particles.

## THE FLOW OF GASES

As it is shown in a flowsheet (Fig. 2), the main component parts of the air-gas flow system are two high pressure fans (5 and 11) working in series in the exhaust system. Data on amount of gas, temperatures, static and velocity pressures are given in Table 1.

The original jet scrubber (a prototype supplied by a Czech drum drier manufacturer) was working very unsatisfactorily. Shortly after its installation it was replaced by a submerged nozzle type scrubber (Fig. 3) with a secondary

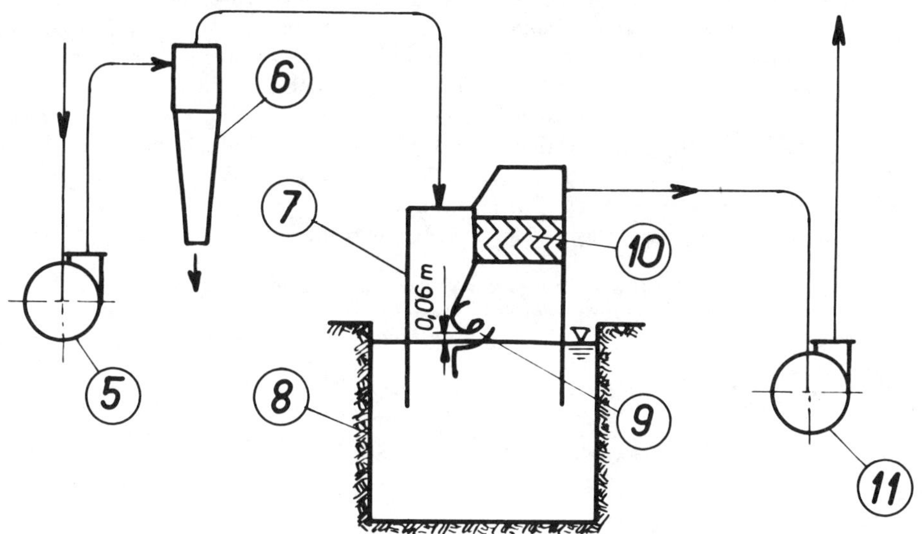


FIG. 3 Secondary cyclone collector and submerged nozzle scrubber — (5) fan (Fig. 2); (6) a pair of cyclones (diameter 1000 mm, length 4188 mm, pressure drop 1080 $N/m^2$); (7) washer; (8) pit (Fig. 2); (9) submerged nozzle (a pair of nozzles with a gap of 60 mm and 1600 mm length each); (10) droplet eliminator; (11) fan (Fig. 2).

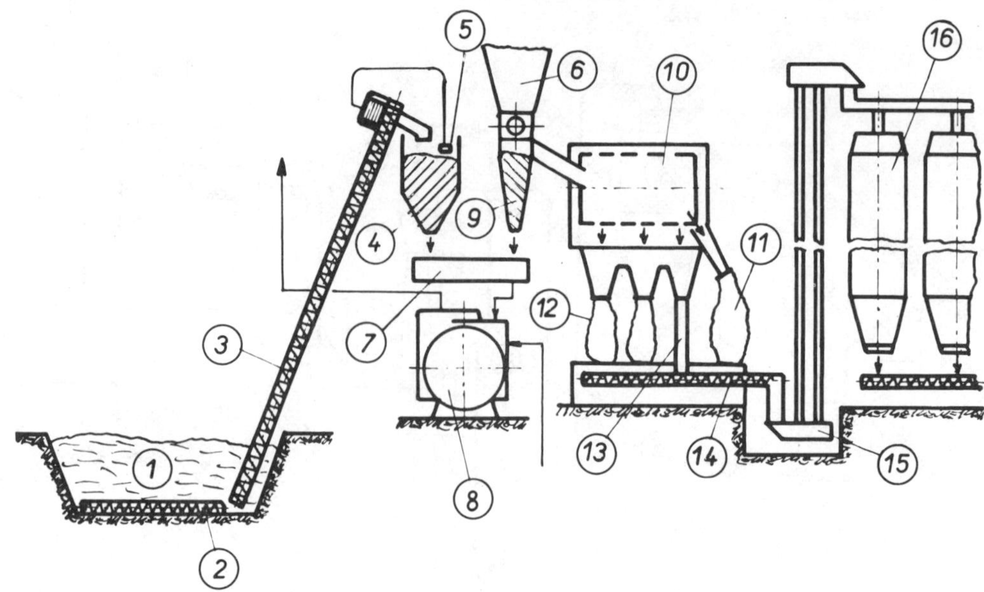

FIG. 4 The flow of material — (1) wet manure pit (25 m$^3$); (2) screw conveyor (diameter 300 mm); (3) screw conveyor (diameter 220 mm, length 9000 mm, motor 7.5 kW); (4) fresh manure hopper (2.2 m$^3$); (5) level controlling switch; (6) bottom of the cyclone with rotary valve; (7) twin intermeshing paddle backmixer (diameter 305 mm, length 2438 mm, 40 rpm, motor 5.5 kW); (8) dryer-pulverizer ATRITOR 18A, motor 120 kW; (9) surge hopper; (10) rotary sieve; (11) bagging of feathers; (12) bagging of DPW; (13) alternative conveying to bulk storage; (14) conveyor; (15) elevator; (16) steel bins (8 bins with total capacity of 100 t).

dust collector at the input and with bent plate droplet eliminators on the discharge end. This one was chosen because of lower air resistance, relatively low water consumption, and simple design.

## THE FLOW OF MATERIAL

The fresh manure (Fig. 4) from the pit (1) is moved by a horizontal screw conveyor to a vertical one (3). The vertical screw conveyor delivers material into the hopper (4). Here the level of manure is controlled by a switch (5) which keeps the motor running for 30 seconds when its sensing element is out of manure. The two controllable screw-type feeders (not shown), one of which delivers the wet material and the other a portion of dry material fed back from the cyclone (6), are arranged to feed into the paddle feeder (7). The paddle feeder then mixes the dry and wet materials together and feeds them into the dryer (8) with suitable moisture content of 38-41 percent w.b.

The feeding back of the required amount of dry material is accomplished by the use of a bifurcated outlet (9) at the base of collecting cyclone, one side feeding the backmixing equipment and the other the simple rotary screen (10). The rotary screen feeds the collected feathers into a bag (11). The dehydrated poultry waste (DPW) can be bagged (12) and/or delivered into the steel bins (16) using horizontal (14) and vertical (15) conveyors.

## SCRUBBING OF FLUE GASES

One of the most difficult problems of the manure drying operation is the odor suppression. The only reliable system at the time seems to be the afterburning. However, this means an additional investment cost due to an expensive heat exchanger and often an increased fuel consumption during operation.

A decision was made to use an air washer as a compromise. It was known that although this type of scrubber was fairly efficient in preventing dust emission, in no case was it able to achieve full odor removal. Since the plant is situated on a hill and the next village is two kilometers away, the district hygienist approved our project without objection. It is clear that if such a plant should be placed in a center of surrounding ridge of hills or in a residential area, air washer scrubbing of smell would not be sufficient. Odor removal depends on the retention time of gases inside the scrubber and this is very short time in this type of scrubber.

We are now considering the replacement of the existing scrubber by a two stage high-rate filter tower (Fig. 5) with plastic filter media. The two settling tanks under the towers

### TABLE 1. DUCT DIMENSIONS, GAS TEMPERATURE, AND FLOW CHARACTERISTICS

| | Duct dimensions | | Gas temperature and flow data (nominal flow of 5.27kg/s) | | | | |
|---|---|---|---|---|---|---|---|
| Point (See Fig. 2) | Dimensions or diameter, mm | Area of cross section, m$^2$ | Temperature,* deg C | Air density, kg/m$^3$ | Air flow volume, m$^3$/s | Air velocity, m/s | Static pressure,* N/m$^3$ |
| A | 762 x 762 | 0.581 | 510 | 0.45 | 11.7 | 20.2 | $\phi$ |
| B | 559 | 0.245 | 120 | 0.88 | 5.99 | 24.4 | −2000 |
| C | 610 | 0.292 | 101 | 0.90 | 5.86 | 20.1 | −5000 |
| D | 508 x 508 | 0.258 | 108 | 0.94 | 5.61 | 21.7 | +1000 |
| E | 800 | 0.503 | 100 | 0.95 | 5.55 | 11.0 | — |
| F | 300 | 0.0707 | 80 | 1.03 | 5.12 | 72.4 | — |
| G | 800 x 650 | 0.52 | 60 | 1.05 | 5.02 | 9.7 | −1000 |
| H | 500 | 0.196 | 50 | 1.06 | 4.97 | 25.4 | −3000 |
| I | 650 x 270 | 0.176 | 40 | 1.14 | 4.62 | 26.3 | + 400 |



MANAGING LIVESTOCK WASTES

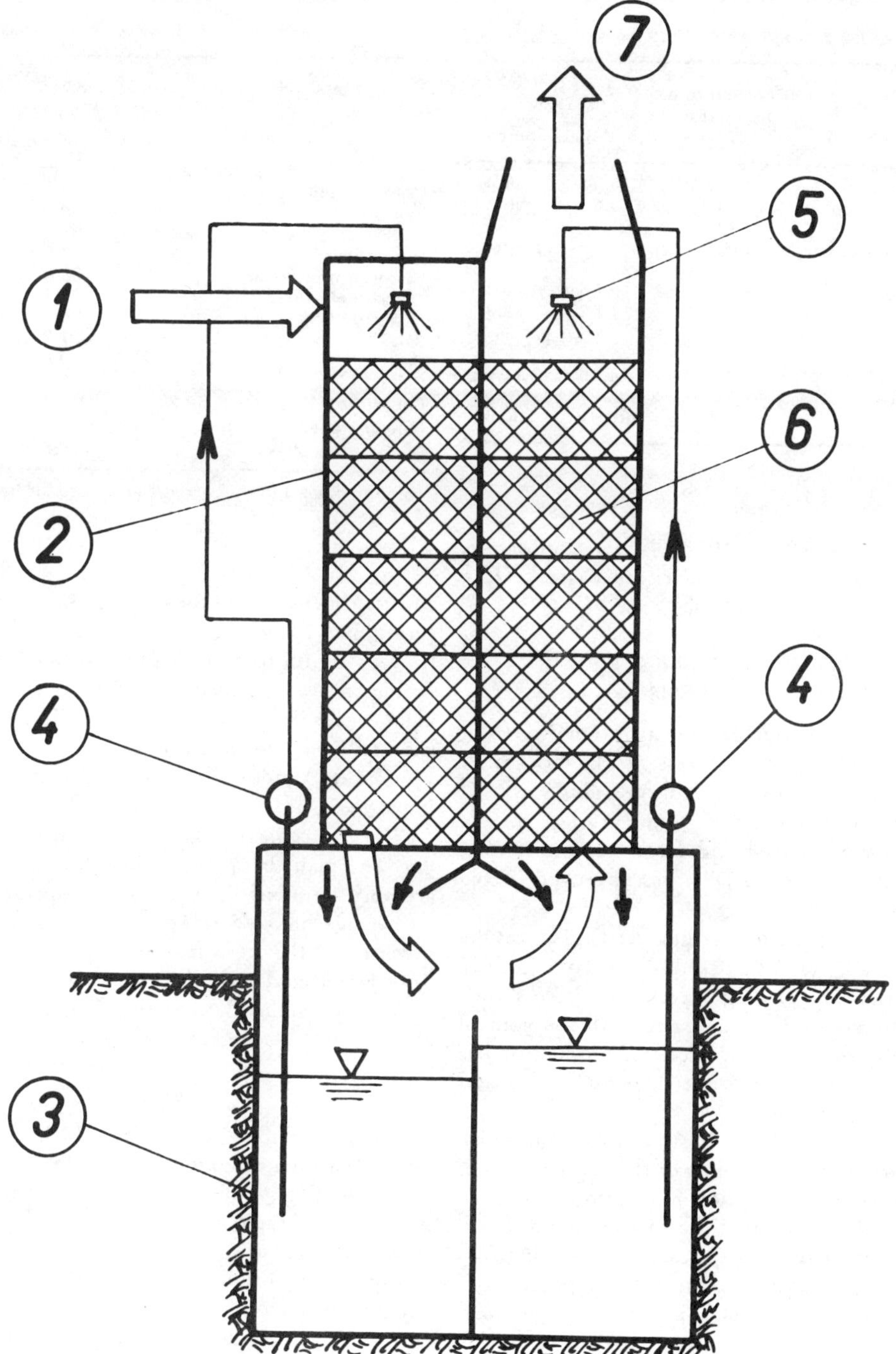


FIG. 5 High-rate biofilter for scrubbing of flue gases — (1) effluent gases; (2) filter tower; (3) pit; (4) pumps; (5) nozzle; (6) modules of plastic filter media; (7) clean gases.

will contain make-up water with different pH (weakly acidic and weakly basic). Plastic media are very light ( a module 1.2 m long, 0.6 m high, and 0.6 m wide weighs 16.8 kg) and can be packed to depths of up to 7.2 m in a simple steel frame or even timber framework clad with PVC sheets. A pilot plant experiment will show if the high temperature of effluent gases will cause problems (temperature of plastics must not be higher than 40 C).

## PLANT PERFORMANCE AND COSTS

The performance of the dryer depends on moisture content of the fresh manure. In our case, this is relatively high, sometimes in excess of 82 percent w.b. (some water is added in houses by waterers, etc.).

The theoretical evaporative capacity of the dryer is 1.7 t/h of water (inlet temperature 510 C, exit temperature 85 C, air flow 4.33 m³/s (STP)). Based on last year's average performance data (output of 281 kg/h of DPW, moisture content of 82 percent of fresh manure and 9 percent of the DPW, average fuel consumption of 95.5 l/h of oil, the effective evaporative capacity was estimated at about 1.14 t/h of water with 12 l of water being evaporated per liter of oil.

One of the reasons for the lower capacity is the fact that the operators prefer the lower oil flow setting of the high-low positions of fuel oil pump of the heater. This makes the plant less sensitive to the possible interruption of wet material feeding and the operation of the plant is therefore

**TABLE 2. COST ANALYSIS**

| Item | Consumption per metric ton of DPW | Costs per metric ton of DPW | |
|---|---|---|---|
| | | U.S. dollars | percent |
| Electric power | 520 kWh | 5.54 | 15.0 |
| Oil | 315 kg | 9.63 | 26.2 |
| Labor—Operation | 6.7 h | | |
| Labor—Repair and Maintenance | 3.0 h | | |
| Labor | 9.7 h | 5.66 | 15.4 |
| Depreciation | | 11.92 | 32.4 |
| Other material costs (spare parts, chemicals, lubrication, oil, etc.) | | 4.06 | 11.0 |
| Total | | 36.81 | 100.0 |

**TABLE 3. ANALYSIS OF DPW**

| | Stored manure* | Fresh manure treated after | | |
|---|---|---|---|---|
| | | Six days | Five days | One day |
| | | Percent moisture free basis | | |
| Crude protein | 21.8 | 30.1 | 31.1 | 35.6 |
| Fat | 5.6 | 2.9 | 2.4 | 2.1 |
| Ash | 32.7 | 21.6 | 22.1 | 21.5 |
| Crude fiber | 15.0 | 13.0 | 12.1 | 10.7 |
| Nitrogen free extract | 24.8 | 23.0 | 23.4 | 21.2 |
| Digestible protein | | 24.7 | 24.5 | 25.2 |
| Digestibility | | 76 | 79 | 71 |
| Salt | | 1.2 | 1.2 | 1.2 |
| $CaCO_3$ | | 6.8 | 6.6 | 6.5 |
| Calcium | 8.2 | 5.6 | 5.5 | 5.2 |
| Phosphorus | 2.6 | 1.8 | 1.5 | 1.5 |
| Moisture content of the sample (w.b.) | | 9.4 | 8.9 | 8.9 |

*Manure stored in the pits at the houses for longer time.

much easier.

The cost analysis taken from 1973-74 operation data is shown in Table 2. Amounts in dollars are based on the exchange rate used in economic calculations.

## SUMMARY AND DISCUSSION OF EXISTING PROBLEMS

In the early period of our drying operation the fresh manure was stored in the 20 m³ pits at the houses. A vacuum tank formerly used for the field spreading was used for hauling of the month old manure to the drying plant. In 1972 the system was changed and fresh manure is now dried. The nutrient value of the DPW has improved (Table 3).

All dried manure is used for feeding cattle. The results with a level of 1-1.5 kg of DPW per head per day in a daily ration are very good. The feeding experiments have shown that the only difference when compared with the control ration was a slightly darker meat. The DPW production has been under steady control of District Hygienst's office and was always found pasteurized.

A serious problem is corrosion. Waste gases are aggressive chemicals. Each stoppage and cooling of the plant results in condensation and formation of acids. In addition there is an erosive action of particles striking the surface of the duckwork and bend. It is therefore necessary to avoid as many as possible of these problems by correct design procedures, including the choice of materials, lining, etc. and eliminating breaks by running on continuous basis and having skilled operators and maintenance staff.

The manure pit (see (1) Fig. 4) has been shown to be too small. It was also necessary to oversize the motors of the wet manure screw conveyors. The DPW screening (recently at the end of process) would be better located adjacent to the cyclone and the screening of feathers combined with the sizing. The coarser part would go back into the dryer and the finer DPW to the output. The dryer itself is running very quietly. Fans are the main source of noise and it would be very useful to furnish the two fans and the part of the piping with sound insulation. The pneumatic system of the plant needs large amounts of air. If the air could be exhausted from the space above wet manure (pit, screw conveyors, etc.), odor would be reduced and the fly control improved.

The full capacity of the dryer has not been fully utilized yet. The plant is operating five days a week and the breaks are very frequent. In 1974 it was operating 223 days with an average daily working time of less than 18 hours. The lowering of the fresh manure moisture content to the usual 75 percent and a transition to a continuous run would make it possible to treat the manure from 400-500 thousand layers with the same equipment.

## SUMMARY

There are some principles which must be considered when designing a poultry farm with DPW as a by-product:

1 A large concentration of layers (e.g. multifloor houses) with rapid manure removal is a big advantage. If some drying occurs during the manure scraping and its transport, the results are lower costs for both investment and operation.

2 The continuous running of the dryer depends on the stockpile in the wet manure pit, reliability of the transport system, and the correct balance among the main system parameters: the number of layers, moisture content of the fresh manure, and effective evaporative capacity.

3 The design must include the effective deodorization of flue gases, corrosion and erosion wear control of the pneumatic system and a reduction of the noise of fans.

# In House Manure Drying—The Slat System

H. A. Elson, A. W. M. King

IN house manure drying systems are of benefit in deep pit (high rise) poultry buildings for several reasons: (a) They reduce the moisture content of the manure, thus reducing its weight, rendering it easier to handle, and enhancing its value. (b) Problems associated with wet pits (ammonia, odours, flies) are avoided. (c) A more amenable environment is provided for staff and stock. (d) Odour emission from buildings is reduced.

Bressler (1969) reported that it is possible to reduce the moisture content in the manure from its original 75 percent down to 30 percent by frequent stirring and continuous aeration in the pit. The weight of the manure is thus reduced to about 1/3 of the original weight and odours during spreading are almost eliminated. Bressler (1970) indicated that the system requires air circulation fans in the pit and harrows which stir the manure daily. Bills (1973) observed that the presence of the histerid Carcinops pumilio can result in local drying of deep pit manure due to an increase in temperature where larval colonies are found.

Whelden (1973) observed a range of 85.5 percent to 92.8 percent moisture in fresh manure samples with a mean of 89.7 percent. He used slats in the pit in an attempt to solve a wet manure problem in a particular deep pit situation. His results show that slats can be used to partially dry manure in deep pits—but that slat position will affect both the amount of manure retained on the slat and the extent of its drying. Using wooden slats in a pit over a short period (7 days) Jaeger, Whelden, Muir and Kittridge (1971) found that moisture content was reduced from 79 percent in the fresh manure to 70.5 percent on 15.2 cm wide slats, to 72.3 percent on 20.3 cm wide slats and only to 78.4 percent on 25.4 cm wide slats.

Elson et al. (1972) reported on trials in which they showed that slats effectively air dried poultry manure under fully stepped cages in deep pit houses. Slats from 5 cm to 61 cm wide were tested. They were able to dry the manure down to 13 percent moisture on 7.6 cm and 10 cm wide slats after 6 months. At the same sampling time manure in a non-slatted area of the pit had a moisture content of 65 percent. This present paper reports and discusses the development of this system in the United Kingdom.

## THE SLAT SYSTEM

The slat system is an efficient and economical method of in house manure drying. The system is used in deep pit houses having positive pressure reverse flow ventilation and fully stepped or flat deck cage configurations.

The technique is to collect manure falling from laying stock directly on wooden slats which retain it in columns subjected to continuous drying. The efficiency of the system derives from the facts that:

1  Fresh manure adheres continuously producing tall columns with a large surface area (see Figs. 1 and 2).

2  The warm ventilation air passes over these columns before being exhausted below the slats through windproof outlets in the pit walls.

3  Heat is provided by stock as they metabolise the energy of the food, and air movement by the existing ventilation system.

## METHOD

Framed timber slats are supported at two levels, the lower row being offset. Gaps about the same width as the slats allow half the manure to pass on to the lower level and provide unrestricted air circulation (see Fig. 1). In the lower row, slats are wider and gaps narrower—in order to retain a large proportion of the manure. One configuration which gives good results is 10 cm wide slats and gaps in the upper row and 12.5 cm wide slats with 7.5 cm wide gaps in the

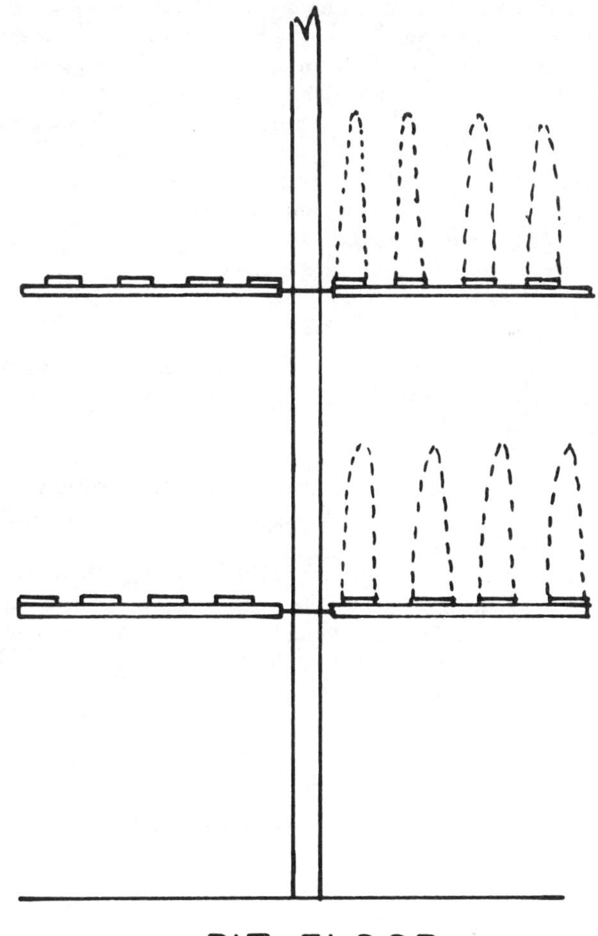



FIG. 1 LEGEND: Tall columns of manure build up on the slats.


The authors H. A. ELSON and A. W. M. KING are employees with Agricultural Development and Advisory Service, Shardlow Hall, Shardlow, Derby, DE7 2GN, England and Kenton Bar, Newcastle-upon-Tyne, NE1 2YA, England, respectively.

FIG. 2 LEGEND: 8 months accumulation of manure on slats.

lower row. Slat frames are pivoted on central posts and supported from the catwalk by cables. At stock depletion the frames are swung down to deposit the manure on the pit floor for removal by tractor and loader, and then returned to a horizontal position.

If the manure starts to "bridge" the drying process slows down. If "bridging" occurs well before stock depletion date the slat frames are swung down to deposit the dried manure on the pit floor, and then returned to the horizontal position. This operation can be carried out from the catwalk in the cage area without entering the pit by the use of extensions to the supporting cables.

A proportion of the manure does not adhere to the slats and is deposited on the pit floor. The proportion on the floor seems to depend on the initial moisture content of droppings, slat size and spacings. However the material on the floor is not wet enough to adhere to the slats and is therefore much lower in moisture content than normal pit manure. It is clearly important that the house be constructed to have a dry pit and that spillage from the cage drinking system is avoided—so that floor manure does not absorb moisture from its surroundings.

## DRYING

Because of the large surface area the manure columns on the slats dry continuously and drying accelerates as the columns become taller. Narrow slats results in quicker drying—especially in the early stages—but because the gaps between the slats are also narrow "bridging" takes place sooner and consequently drying slows down—or slats have to be emptied earlier.

Slats from 5 cm to 61 cm wide have been tried—but the most satisfactory results have been achieved using slats from 10 cm to 15 cm wide with similar gaps. Table 1 gives an indication of how slat width affects drying, and shows that the manures dry continuously.

TABLE 1. THE EFFECT OF SLAT WIDTH
AND TIME ON DRYING RATE.

| Slats in for: | Slat width, cm | | | Pit |
|---|---|---|---|---|
| | 7.6 | 10. | 15.2 | |
| 2 months | 26.0 | 34.4 | 42.5 | 70.9 |
| 4 months | 15.10 | 18.0 | 29.0 | 65.0 |
| 6 months | 13.4 | 12.7 | 15.4 | 65.0 |

If the slats are to be emptied every 6 months, 10 cm wide slats and gaps are satisfactory. However to avoid "bridging" 15 cm wide slats and gaps should be used if the manure is to be allowed to collect on the slats for a whole year.

The dried product can be ground in a hammer mill using a large (as least 2.5 cm diameter) screen and is a pleasant odour free material with a chemical analysis similar to that of machine dried fresh poultry manure. The analysis of a typical sample is shown in Table 2.

TABLE 2. ANALYSIS OF A TYPICAL
SAMPLE OF SLAT DRIED POULTRY
MANURE.

| | |
|---|---|
| Moisture | 15.0 percent |
| Nitrogen | 4.9 percent |
| Phosphorous | 2.1 percent |
| Potassium | 2.3 percent |
| Calcium | 5.2 percent |

## ENVIRONMENT

In order to gain the benefit of feed energy saving by maintaining a minimum house temperature of 21 C, a low ventilation rate (0.9 m$^3$/hr/bird) must be achieved in winter in the UK. With such a low ventilation rate ammonia is often a problem in non-slatted deep pit houses which have wet manure in the pit. Slat manure drying overcomes this problem since the manure is held in a much drier condition—and less ammonia is released. In a series of tests where separate bays of a deep pit house were curtained off, the ammonia level in a slatted bay was half (13 ppm) that of a none slatted bay (26 ppm) at a low ventilation rate (0.9 m$^3$/hr/bird).

Partly as a result of the lower ammonia level in the building, odour emission from slatted houses is lower than that from non-slatted high rise houses, as shown by an assessment of odour at the air outlet by the dynamic dilution method. Mean readings are shown in Table 3.

These odour emission levels are very low when compared with those from many manure drying machines.

TABLE 3. ODOUR ASSESSMENT AT
AIR OUTLET.

| | Threshold dilution, vols/vol sample |
|---|---|
| Slatted houses | 75 |
| Non-slatted houses | 134 |

## COST

In the UK the capital cost of this system is about £5 per m$^2$, i.e. 25 pence per bird. This is no more than a manure drying machine. However, running costs are very low (no fuel required). Total cost per tonne of dried product is about £7.

## SUMMARY

The slat system of in house manure drying has the fol-

(Continued on page 92)

# Control, Collection, and Disposal of Feedlot Runoff


N.P. Swanson, L.N. Mielke, C.L. Linderman

MEMBER          AFFILIATE       ASSOC. MEMBER
ASAE               ASAE               ASAE


FLUSHING the Aegean stables by diverting a river was considered a clever innovation in Greek mythology. However, this technique is hardly applicable to our modern cattle feeding industry in the United States, which was one of the first identified as a threat to water quality. This paper discusses the state of the art and some of the design and management considerations resulting from research on control, collection and disposal of feedlot runoff.

Smith and Miner (1964) reported stream pollution due to feedlot runoff in Kansas. As a result of this and other reports, Congress enacted the Water Quality Act of 1965 and State and Federal regulations were soon initiated for controlling feedlot runoff. Little research had been done at that time on pollution problems related to livestock feeding and the concern to "do something" resulted in some unwarranted regulatory proposals. For example, a bill introduced in the Colorado legislature required frequent complete cleaning of feedlot surfaces to prevent soil and groundwater pollution. However, Stewart et al. (1967) indicate that frequent cleaning of corrals did not alleviate high nitrate accumulations in the groundwater.

Other proposals would have required sealants, concrete surfaces, or plastic barriers to prevent percolation through feedlot surfaces. Mielke et al. (1970) found that a level feedlot on a permeable soil with a shallow, fluctuating water table did not contribute nitrate to the groundwater. The manure surface and the interface layer that formed in the feedlot restricted movement of water and oxygen into the soil profile which limits microbial activity necessary to convert nitrogen to nitrate and leaching by water through the soil profile (Elliott et al. 1972, Mielke et al. 1974). Physical characteristics of the feedlot manure surface cause early runoff with two to five times more runoff than from grassland.

Viets (1971) concluded that groundwater pollution seemed to be mostly a local phenomenon and contamination of aquifers was probably not widespread. However, in regions with sandstone or limestone formations, "sinkholes" may deliver polluted surface flows directly to the groundwater. Smith (1967) attributed nitrate in some Missouri groundwaters to animal wastes.

Animal wastes or products resulting from wastes can be transported by air or water. Air movement transports particulates and/or gaseous compounds. Wind transported particulates can be redeposited outside the feeding area where they are subject to transport by water runoff. Overland flow moves dissolved chemicals and transports microorganisms, organic materials, and soil sediments in suspension and as bedloads.

Dague (1968) stated that hydrology was important in feedlot site selection and in determining the pollution potential and design of runoff control systems. The rate of delivery of runoff is a function of the topography, meteorology, and feedlot hydrology.

The environmental impact of open feedlots is greatest to the immediate watershed. Cattle feedlots in the Corn Belt are an environmental concern, but are not very significant to the total hydrology of the region, since there are about 1000 hectare (ha) of corn for each ha of feedlot. Environmental pollution problems resulting from feedlot materials are intrastate rather than interstate.

Climate is a major factor in designing and managing control, collection, and disposal systems for feedlot runoff. Long-term operations require frequency studies of specific data to determine (a) the probabilities of favorable temperatures, moisture, and sunshine, (b) the probability of coincident soil and weather conditions suitable for timeliness of required operations, and (c) the improbabilities of destructive conditions. Applicable local climatic data must be used in decision making. Current weather data is available from the National Weather Service.

Many valuable articles relating to climate were published in the U.S. Department of Agriculture Yearbook (1941). Wang and Barger (1961) classified 11 000 articles on agricultural meteorology. All mean climatic values must be interpreted relative to rainfall frequencies (Hershfield 1961) and the high variability of daily weather (Brooks 1961).

Solids are transported by overland flow from snowmelt or rainfall. Snowmelt runoff is maximum when the soil surface is frozen. Rain creates an equally or more serious condition for runoff on frozen soils. Where snow can be expected, special design and management considerations are required like shelterbelts or snow fences to control snow accumulation and to provide animal comfort (Martinelli 1973).

Designs for controlling feedlot runoff and disposing of it must include (a) diversion of overland flows from outside the feedlot, (b) control of runoff and soil erosion, (c) adequate control of wind movement and snow accumulation, and (d) maintenance of adequate drainage if high water tables or surface inundation are problems.

Solids transport, initiated by rainfall, results from particles detached by raindrop impact and moved by overland flow. Water erosion of soil by rainfall was quantitatively expressed by Zingg (1940) as a function of land slope and slope length, infiltration rate, and soil physical properties. Ayres (1936) stated the amount, intensity, and duration of rainfall have profound effect on the amount and rate of field runoff, as does the elapsed time since the preceding rain. The rainfall effects on feedlot runoff and solids transport are similar (Swanson et al. 1971, Clark et al. 1975).


Contribution from Animal Waste Management Research Unit, North Central Region, ARS, USDA, in cooperation with the Nebraska Agricultural Experiment Station, Lincoln. A portion of the funds for this research was provided by the Environmental Protection Agency.

The authors are N. P. SWANSON, Supervisory Agricultural Engineer, L. N. Mielke, Soil Scientist, and C. L. LINDERMAN, Agricultural Engineer, ARS, USDA, Lincoln, Nebr.

Snowmelt runoff from feedlots is quite different from rainfall runoff. Sublimation is appreciable on southern exposures. Radiation absorbed by the snow is markedly influenced by any foreign substances such as dust, bark fragments, and twigs (Gartska 1964). Animal traffic rapidly packs snow on the feedlot, causing an ice-like density. Fresh manure deposited on the surface of such a snow pack is particularly vulnerable to transport by overland flow produced by the melting snow and ice. Manure's dark color permits rapid absorption of heat from the sun, and on slopes of 2 or 3 percent or more the mass of supersaturated manure flows readily over the frozen soil surfaces. Lava-like flows of manure and snowmelt water were reported in eastern Nebraska on sloping lots densely stocked at 9.3 and 18.6 m² (100 and 200 ft²) per head (Gilbertson et al. 1970). Swanson (1973a) reported similar but not as heavy manure flows from an eastern Nebraska feedlot with a lower cattle density.

Rapid drainage of a feedlot is very desirable for animal comfort, feeding efficiency, and waste management. Accumulated runoff, moving at relatively high velocities, moves greatly increased quantities of solids, thus creating additional problems. Using mounds and drainways, or terraces and basins limits slope lengths and improves feedlot drainage (Swanson et al. 1973). Using underground drain lines from riser inlets further speeds discharge of runoff (Linderman et al. 1974).

Feedlot drainage is most readily improved by increasing slopes within limits and by decreasing slope lengths. However, steeper slopes cause greater movement of solids. Experiments using a field-plot rainfall simulator on 8 and 13 percent slopes showed that solids eroded from a feedlot may be only half as much as the soil losses from similar slopes on cultivated land (Swanson et al. 1971). If a manure mulch is maintained on the feedlot surface, the erodibility of the soil is not of primary importance. Tolerable soil losses on cultivated land are arbitrarily limited to 11 000 kg/ha (5 ton/ac) per year to conserve productivity. Permissible movement of soil and organic materials from the feedlots may be limited by the frequency and volume of solids removal from settling basins (Swanson et al. 1973). In most feedlots, accumulations of partially decomposed solids must be removed from the feedlot surface mechanically or mounded within the feedlot. Proportion of organic to inorganic material in the mound decreases with time. If a manure mulch is maintained and the redistribution of solids is controlled, many slopes too steep for cultivation can be used for feedlots.

Costs and management requirements are both factors for deciding the permissible slopes and slope lengths for feedlots. Broad-basin terraces in Nebraska accumulated about 1.5 m³ (2 yd³) of solids per head per year on a 100 m(340 ft) long, 15 percent slope as compared with 0.5 m³ (0.6 yd³) of solids on a 50 m (170 ft) long, 7 percent slope (Swanson et al. 1973). Obviously, the longer, steeper slope incurs nearly four times the expense and frequency of solids removal as the lesser slope and also requires increased management. Using diversions to prevent entry of outside runoff onto the feedlot is important in restricting slope length. Both solids transport and runoff volumes are reduced. An equation was developed by Jeschke and Day (1973) to predict the movement of manure solids from unpaved feedlots by rainfall.

The bedload and a part of the suspended solids are deposited rapidly when the velocity of runoff flow is decreased or the runoff is impounded. If uncontrolled quantities of solids are allowed to deposit in a holding pond or a lagoon, three serious problems result. Runoff storage capacity is decreased; drainage of the effluent is more difficult and special equipment such as a dragline may be required for solids removal; and decomposition of organic solids may create odor problems in warm weather.

Separating the settleable solids from the liquid effluent and handling the two materials with separate equipment is usually most economical. Solids may be removed with settling basins either inside or outside the feedlot, with settling channels (Gilbertson et al. 1971) or with flow through solids traps (Swanson and Mielke 1973).

Sumps with float-controlled electric pumps permit the use of underground drains where slope is inadequate for gravity drainage or where runoff must be delivered to a higher elevation (Olson 1972, 1973; Swanson 1973b). About 1/2 hp pumps were adequate and required only modest investment costs. Two such installations were made in eastern Nebraska in 1971. Large lots require installations with multiple sumps and pumps. A 23 m³/hr (100 gpm) capacity, single pump system will remove 16.4 ha-cm (16 ac-in.) of runoff in 72 hr, which is equivalent to the runoff expected from a 10-yr, 24-hr storm on a 1.6 ha (4 ac) feedlot in most of the Corn Belt.

Efforts to drain free water from feedlot surfaces by subsurface drains have usually failed. The formation of a nearly impermeable manure soil interface prevents water movement to conventional tile or other pipe drains. Rapid growth of mycelia on the organic materials soon "plug" gravels or small diameter drains receiving water from the feedlot surface. Unfortunately, those who have experienced such failures have been reluctant to report them and so the mistakes are repeated.

Holding pond storage, adequate for 120 days, is considered desirable in the western Corn Belt and eastern Great Plains. Large storage capacities provide the operator with more options for disposing of effluent.

Feedlot runoff is not a dependable irrigation supply but land application for crop utilization is the most practical disposal method. Runoff effluent was applied for 3 years in quantities exceeding the irrigation requirements of perennial forages in eastern Nebraska (Swanson et al. 1974). Yields were generally increased and forage of excellent quality was produced. In similar studies with corn and forage sorghum, nutrients in the effluent increased yield (Linderman and Swanson 1972, Hinrichs et al. 1974, Sukovaty et al. 1974). Applying effluent at twice the irrigation requirement did not increase nitrates or total salts in the soil profile or cause lasting changes in soil structure. However, available phosphorus increased significantly in the plow layer. Infiltration rates decreased with heavy applications of effluent during the irrigation season, but returned to near normal after winter freezing and thawing.

Runoff effluent can be applied to the land by sprinkler or surface irrigation. Since only larger feeding installations will involve effluent quantities and crop acreages of significant size, disposal system design may be based on convenience and minimum investment rather than on irrigation application and distribution efficiencies. Swanson and Linderman (1974) have reported on low-cost systems for land disposal of runoff.

Under some conditions in many states, distant overland flow of feedlot runoff to the stream is not considered an environmental hazard. In 1973 a land sink was constructed to provide direct land disposal of runoff from a feedlot near Springfield in eastern Nebraska. The level "land sink" provides for impounding and infiltration of feedlot runoff on cropland. A grassed waterway in a serpentine or switchback configuration was installed in 1974 on a hillside below a feedlot located near Gretna, also in eastern Nebraska. The long, serpentine waterway provides longer infiltration opportunity and a long time of concentration, i.e. the time necessary for the feedlot runoff entering at the top of the waterway to reach the receiving stream. Either type of installation can be expected to reduce runoff management problems for small feedlot operators. These installations are fully described in another paper in these proceedings.

In summary, the application of proven soil and water engineering principles assisted in adapting practical structures and facilities into complete systems to control feedlot runoff. Applications include diversions to eliminate surface runoff from the feedlot, debris basins and solids traps for separation of transported solids from runoff water. Riser inlets and underground conduits are used for conveyance of runoff, sumps and pumps provide lift for drainage where gravity flow is not possible, and holding ponds are used for collection of runoff. Effluent disposal requires pumping and distribution equipment and managing of effluent on crops to effectively use nutrients and control pollution.

Usually, bedloads and readily settleable solids should be removed from the runoff in settling basins or debris traps and handled separately. Land application is the most practical method for disposing of effluent. Water and nutrients in the runoff effluent can be beneficial to crops without damaging the soil. Continuously stocked feedlots with the manure-soil interface intact are not a hazard to groundwater quality.

# References


1 Ayres, O. C. 1936. Soil erosion and its control. McGraw-Hill, New York.

2 Brooks, F. A. 1961. Farm climates and solar energy. Chapter 57. Agricultural Enginers Handbook. McGraw-Hill, pp. 817-866.

3 Clark, R. N., A. D. Schneider, and B. A. Stewart. 1975. Analysis of runoff from Southern Great Plains feedlots. TRANSACTIONS of the ASAE 18:319-322.

4 Dague, Richard R. 1968. Animal wastes - a major problem. A paper published in the 50th Anniversary Brochure, Iowa Water Pollution Control Association.

5 Elliott, L. F., T. M. McCalla, L. N. Mielke and T. A. Travis. 1972. Ammonium, nitrate, and total nitrogen in the soil water of feedlot and fieldl soil profiles. Appl. Microbiol. 23:810-813.

6 Gartska, Walter V. 1964. Snow and snow survey. In Handbook of Applied Hydrology. McGraw-Hill. Section 10, pp. 10-29.

7 Gilbertson, C. B., T. M. McCalla, J. R. Ellis, O. E. Cross, and W. R. Woods. 1970. The effect of animal density and surface slope on characteristics of runoff, solid wastes and nitrate movement on unpaved beef feedlots. Nebraska Ag. Exp. Sta. Publ. SB 508.

8 Gilbertson, C. B., T. M. McCalla, J. R. Ellis and W. R. Woods. 1971. Methods of removing settleable solids from outdoor beef cattle feedlot runoff. TRANSACTIONS of the ASAE 14:899-905.

9 Hershfield, David M. 1961. Rainfall frequency atlas of the United States for durations from 30 minutes to 24 hours and return periods from 1 to 100 years. Technical Paper No. 40, U.S. Department of Commerce.

10 Hinrichs, D. G., A. P. Mazurak and N. P. Swanson. 1974. Effect of effluent from beef feedlots on the physical and chemical properties of soil. Soil Sci. Soc. Amer. Proc. 38(4):661-663.

11 Jeschke, J. L. and D. L. Day. 1973. Erosion from feedlots is predicted by adapting the Universal soil loss equation. Illinois Research, University of Illinois Agricultural Experiment Station, Vol. 15, No. 2.

12 Linderman, C. L. and N. P. Swanson. 1972. Disposal of beef feedlot runoff on corn. Agron. Abstr., p. 182.

13 Linderman, C. L., N. P. Swanson and L. N. Mielke. 1974. Riser intake designs for feedlot solids collection basins. Paper No. 74-3030, ASAE, St. Joseph, Mich. 49085.

14 Martinelli, M., Jr. 1973. Snow fences for influencing snow accumulation. In The role of snow and ice in hydrology. Symp. on Measurement and Forecasting, Banff, Alberta, Can., Sept. 1972, p. 1394-1398.

15 Mielke, L. N., J. R. Ellis, N. P. Swanson, J. C. Lorimor and T. M. McCalla. 1970. Groundwater quality and fluctuation in a shallow unconfined aquifer under a level feedlot. In Relationship of Agriculture to Soil and Water Pollution. Proc., Cornell University Conference on Agr. Waste Management, Rochester, N.Y., p. 31-40.

16 Mielke, L. N., N. P Swanson, and T. M. McCalla. 1974. Soil profile conditions of cattle feedlots. Jour. Environ. Quality 3(1):14-17.

17 Olson, E. A. 1972. Control livestock wastes electrically. Rural Electric Nebraska 26(5):12-13.

18 Olson, E. A. 1973. Electric power handles feedlot wastes. Rural Electric Nebraska 27(7):12-14.

19 Smith, G. E. 1967. Contamination of water by nitrates. J. Fert. Solution II(3):8-12, 14-17.

20 Smith, S. M. and J. R. Miner. 1964. Stream pollution from feedlot runoff. Proceedings 14th Sanitary Engineering Conference, University of Kansas.

21 Stewart, B. A., F. G. Viets, Jr., G. L. Hutchinson, and W. D. Kemper. 1967. Nitrate and other water pollutants under fields and feedlots. Environ. Sci. Tech. 1:736-739.

22 Sukovaty, J. E., L. F. Elliott and N. P. Swanson. 1974. Some effects of beef-feedlot effluent applied to forage sorghum grown on a Colo silty clay loam soil. Jour. Environ. Quality 3(4):381-388.

23 Swanson, Norris P. 1973a. Hydrology of open feedlots in the Corn Belt. Proceedings, Midwest Livestock Waste Management Conference, Iowa State University, Ames, November 27-28.

24 Swanson, N. P. 1973b. Feedlots with no room for runoff - do you have to move'em or can you leave'em? Nebraska Farmer 115:13-15.

25 Swanson, N. P. and L. N. Mielke. 1973. Solids trap for feedlot runoff. TRANSACTIONS of the ASAE 16:743-745.

26 Swanson, N. P. and C. L. Linderman. 1974. Low-cost disposal systems for feedlot runoff. AGRICULTURAL ENGINEERING 55(11):20-21.

27 Swanson, N. P., J. C. Lorimor and L. N. Mielke. 1973. Broad basin terraces for sloping cattle feedlots. TRANSACTIONS of the ASAE 16(4):746-749.

28 Swanson, N. P., C. L. Linderman, and J. R. Ellis. 1974. Irrigation of perennial forage crops with feedlot runoff. TRANSACTIONS of the ASAE 17(1):144-147.

29 Swanson, N. P., L. N. Mielke, J. C. Lorimor, T. M. McCalla and J. R. Ellis. 1971. Transport of pollutants from sloping cattle feedlots as affected by rainfall intensity, duration and recurrence. In Livestock Waste Management and Pollution Abatement. ASAE, St. Joseph, Mich. 49085. p. 51-55.

30 U.S. Department of Agriculture. 1941. Climate and man. U.S. Government Printing Office.

31 Viets, Frank G., Jr. 1971. The mounting problem of cattle feedlot pollution. Agricultural Science Review 9(1):1-8.

32 Wang, J. Y. and G. L. Barger. 1961. Bibliography of agricultural meteorology. University of Wisconsin, Madison, Wisconsin.

33 Zingg, A. W. 1940. Degree and length of landslope as it affects soil loss in runoff. AGRICULTURAL ENGINEERING 21:59-64.

# Management of Runoff Water in Relation to Feedlot Operations

H. N. McGill, G. C. Vittetoe

MEMBER
ASAE

MEMBER
ASAE

EXTENSIVE research has been directed at determining the rainfall-runoff relationship from feedlot operations. Though the published results of these studies vary considerably, the engineer is able to make reasonable estimates of runoff that can be expected for his particular project. Also, the chemical composition of feedlot runoff water and its application rate and value in crop production have received considerable attention.

These and other research studies are essential to the engineer if he is to develop a plan for the management of runoff water from feedlot operations.

National, state, and local waste water regulatory agencies have established rules for the disposal of runoff water from feedlots. Generally, these involve storage capacity for runoff from a specified duration and frequency storm.

No consideration is given to hydrologic conditions that might affect the economic utilization or safe disposal of the feedlot runoff water. This requires the engineering application of research information.

To illuminate this requirement, a study was made and is presented in this paper for the wide hydrologic conditions found in Texas.

Presently, the most practical and economical system for controlling runoff from cattle feedlots seems to be one in which (a) as much of the outside drainage as possible is diverted from the feedlot, and (b) the runoff water from the feedlot proper is intercepted and impounded in holding ponds and later disposed of by applying it on agricultural crops. Therefore, systems of retention and irrigation which result in a "no effluent" condition generally will be the type best suited for cattle feedlot settings.

In planning and designing the retention and irrigation type pollution abatement systems, the size of the irrigated area in relation to the area of the feedlot must be considered for the varied conditions encountered. Information on favorable ratios of irrigated areas to feedlot areas is needed by both planners and operators of cattle feedlot pollution abatement systems.

Basic to this study is a simple reservoir from which water is applied to irrigated land when it can be used by the crop and stored in a reservoir when it cannot. Evacuation of the storage reservoir as rapidly as the crop can use the water is the purpose of the operation. Total water need for the crop is not met during periods of extended drought.

In this study evaporation and seepage from the reservoir were not considered because to do so would require area and capacity data for a specific site and would negate the opportunity for developing a generalized criterion. Irrigation efficiencies also were not considered. It is recognized that consideration of evaporation and seepage losses from the reservoir and irrigation efficiencies in the application of the water would materially decrease the storage requirements and increase the irrigation water requirement. The neglect of these adds a safety factor to the operation and simplifies the study.

All values used in the operation study are expressed in inches. This permits its application to any size feedlot.

A discussion of the source of basic data used in the study follows. Its application is illustrated in Table 1.

The authors are: H. N. McGILL, Hydraulic Engineer, and G. C. VITTETOE, State Conservation Engineer, SCS, USDA, Temple, Texas.

**TABLE 1. ILLUSTRATION OF WATER BUDGET PROCEDURE - AREA 7-C.**

| | | | | | One acre irrig./one acre feedlot | | | Two acre irrig./one acre feedlot | | | Four acre irrig./one acre feedlot | | |
|---|---|---|---|---|---|---|---|---|---|---|---|---|---|
| 1 | 2 | 3 | 4 | 5 | 6 | 7 | 8 | 6 | 7 | 8 | 6 | 7 | 8 |
| Month, in. | Precipitation, in. | Runoff, in. | Consumptive use, in. | Effective precipitation, in. | Irrigation water req. 4-5, in. | Storage Begin. month 3+end prev. mo. in. | Storage End month 7-6, in. | Irrigation water req. 4-5, in. | Storage Begin. month 3+end prev. mo. in. | Storage End month 7-6, in. | Irrigation water req. 4-5, in. | Storage Begin. month 3+end prev. mo. in. | Storage End month 7-6, in. |
| 1941 | | | | | | | 0 | | | 0 | | | 0 |
| Jan. | 1.54 | 0.16 | 0.8 | 1.50 | 0.0 | 0.16 | 0.16 | 0.0 | 0.16 | 0.16 | 0 | 0.16 | 0.16 |
| Feb. | 3.57 | 1.30 | 1.2 | 3.05 | 0.0 | 1.46 | 1.46 | 0.0 | 1.46 | 1.46 | 0 | 1.46 | 1.46 |
| Mar. | 4.33 | 1.85 | 2.8 | 3.53 | 0.0 | 3.31 | 3.31 | 0.0 | 3.31 | 3.31 | 0 | 3.31 | 3.31 |
| Apr. | 5.44 | 2.73 | 3.4 | 4.21 | 0.0 | 6.04 | 6.04 | 0.0 | 6.04 | 6.04 | 0 | 6.04 | 6.04 |
| May | 5.52 | 2.80 | 6.1 | 4.26 | 1.8 | 8.84 | 7.04 | 3.6 | 8.84 | 5.24 | 7.2 | 8.84 | 1.64 |
| June | 7.61 | 4.58 | 6.5 | 5.29 | 1.2 | 11.62 | 10.42 | 2.4 | 9.82 | 7.42 | 4.8 | 6.22 | 1.42 |
| July | 3.92 | 0.86 | 6.7 | 3.27 | 3.4 | 11.28 | 7.88 | 6.8 | 8.28 | 1.48 | 13.6 | 2.28 | 0 |
| Aug. | 1.46 | 0.14 | 4.6 | 1.42 | 3.2 | 8.02 | 4.82 | 6.4 | 1.62 | 0 | 12.8 | 0.14 | 0 |
| Sept. | 1.69 | 0.20 | 5.1 | 1.64 | 3.5 | 5.02 | 1.52 | 7.0 | 0.20 | 0 | 14.0 | 0.20 | 0 |
| Oct. | 5.18 | 2.51 | 4.1 | 4.05 | 0 | 4.03 | 4.03 | 0 | 2.51 | 2.51 | 0 | 2.51 | 2.51 |
| Nov. | 1.18 | 0.06 | 2.1 | 1.17 | 0.9 | 4.09 | 3.19 | 1.8 | 2.57 | 0.77 | 3.6 | 2.57 | 0 |
| Dec. | 1.91 | 0.31 | 1.0 | 1.83 | 0.0 | 3.50 | 3.50 | 0 | 1.08 | 1.08 | 0 | 0.31 | 0.31 |
| Maximum required storage | | | | | | | 11.02 | | | 8.62 | | | 6.04 |

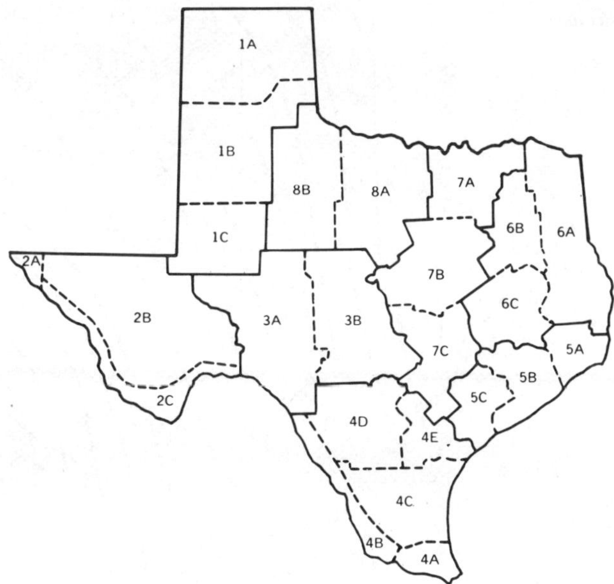

FIG. 1 Subdivisions of irrigated areas.

1 Texas Board of Water Engineers Bulletin 6019, "Consumptive Use of Water by Major Crops in Texas," gives estimated average consumptive use amounts for 12 major crops. Monthly consumptive use was tabulated for each crop in each of 24 areas of major production in the state. Location of these areas is shown on Fig. 1. Consumptive use amounts estimated for perennial pasture used in this study.

2 Monthly precipitation and effective precipitation for the period 1941-1957 for the same area were prepared by the U. S. Study Commission - Texas. Precipitation data for several representative stations (usually five) having complete records were tabulated. These data were averaged to give one set of monthly values for each of the 24 areas of major production in the state. The monthly rainfall-runoff relationship for small watersheds at Tyler, Spur and Riesel Experiment Stations was determined. Effective precipitation data for each subarea were then compiled by modifying the monthly rainfall by the expected monthly runoff.

3 Monthly runoff from the feedlot was determined for each subarea. Soil-cover complex curve number 74 was used with monthly rainfall. This is the equivalent of curve number 90 with individual storm rainfall.

4 Irrigation water requirement is the difference be-

**TABLE 2. FREQUENCY-STORAGE RELATIONSHIPS.**

| Area | 25 Yr 24 Hr rainfall runoff | Ratio irrig. area to feedlot area | Storage requirements, inches | | | | | |
|---|---|---|---|---|---|---|---|---|
| | | | 2 Yr | 5 Yr | 10 Yr | 25 Yr | 50 Yr | 100 Yr |
| 1-A | P = 5.10 | ½:1 | 0.80 | 1.79 | 2.72 | 4.24 | 5.65 | 7.32 |
| | Q = 3.97 | 1:1 | 0.72 | 1.45 | 2.09 | 3.08 | 3.96 | 4.95 |
| | | 2:1 | 0.68 | 1.32 | 1.86 | 2.69 | 3.40 | 4.21 |
| 3-A | P = 6.20 | ½:1 | 0.86 | 1.88 | 2.83 | 4.38 | 5.79 | 7.45 |
| | Q = 5.05 | 1:1 | 0.83 | 1.71 | 2.51 | 3.77 | 4.90 | 6.20 |
| | | 2:1 | 0.79 | 1.53 | 2.17 | 3.14 | 3.99 | 4.94 |
| 4-D | P = 7.60 | 1:1 | 1.60 | 3.77 | 5.89 | 9.49 | 12.92 | 17.03 |
| | Q = 6.41 | 2:1 | 1.35 | 2.97 | 4.49 | 6.98 | 9.28 | 11.98 |
| | | 4:1 | 1.26 | 2.61 | 3.82 | 5.73 | 7.45 | 9.43 |
| 6-B | P = 8.00 | 2:1 | 11.24 | 19.79 | 26.59 | 36.43 | 44.66 | 53.61 |
| | Q = 6.81 | 4:1 | 9.66 | 16.28 | 21.38 | 28.58 | 34.48 | 40.80 |
| 7-A | P = 7.60 | 1:1 | 7.23 | 12.84 | 12.84 | 17.34 | 23.88 | 35.35 |
| | Q = 6.41 | 2:1 | 5.93 | 10.39 | 13.92 | 19.01 | 23.26 | 27.87 |
| | | 4:1 | 4.98 | 8.86 | 11.96 | 16.48 | 20.27 | 24.41 |
| 7-C | P = 8.00 | 1:1 | 4.59 | 9.30 | 13.45 | 19.93 | 25.69 | 32.26 |
| | Q = 6.81 | 2:1 | 3.79 | 7.46 | 10.64 | 15.53 | 19.82 | 24.67 |
| | | 4:1 | 3.38 | 6.52 | 9.18 | 13.23 | 16.75 | 20.71 |
| 8-A | P = 7.10 | 1:1 | 1.99 | 3.89 | 5.53 | 8.02 | 10.21 | 12.68 |
| | Q = 5.92 | 2:1 | 1.91 | 3.66 | 5.13 | 7.37 | 9.31 | 11.49 |
| | | 4:1 | 1.80 | 3.46 | 4.87 | 7.01 | 8.87 | 10.95 |
| 8-B | P = 6.20 | ½:1 | 1.82 | 4.93 | 8.31 | 14.47 | 20.71 | 28.57 |
| | Q = 505 | 1:1 | 1.20 | 2.72 | 4.18 | 6.60 | 8.87 | 11.56 |
| | | 2:1 | 1.12 | 2.42 | 3.60 | 5.51 | 7.26 | 9.29 |

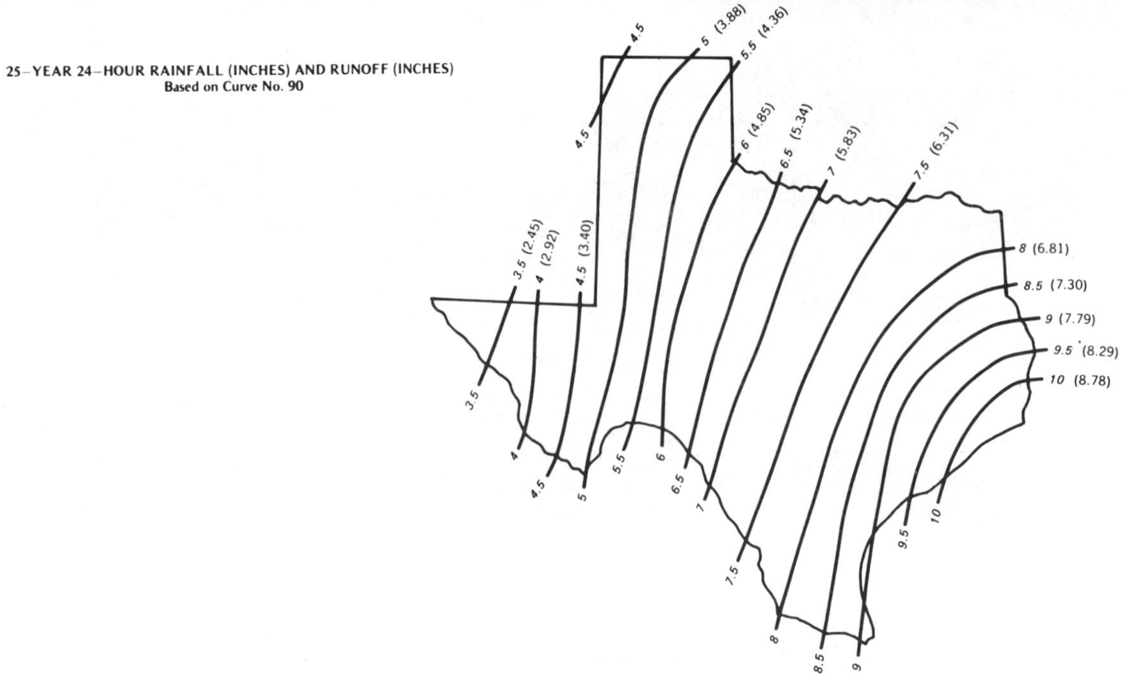

FIG. 3 Annual precipitation—Feedlot storage requirements for 2, 5, 10, and 25 year frequency.

tween consumptive use and effective precipitation. It is the inches of runoff from the feedlot that can be used by the crop when the ratio of the area to be irrigated to the area in the feedlot is one. The irrigation water requirement for other ratios is obtained by multiplying this value by the ratio.

5 The operation is a budget-keeping procedure for the 1941-1957 period. Monthly runoff is added at the beginning of the month, the irrigation water requirement for the month is subtracted, and surplus water is carried in storage to the following month. Thus the storage at the beginning of the month is the sum of the residual from the previous month and the runoff for the current month. Storage at the end of the month is the remaining storage after irrigation water requirements have been satisfied. The maximum amount stored for individual months is the average of the storage at the beginning and end of month.

Operation studies for Areas 1A, 3A, 4D, 6B, 7A, 7C, 8A, and 8B were made. Irrigated area to feedlot area ratios of 0.5, 1, and 2 in western part of the state and 1, 2, and 4 in the eastern part of the state were used.

6 A storage-frequency analysis was made for the 17-year study period using maximum annual storage values for each of the subareas. The results of this analysis are given in Table 2. Also, the table shows the 24-hour, 25-year frequency rainfall and runoff for the center of the areas. Fig. 2 shows the 25-year, 24-hour rainfall and estimated feedlot runoff for the state.

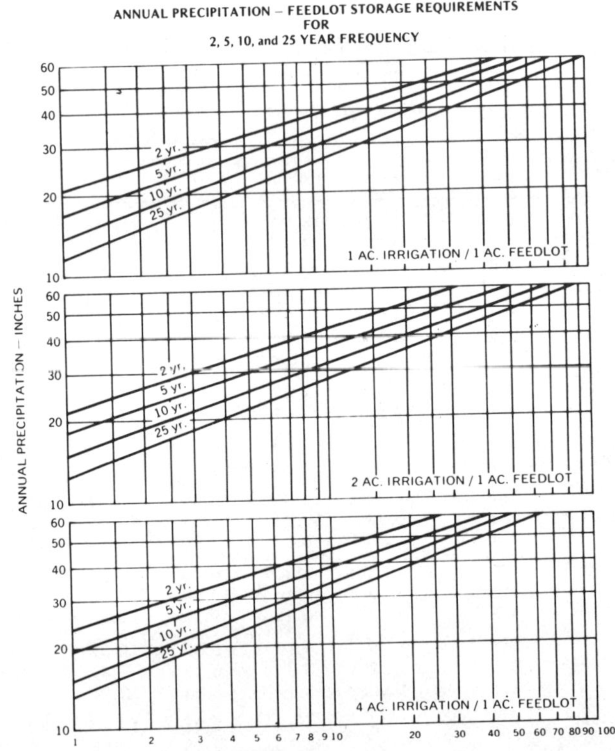


FIG. 2 25-year, 24-hour rainfall (inches) and runoff (inches) based on curve No. 90.



MANAGING LIVESTOCK WASTES

7   The 2-, 5-, 10-, and 25-year frequency storage requirement for the 1, 2, and 4 ratio was related to annual precipitation at the center of the respective area. Fig. 3 shows this relationship. Fig. 4 is a map of the state showing annual normal total precipitation.

The study results in the following conclusions.

1   The required storage capacity varies with the ratio of irrigated land area to feedlot contributing area and the location in the state. The following tabulation from Fig. 3 illustrates this variation for the 25-year frequency.

|  | Storage Required - Inches Runoff | | | |
|---|---|---|---|---|
| Frequency | 20* | 30* | 40* | 50* |
| 2 Year | 1.0 | 3.7 | 10 | 24 |
| 5 Year | 1.8 | 6.3 | 17 | 34 |
| 10 Year | 3.0 | 10.0 | 23 | 44 |
| 25 Year | 4.5 | 14.0 | 31 | 58 |

*Annual rainfall, inches

2   The runoff from the feedlot is not a dependable source of irrigation water supply, but with storage facilities it can be used to supplement other sources. This is illustrated on Table 1. With two acres irrigation per acre of feedlot, there are five months in 1941 when runoff does not equal the irrigation water requirement. Also, from Table 1, it can be seen that with 8.62 inches of storage capacity there are two months in 1941 when the irrigation water requirement would not be available. This deficiency would be increased if evaporation or seepage losses or irrigation efficiencies are considered.

3   The procedures used in this study are reasonable. The addition of refinements such as evaporation and seepage from the storage reservoir or efficiencies in the application will reduce the storage requirement. Irrigation efficiencies can be reasonably estimated and their inclusion in the irrigation water requirement may be desirable. On the other hand, studies have been made that indicate the feedlot effluent would need to be diluted if crop yields are to be maintained. The neglect of these refinements will provide greater opportunity for use of diluting water from other sources. Fig. 5 may be used to adjust the 25-year frequency storage requirement for irrigation efficiency or irrigation water dilution. The ratio of the disposal area to feedlot area is related to the 25-year frequency storage requirement for a range of average annual rainfall amounts.

4   The study shows that a large amount of storage capacity would be needed in the eastern part of the state. This should be expected because in this area there are long periods of time when monthly runoff exceeds irrigation water requirements. It is possible that in this area an overuse of water can be tolerated more often than the 25-year frequency. The following tabulation from Fig. 3 shows the required storage capacity when the ratio of irrigated land area to feedlot area is one.

|  | Storage Required - Inches Runoff | | | |
|---|---|---|---|---|
| Ratio | 20* | 30* | 40* | 50* |
| 1 | 4.5 | 14 | 31 | 58 |
| 2 | 3.7 | 12 | 25 | 47 |
| 4 | 3.2 | 10 | 21 | 39 |

*Annual rainfall, inches

5   Holding ponds with capacity to impound the 25-year, 24-hour runoff from feedlots in the western part of state give operators considerable latitude in dewatering the ponds when the pastures need irrigating. Holding ponds with this same capacity in the eastern part of state do not permit this flexibility of operation. Note on Table 2 that for Area 1-A the 25-year, 24-hour runoff corresponds to a 50-year frequency for a 1:1 ratio of irrigated area to feedlot area; whereas, in Area 7-C the 25-year, 24-hour runoff corresponds to about a 3-year frequency for 1:1 ratio. If holding ponds are dewatered within about two weeks after runoff, the chances are good in the western part of the state

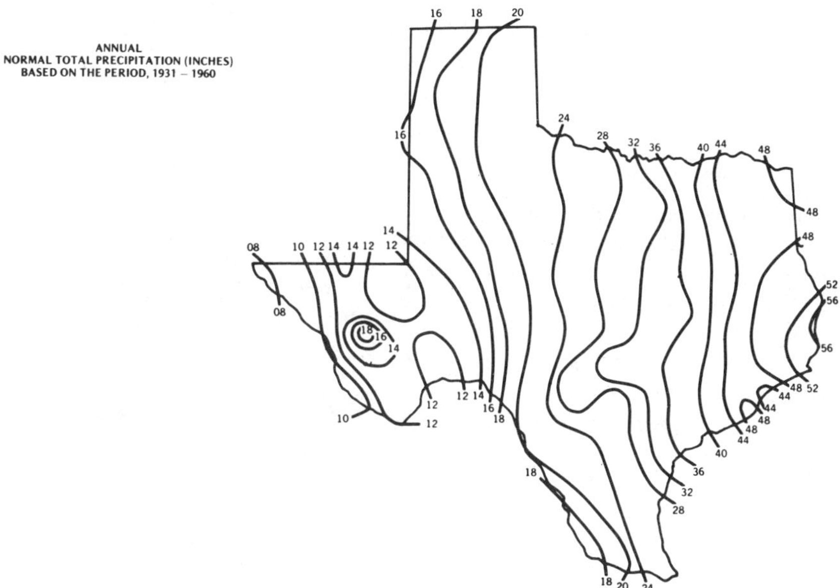


FIG. 4 Annual normal total precipitation (inches) based on period, 1931-1960.

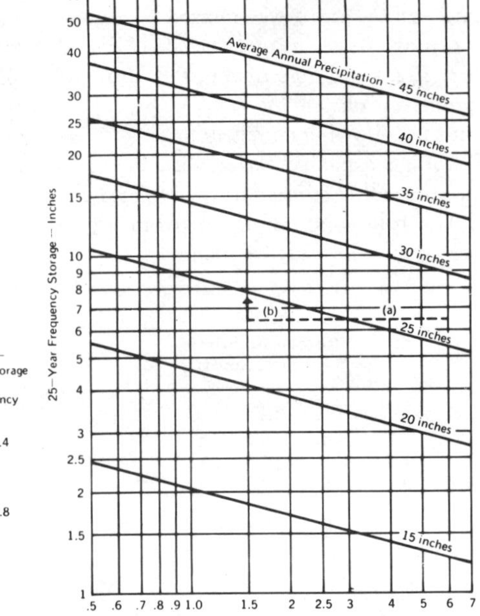


FIG. 5 25-year-frequency storage—ratio is disposal area to feedlot area for annual precipitation amounts.

that the pasture grasses will need additional moisture. The chances are not nearly so good in the eastern part of state. For example, in Area 1-A using a 1:1 ratio of irrigated area to feedlot area, there were only eight months during the 17-year study period where the runoff exceeded the irrigation requirements. However, in Area 7-C the runoff exceeded the irrigation requirements during 81 months.

### References


1  Butchbaker, A. F. 1972. Feedlot runoff disposal on grass or crops. Texas Agricultural Extension Service Leaflet No. 1053.
2  Butchbaker, A. F. 1972. Cattle feedlot hydrology paper presented at Oklahoma Section ASAE Meeting, October 20.
3  Clark, R. N. and B. A. Stewart. 1972. Amounts, composition and management of feedlot runoff. Research Center Technical Report No. 12. Texas Agricultural Experiment Station.
4  Coleman, E. A., Walter Grub, R. C. Albin, G. F. Meenaghan and D. M. Wells. 1971. Cattle feedlot pollution study - interim report no. 2 to Texas Water Quality Board.
5  Kane, J. W. 1967. Monthly reservoir evaporation rates for Texas 1940 through 1965. Texas Water Development Board Report 64.
6  McDaniels, L. L. 1960. Consumptive use of water by major crops in Texas. Texas Board of Water Engineers Bulletin 6019.
7  Pittman, Dwight L. 1971. Guidelines for handling liquid waste from feedlots. Paper presented at SCS - Texas A&M University Workshop.


## In House Manure Drying

*(Continued from page 84)*

lowing benefits:

1  Manure moisture content can be reduced to 12-15 percent, at no fuel cost.

2  The dried material produced is convenient to handle. It can be ground to yield a pleasant odour free product.

3  This product has an enhanced value over wet manure and can be utilised as a fertiliser or feed ingredient.

4  An amenable pollution free house environment is provided for staff and stock.

5  This drying system does not produce offensive odours—indeed the odour emission is lower than from non-slatted wet pit houses.

### References


1  Bressler, G. O. 1969. Solving the poultry manure problem economically through dehydration. Poult. Sci., 48. (5):1789.
2  Bressler, G. O. 1970. Drying poultry manure inside the house. AGRICULTURAL ENGINEERING 51(3):136.
3  Bills, G. T. 1973. Biological fly control in deep pit poultry houses. Br. Poult. Sci., 14. (2):209-212.
4  Elson, H. A. A. W. M. King, and C. L. Benham. 1972. Air drying of poultry manure under full stepped cages in deep pit houses. World's Poult. Sci. Assos. UK Branch Paper March 1972.
5  Jaeger, G. B., H. C. Whelden Jr., F. V. Muir, C. W. Kitteridge. 1972. Manure management in deep pit houses. Res. Life Sci. 20(7):1-4.
6  Whelden, H. C., Jr. 1973. A problem solving study of wet poultry manure in deep pit storage. Personal communication.

# An Illinois Feedlot Runoff Control Project

## Ronald Lawfer

THIS article covers the construction and operation of a "zero runoff" animal waste control system serving an open concrete feedlot on my farm. This system, one of the first of this type constructed in Northern Illinois, has been in use since July, 1973. Since that time it has served to prevent possible feedlot pollution from entering a nearby stream. It also serves as an area demonstration project for other interested feedlot operators.

The feedlot site is situated on the crest of a high ridge and has contained livestock for many years. About five years ago, this feedlot floor was concreted and an auto-mated feed system installed. The capacity of the feedlot is about 100 head of yearling cattle which provides space to feed dairy steers and grow replacement heifers for my dairy operation located at another farmstead. During the summer, some hogs are fattened in this lot as some cattle are moved to pasture.

Before the control system was installed, runoff flowed across the feedlot and off the lower edge of the concreted area as shown in Fig. 1. During seasons when the manure could not be hauled and spread on cropland or pasture due to wet ground conditions, some of the manure washed down the natural waterway below the feedlot. There was a possibility of some polluted water reaching a flowing stream located about 800 ft (244 m) downhill.

Because my farm was one of two host farms for the 1973 Illinois Land Improvement Contractors Association's Soil and Water Conservation Show, it was decided that a "zero runoff" animal waste control system could be both beneficial for my farm and valuable as a demonstration project showing this type of pollution control. At this time the Illinois Environmental Protection Agency and the Illinois Pollution Control Board were considering and studying proposed state regulations to control runoff from farm feedlots. Because the feedlot is similar to many other livestock farms located in rolling terrain characteristic to

The author RONALD LAWFER owns and operates a dairy and livestock farm in Jo Daviess County, Illinois.

Acknowledgements: The author wishes to express appreciation to Jerry Misek, District Conservationist, USDA Soil Conservation Service, Jo Daviess County, Illinois; Herbert Davenport, Area Engineer, USDA Soil Conservation Service for their encouragement, technical aid and assistance of this presentation.

The assistance of Dale Vanderholm, Extension Agricultural Engineer, University of Illinois to interpret initial results of water monitoring was valuable to this author.

Federal, state and local agencies and organizations offering advice and financial assistance for the construction of this research and demonstration project are: Illinois Beef Industry Council, Illinois Institute for Environmental Quality, Illinois Environmental Protection Agency, USDA Soil Conservation Service, USDA Agricultural Stabilization and Conservation Service, University of Illinois Cooperative Extension Service, Illinois Land Improvement Contractors Association.

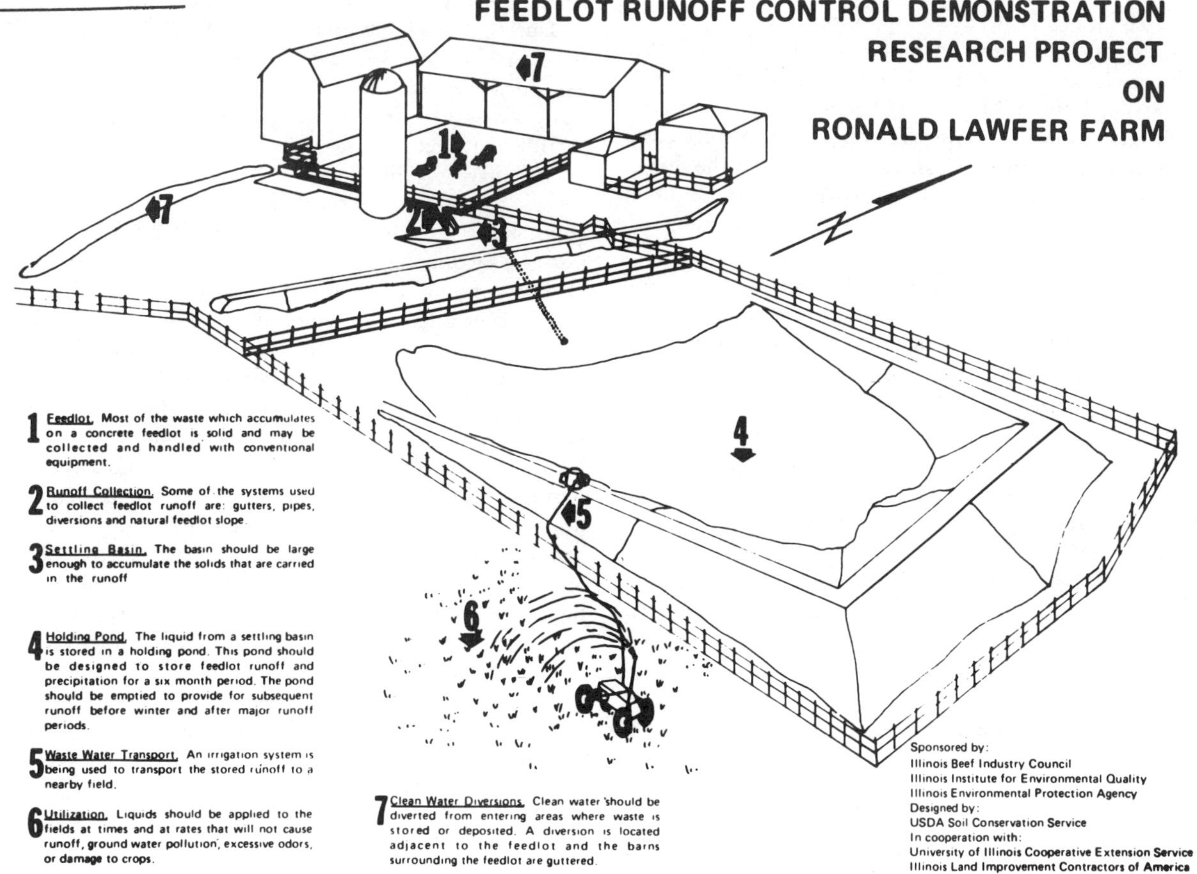



FIG. 1 Feedlot runoff control demonstration research project on Ronald Lawfer Farm.

Northwestern Illinois, EPA officials and other interested agencies, including beef industry officials, felt that this project could serve as a demonstration of runoff control both to area farmers and persons expected to attend the Conservation Show. Soil Conservation Service engineers designed the system and construction was started in June, 1973.

This gravity flow system is designed to collect and store all animal wastes or polluted runoff liquids until they can be applied to the soil.

Feedlot entrances and exits were redesigned to facilitate this system. Fences were removed and replaced as cattle movement routes were changed and animal exclusion from the holding pond was required.

Earth diversions were built above the feedlot to direct unpolluted water away from the system. Spouting was applied to building roofs to direct this water away from the feedlot.

A concrete curb was placed along the lower edge of the feedlot to direct the runoff to a sloped chute. The 2 ft (0.61 m) wide chute carries the runoff by gravity about 30 ft (9.15 m) to a settling basin.

The reinforced concrete basin is designed to collect storm runoff from the cattle feedlot, a nearby hog farrowing house open floor, and the earthen service area below the feedlot and silo. The runoff from about 2 acres (0.81 h) is directed to the settling basin. The basin is 10 by 40 ft (3.05 by 12.2 m) and has a sloped floor that is used to settle out the solids. The liquid then flows out of the basin through a slotted metal outlet into a buried pipe running to the holding pond. At no time is any manure pushed from the feedlot area into the sloped chute or settling basin.

The basin floor slopes from ground level at one end to a depth of 18 in. (45.7 cm) at the opposite end.

Capacity of the basin was calculated on the basis of 2000 cu ft for each acre of feedlot (141 $m^3$ per h). The volume of this settling basin is 417 cu ft (11.8 $m^3$).

Additional capacity above the top of the settling basin was provided by constructing an earthen ridge on the downhill side of the settling basin. This ridge also serves to direct wastes from the farrowing house into the settling basin.

Before construction of the holding pond (located about 100 ft (30.5 m) downgrade from the settling basin) soil borings had shown an impervious subsoil, making an earth dam satisfactory to contain the runoff liquids. This pond was designed for 6 months runoff storage, expecting a 12 in. (30.5 cm) runoff from unpaved areas, and 15 in. (38.1 cm) from the paved areas. Average rainfall in this area is 33 in. (84 cm) per year. The 6 months storage was suggested because this is the period of time (November to May) that field applications may not be practical or possible because of frozen or wet soil conditions.

Construction of the holding pond was much the same as any water containing pond. A "core" trench was dug, backfilled, and compacted with the impervious material excavated from the pond area. Ideal soil and weather conditions enabled the digging of the pond deeper than planned, resulting in a water holding capacity nearly double the planned amount. Total capacity is now 1 250 000 gal (4 731 750 l).

Because of the large amount of liquid runoff, an irrigation system was designed and installed to empty the holding pond as needed. A 7½-hp 3-phase electric centrifugal pump was placed on the earth fill adjacent to a parallel tile outlet terraced cropland field. The intake hose extends from the pump to the center of the holding pond. The hose is suspended from floats that hold the hose inlet near the water surface to prevent plugging.

To provide for the water distribution, a 3 in. (7.6 cm) plastic pipe with three risers spaced 120 ft (37 m) apart was buried in one of the terrace ridges on the adjacent field. Movable lateral lines with three sprinkler heads (with 5/16 in. (8 mm) nozzles) mounted on the aluminum pipe can be attached to these risers to apply the liquid to one acre (0.4 h) of cropland. Each sprinkler covers about a 65 ft (19.8 m) radius at a pressure of 80 psi (5.6 kg per $cm^2$). Complete movement of the lateral lines from each riser results in about 7 acres (2.8 h) irrigated. In addition, a gravity drain pipe was installed through the pond fill. By the use of a perforated pipe, this drain can apply liquids to a permanent pasture located adjacent to the pond. This drain has not been used, but it can draw off the top 4 ft (1.22 m) of liquid, which is about one half the storage volume of the pond.

This "zero runoff" system was completed in August, 1973, and has been in operation since that time. During the first 10 months of operation, rainfall for the area was almost double that of normal, recording about 50 in. (127 cm).

Operation of the feedlot has not been changed since the installation of the system. All manure possible is loaded and spread on cropland when weather conditions permit and labor is available, The manure is handled with a conventional solid manure loader and a spreader with a mechanical endgate. Because of the climate in Northern Illinois, it is not uncommon during the winter months to have 2 to 4 weeks during which the temperature does not rise above freezing. During this period there is no runoff.

During the 18 months that the facility has been in use, its operation has been satisfactory and in some phases has exceeded expectations. Use of the system as a demonstration project to interested persons has resulted in over 10 000 visitors to the site.

The narrow chute that collects the runoff has proven very satisfactory. Debris that does settle or freeze in the chute between rains is readily flushed by the next runoff into the settling basin.

The amount of solids flowing into the settling basin varies according to the amount of solids on the feedlot and the volume of rain that falls. If these solids can be removed mechanically at 2 week intervals, small amounts are washed into the basin. However, if heavy rainfall occurs during a period when a large accumulation of solids are on the feedlot, it is possible for all of this to wash into the settling basin completely filling it. These conditions occur under normal management and weather conditions four to six times a year. Under these conditions, the slotted metal outlet in the corner is not sufficient to handle the volume of liquid. Manure settles at the chute outlet (which is the center of the basin), trapping liquid at the upper end of the basin with solids between the liquids and slotted metal outlet. This makes mechanical loading of the manure difficult as it does not dewater. To solve this problem, a "false wall" was installed the length of the basin on the side of the slotted metal outlet. This wall was blocked 4 in. (10.2 cm) away from the existing concrete wall and covered with ½ in. (1.27 cm) smooth expanded metal. This enables the liquid to dewater from the solids the entire length of the basin. Operation of this addition during the last 9 months has shown it to be very satisfactory. Even during rapid filling of the basin, it dewaters in about 2 days, permitting easy manure loading.

It is important to have the settling basin cleaned before the last fall freeze to have the maximum capacity during a rapid thaw. In the summer, the settling basin needs to be clean whenever it starts to fill up so that the basin capacity is available for subsequent runoff. For my operation, a 50 percent increase in the capacity of the settling basin would allow me to more efficiently utilize labor and equipment.

Because of the capacity of the holding pond, storage was sufficient until July, 1974, at which time the liquid was applied to the adjoining parallel tile outlet terraced corn field. During this period nearly 1 000 000 gal (3 785 400 l) were applied by the overhead sprinklers on 7 acres (2.8 h). Although an early frost prevented accurate yield calculations, the irrigated crop easily doubled the production of the unirrigated area. Nearly 6 in. per acre (15 cm per h) of liquid were applied.

My system is one of several livestock feedlots located in Northern Illinois in which soil and water samples are being monitored by the University of Illinois Agricultural Engineering Department. This research is sponsored in part by the Illinois Institute for Environmental Quality to evaluate runoff control and how different systems function in the prevention of pollution.

This research continues under the direction of an Extension Agricultural Engineer. It shows that about 1/10 of one percent of the liquids in the holding pond is solids, indicating that the settling basin is effectively separating the solids from the liquids. Preliminary studies indicate that the amount of plant nutrients in the irrigation water is relatively small. Sample tests show that about 5 lb of nitrogen, 5 lb of P205 and 20 lb of K20 are applied per acre inch of liquid (2.2 kg of nitrogen, 2.2 kg of P205, 8.8 kg K20 per h cm). Therefore, the 6 in. (15.0 cm) of liquid applied through the irrigation system in 1974 replaced the potash and phosphorus removed by the corn silage crop. Supplemental nitrogen is required to meet 1975 corn crop needs. Sodium and chloride tests on the liquid samples indicate that only under very large application would this be damaging to crops.

As the study of the water and soil samples continues, it is hoped that the results can be used in the future to plan more effective and efficient systems for preventing pollution of the air, water, and soil from livestock feedlots. These systems need to be highly refined and evaluated before their widespread use is required. If this is not accomplished by research and trial by practical applications, both the livestock producer and the general public will suffer, depending on the necessity and efficiency of these costly projects.

Most of these system have been designed to comply with either proposed environmental regulations, general area rules, or what planners or rural development may require. Little of the design thought has taken into consideration possible hinderance of the management and feedlot operation. These systems need to fit into existing feedlots to enable them to maintain their present efficiency and to expand if necessary. Any system that prevents this has not been well designed. My system has been designed to fit well into the feedlot area. It can be easily expanded by using additional settling basins that would empty into the existing holding pond.

It is more important to plan the system according to the wants and needs of the individual operator rather than an arbitrary standard or regulation. The capacity of the holding pond illustrates this point. If it had been necessary to empty liquid during May, 1974, on cropland already too wet to plant, the result would have been most disagreeable. However, this liquid proved very beneficial during a dry August, 1974.

In summary, this "zero runoff" control system fits my need and management requirements very well. I hope that my observations on this system will enable someone to make improvements on future systems.

# Feedlot Waste Recycling with a Flush Cleaning System


C. L. Barth, R. W. Goethe

MEMBER
ASAE


**D**EVELOPMENT of any beef feedlot in South Carolina is the result of much original thinking. Development of the 5000-head feedlot on Walworth Farm with hydraulic cleaning and automatic movement of the liquified waste to the field for irrigation onto cropland is an extraordinary achievement in thinking, planning, design, construction, operation and management.

Walworth Farm, a division of Mower Lumber Company, located at Eutawville, South Carolina, has been in the beef cattle business for over thirty years with 500 to 1000 beef cows with calves grazing the many acres of pasture grasses. Since 1966 there has also been about 1000 feeder cattle on their unsurfaced feedlots as well. Unsurfaced feedlots were inundated in late winter with the heavy precipitation and lack of a frozen surface. In the July and August heat they were dry and dusty.

Waste management on a South Carolina feedlot is a slippery problem. At no time of the year can the manure be allowed to accumulate because of the potential for odor production and feedlot runoff. Winter is cool and rainy, but the waste on the feedlot surface is never frozen. July and August may be hot and somewhat dry, but heavy rainfall is never far away. A nearby State Experiment Station recorded seven days in July and August, 1974, with one or more inches of rainfall. Open feedlots which encourage manure drying are too hot for the animals in summer. Shaded feedlots promote a wet surface and high rates of odor emission.

Environmental impact was a chief consideration in planning the new feedlot. The soil water table fluctuates from three to five feet below the surface and is controlled by Lake Marion, a major recreaction lake, only two miles to the north. A large, busy campground is located two miles to the northeast, downwind.

Planning of the intensive feedlot to replace the "dry" lots began in August, 1971. Construction began in October, 1972. Cattle feeding began September, 1973. This feedlot was to be the answer to South Carolina problems of climate, waste management and practicality.

Schematic layouts of the facilities and the feedlot are shown in Figs. 1 and 2. The building site was flat and fill was required to raise the ends of the feedlot to create the 2-1/2 percent slope toward the center transfer alley. The borrow area, close by, became a three acre, 6-foot deep holding pond to receive rainfall runoff from the feeding pens.

On each side of the center transfer alley the concreted pens slope upward for nearly 500-feet. Four pens, each


The paper has been approved as Technical Contribution No. 1245 by the South Carolina Agricultural Experiment Station, Clemson.

The authors are: C. L. BARTH, Associate Professor, Agricultural Engineering Dept., Clemson University, Clemson, SC; and R. W. GOETHE, Controller, Mower Lumber Co., Eutawville, SC.


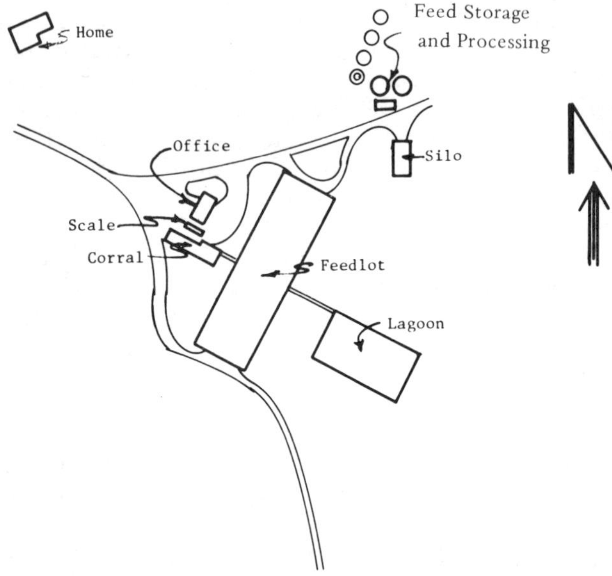



FIG. 1 Plot plan of feedlot and supporting facilities.

120-feet long and 30-feet wide are in each row. Five rows of pens comprise each half of the feedlot. There are a total of 40 pens. Fences are of pipe and cable construction to allow maximum air circulation. At 120 animals per pen each animal has 30 square feet of floor area, 1 foot of feeder space and 15 square feet of shade.

In the northeast half of the feedlot, completed first, the 30-foot wide pens are divided into two 12-foot wide flushing lanes and a 6-foot wide platform, 6-inches high, adjacent to the feed bunk. A six-inch curb separates the two flushing lanes, and the 12-foot high T-shaped posts which support the 15-foot wide shade are installed in the curb at 120 ft spacing. Cables suspended from the posts support the slatted shade material. Each pair of flushing lanes then

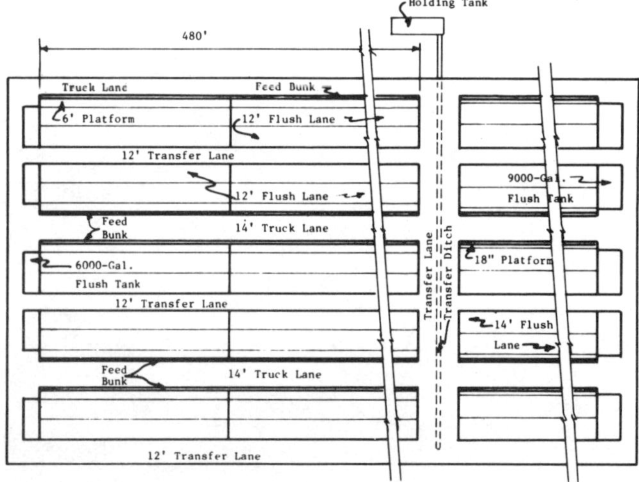



FIG. 2 Schematic layout of feedlot.



**MANAGING LIVESTOCK WASTES**

passes through 4 pens and discharges into a cross channel, located below the center transfer alley, which carries the wastewater to the 30,000-gallon holding tank beside the feedlot.

A 6000-gallon flush tank* is located on the "high" end of each row of feeding pens. Each tank has two flush gates, one for each flushing lane. Regularity of flushing the waste from the feedyard is influenced by two factors — animal density and relative humidity. Greater animal densities require more frequent waste removal and higher humidity levels, which keep the waste in a moist condition, permit longer intervals between waste removal. Experience has shown the need for flushing as often as twice a day or as infrequently as once every three days.

Standard flushing procedure is to release about 1000-gallons of water initially and to wait 5 to 10 minutes for the waste on the floor to "soak". Then the remainder of the water is released to flush the waste from the lot and into the holding tank. The large flush tank is refilled so the second lane can be flushed.

Early experience with the 20 pens in the northeast half of the feedlot showed no advantage for the 6-foot platform adjacent to the bunk. Since it was not flushed, manure accumulated and had to be removed with a tractor scraper. Early experience also showed the need for a greater volume of water for flushing each lane. Due to the tendency for waste to dry and stick to the feedlot surface, unlike many dairy and swine installations with the flush system, greater quantities of water and greater cleaning action is needed.

Changes then incorporated into the construction of the southwest half of the feelot were:

1   Two, 14-foot flushing lanes through each pen with a center curb for the shade-fence posts.

2   An 18-inch wide step adjacent to the fenceline feed bunk.

3   9000-gallon flush tanks at the head of each flush lane with quick recovery filling action so flushing the second lane be executed in a reasonable time.

These changes have improved the effectiveness of the hydraulic manure removal system. Some manure does however, commonly remain clinging to the concrete flushing surface and is attributed to several factors. Constructing the feedyard required transporting and placing huge quantities of fill material. Though increasing the slope of the yard to gain more effective cleaning was desirable it required a substantial added expenditure since an increase of 1/2 percent slope required moving more than 12,000 cubic yards of fill material. The trade-off brought about the concession of slope greater than 2-1/2 percent.

Another factor relating to the incomplete cleaning is the high density and number of animals on the flushing lanes. The removal effect of the wall of water progressing down the slope is dissipated by a large number of animals standing, even lying, in the flushing lanes. There is no real alternative to this on a high density feedlot. Effective cleaning of the concrete surface is also limited in several locations by a small side-to-side slope on the flushing lane which diverts and decreases the cleaning action of the water. This is not an uncommon feature of wide flushing lanes and the importance of a perfectly flat surface, from side to side, cannot be overstressed. The shade installed for the benefit of the animals is also a factor in cleanability of the feedlot, because it is also beneficial in maintaining a moist surface

condition. Shading on the Walworth Farm feedlot covers only 50 percent of the feedlot surface and its favorable effect is limited. Shading and maintenance of a moist feedlot surface can be detrimental from the standpoint of odor production. Dry feedlot surfaces and dry waste accumulation produce less odor than wet conditions, so another trade off may exist between effective cleaning and effective odor control.

Walworth Farm is located in the Coastal Plain on flat Wagram sandy loam soils. Though rainfall is nearly 50 inches per year, field runoff is not a problem. The water table is rather high and in this situation the danger of groundwater pollution from excess nutrients may be the most likely source of trouble.

Cultivated land on Walworth Farm occupies about 300 acres. Pasture and hay land occupy 1500 acres and 1700 acres are wooded. The feedlot and associated facilities occupy 50 acres.

Corn silage and coastal bermudagrass hay, the basis of the feeding program, are produced at Walworth Farm. Additional pruchased feeds are corn, protein suppliment, citrus pulp and bentonite. Corn silage is ensiled in a large pile on a concrete slab. Hay is baled and stacked. Corn is reconstituted, high-moisture grain stored in air-tight silos. The rations are ground and mixed at the central feed processing facility.

Wastewater, after flowing from the feed pens, is held briefly in the 30,000-gallon holding tank where it is blended by a chopper-agitator unit. Then, the waste is quickly pumped to the field for application. A 25-hp electric pump moves the waste through a 6-inch main to the single sprinkler head which covers about one acre per setting. The sprinkler head is moved daily and has been used to cover a total of 200 acres, 24 acres of pasture and the remainder cultivated fields, with the feedlot waste. Liquid manure has been pumped as far as 3/4 mile for application and greater distances will be required in the future as the area of coverage expands. The convenience of the present system would be greatly advanced by the addition of a self-propelled manure gun.

Odor control at the feedlot has been successful. Regular removal of the manure and ready transport to and application on land have kept odor production at the feedlot at a low level. A period of relatively high odor production was coupled with animal digestive disorders at one point. Addition of bentonite to the ration has solved the digestive problem and odor production has similarly beeen controlled.

Confidence in their control of odor pollution of air is demonstrated by the fact that a beef barbeque was served in conjunction with an Open House in October, 1973. Several hundred people dined outdoors only 50 yards from the feedlot. The Governor and a United States Senator of South Carolina and many other dignitaries who were present were unconcerned by the fact that the wind was blowing from the feedlot to the serving area.

The Walworth Farm beef feedlot has been an effective demonstration of a workable system for beef production and waste management in the warm, humid Southeast. Emphasis upon the potentially troublesome subject of waste management is to a high degree responsible for the success. Regular removal of feedlot waste and application to productive cropland in a sensible manner not only maximizes recovery of the nutrients in the waste but also minimizes the potential for pollution from feedlot rainfall runoff, nonpoint source cropland runoff, and odor emission.

---

*Patents for the hydraulic manure removal system are held by Agpro, Inc., Paris, TX.

# Operation of a Beef Manure Flushing System in a Cold Climate


Herman A. Natwick, Philip R. Goodrich
MEMBER
ASAE


THIRTY years of cattle feeding and manure handling stimulated my interest in the possibility of finding a relatively trouble free, semi-automatic manure removal system for a cattle feedlot. Our feedlot is located in northern Minnesota, where temperatures last winter reached -37 F (-38 C) and frequently are at -20 to -30 F (-29 to -35 C). We obviously face differing limitations than feedlots in a moderate climate. All the systems we were aware of, either required too much labor and attention, too much troublesome mechanical equipment, a capital investment which we did not feel could be justified, or would not be operational in our -0 to -30 F (-18 to -35 C) weather.

The idea of a manure flushing system in a confinement barn was first exposed by a group of Iowa and Nebraska cattlemen in early 1973. It immediately seemed to meet our requirements of: (a) low initial cost per head, (b) low operational cost, (c) completely automatic and (d) would function in our extremely cold winters.

It was our observation after having fed cattle in open front buildings with a driveway and feed bunks along the back side for fifteen years, that the manure would move down an incline through the action of the cattle's hoofs. We had also observed that when cattle were concentrated in an open front barn, manure did not freeze except in extreme cases. We had also witnessed that water from overflowing fountains, when given enough volume, would flow a long way in extreme cold without solidifying. These observations, plus the fact that chicken and hog operations had already demonstrated that manure could be flushed, gave me the confidence to proceed without any prior testing of the plan as a complete unit. Several telephone calls to knowledgeable people in the industry led to our starting the planning of a 1,000 head confinement building in July of 1973. Construction was started in August and cattle were placed in the building on December 1, 1973. There were no other such units already constructed and decisions would have to be made as construction progressed, so we decided to act as our own general contractor. This decision was to later save me a great deal of time and money.

The unit we built is a cold confinement pole frame, open ridge roof, metal clad building measuring 50 ft (15.2 m) wide, 416 ft (127 m) long and 13 ft (4 m) high at the eaves. The building runs east and west with the south side completely open for ventilation and air movement. The building has a 12 in. (30 cm) ridge opening the entire length of the roof, and a 10 in. (25 cm) opening under the eave on the back side (north) of the building. Alternate doors and panels make up the entire north side of the building, allowing a good air flow through the building during warm weather.

The building was located on relatively flat ground, with windbreaks of multi-row tree belts on the north, west and south sides. This site was selected because of natural protection from wind, its proximity to our backgrounding lot, working alleys, chutes and sick pens. It was and still is our opinion that cattle should be started for a few weeks in a conventional lot where they are less confined. There seem to be fewer respiratory problems where the cattle are not closely confined in the first few weeks, and the sick cattle are much easier to detect in larger outside lots. A backgrounding lot adjacent to a confinement building also gives us the ability to keep the confinement building filled a much greater percentage of the time.

The manure handling system, which is the unique feature of this barn, is one where the design of the floor, the action of the cattle's hoofs, two electric pumps, flowing water and an aerator do all of the work. The floor in the barn has three gutters or flumes set in concrete below a 2 in. (5.1 cm) wide slot in the floor (Fig. 1). The flumes are 12 in. (30 cm) diameter PVC (poly vinyl chloride) plastic pipes cast in concrete and set 5 in. (12.7 cm) below the top surface of the floor. The barn, floor and flumes all slope toward the east end 1 in. per 16 ft (1 cm/196 cm) totaling 26 in. (66 cm) or approximately 1/2 percent. The barn, which is 50 ft (15.2 m) wide, has 36 ft (11 m) of width in the pen area; the remaining 14 ft (4.3 m) is used for feed bunks and a drive-through alley along the north side of the building. The three flumes are positioned 12 ft (3.6 m) apart in such a manner that there is a 6 ft (1.8 m) drainage area from each side of all three of the flumes. The concrete floor is constructed with a slope of 1 in. per foot (1 cm per 12 cm) toward the slot and flumes. This is steep enough to cause the urine to flow into the slots and the action of the cattle's hoofs to work the solids in, yet it is not steep enough to cause undue slippage by the cattle.

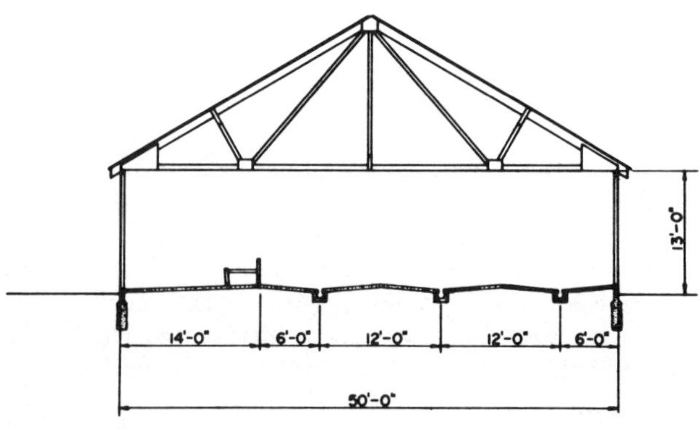


FIG. 1 Cross section of barn and flushing system.


Miscellaneous Journal Series Paper No. 1574, University of Minnesota Agricultural Experiment Station, St. Paul.

The authors are: HERMAN A. NATWICK, Owner and Operator, Natco Inc., Ada, MN; and PHILIP R. GOODRICH, Assistant Professor, Agricultural Engineering Dept., University of Minnesota, St. Paul.

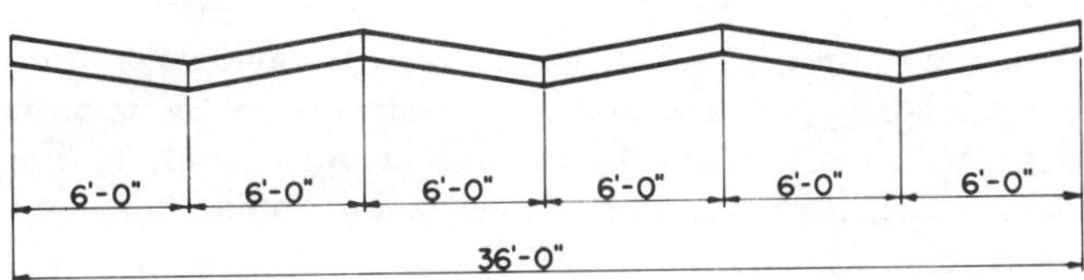

FIG. 2 Special screed used to pour the concrete floor.

Construction of the floor was accomplished by first designing and building a screed using a double row of 12 in. (32 cm) I beams welded together with the proper configurations to pour the floor in one continuous pass (Fig. 2). First, we set the PVC pipe in concrete at the proper elevation. Next, we used the screed to bring the fill under the floor to the desired height and shape.

The next step was to raise the screed 5 in. (12.7 cm), put removeable forms over the PVC pipe for our slots and pour the floor. We used a power winch to propel the 2,800 pound (1265 kg) screed. A vibrator attached to the framework provided uniform compaction to the concrete. A 42 in. (107 cm) wide metal float was used to give the floor a uniform finish. We did not attempt to finish the floor smooth, nor did we leave it excessively rough.

Simultaneously, we constructed a lagoon measuring 350 x 500 x 25 ft (107 x 152 x 7.6 m) deep with a capacity of approximately 15 million gallons (94,500 cubic meters) (Fig. 3). Due to the topography and soil structure, we built it approximately 12 ft (3.7 m) below and 13 ft (4 m) above the ground level. The inside wall of the dike is lined with a minimum of 10 ft (3.1 m) of compacted clay to prevent seepage and erosion. The clay to line the above ground dike was taken from the excavated section which is cut into clay.

It was necessary to reach 10 ft (3.4 m) of water depth in the lagoon in order to become fully operational and recirculating lagoon water for flushing the building. A drilled high capacity well pumped water directly into the high end of the flumes for flushing, enabling us to use the barn while the lagoon was being filled. We began recirculating water from the lagoon through the barn for flushing after the water had reached the required depth. A 12 in. (30 cm) diameter pipe set horizontally 8 ft (2.4 m) above the bottom of the lagoon free flows water into a pumping station, constructed using a 4 ft (1.2 m) diameter culvert set vertically in the center of the lagoon bank. A 4 in. (10 cm) manure pump with a 5 hp (3750 Watt) electric motor is set in the culvert and pumps the water to the barn through an 8 in. (27 cm) PVC pipe.* This pump is controlled by a time clock to allow for the desired pumping frequency. The water flows into a 2 x 3 x 6 ft (0.61 x 0.91 x 1.8 m) high divider box at the head (high end) of the barn where it separates into three short pipes leading to the flumes. The divider box is so constructed that we can direct all of the flow through one or two flumes in the event of a blockage or problem in one. We have found a tapered plastic ice cream pail to be a most effective shutoff to stop the flow of water to one of

---

*Pumps made by Parma Pump Co., Parma ID.

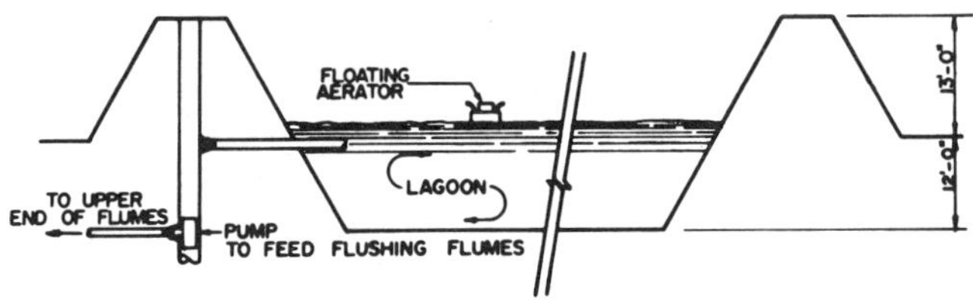



FIG. 3 Cross section of lagoon and pump which feeds the flumes.

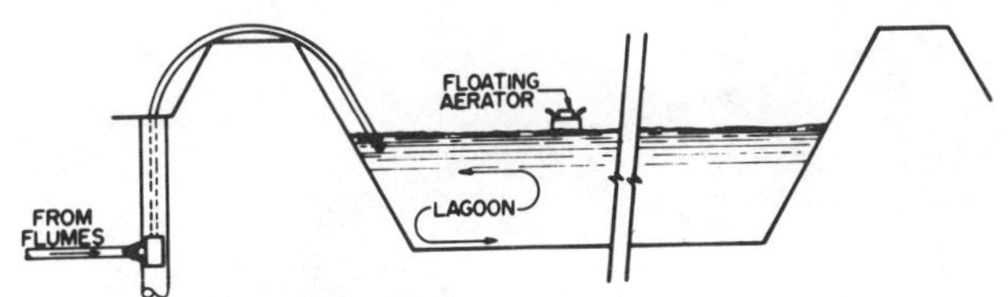



FIG. 4 Lift station collects the waste and pumps into the lagoon.

the pipes leading out of the divider box to a flume.

We estimate we are pumping 70 gallons per minute (0.0044 cubic meters per second) through each of the three flumes. The water flows through the flumes picking up the manure that has accumulated, then flowing into a 15 in. (38 cm) line at the low end of the barn leading back to a waste pumping station adjacent to the lagoon embankment (Fig. 4). Another 4 in. (10 cm) manure pump and a 5 hp (3730 Watt) motor set in a 6 ft (1.8 m) diameter pit 11 ft (3.3 m) deep, lifts the waste material into the lagoon. This pump is controlled by a liquid level float type switch. The system is designed to accomodate equipment to remove the solids at the waste station. As soon as I am satisfied that we can get trouble free equipment, which will reduce the moisture content of the material we remove to 70 percent or under, we plan to install such equipment. I feel we need to make this change for two reasons: (a) to save the material for either its feed or fertilizer value and (b) to reduce the BOD load on the lagoon.

We are presently aerating the water in the lagoon (to control odors) with a 20 hp (15,000 Watt) aerator, with a 10 kW heater for cold weather operation. The cost of electricity for operating the aerator is approximately $4.80 per day in warm weather and $7.20 per day in cold weather. We plan to remove a portion of the material from the lagoon annually and use this to irrigate adjacent corn or alfalfa. By this means, we would expect to prevent a buildup of solids and salts in the lagoon. By the use of constant flow watering fountains and the run-off of precipitation from our conventional lots, we hope to add a considerable amount of water each year. Our experience to date would indicate very little change is noted in the lagoon level, except when we pick up the runoff from several inches of rain on our outside lots.

Our experience after going through one summer and two winters, one which included 17 days of continuous sub-zero weather, has been generally good. We have found that 15 minutes flushing time every four hours seems adequate in warm weather. We flush fifteen minutes each hour in weather down to 0 F (-18 C) and below 0 F (-18 C), we flush continually. When we install the equipment to separate the solids from the liquids, I plan to raise the level of the water to the point where we can eliminate the input pump and free flow the water to the flumes by gravity. We will then attempt to run a much smaller flow of water continuously, perhaps in the order of 20 gallons (0.075 m$^3$) per flume per minute. We have encountered some problems on days when there is a great deal of snow swirling in the air. Our observation has been that the water in the flumes will carry the snow along and it will collect at the point where it leaves the barn and causes a blockage. We are certain we can eliminate this by replacing our pipes and fittings at the lower end of the barn with a 15 in. (38 cm) wide 5 ft (1.5 m) deep concrete trough (Fig. 5).

In sub-zero weather, we do have a build-up of manure on the floor. It is not objectionable when the pens are at the desired density of cattle. We are feeding a ration of one-half

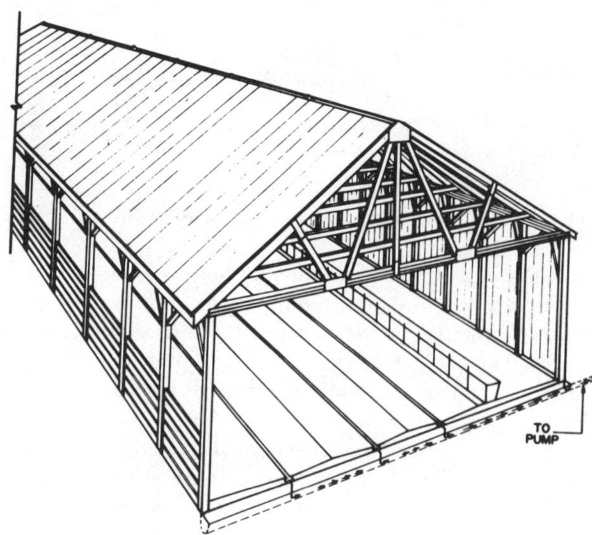

FIG. 5 Proposed trough to collect the outflow from the flumes.

silage and one-half grain with just a little chopped hay. Our experience has shown that, after a prolonged cold spell and when the temperature gets above 0 F (-18 C), the build-up of manure again starts moving into the flumes. If the pens are not at the proper density of cattle, we then seem to have a condition where the build-up is rough and uncomfortable for the cattle. Based on our experience, we would recommend 18 square feet per 1,000 pounds (3.7 square meters per 1000 kg) of animal weight. The cattle naturally have a longer hair coat during cold weather and they do become moderately dirty over their entire body. However, they do not collect as much manure matted in long hair on their flanks and bellies (tagging) as cattle in outside lots, but are comparable to cattle confined in a similar barn on slats. We have found it to be a good practice to cut the switch on cattle's tails to prevent a manure build-up in cold weather.

The location and design, as well as most of the supportive facilities, such as pipes and lagoon pumps, are large enough to accomodate an additional three to four thousand head of cattle. The plan was designed to handle additional buildings of the same type with only minor additions to the pump mechanisms. We feel that by recycling the solids, we could handle a total of at least 5,000 head with this system.

Our total cost on the unit, including fences, bunks, water system, building, floor, flumes, lagoon pumping station, and electrical equipment costs was under $80 per head in late 1973. This cost is only slightly higher than our conventional lots where the cattle have open front housing and concrete lots.

At this point, we would have no reservations about expanding, using the same design with only very minor changes. The advantages of this system over other confinement designs are, in my opinion, very pronounced. In comparison with slats over a pit, our cost is approximately 50 percent less and we do not have to agitate and haul out the waste at certain fixed intervals. In the case of slats over a mechanical scraper device, we do not have the mechanical problems which are so common with these systems, especially in cold weather. I am well pleased with the unit, and we would like to commend and thank the people who first conceived the idea of this system. I feel this will be the basic design of manure handling and housing which will serve the livestock industry in the future.

# Utilization of Beef Cattle Waste from a Slotted-Floor Deep-Pit Barn

Raymond Larson, D. G. Jedele

MEMBER
ASAE

APPROXIMATELY 2 535 000 gallons of liquid manure from the Larson and Taylor Feedlot in DeKalb County, Illinois, is utilized annually as fertilizer on cropland under control of the Feedlot. There are 700 acres of tillable land within 1½ miles of the barns that are available for utilizing manure. The waste management system that has been developed since 1971 makes excellent use of the manure storage capacity available with a minimum of equipment and labor.

Three thousand cattle per year are marketed from two slotted-floor barns with a total capacity of 1340 cattle. This is 2¼ turns of cattle that weigh about 650 lb to start and about 1050 lb when marketed.

Incoming cattle are usually started on feed in a solid-floor barn and lot, although some cattle have been successfully started on slotted floor. But because the solid-floor facilities are available from the previous feeding operation on this farm, they are still used. This means that some manure is handled as a solid.

## MANURE MANAGEMENT SYSTEM

The two slotted-floor barns have 15 manure pits that are 40 ft x 40 ft x 7.5 ft deep. The depth is measured from the bottom of the slats to the floor. Obviously, the pits are never completely empty, and it is not always possible to start pumping when the manure level is exactly at the bottom of the slats. In practice, an average of 75 000 gallons per pit is pumped, so the average storage depth is 6.25 ft.

All the manure is hauled with one 3250-gallon (average load is 3000 gallons) tank wagon towed with a 100-hp tractor. Another 100-hp tractor is used to power the agitator-pump at the barns although a 65-hp tractor could handle it. The average round-trip distance to the fields is one mile, and the average field travel distance per load is one mile. The time required per load is 4 minutes to load, 6 minutes for one mile at 10 mph and 12 minutes for field spreading at 5 mph, or an average time per load of 22 minutes. Theoretically then, 25 loads (75 000 gallons) could be hauled in 9 hours, 10 minutes. In practice, one man can empty one pit (75 000 gallons) in one 10-hour day.

An annual cycle of waste hauling is as follows: About October 1, one man is assigned to feed and observe the cattle and to spend the rest of his day hauling manure. At the start, manure is spread on 150 acres of corn-stubble ground (land from which corn silage was harvested). As the season progresses, corn-stalk ground (land from which shelled corn was harvested) is available to receive the balance of the manure. After the agitator has been started at the barn, the man will chisel plow corn-stubble ground or disk corn-stalk ground where manure was spread the day before. A heavy-duty offset disk with 24-in. blades is used. Because some time is used for feeding and covering the manure spread the previous day, one man takes about two days to empty one pit. With 15 pits, it takes about 30 work days to completely empty all pits.

With pits completely empty in the fall, it would not be necessary to haul again for about 5½ months or about Mid-March. But experience has shown that it is better to haul about five loads out of each pit during January and February on days when the ground is frozen and when snow cover does not interfere with machinery operation. One man will spend about 5 hours a day at manure hauling on these winter days. This manure is spread on corn-stalk ground that will be disked in the spring.

In April, all pits are emptied quickly. The manure is spread on corn-stalk ground near the barns and is disked into the soil just ahead of corn planting. During this season one man is assigned to manure hauling and another to disking down the manure. It is at this time that 75 000 gallons is pumped and spread in one day.

Manure is next hauled in July and August when five 3000-gallon loads (15 000 gallons) or one-fifth the capacity of each pit is removed and chisel plowed into soil from which peas have been harvested.

This annual cycle of manure management is summarized in Table 1.

A total of 845 loads from 15 pits are pumped and hauled annually. This requires about 34 man-days per year. Since the men do some manure hauling and some other work on many days, some attention is given to waste management on about 80 days each year.

## COST FOR HAULING LIQUID MANURE

The manure hauling operation requires one tractor on the pump and one on the tank wagon. Of a total use of 800 hours per year, the tractors are used 309 hours for the manure operation. It is assumed that the average life of the tractors, pump, and tank wagon is eight years, giving an annual fixed cost of 18 percent of the purchase price. The purchase price of the tractors was $10 000 each, the tank wagon $3500, and the pump $1500.

Fixed costs are calculated as follows:

| | | |
|---|---|---|
| Tractors: | $\dfrac{309}{800}$ x $10 000 x 0.18 x 2 = | $1 391.09 |
| Tank Wagon: | $3 500 x 0.18 = | 630.00 |
| Pump: | $1 500 x 0.18 = | 270.00 |
| | Total | $2 291.09 |

The authors are: RAYMOND LARSON, partner and manager of Larson and Taylor Feedlot, DeKalb County, Illinois; and D. G. JEDELE, Professor, Department of Agricultural Engineering, University of Illinois at Urbana-Champaign.

| | Gallons pumped | | Type ground where spread | Acres covered | |
|---|---|---|---|---|---|
| | Per pit | Total | | | |
| Oct.-Nov. | 75,000 | 1,125,000 | Corn-stubble | 150 | Chisel plowed next day |
| | | | Corn-stalk | 130 | Disked next day |
| Jan-Feb. | 15,000 | 225,000 | Corn-stalk | 51 | Disked in spring |
| April | 64,000 | 960,000 | Corn-stalk | 240 | Disked next day |
| July-Aug. | 15,000 | 225,000 | Pea-vine | 35 | Chisel plowed next day |
| Totals | 169,000 | 2,535,000 | | 606 | |

Repair and maintenance costs are assumed to be 85 cents per hour for each tractor, 40 cents per hour for the tank wagon, and 10 cents per hour for the pump. This calculates:

$2.20/hr x 309 hr/yr = $679.80

Operating costs are based on the following assumptions:

| | |
|---|---|
| Weight of tractors | 10 000 lb each |
| Weight of empty tank wagon | 2 000 lb |
| Coefficient of rolling resistance on roads | .05 |
| Coefficient of rolling resistance on fields | .10 |
| Level land | |
| 40-hp pump load | |

Fuel and oil consumption was calculated from formulas and tables by Hunt (1973). The formula for horsepower required is:

$$hp = \frac{\text{coefficient of rolling resistance x weight x mph}}{375}$$

Using this formula, the diesel fuel required for hauling was calculated and tabulated in Table 2.

The hauling tractor idles while the tank is being loaded and this requires 2.5 gal/hr. As stated earlier, the average time per load is 4 minutes to load, 3 minutes to the field, 12 minutes to unload, and 3 minutes to return to the barn. A weighted fuel consumption per hour of hauling is then calculated as follows:

$$\frac{4}{22} \times 2.5 + \frac{3}{22} \times 4.4 + \frac{12}{22} \times 3.6 + \frac{3}{22} \times 3.6 = 3.5$$

The pump load is assumed to require 40 hp with a fuel efficiency of 10 hp-hr per gallon. Therefore, it requires 4.0 gallons per hour.

The tractors have an engine oil capacity of two gallons. Oil is changed at 150-hour intervals; since the tractors are used 309 hours per year for waste hauling, 4 gallons of oil are charged to each tractor per year.

Total operating costs are as follows:

Diesel fuel: 309 hr x 7.5 gal/hr x $0.35/gal = $811.13

Oil: 4 gal x 2 tractors x $2.00 gal = 16.00

Total $827.13

There are more man-hours than machine-hours because of extra operations required of the man. One particular operation found to be necessary to keep things operating smoothly is to wash down the pump each time it is moved from one pit to another. This prevents an excessive build up of hair, etc., on the pump parts.

All costs can now be summarized as follows:

| | |
|---|---|
| Fixed costs | $2 291.09 |
| Repair and maintenance costs | 679.80 |
| Operating costs | 827.13 |
| Labor costs | 1 020.00 |
| Total | $4 818.02 |
| Cost per load | 5.70 |
| Cost per gallon | 0.0019 |

The labor and machine costs for chisel plowing and disking are not charged to manure management because this is the primary tillage operation for crop production.

## BENEFITS FROM HAULING LIQUID MANURE

Five samples of well-mixed liquid manure from pits at the Larson and Taylor Feedlot have been analyzed for nitrogen, phosphorus, and potassium contents. Results are presented in Table 3 and compared with average values from other sources (Vanderholm 1974).

Using our own test averages and current prices for nutrients in commercial fertilizers such as 82-0-0, 0-46-0 and 0-0-60, we have constructed Table 4. We show two columns (A and B) of values of manure based on two assumptions for nutrient losses.

TABLE 3. NUTRIENT CONTENT OF MANURE.

| Sample | Nitrogen, percent | $P_2O_5$, percent | $K_2O$, percent |
|---|---|---|---|
| 12/17/74 | 0.50 | 0.24 | 0.49 |
| 1/10/75 | 0.70 | 0.30 | 0.56 |
| 2/13/75 | 0.64 | 0.28 | 0.36 |
| 2/13/75 | 0.63 | 0.30 | - - - - |
| 2/13/75 | 0.64 | 0.28 | 0.36 |
| Average | 0.62 | 0.28 | 0.44 |
| Published average | 0.55 | 0.40 | 0.47 |

TABLE 2. DIESEL FUEL REQUIRED FOR HAULING LIQUID MANURE

| Load condition | Total weight, lbs | Speed, mph | Power, hp | Fuel efficiency, hp-hrs/gal | Fuel consumption, gal/hr |
|---|---|---|---|---|---|
| Full load on road | 37,000 | 10 | 49.3 | 11.13 | 4.4 |
| Half load in field | 24,500 | 5 | 32.7 | 8.9 | 3.6 |
| Empty on road | 12,000 | 10 | 32 | 8.9 | 3.6 |

TABLE 4. VALUE OF NUTRIENTS IN LIQUID MANURE.

| Nutrient | Pounds per 1,000 gal | Price per pound | Percent loss of nutrients assumptions | | Value per 1,000-gal | | Value per 3,000-gal load | |
|---|---|---|---|---|---|---|---|---|
| | | | A | B | A | B | A | B |
| Nitrogen | 51.6 | $0.20 | 50 | 66 | $5.16 | $3.40 | $15.48 | $10.20 |
| $P_2O_5$ | 23.3 | 0.23 | 0 | 10 | 5.36 | 4.82 | 16.08 | 14.46 |
| $K_2O$ | 36.6 | 0.08 | 0 | 10 | 2.93 | 2.63 | 8.79 | 7.89 |
| | | | | Totals | $13.45 | $10.85 | $40.35 | $32.55 |

## UTILIZATION

Manure is usually spread at the rate of 4000 gallons per acre. Higher rates were tried, but they left the soil surface too slippery for good chisel plowing and disking the following morning. Under most soil conditions the rate of 4000 gallons per acre can be easily incorporated into the soil the day following spreading. This rate amounts to about 103 pounds of nitrogen per acre, 93 pounds of $P_2O_5$, and 146 pounds of $K_2O$. Additional nitrogen is applied to maximize the corn crop, but no additional $P_2O_5$ or $K_2O$ is needed. In fact, soil tests show that phosphorus and potassium levels have increased since the use of liquid manure was started in 1971.

It appears that the crops are not fully utilizing the phosphorus and potassium, but we intend to monitor these elements a while longer before making any changes in our spreading rate. We could use additional information from soil fertility specialists about the maximum level of $P_2O_5$ and $K_2O$ that can be tolerated in soils.

The primary soil types on the farm are Flanagan, Drummer, Saybrook, Catlin, and Octagon.

About 75 loads of liquid manure are spread on frozen ground in January and February and are not incorporated into the soil until spring. We assume that assumption B in Table 4 prevails for these loads. We assume that assumption A in Table 4 prevails for the remaining 770 loads. The total annual value of nutrients in liquid manure from the Larson-Taylor Feedlot is:

```
 75 x $32.55 = $ 2 441.25
770 x $40.35 = $31 069.50
                ----------
       Total    $33 510.75
  Total costs   $ 4 818.02
                ----------
  Net benefit   $28 692.73
```

There is another benefit that is not credited to the beef cattle operation and that is the man and machine time saved by a reduction in the application of commercial fertilizer.

## SUMMARY

Liquid manure hauling as practiced at the Larson and Taylor Feedlot in DeKalb County, Illinois, is profitable. We feel that the management system used on the farm would adapt to many corn-belt cattle feeding operations.

It appears to us that manure from cattle barns with up to 1800-head capacity could be pumped and hauled with one tank wagon, one pump, and two tractors operated by one man. Expansion from the present 1340 capacity to 1800 capacity would add about 12 man- and machine-days to the operation. This might be approaching the limit of availability of favorable weather for manure hauling. Furthermore, more land would be needed to utilize the manure and this would increase manure hauling distances.

For cattle-feeding corn-raising operations, with 400- to 1800-head barn capacity, we would not hesitate to recommend the liquid manure hauling system for utilizing wastes.

**References**


1    Hunt, Donnell. 1973. Farm power and machinery management. Sixth edition. Iowa State University Press, Ames, Iowa.
2    Vanderholm, Dale. 1974. Land applications of manure from modern livestock production facilities. AET Soil and Water Conservation No. 14, 2 p., University of Illinois at Urbana-Champaign.

# Evaluation of Dairy, Beef and Swine Waste Handling Systems

R. L. Maddex, T. L. Loudon, L. R. Prewitt, C. H. Shubert
MEMBER          ASSOC. MEMBER
ASAE                 ASAE

THE design criteria and management schemes for planning, constructing, and operating animal waste handling systems have varied somewhat in Michigan as it has in other states. Although design criteria often appears rather specific for a type of animal or specific system design, most all design values represent an average of data collected in various research trials with adjustment by the planner or decision maker. Observations of waste handling systems in operation on farms show that (a) some systems operate reasonably well; (b) many systems have required adjustment or change in either system components or management schemes and (c) some systems have had serious breakdowns.

In order to evaluate why some waste handling systems performed well and others haven't, an Agricultural Experiment Station project was initiated early in 1974. The objectives are: to evaluate the design criteria and management schemes for existing animal waste systems; to identify and evaluate operational characteristics of various components and management practices; to determine labor requirements, investments and annual costs of systems and components.

## SYSTEM AND COMPONENT IDENTIFICATION

A system to handle animal manure and wastes provides for the organized movement of manure from its source to its utilization or disposal. The handling system includes the machinery, equipment, structures, methods and management required to move manure or waste from the source to the utilization site. Machinery, equipment and structures are components of the handling system. Practices assist or substitute for a component. The management required to make the system work is identified by the system design.

A substantial number of animal waste handling systems have been completed in Michigan during the 1970 to 1974 period. The waste handling systems divide into three general groupings:

1   The addition of system component(s) such as a retention pond for pollution abatement by collecting lot runoff without major changes in livestock numbers. Many installations of this type were made with USDA cost sharing programs.

2   The development of a waste handling system as a part of a remodeling or expansion program which changes the methods and management practices for handling the animal wastes but utilizes much of the existing facilities and waste system components.

3   The development of an animal waste handling system as a part of a new facility.

Waste handling systems are identified as:

1   Liquid—above 82 percent moisture
2   Semi-Solid—72 percent to 82 percent moisture
3   Solid—below 72 percent moisture
4   Polluted Water—wash water and runoff from scraped lots

## SELECTION OF COOPERATORS

Fifteen farms (Table 1) are cooperating in the system evaluation study which will run for one year.

The liquid systems on the three dairy farms differ considerably. The liquid system on Farm No. 1 combines all waste into one holding pond and it is handled as a liquid. Inputs include (a) 100-cow cold free stall with waste scraped to a piston pump and pumped into pond, (b) milking unit wash water piped to pond, (c) lot and cold free stall barn scraped into pond, (d) runoff from approximately 29 000 square feet of lot and roof area, (e) waste feed and (f) bedding and rain and snow falling on pond. The holding pond has a capacity of approximately 800 000 gallons. The liquid system on Farm No. 3 is a cold free stall barn with manure and some bedding and waste feed scraped to a piston pump for underground transfer to the earthen pit. Farm No. 9 is a warm free stall barn with slotted floor alleys and holding pen. Irrigation water can be pumped through the pit to dilute and incorporate liquid waste into the irrigation system.

The two beef liquid systems are both cold units with approximately 36 ft x 160 ft slatted areas, 8 ft deep. The south side of both buildings is open and both have a covered feed drive. The slotted floor area is divided into four pits. Inputs are feces, urine, a very limited amount of waste feed, and water for agitation.

The swine units with liquid systems are all warm units. Slotted areas vary from completely slotted to 30 in. width in one farrowing unit. Depth also varies. Farm No. 1 falls into Group No. 2 which involves the development of a waste handling system as a part of remodeling or expansion with the utilization of existing systems. Farms No. 3, No. 8, No. 9, No. 10, No. 11, No. 14, and No. 15 fall into Group No. 3 which involves the development of a waste handling system as a part of a new unit.

Farm No. 2 is a semi-solid system scraping from the alleys in a new closed cold free stall dairy unit to a concrete bunker at the end of the barn. Inputs are feces, urine, some bedding and waste feed and rain and snow falling on storage area.

Three dairy units (No. 4, No. 6 and No. 7) divert polluted water into separate storage areas and handle the remaining waste as a solid. All are cold free stall units using outside lot feeding. All units had expanded the number of cows milked and two units had added structures. Farm No. 7 follows a daily haul pattern of land application but does

---

The authors are: R. L. MADDEX and T. L. LOUDON, Agricultural Engineering Dept., L. R. PREWITT, Dairy Dept., and C. H. SHUBERT, Agricultural Engineering Dept., Michigan State University, East Lansing.

| Farms | Collection | Storage | Land Application |
|---|---|---|---|
| **Liquid system** | | | |
| 1 (D-176 HD) | Scrape-pump-runoff | Earth pit - 6 months | Tank-broadcast |
| 3 (D-176 HD) | Scrape-pump | Earth pit - 6 months | Tank-broadcast |
| 8 (B-300 HD) | Slotted floor | Concrete pit - 6 months | Tank-broadcast |
| 9 (D-92 HD) | Slotted floor | Concrete pit | Traveling sprinkler |
| 10 (S-1000 HD) | Slotted floor | Concrete pit | Tank-inject |
| 11 (S-1000 HD) | Partial slotted floor | Concrete pit | Tank-broadcast |
| 14 (S-320 HD) | Slotted floor | Concrete pit | Tank-broadcast |
| 15 (S-2000 HD) | Slotted floor | Concrete pit | Tank-broadcast |
| **Semi-solid-system** | | | |
| 2 (D-100 HD) | Scrape | Concrete bunker | Chain-broadcast |
| **Solid system & polluted water system** | | | |
| 4 (D-60 HD) | Scrape-chain-stacker | Stack | Chain-broadcast |
| | Runoff and pipe | Pond | Pump and sprinkler |
| 6 (D-138 HD) | Scrape | Wood bunker | Chain-broadcast |
| | Runoff and pipe | Pond | Pump and sprinkler |
| 7 (D-100 HD) | Scrape | ——— | Chain-broadcast |
| | Runoff and pipe | Pond | Pump and sprinkler |
| **Solid-liquid separation system** | | | |
| 12 (S-1000 HD) | Pit overflow and runoff | Debris basin | Chain-broadcast |
| | Riser pipe | Pond | Pump and sprinkler |
| **Polluted water** | | | |
| 5 (D-115 HD) | Runoff and pipe | Pond | Pump and sprinkler |
| 13 (B-225 HD) | Runoff | Pond | Tank-broadcast |

D = dairy
B = beef
S = swine
HD = head

have a concrete pad that will provide two to three weeks of solid storage. Inputs to the polluted water storage are lot runoff, and milking center waste water. Inputs to solid waste are feces, urine trapped in the feces, waste feed and bedding. All three farms fall into Group No. 2.

Farm No. 12 uses a solid liquid separation system. Waste from two warm slotted swine units overflow from the pits through a pipe into a debris basin. Runoff from an outside feeding lot is flumed to the debris basin. The lot is sloped so that most of the manure flushes into the flume. A riser tube in the debris basin allows liquid to separate and flow into a holding pond. Inputs include feces, urine, waste feed, lot runoff, and rain and snow falling on the outside lot, debris basin and holding pond. This farm falls into Group No. 1—addition of components to an existing facility for pollution abatement.

Polluted water systems were constructed on Farms No. 5 and No. 13. Farm No. 5 uses catch basins to pickup lot runoff. Milking center waste water also goes into the pond. The primary interest is the volume of polluted water and the use of catch basins. Farm No. 13 is a beef feeding unit with a shelter and an outside feeding lot. In both units lots and housing area are scraped and hauled regularly. Inputs include lot runoff and rain and snow falling on pond. Both units fall into Group No. 1.

Eight dairy farm cooperators were chosen for the investigation of component characteristics and management practices. Four have alley scrapers automatically controlled. Three use tractor scrapers and one cooperator has identical barns, one scraped with an alley scraper and one with a tractor and blade. Table 2 gives the information on alley size and scraping time.

## WASTE VOLUME

Waste volume has generally been greater than allowed for in the design criteria. Table 3 shows the design capacity for storages on six cooperating farms and the performance capacity on the same farms with systems in operation. There would appear to have been an error in the design values used in calculating the waste volume on Farm No. 2. Failure to meet design capacity on the other farms could be



| Type | Alley | Frequency | Time per alley | Control | Scraped to |
|---|---|---|---|---|---|
| A (CFS) alley | 184 ft x 9 ft (2) | Continuous morning | 30 min. | Continuous Operator | End-cross conv. slab - |
| 40 HP Tr - 36-in. bucket | 184 ft x 12 ft (2) | | ——— | | |
| B (CFS) alley | 236 ft x 10 ft (4) | 60 min. | 18 min. | Time clock | End-cross conv. |
| C (WFS) alley | 200 ft x 11 ft (2) | 35 min. | 20 min. | Time clock | End tank |
| D (WFS) alley | 176 ft x 11 ft (2) | 40 min. | 23 min. | Time clock | 2 ends-tanks |
| E (WFS) 60 HP Tr - 6-ft blade | ———(2) | ——— | ——— | Operator | Center-cross conv. |
| F (WFS) 14 HP Tr - 42-in. blade | 184 ft x 10 ft (2) | 2 x day | 20 min. | Operator | Center-cross conv. |
| G (WFS) alley | 105 ft x 10 ft (4) | 3 hours | 14 min. | Time clock | End-cross conv. |

CFS = cold free stall
WFS = warm free stall

TABLE 3. EXPERIENCED CAPACITY COMPARED TO DESIGN CAPACITY FOR ANIMAL WASTE STORAGES.

| Farm | Type | Design Capacity | Performance capacity |
|---|---|---|---|
| #11 - Swine finishing - 600 HD | Liquid | 3 months | 2.5 months |
| #11 - Swine nursery - 400 HD | Liquid | 6 weeks | 5 weeks |
| #9 - Dairy - 90 cows | Liquid | 6 months | 4.5 months |
| #2 - Dairy - 120 cows | Semi-solid | 6 months | 3 months |
| #14 - Beef - 320 HD | Liquid | 6 months | 4.6 months |
| #8 - Beef - 300 HD | Liquid | 6 months | 4.5 months |

from errors in estimating animal weights used as a basis for determining manure production, failure to accurately calculate the space available for storage or by using incorrect design data.

Waste accumulation in several pits under slotted floors have been measured through the winter months. Cooperating farmers were asked to measure and record pit levels on a weekly basis. Along with the depth measurements he records the number of animals contributing to the pit and the weights of the animals which is usually a guess. The key to determining manure production accurately has been minimizing measurement uncertainty or error. Considering animals in weight groups, better identification of numbers in pens, using the same measurement site each time and changing measurement scale for easier resolution of weekly increase in depth are helping with data accuracy. The increase in pit depth is converted to pounds per animal unit.

Figs. 1, 2 and 3 show the measured values and a range of manure production value cited in the literature. The dot in the center of the bar indicates calculated values based on information recorded by the farmer. The bar indicates the uncertainty range for measurement experienced by project investigators with the first probe rods used. The shortened error bar results from improved readability of measure probes and measuring techniques.

Measured volumes of waste production in Figs. 1, 2 and 3 are generally greater than figures given in the literature (Loehr 1974 and Miner 1971). However, the values given in the literature do not include spilled water or feed and the measured values do. There is no apparent difference in operation between the two swine finishing units. Estimation of hog weights could be a factor. Waste water goes into the pit in the dairy unit. Measurements in two slotted floor beef units showed essentially the same results of measured values in relation to literature figures for manure production for beef fattening animals.

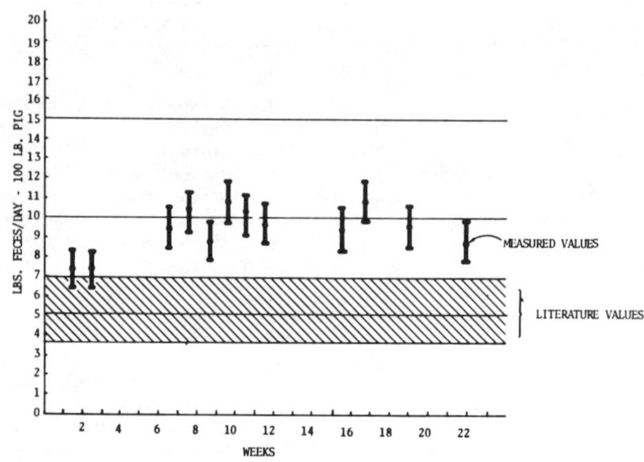



FIG. 2 Swine waste accumulation in slotted floor finishing unit—farm No. 11—10/17/74 to 3/6/75.

Data accumulation is limited but at this point it appears that the criteria of 1/8 cubic foot per day per pig, 1 cubic foot per day per head of beef or 2 cubic feet per day per cow are minimums and possibly below the storage capacity needed unless these values are closely correlated with animal weight and inputs other than feces and urine are minimized.

## SYSTEM MANAGEMENT

The biggest problem for most cooperators is the "how, when and where" of land application. A summary of system planning indicates that:

1    Land area for waste application was not identified except in very general terms.

2    Considerations of slope, soil types, crop growth and cover was not emphasized in relation to problems of system management for land application.

3    Alternate waste handling methods were usually not considered for a breakdown in the management scheme that might occur due to extended poor weather conditions, a breakdown of one or more components, an increase in livestock land or change in the quality and amount of farm labor.

The general attitude of most cooperators in the project places handling of animal waste in the chore classification. The primary goals are meeting regulatory requirements and

(Continued on page 111)

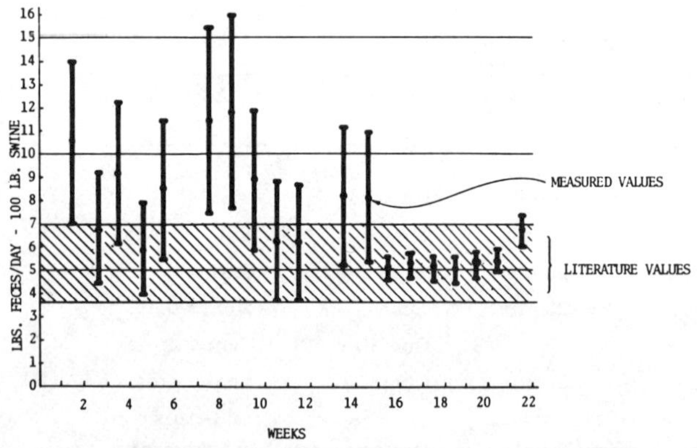



FIG. 1 Swine waste accumulation in slotted floor finishing unit—farm No. 10—10/13/74 to 3/9/75.

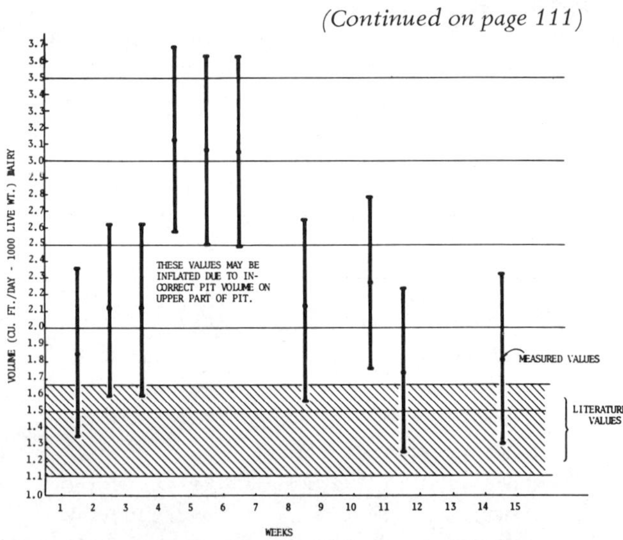



FIG. 3 Dairy waste accumulation in slotted floor dairy free stall barn for milking herd—farm No. 9—10/7/74 to 1/6/75.

# Large Piston Manure Pumps and Outside Manure Storages (Earthen Basins)


R. E. Graves
ASSOC. MEMBER
ASAE


MANY midwest dairymen handle and store manure from free stall dairy barns in a semi-liquid or liquid form. Traditionally a concrete tank under the barn or barnyard has been used. Manure is either scraped to this tank by a tractor scraper or barn cleaner, or in the case of slatted floor barns, it falls directly into the tank after being worked through the slots. At appropriate times the contents of the tank are agitated and pumped out for land spreading.

Problems encountered with handling and storing manure in underground concrete tanks include:

1   High cost of constructing below ground concrete tanks.

2   Size of tanks are usually limited by building configuration and dimensions and available pump lengths (3 m (10 ft) is most common pump length). This limits storage to 6-8 months with most layouts.

3   Gases and odors, especially during mixing and cleaning can be objectionable and sometimes hazardous in livestock barns.

4   Barns with slatted floors over the tanks eliminate manure scraping but are often plagued with ventilation problems.

5   Milking parlor pit drains are often lower than the barn floor. This results in loss of capacity if the drain is connected to the liquid manure tank.

## EARTHEN BASINS

Earthen storage basins have been successfully used for storing liquid manure (Converse et al. 1974). The cost advantage of building these is significant. Costs for basin construction as low as $60 per cow for one year's storage for 100 cows have been reported. Many farmers plan for one year's manure storage. This allows flexibility in planning cleanout with other field operations. On heavier soils many farmers like to spread manure prior to fall plowing to minimize soil compaction from spreader wheels.

It is important to remember that these storage basins, sometimes referred to as storage lagoons, are not designed or managed to provide degradation or treatment of manure. Some separation occurs depending on the amount of waste water added to the basin. This separation usually results in formation of a scum mat which in warm weather will dry on the surface. The scum mat serves to reduce odors, but may be difficult to break up during cleanout.


The author is: R. E. GRAVES, Assistant Professor and Extension Agricultural Engineer, Agricultural Engineering Dept., College of Agricultural and Life Sciences, University of Wisconsin, Madison.


## HOLLOW PISTON PUMP

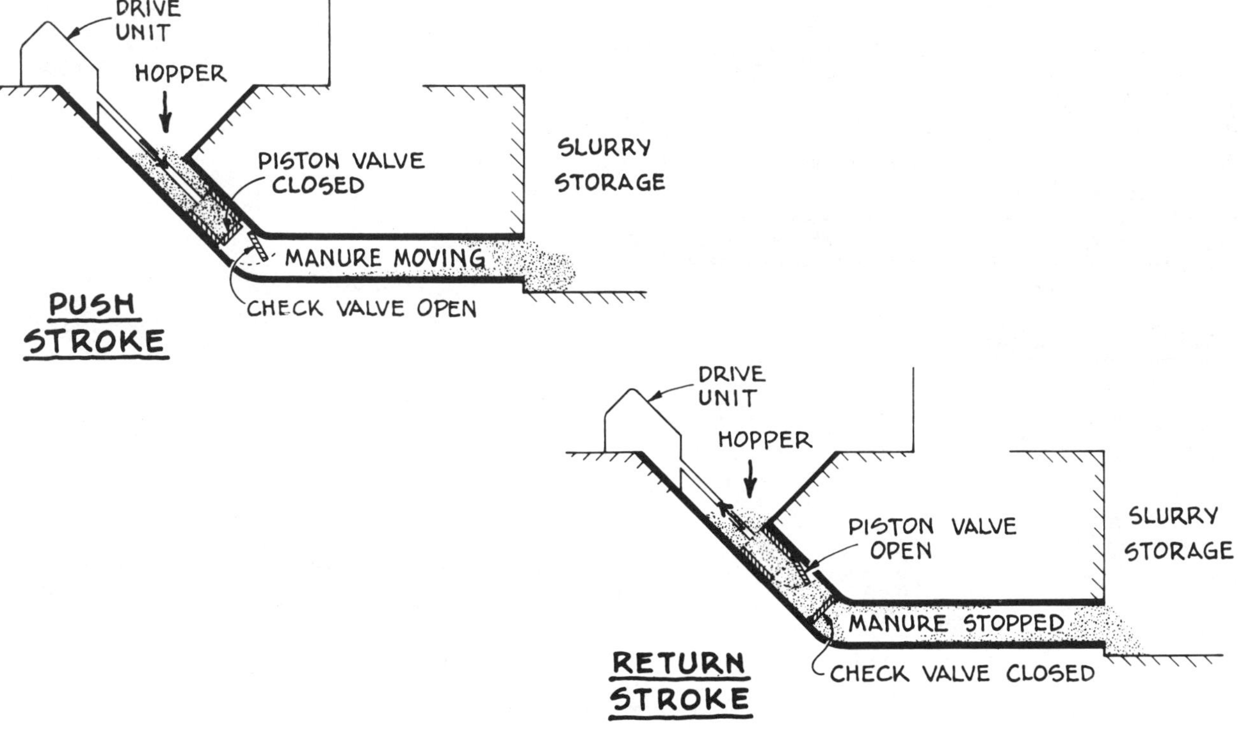



FIG. 1 Hollow piston manure transfer pump. Typical piston size 20 cm square, stroke lengths 30-46 cm, 25-45 strokes per min.

## FILLING STORAGE BASINS

Various methods are used for filling storage basins. Some methods add the manure from the top while others allow the new manure to be added beneath the surface. Many feel that adding manure from the bottom reduces fly and odor problems since the fresh manure is not exposed to the air (Gojmerac 1972). This also reduces problems of freezing. Adding manure from the top often results in a pile up or back up around the loading point. This problem is common during cold weather when the manure is stiffer and does not flow.

Methods used for loading are summarized as follows:

1   Barn cleaner extension - has limited flexibility; storage unit must be adjacent to the barn and requires top loading. Freezing of machinery and manure may be a problem.

2   Tractor scraper or loader - has limited flexibility; storage unit must be adjacent to the barn and requires top loading.

3   Long augers - are used underground for bottom loading. They are not commercially available for this use and limit location of the storage basin.

4   Conventional liquid manure pumps - may be used with a small sump or tank to collect manure. At appropriate intervals the contents are pumped to the storage basin. Top or bottom loading is possible. If the sump is located outside, freezing of the pump may be a problem.

5   Large hollow piston manure pumps - move manure through 23-38 cm (9-15 in.) PVC or steel pipe to the bottom of a storage basin (Fig. 1). These units are commercially available. Manufacturers' recommendations for pumping distance vary from 46 m (150 ft) to 92 m (300 ft). The pumps utilize a hollow piston about 20 cm (8 in.) square, with a 30-46 cm (12-18 in.) stroke.

These pumps are located with the pumping unit at a diagonal pumping down. This allows the drive unit to be located above the floor. The connecting rod runs in the loading hopper and sometimes aids the flow of material to the pump; it also can be fouled by buildup of manure around or under it. Various modifications have been made in the connecting rods to overcome this problem. Both mechanical and hydrualic drive units are used.

Typical pumping rates vary from 25 to 45 strokes per minute. These pumps have relatively loose fitting pistons and flapper type valves on the piston and pump body. Since all material must flow through the pump and valve mechanism, long, stemmy materials such as hay or long straw are not recommended. Installations should include a means of adding water during pumping if manure is not very soupy. One method often employed is to collect milking center waste water. This waste water is pumped into the pump hopper when the manure pump is running.

6   Large solid piston manure pumps - move manure through 23-38 cm (9-15 in.) PVC or steel pipe to the bottom of a storage basin (Fig. 2). These units are commercially available and were originally imported from Europe. The pump has a solid piston about 25 cm (10 in.) in diameter. A hydraulic drive unit is used to operate the pump piston. The pump is installed horizontally at the bottom of a concrete pit with a hopper above it. This pit is typically 1.8-3.0 m (6-10 ft) deep and must be kept dry. The piston is pulled back to allow manure to drop in front of it. On the forward stroke the manure is pushed into the pump cylinder and through a spring loaded check valve. The cylinder opening contains a close fitting die. This die and the edge of the piston serve to shear off any material not completely into the pump. This feature allows the pump to move material with long straw or hay if it can be made to flow down the hopper. The close fitting cylinder creates a vacuum within the cylinder on the return stroke. A phenomenon similar to

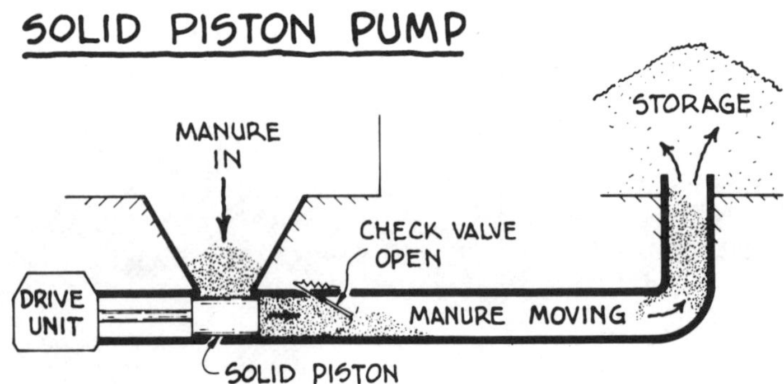


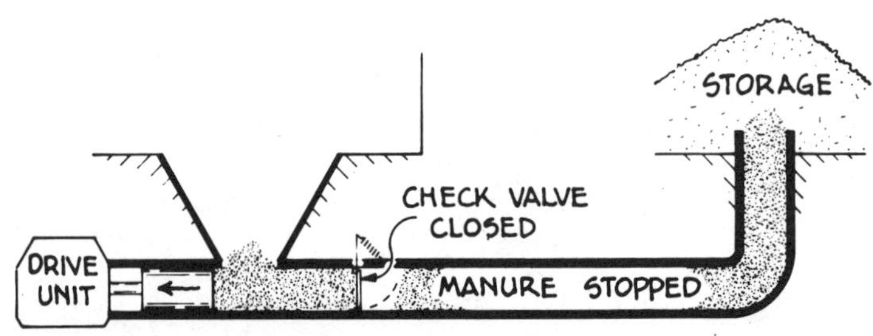


FIG. 2 Solid piston manure transfer pump. Typical piston size 25 cm diameter, stroke lengths 60-90 cm, 5-10 strokes per min.

water hammer sometimes occurs when liquid in the hopper is suddenly pulled into the cylinder.

A more common use of this pump is with stanchion barns where long bedding is used and the manure is stored and handled by traditional solid handling methods.

Both types of large piston pumps have been in use in Wisconsin since 1972. General observations indicate that pump performance improves as manure consistency becomes more liquid. As length of pipe increases, additional water is usually recommended to aid in manure flow and reduce pipe friction. Problems with pipe breakage and joint separation have occurred. Very high pressures can be obtained by these pumps. Information regarding pipe friction or pumping pressures is not readily available if known. Also, many farm construction crews are not experienced in proper installation of large diameter sewer or water pipe.

## BASIN DESIGN

Most storage basins can be classified as one of the three following types:

### Type I

These are rectangular shaped basins with one long vertical wall usually concrete (Fig. 3). a 2.4-3.0 m (8-10 ft) wide strip of the bottom along this wall is also paved. A conventional liquid manure pump can be used from any point along this wall for agitation and pumping (Fig. 4). The remaining sides are usually earthen as is most of the bottom. This is usually the most costly type to build.

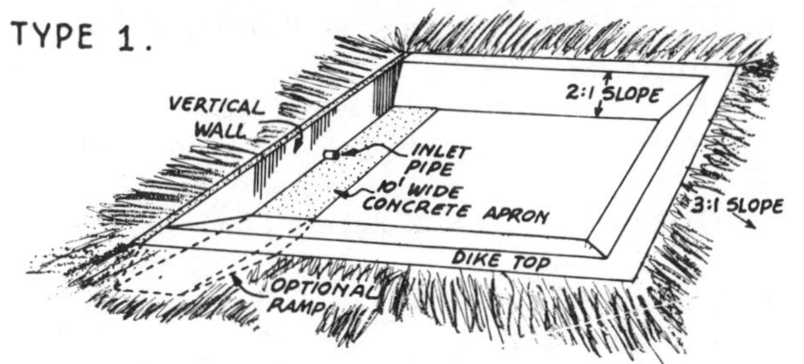


FIG. 3 Type I manure storage basin. Agitate and pump with conventional liquid manure pump along wall.

FIG. 4 Agitating Type I storage basin with conventional liquid manure pump.

### Type II

These are circular or rectangular shaped earthen basins with pumping platform(s) or dock(s) (Fig. 5). Agitation and pumping is done with a conventional liquid manure pump from the platform(s) (Fig. 6).

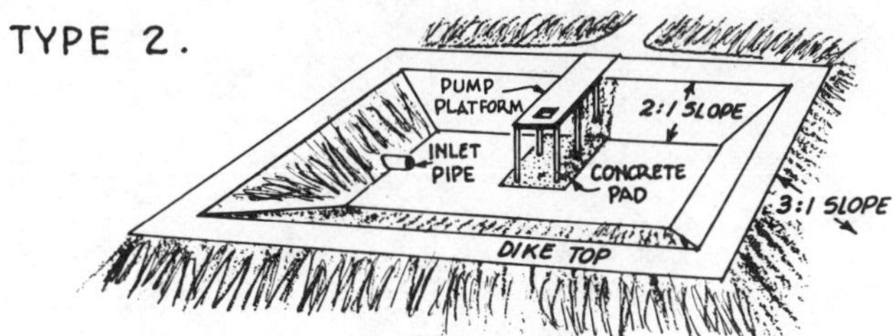


FIG. 5 Type II manure storage basin. Agitate and pump with conventional liquid manure pump from pump platform.

FIG. 6 Type II storage basin with pumping platform and scum mat. The picture is looking down access ramp.

The bottom of the basin may be concrete, although this is probably not necessary. A pad of concrete 3.0-6.0 m (10-20 ft) square where the pump is placed is advisable to protect the bottom from scouring by the pump and keep rocks or debris from being drawn into the pump. The bottom below the pumpout area can also be dug about 0.6 m (2 ft) deeper than the pump.

### Type III

These are circular or rectangular shaped basins with ramp(s) or driveway(s) into them (Fig. 7). Agitation and pumping is done with a modified liquid manure pump (Fig. 8). The pump has no right angle gear box and is mounted horizontally from the three point hitch of a tractor. The tractor is backed down the ramp until the rear wheels of the tractor are just into the manure. The pump extends another 3.0-4.3 m (10-14 ft) into the basin.

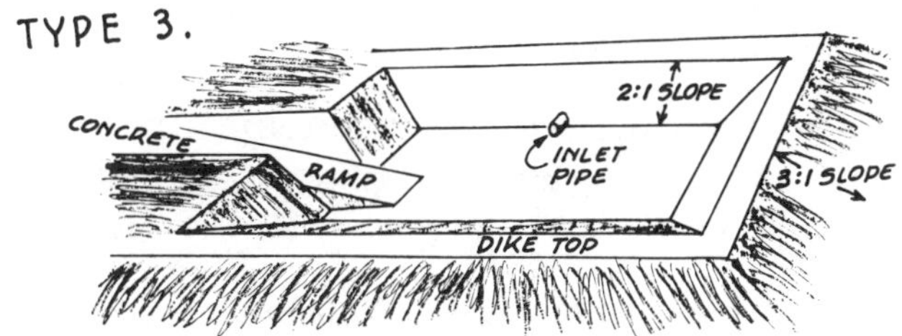


FIG. 7 Type III manure storage basin. Agitate and pump with modified 3-point hitch mounted liquid manure pump backed down ramp.

For loading spreaders, a pipe is extended over the front of the pumping tractor or a long hose is run to a tripod or standpipe located outside the basin. As the manure is pumped from the basin, the pump unit is backed further down the ramp (Fig. 9).

FIG. 8 Modified 3-point hitch mounted liquid manure pump. Note recirculation hood and fill pipe extending over tractor.

FIG. 9 Liquid manure pump on concrete access ramp in storage basin. Note recirculation hood above liquid level and large hose leading to standpipe (out of sight) for filling manure spreader.

If the sides of the basin are not too steep, the pump can be used at other places around the basin for extra agitation. The ramp should be concrete especially if tank wagons are backed down it for filling.

## GENERAL DESIGN RECOMMENDATIONS

1   One year storage capacity provides greatest flexibility in timing field application and may not cost much more to build.

2   Provide extra capacity for rain water and dilution water. Total volumes of 0.07-0.08 m$^3$ (2½-3 cu ft) per 454 kg (1000 lb) cow per day storage are common in Wisconsin. Allow an extra 60 cm (2 ft) of depth for freeboard.

3   Decide on method to be used for emptying storage during design and provide for it. (i.e. dock, ramp, etc.)

4   Evaluate site and soil conditions carefully to assure protection of ground and surface waters.

5   Soil characteristics will affect how steep the banks can be formed. Follow accepted construction methods for sealing, keying dikes, fill lifts, and bank seeding. In general,

steeper banks conserve space and less manure hangs up on them as the basin is emptied. Side slopes of 2:1 to 4:1 (run:rise) are common for the inside banks.

6   Prevent excess surface drainage from entering the basin.

7   Access ramps into ponds should not be too steep, especially if manure spreaders are to be operated on them. Slopes of 8:1 to 10:1 are common. The surface of these ramps should be very rough to provide good traction when wet. Deep groves or ridges can be formed across the ramp with a garden rake or silage fork before the concrete sets.

8   Depth of the basin is affected by site groundwater conditions and emptying equipment. Basin types I and II are usually constructed with a maximum depth of 3.0-3.7 m(10-12 ft). This allows cleanout with readily available liquid manure pumps and leaves some freeboard. Type III basins can be deeper if site conditions permit since the pump is backed down into the basin as cleanout progresses.

9   Length and width of the basin are affected by agitation. 18-24 m (60-80 ft) seem to be distances which can be agitated with minimum trouble. If agitation distances exceed 30 m (100 ft), plan for additional agitation locations.

10   Fence the area to exclude people and livestock when completed.

## LOCATION

Points to keep in mind when planning and locating a manure storage basin include:

1   Distance pump or other loading device can move manure from barn.

2   Required distances from wells (usually 76 m (250 ft)), streams, buildings, highways, property lines and flood plain.

3   Groundwater, bedrock and other unusual soil conditions.

4   Farmstead expansion.

5   Prevailing winds.

6   Access for cleanout operations.

## EMPTYING BASIN

Various problems have been encountered in emptying storage basins. For the most part these have resulted from inadequate agitation. Field experiences vary from very short agitation (less than ½ hour) to several hours of agitating. Careful observation of the basin contents during agitation and pumpout is required. It may be necessary to agitate from several locations if the crust is heavy. In general, once the contents begin to swirl or move it is safe to begin removal of material.

1   If a heavy crust is present be careful not to pump liquid out from under it before sufficient agitation is complete.

2   "Throwing" slurry on top of crust during agitation may be better than agitating below the surface.

3   Agitation which will "start the crust moving or swirling" is helpful in breaking up the crust.

4   If a ramp is used to pump from, consider a stationary standpipe for filling tank wagons. Eliminates backing down ramp.

5   If irrigation is considered allow extra room for additional dilution water.

## CONCLUSION

Experiences with earthen storage basins and large piston manure transfer pumps in Wisconsin have been favorable. Commercially available large area hollow piston and solid piston pumps are used to move manure through underground pipe to storage basins. Careful management and observation allows agitation and pumpout of basins if agitation distances are about 18-24 m (60-80 ft). Properly located and constructed, the basins do not appear to be an undue threat on ground or surface water quality. Controlled studies of pump or basin performance were not made.


**References**

1   Converse, J. C., C. O. Cramer, H. J. Larsen and R. F. Johannes. 1974. Storage lagoon versus underfloor tank for dairy cattle manure. ASAE Paper No. 74-3028, ASAE, St. Joseph, Mich. 49085.

2   Gojmerac, W. L. 1972. Manure handling as related to fly (house and/or stable) control. Final Report - Project 1763. University of Wisconsin, College of Agricultural and Life Sciences.


---

## Evaluation of Waste Handling Systems
*(Continued from page 106)*

labor reduction. This attitude implies that greater importance and emphasis must be given to the management scheme of an animal waste handling system if it is to have the same priority as other production inputs.

### OTHER FINDINGS

The average values from analysis of waste samples taken from liquid storages and ponds have fallen generally within values reported for total solids, volatile solids and total N and P. However, these have considerable variation between samples taken from the same systems and between systems. More work is being done on sampling methods and analysis.

Temperatures of the wastes in storage are being recorded using recording and probe thermometers. Temperatures taken approximately 18 in. from the bottom of the storages were in the mid-thirties during January, February, and early March for the semi-solid storage Farm No. 2 and for the cold slotted floor beef units, Farms No. 8 and No. 14.

The most popular size for liquid tank spreaders is 3000 gallons plus. Farm cooperators use and feel they need a 100 HP plus tractor to handle these spreaders. Time per load has ranged from 8 minutes to 40 minutes. The average is approximately 25 minutes a load.

Component selection is often made on a basis of reducing investment cost or hopefully to reduce labor without carefully considering how this decision might influence or be influenced by the total system operation. The component study of scrapers (Table 2) has identified several problems. Poor construction practices of alley floors and curbs have caused operation problems for alley scrapers. Alley offsets for waterers and stalls increase hand labor. Poor location and design of cross conveyors have resulted in pile up of wastes at transfer points and plugging. Scraping distances of more than 140 ft to 160 ft cause overloading and spillage of waste resulting in poor cleaning.

### SUMMARY

Generally, the design criteria and management schemes used for animal waste handling systems is getting the job done for most farmers. At the same time, most cooperators have gone through a period of trial and adjustment to make the system work. Better planning, better identification of management requirements and more attention to details of components selection and installation could have reduced total labor, and contributed to the installation of handling system more compatible with the total farm operation.


**References**

1   Loehr, R. C. 1974. Agricultural Waste Management, Academic Press, New York, 520 pages.

2   Miner, J. R. (Editor). 1971. Farm Animal-Waste Management, North Central Regional Publication, 12 pages.

# Milking Center Waste Management


H. D. Bartlett, A. E. Branding, L. F. Marriott, M. D. Shaw
MEMBER                                    MEMBER
ASAE                                      ASAE


MILKING center sanitation requires large amounts of water for cleaning the milking equipment and facilities. An irrigation system offers an effective method for disposal of this water with minimum contamination of ground or surface water. However, inherent in the cleaning operation is the removal of substantial amounts of manure. This precludes the use of small high-pressure pumps and relatively small irrigation nozzles which are required for farm application, unless a reliable and practical method is employed for separating the solids from the liquid.

A system for separation of the manure solids and an automatic irrigation system for disposal of the liquid effluent was developed for a 150-cow milking center at The Pennsylvania State University. The milk-room and milking parlor wash water was collected in a sump and pumped into flush tanks by a low pressure, solids handling pump. This water was used to flush the holding area and cow-return alley (Fig. 1). Additional fresh water was available, as needed, for the flush tanks. The flush water and manure were discharged by gravity to a separation bed where the solids were retained, and the liquid portion was released to a float-controlled high pressure pump for irrigation.

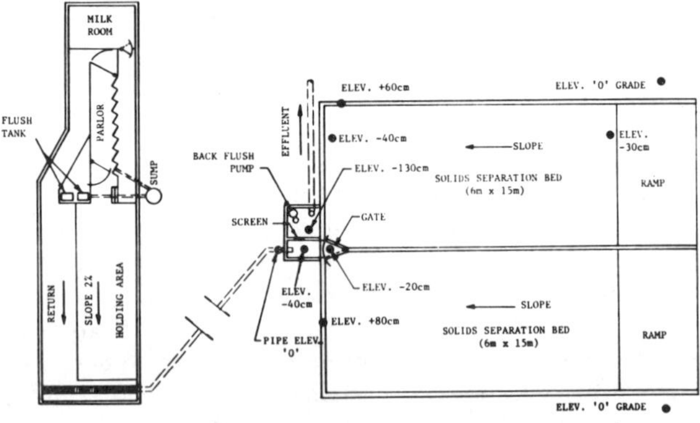


FIG. 1 Layout of milking center and solids separation bed design.

## SOLIDS SEPARATION BED DESIGN

The principle of the separation bed operation was based on distributing the flushed waste over a large area and allowing the liquid to drain off through a screened outlet, leaving the solids distributed relatively uniformly over the bed (Fig. 1). The screened outlet was located near the inlet pipe so that the velocity of incoming waste water flushed the solid materials (manure) away from the screen. The liquid discharged through the screened outlet allowed the

liquid on the bed to drain back from the bed through the screen to a sump.

Preliminary investigation indicated that the flush water would carry the solids a distance of 15 M over the bed. Based on a daily solids production of 0.3 m$^3$ and a 30 cm maximum depth of solids accumulation, a bed of 90 m$^2$ was required to provide 90 days storage.

The bed was constructed in two sections so that one section could be filled while the other side drained. Flow to the respective sections was controlled by a diversion gate, as shown in Fig. 1.

The experimental separation bed was not paved. However, pavement is necessary to prevent seepage, and to facilitate tractor operation for cleaning out the solids.

## SCREEN DESIGN

A screen was required to prevent solids, especially floating material, from being discharged to the irrigation system. The screening material used was galvanized hardware cloth, approximately 15 cm square with 8 mm openings. Frequent removal of solid material which collected against the screen was necessary to prevent blockage. Screen cleaning was accomplished by the use of a time operated pump to backflush the screen at frequent intervals. The pump was a household-type sump pump, installed in the sump at the outlet side of the screen. It was controlled by an interval timer which caused the pump to operate for ten seconds every ten minutes. No freezing problems were encountered with the screen and backflush system when the temperature dropped to -10 C.

## SOLIDS REMOVAL

Accumulated solids in the separation bed were similar in consistency to barnyard manure, and could be handled with a tractor front-end loader. When the solids were allowed to drain for several days during good drying weather they could be handled with a conventional spreader. However, solids which were not well drained were sloppy and require a flail-type tank spreader for satisfactory handling.

Normal operation of the bed did not create fly or odor problems. However, during the cleaning operation the odor was quite strong, but it was not the pungent lingering odor associated with liquid manure pits. When the drained solid material was spread, it had the aroma and consistency of humas.

The repeated flushing of the bed removed most of the soluble material and fine suspended solids, leaving mainly cellulose and fiber. Samples of the accumulated solids showed an analysis of 14.1 percent dry matter with 1.67 percent nitrogen; the effluent analysis was 0.2 percent dry matter with 5.67 percent nitrogen.

## EFFLUENT DISPOSAL

The screened effluent from the separation bed drained


The authors are: H. D. BARTLETT, Professor, Agricultural Engineering Dept., A. E. BRANDING, Assistant Professor, Dairy Science Dept., L. F. MARRIOTT, Associate Professor, Soil Fertility Dept., and M. D. SHAW, Associate Professor, Agricultural Engineering Dept., Pennsylvania State University, University Park.

into a pit, from which it was pumped through a sprinkler irrigation disposal system located in a permanent pasture. A float controlled 3.73 kW, electric pump with a capacity of 2.2 l/s at 4.1 bars operated the irrigation disposal system, consisting of 3 permanently located sprinkler outlets. Only one sprinkler was in use at a time and each of the three sprinklers was operated for approximately the same length of time during each week. Each sprinkler covered a circular area 30 m diameter (approximately 0.07 ha) which resulted in an effluent application of 75 mm per week.

No adverse conditions were noted, except during extremely wet weather, when some water stood temporarily in depressions. Although the area was sloping, there was no perceptable runoff.

The $BOD_5$ and COD of the effluent were approximately 800 mg/l and 3000 mg/l, respectively. This caused no problem in the pasture, but is too high for discharge to a watercourse in Pennsylvania. Soil nitrogen levels in the top 1.2 m of the soil profile increased significantly over the two-year period. This confirms other animal waste disposal studies that annual nitrogen application rates from animal wastes should not exceed the nitrogen uptake by the crop. (Bartlett and Marriott 1970).

## EFFLUENT STORAGE

The effluent disposal system described above requires daily operation of the irrigation equipment throughout the year. In order to avoid the necessity of irrigating during freezing temperatures or other inappropriate times, a holding pond was constructed to provide temporary storage and to observe odor intensity and problems of pumping the effluent after an extended period of storage. The pond, 12 m by 22 m by 1 m deep, had the capacity to store the effluent for about twelve days. Effluent held in the pond for several months did not create objectionable odors, except during brief periods of hot weather, and did not cause any problem with the pump or sprinkler operation. However, solids accumulation can be expected with an ongoing system and some agitation may be required to mix them into suspension.

**Reference**

1    Bartlett, H. D. and L. F. Marriott. 1970. Subsurface disposal of liquid manure. Proceedings of The International Symposium on Livestock Waste, ASAE, pp. 258-260.

This research was supported in part with funds received from the Pennsylvania Fair Fund administered by the Pennsylvania Department of Agriculture.

Authorized for publication as Journal Series Paper No. 4849 of the Pennsylvania Agricultural Experiment Station. Approved April 17, 1975.

# Waste Management at Hall Brothers Dairy


Harold Watson, Harvey E. Hamilton, Donald Hall, Tom McCabe

MEMBER
ASAE


REMOVAL and collection of manure from dairy facilities by hydraulic transport has increased in popularity throughout the South. The system offers several advantages such as low labor, good odor control, and ease of operation but it also poses some new problems such as high water usage, the possibility of excess discharge of effluent to streams, additional facilities for pumping and solids removal, and additional lagoon capacity to maintain an acceptable retention time for treatment.

This study was initiated to evaluate the effectiveness of a flushing system with a screen for solids removal and three lagoons in series which had been in operation for about 2 years. Samples were collected during the fall and winter of 1974-75.

## DESCRIPTION OF FACILITIES

Hall Brothers Dairy constructed a 1200-cow totally confined dairy facility about 10 miles south of Montgomery, Alabama in 1969 in order to consolidate twelve separate company owned dairies. The milk is processed and distributed by the company in the central Alabama area.

The milking parlor, holding pens, milk room, observation room, mechanical room, hospital area, veterinary room and office are located under one roof. Three milking pits, each with four double side opening stalls, in the milking parlor allow 12 cows to be milked simultaneously. A gate arrangement allows the lower end of each of the three holding pens to serve as a 3,000-gallon reservoir for flushing the lanes in the housing area. The parlor is equipped with six 280-gallon flush tanks for cleaning the milking stalls, pits, lanes and holding pens.

The facility is designed to house the cows in groups of one hundred as shown in Fig. 1. The free-stalls are of conventional construction and are covered with a metal roof. Insulation has been added to the underside of some of the roofs. Rain water flows from the roofs onto the cow lanes and must be pumped to the lagoon system. The feed bunks and waterers are designed for flushing. The truck lanes are not flushed but rain water from these lanes must be handled. The 12-cow lanes adjacent to the free-stalls and feed bunk are flushed each time the cows from that lane are milked. The sequence of events is to move the group of

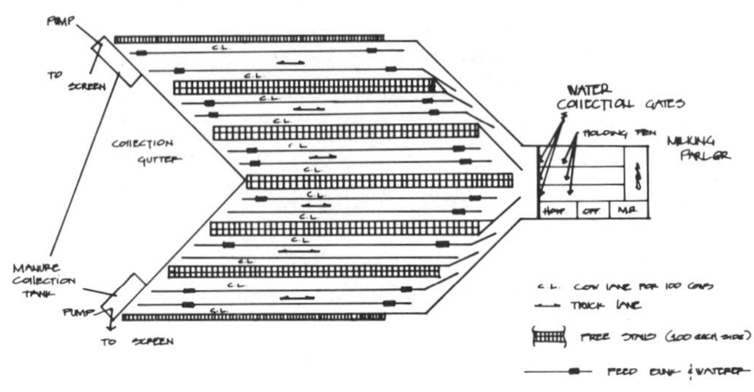


FIG. 1 Milking and housing facility for 1200 cows.

cows to the holding pen, raise the reservoir gate, start water running into the reservoir, milk, flush the parlor and holding pen, and flush the cow lane in the housing area. The slope of the cow and truck lane is 3-percent with no lateral slope.

The manure laden flush and rain waters are collected in two 8,000-gallon tanks located at either side of the housing area. The tanks are equipped with pumps designed to agitate as well as remove the tank contents. The original design included an irrigation system for direct land spreading. This system was not adequate due to the small holding capacity of the tanks (16,000-gallons) in relation to the daily designed consumption for flushing (72,000-gallons) and the high average yearly rainfall rate of 52.60-inches per year. The tanks had to be emptied a minimum of five times daily but the pump ran continuously when rainfall occured. Field application during rainfall events was neither desirable nor practical.

Additional waste treatment facilities were completed and placed into operation in December 1972. These facilities consisted of a 6-foot wide, 300 gpm capacity Bauer Hydrasieve screen with a 0.040-inch openings and three lagoons as shown in Fig. 2. The size and capacity of the lagoons are shown in Table 1. The screen is elevated about 12-feet above a concrete solids collection floor. The manure is pumped at least four times daily from the collection tanks to the screen. The liquid separate flows by gravity to the first lagoon. A sprinkler type irrigation system which will remove effluent from the second or third lagoon cell to surrounding crop land is being planned.

Water is supplied to the facility by two deep wells utilizing 10-horsepower submersible pumps. Presently, all water being used for flushing is fresh water.

Since the lagoon system was installed, a high of 830, low of 608 and an average of 720.6 cows were maintained in the facility. An average of 700 cows were milked during this study.


The authors are: HAROLD WATSON, Specialist, Agricultural Engineering, Alabama Cooperative Extension Service, Auburn University; HARVEY E. HAMILTON, Associate Professor, Agricultural Engineering Dept., College of Agriculture and Agricultural Experiment Station, Auburn University; DONALD HALL, Owner and Manager, Hall Brothers Dairy; and TOM McCABE, County Extension Chairman, Montgomery County, Alabama Cooperative Extension Service, Auburn University.

Acknowledgements: The authors express appreciation to the following for their assistance in this study: Dr. Joseph P. Judkins and Dr. J. M. Morgan of the Civil Engineering Dept., Auburn University, for making their laboratory available for sample analysis; Dr. W. B. Anthony, Agricultural Experiment Station, Auburn University, mineral analysis; Mr. Julius Kendrick and Mr. Cecil Medders, Alabama Power Company, power requirements for water wells; and Mr. Marvin Jones and Mr. Charles Jones, Hall Brothers Dairy, sample collection and other assistance during the course of the study.

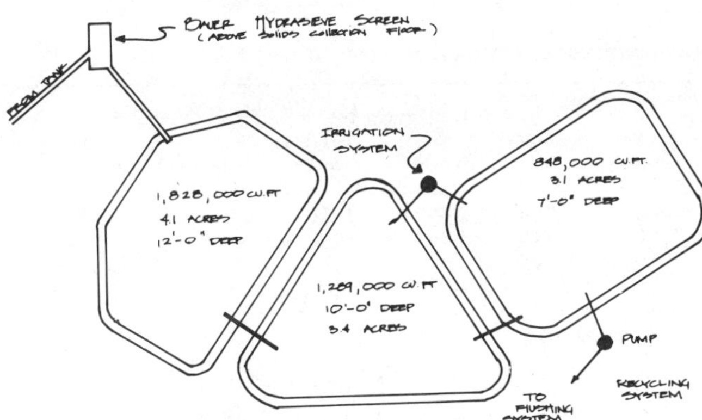


Fig. 2 Screen and three lagoon system.

## SAMPLE COLLECTION AND ANALYSIS

The first and second lagoons were sampled at the overflow well. The effluent flowed from below the surface into a well which discharged below the surface of the next lagoon. Grab samples were taken to accumulate a total sample size of one gallon. The third lagoon was sampled at the surface near the overflow pipe.

Samples of solids removed by the screen were collected as the solids fell from this screen.

Samples were also taken of the manure in the collection tanks prior to being pumped to the screen and another sample was taken of the liquid discharge from the screen as it entered the first lagoon.

Total solids, biochemical oxygen demand (BOD) and chemical oxygen demand (COD) were measured in accordance with Standard Methods (Ref. 4). The suspended solids test by Standard Methods was modified by adding a thick fiberglass pad on top of the fiberglass filter disk in a Gooch crucible. Although the thick pad might trap some dissolved solids, the sample size was increased by about ten times to allow more accuracy in measuring. The volatile solids procedure was modified to allow for the thick pad by increasing the time in the oven to 2 hours for the Gooch crucible and over-night for the solids samples removed from the screen. This procedure assured that a constant weight was reached in the muffle oven.

## RESULTS AND DISCUSSION

### Water Usage

The design water input for manure flushing was 72,000-gallons per day or about 80.4 acre-feet per year. The actual water input was estimated by measuring the total wattage consumed by the pumps over a 1-month period. The measured input was 56,000-gallons per day or about 62.6 acre-feet per year. This reduction is due primarily to the practice of using only well water for flushing and a resulting deterioration of the water system caused by mineral deposits and restriction to water flow.

#### TABLE 1. LAGOON DATA

| Lagoon | 1 | 2 | 3 |
|---|---|---|---|
| Surface area, acres | 4.1 | 3.4 | 3.1 |
| Volume, acre-feet | 42.0 | 29.6 | 19.5 |
| Depth, feet | 12 | 10 | 7 |
| Daily loading rate, lb. VS/1000 cu. ft.* | 1.9 | --- | --- |
| Design retention time, days | 161 | 116 | 77 |
| Actual retention time, days | 198 | 142 | 96 |

*pounds volatile solids per 1000 cubic feet lagoon volume

#### TABLE 2. PRESENT FLUSHING AND LAGOON WATER BALANCE.

| | Acre-feet per year | | |
|---|---|---|---|
| Lagoon | 1 | 2 | 3 |
| Total water input flushing | 62.6 | | |
| Pond evaporation | 14.0 | 11.6 | 10.6 |
| Rainfall on lagoons | 18.0 | 14.9 | 13.6 |
| Feedlot runoff | 16.7 | | |
| Water flowing through lagoons | 83.3 | 86.6 | 89.6 |

The average yearly rainfall of 52.60-inches and pan evaporation of 58.47-inches per year influence the water balance of the lagoon system as shown in Table 2. Pond evaporation is estimated to be 70-percent of pan evaporation. These data are for all fresh water being used for flushing. The water flowing through the third lagoon is available for irrigation. Recycling the water for flushing the housing area will reduce this volume. Another important factor for this system is the large volume of water which must be pumped over the screen. This amounts to the water used for flushing plus the runoff from rain water or 97.3 acre-feet ($25.9 \times 10^6$ gallons) per year.

As a result of this study, a wastewater recycling system for flushing is being planned. The fresh water required by the milking parlor flush tanks is about 13,490-gallons per day or 15.1 acre-feet per year. Assuming the water consumed for flushing to be at the design level, the water available for irrigation from the third lagoon would be the sum of the fresh water from the flush tanks and the runoff less the net evaporation and rainfall on the lagoon or 42.1 acre-feet per year as shown in Table 3. This would be a reduction of 47.5 acre-feet per year when compared to the amount of flow through the third lagoon in Table 2. Operating at design capacity would require pumping the designed water input of 80.4 acre-feet and the runoff of 16.7 acre-feet or a total of 97.1 acre-feet per year. Further evaluation on actual requirements could result in reducing this volume.

The volume of water available for irrigation could be reduced further by diverting the rain water from the roofs over the free-stalls. This is an important factor in the design of a new system.

### Screen Performance

The screen was installed to solve the problem of plugging of pipes and build-up of surface solids in the first lagoon. The residue from the screen contained 13-percent solids which was 95-percent volatile. The COD of the residue was 748 mg per gram (dry basis). Based on reported waste production from an 1100 pound dairy cow of 11.4 pounds total solids, 9.4 pounds of volatile solids and 10.8 pounds COD per day (Jones et al. 1970, Loehr 1974) the screen removed about 42.6-percent of the total solids, 49.1-percent of the volatile solids and 33.7-percent of the

#### TABLE 3. PROPOSED WATER BALANCE.

| | | Acre-feet/year |
|---|---|---|
| (1) | Design water input | 80.4 |
| (2) | Fresh water parlow flush tanks | 15.1 |
| (3) | Amount recycled from 3rd lagoon, (1)-(2) | 65.3 |
| (4) | Lot runoff | 16.7 |
| (5) | Rainfall less pond evaporation from lagoons | 10.3 |
| (6) | Available for irrigation, (2)+(4)+(5) | 42.1 |

TABLE 4. ANALYSIS OF RESIDUE DISCHARGED FROM SCREEN.

|  | Pounds/day/cow |
|---|---|
| Total solids | 4.86 |
| Volatile solids | 4.62 |
| COD | 3.64 |
| Potassium | .00535 |
| Phosphorous | .00146 |
| Nitrogen | .0792 |

TABLE 5. SCREEN AND LAGOON EFFLUENT ANALYSIS.

| Sample location | COD mg/l | BOD mg/l | Solids, mg/l | | | |
|---|---|---|---|---|---|---|
|  |  |  | Total | Total volatile | Suspended | Volatile suspended |
| After screen | 5840 | 2080 | 5780 | 4410 | 1200 | 973 |
| 1st lagoon | 1500 | 146 | 2280 | 1100 | 364 | 314 |
| 2nd lagoon | 865 | 68 | 1690 | 739 | 158 | 124 |
| 3rd lagoon | 667 | 39 | 1520 | 601 | 104 | 85 |

COD. The analysis of residue discharged from the screen indicates a performance comparable to reported laboratory results (Graves et al. 1971). The results of the laboratory analysis of the residue from the screen are shown in Table 4.

Removal of the solids by the screen did eliminate the accumulation of floating solids from the lagoons. The solids are spread on land for disposal.

**Lagoon Performance**

It was physically impossible to collect a representative sample of the manure in the collection tanks without major alterations to the existing system. Several attempts were made but the laboratory results clearly indicated the samples were atypical. The strength of the effluent was estimated by assuming an average weight of 1,100 pounds per cow for the mixed Jersey and Holstein herd. An average of 700 cows were located on the feedlot during this study.

The efficiency of the lagoon system as calculated from the data in Table 5 indicate that the third lagoon may not be needed for a system of this nature especially in view of the low loading rate of the first lagoon. The accumulated COD reduction of 74.3, 85.2 and 88.6 and the accumulated BOD reduction of 93.0, 96.8 and 98.2 in the first, second and third lagoons respectively support this statement. The high removal of volatile solids by the screen would certainly reduce the size requirement of the first lagoon. The size required for the second lagoon would depend on the management of flush and rain waters but would be based on storage requirements prior to irrigation instead of treatment efficiency. The third lagoon would have to be maintained at about one-half capacity to allow for a 10-year 24-hour rainfall event which would result in storage requirements of 8.4 acre-feet of rain water. About 64-percent of this storage requirement is due to rainfall on the lagoons, thus, designing a system with a minimum lagoon surface area becomes an important factor in a no-discharge system.

## SUMMARY

The water usage, solids removal by a hydrasieve screen, and the treatment efficiency of a three cell lagoon system at a 1200-cow dairy facility were studied. It was concluded that wastewater must be recycled from the lagoons for flushing to reduce the irrigation requirements to an acceptable level. The screen removed about 42.6-percent of the total solids, 49.1-percent of the volatile solids and 33.7-percent of the COD from the manure laden flush waters and was effective in eliminating solids accumulation on the surface of the lagoons. The lagoons were lightly loaded resulting in high treatment efficiency. A smaller, two cell lagoon system could be used in a land irrigation system incorporating a screen for solids removal.

**References**

1    Graves, R. E., J. T. Clayton and R. E. Light. 1971. Renovation and Reuse of Water for Dilution and hydraulic Transport of Dairy Cattle Manure. Proceedings of the International Symposium on Livestock Wastes. ASAE, St. Joseph, Mich.
2    Jones, D. D., D. L. Day and A. C. Dale. May 1970. Aerobic Treatment of Livestock Wastes. University of Illinois, Urbana, IL. Bulletin 737. 55 p.
3    Loehr, R. C. 1974. Agricultural Waste Management. Academic Press, New York.
4    13th Edition. 1971. Standard Methods—Water and Wastewater. American Public Health Association, Washington, D.C. 20036.

# Adaptation of a British Waste Management System to the U.S. Environment

Paul Jensen, Gordon Newman, Anthony J. Peters

THE disposition of livestock waste is an outstanding example of the way many things in this country swing like a pendulum. For decades farmers used livestock waste for its plant nutrient value and feeders could sell manure to cash crop farmers. Then chemical fertilizers took over an authority like Ray Loehr (1968) concluded that there was no profitable method of using livestock manure. The P and K were surplus; it was easy and less expensive for the farmer to apply chemical N and feeders had to pay to have manure hauled away.

Now, with N and P costs having tripled, livestock waste is a valuable resource. And while two years ago a two mile trip with a spreader was economically questioned, now we read in the public press and the farm press (Black 1975) about plans to haul it half way around the world to produce blooming deserts in the middle east. Another extreme swing is represented by the magazine Industry Week (1975) which pats itself on the back for having informed its readers about the fortune to be made in meadow muffins — manure as a source of methane or synthetic oil.

A. O. Smith Corporation does not expect to make a fortune from meadow muffins, at least not in this decade, but takes the idea seriously enough to have invested research funds in the operation of a pilot methane generator plant large enough to require a 20 ft diameter Harvestore structure as the containment vessel. This well-engineered chemical processing plant is set up on a 100-cow dairy farm. Its performance will be an appropriate topic at some future meeting on livestock wastes.

Our immediate concern is with a Slurrystore system adapted from British practice, prototyped here in the winter of 1973-74 and now being marketed by A. O. Smith Harvestore Products. In the United Kingdom a Slurrystore system using glass-coated steel Harvestore sheets has been on the market for seven years, in response to the stringent 1963 Water Resources Act (Shattock 1971).

The remainder of this paper is concerned with:

1  Adapting the British system to various United States climates.
2  Variations in system components.
3  Meeting EPA Standards.
4  Cost and benefit aspects of Slurrystore systems.
5  Optimum system operation.
6  From prototypes to production systems.

## CLIMATIC ADAPTATION

The climatic situation for English Slurrystores is predominately wet but mild, so a prime concern was whether severe winter conditions in this country would make it difficult or even impossible to use the above ground storage

The authors are: PAUL JENSEN, A. O. Smith Harvestore Products, Inc., Arlington Heights, IL; GORDON NEWMAN, Agricultural Research Council Institute for Research on Animal Diseases, Compton, Newbury, Berks., England; and ANTHONY J. PETERS, A. O. Smith Harvestore Products, Inc., Arlington Heights, IL.

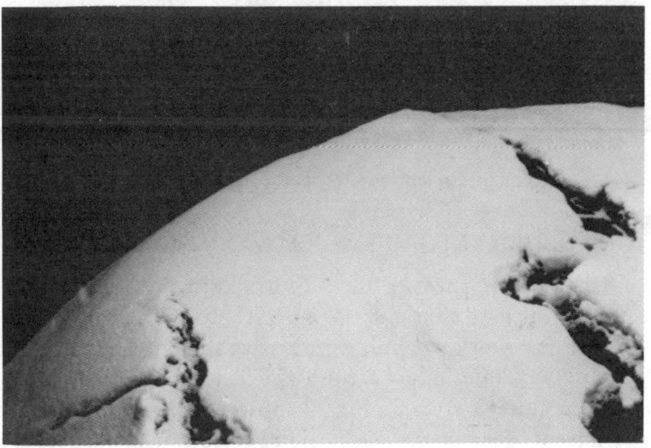

FIG. 1 Frozen crust in bottom-loaded slurrystore.

system. To check it out we built four prototypes last winter. Two are in Wisconsin, to determine feasibility of the system in the northern tier of states. A third is in southern Missouri, where we expected climatic conditions to be near optimum. And the fourth is in southern Texas, to determine the effects of average temperatures much higher than in any British installation.

Of the two Wisconsin installations, one is a 25 ft diameter structure on a hog farm and the other is a 42 ft diameter structure on a 40 cow dairy farm; both are near Greenleaf. For the first winter the hog farm unit was provided with the same auger pump that has been used on all the British Slurrystore installations, and it satisfactorily performed the functions of loading, agitating and unloading the Slurrystore tank. During the winter just ended, that installation has been operated with a Harvestore Model 211 liquanure pump, again with satisfactory results.

The system on the Diny Dairy Farm uses an underground Hedlund pump and 12 in. plastic pipe for forcing the slurry up into the center of the structure. Distance between the pump and the tank is about 100 ft, a convenient arrangement of many farmsteads. As expected and shown in Fig. 1, the material builds up in a mound underneath a frozen crust about 8 in. thick. When it came time to unload this Slurrystore tank in the spring, a conventional centrifugal liquid manure pump was used. The installation at the Williams Dairy Farm, Bolivar, Missouri has a 42 ft diameter Slurrystore tank and the system has functioned in all respects as intended. The installation at the Kettler Farm in Texas is also of the 42 ft size and its operation disclosed during the clean out in the spring of 1974 a kind of problem, easily solved, which would not have occurred with any of the British installations. That is, at 90 F ambient temperatures evaporation was so great as to cause an excessive solids accumulation in the bottom of the tank. The solution was dilution with 3,000 gallons of water dumped into the reception pit, followed by recirculatory agitation to loosen the solids and finally fill a field spreader.

FIG. 2 Site selection—Webber Farms, Kansas.

FIG. 4 Footing concrete poured, anchors placed.

## SLURRYSTORE SYSTEM & COMPONENTS

Selection of the site for a Slurrystore system is important. It must accept all the normal waste, must be convenient for working with pump, spreader, etc., and must prevent any rain-induced runoff around the system which might pollute a stream. Fig. 2 shows a typical site, where the line of trees in the background follows the course of a stream which the Slurrystore system protects. Another typical characteristic of the Slurrystore site is to have the reception pit on the downstream edge of a paved feedlot, for convenient blading of the manure.

Two basic elements of the Slurrystore system are the glass-coated steel slurry tank itself and the reception pit. The third essential is a pump system — one or more pumps; and the fourth is usually a field spreader but could be a piped irrigation system.

### The Slurrystore Tank

We have maximized use of Harvestore design and production know-how in designing Slurrystore tanks. But while buckling is the critical design parameter in the Harvestore forage structure, simple hoop load is the governing factor in the slurry tank. So, in the latter we use closer bolt spacing on the vertical seams. Fig. 3 shows in one graph all structural design factors. Sloping lines represent hydrostatic loads; vertical lines identify allowable tensile and bearing loads for steel sheets of various thicknesses. Continuous horizontal lines locate the horizontal bolt holes (5.022 in. spacing) and the three sets of broken horizontal lines show bolt locations for all vertical seams. One of the vertical lines near the right side indicates the allowable single-shear load for all the 1/2 in. diameter bolts, calculated for a bearing-type connection.

The same mastic sealer tape and seal strip materials which are used in conventional Harvestore structures provide water-tight sealing of the rings of Slurrystore sheets. The rings are mounted on a 15 in. wide concrete footing, the depth of which depends upon soil type and frost line at the building site. Inside the tank there is a 5 in. minimum depth concrete floor incorporating welded wire mesh on top of 3 in. minimum of compacted crushed rock fill. The floor is shaped to drain toward a 12 in. pipe that is cast into it and leads under the tank wall to the reception pit for recirculating the slurry. The 12 in. pipe inlet is fitted with a flapper type emergency valve, operated from the top of the tank by a long rod. A 12 in. irrigation-type gate valve controls slurry recirculation.

### Reception Pit

Our British colleagues offer two sizes of cylindrical glass-coated steel reception pits for sub-grade installation alongside the main Slurrystore tank. We have not yet offered these cylindrical reception pits but may in the future. Thus far all of the reception pits with our system have been custom built, with variations in length, width and depth, depending upon the particular farm site layout. The 6 in. thick walls of these reception pits in order to meet SCS recommendations (Kurtz 1975) are provided with re-bar and the slatted steel grating covering the pit is made more than strong enough to carry the load of a tractor used for scraping manure into the pit. There are also some installations in which stiffened walls of concrete block have been used for the reception pit; it depends upon what is permitted by local codes.

Figs. 4 and 5 illustrate two of the steps involved in con-

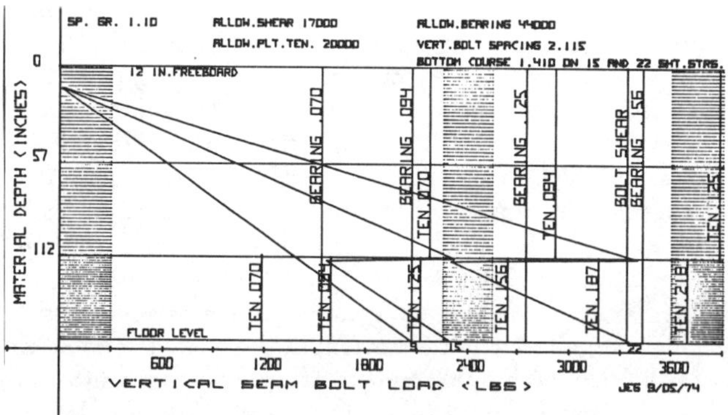


FIG. 3 Slurrystore strength parameters vs. imposed loads.

FIG. 5 Pouring floor.



MANAGING LIVESTOCK WASTES

FIG. 6 Loading over slurrystore rim.

FIG. 8 Jet nozzle manipulation.

structing a 61 ft Slurrystore tank and its reception pit made with concrete block.

## Pumps

Four functions must be performed by the pump or combination of pumps used in Slurrystore systems, as shown in Figs. 6 through 8. These are:

1    Loading the tank (Fig. 6).
2    Agitating by recirculating the slurry from the tank through the reception pit and back.
3    Loading a field spreader from the reception pit.
4    Powering the jet which is used to clean out accumulated solids (Fig. 8).

In most northernly parts of the country a number of operators prefer subsurface loading with a mechanical or hydraulic ram system. This then requires a secondary pump to perform the other three enumerated functions.

Another variation to be introduced will be a 200 or 400 gpm high pressure pump with a 10 horsepower electric motor, actuated by level switches in the reception pit. This would automate the loading of the Slurrystore tank and make it unnecessary to tie up a tractor for its PTO to drive a high capacity (1,500 gpm) centrifual pump.

### Field Spreader Or Irrigation System

Field spreading from a tanker wagon of 2,300 to 3,200 gallon capacity is currently the predominant method of disposing of the slurry when disposal is required and it is possible to work on the fields. As time passes, we expect that more farmers will dispose of slurry through an irrigation system, to save tanker operation labor.

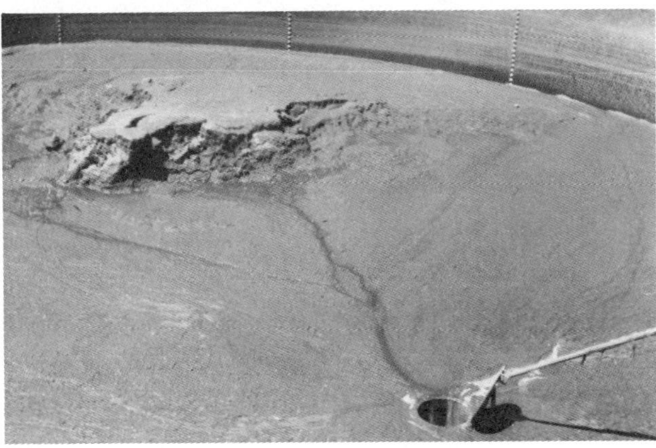

FIG. 7 Typical solids residue.

## MEETING EPA STANDARDS

The Harvestore system, properly sized and managed for the number of animals involved, readily meets all federal EPA regulations (Denit 1974) and the corresponding regulations which have been issued by several states (e.g., Illinois PCB 1974; Minnesota PCA 1974). The design enables the tank to withstand expected pressures and loading. The glass coating of the Slurrystore sheets works against corrosion and the sealed-joint construction enables the tank to meet the requirement of being impermeable. In England concrete stave tanks with steel hoops have proven not to be impermeable. Rather, the slurry leaks through the cracks between the staves, attacking and corroding the hoops, so that the structural strength of the tank itself is ultimately threatened.

With proper sizing of the tank and reception pit, and with removal of slurry frequently enough so that freeboard is available in both the tank and reception pit, the Slurrystore system will meet not only the 1977 requirement for protecting against the effect of a 10-year 24-hour storm, but will also meet the 1983 requirement for protecting against a 25-year 24-hour storm.

## COSTS & BENEFITS OF SLURRYSTORE SYSTEMS

The total installed cost of a Model 6115 Slurrystore tank and its reception pit is about $18,000 to $20,000 but local conditions, concrete cost, etc., can significantly affect this figure. Such a system provides three months storage capacity for the output of 150 dairy cows under ordinary circumstances. A 1,500 gpm centrifugal pump and a 3,200 gallon spreader would add about $8,000 to $10,000 to the cost. This then comes to about $175 per head. The degree of pollution control provided is high and the operating convenience with the Slurrystore system reduces operating and maintenance cost as compared with other systems.

Fortunately the newly increased fertilizer value of liquid manure eases the cost which pollution control measures now impose upon the farmer. But the estimates of slurry value cover an extremely wide range. At the top is a 5 cents or higher per gallon figure given in Time magazine for slurry to be transported by tanker to the middle east. With one dairy cow producing 10 to 12 gallons per day of 10 percent solids waste, her annual manure value at that rate would be $200. On the other end of the scale Gergen (1974) states that most feedlot manure is being sold to contract haulers for about 50 cents to $1 per ton. The dairy cow that produces 88 pounds per day would be producing 16 tons per year, at just $8 to $16. Jim Willrett, however, quoted by

Gergen, estimates that each of his feedlot steers is worth $26 to $28 in manure nutrients and these animals probably average only half the weight of a dairy cow. Rohlf (1975) has given a figure of $60 as the current plant food value of manure from one steer; the corresponding value for a dairy cow output would be $90 to $100. At the $60 per steer rate (about $5 per ton) and anticipating two fills of the 6115 tank annually, we conclude that the first cost of the Slurrystore system is offset in 2 to 3 years by the value of manure retained and utilized.

## OPTIMUM SYSTEM OPERATION

The following procedures are recommended in operation of a Slurrystore system:

1   Avoid excessive evaporation or dilution.

2   Keep long hair or straw out of the Slurrystore, to avoid agitation problems.

3   Don't allow excessive soil to get into the slurry.

4   Mix the material thoroughly before loading the Slurrystore.

5   Agitate once a week when it is not frozen, to keep solids suspended.

6   Don't aerate. Aeration reduces the nitrogen content by 30 to 50 percent (McAllister 1972), depending upon the temperature and losses are mainly of the more soluble nitrogen compounds. Thus aeration reduces the fertilizer value, whereas anaerobic storage actually increases the percentage of soluble nitrogen available for optimum fertilizer use.

7   Spread just before planting and incorporate slurry into the soil as quickly as possible. The tanker with soil injectors is preferred.

8   Don't skimp on horsepower of the slurry pump for agitation.

## FROM PROTOTYPES TO PRODUCTION SYSTEMS

Three of our prototype Slurrystore tanks were 42 ft in diameter, two rings (9 ft-2 in.) high and one was a nominal 25 ft diameter, three rings (13 ft-7 in.) high. From the prototype operations and initial marketing it appears that the most popular size in this country will be the largest, which is 61 ft-7 in. in diameter, 13 ft-7 in. high and identified as the Model 6115. We therefore offer just three sizes: the Models 2515, 4215 and 6115 with nominal capacities of 50,000 140,000 and 300,000 gallons, respectively. From experience to date it appears that certain modifications or options for future systems may be desirable, e.g.:

1   Adding a 12 in. high bottom ring on the 61 ft tank, so that it would be convenient to back concrete trucks over that rim into the interior of the tank to facilitate pouring the floor.

2   Providing a motor driven 400 gpm pump for automatic loading in response to level switches located in the reception pit.

## CONCLUSIONS

We believe the Harvestore Products Slurrystore system provides farmers a good answer to livestock waste management because it:

1   Meets established EPA requirements.

2   Provides a well organized, compact system requiring relatively small space (as compared, for example, with a lagoon system).

3   Maximizes fertilizer value of the liquid manure by allowing aeration to be limited, and spreading to be performed when it can be most beneficial.

4   Reduces the labor required for livestock waste handling.


### References

1   Anon., Industry Week. January 6, 1975. Methane From Feedlots.

2   Black, N. 1975. Crude Oil For Manure, Real Barter Potential. National Hog Farmer January.

3   Denit, J. D. 1974. Effluent regulations for livestock and poultry feedlots. Proc. 1974 Cornell Ag. Waste Mgmt. Conf. March 25-27, 1974 Rochester, N.Y.

4   Gergen, Bill. 1974. Feedlot Manure: Suddenly It's Worth More. Successful Farming, September 1974:24-25.

5   Illinois Pollution Control Board 1974. Livestock Waste Regulations. September 5.

6   Jones, P. W. 1975. Health Hazards Associated With The Disposal Of Cattle Slurry: Research at the Institute For Research On Animal Diseases 1972-1974. Also, abbreviated in Big Farm Management, January:20.

7   Kurtz, D. 1975. Personal Communication with Owen Owens, A. O. Smith Harvestore Products, Inc.

8   Loehr, R. C. 1968. Pollution Implications Of Animal Wastes—A Forward Oriented Review. USDI, Federal Water Pollution Control Administration. 175 p.

9   McAllister, J. S. V. 1972. Problems In The Disposal Of Animal Excreta. Proc. Br. Soc. Animal Prod. 1972:87-92.

10   Minnesota Pollution Control Agency. 1974. Animal Wastes. Manual prepared by Solid Waste Division of MPCA, Roseville, Minn., January.

11   Rohlf, J. A. 1975. Manure Is Worth Spreading. Farm Journal, March.

12   Shattock, G. F. 1971. Manure Disposal From A Dairy Herd. Presented at Autumn National Meeting, Institution of Agricultural Engineers, University of Reading. Whiteknights Park, Reading, England, September 30.

# A Liquid Manure Management System in a Tie Stall Dairy Barn

Gerald S. Meierhofer, Philip R. Goodrich
MEMBER
ASAE

**M**ANURE handling has come a long way since I was a boy on Dad's farm. We handled it by hand with a team of horses and a bobsled. Today, with my system of manure management and a tie stall dairy barn, the cow drops the manure through a series of grates into a pit. After storing the manure for a period of time, we utilize machines to do most of the work of getting this useful manure out onto the crop fields. When we pushed manure out of the cow barn at home with the old carrier and dumped it on a pile, we often noticed that a lot of the juice ran out of the yard. The same yard was a mud problem in the spring and probably the best part of the manure never got back onto the cropland. The runoff did reduce the amount of material that needed to be handled and, at that time, this seemed good. We also used a lot of bedding and that seemed to make it a lot more work to handle that material again. When we hauled it out each day, we seemed to lose very little manure value, but the total work, time and machines required per ton of manure per cow or per year was very high.

Installation of the barn cleaner helped a lot, because it mechanized the process. I thought that this was the best way to handle manure if cost, time and return were considered. I guess that, maybe, it was and may still be for a milking herd of less than 35 cows.

However, when I wanted to expand my herd size, my farm management records showed that I should specialize. They showed that I was making more money and using my crops and time most efficiently in dairy and that the pigs and chickens that I had were holding me down. We were milking 54 cows in the fall of 1971 in a barn originally designed for only 38. It seemed that all my three sons and I really got done was to haul manure. With help from my brother, Will, an Adult Farm Management Instructor at Wadena, we explored the possibility of a new barn. After considerable discussion, looking at various types of barns and talking to other operators, we decided on a tie stall dairy barn with a full manure pit beneath it. The pit was sized to hold manure for at least 6 months.

Our decision to use the tie stall system was based on a number of observations. I felt that I couldn't give the cows enough individual attention and the proper feed in a free stall dairy barn system as I could in a tie stall barn. The parlor system seemed to take a lot of cleanup time. We also were on a herd health program and the Veterinarian felt that herd checks were quite a problem in free stall dairy barns. Experience of many people who were managing and operating dairy farms seemed to indicate clearly to me that an increase in milk production of three to eight percent could be had in a tie stall dairy barn compared to a parlor system, assuming that operators had equal management ability. We therefore, chose to stay with the tie stall dairy

Miscellaneous Journal Series Paper No. 1576. University of Minnesota Agricultural Experiment Station, St. Paul, Minn.
The authors are: GERALD S. MEIERHOFER, Dairyman, Rolling Hills Farm, Watkins, Minn.; and PHILIP R. GOODRICH, Extension Agricultural Engineer, University of Minnesota, St. Paul.

barn with a pipeline milking system. Our Federal Land Bank people felt that a manure pit beneath the barn was too costly and that a pit beside the barn would be better. However, actual cost estimates which included a barn cleaner to move the material from the barn outside to the pit, gave indications that the costs were nearly identical.

We built our new tie stall dairy barn and pit on Rolling Hills Farm, as we proudly call it, in 1972. The barn is 34 feet (10.36 m) wide by 168 feet (51.21 m) long and holds 68 cows, a maternity pen, an office, a bathroom and a milkhouse. We located it 80 feet (24.39 m) south of our old barn. All of the farm buildings are on a hill with a creek at the base of the hill that could potentially be polluted with runoff from our livestock systems. Therefore, we also put a pit beneath a nearby loafing shed to intercept runoff. This pit, along with concrete yards and our new barn pit made our whole operation one with a totally controlled runoff situation. All the stalls have rubber mats imbedded in concrete to eliminate the need for bedding, which would cause problems hanging up on the gutter grates. Also, additional solids in the pit use up manure storage capacity.

Grated openings which allow the manure to fall into the pit replace the gutter used in conventional tie stall barns. The grated openings are 18 inches (45.7 cm) wide with the grates set 3 inches (7.6 cm) below the concrete walkway level. The 3 inch (7.6 cm) depression is important from an aesthetic  point of view in that, as you look down the barn, it gives an impression of having mostly walkway instead of all gutter. Some manure does pile up at times and does not go  through the grates. But, even then, it does not look like it's on the walkway. The grates are constructed of rods 5/8 inch (1.6 cm) diameter and spaced 2.5 inches (6.35 cm) on center. The short cross rods are 3/4 inch (1.9 cm) diameter and spaced 18 inches (45.7 cm) on center. Both edges of the grate are imbedded in concrete. The grate design seems to provide uncomfortable standing for the cows and so it keeps them standing in their stalls. This feature along with the use of electric cow trainers and their perfect positioning is important in getting the manure into the grates. Even though the grates are well designed, a small amount of manure does not go through into the pit. A steel rake with fingers spaced on a cross bar was developed to scrape the grates and push this manure down. It's a very simple job that takes only 5 minutes for the entire barn.

The manure pit is the same size as the barn and 8 feet (2.43 m) deep. However, the whole pit is offset and extends 6 feet (1.83 m) beyond the south wall of the barn. In this offset (see Fig. 1), I have eight large, round pumping ports and two doghouse type pit fans. The pit has four compartments made by divider walls 8 feet (2.43 m) high with a 2 foot (0.61 m) depression in wall for allowing overflow from one pit to the next. The divider walls allow me to pump each pit separately without draining too much liquid off the other pits. The minimum pit depth is 8 feet (2.43 m) with a floor slope of 6 inches (15.2 cm) in the 34 foot

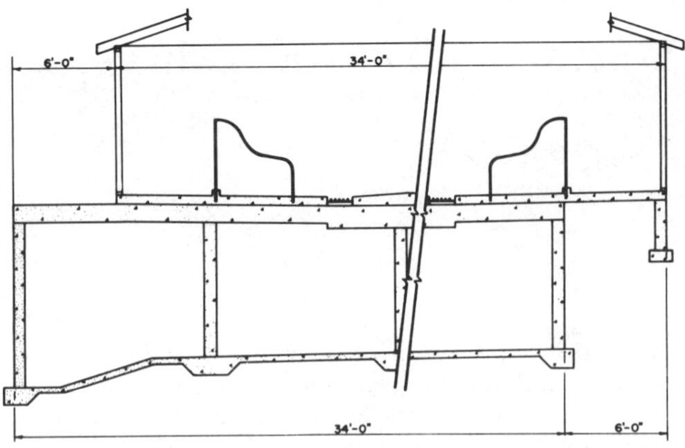


FIG. 1 Cross section of a tie stall dairy barn using manure grates.

(10.36 m) width. I do not feel this is enough slope and that it probably should have been 2 feet (0.61 m) in the 34 feet (10.36 m). In all probability, milking machine parts which are dropped through the grates would then get into the pumps though.

The floor of the pit is depressed 18 inches (45.7 cm) along the south wall to get more complete emptying of the pit. This does help some, but I would make the depression at least 24 inches (61 cm) deep in the new barn. Then we could get more manure out of the pit before the pump quits drawing. A single pumping port on each pit would have been better than the two which I have because a center location between the present 2 would position the pump for better agitation. Also, I am glad that the pit is not wider than 34 feet (10.36 m). I am sure I could not agitate a wider pit. The concrete block columns supporting the beams of the floor are spaced approximately 8 feet (2.43 m) on center. These pilaster blocks used to form the posts must have mortared joints before the cavity within the concrete block is filled with concrete. We tried some without the mortar and split the bottoms open immediately. The six columns in each 34 by 36 foot (10.36 x 10.97 m) pit obstruct the agitation process somewhat, but not enough to justify more expensive clear span concrete beams.

We installed 1¼ inch (3.18 cm) diameter vertical pipes in the barn floor below all drinking cups so that water slop over would drain directly into the pits. Welded catch pans on a ¾ inch (1.90 cm) pipe funnel the water into these pipes from the drinking cups. This works well in helping to keep the animals dry and I highly recommend it.

Managing the collected manure and recycling it to the cropland is a very important part of our total system. Part of our original plan was to pump the liquid manure from the pit and move it to the field through an irrigating system and sprinkling it on the soil. Some initial pumping operations which we had observed had worked quite well. A custom operator was working on a custom basis and pumping liquid manure pits out through aluminum irrigation pipe and spreading it using a large moving gun. It seemed very reasonable that if we could have a custom operator empty our pits, it would save me much work. Therefore, the first time we needed to empty the pit, we got the custom operator to come and work on our pit. However, after the four days of effort, we gave up because of all the difficulty which they had and the high cost. We now pump all our pits twice a year. It takes us about four days each time we pump, working with one 2250 gallon (8517 liter) honeywagon and one pump. We agitate about one hour before we start to haul, and continue the agitation process between loads. Since we are agitating only a 34 by 36 foot (10.36 x 10.97 m) section of the total pit, the agitation process works quite well. The material is hauled out and spread on top of the land. Soon, we hope to inject this material into the soil if the odors become too bothersome to neighbors.

I've heard that some states do not allow dairy cows tied in stalls to be above a pit that is open through gutter grates. There is a fear that the smell and association with the manure in the pit below is detrimental to the dairy herd and would somehow affect the quality of the milk. I feel that this accusation is unfounded and that my barn proves it.

Ventilation is very much a part of the total manure program in moving the air out of the barn through the gutter grates. Air enters the barn through end louvers and flows through the continuous ceiling slot opening on each side of the barn. The air is then pulled down through the gutter grates by pit fans which are mounted in the two doghouse type structures outside the barn on the pit extension. This even flow of air moves odors, and water vapor, out of the barn so that it does not remain near the animals. There are also three wall fans, but they are used only during manure pumping operations and when warm, humid weather conditions necessitate moving more air than the pit fans can pull.

The rolling herd average is now 545 pounds (247 kg) of butterfat and 15 282 pounds (6932 kg) of milk per cow per year. This is proof that animals can do well in a tie stall dairy barn over a manure pit. Handling manure is no longer an unpleasant chore that must be handled every day when many other things that are going to maintain high milk production should be done. I have been able to utilize a lot of information from the University of Minnesota Agricultural Extension Engineers and develop a system which fits my needs very well.

Manure handling has come a long way for me from what we used to do when we handled manure with the pitchfork and the bobsled. Twice a year pumping and recycling of the liquid manure waste on the cropland is an ideal situation for my operation.



MANAGING LIVESTOCK WASTES

# A Complete Dairy Liquid Manure System


W. J. Roberts, M. E. Singley, D. R. Mears

MEMBER ASAE — MEMBER ASAE — ASSOC. MEMBER ASAE


THE Rutgers Circular Dairy Barn was designed and constructed using 3 basic concepts: self-feeding of silage, liquid manure handling and the free stall system of cattle housing. These three were incorporated into a circular building as illustrated in Fig. 1. The design of the building and the performance of the system has been reported earlier by the authors (1966, 1969, 1970 1974).

## FACILITIES

The liquid manure system consists of a circular trench 4 ft deep with an inside radius of 15 ft-9½ in. and an outside radius of 22 ft-5-3/8 in. The surface area is 802 ft$^2$ and the liquid volume when filled to a depth of 1½ ft is 1203 ft$^3$, or 9000 gallons. The trench empties into a holding tank with a capacity of 2373 ft$^3$, or 17 752 gallons, when filled to the level of the trench bottom and 2881 ft$^3$, or 21 560 gallons, when filled to the liquid level in the trench.

The trench is covered by a concrete slotted floor that provides access from the concrete feeding platform to the resting stalls. The slats are 4¼ in. wide at the inside end and 6½ in. wide at the outside. They are 6 in. thick and tapered vertically to assure cleaning. Originally the slats were spaced with 1 in. wide openings between them, but were later separated by 1½ in. openings to reduce the manure accumulation on the slats. The floor is self-cleaning but to improve the rate of disappearance of the manure, the large accumulation next to the curb behind the free stall was often pulled to the center of the floor where cow traffic was heaviest. This was usually done in the morning.

Because of the design of the ventilation system, the slats dried relatively quickly and cow footing was firm. No unusual foot problems occurred except for several heifers that sometimes caught the toe of a hoof in the opening between the slats because of the smaller size of their hoofs. One heifer sprained her front leg in this manner. Though the average slat spacing was 1½ in., the nonuniformity of the slats presented openings of as much as 2 in. Slats should be carefully paired and the slat width for young animals should be less than the 1½ in. spacing, which appears to be optimum for Holstein cows. Though the slats have been used for almost 6 years, the concrete surface remains rough and the original finish is largely intact. There is no sign of corrosion on the concrete surfaces of the manure collection trench.

## MANURE COLLECTION AND PUMPING

Manure which is collected as liquid in the trench is carried to the holding tank by being displaced by liquid pumped into the trench from the tank by a 50 hp electric motor-driven Badger manure pump. The manure is pumped from the holding tank through a 6 in. diameter underground pipe and introduced into the circular trench at a point directly opposite the holding tank (Figs. 2 and 3). The manure enters the trench through a T-fitting 1 ft-3 in. above the floor, which directs the manure to either side of a trench divider. Manure then flows by gravity back to the holding tank over an 18 in. high control gate located tangent to the outer edge of the collection trench as shown in Fig. 3. The pump is operated 2 times a week, usually on Mondays and Fridays, for a period of 20 to 30 min.

When flowing in the right semi-circular path, a floating particle will be carried from the inlet to the dam in about 5.7 min. In the left semi-circular path, it requires about 7.2 min. This indicates that when the pump is operated for 30 min., the volume of liquid in the trench is changed about 4 to 5 times.

Moderate amounts of sawdust are used as bedding on the concrete stall floors. Originally earth-filled stalls were used. Soil which accumulated in the trench and became entrained in the manure could not be handled by the pumping procedure described above. Islands of manure, bedding and soil developed in the trench over a residual layer of soil and manure in the bottom. The trench had to be cleaned by hand after one year's operation using the earth-filled stalls.

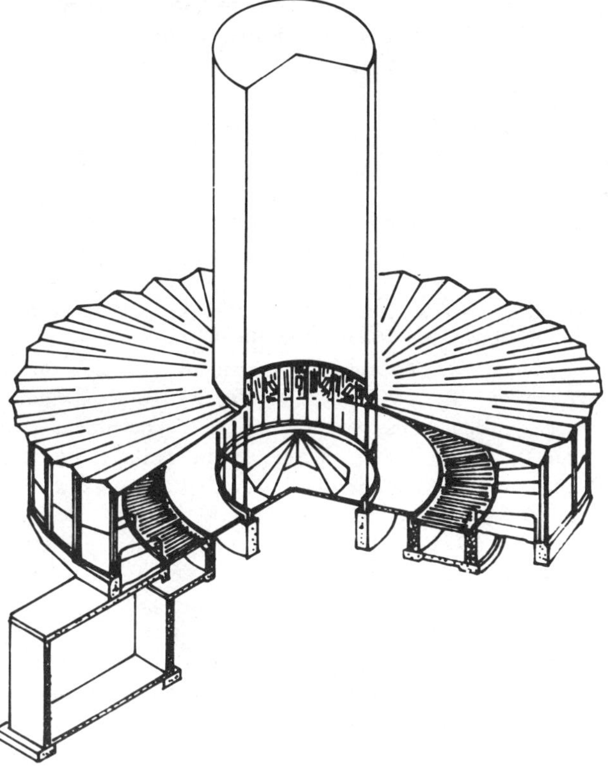

FIG. 1 Cutaway view of circular dairy barn.


Paper of the Journal Series, New Jersey Agricultural Experiment Station, Rutgers University-The State University of New Jersey, Department of Biological and Agricultural Engineering, Cook College, New Brunswick, New Jersey 08903.

The authors are: W. J. ROBERTS, Department Chairman and Specialist in Agricultural Engineering; M. E. SINGLEY, Professor; and D. R. MEARS, Associate Professor, Biological and Agricultural Engineering Department, Cook College, Rutgers University — The State University of New Jersey, New Brunswick, New Jersey.

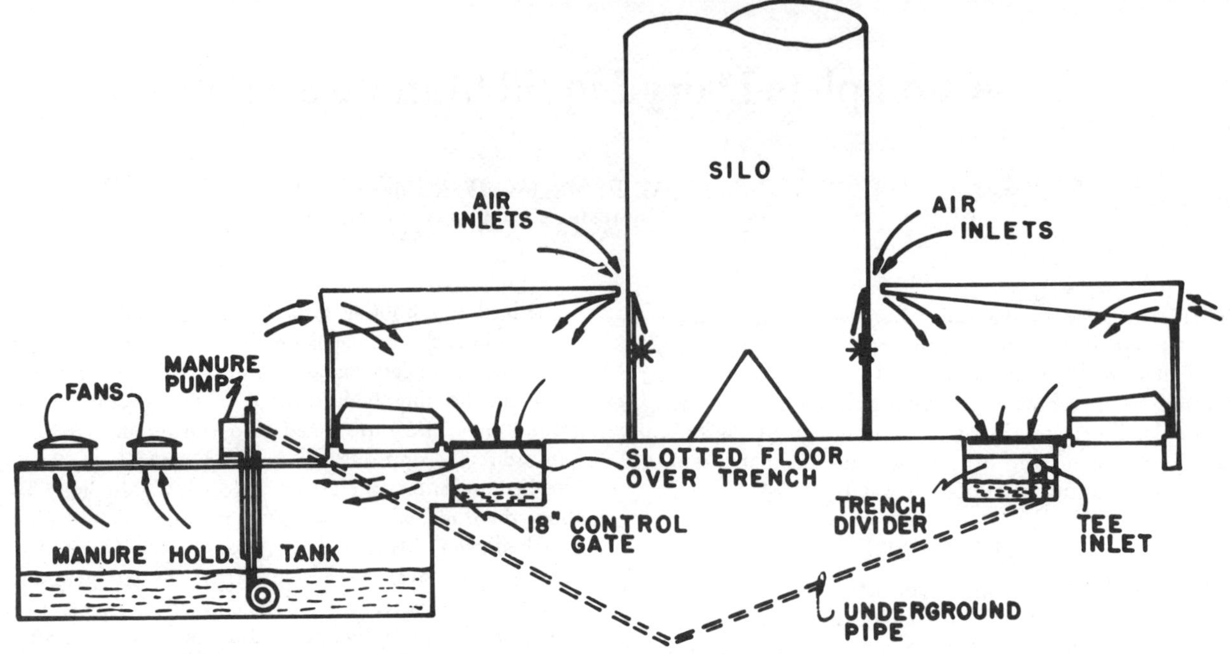


FIG. 2 Schematic cross section of circular dairy barn.

Following paving, no accumulation of manure occurred in the trench. Although the effects were not measured, the soil which flowed through the system undoubtedly adversely affected the pump. Earth-filled stalls should not be used in any slotted floor liquid handling system where gravity flow or recirculation is required.

Several factors in the design and operation of the liquid manure system enhance its operation and are recommended for consideration by others designing liquid systems. Having a flat floor in the collection trench and requiring the recirculated material to flow over rather than under a barrier into the storage tank maintains liquids in the trench at all times. Physically dividing the flow of recirculating manure ensures that each half of the system will be adequately flushed. Management of the herd to limit the amounts of materials other than manure that get into the collection system helps prevent the buildup of deposits of solid materials in the collection trench.

## MANAGEMENT OF COLLECTED MANURE

Manure is pumped by the Badger pump from the holding tank into a transport tanker about every 8 weeks. This is a shorter interval than anticipated. During the operation of the barn, no water was added to the liquid manure system and yet it always seemed to be quite fluid. This coupled with the short time required to fill the tank suggested that groundwater was entering the system. At the time the exercise yard was laid adjacent to the barn and the manure storage tank, a drain for a proposed milking center was laid under the yard and connected to the tank. It has been discovered that groundwater is entering the tank through this pipe at a rate dependent upon the amount of rainfall. During the 1974-1975 mild winter, water entered the tank at the rate of 2.2 in. per week, or 425 gallons per week. At this rate, approximately one quarter of the liquid volume was groundwater. The transport tanker developed by Reed (1970, 1974) is equipped with a mounted plow which places the manure directly into a furrow, covers it and forms a new furrow in one operation. The 800 gallon tank can apply the 800 gallons of material in a 16 in. furrow in about 600 ft. The rate of application of manure at about one gallon per foot of furrow is 130 tons per acre.

This type of system eliminates the field spreading odor problem, which is a significant consideration in suburban New Jersey. Fly problems in the field are also minimized by this technique. The system is only limited by the condition of the soil for plowing; either too dry, too wet, or frozen. New Jersey's climate rarely requires a holding period of more than 10 weeks.

## BARN VENTILATION

One of the key design features of the barn is the interaction between the ventilation system and the manure handling system. Much has been reported on the problem of gas production in storage pits under housed animals, for example Taiganides (1968). The ventilation system is shown in Fig. 2. Two exhaust fans are located in the roof of the concrete holding tank. Slot inlets for the air handling sys-

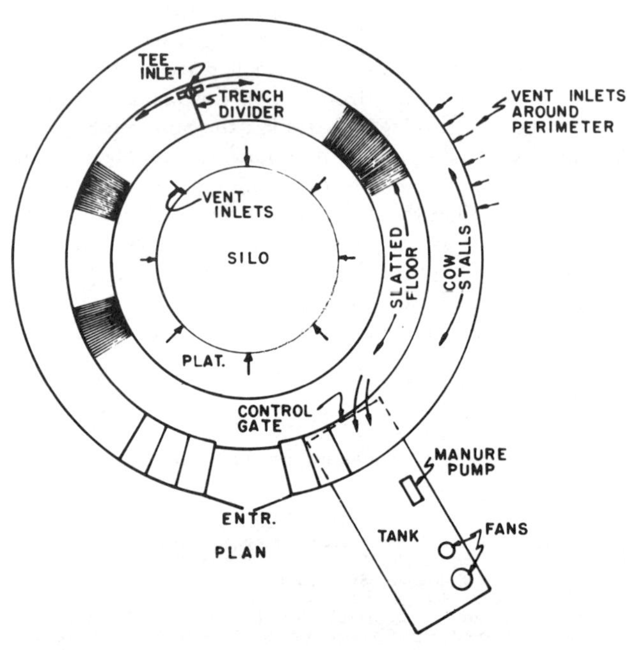


FIG. 3 Plan view of circular dairy barn.



MANAGING LIVESTOCK WASTES

tem are located in the eaves at the perimeter of the building and at the intersection of the silo and roof panels. These slot inlets are manually adjusted 3 times per year to provide an adequate inlet velocity of approximately 800 fpm.

Air is drawn into the building through the inlets and passes down through the area occupied by the animals and through the slotted floor. Once it enters the manure trench, which serves as a plenum, it is drawn into the holding tank and passes out through the exhaust fans.

One of the two exhaust fans is two-speed and runs at one speed or the other continuously. This insures that the air flow pattern will always be as described above. Gases produced by the manure in the trench are always picked up by the ventilation air and exhausted without entering the living area of the animals. Even during pumping and recirculation, noxious gases or odors are not detected within the barn. Moderate to low odors occur at the fan outlet while ventilating. During manure pumping, particularly for recirculation, the odors may be objectionable at the fan outlet. Whether they are objectionable in the vicinity of the building is dependent upon the weather. If the day is foggy or rainy and there is no wind, the odors will not be dispersed and can be detected beyond the vicinity of the building. The manure recirculation system has enough flexibility in the need for recirculation so that pumping on undesirable days can be avoided. At the initiation of the experiment, it was felt that the need for odor control might be the most difficult need to satisfy, particularly since the farm is adjacent to a public road.

Another benefit of the downward flow of air is in moisture control within the building. Outside air, heated as it passes through the building, always has the opportunity to become saturated as it moves through the building and over the surface of the liquid manure in the trench. Maximum moisture can be removed in periods of cold outside temperatures and high inside humidities.

Internal velocities in the barn range from 10 to 50 fpm at low ventilation rates and to 50 to 80 fpm at full ventilation. No 'dead' spots within the barn have been detected.

A series of tests were conducted to attempt to reduce odors created at the exhaust fan ports during the times of pumping and recirculation. Oxidizing agents were added to the manure in the storage tank during pumping. Hollenbach and O'Neill (1972) report, for example, that the addition of 100 ppm of hydrogen peroxide eliminates hydrogen sulphide and reduces the odor of ammonia in the exhausted air.

Most of the literature recommends that some air be exhausted from manure storage areas under animals by drawing outside air through the living area into the manure holding area. Our work indicates that all of the ventilation air should be handled this way. One drawback is that the equipment presently available to ventilate animal structures is not designed for this application. Total failure occurred within 3 years with high quality air-handling equipment. If more animal housing systems are designed with under-floor storage of manure, some manufacturers should investigate using alternate materials for the fan blades, such as fiberglass, or methods of protecting steel from the corrosive atmosphere passing through the fan 24 hours per day.

## ENVIRONMENT AND MILK QUALITY

To determine the influence of the environment within the barn on the quality of the milk produced, both organoleptic and chromatographic evaluations were made by Keller and Kleyn (1970 and 1972). The tests showed that the level of feed flavor was considered to be of slight significance and environmental control in the barn is proving to be very effective in minimizing off-flavors in the milk.

## SUMMARY

The manure handling system and the interaction of the ventilation system as described herein have performed excellently. Minimum daily labor is required; approximately one hour of pumping per week is required except during the field transport period. High capital investments in the floor, trench and holding tank are required, but this cost is offset by the reduction in labor required and the quality of the environment in which the cattle live.


### References

1  Hollenbach, R. C. and E. Y. O'Neill. 1972. Manure odor abatement with peroxygen chemicals. Report No. 5684-R-Inorganic Research and Development. FMC Corporation.

2  Keller, W. and D. H. Kleyn. 1970. The effects of environmental and feeding conditions in a new circular barn in the flavor of milk. Presented at 65th Annual Meeting of the American Dairy Science Association.

3  Keller, W. and D. H. Kleyn. 1972. Head space gas chromatography for objectively determining intensity of haylage flavor in raw milk. Journal of Dairy Science 55(5).

4  Mears, D. R., W. J. Roberts, M. E. Singley and G. H. Nieswand. 1970. Radial folded plate roof for a circular dairy barn. TRANSACTIONS of the ASAE 14(2):377.

5  Mears, D. R., W. J. Roberts and M. E. Singley. 1974. Environmental control in the Rutgers circular dairy barn. Proceedings of the International Livestock Environment Symposium. ASAE, St. Joseph, Michigan 49085.

6  Reed, C. H. 1970. Recycling and utilization of biodegradable waste by plow-furrow-cover and sub-sod-injection. ASAE Paper No. NA 70-206. ASAE, St. Joseph, Michigan 49085.

7  Reed, C. H. 1974. Equipment for incorporating sewage sludge into the soil. Compost Science 15(4):31-32.

8  Roberts, W. J. and D. R. Mears. 1970. The Rutgers round barn, a progress report. Paper No. NA 70-306 presented at ASAE Regional Meeting. ASAE, St. Joseph, Michigan 49085.

9  Singley, M. E. 1966. Space age dairy barn. Paper No. NA 66-306. ASAE, St. Joseph, Michigan 49085.

10  Singley, M. E., W. J. Roberts and D. R. Mears. 1969. Experimental circular dairy barn. Paper No. 69-405. ASAE Annual Meeting. ASAE, St. Joseph, Michigan 49085.

11  Singley, M. E., W. J. Roberts and D. R. Mears. 1970. Experimental circular dairy barn. AGRICULTURAL ENGINEERING 51(2):78.

12  Taiganides, E. P. and R. Waite. 1968. The menace of noxious gases in animal units. ASAE Paper No. 68-557. ASAE, St. Joseph, Michigan 49085.

# A Waste Management System for a 150-Cow Dairy — A 10-Year Case Study

A. C. Dale, J. L. Albright, J. C. Nye, A. L. Sutton
FELLOW
ASAE
ASSOC. MEMBER
ASAE

T HE evolution of a waste management system for the 150-cow herd of the Purdue Dairy Center is presented in this case study. The dairy is similar to many dairies that have evolved from stanchion barns to free-stall barns and milking parlors. The waste from the dairy has been handled as a solid, liquid, and is now treated and handled with a lagoon-irrigation system.

## THE DAIRY FARM

Prior to the summer of 1965 all manure was handled in the solid or semi-solid form. A considerable portion of the liquid escaped the system to drainage ditches. Solids were not removed as rapidly as desired and fly breeding was a problem and created a nuisance.

### The Dairy Farm Layout

At the beginning of the adaptation of a waste management system the dairy consisted fo 140 dairy cows weighing approximately 1300 lb each. The general layout and arrangement are as shown in Fig. 1 and an aerial view is given in Fig. 2. Previous to 1964 two-story stanchion barns were used with the milkhouse and milk handling facilities between them. In 1965 about 44,200 sq ft of outside concrete with four, sixty-cow free-stall housing barns were added. Also a milking parlor and a bulk tank milk storage were added to change to a modern and efficient system of milking.

The drainage pattern is divided with an area of about 20,000 sq ft draining to the west and the remaining area of about 44,200 sq ft draining to the north. A 6-in. curb exists around the entire confinement area to control animal wastes and water runoff.

### Dairy Herd Management

All cows except special research animals are housed in the free-stall barns and fed in outside feed bunks. Feed consists mainly of corn and grass silage with some chopped hay. Although concentrates were at first fed in the milking parlor, this practice was soon stopped. Outside lots were arranged with portable electric fence divisions for experimental nutrition and housing studies.

---

Journal Paper No. 5851, Agricultural Experiment Station, Purdue University, West Lafayette, IN.

Supported in part by grants from Indiana Electric Assn., Indiana Farm Bureau Co-operative Assn., and Grant No. 8R01EC0024-02, Environmental Control Administration, Consumer Protection and Environmental Health Service, Dept. of Health, Education and Welfare.

The authors are A. C. DALE, Professor, Agricultural Engineering Dept., J. L. ALBRIGHT, Professor, Animal Sciences Dept., J. C. NYE, Assistant Professor, Agricultural Engineering Dept., and A. L. SUTTON, Assistant Professor, Animal Sciences Dept., Purdue University, West Lafayette IN.

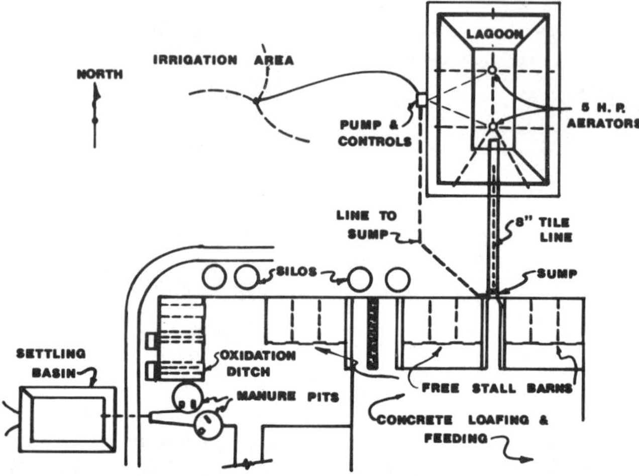


FIG. 1 Layout of deep lagoon-irrigation system at the Purdue Dairy Farm.$

### Liquid Manure System

In 1965 two 24 ft diameter by 10 ft deep 38,000 gallon underground liquid manure tanks were installed for trial and study. These tanks provided storage for the manure from 100-120 cows for 6-8 weeks. The openings into these storage tanks are 8 ft long by 1 ft wide for the tractor scraper and 4 ft long by 2-1/2 ft wide for the pump-agitator to pump into a 1500 gallon wagon for hauling directly to the field.

Solids were scraped directly into the storage tanks. Some runoff was also diverted into the tanks to provide a satisfactory mixture of about 8-10 percent solids for liquid handling. The liquid system reduced fly breeding and allowed more flexibility of labor than the solid waste system by the manure was highly odorous when applied to land. Since the city of West Lafayette is located within

FIG. 2 An aerial view of the Purdue Dairy Farm Center with Lagoon.

1/2 mile of the dairy farm, the odors had to be controlled.

In conjunction with the liquid manure system, a small lagoon was constructed to the west to intercept and treat the feedlot runoff. A three-quarter horsepower centrifugal sewage pump was used to irrigate the effluent onto nearby cropland with little difficulty. The liquid portion of this lagoon was free of any debris or coarse solids as most of the solids in the runoff were deposited as a sediment in the channel or on the concrete lot.

### Other Manure Management Systems

Simultaneously with the installation of the liquid manure tanks one 40 ft x 72 ft free stall barn was modified to study in three separate 24 ft bays (a) a slotted floor over an oxidation ditch, (b) a slotted floor with a pit storage and sump and (c) a concrete floor that was scraped.

**The Oxidation Ditch.** The oxidation ditch under the slotted floor was operated from December, 1965 through June, 1966 and then intermittently after that time. Varying degrees of success were obtained using this oxidation ditch. On several occasions foaming occurred to such an extent that the building became unusable.

In March, 1966 foam reached about 3 ft deep in the free stall bay over the oxidation ditch. This was due to overloading of the ditch during a cold period of 1966 when the rotor aerator inadvertantly stopped. Some extra manure was pushed into the ditch and the layer of ice was broken before the rotor was restarted. Anti-foam agents helped to some extent but did not totally eliminate the problem.

In June, 1966 the oxidation ditch was stopped for cleaning. When the liquid in the ditch was removed an estimate of about 60 percent of the solids were found to have accumulated on the bottom. This accumulation probably contributed to the foaming problem. It was also an indication that the channel was too deep, being 4 ft in depth.

The oxidation ditch was used again in late summer and fall of 1966. Again foaming occurred to such an extent that the system was finally abandoned. However, the ditch was used in 1973 with an Aerob-A-Jet aerator. The results of that study were satisfactory as far as odor and foaming were concerned as reported by Simons, Jones and Dale (1973).

**The Slotted Floor.** A manure movement system was tried in the central part of the free-stall barn. A slotted floor was constructed in the alley over a 4-ft deep pit with a 10-ft deep sump at the end. A sluice gate separated the sump from the pit. When the pit was full, the sluice gate was lifted and the manure was allowed to flow down into the sump. The system was tried on several occasions but the manure seemed to congeal and would not flow to the sump. On one occasion it was flushed down into the pit by fire hoses after allowing the manure to soak for a long period. Even then it flowed quite slowly and did not homogenize itself. It was possible to pump the manure from the sump with an impeller type, chopper pump.

**The Solid Concrete Floor.** The solid concrete floor section of the north end of the barn was used as a control in the dairy housing study. Manure from this area was scraped daily by a tractor mounted scraper to one of the liquid tanks previously described. Although this system required more labor than either of the other two systems during the immediate operation, it has been the preferred system for dairy housing. However, the other

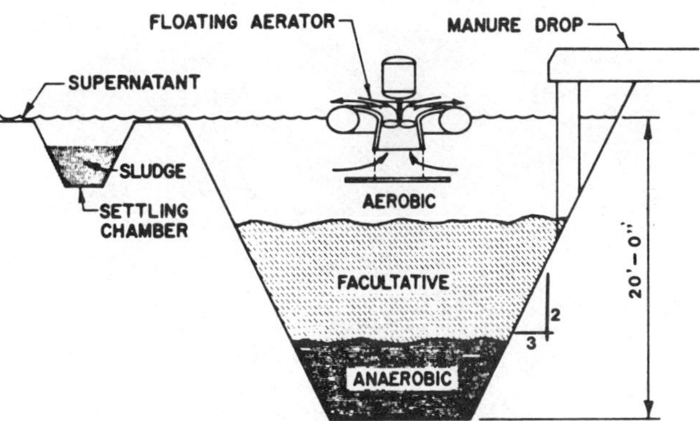


FIG. 3 Combination aerobic-anaerobic lagoon installed at the Purdue Dairy Farm.

two are still used and the liquid manure storage is mixed and pumped and spread on land.

The comparison of the dairy housing system has been reported by Albright and Alliston (1971) and Hill et al. (1973).

## THE DEEP LAGOON AND IRRIGATION SYSTEM

None of the numerous systems tried seemed to improve waste management for the outside floor and free-stalls. The small lagoon to the west had appeared to be a good way of intercepting the wastes for holding until they could be irrigated onto cropland (Dale et al. 1969). Thus an adaptation of this lagoon system was chosen. Further, something had to be done for cows in the two free-stall barns at the east end of the concreted lot. Also, the terrain to the east and north of the barns had a natural slope toward a ditch to the north.

### The Lagoon System

**Deep Lagoon.** A deep, combination aerobic-anaerobic lagoon was designed to fit the dairy farm similar to Fig. 3. Mechanical aerators were used to aerate the lagoon surface. The manure from the feedlot was scraped daily to a sump as shown in Fig. 4. Effluent from the lagoon was recirculated through the sump and the lagoon was loaded almost continuously. The sump was connected to the lagoon with an 8-in. tile line which is shown emptying into the lagoon in Fig. 5. The pump that was used to supply effluent to the sump was also used to irrigate from the lagoon. The irrigation system for the lagoon consisted of two 3/8 in. irrigation nozzles and 500 ft of 1-1/2 in. PVC pipe.

**Size.** The final lagoon size as determined for the operation was 100 ft x 150 ft x 20 ft deep with a side slope of 3 horizontal to 2 vertical. This provided a volume of 174,000 ft$^3$ for storage of the following:

1 Six months runoff of 64,000 cu ft (12 in./ft$^2$ for 64,000 sq ft of feed floor, lagoons, ramps and free-stall barn roofs).

2 Manure produced by 140 cows/six months of 50,000 cu ft.

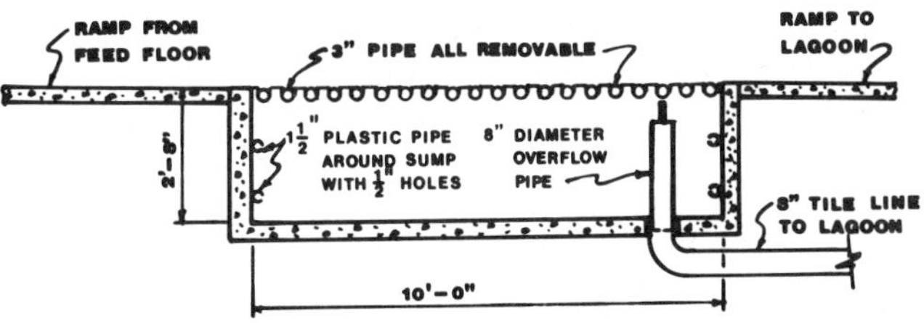


FIG. 4 Cross section through sump and tile line.

FIG. 5 Wastes were flushed through an 8-in. tile onto a ramp leading to the lagoon.

FIG. 6 Light foam occurred on the top of lagoon but foaming has not been a problem.

3 A reserve of 36,000 cu ft for sludge and operation of aerators after irrigation.

4 A free board volume of 24,000 cu ft.

**Aeration Requirements.** Aerator size was based on the maximum oxygen demand as indicated by 1.5 times the $BOD_5$ (5-day Biochemical Oxygen Demand) as follows:

1 One 1300 lb cow produces a 5-day BOD of 2.50 lb/day

2 140 cows produce a BOD of 140 x 2.5 = 350 lb/day

3 1.5 times the 5-day BOD = 1.5 x 350 = 525 lb/day

4 Aeration provided by: 1 horsepower hour of aeration = 3.2 x 0.70 = 2.44 lb $O_2$/hr or 24 hr = 53.7 lb/(horsepower)(day)

5 Horsepower required = 525 - 53.7 = 9.8 hp (10 hp were used.)

**Disposal of Effluent.** The wastes and lagoon contents are irrigated on about 30 acres of land. This is a disposal rate of about 5 cows per acre.

Total water and waste effluent to be irrigated each year on 30 acres:

1 Water from rainfall =

$$\frac{36 \text{ in. (rainfall) - 12 in. (evaporation)}}{12 \text{ in./ft}} \quad (64,000 \text{ ft}^2)$$

$$= 128,000 \text{ cu ft.}$$

2 Manure to be irrigated is 140 x 360 x 2 = 100,000 cu ft/yr.

3 Total (rainfall + manure) = 128,000 + 100,000 = 228,000 cu ft/yr.

4 Depth =

$$\frac{228,000 \text{ ft}^3 \times 12 \text{ in./ft}}{30 \text{ acres} \times 43,560 \text{ ft}^2/\text{acre}} = 2.1 \text{ in./yr}$$

**Summary of Operation.** As the system was placed in operation in August, 1970, a number of factors proved to be in need of modification. Movement of material through the 8 in. tile line from the sump was difficult as considerable amounts of dry corn silage and gravel and chips used for bedding migrated to the sump. Therefore, the manure was not readily washed into the lagoon with the recirculation of effluent and some of it had to be pushed down the ramp.

Several modifications to the water distribution system in the sump were attempted. One potential system used 1-1/2 in. plastic pipe with 3/8 in. and 1/4 in. holes. Turbulence from the recycled effluent steadily washed the manure over the 2 ft high overflow in the sump and

into the 8 in. tile line. Continuous input of manure to the lagoon was achieved to some extent. A complete description was reported by Dale (1971).

Solids in the recycled effluent clogged the pump and pipe creating another problem that resulted in abandoning the sump loading system. After two years of dragging all the manure down the ramp and pushing it into the lagoon the ramp system of loading was also abandoned. One of the liquid tanks is now used for collection of most of the scraped manure and an impeller mixer-pump is installed to mix and pump the manure through a 6 in. plastic pipe to the lagoon. Runoff and liquids enter the lagoon by the concrete ramp.

### The Aeration System and Requirement

The initial selection of the aeration quantity was based on 1.5 times the $BOD_5$ of the 140 dairy cows. This led to the selection of two 5-horsepower floating aerators which were considered to be adequate. Each aerator was rated to supply 3.2 lb of oxygen per horsepower hour at normal test conditions.

During cold weather the aeration system was reduced to one 5-horsepower aerator as shown in Fig. 6. By November 24, 1970, the water temperature was down to 1.5 C. From about December 15, 1970 to March 1, 1971 the surface of the lagoon was completely frozen, except for a small area around the operating aerator. After the first winter the decision was made not to operate the aerators at all from the first of November through March but to operate both aerators from April to November. During these periods odors were controlled by the low temperature.

In the spring as the biological activity increased, both aerators were operated to supply additional oxygen. Some odor was immediately noticed but it soon diminished. Odor has been kept down since, but dissolved oxygen has not been measurable except near the aerators. In the summers one of the floating aerators would overload the rated fuses more often than it should. This particular aerator pulled more than 20 amperes. Finally this aerator was left off and only one 5-horsepower aerator operated. No increase in odor was noted. Thus odor is being controlled by aeration no greater than the equivalent $BOD_5$ or slightly less.

### Irrigation

In the spring of 1971 the irrigation system was started. It consisted simply of a 3-horsepower centrifugal pump,

FIG. 7 A 5/16 in. rotary irrigation nozzle was used for two years to irrigate lagoon contents onto cropland.

1-1/2 in. plastic pipe, and two 3/8 in. revolving sprinkler heads as shown in Fig. 7. The capacity of the pump was about 80 gallons per minute at 40 psi. The sprinklers were placed in the grassland to the west of the lagoon and moved manually each day to new locations as deemed necessary to prevent runoff. Although the nozzles and pump intake plugged occasionally, the system worked fairly well and the lagoon was lowered to a satisfactory level of about 7 ft deep by the fall of 1971.

In the summer of 1972 the same irrigation system was again used, but the suction intake to the pump, the impeller of the pump and the irrigation nozzles continually plugged. Special screened intakes were fabricated to prevent this; however, the screens clogged and collapsed from the outside pressure and the pump cavitated. The nozzles also plugged often. Nevertheless the system was operated on a time clock operating 15 minutes and remaining off 15 minutes through the summer of 1972 with the level of the lagoon finally being lowered to a desirable level for reception of the wastes in the winter of 1972-1973.

The accumulation of solids in the lagoon by the spring of 1973 made the small pump and irrigation system inoperable. Intakes clogged rapidly, nozzles stopped up immediately and operation was impossible. At this time a 40-hp pump and irrigation system which was installed at the Baker-Purdue Animal Sciences Farm was brought to the dairy lagoon to irrigate its contents onto the surrounding land. This irrigation system included a Gorman-Rupp centrifugal pump capable of delivering about 350 gallons per minute at a head of about 80 to 90 psi gage. A big gun-type irrigation nozzle about 1.3 in. in diameter along with 5 in. "quick-coupling" aluminum tubing was used as a main line from the pump to the area to be irrigated with a 4-in. lateral to the nozzle. With a 1 in.-screened intake over the end of a 6-in. suction line the excess accumulation of water and wastes was irrigated in about two weeks onto the disposal area. Large solid particles consisting mainly of small wood chips, shavings, corn silage fibers 1 to 2 in. long, and some grass silage and hay stems did plug the pump occasionally. These materials accumulated over the enclosed impeller of the pump enough to reduce the pressure at the big gun-nozzle and to cause the pump to vibrate excessively. The pump had to be cleaned about once a day to facilitate proper operation.

By the summer of 1974 the average solids accumulation in the lagoon approached 5 percent. A 2-hp sewage open impeller pump was used in conjunction with a 4 in. flexible perforated corrugated pipe to discharge the lagoon liquid onto the soil. By using about 500 ft of this flexible hose and moving it around in a semi-circle it was possible to cover a large area for spreading this 5 to 6 percent solids effluent. The small pump, capable of pumping about 200 to 250 gallons per minute at about 10 to 15 psi, did a very good job for about four weeks at which time one of the motor windings shorted out.

The main difficulty encountered was in moving the irrigation pipe which was full of liquid. However, this can be done with tractors. The lagoon was finally lowered by the Gorman-Rupp pump and irrigation system to a satisfactory level for this year.

### Nutrient Conservation

Results as reported by Jones, Nye and Dale (1973) show that only about 38 percent of the nitrogen is conserved by this combination aerobic-anaerobic lagoon. This is in contrast to the 58 percent that remained in the initial anaerobic lagoon and 92 percent in the liquid manure storages.

## RESULTS AND CONCLUSIONS

The present system is considered to the "best" of all systems tried at the Purdue Dairy Center. Some of the factors in favor of the system are as follows:

1 The system is essentially odorless.

2 A place is provided to put the wastes at all times.

3 Direct runoff into streams is prevented.

4 A relatively small amount of labor is required to operate the system.

5 Aeration for odor control requires only sufficient oxygen to equal the $BOD_5$ during the six warmer months.

6 Irrigation water and about 40 percent of the nitrogen are saved and provided for crops.

Main difficulties with the system were the following:

1 The large silage fibers, hay stems, bedding chips, twine, etc. do not degrade rapidly and thus continually build up over a period of years. Also they do not settle out so a clarified supernatant can be pumped from the top. Thus they often clog up intakes, pumps, and irrigation systems if permitted to enter the storage with the waste materials.

2 Flushing dairy cow manure and bedding through an 8 in. pipe is not easily done.

## RECOMMENDATIONS

To make the system described here work without difficulty, some additional steps must be taken. These could be the following:

1 Install an adequate sedimentation basin between the feed floor and a storage. However, a sedimentation basin would not take out the long fibers of silage and hay that are mixed with the manure which is pumped into the storage.

2 Install a solids-liquid separator between the waste accumulation pits permitting only the liquids to enter the lagoon.

3 Use only finely screened materials for bedding or solids separated by a solids-liquid device for bedding and a finer chopped silage and hay for feed.

References

1 Albright, J. L. and C. W. Alliston. 1971. Effects of varying the

*(Continued on page 131)*

# Self-Unloading Pits in a Dairy Manure Management System

William R. E. Euerle, Gerald O. Euerle, Philip R. Goodrich

MEMBER
ASAE

ONE of our primary goals, when we decided to build a new barn for our dairy operation, was to minimize the time and effort needed to manage the manure from our operation. Time and effort are important because the operation includes 1300 acres (526 ha) of crops in addition to the dairy operation. We looked at a number of systems, including complete pits beneath free stall dairy barns and gutter cleaners emptying into outside pits. These systems did not seem to be just what we wanted.

## BARN DESIGN

Our proposed barn location was on the edge of an area which sloped down into a natural draw, a feature we decided to utilize in planning the building. We constructed a 3/4 acre (0.3 ha) lagoon in the ravine area below the barn site, using cost sharing from the Agricultural Stabilization and Conservation Service and with assistance from the Soil Conservation Service. The lagoon was designed to receive the manure from the storage pit in the new barn and runoff from the outside barn lots. The new barn was placed on the hill above the lagoon with one end facing the lagoon. The cost of the pit in the barn was minimized by designing it for two months storage only. The barn is 148 ft (45.1 m) long and 39 ft (11.9 m) wide. It is a free stall dairy barn with the alleys totally slatted using gang slat sections that are 8 ft (2.4 m) long and made in grids 42 in. (107 cm) wide. Each gang contains six slats with the space between the slats 1 3/4 in. (4.4 cm). Fig. 1 shows the basic layout of the barn. There are 84 free stalls with 39 located along the west side, 33 on the east side and 12 at the north and of the center feed bunk.

There are 4 in. (10.2 cm) of insulation in the walls and 8 in. (20.3 cm) in the ceiling, making it a warm confinement building. To improve the comfort of the cows, indoor-outdoor carpeting is used in the stalls. No other bedding is needed in the free stall area. A maternity area and a milking parlor are adjacent to the barn. The manure from the maternity and sick bay area is scraped into the pit daily. The wash water from the milking parlor drains into the pit near the north end. The pit is 148 ft (45.1 m) long, 24 ft (7.3 m) wide and 6 ft (1.8 m) deep (Fig. 2). It is located below the two 10 ft (3 m) alleys, the feed bunk, and the 12 free stalls at the end of the feed bunk. The pit is not located below the free stalls on the east and west sides of the barn. The two 128 ft (39 m) center support walls run lengthwise each side of the sil-

Miscellaneous Journal Paper No. 5177, University of Minnesota Agricultural Experiment Station.

The authors are: WILLIAM R. E. EUERLE and GERALD O. EUERLE, Owners, Euerle Brothers Dairy Farm, Litchfield, MN; and PHILIP R. GOODRICH, Assistant Professor, Agricultural Engineering Dept., University of Minnesota, St. Paul.

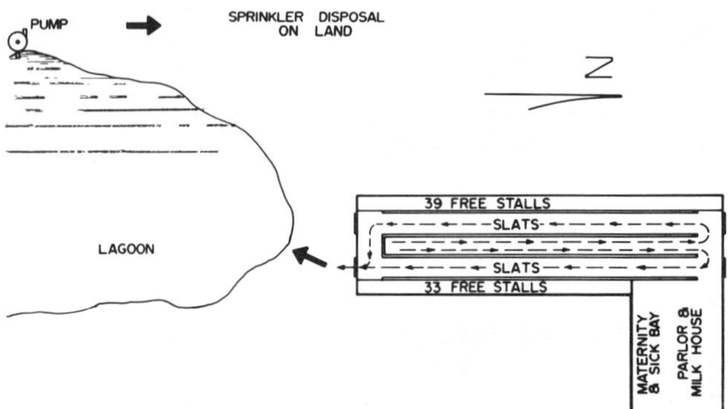


FIG. 1 System layout showing movement of waste to lagoon.

age feed bunk and terminate 10 ft (3 m) from the end walls. An additional wall joins these two center support walls at their southmost point. Thus, the pit is divided into three compartments. The compartment beneath the center feed bunk is totally covered because the feed bunk and 12 free stalls are above it. The two compartments on each side of this central section are covered with slats. The central compartment receives very few solids but fills with liquid which flows into it from the other two compartments.

## CLEANING THE PIT

The wall which was constructed at the south end to join the two center support walls was an afterthought and makes the system work very well. The initial experience with the pit was that some solids tended to remain on the level floor when the liquids all ran out very quickly from the center section directly out the two doors on the south end. By blocking the center compartment at the south end, the liquids are forced to flow out the north end of the central compartment around through the area which receives more solids and then out the door on the southeast corner (Fig. 1). Experience has shown that opening

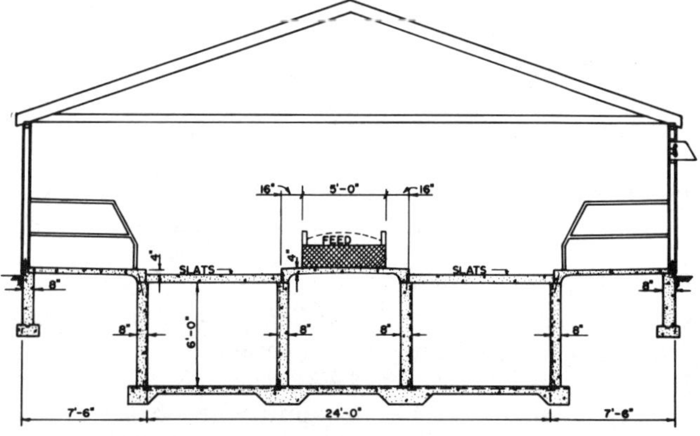


FIG. 2 Cross section of pit beneath free-stall dairy barn.

one door and allowing everything to move out one door is the best method of operating the pit. This flushing-type action moves the solids out very well. We are often asked if the floor should have been sloped from north to south. We feel that the level floor helps to make the system work much better than if the floor was sloped. The level floor allows the system to work when only one door is used and reduces the amount of work involved in emptying the pit.

The doors are 8 ft (2.4 m) wide and 6 ft (1.8 m) high made from double 2 in. (5.1 cm) x 6 in. (15.2 cm) planking. The two layers are built at right angles to each other for strength and sealing. The pits fill up in about two months of operation. We then hook two cable winches on the one door at the southeast corner of the building and lift the door in its slides.

The solids start moving as soon as the door opens and all of the liquid and the solids flush out in about 10 minutes. This mass of slurry moves down the hill and enters the lagoon. Some channeling has occurred in the side slope and a lined flume might be advisable. Because of the ravine, the lagoon is deeper on the south side and the depth of the pond reaches about 14 ft (4.3 m) at its deepest area.

## LAND RECYCLING AND RESIDUE

We usually pump on the deep side in July or August, using a Gorman-Rupp centrifugal pump. We use a 660 ft (201 m) flexible 4 in. (10.1 cm) rubber hose, which connects to a self propelled traveling gun sprinkler. The lagoon and manure are pumped out onto growing crops. We have put the material from the lagoon on alfalfa,

corn, oats, and soybeans and, at no time have we had trouble with burning or hurting the crops.

Since we built this set up in 1970, we have had very good results with the overall system. The manure is very easily managed and returned to the soil. Our location was ideal for this particular type of set up and would not fit some other operations, but for our use, if fits very well. We also return our animal waste to cropland without the compaction that usually accompanies use of the liquid tank wagons. There may be some build-up of solids in the bottom of the lagoon, because we cannot stir and agitate the solids which have settled out. However, we feel that the usefulness of the liquid and the nutrients that we do get out are well worth the $10,000 cost of the irrigation system. We also can use the irrigation system and pump for irrigation on some of our crops from a nearby river. Therefore, the total cost of the system does not have to be charged against the dairy operation. A good sized liquid manure tank and accompanying pit pump would cost nearly as much as our present system and would not provide the alternative use as an irrigation system.

## SUMMARY

Overall, we are very happy with the barn and the manure management system. We don't know how the system could be made any more efficient than it is. Both federal and state inspectors have checked our operation and feel everything is fine. It is very satisfying to have such an easily operated, efficient system to work in with our field operations.

---

## Waste Management for 150 Cows
*(Continued from page 129)*

environment on performance of dairy cows. Symposium on Sound Animal Care—A Prerequisite to Productive Husbandry. Journal of Animal Science 30:566-577.

2 Dale, A. C. 1971. Management of dairy cattle wastes by the deep aerated lagoon and irrigation onto soils and plants. Progress Report, Department of Agricultural Engineering, Purdue University, West Lafayette, IN.

3 Dale, A. C., J. R. Ogilvie, A. C. Chang, M. P. Douglas and J. A. Lindley. 1969. Disposal of dairy cattle wastes by aerated lagoons and irrigation. Animal Waste Management, Cornell Conference on Agricultural Waste Management, Syracuse, NY.

4 Hill, D. L., N. J. Moeller, D H. Yungblut, C. E. Parmelee and J. L. Albright. 1973. Effect of two different free-stall housing systems upon milk production, milk quality, health, and behavior of dairy cows. Journal of Dairy Science 56:668.

5 Jones, Rex E., John C. Nye and Alvin C. Dale. 1973. Forms of nitrogen in animal waste. ASAE Paper No. 73-439. Annual Meeting. ASAE, St. Joseph, Mich. 49085.

6 Simons, Dieter, Don D. Jones and Alvin C. Dale. 1974. Oxidation ditch system analysis and field evaluation of the Aerob-A-Jet. Processing and Management of Agricultural Waste, Proceedings of the 1974 Cornell Agricultural Waste Management Conference, Rochester, NY.

# A Planning Study on Dairy Wastes Management


S. I. Gershon, S. A. Hart,
SENIOR MEMBER
ASAE

A. C. Chang,
ASSOC. MEMBER
ASAE

J. W. Branch, Jr.
ASSOC. MEMBER
ASAE


THE major milkshed for the Los Angeles metropolitan area is the Chino Basin of the Santa Ana River watershed, 35 miles east of Los Angeles (Fig. 1). Approximately 165 000 dairy cows are maintained within a 120-square mile area.

The Santa Ana River watershed is also important as a source of water for the nearby metropolitan area. But as time has gone on, this valuable water supply has been plagued with both quantity and quality problems. As a result, a regional water resources planning organization was created; this is the Santa Ana Watershed Planning Agency.

Concern about pollution of the local water, caused by all land and water users, including dairies, prompted the agency to seek ways to permanently ensure a continuous supply of good quality water. With Federal, State, and local funds, the agency retained consultants to study and recommend various management plans, including the handling of waste from the various contributors of pollutants. Albert A. Webb Associates was selected to make the 8-month study of dairying (Gershon et al. 1974).

## OBJECTIVE

Specifically, the purpose of the study was to determine economically feasible methods by which the dairy industry can reduce by 90 percent the amount of salts it adds to the groundwater basin. This 90 percent of "salts added" amounts to about 50 000 tons of salt (or total dissolved solids) per year.

## CONDUCT OF THE STUDY

To carry out this objective, an extensive data-gathering program was initiated. It included on-farm inspection of manure management and mismanagement practices; in-depth discussions with dairy leaders, practicing dairymen, and representatives of water agencies; evaluations of the research that had been and is being conducted by the University of California; and critical review of and discussions on the proposed and actual regulations being promulgated by the California Regional Water Quality Control Board, Santa Ana Region.

From this information, the extent of the salt problem was determined; alternative means of collecting, treating, and disposing of the dairy waste streams were evaluated; and economic analysis was made of the feasible alternative methods; and a practical plan was recommended.

## DAIRY WASTES AND THE SALT PROBLEM

First of all, we looked into the dairy wastes themselves and determined the extent of the salt problem. Dairy wastes actually consist of three waste streams; these are the manure, the dairy sewage (or wash water). and the runoff water from rain.

Each cow produces about 1.3 cubic feet of fresh manure daily. Of this amount, however, only about 0.5 cubic foot is left in the corral after drying, weathering, and rain (Bishop 1973). The physical characteristics of fresh manure are shown on Table 1.

Sanitation requirements demand that the dairyman wash

**TABLE 1. SUMMARY OF PHYSICAL PROPERTIES**

| Item | Waste production per equivalent cow per day | | |
|---|---|---|---|
| Total wet manure | 85 lb | 1.30 cu ft | 9 gallons |
| Feces | 55 lb | | 5.5 gallons |
| Urine | 30 lb | | 3.5 gallons |
| Wash water | | | 47.0 gallons |
| Corral scrapings at 33 percent moisture conent | 16 lb | 0.5  cu ft | |

his cows before they are milked. He must also thoroughly wash and sanitize the equipment after milking, and he must clean and wash down the milking barn. These processes create the second waste stream, which we call dairy sewage. Our study showed that approximately 47 gallons of wash water are used per cow per day (Chang et al. 1974). And about 10 percent of total manure production is included in the wash water.

The third waste stream is corral runoff water from rain storms. The amount of this water depends, of course, upon the size of the storm. However, the mineral content of this water is generally about that shown in Table 2.

Once these facts were established, we were able to make a determination of the extent of the salt problem. In the area we were studying, we found that, measured as total dissolved minerals, each dairy cow produces about 1321 pounds of salt per year. Assuming the waste from one cow is allowed to stand or is spread over one acre of land, the limitation of 600 pounds per year of salt going into the ground is exceeded (Section 3, Recommended Plan, Santa Ana Watershed Comprehensive Water Quality Management Plan 1974). However, not all salts entering the soil reach the groundwater because soil minerals tend to react with the salts in leaching water. Therefore, some of the salts are removed or are exchanged into other minerals of insoluable form. Groundwater quality investigations in the Chino-Corona Area indicated a localized salinity increase in the existing disposal area (Adriano et al. 1972).

## WASTE MANAGEMENT TECHNIQUES
## NOW BEING USED

Armed with this information, a comprehensive study of the techniques now being used by the dairy industry in the Chino Basin of the Santa Ana River watershed was initiated.


The authors are: S. I. GERSHON, vice president of Albert A. Webb Associates, 3788 McCray Street, Riverside, California 92506; co-authors S. A. HART, A. C. CHANG and J. W. BRANCH, Jr., consultants to Albert A. Webb Associates on the project.

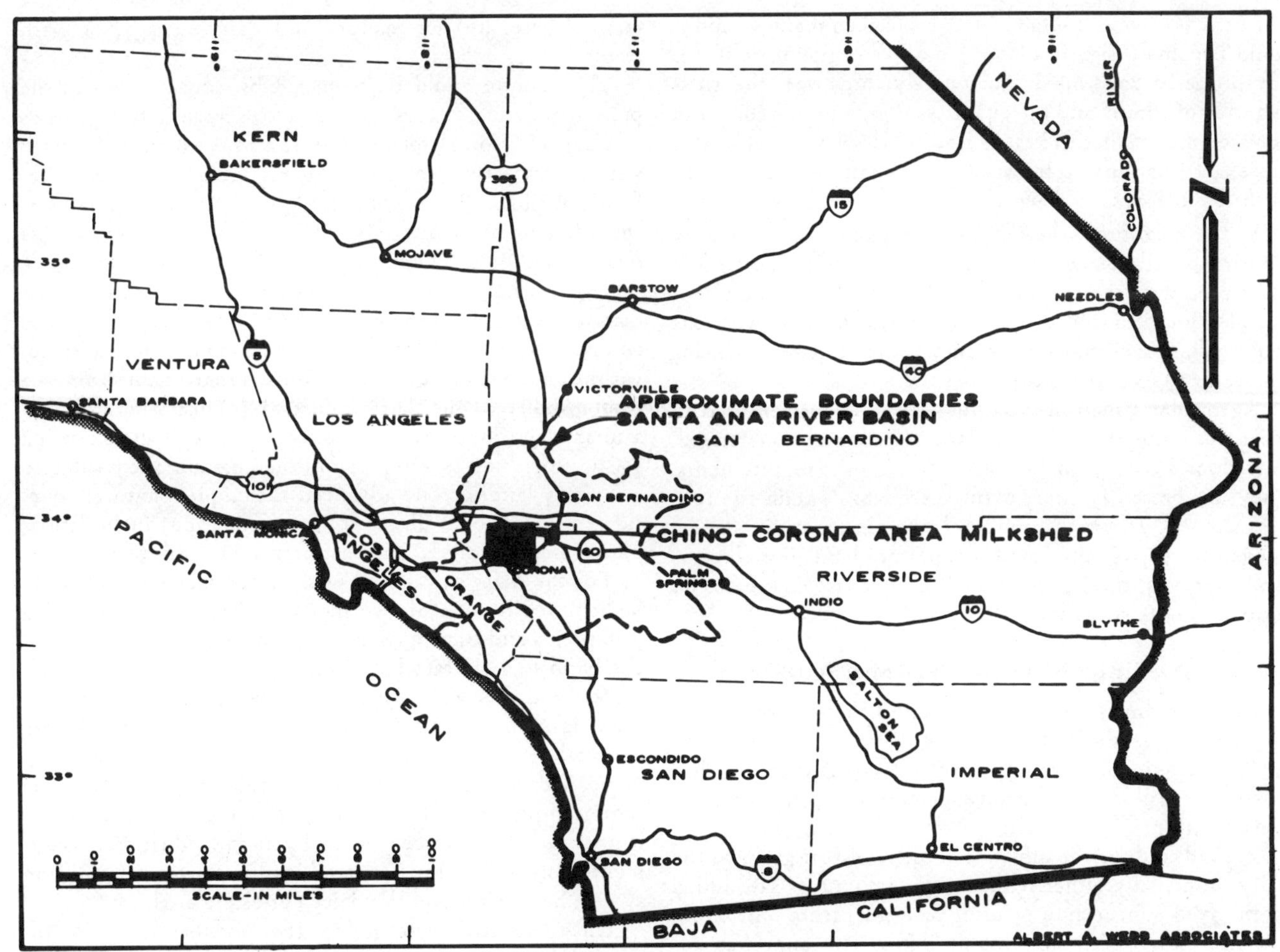



FIG. 1 Study area.

It was found that in most cases the manure is allowed to accumulate for 5 to 6 months on the ground surface or in corrals, where it undergoes weathering. At the end of the long dry summer, and before the winter rains begin, it is scraped out and removed, either to be hauled to cropland or to composting operations. This is repeated in the spring after the corrals have dried out enough to allow vehicular traffic.

Recently, several dairymen have introduced variations designed to help them in handling this waste. These included the utilization of washed manure as bedding in stalls and on-farm composting. A study of each of these was made a part of our overall investigation. All the variations exhibited definite advantages.

The second major waste stream—the wash water, or dairy sewage—is usually collected by troughs from which it flows immediately into sumps. The waste in the sumps may be stored for a short time or it may be pumped to a pasture for disposal. At any rate, the proper collection and disposal of this waste water has been the responsibility of each individual operator, although in some cases, actions have been forced by public health regulations or by court orders.

In the handling of this waste stream also, individual dairymen have demonstrated considerable ingenuity in the development of possible alternative techniques. Most of this waste stream is utilized to irrigate pasture land.

The third and final waste stream is the corral runoff from storms. Surface runoff from uncovered livestock con-

**TABLE 2. QUALITY OF SIMULATED RUNOFF CARRIED WASTE FROM A DAIRY CORRAL***

| Rain no. | Total rain | Amount of rain before runoff | Total runoff | Average rainfall intensity in./hr | pH | E.C. mmho/cm | Percent TS | COD | $NH_4^+$ | $NO_3^-$ | K-N | Cl | $PO_4^{\equiv}$ | $HCO_3$ as $CaCO_3$ | Date of the experiment | Corral condition (days after corral cleaning) |
|---|---|---|---|---|---|---|---|---|---|---|---|---|---|---|---|---|
| | in inches | | | | | | | | | | | | | | | |
| 1 | 1.08 | 0.61 | 0.25 | 0.86 | 7.85 | 6.92 | 0.76 | 3479 | 12.80 | 7.57 | 135.8 | 1408.6 | 0.59 | 922.57 | 5-11-73 | 8 |
| 2 | 1.02 | 0.28 | 0.27 | 0.42 | 8.93 | 5.86 | 0.63 | 3389 | 17.67 | 4.58 | 139.3 | 1054.9 | 0.64 | 870.4 | 5-16-73 | 13 |
| 3 | 0.92 | 0.82 | 0.03 | 0.44 | 8.70 | 6.51 | 0.79 | 4846 | — | — | 411.4 | 1337.1 | — | 934.2 | 5-25-73 | 22 |
| 4 | 0.29 | — | 0 | 0.35 | — | — | — | — | — | — | — | — | — | — | 5-29-73 | 26 |
| 5 | 1.57 | 1.57 | 0 | 0.81 | — | — | — | — | — | — | — | — | — | — | 8-02-73 | 91 |
| 6 | 4.13 | 0.86 | 2.57 | 2.10 | 8.96 | 7.56 | 0.84 | 2862 | 19.78 | 0.93 | 96.5 | 856.0 | — | 1002.3 | 8-08-73 | 97 |
| 7 | 0.77 | — | 0 | 1.55 | — | — | — | — | — | — | — | — | — | — | 8-23-73 | 112 |
| 8 | 4.25 | 0.20 | 3.80 | 1.89 | 9.3 | 3.92 | 0.28 | 1021.5 | 7.6 | 0 | 33.2 | 365.4 | — | — | 9-14-73 | 1 |

*Unpublished data from Dr. A. C. Chang. Department of Soil Science and Agricultural Engineering, University of California, Riverside; Corral slope 3.2 percent, confined 100 milking cows, Corral size 260 ft x 220 ft.

MANAGING LIVESTOCK WASTES

finement is always contaminated by packed manure, and it could become a significant source of water pollution if it is not properly controlled. Fortunately, however, the total quantity of runoff and the quantity of salts in this runoff in our area are small. Studies are now under way to determine the exact relationship between the rainfall and the quality of the runoff water (Chang 1974).

In the meantime, the California Regional Water Quality Control Board has issued requirements that are specifically concerned with the control of the runoff—that is, the use of a collection, transport, and storage system—and with the requirement that the farmer empty the reservoir by using the water to irrigate a field.

Again, dairy men in the Chino Basin have attempted to meet this requirement by a variety of means and, in general, have done a commendable job of meeting the requirements. They are presently constructing basins to contain the runoff from a 10-year storm and the distribution system needed to dispose the runoff on pasture land. For all practical purposes, the handling of this waste stream creates no problem in the study area.

## POSSIBLE ALTERNATIVE METHODS

At this point in our study, we faced the fact that three approaches were possible. These are:

First, eliminate dairying from the basin. This approach is opposed by all interests involved, and it was dismissed from consideration.

Second, reduce the minerals in the feedstuff given to the dairy animals by either reducing the amount of salt added to the feed concentrate or using feed-stuff from only those localities where the salt content is low. This approach has limited possibilities, but does not offer much help.

The third approach open to the dairy industry is that of truly proper management of the wastes to eliminate or to minimize the mineral contribution to the waters of the Santa Ana River Basin. This approach, we contend, is the most logical one.

As pointed out earlier, only two of the waste streams were considered to be significant enough to require manage-ment. These are the manure and dairy sewage, or wash water.

The manure could be managed by continuation of the present practice; that is, continue to scrape it from corrals and spread it on cropland, but at a rate which will reduce salt concentration to meet the requirements of the California Regional Water Quality Control Board. As the average 420-cow dairy has only 40 acres, this won't work. The second method is by exporting it.

Exportation could be achieved by (a) hauling the manure to cropland in another basin, (b) converting it into compost to be used by home gardners and public agencies within and outside the basin (thus, removing it from the responsibility of the dairy industry), (c) disposing of it to a sanitary landfill, or (d) processing it by one of the advanced waste processing techniques and then hauling the residue to a sanitary landfill. The advanced techniques that were considered are aerobic digestion, wet oxidation, incineration, pit burning, pyrolysis, conversion into oil, and refeeding.

For the dairy sewage, or wash water, management could be achieved by one of two methods. These are evaporating it in ponds and transporting the concentrated residue out of the basin to a sanitary landfill or putting it into a collection system and then into a major interceptor line for final disposal into the Pacific Ocean. If this second method were adopted, the dairy sewage would require additional treatment. This could be given by the treatment plants of the sanitation districts of the county, by a regional plant before it goes into the interceptor, or by an individual dairy treatment plant such as the barriered landscape-water renovation system before it leaves the dairy (Erickson et al. 1972).

Once we had determined the possible methods for managing wastes, we undertook to develop cost comparisons (Tables 3 and 4).

## RECOMMENDATIONS

In developing a plan to recommend to the Santa Ana Watershed Planning Agency, we kept in mind that 92 percent of the salts associated with the dairy wastes are in the manure scraped from the corrals. Therefore, it a method

**TABLE 3. COST COMPARISON OF MANURE WASTE MANAGEMENT SYSTEMS
(IN DOLLARS PER EQUIVALENT COW PER YEAR)**

| Alternative | Corral scraping and loading | Trans-portation | Process | Disposal | Total | Remarks |
|---|---|---|---|---|---|---|
| In-basin recycling | ND * | ND | ND | 273.33 | 273.33 | Large capital investmant. |
| Manure export for cropland use | 3.60 | 22.50 | 0 | 3.60 | 30.70 | Disposal on cropland, 100 miles from the Chino-Corona area. |
| Composting | 3.60 | 0 | 2.10 | 0 | 5.70 | Limited market. Income on sale of composted manure reduces cost. |
| Haul to proposed sanitary landfill site in Chino Hills | 3.60 | 2.98 | 0 | 6.42 | 13.00 | Average of 10 and 24 ton trucks. The total cost of disposal in the Badlands for San Jacinto dairies is $14.50. |
| Aerobic digesters | ND | ND | ND | ND | 56.00 | Dairies must be converted to flush out system. Does not include |
| Wet oxidation | ND | ND | ND | ND | 48.12 | cost of removing salts from basin. |
| Incineration | 3.60 | 3.90 | 39.00 | 2.50 | 49.00 | Disposal cost based on haul of residue to Chino Hills site with 80 percent reduction in mass (includes transportation). |
| Pit burning | 3.60 | 2.40 | 16.62 | 2.50 | 25.12 | Air emissions questionable — disposal cost based on haul of residue to Chino Hills site with 80 percent reduction in mass (includes transportation). |
| Pyrolysis | 3.60 | 3.90 | 15.90 | 6.35 | 29.75 | Process cost probably low. Disposal cost based on haul of residue to Chino Hills site with 50 percent reduction in mass (includes transportation). Air emission questionable. |
| Oil conversion | ND | ND | ND | ND | ND | Not enough information available to make estimate. |
| Refeeding (Cerola process) | 20.80 | 6.00 | ND | ND | 26.80 | Cerola process has not been determined to be economically feasible as of this date. |

*ND — Not determined.

| Alternative | Dairy sewage management cost In dollars per milk cow per year | | | | Remarks |
| | Sewer system | Pretreatment | Final treatment | Total | |
| --- | --- | --- | --- | --- | --- |
| Evaporation ponds | 0 | 0 | 0 | 4.88 | Environmental problems-odor, public health, increased humidity |
| Entire treatment at Sanitation Districts of Orange County facilities | 9.84 | 0 | 13.02 | 22.86 | Potential septic problems in interceptor |
| Regional pretreatment plant | 11.25 | 8.11 | 5.42 | 24.78 | |
| BLWRS pretreatment | 8.20 | 14.93 | 3.28 | 26.41 | Not adequately tested |

can be developed for isolating this waste from the basin, only 8 percent of the salts will be left, well within the goal set for our study.

For that reason, we did not recommend management of the dairy sewage at this time. Our recommendations deal instead with management of the manure.

The most economical and practical method we found for managing the manure is composting. However, only about 20 percent of the manure now being produced in the study area is being composted. The lack of a market appears to be the major obstacle to expansion. Therefore, as part of our recommendation, we ask that studies be made on how to expand the market and how to reduce the cost to consumers so that a steady market can be maintained. Perhaps the decline in supplies of commercial fertilizers will provide the solution.

But because of the lack of an immediate market for additional compost, we recommend that a fail-safe backup system for disposing of the manure be provided. For this fail-safe backup system, the method that proved the second most economical is recommended; this is the use of a sanitary landfill. Landfilling is more than fail-safe, it is the system when the market for composted manure runs out. A landfill could also serve as a disposal site for the residue that would result from the various advanced disposal techniques that may in time prove to be economical and feasible.

As a part of our evaluation of the feasibility of our recommended systems, we looked at the facility requirements, their effectiveness in meeting the clean water goals, their costs, their economic feasibility, and the institutional arrangements required.

Additional facilities would, of course, be required for expanded composting; however, we foresee no problems in this connection.

For the required additional sanitary landfill sites, only

(Continued on page 138)

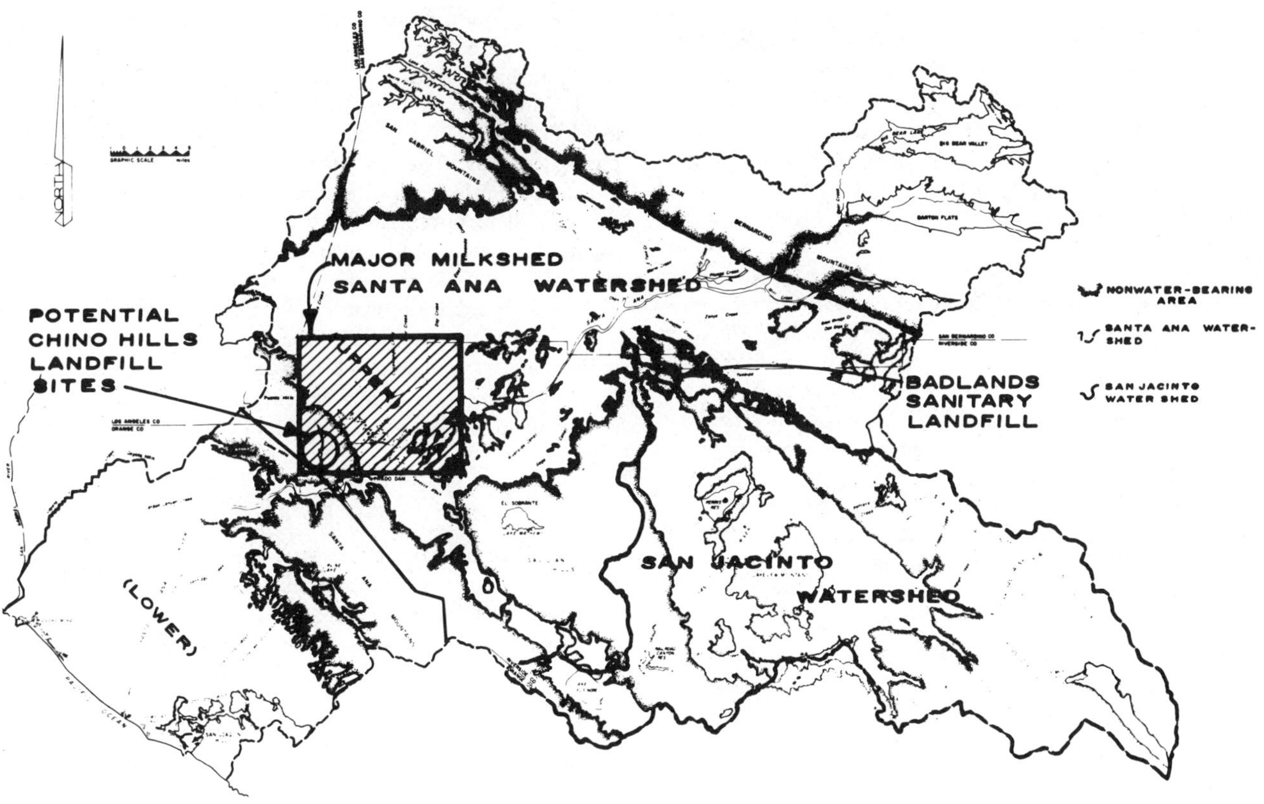


FIG. 2 Sanitary landfill sites in the Santa Ana River Basin.

# Dairy Waste Fiber — A Byproduct With a Future?

W. C. Fairbank, S. E. Bishop, A. C. Chang

MEMBER
ASAE

ASSOC. MEMBER
ASAE

DAIRY waste fiber (DWF) is the screenings extracted from liquefied fresh dairy manure resulting from hydraulically cleaned cow confinement facilities. For clean scouring of the paved milking, traffic, and loafing surfaces, for pipeline conveyance of the total waste stream, and for trouble free pumping and screening, it is desirable to add about ten volumes of new or recycled water for each volume of mixed feces and urine. Several industrial type sewage screens of stainless steel construction have proven their trouble-free ability to remove suspended macro debris which is mostly undigested roughage, seed hulls, and cow hair (plus the unavoidable extraneous trash like barnyard hardware and rocks). Conversely, DWF is the residual fraction of manure after the readily soluble components have been washed out.

Flush-cleaned dairies in California contain 240 to 1000 or more cows in high density confinement. The dairy is constructed to maintain a water flow grade of 1-3 percent from the milking parlor, through the loafing facilities and feed alleys, to the collection sump. Twice each day after milking new water which had only been used to wash cows and equipment and then impounded is released near the parlor. Alternatively, reclaimed water is returned by lift pump from a wastewater storage pond to a release point as close as 30 meters (by State Code) from the milking structure. An open-pipe flow-volume adequate to cause a surface flow of 20 liters per second per meter (100 gpm per foot) of alley width is common. An interceptor sump from 1-3 m deep across the lower end of the alleys is sized to hold a little more than half of the total waste volume per 24 hours, but of no greater capacity lest unwanted decomposition be made possible by less frequent pumping. A float-activated trash pump lifts the waste up to the screen as necessary — usually operating no more than two to four hours per day.

The use of mechanical liquid/solid separators to resolve physical problems of trying to dispose of liquid barn waste through sprinkler irrigation systems was first publicized by Fairbank and Bramhall (1968). By 1971 the acceptance of these factory built devices and their successful performance on California dairies was fait accompli. Bartlett et al. (1973) described methods of separating fibrous material from liquefied bovine manure, characterized particle size and analysis, and alluded to beneficial uses of the solid fraction. Until 1974 most of the interest in liquid/solid separation was as an expedient to waste handling and pollution control. In the same year, however, the shrinking supply and the increasing cost of feed, fiber, fuel and fertilizer stimulated consideration of utilizing or recycling this waste in some beneficial manner.

The authors are: W. C. FAIRBANK, Extension Agricultural Engineer, Cooperative Extension Service, University of California, Riverside; S. E. BISHOP, Farm Advisor, Cooperative Extension Service, Riverside, Calif.; and A. C. CHANG, Assistant Professor and Assistant Agricultural Engineer, Soil Science and Agricultural Engineering Dept., University of California, Riverside.

DWF is a wet byproduct still of little acknowledged value. The "brown water" from which fiber solids are screened is held until irrigation disposal in a wastewater storage pond that has been described as self-sealing by Davis et al. (1972), Meyer et al. (1973), and Oliver et al. (1974). This nutrient enriched water is of constant concern to the operator because it is both the fertilizing and potentially polluting portion of the manure waste stream and is much the greater volume to contain. The innocuous fiber solids, if not immediately used for bedding, are just stockpiled "back of the barn" until Fall plowing when they are broadcast spread and soil incorporated for some marginal benefit.

Economics of DWF are gathering attention as inter-relationships of energy utilization and recovery, feed values, diminishing availability of wood fiber and resource recovery are being recognized concurrent with the dairyman's old antagonist Waste Disposal. For several years it has been casually noted that when DWF is promptly sun-dried, decomposition is arrested and the material takes on the general character of finely chopped straw. The golden fiber certainly suggested a determination of the byproduct potential. Cooperative Extension began field evaluations, to become bolstered in 1974 by investigations in the Agricultural Engineering laboratories at UC Riverside and UC Davis.

The volume of wet DWF per cow per day from high-density confinement ("all concrete" and free-stalls during the lactating period) is approximately 21 l (0.75 ft$^3$). The stack density after free drainage is about 430 kg/m$^3$ (725 lb/yd$^3$). Each 100 cows will produce about 1000 m$^3$ (1300 yd$^3$) per year. The 25 known mechanical separators on California dairies are producing probably 250 m$^3$ (325 yd$^3$) per day. The potential for the entire state at 50 percent usage, however, could approach 3.6 x 10$^6$ m$^3$ per year. Of parallel interest is that the location of DWF would be as diffuse as the dairy industry. This dispersion might improve certain specialty byproducts recovery uses while discouraging other uses that would bear a stronger economy of scale. It is too early to determine marketability of recovered products and thus to project the net effects of dispersal of sources.

## THE PROCESS

Wet fiber falls from the discharge chute of the separator onto a pile, into a dump truck, or into a screw-conveyor with usually little or no attention. Handling and transportation can be by any equipment suitable for handling wet wood shavings — a similar commodity. Most dairies already have suitable equipment.

Storage of wet fiber in dump-truck piles or in windrows on soil is the common situation. During inclement weather decomposition is slow and pollution is nil if runoff water is controlled. During mild weather (5 C+) thermophilic composting commences quickly and within 24 hours the temperature within the pile often exceeds 38 C. Piles can be ignored until ultimate disposal.

The general appearance of fresh fiber varies from that of wet chopped straw if the herd is being fed on alfalfa hay to wet shavings if corn silage is the main roughage. The hulls of barley, cotton seed, and other grains are usually visible in the product. The non-colloidal character of the waste fiber mass allows the surface of the pile to dry quickly in our summer sun and low-humidity, a situation rather limited to the semi-arid Southwest.

Solar drying as the primary means of rapid moisture removal was preordained by energy economics as the only practical form of drying of this heretofore worthless material. In our region where everthing dries if you do nothing, it was only necessary to establish how thick the layer of wet fiber could be before rapid drying gave way to unwanted wet decomposition. It had already been noted that the quickly-formed dry surface became both an effective insulating blanket and a vapor restrictive layer. A rectangular area of 860 m$^2$ (0.2 acres) was graded at a site in the Chino-Corona dairy community which is described by the local flood control agency as having a mean annual rainfall of 33 cm (13 inches); a mean winter temperature of 11 C (51 F); a mean summer temperature of 23 C (74 F); and an annual pan evaporation of 190 cm (75 inches). Five truckloads of 7 m$^3$ each were spread to average depths of about 1-10 cm. Within three days in the summer when midday temperature exceeded 32 C (90 F) the surface layer to about 1 cm reached 25 percent average moisture. Layers thicker than about 3 cm required stirring two or more times per week because the drying action below that depth was so slight the darkening color of wet decomposition degraded the fiber below our acceptance level (we used a flail-type weed chopper but a low velocity tiller would have been the implement of choice). Visual examination of undisturbed dry surfaces revealed that minute fibers filled the voids between large fibers and all were weakly consolidated with soluble solids into a semi-rigid mass with the appearance of low quality paper pulp. This formed a layer impenetrable by sun and air. Approximately 8 cm below the surface of unstirred fiber, drying was nil even during high summer temperatures exceeding 38 C.

Moisture content of 25 percent was excellent for dust-free handling but caused mild heating and mold in 57 l (2 ft$^3$) plastic manure bags. Unheated aeration to a safe moisture of 15 percent after dry screening and before bagging, or within dry storage piles was indicated as probably desirable for a commercial dairy operation.

Classifying of dry fiber into uniform size grades is easily accomplished with conventional equipment. A rotary drum screen of 12 mm (0.5 inch) mesh on a soil shredder worked well for removal of large trash. Small lots run over inclined shaker or gyrating screens indicated that any classifier of woven wire screen used for dry manure, compost or aggregate should work well for DWF. A scalper screen of approximately 8 mm mesh (3 mesh per inch) is desirable to remove such trash as extraneous forage, rocks, beverage can "pop-tops", etc., which find their way in and which would jeopardize acceptance by any would-be user. A main screen of about 4 mesh per cm (10 mesh per inch) would drop out the fines which would not be wanted for most uses of discreet fiber. Air separation is to be considered because we observed amazing size and density classification when one of our screens with a 3 m (10 ft) discharge height was operated during a gentle breeze of about 5 knots.

Pelleting or cubing of DWF to a sufficient bulk density for economic transport and to a form amenable to mechani-cal handling was bench tested in the products packaging laboratory at the Agricultural Engineering Department at UC Davis. No particular limitations were evident. Organic additives are expected to be desirable for certain ultimate uses of the product and their binding effects would further assist compression packaging. Cubes made with water soluble additives quickly dissociate in water — a desirable feature for hydro-seeding uses. Cubes for combustion would obviously suggest use of a hydrocarbon type binder.

Calorific values of DWF dry matter average in excess of 16.5 x 10$^3$ J/kg (3.94 Kcal/gm; 7,100 Btu/lb). Ash is approximately 15 percent. This is comparable to good firewood and clearly suggests heat recovery through municipal type incineration and also interesting possibilities as a solid fuel for consumer uses.

Weed seed content would be a consideration where the produce were to be used in the raw condition for certain sensitive nursery or landscape purposes. Our limited greenhouse tray weed seed germination trials were inconclusive. We have observed cotton seedlings in wet stored fiber which obviously originated from the dairy ration. Some small amount of feed from the bunk or bowl always enters the manure system without being ingested and would be expected to contain some viable seed. Ubiquitous tomato seeds appear in most wastes and need thorough thermophilic composting to be destroyed. Chance of recontamination is unavoidable so it might be inadvisable to suggest "weed-seed-free", yet at this time we do not anticipate problems of this sort.

## USES FOR SUN DRIED DAIRY WASTE FIBER

In general most any non-textile, non-sanitary application for organic fiber or wood shavings has been considered a candidate use for DWF. Organized research on these topics is sparse, but cursory observations have been adequate to stimulate limited commercial evaluation or acceptance while research is being considered.

The shocking cost of feed has led to the consideration of all organic wastes as possible feed ingredients. Most of the nutrients in liquefied dairy waste are solubilized or so finely divided they remain in the effluent and ultimately become plant food. There is no suggestion that DWF need be further examined as a livestock feed ingredient.

Coarse fiber that is slated for use as livestock litter needs to have the "mud" creating silt and fines removed. The fines are impressively uniform and would appear ready for potting soil mixes with no further treatment for the home gardener but with probable steam pasteurization for greenhouse use.

Beef feedlot manure is currently being utilized as a carrier for a paraffin based, colored flame, fireplace log (patent applied for). The marketability of "PRESTO LOG"® is well established. The heat and ash contents of DWF lie between these other two imitation log materials. The possibility of a similar solid fuel consumer product would seem quite natural. There should be no odor generated from burning dry clean manure fiber in a hot fire condition, but the cow hair might create some odor under a smouldering condition. This question is currently unanswered.

Industrial use potential is still under determination. Acceptable product characteristics for any specific use, and therefore the secondary processing and packaging requirements, are not yet confirmed. Energy recovery by direct combustion or pyrolysis would probably favor retention of all of the volatile solids but elimination of as many fixed solids as practical; roofing felts or insulating panels could utilize the tensile properties of the cow hair, expecially, but might favor removal of silt.

Table 1 summarizes our current evaluation of the use potential for sun-dried dairy waste fiber.

**TABLE 1. PROPOSED USES FOR DAIRY WASTE FIBER**

| Use | Evaluation |
|---|---|
| **Livestock litter** | |
| Dairy cow free-stall bedding (wet or damp) | + |
| Dairy maternity pens | ± |
| Calf pens | + |
| Livestock truck and trailer | + |
| Poultry of most classes | + |
| **Horticulture** | |
| Substitute for peat moss as a seed cover | + |
| Hydromulching/hydroseeding | ± |
| Planter Mix (blended with steer manure and soil) | + |
| Mushroom compost | ? |
| Molded fiber planting pots | ? |
| **Industry** | |
| Roofing felt fiber | ? |
| Energy recovery process feedstock | ± |
| Insulating panels | ? |
| **Miscellaneous** | |
| Feed value of the total product or selected fractions | - |
| Fireplace logs (improved buffalo chips) | ? |
| Barbecue briquettes | ? |

+ = Confirmed as generally satisfactory or practical
± = Incomplete evaluation or limited acceptance
- = Negative results or impractical
? = Proposed but untested

## CONCLUSIONS

Sun-dried dairy waste fiber can be substituted for traditional litter in most livestock confinement situations, and for mulch or growth media in many horticultural operations. Industrial or miscellaneous uses appear less certain. Refeeding potential appears nil. All organic waste fibers have traditionally been of low relative cost but now organic carbon and protein from wastes are coming into direct cost comparison with fossil fuel and prime feed. Until now any profit from use of salvaged fiber has been narrow at best. If dairymen will credit some portion of pollution abatement costs to the waste fiber recovery operation the economics could rapidly improve. The general public now accepts and even encourages recycling of wastes which until recently were treated with complete repugnance. The extent to which DWF may become a beneficial byproduct depends first on its established qualities and then on its cost at the point of use relative to alternative products.


### References

1  Bartlett, H. D., R. E. Bos and E. C. Wunz. 1973. Dewatering bovine animal manure. ASAE Paper No. 73-431. 26 p. ASAE, St. Joseph, Mich. 49085.
2  Davis, Sterling, William Fairbank and Herb Weisheit. 1972. Dairy waste ponds effectively self-sealing. ASAE Paper No. 72-222. 10 p. ASAE, St. Joseph, Mich. 49085.
3  Fairbank, W. C. and E. L. Bramhall. 1968. Dairy manure liquid-solids separation. UC Agricultural Extension Service AXT-271. 3 p.
4  Meyer, J. L., et al. 1973. Manure waste ponding and field application rates, part I. UC Agricultural Extension AXT-271. 3 p.
5  Oliver, J. C., et al. 1974. Subfloor monitoring of shady grove dairy liquid manure holding pond. California Agriculture 28(4):6-7.


---

## Planning Study on Dairy Wastes
*(Continued from page 135)*

locations that meet certain specific conditions concerning their geology and location should be considered. A survey of the two counties involved revealed that there are indeed available sites meeting these conditions, and also that the trucking and manpower requirements could be met (Fig. 2).

Composting would cost about $5.70 per cow per year. The cost of disposing into a landfill, on the other hand, is estimated to range from $13.00 to $14.50 per cow per year. Table 5 shows the cost comparisons for composting and for use of the two proposed sanitary landfill sites that would be necessary.

**TABLE 5. BENEFITS — COST ANALYSIS OF MANURE MANAGEMENT (IN DOLLARS PER EQUIVALENT COW PER YEAR)**

| Item | Composting alternative | Manure-hauling alternative | |
|---|---|---|---|
| | | Chino Hills | Badlands |
| **Cost of present system (benefit)** | | | |
| 1. Current annual operating cost for waste disposal | 8.00 | 8.00 | 8.00 |
| 2. Income foregone from disposal land not freed for other beneficial use (Assuming $100 net income per acre per year) | 4.20 | 4.20 | 4.20 |
| 3. Water quality degradation | 8.40 | 8.40 | 8.40 |
| Total benefits | $20.60 | $20.60 | $20.60 |
| **Cost of recommended system (cost)** | 5.70 | 13.00 | 14.50 |
| Cost-benefit ratio | 1/3.61 | 1/1.58 | 1/1.42 |

Landfill disposal of manure will increase the operational cost to the dairymen by $5.00 per cow per year. This would have an impact on the less efficient dairy operators in the Basin. However, the cost of meeting the waste discharge requirements should be passed onto the consumer of milk.

We concluded our study with an environmental impact assessment, which revealed that the major benefit of installation of a management system would be the improvement in the quality of local water in the basin. Conversely, the major adverse effect would be an increase in truck traffic and its related air pollution, plus the adverse effects of creating new landfill sites. In balance, we consider the benefits to outweigh the adverse effects.


### References

1  Adriano, D. C., P. F. Pratt and S. E. Bishop. 1972. Fate of inorganic forms of N and salts from land disposed manure from dairies. Proceedings of the International Symposium on Livestock Wastes, ASAE, St. Joseph, Michigan 49085. pp. 243-246.
2  Bishop, S. E. 1973. Personal communication.
3  Chang, A. C. 1974. 1972-1973 Chino-Corona liquid dairy wastes survey. Final report submitted to California State Water Resources Control Board. May. 72 p.
4  Chang, A. C., G. Yamashita, J. B. Johanson, K. Aref and D. C. Baier. 1974. Quality degradation of dairy washwater. TRANSACTIONS of the ASAE 17(4):757-760.
5  Erickson, A. E., J. Tiedje, B. G. Ellis and C. M. Hansen. Initial observation of several medium sized barriered landscape water renovation systems for animal wastes. Proceedings of 1972 Cornel Agricultural Waste Management Conference. pp. 405-410.
6  Gershon, S. I., S. A. Hart, A. C. Chang and J. W. Branch. 1974. Dairy waste management, Santa Ana Watershed Planning Agency. 177 p.
7  Section 3, Recommended Plan, Santa Ana Watershed Comprehensive Water Quality Management Plan, February 1974. Santa Ana Watershed Planning Agency. 100 p.

# The Dakota System—A Method of Collecting, Storing and Handling Animal Waste

Dennis F. Meyer, P.E.

ASSOC. MEMBER
ASAE

DAIRY cattle in the northern United States must be housed to protect them from the cold winter. In the warmer sections of the country, protection from the heat may be desirable, or even necessary. In either case, a concentration of waste results. One of the more popular housing techniques presently being used is the "Free Stall System".

Design of an animal waste handling system should:

1 Provide sanitary conditions

2 Control odor

3 Provide an efficient and effective method of handling waste

4 Minimize installation and operational costs

5 Maintain high nutrient value of waste.

Nutrient value of animal waste remains high if it is handled in a way that retards biodegradation during the storage period.

The average dairy cow (1200 lb) produces approximately 0.6 pounds of nitrogen (N), 0.18 pounds of phosphate (P), and 0.54 pounds of potassium (K) daily.* Using a value of (current price at Bismarck, N.D. on 1-13-75) 29.12 cents per pound for N, 27.83 cents per pound for P, and 11.38 cents per pound for K, the daily value of the waste produced by one dairy cow is 23.9 cents. A 100-cow herd produces N-P-K valued at $8723.50 per year. In 1973 the Agricultural Census indicated that there were 12 million dairy cows in the United States. These cows produce waste with an N-P-K value of $2 868 000 daily and one billion forty-six million dollars annually.

Mechanical handling of waste in liquid or slurry form enables the operator to virtually eliminate hand labor. Modern liquid manure wagons spread the waste efficiently on the land. A popular size is 3000 gallons which is large enough to contain the daily waste from a 150-cow herd. Irrigation systems may be used to transfer waste to the field, however, they require large amounts of additional water. A dilution ratio of four parts water to one part waste is usually considered to be the minimum. Irrigation guns, 225 gallons per minute or larger, equipped with rubber orifices are recommended. Pipelines and nozzles should be thoroughly flushed at the end of the irrigation operation.

The Dakota System is a slurry or liquid animal waste handling system used primarily in free-stall dairy systems. It has some practical application in swine operations where odors are a problem or the volume of waste to be handled justifies the cost.

The System consists of:

1 A concrete sump (tank) in or adjacent to a confinement area which has a waste storage capacity of 7 days or less

2 A 60-horsepower chopper pump

3 A 10- or 12-inch pipeline

4 An earthen storage pit

5 High-rise trailer.

The waste is pumped from the sump to the pit through the pipeline.

Waste containing more than 96 percent water is considered a liquid and that which contains 80 to 96 percent is considered a slurry when properly agitated. Waste containing less than 80 percent is considered to be a semisolid.

The system enables the operator to maintain sanitary conditions except when excessive freezing in the housing system occurs. The system itself is not affected by adverse weather.

The concrete sump extends across all alleys and is usually charged with a front-end loader or an automatic barn cleaner. The sump depth must be approximately the same depth as the holding pit in order to use the same pump. Usually 8 feet deep with a 1 foot deep pump receptacle in the floor is most economical. When a 7-day capacity concrete holding tank is used, capacity is more closely related to required dimensions to reach all alleys. It allows the operator to empty the tank on a weekly basis which has proven to be a practical interval.

This system is relatively odor free because little or no biodegradation occurs in the sump between pumpings and the earth pit is bottom loaded to prevent surface disturbance.

The same chopper pump may be used to empty the sump and to agitate and pump the waste from the earth pit into liquid manure wagons for field application.

Daily cleaning is a simple indoor operation. Wet field conditions, deep snow, or cold weather do not interfere with barn cleaning operations unless temperatures are extremely low. Freezing in the barn usually does not occur unless the structure has little or no insulation or population in the barn is not maintained at or near capacity. When excessive freezing occurs, the frozen material can be moved outdoors with a front-end loader.

The storage tank should be located as near the milking parlor as possible to reduce plumbing problems.

Milking parlor wastes should be discharged into the sump. Two goals are achieved when this is done: First, a waste disposal problem is solved because detergents and milk products tend to reduce permeability of septic tank fields. Second, a source of water is provided to produce a pumpable slurry. See Fig. 1 for amounts of water needed to produce a resultant liquidity.

Traps must be installed in lines leading from the milking parlor to the concrete tank to prevent odorous gases from

The author is: DENNIS F. MEYER, P.E., Area Engineer, Soil Conservation Service, USDA, Bismarck, ND.

Acknowledgement: The author acknowledges and appreciates the data furnished by the Badger Co., indicating the magnitude of pump forces developed during the agitation process.

*Dairy Handbook, Housing and Equipment, North Dakota State University, Fargo.

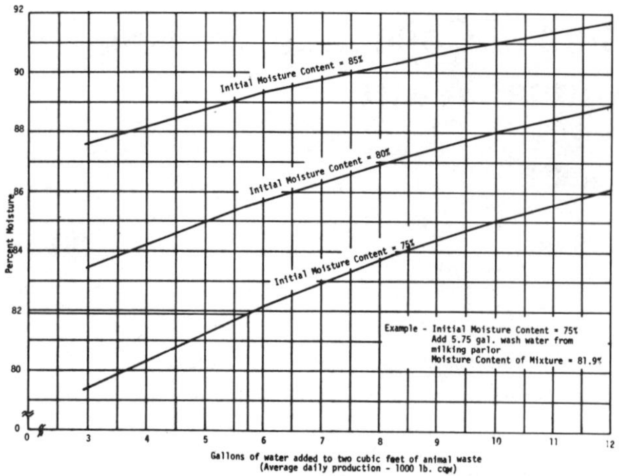


FIG. 1 Moisture waste relationship.

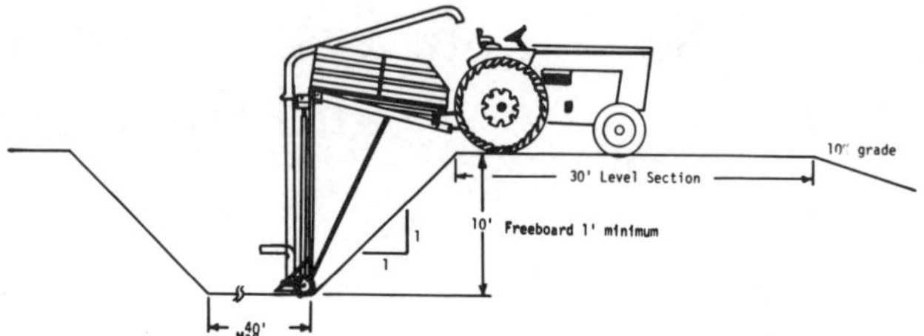


FIG. 3 Cross section—typical holding pond.

escaping into the parlor area. Gases, which are produced in varying amounts, make it unsafe to enter these tanks without a safety rope fastened to the waist and an adult attending the rope.

Installation costs are usually low because one chopper type P.T.O. pump performs all pumping needs. Figs. 2 and 3. Chopper type pumps are needed to handle (condition) straw and wasted roughage during agitation. They should be large enough to require a 60-horsepower tractor. Rating the capacity of these pumps in gallons per minute is usually not possible because the condition of the pumped material varies widely.

A 12-inch P.V.C. class 50 PSI pipeline is recommended for distances up to 500 feet on positive slopes of 2 percent or greater. Smaller diameter pipes on steep slopes and distances of less than 250 feet have been used successfully. Pipelines must be placed below frost zones. In North Dakota 7 feet of cover is required except for small areas near the Canadian border where 8 feet of cover is recommended.

Quick-attach coupling should be used to attach the pump to the pipeline or to the high-rise trailer.

Earth holding ponds (Fig. 4) should be located in areas that will not be needed for feeding, feed storage, or barn expansion. Side slopes on earthen pits are restricted to 1 foot horizontal to 1 foot vertical. End slopes can be equal to side slopes, but flatter slopes such as 4 feet horizontal to 1 foot vertical allows earth moving equipment to remove excavated material more easily. The top width of the dike on the side from which agitation and pumping is to be done should be 30 feet.

The approach slope to the top of this dike should be 10 to 1 or flatter. Dikes on the ends and opposite side should have a top width of 8 feet. Outside slopes for the dikes with the exception of the approach slope can be 3 feet horizontal to 1 foot vertical, although flatter slopes facilitate mowing.

Bottom width of these ponds cannot exceed 40 feet because pumps now available will not effectively agitate greater widths. The bottom length of the ponds is a function of the required capacity. As length increases, the number of pump moves along the side must be increased to effectively agitate the waste material.

Presently, we recommend constructing ponds in relatively impermeable soils. These ponds should be fenced for the safety of children and livestock.

Freezing of the surface during the winter helps maintain lower temperatures for several days to a few weeks into the spring. Favorable field conditions for hauling normally develop during this period.

The ability to move chopper pumps from place to place on one side of the pit is required for complete agitation.

Docks can be constructed to provide an area from which a chopper pump can be operated. One dock is needed for every 75 cows in order to keep all portions of the pit within reach of the pump. High-rise trailers on which chopper pumps are mounted provide a desirable alternative to docks. They cost less than one dock, save additional dock costs for herds larger than 75 cows, and enable the operator to move conveniently from one location to another on the pit perimeter.

Piston pumps are sometimes used in lieu of the concrete sump. One advantage of this system is to be able to pump waste from the barn to the holding pit without the weekly use of a tractor P.T.O. When automatic barn cleaners are used, an additional gathering chain is needed to move the waste to the piston sump. When loaders are used to clean

(Continued on page 143)

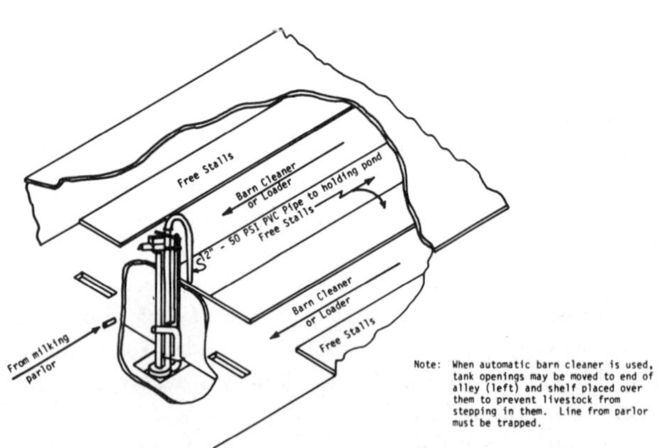


FIG. 2 Typical free stall installation.

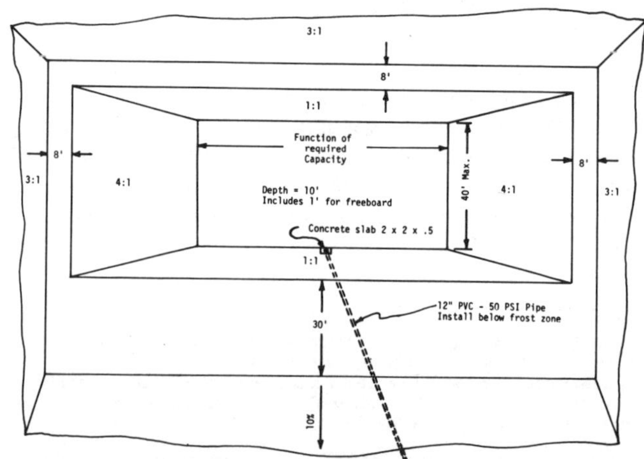


FIG. 4 Top view—typical holding pond.

# Manure Ponds for Minimizing Pollution

A. C. Marini, O. J. Berry, M. L. Knabach
MEMBER
ASAE

A GENERAL definition for pollution is "the addition of something to the water, air, or soil that reduces its quality". The degree of pollution varies with the type and amount of pollutants added to the water, air, or soil.

What to do about the pollution which animal waste creates is important to farmers, particularly dairy farmers. Dairy herds are getting larger. Fewer cows are being pastured, and more are being raised in confinement. Farmers are concerned from the standpoint of waste produced, disposal of waste, and potential effect on land and water quality.

Basic concepts of animal waste management are essentially the same throughout the United States:

1  Divert clean water from the feedlot to prevent its pollution.

2  Grade the feedlot to improve drainage and promote drying.

3  Collect runoff water and waste from the feedlot and manure storage areas.

4  Evaporate or recycle contaminated runoff to the fields for crop fertilization.

5  Collect and store manure solids and liquid manure until they can be safety spread on the land.

Application of animal waste to the land has been an accepted and common practice of disposal for many years. Land spreading is probably the most convenient and economical method of land disposal available at the present time. This method solves the disposal problem for the huge amounts of unwanted waste and recycles the useful nutrients contained in the manure for crop fertilization.

Probably no soil is perfect for waste disposal. The selection of a suitable disposal site requires analysis of waste quantities and the ability of the soil, climate, and crops to recycle the waste. The site selected should contribute minimum damage to the environment for a particular disposal system. On the other hand, there can be no compromise on two important considerations in selecting a disposal site. The site must not be subject to flooding, and soil loss, including absorbed pollutants from the disposal site, must be kept to an acceptable minimum.

## FARM OPERATING UNIT

The Ray Michalski farm (237 acres) is representative of the farms in the Green Bay area of Wisconsin. The soils range from a well-drained sandy loam to poorly-drained silty clay loams. The slope of the land ranges from 1 to 7 percent. The soil is well suited for raising corn, small grain, and hay. The crops are fed to 170 head of livestock that includes a 100-head dairy herd.

The application rate on crop fields to be planted to corn is about 25 tons per acre per year. This is equivalent to

about 250 pounds of nitrogen, 50 pounds of $P_2O_5$, and 200 pounds of $K_2O$.

In the fall of 1974, the corn averaged 150 bushels per acre without the addition of any commercial fertilizer, except starter. Mr. Michalski has not used commercial fertilizer since construction of the waste disposal system 4 years ago. Before this, he used $4500 worth of commercial fertilizer per year on farm land.

## DESIGN AND CONSTRUCTION OF EARTH MANURE POND

The pond to contain the livestock waste was constructed in the earth with a bottom size of 80 feet x 80 feet. The side slope next to the barn was constructed with a slope of 2 horizontal to 1 vertical (2:1). The other slopes were constructed 3:1. The pond was constructed with the excavated material being used to construct dikes along the outside of the pond. By utilizing this method of construction, costs were held to a minimum. The dike around the pond has a top width of 8 feet. The outside slopes of the dike were constructed 3:1, flat enough to establish and maintain vegetation.

The inside slopes in cut and fill as constructed were continuous. Slopes from excavated material to the fill material were not broken. On one end, a ramp was constructed on an 8:1 slope to allow equipment to enter the pond. This ramp is located at the point where the cleanout operations can be most efficient. The bottom of the pond is sloped with one side 2 feet lower than the opposite side.

The bottom was shaped to grade and then covered with 6 inches of well graded, pit-run sand and gravel. Concrete, reinforced with 6 x 6, six-gauge welded wire fabric, was then placed over the entire bottom of the pond. Slab thickness was 5 inches. The ramp also was surfaced with concrete. This prevents equipment from miring in soft mud when emptying the pond.

After construction, the side slopes of the pond were seeded. This reduces soil erosion and provides an environmentally pleasing site for the storage area.

The size of the pond was based on the size of the dairy herd, the amount of time between periods of emptying, and the amount of waste estimated to be produced per day per cow. In this instance, the farmer plans to empty the pond twice a year. Therefore, a storage period of 180 days was used in determining the necessary volume. A 1400 pound dairy cow produces approximately 2.2 cubic feet of waste per day in free-stall conditions with no bedding. For dairy cows weighing approximately 1200 pounds, 1.9 cubic feet per day per cow is produced. In addition, capacity was provided in the pond to handle the milkroom waste water, which is piped directly into the pond. Not only is this an excellent method of disposing of milkroom waste, but it also adds liquid to the pond, facilitating the agitation of the manure at the time of removal.

The authors are: A. C. MARINI, District Conservationist, Soil Conservation Service, Green Bay, WI; O. J. BERRY, State Resource Conservationist, and M. L. KNABACH, State Conservation Engineer, Soil Conservation Service, Madison, WI.

FIG. 1 Pump for emptying the pond is mounted by three-point hitch.

## TRANSPORTING THE ANIMAL WASTE FROM THE BARN TO THE POND

The farmer purchased a commercially fabricated pump for removing the waste from the barn and transmitting it to the pond. The pump is located in a pit inside the barn. The manure from the gutter is scraped into a hopper at the pump. A large piston pushes the manure through a 9-inch plastic pipe into the pond. The outlet for this pipe is located on the bottom of the pond and outlets in an upward direction. This allows the outlet to be immediately covered with manure and will prevent any freezing problem that might occur during the winter. Check valves located at the pump prevent any backup into the system. The pump is driven with an electric motor and needs to run for only a few minutes each day. The purpose of the pump is to move the manure into the pond. It is not used for agitating any of the liquid at the time the pond is emptied.

## EQUIPMENT FOR AGITATING AND EMPTYING THE POND

Mr. Michalski uses a low-head, high-volume manure pump mounted on the three-point hitch of his tractor for agitating and emptying the manure pond. (Fig. 1). The pump is located on a long hitch several feet behind the tractor, allowing the pump to be lowered into the pond while the tractor remains higher on the ramp. The tractor is backed down the ramp. The pump lowered into the manure is run for several minutes to agitate the pond. In agitating the pond, the manure is pumped up and deflected onto the surface of the pond. The pumped slurry assists in breaking up the crust that forms on the surface. Initially, the pump has to run approximately 1 hour before the pond can be emptied. After the initial agitation, the pond is agitated twice a day during emptying operations, each time for approximately 15 minutes. This prevents separating of the liquids and solids.

The pump is a high-volume, low-pressure pump and closely resembles the blower located on field cutters. The propeller pump allows the manure to enter near the center on the back side. A simple deflector valve is used to divert the outflow from the pump into the pond or allows it to continue in a pipe that outlets into a tank used to haul the manure to the fields.

## TANK WAGON USED TO TRANSPORT MANURE TO DISPOSAL FIELDS

The farmer's equipment for hauling the liquid manure is

FIG. 2 A deflector plate on the tank outlet distributes the liquid slurry.

a 3600-gallon tank mounted on a tandem dual truck. The system has worked very well in that it provides a self-propelled tank that fits into the farmer's operational plan. The farmer constructed his own tank outlet that performs the spreading. It is a pipe turned onto a flat deflector plate that serves as the spreading mechanism (Fig. 2). It is relatively simple but has been very effective in distributing the liquid manure. Having the tank on a separate truck provides additional mobility not always available when using a large tractor and trailer unit.

## SPREADING THE LIQUID MANURE

The liquid manure is emptied from the pond twice a year. It takes approximately 4 days to empty the pond, using the 3600-gallon tank. Each load averages approximately 3200 gallons, and the farmer hauls approximately 60 loads per day.

After the liquid has been applied to the land, the ground is disked the same day. This not only conserves the nutrient value of the manure but also reduces the odor that may be present and reduces polluted runoff from occurring. The fall application is plowed under immediately. The odors seem to be greater when the pond is emptied in the fall than when emptied in the spring.

## CONCLUSION

The system of storing liquid manure in open ponds is expected to have increasing use in the future. This type of storage structure has the lowest cost of any structure used to date. It would be relatively easy to regrade the storage area and restore the original site conditions should the system become obsolete.

There are various methods that can be used to remove the liquid manure. The system described in this paper uses a pump mounted on a three-point hitch and the tractor is backed down a ramp. There are other types of pumps that may be mounted on a ramp built out over the pond or that may be dropped over the side of a vertical wall on one side of the pond.

The pumps used to push the manure from the pit in the barn to the pond are being improved each year. However, sufficient use of these pumps has been made to determine this to be a successful method for transporting or transmitting liquid manure. This method completely eliminates spreading manure on frozen ground or the need to spread when undesirable conditions exist. Not only is labor reduced, but venturing forth during adverse winter weather is unnecessary. The capacity of these systems allows the operation to be delayed until proper field conditions or weather conditions exist for the spreading of animal waste. Since little equipment is required, breakdowns also are minimal. Financial requirements areminimum for this type of system.

Experience has shown that 2:1 slopes, used on many of these ponds, have a tendency to clean better, reducing the fly problem, than flatter slopes such as the 3:1 used on this project. Although the amount of manure retained on the slopes was not a problem of great concern, it has been demonstrated that the steeper slopes are advisable.

An entrance ramp is recommended for all installations even though the pump used to empty the pond may be mounted on a catwalk or be dropped over a vertical wall. It should be assumed that sometime in the future, some clean-out with a front-end loader will be necessary. Proper design dictates that this ramp be installed initially to avoid future access problems.

The bottom or floor of the manure pond should slope approximately 2 percent toward the ramp to facilitate cleanout. This is mainly true when the ramp is used to back pumping equipment into the pond area.

The type of surfacing used on the bottom of these ponds may vary. Naturally concrete is the most durable. Considerable savings can be made by using materials such as crushed limestone. The material installed should depend on the type of pump used and the individual's operation and management plan for the system.

If Mr. Michalski were to reconstruct this manure storage facility, he would design for cleanout on a once-a-year basis. This would allow all spreading in the fields to be done in the fall. Not only would the compaction of the field soils be reduced, but he has noted that fuel consumption is reduced by 50 percent when spreading in the fall on these soils due to better traction and less rolling friction encountered by the tank truck.

Another change would be to use a 12-inch pipe between the manure pump in the barn and the pond. This would allow the manure to be pumped with less friction loss in the pipe.

At the present time, the University of Wisconsin-Green Bay is performing tests relative to manure handling facilities. The study is attempting to determine amount of nutrients returned to the soils, using various manure handling methods. It also is studying how soil tilth, friability, moisture-holding capacity, and aeration are affected by manure applications. Mr. Michalski's farm is involved in these studies. The information is not yet available.

---

## The Dakota System
*(Continued from page 140)*

most conventional barns, waste must be moved around corners. One hundred PSI or stronger pipe is needed between the piston sump and holding pit. During periods of high evaporation supplemental water must be added to the waste before pumping to the holding pit.

A chopper pump and high-rise trailer effectively agitate and remove waste from the pit when piston pump systems are used although more agitation is required because of the unchopped straw and roughage in the pit. Surface crusts that form on the pit also are more difficult to break up because of the unchopped straw and roughage.

The Dakota System permits livestock operators to collect, store, and remove animal waste with a minimum of biodegradation. Collection is convenient when the holding area, concrete tank, and associated appurtenances are properly designed. A single pump transfers and loads all wastes to and from the earthen holding pond. Expansion of the system for larger herds is inexpensive because expansion of earthen holding ponds are usually the only modifications required. Geministic (doubling) Viafication (use of living material) is made convenient and economically feasible.

# A Total Recycle Unit System for Dairy Manure Management

A. C. Dale, Roger Swanson

FELLOW
ASAE

IN general the dairy farm has not needed as large a system for management of wastes as beef, swine or poultry production units. Dairy sizes in terms of numbers of animals per unit are not as great. It is difficult to find a 5000 head dairy operation while there are many such beef-cattle operations and a number of equivalent swine and poultry operations. Although there are indications that the dairy herd size is increasing, the average dairy herd is still less than fifty (50) cows (Dale 1971).

Dairy operations have also generally been connected with a farming operation of some 160 to 320 acres of land. Thus, the ratio of animal units per acre has been generally quite low and it has been a matter of returning the wastes to the land in an acceptable concentration to control pollution. Nevertheless, dairy farmers would like to have an improved waste management system.

## NEW METHOD OF HANDLING DAIRY COW MANURE

Waste disposal is a very important aspect of commercial dairy farming. A cow produces several pounds of manure per pound of milk and this waste must be handled in a safe, inoffensive manner. At present there are numerous treatment and disposal methods being used, each with its own advantages and disadvantages. One modification that may improve many of these methods and which may also provide return on the investment is solid-liquid separation with the solid being used for bedding and liquids being pumped to nearby land.

## SOLIDS-LIQUID SEPARATION

Solids-liquid separation is not an entirely new step in a waste handling system, but it appears to be a procedure that could improve many of the present ones (Dale 1972). Thus, it is receiving considerable attention. In this process the large particles end up in a separate container or pile with the liquids going into a storage for other handling or, additional treatment. Essentially, an "unmanageable" product is divided into two by-products which are manageable using conventional pumps and solids conveying systems.

### Separated Solids

The separated solids are the undigested feed particles, straw, hay stems, silage, etc. that add bulk to manure. They make pumping difficult and tend to plug irrigation

systems. Whole cow manure is extremely difficult to irrigate, but when the solids are removed from the mixture the liquid is easily managed. The solids have several possible uses. They may be readily used for bedding in free stalls or loafing areas. Also they make a good odorless soil conditioner. However, their vlaue to the soil is limited to the organic matter which is probably not a big factor in soils where roots and stems are left on the land. Although the status is not clear about refeeding, the solids appear to be a material that may be refed to certain ruminant animals for bulk or roughage feed.

### Liquid

After the large solids have been removed, the liquid portion of the manure becomes easy to treat and handle. The liquid contains about 85 to 90 percent of the five-day BOD, 80 to 95 percent of the minerals and nutrients and most of the dissolved organics along with the finely suspended solids. It may be aerated easily or irrigated directly onto cropland. A point that should be considered is that of keeping the volume of the water added as low as possible. When the volume of the dilution water is minimal (about 1 part water to 2 parts manure) the final separated liquid is a much improved fertilizer and there is less volume to handle. Another factor is that more of the nutrients, particularly nitrogen are saved if the separation is performed each day and anaerobic conditions are maintained. Actually, solids-liquid separation also leads to other new methods of treating and handling wastes.

## THE TOTAL RECYCLE UNIT (TRU) SYSTEM

A system that appears to have much merit in farm animal waste management is the TRU (Total Recycle Unit) developed by Babson Bros. Co., Oak Brook, Illinois. Manure enters this system in the raw state consisting of the mixture of undigested solids, minerals, bacteria, and water. It leaves the system separated into (a) washed undigested solids and (b) a light slurry of concentrated minerals and suspended matter. Other than the fresh, raw manure odor of the wastes entering the system it is practically odorless if processed daily.

The final products produced by the system are usable in a number of ways. The liquid removed in the solids-liquid separation process still contains many small suspended solids, dissolved solids, and most of the oxygen demanding material. However, it is a liquid that works extremely well in an aeration system or it may be easily stored and returned to the land by irrigation or other suitable methods. It lends itself readily to earthen storage techniques and can be easily removed from a large detention pond without ramps and other special pumping aids. The detention pond aspect allows the

Agricultural Experiment Paper No. 5880, Agricultural Experiment Station, Purdue University, West Lafayette, IN.

The authors are: A. C. DALE, Professor, Agricultural Engineering Dept., Purdue University, West Lafayette, IN; and ROGER SWANSON, Babson Bros. Co., Oak Brook, IL.

operator to store nutrients for the most convenient period (180 to 360 days) in a practical, inexpensive structure.

## Concepts and Processes of the TRU System

With the idea that a new and better system could be developed, Babson Bros. Co. engineers set up some rather rigid requirements for such a system. They visualized that it should have the following requirements:

1 Be essentially odorless.

2 Not require an excess of water.

3 Be nearly automatic, requiring as little labor as practical.

4 Be self contained (electric motor driven).

5 Produce products that are useful or easily managed on the dairy farm.

6 Designed to fit a 100-cow dairy unit (this objective was changed as the development progressed).

The objective of the development was to come up with a system that would prevent pollution of both air and water. It was hoped that almost straight, undiluted, non-dried manure could be used by the system. Thus it would not be necessary to add a large quantity of dilution water or to dry the manure before processing.

## Mineral Removal and the Liquid

Recognizing the need to remove bacteria, nutrients and other solids from the liquid to produce a reusable water, attention was turned to this problem. Certain flocculants were found to do a relatively good job of clarifying the liquids but they left many nutrients in the solution. Chemical precipitants were not adequate and they were quite expensive and difficult to use. Finally, consideration was given to electrocoagulation. Rationalizing that the quantity of water was going to be low in the system, and that the rate of flow would be low for a 100-cow dairy, it was suspected that electrocoagulation could work. To date this method has not been entirely satisfactory, but it does appear to offer some possibilities. Thus, at the present time the liquid is being returned directly to the soil.

## Aeration System

With the criteria of minimum odor in mind, the first effort was devoted to the development of some type of aeration system that would insure aerobic conditions throughout the process. Not being satisfied with the systems already known, since most of them were not adaptable to the package they had in mind, a new idea was conceived and developed. This development was a variation of the tower system whereby the material falls through the air and is brought in contact with oxygen in this manner. This tower consisted of perforated disk plates placed in a vertical cyinder with air being drawn along with the material. It did an excellent job with a small amount of expended energy. However, the large amount of solids in the form of fibers, stems, stalks, etc. caused sever problems. Thus the final development was delayed until a solids-liquid separator was developed.

## The Solids-Liquid Separator

As is often the case, one development leads to another or shows the need for another facet. Such was the above. The large solid particles in the manure tended to clog the system, reduced the oxygen transfer and led to number of difficulties. Thus solids separation from the liquids became a clear cut need.

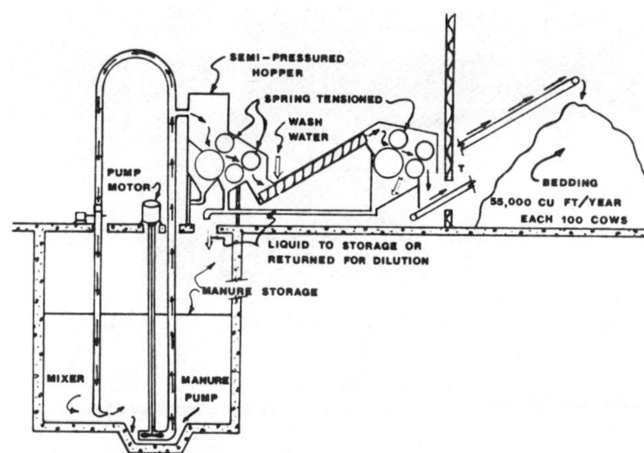


FIG. 1 Schematic of solid-liquid separator.

Utilizing available knowledge and adding new ideas of their own, a progression of solids-liquid separators were developed. Recognizing the need for a special separator, and one that would do the best job possible, screens and roll presses were adapted and developed that would not only separate the solids from the liquids but which would produce solids containing only 65 to 70 percent moisture content by weight. Therefore, they ended up with a solids-liquid separator that far exceeded others in use at that time. However, because of the lack of ruggedness of the screen belts and roller system, the final solids-liquid separator which is being placed in production consists of rollers and revolving screens. A schematic of the system is shown in Fig. 1. The solids-liquid separator is shown in Fig. 2 with a stack of the separated solids in Fig. 3.

## ADAPTATION OF THE SOLIDS-LIQUID SEPARATOR TO A 300-COW DAIRY

As developmental work neared completion, the decision was reached to install a complete solids-liquid separation system on a modern 300-cow dairy. As a progressive farm, the dairy complex is in a constant state of change and new ideas are tried frequently. Thus this equipment was readily accepted.

## The Free Stall Layout and Manure Collection

Fig. 4 shows a layout of the dairy production complex.

FIG. 2 The developed manure processing machine in operation at the 300-cow dairy.

FIG. 3 Washed solid from the solids-liquid separator prior to being spread as bedding.

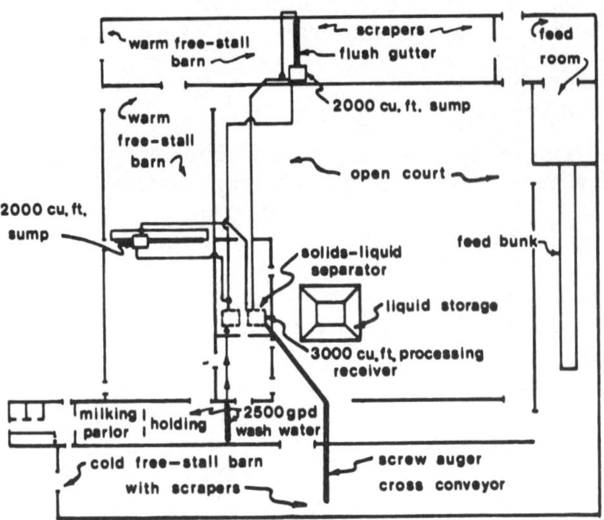


FIG. 4 Layout of manure handling system with TRU.

It consists of (starting at bottom) a 260 stall cold environmental barn, a 160 stall warm environmental barn (left center) which includes maternity, nursery and young stock stalls, and a 250 stall warm environmental barn (top) for dry cows and heifers. Feed storage is in seven sealed structures. Cows are milked in a "Double-3" diagonal (side opening) stall parlor with two prep stalls. Milk is stored in a 11.3 cu meter (400 cu ft) bulk tank.

The manure generated in the cold barn is removed by scraping it to an auger type cross conveyor using a tractor mounted rear blade scraper. It is then conveyed to the 76 cu meter (2700 cu ft) processing receiver pit.

Manure is collected in the other two barns by automatic alley scrapers using a hydraulic cross conveyor system to flush the manure into one pit (per barn). Approximately 9.5 cu meter (335 cu ft)/day of milking parlor wastewater (including cleanup of holding area) is collected in a 76 cu meter (2700 cu ft) gallon receiver. This water is pumped in equal portions to the two 56.8 cu meter (2000 cu ft) cumps. This water is then used to flush the gutters (two in the center barn, one in the top barn). An electrically powered ten horsepower chopper pump is automatically energized to flush the gutter(s) as the scrapers approach the trenches.

These same two pumps are also used to transfer this slurry to the 76 cu meter (2700 cu ft) processing receiver pit on alternate days.

A third ten horsepower chopper pump is used in the 76 cu meter (2700 cu ft) gallon processing receiver pit to homogenize the slurry and to charge the solids-liquid separator. Manure is processed daily at the rate of about 75 cow-days per hour. Thus the manure from 300 cows could be processed daily in 4 hours by one machine.

Solids are removed and used as bedding at approximately 70 percent moisture content. About 1-1/4 ft$^3$/day of solids are produced per 100 lb of animal weight. At this rate there is an excess of bedding (estimated at about 40 percent of production). Transport of bedding is done using a "Hydracat" type vehicle.

Liquids are stored in a 568 cu meter (20.000 cu ft) detention pond. They are removed and transported to the cropland using a 8.3 cu meter (267 cu ft) vacuum tank wagon. Most of the cropland averages 1 kilometer away from the dairy. Future plans include a remote pond nearer the land in use and some type of irrigation system.

### The Solids-Liquid Separator

Manure is fed to the hopper of a perforated metal drum with a radially spring-loaded elastomer covered roll. This produces liquid which can be discharged or proportioned (recycled to the manure pit for dilution).

The solids are further squeezed and then augered into a double squeezing device similar to the others. Further water may be added into the auger to wash solids to remove more of the mucous, dissolved solids, and suspended solids, if desired.

Solids are conveyed from the last squeezer using a conventional solids conveyor to a stockpile. A covered area to keep rain off the dry bedding is recommended (Fig. 3).

## ECONOMICS OF THE BY-PRODUCTS—100 COWS

### Fertilizer Value of Liquid Effluent

It has been found that the nutrients are proportioned roughly 85 to 90 percent in the liquid and 10 to 15 percent in the solids. Values of these are estimated as follows (100 percent availability assumed) for the liquid portion of the waste from 100 cows/day:

1 Nitrogen = 36 lb/day at a value of $0.19/lb is $6.84/day.

2 Phosphorous = 46 lb/day at a value of $193/ton of 46 percent P is $9.66/day.

3 Potassium = 26 lb/day at a value of $85/ton of 60 percent K is $1.82/day.

Therefore the fertilizer value in the liquid effluent is about $18.32/day ($6,687/yr).

Liquid effluent delivered to the detention pond is sufficiently fluid to be pumped by an irrigation system or handled by vacuum tank vehicles. Due to long term inexpensive earthen storage, runoff problems can be minimized by land application at proper times.

### Value of Bedding

Approximately 208 cu meter (54,750 cu ft) of bedding are produced yearly per 100 cows. It consists of coarse undigested solids washed and dried to about 70 percent moisture content. When used in stalls, it packs down, but remains resilient and springy. It does not cling to the animals like sawdust and other types of bedding. Although bacteria counts are positive, in two years use,

*(Continued on page 149)*

# Successful Manure Management System for a Large Commercial Hog Operation

G. D. Gehlbach, A. J. Muehling
MEMBER
ASAE

GEHLBACH Pork Farms, Inc., is a family corporation that currently produces about 8500 market hogs annually in a farrow-to-finish confinement system. This level of production results in about 1.5 million pounds of pork produced or enough for 21 000 people per year. In addition to the hog enterprise, the corporation also farms 680 contiguous acres of highly productive soil ideally suited for corn production. The farm is managed by the father, Albert, and two sons, Gerald and Donald. A total of five full-time workers plus some part-time help operate this farm which specializes in hogs and continuous corn.

Hogs haven't always been produced in confinement on this farm. For about 15 years prior to 1961, a pasture system was used to raise 1500 hogs per year. In 1961 a switch to confinement was begun, using slotted-floor buildings with pits beneath the slats. When the first confinement building was constructed, a 0.4-acre lagoon was constructed to handle the wastes. It was quickly learned that this size of lagoon was inadequate if one attempted to keep the odor level to a minimum. Therefore, field application was necessary. A vacuum tank wagon was built, and a portion of the liquid manure was surface applied.

Since 1961 many changes have taken place. In 1964 and 1966 Gerald and Donald graduated from the University of Illinois and returned to the farm. Since 1961 the hog operation has grown over fivefold. All market hogs are housed in environmentally controlled, totally slotted floor buildings with liquid manure pits below the slats. A good view of the farmstead can be seen in Fig. 1. The pits range in depth from 4 to 6 feet, providing about 3 to 5 months storage in the totally slotted floor buildings. Total capacity of all the pits below the slats is about 864 000 gallons. In the confinement buildings approximately 5000 gallons of liquid manure are produced daily. In a year this amounts to 1.8 million gallons.

Also during this same period, the crop farming acreage has more than doubled. Since land is currently valued near $2000 per acre, every attempt is made to keep all available acres in corn production. Assuming normal corn yields, approximately two-thirds of the total corn requirement for the hogs is produced on the farm. The remaining corn supply plus all soybean meal must be purchased on the market.

Throughout this period of expansion since 1961, the handling of wastes has grown from a minor problem to one that requires the utmost in management to keep air and water pollution to a minimum. In looking back on this expansionary period, it appears that the process of developing an acceptable waste handling procedure was the primary governor on the growth of the hog business. As new manure handling technology and equipment became available, the hog enterprise expanded step by step.

Manure is disposed of in two ways on this farm: (a) hauling from the pits with a vacuum tank wagon and injecting into the soil for fertilizer, and (b) lagooning. Such a combination system results in an almost odor-free disposal of the manure. Eleven residences, including those of some of the operators, exist within a half mile of the farmstead. Such a situation necessitates a system with relatively low odor in order to maintain a pleasant relationship with others in the community.

With hauling to the fields as the primary system of disposal, every attempt is made to pump empty all of the pits below the slats after corn harvest prior to fall plowing. Hauling is again resumed in the spring when applications can be made on the remaining unplowed fields. During the summer the manure is again knifed into the soil on 20 or 30 acres of cropland set aside from corn production.

Manure is currently handled by a 2100-gallon vacuum tank wagon with a soil injection attachment. One load is applied in one round on a quarter-mile length field. The soil injection attachment has two knives on a 60-inch spacing. A single application of manure results in about 3500 gallons of manure applied per acre. The fertilizer content of the manure added per acre with a single application results in about 95 lb of nitrogen (N), 52 lb of phosphorus (P), and 105 lb of potassium (K). The above figures allow for a 40 percent nutrient loss for N and negligible loss for P and K. Most of this loss would come during anaerobic storage, for very little loss would occur when injected into the soil. When manure is applied in the fall after corn harvest, as many acres as possible receive either a single or double application of manure, depending on the extent of depletion of the soil.

During the summer the area left out of corn production receives a heavier application of manure. This 20- to 30-arce tract typically receives repeated applications that approximate six times that of a single application. This tract of land then receives no P or K for several years following this heavier application. Much higher fertilizer prices have caused an acceleration of research to determine when additional nutrients are needed.

The main disadvantages of hauling manure to cropland are: (a) increased soil compaction, and (b) additional expense required to radiate further from the loading point because of the size of the operation.

It is difficult to estimate the amount of manure hauled to the fields and the amount lagooned. An educated guess would be that only 15 to 20 percent of the total manure production is discharged to the 7.5-acre lagoon. The lagoon is operated with a water depth of 5 feet with 4 feet of freeboard above the water level. Odors from the lagoon have been minimal. Wastes from three small buildings are routinely discharged into the lagoon. Additional manure from the large buildings is added whenever weather conditions do not permit hauling. Excess lagoon water is pumped from the lagoon into irrigation lines to maintain an acceptable level. Because of a low loading rate relative to the lagoon surface area, the pumping has been minimal.

The authors are: G. D. GEHLBACH, Gehlbach Pork Farms, Inc., Lincoln, Illinois, 62656; and A. J. MUEHLING, Professor and Extension Agricultural Engineer, University of Illinois, Urbana, Illinois, 61801.

FIG. 1 Aerial view of confinement facilities at Gehlbach Pork Farms, Inc., Lincoln, Illinois.

During 1973 and 1974, soil and plant analyses were made on a tract of land where heavy applications of liquid manure were applied during the summer of 1972. Using Vanderholm's data (1974) to calculate the amount of nutrients applied, approximately 570 lb of N, 312 lb of P, and 630 lb of K were applied per acre. The only additional fertilizer that was applied for the 1973 and 1974 crops was 80 lb of N annually in the form of anhydrous ammonia. Table 1 shows high P and very high K levels on leaf analysis when compared with levels reported by Walker and Peck (1967). These levels are higher than we normally see but are not extremely excessive and would probably not cause yield reductions. More research is needed to determine if these high potassium levels may induce certain deficiencies.

The soil report shown in Table 2 demonstrates again how the liquid manure has precipitated extremely high P and K levels as compared to those of Walker et al. (1968). Despite areas of concern in these reports, the corn plant certainly responds favorably to these heavy applications. Usually the best yielding corn occurs on such tracts the first year following the heavy application.

The total manure handling system has worked well for this large commercial hog farm. The primary disposal system of hauling manure to the fields and knifing it into the soil creates minimal odors and contributes greatly in reducing the amount of commercial fertilizer that must be purchased for corn production. The secondary system of lagooning serves as an excellent relief valve to handle wastes

### TABLE 1. CORN LEAF NUTRIENT LEVELS RESULTING FROM HEAVY APPLICATIONS OF LIQUID MANURE*

| Time of sampling | Nutrients | | | | | | | | | | |
| --- | --- | --- | --- | --- | --- | --- | --- | --- | --- | --- | --- |
| | N | P | K | Ca | Mg | S | Zn | Mn | Fe | B | Cu |
| | Percent | | | | | | Parts per million | | | | |
| At early growth (6th leaf stage) | | | | | | | | | | | |
| 1973 sample | 4.1 | 0.6 | 7.1 | 0.4 | 0.2 | -- | 51 | 53 | 323 | 6 | 10 |
| Sufficiency range (Lockman 1969) | 3.5-5.0 | 0.4-0.8 | 3.0-5.0 | 0.9-1.6 | 0.3-0.8 | 0.2-0.3 | 20-50 | 50-160- | 50-300 | 7-25 | 7-20 |
| At late growth (silking stage) | | | | | | | | | | | |
| 1973 sample | 3.2 | 0.37 | 3.2 | 0.44 | 0.15 | 0.20 | 33 | 61 | 99 | 6 | 4 |
| 1974 sample | 2.9 | 0.27 | 3.1 | 0.53 | 0.20 | 0.16 | 21 | 49 | 91 | 16 | 5 |
| Sufficiency range (Neubert et al. 1969) | 2.6-4.0 | 0.25-0.50 | 1.70-3.00 | 0.21-1.00 | 0.31-0.50 | 0.21-0.50 | 50-150 | 34-200 | 21-250 | 15-90 | 8-20 |

*Approximately 570 lb N, 312 lb P and 630 lb K were applied per acre during the summer of 1972. Application rates were calculated amounts using Vanderholm's data cited earlier. No additional fertilizer was applied for 1973 and 1974 crops other than 80 lb N annually.



MANAGING LIVESTOCK WASTES

TABLE 2. SOIL NUTRIENT LEVELS RESULTING FROM HEAVY
APPLICATIONS OF LIQUID MANURE*

| | pH | OM | P | K | Nutrients | | | | | | | |
| | | | | | Ca | Mg | S | Zn | Mn | Fe | B | Cu |
| | | Percent | lb/ac $P_2O_5$ | lb/ac $K_2O$ | Parts per million | | | | | | | |
|---|---|---|---|---|---|---|---|---|---|---|---|---|
| Late 1973 sample | 6.5 | 3.1 | 200+ | 740 | 5,500 | 810 | 9 | 8 | 142 | 100 | 3.3 | 12 |
| Rating | | | very high | very high | medium | high | medium | medium | very high | medium | high | medium |

*Approximately 570 lb N, 312 lb P and 630 lb K were applied per acre during the summer of 1972. Application rates were calculated amounts using Vanderholm's data cited earlier. No additional fertilizer was applied for 1973 and 1974 crops other than 80 lb N annually.

whenever soil conditions do not permit application. The total system requires no mechanical equipment that must function daily to perform successfully. Undesirable odors around the farmstead are minimal. This could be classified as one successful method of handling swine wastes on a large commercial farm.

### References


1   Lockman, R. B. 1969. Soil testing and plant analysis (1973), pg. 358, edited by L. M. Walsh and J. D. Beaton, Soil Science Society of America, Inc., Madison, Wis.

2   Neubert, P. et al. 1969. Soil testing and plant analysis (1973), pg. 358, edited by L. M. Walsh and J. D. Beaton, Soil Science Society of America, Inc., Madison, Wis.

3   Vanderholm, Dale. 1974. Land application of manure from modern livestock production facilities. Soil and Water Conservation, No. 14, 2 p. Agr. Eng. Dept., Univ. of Ill.

4   Walker, W. M. and T. R. Peck. 1967. Chemical analysis of corn leaves. In: Illinois Research. College of Agr., Univ. of Ill. 9(4):18-19.

5   Walker, W. M., T. R. Peck, S. R. Aldrich, and W. R. Oschwald. 1968. Nutrient levels in Illinois soils. In: Illinois Research, College of Agr., Univ. of Ill. 10(3):12-13.


## Recycle System for Dairy

*(Continued from page 146)*

no animal illness has been attributed to the bedding. Klebsiellae and Salmonellae bacteria counts have always been negative. The bedding also provides good insulation, and dries out in the stalls after a relatively short time. Estimates of the value of good bedding vary, but a conservative estimate would be $2.00/cow/month or $2,400/yr. However, the following additional advantages should be considered:

1 Bedding recovered is in abundant supply and readily available to keep free-stalls well sloped to allow proper urinal drainage.

2 Bedding, by its very nature, "fits" the system and does not "dry up" the slurry as do other types in liquid manure systems.

3 The excess bedding can be used as a soil conditioner or sold commercially as a mulch or bedding.

### OPERATING COSTS—100 COWS

#### Power

The separator requires 14.2 kW for 1 hr, 20 min/day. This is 18.85 kW-hr/day. At $0.025/kW-hr this amounts to $0.47/day ($172/yr).

#### Maintenance

These costs cannot be properly estimated at this time but a conservative estimate would be less than $1,000/yr.

#### Support

Building must be kept above freezing. This varies throughout the country. A nominal assignment is $750/yr.

#### References


1 Dale, A. C. 1971. Status of dairy cattle waste treatment and management research. Animal Waste Management. Proceedings of National Symposium on Animal Waste Management. Council of State Governments, Washington, D.C.

2 Dale, A. C. 1973. Solids-liquid separation—an important step in the recycling of dairy cow wastes. Journal of Milk and Food Technology 36:289-296, Shelbyville, Ind.

# Experience with Open Gutter Flush Systems for Swine Manure Management

Herbert L. Brodie

ASSOC. MEMBER
ASAE

THE open gutter flush system for manure removal from confinement swine buildings has received considerable interest from Maryland swine producers. The system as designed for Maryland farms is more economical to construct, manage and maintain than other types of manure removal systems.

The basic component of the system is a gutter from which manure is periodically flushed utilizing the energy stored water. Animals have direct access to the gutter and use the gutter area for defecation. The periodic flushing action of the water through the gutter reinforces animal activity such that they become trained to defecate in the gutter leaving a majority of the pen area clean.

The design for flush gutters used in Maryland was heavily influenced by the preference of Maryland swine producers. Experience with systems presently in operation has shown that flushing swine manure from open gutters with water is both practical and desirable. Two operating systems have been selected for comparison. A summary of these two systems is shown in Table 1.

System A consists of a finishing building and a gestation building from which waste is flushed to a lagoon. Water for flushing is recirculated from the lagoon. Excess lagoon water is spray irrigated on cropland. System B is a finishing building from which wastes are flushed to an underground concrete storage tank. Collected waste is then spray irrigated onto grassland. Water for flushing is obtained from a well.

## BUILDINGS

Finishing building A, Fig. 1, is of open pole and block construction 9 meters wide by 73 meters long with an average capacity of 600 head. The pen floor slopes from one side of the building to an open gutter. An alley parallels the gutter at the open side of the building. The gutters and floor slope in two sections from each end toward the center of the building on the longer dimension. The structure is not insulated and relies on natural air ventilation.

Gestation building A, Fig. 2, is a totally enclosed block and metal clad wood frame building 11 meters wide by 30 meters long with a capacity of 120 head. The pen floor slopes from either side of a central alley to an open gutter. The gutters and floor slope in two sections to one end of the building. The structure is insulated and mechanically ventilated with thermostatic control.

Finishing building B, Fig. 3, is a partially enclosed block and metal clad wood frame building 7.3 meters wide by 76 meters long with a capacity of 500 head. The pen floor slopes from an alley on one side of the building to an open gutter on the opposite side. The gutters and floor slope in two sections from each end toward the center of the building on the longer dimension. The structure ceiling is insulated. One partially open long wall can be closed with a fabric curtain. Mechanical ventilation is thermostatically controlled.

The author is: HERBERT L. BRODIE, Extension Agricultural Engineer, Agricultural Engineering Department, University of Maryland, College Park, Maryland.

**TABLE 1. SUMMARY OF FLUSH SYSTEMS FOR SWINE**

STRUCTURAL COMPOSITION

Gutter dimensions, m

| | House size, m | Capacity head | Length | Width | Depth | Slope | Pen size, m | Floor slope, percent | Investment, dollars |
|---|---|---|---|---|---|---|---|---|---|
| Finishing A | 73 x 9 | 600 | 36 | 1.8 | 0.1 | 1% for | 3.1 x 7.6 | 6.25 | 27,563 |
| Gestation | 30 x 11 | 120 | 30 | 1.5 | 0.1 | 18 m; | 2.1 x 4.7 | 6.25 | 17,286 |
| Finishing B | 76 x 7.3 | 500 | 37 | 1.2 | 0.1 | 2% for | 3.1 x 6.0 | 4.2 | 34,816 |

FLUSH SYSTEM

Investment, dollars

| | Tank volume, $m^3$ | Dumps per day | Flow, 1/head/day | Floor | Flush | Total | Energy kW·h/day | Labor, hr/yr |
|---|---|---|---|---|---|---|---|---|
| Finishing A | 1.10 | 24 | 88 | 5,651 | 638 | 6,289 | 3.6 | 7 |
| Gestation | 0.94 | 6 | 94 | 2,783 | 515 | 3,298 | 0.89 | 7 |
| Finishing B | 0.72 | 8 | 23 | 4,838 | 315 | 5,153 | 0.76 | 9 |

DISPOSAL SYSTEM

| | Storage | | | Disposal | | | | |
|---|---|---|---|---|---|---|---|---|
| | Type | Volume, $km^3$ | Investment, dollars | Type | Frequency | Investment, dollars | Labor, hr/yr | Energy *, kW·h/day |
| System A | Lagoon | 11.4 | 7,700 | Irrigation | Annually | 5,000† | 18 | 26.5 |
| System B | Concrete tank | 0.13 | 5,000 | Irrigation | Weekly | 6,600‡ | 26 | 25.9 |

*IC engine fuel input @20 percent conversion efficiency.

†Equipment on hand.

‡Does not include tractor.

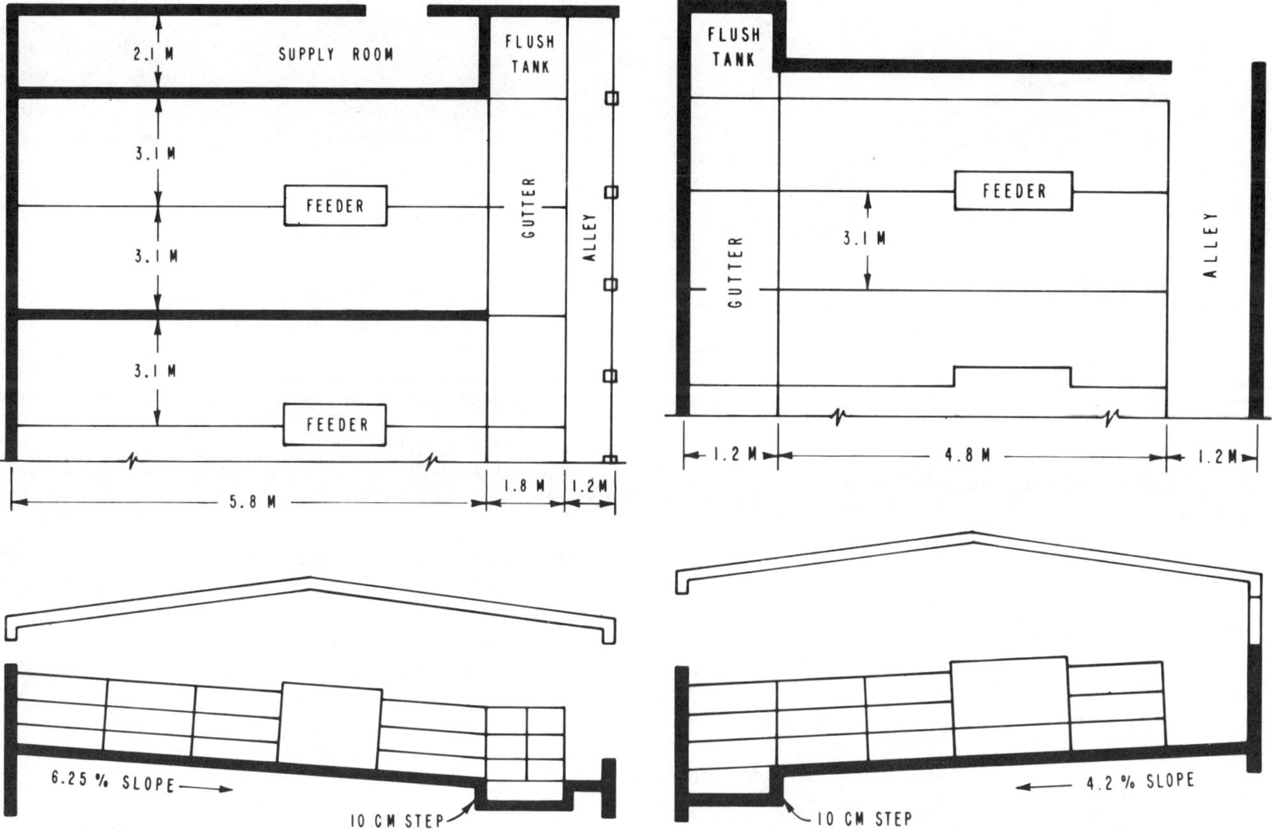

FIG. 1 Layout of swine finishing building, System A.

FIG. 3 Layout of swine finishing building, System B.

## FLUSH GUTTERS

The gutters are rectangular, constant depth, constant width open channels in two sections. A floor slope of 1 percent is used for the first half of the length. The remaining length is at a slope of 2 percent as shown in Fig. 4. The building floor must also be at these slopes to maintain a constant depth gutter.

The maximum gutter length is 38 meters. The finishing buildings are flushed from each end toward the center allowing a building length up to 76 meters. The gutters end at a catch basin from which waste water is transported to storage through a pipe approximately 20 cm in diameter. The size of the catch basin must be sufficient such that water flow is not restricted in the gutter.

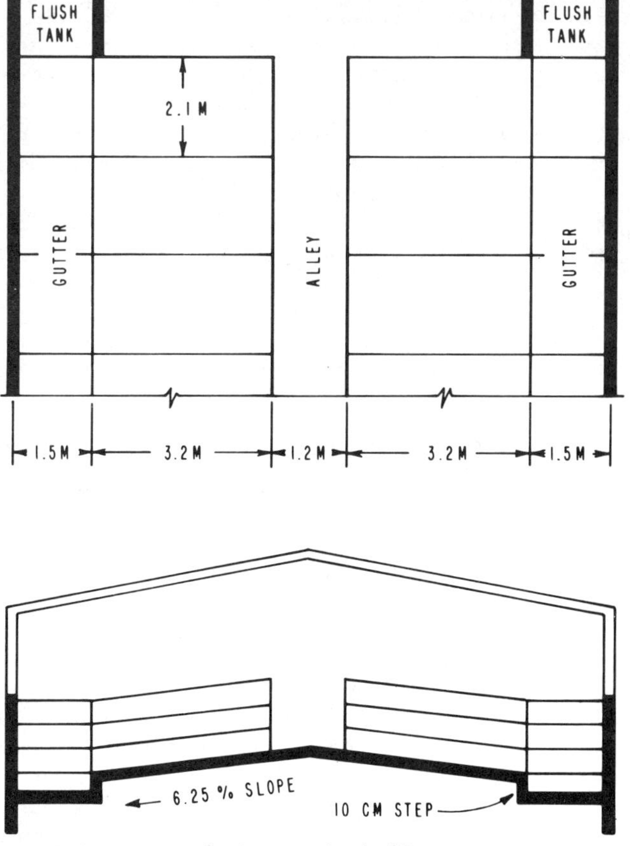

FIG. 2 Layout of swine gestation building, System A.

## FLUSH TANKS

Flush tanks are located at the high end of each gutter. The purpose of these tanks is to store water over a period of time and to release this water at a high rate of flow to form a flushing action. Self dumping tanks are used.

The tank is constructed in such a manner that as it is filled with water, a point of imbalance is reached causing the tank to rotate on an axis and spill out the accumulated water. Once the water is removed the tank rotates back to its upright balanced position. This action requires no external energy input. The frequency of the dumping can be controlled by controlling the rate at which water is allowed to enter the tank.

Flush tanks used in System A are rectangular steel tanks that dump backwards against a contoured wall surface similar to earlier designs shown by Miller and Hansen (1973). Flush tanks used in System B are cylindrical steel tanks that dump backwards against a contoured steel chute as shown in Fig. 5. The dumped water follows the contoured surface to flow under the dump tank and down the gutter. Flow in this manner reduces the amount of splash and impact that might occur if the tank dumped directly on the floor. The

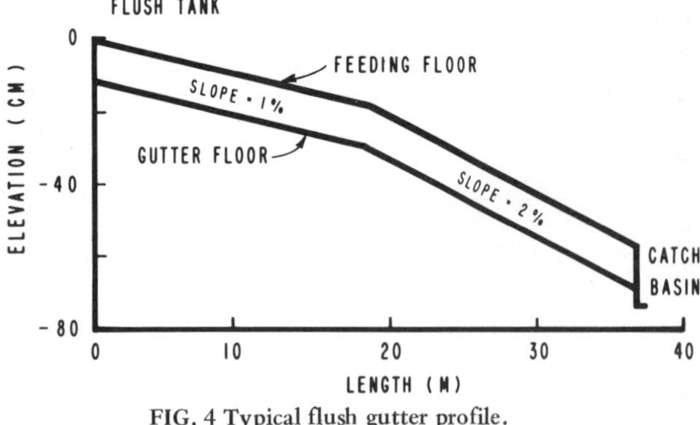

FIG. 4 Typical flush gutter profile.

control of splash allows the tank to be dumped from a high elevation increasing the potential energy of the stored water and reducing the energy loss when the water is introduced into the gutter.

## STORAGE

All waste water from System A is collected in a lagoon with surface dimensions of 95 meters by 91 meters. The depth ranges from 1.2 to 1.8 meters depending on the volume of storage. Ground water conditions prevent the use of a greater lagoon depth at this location.

Lagoon water is returned to the flush tanks by a 70 liter per minute centrifugal pump. The pump is oversized to allow for expension of the confinement system. Presently the pump is operated continuously with a valved by-pass to allow control of the return flow rate. The return flow rate controls the rate of flush tank filling and, thus, the frequency of flushing.

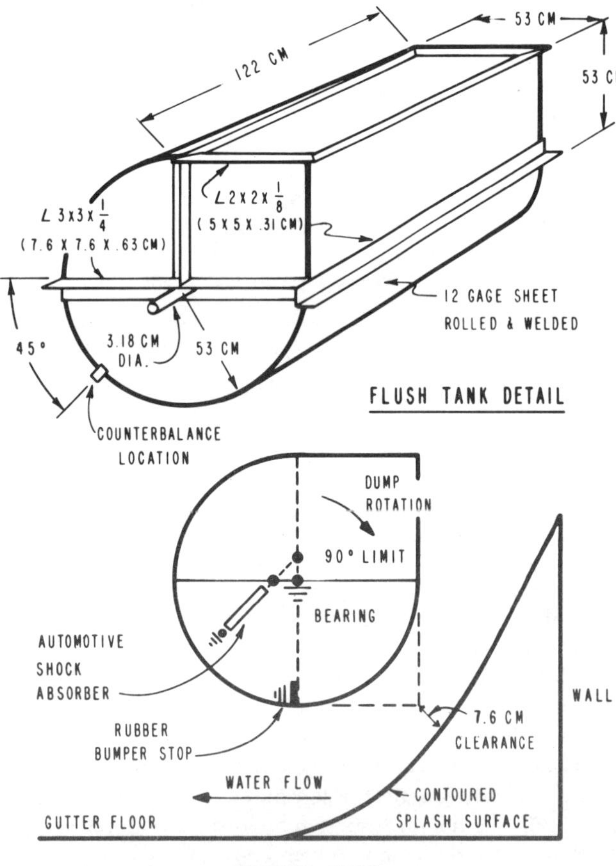


FIG. 5 Cylindrical self dumping flush tank.

Waste water from System B is collected in an underground tank constructed of concrete block. The tank has four interconnected cells and is covered with a concrete slab. The tank dimensions are approximately 3.7 meters by 9.4 meters by 3.7 meters deep. Four openings in the tank top allow access to each cell for cleaning.

## DISPOSAL

Spray irrigation equipment is used for waste water disposal from both systems. The removal of water from the lagoon of System A is performed annually. A portable big gun irrigation system operated by an engine driven centrifugal pump is used to distribute lagoon water on 12 hectares of cropland. An estimated 6.4 megaliters of waste water are disposed annually. Rainwater collected on the lagoon surface accounts for approximately 72 percent of this volume.

Waste water removal from the holding tank of System B is performed once a week. A big gun irrigation system operated by a tractor pto driven progressing cavity pump is used to distribute waste water on four hectares of grassland. An estimated 5.6 megaliters of waste water are disposed annually. Well water used for flushing accounts for approximately 75 percent of this volume.

## OPERATION

### System A

The finishing building operation is extremely satisfactory. Pens are clean and animals are quick to learn defecation patterns. The pen floors at a 6.25 percent slope stay dry even close to the gutter.

The flush tanks when dumping produce considerable splash. The wall surface was constructed of angular sections of roofing metal rather than a continuously curved contour. A baffle placed just above the gutter at the dump location assisted in splash control. The problem could be solved by construction of a properly contoured surface.

The initial wave from a flush overflows the gutter for the first three or four meters of length. The overflow is caused by the wave front hitting the animals in the gutter producing considerable splash. This problem could be controlled with the construction of the contoured surface and by reducing the rotational velocity of the dump tank to lessen the initial volume flow of water.

The gutters clean exceptionally well with a flush frequency of one dump per hour. The volume of water used is 88 liters per head per day. George (1972), recommended 60 gallons per 1000 pounds of animal weight (50 liters per 100 kilograms) which would be approximately 34 liters per head per day. In this finishing building water use efficiency could be increased by reducing the frequency of dumping.

This gestation building operates equally well. Some sows fail to learn proper habits and hand cleaning must occur in their pens. Splash is a problem here also but not to the degree found in the finishing gutter because the concentration of animals is less.

The return rotation of the dump tanks caused considerable noise upon impact with the supports. Automotive shock absorbers were mounted on the tanks to slow the rotation and prevent impact damage to the tank supporters.

The gutters clean exceptionally well. An initial dump frequency of once every 90 minutes was changed to once every four hours with little reduction of cleaning efficiency.



MANAGING LIVESTOCK WASTES

The volume of water used is 94 liters per head per day. George (1972), recommended 40 gallons per 1000 pound of animal weight for sows (33 liters per 100 kilograms) which would be approximately 55 liters per sow per day in this building.

The reduction of volume flow in System A is not of great concern to the operator because water is recycled from the lagoon. Excess volume requires an additional energy cost due to the operation of the return flow pump. However, the operator believes this to be such a small demand as compared to his overall farm power requirements that further reduction in flush volume is not necessary.

**System B**

In this system also the operation is extremely satisfactory. However, the combination of the low pen floor slope of 4.2 percent and the narrow 1.2 meter wide gutter produces a wet floor condition within 2 meters of the gutter edge.

The cylindrical flush tanks are damped with automotive shock absorbers to prevent excessively rapid dumping. The tank rotation is stopped by supports such that all water flows down the chute and no splash can occur. This control prevents excessive splashing even as the wave front moves into the first group of pigs in the gutter.

The gutters clean well with a flush frequency of one dump per four hours. The volume of water used is 23 liters per head per day. This volume is considerably less than that used in System A indicating how much more efficiently System A might be operated. The low water volume results from strict conservation demanded because of the limited capacity of the receiving tank.

## LABOR AND ENERGY

Both systems require a minimal amount of labor for waste handling. Each system requires a small amount of hand cleaning when some animals refuse to become trained to use the gutter. System A requires approximately 32 hours per year of labor including operation of the irrigation system. This is less than an average of 6 minutes per day to handle waste from 720 head of animals.

System B requires more labor for the frequent operation of the irrigation system and because the pen floors near the gutter are occasionally washed with a hose. The total labor input is estimated at 35 hours per year or slightly less than an average of 6 minutes per day to handle waste from 500 head of animals.

The total energy consumption of these systems is also low. The greatest amount of energy is utilized for operating the irrigation systems. The operational energy input to System A is estimated as equivalent to 31 kilowatt hours per day or 15.7 kilowatt hours per animal capacity per year. The operational energy input to System B is estimated as equivalent to 27 kilowatt hours per day or 19.5 kilowatt hours per animal capacity per year.

## DOLLAR COSTS

The capital investment in the flush system including the building floor is low as shown in Table 2. The cost of the storage lagoon or tank and the irrigation equipment represents the major portion of the cost.

The operational costs also shown in Table 2 are primarily those associated with the power requirements of the irrigation system. Labor costs and flush water supply costs are less than 25 percent of the total operating cost.

The total annual cost for waste handling in both the finishing building and the gestation building of System A is approximately $4,337 or $6.02 per head capacity. The total annual cost for waste handling in the finishing building of System B is approximately $3,347 or $6.69 per head capacity.

**TABLE 2. FLUSH SYSTEM COSTS.**

| | |
|---|---:|
| System A. 600 finishing 120 gestating | |
| Investment | |
|     Flush floors and equipment | $ 9,587 |
|     Storage lagoon and irrigation | 12,700 |
| Total waste handling investment | $22,287 |
|     Dollar per animal capacity = $30.95 | |
| Annual fixed cost 17 percent of investment | 3,788 |
| Annual labor @ $3.00 per hour | 96 |
| Annual energy @ $0.04 per kW.h | 453 |
| Total annual cost | $ 4,337 |
|     Dollar per animal capacity = $6.02 | |
| System B. 500 finishing | |
| Investment | |
|     Flush floors and equipment | $ 5,153 |
|     Storage tank and irrigation | 11,600 |
| Total waste handling investment | $16,753 |
|     Dollar per animal capacity = $33.50 | |
| Annual fixed cost 17 percent of investment | 2,848 |
| Annual labor @ $3.00 per hour | 105 |
| Annual Power @ $0.04 per kW.h | 394 |
| Total annual cost | $ 3,347 |
|     Dollar per animal capacity = $6.69 | |

## SUMMARY

Flush systems for hydraulic waste removal from swine buildings result in less capital invested and lower operating costs. Within these savings the overall use of energy is conserved. The environment provided for the animals is improved and operational labor is considerably lessened.


**References**

1  George, R. M. 1972. Designing gutter flushing systems, unpublished paper, Agricultural Engineering Department, University of Missouri, Columbia.

2  Jones, E. E., Jr., George B. Willson and William F. Schwiesow. 1971. Improving water utilization in automatic hydraulic waste removal, Proceedings International Symposium on Livestock Wastes, pp. 154-158.

3  Koelliker, J. K., J. R. Miner, T. E. Hazen, H. L. Person and R. J. Smith. 1972. Automated hydraulic waste-handling system for a 700-head swine facility using recirculated water, Proceedings of the 1972 Cornell Agricultural Waste Management Conference, pp. 249-261.

4  Miller, E. C. and C. M. Hansen. 1973. Flushing, AHSW 7330, Cooperative Extension Service, Michigan State University.

5  Person, H. L. and J. R. Miner. 1972. An evaluation of three hydraulic manure transport treatment systems, including a rotating biological contractor, lagoons, and surface aerators. Proceedings of the 1972 Cornell Agricultural Waste Management Conference. pp. 271-288.

# Swine Production and Waste Management: State-of-the-Art

M. R. Overcash, F. J. Humenik, L. B. Driggers

ASSOC. MEMBER
ASAE

MEMBER
ASAE

MEMBER
ASAE

SWINE on United States farms in December 1974 were estimated at 55.1 million, 10 percent less than a year earlier, and 7 percent below December 1, 1972. This was the lowest December population since 1965. Hogs and pigs kept for breeding were 7.4 million which is down 15 percent from December 1973 and 14 percent less than in December 1972. Market hogs and pigs were 47.6 million which is 9 percent below that of December 1, 1973, and down 6 percent from December 1, 1972. The value of all hogs and pigs on farms December 1, 1974, was $44.00 per head compared with $58.50 on December 1, 1973. The value of all hogs and pigs on farms totaled $83.2 million, down 27 percent from the 1973 value of $114.1 million.

Data collected by personal contacts with individuals in the top ten swine producing states, which account for about 75 percent of the total United States hog production, and national mail surveys indicate that about 60 percent of the swine are grown in pasture or dirt lot conditions, 25 percent in solid concrete floor units, and 15 percent in buildings with slotted floors. The solid concrete floor units are either partially or totally roofed houses or the concrete pad type facilities. The majority of the slotted floor houses have partial rather than total slotted floors. The trend toward enclosed housing is indicated in survey responses.

## SWINE WASTE

The waste load produced in any swine production unit has a single common basis which is derived from the magnitude of animals present expressed in total live weight. The properties of the raw waste may be affected by the multiplicity of factors which determine the animal physiology, but these are basically affected by feed ingredients either in their original form or chemically modified forms.

## NON-DISTINGUISHABLE FACTORS

### Type and Size of Animal

Data on the quantity and quality of waste from different types and sizes of hogs are summarized in Tables 1 and 2. Quantity data of Ngoddy in Table 1 are for swine in environmental chambers which had reduced water consumption because of temperature control and thus are flagged as not representative of most production conditions. Generally data in Table 1 show that the chemical content of waste as reflected in the $BOD_5$ of raw waste is not greatly affected by swine age when normalized on a pound of animal basis.

The authors are: M. R. OVERCASH, Assistant Professor, F. J. HUMENIK and L. B. DRIGGERS, Associate Professors, Biological and Agricultural Engineering Dept., North Carolina State University, Raleigh.

Acknowledgement: This work was supported by the North Carolina Agricultural Experiment Station and Extension Service with supplemental funds from EPA Project 1BB039, Swine Waste—State-of-the-Art and Runoff Research.

**TABLE 1. RAW WASTE CHARACTERIZATION FOR DIFFERENT TYPES OF SWINE**

| Pig wt, lb | Age weeks | Feces | Urine | Total | | Total volume gal |
|---|---|---|---|---|---|---|
| | | ---- lb/day /lb of animal ----- | | | | |
| 12-40 | 3-8 | 0.072 | 0.058 | 0.085 | | 0.0102 |
| 40-80 | 8-12 | 0.043 | 0.048 | 0.091 | 0.07 | 0.0109 |
| 80-120 | 12-16 | 0.054 | 0.061 | 0.115 | 0.05 | 0.0138 |
| 120-160 | 16-20 | 0.046 | 0.058 | 0.104 | 0.042 | 0.0125 |
| 160-200 | 20-23 | 0.047 | 0.051 | 0.098 | 0.034 | 0.0117 |
| Reference | | (1) | (1) | (1) | (2) | (1) |

| Animal | Weight/ animal, lb | $BOD_5$ lb/day/animal | $BOD_5$ lb/day/lb of animal |
|---|---|---|---|
| Boar | 350 | 0.40 | 0.00114 |
| Gestating sow | 275 | 0.40 | 0.00145 |
| Sow with litter | 374 | 0.75 | 0.002 |
| Nursery pig | 35 | 0.07 | 0.002 |
| Growing pig | 65 | 0.13 | 0.002 |
| Finishing pig | 150 | 0.30 | 0.002 |
| Reference | | (3) | (3) |

Reference:
(1) Calculated from data presented by Irgens and Day (1965).
(2) Ngoddy et al. (1971)
(3) Day et al. (1969) and Irgens and Day (1965).

Gestating sows and boars appear to produce a lower $BOD_5$ based upon pound per pound of animal weight and thus have a slightly lower environmental impact.

**TABLE 2. ASAE FACT SHEET, 1972.**

AW-D-1 Fresh manure production and characteristics per 1000 pounds live weight.

| Characteristic | Units | Swine | |
|---|---|---|---|
| | | Feeder | Breeder |
| Raw manure* | lb/day | 75.0 | 50.0 |
| Volume* | Gallons | 9.0 | 6.0 |
| Total solids | lb/day | 6.0 | 4.3 |
| | percent of raw manure | 8.0 | 8.6 |
| Volatile solids | lb/day | 4.8 | 3.2 |
| | percent of total solids | 80.0 | 75.0 |
| $BOD_5$ | lb/day | 2.0 | 1.3 |
| | percent of total solids | 33.0 | 30.0 |
| COD | lb/day | 5.7 | 5.2 |
| | percent of total solids | 95.0 | 90.0 |
| Nitrogen | lb/day as N | 0.5 | ND |
| | percent of total solids | 8.3 | |
| Phosphorus | lb/day as P | 0.15 | ND |
| | percent of total solids | 2.5 | |
| Potassium | lb/day as K | 0.30 | ND |
| | percent of total solids | 5.0 | ND |

References:
Barth, Clyde. (1972) Unpublished data drawn from over 50 references on animal waste, Agr. Engr. Dept., Clemson University.
Miner, J. Ronald, Ed. (1971) Farm Animal-Waste Management. North Central Regional Publications No. 206, Agr. Exp. Sta., Iowa State Univ., Ames, Iowa.
Jones, D. D., Et al. (1971) Aerobic Treatment of Livestock Wastes. Agr. Exp. Sta. Bull. 737, Univ. of Illinois, Urbana, Illinois.

TABLE 3. RAW SWINE WASTE QUANTITIES AND SOLIDS FOR MANURE PLUS URINE.

| Reference | lb | gal | Percent moisture | Total solids | Volatile solids | VS/TS |
|---|---|---|---|---|---|---|
| | | | | - lb/day/100 lb hog- - | | |
| Geldreich (1962) | 9.5 | 1.14 | — | — | — | — |
| Taiganides (1964, 1964a) | 8.4 | 1.0 | 87 | 0.97-1.1 | 0.8-0.91 | 0.83 |
| Clark (1965) | — | — | — | 0.50 | 0.35 | 0.7 |
| Hart and Turner (1965) | — | — | — | 0.8 | 0.62 | — |
| Irgen and Day (1965) | 9.9 | 1.19 | — | — | — | — |
| Kesler (1966) | 9.3 | 1.15 | 94 | 0.56 | — | — |
| Miner, Willrich (1966) | — | — | 92 | — | 0.42 | — |
| Webber (1968) | 8.6 | — | — | — | — | — |
| Schmid and Lipper (1969) | 7.5 | 0.9 | 93 | 0.59 | 0.47 | 0.89 |
| Townsend et al. (1969) | — | — | — | 0.23 | — | — |
| Converse (1970) | 5.2 | 0.63 | 90 | 0.55 | — | — |
| Weller (1970) | — | — | — | 0.47-0.86 | — | — |
| Muehling (1971) and Midwest Plan Service (1969) | 9.2 | 1.1 | — | — | — | — |
| Ngoddy (1971) | — | — | — | 0.50 | 0.41 | 0.83 |
| Scholz (1971) | 8.0 | 0.96 | 90 | 0.8 | — | — |
| Humenik (1972) | — | — | — | 0.57 | 0.47 | 0.83 |
| Value judged most reliable | 8.4 | 1.0 | 92 | 0.64 | 0.47 | 0.83 |

## Waste Constituents

It is intended to include only experimentally measured values obtained from the original author articles in this paper. Review summaries by Middlebrooks (1974), Loehr (1974), ASAE (1972, Table 2), and Muehling (1969) have been helpful in locating original data. The values judged most reliable are not averages of all data available from literature references, but these have been derived from evaluation of information given in the original articles, particularly sample collection and laboratory techniques.

Some variations in average weight and volume data in Table 3 may be due to different levels of water consumption associated with environmental conditions and waterer type. However, no significant trend in swine waste characteristics is apparent for data reported since the mid-1960's; therefore, these values may be taken as representative of swine waste until any significant changes resulting from current nutrition and housing and thus environmental conditions are documented.

Total and volatile solids data show good consistency although solids data may vary to a great degree depending upon the analysis technique used, Prakasam (1972), Humenik and Overcash (1972). The matting and clogging effect of hair has been reported by Schwiesow (1969) and Smith et al. (1971). The urine content of raw waste is about 40 percent on an as-is basis.

The oxygen demand, organic carbon, and elemental characterization data for swine manure plus urine are shown in Table 4. The average daily oxygen demand output per 100-lb hog is 0.71 lb chemical oxygen demand (COD) and 0.28 lb 5-day biochemical oxygen demand (BOD$_5$) for a COD:BOD$_5$ of 2.5. Total organic carbon (TOC) is about 0.2 lb/100-lb hog/day (COD:TOC = 3.5). The range for total nitrogen, potassium and phosphorus is relatively large with daily 100-lb hog outputs taken to be 0.048 lb N, 0.021 lb K, and 0.014 lb P. Total nitrogen data variation was probably due to the non-conservative nature of excreted nitrogen and differential degradation or volatilization before analysis.

Characterization data in mg/l of waste of percent on a dry solids basis are given in Table 5. The range in values for

TABLE 4. RAW SWINE WASTE CHEMICAL CHARACTERISTICS FOR MANURE PLUS URINE.

| Reference | COD | BOD$_5$ | total N | P | K | TOC |
|---|---|---|---|---|---|---|
| | | | - - - lb/day/100-lb hog - - - | | | |
| Hart (1960) | — | — | 0.053 | — | — | — |
| Geldreich (1962) | — | 0.44 | — | — | — | — |
| Taiganides (1964, 1964a) | 0.77-0.96 | 0.35-0.43 | 0.064 | 0.013 | 0.047 | — |
| Clark (1965) | 0.47 | — | — | — | — | — |
| Hart and Turner (1965) | 0.75 | 0.20 | 0.032 | 0.011 | 0.011 | — |
| Kesler (1966, 1966a) | — | — | 0.052 | 0.012 | 0.019 | — |
| Poelma (1966) | — | 0.22 | 0.070 | — | — | — |
| Taiganides and Hazen (1966) | — | — | 0.042-0.060 | 0.012-0.014 | 0.021-0.05 | — |
| Miner, Willrich, Hazen and Pork Producers | 0.70 | 0.29 | 0.04-0.07 | 0.013 | 0.025-0.050 | — |
| McKinney (1967) | — | 0.35 | — | — | — | — |
| Vollenweider (1968) | — | — | 0.041 | 0.024 | — | — |
| Webber (1968) | — | — | 0.058 | 0.014 | 0.016 | 0.21 |
| Day (1969) | — | 0.20 | — | — | — | — |
| Moore (1969) | — | — | 0.051 | 0.014 | — | — |
| Schmid and Lipper (1969) | 0.52 | 0.20 | — | — | — | — |
| Townsend (1969) | 0.55 | 0.33 | 0.043 | 0.02 | 0.01 | 0.16 |
| Converse (1970) | 0.81 | 0.31 | 0.019 | 0.012 | 0.008 | — |
| Weller (1970) | — | 0.19-0.34 | — | — | — | — |
| Ngoddy (1971) | 0.41 | 0.18 | 0.035 | 0.018 | 0.017 | — |
| Humenik (1972) | 0.64 | 0.31 | — | — | — | 0.21 |
| Howell (1963) | 0.69 | 0.26 | 0.053 | 0.015 | — | 0.24 |
| Value judged most reliable | 0.71 | 0.28 | 0.048 | 0.014 | 0.021 | 0.20 |

TABLE 5. RAW SWINE WASTE CONCENTRATIONS.

| References | COD mg/l | BOD$_5$, mg/l | Total Nitrogen N, percent dry basis | Phosphorus, P, percent dry basis | Potassium, K |
|---|---|---|---|---|---|
| Geldreich (1962) | — | 25,000 | — | — | — |
| Taiganides (1964, 1964a) | 160,000 | 68,000 | 4.5 | 1.2 | 4.3 |
| Iregen and Day (1965) | — | 36,000 | — | — | — |
| Hart and Turner (1965) | — | — | 4.0 | 1.4 | 1.4 |
| Kesler (1966, 1966a) | — | — | 5.6 | 1.3 | 2.1 |
| Poelma (1966) | 30,000 | — | — | — | — |
| Scheltinga (1966) | 80,000 | 30,000 | — | — | — |
| Guidi (1967) | 110,000 | 44,000 | — | — | — |
| McKinney (1967) | 80,000 | 30,000 | — | — | — |
| Potin (1968) | — | 19,000 | — | — | — |
| Webber (1968) | — | — | 6.6 | 1.6 | 1.8 |
| Schmid and Lipper (1969) | 70,000 | 20,000 | — | — | — |
| Townsend (1969) | 71,000 | 31,000 | — | — | — |
| Day et al. (1969) and Irgen and Day (1966) | 75,000 | 30,000 | — | — | — |
| Converse (1970) | 155,000 | 60,000 | 3.4 | 2.2 | 1.4 |
| Scheltinga and Poelma (1970) | — | 25,000 | — | — | — |
| Muehling (1971) and Midwest Plan Service (1969) | — | 35,000 | — | — | — |
| Ngoddy (1971) | 80,000 | 43,000 | 7.0 | 3.7 | 3.5 |
| O'Callaghan (1971) | 61,000 | 19,000 | — | — | — |
| Robinson (1971) | — | — | — | 1.8 | 1.0 |
| Calculated value based on data judged most reliable from previous tables | 86,000 | 34,000 | 7.5 | 2.2 | 3.3 |
| Measured value judged most reliable | 80,000 | 35,000 | 5.0 | 1.7 | 2.0 |

mg/l waste characterization data is partially due to water usages and difficulty of obtaining both urine and feces as the total waste.

Trace element concentrations are listed in Table 6 as both mg/g total dry solids and lb/100-lb hog/day assuming a total solids output of 0.64 lb/day/100-lb hog. Loehr's data were converted from mg/g volatile solids to mg/g total solids assuming a volatile solids/total solids ratio of 0.83. Original data of Ngoddy were altered according to a formula accounting for moisture content forwarded by these workers. Copper data are given for trace feed levels and feed additive concentrations of about 125 ppm. Feed additive antibiotic chlorotetracycline concentration in waste is also included in Table 6 for feed levels of about 60 ppm. These data represent first retrieval approximations, and more work would be required to further substantiate these concentrations especially in relation to feed intake.

## PRODUCTION UNIT EFFECT ON DEFECATED WASTE LOAD

### Concrete Floor Units

The amount and concentration of the waste load leaving a partially or completely covered concrete floor production unit is primarily dependent on the total live weight of animals present and frequency and amount of washwater per cleaning event. Water addition from spillage, fogging for animal comfort in hot weather, and rainfall and roof runoff may also contribute to the volume of the production unit waste load.

The waste load expressed in pounds of chemical and biological constituents per 100-lb hog per day is generally reduced in comparison to raw waste defecated within the unit. Decomposition takes place rapidly Grub et al. (1969), but quantitative data on actual losses for concrete floor units are not available. McCalla, et al. (in Gilbertson et al. 1970) using growth chambers projected that 55 percent of the fecal organic matter can be biologically degraded on beef feedlots. Laboratory data of McCalla (1969) and Stewart (1970) show that up to 90 percent of urine nitrogen can be converted to ammonia during dry conditions. Calculations for Gilbertson's (1970, 1971) data show annual beef feedlot losses of 40 percent volatile solids and 75 percent nitrogen for 200 ft$^2$/animal densities.

### Scraping

The concrete pad type units common in cold, dry

TABLE 6. RAW SWINE WASTE TRACE ELEMENTS.

| Reference | Ca | Mg | S | Fe | Zn | B | Mn | Na | fat | Trace Cu | Additive Cu | Chloro-tetracycline |
|---|---|---|---|---|---|---|---|---|---|---|---|---|
| | | | | | | mg/g total dry solids | | | | | | |
| Taiganides and Hazen (1966) | 36 | 5.1 | 9.2 | 1.8 | 0.38 | 0.27 | — | — | — | 0.10 | — | — |
| Loehr (1968) | 24 | 3.3 | 5.6 | 1.2 | — | — | — | — | 18.0 | — | — | — |
| Ngoddy (1971) | 25 | 12.0 | 2.9 | 0.6 | 0.50 | — | 0.2 | 6.3 | — | 0.50 | — | — |
| Robinson (1971) | — | — | — | — | 1.8 | — | — | — | — | — | 1.4 | — |
| Humenik (1972) | — | — | — | — | 0.62 | — | — | — | — | — | 0.98 | 0.15 |
| Value judged most reliable | 29 | 7.0 | 7.0 | 1.2 | 0.50 | 0.27 | 0.2 | 6.3 | 18.0 | 0.10 | 1.2 | 0.15 |
| lb/100 lb/day X 10$^3$ | 18.5 | 4.5 | 4.5 | 0.8 | 0.32 | 0.17 | 0.13 | 4.0 | 11.5 | 0.06 | 0.8 | 0.1 |



MANAGING LIVESTOCK WASTES

regions are generally scraped rather than washed to remove accumulated manure. Waste decomposition prior to scraping or runoff carriage can reduce the waste load from this type of production unit by as much as 50 percent. The pads are sloped to allow drainage and urine removal.

## Waterwash

Waste concentrations from concrete floor-guttered units are reduced because of dilution from washwater and degradation before washing. Water use for hose cleaning is up to about 10 gal/hog/wash with water pressures of 20-40 psi, Taiganides and White (1971) and Humenik (1972). Recycled washwater usage for flush gutters has been reduced from 18 to 6 gal/day/hog while maintaining satisfactory cleaning, Smith et al. (1971).

## Feeding Procedure

In concrete floor and partially slotted units, feeding procedures can influence the unit waste load, Willrich (1966). Floor feeding does not make excess feed available as does self-feeding operations. Hence hogs clean up the floor and less feed becomes part of the production unit waste load. Willrich (1966) reported a reduction in the waste load from 0.6 to 0.4 lb VS/day/100-lb hog associated with conversion from self-feed to floor feeding.

## Summary

The waste load from concrete floor production units is lower than the defecated raw waste load due to volatilization and decomposition prior to washing or scraping, but the percent reduction cannot be accurately estimated for all the particular units at this time. However, waste load concentrations in a housing unit liquid effluent are somewhat proportional to amount of washwater used and thus are about 10 percent of raw waste levels when about 10 gal/day/hog of washwater are used.

## PARTIALLY OR TOTALLY SLOTTED FLOORS OVER PITS

Manure storage pits become full in about 14 to 28 days for partial slats and 60 to 120 days with total slats, based on common storage volumes and assuming an average water precharge depth of 6 inches and wastewater input of 2 gal/day/100-lb hog. Pits may be partially or totally emptied when full or have continuous overflow.

Fully slotted floors require no washing, whereas the amount of washwater for partially slotted floors is dependent upon amount of slats and producer technique. Wastewater from washing partially slotted units will vary about 2 to 7 gal/day/hog depending upon the area ratio of slat to concrete. Wastewater derived from drinking spillage will vary from about 0.5 to 1.0 gal/day/hog for units with automatic waterers. The total waste volume to be handled will approach the defecated manure plus urine volume in totally slatted houses when nipple waterers are used and fogging is not employed.

Data on pit overflow quantity and concentration were not located. A first approximation may be derived from settling characteristics of swine waste and the average daily input of water. From data presented by Willrich (1966) and Howell (1973), it is estimated that about 40 percent to 60 percent of the waste parameters such as TOC, $BOD_5$ COD, N and P remain in the supernatant after settling. However, biological activity could further reduce supernatant concentrations. Assuming no degradation and a waste volume of 1 gal/day/100-lb hog, the pit supernatant concentration would be about 50 percent of the raw waste strength due to settling or about 40,000 mg COD/l. Values of 30,000 mg COD/l have been measured for this condition by Overcash (1975).

Waste load concentrations for manure storage pit effluent are greatly influenced by management techniques. If the pits are emptied upon filling, the amount of chemical

## TABLE 7. UNAERATED SWINE WASTE LAGOON PERFORMANCE

| | Loading rate | | Supernatant or Overflow | |
| | | | COD | BOD₅ |
| Reference | lb VS/day/100 cu ft | cu ft/100 lb | mg/l | |
|---|---|---|---|---|
| Clark (1965) | 4.9 | 92 | 8,560 | 3,600 |
| | 0.36-3.9 | 1250-115 | 1000-4000 | (winter) |
| | 0.36-0.66 | 1250-680 | 1000-3000 | (winter) |
| Hart (1965) | 5 | 90 | — | 500 |
| | 10 | 45 | — | 500 |
| | 14 | 32 | — | 1,000 |
| Curtis (1966) | 13 | 35 | 18,600 | 14,400 |
| | 17.5 | 26 | 3,050 | 1,180 |
| | 12 | 38 | 4,860 | 3,060 |
| | 17.5 | 26 | 69,080 | 43,700 |
| | 11 | 41 | 32,900 | — |
| Willrich (1966) | 3.5-5 | 130-90 | — | 1100-1300 |
| Lynn (1968) | 12.3 | 37 | — | 21,000 |
| | 3.1 | 145 | — | 2,600 |
| Townsend (1969) | 3.7 | 120 | 690 | 450 |
| | 3.7 | 120 | 50,000 | 42,000 |
| Koon (1970) | 2.8 | 160 | 9,342 | 692 |
| | 2.3 | 195 | 1,907 | 284 |
| | 2.7 | 185 | 3,698 | 590 |
| | 2.7 | 185 | 4,950 | 656 |
| Allen (1971) | 1.3 | 350 | 232 | 92 |
| | 7.5 | 60 | 1,688 | 393 |
| Humenik (1972) | 1.2 | 380 | 1,166 | 306 |
| Overcash (1973) | 5 | 90 | 2400-3000 | — |
| Range of concentrations for lagoons judged to be functioning | | | 1000-10,000 | 300-3600 |

and biological waste constituents from the units is somewhat less than the raw waste load since restricted biological activity occurs in pits. Measurements of pit contents after emptying and decomposition studies for raw waste slurries (Overcash 1975) indicate that 80-90 percent of the raw waste remains for pit emptying. The concentration of the defecated raw waste is generally reduced by at least a factor of two because the volume of waste plus water spillage is about 2 gal/day/100-lb hog. Therefore, as a conservative approximation, the waste load volume from pits with no overflow is approximately 2 gal/day/100-lb hog and parameter concentrations are about one-half the level of the defecated raw waste while total constituent amounts are about the same as raw waste defecated.

## LIQUID PRETREATMENT SYSTEMS

The major thrust of in-house of extramural liquid pretreatment systems is management ease and biological degradation of waste constituents that may limit land application. Underfloor treatment units include oxidation ditches, mixing-aeration of pit contents, and slats over lagoons. At present the number of hogs raised in production units with underfloor mechanical treatment represents only a small fraction of the total swine population, although high interest in this approach persists. The operational energy and potential toxic aspects of in-house treatment units are mitigated by utilization of outside liquid treatment units. Another solution for in-house nuisances is the recently developed underfloor ventilation (UFV) system for units with underfloor manure storage pits pursuant to extramural treatment strategies. Fresh air enters from the roof area and passes through floor slats and over storage pit waste just prior to exhaust by a central underfloor plenum.

### Unaerated Lagoon

Loading rates and effluent or supernatant data reported for unaerated swine waste lagoons are summarized in Table 7. At first glance there does not appear to be a relationship between performance and level of waste loading. However, as Willrich (1966) observed, intermittently charged lagoons with loading rates of 5 lb VS/day/1000 ft$^3$ (90 ft$^3$/100-lb hog) or greater were more subject to malfunction. Additionally, examination of data presented in Table 7 indicates that intermittently loaded lagoons with a capacity of less than 90 ft$^3$/100-lb hog have a greater possibility of malfunctioning.

The loading rate for continuously charged lagoons characteristic of underhouse units or frequently cleaned, concrete floored, units can be increased to about 10 lb VS/day/1000 ft$^3$ or 45 ft$^3$/hog without severe hazard of malfunction (Willrich 1966). The overflow concentrations from a lagoon providing good waste degradation as estimated from Table 7 are somewhere in the area of about 3000 mg COD/l and 1000 mg BOD$_5$/l. The variability in loading, climate, and operation preclude a further refinement based on current literature.

### Oxidation Ditch

Production units with an oxidation ditch beneath slotted floors have a treated effluent and residual solids as the characteristic waste load. Research findings reported are for a variety of oxidation ditches with different loading rates, inflow and outflow patterns, and provisions for clarifying final effluent. There appears to be an increasing degree of

### TABLE 8. UNDERHOUSE OXIDATION DITCH.

**Ditch mixed liquor losses in percent.**

| Reference | BOD$_5$ | COD | VS | TS | N | cu ft/hog | retention time, days |
|---|---|---|---|---|---|---|---|
| Scheltinga (1966, 1966a) | 96 | — | — | — | — | 10 | 50 |
| Foree (1969) | 93 | — | — | — | — | 30 | 150 |
| Jones (1969) | — | — | 50-55 | — | — | 08.4 | 42 |
| Smith (1971) | — | 75 | — | 80 | 80 | 12 | 60 |
| Taiganides (1971a, 1972) | 80 | — | — | — | — | 1 | 4.7 |
| Windt (1971) | 90 | 60 | — | 53 | — | 10 | 50 |
| Jones (1972) | 100 | 90 | — | — | 80 | 29 | 145 |
| Milligan (1972) | 55 | 59 | 78 | — | — | 12 | 60 |
| Estimated average reduction at recommended loading of 10 cu ft/hog | 90 | 75 | 80 | 50 | 80 | 10 | 50 |

**Ditch mixed liquor concentration.**

| | BOD$_5$ | COD | TS |
|---|---|---|---|
| | | ---- mg/l ---- | |
| Windt (1969) | 1500 | 15,000 | 15,000 |
| Taiganides (1971a, 1972) | 1000-4000 | 6,000 - 16,000 | 8,000 |
| Jones (1972) | 3000 | 12,000 | 12,000 |
| Estimated typical values | 2500 | 12,000 | 12,000 |

**Settled mixed liquor supernatant.**

| | BOD$_5$ | COD |
|---|---|---|
| | ---- mg/l ---- | |
| Scheltinga (1966, 1966a) | 15 | 570 |
| Smith (1971) | 50-150 | 500-1000 |
| Jones (1972) | — | 2750 |
| Value ranges | 10-150 | 500-2500 |



MANAGING LIVESTOCK WASTES

treatment with increasing ditch residence time or ditch volume per hog, Table 8. However, most researchers simply recommend about 10 ft³/hog for effective aerobic treatment and to reduce foaming, Jones (1969); and at this loading rate, the percent reduction of total input $BOD_5$, COD, VS, TS, and N are approximately 90 percent, 75 percent, 80 percent, 50 percent and 80 percent respectively, based on literature evaluations. Thus the approximate waste load would be 0.32, 0.1, 0.03, 0.18, and 0.01 lb/day/100-lb hog for TS, VS, $BOD_5$, COD, and N, respectively. As pointed out by many researchers, gradual start-up and continual management are essential.

The average effluent concentration from this production unit is difficult to predict. Since the oxidation ditch is nearly a completely mixed unit, the overflow and the pit contents when emptied without solids separation have about the same concentration. In steady-state operation, the mixed-liquor concentrations for an oxidation ditch are about 12,000 mg/l TS and 12,000 mg/l COD or about 15 percent of the raw waste defecated. Clarification or settling of oxidation ditch mixed-liquor provides further reduction of overflow effluent concentrations as indicated by mixed-liquor and supernatant data presented in Table 8. However, because the settled solids are returned to the ditch or remain in the clarifier, the amount of solids is continually increasing. Hence the final concentration of the contents emptied from this unit can be predicted only if the management scheme is known.

Research is being conducted to evaluate the feasibility of oxidation of ditch solids or supernatant refeeding strategies for recycling to swine (Day 1972).

### Aerated Lagoon

The aerated lagoon is similar to the oxidation ditch in that aerobic treatment is provided. However, the oxidation ditch is completely mixed or has continuous velocity gradients whereas the aerated lagoon may have quiescent settling regions. Aeration can be employed for odor control and varying levels of nitrogen reduction by ammonia volatilization and nitrification-denitrification depending upon unit design and management.

Organic removals will be dependent upon amount of aeration and mixing provided. Generally about 90 percent removal of organics and COD and about 80 percent of TKN can be expected in aerated lagoon overflow when loaded at about 10-100 ft³ volume/100-lb hog and provided with aeration ranging from satisfying 50-100 percent of input $BOD_5$ (Dale 1965, Irgens and Day 1969, Humenik et al. 1975).

## RUNOFF FROM HOGS ON PASTURE, DIRT OR CONCRETE PAD TYPE UNITS

Miner and Willrich (1970) concluded several years ago that "although runoff from feeding areas confining animals other than cattle may be expected to be high in organic matter, no data are currently available concerning these sources." At present still only minimal information is available on runoff from production units where hogs are grown on ground. Robbins et al. (1971, 1972) studied the pollutional potential of runoff from watersheds where swine waste was land spread, hogs were on drylot, and even where swine had stream access by comparison with the natural load on streams draining agricultural lands devoid of farm animals in similar and adjacent watersheds. Mass balances for the watershed with 200 sows on 3 acres of drylot plus wastes from 300 confined hogs spread on 5 acres show that 0.69 percent, 1.66 percent, and 3.0 percent of the defecated $BOD_5$, TOC and nitrogen, respectively, was present in the stream draining that watershed. Their conclusions were that even in cases when disposal sites are poorly located or where swine are grown on drylot the amount of pollutants (natural plus animal wastes) that reach streams is less than 10 percent of the raw waste deposited in the watersheds, and receiving water quality is about the same as background conditions recorded for similar streams draining natural lands.

Data assessment, especially mass balances for the 5 watersheds studied by Robbins et al. (1971, 1972), indicates that the pollutional strength of waste load from hogs on drylot at densities between 33 to 67 animals/acre is less than 10 percent of the raw waste defecated. Generally less than 5 percent of waste applied to land and from properly managed and located swine drylots with densities below 67 animals/acre reaches receiving streams. These figures only hold for animal densities that do not result in manure slurry transport during runoff or manure packed in the drylot.

The estimate that less than 10 percent of the defecated swine waste leaves the pasture or dirt lot in surface runoff is corroborated by beef feedlot investigations which either quantitate or estimate the fraction of waste load lost in winter thaw or rainfall-runoff and reports by Miner et al. (1966), Gilbertson et al. (1970), Madden and Dornbush (1971), White and Edwards (1972), McCalla et al. (1972), and Wells et al. (1972).

Gilbertson et al. (1971) and McCalla et al. (1972) in subsequent articles continue to reference previously reported findings (Gilbertson et al. 1970) that only 3-6 percent of the material deposited on beef feedlots will be transported in rainfall runoff. Conclusions presented by Gilbertson et al. (1970) included that about 6 percent of the volatile solids deposited on unpaved beef feedlot surfaces was transported in runoff for lots with stocking rates of 200 ft²/head and 3 percent for stocking rates of 100 ft²/head. If winter runoff is included, the corresponding percentage values were 14.0 and 26.8 percent.

## CURRENT TRENDS

A summary of the annual survey conducted by the National Pork Producers Council initiated in 1970 shows that those who felt waste management would limit production have varied from 35 percent in 1971 to 47 percent in 1973 and 36 percent in 1974. Use of an open wagon manure spreader has remained at about 70 percent, while employment of a liquid manure spreader has increased from 14 percent to 21 percent. Lagoon usage maintains at about 5 percent, and oxidation ditches at about 0.3 percent. Gratifyingly, only about 5 percent rely on natural drainage, indicating that at least the respondents to this poll representing producers of only about 2 percent of all hogs marketed are demonstrating a trend to the recommended or most suitable terminal disposal scheme of land application for utilization and elimination of point source discharges.

More work needs to be conducted on the runoff from swine drylot and pasture areas in response to pending non-point source and feedlot runoff criteria, particularly since most of the swine in the United States are produced under such growing conditions. The trend to enclosed housing will eliminate the need to collect rainfall runoff from growing areas and allow reduction in the amount of manure to be handled if good water conservation practices are employed.

*(Continued on page 163)*

# Swine Waste Nutrient Recovery System Based On the Use of Thermal Discharges


J. R. Miner, L. Boersma, J. E. Oldfield, and H. K. Phinney

MEMBER
ASAE


TWO major concerns of contemporary society are environmental degradation and the consumption of non-renewable natural resources. An integrated swine production system in which waste heat, as is available from a thermal power plant, is used as an aid in manure processing, offers potential to meet these concerns. The Oregon State University swine waste heat nutrient recovery project demonstrates one alternative for application of this approach. The goal is a recirculating hydraulic manure transport system in which the solids are anaerobically digested for methane recovery and nutrient release and algae are grown to capture the soluble plant nutrients.

The experimental facility as built consists of six pens, sufficient for approximately fifty growing finishing hogs, from which the manure is flushed frequently by discharge from an overhead siphon tank as described earlier (Koelliker et al. 1972). The gutter is 75 cm wide, 5 cm deep, and slopes at a rate of one percent. From the gutter, flush water and manure drop into a pump pit designed to accommodate the use of an air lift pump. Slurry is then lifted at a rate of 2 to 10 liters per minute to a rotating flighted cylinder for solid-liquid separation. The solids-rich stream from the separator drops through a metal chute into the anaerobic digester just beneath it. Effluent from the anaerobic digester is combined with liquid effluent from the solid-liquid separator and pumped to the algal growth basins. Effluent from the growth basins is centrifuged for algae harvesting. The liquid fraction is pumped to the flush tanks for reuse. The overall process is shown schematically in Fig. 1.

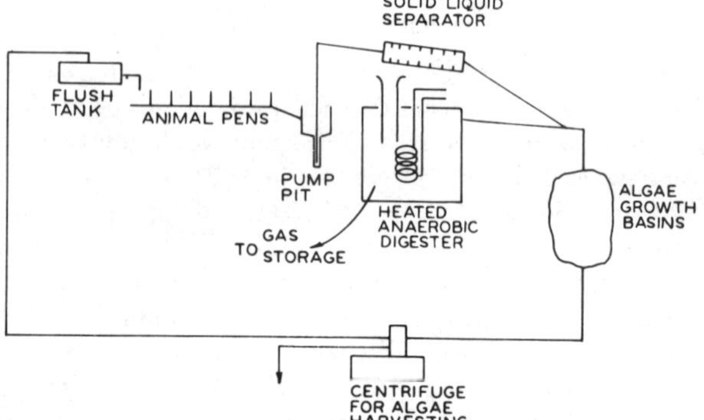


FIG. 1 Schematic diagram of the Oregon State University swine waste management system with nutrient recovery


The authors are J. RONALD MINER, Associate Professor of Agricultural Engineering; LARRY BOERSMA, Professor of Soil Science; JAMES E. OLDFIELD, Professor and Head of Animal Science; and HARRY K. PHINNEY, Professor of Botany and Plant Pathology, Oregon State University, Corvallis 97331. Technical Paper No. 3964. Oregon Agricultural Experiment Station.

Acknowledgments: In addition to the authors, E. W. R. Barlow, Ekkehart Gasper, P. R. Cheeke, and Will Fukui contributed significantly to the project. Financial support was provided in part by the Office of Water Resources Research, through the Water Resources Research Institute of Oregon State University, Project No. B-039-ORE and by the National Science Foundation, Grant No. GI-43681.


## DESIGN RATIONALE

The purpose of this project is to design, construct, and operate a swine production system which will effectively reclaim manure nutrients in a more energy efficient method than application to crop land. To be successful, the system must be compatible with modern swine production techniques, avoid offensive odor production, prevent surface and groundwater pollution, and be sufficiently reliable to be managed by a competent livestock producer under a variety of climatic conditions. This concept is facilitated by the use of waste heat as might be available in the form of warm water from a thermal power plant. If it works well, other heat sources might be considered. A 1,000 megawatt power plant operating with a thermal efficiency of 34 percent will reject 1,940 MW or 462 x $10^6$ cal/sec. Cooling tower water at Portland, Oregon has an annual range in water temperature of 41 to 48 C (Boersma et al. 1974). Such heat emission would be sufficient to maintain up to 150 hectares of water surface at 25 C in the Corvallis area.

### Building Design

An existing swine finishing building at the OSU Swine Center was modified to accommodate this project. Pen partitions were installed to reduce pen widths to 1.5 m and separate feeders and waterers installed. The existing gutter was modified so wastes generated by pigs in the six pens of this project could be kept separate from that generated by the remainder of the Center. A platform supports the water tank fitted with an automatic siphon over the upper end of the gutter. Flushing frequency is controlled by the rate at which returned water is pumped from the centrifuge water storage tank to the flush tank. At the lower end, the gutter was modified to divert the flush water and manure to the wet well for pumping to the treatment system.

### Hydraulic System

The wet well is located immediately adjacent to the swine building and was constructed with the use of a 1.3 m diameter concrete culvert set vertically in the ground to provide a pit 2 m deep. A 20 cm plastic pipe was installed in the center of the pit extending down an additional 5 m. A two-inch plastic pipe installed in this lower section provides an air lift pump to feed the manure slurry to the solid-liquid separator. The volume of the wet well was selected to provide storage for two flushes without overflowing. The air lift pump discharge is a function of the liquid level in the wet well. Flow rates vary from 0 to 30 liters per minute. A low capacity air compressor with a large (130 l) storage tank was selected to provide a continuous flow of air, hence manure slurry to the separator. The wet well and associated piping are shown in Fig. 2.

The solid-liquid separator used in this system is a rotating flighted cylinder that has been previously evaluated on swine manure slurries from this Center. It is a 61 cm diameter metal cylinder with a 15.3 cm helically wound fin on

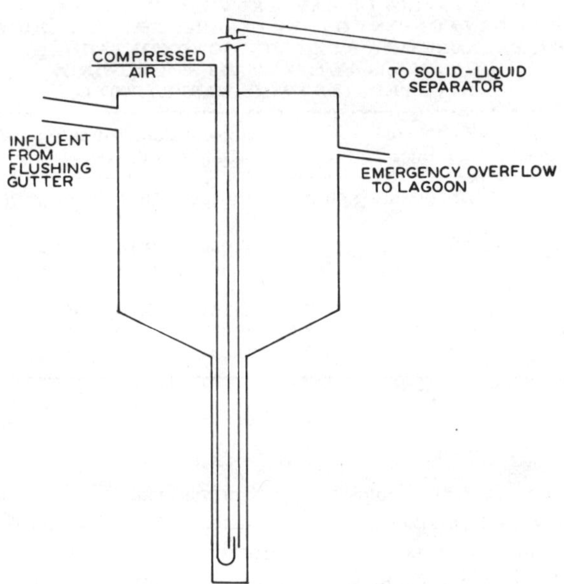

FIG. 2 Wet well and air lift pump

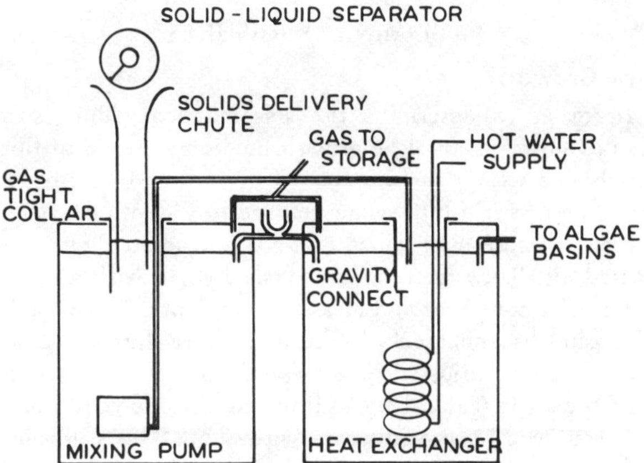

FIG. 4 Anaerobic digesters and associated piping as used in the swine waste management system

the interior surface. There is a 10.2 cm spacing between wraps of the fin. In previous studies, this device, as shown in Fig. 3, was able to remove all the solids retained on a 1.19 mm screen when receiving manure slurry at rates from 17 to 26 liters per minute. This separation device was more completely described by Verley and Miner (1974). In this system, the solid-liquid separator is located immediately over the anaerobic digester so the solids fraction can be dropped into the digester without further handling. Liquid

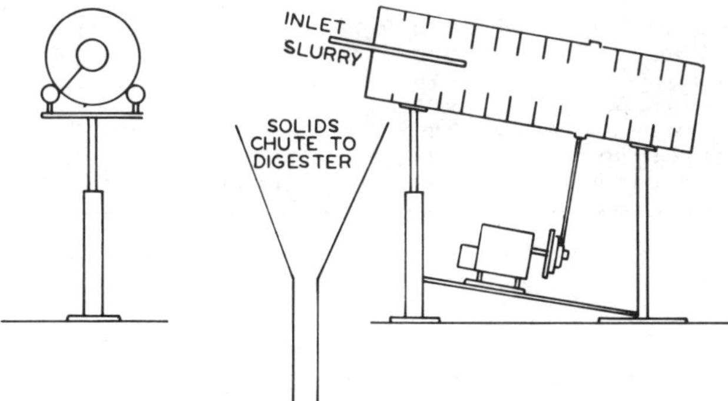

FIG. 3 Rotating flighted cylinder as used in the swine waste management system

from the separator is combined with effluent from the anaerobic digester prior to addition to the algae growth basins.

## Anaerobic Digester

With fifty finishing hogs in the pens served by this system, 16 kg of volatile solids will enter the system daily along with 1.5 kg of nitrogen and 0.22 kg of phosphorus. Based upon a proposed loading rate of 2 kg VS/day m³ (Miner and Smith, 1975), a heated, stirred anaerobic digestion volume of 8 m³ would be required. In order to provide this volume at a minimal cost, two standard septic tanks of 1,800 gallons (6.8 m³) capacity were installed adjacent to

one another so that they can be operated as a single unit by the use of a mixing pump. The digester piping is shown schematically in Fig. 4

The digestion tanks are maintained at 35 C by stainless steel heat exchangers submerged in the tanks. Heat is supplied by a recirculating hot water system. In order to have access to the digester contents and maintain an anaerobic environment within the tank, a cylindrical metal collar was fabricated to fit through the manhole openings of the digestion tanks. All submerged pumps, heat exchangers, and samples are removable through these access ports.

Gas produced by the anaerobic digestion is measured and sampled routinely. In a full scale system, this gas would be available as a heat source within the livestock system or for return to the power plant supplying the warm water.

## Algal Basins

The design of the algae basins was based on achieving an algae growth rate of 20 g/m²/day (Boersma et al. 1974). On this basis, a surface area of 75 m² would be indicated if all the nitrogen were to be converted to algal mass. Initial results have suggested that approximately 30 percent of the dissolved nitrogen can be recovered in the algae; therefore, twelve algae basins providing 2 m² of surface area each have been provided. The algae basins were provided with equipment for mixing and controlling minimum water temperatures from 20 to 35 C.

The basin area was placed in an adjacent pasture about 50 m from the hydraulic system in order to avoid shadows from nearby buildings. Construction cost and the desire to disturb the natural setting as little as possible indicated the use of a wooden deck as the best solution to space and level the algae basins. Six individual platforms were built, in two rows of three, with walkways between. Each platform will support two oval basins placed side by side, which are constructed from fiberglass (30 cm x 91 cm x 210 cm). All twelve basins will have the same elevation.

A heat exchanger system was built for each basin from stainless steel tubing. Four tubes, connected at the ends by U-joints, will run 20 cm apart, parallel to the long axis of a basin.

Nutrient substrate will be supplied on a continuous basis by gravity flow from a feed tank above the basins. A standpipe inside each basin will remove an equal flow of effluent to a two-inch PVC collecting pipe common to all basins and leading to a holding tank for storage prior to centrifugation.

PRELIMINARY RESULTS

**Algae Growth**

In order to determine the suitability of dilute swine manure slurry as an algae growth medium, a series of fiberglass basins with a surface area of $0.1 m^2$ and 12 cm deep were filled with dilute swine manure to a depth of 10 cm. The basins were maintained at 37 C and aerated with compressed air. They were continuously lighted with a combination of fluorescent and incandescent lights. Waste used in the basin was a cheesecloth-filtered manure slurry. The concentration was adjusted to achieve the ammonia concentrations shown in Table 1 by addition of distilled water. Algae were removed by precipitation with alum. This study indicated that the swine manure slurry would provide an adequate growth medium without intensive filtration or chemical clarification.

**TABLE 1. EFFECT OF SWINE WASTE CONCENTRATION ON GROWTH OF *CHLORELLA VULGARIS* 211-8K IN 0.1 m² MINI-PONDS AT 37 C**

| $NH_4^+$ nitrogen concentration, mg/l | Dry matter production, $g/m^2/24$ hr |
|---|---|
| 250 | 56.0 |
| 190 | 44.2 |
| 125 | 27.5 |
| 63 | 15.6 |

Total amino acids comprised about 75 percent of the crude protein in the algae grown in this study. This is typical in microorganisms where such nitrogen is present as nucleic acids and other non-protein compounds. The essential amino acid content of *Chlorella vulgaris* 211-8K grown on swine manure compares very favorably with *Scenedesmus*, *Spirulina*, and soybeans.

Experiments were conducted in which different volumes were withdrawn from the 10 liter basins daily. Harvest rate determines the mean residence time of waste in a basin of constant volume. Thus, a withdrawal rate of one liter per day corresponds to a mean residence time of ten days. Experiments show increased production with decreasing mean residence time (Table 2). The culture volume harvested each day from a fixed pond volume affects algal yield per unit area. Residence time of the waste determines the density of the culture within the pond and consequently, the cost per pound of algae harvested by centrifugation. Therefore, there will be economic trade-offs between maximum product yield and maximum economic yield.

An encouraging aspect of these studies has been the stability of the *Chlorella vulgaris* 211-8K cultures. To date, the cultures have grown at 37 C on at least ten different samples of untreated swine waste for a period of one year, without upset or a significant drop in productivity. During this period, the algae have been subjected to a wide range of environmental conditions, including neglect for brief time intervals.

**Feed Value of Algae**

In preliminary feeding trials, several protein sources were included to provide 12 percent protein in rations fed to laboratory rats (Table 3). The algal material was dried at 80 C for 24 hours. The Protein Efficiency Ratio (PER) is the gain in weight per gram of protein consumed. Casein is a high quality protein while cottonseed meal is of relatively poor quality. In considering both the PER and the growth rate, the centrifuged algae appear to be a relatively good protein source. The algae harvested by alum precipitation did not give good results. It is clear that centrifugation should be used for harvesting, or an alternate resin or polymer for precipitation must be found.

**TABLE 2. EFFECT OF HARVEST VOLUME ON THE $NH_4^+$ NITROGEN CONTENT OF THE EFFLUENT AND NITROGEN BALANCE OF A 0.1 m² MINI-POND CONTAINING *CHLORELLA VULGARIS* 211-8K GROWING ON UNTREATED SWINE WASTE AT 37 C**

| Harvest volume | $NH_4^+$ N content, mg/l | | Daily nitrogen balance of mini-pond, grams of nitrogen | | | |
|---|---|---|---|---|---|---|
| | Waste | Basin | Inflow | Outflow | Algal N Out | $NH_4^+$ N lost percent |
| 1.5 | 250.0 | 18.5 | 0.38 | 0.03 | 0.18 | 0.22 | 58 |
| 2.0 | 250.0 | 30.2 | 0.50 | 0.06 | 0.17 | 0.31 | 61 |
| 2.5 | 250.0 | 43.8 | 0.63 | 0.11 | 0.16 | 0.36 | 58 |
| 3.0 | 250.0 | 49.8 | 0.75 | 0.15 | 0.15 | 0.44 | 59 |
| 3.5 | 250.0 | 50.2 | 0.88 | 0.21 | 0.13 | 0.50 | 57 |
| 4.0 | 250.0 | 71.3 | 1.00 | 0.29 | 0.13 | 0.54 | 54 |

**TABLE 3. RESULTS OF PRELIMINARY FEEDING TRIALS WITH RATS**

| Protein source | Average daily gain, g | Average daily feed intake, g | Protein efficiency ratio, PER |
|---|---|---|---|
| Casein | 3.70 | 13.37 | 2.30 |
| Fungus | 1.96 | 13.17 | 1.44 |
| Torula yeast | 1.76 | 12.29 | 1.17 |
| Brewers yeast | 1.94 | 12.94 | 1.40 |
| Cottonseed meal | 2.23 | 16.26 | 1.13 |
| Algae (centrifuged) | 2.29 | 13.29 | 1.44 |
| Algae (alum pptd) | 1.80 | 16.20 | 0.91 |

CONCLUSIONS

The consistently high yields of dry matter and protein obtained in these studies were most encouraging. The mean dry matter yield of all experiments is equivalent to a production rate of 121.5 tons/ha/year. This corresponds to yields of crude protein of 55 tons/ha/year and total amino acid or true protein yield of 47.6 tons/ha/year. The mean culture density was 1.183 g/l, indicating that culture densities would mean a considerable saving in harvesting costs.

The organic matter in the untreated swine waste appears to provide some $CO_2$ for the photosynthetic algae, as evidenced by the small increase in growth rate when additional $CO_2$ was added to the concentrated waste. Neither the nitrogen or phosphorus content of the waste appears to be limiting algal production. There was no response to nitrogen levels in any experiment. The algal pond system reduced the total nitrogen content of the waste by about 90 percent. The $NH_4^+$ nitrogen level was reduced by about 90 percent also. Nitrate did not accumulate in the mini-basins, and in all cases was less than one mg/l in the outflow.

However, the algae only accounted for 20-40 percent of the nitrogen taken from the waste media. Between 60 and 80 percent of the nitrogen was lost from the system, probably as ammonia gas. The aerobic algal pond system quickly dissipated the odor from the swine waste. Neither the algae product nor the pond system had the odor usually associated with swine waste.

**References**

1   Boersma, L., E. W. R. Barlow, J. R. Miner, and H. K. Phinney. 1974. Animal waste conversion systems based on thermal discharges. Special Report 416, Oregon Agricultural Experiment Station, Corvallis. 54 pp.

2   Koelliker, J. K., J. R. Miner, T. E. Hazen, H. L. Person, and R. J. Smith. 1972. Automated hydraulic waste-handling system for a 700-head swine facility using recirculated water. pp. 249-261. In: Waste Management Research. Proceedings of the 1972 Cornell Agricultural Waste Management Conference.

3   Miner, J. R. and R. J. Smith, editors. 1975. Livestock waste management with pollution control. NC-93 Regional Research Committee Publication. Midwest Plan Service, Ames, Iowa.

4   Verley, W. E. and J. R. Miner. 1974. A rotating flighted cylinder to separate manure solids from water. TRANSACTIONS of the ASAE 17(3):518-520, 525.

---

## Swine Production and Waste Management

*(Continued from page 159)*

The most feasible longterm approach for swine waste management from both a technical and economic basis is to minimize the volume requiring management and recycle it to agricultural lands for utilization and elimination of point source discharges.

**References**

A complete listing of references will be included in the final report currently under preparation for EPA Project 1BB039 entitled Swine Waste Management; State-of-the-Art.

# Managing a Successful Liquid Swine Manure Management System


P. R. George, J. M. Sweeten, S. J. Buchanan
MEMBER
ASAE


LEANCO Corporation, a 600-sow total confinement farrow-to-finish swine operation at Brownwood, Texas, employs a unique three-phase system of liquid manure management that meets or exceeds pollution abatement requirements of three state and federal agencies. The Leanco swine waste management system recycles all manure and wastewater through land disposal; controls both point and nonpoint source pollution through a "no discharge" manure and runoff collection system; and minimizes odor and fly production. A rigorous herd health program prevents a "fourth type of pollution" — spread of communicable swine diseases.

The waste management system described herein consists of these elements: (a) underfloor liquid manure storage pits; (b) soil injection of liquid manure and sludge; (c) retention pond for storage of liquid manure overflow and runoff from manure disosal areas; and (d) irrigation system for dewatering the retention pond. Liquid manure characteristics, measured over a 5-month period in connection with odor control experiments, are also reported.

## OVERVIEW OF MANURE MANAGEMENT SYSTEM

Manure and wastewater, estimated at 27,000 gallons per day (102,000 l/day) are collected in storage pits beneath partially slatted floors (Buchanan 1972). Settleable solids undergo initial digestion during storage. Solids are periodically removed by tank wagon and injected beneath the soil surface.

The supernatant liquid and lighter elements of solid manure flow from the storage pits through overflow pipes and sanitary sewers to a settling pit (Fig. 1). From the settling pit, supernatant liquids are conveyed to the retention pond from which it is terminally disposed of through irrigation.

Surface runoff from pasture and cropland used for disposal of solid and liquid wastes is also collected in the retention pond. This form of nonpoint source water pollution control, i.e. collection of runoff from manure disposal areas, is not generally required by state and federal water pollution control agencies. Liquid wastes collected in the retention pond are disposed of by irrigation onto the adjacent drainage area.

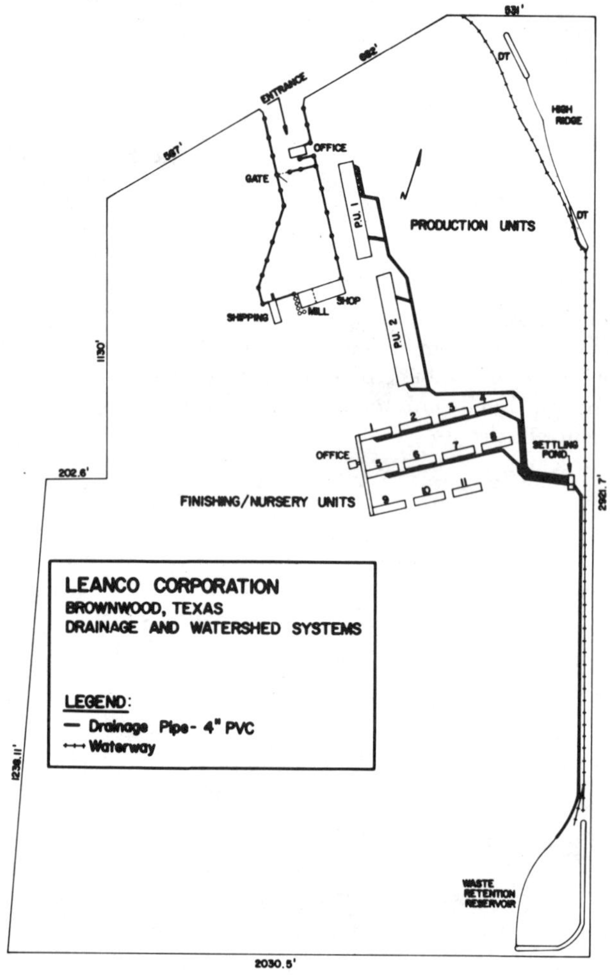

The authors are: P. R. GEORGE, President, Leanco Corp., Brownwood, Texas; J. M. SWEETEN, Agricultural Engineer, Animal Waste Management, Texas Agricultural Extension Service, Texas A&M University, College Station; and S. J. BUCHANAN, Principal Consulting Engineer, Spencer J. Buchanan and Associates, Bryan, Texas.




FIG. 1 Map of Leanco Corp.'s 600 sow farrow-to-finish swine operation showing liquid manure drainage and runoff retention system.

## FEEDING FACILITIES

### Gestation and Farrowing Operations

Two "production units" which contain breeding, gestation, farrowing and nursery sections are the heart of the Leanco operation. Production Units (PU) 1 and 2 occupy 16,250 ft² (1510 m²) and 18,750 ft² (1740 m²) of floor space, respectively, including alleys, offices and feed storage areas. Total capacities of PU-1 and PU-2 are 231 and 336 sows, respectively.

Breeding and gestation is conducted in a clear-span, curtain-walled section of each production unit. Space allotment is approximately 20 ft² per sow. Liquid manure and spilled drinking water are collected in

| Building ident. | Function | Adult hogs served—head | Effective pit storage capacity: | |
|---|---|---|---|---|
| | | | gal/hd | l/hd |
| PU 1 | Breeding and gestation | 150 | 220 | 830 |
| PU 1 | Farrowing and nursery | 81 | 730 | 2770 |
| PU 2 | Breeding and gestation | 216 | 500 | 1900 |
| PU 2 | Farrowing and nursery | 120 | 1180 | 4500 |
| FU 1-4 | Finishing | 400 (ea.) | 75 | 280 |
| FU 5-11 | Finishing | 400 (ea.) | 60 | 230 |

storage pits under partially slatted floors which occupy approximately 1/3 of the total floor area. These pits have a longitudinal bottom slope of 1.0 percent and average depths of 3.5 ft (1.07 m) to 4 ft (1.22 m).

The totally enclosed farrowing and nursery sections of PU-1 and PU-2 are divided into separate, enclosed rooms each having crates for 9 to 15 sows and litters. Each farrowing room and nursery in PU-1 and PU-2 has its own compartmentalized underfloor liquid manure storage pit with an average depth of 3.0 and 4.25 ft (0.92 and 1.3 m), respectively, and 1 percent slope toward the overflow/pump outlet. Pit storage capacities at the operating liquid level (46 cm below the top of the slats) are shown in Table 1.

### Finishing/Nursery Buildings

The finishing/nursery system consists of 11 curtain-walled individual buildings. Each unit can house up to 400 head of market weight hogs. These buildings are roofed to prevent entry of rainwater. Manure and waste-water are collected in liquid manure storage pits under partially-slatted floors (Fig. 2). These pits are 10 to 12 ft (3.1 to 3.7 m) wide and have an average depth of 4 ft (1.2 m). Each 130-ft (40 m) long pit slopes at 1.0 percent toward the center where manure removal takes place. Unit capacities are given in Table 1.

The solid-to-slat relationship in the pen area is 40 to 50 percent slatted. The solid floor section has a 4 percent cross slope. Both solid and slatted floor sections remain essentially free of manure at all times.

Animal liveweight in the finishing/nursery buildings varies from 25 to 200 lb per head (11 to 91 kg). Numbers on feed vary also depending on several management factors. This results in a wide variation in daily manure production between buildings.

## MANAGEMENT OF LIQUID MANURE PITS

### Liquid Overflow System

The underfloor liquid manure pits provide first-stage digestion and temporary storage. Liquid level in the pits is maintained at 18 in. (46 cm) below the top of the slats by the specially-developed liquid overflow drain system shown in Figs. 3 and 4. This system, in effect, separates liquids from settleable solids for separate collection and handling.

Liquid overflow is conveyed through 4 in. (10.2 cm) diameter PVC sewer pipes into a 0.2 ac ft (250 m³) capacity settling basin (Fig. 1). The sewer pipes are placed on a 2 to 3 percent slope.

Solids are removed periodically from the settling basin with a tank wagon. Liquids flow by gravity nearly 1000 ft (300 m) through a 4 in. (10 cm) diameter PVC pipe to the retention pond at the southeast corner of property. A solids cleanout trap is located near the midpoint of this sewer pipe.

Occasionaly, the 4 in. (10 cm) PVC sewer lines become clogged as evidenced by daily inspection of visible inlets and outlets of the system. Stoppage is easily corrected by flushing the pipes using a commercially-available garden hose attachment operated at 50 psi (3.5 kg/cm²) pressure.

Spilled drinking water enters the storage pits in relatively large amounts. By monitoring the removal of liquid manure by tank wagon over a 5 month period in relation to estimated manure production rate, we found that 2.3 to 7.6 gallons (8.7 to 28.8 l) of spilled drinking water is generated for every gallon of manure produced. Buildings containing the smallest hogs produced the most water wasteage, and vice versa. While water spillage increases handling requirements, it is probably beneficial in reducing odors, flies and sewer maintenance problems.

### Management of Solids in Underfloor Storage Pits

Many other confinement swine operations in Texas have underfloor storage pits connected to a lagoon. Frequently, they experience problems with odors, flies and/or overflow line maintenance. These problems are generally caused by infrequent removal of manure from the pits.

Leanco has developed a program that minimizes these

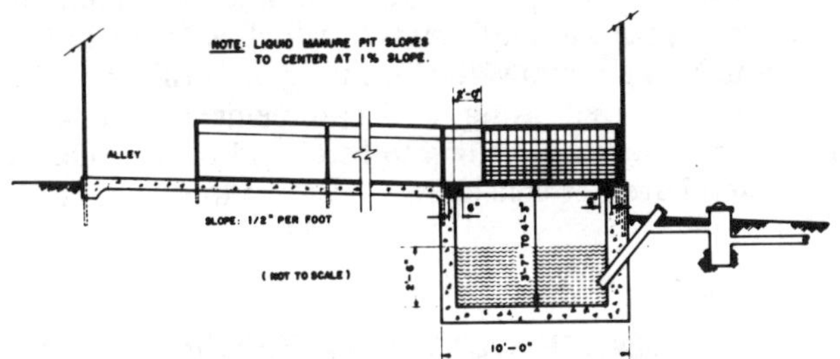


FIG. 2 Cross-section of typical 400 hog finishing/nursery unit depicting liquid manure storage pit configuration.

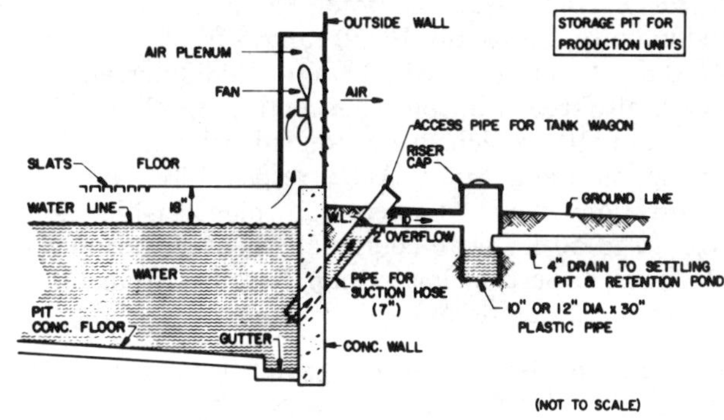


FIG. 3 The Parker George liquid overflow system of waste handling for farrowing and nursery section of Production Units 1 and 2.

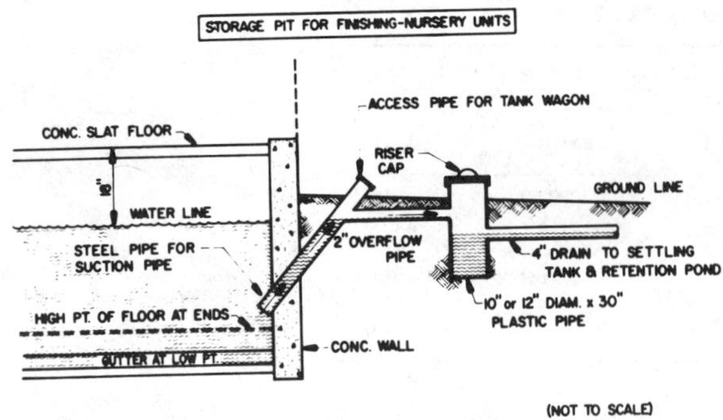


**FIG. 4 The Parker George liquid overflow system of waste handling for finishing/nursery units and breeding and gestation sections of production units.**

problems through frequent withdrawal of settled solids to maintain low solids concentrations. Manure is removed at intervals ranging from 1 day to 21 days, depending upon defecation rates and other factors.

The sludge is vacuum-loaded into a conventional tank wagon through the 7 in. (18 cm) diameter access pipe of the liquid overflow drain system (Figs. 3 and 4). Despite the gentle (1 percent) slopes of the pits in each building, solids flow toward the pumping station without agitation. Accumulation of solids at the distant end of the pits 65 ft (20 m) away has never become a problem.

## STORAGE PIT PERFORMANCE: ODORS AND CHEMICAL PROPERTIES

Storage pit performance was monitored in four finishing/nursery buildings (FU 8, 9, 10 and 11) over a 5-month period (1974-75) by measuring solids concentration (total and volatile), nitrogen (total N and ammonia), pH, chemical oxygen demand (COD) and odors. This work was done in connection with an experiment involving chemicals/biochemicals for odor control (Reddell and Sweeten 1975). Manure samples were extracted from the storage pits on 19 occasions.

The *total solids concentration* in FU 9 (untreated) averaged 5.4 percent wet basis (54,000 mg/l) over a 4 1/2 month period, ranging from 26,000 to 126,000 mg/l. Two of the chemical treatments (potassium permanganate and Clew II*) may have improved solids digestion slightly, reducing the ratio of volatile solids to total solids from 80 percent (control) down to 73 percent (treated).

The *chemical oxygen demand* (COD) stabilized rather quickly and averaged 79,000 mg/l in the untreated building. The chemical treatments showed little or no discernable effect on COD values.

*Nitrogen concentrations* (total N) in the untreated liquid manure storage pit (FU 9) ranged from 2800 mg/l to 4200 mg/l and averaged 3300 mg/l (0.33 percent N). This was also true of nitrogen concentrations observed in FU 8 and 10, which were treated with potassium permanganate and Micro Aid*, respectively. Nitrogen concentrations in FU 11 where baby pigs weighing 9 to 23 kg (20 to 50 lb) were continually housed during the study were 30 percent lower than N values from other buildings both before and after treatment.

Remarkably, 90 percent of the nitrogen was in the

---

*Mention of trade names does not constitute endorsement by the authors or their respective organizations.

ammonia form regardless of treatment. This indicated (a) good digestion in the storage pits, and (b) high nitrogen availability to crops receiving both the manure solids and liquids.

Values of pH were similar for all four buildings. The pH levels decreased from the range of 6.8 to 7.4 in early fall to as low as 5.8 in mid-winter, reflecting a slowdown in digestion with cold weather.

*Odor intensities* were measured using a Scentometer (vapor dilution device) at the downwind side of each building. An average of odor readings taken on 18 occasions between Septemter 25 and November 25 for FU 8, 9, 10 and 11 were, respectively, 5.9, 8.3, 6.7, and 3.1 dilutions to threshold ($D_t$). During this time, average manure loading rates into the pits ranged from 60 gal/day (230 l/day) in FU11 to 440 gal/day (1700 l/day) in FU 9.

After treatments were started (November 25 to February 21), odors were 7 dilutions to threshold ($D_t$) or less. Odor intensities averaged 3.0, 5.9, 2.7, and 5.2 $D_t$, respectively, for FU 8 ($KMnO_4$), FU 9 (Control), FU 10 (Micro Aid) and FU 11 (Clew II). While consistently lower odors were recorded in buildings treated with potassium permanganate and Micro Aid, treatment differences were relatively small. Experiments during summer months are needed to substantiate these trends.

Severe corrosion of steel mixing barrels and pump bearings resulted from use of potassium permanganate. However, further investigation revealed the corrosion problems were caused by chemical interaction of the $KMnO_4$ with the high chloride content (1200 mg/l) of the water supply. According to the manufacturer, the chloride concentration should not exceed 200 mg/l to avoid corrosion with $KMnO_4$.

## LAND DISPOSAL OPERATION

Settled solids from the liquid manure storage pits are collected and hauled to adjacent pasture and cropland using a 1000 gal (3780 l) tank wagon. This sludge is injected 10 in. (25 cm) below the soil surface by a two-row plowdown attachment.

The disposal area consists of 46 acres (19 ha) of coastal bermudagrass and 50 acres (20 ha) of annual hay crops. Hydraulic and nitrogen loading rates are estimated at 22,000 gal/ac/yr (210 m³/ha/yr) and 600 lb/ac/yr (670 kg/ha/yr), respectively. Winter forage crop is sown oats, while hybrid sudangrass is grown in the summer. Top soils are fine sand and loamy fine sand with clay subsoil.

Equipment used for soil injection includes a 1000 gallon (3780 l) tank wagon with plowdown attachments (total cost $2500) and a 94 hp tractor (1971 model). Total cost of the soil injection operation, including labor, equipment, depreciation, repairs, and fuel is about $7.67 per hr. At a hauling rate of three 960-gal (3640 l) loads per hr, the cost of liquid manure soil injection amounts to 0.27 cents per gal (0.07 cents/l) which is about 1/3 of the manure's apparent value for nitrogen fertilizer alone.

Odors measured using a Scentometer at the land disposal site were 2 dilutions to threshold or less. Odors of this level are not considered offensive (Sweeten et al. 1974).

## RUNOFF AND WASTEWATER RETENTION SYSTEM

Leanco Corporation has constructed a large retention

pond at the southeast corner of the property (Fig. 1) to provide limited storage capacity for liquid wastes (manure and wastewater) plus runoff from the 25-yr frequency, 24-hr duration rainfall event. In the Brownwood, Texas area, this design rainfall is 7.1 in. (18 cm). A rainfall of this magnitude is most likely to occur during May or June (Griffiths and Orton 1968).

The retention pond collects runoff from 43 ac (17 ha), including gravel-surfaced parking lot (4 ac), roofed areas (1 ac), and permanent and annual pastureland (38 ac). Runoff coefficients were estimated at 75, 100, and 50 percent, respectively, with a weighted average of 53.5 percent. Hence, runoff resulting from the 25-yr frequency, 24-hr duration rainfall is estimated at 13.6 ac ft or 16,800 m³ (Buchanan 1972).

The retention pond also provides storage for 2.5 ac ft (3100 m³) of manure and wastewater (30-day accumulation) and 1.1 ac ft (1360 m³) of solids storage. Hence, the total storage capacity is 17.2 ac ft (21,200 m³). The maximum depth and water surface area are 18.0 ft (5.5 m) and 2.15 ac (0.87 ha), respectively. Construction cost for the 17.2 ac ft (21,200 m³) retention pond and the 2800 ft (850 m) long diversion terrace along the east boundary of the property was $4000.

Subsurface conditions were explored by five core borings taken at the center and along the south and east sides of the retention pond (Buchanan 1972). Subsoil conditions were found to be very slowly permeable. Piezometers inserted in the borings did not detect seepage after the retention pond was placed in operation. It was concluded that the retention pond meets the 1 X $10^{-7}$ cm/sec (0.1 ft/yr) permeability requirement of the Texas Water Quality Board.

Effluent from the retention pond is applied to 38 ac (15 ha) of pastureland in the drainage area. An irrigation application of 4.5 in. (11 cm) of water during the two-week dewatering period is sufficient to dispose of the 13.6 ac ft (16,800 m³) of runoff from the design storm. The irrigation system consists of a 300 hp pump, 1/2 mile (800 m) of 4 in. (10 cm) and 6 in. (15 cm) diameter aluminum irrigation pipe, and sprinkler nozzles.

## FLY CONTROL PROGRAM

An all-out effort is expended to control flies. Fly breeding prevention efforts are directed primarily toward the finishing/nursery building and the gestation sections of Production Units 1 and 2. The totally enclosed farrowing and nursery of the production units have not experienced fly problems.

The Leanco fly control program consists of several elements. Solids concentrations in the manure pits are kept as low as possible (as discussed previously) to avoid floating material on which flies can breed. The slats have a 6 in. (15 cm) wide solid area at each end so that fresh manure does not adhere to pit walls above the liquid level. Buildings and surrounding areas are kept free of spilled or wasted feed, exposed manure and tall weeds. Soil injection controls fly breeding and emergence from manure disposal areas. Dead animals are promptly disposed of by burial. Various granulated and spray insecticides are used as needed, particularly during prolonged wet weather which is conducive to fly breeding.

## SUMMARY

A 600-sow confinement farrow to finish operation in Central Texas operates a "no discharge" system of contolling point and nonpoint source water pollution; recycles liquid manure through soil injection; and recycles wastewater and runoff onto cropland through irrigation. Through conscientious management of underfloor liquid manure pits, odors, flies and maintenance problems are kept to a minimum.


**References**

1 Buchanan, S. J. 1972. Report of engineering examination of swine wastes and their disposal. Unpublished report, Spencer J. Buchanan and Associates, Bryan, Texas. 12 p.

2 Griffiths, J. F. and R. Orton. 1968. Agroclimatic atlas of Texas, Part 1 - precipitation probabilities. MP-888, Texas Agricultural Experiment Station, Texas A&M University, College Station, Texas.

3 Reddell, D. L. and J. M. Sweeten. 1975. Evaluation of odor intensities at livestock feeding operations in Texas. Proceedings, 3rd International Symposium on Livestock Wastes, Urbana-Champaign, Ill, April 21-24.

4 Sweeten, J. M., D. L. Reddell and H. B. H. Cooper. 1974. Odor measurement for livestock feeding operations. Proceedings, Specialty Conference on Control Technology for Agricultural Air Pollutants, Air Pollution Control Association, Pittsburg, Pa. pp. 55-80.

# Total Waste Management for a Large Swine Production Facility

F. J. Humenik, R. E. Sneed, M. R. Overcash, J. C. Barker, G. D. Wetherill

MEMBER     MEMBER    ASSOC. MEMBER
ASAE       ASAE         ASAE

LEXINGTON Swine Breeders, Inc., operate a 300-sow total confinement facility a few miles south of Lexington, N.C. The total acreage owned, including that covered by buildings, roads, and waste disposal units is about 20 acres. The site is bounded by creeks on two sides with the major one discharging into a recreational lake about two miles downstream. The village of Linwood is about one-half mile southeast of the buildings, and a large furniture plant has been constructed about 400 yards from the property line.

The four major buildings of frame and aluminum sheet construction are: (a) gestation barn, 30 ft x 270 ft, designed to house about 250 sows; (b) farrowing house, 30 ft x 200 ft, for 64 sows in crates; (c) nursery, 25 ft x 90 ft, for 640 pigs up to 30 lb;and (d) finishing barn, 37 ft x 150 ft, for 500, 200-lb hogs or their equivalent. All buildings have partially slatted concrete floors, and the waste is collected in 2 ft deep underfloor pits.

This farm was purchased as an operating enterprise in June 1971. At that time the 1½-acre unaerated lagoon had a liquid depth of only about 4 ft and was considered an adequate waste disposal facility. However, about one year later the lagoon liquid depth had risen to about 8 ft and on occasions overflowed the bank, while some complaints concerning offensive odors were being received from local residents. Land available for the terminal waste receiver system was limited, and additional land acquisition was not possible. Therefore water conservation and extensive pretreatment were necessary to reduce nitrogen to a level which could be accommodated by the available plant-soil receiver.

The pretreatment system consists of series lagoons with a new primary aerated unit constructed ahead of the original unaerated lagoon and an overland flow area utilizing waste irrigation from and return to the 1½-acre unaerated lagoon. Additional nitrogen is lost during sprinkler irrigation and uptake by a continuously harvested cover crop. A schematic diagram of this total waste management system which includes wastewater recycling for manure storage pit cleaning, to minimize wastewater volume, is shown in Fig. 1. Gravity waste flow from the swine buildings can be directed to either the aerated or unaerated lagoon. Surface aeration is provided in the primary lagoon to control odor by reducing the oxygen demand imposed upon both lagoons, and the much larger second lagoon provides the bulk of the size and retention time required for nitrogen volatilization. However, as shown in pilot field projects, surface aeration augments nitrogen volatilization in addition to promoting partial stabilization of organics. The potential for nitrate formation in the aerobic zone and biological denitrification into nitrogen gas in associated anaerobic zones has been demonstrated. At this full-scale installation these separate nitrogen loss mechanisms are compounded with only the overall reduction measurable.

The aeration strategy is to provide surface pumpage to maintain a top aerobic zone with the rest of the lagoon contents relatively undisturbed and anaerobic. This two-phase lagoon design allows for minimum energy input to maintain aerobic surface conditions attendant to organic stabilization and thus odor reduction, while also providing the aerobic-anaerobic interface necessary for biological denitrification. The maintenance of the bottom anaerobic zone by selection of an aerator which promotes surface pumpage and equipped with an anti-erosion shield to eliminate vertical draft allows continuance of anaerobic fermentation of digestion or bottom sludge to gaseous end products which are volatilized.

The new primary lagoon was sized to accommodate two 3-h.p. floating aerators selected to satisfy about 50 percent of the raw waste five day BOD load. Assuming a $BOD_5$ production of 0.32 lb per 100-lb hog per day, 2100 100-lb hog equivalents produce 672 lb BOD per day and thus about 336 lb of oxygen per day would have to be provided. About 5 h.p. of aeration would be required to satisfy this BOD at an oxygen transfer rate of 3 lb of oxygen per h.p.-hr. Correspondingly, two 3-h.p. aerators were selected to give maximum versatility and secure oxygenation capabilities for slightly increased loading. The lagoon size was chosen to be between the complete mixed diameter criteria of 35 to 55 ft and total horizontal effect of 125 ft to 145 ft specified for the purchased oxygenators. Therefore the lagoon surface size was 90 ft by 140 ft with a depth of 8 ft. Side slopes of 1:1 were used attendant to a total volume of about 80,000 ft³. The height of the overflow drain from the first to the second lagoon can be varied with multi-level risers, thus allowing the temporary raising and lowering of lagoon level to kill edge vegetation.

Presently two 3-h.p. units and one 5-h.p. aerator are being used because two 3 h.p. aerators did not result in complete surface mixing. The unmixed areas near the banks became odorous and covered with scum. Data obtained from several field units employing this

**Acknowledgement:** This project was supported by the North Carolina Extension Service and Agricultural Experiment Station and was possible because of the excellent cooperation of Lexington Swine Breeders, particularly Dr. George Wetherill.

The authors are: F. J. HUMENIK, Associate Professor, R. E. SNEED, M. R. OVERCASH and J. C. BARKER, Assistant Professors, Biological and Agricultural Engineering Dept., North Carolina State University, Raleigh; and G. D. WETHERILL, Manager, Lexington Swine Breeders, Inc., Lexington, NC.

The use of trade names in this publication does not imply endorsement by the North Carolina Agricultural Experiment Station of the products named nor criticism of similar ones not mentioned.

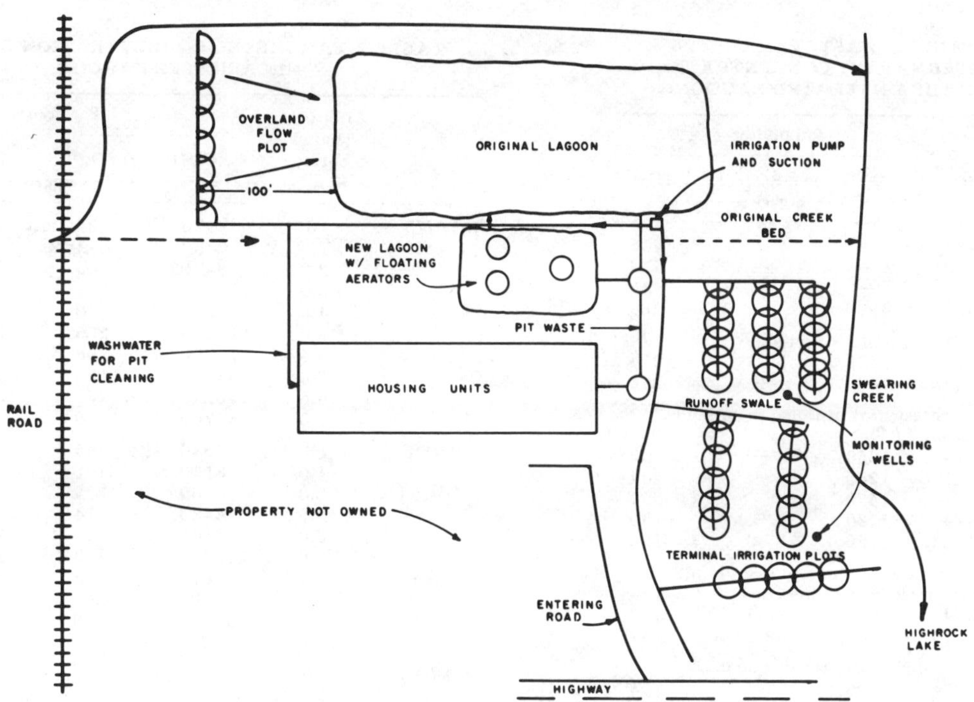


**FIG. 1 Schematic of waste management system at Lexington Swine Breeders.**

oxygenation equipment manufactured by Sydnor Hydro-dynamics, Inc., Richmond, Virginia, indicate that the surface area rating is more effective in determining lagoon size for either complete mixing or total surface agitation. Currently our recommendations for this equipment for high rate mixing are about 450 ft$^2$ per h.p., and about 1000 ft$^2$ per h.p. for complete surface agitation. The surface loading rate for the 11 h.p. of aeration being used at Lexington Swine Breeders is 1145 ft$^2$ per h.p. Visual observations indicate that the effect of a 5-h.p. unit is somewhat less than two 3-h.p. oxygenators of the same manufacturer design series.

Pretreatment waste constituent reduction is also obtained by overland flow of wastewater pumped from the second lagoon to a land area which slopes towards this lagoon. The overland flow area is about 100 ft x 280 ft (0.6 acre) with a uniform slope of approximately 5 percent. Eighty-foot diameter rotary impact, part circle sprinklers spaced 40 ft apart on 2-ft risers are used to apply wastewater from the secondary lagoon to the upper portion of the overland flow area.

Mechanisms for nitrogen conversion in overland flow are organic and ammonia nitrogen conversion to nitrate by soil microorganisms and loss of volatile nitrogen compounds. A cyclic application-recovery period operation has proved to be most desirable on a treatment and operational basis. Recommended loading during the cool season is 1.6 inch/acre, requiring 8 hours of application every fourth day. The warm season loading rate is 2.5 inch/acre requiring 12.5 hours of application every fourth day. The corresponding nitrogen loading rates are 2400 to 3800 lb of nitrogen/acre/year which greatly exceeds uptake capacity. However, nitrates formed but not utilized by the cover crop are returned by surface drainage and soil water interflow to the anaerobic lagoon where denitrification can occur. This recycle aspect allows heavier application rates.

The plant-soil receiver designed as the terminal acceptor system is 3.3 acres of a Kentucky 31 fescue-Johnson grass mixture. A manually activated permanent

irrigation system, utilizing 100 ft diameter, single nozzle rotary impact sprinklers spaced 60 ft apart on 2 ft risers is used to apply waste effluent at the rate of 0.25 inch per hour. Topography is suitable for grazing or hay cropping. A grassed waterway portion of the receiver plot was not irrigated because it would potentially result in direct discharge to receiving waters during runoff periods. Soil infiltration tests disclose that irrigation applications of 0.25 to 0.5 inch per event would be possible without surface runoff. System operation requires that a single lateral be operated for no more than a two-hour irrigation duration to prevent surface runoff as a direct result of any wastewater irrigation event. A second management factor is to forego or discontinue irrigation when wind speeds exceed 5 to 7.5 miles per hour to eliminate drift beyond property boundaries. The irrigation plot design contains a 50 ft

TABLE 1. WASTE PARAMETER SUPERNATANT CONCENTRATION OF THE AERATED LAGOON.

| Date | COD | o-PO$_4$-P | TKN |
|---|---|---|---|
| | | mg/l | |
| Lagoon constructed and filled 9/73 | | | |
| Two 3-hp aerators installed 11/73 | | | |
| 2/17/74 | 5580 | — | 713 |
| 4/23/74 | 10700 | — | 1370 |
| 4/25/74 | 9560 | — | 1220 |
| 5/14/74 | 9370 | — | 1390 |
| 10/17/74 | 4130 | 80 | 756 |
| Overland flow system and manure pit | | | |
| waterwash system installed. | | | |
| 11/7/74 | 4270 | 110 | 958 |
| 11/22/74 | 13500 | 372 | 1310 |
| Additional 5-hp aerator installed | | | |
| 12/12/74 | 10100 | 280 | 1190 |
| 12/19/74 | 10800 | 300 | 1480 |
| 1/2/75 | 9230 | 303 | 980 |
| 1/16/75 | 7330 | 173 | 938 |
| 2/10/75 | 8670 | 215 | 834 |
| 2/18/75 | 8920 | 192 | 918 |
| 2/26/75 | 9350 | 231 | 989 |
| 3/12/75 | 9960 | 258 | 994 |

TABLE 2. WASTE PARAMETER
SUPERNATANT CONCENTRATION
OF THE UNAERATED LAGOON.

| Date | COD | o-PO$_4$-P | TKN |
|---|---|---|---|
| | | - - - - - - - mg/l - - - - - - - - - - - | |
| 5/7/73 | — | — | 465 |
| 5/26/73 | — | — | 473 |
| 6/4/73 | — | — | 435 |
| 8/24/73 | 4370 | — | 286 |
| Aerated lagoon constructed | | | |
| 10/4/73 | 1580 | 53.4 | 274 |
| 11/13/73 | 1820 | — | 296 |
| Two 3-hp aerators installed in aerated lagoon | | | |
| 2/17/74 | 1280 | — | 205 |
| 4/23/74 | 1260 | — | 204 |
| 4/25/74 | 1470 | — | 207 |
| 5/14/74 | 1280 | — | 190 |
| 10/17/74 | 726 | 37.0 | 123 |
| Overland flow system and manure pit waterwash system installed | | | |
| 11/7/74 | 668 | 36.2 | 137 |
| 11/22/74 | 791 | 38.0 | 157 |
| Additional 5-hp aerator installed | | | |
| 12/12/74 | 496 | 17.0 | 134 |
| 12/19/74 | 515 | 20.1 | 134 |
| 1/2/75 | 594 | 21.5 | 151 |
| 1/16/75 | 644 | 19.7 | 155 |
| 2/10/75 | 769 | 20.5 | 185 |
| 2/18/75 | 797 | 18.1 | 176 |
| 2/26/75 | 745 | 22.1 | 209 |
| 3/12/75 | 869 | 29.0 | 228 |

TABLE 3. PARAMETER CONCENTRATION GRADIENT
IN AERATED LAGOON.

| Date | Depth, ft below surface | COD | o-PO$_4$-P | TKN | D.O. |
|---|---|---|---|---|---|
| | | | | - - - - - - - - - - - mg/l - - - - - - - - - - - | |
| 2/10/75 | 0 | 8920 | 224 | 1020 | 2.0 |
| | 1.0 | 7450 | 189 | 933 | 1.5 |
| | 2.0 | 8470 | 209 | 971 | 0.9 |
| | 3.0 | 8710 | 226 | 989 | 0 |
| | 4.0 | 9960 | 240 | 1040 | |
| | 5.0 | 10700 | 255 | 1080 | |
| | 5.5 | 11900 | 288 | 1160 | |
| | 6.0 | 67700 | 1120 | 2900 | |
| | 6.5 | 56200 | 1260 | 2400 | |
| | 7.0 | 80700 | 1170 | 2900 | |
| 2/18/75 | 0 | 5540 | 95 | 784 | 0.2 |
| | 1.0 | 8170 | 201 | 915 | 0.3 |
| | 2.0 | 8560 | 183 | 933 | 0.3 |
| | 3.0 | 8480 | 214 | 915 | 0.2 |
| | 4.0 | 12100 | 256 | 1170 | 0 |
| | 5.0 | 9900 | 278 | 1040 | |
| | 5.5 | 7340 | 145 | 859 | |
| | 6.0 | 12580 | 261 | 1100 | |
| | 6.5 | 57570 | 610 | 2100 | |
| | 7.0 | 64590 | 1100 | 2560 | |
| 2/26/75 | 0 | - | 175 | 840 | 0.8 |
| | 1.0 | - | 230 | 933 | 0.6 |
| | 2.0 | - | 264 | 1440 | 0.5 |
| | 3.0 | - | 215 | 971 | 0.5 |
| | 4.0 | - | 231 | 1010 | 0 |
| | 5.0 | - | 253 | 1040 | |
| | 5.5 | - | 277 | 1230 | |
| | 6.0 | - | 295 | 1160 | |
| | 6.5 | - | 1030 | 2610 | |
| | 7.0 | - | 842 | 2040 | |

grass buffer which safeguards surrounding creeks from direct irrigation or rainfall runoff.

Grass management for net nitrogen removal includes harvesting hay when the grass reaches a height of 12 to 18 inches or continued direct grazing. Emphasis is placed on this operational step as it is the primary nitrogen outlet. Beef cattle have grazed this for about six months with no adverse effects.

## OPERATIONS AND MONITORING

The waste parameter concentrations of the lagoon supernatants have been monitored for approximately two years in the unaerated lagoon and since initiation of the aerated lagoon. The results in Tables 1 and 2 represent composite samples from several locations over the surface area of the lagoons just below the liquid surface. The aerated lagoon supernatant concentrations seem to be influenced only by extensive or shock manure pit dumping events. Reduction in oxygen demand and nitrogen content in this unit is dramatic, based on raw waste inputs of 40,000-80,000 mg/l COD and 2500-5700 mg/l TKN for characteristic waste volumes of 1-2 gallons per 100-lb hog per day as measured at this farm.

Excellent settling occurs in the aerated unit because of the two-zone design possible with the utilized aerators. During 1½ years of operation approximately 1 ft of sludge, with an additional 6 in. of settled solids, have accumulated in this lagoon. The parameter concentration profile (Table 3) shows a slight increase in COD, TKN, and orthophosphate concentrations from the supernatant surface to bottom with a marked increase in the sludge zone. These data on composite samples from several locations obtained while aerators were off verify good surface agitation but not complete unit mixing. Dissolved oxygen levels in the aerated lagoon again depend on pit dumping events, but generally range from 1-2 ppm in the aerated zone decreasing to near zero

approximately 3 ft below the surface. Since installation of the aerators, surface scum formation has been restricted to unagitated areas near the banks. Most odor emission is from these areas, but nuisance levels are confined to the immediate lagoon vicinity.

Sludge zone nitrogen removals of about 25 percent have been recorded for similar aerated units (Humenik and Overcash 1975). Additional reduction of nitrogen is achieved in the unaerated lagoon because total Kjeldahl nitrogen supernatant concentrations (Table 2) are only about 15 percent of the values recorded for the aerated unit. Since construction of the aerated lagoon, the nitrogen content of the unaerated lagoon supernatant has been reduced by 50 percent, while a 75 percent reduction in oxygen demand has been achieved. A depth-concentration profile similar to the aerated unit exists in the unaerated lagoon as indicated by data in Table 4. Approximately 6-12 in. of sludge have accumulated in the unaerated unit.

The results of a limited number of analyses for runoff from the overland flow plot are presented in Table 5. Preliminary indications reveal a 10 percent reduction in the oxygen demand of the applied unaerated lagoon supernatant and a 25 percent reduction in total Kjeldahl nitrogen by overland flow. The sequential application-recovery period management plan already described has been adhered to except during days when the ground was frozen and during very wet periods. A lush, green growth of fescue grass flourishes on this plot.

Two test wells were installed in the terminal irrigation receiver lot to monitor the impact on groundwater quality of continuous surface irrigation of lagoon effluent at heavy nitrogen loading rates. One well was bored on the lower side of the receiver plot through heavy clay soil and cased with 2-in. PVC pipe to a depth of 6½ ft. The groundwater table at this well averages 4-5 ft below the

TABLE 4. PARAMETER CONCENTRATION GRADIENT
IN AERATED LAGOON.

| Date | Depth, ft below surface | Parameter | | | |
|---|---|---|---|---|---|
| | | COD | o-PO$_4$-P | TKN | D.O. |
| | | --- mg/l --- | | | |
| 2/18/75 | 0 | 726 | 19.8 | 177 | 2.7 |
| | 1 | 679 | 18.5 | 179 | 0.6 |
| | 2 | 714 | 19.3 | 190 | 0.4 |
| | 3 | 871 | 24.0 | 255 | 0.4 |
| | 4 | 1390 | 33.9 | 329 | 0 |
| | 5 | 1380 | 45.5 | 370 | |
| | 5.5 | 1700 | 47.5 | 407 | |
| | 6.0 | 4990 | 117.0 | 616 | |
| | 6.5 | 61120 | 1176.0 | 2360 | |
| 2/26/75 | 0 | 747 | 23.1 | 209 | 1.0 |
| | 1 | 756 | 22.2 | 181 | 0.7 |
| | 2 | 719 | 23.5 | 205 | 0.6 |
| | 3 | 739 | 25.4 | 209 | 0.6 |
| | 4 | 853 | 26.5 | 228 | 0 |
| | 5 | 1383 | 44.3 | 379 | |
| | 5.5 | 1375 | 46.5 | 377 | |
| | 6.0 | 1371 | 47.2 | 370 | |
| | 6.5 | 43480 | 870.0 | 2435 | |

TABLE 6. CHEMICAL QUALITY OF GROUNDWATER
BENEATH WASTEWATER IRRIGATION SITE.

| Well | Date | Parameter | | | |
|---|---|---|---|---|---|
| | | COD | o-PO$_4$-P | TKN | NO$_3$-N |
| | | --- mg/l --- | | | |
| Test well 6.5 ft deep (lower side) | 10/17/74 | 36.4 | 8.70 | 3.4 | 0.15 |
| | 11/7/74 | 3.23 | 1.08 | 2.2 | 0.47 |
| | 11/22/74 | 7.87 | 0.46 | 0 | 0.50 |
| | 12/12/74 | 29.4 | 4.38 | 3.2 | 0.45 |
| | 1/16/75 | 18.5 | 2.97 | 2.8 | 0.49 |
| | 2/10/75 | 27.0 | 0.30 | 2.2 | 0.82 |
| | 2/18/75 | 13.0 | 0.24 | 1.1 | 0.71 |
| | 2/26/75 | 41.0 | 0.54 | 7.5 | 0.65 |
| | 3/12/75 | 7.62 | 0.36 | 0 | 0.75 |
| Test well 13 ft deep (upper side) | 10/17/74 | 617 | 10.7 | 44.8 | 0.24 |
| | 11/7/74 | 245 | 4.20 | 26.1 | - |
| | 11/22/74 | 138 | 1.24 | 6.7 | 26.85 |
| | 12/12/74 | 287 | 5.38 | 19.0 | 14.00 |
| | 1/16/75 | 349 | 3.79 | 16.7 | 15.17 |
| | 2/10/75 | 63 | 1.37 | 5.0 | 25.00 |
| | 2/18/75 | 85 | 1.40 | 8.4 | 19.13 |
| | 2/26/75 | 104 | 2.40 | 9.3 | - |
| | 3/12/75 | 170 | 2.46 | 16.2 | 16.25 |
| Water supply | 10/17/74 | 4.05 | 0.30 | 0 | 0.86 |
| | 11/7/74 | 1.61 | 0.15 | 0 | 0.93 |
| | 11/22/74 | 11.8 | 0.08 | 1.7 | 0.89 |
| | 12/12/74 | 13.9 | 0.05 | 1.7 | 0.89 |
| | 1/16/75 | 49.6 | 0.04 | 1.4 | 0.96 |
| | 2/10/75 | 13.0 | 0.06 | 1.1 | 1.05 |
| | 2/18/75 | 3.13 | 0.05 | 0.8 | 1.31 |
| | 2/26/75 | 1.90 | 0.17 | 1.9 | 1.09 |
| | 3/12/75 | 1.90 | 0.18 | 0 | 1.50 |

surface. Nitrate nitrogen values presented in Table 6 have ranged from 0.15-0.82 mg/l indicating no signs of groundwater contamination at this location. The second well was bored on the upslope side of the receiver plot reasonably close to an irrigation lateral and cased through silty clay soil to a depth of 13 ft. The water table ranges from 10-12 ft below the surface. As can be seen from Table 6, this well is obviously contaminated because of the high oxygen demand and nitrate nitrogen values recorded. Cattle grazing in the receiver plot area have been noticed rubbing against the well casing at this site making it difficult to maintain a surface seal. This has probably allowed leakage or irrigated lagoon effluent around the casing into the groundwater. The other well location in Table 6 represents the water supply well adjacent to one of the swine barns which is cased to 100 ft below the ground surface.

The manure pit dumping schedule is managed such that one pit is emptied into the aerated lagoon each day. The pits are flushed once per two emptyings with unaerated lagoon supernatant recycled through the waterwash system. This waterwash system has proven very satisfactory since all solids and sludge are flushed out of the pit, and odor levels within the houses have noticeably decreased. Very little lagoon precharge water is added to the pits after flushing, and only small quantities of washwater are used to clean pens between emptyings. Dissolved oxygen decay rate investigations indicate that reaeration of the recycled lagoon liquid occurs as a result of pit flushing. Typically pit influent has about a 5 ppm D.O. and pit effluent a D.O. of 4 ppm which goes to zero in about 30 minutes.

Excavation and equipment costs of the waste management system excluding the original unaerated lagoon are presented in Table 7. Materials, equipment, excavation, and electric wiring costs are included. Operating costs are judged to be about $250.00 per month during 1975, or about a penny per pound of feeder pig sold.

### References


1  Humenik, F. J. and M. R. Overcash. 1975. Design criteria for swine waste treatment systems. EPA Project R 802203. Final Report in Press.


TABLE 7. EXCAVATION AND EQUIPMENT COSTS (1974)
FOR WASTE MANAGEMENT SYSTEM.

| Item | Cost |
|---|---|
| Lagoon construction and stream routing | $ 4,125 |
| Aerator installation | 2,980 |
| Pumping and irrigation system | |
| *Irrigation system* | |
| 5-hp centrifugal pumps and electric wiring | 775 |
| Sprinklers and risers | 373 |
| Pipe and fittings | 989 |
| Trenching | 453 |
| *Overland flow system* | |
| Sprinklers and risers | 92 |
| Pipe and fittings | 238 |
| Trenching | 70 |
| *Waterwash system* | |
| Pipe and fittings | 330 |
| Trenching and backhoe excavation | 150 |
| Total | $10,575 |

TABLE 5. CHEMICAL QUALITY OF OVERLAND FLOW.

| Location | Distance ft | Date | Parameter | | |
|---|---|---|---|---|---|
| | | | COD | o-PO$_4$-P | TKN |
| | | | --- mg/l --- | | |
| Sprinkler | 0 | 2/18/75 | 692 | 19.4 | 182 |
| | | 2/26/75 | 740 | 20.4 | 205 |
| | | 3/12/75 | 845 | 27.1 | 218 |
| Mid-terrace | 50 | 2/18/75 | 688 | 16.3 | 149 |
| | | 2/26/75 | 720 | 19.5 | 161 |
| | | 3/12/75 | 792 | 23.6 | 87 |
| Effluent | 100 | 2/18/75 | 672 | 14.4 | 131 |
| | | 2/26/75 | 650 | 17.5 | 149 |
| | | 3/12/75 | 739 | 20.0 | 111 |

# Simplifying Manure Handling in a Solid-Floor Swine Housing System


Daniel J. Meyer
ASSOC. MEMBER
ASAE


## BUILDING AND SYSTEM LAYOUT*

A "ROUND" building of 76-foot diameter was constructed which has a feeding-sleeping area toward the outer part of the circular plan, and a manuring-watering area toward the center (Fig. 1). The feeding-sleeping area floor slopes 1/2-inch per foot to a 4-inch step. The step separates the feeding-sleeping area from the manuring-watering area. The manuring-watering area floor slopes 3/4-inch per foot from the step to a central collection pit. The pit is 6 foot in diameter and 4 foot deep. The capacity of the pit is 800 gallons (equivalent to 500 pigs' wastes per day at 140 pounds weight). A paddle-type agitator mixes the solids and liquids prior to removal. The cone-shaped bottom of the pit contains a four-inch PVC pipe for drainage. This drain pipe extends 250 feet at a 3-1/3 percent slope to a concrete storage pit (8 feet x 70 feet x 32 feet) with a six-month capacity.

The 10 pens are wedge-shaped and capable of holding 50 pigs each. The roof over the pens was built in 2 sections and joined together. The wood frame over the feeding-sleeping area was construced first. Then a 48-foot diameter grain bin roof was placed over the manuring-watering area. A ceiling at a 9-foot height in the manuring-watering area was constructed of rigid insulation board supported with brace wire. The thickness of insulation in the ceiling and walls except for a 3-foot high concrete perimeter wall varied between 1 and 1-1/2 inches of urethane. The concrete perimeter wall was not insulated.

A 2-foot wide alley leads from the building entrance to an 8-foot diameter platform which is suspended by 10 cables to allow the mechanical scraper to pass underneath. The main purpose of the platform is to act as a central sorting area. The alley over the manuring-watering area is attached to the platform and therefore, indirectly supported by the cables.

The pen dividers consist of 32-inch high wooden gates covered with a thin gage sheet steel. The lower end of the dividers are attached to the suspended platform and upper end rests on the 4-inch step.

A network of 3 nipple waterers on both sides of alternate pen dividers is used to supply drinking water. The nipples are located 5 feet from the pit.

Initially, several 3 feet x 4 feet ventilation panels located around the perimeter of the building were used for ventilation. Later, 10 single-speed reversible fans with individual thermostats were placed 1-to-a-pen around the outer perimeter of the building providing 51 cfm per pig at 1/8-inch

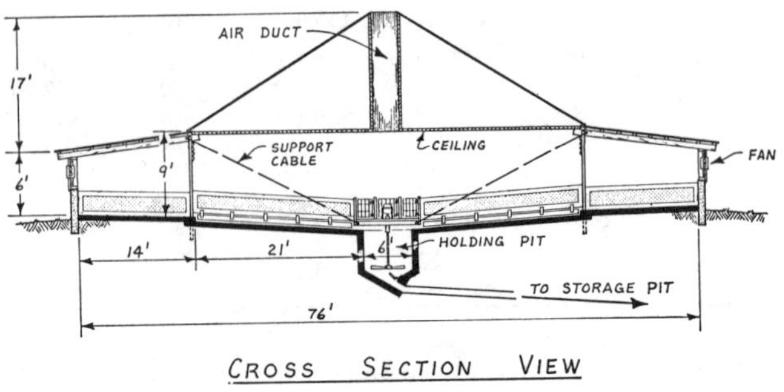

CROSS SECTION VIEW

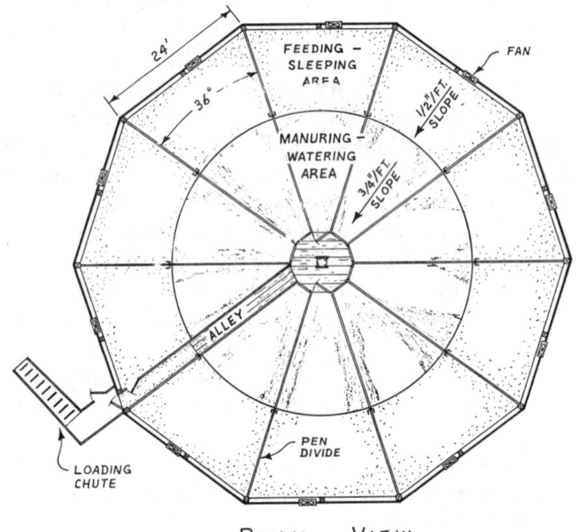

PLAN VIEW

FIG. 1 Building and system layout.

static pressure. The 5-foot diameter hole at the peak of the bin roof is connected to the ceiling by a 14-foot long circular duct. This duct serves as an outlet in the warm season and serves as an inlet in the cold season as the fans pull air into the building in the warm season and exhaust air from the building in the cold season. Additional cooling action is provided with low-volume (1-2 gph) sprinklers located directly above the 4-inch step.

An auger system is used for filling the self-feeders.

The feeding-sleeping area and manuring-watering area comprise approximately 57 percent and 36 percent, respectively, of the total area. These percentages were based on 4.9 square feet per pig in the feeding-sleeping area and 3.1 square feet per pig in the manuring-watering area.

## MANURE HANDLING SYSTEM

The 800-gallon collection pit presently fills nearly full each day with liquids and solids. The mixture is agitated 30-60 seconds before the stopper valve is opened allowing gravity flushing. The flushing takes from 2-4 minutes to

---


The author is: DANIEL J. MEYER, Extension Agricultural Engineer, Agricultural Engineering Dept., University of Wisconsin, Madison.

Acknowledgment: The author acknowledges the research assistance provided by his father, Clarence Meyer.

*The building is in a patent-pending state.

complete. The 4-inch line is being rinsed with clean water after each flushing to avoid solids plugging.

Four features were initially planned into the building to confine the wastes to the manuring-watering area. They were: (a) keeping the sleeping area dry by feeding in that area, (b) locating a 4-inch step between the sleeping-feeding area and the manuring-watering area, (c) bringing cold air in through the air duct and over the manuring-watering area to warm the air before it reaches the sleeping-feeding area, and (d) leaving the air duct open at the top for light to enter and illuminate the manuring-watering area.

Three added features after construction to confine the wastes to the manuring-watering area were: (a) use of crowding gates for newly weaned pigs, (b) use of steel wire pen divides instead of solid pen divides, and (c) changing the location of the nipple waterers from near the pit to over the step (used to keep manuring-watering area wet).

Three different methods were evaluated for moving the manure from the manuring-watering area to the pit. They were: (a) hydraulic flushing, (b) manual scraping, and (c) mechanical scraping. The hydraulic flushing was ruled out in the planning stages of the building because of filtering problems in recycling the large volume of water needed and the excess humidity problem in winter. The manual scraping has been used for 11 months. The average daily time needed to scrape all 10 pens ranges between 10-18 minutes. It was decided to try and reduce the labor requirement by automating the scraping process. Since most of the manure concentrates in the inner portion of the manuring-watering area (roughly 1/8 of total building area), a mechanical scraper was designed to sweep only the inside 30 feet diameter. The scraper unit consists of an endless chain conveyor with a succession of paddles projecting outwardly from the chain. A positive drive unit is being tested since a friction drive unit was not successful.

### GENERAL OBSERVATIONS

1   Opening ventilation panels in the summertime tends to cause dunging in the sleeping-feeding area.

2   When using solid and steel wire pen divides, manure tends to accumulate 1-3 feet from the divider. The feed wastage from improperly adjusted self-feeders works down off the sleeping-feeding area and onto the upper end of the manuring-watering area which a few pigs tend to sleep on.

3   The manuring-watering area was covered twice with water because of water line problems. Since the power unit was elevated on the platform, no problems occurred.

4   Solids plugging has occurred on occasion in the 4-inch line because of (a) the accumulation of solids directly below the inlet in the large storage pit (inlet pipe is 4 foot above the storage pit floor), (b) the accumulation in the storage pit of nearly 1 foot of ice and 1 foot of snow, and (c) inadequate agitating time in the collection pit for proper mixing.

5   A slight dust problem arises when wastage from self-feeders becomes excessive.

6   Condensation did occur on the uninsulated concrete perimeter wall in severe cold weather. This was corrected by insulating the wall with 1-inch rigid urethane board covered with plywood.

7   The open air duct exhausts hot air naturally (chimney effect) which is visibly observed when outside air temperatures drop below 0 F. No moisture condensation occurs on the walls or ceiling even when the fans are not operating. The air inlet for natural ventilation is the un-sealed roof joint.

### SUMMARY

A 10-sided building with solid floor and central manure collection pit with outside manure storage for 500 hogs was winter tested. Manure is scraped manually into the 800 gallon central pit which flows by gravity daily into a large outside concrete storage pit. It takes approximately 10-18 minutes to manually scrape the manure into the collection pit. An agitation unit mixes the slurry before discharge. A new mechanical floor scrape system is being tested to reduce labor requirements. The mechanical ventilation system consists of 10 reversible fans located in the sides with a 5-foot duct in the center. The ventilation system has worked well but solids plugging of the 4-inch line to the storage pit has occurred on occasion.

**References**

1   Meyer, Daniel J. 1974. A new approach in non-slatted liquid waste swine housing systems, ASAE Paper No. 74-4533. St. Joseph, Mich. 49085.

# Double E Farms — Swine Installation


**T. W. Eisenman, R. K. White**
ASSOC. MEMBER
ASAE


THE swine installation is located approximately 18 miles southwest of Columbus, Ohio. The installation is designed for 300 000 lb live weight of hogs (about 510 finishing hogs at any one time). Pigs were put into the facility in 1971. There are three buildings and several outdoor breeding pens. The gestation building, 23 ft x 292 ft, contains 260 tether stalls and 14 farrowing crates/pens. The farrowing/nursery building, 21 ft x 267 ft, has 60 farrowing crates/pens and 60 nursery pens. The finishing building 37 ft x 271 ft, is sectioned to handle pigs in three weight ranges. The pens are sized according to weights; there are 14 pens 7 ft wide for small pigs, 14 pens 12 ft wide for medium sized pigs, and 14 pens 16 ft wide for pigs up to market size. There are two smaller pens in each section adjacent to the oxidation ditch rotors to utilize the space. There are slatted, concrete floors throughout all buildings. Figs. 1 and 2 show the buildings and adjacent areas.

The numbers of animals in each building are governed by the farrowing procedure rather than the capacity of the buildling on a square foot basis. Farrowing has been on a continuous basis (rather than batch) at approximately 16 sows per week. The average swine inventory for 1974 is presented in Table 1. The designed animal weight loading of the installation has never been fully reached.

## WASTE SYSTEM

A flow diagram of the waste handling and treatment system is shown in Fig. 3. The oxidation ditch is used for treatment and odor control. The gestation building has two rotors, the farrowing/nursery has two rotors and the finishing building has eight rotors. The rotors or aerators are each 6 ft long by 16 inch diameter. Design capacity of each aerator is 9.0 lb oxygen per hour in clear water.

The oxidation ditches extend approximately the full length and width of the buildings with only separator walls


The authors are: T.W. EISENMAN, Owner and Operator, Double E Farms, Columbus, Ohio and R. K. WHITE, Assistant Professor, Ohio State University, Columbus.

FIG. 1 Looking across 4.6 acre lagoon — outdoor breeding pens are in the middle, finishing building on the left, farrowing building in the center, and gestation building on the right.

FIG. 2 General view showing handling alleys between farrowing building on the left and gestation building on the right.

to define the channels. Fig. 4 shows a construction view where the two inside and the two outside ditches pass in the finishing building. Concrete slats were poured in place over the oxidation ditches. Fig. 5 shows the slatted floor of the farrowing building. Openings were left in the floor for the aerators as shown in Fig. 6. The pigs are fenced away

### TABLE 1. AVERAGE SWINE INVENTORY FOR 1974.

| Building - section | Design capacity | Average inventory | | | | | Area per animal, sq ft |
|---|---|---|---|---|---|---|---|
| | | Number | | Average weight, lbs | | Total weight, lbs | |
| Gestation | 260 | 218 | x | 340 | = | 74 100 | 13 |
| Farrowing | 74 | 76 | x | 340 | = | 25 800 | 35 |
| Pigs- farrowing | 510 | 375 | x | 7 | = | 2 600 | |
| Pigs - nursery | 510 | 730 | x | 26 | = | 19 000 | 3 |
| Pigs - finish #1 | 510 | 387 | x | 64 | = | 25 800 | 4 |
| Pigs - finish #2 | 510 | 425 | x | 125 | = | 53 000 | 6 |
| Pigs - finish #3 | 510 | 427 | x | 189 | = | 80 700 | 8 |
| | 2884 | 2638 | | | | 281 000 | |

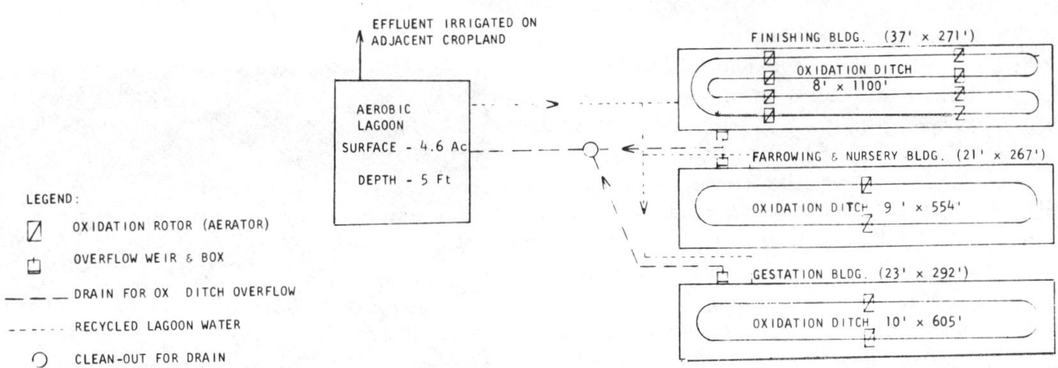


FIG. 3 Flow diagram of Double E Farms swine waste handling and treatment system (designed for 300 000 lb live weight).

from the aerators. The liquid depth in the ditches is maintained at 12 inches by means of a horizontal weir in an outlet box.

Overflow from the oxidation ditches is drained to an aerobic lagoon. The drain from the buildings to the lagoon is an eight inch steel pipe on 0.6 percent slope. A clean-out was provided in the drain but it has not been used. All surface water is diverted from the lagoon but provision is

FIG. 4 Finishing building — oxidation ditch return showing the passing of the two outside ditches and the two inside ditches prior to slat installation.

FIG. 5 Farrowing building — slats have been poured — anchor rings for tethering of sows will be noted.

made to divert roof water to the lagoon if needed. The suspended solids content in the ditches is controlled by a flow of recycle water from the lagoon. Weekly checks of

FIG. 6 Finishing house — aeration wheels have been installed in the 4 oxidation ditches — two sets of four wheels are provided for the finishing house.

the concentration of the oxidation ditch mixed liquor are made by filling a small jar with liquid from the ditch. If after one hour the layer of settled solids was 25 percent of the height, this was considered acceptable. If the percent of settled solids increased, the flow of recycle water was increased until the percent dropped.

The lagoon was sized to receive approximately 100 lb of BOD per day at a loading rate of 21.7 lb of BOD per surface acre. It was assumed that the BOD reduction in the oxidation ditch was 90 percent, because of a reasonably long detention time. Provisions are made for spray irrigation of effluent to control lagoon level. There is no direct outlet for the lagoon.

## OPERATION AND MANAGEMENT—PROBLEMS AND COSTS

The original commercial aerators (rotors) were constructed of too light a material and problems were encountered after only a few weeks of operation. Also, belts broke and bearings failed. Homemade rotors were constructed of ¼ x 2 inch angle iron for fins and ¼ inch iron plate for end and center support. The shaft was a 2 inch pipe with bars of steel welded into each end. Larger bearings were used. After rebuilding the rotors and enlarging the bearings, the cutting of the bearings with hog hair does not seem to be a problem. Single-stage belt drive reductions are used to turn the rotors at about 190 rpm. The drives are much too light and will be replaced with two-stage reductions to assure continuous operation of the rotors.

The replacement rotor accumulated a build-up of white crystaline material, possibly magnesium ammonium phosphate as reported by Iowa State University on pumps used to recycle lagoon water. The accumulation of this crystal was stopped by painting the rotors with Galvanox, a zinc paint. Also, a white crystal builds up in the recycle pump. The initial recycle pump motor was burned out. Now the pump impeller casing is cleaned periodically.

When oats are fed to gestating sows, undigested oat husks build up as a floating mass on the water and must be manually removed. Sand from premixed feed builds up in the bottoms of the ditches and must be manually removed.

Dissolved oxygen readings taken immediately upstream from each rotor in all buildings indicated adequate oxygenation. Table 2 gives data on solids content in the oxidation ditch. Solids and BOD data for influent to lagoon and recycle water were taken shortly after the facility was fully occupied. The quality of recycled water at this early date was equivalent to potable water.

There is a minimum of offensive odors. Visitors have remarked that there is not the characteristic smell of hog buildings. The odor in the finishing building which has the larger weight density of animals had an odor associated with feed as opposed to a swine manure odor. Fig. 7 pictures pigs and slatted areas in the finishing building. Cleanliness of the pigs and slatted floor is evident.

The rotors splashing the liquid in the ditch increases the moisture in the air of the buildings. This moisture, combined with feed dusts, causes metal surfaces in the buildings to corrode and interferes with electrical controls. It will be necessary to replace original electrical switches/controls and their boxes with moisture proof devices.

Due to the geographic location of the installation, only single phase, 240 volt electric power is available. Also, voltage stabilization is very poor. Consequently, many motors burned out and have had to be replaced. Voltage sensing equipment has been installed which cuts the motors off at a

FIG. 7 Finishing building — market size hogs accept the concrete slatted floor and keep clean — absence of tail biting will be noted.

low voltage and has reduced the loss from burned out motors. Signal lights were installed on each aerator so that the herdsman can easily detect when a rotor is not rotating and correct the problem. The three hp electric motors driving each rotor use about 3 kWh. Power costs at the time of the initial installation were just over 1 cent per kWh, but as of December, 1974, had risen to 2.43 cents per kWh.

While actual costs for maintenance of the disposal system are not separated from the general maintenance of the installation, it is estimated that 500 man hours were required during the last year for its maintenance (about 10 hours per week). During the year 1974, approximately $2.25 per marketed hog was attributable to waste control. This amount will become greater as the installation becomes older and electric power rates continue to escalate.

## SUMMARY

While the system appears to offer satisfactory treatment of hog wastes, it has been plagued with many mechanical and electrical breakdown. These problems could probably have been minimized if the state of design had been more fully developed at the time of construction. The 1974 cost of $2.25 for each marketed hog for waste management is too large.

It is our opinion that the installation and operating costs of a waste system as described herein make it prohibitive for most swine operations.

**TABLE 2. ANALYSIS OF OXIDATION DITCH MIXED LIQUOR, INFLUENT TO LAGOON AND RECYCLE WATER.**

| Sample source and date | Total solids, percent | Volatile solids, percent of TS | Suspended solids, percent | BOD, mgO$_2$/l |
|---|---|---|---|---|
| November, 1971 | | | | |
| Farrowing/nursery | 0.81 | 74 | 0.58 | - |
| Gestation | 0.43 | 73 | 0.35 | - |
| Finishing | 1.04 | 76 | 0.77 | - |
| Influent to lagoon | 0.59 | 72 | 0.49 | 2400 |
| Recycle water | 0.05 | 29 | 0.015 | 7 |



MANAGING LIVESTOCK WASTES

# A Waste Management System for a 2500-Head Swine Operation — A Case Study


A. L. Sutton, D. H. Bache, J. C. Nye, A. C. Dale, D. D. Jones

ASSOC. MEMBER ASAE     FELLOW ASAE     ASSOC. MEMBER ASAE

V. B. Mayrose, M. P. Plumlee, L. B. Underwood


ECONOMIC pressures and labor requirements have encouraged the trend to large confinement swine production operations. A result of this trend is the increased use of liquid waste handling. The initial investment, annual operating costs, managerial inputs and requirements for labor, power, and equipment of the various liquid waste systems differ considerably.

The purpose of this paper is to (a) describe the design and performance of a liquid waste management system for a 2500-head swine operation, and (b) discuss some of the advantages and disadvantages of liquid tanker wagon and lagoon-irrigation disposal systems based on experiences from this study.

## HOUSING FACILITIES

The 2500-head per year swine operation utilized in this study is located at the Baker-Purdue Animal Sciences Center, Purdue University, West Lafayette, Indiana (Fig. 1). The farrow-to-finish operation which is used for research in nutrition, breeding, meats and management includes: (a) two 800-head enclosed confinement growing-finishing houses, (b) two 48-sow enclosed confinement farrowing houses, and (c) two 120-head open-front confined gestation houses (Fig. 2). The growing-finishing houses have fully slotted floors over four oxidation ditches each of equal size (Fig. 3). The farrowing houses are in the form of a block H with 24 farrowing crates in each wing on completely slotted floors over oxidation ditches (Fig. 4).

Overflow from the ditches, which have a sluice gate for liquid level control, collects in a sump at the end of the houses. Manure in the sump can be removed by a vacuum liquid tanker wagon or transported through a tile system to a lagoon. The aerator for the oxidation ditch is a 2.3 m (8 ft) paddle wheel which is belt driven by a 5-hp electric motor.

Each gestation house has a partially slotted floor over a 1.2 m (4 ft) deep anaerobic pit (Fig. 3). Perpendicular to the length of the pit is a concrete wall at each pen division (every 3.6 m; 12 ft) with a 25.4 cm (10 in.) square opening at the bottom of the dividing walls. Liquid manure is

drained by a tile system to the sump and pumped into the lagoon. In addition, outlets at each end of the pits permit removal of manure by a vacuum tank wagon. Additional details of the housing facilities have been reported by Mayrose et al. (1974).

**Lagoon**

An anaerobic deep lagoon was excavated in 1971 (Fig. 2). The lagoon was dug 4 m (13 ft) deep, and the excavated dirt was used to form a 2.4 m (8 ft) dike around the lagoon. The lagoon is 45.7 m (150 ft) wide and 76.2 m (250 ft)

FIG. 1 Aerial photograph of the Baker-Purdue Animal Sciences Center swine complex.

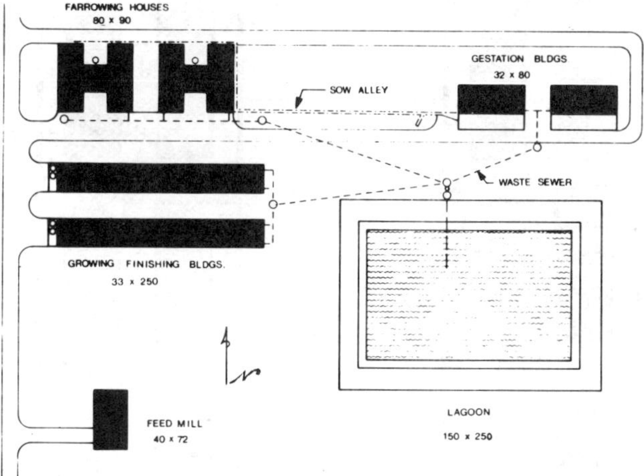


FIG. 2 Layout of the Baker-Purdue Animal Sciences Center swine complex.


Department of Animal Sciences Journal Paper No. 5835, Purdue University Agricultural Experiment Station.

Supported in part by Grant No. 12-14-100-9961 (44) from the Agricultural Research Service, United States Department of Agriculture.

Acknowledgment: The authors acknowledge the assistance of Mr. Roland Halleck and Mr. Frank Grove for providing labor and cost data; Mr. Dan Kelly, Mr. Michael Brumm, Mr. Rickie Chastain and Mrs. Sylvia Shepler with chemical analyses and sample collection; and Prof. William Friday, Agricultural Engineering Department, for illustrations.

The authors are: A. L. SUTTON, Animal Sciences Dept., D. H. BACHE, Agricultural Economics Dept., J. C. NYE, A. C. DALE and D. D. JONES, Agricultural Engineering Dept., V. B. MAYROSE, M. P. PLUMLEE and L. B. UNDERWOOD, Animal Sciences Dept., Purdue University, West Lafayette, Ind.

long at the top of the berm, with about 2:1 side slopes on the inside. The volume is 11 674.9 m$^3$ (15 268.5 yd$^3$) allowing for a 0.61 m (2 ft) free board. Two acres of land were used for lagoon construction.

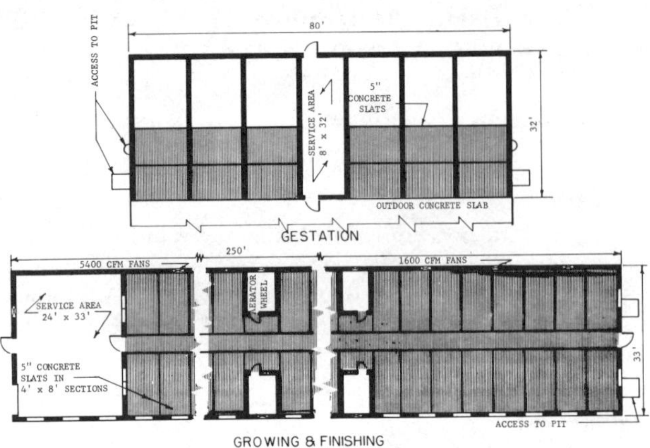

FIG. 3 Floor plan of the gestation and growing-finishing houses.

### Tile and Sump System

Fifteen cm (6 in.) tile (vitrified, clay except cast iron under roadways) drains the waste from all pits to a 5 678 liter (1 500 gal) sump system where it is then pumped into the lagoon (Fig. 2). The slope of the tile line is 0.4 percent from the manure pits to the sump system at the lagoon. The sump system includes a collection pit 1.2 m (4 ft) in diameter by 1.5 m (5 ft) deep and a circular pit housing the sump pump which is 1.5 m (5 ft) in diameter and 2.4 m (8 ft) deep. A gate valve is placed between these two compartments to shut off waste flow into the sump to facilitate servicing of the pump (Fig. 2).

A submersible 3-hp electric vertical cantilever high solids pump which is in the circular pit loads the anaerobic lagoon. A mercury switch with a float automatically starts and stops the pump depending upon the level of liquid waste in the sump. The pump lifts the waste from a 7.6 cm (3 in.) opening into 10.2 cm (4 in.) corrugated plastic tile located about 0.91 m (3 ft) below the top of the lagoon.

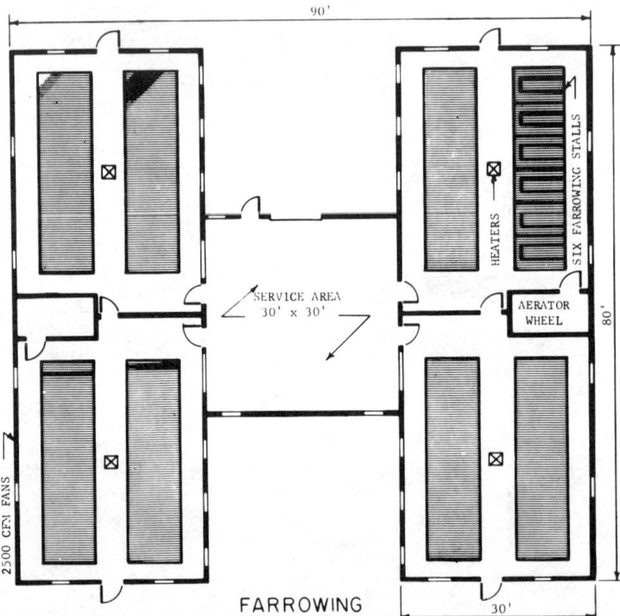

FIG. 4 Floor plan of the farrowing houses.

The lagoon inlet is 0.61 m (2 ft) below the top of the lagoon.

The capcity of the sump pump is about 833 liters (220 gal) per minute with a 4.3 m (14 ft) head.

### Liquid Haul and Irrigation Equipment

A 5678 liter (1500 gal) vacuum liquid tanker wagon pulled by a 50-hp tractor was used to haul and spread waste prior to construction and use of the lagoon-irrigation system. The irrigation equipment used is a hand-carry system. It includes an enclosed impeller centrifugal pump capable of pumping 1363 liters (360 gal) per minute at 7 kg per cm$^2$ (100 psi), 12.7 cm (5 in.) main, and 10.2 cm (4 in.) lateral aluminum irrigation pipe, six 1.3 cm (1/2 in.) Rain Bird EW-TNT nozzles, and an irrigation pipe carrier. The pump is driven by a 40-hp 3-phase electric motor.

### Investment and Operating Costs

Tables 1 and 2 list the initial investment and annual operating costs, respectively, of various components of the swine waste disposal system. These data do not include the cost of slotted floors and oxidation ditch equipment in the confinement buildings.

Construction of the lagoon (Table 1) plus the investment in land taken out of production based on 1975 price estimates is about $6.00 to $7.00 per pig produced yearly. The initial investment per pig for transport and disposal of the waste is quite similar for the irrigation equipment ($2.18) and the liquid tanker wagon ($1.73). The sump and tiling system appears to be a useful alternative for transporting waste from confinement buildings to a lagoon that is at the same grade level or higher than the confinement buildings. The investment is about $2.00/pig produced yearly.

### TABLE 1. INVESTMENT COST OF COMPONENTS FOR THE BAKER-PURDUE WASTE DISPOSAL SYSTEM

| Item | Year purchased | Actual cost | | 1975 estimate | |
|---|---|---|---|---|---|
| | | Total | Per head* | Total | Per head* |
| | | ———————— $ ———————— | | | |
| **Lagoon-irrigation system** | | | | | |
| Liquid pit (0.6 m) | | — | — | 32 832 | 13.13 |
| Tiling system | 1972 | | | | |
|   Tile, excavation | | 2 113 | 0.84 | 2 747 | 1.10 |
|   Labor | | 350 | 0.14 | 402 | 0.16 |
| Sump system | 1972 | | | | |
|   Pump w/controls | | 969 | 0.39 | 1 138 | 0.46 |
|   Sump | | 175 | 0.07 | 275 | 0.11 |
|   Labor | | 350 | 0.14 | 402 | 0.16 |
| Lagoon system | 1971 | | | | |
|   Excavation | | 9 615 | 3.85 | 12 019 | 4.81 |
|   Land (2 acres) | | 1 600 | 0.64 | 2 800 | 1.12 |
|   Electricity installed | | 750 | 0.30 | 1 000 | 0.40 |
| Irrigation system | 1973 | | | | |
|   Pump, pipe, nozzles and trailer | | 4,535 | 1.81 | 5 442 | 2.18 |
| Total (lagoon and irrigation) | | 20 457 | 8.18 | 59 057 | 23.63 |
| **Liquid haul system** | | | | | |
| Liquid pit (1.2 m) (estimated)† | 1969 | 39 398 | 15.76 | 65 664 | 26.26 |
| Liquid vacuum wagon | 1969 | 2 320 | 0.93 | 4 315 | 1.73 |
| Total (liquid pit and liquid haul) | | 41 718 | 16.69 | 69 979 | 27.99 |

*Per market pig sold at average annual capacity.
†Based on 1969 price of 1.6 cents/liter (6 cents/gal).



MANAGING LIVESTOCK WASTES

The higher investment for the liquid haul system compared to the lagoon-irrigation system is mainly due to increased requirements for waste storage.

The total annual operating cost (Table 2) for the lagoon-irrigation system was lower than the liquid hauling system. This was mainly due to an increased cost for labor and tractor operation in the liquid haul system.

Prior to the construction of the lagoon (1971) and tiling system (1972) and the purchase of irrigation equipment, approximately 15 hours per week was devoted to using the liquid tanker wagon to empty anaerobic pits and maintain the liquid level on oxidation ditches (when in operation). Waterer wastage and water for cleaning pens added additional liquid that had to be hauled. Therefore, the cost of hauling the waste may be high as reported in Table 2.

The cup waterers were replaced by adjustable nipple waterers which reduced water wastage.

### TABLE 2. ANNUAL OPERATING COSTS FOR THE BAKER-PURDUE WASTE DISPOSAL SYSTEM

| Item | Total | | Per head* | |
|---|---|---|---|---|
| | Liquid haul | Lagoon-irrigation | Liquid haul | Lagoon-irrigation |
| | | | $ | |
| Depreciation† | 5 459 | 4 793 | 2.18 | 1.92 |
| Interest‡ | 3 149 | 2 658 | 1.26 | 1.06 |
| Insurance§ | 262 | 221 | 0.10 | 0.09 |
| Taxes‖ | 1 050 | 886 | 0.42 | 0.35 |
| Repairs and maintenance# | 200 | 100 | 0.08 | 0.04 |
| Tractor rental** | 3 900 | - - | 1.56 | - - |
| Electric motor depreciation†† | - - | 59 | - - | 0.02 |
| Fuel (780 hr) or electricity (160 hr)‡‡ | 1 061 | 64 | 0.42 | 0.03 |
| Labor§§ | 2 730 | 280 | 1.09 | 0.11 |
| Total | 17 811 | 9 061 | 7.11 | 3.62 |

*Per market pig sold at average annual capacity.
†Depreciation calculated as: 20 percent for vacuum wagon, irrigation system and sump pump; 7 percent for tiling system, sump, lagoon and concrete pit.
‡Interest charged at 9.0 percent on the average investment.
§Insurance charged at 0.375 percent.
‖Taxes charged at 1.5 percent.
#Experience on case farm.
**Based on the custom rate of $5.00/tractor hour.
†† Ten percent of 1975 purchase price ($588).
‡‡Tractor gasoline consumption = 15 liters/hour (4 gallon/hour); gasoline cost = 9 cents/liter (34 cents/gallon); electricity cost = 1.6 cents/kW-hr x 25 hp x hours.
§§Annual - liquid haul = 780 hr at $3.50/hr.
      - irrigation = 80 hr at $3.50/hr.

The high fuel cost for the liquid haul system was due to the large number of operating hours and the high fuel price. Repairs for the liquid tanker wagon during the 5-year period included replacing the vacuum pump once and rebuilding it twice in addition to replacing other minor parts. Maintenance and repair of the irrigation system for the two years of use included replacing bent irrigation pipe, parts to sprinkler heads and other miscellaneous items. A higher maintenance cost may be incurred for the irrigation system after additional years of use.

### Oxidation Ditch Operation

Oxidation wheels were operated from 1969 until the fall of 1971 with liquid level controlled by hauling directly to the field. By 1971, the swine herd had increased and suffi-cient labor was not available to haul waste to maintain the optimum liquid level for good oxidation ditch performance. For a time, the oxidation wheels were operated at high liquid levels contrary to good management techniques for this device. Operation of the oxidation wheels began again in the fall, 1972, with the use of the sluice gate, tile-sump system, and lagoon for liquid level control and storage of excess liquid overflow.

Foaming became a problem when restarting the oxidation wheels in the fall, 1972. Water dilution with some addition of fuel oil and removal of solids from the pit seemed to solve this problem. Foaming has occasionally been a problem during subsequent operation.

The depth of liquid on the rotor is from 5.1 cm (2 in.) minimum to 15.2 cm (6 in.) maximum with a dry matter concentration from 1.7 to 4.6 percent in the ditch.

Maintenance and repair of 10 oxidation wheels averages about two hours per week. This includes lubrication, checking liquid level in the ditch, adjustment of belts and replacement of minor parts. End rotor bearings, seals and belts are the most common parts replaced on a routine basis. In addition, reduction gear coupling spiders, hoods, and rotor paddles have been replaced or rebuilt. An average of $80 cost per year per oxidation wheel was incurred for repairs during the five years of operation. Additional annual costs for the operation of the oxidation wheels were $30 per wheel for maintenance labor and $462 per wheel for electricity (based on 3.3 horsepower demand and 1.6 cents/kw hr).

The oxidation wheels decreased odors in the confinement buildings compared to anaerobic pit conditions. The oxidation ditches performed with less odor, less repair and maintenance problems in the farrowing houses than the finishing houses. This may be due to a lower waste loading rate and greater dilution due to more frequent cleaning in the farrowing houses.

### ANAEROBIC PIT STORAGE

Sludge buildup, partition walls in the pits, and minimal slope on the pit floor in the anaerobic pits in the gestation houses have resulted in plugging of the tile line thus causing greater difficulty in emptying these pits. Some means of agitation and/or flushing with a sloping floor might be advantageous if using the tile system with an anaerobic pit. Perhaps a 20.3 cm (8 in.) tile and greater slope in the tile (1-2 percent) would also be desirable.

Occasionally, an agitator pump and the liquid vacuum wagon have been used to remove the solids from the gestation house pit. On one occasion, the solids had to be removed by hand.

### LAGOON AND IRRIGATION MANAGEMENT

Although the lagoon has been in use only two years, labor required for the management of the lagoon has been minimal. Loading the lagoon with the high solids sump pump has worked quite satisfactorily. It was necessary to enclose the float controlled mercury switch in a weatherproof box to minimize corrosion.

The lagoon is loaded periodically as the sump fills from the overflow of the oxidation ditches and when emptying the anaeroibc pits (gestation units). Odor from the lagoon has been minimal. The lagoon is designed at 0.12 m$^3$ of liquid volume per kg (2 ft$^3$ per lb) of average liveweight for 2000 head averaging 46 kg (100 lb). When the lagoon was initially loaded, it was approximately 1/5 full of water. Fourteen months later the lagoon was full and had to be dewatered.

The lagoon is dewatered to one-half of its volume twice each year. The waste effluent is applied at the rate of 12.6 cm (2 in.) to 48 ha (120 acres) to cropped land (Brookston silty clay loam) adjacent to the lagoon. Labor for waste disposal using this irrigation system averages about 1 man hour for each 2 hours of irrigation time. This included labor needed for set-up and take-down of the irrigation system. The lagoon-irrigation system has greatly reduced labor requirements and has added flexibility for waste disposal as compared to the liquid haul system. The lagoon-irrigation system eliminated, as a problem on the case farm, the necessity to spread waste from full pits during periods of rainy weather or when land was unavailable during the cropping season.

## NUTRIENT (FERTILIZER) VALUE

The nutrient composition of swine waste was obtained (Table 3) from anaerobic pits and oxidation ditches in one of the growing-finishing houses for a 3-year period and from the anaerobic lagoon for a 2-year period. Each month random samples were obtained at different locations in the pits and lagoon, were composited, frozen and stored for analyses. Application of similar rates of the anaerobic and aerobic (oxidation ditch) waste to land cropped to corn resulted in higher corn yields from the anaerobically treated waste compared to aerobically treated waste (Sutton et al. 1973, 1974). This was probably due to the differences in nitrogen content of the wastes.

The fertilizer value of lagoon effluent is much less than liquid pit waste due to losses of nitrogen and the settling of phosphorus in the sludge layer. Fertilizer value per 1000 liters for swine waste from anaerobic pits, the oxidation ditch and the anaerobic lagoon was $3.64, $2.41, and $0.29, respectively, based on fertilizer prices of 18 cents for N, 25 cents for $P_2O_5$, and 9 cents for $K_2O$.

The fertilizer value can help decrease some of the expense of the waste disposal system. However, the system which maximizes fertilizer value may be less desirable from the standpoint of labor flexibility, labor requirements and annual operating costs.

## SUMMARY AND CONCLUSIONS

This paper reports a case study on the design and performance characteristics of a waste disposal system for the 2500-head swine operation at the Baker-Purdue Animal Sciences Center, Purdue University, West Lafayette, Indiana. Components of the waste disposal system include a 15 cm (6 in.) tile line (0.4 percent slope) from pits of the confinement buildings to a 5678 liter (1500 gal) sump which loads an anaerobic lagoon for waste storage and treatment. A hand-carry irrigation system is used for application of the lagoon effluent to adjacent land cropped with corn. Prior to construction and use of the lagoon-irrigation system, a 5678 liter (1500 gal) vacuum liquid tanker wagon and 50-hp tractor were used to haul and spread waste from anaerobic pits and oxidation ditches.

Based on 1975 prices for the waste system described in the study, the initial investment for the lagoon-irrigation system (including tile, sump and pits) was $23.63 per pig marketed annually compared to $27.99 per pig marketed annually for the liquid haul system (including the concrete storage pit). The total annual operating cost, including depreciation, interest, taxes, insurance, repairs, labor fuel, etc., for the lagoon-irrigation system was $3.62 per pig marketed annually compared to $7.11 per pig for the liquid haul system. Labor, fuel, and power requirements were considerably higher for the liquid haul system ($3.07 per pig) compared to the lagoon-irrigation system ($ 0.16 per pig). The annual labor requirement for the liquid haul system was 780 hours compared to 80 hours for the lagoon-irrigation system.

Based on current prices, the estimated fertilizer value of anaerobic pit waste, oxidation ditch liquor, and lagoon effluent was $6.59, $4.36, and $1.52 per pig, respectively. Even though fertilizer value is sacrificed, it was concluded from this case experience that the lagoon-irrigation waste disposal system is an attractive alternative for a large swine operation with limited labor. In addition, the investment in the shallow pit associated with the lagoon-irrigation system is less than the deep pit essential for the liquid haul system.

TABLE 3. AVERAGE NUTRIENT COMPOSITION OF SWINE WASTE FROM VARIOUS WASTE HANDLING AND STORAGE SYSTEMS*

| Analyses | Anaerobic pit[†] | Oxidation ditch[†] | Anaerobic lagoon[‡] |
|---|---|---|---|
| Dry matter, percent | 4.3 | 2.6 | 0.18 |
| Kjeldahl nitrogen, percent | 0.38 | 0.23 | 0.039 |
| Ammonium nitrogen, percent | 0.19 | 0.10 | 0.033 |
| Nitrate nitrogen, mg/l | - - | 11.0 | 3.1 |
| Phosphorus, percent | 0.16 | 0.10 | 0.007 |
| Potassium, percent | 0.13 | 0.14 | 0.03 |
| Sodium, percent | 0.03 | 0.03 | 0.008 |
| Copper, mg/l | 7.0 | 4.0 | 0.44 |
| pH | 6.96 | 7.54 | 7.72 |
| D.O., mg/l | 0.39 | 1.95 | |
| Fertilizer value[§], $/1000 liters | 3.64 | 2.41 | 0.29 |
| Fertilizer value[‖], $/pig sold | 6.59 | 4.36 | 1.52 |

*Wet basis.

[†]Averages of monthly samples for a 3-year period from pits in a growing-finishing house only.

[‡]Averages of monthly samples for a 2-year period.

[§]Fertilizer nutrients are valued at 18 cents, 25 cents, and 9 cents per lb, respectively, for total N, $P_2O_5$ and $K_2O$.

[‖]Estimated fertilizer value based on average annual waste production and average annual pig capacity.

**References**

1 Mayrose, V. B., J. R. Foster, H. W. Jones, M. P. Plumlee, A. L. Sutton and L. B. Underwood. 1974. Baker-Purdue Animal Sciences Center Swine Facilities. Purdue University Cooperative Extension Mimeo AS-390, West Lafayette, Indiana.

2 Sutton, A. L., V. B. Mayrose, D. W. Nelson and J. C. Nye. 1973. Effect of swine waste on corn production. J. Anim. Sci. 37(1):251 (Abstr.).

3 Sutton, A. L., D. W. Nelson, V. B. Mayrose and J. C. Nye. 1974. Effect of liquid swine waste application on soil chemical composition. Proc. 1974 Cornell Ag. Waste Mgt. Conf., New York, p. 503-514.

# KSU Aerobic Swine Waste Handling System —
# Six Years of Problems and Progress


B. A. Koch, R. H. Hines, G. L. Allee, R. I. Lipper

MEMBER
ASAE


NEW confinement, swine-production research facilities which were built at Kansas State University in 1968 were designed specifically to accomodate an aerobic waste-oxidation system. Structural details of waste pits and related components are outlined in Fig. 1 as information only, not recommendations.

## FACILITIES AND EQUIPMENT

Fairfield, pour-in-place concrete slats are used for flooring in all buildings (total slats). Buildings (Butler) are steel and insulated. The farrowing and nursery buildings are completely enclosed. Each is equipped with an overhead furnace bringing in variable amounts (manual control) of outside air. Fans on the furnaces run constantly. An exhaust fan (400 cfm) in each "waste unit" room runs constantly. The ceiling of each unit has four intake fans (1000 cfm each). A separate thermostatic control on each one is manually adjusted. The farrowing building is equipped with an air-conditioning unit bringing in variable amounts of outside air (manual control) when used. Finishing buildings are open to the south in summer (closed in winter) and each has two exhaust fans on the north wall (800 cfm). Each "waste unit" room has an exhaust fan in the ceiling (3400 cfm). Sixteen small (6000 btu/hr.) propane heaters, (C.S.I. Pro-Pig), one over each pen, supply heat in each finishing unit.

Each pit overflows through a standpipe into a 1500-gallon, outside holding tank. Fluid from the tanks is hauled to nearby fields in a tractor-towed, liquid-manure wagon (Hawkeye Steel) (4163 liters - 1100 gallons).

Fairfield "paddle-wheels", originally were used to circulate liquid waste, mix oxygen into the liquid, and keep solids suspended in five separate pits under farrowing, nursery, and finishing swine pens. They were replaced in 1971-72. Their open motors could not withstand high dust and moisture. Their bearings were close to the liquid in pits where foaming was a recurring problem, and odor control was not satisfactory.

All original "paddle-wheels" were replaced by "Aerob-A-Jets" (Fairfield). Each unit consists of a 9-inch diameter, three-blade propeller, mounted on a four-foot hollow shaft driven by a three-horsepower sealed motor. The pit under the number 5 finishing barn was drained and cleaned because it had been used as an anaerobic pit several months. The water level in all pits was raised to accomodate the "Aerob-A-Jets" by extending the overflow standpipes.


Article is contribution KSU No. 484 of the Animal Science and Industry Dept. and contribution No. 213 of the Agricultural Engineering Dept., Kansas State University, Manhattan.

The authors are: B. A. KOCH, R. H. HINES and G. L. ALLEE, Animal Science and Industry Dept., and R. I. LIPPER, Agricultural Engineering Dept., Kansas State University, Manhattan.

Company names and product names are used only for easier communication. No preference or endorsement is implied.


Maintenance problems and labor requirements immediately dropped to acceptable levels. Foaming problems were greatly reduced.

All data and information reported here, except for reference to foam control, have been gathered under "Aerob-A-Jet" conditions.

## FOAMING, DUST, AND ODOR PROBLEMS

We acquired considerable practical experience with foam-control agents while we used the "paddle-wheels." Foaming problems were less serious than those described by others (Jones, Day and Dale)*. Foam never reached floor-level in any of our pits. The most common problem was a 10 to 20 centimeters (4 to 8 inches) layer of foam that did not circulate and, therefore, dried into a thick crusty layer on the surface of the liquid.

Most of the additives we used to successfully break up the foam alleviated the immediate problem but none seemed to prevent more foam from forming. Diesel fuel or kerosene reduced surface foam, but did not prevent a thick unmanageable scum from forming on the liquid surface. Effects of addititves on aerobic bacteria were not investigated, but high odor levels indicated poor results. Commercial defoamers were funished to us by Kemin Industries, Dow-Corning, Rohm & Haas, Feed Flavors, Inc., Arizona Chemicals, and Kalo Company. All effectively reduced the amount of foam on the liquid surface for a short time. None seemed to have any long term effect.

By accident, we found that Foremost-Soweena (a milk replacer for baby pigs) was an effective foam disperser, as was Ferma-Grow (a fermentation feed additive). Poloxalene (active ingredient in Bloat Guard), was also an effective foam-control agent.

The foam dispersers did not overcome a noticeable odor problem (probably hydrogen sulfide), especially in the nursery building. It has been controlled and almost eliminated since April, 1974, by regularly adding small amounts of Puritan Liquid Live Micro-Organisms† to each pit. In November, 1973, a negative ion dispersion unit (Clean Air Systems, Inc.) was installed in the farrowing house and the nursery. It reduced dust so well that part of the air in each building is now being recirculated through the furnace and the air conditioner.

## PROCEDURES

The swine research unit is used primarily for nutritional and management studies with pigs of various ages. The manure disposal system is secondary to overall research


*Aerobic treatment of livestock wastes — U. S. Environmental Protection Agency, 1972.

†Commercially available from Puritan Chemical Co.

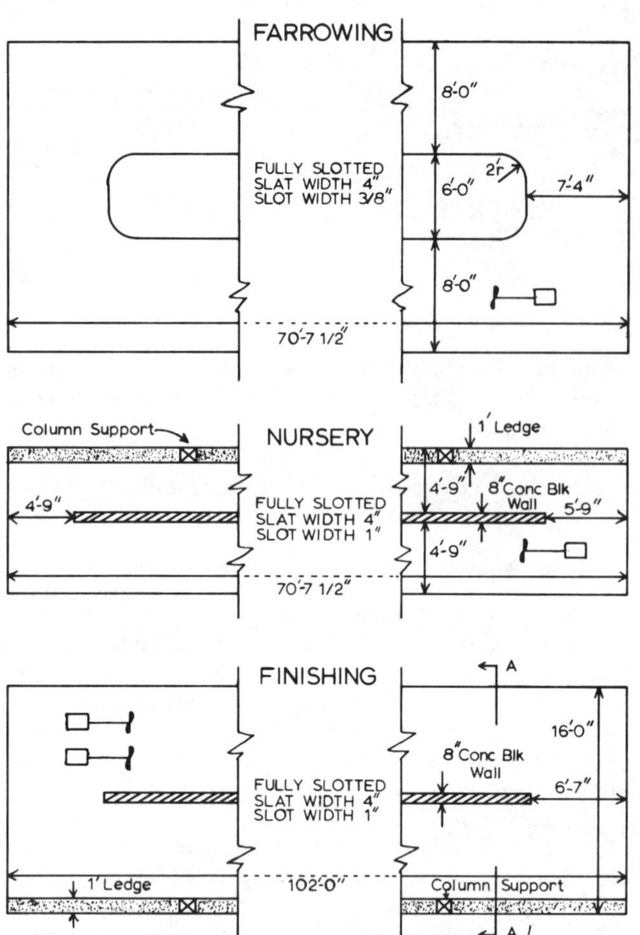


FIG. 1a Plan view of waste pits.

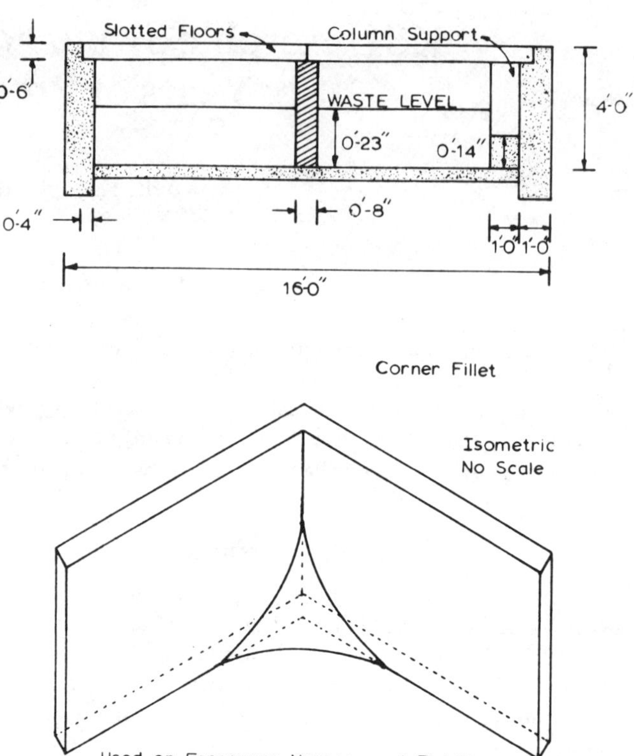


FIG. 1b Typical cross section of waste pits with corner detail.

5   Reduce the "fly" problem.
6   Improve general sanitation.
7   Determine cost of the operation.

planning. Thus pit loadings fluctuate from month to month.

The farrowing house has crates for 29 sows. We have sows farrow every other month and wean pigs at 4 to 6 weeks of age, so the average loading of the farrowing house is approximately 25 sows plus their pigs. All tissue and fluids related to birth drop into the pit (no dead pigs). The septic tank from the headquarters building also drains directly into the pit.

Approximately 100 pigs, averaging 20 kilograms are over each nursery pit. Pigs go into the nursery averaging approximately 10 kilograms each and are moved out at an average weight of 30 to 35 kilograms.

Most pigs are moved directly from the nursery to the finishing barn, where typical loading over each waste pit is 100 pigs averaging 70 kilograms. Pigs are brought into the barn weighing 30 to 35 kilograms and usually removed when they weight approximately 100 kilograms. Time lapse between groups of pigs usually is 2 to 5 days in any house. Floors and walls of pens in all units are cleaned with high-pressure cold water and disinfected between groups of pigs. All cleaning fluid falls into the pits.

## RESULTS AND DISCUSSION

Data collected during the past year are summarized in tables 1-3. Objectives outlined in 1968 when the waste control system was built are being met.

1   Reduce labor requirements to a minimum.
2   Reduce complaints from neighbors to a minimum.
3   Reduce odors in buildings.
4   Furnish pigs a suitable environment.

Labor related to waste handling is slight. A man with a scraper goes through each farrowing crate daily to see that all fecal material falls into the pits. Other pens are scraped when necessary (related to number of pigs in a pen). Labor required to pump overflow tanks and haul liquid to fields could be eliminated by having pits drain directly into a lagoon. Cleaning of pens or buildings between groups of pigs is fast and easy as all wash water falls into the pits. Labor required for Aerob-A-Jet maintenance has been low. A new propeller can be installed on a unit in an hour or less. The complete unit can be replaced in less time. Entries on the maintenance record are listed below.

A new propeller was put on the farrowing house unit March 3, 1974.

August 19, 1974, a new propeller was put on the south nursery unit, that propeller was replaced January 23, 1975, after the flow rate of fluid in the pit slowed noticeably.

Pigs broke a small wood slat off a feeder over the north nursery pit January 24, 1975. The slat fell into the pit and broke the propeller shear pin. A new propeller was installed when the pin was replaced.

In the south finishing barn pit, a new propeller was put on one unit March 27, 1974, and on the other unit, June 25, 1974. August 6, 1974, one motor burned out and was replaced. Liquid flow was slowing in January so both propellers were replaced January 25, 1975.

New propellers were put on both units in the north finishing barn March 3, 1974. One propeller broke and was lost in the pit May 10, 1974. Both propellers were replaced January 3, 1975, after liquid flow slowed.

The "complaint" problem has been almost completely eliminated. Facilities and fields are close to housing areas



MANAGING LIVESTOCK WASTES

TABLE 1. SELECTED DATA FROM AEROBIC OXIDATION PITS AT K.S.U.
SWINE RESEARCH UNIT.

| Item | January, 1975 | | | | | January, 1974 | | | | |
|---|---|---|---|---|---|---|---|---|---|---|
| | Pit No.* | | | | | Pit No.* | | | | |
| | 1 | 2 | 3 | 4 | 5 | 1 | 2 | 3 | 4 | 5 |
| Energy, kW/hr | 2.52 | 1.36 | — | 2.90 | — | 2.92 | 1.75 | — | 2.94 | — |
| Fluid temp., deg C | 20.0 | 20.0 | 20.0 | 18.0 | 20.0 | 19.0 | 21.0 | 20.0 | 17.0 | 20.0 |
| Fluid pH, paper | 5.0 | 8.2 | 8.2 | 7.8 | 8.5 | 6.5 | 8.0 | 8.0 | 8.0 | 8.0 |
| Fluid D.M.†, per-cent | 0.55 | 1.34 | 1.05 | 2.15 | 2.81 | 0.62 | 2.21 | 1.58 | 2.58 | 1.86 |
| Crude protein in D.M., per-cent | 27.1 | 39.6 | 30.4 | 31.7 | 32.8 | 30.2 | 35.6 | 33.8 | 27.1 | 24.1 |
| Crude fiber in D.M., per-cent | 7.5 | 3.2 | 2.4 | 1.9 | 4.5 | 3.9 | 5.9 | 2.6 | 7.3 | 6.7 |
| Total ash in D.M., percent | 35.2 | 31.9 | 42.4 | 42.6 | 32.7 | 39.9 | 30.3 | 38.8 | 38.0 | 43.7 |

*Pit no. 1 = farrowing barn; 2 = south nursery barn; 3 = north nursery barn; 4 = south finishing barn; 5 = north finishing barn.

†D.M. = Dry matter.

‡Temperature of water supply = 12.0 C in January, 1974; 13.0 C in January, 1975.

TABLE 2. PARTIAL COMPOSITION OF DRY MATTER FROM PITS AT K.S.U.
SWINE RESEARCH UNIT.

| Mineral† | January, 1975 | | | | | January, 1974 | | | | |
|---|---|---|---|---|---|---|---|---|---|---|
| | 1 | 2 | 3 | 4 | 5 | 1 | 2 | 3 | 4 | 5 |
| | PPM | | | | | | | | | |
| Phosphorus | 21,590 | 30,120 | 36,960 | 39,690 | 29,430 | 27,900 | 27,400 | 32,200 | 35,400 | 33,700 |
| Calcium | 39,950 | 40,010 | 37,450 | 38,010 | 34,120 | 46,000 | 42,100 | 46,000 | 41,000 | 39,200 |
| Iron | 4,117 | 2,930 | 6,010 | 3,170 | 2,750 | 3,820 | 1,980 | 1,810 | 2,160 | 1,940 |
| Magnesium | 11,290 | 11,090 | 15,090 | 15,960 | 12,050 | 13,373 | 11,660 | 14,245 | 13,663 | 15,388 |
| Manganese | 459 | 798 | 1,074 | 773 | 572 | 303 | 400 | 492 | 411 | 378 |
| Zinc | 445 | 683 | 657 | 556 | 600 | 384 | 694 | 536 | 495 | 339 |
| Copper | 92 | 118 | 92 | 102 | 107 | 46 | 115 | 100 | 82 | 68 |
| Potassium | 44,290 | 46,310 | 75,560 | 75,530 | 59,280 | 56,164 | 52,942 | 71,512 | 61,404 | 85,708 |
| Sodium | 22,250 | 19,601 | 25,950 | 24,150 | 19,960 | 19,323 | 11,749 | 19,996 | 20,920 | 27,194 |

*Pit no. 1 = farrowing barn; no. 2 = south nursery barn; no. 3 = north nursery barn; no. 4 = south finishing barn; no. 5 = north finishing barn.

†Dry matter basis.

but odor from fluid is only slightly noticeable. Fluid is hauled to the fields whenever soil is dry enough regardless of wind direction or humidity.

The overall environment furnished the pigs is conducive to good performance. Twenty-eight sows that farrowed in January, 1975, raised 240 pigs to weaning age. One outstanding pig farrowed October 3, 1974, weighed 98 kilograms (216 pounds) 131 days later (February 11, 1975). Others of similar age exceeded 90 kilograms (200 pounds) each.

Flies are easily controlled. Reproduction of flies in the unit is essentially zero. Flies that migrate into the buildings are easily destroyed by spraying or fogging.

Health problems have been minimal. Facilities are easily cleaned. There has never been a longstanding siege of illness or infection in the facilities. Baby pig diarrhea, a problem at times, is not chronic.

Operational costs have been higher than desired. Table 1 shows electrical energy consumed in three different pits. Differences in consumption in the same pit probably reflect differences in condition or location of the propellers. Face areas of some propeller baldes which were removed after long usage were 28 to 43 percent less than when new. Also, a slight change in location or angle of a propeller will change apparent surface velocity of the fluid in a pit.

Items measured indicate that pits have attained a rather stable state under our system of management (Tables 1 and 2). Measurements made in January 1975 are similar to the same measurements made in January 1974. All items were measured either weekly or monthly throughout the year.

Water removed from overflow pits seems excessive at times. Much of it is apparently water wasted by pigs playing. Overflow from the farrowing house pit has always exceeded that from other pits (Table 3) but the headquarters building septic tank drains into the farrowing house pit. We routinely remove at least one load (4136 liters) from each overflow pit once each week. Water is added to the manure

(Continued on page 185)

TABLE 3. FLUID REMOVED FROM OVERFLOW TANKS
AT K.S.U. SWINE RESEARCH UNIT.

| Month | Farrow | South nursery | North nursery | South finish | North finish |
|---|---|---|---|---|---|
| | | | Liters X $10^3$ | | |
| April '74 | 91.6 | 37.5 | 20.8 | 44.3 | 16.6 |
| May | 60.2 | 34.4 | 23.7 | 54.1 | 16.3 |
| June | 53.7 | 34.3 | 34.3 | 18.5 | 30.1 |
| July | 56.8 | 41.3 | 47.3 | 25.9 | 15.3 |
| Aug. | 80.2 | 31.0 | 45.4 | 9.3 | 13.4 |
| Sept. | 42.6 | 18.5 | 17.2 | 40.3 | 31.0 |
| Oct. | 93.5 | 30.3 | 39.7 | 14.2 | 8.3 |
| Nov. | 17.6 | 8.3 | 4.2 | 33.3 | 39.4 |
| Dec. | 13.8 | 28.8 | 20.8 | 25.9 | 30.1 |
| Jan. '75 | 30.1 | 16.7 | 20.8 | 33.3 | 49.2 |
| Feb. | 71.0 | 13.4 | 20.4 | 20.8 | 0 |
| Mar. | 34.3 | 13.3 | 20.6 | 23.7 | 36.1 |
| Av. | 53.8 | 25.6 | 26.3 | 28.6 | 23.8 |

# Two-Stage Activated Sludge Treatment of
# Effluent from Industrial Hog Breeding Farms

Corneliu A. L. Negulescu

ONE of the problems which may be solved by the activated sludge process is the treatment of wastewaters produced by animal breeding farms. In Romania, during recent years, a system has been applied to breeding hogs concentrated in very limited areas. The agro-industrial type buildings used in this system house 100,000-150,000 hogs per site. In some parts of the country, about 300,000 hogs are produced annually at a single production site. Since waste disposal by land spreading requires a large agricultural area (Laak (1970) and Webber (1971) indicated the minimal land area per animal for corn production is 335 m²/animal), the possibility of treatment and disposal of the waste has to be taken into consideration.

When stables are washed with water under pressure, the manure is flushed out in the sewerage system as wastewater, with characteristics shown in Table 1.

**TABLE 1. CHARACTERISTICS OF RAW WASTEWATERS.**

| Indicator | Units | Range | Mean value |
|---|---|---|---|
| Suspended solids | mg/l | 5,990-7,220 | 6.600 |
| pH | | 7.7-8.5 | 8.0 |
| COD | mg $O_2$/l | 13,157-14,039 | 13,600.0 |
| $BOD_5$ | mg $O_2$/l | 4,705-5,018 | 4.850 |
| Total solids | mg/l | 6,947-8,263 | 7,600.0 |
| Total nitrogen | mg N/l | 825-857 | 840.0 |
| Phosphorus | mg P/l | 33.5-41.5 | 37.0 |

In addition, the daily waste production characteristics were determined: $BOD_5$ = 105 g/animal-day; COD = 240 g/animal-day; N = 17.5 g/animal-day; and P = 1.2 g/animal-day.

The first treatment plants for these complexes were designed according to data found in the literature and had the following flow chart (Negulescu 1970):

1　Mechanical treatment (bar screens, horizontal settling basins)

2　Natural biological treatment (Oxidation ponds, infiltration into the soil)

3　Sludge handling (open digestion tanks and drying beds)

4　Wastewater disinfection (chlorination)

In settling tanks with 2-4 hours detention time, the suspended solids were lowered by 70 percent, the COD by 60 percent, and the $BOD_5$ by 40 percent.

Natural biological treatment by infiltration into the soil did not give the expected results; instead, the soil became clogged by the remaining suspended solids. Nor did the oxidation ponds function as expected; the majority became overloaded, and consequently, operated under anaerobic conditions. The settled sludge in the basins was sent to open digestion tanks, where it remained for 45 days. The

operation of these basins without agitation and heating proved to be inefficient.

From the open digestion tanks, the sludge was sent to drying beds, where it was dewatered for 30-60 days (in the warm season) with the necessary hygienic-sanitary precautions, for potential agricultural use. With this initial arrangement, the realized efficiencies were about 90 percent for $BOD_5$ and about 97 percent for suspended solids. The effluent values were 493 and 204 mg/l, respectively.

Because discharge conditions into streams are generally more restrictive, it became necessary to improve the treatment efficiency. In this respect, an activated sludge — biological treatment stage was introduced after the mechanical treatment stage.

Biological treatment by the activated sludge process was explored in 1970 in a laboratory installation (Negulescu 1971). This study showed that the biological treatment process can be directly initiated with settled water, without additional dilution. The activated sludge growth is quick, due to the bacteria already present in the raw water and the existence of the necessary substrate and nutrients.

Activated sludge concentrations up to 10-12 g/l were used without disturbing the secondary settling process. The oxidation achieved was 0.3-0.8 kg 5-day BOD/kg MLSS per day, the oxygen consumption 1.0 kg $O_2$/kg 5-day BOD removed, and the sludge production 0.65 kg/kg 5-day BOD removed. The detention time of 24 hours and an organic loading of 0.6-1.2 kg COD/kg MLSS per day (respectively 0.3-0.8 kg 5-day BOD/kg MLSS per day) gave treatment efficiencies of 99 percent for 5-day BOD, with an effluent having the 5-day BOD value of 10-40 mg $O_2$/l.

Based upon these results, several treatment plants were designed in 1971 and then partially built according to the following modernized flow chart:

1　Mechanical treatment (bar screens, horizontal settling basins)

2　Biological treatment by two-stage activated sludge process

3　Excess sludge aerobic stabilization

4　Sludge dewatering on concrete sludge beds with central drain

5　Water disinfection by chlorination

The necessary aeration and turbulence were provided by mechanical aerators with a rotor diameter of 1,000 mm, made in Romania. The first results obtained from the biological stage confirmed the laboratory study. The organic matter oxidation as a function of sludge load had very close values, both in laboratory and full-scale tests.

Experimentation began at the start of the summer of 1972; shortly thereafter, some of the equipment broke down. However, by that time, sufficient data had been collected to indicate process reliability. For example, the $BOD_5$ was 2,550 mg $O_2$/l at the aeration tank inlet and 162 mg $O_2$/l at its outlet; the resulting efficiency was

The author is: CORNELIU A. L. NEGULESCU, Senior Research Officer, Institute for Water Management, Bucharest, Romania.

approximately 94 percent. This value was obtained with an insufficient oxygen supply; we hope to achieve better efficiency in the future. The results of the experiments carried on in Romania in laboratory and full-scale tests are shown in Table 2.

**TABLE 2. LABORATORY AND FULL-SCALE EXPERIMENT DATA ON HOG WASTE TREATMENT BY THE ACTIVATED SLUDGE PROCESS.**

| Parameter | Units | Laboratory scale (1970) | Full-scale (1972-1974) |
|---|---|---|---|
| Organic load | kg $BOD_5$/m³-day | — | 1.3-1.5 |
| Sludge load | kg $BOD_5$/kg MLSS, day | 0.3-0.8 | 0.2-0.5 |
| MLSS | g/l | 10-12 | 5.8-10.7 |
| Power consumption | kWh/cap-day | — | 0.08-0.09 |
| $BOD_5$ efficiency | percent | 99.9 | 94-96 |
| in effluent | mg/l | 10-40 | 60-160 |
| Hydraulic load final settling | m³/m²-h | — | 0.5-0.8 |

At low temperatures the biological activity of the microorganisms becomes slower. One of our concerns was to check, at full-scale, some small-scale studies carried on in the open air during the winter of 1971-72. On this occasion, it had been observed that the biological treatment processes may be initiated even at water temperatures of +5 C. The treatment plant which began functioning during the summer of 1972 continued to give satisfactory results during the winter of 1973-74 with efficiencies of 90 percent in removing the organics and suspended solids.

The first results of our research work (treatment plant on full-scale operation) have conclusively demonstrated that the activated sludge process is indicated for this type of wastewater. Our method yields a biologically stabilized effluent and a dry sludge with a potential use as fertilizer, without the release of disagreeable odors.

The research work will continue in 1975. We plan to examine the possibilities of destroying pathogenic germs by disinfection, and to work toward aerobic stabilization of excess sludge. In addition, we soon hope to be able to reduce the amount of nitrogen and phosphorus in the wastewater after biological treatment.

### References


1   Chiriac, V., I. Gueron, and C. A. L. Negulescu. 1974. Le traitement des eaux usées provenant des fermes pour l'élevage intensif des porcs. Journal de CEBEDEAUX. Novembre, 1974. Liège.

2   Hart, S. A. and M. E. Turner. 1965. Lagoon for livestock manure. Journal Water Poll. Control Fed. 37:1578-1596. Washington, D. C.

3   Laak, R. 1970. Cattle, swine and chicken manure challenges waste disposal methods. Water and Sewage Works. pp. 134-138.

4   Loehr, R. C. 1969. Animal wastes — a national problem. Journal of the Sanitary Engineering Division, ASCE. 95(SA2). April, 1969.

5   Negulescu, C. A. L. 1970. Current problems of wastewater treatment in the great industrial units of hog breeding. International Congress on Industrial Waste Water, Stockholm, November.

6   Negulescu, C. A. L. 1971. Probleme actuale ale epurării apelor uzate de la marile complexe industriale de creștere a porcilor. Hidrotehnica, Imbunătățiri Funciare, Construcții Hidrotehnice, Gospodărirea Apelor, Meteorologie. nr. 3/1971, vol. 16, București.

7   Negulescu, C. A. L. and E. Cute. 1973. Quel que résultats obtenus dans le traitement aux boues activées des eaux résiduaires evacuées par les grandes fermes pour l'élevage des porcs. Le Seminaire pour la pollution des eaux par l'agriculture, La Commission Economique ONU pour l'Europe, Vienne. Octobre, 1973.

8   Negulescu, C. A. L. and E. Juriari. 1971. Cercetări de laborator privind epurarea biologică a apelor uzate de la crescătoriile industriale de porci. Studii de protecția și epurarea apelor, vol. XV, pp. 200-230. București, 1971.

9   Taiganides, E. P., T. E. Hazen, E. R. Baumann, and D. Johnson. 1964. Properties and pumping characteristics of hog wastes. TRANSACTIONS of the ASAE. 7:123.

10   Webber, L. R. 1971. Animal wastes. Journal Soil and Water Conservation. 26(2):47-50.


---

## KSU Aerobic Swine Waste Handling

*(Continued from page 183)*

pit if necessary to force that much fluid into the overflow pit.

### SUMMARY

The Kansas State University Swine Research Unit's aerobic oxidation waste disposal system is functioning satisfactorily with minimum maintenance and supervision. Labor requirements are low. Odor is no problem. Waste fluid can be spread on the fields any time soil is not too wet (no complaints from neighbors regardless of wind direction or humidity conditions). Pigs are performing satisfactorily in the unit. Flies are easily controlled and suitable sanitation is easily maintained.

# A Pig Slurry Treatment System Based on Separation Before Aerobic Treatment and Sludge De-Watering

R. Q. Hepherd and L. E. Osborne

ABOUT 24 percent of the pigs in England and Wales are housed on farms with too little land to utilize the plant nutrients in the feces and urine, or to dispose of them without risk of water pollution (Richardson 1974). Many of these farmers have informal arrangements with neighbors willing to accept manures, but if these are in slurry form, the costs of storage until the neighbors are willing to accept them, and of transport to the spreading site, normally fall on the producer. Long term storage (over 150 days) is often necessary but is normally costly, and the emptying of large slurry stores raises handling and management problems.

At present, legal action against farmers arising from pollution is relatively rare (about 30 prosecutions in England and Wales in 1973), the authorities relying more on persuasion than litigation, but possible point sources of pollution will clearly be more and more vigorously checked in future, particularly by the regional water authorities, who recently also became responsible for sewage and industrial waste disposal.

Few storage problems or water pollution hazards arise where manures are handled in solid form, although air pollution nuisances can arise. However, traditional straw-bedded systems of housing are no longer economically viable in many areas, because they are labor consuming or because no cheap straw is available, and the overall cost of manure disposal to land per pig fattened is about double that for a bedding-less system producing slurry.

Following a series of experiments into slurry separation, treatment and sludge dewatering (Hepherd 1975), an experimental process has been evolved, the objectives being to convert as much as possible of the pig slurry into smell-free solids, and to produce liquids that were not sufficiently clean to be run into a watercourse but which could easily be stored and applied to land without much risk of air or water pollution.

This paper summarizes the results of the first year's work on the process, which is described in detail elsewhere (Osborne and Sneath 1974a and b).

## EQUIPMENT AND METHOD

The experiment was designed to handle the daily slurry output from about 400 fattening pigs (1.8 m$^3$/day). Fig. 1 illustrated the process as it would appear if installed adjacent to a pig house. During the experiment, the source of slurry was in fact some miles from the plant and a slurry storage tank on site was necessary.

### Slurry Source

The slurry was taken by tanker from storage channels beneath the partially-slatted floors of a number of buildings housing 1400 pigs, transported to the experimental site, and stored in a 36 m$^3$ holding tank. The slurry was therefore stored for some time (2 to 9 weeks) under anaerobic conditions before processing.

### Separation

A rotating drum and press roller type of separator (Item 3, Fig. 1) was used, the separated liquid being stored in a holding tank and the fiber in a field heap. The latter heated to about 50 C within 5 days of separation, very little seepage occurring at dry matter contents down to 17 percent. The separator performance is summarized in Table 1.

About 90 percent by weight of the particles in the slurry were under 1000 microns in diameter but after separation 90 percent of the particles in the liquid were under 200 microns in diameter. They presumably included the more biodegradable fractions of the solids.

The separator was switched on manually to refill the liquid holding tank (Item 4, Fig. 1) whenever necessary, the machine's capacity being about 1 m$^3$/h.

### Treatment

A high-rate filter delivering into a settling tank was used (Items 8 and 15, Fig. 1). The filter contained 5.2 m$^3$ of plastic sheet medium, arranged in three 1.2 m cubes one above the other, and was clad in 18 mm exterior grade plywood glue-nailed to a softwood frame. The spreading mechanism was a mechanically-driven trough with a series of V-notches along its edge which travelled horizontally to and fro across the top of the medium (Item 7, Fig. 1). Two plants were in fact built. One was fan-ventilated from the bottom at 0.16 m$^3$/sec with warmed air, so that the liquid temperature was maintained at 16 to 18 C: the other was naturally ventilated, the mean liquid temperature (Jan.–May 1974) being 5.7 C. The plants were operated at constant volumetric load for about 5 months before attempting

**TABLE 1. SEPARATOR PERFORMANCE**

| Diameter of drum perforations, mm | Slurry dry matter, percent | Fiber from separator | | Liquid from separator | | Percentage (w/w) of slurry dry matter discharged as fiber |
| --- | --- | --- | --- | --- | --- | --- |
| | | Dry matter, percent | Output as percent of slurry input | Dry matter, percent | Output as percent of slurry input | |
| 1.6 | 7.1 | 18.5 | 20.3 | 4.1 | 79.7 | 53.3 |
| 3.2 | 5.9 | 18.6 | 14.5 | 3.8 | 85.5 | 45.6 |
| 3.2 | 6.1 | 18.1 | 16.0 | 4.2 | 84.0 | 48.6 |
| 3.2 | 6.9 | 16.8 | 20.5 | 5.1 | 79.5 | 49.8 |

The authors are: R. Q. Hepherd and L. E. Osborne, of the Farm Buildings Department, National Institute of Agricultural Engineering, Silsoe, Bedford, England.

to obtain full details of their performance. The operating conditions for each filter during the period recorded in detail (14 Jan. to 27 March 1974) were as follows:

| | |
|---|---|
| Feed rate to filter (separated liquid) - total | 1460 kg/day |
| Frequency of feeding | every 20 min |
| Dose per feed | 20.3 kg |
| Mean $BOD_5$ load - total | 51 kg/day |
|     - per $m^3$ of medium | 9.8 kg/day |
| Mean COD load - total | 118 kg/day |
|     - per $m^3$ of medium | 27.3 kg/day |
| Recirculation rate | 2.7 $m^3$/hr |
| Hydraulic loading on top of medium | 1.9 $m^3/m^2$ per hr |

The settling tanks were pyramid-shaped, 2.7 m square at the top and 1.9 m deep (approx. capacity 4.5 $m^3$).

The volumetric loading on the towers was virtually constant, but because of dilution and some settlement at source, the $BOD_5$, COD and dry matter varied considerably. For example, the $BOD_5$ of separated liquid varied from 25 to 55 g/l, usually being within the range 30 to 40 g/l.

### Sludge Dewatering

The sludge was drawn at 20 min intervals from the bottom of the settling tanks in 30 kg "doses", and was the sole discharge from the treatment stages. The desludging rate was higher than the feed rate in order to give adequate control of the total solids in the treatment stages, the filtrate from the gravity filters used for dewatering being returned to the settling tanks to control solids levels. At the loading rates used, the sludge was difficult to dewater sufficiently rapidly without adding a flocculant. Alum (aluminum chlorohydrate, a commercial grade) was used throughout the experiment because it was easy to meter into the sludge and generally easy to use. It was metered by a small peristaltic pump into a side tube on the outlet nozzle of the sludge discharge pump (Item 11, Fig. 1). The gravity filters used for sludge dewatering were of two types: the first was a rectangular filter tank with straw bale walls two bales high and supported by a frame of 50 mm dia steel tubes. An 8.0 m by 6.3 m tank of this type had a capacity of about 40 $m^3$, or the sludge output from 400 pigs for 6 weeks. When full the tank took about 4 to 6 weeks to drain to 10 percent dm. One wall could then be removed to allow loading to trailers for field stacking. The second filter was a slatted-sided wooden pallet box about 1 m square by 0.75 m high and lined with hessian cloth ($340g/m^2$ grade) retained in place by battens top and bottom (Item 10, Fig. 1). The boxes were handled by fork lift or potato box tippler attachments for a tractor, a box being filled one day with sludge at about 3 percent dm, allowed to stand for 4 days, and then emptied into a trailer at about 10 percent dry matter for stacking in a field heap.

The sludge could easily be retained in a heap about 1.5 m high by a single row of bales, but if loaded into the same trailer as the fiber from the separator, the stack could be formed to any convenient height without any form of retaining wall.

Both types of filter were placed on concrete with sufficient slope to direct the filtrate to a channel and then into a small sump. Much of the filtrate was returned by pump (Item 12, Fig. 1) to the treatment stage to maintain liquid levels, the surplus being spread on land. About 60 to 65 percent of the initial weight left a box within 40 hr of filling, when using a flocculant dosage of 1.5 percent by volume on sludges at about 3 percent dm.

### Plant Control

All pumps, etc could have been automatically controlled either by variable timers or level probes, except the spreader drive and recirculation pump (Item 6, Fig. 1) which ran continuously. However, for convenience of measurement, the separator and filtrate pump were manually operated. Standard equipment was used throughout, and ran trouble-free for 12 months, with the exception of the slurry pump, the rotor and stator of which required changing every 500 running hours (about 2 changes a year) and the spreader drive mechanism on which some worn components had to be replaced. All pumps were of the type with a helical rotor within a rubber stator, except for the small peristaltic pump supplying the flocculant from 250 kg drums. Once set, the sludge removal and flocculant dosing rates were not adjusted.

### RESULTS AND DISCUSSION

#### Mass Balance - Liquids and Solids

The mean daily quantities passing through the plant are shown in Table 2. It was not practicable to measure the total evaporation rate or the weight of dewatered sludge accurately. The dry matter content of three sample batches taken during the recorded period was 10.5 percent. This estimate of average sludge dry matter content was clearly too high. The total evaporation rate from the site of the experiment would have to be 260 kg/day to achieve this average figure, whereas the calculated evaporation rate was 40 to 50 kg/day. If no evaporation had occurred, the sludge dry matter would have been 7.4 percent. The true value must have been between 7.4 and 10.5 percent. The other data were obtained by direct measurement.

The most important result was the reduction in the proportion of the slurry which had to be applied to land as a

TABLE 2. MEAN DAILY MASS BALANCE — LIQUIDS AND SOLIDS

| Input to separator | | Fiber from separator | | | Liquid from separator (feed to filter) | | | Wet sludge from settling tank | | Filtrate return to settling tank | | De-watered sludge** | | | Filtrate to land | | |
|---|---|---|---|---|---|---|---|---|---|---|---|---|---|---|---|---|---|
| Kg | dm*, percent | Kg | dm*, percent | Percent slurry input | Kg | dm*, percent | Percent slurry input | Kg | dm*, percent | Kg | dm*, percent | Kg | dm*, percent | Percent slurry input | Kg | dm*, percent | Percent slurry input |
| 1825 | 7.6 | 365 | 17.0 | 20 | 1460 | 5.2 | 80 | 2160 | 4.1 | 700 | 1.7 | 640 to 900 | 10.5 to 7.4 | 35 49 | 560 | 1.7 | 31 |

*dm = dry matter
**Range estimated from wet sludge weight and total filtrate collected, and from dry matter content of three samples - see text.

| | $BOD_5$, g/l | COD, g/l | SS, g/l | DM, g/l | pH | Correlations |
|---|---|---|---|---|---|---|
| Separated liquid | 35 | 81 | 34 | 5.2 | 7.9 | COD, g/l = 1.15 $BOD_5$, g/l - 42.9* |
| Liquid entering filter | 25 | 48 | 18 | 3.3 | 8.4 | COD, g/l = 1.99 $BOD_5$, g/l - 3.8** |
| | | | | | | DM, % = 0.13 $BOD_5$, g/l - 0.01** |
| | | | | | | SS, g/l = 1.13 $BOD_5$, g/l - 10.38** |
| Wet sludge | 23 | 66 | 31 | 4.1 | 8.4 | No significant relationships |
| Filtrate | 10 | 14 | 0.9 | 1.7 | 7.4 | COD, g/l = 0.92 $BOD_5$ + 4.45** |
| | | | | | | SS, g/l = 0.13 $BOD_5$, g/l - 0.51* |

**Significance at 0.1 percent level
*Significance at 0.5 percent level

liquid (30 percent by weight of the slurry input to the system as opposed to 80 percent at the end of the separation stage). Also this liquid could clearly be applied to land with much less risk of pollution, even if applied at higher rates than would be recommended for slurry (See Table 3).

A mathematical model of the system when integrated into a piggery has been developed and this indicates that the filtrate discharge is likely to be between 0 and 30 percent of the feces and urine input, depending on the time of year and the climatic conditions at the experimental site.

### Plant Nutrients

It was not possible to analyze samples regularly, and the results in Table 4 are very approximate.

### Effects of Temperature

The temperature-controlled tower reduced the soluble $BOD_5$ and COD by 35 percent and 32 percent respectively, whereas for the tower running at near-ambient temperatures the reductions were 17 percent and 22 percent respectively. The difference in $BOD_5$ reduction between the two towers was significant at the 5 percent level, and for COD, at 10 percent. There was no difference between towers, however, in the quality of filtrate produced after dewatering the sludges. The main practical advantage of maintaining liquid temperature is that the liquids will be prevented from freezing.

### Smell

The slurry, separated liquid and freshly separated fiber smell strongly. After stacking, both the fiber and dewatered sludge were virtually odorless, as was the small quantities of filtrate stored at any one time (about 1 m$^3$).

### Handling

The fiber and dewatered sludge were easy to store, handle and spread with conventional manure handling equipment, and both were easier to incorporate into soil by cultivation than straw-based manures.

The filtrate (and wet sludges or separated liquid) were easy to apply to land through sprinkler systems or any other conventional means.

### Costs

The total cost of slurry disposal to land/pig produced, of a system of the type shown in Fig. 1 has been calculated to be double that of the simplest conventional system in use in the U.K. (transport by tanker from channels beneath slats for short distances to land at relatively frequent intervals). The cost was approximately the same as for straw-bedded housing systems but much less labor and no straw were required.

Further work on methods of increasing separation efficiency and reducing the cost of sludge dewatering is currently in progress.

A 500-pig house with the waste treatment system integrated into the structure, and with provision for exhaust air scrubbing, has been designed. It is hoped to construct this unit on a neighboring farm using mainly standard building and other components for field trials of the process when treating freshly-produced dung and urine, to establish performance and costs under typical farm conditions.

### GENERAL CONCLUSIONS

1. Up to 30 percent of a pig slurry which has been stored under anaerobic conditions can be discharged as a relatively clean liquid when using a process of this general type. For cattle and poultry slurries the proportion is likely to be lower, because a higher proportion of the slurry will be discharged as solids during the separation stage (20 to 40 percent has been removed when separating cattle slurry: Hepherd 1975).

TABLE 4. PLANT NUTRIENT ANALYSES

| | Dry matter, percent | | Composition of sample, percent w/w | | | | | |
|---|---|---|---|---|---|---|---|---|
| Source of sample | | Total N | P | K | Ca | Mg | Na |
| Slurry | 5.2 | 0.25 | 0.10 | 0.18 | 0.12 | 0.04 | 0.03 |
| Separated liquid | 3.9 | 0.23 | 0.09 | 0.19 | 0.11 | 0.04 | 0.03 |
| Wet sludge | 2.0 | 0.16 | 0.03 | 0.22 | 0.04 | 0.01 | 0.04 |
| De-watered sludge | 13.7 | 0.17 | 0.23 | 0.32 | 0.18 | 0.06 | 0.05 |
| De-watered sludge after 3 mo. storage in open(Jan-May) | 9.7 | 0.13 | 0.04 | 0.08 | 0.06 | 0.02 | 0.01 |
| Fiber | 21.5 | 0.10 | 0.49 | 0.33 | 0.44 | 0.15 | 0.06 |
| Fiber after 3 mo. storage in open(Jan-May) | 22.1 | 0.09 | 0.21 | 0.21 | 0.33 | 0.06 | 0.03 |
| Filtrate | 1.9 | 0.27 | Trace | 0.22 | 0.04 | 0.04 | 0.04 |

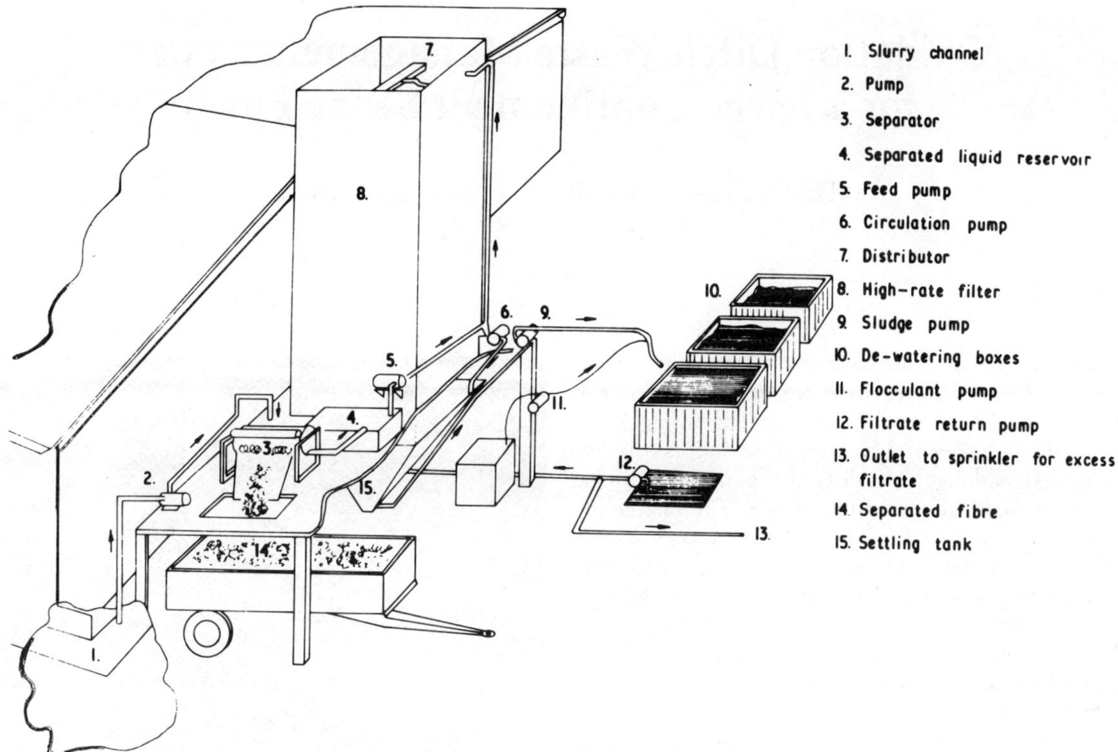


FIG. 1. Diagram of separation, treatment and sludge dewatering process.

2. The process is likely to perform more efficiently if integrated into a piggery so that the feces and urine are processed as soon as possible after leaving the animal.

3. If integrated into a mechanically ventilated piggery, further reductions in the quantity of filtrate produced are possible, particularly if part or all of the warm exhaust air is passed through the filter medium. Freezing risks will also be reduced or eliminated.

4. Exhausting air through the filter tower may also prove to be a fairly efficient means of scrubbing dust and smell from the exhaust air.

5. Since the major part of the output of the treatment system is smell-free solids, storage and handling problems are greatly eased, as are the problems of storage and disposal on land belonging to other farmers.

6. The cost of flocculant was the major component of the running cost. This might be reduced by the modification suggested under (2) above, by the use of cheaper flocculants, by the use of a different treatment stage (e.g. an activated sludge plant) or by a combination of all these factors.

7. In addition to its potential as an air scrubber, a heavily-loaded high-rate filter has other advantages, in that it appears relatively insensitive to changes in loading rate, is very reliable, is easy to operate and install and takes up relatively little space.

8. Much more work is required to establish a plant nutrient flow sheet from the pigs to the land, but the initial results suggest that about 25 percent of the nitrogen is lost, presumably as ammonia from the top of the filter tower and elsewhere.

### References


1    Hepherd, R. Q. 1975. Experiments on slurry handling, treatment and land application at the N.I.A.E., 1968-1974. Report No. 15. National Institute of Agricultural Engineering, Silsoe, Bedford, England.

2    O'Callaghan, J. R., V. A. Dodd and K. A. Pollock. 1973. Long term management of animal manures. Journal of Agricultural Engineering Research, Silsoe, Bedford, England. 18(1):1-12.

3    Osborne, L. E. and R. W. Sneath. 1974a. An integrated separation, aerobic treatment and sludge dewatering system for pig slurry. Dept. Note. No. DN/FB/481/3000. National Institute of Agricultural Engineering, Silsoe, Bedford, England. (Unpublished)

4    Osborne, L. E. and R. W. Sneath. 1974b. The effect of temperature on the performance of high-rate filter towers treating pig slurry. Dept. Note No. DN/FB/499/3000. National Institute of Agricultural Engineering, Silsoe, Bedford, England. (Unpublished)

5    Richardson, S. J. 1974. Animal manures as potential pollutants. Proceedings A.D.A.S./A.R.C. Conference on Agriculture and Water Quality. Technical Bulletin No. 32, Ministry of Agriculture, Fisheries and Food, London. (In press)

# Oxidation Ditch Waste Management System for a Large Confinement Swine Farm

Paul Smart, Floyd McCain, D. L. Day, B. G. Harmon
MEMBER
ASAE

THIS is a case study of the complete waste management system of the Paul Smart confinement swine farm near Lawrence, Kansas. This intensive confinement enterprise is located on only 50 acres of which less than half is suitable for building sites and lagoons. All feed is purchased from a nearby commercial feed processing center. The farrow-to-finish enterprise was started in 1965 and has grown to the present size of 150 farrowing stalls producing 12 000 to 14 000 market hogs per year. An average of 18 pigs per sow per year are marketed from 2.2 litters per sow per year, and the death loss is less than 10 percent of pigs farrowed. Five full-time workers are involved in the husbandry-management of the total enterprise.

This farm demonstrated the first large-scale use of swine oxidation ditches in the U.S.A. Numerous publications have described the waste management system (see references included at the end of this paper).

Experience in the design of total hog confinement systems with complete waste management facilities was lacking in 1965 when the operation began so the owner and his manager travelled extensively to study the best technology available. Many new ideas were generated from trips not only throughout the United States and Canada, but also to several countries in Europe. One thing was apparent, no one had fully solved the problem of manure management.

The site selected for the Paul Smart Hog Farm near Lawrence, Kansas did not include farm land and was near a major highway for easy access to market. Odor control was an important feature of the waste management system to be selected.

Oxidation ditches beneath slotted floors were selected and have been used from the beginning. Surplus waste materials overflow from the oxidation ditches into evaporation lagoons. No other waste management practices such as scraping, scooping, hauling, etc are used. A photograph of the overall farm is shown in Fig. 1 and information about the 11 buildings and 37 rotors is presented in Table 1. Nine evaporation lagoons with a total of 10 to 12 surface acres are used. No hauling or irrigating of lagoon contents has been necessary. Normal annual rainfall is 36 inches, and evaporation is 44 inches.

Several rotor designs and other aeration devices have been tried and rejected. The present design, however, is working satisfactorily, giving good performance and few maintenance problems. The rotors, developed in cooperation with Ross E. McKinney of the University of Kansas Civil Engineering Department, are 32 inches wide by 60 inches in diameter and are powered by 5-hp motors (Fig. 2). This size rotor will pass through ordinary doors in the

The authors are: PAUL SMART, Farm Owner, and FLOYD McCAIN, Farm Manager, Lawrence, Kans.; D. L. DAY, Agricultural Engineering Dept., and B. G. HARMON, Animal Science Dept., University of Illinois, Urbana.

FIG. 1 Aerial view of swine farm. (Photo courtesy of Farm Journal)

TABLE 1. BUILDING SIZES AND CAPACITIES.

| Building | Size, ft | Animal capacity | Number of rotors |
|---|---|---|---|
| Gestation —1 | 32 x 166 | 250 sows | 2 |
| —2 | 36 x 160 | 250 sows | 2 |
| —3 | 36 x 240 | 410 sows | 2 |
| Farrowing—1 | 22 x 140 | 50 sows and 450 piglets | 1 |
| —2 | 22 x 140 | 50 sows and 450 piglets | 1 |
| —3 | 26 x 148 | 50 sows and 450 piglets | 1 |
| Nursery —1 | 32 x 140 | 1100 pigs | 2 |
| —2 | 32 x 140 | 1100 pigs | 2 |
| Growing —1 | 36 x 200 | 800 pigs | 4 |
| Finishing —1 | 36 x 400 | 1500 hogs | 8 |
| —2 | 36 x 500 | 1800 hogs | 12 |

confinement buildings. Bearings of the large diameter rotors are located on the walls of the pit away from the slurry. This along with a 2-7/16 inch diameter shaft practically eliminates bearing problems and allows easy access for occasional maintenance. The bearings are greased only once each month compared to the daily greasing required by many rotors. These large wheels rotate at 100 rpm and typically run for a year or more without bearing replacements. In tests at the University of Kansas 4 pounds of oxygen per horsepower-hour was delivered in clean water at standard conditions.

Experience has shown that the rotors should be located in the ditch straight-of-way instead of at the ends but due to building interior arrangements, this was not possible in all cases. Locating two rotors near each other where possible in adjacent ditches facilitates central locating of electrical supplies and controls.



"

FIG. 2 A rotor inside a finishing building (cover removed). (Photo courtesy of Farm Journal)

Liquid depth in the oxidation ditches is maintained at 12 to 13 inches by using overflow weirs. No extra water is added during normal operation except as wastage from cup-type drinking waterers. Feeders are carefully adjusted to avoid wastage into the ditches, but afterbirths fall directly into the ditches in the farrowing buildings.

A method of refeeding aerobically processed wastes is being tested in a few pens in a finishing building. Oxidation ditch mixed liquor (ODML), lifted by the rotor, flows by gravity down a trough and back into the oxidation ditch. This provides a continuous supply of fresh ODML to the hogs. No other water is provided for these hogs. Results of this study on the benefits of the enriched drinking water are not yet conclusive.

Kansas swine waste management license number H-001 has been assigned to the farm. Several pollution control measurements and observations are regularly required to operate under this license.

The total cost of electricity including that for rotors, feeders, and lights has been about $1.00 per hog marketed at 1 cent per KWH. The farm management is pleased with this low-odor, low-labor method of swine waste management made possible by the use of oxidation ditches and evaporation lagoons.

### References


1    Bella, R. A. 1968. Confined swine manure disposal. M.S. Thesis, Univ. of Kan.

2    Coffman, B. 1974. Manure refeeding cuts odor, solves disposal problems for this hogman. Farm Journal/Hog Extra, Nov. p. H-6.

3    Harvey, J. (Ed.). 1966. An odorless hog house. Successful Farming, June. p. 23.

4    McKinney, R. E. and R. Bella. 1967. Water quality changes in confined hog waste treatment. Project Completion Report. Kansas Water Resources Res. Inst., Univ. of Kan., 88 p.

5    Montgomery, G. A. 1966. Odor-free manure: charge with oxygen and paddle. Hog Farm Management, July. p. 8-9.

6    Smart, P. and R. E. McKinney. 1970. Odorless pork production: from conception to market. Proc., 25th Industrial Waste Conference, Purdue University, Lafayette, Ind. May 5-6. p. 757-760.

7    Trnovsky, Martin. 1974. Characteristics and soil treatment of biologically treated swine wastes. Ph.D. Thesis, Univ. of Kan.

8    Wilmore, R. 1971. An oxidation wheel that works. Farm Journal, Sept. p. 14-15.

# Recovery of Nutrients from Animal Wastes—An Overview of Existing Options and Potentials for Use in Feed


Robert G. Yeck, L. W. Smith, C. C. Calvert
FELLOW
ASAE


THE recovery of feed nutrients from animal wastes can alleviate the waste disposal problem, offset the costs of air and water pollution control for livestock production systems, lower costs to produce animal products (and ultimately the cost to the consumer), increase supplies of available proteins, and conserve natural resources. The current economics of livestock production provides an immediate incentive for the adoption of such processes by some livestock producers. The concept is also ecologically timely. Traditionally, land application of animal wastes results in recovery of nutrients in crops. This presentation reports processes by which nutrients for animal feeding can be recovered more directly and immediately.

The scavaging of nutrients from animal excreta is natural among birds and wild animals and frequently among domestic animals, too. Placing hogs in pens with steers to recover nutrients from beef wastes was once standard practice. Beginning in the early 1940's, researchers began reporting evaluations of nutrients in manure. Smith et al. (1971) and Smith (1973) recently reviewed the nutritive value of animal excreta processed for livestock feed. Whetstone et al. (1974) compiled a comprehensive annotated bibliographic review that provided an insight for alternative processes for utilizing animal wastes, including those for the recovery of energy. Of all concepts for utilization of animal wastes, the recovery of nutrients for feeding to animals appears to be the highest value use for which a practicable technology exists. The nutrient recovery process should result in a net energy saving insofar as it reduces the need for feed grains that require energy for fertilizer, tillage, and transport.

## FEEDING POTENTIAL

The feeding potential of processed animal wastes depends on: Source; conversion process; target species; and the function that is to be served by the ration in which incorporated (i.e., maintenance, reproduction or meat, milk, and egg production).

The crude protein potential of animal wastes can be estimated from the total nitrogen (N) excreted from animals in the United States for 1972 (Tables 1 and 2). In Table 1, excreta N equals the feed N minus the N in body tissue and products. These calculations circumvent the problem of conflicting published data on animal waste composition. Table 1 also shows the total collectable N in the United States to be 2,265,822,000 kg per year. This is equivalent to the total N in the 1972 United States soybean crop.

Table 2 shows another approach for estimating nitrogen in excreta. This approach is based on the amount of protein in the feed as compared with the protein in the products that are derived therefrom. The remainder is assumed to be in the excreta. Nitrogen equivalency was determined by dividing the protein values by 6.25. These conversions do not imply a low nitrogen conversion efficiency of livestock because excreta contains byproducts of digestion, sloughed tissue, and microbial cells—all of which are high in N. The close agreement of the two tables adds credence to the method of calculation that was used for each. The quantity of total excreta N of the tables ($5 \times 10^9$ kg/yr) is about half that previously estimated by Taiganides and Stroshine (1971). They reported excreta N values for beef that are considerably higher than shown in Tables 1 and 2. These tables do not include bedding which may also be a factor. Excreta from poultry, swine, and finishing cattle usually have the highest digestible nutrient content and are, therefore, the most promising resource materials for nutrient recovery. Ruminants are considered the most effective utilizer of such processed excreta due to the microbial population in the rumen that permits efficient utilization of nonprotein nitrogen and fiber (Moore et al. 1967).

## CONSTRAINTS ON USE

As with any feedstuff, good process control and proper handling of the product after processing are needed to assure a high-quality product. There are, however, potential adverse effects from feed additives and inadvertent contaminants that must be recognized (Smith et al. 1971, Calvert 1973, and Fontenot and Webb 1975).

Practically no problems should exist with respect to microorganisms, feed additives (growth promotants and medicants), residues or other chemicals when the user of animal waste-derived feeds has full knowledge of the origin of the waste and wisely uses such knowledge in his decisions as to the circumstances under which it will be fed. When such knowledge is not available, processing technologies based on established guidelines or monitoring criteria will be needed to assure a safe final product. Actually, additives are not routinely fed to dairy cattle and caged layers—two major sources of animal wastes. Researchers and innovative farmers have fed many forms of animal waste to different species of farm animals without adverse effects on the animal or edible animal products—indicating that safety and product quality constraints can be met.

## AVAILABLE PROCESSES

Several processes are now available and have been successfully used for treatment of animal wastes for feed.


The authors are: ROBERT G. YECK, Staff Scientist, National Program Staff, L. W. SMITH and C. C. CALVERT, Research Animal Scientists, Biological Waste Management Lab., Agricultural Environmental Quality Institute, ARS, USDA, Beltsville, MD.

**TABLE 1. ESTIMATED ANNUAL NITROGEN (N) EXCRETION BY US LIVESTOCK — BASED ON NUMBER OF FARM ANIMALS AND THEORETICAL DIFFERENCES IN N BETWEEN FEED CONSUMED AND THAT REPRESENTED BY THE PRODUCT.**

| Class of livestock | Feed N,* | Product N,† | Excreta N,‡ | Head § | Total excreta N, | Collectable N∥ | |
|---|---|---|---|---|---|---|---|
| | kg/head | kg/head | kg/head | $10^3$ | kg $10^3$ | Percent | kg $10^3$ |
| Beef cattle | | | | | | | |
| Beef cows | 34.4 | 6.5 | 27.9 | 41,102 | 1,146,746 | 0 | |
| Cow replacements | 47.9 | 7.2 | 40.7 | 7,470 | 304,029 | 0 | |
| Other  >227 kg | 27.1 | 3.4 | 23.7 | 25,550 | 605,535 | 90 | 544,982 |
| <227 kg | 6.8 | 1.8 | 5.0 | 32,342 | 161,710 | 30 | 48,513 |
| Dairy cattle | | | | | | | |
| Dairy cows | 128.0 | 28.0 | 100.0 | 11,651 | 1,165,100 | 50 | 582,550 |
| Replacement heifers | 39.4 | 7.2 | 32.2 | 3,875 | 124,775 | 20 | 24,955 |
| Swine | | | | | | | |
| Breeding | 44.50 | 0.26 | 44.24 | 9,094 | 402,318 | 10 | 40,232 |
| Market hogs | 8.68 | 2.62 | 6.06 | 104,126 | 631,004 | 80 | 504,803 |
| Sheep | | | | | | | |
| Stock sheep | 10.51 | 0.58 | 9.93 | 14,853 | 147,490 | 0 | |
| Sheep and lambs on feed | 1.21 | 0.24 | 0.97 | 2,873 | 2,787 | 90 | 2,508 |
| Poultry | | | | | | | |
| Laying hens and pullets | 0.964 | 0.217 | 0.747 | 302,120 | 225,684 | 100 | 225,684 |
| Pullets and other chickens | 0.176 | 0.047 | 0.129 | 104,408 | 13,469 | 100 | 13,469 |
| Broilers | 0.130 | 0.058 | 0.072 | 3,074,698 | 221,378 | 100 | 221,378 |
| Turkey (growing) | 1.233 | 0.400 | 0.833 | 128,808 | 107,297 | 50 | 53,649 |
| Turkey (breeder) | 1.964 | 0.169 | 1.795 | 3,453 | 6,198 | 50 | 3,099 |
| GRAND TOTAL | | | | 5,265,520 | | | 2,265,822 |

*Using recommended levels of protein intake. Source: National Research Council (1968, 1970, 1970a, 1970b, 1973).

†Authors' estimates of N accumulated in body tissue and that removed as product during feeding period (fetus growth is included).

‡Excreta N = N in feed minus N in product.

§Number of animals on farm from Tables 439, 443, 460, 478, 574, 575, 583, 593, and 595, US Department of Agriculture, Agricultural Statistics (1973). Cattle and sheep are 1972 estimates, others are 1973. Broilers are total number produced and market hogs are the sum of June 1 and December 1 inventories.

∥Collectability was the judgment by the authors and was based on the practicality of collecting and processing as feed at the farmstead or at a centrally located processing station.

Some of the processes are now in use, and others are in only exploratory stages. Practically all need further improvement. Optimization of processes with relation to nutritional quality and operating efficiencies are particularly important. Of all processes, the following are considered practicable for immediate use:

1   Dehydrating caged layer and other poultry wastes at drying temperatures in excess of 60 C. for use as feed for ruminants and poultry.

2   Ensiling beef cattle wastes and poultry litter for use as feed for ruminants.

3   Supplying litter from broiler houses for use as feed for ruminants.

To this list can be added the following processes that have proved potentially successful but have not been as extensively tested:

1   Physical separation of solids from liquids and further treatment from beef and dairy cattle wastes for use as feed for the same or other species.

2   Recovery of swine oxidation ditch liquor for use as

**TABLE 2. ESTIMATED EXCRETA N FROM US LIVESTOCK IN 1972 — BASED UPON LIVESTOCK PRODUCTS MARKETED, THEIR PROTEIN CONTENT, AND FEED-TO-PRODUCT PROTEIN CONVERSION EFFICIENCY.**

| Product | Amount,* lbs $10^3$ | Product protein, percent | Conversion† efficiency, percent | Excreta protein factor‡ | Excreta N § kg $10^3$ |
|---|---|---|---|---|---|
| Milk | 120,278,000 | 3.5 | 30 | 2.3 | 704,173 |
| Chicken | 1,118,723 | 21.0 | 25 | 3.0 | 51,259 |
| Broiler | 11,477,931 | 21.0 | 25 | 3.0 | 525,898 |
| Turkey | 2,424,145 | 21.0 | 20 | 4.0 | 148,093 |
| Eggs∥ | 8,725,500 | 10.0 | 20 | 4.0 | 253,833 |
| Sheep and lamb | 1,081,254 | 16.0 | 10 | 9.0 | 113,237 |
| Hogs | 20,258,557 | 15.0 | 20 | 4.0 | 884,010 |
| Cattle | 37,185,295 | 15.0 | 15 | 5.6 | 2,271,683 |
| Calves | 767,496 | 15.0 | 15 | 5.6 | 46,887 |
| Total | | | | | 4,999,073 |

*Production and slaughter estimates from US Department of Agriculture, Agricultural Statistics, 1973, Tables 536, 579, 582, 595, 602, 485, 467, and 453.

†Source: Phillips, R. W., 1967.

$$\ddagger \text{Excreta protein factor} = \frac{\text{kg of excreta protein}}{\text{kg of product protein}} = \left(\frac{100}{\text{conversion efficiency \%}}\right) - 1$$

§ Excreta N(kg) = lb of product x percent product protein x excreta protein factor x 0.453 kg/lb x 0.16 kg N/kg protein.

∥Eight eggs, assumed to weigh 1 lb.

feed for swine.

3   Dehydration (in excess of 60 C.) of cattle wastes for use as feed for ruminants.

4   Composting cattle wastes for use as feed for ruminants.

Physical separation can be either an adjunct to some other process or a process in itself. Whole kernel grain recovery from an oxidation ditch is an example of a process in itself (Hegg et al. 1974). Several other physical separation systems have been developed in the private research sector (Ward et al. 1974, Gross 1975).

Waste dehydration technologies that have been used for producing garden fertilizers have been successfully applied to processing dairy cattle excreta for incorporation into ruminant rations (Smith et al. 1974). Although the nutrient content of dairy excreta is relatively low, the large volume that is produced makes it worthy for study. The digestibility of such wastes has been improved by chemical treatment with sodium hydroxide (Smith et al. 1969, 1970). The swine oxidation ditch nutrient recovery system (Harmon et al. 1973) is of particular interest because it represents about the only system that has been tested to date for feeding swine excreta back to swine. Of the foregoing list, composting cattle waste for use as feed for ruminants (Westing and Brandenberg 1974) has been the least tested.

Several more complex conversion systems were reviewed by Calvert (1974). They involve conversion through action of lower order life forms such as earthworms, fly larvae, bacteria, and algae. They would be of particular value where a more distinct break in the nutrient recovery is desired or where nutrient concentration is sought. The nutrient concentration concept could be substantially advanced by recent research on the use of enzymes from *Trichoderma viride* (Griffin et al. 1974). Beef feedlot manure was used to grow *T. viride* for cellulase production; this process may have application in hydrolysis of hemicellulose and cellulose. Nolan and Shull (1973) successfully designed an aerobic thermophylic nutrient reclamation facility in which microorganisms of the order *Actinomycetales* were used. Various environmental factors interfered with performance, indicating a need for further development. In "single cell" conversion processes, we may need to differentiate between naturally occurring wild cultures and those that are selected, upgraded and optimally managed to perform a specific function. The feeding value of residues from anaerobic methane generation systems remains to be proved.

## SELECTING A SYSTEM

A system is composed of resource material, process, and intended use. Guidelines to assist the producer in choosing a system are generally lacking. Careful study of the management practices and intended uses are essential in the selection of the system. Such judgments are the responsibility of the producer. Protection of animal health and quality of animal products without sacrifice to animal productivity must be placed ahead of investment and operating costs of a system. Options will also be limited by applicable government regulations.

The characteristics of the resource material, excreta, are the prime determinents in the choice of a process for a system. If excreta are derived from healthy animals and toxic or residue creating chemicals are not involved in managing the animals or their wastes, the least costly proven processing system would be the choice. If the history of wastes is

### TABLE 3. DIGESTIBILITIES OF ANIMAL EXCRETA BY SHEEP.

| Source | Digestibility | | Authors |
| --- | --- | --- | --- |
| | Dry matter, percent | N, percent | |
| Caged layer | 57 | 77 | Lowman and Knight (1970) |
| Poultry litter | 63 | 68 | Fontenot et al. (1966) |
| Beef feedlot | 42 | 66 | Johnson (1972) |
| Dairy cattle (50 percent forage fed) | 35 | 49 | Smith et al. (1974) |
| Swine | - | 57 | Hennig et al. (1972/73) |

less known, a more complex process will be required. Careful monitoring of processed excreta for undesirable constituents could reduce the need for a more complex process but a process that is adequate to assure safety without monitoring is preferred.

Once safety is assured, optimization of nutrient utilization of excreta is needed. Traditional methods of managing wastes can result in large N losses by volatilization prior to processing. Processing as soon as possible after defecation is one way to reduce N losses. The process, of course, may also alter the nutritive value of the product. The selection of recovery processes will depend on the degree to which a producer desires to conserve or concentrate nutrients. Economic analyses of the costs and benefits of full-scale recovery systems are very limited. Forsht et al. (1974) determined that processes that dried (above 93 C.) layer wastes would be economically feasible for the larger caged flocks (more than 10,000 layers).

The third part of a system, intended use, requires knowledge of source and chemical composition of processed excreta and formulation into rations for maximum utilization. Approximations of crude protein content of various wastes can be made by multiplying the N values of Table 1 by 6.25 but it would be better to obtain an actual analysis. As with any feed, the digestibility of the protein should also be considered. Data in Table 3 show the digestibility by sheep of dry matter and the N in different livestock excreta. All sources of excreta N appear digestible by sheep. These N digestibilities are in the same range of values as those observed for some conventional feeds. Dry matter is less digestible in excreta that comes from animals that are fed the larger percentages of forage. The digestibilities of dry matter for these sources of excreta would be expected to be lower if fed to nonruminants. Although various compositional analyses of processed excreta reveal a variety of other nutritional potentials such as energy and phosphorous, published data on the effectiveness with which the animals utilize them are very limited.

## SUMMARY

Processes for the recovery of nutrients from animal wastes can contribute to reduction of waste disposal problems, reduction of livestock production and consumer product costs, increased feed supplies and conservation of natural resources. Several systems have successfully fulfilled the intended use and met constraints. A producer, however, must exercise good judgment in his choice and management of recovery systems for his enterprise.

### References

1   Calvert, C. C. 1973. Feed additive residue in animal manure processed for feed. Feedstuffs 45(27):32.

2   Calvert, C. C. 1974. Animal wastes as substrates for protein

*(Continued on page 196)*



MANAGING LIVESTOCK WASTES

# Ensiling Poultry Floor Litter and Cage Layer Manure

Stanley A. Vezey, Charles N. Dobbins, Jr.

## SUMMARY

RESULTS of this research indicate that 30 percent cage layer manure can be incorporated into ensilage consisting of 50 percent ground corn, and 20 percent poultry floor litter. 7.5 lb molasses and *Lactobacillus acidophilus* culture were added to accelerate fermentation. Optimum moisture content should be 40-50 percent, adding water, quantity sufficient, to attain this level. *L. acidophilus* fermentation or ensiling provides a method other than dehydrating for processing cage layer manure for utilization as a ruminant feed ingredient.

## INTRODUCTION

There have been numerous published reports on ensiling poultry floor litter as a method of utilizing it as a ruminant feed. Dehydrated cage layer manure and poultry floor litter have also been used as an ingredient of ruminant and poultry feeds. There have been no published reports on ensiling cage layer manure.

This research was done to develop an economical method of utilizing cage layer manure as a ruminant feed ingredient. The physical characteristics of cage layer manure are such that the addition of other matherials are essential for mechanical management of large quantities of the finished product.

E    70.1 percent ground corn, 28.5 percent floor litter, and 20 lb water†.

F    50 lb floor litter, 3.5 lb dried molasses and 2 lb *L. acidophilus* culture.

The above materials were mixed, placed in 30 gallon plastic containers, and stored in an unheated room. The mean temperature was 60 F. Temperatures for optimum growth of *L. acidophilus* should be above 90 F. Optimum temperatures were obtained by placing the containers in a heated room. *L. acidophilus* fermentation was accelerated and minimum pH was attained in less than 7 days. Table 1 lists pH of each mixture.

### TABLE 1. pH OF CORN, LITTER AND CAGE LAYER MANURE AFTER FERMENTATION

| Mixture | pH |
|---|---|
| A | 4.7 |
| B | 4.8 |
| C | 4.6 |
| D | 5.0 |
| E | 6.2 |
| F | 6.1 |

### TABLE 2. CHEMICAL ANALYSIS (Dry Weight)

| Mixture | Percentage | | | | | | Parts per million | | | | | |
|---|---|---|---|---|---|---|---|---|---|---|---|---|
| Mixture | N | P | K | Ca | Mg | Mn | Fe | B | Cu | Zn | Al | Mo |
| A | 2.49 | 1.04 | 1.75 | 1.93 | 0.28 | 165 | 675 | 20 | 166 | 820 | 172 | 3.5 |
| B | 2.41 | 0.99 | 1.76 | 1.80 | 0.20 | 155 | 675 | 20 | 160 | 167 | 776 | 3.5 |
| C | 2.47 | 1.10 | 1.86 | 1.96 | 0.32 | 177 | 723 | 19 | 162 | 166 | 917 | 3.3 |
| D | 2.48 | 0.96 | 1.81 | 1.93 | 0.27 | 155 | 607 | 20 | 146 | 170 | 802 | 3.5 |
| E | 2.63 | 1.44 | 1.82 | 2.80 | 0.37 | 1.78 | 649 | 21 | 180 | 226 | 992 | 4.9 |
| F | 3.43 | 1.48 | 0.283 | 1.76 | 0.29 | 306 | 1124 | 690 | 216 | 216 | 2770 | 6;3 |

## MATERIALS AND METHODS

### Mixtures

Ground corn, poultry floor litter, cage layer manure, dried molasses, water, and *Lactobacillus acidophilus* culture were mixed in the ratios listed below:

A    50 percent ground corn, 20 percent floor litter, 30 percent cage layer manure plus 7.5 lb dried molasses*, 20 lb water†, and 1 lb *Lactobacillus acidophilus*‡ were added to the above.

B    Same as A except 2 lb *L. acidophilus* culture.

C    Same as A except 4 lb *L. acidophilus* culture.

D    Same as A except 0.0 lb *L. acidophilus* culture.

Chemical analyses were conducted on all mixtures and are recorded in Table 2. Table 3 lists the percentage of crude fiber, protein, fat, and moisture. An amino acid profile was conducted on a pooled sample of mixtures A, B and C. The results are in Table 4.

### TABLE 3. NUTRIENT VALUES

| Mixture | C.F.* | Protein percent† | Fat percent | Moisture percent |
|---|---|---|---|---|
| A | 11.0 | 15.56 | 1.40 | 44.10 |
| B | 10.5 | 15.06 | 1.40 | 44.26 |
| C | 10.02 | 15.43 | 1.32 | 45.73 |
| D | 10.0 | 15.50 | 1.14 | 49.76 |
| E | 10.0 | 15.50 | 1.49 | 50.25 |
| F | 25.5 | 21.43 | 1.36 | 47.05 |

*Crude fiber
†6.25 x percent N

---

The authors are STANLEY A. VEZEY and CHARLES N. DOBBINS, JR., Extension Veterinary Department, College of Agriculture, University of Georgia, Athens, Georgia.

*42-45 percent cane sugar absorbed on soybean mill feed.
†Water added to increase moisture content to 40-45 percent.
‡*L. acidophilus* cultured on reconstituted dried non-fat milk.

### TABLE 4. AMINO ACID COMPOSITION*

| gm/100 gm of crude material | |
| --- | --- |
| Aspartic acid | 0.806 |
| Threonine | 0.369 |
| Serine | 0.443 |
| Glutamic acid | 1.558 |
| Proline | 0.750 |
| Glycine | 0.662 |
| Alanine | 1.139 |
| Cysteine | D† |
| Valine | 0.519 |
| Methionine | 0.192 |
| Isoleucine | 0.411 |
| Leucine | 0.917 |
| Tyrosine | 0.352 |
| Phenylalanine | 0.577 |
| Histidine | 0.133 |
| Lysine | 0.40 |
| Ammonia | 0.542 |
| Arginine | 0.009 |
| Unknown | 0.141 |
| TOTAL | 9.560 |

*Clyde T. Young, Ph.D., Dept. of Food Science, Experiment, Georgia
†D = Detectable, but too small to measure

### Bacteriological Results

All samples were cultured by methods described in National Academy of Science publication 1038, 1963 to determine the presence or absence of pathogenic bacteria. Mixtures A, B, C and D were negative for aerobic or anaerobic pathogenic bacteria. Numerous *Streptococci, Staphylococci, Pseudomonas* and *Proteus, sp.* were isolated from mixtures E and F. No *Salmonella sp.* were isolated from any of the mixtures.

Mold growth was prolific on surfaces exposed to the air on all mixtures. No effort was made to identify the different species of mold.

### Palatability

Controlled feeding studies were not conducted, but uncontrolled tests to determine palatability were done. Holstien steers were fed all mixtures *ad libitum*. Regardless of the respective pH, all mixtures appeared to be equally palatable.

### DISCUSSIONS AND CONCLUSIONS

1  It is difficult to obtain optimum *L. acidophilus* growth in small containers unless the surrounding air temperature is heated to a minimum 70 F. This should not be a serious problem in large air-tight silos.

2  Ensiling time can be reduced from 6-8 weeks to less than 3 weeks by adding molasses and *L. acidophilus* culture.

3  Ensiling provides a procedure other than dehydrating for the utilization of cage layer manure as a ruminant feed.

4  Ensiling is economical and uses less fuel than dehydration.

5  Utilizing poultry waste as an animal feed is an efficient and environmentally acceptable method of disposal.

---

## Recovery of Nutrients from Animal Wastes

(Continued from page 194)


production. Fed. Proc. 33(8):1938-1939.

3  Fontenot, J. P., A. N. Bhattacharya, C. L. Drake and W. H. McClure. 1966. Value of broiler litter for ruminants. Proc. Nat. Symp. Anim. Waste Mgmt. ASAE Pub. SP-0366:105-108. St. Joseph, Mich. 49085.

4  Fontenot, J. P. and K. E. Webb, Jr. 1975. Health Aspects of Recycling Animal Wastes by Feeding. J. Anim.Sci. 40: 1267-1277.

5  Forsht, R. Gar, C. R. Burbee and W. M. Crosswhite. 1974. Recycling poultry waste as feed: Will it pay? U.S. Department of Agriculture, Agr. Econ. Rept. No. 254. 51 pp.

6  Gross, C. 1975. Closed system-waste recovery: insulated. Calf News 13(2):36-37.

7  Griffin, H. L., J. H. Sloneker and G. E. Inglett. 1974. Cellulose production by *Trichoderma viride* on feedlot waste. Appl. Microbiol. 27(6):1061-1066.

8  Harmon, B. G., D. L. Day, D. H. Baker and A. H. Jensen. 1973. Nutritive value of aerobically or anaerobically processed swine waste. J. Anim. Sci. 37(2):510-513.

9  Hegg, R. O., J. C. Meiske, R. E. Larson and J. A. Moore. 1974. Beef oxidation ditch settled solids fed to steers. Cornell Agr. Waste Mgmt. Conf. Proc. pp. 382-386.

10  Hennig, A., D. Schüber, H. H. Freytag, C. Voigt, K. Gruhn and H. Jeroch. 1972-1973. Erste Untersuchungen über den Einsatz von Schweinekot in der Rindermast Vorläufige Mitteilung. Jahrbuch für Tierrernahrung und Futterung 8:226-234.

11  Johnson, R. R. 1972. Digestibility of feedlot waste. Okla. Agr. Expt. Sta., Misc. Pub. 87:62-65.

12  Lowman, B. G. and D. W. Knight. 1970. A note on the apparent digestibility of energy and protein in dried poultry excreta. Anim. Prod. 12:525-552.

13  Moore, L. A., P. A. Putnam and N. D. Bayley. 1967. Ruminant livestock: Their role in the world protein deficit. Agr. Sci. Rev. 5:1-8.

14  National Research Council. 1968. Nutrient requirements of sheep. 4th Ed. National Academy of Sciences, Washington, D. C.

15  National Research Council. 1970. Nutrient requirements of beef cattle. 4th Ed. National Academy of Sciences, Washington, D. C.

16  National Research Council. 1971a. Nutrient requirements of dairy cattle. 4th Ed. National Academy of Sciences, Washington, D. C.

17  National Research Council. 1971b. Nutrient requirements of poultry. 6th Ed. National Academy of Sciences, Washington, D. C.

18  National Research Council. 1973. Nutrient requirements of swine. 7th Ed. National Academy of Sciences, Washington, D. C.

19  Nolan, E. J. and J. J. Shull. 1973. Engineering experiences during the demonstration phase of a nutrient reclamation plant. Pres. 75th Nat. Mtg. Am. Inst. Chem. Eng., Detroit, Mich.

20  Phillips, R. W. 1967. Efficiency of various classes of animals as feed converters. In FAO, Inter-American Conf. Anim. Prod. and Health, Gainesville, Fla. p. 36.

21  Smith, L. W. 1973. Recycling animal wastes as protein sources, In Alternative Sources of Protein for Animal Production. National Academy of Sciences, Washington, D. C. pp. 146-173.

22  Smith, L. W., C. C. Calvert, L. T. Frobish, D. A. Dinius and R. W. Miller. 1971. Animal Waste Reuse—Nutritive Value and Potential Problems from Feed Additives: A Review. U.S. Department of Agriculture, ARS 44-224. 56 pp.

23  Smith, L. W., C. C. Calvert and I. L. Lindahl. 1974. Dried manure from lactating dairy cattle as feed for ewes. J. Anim. Sci. 39:1001.

24  Smith, L. W., H. K. Goering and C. H. Gordon. 1969. Influence of chemical treatments upon digestibility of ruminant feces. Cornell Agr. Waste Mgmt. Conf. Proc. pp. 88-97. Cornell University, Ithaca, N.Y.

25  Smith, L. W., H. K. Goering and C. H. Gordon. 1970. In vitro digestibility of chemically-treated feces. J. Anim. Sci. 31:1205-1209.

26  Taiganides, E. P. and R. L. Stroshine. 1971. Impact of Farm Animal Production and Processing on the Total Environment. Proc. Int. Symp. Livestock Wastes. pp. 95-98. St. Joseph, Mich. 49085.

27  U.S. Department of Agriculture. 1973. Agricultural Statistics. Superintendent of Documents, U.S. Government Printing Office, Washington, D. C.

28  Ward, G. M., D. E. Johnson and R. D. Boyd. 1974. Digestibility of steers of cereco sludge produced from feedlot manure. J. Anim. Sci. 39:140.

29  Westing, T. W. and B. Brandenberg. 1974. Beef feedlot waste in rations for beef cattle. Cornell Agr. Waste Mgmt. Conf. Proc. pp. 336-341. Cornell University, Ithaca, N.Y.

30  Whetsone, G. A., H. W. Parker and D. M. Wells. 1974. Study of current and proposed practices in animal waste management. EPA 430/9-74-003. Superintendent of Documents, U.S. Government Printing Office, Washington, D. C.




MANAGING LIVESTOCK WASTES

# Recycling Solids from an Aerated Beef Slurry for Feed


R. O. Hegg, R. E. Larson, J. A. Moore, R. D. Goodrich, J. C. Meiske

ASSOC. MEMBER    MEMBER    MEMBER
ASAE          ASAE       ASAE


RECLAIMED solids from a beef waste oxidation ditch, were fed to finishing steers at three rates (5, 15 and 25 percent of a high energy ration on a dry matter basis (DMB)) to evaluate its value as a ration component.

## DISCUSSION

The source of recycled solids for feed was an oxidation ditch at the Agricultural Engineering Farmstead on the University of Minnesota Agricultural Experiment Station, Rosemount, Minnesota. Thirty-six steers, housed over the oxidation ditch produced the waste that was treated in the ditch. These steers were fed a standard high energy ration of 90 percent corn, 7.5 percent hay (dehydrated alfalfa), and 2.5 percent supplement, mineral, and vitamins (Table 1). A stationary, sloping screen (22 mesh) was used to separate solids from the oxidation ditch slurry. Scrapers at the bottom of the ditch were used to move settled solids to a sump where they could be pumped to the separating screen. The oxidation ditch rotor maintained aerobic conditions in the slurry during the test period. The dewatered solids had a slight, but not unpleasant, odor. The screen did not remove sufficient moisture to produce a material with the desired 20 percent dry matter level, therefore, the solids were pressed to remove additional moisture.

A four month feeding trial was initiated to evaluate these reclaimed solids as a part of a finishing ratin. A group of 20 finishing steers averaging 295 kg (650 lb) each were divided into 4 open lot pens of 5 steers each. Pen one (the control) was fed the high energy ration containing 90 percent corn, 7.5 percent hay, and 2.5 percent supplement (Table 1). The remaining three pens received experimental rations containing 3 levels of reclaimed solids. The experimental rations were developed by replacing a portion of the corn in the control ration with reclaimed solids at levels of 5, 15 and 25 percent DMB.

The solids collected from the oxidation ditch were mixed with feed at the prescribed rates and fed to steers in the respective pens on the same day that wastes were collected from the ditch. The animals were fed once daily as much as they would consume. To accustom steers to the experimental rations, diminishing quantities of hay were fed with the ration during the first 2 weeks. Each

---


Approved for publication as Journal Article No. 9012 of the Minnesota Agricultural Experiment Station.

The authors are: R. O. HEGG and R. E. LARSON, Agricultural Engineers, NCR, ARS, USDA, J. A. MOORE, Instructor, Agricultural Engineering Dept., R. D. GOODRICH and J. C. MEISKE, Professors, Animal Science Dept., University of Minnesota, St. Paul.


### TABLE 1. COMPOSITION OF THE CONTROL RATION.

| Ingredients | Percent |
|---|---|
| Supplement (pelleted) | |
| ground alfalfa hay | 7.5 |
| liquid molasses | 0.657 |
| urea | 0.73 |
| limestone (ground) | 0.565 |
| trace mineralized salt | 0.44 |
| sodium sulfate | 0.075* |
| stilbestrol premix | 0.022† |
| vitamin A and D premix | 0.0105‡ |
| Whole shelled corn | 90.00 |
| total | 100.00 |

*Contained 22 percent sulfur.
†Contained 4400 mg/kg premix.
‡Contained 25,143 IU vitamin A and 4,370 IU vitamin D/gm premix.

animal was weighed once monthly. The animals were taken off feed and water approximately 16 hours before the monthly weighing (Table 2).

The steers receiving the reclaimed solids as part of their ration were hesitant in consuming the feed at the beginning of the experiment and, therefore, for the first (25-day) period lost weight (Table 2). The animals (5) in the control group broke out of their pen early in the experiment. It was several weeks before all the animals were recovered, therefore no results on this pen are available. Two of the five steers fed the 25 percent solids rate, continued to lose weight well into the second month, so they were removed from the pen. For the remainder of the experiment only three steers were fed the ration containing 25 percent solids DMB.

The average daily gain (ADG) for the last three feeding periods were highest for steers fed the 5 percent solids ration and lowest for those fed the 25 percent solids ration. The overall ADG for the 110 day trial for all 3 pens was 0.95 kg/day for the 5 percent level, 0.73 kg/day for the 15 percent solids, and 0.60 kg/day for the 25 percent solids which was below the normal expected gain for finishing steers on the standard high energy ration. Feed consumption was highest for steers fed the 5 percent recycled solids ration (6.59 kg/day) and lowest for those fed the 25 percent solids (4.82 kg/day)(Table 3).

The feed efficiency values are the amounts of feed required to produce a unit of gain (kg of feed/kg of gain). The steers fed the 5 percent recycled solids had the highest feed efficiency (7.13 kg/kg gain), and those fed the 15 percent solids had the lowest (8.35 kg/kg gain). Earlier work at the Agricultural Engineering Farmstead, based on 5 years of feeding steers with the standard

**TABLE 2. AVERAGE DAILY GAINS (ADG) AND FEED EFFICIENCY (FEED/GAIN).**

| time, days | 5 percent solids | | 15 percent solids | | 25 percent solids | |
|---|---|---|---|---|---|---|
| | ADG, kg | Feed/gain | ADG, kg | Feed/gain | ADG, kg | Feed/gain |
| 0-25 | -0.21 kg (-0.46 lb)* | | -0.52 (-1.14) | | -0.54 (-1.19) | |
| 26-54 | 1.36 ( 3.01) | 5.10 | 1.13 ( 2.59) | 5.54 | 1.21 ( 2.66) | |
| 58-83 | 1.03 ( 2.28) | 7.09 | 0.91 ( 2.01) | 7.60 | 0.72 ( 1.60) | 8.52 |
| 84-110 | 1.48 ( 3.27) | 5.47 | 1.27 ( 2.81) | 5.65 | 0.89 ( 1.96) | 7.08 |
| 0-110 | 0.95 ( 2.09) | 7.13 | 0.73 ( 1.62) | 8.35 | 0.60 ( 1.33) | 8.15 |

*Numbers in parenthesis indicate ADG in pounds per day.

ration, showed feed efficiencies ranging from 7.3 to 8.2 kg/kg (Hegg et al. 1974) which suggests that the results reported here compare favorably with those from the earlier work.

The Total Digestible Nutrients (TDN) of the reclaimed solids, were evaluated using the equations of Garret et al. (1969) to predict the amount of TDN/day required by steers, based on their weight and rate of gain. Calculations were made to determine the amounts of TDN/day actually consumed in the form of corn and hay (assuming a TDN for corn of 0.9 and hay of 0.55). The difference between the amount of TDN required by steers and that actually in the form of corn and hay was assumed to be the amount derived from the recycled solids. This difference was divided by the average, daily dry weight of solids consumed to determine the TDN content of the solids. TDN values were determined for the entire test period (110 days) and for only the last 56 days since TDN values for the entire period would include those for the beginning of the experiment when the animals lost weight because they were reluctant to consume the feed. The TDN value of the solids for the last 56 days are probably the more representative estimates of the feed value, because during this time the steers were adapted to the feed and ate it readily. The data indicated that the reclaimed solids ranged in TDN from 33 to 65 percent of the TDN value of corn. The solids were not as valuable as corn because part of the feed value had already been removed by steers over the oxidation ditch during digestion when the corn was initially fed.

A second method used to estimate the feed value of the solids was the Van Soest Summative equation, which is based on two constants: percent CW (cell wall) and the log of the lignin as a percent of ADF (acid detergent fiber). To estimate true digestibility (TD) of the feed: TD = 0.98 (100-percent (CW)) + percent CW [1.473-0.789 (log lignin as percent of ADF)].

To determine the apparent digestibility (an estimate of the TDN) a value of 12 is subtracted from the TD. The TD was 60.5 percent and the apparent digestibility (TDN) was 48.5 percent which is approximately 53 percent  for that of corn.

Both methods for estimating the TDN of the solids gave values of about 50 percent for that of corn. Previously reported work (Hegg et al. 1974) showed that the feed value of solids ranged from 63 to 85 percent for that of corn, but the recycled solids in that study consisted mainly of whole and cracked corn particles.

Chemical analyses of the recycled solids showed that they were high in cellulose and lignin contents, like alfalfa (Table 5). Thus, these solids may be more suitable for a maintenance ration for cattle than in a finishing ration, because of the low energy value of the recycled solids. The solids were analyzed weekly for dry matter content and crude protein (CP) (by the Kjeldahl method). The solids averaged 15.2 percent CP (DMB). further research is needed to evaluate the disease hazard of refeeding recycled solids directly to other animals without pretreatment or processing. When the steers were slaughtered, their carcass characteristics, as

(Continued on page 202)

**TABLE 3. AVERAGE DAILY FEED CONSUMPTION (DRY MATTER) OF STEERS FED RECLAIMED SOLIDS.**

| Solids in ration, percent of DM | Corn, kg | Supp., kg | Solids wet, kg | Solids dry, kg | Hay, kg | Total (dry), kg |
|---|---|---|---|---|---|---|
| 5 | 5.64 | 0.62 | 1.63 | 0.33 | 0.17 | 6.76 |
| 15 | 4.54 | 0.50 | 4.42 | 0.88 | 0.18 | 6.10 |
| 25 | 3.15 | 0.35 | 6.04 | 1.21 | 0.20 | 4.91 |

**TABLE 4. CALCULATED TDN (TOTAL DIGESTIBLE NUTRIENTS) OF RECLAIMED SOLIDS FROM AN OXIDATION DITCH. FED AT 5 PERCENT, 15 PERCENT, AND 25 PERCENT OF THE RATION.**

| Periods | Levels of solids in ration dry matter, percent | | | | | |
|---|---|---|---|---|---|---|
| | 5 | | 15 | | 25 | |
| | TDN | Percent* | TDN | Percent* | TDN | Percent* |
| 110 days | 0 | 0 | 16 | 18 | 87 | 97 |
| last 56 days | 30 | 33 | 56 | 62 | 58 | 65 |

*These values are expressed as a percent of the TDN of corn, using a TDN value for corn of 90 percent.



MANAGING LIVESTOCK WASTES

# Nutrient Availability From Oxidation Ditches

B. G. Harmon, D. L. Day

MEMBER
ASAE

**R**EFEEDING of swine waste or a product derived therefrom has advanced from a hypothesis based on chemical analysis to an acceptable practice on some swine farms. Swine excreta has emerged from a position as a waste to be disposed of as a fertilizer, minimizing the odor nuisance as possible, to considerations of other possibilities for utilization of this biomass. Use as fertilizer may still be the most advantageous, but now the decision can be based on choices among alternatives rather than default.

## REVIEW OF LITERATURE

With the ingredients currently in use, swine diets are approximately 85 percent digestible. The remaining 15 percent is available as an asset or liability. It is readily accessible as a substrate for microbial action. The modification in chemical composition commences with secretions into the intestine and continues postexcretion as the organic mass of the waste supports this biological change. Microflora markedly change the nitrogenous components in fecal material. (Harmon et al. 1970). The amino acid fraction (Table 1) is much greater in feces from conventional rats containing the microflora as compared to that of germ-free rats. Conversely, fecal urea is extremely minute in feces from conventional rats and a major fraction in germ-free rats. The disappearance of urea in conventional rats reflects microbial use of urea.

TABLE 1. AMINO ACIDS AND UREA AS
A PERCENT OF FECAL NITROGEN IN
GERM-FREE AND CONVENTIONAL RATS.

| Diet protein, percent | Forms of fecal nitrogen | | | |
|---|---|---|---|---|
| | Amino acid | | Urea | |
| | Axenic | conv. | Axenic | conv. |
| 0.42 | 30.7 | 55.0* | 44.8* | 6.5 |
| 8.2 | 51.6 | 77.4* | 31.8 | 6.8 |
| 12.2 | 45.2 | 62.7* | 35.3* | 2.2 |
| 10.4 | 35.8 | 50.5* | 17.2* | 5.2 |

*(P<0.01)
Harmon, Jensen & Baker, 1970

The continued microbial action in swine waste has been studied extensively by Day et al. (1969). The oxidation ditch was developed for waste management to utilize aerobic microbiota in reducing the biomass and minimizing odors emitting from the waste. Recycling research at the University of Illinois has been based on utilization of products of this low odor, high microbial-activity system.

The refeeding of swine waste has not progressed with the same speed as the feeding of poultry (Fontenot and Webb 1974) and cattle waste (Anthony 1974) to ruminants. The nitrogen components are the most valuable in waste. While ruminants can use the more simple products like urea and uric acid, swine must have amino acids preformed in the diet. The thrust of our research is to use the single-cell protein resulting from microbial action in the oxidation ditch. The major components of oxidation ditch mixed liquor are amino acids and minerals. In combination they often exceed 90 percent of the dry matter. The carbohydrate level is low suggesting that energy may limit microbial growth within the oxidation ditch.

## LABORATORY STUDIES

In early studies, material was collected for feeding by slowly draining the oxidation ditch and recovering solids from the bottom. Quantities adequate for rat studies were obtained by this method (Harmon, Jensen and Baker 1969). The product containing 27 percent protein could be substituted for one half of the soybean protein in a corn-soybean diet while maintaining gain and efficiency equal to the control diet (Harmon et al. 1972). As the only protein supplement for corn, this material failed to support normal performance (P<0.01).

Subsequently we determined that the fluids that were drained away were rich in amino acids. Holmes et al. (1971) demonstrated, by screening contents of the oxidation ditch, that the percent dry matter and protein in the dry matter increased as the particle sizes decreased. The fraction that passed through the smallest screen openings (200 mesh) had the highest percent dry matter and protein of the dry matter. An analysis of the various fractions (Harmon 1972) showed that individual amino acids increased significantly (P<0.01) as the particle size decreased (Table 2). These findings support the thesis that single-cell protein is being built from the swine waste. The values also support the suggestion that the waste must be biologically enhanced to be of most value to nonruminants.

Further evidence for the bio-upgrading of swine waste can be demonstrated by comparing the amino acid concentration in products of oxidation ditches (Harmon 1972, Orr et al. 1973) and fresh waste (Gouwens 1966) as shown in Table 3.

A more complete nutrient analysis of oxidation ditch mixed liquor (ODML) is shown in Table 4. The values represent a mean of 13 weekly samples (Harmon 1972). In addition to the essential amino acids, ODML is rich in calcium and phosphorus, containing more than 3 percent of each, and the trace elements added to swine diets.

Drying ODML constitutes problems in handling, preserving, and cost that appeared to be prohibitive. Unfortunately, the procedures that separate by sieving or centrifugation are inefficient as the particles that are removed last are the richest in amino acids. Being unable to economically reconcile the problem, ODML was fed as a nutrient solution in subsequent studies. This procedure has practical applica-

The authors are: B. G. HARMON, Animal Science Department, University of Illinois, Urbana, Illinois; and D. L. DAY, Agricultural Engineering Department, University of Illinois, Urbana, Ill.

TABLE 2. ANALYSES OF VARIOUS SIEVED PARTICLES OF
ODML (PERCENT OF DRY MATTER).

| Screen | Size opening, inch | Protein | Lysine | Histoidine | Threonine | Methionine | Isolencine |
|---|---|---|---|---|---|---|---|
| 20 mesh | 0.0328 | 8.4 | 0.38 | 0.24 | 0.98 | 0.32 | 0.81 |
| 50 mesh | 0.0117 | 11.1 | 0.92 | 0.52 | 1.57 | 0.42 | 1.13 |
| 100 mesh | 0.0058 | 20.0 | 1.36 | 0.98 | 2.01 | 0.62 | 1.66 |
| >200 mesh* | 0.0029 | 75.6 | 3.08 | 1.32 | 2.44 | 1.27 | 2.20 |

*Material passing through that screen, remaining in solution. (Harmon 1972)

tion since the pig will normally consume two parts of water per part of dry feed.

In the initial swine studies, ODML was mixed with dry feed (two parts water or ODML per part of dry feed) in a twice-a-day feeding program. The ODML contained from 2-4 percent dry matter. A marginally deficient 12 percent protein corn-soybean meal diet was fed to each group. Pigs eating this diet and drinking ODML could correct amino acid deficiencies in the dry diet. Protein intake is increased by 2 percent and lysine by 0.1 percent. Finishing pigs weighing 50 kg initially and continued to 100 kg gained significantly faster and more efficiently when fed ODML than the pigs offered only water (Harmon et al. 1973).

A second feeding method was devised to simplify the feeding procedure (Fig. 1). The same 12 percent protein diet was offered *ad libitum* in self-feeders and all water was provided from the ditch by constant or intermittent pumping of ODML through a trough in each pen. Control pens received water from a tap. Pigs offered ODML as the only source of water from 40 kg to market weight gained more rapidly and efficiently (Table 5) than pigs given tap water (Harmon and Day 1974). The use of ODML enables the producer to reduce the amount of protein supplement while making lesser savings in calcium, phosphorus, trace minerals and water-soluble vitamins. Parker et al. (1959) reported that the phosphorus in corn and soybean meal that is less than 33 percent available to swine is highly available in the waste.

In each of these studies, it was necessary to add water to the ditch periodically to maintain a constant level of ODML in the ditch. Control and experimental pigs were contributing waste to the ditch, but only the experimental pigs were fed ODML, and yet water additions were still necessary. In spite of these additions, dry matter and ash have built up necessitating emptying every 2-3 years.

TABLE 3. AMINO ACID COMPOSITION OF SWINE FECES
AS DESCRIBED (PERCENT DRY MATTER)

| | Fresh swine feces* | Dried swine feces† | Oxidation ditch mixed liquor* | Oxidation ditch liquor† |
|---|---|---|---|---|
| Phenylalanine | 0.81 | 0.87 | 1.48 | 1.66 |
| Lysine | 0.60 | 1.11 | 1.42 | 1.60 |
| Arginine | 0.44 | 0.67 | 1.28 | 1.45 |
| Threonine | 0.53 | 0.80 | 1.96 | 1.22 |
| Methionine | - - - | 0.58 | 0.77 | 0.60 |
| Isoleucine | 0.52 | 1.03 | 1.49 | 1.54 |
| Leucine | 0.92 | 1.57 | 2.79 | 2.13 |

*Gouwens 1966
†Orr et al. 1973
‡Harmon et al. 1972

TABLE 5. SOURCE OF WATER
AND NUTRIENTS FOR SWINE.

| | Daily gain, kg | | Gain/feed | |
|---|---|---|---|---|
| | Tap H$_2$O | ODML | Tap H$_2$O | ODML |
| Rep. | | | | |
| 1 | 0.61 | 0.63 | 0.303 | 0.308 |
| 2 | 0.66 | 0.70 | 0.296 | 0.297 |
| 3 | 0.77 | 0.90 | 0.244 | 0.343 |
| 4 | 0.60 | 0.72 | 0.230 | 0.278 |
| 5 | 0.58 | 0.66 | 0.260 | 0.263 |
| 6 | 0.71 | 0.79 | 0.309 | 0.338 |
| | 0.66 | 0.73 | 0.274 | 0.304 |

Each value represents 10 pigs (Harmon and Day 1974).

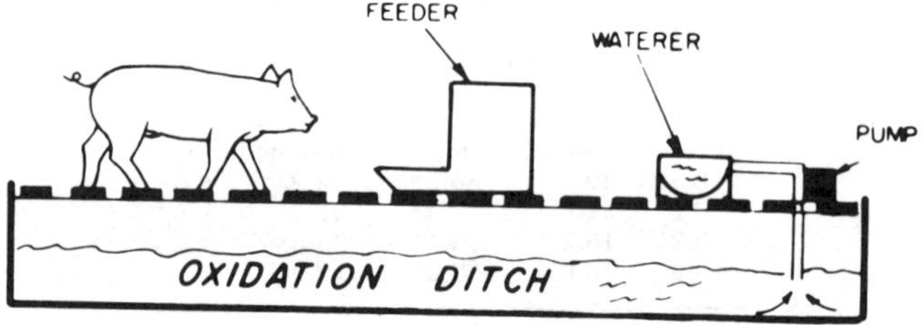


FIG. 1 ODML is pumped from the oxidation ditch directly into a watering trough. No other water is provided.

TABLE 4. NUTRIENT CONTENT OF SWINE ODML.

| | Percent dry matter* | | Percent dry matter* | | Percent dry matter† |
|---|---|---|---|---|---|
| Phenylalanine | 1.48 | Aspartic | 3.73 | Calcium | 3.33 |
| Lysine | 1.42 | Serine | 2.55 | Phosphorus | 3.83 |
| Histidine | 0.47 | Glutamic | 5.06 | Magnesium | 1.49 |
| Arginine | 1.28 | Proline | 1.29 | Sodium | 2.75 |
| Threonine | 1.96 | Glycine | 2.29 | Potassium | 4.04 |
| Valine | 2.06 | Alanine | 2.83 | Iron | 0.5507 |
| Methionine | 0.77 | Tyrosine | 1.17 | Copper | 0.0071 |
| Isoleucine | 1.49 | Tryptophan | 0.28 | Zinc | 0.1148 |
| Leucine | 2.79 | | | | |

*Means of 13 weekly analyses of amino acids except for tryptophan (one analysis)
†Mean of six weekly analyses of minerals, (Harmon 1972)

Studies have been undertaken to measure the balance of water and dry matter in a ditch when all pigs (56, expt. 1; 64, expt. 2) in the partially slotted floor building were given ODML as their sole source of water for feeding periods of 56 and 77 days.

In the first experiment, the average dry matter of the ditch contents was 4.7 percent but increased to 5.3 percent during the progress of the experiment (Table 6). This ditch had operated for 1.5 years prior to the start of this experiment. To maintain volume in the ditch, it was necessary to add 124 gallons daily or 2.2 gallons per pig daily. This experiment was conducted in August and September (Harmon 1975).

TABLE 6. ODML FEEDING TO FINISHING PIGS.

| Exp. no. | 1* | 2† |
|---|---|---|
| Av. daily gain, kg | 0.68 | 0.70 |
| Gain/feed | 0.280 | 0.321 |
| Dry matter, percent | 4.7 | 2.8 |
| $NO_3$, ppm | 160.0 | 252.0 |
| Daily $H_2O$ added, gal. | 124.0 | 83.0 |

*56 pigs; 45 kg initially, 7 pigs per pen, fed for 56 days.
†64 pigs, 26 kg initially, 8 pigs per pen, fed for 77 days.

The second experiment was conducted in November-January in the same building after the ditch was emptied and refilled with water. The average dry matter during the 77-day feeding period was 2.8 percent. For this experiment conducted in the winter, 83 gallons of water were needed daily or 1.3 gallons per pig daily to maintain a constant volume of liquid in the ditch. Based on feed consumption, water intake should approximate 1.25 gallons per pig daily. Remaining losses would be primarily from evaporation.

More information is needed on the production of nitrates in the oxidation ditch. In previous studies (Harmon and Day 1974) we have reported a normal range of 70-225 ppm nitrates. Possibly the upper range of nitrate is detrimental to good performance. In the experiments where all pigs in the building were fed ODML and the feeding trial was broken up into 2-3 week periods, a negative correlation was found between nitrate content in ODML and daily gain. The highest value reported during any of the periods was 370 ppm. However, a cause-and-effect can not be assigned with the data available at the present time. In an early study, an experiment was terminated because of lethal levels of nitrate (5000 ppm) in ODML. This level followed a period when a ditch had been inoperative for several weeks. The paddle wheel was started without emptying and without adding substrate for an extended period. Nitrate values dropped precipitously when carbohydrate was added to the ditch.

High nitrates are associated with a high level of dissolved oxygen. On the other extreme, rat and swine performance are reduced when swine waste is maintained in an anaerobic environment (Harmon et al. 1973). Any substitution of the anaerobic product for corn or soybean meal resulted in reduced gain and feed efficiency. The components that caused reduced performance have not been determined. Since feed consumption was not always reduced, impalatability is not the entire answer.

In a more recent study, the effect of inadequate aeration was demonstrated (Harmon 1975). Sixty-four finishing swine were fed the typical 12 percent protein diet and given either tap water or ODML. Through a series of problems the ditch was operating at less-than-full speed such that the ditch did not remain aerobic. This was clearly detected by odor and color of the material. Performance (Table 7) was reduced for ODML-fed pigs compared to those receiving water. These studies confirm the hazards of using anaerobic or overly aerated products. They also demonstrate the hazards of refeeding without appropriate monitoring of the ditch. Quality control is as critical here as any other part of the feeding program. As an example, Miller et al. (1974) had less-than-optimum performance when feeding the aerated liquid product. They reported that as much as one third of the nitrogen in the ditch was present as ammonia nitrogen. This would suggest that there was insufficient oxygen present to maintain products of an aerobic system.

TABLE 7. PERFORMANCE OF SWINE FED ODML LIMITED IN AERATION.

| | Daily gain, kg | | Gain/feed | |
|---|---|---|---|---|
| | ODML | Tap $H_2O$ | ODML | Tap $H_2O$ |
| Rep 1 | 0.30 | 0.66 | 0.194 | 0.252 |
| 2 | 0.34 | 0.55 | 0.220 | 0.265 |
| 3 | 0.33 | 0.55 | 0.198 | 0.250 |
| 4 | 0.26 | 0.49 | 0.227 | 0.286 |
| Av | 0.31 | 0.56 | 0.210 | 0.263 |

Liquor was not continuously aerobic.

## SUMMARY

A system has been developed by which swine waste can be biologically enhanced forming single-cell protein in a near odor-free aerobic environment. The amino acids can be substituted into the basal diet. The oxidation ditch mixed liquor also provides minerals and vitamins. Because of the low percent dry matter, 2-4 percent, the product is most economically fed in a liquid form.

When all pigs are receiving the liquid, no effluent need leave the building as water must be added to the ditch to maintain a constant volume of liquid in the ditch. Periodic emptying (every 2-3 years) may be necessary when the ash content has reached a very high percent of the total dry matter.

Care must be taken to maintain a constant aerobic environment in the ditch.

**References**

1    Anthony, W. B. 1974. Nutritional value of cattle waste for cattle. Federation Proceedings 33:1939-1941.
2    Day, D. L., D. D. Jones, J. C. Converse, A. H. Jensen and E. L. Hansen. 1969. Oxidation ditch treatment of swine waste. ASAE paper No. 69-924, ASAE, St. Joseph, Mich. 49085.

3   Fontenot, J. P. and K. E. Webb, Jr. 1974. Poultry wastes as feedstuffs for ruminants. Federation Proceedings 33:1936-1937.

4   Gouwens, D. W. 1966. Influence of dietary protein and fiber on fecal amino acid excretion of swine. M.S. thesis. University of Illinois, Urbana, Illinois.

5   Harmon, B. G. 1972. Harvesting nutrients from swine wastes. Proceedings 23rd Minnesota Nutrition Conference. University of Minnesota. p. 1-8.

6   Harmon, B. G. 1975. Recycling swine waste. Proceedings National Feed Ingredients Conference, January 17, 1975. Des Moines, Iowa, p. 167-180.

7   Harmon, B. G., A. H. Jensen and D. H. Baker. 1969. Nutritive value of oxidation ditch residues. J. Anim. Sci. 29:136 (Abstr.).

8   Harmon, B. G., A. H. Jensen and D. H. Baker. 1970. The influence of microbial environment on fecal nitrogen excretion products. Proceedings 9th Annual Symposium on Gnotobiotic Research. Assn. of Applied Gnotobiotics 9:16 (D-2).

9   Harmon, B. G. and D. L. Day. 1974. Nutrition value of amino acids produced in an oxidation ditch from waste. Processing and Management of Agricultural Waste. Proceedings 1974 Cornell Agricultural Waste Management Conference, p. 375-381.

10   Harmon, B. G., D. L. Day, A. H. Jensen and D. H. Baker. 1972. Nutritive value of aerobically sustained swine excrement. J. Anim. Sci. 34:403-407.

11   Harmon, B. G., D. L. Day, D. H. Baker and A. H. Jensen. 1973. Nutritive value of aerobically or anaerobically processed swine waste. J. Anim. Sci. 37:510-513.

12   Holmes, L. W. J., D. L. Day and J. T. Pfeffer. 1971. Concentration of proteinaceous solids from oxidation ditch mixed liquor. Proc. Int. Symp. on Livestock Waste. ASAE. Publication Proc. 271:351-354.

13   Miller, E. R., E. C. Miller, F. F. Green, P. K. Ku, P. A. Whetter and D. E. Ullrey. 1974. Studies with a swine waste recycling system. J. Anim. Sci. 39:186 (Abstr.).

14   Orr, D. E., E. R. Miller, P. K. Ku, W. G. Bergen, D. E. Ullrey and E. C. Miller. 1973. Swine waste as a nutrient source for finishing pigs. Michigan State University, Report on Swine Research 232:AH-SW-7319, p. 81-87.

15   Parker, M. B., H. F. Perkins and H. L. Fuller. 1959. Nitrogen, phosphorus and potassium content of poultry manure and some factors influencing its composition. Poultry Sci. 38:1154-1158.

---

## Recycling Solids from Beef Slurry

*(Continued from page 198)*

**TABLE 5. FIBER ANALYSIS OF SOLIDS RECLAIMED FROM THE OXIDATION DITCH SLURRY COMPARED WITH ALFALFA.**

| Fiber component | | Relcaimed solids, percent | alfalfa, percent |
|---|---|---|---|
| cell wall content | (NDF)* | 56.1 | 36.3 - 42.7 |
| ligno cellulose | (ADF)† | 21.6 | 25.2 - 33.9 |
| lignin | (ADL)‡ | 6.3 | 5.3 - 8.5 |
| cellulose | (ADF-ADL) | 15.3 | |
| hemicellulose | (NDF-ADF) | 34.4 | |

*NDF (neutral detergent fiber)   (Van Soest and Wine 1967)
†ADF (acid detergent fiber)   (Van Soest 1963)
‡ADL (acid detergent lignin)   (Van Soest 1963)

influenced by the rations, was not investigated.

The results have shown the value of this type of material when compared with corn. However, the reclaimed solids used in this study were from waste produced by steers fed a high energy ration (Table 1). Solids reclaimed and recycled from animals fed a ration with more roughage may likely be lower in nutritional value. To evaluate whether a producer or feedlot operator should use reclaimed soilds as part of the ration, considerations will need to be made on the cost of recovering the processing the solids as compared with the costs of other feeds.

### CONCLUSIONS

1 The reclaimed solids collected had approximately 50 percent of the feeding value of corn (DMB).

2 Feed consumption decreased as the percentage of reclaimed solids in the feed increased.

3 The steers needed a period of several days to adjust to the ration before they would readily consume reclaimed animal waste solids. Further work should be done in processing or pretreating the solids to decrease or eliminate this adjustment period.

4 Because of the lower energy in these reclaimed solids they may be more suitable as maintenance rations rather than as finishing rations.

### References

1 Garret, W. N., V. H Meyer and G. P. Lofgren. 1959. The comparative energy requirements of sheep and cattle for maintenance and gain. Journal of Animal Science. 18:528.

2 Hegg, R. O., R. E. Larson, J. A. Moore, J. C. Meiske and R. D. Goodrich. 1974. Five-year beef animal environment study in Minnesota. Proceedings of International Livestock Environment Symposium. Lincoln, NE. 316-326.

3 Hegg, R. O., J. C. Meiske, R. E. Larson and J. A. Moore. 1974. Beef oxidation ditch settled solids fed to steers. Proceedings of the 1974 Cornell Agricultural Waste Management Conference. Rochester, NY. 405-414.

4 Van Soest, P. J. 1963. Use of detergents in the analysis of fibrous feeds. II. a rapid method for the determination of fiber and lignin. Association of Official Analytical Chemists Journal. 46:829-835.

5 Van Soest, P. J. and R. H. Wine. 1967. Use of detergents in the analysis of fibrous feeds. II. determination of the plant cell-wall constituents. Association of Official Analytical Chemists Journal. 50:50-55.

# Nutritional, Pathological and Parasitological Effects of Feeding Feedlot Waste to Beef Cattle

R. R. Johnson, R. Panciera, H. Jordan, L. R. Shuyler

ASSOC. MEMBER
ASAE

THE potential pollutant properties of animal wastes, especially those emanating from the large intensive feedlot operations, have caused considerable concern to scientists in recent years. Numerous methods of handling and disposal have been proposed and tested by which the animal wastes might be removed and rendered innocuous to the environment. Many of these involved financial input with little chance for financial return. Since there is some nutritive value remaining in animal wastes from animals on high concentrate rations, the potential of recycling these wastes through other animlas has been proposed and tested to some extent. The nature of the animal wastes themselves suggest that they might serve a role as either a nitrogen source or a source of roughage factor.

Although investigations are under way at several institutions on the nutritive value of the feedlot wastes, practically no information is available on the potential pathological problems which might arise from long term feeding of animal wastes or on the potential parasite problems resulting from recycling of fecal materials. It was the intent of this research to investigate the effects of including livestock wastes in the feed on the pathology of certain target tissues in finishing beef cattle and the effect on the parasite load after a lengthy period of feeding. Information on the nutritive value and roughage factor value of the animal waste was also obtained.

## EXPERIMENTAL PROCEDURES

Twenty-four yearling beef calves were alloted by sex and weight to eight pens of three animals each. One pen of three heifers and one pen of two steers plus one heifer were assigned to each of the four rations shown in Table 1. The pens were made with concrete slotted floors to allow for manure disposal. All pens were provided with self-feeders and automatic waterers. Animals were weighed at two week intervals and refused feed was weighed back at the same time.

Ration 1 was a standard high concentrate feedlot type ration and Ration 2 was similar except for substitution of feedlot waste (FLW) for the roughage source (cottonseed hulls-CSH). Ration 3 contained both FLW and CSH since the ability of FLW to satisfy the entire roughage requirement was not known. In Ration 4, the FLW was substituted for the majority of the supplementary nitrogen sources on an isonitrogenous basis. By using this design, eighteen animals fed FLW could be compared to the six controls for the pathological and parasitological observations while still comparing FLW as a roughage and N-source. The FLW was

taken from feedlot waste harvested in the spring of 1973 from the Hitch Feedlot, Guymon, Oklahoma. The material was dried and ground prior to incorporation into the rations. All rations were fed for ninety-one days.

At the termination of the feeding trial, the calves were sacrificed in the necropsy room in the College of Veterinary Medicine under the supervision of a Veterinary Pathologist. Samples of rumen walls, abomasal wall, intestinal wall, lung, spleen, liver, trachea, kidney, heart and adrenal tissue were removed from each animal and fixed in formalin for histological examination. The dissected carcasses were observed for any gross pathology. The abomasum and small intestine were removed intact for parasitological studies.

**TABLE 1. RATION COMPOSITION FOR FEEDLOT WASTE TRIAL.**

| | Percentage composition, as is basis | | | |
| Ingredient | 1<br>Control<br>CSH | 2<br>FLW—R | 3<br>FLW—<br>CSH | 4<br>FLW—CSH<br>No SBM |
|---|---|---|---|---|
| Dry rolled sorghum | 74.7 | 74.7 | 74.7 | 73.9 |
| Cottonseed hulls | 15.0 | - | 5.0 | 5.0 |
| Feedlot waste, dried | - | 15.0 | 10.0 | 15.0 |
| Dried molasses | 4.0 | 4.0 | 4.0 | 4.0 |
| Soybean meal, 44 percent | 4.0 | 4.0 | 4.0 | 0.6 |
| Urea | 1.0 | 1.0 | 1.0 | 0.2 |
| Salt, trace mineral | 0.5 | 0.5 | 0.5 | 0.5 |
| Dicalcium phosphate | 0.4 | 0.4 | 0.4 | 0.4 |
| Calcium carbonate | 0.4 | 0.4 | 0.4 | 0.4 |
| Vitamin A 30000 IU/gm | 200gm/T | 200gm | 200gm | 200gm |

## RESULTS AND DISCUSSION

### Animal Performance

Performance of the animals is shown in Table 2. The animals fed the control ration (no FLW) gained faster than any of the other groups. They also were more efficient than treatment 2 (FLW-Roughage) and 4 (no SBM). Apparently, the FLW did not serve well as a roughage source (treatment 2) in this study and as a result, performance was adversely affected. Similarly, this FLW was not satisfactory as an N-source (treatment 4) although that result is confounded with lower consumption figures.

All rations containing FLW were consumed to a lesser extent than the control ration. This was especially evident during the first two to three weeks on test (not delineated in this table) in which consumption of the FLW rations was exceptionally poor. In the later portion of the trial, the intake figures equalized somewhat but by that time the FLW animals were considerably behind in performance. Apparently the dried FLW was objectionable to the cattle even at the 10 and 15 percent levels at the onset. The winter and spring of 1973 were exceptionally wet and pens in most feedlots were cleaned wet and the material taken from the

Work conducted under EPA Contract No. 68-03-0265.

The authors are: R. R. JOHNSON, Professor and Head, Animal Science Dept., University of Tennessee, Knoxville (Former Professor, Animal Nutrition and Biochemistry, Oklahoma State University); R. PANCIERA, Professor of Pathology, H. JORDAN, Professor of Parasitology, College of Veterinary Medicine, Oklahoma State University, Stillwater; and L. R. SHUYLER, Robert S. Kerr Environmental Research Lab., Ada, Okla.

## TABLE 2. PERFORMANCE OF ANIMALS ON FEEDLOT WASTE RATIONS.

| Item | 1 Control CSH | | 2 FLW—R | | 3 FLW— CSH | | 4 FLW– CSH No SBM | |
|---|---|---|---|---|---|---|---|---|
| Pen | 1 | 2 | 1 | 2 | 1 | 2 | 1 | 2 |
| Ave. daily gain, lb | 2.54 | 2.79 | 1.48 | 1.72 | 1.91 | 2.47 | 1.67 | 2.04 |
| Treatment average | 2.66 | | 1.60 | | 2.19 | | 1.86 | |
| Ave. daily feed, lb | 19.1 | | 18.0 | | 17.2 | | 17.8 | |
| Feed per lb gain, lb | 7.58 | 6.73 | 11.21 | 11.25 | 8.26 | 7.50 | 10.31 | 8.97 |
| Treatment average | 7.16 | | 11.23 | | 7.88 | | 9.64 | |

## TABLE 3. INCIDENCE OF GROSS PATHOLOGICAL AND HISTOLOGICAL OBSERVATIONS ON TISSUES.

| Observation | Control | FLW—R | FLW— CSH | FLW— CSH NP |
|---|---|---|---|---|
| **Gross** | | | | |
| Rumen-papillae clumping | 3 | 4 | 5 | 2 |
| Liver-abscess | 1 | 0 | 0 | 0 |
|   foci | 0 | 0 | 1 | 0 |
| Lung-abscess | 0 | 0 | 1 | 0 |
|   pneumonia evidence | 1 | 2 | 0 | 2 |
| Heart | 0 | 0 | 0 | 0 |
| Adrenal | 0 | 0 | 0 | 0 |
| Kidney-gray or white foci | 1 | 3 | 3 | 4 |
| Spleen | 0 | 0 | 0 | 0 |
| **Histological** | | | | |
| Rumen-debris in crypts | 2 | 3 | 4 | 5 |
|   Inflammation | 0 | 2 | 4 | 3 |
|   Keratinization | 4 | 3 | 3 | 3 |
| Abomasum | | | | |
|   Lymphoid aggregates | 5 | 5 | 1 | 2 |
|   Lymphocytes | 0 | 0 | 1 | 1 |
| Jejunum | | | | |
|   Lymphoid aggregates | 2 | 0 | 2 | 1 |
|   Lymphocytes and eosinophils | 6 | 5 | 6 | 6 |
| Ileum | | | | |
|   Lymphocytes and eosinophils | 2 | 5 | 2 | 5 |
| Liver | | | | |
|   Lymphoid aggregates | 1 | 1 | 2 | 2 |
|   Lymphoid nodules | 0 | 0 | 1 | 0 |
|   Fibrosis areas | 2 | 0 | 0 | 0 |
|   Lymphocyte infiltration | 2 | 1 | 1 | 4 |
|   Abscess | 0 | 0 | 1 | 0 |
| Kidney | | | | |
|   Lymphoid aggregates | 1 | 2 | 3 | 2 |
|   Dilated tubules | 1 | 1 | 1 | 0 |
|   Fibrosis areas | 1 | 1 | 0 | 0 |
|   Lymphocytes | 0 | 1 | 0 | 4 |
| Lung | | | | |
|   Lymphocytes | 1 | 0 | 0 | 1 |
|   Inflammation | 1 | 0 | 0 | 0 |
|   Pneumonia lesions | 1 | 3 | 1 | 0 |
| Trachea | | | | |
|   Lymphoid aggregates | 2 | 2 | 1 | 0 |
|   Lymphocytes | 1 | 0 | 0 | 1 |
|   Inflammation | 0 | 2 | 0 | 1 |
| Heart-sarcocysts | 5 | 3 | 3 | 1 |
| Gall bladder | | | | |
|   Salmonella cultured | all negative | all negative | all negative | all negative |

pens required storage in that form. This may have caused an undesirable characteristic in the FLW used in comparison to material tested earlier which was more palatable (Johnson 1972, Johnson, Kropp and McGeehon 1973).

### Pathology Results

Table 3 presents the incidence of gross and histological findings on observation of the tissues. It should be noted that usually only one section of a tissue was preserved for histological examination. Thus, some microscopic lesions might have been missed. However, the whole organ or tissue was scrutinized for gross lesions. The lesions will be discussed in anatomical groups with appropriate interpretation to the extent possible.

Rumen. The rumen epithelium was characterized in both the gross and histological observations by numerous incidences of papillae clumping, debris in the crypts and keratinization of the epithelial cells. These observations are characteristic of rumens from animals that have received a high concentrate ration with very little roughage. They may culminate in a more acute condition called rumen parakeratosis but this degree was not observed in these animals. Furthermore, these signs were random throughout the animals regardless of treatment and therefore do not represent an effect of FLW. Signs of inflammation were more prevalent in those animals receiving FLW.

Abomasum, Jejunum and Ileum. Lymphocyte infiltration and the presence of lymphoid aggregates are generally distributed observations in these animals. These are not considered to be abnormal and would appear to have no relation to treatment.

Liver. Lymphoid aggregates, lymphoid nodules and lymphocyte infiltration were occasional observations in all groups and unrelated to treatment.

Lung. Pneumonia related lesions were also noted in several animals but could be related to a course of pneumonia that progressed through these animals several weeks prior to slaughter. This was apparently brought in by a new group of animals in the same barn and it responded quickly to antibiotic therapy.

Kidney. The most notable observations were lymphoid aggregates, lymphocyte infiltration and white or gray foci. Although the incidence of these observations was greater in the tissue of animals consuming FLW, some also occurred in the controls and an association with theatment is doubtful.

Trachea. The occurrence of lymphoid aggregates, lymphocytes and inflammation was scattered and not related to treatment.

Heart. Sarcocysts appeared in the heart muscle of numerous animals. Although these bodies are not well explained, recent evidence suggests they may be a stage in the life cycle of coccidia. Since coccidia could be transmitted in animal wastes, it was interesting to note that the sarcocysts appeared in least numbers on the FLW rations, thus suggesting no contamination due to this ingredient.

Gall bladder. Since Salmonella represents one of the most important groups to monitor for public health purposes, it is important to note that the bile of all animals was negative for Salmonella.

Conclusions from Pathological examinations. The data presented here suggest no causal relationship or correlation between feeding of FLW and existence of pathological lesions or signs.

### Parasitological Results.

Contents from the abomasum and intestine were taken at slaughter for purposes of counting stomach worms (abomasum) and intestinal parasites. Table 4 presents the

| Animal No. | Abomasum | | Small intestine | |
| --- | --- | --- | --- | --- |
| | No. recovered | Genus | No. recovered | Genus |
| Control | | | | |
| 1 | 1 | *Ostertagia* | 0 | |
| 2 | 1 | *Ostertagia* | 1 | *Cooperia* |
| 3 | 0 | | 0 | |
| 4 | 42 | *Ostertagia* | 2 | *Cooperia* |
| 5 | 1 | *Ostertagia* | 66 | *Cooperia* |
| 6 | 0 | *Ostertagia* | 1 | *Cooperia* |
| FLW-R | | | | |
| 7 | 1 | *Ostertagia* | 1 | *Cooperia* |
| 8 | 0 | | 0 | |
| 9 | 0 | | 0 | |
| 10 | 0 | | 0 | |
| 11 | 0 | | 4 | *Cooperia* |
| 12 | 0 | | 2 | *Cooperia* |
| FLW-CSH | | | | |
| 13 | 116 | *Ostertagia* | 7 | *Cooperia* |
| | 24 | *Cooperia* | | |
| 14 | 0 | | 0 | |
| 15 | 4 | *Ostertagia* | 13 | *Nematodirus* |
| 16 | 1 | *Ostertagia* | 0 | |
| 17 | 0 | | 0 | |
| 18 | 0 | | 0 | |
| FLW-CSH-NP | | | | |
| 19 | 0 | | 1 | *Cooperia* |
| 20 | 0 | | 0 | |
| 21 | 33 | *Ostertagia* | 3 | *Cooperia* |
| 22 | 1 | *Ostertagia* | 3 | *Cooperia* |
| 23 | 0 | | 0 | |
| 24 | 0 | | 0 | |

findings from the parasitological examinations. The number of parasites recovered are reported in addition to the genus of parasite. In the abomasum, *Ostertagia* were the predominant forms observed while Cooperia were the predominant forms in the small intestine. It is readily seen that significant numbers of parasites were recovered from only a few animals and their distribution suggest no association with the consumption of FLW.

Twenty-four beef calves were alloted to four treatments by weight and fed high grain finishing rations containing (a) 15 percent cottonseed hulls (CSH) - control, (b) 15 percent dried feedlot waste (FLW), (c) 5 percent CSH - 10 percent FLW, or (d) 5 percent CSH - 15 percent FLW - no soybean meat supplement. After ninety-one days on feed, all animals were slaughtered and gross and microscopic pathological conditions were observed. Parasite burdens were determined for the abomasum and small intestine. FLW did not serve well as a roughage or protein source. Animals fed FLW rations consumed considerably less feed during the first three weeks of trial, presumably due to an objectionable characteristic of the FLW. The FLW was taken from moist material removed from commercial earthen floor feedlots after a wet winter and spring. No pathological signs could be observed in any distribution associated with FLW feeding. Only a few animals carried detectable parasite burdens and their incidence was not associated with FLW feeding. Although this FLW did not appear to make a major nutritional contribution, its presence in the ration at levels of 15 percent did not appear to have a measurable effect on animal health.

## References

1    Johnson, R. R. 1972. Digestibility of feedlot waste. Oklahoma Agricultural Experiment Station. Miscellaneous Publication No. 87:62-65.
2    Johnson, R. R., J. R. Kropp and M. McGeehon. 1973. Animal wastes as protein sources for ruminants. Oklahoma Agricultural Experiment Station. Miscellaneous Publication No. 90:94-101.

# Microbiological and Chemical Analyses of Anaphage in a Complete Layer Excreta In-House Drying System


**T. S. Chang, J. E. Dixon, M. L. Esmay, C. J. Flegal**
MEMBER    MEMBER
ASAE       ASAE

**J. B. Gerrish, C. C. Sheppard, H. C. Zindel**
ASSOC. MEMBER    EXEC. AFFILIATE
ASAE          ASAE


THE problem of animal waste disposal has become critical because animal production is more concentrated in a confined establishment and farm land is being gradually invaded by residential developments. Furthermore, the increased cost of feed ingredients for animal production paved the way for the utilization of animal waste to reduce the cost of production.

"Anaphage", dehydrated cage layer excreta, has been experimentally used as an ingredient in animal feeds (Thomas 1970, Flegal and Zindel 1970, Zindel 1972) and results were very promising. Handling and drying animal waste have been studied by many researchers (Bressler and Bergman 1971; Surbrook et al. 1971, Chang et al. 1974b, Chang et al. 1975a). A complete in-house automatic manure handling and drying system project was initiated at Michigan State University in 1973 and the description of the project has been reported (Flegal et al. 1974). The importance of a chemical and microbiological analysis of the anaphage in the dehydration of poultry excreta has been reported (Chang et al. 1974a; Chang et al. 1975b). The purpose of this paper is to demonstrate that the chemical and microbiological analyses can be used as guidelines to indicate the efficiency of the dehydration process.

## EXPERIMENTAL PROCEDURE

Anaphage samples were taken directly from the dehydrator aseptically in sterilized glass jars and immediately placed in a refrigerator at the Avian Microbiology Laboratory until chemical and microbiological analyses were conducted.

A.O.A.C. analytical methods for moisture, calcium, phosphorus, ash, crude fiber, ether extract, Kjeldahl Nitrogen and non-protein nitrogen were used for chemical analyses of anaphage.

The microbiological analytical method described by Chang et al. (1974a) was followed with some modifications. A portion of the anaphage sample was saved for chemical analysis. Another portion of the sample was aseptically transferred to a tared poly bottle that contained sterile distilled water to form an approximate 1:10 dilution. The exact dilution was calculated from the weight of the water and sample used.

At the same time 20-25 grams of the anaphage were placed into each of three tared petri dish bottoms. The samples were weighed, placed into a 100 C oven and weighed again after 24 hours to determine percent moisture. The results of the three replicates were averaged to give the final moisture content.

The 1:10 dilution of anaphage was placed into a sterile blender jar and emulsified at high speed for two minutes. The resulting emulsion was used for both bacterial population counts and bacterial identifications. For the population counts, one ml. of the emulsion was transferred into 9 ml. of sterile distilled water, and a serial ten-fold dilution was carried out through six tubes. The procedure was repeated in triplicate.

The Drop-plate method (Miles and Misra 1938) was used for bacterial population counts. One drop from each dilution was transferred to two poured brain-heart infusion plates, using a disposable Pasteur pipette. Each drop was calibrated at 1/35 ml. Again, the procedure was performed in triplicate and one complete set of each dilution series was incubated aerobically while the other was incubated anaerobically. After 24 hours incubation at 37 C, the plates were counted. The first dilution drop that gave definite isolated colonies (generally between 10 and 50 colonies) was counted. The number of colonies was multiplied by the original dilution factor, the inverse of the dilution counted, and 35, and the results of the three replicates were averaged to give total aerobic and total anaerobic counts in bacteria per gram sample.

The emulsion used for the population counts was also used for bacterial identification. One ml. amounts of the anaphage emulsion were placed into 50 ml. of each of the following media: (a) Brain-heart infusion broth (BHI) for general enrichment of all aerobic bacteria, (b) Selenite broth (SB) for enrichment of *Salmonella* spp., (c) Ethylene violet azide (EVA) broth for enrichment of fecal *Streptococcus* spp., (d) EE broth for enrichment of coli-aerogenes group, (e) Rogosa's SL broth (RSL) for enrichment of *Lactobacillus* spp. After 24 hours of incubation at 37 C (all cultures were incubated aerobically except Rogosa's SL), the broth cultures were streaked onto agar plates (BHI broth to BHI agar; SB to Brilliant Green agar; EVA broth to EVA agar; EE broth to EMB agar; RSL broth to RSL agar). After 24 hours incubation at 37 C (RSL agar plates incubated anaerobically) the colonies from these plates were identified morphologically and biochemically.

One ml. of the emulsion was also transferred to each of three SPS agar pour plates to identify and enumerate sulfite-reducing clostridia. These plates were incubated anaerobically for 48 hours, and if clostridial growth was present, the colonies were counted at this time.


Michigan Agr. Expt. Sta. Journal Article No. 7169.

The authors are: T. S. CHANG, Poultry Science Dept., Michigan State University, East Lansing; J. E. DIXON, Agricultural Engineering Dept., University of Idaho, Moscow; M. L. ESMAY, Agricultural Engineering Dept., C. J. FLEGAL, Poultry Science Dept., J. B. GERRISH, Agricultural Engineering Dept., C. C. SHEPPARD and H. C. ZINDEL, Poultry Science Dept., Michigan State University, East Lansing.

**Acknowledgment:** The authors wish to express their appreciation to Mrs. Deborah Flint and Mrs. Elizabeth Linden for their assistance in microbiological and chemical analyses.


## RESULTS AND DISCUSSION

The crude protein (an average of nineteen anaphage samples) was 35.64 percent with a range of 27.46 - 41.82

percent (Table 1). The corrected protein averaged 14.32 percent with a range of 12.04 - 21.45 percent. "Corrected protein" was derived from the subtraction of the percent of non-protein nitrogen from the percent of Kjeldahl nitrogen and then multiplied by 6.25. The protein level of anaphage from this in-house dying system (crude and corrected) was considered good since the protein level in the laying feed was about 16 percent. The calcium and phosphorus contents averaged 8.51 percent and 2.23 percent, respectively. Ash, crude fiber and ether extract averaged 25.72 percent, 11.12 percent and 2.02 percent, respectively. Calcium phosphorus, ash, crude fiber and ether extract were comparable to the result of other anaphage analyses (Chang et al. 1075b).

**TABLE 1. CHEMICAL COMPONENTS OF ANAPHAGE**

| Chemical components | No. samples | Percent* | Range |
|---|---|---|---|
| Calcium | 19 | 8.51 | 6.68 - 10.70 |
| Phosphorus | 19 | 2.23 | 2.07 - 2.57 |
| Ash | 19 | 25.72 | 22.57 - 29.10 |
| Crude fiber | 19 | 11.12 | 9.84 - 12.92 |
| Ether extract | 19 | 2.02 | 1.14 - 2.80 |
| Kjeldahl nitrogen | 19 | 5.81 | 4.39 - 7.71 |
| Crude protein | 19 | 35.64 | 27.46 - 41.82 |
| Non-protein nitrogen | 19 | 3.41 | 2.24 - 4.26 |
| Corrected protein | 19 | 14.32 | 12.04 - 21.45 |

*Dried weight basis.

The microbial count was directly related to the moisture content of the sample (Table 2). There were only 621 aerobic organisms in a gram of anaphage sample when the moisture content was below 1 percent. An average of 290,286 and 702,602 organisms per gram of anaphage were counted when the sample moisture was 1-4 percent and 5-10 percent, respectively. When the moisture of the sample was up to 11-20 percent or higher than 20 percent, the microbial counts averaged 16 and 30 millions per gram sample, respectively. Anaerobic count followed the same pattern when compared to the moisture content of the sample (Table 2).

**TABLE 2. AEROBIC AND ANAEROBIC MICROBIAL COUNT OF ANAPHAGE RELATIVE TO MOISTURE CONTENT**

| Microbes | Moisture, percent | No. of sample | Average microbial count per gram of anaphage |
|---|---|---|---|
| Aerobic | <1 | 2 | 621 |
| | 1-4 | 8 | 290 286 |
| | 5-10 | 8 | 702 602 |
| | 11-20 | 9 | 16 167 511 |
| | >20 | 19 | 30 246 316 |
| Anaerobic | <1 | 2 | 1 030 |
| | 1-4 | 8 | 128 857 |
| | 5-10 | 8 | 667 116 |
| | 11-20 | 9 | 5 579 183 |
| | >20 | 18 | 15 616 666 |

An inverse relation of dehydration temperature and microbial count was apparent from the results (Table 3). Lowering the dehydration temperature resulted in a higher microbial count. The average of aerobic microbial count was decreased from 69,500,000 to 1,739,616 when the dehydration temperature was increased from 350 F to higher than 500 F. The average anaerobic microbial count was decreased from 26,500,000 to 359,167 when the dehydration temperature was increased higher than 500 F. It must be noted that the dehydration temperature in the drying chamber was higher than recorded in the table. The dehydration temperatures in Table 3 were merely the setting indicators on this particular dryer. The temperature varies in different locations of the drying chamber and normally is higher than its setting (Surbrook 1971).

**TABLE 3. THE EFFECT OF DEHYDRATION TEMPERATURE ON AEROBIC AND ANAEROBIC MICROBIAL COUNTS**

| Microbes | Dehydration temperature | Samples analyzed | Average microbial count per gram of anaphage |
|---|---|---|---|
| Aerobic | 350 F | 2 | 69 500 000 |
| | 400 F | 8 | 35 382 000 |
| | 450 F | 8 | 18 392 500 |
| | 500 F | 7 | 17 485 714 |
| | >500 F | 21 | 1 739 616 |
| Anaerobic | 350 F | 2 | 26 500 000 |
| | 400 F | 8 | 20 625 000 |
| | 450 F | 7 | 9 951 429 |
| | 500 F | 7 | 5 925 713 |
| | >500 F | 21 | 359 167 |

It is also interesting to note that the aerobic and anaerobic counts were significantly reduced when the moisture of the sample was reduced to less than 10 percent although the dehydration temperature remained at the same setting (Table 4). Furthermore, only four groups of bacteria (*Bacillus, Clostridium, Lactobacillus,* and fecal *Streptococcus*) were recovered when the dehydration temperature was set at 500 F or higher (Table 5).

*(Continued on page 210)*

**TABLE 4. THE EFFECT OF TEMPERATURE, MOISTURE ON MICROBIAL COUNTS OF ANAPHAGE**

| Microbes | Dehydration temperature | Sample moisture, percent | Average microbial count per gram anaphage |
|---|---|---|---|
| Aerobic | 500 F | >10 | 20 281 666 |
| | | <10 | 710 000 |
| | Over 500 F | >10 | 6 719 520 |
| | | <10 | 183 396 |
| Anaerobic | 500 F | >10 | 6 958 333 |
| | | <10 | 730 000 |
| | Over 500 F | >10 | 1 360 530 |
| | | <10 | 46 241 |

**TABLE 5. MICROORGANISMS RECOVERED FROM ANAPHAGE SAMPLES**

| Microorganisms recovered | Note |
|---|---|
| *Aerobacter aerogenes* | |
| *Alkaligenes faecales* | |
| *Bacillus* spp. | † |
| *Clostridium* spp. | † |
| *Corynebacterium* spp. | |
| *Enterobacter* spp. | |
| *Escherichia coli* | * |
| *Lactobacillus* spp. | † |
| *Proteus* spp' | |
| *Streptococcus* spp., fecal | † |

*Only recovered in high moisture samples.

†Only bacteria recovered when dehydration temperature was 500 F of higher.

# Nutritional Properties of Feedlot Manure Fractionated by Cereco Process

G. M. Ward, D. E. Johnson, E. W. Kienholz

ERES Ecology Corporation of Sterling, Colorado, has developed equipment which produces three fractions from feedlot manure: a high fiber silage (C-I), a dried protein product (C-II–, and a high ash residue. Nutritional studies of the first two products were conducted with cattle, poultry and fish at Colorado State University.

The Cereco silage (C-I) used in this experiment was produced from feedlot manure of steers fed a ration containing 70 percent rolled corn, 18 percent corn silage, and about 4 percent each of alfalfa hay, molasses, and protein supplement. Dried molasses was added as 5 percent of the dry matter to the silage and this was stored in large air-tight plastic bags and allowed to ferment for a least six days before feeding. The Cereco silage was somewhat higher in dry matter, protein and fiber than the corn silage used in this study, but in general, the two feeds were of quite similar composition (Table 1).

Ten Hereford steers averaging 392 kg at the start of the experiment were divided into two groups. One group was fed Cereco silage and the other group was fed corn silage. During a two-week adjustment period, the steers were accustomed to collection stalls and the level of intake was established for each steer. Each group also received 240 g per day of a protein-vitamin supplement. Average daily intake was 11.5 kg for Cereco and 17.0 kg for corn silage. Total collections of feces and urine were made for seven days in collection stalls. Heat and gaseous energy losses were monitored on 4 steers per diet by two 12-hour respiratory exchange measurements using a head-stall open-circuit calorimetry system. After the first seven-day collection, the feeding level was changed; two weeks were allowed for adjustment and the total energy and nitrogen balances were repeated. Feed and faces samples from each animal were composited for seven days for analysis.

The apparent digestibility of nutrients is presented in Table 1. Differences were small in the digestibility of the various nutrients, but NFE digestibility was slightly higher for corn silage resulting in a higher TDN value. Organic matter is presented instead of dry matter digestibility because of differences between lots of corn silage and differences between corn and Cereco silage in their ash content.

The digestible and metabolizable energy values were slightly higher for the corn silage (Table 1). The net energy value for maintenance (Calculated as: NRC maintenance requirement ÷ kg feed for maintenance) was essentially the same for both feeds (1.78 and 1.80 mcal/kg). The most interesting conclusion was that the efficiency of metabolizable energy utilization for gain was somewhat higher for the Cereco silage.

It should be pointed out that Cereco silage was fed as the sole feed only to facilitate determination of its feeding value. We do not suggest it as the only feed for a feedlot

## TABLE 1. FEED COMPOSITION, DIGESTIBILITY AND ENERGY VALUES OF CERECO AND CORN SILAGE

|  | Cereco silage | | Corn silage | |
|---|---|---|---|---|
| Dry matter | 34.4 | | 28.0 | |
| Protein* | 9.0 | | 7.3 | |
| Crude fiber* | 27.5 | | 23.0 | |
| Ether extract* | 1.8 | | 2.1 | |
| NFE* | 50.4 | | 52.2 | |
| Intake head/day (kg) | 11.5 | | 17.0 | |
| *Percent of dry matter | | | | |
| Digestibility of nutrients | | | | |
| Protein | 55.2 | ±2.0† | 51.0 | ±1.9† |
| Crude fiber | 65.3 | ±2.4 | 66.9 | ±2.0 |
| Ether extract | 90.6 | ±1.7 | 84.2 | ±1.4 |
| NFE | 67.1 | ±1.2 | 71.5 | ±0.7 |
| Organic matter | 65.6 | ±1.1 | 68.5 | ±1.1 |
| TDN | 60.2 | ±1.1 | 65.2 | ±1.0 |
| Energy value (mcal/kg) | | | | |
| Digestible energy | 2.74 | ±0.05 | 2.94 | ±0.07 |
| Metabolizable energy | 2.22 | ±0.04 | 2.49 | ±0.07 |
| Net energy-maint. | 1.78 | | 1.80 | |
| Net energy-gain | 1.51 | | 1.37 | |

†Standard error of the mean.

ration. Consumption although lower in this trial has been equal to corn silage when fed as 15-25 percent of the ration. The maximum levels fed (15 kg/day) would, however, be sufficient to meet the maintenance requirements of most cows.

## CERECO PROTEIN SUPPLEMENT

Cereco fraction two (C-II) was prepared from feedlot manure produced by steers fed the same ration as those steers from which the C-I fraction was produced. The chemical composition of the C-II is shown in Table 2. It contained 23.5 percent crude protein and 33 percent ash. The same ten Hereford steers as used for the C-I experiment were changed to a high concentrate ration over a period of 30 days. At the start of the trial, their average weight was 810 pounds. The basal ration consisted of 6 kg of corn silage (28 percent d.m.) and 6.5 kg cracked corn. A protein supplement was fed of either 800 g of soybean meal per steer per day or 1600 g of C-II to produce an approximately isonitrogenous ration. Five steers received the soybean meal supplement and five the C-II supplement. After a two-week adaptation period, a seven-day total collection of feces and urine was made. Following this trial the two groups of steers were reversed and allowed two weeks for

## TABLE 2. ANALYSIS OF C-II FED TO STEERS (PERCENT OF DRY MATTER)

|  | Trial 1 | Trial 2 |
|---|---|---|
| Dry matter | 91.6 | 94.1 |
| Crude protein | 23.8 | 23.1 |
| Fat | 3.0 | 2.8 |
| Fiber | 0.7 | 0.6 |
| Ash | 33.5 | 33.6 |
| NFE | 39.0 | 39.6 |

The authors are: G. M. WARD, D. E. JOHNSON, E. W. KIENHOLZ, Animal Sciences Dept., Colorado State University, Fort Collins.

**TABLE 3. DIGESTION AND METABOLISM OF DIETS CONTAINING C-II OR SOYBEAN MEAL AS PROTEIN SUPPLEMENT**

| | Protein Supplement C-II | | | Soybean Meal | | |
| | Steer group | | | Steer group | | |
| | A | B | Mean | A | B | Mean |
|---|---|---|---|---|---|---|
| D. M. Digest., Percent | 64.7 | 65.3 | 65.0 | 70.6 | 75.2 | 72.9 |
| O. M. Digest., Percent | 68.6 | 68.6 | 68.6 | 72.3 | 76.8 | 74.5 |
| Protein Digest., Percent | 51.7 | 48.2 | 50.0 | 59.1 | 62.6 | 60.9 |
| N Balance (g/day) | 27.3 | 15.2 | 21.2 | 35.1 | 42.8 | 39.0 |
| N Balance — Coefficient of Variation, Percent | 36.8 | 42.8 | 48.1 | 40.5 | 40.0 | 37.3 |
| Percent N retained | 20.8 | 11.6 | 16.2 | 22.7 | 27.9 | 25.3 |
| Percent Digest N retained | 39.7 | 24.0 | 31.8 | 38.4 | 44.6 | 41.5 |

adaptation and another total collection was made over a seven-day period.

The results of both trials are presented in Table 3. Feed intake was slightly higher for the groups fed the soybean meal supplement, 7.7 vs 7.2 kg per head; consequently, protein intake was higher, 962 vs 825 g per day. Protein digestibility, dry matter and organic matter digestibility were similar for both groups of steers when they were fed C-II. However, the steers in Trial 2 had a lower nitrogen balance and less nitrogen retained (Table 3). This difference may be partly an animal effect for it can be seen that steer group A had a higher nitrogen retention than group B and the same trend is seen when soybean meal was the protein supplement for the two groups.

The average result is that the protein of the ration containing C-II (which represented 32.8 percent of the total protein) was digested about 80 percent as well as soybean meal and that nitrogen retention was either about equal (Trial 1) or about one-half (Trial 2) of the value of the soybean meal containing ration.

Table 4 presents the composition of the feces from animals fed the two supplements. The feces produced by steers fed C-II was noticeably higher in ash and lower in fiber than the SBM diet. The ash results from the high intake in C-II. The ash might pose a problem in a continuously recycled system unless a part of the ash is removed as it is in the Cereco process in the form of a high ash fraction.

## POULTRY FEEDING TRIALS

The ruminant is primarily dependent for protein upon microbial cells and it is presumed that their digestive systems are adapted for the digestion of bacterial cells. Whether this is true for non-ruminants is not certain since they are not normally presented with large numbers of microbial cells. The value of single cell protein for animals is a controversial subject. For these reasons and because microbial protein of high quality is potentially of greater value to monogastrics than for ruminants a series of trials were initiated with broilers and layers to evaluate this product.

The first trial was conducted with three-day-old male broilers. These broilers were divided into six treatments with three replicates of 10 birds each. The trial was terminated after 28 days. A low protein diet was chosen to allow a growth response from more protein.

The C-II as 5 percent of the ration supported somewhat better growth than the basal diet while 10 or 15 percent C-II reduced growth rate slightly. The energy density of the ration decreased with increasing amounts of C-II because of its high percentage of ash (40.4 percent). Nevertheless, all levels of C-II produced more efficient gains than the control ration. No diarrhea was encountered. Only one chick died and it was eating the basal diet. The fact that this C-II preparation with its high ash content could substitute for some of the protein supplied in the basal diet by soybean meal was very encouraging.

In an attempt to eliminate the effect of energy dilution caused by the high content of ash, a second experiment was designed to compare C-II with a simulated or substitute C-II also containing a high level of mineral and fiber made up of 25 percent dehydrated alfalfa together with limestone and potassium phosphate. The C-II used in this study was a spray dried product containing 18 percent crude protein and 30 percent ash. The birds were five weeks old at the start of this experiment and the trial continued for 21 days.

The treatments are indicated in Table 5 and consisted of the basal diet as 80 percent of the ration and 20 percent of either C-II, or the simulated C-II, one-half of each or 5 percent C-II and 15 percent substitute. Each treatment was replicated three times with six birds per treatment.

When C-II completely replaced the substitute it supported similar weight gains, but when either 5 or 10 percent C-II was combined with the simulated ration, growth improved but not significantly. This may suggest a complementary combination of essential amino acids from the two protein sources. Note that the same effect may have occurred in the first experiment when 5 percent C-II was added to the basal diet.

Calculations indicate that 20 pounds of C-II could replace 4 pounds of soybean oil meal, 5 pounds of dehydrated alfalfa meal, and 2 pounds of meat and bone meal in 100 pounds of feed for growing broilers.

There were no death losses attributed to either type of ration. Necropsy of selected birds indicated no differences between treatments in the condition of liver or kidneys. Some birds from each treatment were cooked and a group of students asked to compare taste without prior knowledge of the treatments. Taste rankings were apparently random and unrelated to treatments.

The same batch of C-II as fed to steers (Table 2) was incorporated as 0, 15 or 30 percent of a layer ration for 46 hens over a six-week period.

All of the hens, after moving to new cages, were fed the control diet for the first week so that their basal egg quality

**TABLE 4. AVERAGE COMPOSITION OF FECES (PERCENT OF DRY MATTER)**

| Protein supplement | Soybean supplement | C-II supplement |
|---|---|---|
| Dry matter | 26.7 | 27.4 |
| Protein | 16.7 | 16.3 |
| Fat | 5.2 | 2.7 |
| Crude fiber | 12.6 | 7.8 |
| Ash | 9.2 | 17.6 |
| NFE | 56.3 | 55.4 |

**TABLE 5. THE EFFECT OF C-II AND SIMULATED C-II UPON WEIGHT GAIN (5-8 WKS.) AND FEED EFFICIENCY OF BROILERS**

| Diet | | | | |
| Basal, percent | C-II percent | Simulated C-II | Gain, g* | Feed/ gain |
|---|---|---|---|---|
| 80 | 20 | | 865 | 3.08 |
| 80 | 10 | 10 | 933 | 2.90 |
| 80 | 5 | 15 | 932 | 2.82 |
| 80 | | 20 | 868 | 3.01 |

*No differences at 5 percent level of significance.

| Diet | Consumption per bird | Eggs laid | Pounds feed per dozen eggs |
|---|---|---|---|
| Control | 12.05 | 27.92 | 4.97 |
| 15 percent C-II | 12.26 | 29.16 | 5.22 |
| 30 percent C-II A | 12.14 | 28.41 | 5.13 |
| 30 percent C-II B | 12.90 | 29.75 | 5.69 |

and production levels could be obtained. Control hens were laying at 70 percent and the younger Hy-line birds were at 60 percent production. Other hens were at 65 percent egg production. During the next week the first twelve birds were kept on the control diet, twelve were introduced to the 15 percent C-II diet and twenty-two to the 30 percent C-II diet by mixing increasing quantities of the new diet into the feeders. The next four weeks were the main experiment.

Feed consumption was calculated for the entire six-week period. Hen body weights were determined at the start of week 1 and the end of week 4. Egg production records were kept for each bird during the course of the experiment.

Egg quality values were determined daily on every egg laid. Egg shell thickness was measured with a hand micrometer, and albumin thickness with an Ames tripod micrometer. Egg weights were measured; the Haugh units were calculated.

Egg production was uniformly low during the course of the experiment due to the age of the birds. Hens on the 15 percent and 30 percent C-II diets declined slightly in production; however, the 30 percent B hens and controls remained stable but there were no significant differences between groups. Feed consumption was uniform between the diets.

There were no significant differences in the egg quality values during the course of the experiment and egg weights and shell thicknesses showed no change.

None of the birds on experiment had a observable change in appearance. The feces of the 15 percent C-II hens remained firm but black in color; whereas, that of the 30 percent Ceres hens became very loose and black. As compared to normal feeds, the 30 percent diets were "dusty". Small accumulations of the high level diets tended to adhere to the water cups and at times to the bird's beaks, but this was no problem.

All groups of birds gained weight during the experiment. The differences in gain were not statistically significant.

The energy value of C-II product was calculated to be roughly 1100 cal. M.E./lb (2300 kcal/kg). This is similar to the energy in West Coast Oats, Brewer's Dried Grains, and dehulled soybean meal (50 percent protein. Corn is 1560 and soybean is 1020 M.E./lb by comparison.

The replacement value for the C-II in the diets for these hens was calculated to be approximately $150 per ton when corn was selling (July 24, 1974) at $130 and soybean meal at $190 per ton.

---

## Anaphage in Layer Excreta Drying

*(Continued from page 207)*

## CONCLUSIONS

Anaphage contains several nutrient ingredients which can possibly be utilized as ingredients in animal feeds. According to these experimental results, it is concluded that the dehydration temperature for layer excreta must be 500 F or higher with this type of dryer. The moisture content of anaphage should not be higher than 10 percent to maintain the microbial count under one million per gram anaphage. These factors, the microbial count can be used as a guideline for measuring the efficiency of the poultry excreta dehydration process.

### References

1 Bressler, G. O. and E. L. Bergman. 1971. Solving the poultry manure problem economically through dehydration. Livestock Waste Management and Pollution Abatement, ASAE, St. Joseph, Mich. 49085.

2 Chang, T. S., D. J. Currigan, D. W. Murphy and H. C. Zindel. 1974a. Microbiological analysis of poultry anaphage. Poultry Science 53(3):1242-1244. (Mich. Agr. Exp. Sta. Jour. Art. No. 6601).

3 Chang, Timothy S., David Dorn and H. C. Zindel. 1974b. Stability of poultry anaphage. Poultry Science, 53(6). (Mich. Agr. Exp. Sta. Jour. Art. No. 6775).

4 Chang, Timothy S., David Dorn and Elizabeth Linden. 1975a. The effect of dehydrators on the nutrients of poultry anaphage. Mich. Agr. Exp. Sta. Research Report (in press).

5 Chang, T. S., D. J.Currigan, J. E. Dixon, M. L. Esmay, C. J. Flegal, J. B. Gerrish, C. C. Sheppard and H. C. Zindel. 1975b. Preliminary report on the microbiological and chemical analyses of anaphage from a complete in-house drying system. Mich. Agr. Exp. Sta. Research Report (in press).

6 Flegal, C. J. and h. C. Zindel. 1970. The result of feeding dried poultry waste to laying hens on egg production and feed conversion. Mich. Agr. Exp. Sta. Research Report #117:29.

7 Flegal, C. J., M. L. Esmay, J. B. Gerrish, J. E. Dixon, C. C. Sheppard, H. C.Zindel and T. S. Chang. 1974. A complete system for poultry manure in a caged layer production unit. Proc. 1974. National Agricultural Waste Conference, Rochester, N.Y., March 1974. (Mich. Agr. Exp. Sta. Jour. Art. No. 6734).

8 Miles, A. A. and S. S. Misra. 1938. The estimation of the bactericidal power of the blood. Jour. Hygiene 38:732.

9 Surbrook, T. C. 1971. Personal communication.

10 Surbrook, T. C., C. C. Sheppard, J. S. Boyd, H. C. Zindel and C. J. Flegal. 1971. Drying poultry waste. Livestock Waste Management and Pollution Abatement, ASAE, St. Joseph, Mich. 49085. p. 192.

11 Thomas, J. W. 1970. Acceptability and digestibility of poultry and dairy wastes by sheep. Farm Science: Poultry Pollution Problems and Solutions. Mich. Agr. Exp. Sta. Research Report #117:42.

12 Zindel, H. C. 1972. Recycling animal wastes. Proc. Iowa State University Nutrition Symposium on Proteins, June 7-8, Ames, Iowa.

# Nutritional Value of Cattle Feedlot Waste
# For Growing—Finishing Beef Cattle

R. C. Albin, L. B. Sherrod

## INTRODUCTION

NUTRIENT recycling offers a possible approach to feedlot waste management. This concept involves the utilization of feedlot waste as a feedstuff, with reutilization of nutrients as in natural ecosystems (Albin, 1971). A recent review by Anthony (1971) summarized this area of research. One of the striking features of his paper was that no work had been reported in the literature concerning the recycling of waste from southwestern cattle feedlots, where new and improved grain processing techniques are employed. One report from Texas Tech University was included in Anthony's summary, but it involved feedlot waste in which dry-rolled grain sorghum had been used in the cattle rations (Durham et al. 1966). The Environmental Protection Agency has stressed that more information is needed on the practice of nutrient recycling with particular reference to its safety, practicability, and usefulness (Anonymous 1971). Johnson (1972) has reported digestibility values for beef cattle feedlot waste when included at 25 percent and 40 percent levels in high roughage rations.

This study was conducted to determine the nutritive value of feedlot waste from southwestern cattle feedlots where improved grain processing techniques and low levels of roughage are being used.

The specific objectives of this research were:

1 to determine the effect of feeding different levels of feedlot waste to beef cattle upon acceptability, consumption levels (or palatability) and digestibility; and

2 to determine the nutritive value for beef cattle of composted feedlot waste and dry, ground uncomposted feedlot waste.

## EXPERIMENTAL PROCEDURE

Waste that had been scraped from the soil surface of a commercial beef cattle feedlot and stockpiled for approximately 1 month was ground through a hammer mill over a 0.25-in. screen. Rations containing feedlot waste were offered to feeder steers in three, 28-day total collection digestion trials. The steers were tied individually in stalls which sloped toward the feed trough. The animals were fed twice daily, allowed free access to water, and checked daily for health and stress symptoms. Fecal samples were collected daily and sampled for laboratory analyses during the last 7 days of each trial.

Trial I involved including ground, but otherwise unaltered, feedlot waste in four concentrations in a high-energy finishing ration containing adequate protein. The four mixed rations are shown in Table 1 and their chemical and

energy composition are presented in Table 2. Each ration was offered to five steers.

Trial II involved composting the feedlot waste, then using similar percentages of this material as in Trial I (Table 1). The chemical and energy contents are shown in Table 2. For composting, 4000 pounds of ground feedlot waste were piled in a cone shape after mixing enough water to increase the moisture content to near 40 percent, and allowed to compost for five days with daily stirring and mixing. Each ration was offered to five steers.

**TABLE 1. RATION COMPOSITION FOR TRIALS I AND II (PERCENT, AS FED BASIS)**

| | | Rations* | | |
|---|---|---|---|---|
| Ingredients | Control | 20 percent waste | 40 percent waste | 60 percent waste |
| Sorghum grain, dry rolled | 76.0 | 60.8 | 45.6 | 30.4 |
| Cottonseed hulls | 4.0 | 3.2 | 2.4 | 1.6 |
| Alfalfa hay, chopped | 4.0 | 3.2 | 2.4 | 1.6 |
| Cottonseed meal | 3.0 | 2.4 | 1.8 | 1.2 |
| Molasses | 4.0 | 3.2 | 2.4 | 1.6 |
| Supplement (TTA-3)* | 8.0 | 6.4 | 4.8 | 3.2 |
| Ammonium sulfate | 1.0 | 0.8 | 0.6 | 0.4 |
| Feedlot waste, ground† | 0.0 | 20.0 | 40.0 | 60.0 |
| TOTAL | 100.0 | 100.0 | 100.0 | 100.0 |

*Provided 45 mg chlortetracycline and 30,000 IU of vitamin A per pound, with 3.4 percent Ca, 1.1 percent P, 40 percent crude protein, 5 percent salt, 8.5 percent urea, 0.5 percent trace mineral premix, 10 percent dehy alfalfa and ground sorghum.
†Unaltered in Trial I; composted in Trial II.

**TABLE 2. ANALYTICAL COMPONENTS OF RATIONS IN TRIALS I AND II (DM BASIS)**

| | Trial I | | Trial II | |
|---|---|---|---|---|
| Item | Unaltered waste | Control ration | Composted waste | Control ration |
| Dry matter, percent | 82.55 | 87.12 | 78.95 | 86.31 |
| Crude protein, percent | 21.6 | 16.5 | 20.3 | 15.2 |
| Ash, percent | 31.7 | 3.82 | 33.60 | 4.20 |
| Gross energy, kcal/g | 3.546 | 4.286 | 3.390 | 4.424 |
| Ca, percent | 2.35 | 0.52 | 2.55 | 0.50 |
| P, percent | 0.72 | 0.33 | 0.93 | 0.46 |
| Cell soluble material, percent | 49.79 | 76.28 | 44.88 | 78.18 |
| Cell wall material, percent* | 33.95 | 11.17 | 34.52 | 9.88 |

*Corrected for insoluble ash.

This research was supported in-part by the Houston Livestock Show and Rodeo.

Approved by the Dean as College of Agricultural Sciences Publication No. T-5-109.

The authors are: R. C. ALBIN and L. B. SHERROD, Animal Science Department, Texas Tech University, Lubbock and the Texas Tech University Center at Amarillo, Pantex.

Trial III involved feeding unaltered and composted feedlot waste in a low-energy, low protein ration, resembling a high-roughage growing ration (Table 3). The chemical and energy compositions of the rations are shown in Table 4. Each ration was offered to five steers.

**TABLE 3. RATION COMPOSITION FOR TRIAL III USING LOW ENERGY — LOW PROTEIN RATIONS (PERCENT, AS FED)**

| Ingredient | Control ration | Unaltered waste | Composted waste |
|---|---|---|---|
| Sorghum grain, dry-rolled | 17.1 | 6.7 | 7.0 |
| Cottonseed hulls | 60.0 | 48.8 | 47.0 |
| Cottonseed meal | 17.5 | 0.0 | 1.5 |
| Salt | 0.25 | 0.0 | 0.0 |
| Limestone | 0.3 | 0.0 | 0.0 |
| Rock phosphate, defl. | 0.3 | 0.0 | 0.0 |
| Molasses | 4.0 | 4.0 | 4.0 |
| Aureomycin (50 g/lb) | 0.025 | 0.025 | 0.025 |
| Vitamin A (30,000 IU/g) | 0.5 | 0.5 | 0.5 |
| Unaltered feedlot waste | 0.0 | 0.0 | 0.0 |
| Composted feedlot waste | 0.0 | 0.0 | 40.0 |
| TOTAL | 100.0 | 100.0 | 100.0 |

**TABLE 4. ANALYTICAL COMPONENTS OF RATIONS IN TRIAL III (DRY BASIS FOR LOW ENERGY — LOW PROTEIN RATIONS**

| Item | Control ration | Unaltered waste | Composted waste |
|---|---|---|---|
| Dry matter, percent | 87.59 | 87.67 | 85.10 |
| Crude protein, percent | 11.3 | 11.2 | 11.7 |
| Ash, percent | 3.76 | 13.46 | 13.85 |
| Gross energy, kcal/g | 4.584 | 4.217 | 4.102 |
| Ca, percent | 0.34 | 0.97 | 1.05 |
| P, percent | 0.33 | 0.39 | 0.34 |
| Cell soluble material, percent | 35.59 | 34.72 | 35.77 |
| Cell wall material, percent* | 44.95 | 43.44 | 43.87 |

*Corrected for insoluble ash.

**TABLE 5. APPARENT DIGESTION COEFFICIENTS OF RATIONS FOR TRIAL I USING UNALTERED CATTLE FEEDLOT WASTE (PERCENT)**

| Item | Rations* | | | |
|---|---|---|---|---|
| | Control | 20 percent waste | 40 percent waste | 60 percent waste |
| Dry matter | 80.43 | 68.39 | 64.08 | 58.23 |
| Organic matter | 81.64 | 71.16 | 67.49 | 61.97 |
| Gross energy | 79.30 | 67.92 | 63.81 | 61.82 |
| Crude protein, apparent | 72.10 | 61.30 | 54.76 | 52.83 |
| Crude protein, true | 89.15 | 76.87 | 69.66 | 67.49 |
| Cell soluble material | 85.25 | 76.65 | 70.61 | 67.88 |
| Cell wall material | 55.01 | 51.84 | 44.52 | 41.26 |

*Treatment means on the same line are significantly different (P<0.01) in a linear pattern.

## RESULTS AND DISCUSSION

Apparent digestion coefficients of rations in Trial I are presented in Table 5. As the percentage of feedlot waste increased in the ration at levels of 0, 20, 40 and 60 percent, the apparent digestibility of all ration nutrient and energy components decreased in a significant (P<0.01) linear pattern. The apparent digestibility of crude protein decreased from 72.1 percent in the control ration (0 percent feedlot waste) to 52.8 percent in the 60 percent feedlot waste ration, and gross energy apparent digestibility decreased from 79.3 percent to 61.8 percent. The same significant trends were observed in Trial II (Table 6), using composted feedlot waste, except for the apparent digestibility of cell wall material which increased significantly (P<0.05) from 31.0 percent and 29.8 percent in the control and 20 percent feedlot waste rations, respectively, to 41.7 percent and 39.6 percent in the 40 percent and 60 percent feedlot waste rations, respectively. One explanation for this unusual pattern is that when the feedlot waste was composted, apparent digestibility of the feedlot waste was improved from 29.3 percent in unaltered feedlot waste to 42.0 percent in the composted feedlot waste (Table 8).

**TABLE 6. APPARENT DIGESTION COEFFICIENTS OF RATIONS FOR TRIAL II USING UNALTERED CATTLE FEEDLOT WASTE (PERCENT)**

| Item | Rations* | | | |
|---|---|---|---|---|
| | Control | 20 percent waste | 40 percent waste | 60 percent waste |
| Dry matter | 74.29 | 63.35 | 59.87 | 51.66 |
| Organic matter | 75.55 | 66.52 | 63.81 | 56.11 |
| Gross energy | 73.58 | 64.07 | 60.81 | 54.22 |
| Crude protein, apparent | 65.05 | 53.05 | 48.57 | 44.13 |
| Crude protein, true | 83.57 | 70.91 | 65.42 | 59.65 |
| Cell soluble material | 83.57 | 78.76 | 72.81 | 68.07 |
| Cell wall material | 31.03† | 29.84† | 41.72‡ | 39.60‡ |

*Treatment means on the same line are significantly different (P<0.01), except for cell wall material, in a linear pattern.
†,‡Means on the same line having different superscripts are significantly different (P<0.05).

Table 7 portrays apparent digestion coefficients for the rations used in Trial III. Little difference in digestibility of ration contents was observed between the rations containing unaltered or composted waste (low protein - low energy) for dry matter, gross energy, and cell soluble material; but these apparent digestion coefficients were significantly (P<0.05) lower than those for the control ration. Apparent digestion coefficients for organic matter and crude protein were significantly lower (P<0.05) for the composted feedlot waste ration than for the unaltered feedlot waste ration, and these apparent digestibilities were significantly lower (P<0.05) than those for the control ration. Cell wall material apparent digestibility was significantly lower (P<0.05) for the unaltered feedlot waste ration (38.0 percent) than for the control ration (41.4 percent) and the composted waste ration (42.0 percent). This observation would agree with data in Table 6 showing improved cell wall digestibility in the 40 percent composted waste ration.

**TABLE 7. APPARENT DIGESTION COEFFICIENTS OF RATIONS FOR TRIAL III USING LOW — ENERGY LOW PROTEIN CATTLE RATIONS (PERCENT)**

| Item | Rations | | |
|---|---|---|---|
| | Control | Unaltered 40 percent waste | Composted 40 percent waste |
| Dry matter | 52.27* | 44.36† | 42.04† |
| Organic matter | 53.99* | 49.48† | 44.93‡ |
| Gross energy | 52.47* | 46.67† | 43.95† |
| Crude protein, apparent | 23.02* | 17.87† | 13.24‡ |
| Crude protein, true | 48.16* | 41.98† | 38.52† |
| Cell soluble material | 61.59* | 56.64* | 57.11† |
| Cell wall material | 41.40* | 38.00† | 41.98* |

*, †, ‡ Means on the same line having different superscripts are significantly different (P<0.05).



MANAGING LIVESTOCK WASTES

By using the difference method for calculating the digestibility of ration components, an overall summary of the apparent digestibility of cattle feedlot waste when included at a level of 40 percent in either high grain or high roughage rations is presented in Table 8. These data suggest that when feedlot waste is composted and fed in high concentrate — adequate protein rations at a level of 40 percent, the apparent digestibility of energy-yielding components (dry matter, organic matter and gross energy) would be improved 10-15 percent; whereas crude protein, cell soluble material and cell wall material digestibilities would be improved 20-25 percent. The data in Table 8 also indicate that when feedlot waste is included in a low energy - low protein ration, the feedlot waste should be digested to a greater degree than when fed in a high energy - adequate protein ration. These data further suggest that when feedlot waste is composted and fed in low energy - low protein rations at a level of 40 percent, the apparent digestibility of dry matter and organic matter would be unchanged, gross energy would be decreased by about 10 percent, crude protein decreased by 25 percent, and cell soluble material should be decreased by about 8 percent. The apparent digestibility for cell wall material was increased by nearly 100 percent due to composting.

The apparent digestion coefficients reported in Table 8 for unaltered feedlot waste are higher than values reported by Lucas et al. (1974) using 20 percent cattle fecal waste in a 50 percent roughage ration. The difference can be explained by the lower level of fecal waste used by the latter researchers. In the present study, apparent digestion coefficients were calculated by the difference method for cattle feedlot waste when fed at 20 percent, 40 percent and 60 percent levels in cattle rations. Apparent digestion coefficients for feedlot waste at the 40 percent level are presented in Table 8. When compared at the 40 percent level of feeding, apparent digestion coefficients for feedlot waste were lower when fed at a 20 percent level, but higher when fed at a 60 percent level. This linear expression of increasing values for apparent digestibility was unexpected, but could indicate an associative effect of feedstuffs. This observation was reported by Anthony in waste feeding trials with cattle at Auburn University (personal communication).

**TABLE 8. SUMMARY OF APPARENT DIGESTION COEFFICIENTS FOR CATTLE FEEDLOT WASTE WHEN FED AT A 40 PERCENT LEVEL IN CATTLE RATIONS (PERCENT)***

| | | High energy Adequate protein | | Low energy Low protein | |
|---|---|---|---|---|---|
| Item | Type of ration: | | | | |
| | Type of waste: | Unaltered | Composted | Unaltered | Composted |
| Dry matter | | 33.26 | 38.38 | 37.59 | 37.91 |
| Organic matter | | 34.76 | 38.08 | 38.85 | 38.77 |
| Gross energy | | 34.30 | 37.79 | 38.82 | 35.03 |
| Crude protein, apparent | | 26.17 | 32.96 | 58.04 | 45.83 |
| Crude protein, true | | 29.27 | 36.03 | 61.09 | 49.02 |
| Cell soluble material | | 40.83 | 50.25 | 65.32 | 59.53 |
| Cell wall material | | 29.31 | 42.00 | 16.53 | 33.19 |

*Calculated by the difference method.

Johnson (1972) used sheep to determine the digestibility of cattle feedlot waste when included at 25 percent and 40 percent levels in low energy rations with cottonseed hulls as the roughage. The rations he used for the 40 percent levels would be comparable to the 40 percent ration of this study. The digestibility values for dry matter and organic matter of waste alone were about 10 percent higher than those in the present study (about 38 percent, Table 8). The cell wall material digestion coefficient for the waste in this study was near 17 percent. Johnson's crude protein values were higher than the coefficients for this study. In experiments to more closely evaluate beef cattle feedlot waste as a protein source for growing lambs, Johnson et al. (1973) found that the feedlot waste samples were much lower in crude protein values ($<$12 percent) than those reported by Johnson (1972) and that its digestibility was lower ($<$40 percent). Their data suggested that composting may decrease the crude protein value of beef cattle feedlot waste. Our data (Table 8) indicate a similar trend for low energy rations, but that little effect of composting might be detected in high energy rations with lower overall digestibility values.

Ferrell and Garrett (1973) studied the effect of continued recycling of cattle feedlot manure with special interest in the possible buildup of substances in the manure. Their results suggest that there is some nutritive value in cattle manure and that recycling would not create a buildup of minerals or lignin. They reported that only about 26 percent of the dry matter in manure disappeared during the recycling. With only one recycling, the data in Table 8 would suggest about 33-35 percent disappearance of dry matter.

There were no problems with feed consumption. The steers readily consumed all rations, regardless of the level of feedlot waste. No animal health problems were observed.

SUMMARY

This study was conducted to determine the nutritive value of feedlot waste from southwestern cattle feedlots where improved grain processing techniques and low levels of roughage are being used. Waste that had been scraped from the soil surface of a commercial beef cattle feedlot and stockpiled for approximately 1 month was ground through a hammer mill over a 0.25-in. screen. Rations containing feedlot waste were offered to feeder steers in three, 28-day total collection trials. Unaltered or composted feedlot waste was mixed into high energy - adequate protein and low energy - low protein rations at concentrations of 0, 20, 40 and 60 percent. As the percentage of unaltered feedlot waste increased in high energy - adequate protein rations, the apparent digestibility of nutrient and energy components decreased in a significant (P$<$0.01) linear pattern. The same trends were observed when composted waste was used except for the apparent digestibility of cell wall material which increased significantly (P$<$0.05). In low energy - low protein rations, apparent digestion coefficients were significantly lower (P$<$0.05) for rations containing feedlot waste.

Composting feedlot waste lowered the apparent digestibility of rations for organic matter and crude protein, but cell wall digestibility was lower for rations containing unaltered feedlot waste. Apparent digestion coefficients for feedlot waste were higher when the waste was included in low energy - low protein rations than for high energy - adequate protein rations. Composting decreased the apparent digestibility of feedlot waste for all components except cell wall material, which was significantly increased, when included in low energy - low protein rations. The apparent digestibilities for feedlot waste increased at each

*(Continued on page 217)*

# Nutritive Value of Swine Feces for Swine

M. R. Holland, E. T. Kornegay, J. D. Hedges

THE use of livestock and poultry waste as a feed ingredient offers potential for salvaging some of the nutrients and for minimizing the waste disposal problem. A considerable amount of research has been done with processing methods and the feeding of broiler litter to ruminants (Fontenot and Webb 1974) and cattle manure to ruminants (Anthony 1974). Little research has dealt with the refeeding of swine feces, although, dried swine feces (Diggs et al. 1965a, Orr 1973, Eggum and Christensen 1973), oxidation ditch liquor (Orr 1973), and oxidation ditch mixed liquor (Harmon 1971) have been tested as feed ingredients.

The objectives of this study were to characterize wet and dried swine feces as to its proximate, energy and mineral composition and to determine digestibilities and retention values for the proximate, energy and mineral components.

## PROCEDURE

Forty-eight crossbread gilts averaging 125 kg were used in two trials of two replications each. Twelve gilts in each replication were randomly allotted from weight outcome groups to three rations. Stainless steel cages described by Baker et al. (1967) were used for these trials. The gilts were fed twice daily and an excess of water was placed in the feeding trough after the gilts were allowed one hour to eat.

In the first trial (wet) fresh swine feces were incorporated into the rations at the time of feeding. The three test rations were: (1) basal (15 percent crude protein fortified corn-soybean meal*), (2) basal substituted with 22.90 percent swine feces, and (3) basal substituted with 33.83 percent swine feces. The feces were collected from finishing hogs that were housed on slotted floors consuming a similar corn-soybean meal ration. The feces dropped through the slots and were collected from boards that were 0.91 by 1.21 m and sloped 2.54 cm per 0.3 m towards the rear of the pen. The sloped board facilitated the runoff of excess urine so that urine contamination could be kept at a low level. The feces were weighed and incorporated into the basal ration at a level based on its estimated dry matter content. Sufficient water was blended into rations to hold dry matter at a constant level.

In the second trial (dry) fresh feces collected in the same manner as previously described was then placed in large pans lined with wire mesh screen at a depth of approximately 3.0 cm and dried for 24 hours in a forced draft oven at 75 C. The dried swine feces were then ground through a hammer mill with a 60 mesh (0.6 cm X 0.6 cm square) screen. The test diets were blended in a horizontal mixer.

Dried feces were substituted for the 15 percent basal corn-soybean ration previously used at levels of 20.50 percent and 40.71 percent of the dry matter of the basal rations.

In both trials the gilts were placed in the stainless steel cages for a minimum of five to seven days before adjustment to test rations was begun. The level of feces fed in the test ration was gradually increased over a five day period. When the maximum level was reached the gilts were fed for three days and then a five day total collection was carried out. Feces were collected from the gilts twice daily and the pH of the urine was adjusted as needed to prevent the loss of ammonia. Feces collected from the gilts were dried for twenty-four hours at 60 C in a forced air oven at the time of collection and redried in the same manner at the end of the trial.

The feces were then ground through a Wiley Mill equipped with a 1 mm screen and mixed thoroughly. Samples were then taken and packed tightly in feed cups to prevent layering or shifting of the particulate matter. Urine was sampled at the end of the trial and frozen until analysis could be performed.

Crude fiber determinations were made by the procedure of Whitehouse et al. (1945). Other proximate components and total nitrogen were determined by A.O.A.C. (1970) procedures. Phophorus determinations were made by the method of Fisher and Subbarrow (1925). Other minerals were determined by atomic absorption spectrophotometry, after appropriate digestion and dilution.

Statistical analysis was performed on the VPI & SU computer network by the Statistical Analysis System as implemented by Barr and Goodnight (1972). Regressions were calculated by the methods of Snedecor and Cockran (1967).

## RESULTS AND DISCUSSION

There was a wide variation between animals in their acceptance of the feces blended rations with less acceptance of the wet feces blended rations.

Ash, crude fiber, ether extract, and crude protein content of rations fed was increased as feces were substituted for the basal ration (Table 1). Gross energy was about the same and nitrogen free extract (NFE) was lowered as feces was substituted. The concentration of all minerals analyzed was increased as feces were substituted for the basal ration. The composition of feces substituted in these trials was similar to that reported by Orr (1973), Eggum and Christensen (1974) and Kornegay (1974).

The fecal content of crude fiber, ether extract, crude protein, magnesium, copper and zinc increased as the amount of feces substituted was increased (Tables 2 & 3). The fecal content of ash, NFE, calcium, phosphorus and potassium tended to be constant. There was a large increase in the amount of feces excreted by the gilts when feces were substituted for the basal ration with no differences in urinary output. The significant difference in fecal excretion

The Virginia Agricultural Foundation supplied partial funding for this research.

The authors are: M. R. HOLLAND, E. T. KORNEGAY and J. D. HEDGES, Department of Animal Science, Virginia Polytechnic Institute and State University, Blacksburg, Virginia 24061.

*Percentage composition of basal ration: ground yellow corn, 78.5; solvent extracted soybean meal (mx 7 fbr.), 18.3; defluorinated phosphate, 1.2; limestone, 1.0; swine trace mineral salt (high zinc), 0.6; VPI vitamin premix, 0.4; and ferrous sulfate, 0.02.

TABLE 1. AVERAGE PROXIMATE, ENERGY AND MINERAL COMPOSITION OF TEST RATIONS AND FECES SUBSTITUTED IN WET AND DRY TRIALS (100 PERCENT DM BASIS)

| Components | Rations | | | Feces substituted |
|---|---|---|---|---|
| | 1* | 2† | 3‡ | |
| Ash, percent | 4.56 | 6.92 | 8.58 | 15.32 |
| Crude fiber, percent | 4.88 | 7.08 | 8.57 | 14.78 |
| Ether extract, percent | 3.02 | 4.11 | 4.88 | 8.02 |
| Crude protein, percent | 16.65 | 18.19 | 19.21 | 23.50 |
| Nitrogen free extract, percent | 70.91 | 63.71 | 58.76 | 38.28 |
| Gross energy, Kcal/g | 4.41 | 4.45 | 4.53 | 4.57 |
| Calcium, percent | 0.74 | 1.19 | 1.49 | 2.72 |
| Phosphorus, percent | 0.56 | 0.86 | 1.11 | 2.13 |
| Magnesium, percent | 0.18 | 0.33 | 0.44 | 0.93 |
| Potassium, percent | 0.72 | 0.85 | 0.95 | 1.34 |
| Zinc, ppm | 82.43 | 168.41 | 249.65 | 530.37 |
| Copper, ppm | 8.62 | 20.32 | 28.33 | 62.83 |

*Basal corn-soybean ration.
†Feces substituted for 21.74 percent on DM basis.
‡Feces substituted for 37.27 percent on DM basis.

between trials (Table 2) reflects primarily the difference in dry matter intake which was greater in the wet trial, although, there was slightly higher digestibility of dry matter in the wet trial as compared to the dry trial (50.44 vs. 47.71 percent).

TABLE 2. AVERAGE AMOUNT OF FEED, FECES AND URINE AND PROXIMATE COMPONENTS OF FECES AND URINE FROM GILTS ON TEST RATIONS IN WET AND DRY TRIALS (100 PERCENT DM BASIS)

| Components | Rations | | | Sig.† |
|---|---|---|---|---|
| | 1‡ | 2§ | 3‖ | level |
| *Amounts of feed, feces and urine* | | | | |
| Feed intake, g/day | 1302.25 | 1352.71 | 1348.50 | 0.0001# |
| Fecal output, g/day | 136.69 | 268.79 | 348.12 | 0.0001# |
| Urinary output, l/day | 5.43 | 5.01 | 4.88 | 0.8378 |
| *Proximate components and energy of feces* | | | | |
| Ash, percent | 20.51 | 19.99 | 20.40 | 0.7304 |
| Crude fiber, percent | 15.20 | 15.73 | 16.47 | 0.0162# |
| Ether extract, percent | 6.45 | 7.01 | 6.95 | 0.0128# |
| Crude protein, percent | 20.45 | 19.36 | 18.37 | 0.0008 |
| Nitrogen free extract, percent | 37.40 | 37.91 | 37.82 | 0.8412# |
| Gross energy, Kcal/g | 4.62 | 4.70 | 4.72 | 0.0845 |

†Probability of difference between rations.
‡Basal corn-soybean ration.
§Feces substituted for 21.74 percent on DM basis.
‖Feces substituted for 37.27 percent on DM basis.
#Significant (P<0.05) difference between wet and dry trials.

The crude fiber, ether extract, NFE and gross energy values were different for feces collected from gilts on the wet trial versus the dry trial. In the case of crude fiber, the difference was due to a lower content (16.25 vs. 20.46 percent) and higher digestibility (49.25 vs. 29.67 percent) of crude fiber in the rations of the wet trial as compared to the dry trial. For ether extract, NFE and gross energy the differences are due primarily to differences in the rations fed as the digestion coefficients were not different between trials. There was a significant difference between the wet and dry trials for the fecal concentrations of calcium,

magnesium and zinc which was probably due to a higher level being fed as there was no difference in the digestion coefficients (Tables 3 & 6). Phosphorus was the only urinary component that was significantly increased as the level of feces substituted for the basal ration was increased.

TABLE 3. AVERAGE MINERAL COMPONENTS OF FECES AND URINE, AND URINARY NITROGEN AND ENERGY OF URINE AND FECES FROM GILTS ON TEST RATIONS IN WET AND DRY TRIALS (100 PERCENT DM BASIS)

| Components | Rations | | | Sig.† |
|---|---|---|---|---|
| | 1‡ | 2§ | 3‖ | level |
| *Mineral components of feces* | | | | |
| Calcium, percent | 4.39 | 4.18 | 4.27 | 0.3486# |
| Phosphorus, percent | 3.04 | 2.84 | 2.97 | 0.3153 |
| Magnesium, percent | 1.17 | 1.31 | 1.34 | 0.0015# |
| Potassium, percent | 0.91 | 0.92 | 0.93 | 0.0004 |
| Copper, ppm | 73.93 | 84.86 | 92.18 | 0.0004 |
| Zinc, ppm | 633.25 | 745.21 | 805.25 | 0.0001# |
| *Urinary components* | | | | |
| Calcium, ppm | 78.19 | 69.59 | 66.63 | 0.6971 |
| Phosphorus, ppm | 79.99 | 162.60 | 357.22 | 0.0001 |
| Magnesium, ppm | 109.77 | 162.60 | 131.03 | 0.1700 |
| Potassium, ppm | 1725.94 | 1880.71 | 1982.31 | 0.3260 |
| Urinary N, g/l | 4.69 | 5.05 | 5.13 | 0.8484 |
| Urinary energy, Kcal/g** | 2.05 | 2.16 | 2.13 | 0.8081 |

†Probability of difference between rations.
‡Basal corn-soybean ration.
§Feces substituted for 21.74 percent on DM basis.
‖Feces substituted for 37.27 percent on DM basis.
#Significant (P<0.05) difference between wet and dry trials.
**Freeze dried sample.

The apparent digestibilities of the proximate components, apparent net protein utilization, and apparent biological value of the test rations are given in Table 4. Digestion coefficients, for ash crude fiber, dry matter, ether extract, crude protein and NFE were all significantly lowered as feces were substituted for the basal ration with regressed values of 39.49, 39.53, 48.58, 54.23, 58.47 and 37.45 percent, respectively. Orr (1973) reported lower dry matter and protein digestibilities of a corn-dried swine feces ration as compared to a corn-soybean meal ration. The digestibilities of dry matter, crude protein and energy of the basal ration were almost identical to the values obtained by Orr (1973) for a similar basal ration. The digestibilities of dry matter and energy for the corn-dried swine feces ration reported by Orr (1973) were intermediate between rations 1 and 2 in this study; whereas, the crude protein digestibility was higher in this study than reported by Orr (1973).

Orr (1973) reported lower apparent NPU and BV values for a corn-dried swine feces ration as compared to a corn-soybean meal ration which is in contrast to results obtained in this study in which NPU and BV values were not significantly affected. Eggum and Christensen (1974) feeding rats found that NPU and BV values were significantly reduced when pig feces were substituted for casein, but when pig feces were substituted for barley the BV was increased at

TABLE 4. AVERAGE APPARENT DIGESTIBILITIES OF PEOXIMATE COMPONENTS, NET PROTEIN UTILIZATION AND BIOLOGICAL VALUE DETERMINATIONS FOR WET AND DRY TRIALS

| Item | Rations | | | Regressed* | Sig.† |
|---|---|---|---|---|---|
| | 1‡ | 2≠ | 3‖ | 100 percent | level |
| Digestibility of | | | | | |
| Ash, | | | | | |
| percent | 52.94 | 42.27 | 39.33 | 39.49 | 0.0015 |
| Crude fiber, | | | | | |
| percent | 67.22 | 55.03 | 50.47 | 39.53 | 0.0001# |
| Dry matter, | | | | | |
| percent | 89.46 | 80.11 | 74.34 | 48.58 | 0.0001# |
| Ether extract, | | | | | |
| percent | 77.02 | 66.20 | 64.36 | 54.23 | 0.0001 |
| Crude protein, | | | | | |
| percent | 86.95 | 78.76 | 75.60 | 58.47 | 0.0001# |
| Nitrogen free ex- | | | | | |
| tract, percent | 94.43 | 88.12 | 83.26 | 37.45 | 0.0001 |
| Net protein | | | | | |
| utilization, | | | | | |
| percent** | 31.05 | 29.26 | 26.52 | 20.90 | 0.3636 |
| Biological value, | | | | | |
| percent†† | 35.75 | 37.05 | 35.20 | 35.77 | 0.9520 |

*Obtained by regression analysis.
†Probability of difference between rations.
‡Basal corn-soybean ration.
§Feces substituted for 21.74 percent on DM basis.
‖Feces substituted for 37.27 percent on DM basis.
#Significant difference between wet and dry trials.
**Apparent net protein utilization equals retained protein as a percent of protein intake.
††Apparent biological value equals retained protein as a percent of absorbed protein.

TABLE 5. AVERAGE DIGESTIBLE ENERGY, METABOLIZABLE ENERGY AND METABOLIZABLE ENERGY CORRECTED FOR NITROGEN RETENTION FOR WET AND DRY TRIALS (100 PERCENT DM BASIS)

| Components | Rations | | | Regressed* | Sig.† |
|---|---|---|---|---|---|
| | 1‡ | 2§ | 3‖ | 100 percent | level |
| Energy per gram of | | | | | |
| test ration, Kcal | | | | | |
| Digested | 3.92 | 3.51 | 3.32 | 2.27 | 0.0001 |
| Metabolized | 3.76 | 3.33 | 3.15 | 2.07 | 0.0001 |
| Metabolized, cor- | | | | | |
| rected for | | | | | |
| nitrogen re- | | | | | |
| tention | 3.67 | 3.24 | 3.06 | 1.98 | 0.0001 |
| Digestible, | | | | | |
| percent | 88.96 | 78.89 | 73.32 | 46.45 | 0.0001 |
| Metabolizable, | | | | | |
| percent | 85.39 | 74.88 | 69.55 | 42.15 | 0.0001 |
| Metabolizable, cor- | | | | | |
| rected for nitrogen | | | | | |
| retention, | | | | | |
| percent | 83.30 | 72.75 | 67.54 | 40.23 | 0.0001 |

*Obtained by regression analysis.
†Probability of difference between rations.
‡Basal corn-soybean ration.
§Feces substituted for 21.74 percent on DM basis.
‖Feces substituted for 37.27 percent on DM basis.

TABLE 6. AVERAGE MINERAL DIGESTIBILITIES AND RETENTIONS FOR WET AND DRY TRIALS

| Item | Rations | | | Regressed* | Sig.† |
|---|---|---|---|---|---|
| | 1‡ | 2§ | 3‖ | 100 percent | level |
| Absorbed Ca, percent | 38.62 | 30.09 | 27.02 | 16.92 | 0.0276 |
| Retained Ca, percent | 34.91 | 28.11 | 25.40 | 20.13 | 0.0788 |
| Retained Ca, percent | | | | | |
| of Ab. | 89.36 | 92.05 | 92.79 | 95.06 | 0.2362 |
| Absorbed P, percent | 42.59 | 33.99 | 31.49 | 26.35 | 0.0361 |
| Retained P, percent | 37.28 | 23.25 | 21.35 | 11.79 | 0.0013 |
| Retained P, percent | | | | | |
| of Ab. | 86.55 | 64.41 | 61.23 | 46.15 | 0.0031 |
| Absorbed K, percent | 86.42 | 78.38 | 74.88 | 61.75 | 0.0001 |
| Retained K, percent | 13.68 | 17.99 | 13.56 | 16.87 | 0.5200 |
| Retained K, percent | | | | | |
| of Ab. | 16.01 | 22.31 | 18.61 | 25.38 | 0.5472 |
| Absorbed Mg, percent | 30.36 | 21.32 | 21.96 | 17.54 | 0.1435 |
| Retained Mg, percent | 16.35 | 14.76 | 17.98 | 16.73 | 0.0061 |
| Retained Mg, percent | | | | | |
| of Ab. | 40.12 | 50.12 | 59.84 | 63.13 | 0.0061# |
| Absorbed Cu, percent | 17.92 | 17.16 | 18.51 | 17.92 | 0.8432 |
| Absorbed Zn, percent | 20.19 | 15.54 | 18.71 | 15.98 | 0.5390 |

*Obtained by regression analysis.
†Probability of difference between rations.
‡Basal corn-soybean ration.
§Feces substituted for 21.74 percent on DM basis.
‖Feces substituted for 37.27 percent on DM basis.
#Significant difference between wet and dry trials.

lower levels of addition and NPU decreased at a much lower rate. Values of NPU and BV for rations including swine feces were similar in this study to values reported by Orr (1973); however, values for the basal ration were lower in this study as compared to values for the basal ration in the study by Orr (1973). NPU and BV values of the basal ration in this study are somewhat lower than reported previously for smaller hogs (58 kg) fed a similar basal ration (Kornegay 1973).

The predicted digestibility of the fecal crude protein of 58.47 percent in this study compares closely with that of 60 percent determined using rats (Eggum and Christensen 1974). The digestible energy, metabolized energy and metabolizable energy corrected for nitrogen retention all decreased significantly as the level of substituted feces was increased (Table 5). Orr (1973) also reported a decrease in digestible energy when a corn-dried swine feces ration was compared to a corn-soybean meal ration. The energy values obtained for the basal ration in this study are comparable to those reported by Diggs et al. (1965b). The predicted values of digestible energy, metabolized energy and metabolizable energy corrected for nitrogen retention for swine feces are all substantially lower than those for the basal corn-soybean meal ration. This is in agreement with the prediction of Eggum and Christensen (1974), that the "digestible energy in pig feces is much lower than our traditional feed mixtures."

Mineral retention and digestibility data given in Table 6 show that there was a significant decrease in the percentage of absorbed calcium, phosphorus and potassium as feces were substituted for the basal ration. Retained phosphorus as a percent of intake and as a percent of absorbed phosphorus was also significantly decreased. There was a significant increase in retained magnesium as a percent of absorbed magnesium as feces were substituted with the value for the wet trial being significantly higher. Absorbed and retained values for both calcium and phosphorus agree closely with values reported by Kite (1974) for gestating sows.

## SUMMARY

Swine feces collected from finishing hogs was fed to 48 crossbred gilts averaging 125 kg body weight in two total collection metabolism trials to characterize wet and dried swine feces as to its proximate, energy and mineral composition and to determine digestibilities and retention values for the proximate, energy and mineral components. One trial was conducted using unprocessed feces (fresh) and another trial was conducted using dried feces.



MANAGING LIVESTOCK WASTES

Fecal content of crude fiber, ether extract, crude protein, magnesium, copper and zinc increased as the amount of feces substituted for the basal ration was increased. The fecal content of ash, NFE, calcium, phosphorus and potassium tended to be constant. Only urinary phosphorus was found to be significantly increased when feces were substituted for the basal ration. The amount of feces excreted increased as the amount of feces substituted for the basal ration was increased with no differences in urinary output.

Swine feces was found to be of less nutritive value than a basal corn-soybean meal ration, but nutrients were available. Digestion of ash, crude fiber, dry matter, ether extract, crude protein and nitrogen free extract were all significantly reduced when swine feces was substituted for the basal ration. Net protein utilization and biological value were not significantly changed. There were significant decreases in digestible energy, metabolized energy and metabolizable energy corrected for nitrogen retention as feces were substituted for the basal ration. Absorbed calcium, phosphorus and potassium were significantly decreased as feces were substituted for the basal ration with no changes in absorbed copper and zinc. Retained phosphorus as a percent of gross intake and as a percent of absorbed phosphorus was significantly lowered as feces was substituted for the basal ration.

Digestibilties and retention values were similar between the wet and dry trials.


## References

1   Anthony, W. B. 1974. The nutritional value of cattle waste for cattle. Federation Proceedings. 33:1939.

2   A.O.A.C. 1970. Official methods of analysis. (11th Ed.) Association of Official Agricultural Chemists, Washington, D.C.

3   Baker, D. H., W. H. Hiott, H. W. Davis and C. E. Jordan. 1967. A swine metabolism unit. Lab. Practice. 16:1385.

4   Barr, J. A. and J. H. Goodnight. 1972. A user's guide to the statistical analysis system, Student Supply Store, N.C. State Univ., Raleigh, N.C.

5   Diggs, B. G., B. Baker, Jr. and F. G. James. 1965a. The value of pig feces in swine finishing rations. J. Anim. Sci. 24:291, (Abstr).

6   Diggs, B. G., D. E. Becker, A. H. Jensen and H. W. Norton. 1965b. Energy values of various feeds fed young pigs. J. Anim. Sci. 24:555.

7   Eggum, B. D. and K. D. Christensen. 1974. Utilization by rats of proteins in pig feces relative to barley and fortified casein proteins. Agric. Environ. 1:59.

8   Fiske, C. H. and T. Subbarrow. 1925. The colorometric determination of phosphorus. J. Biol. Chem. 66:375.

9   Fontenot, J. P. and K. E. Webb, Jr. 1974. Poultry wastes as feedstuff for ruminants. Federation Proceedings. 33:1936.

10   Harmon, B. G., D. L. Day, A. H. Jensen and D. H. Baker. 1971. Liquid feeding of oxidation ditch mixed liquor to swine. J. Anim. Sci. 33:1149.

11   Kite, B. W. 1974. Evaluation of low and high levels of phosphorus for gestating and lactating gilts and sows. M.S. Thesis, VPI and SU, Blacksburg, Va; p. 56.

12   Kornegay, E. T. 1974. Swine manure as a feed and fertilizer. Eighth Va. Pork Industry Conf., December 3-5.

13   Kornegay, E. T. 1973. Digestible and metabolizable energy and protein utilization values of brewers dried by-products for swine. J. Anim. Sci. 37:479.

14   Orr, D. E. 1973. Swine waste as a nutrient source for finishing pigs. Rpt. Swine Research, Mich. State Univ., E. Lansing, AH-SW-7319, p. 81.

15   Snedecor, G. W. and W. G. Cockran. 1967. Statistical methods. Iowa State University Press. 593 p.

16   Whitehouse, K. A., A. Zarrow and H. Shay. 1945. Rapid method of determining "crude fiber" in distillers' dried grain. J. Assoc. Official Agr. Chem. 28:147.


---

## Nutritional Value of Cattle Feedlot Waste

*(Continued from page 213)*

higher percentage of use.


## References

1   Albin, Robert C. 1971. Handling and disposal of cattle feedlot waste. J. Anim. Sci. 32:803-810.

2   Anonymous. July 24, 1971. EPA takes "lenient" stand on feedlot waste disposal. Feedstuffs, p. 4.

3   Anthony, W. Brady. 1971. Animal waste value - nutrient recovery and utilization. J. Anim. Sci. 32:799-802.

4   Durham, R. M., G. W. Thomas, R. C. Albin, L. G. Howe, S. E. Curl and T. W. Box. 1966. Coprophagy and use of animal waste in livestock feeds. Proc. Natl. Symp. Animal Waste Mgt. ASAE Publ., SP-0366:112-114.

5   Ferrell, C. L. and W. N. Garrett. 1973. Observations concerning the use of cattle manures in drylot feeding. Proc. 12th Ann. Calif. Feeder's Day. p. 4.

6   Johnson, R. R. 1972. Digestibility of feedlot waste. Okla. State Univ. Agr. Exp. Sta. Misc. Publ. 87:62.

7   Johnson, R. R., J. R. Kropp and M. McGeehon. 1973. Animal wastes as protein sources for ruminants. Okla State Univ. Agr. Exp. Sta. Misc. Publ. 90:94.

8   Lucas, D. M., J. P Fontenot and K. E. Webb, Jr. 1974. Digestibility of cattle fecal waste. J. Anim. Sci. 38:231.

# The Inclusion of Pig Manure in Ruminant Diets

## G. R. Pearce

IN Australia, relatively little attention has been given so far to the problems of livestock waste disposal. Pig and poultry industries are generally highly intensified and since many units are located close to urban areas waste disposal is emerging as a major concern. On the other hand, beef cattle and dairy cattle feedlots are few, and except in specific instances where runoff may cause pollution of waterways, their problems are those associated with the mechanics of waste removal and distribution.

In the past, organized disposal of livestock wastes in Australia has been exclusively by application to land as a fertilizer. More recently, the possibilities of recycling back through animals has attracted some attention. Reports of experiments in which poultry manure and poultry litter have been fed to sheep have been made by McInnes, Austin and Jenkins (1968), Leibholz (1969) and Jacobs and Leibholz (1974).

In this paper, studies at the University of Melbourne with cattle and sheep fed diets containing pig manure are summarized.

### Variation in the Composition of Pig Manure

Samples of fresh, uncontaminated pig manure were collected from grower-finisher pigs in 20 piggeries. Each sample was immediately deep-frozen and subsequently freeze-dried and ground with a pestle and mortar. Results of analyses are shown in Table 1. Values for mercury, fluorine and selenium and information on antibiotic and pesticide residues are pending.

### Digestibility Studies with Cattle

Four steers, weighing 350-380 kg, were fed diets in which pig manure dried at 110 C for 20 hours replaced hay in proportions of 0, 15, 30 and 45 percent; the only other ingredient was molasses at 3.0 percent of the diet. Composition of the manure and of the diets is shown in Table 2.

Acknowledgments: This work is being financed by the Australian Meat Research Committee. The cattle and sheep experiments were conducted by G. Stanogias and S. Hendrosoekarjo with the assistance of technical staff of the Animal Production Section, University of Melbourne. Most of the chemical analyses were performed by Laboratory staff of the Animal Production Section, University of Melbourne.

The author is: G. R. PEARCE, Senior Lecturer in Animal Nutrition, School of Agriculture & Forestry, University of Melbourne, Parkville, Victoria, Australia.

**TABLE 1. COMPOSITION OF PIG MANURE FROM 20 DIFFERENT SOURCES (DRY MATTER BASIS)**

| Constituent | Mean | Range | |
|---|---|---|---|
| Crude protein, percent | 18.9 | 12.7 | - 30.0 |
| Ether extract, percent | 6.6 | 4.0 | - 10.9 |
| Acid detergent fiber, percent | 26.5 | 20.7 | - 32.2 |
| Gross energy, kJ/kg | 18 100 | 16 500 | - 20 100 |
| Ash, percent | 18.1 | 13.6 | - 23.6 |
| Calcium, percent | 3.2 | 1.1 | - 5.9 |
| Phosphorus, percent | 2.5 | 1.2 | - 3.4 |
| Potassium, percent | 1.2 | 0.7 | - 2.5 |
| Magnesium, percent | 0.8 | 0.4 | - 1.3 |
| Sulphur, percent | 0.3 | 0.2 | - 0.4 |
| Copper, mg/kg | 249 | 22 | - 636 |
| Zinc, mg/kg | 526 | 128 | - 981 |
| Iron, mg/kg | 1940 | 764 | -4700 |
| Aluminium, mg/kg | 544 | 280 | - 896 |
| Manganese, mg/kg | 342 | 114 | - 561 |
| Cobalt, mg/kg | 6.1 | 2.2 | - 15.2 |
| Molybdenum, mg/kg | 0.3 | 0.2 | - 0.5 |
| Cadmium, mg/kg | 1.0 | 0.3 | - 2.4 |
| Lead, mg/kg | 12.1 | 4.0 | - 42.0 |

The diets were pelleted and fed at the rate of 6.5 kg dry matter/day. A 4 x 4 Latin square design was used under conventional conditions for the collection of feces.

Dry matter digestibility values were $57.9 \pm 2.33$, $52.3 \pm 4.05$, $50.0 \pm 2.09$ and $44.4 \pm 1.32$ for the diets containing 0, 15, 30 and 45 percent manure, respectively. Extrapolating these values linearly suggested that the manure itself had a dry matter digestibility of 29 percent. There was no evidence of associative effects.

### Digestibility Studies with Sheep

Three groups of three wethers were fed diets in which pig manure dried at 100 C for 24 hours replaced hay in the proportions 0, 15, and 30 percent; molasses was included at 3.0 percent of the diet. The composition of the manure and of the diets is shown in Table 3. The diets were pelleted and fed at the rate of 1200 g/day. Each group was fed one diet continuously for 80 days. Over the last 70 days, samples of daily collections of feces and urine were bulked over 7-day periods.

Mean dry matter digestibility values over the 10 7-day periods were $49.2 \pm 0.3$, $45.2 \pm 0.4$ and $41.5 \pm 0.6$ for the diets containing 0, 15 and 30 percent pig manure, respectively. Depression in digestibility with increasing proportions of manure in the diets was of the same order as in the cattle experiment, indicating again a poor digestibility of the manure itself.

**TABLE 2. CHEMICAL COMPOSITION OF MANURE AND DIETS USED IN DIGESTIBILITY STUDIES WITH CATTLE (DRY MATTER BASIS)**

| Constituent | Manure | Diet 1 no manure | Diet 2 15 percent manure | Diet 3 30 percent manure | Diet 4 45 percent manure |
|---|---|---|---|---|---|
| Crude protein, percent | 16.7 | 17.4 | 17.3 | 16.2 | 17.1 |
| Acid detergent fiber, percent | 23.5 | 30.2 | 29.2 | 27.8 | 27.0 |
| Ether extract, percent | 5.9 | 5.6 | 5.1 | 4.5 | 5.9 |
| Ash, percent | 20.4 | 8.2 | 9.8 | 11.2 | 13.2 |
| Gross energy, kJ/kg | 18060 | 24830 | 20400 | 20360 | 19230 |

| Constituent | Manure | Diet 1 no manure | Diet 2 15 percent manure | Diet 3 30 percent manure |
|---|---|---|---|---|
| Crude protein, percent | 19.4 | 13.0 | 13.0 | 13.6 |
| Crude fiber, percent | 25.4 | 31.5 | 30.4 | 29.6 |
| Ether extract, percent | 5.1 | 2.1 | 2.4 | 2.5 |
| Ash, percent | 21.1 | 7.4 | 9.4 | 11.3 |
| Nitrogen-free extract, percent | 29.0 | 46.0 | 44.8 | 43.0 |
| Gross energy, kJ/kg | - | 17050 | 16840 | 16750 |

## Effect of Drying Temperature on the Digestibility of Pig Manure

In the digestibility experiments with cattle and sheep, manure from the same source was used and the temperatures at which the manures were dried were relatively high (110 C and 100 C, respectively). Hay components were different in each experiment. To investigate whether drying temperature was significant with respect to the low digestibility of the manure, batches were dried (i) at 110 C and (ii) at 65 C. Two groups each of 6 sheep were fed 1250 g/day of diets containing 30 percent pig manure. Conventional digestibility trial conditions were applied.

Dry matter digestibilities were 41.7 ± 1.9 percent for the diet containing manure dried at the high temperature and 42.3 ± 1.4 percent for the diet containing manure dried at the low temperature.

Drying temperature did not, therefore, have any effect on digestibility.

## Effect of Copper Content of Pig Manure on Its Digestibility

In the digestibility experiments with cattle and sheep described above, the manure used was derived from pigs which had been fed diets containing high levels of copper (150 ppm) as a growth promotant. The resulting manure had a copper content of approximately 550 ppm and treatments containing 15, 30 and 45 percent had copper contents of 89, 168 and 245 compared with about 5 ppm in the diets of hay alone. To investigate whether the high levels of copper were inhibiting microbial activity, and hence digestion, in the recipient animals manure was collected from pigs fed the same diet as previously except that no added copper was used as a growth promotant. Copper content of this batch of manure was around 25 ppm.

Two sheep were used in a cross-over design to compare the high and low copper manures in diets containing 30 percent manure, with hay. Dry matter digestibilities were similar for both diets, namely around 43 percent, indicating that the high amounts of copper in the previous experiments had not caused depression of digestibility.

## Effect of Continuous Feeding of Dried Pig Manure to Cattle

Two steers were fed a pelleted diet containing 30 percent dried pig manure, 67 percent hay and 3 percent molasses continuously for 92 days at the rate of 7.0 kg/day. Feces and urine were collected daily and aliquot samples bulked over 4-day periods. The animals were slaughtered at the completion of the experiment.

Apart from general effects, specific interest was directed to the effect of the ingestion of the relatively high amounts of copper contained in the manure; this amounted to 1.15 g/day. To this end, samples of blood from the jugular vein

of each animal were taken at weekly intervals and analysed for plasma glutamate oxaloacetate thiaminase (G.O.T.) by a modification of the method of Reitman and Frankel (1957). Plasma G.O.T. levels normally range from about 70 - 140 units and rise to higher levels with liver degeneration induced by copper toxicity. A level of about 200 units may signify the approach of a haemolytic crisis.

In the first 8 weeks of feeding, both steers behaved normally and consumed their diets readily. However, at this point steer 1 suffered digestive upset accompanied by loss of appetite, weakness, and inability to stand. Its condition was rectified by the removal of the pelleted diet and the feeding of long hay. This upset recurred several times when pellets were re-introduced. Until the time of the initial upset the plasma G.O.T. levels of steer 1 fluctuated between 109 and 153 units and with the sickness rose to 182 and 185 units. It was not clear, however, whether its condition was caused by copper toxicity, the prolonged feeding of the pelleted diet or other effects. Steer 2 behaved normally throughout and its plasma G.O.T. levels did not rise above 96 units.

At slaughter, no obvious abnormalities appeared in the tissues of either animal, except for a hardening of the rumen epithelium of steer 1.

## Effect of Continuous Feeding of Dried Pig Manure to Sheep

Sheep are known to be less tolerant than cattle to the ingestion of high amounts of copper. It was previously stated that sheep fed diets containing 0, 15 and 30 percent dried pig manure (with hay) maintained dry matter digestibilities over 80 days of continuous feeding. Associated with these treatments were additional treatments in which molybdenum (as ammonium molybdate) was fed at two levels; sulphate (as sodium sulphate) was included in all diets. The complete range of treatments was as follows:

A   97 percent hay, 3 percent molasses

B   82 percent hay, 3 percent molasses, 15 percent dried pig manure

C   82 percent hay, 3 percent molasses, 15 percent dried pig manure, 90 mg molybdenum/kg dry matter

D   82 percent hay, 3 percent molasses, 15 percent dried pig manure, 176 mg molybdenum/kg dry matter

E   67 percent hay, 3 percent molasses, 30 percent dried pig manure

F   67 percent hay, 3 percent molasses, 30 percent dried pig manure, 90 mg molybdenum/kg dry matter

G   67 percent hay, 3 percent molasses, 30 percent dried pig manure, 176 mg molybdenum/kg dry matter.

All groups included 10.8 g sulphate/kg dry matter. Molybdenum was added to treatments C and G in proportions designed to achieve a copper:molybdenum ratio of 4:3 which was suggested by Matrone (1970) to offer the best opportunity for the alleviation of toxicity effects of dietary copper. The hay contained 10.3 mg copper and 0.24 mg molybdenum/kg. Treatments B, C, D contained 101.3 mg copper/kg, and treatments E, F and G contained 192.5 mg copper/kg.

Each group included three sheep fed 1200 g/day. After a preliminary period of 10 days feces and urine were collected daily and bulked over 7-day periods for 10 weeks. Blood samples were taken weekly from the jugular vein of each sheep for the determination of plasma G.O.T. At the end of the experiment the sheep were slaughtered, examined, and samples taken for liver copper analyses.

*(Continued on page 221)*

# A Summary of Refeeding of Poultry Anaphage, Mortality, Recycling Hens, and Egg Production


C. J. Flegal, H. C. Zindel, C. C. Sheppard, T. S. Chang,    J. E. Dixon, J. B. Gerrish, M. L. Esmay

AFFIIATE
ASAE

MEMBER        ASSOC. MEMBER        MEMBER
ASAE              ASAE                ASAE


COPROPHAGY, dung eating, has been observed in animals and poultry for many centuries. The custom of following beef cattle with swine and poultry in the same pasture was a common practice with diversified farming. However, the intentional inclusion of poultry excrement in the diet of growing or laying chickens has been a rather recent innovation. This technique was an outgrowth of the development of new techniques for manure disposal and/or utilization for the large poultry complex.

The purpose of this experiment was to test a complete system for collecting, handling, air-drying, machine dehydration, and utilization of poultry excreta in a caged layer production unit. This paper will emphasize the utilization of the dried product by laying hens.

## REVIEW OF LITERATURE

The use of 15 percent dried cow manure in laying diets (Palafox and Rosenburg 1951) resulted in a significant reduction of egg production. In this same trial, 5 and 10 percent dried cow manure did not significantly reduce egg production. Durham et al. (1966) fed dried feedlot manure to growing pullets and laying hens at 10, 25 and 40 percent of the diet. No differences in growth rate were observed due to dietary treatment; however, feed efficiency was inversely proportional to the level of dried manure in the diet. In this same trial, lower egg production was observed in the birds that consumed the diets that contained manure but the difference was not significant.

The use of dried poultry excreta in the diet of growing chicks and laying hens has been reported by several investigators. Flegal and Zindel (1971) reported no significant difference in growth rate of Leghorn type chicks (0-28 days) fed diets that contained up to 20 percent dried poultry excreta (DPW). However, when dried poultry excreta was included at 10 or 20 percent of the diet of broiler chicks, a significant reduction in growth rate resulted. Feed efficiency was inversely related to the amount of dried excreta in the diet. Biely et al. (1972) also reported that feeding 5-15 percent dried poultry excreta from 0-30 days to Leghorn chicks did not influence weight gain; but, again feed efficiency was inversely proportional to the amount of DPW in the ration. Other trials (Calvert et al. 1971, Blair and Knight 1973) have reported similar results from feeding DPW to growing chicks.

Quisenberry and Bradley (1968) reported no significant difference in egg production of laying hens fed 20 percent dried hen droppings when compared to the egg production of hens fed a control diet. In the same trial, hens fed 10 percent dried hen droppings produced at a higher rate of egg production than the control fed birds. Flegal and Zindel (1971) reported that the incorporation of up to 20 percent DPW in laying rations had no influence on egg production, Haugh values or shell thickness. Hodgetts (1971) observed that laying hens fed a diet containing 10 percent DPW performed slightly better than hens fed a control ration. Biely et al. (1972) reported that layings hens fed a diet that contained 25 percent DPW produced slightly less eggs than birds fed a control ration. In a trial utilizing continuous recycling of dried poultry excreta, Flegal et al. (1972) reported that birds fed a diet containing 12.5 percent DPW produced slightly more eggs than birds fed a control ration and birds fed a diet that contained 25 percent DPW produced slightly less eggs than the birds fed a control ration.

## MATERIALS AND METHODS

This experiment was conducted in a house described by Sheppard et al. (1974). Four rows of wire cages (stacked triple deck with dropping boards on the two top decks), 21.95 meters long, housed the birds. The individual cages were 30.48 x 40.64 cm. One half of the cages contained three birds per cage and one half of the cages contained four birds per cage.

The entire flock was fed a standard cage layer ration for the first 6½ months in the laying house. After that time, one half of the house was fed each of the diets listed in Table 1. The diets were formulated to be isocaloric and contain the same level of calcium and phosphorus. Feed and water supplied *ad libitum*. A 14-hour light day was provided for the first six months and then a 16-hour light-day for the remainder of the production period. After 13, 28-day laying periods of production, the entire flock was force molted as follows:

Day 1-3 — 24 hours darkness; no feed; no water

Day 4-6 — Water, *ad libitum*

Day 7-28 — Pullet developer, *ad libitum;* 8 hours light

Day 28 + — Placed back on experimental diets; 16 hours light

(Zero production was obtained on day 12, after starting the molt)

The post molt egg production cycle was then continued for 7.5 additional 28-day periods.

Eggs were gathered once daily and recorded according to dietary treatment. All mortality was recorded and veterinary laboratory examination was obtained for all mortality during the trial.

---


The authors are: C. J. FLEGAL, H. C. ZINDEL, C. C. SHEPPARD and T. S. CHANG, Poultry Science Dept., Michigan State University, East Lansing; J. E. DIXON, Agricultural Engineering Dept.,University of Idaho, Moscow; J. B. GERRISH and M. L. ESMAY, Agricultural Engineering Dept., Michigan State University, East Lansing.



Michigan Agricultural Experiment Station Journal Article No. 7173.




MANAGING LIVESTOCK WASTES

| Ingredient | Percent of diet | |
|---|---|---|
| | Diet 1 | Diet 2 |
| Corn, ground | 68.95 | 58.95 |
| Soybean meal, 49 percent | 16.50 | 16.10 |
| Alfalfa, 17 percent | 2.00 | 2.00 |
| Meat and bone meal, 50 percent | 3.00 | 3.00 |
| Limestone, ground | 7.00 | 5.70 |
| Dicalcium phosphate | 1.25 | 0.025 |
| Methionine, DL | 0.05 | 0.075 |
| Salt | 0.25 | 0.25 |
| Fat, A-V | 0.50 | 3.40 |
| Layer premix | 0.50 | 0.50 |
| Anaphage | - - | 10.00 |
| Total | 100.00 | 100.00 |
| **Calculated analysis** | | |
| Crude protein, percent | 16.00 | |
| Fat, percent | 3.54 | |
| Fiber, percent | 2.95 | |
| Calcium, percent | 3.30 | 3.30 |
| Phosphorus, percent (avail.) | 0.48 | 0.48 |
| Metabolizable energy, cal/lb | 1311.0 | 1311.0 |

## RESULTS

Egg production followed a typical curve for the first 75 days after housing. At that time, excessive mortality (due to lymphoid leukosis and hepatitis) occurred. The excessive mortality limits the validity of subsequent egg production and feed efficiency. However, neither egg production or mortality was influenced by dietary treatment before or after molting. The results of this trial are in agreement with those reported by Quisenberry and Bradley (1968), Flegal and Zindel (1971), Hodgetts (1971) and Biely et al. (1972).

### References

1    Biely, J., R. Soong, L. Seier, and W. H. Pope. 1972. Dehydrated poultry waste in poultry rations. Poultry Sci. 51:1502-1511.
2    Blair, R. and D. W. Knight. 1973. Recycling animal wastes. Feedstuffs 45(12):31,35,36,55.
3    Calvert, C. C., N. O. Morgan, and H. J. Eby. 1971. Biodegraded hen manure and adult house flies: their nutritional value to the growing chick. Proc. Intl. Symp. on Livestock Waste, Ohio State University, Columbus, Ohio, pp. 319-320.
4    Durham, R. M., G. W. Thomas, R. C. Albin, L. G. How, S. E. Curl, and T. W. Box. 1966. Coprophagy and use of animal waste in livestock feeds. ASAE Publication No. SP-0366:112-114, ASAE, St. Joseph, Mich. 49083.
5    Flegal, C. J. and H. C. Zindel. 1971. Dehydrated poultry waste (DPW) as a feedstuff in poultry rations. Proc. Intl. Symp. on Livestock Waste, Ohio State University, Columbus, Ohio, pp. 305-307.
6    Hodgetts, B. 1971. The effects of including dried poultry waste in the feed of laying hens. Proc. Intl. Symp. on Livestock Waste, Ohio State University, Columbus, Ohio, pp. 311-313.
7    Palafox, A. L. and M. M. Rosenberg. 1951. Dried cow manure as a supplement in a layer and breeder ration. Poultry Sci. 30:136-142.
8    Quisenberry, J. H. and J. W. Bradley. 1968. Nutrient recycling. Second Natl. Poultry Litter and Waste Mgt. Seminar, College Station, Texas, pp. 96-106.
9    Sheppard, C. C., C. J. Flegal, M. L. Esmay, J. B. Gerrish, J. E. Dixon, H. C. Zindel, and T. S. Chang. 1974. A complete system for collecting, handling, air-drying and machine dehydration of poultry manure in a caged layer production unit. Proc. XV World's Poultry Congress and Exposition, pp. 223-225.

# Inclusion of Pig Manure in Ruminant Diets

*(Continued from page 219)*

Dry matter digestibilities were affected, as previously, by the proportion of pig manure in the diet but not by the amount of molybdenum.

Plasma G.O.T. levels generally fluctuated between the normal range of 70 to 140 units. Nine of the sheep exceeded 140 units at some time (to a maximum of 180 units) but this was usually transitory and no values reached 200 units. Some sheep displayed signs of weakness and trembling but no diagnosis of the cause could be made. Mean liver copper levels at slaughter were A 718, B 1186, C 1440, D 1522, E 1725, F 1560 and G 1703 mg/kg dry matter. Increasing levels of manure in the diet were responsible, therefore, for higher liver copper concentrations and there was no clear effect of the molybdenum additions.

Copper balance estimations can only be reported tentatively, pending completion of analyses. On diet A (no manure), copper retention was low and was stable with time. The inclusion of manure without molybdenum resulted in initially high copper retentions (up to 15 mg/day for diet E in period 1) followed eventually by negative copper balances. Added molybdenum initially had no effect in preventing significant retentions of copper, with positive balances up to 27 mg/day (diet F), except where the higher level of molybdenum was included in the 15 percent manure diet (diet D); in this case only about 1 mg/day was retained during period 1. Thereafter, the effect of molybdenum was variable although some significant negative balances were estimated.

## CONCLUSIONS

Except for the mineral content, the chemical composition of dried pig manure might suggest that it would be well utilized by ruminants. However, dry matter digestibility has been shown to be less than 30 percent. Reasons for this are not clear. If pig manure with a high copper content is fed continuously over a long period of time there may be adverse effects on the recipient animals.

### References

1    Jacobs, G. J. L. and Jane Leibholz. 1974. The digestibility of ensiled broiler-house litter/barley rations for sheep. Proc. Aust. Soc. Anim. Prod. 10:400.
2    Leibholz, Jane. 1969. Poultry manure and meat meal as a source of dietary nitrogen for sheep. Aust. J. Exp. Agric. Anim. Husb. 9(41):589-593.
3    Matrone, G. 1970. Studies on copper-molybdenum-sulphate interrelationships, p. 354-361. In: C. F. Mills, (ed.) Trace element metabolism in animals. E. & S. Livingstone, Edinburgh and London.
4    McInnes, P., P. J. Austin and D. L. Jenkins. 1968. The value of poultry litter and wheat mixture in the drought feeding of weaner sheep Aust. J. Exp. Agric. Anim. Husb. 8(33):401-408.
5    Reitman A. and S Frankel. 1957. A calorimetric method for the determination of serum glutamic oxaloacetic and glutamic pyruvic transaminases. Amer. J. Clin. Path. 28(1):56-63.

# Ensiling Broiler Litter with Corn Forage, Corn Grain and Water

J. P. Fontenot, L. F. Caswell, B. W. Harmon, K. E. Webb, Jr.

$B$ROILER litter, an accumulation of poultry excreta, bedding, wasted feed and feathers has been shown to be a valuable feedstuff for ruminants (Noland et al. 1955, Southwell et al. 1958, Fontenot et al. 1966, 1971). The nutritional value of this product, especially the high nitrogen content, indicates that it could be a very useful feed ingredient for ruminants. In fact, the protein and energy have been shown to be efficiently utilized by ruminants (Bhattacharya and Fontenot 1965, 1966).

Recycling of broiler litter by feeding is not sanctioned by the Food and Drug Administration (Kirk 1967, Taylor 1971). No serious health problems have resulted from feeding the material, but there is apprehension concerning dangers of pathogens in litter fed to animals. It has been shown that broiler litter can be rendered free of pathogens by autoclaving, fumigation and dry heat alone or in combination with paraformaldehyde (Caswell et al. 1975b). Perhaps, use of a natural biological method of processing such as fermentation to destroy microbes may provide an economical and convenient alternative approach.

The ensiling process is characterized by the production of heat and organic acids, followed by quiescence at which time pH of the fermented mass becomes stable at about 4 (Barnett 1954). Ensiling with additional water, forage or grain may offer other advantages such as enhancing the nutritive value of normal feedstuffs, reducing dustiness and improving palatability. Indeed, a more complete feed may result.

## ENSILING WITH CORN FORAGE

The feasibility of ensiling broiler litter with corn forage was studied (Harmon et al. 1975a, 1975b). Wood-shaving broiler litter, collected from a commercial broiler house, in which one group of broilers had been produced, was used. Corn forage, harvested at two stages of maturity (30 and 40 percent dry matter, by field sampling) was ensiled with the litter. Replicated 2 kg mixtures (small-bag silages) were ensiled to study fermentation characteristics and 114 kg mixtures (large-bag silages) were ensiled for feeding sheep. The small-bag silages were firmly packed into doubled polyethylene bags supported by cardboard food containers. The large-bag silages were packed into doubled 0.004 mm polyethylene bags. Treatments for the small-bag silages were: control, untreated; urea, 0.5 percent, wet basis; litter to supply 15, 30 and 45 percent of the total dry matter. The treatments for the large bags were the same as for the small-bag silages, except the 45 percent litter treatment was not used.

Forages, broiler litter, initial mixtures and silages were analyzed for total and ammonia nitrogen, proximate components, volatile fatty acids, lactic acid and water soluble carbohydrate content. Methods were outlined by Harmon et al. (1975a). The pH of the forage mixture and silage samples was determined. Counts for total bacteria and coliforms were determined on asceptically prepared water extracts plated and cultured according to the standard method for testing pasteurized milk (Anonymous 1967).

Digestibility, nitrogen balance and palatability studies were conducted with wethers fed the ensiled material from the large bags. For disgestibility and nitrogen balance, 24 wethers were used in each of two metabolism trials. In conjunction with the metabolism trial, two palatability trials were conducted with 24 sheep 1 to 3 years of age. The digestion and metabolism trials consisted of a 10-day preliminary period followed by a 10-day period during which total fecal and urinary collections were made. The palatability trials consisted of a 5-day period during which the experimental silages were introduced followed by a 15-day experimental period. Dry matter intake measurements were made only during the last 10 days. In the palatability trials the silages were fed in excess (about 5 to 10 percent) of the amount consumed.

The broiler litter contained approximately 83 percent dry matter and 27 percent crude protein, dry basis. The corn forage contained approximately 26 percent dry matter for maturity 1 and 38 percent dry matter for maturity 2. The compositions of the small-bag silages are shown in Table 1. Addition of 15 percent litter, dry basis, raised the crude protein level to 12.5 percent for maturity 1 silages and 10.6 percent for the maturity 2 silages. Increasing the level of litter resulted in further increases in crude protein content. The significant increase in crude protein resulting from litter additions could provide considerable savings in feed costs in high silage feeding operations, since corn silage is deficient in crude protein. The crude fiber levels of the silages containing corn forage alone, or with various additions, were quite similar. The composition of the comparable large-bag silages was quite similar to that of the small-bag silages given in Table 1. The addition of urea and each level of litter resulted in an increase in the proportion of ammonia nitrogen in the fermented mixtures.

The pH and concentrations of lactic and acetic acids for the small-bag silages are given in Table 2. Addition of urea or different levels of litter resulted in increases in the pH of the silages. However, the pH values were certainly in the normal range, which would indicate that normal ensiling occurred. The values for the large-bag silages were similar to those of the corresponding small-bag silages. All silages had accumulated a considerable amount of lactic acid. In maturity 1 silages, lactic acid was generally higher in silages containing urea or litter than in those containing no additives, indicating that perhaps the buffering activity of the am-

The authors are: J. P. FONTENOT, professor, Animal Science Dept., L. F. CASWELL, graduate assistant, Animal Science Dept., K. E. WEBB, JR., assistant professor, Animal Science Dept., Virginia Polytechnic Institute and State University, Blacksburg; B. W. HARMON, assistant professor, Dept. of Agriculture, University of Maryland, Eastern Shore, Md. (formerly graduate assistant, Animal Science Dept., V.P.I. & S.U.).

**TABLE 1. COMPOSITION OF CORN FORAGE ENSILED ALONE AND WITH UREA OR BROILER LITTER**

| Corn maturity | Additive | Dry matter[a], percent | Composition of dry matter | | | | | |
| | | | Crude protein[a], percent | Ether extract, percent | Crude fiber[a], percent | NFE[a], percent | Ash[a], percent | Water soluble carbohydrate, [a,b] percent |
|---|---|---|---|---|---|---|---|---|
| 1 | none | 24.18[d] | 8.84[d] | 1.75[c] | 23.95[f] | 60.92[i] | 4.54[d] | 11.76[f] |
| 1 | 0.5 percent urea* | 23.45[c] | 14.56[h] | 1.94[c] | 24.25[f] | 54.06[f,g] | 5.19[e] | 12.25[f] |
| 1 | 15.0 percent litter† | 27.07[e] | 12.47[g] | 2.15[e,f] | 24.18[f] | 54.69[g] | 6.52[g] | 10.30[e] |
| 1 | 30.0 percent litter† | 31.63[f] | 15.79[i] | 2.64[g] | 23.75[f] | 49.19[e] | 8.63[i] | 6.95[d] |
| 1 | 45.0 percent letter† | 37.07[g] | 18.81[k] | 2.70[g] | 24.15[f] | 43.68[c] | 10.66[k] | 4.11[c] |
| 2 | none | 36.60[g] | 7.59[c] | 2.05[d,e,f] | 21.15[c] | 65.82[i] | 3.39[c] | 14.90[g] |
| 2 | 0.5 percent urea* | 36.66[g] | 11.46[f] | 1.81[c,d] | 21.54[c,d] | 61.73[i] | 3.46[c] | 14.82[g] |
| 2 | 15.0 percent litter† | 40.51[h] | 10.64[e] | 2.23[f] | 22.43[d,e] | 59.23[h] | 5.47[f] | 12.07[f] |
| 2 | 30.0 percent litter† | 44.16[i] | 14.46[h] | 2.16[e,f] | 22.22[d,e] | 53.11[f] | 8.05[h] | 6.82[d] |
| 2 | 45.0 percent litter† | 50.26[j] | 17.50[j] | 2.64[g] | 22.71[e] | 47.89[d] | 9.26[j] | 5.91[d] |

[a]When statistically analyzed over both stages of maturity, values were significantly affected (P<0.01) by maturity.

[b]The interaction of maturity x treatment was significant (P<0.05).

[c,d,e,f,g,h,i,j,k]Means in same column having different superscripts were significantly different (P<0.01).

*Wet basis

†Dry basis

monia in the litter and from urea hydrolysis allowed extended fermentation. Generally, lactic acid values were somewhat higher for the large-bag than small-bag silages, indicating that the larger fermentation mass plus the longer period between preparation and opening of the bags were responsible.

Total bacteria counts and coliform counts for the small-bag silages are given in Table 3. The total counts were quite high, but only the maturity 1, 45 percent litter silage and maturity 2, 30 and 45 percent litter silages had significantly greater numbers than the control silages. Coliform numbers were generally less than 150 per gram and silages containing litter usually had lower coliform counts than the control silages. The differences for the maturity 2 silages were significant. The trend for lower coliform numbers in litter-containing silages than controls suggests ensiling may be a practical means of eliminating potential hazards from presence of pathogens in litter.

Digestibility of dry matter was not significantly different among the various silages. The average digestion coefficient was 64 percent. Apparent digestibility of crude protein was generally higher for the silages treated with urea or litter. As shown in Table 4 the nitrogen intake was higher for the treated than the control silages. The addition of 0.5 percent urea did not improve nitrogen retention but retention was generally higher for the sheep fed the litter treated silages. When expressed as grams per day, the differences were significant and were generally significant when expressed as percent of intake. By assuming that the retention of corn silage nitrogen was constant, it was calculated that 33 percent of the litter nitrogen was retained by the animals on the 15 percent litter and 20 percent by the animals fed the 30 percent litter silages. It is obvious that the sheep were able to utilize the nitrogen from litter efficiently.

In the palatability studies, the consumption of dry matter, expressed as grams per day or per unit of metabolic size per day, was significantly higher for the animals fed the silages containing 15 or 30 percent litter than for the control and urea treated silages. Similar differences existed when the results were expressed in terms of grams of diges-

**TABLE 2. FERMENTATION CHARACTERISTICS OF CORN FORAGE ENSILED ALONE AND WITH UREA OR BROILER LITTER**

| Corn maturity | Additive | pH[a,b] | Percent of dry matter | |
| | | | Lactic acid [a,b] | Acetic acid[a] |
|---|---|---|---|---|
| 1 | none | 3.62[c] | 5.72[e,f] | 1.68[c,d,e] |
| 1 | 0.5 percent urea* | 3.69[d] | 6.62[f,g] | 1.74[d,e] |
| 1 | 15.0 percent litter† | 3.67[c,d,] | 8.59[h] | 1.50[c,d,e] |
| 1 | 30.0 percent litter† | 3.94[f] | 7.63[g,h] | 1.72[d,e] |
| 1 | 45.0 percent litter† | 4.28[g] | 6.38[f] | 1.92[e] |
| 2 | none | 3.71[d] | 4.48[c,d] | 1.18[c] |
| 2 | 0.5 percent urea* | 3.85[e] | 5.64[e,f] | 1.24[c,d] |
| 2 | 15.0 percent litter† | 4.00[f] | 5.01[d,e] | 1.30[c,d] |
| 2 | 30.0 percent litter† | 4.38[h] | 4.19[c,d] | 1.49[c,d,e] |
| 2 | 45.0 percent litter† | 4.68[i] | 3.68[c] | 1.51[c,d,e] |

[a]When statistically analyzed over both stages of maturity, values were significantly affected (P<0.01) by maturity.

[b]The interaction of treatment x maturity was significant (P<0.01).

[c,d,e,f,g,h,i]Means in same column having different superscripts were significantly different (P<0.05).

*Wet basis

†Dry basis

**TABLE 3. TOTAL BACTERIA AND COLIFORMS OF CORN FORAGE ENSILED ALONE AND WITH UREA OR BROILER LITTER**

| Corn maturity | Additive | Total bacteria[a,b] | Coliforms per gram |
|---|---|---|---|
| | | $10^6$ per gram | |
| 1 | none | 7.1[c] | 150[c,d] |
| 1 | 0.5 percent urea* | 5.2[c] | 171[d] |
| 1 | 15.0 percent litter† | 3.8[c] | 100[c,d] |
| 1 | 30.0 percent litter† | 9.4[c] | 84[c,d] |
| 1 | 45.0 percent litter† | 265.3[d] | 170[d] |
| 2 | none | 40.2[c] | 361[e] |
| 2 | 0.5 percent urea* | 30.4[c] | 116[c,d] |
| 2 | 15.0 percent litter† | 79.1[c] | 85[c,d] |
| 2 | 30.0 percent litter† | 458.1[e] | 12[c] |
| 2 | 45.0 percent litter† | 837.1[f] | 106[c,d] |

[a]When statistically analyzed over both maturities, values were significantly affected (P<0.01) by maturity.

[b]The interaction of treatment x maturity was significant (P<0.01).

[c,d,e,f]Means in same column having different superscripts were significantly different (P<0.01).

*Wet basis

†Dry basis

TABLE 4. NITROGEN UTILIZATION BY SHEEP FED SILAGE OF CORN FORAGE ENSILED ALONE AND WITH UREA OR BROILER LITTER

| Corn maturity | Additive | Intake, grams/day | Nitrogen retention | | |
|---|---|---|---|---|---|
| | | | Grams/day[a] | Percent of intake[a] | Percent of absorbed[a] |
| 1 | none | 12.47[b,c] | 1.07[b] | 8.5[b,c] | 14.8[b,c] |
| 1 | 0.5 percent urea* | 15.14[b,c,d] | 1.14[b] | 7.1[b] | 9.9[b] |
| 1 | 15.0 percent litter† | 19.58[d,e] | 3.17[c] | 15.6[c,d,e] | 25.7[c,d,e] |
| 1 | 30.0 percent litter† | 24.18[e,f] | 3.38[c] | 12.4[b,c,d] | 18.6[b,c,d] |
| 2 | none | 10.93[b] | 1.34[b] | 11.3[b,c,d] | 23.1[c,d] |
| 2 | 0.5 percent urea* | 15.74[c,d] | 1.48[b] | 8.7[b,c] | 13.8[b,c] |
| 2 | 15.0 percent litter† | 20.78[e] | 4.47[c,d] | 20.8[e] | 37.4[e] |
| 2 | 30.0 percent litter† | 28.07[f] | 5.18[d] | 18.5[d,e] | 28.8[d,e] |

[a]When statistically analyzed over both maturities, values were significantly affected (P<0.01) by maturity.

[b,c,d,e,f]Means in same column having different superscripts were significantly different (P<0.05).

*Wet basis

†Dry basis

tible dry matter per day. The reason for the increase in dry matter intake for the litter containing silages may have been the bulk density in the material, since the silages were more compact than the control silages, especially for the maturity 2 forage.

## ENSILING BROILER LITTER WITH DIFFERENT LEVELS OF MOISTURE

These studies were conducted to evaluate the feasibility of ensiling broiler litter alone and with added moisture on fermentation characteristics and nitrogen utilization, digestibility and palatability when fed to ruminants (Caswell et al. 1975a). In the initial study, broiler litter was ensiled with moisture levels of 15.6, 20, 30, 40 and 50 percent in small laboratory silos. The litter was ensiled by placing 2 kg of the material in double thickness polyethylene bags supported by cardboard food containers. The silages were tested for the presence of salmonella, shigella and proteus. They were subjected to total bacteria and coliform counts using the procedures described above. The initial mixtures and the silages were analyzed for total, protein, uric acid, ammonia and non-protein nitrogen, proximate components, volatile fatty acids, lactic acid, and water soluble carbohydrates. The pH was determined on water extracts of the materials. Procedures are given by Caswell (1975).

The litter contained approximately 31 percent crude protein, dry basis, and was not changed after ensiling with different levels of moisture. The fermentation characteristics of the initial mixtures and silages are given in Table 5. The initial mixture had a pH of approximately 7.7 and was void of lactic acid. The pH significantly decreased in the silages with increased moisture level, up to 40 percent moisture. Lactic acid followed a similar trend. Although there was a significant difference between the two highest moisture levels, the difference was small. These results indicate that fermentation was approaching maximum with litter ensiled at 40 percent moisture. Generally, ensiling with different levels of moisture did not greatly change the nitrogen components of the silages. There were trends for decreases in uric acid and increases in ammonia nitrogen at the high moisture levels.

The bacterial counts of the initial mixtures and silages are shown in Table 6. Ensiling resulted in large depressions in total bacteria, especially when the moisture level was 30 percent or higher. Coliforms were eliminated by ensiling with moisture levels of 20 percent or higher.

A study was conducted to investigate the apparent digestibility, nitrogen utilization and palatability by ruminants fed rations containing ensiled broiler litter with 20 or 40 percent moisture. The litter, which contained 20 percent moisture, was ensiled alone, and with added water to a final moisture level of approximately 40 percent. A total of 545 kg of the 20 percent moisture litter was ensiled in double

TABLE 5. FERMENTATION CHARACTERISTICS OF INITIAL MIXTURES AND SILAGES OF BROILER LITTER WITH DIFFERENT LEVELS OF MOISTURE

| Type of material | Estimated moisture, percent | pH | Level, percent of dry matter | | |
|---|---|---|---|---|---|
| | | | Lactic acid | Acetic acid | Water soluble carbohydrates |
| Initial mixtures | 15.6 | 7.78 | 0 | 0.83 | 2.65 |
| | 20.0 | 7.74 | 0 | 0.91 | 2.67 |
| | 30.0 | 7.77 | 0 | 1.01 | 2.70 |
| | 40.0 | 7.72 | 0 | 1.20 | 2.62 |
| | 50.0 | 7.68 | 0 | 1.28 | 2.60 |
| Silages | 15.6 | 7.81[a] | 0[a] | 0.99[a] | 2.61[a] |
| | 20.0 | 7.63[b] | 0.24[b] | 1.15[a] | 2.66[a] |
| | 30.0 | 6.12[c] | 2.21[c] | 2.33[b] | 1.89[b] |
| | 40.0 | 5.39[d] | 4.58[d] | 3.14[c] | 1.75[b] |
| | 50.0 | 5.43[d] | 5.14[e] | 4.17[d] | 1.73[b] |

[a,b,c,d,e]Means for silages having different superscripts are significantly (P<0.01) different.

TABLE 6. BACTERIA COUNTS[a] OF INITIAL MIXTURES AND SILAGES OF BROILER LITTER WITH DIFFERENT LEVELS OF MOISTURE

| Type of material | Estimated moisture | Total bacteria | Coliforms |
|---|---|---|---|
| | Percent | $10^6$/gram | $10^1$/gram |
| Initial mixtures | 15.6 | 1750 | 12400[b] |
| | 20.0 | 2240 | 9700[c] |
| | 30.0 | 2000 | 9200[c,d] |
| | 40.0 | 2530 | 8400[c,d] |
| | 50.0 | 2330 | 7000[c] |
| Silages | 15.6 | 209[e] | 1[e] |
| | 20.0 | 187[e] | 0[f] |
| | 30.0 | 24[f] | 0[f] |
| | 40.0 | 7[f] | 0[f] |
| | 50.0 | 24[f] | 0[f] |

[a]Salmonella and shigella were not detected in initial mixtures or silages. Proteus was present in initial mixtures, but was destroyed by ensiling.

[b,c,d]Means for initial mixtures having different superscripts are significantly (P<0.01) different.

[e,f]Means for silages having different superscripts are significantly (P<0.01) different.



MANAGING LIVESTOCK WASTES

| | | Nitrogen supplements | | |
| | | Litter silage | | |
| Item | Processed litter | 20 percent moisture | 40 percent moisture | Soybean meal |
|---|---|---|---|---|
| Nitrogen utilization | | | | |
| Nitrogen intake, g/day | 12.04 | 12.20 | 12.87 | 12.34 |
| Nitrogen excretion, g/day | | | | |
| Fecal | 4.40 | 4.47 | 4.47 | 4.55 |
| Urinary | 7.15[a] | 6.23[a,b] | 6.74[a,b] | 5.79[b] |
| Total | 11.55[a] | 10.70[a,b] | 11.21[a] | 10.34[b] |
| Nitrogen retention | | | | |
| Grams per day | 0.49[a] | 1.49[b] | 1.66[b] | 2.01[b] |
| Percent of intake | 4.10[a] | 12.20[b] | 12.90[b] | 16.30[b] |
| Percent of absorbed | 6.50[a] | 19.50[b] | 19.90[b] | 25.60[b] |
| Apparent digestibility, percent | | | | |
| Dry matter | 74.30[c] | 73.90[c] | 74.30[c] | 69.90[c] |
| Crude protein | 63.50 | 63.30 | 65.30 | 63.20 |
| Ether extract | 68.60[c,d] | 69.50[d] | 71.40[d] | 64.20[c] |
| Crude fiber | 64.20[a] | 64.00[a] | 63.80[a] | 54.80[b] |
| NFE | 80.50 | 79.80 | 80.10 | 76.80 |

[a,b]Means having different superscripts are significantly (P<0.01) different.

[c,d]Means having different superscripts are significantly (P<0.05) different.

thickness 0.004 mm polyethylene bags inside a plywood box with cubic dimensions of 1.2 m. The same amount of dry matter plus the added water were ensiled in similar structures for the 40 percent moisture silage.

A nitrogen balance and digestibility trial was conducted with 24 wethers. The following supplements supplied 50 percent of the dietary nitrogen for the respective rations: (a) dry-heat processed litter; (b) 20 percent moisture ensiled litter; (c) 40 percent moisture ensiled litter and (d) soybean meal. The dry-heat processed litter was heated in a forced draft oven at 260 C for 30 minutes at a litter depth of 1.3 cm.

During the metabolism trial no palatability problems were observed for the rations containing broiler litter. Nitrogen utilization and apparent digestibility for the sheep fed the rations containing processed litter, litter silages and soybean meal are shown in Table 7. Nitrogen retention for the wethers fed the litter silages was not significantly different from that of the wethers fed soybean meal as the source of supplemental nitrogen. Fecal excretion was similar and urinary excretion tended to be lower for the soybean meal than for the litter silage supplemented rations. Supplemental nitrogen source did not significantly affect apparent digestibility of crude protein or NFE. Apparent digestibility of crude fiber was significantly lower for the sheep fed the ration containing the soybean meal. There was also a significant depression in dry matter digestibility for the soybean meal supplemented diet.

The palatability of litter silages was evaluated in two trials with 12 yearling steers. The rations were calculated to contain equal levels of TDN. For the three litter containing rations, litter supplied 50 percent of the total ration dry matter. Each trial consisted of a 10-day adjustment period and a 15-day measurement period. The cattle were individually fed an excess of the individual rations.

Intake was significantly higher for the soybean than for the litter silage supplemented rations, regardless of whether the results were expressed as kilograms per day or grams per unit of metabolic size per day. Perhaps, the feeding period was too short and the animals did not have time to adjust to the rations. In previous experiments (Fontenot et al. 1971), an adjustment was noted by cattle fed rations containing 25 or 50 percent processed broiler litter.

## ENSILING BROILER LITTER WITH HIGH MOISTURE GRAIN

This experiment was conducted to study fermentation characteristics of an ensiled mixture of broiler litter and high moisture corn grain, and to assess digestibility, nitrogen utilization and palatability of rations containing the ensiled mixtures (Caswell et al. 1974). High moisture corn grain containing 26 percent moisture was harvested and ensiled alone and in a ratio of 2:1 with wood-shaving based broiler litter containing 19 percent moisture. One group of birds had been reared on the litter obtained from a commercial, insulated, humidity controlled broiler house. Approximately one metric ton of the high moisture grain and the high moisture grain-litter mixture were ensiled in silos identical in construction, and equipped and sealed as those described for the experiment on ensiling of litter with different levels of moisture.

Ensiling the corn alone or with litter resulted in large decreases in total bacteria and coliforms. The number of coliforms were 199 per gram for the ensiled high moisture corn and 149 for the ensiled corn-litter. Salmonella was not observed in any of the silages or initial materials.

The ensiled corn contained 9.4 percent crude protein on a dry basis. Addition of one-third broiler litter prior to ensiling increased the protein content to 20 percent, dry basis.

A metabolism trial was conducted with 24 wethers. The four rations used were: control, low protein; ensiled corn-litter; ensiled corn and processed litter, and ensiled corn and soybean meal. The total supplementary nitrogen from litter or soybean meal was 57, 64 and 59 percent for the ensiled corn-litter, ensiled corn and processed litter, and ensiled corn and soybean meal rations, respectively.

Data on nitrogen utilization and apparent digestibility are given in Table 8. Nitrogen retention was significantly lower for the unsupplemented ration than for the rations

**TABLE 8. NITROGEN UTILIZATION AND APPARENT DIGESTIBILITY BY SHEEP FED HIGH MOISTURE CORN ENSILED ALONE OR WITH BROILER LITTER**

| Item | Ration designation | | | |
|---|---|---|---|---|
|  | Ensiled corn | Ensiled corn-litter | Ensiled corn + proc. litter | Ensiled corn + soybean meal |
| Nitrogen utilization |  |  |  |  |
| Nitrogen intake, g/day | 5.63 | 10.22 | 10.35 | 10.97 |
| Nitrogen excretion, g/day |  |  |  |  |
| Fecal | 3.53[a] | 4.02[b] | 4.26[b] | 3.60[a] |
| Urinary | 2.47[a] | 5.01[b] | 5.44[b] | 5.10[b] |
| Total | 6.00[a] | 9.03[b,c] | 9.70[b] | 8.70[c] |
| Nitrogen retention |  |  |  |  |
| Grams per day | -0.37[a] | 1.19[b] | 0.65[b] | 2.28[c] |
| Percent of intake | -6.90[a] | 11.70[b] | 6.60[b] | 20.70[c] |
| Percent of absorbed | -19.40[a] | 19.30[b,c] | 11.10[b] | 30.80[c] |
| Apparent digestibility, percent |  |  |  |  |
| Dry matter | 53.60[a] | 68.70[b] | 65.80[b] | 68.80[b] |
| Crude protein | 37.20[a] | 60.70[b] | 58.90[b] | 67.20[c] |
| Ether extract | 76.60[a] | 62.20[b] | 63.50[b,c] | 69.3[c] |
| Crude fiber | 30.30[a] | 67.90[b] | 65.40[b] | 62.80[b] |
| NFE | 64.50[a] | 73.20[b] | 68.90[a,b] | 72.90[b] |

[a,b,c] Means having different superscripts are significantly (P<0.01) different.

containing litter or soybean meal. Retention, expressed as grams per day or percent of intake, was significantly higher for the ration supplemented with soybean meal than for either of the rations containing litter. This was due primarily to lower fecal nitrogen excretion by sheep fed the soybean meal supplemented ration. Expressed as percent of absorbed nitrogen, retention for the ration containing the ensiled corn-litter mixture was not significantly different from the soy supplemented diet. Although retention was lower for the litter fed than the soy fed sheep, it is clear that the litter nitrogen was utilized quite efficiently if the results are compared to those of the sheep fed the low-protein ration.

Apparent digestion coefficients were usually significantly lower for the low protein than for the supplemented rations. Apparent digestibility of crude protein was lower for both the litter containing rations than for the soybean meal ration. Coefficients for other components were not significantly different among the litter and soybean meal treatments.

Two palatability trials were conducted with 12 yearling steers. The calculated crude fiber levels were equalized among the rations. For the two rations containing litter, 27.5 percent of the total ration dry matter was contributed by the waste. Dry matter intakes were 5.9, 7.3, 7.2 and 7.0 kg per day for the cattle fed ensiled corn, ensiled corn-litter, ensiled corn and processed litter, and ensiled corn and soy rations, respectively. All three supplemented rations were consumed in significantly greater amounts than the unsupplemented ration. There was a trend for the consumption to be slightly higher for the rations containing litter than for the ration containing soybean meal. When the results are expressed in terms of grams per unit of metabolic size the differences and trends were similar as when expressed in absolute amounts per day. Results of these studies indicate that litter can be successfully ensiled if mixed with high moisture corn.

## SUMMARY

Ensiling of broiler litter alone, with optimum moisture level, or with low protein feedstuffs appears to be a feasible processing method. The process lowers or eliminates coliform counts. The digestibility of the ensiled material was very good and the nitrogen from the waste was efficiently utilized. Addition of broiler litter to corn forage or high moisture corn grain improved the palatability of the ensiled material.

**References**

1   Anonymous. 1967. Standard methods for the examination of dairy products (12th Ed.). Amer. Public Health Ass., Inc., New York. 438 p.

2   Barnett, A. J. G. 1954. Silage fermentation. Academic Press, New York. 208 p.

3   Bhattacharya, A. N. and J. P. Fontenot. 1965. Utilization of different levels of poultry litter nitrogen by sheep. J. Anim. Sci. 24(4):1174-1178.

4   Bhattacharya, A. N. and J. P. Fontenot. 1966. Protein and energy value of peanut hull and wood shaving poultry litters. J. Anim. Sci. 25(2):367-371.

5   Caswell, L. F. 1975. Fermentation, utilization and palatability of broiler litter ensiled at different moisture levels and with high moisture corn grain. Ph.D. Thesis. Virginia Polytechnic Institute and State University, Blacksburg, Virginia. 127 p.

6   Caswell, L. F., J. P. Fontenot and K. E. Webb, Jr. 1974. Ensiled high moisture corn grain and broiler litter. J. Anim. Sci. 39(1):138. (Abstr.).

7   Caswell, L. F., J. P. Fontenot and K. E. Webb., Jr. 1975a. Ensiled broiler litter with different moisture levels. J. Anim. Sci. 40(1):200. (Abstr.).

8   Caswell, L. F., J. P. Fontenot and K. E. Webb, Jr. 1975b. Effect of processing method on pasteurization and nitrogen components of broiler litter and on nitrogen utilization by sheep. J. Anim. Sci. 40(4):750-759.

9   Fontenot, J. P., A. N. Bhattacharya, C. L. Drake and W. H. McClure. 1966. Value of broiler litter as a feed for ruminants. Proc Natl. Sym. on Amin. Waste Management. ASAE Publ. SP-0366:105-108, ASAE, St. Joseph, Mich. 49085.

10   Fontenot, J. P., K. E. Webb, Jr., B. W. Harmon, R. E. Tucker and W. E. C. Moore. 1971. Studies of processing, nutritional value and palatability of broiler litter for ruminants. Proc. Internatl. Sym. on Livestock Wastes. ASAE Publ. PROC-271:301-304, ASAE, St. Joseph, Mich. 49085.

11   Harmon, B. W., J. P. Fontenot and K. E. Webb, Jr. 1975a. Ensiled broiler litter and corn forage. I. Fermentation characteristics. J. Anim. Sci. 40(1):144-155.

12   Harmon, B. W., J. P. Fontenot and K. E. Webb, Jr. 1975b. Ensiled broiler litter and corn forage. II. Digestibility, nitrogen utilization and palatability by sheep. J. Anim. Sci. 40(1):156-160.

13   Kirk, J. K. 1967. Part 3 - Statements of general policy or interpretation. Use of poultry litter as animal feed. Federal Register 32(171):12714-12715.

14   Noland, P. R., B. F. Ford and M. L. Ray. 1955. The use of ground chicken litter as a source of nitrogen for gestating-lactating ewes and fattening steers. J. Anim. Sci. 14(3):860-865.

15   Southwell, B. L., O. M. Hale and W. C. McCormick. 1958. Poultry house litter as a protein supplement in steer fattening rations. Ga. Agr. Exp. St. Mimeo Ser. n.s. 55.

16   Taylor, J. C. 1971. Regulatory aspects of recycled livestock and poultry wastes. Proc. Internatl. Sym. on Livestock Wastes. ASAE Publ. PROC-271:291-292, ASAE, St. Joseph, Mich. 49085.



MANAGING LIVESTOCK WASTES

# Conversion of Animal Wastes to Feed Supplements Via the Organiform Process

C. K. Davies, G. A. Varga, R. S. Hinkson

THE Organiform Process,* developed by Orgonics, Inc. of Rhode Island, has been used to convert organic waste to high nitrogen organic fertilizers called Organiforms. Two commercial plants are in operation with several more in planning. The first plant in Forestdale, Rhode Island, utilizes leather tankage (L.T.) as the organic waste, while the newest plant in Winston-Salem, North Carolina, processes the cities sewage sludge. Both products, Organiform L.T. and Organiform S.S. are in particular demand for the professional turf grass and home and garden markets. These two commercial products are granular, sterile, and are exceptionally high in nitrogen (20 - 24 percent) of which a large portion is water insoluble, yet available to soil microorganisms. Previous research has shown that the Organiform products provide a slow release of nitrogen to the soil over a period of months.

The Organiform technology has also been applied to raw sewage, slaughter-house wastes, tobacco dusts, mycellium and brewery wastes. In each case, the resulting experimental products were sterile, odorless, granular and possessed a nitrogen content ranging from three to ten times that of the original waste material. The Organiform process is based on urea-formaldehyde technology.

Urea has long been recognized as a valuable supplement to natural proteins (Briggs 1967). However, there is a practical limit regarding the amount of urea that should be added to the ruminants diet. Various attempts have been made to extend this utilization by substituting the dimer, biuret, for urea (Anderson 1959).

More recently attempts have been made to protect dietary proteins from microbial attack in the rumen by reacting soluble protein (casein) with formaldehyde (Ferguson 1967).

In spite of this work, little has been done to control or tailor solubility patterns to protein protection in the rumen as well as availability in the duodenum.

Until 1936, urea-formaldehyde condensed products were generally considered cross-linked high molecular weight compounds suitable for use only in he fields of plastics and synthetic resins. Such compounds were reported as early as 1884 by Tollens (1884).

In 1936 Kadawaki reported thirty-two new compounds derived from the condensation of urea or urea derivatives with formaldehyde. Among these were three compounds which bear a direct relationship to our present day research and development. They are Methylene-diurea, Trimethylene-tetraurea, and Pentamethylene-hexaurea. They are represented by the constitutional formulas shown in Fig. 1.

Utilitarian value of these compounds as a controlled re-

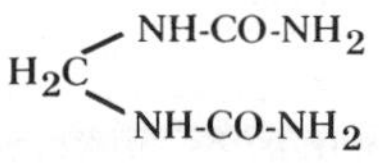

Methylene diurea

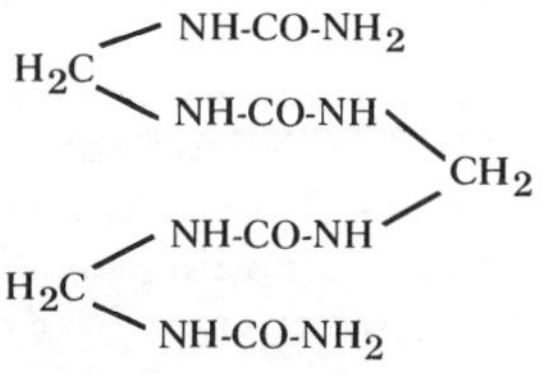

Trimethylene tetraurea

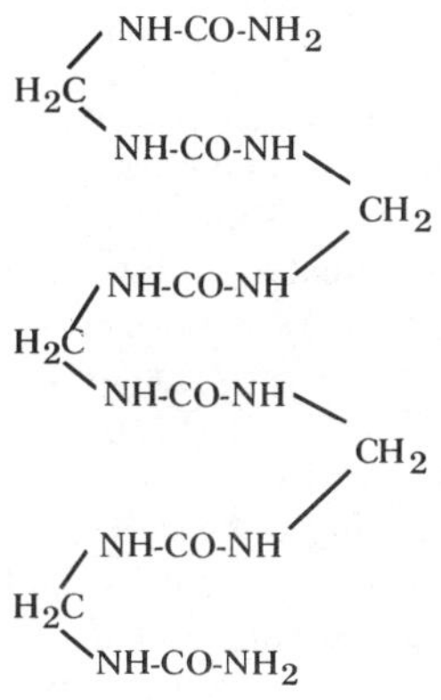

Pentamethylene hexaurea

FIG. 1 Representative methylene ureas derived from the condensation of urea-formaldehyde reactions.

lease nitrogen fertilizer was reported by Clark (1948, 1951, 1952). Strict adherence to reaction parameters was stressed in order to maintain the quality and integrity of the nitrogen compounds.

Kravolic (1954) developed an empirical procedure for determining the quality of the insoluble nitrogen produced under Clark's parameters. The method relies principally on solubility characteristics to indicate the extent of polymerization and therefore molecular weight of the methylene ureas.

Certain criteria were developed as a means of defining the compounds which were given the generic name of "Ureaform" (Smith, 1955).

1    Total nitrogen should be in the range of 36 - 40 percent.

2    Insoluble nitrogen should be at least 60 percent of the total nitrogen (A.O.A.C. - 7th Ed.).

3    At least 40 percent of the insoluble nitrogen should be soluble in the hot phosphate buffered solution as described by Kravolic (1954).

While this method of quality control is generally acceptable when methylene ureas are to be used for fertilizer purposes, it leaves much to be desired when more sophisti-

---

*Patent Pending.

The authors are: C. K. DAVIES, Orgonics, Inc., P. O. Box 543, Slatersville, R. I.; and G. A. VARGA and R. S. HINKSON, Aminal Science Dept., University of Rhode Island, Kingston.

TABLE 1. WATER SOLUBILITY PATTERNS OF
UREAFORM (38 PERCENT N) FRACTIONS.

| | | Percent |
|---|---|---|
| Water soluble fraction: | Urea | 10 |
| (40 percent of the total) | Methylene diurea | 92 |
| | Dimethylene triurea | 8 |
| Hot water soluble fraction: | Dimethylene triurea | 30 |
| (35 percent of the total) | Trimethylene tetraurea | 65 |
| | Tetramethylene pentaurea | 5 |
| Hot water insoluble fraction: | tetramethylene pentaurea | 42 |
| (25 percent of the total) | Pentamethylene hexaurea | 58 |

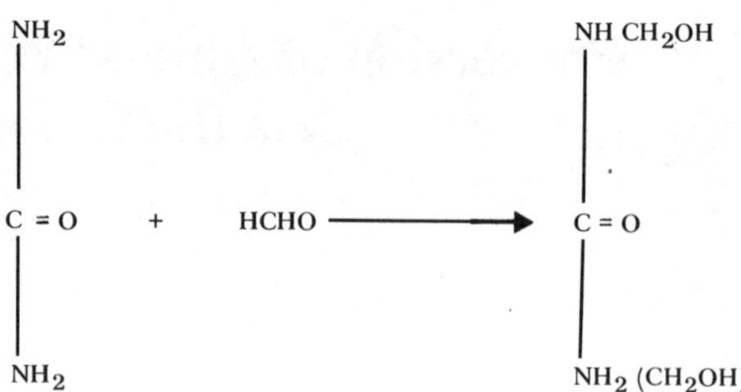



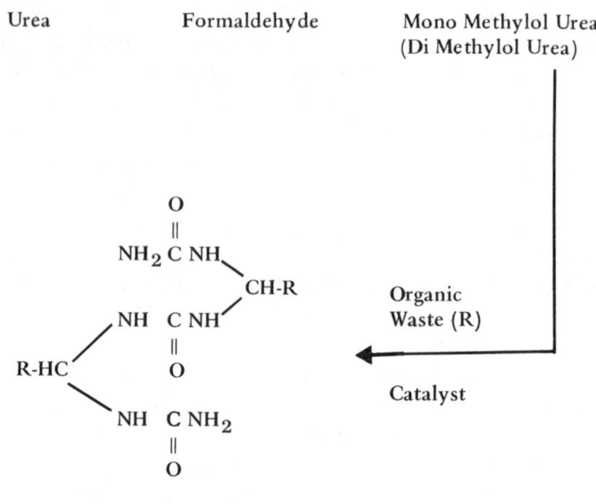



(Tentative Structure)
Dimethylene triurea type

FIG. 2 The organiform reaction.

cated techniques of precondensate preparations are employed. This is especially true when these precondensates are reacted with proteins prior to condensation.

For example, the Kravolic-Morgan test assumes that the water soluble nitrogen is similar to urea or other mineral nitrogen sources. It further assumes that the hot water insoluble fraction is comprised of highly cross-linked polymers which would show inordinate resistance to microbial breakdown. Neither assumption is consistent with the facts.

Work conducted at our laboratories in 1964 clearly indicated that each fraction, that is, water soluble, hot water soluble and hot water insoluble contained individually definable types of methyleneureas as shown in Table 1.

*In vitro* nitrification studies were conducted on the three fractions of ureaform using the techniques developed by Clark (1951). Table 2 shows the percent nitrogen of the faractions converted to nitrate.

While rumen bacteria and nitrobacter bear little or no similarity, the data developed is interesting as a comparison of the microbial degradation between urea, biuret and specific methylene urea polymers. For example, the nitrification pattern or biuret and the soluble fraction or ureaform (comprised mainly of methylene diurea) is almost identical following the initial three-week incubation period. Would it be reasonable to expect that same patterns would hold true for rumen bacteria?

Heretofore, techniques for the production of linear methylene ureas have been restricted to ureaform technology. Thus the aim and objective was the production of a wide range of methylene ureas rather than any specific molecular weight.

The technique of methylolozation prior to condensation was first disclosed by O'Donnell, (1958, 1959, 1966). Essentially, the technique involves the precondensing of urea and formaldehyde to form mono and dimethyol urea prior to condensation.

The method has been further expanded to include a preliminary reaction with proteins. While the exact nature of the reaction is not clearly understood, its existence is amply demonstrated by Walker (1967) and Fraenkel-Conrat (1945). The methylol or $CH_2OH$ group has proven bactericidal properties (Christian 1913) and has even been shown to deactivate viruses (Fraenkel-Conrat 1954).

Following the methylolization step condensation is induced by the use of certain catalysts. The highly bactericidal methylol group is condensed to a methylene ($-CH_2$) bridge. This is shown in Fig. 2.

In 1974, Orgonics, Inc. began pilot plant studies to process manure. The Department of Animal Science at the University of Rhode Island supplied the animals for feeding trials and rumen analysis. This was a pilot study undertaken to determine the feasibility of converting manure to a feed supplement.

Since the results of the pilot study indicated that the animals could utilize the product, and in view of the impending EPA feedlot datelines, Orgonics, Inc. decided to release the findings.

TABLE 2. COMPARISON OF NITRIFICATION VALUES OF UREAFORM FRACTIONS TO
UREA, BIURET, AND UREAFORM OVER A 24 WEEK PERIOD.

| Nitrogen source | Percent nitrogen converted to $NO_3$ | | | | | |
|---|---|---|---|---|---|---|
| | 3 weeks | 6 weeks | 9 weeks | 12 weeks | 15 weeks | 24 weeks |
| Urea | 90 | 96 | 92 | 93 | 95 | 92 |
| Biuret | 50 | 78 | 91 | 92 | 91 | 90 |
| Ureaform — water soluble fraction | 20 | 65 | 90 | 91 | 93 | 91 |
| Ureaform — hot water soluble fraction | 2 | 10 | 23 | 60 | 90 | 91 |
| Ureaform — hot water insoluble fraction | 0 | 2 | 4 | 5 | 40 | 75 |
| Ureaform | 10 | 30 | 34 | 44 | 49 | 76 |

## EXPERIMENTAL

### Manure

10 kg of manure (0 - 1 days old) were blended with 18 - 20 kg of water at a neutral pH. The slurry was heated to 50 - 60 C before addition of the urea-formaldehyde reactant. Manure solids ranged from 15 - 25 percent.

### U/F

Urea and formaldehyde were reacted at a urea to formaldehyde mole ratio of 1.5 to 1 for 20 minutes at 60° C in order to insure complete methylolization of the formaldehyde. Two kg of the final reaction mixture were added to the manure slurry and allowed to stand for 20 minutes with stirring. Condensation to a linear methylene urea — waste product "copolymer" was allowed to proceed overnight.

### Drying

The resulting wet (10.2 percent solids) paste which represented about 3.4 kg of dry Organiformed manure was added to 41.8 kg of ground corn in a vertical mixer. This mix (approximately 50 - 60 percent solids) was air dried, bagged and held until mixed with the final diet ingredients.

### Feeding trials

Two rumen fistulated steers and four dairy heifers were placed in separate pens during the feeding trials. The control and experimental diets are shown in Table 3. The diets were calculated to be isonitrogeneous and isocaloric. The steers were fed 3.5 kg of concentrate (containing Organiformed manure which represented 32.5 percent of the crude protein) and 0.25 kg low quality hay at 8 a.m. and 4 p.m. daily.

Each heifer was fed the experimental diet for 14 weeks. Each steer was fed the control and the experimental diets in a switchback type design for a period of 32 days per dietary treatment period. Rumen content samples were taken just prior to feeding and at two and six hours after feeding on the first experimental day and every seven days thereafter.

In addition to the animal trials, artificial rumen studies are being conducted using Organiformed manures as a potential protein source for rumen microorganisms. The Organiformed manure in the semipurified diets fed to the artificial rumens represented 0.0 to 100 percent of the crude protein contents.

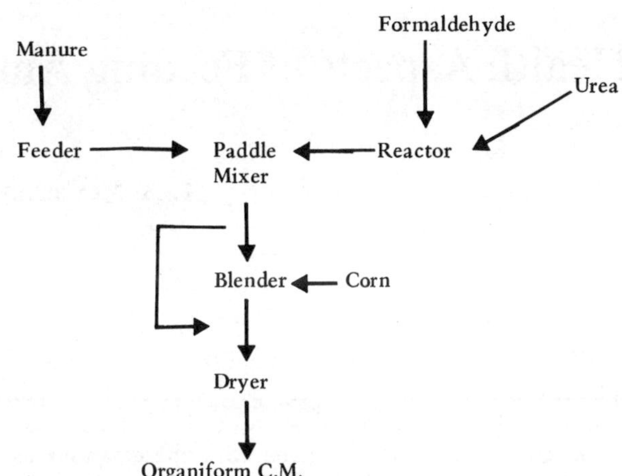



FIG. 3 Organiform process.

## RESULTS AND DISCUSSION

Processing of the manure increased the nitrogen content from 2 percent to 14 percent on a dry matter basis allowing the product to be used in place of soybean meal in the experimental diet. Approximately 30 batches of Organiform C.M. were prepared in the manner described above.

The experimental diet was readily accepted by the animals and no palatability problems were encountered. Both the heifers and the steers showed normal to good body weight gains and no adverse physical effects were noted. There was no significant difference in the bacterial population in the rumen contents of the steers fed either diet. Certain intraruminal results along with the body weight gains of the animals indicate that the Organiformed manure was utilized as a source of protein in the digestive tract of the animals.

Fig. 3 shows a suggested processing scheme to convert animal manure to a feed supplement. The Organiform reaction can be either batch or continuous. The dried Organiformed manure is stable and can be stored indefinitely.

### TABLE 3. CONTROL AND EXPERIMENTAL DIETS FED TO STEERS.

| Ingredient | Control | Organiform |
|---|---|---|
| Corn meal | 773 | 206 |
| Salt (NaCl) | 5 | 5 |
| Limestone ($CaCO_3$) | 15 | 15 |
| Biofos (Ca, P) | 19.5 | 20.0 |
| Delamix | 1.0 | 1.0 |
| Vitamin A (Grams) | 70 | 70 |
| Vitamin D (Grams) | 2 | 2 |
| Oats | 50 | 50 |
| Corn cobs (Ground) | 54.5 | 45 |
| Soybean meal | 82 | 0 |
| Organiform (Premix)* | 0 | 658 |
| Total | 1072 | 1072 |

*Premix contains about 8 percent Organiform C.M. (32.5 percent of the total crude protein) and 92 percent corn meal.


### References

1 Anderson, G. C., G. A. McLaren, J. A. Welch, C. D. Campbell and G. S. Smith. 1959. Comparative effects of urea, uramite, biuret, soybean, protein, creatine on digestion and nitrogen metabolism. J. Animal Science 18:134.

2 Briggs, M. H. 1967. Urea as a protein supplement. Pergamon Press, N.Y.

3 Christian, M. 1913. Disinfection and Disinfectants. Scott, Greenwood & Son. London:74-104.

4 Clark, K. G., J. Y. Yee and K. S. Loue. 1948. New synthetic nitrogen fertilizer. Ind. and Eng. Chem. 40(7):1178-1183.

5 Clark, K. G., J. Y. Yee and K. S. Loue. 1951. Preparation and properties of ureaform. Ind. and Eng. Chem. 43:871-875.

6 Clark, K. G., J. Y. Yee and K. S. Loue. 1952. Ureaform — New nitrogen fertilizer. Crops and soils, 4 (8).

7 Ferguson, K. A., J. A. Hemsley and P. J. Reis. 1967. The effect of protecting dietary protein from microbial degradation in the rumen. J. Science (Australia) 30 (215).

8 Fraenkel-Conrat, H. and H. S. Olcott. 1945. The reaction of formaldehyde with proteins. J. Am. Chem. Soc. 67:950-954.

9 Fraenkel-Conrat, H. 1954. Reaction of nucleic acid with formaldehyde. Biochim. Biophys. Acta. 15:307-309.

10 Kadawaki, I. 1936. New compounds of urea-formaldehyde condensation products. Repts. Imp. Ind. Res. Lab., Osaha, Japan. 6:15-41.

11 O'Donnell, J. U.S. Patent. 2,830,036, 4/8/58.

12 O'Donnell, J. U.S. Patent. 2,882,141, 3/59.

13 O'Donnell, J. U.S. Patent. 3,227,543, 1/4/66.

14 Smith, J. B. 1955. Report of the subcommittee recommendations of the refereenitrogen activity index. J.A.O.A.C. 38 (1):64.

15 Tollens, B. 1884. Ueber einige derivate des formaldehyds, Chemische Berichta. 17:653A-659A.

16 Walker, J. F. 1967. Formaldehyde A.C.S. minograph series, 3rd Ed., Reinhold, N.Y. 399-400.

17 Official Methods of Analysis. (1950). A.O.A.C. Test 2.34, 7th Ed., 15.


MANAGING LIVESTOCK WASTES

# Health Aspects of Feeding Animal Waste Conserved in Silage

T. A. McCaskey, W. B. Anthony

NUMEROUS reports have been published on the use of poultry and animal wastes as sources of animal feed. The incentive for use of animal wastes as feed sources has been the greater economy involved with these rations as compared to conventional rations. However, the use of poultry and animal wastes as feeds or feed ingredients is not sanctioned by the Food and Drug Administration since these feeds are considered adulterated under Section 402 of the Federal Food, Drug and Cosmetic Act. Section 135.104 of the Code of Federal Regulations, revised April 1, 1974, states "the Food and Drug Administration has not sanctioned and does not sanction the use of poultry litter as a feed or component of feed for animals." Although only poultry litter is mentioned in the Code the statement applies to other animal wastes as well (Taylor et al. 1974).

Before the FDA will consider sanctioning the use of wastes as feed, more information is needed to demonstrate their safety for consumption by animals, and that meat and by-products from these animals are safe for human consumption (Taylor 1971). In the event that animal manures are sanctioned for feeding, regulations governing their use and safety will probably be as stringent or perhaps more stringent than those currently in effect for the feeding of meat meal tankage, blood meal, poultry by-product meal, hydrolyzed poultry feathers, and dried paunch contents. Regulations pertaining to microbial safety of these animal by-product feeds currently are covered under Section 135.105 of the Code of Federal Regulations, revised April 1, 1974. Under this Section approved animal by-product feeds shipped interstate will be regarded as adulterated, as within the meaning of Section 402 (a) of the Federal Food, Drug, and Cosmetic Act, when found contaminated with *Salmonella.*

Salmonellae are frequently found in poultry and animal wastes. The potential for health problems associated with salmonellae is perhaps one of the greatest since all serotypes are considered pathogenic and infection can be spread to other animal species as well as humans. Among farm animals, salmonellosis is most prevalent in turkeys, followed by chickens, cattle and swine (Morse and Duncan 1974, Steele 1969). A higher incidence of infection among cattle has been reported to occur during the winter months probably because exposure is greatest when the animals are housed during the winter months (Edwards and Galton 1967).

The salmonellae most often associated with infection in cattle reported by British researchers are *S. dublin* and *S. typhimurium* (Stevens et al. 1967, Lawson et al. 1974, Tutt and Hoare 1974). In the United States *S. typhimurium* is the predominant serotype both in cattle and in humans

(Gibson 1965, Moran 1962). This is also the most frequently isolated serotype, reported by the Center for Disease Control (1974) for both human and non-human sources.

Due to the prevalence of salmonellae in animals and the potential for health problems associated with this group of enteric pathogens in animal wastes, a study was initiated to evaluate the survival of salmonellae in an ensiled, manure-formulated feed.

## PROCEDURES

The health aspects of pathogen survival in an ensiled manure-feed mixture was evaluated with a product containing 45 parts ground shelled corn, 15 parts corn silage, and 40 parts of manure from beef steers confined on concrete. The manure used in formulating the 45-15-40 ration was scraped daily into a manure pit and twice weekly the manure (20-25 percent solids) from the pit was blended with ground shelled corn and corn silage, and ensiled for beef cattle feeding trials. Following blending of the product at the farm a portion of the product was sent to the laboratory for proximate analyses and for pathogen survival studies. The blended product contained about 50 percent solids and 11-12 percent crude protein (dry basis).

The effect of the ensiling process on the survival of enteric pathogens was evaluated with models using several *Salmonella* serotypes. The salmonellae used in the study are listed in Table 1. All cultures were grown in Brain-Heart Infusion broth (BHI) at 37 C for 24 hr prior to their use in the study.

To facilitate recovery of the salmonellae from the manure-blended ration, a 47 mm diameter by 0.8 mm deep plastic disposable dish was packed with the ration (about 16 g) and the ration was seeded with 1 ml of BHI broth culture. Each *Salmonella* culture was inoculated individually in a dish. The seeded dish was covered with two layers of tissue paper, placed in a gas impermeable plastic bag, and covered with the manure-blended ration. Each bag was packed with about 600 g of ration and up to six dishes were placed in each bag. The bags were heat sealed and placed in an incubator to simulate conditions of ensiling.

Following the ensiling process the plates were removed from the bags and the contents of the plates were transferred to 200 ml of Tetrathionate broth (TET). The broth was incubated at 41 C for 24 hr and then streaked on Brilliant Green Sulfa agar (Difco). Suspected salmonellae colonies were picked from the agar and the isolates were confirmed biochemically and serologically by the procedures of Edwards and Ewing (1962). Pre-enrichment of manure and manure-blended ration in Lactose broth for recovery of salmonellae was not used. Preliminary studies indicated that recovery from these substrates was better when pre-enrichment was omitted and the substrates were directly placed into TET broth.

The authors are: T. A. McCASKEY, Associate Professor, and W. B. ANTHONY, Professor, Animal and Dairy Sciences Dept., Alabama Agricultural Experiment Station, Auburn, AL.

| | Serotype | number |
|---|---|---|
| 1 | S. typhimurium | 3124 |
| 2 | S. typhimurium | 3670 |
| 3 | S. typhimurium var. Copenhagen | 3408 |
| 4 | S. typhimurium var. Copenhagen | 3141 |
| 5 | S. heidelberg | 3769 |
| 6 | S. heidelberg | 3758 |
| 7 | S. saint-paul | 3629 |
| 8 | S. saint-paul | 3357 |
| 9 | S. infantis | 3808 |
| 10 | S. infantis | 3531 |
| 11 | S. chester | 3804 |
| 12 | S. chester | 2492 |
| 13 | S. blockley | 3675 |
| 14 | S. blockley | 3672 |
| 15 | S. montevideo | 1718 |
| 16 | S. montevideo | 1666 |
| 17 | S. anatum | 3682 |
| 18 | S. anatum | 3536 |
| 19 | S. typhimurium | 3857 |
| 20 | S. thompson | 3632 |
| 21 | S. thompson | 3611 |
| 22 | S. senftenberg | 3447 |
| 23 | S. senftenberg | 2708 |
| 24 | S. tennessee (lactose fermenting) | --- |
| 25 | S. typhimurium ATCC | 13311 |
| 26 | S. typhimurium ATCC | 14028 |
| 27 | Salmonella sp. | 1974 |

Cultures 1 through 24 were obtained
through Dr. J. E. Williams, Southeast
Poultry, Research Laboratory, Athens,
Georgia. Cultures 25 and 26 were obtained
from the American Type Culture Collection;
and culture 27 was isolated in our laboratory
from a rectal swab of a pig.

Enumeration of selected groups of indigenous micro-organisms in the ration was accomplished with selective media. Coliforms were enumerated on Violet Red Bile agar, yeasts and molds on Potato Dextrose agar acidified with tartaric acid, and acid-producing bacteria were enumerated on Tomato Juice agar with 0.75 percent added calcium carbonate. Spore-forming bacteria in the ration were enumerated by heating a 1:10 dilution of the ration in distilled water for 10 min at 80 C. The heated sample was plated in duplicate on BHI agar and incubated aerobically and anerobically.

Assay for inhibition of growth of salmonellae by bacterial isolates from the manure-blended ration was accomplished by streaking the salmonellae at right angles to the isolates on Phenol Red Dextrose agar. Inhibition was noted where the salmonellae failed to grow near the isolates.

## RESULTS AND DISCUSSION

The survival of the 27 salmonellae in the 45-15-40 blended ration and in the manure used to prepare the blended ration following 3 days of storage at 25 C is shown in Table 2. Twenty-five of the 27 salmonellae survived in the manure; whereas, none of the cultures survived in the manure-blended ration. The higher acidity of the blended ration appeared to be the principal factor accounting for the difference in survival of the salmonellae in the two substrates.

The temperature at which the ensiled ration was held was demonstrated to have a significant effect on the sur-

TABLE 2. SURVIVAL OF SALMONELLAE IN MANURE AND
IN AN ENSILED MANURE-FEED MIXTURE.

| | Recovery after 3 days incubation at 25 C | |
|---|---|---|
| | Manure* | Manure-feed† |
| No. cultures‡ survived | 25 | 0 |
| Percent recovered | 92 | 0 |
| pH, Initial | 6.8 | 6.5 |
| pH, 3-day | 6.0 | 4.5 |

*25 g portions of manure slurry were inoculated with 1 ml of
Salmonella broth culture yielding 10 million or more organisms
per gram of manure in tubes.
†16 g portions of manure-feed mixture were inoculated with
1 ml of culture in dishes yielding 19 million or more organisms
per gram feed.
‡Twenty-seven salmonellae cultures were used.

vival of salmonellae (Table 3). At 5 C, 21 of the 27 salmonellae survived and at 15 C, 25 survived following 4 days of ensiling the ration. Only one culture survived in the ration held at 25 C and none survived following ensiling at 35 C. The freshly blended ration used in this study was considerably more acidic than those previously used; however, the survival of the cultures in the ration was not appreciably affected by the acidic environment at 5 or 15 C. This indicates that there are factors in addition to acid development in the ensiled 45-15-40 ration that affect salmonellae survival. Acid development was greater at 25 C than at 15 C; however, the relatively small difference in pH of the two ensiled rations does not appear to be sufficient to account for the greater survival in the ration held at 15 C.

Prior to ensiling the ration and following 4 days of ensiling at each of the four temperatures, the ration was analyzed to determine the effect of ensiling temperature on the growth of selected groups of indigenous microorganisms (Table 4). In the freshly blended ration yeasts and molds (primarily molds) made up the major group of indigenous organisms (119 million/g) followed by acid-producing bacteria (60 million/g), coliform bacteria (9.7 million/g), and anaerobic spore-forming organisms (6.3 million/g).

Following 4 days of ensiling at 5 C all the microbial groups remained at about the same population levels or higher except for the coliforms. Although the temperature of the ration is not likely to reach 5 C in large quantities of ensiled ration, the temperature at the periphery of the ensiled ration might be expected to fluctuate with climatic conditions. At 15 C and higher the population levels of each of the groups of indigenous organisms decreased as the temperature of ensiling increased. At 35 C the coliform count decreased to less than 1000 per gram of ration. Perhaps the count was less since coliforms were not detected in a 1:1000 dilution of the ration. The yeast and mold population after 4 days of ensiling at 35 C was the major microbial group comprising 8.1 million organisms per gram of ration. The effect of longer ensiling periods on the number of viable indigenous organisms in ensiled manure-blended rations

TABLE 3. EFFECT OF TEMPERATURE ON
SURVIVAL OF SALMONELLAE IN THE
ENSILED MANURE-FEED MIXTURE.

| Survival after 4 days ensiling | Temperature C | | | |
|---|---|---|---|---|
| | 5 | 15 | 25 | 35 |
| No. cultures* | 21 | 25 | 1 | 0 |
| Percent cultures | 78 | 93 | 4 | 0 |
| pH, initial | 4.8 | 4.8 | 4.8 | 4.8 |
| pH, 4-day | 4.6 | 4.4 | 4.0 | 4.1 |

*Twenty-seven salmonellae cultures were used.

| Time and temperature | Yeasts & molds | Coliforms | Acid-producing bacteria | Aerobic spores | Anaerobic spores |
|---|---|---|---|---|---|
| | ---------------- Number/g $(\times 10^{-4})$ (Wet wt.) ---------------- | | | | |
| 0-day | 11900 | 970 | 6000 | 500 | 630 |
| 4-day | | | | | |
| 5 C | 14600 | 110 | 14200 | 2500 | 770 |
| 15 C | 4600 | 25 | 2700 | 740 | 290 |
| 25 C | 630 | 0.12 | 1500 | 700 | 180 |
| 35 C | 810 | < 0.10 | 500 | 650 | 260 |

needs to be evaluated since many of the spore-forming molds and bacteria are better able to survive than are the coliforms.

To determine the effect of the indigenous microflora on the survival of salmonellae, several bacterial colonies were isolated from the 45-15-40 ration after ensiling for 3 days at 25 C. The colonies were selected from plates of Tomato Juice agar following enumeration of acid-producing bacteria in the ensiled ration. The isolates were screened for ability to inhibit growth of the 27 salmonellae. Each of six selected isolates were streaked across the center of a Phenol Red Dextrose agar plate and also on the same medium with 1 percent of added calcium carbonate. After the isolates were incubated at 37 C for 24 hr the salmonellae were streaked at right angles to the isolates. The plates were incubated an additional 24 hr and examined for inhibition of salmonellae growth. The results of this study indicated that the basis of salmonellae inhibition by the six isolates was due principally to acid production (Table 5). All the isolates inhibited growth of the 27 salmonellae on Phenol Red Dextrose agar; whereas, there was no apparent inhibition of a few salmonellae by three of the isolates on the same medium with 1 percent added calcium carbonate.

The critical level of acid development by the isolates from the ensiled 45-15-40 ration on the inhibition of sal-

monellae growth was determined. Isolate Number 16 was grown in Phenol Red Dextrose broth for 48 hr at 37 C. After incubation the pH of the broth culture decreased to 4.6 and a cell-free filtrate of the culture was prepared and tested for sterility. The filtrate was divided into five aliquots and adjusted to pH 4.0, 5.0, 6.0, and 7.0 with 0.1N NaOH or 0.1N HC1. The pH of one aliquot was not adjusted (pH 4.6). The five aliquots were tested again for sterility and 3 ml were dispensed aseptically per sterile 13 x 100 mm screw-capped tube. As a control freshly prepared Phenol Red Dextrose broth was used. Five salmonellae cultures were inoculated (1 loop) into each of the sets of broth and incubated at 37 C for 24 hr.

The salmonellae did not grow in the cell-free filtrates adjusted to pH 4.0 nor in the unadjusted filtrate with a pH of 4.6. All salmonellae grew in the filtrates adjusted to pH 6.0 and 7.0 (Table 6). From these data it would appear that the minimum pH for the growth of salmonellae is in the range of 4.6 to 5.0. Cell-free filtrates prepared from all six of the isolates from the ensiled 45-15-40 ration showed similar pH effects on the growth of the salmonellae.

The effect of different acids on the growth of salmonellae is shown in Table 7. Phenol Red Dextrose broth was adjusted to pH 4.0 through 7.0 with 0.1N HC1, 0.1M lactic and 0.1M acetic acids. The three sets of broth were inoculated with one loop of a 24 hr BHI culture of *Salmonella* and incubated at 37 C for 24 hr. Following incubation, growth at each of the pH levels was determined visually. The pH effects of HC1 and lactic acids were identical. The salmonella did not grow in the broth adjusted to pH 4.0 with either HC1 or lactic acid; whereas, the cultures grew in the broth adjusted to pH 5.0 and higher. Acetic acid was more inhibitory than were HC1 and lactic acid. The salmonellae failed to grow in broth adjusted to either pH 4.0 or 5.0 with acetic acid; however, growth occurred in the broth adjusted to pH 6.0 and 7.0 with acetic acid.

TABLE 5. GROWTH INHIBITION OF SALMONELLAE BY BACTERIA ISOLATED FROM THE ENSILED MANURE-FEED MIXTURE.

| Isolate number | Number of Salmonellae inhibited | |
|---|---|---|
| | PRD agar | PRD agar (1 percent $CaCO_3$) |
| 16 | 27* | 0[a] |
| 17 | 27 | 0 |
| 19 | 27 | 0 |
| 24 | 27 | 5 |
| 35 | 27 | 5 |
| 36 | 27 | 1 |

*Total of 27 salmonellae assayed.

TABLE 6. EFFECT OF pH ON GROWTH OF SALMONELLAE IN CELL-FREE FILTRATE PREPARED FROM GROWTH MEDIUM OF BACTERIAL ISOLATE NUMBER 16.

| Salmonellae | pH of cell-free filtrate* | | | | | |
|---|---|---|---|---|---|---|
| | Fresh broth | No adj (4.6) | 4.0 | 5.0 | 6.0 | 7.0 |
| | ---------- Growth response† ---------- | | | | | |
| Not inoculated | — | — | — | — | — | — |
| S. typhimurium (14028) | + | — | — | — | + | + |
| S. typhimurium (13311) | + | — | — | — | + | + |
| S. heidelberg (3769) | + | — | — | — | + | + |
| S. montevideo (1666) | + | — | — | — | + | + |
| S. montevido (1718) | + | — | — | + | + | + |

*Filtrate was adjusted with 0.1 N NaOH or 0.1 N HC1.
† —, no growth; +, growth.

TABLE 7. EFFECT OF pH ON GROWTH OF SALMONELLAE.

| Salmonellae | Phenol Red Dextrose broth pH adjusted with 0.1 N HCl | | | | |
|---|---|---|---|---|---|
| | 7.4* | 4.0 | 5.0 | 6.0 | 7.0 |
| Not inoculated | — | — | — | — | — |
| S. typhimurium (14028) | + | — | + | + | + |
| S. typhimurium (13311) | + | — | + | + | + |
| S. heidelberg (3769) | + | — | + | + | + |
| S. montevideo (1666) | + | — | + | + | + |
| S. montevido (1718) | + | — | + | + | + |

Results for lactic acid identical to HCl; whereas, acetic acid inhibited growth at pH 4 & 5.
*PRD broth not adjusted.

## CONCLUSIONS

These studies showed that salmonellae do not survive in a ration containing 45 parts ground shelled corn, 15 parts corn silage and 40 parts of bovine manure after ensiling for 3 or 4 days. During ensiling acid must be produced to effect pH of the ration in the range of 4.0 to 4.5 or lower. Development of sufficient acid to kill salmonellae was dependent to some degree upon an ensiling temperature greater than 25 C. Acid development alone was not entirely responsible for death of salmonellae. The manure-blended ration with a pH of 4.0 to 4.5 was sufficient to kill salmonellae in 4 days at 35 C; whereas, only 7 percent of the salmonellae cultures were killed in the same ration ensiled at 15 C. Although the temperature of the ration will likely be higher than 15 C during ensiling, the temperature at the periphery of the ensiled ration will vary with atmospheric temperatures and may affect survival of salmonellae.

Further studies have shown that some acid-producing bacteria in the ensiled manure-blended ration were highly inhibitory to the growth of salmonellae. The mechanism of this inhibition was demonstrated to be due principally to acid development. Inhibition of salmonellae by acids was shown to be affected by kind and quantity of acid. In a medium adjusted to various pH levels with lactic acid, salmonellae did not grow at pH 4.0, whereas growth occurred at pH 5.0 and higher. Acetic acid was more inhibitory than lactic acid preventing growth of salmonellae at pH 4.0 and 5.0.


### References

1   Center for Disease Control. 1974. Salmonella Surveillance Annual Summary 1973, issued December 1974.

2   Edwards, P. R. and W. H. Ewing. 1962. Identification of enterobacteriaceae. Burgess Publ. Co., Minneapolis, Minn.

3   Edwards, P. R. and M. M. Galton. 1967. Salmonellosis. Adv. in Vet. Sci. 11:1-63.

4   Gibson, E. A. 1965. Reviews of the progress of dairy science: Section E. Diseases of dairy cattle. Salmonella infection in cattle. J. Dairy Res. 32:97-134.

5   Lawson, G. H. K., E. A. McPherson, A. H. Laing, and P. Wooding. 1974. The Epidemiology of Salmonella dublin infection in a dairy herd. J. Hyg. Camb. 72:311-327.

6   Morse, E. V. and M. A. Duncan. 1974. Salmonellosis—an environmental health problem. J. Amer. Vet. Med. Assoc. 165:1015-1019.

7   Moran, A. B. 1962. Occurrence and distribution of Salmonella in animals in the U.S. Proc. 65th Ann. Meet. U.S. Livestock Sanitary Assoc., Minneapolis, Minn. p. 441-448.

8   Steele, J. H. 1969. Salmonellosis. A major zoonosis. Arch. Environ. Health. 19:871-875.

9   Stevens, A. J., E. A. Gibson and L. E. Hughs. 1967. Salmonellosis III. Recent observations on field aspects. Vet. Record. 80:154-161.

10   Taylor, J. C. 1971. Regulatory aspects of recycled livestock and poultry wastes. Proc. Intern. Symp. on Livestock Wastes, April 19-22, 1971, Ohio State University. p 291-292.

11   Taylor, J. C., D. A. Gable, G. Graber and E. W. Lucas. 1974. Health criteria for processed wastes. Federation Proceedings 33(8) 1945-1946.

12   Tutt, J. B. and D. I. B. Hoare. 1974. Disease associated with S. typhimurium in cattle. Vet. Record. 95:334-337.

# Start-Up of Pilot Scale Swine Manure Digesters
## for Methane Production


H. M. Lapp, D. D. Schulte, E. J. Kroeker, A. B. Sparling, B. H. Topnik

MEMBER  ASSOC. MEMBER  ASSOC. MEMBER
ASAE  ASAE  ASAE


**P**RODUCTION of methane gas from animal manure has received increased attention throughout the world as efforts to find alternate energy sources intensify. Animal manure contains large quantities of organic matter which, if broken down by anaerobic bacteria in a controlled environment, may produce significant quantities of methane gas. The potential for energy recovery from animal waste is enhanced by modern systems of confinement housing where large quantities of manure are rapidly accumulated.

Farm scale digesters have been operated in Europe,


The authors are: H. M. LAPP, Professor, D. D. SCHULTE, Assistant Professor, and E. J. KROEKER, Research Engineer, Agricultural Engineering Dept., A. B. SPARLING, Associate Professor and B. H. TOPNIK, Research Fellow, Civil Engineering Dept., University of Manitoba, Winnipeg, Canada.

**Acknowledgements:** The authors wish to acknowledge the Biomass Energy Institute Inc., the Faculty of Agriculture at the University of Manitoba, Agriculture Canada and Shell Canada Ltd. for financial support of the project. Special thanks are due to Professor L. C. Buchanan, J. D. Haliburton, N. Parreno and R. H. Mogan for technical assistance on the project.


Asia, Africa and India (Imhoff 1946, Fry 1973, Singh 1973). However, the technical and economic feasibility of large-scale anaerobic digestion of animal manure has not been assessed in a cold climate characteristic of Canada. Objectives of the project described in this paper include: (a) determination of design parameters for methane production from animal manure in cold climates; (b) determination of the efficiency of the anaerobic digestion process in recovering energy from animal wastes; (c) development of safe and economical methods of collecting, scrubbing, storing and utilizing digester gas on farms; (d) analysis of the effluent from anaerobic digestion and assessment of its value as a fertilizer; and (e) assessment of the environmental impact, if any, of the anaerobic digestion process.

## EXPERIMENTAL PROCEDURE
### Pilot-Plant

Intensive pilot-scale studies began in 1974. The pilot-plant is located adjacent to a four-building swine complex at the University of Manitoba Glenlea Research

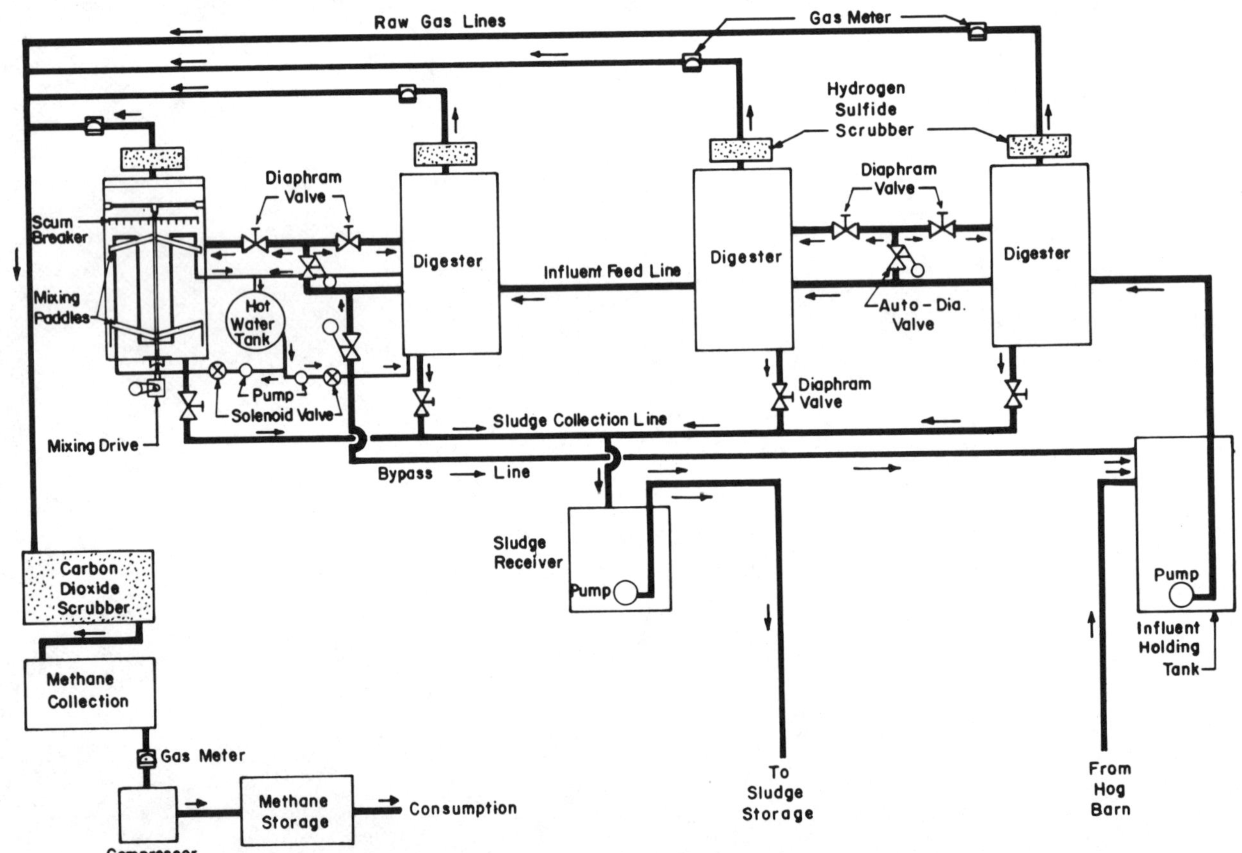


**FIG. 1 Schematic diagram of pilot-plant.**

| Parameters | Digester | | | |
| --- | --- | --- | --- | --- |
| | A | B | C | D |
| | Before | Before | Before | Before |
| Volatile acids, mg/l HAc | 200 | 200 | 300 | 300 |
| Alkalinity, mg/l $CaCO_3$ | 4880 | 4880 | 5640 | 5640 |
| pH | 7.4 | 7.4 | 7.2 | 7.2 |
| Ammonia, mg/l N | 850 | 850 | 950 | 950 |
| $CO_2/CH_4$ | — | — | — | — |

Station and is housed in a 7.3 x 7.9 m (24 x 26 ft) steel building. All electrical devices in the building are explosion-proof. A small steel building adjacent to the pilot-plant houses auxiliary non-explosion proof equipment.

The pilot-plant, shown schematically in Fig. 1, consists of four single-stage 3194 liter (700 Imp gal) fiberglass digestion tanks. The pilot-plant is serviced by a 91.3 m³ (20 000 Imp gal) influent holding tank located between the steel building and the swine barns. The influent holding tank is filled periodically by an overhead liquid manure transfer system. The influent holding tank and overhead transfer system were placed into operation during the course of the experiment reported in this paper. Digester gas is collected over water where it is held temporarily prior to processing, handling and utilization.

### Start-Up Procedure

Initially, two of the four digester (A and B) were each seeded with 2000 liters (440 Imp gal) of digesting sludge from the Winnipeg North End Municipal Sewage Treatment Plant. Table 1 summarizes the condition of the seed at the initiation of the start-up procedure. The following day, digesters A and B were fed 23 liters (5 Imp gal) of raw liquid swine manure. This loading rate was increased 23 liters each day for four days, bringing the mixed liquor volume in each to operational capacity (2300 liters or 505 Imp gal). Both digesters A and B were then operated at a solids retention time (SRT) of 20 days.

One week after seeding digesters A and B, digesters C and D were seeded to capacity (2300 liters or 505 Imp gal) with digesting sludge from the North End plant. Conditions of the seed in digesters C and D are also shown in Table 1. The following day, digesters C and D were started at a SRT of 50 days by adding 46 liters of raw liquid swine manure. The feed rate was increased over the next eight days to a SRT of 10 days.

### Analytical Procedures

Gas analyses were performed by gas chromotography on a Fisher-Hamilton gas partitioner using a dual column arrangement of 30 percent DEHS on 60/80-mesh Chromosorb P and a 40/60-mesh Molecular Sieve (14X). Volatile acids were determined spectrophotometrically at 490 m$\mu$ using procedures outlined by the Hach Chemical Corporation (1971). Alkalinity was determined potentiometrically according to Standard Methods (1971). Ammonia was also determined according to Standard Methods (1971).

## RESULTS

Due to the accelerated start-up discussed earlier, gas production rates in digesters C and D began decreasing during the latter part of the start-up period.

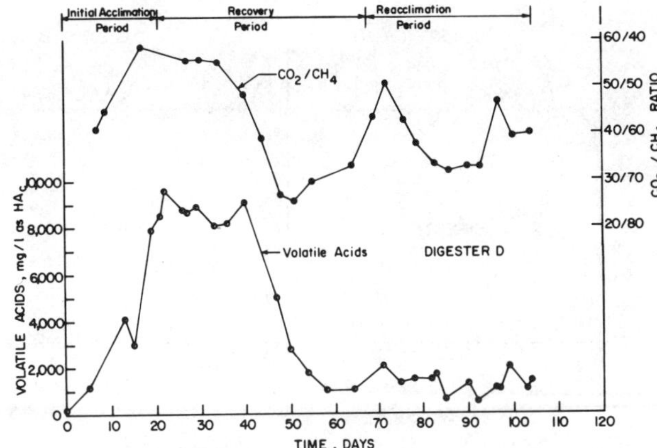


FIG. 2 Volatile acids content and $CO_2/CH_4$ ratio in digester "D" during start-up, recovery and reacclimation.

Consequently, the SRT's were adjusted to 20 days so that all digesters were fed at a rate of 115 liters per day (25 Imp gal/day). During the following six days gas production also began dropping in digesters A and B. Volatile acids, alkalinity and ammonia continued to rise in all digesters. All digesters were then adjusted to a SRT of 40 days in an effort to stabilize the process. During an additional six day period gas production rates continued to decrease. Feeding was then stopped in all digesters and an extended recovery period (47 days) ensued. During this time the influent holding tank and automatic manure transfer system were placed into operation to improve control of the loading rates.

Fig. 2 illustrates the levels of volatile acids and the $CO_2/CH_4$ ratio in the gas from digester D during start-up, recovery and reacclimation. Table 2 shows the condition of the digesters at the beginning of the recovery period.

In an attempt to reduce the mixed liquor ammonia concentration, digesters A and B were diluted during the recovery period by the removal of 550 liters (120 Imp gal) of mixed liquor from each digester and the addition of an equal volume of non-brackish water. Table 2 also shows the condition of each digester at the end of the recovery period.

TABLE 2. PERFORMANCE INDICATORS AT BEGINNING
AND END OF DIGESTION RECOVERY PERIOD.

Beginning*

| Parameters | Digester | | | |
| --- | --- | --- | --- | --- |
| | A | B | C | D |
| Volatile acids, mg/l HAc | 8250 | 7350 | 8000 | 8550 |
| Alkalinity, mg/l $CaCO_3$ | 10180 | 10160 | 10240 | 10680 |
| pH | 7.5 | 7.4 | 7.4 | 7.3 |
| Ammonia, mg/l N | 2310 | 2280 | 2320 | 2270 |
| $CO_2/CH_4$ | 52/48 | 48/52 | 54/46 | 56/44 |

End†

| Parameters | Digester | | | |
| --- | --- | --- | --- | --- |
| | A | B | C | D |
| Volatile acids, mg/l HAc | 1000 | 940 | 1080 | 1050 |
| Alkalinity, mg/l $CaCO_3$ | 9560 | 9400 | 12760 | 14400 |
| pH | 8.2 | 8.1 | 8.3 | 8.3 |
| Ammonia, mg/l N | 1720 | 1790 | 2600 | 2740 |
| $CO_2/CH_4$ | 38/62 | 45/55 | 47/53 | 33/67 |

*Digesters A and B-27 days from start-up; Digesters C and D-20 days from start-up.
†Digesters A and B-74 days from start-up; Digesters C and D-67 days from start-up.

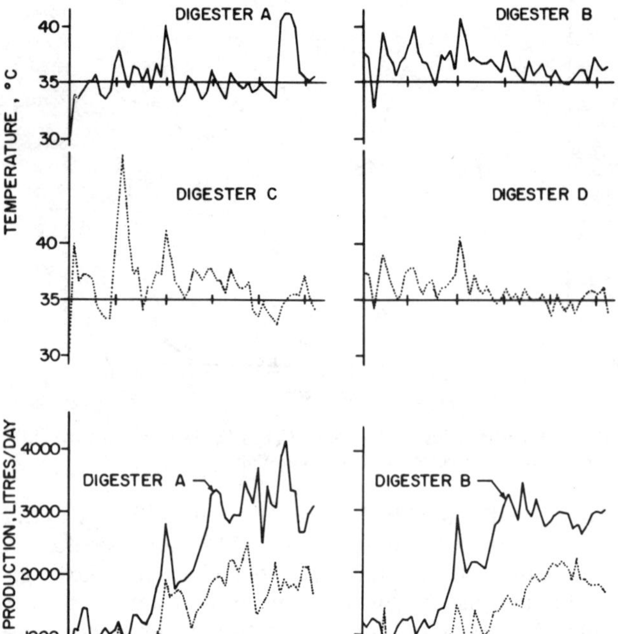

FIG. 3 Gas production and temperature during reacclimation.

**TABLE 3. PERFORMANCE INDICATORS AT END OF REACCLIMATION PERIOD.***

| Parameters | Digester | | | |
| --- | --- | --- | --- | --- |
| | A | B | C | D |
| Volatile acids, mg/l HAc | 1330 | 1330 | 1500 | 1450 |
| Alkalinity, mg/l $CaCO_3$ | 14480 | 14960 | 14640 | 14340 |
| pH | 8.4 | 8.4 | 8.2 | 8.5 |
| Ammonia, mg/l N | 2930 | 2930 | 2870 | 2840 |
| $CO_2/CH_4$ | 34/66 | 35/65 | 33/67 | 39/61 |

*Digesters A and B-111 days from start-up; Digesters C and D-104 days from start-up.

Gas production rates had fluctuated at low levels throughtout the recovery period but the $CO_2/CH_4$ ratio decreased from those shown in Table 2. Reacclimation was accomplished over an additional 37-day time interval after which a 15-day SRT and a 30-day SRT had been reached for digesters A and B and C and D, respectively. Table 3 summarizes the data after reacclimation.

Before installation of the influent holding tank and manure transfer system, solids concentration in the feed material fluctuated widely (2.5 to 7.8 percent total solids) making maintenance of a consistent solids loading rate difficult at a given SRT. This variation in solids loading rate in the initial acclimation period may have been partially responsible for digester instability. The utilization of SRT as the controlled loading parameter is different from many laboratory studies where volatile solids loading rates have generally been the controlled parameter. However, SRT has proven to be a more practical parameter in terms of pilot-scale operation where influent solids content is variable. This may also prove to be the case for farm application. At the end of the 37-day reacclimation period the 30-day SRT of digesters C and D corresponded to a volatile solids loading rate of approximatley 1.3 gm/liter/day (0.08 lb/ft³/day) and the 15-day SRT of digesters A and B corresponded to approximately 2.6 gm/liter/day (0.17 lb/ft³/day).

The manure from the influent holding tank was fed to the digesters once daily at a temperature of approximately -1 C (30 F) during mid-winter. The mixed liquor temperature at loading was reduced by as much as 3 C (5.5 F) depending on the SRT. Such temperature drops and occasional sudden temperature increases (Fig. 3) did not seriously reduce gas production rates. Gas production rates and digester mixed liquor temperatures during reacclimation and beyond are shown in Fig. 3.

## DISCUSSION

Several researchers (Anthonisen and Cassell 1966, Schmid and Lipper 1969 and McKinney 1962) have noted that the ammonia nitrogen in liquid manure may occur at concentrations above the limiting level and thus may retard methane production. According to McCarty (1964), ammonia nitrogen concentrations between 1500 and 3000 mg/liter are inhibitory and those above 3000 mg/liter are completely toxic to methane forming bacteria. Yet, Hart (1963) reported data translating to concentrations of 4260 and 4140 mg/liter of ammonia nitrogen in digesting chicken manure. Methane production was depressed but not eliminated. Tables 2 and 3 show that all ammonia nitrogen concentrations were in the inhibitory range. At the end of the reacclimation period (Table 3) near toxic conditions supposedly existed but gas production and quality (Fig. 3 and Table 3) reached their highest levels.

Two factors may, alone or in combination, explain the tolerance to high ammonia levels. First, the methane formers may have acclimated to the high ammonia concentrations. Kotze et al. (1968) observed that adaptation periods in the laboratory studies of anaerobic digestion are usually not long enough to take advantage of increases in enzymatic acitvity that occur when switching from digesting sewage slude to synthetic or other substrates. In reports of digestion failure (Hart 1963, Schmid and Lipper 1969), gas quality measured as percent $CH_4$ was low at the onset of acclimation to manure loading, making adaptation of the methane formers to the high ammonia levels increasingly difficult. In this experiemnt, good digestion characteristics were present initially (Table 1) but the loading rate at start-up may have been increased too rapidly, preventing acclimation to the ammonia levels shown in Table 2. Gramms et al. (1971) reported poor gas production from swine and poultry manure when 35-day acclimation periods were employed. Melbinger and Connellon (1971) demonstrated that even longer acclimation periods may be necessary when high ammonia levels occur. Their experience, in which full-scale sewage sludge digesters were reacclimated over a three month period after initial failure due to ammonia toxicity, showed that ammonia concentrations above normally reported threshold levels are not necessarily toxic if the rate of increase of the ammonia is less rapid than acclimation of the methane formers.

The form and concentration of ammonia which is toxic to the methane formers is pH dependent (McCarty and McKinney 1961). Fig. 4 illustrates the ammonia nitrogen conditions at which digesters have been operated successfully and at which several have failed. Successful reacclimation of the digesters in this study (Fig. 3) at high free ammonia concentrations (Fig. 4) may be partially explained by adaptation of the methane formers to the supposedly toxic conditions.

Secondly, cation antagonism may have offset the



MANAGING LIVESTOCK WASTES

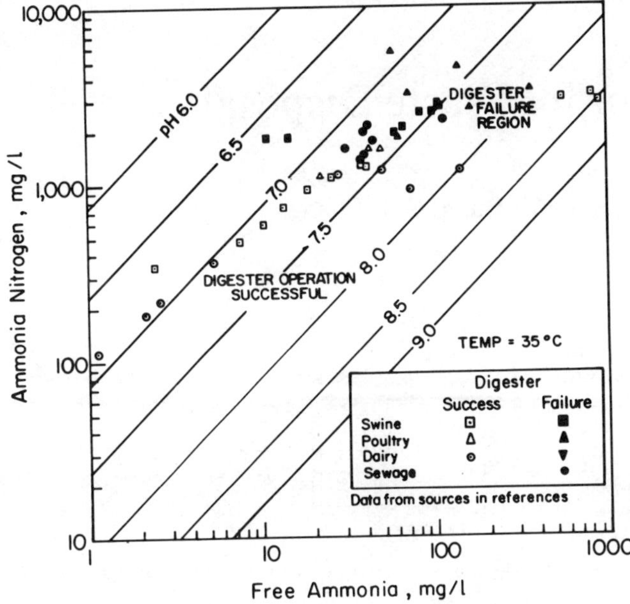

FIG. 4 Relationship of digester failure to ammonia nitrogen and pH [After Jewell et al. 1974].

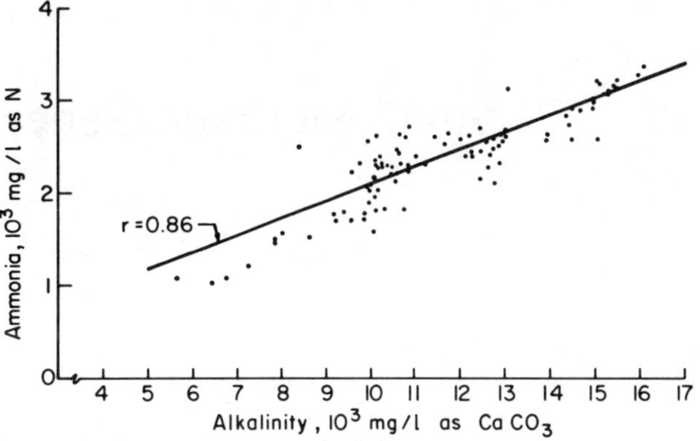

FIG. 5 Ammonia nitrogen and alkalinity in swine manure digesters A, B, C and D.

secondly, cation antagonism may have offset the potential ammonia toxicity. Kugelman and Chin (1971) outlined experiments and evidence supporting this theory. Anthonisen and Cassell (1966) used cation antagonism successfully when they added sodium chloride to a poultry manure anaerobic digester. They reported significantly higher gas production in a digester receiving the sodium antagonist than in one without. The water consumed by the swine and used for washing pens at the Glenlea Research Station is brackish. The sodium content of the water is approximately 900 mg/liter while sulfates and chlorides occur in concentrations of 700 and 1300 mg/liter, respectively. It is possible that cation antagonism reduced the toxic potential of the ammonia in this study.

Efforts have been made in this experiment to find suitable relationships for monitoring digester performance. Fig. 2, for example, illustrates a correlation between volatile acids concentration and the $CO_2/CH_4$ ratio of the gas. Kotze et al. (1972) suggested that this ratio, rather than the total volume of gas produced, could be used as an indicator of unbalanced conditions. Volatile acids in themselves are not toxic to methane bacteria (McKinney 1962) but are useful indicators of balanced digestion. High concentration of volatile acids and a declining pH were reflected in increased $CO_2$ content of the digester gas (Table 2).

A resistance to the decline of pH, in spite of high volatile acids concentrations, was noted throughout this experiment. The pH of digester D, for example, remained between 7.6 and 8.0 until a volatile acids concentration of 8550 mg/liter had been reached. The pH then dropped to 7.3 before loading was stopped to allow recovery. Schmid and Lipper (1969), attributed the buffering capacity of swine manure digesters to ammonium bicarbonate. Melbinger and Donnellon (1971) stated that about 23 percent of total alkalinity seemed to be in the form of ammonia nitrogen. A similar interpretation of Fig. 5 shows that approximately 18 percent of the total alkalinity in this experiment was associated with ammonia nitrogen. From Fig. 5 it appears that it may be possible to monitor the rate of increases of ammonia nitrogen through changes in alkalinity. Interpretation of the toxicity of such changes in ammonia nitrogen bears further investigation as it relates to the effectiveness of dilution, loading rates, SRT, salt additions and other control procedures.

## SUMMARY AND CONCLUSIONS

Pilot-scale energy recovery from swine manure in four single-stage 3194 liter anaerobic digesters indicated that fundamental considerations must be given to proper start-up procedures. Swine manure may be digested successfully at ammonia levels in excess of 3000 mg/liter if the methane forming bacteria are acclimated for time periods exceeding those normally reported for animal manure digestion. Cation antagonism may also account for the abnormal tolerance to ammonia found in this study.


### References

1 Albertson, O. E. 1961. Ammonia nitrogen and the anaerobic environment. Jour. Water Poll. Control Fed. 33:978-995.

2 Anthonisen, A. A. and E. A. Cassell. 1966. Studies on chicken manure disposed: Part 1 Laboratory studies. Research Report No. 12. New York State Department of Health, Albany, N.Y. 128 p.

3 Fry, J. L. 1973. Methane digestion for fuel gas and fertilizer. The New Alchemy Institute, Woods Hole, Mass.

4 Gramms, L. C., L. B. Polkowski and S. A. Witzel. 1971. Anaerobic digestion of farm animal wastes (dairy bull, swine and poultry). TRANSACTIONS of the ASAE 14(1):7-11, 13.

5 Hach "Colorimeter" Methods Manual. 1971. 7th Edition. Hach Chemical Corporation. Ames, Iowa. p. 124-125.

6 Hart, S. A. 1963. Digestion tests of livestock wastes. Jour. Water Poll. Control Fed. 35:748-757.

7 Imhoff, K. 1946. Digester gas for automobiles. Sewage Works Jour. 18:17-25.

8 Jewell, W. J., G. R. Morris, D. R. Price, W. W. Gunkel, D. W. Williams and R. C. Loehr. 1974. Methane generation from agricultural wastes: Review of concept and future applications. ASAE Paper No. NA74-107. ASAE, St. Joseph, Mich. 49085.

9 Kotze, J. P., P. G. Thiel, D. F. Toerien, W. H. J. Hattingh and M. L. Siebert. 1968. A biological and chemical study of several anaerobic digesters. Water Research 2:195-213.

10 Kugelman, I. J. and K. K. Chin. 1971. Toxicity, synergism and antagonism in anaerobic waste treatment processes. pp. 55-90. In: F. G. Pohland (ed.) Anaerobic Biological Treatment Processes. Advances in Chemistry Series 105. American Chemical Society. Washington, D.C.

11 McCarty, P. L. 1964. Anaerobic waste treatment fundamentals Part III: Toxic materials and their control. Public Works. 95:91-94.

12 McCarty, P. L. and R. E. McKinney. 1961. Salt toxicity in anaerobic digestion. Jour. Water Poll. Control Fed. 33:399-415.

13 McKinney, R. E. 1962. Microbiology for Sanitary Engineers. McGraw-Hill, New York, N.Y.

14 Melbinger, N. R. and J. Connellon. 1971. Toxic effects of ammonia nitrogen on high-rate digestion. Jour. Water Poll. Control Fed. 43:1658-1670.

15 Schmid, L. A. and R. I. Lipper. 1969. Swine wastes, characteri-


(Continued on page 243)

# Small Methane Generator for Waste Disposal

Chung Po, H. H. Wang, S. K. Chen, C. M. Hung, C. I. Chang

TAIWAN produces about six million hogs a year, most of which are kept in small "family" units, and frequently "Manure Credit" is the only profit in pig raising. The floors of hog pens are now most commonly of concrete, and the feces and urine can be easily washed into the gutter leading to a pit in which manure is stored. Manure is traditionally deposited on crop land. Here, the decomposing manure serves as a source of plant nutrients and as organic humus to improve the soil. In some estuary or coastal villages, pigs are housed near or over fish ponds and their waste provides a constant source of nutriment for the fish.

The Taiwan Sugar Corporation has a large pig production program which now markets 400 000 hogs a year. When the program was initiated in 1953, the main objective was to obtain hog manure for their cane fields. Liu (1962) indicated that the original plan was to raise 100 000 hogs for their 20 000 hectares of sugar cane, and as a result, the yield of cane increased by 7-10 percent.

Yeh (1969) reported that the application of 30-40 tons of compost, 20 tons of solid manure of 100 tons of liquid manure in addition to mineral fertilizers (200 kg N, 100 kg $P_2O_5$, and 100 kg $K_2O$ per hectare) increased the length of cane stalks by 10 cm, the amount of millable cane be 2400-6000 (an increase of 2-5 percent) per hectare, and the sugar yield by 1.3-1.9 tons (an increase of 12-18 percent) per hectare as compared to mineral fertilizers alone.

In recent years, research in breeding, nutrition, and disease control in swine production has provided great advances in reducing the feed cost per unit of production. Because of increased demand for better food, large quantities of feed grains and soybeans are imported each year to feed livestock and poultry. In the old days, a farmer fed three or four pigs in his backyard as scavengers. Today a farm contianing 10 000 pigs is not uncommon. Both operations pose the formidable problem of disposal of vast quantities of wastes which are far beyond the capacity of the nearby land to absorb.

In Taiwan, methane gas was generated and used during World War II by the Japanese. The technical know-how died, however, when Taiwan was restored to China. Mr. W. H. Horng (1956) of JCRR obtained a copy of an article entitled "Gaz du Fumier a la Ferme" par F. Migotte, La Maison Rustique, Paris (1952), and in cooperation with Mr. P. N. Soong, Research Chemist of Taiwan Sugar Experiment Station, the first batch-type digester was constructed in Tainan.

Kuan and Chou (1963) reported that a continuous type methane generator was successfully developed, and that methane was produced from fermenting hog waste and bagasse. In the same year, the Taiwan Provincial Research

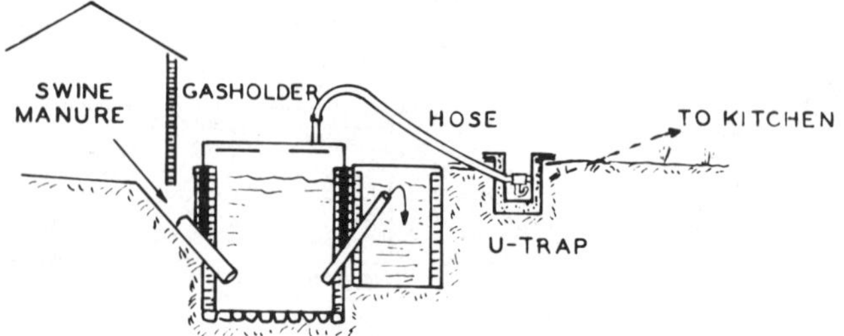



FIG. 1 Diagrammic illustration of the small methane generator.

Institute for Environmental Sanitation developed a concrete digester of the continuous type for methane production. However, the cost was prohibitively high for farmers. P. C. Chen of Technology Institute of Taipei then established a methane generation device which consisted of an excavated tank and a gas holder made of galvanized iron sheet mounted on a wooden frame. This work was continued by J. T. Yu and N. L. Tai of JCRR under a swine project. The device gave little attention to sanitation, and difficulties were experienced in keeping the gas holder air tight. Chung (1965) modified the fermenter into the model shown in Figs. 1 and 2, which was widely duplicated over the island.

## SYSTEM DESCRIPTION

The generator consists of an excavated brick-lined digester, 1.5 x 1.5 x 1.8 m, and an inverted steel gas holder, 1.8 x 1.8 x 0.9 m, resting in the water seal (Fig. 1). The digester is connected to the pigsty by a cement pipe through which wastes and wash water from 10-15 hogs are loaded daily, to assure continuous production of methane gas. The effluent is temporarily stored in a chamber next to the digester. A U-trap is installed to catch any water condensations in the gas delivery line.

In order to reduce the cost of construction, and to make it possible to build larger digesters for bigger hog farms, the authors have developed a rubber gas holder of 0.55 mm thick Hypalon laminated with Neoprene (Du Pont prod-

FIG. 2 Methane generator in operation.

The authors are: CHUNG PO, Specialist, Joint Commission on Rural Reconstruction, Taipei, Taiwan; H. H. WANG, Professor, and S. K. CHEN, Research Assistant, Laboratory of Applied Microbiology, National Taiwan University, Taipei, Taiwan; C. M. HUNG, Specialist, Taiwan Livestock Research Institute, Hsinhua, Tainan, Taiwan; and C. I. CHANG, Head, Rubber Laboratory, Union Industrial Research Laboratories, Hsinchu, Taiwan.

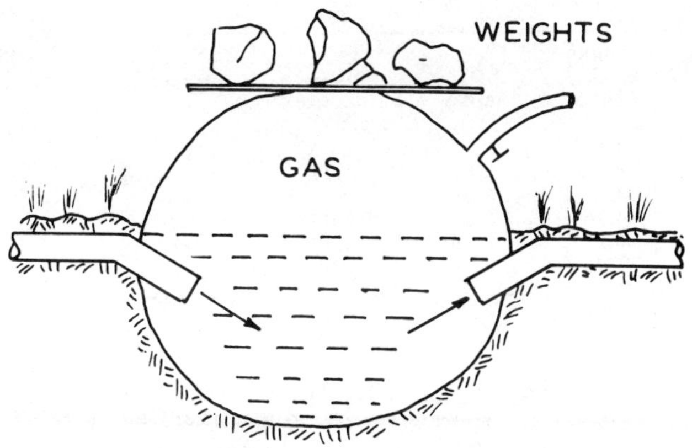

FIG. 3 Rubber-bag fermentor for methane generation.

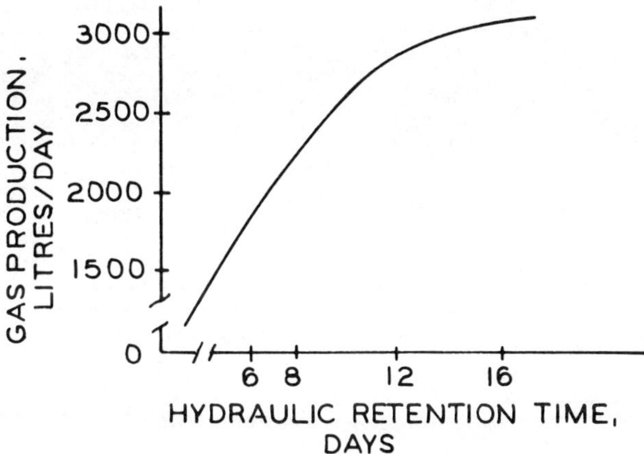


FIG. 4 Gas production at different HRT.

ucts) and reinforced with 420D nylon sheet. Weights are put on top to increase the pressure. In fact, the digester and the gas holder can be combined in one bag, which can be made as large as necessary at very low cost (Fig. 3).

In order to increase the efficiency of the methane generators, all the operating parameters must be determined and measured. The authors determined the characteristics of swine wastes and measured the influent and effluent at different hydraulic retention times, the removal of BOD, and gas production, according to Standard Methods (APHA 1971). A 3-hp Kohler gasoline engine was modified and connected to an electrical generator and run with biogas as the fuel.

## RESULTS AND DISCUSSION

### Hydraulic Retention Time (HRT)

The volume of the digester is 5530 liters. Each day, feces and urine from 10-15 hogs and some wash water enter the digester, and an equal amount of liquid (effluent) is expelled. Because loading of wastes is continuous (Tables 1 and 2), the length of time the wastes remain in the digester is determined by the amount of water which enters with the wastes. The more wash water, the shorter the retention time. From Table 3, it appears that 12-16 days HRT is optimal for maximum gas production and BOD removal. If the amount of wash water is fixed, then the digester must be designed to provide the adequate retention time as shown in the following equation:

$$\text{Hydraulic retention time} = \frac{\text{volume of digester}}{\text{wash water + wastes}}, \text{ OR}$$

Volume of digester = hydraulic retention time x (wash water + wastes)

### Temperature

Contrary to general belief, the difference between the environmental temperature and the digester temperature is very small. During the day, digester temperatures is 1-1.5 C lower; at night, it is 1-1.5 C higher. The heat generated by fermentation is absorbed by the vast amount of water as latent heat. However, environmental temperature is an important factor in gas production, and considerable difference in gas generation between winter and summer as well as north and south is noted.

### Gas Production

Gas production depends upon many factors — loading of volatile solids, temperature, hydraulic retention time, pH, ammonia toxicity, etc. In one case, a 16-day HRT proved best. Above this level, the pH and $NH_3$-N rose, and gas production and volatile solids destruction declined (Fig. 4). The gas contained 63-65 percent methane.

The authors observed that in one case, the gas produced at 5-10 days HRT was more than at 22 days and suggested that ammonia toxicity was the cause. At 5-10 days, the contents of the digester were more dilute, and inhibition of methanogenic bacteria did not occur.

In another observation, where only feces and water were loaded into the digester, gas production proceeded steadily up to 55 days HRT (Fig. 5). This may be explained by the fact that without urine, the nitrogen content was lower and less inhibition was caused by ammonia toxicity.

For practical purposes, it is better to flush both feces and urine into the digester with water. Thus, a shorter retention time, say 10-12 days, should be adequate.

### Destruction of Volatile Solids

The methane generator also serves the purpose of BOD (Biochemical Oxygen Demand) removal. The near complete removal of BOD would require a very long time, possible 60 days. Therefore, the 86 percent removal of 16-day HRT, when gas production is at its maximum, should be satisfactory.

### Utilization of Methane

Methane gas generated from swine wastes has already been used for cooking purposes. It was determined that for a family of five, 500 liters of gas were required to cook a meal (Wang et al. 1974); hence, the average daily production of 2500 to 3000 liters of gas should be adequate for

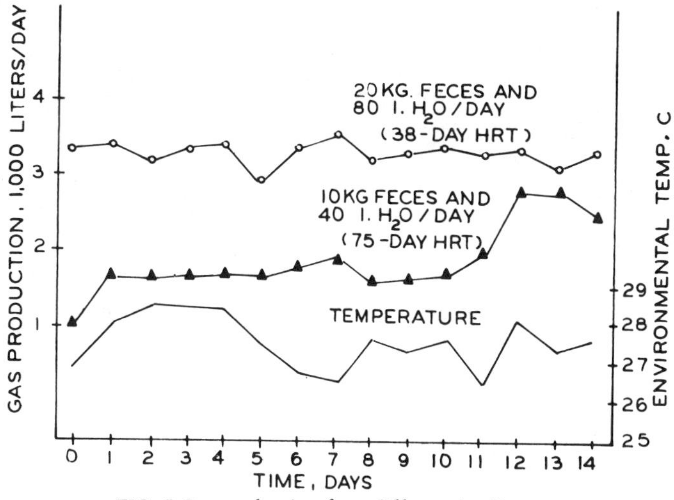


FIG. 5 Gas production from different loadings.

TABLE 1. CHARACTERISTICS OF SWINE FECES

| Sample | Feed used kg/day | Feces excreted kg/day | TS/ feces percent | Volatile solids, kg/day | COD kg/day | BOD kg/day | Total nitrogen g/day |
|---|---|---|---|---|---|---|---|
| 1 | 22.6 | 19.28 | 29.3 | 4.65 | 3.39 | 1.02 | 203.8 |
| 2 | 22.2 | 15.76 | 30.6 | 4.08 | 2.89 | 1.07 | 173.5 |
| 3 | 23.0 | 18.33 | 31.6 | 4.25 | 3.06 | 0.89 | 183.8 |
| 4 | 22.2 | 16.27 | 31.3 | 4.18 | 3.05 | 1.00 | --- |
| 5 | 22.2 | 17.24 | 28.8 | 4.12 | 2.98 | 0.93 | --- |
| Average | 22.4 | 17.38 | 30.3 | 4.25 | 3.08 | 0.98 | 187.0 |

a. There were six sows of about 200 kg and five hogs of about 90 kg.
b. With most care to avoid sampling urine.

kitchen use. The gas was also used to run an engine. The following shows the length of time the engine should run:

A 90 kg hog (3.0 kg feed/day)→3.0 kg feces→0.3M$^3$ gas/day

0.3M$^3$ x 65% x 38 000 BTU/M$^3$ = 7220 BTU

7220 BTU + 42 BTU/HP/min x 15% = 26 min/HP/pig

## Utilization of the Effluent

For years, the effluent from the digester has been successfully used by Taiwan farmers for growing crops and fruits. In some cases, it has also been used for fish culture, since the nutrients are good fertilizer for the plankton in the ponds which provide food for fish.

The effluent has also been used on a trial basis to grow algae by circulating the liquid media from receiving pond to a roof. In addition, three causes of the naturally existing photopseudomona, a purple photosynthetic bacteria, have been found in Taiwan (Wang 1974). It is promising that this bacteria could be used for waste disposal.

At shorter hydraulic retention times, certain organic acids would not be converted to $CH_4$ and $CO_2$. In addition, the ammonia in the effluent is a source of nitrogen for the algae and the photosynthetic bacteria. Meyer et al. (1971) showed a reduction and loss of infectivity of many pathogens in the wastes when exposed to anaerobic digestion. Forstner (1970) indicated that eggs of some parasites were incapable of further development after the fermentation.

Davis and Gloyna (1970, 1972) measured the bactericidal effects of algae on enteric pathogenic organisms in a series of laboratory and field tests and showed that bacterial die-off rates increased when mixed cultures of algae were present. Therefore, algae can be safely fed *per se* to the animals.

## CONCLUSION

A large hog fed three kg of feed will void about three kg of feces containing 0.5 kg of total solids. This requires 0.5 M$^3$ digester capacity to produce about 300 liters of gas a day at a 12-day hydraulic retention time.

As some 80 percent of the BOD in the wastes has been removed by anaerobic fermentation, the remaining 20 percent may be further treated by loading it into an oxidation ditch. This is currently under study at the Taiwan Provincial Livestock Research Institute.

## References


1  Amer. Publ. Health Assn. 1971. Std. methods for the exam. of water and waste water. 13 ed.
2  Chung, Po. 1965. The animal-methane-chlorella cycle. JCRR brochure.
3  Davis, E. M and E. F. Gloyna. 1970. Bactericidal effects of algae on enteric organisms. FWOA Res. Ser. 18050 DOL, Washington, D.C.
4  Davis, E. M. and E. F. Gloyna. 1972. Bacterial dieoff in ponds. Jour. San. Eng. Div., Proc. Amer. Soc. Civil Engr., 98, 59.
5  Forstner, M. J. Z 1970. Investigations on the effect of anaerobic sludge digestion in Emsher Tanks at main sewage works on the viability of worm eggs. Wass. Abwass. Forsch. 3. 57; Water Poll. Abs., 44, 12,2741 (1971).
6  Horng, W. H. 1956. Demonstration manure digestion plant for sugar cane growers. JCRR Project TW-K-79.
7  Kuan, S. S. and J. C. Chou. 1963. Methane gas production. Taiwan Sugar X(4):17.
8  Liu, K. C. 1962. The agricultural review of the sugar industry in Taiwan. Taiwan Sugar IX(4):15.
9  Meyer, R. C., et al. 1971. Porcine enterovirus survival and anaerobic sludge digestion. Livestock waste management and poll. Abatement, Intl. Symp. Proc., Columbus, Ohio, 183.
10  NTU Civil Engineering Research Institute: Swine waste disposal. 1973.
11  Wang, H. H. 1974. Unpublished data.
12  Yeh, T. P. 1969. Comparison of the effect of compost, solid and liquid manure on sugar cane yields and on physico-chemical properties of the soils. Proc. Intern. Soc. of Sugar Cane Tech. XIII Cong. 809-813.


TABLE 2a. CALCULATED CHARACTERISTICS OF INFLUENT AT DIFFERENT HYDRAULIC RETENTION TIME*

| Retention time, days | 6 | 8 | 12 | 16 |
|---|---|---|---|---|
| Total solid, g/l | 5.57 | 7.42 | 11.13 | 14.84 |
| Volatile solid, g/l | 4.61 | 6.14 | 9.21 | 12.29 |
| VS/TS,% | 82.9 | 82.9 | 82.9 | 82.9 |
| COD, mg/l | 3,345 | 4,460 | 6,690 | 8,920 |
| BOD, mg/l | 1,070 | 1,427 | 2,140 | 2,854 |
| BOD/COD% | 32.0 | 32.0 | 32.0 | 32.0 |

*Values are calculated on the basis of data given in Table 1.

TABLE 2b. MEASURED CHARACTERISTICS OF EFFLUENT AT DIFFERENT HYDRAULIC RETENTION TIME

| Retention time, days | 6 | 8 | 12 | 16 |
|---|---|---|---|---|
| pH | 7.2 | 7.1-7.2 | 7.2-7.4 | 7.3-7.5 |
| Total solid, g/l | | 3.218 | 4.338 | 4.938 |
| Volatile solid, g/l | | 2.032 | 2.622 | 2.638 |
| VS/VT, % | | 62.2 | 60.4 | 534 |
| COD, mg/l | 1,700 | 1,900 | 2,160 | 2,160 |
| BOD, mg/l | 400 | 380 | 410 | 380 |
| BOD/COD, % | 23.5 | 20.5 | 18.9 | 17.2 |
| Total nitrogen, mg/l | 361 | 476 | 710 | 921 |
| Ammonium nitrogen, mg/l | 269 | 346 | 529 | 808 |
| An/TN | 74.5 | 76.1 | 75.0 | 87.8 |

TABLE 3. PROCESS OPERATING CHARACTERISTICS

| Retention time, days | 6 | 8 | 12 | 16 |
|---|---|---|---|---|
| a. Efficiency of gas production | | | | |
| liter (STP)/kg TS destroyed | --- | 803 | 1001 | 993 |
| liter (STP)/kg VS destroyed | --- | 822 | 1032 | 1019 |
| liter (STP)/kg COD reduced | 1201 | 1318 | 1501 | 1455 |
| liter (STP)/kg BOD removed | 2944 | 3222 | 3930 | 3890 |
| b. Efficiency of treatment | | | | |
| TS reduction percent | --- | 56.6 | 61.1 | 66.7 |
| VS reduction percent | --- | 66.9 | 71.5 | 78.5 |
| COD reduction percent | 49.2 | 57.4 | 67.7 | 75.8 |
| BOD removal percent | 62.6 | 73.4 | 80.4 | 86.7 |

| Loadings: | | |
|---|---|---|
| Total solids | 0.927 | g/l/day |
| Volatile solids | 0.768 | g/l/day |
| COD | 557 | mg/l/day |
| BOD | 178 | mg/l/day |

# Product Applications of Treated Livestock Wastes

C. Corvino, B. Dunn, E. Tseng, J. D. Mackenzie

IN normal feedlot operation, large quantities of manure are generated in a confined area. This manure poses a serious solid waste disposal problem as well as a source of water and air pollution. The present research is aimed at converting the waste material into a useful raw material with subsequent product application. The products must adhere to the following points; (a) They must be capable of incorporating large quantities of the manure. (b) They should be of comparable quality and value to existing products. (c) Production methods and costs must be economically feasible.

The conversion process involves low temperature pyrolysis of the waste material. The resulting by-products of this treatment, the condensed vapor and the solid residue, are then added to other materials and synthesized to form useful products. Although the pyrolysis treatment and product applications of cattle wastes are considered in this paper, similar results have been obtained from other livestock wastes.

## PYROLYSIS

The pyrolysis of livestock wastes has been previously reported (White and Taiganides 1971). The treatment described in the present paper is rather difficult because it is addressed to commercial applications as opposed to experimental interests. Considerably larger loads are used (1000 g) at lower temperatures (200-400 C).

A flowchart of the manure treatment operation is shown in Fig. 1. A semi-continuous process has been developed for both wet and dry wastes. The pyrolyzer consists of an inner rotating cylinder (1 ft diameter by 3 ft long) housed within a rectangular enclosure. The inner cylinder has walls of stainless steel mesh so that once the manure is reacted, it reduces in size and drops through the perforations in the vessel walls to a storage container. The vessel is mounted on an axle which serves as the opening through which the material enters the chamber on a continuous basis. Concurrently, the volatiles driven off are condensed in a water cooled unit. The pyrolyzer is gas fired and enclosed in refractory bricks to prevent unnecessary heat loss. Convective heat flow from the unit is adequate to pre-dry the wet manure in a small chamber placed on top of the pyrolyzer.

The condensate may be readily separated into two liquids — an aqueous fraction and an oil fraction. The ash is a black, odorless solid. A range of yields for these components is shown in Table 1. This variation results from several different contributions; pyrolysis conditions (temperature, time), cattle feed, manure source and the presence of soil mixed in with the manure. The chemical analyses indicated in the table are also sensitive to these conditions. The relationship between ash content and pyrolysis temperature

is shown in Fig. 2. The decrease in ash at greater temperatures is expected because less carbon is present. On the other hand, more oil is extracted at these temperatures so that the carbon is being transferred between fractions. The carbon content is an important consideration for product applications.

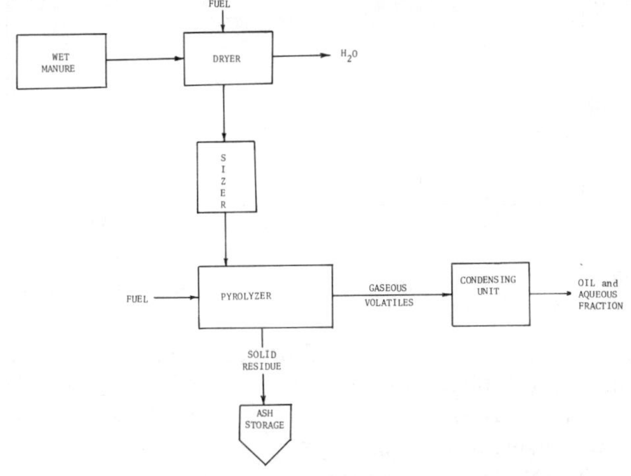



FIG. 1. Flowchart of manure treatment operation.

**TABLE 1. YIELDS AND TYPICAL CHEMICAL ANALYSIS OF PYROLYSIS PRODUCTS**

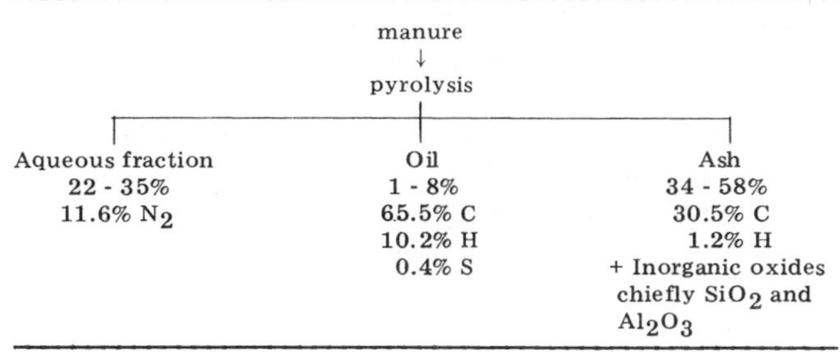

| | manure ↓ pyrolysis | |
|---|---|---|
| Aqueous fraction | Oil | Ash |
| 22 - 35% | 1 - 8% | 34 - 58% |
| 11.6% $N_2$ | 65.5% C | 30.5% C |
| | 10.2% H | 1.2% H |
| | 0.4% S | + Inorganic oxides chiefly $SiO_2$ and $Al_2O_3$ |

## PRODUCT APPLICATIONS

The pyrolysis by-products and their applications are catagorized in Fig. 3. The liquid separates into two fractions. The aqueous portion is rich in nitrogen and holds promise as a fertilizer. The oil is of obvious importance. Chevron Oil Company has evaluated the crude as a good low sulfur fuel oil. In our own laboratories, we have produced carbon black by a decomposition method commonly used in industry (Schubert, Ford, Lyon 1969). The quality of the resulting material was equal to that of commercial carbon black.

The solid residue is a carbonaceous alumino-silicate. It may be used as a carbon black substitute because of its high carbon content or as an inexpensive filler material. In the


The authors are: C. Corvino, Laboratory Assistant; B. Dunn, Assistant Professor; E. Tseng, Graduate Student; and J. D. Mackenzie, Professor, Materials Department, University of California, Los Angeles.

Acknowledgment: We are grateful for the support of this project by the Environmental Protection Agency.

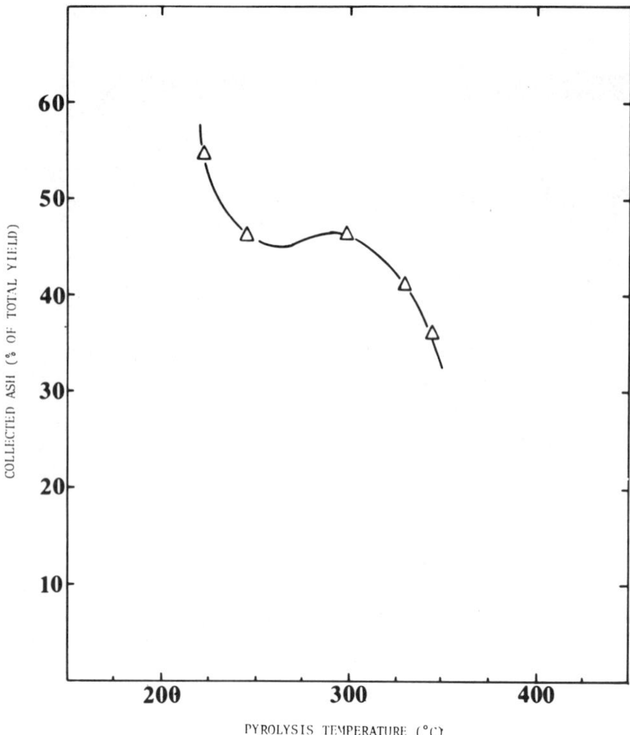



FIG. 2 Variation in ash content as a function of pyrolysis temperature. Treatment time is 90 minutes.

former, the ash may be used in ink, paint pigments and rubber. As a filler material, uses include rubber, bricks, tiles and charcoal briquettes. Figs. 4 and 5 are photos of products actually made using the pyrolysis residue.

The printing ink was prepared according to methods normally used in the industry (Printing Ink Manufacturers 1967), with the treated ash substituting for 50 percent of the carbon black. Although satisfactory ink has been fabricated on a laboratory scale, certain problems remain for commercial applications. The difficulties involve the particle size and grittiness due to the incorporation of sand particles present in the pen. The carbon black prepared from the condensed oil fraction is suitable for commercial processes.

Carbon black is frequently used in rubber tires as a filler (Burgess, Scott, Hess 1970). The finer particles are used in tire treads while coarser sizes find application in tire carcasses. In this latter capacity, pyrolyzed ash has been completely substituted for carbon black.

The ash has been successfully substituted as a filler material in ceramic tiles. The residue is reduced in size and

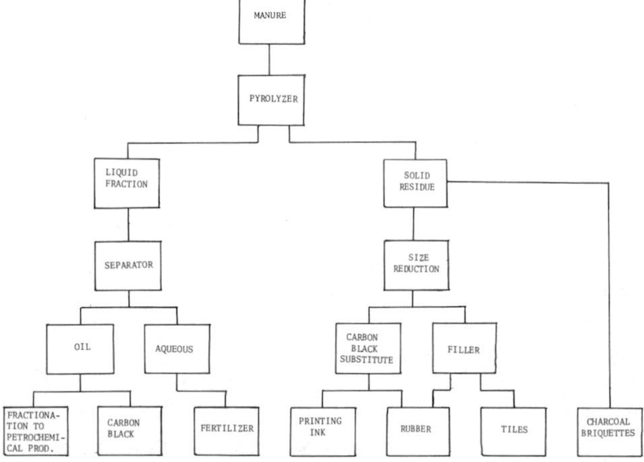



FIG. 3 Flowchart of the various applications for by-products of pyrolyzed cattle wastes.

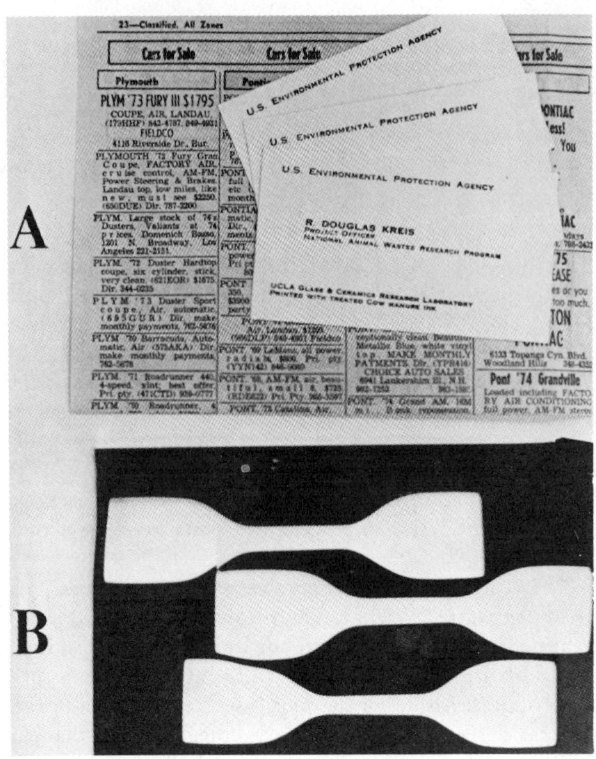



FIG. 4 Product applications of solid residue as a carbon black substitute. (a) Items printed from an ink incorporating the residue (b) Rubber sample prepared using the residue.

combined in equal amounts with ground glass bottles — another bothersome solid waste — and prepared according to conventional ceramic tile techniques. During the firing process, the glass particles fuse together to form a continu-

FIG. 5. Product applications of solid residue as filler material. (a) Charcoal briquettes made using the residue (b) Tile prepared from residue mixed with ground glass bottles.

ous matrix containing the dispersed ash particles. The resulting tile possesses excellent properties which are, in many regards, superior to existing products on the market. Its performance and cost are economically competitive. Table 2 lists the pertinent properties. The tiles are suitable for roof, floor and wall applications.

### TABLE 2. PROPERTIES OF HOT-PRESSED TILES

| | |
|---|---|
| Density, controllable | 1.8 to 2.4 $g/cm^3$ |
| Incombustible | class A |
| Flexural strengths | 6000 to 8000 psi |
| Apparent impact strength | 95 ft lb at 1.8 $g/cm^3$ density |
| Abrasion wear index (Taber) | 55 to 130 (min. acceptable is 35) |
| Moisture absorption as per UBC Standard 32-12 | 2.2 percent at 1.8 $g/cm^3$ density |
| Wt. per area | 4 lb per sq ft at 3/8 in. thickness |
| Hardness, Moh scale | 6 |
| Decoration | bulk and surface colors possible; glazed easily |
| Miscellaneous | can be painted, glued, nontoxic and odorless |

Charcoal briquettes have been made using the pyrolyzed residue. This product does not require any additional size reduction beyond that occurring in the pyrolysis chamber. A mixture of corn starch and water serves as a binder and samples are formed with a commercial briquetting press.

Field tests indicate comparable performance to existing charcoal products. Another possible application of briquettes is to burn the remaining carbon. In this way the briquettes may be used as a fuel supplement to coal fired utilities. Their form is convenient for handling and transportation (Gillmore 1975).

The cost of producing pyrolyzed ash has been estimated. For a plant to process 30,000 tons of manure (25 percent water) per year, the cost of the ash is on the order of $3.50/ton. The crude oil is taken as a credit against costs. For most of the product applications, some size reduction is necessary. To produce this highly carbonaceous residue in a form as fine as carbon black, a total cost of $10 per ton is estimated. Our industrial contacts have indicated considerable interest in obtaining carbon black substitutes at this price.

### References


1  Burgess, K. A., C. E. Stott and W. M. Hess. 1970. Vulcanizate performance as function of carbon black morphology. Rubber Chem. Technol. 43:230-248.

2  Gillmore, D. W. 1974. Personal communication.

3  National Association of Printing Ink Manufacturers. 1967. Printing ink handbook. National Association of Printing Ink Manufacturers, New York.

4  Schubert, B., F. P. Ford and F. Lyon. 1969. Analysis of carbon black. p. 179-243. In: Encyclopedia of industrial chemical analysis. Vol. 8. John Wiley and Sons, New York.

5  White, R. K. and E. P. Taiganides. 1971. Pyrolysis of livestock wastes. p. 190-194. In: Proceedings of international symposium on livestock wastes. ASAE, St. Joseph, Mich. 49085.


---

## Pilot Scale Swine Manure Digesters for Methane

*(Continued from page 237)*


zation and anaerobic digestion. Proc. Conf. on Agric. Waste Mgt., Cornell Univ., Ithaca, N.Y. pp. 50-57.

16 Singh, Ram Bux. 1973. Bio-gas-plant generating methane from organic wastes. Gobar Gas Research Station, Ajitmal, Etawah (U.P.) India.

17 Standard Methods for the Examination of Water and Wastewater. 1971. 13th Edition. American Public Health Association. New York, N.Y.

# Characterization of Methane Production from Poultry Manure


H. Moustafa Hassan, David A. Belyea, Awatif El-Domiaty Hassan

ASSOC. MEMBER      MEMBER
ASAE         ASAE


**M**ETHANE production through anaerobic decomposition of organic wastes has long been known, but the commercial feasibility of such a process has remained uncertain. However, the shortage and increased cost of energy have stimulated the interest of economists and researchers to reinvestigate the economic feasibility of anaerobic digestion to produce methane gas.

Recent reports have dealt with the biochemistry and kinetics of anaerobic treatment. Lawrence and McCarty (1969) have studied the kinetics of such fermentation under idealized conditions, where pure substrates (volatile acids) were supplied directly to the culture. These idealized conditions do not normally occur under natural conditions. The rate of methane production from organic wastes could be limited by the rate of hydrolysis of the complex organic compounds in the waste and the rate of organic acid production.

Chicken manure contains a significant quantity of energy which can be reclaimed through anaerobic conversion to methane. Therefore, the objectives of the present study were to experimentally establish the optimum conditions for the anaerobic digestion of poultry manure, and to develop the criteria for designing a digestion system for commerical application.

This paper reports the overall kinetics of the anaerobic fermentation of poultry manure to methane under laboratory and field conditions.

## EXPERIMENTAL PROCEDURES

### Manure

The manure used in this investigation was collected from a pit under caged Black Cross laying hens at the University of Maine Farm. The solids concentrations of the fresh manure ranged from 25 percent to 28 percent during the spring, fall, and winter months, and from 35 percent to 40 percent during the summer months.

### Digesters

The laboratory scale digesters consisted of different size bottles connected to gas collectors as shown schematically in Fig. 1. Eight 8-liter bottles filled with diluted manure (1 manure:2 water by weight) were placed in a chamber maintained at 35 C (95 F) and mechanically mixed for 30 seconds every hour by a timed shaking device. Similarly, twenty 1-liter boiling flask digesters were maintained at the specified temperatures and mixed once daily by manual shaking. The accumulated gas volume was recorded at atmospheric pressure and room temperature (25 C), unless indicated otherwise. The gas was then removed and analyzed for composition.

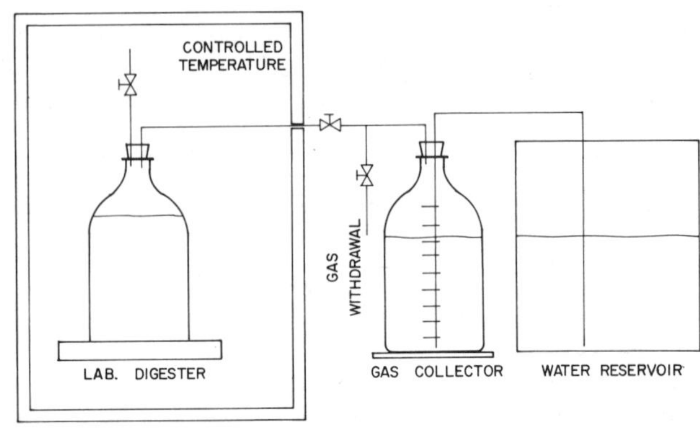


FIG. 1 Schematic diagram of a laboratory digestion unit.

The field digester unit is shown schematically in Fig. 2. The digester is an insulated 2.63 m³ (93 ft³) cylindrical steel tank. The digester was filled with poultry manure diluted to a solids concentration of 7 percent for a total volume of 2.07 m³ (73 ft³). The waste was mixed in an external tank and fed into the digester by a grinder pump which was also used for mixing the sludge during digestion. The mixing period was set for 15 minutes every three hours. A scum breaker was installed at the top of the tank and operated by a timer for 15 minutes every two hours. Samples of the digester contents were taken both from the tank side taps and the discharge side of the circulation piping.

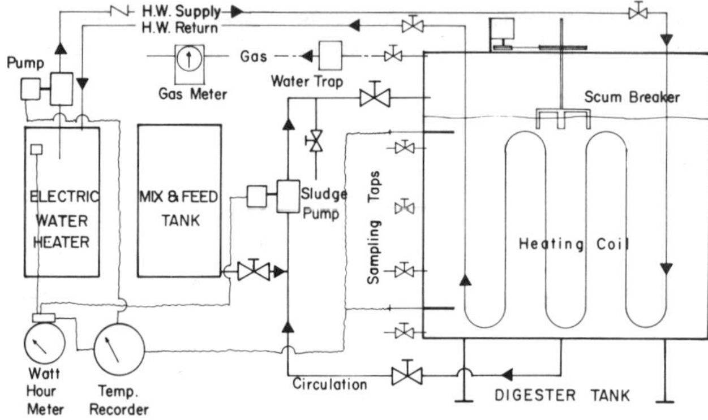


FIG. 2 Schematic diagram of the field digester (2.63 m³).

The digester was heated by circulating hot water into internal coils, from an electric water heater. The digester temperature was maintained at 35 C (95 F) by a dual element temperature recorder/controller wired to a circulating pump within the heating system. The energy input to


Acknowledgement: The continued and generous support of the New England Regional Commission is gratefully acknowledged.

The authors are: H. MOUSTAFA HASSAN, Microbiologist, Microbiology Dept., DAVID A. BELYEA, Graduate Assistant, and AWATIF EL-DOMIATY HASSAN, Assistant Professor, Agricultural Engineering Dept., University of Maine, Orono, ME 04473.

the system was measured by watt-hour meters installed in the electrical supplies to the grinder pump and water heater. The evolved gas was measured by a calibrated commercial gas meter.

Temperature fluctuations in all laboratory and field experiments were within ± 1 C (~2 F).

## Gas Analysis

Gas analyses were performed at 1- to 3-day intervals with a Fisher-Hamilton Model 29 Gas Partitioner. This gas chromatograph detected $CO_2$, $O_2$, $N_2$, $CH_4$, and CO. A laboratory recorder was used with the chromatograph to record the percentages of the respective gases in the collected samples.

The hydrogen sulfide content of the gas was determined by bubbling a known volume of gas through 100 ml of 0.1 N zinc acetate solution, and the absorbed sulfur was determined by the Standard Methods procedure (1965), and calculated as ppm total sulfide.

## Solids Concentration

Solids concentration analyses were performed by oven drying 0.1-0.2 kg samples at 75 C (170 F) to a constant weight.

## pH

All pH measurements were made with a Corning Model 10 temperature compensated pH meter.

RESULTS AND DISCUSSION

## Effect of Initial Solids Concentration

Anaerobic digestion is directly related to the activity of "acid forming" and "methane producing" bacteria. It is a well recognized fact that the growth rates, metabolic activity, and the total cell population of bacterial cultures are functions of the initial concentration of organic nutrients. Therefore, a kinetic study on the effect of the initial concentration of chicken manure solids on the rate and extent of methane production was needed in order to establish the optimum concentration. Fig. 3 shows typical time progression curves for three different solids concentrations. Table 1 shows the effect of initial solids concentrations on the maximum yield of methane attainable after 110 days. It is evident that concentrations higher than 8.2 percent total solids resulted in fast decline in the rate and extent of methane production, while at concentrations between 4.6 and 7.6 percent a plateau was reached. These data suggest that operating anaerobic digesters at concentrations of 7.0-7.5 percent will be the most economical for poultry manure.

The percent reduction in total solids was determined by drying the samples to a constant weight before and after digestion. The data (Table 1) indicate that the high initial solids concentration greatly reduced the efficiency of the anaerobic destruction of organic solids. Thus, it is evident that too high substrate concentration inhibits the growth and the metabolic acitivity of the organisms involved in methane production. This type of substrate inhibition is similar to that reported by Edwards (1970) for different organisms and different substrates. This inhibition might have been also due to the presence of some toxic compounds (e.g. $NH_3$) in the fresh manure. Recently, Gramms et al. (1971) demonstrated that high loading rates of chicken manure resulted in a decrease in digestion effi-

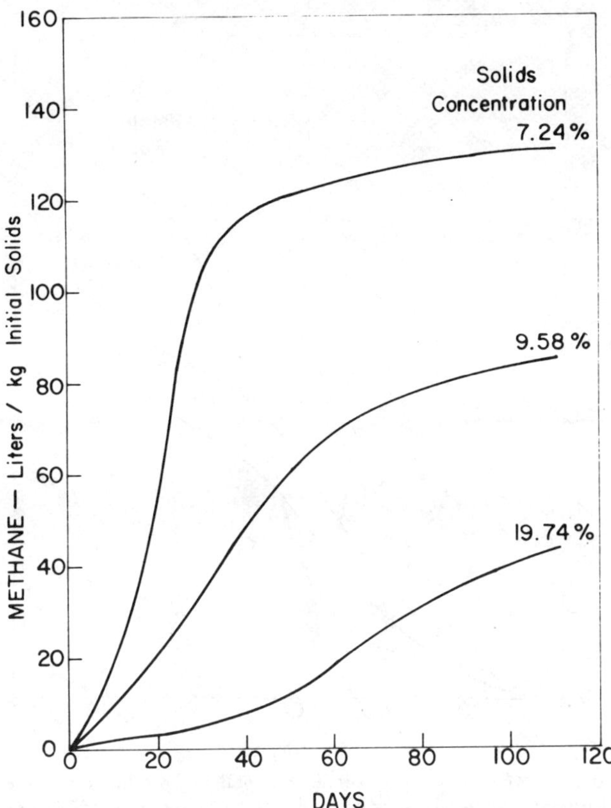

FIG. 3 Effect of initial solids concentrations on methane production. The fresh manure was mixed with water to the specified solids concentration. 1-liter digesters were used; incubation temperature was 25 C (78 F).

TABLE 1. EFFECT OF INITIAL SOLIDS CONCENTRATION ON METHANE PRODUCTION AND SOLIDS REDUCTION AFTER 110 DAYS OF ANAEROBIC DIGESTION.

| Initial solids concentration, percent | $CH_4$*, liters/kg of initial solids | Solids reduction, percent |
|---|---|---|
| 1.9 | 92 | 28.0 |
| 4.6 | 146 | 39.4 |
| 5.6 | 137 | 39.0 |
| 6.6 | 146 | 37.7 |
| 7.2 | 129 | 45.3 |
| 7.6 | 149 | 34.0 |
| 8.2 | 106 | 30.4 |
| 9.6 | 85 | 29.2 |
| 11.5 | 62 | 26.3 |
| 19.7 | 41 | 18.3 |

*The gas was collected at atmospheric pressure and 25 C.

ciency, which was related to the high concentration of ammonia. Furthermore, Albertson (1961), and Anthonisen and Cassell (1974) have demonstrated that high concentrations of ammonia were toxic to the growth of methanobacteria. Thus, the recommended range of solids concentrations (7-7.5 percent) will also result in lower ammonia concentrations.

## Effect of Carbon Sources

Chicken manure contains a high percentage of nitrogen. A delicate balance between C:N ratio is required for the anaerobic methane fermentation. Therefore, a study on the use of different carbon wastes was undertaken. Fig. 4 shows the effect of adding potato wastes and wood sawdust on the kinetics of methane production. It is evident that a longer lag period was observed when potato waste was used, but it resulted in the highest rate of methane production

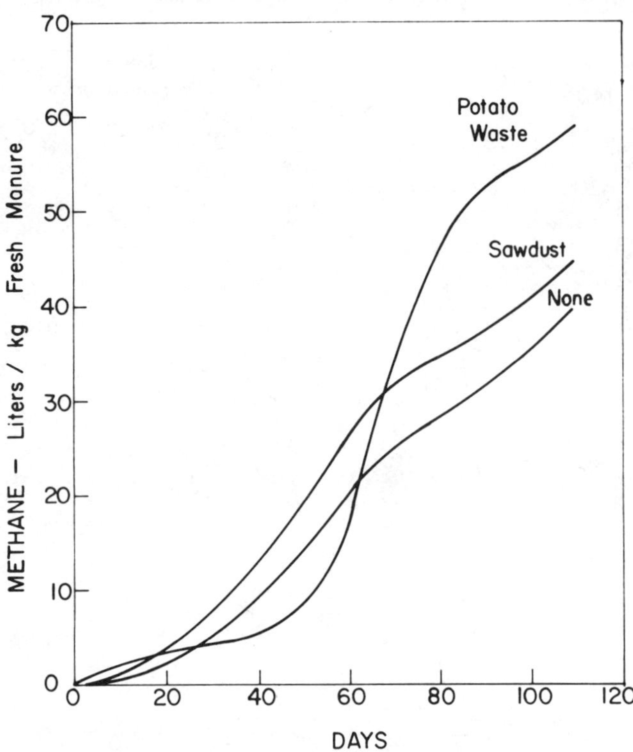

FIG. 4 Effect of carbon source on methane production. 8-liter digesters were used. Sawdust and potato waste were added at 4 percent by weight of fresh manure. A digester with no additional carbon source was included as a control. Incubation temperature was 35 C (95 F).

after about 60 days. Sawdust resulted in slight but definite stimulation in the rate of methane production. However, because sawdust is normally used in the broiler industry, it was of interest to study the effect of different concentrations of sawdust on methane production. Table 2 shows that adding sawdust at a concentration of 2-4 percent resulted in the same stimulatory response. However, at 0 percent and 8 percent sawdust, the amount of methane produced was significantly lower.

**TABLE 2. EFFECT OF SAWDUST ON THE METHANE PRODUCED AFTER 110 DAYS AT 25 C.**

| Sawdust*, percent | $CH_4$, liters/kg solids† |
|---|---|
| 0 | 89 |
| 2 | 109 |
| 4 | 106 |
| 8 | 93 |

*Sawdust added as percent of wet manure.
†Total solids concentration was 8.2 percent.

## Effect of Inoculum

The eight-liter digesters were seeded with an inoculum from a batch of previously digested chicken manure at 0, 20, and 50 percent by volume, and incubated at 35 C. The results (Fig. 5) clearly illustrate that the amount of inoculum significantly affected the rate of methane production, and the length of the lag period. From Fig. 5, the production of 40 liters of methane per kg of fresh manure was achieved after 29, 74, and 96 days when 50, 20, and 0 percent inoculum were used, respectively.

The absence of lag period when 50 percent inoculum was used may be due to the presence of some essential substances carried over from the previous digestion, and an actively growing methanogenic culture.

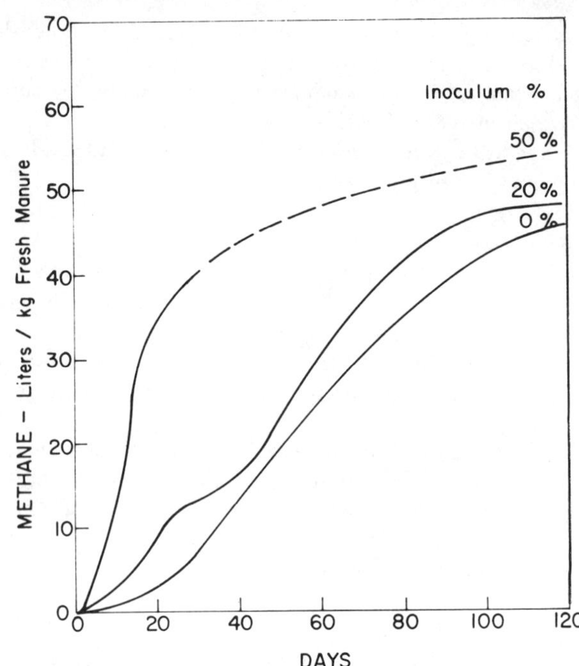

FIG. 5 Effect of inoculum size on the rate of methane production. 8-liter digesters were used. Sawdust was added at 4 percent by weight of the fresh manure. Incubation temperature was 35 C (95 F).

## Effect of Temperature

The effect of temperature on methane production from chicken manure is shown in Fig. 6. The results show a bell-shaped curve. It is evident that 35 C is the optimum temperature for the production of methane from chicken manure under the present conditions. This might indicate that the natural microflora of the chicken manure available in Maine does not contain the thermophilic methanobacteria. However, Savery and Cruzan (1972) reported a greater methane production from chicken manure at the

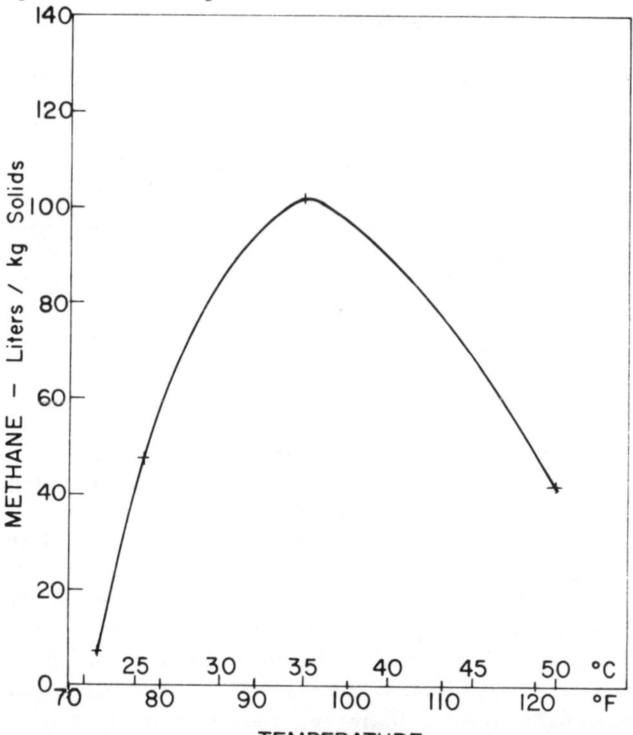

FIG. 6 Effect of incubation temperature on methane production after 25 days of digestion. 1-liter digesters were used; inoculum size was 25 percent; and no additional carbon source was added.



MANAGING LIVESTOCK WASTES

thermophilic range (51 C) than at the mesophilic range (29 C), when using a batch reactor, but this was not the case when the digester was operated continuously.

Thus, low temperatures (23 C or less) will decrease the efficiency of the system and will require longer retention periods, while temperatures in the thermophilic range (50 C) seem to be uneconomical.

### Isolation and Identification of the Methanobacteria Present

The methanobacteria predominantly present in the anaerobic digesters were isolated by the method of Barker (1936) using acetate and butyrate as organic substrates at a concentration of 1 percent. The culture was identified as a mixture of *Methanobacterium söhngenii* and *Methanobacterium omelianskii* (Fig. 7).

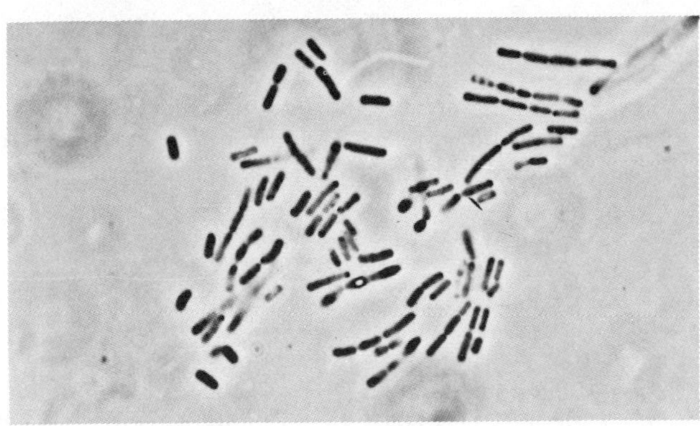

FIG. 7 A phase micrograph of the methanobacteria isolated from an anaerobically digested chicken manure (Magnification = 3500 x).

### Field Studies

Field trials were necessary for a realistic evaluation of the reliability and economy of an anaerobic digester under farm conditions. Based on the laboratory findings of this study, the field digester (2.63 m³) was operated under the following optimum conditions: sludge solids concentration was maintained at approximately 7 percent; digester temperature was set at 35 C (95 F); the digester was recharged at an interval of 21 days by removing 25 percent of the tank contents thus providing a 75 percent inoculum for the following batch.

Fig. 8 shows the accumulative methane production from the field digester and its methane percentage over the period November, 1974, to February, 1975. A summary of the results of a year's operation of the field digester is presented in Table 3. The gas volumes were measured at atmospheric pressure and 18 C. The gas produced contained considerable quantities of water vapor that was partially removed by a water trap, and hydrogen sulfide (4-7 ppm of sulfide). This combination is extremely corrosive and must be removed before being used in any commercial device. Feathers in the manure, especially with a flock of molting laying hens, caused a troublesome pump clogging problem in this experiment and should be considered in the design of a commercial digester.

The values of methane contents (column 2 of Table 3) represent the calculated averages for each entire trial period. Methane production in the range of 133 to 161 liters/kg dry manure and a net output of 3489 to 4187 kJ/kg dry manure were obtained. Using a conservative value of 3500 kJ/kg dry manure for year around operation and daily fresh

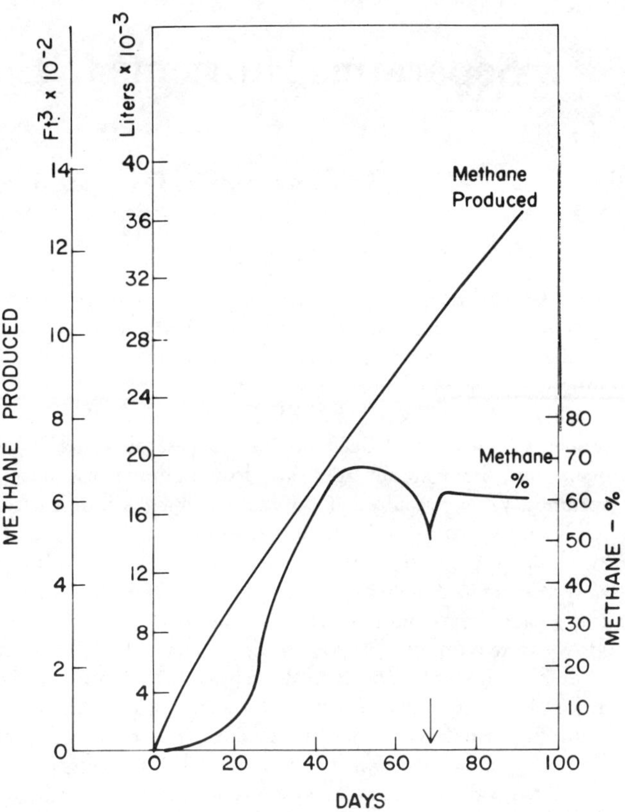


FIG. 8 Kinetics of methane production from the field digester (accumulative methane volume and methane percentage). Arrow indicates the time of recharging.

TABLE 3. SUMMARY OF RESULTS OF FILED TRIALS AFTER 90 DAYS OF OPERATION BASED ON DRY WEIGHT OF MANURE.

| Trial no. | Average methane content, percent | Methane production* | | Calculated energy output† | | | |
|---|---|---|---|---|---|---|---|
| | | liter/kg | ft³/lb | Gross | | Net | |
| | | | | kJ‡/kg | Btu/lb | kJ/kg | Btu/lb |
| 1 | 59.0 | 133 | 2.13 | 4954 | 2130 | 4187 § | 1810 § |
| 2 | 52.8 | 161 | 2.59 | 6024 | 2590 | 3489 | 1500 |

*Gas collected at atmospheric pressure and 18 C.
†Calculated using methane at 37.25 kJ/liter (1000 Btu/ft³).
‡kJ = 1000 joules.
§No heating input required during summer trial.

manure production of 0.109 kg (3.84 oz)/bird (Graves and Massie 1972), a 50 000 bird laying flock would have a net energy production potential of 4.75 x 10⁶ kJ/day. This represents the energy equivalent of 32.5 gallons (123 liters) of fuel oil per day (based on 139 000 Btu/gal for #2 fuel oil) or 11 000 gallons (41 624 liters) per year, for 350 days of operation per year.

### CONCLUSIONS

The following conclusions apply to the anaerobic digestion of chicken manure.

1  Initial solids concentrations in the range of 4.6 percent to 7.6 percent resulted in the maximum yield of methane with the most economical operation being at the higher end of the range.

2  An added carbon source stimulated methane production in the digester provided that the carbon source is easily degradable and that the concentration is not too high.

*(Continued on page 251)*

# Separating Nutrients to Enhance Swine-Waste Digestion


L. A. Schmid, R. I. Lipper, J. K. Koelliker, C. A. Cate, J. W. Daber

MEMBER    ASSOC. MEMBER
ASAE        ASAE


$\mathbf{I}$NCREASED environmental concerns, energy supply and costs, and the availability and costs of commercial fertilizer have directed engineers' attention toward maximum use of a common resource, animal waste.

This paper concerns anaerobic digestion of agricultural wastes which has not been entirely successful where it has been tried. Poor stabilization and low methane production commonly have resulted. To obtain good digestion, it often has been necessary to dilute the waste concentration, resulting in increased digester capacity to handle the dilution water, increased volumes of wastes for ultimate disposal, and greater costs. Basic research on anaerobic digestion has shown that ammonia is quite toxic to anaerobic organisms when the concentration is in the range of 2000 mg/l. Waste disposal systems that capture the entire waste discharge, including urine, are especially susceptible in that undiluted ammonia concentrations often exceed 3000 mg/l. The trend toward confinement systems means that producing concentrated wastes will continue to increase. Hence, any procedure developed to handle such wastes successfully in their concentrated form should offer an economic advantage.

In this project, we explored a novel method of waste treatment to remove toxic ammonia, increase waste stabilization and methane-gas production, produce a purified methane gas, eliminate the need for dilution water, reduce the volume for ultimate disposal to the land, and in the process capture and store nitrogen as a clean liquid fertilizer which can be applied to the land more efficiently than can raw liquid swine wastes. Heating, mixing, and gas collection become a necessity once the decision has been made to optimally digest the waste. The extra facilities to accomplish those operations become a smaller part of overall project costs when waste volume is minimized.

## PROCESS DESCRIPTION

If ammonia can be removed as a gas from the waste and captured for later use, ammonia toxicity and dilution requirements can be eliminated. See Srinath and Loehr (1974) for a comprehensive article on ammonia desorption from animal wastes.

Ammonia exists in equilibrium with ammonium ion in water,

$$NH_4^+ + H_2O \rightleftharpoons NH_3 + H_2O + H^+ \quad \dots \dots \dots \dots \dots \dots \dots [1]$$

Undissociated ammonia ($NH_3$-N) is the only form that can come off as a gas, and the amount available is described by the equation:

$$NH_3\text{-}N = (NH_4\text{-}N) \cdot \cfrac{10\text{pH}}{\cfrac{k_b}{k_w} + 10\text{pH}} \quad \dots \dots \dots \dots \dots \dots [2]$$

where $k_b$ and $k_w$ are the ionization constants of aqueous ammonia and water, respectively. The temperature effect can be calculated from the equation;

$$k_b/k_w = [\text{-}3.398 \ln (0.0241 \cdot \quad \text{deg C})] \cdot 10^9 \dots \dots \dots \dots \dots [3]$$

Solving Equation [3] at an ($NH_4$-N) concentration of 2000 mg/l results in a free $NH_3$-N concentration at pH 7 and 10 C of 4.1 mg/l; at 35 C the concentration would be 34 mg/l. The $NH_3$-N concentration would be 292 mg/l at pH 8 and 35 C, resulting in a desorption driving force 72 times greater than at pH 7 and 10 C.

Thus, one would not expect ammonia to be volatilized to any great extent in the pH range near 7 or less, where most anaerobic digesters operate. However, two factors combine to reduce the problem of low theoretical removal of ammonia: (a) The digester gas from which ammonia is to be removed can be recirculated many times through the digester if the gas is used for mixing; and (b) A long retention time, 15 to 20 days, is allowed for digestion. Therefore, the volume of gas available for stripping to the volume of liquid becomes high. A heated digester can operate at near 35 C, at which temperature the fraction of dissociated ammonia available for stripping is nearly 8.5 times greater than it would be at 10 C. So, given a high gas to liquid ratio and high temperature, it becomes possible to remove ammonia at the low pH.

To further increase the efficiency of desorption, the pH can be raised to near 8, which should not greatly affect the organisms' metabolic activity. Two ways to raise the pH of the digester are: (a) add a base continuously to the digester, which (with the high buffering capacity of high-strength digester liquor) would result in a large chemical demand and more salts in the liquid effluent; and (b) remove the $CO_2$ from the gas stream and use the $CO_2$-free gas for mixing and further stripping of $CO_2$, which would result in raising the equilibrium pH of the digester. A basic solution can be used for that stripping, but it results in a chemical demand as in the other system. (Other processes available will be discussed later in the paper.)

Ammonia could be captured by stripping it from the gas by an acid solution, ideally phosphoric in that it would be desirable to keep the nitrogen to use later as a fertilizer. The ammonia would neutralize the acid, resulting in liquid


**Acknowledgments:** The research reported in this paper was supported by a grant from the Kansas Water Resources Research Institute and by the Kansas Agricultural Experiment Station. The valuable assistance of Dr. Berl Koch, Animal Science and Industry Dept., and others at the Kansas Boar Testing Station, Kansas State University, is appreciated. The assistance of John Anchutz in developing and constructing the research apparatus is gratefully acknowledged.

The authors are: L. A. SCHMID, Associate Professor, Civil Engineering Dept., R. I. LIPPER, Professor, J. K. KOELLIKER, Assistant Professor, Agricultural Engineering Dept., C. A. CATE and J. W. DABER, Graduate Assistants, Civil Engineering Dept., Kansas State University, Manhattan.

ammonium phosphate, which would be stable and storable for use as a fertilizer.

## FIELD STUDY

In a field study at the Kansas State University Boar Testing Station the feasibility of stripping ammonia from the wastes was tested while enhancing the overall anaerobic digestion process. Information from previous research on anaerobic digestion by Schmid and Lipper (1969) was used to characterize the wastes. Figs. 1 and 2 show, respectively, diagrams of the station and waste disposal facilities and of the field experimental setup.

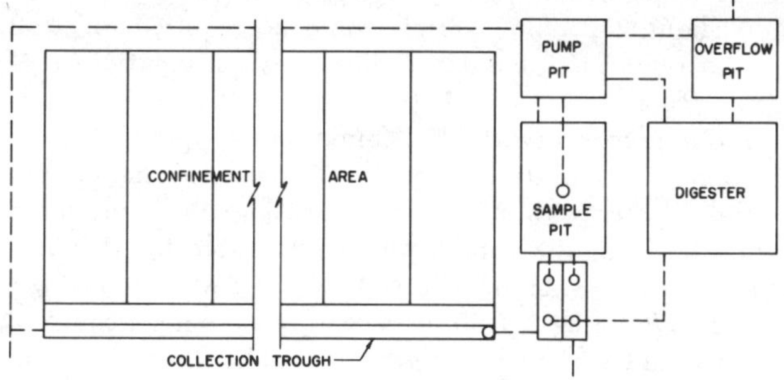


FIG. 1 Schematic of waste collection and treatment facilities.

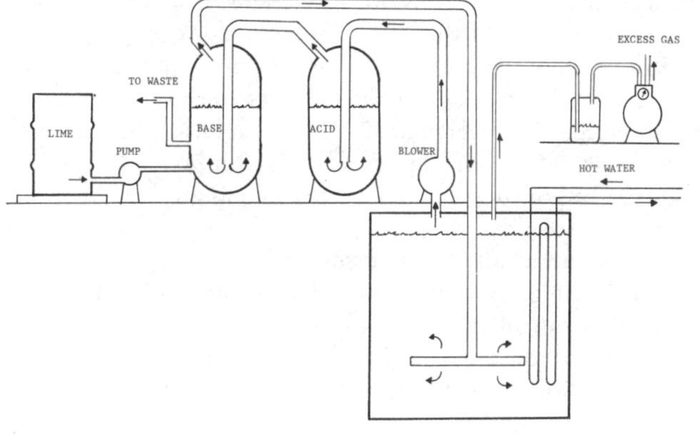


FIG. 2 Schematic of gas purification units.

Two rotary lobe compressors in series provide for the gas mixing and pumping. The gas diffuser within the digester is a PVC pipe with drilled holes. The ammonia stripper tank contains a solution of 85 percent phosphoric acid, and gas is discharged into it through a plastic inlet pipe with drilled holes. The gas then enters a basic solution in the next tank, where $CO_2$ is removed, using lime as the base. The gas, now stripped of ammonia and $CO_2$, is recycled back to the digester for mixing and further stripping. The digester is heated by recirculating water from a 40-gallon, LP gas-fired water heater through copper coils within the digester.

The liquid volume of the digester is 450 cf (12.7 m$^3$). One severe problem in using the $CO_2$ stripping unit is that feeding of lime results in deposition and encrustation of calcium carbonate sludge, causing the diffuser to plug. That leads to line leaks, mechanical breakdowns, freezing lines, and other problems. We have not had long runs to generate good data to date. Difficulties with hardware had been expected. The primary objective was to determine whether removing $CO_2$ from the gas would raise the digester pH sufficiently to result in significant ammonia desorption. Results from two periods during which the system functioned mechanically are reported in Table 1.

The first phase, without $CO_2$ stripping, showed a pH near 7 and nearly equal quantities of methane and carbon dioxide in the gas. The ammonia in the digester at that time was low because of dilution water from start-up and leaky waterers. Removal was slight, as evidenced by the small change in ammonia concentration in the ammonia stripper. The acid tank contained 40 gallons (151 l) of acid, resulting for the first phase in an uptake of 116 grams, or 6.5 grams per day.

During the second phase, when the $CO_2$ stripping unit was active, $CO_2$ essentially had been removed on January 6 and 9, and the pH of the digester had increased from near 7 to more than 8. By January 14, $CO_2$ had increased because of lime loss and a drop of pH in the stripper, resulting in a drop to 7.6 pH in the digester. Uptake during the 3 day period, January 6 to 9 was 123 grams per day.

## DISCUSSION OF RESULTS

Desorption of ammonia from an aqueous solution can be described by this equation as presented by Srinath and Loehr (1974) for a continuous-flow system:

$$\frac{C_1 - C_2}{C_2} = K_D \cdot F \cdot t_{HR}, \quad \dots\dots\dots\dots\dots\dots\dots [4]$$

**TABLE 1. RESULTS OF FIELD TESTS TO REMOVE AMMONIA FROM SWINE WASTES TO ENHANCE ANAEROBIC DIGESTION.**

| Date | Digester | | | Waste gas | | NH$_4$ stripper | CO$_2$ stripper |
| | Temp., deg C | pH | NH$_4$-N, mg/l | Percent CH$_4$ | Percent CO$_2$ | NH$_4$-N, mg/l | pH |
| --- | --- | --- | --- | --- | --- | --- | --- |
| **Phase I** | | | | | | | |
| 9/12/74 | 26.0 | 7.25 | 1360 | 46 | 45 | 1156 | * |
| 9/16/74 | 26.0 | 7.15 | - | 46 | 48 | - | * |
| 9/20/74 | 27.0 | 7.05 | 715 | 48 | 45 | 1645 | * |
| 9/24/74 | 27.0 | 7.15 | 795 | - | - | 1825 | * |
| 9/26/74 | 26.0 | 7.15 | 750 | 49 | 41 | 1740 | * |
| 9/30/74 | 24.5 | 7.15 | 760 | - | - | 1925 | * |
| **Phase II** | | | | | | | |
| 1/6/75 | 28.5 | 7.55 | 1290 | 86 | 1 | 6000 | 11.2 |
| 1/9/75 | 32.0 | 8.08 | 1400 | 87 | 0.3 | 8450 | 12.6 |
| 1/19/75 | 27.0 | 7.63 | 1325 | 58 | 32 | 9390 | 9.0 |

*$CO_2$ stripper not in use during this period.

where

$C_1$ = the influent $NH_4$-N concentration, mg/l
$C_2$ = the effluent $NH_4$-N concentration, mg/l
$K_D$ = a desorption coefficient
$F$ = ratio of undissociated (free) ammonia to the total ammoniacal nitrogen
$t_{HR}$ = total time in hours over which desorption occurs.

$C_1$ and $C_2$ are measured values and F is calculated from Equation [2]. Once $K_D$ is determined, it can be corrected to other temperatures by:

$$K_{DZ} = K_{D1} \, 1.063^{(\deg C_2 - \deg C_1)} \quad \dots \dots \dots \dots \dots \quad [5]$$

The $K_D$ values were calculated for the two phases of results presented in Table 1 and corrected to 35 C, with the following results:

Phase 1, $K_d$ = 5.85 x 10$^{-3}$
Phase 2, $K_D$ = 7.95 x 10$^{-3}$.

Most of the work done by Srinath and Loehr (1974) was for air rates to liquid volume ratios many times greater than those used in this study, in which gas rate was approximately 50 cfm/1000 cf of digester capacity (sufficient to provide mixing within the tank and what would probably be provided if mixing were the only criterion). The formulas presented by Srinath and Loehr (1974), which give values of $K_D$ about 20 times greater than the experimental values for Phase 1 and 2, should be considered valid only for the higher aeration rates used by them.

Henry's Law can also be used to estimate a value of $K_D$, arranged to give the equation:

$$K_D = \frac{A \cdot \rho a}{V \cdot H},$$

where

A = volume of air, or gas applied per hour
V = volume of liquid containing ammonium
$\rho a$ = density of ammonia
H = Henry's constant for solubility (mg/l per atmosphere).

Solving for an air flow rate of 50 cfm/1000 cf, a value for $K_D$ of 4.55 x 10$^{-3}$ at 35 C is obtained. Because $K_D$ is directly related to A/V, the experimental data of Srinath and Loehr (1974) can be used by multiplying their $K_D$ values by the inverse ratio of A/V of their work to the A/V ratios used in this study:

$$0.52 \cdot \frac{50 \text{ cfm}/100 \text{ cf}}{5660 \text{ cfm}/100 \text{ cf}} = 4.59 \times 10^{-3}.$$

These values are compared with the research results in Table 2 and they are all in general agreement. With a value $K_D$ known, estimates of ammonia removals can be made for other conditions by using an average value for $K_D$ of 5.7 x 10$^{-3}$ at 35 C and a gas flow rate of 50 cfm/1000 cf. For greater gas flows, we proportionately increased $K_D$, obtaining results presented in Table 3.

Table 3 shows that for significant ammonia removals, air-flow rates in the range of 200 to 400 cfm/1000 cf or greater would be required. However, to reduce the ammonia concentration for good digestion, gas-flow rates of only about 200 cfm/1000 cf would be required for a 50 percent removal. Table 3 also illustrates the significance of temperature.

From previous work by Schmid and Lipper (1969) at the same location, the ammonia concentrations that existed in the digester without stripping, are known. Using those concentrations, the total ammonia available for stripping was estimated to be 0.04 lb $NH_3$-N per 150 lb animal per day (18 g/67.5 kg). If 50 percent of the ammonia could be stripped and captured, the system could collect 14 600 lb (6570 kg) of nitrogen from a 2000-head operation per year.

Other $CO_2$ removal schemes are being investigated. One consists of stripping both gases (ammonia and $CO_2$) in one unit by passing the gas into a packed tower under several atmospheres of pressure with phosphoric acid as the liquid media. Ammonia is readily adsorbed, and much of the $CO_2$ goes into solution in the acid. The acid is recycled around the tower, where exposing it to atmospheric pressure prior to pumping it back releases the $CO_2$ gas to the atmosphere; the ammonia, however, remains in the solution because of its low pH. Much of the power for compression can be recaptured upon pressure release, reducing greatly the cost of compression.

How economically feasible such schemes are, cannot yet be said. If a high-rate, heated, and mixed anaerobic digester is feasible from an energy standpoint, then a system such as this one incorporating additional gas cleaning may very well be feasible. Studies will be continued.

**TABLE 2. VALUES OF $K_D$ FOR GAS-FLOW RATE OF 50 CFM PER 1000 CF OF LIQUID AT 35 C.**

| | |
|---|---|
| KSU, Phase 1, pH 7.1 | 5.85 x 10$^{-3}$ |
| KSU, Phase 2, pH 7.8 | 7.95 x 10$^{-3}$ |
| Calculated from Henry's Law | 4.55 x 10$^{-3}$ |
| Calculated from data of Srinath and Loehr (1974) | 4.59 x 10$^{-3}$ |

**TABLE 3. PERCENTAGE OF AMMONIA REMOVAL FOR A 15-DAY DIGESTER RETENTION TIME.**

| Temp, deg C | pH | Percent removal 50 cfm/1000cf | Percent removal 100 cfm/1000cf | Percent removal 200 cfm/100cf | Percent removal 400 cfm/1000cf |
|---|---|---|---|---|---|
| 15 C | 7.0 | 0.2 | 0.4 | 0.7 | 1.4 |
| 15 C | 8.0 | 1.7 | 3.3 | 6.3 | 11.9 |
| 25 C | 7.0 | 0.5 | 1.1 | 2.2 | 4.2 |
| 25 C | 8.0 | 5.5 | 10.4 | 18.8 | 31.7 |
| 35 C | 7.0 | 2.8 | 5.4 | 10.3 | 18.7 |
| 35 C | 8.0 | 20.4 | 33.9 | 50.6 | 67.2 |



MANAGING LIVESTOCK WASTES

## CONCLUSIONS

1   Removing carbon dioxide from the gas stream of a digester, with recycling of the $CO_2$-free gas back for mixing, can increase digester pH to near 8,

2   At digestion temperatures of 35 C and pH 8, significant amounts of ammonia can be stripped from an anaerobic digester by gas stripping with ammonia free gas.

3   Removing carbon dioxide for pH control has an added advantage of producing a digester waste gas of nearly 100 percent methane.

4   Ammonia gas is readily adsorbed into solutions of phosphoric acid, resulting in ammonium phosphate, a valuable fertilizer.

### References


1   Schmid, L. A. and R. I. Lipper. 1969. Swine wastes, characterization and anaerobic digestion. Proceedings, Cornell Agricultural Animal Waste Conference, N.Y. 50-57.

2   Srinath, E. G. and R. C. Loehr. 1974. Ammonia desorption by diffused aeration. JWPCF. 46(8):1939-1957.


---

## Methane Production from Poultry Manure

*(Continued from page 247)*

3   Using inoculum or seed culture greatly reduced the time lag in starting a digester, and suggested use of a continuously fed system or a batch system where 50 percent or less of the digester volume is emptied at one time.

4   The optimum temperature for methane production was 35 C (95 F).

5   Water vapor and hydrogen sulfide are present in the gas and must be scrubbed out.

6   Methane recovery of 130 to 160 liters/kg (2 - 2.5 ft$^3$/lb) dry manure and a net energy gain of 3500 kJ/kg dry manure was obtained from the full scale operation.

7   Application of the laboratory findings to the field digester proved to be successful.

### References


1   Albertson, O. E. 1961. Ammonia nitrogen and the anaerobic environment. J. Water Poll. Control Fed., 33:978-995.

2   Anthonisen, A. C. and E. A. Cassell. 1974. Methane recovery from poultry waste. North Atlantic Region of ASAE, Paper No. NA74-108. 28p.

3   Barker, H. A. 1936. Studies upon the methane-producing bacteria. Archiv für Mikrobiol. 7:420-438.

4   Edwards, V. H. 1970. The influence of high substrate concentrations on microbial kinetics. Biotechnol. Bioeng. 12:679-712.

5   Gramms, L. C., L. B. Polkowski and S. A. Witzel. 1971. Anaerobic digestion of farm animal wastes (dairy bull, swine, and poultry). TRANSACTIONS of the ASAE 14(1):7-11.

6   Graves, R. E. and L. R. Massie. 1972. Manure facts and figures. University of Wisconsin Cooperative Extension Programs, Madison, Wisconsin. 16p.

7   Lawrence, A. W. and P. L. McCarty. 1969. Kinetics of methane fermentation in anaerobic treatment. J. Water Poll. Control Fed. 41:R1-R17.

8   Savery, C. W. and D. C. Cruzan. 1972. Methane recovery from chicken manure digestion. J. Water Poll. Control Fed. 44:2349-2354.

9   Standard Methods for the Examination of Water and Wastewater. 1965. p. 427-428. 12th Ed. American Public Health Association, Inc., New York.

# Residual and Annual Rate Effects of Manure on Grain Sorghum Yields

A. C. Mathers, B. A. Stewart, J. D. Thomas

RECENT energy and fertilizer shortages have changed the emphasis from disposal to use of feedlot wastes. With this emphasis, pollution problems such as $NO_3^-$ accumulations in soil or groundwater will be minimized. Numerous authors have indicated that nitrate accumulates in the soil when large amounts of feedlot manure are applied to soil (Adriano, Pratt and Bishop 1971; Mathers and Stewart 1971, 1974; Meek et al. 1974; Murphy et al. 1972; and Olsen, Hensler and Attoe 1970). Because nitrate moves with water percolating through the soil, it may be lost as a plant nutrient and become a problem in groundwater. However, when rainfall is low or irrigation is controlled so that nitrate remains in the root zone, crops can utilize it.

Man has long realized the fertilizer value of manure. When commercial nitrogen fertilizer was cheap, however, the fertilizer value of manure would not pay the hauling cost. Now, the increased cost of commercial fertilizer has again made it profitable to return manure to the land for fertilizer. Herron and Erhart (1965) found that a metric ton of high-quality feedlot manure was equal to 11 kg of N supplied as ammonium nitrate. Mathers and Stewart (1971, 1974) and Vitosh, Davis and Knezek (1973) found that 22.4 metric tons/ha supplied the nutrients needed to produce excellent yields of corn (Zea mays) or grain sorghum (Sorghum bicolor). When larger amounts of manure were applied, potassium and other salts built up in the soil (Cross, Mazurak and Chesnin 1971; Mathers and Stewart 1971, 1974; Vitosh, Davis and Knezek 1973). High concentrations of nitrate and salts in the soil cause inefficient use of all elements and reduce yields.

In this experiment we determined the effect of various rates of manure and the residual effects from large applications of manure on grain sorghum yields.

The experiment was started in April 1969 on Pullman clay loam. Manure was applied annually at 0, 22, 67, 134 and 268 metric tons per hectare (T/ha) (wet weight, approximately 50 percent water). Other treatments were 538 T/ha applied for 1 year only and annually for 3 years. Application dates and chemical properties of the manure are shown in Table 1. Commercial fertilizer treatments (N and NPK) were included for comparison at rates of 268-0-0 and 268-56-56 kg of element per hectare for the first 2 years and 134-0-0 and 134-56-56 the next 3 years. The sources of NPK were ammonium nitrate, concentrated superphosphate and potassium chloride. Manure and fertilizer were plowed down (about 20 cm) as soon after application as possible.

Hybrid grain sorghum (Sorghum bicolor L. Moench 'RS-671' for the first 3 years and 'DeKalb E-59' the last 2 years) was planted in level borders about the first of June each year in rows 75 cm apart. A preplant irrigation was applied every year except the first. After stand establishment, irrigation was applied as needed to maintain good plant growth.

Soil samples were taken to a depth of 3.6 m before applying initial treatments in 1969. Samples were taken to 15 cm after stand establishment early in June each year. Following harvest, soil samples were taken to 1.8, 3.6, 1.8, 3.6 and 6 m in 1969, 1970, 1971, 1972 and 1973, respectively. These samples were analyzed for nitrate as described by Kamphake, Hannah and Cohen (1967) and for soluble salts as described by the U.S. Salinity Laboratory Staff (1954).

**TABLE 1. APPLICATION DATES AND CHEMICAL PROPERTIES OF MANURE APPLIED TO IRRIGATED GRAIN SORGHUM PLOTS (PERCENT WET WEIGHT)**

| Date applied | $H_2O$, percent | N, percent | P, percent | K, percent | Na, percent |
|---|---|---|---|---|---|
| 4-22-1969 | 50 | 1.78 | 0.31 | 1.12 | 0.38 |
| 2-26-1970 | 48 | 1.32 | 0.34 | 0.91 | 0.48 |
| 12-15-1970 | 35 | 1.86 | 0.40 | 1.32 | 0.55 |
| 3-12-1972 | 55 | 1.30 | 0.22 | 0.80 | 0.33 |
| 4-17-1973 | 54 | 1.19 | 0.29 | 0.78 | 0.33 |

The analyses were made on oven dry (70 C) manure.

Grain yields were determined by hand harvesting four rows 4.65 m long from each plot when the grain was mature in October. The samples were threshed, moisture content of the grain was determined gravimetrically, and yields were calculated at 14 percent moisture in the grain.

## RESULTS AND DISCUSSION

Either 22 T manure/ha or 134 kg N/ha was adequate to produce high grain sorghum yields (Table 2). Larger amounts of manure did not further increase yields. Because the experiment was established on an area containing a sig-

**TABLE 2. GRAIN SORGHUM YIELDS AS AFFECTED BY RESIDUAL AND ANNUAL MANURE APPLICATIONS**

| Treatment | Year | | | | |
|---|---|---|---|---|---|
| | 1969, kg/ha | 1970, kg/ha | 1971, kg/ha | 1972, kg/ha | 1973, kg/ha |
| Check | 5,165a* | 5,341abc | 5,285b | 5,424d | 3,963c |
| N† | 5,311a | 6,031a | 8,351a | 7,756ab | 8,620a |
| NPK† | 5,132a | 5,174abc | 8,901a | 8,375a | 8,369ab |
| Manure-T/ha‡ | | | | | |
| 22 | 5,378a | 5,962a | 8,863a | 8,318a | 8,708a |
| 67 | 5,535a | 5,463abc | 8,721a | 8,204a | 8,448ab |
| 134 | 5,378a | 5,692ab | 5,692ab | 7,836ab | 8,335ab |
| 268 | 1,470b | 4,275c | 7,993a | 7,308bc | 7,640b |
| 536-3 | 375b | 1,237d | 1,403c | 6,847c | 8,386ab |
| 536-1 | 375b | 4,584bc | 8,682a | 8,421a | 8,594a |

*Values in each column followed by the same letter are not significantly different at the 5 percent level.
†N was applied at 268 kg/ha in 1969, 1970 and at 134 kg/ha in 1971, 1972 and 1973.
‡Wet weight (approximately 50 percent water).

Contribution from the Soil, Water, and Air Sciences, Southern Region, Agricultural Research Service, USDA, in cooperation with the Texas Agricultural Experiment Station, Texas A&M University.

The authors are A. C. MATHERS and B. A. STEWART, Soil Scientists; and J. D. THOMAS, Agricultural Research Technician, USDA Southwestern Great Plains Research Center, Bushland, Texas 79012.

TABLE 3. CONDUCTANCE OF SATURATED PASTE EXTRACTS FROM SEED ZONE SOIL SAMPLES AT PLANTING TIME AND AFTER HARVEST WHERE VARIOUS RATES OF MANURE OR COMMERCIAL FERTILIZER WERE APPLIED

| | Year | | | | | | | | | |
|---|---|---|---|---|---|---|---|---|---|---|
| | 1969 | | 1970 | | 1971 | | 1972 | | 1973 | |
| Treatment | Planting, mmhos/cm | Harvest, mmhos/cm | Planting, mmhos/cm | Harvest, mmhos/cm | Planting, mmhos/cm | Harvest, mmhos/cm | Planting, mmhos/cm | Harvest, mmhos/cm | Planting, mmhos/cm | Harvest, mmhos/cm |
| Check | 1.27f* | 0.45c | 0.60b | 0.84c | 0.85c | 0.49b | 0.51e | 0.61e | 0.49d | 0.48c |
| N† | 2.48e | 0.44c | 0.67b | 0.93c | 0.54c | 0.55b | 0.77de | 0.65e | 1.40cd | 0.51c |
| NPK† | 3.43de | 0.44c | 0.73b | 0.95c | 0.64c | 0.45b | 1.18de | 0.66e | 1.02cd | 0.51c |
| Manure-T/ha‡ | | | | | | | | | | |
| 22 | 2.65de | 0.50c | 0.71b | 1.06c | 1.03c | 0.56b | 1.41d | 1.21de | 1.50cd | 0.64bc |
| 67 | 3.63d | 0.49c | 0.87b | 1.10c | 1.51c | 0.69b | 2.32c | 2.03cd | 2.41bc | 1.00bc |
| 134 | 6.27c | 1.31b | 1.47b | 1.25c | 1.90c | 0.84b | 2.37c | 2;06c | 3.36b | 1.32bc |
| 268 | 8.33b | 1.67b | 3.48b | 3.38b | 3.81b | 1.47b | 5.38z | 3.14b | 8.67a | 3.68a |
| 536-3 § | 11.73a | 3.83a | 10.60a | 9.43a | 9.68a | 5.90a | 3.72b | 3.93a | 1.78cd | 1.43b |
| 536-1‖ | 11.03a | 4.11a | 3.00b | 2.41bc | 1.23c | 0.93b | 1.00de | 1.39cde | 0.93d | 0.62bc |

*Values in each column followed by the same letter are not significantly different at the 5 percent level.
†N applied at 268 kg/ha in 1969 and 1970 and at 134 kg/ha in 1971, 1972 and 1973.
‡ Wet weight (approximately 50 percent water).
§ Manure applied for first 3 years only.
‖Manure applied for first year only.

nificant amount of residual fertilizer nitrogen, the check plots had adequate nitrogen for maximum yields the first year. The check plots did not become severely nitrogen deficient until the third crop. Therefore, the only significant changes in yield the first and second years were reductions caused by high rates of manure. For the other 3 years, there were large yield increases over the check. The increased yields are attributed to nitrogen because there was no significant difference between yields on plots receiving nitrogen alone and plots receiving nitrogen, phosphorus, and potassium or manure at rates of 134 T/ha or less. Nitrogen fertilizer rates were reduced from 268 to 134 kg N/ha after the second crop without causing any reduction in yield. Phosphorus and potassium did not cause a significant change in yields.

Yields were reduced when 268 or 536 T manure/ha were applied. The yield reduction was greater the first year when a preplant irrigation was not applied (Table 2). Also, the conductance of saturated paste extracts was higher the first year (Table 3). In subsequent years, when manure was applied earlier in the season and preplant irrigations were applied, the salt content of the seed zone was lower. Seedlings did not show as much stress, because the irrigation water leached the salt. However, stress increased with increasing manure rates. Conductance values in the seed zone were always higher than 3 mmhos where manure was applied at

268 T/ha or more. This salt, combined with high ammonium levels in the soil, caused severe stress early in the season and reduced yields.

The harmful salinity effects of high manure rates generally do not persist. The first year after high rates were discontinued, yield markedly increased (Fig. 1). The second year, yields were near maximum for the variety and crop management system. Thus, when the ammonium was nitrified and sufficient water was applied to disperse the salt, sorghum yields were not reduced. Adequate nutrients remained in the soil to produce good crop growth.

When manure is mixed into soil, the organic forms of nitrogen are converted to ammonia. The ammonia is then nitrified under aerobic conditions. Thus, nitrate forms in the soil and is used by plants, accumulated, or leached with water from irrigation or rainfall. If conditions become anaerobic, the nitrate may be denitrified and lost to the air as $N_2$. Where manure was applied, nitrate accumulated and moved down in the soil profile as shown in Fig. 2. As the ap-

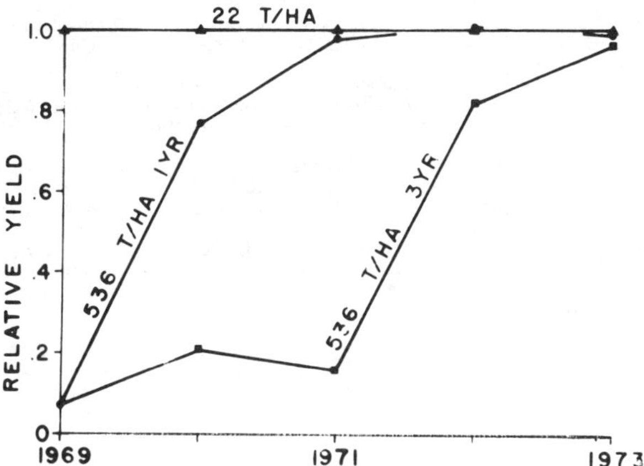


FIG. 1 Yield of grain sorghum as affected by manure rates showing recovery from 536 tons per hectare applied in 1969 only and applied in 1969, 1970 and 1971.

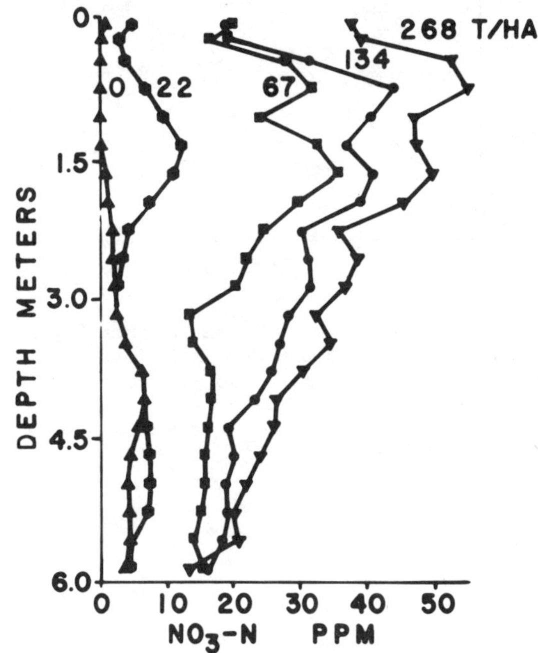


FIG. 2 Nitrate-N in Pullman soil after five annual applications of manure at indicated rates to irrigated grain sorghum.

MANAGING LIVESTOCK WASTES

plication rates were increased, the nitrate accumulation and depth of movement increased. Where manure was applied at 22 T/ha, only a small amount of nitrate accumulated and none moved below 2 m. However, where higher rates were applied, large quantities of nitrate accumulated in the top 2 m and some moved to 6 m.

Nitrate accumulated in the surface 3 m of soil increased for 4 years, then decreased the fifth year (Fig. 3). This decrease can be attributed to denitrification and leaching because water was ponded on the level borders for some time in 1973. Nitrate decrease was greatest in the top meter of soil.

The nitrate-nitrogen accumulated to 3 m where 536 T manure/ha was applied one time on April 22, 1969, or three times on April 22, 1969, February 26, 1970, and December 15, 1970, is shown in Fig. 4. When the experiment was started about 170 kg N/ha was present in the top 3 m of soil. After the 536-T/ha applications of manure were discontinued, nitrate continued to increase in the soil the

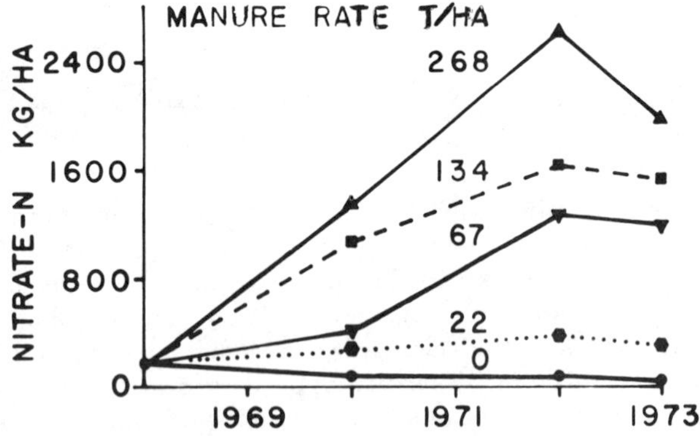


FIG. 3 Nitrate-N accumulation to 3 meters as affected by annual manure applications.

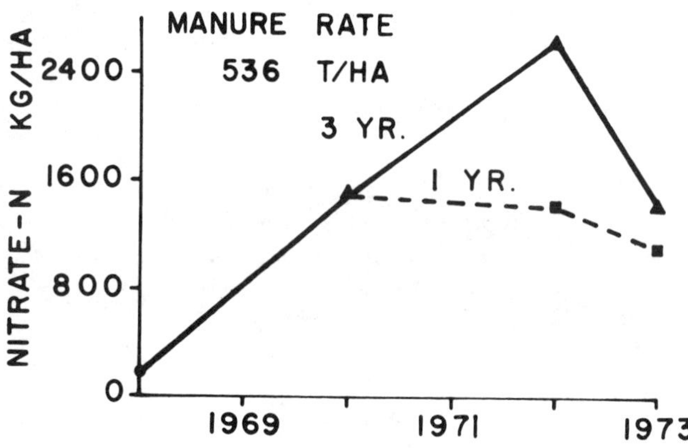


FIG. 4 Nitrate-N accumulation to 3 meters as affected by one or three manure applications of 536 tons per hectare. Manure was applied April 22, 1969, on all plots. Plots receiving manure 3 years received extra manure February 26, 1970, and December 15, 1970. Soil samples were taken to 3 meters April 1969, October 1970, February 1973 and November 1973.

following year. Residual manure was still producing nitrate faster than nitrate was taken up by the sorghum crop. However, the second year after the last application, nitrification decreased. The combined plant uptake, denitrification, and leaching losses exceeded nitrification. Thus, the amount of nitrate remaining in the top 3 m of soil decreased.

## SUMMARY


During the 5 years of 1969 through 1973, annual applications of manure at 22 T/ha produced maximum yields of grain sorghum. Grain sorghum yields were decreased by 268- or 536-T/ha applications of manure as a result of high salt and ammonia concentrations in the root zone. However, these high concentrations were temporary, and yields increased the first year after the heavy applications of manure were discontinued. Leaching of salts continued the second year and yields increased to a maximum on these plots.

Nitrate did not accumulate where manure was applied at 22 T/ha. However, when 67 T manure/ha or more was applied, nitrate accumulated in the top 2 m of soil. Some nitrate moved to a depth of 6 m.



**References**

1    Adriano, D. C., P. F. Pratt and S. E. Bishop. 1971. Fate of inorganic forms of N and salt from land-disposed manures from dairies, p. 243-246. In: Livestock waste management and pollution abatement. ASAE, St. Joseph, Michigan 49085.

2    Cross, O. E., A. P. Mazurak and L. Chesnin. 1971. Animal waste utilization for pollution abatement. TRANSACTIONS of the ASAE 16(1):160-163.

3    Herron, G. M. and A. B. Erhart. 1965. Value of manure on an irrigated calcareous soil. Soil Sci. Soc. Amer. Proc. 29(3):278-281.

4    Kamphake, L. J., S. A. Hannah and J. M. Cohen. 1967. Automated analysis for nitrate by hydrazine reduction. Water Research 1:205-216.

5    Mathers, A. C. and B. A. Stewart. 1971. Crop production and soil analyses as affected by applications of cattle feedlot waste, p. 229-231, 234. In: Livestock waste management and pollution abatement. ASAE, St. Joseph, Michigan 49085.

6    Mathers, A. C and B. A. Stewart. 1974. Corn silage yield and soil chemical properties as affected by cattle feedlot manure. J. Environ. Quality 3(2):143-147.

7    Meek, B. D., A. J. MacKenzie, T. J. Donovan and W. F. Spencer. 1974. The effect of large applications of manure on movement of nitrate and carbon in an irrigated desert soil. J. Environ. Quality 3(3):253-258.

8    Murphy, L. S., G. W. Wallingford, W. L. Powers and H. L. Manges. 1972. Effects of solid beef feedlot wastes on soil conditions and plant growth. Proc. Cornell Agr. Waste Mgmt. Conf., p. 449-464.

9    Olsen, R. J., R. F. Hensler and O. J. Attoe. 1970. Effect of manure application, aeration and soil pH on soil nitrogen transformations and on certain soil test values. Soil Sci. Soc. Amer. Proc. 34(2):222-225.

10    U. S. Salinity Laboratory Staff. 1954. Determination of the properties of saline and alkali soils, p. 7-33. In: L. A. Richards (ed.) Diagnosis and improvement of saline and alkali soils. USDA Agriculture Handbook No. 60.

11    Vitosh, M. L., J. F. Davis and B. D. Knezek. 1973. Long-term effects of manure, fertilizer and plow depth on chemical properties of soils and nutrient movement in a monoculture corn system. Jour. Environ. Quality 2(2):296-299.




MANAGING LIVESTOCK WASTES

# Direct Land Disposal of Feedlot Runoff


N. P. Swanson, C. L. Linderman, L. N. Mielke

MEMBER ASAE     ASSOC. MEMBER ASAE     AFFILIATE ASAE


THE pollution potential of uncontrolled feedlot runoff is aggravated by the chemical characteristics of the first runoff leaving the feedlot and by the greater amount and frequency of runoff from feedlots as compared with runoff from surrounding lands. Under conditions in the western Corn Belt, feedlots have several more runoff events per year and 2 to 3 times more runoff per unit area than cropland or pasture (Swanson 1973). During a rainfall event, runoff will begin sooner from a feedlot than from adjacent cropland. Ammonium-nitrogen ($NH_4$-N) and nitrate-nitrogen ($NO_3$-N) leach rapidly from the feedlot surface and are transported in the initial runoff (Swanson et al. 1971). The high concentrations of $NH_4$-N and $NO_3$-N in the runoff decrease as precipitation continues, but the initial feedlot runoff adds seriously to the pollution problem. If the receiving stream is at a low or normal flow, a minimum of dilution is provided and the feedlot runoff may move downstream as a "slug."

Runoff control systems are required for many feedlots to comply with discharge elimination laws. Most systems use a debris basin for separating solids and a holding pond for storing effluent for later disposal. Holding ponds, however, are often a source of complaint from feedlot operators. They sometimes cause odor in warm weather, are a safety hazard, and foster weed growth. Equipment is required for pumping effluent and distributing it on cropland. Direct land disposal of runoff effluent by gravity flow from the debris basin can eliminate the need for a holding pond and pumping equipment.

Two alternate systems for direct land disposal were constructed in eastern Nebraska. One system uses a "field sink," a level basin surrounded by an earth dike, to impound and infiltrate feedlot runoff. The other uses a grassed waterway consisting of bench terraces connected in series in a serpentine or switchback configuration on a hillside. Feedlot runoff from this system will not reach the receiving stream unless the runoff rate and duration are sufficient for runoff to travel the length of the vegetated waterway. The two systems are discussed in this paper.

## DIRECT LAND DISPOSAL

Feedlot runoff which flows a relatively long distance overland toward a stream is not considered an environmental hazard in many states since overland flow permits the highly polluted initial runoff to infiltrate the soil. Also, overland flow of runoff greatly increases the time for delivery to a receiving stream. This delay increases the opportunity for runoff from the surrounding watershed to dilute the feedlot runoff. Direct land disposal on cropland or pasture is possible on many sites by installing structures, such as field sinks and serpentine waterways, to increase the infiltration opportunity and the time necessary to deliver feedlot runoff to a stream.

Early civilizations conceived the practice of storing runoff water for crop use by providing for infiltration and storage as soil water. Matson (1961) described "syrup-pan" and basin methods as recent practices for water spreading. Runoff water was utilized in terrace benches (Zingg and Hauser 1959) and level pan basins (Mickelson 1966) in semi-arid regions. Wittmuss and Ehlers (1968) measured the crop response on level benches receiving runoff water.

A water-spreading system of basins or terraces, alone or in combination, can be used for infiltration of feedlot runoff. Basin systems, like the field sink, are possible when nearly level land, not subject to frequent overflow, separates the feedlot from the receiving stream. The "syrup-pan" system with spreader dikes is readily adaptable to sites with deep permeable soils and up to 1 percent slopes. Slopes of 2 percent or more will require construction of a bench terrace cross section to adequately control the runoff. Low-gradient terraces, connected in series, like the serpentine waterway, will provide in-transit storage and increase the infiltration opportunity time. The serpentine waterway zigzags across the slope with "hair-pin turns" at each side of the traversed area.

Any runoff control system, including direct land disposal systems, should provide for the collection of bedloads and readily settleable solids from the runoff. Solids collecting in the land disposal area could damage crops and interfere with water spreading. Solids can be collected by terrace basins (Swanson et al. 1973) or solids traps (Swanson and Mielke 1973) inside or outside the feedlot. Discharge from a basin may be through riser intakes discharging into underground pipe (Linderman et al. 1974). Delivery from a solids trap may be by either underground pipe or open channel.

## FIELD SINK

The climates of the western Corn Belt and the Great Plains are well suited to the operation of field sinks. Anticipated runoff and the ratio of the contributing feedlot area to the field sink area are the major design considerations. As the ratio of the field sink area to the feedlot area increases, management problems decrease, but the land requirement and construction costs increase. The anticipated management problems are possible interference with timely field operations and decreased yields because of wet soil. Discharge of runoff from a field sink should occur only under extraordinary conditions.

Installation of a field sink was completed in July 1973,


Contribution from Animal Waste Management Research Unit, North Central Region, ARS, USDA, in cooperation with the Nebraska Agricultural Experiment Station, Lincoln. A portion of the funds for this research was provided by the Environmental Protection Agency. Prepared for presentation and publication - Proceedings, International Symposium on Livestock Wastes, University of Illinois, Urbana-Champaign, April 21-24, 1975.

The authors are: N. P. SWANSON, Supervisory Agricultural Engineer; C. L. LINDERMAN, Agricultural Engineer; and L. N. MIELKE, Soil Scientist, ARS, USDA, Lincoln, Neb.

Acknowledgment: Marvin D. Keebler, Area Engineer, and Lester D. Hovorka, Civil Engineering Technician, Soil Conservation Service, Lincoln, Nebraska, provided consultation for design of the cross section of the serpentine waterway and completed the field plan and staking for construction. The authors are grateful for this engineering assistance and other valuable suggestions made in completing the installation and instrumentation on the Harland Krambeck farm, Gretna, Nebraska.

at Springfield, near Omaha, Nebraska. The site is adjacent to a creek, but is not subject to flooding by the design storm. The sink, 50 m (160 ft) wide and 100 m (325 ft) long, is located on a leveled area between the feedlot and the creek and is surrounded by a 30 cm (1-ft) high dike. The leveled area slopes 0.1 percent to aid runoff distribution from the smaller events which would not completely cover a level sink. Feedlot runoff collects in a broad-basin terrace and discharges to the field sink via a riser intake and underground pipe. The feedlot area is 1.2 times the field sink area. Therefore, the average annual feedlot runoff of 23 cm (9 in.) (Swanson et al. 1971) will contribute 28 cm (10.8 in.) of added water to the field sink. Researchers have shown that 25 to 35 cm (10 to 15 in.) of feedlot runoff per year used as irrigation water is beneficial to corn, perennial forages, and forage sorghum (Linderman and Swanson 1972, Swanson et al. 1974, Sukovaty et al. 1974).

The field sink was designed for a 10-year, 24-hour storm of 11.9 cm (4.7 in.) precipitation which is expected to produce 10.7 cm (4.2 in.) of runoff from the feedlot (Table 1). This will add 24.7 cm of direct rainfall and feedlot runoff to the field sink. The soil is a silty clay loam with an expected base infiltration rate of 0.50 cm (0.20 in.) per hour. Thus, the 12.7 cm (5.0 in.) of impounded water remaining after the storm will be infiltrated within about 25 hours.

This installation functioned well with above normal rainfall in the fall of 1973, snowmelt in December 1973 and January 1974, and above normal rainfall in late May and early June 1974. A corn crop estimated at 3800 kg/ha (60 bu/ac) was grown on the field sink in 1974 despite severe drought and high temperatures from mid-June to mid-August. Weather conditions caused failure of much of the local corn crop.

## SERPENTINE OR SWITCHBACK WATERWAY

Runoff begins earlier on the feedlot than on adjacent cultivated or grassed land, flushing off $NH_4$-N and $NO_3$-N and reaching the waterway which still has a capacity for water intake. The low gradient of the waterway increases the time for feedlot runoff to reach the end of the waterway, thereby increasing the opportunity for infiltration. Thus, a major runoff event would be required for delivery of an appreciable discharge of effluent from the feedlot to the receiving stream. When discharges do occur, they will (a) have low COD and very low $NH_4$-N and $NO_3$-N contents as compared with the initial feedlot runoff, (b) have increased oxygen contents due to aeration caused by overland flow and raindrop addition and splashing, (c) be reduced in volume by continuing infiltration on the water-

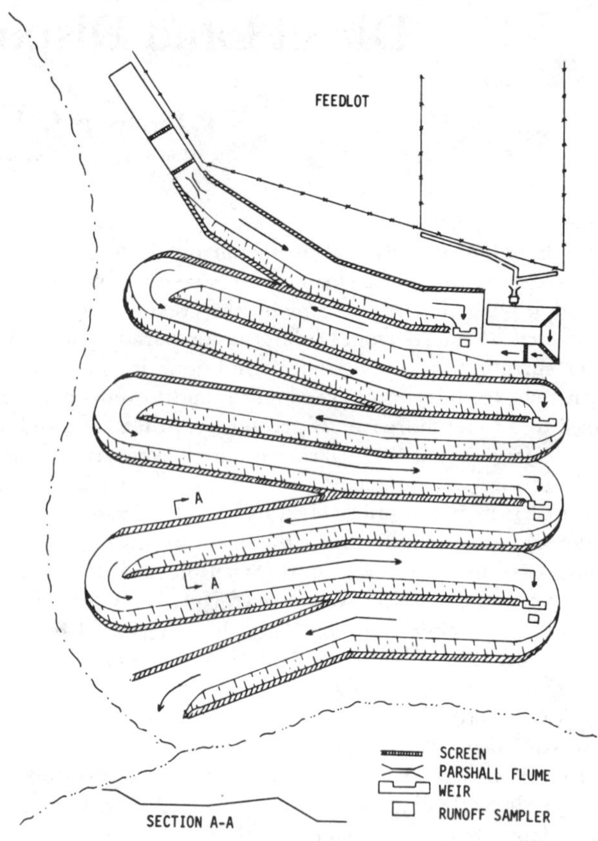


FIG. 1 Serpentine waterway for feedlot runoff disposal, Gretna, NE.

way, (d) be diluted by rainfall, and (e) enter the receiving stream with peak discharge from the total watershed diluting the feedlot runoff.

The design of the serpentine waterway should include (a) the maximum wetted perimeter permitted by the cross slope to promote infiltration, (b) a minimum gradient to prevent "ponding" due to inaccuracies in construction, (c) vegetal retardance to provide the maximum feasible roughness coefficient, and (d) a length adequate to provide a comparatively long time of concentration. The waterway should provide complete infiltration of feedlot runoff, except for those events which are accompanied by diluting runoff from the surrounding watershed. Even when runoff discharges from the waterway, the initial runoff with high $NH_4$-N and $NO_3$-N will have been infiltrated.

A switchback waterway was constructed below a 1.2 ha (3 ac) feedlot on a research site at Gretna, near Omaha, Nebraska, in May 1974. The waterway was on a formerly terraced hillside with 4 to 8 percent slopes on a deep silty clay loam soil. A sketch and typical cross section is shown in Fig. 1. There are eight traverses of the waterway on the hillside for a total length of 790 m (2600 ft). The gradient of the channel is 0.0025 with much steeper slopes (4 to 7 percent) on the "hair-pin" turns. The channel has a 4.1 m (13.5 ft) wide flat bottom, a 4.1 m (13.5 ft) wide, 13.5:1 front slope, and a steep backslope. On the 4 percent slopes near the bottom of the hill, the flat bottom of the waterway is doubled to 8.2 m (27 ft) wide. The 4.1 m (13.5 ft) widths will permit row crop production, using four 102 cm (40 in.) rows. The hair-pin turns must, of course, be kept in grass to prevent erosion.

**TABLE 1. DESIGN OF FIELD SINK FOR 10-YEAR, 24-HOUR STORM, SPRINGFIELD, NE.**

| | |
|---|---:|
| Direct rainfall on field sink, cm* | +11.9 |
| Added feedlot runoff (1.2 x 10.7), cm† | 12.8 |
| Total water on field sink, cm | 24.7 |
| Infiltration during 24-hr storm (0.50 x 24), cm‡ | 12.0 |
| Depth ponded after 24-hr storm, cm | +12.7 |
| Freeboard, cm | 15.0 |
| Impoundment depth for field sink, cm | 27.7 |

*10-year, 24-hour rainfall = 11.9 cm.
†Expected runoff is 90 percent of rainfall and the feedlot area is 1.2 times the field sink area.
‡Base infiltration rate = 0.50 cm/hr.



MANAGING LIVESTOCK WASTES

| Depth of feedlot runoff, cm | Frequency | |
|---|---|---|
| | No. of events | Cumulative percentage |
| 0.00-0.24 | 27 | 30 |
| 0.25-0.49 | 8 | 39 |
| 0.50-0.74 | 10 | 50 |
| 0.75-0.99 | 11 | 63 |
| 1.00-1.24 | 5 | 68 |
| 1.25-1.49 | 3 | 72 |
| 1.50-1.74 | 6 | 79 |
| 1.75-1.99 | 2 | 81 |
| 2.00-2.99 | 7 | 89 |
| 3.00-3.99 | 5 | 94 |
| 4.00-4.99 | 2 | 97 |
| 5.00-5.99 | 1 | 98 |
| 6.00-6.99 | 1 | 99 |
| 7.00-7.99 | 1 | 100 |
| Total | 89 | |

The total waterway area is about the same as the contributing area of the feedlot. Solids traps collect transported solids from the runoff before it enters the waterway. A cover of smooth bromegrass with alfalfa has been established. A 20-year frequency summer drought in 1974 prevented establishment of vegetation and precluded runoff until August.

Anticipated performance of the waterway can be calculated with the Manning open channel flow equation. The retardance coefficient, n, for short grass is assumed to be 0.04 (Chow 1959). The minimum depth for appreciable flow is probably about 3.0 cm (0.1 ft). This requires 58.1 $m^3$/hr (0.57 cfs) with a velocity of 7.1 m/min (23.3 ft/min). The time of concentration will be 112 min (790 m ÷ 7.1 m/min) or 1.86 hours. Feedlot runoff from 1.2 ha (3 ac) must be 0.48 cm/hr (0.25 in./hr) to maintain 58.1 $m^3$/hr (0.57 cfs). If the runoff continued long enough to discharge from the waterway, it would be diluted with 0.9 cm (0.48 x 1.86) of rainfall. If infiltration in the channel and runoff from the uninundated portion of the waterway is ignored, the waterway discharge would be 25 percent rainfall. As the soil becomes saturated and the infiltration rate approaches zero, the waterway discharge will approach 50 percent rainfall since the feedlot and waterway areas are equal.

A flow depth of 3.0 cm (0.1 ft) provides over 1.2 ha-cm (1.2 ac-in.) of in-transit storage. Greater flow depths increase the wetted perimeter and thus increase infiltration of runoff while increasing the in-transit storage.

On bromegrass pasture in western Iowa, rainfall runoff of 0.10 cm or more occurred only 4 times annually (Saxton 1974, Saxton et al. 1971). Runoff less than 0.10 cm would not cause appreciable dilution of feedlot runoff. Table 2 summarizes 89 feedlot runoff events numbering about 20 annually. If only four of these feedlot runoff events are diluted by cropland and pasture runoff, then about 16 events or 80 percent should be controlled by the waterway. In Table 2, 72 of the 89 feedlot runoff events, or 81 percent, had runoff depths of 2.00 cm (0.75 in.) or less.

Ordinarly, rainfall of 2.00 cm or more is required to cause appreciable runoff and no runoff is expected from eastern Nebraska feedlots with rainfall of less than 1.25 cm (0.5 in.) unless precipitation has occurred during the previous 72 hours. Rains of 2.5 to 3.8 cm (1.0 to 1.5 in.) have caused runoff of 0.75 to 3.0 cm (0.3 to 1.2 in.). If the level part of the waterway, 0.41 ha (1.01 ac), can infiltrate 9.0 cm (3.5 in.), it would absorb 3.0 cm (1.2 in.) of rainfall and 2.0 cm (0.75 in.) of runoff from the 1.2 ha (3 ac) feedlot. This would control about 80 percent of the runoff events expected from the feedlot (Table 2).

## Conclusion

Further experience and data from the research installations will provide better criteria for future designs. However, runoff information now available for small watersheds and feedlots permits design for direct land disposal commensurate with pollution control requirements.

Very gently sloping land will permit the installation of either a field sink or a serpentine waterway. Where the topography permits, it is possible to discharge a waterway into a field sink.

Land used for direct feedlot runoff disposal can be cropped successfully with row crops, close growing crops, or perennial forage. Direct land application can save investment and time for many feeders with average or smaller size feedlots, and provide adequate protection for the environment.


## References

1   Chow, Ven Te. 1959. Open-channel hydraulics. McGraw-Hill Book Co., Inc., New York, p. 98-114.

2   Linderman, C. L. and N. P. Swanson. 1972. Disposal of beef feedlot runoff on corn. Agron. Abstr. p. 182.

3   Linderman, C. L., N. P. Swanson and L. N. Mielke. 1974. Riser intake designs for feedlot solids collection basins. ASAE Paper No. 74-3030, ASAE, St. Joseph, Mich. 49085.

4   Matson, Howard W. 1961. Water management, conservation use, and legal aspects. Chapter 42, Agricultural Engineers Handbook. pp. 495-497.

5   Mickelson, Rome. 1966. Level pan system for spreading and storing watershed runoff. Soil Sci. Soc. Am. Proc. 30:388-392.

6   Saxton, Keith. 1974. Unpublished data, ARS-USDA, Columbia, Missouri.

7   Saxton, Keith E., Ralph G. Spomer and Larry A. Kramer. 1971. Hydrology and erosion of loessial watersheds. Journal of the Hydraulics Division, ASCE, 97(HY 11), Proc. Paper 8523:1835-1851.

8   Sukovaty, J. E., L. F. Elliott and N. P. Swanson. 1974. Some effects of beef-feedlot effluent applied to forage sorghum grown on a Colo silty clay loam soil. J. Environ. Quality 3(4):381-388.

9   Swanson, N. P., L. N. Mielke, J. C. Lorimor, T. M. McCalla, and J. R. Ellis. 1971. Transport of pollutants from sloping cattle feedlots as affected by rainfall intensity, duration, and recurrence. In: Livestock Waste Management and Pollution Abatement, p. 51-55. ASAE, St. Joseph, Mich. 49085.

10   Swanson, N. P., J. C. Lorimor, and L. N. Mielke. 1973. Broad basin terraces for sloping cattle feedlots. TRANSACTIONS of the ASAE 16(4):746-749.

11   Swanson, Norris P. 1973. Hydrology of open feedlots in the Corn Belt. Conference Proceedings, Midwest Livestock Waste Management, Iowa State University, Ames, November 27-28, 1973.

12   Swanson, N. P. and L. N. Mielke. 1973. Solids trap for beef cattle feedlot runoff. TRANSACTIONS of the ASAE 16(4):743-745.

13   Swanson, N. P., C. L. Linderman and J. R. Ellis. 1974. Irrigation of perennial forage crops with feedlot runoff. TRANSACTIONS of the ASAE 17(1):144-147.

14   Wittmuss, H. D. and P. L. Ehlers. 1968. Crop responses to rainfall multipliers. TRANSACTIONS of the ASAE 11(4):484-486.

15   Zingg, A. W. and V. L. Hauser. 1959. Terrace benching to save potential runoff for semiarid land. Agron. J. 51:289-292.

# Land Disposal of Beef Wastes: Climate, Rates, Salinity, and Soil


M. L. Horton, J. L. Halbeisen, J. L. Wiersma, A. C. Dittman, R. M. Luther

AFFILIATE ASAE     MEMBER ASAE


ANIMAL wastes can be considered a disposal problem or a source of nutrients. In either case, the soil is the recipient. Soils vary in their chemical and physical make-up; therefore, the effect of a given input of wastes differs from soil to soil.

Variation in climate influences the amount of water and the temperature, both of which affect chemical reaction and movement of waste components in the soil.

Due to the non-uniformity of wastes, the variability in soils, and climatic factors, a single guideline for application of wastes to agricultural lands is not possible.

A recent publication by Powers et al. (1974) discusses guidelines for applying beef wastes to fields. The circular emphasizes the need for knowledge about the composition of the wastes, the characteristics of the soil at the disposal site, and the water supply from either rainfall or irrigation. Two important hazards from wastes recognized by Powers et al. (1974) were dispersion of the soil and build-up of salts. Sodium and potassium are primary constituents of wastes that contribute to the dispersion hazard. The salts of Na, K, Ca, and Mg are the primary contributors to the salinity hazard.

Reddel et al. (1974) have demonstrated a direct relationship between the electrical conductivity (EC) of a saturated extract from surface soil and the rate of applied beef waste. Also, the water soluble Na and K increased directly in the surface of the soil with the amount of waste applied.

Frey et al. (1972) have shown an almost linear relationship between salt level of the ration and the salinity of wastes and runoff waters from beef cattle feedlots.

Vitosh, et al. (1972) suggested that forage grown on soils receiving large amounts of beef waste may have a high enough K to Mg ratio to warrant an addition of Mg to the feed if the forage is fed to beef or dairy cattle.

The Northern Great Plains with its continental-type climate characterized by temperature and moisture extremes, with its montmorillonite-type clay soils generally high in salts, and with a high population of beef cattle presents a broad range of animal waste management and disposal problems. The lack of sufficient leaching waters and the susceptibility of the soils to dispersants and excess salts are sufficient reasons for waste management studies.

The purpose of this research is to establish some basic quidelines for disposal of beef wastes on agricultural lands in the Northern Great Plains. The specific objectives of this study were to determine the influence of salt in the ration, rates of waste applied, and climate upon the concentration and movement of chemical constituents in the soil, composition of runoff waters, and growth and yield of crops produced on disposal plots. The results reported here are preliminary due to only one year of data.

## MATERIALS AND METHODS

The research was initiated in August, 1973 with 16 pens of beef steers established at the Southeast South Dakota Experiment Farm. The experiment consisted of eight pens in cold confinement in a barn and eight pens in the open with no shelter. Each pen was initially established with 11 steers. Each pen was concrete surfaced and lined with retaining boards to prevent mixing or loss of wastes.

A common basic ration was used with four levels of added salt (NaCl) which were 0.00 percent, 0.25 percent, 0.50 percent, and 0.75 percent of the ration on a dry weight basis. During the first four months of the feeding trial, corn silage with appropriate supplements was used as the basic ration. During the last four months of the feeding trial, the silage was replaced with corn and alfalfa to increase the nutritional needs of the steers during finishing for market.

All wastes accumulated in the pens were sampled, weighed, and applied to the field disposal plots.

The wastes applied to the field plots consisted of two salinity levels—low (0.0 percent and 0.25 percent combined) and high (0.50 percent and 0.75 percent combined). Four rates of wastes were applied in addition to a check plot—44.8, 89.6, 134.4, and 179.2 MT/ha; and four replications of each treatment were used. Each field plot was 6.1 m by 36.6 m in size. Applied wastes were incorporated with a chisel plow soon after application.

Waste samples were collected at the time of field application for water content and chemical analyses.

Soil samples were collected to a depth of 150 cm using a power probe in October 1973, May 1974, and September 1974. Soil samples were air dried for 72 hours before being ground and passed through a 2 mm screen in preparation for analyses. Water soluble and extractable Ca, Mg, Na, and K contents of the soil samples were determined by atomic adsorption (Isaac and Kerber 1971). Electrical conductivities were determined from saturated extract from the soil according to U.S. Salinity Laboratory (1954) procedures.

After plowing and seedbed preparation, all plots were planted to corn on May 24, 1974, at a population of 47,000 plants per hectare in 76.2 cm rows. Leaf samples were collected from all plots during the final week of July 1974. The dried and ground leaf samples were analyzed for major ions by spark-emission spectroscopy.

A portion of each plot was harvested for silage yields on September 9, 1974, and a portion harvested for corn grain yields on September 18, 1974.


The authors are: M. L. HORTON, Professor, J. L. HALBEISEN, Research Assistant, Plant Science Dept., and J. L. WIERSMA, Director, Water Resources Institute, A. C. DITTMAN, Research Associate, Agricultural Engineering Dept., R. M. LUTHER, Professor, Animal Science Dept., South Dakota State University, Brookings.

This study was supported through EPA Demonstration Grant S-802532.

Approved for publication by the Director of the South Dakota Agricultural Experiment Station, Brookings, South Dakota, as Journal Paper No. 1350.

The initial soil analyses revealed extremely high values for extractable Ca and Mg for most plots below the 30 cm depth and for some plots at the 0 to 30 cm depth. It was concluded that free $CaCO_3$ and related salts were responsible for the high and erroneous values.

## RESULTS AND DISCUSSION

The chemical analyses for the wastes of animals consuming different amounts of added salt (NaCl) in the ration for two environments are given in Table 1. The values given are averages over six sampling dates. These analyses show that while the other elements remain constant, Na varies according to the amount of NaCl in the ration. Wastes produced in the open environment tend to have a lower cation concentration. Thus, waste reflects the different constituents of the ration used by the feedlot operation and the environment under which they are produced.

The ear corn yields and silage yields are presented in Table 2. The silage yields are not significantly different for the various treatments, while the ear corn yields are significantly different at the 0.05 level.

The leaf analyses data are presented in Table 3. The level of N in the leaves increased as waste rate increased. The element Mg shows a definite inverse relationship with the amount of waste applied to the soil. The treatments are significantly different for the Mg content found in the leaves. The reduction of Mg in the leaf indicates a possibility of decreased Mg content in the total forage. If this assumption proves true and if the same trend occurs in forage grasses, hypomagnesemia could become a problem when these forages are grazed by animals not receiving a Mg enriched supplement.

**TABLE 3. LEAF ANALYSES FROM CORN PLANTS GROWN ON PLOTS RECEIVING THE INDICATED AMOUNTS OF WASTE.***

| Waste rate, MT/ha | Salt level | N† | P | K | Ca | Mg |
|---|---|---|---|---|---|---|
| | | | | Percent | | |
| Check | | 2.70 | 0.27 | 2.61 | 0.55 | 0.54 |
| 38.5 | Low | 2.97 | 0.35 | 2.61 | 0.53 | 0.39 |
| 26.9 | High | 2.80 | 0.34 | 2.59 | 0.55 | 0.44 |
| 101.5 | Low | 3.05 | 0.35 | 2.58 | 0.51 | 0.36 |
| 85.3 | High | 3.06 | 0.35 | 2.68 | 0.53 | 0.34 |
| 135.2 | Low | 3.18 | 0.36 | 2.61 | 0.48 | 0.32 |
| 119.7 | High | 3.08 | 0.36 | 2.65 | 0.52 | 0.30 |
| 169.6 | Low | 3.25 | 0.35 | 2.69 | 0.47 | 0.32 |
| 172.7 | High | 3.12 | 0.35 | 2.52 | 0.46 | 0.26 |

*All Na analyses were less than 0.01 percent
†N was determined by Kjeldahl

**TABLE 1. AVERAGE CATION COMPOSITION OF WASTE OVER ENVIRONMENTAL AND RATION SALT LEVELS.**

| Percent NaCl in the feed | Environment | Pen no. | Ca | Mg | Na | K | Total of cations |
|---|---|---|---|---|---|---|---|
| | | | | | Percent | | |
| 0.00 | Confinement | 4 | 1.19 | 0.76 | 0.49 | 3.22 | 5.66 |
| | | 9 | 1.16 | 0.75 | 0.41 | 3.19 | 5.51 |
| | Open | 6 | 1.16 | 0.76 | 0.38 | 3.01 | 5.31 |
| | | 13 | 1.14 | 0.76 | 0.35 | 3.02 | 5.27 |
| 0.25 | Confinement | 1 | 1.25 | 0.77 | 0.73 | 3.14 | 5.89 |
| | | 12 | 1.23 | 0.75 | 0.84 | 3.15 | 5.97 |
| | Open | 7 | 1.09 | 0.76 | 0.65 | 3.16 | 5.66 |
| | | 16 | 1.10 | 0.72 | 0.75 | 3.05 | 5.62 |
| 0.50 | Confinement | 2 | 1.22 | 0.76 | 1.10 | 3.26 | 6.34 |
| | | 10 | 1.29 | 0.79 | 1.15 | 3.46 | 6.69 |
| | Open | 8 | 1.20 | 0.78 | 1.04 | 3.28 | 6.30 |
| | | 15 | 1.04 | 0.69 | 0.97 | 2.85 | 5.55 |
| 0.75 | Confinement | 3 | 1.17 | 0.74 | 1.58 | 3.20 | 6.69 |
| | | 11 | 1.15 | 0.75 | 1.49 | 3.23 | 6.62 |
| | Open | 5 | 1.11 | 0.71 | 1.22 | 3.11 | 6.15 |
| | | 14 | 1.06 | 0.71 | 1.36 | 3.24 | 6.37 |

**TABLE 2. YIELD OF EAR CORN AND SILAGE FROM PLOTS RECEIVING FOUR RATES OF APPLIED BEEF WASTES.**

| Waste rate, MT/ha | NaCl treatment | Ear corn yield (15.5 percent water) | | Silage yield (dry weight) | |
|---|---|---|---|---|---|
| | | hl/ha | bu/a | MT/ha | t/a |
| Check | | 39.22 | 45.08 | 5.67 | 2.53 |
| 38.5 | Low | 55.19 | 63.43 | 8.47 | 3.78 |
| 26.9 | High | 35.27 | 40.54 | 6.72 | 3.00 |
| 101.5 | Low | 62.32 | 71.64 | 7.54 | 3.37 |
| 85.3 | High | 54.50 | 62.65 | 7.86 | 3.51 |
| 135.2 | Low | 52.77 | 60.65 | 5.98 | 2.67 |
| 119.7 | High | 61.00 | 70.11 | 6.68 | 2.98 |
| 169.6 | Low | 62.19 | 71.48 | 7.68 | 3.43 |
| 172.7 | High | 53.89 | 61.94 | 6.37 | 2.85 |

Examination of the soils data indicates that the chemical effects of the waste remained in the 0 to 30 cm depth. This is reasonable since the rainfall for the year from October 1973 to September 1974 was only 44.2 cm which is 18.8 cm less than the normal rainfall for the same period.

A recent publication (Powers et al. 1974) examined the cationic ratios present in the various types of waste as a method of predicting the possible dispersion hazard in the soils receiving these wastes. The cations listed by Powers et al. (1974) included Ca, Mg, Na, and K since these are the major cations absorbed on the exchange complex of the soils in the Great Plains area. However, in South Dakota many of the soils contain free Ca and Mg related salts throughout the entire profile, while some soils have an accumulation of these salts at the surface. With the presence of these free salts, it is difficult to measure the exchangeable (Exc) Ca and Exc Mg in the soil. These accumulations are related to the characteristic low rainfall of the sub-humid climate.

This characteristic low rainfall also increases the importance of the Na ion since the amount of leaching water is small. Therefore, Exc Na is used as a parameter to predict any future soil problems related to dispersion.

Exchangeable Na for the various treatments is given in Table 4. A factorial design was used in detecting the difference among the treatments. Since the season effect reflects the application of waste, it must be present in an interaction or main effect before that parameter has any meaning. Table 5 shows that the season by waste rate and season by NaCl treatments are significantly different at the 0.01 and 0.05 levels respectively.

Since the data showed the significant difference in Exc Na with an increase in waste rate and NaCl treatment, a multiple regression analysis was completed to derive an

TABLE 4. SOIL ANALYSES (0-30 cm DEPTH) FOR PLOTS RECEIVING THE
INDICATED AMOUNTS OF WASTE.

| Waste rate, MT/ha | Salt level | Fall 1973 | | | Fall 1974 | | |
|---|---|---|---|---|---|---|---|
| | | Exc Na, meq/100g | Exc K, meq/100g | EC, $\mu$mhos/cm | Exc Na, meq/100g | Exc K, meq/100g | EC, $\mu$mhos/cm |
| 38.5 | Low | 0.25 | 0.48 | 682 | 0.15 | 0.85 | 2418 |
| 26.9 | High | 0.08 | 0.46 | 460 | 0.18 | 0.88 | 1928 |
| 101.5 | Low | 0.06 | 0.82 | 724 | 0.27 | 1.86 | 4873 |
| 85.3 | High | 0.04 | 0.76 | 713 | 0.36 | 1.47 | 3956 |
| 135.2 | Low | 0.10 | 0.80 | 683 | 0.38 | 2.46 | 5886 |
| 119.7 | High | 0.06 | 0.77 | 745 | 0.54 | 1.76 | 5456 |
| 169.6 | Low | 0.08 | 0.84 | 923 | 0.33 | 2.18 | 4903 |
| 172.7 | High | 0.06 | 0.71 | 830 | 0.56 | 1.93 | 6016 |

equation which would predict a change in Exc Na. The dependent variable was the change in Exc Na from the fall 1973 sampling to the Fall 1974 sampling. The independent variables were waste rate and amount of applied Na which was calculated from the chemical analyses given in Table 4.

A simple linear regression equation [1] was derived from the data; 70 percent of the data variation was explained by this equation.

$$Y = 0.05007 + 0.00024 X_i \dots \dots \quad [1]$$

where Y is the change in Exc Na and $X_i$ is the kilograms per hectare of applied Na.

Using this equation it is possible to predict the potential for soil dispersion from chemical analyses of the waste and chemical analyses of the soil. It is necessary to make some assumptions concerning the types of clay present in the soil before predicting the possibility of soil dispersion. The critical value of 15 percent Exc Na for dispersion of soil is open to question. Some soil scientists believe that this critical value changes with the types of clay that are present in the soil.

A similar approach was taken for Exc K as that taken for Exc Na. However, the variability of the data prohibited the establishment of any significant relationships and a suitable prediction equation. This data variability for Exc K is related to the amount of illite present in South Dakota soils.

The analyses of variance for electrical conductivity (EC) showed a highly significant difference in the treatments for the season by waste rate interaction. The increase in EC is related to an increase in the amount of applied waste. Since the rainfall of South Dakota is relatively low which reduces the amount of leaching water, this significant increase in EC could pose a serious problem after several applications of waste.


## SUMMARY

1. The amount of salt (NaCl) in the beef ration directly affects the sodium level of the waste which in turn affects the exchangeable sodium in the soil.

2. The nitrogen level of corn leaf samples increased and the magnesium level decreased as the rate of waste applied to the plots increased.

3. Corn silage and grain yields were variable with rate of applied waste. The below normal rainfall may have contributed to the variable results.

4. Wastes applied to the soil in sub-humid regions where irrigation is not available remain near the soil surface and may greatly affect the chemical and physical characteristics of the soil.


TABLE 5. ANALYSIS OF VARIANCE FOR EXC NA IN SOIL.

| Source | | df | Mean square |
|---|---|---|---|
| Season | (S) | 1 | 1.0686** |
| Waste rate | (W) | 3 | 0.4576 NS |
| NaCl treatment | (N) | 1 | 0.1723 NS |
| Replication | (R) | 3 | 0.0126 |
| S x W | | 3 | 0.1281** |
| S x N | | 1 | 0.1434* |
| W x N | | 3 | 0.0211 NS |
| S x W x N | | 3 | 0.0043 NS |
| S x R | | 3 | 0.0162 |
| W x R | | 9 | 0.0186 |
| S x W x R | | 9 | 0.0083 |
| N x R | | 3 | 0.0186 |
| S x N x R | | 3 | 0.0076 |
| W x N x R | | 9 | 0.0137 |
| Error | | 9 | 0.0112 |

**is significant at the 0.01 level.
* is significant at the 0.05 level.
NS is nonsignificant at the 0.05 level.


## References

1 Frye, A. L., R. W. Hansen, M. G. Petit, R. P. Martin, J. K. Matsushima, S. M. Morrison, B. R. Sabey, J. L. Smith, J. C. Ward, and R. C. Ward. 1972. Animal waste management with pollution control. Annual Report of Colorado Contributing Project to NC 93. Colorado State University, Fort Collins.

2 Isaac, R. A. and J. D. Kerber. 1971. Atomic absorption and flame photometry: techniques and uses in soil, plant, and water analysis. pp. 17-37. In: L. M. Walsh (ed.) Instrumental methods for analysis of soils and plant tissue. Soil Science Society of America, Madison, Wisconsin.

3 Jones, J. B. 1967. Interpretation of plant analysis for several agronomic crops. pp. 49-58. In: Soil testing and plant analysis, plant analysis part II. Soil Science Society of America, Madison, Wisconsin.

4 Powers, W. L., G. W. Wallingford, L. S. Murphy, D. A. Whitney, H. L. Manges, and H. E. Jones. 1974. Guidelines for applying beef feedlot manure to fields. Kansas State University, Cooperative Extension Service Bulletin C-502.

5 Reddell, D. L., R. C. Egg, and V. L. Smith. 1974. Chemical changes in soils used for beef manure disposal. ASAE Paper No. 74-4060. ASAE, St. Joseph, Mich. 49085.

6 U.S. Salinity Laboratory Staff. 1954. Methods for soil characterization. Chapter 6 in USDA Agricultural Handbook No. 60. L. A. Richards (ed.) pp. 83-126.

7 Vitosh, M. L., J. F. Davis, and B. D. Knezek. 1972. Long-term effects of fertilizer, manure, and plowing depth on corn. Michigan State University, Agricultural Experiment Station Research Report No. 198.

# Disposal of Beef Feedlot Wastes onto Land

H. L. Manges, R. I. Lipper, L. S. Murphy, W. L. Powers

MEMBER ASAE     MEMBER ASAE

WASTES from beef feedlots are in both the liquid and solid form. Liquids are the result of stormwater runoff from the feedlot surface. Runoff is caught and stored in storage reservoirs until it can be applied to land. Solids consist of manure and accompanying soil removed from the pens during cleaning. Manure is normally taken from the pens and stored in stockpiles until cropland is available for manure spreading. As the wastes are kept separate during storage, they are spread on different land areas.

We have postulated that pollution from beef feedlots can be minimized by the orderly disposal of wastes onto agricultural lands. The objectives of the reported research were to determine the optimum feedlot waste application rates onto land from economical and pollutional considerations.

## METHODS AND PROCEDURES

This research was conducted in cooperation with the Pratt Feedlot, Inc., a 35 000-head commercial feedlot located near Pratt in southcentral Kansas. Manges et al. (1971) have described the feedlot and waste-handling facilities.

Manure and effluent from runoff-storage reservoirs were applied to field plots, 9 m wide and 61 m long, which were located on a silty clay loam soil. Plots were laid out in a randomized complete block design with four replications. Treatments of 0, 22, 45, 90, 180, 360 and 720 metric ton/ha of dry manure annually and 180, 360 and 720 metric ton/ha the first year only were selected for the manure plots. Annual treatments of 0, 5, 10, 20 and 40 cm were selected for the effluent plots.

As it was difficult to control waste-application rates onto the plots, actual waste treatments were measured. Manure applications were determined by weighing the manure deposited on a plastic sheet spread across the plot. Depths of effluent applied were determined from inflow and outflow measurements. As actual waste treatments varied from desired treatments, results were analyzed by regression techniques.

Furrow-irrigated corn was grown on the waste-disposal plots. Manure was spread after corn harvest and immediately plowed under with a moldboard plow. Effluent was applied during the irrigation season. Well water was applied to all plots as needed for high corn yields. Corn forage yields were determined by weighing forage mechanically harvested from test rows. Soil cores were taken after corn harvest and prior to manure spreading for analyses to determine changes in soil chemical properties.

Properties of wastes applied varied widely as shown in Tables 1 and 2. The irrigation water applied had an electrical conductivity of 0.55 millimhos/cm and soluble sodium percentage of 40.7. Additional details on the waste disposal plots, characteristics of wastes and laboratory procedures have been given by Manges et al. (1971), Murphy et al. (1972) and Wallingford et al. (1974).

Feedlot wastes may be applied to land with waste disposal as the main objective or they may be applied as a fertilizer to increase crop yields. Equations were developed to determine annual net income from crop production when wastes were applied for disposal or as fertilizer. Most profitable waste-application rates were determined from net-income equations using yield-prediction equations and 1974 prices.

## RESULTS

### Effluent Disposal

Assuming unlimited land area for disposal of effluent, annual crop income attributed to the effluent of the Pratt Feedlot, Inc., can be expressed as:

$$ I_e = \frac{P_f Y_e T_e}{E} - \frac{T_e}{E}(C_f + C_1 + C_w - C_w E) - C_e T_e \dots \dots [1] $$

where

$I_e$ = Annual net income, \$

$P_f$ = Selling price of corn forage, \$/metric ton

$Y_e$ = Corn forage yield, metric ton/ha

**TABLE 1. PROPERTIES OF EFFLUENT APPLIED TO DISPOSAL PLOTS.**

| | Electrical conductivity, mmhos/cm | Na | K | Ca | Mg | $NH_4^+$-N | $NO_3^-$-N |
|---|---|---|---|---|---|---|---|
| | | | | parts per million | | | |
| High | 7.6 | 660 | 1840 | 615 | 239 | 179 | 63 |
| Low | 1.6 | 112 | 259 | 83 | 35 | 4 | 1 |

**TABLE 2. PROPERTIES OF MANURE APPLIED TO DISPOSAL PLOTS.**

| Year | N | P | K | Ca | Mg | Na |
|---|---|---|---|---|---|---|
| | | | Percent of dry weights | | | |
| 1969 | 1.04 | 0.41 | 1.09 | 0.78 | 0.39 | 0.23 |
| 1970 | 3.12 | 0.71 | 0.39 | 0.56 | 0.44 | 0.17 |
| 1971 | 0.89 | 0.57 | 0.97 | 0.98 | 0.42 | 0.25 |
| 1972 | 0.84 | 0.57 | 1.37 | 0.99 | 0.42 | 0.29 |

Paper is contribution No. 217, Dept. of Agricultural Engineering and contribution No. 1513, Dept. of Agronomy, Kansas Agricultural Experiment Station, Manhattan.

Partial support for this research was provided by the Environmental Protection Agency, Water Quality Office, under Grants Nos. 13040 DAT and S800923 and by the Kansas State Board of Animal Health.

The authors are: H. L. MANGES and R. I. LIPPER, Research Agricultural Engineers, L. S. MURPHY and W. L. POWERS, Research Agronomists, Kansas Agricultural Experiment Station, Manhattan.

$T_e$ = Annual effluent from storage reservoirs, ha-cm
$E$ = Annual effluent application rate, cm
$C_f$ = Cost of field operations, \$/ha
$C_l$ = Cost of land, \$/ha
$C_w$ = Cost of applying well water, \$/ha-cm
$W$ = Depth of well water applied annually, cm
$C_e$ = Cost of applying effluent, \$/ha-cm

Selling price of corn forage was taken as \$15 per metric ton standing in the field. The following corn-forage prediction equations were developed from yield data:

$$Y_{e-70} = 0.498\ E + 41.5, \qquad R^2 = 0.32 \quad \ldots\ldots [2]$$

$$Y_{e-71} = -0.0164\ E^2 + 0.85\ E + 36.3, \qquad R^2 = 0.37 \quad \ldots\ldots [3]$$

$$Y_{e-72} = -0.0290\ E^2 + 1.28\ E + 49.2, \qquad R^2 = 0.27 \quad \ldots\ldots [4]$$

$$Y_{e-73} = -0.0170\ E^2 + 0.98\ E + 33.8, \qquad R^2 = 0.33 \quad \ldots\ldots [5]$$

where

$Y_{e-70}$ = Corn forage yield in year of subscript, Metric ton/ha.

$R$ = Coefficient of multiple regression.

Cost of field operations which includes tillage, planting and weed control was taken as \$125 per ha. Cost of land which includes interest on the investment and property taxes was taken as \$200 per ha. Although cost of irrigation, including depreciation on the system, interest on the investment and operating costs is quite variable, it was taken as \$4 per ha-cm. Average annual irrigation requirement for the Pratt area was taken as 40 cm and it was assumed that irrigation requirement was reduced by the amount of effluent applied. As effluent was applied through the irrigation system using an additional pump, cost of applying effluent was assumed to be the same as for irrigation water. Effluent available for disposal was considered to be a constant amount annually for a feedlot.

For each of the four years, maximum net income resulted from the application of the first unit of effluent. Estimated cost of corn production would have had to exceed \$622, \$544, \$738, and \$507 per ha for 1970 through 1973, respectively, before it would have been profitable to apply effluent with disposal as the main objective rather than using it as fertilizer.

Maximum net income results when the return from the last unit of fertilizer added is equal to its cost. As a larger volume of material must be applied to supply a unit of plant nutrient through feedlot wastes than through commercial fertilizer, it was assumed that wastes would be spread onto land near the feedlot to keep transportation costs to a minimum and that wastes would be the only fertilizer used.

When effluent is used as a fertilizer, net income per ha can be expressed as:

$$i_e = P_f Y_e - C_f - C_l - C_w W + C_w E - C_e E \quad \ldots\ldots\ldots\ldots [6]$$

where

$i_e$ = Annual net income, \$/ha

Maximum net income resulted from the effluent application rate which gave highest yields. These rates were 49 cm (highest amount applied) in 1970, 26 cm in 1971, 22 cm in 1972 and 29 cm in 1973.

## Manure Disposal-Annual Application

Annual net income from farming operations to dispose of feedlot manure where no additional fertilizer is added can be expressed as:

$$I_m = \frac{P_f Y_m T_m}{M} - \frac{T_m}{M} (C_f + C_l + C_i) - C_m T_m \quad \ldots\ldots\ldots\ldots [7]$$

where

$I_m$ = Annual net income, \$
$Y_m$ = Corn forage yield, metric ton/ha
$T_m$ = Annual manure from feedlot, metric ton
$M$ = Annual manure application (dry weight), metric ton/ha
$C_i$ = Cost of irrigation, \$/ha
$C_m$ = Cost of applying manure, \$/metric ton

The following corn forage-yield prediction equations were developed from yield data:

$$Y_{m-70} = -0.000127\ M^2 + 0.0583\ M + 45.4, \qquad R^2 = 0.25 \ldots [8]$$

$$Y_{m-71} = -0.000178\ M^2 + 0.0536\ M + 37.5, \qquad R^2 = 0.49 \ldots [9]$$

$$Y_{m-72} = -0.0459\ M + 60.2, \qquad R^2 = 0.48 \ldots [10]$$

$$Y_{m-73} = -0.000149\ M^2 + 0.0556\ M + 42.4, \qquad R^2 = 0.51 \ldots [11]$$

$$Y_{m-74} = -0.000526\ M^2 + 0.167\ M + 42.5, \qquad R^2 = 0.21 \ldots [12]$$

where

$Y_{m-70}$ = Corn forage yield in year of subscript, metric ton/ha

Cost of applying irrigation water was taken as \$160 per ha. Cost of applying manure to land which includes cleaning of pens, hauling to the stockpile, loading from the stockpile, hauling to fields and spreading onto land was taken as \$4.95 per metric ton. (Manure-application costs were based upon measurements which show the average moisture content of manure taken from the stockpile was about 33 percent and current manure-handling costs of \$3.30 per metric ton of material.) Other prices and costs were the same as for the effluent study.

As for effluent, maximum net income resulted from the application of the first unit of manure. Cost of corn production which was \$485 per ha would have had to exceed \$681, \$562, \$903, \$636 and \$637 per ha for 1970 through 1974, respectively, before it would have been profitable to apply manure with disposal as the main objective rather than using it as a fertilizer.

When manure is used as a fertilizer, net income per ha becomes:

$$i_m = P_f Y_m - C_f - C_l - C_i - C_m M \quad \ldots\ldots\ldots\ldots [13]$$

where

$i_m$ = Annual net income, \$/ha

For each of the five years, cost of applying manure exceeded income from manure. Cost of applying manure would have had to have been \$0.87, \$0.80, \$0.69, \$0.83 and \$2.51 per metric ton for 1970 through 1974, respectively, before it would have been profitable to apply the first unit.

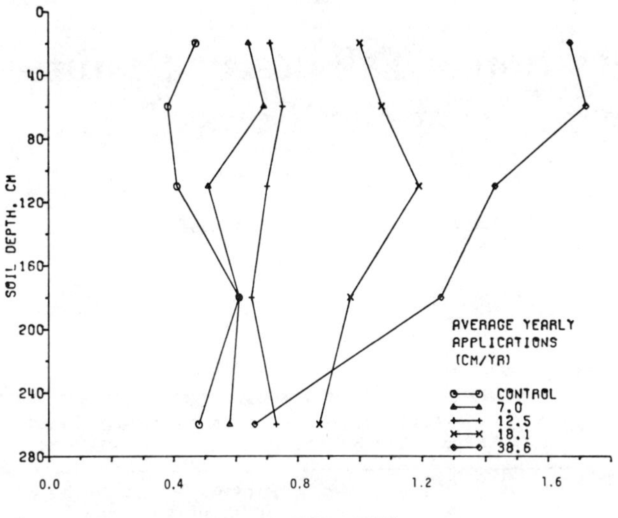

FIG. 1 Effects of effluent applications on the electrical conductivity of soil-saturation extracts, fall 1973.

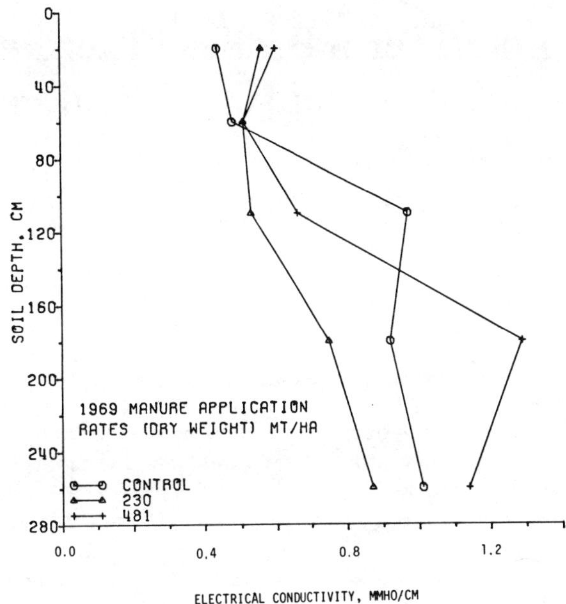

FIG. 3 Effects of residual manure treatments on electrical conductivity of soil-saturation extracts, fall 1973.

## Manure Disposal-First Year Only

When manure was applied the first year with no other fertilizer added in subsequent years, increases in corn forage yields during the study were attributed to the first year's application. The following corn forage-yield prediction equations were developed from the yield data:

$$Y_{m\text{-}70} = -0.000127\,M^2 + 0.0583\,M + 45.4, \qquad R^2 = 0.25 \quad [14]$$

$$Y_{m\text{-}71} = -0.0000991\,M^2 + 0.0638\,M + 34.1, \qquad R^2 = 0.21 \quad [15]$$

$$Y_{m\text{-}72} = -0.0000561\,M^2 + 0.0364\,M + 54.8, \qquad R^2 = 0.16 \quad [16]$$

$$Y_{m\text{-}73} = -0.000106\,M^2 + 0.0947\,M + 32.0, \qquad R^2 = 0.69 \quad [17]$$

$$Y_{m\text{-}74} = 0.0361\,M + 42.4, \qquad R^2 = 0.16 \quad [18]$$

Net income was determined from Equation [13] using the same prices and costs as for annual manure applications and the sum of $Y_m$'s since manure was applied for the yield prediction equations. Cost of applying manure would have had to have been $0.87, $1.83, $2.38, $3.80 and $4.34 per metric ton from 1970 through 1974, respectively, before it would have been profitable to apply the first unit.

## Chemical Properties of Soil

In the fall of 1973 after four years of the study, nitrate-nitrogen in the soil increased with increasing feedlot waste-loading rates. However, only where average annual manure applications were greater than 311 metric ton/ha did the nitrate-nitrogen concentration in soil-saturation extracts exceed 40 ppm in the top 280 cm.

Figs. 1, 2 and 3 give the electrical conductivity of soil-saturation extracts under the waste plots in the fall of 1973. In general, electrical conductivity increased with waste-loading rate. However, saline conditions were found only under the plots which received average annual manure applications of 601 metric ton/ha.

## DISCUSSION

Maximum profit resulted from effluent application rates which gave highest yield. If supplemental irrigation is not practiced, maximum net profit would result from somewhat lower effluent application rates.

There are two reasons why returns from annual manure applications were less than costs of applying manure. First, the soil had a high initial fertility level as evidenced by high yields from the check plots throughout the study. Secondly, available nitrogen was limiting under low annual manure applications. Herron and Erhart (1965) found that about 50 percent of the nitrogen in feedlot manure was removed by irrigated sorghum the first year after it was applied. Manure at low annual application rates didn't provide sufficient available nitrogen for optimum corn growth during the initial years of the study.

Nitrate-nitrogen concentrations and electrical conductivity of soil-saturation extracts from waste disposal plots were at acceptable levels except where 601 metric ton/ha of manure were applied annually. These relatively low concentrations were due to leaching from precipitation which exceeded the normal of 62 cm per yr during 3 of the 5 years of the study along with the average of 40 cm per yr added by irrigation.

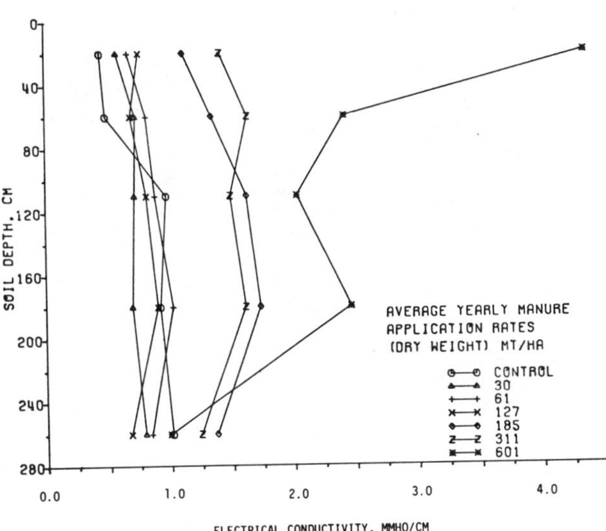

FIG. 2 Effects of yearly manure applications on electrical conductivity of soil-saturation extracts, fall 1973.

*(Continued on page 268)*

# Long-Term Broiler Litter Fertilization of Tall Fescue Pastures and Health and Performance of Beef Cows

J. A. Stuedemann, S. R. Wilkinson, D. J. Williams, H. Ciordia, J. V. Ernst,

W. A. Jackson and J. B. Jones, Jr.

INTENSIVE poultry production areas tend to be concentrated in various geographical regions including the Southern Piedmont, the Boston Mountains, and the Appalachian and Ozark highland regions of the Southeastern United States. Tall fescue is a major pasture species in all of these regions.

Convenience, limited pasture acreage, and economic factors have resulted in heavy use of broiler litter as a fertilizer in certain areas. This practice has been associated with various beef cattle health problems (Stuedemann et al. 1973, Wilkinson et al. 1971, Williams et al. 1969), including grass tetany, fat necrosis, and nitrate toxicity.

Studies were initiated to assess the cumulative, long-term effects of a high rate of broiler litter fertilization of fescue on various soil, plant, and animal parameters in a grazing ecosystem. This paper presents grazing systems research results related to beef cattle health problems.

### EXPERIMENTAL PROCEDURE

Three established Kentucky-31 tall fescue *(Festuca arundinacae,* Schreb) pastures fertilized at three different N levels were used (Table 1). All pastures were stocked with Angus brood cows of similar genetic background.

An 8.5 ha fescue pasture, with no history of poultry litter fertilization was fertilized with broiler litter (high N) at the rate of 24.08, 22.26, 19.08, 22.40, 17.29, 10.68, and 12.36 metric tons/ha (dry matter basis) in 1968-69, 1969-70, 1970-71, 1971-72, 1972-73, 1973-74, and 1974-75, respectively. From 1968-69 through 1971-72 approximately equal applications were made in August, November, February, and May of each year. From 1972-73 through 1974-75 several applications were omitted due to the unavailability of broiler litter. Samples of litter from each application were analyzed for P, K, Ca, Mg, Mn, Fe, B, Cu, Zn, and Mo by direct reading emission spectroscopy (Jones and Warner 1969), N by Kjeldahl digestion (Horowitz 1970) and S (Jones and Isaac 1972). Nutrient concentrations and inputs for N, P, K, Ca, Mg, S, Mn, B, Cu, Zn, and Mo are presented in Table 2.

A second fescue pasture (moderate N) with no history of poultry litter fertilization was fertilized with 134 kg N/ha as $NH_4NO_3$ in 1969 and with 224 kg N/ha per year in 1970 through 1974. Equal N applications were made in February, July, and September of each year. To maintain a "medium" soil test level of P and K, 49.3 kg of $P_2O_5$ and 98.5 kg of $K_2O$ were applied per ha in 1971. The number of cows and the size of the area grazed were adjusted on several occasions until a 24-cow herd stocked at a rate of 0.4 ha/cow was achieved in February 1971 (Table 1).

A third fescue pasture (low N) with no history of poultry litter fertilization was added to the experiment in 1972. It was fertilized with 84 kg N/ha as $NH_4NO_3$ in 1972 and 74 kg N/ha per year in 1973 and 1974. To maintain a "medium" soil test level of P and K, 50 kg $P_2O_5$ and 113

**TABLE 1. FERTILIZATION LEVELS AND STOCKING RATES OF THREE KENTUCKY-31 TALL FESCUE PASTURES.**

| Pasture fertilization | Year | Level of fertilization | No. cows | Hectares per cow |
|---|---|---|---|---|
| | | kg/ha/yr | | |
| Broiler litter | 1968-69 | 24,100 | 20 | 0.42 |
| (High N) | 1969-73 | 20,200 | 21 | 0.40 |
| | 1973-75 | 11,500 | 21 | 0.40 |
| | | Nitrogen | | |
| Moderate $NH_4NO_3$ | 1969 | 134 | 34 | 0.64 |
| (Moderate N) | 1970 | 224 | 24 | 0.61 |
| | 1971-75 | 224 | 24 | 0.40 |
| Low $NH_4NO_3$ | 1972 | 84 | 15 | 0.43 |
| (Low N) | 1973-75 | 74 | 16 | 0.40 |

**TABLE 2. PLANT NUTRIENT CONCENTRATION OF BROILER LITTER APPLIED AND AVERAGE ANNUAL PLANT NUTRIENTS SUPPLIED FROM AUGUST 1968 TO AUGUST 1972*.**

| Plant nutrient | Concentration† percent | Amount supplied, kg/ha/yr |
|---|---|---|
| N | 3.58 | 789 |
| P | 1.53 | 335 |
| K | 1.88 | 413 |
| Ca | 1.68 | 375 |
| Mg | 0.42 | 93 |
| S‡ | 0.45 | 84 |
| | ppm | |
| Mn | 321 | 7.00 |
| B | 36 | 0.80 |
| Cu | 127 | 2.82 |
| Zn | 272 | 6.03 |
| Mo | 8 | 0.18 |

*Mean annual dry matter input was 22.0 metric tons/ha.
†Mean of 164 determinations on separate manure samples.
‡Sulfur analyzed only from August 1970 through August 1971.

Contribution from Soil, Water and Air Sciences, Southern Region, ARS, USDA, in cooperation with the University of Georgia Agricultural Experiment Station.

The authors are: J. A. STUEDEMANN, Animal Scientist, and S. R. WILKINSON, Supervisory Soil Scientist, ARS, USDA, Watkinsville, GA; D. J. WILLIAMS, Professor, Veterinary Medicine and Surgery, University of Georgia, Athens; H. CIORDIA, Microbiologist, ARS, USDA, Experiment, GA; J. V. ERNST, Microbiologist, ARS, USDA, Auburn, AL; W. A. JACKSON, Chemist, ARS, USDA, Watkinsville, GA; and J. B. JONES, JR., Chairman, Horticulture Division, University of Georgia, Athens.

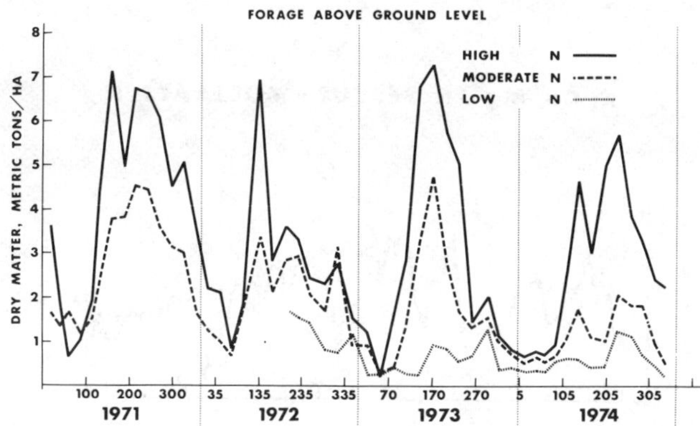

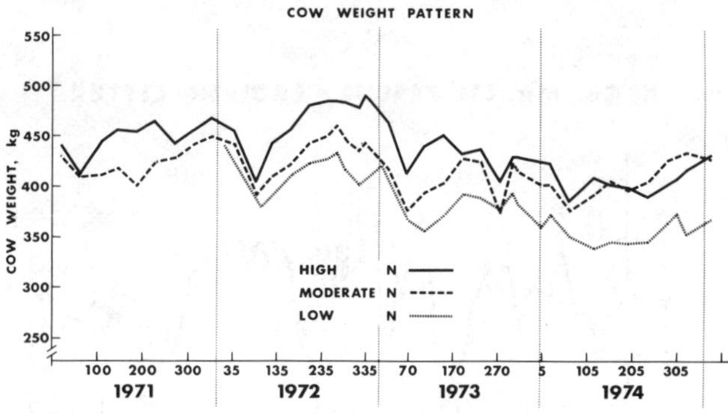

FIG. 1 The quantity of forage above ground level in the three fescue pastures fertilized with high (broiler litter), moderate, and low levels of nitrogen.

FIG. 2 Weight pattern of cows grazing the three fescue pastures fertilized with high (broiler litter), moderate, and low levels of nitrogen.

kg $K_2O$ were applied per ha in 1972. The pasture was stocked at 0.43 ha/cow in 1972 and adjusted to 0.40 ha/cow in 1973 and 1974 (Table 1).

All cattle were managed in a similar manner. A 90-day breeding season beginning in mid-April was used. Beginning in 1971, bulls were rotated at least once during the breeding season. Except for cows grazing the broiler littered fescue through 1971, all open cows, as judged by rectal palpation, were routinely replaced with pregnant cows that had calved at least twice. Calf production was estimated by Beef Cattle Improvement Association procedures.

During periods of insufficient forage, cows grazing the broiler littered and moderate N fertilized fescue received hay harvested from their respective pastures. No hay was harvested from the low N fertilized fescue. In 1972, cows grazing the low N fertilized fescue were fed Coastal bermudagrass hay during periods of insufficient forage. In subsequent years, they received fescue hay harvested from a field fertilized at the same rate as the pasture.

Beginning in 1971, blood serum samples were obtained from all cows at approximately 6-week intervals and more often during the grass tetany season. Before 1971, blood serum samples were obtained only during the tetany season. Samples were analyzed for Ca, Mg, and K by atomic absorption (Anon. 1971).

Available forage was sampled in the broiler littered and moderate N fertilized fescue pastures from late 1968 through the present. Beginning in July 1972, forage was sampled from the low N fertilized fescue. Herbage samples were freeze dried, ground through a 30-mesh screen, and analyzed for total N by Kjeldahl digestion (Horowitz 1970) and for P, K, Ca, Mg, Mn, Cu, Zn, Fe, B, and Mo content by direct reading emission spectroscopy (Jones and Warner 1969). Nitrate-N was determined by the Technicon Cadmium Reduction method (Anon. 1969).

Nematode and coccidia parasitism was monitored in cattle on all three pastures. Salmonellae determinations were made on litter, soil, and cattle feces from the broiler littered pasture from February 1970 through June 1972. Salmonellae, coliform bacteria, nitrate, and phosphorus were determined on the water samples from a small pond located immediately adjacent to the littered pasture during the February 1971 to October 1972 period. The watershed of the pond was essentially the littered pasture.

## RESULTS AND DISCUSSION

As the level of fertilization increased, the quantity of available forage increased, particularly during periods of active growth (Fig. 1). Hay was harvested from the littered and moderate N fertilized pasture in June of each year. During spring and summer, forage levels usually exceeded 3000 to 4000 kg/ha on the littered or moderate N fertilized fescue. However, the quantity of forage available on the low N pasture rarely exceeded 1200 kg/ha.

Mean cow weights were directly related to the level of N fertilization (Fig. 2). However, these differences were not reflected in calf performance. Adjusted 205-day calf weights were similar (Table 3) and seemed to stabilize near 165 kg for the littered and moderate N fertilized pasture.

TABLE 3. PERFORMANCE OF CALVES PRODUCED ON FESCUE PASTURES FERTILIZED WITH BROILER LITTER AND MODERATE AND LOW LEVELS OF $NH_4NO_3$.

| Pasture fertilization | Year calved | No. cows | Ha per cow | No. calves born | No. calves marketed | Adj 205-day wt., kg | Adj kg calf per ha |
|---|---|---|---|---|---|---|---|
| Broiler litter | 1969 | 20 | 0.42 | 20 | 17 | 158 | 316 |
| | 1970 | 21 | 0.40 | 21 | 17 | 163 | 326 |
| | 1971 | 21 | 0.40 | 9 | 7 | 178 | 146 |
| | 1972 | 21 | 0.40 | 18 | 15 | 168 | 295 |
| | 1973 | 21 | 0.40 | 20 | 19 | 167 | 373 |
| | 1974 | 21 | 0.40 | 21 | 21 | 163 | 403 |
| Moderate $NH_4NO_3$ | 1969 | 34 | 0.64 | 27 | 23 | 183 | 193 |
| | 1970 | 24 | 0.61 | 20 | 18 | 188 | 232 |
| | 1971 | 24 | 0.40 | 21 | 21 | 176 | 380 |
| | 1972 | 24 | 0.40 | 21 | 20 | 173 | 355 |
| | 1973 | 24 | 0.40 | 23 | 22 | 165 | 373 |
| | 1974 | 24 | 0.40 | 24 | 22 | 165 | 373 |
| Low $NH_4NO_3$ | 1972 | 15 | 0.43 | 15 | 15 | 198 | 458 |
| | 1973 | 16 | 0.40 | 16 | 16 | 166 | 410 |
| | 1974 | 16 | 0.40 | 16 | 15 | 138 | 318 |

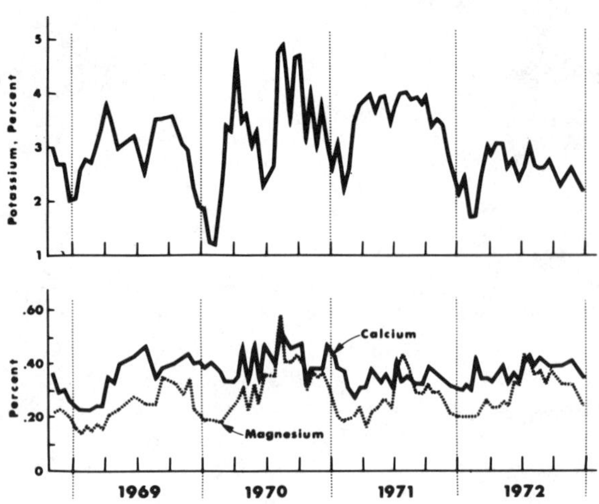


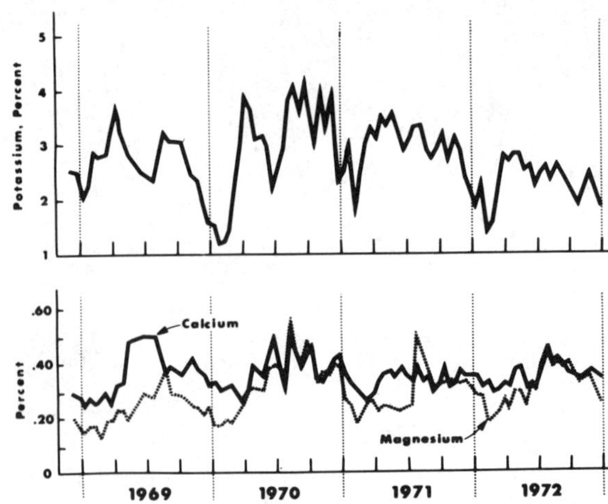


FIG. 3 Percent potassium, calcium, and magnesium in fescue fertilized with high (broiler litter) and moderate levels of nitrogen.

However, calf weights have decreased progressively each year on the low N pasture.

Cow numbers were inadequate to detect differences in reproductive rates. Except for a venereal disease problem identified in the cows grazing the broiler littered fescue in 1970, conception rates were acceptable.

### Grass Tetany

From 1969 through 1972, 11 of 21 cows grazing the littered fescue were given a form of Mg supplementation; in 1973 and 1974, 14 of 21 received Mg supplementation. From 1970 through 1974, 16 of 24 cows grazing the moderate N fescue received Mg supplementation. Through 1974, one-half of the cows grazing the low N fescue received Mg. No cows were supplemented with Mg in 1975. A discussion of the tetany prevention research is beyond the scope of this paper.

The number of cases and/or deaths as a result of grass tetany is direct evidence for evaluating the effect of pasture fertilization on the tetanigenicity of a given pasture. No tetany occurred in cows receiving Mg supplementation. All tetany occurred either in control cows or before the Mg supplementation was implemented. From 1970-74, ten, seven, and zero cases of tetany occurred on the broiler littered, moderate N, and low N fertilized fescue pastures, respectively. The broiler littered and moderate N fertilized pastures had four and five deaths, respectively. In 1975, when no cows were supplemented with Mg, all tetany cases resulted in deaths with four and one tetany deaths on the littered and moderate N fertilized fescue, respectively. Although these results are not conclusive, the broiler littered pasture seemed more tetanigenic than the moderate N fertilized fescue.

Blood serum Mg levels of control animals grazing fescue at each fertility level indicated that blood serum Mg levels generally reflected the tetanigenicity of a given pasture. Serum Mg levels were lower in cows on all pastures from January through March than at other times of the year. During this period, levels for cows grazing the littered and moderate N fertilized fescue were generally lower than those among cows grazing the low N pasture.

Herbage Mg levels were lowest in both the broiler littered or moderate N fertilized pastures from January through March (Fig. 3). Herbage K levels tended to reach their lowest concentrations in January or February and then increased rapidly before the rise in plant Mg levels. Herbage K levels appeared to peak earlier in the broiler littered fescue than in the moderate N fertilized fescue.

There was little difference in plant N levels between the littered and moderate N fertilized pastures, particularly after the first 2 years (Fig. 4). Nitrogen concentrations rose rapidly each spring in a manner similar to K levels. High herbage levels of K and N have been associated with increased incidence of tetany (Kemp 1971, Grunes et al. 1970, Wilkinson et al. 1972). However, in this experiment, little difference in herbage N and K were observed in the littered and moderate N fertilized pastures, though there appeared to be a tendency for greater incidence of tetany in the littered pasture.

### Fat Necrosis

The incidence of fat necrosis (hard fat), as determined by rectal palpation among cows grazing the various fescue pastures, is shown in Table 4. The incidence of fat necrosis increased as N levels increased. No fat necrosis was detected in cows grazing fescue fertilized at the moderate N level until approximately 6 months after the N fertilization level

### NITROGEN IN FESCUE

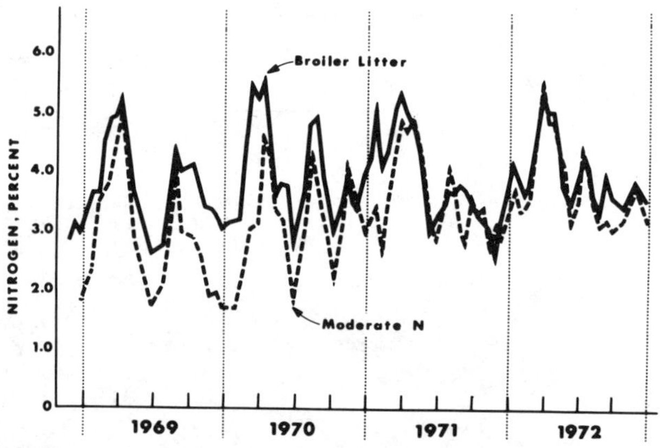


FIG. 4 Percent nitrogen in fescue fertilized with high (broiler litter) and moderate levels of nitrogen.



MANAGING LIVESTOCK WASTES

**TABLE 4. INCIDENCE OF FAT NECROSIS (FN) IN COWS GRAZING FESCUE PASTURES FERTILIZED WITH BROILER LITTER, MODERATE AND LOW LEVELS OF $NH_4NO_3$*.**

| Pasture fertilization | Palpation date | No. cows | No. cows with FN | No. palpable lesions | Estimated size range of palpable lesions |
|---|---|---|---|---|---|
| Broiler litter | 10-02-69 | 20 | 2 | 2 | 0.6 cm dia. to 1.3 cm dia. |
| | 09-09-70 | 21 | 7 | 8 | 1.9 cm dia. to 10.2 x 7.6 x 2.5 cm |
| | 09-30-71 | 21 | 14 | 32 | 0.3 cm dia. to 25.4 x 20.3 x ____ cm† |
| | 09-27-72 | 21 | 15 | 48 | 0.3 cm dia. to 25.4 cm dia. |
| | 10-03-73 | 21 | 14 | 52 | 0.3 cm dia. to 12.7 x 7.6 x ____ cm† |
| | 10-09-74 | 21 | 10 | 33 | 0.3 cm dia. to 12.7 x 10.2 x 3.8 cm |
| Moderate $NH_4NO_3$ | 10-02-69 | 34 | 0 | 0 | |
| | 09-09-70 | 23 | 1 | 1 | 7.65 x 5.1 x 2.5 cm |
| | 09-30-71 | 24 | 6 | 17 | 0.3 cm dia. to 4.4 x 2.5 x 1.3 cm |
| | 09-27-72 | 24 | 5 | 10 | 0.6 cm dia. to 6.4 x 2.4 x 2.5 cm |
| | 10-03-73 | 24 | 4 | 8 | 0.6 cm dia. to 22.9 cm dia. |
| | 10-09-74 | 24 | 0 | 0 | |
| Low $NH_4NO_3$ | 09-26-72 | 15 | 1 | 1 | 2.54 x 1.3 x 1.3 cm |
| | 10-03-73 | 16 | 1 | 1 | 0.6 cm dia. |
| | 10-09-74 | 15 | 0 | 0 | |

*All palpations were done by the same person.
†Unable to estimate all dimensions.

had been increased to 224 kg/ha per year. At any time no more than one cow grazing the low N fescue was detected to have fat necrosis. Data in Table 4 only reflect the incidence of fat necrosis at the time of palpation; it does not account for cows that died or were replaced. From 1968 through 1974, five cows that had grazed the broiler littered fescue died as a direct result of intestinal constriction by hard fat. Another cow grazing the littered pasture died as a result of hard fat which almost completely encased the omasum.

These results indicate that the etiology of fat necrosis is associated with the level of plant nutrient inputs to fescue. Moreover, fat necrosis should not be associated with any broiler litter effect other than its plant nutrient input. As illustrated in Fig. 2, mean cow weights, like the occurrence of fat necrosis, were directly related to the level of N fertilization. Thus, the incidence of fat necrosis and general cow condition may be confounded in this experiment.

## Parasitism

No difference and generally low levels of nematode eggs and coccidia oocysts were observed among cows grazing the different pastures. However, calves were more parasitized than cows. The number of nematode eggs recovered from the calves increased as the level of fertilization decreased. In general, calves grazing the broiler littered fescue had the lowest egg counts.

The average numbers of coccidia oocysts from the three pastures did not exhibit clear cut differences among pastures. Twelve species of coccidia were identified from calves grazing each pasture, including two highly pathogenic species, *Eimeria bovis* and *E. zurnii*. No detectable health problems were associated with the coccidia.

When initially parasite-free tracer calves were slaughtered after grazing either the broiler littered or moderate N fertilized fescue, the number of gastrointestinal parasites recovered were not significantly different. In 1973 and 1974, two calves from each pasture were slaughtered at weaning to obtain actual worm counts. Average numbers of worms/ calf for 1973 and 1974 were 6,193, 17,773, and 37,197 for the broiler littered, moderate N, and low N fertilized fescue, respectively. These results follow the same trends obtained from nematode egg counts.

Apparently, the pastures' condition as affected by vari-

ous factors, such as grazing intensity, may be more closely associated with the level of parasitism than with the use of broiler litter. The level of parasitism was inversely related to the quantity of available forage which in turn was directly related to the level of fertilization.

## Nitrate Toxicity

Wilkinson et al. (1971) reported that plant levels of $NO_3$-N reached concentrations of 3,300 and 2,000 ug/g on the broiler littered and moderate N fertilized fescue in August 1969. Nitrate-N levels were lower in 1970 and 1971. No direct evidence of an effect of these $NO_3$-N levels on animal health was observed.

## Salmonella

From February 1970 through March 1971, salmonellae were isolated from chicken litter only one time (*Salmonellae thompson*). Salmonellae were recovered from a pond adjacent to the littered pasture, but at no time were the same serotype found in the pond and litter (courtesy Enteric Diseases Section, Center for Disease Control, Atlanta, Georgia). Inasmuch as the litter was essentially free of salmonella, the potential hazard of salmonella infection to cattle grazing the broiler littered fescue could not be evaluated.

## Pond Coliform Bacteria

Coliform bacteria levels of water from a pond adjacent to the littered pasture from February 1971 through October 1972 showed that coliform bacteria were low during periods of no or low rainfall and increased with rainfall. The bacteria could have originated from the broiler litter or from the cattle grazing the pasture. It was not determined whether any of the coliform bacteria were pathogens. Intermittant analyses of pond water from littered pasture runoff indicated a maximum $NO_3$-N content of 5.0 ppm (Giddens et al. 1973).

## Arsenic in Broiler Litter

Mean As content of litter was 23.8 with a range of from 3.5 to 41 ppm. Soil samples contained mean extractable As levels of 2.2, 3.6, and 1.9 ppm from the littered, moderate N, and low N fertilized fescue, respectively. Cow hair samples contained mean As levels of 0.99, 0.35, and 0.48 ppm from the littered, moderate N, and low N fertilized pastures, respectively. The significance of this accumulation of As in hair of cows grazing the littered fescue is unknown.

## Selenium

Mean Se levels of fescue fertilized with broiler litter and moderate N were 0.042 and 0.053 ppm, respectively. These samples were obtained from November 1968 through May 1970. Though no white muscle disease was evident, in 1970 several calves were born weak and subsequently died. From January 1971 through October 1972, blood Se levels were consistently lower in cows grazing the broiler littered fescue. Overall mean blood serum levels were 0.054 and 0.068 ppm for cows grazing the littered and moderate N fertilized fescue, respectively (Se analyses by E. E. Cary, U.S. Plant, Soil, and Nutrition Laboratory, Ithaca, New York). However, these plant and blood levels were above Se responsive conditions (Andrews et al. 1968).

## SUMMARY

Animal health problems of grass tetany, fat necrosis, and nitrate toxicity were studied in cattle grazing three Kentucky-31 tall fescue pastures fertilized at a high N level (from broiler litter), and at moderate and low N levels (from ammonium nitrate) from 1968 to 1975. Cow and calf performance was monitored during this time interval. At a stocking rate of 0.4 ha/cow, little difference was observed between calf production at the high and moderate N levels, with a progressive trend toward lower calf production at the lower N level. Grass tetany was a problem on the broiler littered and moderate N fertilized fescue, with an indication of a greater problem on the littered fescue, although herbage levels of N, K, Ca, and Mg were not greatly different between the two pastures. The incidence of fat necrosis increased with increasing levels of N fertilization. Fat necrosis was not associated with any broiler litter effect other than its plant nutrient input. Pasture condition, as affected by various factors, appeared more closely associated with the level of parasitism than with the use of broiler litter. No direct evidence of nitrate toxicity in cattle was observed. On the basis of the demonstrated potential for increased incidence of grass tetany and fat necrosis, broiler litter fertilization is recommended at rates of 9 metric tons/ha per year or less on tall fescue pastures.


## References

1   Andrews, E. D., W. J. Hartley, and A. B. Grant. 1968. Selenium-responsive diseases of animals in New Zealand. New Zealand Vet. J. 16:3-17.

2   Anonymous. 1969. Technicon-Cadimum reduction. FWPCA methods for chemical analysis of water and wastes. USDI-FWPCA, Div. Water Qual. Res. Anal. Qual. Control Laboratory. 1014 Broadway, Cincinnati, OH.

3   Anonymous. 1971. Analytical methods for atomic absorption spectrophotometry. Perkin-Elmer, Norwalk, CT.

4   Giddens, J. A., A. M. Rao, and H. W. Fordham. 1973. Microbial changes and possible ground-water pollution from poultry manure and beef cattle feedlots in Georgia. Environmental Resources Center, Atlanta, GA. No. 0573 p. 57.

5   Grunes, D. L., P. R. Stout, and J. R. Brownell. 1970. Grass tetany of ruminants. Advance. Agron. 22:331-374.

6   Horowitz, W. (ed.) 1970. Official methods of analysis. 11th Ed. Assoc. Official Analytical Chemists, Washington, D.C. p. 17.

7   Jones, J. B., Jr. and R. A. Isaac. 1972. Determination of sulfur in plant material using a Leco Sulfur Analyzer. Agr. Food Chem. 20:1292-1294.

8   Jones, J. B., Jr. and M. H. Warner. 1969. Analysis of plant-ash solutions by spark-emission spectroscopy. In: E. L. Grove and H. D. Perkins, (ed.) Developments in applied spectroscopy. 7A:152-160. Plenum Press, NY.

9   Kemp, A. 1971. The effects of K and N dressings on the mineral supply of grazing animals. p. 1-14. In: Potassium and systems of grassland farming. Proc. 1st Colloq. Potass. Inst., Ltd. Higgs and Co., Caxton House, Henley-on-Thames.

10   Stuedemann, J. A., S. R. Wilkinson, D. J. Williams, W. A. Jackson and J. B. Jones, Jr. 1973. The association of fat necrosis in beef cattle with heavily fertilized fescue pastures. Proc. Fescue Toxicity Conference, University of Kentucky, Lexington, Ky., May 31 and June 1, 1973. Published by University of Missouri, Columbia, MO. p. 9-23.

11   Wilkinson, S. R., J. A. Stuedemann, J. B. Jones, Jr., W. A. Jackson and J. W. Dobson. 1972. Environmental factors affecting magnesium concentrations and tetanigenicity of pastures. p. 153-175. In: J. B. Jones, Jr., M. Blount and S. R. Wilkinson (ed.) Magnesium in the environment, soils, crops, animals and man.

12   Wilkinson, S. R., J. A. Stuedemann, D. J. Williams, J. B. Jones, Jr., R. N. Dawson and W. A. Jackson. 1971. Recycling broiler house litter on tall fescue pastures at disposal rates and evidence of beef cow health problems, p. 321-324, 328. In: Livestock Waste Management and Pollution Abatement, PROC-271, ASAE, St. Joseph, MI. 49085.

13   Williams, D. J., D. E. Tyler and E. Papp. 1969. Abdominal fat necrosis as a herd problem in Georgia cattle. J. Amer. Vet. Med. Assoc. 154:1018-1021.


---

## Disposal of Beef Feedlot Wastes

*(Continued from page 263)*

## SUMMARY

Effects of feedlot wastes on corn forage yields and soil chemical properties have been measured for five years on an irrigated silty clay loam soil in southcentral Kansas. It was not profitable to apply feedlot wastes with disposal as the main objective.

Most profitable effluent application rate averaged 26 cm per yr for the last three years of the study. Manure applications didn't return a profit in any of the study years. However, a large first year manure application followed by no fertilizer in subsequent years came closest to returning a profit.

Excessive nitrate-nitrogen concentrations and electrical conductivities of soil-saturation extracts existed only where annual manure applications exceeded 311 metric ton/ha.


## References

1   Herron, G. M. and A. B. Erhart. 1965. Value of manure on an irrigated calcareous soil. Soil Sci. Soc. Amer. Proc. 29:278-281.

2   Manges, H. L., L. A. Schmid and L. S. Murphy. 1971. Land disposal of cattle feedlot wastes, pp. 62-65. In: Livestock Waste Management and Pollution Abatement. American Society of Agricultural Engineers. St. Joseph, Mich. 49085.

3   Murphy, L. S., G. W. Wallingford, W. L. Powers and H. L. Manges. 1972. Effects of solid beef feedlot wastes on soil conditions and plant growth, pp. 449-464. In: Waste Management Research, Proceedings of the 1972 Cornell Agricultural Waste Management Conference. Graphics Management Corporation. Washington, D. C.

4   Wallingford, G. W., L. S. Murphy, W. L. Powers and H. L. Manges. 1974. Effect of beef-feedlot-lagoon water on soil chemical properties and growth and composition of corn forage. Journal of Environmental Quality 3(1):74-78.

# Manure from Caged Hens Evaluated on Fescue Pasture

J. M. Vandepopuliere, C. J. Johannsen, H. N. Wheaton

AN increasing number of poultry producers have concerns about the maximum level of cage hen manure that can be applied to pasture without detrimental effects. Questions related to the deposition of large quantities of manure or repeated applications on the same field and their potential effects on plant kill, dry matter yield and pollution of the environment are major concerns. These questions led to the initiation of a project to study the effects of applications of up to 89.8 tons of poultry manure per hectare on established fescue pasture.

Experimental field plots were located on three poultry farms near Ste. Genevieve, Missouri. Plots with dimensions of 6.1 x 54.9 meters were established to accomodate application with flail spreaders and 12.2 x 27.4 meters for tank wagon application. Thirty-six plots, (1.8 x 3.6 meters) were established on the poultry farm at the University of Missouri-Columbia. Manure applied to the University plots was weighed and spread manually.

The manure was applied to the test plots in February, 1973. The delivery rate of the manure spreaders was calibrated using plastic sheets or shallow containers. Plot treatment levels of 22.5, 44.9, 67.4 and 89.8 tons manure/hectare were attempted on farms A and B (flail spreader). A honey wagon was used on farm C and approximately twice as much liquid manure was delivered to duplicate the level of nutrients applied on farms A and B. Chemical fertilizer was used as a positive control. Check plots did not receive any type of fertilizer treatment.

Plastic sheets or shallow containers were placed in the field plots to determine the amount of manure applied. Table 1 describes the quantity of manure applied per treatment at all locations. Samples of manure were obtained for analyses at the time of application. The nitrogen level applied per treatment is listed in Table 2.

The soil on farm A was typed as Lindley silt loam and farms B and C were Menfro silt loam. All soils were thick loess deposits derived from the Missouri River bottoms.

The assay results of the soil sample taken at the May harvest, Table 3, show increased calcium, phosphorus and potassium values. There was an unexplainable calcium level decrease with increased manure application on plots at the UM poultry farm. The soil pH was not affected by the manure application.

The plots were harvested on May 9, August 1, and October 15, 1973 by mowing random strips 3.05 meters long. Samples were taken from the grass catcher for moisture and nutrient composition assay.

May forage samples were analyzed for 14 elements (Table 4). Nitrogen, sodium, calcium, magnesium and potassium levels showed increases due to the effect of the manure application. The phosphorus values were variable which is probably due to the complexing with the calcium found in the soil and the minerals supplied by the manure. Other elements showing increased levels with manure application were Cu, Fe, Zn, B, Mn, Al, and Ba, however they did not reach levels that are considered toxic to animals.

**TABLE 1. MANURE APPLICATION ON THE FIELD PLOTS, (t/ha).**

| Design rate | Farm | | | |
| --- | --- | --- | --- | --- |
| | A* | B* | C† | UM |
| Manure | | | | |
| 22.5 | 24.4 | 16.7 | 52.8 | 22.5 |
| 44.9 | 52.8 | 44.0 | 89.5 | 44.9 |
| 67.4 | 76.8 | 59.2 | 172.1 | 67.4 |
| 89.8 | 94.9 | 114.9 | 220.0 | 89.8 |

*Flail spreader
†Tank wagon

**TABLE 2. NITROGEN APPLICATION ON THE FIELD PLOTS.**

| | Farm | | | |
| --- | --- | --- | --- | --- |
| | A* | B* | C† | UM |
| | - - - Nitrogen, kg/ha - - - | | | |
| Chemical fertilizer | 339 | 339 | 339 | 339 |
| Manure, t/ha | | | | |
| 22.5 | 329 | 269 | 465 | 216 |
| 44.9 | 707 | 713 | 787 | 431 |
| 67.4 | 1028 | 959 | 1515 | 647 |
| 89.8 | 1271 | 1861 | 1937 | 862 |

*Flail spreader
†Tank wagon

**TABLE 3. SOIL ANALYSES TAKEN FOUR MONTHS AFTER MANURE APPLICATION (0-7.6 cm depth)**

| Nutrient | Manure, t/ha | Farm | | | |
| --- | --- | --- | --- | --- | --- |
| | | A | B | C | UM |
| Organic matter, percent | Check | 3.1 | 2.0 | 2.5 | 3.5 |
| | 89.8 | 3.1 | 2.4 | 3.2 | 3.9 |
| pH | Check | 6.5 | 6.6 | 6.6 | 6.7 |
| | 89.8 | 6.5 | 6.5 | 6.4 | 6.2 |
| Ca, ppm | Check | 3800 | 3200 | 3200 | 5600 |
| | 89.8 | 4900 | 5400 | 5600 | 3400 |
| $P_2O_5$, ppm | Check | 448 | 372 | 330 | 302 |
| | 89.8 | 448 | 448 | 448 | 448 |
| K, ppm | Check | 410 | 210 | 190 | 350 |
| | 89.8 | 500 | 390 | 390 | 340 |
| Mg, ppm | Check | 280 | 460 | 300 | 280 |
| | 89.8 | 300 | 360 | 380 | 300 |

The authors are: J.M. VANDEPOPULIERE, Associate Professor, Poultry Husbandry Dept., C.J. JOHANNSEN, Associate Professor, and H.N. WHEATON, Professor, Agronomy Dept., University of Missouri, Columbia.

TABLE 4. MINERAL ANALYSIS OF PLANT TISSUES.*

| Mineral | Manure (t/ha) | Farm A | B | C | UM |
|---|---|---|---|---|---|
| | | | Percent | | |
| N | Check | 2.02 | 2.32 | 1.58 | 1.65 |
| | 89.8 | 2.47 | 2.75 | 3.66 | 1.89 |
| Ca | Check | 0.56 | 0.51 | 0.34 | 0.49 |
| | 89.8 | 0.56 | 0.63 | 0.45 | 0.51 |
| P | Check | 0.31 | 0.21 | 0.27 | 0.26 |
| | 89.8 | 0.30 | 0.56 | 0.38 | 0.27 |
| K | Check | 2.06 | 2.41 | 1.86 | 2.00 |
| | 89.8 | 3.28 | 4.35 | 4.50 | 2.86 |
| Na | Check | 0.06 | 0.04 | 0.04 | 0.05 |
| | 89.8 | 0.11 | 0.13 | 0.19 | 0.14 |
| Mg | Check | 0.15 | 0.15 | 0.14 | 0.14 |
| | 89.8 | 0.22 | 0.30 | 0.22 | 0.19 |
| | | | ppm | | |
| Cu | Check | 9 | 8 | 8 | 8 |
| | 89.8 | 11 | 15 | 13 | 8 |
| Fe | Check | 100 | 45 | 54 | 54 |
| | 89.8 | 86 | 120 | 155 | 72 |
| Zn | Check | 17 | 21 | 20 | 15 |
| | 89.8 | 26 | 38 | 30 | 21 |
| B | Check | 2 | 2 | 4 | 2 |
| | 89.8 | 4 | 7 | 4 | 4 |
| Mn | Check | 40 | 51 | 45 | 45 |
| | 89.8 | 45 | 70 | 75 | 40 |
| Al | Check | 152 | 46 | 40 | 51 |
| | 89.8 | 85 | 85 | 101 | 73 |
| Ba | Check | 11 | 10 | 5 | 12 |
| | 89.8 | 18 | 16 | 11 | 14 |
| Mo | Check | 1.1 | 0.5 | 0.9 | 1.2 |
| | 89.8 | 0.8 | 1.5 | 1.3 | 0.9 |

*Samples taken May 9, 1973.

TABLE 5. SUMMARY OF TOTAL FORAGE YIELDS, (t/ha).

| | Farm A* | B* | C† | UM* | UM‡ |
|---|---|---|---|---|---|
| Check (avg.) | 8.3 | 3.8 | 3.1 | 11.9 | 10.0 |
| Chemical fertilizer | 11.3 | 8.4 | 6.0 | 16.9 | 10.7 |
| Manure | | | | | |
| 22.5 | 9.9 | 7.7 | 7.0 | 14.4 | 10.8 |
| 44.9 | 12.5 | 9.4 | 8.0 | 16.1 | 11.9 |
| 67.4 | 15.1 | 11.0 | 7.5 | 17.7 | 11.8 |
| 89.8 | 16.4 | 11.2 | 6.6 | 17.8 | 11.3 |

*Three harvests - May, August, October, 1973
†Two harvests - May, August, 1973
‡Two harvests - June, October, 1974

The forage yield increased consistently at all locations as the level of manure increased from 0 to 44.9 tons/hectare (Table 5). Higher rates produced some increase at two locations but resulted in a decrease on one farm and on the UM plots. This decrease was probably due to killing or stunting of plant growth.

In order to study the carry over of nutrients, UM plots were harvested one year later without any additional manure application. There appeared to be a 10-20 percent improvement in yield indicating limited nutrient availability the second year. When considering the amount of nutrient uptake and removal by harvesting as compared to the amount applied, it is doubtful that there was much loss of nitrogen by leaching up to the 44.9 tons/hectare rate. Future recommendations will encourage producers not to apply rates over 44.9 tons/hectare on deep loess soils.

These data indicate that manure from cage hens should be spread as thinly as possible on fescue pasture in order to obtain maximum benefit. The maximum use level should not exceed 44.9 tons/hectare. Environmental pollution due to surface runoff was minimal but was not monitored other than visually.



MANAGING LIVESTOCK WASTES

# The Efficiency of Using Sludge from Pig Growing Complexes as Organic Fertilizer

Vl. Ionescu-Sisesti, I. Jinga, Gh. Roman, Gh. Pricop

ONE of the major problems generated by the concentration of livestock breeding in large industrial units is that of the disposal and treatment of effluents - two operations considered today as component parts of the production process (Ionescu-Sisesti 1973, Roman 1974).

The volume of waste waters to be disposed of is enormous. For instance, in pig complexes with capacities of 100 000 to 150 000 head, the daily volume to be discharged ranges from 2000 to 4000 m$^3$. Due to their high content of impurities, waste waters have detrimental polluting effects on the environment, such as water courses, soil, and air, and present health hazards for both humans and animals (Ionescu-Sisesti 1973, Jinga 1971, Minciuna 1973, Miner 1971, Roman 1974).

It is, therefore, essential to equip such complexes with treatment installations where impurities are retained, thus allowing large volumes of conventionally clear waters to be either recycled or discharged into neighboring water courses, with practically no pollution risks.

Impurities retained in sludge beds amount to 10.5 to 15 tons daily (3000-5000 tons yearly) (Ionescu-Sisesti, Roman and Jinga 1973). They form a "raw sludge", i.e., a fluid with 90-92 percent moisture content and a high level of fertilizing elements. To be used as organic fertilizer, this sludge requires processing to decrease its moisture content and noxiousness and to allow concentration of valuable substances (Ionescu-Sisesti, Roman and Jinga 1973, Roman 1974).

In practice, these operations include dewatering on drying beds down to 60-75 percent moisture content i.e., a consistency allowing easy handling and storage for further drying and subsequent digestion.

Digested sludge is an organic fertilizer exhibiting the following characteristics (Ionescu-Sisesti, Roman and Jinga 1973): pH, 4.8-6.4; mineral residue, 5.91-10.24 percent; total nitrogen, 2.31-3.05 percent; total phosphorus, 0.34-0.63 percent; and total potassium, 0.09-0.17 percent, all on a dry matter basis.

Experiments carried out in various countries show that "the fertilizing effect of pig sludge is similar to that of manure, perhaps even more marked in the first year of application because of higher availability of nutrients" (O'Callaghan et al. 1971, Ionescu-Sisesti, Roman and Jinga 1973, Miner 1971). The use of sludge may be detrimental, however, because of environmental contamination hazards by pathogenic agents. Therefore, all sludge processing operations should aim towards the destruction of such agents (eggs of parasites, *Salmonella*, etc.). Certain rules concerning rates, time of application, and crop should also be kept in mind when applying sludge (O'Callaghan et al. 1971, Ionescu-Sisesti 1973, Jinga 1971, Loehr 1971, Minciuna

1973, Miner 1971, Roman 1974, Taiganides 1970).

Considering the existing legal regulations concerning the compulsory treatment of pig-breeding complex effluents, and the subsequent storage of high volumes of sludge, which may provide a useful material for agriculture, the Academy for Agricultural and Forestry Sciences has initiated a complex research program. Among the cooperators is the Chair of Phytotechnics of the Institute of Agronomy, Nicolae Balcescu of Bucharest, Romania.

This paper presents the first results obtained in Romania on the use of sludge as organic fertilizer. Experiments were carried out on the Experimental Field of Cazanesti, county of Ialomitza.

## Natural Conditions and Method of Work

Trials were conducted on two neighboring fields: one located on the terrace of the Ialomitza River, a slightly leached chernozem soil, well supplied in humus; the second in the river valley on an alluvial carbonatic soil, with a medium humus supply, but well supplied in phosphorus and potassium.

On the terrace, the influence on yields of several crops was studied, as compared to the rates recommended for the area. Two experiments were undertaken in the valley on corn and mangold, with various rates of sludge and with the addition of chemical fertilizers (nitrogen, phosphorus, and potassium).

Sludge was supplied by the treatment plant of the pig complex of Cazanesti, adjacent to the experimental fields. Applications were made in the fall by ploughing under after all corn crop residues had been removed.

## RESULTS

### Effect of Sludge Fertilization on Yields of Various Field Crops

Table 1 presents the results obtained. It appears that an organic-mineral sludge fertilization of various crops does not result in higher yields as compared with normal chemical or organic-chemical and manure applications (to alfalfa). The only exception was for soybeans, where significant but small yield increases of 200 kg/ha were recorded.

Economic indexes (net returns, cost price, profitability rates) were lower for most crops in the case of organic-chemical plus sludge fertilization, because of high transport and application costs.

These results show that on soil with a good humus supply, the addition of sludge to a well balanced chemical fertilization not only does not bring about a significant yield increase in the first year of application, but even induces a decrease of the crop's economic efficiency. The explanation lies in the "resistance to mineralization" of sludge in the first year, a behavior similar to that of undecomposed ploughed-in plant residues.

The authors are: VL. IONESCU-SISESTI, I. JINGA, GH. ROMAN, GH. PRICOP, Research Irrigation Dept., Land Reclamation Research Institute, ICIF, Bucharest, Romania.

TABLE 1. SLUDGE FERTILIZATION ON YIELDS OF VARIOUS FIELD CROPS
GROWN ON A SLIGHTLY LEACHED CHERNOZEM SOIL — CAZANESTI.

| Crop | Basal dressing | Yields, t/ha | D.L., 5 percent | Difference, t/ha | Net return 1000 lei/ha | Cost price, lei/t | Profitability rate, percent |
|---|---|---|---|---|---|---|---|
| | | | | | Economic efficiency indexes | | |
| Corn (grain) | $N_{50}P_{35}$ | 10.33 | 0.66 | — | 6.6 | 194 | 311.1 |
| | 30 t sludge + $N_{50}P_{70}$ | 10.27 | — | -0.06 | 5.0 | 344 | 132.8 |
| Soybean (grain) | $N_{50}P_{50}$ | 2.22 | 0.08 | — | 4.1 | 1 118 | 165.1 |
| | 30 t sludge + $N_{50}P_{70}$ | 2.51 | — | 0.29 | 3.4 | 1 582 | 86.9 |
| Sunflower | $N_{50}P_{70}$ | 2.55 | 0.75 | — | 3.2 | 764 | 165.8 |
| | 30 t sludge + $N_{50}P_{70}$ | 2.64 | — | 0.09 | 2.0 | 1 261 | 60.9 |
| Sugarbeet | $N_{100}P_{100}K_{40}$ | 49.9 | 4.96 | — | 6.2 | 125 | 100.1 |
| | 30 t sludge + $N_{50}P_{70}$ | 54.7 | — | 4.7 | 6.3 | 135 | 84.8 |
| Potatoes | $N_{150}P_{70}K_{80}$ | 20.0 | 1.43 | — | 12.6 | 371 | 169.4 |
| | 30 t sludge + $N_{50}P_{70}$ | 21.4 | — | 1.4 | 3.2 | 383 | 161.3 |
| Alfalfa (dry matter) | 30 t manure + $N_{50}P_{50}$ | 6.56 | 1.25 | — | 0.4 | 447 | 11.9 |
| | 30 t sludge + $N_{50}P_{70}$ | 6.67 | — | 0.11 | 0.8 | 399 | 25.2 |
| Mangold | $N_{150}P_{100}K_{40}$ | 98.7 | 5.46 | — | 5.1 | 99 | 51.4 |
| | 30 t sludge + $N_{50}P_{70}$ | 101.3 | — | 2.6 | 4.7 | 104 | 44.1 |
| Hemp | $N_{150}P_{70}$ | 7.99 | 0.96 | — | 4.3 | 456 | 119.1 |
| | 30 t sludge + $N_{50}P_{70}$ | 8.36 | — | 0.4 | 3.8 | 545 | 83.4 |

## Various Rates of Sludge, Applied Either Alone or in Association with Chemical Fertilizers on Grain Corn Yields

Table 2 presents the results, an average for two years, obtained on an alluvial soil with a medium humus supply. In general, yields were lower than on the terrace, because of excess moisture in both experimental years.

Yields in plots with various sludge application rates were significantly higher than those of checks. The highest were obtained in the plot with 20 ton/ha sludge. The addition of chemicals did not cause yield increases. On the other hand, yields of plots with sludge plus chemical applications exceeded by 430-1140 kg/ha those of manure fertilized plots.

It thus appeared that the highest returns were obtained with the 20 tons sludge alternative. However, cost prices and profitability rates are more favorable for check plots as production costs are lower, and for the 10 and 20 tons sludge alternatives.

## Various Rates of Sludge Applied Either Alone or in Association with Chemicals to Mangold Crops

Table 3 presents the average results for two years, obtained on the same soil as above.

In general, results are similar to those obtained for corn with the difference that the highest yields were obtained on the 30 tons/ha sludge alternative and that chemical plus sludge organic fertilization yields more only with rates of 30 tons.

Yields obtained on the 30 tons sludge alternative exceeded by 36.6 tons/ha and 14.2 tons/ha those obtained on the check plots and the 20 tons/ha alternative, respectively.

On the other hand, the most favorable economic indexes were obtained when sludge was applied alone.

## CONCLUSIONS

1   Sludge obtained by treatment of waste waters from pig breeding complexes may be successfully used as organic fertilizer.

2   Behavior of ploughed-in sludge is similar to that of crop residues, i.e., the second year after application and on soils with good humus supply, sludge fertilization provides yield increases.

3   As an organic fertilizer, sludge may be applied to any commercial or forage crop.

4   Organic sludge fertilization may be used instead of chemical or organic-chemical with manure fertilizations. Sludge should, therefore, be considered as an important alternative for organic fertilization.

5   In conditions similar to those of the alluvial soils of Cazanesti, addition of chemical to sludge fertilization is unnecessary.

6   On similar soils recommended application rates range from 15-20 tons/ha for corn (grain) to 25-30 tons/ha for mangold.

7   Sludge fertilization may bring about potential contamination hazards. Preventive sanitary and veterinary

TABLE 2 . VARIOUS SLUDGE RATES, APPLIED EITHER SINGLE OR IN ASSOCIATION
WITH CHEMICALS ON CORN YIELDS (ALLUVIAL SOIL — CAZANESTI).

| Alternative | Grain yield, t/ha | Difference as compared to: Check, t/ha | Organic fertilizer manure, t/ha | Net return, 1000 lei/ha | Cost price, lei/t | Profitability rate, percent |
|---|---|---|---|---|---|---|
| | | | | Economic efficiency indexes | | |
| Check | 6.98 | — | -0.30 | 3.8 | 328 | 166.7 |
| $N_{150}P_{50}$ | 8.25 | 1.27 | 0.97 | 3.8 | 414 | 111.8 |
| Manure 20 t | 7.28 | 0.30 | — | 2.8 | 495 | 76.7 |
| Sludge 10 t | 7.71 | 0.73 | 0.43 | 3.8 | 387 | 126.0 |
| Sludge 20 t | 8.42 | 1.44 | 1.14 | 4.0 | 394 | 122.0 |
| Sludge 30 t | 7.78 | 0.80 | 0.50 | 3.3 | 456 | 91.8 |
| Sludge 10 t + $N_{50}P_{35}$ | 7.68 | 0.70 | 0.40 | 3.3 | 447 | 95.5 |
| Sludge 20 t + $N_{50}P_{35}$ | 8.06 | 1.08 | 0.78 | 3.3 | 464 | 88.4 |
| Sludge 30 t + $N_{50}P_{35}$ | 8.37 | 1.39 | 1.09 | 3.3 | 483 | 81.0 |
| DL 5 percent | 0.15 | | | | | |



MANAGING LIVESTOCK WASTES

| Alternatives | Yields — roots, t/ha | Differences as compared to: | | Economic efficiency indexes | | |
| --- | --- | --- | --- | --- | --- | --- |
| | | Check t/ha | Manure organic fertilization t/ha | Net return 1000 lei/ha | Cost price, lei/t | Profitability rates, percent |
| Check | 116.5 | — | -17.4 | 6.7 | 93 | 60.9 |
| $N_{150}P_{50}K_{40}$ | 133.9 | 17.4 | 0 | 7.0 | 98 | 52.4 |
| Manure 20 t | 133.9 | 17.4 | — | 7.2 | 96 | 56.2 |
| Sludge 10 t | 134.2 | 17.7 | 0.3 | 7.4 | 95 | 58.6 |
| Sludge 20 t | 141.8 | 25.3 | 7.9 | 7.8 | 95 | 57.7 |
| Sludge 30 t | 148.1 | 31.6 | 14.2 | 8.1 | 96 | 56.5 |
| Sludge 10 t + $N_{50}P_{35}K_{40}$ | 136.5 | 20.0 | 2.6 | 7.1 | 98 | 52.6 |
| Sludge 20 t + $N_{50}P_{35}K_{40}$ | 140.4 | 23.9 | 6.5 | 7.1 | 99 | 50.9 |
| Sludge 30 t + $N_{50}P_{35}K_{40}$ | 143.4 | 26.9 | 9.5 | 7.1 | 101 | 48.9 |
| DL 5 percent | 7.31 | | | | | |

measures are, therefore, compulsory both in the treatment plants and drying and digestion beds.

8  Pig sludge represents not only an important source as organic fertilizer, but the extension of its use represents rational means for the disposal of highly polluting residues. Further experiments on various soils and crops, as well as research on new decontamination methods are, therefore, essential.

## References

1   Ionescu-Sisesti, Vl. 1973. Tehnologii combinate in vederea epurarii apelor uzate (Combined waste waters treatment technologies). In: Viata economica, No. 9.

2   Ionescu-Sisesti, VI., I. Jinga. Gh. Roman, Gh. Pricop. 1973. Eficienta utilizarii ca ingrasamint organic a namolului provenit de la epurarea apelor uzate din complexele de crestere a porcilor (The efficiency of the utilization as organic fertilizer of the sludge resulting from the pig growing complexes). In: Probleme agricole, No. 9.

3   Ionescu-Sisesti, VI., Gh. Roman, I. Jinga. 1973. Ricerche sull 'utilizzazione in agricoltura dei faughi risultanti dall 'epurazioue dei camplessi di alleva menti dei maiali in Romania. In: Convegno Agrobidlgia, Bologna, October.

4   Jinga, I. 1971. Cercetari privind valorificarea prin irigatie a apelor reziduale provenite de la complexele pentru cresterea si ingrasarea porcilor (Researches on the use for irrigation purposes of pig-breeding complexes effluents). Teza de doctorat, Institutul ag ronomic, Bucuresti.

5   Loehr, R. C. 1971. Alternatives for the treatment and disposal of animal wastes. In: J. of Water Pollut. Control Fed., 4, 43, No. 4.

6   Minciuna, V. 1973. Potentialul epizootologic al apelor reziduale provenite din complexele zootehnice (Epizootological potential of livestock complexes effluents). In: Stiinta solului, No. 4.

7   Miner, J. R. 1971. Farm animal-waste treatment. In: Iowa Agr. Exp. Sta. Spec. Rep. 67.

8   O'Callaghan, J. R. et al. 1971. Land spreading of manure from animal production units. In: J. Agric. Engng. Res. 16(3).

9   Pricop, Gh. and G. Salay. 1968. Irigatii cu ape freatice si uzate (Ground and waste waters for irrigation purposes). Bucuresti.

10   Roman, Gh. 1974. Contributii la studiul epurarii si folosirii apelor uzate provenite de la complexele industriale de cresterea porcilor (Contributions to the study of the treatment and use of industrial pig-breeding complexes effluents). Teza de doctorat, Institutul Agronomic, Bucuresti.

11   Taiganides, P. 1970. Agricultural wastes and the environment. In: AGRICULTURAL ENGINEERING, June.

12   xxxxx 1971. International Symposium on Livestock Wastes. In: AGRICULTURAL ENGINEERING, May.

# The Yield Response of Grass to Aerobically Stabilized Swine Waste

S. M. Mutlak, A. D. McKelvie, K. Robinson

RATES of application to land of raw animal wastes on the basis of their N, P, K, content are reasonably established and on large arable farms can be used without causing undue concern about environmental problems. In the case of an intensive animal unit with a restricted area of land available the dilemma is the disposal or utilization of a given volume of waste per day. Adequate storage in a lagoon is one way of avoiding discharge during the winter but results in the need to apply larger quantities at other times of the year. It was considered that aerobic stabilization of a waste, by removal of most of the readily available biodegradable material and the elimination of odor, might be useful as a means of increasing the hydraulic loading rate to land, preserving most of the fertilizer value and reducing the immediate biological load on the soil microflora. A further advantage would be the predictable composition of the effluent from an oxidation ditch compared with that from a manure storage system.

There are three possibilities for the land application of oxidation ditch effluent, either as the mixed liquor, or the settled supernatant or the separated suspended solids. The purpose of the work described here was to examine the more immediate effects of aerobically stabilized swine waste on the growth and composition of grass at high hydraulic loading rates, to identify the major factors limiting application rate and to establish guidelines for similar long-term work. In view of the use of anaerobic lagoon supernatant for land application it was considered convenient to include an examination of this material in the experiment.

## MATERIALS AND METHODS

The following materials were applied, at the rates shown in Table 1, at two-week intervals to l/m² plots in a randomized block design with two replications.

The authors are: S. M, MUTLAK, Institute for Applied Research on Natural Resources, Abu-Ghraib, Bagdad, Iraq; A. D. McKELVIE and K. ROBINSON, School of Agriculture, Aberdeen, Scotland.

1   Oxidation ditch supernatant liquor (ODL). This was obtained from an oxidation ditch treating swine waste after removal of the suspended solids by rotary screen, cyclone and gravity settling.

2   Suspended solids (S). These comprised suspended solids removed from an oxidation ditch by rotary screen and cyclone mixed in the proportions 3:1.

3   Slurry (SL). This consisted of a mixture of ODL and S to give a suspended solids concentration of 8 percent.

4   Anaerobic lagoon supernatant (LS).

The facilities for stabilization have been described elsewhere (Report 1974, Robinson 1974).

The plots were arranged in a field with a sandy loam soil and an established sward of perennial rye-grass with some white clover. The effects of applying waste were compared with the results obtained from control plots receiving no treatment.

**TABLE 1. RATES OF APPLICATION OF AEROBICALLY AND ANAEROBICALLY STABILIZED SWINE WASTE.**

| Waste | Rate | Rate of application Liquid, $m^3$/ha | Solid t/ha |
|---|---|---|---|
| Oxidation ditch | 1 | 12.45 | |
| supernatant | 2 | 24.90 | |
| (ODL) | 3 | 37.35 | |
| | 4 | 49.80 | |
| Suspended solids | 1 | | 0.99 |
| (S) | 2 | | 1.98 |
| | 3 | | 2.97 |
| | 4 | | 3.96 |
| Slurry | 1 | 12.45 | 0.99 |
| (SL) | 2 | 24.90 | 1.98 |
| | 3 | 37.35 | 2.97 |
| | 4 | 49.80 | 3.96 |
| Anaerobic lagoon | 1 | - | |
| supernatant | 2 | 24.90 | |
| (SL) | 3 | - | |
| | 4 | 49.80 | |

**TABLE 2. THE COMPOSITION OF AEROBICALLY AND ANAEROBICALLY STABILIZED SWINE WASTE.**

| Waste | Month | $NH_4$-N ppm | $NO_2$-N ppm | $NO_3$-N ppm | $K_j$-N ppm | Extractable N, ppm | Dry matter percent | COD ppm |
|---|---|---|---|---|---|---|---|---|
| Oxidation ditch supernatant | July | 70 | 124 | 74 | 175 | - | 0.5 | 1800 |
| | August | 294 | 455 | 40 | 784 | - | 0.5 | 1750 |
| | September | 341 | 374 | 35 | 855 | - | 0.5 | 1600 |
| Suspended solids | July | - | - | - | 2400 | 424 | 20 | - |
| | August | - | - | - | 2600 | 502 | 21.2 | - |
| | September | - | - | - | 2700 | 508 | 21.2 | - |
| Anaerobic lagoon supernatant | July | 2224 | 137 | 78 | 4180 | - | 3.5 | 5500 |
| | August | 1255 | - | 30 | 1820 | - | 0.4 | 4700 |
| | September | 1288 | 21 | 27 | 1773 | - | 0.4 | 4050 |

TABLE 3. SOME MAJOR AND MINOR ELEMENT CONCENTRATIONS OF AEROBICALLY AND ANAEROBICALLY STABILIZED SWINE WASTE.

| | N | K | Ca | P | Mg | Cu | Mn | Zn |
|---|---|---|---|---|---|---|---|---|
| | | | | | Element, ppm | | | |
| Oxidation ditch supernatant | 565 | 1200 | 60 | 53 | 34 | 1 | 1 | 1 |
| Suspended solids | 4730 | 3500 | 18 400 | 12 200 | 2530 | 494 | 341 | 121 |
| Anaerobic lagoon supernatant | 300 | 700 | 115 | 60 | 58 | 2 | 2 | 0.8 |

Grass was harvested from the plots at intervals of one month commencing one month after the first application of waste. The methods for the extraction and analysis of nitrogen such as those of Bremner (1965) and Tinsley (1970) have been described in detail elsewhere (Mutlak 1974).

## RESULTS

The composition of ODL, S, SL and LS during the period of the experiment are shown in Table 2. It can be seen that LS has a much higher ammonia, $K_j$-N and COD content than ODL and that most of the nitrogen in the latter is present as oxidized nitrogen (mainly nitrite). In the case of S most of the nitrogen was present in the organic form and the extractable nitrogen was low compared with the $K_j$-N. A reduction in the concentration of components in LS occurred after the first month due to rapid extraction of supernatant from the lagoon and a much reduced retention time for the waste supply to the lagoon. The concentrations of some of the major and minor elements of the aerobically and anaerobically stabilized wastes are shown in Table 3. The distribution of copper is of particular interest in view of its potential role as a toxic element in the soil and the possibility of its uptake by growing crops.

Compared with the control S gave no increase in yield whereas the other treatments produced yields increasing in the order SL, ODL, and LS (Table 4). The percentage increases in yield averaged over the three cuts for S, SL, ODL, and LS respectively were 0, 22, 31 and 64. There was no clear trend of increased yields as the rate of application was raised nor with successive cuts.

All treatments except S gave increased nitrogen content compared with the control. The pattern of increases is shown in Table 5. There was no clear trend of nitrogen content increasing with rate of application.

An increase in the nitrate content of the grass was observed with all four waste materials and as shown in Table 6 was particularly large in those plots receiving the highest rate of LS. These increases are undoubtedly a reflection of the increases in the nitrate content of the soil in the plots. As shown in Table 7 the levels in those plots receiving LS were, compared to the control, extremely high and at the LS2 rate were accompanied by a marked fall in soil pH.

These effects must restrict the rate at which the super-

TABLE 4. THE EFFECT OF STABILIZED SWINE WASTE ON THE DRY YIELD OF GRASS.

| Cut | Oxidation ditch supernatant ODL | | | | Suspended solids, S | | | | Slurry, SL | | | | Anaerobic lagoon supernatant LS | | Control |
|---|---|---|---|---|---|---|---|---|---|---|---|---|---|---|---|
| | Rate* | | | | Rate | | | | Rate | | | | Rate | | |
| | 1 | 2 | 3 | 4 | 1 | 2 | 3 | 4 | 1 | 2 | 3 | 4 | 2 | 4 | |
| | | | | | | | | | | | | | | | Yield, t/ha |
| 1 | 2.80 | 2.10 | 2.20 | 1.60 | 1.45 | 1.30 | 1.35 | 1.85 | 2.0 | 1.80 | 1.85 | 2.50 | 2.90 | 3.45 | 1.45 |
| 2 | 1.75 | 1.75 | 1.95 | 2.10 | 1.35 | 1.15 | 1.30 | 1.75 | 1.50 | 1.80 | 1.20 | 1.65 | 2.30 | 2.40 | 1.35 |
| 3 | 1.50 | 1.65 | 1.95 | 1.85 | 1.65 | 1.65 | 1.65 | 1.30 | 2.0 | 1.85 | 1.80 | 1.65 | 1.65 | 1.90 | 1.48 |
| Treatment Mean | 1.93 ±0.10 | | | | 1.48 ±0.10 | | | | 1.80 ±0.10 | | | | 2.43 ±0.14 | | 1.48 ±0.20 |

*See Table 1.

TABLE 5. THE EFFECT OF STABILIZED SWINE WASTE ON THE NITROGEN CONTENT OF GRASS.

| Cut | Oxidation ditch supernatant | | | | Suspended solids | | | | Slurry | | | | Anaerobic lagoon supernatant | | Control |
|---|---|---|---|---|---|---|---|---|---|---|---|---|---|---|---|
| | Rate | | | | Rate | | | | Rate | | | | Rate | | |
| | 1 | 2 | 3 | 4 | 1 | 2 | 3 | 4 | 1 | 2 | 3 | 4 | 2 | 4 | |
| | | | | | | | | | | | | | | | Total nitrogen, mg/g |
| 1 | 26.0 | 31.6 | 29.4 | 31.0 | 28.4 | 29.8 | 30.0 | 27.0 | 33.2 | 36.2 | 32.2 | 34.4 | 39.8 | 40.4 | 28.4 |
| 2 | 31.0 | 26.6 | 28.8 | 31.8 | 28.6 | 27.2 | 27.8 | 28.0 | 33.2 | 30.4 | 36.2 | 35.8 | 38.2 | 36.4 | 26.0 |
| 3 | 42.1 | 39.6 | 40.4 | 42.2 | 33.2 | 34.2 | 37.0 | 39.3 | 39.3 | 40.4 | 39.7 | 39.7 | 43.6 | 44.1 | 34.0 |
| Treatment Mean | 33.4 ±0.97 | | | | 30.9 ±0.97 | | | | 35.9 ±0.97 | | | | 40.4 ±1.37 | | 29.5 ±1.93 |

TABLE 6. THE EFFECT OF THE HIGHEST RATE OF APPLICATION OF STABILIZED SWINE WASTE ON THE NITRATE CONTENT OF GRASS.

| | Oxidation ditch supernatant | Suspended solids | Slurry | Anaerobic lagoon supernatant | Control |
|---|---|---|---|---|---|
| $NO_3$-N (ppm dry matter) | 1.8 | 2.0 | 3.0 | 6.0 | 0.5 |

TABLE 7. THE EFFECT OF THE LOWEST AND HIGHEST RATES OF APPLICATION OF STABILIZED SWINE WASTE ON SOIL pH AND NITRATE CONCENTRATION.

| | Oxidation ditch supernatant | | Suspended solids | | Slurry | | Anaerobic lagoon supernatant | | Control |
|---|---|---|---|---|---|---|---|---|---|
| | Rate | | Rate | | Rate | | Rate | | |
| | 1 | 4 | 1 | 4 | 1 | 4 | 1 | 4 | |
| pH | 5.45 | 6.2 | 6.0 | 5.9 | 6.0 | 5.9 | 6.0 | 5.4 | 6.2 |
| $NO_3$-N | 4.8 | 10.0 | 7.4 | 14.7 | 20.4 | 28.6 | 47.0 | 96.0 | 3.8 |

natant liquor from a lagoon can be applied to land since despite the higher yield the increases in nitrate content of the grass may give rise to toxicity in animals and the increased nitrate concentration in the soil to contamination of groundwater supplies.

Copper is widely used as a component of swine rations and the land application of copper-containing wastes is of some concern. It can be seen from Table 8 that copper uptake by grass does occur and that the concentration in the grass increases with rate of application for S and SL. Copper uptake was high with suspended solids and it appears that when it is applied together with oxidation ditch liquor even more copper is made available to the grass. There was no clear pattern of uptake and there were large fluctuations in copper content for different cuts.

## DISCUSSION

In general the rates of aerobically and anaerobically stabilized wastes applied to the field plots produced an increase in yield. The increases occurred mainly with supernatant liquors and could perhaps be attributed to the high concentrations of available nitrogen and potassium in these wastes. The size of the yield increases was, however, dependent on the type of waste and on the time of application and cutting. The suspended solids did not encourage increased growth probably because of the low availability of plant nutrients and because in the period under examination there had been insufficient time for its incorporation in the soil and breakdown by microorganisms. The fact that the combination of the suspended solids with oxidation ditch supernatant produced smaller increases in yield than the latter alone suggests that the suspended solids have a toxic effect. Despite the fact that no increases in yield were obtained with the slurry the nitrogen content of the grass, except for the third cut, was higher than that of the oxidation ditch liquor alone. A disturbing aspect of the increased nitrogen content relates to the nitrate concentration in the grass which was markedly higher than that of the grass from the control plots particularly at the highest rate of application of anaerobically stabilized supernatant. The results indicate that the form of inorganic nitrogen at the time of application is important in the accumulation of nitrogen in the soil and plant but has not much to do with the form of the accumulated nitrogen. For instance, most of the inorganic nitrogen in the aerobically stabilized waste was present as nitrite or nitrate whereas the anaerobic supernatant contained mainly ammonium nitrogen; nevertheless the nitrate content of the soil and herbage receiving the anaerobic supernatant was higher. It can be concluded that if effects such as the accumulation of concentrations of nitrate in grass which are toxic to animals, increased possibility of nitrate leaching and undesirable soil pH changes, are to be avoided it is preferable that nitrification whould take place in an aerobic stabilization process rather than in the soil.

Although increases in the copper concentration of the grass were observed when copper containing wastes were applied it remains to be shown whether the form of copper accumulated in the crop would be toxic to animals. Since the suspended solids contain the highest concentrations of copper it will obviously take some time for this copper to be released into the soil. The rate and extent of this release and its effect on the crop and soil microflora have yet to be established.

General observations of the field plots during the experiment showed that there was an increased activity and number of earthworms, flies were very evident on those plots treated with suspended solids; in some plots the grass was weak-rooted probably because of nutrient accumulation and increased surface soil friability; suspended solids accumulated on the surface of the plots; and there was no apparent smell problem.

(Continued on page 281)

TABLE 8. THE EFFECT OF STABILIZED SWINE WASTE ON THE COPPER CONTENT OF GRASS.

| | Copper concentration, ppm | | | | | | | | | | | | | | |
|---|---|---|---|---|---|---|---|---|---|---|---|---|---|---|---|
| | Oxidation ditch supernatant | | | | Suspended solids | | | | Slurry | | | | Anaerobic lagoon supernatant | | Control |
| Cut | Rate | | | | Rate | | | | Rate | | | | Rate | | |
| | 1 | 2 | 3 | 4 | 1 | 2 | 3 | 4 | 1 | 2 | 3 | 4 | 2 | 4 | |
| 1 | 11.6 | 15.1 | 11.9 | 13.3 | 17.8 | 19.1 | 14.5 | 46.4 | 33.3 | 30.3 | 31.3 | 43.8 | 28.3 | 35.9 | 12.0 |
| 2 | 12.5 | 14.1 | 12.0 | 11.5 | 22.4 | 35.4 | 43.8 | 41.3 | 36.3 | 36.8 | 35.1 | 38.3 | 18.4 | 18.0 | 13.4 |
| 3 | 12.1 | 11.4 | 16.0 | 13.2 | 27.9 | 25.0 | 30.4 | 47.5 | 15.0 | 15.4 | 36.7 | 35.5 | 21.3 | 27.9 | 12.4 |
| Treatment Mean | 12.9 ±2.02 | | | | 28.5 ±2.02 | | | | 32.3 ±2.02 | | | | 25.0 ±2.86 | | 12.6 ±4.04 |

# A Practical Management System for Pollution-Free Land Spreading of Animal Wastes

K. A. Pollock, J. R. O'Callaghan

UNDER the 1974 Control of Pollution Act, British farmers face new stringent restrictions upon the disposal of livestock wastes. The chief restriction to land spreading of farmyard manure and slurry has, under normal practice, been trafficability on soft ground in winter, and possible crop damage through poaching and smothering. Increased fertilizer prices have further stimulated the change of attitude from the cheapest disposal of such wastes, to the maximum economic utilization of their manurial nutrient contents.

Previous work at Newcastle (O'Callaghan et al. 1971 (a), 1973), proposed methods of calculating the maximum application rates for livestock wastes to agricultural land, based on the avoidance of water pollution and crop damage. Such methods require costly winter storage of manure, and planned management of application during the growing season. It was appreciated that to be widely adopted, their validity and practicability would need to be demonstrated.

A two-year field trial was planned in conjunction with the advisory service, ADAS, to investigate these two points, and this paper summarizes the results, which are more fully reported elsewhere, (O'Callaghan and Pollock 1975, Pollock et al. 1974).

## THE FIELD TRIAL

The proposed hypotheses were based upon two principles of limitation to application rates.

1    Slurry should be applied to satisfy, but not exceed, an existing soil moisture deficit, so that all liquid is retained within the soil matrix, and organic matter in the slurry is broken down by soil organisms.

2    The total available nutrients applied in a season should not exceed the amount that can be removed in the crop to which it is applied in that season. If applied to a soil in deficit, such nutrients should be absorbed and retained by the soil until required by the crop.

The field trial was planned to investigate the accuracy of the evapotranspiration estimates in assessing the amounts of liquid that should be applied, and also the management implications of implementing the scheme. Two sites with different soil types were selected on widely spaced commercial farms, one at Darlington, C. Durham and the other at Acklington, Northumberland. On each site, the well-established tile drain system enabled the isolation of a plot of approximately 0.4 ha,

so that total drain flow could be intercepted by a 400 gallon tank under a single main drain. At the same point a 200 gallon tank collected water from an adjacent lateral, which drained the control area, (Figs. 1, 2).

Pig slurry was spread on the main plots at both sites, and the amounts applied, and rainfall received were measured. The accuracy of the evapotranspiration estimates in predicting soil moisture deficit could be assessed by monitoring drain flow at frequent intervals, during visits to the sites.

The quantity of pig slurry that could be applied to grassland grown for conservation in the season was estimated in advance by examination of 15 years of rainfall records for the two sites, and the appropriate county average values for evapotranspiration. Single dosage rates were always below deficit estimates, but a further limitation was imposed to minimize crop damage. Initially, 12.5 mm/dressing was chosen, but over the first season, the danger of severe sward kill

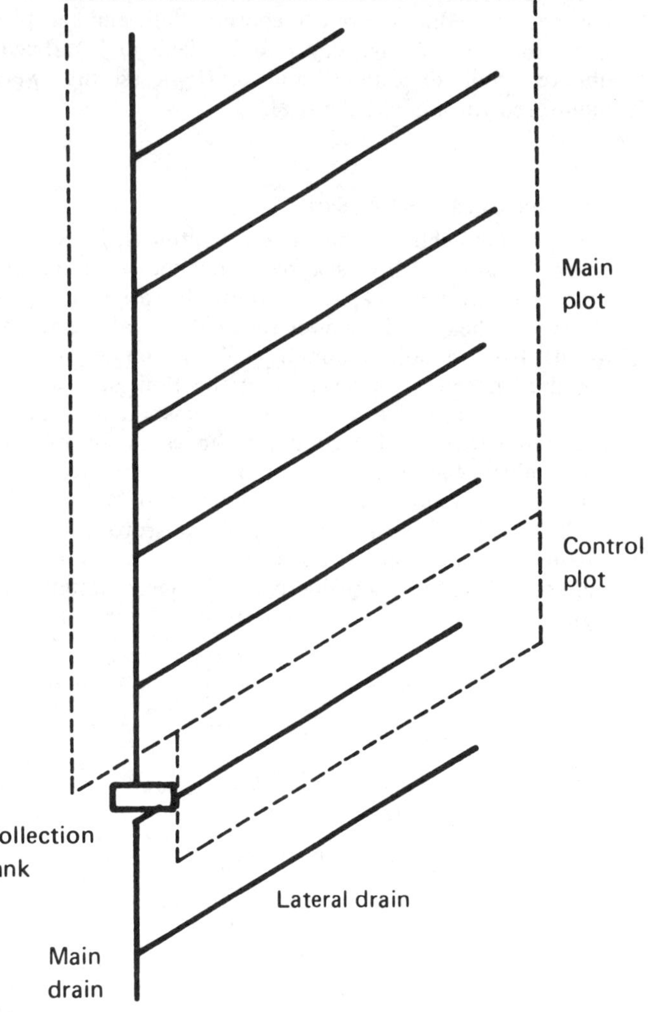



**FIG. 1 Schematic layout of the drainage system and position of the collecting tank [after O'Callaghan and Pollock 1975].**


The authors are: K.A. POLLOCK, Lecturer and J.R. O'CALLAGHAN, Professor, Agricultural Engineering Dept., Newcastle University, England.

**Acknowledgement:** The authors wish to acknowledge the help of the Agricultural Development and Advisory Service, Northern Region, for assistance in planning and analysis of samples; of the Agricultural Research Council for financial support; of T. Robertson for carrying out the field observations; and of C. Pickering, J. Moore, W.E. Manners and R. Manners, the farmers who provided the necessary facilities.

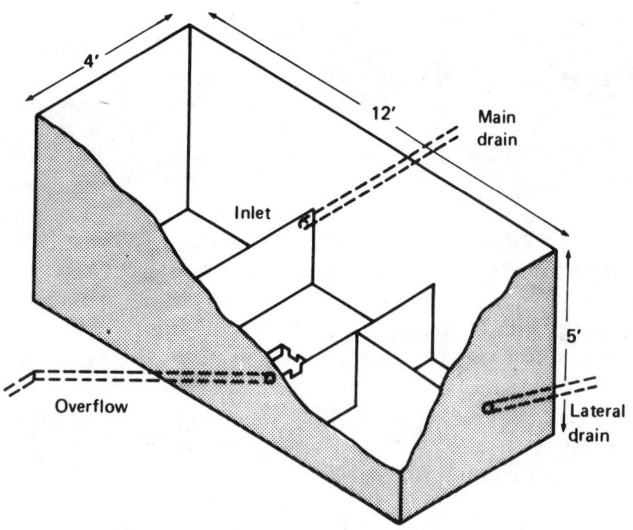



**FIG. 2 Cutaway drawing of the collecting tanks [after O'Callaghan and Pollock 1975].**

forced a further lowering of rates closer to crop requirements.

Slurry applied was analyzed for pollutant load and nutrient content, as was drainage flow collected in the tanks from the treated and control plots, Table 1. Grass yield was estimated from controlled cutting of strips of the crop immediately prior to conservation, and samples were analyzed to give dry matter yield and nutrient uptake. Soil conditions and nutrient status were monitored throughout the trial.

## RESULTS

### Water Movement and Analysis

Experience showed that at Darlington there was no sharp cessation of flow at a particular moisture content equivalent to field capacity. Similarly, application of slurry or a heavy rain shower resulted in leakage to the drains from a soil nominally still in deficit. It was possible, however, on examining the flow of drainage water to identify the pattern of sustained leakage associated with field capacity. Histograms for rainfall and drainflow are shown in Fig. 3, along with values for soil moisture content in the top 150 mm of soil.

The cumulative balance of soil moisture deficit as calculated from rainfall and evapotranspiration estimates is also given, and indicates the large deficit that built up during 1972, when rainfall was 25 percent below

the 30 year average, Table 2. In consequence of this, it was not possible to return the soil profile to field capacity, during the season, but the estimates accurately predicted the time during the following winter when sustained drainflow occurred. During the 1973 spreading season, when rainfall was higher than average, futher confirmation of this accuracy of estimation was obtained on two occasions, and field capacity was again predicted at the time of sustained drainflow in the late autumn. The dry weather at the Acklington site over the whole of the trial, (31 percent below average) resulted in no drainflow from this freely drained soil at the levels of slurry application permissible to ensure good sward growth.

**TABLE 2. ACTUAL AND AVERAGE RAINFALL.**

| | | April–Sept. '72 | Oct. '72–Mar. '73 | April–Sept. '73 | Oct. '73–Mar. '74 |
|---|---|---|---|---|---|
| Darlington trial site Hartburn weather station | actual | 232 | 153 | 376 | 267 |
| | average | 300 | 268 | 300 | 268 |
| Acklington trial site Airport weather station | actual | 248 | 180 | 227 | 243 |
| | average | 325 | 323 | 325 | 323 |

It will be seen from Fig. 3 that high peaks of drainflow occurred after rain or slurry spreading, even when the soil was nominally in deficit according to the estimates and soil moisture content measurements. There was found to be rapid run through of applied liquid to the drains under these conditions, but the volume was in total small, and the flow not sustained. On analysis, however, it was clear that slurry applied was not completely purified by its filtration through the soil, and the BOD of the drainage water rose to about one tenth of the value in the pure slurry. This pollution was short lived, and, in all cases, had declined to background levels in about 36 hours. Typical values for the response to a single spreading are given in Table 3.

A similar pattern of temporarily increased concentration was found with nutrient levels. Ammonium nitrogen increased greatly, although nitrate to a much smaller extent, as would be expected from the form of nitrogen in slurry. Phosphorus and potassium levels also increased, the former probably in association with the increased suspended solids levels. Winter rainfall caused leaching of nitrate nitrogen, as opposed to $NH_4$-N, with sustained

**TABLE 1. SLURRY ANALYSIS FOR DARLINGTON AND ACKLINGTON SITES.**

| | Darlington | | | | Acklington | | | |
|---|---|---|---|---|---|---|---|---|
| | Volume l/ha x $10^{-3}$ | N | P | K | Volume l/ha x $10^{-3}$ | N | P | K |
| **1972** | | | | | | | | |
| 1 | 64.6 | 138 | 39.2 | 38.1 | 124 | 402 | 135 | 242 |
| 2 | 121 | 568 | 194 | 194 | 119 | 919 | 177 | 235 |
| 3 | 121 | 534 | 194 | 145 | 55.0 | 286 | 83.0 | 109 |
| Total | 307 | 1240 | 427 | 377 | 298 | 1610 | 395 | 591 |
| **1973** | | | | | | | | |
| 1 | 60.7 | 260 | 54.5 | 138 | - | - | - | - |
| 2 | 121 | 411 | 228 | 390 | 28.1 | 118 | 22.4 | 112 |
| 3 | 121 | 472 | 157 | 429 | - | - | - | - |
| Total | 303 | 1140 | 439 | 952 | 28.1 | 118 | 22.4 | 112 |

nutrient levels in kg/ha



MANAGING LIVESTOCK WASTES

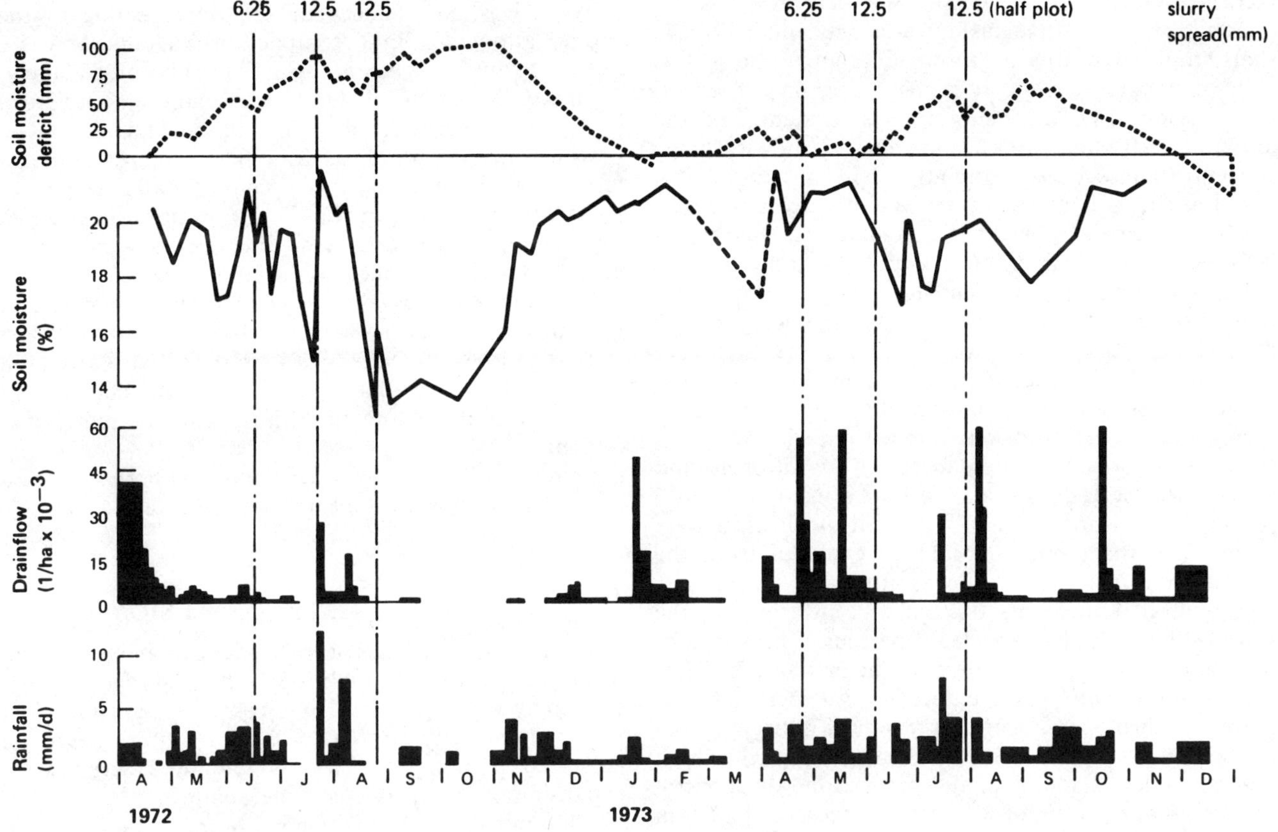


FIG. 3 Water movement chart for Darlington site [after O'Callaghan and Pollock 1975].

readings of around 15 mg/l, Table 3.

The total quantities of pollutants leached were very low under this summer spreading regime, and could be much lower again if nutrients were applied only to meet crop requirements. The total efficiency of purification of the slurry amounted to approximately 98 percent, when the flow and concentration reductions were combined.

Rainfall was commonly read at weekly intervals between spreading, and the calculation of soil moisture deficit proved easy to use under British conditions.

During the two years of the experiment, it was necessary to reduce the level of weekly potential evapotranspiration to take account of the high pre-existent soil moisture deficit on only five occasions, and these were all during the dry first year. It might be possible to ignore this adjustment in some of the cloudier, wetter areas of Great Britain, and retain acceptable accuracy.

### Grass

It was accepted that slurry was to be applied at higher

TABLE 3. DRAINAGE ANALYSIS FOR A SINGLE SLURRY SPREADING ON
DARLINGTON SITE (ALL CONCENTRATIONS IN mg/l).

| Date | BOD$_5$ | COD | NO$_3$-N | NH$_4$-N | Total N | Total P | Total K | Total Ca | S solids |
|---|---|---|---|---|---|---|---|---|---|
| Background average | <1 | 2.0 | 6.0 | <0.2 | 5.9 | <0.03 | <2.0 | 132 | 23.9 |
| 14 Jun 72; 14.00 | 1.0 | <1 | 4.0 | 0.1 | 4.6 | 0.09 | 1.8 | 156 | 44 |
| 14 Jun 72; 14.30 | (6.25 mm slurry spread; estimated soil moisture deficit 50 mm). | | | | | | | | |
| 14 Jun 72; 16.00 | 482 | 1260 | 11.2 | 60.6 | 125 | 16.40 | 70.0 | 135 | 860 |
| 14 Jun 72; 18.00 | 630 | 2560 | 11.2 | 108.0 | 203 | 19.80 | 123.0 | 151 | 1010 |
| 14 Jun 72; 20.00 | 835 | 2300 | 11.2 | 97.0 | 189 | 16.00 | 105.0 | 158 | 1010 |
| 15 Jun 72; overnight | 655 | 1460 | 11.2 | 70.3 | 131 | 9.80 | 70.0 | 161 | 770 |
| 15 Jun 72; 12.00 | 58 | <1 | 0.9 | 5.0 | 14.1 | 1.76 | 7.5 | 169 | 58 |
| 15 Jun 72; control | 34 | <1 | Nil | 0.1 | 12.4 | 0.64 | 3.7 | 151 | 170 |
| 16 Jun 72; 12.00 | 8 | 21 | 3.7 | 2.8 | 9.9 | 0.56 | 6.1 | 156 | 40 |
| 18 Jun 72; 13.30 | 5 | <1 | 4.2 | 0.3 | 5.2 | 0.10 | 1.9 | 156 | 26 |
| Winter average | 4.8 | 17.5 | 14.8 | 0.2 | 15.4 | <1 | 3.1 | 92 | 22 |

levels than desirable for maximum grass growth, but in fact the grass yielded well despite the occurance of some scorch, Table 4. Although slurry was only applied to the short stubble left after a conservation cut, some of this effect may have been due to surface scorching of leaves. On occasions physical smothering was evident, and the levels of nutrients applied were consistent with toxic concentrations occuring within the plants although levels found at subsequent harvesting were not excessive.

Where slurry is applied under commercial conditions, it is clear that very often the dosage levels will be determined by crop management considerations, rather than concern for water pollution. The chemical nutrient demands of the crop impose a far more stringent limitation upon application than the hydraulic loading capacity of the soil under most British conditions.

Analyses of the crop were carried out at each grass cutting, and demonstrated increases of nitrogen and phosphorus levels despite a low overall utilization. Only 9 percent of phosphorus and 25 percent of nitrogen were removed in the crop, as would be expected from the excessive applications. Potassium, however, was recovered at 90 percent, and the concentrations in the leaf dry matter often exceeded 3 percent. These figures have to be viewed in the context of the previous fertilizer history of the plots, and the dry winters experienced, as there was clearly significant carry-over of nutrients from previous years. This was demonstrated by the high potassium levels found in the grass from the control plots, so that the ratio of K:Mg, while exceeding 20:1 on occasion in the treated grass, was simlar to that from the control plots, Table 5.

## Soil

There was no evidence of soil damage due to the slurry, although slight compaction was noted where repeated runs had been made over freshly spread slurry. High $NH_4$-N and $NO_3$-N levels were found on both sites, with the former declining in winter, and the latter moving down the profile as anticipated. Increases in soil levels would be insufficient to account for all nitrogen not leached or recovered in the crop, and atmospheric loss through ammonia and denitrification is expected to account for the balance.

High potassium levels existed before treatment, and were increased, but levels differed significantly between sites. The level associated with sward kill at Darlington was high, but lower than that found in the control and treated plots at Acklington at times when no scorch was evident.

Increases of phosphorus were noted on both sites, and no movement down the profile was detected.

## DISCUSSION AND CONCLUSION

Waste from all classes of livestock is commonly spread on land under British conditions. The trial reported here dealt only with pig slurry, but contains principles that may be applied to any waste. They require a degree of advanced planning not always employed, even when expansion of enterprises is contemplated.

The limited results for water movement indicate that the method of calculating soil moisture deficit from actual rainfall and average evapotranspiration is reasonably accurate, and may be usefully employed to assist management. Historical rainfall values may be used to calculate average hydraulic loading rates for different crops, so that a maximum absorptive capacity of the farm to avoid organic water pollution may be deduced. In any one year, weekly rainfall figures, collected on site, may be used in conjunction with average evapotranspiration to give an estimate of soil moisture deficit. This should be calculated from the observed occurance of field capacity, i.e. when drainflow effectively ceases, or alternately by assuming field capacity on April 1st. The resulting deficit sets a limit to slurry application rate at the appropriate time, be it early in the spring before grass growth, or later after conservation and grazing.

A tighter control is imposed by the nutrient requirements of the crop, and the principle of economic utilization should dictate the level of application below the maximum that can be safely removed by the crop. In the field trial, excessive levels of nitrate were not found, despite the high loading rates, and the highest values were similar to those found following artificial nitrogen application (Hewgill and Le Grice 1975).

Anticipating the pattern of cutting of the crop, the manager may apportion his slurry application to achieve the desired yield without supplying excess nutrients. This will be dependent upon prior soil analysis to detect possible deficiencies, and slurry analysis to determine application rates. The latter may well indicate the need for applications of supplementary K and P fertilizers, although these would not need to take place at the same time. Accurate slurry analysis does present a problem under variable production conditions, and presupposes homogenous material on application. These conditions

### TABLE 4. GRASS YIELD AND ANALYSIS.

|  |  | Darlington | | Acklington | |
|---|---|---|---|---|---|
|  |  | Treated | Control | Treated | Control |
| 1972 | Dry matter, kg/ha | 11625 | 8360 | 11154 | 2780 |
|  | Nitrogen, kg/ha | 297 | 152 | 407 | 62.4 |
|  | Phosphorus, kg/ha | 41.0 | 27.0 | 47.5 | 8.9 |
|  | Potassium, kg/ha | 349 | 216 | 355 | 85.4 |
| 1973 | Dry matter, kg/ha | 8186 | 9290 | 4944 | - |
|  | Nitrogen, kg/ha | 236 | 234 | 148 | - |
|  | Phosphorus, kg/ha | 33.7 | 30.5 | 18.5 | - |
|  | Potassium, kg/ha | 297 | 296 | 145 | - |
|  | Average concentration, percent |  |  |  |  |
|  | N | 2.90 | 2.19 | 3.53 | 2.46 |
|  | P | 0.38 | 0.33 | 0.42 | 0.37 |
|  | K | 3.34 | 2.93 | 3.12 | 3.19 |

### TABLE 5. RATIO OF K:Mg IN GRASS CUT FOR CONSERVATION.

|  |  | Treated | Control |
|---|---|---|---|
| Darlington | 1972 | 19.5 | 19.5 |
|  |  | 16.6 | 17.5 |
|  |  | 14.4 | 12.0 |
|  | 1973 | 25.2 | 23.4 |
|  |  | 13.4 | 14.1 |
|  |  | 22.8 | 15.5 |
| Acklington | 1972 | 18.0 | 19.3 |
|  |  | 16.0 | - |
|  |  | 14.9 | 15.0 |
|  | 1973 | 16.9 | - |
|  |  | 10.9 | - |
|  |  | 16.6 | 15.5 |

probably only prevail in controlled environment pig and poultry houses, where output may be related to published figures for animals on similar diets, (O'Callaghan et al. 1971 (b)).

Homogeneity can only be achieved by stirring and aeration, at the possible cost of nitrogen loss to the atmosphere. It would be feasible but expensive to analyze each load at spreading, and adjust future rates accordingly. It is more realistic to rely upon compensatory growth between applications, at least when dealing with a crop such as grass, which is harvested several times in a single season.

By the adoption of these precautions, adequate land may be retained to accept the expected slurry output, and a pattern of disposal may be planned in advance. This would be adjusted where necessary during the season, according to weather conditions, as these might affect the soil moisture deficit, and the yield of the crop. In consequence, satisfactory utilization of the fertilizer nutrients in the livestock wastes, without causing water pollution or crop damage, may be achieved.

**References**

1 Hewgill, D. and S. LeGrice. 1975. A lysimeter study with pig slurry. In: Agriculture and Water Quality, MAFF Technical Bulletin No. 32 (In preparation).

2 O'Callaghan, J.R., K.A. Pollock and V.A. Dodd. 1971 (a). Land Spreading of manures from animal production units. J. Agric. Engng. Res. 16 (2):280-300.

3 O'Callaghan, J.R., V.A. Dodd, P.A.J. O'Donoghue and K.A. Pollock. 1971 (b). Characterization of waste treatment properties of pig manure. J. Agric. Engng. Res. 16 (4):399-419.

4 O'Callaghan, J.R., V.A. Dodd and K.A. Pollock. 1973. The long term management of animal manures. J. Agric. Engng. Res. 18(1):1-12.

5 O'Callaghan, J. R. and K. A. Pollock. 1975. A land spreading trial with pig slurry. In: Agriculture and Water Quality, MAFF Technical Bulletin No. 32 (In preparation).

6 Pollock, K. A., J. R. O'Callaghan and T. W. Robertson. 1974. Joint Land Spreading Trial Final Report. Department of Agricultural Engineering and Agricultural Development and Advisory Service, Northern Region.

---

## Yield Response of Grass to Swine Waste

*(Continued from page 276)*

### SUMMARY

1   Application of oxidation ditch supernatant increases both the yield and nitrogen content of grass and on the basis of the results it would appear that application rates could exceed 49.8 m$^3$/ha.

2   The suspended solids harvested after aerobic stabilization have no immediate beneficial or harmful effects on crop yield or nitrogen composition. The rate of application could be limited by copper concentration either because of copper accumulation in the soil or because of uptake by the plant.

3   The combined effect of oxidation ditch supernatant and suspended solids is to produce an increase in the nitrogen content of grass but only a small increase in yield. Factors restricting the application of this type of slurry would be its influence on the nitrate concentration in or copper uptake by the grass.

4   The supernatant liquor from an anaerobic lagoon stimulates the production of the largest yields of grass and with the highest nitrogen content. The major factors restricting application rate would be the increase in nitrate content of the grass and also the soil.

**References**

1   Bremner, J. M. 1965. Total nitrogen. In: Methods of soil analysis. Agron. Monograph Pt. 2. p. 1149. Amer. Soc. Agron., Madison, USA.

2   Mutlak, S. M. 1974. Effect of treated animal wastes on soil structure and fertility with respect to environmental pollution. Ph.D. thesis. Univ. of Aberdeen, Scotland.

3   Report. 1974. Treatment of piggery wastes. North of Scotland College of Agric., Aberdeen, Scotland.

4   Robinson, K. 1974. Waste treatment with a protein bonus. Proc. Cornell Agric. Waste Management Conf., 'Processing and management of Agricultural Waste'. Cornell Univ., Ithaca, NY 14890 USA.

5   Tinsley, J. 1970. A manual of experiments, 2nd edition. Dept. of Soil Science Univ. of Aberdeen, Scotland.

# Nutrient Losses From Livestock Waste During Storage, Treatment, and Handling

Dale H. Vanderholm

THIS paper summarizes information on losses of fertilizer nutrients from livestock wastes from the very early research up to the most recent information available for modern livestock production systems. Utilization of manure as a fertilizer for agricultural crops is a very old and well-established practice, dating back perhaps thousands of years. The scientific study of manure as a fertilizer is somewhat more recent. After the fertilizer properties of manure began to be reasonably well understood and the economic aspects of using manure as fertilizer were realized, researchers began to look at the problems of nutrient losses from manure before it was spread on land. As early as 1897, Aeby et al. (1897) wrote about the use of chemicals to prevent nitrogen losses from stored manure. Midgely and Weiser (1937) discuss the use of superphosphate as well as several other chemical and physical methods for preventing nitrogen losses. Since that time, a great deal of research dealing with the loss and preservation of fertilizer nutrients in manure has been conducted and reported.

Currently, recycling of manure as a fertilizer, as a protein source in feedstuffs, or in many other ways has become the main thrust of much research effort. Most people in the waste management area have come to the opinion that treatment of livestock waste to a quality suitable for discharge is economically impractical, and land application has again emerged as the primary ultimate disposal method. Nutrient losses have renewed importance, but livestock characteristics, rations, and housing systems are changed as well as the storage, treatment, and handling of manure, so once again the processes involved and the levels of nutrient losses to be expected are under study.

## NUTRIENT LOSS PROCESSES

This paper will deal only with the major nutrients: nitrogen (N), phosphorus (P), and potassium (K).

Up to 50 percent or more of the nitrogen in fresh manure may be in ammonia form or be converted to ammonia form in a very short time following excretion. Papanos and Brown (1950) note that the resulting free ammonia is very volatile, and unless it is absorbed by or reacts chemically with some substance, much of it volatilizes into the air. In addition to ammonia, Elliott et al. (1971) found organic N compounds entering the atmosphere from beef feedlot surfaces.

The ammonia in manure may oxidize to nitrites and nitrates under aerated conditions. If these nitrites and nitrates then encounter an anaerobic or reducing environment, gaseous nitrogen may be produced which is likely to escape into the air. This process, commonly referred to as denitrification, was demonstrated in stored manure as early as 1917 by Russell and Richards (1917). Soluble forms of nitrogen may be leached by water passing through manure stored in outside sites on uncovered feedlot surfaces.

Losses of phosphorus and potassium are not nearly as well documented as those of nitrogen and not nearly as significant. Phosphorus may be lost by the formation of phosphorus-containing precipitates in anaerobic treatment or storage facilities which then settle to the bottom sludges. The precipitation of magnesium ammonium phosphate in anaerobic swine lagoons as described by Booram et al. (1973) is an example of this. Unpublished data collected by Vanderholm (1975) show losses of potassium in the supernatant of lagoons and holding ponds and high potassium contents in the bottom sludges, indicating a process similar to phosphorus precipitation is occurring with potassium. Information on the process or compounds involved has not been found.

Both phosphorus and potassium are subject to a certain amount of leaching loss due to precipitation on outdoor storage facilities or feedlot surfaces.

## NITROGEN LOSSES

### Aerated Systems

A great deal of research effort has been devoted to the study of aerated manure treatment systems and a side benefit to this is the availability of much information on the nitrogen losses in these systems. Ludington et al. (1972) found N losses from poultry manure of 30 percent during treatment in an oxidation ditch and 50 percent to 60 percent in storage facilities using diffused aeration. The losses were attributed both to ammonia stripping and denitrification. Loehr et al. (1971) observed an N loss of 31 percent during 9 months of operation of an oxidation ditch treating poultry manure. Dunn and Robinson (1972) found N losses up to 81 percent in treating poultry manure by oxidation ditches, while Hegg et al. (1974) found N losses from beef wastes in a batch operated oxidation ditch ranged from 50 percent to 75 percent with temperature as a major variable affecting loss. Edwards and Robinson (1969) reported oxidation ditch N losses from 12.3 percent to 54.4 percent, depending on length of time the ditch operated. Stewart and McIllwain (1971) were unable to account for 66 percent of the nitrogen added during operation of an oxidation ditch with poultry manure for a year. Vasseur (1974) also reported a 66 percent N loss from an oxidation ditch treating swine wastes. Prakasam et al. (1974) describe attempts to maximize or minimize N loss from oxidation ditches by modifying operating procedures.

In aerated storage of dairy manure, Bloodgood and Robson (1969) observed Kjeldahl N losses from 16 percent to 43 percent with temperature as the major variable. Hashimoto (1971) found N losses of 60 percent to 70 percent in a diffused-air aeration system under caged laying hens, primarily attributed to denitrification. Chang et al. (1971) studied N losses during aerobic digestion of dairy

The author is: DALE H. VANDERHOLM, Assistant Professor, Department of Agricultural Engineering, University of Illinois at Urbana-Champaign.

wastes, finding a 90 percent N loss when denitrification was complete. Poelma (1974) reported nitrogen losses of over half when using a floating aerator in under-floor manure storage. However, only a 10 percent N loss was found by Zeisig (1974) when using intermittent aeration of liquid manure to reduce odors.

### Anaerobic Lagoons

Willrich (1966) found losses from an anaerobic swine waste lagoon of 45 percent to 50 percent over a 30-month period. Koelliker and Miner (1971) describe the N transformation within anaerobic lagoons as well as the desorption process by which large amounts of N can be lost. Their data indicate N losses up to 65 percent of the total entering the lagoons annually. Smith et al. (1971) report a 67 percent loss of N in the combination lagoon-oxidation ditch system. In a California study, Meyer (1973) found N losses from dairy waste holding ponds from 20 percent to 50 percent.

### Anaerobic Storage Facilities

Papanos and Brown (1950) indicate N losses from poultry manure ranging from 62 percent for slurries stored in pots at room temperature for 4 weeks to 77 percent from dropping pits over a 20-week period. Eno (1962) states that 50 percent to 75 percent of nitrogen can be lost from poultry dropping pits in 4 to 10 weeks and describes a process known as "fire-fanging" whereby almost all the N is converted to ammonium carbonate and lost by volatilization. In laboratory studies, McAllister (1964) found N losses of 42 percent from cattle waste and 56 percent from swine waste during a 12-week storage of the slurry in open pots. Vanderholm et al. (1974) collected samples from liquid beef manure pits and found N losses to be small. Jones et al. (1973) also found low N losses from liquid dairy manure in pits, around 8 percent. Their observations of liquid swine manure, however, showed losses of N during storage of nearly 60 percent. Hashimoto and Ludington (1971) report a 60 percent N loss from anaerobic poultry manure slurry which they attributed to ammonia desorption.

### Composting

Wells et al. (1969) report that considerable N is lost as ammonia during composting, a find also reported by Martin et al. (1972) who indicate that the carbon-nitrogen ratio of the waste affected the amount of loss. Loehr (1974), however, states that composting conserves much of the nutrient content, including N.

### Feedlot Surfaces

In a laboratory study simulating cattle feedlot surface conditions, Stewart (1970) found N losses from urine ranging from 25 percent to 90 percent largely due to ammonia volatilization. McCalla and Viets (1969) also simulated feedlot surfaces at different stocking densities and found as much as 90 percent N loss from urine added. Adriano et al. (1971) found volatilization losses from urine on simulated feedlot surfaces of up to 99 percent of urine N or nearly 50 percent of total N and pointed out this was consistent with their field measurements which indicated N losses of 40 percent from corral surfaces. In studying solid wastes from feedlot surfaces, Gilbertson et al. (1971) recovered only 42 percent to 55 percent of the estimated excreted N, indicating the rest was lost. The N escaping from feedlot surfaces can eventually find its way into nearby surface waters, according to Hutchinson and Viets (1969).

### During and After Land Application

Lagoon and holding pond effluents are commonly applied to land by sprinkler irrigation systems. Koelliker and Miner (1971) report N losses of 15 percent to 30 percent between the lagoon and the ground surface during sprinkling. In contrast, Welsh and Goodrich (1974) found very little change in ammonia content of liquid waste during sprinkling, although much of their data was collected in cold weather.

Several researchers have reported losses of N from surface-spread manure before incorporation. In early Wisconsin experiments with fermented manure, Heck (1931) measured N losses of 25 percent to 33 percent within periods of 12 hours to 7 days after spreading. Walsh and Hensler (1971) state the average N loss from fermented manure between spreading and incorporation is 20 percent after 6 hours, 25 percent after 24 hours, and 45 percent after 96 hours.

In greenhouse studies, Adriano et al. (1974) found N losses from applied manure of 26 percent to 45 percent depending on soil moisture content and attribute the losses primarily to ammonia volatilization. Working with sewage sludge rather than manure, King (1973) found N losses of 21 percent to 36 percent for surface-applied sludge and 16 percent to 22 percent for incorporated sludge and cited denitrification as the primary loss mechanism. Additional information on losses of N from land-applied manure as a function of soil pH, application rate, and other factors is given by Peters and Reddell (1973), and Mathers and Stewart (1970), and Webber and Lane (1969).

### Outdoor Storage

In addition to the many types of N losses already discussed, Salter and Schollenberger (1938) point out the total soluble nutrients in stored manure are subject to loss by rainfall leaching which might be as much as 50 percent of the total N.

## PHOSPHORUS AND POTASSIUM LOSSES

Salter and Schollenberger (1938) state that potential leaching losses are up to 58 percent of the P and 97 percent of the K in uncovered outdoor storages. Booram et al. (1973) discuss the process of magnesium ammonium phosphate precipitation in anaerobic lagoons, but do not specify any loss levels. Norstedt et al. (1971) found a 33 percent loss of ortho-phosphate in anaerobic lagoons. Of course, these precipitation losses are not true losses as the nutrients can be recovered if the lagoon bottom sludges are removed.

## SUMMARY

Many examples of nutrient losses from system components have been cited, and a summary of the nitrogen losses is given in Table 1. To be most usable, these losses need to be combined to estimate losses for a total system. Examples of attempts to do this are given by Willrich et al. (1974), Vanderholm (1973), and Sutton et al. (1974). While making accurate estimates of system nutrient losses is nearly impossible using information obtained from a wide variety of experiments under a wide variety of conditions, the numbers do seem to show enough consistency

### TABLE 1. NITROGEN LOSSES FROM MANURE DURING STORAGE, TREATMENT, AND HANDLING

| Type of facility | Observed nitrogen loss, percent | Source |
|---|---|---|
| Oxidation ditch | 30 | Ludington et al. (1972) |
| Oxidation ditch | 31 | Loehr et al. (1971) |
| Oxidation ditch | 81 | Dunn and Robinson (1972) |
| Oxidation ditch | 50-75 | Hegg et al. (1974) |
| Oxidation ditch | 12.3-54.5 | Edwards and Robinson (1969) |
| Oxidation ditch | 66 | Stewart and McIlwain (1971) |
| Oxidation ditch | 66 | Vasseur (1974) |
| Aerated storage | 50-60 | Ludington et al. (1972) |
| Aerated storage | 16-43 (Kjeldahl N) | Bloodgood and Robson (1969) |
| Aerated storage | 60-70 | Hashimoto (1971) |
| Aerated storage | 50 | Poelma (1974) |
| Aerated storage | 10 | Zeisig (1974) |
| Aerobic digester | 90 | Chang et al. (1971) |
| Anaerobic lagoon | 65 | Koelliker and Miner (1971) |
| Anaerobic lagoon | 45-50 | Willrich (1966) |
| Anaerobic lagoon | 67 | Smith et al. (1971) |
| Holding pond | 20-50 | Meyer (1973) |
| Anaerobic storage | 42-56 | McAllister (1964) |
| Anaerobic storage | 8-60 | Jones et al. (1973) |
| Anaerobic storage | 60 | Hashimoto and Ludington (1971) |
| Anaerobic storage | 62 | Papanos and Brown (1950) |
| Anaerobic storage | 50-75 | Eno (1962) |
| Feedlot surfaces | 25-90 (urine N) | Stewart (1970) |
| Feedlot surfaces | 90 (urine N) | McCalla and Viets (1969) |
| Feedlot surfaces | 45-58 | Gilbertson et al. (1971) |
| Feedlot surfaces | 40-50 | Adriano et al. (1971) |
| Sprinkler system | 15-30 | Koelliker and Miner (1971) |
| Sprinkler system | negligible | Welsh and Goodrich (1974) |
| Disposal land surface | 20-45 | Walsh and Hensler (1971) |
| Disposal land surface | 25-33 | Heck (1931) |
| Disposal land surface | 26-45 | Adriano et al. (1974) |
| Disposal land surface | 16-36 | King (1973) |

to be useful. Table 2 has been included to illustrate some expected losses on a system basis. Their primary value should be in planning land application requirements for wastes from specific livestock production systems.

### TABLE 2. ESTIMATED NUTRIENT LOSSES DURING STORAGE, TREATMENT, AND HANDLING FOR VARIOUS WASTE-MANAGEMENT SYSTEMS

| System | Nutrient loss, percent | | |
|---|---|---|---|
| | N* | $P_2O_5$ | $K_2O$ |
| Deep pit storage, liquid spreading | 30 to 65 | — | — |
| Anaerobic lagoon, irrigation or liquid spreading | 60 to 80 | 30 to 50 | † |
| Oxidation ditch, anaerobic lagoon, irrigation or liquid spreading | 70 to 90 | † | † |
| Bedded confinement, solid spreading | 30 to 40 | — | — |
| Open lot, solid spreading, runoff collected and irrigated | 50 to 60 | † | † |

*Nitrogen loss values assume that manure is applied to the ground surface and is incorporated within a few hours. If soil injection is used, losses may be reduced; or if incorporation is delayed, losses will increase.

†Losses are likely, but not enough data are available to estimate.

## References


1   Adriano, D. C., P. F. Pratt, and S. E. Bishop. 1971. Fate of inorganic forms of N and salt from land disposed manures from dairies. Proc., Int. Symposium on Livestock Wastes, ASAE, St. Joseph, Mich. 49085. p. 243-246.

2   Adriano, D. C., A. C. Chang, and R. Sharpless. 1974. Nitrogen loss from manure as influenced by moisture and temperature. J. Environ. Quality 3:258-261.

3   Aeby, J. et al. 1897. The fertilizing value and preservation of nitrogen of barnyard manure (Trans. title) Landw. Vers. Stat. 48(2-3):247-360. Abs. in Exp. Sta. Red. 8:873-875.

4   Bloodgood, D. E. and C. M. Robson. 1969. Aerobic storage of dairy cattle manure. Proc. Cornell Univ. Conf. on Agr. Waste Mgt. p. 76-80.

5   Booram, C. V., R. J. Smith and T. E. Hazen. 1973. Some chemical and physical aspects of phosphate precipitation from anaerobic liquors derived from animal waste lagoons. ASAE Paper No. 73-4522, ASAE, St. Joseph, Mich. 49085.

6   Chang, A. C., A. C. Dale and J. M. Bell. 1971. Nitrogen transformation during aerobic digestion and denitrification of dairy cattle wastes. Proc., Int. Symposium on Livestock Wastes, ASAE, St. Joseph, Mich. 49085. p. 272-274.

7   Dunn, G. G. and J. B. Robinson. 1972. Nitrogen losses through denitrification and other changes in continuously aerated poultry manure. Proc., 1972 Cornell Univ. Conf. on Agr. Waste Mgt., p. 545-554.

8   Edwards, J. B. and J. B. Robinson. 1969. Changes in composition of continuously aerated poultry manure with special reference to nitrogen. Proc., 1969 Cornell Univ. Conf. on Agr. Waste Mgt., p. 178-184.

9   Elliott, L. F., G. E. Schuman and F. G. Viets, Jr. 1971. Volatilization of nitrogen containing compounds from beef cattle area. Soil Sci. Soc. Amer., Proc. 35:752-755.

10   Eno, C. F. 1962. Chicken manure, its production, value, preservation and disposition. Univ. of Florida Agr. Exp. Sta. Circ. S-140. 48 p.

11   Gilbertson, C. B., T. M. McCalla, J. R. Ellis and W. R. Woods. 1971. Characteristics of manure accumulations removed from outdoor, unpaved, beef cattle feedlots. Proc., Int. Symposium on Livestock Wastes, ASAE, St. Joseph, Mich. 49085. p. 56-59.

12   Hashimoto, A. G. 1971. An analysis of a diffused air aeration system under caged laying hens. ASAE North Atlantic Region Paper No. NAR 71-428, ASAE, St. Joseph, Mich. 49085.

13   Hashimoto, A. G. and D. C. Ludington. 1971. Ammonia desorption from concentrated chicken manure slurries. Proc., Int. Symposium on Livestock Wastes, ASAE, St. Joseph, Mich. 49085. p. 117-121.

14   Heck, A. F. 1931. Conservation and the availability of nitrogen in farm manure. Soil Sci. 31:335-363.

15   Hegg, R. O., E. R. Allred, W. J. Maier and E. L. Schmidt. 1974. Nitrification of aerated beef wastes at reduced temperatures. Paper 74-II-105 presented at the VIIIth International Congress of Agr. Eng., Elberg, The Netherlands, Sept. 22-28.

16   Hutchinson, G. L. and F. G. Viets, Jr. 1969. Nitrogen enrichment of surface water by absorption of ammonia volatilized from cattle feedlots. Science 166:514-515.

17   Jones, R. E., J. C. Nye and A. C. Nye. 1973. Forms of nitrogen in animal waste. ASAE Paper No. 73-439, ASAE, St. Joseph, Mich. 49085.

18   King, L. D. 1973. Mineralization and gaseous loss of nitrogen in soil-applied liquid sewage sludge. J. Environ. Quality 2(3):356-358.

19   Koelliker, J. K. and J. R. Miner. 1971. Desorption of ammonia from anaerobic lagoons. ASAE Mid-Central Region Paper No. MC 71-804, ASAE, St. Joseph, Mich. 49085.

20   Loehr, R. C. 1974. Agricultural waste management problems, processes, and approaches. Academic Press. New York. 576 p.

21   Loehr, R. C., D. F. Anderson and A. C. Anthonisen. 1971. An oxidation ditch for the handling and treatment of poultry wastes. Proc., Int. Symposium on Livestock Wastes, ASAE, St. Joseph, Mich. 49085. p. 209-212.

22   Ludington, D. C., A. T. Sobel, R. C. Loehr and A. G. Hashimoto. 1972. Pilot plant comparison of liquid and dry waste management systems for poultry manure. Proc. 1972 Cornell Univ. Conf. on Agr. Waste Mgt., p. 569-580.

23   Martin, J. H. Jr., M. Decker, Jr. and K. C. Das. 1972. Windrow composting of swine wastes. Proc. 1972 Cornell Univ. Conf. on Agr. Waste Mgt., p. 159-172.

24   Mathers, A. C. and B. A. Stewart. 1970. Nitrogen transformations and plant growth as affected by applying large amounts of cattle feedlot wastes to soil. Proc. 1970 Cornell Univ. Conf. on Agr. Waste Mgt., p. 207-214.

25   McCalla, T. M. and F. G. Viets, Jr. 1969. Chemical and microbial studies of wastes from beef cattle feedlots. Proc., Pollution Research Symposium, Univ. of Nebraska, Lincoln, Neb., May 23.

26   McAllister, J. S. V. 1964. Investigations into storage and use of slurries II. Changes in the dry matter and N content of slurry during storage. Res. and Exp. Record, Min. of Agr., Northern Ireland 12, part 2:135-142.

27   Meyer, J. L. 1973. Manure waste ponding and field application rates. Unpublished mineo, Univ. of California Agricultural Extension Service, Stanislaus, San Joaquin, and Merced Counties, Calif.

28   Midgley, A. R. and V. L. Weiser. 1937. Effect of superphosphates in conserving nitrogen in cow manure. Univ. of Vermont, Vermont Agr. Exp. Sta. Bull. 419. 23 p.

29   Norstedt, R. A., L. B. Baldwin and C. C. Hortenstine. 1971. Multistage lagoon systems for treatment of dairy farm waste. Proc., Int. Symposium on Livestock Wastes, ASAE, St. Joseph, Mich. 49085. p. 77-80.

30   Papanos, S. and B. A. Brown. 1950. Poultry manure, its nature, care and use. Univ. of Conn., Storrs Agr. Exp. Sta. Bull. 272. 51 p.

31   Peters, R. E. and D. L. Reddell. 1973. Ammonia volatilization and nitrogen transformations in high pH soils used for beef manure disposal. ASAE Paper No. 73-428, ASAE, St. Joseph, Mich. 49085.

32   Poelma, H. R. 1974. Aeration of liquid manure and forced recirculation of aerated manure. Paper 74-II-112 presented at the VIIIth International Congress of Agr. Engr., Elberg, The Nether-




MANAGING LIVESTOCK WASTES

lands, Sept. 22-28.

33    Prakasam, T. B. S., E. G. Srinath, A. C. Anthonisen, J. H. Martin, Jr. and R. C. Loehr. 1974. Approaches for the control of nitrogen with an oxidation ditch. Proc. 1974 Cornell Univ. Conf. on Agr. Waste Mgt., p. 421-435.

34    Russell, E. J. and E. H. Richards. 1917. The changes taking place during the storage of feedlot manure. J. Royal Agr. Soc. 77:1-36.

35    Salter, R. M. and C. J. Schollenberger. 1938. Farm manure. Soils and Men, 1938 Yearbook of Agriculture, USDA, Wash., D.C. p. 445-461.

36    Smith, R. J., T. E. Hazen and J. R. Miner. 1971. Manure management in a 700-head swine finishing building; two approaches using renovated waste water. Proc., Int. Symposium on Livestock Wastes, ASAE, St. Joseph, Mich. 49085. p. 149-153.

37    Stewart, B. A. 1970. Volatilization and nitrification of nitrogen from urine under simulated cattle feedlot conditions. Environ. Sci. and Tech. 4(7):579.

38    Stewart, T. A. and R. McIlwain. 1971. Aerobic storage of poultry manure. Proc., Int. Symposium on Livestock Wastes, ASAE, St. Joseph, Mich. 49085. p. 261-262, 266.

39    Sutton, A. L., J. V. Mannering, D. H. Bache, J. F. Marten, and D. D. Jones. 1974. The fertilizer value of animal wastes, Proc., Purdue Univ., Agr. Waste Conf., Weste Lafayette, Ind. Nov. 7, 1974.

40    Vanderholm, D. H. 1973. Area needed for land disposal of beef and swine wastes. Iowa State Univ. Coop. Ext. Service Pub. Pm-552 (Rev.). 2 p.

41    Vanderholm, D. H. 1975. Unpublished data. Agr. Engr. Dept., Univ. of Illinois at Urbana, Champaign.

42    Vanderholm, D. H., J. C. Lorimer and S. W. Melvin. 1974. Field performance of selected beef feedlot waste handling systems. ASAE Paper No. 74-4015, ASAE, St. Joseph, Mich. 49085.

43    Vasseur, J. 1974. Lowering the cost of deodorizing liquid manure by aerated storage. Paper 74-II-120 presented at the VIIIth International Congress of Agr. Engr., Elberg, The Netherlands, Sept. 22-24.

44    Walsh, L. M. and R. F. Hensler. 1971. Manage manure for its value. Univ. of Wisconsin Coop. Ext. Circ. 550. 7 p.

45    Webber, L. R. and T. H. Lane. 1969. The nitrogen problem in the land disposal of liquid manure. Proc. 1969 Cornell Univ. Conf. on Agr. Waste Mgt., p. 124-130.

46    Wells, D. M., R. C. Albin, W. Grub, and R. Z. Wheaton. 1969. Aerobic decomposition of solid wastes from cattle feedlot. Proc. 1969 Cornell Univ. Conf. on Agr. Waste Mgt., p. 58-62.

47    Welsh, S. K. and P. R. Goodrich. 1974. Effect of sprinkling on liquid animal waste properties. ASAE Paper No. 74-4034, ASAE, St. Joseph, Mich. 49085.

48    Willrich, T. L. 1966. Primary treatment of swine wastes by lagooning. Proc. National Symposium on Mgt. of Farm Animal Wastes, ASAE, St. Joseph, Mich. 49085. p. 70-74.

49    Willrich, T. L., D. O. Turner and V. V. Volk. 1974. Manure application guidelines for the Pacific Northwest. ASAE Paper No. 74-4061, ASAE, St. Joseph, Mich. 49085.

50    Zeisig, H. D. 1974. Surface ventilation of liquid manure for the purpose of limiting emissions of smell. Paper 74-II-119 presented at the VIIIth International Congress of Agr. Engr., Elberg, The Netherlands, Sept. 22-24.

# Dairy Lagoon System and Groundwater Quality

John I. Sewell, James A. Mullins, H. O. Vaigneur

MEMBER      MEMBER      MEMBER
ASAE        ASAE        ASAE

IN 1973 a primary anaerobic lagoon and holding pond system was constructed (Fig.1) to receive and treat the waste from the new West Tennessee Experiment Station Dairy at Jackson. The dairy milks 100 cows and is equipped with an alley flush system utilizing recycled water from the holding pond (reservoir for holding effluent from lagoon). The objective of this study was to evaluate the effects of the lagoon system on nearby groundwater quality.

## SITE DESCRIPTION

The primary lagoon and wells 1-3 (Fig. 1) were constructed near the shoulder of a geologic terrace formation, and the holding pond with wells 4-7 were constructed just off the terrace in an alluvial bottom land. The soil of the terrace formation was investigated and found to be silt loam and sandy loam, 0- to 1-m depth; clean white sand, 1- to 4-m depth; a 0.3- to 0.5-m stratum of black sandy clay at 4- to 5-m depth; and below this, clean white sand. The water table varies from about 3 to 10 m below the surface. The alluvial bottom-land soil in which the holding pond was built is about 12 m lower than the terrace and consists of sandy loam topsoil, 0 to 1 m; sands with some clay, 1 to 2 m; and clean white sand, below 2 m. The water table varies from 1 to 3 m below the soil surface. The average annual rainfall is 125 cm.

## SYSTEM DESCRIPTION AND OPERATION

The primary lagoon equipped with a float-level recorder was filled with water from a nearby stream before being loaded with manure slurry. The lagoon level dropped from 2.8 to 1.5 m in depth within two weeks after it was first filled with water. On May 15, 1974 when the primary lagoon depth was stable at 1.5 m, loading with manure slurry flushed from the dairy was begun. The lagoon then held 1650 m$^3$ water, and an additional 1800 m$^3$ were needed to fill the lagoon. After 30 days, the primary lagoon began overflowing into the holding pond. During this period, three flush tanks were each emptied one or two times per day, and the average total daily tank emptyings was five. Flush water plus rainfall from the unroofed portion of the slab discharged 2340 m$^3$ into the lagoon. The facility is equipped with roof gutters to divert most of the rainfall from the system. Rainfall of 31 cm into the lagoon contributed an additional 1140 m$^3$, making a total inflow of 3480 m$^3$ during the period. Considering surface evaporation of 170 m$^3$ from the lagoon surface (no scum mat had formed), seepage losses were approximately 1510 m$^3$ or 46 percent of inflow less evaporation.

Water for the flushing operations started on May 15 was obtained through October from the station water supply well. Then the holding pond was approximately two-thirds full, and recycling of holding pond water for flushing was started. In November, the recycle pump hose failed and fresh water was again used for flushing until repairs were made in early January. By then, the holding pond had completely filled; and its level was lowered by furrow irrigation on crop land. Recycled water has been continuously used for flushing since January, 1975.

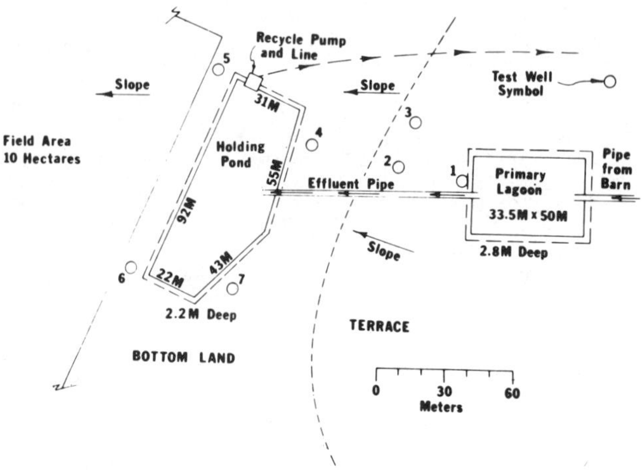

In June, 1973, test wells of 10-cm ID polyvinyl chloride pipe were installed (Fig. 1) to depths of at least 6 m to obtain background levels of water quality parameters before lagoon system operation. The wells were located from 5 to 15 m from the lagoon or pond. Fine white sand rose rapidly inside the well casings, especially in wells 1-3. In September, 1973 the sand was removed from the wells; but it again rapidly rose with the groundwater table to near the levels reported in Table 1.

## WATER SAMPLING PROCEDURE

Water samples were collected and well-level measure-



FIG. 1 Sketch of the West Tennessee Experiment Station dairy lagoon system layout.

TABLE 1. WATER DEPTHS OF TEST WELLS IN METERS BEFORE AND AFTER LOADING SYSTEM.

| Well | Well* depth | Water depth in well before loading† | | Water depth in well after loading‡ | |
|---|---|---|---|---|---|
| | | Median | Range | Median | Range |
| 1 | 3.1 | 0 | 0 | 0 | 0 - 0.5 |
| 2 | 3.0 | 0 | 0 - 0.8 | 0.6 | 0 - 1.1 |
| 3 | 3.8 | 0 | 0 | 0 | 0 - 0.4 |
| 4 | 3.4 | 1.2 | 1.1 - 1.3 | 1.4 | 1.1 - 1.5 |
| 5 | 2.6 | 0.2 | 0.1 - 1.2 | 0.7 | 0.1 - 1.5 |
| 6 | 1.8 | 0.5 | 0.2 - 1.1 | 0.6 | 0.3 - 1.4 |
| 7 | 2.7 | 0.2 | 0 - 0.8 | 1.0 | 0.2 - 1.7 |
| 8§ | 42.7 | ---- | --- | ----- | ---- |

*Depth from gound surface to bottom of well on February 5, 1975.
†System was loaded in May and June, 1974; 5 observations.
‡12 Observations through February 5, 1975.
§Experiment station water supply.

The authors are: JOHN I. SEWELL, Professor and Associate Head, Agricultural Engineering Dept., University of Tennessee, Knoxville; JAMES A. MULLINS, Professor, Agricultural Engineering Dept., and H. O. VAIGNEUR, Associate Professor, Agricultural Engineering Extension, University of Tennessee, Jackson.



MANAGING LIVESTOCK WASTES

TABLE 2. CHEMICAL CONCENTRATIONS IN MG/L OF TEST WELL SAMPLES BEFORE
AND AFTER LOADING SYSTEM IN MAY AND JUNE, 1974

| | Before Loading | | | | | | After Loading | | | | | |
| | $NO_3$-N | | | Chloride | | | $NO_3$-N | | | Chloride | | |
| Well | Median | Range | No.* | Median | Range | No.* | Median | Range | No.* | Median | Range | No.* |
|---|---|---|---|---|---|---|---|---|---|---|---|---|
| 1 | 2 | † | 1 | † | 2-5 | 2 | 10.5 | † | 1 | 12 | † | 1 |
| 2 | † | 1.5-2.4 | 2 | 10 | 7-11 | 3 | 10.8 | 1.4-14.5 | 8 | 17.5 | 5-26 | 9 |
| 3 | † | † | † | † | † | † | † | 2.2-18 | 2 | † | <1-8.5 | 2 |
| 4 | 10 | 4.8-16 | 4 | 11 | 3.0-12.5 | 5 | 11 | 4.0-2.2 | 7 | 13.3 | 8.7-15 | 7 |
| 5 | 9.1 | 6.0-15.5 | 4 | 12 | 5.0-18 | 5 | 13.5 | 2.0-16.5 | 8 | 16.5 | 12.5-20.5 | 9 |
| 6 | 7.0 | 0.8-14.5 | 3 | 10 | 8.5-21 | 4 | 12.5 | 1.0-18 | 9 | 12.0 | 7.4-19.8 | 9 |
| 7 | 6.0 | 1.5-14 | 3 | 10 | 8.0-13 | 4 | 11 | 2.2-17 | 9 | 8.7 | 2.8-10 | 9 |
| 8‡ | | | | | | | 8 | 4.0-16.5 | 9 | 7.5 | 5.0-9.5 | 9 |

*Number of observations
†Insufficient water for testing
‡Experiment station water supply

ments were made 17 times between September, 1973 and February, 1975. Background levels of water-quality parameters were evaluated before lagoon system operation. Some wells contained no water on many dates, and all parameters were not evaluated for every date on which samples were available.

The samples were collected in Jackson soon after 8:00 A.M., immediately placed in ice, and shipped by air to Knoxville where they arrived in time to make chemical analyses and prepare bacterial cultures the same day. Most of the samples were analyzed in the Agricultural Engineering Laboratory; however, three sets were analyzed by a commercial laboratory. All bacterial tests were made by the Millipore Filter technique.

### RESULTS

Test-well water depths were affected by local groundwater conditions. The groundwater normally recedes after May and June, the period when the lagoon system was being initially loaded. However, Table 1 suggests that loading the lagoon system increased the median level of the water table by about 0.5 m near the lagoon system. Holding-pond water has averaged 115 ppm total Kjeldahl N, 8 ppm $NO_3$-N, and 74 ppm chloride.

Wells 2, 4, and 5 (Tables 2 and 3, Figs. 2 and 3) were representative of all wells. Nitrate-nitrogen and chloride concentrations varied widely with time. The variation in $NO_3$-N determinations was greater after system loading than before; however, more laboratory analyses were made after loading. Median $NO_3$-N determinations averaged 6.0 and 11.5 ppm before and after loading, respectively. Median chloride determinations averaged 9.3 and 13.3 ppm before and after loading, respectively.

As loading of the holding pond progressed, $NO_3$-N concentrations (Fig. 2) tended to increase markedly for two months, after which $NO_3$-N began to decrease to levels near those before loading. Also, since two months after loading the holding pond, chloride concentrations (Fig. 3) have

TABLE 3. BACTERIAL CONCENTRATIONS IN COLONIES PER 100 ML
OF TEST WELL SAMPLES BEFORE AND AFTER LOADING

| | Before Loading | | | | | After Loading | | | | | |
| | Total Coliform | | Fecal Coliform | | Fecal Strep. | Total Coliform | Fecal Coliform | | | Fecal Streptococci | | |
| Well | Median* | Range | Median* | Range | Median† | Range‡ | Median | Range | No.§ | Median | Range | No.§ |
|---|---|---|---|---|---|---|---|---|---|---|---|---|
| 1 | 1 200 | 480-8 400 | 10 | <10-20 | # | 1 000-2 500 | 50 | * | 1 | 10 | * | 1 |
| 2 | 800 | 200-2 400 | <10 | 0-<10 | | 2 000-6 000 | 10 | 0-50 | 8 | 15 | 0-25 | 9 |
| 3 | # | # | # | # | # | # | # | 50-110 | 2 | # | 7-2 000 | 2 |
| 4 | 2 000 | 10-330 000 | <10 | 0-20 | # | 1-20 | 2 | 0-600 | 7 | 1 | 0-120 | 8 |
| 5 | 80 | 60-200 | <10 | 0-10 | 0 | 1-10 | 10 | 0-1 500 | 9 | 5 | 1-12 | 9 |
| 6 | 850 | 10-1 400 | <10 | 0-10 | <5 | 1-4 000 | 40 | 0-20 000 | 10 | 50 | 1-600 | 10 |
| 7 | 1 000 | 50-9 900 | 10 | <10-30 | 0 | 1-50 | 5 | 0-1 000 | 10 | 5 | 0-180 | 10 |
| 8‖ | All bacterial tests were negative. | | | | | | | | | | | |

*Three observations
†One observation
‡Two observations
§Number of observations
#Insufficient water for testing
‖Experiment station water supply

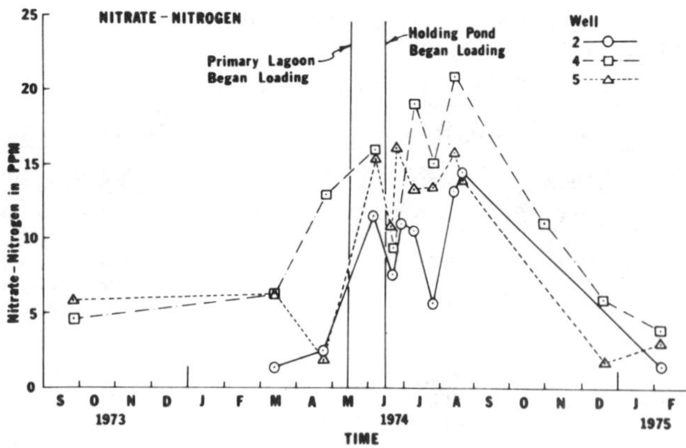


FIG. 2 Variation of nitrate-nitrogen with time in representative test wells.

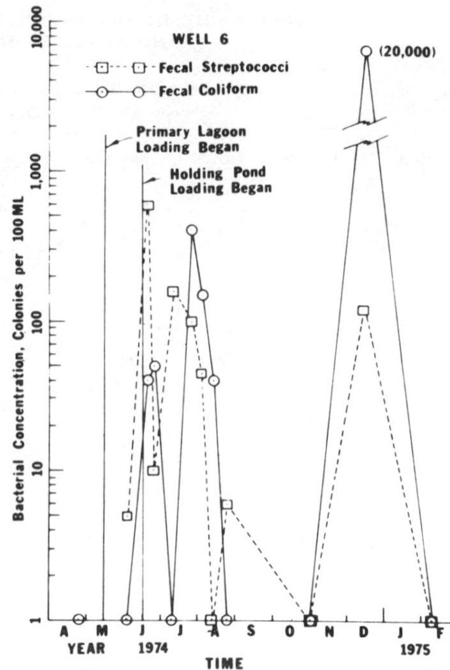


FIG. 4 Variation of fecal coliform and fecal streptococci with time in test well 6.

tended to gradually decrease to levels near those prior to loading. Thus, the previous analysis of median parameter concentrations may erroneously suggest that $NO_3$-N and chloride concentrations have increased markedly because of lagoon system operation.

Soil microbial activity and nitrification increased with increasing temperature; therefore, a higher nitrate level might be expected in the soil water during summer than in winter. Thus, some of the elevation of the nitrate-nitrogen levels may be attributed to increasing temperatures during the summer lagoon loading period. Potassium chloride fertilizer applied in the spring to nearby fields may have been the source of some chloride elevations detected during the summer.

After the initial filling of the primary lagoon, outflow into the holding pond has been continuous and seepage has decreased. This same type of sealing has been observed by other investigators. Decreases of groundwater contamination associated with lagoon sealing were reported by Oliver et al. (1974) who concluded that a California lagoon became sealed in 55 days after initial loading. Meyer et al. (1972) reported that California manure holding ponds were self sealing. Nordstedt et al. (1971) found that a dairy lagoon system in Florida had little adverse effect on water quality beyond 30 meters from the lagoon system.

Determinations of bacterial parameters (Table 3 and Fig. 4) were infrequent prior to system loading; however, differences in concentrations before and after loading appear small. Well 6 (Fig. 4), which is typical of all wells, shows high concentrations of fecal coliform and fecal streptococci in mid-December, 1974. This is the period during which the

holding pond was completely filled and, presumably, the higher hydrostatic pressure forced contaminated holding water into the surrounding groundwater.

Bacterial tests of samples from the station water supply well were negative. Chloride and $NO_3$-N determinations varied erratically, and apparently they were not related to lagoon system operation. This would be expected since the station water supply well is 43 m deep and 150 m away from the primary lagoon.

## CONCLUSIONS

Operation of the lagoon system caused nearby test-well levels to increase by approximately 0.5 m. Initial lagoon seepage markedly decreased within two months after loading. Initial lagoon loading was associated with increased chloride and $NO_3$-N in nearby test wells; however, after two months, these concentrations began decreasing to levels near those of the period before loading.

Fecal coliform and fecal streptococci concentrations were highly variable before and after lagoon loading. During one period when the holding pond was completely filled, high hydrostatic pressure apparently caused seepage of contaminated water into nearby groundwater.

After the initial period of operation, the lagoon system apparently has not caused a marked deterioration in nearby groundwater quality except during one period when the holding pond completely filled. Excessive lagoon seepage which could be caused by high lagoon and holding pond levels can adversely affect groundwater quality. Because of the short duration of this study, no conclusions about the ultimate effects of the lagoon system on groundwater can yet be made.

### References

1  Meyer, J. L., E. Olson and D. Baire. 1972. Manure holding ponds found self-sealing. California Agriculture. May, pp. 14-15.
2  Nordstedt, R. A., L. B. Baldwin and C. C. Hornstein. 1971. Multistage lagoon systems for treatment of dairy farm waste, pp. 77-80. In: Livestock Waste Management, ASAE, St. Joseph, Mich. 49085.
3  Oliver, J. D., W. C. Fairbank, J. L. Meyer and J. M. Rible. 1974. Subfloor monitoring of Shady Grove Dairy liquid manure holding pond. California Agriculture. April, pp. 6-7.

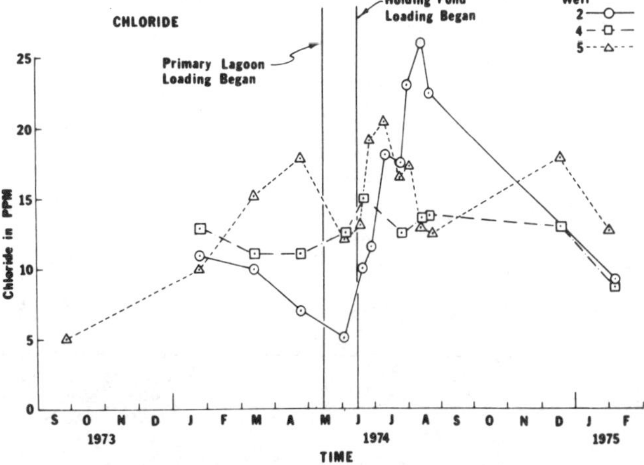


FIG. 3 Variation of chloride with time in representative test wells.



MANAGING LIVESTOCK WASTES

# Seepage Beneath Feedyard Runoff Catchments

R. N. Clark
MEMBER
ASAE

S TATE and Federal regulations for control of surface water pollution require that storm runoff from feedyards be caught and prevented from flowing into natural drainageways. However, seepage from these catchments could pollute the ground water. Several investigators have shown that feedyard runoff effectively reduces seepage from earthen catchments. Meyer et al. (1972) reported seepage of 0.1 cm/day from manure-laden holding ponds in soils ranging from clay loam to sand. They did not discuss the mechanisms of sealing, but did say that artificial seals were not required. Davis et al. (1973) found that when a sandy loam was inundated with liquid dairy cattle wastes for 4 months, infiltration decreased from 122 cm/day to only 0.5 cm/day. They concluded that the primary mechanism of sealing was biological activity in the presence of high organic matter concentrations. Chang et al. (1974) reported that soil hydraulic conductivity decreased quickly in soil columns that had been recovered from the bottom of a holding pond. They stated that initial sealing was caused by physical entrapment of suspended particles in soil, followed by microbial growth that completely sealed the soil from water movement. Travis et al. (1971) found that feedyard runoff effectively sealed undisturbed soil cores, ranging in texture from loam to clay loam. Sealing was attributed to dispersion caused by Na, $NH_4$, and K ions.

Research studies were begun in 1969 at Bushland, Texas, to evaluate the seepage and sealing effects of impounded feedyard runoff. This paper presents results from two types of runoff catchments, naturally occurring playa lakes and man-made holding ponds.

## PLAYA LAKES

Numerous playa lakes dot the Southern High Plains. Most are circular to oval with similar watershed configurations, and occur as single basins. They range from a few hectares to several hundred hectares in size and from 1 to 10 meters deep (Clyma and Lotspeich 1966). These playas are usually dry, but they do collect most of the surface runoff from rainstorms. Most playa bottoms contain very slowly permeable Randall clay. Permeability rates are estimated at about 0.1 cm/hour (Jacquot et al. 1970). Because the playas do not drain into streams or rivers, they have been attractive as locations for feedyards. By placing the feedpens on the playa slopes, pen drainage is adequate and all runoff is impounded.

In 1969, Lehman et al. (1970) collected cores from two playas near Bushland, Texas. One playa collected runoff from an adjacent feedyard and the other collected runoff

from cultivated agricultural land. The feedyard playa had been inundated for 4 months before sampling. Eleven locations across the playa bottom were cored to a depth of 3.5 meters. These cores were analyzed for chloride, nitrate, nitrite, and water contents. Again in 1974, the feedyard playa was cored at approximately the same locations and depth. Figs. 1, 2, and 3 show the comparisons between the cores of 1969 and 1974.

The water contents (Fig. 1) differed little between the years even though the playa contained water most of the

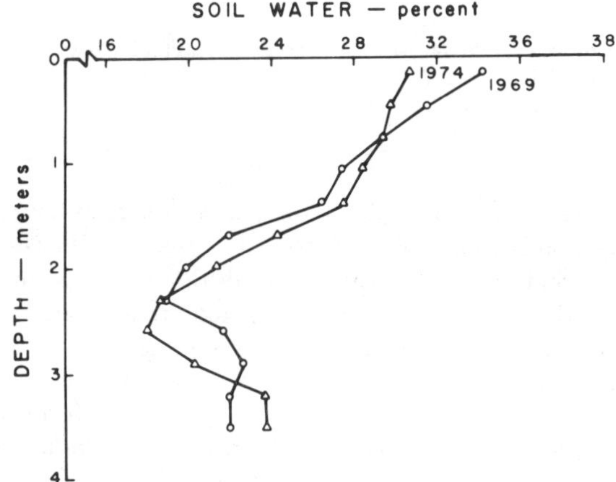


FIG. 1 Soil water content, dry-weight basis (110 C) beneath a feedyard playa, 1969 and 1974.

time from 1969 to 1974. The 2- to 3-m depth was below field capacity, indicating that water had not moved through this zone this rapidly enough to keep the soil saturated.

The nitrate contents (Fig. 2) showed no significant increase during the 5 years. The values remained less than 2 ppm below the 1.5-m depth. The reduction in nitrate in the surface 15 cm indicates considerable denitrification. Nitrite

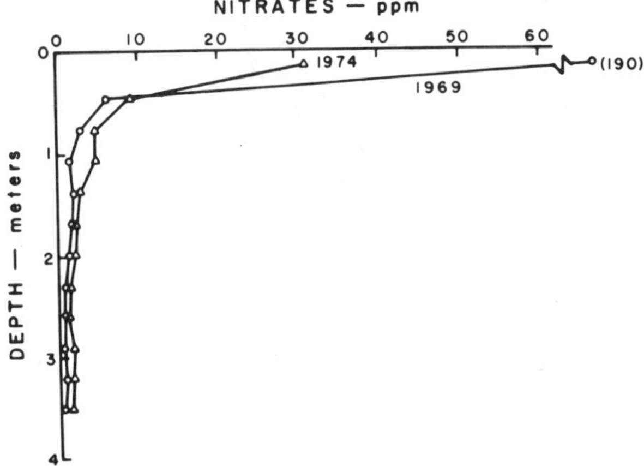


FIG. 2 Nitrates ($NO_3$-N), dry-weight basis (110 C) beneath a feedyard playa, 1969 and 1974.


Contribution from the Soil, Water, and Air Sciences, Southern Region, Agricultural Research Service, USDA, in cooperation with the Texas Agricultural Experiment Station, Texas A&M University.
The author is R. N. Clark, Agricultural Engineer, USDA Southwestern Great Plains Research Center, Bushland, Texas 79012.

values are not shown, but the distribution with depth was similar to that of nitrate and the average value was about 50 ppb.

The only significant change was the chloride concentration, which showed an average increase of about 100 ppm at each depth (Fig. 3). This increase clearly shows that water has moved through the bottom of the playa; however, chlorides are not hazardous at these levels.

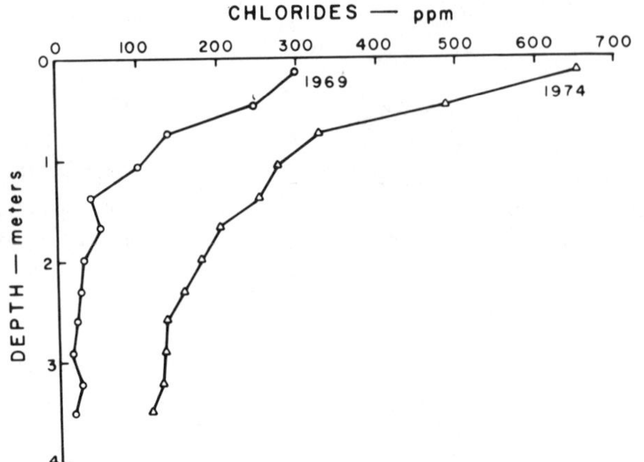



FIG. 3 Chlorides, dry-weight basis (110 C) beneath a feedyard playa, 1969 and 1974.

## HOLDING PONDS

For feedyards not located on playa slopes, excavated holding ponds must be used to catch the runoff. These holding ponds normally expose subsoils that have a permeability of about 4 cm/hr. Three soil sealing treatments were compared in three newly constructed holding basins at the USDA Southwestern Great Plains Research Center. One basin was covered with a 10-cm-thick layer of clay from the bottom of a playa. The second basin had 5 kg/m$^2$ of bentonite rototilled to a depth of 10 cm. The third basin, the check treatment, was not disturbed or packed. Each basin was 4.25 m square and 2 m deep with vertical sides. The sides were lined with plywood to prevent sloughing. Each basin was instrumented with a water level recorder and a neutron access tube for soil water content measurements. Runoff was divided equally into the three basins as the water was discharged from a measuring flume.

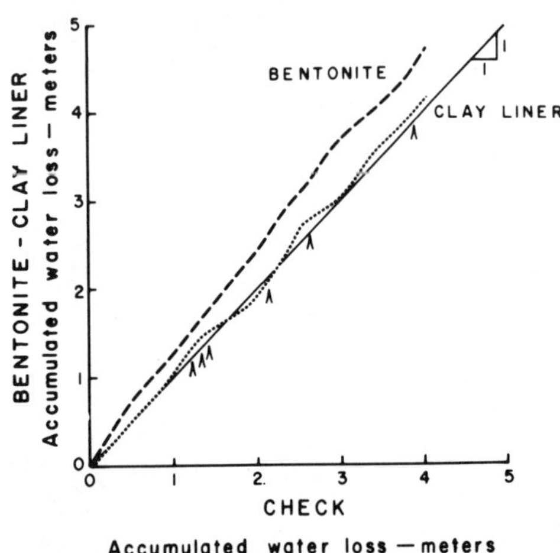



FIG. 4 Comparison of accumulated water loss from the bentonite and clay liner treatments with that from the check treatment, 1973. (Arrows denote increases in head.)

Figs. 4 and 5 are double-mass plots of the accumulated water loss from the two treated basins against that from the check basin in 1973 and 1974, respectively. In 1973, the water loss rate was greater for the bentonite basin than for the check basin until about 2.5 m (check treatment) had been lost, then the loss rate was the same for all treatments. This higher initial water loss rate was probably due to rototilling that disturbed the surface structure of the soil in this treatment. In 1974 (Fig. 5), the water loss rates were similar until the head was increased to 2 m, after which the check treatment lost water faster for 2 to 3 days. Thereafter, the water loss rates were the same.

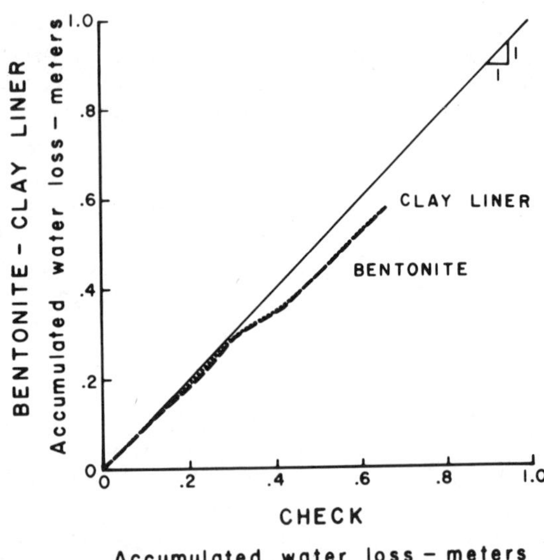



FIG. 5 Comparison of accumulated water loss from the bentonite and clay liner treatments with that from the check treatment, 1974. (Arrow denotes an increase in head.)

In Fig. 6, the accumulated water loss from the clay liner treatment is shown along with accumulated pan evaporation. The evaporation was taken from a Young screen pan and has not been adjusted to represent lake evaporation. These data show that water loss rate increased whenever head increased; however, this change in rate diminished with time. Also, these data show that water is moving into the soil and that about 30 percent of the water lost could be attributed to seepage. Over the 110-day period, the average seepage rate was estimated as 0.015 cm/hr. This seepage rate was based entirely on the differences in total water loss from the basin and Young screen evaporation pan.

*(Continued on page 295)*

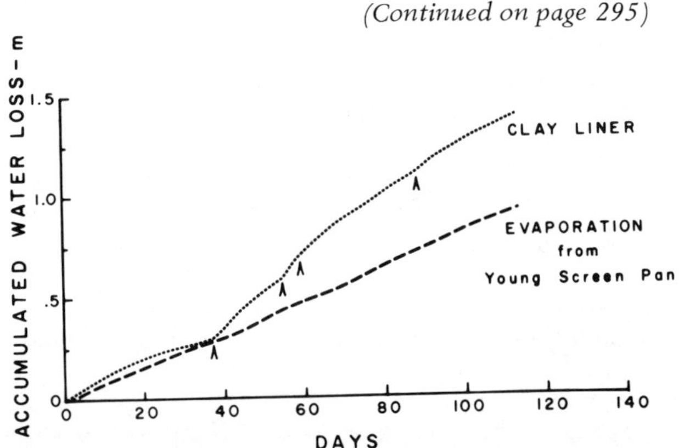



FIG. 6 Evaporation plus seepage from the clay liner treatment and estimated evaporation loss, 1974. (Arrows denote increases in head.)



MANAGING LIVESTOCK WASTES

# Nutrient Losses from Manure under Simulated Winter Conditions

T. S. Steenhuis, G. D. Bubenzer, J. C. Converse

MEMBER          MEMBER
ASAE            ASAE

THERE have been many claims but little factual information concerning the effect of winter spreading of manure. Detailed knowledge of the effects of land application of manure is, however, necessary in order to maximize nutrient usage and minimize its pollution potential while retaining as many management alternatives as possible. Research is presently being conducted on manurial nitrogen transformation and transport processes under late fall, winter and early spring climatic conditions. The objective of this research is the development of quantitative mathematical models which predict nitrogen transformations and movement from land applications of livestock waste. Within this framework processes associated with subsurface nitrogen transformations and movement in a non-cracking soil have been developed (Walter et al. 1974a, b, 1975). The present paper describes the variables for overland flow associated with transport processes of manurial nitrogen by snow melt and rainfall.

## LABORATORY FACILITIES AND PROCEDURES

Biotron facilities (Shove and Bubenzer 1974) were used to produce snow and to control snow melt rates from runoff boxes (30 x 45 cm). One-half of the boxes contained 7.5 cm of a Miami silt loam soil. The remaining boxes contained no soil. All boxes were insulated with 1 cm molded expanded polystyrene on all inside surfaces to prevent rapid heat transfer through the walls and bottom. The soil was frozen before the snow or manure was applied.

Snow used in the study was produced using two fog nozzles. A mixture of demineralized and tap water was used for the first trial. Remaining trials used only demineralized water. Snow of the first run was slightly more dense than subsequent runs.

Urine and fecal wastes were investigated independently to determine the effect of snow melt on their nutrient release pattern. Forty-seven g urine were mixed with one g of anaerobically stored liquid manure and absorbed into 25 g straw before being placed in one-half of the experimental boxes. The remaining boxes received 150 g of feces to which 3.5 g of liquid manure had been added. Application rates were comparable to 6 and 12 metric tons/h, respectively. Both the feces and the straw-urine mixtures were placed at the base, center or top of a 20 cm snow pack in those boxes which did not have the soil layer and at the base or top of the snow pack over the frozen soil.

Immediately after application of the snow pack, the runoff boxes were placed at a 30 percent slope in climatic chambers. Daily sinusoidal temperature cycles of either -8 to 0, -4 to 4, 0 to 8 or 4 to 12 C were maintained until all the snow had melted or for a maximum period of 10 days (Table 1). A daily light and dark period was maintained in the chambers. Light intensity was comparable to a cloudy January day in Wisconsin. In cases where all the snow did not melt within 10 days, the temperature was increased to, and maintained at, 12 C until all snow melted. As soon as the snow melted a portable rainfall simulator (Bubenzer and Meyer 1965) was used to apply 0.75 cm of rainfall to the runoff boxes.

Runoff samples from both the snow melt and the rain were analyzed for total nitrogen, ammonium, nitrate and

Research supported by the Office of Water Resources Research, US Department of Interior (OWRR-B-076-WIS) and by the College of Agricultural and Life Sciences, University of Wisconsin, Madison.

The authors are: T. S. STEENHUIS, Research Assistant, G. D. BUBENZER, Associate Professor, and J. C. CONVERSE, Assistant Professor, Agricultural Engineering Dept., University of Wisconsin, Madison.

TABLE 1. SOME CHARACTERISTICS OF MANURE AND SNOW

| Run | Density snow, g/c$^3$ | Temp. first 10 days, C | Temp. melt period, C | Average melt rate, cm/day | Straw-urine Mixture ‡ Ammonia mg/g | mg/u† | Total nitrogen mg/g | mg/u | Feces § Ammonia mg/g | mg/u | Total nitrogen mg/g | mg/u |
|---|---|---|---|---|---|---|---|---|---|---|---|---|
| 1 | 0.32 | -8 to 0<br>-4 to 4 | 12<br>12 | 0.0/6.0*<br>0.8/6.0 | 0.53 | 38 | 7.14 | 514 | 0.14 | 20.3 | 4.05 | 608 |
| 2 | 0.26 | 0 to 8<br>4 to 12 | —<br>— | 2.5<br>4.0 | 0.92 | 66 | 7.01 | 505 | 0.13 | 20.0 | 4.12 | 617 |
| 3 | 0.28 | -4 to 4<br>0 to 8 | 12<br>— | 0.8/6.0<br>2.5 | 0.59 | 42 | 8.84 | 636 | 0.11 | 16.3 | 3.77 | 565 |
| 4 | 0.18 | -8 to 0<br>4 to 12 | 12<br>— | 0.0/6.0<br>4.0 | 0.96 | 69 | 8.86 | 638 | 0.11 | 16.3 | 4.28 | 641 |

*  Snow melt during first 10 days/snow melt during 12 C period.

†  mg of $NH_3$, TN as added per experimental unit.

‡  25 g straw 47 g urine per runoff box.

§  150 g feces per runoff box

TABLE 2. PARTITIONING OF TOTAL NITROGEN LOSS FOR STRAW-URINE MIXTURE SUBJECTED TO VARIOUS TREATMENTS AND NO HEAT LOSS TO UNDERLAYING SOIL.

| Temperature cycle, C | Placing urine straw mixture | Percent of total N lost in melt of 20 cm pack | Percent of total N lost in 0.75 cm rain | Percent of total N volatilized per total period | Percent of total N in straw at end experiment | Percent of total N volatilized per day |
|---|---|---|---|---|---|---|
| -8 to 0 | Base | 75.5 | 2.8 | 7.7 | 14.0 | 0.6 |
| | Midpoint | 51.9 | 4.0 | 21.9 | 22.2 | 1.8 |
| | Top | 14.2 | 27.2 | 22.2 | 36.4 | 1.9 |
| -4 to 4 | Base | 86.1 | 0.8 | 4.4 | 8.7 | 0.4 |
| | Midpoint | 85.3 | 2.6 | 2.2 | 9.9 | 0.2 |
| | Top | 16.9 | 23.1 | 18.2 | 41.9 | 1.5 |
| 0 to 8 | Base | 66.6 | 3.3 | 18.2 | 11.9 | 3.6 |
| | Midpoint | 70.2 | 6.1 | 7.4 | 16.3 | 1.5 |
| | Top | 8.3 | 27.7 | 26.5 | 37.5 | 5.3 |
| 4 to 12 | Base | 76.9 | 3.9 | 5.5 | 13.7 | 1.8 |
| | Midpoint | 70.2 | 7.6 | 4.3 | 17.9 | 1.4 |
| | Top | 12.0 | 25.4 | 35.1 | 27.5 | 11.8 |

nitrite. Total nitrogen was determined by the semi-micro Kjeldahl procedure with pretreatment with $KMnO_4$ and reduced iron (Bremner 1965). Determination of ammonium, nitrite and nitrate were made using the method developed by Keeney and Bremner (1967).

## RESULTS AND DISCUSSION

The pollution potential of winter spread manure depends to a large part upon the temperature of the soil. When the soil is not frozen the water may infiltrate into the soil where the ammonia is absorbed on the soil particles and is not readily converted to nitrate as long as the soil temperature remains cold (Walter et al. 1974a). The potential for surface runoff from snow melt is greatly increased for frozen soils. Midgley and Dunklee (1945) reported the amount of snow to be more important than the physical characteristics of the land in determining nutrient losses from manure applied to snow covered fields.

A distinction was made between two processes which govern the loss of manurial nitrogen in the runoff water from snow melt or rain. The first process is the loss of water soluble nitrogen. It is most distinct for urinal nitrogen losses where all of the nitrogen in cattle urine is water soluble and less important for fecal nitrogen losses where the ratio of water soluble to non-water soluble nitrogen is 1:5 (Scheffer 1961). The second process is the detachment of particles from the feces which contain non water soluble nitrogen. An outside energy source in the form of freezing and thawing or raindrop impact is required to detach and transport the manure flakes.

### Straw-Urine Mixture

Total nitrogen (TN) was used as an indicator of future inorganic nitrogen for the straw-urine mixture. The mass balance for the total nitrogen was:

$$M_1 = M_2 + L_s + L_r + V \dots\dots\dots\dots\dots\dots\dots [1]$$

where

$M_1$ = the initial nitrogen content of the manure (g)

$M_2$ = the residual nitrogen content of the manure after leaching by snow melt and rain (g)

$L_s$ = the nitrogen lost in snow melt (g)

$L_r$ = the nitrogen lost during rainfall (g)

$V$ = the nitrogen lost through volatilization (g)

The mass balance is first discussed for non soil conditions where there was no refreezing of water at the base of the snow pack and then the effect of refreezing due to the cold soil is described.

The placement height of the straw-urine mixture in the snow significantly affected the relation between cumulative nitrogen loss and cumulative runoff. Straw-urine mixture which was placed at base and midpoint of snow had an average loss of nitrogen of 73 percent during the snow melt process in the non-soil boxes. When placed at the top there was less contact with free water and, consequently, on the average, only 13 percent of the total nitrogen was lost in the snow melt process (Table 2). This effect was significant at the 1 percent level. Cyclic temperatures averaged over placings did not cause significant differences. For placement at the top, the concentration of total nitrogen in the first cm snow melt varied inconsistently among treatments between 10 and 75 ppm. The concentration of nitrogen in snow melt in excess of first cm was between 5 and 10 ppm for all treatments (Table 3). Volatilization losses were not measured directly but were obtained by difference in the

TABLE 3. CONCENTRATION OF TOTAL NITROGEN IN RUNOFF WATER FROM SNOW MELT AND RAIN FOR STRAW-URINE MIXTURE PLACED AT TOP OF SNOW PACK.

| Run | Temp. first 10 days, C | Temp. melting period, C | Melt rate cm/day | Concentration first cm snow melt, µg/g | Concentration snow melt water in excess of 1 cm, µg/g | Concentration runoff water rain, µg/g |
|---|---|---|---|---|---|---|
| 1 | -8 to 0 | 12 | 0.0/5.5* | 16 | 6 | 113 |
| | -4 to 4 | 12 | 0.2/5.5 | 75 | 5 | 88 |
| 2 | 0 to 8 | ---- | 0.9 | 10 | 6 | 150 |
| | 4 to 12 | ---- | 3.4 | 17 | 6 | 128 |
| 3 | -4 to 4 | 12 | 0.4/5.5 | 26 | 6 | 184 |
| | 0 to 8 | ---- | 1.8 | 12 | 5 | 143 |
| 4 | -8 to 0 | 12 | 0.0/5.5 | 51 | 8 | 149 |
| | 4 to 12 | ---- | 3.5 | 20 | 10 | 122 |

* Snow melt during first 10 days/snow melt during 12 C period.



MANAGING LIVESTOCK WASTES

| Run | Density snow, $g/cm^3$ | Melt rate, cm/day | d, $day^{-1}$ | Temperature cycle C |
|---|---|---|---|---|
| 1 | 0.32* | 6.0 | 4.29 | -8 to 0/12 |
|   | 0.32 | 0.8 | 0.88 | -4 to 4/12 |
| 2 | 0.26 | 2.5 | 3.18 | 0 to 8 |
|   | 0.26 | 4.0 | 3.20 | 4 to 12 |
| 3 | 0.28 | 0.8 | 1.90 | -4 to 4/12 |
|   | 0.28 | 2.5 | 2.15 | 0 to 8 |
| 4 | 0.18* | 6.0 | 15.0 | -8 to 0/12 |
|   | 0.18 | 4.0 | 5.71 | 4 to 12 |

*Snow type changed at time of melting due to delay period of 10 days in melting.

mass balance. Volatilization was small when straw-urine mixture was placed at the base or midpoint but was as much as 35 percent per melting period or up to 12 percent per day when it was on top (Table 2). Additional experimentation proved that the total nitrogen retained by the straw after treatment was equal to original nitrogen content of the straw plus 1 to 2 mg urinal nitrogen per g of straw.

The cumulative nitrogen loss (L) expressed as portion or total nitrogen lost had the following exponential relation for the straw-urine placed at the base for non-soil boxes.

$$L = \left[1 - \exp(-Yd/R)\right] *100 \quad \dots \dots \dots \dots \dots \dots [2]$$

where

L = the percent of total nitrogen lost at runoff depth, percent

Y = the cumulative runoff, cm

R = the melting rate, cm/day

d = a constant dependent upon snow type, $day^{-1}$

The fitted values for d are shown in Table 4. Runs 1 and 4 showed a wide discrepancy in the value of d between the two treatments. This discrepancy was attributed to the changes in snow texture during the 10 day delay period before melting for the -8 to 0 C temperature cycle. This change from fresh snow to coarser grained snow has also been observed by many Japanese workers (Shimuzu 1970). The d values for runs 2 and 3 were very similar for the two different temperature cycles for the same snow type. Predicted curve and observed values for the second run are shown in Fig. 1 as an example.

Nitrogen loss from the straw-urine mixture at midpoint of the 20 cm snow pack was initially non-exponential. After approximately one cm of runoff the total nitrogen loss in the snow melt was exponentially related to cumulative runoff with a value of d as computed for the nitrogen runoff when straw-urine was placed at base. Equation [2] had to be adjusted for the initial non-exponential nitrogen loss:

$$L = \left[1 - \exp\left[\frac{-(Y - Y^*)d}{R}\right] + \frac{M_3^*}{M_1}\right] *100$$

$$\dots \dots \dots \dots \dots \dots \dots \dots \dots \dots \dots \dots \dots \dots [3]$$

where

L. Y, $R^d$, = as in Equation [2]

$M_3^*$ = total loss at $t^*$, mg

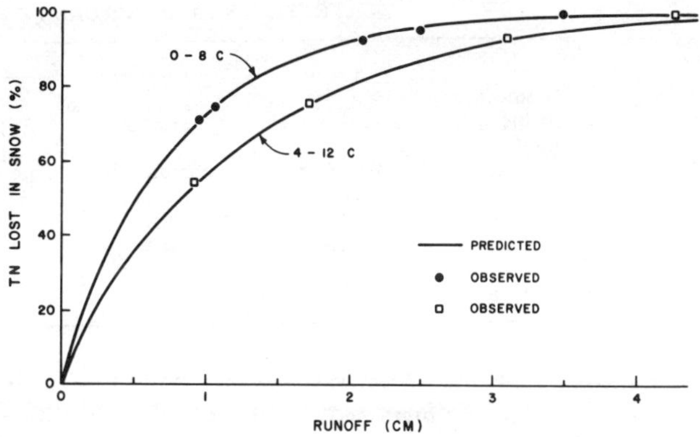

FIG. 1 Total nitrogen (TN) lost in snow melt water for straw-urine mixture placed at base.

$Y^*$ = cumulative loss at $t^*$, cm

$t^*$ = a time at initial point of the exponential curve.

The predicted curve and measured values for the second run for the straw-urine mixture placed in the middle are shown in Fig. 2.

For freshly fallen snow decreasing values of d were related to increasing densities. Fresh fallen snow also had a lower d value than the mature snow pack (Table 4).

Nitrogen concentration in runoff from simulated rain was significantly higher (1 percent level) for straw-urine mixture which had been on top of the snow pack compared with placement at base and midpoint, where most of the nitrogen had been lost in the snow melt (Table 2).

Ammonia percentage in the runoff water of the straw-urine mixture varied between 10 and 20 percent of the total nitrogen concentration. The initial ammonia concentration was dependent upon the time between collection and mixing of the manure and the application on the snow pack. The percentage increased during the course of the melting period due to the continued formation of ammonia from the urea. Similar observations have been reported by Chin and Kroontje (1963).

## Feces

The feces in the non-soil boxes exhibited a linear relationship between the cumulative nitrogen loss and the cumulative snow melt for all placements as long as the daily melt rate was constant. Cumulative nitrogen losses from the feces samples were much lower than those from the straw-urine mixture during the snow melt period. When the snow melt rate exceeded 2.5 cm/day, less than 3 percent of the

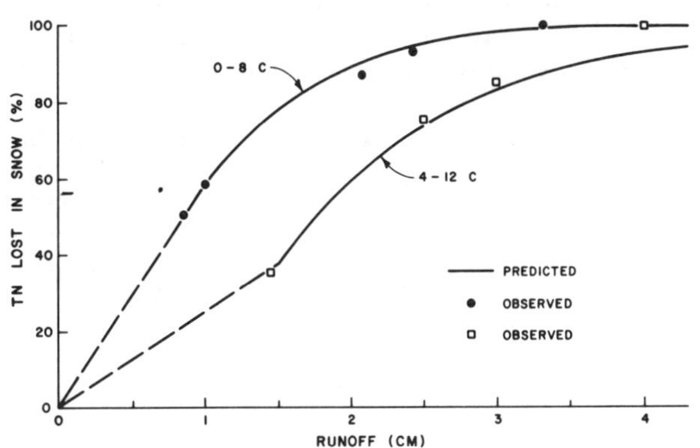

FIG. 2 Total nitrogen (TN) lost in snow melt for straw-urine mixture placed at midpoint of pack.

**TABLE 5. AVERAGE TOTAL NITROGEN LOSS PER CM SNOW MELT AND PER CM SIMULATED RAINFALL IN NON-SOIL CONTAINING RUNOFF UNITS. TOTAL NITROGEN LOSS IS EXPRESSED AS PERCENT OF THE TOTAL NITROGEN ADDED.**

| Temperature during snow melt period C | Base | | Midpoint | | Top | |
|---|---|---|---|---|---|---|
| | Snow melt | Rainfall | Snow melt | Rainfall | Snow melt | Rainfall |
| | percent/cm | percent/cm | percent/cm | percent/cm | percent/cm | percent/cm |
| -8 to 0 | 1.8 | 3.1 | 2.9 | 5.8 | 2.0 | 5.7 |
| -4 to 4 | 1.6 | 4.1 | 4.8 | 10.5 | 6.4 | 7.1 |
| 0 to 8 | 0.9 | 2.8 | 1.7 | 3.0 | 1.3 | 3.3 |
| 4 to 12 | 1.6 | 7.9 | 1.5 | 3.0 | 1.1 | 3.7 |

total nitrogen of the feces was lost per cm snow melt. Freezing and thawing increased significantly the percentage total nitrogen lost per cm snow melt. The percentage total nitrogen loss per cm snow melt averaged 5.6 percent for the -4 to 4 C cycle when the feces was placed at midpoint and top of snow (Table 5). Temperature variations in the upper half of the snow pack were greater than at base, causing a flaking of the feces, due to freezing and thawing action.

The total nitrogen concentration in runoff from the simulated rain was up to 494 percent higher than in snow melt. Placing and temperature did not significantly influence the total nitrogen loss in the rain water (Table 5).

Ammonia concentrations from plots spread with feces ranged from 0.5 to 2 ppm in the melt water, except for plots with very slow melt rates where an average concentration of 4 ppm was observed. The concentrations may be higher than those expected from pure feces due to addition of 3.5 g anaerobic stored manure per 150 g feces.

## Effects of Soil Layer

The soil acted primarily as a heat sink. Melt water from the top which refroze at the base caused an inclusion of the solutes in the ice layer. The solutes were released only after the ice layer began melting.

Nitrates in the overland flow were negligible. However, nitrate content in the water that percolated through the soil was high. In some cases nitrate levels in excess of 100 ppm were observed. These nitrates were leached from the soil layer.

Total nitrogen losses for the runoff boxes containing soil were comparable to experimental units without the soil layer. Fecal nitrogen losses in snow melt and rainfall runoff were much lower than urinal nitrogen losses. Straw-urine mixture placed at base of snow pack resulted in higher total nitrogen losses in snow melt than for straw-urine mixture at top of the pack. No exponential relation was observed between cumulative urinal nitrogen losses and cumulative runoff due to the trapping of solutes in the ice layer. Urinal nitrogen losses in rain were inversely related to the nitrogen loss during the snow melt period (Table 6).

Total nitrogen losses in 0.75 cm rain for runoff boxes with both feces and soil were slightly higher than those with only soil (Table 6).

Additional experimentation showed that when the straw-urine mixture was not exposed to snow melt, nitrogen losses in rain were as high as when the mixture was placed at the top of the pack. Nitrogen losses of feces in rain were slightly lower when not exposed to snow melt due to formation of a feces crust.

## SUMMARY

Pollution potential of land application of manure was studied under simulated winter conditions. Urinal and fecal waste separately were placed in runoff units at base, midpoint or top of a 20 cm snow pack. The units were exposed to a daily light and dark cycle and a sinusoidal fluctuating temperature ranging from -8 to 12 C.

Urinal losses were determined primarily by the quantity of water which passed through. Consequently, for placement at base and midpoint on the average 73 percent of the total nitrogen added was lost in the snow melt. Only 13 percent of the total nitrogen was lost added for placement at top. Urinal nitrogen loss in rain was inversely related to the loss in the snow melt period.

Fecal nitrogen losses were much lower than urinal nitrogen losses. Raindrop impact and freezing and thawing action were important factors which influenced the fecal nitrogen losses.

### References


1 Bremner, J. M. 1965. Total nitrogen, p. 1174-1175. In: C.A. Black et al. (ed) Methods of soil analysis, Part 2, Chemical and microbiological properties. Agronomy Series. Amer. Soc. of Agron. Inc., Madison, Wisc.

2 Bubenzer, G. D. and L. D. Meyer. 1965. Simulation of rainfall and soils for laboratory research. TRANSACTIONS of the ASAE 8:73-75.

3 Chin, W. and W. Kroontje. 1963. Urea hydrolysis and subsequent loss of ammonia. Soil Sci. Soc. Amer. Proc. 27:316-318.

4 Colbeck, S. C. 1974. Water flow through snow overlying an impermeable boundary. Water Resources Res. 10:119-123.

5 Keeney, D. R. and J. M. Bremner. 1967. Determination and isotope-ratio analysis of different forms of nitrogen in soils: Mineralizable nitrogen. Soil Sci. Soc. Amer. Proc. 31:34-39.

6 Midgley, A. R. and D. E. Dunklee. 1945. Fertility runoff losses from manure spread during the winter. Vermont Agr. Exp. St. Bul. 523:29p.

7 Scheffer, F. 1961. Stickstoff und Pflanze: 121-170. In: Der Stickstoff, seine Bedeutung fuer die Landwirtschaft und die Ernaehrung der Welt. Fachverband StickstoffIndustie e.V. Duesseldorf.


**TABLE 6. LOSS OF TOTAL NITROGEN FOR EXPERIMENTAL UNITS WITH AN UNDERLAYING SOIL LAYER.**

| Temperature C | Straw-urine mixture | | | | Feces | | | | No manure | |
|---|---|---|---|---|---|---|---|---|---|---|
| | Base | | Top | | Base | | Top | | Base | Top |
| | Snow melt, mg/cm | Rainfall, mg/cm | Snow melt, mg/cm | Rainfall, mg/cm | Snow melt, mg/cm | Rainfall mg/cm | Snow melt mg/cm | Rainfall, mg/cm | Snow melt, mg/cm | Rainfall, mg/cm |
| -8 to 0 | 26 | 52 | 13 | 164 | 6 | 81 | 14 | 79 | 1 | 49 |
| -4 to 4 | 27 | 37 | 34 | 231 | 6 | 76 | 12 | 111 | 1 | 77 |
| 0 to 8 | 27 | 56 | 8 | 200 | 2 | 80 | 8 | 57 | 1 | 49 |
| 4 to 12 | 72 | 79 | 5 | 201 | 5 | 103 | 6 | 105 | 1 | 49 |

8   Shimuzu, H. 1970. Air permeability of deposited snow. Contributions from the Institute of Low Temperature Science. Series A, 22:1-32.

9   Shove, G. C. and G. D. Bubenzer. 1974. The Biotron. AGRICULTURAL ENGINEERING 55(10):75-76.

10   Walter, M. F., et al. 1974a. Movement and transformation of manurial nitrogen through soils at low temperatures. Sixth National Agricultural Waste Management Conference. Rochester, NY, March 25-27, 1974.

11   Walter, M. F., et al. 1974b. Evaluation of one soil nitrate transport model. ASAE Paper No. 74-2557. ASAE, St. Joseph, Mich. 49085.

12   Walter, M. F., et al. 1975. Movement of manurial nitrogen in cool, humid climates. TRANSACTIONS of the ASAE 18:100-105.

---

## Seepage Beneath Feedyard

*(Continued from page 290)*

### DISCUSSION

Feedyard catchments do lose water by seepage. Chemical analyses of cores from naturally occurring playa lakes indicated water had moved past the 3.5-m depth. However, this was indicated only by an increase in the chloride content. The chloride content increased from 25 to 125 ppm at the 3-m depth when the surface was inundated with water containing over 1000 ppm chloride for 5 years. Even though chloride increased, little nitrate or nitrite moved below 1 m. Also, the water contents between 2 and 3 m were below field capacity, indicating that water moved slowly. Therefore, little hazard to groundwater is likely from feedyard runoff caught in playa lakes.

Water moved through the holding ponds independently of the soil lining treatments. Water loss rates were the same in the three treatments for 2 years, except for short periods after head increased.

The seepage rate was difficult to measure because it was too slow. For field studies, evaporation has not been measured but has been estimated from evaporation pans that contain clear water instead of colored, manure-laden water. The results of this study and others described earlier clearly show that feedyard runoff greatly reduces the seepage rate, but does not completely seal the bottom of holding ponds. Small amounts of water do seep through these ponds into the groundwater; however, these waters are low in nitrate and nitrite. The only increase seems to be in the chloride content.

### References

1   Chang, A. C., W. R. Olmstead, J. B. Johanson, and G. Yamashita. 1974. The sealing mechanism of wastewater ponds. J. Water Pollution Control Fed. 46(7):1715-1721.

2   Clyma, W. and F. B. Lotspeich. 1966. Water resources in the High Plains of Texas and New Mexico. USDA, ARS 41-114.

3   Davis, S., W. Fairbank, and H. Weisheit. 1973. Dairy waste ponds effectively self-sealing. TRANSACTIONS of the ASAE 16(1):69-71.

4   Jacquot, L., L. C. Geiger, B. R. Chance, and W. Tripp. 1970. Soil Survey Randall County, Texas. USDA-SCS.

5   Lehman, O. R., B. A. Stewart, and A. C. Mathers. 1970. Seepage of feedyard runoff water impounded in playas. Tex. Agr. Exp. Sta. MP944.

6   Meyer, J. L., E. Olsen, and D. Baier. 1972. Manure holding ponds found self-sealing. Calif. Agr. 26(5):14-15.

7   Travis, D. O., W. L. Powers, L. S. Murphy, and R. K. Lipper. 1971. Effect of feedlot lagoon water on some physical and chemical properties of soils. Soil Sci. Soc. Amer. Proc. 35:122-126.

# Animal Waste Contribution to Nitrate Nitrogen in Soil


L. F. Marriott, H. D. Bartlett

MEMBER
ASAE


THE use of animal manure for crop production is a recommended practice which is becoming more important in these times when fertilizers are scarce and costly. At the same time, concern about pollution dictates that judgment must be exercised in the use of this waste material. Since the nitrogen content is of prime importance both in crop requirements and as a potential source of pollution, this investigation was concerned with management of the manure and with effects on soil and crops. Subsurface injection of the manure was used because it was effective in odor control in the field, reduced chance of surface runoff, and conserved nitrogen by reducing loss by volatilization following field application. The main objective was to determine the maximum amount of manure nitrogen that can be applied to an existing sod and still have minimum nitrate pollution hazard to groundwater.

The initial concept of the investigation was the use of land application as a means of manure disposal for operators with a concentrated livestock operation and limited land. As a result, the rates chosen were much higher than crop requirements would justify. Briefly, dairy manure slurry was injected beneath the surface of an orchardgrass sod at annual rates to supply from 784 to 3920 kg N/ha for three consecutive years. There was considerable mechanical damage to the sod by the injection equipment, with greater damage as the rate increased. Nitrate nitrogen was determined in soil water sampled by means of suction lysimeters using a vacuum of 40 to 50 centibars. Soil and crop samples were taken for analysis. More detailed procedures and initial results are reported in Bartlett and Marriott (1971).

The second phase of the study was designed to determine whether the application of manure nitrogen at twice the estimated crop requirement would contribute appreciably to nitrate nitrogen in soil and soil water below normal rooting depth. Dairy manure slurry was applied with commercially manufactured injection equipment for liquid manure tank spreaders. The equipment was modified to locate the injectors behind the wheels with lateral adjustment as required. The manure was injected under bluegrass at 300, 400 and 500 kg N/ha in April 1973. In a second experiment, manure was injected under orchardgrass at 340, 450 and 560 kg N/ha in November 1973 and in April 1974. A control treatment of urea at 112 kg N/ha was also applied in April. Nitrate nitrogen monitoring was similar to that described previously. The soil in these studies was a Hagerstown silt loam, a well-drained soil with silty clay to clay B horizon.

## RESULTS AND DISCUSSION

For purposes of this discussion, potential nitrate nitro-

gen pollution is assumed to exist when the concentration in soil water at 90 to 120 cm exceeds 10 mg/l, the potable water standard. This is not a strictly valid assumption without other supporting evidence, but does represent a goal in this study.

## Manure Disposal on Orchardgrass

Manure rates of 1568 kg/ha/yr or higher produced excessive levels of nitrate nitrogen in soil water at all depths of sampling in 1971, the last year of application (Fig. 1). With-

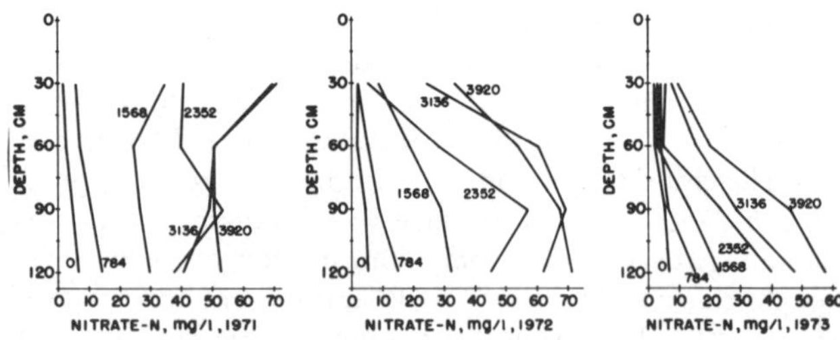

FIG. 1 Average annual concentration of nitrate nitrogen in soil water sampled with suction lysimeters. Dairy manure slurry was applied in 1969, 1970 and 1971 to supply the indicated annual rates of N in kg/ha.

in two years there was a marked reduction in concentration at 30 to 60 cm. Apparently by 1973, the utilization of available N by the grass was equal to or greater than the rate of nitrification except at the two highest rates of manure. However, the 1971 data demonstrated that annual applications of manure supplying 1568 kg N/ha or more far exceeded the capacity of the grass to utilize it. The lowest rate of manure (784 kg N/ha/yr) contributed the least to pollution. However, the increase in soil nitrogen resulting from three years of applications (Table 1) can be expected to continue with further applications and to release greater amounts of N to a crop already over supplied. Thus the annual rate of 784 kg N/ha would be excessive for this crop.

TABLE 1. NITROGEN CONTENT OF MANURE PLOT SOILS IN OCTOBER, 1972 AFTER 1969, 1970 and 1971 DAIRY MANURE APPLICATIONS SUPPLYING THE INDICATED ANNUAL RATES OF N

| Depth, cm | Annual rate of nitrogen supplied by manure, kg/ha | | | | | |
|---|---|---|---|---|---|---|
| | None | 784 | 1568 | 2352 | 3136 | 3920 |
| 0- 15 | 0.139 | 0.197 | 0.244 | 0.264 | 0.285 | 0.359 |
| 15- 30 | 0.051 | 0.074 | 0.100 | 0.113 | 0.122 | 0.141 |
| 30- 45 | 0.029 | 0.035 | 0.047 | 0.038 | 0.031 | 0.059 |
| 45- 60 | 0.026 | 0.026 | 0.030 | 0.036 | 0.032 | 0.033 |
| 60- 75 | 0.024 | 0.024 | 0.030 | 0.031 | 0.030 | 0.029 |
| 75- 90 | 0.023 | 0.026 | 0.028 | 0.024 | 0.029 | 0.030 |
| 90-105 | 0.022 | 0.024 | 0.026 | 0.024 | 0.028 | 0.027 |
| 105-120 | 0.024 | 0.022 | 0.028 | 0.027 | 0.024 | 0.025 |


Authorized for publication as Journal Series Paper No. 4837 of the Pennsylvania Agricultural Experiment Station. Authorized March 17, 1975.

The authors are: L. F. MARRIOTT, Associate Professor, Agronomy Dept., H. D. BARTLETT, Professor, Agricultural Engineering Dept., Pennsylvania State University, University Park.

TABLE 2. DRY MATTER YIELD AND NITROGEN REMOVAL IN ORCHARDGRASS FORAGE AS AFFECTED BY
SUBSURFACE APPLICATIONS IN 1969, 1970 AND 1971 OF DAIRY MANURE SLURRY TO SUPPLY
THE INDICATED ANNUAL RATES OF N

| Nitrogen, kg/ha | 1971 | | | 1972 | | | 1973 | | |
|---|---|---|---|---|---|---|---|---|---|
| | ton/ha | percent N* | N, kg/ha | ton/ha | percent N | N, kg/ha | ton/ha | percent N | N, kg/ha |
| 0 | 1.86 | 1.67 | 31 | 3.43 | 1.83 | 63 | 2.49 | 1.72 | 43 |
| 784 | 5.85 | 2.68 | 157 | 6.72 | 2.37 | 159 | 5.42 | 1.98 | 107 |
| 1568 | 6.41 | 2.99 | 192 | 7.42 | 2.68 | 194 | 6.45 | 2.00 | 129 |
| 2352 | 5.53 | 2.96 | 164 | 8.18 | 2.98 | 244 | 7.46 | 2.29 | 171 |
| 3136 | 5.26 | 2.85 | 150 | 8.13 | 3.25 | 264 | 7.37 | 2.67 | 197 |
| 3920 | 5.42 | 2.91 | 158 | 8.06 | 3.33 | 268 | 8.06 | 2.93 | 236 |
| LSD (0.05) | 0.52 | | | 0.88 | | | 1.15 | | |
| CV, percent | 6.6 | | | 8.4 | | | 12.3 | | |

*Average percent N = total N in forage/total dry matter yield in three cuttings.

Yields and nitrogen content of the forage (Table 2) support the above observations. Yields in 1971 were depressed to some extent by the physical damage to the sod by the injection equipment, particularly at the higher rates of manure. The 1972 data reflect the recovery of the sod along with some indication of declining nitrogen availability at the two lower rates of manure. In 1973, the general decline in yields and nitrogen content further indicate less available nitrogen.

The nitrate nitrogen content of soil samples taken October 1972 (Fig. 2) shows the same general relationship as the nitrate concentrations in 1972 water samples. The close agreement of the curves for the check and the lowest manure rate suggest that the nitrogen effects of the low rate of manure had disappeared except in the surface 15 cm. The higher values at the 0-15 cm depth resulted from the accumulation of nitrate nitrogen following the last harvest in late August. The magnitude of the increase is closely related to the amount of nitrogen in the soil (Table 1).

Although excessive amounts of nitrogen were applied, the first cut forage in 1971 was considered safe for feeding since it contained less than 0.2 percent nitrate nitrogen. The respective values for June 1 harvested grass were 0.0147, 0.0999, 0.1235, 0.1705, 0.1867 and 0.1470 percent. The decrease in available nitrogen over the next two years resulted in May 30, 1973 values of 0.0139, 0.0156, 0.0255, 0.0286, 0.0345 and 0.1870 percent, respectively. Forage containing over 0.2 percent nitrate nitrogen should make up only part of the ration while material containing over 0.4 percent is considered potentially toxic and may result in nitrate poisoning.

Under the conditions of this study, continued annual application of the lowest rate of manure (784 kg N/ha/yr) exceeded the capacity of orchardgrass to utilize the available nitrogen. Soil water data indicated that nitrate nitrogen exceeded 10 mg/l at 120 cm and that potential nitrate pollution of water existed below that depth. The extent of such further movement or the amount of nitrogen involved is not known.

## Manure Treatments on Bluegrass

Application of manure to supply 300, 400 and 500 kg N/ha to bluegrass apparently contributed little to nitrate nitrogen in groundwater. This is based in part on data from soil samples taken to a depth of 120 cm in 15 cm increments in May and again in October. The average nitrate nitrogen contents of the May samples, from 0 to 120 cm, were 20, 17, 7, 6, 8, 11, 15 and 19 ppm, respectively. The corresponding values in October were 17, 13, 6, 6, 7, 9, 11

and 13 ppm. It is extremely unlikely that the nitrate load at the 90 to 120 cm depths in the spring samples came from the manure. A similar spring manure application in an adjoining field resulted in very little nitrate movement until more time had elapsed. Also there was no indication of an increase in nitrate nitrogen in the zone from 30 to 75 cm.

The concentration of nitrate nitrogen in soil water samples generally followed the same pattern as in the soil, with values in excess of 10 mg/l at 30, 90 and 120 cm, and never in excess of that level at 60 cm. These findings confirmed the soil data and further indicated that nitrate nitrogen below 60 cm was not derived from the manure. In 1974, nitrate nitrogen concentration decreased to less than 10 mg/l at all sampling depths. Soil data from plots receiving urea were similar to the data from the manured plots. The water data did differ in that the concentration of nitrate nitrogen at 30, 60 and 90 cm never exceeded 10 mg/l in 1973, and was less than 6 mg/l at all depths in 1974.

Yields of bluegrass where manure was applied were 60 to 80 percent of that obtained where urea was applied (4.2 ton/ha). Mechanical damage to the sod was partly responsible for this. In 1974, the residual effect of the manure resulted in yields 80 to 107 percent of 4.2 ton/ha, the yield with urea. Average annual removal of nitrogen in the forage was 115 kg/ha (urea) and 73 to 95 kg/ha (manure). After correction for estimated removal where no nitrogen was applied, there was an apparent recovery of about 10 percent of the manure nitrogen per year.

Very likely some nitrate nitrogen from the upper soil

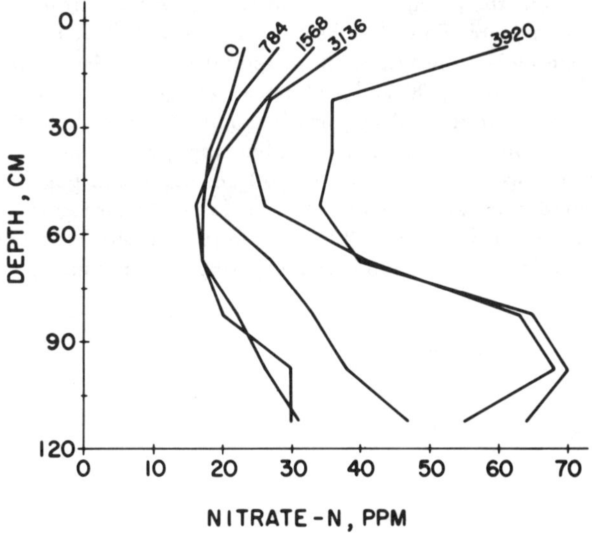


FIG. 2 Nitrate nitrogen in soil sampled in October 1972. Dairy manure slurry was applied in 1969, 1970 and 1971 to supply the indicated annual rates of N in kg/ha.

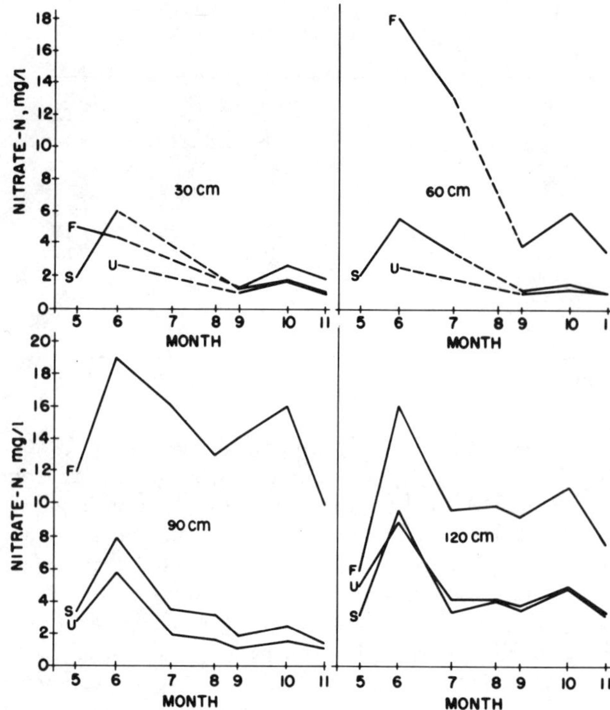

FIG. 3 Average concentration of nitrate nitrogen in soil water sampled with suction lysimeters. Dairy manure slurry was applied in November 1973 (F) and April 1974 (S) at rates to supply N at 340, 450 and 560 kg/ha. Urea (U) was applied in April at the rate of 112 kg N/ha.

layers was leached from the profile during the winter months, since the water data indicate that there was a loss from the 120 cm depth. However, it is believed the loss from manure nitrogen was quite low.

### Fall-Spring Manure on Orchardgrass

Manure was injected under orchardgrass at 340, 450 and 560 kg N/ha in the fall of 1973 or in April 1974, but suction lysimeters were not installed until after the April manure application. Data from the water samples were averaged over the three manure rates because of no significant difference between rates. However, there was a trend toward higher nitrate levels with higher manure rates. The graphs in Fig. 3 show that nitrate nitrogen from the fall manure had moved down to 90 cm prior to the May sampling, but there had been no apparent movement from the spring manure. Four days before the June sampling, 4.5 cm of rain moved the nitrate down as indicated by the greater concentration at 90 and 120 cm. Precipitation amounting to 5.5 cm after the July sampling, and 11 cm in the eight days before the September sampling apparently kept some nitrate moving down from the 60 cm depth and possibly beyond the 120 cm level. However, the concentration at 120 cm remained near 10 mg/l after July, so water moving out of this zone should have contributed a minimum of

pollution to groundwater.

Soil samples were taken in May and October. There were no significant differences in nitrate nitrogen or in total nitrogen content of the samples related to rate or time of application of the manure.

Yields are shown in Table 3. The low first harvest yield for spring manure was caused by root disruption by the injection equipment which apparently contributed to the nitrogen deficient appearance of the grass until near harvest. There was an approximate 25 to 30 percent net recovery of the manure nitrogen.

TABLE 3. DRY MATTER YIELD AND NITROGEN REMOVAL IN ORCHARDGRASS FORAGE AS AFFECTED BY SUBSURFACE APPLICATIONS OF DAIRY MANURE SLURRY IN NOVEMBER 1973 (F) OR APRIL 1974 (S) AND SPRING SUPPLIED UREA (U) SUPPLYING THE INDICATED AMOUNTS OF N. 1974

| Nitrogen, kg/ha | | Dry matter, ton/ha | | | | Nitrogen in forage, kg/ha | | | |
|---|---|---|---|---|---|---|---|---|---|
| | | 1st | 2nd | 3rd | Total | 1st | 2nd | 3rd | Total |
| (F) | 340 | 2.93 | 1.30 | 1.81 | 6.04 | 84 | 35 | 33 | 152 |
| | 450 | 3.20 | 1.50 | 1.95 | 6.65 | 98 | 45 | 42 | 185 |
| | 560 | 3.25 | 1.50 | 2.08 | 6.83 | 100 | 49 | 43 | 192 |
| (S) | 340 | 1.66 | 1.86 | 1.68 | 5.20 | 48 | 53 | 32 | 133 |
| | 450 | 1.84 | 2.15 | 2.15 | 6.14 | 57 | 66 | 43 | 166 |
| | 560 | 1.41 | 2.22 | 1.99 | 5.62 | 49 | 75 | 45 | 169 |
| (U) | 112 | 3.25 | 0.85 | 1.03 | 5.13 | 92 | 17 | 17 | 126 |
| LSD (0.05) | | 0.56 | 0.47 | 0.61 | | 21 | 16 | 14 | |
| CV, percent | | 15.0 | 19.3 | 22.1 | | 18.3 | 22.1 | 24.5 | |

## CONCLUSION

Manure applications supplying available nitrogen in excess of crop requirements may be sources of potential nitrate pollution of groundwater. Data presented here lend support to the conclusion that for harvested grass swards on a deep well drained soil, manure supplying total nitrogen at the rate of approximately two times the crop requirement will contribute minor amounts of nitrate nitrogen to groundwater supplies. Continued annual applications at this rate are planned to determine whether the evidence will indicate a significant loss of nitrate nitrogen from the rooting zone.


**Reference**

1 Bartlett, H. D. and L. F. Marriott. 1971. Subsurface disposal of liquid manure. Proc. Int. Sym. on Livestock Wastes. ASAE, St. Joseph, Mich. 49085

# Effectiveness of Forest Buffer Strips in Improving the Water Quality of Manure Polluted Runoff

R. C. Doyle, D. C. Wolf, D. F. Bezdicek

W ITH today's environmental concerns, it is necessary to evaluate the pollution potential of the traditional method of agricultural manure disposal, namely land application. The economics of returning manure to the land as a method of crop fertilization has become increasingly desirable during the past two years because the price of nitrogen alone has increased by a factor of three and supplies are limited (Humenik et al. 1974).

The objectives of this research were to determine the movement of fecal coliform, fecal streptococcus, total soluble P, K, Na, $NO_3$-N, $NH_4$-N, and organic-N in runoff water from manured land and to evaluate the effectiveness of forest buffer zones in improving the water quality of manure-polluted runoff.

## MATERIALS AND METHODS

Dairy manure (69 percent moisture) was applied at the rate of 90 metric tons per hectare on 0.19 hectare of a Chester gravelly silt loam soil (Typic Hapludult; fine loamy, mixed, mesic) having a 4 percent slope and planted in alfalfa *(Medicago sativa)*. The manure-treated area was divided into three triangularly shaped areas being 45 m in length and 20 m across the lower perimeter. The forest buffer area was sectioned into three 20 x 30.5 m plots for runoff collection. The plots had a slope of 35-40 percent and consisted of a mixed population of deciduous trees and

a uniform ground cover of honey suckle (*Lonicera japonica*).

Runoff was collected by means of dust pan lysimeters installed at 0.0, 3.8, 7.6, 15.2 and 30.5 m intervals from the manured area. Four lysimeters were placed at each distance in each plot. Runoff samples were collected from four natural rainfall events after an initial August 1973 manure application. A second 90 metric tons per hectare of manure was spread in November 1973 and runoff from three subsequent rains was collected. Temperature and rainfall data are presented in Table 1. Analysis of composite manure samples is presented in Table 2. Approximately 30 m west of the manure plots, a check plot was established in an isolated waterway. The alfalfa field portion of the check was similar to the manured area of the treated plots with respect to vegetation. The check plot was steeper in slope (6 percent slope), but the wooded area had only a 5 to 8 percent slope and had a predominance of herbaceous annuals with a sparse honey suckle cover and an intermittent canopy.

### TABLE 2. CHEMICAL AND BIOLOGICAL ANALYSES OF THE MANURE APPLIED ON THE FOREST BUFFER PLOTS*

| Date applied | FC/g x $10^4$ | FS/g x $10^4$ | Total-N percent | Total-P percent | Total-K percent |
|---|---|---|---|---|---|
| August 8, 1973 | 4670 | 2790 | 3.1 | 0.7 | 3.7 |
| November 13, 1973 | 73 | 1014 | 2.6 | 0.5 | 1.7 |

*All values are expressed on a dry weight basis.

Acknowledgment: The research reported here was supported in part by funds from the McIntire-Stennis Cooperative Forestry Research Program. The authors also acknowledge the very capable technical advice of Dr. Ernest Rebuck and Dr. Larry Douglass.

Contribution no. 5040 and Scientific Article no. A2086 of the Maryland Agr. Exp. Sta., Dept. of Agronomy.

The authors are: R. C. DOYLE, Graduate Research Assistant; D. C. WOLF, Assistant Professor, Agronomy Dept., University of Maryland, College Park, 20742; and D. F. BEZDICEK, Associate Professor, Agronomy Dept., Washington State University, Pullman, (formerly Assistant Professor, Agronomy Dept., University of Maryland.) The research work reported here is part of the M.S. thesis submitted by the senior author in partial fulfillment of the requirements for the M.S. degree.

Runoff samples were transferred to sterile urine cups after each rain and returned immediately to the laboratory for analysis. Microbiological analysis for fecal coliform (FC) was carried out according to standard methods utilizing the most-probable-number (MPN) technique (A.P.H.A. 1971). Fecal streptococci (FS) were determined by the dilution plate method (A.P.H.A. 1971). Analysis for $NH_4$-N, $NO_3$-N

### TABLE 1. THE TOTAL RAINFALL, DURATION OF RAIN, AND AVERAGE RUNOFF COLLECTED FOR ALL RUNOFF PRODUCING RAINS ON THE FOREST BUFFER PLOTS. INCLUDED IS THE TOTAL RAINFALL AND AVERAGE TEMPERATURE FOR TEN DAYS PRIOR TO RUNOFF EVENTS AND TOTAL SNOWFALL FOR FIVE DAYS PRIOR TO RUNOFF EVENTS

| Rain number | Date | Total rainfall, cm | Duration, hour | Total rainfall 10 days prior, cm | Mean temperature 10 days prior, C | Snowfall 5 days prior, cm | Mean runoff collected per lysimeter, ml |
|---|---|---|---|---|---|---|---|
| 1 | Aug. 14 | 3.03 | 1.5 | 0.05* | 25.5* | 0.00 | 319 |
| 2 | Sept. 14 | 5.70 | 14.0 | 0.12 | 21.2 | 0.00 | 376 |
| 3 | Oct. 2 | 3.00 | 7.0 | 0.65 | 18.1 | 0.00 | 193 |
| 4 | Oct. 29 | 5.05 | 13.0 | 0.00 | 11.1 | 0.00 | 301 |
| 5 | Dec. 5 | 2.80 | 11.5 | 0.62 | 9.2 | 0.00 | 95 |
| 6 | Dec. 9 | 3.16 | 17.5 | 3.02 | 6.5 | 0.00 | 233 |
| 7 | Dec. 21 | 7.98† | 18.5 | 2.63 | -0.9 | 18.75 | 257 |

*Values were obtained from 8 days prior to rain number 1.
†Total rainfall includes 4.23 cm rain and 3.75 cm snow melt.

and organic-N were performed by semimicro-Kjeldahl procedures (Bremner 1965a, 1965b). Total soluble P was determined according to Murphy and Riley (1962). Analysis for K and Na in the runoff was performed by the University of Maryland Soil Testing Laboratory (Bandel et al. 1969).

Runoff data within each rain were evaluated by means of a least squares analysis of variance. The design was a randomized complete block with the check plot excluded from the analysis. The Duncan Multiple Range comparison of means was used when significance was indicated in the analysis of variance. Significance testing was conducted at the 5 percent level of probability. All data were evaluated in log form. The means reported herein are geometric averages.

## RESULTS AND DISCUSSION

### Runoff Analysis of Manure Plots

Only data from runoff collected at the 0.0 m distance from the manured area is presented for the first manure application. This corresponds to runoff from rain numbers 1, 2, 3 and 4 (Table 1). The runoff data from the second manure application is presented for all distances and corresponds to rain numbers 5, 6 and 7.

Runoff collected at the 0.0 m distance from the manured area generally had high concentrations of all chemical parameters measured (Table 3). The highest concentrations were generally found in the first rain after manure application. There was a significant decrease in N, P and K concentrations in runoff from successive rains at the 0.0 m distance. All chemical parameters exhibited concentration increases at 0.0 m in runoff from rain 4. The reason for this is not known.

**TABLE 3. CHEMICAL AND MICROBIOLOGICAL ANALYSES OF RUNOFF COLLECTED AT 0.0 m DISTANCE FROM THE MANURED AREA FOR ALL RAINS**

| Rain number | N | P | K | Na | FC x $10^4$/ 100 ml | FS x $10^4$/ 100 ml |
|---|---|---|---|---|---|---|
| | | | ppm | | | |
| 1* | 21.3a† | 8.68ab | 298c | 32.1a | N.S.‡ | 347a |
| 2 | 22.4a | 6.55a | 100a | 11.3a | 1.3bc | 126a |
| 3 | 10.5a | 11.47ab | 23a | 6.4a | 0.7b | 3a |
| 4 | 27.9ab | 14.17ab | 133a | 26.7a | < 0.1a | 1a |
| 5§ | 175.2d | 56.81c | 654d | 20.2a | 29.1d | 5843a |
| 6 | 69.3c | 23.60ab | 305bc | 38.9a | 22.9d | 580a |
| 7 | 47.0bc | 21.23b | 141ab | 25.4a | 4.2c | 135a |

*An initial 90 metric tons per hectare of manure was applied before rain number 1.

†Means in a column that do not have a letter in common differ significantly at the 0.05 level.

‡N.S. indicates that no data were obtained due to experimental error.

§A second 90 metric tons per hectare of manure was applied before rain number 5.

The initial runoff at 0.0 m after the second manure application contained concentrations of N, P and K five times greater than those of any of the four previous rains. Increased N, P and electrical conductivity levels in winter runoff from manure have been reported by other researchers and they attributed this to reduced infiltration rates encountered during the winter months when the soil had a higher water content (Gilbertson et al. 1971). The low N values observed in runoff collected during the late summer and fall could be related to the volatilization rate of $NH_3$-N (Adriano et al. 1974). Edwards et al. (1972) postulated that reductions of N in runoff from manure during the warmer seasons were related to gaseous losses of N or reduction of soluble forms of N through immobilization. The overall increased pollution load in the December runoff can be related to the increased solubility of the manure when maintained at a high moisture level, as reported by Miner et al. (1966). However, the December 21 runoff exhibited lower pollution than the two preceeding rains although the manure was moist from light rainfall and snow for four days prior, and the ground was partially frozen. The additive effect of the two manure applications cannot be discounted as an explanation for the increased pollution of the winter runoff.

Examination of the runoff data shows that the high chemical pollutant concentrations found at 0.0 m had no effect on the runoff collected at other distances (Tables 4 and 5). In only two instances were statistically significant differences found between runoff collection intervals other than 0.0 m. During rain 5 the $NO_3$-N value at the 3.8 m distance was greater than the 15.2 and 30.5 m intervals and the $NH_4$-N value from rain 7 at 3.8 m was significantly greater than that observed at 30.5 m. These higher values, however, were 5.0 and 0.7 ppm for the $NO_3$-N and $NH_4$-N respectively, which are not exceptionally large when compared with background values for runoff.

**TABLE 4. CONCENTRATION OF TOTAL SOLUBLE NITROGEN ($NH_4$-N + $NO_3$-N + ORGANIC-N) AND PHOSPHORUS IN RUNOFF COLLECTED FROM THREE RAINS AFTER THE SECOND APPLICATION OF 90 METRIC TONS PER HECTARE OF DAIRY MANURE**

| | Rain number | | | | | |
|---|---|---|---|---|---|---|
| | 5 | | 6 | | 7 | |
| Distance, meters | N | P | N | P | N | P |
| | | | ppm | | | |
| 0.0 | 175.2b* | 56.81b | 69.3b | 23.60b | 47.0b | 21.23b |
| 3.8 | 9.7a | 0.10a | 3.0a | 0.03a | 2.9a | 0.10a |
| 7.6 | 6.9a | 0.21a | 3.3a | 0.14a | 2.3a | 0.08a |
| 15.2 | 3.7a | 0.15a | 2.6a | 0.11a | 1.6a | 0.07a |
| 30.5 | 3.6a | 0.13a | 1.1a | 0.07a | 1.3a | 0.06a |

*Means in a column that do not have a letter in common differ significantly at the 0.05 level.

**TABLE 5. CONCENTRATION OF TOTAL SOLUBLE K AND Na IN RUNOFF COLLECTED FROM THREE RAINS AFTER THE SECOND APPLICATION OF 90 METRIC TONS PER HECTARE OF DAIRY MANURE**

| | Rain number | | | | | |
|---|---|---|---|---|---|---|
| | 5 | | 6 | | 7 | |
| Distance, meters | K | Na | K | Na | K | Na |
| | | | ppm | | | |
| 0.0 | 654b* | 20.2b | 305b | 38.9b | 141b | 25.4b |
| 3.8 | 43a | 3.1a | 7a | 0.6a | 8a | 0.8a |
| 7.6 | 47a | 2.9a | 25a | 0.8a | 11a | 0.7a |
| 15.2 | 38a | 2.9a | 13a | 0.8a | 6a | 0.8a |
| 30.5 | 25a | 2.3a | 12a | 1.0a | 5a | 0.7a |

*Means in a column that do not have a letter in common differ significantly at the 0.05 level.

No FC data were obtained for the runoff from the first rain and the period before the second rain was exceptionally hot and dry. For a seven day interval the high temperature ranged from 35.0 to 36.7 C and rainfall totaled 1.12

cm for the entire 31 day period. Three days prior to the second rain, the FC population was less than 23/g dry manure and the FS density was less than 13/g dry manure. Such harsh environmental conditions were not encountered after the second manure application. The FC and FS numbers in the second manure application were not reduced as rapidly. Fecal streptococci in the manure increased over ten fold in the first 20 days after the manure was spread, and fell to roughly half of the initial value after 30 days. Van Donsel et al. (1967) also reported the favored survival of FS over FC in the soil during the winter, but found a steady decrease in the FS population with 90 percent reduction occurring at 20 days.

Fecal streptococci in the runoff at 0.0 m from the manure were noticeably higher than values obtained at other collection intervals further down the slope for all three rains after the second manure application, and exhibited a decrease after each successive rain (Table 6). The FC density showed a significant reduction at 0.0 m with successive rains after manure application, and although FS numbers at 0.0 m were not significant between rains, they did indicate a reduction with time (Table 3).

TABLE 6. CONCENTRATION OF FECAL COLIFORM (FC) AND FECAL STREPTOCOCCI (FS) IN RUNOFF COLLECTED FROM THREE RAINS AFTER THE SECOND APPLICATION OF 90 METRIC TONS PER HECTARE OF DAIRY MANURE

| | Rain number | | | | | |
| | 5 | | 6 | | 7 | |
| Distance, meters | FC, $10^4$/ 100 ml | FS, $10^4$/ 100 ml | FC, $10^4$/ 100 ml | FS, $10^4$/ 100 ml | FC, $10^4$/ 100 ml | FS, $10^4$/ 100 ml |
|---|---|---|---|---|---|---|
| 0.0 | 29b* | 5843b | 23a | 580b | 4a | 135a |
| 3.8 | 0a | 16a | 0a | 1a | 1a | 0a |
| 7.6 | 0a | 6a | 31a | 3a | 3a | 1a |
| 15.2 | 0a | 1a | 0a | 3a | 0a | 0a |
| 30.5 | 0a | 10a | 3a | 1a | 0a | 0a |

*Means in a column that do not have a letter in common differ significantly at the 0.05 level.

The overall quality of the runoff water did not appear to be related to the amount of rainfall or the quantity of runoff collected, but was dependent on the number of rains previously leaching the manure, and the temperature and moisture condition of the manure and soil. From the variable nature of the results, it is difficult to assign the relative importance of these factors to the transport of fecal wastes in runoff. Rain intensity is a decisive factor in the movement of manure pollutants in runoff as has been observed by several researchers (Miner et al. 1966 and Swanson et al. 1971). The average intensity of each rain may be calculated from the data in Table 1; however, this average intensity does not indicate short periods of very intense rainfall which might account for significant manure contamination in the runoff.

It should be noted that the buffer area was completely open for bird and animal habitation, and the infrequent values that suggested some minor movement of fecal waste may be of wild animal origin.There is no substantial evidence that any components of the manure reached the 3.8 m distance in sufficient quantity to represent a pollution hazard.

## Runoff Analysis of Check Plot

After initiation of the project, it was discovered that the check plot was frequented by widlife from the surrounding area. The plot was longitudinally bisected by a trail polluted with animal droppings which resulted in inflated biological and chemical indicator values in the runoff. Other authors have reported that the soil can act as a reservoir for FC and FS deposited by wild animals and birds (Geldreich et al. 1968 and Presnell and Miescier 1971). Because of the presence and activity of workers and the use of an animal repellant, nicotine sulfate, the use of the trail by animals was apparently reduced and the quality of the runoff water showed substantial improvement with time. Fecal coliform values were reduced 92 percent from the second to the seventh rain and FS decreased 99.8 percent from the first to the seventh rain (Fig. 1). Total N and P levels dropped in an approximately linear manner with respect to runoff events (Fig. 2). The coefficients of determination were 0.865 and 0.807 for N and P, respectively. Sodium and K values did not display any consistently increasing or decreasing trend and tended to remain low.

The wide range of both chemical and bacteriological pollutant values in the check plot runoff appeared to be largely determined by wildlife activity. This so called "natural" fecal pollution is quite variable as indicated in Figs. 1 and 2 and may reach levels as high as those found in runoff from manured land.

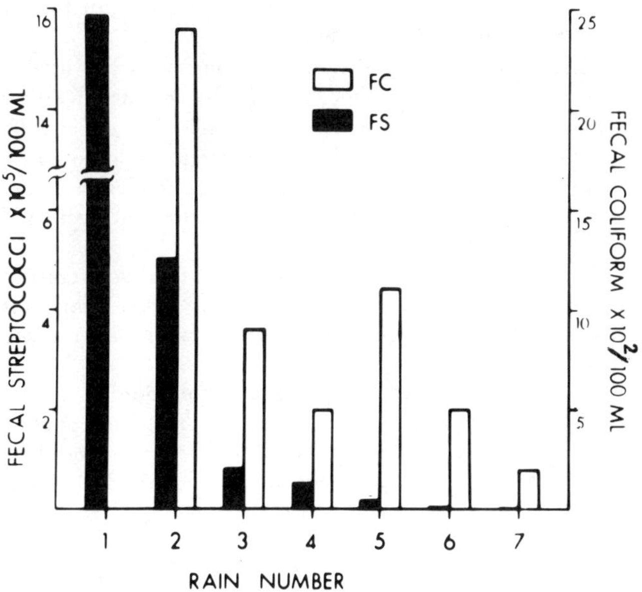


FIG. 1 Fecal coliform (FC) and fecal streptococci (FS) analyses of runoff from the check plot. Values reported are geometric means of all collection intervals within the plot.

## CONCLUSIONS

Application of 90 metric tons of dairy manure per hectare elevated levels of N, P, K and Na in runoff water at the 0.0 m distance from the treated area. The concentrations of these nutrients were dependent on the number of rains previously leaching the manure, but was independent of the total rainfall and the amount of runoff collected. Runoff at the 0.0 m distance from the manure during the winter contained concentrations of chemical pollutants over four times greater than those in the early fall. It appears probable that the temperature and moisture condi-

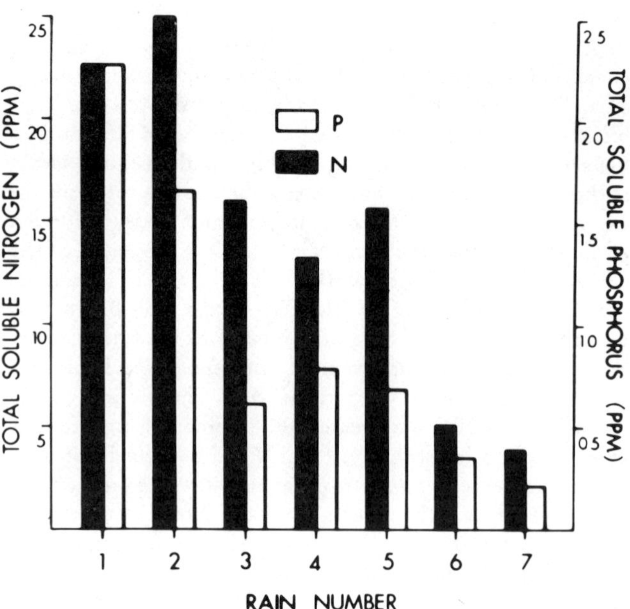

FIG. 2 Total soluble nitrogen ($NH_4$-N + $NO_3$-N + organic-N) and phosphorus of runoff from the check plot. Values reported are geometric means of all collection intervals within the plot.

tions in the manure and soil have a strong influence on the release of nutrients into the runoff, but the accumulation of manure from the autumn and summer applications may also be a significant factor affecting the water quality of runoff from the manured area.

The FC and FS populations in the manure were variable and influenced by climatic conditions. Dry hot weather rapidly reduced microbiological densities, but cool moist conditions prolonged coliform and streptococci survival. Both indicators showed increases over natural levels in runoff at the 0.0 m distance from the manure, but decreased with time and each subsequent rain.

Despite the elevated concentrations of fecal pollutants in runoff from manure collected at 0.0 m, no significant movement of these contaminants 3.8 m or beyond was observed.

Runoff from an area receiving fecal pollution from wildlife showed improvement in quality as the animal population was reduced. The FC and FS levels were reduced 92 and 99.8 percent, respectively in runoff within 129 days.

Nitrogen and P also showed reductions of 83 and 91 per cent in the same area.

On the basis of these data, it could be concluded that under the conditions of this experiment, a forest buffer strip of 7.6 m is sufficient to prevent stream pollution from animal wastes. It should also be noted that manure application in the winter months presents a greater pollution potential than summer application due to decreased infiltration rates and longer FC and FS survival during the winter months.


### References

1  Adriano, D. C., A. C. Chang and R. Sharpless. 1974. Nitrogen loss from manure as influenced by moisture and temperature. J. Environ. Quality 3(3):258-261.

2  American Public Health Association. 1971. Standard methods for the examination of water and wastewater, 13th ed. American Public Health Assoc., Washington, D. C. pp 635-693.

3  Bandel, V. A., K. C. Stottlemyer and C. E. Rivard. 1969. Soil testing methods at the University of Maryland Soil Testing Laboratory. Univ. of Md. Agron. Mimeo No. 37.

4  Bremner, J. M. 1965a. Total nitrogen. In: C. A. Black (ed.) Methods of soil analysis. Agronomy 9: 1172-1175. Amer. Soc. Agron., Madison, Wis.

5  Bremner, J. M. 1965b. Inorganic forms of nitrogen. In: C. A. Black (ed.) Methods of soil analysis. Agronomy 9: 1179-1206. Amer. Soc. Agron., Madison, Wis.

6  Edwards, W. M., E. C. Simpson, M. H. Frere. 1972. Nutrient content of barnyard runoff water. J. Environ. Quality 1(4):401-405.

7  Geldreich, E. C., L. C. Best, B. A. Kenner and D. J. Van Donsel. 1968. The bacteriological aspects of storm-water pollution. J. Water Pollut. Control Fed. 40(11):1861-1872.

8  Gilbertson, C. B., T. M. McCalla, J. R. Ellis, O. E. Cross and W. R. Woods. 1971. Runoff, solid wastes and nitrate movement on beef feedlots. J. Water Pollut. Control Fed. 43(3):483-493.

9  Humenik, F. J., M. R. Overcash, L. B. Driggers and G. J. Kriz. 1974. Cleaning the animal farm environment. Environ. Sci. Tech. 8(12):984-989.

10  Miner, J. R., R. I. Lipper, L. R. Fina and I. W. Funk. 1966. Cattle feedlot runoff—Its nature and variation. J. Water Pollut. Control Fed. 38(10):1582-1591.

11  Murphy, J. and J. R. Riley. 1962. A modified single solution method for the determination of phosphate in nautral waters. Anal. Chim. Acta. 27(1):31-36.

12  Presnell, M. W. and J. J. Miescier. 1971. Coliforms and fecal coliforms in an oyster-growing area. J. Water Pollut. Control Fed. 43(3):407-416.

13  Swanson, N. P., L. N. Mielke, J. C. Lorimor, T. M. McCalla and J. R. Ellis. 1971. Transport of pollutants from sloping cattle feedlots as affected by rainfall intensity, duration and reoccurrance. Livestock Waste Management and Pollution Abatement (Proceedings, International Symposium on Livestock Wastes), ASAE, St. Joseph, Michigan 49085. pp. 51-55.

14  Van Donsel, D. J., E. E. Geldreich and N. A. Clark. 1967. Seasonal variations in survival of indicator bacteria in soil and their contributions to storm-water pollution. Appl. Microbiol. 15(6):1362-1370.

# Effect of Anaerobic Swine Lagoons on Groundwater Quality in High Water Table Soils

E. R. Collins, Jr., T. G. Ciravolo, D. L. Hallock
MEMBER
ASAE

D. C. Martens, H. R. Thomas, E. T. Kornegay

ALTHOUGH not classed as a major swine-producing state, the gross value of Virginia swine products was over $3.5 million in 1973. Most production is concentrated in the Tidewater area of the state where water tables are frequently 1.8 m or less below the surface and where soils are often poorly drained. Many production units rely on anaerobic lagoons for waste holding and partial treatment prior to land disposal of effluent. Numerous units have been in operation for long periods of time with few problems of rapid sludge accumulation or severe odors.

In 1973 Virginia environmental authorities became concerned that seepage from animal waste lagoons in high water table soils constituted an environmental hazard and suggested that use of such facilities should be prohibited. These measures would cause considerable financial and managerial hardships on many Virginia swine producers, and would be especially undesirable without evidence to prove that lagoons are significant groundwater pollution sources.

The Departments of Agricultural Engineering, Agronomy, Animal Science, and the Tidewater Research and Continuing Education Center of the Virginia Polytechnic Institute and State University mounted a joint research effort to provide sound answers upon which producers and regulatory agencies could base decisions. Objectives of the study were: (a) to determine the amount, distance, and rate of movement of biological and chemical constituents from swine waste lagoons into groundwater; and (b) to measure contamination variability in groundwater as affected by inherent soil drainage characteristics. This discussion will be limited to results obtained thus far.

## LITERATURE REVIEW

Considerable research has been reported concerning lagoon treatment of livestock wastes. However, Loehr (1971) has observed that current liquid treatment systems for livestock wastes will not likely produce effluents suitable for discharge into receiving streams. It follows, then, that significant leakage from lagoons into groundwater may be equally undesirable.

Koelliker et al. (1971) conducted studies to determine optimum land application rates of anaerobic swine lagoon effluent. Tile drainage was used beneath disposal plots to measure infiltration and renovation of effluent through the

soil profile. Almost all of the nitrogen reaching the tile drainage was of the $NO_3^-$ form. Soil filtration decreased nitrogen concentration by nearly 50 percent for the heaviest effluent application rate. Since soil-moisture content was usually high and considerable organic material was added in the lagoon effluent, the decrease in N was attributed to denitrification. Chloride concentration was used to follow water movement through the soil.

Few studies have been reported concerning seepage from livestock waste lagoons. Nordstedt et al. (1971) studied a multistage dairy lagoon system in a sandy, high-water table soil. Lagoons were located above a clay layer so vertical seepage was not considered likely. Test wells 2.4 - 3 meters deep were driven at distances of 4.6, 15, and 30 m from both an anaerobic and an aerobic lagoon. Total salts, BOD, and $NO_3^-$ values from test well samples indicated some leakage from the anaerobic lagoon, but indicator values did not vary significantly from reference well samples at the 30-meter distance.

Davis et al. (1972) measured the infiltration rate of a dairy manure waste pond. Infiltration rates of a newly-constructed pond in a sandy loam soil were measured with fresh water and thereafter with manure water. Infiltration decreased from 122 cm per day with fresh water to 0.5 cm per day with manure water after 4 months. Concurrent evaporation measurements averaged 0.4 cm per day during the 4-month period. They concluded that the manure slime and biological sludge developed on the bottom of a manure pond were forced into soil pores by hydrostatic pressure, effectively blocking further infiltration.

Humenik (1972) evaluated a single unaerated lagoon, two unaerated series lagoons, and packed soil lysimeters for swine lagoon effluent application. He observed that poor effluent quality and potential for odor and leakage are major constraints associated with the excellent waste reduction by the lagoons under study. Leachate $NO_3^-$ from lysimeters loaded with effluent was shown to be removable without supplemental organic material by providing a controlled water table and promoting biological denitrification in the anaerobic zone.

## EXPERIMENTAL PROCEDURE

Three swine waste lagoons in the Coastal Plains Region of Virginia were initially selected to investigate possible groundwater contamination. One lagoon was located at the Virginia Swine Evaluation Station and one at the VPI & SU Tidewater Research and Continuing Education Center (TRACEC) near Holland, Virginia. The lagoon at TRACEC was abandoned in August, 1974, when it was replaced by a new, two-stage lagoon system. The new lagoon system was added to the study in September, 1974. The third lagoon was located on a private farm and was incorporated into the study in September, 1974. Soils at all locations ranged from sandy loam to fine sandy loam.

Based on work supported in part by the Virginia Agricultural Foundation and in part by the United States Department of the Interior, Office of Water Research and Technology, administered by the Virginia Water Resources Research Center as Project B-068-VA.

The authors aare E. R. COLLINS, Assistant Professor, Department of Agricultural Engineering; T. G. CIRAVOLO, Graduate Research Assistant and D. L. HALLOCK and D. C. MARTENS, Associate Professors, Department of Agronomy; H. R. THOMAS, Associate Professor and E. T. KORNEGAY, Professor, Department of Animal Science; Virginia Polytechnic Institute and State University, Blacksburg, Virginia 24061.

Test wells were installed at each lagoon site to depths of 3, 4.6, and 6 m and at distances of 3, 15, and 30 m from the lagoon edge (Fig. 1). Wells were constructed of 5-cm diameter PVC pipe which was water jetted to the specified depth. Each well was constructed with a permanently installed sampling tube to simplify sampling procedures, and was capped to prevent contamination from external sources. Reference wells have been established in fields some distance from lagoon sites, but sample data were not available for this report.

FIG. 1 Test well installations with sampling in progress.

Wells were pumped out 24 hours prior to sample collection to assure that fresh groundwater would be in the wells. Samples were drawn using a shallow-well jet pump which supplied ample pumping volume when collecting from deeper wells. A schematic of the sample collection procedure is shown in Fig. 2. Samples were collected monthly beginning in August, 1973, from wells at the Virginia Swine Evaluation Station. Wells at TRACEC and the private farm were sampled monthly beginning in September, 1974.

Groundwater constituents initially determined were COD, $NH_4^+$, $NO_3^-$, $PO_4^{3-}$, $Cl^-$, $Cu^{2+}$, $Zn^{2+}$, $Mg^{2+}$, $Mn^{2+}$, $Na^+$, $K^+$, $Ca^{2+}$, and fecal coliform bacteria. Because of the large number of samples collected and limitations of laboratory space and technician service, the constituents presently being determined are $NH_4^+$, $NO_3^-$, $PO_4^{3-}$, $Cl^-$, $Cu^{2+}$, $Zn^{2+}$, $Mn^{2+}$, and fecal coliform. Ammonia and $Cl^-$ were determined using electrode methods; $Ca^{2+}$, $Mg^{2+}$, $K^+$, $Na^+$, $Cu^{2+}$, $Zn^{2+}$, and $Mn^{2+}$ using atomic absorption spectrophotometry; $NO_3^-$ by the phenoldisulphonic acid method;

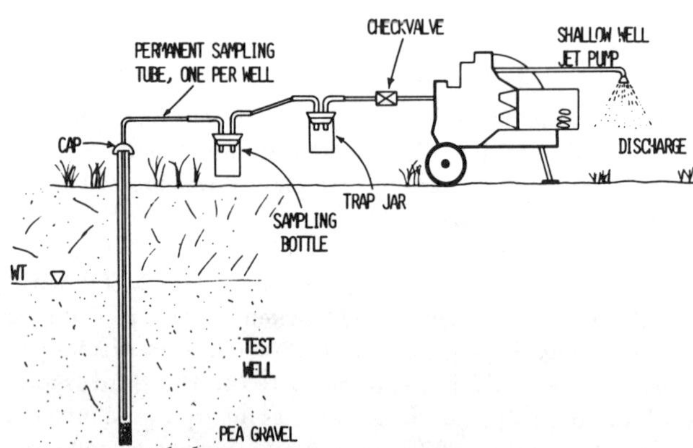


FIG. 2 Test well and sampling system.

$PO_4^{3-}$ by the ascorbic acid method; and fecal coliform bacteria by the millipore membrane filter technique.

## RESULTS AND DISCUSSION

Studies cited previously (Davis et al. 1972, Humenik 1972) have raised the spectre of seepage and consequent groundwater pollution from livestock waste lagoons. Also mentioned (Davis et al. 1972) was the sealing by biological materials under hydrostatic pressure. The lagoons reported herein are not likely to be subject to the condition of sludge being forced into the soil under hydrostatic pressure. On the contrary, much of the time there is an influent condition of boundary flow of groundwater into lagoons. At best, sealing may take place as a sludge blanket settles on the bottom and sides.

Even if effective biological sealing does not occur in high-water table lagoons, a hydraulic gradient for water movement through the soil must exist if significant amounts of seepage are to occur. As long as the soil remains saturated by a water table at, or near, the lagoon level, the most likely mode of pollutant leakage is by diffusion through the soil. We have noted that the water table near the studied lagoons has ranged between 0.3 and 1.8 m below the soil surface throughout the period wells have been monitored.

Data collected during the first year included determination of $Ca^{2+}$, $Mg^{2+}$, $K^+$, and $Na^+$. Levels of these materials in groundwater samples were consistently low and are no longer being measured. Total nitrogen measurements have not been made because time and laboratory space have been limited. The approach taken initially was to measure relative intensity of groundwater pollution indicators at various depths and distances around lagoon sites. More recently, we have recorded levels of pollutants in the lagoon liquids concurrently with groundwater sampling (Table 1).

**TABLE 1. POLLUTANT CONCENTRATIONS ANAEROBIC SWINE WASTE LAGOONS**

| Date | $NH_4^+$ | $NO_3^-$ | $Cl^-$ | Fecal coliform |
|---|---|---|---|---|
|  | $-----$ mg/l $-----$ |  |  | no./100 ml |
| *Virginia Swine Evaluation Station* |  |  |  |  |
| April 1974 | — | — | — | 44,800 |
| July 1974 | — | — | — | 206,400 |
| Sept. 1974 | — | — | — | 400,000 |
| Oct. 1974 | 132 | <0.20 | 415 | — |
| Nov. 1974 | 86 | 0.20 | 145 | 43,800 |
| Jan. 1975 | 69 | 0.36 | 140 | — |
| *Tidewater Research and Continuing Education Center* |  |  |  |  |
| Sept. 1974 | — | — | — | 172,800 |
| Oct. 1974 | 115 | <0.20 | 209 | 117,000 |
| Nov. 1974 | 272 | 0.36 | 266 | 397,000 |
| Dec. 1974 | — | — | — | 365,000 |
| Jan. 1975 | 266 | — | 227 | — |
| *Private farm, Suffolk, Virginia* |  |  |  |  |
| Sept. 1974 | — | — | — | 156,000 |
| Oct. 1974 | 294 | <0.20 | 213 | 313,000 |
| Nov. 1974 | 205 | 0.30 | 346 | 223,000 |
| Dec. 1974 | 382 | 0.43 | 289 | 284,000 |
| Jan. 1975 | 486 | <0.20 | 367 | — |

### Swine Evaluation Station

The most complete data set in this study is from the lagoon located at the Virginia Swine Evaluation Station. One series of wells at this site has been sampled since August, 1973. Another full series and two partial series were added in Summer, 1974, and have been sampled since September, 1974. Levels of $Mn^{2+}$, $Zn^{2+}$, $Cu^{2+}$, and $PO_4^{3-}$



MANAGING LIVESTOCK WASTES

in well samples have remained below recommended standards for U. S. public water supplies. (U. S. Environmental Protection Agency 1973).

Ammonia levels in all series of wells at this site show a decreasing concentration of $NH_4^+$ with increasing distance from the lagoon and increasing well depth (Table 2). The $NH_4^+$ conecntration was much lower in the 15- and 30-m wells than in the 3-m wells. Apparently leakage of effluent high in $NH_4^+$ moved as far as 3 m from the lagoon but no more than 15 m. Since the groundwater level was usually high, much of this movement was probably by diffusion.

Nitrate levels at all depths, distances, and depth-distance combinations were not statistically significant, and averaged 0.22 mg/l. There are several reasons to expect $NO_3^-$ levels to be low under the study conditions. First, $NO_3^-$ levels in the anaerobic lagoon were low (Table 1), and if leakage occurred, groundwater $NO_3^-$ levels would not be appreciably increased. Secondly, the poor soil drainage characteristics are generally unsuitable for nitrification of $NH_4^+$, but conditions are ideal for denitrification of $NO_3^-$.

Chloride levels in lagoon fluids were expectedly high (Table 1). Due to its mobility, $Cl^-$ was measured in groundwater samples as an indicator of degree of leakage from lagoons. Since the addition of more well series in September, 1974, only two observations for $Cl^-$ have been available because of a laboratory error. Data on a single series of wells collected from August, 1973, through January, 1975, indicate a highly significant difference in levels of $Cl^-$ in wells at different depths and distances (Table 2). Although more data would be desirable for good statistical analysis, a distinct trend has developed showing high concentration of $Cl^-$ in 3-m distant wells with much lower concentrations in the 15- and 30-m wells. Again, there is little apparent movement of lagoon pollutants beyond the 3-m distance.

**TABLE 2. MEAN POLLUTANT CONCENTRATIONS IN GROUNDWATER, VIRGINIA SWINE EVALUATION STATION***

| Well depth, meters | Distance from lagoon, meters | | |
|---|---|---|---|
| | 3 | 15 | 30 |
| | — — — — $NH_4^+$, mg/l — — — — | | |
| 3.0 | 26.68 | 0.23 | 0.17 |
| 4.6 | 0.19 | 0.14 | 0.21 |
| 6.0 | 10.83 | 0.17 | 0.15 |
| | — — — — $Cl^-$, mg/l — — — — | | |
| 3.0 | 70.2 | 9.9 | 10.7 |
| 4.6 | 18.9 | 7.7 | 10.7 |
| 6.0 | 7.1 | 12.6 | 10.6 |
| | — fecal coliform, no./100 ml — | | |
| 3.0 | 1093.0 | 0.1 | 0.0 |
| 4.6 | 0.0 | 0.0 | 0.0 |
| 6.0 | 4.0 | 0.6 | 0.0 |

*Mean values of $NH_4^+$ for monthly samples, Sept. 1974 - Jan. 1975. Mean values of $Cl^-$ and fecal coliform for monthly samples, Aug. 1973 - Jan. 1975.

Fecal coliform determinations were made on samples to measure bacteriological contamination of groundwater. Table 2 shows the same pattern established by analysis of $NH_4^+$ and $Cl^-$ data. Although fecal coliform levels in the lagoon liquids were always relatively high, contamination in groundwater wells was only detectable at the 3-m distant wells, and at the 3-m depth.

### Tidewater Research and Continuing Education Center

Wells at TRACEC were installed around an existing lagoon when the study was initiated. The decision was made to improve waste handling facilities in 1974, a new two-stage lagoon system was built, and the old lagoon filled with soil.

Well series were installed around the new anaerobic lagoon. At this location we were unable to extend wells beyond 15 m from the lagoon because of space limitations imposed by buildings and other facilities. Therefore, sampling was limited to 3 series of wells installed at depths of 3, 4.6, and 6 m and at distances of 3 and 15 m from the lagoon. Data reported here is limited to samples taken beginning in September, 1974, around the new lagoon.

It has been difficult to draw conclusions thus far about pollutant movement from the above system. Data have been quite variable, and an additional year of sampling will be needed before firm conclusions can be made. The first two months of data indicated high levels of fecal coliform organisms in most of the wells at both distances. However, no fecal organisms have been recorded since October, 1974, in any of the wells. Earlier high levels may have been due to contamination during well installation or could have occurred because of leakage from the lagoon before biological sealing. Ammonia levels in all wells have been relatively low compared with levels within the lagoon, and compared to levels in wells at the Swine Evaluation Station. In contrast, $NO_3^-$ levels have been relatively high. No satisfactory explanation can be advanced at this time. Because of the location, the section of lagoon near the test wells was constructed mostly above original ground level using an embankment. If leakage had occurred, it could have moved 1.8 - 3 m through an unsaturated profile before reaching the water table from which samples were taken. Some nitrification may have occurred under these conditions but supporting data is not available.

### Private Farm

Two series of wells were installed around an existing lagoon. The first series was in a low, poorly-drained area which at times contained surface water. The second series was in an area with better surface and internal drainage.

Data collected beginning in September, 1974, are inconclusive. As with the TRACEC location, more data must be obtained before effects can be clearly identified. A trend is developing indicating decreasing $NH_4^+$ and $Cl^-$ concentrations with increasing distance of wells from the lagoon. An interesting pattern of high $NH_4^+$ - low $NO_3^-$ and low $NH_4^+$ - high $NO_3^-$ for the first and second well series, respectively, is developing. This pattern may be related to drainage characteristics and, perhaps, direction of groundwater flow. Fecal coliform levels have been consistently low in all wells except for erratic values at the initial sampling. These may have been caused by external contamination when wells were installed.

## SUMMARY AND CONCLUSIONS

Studies are continuing in the Coastal Plain of Virginia to determine the degree of pollution caused by three swine waste lagoons on groundwater in poorly drained, high water table soils. Main indicators of pollutant movement are $NH_4^+$, $NO_3^-$, $Cl^-$, and fecal coliform bacteria.

Data collected for 16 months on one lagoon are unable to indicate significant effects on groundwater beyond 3 m from the lagoon edge. Only five months of data are available for two other lagoons, but a trend similar to that of the first lagoon seems to be developing.

*(Continued on page 313)*

# Nutrient Characteristics of Wastes from Deep Pits and Anaerobic Lagoons


J. C. Lorimor, S. W. Melvin, B. M. Leu

ASSOC. MEMBER    MEMBER
ASAE           ASAE


**W**ASTE management problems are very real to the producer who has committed himself to a new system or concept which has no record of failure or success. With today's rapid changes in livestock production, research has not been able to keep ahead of the industry in answering immediate waste management needs. To help bridge this gap, we initiated on-farm monitoring of innovative waste management systems in 1970 (Melvin et al. 1973, Vanderholm et al. 1974). Since 1973 anaerobic lagoons for large beef confinement systems have been installed in many areas of the Midwest in conjunction with the recent solid-floor, flume flushing confinement units. Our recent work, therefore, has been concentrated on liquid manure in deep pits and anaerobic lagoons.

High fertilizer prices have created a need to better characterize the nutrient concentrations and nutrient retention in various liquid manure systems. Since deep pits and anaerobic lagoons are the most common liquid manure systems, our monitoring efforts concentrated on the comparison between these two types.

## OBJECTIVES

Our objectives were to (a) characterize waste nutrient concentrations in the liquid manure generated from various commercial beef and swine operations; (b) to determine the system parameters affecting these concentrations; and (c) to obtain waste volume generation data.

Even though much information is already published regarding the above objectives, the data has not always reflected waste management practices used by commercial producers. This has been especially true for conditions describing transient conditions such as the start-up of anaerobic beef lagoons.

## SYSTEM DESCRIPTION

This report covers results obtained from four anaerobic beef lagoon systems, three anaerobic swine lagoon systems, and two beef deep pit systems. Our monitoring program included swine deep pits, but this data is omitted as a result of poor control of the systems monitored. Unmetered water additions to the pits and nonuniformity in pen weights and numbers between monitoring dates resulted in data with limited value.

Beef systems are described in Table 1. All systems were totally roofed confinement buildings open to the


Contribution by Iowa State University Extension Pilot Project, Disposal of Solid Animal Wastes, funded by ES-USDA.

The authors are: J. C. LORIMOR, Former Assistant Professor, and S. W. MELVIN, Associate Professor, Agricultural Engineering Dept., and B. M. LEU, Graduate Assistant, Animal Science Dept., Iowa State University, Ames.


TABLE 1. BASIC DESCRIPTION OF BEEF FINISHING SYSTEMS.

| System | Waste system | Capacity* | Initial use | Lagoon or pit size $m^3$/kg | Initial lagoon dilution $m^3$/kg |
|---|---|---|---|---|---|
| A | Lagoon | 640 | 3-74 | 23.4 | 0.029 |
| B | Lagoon | 540 | 12-73 | 23.0 | 0.0075 |
| C | Lagoon | 1900 | 10-73 | 44.6 | 0.024 |
| D | Lagoon | 600 | 11-73 | 36.2 | ---† |
| F | Pit | 200 | 1971 | 4.08 | |
| G | Pit | 500 | 1970 | 2.66 | |

*One time capacity.
†Diluted with feedlot runoff.

south. Even though rations were varied, the major ingredients were corn grain and corn silage. Corn grain in the ration was increased proportionately throughout the feeding period. Supplemental protein was supplied by either urea or soybean oil meal. The range in weights of cattle in the buildings was 600-1100 pounds.

All beef lagoon systems were initially stocked either in the fall of 1973 or in the spring of 1974 and all are recycling lagoon effluent for manure removal. Initial dilution water in the lagoons varied from 0.00749 $m^3$/kg liveweight in System B to 0.0293 $m^3$/ kg liveweight in System A. the two deep pit buildings have been operated for 3 years or more.

Swine systems are described in Table 2. Systems H and I are finishing operations, while J is a nursery-finishing unit. All systems used flushing gutters for manure removal from the building. Corn is the basic grain in the ration. Soybean oil meal is used for supplemental protein. System J differs from the other two systems in that a settling tank with a detention time of approximately 50 days of waste storage is used prior to discharge into the lagoon. All three systems are recyling lagoon effluent for manure removal. Initial dilution of these lagoons was not known. The effect of initial dilution is slight since the facilities have been in use for several years. The lagoon volumes were designed for approximately 0.0624 $m^3$/kg liveweight of the animals in all three system. Systems H and I have had waste disposed from them for several years, but System J has never had liquid removed except through the cleaning of the settling tank with a vacuum loader. The lagoon is nearing capacity after 3 years of operation.

TABLE 2. BASIC DESCRIPTION OF SWINE PRODUCTION SYSTEMS.

| System | Type | Capacity | Initial use | Lagoon volume* $m^3$/kg |
|---|---|---|---|---|
| H | Finishing | 500 | 1968 | 0.062 |
| I | Finishing | 700 | 1962 | 0.062 |
| J | Nursery-finishing | 240/480 | 1971 | 0.062 |

*Approximate

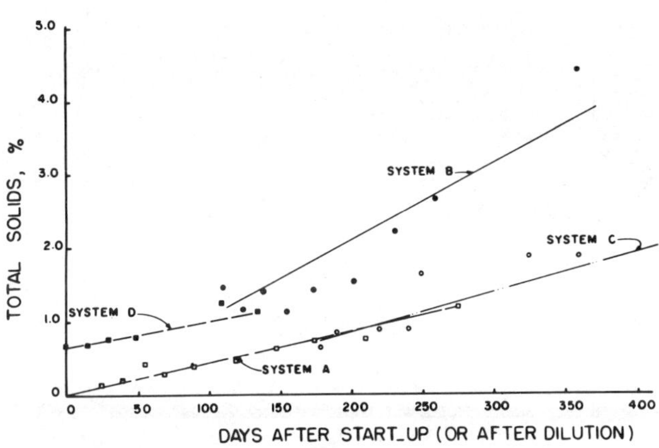

FIG. 1 **Relationship between total solids buildup and days after start-up or dilution for beef lagoons.**

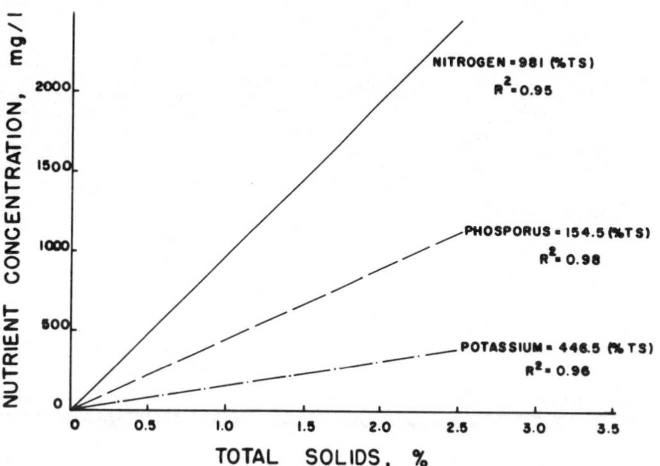

FIG. 3 **Relationship between nutrient concentrations and total solids in beef lagoons.**

## METHODS AND PROCEDURES

During 1974, each site was visited twice a month in April and May, and once a month thereafter through mid-October. During each visit, observations concerning animal numbers, weight, and rations were obtained. Samples of liquid manure were obtained from the lagoon system with a large drenching syringe from the recyling water if possible and from the lagoon if not. Samples of deep pit manure were obtained using a 3.05 m by 3.81 cm diameter PVC tube with a manually-operated ball valve on the bottom. This technique provided a vertically integrated sample of the layered pit contents.

Composite samples were obtained at each site. Samples were placed in an ice chest until delivery to the Iowa State University Engineering Research Institute Analytical Laboratory.

Manure accumulation observations were obtained by measuring liquid levels from an established reference. Depth data was converted to volume data from measured dimensions of the pits and the lagoon surfaces.

Chemical analyses performed at the laboratory included total Kjeldahl nitrogen, total phosphorus on an autoanalyzer, potassium on a flame emission spectrophotometer, and solids and COD tests using standard methods.

## RESULTS AND DISCUSSION

### Beef Anaerobic Lagoon System

Data obtained from beef anaerobic lagoons indicate a linearly increasing total solids level with time. This information is illustrated in Fig. 1. The rate of increase

of solids with time is inversely related to the initial dilution water used. Initial dilution information for System D was not available since runoff from an open feedlot was used to initially dilute the lagoon. Recycling of lagoon effluent has not been a problem except for System B where the total solids level is approaching 4.0 percent. Additional dilution water will probably be needed to control the total solids level in each of these lagoons.

Fig. 2 illustrates the rate of increase of one nutrient (total nitrogen) measured during 1974 in each of the four lagoons. Notice again the linearity with time in the range of the variables studied. This rate of increase would be expected to decrease with time as a higher percentage of the nitrogen is lost through volatilization of ammonia. An explanation for the extremely high initial level of nitrogen in System B is not known.

Since total solids, nutrient level and time appear to be related in the range of variables studied, nutrient concentrations were plotted against total solids. Fig. 3 shows the pooled nutrient information for the four lagoons plotted as a function of total solids. A regression forcing the line through the origin was run for each data set. In the range of variables studied, a good predictive relationship exists between each of the three major nutrients and total solids.

During the monitoring period of 1974, accumulations of waste volume were measured. These values include net rainfall accumulations in each lagoon during the summer of 1974. Total accumulations average 0.033 $m^3$/hd-day for System A; 0.027 $m^3$/hd-day for System B; and 0.020 $m^3$/hd-day for System C. By combining liquid accumulation data with nutrient concentrations and adjusting for animal weights, an estimated daily nutrient production rate was calculated. The values listed in Table 3 are based on an average weight of 363 kg and represent the nutrients expected in the liquid

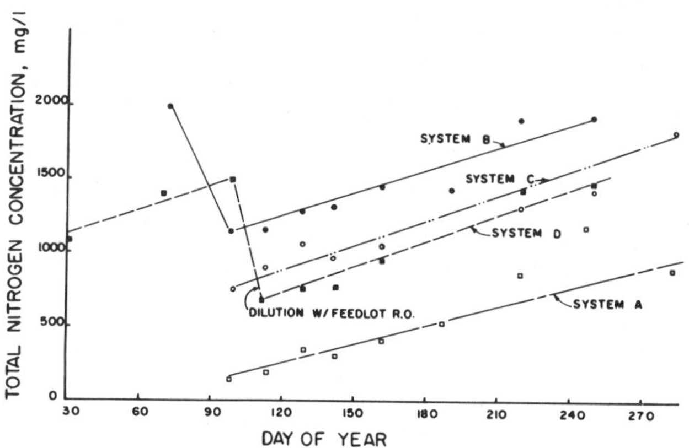

FIG. 2 **Total nitrogen concentrations in beef lagoons in 1974.**

TABLE 3. DAILY NUTRIENT ACCUMULATION IN
BEEF LAGOONS — 1974.

| System | Nitrogen kg/day | Phosphorus kg/day | Potassium kg/day |
|---|---|---|---|
| A | 0.113 | 0.012 | 0.033 |
| B | 0.077 | 0.014 | 0.044 |
| C | 0.095 | 0.015 | 0.063 |
| Avg. for lagoon systems | 0.095 | 0.014 | 0.047 |
| F | 0.113 | 0.018 | 0.035 |
| G | 0.150 | 0.032 | 0.045 |
| Avg. for pit systems | 0.127 | 0.026 | 0.040 |

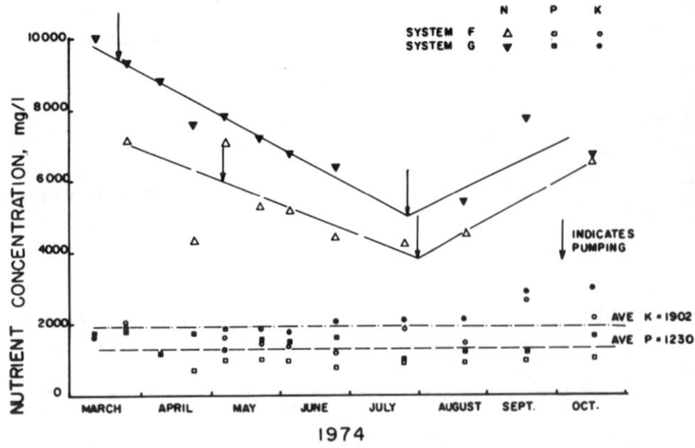

FIG. 4 Nutrient concentration levels in beef deep pit manure in 1974.

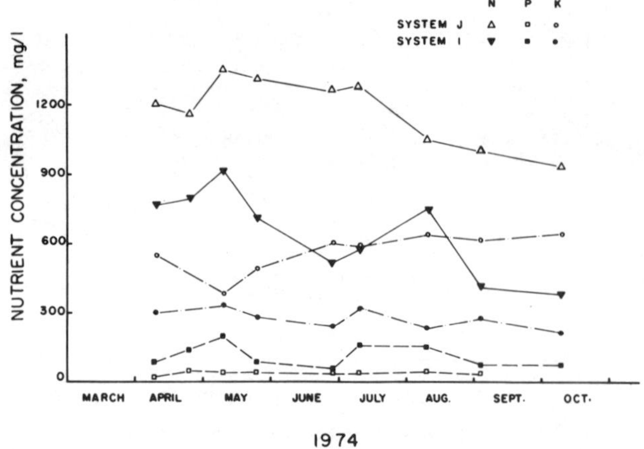

FIG. 5 Nutrient concentration levels in swine anaerobic lagoons in 1974.

effluent. The average values shown represent a higher percentage nitrogen retained than has been previously reported (Vanderholm 1973, Jones et al. 1973). Nutrients in the sludge layer are not included.

### Beef Deep Pit Systems

Deep pit beef manure samples had an average total solids level of 11.7 percent with a range from 7 to 15 percent. Nutrient concentrations for nitrogen were highly variable while phosphorus and potassium were relatively constant throughout the season. The pooled averages were as follows: nitrogen, 6992 mg/l; phosphorus, 1230 mg/l; and potassium, 1902 mg/l. Nutrient concentrations were plotted against time in Fig. 4. From this data, it appears that the concentration of nitrogen may decrease by as much as 50 percent during the summer period. Since dilution and rations were relatively unchanged, we would attribute this nitrogen decrease to volatilization of ammonia.

Measurements of deep pit depths provided information suggesting less manure volume production than commonly accepted (ASAE 1973). When animal weights were corrected to 1000 pounds, the average manure production rate in System F was 0.018 m$^3$/animal unit-day and 0.022 m$^3$/animal unit-day in System G.

Combining nutrient concentration with waste production rate provides an estimate of daily nutrient generation for an 800-pound animal in a deep pit system. These values are listed in Table 3 and again show high nitrogen retention.

### Swine Anaerobic Lagoons

The swine lagoons studied were older, more seasoned lagoons. The data obtained should represent a more stable operating condition than the beef lagoons during start-up. Fig. 5 illustrates the comparison of Systems I and J. For sake of clarity, data for System H is not plotted. It showed close relationship with System I. The pooled nutrient averages for the three lagoons were 804 mg/l; 73 mg/l; and 416 mg/l for nitrogen, phosphorus, and potassium, respectively. Total solids averaged 0.49 percent for the three lagoons. Nitrogen concentration tended to decrease over the summer period as a result of ammonia volatilization. Less dilution water was added to System J, resulting in higher nitrogen and potassium levels. The phosphorus in System J was only a third as

high as System I as a result of the primary sedimentation unit removing phosphorus-rich settleable solids.

## SUMMARY AND CONCLUSIONS

During 1973 and 1974, four newly established beef anaerobic lagoons, two well established beef deep pits, and three well established swine anaerobic lagoons were monitored. The following conclusions were drawn:

1 During start-up, beef anaerobic lagoon total solids levels increased linearly with time. The rate of increase was related inversely to initial dilusion; i.e., initial total liquid volume.

2 Major nutrient concentrations in beef lagoons during start-up are linearly related to total solids, and can be quite accurately predicted until the total solids levels exceed 2.5 percent.

3 Volatilization and other losses might slow the rate of nitrogen incrase in beef lagoons as higher nutrient levels are reached. For the range studied, however, no signs of leveling off have occurred.

4 Nitrogen losses occur through the summer in beef deep pits and seasoned swine lagoons.

5 Settling solids prior to discharging swine wastes to a lagoon significantly decreases phosphorus levels in the lagoon.

6 In beef deep pits, waste accumulations were less than commonly accepted excretion values. Nitrogen retention was high.

### References


1 American Society of Agricultural Engineers. Revised June 1973. Structures and Environment Committee 412, Report AW-D-1.

2 Jones, R. E., J. C. Nye and A. C. Dale. 1973. Forms of nitrogen in animal waste. ASAE Paper No. 73-439. ASAE, St. Joseph, Mich. 49085.

3 Melvin, S. W., D. H. Vanderholm and J. C. Lorimor. 1973. Monitoring on-farm waste management systems. ASAE Paper No. 73-5542. ASAE, St. Joseph, Mich. 49085.

4 Standard Methods for the Examination of Water and Wastewater. 1965. American Public Health Association, Inc. New York. Twelfth Ed.

5 Vanderholm, D. H. 1973. Area needed for land disposal of beef and swine wastes. Iowa State University. Pamphlet Pm-552 (Rev.).

6 Vanderholm, D. H., J. C. Lorimor and S. W. Melvin. 1974. Field performance of selected beef feedlot waste-handling system. ASAE Paper No. 74-4015. ASAE, St. Joseph, Mich. 49085.




MANAGING LIVESTOCK WASTES

# Nitrogen Removal and Recovery from Poultry Wastewater by Ion Exchange


**Lee A. Mulkey**
Assoc. Member
ASAE


ION exchange has been widely used to separate and recover certain types of chemical constituents from liquid or gas streams. Application of the principle to industrial and municipal wastewater pollution abatement problems includes removal of various inorganic contaminants such as nitrogen compounds (Sanks 1973, Kreusch and Schmidt 1971, Robinson et al. 1974, Dow Chemical Co. 1971, University of California 1971). Some systems that include recovery of the pollutants in the form of useful by-products have been demonstrated in commercial installations (Higgins and Chopra 1970).

Ion exchange treatment of animal wastewaters, however, has not been previously considered. Laboratory studies were therefore carried out to determine the feasibility of various possible exchange treatment alternatives for removal and recovery of nitrogen from poultry wastewaters. Laboratory data provided the basis for examining scale-up possibilities.

## THEORY

Ion exchange is a reversible process during which insoluble high molecular weight polyelectrolytes exchange their mobile ions for other ions in a surrounding medium (Dorfner 1972). This reversible reaction may be represented by the equilibrium equation:

$$aR\text{-}A + B^a \quad (R)_a\text{-}B + aA \quad \dots\dots\dots\dots\dots [1]$$

where

R- = resin exchange site
B = ion to be removed from solution
A = ion on resin available for exchange
a = valence of ion B, expressed as positive number

Partitioning of ions between resin-bound and dissolved phases determines ion exchange behavior. Both the extent and time rate of partitioning are important. The extent of separation for simple exchange can be measured by the equilibrium constant, $K_s$, (also called the selectivity coefficient (Dorfer 1972)) for Equation [1] expressed as

$$K_s = \frac{Y_B / Y_A}{X_B / X_A} \quad \dots\dots\dots\dots\dots\dots\dots [2]$$

where

Y = concentration of ions A or B on resin and
X = concentration of ions A or B in solution.

A more useful partitioning term, the separation factor (r), accommodates ion exchange and other parallel reactions (Heister et al. 1969) and can be written as

$$r = \frac{x\,(1-y)}{y\,(1-x)} \quad \dots\dots\dots\dots\dots\dots [3]$$

where

x = concentration of an ion in solution expressed as fraction of its initial concentration in solution, and
y = concentration of the same ion bound to the resin expressed as fraction of total resin capacity.

For a system at equilibrium, $r<1$, indicates that more of the waste stream ion is associated with the resin than with the solution and "favorable" exchange is said to exist. For $r>1$, the opposite is true and "unfavorable" exchange prevails. For cases where only the reactions of Equation [1] are involved, i.e., no parallel reactions occur, $r = \frac{1}{K_s}$.

The reaction rate of Equation [1] can be described by various kinetic equations (Heister et al. 1969). For the fixed- and fluidized-bed reactors investigated in this study the ion exchange can be described by

$$\ln(x/(1-x)) = \frac{(1-r)\,k\bar{t}\,C_i\,(Ft - VE)}{QdV} \quad \dots\dots\dots\dots [4]$$

where

k = mass transfer coefficient
$\bar{t}$ = mean contact time
$C_i$ = feed concentration
F = feed flow rate
t = time
V = resin volume
E = void volume expressed as a fraction of total volume
Q = resin exchange capacity
d = resin density

Equation [4] can be used to calculate resin volumes, flow volumes, or run time given the separation factor, r, and mass transfer coefficient, k. When r approaches 1 an alternate solution to Equation [4] must be used (Heister et al. 1969). Equations [1], [3], and [4] can be used to describe both the resin loading and regeneration cycles.

## WASTEWATER CHARACTERISTICS

The chemical composition of the influent largely determines the efficacy of treatment. Inorganic nitrogen in


The author is: LEE A. MULKEY, Agricultural Engineer, Agro-Environmental Systems Branch, Southeast Environmental Research Laboratory, Environmental Protection Agency, Athens, GA.

TABLE 1. WASTEWATER CHARACTERISTICS

| Cation | Relative equivalent concentrations | Percent of total |
|---|---|---|
| $NH_4^+$ | 1.0 | 69.2 |
| $K^+$ | 0.156 | 10.8 |
| $Na^+$ | 0.112 | 7.7 |
| $Mg^{++}$ | 0.029 | 2.0 |
| $Ca^{++}$ | 0.149 | 10.3 |

poultry wastewater is present largely in the form of $NH_3$ and $NH_4^+$. Conversion to $NO_2^-$ and $NO_3^-$ could be accomplished if removal in the anionic form were to prove more practical. However, because anionic resins are more easily fouled than cationic resins, and the cationic form of nitrogen is more easily maintained than the anionic, nitrogen removal is more practical in the $NH_4^+$ form. Since the strong acid resin used for removal is not ion specific, competing ion effects also had to be considered. Table 1 gives the relative cationic composition for the poultry wastewaters that were subjected to aerobic stabilization.

Two specific reactions of interest are

$$R\text{-}H^+ + NH_4^+ \rightleftharpoons R\text{-}NH_4^+ + H^+ \dots \dots \dots [5]$$

$$H^+ + HCO_3^- \rightleftharpoons H_2O + CO_2\uparrow \dots \dots \dots [6]$$

The selectivity coefficient for the reaction shown in Equation [5] is

$$K_s = \frac{[R\text{-}NH_4^+]\,[H^+]}{[R\text{-}H^+]\,[NH_4^+]} \dots \dots \dots [7]$$

If Equation [5] represents the ion exchange treatment, the resin utilization is determined largely by $K_s$, i.e.,

$$[R\text{-}NH_4^+] / [R\text{-}H^+]$$

must adjust to satisfy Equation [7] as the reaction in Equation [5] goes from left to right. If significant amounts of bicarbonate are present, $H^+$ is depleted, and utilization will be complete. If only part of the $H^+$ present is depleted resin utilization will be determined by the equilibrium constant of Equation [6], the value of $K_s$ in Equation [7], and the amount of $CO_2$ removed from the system.

The $HCO_3^-$ ion in Equation [6] was chosen to illustrate the buffering effect because it has been shown to be present in poultry wastewater. In principle the same analysis holds for any buffering agent. A rigorous mathematical analysis of exchange equilibrium using values of $K_s$ requires information on the exact nature and quantity of buffering agents and the associated reactions. Use of the separation factor, r, is much easier and more reliable.

## RESULTS AND DISCUSSION

Treatment of a solution with a strong acid resin is similar to titrating it with a strong acid. Fig. 1 shows a standard titration curve obtained by titrating a sample of poultry waste with 0.1N $H_2SO_4$. The wastewater is highly buffered, as illustrated by the long flat portion of the titration curve. The large quantities of $CO_2$ released during this part of the titration as indicated by severe foaming suggests that the major buffering agents are $HCO_3^-$ and $CO_3^=$. If the reaction of Equation [6] is the source of $CO_2$, resin utilization

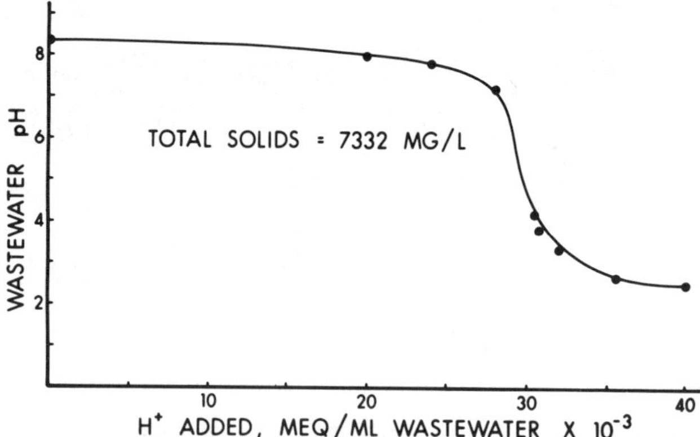

FIG. 1 Poultry wastewater response to addition of strong acid (0.1N $H_2SO_4$.

should be higher than that indicated by the selectivity values based on Equation [2]. It should be noted that the buffering capacity of poultry waste is not unlimited, i.e., column effluents were typically in the pH range of 2-3.

### Loading Isotherm and Flow Directional Studies

Loading isotherms over a range of initial $NH_4^+$ concentrations were generated for both fixed and fluidized beds. An unbuffered synthetic waste made from chloride salts having the cationic profile of Table 1 was also used to determine the relative influence of Equations [5] and [6]. Typical results are shown in Fig. 2. The poor operating characteristics of fixed beds (low resin utilization values) were caused by solids clogging and formation of gas pockets and consequent channeling. The $CO_2$ released, denoted by Equation [6], cannot escape readily from down-flow fixed beds and severely limits operation.

Resin utilization values for fluidized beds are very high and leakage is minimal. Fluid-bed performance for the synthetic waste run clearly shows the beneficial role of buffering agents like $HCO_3^-$ present in the poultry waste. All resin utilization values are corrected for total cations according to the data in Table 1.

### Regeneration

The objective of recovering the $NH_4^+$ as a useful by-product led to the use of $HNO_3$ as a regenerant. Regeneration can be represented by the following equation:

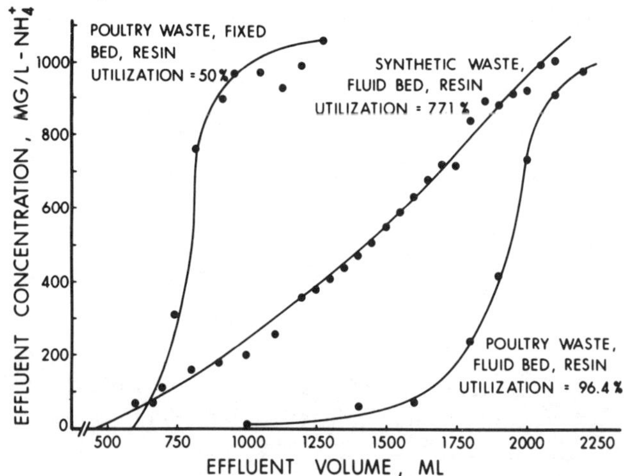

FIG. 2 Influence of flow direction and wastewater characteristics on ion exchange isotherms.



MANAGING LIVESTOCK WASTES

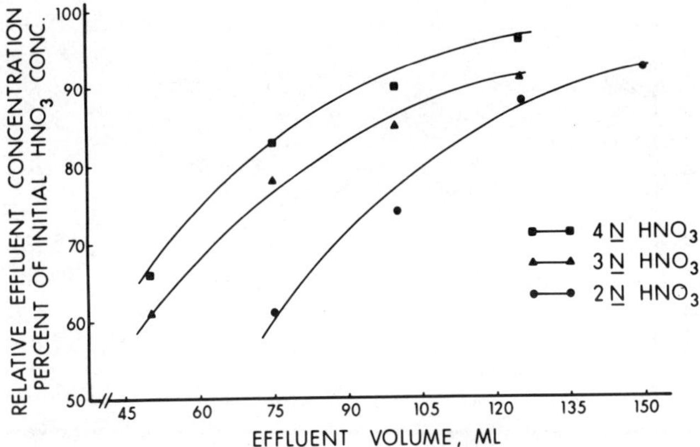

FIG. 3 Regeneration isotherms for different regenerant concentration.

$$R - NH_4^+ + HNO_3 \rightleftharpoons R - H^+ + NH_4NO_3 \ldots \ldots \ldots [8]$$

Ammonium and other nitrate salts, e.g., $Ca^{++}$, $Mg^{++}$, $Na^+$, and $K^+$, are formed yielding a high quality plant nutrient mixture.

Regeneration with 0.5, 1, 2, 3, and 4N $HNO_3$ was accomplished to investigate the influence of acid strength. The 0.5 and 1N $HNO_3$ did not adequately strip the resin of the divalent cations, $Ca^{++}$ and $Mg^{++}$. Typical isotherms for 2, 3 and 4 N $HNO_3$ are shown in Fig. 3. Regeneration is much less efficient than loading. The acid leakage (the area under each curve) is significant in all cases, i.e., considerably more than the stoichiometric amount of acid is required for resin regeneration.

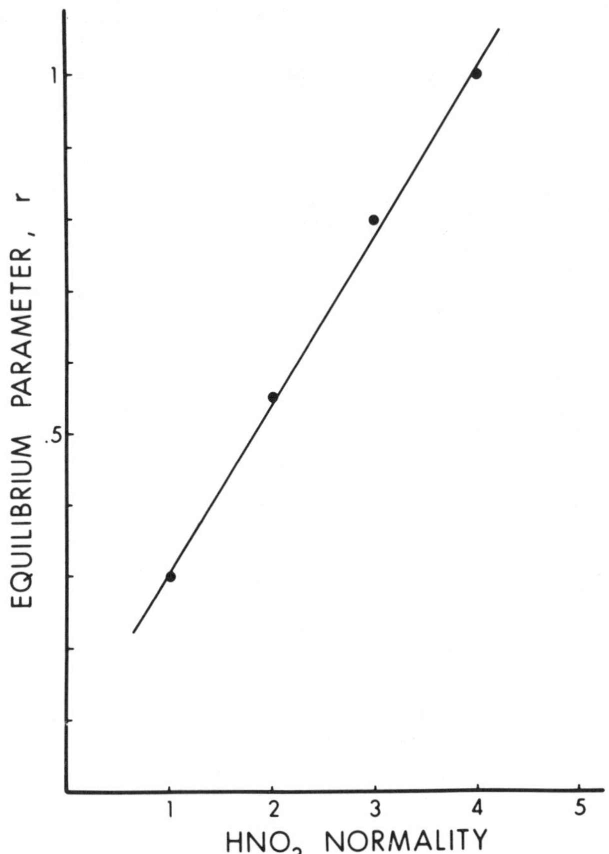

FIG. 4 Influence of regenerant concentration on exchange equilibrium.

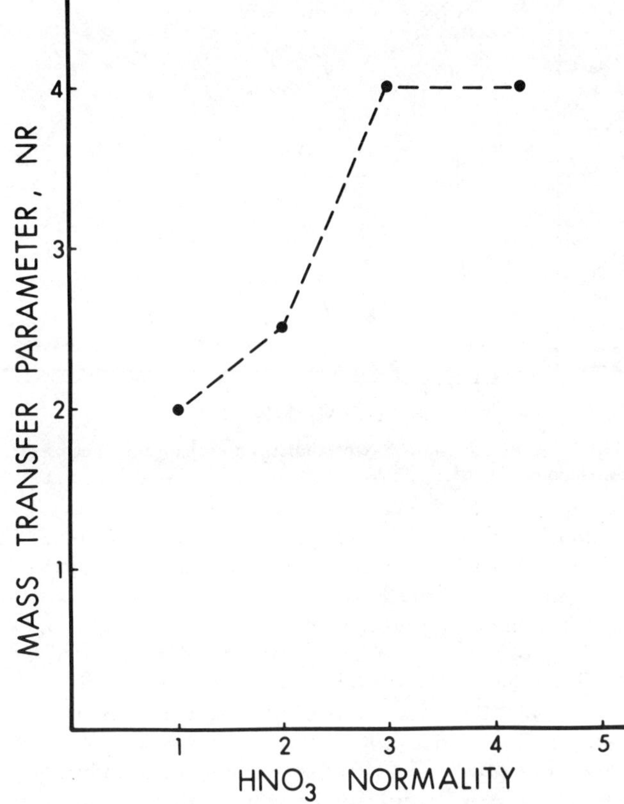

FIG. 5 Influence of regenerant concentration on mass transfer rates.

## Evaluation of Separation and Mass Transfer Parameters

Data from several loading and regeneration isotherms were analyzed to determine the r and k values of Equation [4]. Mean contact time was kept at three minutes for all runs, enabling definition of a new mass transfer term called the column capacity parameter (NR), the product at k and $\bar{t}$ (Heister et al. 1969). If the distribution of cations in Table 1 is assumed to be constant, values developed with $NH_4^+$ data will include the competing ion and parallel reaction effects and accurately represent the system behavior.

Equation [3] was curve fitted to x-y plots from loading and regeneration isotherms. For the $NH_4^+$ concentration range investigated during loading (459-2200 mg/l) the resulting r is independent of concentration (r=0.07). For regeneration at varying acid strengths, r varied according to Fig. 4.

Equation [4] was curve fitted to loading and regeneration isotherms given the previously determined r values and the conditions of the test. For the $NH_4^+$ concentrations during loading, NR was independent of concentration (NR=30.0). For regeneration at varying acid strengths NR varied according to Fig. 5. In the case of the 4N $HNO_3$, an alternate solution to Equation [4] was used (Heister et al. 1969).

Figs. 6 and 7 show the ability of Equation [4] to describe ion exchange treatment of poultry wastewater. Predicted values compare quite well with observed data.

### SCALE-UP DESIGN

Ion exchange must be preceded by a biological treatment process if maximum amounts of $NH_4^+$ are to be removed. Fortunately, conversion of organic nitrogen to

MANAGING LIVESTOCK WASTES

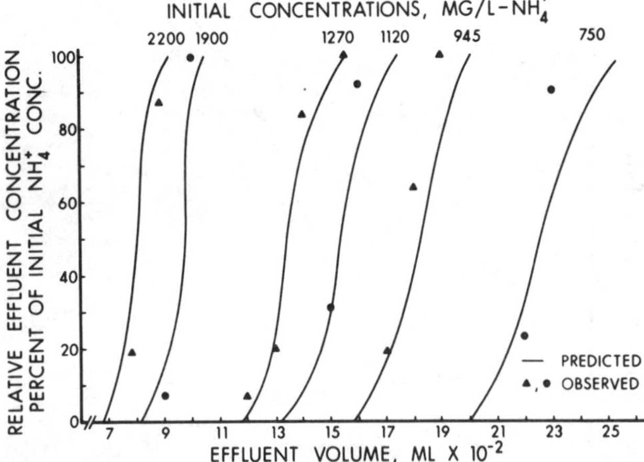

FIG. 6 Use of Equation [4] to predict ion exchange isotherms for resin loading.

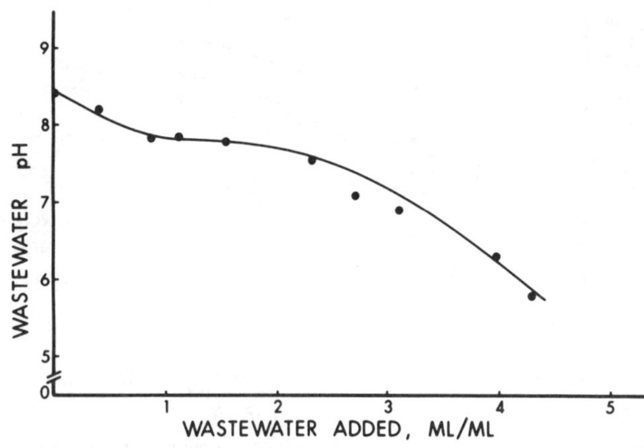

FIG. 8 Poultry wastewater response to addition of acidic, ion-exchange treated wastewater.

ammoniacal forms is a result of biological treatment and can be achieved in either aerobic or anaerobic systems. Nitrogen losses through volatilization and denitrification must be minimized to recover maximum amounts of $NH_4^+$.

The properties of poultry wastewater demonstrated by Fig. 1 can be used to show how ion exchange is compatible with existing biological waste management systems and can improve their efficiency. Oxidation ditches under caged layers are known to be effective in odor control and stabilization, but mass balance studies indicate that large quantities of $NH_3$ can be lost via stripping at the rotor where a large surface area of wastewater is exposed to air (Prakasam et al. 1974). If the pH of the oxidation ditch mixed liquor were lowered just prior to rotor contact, free $NH_3$ would be converted to $NH_4^+$ and be rendered unavailable for air stripping. In the wastewater acid addition studies demonstrated by Fig. 1 each acid increment caused in immediate, almost instantaneous, drop of 2 to 3 pH units, which was followed by a gradual increase in pH to an equilibrium value over a period of 15-20 minutes. Therefore with acid addition to the oxidation ditch, the pH, after dropping sharply at the rotor, would gradually increase as the mixed liquor moves and mixes downstream where $NH_3$ stripping is not a problem.

Acid addition to oxidation ditches for pH control is not practicable, but effluent from a strong acid ion exchange column treating poultry waste is acidic (pH 2-3) and could accomplish pH control in the same manner as pure acid. To test the utility of the acidic, treated poultry wastewater

another titration was performed using it as the titrant. The resulting titration curve is shown in Fig. 8. The abscissa indicates the quantity of recycled wastewater added as the ratio of recycled water volume to fresh wastewater volume. As shown by the curve, four times the original volume can be added, i.e., a recycle rate of 400 percent can be achieved, while depressing the equilibrium pH to no lower than 6. Since at most 100 percent recycle can be achieved during continuous flow operation, and wastes (the source of buffering agents) are being added at the rate of ion exchange removal, the need for land disposal of the acidic wastewater is eliminated. These data suggest the treatment scheme given in Fig. 9. The addition of $NH_3$ or other neutralizers to the regenerant shown in the figure is an option if the excess acid (determined by the area under the curves of Fig. 3) is undesirable.

A system similar to that of Fig. 9 was designed via Equation [4] and waste load data (Mulkey 1975). The design assumptions are as follows:

1   number of birds: 50,000

2   nitrogen loading rate: 3.154 grams Total Kjeldahl Nitrogen (TKN)/bird/day

3   nitrogen wastewater conversion: $NH_4^+$/TKN = 0.8, $NH_3$ losses 10 percent

4   operational mode: up-flow, fluid bed

5   regenerant: 3N $HNO_3$

6   design $NH_4^+$ concentration: $C_i = 1000$ mg/l $NH_4^+$

7   regeneration efficiency: 85 percent

8   resin type: $H^+$ form strong acid, 20-50 mesh with E = 0.4, Q = 2.3 meq/gm, wet basis

9   breakthrough point for loading exhaustion: x = 0.1

The design summary is as follows:

1   Resin Volume: 608.4 liters

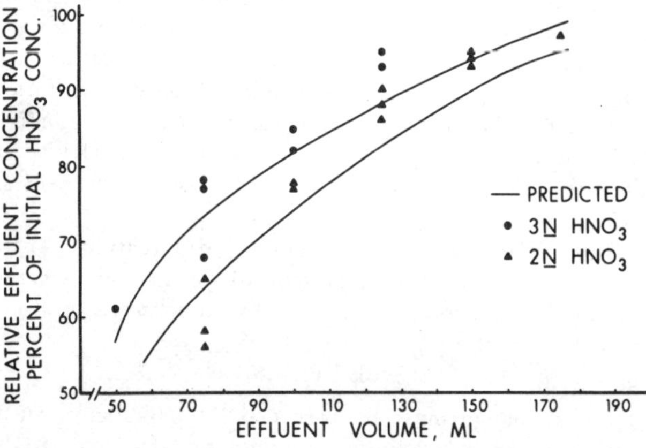

FIG. 7 Use of Equation [4] to predict ion exchange isotherms for resin regeneration.

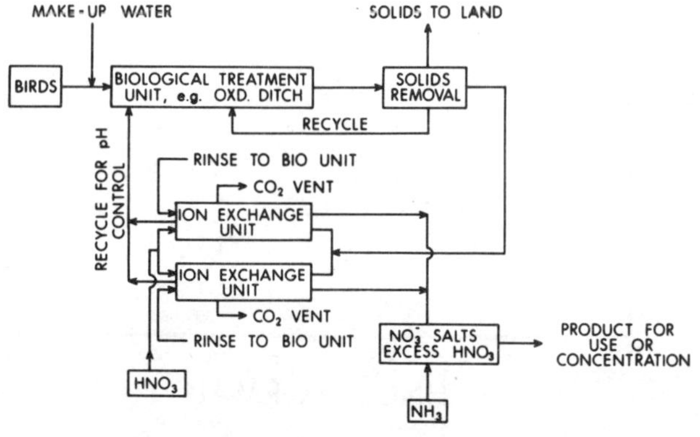


FIG. 9 Process flow chart for ion exchange treatment of poultry wastewaters.

2 Flow Rate: 101.2 liters/min

3 Column Specifications: (a) number, 2 (b) volume, 528.5 liters (c) height, 189.6 cm (d) diameter, 59.6 cm.

4 Operating Characteristics: (a) loading cycle time, 60 minutes (b) regeneration cycle time, 4.21 minutes (c) regenerant volume, 427 liters per cycle; 10,248 liters per day (d) neutralization volume, 11.611 kg $NH_3$ per cycle; 278.7 kg $NH_3$ day.

## CONCLUSION

Ion exchange treatment of poultry wastes with a strong acid $H^+$ form resin regenerated by $HNO_3$ provides an efficient method to remove and recover $NH_4^+$ and $NH_3$ from wastewaters. Fixed-bed columns operate poorly because the $CO_2$ released during exchange reactions and the high suspended solids loading result in channeling, clogging, and poor resin utilization. Resin utilization is essentially complete, however, for fluidized bed systems and regeneration results in formation of a concentrated nitrate salt solution $(NH_4^+, Ca^{++}, Mg^{++}, Na^+, K^+)$ for use as a nitrogenous fertilizer source.

Ion exchange units are compatible with conventional waste management systems and can be designed to enhance the overall performance by preventing $NH_3$ losses through air stripping. No special land disposal for acidic waste effluents is required. Design projections for a 50,000 bird operation indicate complete treatment can be achieved with a modest resin requirement of only 9.76 grams resin/bird.


**References**

1 Dorfner, Konrad. 1972. Ion exchangers, properties and applications. Ann Arbor Science Publishers, Inc. 316 p.

2 Dow Chemical Company, Walnut Creek, Calif. 1971. Nitrate removal from wastewaters by ion exchange. EPA-17010FS01/71. U.S. Environmental Prot DC.

3 Hiester, N. K., Vermeulen, T. and C. Klein. 1969. Adsorption and ion exchange. Section 16, p. 1-40. In: J. H. Perry, (ed.) Chemical engineer's handbook. 4th edition. McGraw-Hill Inc. New York.

4 Higgins, I. R. and R. C. Chopra. 1970. Chem-Seps continuous countercurrent ion exchange for removal and recovery of ammonium nitrate. Preliminary Report, EPA Project 12020EGM. 25 p.

5 Kreusch, E. and K. Schmidt. 1971. Wastewater demineralization by ion exchange. EPA-17040EEE 12/71. U.S. Environmental Protection Agency, Washington, DC. 129 p.

6 Mulkey, L. A. 1975. Nitrogen control in animal wastewaters by ion exchange. U.S. Environmental Protection Agency Report in preparation.

7 Prakasam, T. B. S., R. C. Loehr, P. Y. Yang, T. W. Scott, and T. W. Bateman. 1974. Design parameters for animal waste treatment systems. EPA-660/2-74-063. U.S. Environmental Protection Agency, Washington, DC. 218 p.

8 Robinson, D. J., H. E. Weisberg, G. I. Chase, R. L. Libby, and J. L. Capper. 1974. An ion-exchange process for recovery of chromate from pigment manufacturing. EPA-670/2-74/044. U.S. Environmental Protection Agency, Washington, DC. 91 p.

9 Sanks, R. L. 1973. Ion exchange color and mineral removal from kraft bleach wastes. EPA R2-73-255. U.S. Environmental Protection Agency, Washington, D.C. 189 p.

10 University of California, Berkeley. 1971. Optimization of ammonia removal by ion exchange using clinoptilolite. EPA-17010FS01/71. U.S. Environmental Protection Agency, Washington, D.C. 189 p.


---

# Effect of Swine Lagoons on Groundwater

*(Continued from page 305)*


**References**

1 Davis, S., W. Fairbank and H. Weisheit. 1973. Dairy waste ponds effectively self-sealing. TRANSACTIONS of the ASAE 16(1):69-71.

2 Humenik, F. J. 1972. Swine waste characterization and evaluation of animal waste treatment alternatives. Report No. 61, Water Resources Research Institute, University of North Carolina.

3 Koelliker, J. K., et al. 1971. Treatment of livestock - lagoon effluent by soil filtration. Proceedings of International Symposium on Livestock Wastes: 329-333. Columbus, Ohio.

4 Loehr, R. C. 1971. Alternatives for treatment and disposal of animal wastes. Journal of the Water Pollution Control Federation 43: 668-678.

5 Nordstedt, R. A., L. B. Baldwin and C. C. Hortenstine. 1971. Multistage lagoon systems for treatment of dairy farm waste. Proceedings of the International Symposium on Livestock Wastes: 77-80. Columbus, Ohio.

6 U. S. Environmental Protection Agency. 1973. Water quality criteria - 1972. Report EPA - R3-73-033: 48-104.

# Oxidation-Nitrification and Denitrification of Veal Calf Manure

H. G. van Faassen, H. van Dijk

AT the ASAE International Symposium on Livestock Wastes in Ohio in 1971, Ten Have (1971) reported on aerobic decomposition of veal calf and pig liquid manure and pig urine in 35 to 1000 m³ aeration basins and oxidation ditches.

To get a more flexible approach we started experiments in 2 to 20 liter fermentors. In this way it is much easier to investigate the influence of varying conditions and process operation on purification results. We studied the following questions:

Why is there an incomplete oxidation of ammonium and an accumulation of nitrite? Is this inherent to the procedure applied, to substrate composition or to environmental factors?

How can biological treatment be optimized to get an acceptable effluent quality regarding BOD, N and P and a small volume of surplus sludge?

## THEORY OF OXIDATION-NITRIFICATION AND DENITRIFICATION

What happens when we add manure, containing BOD and ammonia (= NOD), to an aerated activated sludge suspension (Fig. 1A)? A rapid substrate respiration will occur, leading to a low level of dissolved oxygen, thus inhibiting nitrification and even bringing along denitrification of part of the nitrite or nitrate accumulated in the foregoing period. After some time the BOD cannot be met and respiration will slow down to endogenous respiration. The level of dissolved oxygen will then increase, resulting in the inhibition of denitrification and stimulating nitrification (nitrifiers use carbondioxide as a carbon source). When the latter is completed, the dissolved oxygen level will further increase and a new load may be added.

However, the "over-all" removal of nitrogen via nitrification and denitrification has certain limits. The substrate has to supply enough BOD for the denitrification if no additional BOD, as for example methanol, is added or otherwise there will be an accumulation of nitrite or nitrate. It is also known that denitrification of nitrate demands more BOD than denitrification of nitrite. Theoretically, complete denitrification demands at least 1.7 g of substrate-BOD per g

of nitrite-nitrogen and 2.9 g per g of nitrate-nitrogen. In practice complete denitrification requires about twice the theoretical amount of BOD (some BOD is used for cell synthesis). The overdose of BOD should be removed afterwards via respiration with oxygen.

It will be clear that the largest possible denitrification will be achieved when we inhibit the substrate respiration for some time after the addition of fresh substrate, which can be done by stopping the aeration. This procedure is outlined in Fig. 1B.

It may be mentioned that in certain cases nitrification and denitrification can also occur simultaneously when an oxygen gradient exists in the sludge flocs. We met this situation under certain conditions in experiments with pig manure.

For a more detailed discussion of the theory we refer to McCarthy and Haug (1971).

## MATERIALS AND METHODS

Well defined liquid manure was prepared by mixing fresh urine and feces, in the ratio as excreted, sieving coarsely (0.6 mm) and storing at 4 C. Difficulties with highly variable samples, taken from manure pits, were thus eliminated. Once a day or more the activated sludge was loaded with liquid manure, after removal of effluent (further on called primary effluent) or mixed liquor. Primary effluent or mixed liquor was drawn off when ammonium-nitrogen was almost completely nitrified during aeration. As in Fig. 1B, aeration was stopped after addition of a load to denitrify nitrite and/or nitrate present in the mixed liquor. After more or less complete denitrification, aeration was resumed. More than 90 percent of the denitrification took place in the mixed liquor, since less then 10 percent of the nitrite was removed with primary effluent. This 10 percent was lost by denitrification during storage; to increase that denitrification rate some methanol was added. A "secondary denitrified" effluent could then be separated from "denitrifying sludge" by gravity settling after stirring.

Three different veal calf manures have been used. Their composition and the percentages of the components derived from the urine are shown in Table 1. Samples A, B and C were from 2, 6 and 6 weeks old calves. In case of sample B the calves had been fed with soya as part of their

The authors are: H. G. VAN FAASSEN and H. VAN DIJK, Institute for Soil Fertility, Haren (Gr.), The Netherlands.

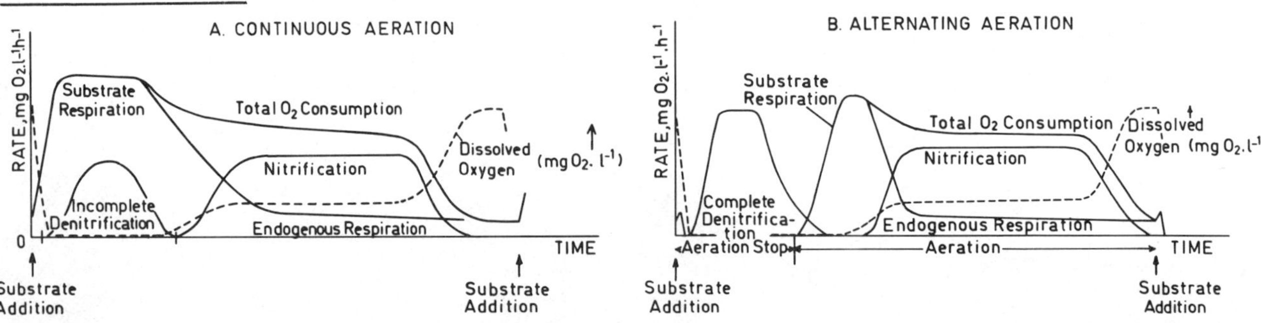


FIG. 1 Dissolved oxygen and rates of respiration, nitrification and denitrification in an activated sludge system with continuous or alternating aeration.

**TABLE 1. COMPOSITION OF THREE VEAL CALF MANURES USED IN THE EXPERIMENTS AND PERCENTAGES OF THE COMPONENTS CONTRIBUTED BY URINE.**

| Components | g per liter in sample | | | Percent contributed by urine to sample | | |
|---|---|---|---|---|---|---|
| | A | B | C | A | B | C |
| COD | 24 | 38 | 24 | 36 | 20 | 41 |
| Volatile solids | 12 | 23 | 11 | 42 | 24 | 47 |
| Ash | 12 | 13 | 12 | 89 | 64 | 81 |
| Kjeldahl-N | 4.4 | 6.0 | 7.4 | 86 | 74 | 91 |
| $NH_4$-N | 3.5 | 4.4 | 6.6 | 97 | 97 | 98 |
| Phosphorus (P) | 0.8 | 1.3 | 1.5 | 91 | 52 | 77 |

ration and therefore produced relatively more feces. As can be seen from Table 1 most of the nitrogen, phosphorus and ash, but less than half of the Chemical Oxygen Demand (COD) and volatile solids, came from the urine. Liquid manure, mixed liquor and effluent were analysed for COD, Kjeldahl-nitrogen, ammonium-nitrogen, total and volatile solids (ash) and phosphorus in every experiment; thus, material balances could be calculated. Once a day or more ammonium-, nitrite-, and nitrate-levels were estimated by color-reactions and teststrips; besides pH and redox potential were measured. Dissolved oxygen in the mixed liquor and the oxygen pressure in the air after passage through the mixed liquor were measured with temperature compensated oxygen electrodes. From the recorded value of the oxygen pressure after passage through the mixed liquor and the aeration rate, the amount of oxygen consumed could be calculated. Phosphates were separated from the effluent by flocculation with iron-aluminum-sulphate or lime, followed by settling and filtration; by adding lime to the mixed liquor more phosphate could be discharged with the surplus sludge.

The series of experiments described here was performed in a 2 liter fermentor with 1.2 liter (number 1-12) or 1.6 liter (number 12-16) activated sludge, during 7 months.

## CONTROL OF PROCESS CONDITIONS

We tried to control the process conditions so that no bad smell was noticeable and a maximum loss of nitrogen was achieved. Air supply during aeration was amply sufficient in all cases. An aeration scheme was started with experiment 13 in which aeration was alternated with non-aeration 4-6 times per 24 hours. This appeared to be no success because the nitrification of the same load was no longer completed in 24 hours and too much BOD was lost during aeration periods, thus leading to incomplete denitrification. This schedule may be more successful, if the load is also split up, which is being investigated at the moment. Aeration rate and time, temperature and sludge concentration (estimated from analyses) are given in Table 2. Stirring rate was around 450 ppm. This rate influenced the oxygen transfer to the solution and therefore oxygen consumption and dissolved oxygen level. Thus, varying this rate occasionally was used to lengthen either the denitrification time or the nitrification time (Fig. 1B). The pH of the mixed liquor was sometimes adjusted by adding lime, if pH went below 7.0. The sludge load varied with variations in sludge level and loading. The sludge load with COD and with NOD (calculated as Kjeldahl-N times 3.4) is given in Table 3. The removal of COD and NOD in our system proved to be about 50 percent and more than 80 percent, respectively (Table 4). The average daily load with COD and NOD (Table 3) was about 2.0 and 1.2, respectively. Thus, the amount of oxygen used for the oxidation of nitrogen was about equal to the amount used for the oxidation of organic carbon. During part of experiment 15 and during experiment 16 the BOD-load was increased by addition of methanol to the mixed liquor together with the manure load. This was done to

**TABLE 2. AERATION RATE, AERATION TIME, TEMPERATURE AND ESTIMATED SLUDGE CONCENTRATION**

| Experiment number | Aeration rate, 1 air $.h^{-1}$ | Aeration time, Percent of time | Temperature deg C | Sludge concentration g MLVSS $.1^{-1}$† |
|---|---|---|---|---|
| 1 - 5 | 2 | 85 - 94 | 16 - 29 | 5 - 12 |
| 6 | 2 | 90 | 14.5 | 7 - 13 |
| 7* | 2 | 70 | 14.5 | 8 - 15 |
| 8 | 1 - 2 | 82 | 14.5 | 10 - 28 |
| 9 - 11 | 0.5 - 1 | 82 - 92 | 14.5 | 7 - 15 |
| 12 - 14 | 1 | 71 - 90 | 14.5 | 7 - 10 |
| 15* | 1 | 54 | 14.5 | 6 - 7 |
| 16 | 1 | 50 | 14.5 | 6 - 7 |

*Transition to sample B and C, respectively
†MLVSS = Mixed Liquor Volatile Suspended Solids

**TABLE 3. DAILY LOAD, SLUDGE LOAD AND VOLUME LOAD WITH OXIDIZABLE CARBON (COD) AND NITROGEN COMPOUNDS (NOD) FROM MANURE.**

| Experiment number | Daily load | | Sludge load | | Volume load |
|---|---|---|---|---|---|
| | COD | NOD | COD | NOD | COD + NOD, kg $0_2.m^{-3}.day^{-1}$ |
| | g $0_2.day^{-1}$ | | g $0_2.$g MLVSS$^{-1}.day^{-1}$ | | |
| 1 - 5 | 2.4 | 1.4 | 0.20 | 0.12 | 3.1 |
| 6 | 1.9 | 1.2 | 0.14 | 0.09 | 2.6 |
| 7 | 3.0 | 1.5 | 0.19 | 0.10 | 3.7 |
| 8 | 3.0 | 1.6 | 0.13 | 0.06 | 3.8 |
| 9 - 11 | 1.6 | 0.9 | 0.10 | 0.05 | 1.8 |
| 12 - 14 | 2.1 | 1.2 | 0.14 | 0.08 | 2.1 |
| 15 | 0.85 (0.90*) | 0.9 | (0.08*) | 0.08 | 1.1 |
| 16 | 0.65 (1.0 *) | 0.7 | (0.09*) | 0.07 | 1.1 |

*Addition of added methanol included

TABLE 4. REMOVAL OR TRANSFORMATION; DISCHARGE OF NITROGEN COMPOUNDS WITH SURPLUS SLUDGE AND EFFLUENT.

| Experiment number | Removal or transformation | | | | Discharge of nitrogen compounds | |
|---|---|---|---|---|---|---|
| | COD | TVS† | Kjeldahl-N | $NH_4$-N | with surplus sludge | with effluent |
| | Percent of influent | | | | Percent of influent | |
| 1 - 5 | 60 | 36 | 86 | 97 - 100 | 8 - 13 | 1 - 6 |
| 6 | 65 | 21 | 83 | 97 | 14 | 4 |
| 7 | 51 | 32 | 80 | 99 | 15 | 4 |
| 8 | 35 | 30 | 77 | 97 | 21 | 2 |
| 9 - 11 | 41 | 24 | 76 | 95 - 103 | 15 - 22 | 2 - 4 |
| 12 - 14 | 48 | 36 | 81 | 96 - 102 | 16 - 18 | 0 - 3 |
| 15 | 51(53*) | 25 | 88 | 96 | 8 | 4 |
| 16 | 58(70*) | 40 | 85 | 95 | 10 | 5 |

*Removal of added methanol included
†TVS = Total Volatile Solids

improve the very low BOD/N ratio in manure sample C, and thus make complete denitrification possible. We also adjusted the time during which the mixed liquor was not aerated, so that maximum use was made of the carbon compounds in the substrate for denitrification. As the denitrification time was increased we had to lower the daily load since then less time was available for nitrification and removal of the overdose of BOD. The sludge concentration was controlled by removal of mixed liquor at the end of each experiment. Because of the presence of sludge, discharged mixed liquor gave a more rapid denitrification than primary effluent; no methanol had to be added to achieve this.

## RESULTS AND DISCUSSION

The effluent quality is shown in Table 5. Ammonium- and Kjeldahl-nitrogen levels were still higher than desirable, but much lower than the values given by Ten Have (1971).

Remarkably even in the primary effluent hardly any nitrate could be detected. This was in contrast with preliminary experiments at a temperature of about 30 C, where nitrate dominated. The absence of nitrate might be explained from the fact that dissolved oxygen levels were more frequently and for a longer period zero or very low due to higher loading and longer periods without aeration. Nitrification to nitrite only means a more economic use of BOD as explained under "Theory". The remainder of nitrite in the denitrified effluent was caused by too small a dose of methanol or too short a denitrification time. The effluent still had a considerable COD-value, even after chemical flocculation. It is interesting that this did not mean that the $BOD_5$-level too was very high. For some effluents the COD/$BOD_5$ ratio was determined and appeared to be about 20. For the influent this ratio was about 2. Without lime addition the phosphate levels of the effluent were high, in spite of a large decrease compared to those of the influent. The lower phosphate levels during experiments 8-14 might be explained, at least partly, from the lime additions to the mixed liquor.

### Decomposition, Transformation and Discharge

Mostly somewhat more of the influent Kjeldahl-nitrogen was lost than initially present as ammonium-nitrogen. This means that part of the organic nitrogen of the influent has been ammonified and afterwards nitrified and denitrified. The transformation of ammonium-nitrogen thus almost always was more than 100 percent when related to the amount initially added.

Table 4 shows the calculated discharge of nitrogen compounds with surplus sludge and effluent; in the calculation all $NH_4$-N present in the mixed liquor at the end of the experiments has been included in that discharged with the effluent. Thus, the discharge of nitrogen with effluent as shown in Table 4 is less favorable than in reality. A discharge of 1 percent of influent nitrogen with the effluent seems real. The discharge of nitrogen compounds with the surplus sludge may depend both on temperature and influent composition. The larger decrease of the COD (Table 4) in the experiments 1-6 compared to 8-14 may be caused by differences in temperature, substrate and/or other conditions.

### Oxygen Consumption

The rate of oxygen consumption was higher at the higher temperatures. The oxygen consumed per kg of manure has been calculated from direct measurements and indirectly from material balances. From the theory the "overall" decrease of COD plus NOD should be equal to the oxygen consumed. The decrease of NOD is calculated as 1.7 times the nitrogen loss. In Table 6 the amount of oxygen consumed is compared with the calculated amount for some experiments; the agreement is good. From the values in Table 6 it can be calculated that 42, 49 and 84 percent, respectively, of the oxygen used for the oxidation of carbon compounds, from the samples A, B and C, is delivered by nitrite during denitrification. However, with sample C methanol was also added to the mixed liquor. If the oxidation of methanol is not included in the calculation for sample C, then 45 percent of the carbon oxidation took place by denitrification, which fits in rather well with the values of 42 and 49 percent. These percentages again show that denitrification demands an overdose of BOD. This can also be seen from the ratio of COD-removal and nitrogen

TABLE 5. EFFLUENT QUALITY OF SECONDARY (DENITRIFIED) EFFLUENT

| Experiment number | COD, g $O_2.1^{-1}$ | Kjeldahl-N | $NH_4$-N | $NO_2$-N | P |
|---|---|---|---|---|---|
| | | mg.$1^{-1}$ | | | |
| 1 - 5 | 1.5 - 2.3 | 70 - 220 | 35 - 70 | 2 - 20 | 220 - 380 |
| 5 | 1.2 | 100 | 40 | 25 | 150 |
| 7 | 2.2 | 190 | 100 | 5 | 280 |
| 8 | 1.7 | 100 | 50 | 30 | 25 |
| 9 - 11 | 1.5 | 100 | 85 | 0 - 10 | 15 - 120 |
| 12 - 14 | 1.8 - 2.2 | 60 - 140 | 10 - 25 | 0 - 40 | 40 |
| 15 | 1.6 | 40 | 35 | 0 | 35 |
| 16 | 2.4 | 130 | 30 | 5 | 70 |

**TABLE 6. COMPARISON OF CALCULATED AND MEASURED OVER-ALL OXYGEN CONSUMPTION: RATIO BETWEEN DECREASE OF COD AND NITROGEN LOSS.**

| Experiment number | Decrease of COD | + Decrease of NOD | = | Calculated $O_2$-consumption | Measured $O_2$-consumption | $\dfrac{\text{Decrease of COD}}{\text{Nitrogen loss}}$ g $O_2$/g $N_2$ |
|---|---|---|---|---|---|---|
| | | "overall" values expressed as g $O_2$/kg veal calf manure | | | | |
| 1 - 5 | 15.5 | 6.5 | | 22.0 | 21.5 ± 2 | 4.1 |
| 6 | 14.5 | 6.0 | | 20.5 | 22 ± 2 | 4.0 |
| 7 | 18.5 | 7.5 | | 26.0 | --- | 4.2 |
| 8 | 13.5 | 7.5 | | 21.0 | 24 ± 2 | 3.1 |
| 9 - 11 | 16.0 | 8.0 | | 24.0 | — | 3.3 |
| 12 - 14 | 17.5 | 9.0 | | 26.5 | — | 3.3 |
| 15 | 12.0 | 11.0 | | 23.0 | — | 1.9(2.1*) |
| 16 | 13.0 | 11.0 | | 24.0 | — | 2.1(3.7*) |

*Removal of added methanol included

loss (Table 6); during the experiments 15 and 16 with manure sample C, this ratio had to be increased by methanol addition, to get complete denitrification.

## CONTINUATION OF THE PROJECT

We are now automating the procedure, in order to investigate the results with more frequent loading split up over 24 hours. Experiments with larger scale plants have recently been started in The Netherlands in Wageningen. Efforts will be made to improve the effluent quality still more. Parallel, experiments with pig manure are also going on. Because of the higher BOD to nitrogen ratio in that substrate, the overdose of nitrogen is eliminated by simultaneous nitrification and denitrification. Our aim here is to produce biomass for animal feed combined with an acceptable effluent.

## CONCLUSIONS

1 The nitrite accumulations found in practice during aerobic treatment of veal calf manure are caused by incomplete denitrification, during the time that dissolved oxygen is zero (Fig. 1A). Because of continued aeration after loading, maximum use of BOD for denitrification is not made (Fig. 1A vs. 1B). Over-aeration will lead to high dissolved oxygen levels and a high redox potential, further inhibiting denitrification. However, even with an optimal process, the BOD/N ratio of the manure may be too low for complete denitrification without addition of e.g. methanol.

2 The nitrite accumulation, accompanied by a low pH will inhibit further nitrification. This explains the accumulation of ammonia.

3 For optimal biological treatment, oxidation-nitrification should be alternated with denitrification, in such a way that in each cycle the added ammonia is fully nitrified and nitrite is kept below 500 mg/l $NO_2$ by denitrification. Thus, per cycle about 150 mg of ammonifiable nitrogen can be added per liter mixed liquor. How long aeration must be stopped per cycle will depend on the denitrification rate at the particular temperature and substrate composition. The pH should preferably be kept between 7 and 8, e.g. by adding lime.

4 Nitrification should be limited to nitrite formation in the case of wastes with a low BOD/N ratio; this can be done by working at low dissolved oxygen levels and a redox potential ($E_{cal}$) below + 100 mV. In this way complete denitrification can be achieved with the smallest amount of BOD.


**References**

1 McCarthy, P. L. and R. T Haug. 1971. Nitrogen removal from wastewaters by biological nitrification and denitrification p. 215-232. In: G. Sykes and F. A. Skinner (eds.), Microbial aspects of pollution. Soc. Appl. Bacteriol. Symp. Ser. 1. Academic Press, New York.

2 Ten Have, P. 1971. Aerobic biological breakdown of farm waste, p. 275-278. In: Livestock waste management and pollution abatement. Proc. Int. Symp. Livestock Wastes. ASAE, St. Joseph, Mich. 49085.

# Bacterial Analysis and Land Disposal of Farm Waste Lagoon Waters

D. R. Smallbeck, M. C. Bromel

THE combination of expanding populations and dwindling land resources has resulted in the confinement of large cattle herds on a minimum amount of land. In order for such a system to succeed, it must not create a public health hazard that may increase the likelihood of disease transmission.

Van Donsel and Geldreich (1971) have reported that analysis of bottom sediments from freshwater environments may furnish a concentrated and stable index of the quality of the overlying water, especially where there is a great variability in the bacterial quality of the water. Rittenberg et al. (1958) observed higher densities of fecal coliforms in the mud than in the water around California marine sewage outfalls, suggesting prolonged bacterial survival after sedimentation. Van Donsel and Geldreich (1971) observed a positive correlation between fecal coliform densities and salmonellae occurrence in bottom sediments of rivers and streams.

The survival capability of potentially pathogenic organisms must be considered whenever discharge of wastewater into streams or on land is a common practice. Mair and Ross (1960) reported survival of salmonellae in soil for up to 307 days. Dazzo et al. (1973) examined the influence of frequent recontamination of soil with manure slurry on the survival rate of *Salmonella enteritidis* and found that repeated applications of the slurry favored the extended survival of *S. enteritidis.*

This study was conducted to establish the following: the relationships between the lagoon sediments and the overlying water of certain groups of bacteria, the isolation of pathogenic organisms from the lagoon sediment and water, and the effect of land disposal of wastewater on the survival of certain indicator bacteria.

## MATERIALS AND METHODS

Two animal waste lagoons located at North Dakota State University were the source of the water and sediment samples which were collected bimonthly. The locations chosen for wastewater disposal received extremes of environmental exposure. One site was located in a wooded area, with limited human and animal activity. The other site, located in a pasture area with 4 to 18 inches of grassy vegetation was subjected to considerable animal activity. The soil in both sites was a coarse Fargo clay, rich in organic material.

Organisms of the genus *Bacillus*, were enumerated by heat shocking 5 ml quantities of 10-fold serial dilutions of water and sediment samples with subsequent plating on trypticase Soy Agar (TSA) (Baltimore Biological Laboratories) (BBL). The coliform and fecal coliform groups were enumerated using the multiple tube fermentation technique, presumptive, confirmed, completed, and elevated temperature tests as described in Standard Methods for the Examination of Water and

Wastewater (APHA 1971). The fecal streptococci were enumerated using KF Streptococcus Agar (BBL) and the pour plate technique in accordance with Standard Methods (APHA 1971).

Ten milliliters of each water and sediment sample were enriched for *Salmonella* in Selenite Cystine Broth (BBL). Positive enrichment tubes were verified by streaking on Hektoen Enteric Agar (Difco) from which subsequent isolates were subjected to biochemical tests including: Triple Sugar Iron Agar (Difco), urease, indole, lysine decarboxylase, methyl red, Voges Proskauer, and rhamnose fermentation. Biochemically positive cultures were confirmed by serological typing with *Salmonella* Polyvalent O Antiserum (BBL).

The isolation and characterization of coagulase-positive staphylococci was achieved by employing spread plates of Vogel Johnson Agar (BBL) plus 40 ml of a 1 percent potassium tellurite solution per liter. Suspect colonies were gram stained and subjected to the coagulase plasma test (Difco) by the tube method.

*Salmonella typhimurium*, ATCC number 13311, and *Escherichia coli*, ATCC number 11775, were the control organisms used in the survival study.

Soil inoculum was prepared by inoculating Trypticase Soy Broth (BBL) with the appropriate organism and incubated at 37 C for 24 hours after which the 50 ml portion was added to a flask containing 450 ml of filter sterilized lagoon water. A 2 ft by 1 ft soil area duplicating as closely as possible the disposal area was inoculated by evenly pouring the 500 ml of the seeded lagoon water on the soil surface.

Initial soil counts were made 24 hours after inoculation, then daily for 2 weeks and finally, three times a week until the organism could no longer be detected. Fecal coliforms were enumerated as previously described. Salmonellae were enumerated on 150 x 15 mm spread plates of Hektoen Enteric Agar receiving 1 ml aliquots of the sample and incubated at 43 C with subsequent identification according to biochemical and serological criteria. Uninoculated soil controls were examined and found to be free of salmonellae.

## RESULTS AND DISCUSSION

The relationship of the fecal coliform group and the fecal streptococcus group in lagoon sediments and overlying water is shown in Fig. 1. The fecal coliform populations in the sediments were 8 to 10 times higher than in the overlying water. The fecal streptococcus population in the sediments versus the overlying water did not vary appreciably during the winter months; however, a one-to-three-fold streptococcal population increase in the sediment was observed during the summer months. The decrease in fecal coliform and fecal streptococcus population densities in the water during the summer months may possibly be attributed to such environmental factors as increased ultraviolet radiation and temperature increase. The increased population

The authors are: D.R. SMALLBECK and M.C. BROMEL, Bacteriology Dept., North Dakota State University, Fargo.



MANAGING LIVESTOCK WASTES

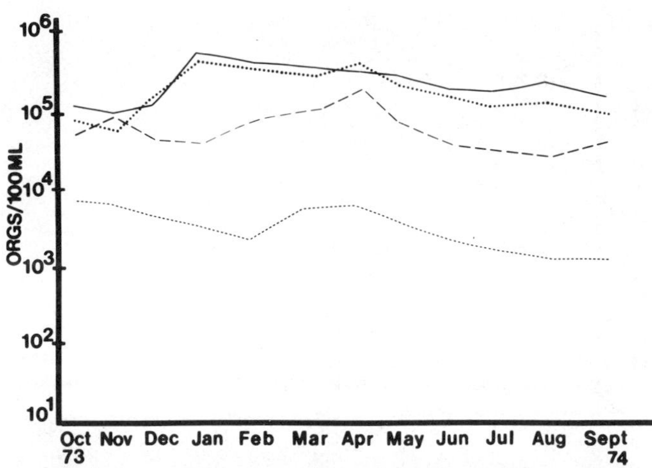

FIG. 1 FC-FS populations in lagoon sediment versus overlying water. -----FC water. — —FC sediment. ·····FS water. ——FS sediment.

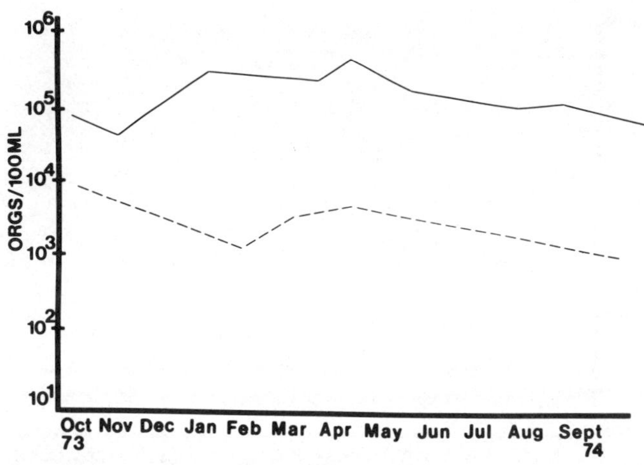

FIG. 3 Comparison of FC-FS populations in lagoon water. — —FC. ——FS.

densities of the fecal coliform group in the sediment are in agreement with Van Donsel and Geldreich (1971) and Rittenberg et al. (1958).

The relationship of the aerobic sporeforming *Bacillus* population densities in the lagoon sediment and the overlying water is shown in Fig. 2. Aerobic sporeforming *Bacillus* populations in the sediments were 10 to 100 times greater than in the overlying water. This group of microorganisms exhibited a trend opposite to that of the fecal coliform and fecal streptococcal group since an increase in the *Bacillus* population of the overlying water occurred during the summer months. The increase in the numbers of the aerobic sporeformers are more resistant to environmental conditions deleterious to the fecal coliform and fecal streptococcus groups. It is also possible that the increase in the sporeforming bacilli could be due to the influx of proteinaceous nutrients which would promote increased growth.

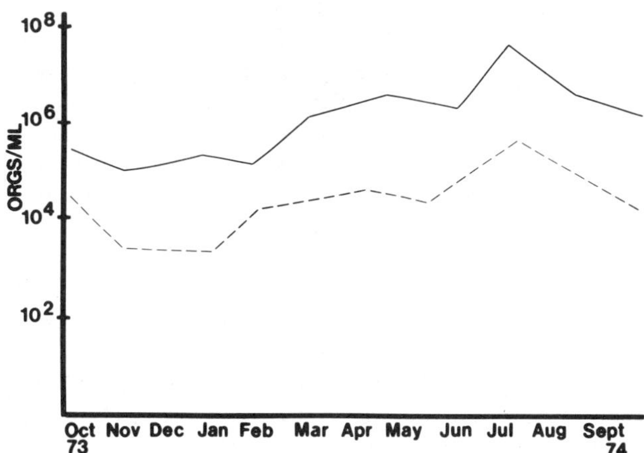

FIG. 2 Aerobic sporeforming bacilli populations in lagoon sediment versus overlying water. — —water. ——sediment.

The average population densities of fecal coliforms and fecal streptococci in lagoon water are depicted in Fig. 3. During the first 5 months the fecal coliform densities decreased slowly while in contrast the fecal streptococcus densities dropped initially and then rose and remained at the higher level throughout the

sampling period. Blannon and Peterson (1974) have reported that the fecal streptococcus group survived longer than the fecal coliform group in a sanitary landfill. Geldreich (1970) has observed that the fecal streptococci survive longer than the fecal coliforms in irrigation waters and that they should be considered as another parameter for determining fecal contamination in water.

The isolation frequency of pathogenic organisms is shown in Table 1. Seven of eight sediment samples were positive for *Salmonella* as compared to four of eight water samples. The isolation frequency of coagulase-positive staphylococci was also observed to be higher in the sediment than in the overlying water. Van Donsel and Geldreich (1971) has reported an isolation frequency for salmonellae of 80 percent in sediments when fecal coliform population densities are in excess of 2000 per 100 ml of water. Our results indicate a *Salmonella* isolation frequency of 87.5 percent which may be explained by the closed environment and rich nutrient loading afforded by the lagoon studied. Also *Salmonella* infection rates in animal populations are generally higher than in humans. In clinically healthy cattle, 13 percent have been found to be carriers of salmonellae (Prost and Rieman 1967).

TABLE 1. RECOVERY OF SALMONELLAE AND STAPHYLOCOCCI

| Groups of microorganisms | No. of samples | Percent of positive water samples | Percent of positive sediment samples |
|---|---|---|---|
| Salmonellae (sero-typed) | 8 | 50 | 87.5 |
| Staphylococci (coagulase test) | 3 | 33 | 67 |

The fecal coliform and fecal streptococcus groups were employed to study the effects of wastewater disposal in a free area with the results depicted in Fig. 4. The initial populations of the indicator organisms obtained on June 10 were found to be approximately 200 per 10 g soil for the fecal coliform group and 700 per 10 g soil for the fecal streptococcus group. A steady increase in the fecal coliform population as well as the fecal streptococcus population occurred during the disposal period. These

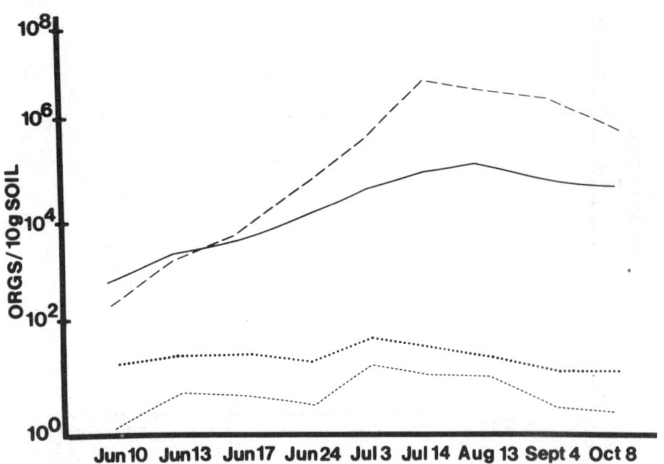

FIG. 4 FC-FS populations during land disposal of wastewater. -----FC control. ·····FS control. ———FS disposal area. — —FC disposal area.

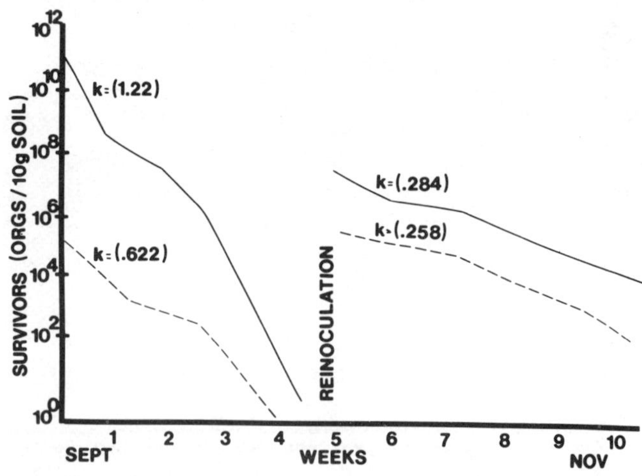

FIG. 6 Survival curves of *E. coli* and *S. typhimurium* in soil. ———*E. coli.* — —*S. Typhimurium.*

results indicate that these fecal bacteria can withstand the competition of soil organisms. The schedule of waste water disposal is shown in Table 2. The large amount of wastewater released on the disposal area may have contributed to the increase in fecal bacteria by modifying the soil environment with high nutrient loading and in keeping the soil in a saturated state which may have promoted the survival of these organisms.

TABLE 2. SCHEDULE OF PUMPING AT
RESEARCH CENTER.

| Month | Total gallons | Equivalent rainfall, inches of water |
|---|---|---|
| June | 66 300 | 7.4 |
| July | 99 300 | 11.1 |
| August | 93 000 | 7.9 |
| September | 317 929 | 16.7 |
| Total | 576 529 | 42.1 |

The average population densities of fecal coliforms and fecal streptococci in soil are shown in Fig. 5. Van Donsel et al. (1967) has reported that the fecal coliform group survived slightly longer than the fecal streptococcus group during the summer months. Our results indicate that the fecal streptococcus group survived slightly longer than the fecal coliform group which may have been due to a variety of environmental conditions, i.e. temperature and pH, capable of extending the survival period of the fecal streptococci.

The survival curves of *Escherichia coli* and *Salmonella typhimurium* in Fargo clay soil are shown in Fig. 6. The death constants were calculated for the organisms according to the formula of Chick (1908) as modified by Klock (1971):

$$\log \frac{n}{n_0} = -kt$$

where $n_0$ is the number of organisms initially, n is the number of organisms at the completion of the experiment, t is time, and k is the constant for the death rate. These constants, given in Fig. 6, were determined after the slope had been established and before it began to drop off. The higher the death constant (k), the faster the death rate. Both organisms exhibited a rapid decline during the first 2 weeks of incubation and were undetectable by the 4th week of the study. It may be seen that after the initial inoculation the death rate for *E. coli* was twice as high as that of *S. typhimurium*. A reinoculation of the soil area with the appropriate organism greatly extended the survival of both organisms as evidenced by the death constants. Our results are supported by Dazzo et al. (1973) who have shown that *S. enteritidis* and fecal coliform survival in soil was extended by repeated applications of manure slurry on the soil.

## SUMMARY


Fecal coliform populations in the sediment were 10 times higher than in the overlying water. In contrast, the fecal streptococcus populations did not vary appreciably during the winter months; however, a one- to three-fold increase of the fecal streptococci in the sediment versus the overlying water was observed during the summer months. Aerobic sporeforming *Bacillus* populations in the sediment were approximately 100 times higher than in the overlying water.

Salmonellae, as well as coagulase-positive staphylococci, were isolated more frequently from the sediments than from the overlying water. In view of the fact that the sediment-water interface is not a static system, any disturbance such as too vigorous pumping of wastewater may cause resuspension of bottom sediments thus


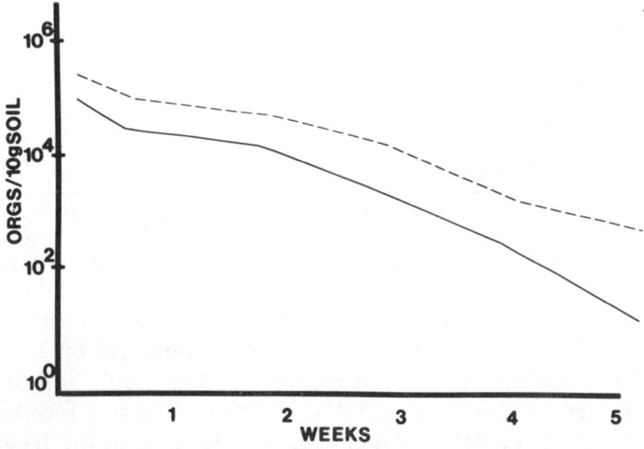

FIG. 5 Comparison of FC-FS populations in soil. ———FC. — —FS.



MANAGING LIVESTOCK WASTES

increasing the potential health hazard associated with the microbial populations in the sediment.

The survival of pathogens as well as indicator bacteria in the soil indicates that these organisms can survive competition by soil organisms if adequate nutrients are available. Therefore, the amount and frequency of waste water applications to soil should be carefully calculated in order for land disposal systems to succeed in safety.

## References


1  American Public Health Association. 1971. Standard methods for the examination of water and wastewater. 13th ed. American Public Health Association, Inc., New York.

2  Blannon, J.C., and M.L. Peterson. April 12, 1974. Survival of fecal coliforms and fecal streptococci in a sanitary landfill. U.S. Environmental Protection Agency. National Environmental Research Center. Cincinnati, Ohio.

3  Chick, H. 1908. Investigation of the laws of disinfection. In Klock, J.W. 1971. Survival of coliform bacteria in wastewater treatment lagoons. J. Wat. Pollut. Contr. Fed. 43:2071-2083.

4  Dazzo, F., P. Smith, and D. Hubbell. 1973. The influence of manure slurry irrigation on the survival of fecal organisms in scranton fine sand. J. Environ. Quality. 2(4):470-473.

5  Geldreich, E.E. 1970. Applying bacteriological parameters to recreational water quality. J. Am. Wat. Wks. Assn. 62:113-120.

6  Klock, J.W. 1971. Survival of coliform bacteria in wastewater treatment lagoons. J. Wat. Pollut. Contr. Fed. 43:2071-2083.

7  Mair, N.S., and A.I. Ross. 1960. Mon. Bull. Min. Hlth. Lab. Serv. 19:39.

8  Prost, E., and H. Riemann. 1967. Food-borne salmonellosis. Ann. Rev. Microbiol. 21:495-528.

9  Rittenberg, S.C., T. Mittwer, and D. Ivler. 1958. Coliform bacteria around three marine sewage outfalls. Limnol. Oceanagr. 3:101-108.

10  Van Donsel, D.J., and E.E. Geldreich. 1971. Relationships of salmonellae to fecal coliforms in bottom sediments. Wat. Res. 5:1079-1087.

11  Van Donsel, D.J., E.E. Geldreich, and N.A. Clarke. 1967. Seasonal variations in survival of indicator bacteria in soil and their contribution to storm-water pollution. Appl. Microbiol. 15(6):1362-1370.

# A Mycological Investigation of Beef Feedlot Manure in a Semiarid Temperate Climate

## R. G. Bell

**M**ANURES from herbivorous animals support the development of a large and diverse fungal population. These coprophilous fungi have been investigated extensively to discover their sequence of colonization (Harper and Webster 1964; Webster 1970; Wicklow and Moore 1974). The spores, which are ingested with herbage, resist breakdown in the digestive system and then germinate in the feces.

After the discovery in the early 1960's of the aflatoxins, toxic metabolites produced by some strains of *Aspergillus flavus* (Sargeant et al. 1961), concern has been expressed about mycotoxins in animal feedstuffs. Hendrickson and Grant (1971) demonstrated that aflatoxins were produced in sterilized feedlot manure. This finding, combined with that of Robert et al. (1962) that some *A. flavus* strains produced the wilt-inducing phytotoxins Aspergillomarasmines A and B, is particularly disturbing in view of the present interest in feeding with manure and applying manure at high rates to agricultural land.

This study was undertaken to identify and determine the relative frequency and possible significance, to both animals and plants, of fungi in manure accumulating on the surface of an unpaved feedlot under semiarid conditions.

## MATERIALS AND METHODS

Surface manure and freshly voided feces were sampled randomly at weekly intervals in the summer of 1974 from the surface of a single unpaved pen 75 x 50 m at a commercial feedlot in Lethbridge, Alberta, that contained 68 finishing cattle on a high-energy diet of barley, roughage, and concentrate. Samples were examined either upon return to the laboratory or, when stated, after storage at room temperature in closed polyethylene bags.

Serial dilutions of the manure were prepared in sterile distilled water; 0.5 ml aliquots of appropriate dilutions were spread onto the surface of petri dishes containing Sabourard's (Difco), Czapek-Dox (Difco), or manure extract* agars supplemented with 0.675 g rose-bengal and 0.03 g streptomycin sulface per liter. The fungal colonies that developed after incubation at 25, 37, or 55 C for 2 to 7 days were counted and identified to order or family and, when possible, to genus or species (Gilman 1957, Cooney and Emerson 1964, Barron 1968, Ellis 1971, Hesseltine and Ellis 1973). *A. flavus* isolates were checked for aflatoxin production potential using the Aspergillus Differential Medium (ADM) of Bothast and Fennell (1974).

## RESULTS AND DISCUSSION

*Thermophilic fungi.*—Four thermophilic fungi, *Mucor pusillus, Thermomyces lanuginosa, Talaromyces thermophilus,* and *Chaetomium thermophile,* were found in manure collected from the surface of the feedlot (Table 1). The first three were also isolated from freshly voided feces and from the feed. Minimum and maximum temperatures for growth of these species are: *T. lanuginosa* 30-60 C, *T. thermophilus* 27-59 C, *C. thermophile* 27-58 C, and *M. pusillus* 20-55 C (Cooney and Emerson 1964).

The population remained almost unchanged throughout the study, suggesting that the fungal propagules were present as spores that were unable to germinate because of a combination of subliminal temperatures and the low water content of the manure. Loss in spore viability under these conditions probably explains the differences between the *M. pusillus* population in fresh feces and in feedlot

The author R. G. BELL is Research Scientist, Research Station, Agriculture Canada, Lethbridge, Alberta, Canada T1J 4B1.

Acknowledgements: I wish to thank Dr. D. R. Fennell, NRRL, Peoria, Illinois, for identification of the *Aspergillus flavus*; the staff of the Mycology Section of the Biosystematics Institute, Agriculture Canada, Ottawa, Ontario, for confirming the identity of the other fungi identified to species level; Dr. M. G. Prior for the toxicological investigation of the *Fusarium solani*; and W. T. Hill Farms Limited, Lethbridge, Alberta, for access to their feedlot.

*Manure extract agar: Mix 250 g manure with 1 liter of distilled water. Autoclave mixture at 103.4 k pascals for 15 min. Cool and centrifuge. Add distilled water to bring volume of supernatant up to 1 liter. Adjust pH to 5. Add 1 g glucose and 20 g agar; autoclave at 103.4 k pascals for 15 min.

TABLE 1. FUNGI IN BEEF FEEDLOT SURFACE MANURE AS ENUMERATED ON MANURE EXTRACT AGAR INCUBATED AT 55 C FOR 48 HR.

| Date sampled | Moisture, percent* | Propagules per g dry manure | | | |
| --- | --- | --- | --- | --- | --- |
| | | Mucor pusillus | Thermomyces lanuginosa | Talaromyces thermophilus | Chaetomium thermophile |
| June 24 | 22.1 | 1,140 | 229 | 231 | 46 |
| July 2 | 55.2† | 1,030 | 206 | 205 | 41 |
| July 8 | 44.6† | 855 | 171 | 167 | 34 |
| July 15 | 12.3 | 1,230 | 247 | 247 | 50 |
| July 22 | 10.5 | 1,180 | 236 | 249 | 48 |
| July 29 | 14.2 | 1,100 | 220 | 218 | 44 |
| Aug. 6 | 13.3 | 940 | 188 | 192 | 38 |
| Fresh manure | 77.4 | 2,640 | 273 | 264 | - --‡ |

*Wet weight basis.

†Sample taken on morning after overnight rain.

‡<20 per g dry manure.

TABLE 2. FUNGI IN BEEF FEEDLOT SURFACE MANURE AS ENUMERATED ON
SABOURARD'S AGAR INCUBATED AT 25 C FOR 48 HR.

| Date sampled | Moisture, percent* | Propagules per g dry manure | | | | |
|---|---|---|---|---|---|---|
| | | Mucorales | Aspergillus flavus | Aspergillus niger | Penicillium spp. | Geotrichum spp. |
| June 24 | 22.1 | 1,590 | 120 | 120 | 120 | - - -‡ |
| July 2 | 55.2† | 9,700 | 400 | - - - | 200 | - - - |
| July 8 | 44.6† | 21,000 | 200 | - - - | - - - | - - - |
| July 15 | 12.3 | 21,000 | 80 | - - - | - - - | - - - |
| July 22 | 10.5 | 246 | 246 | - - - | - - - | - - - |
| July 29 | 14.2 | 814 | 80 | - - - | - - - | - - - |
| Aug. 6 | 13.3 | 2,130 | 150 | - - - | - - - | 500 |
| Fresh manure | 77.4 | 115,000 | - - - | - - - | - - - | - - - |

*Wet weight basis.
†Sample taken on morning after overnight rain.
‡<40 per g dry manure.

surface manure.

The temperature used to grow the thermophilic population (55 C) was near the maximum for *M. pusillus*, which grew poorly and produced neither aerial mycelium nor sporangia. When these plates were reincubated at 45 C, however, the stunted *M. pusillus* colonies developed normally. *M. pusillus* is a causative organism of mucormycosis and could be injurious to animal health. Since conditions in the manure pack are unfavorable for its growth and sporulation, inhalation of an infective dose of this low-virulence pathogen is unlikely.

*Mesophilic fungi.*—Mucorales represented by the genera *Mucor, Rhizopus, Absidia,* and *Mortierella* predominated among fungi isolated at 25 C from the manure and the freshly voided feces (Table 2). These fungi, which can rapidly utilize simple substances, are typical of the early stages of organic matter decomposition. The appearance of mucorateous sporangia on the surface of manure 2 days after the moisture content was increased to 70 percent clearly demonstrated the continued presence of suitable substrates and indicated that these fungi did not represent spores left by the initial colonizers after substrate exhaustion.

Moisture content and the size of fungal populations, especially the phycomycetes, were positively related (Table 2). The highest total population (21 200) was obtained from a manure sample taken 1 day after the second rainfall within 1 week. I suspect that the rain which fell on July 1 and on July 7 stimulated a flush of fungal growth that was detected in samples taken on July 8 and on July 15. This population declined rapidly as the manure dried out again. The flush of growth after the rainfall resulted almost entirely in the development of Mucorales while numbers of the slower-growing *A. flavus* remained almost unchanged.

Rapid colonization of the sugar-rich (4 percent maltose) Sabourard's agar by Mucorales obscured slower-growing fungi. Numbers of *Aspergillus* spp. and *Geotrichum* spp. colonies could not be estimated accurately (Table 3). On Czapek-Dox agar, which contains 3 percent sucrose, the phycomycete colonies developed slowly so that even after incubation for 7 days at 25 C the hyphomycete colonies could be counted accurately. Because the slower-growing hyphomycetes are masked, Sabourard's agar is unsuitable for enumerating fungal populations in manure, especially when species such as *A. flavus* with an aflatoxic potential may not be accurately enumerated. Furthermore, three numerically important hyphomycetes were not detected at all on Sabourard's agar (Table 3). Some species of one of these, *Scopulariopsis,* are pathogenic to houseflies (Domsch and Gams 1972) and could play a role in limiting fly populations associated with decomposing manure.

Although dematiaceous hyphomycetes are common spoilage organisms associated with stored forage and feed grains, they were, with the exception of an *Aureobasidium* sp., absent from the feedlot manure while *A. flavus,*

TABLE 3. FUNGI IN BEEF FEEDLOT
SURFACE MANURE SAMPLED ON AUG.
6 AS ENUMERATED ON SABOURARD'S
AND CZAPEK-DOX AGARS AFTER
INCUBATION AT 25 C FOR 2 AND 7
DAYS, RESPECTIVELY.

| Fungi | Propagules per g dry manure | |
|---|---|---|
| | Sabourard's agar | Czapek-Dox agar |
| Mucorales | $2.13 \times 10^3$ | $1.74 \times 10^3$ |
| Aspergillus flavus | $1.50 \times 10^2$ | $7.68 \times 10^3$ |
| Geotrichum spp. | $5.00 \times 10^2$ | $7.68 \times 10^3$ |
| Scopulariopsis spp. | - - -* | $1.47 \times 10^4$ |
| Fusarium spp. | - - - | $7.80 \times 10^3$ |
| Aureobasidium spp. | - - - | $1.76 \times 10^4$ |
| Gliocladium spp. | - - - | $4.61 \times 10^2$ |

*<40 per g dry manure.

another common cause of spoilage, was always a major component of the mycoflora. Of the *A. flavus* isolates, 90 percent produced cadmium-yellow undersided, asporogenous colonies that are typical of potentially aflatoxic strains when cultured on ADM (Bothast and Fennell 1974).

Fungal counts on both Sabourard's and Czapek-Dox media incubated at 37 C were about five times those observed at 25 C. A mixture of *Absidia* spp. and *Mucor* spp. predominated on these plates. Yeasts (*Candida* spp.) and the known pathogen *Aspergillus fumigatus* were also present. The thermophilic fungi *Thermomyces lanuginosa* and *Malbranchea pulchella* var. *sulphurea* were also recorded. Incubation at 37 C obviously allowed both mesophiles and thermophiles to grow.

### Influence of Manure Moisture on Fungal Population

The flush of mucor growth after the addition of water suggests that the development of the fungi was restricted by low moisture content of the manure. In support of this hypothesis, consider that, if a beef animal excretes 28.3 liters of urine plus feces each day (Hart 1960), then the 68 cattle confined in the pen would excrete $2.89 \times 10^5$ liters

(68 cattle x 150 days x 28.3 liters) from May to September. This volume spread over the pen of area 3750 m$^2$ would be 8 cm deep. In the Lethbridge area during this period, the deficit of precipitation to evaporation is normally 36 cm (Hobbs 1970), i.e., more than adequate to dry the wastes accumulating on the feedlot surface.

To test the hypothesis that water is limiting, part of the manure sample collected on July 8 was kept at room temperature and maintained at about 55 percent moisture for 4 wk. After this period, the fungal population, moisture content, and ash content were compared with those of manure left on the feedlot surface (Table 4). Mucorales had disappeared completely and numbers of both *A. flavus* and *Fusarium solani* had increased drastically in the moist manure compared with that under feedlot conditions. The moist manure also had a greater ash content, reflecting decomposition of the organic matter. The phycomycetes probably disappeared because of substrate exhaustion and loss of spore viability; however, the effect of dormant zygospore production, direct parasitism, and accumulation of toxic by-products, possibly from *A. flavus* and *F. solani,* cannot be excluded. The *A. flavus* was shown to be potentially aflatoxic using the ADM technique (Bothast and Fennell 1974). Aspergillomarasmine production was not investigated. A toxicological examination of the *F. solani* —often a plant pathogen which produces lycomarasmine, a wilt-inducing phytotoxin of similar structure to the aspergillomarasmines (Robert et al. 1962)—revealed the presence of at least two mammalian mycotoxins (M. G. Prior, personal communication).

Surface manure contained almost twice as much ash (26 vs. 15 percent) as freshly voided feces, which could indicate either decomposition or mixing with soil. The presence of small stones and red, burnt clay aggregates in the ash of surface manure supports the latter explanation. When freshly voided feces were left exposed to the prevailing climatic conditions, however, the ash content increased from 15 to 22 percent (moisture content decreased from 77 to 10 percent) in 7 days and then remained almost unchanged for 4 months (unpublished data). It is concluded that decomposition of manure on the feedlot surface is limited to the first week after excretion or after rainfall, i.e., at times when moisture content is not limiting.

## Pathogenic and Toxigenic Fungi in Manure

Potentially pathogenic and toxigenic fungi were isolated from feedlot surface manure. Large populations of these organisms are unlikely to develop within feedlots in semiarid regions because, in summer when the temperature is favorable for rapid fungal growth, such development is normally prevented by the low moisture content of the manure. In contrast, if the manure were incorporated at high rates in agricultural land, the more favorable moisture regime in the soil and the relative abundance of readily utilizable organic matter could permit growth of toxigenic fungi which usually, under the conditions of nutritional austerity in the soil, could not compete with normal soil microflora. However, the diverse and numerous microorganisms in agricultural soils would probably compete strongly for the added organic matter with those introduced with the manure. The possibility of aflatoxin production in manure should not be overlooked as *A. flavus* is clearly an effective competitor in that medium. Recycling by refeeding could be an unsafe procedure if other than fresh manure were used.



### SUMMARY

In semiarid climates, fungal development and organic matter decomposition are definitely limited in manure on the feedlot surface. Field observations and laboratory experiments show that low water content is the main limiting factor.

Fungi that are pathogenic or toxigenic to animals and plants are normally present in feedlot surface manure. The health hazard posed by these organisms within the feedlot is probably minimal because the adverse water regime restricts their development. When this relatively undecomposed, dehydrated manure is ultimately disposed of, however, moisture may no longer be limiting and rapid growth of these pathogenic and toxigenic fungi may occur.




### References

1   Barron, G. L. 1968. The genera of Hyphomycetes from soil. Williams and Wilkins, Baltimore. 364 p.

2   Bothast, R. J., and D. R. Fennell. 1974. A medium for rapid identification and enumeration of *Aspergillus flavus* and related organisms. Mycologia 66:365-369.

3   Cooney, D. G., and R. Emerson. 1964. Thermophilic fungi: an account of their biology, activities, and classification. W. H. Freeman & Co., San Francisco. 188 p.

4   Ellis, M. B. 1971. Dematiaceous Hyphomycetes. Commonwealth Mycological Institute, Kew, England. 608 p.

5   Domsch, K. H., and W. Gams. 1972. Fungi in agricultural soils. John Wiley and Sons, New York. 290 p.

6   Gilman, J. C. 1957. A manual of soil fungi. Iowa State College Press, Ames. 450 p.

7   Harper, J. E., and J. Webster. 1964. An experimental analysis of the coprophilous fungus succession. Trans. Brit. Mycol. Soc. 47:511-530.

8   Hart, S. A. 1960. The management of livestock manure. TRANSACTIONS of the ASAE 3(2):78-80.

9   Hendrickson, D. A., and D. W. Grant. 1971. Aflatoxin formation in sterilized feedlot manure and fate during simulated water treatment procedures. Bull. Environ. Contamination Toxicol. 6:525-531.

10   Hesseltine, C. W., and J. J. Ellis. 1973. Mucorales, p. 187-217. In: G. C. Ainsworth and A. S. Sussman (eds.) The fungi—an advanced treatise. Vol. 4B. Academic Press, New York.

11   Hobbs, E. H. 1970. The agricultural climate of the Lethbridge area, 1902-1969. Can. Dep. Agr. Res. Sta., Lethbridge, Alta. Agrometeorol. Publ. 1. 13 p.

12   Robert, M., M. Barkier, E. Lederer, L. Roux, K. Biemann, and W. Vetter. 1962. (Two new natural phytotoxins, Aspergillomarasmines A and B, and their identity to lycomarasmine and its derivatives.) Bull. Soc. Chim. Franc. (1962):187-188.

13   Sargeant, K., A. Sheridan, J. Kelly, and R. B. A. Carnaghan. 1961. Toxicity associated with certain samples of groundnuts. Nature (Lond.) 192:1096-1097.

14   Webster, J. 1970. Coprophilous fungi. Trans. Brit. Mycol. Soc. 54:161-180.

15   Wicklow, D. T., and V. Moore. 1974. Effect of incubation temperature on the coprophilous fungal succession. Trans. Brit. Mycol. Soc. 62:411-415.


TABLE 4. FUNGAL POPULATION AS ENUMERATED ON SABOURARD'S MEDIUM INCUBATED AT 25 C FOR 48 HR., MOISTURE CONTENT, AND ASH CONTENT OF (A) FEEDLOT SURFACE MANURE COLLECTED ON JULY 8, (B) FEEDLOT SURFACE MANURE COLLECTED ON AUG. 6, AND (C) JULY 8 FEEDLOT MANURE SAMPLE MAINTAINED AT ABOUT 50 PERCENT MOISTURE CONTENT AND INCUBATED AT 25 C FOR 4 WK (JULY 8 TO AUG. 6).

| Fungi | Propagules per g dry manure | | |
|---|---|---|---|
| | A | B | C |
| Mucorales | $2.10 \times 10^4$ | $2.13 \times 10^3$ | - - -* |
| Aspergillus flavus | $2.00 \times 10^2$ | $1.50 \times 10^2$ | $7.88 \times 10^6$ |
| Fusarium solani | - - - | - - - | $1.97 \times 10^6$ |
| Moisture content† | 44.6 | 13.3 | 54.0 |
| Ash content‡ | 25.7 | 26.1 | 43.7 |

*<40 per g dry manure.
†Percent of wet weight.
‡Percent of dry weight.

# Modification and Enzymatic Hydrolysis of Feedlot Waste

G. Keith Elmund, Dale W. Grant, S. M. Morrison

FEEDLOT waste (FLW) contains significant quantities of partially digested plant residues including cellulose, hemicelluloses, pectins and lignin. McCalla et al. (1970) determined that fresh FLW contains 60 to 75 percent biodegradable materials. During decomposition, some of the hemicelluloses, starch and manure proteins are readily attacked by microorganisms. Plant cellulose, however, is relatively resistant to microbial enzymes (Waksman 1952). The resistance of native cellulose to enzymatic hydrolysis is related to crystalline regions in its structure, particle size and the association of lignin with the polysaccharide matrix. Modification of plant cellulose to enhance its susceptibility to enzymatic hydrolysis would be advantageous for refeeding FLW and for other recycling procedures such as the production of single cell protein. Techniques for enhancing the digestibility of cellulosic materials include reaction with dilute solutions of sodium hydroxide and ball milling (Mandels et al. 1974). One phase of our research project is the evaluation of hydrogen peroxide and ferrous sulfate (HPFS) for effecting partial depolymerization of the cellulosic fraction of FLW. Our goal is to enhance the susceptibility of cellulose to cellulase and to maximize its conversion to hydrolytic products which may subsequently serve as substrates for the production of microbial biomass, thereby upgrading the nutritional value of FLW for refeeding.

Mixtures of hydrogen peroxide and ferrous sulfate affect the chemical reactivity and physical nature of plant cellulose (Bains 1972) and other carbohydrates (Moody 1963) by the generation of free radicals on treated substrates. Relatively low concentrations of HPFS partially depolymerized and solubilized cotton fibers, sawdust and other plant materials (Halliwell 1965). The rate of fiber disintegration was dependent on the relative concentrations of HPFS reactants as well as pH and temperature. Treatment with HPFS has also been reported to increase the alkali solubility and to reduce the strength and degree of polymerization of cotton fibers (Koenigs 1972a). In addition, treated cotton fibers showed a 2.5-fold increase in susceptibility to cellulase as measured by weight loss compared with controls treated with heat-inactivated enzyme. The enhanced susceptibility to cellulase was attributed to swelling and decreased degree of polymerization of the cotton fibers.

In preliminary optimization studies, 250 mg aliquots of the cellulosic substrates were weighed into 250 ml Erlenmeyer flasks and suspended in 40 ml deionized water containing initial hydrogen peroxide and ferrous sulfate concentrations ranging from 30 to 1500 mM and 0.2 to 22 mM, respectively, The reaction mixtures were subsequently analyzed for remaining dry weight, cellulose (Updegraff 1969), reducing sugars (Miller 1959) and susceptibility to cellulase (Mandels and Weber 1969). The results indicated that the concentrations of HPFS reactants were critical in determining the extent of depolymerization and oxidation of both cotton and lyophilized fresh manure substrates. Increasing the initial concentration of HP above 600 mM relative to constant initial levels of substrate plus FS resulted in increased conversion of the substrate to filterable materials; in some samples, dry weight, cellulose, and reducing sugar determinations did not account for over half of the initial substrate. It was subsequently found that a significant quantity of the organic material was converted to carbon dioxide. Table 1 shows the percent of cotton and manure substrates converted to carbon dioxide during two and four days incubation with HPFS. At an initial concentration of 1500 mM HP, 65 and 66 percent of the cotton and manure substrates, respectively, were converted to glucose equivalents of carbon dioxide during four days incubation. Halliwell (1965) reported that although small quantities of soluble carbonyl compounds were found when cotton fibers were treated with HPFS, organic matter "disappeared" from solution as the reaction progressed. Moody (1963) established that carbon dioxide and other oxidation products are generated when various carbohydrates, including glucose and cellobiose, are treated with HPFS.

TABLE 1. PERCENT SUBSTRATE CONVERTED TO GLUCOSE EQUIVALENTS OF CARBON DIOXIDE DURING TREATMENT WITH HYDROGEN PEROXIDE AND FERROUS SULFATE*

| | Percent Conversion | | | |
| | Cotton | | Feedlot waste | |
| Initial $H_2O_2$ concentration | Day 2 | Day 4 | Day 2 | Day 4 |
|---|---|---|---|---|
| None, control | 1 | 2 | 2 | 2 |
| 150 mM $H_2O_2$ | 7 | 7 | 6 | 6 |
| 300 mM $H_2O_2$ | 14 | 16 | 15 | 16 |
| 600 mM $H_2O_2$ | 32 | 32 | 35 | 47 |
| 1500 mM $H_2O_2$ | 62 | 65 | 56 | 66 |

*Samples reacted in 2.2 mM ferrous sulfate. Carbon dioxide captured in barium hydroxide traps.

Cotton, sawdust and manure substrates, reacted with relatively low concentrations of HPFS, also showed enhanced susceptibility to hydrolysis with cellulase. Table 2 shows the percent hydrolysis after one hour and 24 hour reaction with cellulase of sawdust pre-treated with HPFS. After 24 hour hydrolysis, sawdust pre-treated with 150 mM HP and 2.2 mM FS released 55 percent more glucose equivalents of reducing sugars than controls treated with cellulase alone. However, samples reacted with 600 mM HP and 2.2 mM FS showed an apparent decrease in susceptibility to cellulase compared with either controls or samples receiving less initial hydrogen peroxide. The decreased enzyme susceptibility of samples receiving excess HP may be due to excessive oxidation of the cellulosic constituents or the solubilization of inhibitory substances.

The authors are: G. KEITH ELMUND, DALE W. GRANT and S. M. MORRISON, Department of Microbiology. Colorado State University. Ft. Collins, CO 80523.

**TABLE 2. PERCENT HYDROLYSIS OF SAWDUST TREATED WITH HYDROGEN PEROXIDE AND FERROUS SULFATE AND SUBSEQUENTLY HYDROLYZED WITH CELLULASE***

| Initial $H_2O_2$ concentration | Percent hydrolysis at one hour | Percent hydrolysis at 24 hours |
|---|---|---|
| None, control | 7 | 11 |
| 80 mM | 11 | 17 |
| 150 mM | 12 | 17 |
| 300 mM | 11 | 17 |
| 600 mM | 1 | 8 |

*Samples reacted in 2.2 mM ferrous sulfate.

Experiments were also conducted to determine the effect of incubation temperature during reaction with HPFS on the susceptibility of FLW to cellulase. Three different levels of FLW were treated with 2.2 mM FS and 150 mM HP and incubated four days at temperatures ranging from 5 to 50 C. Subsequent to HPFS treatment, the samples were adjusted to pH 4.8 with citrate buffer containing 0.005 percent merthiolate and cellulase (Nutritional Biochemicals Co) at a substrate to enzyme ratio of 10 to 1. After incubation with cellulase at 50 C for one hour and 24 hours, filtrates were assayed for reducing sugar (Miller 1959). The results shown in Table 3 suggest a four-fold increase in susceptibility to cellulase in samples treated with HPFS compared with controls treated with cellulase alone. At low substrate concentrations (6 g/l), enhanced susceptibility to cellulase was most evident in samples treated with HPFS at relatively low temperatures (5 C). Conversely, at higher substrate concentrations (25 g/l), pre-incubation at moderately higher temperatures (30 C) was required for optimal enzymatic hydrolysis. These optimization studies established that treatment of FLW and other cellulosic materials for enhanced susceptibility to cellulase is dependent not only on the concentrations of HPFS reactants but upon the concentration of substrate and incubation temperature as well. In addition, pH may also affect the depolymerization reaction. In our initial experiments, the acetate buffer system (pH 4.2) described by Halliwell (1965) was used until recovery balances suggested that a portion of the acetate was being converted to carbon dioxide. Larsen and Smidsrod (1967) reported that buffer ions compete with the substrate for $FE^{2+}$ and HP. Subsequently, unbuffered reactions were used during HPFS treatment of FLW, sawdust and cotton fibers. Under these conditions, acidic products accumulated.

**TABLE 3. PERCENT HYDROLYSIS OF FEEDLOT WASTE TREATED WITH HYDROGEN PEROXIDE AND FERROUS SULFATE AT DIFFERENT TEMPERATURES***

| Temperature | Feedlot waste, gm/l | | |
|---|---|---|---|
| | 6 | 13 | 25 |
| Control | 8 | 7 | 9 |
| 5 C | 33 | 28 | 26 |
| 10 C | 27 | 33 | 27 |
| 20 C | 24 | 30 | 30 |
| 30 C | 18 | 26 | 32 |
| 40 C | 15 | 21 | 29 |
| 50 C | 12 | 21 | 24 |

*Samples reacted with 150 mM hydrogen peroxide and 2.2 mM ferrous sulfate.

The role of lignin in limiting the accessibility of FLW and other cellulosic materials to microbial enzymes is critical for developing recycling processes. The free radical reactions initiated by mixtures of HPFS may be similar to the chemical and physical changes observed in $\gamma$-irradiated cellulosic substrates. For example, $\gamma$-irradiation of wheat straw and sheep feces increased *in vitro* dry matter digestibility by apparently disrupting the lignin-cellulose complex (Pigden and Heaney 1969). In addition, the oxidases of certain wood-rotting fungi may generate sufficient quantities of hydrogen peroxide to disrupt the lignin-cellulose complex (Koenigs 1972b).

Lignin assays (AOAC 9th ed) were conducted on fresh lyophilized FLW and sawdust treated with HPFS and on untreated controls (Table 4). Substrates treated with 2.2 mM FS and 150 mM HP showed a 33 percent decrease in lignin content compared with untreated controls. During treatment of sawdust with HPFS, phenolic materials were solubilized and detected by reaction with Folin-Phenol reagent (Lowry et al. 1951). The acid-detergent fiber (ADF) analysis is another parameter suggested for evaluating the potential digestibility of cellulosic materials (Van Soest 1973). A 400 g (dw) sample of fresh FLW was reacted in 2.2 mM FS and 150 mM HP. After four days incubation at 40 C, the sample was lyophilized and aliquots were assayed for remaining ADF and lignin. Aliquots of FLW reacted with HPFS showed 21 and 23 percent decreases in ADF and lignin content, respectively, compared with untreated controls. The reduced lignin and ADF content of FLW treated with HPFS confirmed that cellulosic materials processed in this manner may serve as nutritionally enhanced supplements to feedlot rations.

**TABLE 4. LIGNIN CONTENT OF FEEDLOT WASTE AND SAWDUST TREATED WITH HYDROGEN PEROXIDE AND FERROUS SULFATE**

| Substrate | Lignin recovered | |
|---|---|---|
| | Control, mg/gm | Treated, mg/gm |
| Feedlot waste | 72 | 48 |
| Sawdust | 240 | 160 |

The production of cellulases and the enzymatic hydrolysis of various cellulosic materials have received extensive investigation (Pathak and Ghose 1973). *Trichoderma viride* mutant strains QM 9123 and QM 9414 (Mandels et al. 1974) give relatively high yields of cellulase when grown on various cellulosic substrates including FLW (Griffin et al. 1974).

Studies were conducted to evaluate cellulase production during growth of QM 9123 and QM 9414 on fresh FLW treated with HPFS and untreated controls. The cultural conditions and cellulase assay procedures were those of Mandels et al. (1971). Cellulase yields in replicate samples within an experiment were relatively consistent; however, cellulase yields from replicate experiments were sufficiently variable to allow only qualitative discussion of the results. Cellulase ($C_1$ and $C_x$) activities in culture filtrates of QM 9414 grown on treated and control substrates were higher than activities in filtrates from QM 9123. In addition, cellulase yields from both QM 9123 and QM 9414 grown on HPFS-treated fresh FLW were higher than yields from untreated controls. The initial concentration of substrate in the growth medium also affected cellulase yields; preliminary results suggest a fresh FLW concentration between 4 and 10 percent (dry wt/vol) may be optimal.

Filtrates from manure and cotton substrates after treatment with HPFS and cellulase were evaluated for their ability to serve as a carbon source for the growth of the feed yeast *Candida utilis*. A basal medium was added to diluted filtrates, filter sterilized, and dispensed into replicate test tubes with Morton closures. The tubes received a standardized inoculum and were incubated at 30 C on a rotary shaker. Growth was monitored spectrophotometrically at 525 nm. The results, shown in Table 5, indicate that the hydrolytic products of FLW treated with HPFS and cellulase are suitable substrates for the growth of *C. utilis*. For example, filtrates from FLW reacted with 2.2 mM FS and 150 mM HP and cellulase produced five times more biomass of yeast cells than controls treated with cellulase alone.

addition, the breakdown products of FLW treated with HPFS and cellulase were suitable substrates for the growth of *Candida utilis*. The production of *T. viride* and *C. utilis* biomass on HPFS-treated substrates may increase the nutritional value of FLW and other cellulosic materials as supplements to feedlot rations.

Additional research is needed to evaluate other parameters which may affect optimization of HPFS-treatment of FLW and other cellulosic substrates for facilitating their enzymatic hydrolysis with cellulase. The studies should include evaluation of the effectiveness of large-scale processing under HP-limited conditions as well as the evaluation of the digestibility of processed substrates *in vitro* and *in vivo*.

**TABLE 5. GROWTH OF *CANDIDA UTILIS* ON FILTRATES FROM FEEDLOT WASTE TREATED WITH HPFS* AND CELLULASE**

| Treatment | mg dry wt cells†/gm FLW, w/w | | |
|---|---|---|---|
| | Day 2 | Day 4 | Day 4 + cellulase |
| Control, no. HPFS | 32 | 32 | 76 |
| 0.22 mM FeSO$_4$ | 48 | 32 | 88 |
| 1.1 mM FeSO$_4$ | 36 | 32 | 370 |
| 2.2 mM FeSO$_4$ | 64 | 48 | 410 |
| 3.3 mM FeSO$_4$ | 48 | 36 | 300 |
| 22 mM FeSO$_4$ | 36 | 36 | 230 |

*Samples reacted in 150 mM hydrogen peroxide.

†Calculated from dry weight determinations on cells grown in basal medium plus dextrose and collected on tared 0.45 $\mu$ Millipore filters.

## SUMMARY

Feedlot waste modified with HPFS had reduced acid-detergent fiber and lignin contents as well as increased susceptibility to enzymatic hydrolysis with cellulase. These results indicate that HPFS-treatment may enhance the digestibility of FLW and other cellulosic materials. The extent of modification depended on various factors including the substrate concentration, the concentrations of reactants, and the incubation temperature; research of others suggests that the reaction of carbohydrates with HPFS also depends on pH (Halliwell 1965, Larsen and Smidsrod 1967) and oxygen tension (Ingles 1972). Fresh FLW treated with HPFS and untreated controls also served as substrates for cellulase formation during the growth of *Trichoderma viride* mutant strains QM 9123 and QM 9414. With both treated and control substrates, cellulase yields depended on the initial concentration of FLW in the growth medium. In


### References

1   Association of Official Agricultural Chemists, Official Methods of Analysis 9th ed. Washington, D.C.

2   Bains, M. S. 1972. Inorganic redox systems in graft polymerization onto cellulosic materials. J. Polymer Sci.: Part C 37:125-151.

3   Griffin, H. L., J. L. Sloneker and G. E. Inglett. 1974. Cellulase production by *Trichoderma viride* on feedlot waste. Appl. Microbiol. 27:1061-1066.

4   Halliwell, G. 1965. Catalytic decomposition of cellulose under biological conditions. Biochem. J. 95:35-40.

5   Ingles, D. L. 1972. Studies of oxidations by Fenton's reagent using redox titration. II. Effect of oxygen. Aust. J. Chem. 25:97-104.

6   Koenigs, J. W. 1972a. Effects of hydrogen peroxide on cellulose and on its susceptibility to cellulase. Mater. und Organism 7(2):133-147.

7   Koenigs, J. W. 1972b. Production of extracellular hydrogen peroxide and peroxidase by wood-rotting fungi. Phytopathol. 62:100-110.

8   Larsen, B. and O. Smidsrod. 1967. The effect of pH and buffer ions on the degradation of carbohydrates by Fenton's reagent. Acta Chem. Scand. 21:552-564.

9   Lowry, D. H., N. J. Rosebrough, A. L. Farr and R. J. Randall. 1951. Protein measurement with the Folin phenol method. J. Biol. Chem. 193(1):265-275.

10   Mandels, M. and J. Weber. 1969. The production of cellulases. Advan. Chem. Ser. 95:391-414.

11   Mandels, M., J. Weber and R. Parizek. 1971. Enhanced cellulase production by a mutant of *Trichoderma viride*. Appl. Microbiol. 21(1):152-154.

12   Mandels, M., L. Hontz and J. Nystrom. 1974. Enzymatic hydrolysis of waste cellulose. Biotechnol. Bioeng. 16:1471-1493.

13   McCalla, T. M., L. R. Frederick and G. L. Palmer. 1970. Manure decomposition and fate of breakdown products in soil. pp. 241-255. In: Agricultural practices and water quality, eds. T. L. Willrich and G. E. Smith. Iowa State University Press. Ames, Iowa.

14   Miller, G. L. 1959. Use of dinitrosalicylic acid reagent for determination of reducing sugar. Anal. Chem. 31(3):426-428.

15   Pathak, A. N. and T. K. Ghose. 1973. Cellulases-1: sources, technology. Process Biochem. 8(4):35-38.

16   Pigden, W. J. and D. P. Heaney. 1969. Lignocellulose in ruminant nutrition. Advan. Chem. Ser. 95:245-261.

17   Updegraff, D. M. 1969. Semimicro determination of cellulose in biological materials. Anal. Biochem. 32(3):420-424.

18   Van Soest, P. J. 1973. Collaborative study of acid-detergent fiber and lignin. j. Assoc. Official Anal. Chem. 56(4):781-784.

19   Waksman, S. A. 1952. Soil Microbiology. John Wiley and Sons, Inc. New York.

# Influence of Antibiotics and Growth Promoting Feed Additives on the Manuring Effect of Animal Excrements in Pot Experiments with Oats

## Cord Tietjen

SUPPLEMENTATION of animal feed by antibiotics and other additives is aimed at increasing weight gains, increasing feed conversion efficiency, and the maintenance or the restoration of animal health. A lasting effect in the excrement after defecation cannot be excluded. Seventy-five percent of the dietary chlortetracycline was excreted by yearling steers in experiments of Elmund et al. (1971). Antibiotic supplementation apparently altered the digestive processes in the digestive tract, resulting in feces which were less biodegradable. As a consequence, supplementation of animal feed by additives of high stability might need consideration in waste disposal as well as in animal manure utilization in crop production (Calvert et al. 1971).

### EXPERIMENTS WITH PIG MANURE

Different feed additives were applied under comparable conditions for pig fattening in an experiment carried out in the Institute of Animal Nutrition of the Agricultural Research Center, Braunschweig-Völkenrode, for investigations on weight gains, residue formation, and resistance. Urine and feces were collected and applied in aliquot amounts in pot experiments with oats to investigate the effect on crop yield and the uptake of nitrogen. The basis of comparison was the nitrogen content. In Table 1 the applied feed additives are listed and the dosage of 'gülle' (feces + urine in aliquot amounts) is noted with the nitrogen content and quantity per pot. In order to reach maximum yield levels, increasing rates of nitrogen fertilizer from 0 to 2.5 g per pot filled with 6 kg soil were applied additional to the gülle doses.

### RESULTS

#### Crop Yield

Fig. 1 shows the influence of nitrogen application on the yield increase of grain and straw dry matter. The yield level rises by gülle in the average of the eight variants about seven times, from 8 to 57 g per pot, in the series without nitrogen fertilizer. This increase corresponds to the yield effect of 0,4 g mineral nitrogen fertilizer. In Fig. 1, the

The author is: CORD TIETJEN, Institut fur Pflanzenbau und Saatgutforschung der Forschungsanstalt fur Landwirtschaft, Braunschweig-Volkenrode, Braunschweig, Bundesallee, Federal Republic of Germany.

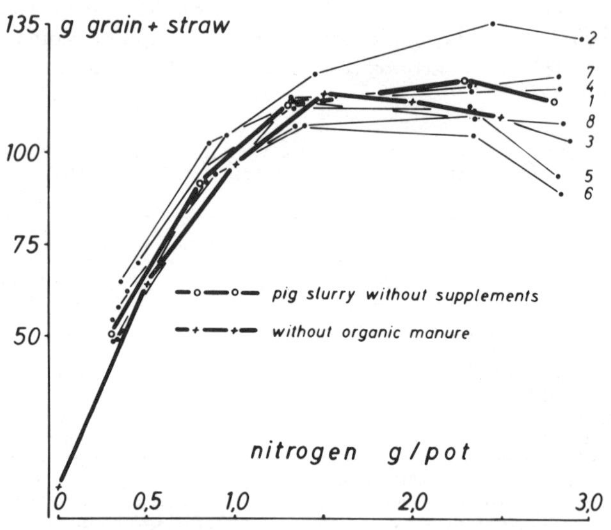


FIG. 1 Crop yield effect of increasing nitrogen fertilizer rates and gulle from pig feeding experiments with additives

| | |
|---|---|
| 1 gülle (no feed additive) | 5 gülle (oxytetracycline) |
| 2 gülle (carbadox) | 6 gülle (flavophospholipol) |
| 3 gülle (oleandomycin 1) | 7 gülle (virginiamycin) |
| 4 gülle (oleandomycin 2) | 8 gülle (zinc bacitracin). |

different curves are adjusted to this result.

There is a rather good conformity between the yield curves obtained 'without organic manure' and with 'pig gülle without feed supplement'. This demonstrates distinctly the varying effect of some gülle variants originating from animals fed with additives. At a higher yield level, as an interaction with fertilizer nitrogen, the gülle variant 6 (flavophospholipol) caused depressions, while the gülle variant 2 (carbadox) brought higher yields with each nitrogen fertilizer rate. The highest yield of 135 g is 13 percent above the level obtained by gülle without additive. In comparison with the gülle variant 1 (no additive), the depressive effect of variants 8 (bacitracin), 5 (oxytetracyclin), and 3 (oleandomycin 1) together with the two highest nitrogen rates are remarkable.

### Nitrogen Contents in Grain and Straw

The nitrogen content in grain and in straw increases with higher fertilizer nitrogen rates (Table 2). Most of the values

**TABLE 1. EXPERIMENTAL VARIANTS "FEED ADDITIVE" AND "GÜLLE".**

| Group<br>Feed additive | | 1<br>None | 2<br>Carbadox | 3<br>Oleando-<br>mycin 1 | 4<br>Oleando-<br>mycin 2 | 5<br>Oxytetra-<br>cycline | 6<br>Flavo-<br>phospholi-<br>pol | 7<br>Virginia-<br>mycin | 8<br>Zinc-<br>Bacitra-<br>cin |
|---|---|---|---|---|---|---|---|---|---|
| Additive dosage | mg/kg feed | | 50 | 20 | 10 | 80 | 15 | 50 | 80 |
| Gülle | g/pot | 40 | 60 | 55 | 48 | 45 | 45 | 45 | 45 |
| Nitrogen | percent | 0.76 | 0.76 | 0.71 | 0.69 | 0.71 | 0.76 | 0.71 | 0.79 |
| in Gülle | g/pot | 0.30 | 0.45 | 0.39 | 0.33 | 0.32 | 0.34 | 0.32 | 0.35 |

TABLE 2. NITROGEN IN GRAIN AND
STRAW, PERCENT IN DRY MATTER.

| | | Fertilizer N, g/pot | | |
|---|---|---|---|---|
| Group | Feed additive | 0 | 1,0 | 2,5 |
| | | Nitrogen in grain | | |
| 1 | None | 1,52 | 1,87 | 2,74 |
| 2 | Carbadox | 1,59 | 2,12 | 2,62 |
| 3 | Oleandomycin 1 | 1,44 | 2,11 | 2,80 |
| 4 | Oleandomycin 2 | 1,86 | 2,46 | 3,71 |
| 5 | Oxytetracycline | 1,71 | 2,25 | 3,82 |
| 6 | Flavophospholipol | 1,78 | 2,86 | 3,74 |
| 7 | Virginiamycin | 1,35 | 1,74 | 2,62 |
| 8 | Zinc Bacitracin | 1,38 | 1,78 | 3,05 |
| 0 | (No gülle) | 1,63 | 1,76 | 2,52 |
| | | Nitrogen in straw | | |
| 1 | None | 0,36 | 0.47 | 1,43 |
| 2 | Carbodox | 0,45 | 0.46 | 1,24 |
| 3 | Oleandomycin 1 | 0,39 | 0.46 | 1,24 |
| 4 | Oleandomycin 2 | 0,37 | 0.47 | 1,29 |
| 5 | Oxytetracycline | 0,35 | 0.53 | 1,79 |
| 6 | Flavophospholipol | 0,41 | 0.69 | 2,14 |
| 7 | Virginiamycin | 0,35 | 0,50 | 1,26 |
| 8 | Zinc Bacitracin | 0,36 | 0,50 | 1,92 |
| 0 | (No gülle) | 0,46 | 0.38 | 1,17 |

TABLE 3. EFFECT OF CHLORTETRACYCLIN, ZINC
BACITRACIN AND STREPTOMYCIN APPLIED TO THE
SOIL IN POT EXPERIMENTS WITH OATS, ON CROP
YIELD AND NITROGEN CONTENTS (dm)*.

| | Grain | | Straw | |
|---|---|---|---|---|
| | Yield, g | Nitrogen, percent | Yield, g | Nitrogen, percent |
| No antibiotics | | | | |
| No organic manure | 63,9 | 1,36 | 49,6 | 0,34 |
| With organic manure | 85,8 | 1,66 | 68,0 | 0,38 |
| With antibiotics | | | | |
| No organic manure | 63,8 | 1,39 | 51,2 | 0,32 |
| With organic manure | 72,1 | 2,83 | 64,5 | 0,73 |

*Mean values of two application rates (0.8 and 4.2 mg/kg soil) of the
three antibiotics and two varieties of chicken manure.

obtained by gülle application are higher than those in the series without gülle. Especially high nigrogen contents in grain were found in the series variant 4 (oxytetracycline), and 6 (flavophospholipol), while several low values were caused by variants 7 (virginiamycin) and 8 (zinc bacitracin).

Similar results, but not so pronounced were obtained with straw. A few striking increases in the nitrogen content were caused by the gülle variants 6 (flavophospholipol), 5 (oxytetracycline), and 7 (virginiamycin).

The results show that feed supplementation may influence the efficiency of pig manure on the crop yield and on the nitrogen contents in grain and in straw, especially when applied together with nitrogen fertilizer on a higher level. Hindering as well as furthering seems to depend on the specific additive and its concentration.

## EXPERIMENTS WITH BROILER DROPPINGS

Broiler feed was supplemented with quindoxin, 20mg/kg, for the one group, and with payzone, 12 mg/kg, for the other group. The collected droppings showed a remarkable difference in their consistency and in the nitrogen contents: the payzone group droppings were high in water content with 64 percent against 45 percent in the droppings of the control group and the quindoxin group. The corresponding nitrogen values were 7,8 percent in dry matter and 5,4 percent respectively.

Fig. 2 shows the results obtained with broiler droppings in pot experiments with oats. The feed supplementation with quindoxin caused a significant increase in yield at all nitrogen fertilizer levels, while droppings from the payzone group differed from the control group only in the series with the highest nitrogen rate.

## EXPERIMENTS WITH SOIL APPLICATION OF ANTIBIOTICS

Chlortetracyclin, zinc bacitracin and streptomycin were applied in pot experiments to the soil; the doses were 0,8 and 4,2 mg/kg soil. Two different organic manure variants were added: powdery dry manure from laying hens in deep litter management, and compost prepared with peat and wet droppings from laying hens in battery cage management. About 2.5 g nitrogen were contained in the manure doses per pot; 1 g nitrogen was added as chemical, $NH_4NO_3$. Table 3 shows the summarized results. The antibiotics exercised nearly no influence on the dry matter production and on the nitrogen content of the crop in the series without organic manures. Streptomycin was an exception: the small dose increased the nitrogen content in the grain remarkably, the great dose less pronounced. However, when applied together with the chicken manures, the antibiotics caused a depression in the yield of plant material and an essential increase of the nitrogen content (Table 3).

Flavophospholipol was applied as a chemical in pot experiments without organic manure or together with wet chicken manure added. In two series straw was added to the chicken manure. Fig. 3 shows the results. Flavophospholipol did not exercise a yield increasing effect in any case; on the contrary, together with higher nitrogen fertilizer rates the yield was decreased.

## DISCUSSION

Antibiotics applied in plant nutrition and in plant pro-

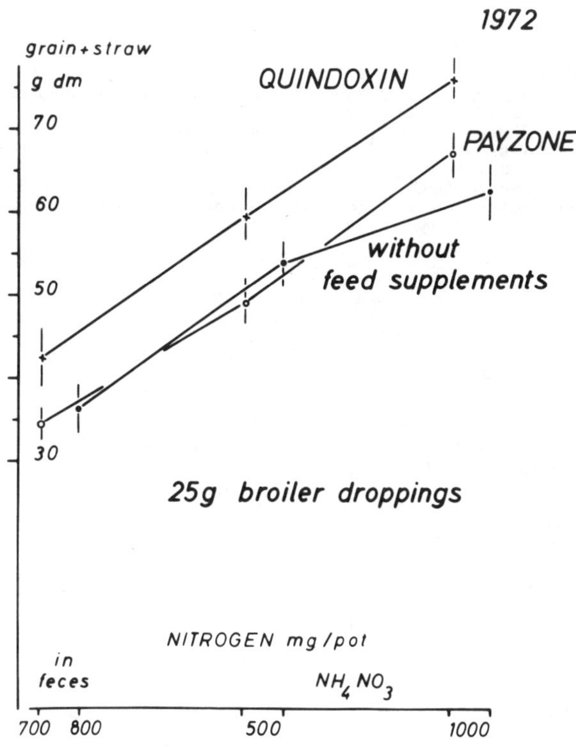


FIG. 2 Crop yield effect of broiler manure from feeding experiments with quindoxin and payzone (700 mg nitrogen in the droppings from the quindoxin and the payzone groups, and 800 mg nitrogen in the droppings without feed supplements; 500 and 1000 mg N were applied additionally in the form of $NH_4$-$NO_3$).

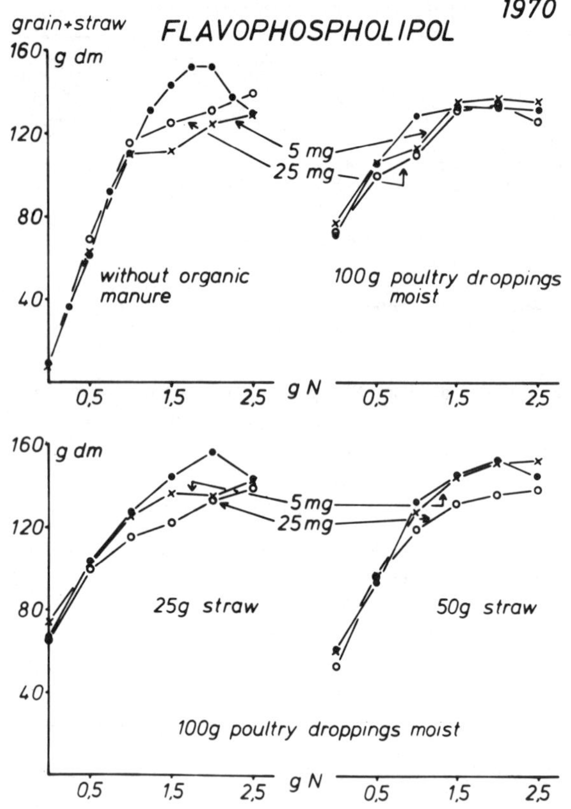



FIG. 3 Effect of flavophospholipol applied to the soil in pot experiments with chicken manure and straw.

the soil, and in the nutrition of the crop. Adsorption of antibiotics by clay minerals is different depending on the kind of mineral and on chemical properties of the antibiotic (Allison 1973). Desorption is possible, that means lasting efficiency of still unknown degree. We observed in our experiments hindering as well as furthering of crop mass production, and also a remarkable increase in the nitrogen content of the grain by the application of excrements resulting from feed supplemented by additives, as shown in Table 4. We wish to know the duration of effect after the defecation in the gülle pit, in the soil, and perhaps also in the crop.

## SUMMARY


In pot experiments with oats, pooled fresh excrements from pigs and from broilers were used which were collected during investigations on the nutritive effect of feed supplementation with carbadox, oleandomycin, oxytetracycline, flavophospholipol, virginiamycin, zinc bacitracin, payzone, and quindoxin. As interaction with increasing rates of nitrogen fertilizer, we observed hindering as well as furthering of crop growth. Remarkable increase of dry matter production was caused by carbadox; higher nitrogen contents were related to flavophospholipol, oxytetracycline, and oleandomycin.

Soil application of chlortetracyclin, zinc bacitracin, and streptomycin did not affect the production of dry matter and the nitrogen content, but applied together with two varieties of chicken manure, dry matter production was decreased, and nitrogen content was increased to a remarkable extent.

Dietary supplementation by antibiotics and other kinds of additives may modify the biodegradation of the excrements as well as their manuring effect in crop production.


tection exercised manifold effects: furthering and hindering, retardation, and inhibition in the growth of shoot and root, mass production, and flowering; necrosis, chlorosis, and morphological deformity; interactions with Mg, Ca, Mn, and Fe; changing cell respiration, and other physiological processes, with mode of action mostly unknown (Brandl 1965, Leh 1960). A systematic application of antibiotics in the field of crop production is unusual.

When animal excrements containing residues of feed additives are utilized as manure, nothing is known of the effective dose and its orientation. High stability may be an advantage for the passage through the digestive tract, but it raises the question of the continuing activity in the manure,


### References

1    Allison, F. E. 1973. Soil organic matter and its role in crop production. Elsevier Sc. Publ. Cy. p. 368.

2    Brandl, E. 1965. Antibiotica im Pflanzenschutz, p. 816-887. In: R. Brunner; G. Machek, (ed.) Die Antibiotica, II. Hans Carl, Nürnberg, 992 p.

3    Calvert, C. C., L. T. Frobish, D. A. Dinius. 1971. Animal waste reuse — nutritive value and potential problems from feed additives. United States Department of Agriculture, ARS 44-224, 56 p.

4    Elmund, G. K., S. M. Morrison, D. W. Grant and Sr. M. P. Nevins. 1971. Role of excreted chlortetracycline in modifying the decomposition process in feedlot waste. Bulletin of Environmental Contamination & Toxicology 6 (2):129-132.

5    Leh, H. O. 1960. Die Wirkung von Streptomycin auf das Wachstum einiger Kulturpflanzen. Pflanzenernähr., Düngung, Bodenkde. 88:129-148.


TABLE 4. SOME SPECIFICATIONS OF THE FEED ADDITIVES APPLIED IN THE EXPERIMENTS.

| | | | | |
|---|---|---|---|---|
| Carbadox | Fortigro | Methyl-3-(2-Chinoxalinyl-Methylen) | $C_{11}H_{10}N_4O_4$ | Pfizer |
| Chlortetracyclin | Aureomycin | 7-Chloro-4-Dimethylamino-1,4,4a,5, 5a,6,11,12a-Octohydro-3,6,10,12, 12a-Pentahydroxy-6-Methyl-1,11- Dioxo-2,Naphtacenecarboxamide | $C_{22}H_{23}O_8N_2Cl.HCl$ | Cyanamid |
| Flavophospholipol | Flavomycin, Moenomycin | | $C_{70}H_{124}N_6O_{40}P$ | Hoechst |
| Oleandomycin | | | $C_{35}H_{61}O_{12}N$ | Pfizer |
| Oxytetracycline | Terramycin | | $C_{22}H_{24}O_9N_2HCl$ | Pfizer |
| Payzone-Nitrovin | Peson | 1,5-di(5-nitr-2-furyl)-1,4-penta- dien-3-1-amidinohydrazon hydro- chlorid | | Cyanamid |
| Quindoxin | Grofas | Quinoxaline-1,4-dioxid | $C_8H_6N_2O_2$ | ICI |
| Streptomycin | | | $C_{21}H_{39}O_{12}N_7$ | |
| Virginiamycin | | | $C_{28}H_{35}N_3O_7$ | Lohmann |
| | | | $C_{43}H_{49}N_7O_{10}$ | |
| Zinc Bacitracin | | Polypeptides, Zn | | |

# Optimum Dilution of Swine Wastes for Growth of *Lemna Minor* L. and *Euglena* sp.

Ronald A. Stanley, Carl E. Madewell

PROPER dilution rates of swine waste water in northwest Alabama for growth of *Lemna minor* L. and *Euglena* sp. were determined as the initial step in development of a biological treatment system from which protein could be recovered. For *Lemna minor*, optimum growth was obtained with 19 ml/l/wk. At this loading rate, oxygen was saturated during the day and half saturated at night. For *Euglena*, optimum growth was obtained with the highest loading rate tested, 150 ml/l/wk. At this loading rate, oxygen was below 2 ppm both day and night. In growth room tests, chemical constituents from a standard duckweed medium gave no additional benefit in optimally diluted swine waste water.

## INTRODUCTION

The proper disposal of agricultural wastes is a problem because of the increase in quantity and the increase in concern for the environment. One beneficial method of disposal is the use of wastes to enrich enclosed ponds for increased production of fish or shellfish. Though biological treatment in itself is not a particularly efficient means of treatment, coupled with protein production it might be worthwhile (Middlebrooks et al. 1974). This approach is being used at Woods Hole Oceanographic Institute for marine shellfish (Douglas 1974, Dunstan and Tenore 1974, Tenore, Goldman and Clarner 1973). Such a system should work for freshwater organisms also. Three possible systems that the Tennessee Valley Authority intends to test are: (a) *Lemna minor* L. (or some other duckweed) consumed by White Amur, (b) algae consumed by a freshwater shellfish such as Asiatic clams, and (c) algae consumed by a phytoplanktivorous fish such as Silver Amur (Stanley 1974). Duckweed growth is supported by waste water (Culley and Epps 1973, Harvey and Fox 1973), and White Amur has a high preference for duckweed (Blackburn and Sutton 1971). Similar support can be given for other possible combinations.

## METHODS AND MATERIALS

Swine waste fluid was obtained from two primary treatment lagoons for swine feedlot wastes near Muscle Shoals, Alabama. Chemical analyses were run on fluid from one lagoon and on local tap water using standardized accepted techniques, including flame absorption and emission spectroscopy. Results (Table 1) are compared to concentrations in Hutner's (1953) standard duckweed nutrient solution.

In greenhouse dilution tests, swine waste fluid was added three times a week to 6000 ml glazed ceramic vessels with a surface area of 350 cm$^2$. Additions gave a twofold dilution series from 1.25 percent per week to 20 percent per week.

Innoculum for the *Euglena* series was about 2 grams fresh weight of cultures grown from wild, mixed populations of euglenoids obtained near several local swine waste treatment lagoons. Innoculum for the *Lemna minor* L. series was 10 grams fresh weight of a culture from an isolated pond in Florence, Alabama. Plants were harvested weekly for a period of 8 weeks during July and August leaving enough to reinnoculate vessels for the following week's growth. Harvested plants were dried overnight at 70 C and weighed. Oxygen was determined with a YSI Model 53 oxygen monitor. Growth room dilution tests were conducted in 100 ml bottles at 20 C, with about 3000 lux continuous light. In growth room tests, a single addition of swine waste water was made at the beginning of the experiment, and results were evaluated by counting fronds after one week.

TABLE 1. CHEMICAL ANALYSIS OF SWINE WASTE WATER AND TAP WATER AND CONSTITUTION OF HUTNER'S (1953) MEDIUM.

| Chemical | Swine waste water | Tap water | Hutner's (1953) medium |
|---|---|---|---|
| | Concentration of chemical in: | | |
| Total N, mM | 12.7 | 0.43 | 0.50 |
| $NH_4^+$, mM | 3.8 | <0.06 | 0.25 |
| $NO_3^-$, mM | - | - | 0.25 |
| K, mM | 10.0 | 0.05 | 0.46 |
| P, mM | 3.2 | <0.04 | 0.23 |
| Ca, mM | 1.2 | 0.65 | 0.15 |
| Mg, mM | 1.6 | 0.17 | 0.20 |
| S, mM | 0.7 | - | 0.20 |
| Fe, $\mu$M | 72.0 | <2.0 | 9.0 |
| Zn, $\mu$M | 14.0 | 1.5 | 23.0 |
| Mn, $\mu$M | 13.0 | <2.0 | 9.1 |
| Cu, $\mu$M | 2.0 | <2.0 | 1.6 |
| B, $\mu$M | - | - | 93.0 |
| EDTA, mM | - | - | 0.17 |
| Solids, percent | 0.060 | <0.001 | - |
| pH | 7.70 | 8.08 | 6.0 |

## RESULTS AND DISCUSSION

In the greenhouse tests, loading rate for optimum growth of *Lemna minor* L. was 19 ml/l/wk (Fig. 1). Growth at this dilution was about twofold more than at higher and lower concentrations. The duckweed fronds at the next higher concentration were particularly susceptible to a bleaching condition in which fronds became devoid of chlorophyll and growth was greatly retarded. Dissolved oxygen during the day at the optimum loading rate was normal (saturated) while at night oxygen was about one-half saturated (4.6 ± 0.8 ppm) (Fig. 2).

Maximum growth of *Euglena* sp. in the greenhouse was obtained at the highest loading rate tested, 150 ml/l/wk

The authors are: RONALD A. STANLEY, Biologist, University of South Dakota at Springfield, Springfield, South Dakota 57062 (formerly with TVA's Environmental Biology Branch); and CARL E. MADEWELL, Agricultural Economist, Division of Agricultural Development, TVA, Muscle Shoals, Ala.

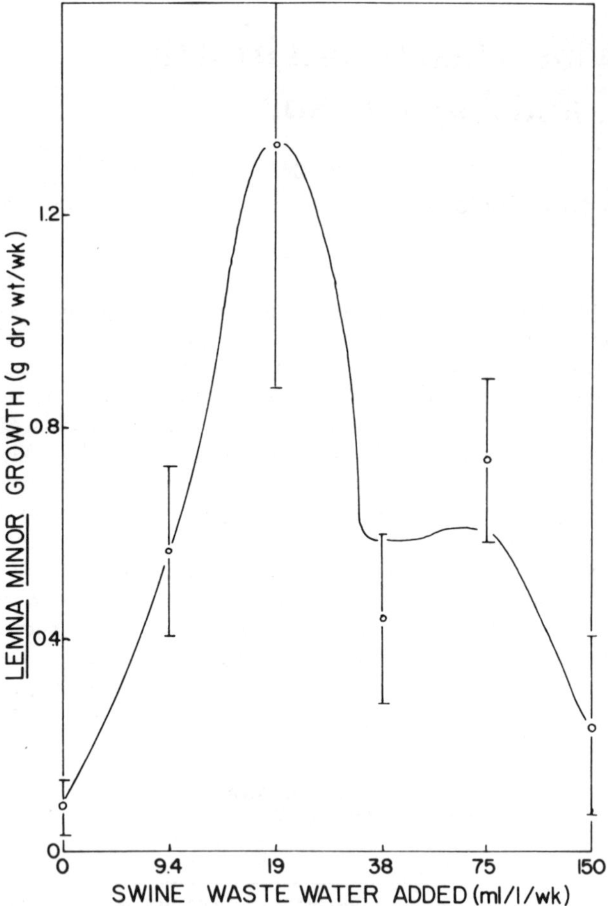

FIG. 1 Growth in the greenhouse of *Lemna minor* L. with various levels of swine waste water enrichment. Average of 8 weeks.

(Fig. 3). Day and night oxygen concentrations at this loading rate were below 2 ppm (Fig. 4). The highest loading rate at which dissolved oxygen remained above 2 ppm both day and night was 38 ml/l/wk.

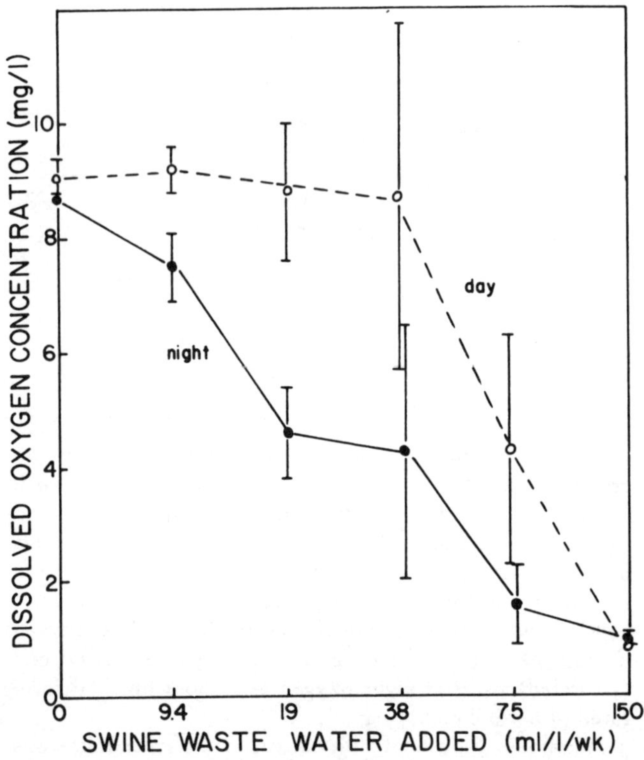

FIG. 2 Oxygen concentration in containers with *Lemna minor* L.

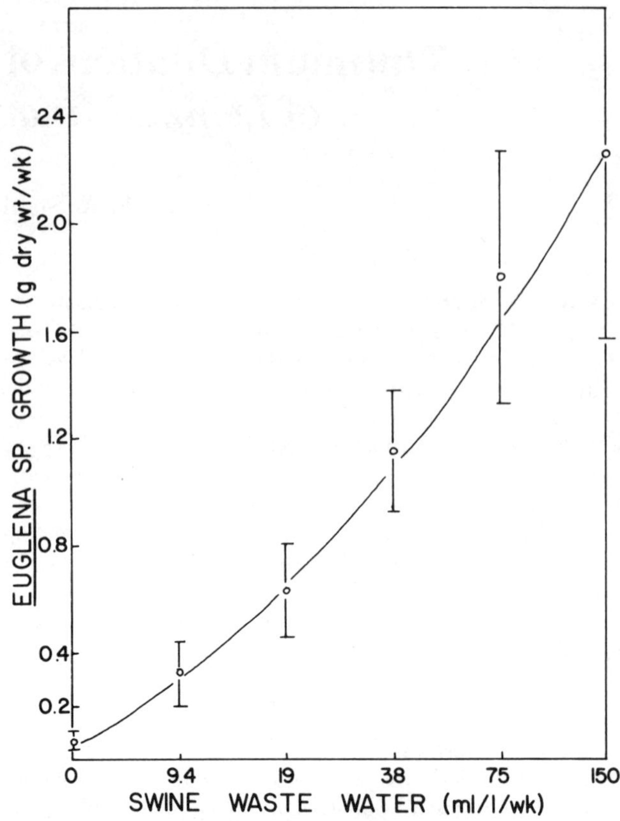

FIG. 3 Growth in the greenhouse of *Euglena* sp. with various levels of swine waste water enrichment. Average of 8 weeks.

Optimum loading rate can be converted into rate of enrichment by specific chemicals (Table 2) by multiplying by each chemical concentration (Table 1). Comparison of these enrichment rates with the concentrations of each chemical concentration in tap water (and in Hutner's (1953) Medium) will give clues about possible controlling nutrients. For *Lemna minor* L. the optimum loading rate gives a weekly addition of ammonium, potassium, and phosphate greater than the concentration in tap water. These are possible limiting nutrients below 19 ml/l/wk and possible toxic nutrients above 19 ml/l/wk. For *Euglena* sp. only calcium is present in tap water at a higher concentration than is being added at the optimum loading rate.

In growth room tests, optimum growth of *Lemna minor* was obtained at a similar dilution rate, but at higher and lower concentrations *Lemna minor* showed a better relative

TABLE 2. ADDITION RATE OF VARIOUS CHEMICALS IN SWINE WASTE WATER FOR OPTIMUM GROWTH OF *LEMNA MINOR L.* AND *EUGLENA* sp. IN THE GREENHOUSE.

| Chemical | Optimum enrichment rate ($\mu$M/wk) for: | |
|---|---|---|
|  | *Lemna minor L.* | *Euglena sp.* |
| Total N | 240 | 1900 |
| $NH_4$ | 72 | 570 |
| K | 190 | 1500 |
| P | 61 | 480 |
| Ca | 23 | 180 |
| Mg | 30 | 240 |
| S | 13 | 105 |
| Fe | 1.4 | 11 |
| Zn | 0.27 | 2.1 |
| Mn | 0.25 | 2.0 |
| Cu | 0.04 | 0.3 |



MANAGING LIVESTOCK WASTES

growth response than in the greenhouse (Fig. 5). Addition of $NaHCO_3$ at 0.3 and 1.0 mM was beneficial in the absence of swine waste enrichment but was invariably detrimental in enriched media (Fig. 6). Tests with other constituents of a standard duckweed medium (Hutner 1953) gave similar results with $NaHPO_4$, but even in the absence of enrichment, *Lemna minor* showed no beneficial response to $NH_4NO_3$, $Ca(NO_3)_2$, or $MgSO_4$. These results suggest that potassium, which was not tested, is the most likely governing chemical factor. Only complete Hutner's medium was as effective in promoting growth of *Lemna minor* as optimally diluted swine waste water.

In these tests, *Lemna minor*, at optimum dilution, produced an average of 37 g/m$^2$/wk compared to 3.4 kg/m$^2$/yr found by previous workers (Cully and Epps 1973). Growth of *Euglena* was also less than previously reported productivity for algae (Dugan, Golueke and Oswald 1972, McGarry and Tongkasame 1971). Increased productivity

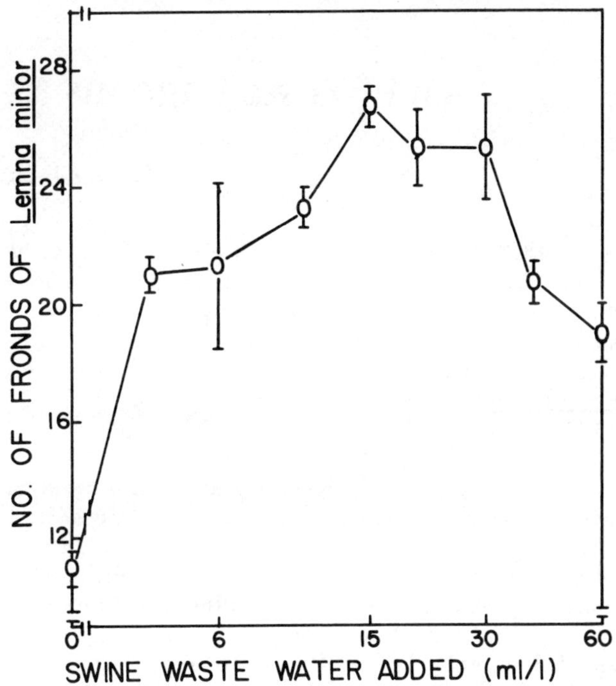

FIG. 5 Growth of *Lemna minor* L. in the controlled environment room with various levels of swine waste water enrichment.

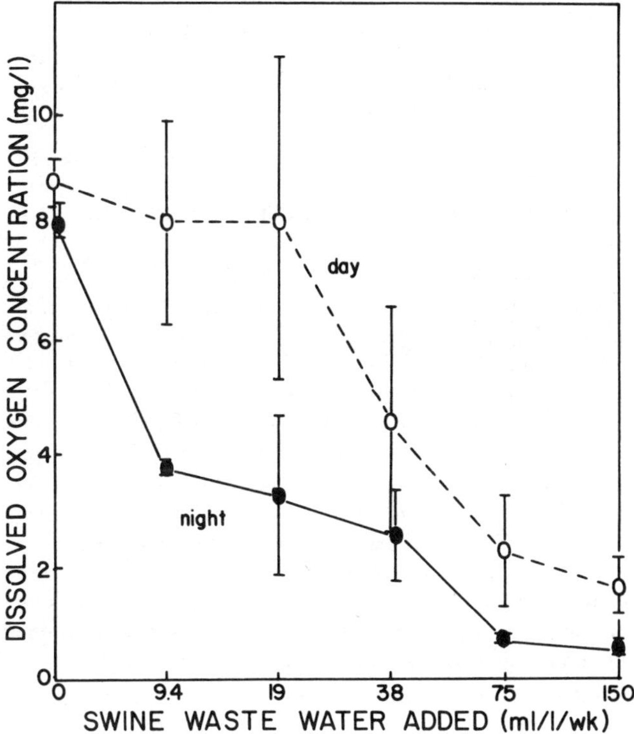

FIG. 4 Oxygen concentration in containers with *Euglena* sp.

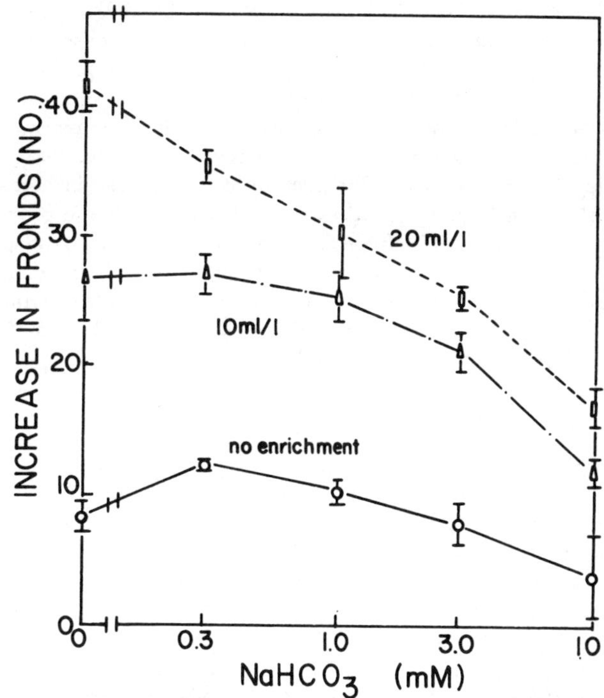

FIG. 6 Growth of *Lemna minor* L. in the controlled environment room with combinations of swine waste water enrichment and $NaHCO_3$.

should be achieved by more frequent harvesting so that self-competition is minimized. Another possible limiting factor in these tests is micronutrients, which were not studied. In seaweeds, micronutrients but not macronutrients were stimulatory as additives to secondary sewage treatment effluents (Prince 1974).


## References

1   Blackburn, R. and D. Sutton. 1971. Growth of the White Amur (Ctenopharyngodon idella Val.) on selected species of aquatic plants. Proc. Eur. Weed Res. Council 3:87-93.

2   Culley, D. and E. Epps. 1973. Potential usefulness of duckweeds in wastewater treatment and animal feeds. Jour. Water Poll. Contr. Fed. 45:337-347.

3   Douglas, John H. 1974. Oysters, algae, and sewage. Science News 106(11):170-171.

4   Dugan, G. L., C. C. Golueke and W. J. Oswald. 1972. Recycling system for poultry wastes. Jour. Water Poll. Contr. Fed. 44:432-440.

5   Dunstan, W. M. and K. R. Tenore. 1974. Control of species composition in enriched mass cultures of natural phytoplankton populations. Jour. Applied Ecol. 11:529-536.

6   Harvey, Richard M. and J. L. Fox. 1973. Nutrient removal using *Lemna minor*. Jour. Water Poll. Contr. Fed. 45:1928-1938.

7   Hutner, S. H. 1953. Comparative Physiology of heterotrophic growth in plants. In: W. E. Loomis, (ed.) Growth and differentiation in plants. Iowa State College Press, pp. 417-446.

8   McGarry, M. G and C. Tongkasame. 1971. Water reclamation and algae harvesting. Jour. Water Poll. Contr. Fed. 43:824-835.

9   Middlebrooks. E. J., D. B. Porcella, R. A. Gearheart, G. R. Marshall, J. H. Reynolds and W. J. Grenney. "Techniques for Algae Removal from Wastewater Stabilization Ponds," 1974. Jour. Water Poll. Contr. Fed. 46:2676-2695.

10   Prince, Jeffery S. 1974. Nutrient assimilation and growth of some seaweeds in mixtures of sea water and secondary sewage treatment effluents. Aquaculture 4:69-79.

11   Stanley, R. A. 1974. Methods of biological recycling of nutrients from livestock waste: a literature review and systems analysis. Bulletin Y-80, Agricultural Resource Development Branch, Tennessee Valley Authority, Muscle Shoals, Alabama.

12   Tenore, K. R., J. C. Goldman and J. P. Clarner. 1973. The food chain dynamics of the oyster, clam, and mussel in an aquaculture food chain. Jour. Exp. Mar. Biol. Ecol. 12:157-165.

# Swine Waste Lagoons as Potential Disease Reservoirs

## R. D. Glock, K. J. Schwartz

ANAEROBIC lagoons and pits have been shown to be practical and efficient for the storage of swine wastes (Stewart and Sweeten 1973). Increasingly common use of these systems has raised the question of whether they might serve as potential reservoirs for infectious organisms. An example of the potential hazard was reported by Findlay (1972) who detected survival of *Salmonella dublin* in cattle waste slurry.

Schmitt et al. (1974) reported that effluent from an open-channel flushing gutter system produced no adverse effect on performance when fed to growing swine, but they did note unexplained enlargement of lymph nodes in some animals. Glock et al. (1975) reported the establishment of *Salmonella saint paul* and swine dysentery in pigs which were fed anaerobic lagoon effluent from an infected herd. The pathogenic potential of salmonella is well established. Recent work by Harris et al. (1974) and Taylor and Alexander (1971) has shown that a spirochete, *Treponema hyodysenteriae*, is the primary cause of swine dysentery. This organism is anaerobic and its survival or even proliferation in the anaerobic lagoon would seem feasible.

A survey of anaerobic lagoons and underfloor pits was conducted to determine the frequency with which *Salmonella spp.* and *T. hyodysenteriae* could be detected in herds with or without a specific history of enteric disease problems. Cultures of *Salmonella spp.* isolated in the studies were inoculated intragastrically into pigs to determine their pathogenicity. Lagoon effluent from herds with swine dysentery was also fed to pigs to determine whether *T. hyodysenteriae* infections would result.

## EXPERIMENTAL DESIGN

### Survey of Swine Farms

Multiple specimens were collected from 17 farms in the central United States. Fourteen of the facilities had anaerobic lagoons. Eight of these 14 used recycled effluent in flushing-gutter systems and four used fresh water for flushing. Two were using systems not utilizing flushing of gutters. One of the farms had a history of swine dysentery and salmonellosis and two others had a history of swine dydentery. In addition to these 14 herds, specimens were obtained from three herds housed in buildings with anaerobic pits under slatted floors and in which dysentery was present.

Effluent was collected in sterile bottles and refrigerated for transport to the laboratory. Three drops of effluent from each sample were placed on glass slides and examined by phase microscopy for the presence of *T. hyodysenteriae* (Glock et al. 1974).

An aliquot was also centrifuged at 2000 rpm for 10 minutes to remove large particulate material. The supernate was placed in a dialysis tube and suspended in polyvinylpyrrolidinone to dehydrate it to approximately 5 percent of its original volume. Additional phase microscopic examinations were then performed.

Approximately 0.1 ml of each sample was streaked on turgitol-7 agar and 2 ml was added to 10 ml of tetrathionate broth and selenite-F broth. All of these were incubated at 37 C for 36 hours and subcultured on brilliant green agar with sulfa and salmonella-shigella agar. An additional tube of selenite-F broth was incubated at 42 C and subcultured on desoxycholate citrate agar. Colonies with no lactose fermentation and with morphology typical of *Salmonella spp.* were selected and biochemical tests were conducted as described by Oetjen and Harris (1973). Those with biochemical reactions suggestive of *Salmonella spp.* were submitted for serotyping.

Fecal specimens from at least six pigs on each farm were examined by phase microscopy. Culture techniques on each sample were similar to those described for effluent.

### Effluent Feeding Trials

Approximately 80 l of effluent was collected from anaerobic lagoons on two farms and underfloor pits on two other farms. Each of the farms had pigs with clinical swine dysentery at the time of effluent collection.

The effluent from each source was offered as the only source of water to two pigs weighing 12-18 kg. The pigs were obtained from a herd with no history of swine dysentery. *Treponema hyodysenteriae* and *Salmonella spp.* were not detected in feces collected before the trials began. The pigs were housed in strict isolation and were fed a ration containing 16 percent protein and free of antibacterial compounds. Feces collected at 2-day intervals were examined by phase microscopy and isolation of salmonella was attempted as previously described.

### Salmonella Pathogenicity Studies

Four groups of three pigs each were housed and fed as previously described. Each of the groups was inoculated with a different species of salmonella isolated from an anaerobic lagoon or pit. A fifth group served as controls.

Each of the salmonella isolates was grown with agitation in tryptose broth at 37 C for 4 to 5 hours. The culture was then centrifuged to sediment the bacteria which were resuspended in phosphate-buffered saline (pH 7.4) to a concentration of approximately $2 \times 10^9$ organisms per ml. Each pig was given 30 ml of this inoculum via stomach tube after a 24-hour period of starvation.

Pigs were observed daily for evidence of diarrhea. Feces were examined for *T. hyodysenteriae* and salmonella at 2-day intervals. After 7 days one pig from each group was killed and isolation of salmonella was attempted from liver, spleen, and colon.

## RESULTS

### Survey of Swine Farms

Four of 14 anaerobic lagoons contained salmonella.

The described work was supported in part by a grant from the National Pork Producers Council and USDA-ARS grant No. 12-14-100-11187.

Acknowledgment: The authors thank Dr. Billie Blackburn, Diagnostic Bacteriology, Veterinary Services Laboratories, APHIS, USDA, Ames, Iowa, for salmonella species typing.

The authors are: R. D. GLOCK and K. J. SCHWARTZ, Veterinary Pathology Dept., Iowa State University, Ames



MANAGING LIVESTOCK WASTES

**TABLE 1. INCIDENCE OF DETACTABLE SALMONELLS AND
*TREPONEMA HYODYSENTERIAE* IN EFFLUENT AND
RECTAL SWABS.**

| Type system | No. of herds | Salmonella spp. | | T. hyodysenteriae | |
|---|---|---|---|---|---|
| | | Effluent | Rectal swabs | Effluent | Rectal swabs |
| Effluent flush | 8 | 3 | 1 | 1 | 2 |
| Water flush | 4 | 0 | 0 | 0 | 0 |
| Lagoon—no flush | 2 | 1 | 1 | 0 | 1 |
| Slats and pit | 3 | 1 | 0 | 0 | 2 |
| Total | 17 | 5 | 2 | 1 | 5 |

(Table 1). The four serotypes isolated were *S. saint paul, S. typhimurium, S. molade,* and *S. manhattan*. Three of the isolates were from systems using recycling flushing-gutters and one was from a facility utilizing shallow subfloor pits with overflow into an anaerobic lagoon. No salmonella were isolated from four facilities which were relatively newly established and where fresh water was used for flushing of gutters. In two herds the same serotypes of salmonella were isolated on repeated sampling of the lagoons and rectal swabs from pigs. Three of the four systems with salmonella present also had herd histories of severe enteric disease diagnosed as swine dysentery. One of these herds had a history of salmonellosis but no conclusions could be made on the other two.

*Treponema-hyodysenteriae*-like organisms were observed in rectal swabs of pigs from the three herds with active swine dysentery but were found in very low numbers in effluent from only one.

Effluent from one of three farms with pits under slatted floors and active swine dysentery contained *S. agona*, but no salmonella were isolated from rectal swabs of pigs. Numerous *T. hydysenteriae*-like organisms could be demonstrated in rectal swabs from two herds but none were found in pit effluent.

### Effluent Feeding Trials

Effluent feeding trials were conducted to determine whether swine dysentery could be transmitted through effluent from four herds with known infection. It was suspected that lack of identification of *T. hyodysenteriae* in effluent might be the result of insensitive methods for detection rather than lack of its presence.

Pigs which were fed effluent from two of the four herds developed diarrhea with blood, mucus, and numerous *T. hyodysenteriae* in their feces. The pigs from each of the two groups were killed when moribund. There were typical lesions of swine dysentery in their colons and *T. hyodysenteriae* was isolated.

### Salmonella Pathogenicity Studies

Each group of three pigs was inoculated with one of four serotypes of salmonella from anaerobic storage systems. The serotypes used in the four groups were *S. molade, S. agona, S. typhimurium,* and *S. saint paul.*

Pigs in all groups had diarrhea, partial anorexia, and mild depression beginning 2 to 3 days post-inoculation and lasting an average of 3 days. Salmonella were isolated from their feces and from liver, spleen, and colon of one pig in each group which was killed at day 7. No salmonella were detected in control pigs.

## DISCUSSION

These studies indicate that two types of enteric pathogens, salmonella and *T. hyodysenteriae,* can survive for a significant, undetermined time in anaerobic pits and lagoons used for storage of swine wastes.

Salmonella were found in a significant number of effluent samples and in two cases the same serotype was isolated from rectal swabs collected on the same farm. This suggests the establishment and perhaps the recycling of dominant serotypes on individual farms. Pathogenicity studies showed that the salmonella were pathogenic at least to the extent that they produced diarrhea and systemic infection in experimental pigs.

Methods of detecting *T. hyodysenteriae* are very insensitive as indicated by the low number of cases in which they were detected in effluent from herds with swine dysentery. Feeding trials indicated that these organisms were present and infectious in effluent from two of four sources. In one case the effluent was taken from storage tanks used for flushing gutters so they apparently survived circulation through a lagoon.

These results should not be construed to indicate that anaerobic lagoons or recycling flushing-gutter systems are undesirable as supported by the fact that no significant disease probelms were encountered in most cases. However, design of these facilities should include consideration of the potential hazard. For instance, it would seem appropriate in recycling of effluent to place outlets and intakes as far apart as possible to allow maximum transit time. A few facilities were found to have the two openings nearly side by side. Design of gutters should permit installation of slats to prevent pigs from ingesting effluent if disease problems arise.

More study is needed to develop techniques sufficiently sensitive to determine actual survival times of various pathogens. A report by Stadler (1974) suggests the use of calcium cyanamide to decontaminate swine waste lagoons or pits, but more information is needed before general recommendations can be made for use in situations where persistent disease problems exist.


### References

1   Findlay, C. R. 1972. The persistence of *Salmonella dublin* in slurry tanks and on pasture. Vet. Rec. 91:233-235.

2   Glock, R. D., D. L. Harris and J. P. Kluge. 1974. Localization of spirochetes with the structural characteristics of *Treponema hyodysenteriae* in the lesions of swine dysentery. Infect. Immun. 9:167-178.

3   Glock, R. D., K. J. Vanderloo and J. M. Kinyon. 1975. Survival of certain pathogenic organisms in swine lagoon effluent. J. Amer. Vet. Med. Ass. 166:273-275.

4   Harris, D. L., R. D. Glock, C. R. Christensen and J. M. Kinyon. 1972. Swine dysentery I: inoculation of pigs with Treponema hyodysenteriae (new species) and reproduction of the disease. Vet. Med./Small Anim. Clin. 67:61-64.

5   Oetjen, K. A. B. and D. L. Harris. 1973. Scheme for systematic identification of aerobic pathogenic bacteria. J. Amer. Vet. Med. Ass. 163:169-174.

6   Schmitt, L, W., T. E. Hazen and R. J. Smith. 1974. In: Processing and management of agricultural wastes. Proceedings, Cornell agricultural waste management conference. 356-374.

7   Stadler, E. 1974. Die schweinedysenterie als umweltbedingte infektionskrankheit-die gulle als erregerreservoir und ihre behandlung mit alzogur. Tierarztliche Umschaue. 11:592-603.

8   Stewart, B. R. and J. M. Sweeten. 1973. Liquid manure management for swine. Texas Agr. Ex. Serv. Bull. 26 p.

9   Taylor, D. J. and T. J. L. Alexander. 1971. The production of dysentery in swine by feeding cultures containing a spirochaete. Brit. Vet. J. 127:58-61.

# Excretion of Salts by Feedlot Cattle in Response to Variations in Concentrations of Sodium Chloride Added to Their Ration


L. R. Shuyler, D. A. Clark, J. Barth, D. D. Smith
ASSOC.
MEMBER


AGRICULTURAL practices such as addition of fertilizers, confined feeding of livestock, and large applications of pesticides have been major contributors to percolating groundwater contamination and surface water pollution. In areas of intensive land use, tremendous quantities of nutrients, salts, and organic wastes have entered the ground and surface waters and seriously degraded their quality. During the decade ending in 1972, the cattle feeding industry expanded continuously and the number of fed cattle annually marketed in the United States increased nearly 40 percent, thus increasing the problem of surface and groundwater contamination from feedlot runoff. A complicating factor may have been the fact that feedlot operators have generally supplemented their feeds with varying quantities of salt (sodium chloride) during the 90 to 150 days that beef cattle are on full feed immediately prior to slaughter.

Although the salt metabolism of domestic animals has been studied in the past (Aines and Smith 1957, Dyer 1969, Morrison 1959, and Lomba et al. 1969), previous efforts have not addressed the potential long-term environmental salt accumulation resulting from feedlot waste production. Therefore, a study was designed by the National Environmental Research Center (NERC-LV), Las Vegas, Nevada, and the Robert S. Kerr Environmental Research Laboratory (RSKERL), Ada, Oklahoma, to determine whether the excretion of salts in the wastes (urine and feces) of feedlot beef animals may be significantly reduced by varying the concentration of added salt in their ration from 0.5 percent (National Research Council accepted level) to 0 percent without affecting the rate of gain. With the concurrence of the Nevada Operations Office (NV) of the Energy Research and Development Administration (ERDA), the study was conducted on the Nevada Test Site (NTS) experimental farm located near Las Vegas, Nevada, in a desert environment.

## BACKGROUND INFORMATION

The rationale for a salt supplement for cattle is to furnish additional sodium since most rations supply adequate chlorine. Aines and Smith (1957) report that a comparison of sodium and chloride ions for the treatment of salt-deficient cows showed the sodium ions but not chloride ions to be of therapeutic value. When sodium chloride was fed to salt-deficient cows, increases in milk production, body weight, and appetite occurred. When sodium bicarbonate was fed, similar increases occurred.

According to Baumgart (1969) all animals require salt and have been credited with the ability to select and regulate salt intake. Most of the sodium and chlorine included in the diet of domestic animals is excreted. When the intake is at a minimum, the body makes an adjustment whereby the output of sodium and chlorine in the urine nearly ceases. Large amounts of sodium are conserved and recycled in ruminants via the saliva. Morrison (1959) stated that the amount of salt supplement needed may vary depending on the salt content of the feed and water.

## GENERAL PROCEDURES

Nine, 10 ft x 10 ft, open-air, concrete-floored, individual feeding pens were prepared at the experimental farm. These pens were provided with a sun shade, individual feeders, and metered waterers. Modifications were made to allow conversion of the pens to temporary metabolism stalls.

Nine, grade, yearling Hereford heifers were selected from the NTS beef herd and divided into three study groups. Criteria for selection included temperament and weight. Individual animals making up each group were as similar as possible and the total weight of each group was nearly identical.

Throughout the 18-week project, each group was handled identically and received the same basal rations which were prepared by RSKERL. The only treatment difference was the percentage of salt added to the basal ration, i.e., Group I received 0 percent; Group II, 0.25 percent; and Group III, 0.5 percent. To simulate actual feedlot conditions each regime was divided into a series of three, six-week rations—a starter ration (Phase 1), an intermediate ration (Phase 2), and a finishing ration (Phase 3). The composition of each ration is presented in Table 1, and the


The authors are: L. R. SHUYLER and D. A. CLARK, Robert S. Kerr Environmental Research Laboratory, US Environmental Protection Agency, Ada, OK; J. BARTH and D. D. SMITH, National Research Center, US Environmental Protection Agency, Las Vegas, NV.


**TABLE 1. COMPOSITION OF FEED RATION.**
**(PERCENT WET WEIGHT)**

| | Phase 1 Percent | Phase 2 Percent | Phase 3 Percent |
|---|---|---|---|
| Hay (ground alfalfa) | 40.5 | 18.9 | 10.5 |
| Grain (corn, sorghum, barley) | 51.0 | 73.3 | 81.0 |
| Cottonseed meal | 8.0 | 7.4 | 8.0 |
| $CaCO_3$ (or limestone) | 0.5 | 0.46 | 0.5 |
| Trace minerals (0.5 lb/ton) | 0.025 | 0.023 | 0.025 |
| Sodium Chloride added: | | | |
| Group I | 0.0 | 0.0 | 0.0 |
| Group II | 0.25 | 0.25 | 0.25 |
| Group III | 0.5 | 0.5 | 0.5 |

| Parameter | Starter ration (Phase 1) | | | Intermediate ration (Phase 2) | | | Finishing ration (Phase 3) | | |
|---|---|---|---|---|---|---|---|---|---|
| | Group I Percent | Group II Percent | Group III Percent | Group I Percent | Group II Percent | Group III Percent | Group I Percent | Group II Percent | Group III Percent |
| Sodium | 0.079 | 0.2 | 0.38 | 0.047 | 0.15 | 0.23 | 0.016 | 0.093 | 0.2 |
| Potassium | 1.3 | 1.4 | 1.3 | 0.85 | 0.71 | 0.69 | 0.69 | 0.68 | 0.6 |
| Total Kjeldahl nitrogen | 2.7 | 2.7 | 2.7 | 2.5 | 2.4 | 2.5 | 2.4 | 2.3 | 2.4 |

analyses are shown in Table 2. All rations contained about 12.5 percent moisture.

Once daily each animal was free fed. Net feed intake weights, water consumption, and any notes on the animal's health or behavior were entered daily on each animal's record sheet. At the end of the study the animals were sacrificed and the carcass grade was determined.

### SAMPLING PROCEDURES

Grab urine and fecal samples were collected from each animal at the end of weeks 2, 3, 4, 5, 8, 9, 10, 11, 14, 15, 16, and 17. Twenty-four-hour total urine and fecal collections were made at the end of weeks 6, 12, and 18. Samples were not collected until the end of the second week of each feeding regime to allow a period of adjustment to each ration.

During the weekly collections, urination was stimulated by stroking the ventral commissure of the vulva. Approximately 200 cc of urine was collected from the resultant urine flow. The feces samples were collected by removing 200 to 300 grams from the rectum with a gloved hand.

Each six weeks the animals were weighed and the individual pens were modified into temporary metabolism stalls. The animals were haltered with access to feed and water. At approximately 9 a.m., an indwelling latex catheter (16 Fr) with a 30 cc balloon was placed in the bladder. Tygon tubing was run over a pulley arrangement to a receptacle. At 1 p.m., after allowing 4 hours for seating of the catheter, the 24-hour urine collection began and plexiglass sheets were placed on the floor immediately behind each animal. A 24-hour watch was maintained on each animal. Immediately after defecation, the feces were scraped from the plastic sheet and placed in a plastic bag.

At the end of the 24-hour sampling period, feces and urine were weighed, mixed, and aliquots removed for analysis. The pH of each urine sample was determined.

Following the collection, 30 cc of blood was removed from the jugular vein of each animal via a heparinized hypodermic syringe. This blood was centrifuged and the serum collected for analysis.

All samples collected were packaged and shipped in a refrigerated chest by air freight to RSKERL for analysis.

### PARAMETERS OF ANALYSIS

The weekly grab samples of feces were analyzed as follows: sodium, potassium, chloride, nitrate, nitrite, ammonia nitrogen, total Kjeldahl nitrogen, and moisture. In addition to the analyses performed on weekly grab samples, the 24-hour fecal composite samples were analyzed as follows: proximate analysis including crude protein, ether extract, ash, and fiber; and calcium, magnesium, total phosphorus, iron, copper, zinc, manganese, and cobalt.

Weekly grab samples of urine were analyzed as follows: sodium, potassium, chloride, carbonate and bicarbonate, nitrate, nitrite, ammonia nitrogen, total Kjeldahl nitrogen, pH, and specific gravity. In addition to the analyses performed on weekly grab samples, the 24-hour urine composite samples were analyzed as follows: calcium, magnesium, total phosphorus, iron, copper, zinc, manganese, cobalt, moisture, and ash.

The blood serum samples were analyzed as follows: sodium, potassium, chloride, calcium, magnesium, and total phosphorus.

Feed samples were analyzed as follows: sodium, potassium, chloride, calcium, magnesium, total phosphorus, iron, copper, zinc, manganese, cobalt, moisture, total Kjeldahl nitrogen, crude protein, ether extract, ash, and fiber.

Water samples were analyzed as follows: sodium, potassium, chloride, carbonate and bicarbonate, calcium, magnesi-

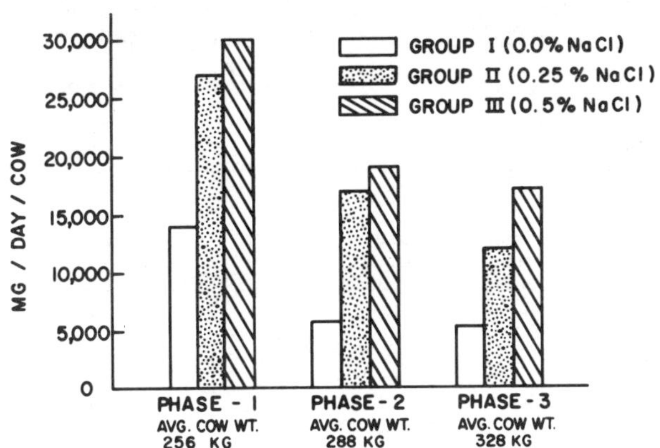


FIG. 1 Sodium in total waste output.

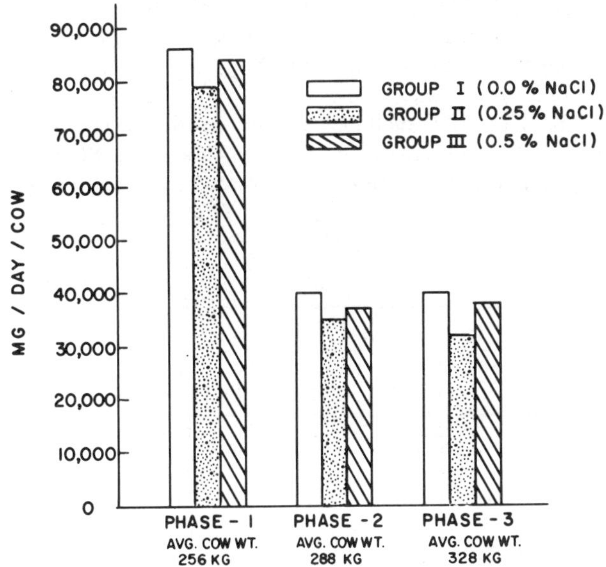


FIG. 2 Potassium in total waste output.

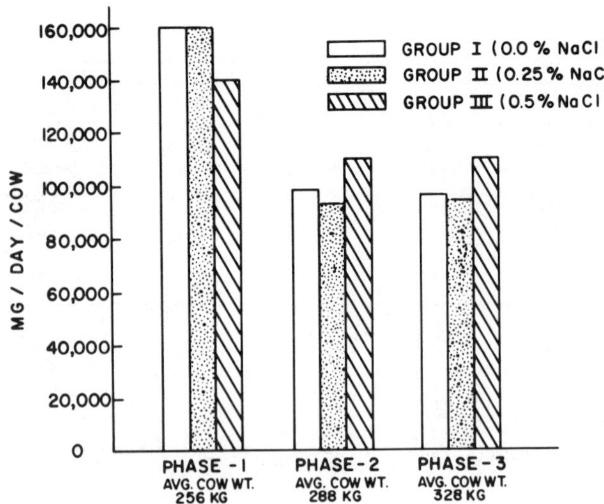



FIG. 3 Total Kjeldahl nitrogen in total waste output.

um, total phosphorus, iron, copper, zinc, manganese, and cobalt.

## ANALYTICAL PROCEDURES

The urine pH analysis was performed immediately after compositing the sample. The Department of Animal Sciences and Industry of Oklahoma State University analyzed the 24-hour fecal composite samples and the feed samples for crude protein, ether extract, ash, and fiber. The remaining analyses were done in the laboratory of the RSKERL.

Urine and blood samples were analyzed in accordance with EPA's "Manual of Methods for Chemical Analysis of Water and Wastes." Methods of analysis for fecal and feed samples were obtained from EPA methods and "Official Methods of Analysis of the Association of Official Agricultural Chemists, Tenth Edition, 1965."

Fecal and feed samples were prepared for analysis as follows: a representative sample was digested with distilled water at 80 C for one hour, cooled, and diluted to volume for nitrate and nitrite analysis by EPA methods. Ammonia nitrogen and total Kjeldahl nitrogen analyses were performed on the feces directly. The moisture content was determined on 10 to 30 grams of material dried at 105 C for 16 hours. The dried samples were subsequently ashed at 450 C, dissolved in approximately 5 percent nitric acid, and diluted to volume for the remaining analyses performed at the EPA laboratory.

Standard EPA analytical quality control tests from the "Handbook for Analytical Quality Control in Water and Wastewater Laboratories" were conducted to assure precision and accuracy of all data. Analytical quality control reference samples from the EPA Analytical Quality Control Laboratory were analyzed to check analytical techniques and standards utilized in computing standard curves. The accuracy throughout sample preparation and analysis was $\leq$ 20 percent.

## DISCUSSION OF DATA

Only the relationship between sodium, potassium, and total Kjeldahl nitrogen will be discussed in this paper. A final report that will present and interpret the remaining data obtained from the study is forthcoming.

Relative to Group III's sodium output, Groups I and II, respectively, of Phases 1, 2, and 3 produced the following percentage of sodium: Phase 1—46 percent and 90 percent, Phase 2—30 percent and 89 percent, and Phase 3—31 percent and 70 percent (Figs. 1, 2, and 3). Potassium and total Kjeldahl nitrogen levels remained relatively constant for all three groups in each phase. The maximum difference between groups was 8 percent, 12 percent, and 20 percent in Phases 1, 2, and 3, respectively, for potassium output and 12 percent, 15 percent, and 15 percent for total Kjeldahl nitrogen output in Phases 1, 2, and 3, respectively. These differences were within the range of accuracy for the analysis.

The figures also show that the total waste output of sodium for each group decreased in each subsequent phase of the study. Group I produced 48 percent (Phase 2) and 39 percent (Phase 3) of Phase 1 total waste output; Group II produced 63 percent (Phase 2) and 44 percent (Phase 3) of Phase 1; and Group III produced 63 percent (Phase 2) and 57 percent (Phase 3) of Phase 1. This downward trend is explained in part by observing the feed analysis in Table 2. For example, the sodium content of the ration fed to Group I decreased from 0.079 percent (Phase I) to 0.04 percent (Phase 2) to 0.016 percent (Phase 3). The hay used in preparing the rations contributed significantly to the total sodium content of the ration. The compositions of the rations given in Table 1 show that the 40 percent hay content in Phase 1 was reduced to 19 percent in Phase 2 and 10.5 percent in Phase 3. The sodium in the waste decreased approximately the same. The potassium and total Kjeldahl nitrogen in the waste decreased in the same manner as sodium and may be explained by their concentrations in the rations for the three phases.

The total waste output was calculated from the average feces and urine outputs on the three 24-hour composite collection periods. An average of a larger number of data points would give a more accurate total waste output.

No difference in the health of the groups of heifers was observed during the study. The average daily gain and

(Continued on page 343)

**TABLE 3. RATE OF GAIN AND SLAUGHTER DATA**

| Animal no. | NaCl percent | Average daily gain for study, kg | lb feed/lb gain | Live wt., kg | Hot carcass wt., kg | Dressing percent | Quality grade | Yield grade | Percent cutability |
|---|---|---|---|---|---|---|---|---|---|
| I-1 | 0.0 | 0.91 | 7.16 | 361.4 | 197.7 | 54.7 | Choice | 2.5 | 51.2 |
| I-2 | 0.0 | 0.60 | 9.90 | 306.4 | 181.4 | 59.2 | Choice − | 2.7 | 50.7 |
| I-3 | 0.0 | 0.40 | 12.80 | 271.8 | 150.5 | 55.4 | Good − | 1.6 | 53.3 |
| II-1 | 0.25 | 0.78 | 7.59 | 343.2 | 183.2 | 53.4 | Good | 2.3 | 51.6 |
| II-2 | 0.25 | 0.93 | 6.67 | 350.9 | 194.1 | 55.3 | Good + | 1.5 | 53.3 |
| II-3 | 0.25 | 0.78 | 8.09 | 321.8 | 177.7 | 55.2 | Choice − | 2 | 52.3 |
| III-1 | 0.5 | 0.60 | 10.12 | 315.9 | 174.1 | 55.1 | Good − | 2 | 52.3 |
| III-2 | 0.5 | 0.72 | 9.10 | 329.1 | 189.5 | 57.6 | Good | 2 | 52.3 |
| III-3 | 0.5 | 1.0 | 7.98 | 350.9 | 205 | 58.4 | Choice | 2.5 | 51.2 |

# Particle-Size Distribution of Livestock Wastes


A. C. Chang, J. M. Rible
ASSOC. MEMBER
ASAE


RECENT animal waste management research has resulted in three potential solutions for waste disposal:

1   Land disposal of animal wastes, returning plant nutrients to cropland;

2   Refeeding of wastes to livestock, salvaging residue feed value;

3   Severe destruction of wastes, recovering heat energy. However, the value of animal wastes as low quality material for any of these suggested uses has always been marginal.

The practice of returning animal wastes to cultivated land is neither the most efficient or the most economical one (Gershon et al. 1975, Chang et al. 1974, Pratt et al. 1974). Reusing defecated wastes as feed supplement at best accounts for only a small percentage of the disposal problem (Anthony 1971, Bull and Reid 1971, Westing and Bradenberg 1974). Finally, fuel recovery is made difficult by waste's high ash and/or moisture content (Garner et al. 1972).

Wastes are now causing great environment concerns, but wastes from a concentrated animal production area can also be viewed as raw materials if useful end products could be developed. Lindley (1970) studied the particle-size distribution of freshly collected dairy wastes and determined the effect of particle size on the aerobic stabilization of these wastes. It was shown that the finest fraction of wastes was decomposed to the greatest extent and that the coarsest fraction was the least degraded. This suggested that the chemical and biological properties of each fraction of waste were different due to differential particle size. This study's objective is to determine the particle-size distribution pattern of livestock wastes and to characterize each waste size class in terms of its value as fertilizer, feed supplement and fuel.

## PROCEDURE

Animal waste samples from livestock operations in southern California were selected at random for laboratory analysis. A total of 24 sets of waste samples from dairy cattle, beef cattle and poultry were analyzed. These represented freshly defecated wastes, composted wastes and wastes that had accumulated on the ground surface. One poultry waste sample from a rotary drier was included and labelled "fresh dried."

### Particle-Size Classification

Depending upon the nature of the waste sample, either a dry-sieving (for deposited and composted wastes) or a wet sieving (for freshly-collected wastes) was employed. The

size classification was accomplished by using a set of ASTM standard 12.7 cm-diameter (5 in.) wire screen sieves of 18, 35, 60, 120, and 270 mesh openings to separate waste samples into six size fractions.

In dry sieving, a 50 g portion of the sample was placed in the top sieve of the stack and shaken by a mechanical vibrating shaker for 10 minutes. Each fraction was then decanted and weighed to determine its percentage of the total. Fifteen portions of each collected waste sample were sieved, and their average represents the particle-size distribution of that sample. The vibration shaker was fabricated by mounting the vibrator (powered by 1/12 hp Dayton Shaded Pole Motor with two rotating off-center weights mounted on the motor shaft) on a fixed-spring seat. Lindley's (1970) wet-sieving procedure was followed to determine particle-size distribution of freshly-collected waste samples.

### Method of Analysis

Standard animal feed analysis procedures were used to determine crude fiber, crude protein, ether-extracted fat, moisture, and ash contents of each fraction of the wastes (Horwitz 1970). Nitrogen-free extract was calculated by difference. The only deviation from the standard procedures was in the determination of crude fiber. Since the filtration of acid and base digested liquor through an Oklahoma State filter screen or a modified California State buchner funnel was very slow, centrifugation was used to separate solids from liquid. A check using two standard feed samples showed that such deviation from standard procedure did not produce any significant difference in results.

To determine fertilizer value of each particle-size class, total Kjeldahl nitrogen and total phosphorus were determined (Bremner 1965, Olsen and Dean 1965). Heat of combustion was determined using a Parr Model 1341 oxygen bomb calorimeter.

### RESULTS AND DISCUSSION

The particle-size distribution patterns of the analyzed livestock waste samples are summarized in Table 1. The solid (composted and deposited) wastes follow a distribution pattern that is completely different from that of the freshly-collected wastes. More than 70 percent of the particles in dried wastes are greater than 0.500 mm. Particles smaller than 0.105 mm are an insignificant fraction of the total wastes. In the freshly-collected wastes, particles are almost equally divided between sizes greater than 1.000 mm and sizes smaller than 0.053 mm, with usually less than 25 percent of the total particles distributed in the intermediate ranges. This difference in particle-size distribution undoubtedly will affect the decision about how each size fraction of waste can be utilized.


The authors are: A. C. CHANG, Assistant Professor and Assistant Agricultural Engineer, Soil Science and Agricultural Engineering Dept., and J. M. RIBLE, Area Soil and Water Specialist, Cooperative Extension, University of California, Riverside.

Tables 2, 3, and 4 summarize the feed value analysis of each size class of wastes of dairy cattle, beef cattle, and poultry. There is a consistent trend of declining crude fiber content and increasing crude protein and ash content with the increasing fineness of waste particles. When the size distribution pattern is considered, crude fiber can be effectively concentrated into particles greater than 1.000 mm and crude protein into particles smaller than 0.053 mm

**TABLE 1. PARTICLE-SIZE CLASSIFICATION OF LIVESTOCK WASTES**

| Livestock | Nature of the wastes | No. of samples analyzed | Particle-size Distribution (percent total) | | | | | |
|---|---|---|---|---|---|---|---|---|
| | | | Greater than 1.000 mm | 1.000-0.500 mm | 0.500-0.250 mm | 0.250-0.105 mm | 0.105-0.053 mm | Less than 0.053 mm |
| Dairy | Deposited | 5 | 56.9 | 19.9 | 13.8 | 5.2 | 2.3 | 0.9 |
| | Composted | 2 | 67.7 | 12.4 | 10.8 | 6.6 | 2.1 | 0.4 |
| | Collected fresh | 3 | 41.8 | 7.1 | 7.2 | 3.9 | 2.0 | 38.0 |
| Beef cattle | Deposited | 3 | 65.4 | 16.1 | 10.1 | 5.3 | 1.6 | 0.5 |
| | Collected fresh | 3 | 30.7 | 9.0 | 6.7 | 6.1 | 3.6 | 43.9 |
| Poultry | Deposited | 3 | 15.2 | 29.3 | 25.0 | 16.2 | 8.5 | 6.5 |
| | Composted | 1 | 48.9 | 37.7 | 9.6 | 2.5 | 8.5 | 6.5 |
| | Collected fresh | 3 | 23.6 | 11.6 | 16.3 | 8.3 | 4.8 | 35.6 |
| | Fresh dried | 1 | 52.7 | 21.0 | 15.1 | 8.1 | 2.6 | 0.4 |

**TABLE 2. FEED VALUES OF DAIRY CATTLE WASTES**

| Nature of the waste | Particle size, mm | Moisture content, percent | Crude fiber, percent | Crude protein, percent | Fat, percent | Nitrogen free extract, percent | Ash, percent | Estimated gross energy† kcal/gm |
|---|---|---|---|---|---|---|---|---|
| | >1.000 | 12.0 | 23.8 | 12.5 | 1.39 | 21.4 | 28.9 | 3.2 |
| | 1.000-0.500 | 10.9 | 25.7 | 12.5 | 0.65 | 22.4 | 27.6 | 3.1 |
| Deposited | 0.500-0.250 | 8.2 | 19.5 | 12.6 | 0.65 | 27.3 | 31.8 | 3.0 |
| | 0.250-0.105 | 9.8 | 12.4 | 14.0 | 0.67 | 25.7 | 37.5 | 2.7 |
| | 0.105-0.053 | 8.2 | 10.5 | 14.6 | 1.35 | 21.4 | 43.9 | 2.5 |
| | <0.053 | 8.4 | 8.6 | 15.1 | 0.69 | 26.0 | 41.2 | 2.6 |
| | >1.000 | 12.3 | 14.4 | 7.3 | 0.71 | 11.3 | 54.0 | 1.8 |
| | 1.000-0.500 | 10.7 | 13.2 | 7.9 | 0.69 | 13.3 | 54.2 | 1.8 |
| Composted | 0.500-0.250 | 7.4 | 8.6 | 5.3 | 0.59 | 12.7 | 65.4 | 1.3 |
| | 0.250-0.105 | 5.0 | 6.1 | 4.7 | 0.44 | 10.3 | 73.3 | 1.1 |
| | 0.105-0.053 | 3.2 | 5.4 | 5.1 | 0.36 | 8.7 | 77.2 | 1.0 |
| | <0.053 | 3.4 | 6.6 | 6.7 | 0.52 | 8.2 | 74.6 | 1.1 |
| | >1.000 | * | 41.8 | 11.4 | 3.15 | 33.6 | 10.0 | 4.2 |
| | 1.000-0.500 | * | 33.8 | 12.4 | — | — | 8.1 | — |
| Fresh collected | 0.500-0.250 | * | 35.4 | 10.2 | 3.25 | 40.9 | 10.2 | 4.2 |
| | 0.250-0.105 | * | 27.7 | 17.2 | 3.82 | 40.0 | 11.3 | 4.2 |
| | 0.105-0.053 | * | 11.7 | 23.9 | 7.41 | 37.1 | 19.9 | 4.1 |
| | <0.053 | * | 4.4 | 29.0 | 12.92 | 16.7 | 37.0 | 3.8 |

* On dry weight basis.
† Calculated on dry weight basis by assuming 4.3 kcal/gm for crude fiber and nitrogen free extract,
5.6 kcal/gm for crude protein,
9.3 kcal/gm for fat.

**TABLE 3. FEED VALUE OF BEEF CATTLE WASTES**

| Nature of the waste | Particle size, mm | Moisture content, percent | Crude fiber, percent | Crude protein, percent | Fat, percent | Nitrogen free extract, percent | Ash, percent | Estimated gross energy† kcal/gm |
|---|---|---|---|---|---|---|---|---|
| | 1.00- | 9.1 | 17.2 | 12.2 | 2.33 | 31.8 | 28.4 | 3.3 |
| | 1.000-0.500 | 9.3 | 18.3 | 12.3 | 1.70 | 31.9 | 27.5 | 3.2 |
| Deposited | 0.500-0.250 | 8.5 | 16.3 | 12.9 | 1.88 | 30.0 | 30.4 | 3.2 |
| | 0.250-0.105 | 6.6 | 13.3 | 13.7 | 2.29 | 30.7 | 33.4 | 3.1 |
| | 0.105-0.053 | 5.7 | 11.3 | 14.3 | 2.16 | 23.4 | 43.1 | 2.6 |
| | 0.053 | 7.5 | 11.1 | 16.4 | 2.84 | 14.6 | 47.5 | 2.5 |
| | 1.000 | * | 43.7 | 10.6 | 4.76 | 32.2 | 8.7 | 4.3 |
| | 1.000-0.500 | * | 58.7 | 13.7 | 2.93 | 11.8 | 12.9 | 4.1 |
| Fresh collected | 0.500-0.250 | * | 32.8 | 15.8 | 3.50 | 33.9 | 14.0 | 4.1 |
| | 0.250-0.105 | * | 27.2 | 16.6 | 4.13 | 41.0 | 11.1 | 4.2 |
| | 0.105-0.053 | * | 16.6 | 17.5 | 3.39 | 46.2 | 16.3 | 4.0 |
| | 0.053 | * | 10.2 | 30.6 | 3.11 | 34.4 | 21.7 | 3.9 |

* On dry weight basis.
† Calculated on the dry weight basis by assuming 4.3 kcal/gm for crude fiber and nitrogen free extract,
5.6 kcal/gm for crude protein,
9.3 kcal/gm for fat.



MANAGING LIVESTOCK WASTES

from freshly-collected wastes. Composted wastes, which are most stable biologically, always have lower nutrient values.

The value of livestock wastes as feed supplements depends heavily on the digestibility of each analyzed food component. At the present time, their digestibilities are not known. By assuming that all components except ash and moisture are potential energy sources (carbohydrate, protein, and fat yield approximately 4.3, 5.6, and 9.3 kcal gross energy per gram, respectively), we calculated gross energy for each particle-size class (Tables 2, 3, and 4) (Crampton and Harris 1969). Based on estimated gross energy, fresh wastes appear to have the highest nutrient value for refeeding. Particle size does not have any significant effect on the calculated gross energy value. If protein content is used as a criterion to select feed supplements, finer fractions with high protein content will be a better choice. Because of their higher protein content, poultry wastes probably have a greater value as feed supplements than do beef and dairy cattle wastes. However, their suitability for refeeding should be further analyzed by determining their amino acid content, something not determined in the present study.

Total Kjeldahl nitrogen, total phosphorus and heat of combustion of each waste fraction were analyzed to determine the fraction's value as fertilizer and as fuel (Tables 5, 6, and 7). The coarse fractions are higher in heating value and much lower in ash content. This suggests their greater value as fuel after removal of moisture. The nitrogen-enriched fine fraction undoubtedly would be best suited for fertilizer use. The total phosphorus content does not seem to vary with particle size. The possibility of using a chemical agent to induce a more effective separation of nutrients

### TABLE 4. FEED VALUE OF POULTRY WASTES

| Nature of the waste | Particle size, mm | Moisture content, percent | Crude fiber, percent | Crude protein, percent | Fat, percent | Nitrogen free extract, percent | Ash, percent | Estimated gross energy† kcal/gm |
|---|---|---|---|---|---|---|---|---|
| Deposited | >1.000 | 10.2 | 12.4 | 26.2 | 1.58 | 17.8 | 31.9 | 3.2 |
| | 1.000-0.500 | 9.8 | 15.4 | 23.4 | 1.85 | 25.0 | 24.5 | 3.6 |
| | 0.500-0.250 | 8.8 | 15.7 | 24.1 | 1.17 | 24.5 | 25.7 | 3.5 |
| | 0.250-0.105 | 8.8 | 13.7 | 30.6 | 2.03 | 16.2 | 28.7 | 3.5 |
| | 0.105-0.053 | 8.9 | 8.8 | 27.7 | 2.15 | 9.0 | 33.5 | 3.9 |
| | <0.053 | 7.3 | 7.3 | 45.1 | 3.10 | — | 34.7 | 3.4 |
| Composted | >1.000 | 11.6 | 17.9 | 19.2 | 0.89 | 10.3 | 40.1 | 2.7 |
| | 1.000-0.500 | 11.7 | 17.3 | 19.1 | 0.56 | 7.4 | 43.9 | 2.5 |
| | 0.500-0.250 | 11.4 | 16.1 | 19.2 | 0.86 | 4.0 | 48.4 | 2.3 |
| | 0.250-0.105 | 8.5 | 12.9 | 20.4 | 1.25 | 1.0 | 55.9 | 2.0 |
| | 0.105-0.053 | 7.4 | 10.4 | 23.9 | 0.65 | 0.5 | 57.1 | 2.0 |
| | <0.053 | 10.3 | 9.4 | 29.2 | 0.75 | — | 53.5 | 2.4 |
| Fresh collected | >1.000 | * | 29.9 | 27.9 | 0.97 | 14.4 | 26.8 | 3.6 |
| | 1.000-0.500 | * | 24.1 | 22.0 | 1.24 | 33.6 | 19.0 | 3.8 |
| | 0.500-0.250 | * | 23.7 | 27.8 | 1.55 | 30.6 | 16.7 | 4.0 |
| | 0.250-0.105 | * | 18.8 | 34.3 | 2.58 | 23.7 | 20.6 | 4.0 |
| | 0.105-0.053 | * | 17.4 | 29.5 | — | — | 16.0 | — |
| | <0.053 | * | 1.7 | 56.7 | 3.20 | 8.1 | 30.3 | 3.9 |
| Fresh dried | >1.000 | 3.3 | 12.0 | 33.3 | 2.44 | 19.4 | 29.6 | 3.4 |
| | 1.000-0.500 | 2.8 | 13.1 | 30.7 | 2.25 | 15.4 | 35.7 | 3.2 |
| | 0.500-0.250 | 2.2 | 10.7 | 25.7 | 2.96 | 15.5 | 42.9 | 2.8 |
| | <0.250 | 2.7 | 11.2 | 38.2 | 3.22 | 5.8 | 38.9 | 3.2 |

*   On dry weight basis.
†   Calculated on dry weight basis by assuming 4.3 kcal/gm for crude fiber and nitrogen free extract,
         5.6 kcal/gm for protein,
         9.3 kcal/gm for fat.

### TABLE 5. PLANT NUTRIENTS AND HEAT OF COMBUSTION VALUES OF DAIRY WASTES

| Nature of the waste | Particle size, mm | Nitrogen, percent d.w. | Phosphorus, percent d.w. | Heat of combustion kcal/gm ---------- d.w. | (BTU/lb) |
|---|---|---|---|---|---|
| Deposited | >1.000 | 2.0 | 0.52 | 3.3 | (5924) |
| | 1.000-0.500 | 2.0 | 0.47 | 3.4 | (6103) |
| | 0.500-0.250 | 2.0 | 0.38 | 3.1 | (5562) |
| | 0.250-0.105 | 2.2 | 0.40 | 2.8 | (4980) |
| | 0.105-0.053 | 2.3 | 0.42 | 2.5 | (4525) |
| | <0.053 | 2.4 | 0.32 | 2.5 | (4414) |
| Composted | >1.000 | 1.2 | 0.31 | 1.9 | (3371) |
| | 1.000-0.500 | 1.3 | 0.27 | 2.2 | (3877) |
| | 0.500-0.250 | 0.8 | 0.27 | 1.6 | (2965) |
| | 0.250-0.105 | 0.8 | 0.15 | 0.9 | (1562) |
| | 0.105-0.053 | 0.8 | 0.12 | 0.8 | (1562) |
| | <0.053 | 1.1 | 0.13 | 0.8 | (1435) |
| Fresh collected | >1.000 | 1.8 | 0.51 | * | |
| | 1.000-0.500 | 2.0 | 0.63 | * | |
| | 0.500-0.250 | 1.7 | 0.62 | * | |
| | 0.250-0.105 | 2.8 | 1.20 | * | |
| | 0.105-0.053 | 3.8 | 0.76 | * | |
| | <0.053 | 4.6 | 0.98 | * | |

* Not analyzed.

in the wastes was not investigated. The heat of combustion of each fraction agrees closely with calculated gross energy in Tables 2, 3, and 4.

It is obvious from these analyses that fresh wastes have the highest value for re-utilization as fertilizer, feed supplements or fuel. The application of size classification techniques suggests separating wastes into a least two major fractions for possible separate utilization. Techniques for more effective nutrient separation and possibilities of further processing are under investigation. Nutrient values in the wastes are lost through weathering and decomposition. Therefore, deposited wastes and composted wastes generally have lower values for utilization, and particle-size separation did not appear effective as a means of concentrating nutrient value for more efficient utilization of these wastes.

## SUMMARY

1   Livestock waste samples of various nature were analyzed to determine their particle-size distribution and to characterize each size fraction in terms of its value as fertilizer, feed supplement, and fuel.

2   The particle-size distribution pattern of fresh livestock wastes and the effectiveness of using size classification techniques to concentrate nutrient components from these wastes warrant their separation into at least two size fractions for separate reutilization.

3   The value of deposited wastes and composted wastes was much lower due to loss of nutrient value during weathering and decomposition. Heavy concentration of particles in coarse fractions from these wastes make separate utilization by size classification not very promising.

4   Further efforts to improve the effectiveness of particle separation and nutrient concentration, with the possibilities of using these fractions as raw materials for further processing are presently under investigation.

## References


1   Adriano, D. C., P. F. Pratt and S. E. Bishop. 1971. Fate of inorganic forms of N and salts from land-disposed manure from dairies. In: Livestock Waste Management and Pollution Control Abatement, PROC-271, ASAE, St. Joseph, Mich. 49085. pp. 243-246.

2   Anthony, W. B. 1971. Cattle manure as feed for cattle. In: Livestock Waste Management and Pollution Control Abatement, PROC-271, ASAE, St. Joseph, Mich. 49085. pp. 293-296.

3   Bremner, J. M. 1965. Total nitrogen. In: C. A. Black (Editor-in-Chief) Methods of Soil Analysis. Amer. Soc. Agronomy, Inc., Madison, Wisconsin. pp. 1149-1178.

4   Bull, L. S. and J. J. Reid. 1971. Nutritive value of chicken manure for cattle. In: Livestock Waste Management and Pollution


### TABLE 6. PLANT NUTRIENTS AND HEAT OF COMBUSTION VALUES OF BEEF CATTLE WASTES

| Nature of the waste | Particle size, mm | Nitrogen, percent d.w. | Phosphorus, percent d.w. | Heat of combustion kcal/gm d.w. | Heat of combustion (BTU/lb) d.w. |
|---|---|---|---|---|---|
| Deposited | >1.000 | 1.9 | 0.36 | 3.4 | (6195) |
|  | 1.000-0.500 | 1.9 | 0.41 | 3.16 | (6440) |
|  | 0.500-0.250 | 2.1 | 0.55 | 3.43 | (5873) |
|  | 0.250-0.105 | 2.2 | 0.32 | 2.9 | (5298) |
|  | 0.105-0.053 | 2.3 | 0.47 | 2.16 | (4615) |
|  | <0.053 | 2.6 | 0.60 | 2.0 | (5019) |
| Fresh collected | >1.000 | 1.7 | 0.83 | * |  |
|  | 1.000-0.500 | 2.2 | 0.39 | * |  |
|  | 0.500-0.250 | 2.5 | 0.42 | * |  |
|  | 0.250-0.105 | 2.7 | 0.73 | * |  |
|  | 0.105-0.053 | 2.8 | — | * |  |
|  | <0.053 | 4.9 | 1.42 | * |  |

* Not analyzed.

### TABLE 7. PLANT NUTRIENTS AND HEAT OF COMBUSTION VALUES OF POULTRY WASTES

| Nature of the waste | Particle size, mm | Nitrogen, percent d.w. | Phosphorus, percent d.w. | Heat of combustion kcal/gm d.w. | Heat of combustion (BTU/lb) d.w. |
|---|---|---|---|---|---|
| Deposited | >1.000 | 4.2 | 0.26 | 3.5 | (6339) |
|  | 1.000-0.500 | 3.7 | 0.25 | 3.4 | (6136) |
|  | 0.500-0.250 | 3.9 | 0.81 | 3.3 | (5982) |
|  | 0.250-0.105 | 4.9 | 0.45 | 2.8 | (5077) |
|  | 0.105-0.053 | 6.0 | 0.54 | 2.5 | (4505) |
|  | <0.053 | 7.2 | 0.29 | 2.5 | (4441) |
| Composted | >1.000 | 3.1 | 0.75 | 2.4 | (4408) |
|  | 1.000-0.500 | 3.1 | 0.79 | 2.5 | (4582) |
|  | 0.500-0.250 | 3.1 | 0.55 | 1.7 | (3187) |
|  | 0.250-0.105 | 3.3 | 0.60 | 1.8 | (3177) |
|  | 0.105-0.053 | 3.8 | 0.48 | 1.0 | (1880) |
|  | <0.053 | 4.7 | 0.79 | 1.9 | (3383) |
| Fresh collected | >1.000 | 4.4 | 2.15 | * |  |
|  | 1.000-0.500 | 3.5 | 2.19 | * |  |
|  | 0.500-0.250 | 4.4 | 1.91 | * |  |
|  | 0.250-0.105 | 7.1 | — | * |  |
|  | 0.105-0.053 | 4.7 | 2.54 | * |  |
|  | <0.053 | 9.1 | 3.35 | * |  |
| Fresh dried | >1.000 | 5.3 | 0.55 | 3.1 | (5565) |
|  | 1.000-0.500 | 4.9 | 0.85 | 2.8 | (5082) |
|  | 0.500-0.250 | 4.1 | 0.71 | 2.2 | (4920) |
|  | <0.250 | 4.8 | 1.15 | 2.5 | (4444) |

* Not analyzed.



MANAGING LIVESTOCK WASTES

Control Abatement, PROC-271, ASAE, St. Joseph, Mich. 49085. pp. 297-300.

5   Chang, A. C., P. F. Pratt, K. Aref and D. C. Baier. 1974. Soil modification for the disposal of dairy cattle wastes. In: Processing and Management of Agricultural Wastes. Proc. 1974 Cornell Agricultural Waste Management Conf. pp. 522-532.

6   Crampton, E. W. and L. E. Harris. 1969. Applied Animal Nutrition. W. H. Freeman and Co., San Francisco. 2nd Edition. 753 pp.

7   Garner, W., C. E. Bricker, T. L. Ferguson, C. J. W. Wiegland and A. D. McElroy. 1972. Pyrolysis as a method of disposal of cattle feedlot wastes. In: Processing and Management of Agricultural Wastes. Proc. 1972 Cornell Agricultural Waste Management Conf. pp. 101-123.

8   Gershon, S. I., S. A. Hart, A. C. Chang and J. W. Branch. 1975. A planning study of dairy waste management. Paper to be presented at Int'l. Symp. Livestock Wastes, 1975, Urban-Champaign, Ill., April 21-24.

9   Horwitz, W. (ed.) 1970. Official methods of analysis of the Association of Official Analytical Chemists, 11th Ed. The Association of Official Analytical Chemists, Washington, D.C. 1015 pp.

10   Lindley, J. A. 1970. Effects of particle size on the aerobic treatment of animal waste. Unpublished M.S. thesis, Purdue University, August. 113 pp.

11   Olsen, S. R. and L. A. Dean. 1965. Phosphorus. In: C. A. Black (Editor-in-Chief) Method of Soil Analysis. Amer. Soc. Agronomy, Inc., Madison, Wisconsin. pp. 1035-1049.

12   Pratt, P. F., S. Davis, R. G. Sharpless, W. J. Pugh and S. E. Bishop. 1974. Nitrate contents of sudangrass and barley forage grown on plots treated with animal manures. Soil Sci. Soc. Amer. Proc. (In Press).

13   Westing, T. W. and B. Brandenberg. 1974. Beef feedlot waste in rations for beef cattle. In: Processing and Management of Agricultural Wastes, Proc. 1974 Cornell Agricultural Waste Management Conf. pp. 336-341.

---

## Excretion of Salts

*(Continued from page 338)*

pounds of feed per pound of gain are shown in Table 3. The data does not indicate that decreasing the salt ration affected these parameters.

### CONCLUSIONS

1   No differences in growth rate or quantity and quality of beef production were observed between the groups.

2   A significant decrease in the sodium waste discharged to the environment is obtained by reducing the salt added to the ration.

3   Some feed rations may have adequate sodium without supplementing the ration with salt.

### References

1   Aines, P. D. and S. E. Smith. 1957. Sodium versus chloride for the therapy of salt-deficient dairy cows. J. Dairy Sci. 40:682-688.

2   Baumgart, B. R. 1969. Voluntary feed intake. In: Animal growth and nutrition, E.S.E. Hafez and I.A. Dyer. Lea and Febiger, Philadelphia. pp. 121-137.

3   Dyer, I. A. 1969. Mineral requirements. In: Animal growth and nutrition, E.S.E. Hafez and I.A. Dyer. Lea and Febiger, Philadelphia. pp. 312-331.

4   Handbook for analytical quality control in water and waste-water laboratories. June 1972. U.S. Environmental Protection Agency, Cincinnati.

5   Lomba, R., R. Paquay, V. Bienfet, and A. Lousse. 1969. Statistical research on the fate of dietary mineral elements in dry and lactating cows. VI. Sodium. J. Agri. Sci. 73:453-458.

6   Manual of methods for chemical analysis of water and wastes. 1974. EPA-625/6-74-003. U.S. Environmental Protection Agency, Cincinnati.

7   Morrison, Frank B. 1959. Feeds and feeding. Twenty-second Edition. The Morrison Publishing Co., Clinton, Iowa.

8   Official methods of analysis of the association of official agricultural chemists. 1965. Tenth Edition. Association of Official Agricultural Chemists, Washington, D.C.

# Decomposition Rates of Beef Cattle Wastes

Marvin L. Stone, Judson M. Harper, Ralph W. Hansen

MEMBER
ASAE

MEMBER
ASAE

INCREASED interest in utilizing manure as a fertilizer, fuel, and feed source has focused attention on manure qualities. The reuse of manure emphasizes the need for proper management of the manure to retain its utilizable components. Many components of the manure can be lost or reduced through decomposition on the feedlot surface and during storage. Losses of up to 90 percent of the nitrogen in manure have been reported while the manure was left on the feedlot surface (McCalla et al. 1971). Data are needed that can be used as a basis to determine the best manure management and harvesting practices. Component losses may then be reduced to a minimum and harvesting practices used to produce optimal manure utilization techniques.

Decomposition of manure occurs through biological action and spontaneous chemical reactions. These processes are affected by the microclimate surrounding the manure, the chemical composition of the manure, and the microbiological populations existing in the manure. The initial chemical and biological composition of manure is a function of the animal's feed, age, and other factors. For a given manure, the microclimate then controls decomposition. The purpose of this study was to examine the effects of microclimate on decomposition rates of representative samples of beef cattle wastes.

## PROCEDURES

Temperature and water availability in manure are the major influences on the growth of bacteria and on chemical reaction rates and were chosen as the major independent variables in the study. The levels chosen for these variables reflect values comparable to those which might be found in the field. Temperature levels selected were 40 F (4.4 C), 80 F (26.7 C), and 120 F (48.8 C), and moisture levels were 30, 50, and 70 percent on a wet basis.

Decomposition level of the manure was measured as changes in the chemical and physical properties. Properties measured were those physical and chemical properties important when reusing the manure, specifically, viscosity, bulk density, particle size distribution, squeezability, and odor.

A Brookfield RVT viscometer using a no. 3 spindle was used to measure the viscosity of a 15 percent solids slurry of the manure. This viscometer rotates a spindle in the manure and measures the resulting torque which can be related to viscosity. Shear rate then is directly proportional to spindle speed. The power law was used to model viscosity as:

$$\mu = K(SS)^n$$

when

K,n = constants
$\mu$ = viscosity in centipoise
SS = spindle speed of viscometer in rpm

A wet sieve analysis (Frecks and Gilbertson 1973) was used to determine particle size distribution. Squeezability was developed as a measure of the percent of liquid from a solids slurry that would pass through a hand squeeze press. It is an index of the ease with which the liquid and fiber portions of the manure slurry can be separated.

The hand press used for the squeezability test had a 3 in. diameter perforated cup in which the sample was pressed with a piston. The sample was stirred and pressed until no more liquid would pass.

Chemical properties measured included nitrogen, ash, fiber, and pH. Nitrogen was separated into protein and non-protein nitrogen and ammonia nitrogen. Total nitrogen was analyzed by the Kjeldahl method. Protein nitrogen was measured as the Kjeldahl nitrogen content of a trichloroacetic acid precipitate. Ammonia nitrogen was found using the method of Bremmer and Keeney (1965).

Acid detergent fiber was measured using the method described in "Forage Fiber Analysis" (Goering and Van Soest 1970).

To establish the initial composition of the manure, a composite urine-free sample was taken from a feedlot. Fresh feces samples were collected in equal parts from pens with animals weighing approximately 500, 750, and 1000 pounds. After collection the manure was placed in a large container, mixed, and placed on a plastic sheet in a thin layer and dried. One third of the manure sample was removed when the manure reached each of the desired moisture levels.

Each third of the manure was again divided into thirds, packed to maximum density in brick form and frozen. This provided nine bricks of manure, one for each temperature and moisture combination.

A constant temperature/humidity chamber, which could control the temperature of the air and humidity over all feasible ranges, was used to age the manure at the desired levels of moisture content and temperature.

During the aging process the manure was sampled every other day for a ten-day period, with the exception of the samples aged at 40 F (4.4 C). Manure at this temperature was expected to decompose slowly and was, therefore, sampled every other day for the first and last four days of a twenty-day period and every four days for the remaining eight days.

The data collected was analyzed statistically. A

The authors are: MARVIN L. STONE, JUDSON M. HARPER and RALPH W. HANSEN, Agricultural Engineering Dept., Colorado State University, Fort Collins.

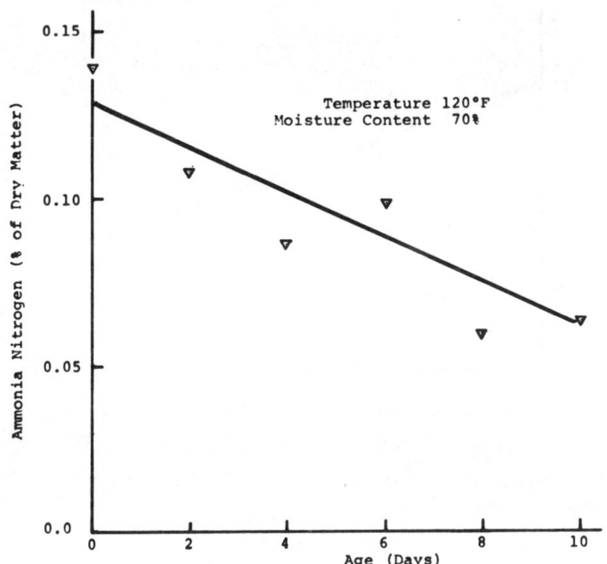

FIG. 1 Ammonia nitrogen content of manure aged at 120 F temperature and 70 percent moisture content.

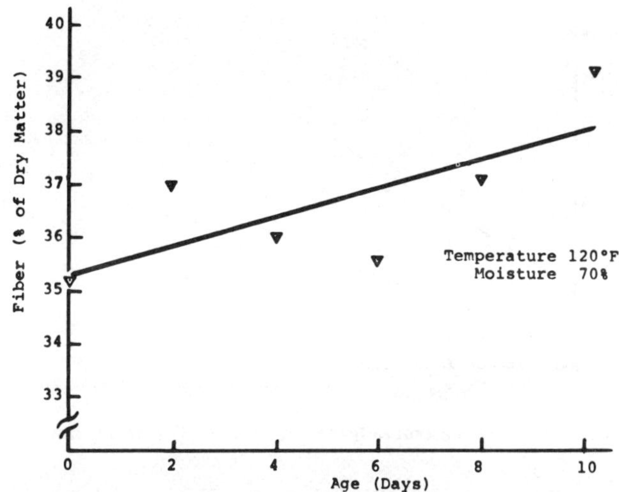

FIG. 3 Fiber content of manure aged at 120 F temperature and 70 percent moisture content.

two-way analysis of variance was performed with respect to temperature and moisture content, with time included as a covariate. This analysis eliminated parameters that had no significant overall effect. Variables that had significant effects were analyzed further for rate of change for each parameter for each temperature and moisture combination. The rates of change for each parameter were compared.

## RESULTS AND CONCLUSIONS

Significant changes with respect to time were found in five parameters. Chemical parameters undergoing changes were ammonia, acid detergent fiber, and ash, and physical properties undergoing changes were viscosity and squeezability.

Of the chemical parameters, ammonia underwent the greatest overall change with respect to time, a decrease of 35 percent. This is relatively unimportant, however, as ammonia is only 3 to 4 percent of the total nitrogen and 0.05 to 0.1 percent of the total dry matter. The greatest change in ammonia occurred at a temperature of 120 F (48.8 C) and 70 percent moisture content as shown in Fig. 1. Further analysis revealed the mean rate of change

of ammonia to increase with temperature and moisture content. To conserve this nitrogen, keep the manure dry; but since ammonia is such a small part of the total nitrogen, its loss should not be considered significant.

The changes in acid detergent fiber (ADF) and ash were significant, but produced less change on a percentage basis than ammonia. Ash content increased 4 percent overall with a mean content of 25.6 percent of the dry matter. Fig. 2 illustrates the change in ash at 120 F (48.8 C) and 70 percent moisture content. ADF increased 3 percent with a mean content of 34.1 percent of the dry matter. The change that occurred at 120 F (48.8 C) can be seen in Fig. 3. Increase in ash content is expected as bacterial action converts volatile solids to non-volatile solids or ash. Increase in fiber percentage may be attributed to the fact that it decomposed very slowly and would, therefore, tend to increase as the total manure solids decrease with time.

Ash and ADF were major chemical constituents in manure. The increase in ash means conversely a decrease in organic matter and suggests shorter harvesting periods, since the 4 percent increase in this study occurred in ten to twenty days.

No significant linear trends with time were found in pH, but a pattern in its change occurred. During the 120

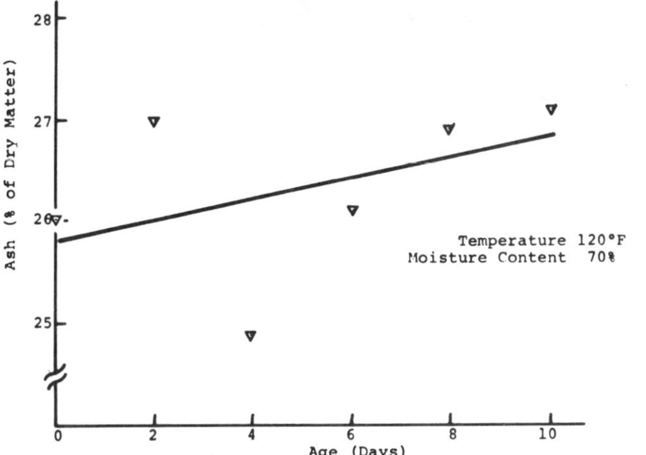

FIG. 2 Ash content of manure aged at 120 F temperature and 70 percent moisture content.

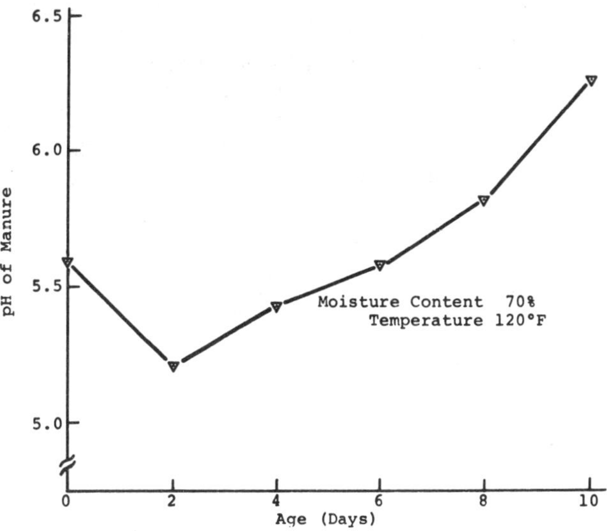

FIG. 4 pH of beef feces aged at 120 F and 70 percent moisture content versus time.

MANAGING LIVESTOCK WASTES

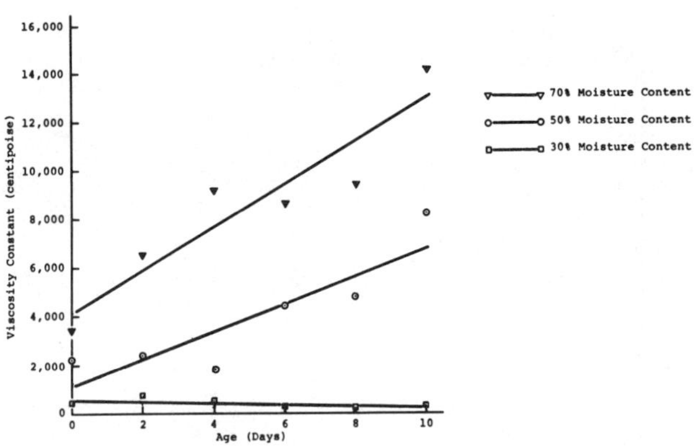

FIG. 5 Viscosity constant of beef feces aged at 120 F versus time.

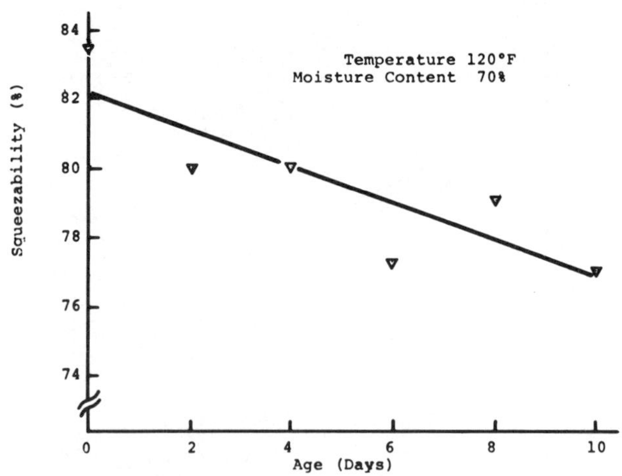

FIG. 6 Squeezability of manure aged at 120 F temperature and 70 percent moisture content.

F (48.8 C) and 80 F (26.7 C) tests, pH dropped for the first several days, then slowly increased for the duration of the test as shown in Fig. 4. This pattern was also observed to correlate with changes in the odor of the manure. During the first several days, a strong silage odor developed and thereafter that odor decreased and was replaced with a strong moldy odor. Mold-like growths were observed throughout the manure during the latter periods of these tests.

Viscosity of the aged manure slurry increased with time. These changes in viscosity can be portrayed as a change in the constant in the viscosity model with time. During the ten- and twenty-day tests the overall change was an increase of 50 percent. Significant differences were found in rates of change of this constant with different treatments. Higher rates of change were associated with higher moisture contents and temperatures. Fig. 5 illustrates the absolute value of the viscosity constant for different moisture contents aged at 120 F (48.8 C). The viscosity constant is the actual viscosity of the manure at a particular spindle speed of one on the viscometer. An increase was found in the rate of change of the constant with increasing moisture.

These viscosity changes may have a significant effect on manure reuse. Any process that uses slurried manure could develop large increases in viscosity over relatively short periods of storage. To maintain a lower viscosity, the use of shorter harvesting periods, while keeping the manure as dry as possible, would be the appropriate management practice.

Squeezability exhibited a 4 percent overall decrease with time, with a mean of 74 percent passing through the press. The greatest change in squeezability occurred during the 120 F (48.8 C) and 70 percent moisture content tests as shown in Fig. 6. These changes were negatively correlated with viscosity. An increase in viscosity decreases the amount passing through the press.

## SUMMARY

No significant changes occurred in total or protein nitrogen for the ten- to twenty-day test periods, although a 35 percent drop in ammonia nitrogen occurred. Ash and fiber content both increased slightly with time as total solids decreased. Viscosity increased rapidly with a corresponding decrease in squeezability during time.

Manure may be easier to process and additional constituents conserved if management practices provide for more frequent harvesting and adequate storage. By determining the physical and chemical changes in the manure which significantly affect a given recycling process and applying available management practices to minimize the level of these changes, the final yield of product from the harvested manure can be optimized.


### References

1 Bremmer, J. M. and D. R. Keeney. 1965. Steam distillation methods for determinating ammonium, nitrate and nitrite. Analytica Chimica Acta, 32:485-495.

2 Frecks, G. A. and C. B. Gilbertson. 1973. The effects of ration on the engineering properties of beef cattle manure. ASAE Paper No. 73-442. ASAE, St. Joseph, Mich. 49085.

3 Goering, H. K. and P. J. Van Soest. 1970. Forage fiber analysis. Agricultural Handbook No. 379. ARS-USDA.

4 McCalla, T. M., L. R. Federick and G. L. Palmer. 1970. Manure decomposition and fate of breakdown products in the soil. In: Agricultural Practices and Water Quality, Iowa State University Press, Ames, Iowa.

# Chemical Characteristics of Beef Feedlot Manures as Influenced by Housing Type

D. C. Adriano

ANIMAL wastes, fertilizers, and sediments have been identified as major pollutants to our water courses and/or groundwaters. In the past, animal wastes were almost totally neglected resulting, in some areas, to extensive environmental contamination. That was before the United States Environmental Protection Agency (EPA) was created and when commercial fertilizers were cheap. But now EPA and many state governments have enforced stringent discharge regulations and fertilizers' prices have skyrocketed. These are probably two main reasons why manures are now being discharged with a great deal of caution.

Farmers have been increasingly returning manures to the farms because of their high fertilizer values. The amounts of fertilizing nutrients that these manures can provide depend upon the volumes to be applied and the nutrient concentrations in them. Therefore, for a certain application rate, the concentration of nutrients, or elemental composition, must be known prior to aplication.

The nutrients in manure are not stable. They undergo changes during decomposition, a process influenced by climate. Of the various climatic parameters affecting this process, perhaps the most important are temperature and moisture. Various housing types of feedlots modify this climatic influence. Thus, various housing types could influence the decomposition of manures and transformation of its constituents.

This study was undertaken to determine the chemical composition of beef feedlot wastes collected from various housing systems.

## FEEDLOT DESCRIPTION AND STUDY SITES

In Michigan, as in other areas in the north central region of the United States, there are several feedlot housing types which provide varying degrees of protection of animals from weather exposure. The diagramatic representation of the five basic housing types in Michigan are depicted in Fig. 1. In the open lot, no shelter is provided for the animals and the facility solely consists of a dirt lot. The drylot unpaved feedlot has a shelter area with an unpaved outside lot, whereas, the drylot paved feedlot consists of a shelter with an open

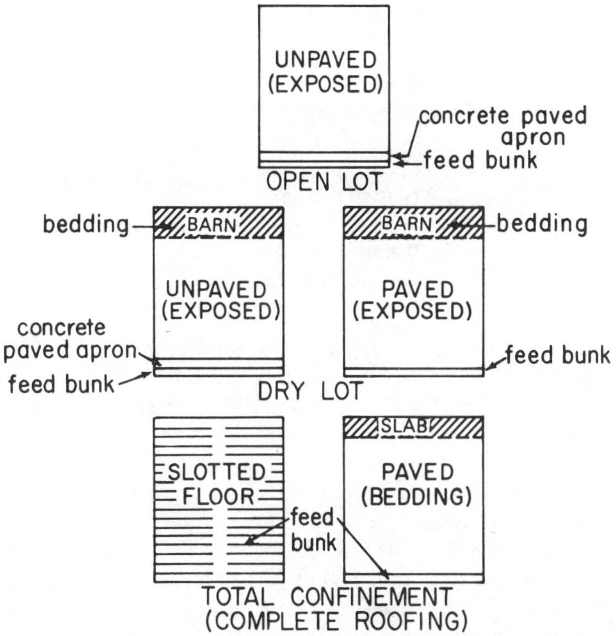

FIG. 1 Common beef housing systems in Michigan.

front allowing access to a paved outside lot. Total confinement (cold) housing (TCH) may be one of two types. In the first type, feeders are completely confined in a shelter with a solid concrete floor. Manure is scraped from the floor regularly and either stored to spread later or spread immediately. The other type may have a slotted floor provided with a pit below the floor for storage purposes. The waste is pumped into wagons and spread on fields when the pits are full.

Scraping and hauling of manures is required more frequently in drylot and total confinement systems because of the smaller areas per animal they provide. Paved drylot housing systems generally require a much smaller exposed area per animal than open lots and unpaved drylot systems. Bedding is often used in drylot and total confinement systems.

As of 1969, drylot unpaved comprised about 53 percent of the total (1,686) feedlots in Michigan. About 40 percent was drylot paved and 4 percent each was open lot and completely covered lot. Most of the feedlots were in the < 100 to 500-head capacity (Johnson and Davis 1974).

The extent of nutrient losses via runoff, leaching, or volatilization should vary in these various housing types. The greatest runoff problems may be expected in an open lot system because of lack of roof covering, whereas, in total confinements this problem is nonexistent. Leaching could occur in open lot and drylot unpaved systems. Volatilization losses would be expected to be high in systems with open areas.

Six feedlots, located in the same geographic area in


Contribution from the Crop and Soil Sciences Dept., Michigan Agricultural Experiment Station, Michigan State University, East Lansing, Mich. 48824. Financial support from NSF-RANN through Grant GI-20 is gratefully acknowledged.

Michigan Agricultural Experiment Station Journal Article No. 7265.

**Acknowledgments;** The assistance of L. Ameen in sampling and analyses of samples and Dr. J. Johnson of the Agricultural Economics Dept. and Dr. R. Maddex of the Agricultural Engineering Dept. in feedlot site selection are gratefully acknowledged.

The author is: D. C. ADRIANO, Soil Scientist, Savannah River Ecology Laboratory, Institute of Ecology, University of Georgia and the US Energy Research and Development Administration, Aiken, South Carolina (formerly Michigan State Universtiy, East Lansing).

| Feedlot no. | Housing System | Approximate animal size, pounds | Approximate animal density, sq ft/head | Approximate scraping frequency | Type of ration |
|---|---|---|---|---|---|
| 1 | Open lot (unpaved) | 600-900 | 870 | once in 2-3 years | corn silage |
|  | Total confinement (paved bedding) | 800-1200 | 50 | twice a year | corn silage |
| 2 | Open lot (unpaved) | 450-700 | 350 | none | corn silage & grain |
|  | Drylot (paved— no bedding) | 700-1100 | 60 | once in 2 weeks | corn silage & grain |
| 3 | Drylot (paved— bedding) | 1000-1200 | 60 | twice a year | corn silage & grain |
| 4 | Drylot (paved— bedding) | 500-1200 | 30 | once a month | corn silage, grain & hay |
| 5 | Total confinement (slotted floor) | 700-1200 | 20 | twice a year | corn silage & grain |
| 6 | Total confinement (paved bedding) | 600-1100 | 60 | once in 2 weeks | corn silage |
|  | Open lot (unpaved) | 800-900 | 730 | none | corn silage |

southern Michigan to achieve similarity in climatic conditions, were selected in this study. In Table 1 are presented the physical characteristics of these feedlots, rations fed, and approximate manure management. At sampling time, Feedlot 1 had a total of about 1,000 head; 820 were in an open lot and the rest in the total confinement system. The open lot system was fenced into four sections, with 600 to 700-lb animals confined in two adjacent sections, and 800 to 900-lb animals confined in the other two. Manure piles had formed in these sections. In the total confinement, animals were in the 800 to 1200-lb weight range. Feedlot 2 had about 1400 head with about 350 head in each of two adjacent open lots, and the remainder in a paved drylot. One open lot was unpaved, and the other paved. The animals in open lots weighed between 450 and 700 lb, and those in the drylot weighed between 700 and 1100 lb. Feedlot 3 had about 800 head, weighing between 900 and 1200 lb, housed in paved drylots. In this feedlot manure in the exposed areas was being flushed with water and collected in a concrete sump. There was about 400 head in Feedlot 4 in a paved drylot housing system. The animals weighed from 500 to 1200 lb. Feedlot 5 had about 400 head housed in total confinement with a slotted floor. The animals weighed from 500 to 1150 lb and were housed in five sections. Waste were collected in concrete pits about 8 feet deep. Feedlot 6 had about 3000 head, with about 675 animals housed in paved total confinement, and the balance in open lots. The confined animals weighed from 600 to 1000 lb.

The total confinement systems provide the least footage areas, ranging from 20 to 60 ft² per head. Open lots provided the largest areas per animal, ranging from 350 to 870 ft². The scraping frequency varied between the various housing systems, from as often as once in 2 weeks in a paved total confinement (Feedlot 2) to almost none in two open lots, in which exposed areas were not surfaced.

## EXPERIMENTAL PROCEDURES

Manure and ration samples were collected during spring to fall of 1973. These samples were collected four times at bi-monthly intervals, starting in the last week of March. The composite samples for old manures were collected by digging manure packs in three to four different areas in a pen. The old manure consisted of feces, urine, and bedding materials, generally straw. From each pen, fractions of six to eight freshly defecated fecal mounds were composited to represent fresh fecal samples. Caution was observed to make sure that fresh fecal samples were not urine contaminated. Ration samples were collected from the feed bunks. Manure, fecal and ration samples were placed in plastic bags and frozen until chemical analysis could be conducted. Prior to freezing, fractions were separated for air drying to determine moisture content gravimetrically for calculating the total dry solid contents.

Wet manure and fecal samples were analyzed for total Kjeldahl N using a semi-micro acid digestion and distillation technique (Bremner 1965). Ration samples were first air dried and ground before the Kjeldahl N analysis.

For total elemental analysis, dried and ground samples were predigested with nitric acid using the micro-Kjeldahl apparatus prior to digestion with perchloric acid. Phosphorus was colorimetrically analyzed (Jackson 1958). Potassium and Na were determined using a flame photometer. Calcium, Mg, Fe, Mn, Zn, and Cu were determined using an atomic absorption spectrophotometer. Electrical conductivity and pH were determined from extracts of wet manure and feces using a conductivity bridge and pH electrode, respectively.

## RESULTS AND DISCUSSION

Nitrogen excreted by cattle is primarily in the organic form. Approximately 50 percent of this N is in the urine as urea, $CO(NH_2)_2$, and 50 percent is in the feces as proteinaceous N. Urea in urine is quickly hydrolyzed to ammonium carbonate, $(NH_4)_2CO_3$, and ammonia, $NH_3$, which can either be adsorbed by the clay particles or evolved to the atmosphere. High temperature, low or high moisture contents of the manure and soil, and soils with low cation exchange capacities (i.e., sandy soils) would promote large N losses through $NH_3$ volatilization. Large losses of N through this pathway have been reported by several investigators (Adriano et al. 1973, Adriano et al. 1974, Stewart 1970). Increased volatilization of $NH_3$ with increasing evaporation from a wet soil surface has been demonstrated to occur in the laboratory (Stewart 1970), in the greenhouse (Adriano et al. 1974) and in a feedlot (Elliot 1971). Evidence indicates that a

TABLE 2.
SOME CHEMICAL CHARACTERISTICS OF MANURE FROM FEEDLOTS IN SOUTHERN MICHIGAN. VALUES FOR THE ELEMENTS ARE ON DRY WEIGHT BASIS

| | Number of samples | Total dried solids | Kjeldahl N | P | K | Ca | Mg | Na | Fe | Mn | Zn | Cu | EC mmho/cm | pH |
|---|---|---|---|---|---|---|---|---|---|---|---|---|---|---|
| | | | | | | | Total | | | | | | | |
| | | | | | | percent | | | | | ppm | | | |
| **Openlot:** | | | | | | | | | | | | | | |
| Old Manure: | 18 | | | | | | | | | | | | | |
| Unpaved: | | | | | | | | | | | | | | |
| Range | | 32-71 | 0.53-2.12 | 0.11-0.79 | 0.50-1.58 | 0.91-9.33 | 0.42-2.62 | 0.07-0.41 | 0.72-3.05 | 332-1275 | 63-149 | 13-39 | 1.10-6.20 | 7.2-8.9 |
| Average | | 51 | 1.28 | 0.40 | 0.90 | 3.55 | 1.07 | 0.21 | 1.25 | 586 | 103 | 26 | 3.41 | 7.9 |
| C.V. percent | | 21 | 40 | 52 | 31 | 76 | 67 | 52 | 31 | 42 | 30 | 31 | 39 | 7 |
| Piled: | 6 | | | | | | | | | | | | | |
| Range | | 42-61 | 0.55-1.33 | 0.24-0.85 | 0.41-1.07 | 1.57-2.52 | 0.60-0.93 | 0.11-0.18 | 0.99-1.29 | 404-685 | 91-140 | 18-33 | 2.51-3.80 | 8.0-8.8 |
| Average | | 52 | 0.94 | 0.49 | 0.85 | 2.03 | 0.77 | 0.16 | 1.17 | 494 | 119 | 27 | 3.10 | 8.4 |
| C.V., percent | | 15 | 27 | 55 | 28 | 21 | 18 | 26 | 110 | 21 | 15 | 22 | 18 | 4 |
| Fresh Feces: | 18 | | | | | | | | | | | | | |
| Range | | 17-28 | 1.37-2.95 | 0.21-1.18 | 0.44-1.49 | 0.44-2.02 | 0.28-0.56 | 0.10-0.61 | 0.080--0.97 | 156-318 | 80-251 | 13-133 | 1.60-3.90 | 5.0-8.2 |
| Average | | 21 | 2.05 | 0.65 | 0.95 | 1.19 | 0.38 | 0.30 | 0.30 | 222 | 150 | 40 | 2.60 | 6.8 |
| C.V., percent | | 15 | 23 | 33 | 28 | 36 | 20 | 51 | 123 | 20 | 35 | 72 | 20 | 17 |
| **Drylot:** | | | | | | | | | | | | | | |
| Old Manure: | 18 | | | | | | | | | | | | | |
| Range | | 17-52 | 1.56-3.19 | 0.30-1.31 | 1.28-3.81 | 0.73-1.93 | 0.41-0.65 | 0.16-0.64 | 0.049-0.76 | 60-251 | 65-283 | 12-39 | 3.10-10.0 | 5.4-9.0 |
| Average | | 30 | 2.47 | 0.67 | 2.18 | 1.10 | 0.53 | 0.38 | 0.26 | 144 | 104 | 23 | 6.40 | 7.3 |
| C.V., percent | | 31 | 17 | 35 | 32 | 27 | 13 | 38 | 138 | 46 | 46 | 33 | 32 | 17 |
| Fresh feces: | 18 | | | | | | | | | | | | | |
| Range | | 14-31 | 1.40-3.06 | 0.56-1.53 | 0.76-1.65 | 0.58-1.40 | 0.26-0.61 | 0.12-0.41 | 0.052-0.15 | 85-213 | 72-401 | 12-100 | 2.30-4.40 | 4.2-8.4 |
| Average | | 22 | 2.08 | 0.91 | 1.20 | 0.98 | 0.40 | 0.24 | 0.089 | 134 | 131 | 33 | 3.40 | 6.4 |
| C.V., percent | | 20 | 18 | 27 | 20 | 25 | 24 | 31 | 37 | 31 | 64 | 58 | 32 | 26 |
| **Total Confinement:** | | | | | | | | | | | | | | |
| Old Manure: | | | | | | | | | | | | | | |
| Slotted floor: | 4 | | | | | | | | | | | | | |
| Range | | 15-36 | 2.22-2.83 | 0.69-0.89 | 1.93-3.41 | 0.60-1.04 | 0.48-0.63 | 0.09-0.24 | 0.066-0.071 | 64-75 | 65-94 | 14-17 | 5.40-6.90 | 6.9-8.0 |
| Average | | 22 | 2.56 | 0.75 | 2.57 | 0.81 | 0.54 | 0.17 | 0.069 | 70 | 83 | 16 | 6.30 | 7.6 |
| C.V., percent | | 43 | 10 | 13 | 28 | 27 | 15 | 44 | 3 | 8 | 19 | 10 | 10 | 7 |
| Paved (beddings) | 23 | | | | | | | | | | | | | |
| Range | | 18-49 | 1.36-4.39 | 0.15-1.08 | 1.55-4.02 | 0.98-6.50 | 0.37-2.47 | 0.05-0.69 | 0.038-3.31 | 68-4240 | 41-310 | 15-107 | 3.80-12.0 | 6 5-9.3 |
| Average | | 28 | 2.77 | 0.63 | 2.56 | 1.74 | 0.70 | 0.33 | 0.30 | 418 | 130 | 32 | 6.43 | 8.4 |
| C.V., percent | | 26 | 30 | 48 | 29 | 67 | 67 | 53 | 86 | 216 | 44 | 68 | 30 | 8 |
| Fresh feces: | 23 | | | | | | | | | | | | | |
| Range | | 16-30 | 1.31-3.14 | 0.26-1.10 | 0.55-1.68 | 0.71-2.86 | 0.21-0.72 | 0.08-0.51 | 0.036-0.22 | 52-336 | 86-291 | 13-83 | 2.20-8.20 | 4.2-8.3 |
| Average | | 20 | 2.25 | 0.68 | 1.15 | 1.45 | 0.43 | 0.24 | 0.087 | 144 | 144 | 50 | 3.20 | 7.1 |
| C.V., percent | | 19 | 25 | 42 | 27 | 36 | 31 | 49 | 47 | 46 | 51 | 55 | 40 | 16 |

large percentage of evolved $NH_3$ came from the urine fraction of the manure (Adriano et al. 1971, Adriano et al. 1973, Adriano et al. 1974, Stewart 1970).

Two other modes of N loss from manures in feedlots are possible. First, some of the $NO_3$ that is not denitrified or absorbed by the plants can leach through the soil profile and contaminate the groundwater. Several studies in southern California indicated that $NO_3$ leached through the profiles at high concentrations beneath some dairy corrals, beef feedlots, and croplands receiving dairy and feedlot manures (Adriano et al. 1971, Chang et al. 1973). The other mode of N loss is by surface runoff. Organic and inorganic forms of N present in feedlot manures can be subject to this type of loss and were detected in barnlot runoffs in Ohio (Edwards et al. 1972).

Data in Table 2 indicate that the total dried solids (TDS) in fresh feces collected from the various housing systems were about equal (21 percent). Old manure from unpaved and piled areas in open lots had about equal TDS contents (51 percent). Manure from paved drylots and the paved floor TCH were about 30 percent TDS. Manure from the slotted floor TCH contained 22 percent TDS. The range (minimum to maximum) values were about equal for the feces, but these values were wider for the manures. No attempt was made to determine the extent of soil-particle contamination of manure from the open lots. It should be pointed out that no sampling was done right after a rainfall event.

The N content of manure from open lots averaged 1.1 percent (Table 2). Manure from drylot and TCH systems had more than twice the N of manure from open lots.

This pattern suggests that larger amounts of this nutrient were lost from open lots, possibly by $NH_3$ volatilization. Fresh feces from various lots had about equal N contents (2.1 percent), about double that of manure from open lots but slightly lower than in manure from drylot and TCH systems. Bedding or litter, usually straw or corn cobs, is commonly used in drylots in Michigan. These materials are low in nutrient contents and will dilute the nutrient contents of manure. However, bedding or litter would prevent large losses of N primarily from the urine fractions by volatilization. Assuming minimal losses of N from TCH systems (average of 2.7 percent), more than 50 percent of total N defecated in open lots were lost, as indicated by N data for open lots (average of 1.1 percent). Nitrogen content is least variable in TCH systems with slotted floors.

Phosphorus is mainly excreted through the fecal fractions (Azevedo and Stout). In open lots, the P content of manure tended to be lower than in fresh feces, probably caused by runoff or leaching losses. However, in TCH systems, P content of manure was about equal to that in fresh feces. The average P content of manure was 0.40 percent for the unpaved lot, compared to 0.75 percent for the TCH system with slotted floor. Phosphorus contents in open lots was quite variable, with the least variability observed in slotted floor TCH systems. Potassium is excreted mainly through the urinary fraction, which accounts for about 70 percent of the total excreta (Azevedo and Stout). This explains the tendency for fresh feces to have lower K than manure. The modes of losses for this nutrient can be expected to be quite similar to those for P, i.e. mainly by runoff and to some extent by leaching. The K content of manure was

MANAGING LIVESTOCK WASTES

## TABLE 3. AVERAGE CHEMICAL COMPOSITION OF RATION AND MANURE SAMPLES TAKEN FROM SIX SOUTHERN MICHIGAN FEEDLOTS*

| | No. of samples | TDS | Kjel. N | P | K | Total Ca | Mg | Na | Fe | Mn | Total Zn | Cu | EC mmho/cm | pH |
|---|---|---|---|---|---|---|---|---|---|---|---|---|---|---|
| | | | | | | Percent | | | | | ppm | | | |
| Ration | 50 | 40 | 1.65 | 0.40 | 1.02 | 0.53 | 0.22 | 0.15 | 0.034 | 43 | 58 | 21 | — | — |
| Fresh feces | 59 | 21 | 2.13 | 0.75 | 1.10 | 1.21 | 0.40 | 0.26 | 0.16 | 185 | 142 | 34 | 3.07 | 6.8 |
| Old manure | 73 | 39 | 1.84 | 0.61 | 1.63 | 1.69 | 0.66 | 0.23 | 0.67 | 369 | 113 | 30 | 5.02 | 7.7 |

* All chemical values are expressed on dry-weight basis. These values were obtained by combining data from the various housing types.

generally low in open lots, with an average of 0.82 percent. Manure from drylot and TCH systems contained more than 2 percent K.

The percentages of Ca and Mg in manure from open lots were generally high (Table 2). But these nutrients were quite variable in open lots and also in paved TCH systems. The Na content was about equal in unpaved and piled areas in open lots. The Fe content of manure in open lots averaged 1.2 percent as compared to 0.26 percent in drylot, and 0.18 percent in TCH systems. The Mn, Zn and Cu portions of manure were also generally high in open lots. Zinc tended to be lower in manure than in fresh feces. Electrical conductivity, a measure of the soluble salt concentration in manure, was generally high in drylot and TCH systems. No discernible pattern could be detected for manure pH from the various housing types. The pH levels ranged from 6.4 in the paved open lot to 8.4 in piled open lot and paved total confinement.

The average values for the various parameters for the ration, manure and fecal samples from all lots are presented in Table 3. Fresh feces contained larger percentages of N and P than was contained in the rations. Conceivably, these differences are attributable to the animal digestion process. Cattle are inefficient in utilizing the nutrients in feeds. It has been estimated that about 75 to 89 percent of the N, 70 to 85 percent of P, and 80 to 90 percent of K fed in ration is excreted (Azevedo and Stout). On a dry weight basis, the amount of organic matter defecated is less than that fed. This is probably one factor causing the higher nutrient concentrations in feces, i.e. by concentrating nutrients in the smaller amount of defecated organic matter.

There are other elements which are generally highly concentrated in manure. Iron is being added as a diet supplement but only a small fraction is retained in the animals's bodies, which could explain the unusually high concentration of this element in manure. Sodium, Zn, and Cu concentrations in feces and old manure were about equal on the average. The electrical conductivity of 5.02 mmho/cm for manure was markedly higher than the 3.07 mmho/cm for fresh feces. Manure generally had higher pH than fresh feces.

## CONCLUSIONS

Climate influences the decomposition of manure and the transformation of its constituents. Temperature and moisture are perhaps the most important climatic variables affecting these processes. The type of feedlot housing employed modified climatic influences; and thus, the type of housing in use affected the decomposition and composition of manure.

The total dried solid (TDS) content of manure probably indicates its degree of exposure to climate as affected by housing type. In lots with more favorable evaporative conditions, TDS was high. The TDS content of manure was highest for open lots, followed by drylots and total confinement systems.

The N content of manure from open lots was generally low. Manure from drylot and total confinement systems had more than twice the N of open lot manure. This suggests that larger amounts of N were lost from open lots, possibly mainly by ammonia volatilization. The N content of fresh feces averaged 2.1 percent, about double that for manure in open lots but slightly less than the manure in drylot and total confinement systems.

Phosphorus content of manure in open lots and drylots is less than that in fresh feces, caused perhaps by runoff or leaching losses. However, in total confinement systems, the P contents of fresh feces and manure were about equal. Potassium content of open lot manure was generally low with an average of 0.82 percent, whereas manure from drylot and total confinement systems contained more than 2 percent K.

Calcium and Mg contents of manure were generally high in open lots. Sodium content of manure was about equal in unpaved and piled areas in open lots with a tendency for the Na content of manure to increase in paved drylot and total confinement systems. Iron, Mn, Zn, and Cu portions of manure were generally high in open lots.


### References

1 Adriano, D. C., P. F. Pratt and S. E. Bishop. 1971. Fate of inorganic forms of N and salt from land-disposed manures from dairies. Proc. Intl. Symposium on Livestock Wastes. ASAE, St. Joseph, Mich. p. 234-246.

2 Adriano, D. C., A. C. Chang, P. F. Pratt and R. Sharpless. 1973. Effect of soil application of dairy manure on germination and emergence of some selected crops. J. Environ. Quality 2:396-399.

3 Adriano, D. C., A. C. Chang and R. Sharpless. 1974. Nitrogen loss from manure as influenced by moisture and temperature. J. Environ. Quality 3:258-261.

4 Azevedo, J. and P. R. Stout. Animal manures. University of California Agricultural Publications, Berkeley, Calif. (In press).

5 Bremner, J. M. 1965. Total nitrogen. In. C. A. Black (ed). Methods of soil analysis (Part 2). Chemical and microbiological properties. Agronomy 9:1149-1178.

6 Chang, A. C., D. C. Adriano and P. F. Pratt. 1973. Waste accumulation on a selected dariy corral and its effect on the nitrate and salt of the underlying soil strata. J. Environ. Quality 2:233-237.

7 Edwards, W. M., E. C. Simpson and M. H. Frere. 1972. Nutrient content of barnlot runoff water. J. Environ. Quality 1:401-405.

8 Elliot, L. F., G. E. Schuman and F. G. Viets, Jr. 1971. Volatilization of nitrogen containing compounds from beef cattle areas. Soil Sci. Soc. Amer. Proc. 35:752-755.

9 Jackson, M. L. 1958. Soil chemical analysis. Prentice-Hall, Inc., Englewood Cliffs, NJ. p. 134-182.

10 Johnson, J. B. and G. A. Davis. 1974. Economic effects of surface water runoff controls on Michigan beef feedlots. Michigan Farm Economics. No. 374. Mich. State Univ. March Issue.

11 Stewart, B. A. 1970. Volatilization and nitrification of nitrogen from urine under simulated cattle feedlot conditions. Environ. Sci. Technol. 4:579-582.

# Identification and Measurement of Volatile Compounds Within a Swine Building and Measurement of Ammonia Evolution Rates From Manure-Covered Surfaces

J. Ronald Miner, Michael D. Kelly, A. W. Anderson
MEMBER
ASAE

ASSOC. MEMBER
ASAE

IN an effort to devise a field technique for sampling and measuring airborne volatile organic compounds in the vicinity of livestock production facilities, a trapping procedure has been developed. The traps consist of a short length of tubing fitted with a packing of high retention capacity. After having trapped the organic compounds from a measured quantity of air, the traps are returned to the laboratory, their contents transferred to an injector tube under cryogenic conditions, and subjected to chromatographic and mass spectrographic analyses. Thirty compounds were identified by this procedure and the concentrations of eight of these determined in a swine building atmosphere.

In related works, a sampling box was designed and built which permits the measurement of ammonia generation rates from earth, building, and treatment system surfaces. The box allows capture of the total gaseous emission from a defined area during the sampling program. These measurements quantify the rate of ammonia release from dairy and swine housing areas, manure storage facilities, and grassland used for manure disposal.

## VOLATILE COMPOUND IDENTIFICATION

Air samples were collected 2.7 m above the floor of the OSU Swine Center partially slotted floor finishing building. Samples were pumped for 24 hours through either Poropak Q (Waters Assoc., Farmingham, Massachusetts; 80/100 mesh ethylvinylbenzenedivinylbenzene polymers) or Tenax GC (Applied Science Laboratory, State College, Pennsylvania; 35/60 2,6-diphenyl p-phenylene oxide polymers) packed in stainless steel traps 10.3 cm long, 3 mm inside diameter. After loading, the traps were capped and refrigerated prior to analysis. Pumping rate for the Poropak Q traps was 20 liters per hour; for the Tenax traps it was 30 liters per hour. Samples were also collected in a similar manner on the OSU campus.

The first step in processing the traps was to purge them with nitrogen at 55 C for one hour to remove water. The traps were next heated, Poropak Q to 150 C, and Tenax GC to 200 C, and purged with nitrogen to transfer the trapped volatiles to a stainless steel trap immersed in dry ice. The cold traps were transferred to the inlet system of either the gas chromatograph or gas chromatograph - mass spectrometer combination.

Chromatographic analyses were made on a F & M Model 402 gas chromatograph fitted with dual flame ionization detectors, a strip chart recorder and a Hewlett-Packard

3370 A integrator. A Finnigan 1015C mass spectrometer in combination with a Varian Aerograph series 1440 chromatograph was used for mass spectral analyses.

### Columns

The following columns were used: a capillary column 30 m x 0.75 mm I.D. stainless steel coated with five percent Ethylene Glycol Succinate (EGS), programmed to run at 110 C for four minutes, then raised to 175 C at 4 C/min and held; a 61 m x 0.75 mm I.D. stainless steel column coated with five percent Triton X305, programmed to operate at 60 C for 5 minutes and raised to 150 C at 4 C/min and held; a 153.8 m x 0.75 mm I.D. stainless steel column coated with eight percent Carbowax 20M, operated at 70 C for 5 minutes and then temperature programmed to 150 C at 2 C/min; and a 2 m x 3.0 mm I.D. stainless steel tube packed with four percent Carbowax 20M +0.8 percent KOH on Carbopack B (Supelco, Incorporated, Bellefonte, Pennsylvania), run isothermally at 90 C. Carrier gas flow rates were 20 ml/min of helium for the 3.0 mm columns and 12-15 ml/min of helium for the capillary columns.

### Results

Over thirty Swine Center samples were studied. Before running the cold traps on the chromatograph, they were checked in the labs for odor retention. The escaping volatiles from representative traps were smelled in the lab to get a comparison of the odors. The trap from the swine barn smelled like manure and the one from the campus had a slightly musty smell.

Table 1 shows the compounds identified from the swine building atmosphere, the traps and columns used, and the retention times. Many compounds were detected in more than one column, but for convenience, only listed once. There were two xylene isomers and several alkyl benzene isomers, hence the variable retention times.

Compounds identified on the Triton X305 column from Tenax traps exposed on the main campus were similar to those from the swine barn. The alkyl benzene isomers were common in both locations but the concentrations were higher in the Swine Center. However, the chromatographic results on the EGS column were very different. The acids and phenolic compounds were absent from the campus traps. The chromatographic results from a Carbopack B column which separated amines showed more peaks from the campus traps than from the Swine Center. Trimethyl amine was more concentrated in the Swine Center than on campus. Isopropyl amine was tentatively identified in both places.

By using a gas chromatograph equipped with an integrator, a quantitative check could be made on various compounds. A standard solution was injected into the chromatograph to establish an area-quantity relationship which

The authors are: Associate Professor, Department of Agricultural Engineering; Graduate Research Assistant, Department of Microbiology; and Professor, Department of Microbiology, Oregon State University, Corvallis, Oregon 97331. Technical Paper No. 3972. Oregon Agricultural Experiment Station.

Acknowledgments: In addition to the authors, the laboratory assistance of Cheryl Gould, James Christian, and Kris Marvell was instrumental in the success of this project. Financial support was provided in part by Grant No. S-802009, Office of Research and Monitoring, U.S. Environmental Protection Agency.

**TABLE 1. VOLATILES IDENTIFIED FROM THE OSU SWINE CENTER ATMOSPHERE USING THE TRAP METHOD AND COMBINED GLC MASS SPECTRAL ANALYSIS.**

| Compound | Column | Trap* | Retention time, seconds |
|---|---|---|---|
| 2- butanol | Triton | B | 75 |
| Sec-butanol | ,, | B | 81 |
| Hexanal | ,, | B | 97.5 |
| Dimethyl disulfide (DMDS) | ,, | B | 105 |
| 3-amino pyridine | ,, | T | 120 |
| n-butanol | ,, | B | 140 |
| Dimethyl Trisulfide (DMTS) | ,, | T | 450 |
| Toluene | ,, | B | 130 |
| Xylene | ,, | B | varies |
| Alkyl benzenes | ,, | B | varies |
| 2, 3-butanediol | ,, | B | 170 |
| Acetoin | ,, | B | 180 |
| Indane | ,, | T | 345 |
| Benzaldehyde | ,, | B | 540 |
| Me-naphthalene | ,, | T | 1,440 |
| Diacetyl | Carbowax | B | 240 |
| 2-octanone | ,, | T | 210 |
| Acetic acid | EGS | B | 85 |
| Propionic acid | ,, | B | 115 |
| N-butyric acid | ,, | B | 150 |
| Valeric acid | ,, | B | 210 |
| Acetophenone | ,, | T | 240 |
| Caproic acid | ,, | T | 275 |
| Enanthic acid | ,, | T | 300 |
| Phenol | ,, | B | 455 |
| P-cresol | ,, | B | 515 |
| 2-ethoxy-1-propanol | ,, | P | 195 |
| Et-phenol | ,, | P | 580 |
| Benzoic acid | ,, | P | 645 |
| Trimethyl amine (TMA) | Carbopack B | B | 75 |

*T, Tenax; P, Poropak; B, Both.

could then be compared with the chromatogram of swine building air samples. The concentrations listed in Table 2 were determined in this manner. The concentrations estimated in this work were all below the individual compound threshold odor concentrations.

**TABLE 2. AVERAGE CONCENTRATIONS OF VOLATILES IN SWINE CENTER AIR PASSED THROUGH POROPAK Q TRAPS, MAY 1974.**

| Compound | $10^{-4} \mu g/m^3$ |
|---|---|
| Acetic acid | 15 |
| Propionic acid | 26 |
| Butyric acid | 10 |
| Valeric acid | 12 |
| Phenol | 25 |
| Cresol | 45 |
| Dimethyl disulfide | 22 |
| Xylene | 45 |

## AMMONIA RELEASE RATES

Ammonia release from manure-covered surfaces which are in the immediate proximity of livestock production facilities has been demonstrated. Koelliker and Miner (1973) documented the release of ammonia from an anaerobic swine manure lagoon surface. Ammonia concentrations in air near livestock feeding operations have been measured as significantly higher than those in other agricultural areas (Hutchinson and Viets 1969, and Luebs, Laag, and Davis 1973). Due to the solubility of ammonia in water, the potential exists for livestock production enterprises to make significant contribution to the nitrogen content of surface impoundments, thereby contributing to enrichment.

In order to quantify the rate of ammonia release from surfaces associated with livestock production systems, the

sampling box shown schematically in Fig. 1 and photographed in Fig. 2 was constructed. This box covers a square area 0.61 m on a side. There is a plywood deck 0.3 m from the bottom of the box. A diaphragm pump pulls air from beneath the deck through absorption tubes and finally through a wet test meter for air volume measurement. Air admitted to the space beneath the deck through a copper tube passes through a can filled with activated carbon, which insures that ammonia-free air enters the system. A metal strip attached to the lower edge of the sampling box pressed into the surface prevents the entrance of unfiltered air. The air pump was driven by a twelve volt battery and DC-AC converter when other electrical connections were not accessible.

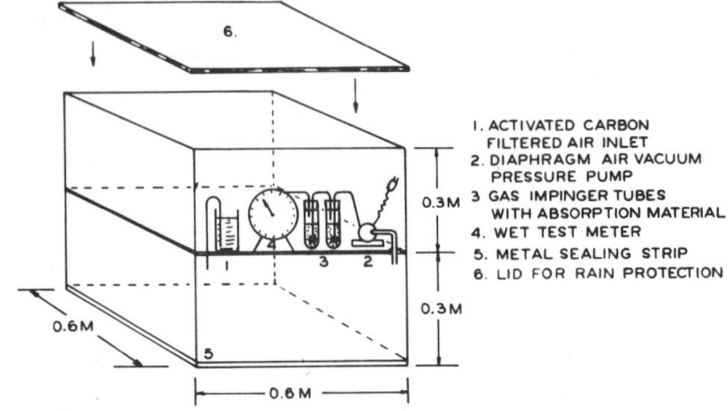


FIG. 1 Construction of the sampling box to capture the released volatile compounds from a soil surface previously exposed to animal manures.

Ammonia was collected from air pulled through the box by a one percent sulfuric acid solution and porous ceramic diffuser stone in a large test tube. After a one hour sampling period, the absorbing solution was returned to the laboratory and split. One half was used for ammonia analysis with Nesslers reagent (APHA, 1971). The other half was subjected to Kjeldahl digestion (APHA, 1971) prior to Nesslerization. The difference between these two values is called non-ammonia nitrogen and is considered to be principally amines.

FIG. 2 Photograph of the sampling box in operating position.

### Results

The sampling box was used to measure the ammonia release rates from a variety of surfaces associated with the

OSU Dairy, Swine Center, and campus. Each of the sampling locations was subject to a variety of short term variations and precise values were not reproducible. By collecting five or more samples from forty locations, useful information was obtained. The results are summarized in Table 3. These values were measured under summer conditions of 30 C daytime temperatures and during a period without rainfall.

In addition to the ammonia evolution, these studies indicated a non-ammonia nitrogen evolution rate ranging from 0.25 to 0.75 of the ammonia. No correlation is evident from these data between age of manure and non-ammonia nitrogen-ammonia release rates. Non-ammonia nitrogen values were consistently low from the swine manure lagoon surface.

A lagoon surface releasing 25 mg/m$^2$-day would release approximately 90 kg/ha (80 lb/acre) annually. This value is considerably smaller than anticipated, based upon previous work (Koelliker and Miner, 1973). This is best explained by pH (7.9) of this lagoon, which was relatively low for ammonia desorption. At this pH, less than four percent of the ammonia is present as $NH_3$ and exhibiting a vapor pressure (Miner 1974). The same explanation—low soil pH—also explains the low ammonia release rates measured in this study compared to those reported by Denmead, Simpson, and Freney (1974). They reported nitrogen flux rates ranging from 25 to 80 mg/m$^2$-day from a grazed alfalfa pasture.

This technique offers a simple quantitative technique for the measurement of ammonia release rates from surfaces associated with livestock production. The values measured correlate well with observed odor release and lead to a prediction of the potential contribution of livestock feeding to airborne plant nutrients which can be absorbed by nearby surface waters.

## Acknowledgments

In addition to the authors, the laboratory assistance of Cheryl Gould, James Christian, and Kris Marvell was instrumental in the success of this project. Financial support was provided in part by Grant No. S-802009, Office of Research and Monitoring, U.S. Environmental Protection Agency.

**TABLE 3. EVOLUTION RATE OF AMMONIA FROM SEVERAL DIFFERENT SURFACES IN THE VICINITY OF LIVESTOCK PRODUCTION FACILITIES.**

| Surface description | Evolution rate, mg/day -m$^2$ |
| --- | --- |
| On pasture grass and bare soil more than 30 m from dairy barn | 1 - 2 |
| On pasture with dried manure and on manure-free dairy barn surfaces | 2 - 5 |
| Pasture land after recent liquid dairy manure application | 5 - 20 |
| Manure-covered aisle in freestall dairy barn | 50 - 100 |
| Grassland near swine barn with no direct manure contact | 2 - 3 |
| Soil and grass with some previous manure application | 2 - 5 |
| Lagoon water | 20 - 40 |
| Campus sidewalk and lawn surfaces | 0.5 - 1.5 |

## References

1   APHA. 1971. Standard methods for the examination of water and wastewater. Thirteenth edition. American Public Health Association. Washington, D.C.

2   Denmead, O. T., J. R. Simpson, and J. R. Freney. 1974. Ammonia flux into the atmosphere from a grazed pasture. Science 185:609-610.

3   Hutchinson, G. L. and F. G. Viets, Jr. 1969. Nitrogen enrichment of surface water by absorption of ammonia volatilized from cattle feedlots. Science 166:515.

4   Koelliker, J. K. and J. R. Miner. 1973. Desorption of ammonia from anaerobic lagoons. TRANSACTIONS of the ASAE 16(1):148-151.

5   Luebs, R. E., A. E. Laag, and K. R. Davis. 1973. Ammonia and related gases emanating from a large dairy area. Cal. Agr. 27(2):10-12.

6   Miner, J. R. 1974. Odors from confined livestock production. Environmental Protection Agency Technology Series. EPA-660/2-74-023. 125 pp.

# Quantitative Measurement and Sensory Evaluation of Dairy Waste Odor


C. N. Ifeadi, E. P. Taiganides, R. K. White

ASSOC. MEMBER
ASAE

MEMBER
ASAE

ASSOC. MEMBER
ASAE


**A** MAJOR environmental concern of any livestock production is the odor problem. Offensive odors from animal production facilities have prompted neighbors to legal actions which have limited and even closed some animal feedlots (Willrich and Miner 1971).

The study of odors from animal wastes has been investigated from different approaches. Taiganides and White (1969) reviewed the effects of odorous gases on animals raised in confinement. Others (Burnett and Dondevo 1969, Merkel et. al. 1969, and White et. al. 1971) have identified and characterized qualitatively odorous compounds in the volatile gases released from animal wastes.

Quantification is the next most important step, particularly from the standpoint of establishing odor standards and developing engineering systems for odor management and control.

The objectives of this study were to develop research instrumentation capable of characterizing odors and to evaluate odors from dairy animal wastes.

## INSTRUMENTATION AND PROCEDURE

Proper characterization of an odor requires two types of analysis, as pointed out by several investigators (Wright 1964, Dravnieks 1968, Lindvall 1970): an objective or analytical measurement, and an organoleptic or sensory evaluation of the odor.

Dairy waste odors were generated in concentrations and manner suitable for analysis by objective and organoleptic measurement systems. The instrumentation system for the objective odor measurement consisted of: (a) dairy waste odor generator and diffusion cell for pure ocorous compounds, (b) sample odor collector, (c) odor transfer system, (d) injection system, (e) gas chromatograph (GC), and (f) chemical ionization (CI) mass spectrometer. Two additional instruments: (a) a dilution train, and (b) a sniffing hood were developed for sensory evaluation (Ifeadi 1972).

### Waste Odor Generator

Dairy waste odor was generated in 5-liter flasks held at 20 C in a water bath. The flasks contained initially 600 g of dairy waste (mixture of 70 percent feces and 30 percent urine), were fed daily with 50 g of fresh waste. In one series of tests the wastes were diluted with 3000 ml of distilled water. In others, the waste remained in the flasks undiluted.


Approved for publication as Journal Article No. 47-75 of the Ohio Agricultural Research and Development Center (OARDC). Supported in part by funds from the Agricultural Research Service, USDA.

The authors are: C. N. IFEADI, Research Scientist, Battelle Laboratories, Columbus, Ohio (Former Research Associate, OARDC); E. P. TAIGANIDES, Professor, and R. K. WHITE, Associate Professor, Agricultural Engineering Dept., OARDC and Ohio State University, Columbus.


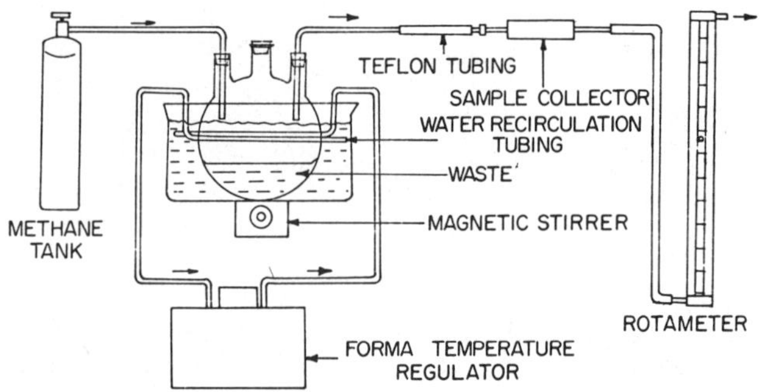



FIG. 1 Instrumentation for odor generation and collection.

Tests were conducted on samples taken from the flask head space gases. Three series of test were conducted lasting respectively, 5, 6, and 12 days.

### Odor Collector

Concentrations of the odorous compounds in the head space of the flasks were too low for direct gas chromatographic identification and quantification (White 1969). A device to concentrate the collected gases consisted of a cylindrical collector packed with Chromosorb 102 (Dravnieks et al. 1970).

Fig. 1 illustrates the set up for collecting odorous gases from the decomposing dairy waste, while Fig. 2 shows a sectional view of the collector.

### Sample Transfer and Injection

Samples in the collector were transferred to a specially designed injection needle by placing the collector in an oven maintained at a constant temperature of 120 C, and flushing the desorbed samples with nitrogen at the rate of 20 cc/min for 15 minutes (Ifeadi 1972).

Fig. 3 shows the injection needle supported between the cold/hot block scissor mechanism. During sample transfer to the injection needle the long legs of the cold blocks were immersed into liquid nitrogen in order to cool the injection needle and to trap the volatiles as they were desorbed from the collector. During injection of the sample into the gas chromatograph (GC) for analysis, the cold/hot block scissor mechanism was operated by a solenoid to bring the hot copper blocks, maintained at a constant temperature of 200 C, into contact with the injection needle. The high temperature of the hot blocks caused the sample to volatilize. The injection into the GC was completely by means of a bellow mechanism (White 1969);

### Gas Chromatographic Analysis

The sample was injected into a GC column, packed with 10 percent Carbowax 20 M on Chromosorb P, AW-DMCS,

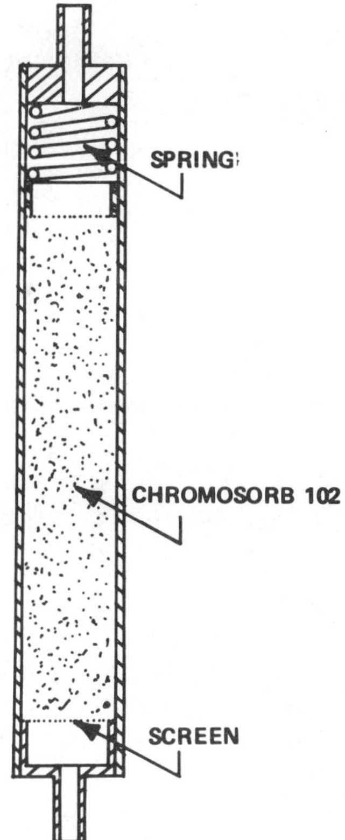

FIG. 2 Sample collector.

FIG. 4 Diffusion cell.

80/100 mesh. The carrier gas was helium, and flow rates for helium, hydrogen and air were 30, 30 and 300 cm/min respectively. The GC temperature was programmed at the rate of 4 C/min.

Certain compounds in the dairy waste volatiles had been tentatively identified previously (White et al. 1971). The presence of dimeythyl sulfide and diethyl sulfide was confirmed for the condition in this study. To analyze these compounds, which in pure forms are liquid at room temperature and pressure, they were metered from a diffusion cell shown in Fig. 4. The cells enabled the diffusion of known concentration of dimethyl and diethyl sulfides into the collector for the GC analysis and development of calibration curves.

## Chemical Ionization Spectra

The presence of dimethyl and diethyl sulfides within the dairy waste volatile was also confirmed by the combined use of gas chromatography/mass spectroscopy. Chemical ionization (CI) (Munson 1971) was the mass spectrometer used. Pure compounds and the dairy waste volatiles were analyzed by the GC/CI system.

## Instrumentation for Sensory Evaluation

The instrumentation for the odor evaluation is shown in Fig. 5. The essential features for sensory evaluation are the odor source, the dilution train, and the sniffing hood. Odor was metered into the dilution train either from the diffusion cell with the known compound(s) or the dairy odor generator depending on the type of analysis.

The dilution train consisted of an air filter and a series of rotameters. Odorants from either of the sources were diluted with filtered air to desired concentrations for sensory evaluation and GC analysis. The bleeding ports and exhaust ports in the dilution train were connected to carbon filter cannisters which filtered the outgoing odorants, thus preventing the contamination of the air in the room.

The intensity of each stage of the dilution was rated by one man on the basis of a six-point rating system: no odor-0, barely perceptible-1, distinct-2, moderate-3, strong-4, and overpowering-5. Nitrogen gas under pressure

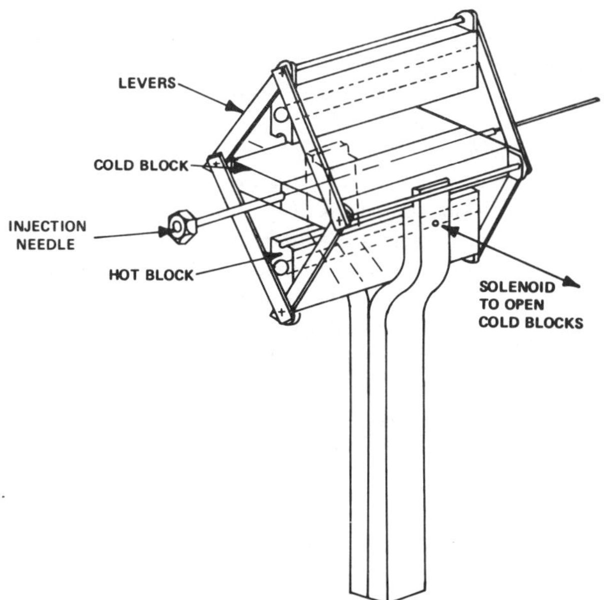



FIG. 3 Cold/hot block scissors mechanism.

**MANAGING LIVESTOCK WASTES**

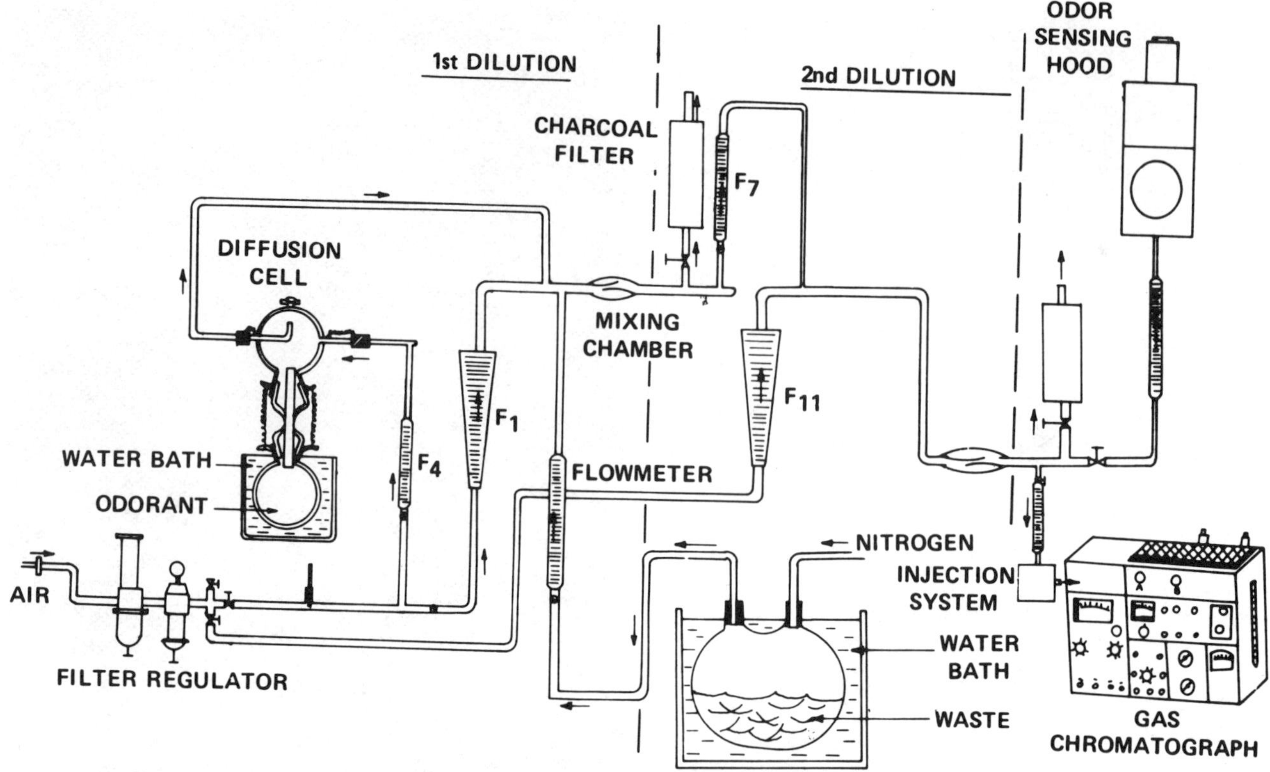


FIG. 5 Odor analysis instrumentation.

was used to flush dairy waste volatiles from the flask. Displaced odorous volatiles were diluted serially, using odor-free air and were sniffed at the sniffing hood.

## EXPERIMENTAL RESULTS AND DISCUSSION

### Confirmation of Dimethyl and Diethyl Sulfides

The GC and chemical ionization mass spectrometer was successfully used to confirm the presence of the dimethyl sulfide and diethyl sulfide in dairy waste volatiles by comparing the chemical ionization spectra of the pure compounds with those obtained from dairy waste volatiles.

The other compounds which were identified as present in the volatiles were pyruvaldehyde, caproaldehyde, benzaldehyde, acetophenone, and penthyl mercaptan.

### Objective Quantitative Results

The first step in the quantification of dimethyl and diethyl sulfides composition of the dairy waste odor was the development of the calibration curves using the pure compounds. Since these compounds are in liquid state at normal room temperature and pressure, it was necessary to determine the diffusion rates from the diffusion cell. Two methods were used (a) direct measurement, and (b) theoretical calculation using diffusion theory (McKelvy and Hoelschen 1957, and Sherwood and Tigford 1952). The measured and theoretical diffusion rates for dimethyl sulfide at 25 C were 0.35 and 2.73 cc/min respectively, while those for diethyl sulfide at 40 C were respectively 0.10 and 1.87 cc/min. The temperatures choice at which to diffuse the liquid compounds from the diffusion cell were just high enough to cause the volitization of the liquids, yet low enough (i.e. below the boiling points) to prevent massive convectional transport of the compounds. The theoretical diffusion rates were higher than the measured values by a

factor of about 10 for both compounds. The difference between the two methods is large but these numbers are the results of a first attempt to quantify individual odorants in a natural odor sample.

With the diffusion rates known, the volatilized pure compounds were serially diluted using the dilution train (Fig. 5). Samples were collected at different concentrations and analyzed with the GC. GC concentration calibration curves were thus developed. The calibration curves were not straight lines but rather, S-shaped, which is a typical shape obtained from the absorption of a sample by the stationary phase (Carbowax 20 M) of the GC column. However, in absolute calibration direct elution is not essential provided standardized procedures are adopted. The calibration curves of the pure compounds were then used to obtain the concentrations of these compounds in the waste volatiles generated.

Table 1 shows the results of three series of tests to determine the concentrations of dimethyl sulfide and diethyl sulfide in the head space gases of the flasks containing dairy wastes. These results indicate that the diluted dairy waste released more dimethyl sulfide than undiluted waste, while the undiluted dairy waste released more diethyl sulfide than the diluted waste. The amount of dimethyl sulfide released from either the diluted or undiluted dairy waste was always higher than the amount of diethyl sulfide. The typical chromatograms obtained for the diluted and undiluted dairy wastes are shown in Fig. 6.

### Sensory Evaluation

Table 2 presents the relationship of quantitative and organoleptic analyses for the odor intensity of diluted and undiluted dairy waste volatiles.

The diluted waste had a lower odor threshold, hence

## AVERAGE DAILY CONCENTRATION OF DIMETHYL AND DIETHYL SULFIDE IN DILUTED AND UNDILUTED DAIRY WASTE

### (CONCENTRATION, ppm BY VOLUME)

| | Dimethyl sulfide | | | | Diethyl sulfide | | | |
| | Diluted waste | | Undiluted waste | | Diluted waste | | Undiluted waste | |
| Day* | Theoretical† | Measured‡ | Theoretical† | Measured‡ | Theoretical† | Measured‡ | Theoretical† | Measured‡ |
|---|---|---|---|---|---|---|---|---|
| 1 | 45.0 | 6.1 | 21.1 | 2.7 | 0.0 | 0.0 | 0.0 | 0.0 |
| 2 | 221.5 | 28.5 | 43.8 | 5.7 | 2.9 | 0.2 | 51.0 | 2.3 |
| 3 | 666.2 | 86.6 | 209.0 | 37.5 | 6.7 | 0.4 | 7.4 | 0.4 |
| 4 | 780.4 | 100.5 | 177.0 | 22.4 | 6.7 | 0.3 | 41.0 | 2.0 |
| 5 | 461.7 | 62.8 | 558.6 | 72.7 | 3.3 | 0.2 | 75.1 | 3.4 |
| 6 | 725.7 | 93.8 | 619.0 | 92.0 | 4.6 | 0.2 | 118.0 | 5.3 |
| 7 | 499.6 | 76.6 | | | 2.7 | 0.1 | | |
| 8 | 767.6 | 99.2 | | | 3.2 | 0.2 | | |
| 9 | 551.3 | 76.2 | | | 4.4 | 0.2 | | |
| 10 | 634.3 | 84.2 | | | 1.9 | 0.1 | | |
| 11 | — | — | | | — | — | | |
| 12 | 728.7 | 95.5 | | | 10.5 | 0.6 | | |
| Average | 507.0 | 67.5 | 271.4 | 38.9 | 3.9 | 0.2 | 48.8 | 2.2 |

* Time after dairy waste was placed in the incubator.

† Concentration obtained from the calibration curve based on the principles of diffusion theory.

‡ Concentration obtained from the measured calibration curve.

requiring a larger volume of air to render the volatiles odorless, than for the solid undiluted waste. The presence of some compounds (not identified) with low threshold levels could have contributed to the low threshold measurements. Whether all the significant odor relevant compounds in an animal waste odor are known or not, such data as presented in Table 2 would be useful in assessing odor emissions from animal waste. If it were that the odor intensity standard were set for less than No. 3 (i.e. moderate), then diluting the dairy waste volatiles down to a concentration of 0.9 ppm for the diluted waste and 1.5 ppm for the undiluted will meet the standard. This is a way of relating control standard to concentration level and odor intensity of particular odorants. Such correlation is an important criterion in establishing and enforcing odor pollution regulations.

## CONCLUSIONS

Instrumentation for quantitative measurement and sensory evaluation which proved useful in analyzing odors has been developed.

The combined use of GC and chemical ionization mass spectrometer to analyze dairy animal waste volatiles confirmed the presence of dimethyl sulfide and diethyl sulfide. Other compounds identified as present in the mixture were pyruvaldehyde, caproaldehyde, benzaldehyde, acetophenone, and pentyl mercaptan.

Quantitative analysis of the diethyl and dimethyl sulfides released from stored dairy waste gave an average daily value of 0.2 ppm for diethyl sulfide and 67.5 ppm for dimethyl sulfide. For undiluted waste, the values were 2.2 ppm for diethyl and 38.9 ppm for dimethyl sulfide.

Sensory evaluation indicated that the diluted dairy waste had lower odor threshold level than the undiluted waste.

### References

1  Burnett, W. E. and N. C. Dondero. 1969. Microbiological and chemical changes in poultry manure associated with decomposition and odor generation. P. 271, In: Proceedings, Cornell Uni-

*(Continued on page 368)*

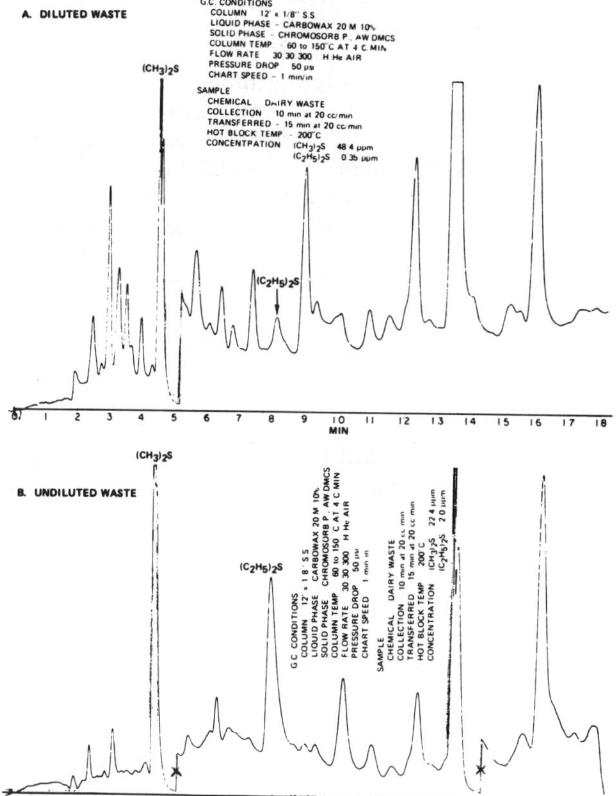



FIG. 6 Chromatograms of samples obtained on the fourth day of incubation.

## TABLE 2. DAIRY ODOR INTENSITY MEASUREMENT RELATED TO MEASURED CONCENTRATIONS OF DIMETHYL AND DIETHYL SULFIDES

| Concentration of $(CH_3)_2S$ plus $(C_2H_5)_2S$ ppm† by volume | Odor intensity* no. | Concentration of $(CH_3)_2S$ plus $(C_2H_5)_2S$ ppm† by volume | Odor intensity* no. |
|---|---|---|---|
| 0.01-0.05 | 0 to 1 | 0.02-0.07 | 0 to 1 |
| 0.05-0.90 | 2 to 3 | 0.07-1.49 | 2 to 3 |
| 0.91-2.0 | 4 | 1.5-5.0 | 4 |
| Over 2.0 | 5 | Over 5.0 | 5 |

* 0 - no odor
1 - Barely perceptible
2 - Distinct
3 - Moderate
4 - Strong
5 - Overpowering odor.

† Concentrations are based on measured calibration curve.

# Evaluation of Odor Intensities at Livestock Feeding Operations in Texas


Donald L. Reddell, John M. Sweeten

MEMBER
ASAE

MEMBER
ASAE


ODOR is a difficult environmental problem associated with livestock operations. Methods of measuring odor are needed to (a) evaluate parameters affecting odor production, (b) determine effective livestock management systems which will reduce odor, (c) evaluate chemical and biological treatment systems for odor reduction, (d) provide guidelines for site selection and land use control near livestock operations, and (e) provide regulatory agencies with a reasonable tool for enforcing odor intensity standards. Probably no feedlot is perennially free of odor.

The gas chromatograph has been successfully used to identify and measure the concentrations of odorous compounds released from animal manure (Miner and Hazen 1969, Burnett and Dondero 1970, and Bethea and Narayan 1972). Identification of the more common and abundant gases released from manure has not satisfactorily explained observed odors. Hanna (1972) indicated that attempts to relate odorant concentrations to realized odor intensities generally have not been successful.

Organoleptic techniques which utilize the human nose to measure odor intensity are the primary means of measuring odor under field conditions. Odor panels of not less than five persons, and preferably 10 or more, are recommended for olfactory measurement of odor (American Public Health Association 1971). Human limitations of odor measurement were reviewed by Barth (1973) and Summer (1971), and include fatigue, adaptation, imprecise resolution, personal habits, physical condition, and admixing of pollutants. Climatic factors such as temperature and moisture content affect odor emissions (Ludington et al. 1971), while wind velocity, atmospheric stability and relative humidity influence odorant transport (Smith 1968). All of these factors and many more can limit the accuracy of organoleptic techniques for odor measurement.

Attempts to correlate odor intensity, as measured with an odor panel, with possible parameters causing or affecting odors has not been successful. A need for greater precision in odor measurement is evident. The use of probability theory and frequency analysis to quantify odor intensities is a possible means of overcoming many of the shortcomings associated with olfactory techniques. The objectives of this study were as follows:

1 Determine if probability distribution functions can be used to quantify odor intensities and thus provide a meaningful odor measurement, and

2 Determine typical odor probability distributions for several livestock feeding operations.

## PROCEDURE

Two olfactory procedures for measuring odor intensity were used in this study. The first was a direct inhalation (dynamic dilution) procedure using a Scentometer* manufactured by Barneby-Cheney in Columbus, Ohio. Known flow rates of odorous and odor-free air were mixed together until an odor threshold level was reached by the panel member. Odor intensities measured with the Scentometer were expressed as dilutions to threshold ($D_T$), which was the number of times that odorous air was diluted with odor-free air to reach the point where odor was barely perceptible. The Scentometer allowed odor readings over the range of 1.5 to 170 $D_T$. Primary readings suggested by the manufacturer are 2, 7, 31, and 170 $D_T$. The large gap between 31 and 170 $D_T$ was significant because the dividing line between reasonable and intolerable odors appeared to lie within this range. Cooper (1973) described the Scentometer and procedures for its use in detail.

The second olfactory measuring procedure used in this study was the liquid dilution procedure described by Sobel (1969) and used by Burnett and Dondero (1969). This procedure involved diluting animal waste with odor-free water until the threshold level was obtained. Odor intensity measured with the liquid dilution technique was expressed as an odor intensity index (OII), which was the number of times the original sample concentration was halved with odor-free dilution water to yield a barely distinguishable odor.

Liquid dilution tests were performed in the laboratory on samples brought from the field and were used to determine the odor potential of manure. The method assumed that odors contained in the soluble waste fraction were representative of total manure odor. The dynamic dilution (Scentometer) procedure was concerned with odors emitted from manure. How well these two procedures correlate with each other has not been established.

## RESULTS AND DISCUSSION

Olfactory techniques for measuring odor are widely used. These techniques require the use of panel members to measure odor intensity. For various reasons, a certain amount of training and screening of panel members should be undertaken.


The authors are: DONALD L. REDDELL, Associate Professor, Agricultural Engineering Dept., Texas Agricultural Experiment Station, and JOHN M. SWEETEN, Agricultural Engineer, Animal Waste Management, Texas Agricultural Extension Service, Texas A&M University, College Station.

Approved as Texas Agricultural Experiment Station Paper No. TA-11883.



*Mention of trade names does not imply endorsement by either the Texas Agricultural Experiment Station or Texas Agricultural Extension Service, Texas A&M University.

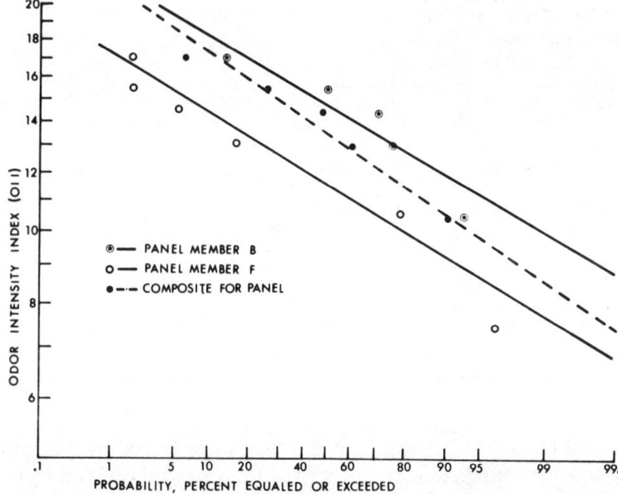



**FIG. 1 Lognormal probability distributions of the odor intensity index [OII] for two panel members and a composite distribution for the entire panel obtained for liquid swine waste using the liquid dilution technique.**

## Variations Among Odor Panel Members

Four samples per week of liquid swine waste were brought into the laboratory over a 5-month period, and the odor evaluated with a 10-member panel using the liquid dilution technique. Because of a few missing weekly samples due to holidays and the occasional absence of panel members, approximately 50 odor readings were made by each panel member during the period of study.

The following four statistics were calculated from these data:

$$X_A = \frac{\sum\limits_{i=1}^{N} X_i}{N} \quad \ldots\ldots\ldots\ldots\ldots\ldots\ldots\ldots\ldots [1]$$

$$X_G = \text{Antilog} \left( \frac{\sum\limits_{i=1}^{N} (\log X_i)}{N} \right) \ldots\ldots\ldots\ldots [2]$$

$$S_A = \left( \frac{\sum\limits_{i=1}^{N} (X_i - X_A)^2}{N-1} \right)^{0.5} \ldots\ldots\ldots\ldots [3]$$

and

$$S_G = \text{Antilog} \left( \frac{\sum\limits_{i=1}^{N} (\log X_i - \log X_G)^2}{N-1} \right) \ldots\ldots\ldots\ldots [4]$$

where $X_A$ = arithmetic mean, $X_G$ = geometric mean, $S_A$ = arithmetic standard deviation, $S_G$ = geometric standard deviation, $X_i$ = value of individual odor intensity measurement, and $N$ = number of odor intensity mesurements made.

A substantial variation among panel members is indicated by these data. To illustrate this variation, the results from six members of the panel are shown in Table 1. The arithmetic means vary from 11.8 to 15 OII. Panel members A, B, and C are similar and panel members C and D are similar. However, panel members E and F indicate a less sensitive sense of smell. Variations in odor intensities of this magnitude make the use of odor panels as a research tool difficult; particularly when some panel members are dropped and new ones added over the life of a research project.

The geometric means shown in Table 1 tend to be slightly less than the arithmetic means; indicating the data are slightly skewed. The arithmetic standard deviations vary from 1.62 to 2.75 OII, which is a large variation. However, the geometric standard deviations only vary from 1.12 to 1.29 OII, which is a much smaller numerical variation.

## Probability Distribution of Odor Intensity

From studying the individual data points obtained by each panel member and the means and standard deviations in Table 1, it became obvious that each panel member had a unique and repeatable odor frequency or odor probability distribution. Because the geometric standard deviation had a small numerical variation among panel members and was calculated using the logarithms of numbers, the odor data for each panel member were tested to see how well they fit a lognormal probability distribution.

A theoretical lognormal plot was obtained by plotting the following three points,

1  $X_G$ at P = 50 percent,
2  $X_G \cdot S_G$ at P = 15.9 percent, and
3  $X_G / S_G$ at P = 84.1 percent,

on log-probability paper. The resulting plot was a straight line. The actual probabilities were calculated from the data using the relationship

**TABLE 1. MEANS AND STANDARD DEVIATIONS OF THE ODOR INTENSITY INDEX (OII) OBTAINED BY SIX PANEL MEMBERS FROM LIQUID SWINE WASTE USING THE LIQUID DILUTION TECHNIQUE.**

| Panel* member | Geometric† mean, OII | Geometric standard deviation, OII | Arithmetic† mean, OII | Arithmetic standard deviation, OII |
|---|---|---|---|---|
| A | 14.91 a | 1.12 | 15.00 a | 1.62 |
| B | 14.76 a | 1.18 | 14.95 a | 2.27 |
| C | 14.32 ab | 1.17 | 14.48 ab | 2.05 |
| D | 13.56 b | 1.20 | 13.78 b | 2.33 |
| E | 12.50 c | 1.29 | 12.86 c | 2.75 |
| F | 11.62 d | 1.19 | 11.79 d | 1.91 |

*The mean and standard deviation for each panel member is computed from approximately 50 different odor readings taken over a 5 month period.
†Means followed by the same letter are statistically the same at the 5 percent level of significance.

TABLE 2. MEANS AND STANDARD DEVIATIONS OF THE ODOR
INTENSITY INDEX (OII) OBTAINED BY TWO DIFFERENT
SIZED ODOR PANELS ON BEEF FEEDLOT WASTE
USING THE LIQUID DILUTION TECHNIQUE.

| Panel*<br>no. | No. of<br>panel<br>members | Geometric†<br>mean,<br>OII | Geometric<br>standard<br>deviation,<br>OII | Arithmetic†<br>mean,<br>OII | Arithmetic<br>standard<br>deviation,<br>OII |
|---|---|---|---|---|---|
| 1 | 17 | 9.66 a | 1.19 | 9.81 a | 1.65 |
| 2 | 5 | 10.12 a | 1.16 | 10.23 a | 1.43 |

*The means and standard deviations for Panel Nos. 1 and 2 were computed
from 161 and 101 odor readings, respectively, taken over a 6 week period.
†Means followed by the same letter are statistically the same at the 5 percent
level of significance.

$$P = \left( \frac{M}{N+1} \right) 100, \quad\dots\dots\dots\dots\dots\dots\dots\dots\dots\dots\dots\dots \quad [5]$$

where P is the probability (percent) that a given odor
level will be equalled or exceeded, M is the cumulative
number of odor readings in a sequence which equalled or
exceeded the given odor level, and N is the total number
of odor readings in the sequence. Thus, if the odor
readings 7, 8, 8, and 9 OII were obtained, then the
probability of an odor reading equaling or exceeding 7.5
would be 60 percent [i.e. $P(X \geqslant 7.5) = 100(3/5) = 60$
percent].

Fig. 1 shows the lognormal probability plots vs. the
odor intensity index for panel members B and F from
Table 1. The solid lines of the figure are the theoretical
plots. The actual probabilities calculated from Equation
(5) are shown as data points. From these data, it is seen
that a lognormal distribution adequately describes the
frequency of odor intensity as measured by individual
panel members. The distributions for the other panel
members in Table 1 lie in between the two given in Fig.
1.

Using Monte Carlo simulation techniques and the
theoretical distributions for each panel member, a
composite theoretical distribution was predicted for the
entire panel, and is shown as the dashed line in Fig. 1.
The actual probabilities calculated for the entire panel
are plotted as solid circles. Again a good fit is obtained.

As a result of this information, it is possible to
establish some lognormal probability distribution as a
standard for a given animal enterprise and odor test
procedure. By screening potential panel members to
establish their lognormal probability curves and using
Monte Carlo techniques, a panel can be selected which
will yield the standard distribution curve within an
allowable error range.

For instance, manure samples were collected from a
beef feedlot and odor intensity measured by the liquid
dilution technique. A 17 member panel was used to
establish a standard distribution for the beef feedlot.
Using the individual probability distributions and Monte
Carlo simulation, a 5-member panel was established that
could essentially yield the standard distribution. A
comparison of the means and standard deviations from
the 17- and 5-member panels is shown in Table 2. As is
readily seen, there is only a slight difference in the means
obtained by the two panels and the geometric standard
deviations are essentially the same. With the standard
distribution established, replacement panel members
can be screened and selected in a manner to yield the
standard distribution. This is a valuable technique for
research efforts lasting over extended periods of time and
requiring some change in panel members during the
research period.

### Odor Readings at Swine and Beef Operations

Odor measurements were made at a 600-sow farrow to
finish operation (Swine BR) and a 30,000 head open beef
feedlot (Feedlot FR). At the swine operation,
Scentometer readings were made approximately twice
weekly from September 1974 until March 1975 by a
single observer. Liquid manure samples were collected
from the deep pit beneath the slatted floors and carried
to the laboratory for a liquid dilution test with 10 panel
members. At the beef feedlot, Scentometer readings were
made twice per week during May and June of 1974 by a
single observer. Manure samples from the feedlot surface
were carried to the laboratory for a liquid dilution test
before an 8-member panel.

The results from the liquid dilution tests are shown in
Fig. 2. The actual probability data gave a good fit to the
theoretical lognormal probability curve (solid lines). The
swine manure has a much greater potential for odor than
the beef feedlot. The 50 percent probability for swine is
approximately 15 OII, and that for beef is approximately
10 OII.

The distribution of odor intensity measurements
obtained with the Scentometer and a single observer are
shown in Fig. 3. The odor readings from the Scentometer
also appear to fit a lognormal distribution very well. In

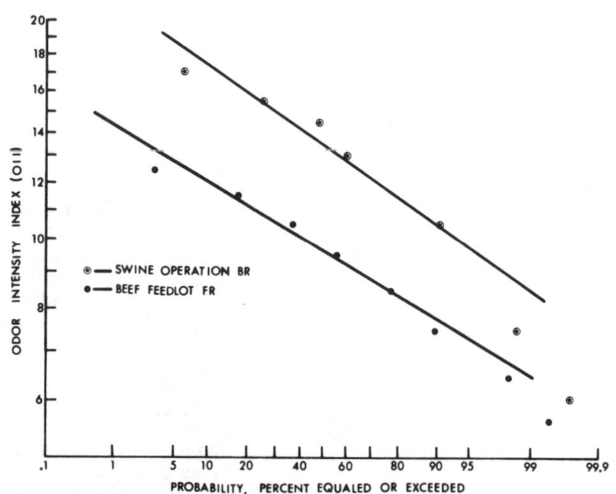


FIG. 2 Lognormal probability distributions of the odor intensity index
[OII] of liquid swine waste and beef feedlot waste using the liquid
dilution technique.



MANAGING LIVESTOCK WASTES

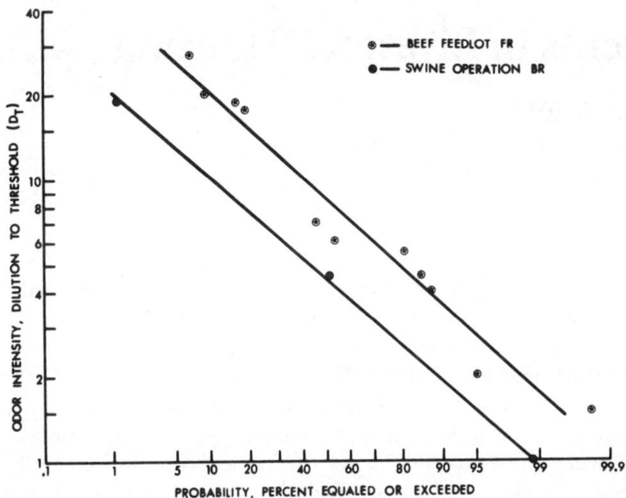

FIG. 3 Lognormal probability distributions of the odor intensity, dilutions to threshold [$D_T$], at a swine operation and a beef feedlot using the Scentometer.

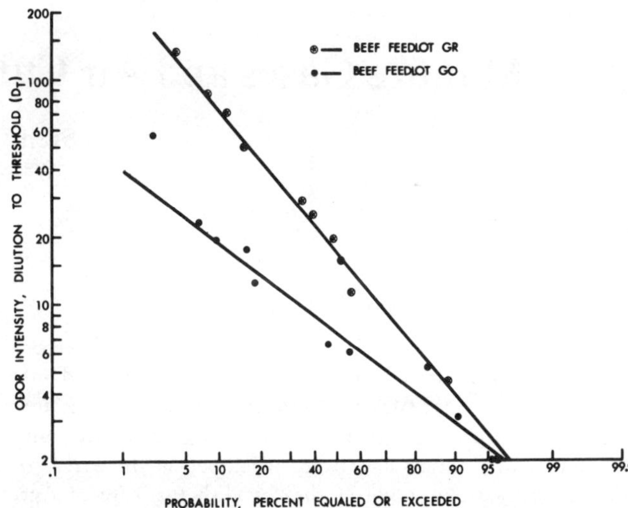

FIG. 4 Lognormal probability distributions of the odor intensity, dilutions to threshold [$D_T$], at two beef feedlots using the Scentometer.

this case however, the beef feedlot indicates more odor than the swine operation. This particular swine operation is well managed and these low odor readings with the Scentometer are typical of the operation.

No panel was used with the Scentometer. Therefore, a comparison between the liquid dilution technique and the Scentometer was impossible. However, no correlation appeared to be evident from these data.

### Odor Readings at Two Beef Feedlots

Odor intensities were measured at a 4000- (Feedlot GR) and a 12,000- (Feedlot GO) head beef feedlot during the spring and summer of 1973 with the Scentometer. At Feedlot GR, 4 pens were randomly selected and odor measurements taken at the center of each pen by a single observer on 8 different dates. At Feedlot GO, odor measurements were made in 3 pens by a single observer on 9 different dates and by 2 or more observers on 5 additional dates.

The frequency of odor readings for Feedlot GR and GO is shown in Fig. 4. Again a good fit with the lognormal distribution is obtained. Several states have enacted odor standards. Fig. 4 shows that the 127 $D_T$ standard for Oklahoma and Colorado was exceeded 0 and 4 percent of the time, respectively, at Feedlots GO and GR. A 7 $D_T$ standard would be exceeded 52 percent of the time at Feedlot GO and about 78 percent of the time at Feedlot GR. Most of the difference between the odor frequency curves for Feedlots GR and GO is probably due to the fact that different observers were used. This is another situation where standardization of the odor panel would be very beneficial.

## CONCLUSIONS

Based on data and analysis from this study, the following conclusions were made:

1 The lognormal probability distribution can be used to describe the frequency of odor occurrence by individual panel members, animal enterprises, or odor measuring techniques.

2 Use of the lognormal distribution in conjunction with Monte Carlo simulation provides a means for standardizing odor panels, so that the addition or deletion of panel members during a research project can be undertaken.

3 The Scentometer has limitations as a research instrument due to imprecise measurement.

4 The minimum odor level that can reasonably be expected inside a feedlot is 7 $D_T$. Readings of 170 $D_T$ are probably unreasonable and correctable.

### References


1 American Public Health Association. 1971. Standard methods for the examination of water and wastewater. Washington, D.C. 874 p.

2 Barth, C. L. 1973. Odor sensation theory and phenomena and their effects on olfactory measurements. TRANSACTIONS of the ASAE 16(2):340-347.

3 Bethea, R. M. and R. S. Narayan. 1972. Identification of beef cattle feedlot odors. TRANSACTIONS of the ASAE 15(6):1135-1137.

4 Burnett, W. E. and N. C. Dondero. 1969. Microbiological and chemical changes in poultry manure associated with decomposition and odor generation. Proceedings, Cornell University Conference on Agricultural Waste Management, Syracuse, NY. pp. 271-291.

5 Brunett, W. E. and N. C. Dondero. 1970. Control of odors from animal wastes. TRANSACTIONS of the ASAE 13(2):221-224, 231.

6 Cooper, H. B. H. 1973. Can you measure odor? Hydrocarbon Processing 52:97-101.

7 Hanna, G. F. 1972. Odor classifications of organic chemicals. Paper presented at the Third Symposium on Hazardous Chemicals Handling and Disposal, Indianapolis, Ind., April 11-13.

8 Ludington, D. C., A. T. Sobel and B. Gormel. 1971. Control of odors through manure management. TRANSACTIONS of the ASAE 14(4):771-774, 780.

9 Miner, J. R. and T. E. Hazen. 1969. Ammonia and amines: components of the swine building odor. TRANSACTIONS of the ASAE 12(6):772-774.

10 Smith, M. (ed.). 1968. Recommended guide for the prediction of the dispersion of airborne effluents. The American Society of Mechanical Engineers. New York, NY. 85 p.

11 Sobel, A. T. 1969. Measurement of the odor strength of animal manures. Proceedings, Cornell University Conference on Agricultural Waste Management, Syracuse, NY. pp. 260-270.

12 Summer, W. 1971. Odor pollution of air-causes and control. The Chemical Rubber Company Press. Cleveland, Ohio. 310 p.

# Manure Gases and Air Currents in Livestock Housing

Sven-Uno Skarp
ASSOC. MEMBER
ASAE

## BACKGROUND

THE method of handling waste products from livestock buildings as liquid manure was first introduced into Sweden in the mid-fifties. It did not, however, come into more general use until systems of cleaning-out by flooding, flushing and floating had been developed in the early sixties. Investigations made at the Swedish Institute of Agricultural Engineering (SIAE), among other places, showed that the most rational and the most favorable method, economically, of handling manure from livestock buildings was to handle it in liquid form together with urine, water and, when relevant, silage juices.

The new system for manure handling met with great interest from the farmers and the number of liquid manure installations increased rapidly. However, in due course it appeared that problems were caused by the formation of gases by anaerobic activity in the liquid manure. In some places these problems were aggravated by unsuitable design of systems for handling the manure and by insufficient ventilation systems. During the severe winter of 1964-65 these deficiencies resulted in a number of serious cases of hydrogen sulphide poisoning, which focussed interest on the problem and required its rapid solution. However, all the known cases of mortality in cattle and pig herds had occurred in connection with agitation, pumping or cleaning-out of the liquid manure, when large volumes of gases were released.

## THE PLANNING AND COMPLETION OF THE INVESTIGATION

Comprehensive studies and measurements of manure gases in livestock buildings were started by SIAE in different parts of Sweden during different times of the year. These studies were intended to reveal which measures were necessary for solving the problem of manure gas poisoning. The livestock buildings were chosen so that manure handling systems for both solid manure and liquid manure were included. Thus, measurements were made in livestock units with: a slatted floor above a manure pit; a system with combined sluice gate and flushing; a floating system; mechanical scraping of the manure; and a system with combined scraping and flushing. Altogether 68 studies were made in 58 livestock houses, of which 18 were pig houses and 40 cattle houses.

The survey of the concentration of manure gases required the development and construction of a mobile laboratory, which was mounted on a truck. The laboratory was equipped with four gas analyzers, namely two infra-red analyzers for determining carbon dioxide ($CO_2$) and methane ($CH_4$), and two ionoflux analyzers for determining hydrogen sulphide ($H_2S$) and ammonia ($NH_3$). The air for analysis was piped to the analyzers via stainless steel

tubes (inner diameter 6 mm). The values given by the analyzers were recorded directly onto a recorder. The same truck also carried ventilation equipment consisting of two propeller fans and ventilation tubes that could be coupled to the livestock buildings and which enabled the volume of exhausted ventilation air to be accurately measured (Fig. 1). During these measurements the normal ventilation equipment in the building was switched off and all the air was drawn out through the experimental equipment.

FIG. 1 The mobile laboratory for analysis of manure gases. The equipment includes fans which are used to obtain the required ventilation in the buildings.

The extent and the movements of the air currents were found to be of great importance and were studied by means of smoke. The measurements were made both during the collection period, i.e. in between cleaning-out times when the manure accumulated in the gutters and channels but remained static, and during cleaning-out or agitation phases.

### GASES DURING THE COLLECTION PERIOD

The concentrations of carbon dioxide and methane varied widely throughout the day according to the activity of the animals. Fig. 2 (a-d) shows the hourly variations in cattle houses, the concentrations being measured in the exhausted ventilation air. The lowest concentrations were in the early hours of the morning when the animals were at rest. The $CO_2$ and $CH_4$ production increased during milking and feeding. The increases in concentration naturally depended on the capacity of the existing ventilation equipment. $CO_2$ concentrations varied from about 800 ppm at ventilation capacities corresponding to maximum ventilation = 250 m³ per head of cattle (500 kg) per hour to about 4000 ppm at ventilation capacities corresponding to minimum ventilation, i.e. 50 m³ per head of cattle per hour. The diagram is composed of representative results from cattle houses and it shows that there were no differences between buildings with different systems of handling manure

The author is: SVEN-UNO SKARP, Swedish Institute of Agricultural Engineering, Uppsala, Sweden.

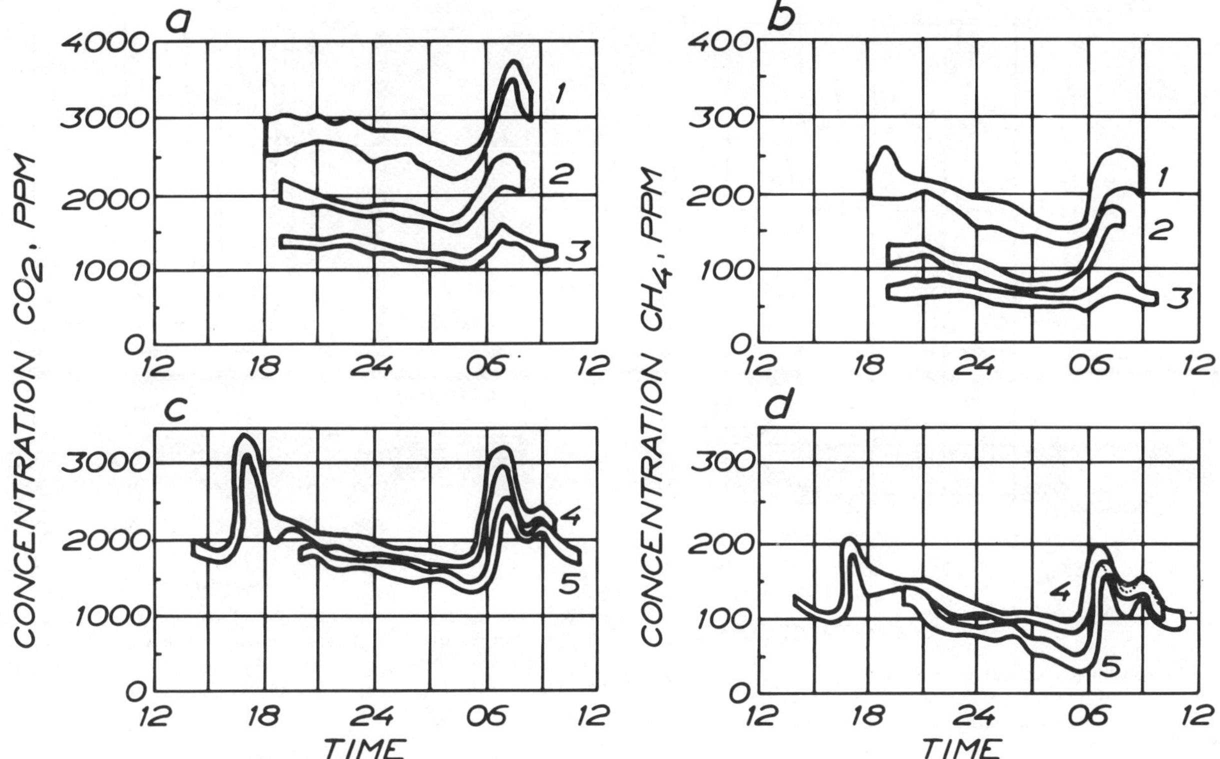



FIG. 2 Variation of $CO_2$ and $CH_4$ concentrations in a cattle shed during 24 hours. The concentrations are measured in the exhausted ventilation air. Bands 1, 2 and 3 in diagrams (a) and (b) represent buildings with liquid manure systems where the indoor storage period was at least one week. The ventilation capacity was 50, 130 and 240 $m^3$ per head of cattle per hour for 1, 2 and 3 respectively. In diagrams (c) and (d) band 4 represents a building where the liquid manure was removed once a week and band 5 a building where solid manure was removed twice a day at about the same ventilation capacities, namely 120 $m^3$/head/hour.

indoors. The variations in the concentrations of $CH_4$ followed the same pattern as those for $CO_2$ but at a level about 10 times lower.

Ammonia ($NH_3$) generally showed no direct variations throughout the day in cattle houses. The concentrations were very low and in some cases they were undetectable. Only three cattle houses had concentrations above 20 ppm and the highest level was 30 ppm.

In pig houses the situation was similar to that in cattle houses. However, here $CH_4$ was very seldom present and did not follow the $CO_2$ pattern. In some cases $CH_4$ was hardly detectable. The highest level measured was 30 ppm. The concentrations of $CO_2$ varied between ca 800 and ca 2000 ppm, according to the ventilation capacity. $NH_3$ varied widely and the values ranged from a few ppm to a maximum of 30 ppm. In pig houses using systems for solid manure the gas concentrations were about the same as in those with liquid manure.

The dispersion of gases in the building depended on the building's shape, the ventilation system's design, the ventilation capacity, the herd size and the associated convection currents etc. The gases mainly followed the air currents, the movements of which are largely determined by the design and location of the air inlets. High concentrations of gases were sometimes measured at places lacking air inlets or with widely spaced air inlets.

The vertical distribution of gases in cattle houses during the winter (Fig. 3) was such that $CO_2$ and $CH_4$ had higher concentrations at roof-level than at floor-level. In these cases the manure temperature was about +10 C or just below. The explanation of this distribution is probably that these gases come mainly from exhaled breath and from the rumen and, being warm, they rise towards the roof. There were few differences in the vertical distribution of $NH_3$,

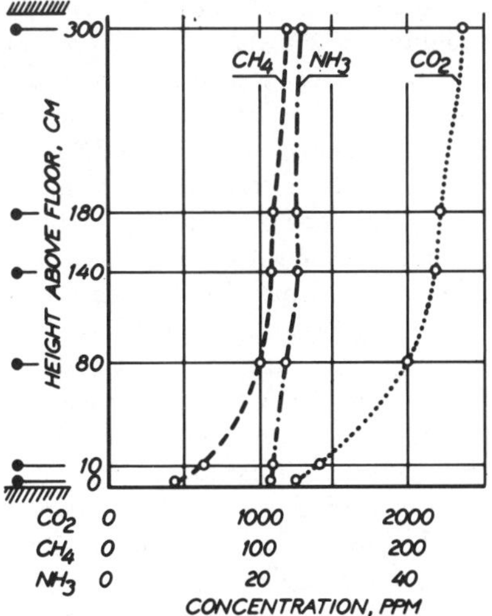



FIG. 3 Distribution of gases between floor and roof in a cattle building using a liquid manure system. The measurements were made in the winter when the building was ventilated at about half the maximum capacity.

but this gas is released by feces and urine and has much the same temperature as the surrounding air.

In the stored manure the temperatures were about +10 C in winter and about +20 C in summer. In summer the stored manure released greater volumes of gases, which resulted in higher concentrations, particularly in pig houses. $H_2S$ also occurred in some cases. The increased ventilation rate during the summer helped to even out the vertical distribution of gases. A diagram showing measurements in a pig house with mechanical scrapers and liquid manure is given in Fig. 4. It shows clearly that the gas concentrations

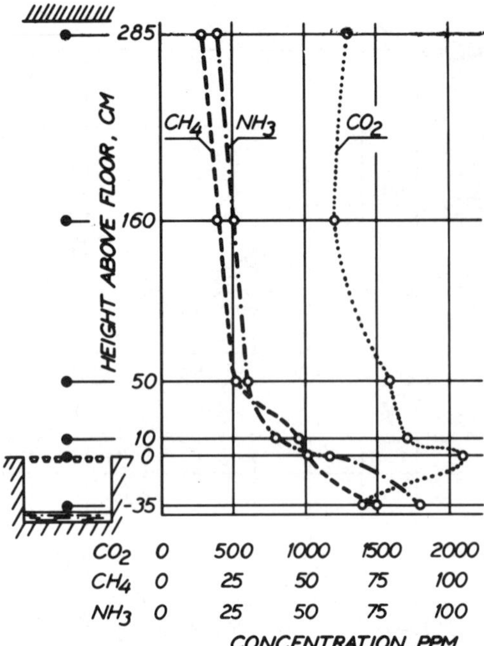



FIG. 4 Distribution of gases between floor and roof above the manure channel in a pig building using a liquid manure system. The measurements were made in the summer when the ventilation was at maximum capacity, which caused a rapid exchange of air above the slatted floor and evened out the differences in the gas concentrations.

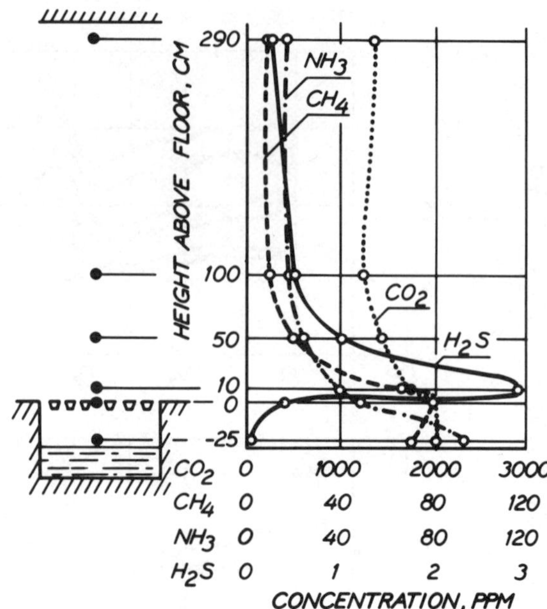



FIG. 5 During the summer when liquid manure stored indoors has a higher temperature than during the winter, the occasional release of H₂S may occur from pig manure. The figure shows the distribution of gases from floor to roof on an occasion when H₂S was being released.

in the space between the surface of the manure and the slatted floor, and just above the slats, were greater than those higher up where the concentrations were more uniform and lower than in seasons with lower ventilation rates. In livestock buildings where the fresh air was effectively spread by means of suitable design and location of air inlets, distribution systems for fresh air and, in some cases, circulation fans, the gases had a more even distribution.

In the breathing zones of cattle, or in manure channels or outlet pipes, $H_2S$ did not occur in detectable quantities as long as the manure remained static during the collection period. In pig houses $H_2S$ may be released spontaneously from liquid manure, especially during the summer when the manure has a higher temperature. $H_2S$ was thus recorded locally in pig houses just above the channels and gutters. However, the volumes were never so large that $H_2S$ could be detected in the exhausted ventilation air. $H_2S$ could also be detected in pig houses using mechanical scrapers and liquid manure, despite the storage period being maximally 12-14 hours. Fig. 5 shows the concentrations on one occasion in a pig house using the above system. The highest concentration of $H_2S$, about 3 ppm, was measured at a height of about 10 cm above the slatted floor, while the highest concentrations of other gases were at lower heights.

## OCCURRENCE DURING CLEANING-OUT AND AGITATION

When liquid manure is set in motion during agitation or cleaning-out, gases are immediately released, $H_2S$ being among them. Solid manure, i.e. manure mixed with bedding which absorbs urine and water, does not release detectable volumes of $H_2S$ during cleaning-out. The release of gases depends on the fluidity of the manure. At higher water contents the manure floats more easily and the movement will be more vigorous during cleaning-out, which leads to the release of large volumes of $H_2S$. The release of gas is

also linked with the manure temperature. When the manure temperature is higher than about +20 C, the normal summer temperature, there is a higher decomposition rate in the manure which results in larger amounts of $H_2S$ being released when the manure is agitated. Methane and carbon dioxide are also released, while ammonia usually decreases in concentration, especially when $H_2S$ is present. Ammonia and hydrogen sulphide probably combine to form a compound, which would explain this decrease. Apart from the manure's temperature and fluidity, the release of gases depends on the length of storage and the type of manure handling system. The gases are mainly released in outlet pipes, pump pits, at flushing nozzles and at sluice gates.

In systems using mechanical scrapers and solid manure, no increases were noticed in the concentrations of gases, or in the presence of $H_2S$ in the breathing zones. In systems using sluice gates and systems using a combination of sluice gates and circuit flushing the volumes of $H_2S$ released in connection with cleaning-out were sometimes so large that they clearly influenced the exhausted ventilation

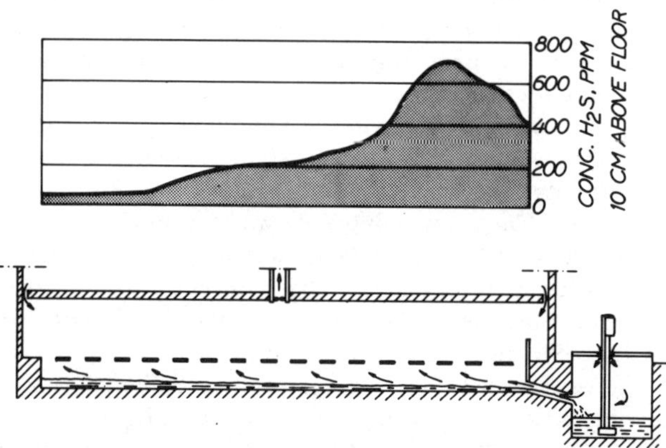



FIG. 6 Lethal concentrations of H₂S have occured when gas traps and exhaust ventilation of outlet pipes are excluded in cleaning systems, particularly in pig buildings with sluice gate systems where H₂S comes into the building from pump pits and outlet pipes. The diagram shows the concentrations of H₂S at a height of 10 cm above the slats over the channel during cleaning.



MANAGING LIVESTOCK WASTES

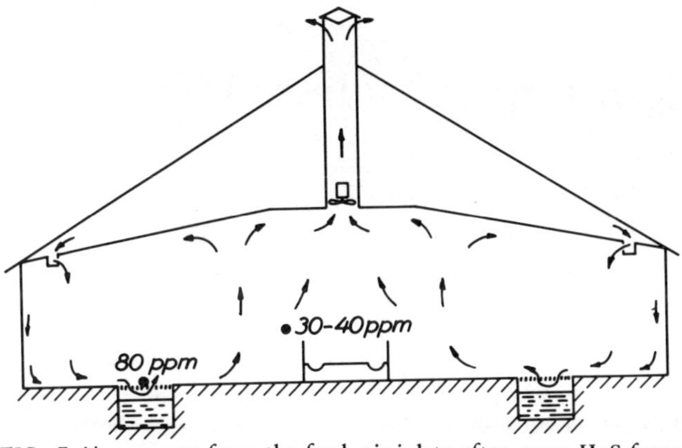

FIG. 7 Air currents from the fresh air inlets often carry H2S from the manure channel into the breathing zone. Here H2S is recorded during cleaning out.

air. However, the buildings were not fitted with gas traps or with ventilated outlet pipes. At some places inside the buildings the gas concentrations were very much higher than in the exhausted air. The $H_2S$ concentrations were particularly high above the manure channels, where the lethal dose of 800 ppm was sometimes exceeded in pig houses. Measurements of $H_2S$ at a height of 10 cm above a slatted floor during cleaning-out are given in Fig. 6. Under corresponding conditions in cattle houses, the $H_2S$ levels were much lower. In buildings with tied-up animals the air currents (Fig. 7) may carry gases to the breathing zones, where 30-40 ppm of $H_2S$ were measured when the concentrations above the manure channels were ca 80 ppm.

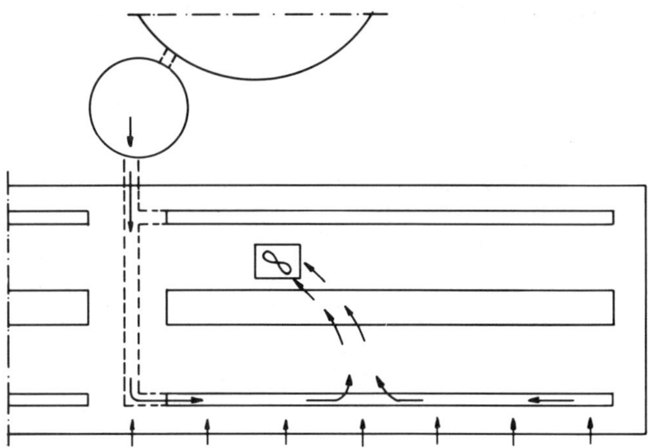

FIG. 8 Smoke is used to study the routes of gas-laden cold air which flows in from pump pits and outlet pipes during cleaning-out in a cattle building. Here there were no gas traps and no ventilation of the outlet pipes.

These conditions emphasize the importance of having gas traps in the cleaning-out system. The high $H_2S$ concentrations in the breathing zones depends, namely, on the large volumes of gases that are released in pits and outlet pipes being drawn into the livestock building if there are no gas traps. When gas traps were installed the $H_2S$ concentrations decreased to a fraction of the earlier values and were only registered around pit covers and flushing nozzles where the manure was in rapid movement. A separate ventilation fan is also needed to remove gases that are released in the outlet pipe. The gas-laden air that penetrates livestock buildings from manure pits and outlet pipes first follows the pipes and channels and then spreads through the building according to the design of the ventilation.

Fig. 8 gives an example of how gases from a pump pit spread throughout a building during cleaning-out. The gas-

laden air moves at low levels due to its lower temperature and, thus, greater weight. Competition for the lowest level between gas-laden cold air from the cleaning-out system and fresh cold air from the air inlets has resulted in the distribution pattern illustrated.

## INFLUENCE OF NORMAL VENTILATION ON AIR CURRENTS AND GAS DISTRIBUTION

Ventilation systems using a slight negative pressure are the most common ones in Sweden. They were also most represented in the investigations reported here. One of the most serious principle errors was that the air inlet area was of the wrong size, which caused incorrect flow speeds of incoming air and consequently unsatisfactory distribution of fresh air. In many buildings the size of the air inlet was too large in relation to the volume of exhausted ventilation air. This resulted in the absence of a stable negative pressure in the building and an uncontrolled intake of fresh air through the inlets. Only inlets near the fans admitted fresh air and the wind conditions outdoors completely decided how the air entered the building. The investigations also showed that the total area of the air inlets must not be greater than that enabling the exhausted ventilation air to give the incoming fresh air a speed of at least 2 m per second in the inlets. The inlet area will then be such that a stable, static, negative pressure is formed throughout the entire building. The air inlets should, thus, be adjustable to suit summer or winter ventilation conditions. The exhaust fans can be placed in walls, the roof, or under the floor. Provided the volume of exhausted ventilation air is the same the distribution patterns of fresh air will be similar.

The study of different types of ventilation systems showed that balanced ventilation, i.e. where both incoming fresh air and exhausted ventilation air are controlled by fans, gave the best ventilation results. The incoming fresh air was distributed via ducts to all parts of the building. However, the systems for incoming air and exhaust air must be controlled by the same thermostat.

### SUMMARY

Results of the investigation into the presence, concentrations and distribution of manure gases in livestock housing can be summarized as follows:

1  Gases released from the manure were mainly $CO_2$, $CH_4$, $NH_3$ and $H_2S$.

2  Solid manure did not release gases in quantities injurious to animals and humans.

3  Static liquid manure released $H_2S$ in measurable quantities only if the manure originated from pigs.

4  Liquid manure set in motion by pumping, mixing or cleaning-out released large amounts of gases, particularly $H_2S$ which sometimes appeared in lethal concentrations.

5  The normal ventilation design was found to have a great influence on the distribution of manure gases. The largest problems were caused by currents of cold air at low heights due to ineffective mixing and distribution of the incoming fresh air from the air inlets.

6  The design, location and area of air inlets determined the way the fresh air was distributed.

7  The design and location of the exhaust fans were of minor importance for the correct control of incoming fresh air.

8  Balanced ventilation systems gave the best conditions compared with systems using slight negative or positive pressures.

# Exhaust Systems for Underfloor Liquid Manure Pits

David S. Ross, Robert A. Aldrich, Dwight E. Younkin, Grant W. Sherritt, Joseph A. McCurdy

ASSOC. MEMBER
ASAE

MEMBER
ASAE

MEMBER
ASAE

S WINE producers are concerned about the gases produced in liquid manure storage pits in enclosed buildings with partially slotted floors. The gases produced in the pit cause unpleasant odors and can be injurious to both people and animals in the buildings.

One method for removing these gases from a manure storage pit is to exhaust air through the pit to the outside of the building. Continuous ventilation should prevent pit gases from moving into the animal occupied area of the building.

Laboratory and field installation studies on two exhaust systems are reported in this paper. Driggers (1971), DeShazer, et al. (1972) and Horsfield (1973) have reported on other studies. The popular press has reported experiences with several systems installed in commercial buildings.

## PERFORATED DUCT DESIGN

Steele and Shove (1969) gave design charts for flow and pressure distribution in perforated air ducts with uniform cross section. Two general cases of flow were presented; these were (a) uniform discharge or intake along the duct, and (b) uniform openings along the duct. This application used ducts with uniform discharge or intake.

The charts expand on the earlier development by Shove and Hukill (1963) of an equation for the cases of uniform intake and discharge of air along a duct.

The design procedure followed that of Steele and Shove (1969); a brief outline of this application was reported by Ross, et al. (1974).

### Laboratory Tests

In a closed laboratory, tests were run using 100 ft (30.5 m) each of 6- and 8-in. (15.2 and 20.3 cm) diameter plastic pipe and a Model 15P ILG centrifugal fan. The 0.25 hp fan discharged 380 cfm (10.8 cu m per min) at 0.25 in. (0.64 cm) water static pressure.

The following duct designs and test combinations were studied. A flow of 4 cfm per ft of duct length was chosen for minimum ventilation; this being adequate for a farrowing house. The design procedure indicated the static pressure at the fan would be too high to put the fan at the end of 100 ft of 6-in. pipe. The fan was relocated to the center of the duct, creating two 50 ft ducts, A and B. The number of orifices placed in these ducts were varied; duct A had one orifice per ft while duct B had one orifice each 6 in. or two orifices per ft. Measurements of air velocity were made at each orifice.

---

The authors are: DAVID S. ROSS, Assistant Professor, Agr. Eng. Dept., University of Maryland, College Park (formerly Graduate Assistant, Agr. Engr. Dept., Penn State); ROBERT A. ALDRICH, Professor, Agr. Engr. Dept.; DWIGHT E. YOUNKIN, Professor, Animal Sci. Ext.; GRANT W. SHERRITT, Associate Professor, Animal Sci. Dept.; and JOSEPH A. MCCURDY, Professor, Agr. Engr. Ext., The Pennsylvania State University, University Park.

Approved for publication as Scientific Article No. A2087, Contribution No. 5041 of the Maryland Agricultural Experiment Station, Department of Agricultural Engineering.

An 8 in. diameter pipe, duct C, was large enough for the fan to operate at the end of 100 ft with an airflow of 4 cfm (0.11 cu m per min) per ft.

In all cases the orifices were drilled using bit sizes reasonably close to the calculated diameters. The air velocities were measured by a hot wire anemometer centered inside a length of 3 in. (7.6 cm) aluminum tube used to straighten the airflow to the orifice. A gasket was used to provide a tight fit against the duct around the orifice.

The data from the tests are shown in Figs. 1 and 2. Two runs were made for each duct. In all ducts the airflow decreased with distance from the fan with the average value above the design value. The standard deviation was about ±15 percent of the mean value.

### Field Installation

Two field installations using the perforated duct system have been observed. Some air distribution data was taken in one farrowing barn where the air volume moved through the pit was based on a minimum continuous flow of 20 cfm (0.57 cu m per min) per pen. Additional ventilation was

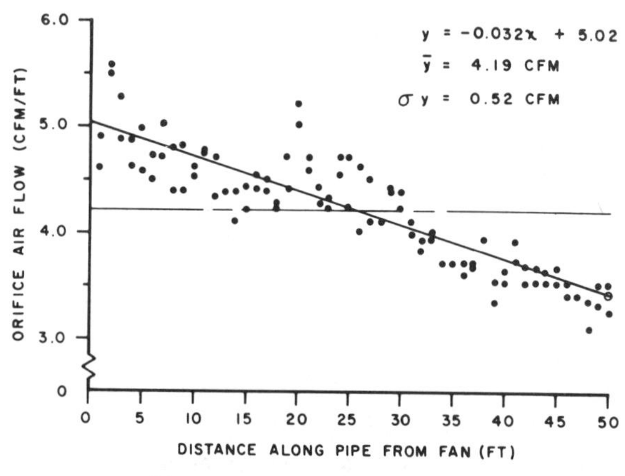

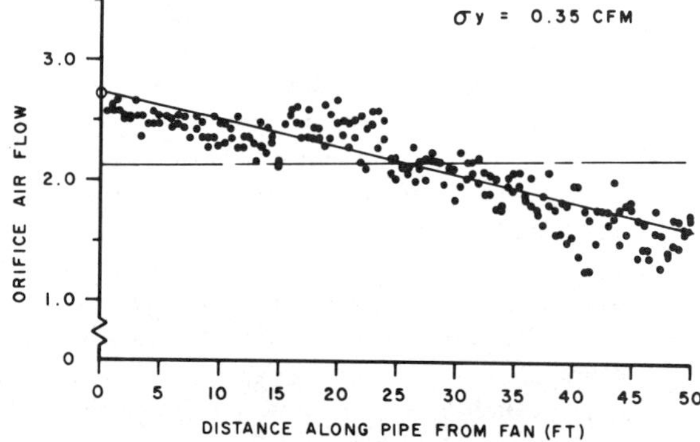


FIG. 1 Airflow into varying diameter orifices along 6 in. diameter plastic pipe. Flow rate designed for 4 cfm/ft.

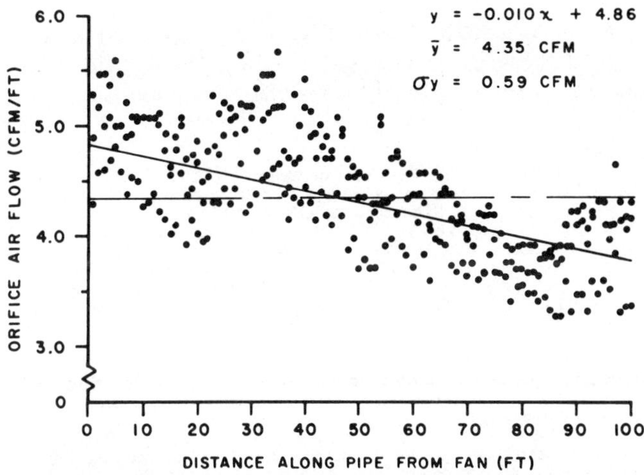



FIG. 2 Airflow into varying diameter orifices along 8 in. diameter plastic pipe. Holes 12 in. apart. Flow rate designed at 4 cfm/ft.

provided by slot inlet and exhaust fans. The operation checked had a dual duct system in the pit with a 6 in. diameter pipe on one side serving as an inlet to provide an air curtain across the top of the pit just below the slats and a second 6 in. diameter pipe on the opposite side of the pit picking up the flow and exhausting it from the building. Holes of varying size were drilled 6 in. on center through the pipe wall. Air velocity in the pit was measured with a hot wire anemometer. The airflow measured was in reasonable agreement with design values. It was concluded from general observations that it was not necessary to provide the air curtain and the inlet duct operation was discontinued. The swine grower was satisfied with the level of odor control obtained by the exhaust system.

## TAPERED DUCT SYSTEM

It is not possible to install a perforated pipe system in all barns so an alternative was designed. This consisted of a weighted curtain inlet, an above slot pickup duct with the weighted curtain controlling flow into a tapered exhaust duct. The tapered duct was designed according to ASHRAE Guide procedures. In this system all the ventilation air passes through the pit. The details of a typical system are shown in Figs. 3 and 4. Design tables developed from ASHRAE procedures are given by McCurdy (1972).

### Field Installation

There have been several installations of this design operating for a year or longer. The operators indicate the systems are providing acceptable odor control and are satisfying the ventilation requirements.

A system was installed in 1973 in a barn at the Swine Research Center in the University Park Campus of The Pennsylvania State University. The building is a farrow-to-finish barn with hot water heating. It has a center slot inlet with side pits. The pickup ducts and tapered exhaust ducts were designed according to ASHRAE (1972). Ventilation volume control is provided through one variable speed fan rated at 1490 cfm (42.2 cu m per min) minimum and 4000 cfm (113.3 cu m per min) maximum, and one single speed fan rated at 8100 cfm (229.4 cu m per min). The barn had been in operation for 6 years prior to the installation of this exhaust system. The system was monitored during the winter of 1973-74 by checking air and temperature distribution, variable speed fan operation, and subjective evaluation of odor control. The hog population varied from 1/3 to 3/4

of capacity based on hog numbers. The barn was never more than one-half full of market weight hogs. The controls were set to maintain a hog height temperature of about 65 F (18 C).

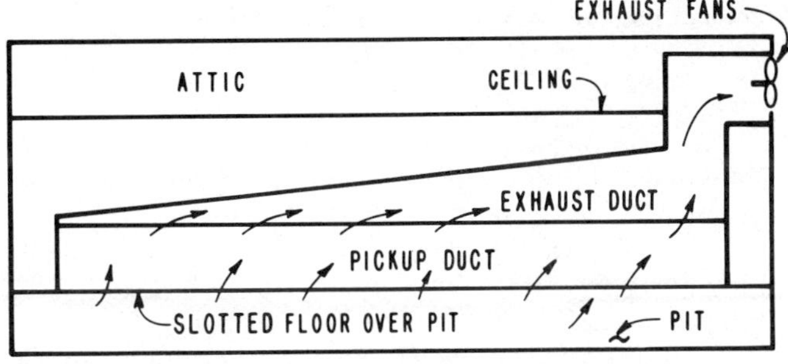



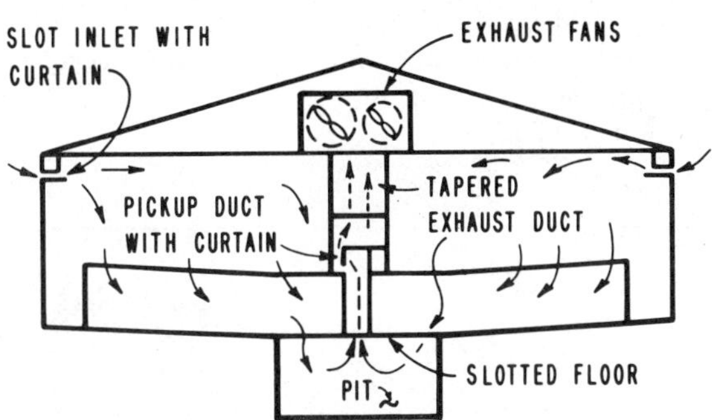



FIG. 3 Cross sections showing pickup and tapered exhaust duct above a center pit for gas and odor control.

### Results

Air distribution and temperature controls were acceptable throughout the winter period. However, it was apparent early in the season that pit odor in the occupied zone was not being satisfactorily controlled. The variable speed fan was dropping to the minimum speed during cold periods and the airflow through the pits was not enough to prevent pit odors from moving into the occupied zone. The set point for variable speed fan control was adjusted to the

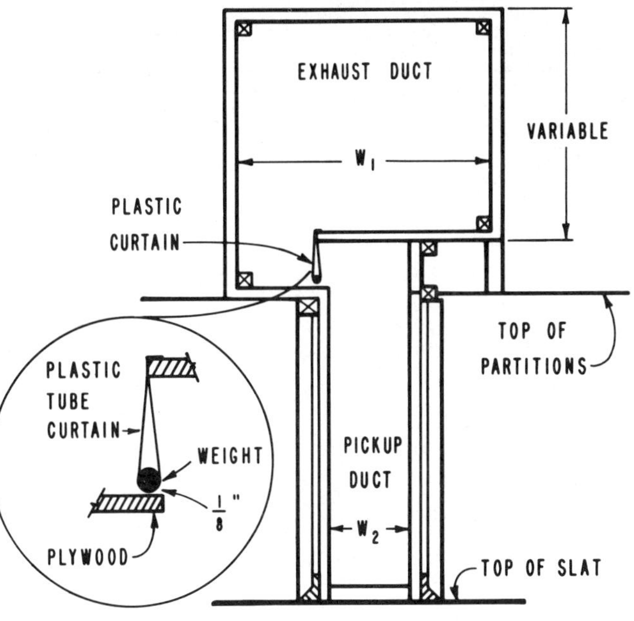



FIG. 4 Expanded view for pickup and tapered exhaust duct for center pit. Dimensions to fit design. Exterior plywood construction.

minimum value (below 45 F or 7 C) for a period of 10 days to see if a higher minimum airflow would provide better pit odor control. This appeared to greatly reduce odors but resulted in higher fuel consumption to maintain acceptable temperature control with the low hog numbers.

## SUMMARY

The following can be stated based on the laboratory study and field observations of the two types of exhaust systems.

1   It is possible to design manure pit exhaust systems using either the Steele and Shove equations for perforated ducts or the ASHRAE equations for a tapered duct.

2   In the buildings monitored, the concentration of pit odors in the occupied zone was reduced; but it was not possible to say they were held to an acceptable level because only subjective evaluations were used.

3   There appears to be a minimum airflow necessary to prevent pit odors from entering the occupied zone. This has not been determined.

4   A system in which all ventilation air must be exhausted through the pits is more easily designed than one that exhausts only part of the ventilation required.


**References**

1   ASHRAE Handbook of Fundamentals. 1972. American Society of Heating, Refrigerating, and Air Conditioning Engineers.

2   DeShazer, J. S. and E. A. Olson. 1970. Gas removal from swine housing. E.C. 70-219. 1970 Nebraska Swine Report, Cooperative Extension Service, University of Nebraska, pp. 25-27.

3   Horsfield, B. C. 1973. Underfloor ventilation for confinement swine. Farm Builders' Conference, Mimeo, Purdue University.

4   Driggers, L. B. 1971. Ventilation of swine buildings. Agricultural Extension Service EBAE-713, North Carolina State University.

5   McCurdy, J. A., R. A. Aldrich and D. E. Younkin. 1972. Pit ventilation for hog barn odor control. Agricultural Engineering Fact Sheet EPP-10, Cooperative Extension Service, The Pennsylvania State University.

6   Ross, D. S., R. A. Aldrich, D. E. Younkin, G. W. Sherritt and J. A. McCurdy. 1974. Exhaust systems for underfloor liquid manure pits. ASAE Paper No. 74-4526, ASAE, St. Joseph, Mich. 49085.

7   Shove, G. C. and W. V. Hukill. 1963. Predicting pressure gradients in perforated grain ventilation ducts. TRANSACTIONS of the ASAE 6(2):115-116, 122.

8   Steele, J. L. and G. C. Shove. 1969. Design charts for flow and pressure distribution in perforated air ducts. TRANSACTIONS of the ASAE 12(2):220-224.


---

# Evaluation of Dairy Waste Odor

*(Continued from page 357)*


versity, 1969 Agricultural Waste Management Conference, Ithica, N.Y.

2   Dravnieks, A. 1968. Approaches to objective olfactometry. P. 2, In: N. Necmi (ed), Theories of odor and odor measurements. Circa Publications, Inc., Pelham, N.Y.

3   Dravnieks, A., B. K. Kratoszynski, Joyce Barton, O'Donnell, A. and T. Burgwald. 1970. Gas chromatographic anlaysis of amino compounds in atmosphere. In: Proceedings, 11th Conference on Methods in Air Pollution and Industrial Hygiene Studies, University of California, Berkeley, Calif.

4   Ifeadi, C. N. 1972. Quantitative measurement and sensory evaluation of dairy waste odor. Unpublished. PhD Dissertation, The Ohio State University, Columbus, Ohio.

5   Lindvall, Thomas. 1970. A sensory evaluation of odorous air pollutants. Nordisk Hygianisk Jidskrift, Stockholm, Sweden.

6   McKelvey, J. M. and H. E. Hoelschen. 1957. Apparatus for preparation of very dilute gas mixtures. Analytical Chemistry, (29):123.

7   Merkel, J. A., T. E. Hazen and J. R. Miner. 1969. Identification of gases in a confinement swine building atmosphere. TRANSACTIONS of the ASAE 12(3):310.

8   Munson, M. S. B. 1971. Chemical ionization mass spectrometry, Analytical Chemistry, 43:284.

9   Sherwood, T. K. and R. L. Tigford. 1952. Absorption and extraction. McGraw-Hill, N.Y.

10   Taiganides, E. P. and R. K. White. 1969. The menace of noxious gases in animal units. TRANSACTIONS of the ASAE 12(3):359-362.

11   Willrich, T. L. and J. R. Miner. 1971. Litization experiences of five livestock and poultry producers. P. 99-101. St. Joseph, Mich. 49085.

12   White, R. K. 1969. Gas chromatographic analysis of odor from dairy animal wastes. Unpublished PhD dissertation, The Ohio State University, Columbus, Ohio.

13   White, R. K., E. P. Taiganides and G. Cole. 1971. Chromatographic identification of malodors from dairy animal waste. P. 110-113. In Livestock waste management and pollution abatement. ASAE Publications. PROC. 271. St. Joseph, Mich. 49085.

# Malodor Reduction in Beef Cattle Feedlots


Willie L. Ulich, John P. Ford

SENIOR MEMBER     ASSOC. MEMBER
ASAE                      ASAE


THE control of malodors is a persistent problem faced by beef cattle feedlot operators. Managers of these vital agricultural industry units are experiencing the ever-tightening economic vs. regulation squeeze. Many of these units contain 20 000 to 40 000 animals and are stocked at rates varying from 80 to 140 sq. ft. per animal. Commercially fed cattle are usually fed approximately 150 days, at an average rate of about 26 pounds of feed per day, and gain about 1 pound of live weight per 8 pounds of feed. a 1 000 pound animal produces approximately 27 pounds of excreta, excluding urine, per day. It is, therefore, evident that the feedlot manager has a tremendous waste control problem.

The organic waste material encountered is such that malodors develop rather rapidly and make up one of the managers major problems. The odors developed have an adverse effect upon the animals, operational labor, and the surrounding neighborhood. Some state air pollution agencies currently have odor control standards and others are formulating such standards. Most current legal cases are prosecuted on the basis of nuisance rather than on odor threshold standards; however, odor control standards are inevitable in the future.

Primary approaches to odor control are good housekeeping, masking agents, oxidation, deodorants, and the destruction or inhibition of bacteria and enzymes which produce odorous compounds. Interference with the olfaction process and reduction in the concentration of the odorant are two basic methods of feedlot odor control. There are currently several types of materials being marketed for this purpose; however, comparative objective evaluations have been limited. Operators have experienced various degrees of success; however, some are plagued by high cost or undesirable side effects. The purpose of the investigations, herein reported, was to provide an objective evaluation of certain materials which may have a potential for open cattle feedlot odor control systems.

## CONTROL AGENTS STUDIED

In 1970, Burnett and Dondero (1970) reported good control of odors from liquid poultry manure by the use of masking agents. Also in 1970, Ludington and Sobel (1970) evaluated some 40 odor control products. While no products was deemed satisfactory, masking and counter acting agents were most effective. Considerable work has also been accomplished in the way of feeding animals certain odor reduction ingredients and with varying degrees of success. Faith (1964) reported satisfactory feedlot odor control by the use of potassium permanganate. Varying success of odor control by use of lime, orothodi-chlorobenzene, hydrogen peroxide, and paraformaldehyde have also been


The authors are: WILLIE L. ULICH, Professor, and JOHN P. FORD, Instructor, Agricultural Engineering Dept., Texas Tech University, Lubbock.


**TABLE 1. GAS ABSORBENT AGENTS.**

| Absorbent | Gases absorbed |
|---|---|
| 1.2 N Hydrochloric acid | Amines, ammonia |
| Cadimum hydroxide suspension | Hydrogen sulfide |
| 3 percent w/v Mercuric chloride | Mercaptans |
| 4 percent w/v Mercuric cyanide | Disulfides |

reported. It is assumed the varying results obtained have been mainly due to the varying rates of application and the age of the fecal material under test.

Based upon prior work and preliminary tests, six materials were selected for further investigation. The criteria of selection was based upon a high probability of modifying the production of odorous gases. The materials included potassium permanganate, potassium nitrate, paraformaldehyde, a formulation of orthodi-chlorobenzene marked under the brand name as Ozene, hydrogen peroxide, and a confidential formulation known as Formula-2. The effect of these materials on beef animal excreta was then subjected to chemical qualative tests and to organoleptic tests.

## LABORATORY TESTS

Fresh manure samples were collected from a minimum of three animals, mixed and slurried at the rate of 2 liters of water per 500 grains of manure, and placed in 5 liter test containers. Nitrogen, a non-reactive gas, was bubbled through this mixture at a low rate for 72 hours. This method also provided a mild stirring action. The volatile gases produced from the manure were swept by the nitrogen through a selective absorption system which trapped and concentrated the odorous gases. The selective absorption system of Merkel et al. (1969) was modified by using a cadium hydroxide suspension for the absorption of hydrogen sulfide (Jacobs et al. 1957). The selective absorbents used and the gases absorbed are given in Table 1 and Fig. 1 attached. The selective absorbents were then colorimetrically analyzed to determine whether the various classes of odorous gases were present.

Preliminary tests showed that all classes of odorous gases, except hydrogen sulfide, were present in detectable

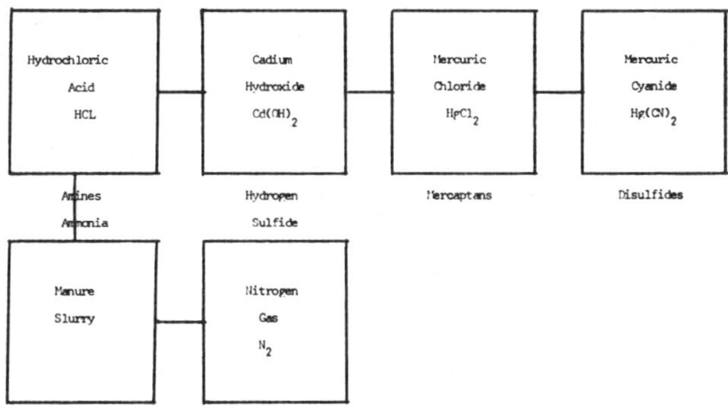


FIG. 1 Malodor absorbent test.

| Material | Quantity per acre | Unit cost | Cost per acre |
|---|---|---|---|
| Potassium permanganate | 20.0 lb | $1.54/lb | $ 30.80 |
| Potassium nitrate | 20.0 lb | 1.54/lb | 30.80 |
| Parafarmaldehyde | 2178.0 lb | 0.50/lb | 1078.11 |
| Ozene | 6.0 gal | 3.38/gal | 20.24 |
| Hydrogen peroxide, 10 percent | 12.4 gal | 2.48/gal | 30.80 |
| Formula 2 | 5.9 gal | 5.95/gal | 35.10 |



| | Test I | | Test II | | | |
|---|---|---|---|---|---|---|
| | P. | O. | P. | O. | P. | O. |
| Potassium permanganate | 3.1 | 3.3 | 3.3 | 3.4 | 2.7 | 2.7 |
| Potassium nitrate | 2.5 | 2.6 | 3.9 | 3.9 | 4.7 | 4.2 |
| Paraformaldehyde | 3.3 | 3.2 | 5.1 | 5.0 | 2.7 | 2.6 |
| Hydrogen peroxide | 3.6 | 3.6 | 3.0 | 3.0 | 4.6 | 4.5 |
| Ozene | 5.4 | 6.0 | 2.9 | 3.4 | 3.0 | 2.9 |
| Formula 2 | 6.2 | 6.2 | 5.1 | 5.1 | 4.4 | 4.4 |
| Control A | 4.1 | 4.5 | 4.0 | 3.8 | 4.4 | 4.4 |
| Control B | 4.7 | 5.1 | 4.9 | 4.2 | 4.2 | 4.2 |

Note: Detection P. = presence and O. = offensiveness

quantities at the end of a 24-hour reaction period. Further tests indicated hydrogen sulfide was produced within 48 hours in about half the tests and was always present within 72 hours.

To verify that the same gases were produced without the addition of moisture, 500 grams of manure was distributed on four pieces of wire in the reaction vessel. Nitrogen gas, saturated with water vapor, was intorduced at the bottom of the reaction chamber and allowed to sweep upward around the manure. The off gases were selectively absorbed as previously outlined. The same gases were found to be present from the solid manure as from the liquid manure slurry.

Following the odor detection procedural tests, repeatable qualification, aand minimum effective control rate tests were performed using each of the materials selected. Where favorable results were obtained in the qualification tests, linear reduction of strength tests were performed to establish control rate parameters. It was recognized, however, that various combinations of the gases might greatly affect a control agent on the human malodorous scale.

## ORGANOLEPTIC TESTS

Since complaints about odors are the results of the sense of smell, the human nose is the ultimate judge as to the effectiveness of an odor control method. For this reason, the organoleptic test described by Sobel (1970) was used for further evaluates fo the odor control materials. This test consists of a 0-10 rating scale for both presence and offensiveness of odors.

Two pound manure samples containing various agent treatments were prepared in perforated top containers and placed in a paper bag to prevent visual prejudice. Each agent was applied at the recommended rate except for hydrogen peroxide and potassium nitrate which were applied at rates equivalent to the cost of potassium permanganate. Excpet for hydrogen peroxide and paraformaldehyde, the odor control materials were applied by spray in a 1% solution. The hydrogen peroxide was applied without dilution, and the paraformaldehyde flakes were simply sprinkled on the surface of the manure. Two samples contained no treatment and one contained wet sand as a check. Application rates and costs are given in attached Table 2. The unit costs of the materials, while believed to be realistic, are subject to price fluctuations. The per acre rates were considered as one treatment unit and are so referred to in further discussion in this report. A minimum of 10 judges participated in each evaluation, and the same people were used as panelists whenever possible.

Following the placement of manure in the containers, 1/4 inch of water and one treatment unit of test materials were applied to the manure. Approximately 24 hours later, the odor panel conducted the first evaluation. After another 24-hour period, the odor panel again conducted an evaluation. Five days later, the manure in each container

was mixed and, except for the paraformaldehyde, three treatment units of test material were applied. The odor panel again conducted an evaluation one day later.

The evaluation of the slurry experiments showed that six of the seven materials suppressed the release of sulfurous compounds at all levels tested. The materials exhibited varying capabilities for suppressing the release of amines and ammonia. The data for amines and ammonia for each of the materials was plotted and the percent transmittance was linearly regressed on the amount of odor control material according to the method of least squares. The equations developed for the materials which suppressed the release of odorous gases were then plotted. The point where the line represented by the equation crossed the 90% transmittance line was the estimate of the average amount of odor control material to totally suppress the release of nitrogenous and sulfurous gases.

The panelists' evaluations for presence and offensiveness were summed and averaged in Table 3. Responses among the panelists were highly variable, however, an individual's responses as to the presence and the offensiveness of odor were generally parallel.

### Potassium Permanganate

Potassium permanganate was found to be the most economical odor control agent of the six materials considered in these tests for totally suppressing the release of the important malodorous gases. Potassium permanganate suppressed the release of sulfurous gases at all levels in the slurry experiments. The quantity of potassium permanganate required to totally suppress the release of the important malodorous gases was estimated to be 14 grams per 500 gram sample or an equivalency of 56 pounds per ton of manure. The cost for this quantity of potassium permanganate was estimated to be $86.24. While this was the least cost of any material considered in these total suppression tests, it was still judged to be excessive. Organoleptic testing showed that potassium permanganate was effective in the reduction of malodors in much lower quantities than required for their elimination. The panelists' average ratings for presence and offensiveness for the manure treated with potassium permanganate was 3.0 and 3.1. These were lower average ratings than achieved by the other test materials. Potassium permanganate was judged to be effective in the control of cattle feedlot odors when applied to the feedlot surface layer at the rate of 20 pounds per acre in a 1 percent water solution. At current prices, the cost for this application was estimated to be $30.80, which is 1.5 times the cost to achieve comparable results with Ozene.

### Potassium Nitrate

Potassium nitrate was estimated to be a much more expensive method of totally suppressing the release of the important malodorous gases than potassium permanganate. Paraformaldehyde suppressed the release of sulfurous gases at all levels in the slurry experiments. The quantity of paraformaldehyde to totally suppress the release of the important malodorous gases was estimated to be 49 grams per 500 grams or 197 pounds per ton of manure. The cost for this amount of paraformaldehyde would be $97.71, which is 1.1 times the cost to achieve equivalent results with potassium permanganate. Organoleptic evaluations were somewhat ambiguous, but did not indicate effectiveness under the conditions of the test.

### Hydrogen Peroxide

Hydrogen peroxide, 3 percent weight per volume, was estimated to be a slightly more expensive means of totally suppressing the release of the important malodorous gases than potassium permanganate. Hydrogen peroxide suppressed the release of sulfurous gases at all levels in the slurry experiments. The quantity of hydrogen peroxide to totally suppress the release of the important malodorous gases was estimated to be 82 milliliters per 500 grams, or 19.3 gallons of hydrogen peroxide per ton of manure. The cost for this amount of hydrogen peroxide would be $97.46 or 1.1 times the cost to attain equivalent results with potassium permanganate. Organoleptic evaluations were somewhat ambiguous, but did not indicate effectiveness at the rates applied in the test.

### Ozene

Ozene, the brand name of a formulation of orthodichlorobenzene, was estimated to be a slightly more expensive method of totally suppressing the re-release of the important malodorous gases than potassium permanganate. Ozene suppressed the release of sulfurous gases at all levels in he slurry experiments. The quantity of Ozene to toally suppress the release of the important malodorous gases was estimated to be 60 milliliters per 500 grams, or 28.8 gallons per tone of manure. The cost for this amount of Ozene would be $97.00, or 1.1 times the cost to attain equivalent results with potassium permanganate. Organoleptic testing showed that Ozene was effective in the reduction of malodors in much lower concentrations than required for their elimination. The panelists's average rating for presence and offensiveness of manure treated with one treatment unit of Ozene did not indicate effectiveness. The addition of a second treatment unit achieved good control. The estimated cost for two treatment units of Ozene was $20.24.

### Formula 2

Formula 2 was estimated to be a very expensive method of odor control. It suppressed the release of sulfurous compounds at all levels in the slurry experiments. The quantity of Formula 2 to totally suppress the release of the important malodorous gases was estimated to 193 milliliters per 500 grams, or 92.5 gallons of Formula 2 per ton of manure. The cost for this quantity of Formula 2 would be $550.38, which is 6.4 times the cost to achieve equivalent results with potassium permanganate. Organoleptic testing did not indicate effectiveness at the levels applied. The panelists' average rating for presence and offensiveness of manure treated with Formula 2 exceeded that for manure receiving no treatment. Panelists' responses, however, indicated a decrease in both presence and offensiveness with an increase in time and the concentration of Formula 2.


## SUMMARY

Sulfurous compounds, amines, and possibly ammonia were shown to be the important components of cattle feedlot odors. The same classes od odorous gases were found to be produced under anaerobic conditions by both liquid and solid manures. Under the restraints of this investigation good housekeeping and good drainage combined with the timely applications of odor control agents, such as Ozene or potassium permanganate, during wet periods was judged to be the least cost methods of cattle feedlot odor control.



### References

1   Allied Chemical Company. Industrial use of ozene, Form no. 52-42, Specialty Chemicals Division, Allied Chemical Company, Morristown, New Jersey. 6 pages.

2   Anthony, W. B. Cattle manure: Re-use through wastelage feeding, Animal Waste Management: Cornell University Conference on Agricultural Waste Management. Syracuse, New York: Cornell University. 105-113.

3   Burnett, W. E. and Dondero, N. C. 1970. Control of odors from animal wastes, TRANSACTIONS of the ASAE 13(2):221.

4   Deibel, R. H. 1967. Biological aspects of the animal waste disposal problem. Agriculture and the quality of our environment, ed. Nyle C. Brady. Washington, D. C.: American association for the advancement of science, Publication No. 85:395-399.

5   Faith, W. L. 1964. Odor control in cattle feed yards. Air Pollution Control Association Journal XIV 459-464.

6   FMC Corporation. Hydrogen peroxide control of sewage hydrogen sulfide. New York: FMC Corporation, Inorganic Chemicals Division (undated bulletin).

7   Grant, D. L. August 15, 1971. Odorless stockyards being eyed. Lubbock (Texas) Avalanche-Journal, p. 5 F.

8   Hammond, W. C.; Day, D. L.; and Hansen, E. L. June, 1968. "Can lime and chlorine suppress odors in liquid hog manure?" AGRICULTURAL ENGINEERING (6)49:340-343.

9   Jacobs, M. B.; Braverman, M. M.; and Hochester, Seymour. September, 1957. "Ultramicrodetermination of sulfides in air." Analytical Chemistry XXIX:1249-1351.

10   Ludington, D. C. and Sobel, A. T. September, 1970. Moisture increases manure odors, Poultry Digest. 445.

11   Merkel, J. A.; Hazen, T. E.; and Miner, J. R. 1969. Identification of gases in a swine building atmosphere. TRANSACTIONS of the ASAE 13(3):310-313, 315.

12   Nakano, N. 1968. Feeds producing odor free excreta. U.S. Patent 3 370 953.

13   Narayan, Raghu S. 1971 Identification and control of beef cattle feedlot odors. Unpublished M.S. thesis, Texas Tech University, Lubbock, Texas.

14   Seltzer, W.; Moum, S. G.; and Goldhaft, T. M. November, 1969. A method for the treatment of animal wastes to control ammonia and other odors. Poultry Science XLVII- (6):1912.

15   Sobel, A. T. 1970. Olfactory measurement of animal manure odor. ASAE Paper No. 70-417, St. Joseph, Mich. 49085.

# The Use of Dried Bacteria Cultures and Enzymes to Control Odor and Liquefy Organic Waste Found in Hog, Dairy and Poultry Producing Units as Well as Lagoons

John F. Bergdoll

FOR years I have conducted extensive tests on different dried bacteria cultures and enzyme products in an effort to find a combination product that would control ammonia and other odors produced by livestock and poultry. Time has also been spent trying to get a product and a system that would break up the solids found in lagoons and beneath slatted floor hog houses. Some of the tests have been quite discouraging; but the latest work has shown very encouraging results.

The bacteria and enzyme products tested have varied greatly in activity. One of the products, worked with initially, was shown in tests conducted by Dr. Timothy S. Chang of Michigan State University to have only 1/20th of the bacterial activity of the products I'll report on here. This was determined by using the standard bacteriological method of counting bacteria. We started with dilution by weight of the products tested and then used the pour plate method to make the count for bacterial activity.

Four separate tests were conducted in different areas with dairy cattle waste, pig waste in deep pit hog houses, laying hen manure, and a lagoon at a by-product rendering plant, respectively.

## Test No. 1:

Testing was run on the manure produced by approximately 550 holstein dairy cows located in a residential area of a large West Coast city. The cows were being kept in a dry lot with a concrete floor. The entire lot was surrounded by middle class homes, and the only thing separating the homes from the cattle lot was a 5 foot high concrete block wall. The nearest homes were within 40 feet of the concrete block wall. The dairyman had received numerous complaints about odors and fly problems. Because of our experience with this type situation, we sprayed 2 oz of Big Dutchman Improved Odor Control* for every 1000 square feet of the lot on a weekly basis for six weeks. After the treatments there was a marked reduction in objectionable odor; and there was not one complaint from neighbors. The manure became much drier after spraying. In fact, before treatment the dry lot was usually damp, but after treatment it became dusty at times. There was also a great reduction in the number of flies in the lot. This was in late August, a time when fly problems might be expected. The owner of the dairy stated about 4 months after testing, "Find me a fly now". After 2 months of spraying, or 8 treatments, we reduced the material to 1 ounce per 1000 square feet. The material was applied with a boom spray on the back of a pickup truck.

Results from this test, and succeeding tests with dry lot dairy cattle, let us conclude that offensive odors can be kept to an absolute minimum and fly problems will be greatly lessened when the treatment is applied regularly and properly. The odor control product used in this test is being mass produced and is stable so far as strength and activity is concerned.

## Test No. 2:

This test was conducted at a Research Farm of a major feed manufacturer where two identical slatted floor hog houses had pits and the level of waste was kept about 3 feet deep. There was agitation by a Fairfield Aeroba-Jet in both pits, and the waste from both hog houses was allowed to overflow continually into a small lagoon. One house was treated and the other used as control; each house was identical in size and each contained 500 hogs of about equal size. Chemical tests were conducted each week, odors were checked, and the color of liquid collected from the pits.

A bacteria and enzyme mixture† designed primarily to control odors was sprayed on the walls, ceiling, and pigs at a rate of 1 ounce material for each 1000 square feet (i.e., 1 ounce material in 2 gallons water). This initial treatment did get some reduction in odor. In the beginning odor levels were determined by a panel of people; later instrumentation was used to determine the ammonia levels. Then a liquid waste control product designed to liquefy waste in deep pits was applied at the rate of 1 pound for each 6000 gallons of waste in the pit.

Before the tests there was about 6 to 10 inches of solids on the pit bottom in each house. After the initial treatment, we sprayed twice a week using 3 1/2 pounds of material for each 60 000 pounds of live animal weight contributing waste to the pit. Effort was made to distribute the material very evenly over the entire surface of the pit. We have learned it is extremely important for best results that no corners or edges of the pit be missed, and that agitation and/or aeration of any type will speed up the reaction. Twenty-four hours after initial treatment, there was a great amount of foaming and general activity in the pit. The odors in the treated house became much more severe for 3 to 5 days, which we expected since the bacteria was loosening the solids on the pit bottom. About 3 weeks after treatment, inspection of the pit showed very little solid on the pit, and that was only along the edges of the wall and in the corners. Temperature in the treated pit averaged 2 deg to 4 deg higher during testing than in the control pit.

The actual chemical results are presented in Table 1.

The analysis was run only on the liquid. Conducting this test again, it may be more accurate to take a cylinder straight down in the pit of the control house to bring up

---

The author is: JOHN F. BERGDOLL, Indianapolis, IN.

*This product now being marketed in the USA by Central Soya, Inc., Fort Wayne, IN, as odor control.

**TABLE 1. ACTUAL CHEMICAL RESULTS FROM TEST NO. 2.**

| | Treated pit | | Control pit | |
| --- | --- | --- | --- | --- |
| | High | Low | High | Low |
| Suspended solids total mg/L | 8 660 | 1 100 | 27 940 | 740 |
| 5 day BOD mg/L | 5 100 | 981 | 3 952 | 240 |
| Chemical oxygen demand mg/L | 16 563 | 3 280 | 17 143 | 2 143 |
| Average pH | 7.4 | | 8.0 | |

| Aerated lagoon | Test start | Test end |
| --- | --- | --- |
| 5 day BOD mg/L | 135 | 56 |
| Suspended solids mg/L | 1 640 | 320 |
| Chemical oxygen demand mg/L | 3 389 | 543 |

the solids from the bottom for analysis.

It was concluded that proper application of the enzyme bacteria product in a hog pit will bring liquefication; and that aeration or agitation, or a combination of both will greatly increase the rapidity of this process. As shown by chemical tests, this has a very beneficial effect on the lagoon in which the waste is being pumped. There was a definite change in color of the material in the pit and the lagoon. Approximately 2 weeks after treatment the color was much darker in the pit and as time went on the color became lighter and acquired a slight purple tint. Odors were reduced but not eliminated, particularly in the farm lagoon. In other tests where the manure from a deep pit hog house was pumped into honeywagons and spread on fields, there was a marked reduction in odor of the material being spread.

## Test No. 3:

This test was conducted in a cage layer house where birds were kept in double deck wire cages. The ventilation was forced air, and mechanical pit cleaners were employed. This house was chosen because it always had high ammonia levels.

The initial application was 1 oz bacteria product soaked in 2 gallons of water for about 2 hours, then sprayed over the droppings of each 1000 laying hens. The same treatment was applied 7 days later. The first application included spraying the entire walls, ceilings, birds, cages and droppings.

The following ammonia level results were obtained from readings taken in several places, and are averaged:

| | |
| --- | --- |
| 249 ppm | Before treatment |
| 100 ppm | 1 hr after 1st treatment |
| 45 ppm | 3 hr after 1st treatment |
| 55 ppm | 7 days after 1st treatment |
| | - - - Second Treatment applied - - - |
| 30 ppm | 3 hr after 2nd treatment |
| 20 ppm | 7 days after 2nd treatment |
| 6 ppm | 14 days after 2nd treatment |

ppm=parts per million

All readings were taken with a Model No. Th 20501 ammonia meter manufactured by Draegerwerk, Luebeck, Germany. The ammonia level stabilized at 6 ppm. Theoretically, one would suspect the bacteria could sustain their levels and that additional applications would not be necessary. However, after 2 weeks without additional spraying, the ammonia levels rose rather dramatically, and after 3 weeks the ammonia levels rose to 50-75 ppm. The temperature and humidity in the buildings was a factor. We noted that when the room temperature was high (80-95 F) and the humidity above 70 percent, the ammonia levels and odors in general tended to be higher. We continued this program for several months. Good control of ammonia odors from cage layer droppings was achieved with proper treatment. Further tests showed that spraying about every 7 days gave the best results, and that spraying walls, ceilings, birds, and cages with the initial application was important to get the quick knockdown of ammonia level.

## Test No. 4:

This test was conducted at the world's largest poultry by-product rendering plant located in Georgia, which produces several hundred tons per day of poultry by-product meal and feather meal. Their waste management program incorporated the use of 5 different connected lagoons. Aerators were used in the primary lagoon. Approximately 3000 gallons of water per minute is used; and because of the volume, the water is taken from the fifth lagoon and reused back through the plant. They incorporate a so-called closed system whereby under normal operation, there would be no effluent escaping from any of the lagoons. Despite a good amount of aeration in the first lagoon, a buildup of fat and blood had reached the point where approximately 2/3 of the surface was covered with this buildup. The crust ranged in depth from less than 1 inch to as much as 7 inches. Our objective was to see whether the bacteria product would break up or consume the organic waste found on top of the water.

One pound of Big Dutchman Improved Liquid Waste Control marketed by Central Soya Inc. of Fort Wayne, Indiana) was added to each 10 000 gallons of water in the lagoon. Approximately 600 pounds of material was spread from a motorboat, the entire lagoon surface was covered. One week later an additional 200 pounds of material was added via the recirculating water that is run through the plant. In about 2 weeks the lagoon changed color completely and became quite purple. (This color change has been reported also by Michigan State researchers, Ted Loudon and John Gerrish, in their work with a swine lagoon.) Over 1/2 of the surface was at this time completely free of any buildup. After 4 weeks approximately 90 percent of the surface of the lagoon was free of any buildup. Additionally, within the plant's recirculation system there was a noticeable decrease in the buildup of organic waste.

This test began in September when the average daily temperature ranged from about 50 deg to 85 deg. It is safe to assume that under such conditions a bacteria product properly applied can greatly reduce organic buildup and that the BOD level in the primary lagoon of a rendering plant will be greatly reduced.

The following conclusions were drawn from all of these tests:

1    The proper distribution of a bacteria product is extremely important whether spraying the product directly on manure, spraying droppings beneath cages, or applying the material to the lagoon surface. All areas must be covered evenly.

2    Water containing chlorine levels of over 3 to 5 ppm will have a detrimental effect on the bacteria effectiveness.

---

†Big Dutchman improved liquid waste control now being marketed in the US by Central Soya Inc., Fort Wayne, IN, as liquid waste control.

MANAGING LIVESTOCK WASTES

# Odor Control of Liquid Dairy and Swine Manure Using Chemical and Biological Treatments

C. A. Cole, H. D. Bartlett, D. H. Buckner, D. E. Younkin

MEMBER
ASAE

THERE are many livestock operations that manage animal wastes as a liquid, storing it in pits until it is convenient to dispose on land. The pits are continuously anaerobic and biological decompositions produce foul decomposition products. It requires only trace amounts of some of these materials to surround exposed waste with foul manure odors for hundreds of yards.

The objective of this research was to comprehensively investigate biological and chemical treatments that could be used to remove the odorous materials from dairy and swine pits. Treatment of the manure pit contents to either prevent formation of odorous materials, or destroy them will effectively eliminate the airborne odor. The short term destruction of the odorous materials is often accomplished by oxidation or alternately the odorous gases may be adsorbed into a material such as activated carbon. However, if the biological decomposition is varied to result in different end products, the odor problem can be controlled. The commercial dried bacteria and enzymes tested in this work were claimed to control odor over a long time because of their specifically cultured properties. Nitrate supplied a hydrogen acceptor so that the system redox potential should not have permitted production of the odorous materials. Orthodichlorobenzene containing preparations and formaldehyde act as a bacteriacide and/or emulsifier.

Limited work has been previously published in the area of chemical treatments to reduce odors. Yushok and Bear (1948) reported that hydrated lime deodorized poultry manure and reduced nitrogen losss. Hammond, Day and Hansen (1968) and Deibel (1967) verified this deodorization. Benham (1967) claimed similar results for NaOH treatment of poultry droppings in slurry pits.

Faith (1964) and Emmanual (1965) reported that the oxidant, $KMnO_4$, was effective for reducing cattle manure odors. Hammond (1968) also using chemical oxidants reported that $Cl_2$ treatment of swine waste reduced $H_2S$, $NH_3$ and $CH_4$ production. However, Deibel (1967) reported that $O_3$ as well as proprietary bacterial compounds failed to control odors.

Kibbel et al. (1972) reported on the successful use of $H_2O_2$ for manure odors. Collins and Eastburn (1973) concluded that $H_2O_2$ effectively eliminated $H_2S$ in dairy manure and also that odor muskants were beneficial for odor control. Burnett and Gormel (1969) also reported that commercial masking agents were effective for odor control in liquid chicken manure pits. Burnett and Dondero (1970) found masking agents to be the most effective and digestive enzymes to be the least effective of 40 commercial materials evaluated for chicken manure. Seltzer et al. (1969) reported that flake paraformaldehyde resulted in reduced $NH_3$ and noxious gas formation.

## EXPERIMENTAL PROCEDURES

Short term controled experiments were made using pit contents in a bench scale jar test apparatus followed by full scale demonstrations. The bench scale jar tests were made using one liter samples of the swine manure pit contents placed in the beakers and the chemicals (1 percent solution) added and stirred to insure complete mixing. Analyses were made to determine soluble ammonia, soluble sulfide, pH and odor quantity and quality. Sulfide and ammonia were measured since they were known to contribute to odor, especially sulfide. (The analyses were performed using specific ion electrodes.) Odor was evaluated by an odor panel rating each treated sample and control sample for quantity and quality of odor on a zero to ten scale with ten being very strong and very obnoxious, and zero being odor-free and pleasant (Collins and Eastborn 1973). The panel consisted of at least five people that were chosen randomly. The arithmetic mean which was very close to the geometric mean was calculated for each sample for odor quantity and quality.

Experiments using pit contents to evaluate long term methods of preventing odor formation (dried bacteria, enzymes, orthodichlorobenzene formaldehyde or sodium nitrate) were made in 208 liter simulated pits. One hundred and thirteen liters of actual pit contents was initially added to each simulated pit. Some vessels were treated and others were used as controls. Approximately four liters of fresh manure was added to each simulated pit weekly to approximate actual conditions. The treatments using the proprietary compounds were made according to the manufacturers' recommendations for the swine manure but doubled for the dairy manure. Analyses for ammonia, sulfide, pH, temperature, dissolved oxygen and odor were made weekly on the drum contents. The odor was compared weekly using an odor panel to rate a sample from each treatment as to odor quantity and quality similar to that for the short term tests. Dissolved oxygen was measured using a membrane probe instrument.

## SWINE MANURE RESULTS

### Short Term

Numerous short term chemical treatments were screened and the oxidants sodium hypochlorite (NaCl), hydrogen peroxide, $(H_2O_2)$ chlorine dioxide $(ClO_2)$ and potassium permanganate $(KMnO_4)$ and the adsorbant, powdered activated carbon, were selected as showing promise and investigated further as reported here.

The authors are: C.A. COLE, Associate Professor, H.D. BARTLETT, Professor, D. H. BUCKNER, Research Assistant, and D. E. YOUNKIN, Professor, Pennsylvania State University, University Park.

Authorized for publication as paper no. 4872 in the Journal Series of the Pennsylvania Agricultural Experiment Station.

TABLE 1. SULFIDE CONCENTRATION IN SWINE PIT WASTE.

| Dose, mg/l | Sulfide Concentrations, mg/l | | | | | | | | |
|---|---|---|---|---|---|---|---|---|---|
|  | 100 | 300 | 500 | 500 | 1000 | 2000 | 5000 | 2000 | 5000 |
| $H_2O_2$ | 4 | <1 | <1 | 6 | 4 | - | - | - | - |
| NaOCl | 16 | <1 | <1 | 8 | 5 | - | - | - | - |
| $ClO_2$ | 4 | <1 | <1 | 6 | 3 | - | - | - | - |
| $KMnO_4$ | 16 | <1 | <1 | 28 | 12 | - | - | - | - |
| Act carbon | - | - | - | - | - | 40 | 1 | 25 | 29 |
| Control | 120 | 120 | 45 | 100 | 105 | 120 | 45 | 100 | 105 |
| Manure solids | 1-2% | 1-2% | 1-2% | 5-7% | 5-7% | 1-2% | 1-2% | 5-7% | 5-7% |

The sulfide concentrations are shown in Table 1 for swine manure solids concentrations of 1-2 percent and 5-7 percent and varying dosages of chemical treatments. Pit contents with only 1-2 percent total solids had sulfide concentrations less than 1 ppm with a dose of oxidant of 300 or 500 mg/l. Activated carbon however, only reduced the sulfide concentration to 1.5 mg/l even at a dose of 5000 mg/l with 1-2 percent solids. It was not possible to reduce sulfide nearly as much at manure pit solids concentrations of 5-7 percent with the oxidants or activated carbon. Doses of hydrogen peroxide, sodium hypochlorite and chlorine dioxide at 500 or 1000 mg/l appeared the most effective and did reduce the sulfides to levels under 10 mg/l, i.e., 90 + percent removal.

Results of experiments measuring sulfide concentration as a function of time showed that the reaction had stopped after 15 minutes time lapse and no further reduction in sulfide level occurred with any of the oxidants or activated carbon.

None of the chemical or physical treatments had any significant effect on the ammonia concentration (generally greater than 1000 ppm). However, it was felt that the ammonia odor was more acceptable than the other odors.

The results of odor measurement are the proof for any successful odor control, Table 2. The most effective treatment was chlorine dioxide at all comparable concentrations. However, all the treatments reduced the odor strength to the definite or lower range from very strong. It was seen that the 5-7 percent solids had more residual odor than the 1-2 percent in all cases. The reduction of odor from very strong to faint to definite was considered substantial. It should be noted that activated carbon was found to be very difficult to handle because of dusting problems.

Field trials were made using swine pit contents and an 800 gallon liquid manure wagon. Observors rated plots treated with sodium hypochlorite, potassium permanganate and hydrogen peroxide at 500 and 1000 ppm dose one-half hour after treatment. Definite improvement in the odor was noted with some individuals preferring one treatment over another. However, it should be noted that the treatments did not eliminate the odor but did reduce it.

The cost of the short term treatments to the swine operator was calculated on a dose that was found to cause a substantial reduction in odor strength and sulfide concentration using current chemical prices. The cost of hydrogen peroxide, potassium permanganate, sodium hypochlorite and chlorine dioxide for a 500 ppm dose was $2.58, $3.12, $4.98 and $72.70 per thousand gallons treated. The cost for 5000 ppm activated carbon was $12.35 per thousand gallons treated. It therefore appeared that the cheapest effective treatments were hydrogen peroxide and potassium permanganate for a substantial odor reduction.

**Long Term Control**

Sulfide level in the simulated pit contents was monitored as a quantitative indicator of the odor producing capacity of the waste. The results for the tested materials were compared with a stirred and unstirred control barrel to which no treatment was applied, Table 3. It can be seen that there is a substantial and consistent reduction in sulfide in the stirred control barrel over the quiescent barrel so there is an apparent advantage of agitating the pit contents periodically in order to prevent sulfide buildup to extremely high levels. This is similar to the conclusions of Breisler (1971). Comparison of the various treatments including dry bacteria, enzymes and $NaNO_3$ showed only the $NaNO_3$ to have a

TABLE 2. ODOR PANEL RESULTS FOR SWINE PIT WASTE.

| Dose, mg/l | Level of Odor Strength* | | | | | | | | |
|---|---|---|---|---|---|---|---|---|---|
|  | 100 | 300 | 500 | 500 | 1000 | 2000 | 5000 | 2000 | 5000 |
| $H_2O_2$ | 5.6 | 5.1 | 4.6 | 5.2 | 4.0 | - | - | - | - |
| HaOCl | 6.8 | 4.6 | 3.8 | 5.2 | 5.8 | - | - | - | - |
| $ClO_2$ | 5.1 | 4.0 | 2.4 | 3.7 | 5.2 | - | - | - | - |
| $KMnO_4$ | 5.6 | 4.3 | 3.3 | 5.2 | 5.2 | - | - | - | - |
| Act carbon | - | - | - | - | - | 6.0 | 5.3 | - | 6.5 |
| Control | 10 | 10 | 10 | 10 | 10 | 10 | 10 | 10 | 10 |
| Manure solids | 1-2% | 1-2% | 1-2% | 5-7% | 5-7% | 1-2% | 1-2% | 5-7% | 5-7% |

*Zero, no odor; two, very faint odor; four, faint odor; six, definite odor; eight, strong odor; ten, very strong odor

TABLE 3. LONG TERM CONTROL SULFIDE CONCENTRATION
FOR SWINE PIT WASTE.

| Treatment | Sulfide concentration, mg/l | | | | | | | |
|---|---|---|---|---|---|---|---|---|
| | Time, days | | | | | | | |
| | 7 | 14 | 21 | 28 | 35 | 42 | 56 | 63 |
| Dried enzyme, A | 85 | 66 | 62 | 48 | 35 | 18 | 76 | 37 |
| Dried bacteria, B | 70 | 130 | 66 | 72 | 50 | 38 | 54 | 89 |
| Dried enzyme, C | 19 | 40 | 32 | 54 | 46 | 19 | 68 | 70 |
| Disinfectant, E | 34 | 35 | 42 | 51 | 68 | 40 | 34 | 29 |
| Disinfectant, F | 37 | 25 | 27 | 27 | 35 | 34 | 42 | 37 |
| NaNO$_3$ | 19 | 5 | 3 | 21 | 19 | 17 | 3 | 3 |
| Stirred control | 55 | 54 | 45 | 78 | 49 | 32 | 32 | 40 |
| Unstirred control | 70 | 62 | 70 | 150 | 170 | 135 | 330 | 93 |

consistent reduction in sulfide below that of the stirred control.

Table 4 shows the results of odor strength measurement. The treatments and stirred control were compared for odor against the unstirred control. The arithmetic mean of all the results is about 6, definite, so there is no substantial difference among the treatments and the stirred control. The spread of the individual results of the panel was very high with some people disliking the NaNO$_3$ treatment odor and others finding it lessened and acceptable.

The treated drums at the end of the test (5-7 percent total solids) were spread on a recently cut hay field. A panel compared the various treatments and controls and concluded that only NaNO$_3$ treated waste had substantially less odor. The experiments did not substantiate the claims made for the proprietary materials for odor control.

TABLE 4. LONG TERM ODOR CONTROL FOR SWINE
PIT WASTE.

| Treatment | Odor strength* | | | | | | | |
|---|---|---|---|---|---|---|---|---|
| | Time, days | | | | | | | |
| | 7 | 14 | 21 | 28 | 35 | 42 | 56 | 63 |
| Dried enzyme A | 7.1 | 6.7 | 7.4 | 6.7 | 6.2 | 6.6 | 6.9 | 7.7 |
| Dried bacteria B | 7.1 | 7.0 | 6.3 | 5.3 | 6.2 | 7.0 | 7.6 | 7.5 |
| Dried enzyme C | 6.2 | 6.1 | 5.2 | 5.0 | 5.5 | 5.9 | 7.5 | 7.5 |
| Disinfectant E | 6.4 | 6.5 | 6.2 | 7.3 | 6.6 | 6.6 | 6.1 | 5.5 |
| Disinfectant F | 6.0 | 5.5 | 5.0 | 5.2 | 6.0 | 5.2 | 5.4 | 6.1 |
| NaNO$_3$ | 7.0 | 50. | 8.5 | 8.2 | 8.9 | 7.1 | 6.6 | 6.6 |
| Stirred control | 6.0 | 5.1 | 6.9 | 6.0 | 5.7 | 5.6 | 6.1 | 7.7 |
| Unstirred control | 10.0 | 10.0 | 10.0 | 10.0 | 10.0 | 10.0 | 10.0 | 10.0 |

*See Table 2 footnote

## DAIRY MANURE RESULTS

Short term chemical treatments similar to the swine manure tests were used on dairy manure pit contents. The solids of the dairy manure was between 11 and 15 percent total solids, the initial sulfide concentration was only in the 40-50 mg/l range, about one-half that of swine manure, and the odor was considered to be much less strong and offensive. For example, an experiment comparing untreated swine and dairy manure pit contents found mean odor numbers of 8.0 (strong) versus 4.8 (faint to definite) respectively. Sulfide levels were reduced only by about one-half by the chemical treatments and no substantial improvement in odor was found, Tables 5 and 6. Field trials verified these results.

TABLE 5. SULFIDE RESIDUAL IN DAIRY PIT
WASTE AFTER TREATMENT.

| Treatment | Sulfide concentration, mg/l | | | | | |
|---|---|---|---|---|---|---|
| | Dose, mg/l | | | | | |
| | 250 | 500 | 1000 | 2000 | 5000 | 10 000 |
| H$_2$O$_2$ | 31 | 23 | 26 | - | - | - |
| NaOCl | 28 | 30 | 33 | - | - | - |
| ClO$_2$ | 19 | 13 | 17 | - | - | - |
| KMnO$_4$ | 35 | 26 | 28 | - | - | - |
| Act. carbon | - | - | - | 38 | 27 | 30 |
| Control | 45 | 39 | 50 | 45 | 39 | 50 |

TABLE 6. ODOR PANEL RESULTS FOR DAIRY PIT
WASTE AFTER TREATMENT.

| Treatment | Level of odor strength* | | | | | |
|---|---|---|---|---|---|---|
| | Dose, mg/l | | | | | |
| | 250 | 500 | 1000 | 2000 | 5000 | 10 000 |
| H$_2$O$_2$ | 7.0 | 7.2 | 4.6 | - | - | - |
| NaOCl | 5.4 | 5.3 | 4.5 | - | - | - |
| ClO$_2$ | 4.9 | 4.3 | 4.6 | - | - | - |
| KMnO$_4$ | 5.7 | 4.9 | 5.0 | - | - | - |
| Act. carbon | - | - | - | 5.0 | 4.1 | 3.5 |
| Control† | ------------------ taken as 5.0 ------------------ | | | | | |

*See Table 2 footnote
†See explanation in text

The dried bacteria, enzymes, orthodichlorobenzene containing material, formaldehyde and sodium nitrate were evaluated for the long term control of dairy manure odor as was evaluated for swine manure, Tables 7 and 8. No consist-

TABLE 7. LONG TERM CONTROL
SULFIDE CONCENTRATION FOR
DAIRY PIT WASTE.

| Treatment | Sulfide concentration, mg/l | | | |
|---|---|---|---|---|
| | Time, days | | | |
| | 14 | 28 | 42 | 56 |
| Dried enzyme, A | 50 | 75 | 58 | 34 |
| Dried bacteria, B | 50 | - | 54 | 58 |
| Dried enzyme, C | 74 | 60 | 32 | 54 |
| Disinfectant, F | 50 | 42 | 43 | 42 |
| Formaldyhyde | 20 | 39 | 17 | 32 |
| NaNO$_3$ | <1 | 17 | 24 | 4 |
| Stirred control | 62 | 90 | 100 | 52 |
| Unstirred control | 44 | 54 | 50 | 44 |

TABLE 8. LONG TERM ODOR CONTROL FOR DAIRY
PIT WASTE.

| Treatment | Odor strength* | | | | | | |
|---|---|---|---|---|---|---|---|
| | Time, days | | | | | | |
| | 14 | 21 | 28 | 35 | 42 | 49 | 56 |
| Dried enzyme, A | 8.7 | 7.0 | 8.5 | 4.5 | 6.2 | 4.1 | 7.9 |
| Dried bacteria B | 7.5 | 7.1 | 7.5 | 5.9 | 6.5 | 6.8 | 9.0 |
| Dried enzyme, C | 8.5 | 7.5 | 7.9 | 7.2 | 8.7 | 8.2 | 8.4 |
| Disinfectant, F | 8.6 | 6.1 | 8.0 | 8.3 | 6.7 | 7.7 | 7.9 |
| Formaldehyde | 7.2 | 6.8 | 6.6 | 6.8 | 6.7 | 7.0 | 7.9 |
| NaNO$_3$ | 6.2 | 6.0 | 8.3 | 6.8 | 5.8 | 5.5 | 5.4 |
| Stirred control | 8.6 | 7.6 | 7.9 | 7.2 | 6.8 | 6.5 | 7.8 |

*See Table 2 footnote



MANAGING LIVESTOCK WASTES

ent substantial improvement was found for any of the materials except sodium nitrate where variable results were again noted. The sulfide level was very low for $NaNO_3$ treatment but odors were not substantially improved.

## SUMMARY

Short term control experiments indicated that hydrogen peroxide, sodium hypochlorite, chlorine dioxide and potassium permanganate dosed at 500 ppm greatly reduced sulfide and odor level in liquid swine manure; however, activated carbon dosed at even 5000 ppm was not as effective. The most cost effective material was hydrogen peroxide. Long term control experiments with swine manure showed that dried enzymes, dried bacteria and orthodichlorobenzene were not effective for reducing odor or sulfide levels. Short term odor control with liquid dairy manure was not successful. The odor of liquid dairy manure was judged by the panel to be much less strong and offensive than swine manure and a substantial reduction by treatment could not be found. Long term control experiments on dairy manure showed that the dried enzymes, dried bacteria and orthodichlorobenzene were again not effective for reducing odor. Sodium nitrate was found to change the odor, reduce sulfide level and cause suspended solids to float for both swine and dairy manures. Formaldehyde appeared to slightly reduce sulfide level but did not reduce the odor.


## References

1   Benham, C. L. 1967. The effectiveness of caustic soda in preventing the development of smell in poultry droppings slurry pits. Report Nat. Agricultural Advisory Service. Shardlow Hall, Shardlow, Derby England.

2   Breisler, G. O. 1971. Why stirring manure reduces odor. Poultry Digest. February. 60 p.

3   Burnett, W. E. and N. C. Dondero. 1970. Control of odors from animal wastes, TRANSACTIONS of the ASAE 13(2):221-224.

4   Burnett, W. E. and B. Gormel. 1969. Odor control by chemical treatment. Gases and particulate matter from high density poultry management systems as they relate to air pollution. Contract C1101. New York Dept. of Health.

5   Collins, N. E. and R. P. Eastburn. 1973. Chemical treatment to reduce the offensiveness of liquid manure odors during handling. ASAE Meeting. North Atlantic Region. Orono, Maine.

6   Deibel, R. H. 1967. Biological aspects of the animal waste disposal problem. p. 395-399. In: N. C. Brady, (ed.) Agriculture and the quality of our environment. AAAS. Pub No. 85 Washington, D. C.

7   Emanuel, A. G. 1965. Potassium permanganate offers new solutions to air pollution control. Air Engineering, September, 19-21.

8   Faith, W. L. 1964. Odor control in cattle feed yards. JAPCA 14(11):459-460.

9   Hammond, W. C., D. L. Day and E. L. Hansen. 1968. Can lime and chlorine suppress odors in liquid hog manure? AGRICULTURAL ENGINEERING 49(6):340-343.

10   Kebbel, Jr., W. H., C. W. Raleigh and J. A. Shepherd. 1972. Hydrogen peroxide for industrial pollution control. 27th Purdue Industrial Waste Conference. Lafayette, Ind. p. 824-829.

11   Seltzer, W., S. G. Moum and T. M. Goldshaft. 1969. A method for treatment of animal wastes to control ammonia and other odors. Poultry Science 48(6):1912.

12   Yushok, W. and F. E. Bear. 1948. Poultry manure — its preservation, deodorization and disinfection. N. J. Ag. Expt. Station. Bull. No. 707. New Brunswick, N. J.


---

## Dried Bacteria to Control Odor

*(Continued from page 373)*

(This was determined in the laboratory by a Quality Control Bacteriologist employed by the basic manufacturer of the bacteria used in the Liquid Waste product.)

3   It is extremely important that all spreading and/or spraying equipment be free from all types of disinfectants, insecticides and other harmful chemicals.

4   When applying the bacteria product to a pit or lagoon, the effectiveness of the bacteria can be greatly accelerated by use of an aerator or some type of agitation. Temperature levels are also critical, particularly when the product is used in a shallow lagoon or pit.

5   When mixed, the bacteria product should not be left in a sprayer or other distributing equipment for many hours or the bacteria count will be greatly reduced.

6   The first few days after treatment of a badly contaminated lagoon or deep pit (particularly true in a deep pit hog house), odors and ammonia level will be higher because the bacteria will be breaking up the accumulation of organic matter on the pit or lagoon bottom. This is not the case with a dry lot cattle operation or droppings beneath poultry cages.

7   Because the bacteria are grown on wheat bran and even though it is extremely fine after processing, some slight problems may develop when spraying the material through some types of sprayers. The reason is that when the bran is wet it will swell in size and some clogging may result. This is why it is important that the material be sprayed immediately after mixing.

From these tests we find that odors and harmful gases can be greatly reduced in hog houses, poultry houses, dairy barns, rendering plants, etc. by the proper use of the right bacteria enzyme product. The total volume of manure can also be reduced by at least one-third and in some cases even more. This was determined by spraying the manure under every other row of cages in a large cage layer house. The height of the manure beneath the cages of both the treated and untreated rows was measured once a week. The manure in a set area was also weighed. If the treatment is started before the fly breeding season begins, fly problems can be greatly reduced.

The reduction in the loss of nitrogen found in the treated manure greatly enhances the fertilizer value. In fact, some tests have shown that the fertilizer value will be enough greater in nitrogen level to more than pay for the cost of material and labor required to apply the product. In all cases the caretakers and operators felt that working conditions were greatly improved with the use of the bacteria enzyme product.

# Management of Odors Associated with Livestock Production


J. Ronald Miner

MEMBER
ASAE


ANIMAL waste management research has been encouraged by a variety of environmental pressures and public concern about the impact of livestock production on environmental quality. Advances in livestock pollution control technology have been accompanied by demands for increasingly stringent environmental protection. Technologies have evolved making water quality protection an achievable goal. Further work is under way to more effectively utilize livestock waste constituents and to devise systems more compatible with sophisticated livestock production techniques.

The discharge of volatile organic compounds into the air from livestock production systems presents a current challenge to livestock production enterprises. Conflicts over the release of volatile organic compounds into the atmosphere have been based predominantly upon odor complaints from nearby residents, commercial operations, and recreational interests. More recently, odor concern has been supplemented by allegations that ammonia and other water pollutants discharged from livestock operations into the air are being transported and later absorbed by nearby surface waters. Increased ammonia levels in the atmosphere near livestock feeding operations have been documented by Hutchinson and Viets (1969) and by Luebs, Laag, and Davis (1973).

Conflicts between livestock producers and the public concerning odor complaints have been documented (Willrich and Miner 1971). These conflicts are now being reflected in rules and regulations designed to protect the public from malodors generated by livestock production enterprises. These rules and regulations are being applied as additional restrictions on the location, design, and operation of commercial livestock and poultry production enterprises (Sweeten 1974).

The most obvious fact concerning livestock production and odor release is the complexity of the release mechanism, transport system, and receptor reaction. The evaluation of odors is a personal response based upon the sensitivity of the person involved, his previous experience, and momentary disposition. Research has identified more than forty organic compounds in the air near manure storage or treatment devices. See Table 1 for a summary of compounds previously identified. Many of these compounds are known to be odorous in trace concentrations. The levels at which some of these compounds are detectable by the human nose are shown in Table 2. Most quantitative measurements of odorant concentrations suggest, however, that the odor perceived from livestock production enterprises is a result of the mixture of odorous compounds, since all of the measured compounds are present in concentrations below their threshold.


The author is J. RONALD MINER, Associate Professor, Agricultural Engineering Department, Oregon State University. Technical Paper No. 3934. Oregon Agricultural Experiment Station. Financial support was provided in part by Grant No. S-802009, Office of Research and Monitoring, U. S. Environmental Protection Agency.


## MEASUREMENT OF ODORS

Various schemes have been proposed for the measurement of odors. The concentration of the individually odorous compounds has been used by some investigators as an

**TABLE 1. COMPOUNDS IDENTIFIED IN THE AIR FROM THE ANAEROBIC DECOMPOSITION OF LIVESTOCK AND POULTRY MANURE**

| Alcohols | Acids | Amines | Fixed gases |
|---|---|---|---|
| Methanol | Butyric | Methylamine | Carbon dioxide |
| Ethanol | Acetic | Ethylamine | Methane |
| 2-Propanol | Propionic | Trimethylamine | Ammonia |
| n-Propanol | iso-Butyric | Triethylamine | Hydrogen sulfide |
| n-Butanol | iso-Valeric | | |
| iso-Butanol | | | |
| iso-Pentanol | | | |

| Carbonyls | Esters | Sulfides |
|---|---|---|
| Acetaldehyde | Methyl formate | Dimethyl sulfide |
| Propronaldehyde | Methyl acetate | Diethyl sulfide |
| iso-Butyraldehyde | iso-Propyl acetate | Disulfides |
| Hexanal | iso-Butyl acetate | |
| Acetone | iso-Propyl propionate | Mercaptans |
| 3-Pentanone | Propyl acetate | |
| Formaldehyde | n-Butyl acetate | Methyl mercaptan |
| Heptaldehyde | | Nitrogen heterocycles |
| Valeraldehyde | | |
| Octaldehyde | | Indole |
| Decaldehyde | | Skatole |

Source: J. R. Miner. Odors from confined livestock production. Environmental protection agency technology series, EPA-660/2-74-023. 125 pp. 1974

**TABLE 2. THRESHOLD LIMIT VALUES FOR VARIOUS GASES ASSOCIATED WITH ANIMAL WASTE ODORS**

| Substance | Concentration in air, $10^{-9}$ g/l |
|---|---|
| Acetaldehyde | 360 |
| Acetic acid | 25 |
| Ammonia | 35 |
| n-Butyl acetate | 710 |
| Butyl mercaptan | 35 |
| Diethylamine | 75 |
| Dimethylamine | 18 |
| Ethylamine | 18 |
| Ethyl mercaptan | 25 |
| Isopropylamine | 12 |
| Methylamine | 12 |
| Methyl mercaptan | 20 |
| Triethylamine | 100 |

Notes: The Threshold Limit Values refer to airborne concentrations under which it is believed that nearly all workers may be repeatedly exposed without adverse effect.

Source: American conference of governmental industrial hygienists. Threshold limit values for 1967.

indication of odor transport and odor pollution. Ammonia and hydrogen sulfide have been widely used for this application because of their ease of measurement and known odorous characteristics. In the concentrations measured, however, they would not be detectable by the human nose were they not accompanied by other odorous compounds.

The measurement of odor intensity based upon the number of dilutions required to reduce the concentration to a barely detectable level has been the most generally accepted method for evaluating odor concentrations. The Scentometer as distributed by Barneby-Cheney and described by Rowe (1963) has been used by researchers (Sweeten et al. 1973) and more recently, regulatory agencies (Sweeten et al. 1974) to evaluate odor control techniques and odor transport. The Scentometer as shown schematically in Fig. 1 is a device for dilution of odorous air with odor-free air to determine the number of dilutions necessary to reach the barely detectable level. Similar measurements have also been made in the laboratory using more sophisticated devices (Moncrieff 1961) and in a mobile unit described by Lindvall et al. (1974).

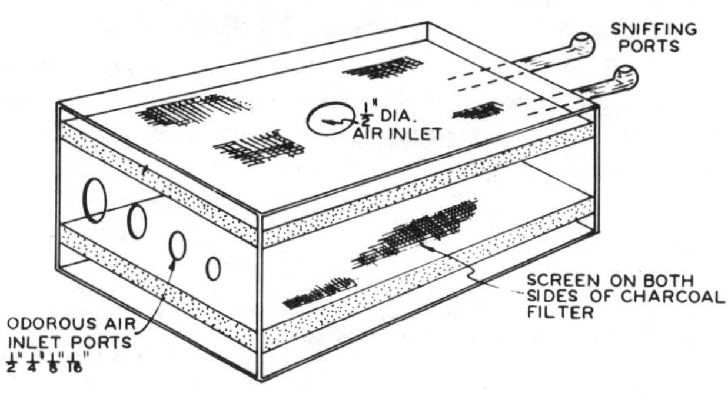


FIG. 1 Sketch of Scentometer.

## ODOR SOURCES

To avoid odor complaints and to minimize the escape of potential water pollutants to the air, livestock producers are faced with the need to control the evolution of organic compounds from manure treatment processes. These gases may arise from several sources; however, they may be grouped in terms of feed materials, fresh manure, and stored or decomposing manure. Greatest research efforts have been directed toward the control of volatile compounds released by the storage and treatment of manures. Feed odors are most commonly associated with the feeding of waste materials or fermented products which have odors that are objectionable. Among the materials most commonly related to feed odors are potato wastes used as feed, other food processing wastes, or fermented feeds.

The odor of fresh manure is generally described as less objectionable than that of anaerobically decomposing manure. Fresh manure does evolve large quantities of ammonia but, in general, this ammonia is not accompanied by other decomposition products which contribute the most objectionable characteristics. The odorous compounds evolved from manure-covered surfaces or treatment facilities are a function of the biological reactions taking place within the material, the nature of the material as excreted, and the physical arrangement of the storage or treatment surface.

## ODOR CONTROL TECHNIQUES

Techniques for the control of odor release are based upon a limited number of rather specific control phenomena. The volatile compound present in the feed, manure, or manure slurry may be converted to a non-volatile form by precipitation or pH control, or by biological conversion to a less odorous or volatile compound. This is most commonly

done by the addition of a pH adjusting chemical, by a controlled biological operation, or by the addition of a chemical known to convert the compound of concern to a less volatile form. Examples of these three approaches would include the addition of lime as a pH adjusting material to control the release of hydrogen sulfide (Day 1966). As indicated in Table 3, the volatility of hydrogen sulfide is a strong function of pH, and if the pH is raised above 9.5, the escape of hydrogen sulfide should be insignificant. Sulfides are oxidized to sulfates under aerobic waste treatment as is practiced in oxidation ditches and aerated lagoons. Paraformaldehyde has been added to manure as a means of converting ammonia to non-volatile hexamethylenetetramine (Seltzer et al. 1969).

A second technique for the control of odor is to inhibit the anaerobic decomposition of manure. This is most frequently accomplished by keeping feedlots sufficiently dry to allow oxygen permeation of the surface. In systems utilizing bedding, the bedding serves a similar role. In liquid systems, odorant-producing decomposition is most often inhibited by maintaining the slurry in an aerobic condition. Oxidation ditches and aerated lagoons are two examples of this practice. Bacterial decomposition of manure is also inhibited by low temperatures, but, except in site selection, little use has been made of this fact.

Physical control of odorant release offers a third potential for odor control. Covers on manure storage tanks and anaerobic manure treatment devices are effective in controlling the escape of odorants. By reducing the air exchange over manure slurries, the volatilization of odorous gases is reduced. Gases escaping from enclosed tanks may be further managed by incineration, liquid scrubbing, or by soil column absorption. Although no entirely satisfactory cover material has been developed, covering of anaerobic lagoons and venting the released gases to a burner or a soil absorption field offer potential for odor control.

TABLE 3. FRACTION OF SULFIDE PRESENT AS $H_2S$ IN
WATER AT 25 C AS A FUNCTION OF pH

| pH | Fraction of sulfide present as $H_2S$ |
|---|---|
| 6.0 | 0.89 |
| 7.0 | 0.44 |
| 7.4 | 0.24 |
| 7.8 | 0.11 |
| 8.2 | 0.046 |
| 8.8 | 0.012 |
| 9.2 | 0.005 |
| 9.6 | 0.0019 |

Source. Standard methods for the examination of water and wastewater. Thirteenth edition. American public health association. 1971.

## ODOR CONTROL CHEMICALS

The addition of odor control chemicals to manure storage tanks or animal feeding areas has attracted widespread interest. Materials having potential for odor control, when used in this manner, are those which can prevent the release of odorous compounds, inhibit their formation, or mask their odor.

Oxidizing agents have potential as odor control chemicals. Faith (1964) proposed the use of potassium permanganate on cattle feedlots in Arizona at a rate of 22 kg/ha (20 lb/acre). Ford and Ulich (1973) conducted a series of tests and concluded that potassium permanganate at a level of 28 g/kg (56 lb/ton) of manure totally suppressed the release of odorous gases. It was also judged effective by their panel in reducing malodors at lower application rates. They reported potassium permanganate was effective in reducing the odor of well-managed feedlot surfaces when applied at the rate of 22 kg/ha (20 lb/acre) as a one percent water solution. Potassium nitrate, paraformaldehyde, hydrogen peroxide, and Ozene (a commercial formulation of orthodichlorobenzene) were also evaluated by Ford and Ulich (1973) but judged less effective than potassium permanganate.

Enzymes and other digestive aids have also been proposed for the control of livestock production odors. These products have been difficult to evaluate under controlled conditions because of the unwillingness of manufacturers to identify their composition. Ford and Ulich (1973) sujected "Formula 2" and a digestive deodorant to their testing program. Their odor panel concluded that neither was effective in reducing the strength or offensiveness of manure odors. This conclusion is similar to the earlier results of Burnett and Dondero (1968).

Feed additives for the modification of manure odors remain of interest to researchers. Although it is known that ration ingredients influence the odor of fresh feces and urine, no usable feed additives have gained widespread application. Matsuhima (Feedlot Management, 1972) used sagebrush as an additive in cattle rations and reported an odor reduction. Kellems (R. J. Kellems, unpublished data, Animal Science Department, Oregon State University, Corvallis, 1974) fed sagebrush, up to two percent of the ration, to beef animals and was unable to detect any influence on the odor of decomposing manure. He did, however, report a significant carryover of mint oil odor from ration to urine. Sagebrush, dry *Lactobacillus acidophilus,* lyophilized yeast culture, and activated charcoal, all at the five percent level, were fed to swine by Ingram et al. (1973) with no significant change in fecal odor, according to an odor panel.

## DESIGN AND MANAGEMENT

Levels of understanding of odor control by chemical feed additives or odor-masking techniques are not sufficient to offer solutions to livestock producer problems. Until advances are made in these areas, site selection, facility design, and careful management offer the greatest potential for avoiding odor conflicts. Sweeten et al. (1974) monitored the decrease in odor intensity downwind on a 4,000-head cattle feedlot in north-central Texas. Although odor intensities were as high as 64 dilutions to threshold at the feedlot, readings were always 1.5 or less one quarter of a mile away. Odors at this intensity are difficult to measure and may be of little societal consequence.

In addition to maintaining livestock production sites remote from potential odor complainants, other management tools to minimize odor release involve inhibition of anaerobic manure decomposition, keeping feedlots dry, and using aerobic treatment and storage devices. Presenting a neat and orderly appearance, although not directly related to odors, has an obvious impact in decreasing odor complaints.

## TECHNOLOGY NEEDED

Conflicts between livestock producers and nearby residents and the general public over odors are based upon the lack of an adequate odor control technology. This vacuum has invited a series of expensive court litigations, and, more recently, governmental regulations for which the livestock and poultry producers have few viable solutions. Further research is needed to evolve odor control systems compatible with intensive livestock production and economically compatible with the problem.

Until such technology evolves, odor control regulations and court decisions must be made with full recognition of the problems involved. Odors, like other nuisances, can best be defined when intensity, duration, and frequency are all given full consideration.


### References

1 Burnett, W. E. and N. C. Dondero. 1968. The control of air pollution (odors) from animal wastes - evaluation of commercial odor control products by an organoleptic test. ASAE Paper No. 68-909, ASAE, St. Joseph, Mich. 49085.

2 Day, D. L. 1966. Liquid hog manure can be deodorized by treatment with chlorine or lime. Illinois Research. (summer) 127:16.

3 Faith, W. L. 1964. Odor control in cattle feed yards. Jour. Air. Poll. Control Assoc. 14:459-460.

4 Feedlot Management. 1972. Sagebrush for odor control: in the feed or in the manure? Feedlot Management 14(5):74.

5 Ford, J. P., and W. L. Ulich. 1973. Odor control for confined beef cattle feedlots. Presented at the First Annual Symposium on Air Pollution Control in the Southwest. Texas A&M University, November 5-7.

6 Hutchinson, G. L., and F. G. Viets, Jr. 1969. Nitrogen enrichment of surface water by absorption of ammonia volatilized from cattle feedlots. Science 166:515.

7 Ingram, S. H., R. C. Albin, C. D. Jones, A. M. Lenon, L. F. Tribble, L. B. Porter, and C. T. Gaskins. 1973. Swine fecal odor as affected by feed additives. (abstr.) J. Animal Sci. 36(1):207.

8 Lindvall, T., O. Norén, and L. Thyselius. 1974. Odor reduction for liquid manure systems. TRANSACTIONS of the ASAE 17(3):508-512.

9 Luebs, R. E., A. E. Laag, and K. R. Davis. 1973. Ammonia and related gases emanating from a large dairy area. Cal. Agr. 27(2):10-12.

10 Miner, J. R., and T. E. Hazen. 1969. Ammonia and amines: components of swine building odor. TRANSACTIONS of the ASAE 12:772-774.

11 Moncrieff, R. W. 1961. An instrument for measuring and classifying odors. Jour. Appl. Physiology. 16:742.

12 Rowe, N. R. 1963. Odor control with activated charcoal. Jour. Air Poll. Control Assoc. 13:150-153.

13 Seltzer, Moum, Goldhaft. 1969. A method for the treatment of animal wastes to control ammonia and other odors. Poultry Sci. 48:1912.

14 Sweeten, J. M. 1974. Environmental protection requirements for swine operations. Fact sheet, Texas Agr. Ext. Serv. L-1302. 4 pp.

15 Sweeten, J. M., D. L. Reddell, L. M. Schake, and B. Garner. 1973. Odor intensity at cattle feedlots. Presented at the First Annual Symposium on Air Pollution Control in the Southwest. Texas A&M University, November 5-7.

16 Sweeten, J. M., D. L. Reddell, and H. B. H. Cooper. 1974. Odor measurement for livestock feeding operations. In: Proceedings of specialty conference on control technology for agricultural air pollutants. Air Poll. Control Assoc. (in press).

17 Willrich, T. L., and J. R. Miner. 1971. Litigation experiences of five livestock and poultry producers. pp. 99-101. In: Livestock waste management and pollution abatement. ASAE publication PROC-271, ASAE, St. Joseph, Mich. 49085.

# Chemical Treatment of Liquid Dairy Manure to Reduce Malodors


W. F. Ritter, N. E. Collins, Jr., R. P. Eastburn

ASSOC. MEMBER   ASSOC. MEMBER   AFFILIATE
ASAE         ASAE         ASAE


IN the dairy and swine industry, liquid manure is often kept in large underground storage tanks until field and climatic conditions permit its disposal. Anaerobic conditions prevail in the storage tank when the manure isn't aerated and odors are emitted. If the liquid manure is left undisturbed, the odors are generally a very localized problem, but if the liquid is disturbed and spread on the land, these odors can become a public nuisance.

Several researchers have reported using chemicals to control odors of livestock wastes (Hollenback 1971, Barnett and Dondero 1970). Hollenback (1971) reported that both 50 and 100 ppm of hydrogen peroxide effectively removed hydrogen sulfide odors from a liquid manure system, but when the manure was agitated 18 hours later, the hydrogen sulfide odor had returned. Hydrogen peroxide also has been used to control odors in municipal wastes (U.S. Environmental Protection Agency 1974). Today, there are also chemical products on the market that are recommended for odor control in livestock facilities.

The objective of this research was to evaluate the effectiveness of hydrogen peroxide and other chemical agents in the control of odors emanating from liquid manure.

## EXPERIMENTAL PROCEDURES

The experimental work covered two separate phases. The first phase involved research to determine the minimum level of hydrogen peroxide required to eliminate hydrogen sulfide odor in liquid manure. Phase two involved the consensus of nine test panels to evaluate the effectiveness of different chemicals to control odors in liquid manure.

In the first phase of the research, liquid dairy manure was taken from a storage tank adjacent to the large animal research laboratory on the Delaware Agricultural Experiment Station Farm and mixed with fresh feces from the dairy loafing shed. Twenty-four 208 liter drums were each loaded daily 5 times per week for 6 weeks with 3.785 liters of slurry of approximately 10 percent total solids to simulate loading and storing of liquid manure.

In the tests to determine the minimum level of hydrogen peroxide required to eliminate the hydrogen sulfide odor in the liquid manure, the manure in a test drum was stirred for 2 minutes with a rotating paddle that was driven by a 6.35 mm electric drill. A specified amount of hydrogen peroxide was added then to the test drum and the contents were stirred for 2 minutes. The hydrogen sulfide and dissolved


The authors are: W. F. RITTER, Assistant Professor; N. E. COLLINS, JR., Assistant Professor; and R. P. EASTBURN, Research Assistant, Agricultural Engineering Department, University of Delaware, Newark, Delaware.

Acknowledgments: Part of the financial support for this research was provided by the E. I. Du Pont de Nemours Company.

Published with the approval of the Director of the Delaware Agricultural Experiment Station as Miscellaneous Paper No. 716.


**TABLE 1.**
**THE FORM USED BY PANELISTS TO EVALUATE THE ODOR**

Date _______________________

Name _______________________

Position _______________________

Please rate the samples as to presence of odor and as to the offensiveness of the odor to you. If the odor brings to mind a descriptive term, please feel free to comment. Thank you for your assistance in this matter.

| | Presence | | Offensiveness |
|---|---|---|---|
| 0 | No odor | 0 | Not offensive |
| 1 | | 1 | |
| 2 | Very faint | 2 | Very faint offensiveness |
| 3 | | 3 | |
| 4 | Faint | 4 | Faint offensiveness |
| 5 | | 5 | |
| 6 | Definite | 6 | Definite offensiveness |
| 7 | | 7 | |
| 8 | | 8 | Strong offensiveness |
| 9 | | 9 | |
| 10 | Very strong | 10 | Very strong offensiveness |

| Sample | Presence | Offensiveness | Comment |
|---|---|---|---|
| 1 | | | |
| 2 | | | |
| 3 | | | |
| 4 | | | |
| 5 | | | |
| Etc. | | | |

oxygen concentrations of the liquid manure slurry were measured before the hydrogen peroxide was added and at various intervals after it was added. Hydrogen sulfide concentrations were measured by the Hach Chemical Company procedure (1968) and the dissolved oxygen content was measured with an oxygen probe. For several of the tests, the hydrogen sulfide concentration in the air above the liquid was measured with a Unico 400 precision gas detector.

The olfactory measurement procedure (Table 1) developed by Sobel (1972) was used to evaluate the effectiveness of the chemicals. Treatments evaluated were as follows:

1 Panel 1 evaluated the effectiveness of pretreatment with hydrogen peroxide and then the addition of varying levels of the masking agent Alamask 520A.

2 Panel 2 evaluated the effectiveness of pretreatment with hydrogen peroxide and different levels of the masking agents Alamask 151A, Alamask 518B and NaOCl.

3 Panel 3 evaluated the effectiveness of pretreatment with hydrogen peroxide.

4 Panel 4 evaluated the acceptability of masking agents Alamask 151A, 520A and 518B, and NaOCl.

5 Panel 5 evaluated the effect of mechanical stirring during storage followed by chemical treatment with Alamask 151A, 510A and 518B.

**TABLE 2. HYDROGEN SULFIDE AND DISSOLVED OXYGEN CONCENTRATIONS AS A FUNCTION OF TIME IN LIQUID MANURE TREATED WITH HYDROGEN PEROXIDE**

| Test No. | $H_2O_2$ added, ppm | Time elapsed, min | $H_2S$, ppm | DO, ppm | Test | $H_2O_2$ added, ppm | Time elapsed, min | $H_2S$, ppm | DO, ppm |
|---|---|---|---|---|---|---|---|---|---|
| 1 | 100 | 0 | 2 | 0.8 | 4 | 25 | 0 | 7 | 0.4 |
| | | 5 | 0 | 10.8 | | | 5 | 0 | — |
| | | 10 | 0 | —* | | | 15 | — | 8.2 |
| | | 25 | — | 15 | | | 35 | 0 | 4.6 |
| | | 70 | — | — | | | 65 | 0 | 0.8 |
| | | 120 | 0 | 15 | | | 90 | 0 | 0.6 |
| | | 450 | 2 | 0.8 | | | 160 | 5 | — |
| 2 | 50 | 0 | 5 | 0.8 | 5 | 12.5 | 0 | 4 | 0.6 |
| | | 5 | 0 | 14.6 | | | 5 | 0 | 3.2 |
| | | 45 | — | 15 | | | 35 | 0 | 0.2 |
| | | 90 | 0 | 10.2 | | | 60 | 0 | 0.2 |
| | | 352 | 2 | 0.8 | | | 129 | 1 | — |
| 3 | 25 | 0 | 3 | 0.8 | | | 152 | 2 | — |
| | | 5 | 0 | 9.6 | 6 | 6.5 | 0 | 6 | 0.4 |
| | | 10 | — | 9.0 | | | 5 | T† | 1.4 |
| | | 25 | — | 5.2 | | | 35 | T | 0.2 |
| | | 360 | 3 | 0.8 | | | 95 | 5 | — |

* No data collected

† Trace

6    Panel 6 evaluated the effectiveness of chemical treatment with Alamask 151A, 510A and 518B, plus lemon and vanilla extract.

7    Panel 7 evaluated the effectiveness of treatment with Agri-Gest, Cairox (potassium permanganate), Roebic, Rid-X, Wham and King Activator.

8    Panel 8 evaluated the effectiveness of treatment with hydrogen peroxide and Cairox or Agri-Gest.

9    Panel 9 evaluated the effectiveness of Cairox with time.

Panels 1-5 and 7-9 were composed of staff and students of the College of Agricultural Sciences. Panel 6 consisted of farmers and Extension personnel who attended the annual dairy meetings in each county.

The liquid manure from the 24 drums was used for the first 5 panels. For panel 1, the liquid manure in the drum was agitated for 2 minutes and the untreated sample was removed. Hydrogen peroxide (12.5 ppm) was added and the liquid agitated for 2 minutes before the hydrogen peroxide treatment sample was removed. The stirring procedure then was repeated as 50 ppm increments of masking agent 520A were added. For panels 2 and 3 each sample was taken from a different drum after the chemicals and manure had been mixed for 2 minutes. Only the diluted masking agent or control materials were placed in the sample for panel 4. All samples were prepared 1 hour before the odor panels began evaluating the treatments.

For panel 6, 18.9 liter samples of liquid dairy manure from the storage tank adjacent to the large animal research laboratory and the chemical agent were mixed at the site of the panel evaluation. Liquid dairy manure from another storage tank on the Delaware Agricultural Experiment Station Farm and fresh feces were mixed in 18.9 liter drums and allowed to stand for several weeks before the odor tests were conducted for test panels 7-9.

In the case of panels 7 and 8, all the chemicals except the hydrogen peroxide were added and mixed 48 hours prior to evaluation. The Cairox treated samples for test panel 9 were the same samples used for panel 8, but the odor tests were conducted 24 hours later to evaluate the effectiveness of Cairox over an extended period of time.

## RESULTS AND DISCUSSION

The results of the tests to determine the minimum amount of hydrogen peroxide required to eliminate the hydrogen sulfide in the liquid manure, are shown in Table 2. The addition of 12.5 ppm or more of hydrogen peroxide quickly reduced the hydrogen sulfide in the liquid manure and increased the dissolved oxygen content. The dissolved oxygen content in the liquid manure was a function of the amount of hydrogen peroxide added to the manure. When 100 and 50 ppm of hydrogen peroxide were added to the manure, respectively, the dissolved oxygen content was increased above saturation levels. Treatment of the liquid manure with 6.5 ppm of hydrogen peroxide was ineffective in suppressing the hydrogen sulfide or increasing the dissolved oxygen content. Since 12.5 ppm of hydrogen peroxide eliminated the hydrogen sulfide in the liquid manure for at least 1 hour, it was selected as the treatment level required for suppression of hydrogen sulfide in the remaining tests where hydrogen peroxide was used.

Measurements of hydrogen sulfide concentrations in the air above the liquid during several of the tests showed the maximum concentration of 10 ppm immediately after agitation. Tests taken 5 minutes later indicated a significant decrease in hydrogen sulfide concentration.

Panels 1 and 2 (Tables 3 and 4) evaluated the effective-

**TABLE 3. THE OFFENSIVENESS OF ODORS EMANATING FROM LIQUID MANURE TREATED WITH DIFFERENT LEVELS OF ALAMASK 520A***

| Treatment | Not offensive | Mildly offensive | Definitely offensive | Very offensive |
|---|---|---|---|---|
| | | (Percent of response) | | |
| Manure, | 4.8 | 33.3 | 23.8 | 38.1 |
| 12.5 ppm $H_2O_2$ | 11.9 | 26.2 | 26.2 | 35.7 |
| 25 ppm 520A† | 21.4 | 45.5 | 16.2 | 16.2 |
| 50 ppm 520A† | 31.0 | 33.3 | 23.8 | 11.9 |
| 100 ppm 520A† | 19.0 | 40.5 | 33.3 | 7.2 |
| 200 ppm 520A† | 28.6 | 31.0 | 33.3 | 7.2 |
| 300 ppm 520A† | 19.0 | 42.9 | 28.6 | 9.5 |

* Panel 1 contained 21 respondents

† Pretreated with 12.5 ppm of $H_2O_2$

TABLE 4. THE OFFENSIVENESS OF ODORS EMANATING
FROM LIQUID MANURE TREATED WITH
DIFFERENT CHEMICAL AGENTS*

| Treatment | Not offensive | Mildly offensive | Definitely offensive | Very offensive |
|---|---|---|---|---|
| | | (Percent of responses) | | |
| Manure | 16.7 | 54.2 | 8.3 | 20.8 |
| 12.5 ppm $H_2O_2$ | 16.7 | 33.3 | 41.7 | 8.3 |
| 50 ppm 518B† | 29.2 | 41.7 | 20.8 | 8.3 |
| 100 ppm 518B† | 41.7 | 41.7 | 12.5 | 4.1 |
| 50 ppm 151A† | 33.3 | 58.4 | 8.3 | 0.0 |
| 100 ppm 151A† | 45.8 | 37.5 | 12.5 | 4.1 |
| 50 ppm NaOCl† | 41.7 | 20.8 | 29.2 | 8.3 |
| 100 ppm NaOCl† | 29.2 | 33.3 | 29.2 | 8.3 |

* Panel 2 contained 12 respondents

† Pretreated with 12.5 ppm of $H_2O_2$

ness of Alamask 520A, 518B, 151A, sodium hypochlorite, and pretreatment with 12.5 ppm hydrogen peroxide in controlling odors. Table 3 shows that over 50 percent of the panelists found the untreated liquid manure as well as the liquid manure treated with 12.5 ppm of hydrogen peroxide to be definitely or very offensive. This suggests that the hydrogen peroxide may have changed the nature of the odor rather than have reduced its offensiveness. One hundred ppm of Alamask 518B and 151A were considered not offensive or mildly offensive. Over 35 percent of the panelists found the sample treated with 300 ppm of Alamask 520A to be definitely offensive or very offensive.

The effect of treating liquid manure with hydrogen peroxide before the masking agents NaOCl, Alamask 151A or 518B were added are shown in Table 5. Alamask 518B and NaOCl seemed to be the least offensive without hydrogen peroxide while Alamask 151A appeared to be the most effective with pretreatment. From the results of Table 5 one may conclude that pretreatment with hydrogen peroxide will be determined by the masking agent used.

Test panel 4 revealed that the masking agent itself may be offensive. Over 20 percent of the panelists rated Alamask 151A, 518B, 520A and NaOCl as definitely or very offensive. This suggests that the substitution of one odor for another may not solve the malodor problem of liquid manure.

Panel 5 evaluated the offensiveness of manure that had been stirred daily and treated with Alamask 510A, 151A, and 518B while panel 6 evaluated the effectiveness of Alamask 151A, 510A and 518B plus lemon and vanilla extracts as masking agents. Both panels found Alamask 151A and 518B to be the most effective agents in reducing malodors. Daily agitation did not consistently improve the effectiveness of the compounds tested in panel 5. Panel 6, which was made up mostly of dairy farmers, found liquid

TABLE 5. THE OFFENSIVENESS OF ODORS EMANATING
FROM LIQUID MANURE PRETREATED
WITH HYDROGEN PEROXIDE*

| Treatment | Not offensive | Mildly offensive | Definitely offensive | Very offensive |
|---|---|---|---|---|
| | | (Percent of responses) | | |
| 100 ppm 518B | 27.8 | 22.2 | 27.8 | 22.2 |
| 100 ppm 518B† | 14.8 | 25.9 | 31.5 | 27.8 |
| 100 ppm 151A | 29.6 | 16.7 | 29.6 | 24.1 |
| 100 ppm 151A† | 25.9 | 33.3 | 22.2 | 18.5 |
| 100 ppm NaOCl | 18.5 | 37.0 | 31.5 | 13.0 |
| 100 ppm NaOCl† | 20.4 | 31.5 | 22.2 | 25.9 |

*Panel 3 contained 27 respondents

† Pretreated with 12.5 ppm of $H_2O_2$

TABLE 6. THE OFFENSIVENESS OF ODORS EMANATING
FROM LIQUID MANURE TREATED WITH
AGRI-GEST AND CAIROX*

| Treatment | Not offensive | Mildly offensive | Definitely offensive | Very offensive |
|---|---|---|---|---|
| | | (Percent of responses) | | |
| 48 ppm Agri-Gest | 17.3 | 40.4 | 25.0 | 17.3 |
| 48 ppm Agri-Gest† | 23.0 | 50.0 | 21.2 | 5.8 |
| 24 ppm Agri-Gest | 65.4 | 25.0 | 3.8 | 5.8 |
| 24 ppm Agri-Gest† | 80.7 | 13.5 | 5.8 | 0.0 |
| 480 ppm Cairox | 84.6 | 13.5 | 1.9 | 0.0 |
| 480 ppm Cairox† | 73.1 | 25.0 | 1.9 | 0.0 |
| 240 ppm Cairox | 71.2 | 26.9 | 1.9 | 0.0 |
| 240 ppm Cairox† | 75.0 | 21.2 | 3.8 | 0.0 |
| 12.5 ppm $H_2O_2$ | 48.0 | 26.9 | 17.3 | 7.7 |
| Manure | 7.7 | 42.3 | 34.6 | 15.4 |

* Panel 8 contained 26 respondents

† Treated with 12.5 ppm $H_2O_2$

manure as offensive as the other test panels which indicates the background of the panelists apparently did not influence the response to the offensive odors emanating from the liquid manure. Based on the results of the first 6 panels, Alamask 518B and 151A would be rated as the most desirable masking agents tested.

Test panel 7 evaluated the commercial compounds Agri-Gest, Wham, Roebic, King Activator, Rid-X and Cairox (potassium permanganate). Agri-Gest, Roebic, King Activator and Rid-X are bacterial or enzyme cultures. Cairox was added to the liquid manure at the rate of 480 ppm while the rest of the compounds were added at the rates recommended by the manufacturer. The results showed that only Agri-Gest and Cairox were less offensive than the raw manure. All the other compounds had odors that were as offensive or more offensive than the liquid manure itself.

Panel 8 evaluated the effect of the addition of 12.5 ppm hydrogen peroxide to Agri-Gest and Cairox. The results in Table 6 indicate that 90 percent or more of the panelists found that the odors emanating from liquid manure treated with 480 ppm or 240 ppm Cairox or 24 ppm Agri-Gest were not offensive or mildly offensive. The addition of hydrogen peroxide to the Cairox or Agri-Gest had little effect. A larger percentage of the panel found the liquid manure treated with 48 ppm Agri-Gest more offensive than the manure treated with 24 ppm Agri-Gest. This indicates possibly that at the higher concentration the odor of the Agri-Gest may become offensive.

The effectiveness of treating liquid manure with Cairox three days before the odor tests were conducted is shown in Table 7. After 24 hours, panel 9 was convened to re-evaluate the samples used by panel 8. Over 90 percent of the panelists still found the odors emanating from the liquid manure treated with Cairox to be not offensive or mildly offensive although a larger percentage of panel 9 found the odors mildly offensive. This suggests that the effectiveness

TABLE 7. THE OFFENSIVENESS OF ODORS EMANATING
FROM LIQUID MANURE AFTER TREATING WITH CAIROX
THREE DAYS BEFORE THE TEST*

| Treatment | Not offensive | Mildly offensive | Definitely offensive | Very offensive |
|---|---|---|---|---|
| | | (Percent of responses) | | |
| 480 ppm Cairox | 53.9 | 40.4 | 3.8 | 1.9 |
| 240 ppm Cairox | 57.7 | 34.7 | 3.8 | 3.8 |
| Manure | 15.4 | 34.6 | 34.6 | 15.4 |

*Panel 9 contained 26 respondents

TABLE 8. THE COST OF TREATING LIQUID MANURE
FOR ODOR CONTROL

| Treatment | Level, ppm | Cost per 10 M$^3$ |
|---|---|---|
| Hydrogen Peroxide | 12.5 | $ 0.06 |
| Alamask 520A | 100 | 5.07 |
| Alamask 518B | 100 | 3.30 |
| Alamask 151A | 100 | 12.12 |
| NaOC1 | 100 | 0.40 |
| Cairox | 240 | 3.33 |
| Agri-Gest | 24 | 3.04 |
| Oxidation Ditch | | 13.98 |

of the Cairox in controlling the odors decreased slightly in the 24-hour period.

The cost of chemical treatment (Table 8) prior to spreading the liquid manure on land varies from $0.06 to $12.12 per 10 cubic meters. The cost of chemical treatment is less than the cost of operating an oxidation ditch, based on a dairy cow producing 1000g $BOD_5$ (5-day biochemical oxygen demand) and 42.5 liters of waste per day and power requirements of 2.2 KWH per kg $BOD_5$ (Jones et al. 1970). Electricity costs were assumed to be 5 cents per KWH.

## CONCLUSIONS

1   Hydrogen peroxide effectively eliminated hydrogen sulfide in liquid dairy manure for a short period of time, but other malodors associated with liquid dairy manure remained.

2   Alamask 518B and 151A, Cairox and Agri-Gest appeared to be the most effective of the compounds tested in reducing the malodors associated with liquid dairy manure.

3   Cairox was effective in reducing the malodors in the liquid dairy manure for at least 72 hours.

4   The estimated cost of a single chemical treatment is less than the cost of odor control by the oxidation ditch.

5   The estimated cost of treating liquid manure with 12.5 ppm hydrogen peroxide is less than the treatment costs of the other chemical agents tested.


## References

1   Barnett, W. E. and N. C. Dondero. 1970. Control of odors from animal wastes. TRANSACTIONS of the ASAE 13(2):221-224, 231.

2   Hach Chemical Company. 1968. Engineer's laboratory methods manual. 6th Edition. Ames, Iowa.

3   Hollenback, R. C. 1971. Manure odor abatement using hydrogen peroxide. Report No. 5638-R Food Machinery Corporation, Princeton, N.J.

4   Jones, D. D., D. L. Day and A. C. Dale. 1970. Aerobic treatment of livestock wastes. University of Illinois Agricultural Experiment Station Bul. 737.

5   Sobel, A. J. 1972. Olfactory measurement of animal manure odor. TRANSACTIONS of the ASAE 15(4):696-699.

6   U.S. Environmental Protection Agency. 1974. Sulfide control in sanitary sewerage systems. EPA Technology Transfer Design Manual.

# Land Application of Manures — Wisconsin's Manure Management Plan


L. R. Massie, R. D. Powell, R. E. Graves

MEMBER
ASAE

ASSOC. MEMBER
ASAE


THE production and related need to dispose of manure from a farm's livestock operation often is not compatible with the crop production schedule. Manure production and utilization situations can be examined to point up and resolve conflicts between the livestock and crop production operations. Each farmer can have a complete manure management program which determines the number of animals the farm can support, based on the imposed limitations. Essentially, the farm has a manure management plan similar to a soil conserving or livestock production plan.

Farmers should be able to demonstrate that the amount applied, timing of applications, and methods of spreading and incorporating the manure agree with the concept of "best practice available" for protecting water quality.

## APPROACHES TO MANURE APPLICATIONS

1   Apply manure based on crop removal of nutrients. Crops with high nitrogen and phosphorus requirements, such as corn and small grains, use manure applications to best advantage.

2   Use fallowed land disposal sites. This type of manure disposal is not recommended because the possible contamination effects have not been completely investigated.

## THE SOILS MAP

A soils map describes the location and areal extent of all soil types, average land slope, and the erosion or deposition that has occurred. Each soil type represents a unique combination of characteristics.

Soil conditions which may affect land application of manure are:

1   Internal Drainage. Excessively well-drained soils (sands) may permit rapid movement of water and/or groundwater contamination. Some soils have seasonally high water tables and may be subject to flooding. Others need artificial drainage to help provide better infiltration, satisfactory percolation, and an aerobic environment for bacterial decomposition.

2   Slow Water Intake Rates. Low infiltration rates produce larger runoff rates with the associated capacity to carry large amounts of manure and soil from the field.

3   Shallow Soils. A shallow soil or one with a restricting layer (Fragipan) has small storage capacities. Cracked or creviced bedrock near the surface may allow contamination of the groundwater.

4   Erosion. The subsoil exposed by erosion is higher in clay content than the original topsoil. It has a lower infiltration rate and increases the runoff problem.

5   Location of the Soil Body on the Landscape. A soil type is always found in about the same position on the landscape; that is, at the top of a hill, at the bottom, or in the middle of the slope. Knowing the soil type provides some insight as to whether the runoff is generated on the soil body, flows down across it from upslope, or ponds upon it.

## LAND CAPABILITY

A land capability map relates the soils information and crop rotations on a field basis.

The land capability classification is based on the combined effects of soil features and climate on the risk of soil damage, on limitations of safe use, and on the difficulty in applying conservation practices when the land is cultivated. Limitations are judged as they relate to current soil-using practices in the United States.

Capability Classes I, II, and III include land that is suitable for regular cultivation. Class IV land has very severe limitations requiring careful management. It is suitable for occasional cultivation.

## THE LAND USE MAP AND CROPPING PLAN

The land use map describes the physical arrangement of the fields, pasture, and woodland. It may also show the location of neighbors' houses in close proximity to the farm. The acreage of each field and the conservation practices are usually shown.

A cropping plan compatible with the land capability classes and the conservation practices is established to maximize feed production.

## MANURE MANAGEMENT RECOMMENDATIONS FOR WISCONSIN

1   Spread and incorporate manure on cropland (0-16 percent slopes) during the growing season when major tillage (plowing) occurs. Manure spread in the fall should be incorporated by plowing or disking to reduce nutrient losses in winter runoff.

2   Allow winter spreading on Class I and Class II land where the slopes do not exceed 6 percent, the soils are moderately to well drained, the runoff is generated locally, and water management practices have been installed.

3   Spread and immediately incorporate (same day) the manure on flood plains and those relatively small areas identified as major contributors of concentrated surface flow in each field after the danger of major runoff events is past.

4   Determine the manure distribution schedule after

---


The authors are: L. R. MASSIE, Extension Agricultural Engineer, R. D. POWELL, Associate Professor, Soil Science Dept., and R. E. GRAVES, Extension Agricultural Engineer, College of Agriculture and Live Sciences, University of Wisconsin, Madison.

considering crop planting and harvesting dates, average snow depths, distance from the farmstead, access, soil compaction, and the kind of manure (bedded-pack, freestall, etc.). The schedule should be developed for each year of the crop rotation.

5   Minimize nitrate-nitrogen accumulations in the soil from manure applications by using rates based on crop removal of nutrients. The amount of nitrogen that can be applied annually without accumulating nitrates is estimated to be 250 pounds per acre. If dairy manure contains 10 pounds of nitrogen per ton, 25 tons of wet manure could be applied per acre. If 25 percent of the nitrogen is lost prior to incorporation, the maximum rate could be 33 tons per acre. Single applications on a land disposal basis might reasonably be two times the annual rate.

6   Application rates should be lower on sandy soils than on fine-textured, poorly-drained soils where denitrification losses are largest.

7   Apply manure to soils planted to non-legume crops (corn or small grain) that use nitrogen to best advantage. Manure should not be applied to alfalfa. The growth of grasses is encouraged, resulting in a possible reduction of forage quality as measured by the protein content.

8   Rates of application to crops that lodge (small grains) will have to be lower than rates for corn. Suggest 100 pounds per acre.

9   If corn is grown for grain, the nitrate buildup in the grain is very small. However, when corn is grown for silage or if cattle are pastured on corn stubble, nitrates may accumulate in the plant, causing nitrate toxicity in the animals.

10   Establish lower levels of manure application on shallow soils over bedrock to reduce possible nutrient movement into the groundwater system.

11   Use runoff control practices on field slopes of 6 percent and over.

12   Divert runoff generated on upslope pastures, woodlots, etc. that presently flows across the farm fields. Use grass waterways to move the runoff through the field.

13   On long slopes (greater than 300 feet), install a system of terraces or surface drains to trap organic debris and sediment and to increase the time needed to concentrate large flows.

14   Use drainage systems (deep and/or surface) on poorly-drained soils to produce desirable aerated soil conditions in cultivated fields.

15   Avoid spreading manure in runoff collection and control facilities (waterways, terraces, surface drains, etc.).

## THE MANURE YEAR

The manure distribution schedule is most easily developed for a year, beginning on some date other than January 1. October 1 works well for Wisconsin dairy farmers. The crops are harvested, stored manure can be emptied and incorporated into the soil during fall plowing, and the cattle are coming off pasture.

The manure year will begin for others when their manure storages are empty, perhaps May 1 or June 1.

## THE DISTRIBUTION SCHEDULE

The distribution schedule relates manure production to the cropland available for spreading. All conditions that affect the application of manure should be considered as the distribution schedule is developed. The crop being grown, snow cover, soils data, topography, conservation practices, type of livestock housing, rate of application, and the location of the field on the farm are factors that determine the time of spreading and the amount of manure

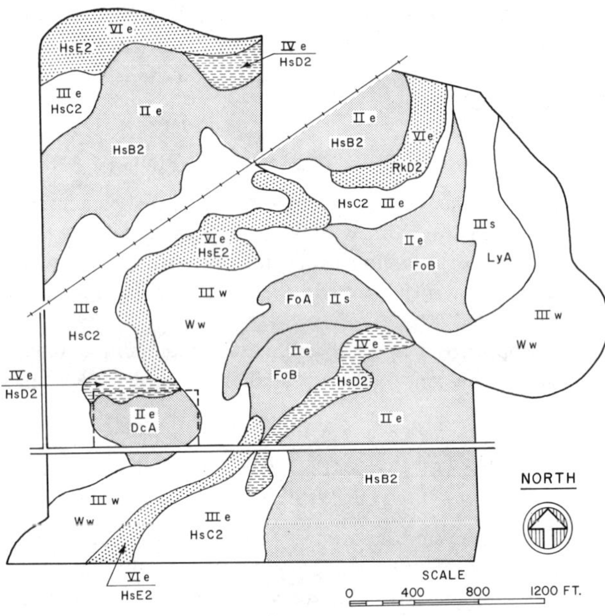

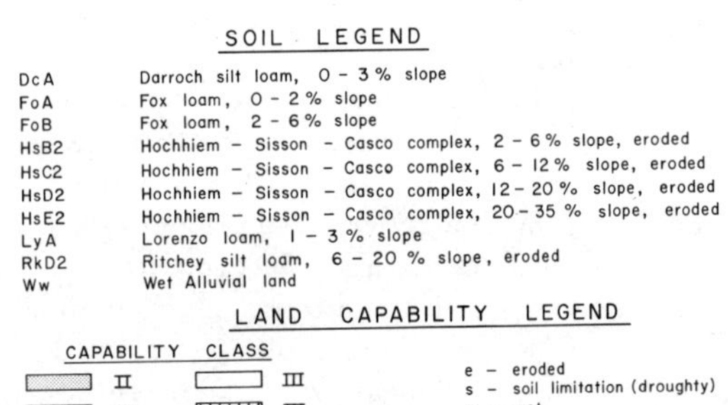

SOIL  LEGEND

| | |
|---|---|
| DcA | Darroch silt loam, 0 - 3 % slope |
| FoA | Fox loam, 0 - 2 % slope |
| FoB | Fox loam, 2 - 6 % slope |
| HsB2 | Hochhiem – Sisson – Casco complex, 2 – 6 % slope, eroded |
| HsC2 | Hochhiem – Sisson – Casco complex, 6 – 12 % slope, eroded |
| HsD2 | Hochhiem – Sisson – Casco complex, 12 – 20 % slope, eroded |
| HsE2 | Hochhiem – Sisson – Casco complex, 20 – 35 % slope, eroded |
| LyA | Lorenzo loam, 1 – 3 % slope |
| RkD2 | Ritchey silt loam, 6 – 20 % slope, eroded |
| Ww | Wet Alluvial land |

LAND  CAPABILITY  LEGEND

CAPABILITY  CLASS

II   III   IV   VI

e – eroded
s – soil limitation (droughty)
w – wet

FIG. 1 Soil and Capability map.

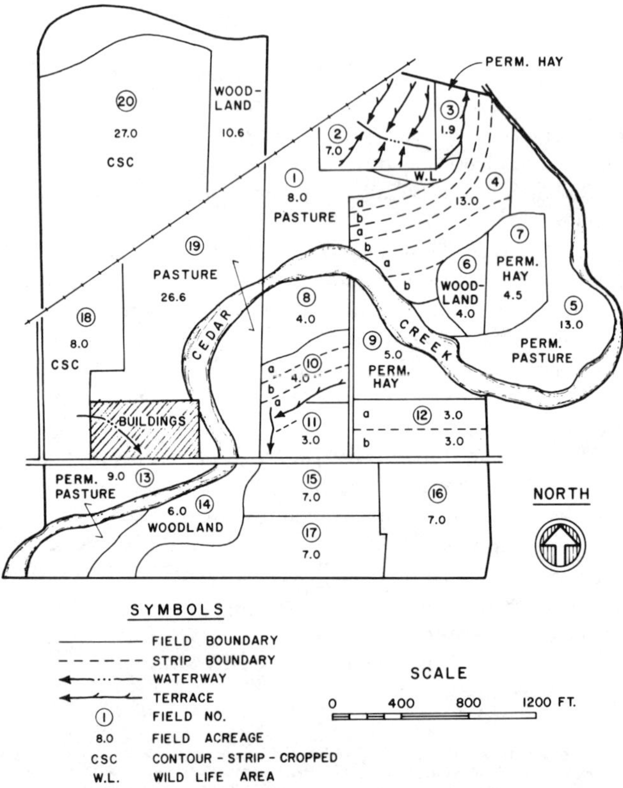


SYMBOLS

| | |
|---|---|
| | FIELD  BOUNDARY |
| | STRIP  BOUNDARY |
| | WATERWAY |
| | TERRACE |
| ① | FIELD  NO. |
| 8.0 | FIELD  ACREAGE |
| CSC | CONTOUR - STRIP - CROPPED |
| W.L. | WILD  LIFE  AREA |

FIG. 2 Land use map.



MANAGING LIVESTOCK WASTES

## TABLE 1. CROPPING PLAN

| Field no. | Conservation practices | Acres | Years 1974 | 1975 | 1976 | 1977 |
|---|---|---|---|---|---|---|
| | | | | | | |
| **UNIT I WITH CROP ROTATION OF C-C-O-H*** | | | | | | |
| 2 | Terraces | 7 | H | C | C | O |
| 11 | None | 3 | C | C | O | H |
| 12A | Field | 3 | C | O | H | C |
| 12B | Strips | 3 | O | H | C | C |
| 15 | Contoured | 7 | C | C | O | H |
| 16 | Contoured | 7 | C | O | H | C |
| 17 | Contoured | 7 | O | H | C | C |
| **UNIT II WITH CROP ROTATION OF C-O-H-H** | | | | | | |
| 4A | Contour strip | 7 | C | O | H | H |
| 4B | | 6 | H | H | C | O |
| 8 | None | 4 | H | H | C | O |
| 10A | Contour strip | 2 | O | H | H | C |
| 10B | | 2 | H | C | O | H |
| 18A | Contour strip | 5 | H | H | C | O |
| 18B | | 3 | C | O | H | H |
| 20A | Contour strip | 13 | O | H | H | C |
| 20B | | 14 | H | C | O | H |
| **UNIT III—PERMANENT HAY AND PASTURE†** | | | | | | |
| 7 | | 4.5 | H | H | H | H |
| 9 | | 5.0 | H | H | H | H |
| 1 | | 8.0 | P | P | P | P |
| 13 | | 9.0 | P | P | P | P |
| 19 | | 26.6 | P | P | P | P |

* C = Corn, O = Oats, H = Hay.

† H = Hay, P = Pasture

applied to each field.

Once developed, the producer will have additional information when considering winter or summer manure storages; changes in crop rotations, livestock housing, and animal concentrations; more intensive conservation practices and obtaining more land.

## MANURE DISTRIBUTION SCHEDULE FOR EXAMPLE FARM

A 36-cow dairy farm is used to demonstrate the basic information, calculations, and decisions needed to develop a manure distribution schedule for the 1975 crop year. The soils map, land use map, and crop rotations are shown in Figs. 1 and 2, and Table 1.

The daily, monthly, and yearly amounts of manure and total fertilizer nutrients produced by each animal age group are shown in Table 2. The type of manure and its location are also indicated.

The amounts of manure to be applied to the cropland (Table 4) were calculated using the total nitrogen values (Table 2 or 3). The amounts can also be determined for another nutrient or by using the available nutrient values (Table 3). The fertilizer equivalent (lb N/acre) is the amount available to the crop, so is credited to the fertilizer budget. The available nitrogen applied to the small grain is about 25 lb/acre, and to the corn is 100 lbs/acre.

The distribution schedule (Table 4) shows manure appli-

## TABLE 2. MANURE AND TOTAL NUTRIENT PRODUCTION SCHEDULE*

| Animal | No. | Average weight lb | Animal units | Manure as voided† lbs/day | tons/mo | Manure produced per year Location | Months | Amount tons | Total manure nutrients produced per year Nitrogen lb | Phosphorus lb | Potassium lb |
|---|---|---|---|---|---|---|---|---|---|---|---|
| Milk cows | 30 | 1420 | 42.6 | 3834 | 57.5 | Stanchion barn | 12 | 690 | 6900 | 1380 | 5520 |
| Dry cows | 6 | 1420 | 8.5 | 765 | 11.5 | Bedded Pack | 7 | 80 | 800 | 160 | 640 |
| | | | | | | Pasture | 5 | 57 | 570 | 114 | 455 |
| Heifers, 10 mo-fresh | 10 | 950 | 9.5 | 855 | 12.8 | Bedded Pack | 7 | 89 | 890 | 178 | 712 |
| | | | | | | Pasture | 5 | 64 | 640 | 128 | 512 |
| Calves, 1½-10 mo | 11 | 450 | 5.0 | 450 | 6.7 | Barn and yard | 12 | 82 | 820 | 163 | 655 |
| Calves, 0-1 ½ mo | 5 | 120 | 0.6 | 54 | 0.8 | Stanchion barn | 12 | 10 | 100 | 20 | 80 |
| Totals | | | 66.2 | 5958 | 89.3 | | | 1072 | 10720 | 2143 | 8574 |

* One animal unit (A.U.) = 1000 lb live weight. Daily manure production = 90 lb/A.U. One month = 30 days. The chemical analysis per ton of dairy cattle manure is 10 lb Nitrogen, 2 lb Phosphorus, and 8 lb Potassium.

† For milk cows, the manure as voided/day = (90 lb/day/A.U.) (42.6 A.U.) = 3834 lb/day. Manure as voided/month = (3834 lb/day) (30 days/mo) ÷ 2000 lb/ton = 57.5 tons/mo.

## TABLE 3. TOTAL AND AVAILABLE MANURE NUTRIENTS PRODUCED PER YEAR

| Animal | No. | Location | Nitrogen Total lb | Available lb | Phosphorus Total lb | Available lb | Potassium Total lb | Available lb |
|---|---|---|---|---|---|---|---|---|
| Milk cows | 30 | Stanchion barn | 6900 | 2760 | 1380 | 828 | 5520 | 3864 |
| Dry cows | 6 | Bedded pack | 800 | 320 | 160 | 96 | 640 | 448 |
| | | Pasture | 570 | 228 | 114 | 68 | 455 | 319 |
| Heifers, over 10 mo | 10 | Bedded pack | 890 | 356 | 178 | 107 | 712 | 498 |
| | | Pasture | 640 | 256 | 128 | 77 | 512 | 358 |
| Calves, 1½-10 mo. | 11 | Barn and yard | 820 | 328 | 163 | 98 | 655 | 459 |
| Calves, 0-1 ½ mo | 5 | Barn | 100 | 40 | 20 | 12 | 80 | 56 |
| Totals | | | 10720 | 4288 | 2143 | 1286 | 8574 | 6002 |

* Available nutrients calculated using Wisconsin Soil Test recommendations that 40 percent N, 60 percent P, and 70 percent K are used during that crop year.

**MANAGING LIVESTOCK WASTES**

| Field | Crop | Acres | Amount of manure applied | | Fertilizer equivalent | Month to spread | Notes |
|---|---|---|---|---|---|---|---|
| | | | Maximum* | Actual | | | |
| | | | Tons/yr | Tons/yr | lbs N/acre | | |
| 2 | Corn | 7.0 | 175 | 170 | 97 | October | Runoff controlled, difficult to reach winter and spring, fall-plowed. |
| 4A | Oats | 7.0 | 70 | 49 | 28 | October | Some Class III land, fall-plowed. |
| 10B | Corn | 2.0 | 50 | 45 | 90 | April | Accessible in winter. |
| 11 | Corn | 3.0 | 75 | 66 | 88 | April | Class IV land. |
| 12A | Oats | 7.0 | 70 | 42 | 24 | October | Class II land, easy access, fall-plowed. |
| 15 | Corn | 7.0 | 175 | 169 | 97 | May | Easy access in winter, some Class III land. |
| 16 | Oats | 7.0 | 70 | 42 | 24 | October | Class II land, easy access, fall-plowed. |
| 18B | Oats | 3.0 | 30 | 18 | 24 | October | Limited runoff control, Class III land, fall-plowed. |
| 20B | Corn | 14.0 | 350 | 350 | 100 | April-May | Some Class III land, limited runoff control, reasonable access. |
| | Totals | | 1065 | 951 | | | |

* The maximum amount of manure applied is calculated using 250 lb total N/acre for corn and 100 lb total N/acre for oats. No manure is applied on continuing alfalfa hay.

cations in the fall and spring to minimize nutrient losses. Summer manure production (May 1 to October 1) totals 326 tons, and winter production, in addition to the bedded pack, in 456 tons. The notes in Table 4 summarize the soil, slope, and runoff control conditions and accessibility that may affect the decision on time of spreading. Winter spreading is possible on Fields 10B, 15, and 20B, so the volume of winter manure storage needed could be small. Short-term manure storage near the farmstead is recommended. The critical period is during the summer months when cropland is not available for spreading. Summer manure storage is a necessity for this farm.

This farm is not over-stocked from a manure utilization viewpoint. It could carry an additional five milk cows. This is typical of Wisconsin farms that produce most of the feed for their livestock on-site.

## SUMMARY

The Wisconsin Manure Management Plan was developed to coordinate the production and handling of manure with maximum utilization and a minimum pollution potential. It is a simple, rational procedure consistent with sound agricultural practices and environmental concerns.

### References


1  Parker, Dale E., Donald C. Kurer and Joseph A. Steingraeber. 1970. Soil survey Ozaukee County Wisconsin. USDA-Soil Conservation Service. 92 p.
2  Walsh, L. M. and R. F. Hensler. 1971. Manage manure for its value. University of Wisconsin Extension Circular 550. 7 p.




MANAGING LIVESTOCK WASTES

# If You Cannot Spread It, Treat It

Peter M. Wilson

THE approach to management of livestock in the last twenty years has led to the operation of large animal units with high density populations, in particular poultry and pig farming. Two main types of intensive units have been developed, (a) where the enterprise is quite separate from the land and (b) where it is an integral part of the farm. This type of livestock management has resulted in a serious problem of waste disposal.

Legislation to control pollution has clamped down on indiscriminate disposal of farm waste. However, the problem is not insoluble and much encouraging research is being carried out. What the farmer requires in most cases is simple practical advice both to identify the problem and to offer the best solution in the particular circumstance. Such advice may be obtained from the agricultural colleges and the services of a number of consulting engineers. These disciplines can also provide advice on negotiation with river authorities, as frequently the problem is aggravated by a lack of understanding between the farmer and authority.

In particular, I have studied disposal of waste from pig fattening units, although many of the principles may be applicable to the disposal of other livestock wastes.

## LAND SPREADING

Traditionally, livestock manure was applied to land. This technique was not only an effective method of disposal but also supplemented soil fertility. However, the use of man-made fertilizers increased on account of their convenience and labor-saving. With the increasing cost of fertilizer a return to the practice of land spreading of manure seems likely.

Land spreading has since been studied with a more scientific approach. The rate and frequency of spreading are not entirely determined by the season and the weather, but also by the agricultural use of the land, soil type, the dangers of cross infection by pathogenic organisms, chemical toxicity, organic and inorganic pollution of watercourses. More than often the problem is the smell and flies associated with spreading of raw manure.

My advice is that wherever possible spreading is the most economic solution and provides a financial saving, or an income, for the farmer. However, for the reasons mentioned above some degree of treatment of livestock waste is necessary.

## DESIGNING A TREATMENT SYSTEM
## INITIAL CONSIDERATIONS

The nature, extent and cost of a treatment system are determined by several factors of which some of the most important are:

1    Size of the pig unit.

2    Location of the pig unit, for example, proximity to urban communities, other farms and watercourses.

3    Availability, characteristics and use of surrounding land.

4    Existing systems for handling, storage and disposal of waste.

5    Properties of the pig waste.

6    Ultimate fate of treatment products.

With these points in mind, three degrees of treatment may be appropriate:

### Odor Control

Elimination of odor is commonly required where slurry is spread on land close to urban communities.

### Partial Treatment

This is necessary to reduce the danger of pollution of watercourses by runoff and leaching of potential pollutants from land on which waste is applied.

### Complete Treatment

Such a system might be required for the intensive pig unit which has little or no associated land for spreading slurry. It would probably involve a costly tertiary treatment process and a wasteful solids disposal system.

Having defined what is required, it is important to design a simple but reliable system that requires minimum supervision and maintenance. Due to tight economic restraints the farmer must be provided with a system which has the lowest capital and running costs. However, the cost of treatment may be offset by utilization of the by-products on land.

## INTRINSIC MEASURES OF CONTROL

Steps should be taken to minimize the waste quantity and quality. Frequently, an improvement in husbandry techniques, change of feeding regime, change of diet, avoidance of overfeeding and wastage of food can substantially reduce the volume and strength of the waste. Indeed, several of these measures will benefit the farmer by optimizing his livestock management.

Extraneous sources of uncontaminated water, if draining into the waste collection tank, should be diverted directly to the nearest watercourse; a significant source is rainwater from roofs. Where an underground collection tank is employed sources of groundwater seepage should be eliminated.

## TREATMENT SYSTEMS

Several treatment methods, based on biological, chemical and physical principles, are being studied. Although en-

Acknowledgments: While the views expressed in this paper are those of the author, I wish to thank my colleagues at Babtie, Shaw & Morton for their invaluable advice and encouragement.

I also wish to thank members of the North of Scotland and West of Scotland Agricultural Colleges, The University of Newcastle and many farmers in providing practical information.

The author is: PETER M. WILSON, Public Health Engineering Div., Babtie, Shaw & Morton, Glasgow, Scotland.

couraging research is continuing, processes such as anaerobic digestion and incineration are economically infeasible at present and would require experienced supervision.

A treatment system may well incorporate a combination of methods as there is no single all-embracing disposal operation. One of the most successful partial treatment systems involves three stages:

1 Mechanical separation of fibrous solids.
2 Aerobic stabilization of liquor.
3 Final clarification.

The products of this system include biologically stabilized solids and sludges, and a clarified liquor with a much reduced oxygen demand and solids content but high in inorganic nitrogen and phosphorus. The liquor can be expected to have a minimum of 50 mg/liter BOD and 50 mg/liter suspended solids and contain about 200 mg/liter nitrate and 50 mg/liter phosphate which is unlikely to be suitable for direct discharge to a watercourse. It may be used for irrigation or washwater in the piggery. Alternatively, it may be possible to discharge into the public sewer. The fibrous solids may be used for soil conditioning and could be marketed. The sludge can be land spread and may be more readily accepted by neighboring farmers.

## MECHANICAL SEPARATION OF FIBROUS SOLIDS

As much as 60 percent total suspended solids can be attributed to relatively inert fibrous solids, for example, pig hairs and cereal fiber. Their removal will subsequently reduce the build up of a biologically inactive proportion of mixed liquor suspended solids in the aerobic treatment.

Vibrating screens have proved to be the most reliable devices for separating out the fibrous solids and their capital and running costs are much less than other devices. A 100-mesh screen made of either nylon or stainless steel is recommended. The screens are usually circular and vibrated by a low horsepower motor, usually ½ to 3 hp.

For optimum performance, slurry should be applied as fresh as possible. Sixty to seventy-five percent solids removal, up to 20 percent reduction in BOD and 5 to 10 percent reduction in volume can be achieved. The solids removed have a minimum dry matter content of 15 percent and will dry further on standing (drainage liquors should be returned to the rest of the treatment system). The dried product can be used for conditioning soil and could be marketed.

The liquor from screening contains a high BOD, 10 000 to 20 000 mg/liter, and up to 1.5 percent suspended solids. Although it could be spread on land it is potentially polluting and requires biological stabilization.

## AEROBIC STABILIZATION OF LIQUOR

Aerobic treatment of the liquor can achieve four objectives:

1 reduction of odor.
2 reduction of soluble BOD and to some extent insoluble BOD.
3 stabilization and bio-flocculation of fine solids.
4 nitrification.

Odor reduction is achieved by a relatively short aeration period of one to three days but to achieve the other three objectives an aeration period of up to 30 days is required as well as low biological loading rates.

Aeration is largely limited by the concentration of mixed liquor suspended solids in order to obtain adequate mixing and efficient oxygen transfer. For maximum BOD reduction a minimum design factor of 2 kg $O_2$ per kg BOD applied is employed.

Oxygen can be supplied by blowing air into the liquor or by vigorous agitation by rotating blades. Air blowing systems often require auxiliary mixing equipment and the compressors are noisy. There are several types of aerating agitators but as a general rule I recommend the following systems:

1 Pig unit less than 1000 pigs — an oxidation ditch utilizing a horizontally mounted rotating paddle.
2 Pig unit more than 1000 pigs — an aeration tank in which aeration is provided by a floating surface aerator.

In some cases it may be possible to install a surface aerator in the existing slurry storage tank.

## CLARIFICATION

Although the aerobically treated liquor could be spread on land, clarification may be necessary to deal with solids and liquids separately. Clarification may be brought about by gravity settlement with regular sludge draw off.

## INORGANIC NITROGEN AND PHOSPHOROUS

At low biological loadings ammoniacal-nitrogen is largely oxidized to nitrite and nitrate nitrogen by nitrifying bacteria. Due to the association of these forms of nitrogen with eutrophication and toxicity the ultimate fate of the clarified liquor must be carefully considered as nitrate is poorly retained in soils.

Intermittent periods of non-aeration will encourage loss of nitrogen, as the gas, by the activity of denitrifying bacteria.

Phosphate reduction is most effectively carried out by application of the liquor to land as phosphate is well retained by soil. Where this is not feasible chemical removal of phosphate is recommended.

## COST OF TREATMENT

It is often misleading to quote costs because of the ability of the farmer to improvise and utilize second-hand equipment and to carry out construction work himself. Overall cost will also depend on many of the factors mentioned above.

An approximate guide to the order of cost for the 3-stage system described, is given in the following example:

5000 pig fattening installation, meal-based diet, waste production, 45 m³ waste per day

| | |
|---|---|
| Mechanical separation using a vibrating screen | £ 5 000 |
| Aerobic stabilization of liquor by mechanical aeration.    (a) Mechanical | £17 000 |
|    (b) Civil | £20 000 |
| Final clarification | £10 000 |
| Total | £52 000 |

These costs include civil costs, pipework, valves and ancillary equipment. On a 15-year amortized basis, plus 15 percent fixed interest, plus running costs this is about £13 000 per year, i.e. approximately £2.60 per pig place or 90p. per pig produced.

*(Continued on page 394)*

# Evaporation of Water from Holding Ponds

G. L. Pratt, A. W. Wieczorek, R. W. Schottman, M. L. Buchanan

MEMBER
ASAE

THE average annual rainfall at Fargo, North Dakota, is about 51 centimeters (cm), and the average annual evaporation from lake surfaces is 76 cm. In western North Dakota the rainfall averages less than 41 cm, and evaporation from lake surfaces averages about 96 cm. This moisture deficit range of from 25 to 56 cm is typical of Great Plains States where evaporation can be used effectively as a partial disposal system for waste water.

Factors that may affect evaporation include temperature, barometric pressure, relative humidity, wind velocity, atmospheric pollution, and water pollution. Data on evaporation have been recorded from clear water lakes. Waste water carries a high solids content and coloration. The research reproted here was designed to determine the effects of contaminants on evaporation of waste water.

## MATERIALS AND METHODS

A barn for beef cattle and an evaporation pond were used for conducting these tests (Fig. 1). The barn houses 20 head of feeder cattle. It has a slatted floor over a pit that is about two feet deep. The concrete floor in the pit has a two percent slope to one end of the building. The sloping floor provides gravity flow separation of urine from the feces. The urine is collected at the low end and pumped to the evaporating pond. A cable type scraper system conveys the solids to the opposite end of the pit. A cross conveyor intercepts the solids and takes them out of the building.

A 10 mil vinyl (PVC) liner was placed on the pond bottom and covered with eight inches of soil to prevent seepage from the pond. The soil cover prevented deterioration of the liner by the sun.

A holding tank was used to measure inflow into the pond from the barn. Water was pumped to the tank. The volume of water in the tank was measured each day and the tank was drained. Approximately 380 cubic meters (m$^3$) of clean water were pumped into the pond at the beginning of the test to provide a base line for measuring evaporation.

Waste water drained from the building at a rate of just over 0.021 liters per day-kilogram (kg) of animal weight. This figure appeared to be fairly constant throughout the test period. The average total production of liquid waste drained from the barn for 400 to 500 kg feeders was 9.45 liters/day/head. In a one-year period this would be 3.45 m$^3$ per head. This amount, plus the precipitation on the pond, must be disposed of each year to prevent the pond from overflowing.

The barn was provided with a tile drainage system around a basement under part of the building. Liquids from the tile line were pumped to the lagoon. Average monthly flow was about 1,590 liters.

Work on which this project is based was supported by the North Dakota Agricultural Experiment Station, North Dakota State University. Project No. 1422.

The authors are: G. L. PRATT, Professor, A. W. WIECZOREK, Former Graduate Research Assistant, R. W. SCHOTTMAN, Assistant Professor, Agricultural Engineering Dept., and M. L. BUCHANAN, Professor and Chairman, Animal Science Dept., North Dakota State University, Fargo.

FIG. 1 Test site. Beef barn on right. Lower pond utilized for evaporation studies.

Table 1 is a record of the rainfall for each month during the summer test period and the corresponding water level increases in the pond. Part of the rain that fell on the banks of the pond flowed into the pond.

Samples of the urine drained from the sloping barn floor were collected once a month throughout the summer months for quality tests. At the same time waste water samples were collected from the evaporation pond for testing.

**TABLE 1. SUMMER RAINFALL RECORD AT EXPERIMENT SITE.**

| Date | | Rainfall, cm | Pond depth increase, cm |
|---|---|---|---|
| June | 3 | 1.02 | 1.52 |
| | 8 | 1.02 | 1.52 |
| | 16 | 1.12 | 1.22 |
| | 18 | 0.76 | 0.91 |
| | 19 | 0.79 | 0.91 |
| | 20 | 0.13 | --- |
| | 27 | 0.41 | -- |
| July | 2 | 0.41 | 0.91 |
| | 9 | 1.27 | 2.13 |
| | 11 | 0.13 | 0.30 |
| | 17 | 0.30 | 0.46 |
| | 18 | 1.78 | 2.74 |
| | 23 | 0.33 | 0.46 |
| | 24 | 0.46 | 0.61 |
| | 26 | 1.22 | 1.83 |
| | 27 | 0.46 | 0.61 |
| | 28 | 0.13 | --- |
| | 29 | 0.25 | 0.46 |
| August | 6 | 1.07 | 1.50 |
| | 7 | 1.37 | 2.01 |
| | 13 | 0.25 | 0.46 |
| | 15 | 0.13 | --- |
| | 21 | 2.72 | 5.49 |
| September | 1 | 4.98 | 14.63 |
| | 3 | 4.62 | 3.66 |
| | 15 | 0.10 | 0.46 |
| | 21 | 1.27 | --- |
| | 22 | 2.54 | 4.27 |
| | 24 | 3.48 | 4.27 |
| | 27 | 0.23 | 0.46 |

| 1973 | Wind movement km/hr | | Percent, relative humidity | | Temperature, deg C | |
|---|---|---|---|---|---|---|
| | Experiment site | Agroclimatic station | Experiment site | Agroclimatic station | Experiment site | Agroclimatic station |
| June | 7.08 | 19.31 | 62.0 | 63.6 | 18.6 | 19.3 |
| July | 7.74 | 7.26 | 65.0 | 67.3 | 20.7 | 20.4 |
| August | 7.88 | 5.66 | 68.3 | 72.3 | 22.3 | 22.6 |
| September | 5.86 | 6.24 | 71.5 | 75.8 | 12.9 | 13.4 |
| Average | 7.14 | 9.62 | 66.7 | 69.7 | 18.6 | 18.9 |

Tests were completed to determine total solids, volatile solids, fixed residue suspended solids, dissolved solids, volatile suspended solids, and volatile dissolved solids. Procedures used are given in Standard Methods. Five-day 20 C BOD tests were conducted using a Hach Manometric BOD Apparatus. Three replications of each sample were tested. Turbidity tests of each sample were conducted using a Hach Laboratory Turbidimeter. Values were recorded in Jackson Turbidity Units.

Urethane foam was sprayed on the outer sides of a standard evaporation pan 25.4 cm deep and 120.6 cm in diameter to cause the pan rim to float 6.4 cm above the pond water level. The water was maintained in the pan at a level of 6.4 cm from the top of the pan rim to keep it equal to that of the pond. The floating pan was centered in the pond and held there by guide ropes anchored to the banks. Other equipment included a water level recorder, a remote reading thermograph, a totalizing anemometer, a weighing type rain gage, and a hygrothermograph.

### Climatic Conditions

Climatic data collected at the lagoon site were compared with the climatic data collected at the North Dakota State University Agroclimatic Station. The lagoon was sheltered on the west by trees. The Agroclimatic Station was located about a mile from the lagoon and was not protected by trees. The climatic variables, air temperature, water temperature, relative humidity, wind movements, and water quality differences were recorded.

Monthly mean values of the climatic variables at the Agroclimatic Station and the pond site are recorded in Table 2. The mean wind velocity for the summer months was 2.65 km/hr higher at the Agroclimatic Station than it

was at the pond site. The monthly mean relative humidities were higher each month at the Agroclimatic Station than at the pond site. The mean at the pond site was 66.7 percent and Agroclimatic Station it was 69.7 percent. The mean air temperature was highest at the pond site in July and at the Agroclimatic Station in the other three months. The overall difference in the mean air temperature at the two sites was 0.3 C. The monthly mean water temperature for the months from June through September was 19.7 C in the Class A land pan at the Agroclimatic Station while the value for the pond surface was 20.1 C. The mean temperature of the water in the floating pan was 20.3 C.

## RESULTS AND DISCUSSION

### Pond Water Level Fluctuation

Fig. 2 is a record of the monthly water level fluctuation of the evaporation pond utilized for the disposal of the animal liquid wastes. Pond depth increased 15.9 cm in the months from November through March. This represents drainage from the barn plus some snow accumulation. The pond level dropped 24.4 cm from March through July. Since the pond was lined with a plastic liner, this loss can be attributed to evaporation. Drainage from the barn continued during the time this evaporation took place.

Fig. 3 gives a series of curves showing the rate at which water was added to and evaporated from the pond during June, July, August, and September. The top curve gives the cumulative flow into the pond from building drainage and

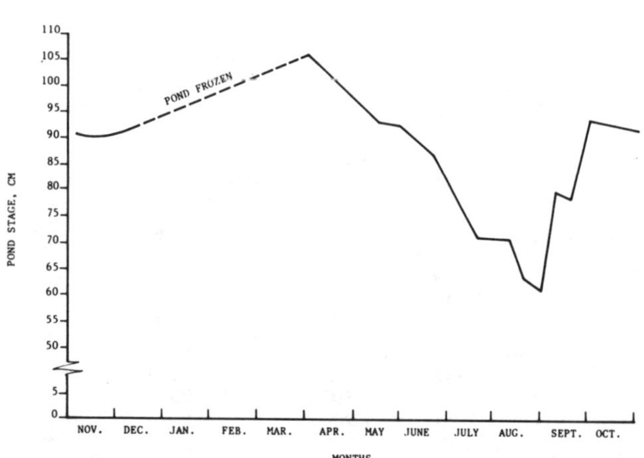


FIG. 2 Observed pond levels for the period from November 1972 to October 1973.

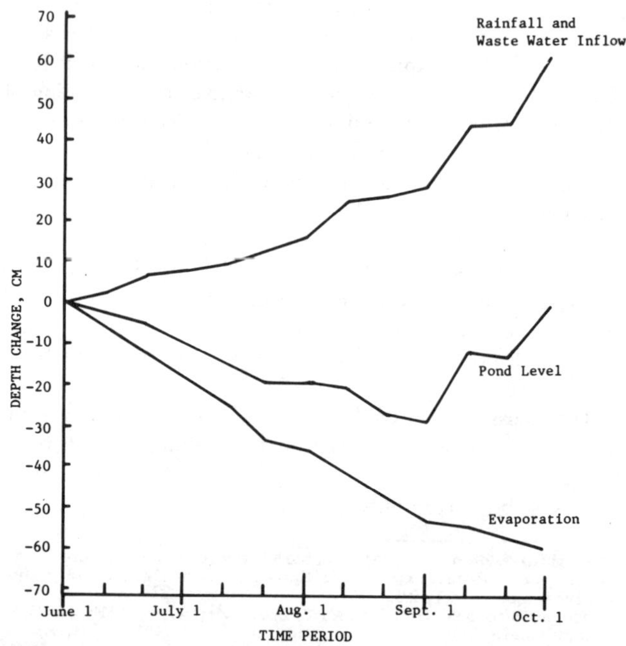


FIG. 3 The combined effects of inflow and evaporation on pond level for the Summer of 1973.



MANAGING LIVESTOCK WASTES

TABLE 3. WATER TEMPERATURES, DEG C

| 1973 | Land pan agroclimatic station | | | Floating pan experiment station | | | Evaporation pond surface | | |
|---|---|---|---|---|---|---|---|---|---|
| | Max. | Min. | Mean | Max. | Min. | Mean | Max. | Min. | Mean |
| June | 26.3 | 12.8 | 19.6 | 25.1 | 16.0 | 20.6 | 23.6 | 15.6 | 19.6 |
| July | 28.6 | 14.8 | 21.7 | 27.1 | 17.9 | 22.5 | 26.9 | 18.6 | 22.8 |
| Aug. | 29.5 | 16.4 | 22.9 | 27.2 | 18.6 | 22.9 | 27.3 | 18.5 | 22.9 |
| Sept. | 20.2 | 8.8 | 14.5 | 19.2 | 10.8 | 15.1 | 17.7 | 12.6 | 15.2 |
| Average | 26.2 | 13.2 | 19.7 | 24.6 | 15.8 | 20.3 | 23.9 | 16.3 | 20.1 |

TABLE 4. MEAN WATER QUALITY VALUES FROM TESTS<br>CONDUCTED FROM JUNE THROUGH AUGUST, 1973.

| | Barn waste water analyses | Evaporation pond waste water analyses |
|---|---|---|
| Temperature deg C | 17.3 | 20.3 |
| BOD, ppm | 3,700.0 | 681.0 |
| Total solids, ppm | 17,496.6 | 3,041.2 |
| Volatile solids, ppm | 7,626.6 | 761.8 |
| Fixed solids, ppm | 9,872.0 | 2,279.4 |
| Suspended solids, ppm | 2,725.3 | 138.6 |
| Dissolved solids, ppm | 14,771.3 | 2,902.6 |
| Volatile S.S., ppm | 1,183.0 | 118.0 |
| Volatile D.S., ppm | 6,441.6 | 643.8 |
| Turbidity, JTU | 64.7 | 32.7 |

precipitation. The lower curve gives the total evaporation as was recorded from the floating evaporation pan. The third curve shows the change in the pond level for the four-month period. A high rainfall rate in September brought the pond level back up to about the level that it was in November of 1972.

## Water Quality

Laboratory tests conducted on the liquid wastes drained from the confinement building and collected from the evaporation pond are recorded in Table 4. These tests indicate that the organic load in the liquids draining from the confinement building was high in comparison with the organic load in the evaporation pond. Since evaporation nearly compensated for the amount of precipitation in the test, the effect of dilution from precipitation was probably negligible. Much of the improvement can be attributed to settling of solids and bacterial decomposition.

## Pan Coefficients

Observed monthly evaporation data and derived coefficients for the evaporation of waste water are presented in Table 5. Evaporation from the Class A land pan was an average of 4.8 cm/month greater than evaporation from the pond for the months of May through October. The highest difference occurred in May and amounted to 6.99 cm. The

TABLE 5. SUMMARY OF OBSERVED PAN AND RECORDED POND<br>EVAPORATION AND MONTHLY PAN COEFFICIENTS.

| Period 1973 | Evaporation, cm | | | Pan coefficients | |
|---|---|---|---|---|---|
| | Land pan agroclimatic station | Floating pan | Evaporation pond | Land pan agroclimatic station | Floating pan |
| May | 18.72 | --- | 11.73 | 0.63 | --- |
| June | 23.14 | 17.20 | 16.84 | 0.73 | 0.98 |
| July | 22.52 | 18.58 | 18.57 | 0.82 | 0.99 |
| August | 19.84 | 16.52 | 16.05 | 0.81 | 0.97 |
| September | 8.56 | 10.02 | 9.49 | 0.67 | 0.95 |
| October | | --- | 5.33 | 0.62 | --- |
| June - September Ave. | | | | 0.76 | 0.98 |
| Totals | 79.63 | 62.31 | 60.95 | | |
| May - October Ave. | | | | 0.73 | --- |
| Totals | 106.91 | | 78.02 | | |

monthly evaporation from the floating pan was an average of 0.43 cm higher than the pond evaporation. The pan coefficient for the floating pan is calculated by dividing evaporation from the pond by the evaporation from the floating pan. The average coefficient for the summer of 1973 was 0.98. This shows the close relationship between evaporation from the pond and from the floating pan. The fact that losses from the floating pan were slightly greater than from the pond would suggest that the pond liner had effectively controlled seepage of water into the soil.

Land pan coefficients were determined by dividing monthly pond evaporation values by monthly observed land pan evaporation values. These calculated coefficients values are low during May and the late summer months of August and September. The remaining months have an average coefficient value of 0.78. The total averaged six-month pan coefficient is 0.73. When multiplying these coefficient values by observed Class A pan evaporation data, a reasonable estimate of waste water evaporation will be provided.

The pan coefficient of 0.73 that was calculated from data recorded from the test pond is similar to the pan coefficient of 0.75 that is generally used for estimating the evaporation of lake water in eastern North Dakota. The water in the pond was made up of urine and rainwater. Most fecal solids were removed from the urine. Dissolved solids in the water averaged 2902.6 ppm. Suspended solids averaged 138.6 ppm. Turbidity averaged 32.7 JTU. Test results indicate that evaporation characteristics of water of this quality are similar to those of lakewater.

Land pans or floating pans may be used to evaluate seepage losses from waste storage ponds. Clay linings in ponds tend to seal when kept wet but often crack when they become dry after they are pumped. Ponds with porous bottoms do not seal well. Seepage from ponds such as these can be estimated by using the land pan or floating pan to measure evaporation losses. Total losses during periods of little or no runoff can be estimated from measurements taken on the pond. A staff gage in a stilling well will provide data on changes in pond dept. Pond surface areas, measured at the beginning and end of a test peiod, can be averaged and combined with the loss in pond depth to estimate total losses. Total losses minus evaporative losses will give the estimate of seepage losses. Floating pans may be preferable to land pans for these measurements. Floating pan coefficients do not appear to vary with the season of the year as do the land pan coefficients.

---

## Spread it or Treat it

*(Continued from page 390)*

### SUMMARY

A 3-stage treatment system for pig waste has been described, which will reduce odor and the pollutional characteristics of the raw waste. A high quality effluent is not expected and it is assumed that the products of treatment will be utilized rather than disposed of. Costly tertiary methods would be required for a higher degree of treatment. An approximate cost of 90p. per pig produced is quoted but the apparent cost depends on several local factors.

# An Economic and Managerial Evaluation of Manure Fluming and Land Application Systems

Paul B. Bohley, C. Robert Near, Dale Rasmussen
EXEC. AFFILIATE
ASAE

E VOLUTION of beef cattle feedlots, in size, location, design, and density has followed a somewhat logical progression. Economics, increased demand for beef, growing competition for land, coupled with rising labor costs and widely fluctuating meat and grain prices, are factors which have influenced the livestock producers during the past decade. The complexity of the situation has been compounded by the American farmers' forced entry into a world market for purchases, as well as sales, and by increased environmental attention.

The energy shortage, real or imagined, has added still a further variable for management to consider. Recent evidence points towards a developing trend of optimum size for feedlots rather than maximum size. In fact, several lots in consideration by this report are maintaining or increasing human input as opposed to more labor saving, but energy using equipment.

Recycling animal waste on land has always been favored, primarily as disposal. The oil shortage resulting, among other things, in the skyrocketing cost of commercial fertilizer, has emphasized the value of land waste recycling and offered potential savings in crop cost inputs. A primary problem then exists of how best to accomplish this re-use program.

One waste management system that has been readily accepted by livestock producers using confinement housing, is manure fluming. There are diverse approaches to the use of waste fluming, depending on livestock type, climate, topography, operator perference, and other variables. During the summer of 1974 the authors visited several beef cattle feedlots in western Iowa and eastern Nebraska. The waste management systems were examined in some detail with emphasis on component parts; i.e., production – collection – holding – transfer – storage and re-use. Two systems are described and evaluated from a management standpoint relating to ease of operation, reliability and completeness. No attempt is made to evaluate or compare the biological processes in various methods of feedlot waste management. Solids separation total application on land and lagoon longevity and activity are considered in relationship to the mechanics and hydraulics of the complete system.

System No. 1 is an 1100-head feelot built in 1971, with cattle being confined on December 12 of that year. An open lot is used for holding approximately 500 head prior to moving into the confinement building.

The barn is 550 ft (168 m) long, oriented east and west, and open to the south. It is 52 ft (16 m) wide and has an 18 in. (0.5 m) slope in the floor from each end to the center.

The operation is located in a township with approximately 500 population density, including a small community of 200 people one mile south of the lagoon location. Prevailing winds in the winter are W-NW and in the summer, W-SW. Land is quite level to gently rolling, with the soil type a light silty clay loam.

The flume in this confinement housing set-up is 7 ft (2.1 m) wide and 2 ft (0.60 m) deep, covered by a slotted floor and located longitudinally in the center of the building. The flushing action is from each end towards the center, through the 7 ft (2.1 m) flume, the manure being evacuated through a gravity system to a lagoon which is approximately 150 ft (46 m) to the south of the barn and at a considerably lower grade. (Fig. 1).

The lagoon is 200 ft by 200 ft (70 m by 70 m), with a maximum depth of 28 ft (8.5 m). It is located 12 ft (3.5 m) above grade and approximately 20 ft (6.1 m) below grade, with sloping sides. The concrete pump house is built into the dike at the west end, and contains a 6 in. (152 mm) pump. The suction pipe enters the lagoon about 7 ft (2.1 m) above the bottom of the lagoon.

Initially, the flushing was conducted under a manual system during the two and one-half to four hour feeding cycle. In the early fall of 1974 the system was automated to operate for six minutes during each hour throughout the year. The discharge of the pump divides with a "Y" pipe fitting at the confinement barn, flushing from each end at the same time. With the advent of the automatic flushing system, the operator indicates almost no trouble at all in handling the waste, and the animals in the barn are very clean.

The 6 in. (152 mm) pump moves approximately 800 to 900 gallons (3028 to 3406 liters) of effluent per minute

FIG. 1 Aerial view of lagoon construction showing relationship to barn.

---

The authors are: PAUL B. BOHLEY, Agricultural Marketing Manager, The Gorman-Rupp Co., Mansfield, OH; C. ROBERT NEAR, Sales Manager, and DALE RASMUSSEN, Animal Waste Specialist, Hastings Irrigation Pipe Co., Hastings, NE.

**MANAGING LIVESTOCK WASTES**

FIG. 2 Lagoon effluent application prior to planting.

FIG. 4 Flume installation showing arrangement of three flumes.

during the six minute flushing cycle. A dishcarge check valve in the line prevents return flow upon shutdown. Since flooded suction is utilized, no air relief valve is needed in the system to assist in priming. Velocity in the discharge pipe is about 2.5 ft (0.8 m) per second. This is considered minimal to prevent solids settling out. However, the static elevation difference and the low solids content, 3 percent, reduce the solids settling problem considerably. A clean-out is provided, and the system can be flushed with fresh water if needed.

It is possible, by valving, to direct out-flow to one or the other end of the barn. A noticeable build-up occurs at the sides of the 7 ft (2.1 m) flume channel, due to he sudden reduction in velocity and force of the flush water as it spreads out in the flume. The frequent sequence of flushing as opposed to the manual flushing technique used originally, and the slope of the flume floor, help control this.

Within the six minute time span, the flumes are cleaned and the manure washed to the lagoon. In four years there has been some build-up around the discharge area of the 24 in. (609 mm) gravity return line. One estimate places this at approximately 70 ft (2.1 m) in diameter, comprising some 760 tons of material with 27 percent total solids. This, together with the 1 1/2 to 4 ft (0.45 to 1.2 m) thick crust, estimated at 1500 to 2000 tons, and the build-up of solids on the bottom, suggests a useful life of the lagoon to be something less than ten years at present loading rates, unless solids reduction of one type or another is practiced.

Irrigation through gated pipes is accomplished during the summer with the manure being applied to growing corn. An 8 in. (203 mm) engine driven pump is used to evacuate

FIG. 3 Effluent application on sod by spray irrigation.

approximately three acre feet of manure water, the suction inlet pipe located 10 ft (3 m) above the lagoon floor. At the same time, an equal amount of fresh water is applied to the lagoon from a deep well, with the discharge being also at a low level of the lagoon. With the effluent being removed from one end of the lagoon and the fresh water applied at the other end, a circulation is set up which helps to remove some of the solids from the sides and bottom, but no attempt is made to break up the crust since odor control is desirable.

The manure lagoon at 22 ft (6.7 m) of depth contained approximately $39,600 worth of nitrogen based on fall 1974 nitrogen prices. However, many factors affect the nitrogen content; therefore, the contents of each lagoon should be analyzed.

The value of the lagoon product is realized on growing corn as an adjunct to the normal fertilizer program. Timing of irrigation with lagoon effluent is important. More research and testing are needed to determine correct parameters for growing crops. Spray irrigation manure lagoons evaluated for this paper were all of sufficient size to hole several years' production. Thus, the need for evacuation at any one time to make additional room has been eliminated and land application can be optimized.

The use of gated pipe-gravity irrigation (Fig. 2), or a high pressure volume gun approach (Fig. 3), eliminates the need for heavy hauling equipment and reduces labor costs considerably. System No. 2 is located in located in northwest Iowa. Topography is quite rolling, and the immediate area has a low population density. Prevailing winds are W-NW in the winter and W-SW in summer.

The 2000-head confinement barn is approximately 900 ft (274 m) long by 58 ft (18 m) wide. A 200 ft by 700 ft (61 m by 213 m) manure lagoon with a maximum depth of 36 ft (11 m) is immediately adjacent. Capacity is approximately 31 million gallons (119.2 million l) when the lagoon is filled to a depth of 30 ft (9 m). A second confinement barn is under construction on the other side of the lagoon. Both buildings open to the south, being oriented on an east-to-west axis. The barn was first occupied in October 1973.

The floor is cleaned by three flumes, 14 in. (355.6 mm) diameter PVC pipe (Fig. 4), supplied by effluent from the lagoon with a 6 in. (152 mm) chopper type trash pump (Fig. 5). The sequence of flushing is ten minutes every hour in winter and fifteen minutes every two hours in summer. During extremely cold weather, flushing may be continuous. Normally the pump is on automatic control.

A butterfly valve in the discharge of the pumping system



MANAGING LIVESTOCK WASTES

FIG. 5 Six-inch chopper pump in pump house.

induces artificial head, reducing flow capacity to 250 GPM (946 L/M). Effluent for flushing is supplied through a channel at the end of the barn and all flumes are supplied evenly. There is approximately two minutes' separation between the actual flushing acitivty of the flumes, with the center flushing clean first, the flume on the open side of the barn flushing next, and the feed bunk flume flushing last due to the heavier build-up of manure in this flume. During flushing, the water flows over the top of the waste, and commences eroding and flushing the manure from the downstream end. The flume discharge is collected by a manifold at the end of the barn, prior to evacuation to the lagoon (Fig. 6).

There is a build-up on part of the surface of the lagoon and a minimal build-up on the bottom, since this lagoon had been in operation for approximately ten months when visited.

Recent tests by the University of Iowa indicated that with the lagoon at a depth of 22 ft (6.7 m), there was an

FIG. 6 Flume collection return trough.

FIG. 7 Flume opening showing relative cleanness of cattle and floor,

average solids content of approximately 2 percent. Calculating the nutrient value indicated that with 21 million gallons (784,850 cu m) in the lagoon, the nitrogen content was approximately 115,000 lb (52,164 kg), the phosphorus content about 35,000 lb (15,876 kg), and the potash content about 69,000 lb (31,298 kg). Based on October commercial fertilizer prices, this would accrue a value of around $35,000. A high pressure irrigation system is programmed for use during the summer of 1975 to take advantage of this valuable product.

The manager of this lot previously operated on open feedlot and his comments are pertinent in comparing the efficiency and economy of a confinement lot with a fluming system to open lot operation. He indicates that his incidence of hoof rot has practically disappeared and he has had no problems with respiratory or other diseases of this type. He also indicates that the considerable saving in labor is a particular benefit to be derived from the use of confinement housing manure lagoon program.

He allows approximately 18 to 20 sq ft (1.67 to 1.85 cm$^2$) of space per animal, and his animals are very clean (Fig. 7). The extremely heavy snowstorm that occurred in northwest Iowa and eastern Nebraska in early February 1975, accounting for some 96,000 cattle deaths in open lots and on the range, pointed up the value of confinement housing where the incidence of loss was extremely low.

In both confinement buildings considered, freezing has been minimal in the flume intakes and outlets, due to the frequent sequence of flushing, and in the one case to continous flushing when the temperature was at minus 20-25 degrees F (minus 4-13 degrees C).

The use of a manure storage lagoon in conjunction with a flushing system is of particular benefit from an environmental standpoint. While there is still much to be learned, it appears that with proper management, including loading rate of lagoon and correct system design, odor can be controlled. The operator of the first system described indicates that during the flushing sequence there can be some objectionable odor and he also indicates that during the gated pipe application to the field there is quite a prevailing odor for a short period of time. It may be that a high pressure irrigation system on this lagoon would have deleterious odor effects were the population density greater than that which exists at the present time. No fly problem has developed with either of these lagoons, the greatest number of

(Continued on page 401)

# Energetics of Alternative Waste Management Systems

H. C. Kim, D. L. Day

MEMBER
ASAE

IT is necessary to assess major waste management systems in the livestock industry and to determine the most appropriate one in terms of monetary budget. For the benefit of national energy supplies, it is also interesting to analyze and compare the major alternatives in livestock waste management in terms of energy budgets.

This paper summarizes a report by Kim and Day (1974).

## OBJECTIVE

The objectives of this work are to evaluate and compare the major systems of livestock waste management and to decide which one is the most economic, in energy as well as monetary budgets.

This report is limited to studying hypothetical swine production facilities that market 4000 hogs per year. These facilities maintain farrow-to-finish production stages, averaging 2000 hogs at a time.

The systems considered in this paper all include some method of utilizing the wastes. They are as follows:

1 Anaerobic storage systems, with contents spread on crop land for fertilizer.

2 Oxidation-ditch, with feeding of oxidation-dtich mixed liquor (ODML).

3 Anaerobic digestion, producing methane gas.

4 Drying and refeeding. Actually, there are some surplus liquid products for fertilizer in all the systems except drying.

## PROCEDURE

The energy expenses include all man-controlled energy inputs—energy in materials, labor force, and equipment and facility expenses averaged over the life of the equipment. Operations that require direct energy such as gasoline, fuel oil, and electricity are applied in terms of energy contents rather than as direct inputs due to the method of production. Energy returns refer to energy savings due to utilization of waste—as fertilizer; replacements of diet (i.e., single-cell protein and dried manure; heat; and electric power.

These energy values are expressed in the average annual expenses of 1974 dollars, then converted into 1963 values to obtain the corresponding energy expenses by referring to an energy input-output matrix (Table 1). The entire input-output analysis as reported in the original bulletin (Herendeen 1973) offers a theoretical framework and large data base well suited to estimate the total energy costs of many classes of products in the United States on a 1963 price structure.

Cost estimations of equipment, such as drier and heat exchanger, etc., were carried out according to the Chemical Engineers Handbook (1963) and Cost Engineering (Morgan 1960). The costs were converted into current costs by referring to Fig. 1, which shows material indexes for metal products group (Zimmerman et al. 1971). The 1974 index was obtained by extrapolating the previous values. Cost conversions for farm machinery and buildings were made according to Fig. 2, adapted from Agricultural Statistics (USDA 1973) and Agricultural Prices (USDA 1974b; 1974c).

The basic requirements of confinement housing facilities were assumed to be the same in all systems, i.e., roofs, walls, ventilation, lighting, and watering systems. The expenses of floors and storage systems were involved in the cost comparisons of building facilities. A minimum of 120 days storage was included in the pit for anaerobic storage and in the lagoon for anaerobic digestion.

Hauling expenses were evaluated as suggested in the AGRICULTURAL ENGINEERS YEARBOOK (1969). Re-

The authors are: H. C. KIM, Research Assistant and D. L. DAY, Professor, Agricultural Engineering, University of Illinois at Urbana-Champaign 61801.

## TABLE 1. THE TOTAL PRIMARY ENERGY PER FINAL DEMAND

| Used in this paper | Code number* | Information given | (Cal/dollars 1963) x $10^{-4}$ |
|---|---|---|---|
| Digester | 13.03 | Tank | 2.2727 |
| Feed | 14.15 | Prepared feeds for animals | 1.8327 |
| Fertilizer | 27.02 | Fertilizer | 4.5322 |
| Concrete material | 36.12 | Ready-mixed concrete | .3606 |
| Oil burner | 40.03 | Heating equipment except electric | 1.8524 |
| Heat exchanger | 42.08 | Pipe | 1.8583 |
| Pipe | 42.08 | Pipe | 1.8583 |
| Tractor and spreader | 44.00 | Farm machinery | 1.9432 |
| Conveyor | 46.02 | Conveyor | 1.6225 |
| Dryer | 48.06 | Special industry machinery | 1.6400 |
| Rotor | 49.01 | Pump or compressor | 1.4680 |
| Pump | 49.01 | Pump or compressor | 1.4680 |
| Blower | 49.03 | Blowers and fans | 1.5960 |
| Motor and generators | 53.04 | Motor and generators | 1.6475 |

* Reference number of corresponding energy value in the original input and output table.

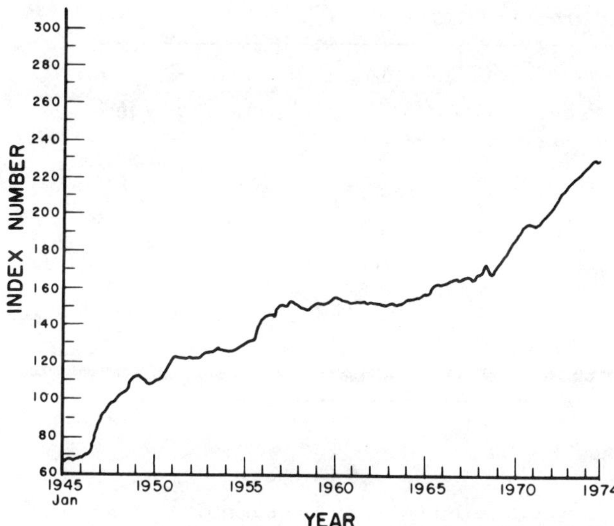

FIG. 1 The wholesale price indexes for metals and metal products group.

turns from hauling were estimated on the basis of replacing fertilizer and on the assumption that the cropland is sufficient to utilize the manure. During the storage of digested sludge in a lagoon, nitrogen losses were considered to be 20 percent (Vanderholm 1974). Some settling losses of P and K may also occur in the lagoon.

The 1963 prices for use in the matrix were N, 15.3 cents/Kg; $P_2O_5$, 20.3 cents/Kg; $K_2O$, 9.8 cents/Kg (USDA 1964). The 1974 prices used in evaluating dollar returns were N, 27.6 cents/Kg; $P_2O_5$ 52 cents/Kg; $K_2O$, 15.2 cents/Kg (USDA 1974a, p. 24). The energy consumption of the labor force was based on the labor of a normal man, who weighs 150 lb, is moderately active, and has a daily energy intake of 3250 Kcal (Swift 1959).

Energy return in the ODML was calculated, based on the limiting component of lysine in the ODML (Day and Harmon 1974) and in dried manure 15 percent replacement of commercial diet (Diggs et al. 1965). Estimation of refeeding benefits was based on the value of commercial feed replaced. The 1963 prices for use in the matrix were soybeans, 11.1 cents/c; corn, 7.2 cents/g (USDA 1964). The 1974 prices used in evaluating dollar returns were soybeans, 22.7 cents/g; corn, 17.2 cents/g (USDA 1974a, p. 18).

In the confinement housing with farrowing, nursery,

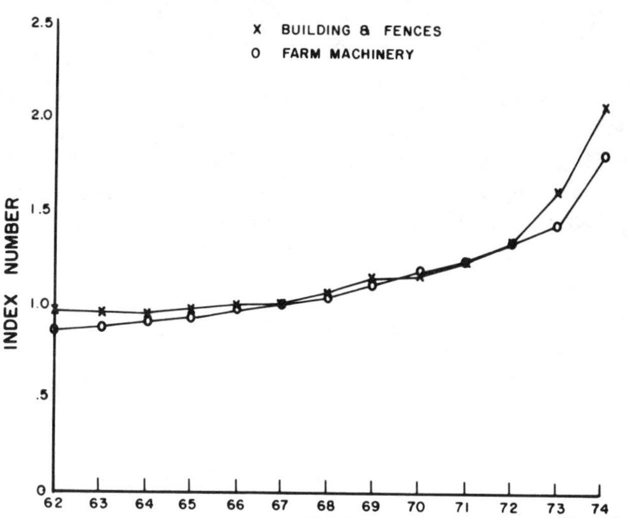


FIG. 2 Price indexes paid by farmers: by group of communities, U.S. average.

growing, and finishing facilities, the average weight of hogs was obtained from Table 2, which was computed from a typical growth curve (Midwest Plan Service 1972).

TABLE 2.

ESTIMATE OF WEIGHTED AVERAGE WEIGHTS OF SWINE

| Growth stage, Kg | Days required | Fraction of total days | Average weight of stage, Kg | Weighted average, Kg |
|---|---|---|---|---|
| 2.27-13.61 | 42 | .23 | 158.76* | 5.22 |
| 13.61-45.36 | 56 | .30 | 29.48 | 8.85 |
| 45.36-68.04 | 35 | .20 | 56.70 | 11.34 |
| 68.04-99.79 | 49 | .27 | 63.92 | 22.68 |
|  | 182 | 1.00 |  | 48.09 |

* 103 Kg for a sow and 55.76 Kg for seven pigs.

## RESULTS AND DISCUSSION

Table 3 shows the comparison of input elements in the average annual costs. The major equipment refers to the five rotors of 5 hp unit with drives and motors (Newton 1970), a digester of 240 $m^3$ made of chromized steel plate (Ross 1974), and a continuous through-circulation conveyor dryer with shelter (Chemical Engineers Handbook 1963). Appurtenance refers to a "V" through-style dual-roll conveyor for fresh manure transportation, a heat exchanger, scrubber, pumps, pipes and fittings, etc.

The percentage of floors and storage expense over the total expense appears to be quite high except for the drying system. This is because the manure is not handled in liquid phase but in solid phase, thereby excluding the bulk manure storage facilities, which are estimated to be quite expensive in the other three systems. The energy expenses of those bulk facilities, however, do not show a high percentage of total expense since concrete has a low energy requirement compared to other inputs.

The operating energy costs in these systems, which require large quantities of heat energy and electric power, are by far the greatest part of energy input.

From Table 3, the energy input in drying operation reaches near 90 percent of the total expense, since a tremendous amount of fuel oil combustion is required. The inconsistency of the energy over dollar value is considered to be caused by the various energy requirements in the different input elements. A comparison of operating costs of the oxidation ditch and anaerobic digestion also shows a discrepancy between the energy and the dollar value. Although a large quantity of heat is required in the operation of the anaerobic digestion, this energy is predominantly from methane gas from within the system at the low cost of $2.98 per million Kcal (Pfeffer 1973). The high energy consumption in the oxidation ditch operation is mainly due to the electric power for running rotors (470 kWh per day), which makes it expensive to run rotors. Based on 2 cents kWh, the ratio of dollars to Kcal turns out to be $23.26/ million Kcal.

As seen in Table 3 the hauling cost increases in the digestion system, since it requires dilution with a water/manure ratio of 1.9 to avoid $N\text{-}NH_3$ toxicity (Miner and Smith 1974); also, a mixed form of digested sludge must be disposed of, rather than the settled sludge only, to prevent loss of N-component in the supernatant.

The energy consumption in the drying system reaches

| | Anaerobic storage | | Oxidation ditch | | Anaerobic digestion | | Drying | |
|---|---|---|---|---|---|---|---|---|
| | Dollars | Cal x $10^{-6}$ | Dollars | Cal x $10^{-6}$ | Dollars | Cal x $10^{-6}$ | Dollars | Cal x $10^{-6}$ |
| Capital costs | | | | | | | | |
| Major equipment | — | — | 2,630 | 25.52 | 2,412 | 35.75 | 7,930 | 114.58 |
| Appurtenance | — | — | — | — | 1,286 | 16.84 | 3,624 | 38.86 |
| Floors and storage | 5,686 | 9.51 | 5,394 | 9.02 | 4,488 | 15.03 | 2,196 | 3.67 |
| Operating costs | | | | | | | | |
| Electricity | — | — | 3,428 | 147.45 | 276 | 11.87 | 1,117 | 48.03 |
| Fuel (or heat) | — | — | — | — | 1,412 | 262.40 | 13,498 | 1,399.35 |
| Labor | — | — | — | — | 1,095 | .05 | 12,529 | .57 |
| Hauling costs | 2,120 | 26.28 | 146 | 1.71 | 4,583 | 62.55 | — | — |
| Total | 7,806 | 36.19 | 11,598 | 183.70 | 15,552 | 404.49 | 49,894 | 1,605.06 |

the highest level among the alternative systems considered in this paper. This may be caused by the fact that the drying is accomplished by the heat from burning the fuel oil, whereas in the oxidation ditch and in the anaerobic digestion the microorganisms are employed free of charge. The latter two systems only supply the proper environments for the worker: aerobic bacteria for the oxidation ditch, and methane bacteria for the anaerobic digestion.

Table 4 shows the dollar and energy returns for the alternative systems. In the drying for refeeding system, 15 percent of the commercial diet is replaced by the dried manure, according to the results of refeeding experiments of Diggs (1965). On the other hand, only the protein portion of the ODML (based on the limiting nutrient of lysine) is counted without any replacement of the remaining diet, which thus replaces only 36.1 g of soybeans by corn per 1 Kg diet, including a slight benefit of phosphorous available in the ODML. There are other nutrients or catalytic elements available in the ODML that could also be accounted for.

Table 5 shows the comparisons of the balanced costs of the alternative systems. The negative sign is considered to be reasonable, since this paper excludes the benefits from the marketing of hogs and the amounts of energy gained in the hogs. In the anaerobic system, however, the balances in the monetary and energy budget are positive, since this system has a quite profitable benefit of fertilizer without any expensive input, compared with other systems.

The anaerobic digestion system appears to be quite beneficial over the ODML feeding system due to the various returns such as heat, electricity, and a great deal of fertilizer. On the other hand, the main benefit in the oxidation ditch operation is the amount of the ODML fed to the hogs, but the energy benefit assumed in the ODML is not as much compared with those returns in the anaerobic digestion, particularly the value of fertilizer.

The comparison of the balanced costs, as shown in Table 5, shows that there is no linear relationship between dollar and energy value with the various inputs and outputs. It does, however, indicate that dollar expenditures increase with energy consumption.

TABLE 5. THE COMPARISONS OF BALANCED COSTS
(RETURN-EXPENSE)/YEAR/HOG

| | Dollars | Cal x $10^{-3}$ |
|---|---|---|
| Anaerobic storage | + .22 | + 33.0 |
| Digestion | −1.40 | − 7.0 |
| Oxidation ditch | −2.35 | − 35.7 |
| Drying | −3.04 | −345.7 |

## SUMMARY

The anaerobic storage system for fertilizer appeared to be the most profitable one among the alternative waste management systems studied in this paper in both the energy and the monetary budgets. This system has, however, a limit in the availability of the cropland as well as the potential odor problem. This system might not be allowed to operate in the vicinity of a community, unless special odor controls were used. In general, odor is controlled by aeration, as with the oxidation ditch operation. Using odor control systems with the anaerobic storage system, however, might upset the budget of this system.

The drying for refeeding system resulted in the most expensive method in both monetary and energy budgets, primarily because of the tremendous amount of fuel consumption. Besides this disadvantage, this system has also a problem in utilizing the amounts of dried manure left over. Deodorizing facilities may be necessary, since, in order to preserve nutrients in the dried feces, the operating tempera-

TABLE 4. THE COMPARISONS OF OUTPUT ELEMENTS (RETURNS)

| | Anaerobic storage | | Oxidation ditch | | Anaerobic digestion | | Drying | |
|---|---|---|---|---|---|---|---|---|
| | Dollars | Cal x $10^{-6}$ | Dollars | Cal x $10^{-6}$ | Dollars | Cal x $10^{-6}$ | Dollars | Cal x $10^{-6}$ |
| Refeed | — | — | 1,934 | 24.29 | — | — | 28,717 | 222.40 |
| Fertilizer | 8,665 | 168.05 | 243 | 16.70 | 8,192 | 154.89 | — | — |
| Heat | — | — | — | — | 500 | 167.88 | — | — |
| Electricity | — | — | — | — | 1,248 | 53.73 | — | — |
| Total | 8,665 | 168.05 | 2,177 | 40.99 | 9,940 | 376.50 | 28,717 | 222.40 |

ture of this system is not high enough to eliminate the odor.

The anaerobic digestion returns various benefits—heat, electric power, and fertilizer. This system, however, requires a delicate operation control, considering the instability of the methane bacteria over a slight change of the environment (temperature, pH, volatile acids, and other toxic materials). Because the fertilizer in the surplus sludge accounts for a great deal of the benefits of this system, the availability of cropland for sludge disposal is also a factor which could shift the balance.

The oxidation ditch operation, in which the ODML is fed, is quite attractive in both the energy and the monetary budgets even with the high energy consumption for running the rotors, since it eliminates the problems of limitations of cropland, objectionable odor, and intricate operations faced in the alternative methods. Estimates of the ODML value might reasonably include other elements available in the ODML as well as comparisons with water consumption in the other systems. In the operation of the oxidation ditch, nitrogen toxicity should be carefully avoided for the sake of animal health.

It is noted that utilization of treatment products is desirable for all systems, for disposals cost a great deal of energy and money. The method of analyzing energy budgets used in this paper also seems appropriate for studying alternatives of other systems.


### References

1   AGRICULTURAL ENGINEERS YEARBOOK. 1969. ASAE, St. Joseph, Mich. 49085.

2   Chemical Engineers Handbook. 1963. 4th Ed. Prepared by a staff of specialists under the editorial direction of R. H. Perry, C. H. Chilton and S. P. Kirkpatrick. New York, McGraw-Hill. pp. 20-12, 13.

3   Day, D. L. and B. G. Harmon. 1974. Nutritive value of aerobically treated livestock and municipal wastes. Proc., The Use of Waste Water in the Production of Food and Fiber. Okla. State Dept. of Health, Okla. City, Mar. 5-7. EPA-660-/2-74-041, June, pp. 240-255.

4   Diggs, B. G., B. Baker, Jr. and F. C. James. 1965. Value of pig feces in swine finishing rations. J. Anim. Sci. 24:291.

5   Herendeen, R. A. 1973. An energy input-output matrix for the U.S., 1963. User's Guide, Center for Advanced Computation, Univ. of Ill. at Urbana-Champaign. pp.48-81.

6   Kim, H. C. and D. L. Day. 1974. Energetics of alternative waste management systems. Progress report, Agri. Engr. Dept., Univ. of Ill. at Urbana-Champaign, Project 31 15 10 375.

7   Midwest Plan Service. 1972. Structures and Environment Handbook, 127 p.

8   Miner, J. R. and R. J. Smith (Ed.). 1974. Anaerobic treatment. In: Livestock Waste Management with Pollution Control, North Central Regional Research Publication (in press).

9   Morgan, J. C., P. E. Sullivan and R. A. Troupe. 1960. Equipment weight as a basis for cost estimation. 6 p. In: O. T. Zimmerman and I. Lavine (Ed.) Cost Engineering 5(4):6. Industrial Research Service, Inc., Dover, New Hampshire.

10   Newtson, K. 1970. Personal communication. Thrive Center, Inc.

11   Pfeffer, J. T. 1973. Reclamation of energy from organic refuse. Prepared for the office of research and monitoring. U.S. EPA National Environmental Research Center, Cincinnati, Ohio. p. 95.

12   Ross, I. J. 1974. Personal communication. Agri. Engr. Dept., Univ. of Kentucky.

13   Swift, R. W. 1959. Food energy. pp. 39-56. In: Food. the Yearbook of Agriculture. USDA, Washington, D.C.

14   USDA. 1964. Agricultural prices annual summary. 1963. Statistical Service Reporting Service Pr1-3(64):167-175.

15   USDA. 1973. Agricultural statistics, U.S. Agriculture Printing Office, Washington, D.C. 1974, p. 461.

16   USDA. 1974a. Agricultural prices, Statistical Reporting Service PR1(9-74).

17   USDA. 1974b. Agricultural prices, Statistical Reporting Service PR1(11-74).

18   USDA. 1974c. Agricultural prices, Annual Summary. 1973. Statistical Reporting Service Pr1-3(74). p. 10.

19   Vanderholm, D. 1974. Land application of manure from modern livestock production facilities, Soil and Water Conservation, No. 14, Agr. Engr. Dept., Univ. of Ill. at Urbana-Champaign.

20   Zimmerman, O. T. and I. Lavine. 1971. Cost engineering, 16(3):13. Industrial Research Service, Inc., Dover, New Hampshire.


---

## Manure Fluming and Land Application

*(Continued from page 397)*

flies being located around the feed bunks — and these are controlled with chemicals.

In the confinement buildings the authors visited, the various aspects of waste managment were being carried out with excellent efficiency. The manure is moved rapidly to the flumes by the action of the animals' feet. Even though the slightly sloping floors are ribbed or brushed to prevent slippage, cattle injury is almost non-existent. The flumes are sized adequately to collect and hold the waste, and the flexibility of the system is such that cleaning presents few problems. The manure is tranferred to the holding, or storage, lagoon, either by gravity or by pumping, and in one particular case, with the lagoon being located at a higher elevation than the barn, the fluming action is done by gravity pressure, the waste being collected in a concrete sump and pumped back to the lagoon. The fluming action does not add new water to the system, and the action of fluming, in effect, aerates the material being pumped. Storage is provided for several years without evacuation, but

land application is easily accomplished through either a gravity irrigation system or a high pressure system.

The volume to be handled makes a field tank distribution system considerably more expensive, from an operating standpoint, although the initial cost of a complete irrigation system may be slightly higher.

Spray or gravity irrigation of manure lagoon product adds to the total efficiency of a confinement feeding system. Hydraulic transfer reduces labor, nutrient utilization, and power cost, can be optimized by trading volume for time; confinement housing reduces land requirements, aids in herd health, and improves feed conversion ratios. Energy conservation and soil improvement are important pluses that accrue to such a program.

Because of the cost effectiveness of the program, the efficiencies gained and the value to be derived at all levels of operation, confinement beef feeding with manure fluming, lagoon storage and irrigation application will increase.

# Field Evaluation of a Settling Chamber for Swine Wastes

E. T. Oatway, D. D. Schulte, L. Shwaluk

ASSOC. MEMBER
ASAE

THE swine manure handling and storage system evaluated in this project was constructed in 1970 by Mr. J. Van Schepdael at Libau, Manitoba. The objectives were to design a system which would (a) use partially slotted floors; (b) detain wastes for minimum periods of time within the barn; (c) provide storage for four to six months; and (d) require minimum capital investment.

The barn was designed for a herd of 50 sows in a farrow to finish enterprise. There are slotted areas 4 ft wide (1.22 m) in the pens used for weaners, feeders and breeding animals. In the farrowing crates the slotted areas are 2 ft wide (0.61 m). The pits under the slotted floors are 1.5 ft deep (0.46 m). Each pit is fitted with a watertight gate to control the flow of waste from the pits to the collection sewer and storage (Fig. 1).

The manure storage system consists of two units connected in series (Fig. 2). The first unit (the settling chamber) is a covered concrete tank, 24 ft x 24 ft x 8 ft (7.3 x 7.3 x 2.4 m) with an effective capacity of 3000 ft$^3$ (84.9 cu m). The second unit is an earthern pit, 80 ft x 80 ft x 5 ft (24.4 x 24.4 x 1.67 m), with a capacity of approximately 25,000 ft$^3$ (707.5 m$^3$). The storage units are connected by a 6 in. diameter (15.2 cm) gravity flow pipe located 6 ft (1.83 m) above the bottom of the concrete tank.

Another 6 in. diameter pipe fitted with a valve, permits the operator to control flow of liquid from the earthen pit to the concrete tank to (a) facilitate emptying the earthen pit, and (b) dilute material in the concrete tank for agitation and pump out.

The operational procedure involves opening the barn pit gates as required, usually every five to ten days, permitting the manure to flow into the concrete tank. When the concrete tank is full and a barn pit is emptied, an equal volume of liquid overflows into the earthen pit. Theoretically, if enough time were available in the concrete tank, the solids would settle out making manure handling somewhat more manageable.

## METHODS AND PROCEDURES

The purpose of the study was to evaluate the system by determining the effectiveness of the concrete tank as a solid-liquid separation device, and to develop design

**Acknowledgments:** The authors wish to thank Mr. and Mrs. J. Van Shepdael for their cooperation and assistance on this project and Miss N. Parreno for laboratory analyses.

The authors are: E. T. OATWAY, Chief, Agricultural Engineering Division, Technical Services Branch, Manitoba Dept. of Agriculture; D. D. SCHULTE, Assistant Professor, Agricultural Engineering Dept., University of Manitoba, Winnipeg, MB, Canada; and L. SHWALUK, Livestock Waste Technician, Manitoba Dept of Agriculture.

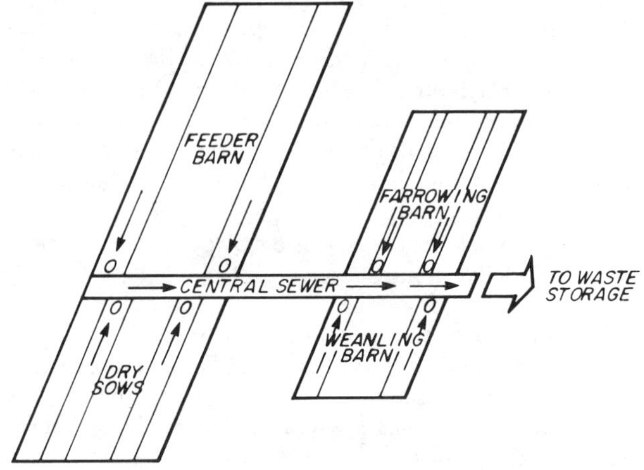



**FIG. 1 Layout of swine barn.**

criteria and operational procedures for such a system.

The study was conducted from July to October of 1974 by the Manitoba Department of Agriculture and the Agricultural Engineering Department of the University of Manitoba. The concrete tank was emptied by Mr. Van Shepdael in late June, 1974. Wastes were accumulated in the concrete tank for approximately four weeks when overflow to the earthen pit began. The first samples were taken July 29, 1974 and thereafter at two to three week intervals until October 17, 1974.

Samples were taken of the waste (a) as it entered the concrete pit; (b) in the concrete tank at the surface and at 1, 2, 3, 4, 5, and 6 ft depths (0.3, 0.6, 0.9, 1.2, 1.5, and 1.8 m); and (c) entering the earthen pit. The samples were analyzed at the University of Manitoba Agricultural Waste Management Laboratory for the content of total solids, filterable solids, total nitrogen and total phosphorus. All analyses were performed according to Standard Methods (1971).

Some difficulties were experienced in obtaining accurate samples. Two sampling devices, a ball and cone and a stoppered bottle, were used (Fig. 3). Samples taken by the ball and cone tended to be higher in solid content while those taken by the stoppered bottle tended to be lower in solids content. The ball and cone was more satisfactory for samples of higher solid contents ($\geqslant 7$

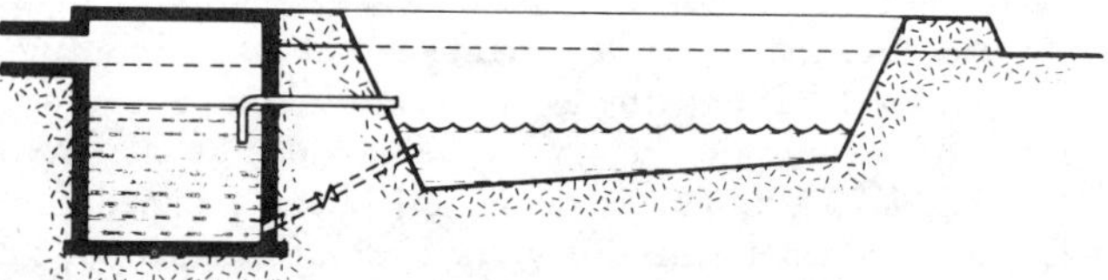

**FIG. 2 Cross-section of concrete tank and earthen pit.**

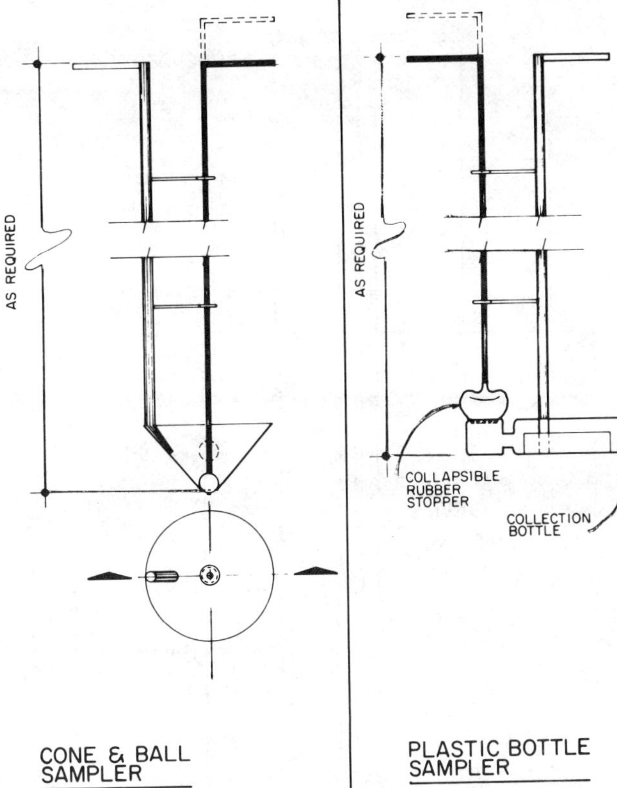



**FIG. 3 Types of samplers.**

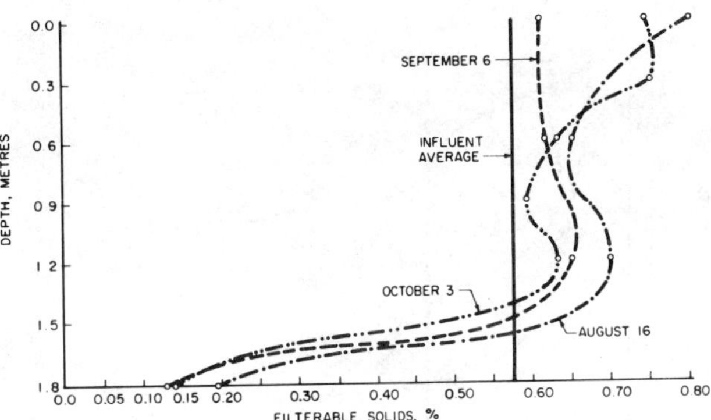



**FIG. 5 Variation of filterable solids during testing period.**

percent) while the stoppered bottle worked best for samples of lower solids content (≤ 7 percent).

## RESULTS AND DISCUSSION

The normal detention time of liquid in the tank was determined by the difference between the periods of inflow and outflow when a barn pit was emptied. For example, one of the barn pits with a capacity of 3000 ft³ (8.49 m³) required 4 minutes to overflow into the concrete tank and the time for an equal volume to overflow into the earthen pit was 34 minutes. In this case, assuming complete mixing, the solids settling time was 30 minutes.

The manure, as produced in the barn pits, had an average total solids content of 5.6 percent with a range of 1.4 to 9.6 percent. Of the total solids, an average of 10 percent was filterable. Fig. 4 illustrates the effectiveness of the concrete settling tank as a solids separator. The verticl/line at the 5.6 percent level in Fig. 4 represents the average total solids content in the liquid manure received from the barns. The tank began filling with

solids rapidly after October 3 and by October 17 had lost its effectiveness as a separation device. The tank had performed well for 14 weeks, discharging an effluent of approximately 1 percent total solids over the entire length of time. The effluent was pumped from the earthen pit and used as irrigation water.

Fig. 5 indicates that before October 3 the solids remaining near the surface of the concrete tank consisted primarily of filterable solids while those at the bottom were largely non-filterable. Good separation was occurring in the tank and a minimum of settling would be expected in the earthen pit.

The nitrogen and phosphorus contents of the manure in the settling chamber are shown in Figs. 6 and 7. The average total nitrogen and total phosphorus concentrations in the liquid manure from the barns were 2751 and 1658 mg/l, respectively. Nitrogen was influenced to a lesser extent by settling than was phosphorus or toal solids. However, as the tank filled, the nitrogen content throughout the tank increased rapidly, approaching the nitrogen concentration of the manure flushed from the barns. Subsequent analyses have shown that the increased nitrogen content at the 6 ft depth (1.8 m) was due to an increased amount of particulate organic nitrogen. Since this study was not intended to comlete a nitrogen balance, factors which may have influenced nitrogen losses, such as ammonia volatilization, were not studied.

Based on the normal liquid discharge rates from the hog barns in this study, approximately four times the liquid capacity of the tank pass through the tank before excess solids accumulated and begin passing out the

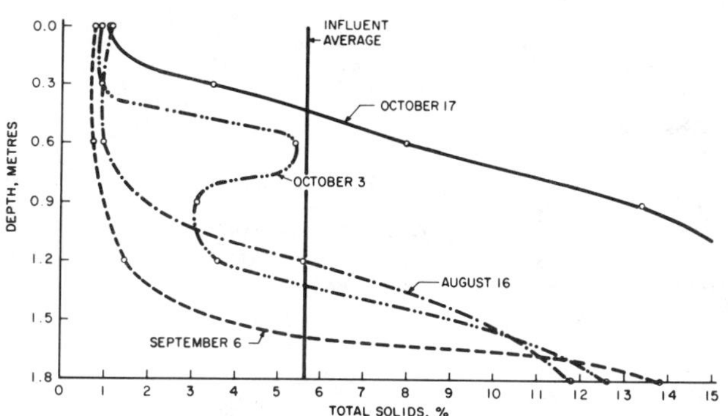



**FIG. 4 Variation of total solids during testing period.**

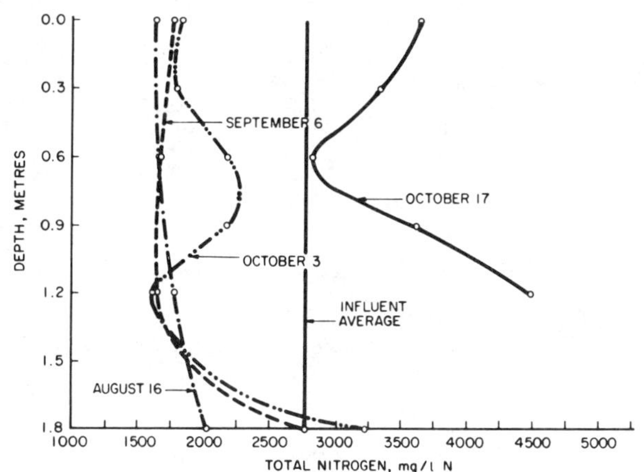



**FIG. 6 Variation of total nitrogen during testing period.**

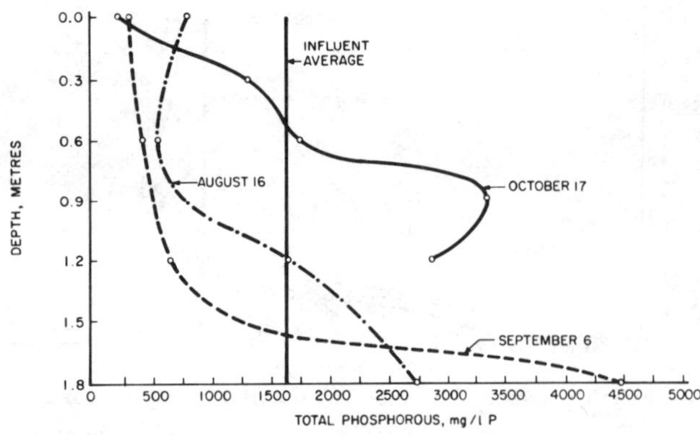

FIG. 7 Variation of total phosphorus during testing period.

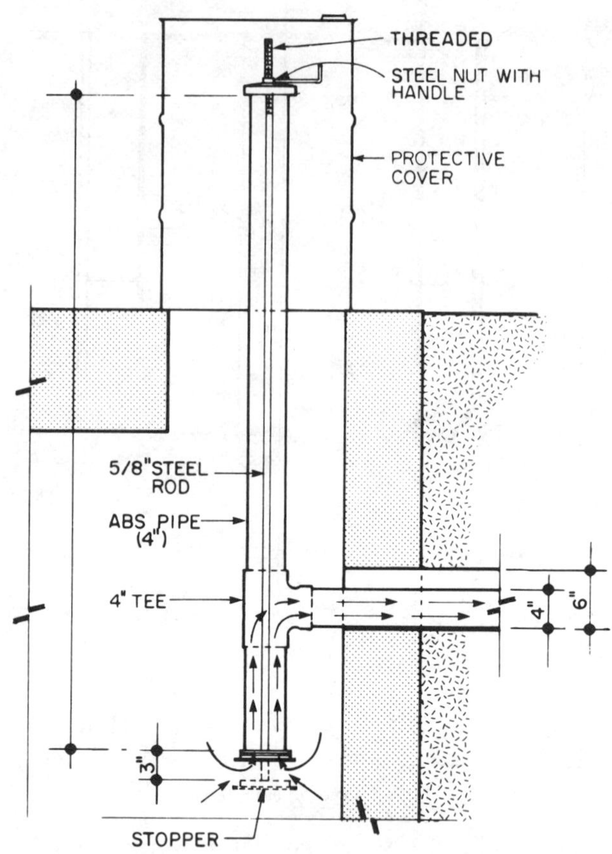


FIG. 8 Detail of detention valve.

gravity discharge pipe. Correspondingly, four times the holding capacity of the concrete tank would be required for an equal hydraulic retention time in the earthen pit. This, of course, could vary depending on the settleability of the solids, the solids content of the manure itself, evaporation and water management in the barn. The material which must periodically be removed from the concrete settling chamber requires the use of an agitation type manure pump for handlig. The system. is relatively odor-free and has worked satisfactorily for several years.

In late November, 1974 a valve (Fig. 8) was placed in the 6 in. (15.2 cm) gravity pipe connecting the concrete tank and the earthen pit. the valve was located approximately 1 ft (30.4 cm) below the surface of the liquid to take advantage of the low solids content at this location. The valve can be closed prior to flushing to permit more settling and is opened 24 hours later for discharge. this also eliminates the possibility of discharge of a higher solids content than necessary due to the turbulence in the tank at the time of flushing. Tables 1 and 2 summarize all of the data from this study.

## CONCLUSION

1 The lowest percentage of total solids in the concrete tank occurs approximately 1 ft (30.5 cm) below the surface of the waste.

2 During and immediately after flushing of barn pits,

(Continued on page 411)

TABLE 1. SUMMARY OF DATA FROM SETTLING CHAMBER.

| | Sampling date | | | | | | |
|---|---|---|---|---|---|---|---|
| Depth, m | July 29 | Aug. 16 | Aug. 29 | Sept. 6 | Sept. 19 | Oct. 3 | Oct 17 |
| | Total solids, percent | | | | | | |
| Surface | 5.316 | 1.118 | 0.736 | 0.773 | 1.061 | 0.906 | 1.085 |
| 0.3 | — | --- | — | --- | 0.845 | 0.895 | 3.491 |
| 0.6 | 0.763 | 0.965 | 0.730 | 0.762 | 0.859 | 5.404 | 7.984 |
| 0.9 | — | --- | — | --- | 1.243 | 3.106 | 13.370 |
| 1.2 | 1.732 | 5.596 | 5.513 | 1.445 | 4.189 | 3.590 | 15.708 |
| 1.8 | 14.107 | 11.813 | 13.244 | 13.829 | 7.747 | 12.610 | 15.708 |
| | Filterable solids, percent | | | | | | |
| Surface | 0.41 | 0.80 | 0.60 | 0.61 | 0.78 | 0.74 | — |
| 0.3 | --- | --- | — | --- | 0.76 | 0.75 | — |
| 0.6 | 0.64 | 0.65 | 0.58 | 0.62 | 0.77 | 0.64 | — |
| 0.9 | — | --- | — | --- | 0.65 | 0.59 | — |
| 1.2 | 0.57 | 0.70 | 0.71 | 0.65 | 0.55 | 0.63 | — |
| 1.8 | 0.16 | 0.19 | 0.21 | 0.13 | 0.43 | 0.14 | — |
| | Total nitrogen, mg/l N | | | | | | |
| Surface | 2,041 | 1,630 | 1,560 | 1,764 | 1,999 | 1,837 | 3,640 |
| 0.3 | — | --- | — | --- | 1,876 | 1,792 | 3,338 |
| 0.6 | 1,652 | 1,658 | 1,548 | 1,671 | 1,926 | 2,173 | 2,811 |
| 0.9 | — | --- | — | --- | 1,971 | 2,173 | 3,606 |
| 1.2 | 1,649 | 1,782 | 1,719 | 1,657 | 2,100 | 1,613 | 4,480 |
| 1.8 | 2,394 | 2,021 | 2,508 | 2,755 | 2,576 | 3,225 | |
| | Total phosphorous, mg/l P | | | | | | |
| Surface | 2,115 | 792 | 392 | 271 | 436 | — | 228 |
| 0.3 | — | --- | — | --- | 184 | — | 1,312 |
| 0.6 | 274 | 544 | 455 | 378 | 380 | — | 1,744 |
| 0.9 | — | --- | — | --- | 760 | — | 3,360 |
| 1.2 | 880 | 1,664 | 1,020 | 648 | 815 | — | 2,880 |
| 1.8 | 1,560 | 2,720 | 2,360 | 4,480 | 1,200 | — | — |

# Livestock Agriculture in the State of Hawaii—
# A Regional Approach to Waste Management

G. M. Wong-Chong, W. I. Hugh, J. H. Koshi, T. Tanaka, C. Schlottfeldt

THE 1973 census showed the State of Hawaii having a population of 832,147 residents and a resident equivalent of 62,396 visitors (Hawaii Data Book 1974). Of this population 81.5 percent resides on Oahu. To feed this population some food is produced locally while a greater proportion is imported from mainland U.S.A. and other countries. Table 1 shows a breakdown of the animal protein supply for 1973 (Statistics of Hawaiian Agriculture 1973) and it could be assumed that a fraction of this supply proportional to the population was consumed on Oahu.

Livestock agriculture in Hawaii is influenced by several factors:

1. almost all feed stuffs are imported;
2. the uneven population distribution, with the major market outlet on Oahu;
3. the high cost and irregularity of inter-island transportation; and
4. increasing urban encroachment on agricultural land.

The first three factors dictate the economic desirability of having livestock production on Oahu. However, increasing urban encroachment is amplifying the problems of animal waste management. The Waianae area (3500 acres) is the livestock producing center on Oahu (and for the state). Actual land utilization for livestock in this area is about 971 acres, and Table 2 gives a breakdown of the production intensity. In this same Waianae area, a 1970 census showed the region to have grown 46.3 percent in the period 1960-1970 and observations on construction activities indicate the area one of the most intense on the island. This combination of increasing urbanization and livestock production will necessitate rigid waste management practice.

At present waste management practice is at an extremely low priority, limited to anaerobic lagoons for swine wastes and disposal of the solid wastes (layers and dairy) wherever and whenever possible. This investigation examined the feasibility of a centralized solid waste management system for the Waianae area. The waste treatment process considered was composting and the market potential for compost was examined as the ultimate disposal. Major concerns of the study were nuisance (fly and odor) control so that urban dwelling and livestock production could coexist in the same region.

Published with the approval of the director of the Hawaii Agricultural Experiment Station as Journal Series Paper No. 1854.

The authors are: G.M. WONG-CHONG, Assistant Professor, Agricultural Engineering Dept., W.I. HUGH, J.H. KOSHI, T. TANAKA, Extension Specialists, Animal Science Dept., and C. SCHLOTTFELDT, Graduate Assistant, Agricultural Economics Dept., College of Tropical Agriculture, University of Hawaii, Honolulu.

## TABLE 1. ANIMAL PROTEIN SUPPLY, STATE OF HAWAII, 1973 (REFERENCE 7).

| Commodity | Supply (1000 lbs) | | | Percent of total produced locally |
|---|---|---|---|---|
| | Total | Import | Local | |
| Milk | 136 400 | | 136 400 | 100.0 |
| Beef & Veal | 71 506 | 40 493 | 31 013 | 43.4 |
| Pork | 23 866 | 16 433 | 7 433 | 31.1 |
| Broilers | 26 515 | 21 488 | 5 027 | 19.0 |
| Eggs* | 18 504 | 1 204 | 17 300 | 93.5 |

*Eggs in 1000 doz.

## TABLE 2. INTENSITY OF LIVESTOCK AGRICULTURE IN WAIANAE.

| Commodity | No. farms | No. animals | Percent supply | | Percent state consumption |
|---|---|---|---|---|---|
| | | | Oahu | State | |
| Broiler | 4 | 272 500 | 69 | 65 | 13 |
| Dairy | 19 | 6 872 | 66 | 55 | 55 |
| Egg | 7 | 460 300 | 50 | 33 | 30 |
| Pork | 41 | 37 032 | 60 | 50 | 15 |

## SOLID WASTE POTENTIAL AND COLLECTION

Based on an animal census, Table 2, the solid waste potential was estimated (Gerry 1968, Sprague et al. 1972, USA-EPA 1973) and is summarized in Table 3. This waste potential comes from 30 different sources, each with its own operational variations because of animal species, size of operation, and general management practice.

In existing operations the manure from the animals is permitted to accumulate for a period of time, which may vary from 3 months (dairy or broilers) to more than a year (layers). Under these circumstances, especially with the open corral dairy cattle operations, there may be significant odor and fly problems after a rainfall event. These nuisance problems may persist, possibly on a lower frequency, with frequent removal of the manure from the farm site. This frequency may be governed by accumulations to meet full truck loads. On the basis of

## TABLE 3. SUMMARY ESTIMATE OF SOLID ANIMAL WASTES GENERATION IN THE WAIANAE AREA.

| Animal | Fresh manure | | Aged manure | |
|---|---|---|---|---|
| | Tons/day | Percent moisture | Tons/day | Percent moisture |
| Broilers and pullets | | | 10.0 | 25 |
| Dairy cattle | 328.0 | 80 | 44.7 | 30 |
| Layers | 36.3 | 75 | 17.5 | 50 |
| | 364.3 | 79.5 ave | 71.2 | 35 ave |
| Volume | 120 000 ft$^3$ (60 #/ft$^3$) | | 4000 ft$^3$ (35#/ft$^3$) | |

having a 15 cubic yard truck capacity, a full load would represent having an accumulation up to about 7 tons. With the small dairy farms this accumulation period could be as long as 10-14 days. On the "average" there will be about 10 truck loads per day. This average collection rate may be somewhat misleading in that scheduling should be made to meet the full load condition (with the dairy and layers) but with the broilers the limitation will be the end of a cycle period, at which time the waste litter will be removed from the pen. Assuming batches of 10,200-20,000 birds every two weeks, then there will be batches of 22.5 to 45.0 tons of manure to be removed at a time.

The nuisance problems could be further reduced by daily collection of manure from the dairy animals. This daily collection could be an on-farm facility, where the manure is removed from the animal area. To achieve this facility the operations will require conversion from open corral to a free stall with paved floors. This change would also facilitate greater land utilization and in times of severe weather (muddy and hot) conditions reduce the stresses on the animals. These reductions of stress could mean greater productivity. However, it is recognized that such a change could require a substantial financial outlay by the producers.

Further consideration in the collection system must be a "sanitation station" where trucks will be washed and disinfected prior to entering a farm. This Station could be located at the exit end of the processing site. Further sanitation and precautions could be achieved by having tire baths (or spray) at the farm site, and the actual manure pick-up location at some remote point on the farm.

Conceptually, the collection system would be two trucks with facilities available at the farm site for loading of the trucks. The operation would be a 5-day per week pick-up schedule.

## WASTE PROCESSING

The waste will be stabilized by aerobic thermophilic composting. Several factors contribute to the rate of stabilization and these include (a) aeration, (b) moisture content, and (c) inoculation. A moisture content of 40 to 60 percent on a wet-weight basis has been reported to be about ideal for composting (Jeris et al. 1973).

Aeration could be achieved by physically turning the compost pile or by forced aeration. Effective inoculation could be achieved by a 10-20 percent recycle of composted material to the raw waste, thus circumventing the lag-time required for establishing the necessary microflora.

Assuming the processing site will be in the Waianae area, odor control will be a necessary criteria. Martin and Das (1972) reported the inability to control odors in simulated windrow composting of swine waste. We were also unable to achieve odor control in windrowing dairy manure, even with daily turning. Odors were most obvious during the initial 7 days. These odors are the result of anaerobic decomposition of certain organic compounds (nitrogenous, sulfur and long chain fatty acids) in the wastes. To minimize anaerobic action forced aeration was examined.

The experimentation was conducted in a 3.0 ft$^3$ cylindrical container which was insulated to reduce heat loss. An air diffuser was located in the bottom of the vessel and on the removable cover there was an air exhaust vent. There were 4 temperature sensing ports, 3 located on the side of the vessel and one at the center of the cover. Fig. 1 presents a typical reaction profile for forced aeration composting. The material composted was poultry (layer) manure with 20 percent inoculation and aeration rates of 1.00 and 1.67 SCFH per ft$^3$ manure. Oxygen utilization during the early stages of composting resulted in 4.0 and 7.0 percent oxygen in the exhaust gas. In a confined space there were some odors for the first two days but with normal ventilation, these odors were reduced to a trace, with no odors after the fourth day, suggesting the near complete degradation of the putrescible components of the wastes.

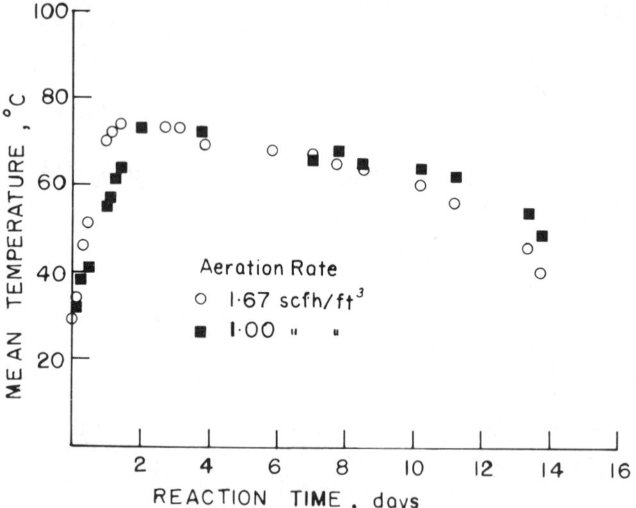


FIG. 1 Typical reaction profile of forced aeration composting with 20 percent inoculation.

Based on the laboratory forced aeration studies, a process was designed for estimation of economic feasibility. Fig. 2 is a schematic diagram of the process. The inoculation would be about a 10-20 percent recycle and the forced aeration would be at a 0.75 to 1.50 SCFH/ft$^3$ manure. The residence time at the forced aeration stage would be about 4-5 days. This aeration could be achieved with a 10-hp rotary compressor which will deliver about 100 cfm at 20 psig. A 50-hp front-end bucket loader tractor would provide the material handling—moving, mixing, and windrow aeration.

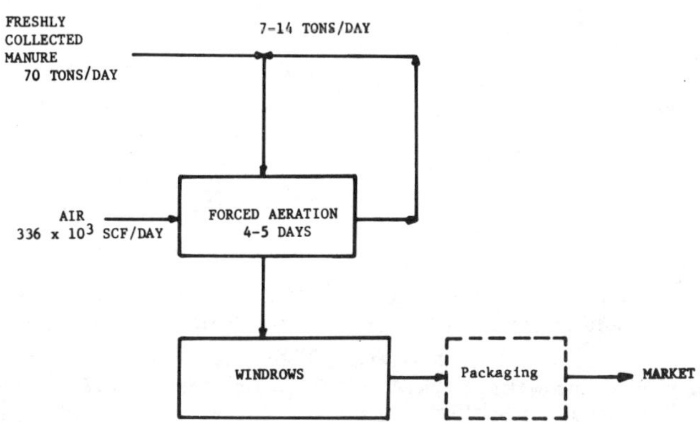


FIG. 2 Schematic flow diagram of the proposed composting process.

**TABLE 4. SYSTEM ECONOMIC ANALYSIS.**

Capital investment

| | | |
|---|---|---|
| 1. | Two manure collection trucks 15 yds$^3$ capacity | $ 60 000 |
| 2. | Manure front-end loader tractor with attachments | 12 000 |
| 3. | Air blower and drive | 5 000 |
| 4. | Piping for aeration grid | 500 |
| 5. | Instrumentation for aeration system | 750 |
| | 30 percent contingency | 23 475 |
| | Total | $101 725 |
| | Annual cost at 20 percent/year amortization | 20 862 |
| | Capital investment/ton of product = $1.60 | |

Operating costs

| | | |
|---|---|---|
| 1. | Manpower (4 persons at $5.00/hr) | $ 40 000 |
| 2. | Electricity [20 000 kw-hr/year at 3 ¢] | 600 |
| 3. | Fuel for trucks and tractor [200 bbl/year at $15/bbl] | 3 000 |
| 4. | Maintenance [tractor, trucks & blower] 15 percent of investment | 11 550 |
| | Total | $ 55 150 |
| | Operating cost/ton of product = $4.25 | |

Total production cost = $5.85/ton of produce

## ECONOMIC ANALYSIS OF CENTRALIZED COMPOSTING PROCESS IN THE WAIANAE REGION

The waste handling and processing system would be essentially that shown in Fig. 2 along with two 15 yd$^3$ trucks for waste collection. Table 4 presents the details of the economic analysis. The analysis was based on the availability of about 70 tons of manure each day which when processed yielded about 36 tons of compost. To the estimated $5.85/ton of compost the cost of (a) packaging and (b) site development structures must also be added.

## MARKET POTENTIAL

The organic matter content of many Hawaiian soils is low and most areas of continuous vegetable crop agriculture require organic matter amendment to the soil. Similar organic matter amendment is required for much of the highway landscaping, recreational areas, and greenhouse operations. This demand has been met mainly by the importation of compost and sewage sludge from the mainland. Imported compost retails at about $200/ton and nitrogen fortified (7 percent N) sewage sludge at about $80/ton. Local steer and chicken manures retail at about $45/ton. The Hawaii State Department of Transportation estimated some 5000 tons of organic soil amendment material were imported to the state in 1972 (reference 1).

The University of Hawaii Extension Service recommends applications of 3-5 tons compost per acre per year for vegetable farming, landscaping, home gardens, and recreational areas. Based on these recommendations the potential market for compost on Oahu was estimated (Table 5). Fifty percent of the maximum estimated could be considered as reasonable, and this amounts to about 26,185 tons per year. From the waste potential for the **Waianae area**, the potential production rate is about 13,000 tons per year, which is about 50 percent of the estimated reasonable market potential. In the worst case the potential production is still less that the most pessimistic (20 percent) estimation of the market potential.

## SUMMARY AND CONCLUSIONS

An economic feasibility study of a centralized solid waste handling and processing system for the livestock industry in the Waianae Region was undertaken. A major concern of the study was the nuisance control to permit compatible coexistence of urban dwelling and the industry. The waste processing considered was combined forced aeration-windrowing composting, with marketing of the compost as ultimate disposal.

The following are the conclusions:

1 For the Waianae area there is a potential of about 70 tons of manure per day.

2 A centralized forced aeration-windrow composting operation could produce at about $5.85 per ton. This processing cost is very small in comparison to some of the current market (retail) prices for compost and similar materials.

3 The market potential for compost was conservatively estimated at about 26,000 tons per year on Oahu, and the compost production rate at about 13,000 tons per year.

*(Continued on page 411)*

**TABLE 5. ESTIMATED MARKET POTENTIAL FOR COMPOST UTILIZATION IN OAHU.**

| | Area, acres | | Estimated potential, tons/year | | | |
|---|---|---|---|---|---|---|
| | Total | Consid. | 100% | 80% | 50% | 20% |
| A. Vegetables and melons | 1600 | 1600 | 6 400 | 5 120 | 3 200 | 1 920 |
| B. Golf courses | 2880 | 2880 | 11 520 | 9 216 | 5 760 | 3 456 |
| C. Recreation area | 2304 | 1612 | 6 448 | 5 160 | 3 224 | 1 936 |
| D. Road protection and landscaping | 1097 | 1097 | | | | |
|   1. Bed formation | | | 12 089 | 9 671 | 6 044 | 3 627 |
|   2. Remaining areas | | | 1 308 | 1 048 | 652 | 392 |
| E. Gardens | | | | | | |
|   1. Civilian area | 3700 | 3700 | 11 100 | 6 660 | 4 440 | 2 220 |
|   2. Military area | 1750 | 1750 | 5 250 | 4 200 | 2 625 | 1 575 |
| F. Greenhouse crops | 5 | 5 | 140 | 140 | 140 | 140 |
| G. Other | -- | -- | 100 | 100 | 100 | 100 |
| Total | | | 54 355 | 41 315 | 26 185 | 15 366 |

# Estimating Quantity and Quality of Runoff from Eastern Beef Barnlots

W. M. Edwards, J. L. McGuinness

RECENTLY economic conditions have given rise to an increase in the numbers and sizes of beef cattle herds in the hill country of the humid eastern third of the United States. Typically, calves from these herds are fed for the beef and feeder-calf markets in many small barnlot feeding operations. Although the individual feeding operations commonly involve fewer than 100 head of cattle, the stocking rate (lot area per animal) is similar to that of the larger western operations, and the concentration of nutrients in runoff from these eastern lots is quite high (Edwards et al. 1972).

Recognizing these potential "hotspots" of downstream water pollution, the United States Environmental Protection Agency proposed feedlot effluent limitations guidelines (EPA 1974), directing that by July 1, 1977, all cattle feeding operations have the "best practicable control technology available" in use to control runoff. By 1983, the runoff control must be the "best available technology economically achievable." These proposed 1977 and 1983 standards require that each operation has sufficient storage capacity to contain waste water plus the runoff from the highest 24-hour storm event likely to occur in 10- and 25-year periods, respectively. Although the current thrust of proposed regulations concerns large feeding operations, effluent limitations for smaller operations are forthcoming (EPA 1974).

This paper describes a procedure to estimate runoff volumes from an unpaved barnlot in eastern Ohio for 10- and 25-year frequency storms. Included also are runoff estimates for storms of various durations (up to 30 days) and the expected nutrient content of the runoff water, both of which are useful in designing runoff control systems. Although the examples presented here describe rainfall-runoff relationships for one specific barnlot, our primary purpose is to present a generally applicable method, not local findings.

## PROCEDURES AND RESULTS

Runoff from a 0.17-ha unpaved barnlot watershed (No 163) at the North Appalachian Experimental Watershed, USDA, ARS, Coshocton, Ohio, was measured for 181 separate runoff events during a 52-month period, 1968 to 1972. Fig. 1 shows the size distribution of these runoff events. Physical characteristics of the watershed and runoff quality have been previously defined (Edwards et al. 1971, 1972).

Since our primary objective was to predict runoff for 24-hour storms, the data were grouped into 130 separate 24-hour rainfall periods. Therefore, runoff data from

two or more distinct storms within a 24-hour period were combined. Runoff was then regressed with:

1 Rainfall Amount:   Total rainfall in 24-hour period (cm);
2 Antecedent Soil Moisture:   ASM-Estimated soil water content of the topsoil when rainfall begins (cm water in top 18 cm of soil);
3 Rainfall Intensity:   Maximum 5-minute intensity in the runoff-producing storms (cm/hr.);
4 Barnlot Usage:   Whether cattle were in the lot when the storm occurred or the lot was idle.

The resulting multiple regression equation explained 87 percent of the total deviations (Table 1). Since neither rainfall intensity nor barnlot usage was statistically significant when the other two parameters (rainfall amount and ASM) were included, they were omitted from the final regression equation. The data were grouped into time periods of 2, 4, 7, 10 and 30 days to define runoff expected for time periods longer than one day. Again, although all four independent variables were considered, rainfall intensity and barnlot usage were not significant and were omitted. The final regression equations are presented in Table 1.

**TABLE 1. REGRESSION EQUATIONS, $R^2$ AND n FOR PREDICTING BARNLOT RUNOFF FOR 1-, 2-, 4-, 7-, 10- AND 30-DAY PERIODS**

| Period, days | Equation | $R^2$* | n† |
|---|---|---|---|
| 1 | Q = 0.0957 ASM‡   + 0.6193 P§   - 0.8915 | 0.873 | 130 |
| 2 | Q = 0.1203 ASM   + 0.6255 P   - 1.0394 | 0.766 | 129 |
| 4 | Q = 0.1432 ASM   + 0.6164 P   - 1.1770 | 0.776 | 114 |
| 7 | Q = 0.1747 ASM   + 0.5851 P   - 1.3406 | 0.755 | 86 |
| 10 | Q = 0.2019 ASM   + 0.6259 P   - 1.6294 | 0.775 | 80 |
| 30 | Q = 0.2998 ASM   + 0.4585 P   - 1.6769 | 0.733 | 38 |

*$R^2$ = Portion of total deviations explained by regression.
†n = Number of observations used in deriving each regression equation
‡ASM = $H_2O$ in top 18 cm of soil, cm.
§P = Rainfall, cm.

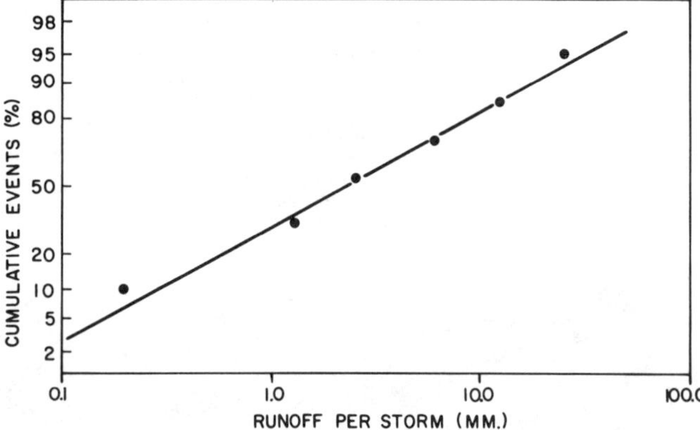



**FIG. 1 Size distribution of measured runoff events.**


Contribution of the USDA, ARS, NCR, North Appalachian Experimental Watershed, Coshocton, Ohio in cooperation with the Ohio Agricultural Research and Development Center, Wooster, Ohio.

The authors are W. M. EDWARDS, Soil Scientist, USDA and Assistant Professor, OARDC; and J. L. McGUINNESS, Statistician, USDA, respectively.




MANAGING LIVESTOCK WASTES

Fig. 2 gives the rainfall expected for a range of probabilities and durations as estimated by Weather Bureau methods (U.S.D.C. 1960, Miller 1964) for the Coshocton area. The return period is the reciprocal of the probability percentage.

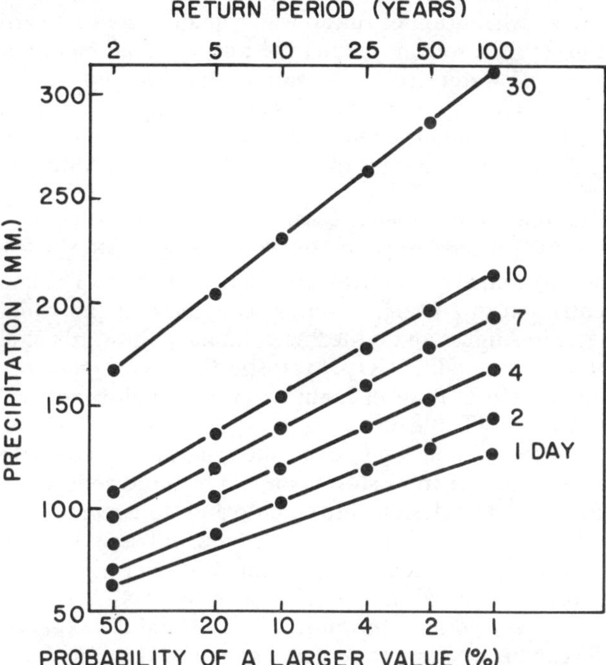


FIG. 2 Precipitation expected at different probabilities for various time periods (Coshocton, Ohio).

The probabilities of antecedent soil moisture (Fig. 3) were developed from long-term precipitation records and a topsoil moisture budgeting procedure (Youker and

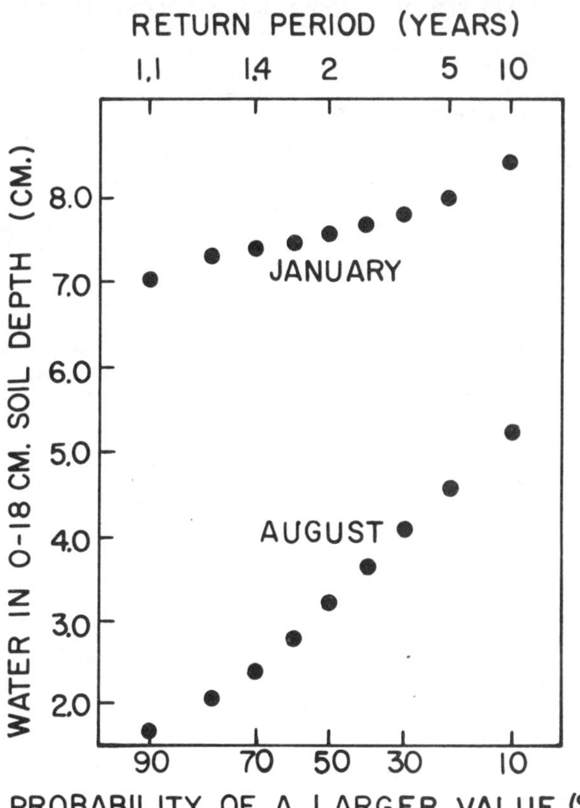


FIG. 3 Antecedent moisture probability.



Edwards 1969) revised to include the dormant season. Since no moisture measurements were made in the barnlot, values estimated for meadow were used as an index of barnlot soil moisture. Values from Figs. 2 and 3 were used in the multiple regression equation for 1-day periods (Table 1) to calculate predicted runoff shown in Table 2.

TABLE 2. PREDICTED BARNLOT RUNOFF (cm) FROM 10- AND 25-YEAR STORMS (24-HOUR) FALLING ON DRY, AVERAGE AND WET ASM IN JANUARY AND AUGUST

| Month | Return* period, years | Runoff | | |
|---|---|---|---|---|
| | | Dry ASM cm | Avg. ASM cm | Wet ASM† cm |
| January | 10 | 5.44 | 5.49 | 5.56 |
| | 25 | 6.38 | 6.43 | 6.50 |
| August | 10 | 4.93 | 5.08 | 5.26 |
| | 25 | 5.87 | 6.02 | 6.22 |

*10-year storm = 9.1 cm rainfall
 25-year storm = 10.7 cm rainfall.
†Jan.:  dry, average, wet = 6.9, 7.4 and 8.4 cm $H_2O$ in top 18 cm soil
 Aug.:  dry, average, wet = 1.7, 3.2 and 5.2 cm $H_2O$ in top 18 cm soil
 The dry, average, and wet values will be exceeded 90, 50 and 10 percent of the time, respectively.

Table 3 gives nutrient transports from the design storms estimated from predicted runoff volumes (Table 2) and previously determined seasonal distributions of $NO_3$-N, total N, P, and K concentrations. Table 4 gives similar information on runoff and nutrient transport for 2-, 4-, 7-, 10- and 30-day periods.

TABLE 3. ESTIMATED RUNOFF AND NUTRIENT TRANSPORT* IN RUNOFF WATER FROM 10- AND 25-YEAR STORMS (24-HOUR) WITH AVERAGE† ASM IN JANUARY AND AUGUST

| Month and return period ‡ | Runoff § 1 x $10^{-3}$ | Nutrient transport | | | |
|---|---|---|---|---|---|
| | | $NO_3$-N g | Total-N g | P g | K g |
| January | | | | | |
| 10-year | 94 | 160 | 6400 | 800 | 31 000 |
| 25-year | 110 | 190 | 7500 | 940 | 36 000 |
| August | | | | | |
| 10-year | 87 | 500 | 700 | 120 | 11 000 |
| 25-year | 103 | 600 | 900 | 140 | 13 000 |

*Concentrations from Edwards et al. 1972.
†Average ASM =   7.4 cm $H_2O$ in top 18 cm soil, January
                 3.2 cm $H_2O$ in top 18 cm soil, August.
‡10-year storm = 9.1 cm rainfall; 25-year storm = 10.7 cm rainfall.
§Barnlot area = 0.17 ha.

## DISCUSSION

The regression equations were developed from the 52-month runoff record from barnlot watershed 163, assuming that the period represented expected weather conditions. Briefly, total runoff during this period was 117 cm, or 28 percent of the 425 cm of rainfall input. Snow added another 50 mm of water, most of which melted slowly enough that it did not produce runoff. The largest 24-hour runoff event accounted for 84 mm of runoff, caused by 138 mm of rain which fell in 24 hours on July 5, 1969. The return period of this event was approximately once in 200 years. The smallest event recorded was 0.02 mm, or 26 liters from the 0.17-ha lot. Only nine of the 181 events (5 percent) produced more than 25 mm of runoff; but they accounted for 35 percent of the total runoff (Fig. 1). Ninety-five events produced 2.5 mm of runoff or less.

The regression equation is a statistical best fit of the predicted and observed runoff. The regression was calculated twice, first with the value from the 200-year storm included, then with this extreme value omitted. Since the regression coefficients did not change significantly, this means that the large event correlated with the others, adding credence to the prediction of runoff from large events.

In the regression equations of Table 1, the amount of rainfall is the most important predictor. The effect of ASM increases with storm duration. ASM and runoff are linked through the annual distribution of rainfall. There will likely be more runoff during a 30-day period in March, when the rainfall averages 9 cm, than during a 30-day period in October, when the average rainfall is 6 cm. ASM for March events, because of unfavorable climatic conditions for drying, are usually quite wet but dry ASM conditions are expected in the late summer and fall months. Therefore, for the longer runoff periods, high runoff amounts should be associated with wet antecedent conditions.

For 1-day periods, however, the above relationship is not true. For short durations, high runoff amounts are caused by high rainfall intensities, which occur most frequently (almost exclusively) in the late spring and summer months when ASM is lower (Harrold 1961).

Under dry ASM conditions, the barnlot surface becomes hard and densely packed. In the wet season, the cattle frequently stand knee-deep in the mixture of mud and manure. Because both surface conditions greatly mask the normal soil profile hydrologic characteristics, only extreme soil type differences should influence the rainfall-runoff relation.

The meadow topsoil moisture estimates are a satisfactory index of barnlot ASM even though absolute values of water content under the two surfaces may be different. When the meadow is wet, the barnlot will be wet; and both surfaces lose water under warm, dry climatic conditions.

When rainfall intensity was tested for use in the multiple regression equation, its statistical effect was very similar to ASM. The $R^2$ values were very nearly the same as those shown in Table 1. Since ASM can be estimated with available models (Youker and Edwards 1969, Saxton et al. 1974), we included ASM instead of maximum 5-minute rainfall intensity. If at other locations the intensity data are more readily available than estimates of ASM, they can be used instead of ASM in developing a runoff prediction.

The effect of ASM on predicted runoff is shown in Table 2. The specific values of ASM were selected so that wetter antecedent conditions would be expected 90, 50, and 10 percent of the time. In January, the spread between wet and dry ASM is only 1.5 cm of water, and only 10 percent of the time is there less than 6.9 cm of water in the 0 to 18 cm surface layer. In August, however, the spread between wet and dry ASM is 3.5 cm and the wet condition is drier than the dry ASM value for January. These figures show that January runoff cannot be influenced much by ASM because ASM is consistently high.

But there can be almost 0.7 cm more runoff from a 25-year storm during January than from the same storm in August because of possible differences in ASM (Table 2).

Using the runoff prediction presented here and the yearly distribution of runoff water quality from Edwards et al. (1972), we can predict the nutrient load contained in runoff water from the barnlot for 10- and 25-year storms. The values in Table 3 were obtained by multiplying average January and August nutrient concentrations by the runoff estimates of Table 2 for average ASM conditions.

January and August periods were shown in Tables 3 and 4 because total N, P, and K concentrations peak in January and are at or near their lows in August. The concentration of $NO_3$-N, however, is highest in the runoff water in August and quite low in January (Edwards et al. 1972). As a result, $NO_3$-N transport can be higher in a summer storm than in January, even though runoff volume is less (Table 3).

Another dimension of controlling barnlot runoff water involves runoff from storms on two or more consecutive days. A system designed to contain and store runoff from a 25-year storm on 1 day could be full and have no capacity for the next day's storm. Considering this, we calculated the rainfall probabilities for 2-, 4-, 7-, 10-, and 30-day periods shown in Fig. 2. Rainfall events for 25-year storms from these graphs were used in regression equations of Table 1 with average ASM values for January and August to estimate runoff for the respective periods (Table 4). Nutrient transports were calculated as for the 1-day periods.

TABLE 4. RAIN, RUNOFF, AND NUTRIENT TRANSPORT* IN RUNOFF WATER FROM 25-YEAR STORMS OF 2-, 4-, 7-, 10- AND 30-DAY PERIODS (AVERAGE ASM)† OCCURRING IN JANUARY AND AUGUST

| Month | Period days | Rain cm | Runoff‡ $1 \times 10^{-3}$ | Nutrient transport | | | |
|---|---|---|---|---|---|---|---|
| | | | | $NO_3$-N g | Total-N g | P g | K g |
| Jan. | 2 | 12 | 120 | 210 | 8 500 | 1100 | 41 000 |
| | 4 | 14 | 150 | 250 | 9 900 | 1200 | 47 000 |
| | 7 | 16 | 160 | 270 | 11 000 | 1400 | 52 000 |
| | 10 | 18 | 190 | 320 | 13 000 | 1600 | 61 000 |
| | 30 | 26 | 220 | 370 | 15 000 | 1900 | 71 000 |
| Aug. | 2 | 12 | 120 | 670 | 1 000 | 160 | 14 000 |
| | 4 | 14 | 130 | 780 | 1 200 | 190 | 16 000 |
| | 7 | 16 | 150 | 850 | 1 300 | 210 | 18 000 |
| | 10 | 18 | 170 | 1000 | 1 500 | 240 | 21 000 |
| | 30 | 26 | 190 | 1100 | 1 700 | 270 | 24 000 |

*Concentrations from Edwards et al. 1972.
†Average ASM =   7.4 cm $H_2O$ in top 18 cm soil, January
      3.2 cm $H_2O$ in top 18 cm soil, August
‡Barnlot area = 0.17 ha

## CONCLUSIONS

A short-term experimental record of the rainfall-runoff relation for an unpaved cattle barnlot watershed was used with precipitation probabilities and the most probable levels of antecedent soil moisture to predict runoff volumes for 10- and 25-year design storms. The method eliminates the need for long-term runoff measurements and shows how short-term experimental results can be combined with available climatic data to form a rational basis for designing systems for handling barnlot effluent.



MANAGING LIVESTOCK WASTES

## References

1 Edwards, W. M., F. W. Chichester and L. L. Harrold. 1971. Management of barnlot runoff to improve downstream water quality. Int. Livestock Waste, Mgmt. Symp. (Columbus, Ohio) Proc.:48-50.

2 Edwards, W. M., E. C. Simpson and M. H. Frere. 1972. Nutrient content of barnlot runoff water. Journ. of Environ. Qual. 1(4):401-405.

3 Environmental Protection Agency. 1974. Feedlots point source category, effluent guidelines and standards. Federal Register 39(32): 5704-5710, Feb. 14.

4 Harrold, L. L. 1961. Hydrologic relationships on watersheds in Ohio. Soil Conservation 26(9):208.

5 Miller, John F. 1964. Two- to ten-day precipitation for return periods of 2 to 100 years in the contiguous United States. U.S. Dept. of Commerce, Weather Bureau Tech. Paper No. 49, Wash., D.C. 29 pp.

6 Saxton, K. E., H. P. Johnson and R. H. Shaw. 1974. Modeling evapotranspiration and soil moisture. TRANSACTIONS of the ASAE 17(4):673-677.

7 U.S. Dept. of Commerce, Weather Bureau. 1960. Rainfall intensity-frequency regime. Part 5 - Great Lakes Region. Tech. Paper No. 29, Wash., D.C. 31 pp.

8 Youker, R. E. and W. M. Edwards. 1969. Simplified budgeting of plow layer moisture under meadow. USDA, ARS 41-154, 11 pp.

## Settling Chamber for Swine Wastes

*(Continued from page 404)*

turbulence increased the total solids content at and near the surface of the concrete tank.

3 The concrete settling tank reduces the total solids content of the overflow liquid to approximately 1 percent and continues to do so for a period of time equal to four times the storage period of the settling tank.

### References

1 Standard Methods for the Analysis of Water and Wastewater. 1971. American Public Health Association. New York, NY.

TABLE 2. SUMMARY OF DATA FROM INFLUENT TO SETTLING CHAMBER.

| Sample date 1974 | Total | Solids Filterable percent | Total nitrogen | Total phosphorous |
|---|---|---|---|---|
| | | | mg/l | |
| July 29 | 1.36 | 0.59 | 1993 | 288 |
| | 5.79 | 0.50 | 2732 | 1392 |
| August 16 | 5.87 | 0.54 | 2341 | 1856 |
| | 7.58 | 0.43 | 2304 | 1760 |
| August 29 | 3.92 | 0.72 | 2791 | 2140 |
| September 6 | 5.34 | 0.81 | 3502 | 572 |
| | 9.61 | 0.43 | 3595 | 3600 |
| Average | 5.64 | 0.58 | 2751 | 1658 |

## Livestock Agriculture in Hawaii

*(Continued from page 407)*

4 For the Waianae region a centralized waste handling and processing system producing compost is economically feasible.

5 The preliminary laboratory studies indicated forced aeration may provide odor control. However, this should be tested at a pilot scale to assess the degree of deviations from the rigid controls achievable in laboratory scale experiments.

### References

1 Communication to Councilwoman Mary George (DEP-L 1.9415) from Mr. E.A. Wright, Deputy Director, State of Hawaii, Department of Transportation, February 12, 1973.

2 Gerry, R.W. 1968. Manure production from broilers. Poultry Science 47:339-340.

3 Hawaii Data Book. 1974. State of Hawaii, Department of Planning and Economic Development.

4 Jeris, S. J., and R. W. Regan. 1973. Controlling environmental parameters for optimum composting. Compost Science, 14(2):8-15.

5 Martin, J. H., Jr. and K. C. Das. 1972. Windrow composting of swine waste. Proc. Cornell University, Agricultural Wastes Management Conference.

6 Sprague, D. C., A. T. Sobel, H. R. Davis, and T. L. Todd. 1972. Performance observations of a high-rise system for collecting, conditioning, handling and utilizing cage layer manure, N.Y.S. College of Agricultural and Life Sciences, Cornell University, AWM 72-06.

7 Statistics of Hawaiian Agriculture. 1973. State of Hawaii, Department of Agriculture.

8 U.S. Environmental Protection Agency. 1973. Development document for proposed effluent limitations guidelines and new source performance standards for the feedlots. EPA 440/1-73/004.

# A Computer Simulation of Storage and Land Disposal of Swine Waste


C. Roland Mote, E. Paul Taiganides

ASSOC. MEMBER    MEMBER
ASAE          ASAE


A DIGITAL computer simulation program that potentially may aid in the development of design and operating criteria for swine waste storage and land disposal systems has been developed and demonstrated. The program simulates the interaction of such factors as variations in animal population, soil trafficability conditions, and crop production stages, and reports the resulting performance and cost of selected system designs and operating criteria.

The program is briefly described and the results from the study of a specific system are presented herein.

## THE COMPUTER SIMULATION PROGRAM

The waste storage and land disposal system simulated by the computer program consists of a confinement swine production unit from which the waste is hydraulically transported into a storage tank. Waste from the storage tank is hauled with tank wagons and applied to cropland. In this system the quantity of waste produced is a function of the number of animals present in the production unit and the per animal waste production rate. Waste can be hauled onto the cropland with tank wagons only when the soil will support the machinery traffic, and the crops will not be damaged by the hauling operation. Application of waste to the cropland may also be limited by the capacity of the crop to utilize the waste. Ground water and soil pollution may result from excessive waste application rates.

With the use of GASP II (Pritsker and Kiviat 1969), a FORTRAN based simulations language, four routines which collectively simulate the above mentioned system characteristics were combined into a computer program that simulates the operation of the complete system. A block diagram of the computer simulation program is shown in Fig. 1.

Two of the four routines, "Swine Population," and "Soil Trafficability", are scheduled independently and serve as sources of information to the other routines. The "Swine Population" routing simulates the changes in the number of animals that occur as pigs are brought in and finished animals are sold. The waste production rate is calculated as a function of the number of animals in the production unit at any given time.

The "Soil Trafficability" routine determines on a daily basis whether or not the soil will support the traffic of the


This article was approved as Journal Article No. 19-75 of the Ohio Agricultural Research and Development Center (OARDC). The study was supported in part by Grant No. R-801125 from the U.S. Environmental Protection Agency.

The authors are: C. ROLAND MOTE, Assistant Professor, Agricultural Engineering Dept., University of Arkansas, Fayetteville (Former Graduate Research Associate, Ohio State University); and E. PAUL TAIGANIDES, Professor, Agricultural Engineering Dept., Ohio Agricultural Research and Development Center and Ohio State University, Columbus.

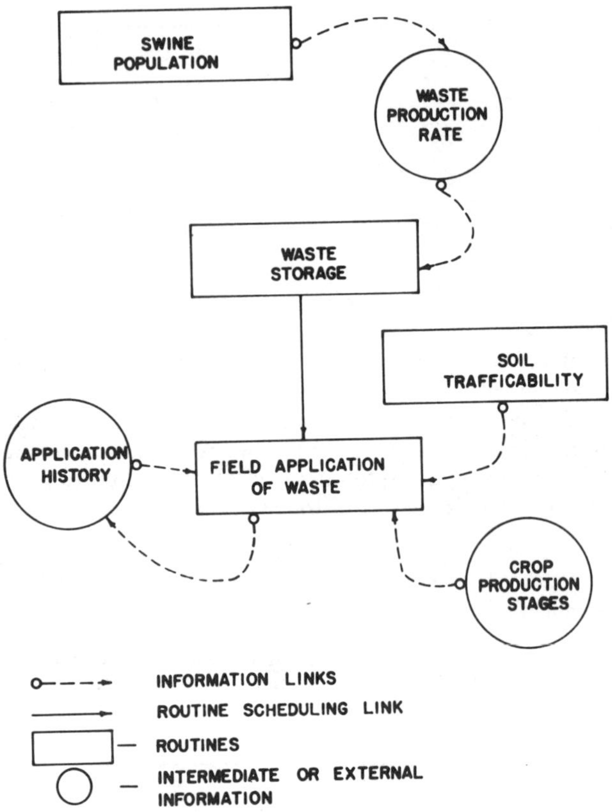

FIG. 1 Block diagram of computer simulation program.

hauling machinery. This routing simulates natural precipitation and maintains a moisture balance on the soil. Trafficability conditions are determined by comparing the soil moisture content with moisture based soil trafficability criteria reported by Thornthwaite et al. (1958). During the winter months when the soil is likely to be frozen some of the time, the soil is considered to be trafficable for a fixed portion of each day regardless of what the soil moisture content is.

The "Waste Storage" routine schedules the "Field Application of Waste" routine when the waste storage tank becomes filled to a level pre-established by the program user. However, before waste is applied to any cropland, the stage of crop production and the amount of waste previously applied during the current production cycle is checked. If the crop is in a growing or maturing stage or if the fertilizer requirements of the crop have already been satisfied by previously applied waste, no waste will be hauled. The soil trafficability condition is also checked before hauling is initiated. If any condition prohibits hauling, nothing is done until the next day when conditions are again checked to determine if hauling operations are possible. If the storage tank is full and waste cannot be hauled, all waste produced in

| Parameter | Value |
|---|---|
| **Constant value parameters:** | |
| Production unit capacity | 1 500 pigs |
| Cropland available for waste disposal | 40.47 ha |
| Number of tank wagons | 1 |
| Tank wagon capacity | 4 163.5 l |
| Soil type | Plastic |
| Soil moisture content at field capacity | 23.6 cm |
| Flush water rate | 6.1 l/pig/day |
| Waste production rate | 6.4 l/pig/day |
| Cost index | 143 |
| Fertilizer value of waste | $0.42/1000 l |
| **Variable parameters:** | |
| Storage capacity | 30 days |
| | 60 days |
| | 90 days |
| | 120 days |
| | 180 days |
| | 240 days |
| | 300 days |
| Percent of storage tank full when hauling is first attempted (hauling criterion) | 20 |
| | 50 |
| | 80 |
| **Cropping programs** | |
| Number 1 | |
| Corn | 40.47 ha |
| Number 2 | |
| Corn | 20.24 ha |
| Wheat | 20.24 ha |
| Number 3 | |
| Corn | 20.24 ha |
| Wheat | 6.07 ha |
| Forage | 10.12 ha |
| Uncultivated land | 4.05 ha |

subsequent time periods is considered as overflow.

The computer program also has calculation procedures for estimating the cost of owning and operating the system. Cost estimates are based on cost-capacity relationships for tank wagons and storage tanks and a fixed percentage for overhead. The fertilizer value of waste applied to the soil is subtracted from the estimated cost to yield an estimated net cost.

A more detailed and complete presentation of the computer program, how it works, and how to use it is given by Mote (1974).

## RESULTS OBTAINED WITH
## THE COMPUTER SIMULATION PROGRAM

A particular waste storage and field disposal system was studied using the computer simulation program to investigate the variations in the average annual storage overflow volume and the annual net cost in response to changes in the design and operating criteria. The criteria that were varied were days of storage capacity, the portion of the storage tank that is full when hauling operations are first attempted (haul criterion), and the cropping program on the cropland available for waste disposal.

The system studied was to handle the waste from a 1500 animal capacity swine production unit located on a 40.45 ha farm in Central Ohio. Program parameter values that were held constant as well as those that were varied during the course of the study are presented in Table 1.

Note in Table 1 that the seven storage capacities, three haul criteria, and three cropping programs combine to provide a total of 63 different situations to be studied.

The operation of the system was simulated for a period of 5 years under each of the 63 conditions.

A multi-year simulation period was desirable for the purpose of taking into account variations in system performance that might occur over a period of time. Variations could be expected due to such things as differences in precipitation magnitudes and occurrences from year to year and the resulting soil trafficability, as well as variations in the animal population that might occur. The 5 year simulation period was arbitrarily selected for this study as a means of comparing the system operating results under the various situations for a reasonable operating period.

### Hauling Criteria

A waste storage tank does not have to be emptied until is is full. However, the simulation studies show that under realistic conditons where there are factors such as crop production stages and soil moisture conditions that limit the periods available for hauling, the average annual overflow volume can be reduced by initiation of hauling operations well before the tank is full. As can be seen from Figs. 2, 3, and 4, for each of the cropping programs and storage capacities the average annual overflow volume declined as the haul criterion was reduced from 80 to 20 percent. For the 180 day storage capacity, for example, the average annual overflow volume for cropping program number 1 (40.47 ha of corn) was reduced 73 percent by lowering the haul criterion from 80 to 20 percent.

### Storage Capacity

An increase in the number of days of storage capacity might be expected to reduce the annual overflow volume. The data presented in Figs. 2, 3, and 4, show that this is

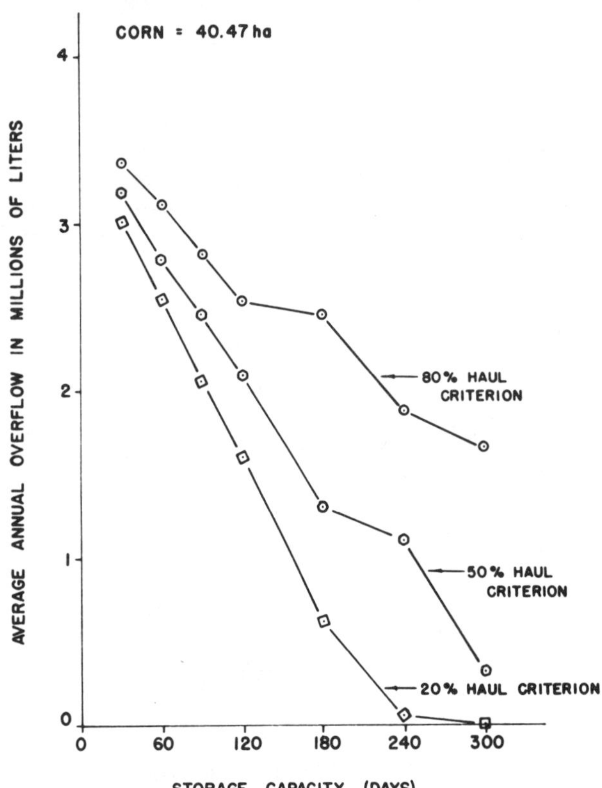


FIG. 2 Relationship of Storage capacity to overflow volume for cropping program Number 1.

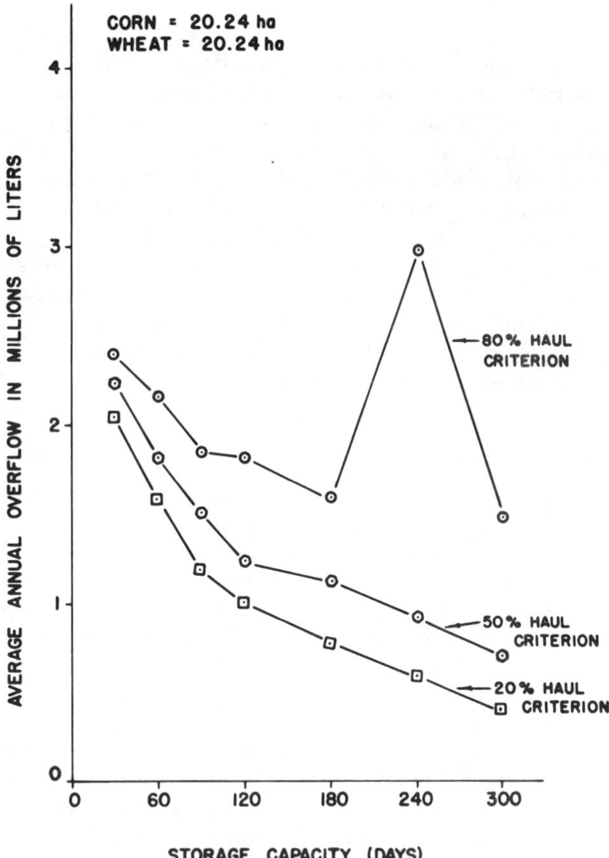



FIG. 3 Relationship of Storage Capacity to overflow volume for cropping program Number 2.

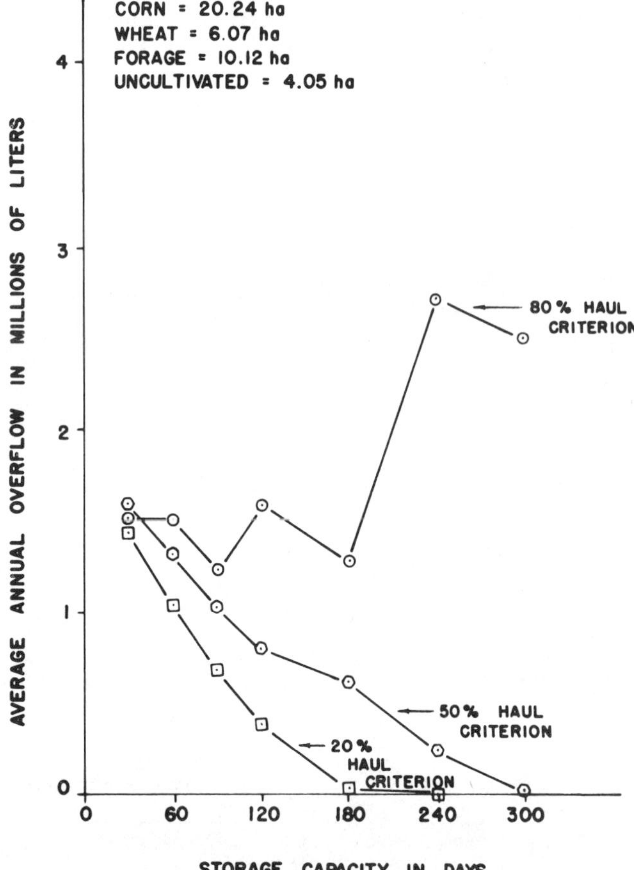



FIG. 4 Relationship of storage capacity to overflow volume for cropping program Number 3.

generally true. In fact for cropping program number 3 and a 20 percent haul criterion the average annual overflow volume was reduced by 98 percent by increasing the storage capacity from 30 to 180 days. However, as shown very clearly in Figs. 3 and 4, if the haul criterion is 80 percent, an increase in storage capacity can actually cause an increase in the annual storage tank overflow rather than a decrease. Such a condition arises when the size of the storage tank and the hauling criterion used cause a large number of the hauling operations to be scheduled at times when either the soil trafficability or crop development stage prohibit hauling. Small storage capacities and lower haul criteria cause hauling operations to be attempted more frequently and thus take advantage of more of the periods during which hauling is possible.

### Cropping Program

The type crops being grown on the land used for waste disposal affect the relationship between storage capacity and overflow volume. A comparison of Figs. 2, 3, and 4 show that the overflow volume from the smallest storage tank was greatest for an all corn cropping program (Cropping program 1), and smallest for cropping program 3 where some of the land was uncultivated. Figs. 2 and 4 show that for cropping programs 1 and 3, it is possible to provide a storage capacity that will completely eliminate storage overflow if hauling operations are attempted when the storage tank is 20 percent filled. However, for cropping program 2, overflow continued to exist even at storage capacities of 300 days. From the results illustrated in Figs. 2, 3, and 4 it appears that cropping programs which provide land with minimum crop restriction to hauling (i.e., cropping program number 3) require the smallest storage capacity to completely eliminate storage tank overflow.

### Net Annual Cost

Cost incurred as the result of waste produced in excess of the storage capacity (overflow) may vary from system to system. The cost may be in the form of expenses for providing temporary storage, clean up costs, or even fines from some environmental control agency. Because it is not known what the exact cost of handling waste overflow should be the net annual cost has been calculated for overflow costs of 0, 1, and 5 cents per liter.

In Figs. 5, 6, and 7 the relationship between storage capacity and relative average annual net cost has been plotted for each of the 3 cropping programs. For each cropping program, the relative net cost curves are for a haul criterion of 20 percent, because, as seen before in Figs. 2, 3, and 4, the 20 percent haul criterion produced the least amount of overflow.

All the relative average annual net costs plotted in Figs. 5, 6 and 7 are relative to the same value. Therefore, cost comparison can be made between the three figures.

As can be seen from Figs. 5, 6, and 7 the necessary storage capacity for minimum annual net cost is 30 days or less if there is no cost for handling overflow. At zero overflow cost, the only incentive to store the waste is its fertilizer value. The curves in Figs. 5, 6, and 7 thus show that fertilizer value alone is insufficient incentive to eliminate overflow. The curves show, however, that there is incentive to store more of the waste if overflow costs as much as 1 cent per liter.



MANAGING LIVESTOCK WASTES

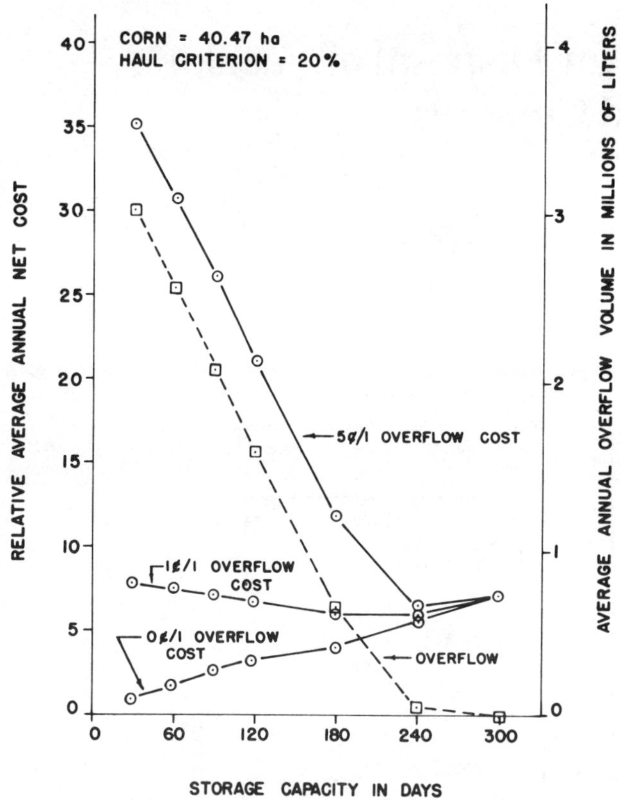



**FIG. 5 Relationship of storage capacity to relative net cost for cropping program Number 1.**

For the all corn cropping program (Fig. 5), the lowest cost is provided by a 180 day storage capacity if the overflow cost is 1 cent per liter, and by a 240 day storage capacity if the overflow cost is as much as 5 cents per liter.

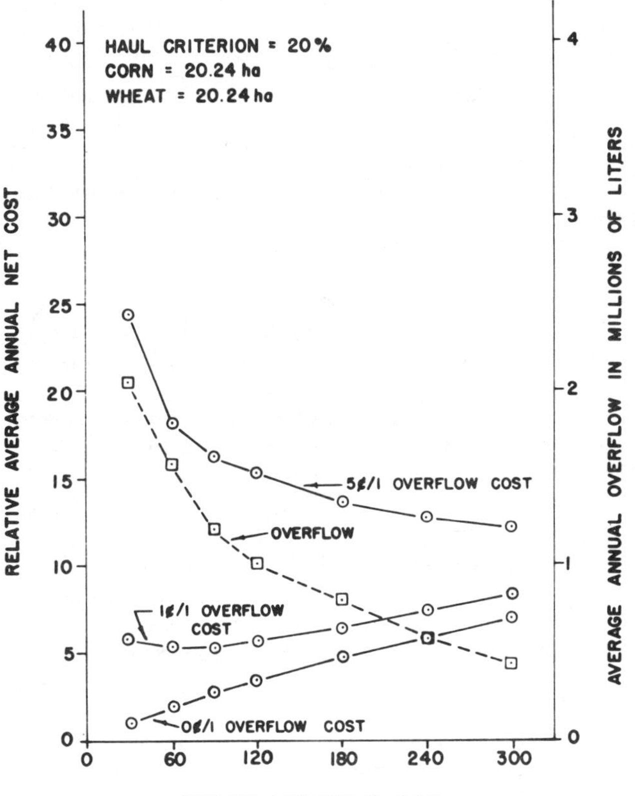



**FIG. 6 Relationship of storage capacity to relative net cost for cropping program Number 2.**

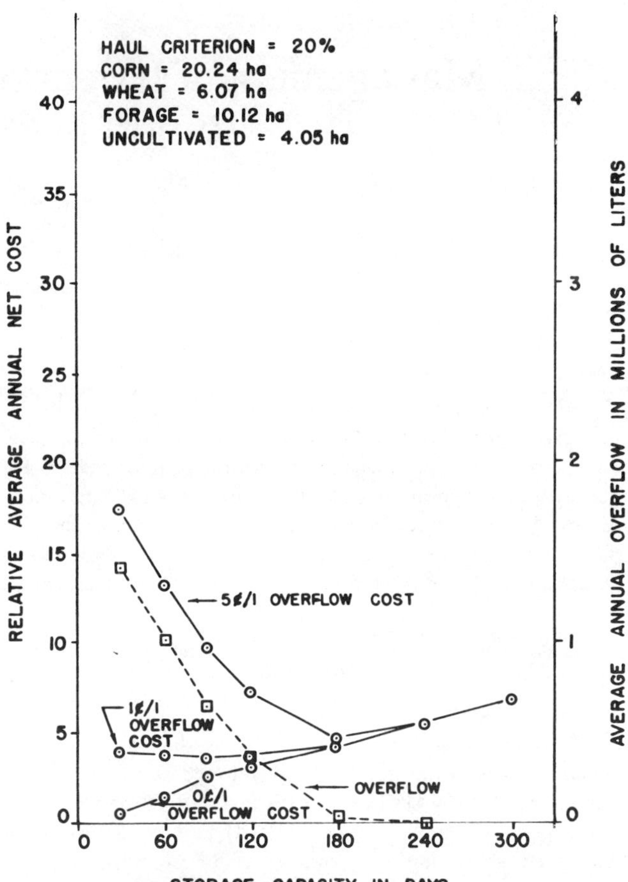



**FIG. 7 Relationship of storage capacity to relative net cost for cropping program Number 3.**

For cropping program 2 (Fig. 6), the lowest net cost for an overflow cost of 1 cent per liter is provided by a storage capacity between 60 and 90 days. If the overflow cost is 5 cents per liter, the net annual cost continues to decline as the storage capacity is increased to 300 days, however, the rate of decline is greatly reduced for increases in storage capacity above 90 days.

For the cropping system that provides some uncultivated acreage (Fig. 7) the lowest net annual cost for an overflow cost of 1 cent per liter is provided by a storage capacity of 90 days, but a 180 day storage capacity provides the minimum net annual cost if the overflow cost is 5 cents per liter. It can also be seen that the minimum costs in Fig. 7 are less than the minimum costs of the other two cropping programs.

## CONCLUSIONS AND RECOMMENDATIONS

Results from the runs made with the computer simulation program illustrate that the incentive to store waste and apply it to cropland is directly related to the cost, other than loss of fertilizer value, incurred by waste accumulations in excess of storage capacity (overflow). In other words, for the conditions investigated, the fertilizer value of the waste was not sufficient to justify storing every drop of the waste for field application.

The computer results also point out that the volume of storage tank overflow can be reduced by, (a) initiation of hauling operations well before the tank is full, and (b) selecting cropping programs that will minimize the amount of time that no land is available for hauling onto. It was also shown that an increase in storage capacity may actually increase the annual waste overflow volume

*(Continued on page 420)*

# Management of Irrigation for Disposal of Feedlot Runoff in Cold Climates

R. W. Schottman, C. W. Thoreson, J. K. Koelliker


ASSOC. MEMBER
ASAE

MEMBER
ASAE


T HE objectives of this study were to develop a computer program to accurately model a feedlot runoff management system to determine a set of management criteria for detention pond pumping and irrigation. It was to be practical for North Dakota feedlots, and would compare the predicted effectiveness of Kansas, Minnesota, and North Dakota management systems in restricting detention pond overflows.

## THE MODEL

### General Description

The runoff model described here is the North Dakota version of the computer model developed at Minnesota by Larson et al. (1974). They, in turn, had modified a model developed by Koelliker et al. (1974) at Kansas. As input data, the three models required daily precipitation and temperature data, as well as several runoff and disposal system parameters. These parameters included the area of the feedlot, the Soil Conservation Service (SCS) runoff curve number for the feedlot surface, the dimensions of the runoff control structure and the pumping rates.

The computer models examined weather records consisting of maximum and minimum temperatures and precipitation on a daily basis to determine values for runoff, evaporation, overflow, pumping and precipitation onto the pond's surface. All quantities of water were expressed in terms of depth in centimeters over the feedlot area. Fig. 1 shows a schematic of the model with the physical facilities being represented by squares and the processes being shown as lines and arrows. The design parameters used in the North Dakota model also are listed there.

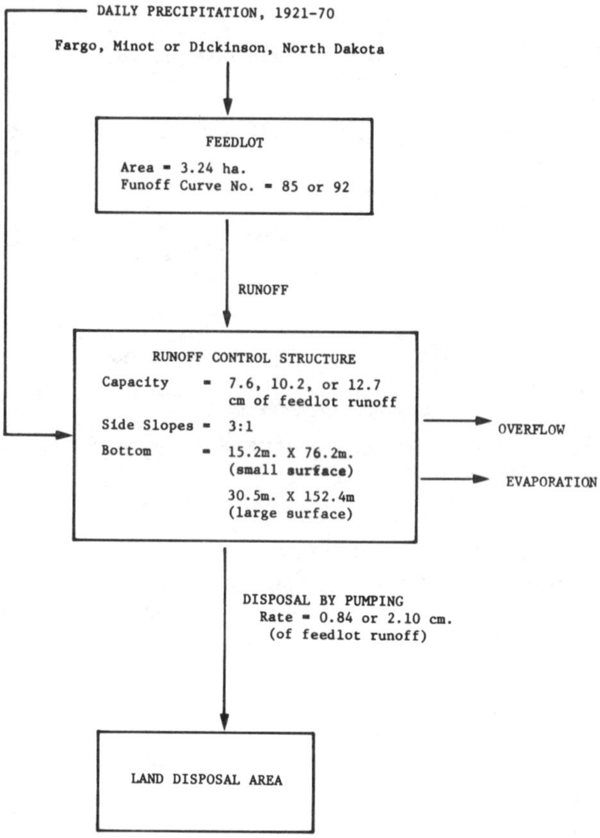

The model was written in the Fortran IV computing language, and all computer operations were made on the IBM 360 computer located at North Dakota State University. The computer program allowed for several management alternatives and conditions to be run at one time.

### Runoff Calculations

The North Dakota model used the Soil Conservation Service method (United States Department of Agriculture 1964) for calculating runoff for the period April 1 to November 20. Runoff during the period November 20 to December 1 was predicted using the SCS method if the weather was warm enough, or was accumulated if temperatures were below -1.1 C. All of the precipitation during the


Work on which this project is based was supported by the North Dakota Agricultural Experiment Station, Project No. 1426.

The authors are: R. W. SCHOTTMAN, Assistant Professor, C. W. THORESON, Graduate Research Assistant, Agricultural Engineering Dept., North Dakota State University, Fargo; and J. K. KOELLIKER, Assistant Professor, Agricultural Engineering Dept., Kansas State University, Manhattan.


DAILY PRECIPITATION, 1921-70

Fargo, Minot or Dickinson, North Dakota

FEEDLOT
Area = 3.24 ha.
Funoff Curve No. = 85 or 92

RUNOFF

RUNOFF CONTROL STRUCTURE
Capacity = 7.6, 10.2, or 12.7 cm of feedlot runoff
Side Slopes = 3:1
Bottom = 15.2m. X 76.2m. (small surface)
30.5m. X 152.4m (large surface)

OVERFLOW

EVAPORATION

DISPOSAL BY PUMPING
Rate = 0.84 or 2.10 cm. (of feedlot runoff)

LAND DISPOSAL AREA

FIG. 1 Elements of the runoff control system with assumed design parameters.

period December 1 to December 10 was assumed to accumulate on the feedlot until the previous 5-day average temperature exceeded -1.1 C. This accumulated quantity of water was then assumed to run off immediately. Precipitation for the remaining days of the year was assumed to accumulate and run off on April 1. The cutoff dates were based on observations of local temperature data.

The assumption of the complete accumulation of snowfall during the winter months was rather arbitrary. Larson et al. (1974) had assumed that 0.63 cm of water would be lost from each precipitation event occurring during freezing conditions and that the remainder would be accumulated as potential runoff. Unpublished results for the runoff for one winter at Fargo as well as observations at other feedlots indicated that the 0.63 cm loss was somewhat high, particularly when drifting occurred within the feedlot. Complete accumulation was therefore assumed for lack of a better estimate. To put this assumption into perspective, one should realize that precipitation for Fargo, North Dakota, for the period December 1 to April 1 is approximately 7.6

TABLE 1. PRECIPITATION, RUNOFF AND PERFORMANCE FOR FEEDLOT CONTROL FACILITIES IN NORTH DAKOTA.

| Location | Average annual precipitation, cm | 25-year 24-hour storm, cm | 25-year* 24-hour runoff, cm | Years with† overflow for smallest structures, percent | Runoff† controlled for smallest structures, percent |
|---|---|---|---|---|---|
| Fargo | 47.3 | 9.2 | 7.0 | 68 | 87.3 |
| Minot | 40.1 | 10.3 | 8.1 | 40 | 92.0 |
| Dickinson | 40.8 | 9.0 | 6.9 | 38 | 93.0 |

*Based on curve number = 92.
†Based on pumping rate = 2.10 cm/day, large surface area, runoff curve number = 92, 30-day pumping model, storage capacity = 7.6 cm of feedlot runoff.

cm (Weather Bureau Climatological Data).

Thoreson (1974) measured rainfall, pan evaporation, pond levels, and pond cross sections for feedlots in eastern North Dakota during the summer of 1974, and used this data as a check on the computer model. Predicted and observed pond levels showed close agreement. Winter and spring runoff was not measured.

## Pumping Schedules

Management schedules indicative of a 3.2 hectare feedlot operation in North Dakota were defined. Other sizes of feedlot might have been used since all water quantities were independent of the absolute size of the feedlot.

The pumping schedules used in the Kansas and Minnesota studies allowed pumping whenever soil and weather conditions permitted. This type of scheduling was not felt to be applicable to labor availability conditions on North Dakota farms. Determination of a feasible time for runoff disposal was necessary. To do this, the work loads and irrigation needs of several North Dakota feedlot operators were examined, and a pumping schedule was determined. The North Dakota Crop and Livestock Reporting Service (1965) reported that at least 90 percent of all small grains have normally been planted by the first of June, and 95 percent of all small grains have been harvested by September 15. April and May were initially ruled out as pumping months based upon this and upon observations of pumping operations of several feedlot operations.

Disposal was therefore scheduled for the first 15 days of June and the last 15 days of September. The selection of these dates also insured that the farmer would attempt to empty the pond early in the summer and again in the fall.

This pumping schedule was designated as the 30-day pumping model, and simulations were run for Fargo, Minot, and Dickinson, North Dakota. These three sites represented the conditions in south-eastern, north-central and south-western North Dakota, respectively.

A further simulation, designated as the 33-day model, showed the effects of three additional pumping days on April 15, May 15, and August 15 at Minot.

The potential pumping days each year were specified in the model. However, the actual number of days when pumping occurred was not necessarily the same because of variations in weather conditions and pumping requirements. Pumping was permitted after checking the mean temperature for the day, the mean temperature for the preceeding five days, the precipitation for the day, the precipitation in the last five days, and the level of the pond. An attempt to pump was made on the following day if weather conditions did not permit pumping on a scheduled day. Further attempts to pump continued on succeeding days until weather conditions permitted pumping or until another scheduled pumping day was encountered.

The percentage of years in which overflow occurred, and the storage efficiencies of the structures are the two measures which were used to illustrate the effectiveness of various management systems. The storage efficiency, as defined in equation [1], represents the percent of runoff that was controlled. It is based on the average annual runoff, RO, and the average annual overflow, OF.

$$\text{Efficiency} = 100\,(RO - OF)/RO \quad \cdots\cdots\cdots\cdots\cdots\cdots\cdots [1]$$

TABLE 2. PERFORMANCE OF 25-YEAR, 24-HOUR SIZED RUNOFF CONTROL FACILITIES IN MINNESOTA AND KANSAS, UNSURFACED FEEDLOT.

| Location | Annual precipitation, cm | 25-yr, 24-hr precipitation, cm | Years with overflow, percent | Runoff controlled, percent |
|---|---|---|---|---|
| **Minnesota (2)** | | | | |
| Bird Island | 71.2 | 12.0 | 20 | 96.0 |
| Crookston | 54.7 | 10.1 | 17 | 98.4 |
| Minneapolis | 66.3 | 12.1 | 6 | 99.3 |
| Morris | 61.3 | 11.3 | 6 | 98.9 |
| Waseca | 75.9 | 12.5 | 10 | 99.5 |
| Worthington | 66.4 | 12.4 | 3 | 98.3 |
| **Kansas (1)** | | | | |
| Colby | 49.3 | 10.9 | 6.7 | 98.6 |
| Garden City | 48.0 | 11.9 | 0.0 | 100.0 |
| Ellsworth | 71.4 | 14.0 | 10.0 | 97.9 |
| Horton | 93.7 | 15.0 | 30.0 | 93.0 |
| Independence | 94.0 | 17.3 | 30.0 | 95.5 |

**TABLE 3. PERCENT OF YEARS THAT DESIGN CAPACITY WAS EXCEEDED FOR MINOT, NORTH DAKOTA. (30 DAY PUMPING MODEL)**

| | | SCS curve number | | | |
| | | CN = 92 | | CN = 85 | |
| | Relative surface | Pumping rate, cm per day | | | |
| Pond capacity, cm | area | 0.84 | 2.1 | 0.84 | 2.1 |
|---|---|---|---|---|---|
| 7.6 | Large | 44 | 40 | 16 | 14 |
| | Small | 56 | 52 | 24 | 22 |
| 10.2 | Large | 34 | 24 | 8 | 6 |
| | Small | 34 | 28 | 8 | 6 |
| 12.7 | Large | 14 | 10 | 2 | 2 |
| | Small | 20 | 14 | 4 | 4 |

## RESULTS

### Geographic Trends of Control Effectiveness

General trends regarding years of overflow for the Fargo and Dickinson locations were similar to those obtained for the Minot location except that the control of the runoff was more effective as one moved from east to west across the state. This can be attributed to the higher annual evaporation deficit in the western part of the state. Table 1, which shows a representative set of results for the three locations, illustrates this trend. A comparison of these results for a structure containing 7.6 cm of feedlot runoff to those summarized for Kansas and Minnesota in Table 2 for 25-year, 24-hour storms show that the frequency of overflow is higher using the North Dakota management model.

This difference is primarily due to the fact that tighter restrictions on pumping were in effect in North Dakota.

### Pumping Capacity, Pond Capacity and Surface Area

Table 3 illustrates the relatively little importance of increasing the pumping capacity of the system for the purpose of controlling the number of overflow events. In all cases, the percentage of years having overflow events declined, but still exceeded 10 percent when a curve number of 92 was specified.

The percentage of years that the design capacity was exceeded always decreased when the capacity of the control structure was increased. The percentage declined to less than four when runoff was low (CN = 85) and declined to less than twenty when higher runoff occurred (CN = 92). The higher runoff value would be more typical of most feedlots in North Dakota (United States Department of Agriculture 1971).

Table 3 also illustrates the effectiveness of a control structure having a relatively large surface area. The percentage of years that the design capacity was exceeded was consistently higher for the pond having a smaller surface area, in some cases the difference being as much as 12 percent.

### Frequency of Pumping

A feedlot operator attempting to manage the runoff would also be interested in estimating the frequency of pumping. Tables 4 and 5 list the average days of pumping

**TABLE 4. CONTROL STRUCTURE EFFICIENCY AND ACTUAL DAYS OF PUMPING FOR MINOT, NORTH DAKOTA. (30-DAY PUMPING MODEL)**

| | Large surface area control structure | | | | | | Small surface area control structure | | | | | |
| | Control structure efficiency | | | Average days of pumping per year | | | Control structure efficiency | | | Average days of pumping per year | | |
| Pumping rate | Capacity, cm | | | Capacity, cm | | | Capacity, cm | | | Capacity, cm | | |
| | 7.6 | 10.2 | 12.7 | 7.6 | 10.2 | 12.7 | 7.6 | 10.2 | 12.7 | 7.6 | 10.2 | 12.7 |
|---|---|---|---|---|---|---|---|---|---|---|---|---|
| **Watershed CN 92** | | | | | | | | | | | | |
| 0.84 cm/day | 90.3 | 94.9 | 97.5 | 9.4 | 10.5 | 11.1 | 86.6 | 93.0 | 96.4 | 13.1 | 14.3 | 15.1 |
| 2.10 cm/day | 92.0 | 96.0 | 98.1 | 5.1 | 5.4 | 5.6 | 88.9 | 94.7 | 97.7 | 7.1 | 7.6 | 7.8 |
| **Watershed CN 85** | | | | | | | | | | | | |
| 0.84 cm/day | 96.2 | 98.3 | 99.3 | 6.3 | 6.8 | 7.0 | 94.6 | 97.7 | 99.2 | 9.6 | 10.3 | 10.8 |
| 2.10 cm/day | 96.6 | 98.5 | 99.3 | 3.6 | 3.8 | 3.9 | 95.2 | 98.1 | 99.2 | 5.3 | 5.7 | 6.1 |

**TABLE 5. CONTROL STRUCTURE EFFICIENCY AND ACTUAL DAYS OF PUMPING FOR MINOT, NORTH DAKOTA. (33-DAY PUMPING MODEL)**

| | Large surface area control structure | | | | | | Small surface area control structure | | | | | |
| | Control structure efficiency | | | Average days of pumping per year | | | Control structure efficiency | | | Average days of pumping per year | | |
| Pumping rate | Capacity, cm | | | Capacity, cm | | | Capacity, cm | | | Capacity, cm | | |
| | 7.6 | 10.2 | 12.7 | 7.6 | 10.2 | 12.7 | 7.6 | 10.2 | 12.7 | 7.6 | 10.2 | 12.7 |
|---|---|---|---|---|---|---|---|---|---|---|---|---|
| **Watershed CN 92** | | | | | | | | | | | | |
| 0.84 cm/day | 92.5 | 96.6 | 98.4 | 10.2 | 11.1 | 11.6 | 90.1 | 95.4 | 97.8 | 13.9 | 15.1 | 15.8 |
| 2.10 cm/day | 95.4 | 97.9 | 99.1 | 5.5 | 5.7 | 6.0 | 93.7 | 97.4 | 99.0 | 7.6 | 8.1 | 8.2 |
| **Watershed CN 85** | | | | | | | | | | | | |
| 0.84 cm/day | 97.3 | 98.8 | 99.5 | 6.7 | 7.1 | 7.4 | 96.3 | 98.5 | 99.5 | 10.1 | 10.7 | 11.1 |
| 2.10 cm/day | 98.1 | 99.2 | 99.8 | 3.8 | 4.0 | 4.1 | 97.7 | 99.0 | 99.9 | 5.5 | 5.9 | 6.3 |

| Pumping rate (cm/day) | Pond capacity (cm) | | |
|---|---|---|---|
| | 7.6 | 10.2 | 12.7 |
| 0.84 | 1970 | 1970 | 1970 |
| | 1969 | 1969 | |
| | 1967 | 1953 | |
| | 1953 | | |
| | 1944 | | |
| | 1943 | | |
| 2.10 | 1970 | 1970 | 1970 |
| | 1969 | 1969 | |
| | 1944 | | |

per year for each of the management schedules. The actual number of pumping days varied from a minimum of about four for a small capacity pond having a large surface area, a low runoff rate, and a large pumping rate, to a maximum of about 16 for a large capacity pond having a small surface area, a high runoff rate, and a low pumping rate. A comparison of the results in Tables 4 and 5 also shows that about one extra pumping day would result when the 33-day model was used instead of the 30-day model.

A tabulation of the number of years that overflow occurred does not completely describe the extent to which the runoff was being controlled. A consideration of the storage efficiencies is also necessary. Tables 4 and 5 show the storage efficiencies for the 30-day and the 33-day models at Minot. All of the efficiencies were between 86 and 99.9. The additional pumping days increased the efficiencies by about one to two percent, but in no case provided 100 percent control for the storms occurring during the 50 year period of record.

### Overflow by Seasons of the Year

The correlation between overflow events and particular seasons of the year was also studied to gain a better understanding of effective management techniques. The following discussion assumes that a feedlot located in Minot, North Dakota had a runoff curve number equalling 85 (a relatively low value). In addition, the strictest set of management techniques was assumed to control overflow. These included a pond having a relatively large surface area and a management schedule using 33 days of pumping. This set of conditions would produce the least runoff, and would dispose of the largest quantity of runoff. If overflow occurred using this model, then it would occur in the other models. The monthly rainfall records for the years in which overflow occurred were examined to determine whether the large runoff events occurred during a particular season of the year.

Table 6 indicates the years in which overflow most frequently occurred, using different management systems. The average monthly rainfall for these years is shown in Table 7. These data indicate that combinations of storms producing large runoff events occurred throughout the spring and summer. Many of the years for which overflow occurred had monthly rainfalls more than 50 percent above the average.

### Environmental Protection Agency (EPA) Regulations

The EPA regulations (1974) require that by 1983 runoff from a 25-year, 24-hour storm must be contained in some type of runoff control structure. Table 1 gives the magnitudes of the 24-hour, 25-year runoff events for each of the North Dakota locations, assuming a curve number of 92 for an unsurfaced feedlot. Predictions of the overflow conditions which would result from structures having these capacities can be estimated using the results shown in Table 3.

Overflow occurred in more than 40 percent of the years when a pond had a capacity of 7.6 cm, and a curve number of 92. A similar number of overflow events would be predicted for a pond built to meet the 25-year, 24 hour criteria.

Another computer simulation using the EPA criteria, a 33-day pumping model, and a curve number of 85 showed that the design capacity was exceeded in 36 percent of the years. Although failing to contain all of the water in over a third of the years, the structures still managed to maintain efficiencies of 95 percent.

## CONCLUSIONS

Several conclusions can be drawn regarding feedlot pumping management for the relatively short pumping season available to feedlot operators in North Dakota. Control of the feedlot runoff was not complete, using the management schedule specified for North Dakota. As an example, a structure meeting the EPA 25-year, 24-hour storm criteria would have at best controlled 95 percent of the runoff and would have overflowed in at least 36 percent of the years studied for north-central North Dakota. In contrast, the management schedules allowing pumping whenever possible in Minnesota and Kansas had fewer years of runoff than did the more restrictive North Dakota schedule which placed its main emphasis on using the water for irrigation during the summer and fall. Nevertheless, storage efficiencies for all three states were of the same order of magnitude.

An increase in the design capacity was not the answer to providing complete runoff control when using the North Dakota schedule. Even a structure having a capacity of 12.7 cm of feedlot runoff (approximately twice as large as the 25-year, 24-hour storm) would have failed to contain all of the runoff during the 50 year period. The efficiency of the

TABLE 7. RAINFALL FOR YEARS WHEN OVERFLOW
OCCURRED, CENTIMETERS.

| Year | Months when rainfall occurred | | | | | | | |
|---|---|---|---|---|---|---|---|---|
| | March | April | May | June | July | August | Sept. | Oct. |
| 1970 | 1.8 | 18.5 | 9.1 | 3.9 | 17.9 | 1.7 | 5.3 | 1.1 |
| 1969 | 1.8 | 2.8 | 3.1 | 12.3 | 7.9 | 2.5 | 0.2 | 4.2 |
| 1967 | 1.0 | 12.1 | 0.6 | 2.5 | 0.2 | 1.7 | 3.7 | 3.8 |
| 1953 | 2.2 | 7.9 | 13.9 | 13.9 | 6.0 | 2.6 | 0.7 | 2.1 |
| 1944 | 1.1 | 2.2 | 4.1 | 27.0 | 3.3 | 9.3 | 3.4 | 0.0 |
| 1943 | 4.3 | 5.9 | 6.8 | 11.9 | 10.3 | 8.8 | 0.8 | 0.8 |
| Average (1931- 1969) | 1.9 | 2.6 | 4.6 | 8.7 | 5.5 | 4.7 | 3.5 | 2.1 |

control structure could be improved by up to five percent by additional pumping during April, May and August. These are months not ordinarily requiring irrigation water.

Another point that emerged from an examination of the monthly rainfall records was that a pumping strategy aimed at preventing overflows due to spring runoff was not wholly successful. Overflow occurred during late summer which is susually considered to be a dry season.

### References


1   Koelliker, J. K., Manges, H. L. and Lipper, R. I. 1974. Performance of feedlot runoff control facilities in Kansas, ASAE Paper No. 74-4012. ASAE, St. Joseph, Mich. 49085.

2   Larson, C. L., James, L. G., Goodrich, P. R. and Bosch, J. 1974. Performance of feedlot runoff control system in Minnesota, ASAE Paper No. 74-4013. ASAE, St. Joseph, Mich. 49085.

3   Thoreson, C. W. 1974. Runoff Disposal Management for Feedlots. Unpublished M.S. thesis, North Dakota State University Library, Fargo.

4   United States Department of Agriculture, Soil Conservation Service 1971. North Dakota Supplement to Engineering Field Manual, Chapter 2.

5   United States Department of Agriculture, Soil Conservation Service, Soil Conservation Service National Engineering Handbook. 1964. Estimation of direct runoff from storm rainfall, Washington, D. C. Section 4, Chapter 10.

6   United States Department of Agriculture Statistical Reporting Service, North Dakota Crop and Livestock Reporting Service. 1965. North Dakota Weather Crop Bulletin 1950-1965.

7   United States Environmental Protection Agency. February 14, 1974. Effluent limitations guidelines for existing sources and standards of performance adn pretreatment standards for new sources, feedlot category. Federal Register 39:5704-5708.

8   Weather Bureau Climatological Data, North Dakota Section, United States Department of Commerce. Volumes 72 through 81, annual summaries.


---

## Computer Simulation of Swine Waste Disposal

*(Continued from page 415)*

if the proper hauling initiation criterion is not being employed.

It is recommended that the computer simulation program be used to expand this study and investigate the effect of the number and size of tank wagons on the overflow volume and the net annual cost of operating the waste storage and land disposal system. Application of the program to various size production units, cropland acreages, and cropping systems could also be used to develop a series of reference curves that would assist swine producers in selecting waste handling systems to fit their conditions.

The computer simulation discussed in this paper could also be incorporated into a simulation program with the broader scope of simulating the entire farming operation. Such a simulation program could take into account the potential for overlapping demands on tractors and personnel by the waste disposal and crop production related activities. Such an expansion of the scope of the simulation program would enhance its usefulness as a management tool.

### References


1   Mote, Charles Roland. 1974. A computer simulation of treatment, storage, and land disposal of swine waste. Unpublished Ph.D. Dissertation, Ohio State University, 1974.

2   Pritsker, A. Alan B. and Philip J. Kiviat. 1969. Simulation with GASP II. Prentice-Hall, Inc., Englewood Cliffs, New Jersey.

3   Thornthwaite, C. W., J. R. Mather, D. B. Carter, and C. E. Molineaux. 1958. Estimating soil moisture and tractionability conditions for strategic planning. Geophysics Research Directorate, United States Air Force, Bedford, Mass., AF CRC-TN-58-201, ASTIA Document No. A. D. 146789, March, 1958.

# Runoff Control Facilities for Beef Cattle Feedlots in Eastern Nebraska


J. A. Nienaber, C. B. Gilbertson, T. E. Bond, J. L. Gartung

ASSOC. MEMBER    MEMBER    FELLOW  ASSOC. MEMBER
ASAE          ASAE      ASAE     ASAE


FEEDLOT runoff pollution potential has been well documented in the United States and the need to control it is evident (Gilbertson and Nienaber 1973a, Madden and Dornbush 1971, Miner et al. 1966, Swanson et al. 1971). Designing runoff control facilities to reduce management problems and provide adequate environmental protection is relatively new. However, methods for controlling runoff are available and in use (Gilbertson et al. 1971, Swanson et al. 1973, Swanson and Mielke 1973).

This study was made to evaluate the performance of runoff control facilities in eastern Nebraska based on select, measured, and observed parameters.

### Sites and Facilities

Feedlot runoff control facilities were constructed on the University of Nebraska Feedlot, near Mead, in 1968, and on cooperator sites at Nebraska City and Oakland, Nebraska, in 1970. In 1971, facilities were installed at US Meat Animal Research Center, Clay Center, Nebraska, and in 1973 Environmental Protection Agency funded a demonstration-research project on a cooperator site near Papillion, Nebraska, (Fig. 1). The 10-yr, 24-hr design storm ranged from 11- to 12-cm for all locations.

Facilities designed to control feedlot runoff were composed of a debris basin, holding pond, and disposal system. Runoff from areas outside the feedlot was diverted to keep the runoff volume at a minimum.

### Debris Basin

Debris basins are designed for separating settleable solids from runoff. Since this separation is the key to managing the runoff, research on the debris basin operation was emphasized.

There are two types of debris basins; the continuous flow and batch type (Figs. 2a and b). Controlled release of liquid from the continuous flow-type basin reduces velocity sufficiently to allow solids to settle. The batch type is designed to retain runoff until liquids are pumped to a holding pond. The combination debris basin-holding pond, a form of the batch type system, stores both liquid and solids until disposal (Fig. 2c).

Debris basins were located inside or outside of the feedlot fenceline, directly downslope from the feedlot, or a diversion channel was used to direct the runoff into the basin (Fig. 3a). Debris basins were higher than, at an equal elevation, or lower than the holding pond (Figs. 3b, c, and d). When the basin elevation was lower than the holding pond, a pumping station was required.

Shallow basins (less than 1-m deep) were required for


Contribution from Animal Waste Management Research Unit, North Central Region, ARS, USDA, in cooperation with the Nebraska Agricultural Experiment Station, Lincoln.

A portion of the funds for this research was provided by the Environmental Protection Agency.

The authors are: J. A. NIENABER, C. B. GILBERTSON, T. E. BOND and J. L. GARTUNG, Agricultural Engineers, ARS, USDA and University of Nebraska, Lincoln.


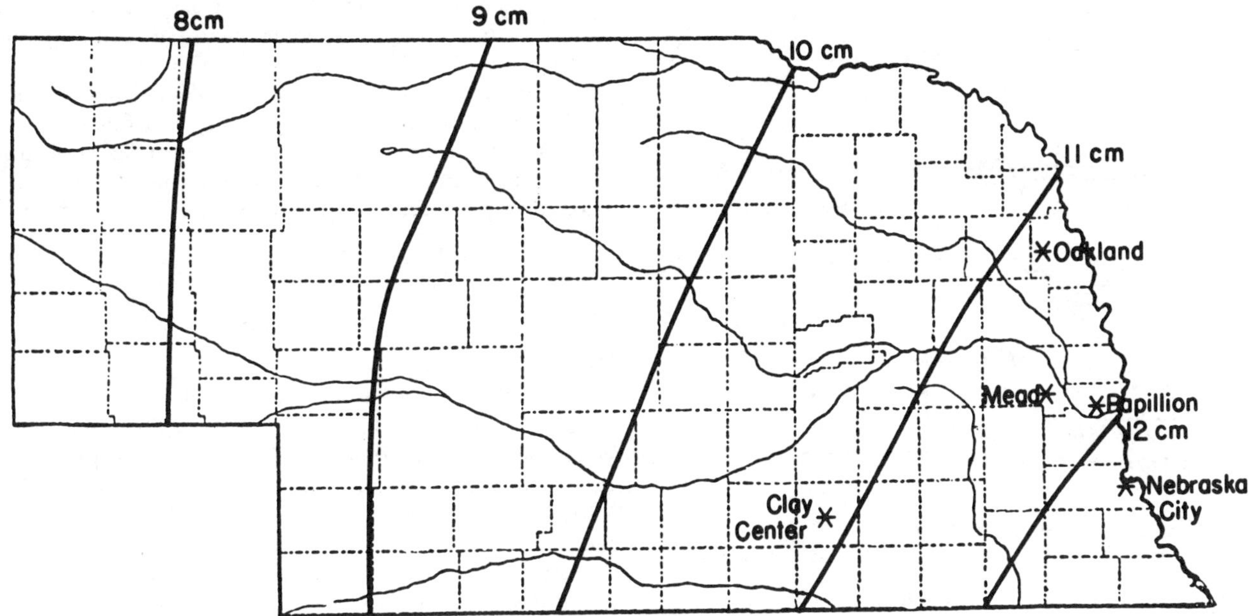


FIG. 1 Feedlot runoff control research sites with 10-yr, 24-hr reoccurance design storms.

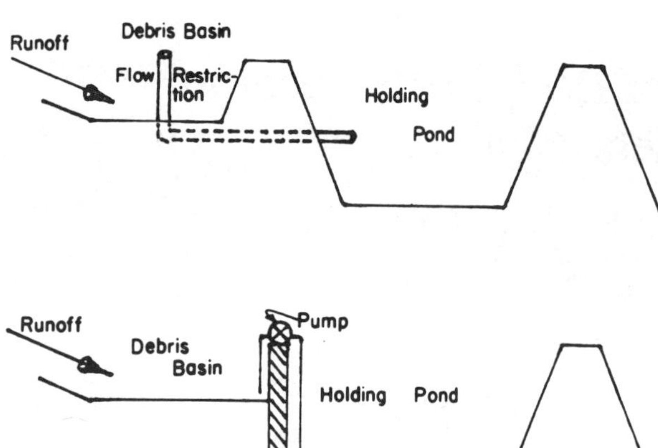

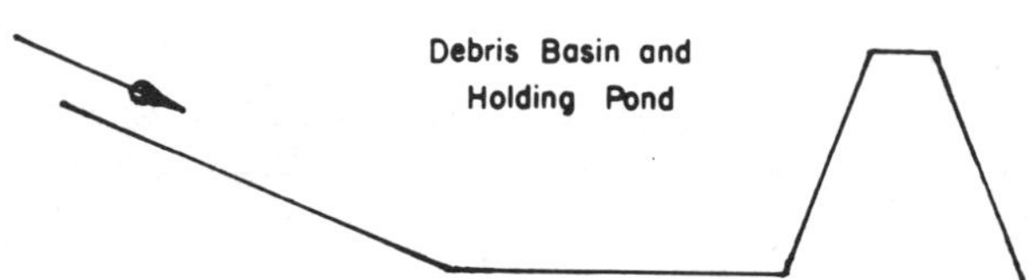

FIG. 2 Types of debris basins

    (a)    continuous flow — continuous liquid release at restricted flow

    (b)    batch system — liquid in static condition until pumped

    (c)    combination — debris basin and holding pond combined.

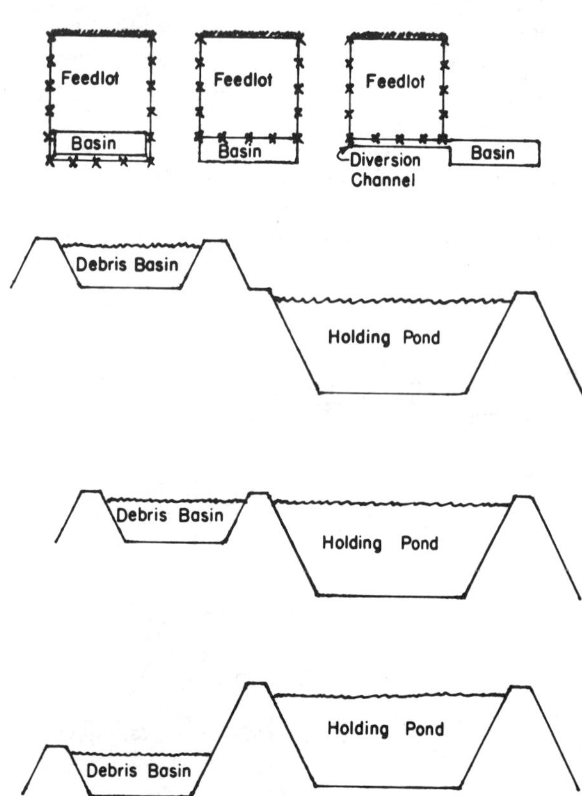

FIG. 3 Debris basin location and elevation

    (a)    debris basins located inside, outside, and remote from the feedlot, respectively

    (b)    debris basin at higher elevation than holding pond

    (c)    debris basin at same elevation as holding pond

    (d)    debris basin at lower elevation than holding pond.

dewatering settled solids and for ease of cleaning. Deep debris basins (more than 1-m deep) required specialized cleaning equipment and were located outside the feedlot to protect animals.

Several types of liquid outlets were utilized to drain the basins: (a) a channel with no flow restrictions, (b) a dam formed with 2.5-cm crushed rock, (c) a slotted wood dam, (d) a perforated riser (16-mm holes) connected to buried drainline, (e) a floating plastic drainline, (f) and a board dam and (g) buried drainline — both backfilled with rock (Figs. 4a through g).

All basins were designed to retain a minimum of 3.2-ha cm/ha of feedlot runoff (Gilbertson and Nienaber 1973b). For basins remote from a holding pond, a safety factor of 2.5 was used.

### Holding Ponds

The holding pond is the ultimate control for storing runoff until it can be disposed of safely or evaporated. Holding ponds were designed to retain 75 to 100 percent of the 10-yr, 24-hr design storm at all research facilities (Gilbertson and Nienaber 1973b).

The elevation of the debris basin will affect the total volume of storage available. When the debris basin and holding pond are at equal top elevations, the net storage volume is effectively increased.

On sandy soils, moisture barriers, including 0.15-mm polyethylene plastic, soil cement, and bentonite, were studied to determine their effectiveness against soil and groundwater contamination. Holding ponds constructed on soils with high clay content were not sealed.

### Disposal Systems

Centrifugal pumps (electric, gas engine, or tractor driven) were used with several types of irrigation distribution equipment, including portable sprinklers, permanent buried sprinkler lines, and a center pivot. Disposal time, as affected by system capacity, ranged from 1 to 13 days for the design storm. Disposal equipment components were selected on the basis of their availability, soil intake rate, type of distribution system, and desired automation.

Minimum disposal area design was based on anticipated runoff volume, crop water requirements, and soil properties (Nienaber et al. 1974).

Crops, including corn, corn slot planted in grass, and grass, were tested for their adaptability to effluent application. The plant material was removed from the corn silage field at harvest, while the corn planted in grass and grass planted in pastures, had ground cover throughout the year.

### RESULTS AND DISCUSSION

### Debris Basins

The solids settling efficiency is defined as the ratio of the solids settled in the debris basins to the solids which would settle out of runoff in 24 hrs, under static conditions. Gilbertson and Nienaber (1973a) suggested a settling efficiency of 85 percent to maintain the design volume in the holding pond for a 10-yr life. The batch system (Fig. 2b) was 100 percent efficient, while basins with continuous liquid discharge varied from 5 to 95 percent efficient. A no restriction flow through channel (Fig. 4a) was only 5 percent efficient in solids removal while 3 rock dams (Fig. 4b),

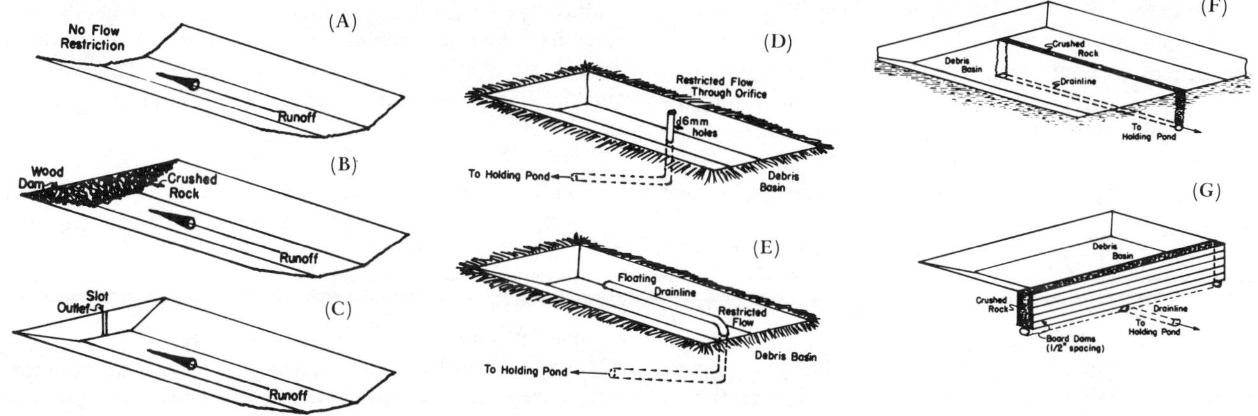


FIG. 4 Types of liquid outlets used

    (a)    flow through channel
    (b)    rock dam
    (c)    slotted-wood dam
    (d)    perforated riser (steel or plastic) and drainline
    (e)    floating plastic drainline and buried drainline
    (f)    buried drainline backfilled with crushed rock
    (g)    board dam backfilled with crushed rock.

installed in series in a 60-m channel averaged 95 percent efficient initially but dropped to 72 percent efficient the fourth year. One slotted-wood dam (Fig. 4c) was 76 percent efficient, and 2 slotted-wood dams in series averaged 80 percent efficient; however, the settled solids load was distributed over a greater area and dried more completely.

As a comparison to similar basin designs, Swanson and Mielke (1973) reported 80 percent settling efficiency of solids using 6.4- and 9.5-mm hardware cloth screens installed across a channel.

Settling efficiency was not measured on the debris basins with perforated drain risers (Fig. 4d). However, based on solids settling characteristics of runoff, the efficiency should be greater than 85 percent, since a flow release time of 2 to 3 days is designed for the 10-yr, 24-hr storm (Gilbertson et al. 1972). The efficiency of the floating plastic drainline (Fig. 4e) was 82 percent.

The solids settling efficiencies of the buried drainline and the double wall board dam — both backfilled with crushed rock (Figs. 4f and g) were 95 percent, however, the liquid outlets both plugged within 1-yr operation.

The type of discharge outlet used also affected the dewatering of the settled solids. The no restriction channel, the rock dam, and the slotted-wood dam (Figs. 4a, b, and c) provided the most complete drainage of liquid. The plastic and steel perforated risers with 16-mm holes (Fig. 4d) became plugged at low water levels which required shovel work to drain the basin. The floating plastic drainline (Fig. 4e) became partially silted in during the second year of operation.

The basin depth directly affected the type of solids removal equipment used. Solids in the 2-m deep, batch-type debris basin (Fig. 2b) were agitated with a diaphragm pump and removed with a liquid manure wagon. A commercial manure pump with mixing capabilities and a high volume gun-type sprinkler were used the last 2 yrs.

Solids removal from shallow basins was influenced by the dewatering provided by the discharge outlet and the depth of solids accumulation. When solids depth was limited to less than 1/2 m, dewatering was sufficient for removal with conventional tractor-type or track-type loaders. Machine operator judgment is critical for proper maintenance of a debris basin. An inexperienced operator may lower the basin bottom and cause puddling.

When the debris basin was located inside the feedlot, the drainage of the feedlot surface was improved. However, drainage of liquid from basins was affected by cattle movement which caused depressions in the basin and retained liquid. When liquid was retained in the basin for more than 1 week, the cattle agitated the liquid and solids into a slurry which plugged 16-mm holes in the perforated riser. When an incompletely drained basin was frozen, it became a hazard to cattle when steers that laid on the ice, melted through.

Weed control has been a continuous problem for basins located outside the feedlot. The fenceline at the lower end of the feedlot generally retained some solids due to a manure dam formed by cattle movement. It was assumed that the manure dam reduced maintenance of the debris basin by retaining solids, however, the fenceline required additional cleaning.

Solids were deposited in diversion channels when debris basins were located remote from the feedlot (Fig. 3a). When the working alley was used as the diversion channel, solids which were deposited in the channel hampered cattle movement during wet periods.

The debris basin elevation as compared to the holding pond also affected the drainage of liquid from the debris basin. At a higher elevation than the holding pond (Fig. 3b), debris basin gravity drainage was not affected by the level of liquid in the holding pond. At the same elevation as the holding pond, debris basins do not drain when the level of liquid in the holding pond is above the bottom elevation of the debris basin (Fig. 3c). If the debris basin is at a lower elevation than the holding pond and drainage is dependent on a sump pump (Fig. 3d), debris basins will not drain if there is equipment failure.

Overflows from the debris basin into the holding pond were expected with the volume of 3.2-ha cm/ha of feedlot. While samples were not taken, information on settling characteristics of feedlot runoff indicated direct release of runoff to the holding pond caused little loss of settling efficiency.

Basins remote from the holding pond were successful in containing runoff from a series of storms within a 3-day period which totaled 20 percent more than the design storm. Although the basins were filled, the loss was estimated at less than 5 percent. Without the 2.5 safety factor, 60 percent of the runoff would have been lost to the environment.

There was snowmelt runoff in 4 of the 6 years at the research sites; however, a significant amount moved off the feedlot in only 2 years. Depth of solids accumulation in the debris basin affected manageability of snowmelt runoff. The 2-m deep basin had to be cleaned with a dragline, and a shallow basin (1-m deep) had to be reconstructed after cleaning. However, basins retaining only 1/2 m of slurry were easily cleaned with a track-type loader. The diversion channel was also effective in spreading out the slurry.

## Holding Ponds

Ponds designed with a volume equal to 100 percent of the design storm were more easily managed because chance of overflow was reduced and disposal operations were flexible. However, the ponds must be emptied after each runoff event to provide capacity for future runoff, unless additional capacity is provided for long-term storage.

Initially, bentonite was an effective sealant, but continued wetting and drying caused the banks to crack and when the pond elevation increased, a significant amount of seepage was measured. The plastic liner remained sealed, when properly installed, throughout the test period. Soil

TABLE 1. ADVANTAGES AND RESTRICTIONS OF ALTERNATIVES<br>FOR USE IN DESIGN OF FEEDLOT RUNOFF CONTROL FACILITIES

| DEBRIS BASINS | Advantages | Restrictions |
|---|---|---|
| Batch (deep) | Conserve space; concentrates solids; high settling efficiency | Special cleaning equipment; pump out liquid; outside feedlot only |
| Combination with holding pond | Eliminates one structure; adaptable to inside feedlot | Special cleaning equipment or large surface area |
| Continuous flow (shallow) | No pumping required; adaptable to inside feedlot; conventional manure handling—usually | |
| Higher elevation | Gravity drainage | Topography |
| Same Elevation | Increases total system capacity | May prevent basin drainage; use only outside feedlot |
| Lower elevation | Special locations; may pump directly to field disposal | Dependent on sump pump for drainage |
| Inside feedlot | Good feedlot drainage; use basin berm as mound; solids pushed back on feedlot without hauling | Cattle churn imponded water and may plug risers; frozen imponded water may be a hazard |
| Outside feedlot | Runoff and animals are separated; solids pushed onto adjacent cropland | Weed control; land utilization is low; fencelines drain poorly |
| Diversion channels | Flexibility in locating facilities; spreads winter runoff | Not recommended for working alley |
| Flor through Outlet | Excellent drainage; winter runoff spread out | Not suitable for solids removal of rainfall runoff |
| Rock dam; Board dam with slot | Good drainage; moderate settling efficiency; winter runoff spread out | Large surface area |
| Rock and Draintile | High settling efficiency | Plugging resulted in poor drainage |
| Plastic or Steel riser | Steel suited to inside feedlots; moderate to good settling efficiency | Requires maintenance for drainage |
| Floating drainline | Moderate settling efficiency; easily moved for solids removal | Drainline may silt in; buckled pipe will retard drainage |
| **HOLDING PONDS** | **Advantages** | **Restrictions** |
| Capacity = 75% design storm | Minimum volume—area restriction | Requires disposal regardless of soil condition |
| Capacity = 100% design storm | Maximum protection; allows flexibility in disposal | |
| Sealant—plastic | Effective unless ruptured | Installation difficult |
| Sealant—bentonite | Effective under continuous wet conditions | Cracking walls permit seepage |
| Sealant—soil cement | Effective | Will not sustain vehicle traffic; high initial cost |
| **DISPOSAL SYSTEMS** | **Advantages** | **Restrictions** |
| Crop | Good fertilizer | Check effluent; dilute when salt concentration is high and pH low |
| Irrigation equipment | Centrifugal pumps, sprinklers, gated pipes, etc., well suited | Spray nozzles plugged; intakes should be screened |
| Size—minimum of ½ ha/ha feedlot | Least area and equipment required | Needs continuous cover, runoff controlled |
| Size—twice feedlot area | Adaptable to disposal under wet conditions; USE AVAILABLE EQUIPMENT | |



MANAGING LIVESTOCK WASTES

cement has remained intact for 5 yrs and has been effective in eliminating erosion on steep slopes.

## Disposal Systems

Irrigation equipment and techniques were well adapted to effluent disposal. Centrifugal pumps, sprinklers, and pipelines have handled liquid from the holding ponds without difficulty because large particles were settled out in the debris basin. The center pivot sprinkler system has functioned well on 15 percent terraced slopes; however, 3.6-mm spray nozzles, installed in one portion of the system, plugged rapidly while 6.3-mm nozzles remained open.

Disposal-area size requirements was dependent on the topography and crop cover. The minimum size of 1/2-ha/ha of feedlot provided adequate area for disposal (Nienaber et al. 1974).

Effluent was beneficial to grass production (Satterwhite and Gilbertson 1972). Corn, corn silage, and grass were harvested from disposal areas. All crops tested seemed well adapted to holding pond effluent application; however, vegetation kill resulted when effluent from a holding pond receiving direct snow-melt runoff was applied (Satterwhite and Gilbertson 1972). Salt crystals were visible on the wilted plants. This was at one site only and could have been prevented by dilution.

## CONCLUSIONS AND RECOMMENDATIONS

The performance of the runoff control facilities studied was dependent on the factors noted in Table 1, with no system being optimum for all locations. For example, the cooperator site at Nebraska City used a debris basin at the same elevation as the holding pond which had to be located outside the feedlot since it did not drain when the holding pond was full. The total storage capacity was increased by the volume of the debris basin which prevented an overflow during a 3-day rainfall event of 16.5 cm.

The cooperator site at Papillion had restricted feedlot space, but was located on a 10 percent slope which made equal pond elevations impractical. Therefore, the debris basins were located inside the feedlot to conserve space. At the same location, an experimental deep narrow basin of minimum area was installed, but a special manure pump was required for solids removal.

Each feedlot site has important physical characteristics that will affect the optimum solution. Table 1 presents information for making initial decisions on the type and location of runoff control structures, and it should be helpful in analyzing alternatives available for controlling feedlot runoff.


## References

1  Gilbertson, C. B., T. M. McCall, J. R. Ellis and W. R. Woods. 1971. Methods of removing settleable solids from outdoor beef cattle feedlot runoff. TRANSACTIONS of the ASAE 14:899-905.

2  Gilbertson, C. B. and J. A. Nienaber. 1973a. Beef cattle feedlot runoff — physical properties. TRANSACTIONS of the ASAE 16(5):997-1001.

3  Gilbertson, C. B. and J. A. Nienaber. 1973b. Feedlot runoff control system design and installation — a case study. TRANSACTIONS of the ASAE 16(3):462-470.

4  Gilbertson, C. B., J. A. Nienaber, T. M. McCalla, J. R. Ellis and W. R. Woods. 1972. Beef cattle feedlot runoff, solids transport and settling characteristics. TRANSACTIONS of the ASAE 15(6):1132-1134.

5  Madden, J. M., J. N. Dornbush. 1971. Pollution potential of runoff from livestock feeding operations. ASAE Paper No. 71-212. ASAE, St. Joseph, Mich. 49085.

6  Miner, J. R., L. R. Fina, J. W. Funk, R. I. Lipper and G. H. Larson. 1966. Stormwater runoff from cattle feedlots. ASAE Publication No. SP-0366, p 23-27. ASAE, St. Joseph, Mich. 49085.

7  Nienaber, J. A., C. B. Gilbertson, T. M. McCalla and F. M. Kestner. 1974. Disposal of effluent from a beef cattle feedlot runoff control holding pond. TRANSACTIONS of the ASAE 17(2):375-378.

8  Satterwhite, M. B. and C. B. Gilbertson. 1972. Grass response to applications of beef cattle feedlot runoff. Proceedings 1972 Cornell Agricultural Waste Management Conference, Cornell University, 465-480.

9  Swanson, N. P., L. N. Mielke, J. C. Lorimor, T. M. McCalla, and J. R. Ellis. 1971. Transport of pollutants from sloping cattle feedlots as affected by rainfall intensity, duration and recurrence. In: ASAE PROC-271, Livestock Waste Management and Pollution Abatement. ASAE, St. Joseph, MI 49085. pp. 51-55.

10  Swanson, N. P., J. C. Lorimor, L. N. Mielke. 1973. Broad basin terraces for sloping cattle feedlots. TRANSACTIONS of the ASAE 16(4):746-749.

11  Swanson, N. P. and L. N. Mielke. 1973. Solids trap for beef cattle feedlot runoff. TRANSACTIONS of the ASAE 16(4):743-745.

# Design Runoff Volume from Feedlots
# in the Southwestern Great Plains


Victor L. Hauser

MEMBER
ASAE


RAINFALL-INDUCED runoff from open cattle feedlots is one of the most concentrated and potentially damaging wastes produced by agricultural operations. Thus, most states have strict regulations governing its movement, storage, and disposal.

Large volumes of runoff from feedlots may accumulate during a few hours of rainfall. The runoff must be stored during the storm and removed from the feedlot area between storms. These conditions require reservoirs, pumps, pipelines, and other equipment, all of which must be big enough to handle the expected runoff volume. Over-design wastes money and under-design may cause water pollution.

When most of the existing regulations and design critera were established, there were few research data available on feedlot runoff volume. There was great urgency in setting up regulations; therefore, existing equations and data were sometimes used with little proof that they applied. However, more research data is now available and should be used to reevaluate design methods and regulations.

The depth of runoff from the lot is determined by 3 design steps: (a) select rainfall return period, (b) select rainfall amount, and (c) compute runoff depth. Both the 10- and 25-year return periods are used for design (Texas Water Quality Board 1972, Shuyler et al. 1973). Storm rainfall amount is obtained from the maps in Hershfield's (1961) Technical Paper No. 40 (TP-40). The Soil Conservation Service (SCS) equation which is usually used to compute runoff depth (Mockus 1964, Ogrosky and Mockus 1964) is,

$$Q = (P - 0.2S)^2/(P + 0.8S) \quad\dots\dots\dots\dots\dots\dots\dots\dots [1]$$

where S is defined by

$$S = (1000/CN) - 10 \quad\dots\dots\dots\dots\dots\dots\dots\dots\dots\dots [2]$$

and Q is runoff in inches, P is rainfall in inches, and CN is a curve number determined from tables. CN is a function of antecedent rainfall, crop, cover, and soil type. The SCS equation was developed for watersheds covered by cultivated crops or grass, not for feedlots. Feedlots differ from farm fields because they develop an impermeable layer between the manure and the underlying soil (Mielke 1974), the water holding capacity of the manure differs from soil, and the hoofs of moving cattle may change the surface detention during a storm. Early work by Miner et al (1966) showed that a CN of 90 or greater should be used in the SCS equation for feedlots. The CN is assumed constant (usually 90) for current feedlot design and variables like slope, stocking density, or manure water content, are ignored.


Contribution from the Water Quality Management Laboratory, USDA, Agricultural Research Service, Durant, OK.
The author is: VICTOR L. HAUSER, Agricultural Engineer, Water Quality Management Lab., USDA, ARS, Durant, OK 74701.


Published feedlot runoff data were studied to learn how closely the SCS equation (CN = 90) predicts runoff from large storms. Runoff events caused by storm rainfall greater than 2.5 cm were studied because large storms are critical in design. The data were from Bellville, Texas (Reddell and Wise 1974), Gretna, Nebraska (Swanson et al. 1971), and Bushland, Texas (Clark and Stewart 1973). The total runoff computed by the SCS equation varied less than 5 percent from that measured for Bellville and Gretna; however, the SCS equation predicted 1.8 times more runoff than was measured at Bushland. The climate at Bellville is humid, and at Gretna it is subhumid, but at Bushland it is semiarid. This study evaluated present design methods for feedlots in the semiarid Texas High Plains.

## DATA AND PROCEDURE

A test of runoff design methods requires reliable return period estimates of rainfall and runoff amounts. Accurate, measured daily rainfall data were available. However, the longest available (published) record of measured runoff data was for 2 years. Therefore, computed values of daily runoff were used to estimate runoff return period.

Feedlot runoff data were available from the Southwestern Great Plains Research Center, USDA, ARS, Bushland, Texas (15 miles west of Amarillo). Daily rainfall data were obtained for Amarillo from the National Oceanic and Atmospheric Administration, National Climatic Center, Asheville, NC, for the entire 82-year record at the station.

Feedlot runoff data and a feedlot runoff equation derived from a 2-year feedlot study at Bushland were obtained from Clark and Stewart (1973). The Bushland runoff equation is:

$$Q = 0.369 R - 0.310 \quad\dots\dots\dots\dots\dots\dots\dots\dots\dots\dots [3]$$

where Q is daily runoff in cm and R is daily rainfall in cm. This equation computes slightly more runoff than a more recent equation derived for the same feedlot by Clark, Schneider and Stewart (1974). The measured feedlot had an average slope of 1.5 percent, stocking density of 13 sq m/animal (300 animals/acre), and is typical of feedlots in the area.

Daily runoff amount was computed using both the SCS (CN = 90) and the Bushland equations from daily rainfall at Amarillo for the entire 82-year record. Sums of both rainfall and runoff were computed for 1, 2, 3, 5, 7, 10, and 14 days, beginning on each day of the 82-year record. From each of these sets of numbers, the largest value for each year (annual series) was selected for frequency analysis.

Return periods for both rainfall and runoff were obtained by the modified log-normal method (McGuinness and Brakensiek 1964). The frequency plotting position for the data was computed by the formula:

$$F = m/(n + 1) \quad \dots\dots\dots\dots\dots\dots\dots\dots\dots\dots [4]$$

where F is frequency; m is the order number; and n is the number of data points (McGuinness and Brakensiek 1964).

Finally, the SCS equation was fitted to the Bushland data by optimizing the value of CN. There were 20 runoff events in the Bushland record where rainfall exceeded 1.3 cm (0.5 in.) (Clark and Stewart 1973). Runoff was computed, for each of the 20 rainstorms, by the SCS equation using CN between 70 and 90. The error of the SCS equation for each CN was estimated by the sum of squares of the difference between measured and SCS computed runoff. In this way, CN between 70 and 90 were compared to find the CN which produced the smallest error in the least squares sense.

## RESULTS

The return period curves and the data for the 1-, 3- and 14-day sums of rainfall derived from the Amarillo data, are shown in Fig. 1. Return period curves for 2-, 5-, 7- and 10-day rainfall amounts are not shown.

The rainfall amounts predicted for 10- and 25-year return periods and sums of 1- to 14-day rainfall are shown in Fig. 2. The straight lines were derived by least squares regression of the 2- to 14-day amounts only, and were extrapolated to the 1-day rainfall amount. The corresponding 1-day rainfall amounts from TP-40 (Hershfield 1961) are also shown. The 25-year rainfall amount from TP-40 is 2.3 cm (0.9 in.) greater than that found in this study. Similarly the 7-day rainfall amount estimated by Miller (1964), (not shown in Fig. 2), is 1.8 cm (0.7 in.) too large. The 10-year, 1-day rainfall amount from TP-40 equals the 10-year, 2-day rainfall amount observed from the Amarillo data.

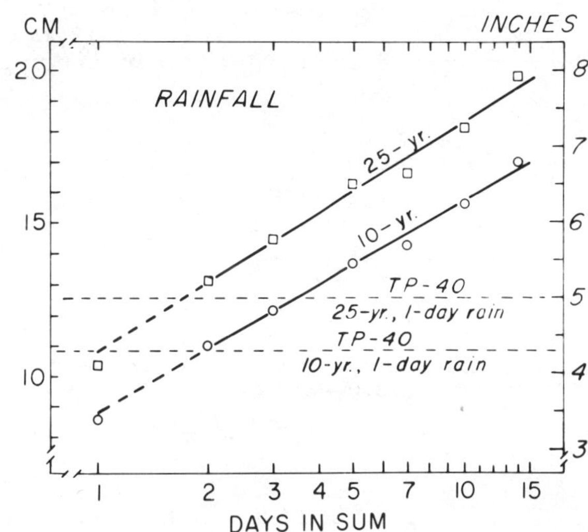



FIG. 2 Rainfall amounts for 1- to 14-day periods, expected to occur once in 10 or 25 years at Amarillo, Texas. One-day rainfall amounts predicted by TP-40 (Hershfield 1961) expected to occur once in 10 or 25 years at Amarillo are also shown.

The results of the runoff study for the 1- and 14-day sums only are shown for both the SCS (CN = 90) and Bushland equations in Fig. 3. As expected, the differences between runoff amounts derived from each of the two equations were large.

The runoff amounts for the 10- and 25-year return period were obtained from each of the seven derived curves (five not shown) for both the SCS (CN = 90) and Bushland equations (Fig. 4). The straight lines were derived by least squares regression of the 2- to 14-day sums only. The design runoff amount for 1 day, according to present design practice, was computed by the SCS equation (CN = 90) with rainfall amounts read from TP-40; these values are also shown on Fig. 4. The SCS equation computed about twice as much runoff for both the 10- and 25-year return frequency, as did the Bushland equation (Fig. 4). The design

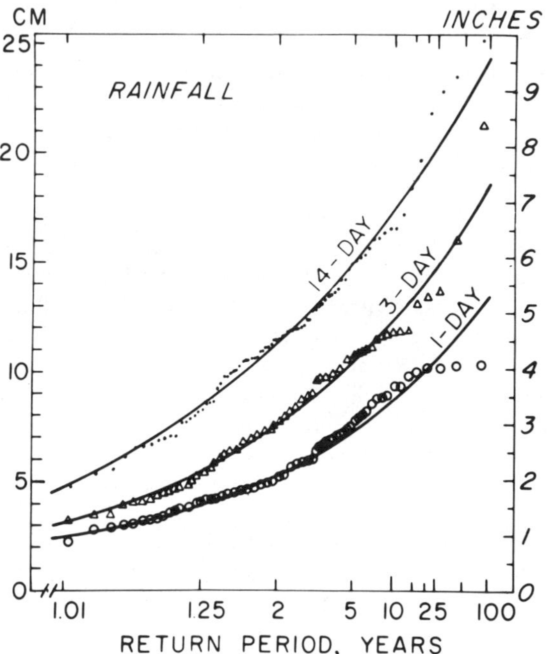



FIG. 1 Return period in years for 1-, 3- and 14-day rainfall amounts at Amarillo, Texas.

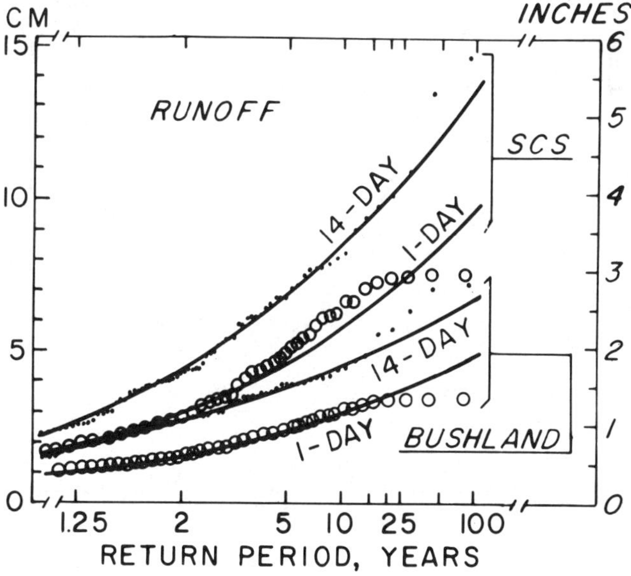



FIG. 3 Return period in years for 1- and 14-day runoff amounts computed by the SCS (CN = 90) and Bushland equations for the Amarillo, Texas, rainfall record. (Several small values omitted for clarity.)

MANAGING LIVESTOCK WASTES

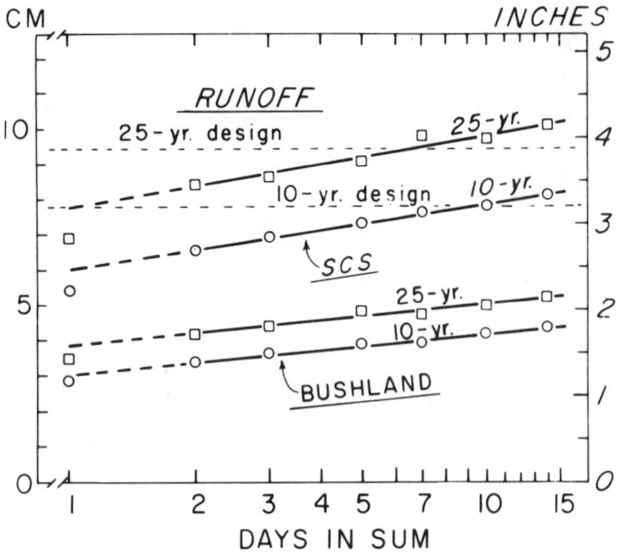


FIG. 4 Runoff amounts for 1- to 14-day periods, expected to occur once in 10 or 25 years, as computed by the SCS (CN = 90) or Bushland equations for Amarillo, Texas. The 1-day, 10- or 25-year design values (by present methods) are shown for comparison.

runoff amount is more than 2 times larger than the 1-day amount obtained from the Bushland curve for either 10- or 25-year return periods. The design runoff amount (using the SCS equation) is even larger than that computed by the SCS equation from the Amarillo rainfall data (Fig. 4). Clearly, the design runoff amount is greater than one can expect to occur in a reasonable number of years.

The Bushland data showed that a CN of 82 provided the best estimate of runoff by the SCS equation (Fig. 5). The Bushland data used in this analysis is shown in Fig. 6, with the Bushland equation and the SCS equation for CN of 82, 85, and 90. The 10- and 25-year rainfall amounts for 1 day

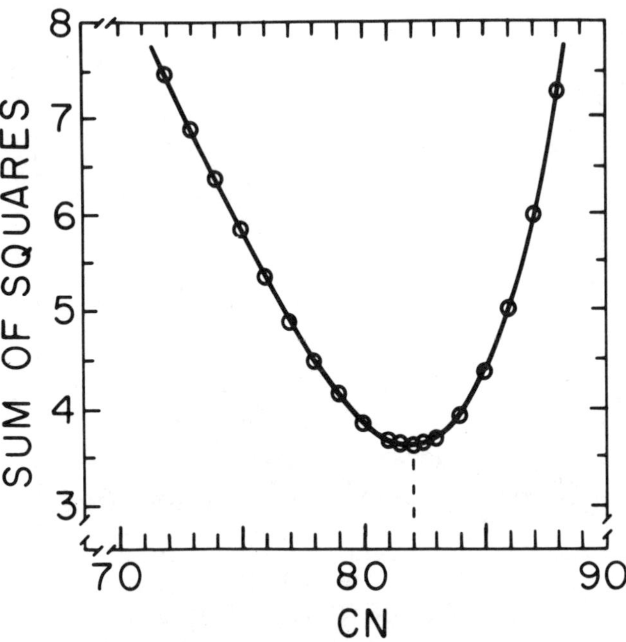


FIG. 5 Sum of squares of the difference between measured runoff and runoff computed by the SCS equation for different CN at Bushland. The computations were for feedlot runoff data published by Clark and Stewart (1973) where rainfall amount was greater than 1.3 cm (0.5 in.).

as estimated from Fig. 1 are also shown. A CN of 82 should be used with the SCS equation if it is used for design in the Texas High Plains. Although this value will compute appreciably more runoff for the 10- or 25-year storms than the Bushland equation, it will compute substantially less runoff than the SCS equation when a CN of 90 is used.

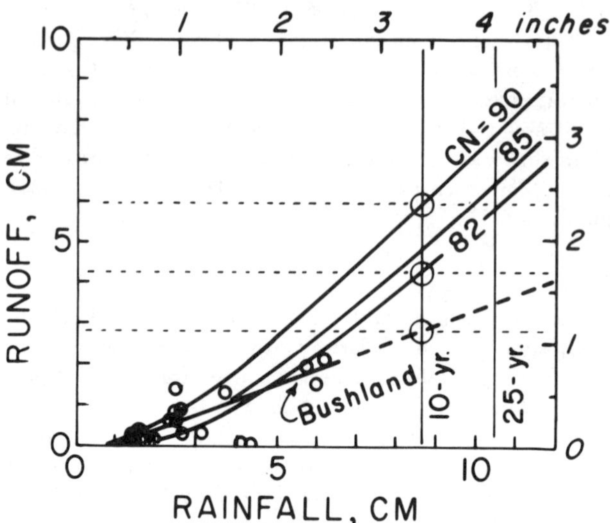


FIG. 6 Feedlot runoff measured at Bushland, where storm rainfall exceeded 1.3 cm (0.5 in.), and the Bushland equation (Clark and Stewart 1973). Lines computed by the SCS equation for Cn = 82, 85 and 90 and the 10- and 25-year, 1-day rainfall predicted from Fig. 1 are also shown.

## DISCUSSION

If the cost of runoff control were low, then overdesign of runoff water holding reservoirs would be of no concern. However, the cost is not negligible (Jones 1974). This study shows that the estimated runoff storage volume may safely be reduced, perhaps as much as 50 percent, in the Texas High Plains.

The errors in current runoff volume estimates are due partly to errors in rainfall amount read from TP-40. Hershfield (1961) showed that observational day rainfall values (used in this study) should be multiplied by 1.13 to produce the equivalent 1440-minute amount. The 1440-minute or 24-hour amount is more appropriate for runoff design and TP-40 shows 1440-minute amounts. Even so, after multiplying the rainfall amounts of Fig. 1 by 1.13, rainfall amounts shown in TP-40 are still 1.3 cm (0.5 in.) and 0.8 cm (0.3 in.) larger than those found in this study for the 10- and 25-year rainfalls, respectively.

Conclusions regarding feedlot runoff estimates must be tempered by the fact that runoff was computed for this study. Nevertheless, it seems safe to say that present design amounts of runoff are substantially larger than can be expected.

Regulations for feedlot runoff control often require that all accumulated runoff be pumped out of the reservoir in a specified number of days, n. Because of this, the runoff holding reservoir should be designed to hold an n-day runoff volume rather than the presently used 1-day runoff amount.

*(Continued on page 436)*



**MANAGING LIVESTOCK WASTES**

# Quantity and Quality of Beef Feedyard Runoff in the Great Plains


R. N. Clark, C B. Gilbertson, H. R. Duke

ASSOC. MEMBER        MEMBER        ASSOC. MEMBER
ASAE                 ASAE                 ASAE


THE Great Plains Region has become the world's largest confined cattle feeding area during the last 10 years. This region is comprised of all or parts of the states of North and South Dakota, Montana, Wyoming, Nebraska, Colorado, Kansas, New Mexico, Oklahoma, and Texas. Fed cattle production increased from 6 million head annually in 1963 to over 14 million in 1973. Many of these cattle are fed in large confined feedyards that are highly mechanized on small areas of land. Normal stocking rates range from 10 to 50 square meters per animal, which allows for feeding up to 25 000 animals on about 30 hectares. Because of these large concentrations of cattle, large amounts of waste accumulate which create a potential for pollution when storm runoff occurs.

After several fish kills were attributed to cattle feedyard runoff, states began establishing water control policies and the U.S. Environmental Protection Agency issued its regulations for controlling feedyard runoff. Most regulations specify no discharge from feedyards to public waters, which means feedyard operators must establish holding ponds capable of collecting and storing all runoff from feeding areas until it can be disposed of properly. Because of the lack of feedyard hydrology data needed to design these facilities, several researchers began investigations to determine the amounts and quality of runoff that could be expected from feedyards. This work began in 1967 in Colorado and Nebraska and is continuing presently in most states.

The objective of this paper is to combine and summarize data from several research studies in the Great Plains. Data from eight feedyards in five states are discussed. General information for each feedyard is given in Table 1.

All feedyards were private commercial yards, except the one at Mead, NE, which was a research unit operated by the University of Nebraska. All feedyards were constructed on soil with only a limited amount of paving around feedbunks and water troughs, except the Pratt, KS, feedyard which has 18 m concrete apron for the feedbunks.

Usually, only a part of the feedyard was instrumented for measuring runoff. These instrumented areas ranged from 0.1 hectare to over 13 hectares. Slopes varied from 1 to 9 percent with an average slope of about 3 percent. A moisture deficit term was calculated for each feedyard (defined as the difference between annual evaporation and annual precipitation). This value gives an indication of the average dryness of the feedyard surface.

## QUANTITY OF RUNOFF

Rainfall runoff accounted for the largest quantities of runoff for all feedyards. A linear relationship between precipitation and runoff was reported by seven of the eight feedyards. Fig. 1 shows the linear relationships


Contribution from the Soil, Water, and Air Sciences, Southern, North Central, and Western Regions, Agricultural Research Service, USDA, in cooperation with Agricultural Experiment Stations—Texas, Nebraska, and Colorado, respectively.

The authors are: R. N. CLARK, Agricultural Engineer, Southwestern Great Plains Research Center, ARS, USDA, Bushland, Texas; C.

The authors are: R. N. CLARK, Agricultural Engineer, Southwestern Great Plains Research Center, ARS, USDA, Bushland, Texas; C. B. GILBERTSON, Agricultural Engineer, ARS, USDA, University of Nebraska, Lincoln; and H. R. DUKE, Agricultural Engineer, ARS, USDA, Colorado State University, Fort Collins.


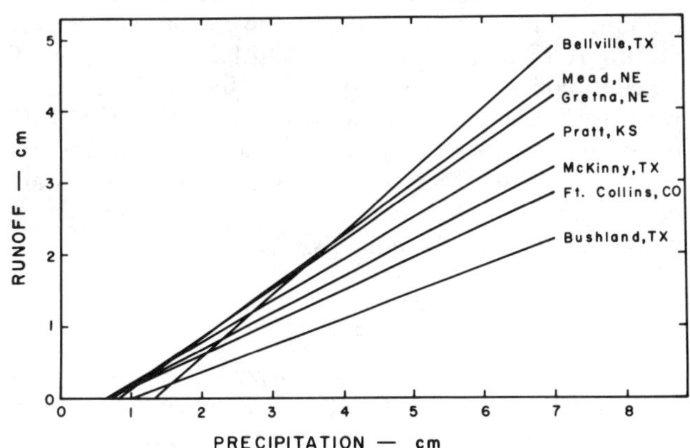


FIG. 1 Precipitation-runoff relationships for beef cattle feedyards at seven locations in the Great Plains.

TABLE 1. GENERAL INFORMATION FOR EACH BEEF CATTLE FEEDYARD STUDIED.

| Location | Researcher | Years of study | Area | Average slope | Stocking rate | Annual precipitation | Annual evaporation | Moisture deficit |
|---|---|---|---|---|---|---|---|---|
| | | | ha | percent | m$^2$/animal | cm | cm | cm |
| Bellville, TX | Reddell | 72 | 7.5 | 3.0 | 35 | 110 | 135 | 25 |
| Bushland, TX | Clark | 71-73 | 4.0 | 1.5 | 12 | 50 | 165 | 115 |
| Ft. Collins, CO | Duke | 69-71 | 0.4 | 6.0 | 19 | 40 | 100 | 60 |
| Gretna, NE | Swanson | 68-72 | 0.3 | 6.0 | 46 | 70 | 110 | 40 |
| McKinney, TX | Kreis | 69-70 | 13.0 | 3.0 | 9 | 90 | 140 | 50 |
| Mead, NE | Gilbertson | 68-72 | 0.1 | 3,6,9 | 19 | 70 | 110 | 40 |
| Pratt, KS | Manges | 69-70 | 13.3 | 1.3 | 33 | 60 | 150 | 90 |
| Sioux Falls, SD | Dornbush | 70-71 | 1.6 | 3.0 | 53 | 60 | 90 | 30 |

developed from individual storms that produced runoff. In all but one feedyard, nearly 1 cm of rainfall was required to produce runoff, even when rainfall had occurred the previous day. Slopes of the regression lines ranged from a high of 0.86 for Bellville, TX, to a low of 0.36 for Bushland, TX. These data show that feedyards in drier areas have less runoff from the same precipitation than feedyards in wetter areas. These regression lines seemed to be proportional to the moisture deficit (annual evaporation minus annual precipitation) for all feedyards except Pratt, KS, which had 20 percent more paved area than the other feedyards which tended to increase runoff.

Because these regression lines seemed proprotional to moisture deficit, Fig. 2 was developed. The numerical regression slope value for each feedyard was compared to the annual moisture deficit. This method was used to combine all feedyard precipitation-runoff data from the Great Plains and to form a regression line as a linear function of annual moisture deficit. Through the use of Fig. 2 and the annual moisture deficit, an estimate could be made of the amount of runoff expected for a particular area. For example, Wichita, KS, has an annual moisture deficit of 60 cm which would give a regression line slope of 0.56 (Fig. 2). Combining this regression line slope with the average y-intercept yields the prediction equation

$$\text{Runoff} = 0.56 \text{ precipitation} - 0.33.$$

This information would be helpful in designing new holding pond facilities for open, unpaved beef feedyards.

Most researchers have concluded that feedyard slope and stocking rates have little influence on runoff amounts. Gilbertson et al. (1970) showed little difference in runoff relationships for feedyards on 3, 6, and 9 percent slopes. Another significant finding from feedyards with predominantly dry manure packs is that runoff volumes are less when rainfall has occurred the previous day. When these lots are wetted, depressions are created in the wet manure, unlike a dry lot which is packed smooth by the animals.

## QUALITY OF RUNOFF

All researchers who have analyzed feedyard runoff have emphasized the variability of its chemical composition. However, for comparison of one location with another, only average values are given in Tables 2 and 3. Table 2 shows total solids concentrations, chemical oxygen demand, electrical conductivity, and concentrations of total nitrogen and phosphorus, which are the

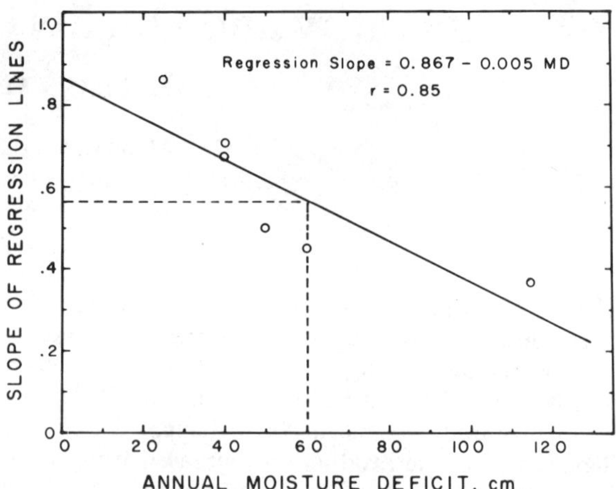


FIG. 2 Slope of the feedyard precipitation-runoff regression line as a linear function of the annual moisture deficit [evaporation-precipitation].

TABLE 3. CONCENTRATIONS OF SELECTED IONS IN RUNOFF FROM BEEF CATTLE FEEDYARDS.

| Location | Sodium, ppm | Calcium, ppm | Magnesium, ppm | Potassium, ppm | Chloride, ppm |
|---|---|---|---|---|---|
| Bellville, TX | 230 | — | — | 340 | 410 |
| Bushland, TX | 588 | 449 | 199 | 1320 | 1729 |
| McKinney, TX | 408 | 698 | 69 | 761 | 450 |
| Mead, NE | 478 | 181 | 146 | 1864 | 700 |
| Pratt, KS | 511 | 166 | 110 | 815 | — |

most commonly measured properties used to indicate the quality of the feedyard runoff. The concentrations of solids transported in the runoff varied from about 3000 ppm at Sioux Falls, SD, to more than 17 000 ppm at Ft. Collins, CO. The solids transported in runoff seemed more closely related to stocking rate than to any other parameter. Between 40 and 50 percent of the total solids removed were classified as volatile. Gilbertson et al. (1972) have reported that settleable solids range from 15 to 65 percent of the total solids with an average of about 37 percent. Most of these settleable solids could be removed in approximately one-half hour of detention time. Allowances for solids should be taken into consideration in designing holding ponds and either the size increased or a settling basin used ahead of the pond.

The chemical oxygen demand ranged from a low of 2160 ppm at Sioux Falls, SD, to a high of 17 800 ppm at Ft. Collins, CO, while total nitrogen ranged from 50 ppm at Bellville, TX, to 1153 ppm at Ft. Collins, CO. These data, except for phosphorus, show the diluting effect of larger annual precipitation and lower stocking rates. For example, the Mead, NE, and Ft. Collins, CO, feedyards had about the same stocking rate, but the electrical

TABLE 2. AVERAGE CHEMICAL CHARACTERISTICS OF RUNOFF FROM BEEF CATTLE FEEDYARDS.

| Location | Total solids, ppm | Electrical conductivity, mmhos/cm | Chemical oxygen demand, ppm | Total nitrogen ppm | Total phosphorus, ppm |
|---|---|---|---|---|---|
| Bellville, TX | 9000 | — | 4000 | 9000 | 85 |
| Bushland, TX | 15 000 | 8.4 | 15 700 | 15 300 | 205 |
| Ft. Collins, CO | 17 500 | 8.6 | 17 800 | 17 500 | 93 |
| McKinney, TX | 11 429 | 6.7 | 7210 | 11 400 | 69 |
| Mead, NE | 15 200 | 3.2 | 3100 | 15 200 | 300 |
| Pratt, KS | 7500 | 5.4 | 5000 | 7500 | 50 |
| Sioux Falls, SD | 2986 | — | 2160 | 3000 | 47 |

                                                                    MANAGING LIVESTOCK WASTES

conductivities were 3.2 and 8.6 mmhos/cm, and annual precipitation was 70 and 40 cm, respectively. Since the electrical conductivity is a measure of total salts, the effect of rainfall dilution can be seen. Other than these comparisons, little continuity of data can be found; again pointing to the high degree of variability.

The level of pollutant concentrations in feedyard runoff is several times greater than untreated municipal sewage; therefore, land application rather than treatment is the recommended practice for disposal. Land application offers an economical procedure to safely remove these wastes from the holding ponds. Also, nutrients contained in the feedyard runoff can be used beneficially. However, the runoff contains harmful salts that can adversely affect soil structure and crop growth.

Table 3 shows concentrations of ions that are important when considering land disposal of the runoff. Calcium, sodium, and magnesium are used in determining the sodium adsorption ratio, which is a measure of the sodium hazard of the water. Waters that contain large concentrations of exchangeable sodium cause the soil to disperse and water intake to be reduced. The sodium concentrations shown in Table 3 are very large as compared with those of calcium and magnesium and may cause problems if applied repeatedly. Also, the effect of potassium on alkalinity is not clearly understood because natural waters do not contain the high concentrations that feedyard runoff contains. Potassium in high enough concentration will probably have effects similar to sodium in causing soil dispersion. Therefore, careful management of the sodium-potassium-calcium-magnesium ratios in the disposal areas is needed. Powers (1973) and Clark et al. (1974) have recommended dilutions of at least four parts irrigation water to one part feedyard runoff to prevent salinizing disposal areas. In areas where annual precipitation exceeds 60 cm, considerable natural dilution occurs, thus reducing the salinity hazard.

## SNOWMELT RUNOFF

Snowmelt runoff is much different than rainfall runoff because the flow rate is slower and runoff usually occurs over several days. Most of the feedyards in Kansas and the South did not have significant snowmelt runoff because of their small snow and ice accumulations. Snowmelt becomes important in the northern states where the average annual snowfall exceeds 50 cm. For example, about 30 percent of the total annual runoff was attributed to snowmelt at Sioux Falls, SD. Data from Mead, NE, showed that snowmelt transports about three times more solids and the electrical conductivity is about twice that of rainfall runoff.

## SUMMARY

Feedyard runoff data from eight Great Plains feedyards in five states have been studied and summarized.

In all cases, the precipitation-runoff relationships were linear; however, the regression slopes varied from 0.36 to 0.86. Runoff did not occur on these feedyards until at least 1 cm of rainfall had fallen. The regression slopes of the precipitation-runoff relationships were found to be proportional to the annual moisture deficit between evaporation and precipitation.

The quality of runoff was quite variable at each location and depended on factors like rainfall intensity and duration, time since last runoff, and stocking rate. However, noticeable quality differences were found between the various research locations. The concentrations of salts were less where the annual moisture deficit was low and increased in areas where the moisture deficit was higher. Runoff normally contained too much sodium to be continually applied to the same land area without dilution.


### References

1 Clark, R. N., A. D. Schneider, and B. A. Stewart. 1974. Analysis of runoff from Southern Great Plains feedlots. ASAE Paper No. 74-4017. ASAE, St. Joseph, Mich. 49085.

2 Dornbush, J. N., and J. M. Madden. 1973. Pollution potential of runoff from production livestock feeding operations in South Dakota. Report No. A-025-SDAK. South Dakota State University, Brooking, SD.

3 Gilbertson, C. B., T. M. McCalla, J. R. Ellis, O. E. Cross, and W. R. Woods. 1970. The effect of animal density and surface slope on characteristics of runoff, solid wastes, and nitrate movement on unpaved beef feedlots. Nebraska Agr. Exp. Sta. Pub. No. SB508.

4 Gilbertson, C. B., J. A. Nienaber, T. M. McCalla, J. R. Ellis, and W. R. Woods. 1972. Beef cattle feedlot runoff, solids transport, and settling characteristics. TRANSACTIONS of the ASAE 15(6):1132-1134.

5 Gilbertson, C. B., J. R. Ellis, J. A. Nienaber, T. M. McCalla and T. Klopfenstein. 1975. Physical and chemical characteristics of outdoor beef cattle feedlot runoff. In Press as a University of Nebraska Research Bulletin.

6 Kreis, R. D., M. R. Scalf, and J. F. McNabb. 1972. Characteristics of rainfall runoff from a beef cattle feedlot. EPA-R2-72-061. Robert S. Kerr Water Research Center, Ada, OK.

7 Manges, H. L., L. A. Schmid, and L. S. Murphy. 1971. Land disposal of cattle feedlot water. In Livestock Waste Management and Pollution Abatement. ASAE, St. Joseph, Mich. 49085. pp. 62-65.

8 Powers, W. L. 1973. Guideline for land disposal of feedlot lagoon water. Kansas Extension Service C-485, Manhattan, Kansas.

9 Reddell, D. L. and G. G. Wise. 1974. Water quality of storm runoff from a Texas beef feedlot. Texas Agr. Expt. Sta. PR-3224. College Station, TX.

10 Swanson, N. P., L. N. Mielke, J. C. Lorimor, T. M. McCalla, and J. R. Ellis. 1971. Transport of pollutants from sloping cattle feedlots as affected by rainfall intensity, duration, and recurrence. In Livestock Waste Management and Pollution Abatement. ASAE, St. Joseph, Mich. 49085. pp. 51-55.

11 Wallingford, G. W., L. S. Murphy, W. L. Powers, and H. L. Manges. 1974. Effect of beef-feedlot-lagoon water on soil chemical properties and growth and composition of corn forage. J. of Environ. Quality 3(1):74-78.

# Properties of Solid and Liquids from Stacked Manure

J. C. Converse, C. O. Cramer, G. H. Tenpas, D. A. Schlough

MEMBER
ASAE

MEMBER
ASAE

ENVIRONMENTAL pressures and the drudgery of daily hauling of manure has prompted many livestock producers, especially dairymen, to look at manure handling alternatives. One such alternative is manure stacking, which is a very viable manure handling system for stanchion type dairy facilities where bedding is used. The majority of the dairy herds are housed in stanchion type barns, thus the importance of manure stacking. Many dairy facilities are located along streams so stacking the manure on the ground outside the barn may cause serious pollution of the stream. The objectives of this research were to: (a) determine the quantity and quality of the seepage from manure stacks, (b) determine quantity and quality of the manure in the stack, and (c) to define design and management criteria for solid manure handling.

## FACILITIES

This research was carried out at the Ashland Experimental Farm and at the Electric Research Farm near Madison. The study covers a three-year period at each location.

### Ashland

The manure from the 32 cow stanchion barn, which also includes 12 floor level calf stalls, 1 calf pen and 2 cow pens, is stored in a bunker type storage unit. The unit is 17 m (56 ft) long, 7.3 m (24 ft) wide with 3.4 m (11 ft) high side walls constructed of tongue and groove planks nailed to treated posts. A trussed roof supports the manure distributing equipment consisting of an extension to the barn cleaner, a cross conveyor and a reversible horizontal distributing conveyor. The base is concrete with two floor drains located along the center line. The drains connect to a 3785 l (1000 gal) sampling tank which collects seepage from the storage unit. Periodically, the seepage is sampled, measured and pumped to an adjacent detention pond. Vertical 10 x 10 cm (4 x 4 in.) timbers located 2.4 m (8 ft) O.C. on the walls and 10-15 cm (4 to 6 in.) of a porous media such as chopped corn stalks, shredded bark or corn cobs spread on the floor facilitate drainage of the liquid from the unit (Cramer et al. 1973, Tenpas et al. 1973). The cows are housed in the barn from early November to late May,

except for about one-half hour outside exercise each day. During the summer they are in the barn only for milking. The manure is cleaned out in mid May and late October. In most years some winter manure is left in the storage unit until early fall before it is removed.

### Electric Research Farm—ERF 1

The manure from 28 cows of a 54 cow stanchion barn is stacked on a concrete platform. A concrete curved wall supports a barn cleaner extension. A pivot placed near the barn allows the extension to swing 60 degrees. Two 2.4 m (8 ft) high walls consisting of pressure treated posts and tongue and groove planks extend out 10.7 m (35 ft) along two sides. The other curved edge of the base has no wall. The concrete floor area is 273 m² (2938 sq ft) and slopes to one corner where a 0.6 x 0.6 m (2 x 2 ft) drain is connected to a 3785 l (1000 gal) sampling tank. Periodically the seepage is sampled, measured and pumped to an adjacent detention pond. Ten to fifteen cm (4 to 6 in.) of shredded corn stalks on the concrete base aid in draining the liquids from the unit. The cows are housed year around in the barn except for approximately 45 minutes each day. The manure is removed in early May and late November. Converse et al. (1973) further describe the system.

### Electric Research Farm—ERF 2

The manure from 26 cows of this 54 cow stanchion barn is stacked on a pie shaped concrete platform with one 3.4 m (11 ft) high side wall and one curved end wall consisting of posts with tongue and groove planks. The other edge of the platform has no wall. A thrower type stacker is placed under the barn cleaner at the narrow end of the platform. It throws the manure into a windrow. A 0.3 m (1 ft) wide by 14 m (46 ft) long gutter extends lengthwise down the center of the storage unit with the floor sloping toward it from both directions. Short 5 x 15 cm (2 x 6 in.) planks spaced 1.3 cm (1/2 in.) apart cover the gutter, which is connected to a 3785 l (1000 gal) sampling tank. The seepage is pumped to an adjacent detention pond. To aid drainage to the gutter, 10 to 15 cm (4 to 6 in.) of shredded corn stalks are placed on the 211 m² (2275 sq ft) platform. The cows are housed year around in the barn except for 45 minutes each day. The manure is removed in early May and late November. Converse et al. (1973) further describe the system.

## RESULTS AND DISCUSSION

### Fresh Manure

Fresh manure was collected from two cows for three consecutive days for each month from December to April at the Ashland site. Five different cows were used during the three year study. Gutter pans collected the manure

Research supported by the College of Agricultural and Life Sciences, University of Wisconsin, Madison, U. S. Department of Agriculture, ARS Contract 12-14-100-10,090(42) and the Wisconsin Electric Utilities Research Foundation.

The authors are: J. C. CONVERSE, Assistant Professor, C. O. CRAMER, Professor, Agricultural Engineering Dept., University of Wisconsin, Madison; G. H. TENPAS, Professor, and D. A. SCHLOUGH, Specialist, Experimental Farm, University of Wisconsin, Ashland.

TABLE 1. AVERAGES FOR PHYSICAL, CHEMICAL AND BIOCHEMICAL
CHARACTERISTICS OF FRESH MANURE FROM COWS AT ASHLAND FOR
THREE YEARS BASED ON 454 KG (1000 LB) LIVE BODY WEIGHT PER DAY.*

| Parameters | Units | Winter 71-72 | Winter 72-73 | Winter 73-74 | Ave. |
|---|---|---|---|---|---|
| Ave. wt. of | Kg | 658 | 627 | 678 | 656 |
| animals | lbs | 1449 | 1381 | 1493 | 1444 |
| Quantity | Kg | 53.1 | 51.3 | 44.0 | 49.5 |
| | lbs | 117 | 113 | 97 | 109 |
| Volume | $m^3$ | 0.054 | 0.053 | 0.046 | 0.051 |
| | $ft^3$ | 1.94 | 1.86 | 1.62 | 1.80 |
| Density | $Kg/m^3$ | 965 | 972 | 960 | 966 |
| | $lb/ft^3$ | 60.3 | 60.7 | 59.9 | 60.3 |
| Total solids | Percent | 17.7 | 16.6 | 16.4 | 16.9 |
| Volatile solids | Percent TS | 89.4 | 88.9 | 88.7 | 89 |
| COD | ppm | 169000 | 182000 | 187000 | 178000 |
| Total nitrogen | ppm as N | 4772 | 5102 | --- | 4937 |
| Ammonia | ppm as N | 1387 | 1378 | --- | 1383 |
| Total phosphorous | ppm as P | 1205 | 1484 | 1396 | 1362 |
| Potassium | ppm as K | 3450 | 2865 | 3616 | 3310 |
| pH | --- | 6.9 | 7.1 | 7.3 | 7.1 |

*Average weight for two cows taken for three consecutive days each month from December to April.

including bedding. The manure from each gutter pan was dumped in a barrel for measuring and sampling. At the ERF 1 and ERF 2, the fresh manure was collected in the manure spreader and weighed for five consecutive days each spring and fall. Some liquid losses occurred between collection and weighing. Laboratory analysis for all manure and seepage samples followed standard procedures as described by Cramer et al. 1973.

Table 1 gives the quantities and characteristics of fresh manure based on 454 kg (1000 lb) live body weight for the three collection periods at Ashland. It shows that the average production of manure was 49.5 kg (109 lb) with an average density of 966 kg/m³ (60.3 lb/ft³) and a total solids of 16.9 percent. Bedding quantities averaged 3.9 kg (8.7 lb) per 454 kg (100 lb) live body weight per day over the three-year period. Bedding material was a combination straw and shavings. These values are very close to what Tenpas et al. (1973) found for the previous two years of study.

Average amount of fresh manure collected from the cows for ERF 1 and ERF 2 was 32.2 kg (71 lb) and 30.9 kg (68.1 lb) per 454 kg (1000 lb) live body weight per day. Average bedding used was 1.8 to 2.3 kg (4 to 5 lb) per 454 kg (1000 lb) live body weight and consisted of baled straw or chopped corn stalks and uneaten hay. These manure quantities are considerably lower than found at Ashland. However, some loss in liquids occurred between collection and weighing as the manure dropped from the barn cleaner into the manure spreader. No analysis was made on the composition of the gutter manure from ERF 1 and ERF 2.

**Stored Manure**

Manure was stored during summer and winter periods in all three units. At the Ashland unit only the winter stored manure was used for evaluation because the cows are in the barn only for milking during the summer. At the Electric Research Farm the animals were kept in the barn throughout the year, except for about 45 minutes per day. Table 2 gives the inplace volume and density of manure for each period and a summer and winter average. The average winter volumes were 0.040 m³ (1.41 ft³), 0.033 m³ (1.16 ft³) and 0.030 m³ (1.07 ft³) for Ashland, ERF 1 and ERF 2, respectively. Summer volume averages for ERF units were about 25 percent less during summer than during the winter period, but the average density was about 64 kg/m³ (4 lb/ft³) higher in the summer than winter, indicating a wetter manure due to precipitation and poor drainage. Bedding usage was about the same during the summer as in the winter.

TABLE 2. INPLACE VOLUME AND WEIGHT OF MANURE FOR THREE STACKS FOR THREE YEARS BASED ON
454 KG (1000 LB) LIVE BODY WEIGHT PER DAY OF STORAGE.

| Period | Ashland Kg (lb) | Ashland $m^3$ ($ft^3$) | Ashland $Kg/m^3$ ($lb/ft^3$) | ERF 1 Kg (lb) | ERF 1 $m^3$ ($ft^3$) | ERF 1 $Kg/m^3$ ($lb/ft^3$) | ERF 2 Kg (lb) | ERF 2 $m^3$ ($ft^3$) | ERF 2 $Kg/m^3$ ($lb/ft^3$) |
|---|---|---|---|---|---|---|---|---|---|
| Summer 71 | --- | --- | --- | 20.6 (45.3) | 0.020 (.7) | 1008 (62.9) | 17.3 (38.0) | 0.021 (.8) | 812 (50.7) |
| Winter 71-72 | 31.9 (70.3)* | .042 (1.5) | 21.6 (47.5)* | 26.7 (58.8) | 0.031 (1.1) | 849 (53.0) | 16.6 (36.6) | 0.028 (1.0) | 593 (37) |
| Summer 72 | --- | --- | --- | 24.9 (54.8) | -- | --- | 22.5 (49.6) | --- | --- |
| Winter 72-73 | 34.1 (75) | .042 (1.5) | 23.5 (51.7) | 29.2 (64.3) | 0.031 (1.1) | 953 (59.5) | 27.6 (60.7) | 0.031 (1.1) | 876 (54.7) |
| Summer 73 | | | | 24.1 (53.1) | 0.027 (1.0) | 895 (55.9) | 23.8 (52.4) | 0.026 (.9) | 932 (58.2) |
| Winter 73-74 | --- | .037 (1.3) | --- | 32.4 (71.3) | 0.037 (1.3) | 878 (54.8) | 28.2 (62.2) | 0.031 (1.1) | 897 (56) |
| Winter Ave. | --- | .040 (1.4) | --- | 29.4 (64.8) | 0.033 (1.2) | 894 (55.8) | 24.1 (53.2) | 0.030 (1.1) | 789 (49.2) |
| Summer Ave. | --- | --- | --- | 23.2 (51.1) | 0.024 (.8) | 952 (59.4) | 21.2 (46.7) | 0.024 (.8) | 872 (54.5) |

*Estimates as not all manure removed at end of winter storage periods.

TABLE 3. AVERAGES FOR PHYSICAL, CHEMICAL AND BIOCHEMICAL CHARACTERISTICS OF MANURE REMOVED FROM STACK FOR THREE WINTER AND SUMMER PERIODS FOR THREE DIFFERENT STACKS.

| Parameter | Units | Winter | | | | Summer | | | |
|---|---|---|---|---|---|---|---|---|---|
| | | Ashland | ERF 1 | ERF 2 | Ave. | Ashland* | ERF 1 | ERF 2 | Ave. |
| Total solids | Percent | 23.9 | 19.9 | 21.0 | 21.6 | --- | 20.8 | 20.7 | 20.8 |
| Volatile solids | Percent TS | 87.7 | 84.0 | 83.3 | 85 | --- | 83.3 | 79.7 | 81.5 |
| COD | PPM | 256000 | 209000 | 221000 | 229000 | --- | 206000 | 228000 | 217000 |
| Total nitrogen | ppm as N | 6356 | 5851 | 6012 | 6073 | --- | 5314 | 6099 | 5707 |
| Ammonia | ppm as N | 2071 | 1749 | 1626 | 1815 | --- | 1751 | 2037 | 1894 |
| Total phosphorus | ppm as P | 1355 | 838 | 1138 | 1110 | --- | 979 | 1238 | 1109 |
| Potassium | ppm as K | 4556 | 4140 | 4443 | 4380 | --- | 3573 | 3914 | 3744 |
| pH | --- | 7.1 | 7.2 | 6.4 | 6.9 | --- | 6.6 | 6.6 | 6.6 |

*Stacking not done during summer.

At Ashland, the average inplace volume was approximately 22 percent less than the average volume of fresh manure collected, thus indicating some shrinkage during the year. However part of the reduction is due to separation of the liquids from the solids. During the 1972 summer period a portion of the winter manure was held in the Ashland storage unit, resulting in a 16 percent volume reduction over a three-month period.

Table 3 gives the characteristics of the stored manure for all three units. The overall average total solids concentration of all three units was 21.6 and 20.8 percent for the winter and summer periods, respectively.

### Seepage

Seepage samples were taken each time the holding tanks were pumped. Table 4 gives the quantity of seepage collected for each storage period for each system and a winter and summer average volume collected. An average of 5.9, 10.9 and 7.5 l (1.6, 2.9 and 2.0 gal) per day per 454 kg (1000 lb) live body weight was collected from Ashland, ERF 1 and ERF 2, respectively. The average volume of seepage collected during the summer was less even though the precipitation was greater. This wa probably due to the larger amount of manure surface area during the summer, as the manure did not stack as easily and flowed more during the summer than the winter, especially in the ERF 1 and ERF 2 systems.

Table 5 gives the average quantity of seepage collected

TABLE 4. QUANTITIES OF SEEPAGE IN LITERS COLLECTED FOR THREE YEARS FROM THREE DIFFERENT STACKS PER 454 KG (1000 LB) LIVE BODY WEIGHT PER DAY.

| Period | Ashland | ERF 1 | ERF 2 |
|---|---|---|---|
| Summer 71 | --- | 8.9 (2.4) | 6.4 (1.7) |
| Winter 71-72 | 5.2 (1.4) | 10.9 (2.9) | 7.7 (2.0) |
| Summer 72 | 5.2 (1.4) | 5.2 (1.4) | 4.4 (1.2) |
| Winter 72-73 | 8.0 (2.1) | 12.5 (3.3) | 8.5 (2.3) |
| Sumer 73 | 2.3 (0.6) | 6.2 (1.6) | 4.2 (1.1) |
| Winter 73-74 | 4.5 (1.2) | 9.4 (2.5) | 6.2 (1.6) |
| Summer 74 | 3.1 (0.8) | --- | --- |
| Winter average | 5.9 (1.6) | 10.9 (2.9) | 7.5 (2.0) |
| Summer average | 3.5 (0.93) | 6.7 (1.8) | 5.0 (1.3) |

along with the average rainfall for the winter and summer periods for each system. The larger amount of seepage resulting from ERF 1 and ERF 2 was due to the larger surface area, and no roof, even though the average rainfall was higher at Ashland. In three out of six storage periods the gutter in ERF 1 plugged which resulted in the lower average rainfall. The seepage portion of the study was terminated and the accumulated rainfall and seepage were used to determine average rainfall and seepage volumes. The large surface area resulted in the large volume of seepage collected. Consequently, smaller surface areas would result in less amounts of seepage collected.

TABLE 5. WEIGHTED AVERAGES FOR PHYSICAL, CHEMICAL AND BIOCHEMICAL CHARACTERISTICS OF SEEPAGE FROM THREE WINTER AND SUMMER PERIODS FOR THREE DIFFERENT STACKING SYSTEMS.

| Parameters | Units | Winter | | | | Summer | | | |
|---|---|---|---|---|---|---|---|---|---|
| | | Ashland | ERF 1 | ERF 2 | Ave. | Ashland | ERF 1 | ERF 2 | Ave. |
| Quantity* | Liters | 53486 | 59476 | 41573 | --- | 23266 | 35328 | 30747 | --- |
| | gal | 14931 | 15714 | 10985 | --- | 6147 | 9334 | 8123 | --- |
| Rainfall** | cm | 29.5 | 20.5 | 26.1 | --- | 49.3 | 24.2 | 47 | --- |
| | in. | 11.6 | 8.1 | 10.3 | --- | 19.4 | 9.5 | 18.5 | --- |
| Total solids | Percent | 2.4 | 1.9 | 2.3 | 2.2 | 1.6 | 2.4 | 2.0 | 2.0 |
| Volatile solids | Percent TS | 53.9 | 62.3 | 61.7 | 59.1 | 49.9 | 57.0 | 55.0 | 54.3 |
| Total suspended solids | Percent | 0.19 | 0.24 | 0.23 | 0.22 | 0.14 | 0.33 | 0.27 | 0.26 |
| Volatile suspended solids | Percent TSS | 63.6 | 76.0 | 70.4 | 69.4 | 61.6 | 70.9 | 68.75 | 67.6 |
| BOD | ppm | 9813 | 9179 | 9169 | 9398 | 6037 | 12798 | 7903 | 9136 |
| COD | ppm | 24000 | 21000 | 24000 | 23000 | 15000 | 29000 | 20000 | 22000 |
| Total nitrogen | ppm as N | 2641 | 1889 | 2363 | 2339 | 1315 | 1930 | 1554 | 1623 |
| Ammonia | ppm as N | 1961 | 1108 | 1495 | 1559 | 931 | 1366 | 1048 | 1119 |
| Total phosphorous | ppm as P | 103 | 110 | 156 | 126 | 51 | 146 | 107 | 106 |
| Potassium | ppm as K | 4171 | 3695 | 3466 | 3804 | 3377 | 3784 | 3752 | 3664 |
| pH | --- | 7.8 | 7.8 | 7.9 | 7.8 | 7.7 | 7.4 | 7.9 | 7.7 |

*Average quantity of seepage collected, which includes urine and precipitation, for three winter and three summer periods.
**Average precipitation for three winter and three summer periods. For winter and summer periods average rainfall was not equal for ERF 1 and ERF 2 because seepage collection periods for ERF 1 were shorter in three of six storage periods than for ERF 2 which was due to plugging of drain.



MANAGING LIVESTOCK WASTES

TABLE 6. RELATIONSHIP BETWEEN SEEPAGE COLLECTED AND PRECIPITATION ON STACKING UNIT (S/P)

| Period | Ashland | ERF 1 | ERF 2 |
|---|---|---|---|
| Summer 71 | --- | 0.66 | 0.56 |
| Winter 71-72 | 1.28 | 1.16 | 1.04 |
| Summer 72 | 0.45 | 0.46 | 0.22 |
| Winter 72-73 | 1.67 | 0.89 | 0.70 |
| Summer 73 | 0.29 | 0.40 | 0.23 |
| Winter 73-74 | 1.37 | 1.39 | 0.60 |
| Summer 74 | 0.41 | --- | --- |
| Winter average | 1.44 | 1.15 | 0.78 |
| Summer average | 0.38 | 0.51 | 0.34 |

Table 5 gives the weighted average values of the seepage quality collected for each system for winter and summer periods. The weighted average values for each system were very close, so they were averaged to give a winter and summer average for each parameter. The average T.S. concentration was 2.2 percent and 2.0 percent for winter and summer, respectively. In each case the average BOD concentration was approximately 9000 ppm for both winter and summer. Ammonia concentration accounted for over 60 percent of the total nitrogen.

### Detention Pond

Sizing the detention pond is difficult due to the many variables involved, such as amount of precipitation, amount of bedding used in the barn, surface area of manure in storage and the ability of the porous media to drain. Table 6 gives the relationship between seepage collected and the amount of precipitation falling on the storage unit. It was determined by measuring the volume of precipitation falling on the storage unit for the storage period and dividing it into the seepage collected. The S/P ratio for winter is 1.44, 1.15 and 0.78 for Ashland, ERF 1 and ERF 2, respectively. The S/P values for summer were less than 50 percent of what they were during the winter. These ratios do not take into account any snow drifting into the storage unit.

Since dentention ponds are relatively inexpensive to construct, it is recommended that the storage volume be twice as large as the volume resulting from the average precipitation falling on the storage unit for the desired storage period.

### Design Recommendations

The recommended sizing for manure storage with liquid separation is 0.042 m³ (1.5 ft³) per 454 kg (1000 lb) live body weight per day. This includes bedding up to about 4 kg (9 lb) per 454 kg (1000 lb) live body weight per day and also includes shrinkage.

A 10 to 15 cm (4 to 6 in.) thickness of porous media such as chopped corn stalks, corn cobs or shredded bark spread on the floor is necessary to obtain adequate drainage. Vertical 10 x 10 cm (4 x 4 in.) timbers are necessary to allow drainage down the sides in the bunker unit.

The open sided manure stack is recommended for winter time storage only if at least 3 kg (6 lb) per 454 kg (1000 lb) live body weight of bedding is used. Summer-time storage requires a totally enclosed structure as the heavy precipitation results in manure flowage, unless larger quantities of bedding are used. The open sided unit is desirable for manure removal.

A hard surface such as concrete or limestone is required as base for the stacking area.

Concrete or tongue and groove plank and post walls are recommended for side walls. If liquid ponds inside the unit, seepage will pass through the side walls of the post and plank.

The storage unit should be so located so as not to distract from the farmstead. If it cannot be so located, consideration should be given to a bunker type storage unit over a platform type unit. A roof on the bunker storage will greatly enhance its appearance.

For a platform storage unit drains should be located so manure does not initially cover the drain. Drain area of 0.1 m² (1 ft²) per 7 m² (75 ft²) of platform area is recommended. Five x fifteen cm (2 x 6 in.) short planks spaced 1 to 2 cm (1/2 to 3/4 in.) apart across the drain are recommended.

Manure that has dried out on the surface is not conducive to fly breeding. Therefore, winter stored manure can be held in the storage unit over summer. Summer-time stacking is conducive to fly breeding. If daily summer-time stacking is anticipated, confining chickens to the stack with a very limited diet will control fly breeding.

There are no odors emitted from the stacks during storage. A fermentation odor is emitted during hauling periods but it is not nearly as offensive as found in liquid manure systems.

Detention ponds should be sized to receive two times the volume of the precipitation falling on the storage area for the storage period desired.

An 1118 W (1-1/2 hp) centrifugal pump and a perforated 3.8 cm (1-1/2 in.) plastic pipe or several sprinkler heads will adequately remove and distribute the contents of the detention pond.

### SUMMARY

Three manure stacking systems were evaluated over a three-year period for quantity and quality of seepage and manure accumulation. Stacking was done for three winter periods and three summer periods for 32, 28 and 26 cow herds. One of the systems was roofed. Liquids were separated by gravity from the solids and stored in detention ponds.

An average of 0.051 m³ (1.8 ft³) per 454 kg (1000 lb) of live body weight of manure, including bedding, was collected at the Ashland Station with an average solids content of 16.9 percent. The inplace volume of storage at Ashland averaged 0.040 m³ (1.4 ft³) per 454 kg (1000 lb) live body weight, thus indicating a 22 percent reduction in volume required. Liquid separation and shrinkage accounted for the volume reduction. Inplace storage was 0.033 m³ (1.2 ft³) and 0.030 m³ (1.1 ft³) per 454 kg (1000 lb) of live body weight for ERF 1 and ERF 2, respectively.

The average quantity of seepage (precipitation and urine) collected during the winter periods was 5.9 l (1.6 gal), 10.9 l (2.9 gal) and 7.5 l (2.0 gal) per 454 kg (1000 lb) live body weight per day at Ashland, ERF 1 and ERF 2. During the summer, these quantities were considerably less. The total solids concentration for all periods averaged 2.1 percent, while the BOD concentration was approximately 9200 ppm.

The ratio of seepage collected to precipitation falling on the storage unit ranged from 0.22 to 1.67. Sizing of a detention pond can be estimated by using twice the volume of precipitation falling on the storage unit over the desired storage period.

**References**

1 Converse, J. C. and C. O. Cramer. 1973. Storage structures for solid manure. Proceedings of 1973 Livestock Waste Management Conference, Ag Eng Dept, University of Illinois, Urbana-Champaign.

2 Cramer, C. O., J. C. Converse, G. H. Tenpas and D. A. Schlough. 1973. Design of solid manure storage for dairy herds. TRANSACTIONS of the ASAE 16(2):354-360.

3 Tenpas, G. H., D. A. Schlough, C. O. Cramer and J. C. Converse. 1972. Roofed vs. unroofed solid manure storage for dairy cattle. ASAE Paper No. 72-949. ASAE, St. Joseph, Mich. 49085.

---

## Design Runoff from Feedlots

*(Continued from page 428)*

### SUMMARY AND CONCLUSIONS

Rainfall-induced runoff from open cattle feedlots is a concentrated and potentially damaging waste. Because of this and the past lack of runoff measurements, current regulations and design criteria predict more runoff than can reasonably be expected in the Texas High Plains.

The analysis of 82 years of observational day rainfall from Amarillo, Texas, show that TP-40 estimates more rainfall for that location than can be expected.

Runoff amounts were computed for each day of the 82-year rainfall record for Amarillo by both the SCS equation and the Bushland equation which was recently derived from measured runoff events. The SCS equation computed substantially more runoff that the Bushland equation. If the SCS equation is used to compute runoff, the CN of 82 should be used instead of the currently used value of 90.

**References**

1 Clark, R. N., A. D. Schneider and B. A. Stewart. 1974. Analysis of runoff from Southern Great Plains feedlots. Presented to the ASAE, June 1974, Stillwater, OK., ASAE paper no. 74-4017, ASAE, St. Joseph, Michigan 49085.

2 Clark, R. N. and B. A. Stewart. 1973. Amounts, composition and management of feedlot runoff. Sw. Great Plains Research Center, Bushland, Texas, 79012, Technical Report No. 12.

3 Hershfield, David M. 1961. Rainfall frequency atlas of the United States for durations from 30 minutes to 24 hours and return periods from 1 to 100 years. U.S. Dept. of Commerce, Weather Bureau, Technical Paper no. 40. 115p.

4 Jones, Bill. 1974. EPA and the livestock feeder. AGRICULTURAL ENGINEERING 55(3):30-31.

5 McGuinness, J. L. and D. L. Brakensiek. 1964. Simplified techniques for fitting frequency distributions to hydrologic data. USDA, Agricultural Research Service, Agricultural Handbook 259. 42p.

6 Mielke, Lloyd N. 1974. Physical characteristics of the surface and interface layers of a level beef cattle feedlot. PhD. dissertation, Dept. of Agronomy, Univ. of Nebraska, Lincoln. 134p.

7 Miller, John F. 1964. Two- to ten-day precipitation for return periods of 2 to 100 years in the contiguous United States. U.S. Dept. of Commerce, Weather Bureau, Technical Paper no. 49 29p.

8 Miner, J. R., L. R. Fina, J. W. Funk, R. I. Lipper and G. H. Larson. 1966. Stormwater runoff from cattle feedlots. Proc. Nat. Symp. on Animal Waste Management, ASAE Pub. no. SP-0366. pp. 23-27. ASAE, St. Joseph, Michigan 49085.

9 Mockus, Victor. 1964. Estimation of direct runoff from storm rainfall, Chapt. 10, In: Soil Conservation Service, National Engineering Handbook, Section 4, Hydrology (revised 1972), U.S. Dept of Agriculture, Washington, D.C. 20250.

10 Ogrosky, Harold O. and Victor Mockus. 1964. Hydrology of agricultural lands, Section 21, In: Ven Te Chow, (ed.) Handbook of applied hydrology. McGraw-Hill Book Co., New York.

11 Shuyler, Lynn R., David M. Farmer, R. Douglas Kreis and Marsha E. Hula. 1973. Environmental protection concepts of beef cattle feedlot wastes management. Report of project no. 21AOY-05, element 1B2039, Environmental Protection Agency, P. O. Box 1198, Ada, OK 74820.

12 Swanson, N. P., L. N. Mielke, J. C. Lorimer, T. M. McCalla and J. R. Ellis. 1971. Transport of pollutants from sloping cattle feedlots as affected by rainfall intensity, duration, and recurrence. Proc. Int. Symp. on Livestock Wastes, ASAE Pub. Proc-271. pp. 51-55. ASAE, St. Joseph, Michigan 49085.

13 Texas Water Quality Board. 1972. Tentative proposed regulation for control of wastes from confined areas for feeding and/or holding of certain animals. Texas Water Quality Board, P.O. Box 13246, Austin, Tex. 78711.

14 Reddell, D. L. and G. G. Wise. 1974. Water quality of storm runoff from a Texas beef feedlot. Texas Agric. Exp. Station, College Station, PR-3224. 7 pp.

# Management of a Flushing-Gutter Manure-Removal System to Improve Atmospheric Quality in Housing for Laying Hens

R. L. Fehr, R. J. Smith

THE use of flushing-gutter manure-removal systems has proved a successful manure-management technique. Types of flushing devices and gutters tried have varied greatly. Tipping buckets, dosing siphons, trapdoor tanks, and large-capacity pumps have been used as flushing devices. Gutter systems have included open-channels, between-slats, under-slats, flumes and tubes. The flushing-gutter system used at the Iowa State University Poultry Science Center is a combination of dosing siphons and open channels under cages. This paper is a report on the effects of management on flushing-water quality and on the building environment.

## REVIEW OF LITERATURE

Airborne ammonia has been associated with several adverse effects on layers. In the early 1950s, $NH_3$, at levels exceeding 86-100 $\mu$g/l, was found the principal cause of keratoconjunctivitis (Faddoul and Ringnose 1950, Valentine 1964). More recently, studies have indicated that $NH_3$ also increases susceptibility to newcastle disease, decreases feed intake, and decreases egg production (Anderson et al. 1964, Charles and Payne 1966). Although there are many odorous compounds in a poultry house, $NH_3$ has a strong correlation with odor offensiveness (Hashimoto 1972).

The level of $NH_3$ concentrations in poultry housing has been closely associated with ventilation rates and manure-management systems (Lampman et al. 1967, Longhouse et al. 1963). A flushing-gutter manure-removal system, however, involves moving large quantities of water through the building daily; thus, the condition of the water also becomes a factor.

## DESCRIPTION OF UNIT D, POULTRY RESEARCH CENTER, IOWA STATE UNIVERSITY

The flushing-gutter manure-removal system at Unit D includes in-house recycling, effluent treatment, and effluent recycling (Fig. 1). Treated effluent is pumped 610 m (2000 ft) to Unit D from an aerobic-lagoon cell. The treated water is used for flushing other buildings at other research facilities.

After entering Unit D, the treated water flows through a ball valve, then into 1.9-cm (3/4-in.) ABS pipe. This pipe runs to the ceiling of Unit D, where it tees and runs to both the east and west sides. Each side of Unit D has its own in-house recycling system (Fig. 2). To control the treated-water flow into either side, solenoid valves and plastic gate

Journal Paper No. J-8143 of the Iowa Agriculture and Home Economics Experiment Station, Ames, IA, Project No. 1842.

The authors are: R. L. FEHR, Graduate Research Assistant, and R. J. SMITH, Assistant Professor, Agricultural Engineering Dept., Iowa State University, Ames.

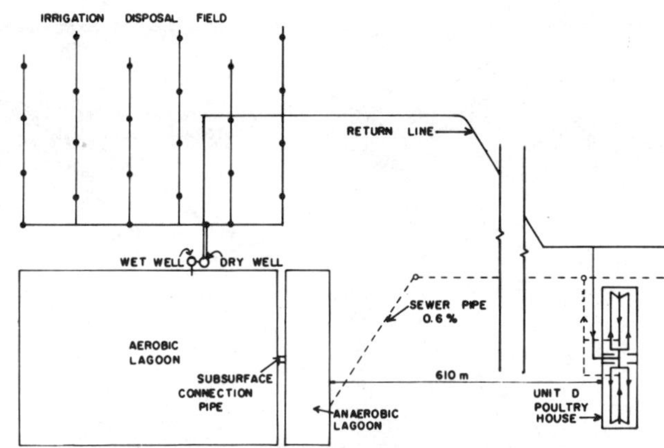


FIG. 1 The waste treatment, recycling, and disposal system for Unit D of the Poultry Research Center, Iowa State University, Ames.

valves were installed in both lines. The gate valve provides a means of complete treated-water shutoff, which is useful when draining the recycling water out of the building. The solenoid valve responds to provide treated water to the in-house system when there is insufficient water to operate the recycling pump.

ABS pipe runs from the solenoid valve to a tee and then into two separate flushing-tank systems, one on each side of a center pit. Each flushing-tank system consists of two 0.57-m³ (150-gal) livestock-watering tanks interconnected by a 20-cm (8-in.) tube. The recycled water is flushed by dosing siphons installed on one of the tanks. Three flush tubes, 7.6-cm (3-in.) PVC pipe, are located under one bell, 76 cm x 15 cm x 20 cm (30 in. x 6 in. x 8 in.), which causes all three tubes to flush simultaneously.

Below each flushing-tank system is an epoxy-coated metal tray. The trays were formed by riveting three prefabricated metal sections, 0.9 mm (20 gage) thick, 54 cm (21.3 in.) wide, and 3 m (120 in.) long. To increase rigidity and to provide flow dividers, which prevent meandering, a 2.5 cm (1 in.) vertical lip was designed for riveting the sections together. Each tray is located below 4 rows of layer cages, which are in a stairstep configuration. When completed, each tray is 162 cm (64 in.) wide and 24 m (80 ft) long and is installed at a 0.625 percent slope.

At the end of each tray is a concrete collection gutter 36 cm (14 in.) long and 162 cm (64 in.) wide. The collection gutters are sloped toward the center of the building to move the recycled water to a central concrete pit.

The central pit is 162 cm (64 in.) wide and 24 m (80 ft) long. Above the pit are 4 rows of layer cages in a stairstep configuration. To provide an area for recycling water storage, the pit was constructed with three different zones. The first 18 m (60 ft) has a slope of 0.42 percent and is cleaned by the recycled water from the trays. Meandering

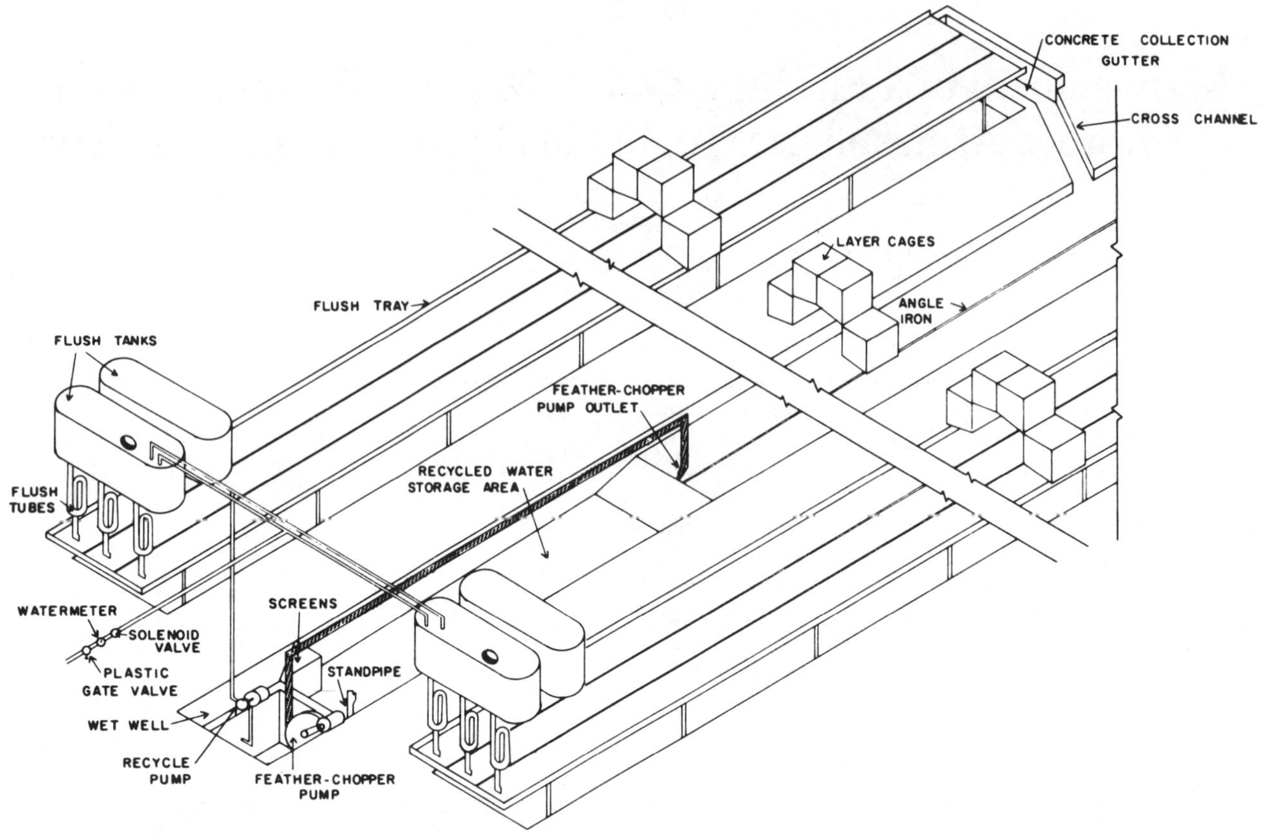



FIG. 2 The Unit D Poultry House recycling system at the Poultry Research Center, Iowa State University, Ames.

was prevented in this section by anchoring an angle iron down the center. The next 60 cm (24 in.) section of pit drops 46 cm (18 in.) to provide 3400 liter (120 ft$^3$) of area for recycling water storage in the final 5.5-m (18-ft) section. The final section has a 1.85 percent slope, which aids in cleaning the recycled water out of the building. The configuration described is not optimum. Internal recycling was judged necessary because of the remote lagoon location and inadequate sewer slope (Fig. 1).

Recycled water must pass through 6.4-mm (1/4-in.) and 3.2-mm (1/8-in.) screens before it enters the wet well to be pumped back to the flush tanks. The screens were installed to prevent feathers from plugging the pump. The recycling pump is a Jabsco* No. 777, plastic-bodied pump with a flexible rubber impeller, powered by a 560 W (3/4 hp) electric motor.

The recycling pump is controlled by a cyclic time switch, which allows a variable number of daily tray flushings. Damage to the rubber impeller is prevented by a float switch that stops the pump if the recycled water level is too low in the storage area. When the cyclic time switch is operating the recycling pump and the float switch stops the pump; then, the solenoid valve opens, allowing treated water to enter the flush tanks.

Recycled water is removed from the storage area by removing a standpipe at the end of the storage area. A feather-chopper pump is used to aid in mixing and to prevent feathers from plugging the sewer. The chopper pump was developed at Iowa State University in 1972 (Van Ee 1974). The chopper pump mixes the poultry manure, which has settled in the storage area, by having its outlet at

the beginning of the storage area and using a high flow rate of 22 liter/sec (350 gal/min).

The effluent from Unit D flows to a two-stage lagoon system. The first stage is an anaerobic lagoon connected by a subsurface pipe to the second stage of treatment, an aerobic lagoon. These lagoons were designed to handle the wastes from other livestock at the research center. The aerobic-lagoon effluent is irrigated on adjacent sheep pasture annually (Fig. 1).

## MANAGEMENT OF THE
## MANURE-HANDLING SYSTEM

Since installation in 1973, the manure-handling system's management has included biweekly cleaning of the recycled water-storage area, monthly cleaning of the flush trays, and occasional cleaning of the flush tanks. These management methods were adopted by the personnel working in Unit D as the best compromise between labor input and proper system operation. Normal operating procedure includes allowing the continuous-flow, bird-watering system to flow into the recycled-water system at the rate of 9690 liter/day (2560 gal/day).

The management of the recycling system within the unit can be controlled, but the management of the effluent-treatment system cannot. The lagoons are subject to slug loading from other sections of the research center as well as to continuous loading from a flushing-gutter, swine-finishing unit. At present, the final disposal system consists of a gasoline-engine-driven irrigation pump and hand-move pipe that is operated when needed.

## LABORATORY ANALYSES

Recycled water in Unit D and treated effluent from the

---

*Mention of commercial products does not imply endorsement by Iowa State University.

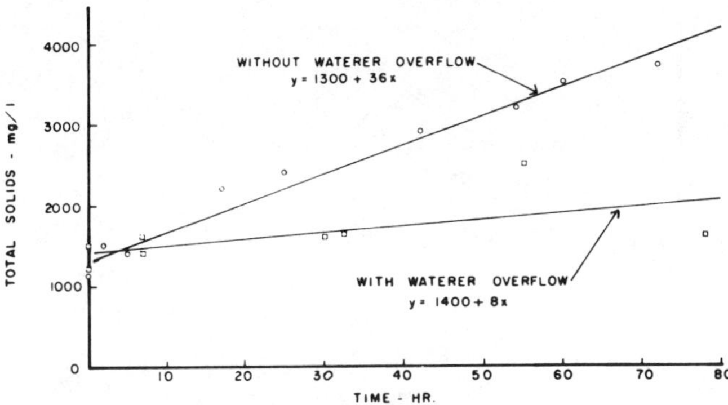


FIG. 3 Variations in total-solids concentrations in the recycled water with and without the overflow from the layer waterers entering the recycled water in the Unit D poultry house, Iowa State University, Ames.

aerobic lagoon cell was sampled for Kjeldahl-N, $NH_3$-N, COD, and total solids. These analyses were done according to Standard Methods (A.P.H.A. 1971).

Ammonia-gas concentrations were determined with a sulfuric acid trap and Nessler's reagent. Air was sampled at the rate of 1.03 liter/min (0.036 $ft^3$/min) by using a calibrated limiting orifice (Lodge et al. 1966). A limiting orifice allows accurate flow rates to be simply maintained. Two midget impingers containing 20 ml of 0.02 N sulfuric acid were used in series to trap virtually all the $NH_3$ present in the air (Leithe 1970). The samples were returned to the laboratory for Nesslerizing, and the percentage transmittance was measured. A table relating transmittance to nitrogen concentration was used to determine the amount of $NH_3$ in the sulfuric acid. The $NH_3$ concentrations reported are the average of two 15-min samples taken consecutively.

## PRELIMINARY RESULTS

The results presented contain measurements taken under two types of management conditions. Overflow from the waterers was allowed to enter the recycling water-storage area for one test and was diverted for the second test. The two sets of conditions caused noticeable differences in the measurements.

The solids buildup in the recycling water was a primary concern when the system began operation. Damage to the recycling pump, flush tank malfunctions, and sewer plugging would be possible consequences of large solids concen-

trations. The general equation that can be used to describe solids buildup in the recycling storage area is:

$$X(t) = C + Ae^{-t/B}$$

where

$X(t)$ = amount of solids at time t, kg

$t$ = elapsed time, day

$$C = \frac{(\text{solids input/time}) \times \text{storage area volume}}{\text{liquid volume added/unit time}},$$

$$\frac{(kg/day) \times liter}{liter/day}$$

$$B = \frac{\text{storage area volume}}{\text{liquid volume added/unit time}},$$

$$\frac{liter}{liter/day}$$

$A$ = constant determined by initial conditions, kg

With the overflow from the waterers entering the storage area, the equation can be evaluated with use of the following conditions: 1120 layers at 1.8 kg/layer; solids input = 13.4 kg/1000 kg liveweight-day; liquid volume added = 39.6 liter/1000 kg liveweight-day (manure) + 9690 liter/day (waterer overflow); storage area volume = 3406 liter; initial conditions = 4.09 kg solids at t = 0. This results in the equation:

$$X(t) = 9.34 - 5.25e^{-t/0.346}$$

When this equation is solved for t = ∞, the concentration of solids approaches 2742 mg/l. After 2 days, the concentration should approach 2737 mg/l. The lower line in Fig. 3 indicates how the total-solids concentration in the recycled liquid varies with time, with the waterer overflow entering the storage area. The slope of the lower line is not significantly different from zero, which would be expected from the general equation. Complete mixing of solids is assumed in the general equation, and this condition does not exist in the storage area. The lack of mixing explains the difference between the average solids concentration in the recycled liquid, 1625 mg/l, and the predicted value, 2742 mg/l.

When the waterer overflow is not allowed to enter the storage area, the general equation, with a different liquid flow condition, is:

$$x(t) = 1144 - 1140e^{-t/42.4}$$

Solved for t = ∞, the concentration of solids approached 335,880 mg/l. After 2 days, the concentration should approach 16,600 mg/l. The upper line in Fig. 3 indicates how the total-solids concentration in the recycled liquid varies with time, with no waterer overflow entering the storage area. The slope of the line is significantly different from zero, which would be expected from the general equation. With the equation for the upper line given in Fig. 3, a solids concentration of 2166 mg/l results after 2 days. More solids settlement is indicated by the large difference between predicted and the actual solids concentration under this condition as compared with the situation when the waterer overflow enters the storage area. This increase in solids settlement results from the decrease in the amount of water, which is turbulent and high in suspended solids,

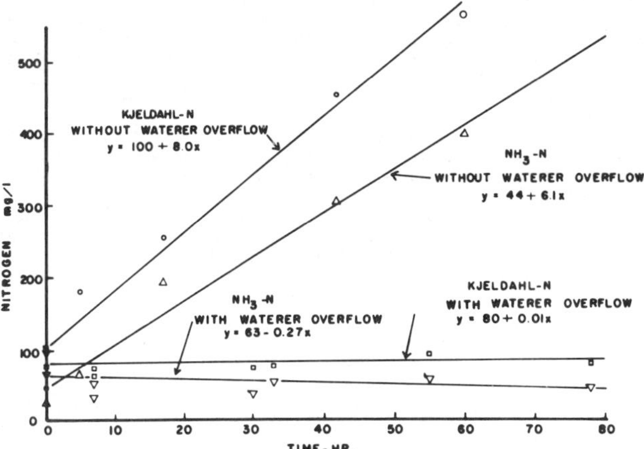


FIG. 4 Variations in Kjeldahl and $NH_3$-N concentrations in the recycled water with and without the overflow from the layer waterers entering the recycled water in the Unit D poultry house, Iowa State University, Ames.

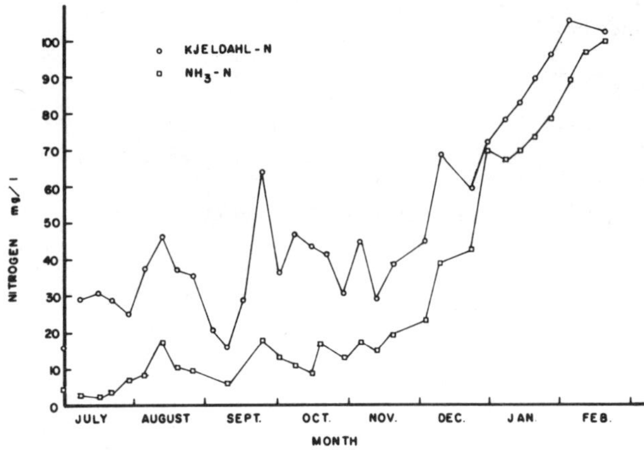

FIG. 5 Variations in Kjeldahl and NH₃-N concentrations in the aerobic lagoon at the Poultry Research Center, Iowa State University, Ames.

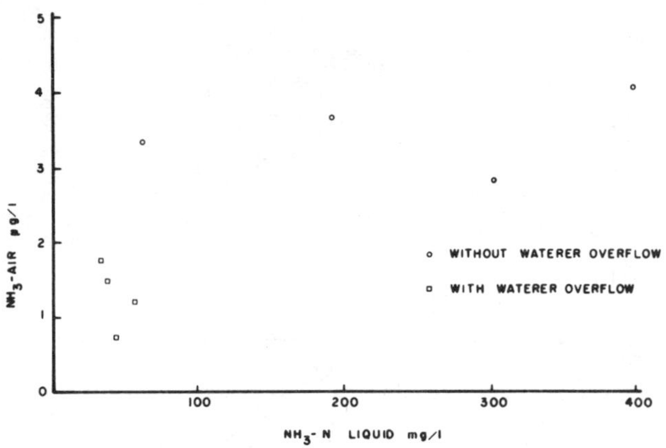

FIG. 6 Ammonia levels in the air at varying NH₃-N levels in the recycled water with and without the overflow from the layer waterers entering the recycled water in the Unit D poultry house, Iowa State University, Ames.

that flows out of the storage area over the standpipe.

The Kjeldahl and $NH_3$-N levels in the recycling water (Fig. 4) did not increase when the waterer overflow was entering the storage area. The slope of both lines under this condition is not significantly different from zero. Because of $NH_3$ volatilization, a general equation is difficult to develop for Kjeldahl or $NH_3$-N levels in the recycling water. Ammonia volatilization in this system would be dependent on many variables, including $NH_3$ levels in the liquid, agitation of liquid, temperature, pH, and $NH_3$ levels in the air. We observed that the $NH_3$-N concentration in the recycling water decreases after entering the building. This decrease is a winter phenomenon resulting from the buildup of $NH_3$-N in the lagoon after it becomes ice-covered (Fig. 5).

When the waterer overflow was prevented from entering the storage area, both Kjeldahl and $NH_3$-N levels increased in the recycling water (Fig. 4). Under this condition, the slopes of both lines are significantly different from zero. Experiments covering extended periods are being conducted to determine if this buildup will continue or if the rate of $NH_3$ volatilization will increase until it equals $NH_3$ production by the accumulated manure in the pit.

Fig. 6 indicates the relationship between $NH_3$-N levels in the recycling water and $NH_3$ concentrations in the air. There seems to be a relationship between them, as would be expected, but more data will be required before this relationship can be defined. The ventilation rate used when allowing the waterer overflow to enter the storage area was approximately half the rate used during the test when the waterer overflow was not allowed to enter the storage area. This indicates that, although the $NH_3$ concentrations are lower with waterer overflow entering the pit, they could have been reduced further by the use of equal ventilation rates during both tests.

## CONCLUSIONS

1   Overflow waterers can add more than 200 times more water to a poultry manure system than the layers excrete in their manure.

2   Continuous-overflow waterers are common in layer houses, and these tests indicate that, with a liquid-manure system, the overflow should be allowed to enter the pits to reduce ammonia-gas concentrations in a building.

3   Ammonia-nitrogen levels in a continuously fed aerobic lagoon increase after they are ice covered to a level at which $NH_3$ gas is released when it is exposed to the air.

4   Ammonia concentrations in the atmosphere of the building were at all times far below levels associated with adverse effects on layers.


## References

1   American Public Health Association. 1971. Standard methods for the examination of water and wastewater. 13th ed. American Public Health Association. Washington, D.C.

2   Anderson, D.P., C. W. Beard, and R. P. Hanson. 1964. The adverse effects of ammonia on chickens including resistance to infection with newcastle disease virus. Avian Diseases 8:369-379.

3   Charles, D. R., and C. G. Payne. 1966. The influence of graded levels of atmospheric ammonia on chickens. I & II. British Poultry Sci. 7:177-187, 189-198.

4   Faddoul, G. P., and R. C. Ringnose. 1950. Avian keratoconjunctivitis. Veterinary Medicine 45:492-493.

5   Hashimoto, A. 1972. Aeration under caged laying hens. TRANSACTIONS of the ASAE 15:119-1123.

6   Lampman, C. E., J. E. Dixon, C. F. Peterson, and R. E. Black. 1967. Environmental control for poultry housing. Idaho Agricultural Experiment Station Bulletin 456. 27 pp.

7   Leithe, W. 1970. The analysis of air pollutants. Ann Arbor-Humphrey Science Publishers, London.

8   Lodge, J. P. Jr., J. B. Pate, B. E. Ammons, and G. A. Swanson. 1966. The use of hypodermic needles as critical orifices in air sampling. Journal of the Air Pollution Control Association 16¾197-200.

9   Longhouse, A. D., C. S. Shaffer, and G. O. Bressler. 1963. Poultry housing—Basic data useful for design purposes in the northeastern states. West Virginia Agricultural Experiment Station Bulletin 486T. 43 pp.

10   Valentine, H. 1964. A study of the effect of different ventilation rates on the ammonia concentrations in the atmosphere of broiler houses. British Poultry Science 5:149-159.

11   Van Ee, G. R. 1974. Chopper pump. U.S. Patent 3,807,644. April 30.

# Performance of Screw Conveyors for Unloading Sludges from Field Transports

Mead Weil, Andrew Higgins

ASSOC. MEMBER
ASAE

VARIOUS alternative means for disposal of biodegradable wastes have been devised in the past few years to minimize air and ground water pollution and maximize the economic and nutritive returns from these wates. One method being investigated is the injection of biodegradable wastes into. the aerobic region of the soil profile (upper 8 in.). Microorganisms present in the soil can then break down these wastes into simpler compounds that can be readily absorbed and utilized by the soil vegetative cover.

Equipment has been developed to incorporate a wide range of biodegradable wastes into the upper 8 in. of the soil by Plow-Furrow-Cover, Sub-Sod-Injection, and Ridge and Furrow. The physical properties of the wastes unloaded from this equipment may vary from thin slurries to caked materials with up to 75 percent solids content (wet basis). A field transport for such material must be water tight and readily unload sludges with a wide range of physical properties. Screw conveyors have been found to be well suited to unloading sludges from field transports, due to their simple operation and water tight capabilities.

A review of literature indicates the difficulty of describing the physical properties of sewage sludges. Percent solids does not adequately indicate the handling characteristics of this material. Tests were conducted on two types of screw conveyors to determine which sewage sludge physical properties and which screw conveyor operational variables were most closely correlated with the total horsepower input and the mass flow rate for the conveyors.

Fig. 1 illustrates the screw conveyor performance testing setup for determining total horsepower and mass flow rate for conveying sewage sludges of varying physical properties. The screw conveyors were powered by a direct drive 7 hp hydraulic motor. Instrumentation for measuring torque included a 1000 in.-lb strain-gage torque transducer coupled to a 10-in. deflection strip-chart recorder. Rotational speed was plotted on another strip-chart recorder as 2 deflections per screw conveyor revolution.

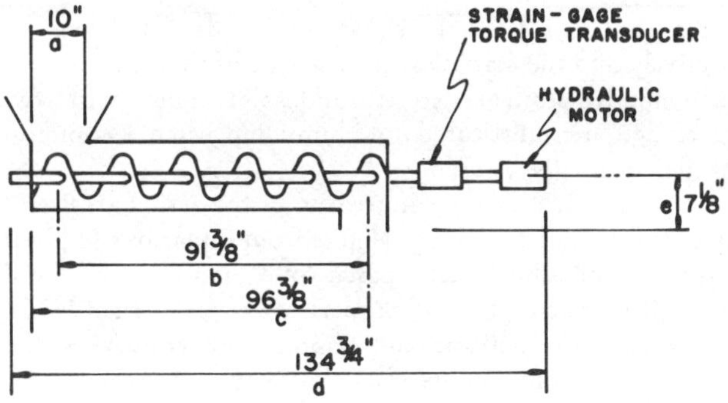

## PROCEDURE

Nine-in. diameter helicoid and ribbon screw conveyors were tested independently, using the same sewage sludge, to determine the total horsepower input and mass flow rate of each conveyor. A complete factorial design was followed (i.e., 36 trails plus 5 repetitions per screw conveyor) using the following variables:

1   Screw Conveyor Type — 9-in. diameter, standard pitch, helicoid and ribbon

2   Screw Conveyor RPM — 50, 100, 150



a – INTAKE LENGTH

b – EFFECTIVE CONVEYING LENGTH

c – INSTALLED CONVEYING LENGTH

d – TOTAL LENGTH

e – DISCHARGE LENGTH

FIG. 1 Screw conveyor performance testing setup.

3   Screw Conveyor Inclination — 0, 30 deg. above horizontal

4   Sewage Sludge Solids Content — 15, 12, 9 percent (wet basis)

Additional tests were conducted on the sewage sludge at each solids content to determine (a) the Portland Cement Association Concrete Slump Test (ASTM designation C143), (b) non-Newtonian viscosity, and (c) mass density. The sewage sludge tested was a domestic, secondary sludge, aerobically treated and mechanically processed.

## RESULTS

A stepwise linear regression analysis (computer program BMD O2R, University of California Press) was used to statistically analyze the 46 sets of data for total horsepower ($hp_T$) and mass flow rate (MFR) for both screw conveyors. The dependent variables $hp_T$ and MFR were plotted against the seven independent variables (4 sewage sludge physical properties and 3 screw conveyor type and operational variables) to determine the best mathematical model for predicting screw conveyor $hp_T$ and MFR. Table 1 lists these equations and their statistical significance.

The single correlation coefficients for the four sewage sludge physical properties alone are listed in Table 2. These coefficients indicate that solids content and non-Newtonian viscosity are very highly correlated with slump and could therefore be substitiued in the $hp_T$ and MFR Equations

The authors are: MEAD WEIL, Graduate Student and ANDREW HIGGINS, Engineering Research Associate, Biological and Agricultural Engineering Dept., Rutgers University, New Burnswick, N. J.

TABLE 1. REGRESSION ANALYSIS WITH 3 VARIABLES FOR THE COMBINED 9-INCH DIAMETER HELOCOID AND RIBBON SCREW CONVEYOR MODEL.

| Equation number | Dependent variable | Multiple regression coefficient (R) | Standard error of estimate (S.E.) | Independent variables | | |
|---|---|---|---|---|---|---|
| | | | | Parameter (P) | Coefficient | (S.E.P.) |
| [1] | MFR | 0.8796 | 4.0181 | C | 4.47581 | |
| | | | | N | 0.14190 | 0.01781 |
| | | | | AM | -0.04734 | 0.00534 |
| [2] | $hp_T$ | 0.9210 | 0.3234 | C | 1.47926 | |
| | | | | N | 0.00948 | 0.00143 |
| | | | | M | -0.21288 | 0.01568 |

TABLE 2. CORRELATION BETWEEN SEWAGE SLUDGE SOLIDS CONTENT AND THREE OTHER SEWAGE SLUDGE PHYSICAL PROPERTIES.

| Sewage sludge physical property | Correlation coefficient |
|---|---|
| Solids content (S) | 1.000 |
| non-Newtonian viscosity (V) | 0.980 |
| Mass density (D) | 0.694 |
| Slump (M) | -0.981 |

[1] and [2] of Table 1 with little loss in confidence.

Tables 3 and 4 list the equations for MFR and $hp_T$ respectively and the statistical significance of their parameters when non-Newtonian viscosity and solids content of sewage sludge are substituted for slump and when a common coefficient for (N) is used.

Table 5 indicates the comparison of the actual MFR and $hp_T$/10 ft values for sewage sludge from Equations [1] and [2] to the handbook data, based on a theoretical material class with similar physical properties (material class H36).

For both the helicoid and ribbon screw conveyors, the horsepower requirements for the empty conveyor was approximately 12 percent of the total horsepower requirement for a full conveyor.

are considerably less reliable than Equations [1] and [2], when in fact they are almost equal in validity as shown by the (S.E.).

It can be seen from the results of Tables 1, 3, and 4 that screw conveyor RPM (N) and one sewage sludge physical property are all that is necessary to adequately describe the horsepower input to the screw conveyor. The mathematical model for MFR is only slightly more complicated in that the angle of screw conveyor operation is required in addition to RPM (N) and one sewage sludge physical property. This simplification offers the designer of sludge field transports the option of choosing whatever sewage sludge physical property is available to him for determining $hp_T$ and MFR without loss in confidence of the equation used.

## DISCUSSION

In comparing the precision of Equations [3] thru [8] with Equations [1] and [2], the reader should give primary emphasis to the (S.E.) values instead of the (R) values. Due to the way in which (R) is calculated, those values for (R) would indicate that Equations [3] thru [5] in particular

## SUMMARY

Forty-six tests were conducted on 9-in. diameter, standard pitch ribbon flight and helicoid flight screw conveyors conveying domestic secondary sewage sludge with solids between 15 percent and 9 percent (wet basis). A complete factorial experimental design was followed using

TABLE 3. FORCED REGRESSION ANALYSIS WITH THREE SEWAGE SLUDGE PHYSICAL PROPERTIES AND A COMMON (N) COEFFICIENT IN THE 9-INCH DIAMETER SCREW CONVEYOR MFR MODEL (RIBBON AND HELICOID SCREW CONVEYORS).

| Equation number | Sewage sludge physical property | Multiple regression coefficient (R) | Standard error of estimate (S.E.) | Independent variables | | |
|---|---|---|---|---|---|---|
| | | | | Parameter (P) | Coefficient | (S.E.P.) |
| [3] | slump (M) | 0.8168 | 3.9766 | C | 5.23577 | |
| | | | | N | 0.13439 | |
| | | | | AM | -0.04719 | 0.00527 |
| [4] | viscosity (V) | 0.5315 | 5.8385 | C | 4.06946 | |
| | | | | N | 0.13439 | |
| | | | | AV | -0.00048 | 0.00012 |
| [5] | solids (S) | 0.6491 | 5.2433 | C | 4.75714 | |
| | | | | N | 0.13439 | |
| | | | | AS | -0.02451 | 0.00454 |

TABLE 4. FORCED REGRESSION ANALYSIS WITH THREE SEWAGE SLUDGE PHYSICAL PROPERTIES AND A COMMON (N) COEFFICIENT IN THE 9-INCH DIAMETER SCREW CONVEYOR TOTAL HORSEPOWER MODEL ($hp_T$ PER 10 FT OF LENGTH, RIBBON AND HELICOID SCREW CONVEYORS).

| Equation number | Sewage sludge physical property | Multiple regression coefficient (R) | Standard error of estimate (S.E.) | Independent variables | | |
|---|---|---|---|---|---|---|
| | | | | Parameter (P) | Coefficient | (S.E.P.) |
| [6] | Slump (M) | 0.9088 | 0.3194 | C | 1.47249 | |
| | | | | N | 0.00955 | |
| | | | | M | -0.21293 | 0.01546 |
| [7] | Viscosity (V) | 0.8249 | 0.4327 | C | -1.72204 | |
| | | | | N | 0.00955 | |
| | | | | V | 0.00330 | 0.00036 |
| [8] | Solids (S) | 0.8845 | 0.3570 | C | -3.84406 | |
| | | | | N | 0.00955 | |
| | | | | S | 0.30212 | 0.02519 |

**TABLE 5. COMPARISON OF ACTUAL MFR AND $hp_T/10$ FT FROM EQUATIONS [1] AND [2] TO HANDBOOK DATA.**

| Sewage sludge physical properties | Screw conveyor speed (rpm) | Percent trough filled | | Percent theoretical* $hp_T/10$ ft |
| --- | --- | --- | --- | --- |
| | | 0 deg | 30 deg | 0 deg |
| M = 4.13 in. | 50 | 75.1 percent | 30.7 | 76.5 percent |
| S = 14.1 percent | 100 | 60.6 | 41.5 | 66.0 |
| | 150 | 55.7 | 43.0 | NA |
| M = 9.44 in. | 50 | 75.1 percent | NA | NA |
| S = 11.1 percent | 100 | 60.6 | 17.1 percent | 38.0 percent |
| | 150 | 55.7 | 26.7 | NA |

*Theoretical horsepower based on calculations made for material class H36 of FMC Book 3089, Screw Conveyors and Screw Feeders, (1969).

the following parameters: (a) type of screw conveyor, (b) screw conveyor RPM, (c) screw conveyor angle of inclination, and (d) sewage sludge solids content. Multiple regression analysis was used to evaluate the performance of the screw conveyors and to determine whichever sewage sludge physical properties and whichever screw conveyor operational variables were most significant in calculating mass flow rate and total horsepower. The resulting mathematical models for MFR and $hp_T/10$ ft are as follows:

$$MFR = 4.48 + 0.14 (N) - 0.047 (A) (M)$$
$$hp_T/10 \text{ ft} = 1.48 + 0.009 (N) - 0.21 (M)$$

If other sewage sludge physical properties are known instead of slump, the following models for MFR and $hp_T/10$ ft can be used:

$$
\begin{aligned}
MFR &= 5.24 + 0.134 (N) - 0.047 (A) (M) \\
&= 4.07 + 0.134 (N) - 0.00048 (A) (V) \\
&= 4.76 + 0.134 (N) - 0.024 (A) (S) \\
hp_T/10 \text{ ft} &= 1.47 + 0.0096 (N) - 0.213 (M) \\
&= 1.72 + 0.0096 (N) + 0.0033 (V) \\
&= 3.84 + 0.0096 (N) + 0.302 (S)
\end{aligned}
$$

Three sewage sludge physical properties were correlated with solids content: (a) non-Newtonian viscosity, (b) mass density, and (c) slump. Of the three, only mass density showed a low correlation coefficient with solids content.

### References


1 FMC Corporation Book 3089. 1969. Screw Conveyors and Screw Feeders, 32-41.

2 W. J. Dixon. 1967. Biomedical Computer Programs, University of California Press, Berkley and Los Angeles, California, Program BMD-O2R, 15-22, 233-247.

3 R. M. Peart, B. A. McKenzie and F. L. Herum. 1967. Dimensional Standards and Performance Procedures for Screw Conveyors, ASAE Paper No. 66-848. TRANSACTIONS of the ASAE, 667-669.

4 Reed, Charles H. 1975. Equipment for incorporating sewage sludges into the soil. Paper presented at Third International Symposium on Livestock Wastes. University of Illinois, Urbana, Illinois.


Nomenclature:

| | | |
| --- | --- | --- |
| MFR | = | mass flow rate ($ft^3$/min) |
| $hp_T/10$ ft | = | total horsepower to screw conveyor drive shaft per 10 ft of length |
| C | = | constant |
| N | = | screw conveyor rpm |
| A | = | screw conveyor angle of inclination (degrees) |
| M | = | sewage sludge slump (in.) ASTM designation C143 |
| D | = | sewage sludge mass density ($lb/ft^3$) |
| V | = | sewage sludge non-Newtonian viscosity (cps) |
| S | = | sewage sludge solids content (percent, wet basis) |

# Equipment for Incorporating
# Animal Manures and Sewage Sludges into the Soil

Charles H. Reed
FELLOW ASAE

THE incorporation of biodegradable wastes into the upper 8 inches (0.203 m) of the aerobic layer of the soil is preferable to burying in landfills, dumping in the surface waters, disposal in the atmosphere by incineration, or surface spreading. Numerous writers have documented this fact, Bohn and Cauthorn (1972) summarizing it, "compared to air and water, the soil has a vastly greater potential for waste disposal and transformation . . . and it has the capacity to absorb far more material (nature's wastes) than it can produce or that is added to it." When biodegradable wastes are incorporated directly into the soil there are no odors, no opportunity for flies or other pests to feed or breed, no runoff or surface erosion of wastes, and the wastes are. placed in the best possible medium for immediate degradation to plant nutrients and utilization by plants. This conforms to the concept of land treatment as defined by Stevens et al. (1972): conveying the idea of reciprocal and beneficial relationships between both the land and the waste.

This paper will describe equipment which has been developed at the N. J. Agricultural Experiment Station for three techniques of incorporating biodegradable wastes into the soil: Plow-Furrow-Cover, Sub-Sod-Injection, and Ridge-and-Furrow.

To Plow-Furrow-Cover (PFC), waste is deposited into a 6-to-8-inch-deep (0.152-0.203 m) plowed furrow. Immediately after deposition, and in the same operation, a plow covers the waste and opens the next furrow. With properly adjusted equipment 1 1/2 to 2 inches (0.038-0.051 m) of waste can be completely covered. This is approximately 170 to 225 tons per acre. (38 000-50 000 kg/hectare). A well-formed furrow, 16 inches (0.41 m) wide, 7 to 8 inches deep (0.177-0.203 m) and 400 feet (122 m) long containing 1 1/2 inches (0.038 m) of waste, amounts to 500 gallons (1890 liters), or approximately 2 tons (1800 kg). PFC leaves the soil well plowed and ready for disking and seeding.

An implement for the Sub-Sod-Injection (SSI) technique has been developed to inject any slurries that will flow through a 6-inch-diameter (0.152 m) hose 2 feet long (0.61 m). Animal manures with up to 20 percent solids and sewage sludge with up to 10 percent solids can be injected into the soil at the rate of 400 gallons (1510 liters) per minute. It is deposited in a band up to 2 inches (0.051 m) thick and 28 inches (0.710 m) wide and from 6 to 8 inches (0.152-0.203 m) beneath the surface without turning over the soil. The injector has a spring-trip release for passing over subsurface objects. It is comparable to a two-bottom plow in weight and durability. This equipment is not yet available commercially.

The Ridge-and-Furrow irrigation technique may be used for applying slurries to the soil and is briefly described by White et al. (1975). The furrows are made on-the-contour or slightly sloping to permit the water to filter into the soil. Aerobic conditions should be maintained in and at the bottom of the furrow. A right-hand and left-hand plow bolted together on one beam is an excellent ridge-and-furrow opener. Unless the waste is covered immediately it might be considered a surface method of application.

For the purpose of this paper, animal manures and sewage sludges will be classified into 3 types on the basis of their physical properties and handling characteristics: (a) slurries, which will flow by gravity through a 6-inch-diameter hose (0.152 m) with less than 5 feet (1.5 m) of head; (b) semisolids, which are difficult or impossible to pump through a 6-inch-diameter (0.152 m) hose and perhaps best described by the amount of slump when tested in the ASTM C143, Concrete Slump Test; and (c) solids which have no flow characteristics.

Table 1 summarizes the different operations which are possible with basic components (2 chassis and multipurpose tank) and a selection of appropriate implements and accessories. The last line indicates that when the tank is mounted on a truck chassis it can be used as a combination highway transport and field application unit. With this the furrows would be previously opened, then filled from the truck tank, and in a third pass over the ground the waste would be covered with a tractor-mounted plow, or blade. The latter implements should be adjusted to keep the tractor out of the furrow.

Standard farm tractors with a 3-point hitch have been used in our R&D. A 50 hp (37 000 W) appears to be very satisfactory for trailers with tanks of up to 1000-gallons capacity (3780 liters); and 75 to 90 hp (56 000-67 000 W) for tanks of up to 1600-gallons capacity (6000 liters).

## THE GOOSENECK TRAILER CHASSIS

The gooseneck trailer chassis as developed at the N. J. Agricultural Experiment Station for PFC and SSI has been described in previous publications, Reed (1974, 1975) and White (1975). The gooseneck may be custom-made as an integral part of the tank or a tank may be mounted on a separate gooseneck trailer chassis. The gooseneck provides unexcelled maneuverability when (a) a 16-inch (0.406 m) single-bottom moldboard plow, (b) a 28-inch (0.710 m) disk plow, (c) a ridge-and-furrow opener, or (d) a sub-sod-injector is mounted on the 3-point hitch of the tractor. The axle of the trailer is offset in the PFC operation permitting the right tractor wheels to run in the previously opened furrow, and the right trailer wheel in the clean furrow that was just opened by the plow which is located between the tractor and trailer wheels. The trailer wheels follow directly

The author is: CHARLES H. REED, Professor, Biological and Agricultural Engineering Dept., Rutgers University, New Brunswick, N.J.

**TABLE 1. COMPONENTS AND ACCESSORIES REQUIRED FOR PFC, SSI, AND RIDGE-AND-FURROW TECHNIQUES.**

| Technique | Type of waste | Location of implements | | Type of chassis (for tank) | Cross conveyor | Longitudinal conveyor | Choice of power for implements & accessories |
|---|---|---|---|---|---|---|---|
| | | Plow | SSI or R & F | | | | |
| PFC | Slurry | Tractor | | Gooseneck | None | Desirable | STHS |
| SSI | Slurry | | Tractor | Gooseneck | None | Desirable | STHS |
| PFC | Slurry | Chassis | | "A" Frame | None | Desirable | STHS or PTO or Auxiliary |
| PFC | Semisolid | Chassis | | "A" Frame | Yes | Yes | STHS or PTO or Auxiliary |
| SSI | Slurry | | Chassis | "A" Frame | None | Desirable | STHS or Auxiliary |
| Transport | { Slurry | None | None | Truck | None | Desirable | Auxiliary or Truck PTO |
| | { Semisolid | None | None | Truck | Yes | Yes | Auxiliary or Truck PTO |

Legend: STHS - A standard tractor hydraulic system.
PTO - Power takeoff on tractor.
Auxiliary engine or hydraulic system.
Ridge-and-furrow opener is the same as SSI.

behind the tractor wheels in the SSI operation.

Satisfactory tanks for slurries have had from 1000- to 1200-gallon (3800-4500 liters) capacity, and 6-inch (0.152 m) or 9-inch (0.228 m) longitudinal screw along the bottom, and a 6-inch-diameter (0.152 m) outlet valve with a 6-inch (0.152 m) flexible hose that is 5 feet (1.5 m) long for PFC, and 2 feet (0.61 m) long for SSI.

The gooseneck chassis with tank is the most simple combination of components to PFC or SSI any uniform slurry which will flow by gravity through 5 feet (1.5 m) of 6-inch-diameter (0.152 m) hose. The hydraulic system within a standard farm tractor is used to control the plow, injector, or ridge-and-furrow opener which is mounted on the tractor 3-point hitch. There are usually two hydraulic power outlets; one may be used for controlling the gate valve on the tank and the other for running the longitudinal conveyor.

The relatively simple systems on the gooseneck tongue to PFC and SSI slurries will not handle semisolid waste. This material must be deposited within a foot (0.3 m) of the tank outlet thus necessitating mounting the plow and beam on an "A" frame chassis to the rear of the tank outlet and in front of the trailer wheel. Also, a considerably more complicated power system is required.

## THE CONVENTIONAL "A" FRAME TRAILER CHASSIS

The first chassis used in this development for PFC was a conventional "A" frame type for one-axle trailer commonly used for farm equipment. A 16-inch (0.406 m) single-bottom moldboard plow was mounted on the right side of the trailer chassis just forward of the wheel. Because of the lines of force of a moldboard plow, it did not function satisfactorily.

Any one of three implements may be mounted on the trailer chassis and controlled by a hydraulic piston: (a) a 28-inch (0.710 m) disk plow may be mounted on the plow beam in front of the right trailer wheel to PFC in one pass over the ground, (b) a sub-sod-injector may be mounted in the front of the "A" frame for SSI, (c) a ridge-and-furrow opener may be mounted in the "A" frame.

The trailer wheels and axle are located so that the maximum static vertical load on the tractor drawbar does not exceed 1200 lb (540 kg) or the tractor manufacturer's recommendation for the tractor being used. A single axle has been preferable to a tandem or bogie axle. The tires should be 15.50 x 20 or equivalent, and not wider than the furrow made by the plow.

The tank-chassis mountings should permit the tank to be turned end-for-end so that the valve assembly will be in the rear to (a) PFC in 2 passes over the ground, (b) fill a previously opened ridge-and-furrow, or (c) surface-spread waste with a spinner attachment.

## A 28-INCH (0.7 m) DISK VS A 16-INCH (0.4 m) MOLDBOARD PLOW)

A 28-inch (0.710 m) disk plow with a 3-point hitch mounted on a tractor and used with the gooseneck tongue trailer has outstanding advantages over the moldboard plow for PFC:

1 There is no compacted plow sole.

2 The semielliptical furrow bottom is about the same shape as the track of a 16-inch-wide (0.406 m) tire used on the tractor and trailer wheels.

3 A disk plow will function under more adverse conditions than a moldboard plow.

4 Adjustments on a disk plow are not as exacting as those on a moldboard plow.

When the disk plow is mounted on an "A" frame chassis, the above advantages are still valid, but there are some disadvantages when compared to the gooseneck trailer:

1 Additional linkage and a hydraulic piston is needed for the plow.

2 There is a limited space for the plow and beam between the outlet of the tank and the trailer wheel. Three schemes are now being field tested: (a) the plow beam to push the plow, (b) the plow beam to pull the plow, and (c) the plow mounted on a vertical beam.

3 The hydraulic power for the piston to close the gate valve on the tank loses its priority. The piston to close the gate valve must have priority for power over the other components.

4 The trailer wheel rolls in the newly formed furrow but because the tongue or axle is not offset the tractor must run entirely on the land; not with the two right wheels in the previously opened furrow as with the gooseneck tongue. If plowing is difficult the tractor wheels may slide into the furrow.

The conveyor system used in the most recently constructed 1600-gallon (6000 liters) tank now being field tested consists of two components:

1 A longitudinal conveyor which is required for semisolids and is desirable for slurries with settleable solids. It may be either slats attached to a pair of drive chains, such as a manure spreader; or screw conveyors.

2 A cross-screw conveyor which is required to convey

*(Continued on page 451)*

# Shortest Path Network Analysis of Manure Handling Systems to Determine Least Cost — Dairy and Swine


J. R. Ogilvie, P. A. Phillips, K. W. Lievers

MEMBER
ASAE

MEMBER
ASAE

MEMBER
ASAE


THE need for careful control of livestock wastes, combined with wide variations in climate, type and size of operations, cropping systems, and soil type, make choosing between different manure handling alternatives a difficult task. Due to these complexities of the crop-livestock industry, the use of a systematic approach is essential to maximize the use of our resources when the costs of farm inputs are steadily rising. People working in support of agriculture, as well as farmers themselves can benefit from systems analysis techniques and results as a valuable aid in the decision making process.

Networks can be used to graphically represent alternative systems in any phase of agricultural production or utilization that involves a number of possible machines or methodologies. Critical path (CPM) type networks have been used successfully in the past for irrigation systems (Mussivand 1966), forage harvesting systems (Coupland and Halyk 1969) and swine manure handling systems (Morris and Nygaard 1964).

The set of possible alternative components can be depicted as a directed network with unique initial and terminal nodes where arcs represent activities or mechanical components and nodes represent events or completed operations. If a value (time, cost, etc.) can be provided for each arc, an optimal system (minimum time, cost etc.) can be found. For solving large networks linear programming (LP) techniques could also be used.

For this problem, an algorithm called Shortest Path Network Analysis (SPNA) was used for the following reasons: (a) the network is itself an integral part of the approach and once constructed, relates directly to the program input, (b) the SPNA program is very easy to use with algebraic relationships input as FORTRAN equations to assign values to each arc of the network, (c) non-linear functions can be used directly without linear approximation. Grasp of mixed integer and separable programming techniques necessary for an LP-type package are not required. Minimum time spent on program familiarization means maximum time on problem definition, (d) initial output of the analysis includes a valuation and rank for all possible system alternatives, not just the optimal or sub-optimal as provided by LP, (e) the network and the SPNA application provide a useful basis for later simulation of selected system alternatives involving stochastic parameters such as weather.

The program was initially formulated by Preston (1967). A description of the program plus details on the modification to give ranked output of all paths in the network can be found in Lievers (1974).

In general, manure handling systems consist of three main stages, collection and transfer to storage, storage (may include processing) and removal from storage for land application. Under Canadian conditions, long term storage is included in every system to prevent spreading manure on frozen soil and subsequent runoff pollution problems. At present there are many types of housing and handling equipment suitable for the various management preferences and climatic conditions. The SPNA program was used to evaluate some acceptable dairy and swine systems to determine least cost.

Careful network construction can lead to significant savings in computational cost and time as well as easing later equation modifications for different measures of effectiveness. The present SPNA program can easily cope with expansions in the network but needs considerable data modification to deal with contractions. In agricultural mechanization problems of this sort, the possibility of multiple-use machines and of variable interdependence exists. The only way to fit such problems into the present SPNA format is to construct the network so as to avoid computational backtracking and loops. Although certain variables used in the equations may have known probability distribution associated with them, SPNA cannot cope with these distributions directly. However, consecutive sets of variables (eg. minimum, maximum and mean) can provide meaningful point comparisons.

## EQUIPMENT AND STRUCTURES

Most of the systems chosen for SPNA analysis were confinement housing units shown in plans published by the Canada Plan Service (CPS). Costs of equipment were solicited from farm machinery distributors in eastern Canada. Structure costs were calculated according to rates supplied in local construction cost manuals. Cost of use of equipment and structures was calculated in accordance with the ASAE Yearbook (1974)*. Field studies were carried out to determine the time requirements for the various manure handling activities.

The dairy plans include tie stall arrangements for 40-80 milking cows with alternative solid, semi-solid, and liquid manure handling facilities. Plans for free stall barns cover a range of herd sizes from 60 to 200 milking cows. Warm barns can use slotted floors or mechanical scraper in conjunction with liquid or semi-solid manure handling. Cold free stall barns require paved alleys and regular tractor scraping in conjunction with alternative liquid and semi-solid handling facilities. In addition, there are 8 CPS manure storage plans for long-term bulk storage of liquid, solid

---


The authors are: J. R. OGILVIE, Agricultural Engineering Dept., McGill University, Ste. Anne de Bellevue, Quebec, Canada; P. A. PHILLIPS and K. W. LIEVERS, Engineering Research Service, Research Branch, Agriculture Canada, Ottawa, Ontario, Canada.


*Fixed cost includes depreciation, interest, taxes, housing and insurance. Variable cost includes repair, maintenance, energy and labor cost.

or semi-solid manure. The complete dairy manure handling network for SPNA consists of over 100 elements such as gutter cleaners and their fixed and variable costs. A much simplified network showing 42 of these elements and the possible alternate systems is shown in Fig. 1.

For dairy operations, the cropping program was assumed to be primarily corn, hay and pasture. It was further assumed that the land cropped would equal 3 acres per cow milked with some additional feed being purchased. Manure was assumed to be spread on 1/3 of the land each year. Only manure handling costs for the milking cows in the herd were considered in this analysis. Costs for handling

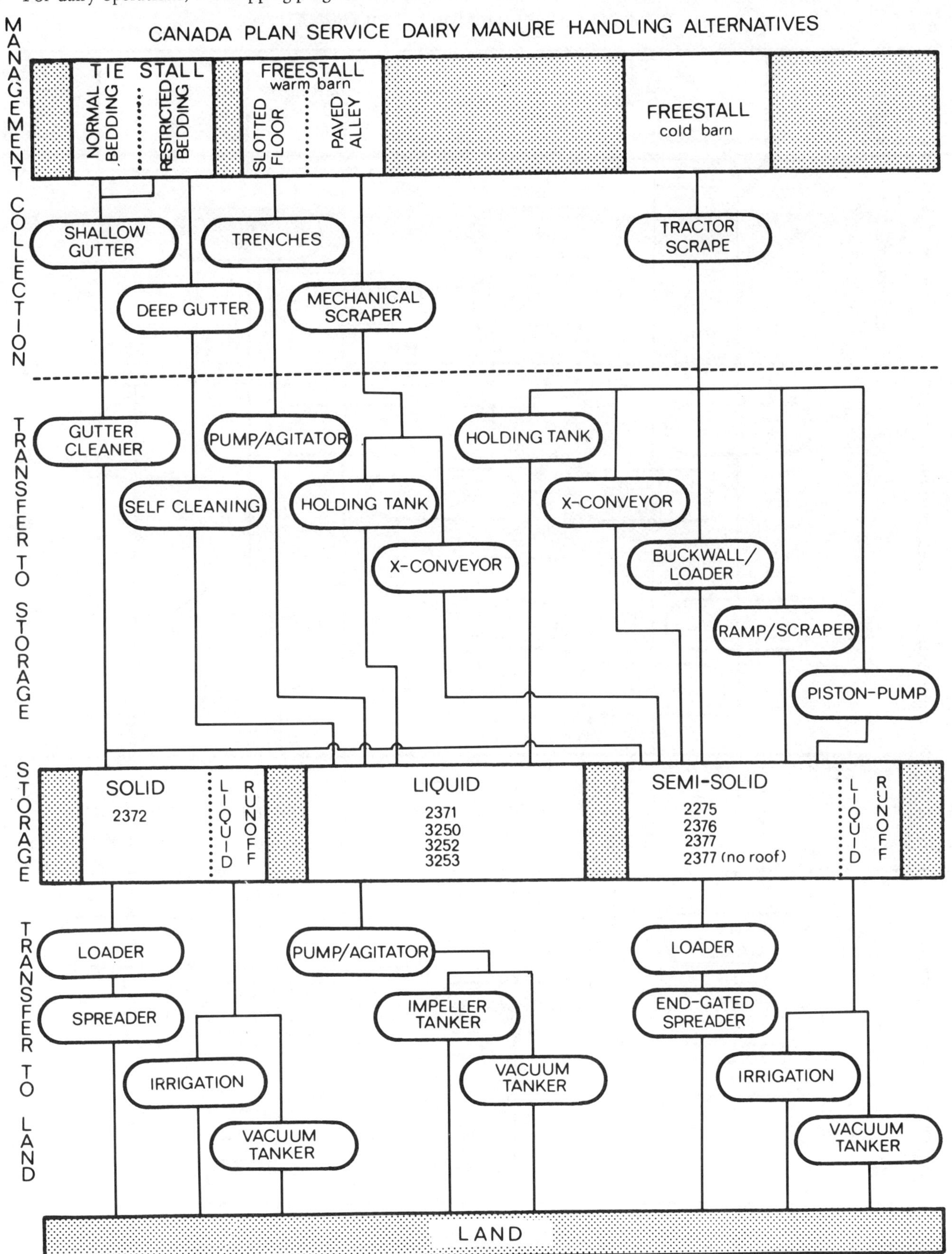


FIG. 1 Simplified network for SPNA dairy analysis.

manure do not include credit for nutrients recovered by the growing crops.

For swine operations, only capital costs were calculated. The herd sizes used (200 to 1000 head) represent commercial size units. Growing-finishing buildings were considered for this example. The variations from a standard gutter cleaner floor were the only costs charged against the system for building construction. A simplified outline of the system alternatives for swine is shown in Fig. 2.

The essential function performed by the SPNA program is the rapid summation and tabulation of cost functions based on given variables. The user is free to change or up-

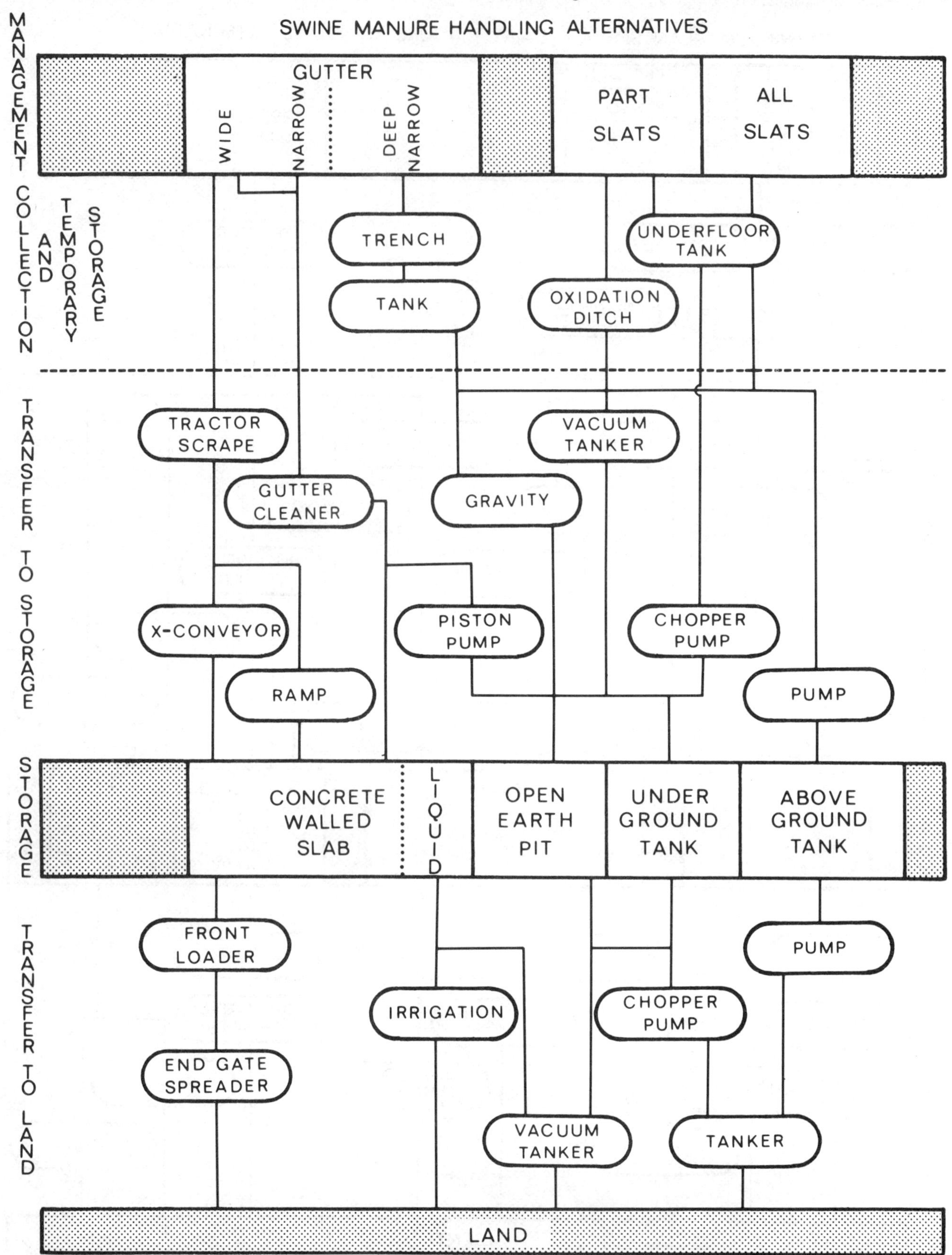


FIG. 2 Simplified network for SPNA swine analysis.

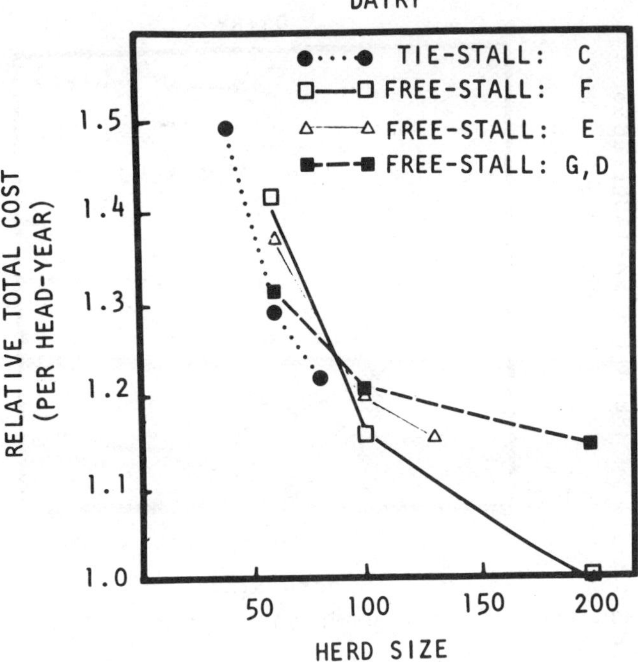

FIG. 3 Some alternate dairy manure systems.

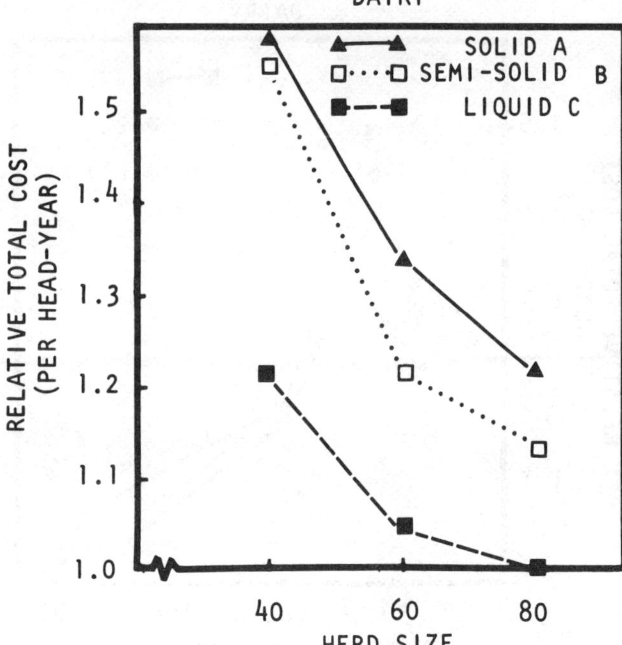

FIG. 4 Alternate tie stall manure systems.

date variables (equipment costs, labor requirements, depreciation rates, repair rates, etc.) at any time. In addition the program allows incrementation of one variable (eg. herd size) to evaluate sensitivity to that variable.

## RESULTS AND DISCUSSION

The analysis of dairy manure handling focused on the relationships:

1   Between different housing systems† for different herd sizes.

2   Within a specific housing system for different herd sizes.

3   Between liquid, semi-solid, and solid handling systems and the effect of storage time.

4   The effect of size of field spreading unit and herd size.

In analysing alternative manure handling systems it is important to remember that in practice each manure handling problem will be unique and no one set of answers can solve all problems. In addition, costs of farm equipment and farm structures vary widely throughout the country. The use of systems analysis techniques therefore requires both good data and uniform methodology to provide meaningful relationships which can help the farmer maximize the use of his dollar.

In Fig. 3 the essential trend is a noticeable decrease in relative total cost‡ as capital costs become defrayed over a larger number of head. Differences which can be attributed to type of housing reflect primarily the management of bedding and choice of manure handling facilities. Variable costs of the warm barn liquid systems range from 50 to 70 percent of the variable costs of other systems over all herd sizes. Higher fixed costs for the liquid systems, however, make total costs less favorable at larger herd sizes.

Within the alternate tie stall housing systems the effects of scale are important (Fig. 4). The high cost of solid system A can be attributed mainly to the high bedding needs. The lower overall total cost of the liquid system C is brought about chiefly by the approximate 50 percent lower variable costs of this system. For the farmer who may plan later herd expansion, or who would prefer to substitute labor and other variable costs for capital, the semi-solid system B offered the lower long term fixed costs at herd sizes from 60 to 80 head.

The variation in slopes in Fig. 5 reflects the degree of sensitivity in cost to additional days of storage for selected systems at different moisture contents. The low response slope for solid system A may be attributed to the efficient

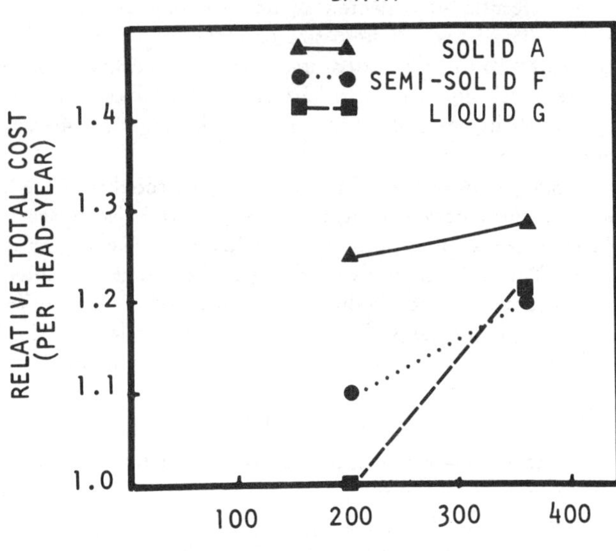

FIG. 5 Effect of manure storage period on selected solid, semi-solid and liquid systems (60 cow herd).

---

†Based upon selected CPS dairy housing plans and manure handling systems: (A) Tie Stall/Normal Bedding/Shallow Gutter/Gutter Cleaner/CPS2372, (B) Tie Stall/Restricted Bedding/Shallow Gutter/Gutter Cleaner/CPS2275, (C) Tie Stall/Restricted Bedding/Deep Gutter/Self-Cleaning/CPS3253, (D) Free Stall/Restricted Bedding/Slotted Floor/Alley Trenches/X Trench/CPS3253, (E) Free Stall/Restricted Bedding/Mechanical Scraper/Holding Tank/CPS3253, (F) Free Stall/Restricted Bedding/Tractor Scrape/Piston-Pump/CPS-2275, (G) Free Stall/Restricted Bedding/Tractor Scrape/Holding Tank/CPS3253. Note: systems A to E = warm barn; systems F, G = cold barn.

‡Sum of fixed and variable costs per head per year.

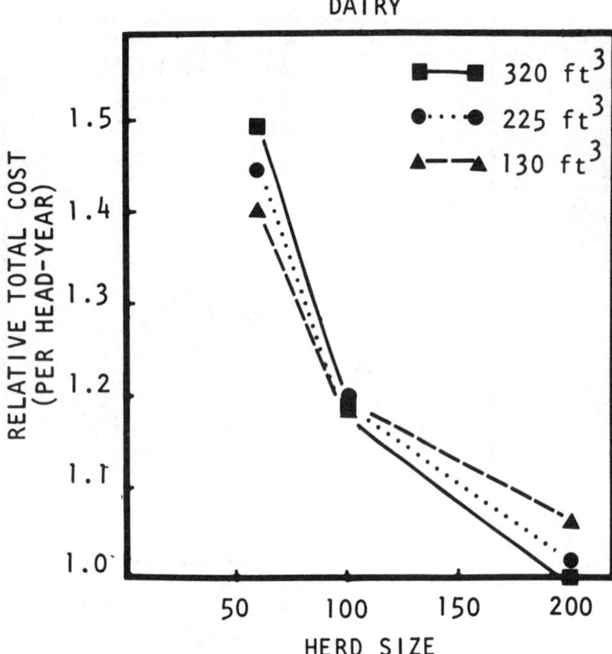



FIG. 6 Effect of spreader capacity on total cost of System F.

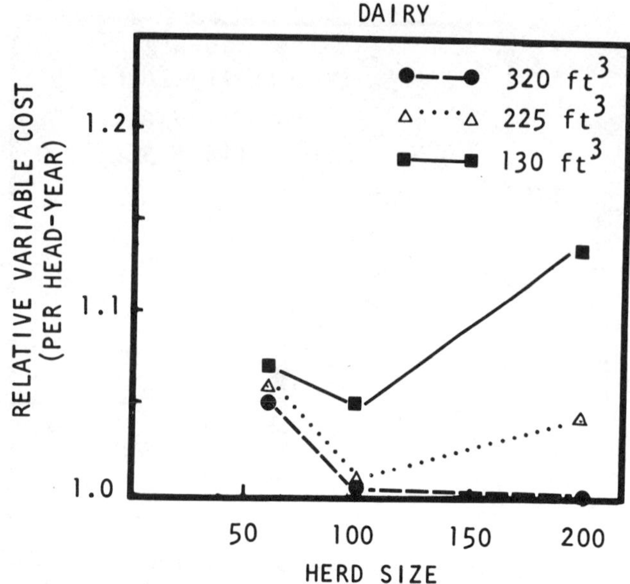



FIG. 7 Effect of spreader capacity on variable cost of System F.

use of enlarged storage facilities while the total system cost is still the most expensive overall. The response of the liquid system G is due to the additional volume required per head to store manure at a higher moisture content. Constant labor costs have been used throughout, so total costs do not attempt to weigh any benefits to the farm operator from improved labor distribution.

The effect of manure spreader capacity on the total and variable costs of the selected system F is shown in Figs. 6 and 7. Differences between spreader sizes are due to a trade-off of the lower variable cost for the larger spreader due to reduced travelling time, versus longer average hauling distance for the larger herd sizes. The differences apparent within spreader sizes in Fig. 7 reflect a general reduction in repair costs per head versus increases in labor and tractor use due to longer trips to the field at larger herd sizes. The overall lower variable cost of the larger spreader may justify its purchase even at smaller herd sizes.

The results for selected swine systems§ are shown in Fig. 8. The effect of herd size or scale has a minor effect compared to the effect of storage type. The costs of storage are relatively insensitive to herd size and mask the effect of spreading the costs of equipment over the larger numbers of animals.

While many farmers may be swayed by the difference in capital costs between swine systems, further analyses of variable costs must be done to evaluate the overall annual costs. This is particularly true when labor costs are added.

For cold climate conditions the major cost is usually for storage and indicates the need to search for lower priced bulk units useable with regular unloading equipment.

## SUMMARY

Network analysis techniques were used to assess alter-natives in manure management systems. The shortest path network analysis (SPNA) was adopted. This modification of CPM and PERT techniques yielded the least cost when the durations of activities were expressed as capital, fixed or variable costs. The objective was to use SPNA to evaluate certain recommended systems to determine least cost to the farmer. The equipment and structures for dairy manure handling comprised 100 elements (gutter cleaners, tractor loaders, etc.) while the swine systems comparised 36 such elements.

SPNA reduced the calculation time and allowed evaluation of many more combinations than would normally be considered by farm advisers. The program allows for regional needs and costs since these are supplied by the user.

In dairy manure systems the most noticeable effect is that of scale or herd size. In tie stall barns, the effect of a change from the solid to the liquid system was just as great.

There was little difference between liquid systems in free

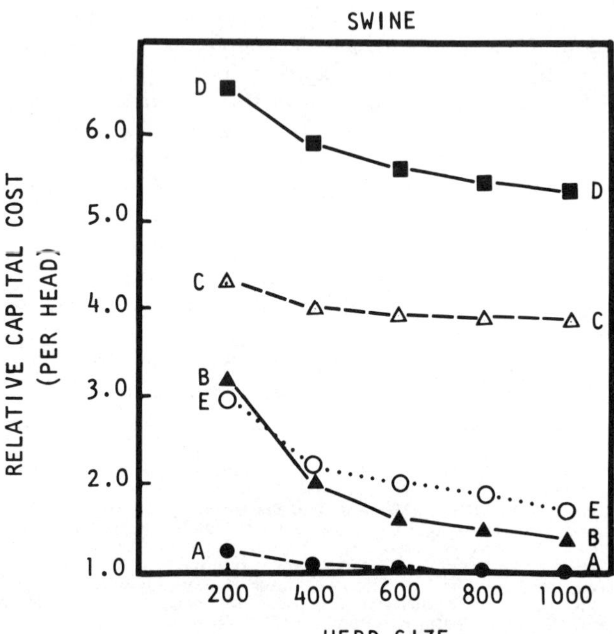



FIG. 8 Effect of herd size on relative capital cost.

---

§ Selected swine manure handling systems: (A) All Slats/Underfloor Tank/Vacuum Tanker/Open Earth Pit, (B) Wide Gutter/Tractor Scrape/Ramp/Conc. Walled Slab/Loader & Irrigation/Spreader, (C) Part Slats/Oxidation Ditch/Vacuum Tanker/Underground Conc. Storage, (D) Deep Narrow Gutter/Trench Storage/Gravity/Underground Tank/Chopper Pump/Tanker, (E) Part Slats/Underfloor Tank/Pump/Silo Storage/Tanker.



MANAGING LIVESTOCK WASTES

or tie stall barns but in total costs, semi-solid systems were favored over liquid for larger than 100 cow herds. Increased manure storage time became less costly in open top semi-solid storages than in liquid units.

The relative capital costs of the selected swine systems was quite wide reflecting the need for built-in concrete structures in some operations.

The proper selection of manure spreader capacity was important. This resulted in savings of 14 percent in variable costs of one complete system.


**References**

1   Coupland, G. A. and R. M. Halyk. 1969. Critical path scheduling of forage harvesting systems in Quebec. Paper No. 69-678, ASAE, St. Joseph, Mich. 49085.

2   Lievers, K. W. 1974. The modified shortest path network analysis (SPNA) program. Report No. 476. Engineering Research Service, Agriculture Canada, Ottawa.

3   Morris, W. H. M. and A. Nygaard. 1964. Application of an optimizing path algorithm in the comparison of farm work methods. Jour. of Farm Econ. 46(2):410-417.

4   Mussivand, T. 1966. The application of methods engineering in irrigation. Unpublished M.Sc. thesis. University of Alberta.

5   Preston, T. A. 1967. A computer programme for the evaluation of alternative methods. Can. Agr. Eng. 9:109-112.


---

## Incorporating Sludges into the Soil

*(Continued from page 445)*

semisolids to the furrow may not be needed with some slurries. The duel conveyor system is designed to unload material at rates ranging from 0 to 60 ft³/min (0.0284 m³/s).

A valve assembly fits either an opening at the center of the front of the tank for SSI or to fill a ridge-and-furrow, or at the outlet end of the cross-conveyor for PFC. The valve has a 12-inch-diameter (0.305 m) opening to be used for wastes with solid contents ranging from 10 percent to 25 percent. A reducer with a 6-inch-diameter (0.152 m) opening for a flexible hose with a quick coupling is used for slurries with solid contents up to 10 percent. The valve opening is regulated by a hydraulic piston attached to the stainless steel blade.

The last column in Table 1 lists the choices of power for running the conveyor systems and operating the blade of the gate valve. The longitudinal and cross conveyors could be driven directly by the tractor PTO except that the PTO shaft would interfere with the SSI or ridge-and-furrow opener which are mounted in the center of the front of the "A frame. A PTO shaft would not interfere with the PFC operation. If the SSI technique was not to be used, an accessory kit to use the PTO of the tractor might be more simple and less expensive for driving the conveyors for the PFC operation than would be hydraulic power. Another alternative would be to use an auxiliary gasoline engine to drive the conveyors if the SSI technique was to be used. Also, the PTO of the tractor might be used to power an auxiliary hydraulic system.

If hydraulic power is to be used for lifting the plow, controlling the gate valve, and running the longitudinal and cross conveyors, the system or systems must: (a) permit independent control of each componenet, including the speed of the conveyors; (b) permit the plow to float when the control valve is in the neutral position; (c) give priority of power to the gate valve over the other 3 components. Also, an independent manual control of the gate valve should be provided in the event the tractor engine or hydraulic system is inoperative.

## ADVANTAGES OF MULTIPURPOSE EQUIPMENT

There would be an economical advantage for the commercial fabricator and ultimately to the public, if basic components of equipment could be mass produced. These basic components such as the tank and chassis could be designed so that different accessories or adaptation kits could be made available for the basic components depending upon the function for which the purchaser needed it.


**References**

1 Bohn, Hinrich L. and R. C. Cauthorn. 1972. Pollution: The problem of misplaced waste. American Scientist 60(5):561-565.

2 Reed, Charles H. 1973. Equipment for incorporating sewage sludge and animal manures into the soil. Proceedings of Conference on Land Disposal of Municipal Effluents and Sludges. USEPA Region II and Rutgers—The State University of New Jersey. Also published in Compost Science 14(6):30-32.

3 Reed, Charles H. 1974. Equipment for incorporating sewage sludges into the soil, Compost Science 15(4):31-32.

4 Stevens, R. Michael, D. J. Elazar, Jeanne Schlesinger, J. F. Lockard and B. A. Stevens. 1972. Green land—clean streams. Center for the Study of Federalism, Philadelphia, Penn.

5 Weil, M., C. H. Reed and A. J. Higgins. 1975. Performance of screw conveyors for unloading sludges from field transports. Paper presented at Third International Symposium on Livestock Wastes. University of Illinois, Urbana, Ill.

6 White, R. K., M. Y. Handy and T. H. Short. 1975. Systems and equipment for disposal of organic wastes on soil. Research Circular 197, Ohio Agricultural Research and Development Center, Wooster, Ohio.

# Application of the Rotating Flighted Cylinder to Livestock Waste Management

J. Ronald Miner, Wayne E. Verley

MEMBER
ASAE

ASSOC. MEMBER
ASAE

SEDIMENTATION is a well established process for separating manure solids from transport liquid. Sedimentation may be accomplished in a variety of physical facilities ranging from simple settling basins to continuous flow centrifuges. Verley and Miner (1974) described a concept for solid-liquid separation involving the flow of manure slurries through an inclined tube fitted with a helically-wound fin attached to the interior surface. As a dilute slurry flows down the open space of the tube, settleable solids are trapped between wraps of the fin. If the tube is slowly rotated, the solids are carried toward the upper end and finally discharged as slurry. The solids concentration of the upper end effluent is a function of the design of the upper wraps of the fin, the tube rotational speed, and the solids content of the influent slurry. The tube in operating position is shown in Fig. 1.

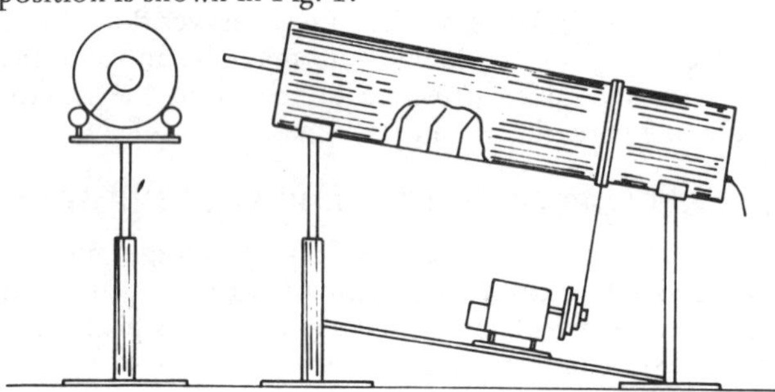

FIG. 1 Schematic diagram of the rotating flighted cylinder solid-liquid separator.

The first tube constructed to demonstrate the performance of a solid-liquid separator of this design was 20 cm in diameter and 145 cm long. Holes were drilled in the upper fins to reduce the amount of liquid discharged from the upper end. This design proved effective in removing beet pulp solids (Verley and Miner 1974) at a flow rate of 2 l/min. This experience demonstrated the potential of the concept to concentrate solids in a manure slurry but the flow rate was too low to allow feeding representative manure slurry samples.

To evaluate the performance of a rotating flighted cylinder for the concentration of manure slurries, a 61 cm diameter model was fabricated of sheet steel. Dimensions of this model are given in Table 1. The height of the fin was reduced (5 cm per wrap from the sixth wrap from the upper end to the upper end) to decrease the amount of water

The authors are J. RONALD MINER, Associate Professor, Agricultural Engineering Department, Oregon State University; and WAYNE E. VERLEY, Sanitary Engineer, Kansas State Department of Health and Environment, Topeka, Kansas, formerly Graduate Research Assistant, Oregon State University. Technical Paper No. 3952. Oregon Agricultural Experiment Station.

Acknowledgment: In addition to the authors, John O'Brien, Kris Marvell, Nora Webb, Ronald Southard, and Tai Hak Chung contributed to collection of the data reported in this paper. Financial support was provided in part by the Oregon Agricultural Engineering Research Foundation, and the Water Resources Research Institute of Oregon State University.

discharged from the upper end and thereby increase the solids concentration. By selection of this rate of decrease and the mounting angle of 17.5 degrees, water overtopping the upper wraps of the fin flowed toward the lower end of the device.

TABLE 1. DIMENSIONS OF THE 61 cm METAL FLIGHTED CYLINDER FOR SOLIDS REMOVAL FROM DILUTE MANURE SLURRIES AND THE 20.3 cm PLASTIC FLIGHTED CYLINDER FOR SOLID-LIQUID SEPARATION AND AEROBIC BIOLOGICAL WASTE TREATMENT

| Parameter | Metal cylinder | Plastic cylinder |
|---|---|---|
| Tube diameter, cm | 61 | 20.3 |
| Tube length, cm | 153 | 122 |
| Fin height, cm | 15.3 | 6.3 |
| Fin spacing, cm | 10.2 | 2.5 |
| Mounting angle of cylinder to horizontal, degrees | 17.5 | 19 |
| Tube rotational speed, rpm | 0.44 | 1.0 |
| Tube peripheral speed, cm/min | 85 | 64 |

## PERFORMANCE — DAIRY MANURE SLURRY SOLIDS CONCENTRATION

Two problems encountered immediately in evaluating the device for removing solids from a dairy manure slurry were: (a) a slurry feeding system, and (b) a sampling technique which adequately reflected its performance. An air lift pumping system was devised which had the capability of providing flow rates of 2 to 60 l/min (Verley and Miner 1974) from a barrel which was continuously receiving and discharging a high flow rate of a well-agitated manure slurry. An alternative that also proved satisfactory was a gasoline engine driven variable speed diaphragm pump. The sampling problem was resolved by the fabrication of a funnel strainer as shown in Fig. 2, which was used to collect

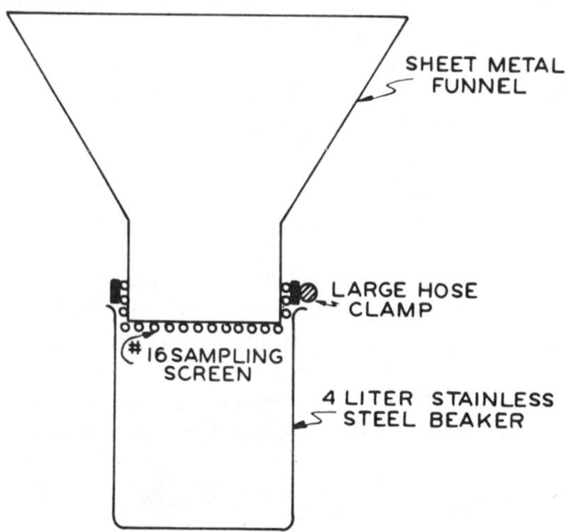


FIG. 2 Cross section of the solids sampling funnel used to evaluate the performance of the 61 cm flighted cylinder for removing solids from dairy and swine manure slurries.

the flow for a specified time. The volume collected was measured, the solids retained on the screen (1.19 mm openings) weighed, and the moisture content determined. From these data it was possible to calculate the retained solids content of the influent and two effluent streams.

When a dilute (0.05 to 1.2 percent retained solids) dairy manure slurry from flushing a free stall dairy barn was passed through the rotating cylinder, retained solids removals ranged from 20 to 80 percent, depending primarily upon the flow rate. These data are summarized in Table 2. Most of the discharged solids greater than 1.19 mm were wood chips, plant stems, and other low density materials not subject to removal by sedimentation. During the testing of this device, power consumption was measured and found to be 0.136 hp, independent of the flow rate. This power consumption does not include lifting the slurry from the storage reservoir to the separator.

## PERFORMANCE — SWINE MANURE SLURRY SOLIDS CONCENTRATION

Utilizing the same separator and air lift pump as previously described, a series of trials were conducted to determine the performance of the unit in removing solids from a dilute swine manure slurry. A gasoline-powered diaphragm pump lifted the slurry from a pit beneath a partially slatted feeding floor to the barrel holding the air lift pump. The swine were fed a finely ground complete ration pelleted for storage and handling ease. There was no floating material present in the slurry. Flow rates to the separator ranged from 17.8 to 26.5 l/min. In every trial, the device removed all the solids retained by a screen of 1.19 mm openings. Concentration factors as defined in Table 2 ranged from 20 to 112, with a median of 62. Solids stream retained solids were as high as 4.3 percent when the feed slurry was 0.04 percent retained solids.

**TABLE 2. SUMMARY OF PERFORMANCE DATA FROM TESTS OF THE 61 cm DIAMETER ROTATING FLIGHTED CYLINDER WITH DILUTE DAIRY MANURE SLURRIES**

| Flow rate range, l/min | Average retained solids removal, percent* | Average concentration factor† |
|---|---|---|
| 4 - 12 | 54.1 | 11.7 |
| 12 - 16 | 46.4 | 12.3 |
| 16 - 20 | 34.6 | 13.9 |
| 20 - 24 | 28.6 | 14.0 |
| 24 - 28 | 10.5 | 7.5 |
| 28 - 36 | 14.3 | 10.2 |
| 36 - 50 | 13.1 | 12.5 |

* Retained solids are defined as those which will not pass through a screen with 1.19 mm openings.

† Concentration factor is the upper retained solids concentration divided by the influent retained solids concentration.

## BIOLOGICAL WASTE TREATMENT

Biological waste treatment systems in which a solid surface is alternately exposed to the atmosphere and to a soluble BOD laden water are commonplace in wastewater treatment technology. Trickling filters, contact beds, and sand filters represent prime examples. Rotating disk processes, which function similarly for the removal of soluble BOD, have been applied to municipal sewage (Autotrol Corporation 1971) and to partially treated livestock wastes (Miner et al. 1973). In the rotating disk process, parallel circular disks mounted on a horizontal shaft are slowly rotated one to five revolutions per minute. The shaft is mounted just above the water surface of a tank so the disk surfaces are alternately exposed to the atmosphere and submerged in the wastewater. A biological growth, similar to that on trickling filter stones, develops on the disk, and feeds upon the dissolved waste materials. As the growth becomes sufficiently heavy, portions are shed, which can be removed from the water by traditional sedimentation processes. BOD removals of up to 90 percent were reported by Autotrol Corporation (1971) for a pilot plant treating municipal sewage. The major advantage of this process over activated sludge is a lower power consumption and less complex mechanical equipment.

The rotating flighted cylinder is envisioned as having the potential to operate in a manner biologically similar to a rotating disk device, but with the additional advantage that a separate sedimentation device would not be required. Solids sloughed from the fin would be worked upward in the tube and returned to the wet-well from which the waste was pumped. Since the device also has solid-liquid separation capabilities, the traditional functions of primary and secondary wastewater treatment might be accomplished in a single unit. If successful, this device would have application in the treatment of domestic wastes from small systems as well as livestock wastes.

To evaluate the waste treatment capabilities of this concept, a 20 cm diameter PVC tube 122 cm long was fitted with a fiberglass fin to allow biological waste treatment studies without metal toxicity problems. The dimensions of this tube are given in Table 1. Holes, 0.5 cm in diameter, were drilled in the upper five wraps of the fin to decrease the quantity of liquid discharged from the upper end.

### OXYGEN TRANSFER CAPACITY

The oxygen transfer capacity of the 20 cm plastic tube was determined as a function of liquid flow rate using the system pictured in Fig. 3 according to the procedure outlined by Eckenfelder and O'Connor (1961). Tap water was deoxygenated by the addition of a sodium sulfite solution with a cobalt catalyst. Various flow rates were achieved using an air lift pump. Oxygen transfer capacity data were

FIG. 3 Rotating flighted cylinder in operating position for oxygen transfer and biological waste treatment studies.

converted to the common basis of 20 C and zero dissolved oxygen. The results are shown graphically in Fig. 4.

## DAIRY MANURE SLURRY TREATMENT

The 20 cm PVC tube previously used for the oxygen transfer studies was used for the treatment of liquid dairy manure at two loading rates. This treatment is shown in Fig. 3. The barrel was used as a storage tank. Effluent was removed from it once a day; feed slurry was added to it after effluent removal. The air lift pump continuously transferred barrel contents from near the bottom of the barrel to the rotating flighted cylinder at a point about 30 cm from the upper end. Both upper and lower effluent streams were discharged to the barrel. This arrangement of equipment is similar to that which would result if a rotating flighted cylinder were placed over or adjacent to an existing manure storage tank receiving flushes of liquid manure on a daily basis.

Samples for these studies were obtained by collecting a supply of fresh dairy manure from the OSU Dairy Barn in a large metal container, mixing to assure homogeneity, and packaging in 100 and 200 g units in plastic freezer bags. The manure samples were stored in a freezer until the day of use.

The treatment system was fed and sampled daily Monday through Friday. Samples of effluent and feed were analyzed immediately using procedures from Standard Methods (APHA 1971), facilitated with supplies from Hach Chemical Company where appropriate.

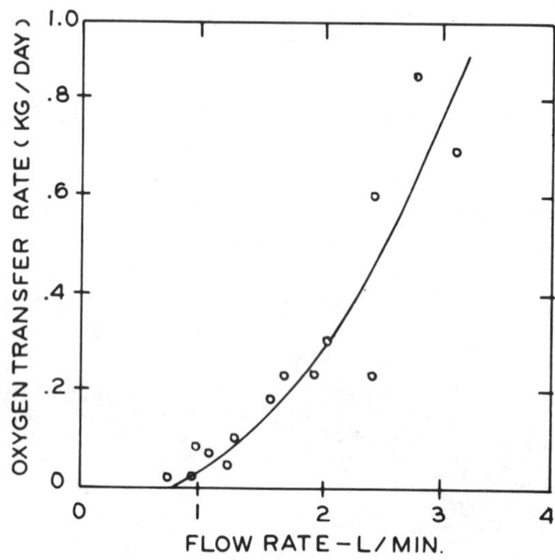

FIG. 4 Oxygen transfer as a function of the flow rate of clean water at 20 C for the 20.3 cm fiberglass separator tube.

### Five Hundred Gram Per Day Studies

In the first dairy manure study, 500 g of manure were fed and 20 l of effluent removed daily. Immediately after removing 20 l of effluent from the top of the barrel, four 100 g manure packets were added. A fifth manure packet was mixed with 4 l of water to provide the feed sample for analyses. After the necessary samples were obtained, the residual was added to the barrel and the liquid volume returned to 175 l by the addition of water.

During this trial, it was evident that the system was operating satisfactorily by the lack of odors and the generally desirable-appearing effluent. The data for this study are summarized in Table 3. At the end of this three week trial, it was concluded that the full capability of the device was not being utilized. A portion of the removal that was being achieved was due to solids settling and accumulating in the barrel.

TABLE 3. SUMMARY OF OPERATING DATA AND PERFOR-MANCE OF THE 20 cm ROTATING FLIGHTED CYLINDER AS A BIOLOGICAL TREATMENT DEVICE WHEN FED 20 l OF DAIRY MANURE SLURRY PER DAY*

| Parameter | Feed† | Effluent† |
|---|---|---|
| Volume, l/day | 20 | 20 |
| COD, mg/l | 10,000 | 400 |
| $BOD_5$, mg/l | 200 | 20 |
| Total solids, mg/l | 1,500 | 600 |
| Volatile solids, mg/l | 1,200 | 300 |
| Dissolved oxygen, mg/l | — | 8 |
| Kjeldahl N, mg/l | 75 | 9 |
| Ammonia N, mg/l | 30 | 4 |

* Flow rate through the tube was 1.2 l/min.

† Data reported represent an average of values determined after a period of adjustment.

### One Thousand Gram Studies

This trial was conducted in a manner similar to the one above, except that 200 g packets were fed and 40 l of effluent were removed daily. In order to prepare a feed sample, a 200 g packet of manure was mixed with 8 l of water. Thus, the system was loaded at twice the previous rate. Results of this trial are summarized in Table 4.

TABLE 4. SUMMARY OF OPERATING DATA AND PERFOR-MANCE OF THE 20 cm ROTATING FLIGHTED CYLINDER AS A BIOLOGICAL WASTE TREATMENT DEVICE WHEN FED 40 l OF DAIRY MANURE SLURRY DAILY*

| Parameter | Concentration† | |
|---|---|---|
| | Feed | Effluent |
| Dissolved oxygen, mg/l | — | 8 |
| $BOD_5$, mg/l | 600 | 50 |
| COD, mg/l | 10,000 | 1,200 |
| Total solids, mg/l | 2,500 | 950 |
| Volatile solids, mg/l | 1,500 | 600 |
| Kjeldahl N, mg/l | 140 | 25 |
| Ammonia N, mg/l | 35 | 5 |

* Flow rate through the tube was 1.5 l/min.

† Data reported represent an average of values determined after a period of adjustment.

Again, the device was able to accept the applied waste load for a four week period, maintain an aerobic environment, and yield an effluent showing 60 to 90 percent pollutant reduction depending upon the constituent of concern. In this trial, the loading rate was near the maximum acceptable if maintaining aerobic conditions in the barrel is essential. Although dissolved oxygen was present in the barrel, concentrations were frequently less than 1.0 mg/l during the study. The thickness of foam on the barrel increased to about 6 cm during this trial but no complications resulted. Heavier solids accumulated in the bottom of the barrel as before.

DOMESTIC SEWAGE STUDY

The experimental equipment described and shown in Fig. 3 was next used for the treatment of raw sewage from the City of Corvallis. Fresh comminuted sewage was used to refill the barrel each day after 120 l of effluent had been removed. The barrel was operated to hold 200 l of liquid. Samples of the feed and effluent were analyzed. The results of those analyses are shown in Table 5.

Under the conditions tested, the device produced an effluent comparable to that from a trickling filter or activated sludge treatment plant. Under these conditions, the device was not loaded to full capacity as indicated by the high dissolved oxygen concentration in the effluent.

**TABLE 5. AVERAGE OF RAW SEWAGE FED AND EFFLUENT ANALYSES WHEN 120 l OF COMMINUTED SEWAGE WAS SLUG FED TO THE ROTATING FLIGHTED CYLINDER DAILY***

| Parameter | Concentration, mg/l | |
|---|---|---|
| | Feed | Effluent |
| Dissolved oxygen | 0 | 5 |
| $BOD_5$ | 260 | 14 |
| COD | 435 | 70 |
| Ammonia N | 10.5 | 2.3 |
| Total solids | 675 | 280 |
| Volatile solids | 385 | 85 |

* Pumping rate through the cylinder was 9 l/min.

CONCLUSION

The rotating flighted cylinder was tested as both a solid-liquid separator and as a biological waste treatment device. As a solid-liquid separator, it was demonstrated to be effective in removing setteable particles from a dilute slurry and concentrating them into a low volume concentrated stream. As a biological waste treatment device, it effectively combined primary and secondary waste treatment into a single unit and produced an effluent comparable to that obtained from conventional secondary sewage treatment devices. The main advantages of this device are its mechanical simplicity, low power consumption, and trouble-free operation.


### References

1  APHA. 1971. Standard methods for the examination of water and wastewater. 13th edition. American Public Health Association. New York.

2  Autotrol Corporation. 1971. Application of rotating disc process to municipal wastewater treatment. U. S. Environmental Protection Agency. 17050 DAM 11/71. 75 pp.

3  Eckenfelder, W. W., and D. J. O'Connor. 1961. Biological waste treatment. Pergamon Press. New York. 209 pp.

4  Miner, J. R., T. E. Hazen, R. J. Smith, and G. B. Parker. 1973. Demonstration of three recirculating swine waste management systems. U. S. Environmental Protection Agency. EPA-660/2-74-009. 148 pp.

5  Verley, W. E., and J. R. Miner. 1974. A rotating flighted cylinder to separate manure solids from water. TRANSACTIONS of the ASAE 17(3):518-520, 525.

# Settling Characteristics of Swine Manure as Related to Digester Loading


J. R. Fischer, D. M. Sievers, C. D. Fulhage

ASSOC. MEMBER    ASSOC. MEMBER    ASSOC. MEMBER
ASAE            ASAE            ASAE


LIVESTOCK wastes are commonly removed from confinement facilities by flushing the facilities with water. Flushing is a convenient method of transporting manure to an anaerobic digester where the manure solids may be used to generate methane. However, two problems are encountered when this method is used: (a) the digester volume must be larger, and therefore more expensive, if all of the flush water is added to the digester; and (b) the manure slurry is diluted below the solids concentration recommended for good digestion and gas production. Preferably, the manure solids should be concentrated after flushing. Gravity settling is a low cost and easily automated means of concentrating the solids.

Previous work on settling has dealt mainly with settling of total solids. Settling rates of swine manure (Jett et al. 1973) were evaluated for depths (0 to 6 feet), solids concentrations (0.5 to 2 percent), and time (0 to 60 minutes). Moore et al. (1973) evaluated the settling of swine manure from the liquid surface over a 1000-minute interval. Settling, with respect to depths, was not reported. Removal of constituents other than total solids, such as volatile solids, nitrogen, and phosphorus, is important when considering anaerobic digestion. Reductions in volatile solids would decrease the gas production. Also, wasting the liquid effluent after settling could result in nutrient deficiencies in the digester. Also, the partial removal of nitrogen, phosphorus, and potassium in the clarified effluent reduces the fertilizer value of the digester effluent. In this paper, we evaluate the effect of gravity settling of swine manure on the organic and nutrient loading of an anaerobic digester.

## EXPERIMENTAL AND ANALYTICAL PROCEDURES

### Experimental Procedures

The settling chamber was a 14.61 cm (5.75-inch) ID plastic column with a 1/2-inch outlet at 0, 2, 4 and 6 ft from the top. A 3.81 cm (1 1/2-inch) hose was attached to the bottom of the settling chamber and connected to the outlet side of the sludge pump. The suction hose of the sludge pump was placed in the top of the settling chamber. Manure used for the settling tests came from several groups of fattening hogs that were fed a 14 percent corn-soybean finishing ration. No attempt was made to use manure from the same group of hogs for each test.


Approved by the Director as a contribution from the Missouri Agricultural Experiment Station (Journal Series Number 7213).

The authors are: J.R. FISCHER, Agricultural Engineer, ARS, USDA, D.M. SIEVERS and C.D. FULHAGE, Assistant Professors, Agricultural Engineering Dept., University of Missouri, Columbia.


The procedure for a given test was as follows: The test weights of manure and water were placed into the settling chamber to obtain the desired solids concentration, and the sludge pump was turned on to mix the slurry. Solids concentrations of 0.05, 0.5, and 5.0 percent were selected for evaluation. However, the actual values of solids tested were 0.04, 0.42, and 4.5 percent. The total dissolved solids in the tap water was determined and the mean value was subtracted from the total solids value obtained for each replication. When the slurry was thoroughly mixed, a sample was taken while the pump was still running. Then the pump was turned off and timing of the experiment began. Samples were collected from each of the four outlets at 1, 10, 100, and 1000 minutes of settling time. The samples were analyzed for total solids (T.S.), volatile solids (V.S.), ammonia, total Kjeldahl nitrogen (TKN), phosphate ($PO_4$), potassium, and COD. The experiment (replicated three times) was a 4 x 4 x 3 split-split plot factorial with depth and time as the subplots.

### Analytical Procedure

The three different types of settling analyses performed on the data were mass balance, type II settling, and zone settling. In type II settling, the particles coalesce, increasing in size and in velocity of settling. In zone settling, the concentration of solids is such that the mass of solids settle as a unit and displace the fluid upwards. The zone settling analysis was performed only with the 5.0 percent solids concentration.

The percent removal of each of the constituents was obtained by the mass balance analysis. In this analysis, the quantity of each constituent (mg) removed was determined by the change in density (from start to time of withdrawal) multiplied by the increment of volume that was represented by each withdrawal port (as measured by one half the distance from the port in question to one half the distance to the next port above it).

Total milligrams removed at time T was calculated by summing the milligrams removed at each depth for a given time. Percent removed was calculated based on the sum of solids for each settling time divided by the total milligrams of a constituent in the chamber at time zero. The percent removed was averaged over the three replications for each time and depth. The standard deviations for these data were obtained.

The percent removal (total solids only) was obtained by type II settling analysis. Percent removed from type II settling analysis was obtained by the method of Clark and Viessman (1969, pages 280-282). The zone settling analysis was performed as outlined in the same reference, pages 282-286. The settling analyses were performed on organic nitrogen, which was obtained by subtracting ammonia from TKN.

## RESULTS AND DISCUSSION

Mean values for each of the constituents for the three solids concentrations are listed in Table 1. Analysis of the first replication showed that potassium concentration did not change with time and/or depth in the settling chamber because the potassium was dissolved. Therefore, data were not collected for potassium in the last two replications.

TABLE 1. MEAN CONCENTRATIONS (MG/LITER) OF EACH OF THE FIVE CONSTITUENTS FOR THE THREE SOLIDS CONCENTRATIONS.

| | Solids concentration, percent | | |
| --- | --- | --- | --- |
| Constituent | 0.04 | 0.42 | 4.5 |
| T.S | 372.0 | 4 254 | 44 583 |
| V.S. | 327.0 | 3 654 | 37 687 |
| Org. N. | 16.9 | 178 | 1 518 |
| PO$_4$ | 8.7 | 86 | 975 |
| COD | 666.0 | 6 293 | 49 286 |

### Settling Data

The 0.04 and 0.42 percent solids concentrations demonstrated type II settling. The change in slope of the curve for the 4.5 percent solids concentration demonstrated zone settling (Fig. 1). The formation of this mass of solids is shown in Fig. 1 as the compaction point. Zone settling analysis can predict the time and depth at which a given solids content will occur, but it cannot predict total solids removal.

The mass balance and type II settling analysis predicted approximately the same total solids removal for the 0.04 and 0.42 percent solids concentrations (Fig. 2). Thus, the mass balance analysis was used in all calculations. Percent removal was higher with a 0.42 percent solids than with 4.5 percent solids for a settling time less than 300 minutes. This is because 0.42 percent solids follows the type II settling and the 4.5 percent solids follows zone settling. Removal of total solids increases as the solids concentration increases from 0.04 to 0.42 percent. When compared with Jett's data (Fig. 3), it seems that settling efficiency increases to about a 1 percent solids concentration and then begins to decrease as solids concentration increases further. This indicates a possible change from type II settling to zone settling between 1 and 2 percent solids concentration.

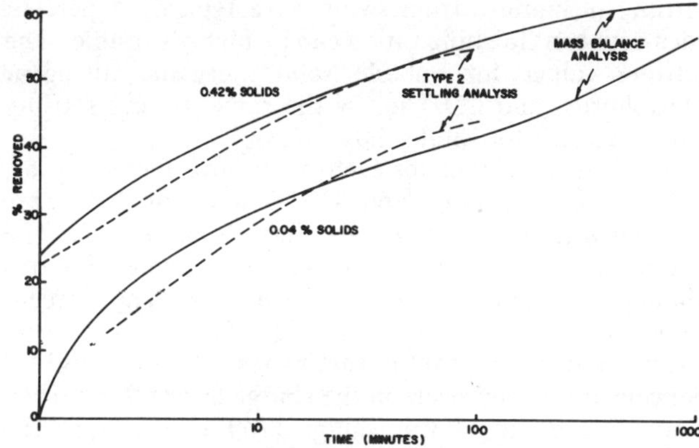


FIG. 2 Percent removal versus time comparing results from mass balance and Type II settling analysis.

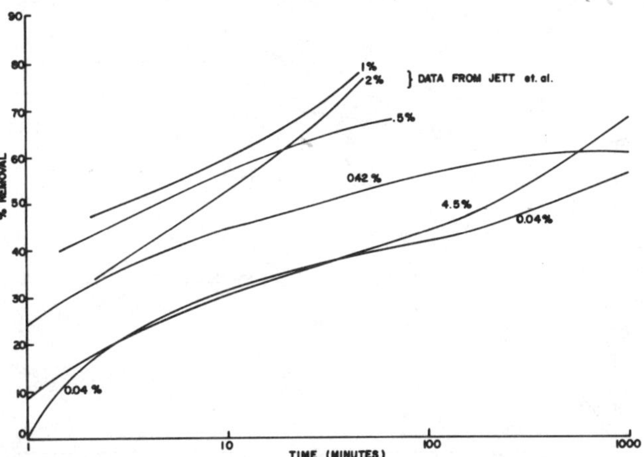


FIG. 3 Comparison of percent removal versus time for solids concentration including data of Jett.

Jett obtained a higher percent removal for 0.5 percent solids than we obtained for 0.42 percent. Two possible reasons for this discrepancy are (a) the difference in rations fed could produce differences in manure, or (b) Jett's data represents one observation, whereas our experiment was replicated three times. One replication of this experiment yielded approximately the same percent removal as Jett obtained. The large standard deviations for removal for the 0.42 percent solids concentration presented in Fig. 4 are typical of those for

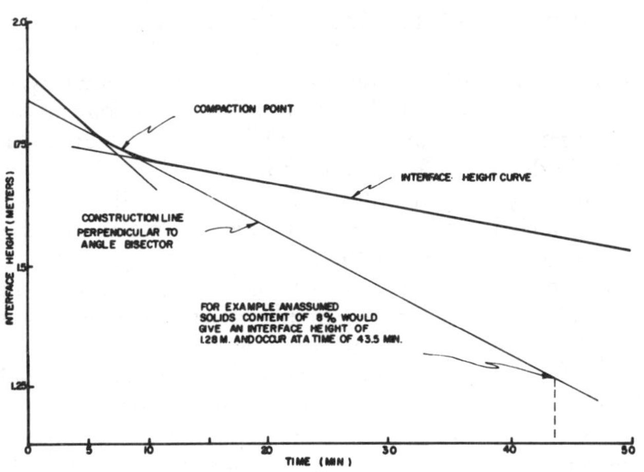


FIG. 1 Interface height versus time used for zone settling analysis for 4.5 percent solids concentration.

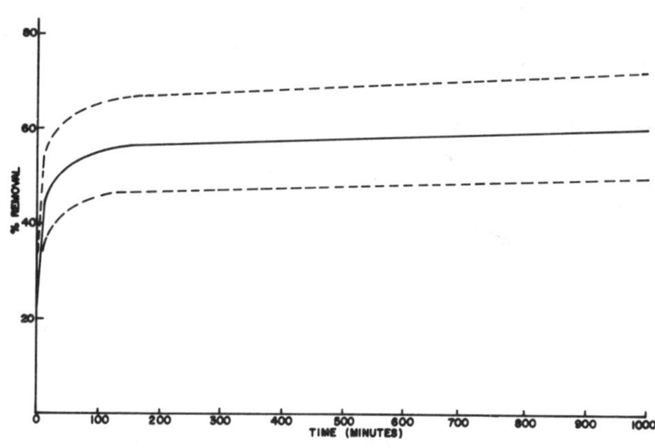


FIG. 4 Percent removal of total solids versus time for 0.42 percent solids—dotted lines are standard deviations.

the other two solids concentration and indicate that the settling of manure from swine on a typical 14 percent corn-soybean finishing ration can be highly variable. The settling values for volatile solids, organic nitrogen, phosphorus, and COD follow the same general settling curves as that for total solids.

The settling curves for each of the five factors (T.S., V.S., Org. N., $PO_4$, and COD) and for the three concentrations are shown in Figs. 5, 6 and 7. The standard deviations shown in Fig. 4 indicate that one should be cautious in using these data. The percent removal values presented in Figs. 5, 6 and 7 were calculated from the percentages of the original concentrations that were in the sludge layer after a given time. The depth of this sludge layer increases as the solids content of the slurry increases. The data showed the remaining solids above the sludge layer are distributed about equally throughout this liquid layer.

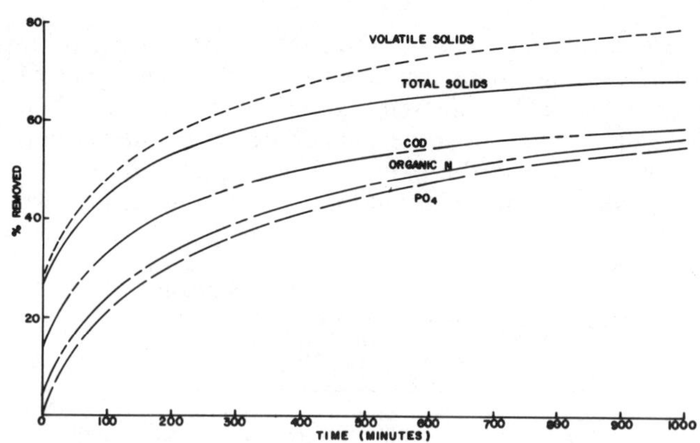

FIG. 7 Percent removed versus time for each parameter for 4.5 percent solids.

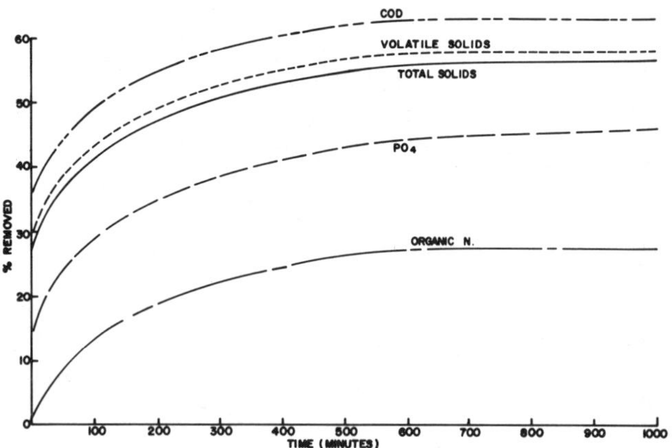

FIG. 5 Percent removed versus time of each characteristic for 0.04 percent solids.

### Effect of Settling on Digester Loading

Volatile solids is the characteristic most often used to determine digester loading rates. For the 0.04 and 0.42 percent solids, approximately 75 and 91 percent, respectively, of the volatile solids that settled in 1000 minutes settled within the first 100 minutes. For 4.5 percent solids, the comparable value was only 55 percent of the volatile solids. Zone settling is more time dependent, than type II settling, but eventually removes

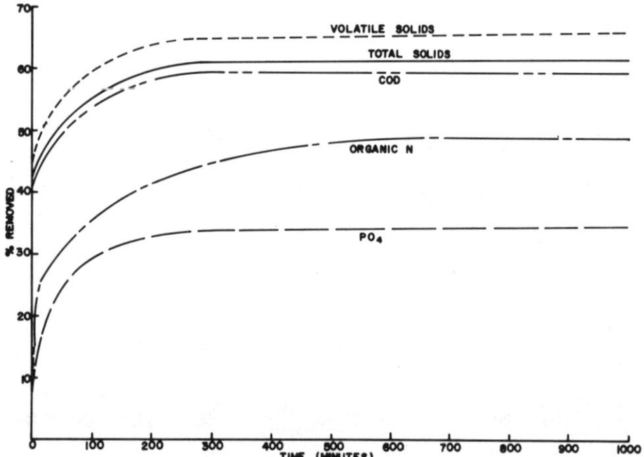

FIG. 6 Percent removed versus time for each characteristic for 0.42 percent solids.

more solids on a mass basis. For the 4.5 percent solids content, approximately 69 percent of the total volatile solids can be expected to settle in 1000 minutes, whereas for 0.04 percent solids content, only 45 percent of the total volatile solids settled in 1000 minutes.

When a settling basin is used with an anaerobic digester, only 45 to 70 percent of the volatile solids will settle in the sludge. If most of the liquid above the sludge is discarded, the digester will be loaded about one-half as much as if the whole volume of manure is added. Thus, gas production from the digester can be reduced by equivalent percentages.

The settling of nitrogen ranged from 27 to 53 percent, as shown in Figs. 5, 6 and 7. This could affect digester loading in two ways. First, the bacteria in the digester could become nitrogen limited and therefore decrease gas production. Second, the nitrogen could be limited to the point where it affects the ammonia concentration in the digester, which could weaken the buffering capacity of the digester. The settling of both nitrogen and phosphorus are important because of their importance as fertilizer elements.

The settling of COD indicates that the liquid from the settling basin has a high pollution potential and should be properly disposed of.

Using these data, a settling basin and digester were designed for a swine confinement building housing 320 hogs which is to be flushed on the average of 2.5 times per day with 6.6 m³ (234 ft³) of water per flush. The digester is to be loaded at a rate of 1.91 Kg V.S./m³ (0.12 lb V.S./ ft³) per day and the detention time is to be 15 days. The settling basin is built into the end of the building and its dimensions are 1.7 m (5.5 ft) wide, 0.8 m (2.5 ft) deep and 7.3 m (24 ft) long.

The digester was designed based on the settling characteristics of V.S. The digester volume required is 132.4 m³ (4677 ft³). If the settling basin were not used prior to the digester, two alternatives are possible: (a) add all the flush water to the digester which would require a digester about twice as large, and (b) add one-half of the unsettled flush water to the digester which would reduce the methane production. The volume of digester influent is 8.8 m³ (312 ft³) per day. Amounts and percentages of the constituents introduced into the digester daily are shown in Table 2.

*(Continued on page 462)*

# A Rotating Conical Screen Separator
# for Liquid-Solid Separation of Beef Waste


Ralph Shirley, Allen Butchbaker
ASSOC. MEMBER    MEMBER
ASAE        ASAE


DEWATERING of animal waste is a problem of interest to many in the field of livestock production. Once the waste has been dewatered, there are several handling and treatment options available. The liquid manure fraction may be used for flushing animal waste in production facilities, for irrigation or for utilization in other ways. The solid fraction may be used as a soil conditioner, as bedding for animals (Jacinto 1973). s a supplement in animal feed (day 1974 and Nye 1974), or in other ways to dispose of it.

Solid-liquid separation techniques may be classified into two broad categories: (a) separation due to density difference, and (b) separation due to size restriction. Separation by density difference follows Stoke's Law at dilute concentrations. Sedimentation does not occur at high concentration such as is found in most animal raw waste slurries. Centrifuges have been used for separating solids from animal wastes, however, most operate as batch systems and high energy input is needed.

Many devices for solid-liquid separation of animal waste based upon size restriction have been used. Most successful have been the stationary or vibrating screen separators. Graves (1972) reported good separation with a stationary sloping screen was achieved with dairy manure when waste was diluted 6:1 or more. Fairbanks (1967) reported that vibrating screens produced waste solids resembling wet saw dust from flushing systems for dairy cattle which can be dried and used as bedding. Rapid wearing due to vibration of the screen occurs if the solids contain much grit or sand (Graves 1972). Bartlett (1974) found that for both stationary and vibrating screen separators a need existed for screen deblinding devices, or a means of dislodging particles, to allow release of free water through screen openings and a means of applying pressure beyond gravitational force.

Vacuum filtration has become a common method of dewatering sewage sludge in areas where climate or space does not permit drying beds (Clark 1971). Glerum (1971) reported that vacuum filtration had a low capacity and also had a low power requirement. Backer (1973) also found a low cake and filtrate yield for vacuum filtration of fresh, untreated cattle manure.

Bartlett (1974) examined pressure filtration methods of dewatering dairy manure slurry. Without removal of the fine particles, compression of the slurry resulted in a completely impervious cake over the screen. A perforated shell screw conveyor separator was also tested but it required extremely close clearances between the auger flight and shell in order to prevent complete clogging of the perforated shell. Microscreening can be used at high flows and low solids concentration to remove microorganisms from water (Clark 1971). The process uses a fine stainless screen instead of a cloth of create a pressure head difference to pull liquid through the screen.

Glerum (1971) tested a centrasieve and found it to yield very good results. The centrasieve is a perforated bowl which spins at a high rate of speed with the slurry introduced in the center. The water passes through the perforated bowl and solids work their way to the outside edge where they are slung off. One drawback of the centrasieve is a high-power requirement. Bartlett (1974) examined a perforated cone centrifuge but found it to be extremely sensitive to rotation speed, slurry consistency, and cone finish.

Holmes (1971) showed that the larger particles of pig oxidation ditch mixed liquor (ODML) contained low percentages of crude protein. Their work with beef ODML showed similar results. Similarly, the soluable protein content of screened or pressed solids from dairy manure was lower than the raw manure slurry (Bartlett 1974).

The purpose of this investigation was to develop a dewatering device for beef animal waste slurry that has low energy requirement and is effective in separating large particles of the beef waste consisting of hair or whole or cracked grain from the remaining liquid slurry. The objective was to improve the pumpability of the slurry and to increase the protein content in the slurry for future refeeding investigations. This paper reports on the development of a dewatering device: a rotating conical screen separator.

## DEVELOPMENT OF THE
## ROTATING CONICAL SCREEN SEPARATOR

After reviewing the various methods that have been used for separating solids from the liquid manure, it was decided that the two most effective methods were: (a) size separation by screening and (b) size separation by centrifugal force using a centrifuge. Screens have a tendency to blind or plug. Centrifuges require high energy and most are batch operated. It was theorized that both methods (screening and centrifuging) could be included in one device. Since energy to a centrifuge was high due to the high rotational speed it was also theorized that by increasing the diameter that the speed of roatation could be reduced.

The fundamental relationship for the forces developed from rotational motion is:

$$\Sigma F_n = \frac{W}{g} \frac{v^2}{r} = \frac{W}{g} r\, \omega^2 = M r \omega^2$$

where

$F_n$ = forces normal to the particle in the direction of the acceleration $a_n$, $lb_f$ (newtons)

$W$ = weight of the particle, $lb_f$ (newtons)

$g$ = acceleration of gravity, ft/sec$^2$ (m/sec$^2$)

$v$ = peripheral velocity ($v = r\,\omega$), ft/sec (m/sec)

$r$ = radius, ft(m)

$\omega$ = angular velocity, radians/sec

$M$ = mass, $lb_m$ (kg)

---


The authors are: RALPH SHIRLEY, former Graduate Student and ALLEN BUTCHBAKER, Associate Professor, Agricultural Engineering Dept., Oklahoma State University, Stillwater.

By increasing the radius r, angular velocity $\omega$ can be reduced to achieve the same force acting normal to the particle towards the center of curvature.

## Preliminary Testing

Model I was a flat screen, 22 in. (55.9 cm) in diameter with a 1/8-in. mesh hardware cloth, mounted horizontally on a wheel with four 0.5-in. (1.25 cm) spokes. A pan with a drain was installed below the screen to catch the fluid passing through the screen and a shield was constructed around the screen to stop the solids that spun off the screen. The solids then fell onto a collection pan and the liquid collected and drained away.

The waste used for testing was collected in the oxidation ditch below an open-type sheltered beef-feeding pen with slotted floors located on the Oklahoma State University campus. The slurry was continually mixed in the oxidation ditch by a rotor. The cattle were fed a high-concentrate ration consisting primarily of corn. A considerable amount of spilled feed was in the pit as well as an accumulation of hair from over approximately a one-year period.

When the waste was pumped to the center of the screen of Model I by a variable flow airlift pump, the solids built up to about 3/4-in. thick on the screen. The solids were not slung off until the speed was increased to 200 rpm. At 200 rpm, both the solids and liquids were slung off the screen, and nothing passed through the screen to the catch pan.

Model II consisted of a conical screen with a 45 deg angle below the horizontal in place of the flat screen. This gave the particles an uplift force off the surface of the screen breaking the surface tension between the particle and the liquid. Some of the liquid got through the screen and fell into the pan, but most of the liquid ran down the underside of the screen held by surface tension and was slung off the outer edge by centrifugal force. The solids slid off the screen at speeds as low as 15 rpm for this 22-in. diameter screen measured at the base.

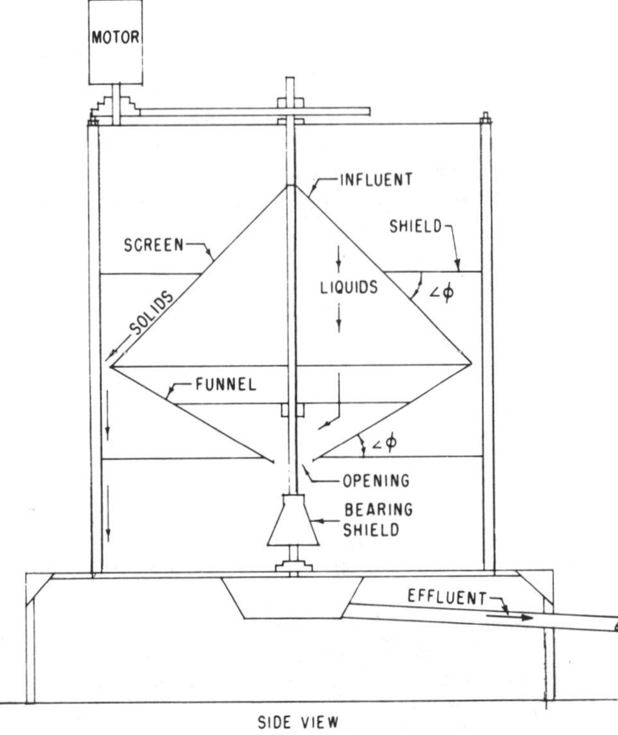



FIG. 1 Schematic of rotating conical screen separator with attached funnel.

## Prototypes

Model III had the same 22-in. (55.9 cm) diameter screen with 1/8-in. mesh hardware cloth as Model II. However, a conical funnel was attached to the underside of the screen (Fig. 1). The liquid then diverted into the funnel which had an opeing at the bottom for directing the liquid into the catch pan. The funnel also eliminated the need for spokes on the screen because the funnel was attached to the shell by a wheel instead of the screen being attached to the shaft. Model III achieved the objectives of the project in effectively separating large solid waste particles and hair from the manure slurry.

Model IV was built geometrically the same as Model III except the size was increased to a 4-ft (121.9 cm) diameter screen to attempt to determine some design criteria for the process (Fig. 2). The shield surrounding the screen was also lowered because of experience from Model III indicated less splash and fewer solids accumulating at the higher levels on the screen than initially anticipated.

FIG. 2 A 4-ft (121.9 cm) diameter rotating conical screen separator.

## RESULTS

For both Models III and IV there were three distinct actions occurring on the screen which are illustrated in Fig. 3. The top third of the screen (Zone 1) was kept free from solids buildup because of the washing effect of the liquid portion of the waste. As the liquid drained through the screen, the transport effect ended and the solids deposited in Zone 2. Here the solids moved slowly down the incline, draining and gently pushed by solids being deposited above. This deposit of solids was about 0.75 in. (1.91 cm) thick. At some point near the edge a combination of gravity and centrifugal forces caused the solids to slough off the screen



MANAGING LIVESTOCK WASTES

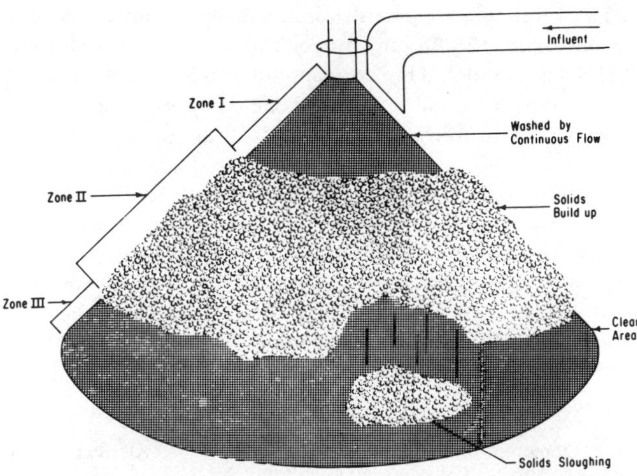



FIG. 3 Action of solid-liquid separation for the rotating conical screen separator.

leaving a clear zone (Zone 3). Some liquid drained over the top of the material in Zone 2 and passed through the screen in Zone 3.

## Model III:

The results of testing the 22-in. (55.88 cm) diameter cone of Model III indicated that the concentration of the screened solids was a function of inflow rate and speed of rotation. The removed solids concentration decreased as influent flow rate increased (Fig. 4). Between 5 and 25 kg/min influent flow rate, the expression relating removed solids concentration versus influent flow rate is

$$C = 100\,000\,e^{-0.054R}$$

where

C = removed solids concentration, mg/l
R = influent flow rate, kg/min

For these tests rotational speed was held constant at 33 rpm and influent concentration was 25 100 mg/l total solids. At an influent flow rate of 60 lb/min (27.3 kg/min) some of the waste splashed off the screen of Model III and into the removed solid area. This was considered an overloading rate of the inflow.

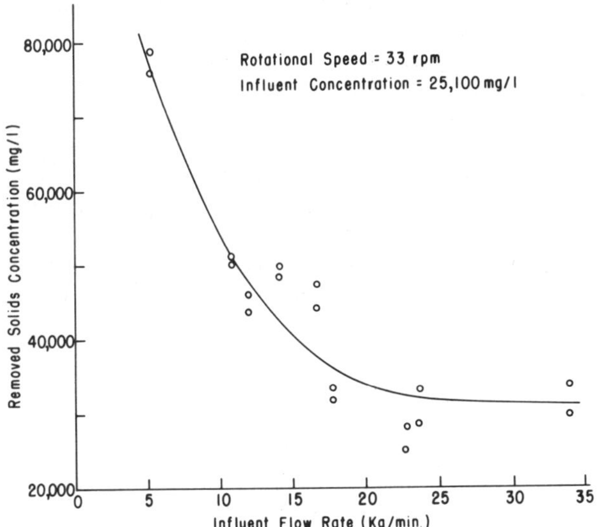



FIG. 4 Removed solids concentration versus influent rate for 22-in. (55.9 cm) diameter rotating conical screen separator.

The effect of rotational speed on Model III is illustrated in Fig. 5. The optimum speed of rotation was 29 rpm. This corresponded to a peripheral speed at the outside diameter of the cone of 167.0 ft/min (50.91 meters per minute). At low rotational speeds (below optimum speed) the solids were not slung from the screen and caused a mat of solids to buildup sealing the screen. At high rotational speeds (above optimum speed) the solids had less time in contact with the screen before being slung off, consequently the removed separated solids had a lower solids concentration. The flow rate for these tests was 11.46 kg/min with an input concentration of 40 960 mg/l.

The removed solids concentration was dependent upon the input waste concentration. Test results showed influent concentrations of 40 960 mg/l and 25 100 mg/l gave a removed solids concentration of 71 000 mg/l and 50 000 mg/l, respectively for the 22-in. diameter screen rotating at 33 rpm with 11.46 kg/min inflow. Thus, as input solids concentration decreases the removed solids concentration increases. This was consistent with the results found by Graves (1972) for stationary screens.

An attempt was made to use a smaller screen size (1/16-in. mesh), but after two hours of operation hair had plugged the screen.

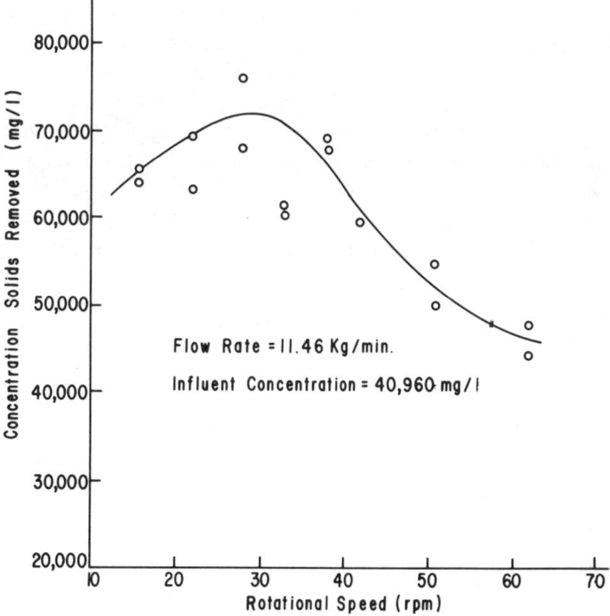



FIG. 5 Concentration of solids removed versus rotational speed for 22-in. (55.9 cm) diameter rotating conical screen separator.

## Model IV:

Model IV was built with a 48-in. (121.9 cm) diameter screen at an angle of 45 deg. It was constructed to check some of the parameters that were considered to be major factors affecting separation of the solids from the liquid slurry. It required a much lower rpm range than the smaller 22-in. diameter model. The shield around the outside was also lowered because of less splashing due to the lower speed of rotation.

The optimum speed of rotation was found to be 13.5 rpm. The action on the screen was similar to Model III with the three clearly defined zones. The peripheral speed at the outside diameter of the cone at optimum separation of 13.5 rpm was 159.9 ft/min (51.71 m/min). The optimum speed of rotation 13.5 rpm was independent of the influent concentration.

**Summary of Results**

Based upon the tests on the rotating conical screen separators, several findings were made:

1　Optimum peripheral speed at the outside diameter of the screen was between 165-170 feet per minute (50.3-51.8 meters/min) for a 45 deg angle screen.

2　Specific influent flow rate (flow per unit area of screen) of 2-5 lb/sec-ft$^2$ (9.8-24.4 kg/sec-m$^2$) gave the best separation efficiency.

3　As peripheral diameter increased the removed solids concentration incarased for the same specific flow rate.

4　As influent flow rate decreased, the removed solids concentration increased.

5　As the influent waste solids concentration decreased, the removed solids concentration increased.

6　Theoretical calculations of the power requirements for rotating the mechanism and slurry, neglecting belt and bearing losses, indicated a maximum power requirement of approximately 0.4 watts.

7　The conical rotating screen was self-cleaning and did not plug after several days of continuous operation.

SUMMARY

A rotating conical screen separator was developed for the separation of larger particles, including spent grain and hair, from liquid manure slurries. The screen was shaped into a cone to give an uplift force to the particles on the screen and thus the screen was self-cleaning. Liquids were collected on the underside of the screen by a funnel and directed toward an effluent collection pan. Two different screen diameters were tested, a 22-in. (55.88 cm) and a 4-ft (121.9 cm) with screen angles of 45 deg below the horizontal.

The total solids concentration was maximum at 29 rpm for the 22-in. (55.88 cm) model and 13.5 rpm for the 4-ft (121.9 cm) model. This corresponded to a peripheral speed at the outside diameter of the cone of approximately 170 ft/min (51.8 m/min). The removed solids concentration increased as input concentration decreased, decreased as influent flow rate increased, and increased as diameter increased for the same specific flow rate. The power requirements were extremely small due to the slow speed of rotation and thus only the power needed to over come the friction losses due to bearings and belt losses need be considered in the design.

**References**

1　Backer, L. F., R. L. Witz, G. L. Pratt and M. L. Buchanan. 1973. Vacuum filtration of cattle manure. ASAE Paper No. 73-4531, ASAE, St. Joseph, Mich. 49085.

2　Bartlett, H. D., R. E. Bos and E. C. Wung. 1974. Dewatering bovine animal manure. TRANSACTIONS of the ASAE 17(5):968-972.

3　Clark, J. W., W. Viessman, Jr. and M. J. Hammer. 1971. Water supply and pollution control. International Textbook Co., Scranton, PA.

4　Day, D. L. and B. G. Harmon. 1974. A recycled feed source from aerobically processed swine wastes. TRANSACTIONS of the ASAE 17(1):82-84.

5　Fairbanks, W. C. and E. L. Bramhall. 1967. Dairy manure liquid-solid separation, University of California Agricultural Extension. Pamphlet No. AXT-27.

6　Glerum, J. C., K. Klomp and H. R. Poelma. 1971. The separation of solid and liquid parts of pig slurry. p. 385-347. International Symposium on Livestock Waste. ASAE, St. Joseph, Mich. 49085.

7　Graves, R. E., J. T. Clayton. 1972. Stationary sloping screen to separate solids from dairy cattle manure slurries. ASAE Paper No. 72-951, ASAE, St. Joseph, Mich. 49085.

8　Holmes, L. W. J., D. L. Day, J. T. Pfeffer. 1971. Concentration of proteinaceous solids from oxidation ditch mixed liquor. p. 351-359. International Symposium on Livestock Waste. ASAE, St. Joseph, Mich. 49085.

9　Jacinto, Frank. 1973. Cows make their bedding. Progressive Farmer, July.

10　Nye, J. C., A. C. Dale, T. Wayne Perry. 1974. Recovering protein from dairy cattle waste. TRANSACTIONS of the ASAE 17(6):1155-1160.

---

Settling of Swine Manure

*(Continued from page 458)*

TABLE 2. THE AMOUNTS AND PERCENTAGES OF THE VARIOUS CONSTITUENTS THAT WILL BE INTRODUCED INTO THE DIGESTER PER DAY.

| Constituent | lb/flush | Percent* |
|---|---|---|
| T.S. | 712.50 | 78 |
| V.S. | 525.00 | 72 |
| N | 18.60 | 55 |
| P | 9.70 | 38 |
| K | 9.60 | 53 |
| COD | 725.00 | 65 |

$$*Percent = \frac{lbs\ added\ to\ digester}{lbs\ removed\ from\ building} (100)$$

SUMMARY

1　Settling removes little potassium, probably because the potassium is dissolves.

2　Swine manure exhibits class II-type settling up to a total solids concentration of about 1 percent. Slurries with higher solids concentrations exhibit zone-type settling.

3　Removal rate of total and volatile solids seems to be highest at a concentration range of 0.5 to 1 percent total solids up to a time of 300 minutes.

4　Most type II settling occurs within the first 100 minutes.

5　Zone settling is more time dependent than is type II settling.

6　The settling of manure from hogs on a typical 14 percent corn-soybean finishing ration is highly variable.

7　Calculations of digester loading based on manure production may lead to over-estimation of gas production if settling is used, because volatile solids are lost in the liquid.

**References**

1　Clark, John W., Warren Viessman, Jr. 1969. Water supply and pollution control. International Textbook Company. Scranton, Pennsylvania. 575 p.

2　Jett, S.C., H.E. Hamilton and I.J. Ross. 1973. Settling characteristics of swine manure. TRANSACTIONS of the ASAE 17(5):965-967.

3　Moore, J.A., R.O. Hegg, D.C. Scholz and Egon Strauman. 1973. Settling solids in animal waste slurries. ASAE Paper No. 73-438. ASAE, St. Joseph, Mich. 49085.

# Evaluation of Solids Separation Devices

J. W. Shutt, R. K. White, E. P. Taiganides, C. R. Mote

ASSOC. MEMBER ASAE    ASSOC. MEMBER ASAE    MEMBER ASAE    ASSOC. MEMBER ASAE

S EPARATION and concentration of the solids fraction from the liquid is frequently necessary, and it may precede or follow another unit operation for the biochemical treatment and/or recycling of livestock wastes. There are several mechanical methods and devices which may be used for this purpose. Devices and methods which are either common or are considered the most promising in terms of practical utility are: stationary and vibrating screens, liquid cyclones and settling tanks. These devices and some laboratory scale devices were evaluated for their efficiency of separation in the removal of solids from streams of untreated and treated wastewater. Separation and concentration performance was measured by changes in total volume, total solids (TS), total volatile solids (TVS), total suspended solids (TSS), 5-day Biochemical Oxygen demand (BOD), and chemical oxygen demand (COD).

## EQUIPMENT PERFORMANCE

### Stationary, Run-down Screen

A stationary screen, 46 cm wide and 76 cm long (surface area = 3496 cm$^2$) of the type shown in Fig. 1 was tested. Testing was done in Botkins, Ohio, where such a screen was installed as part of a complete system of swine wastewater treatment and recycle (Taiganides and White, 1971 and 1974). Controlled tests to determine efficiency of solids separation with various size screen openings and flow rates were also carried out at the Ohio State University Animal Science Livestock Center in Columbus. The wastewaters tested were untreated manure slurries flushed out of pig fattening barns by discharging large quantities of water down the gutters.

Two screen openings sizes (0.10 and 0.15 cm), and four different inflow flow rates (123, 183, 235 and 313 lpm) were tested. The influent wastewater to the screen ranged from 0.2 to 0.7 percent T.S. Table 1 is a summary of average solids removal and concentration performance. There was a great variability in the data. The standard deviation was highest for the COD results which ranged from 14 percent to 43 percent of the mean. For the rest of the parameters the standard deviation was mostly below 10 percent of the mean.

The data in Table 1 indicate that the screen with

Approved as Journal Article No. 48-75 of the Ohio Agricultural Research and Development Center (OARDC).

These studies were supported in part by Grant No. R-801125 from the US Environmental Protection Agency.

Commercial equipment used in these studies were donated by the manufacturers, Bauer Bros. Co., Springfield, OH and Sweco, Inc., Los Angeles, CA.

The authors are: J. W. SHUTT, Associate Engineer, Floyd G. Browne and Associates, Ltd.; R. K. WHITE, Associate Professor, E. P. TAIGANIDES, Professor, and C. R. MOTE, Former Research Associate, Agricultural Engineering Dept., OARDC and Ohio State University, Columbus.

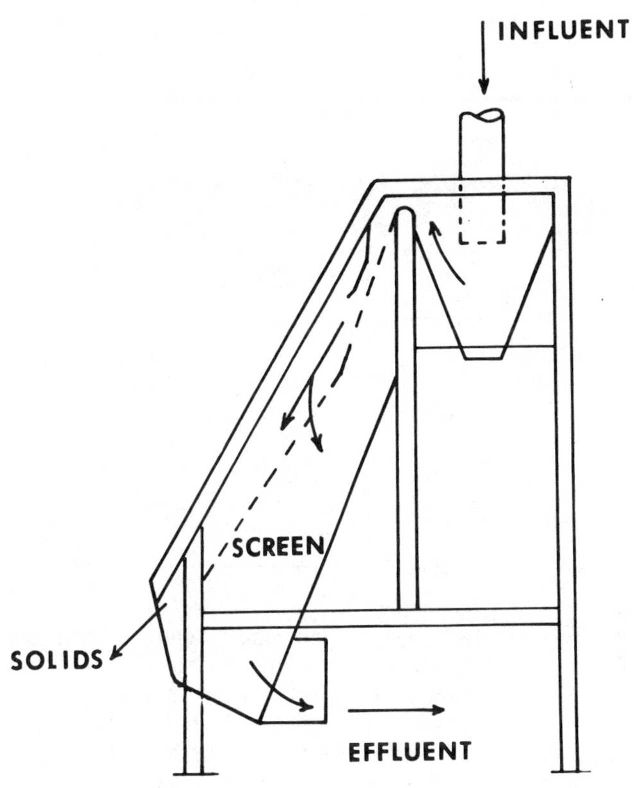



FIG. 1 Stationary, rundown screen.

smaller openings (0.10 cm) gave consistently better performance at low loading rates of about 120 lpm, which is equivalent to 3 lpm per cm of screen width. At this inflow rate, 35 percent of the solids, 62 percent of the BOD and 69 percent of the COD in the influent wastewater were removed by the screen, even though only 2 percent of the total flow was diverted. At the high flow rate of 313 lpm, the flow volume diverted by the stationary screen was 0.6 percent and 0.4 percent for the 0.10 cm and 0.15 cm screen openings, respectfully. BOD and TS removal was several times lower than that achieved with lower inflow rates. The screen with 0.15 cm openings produced the best results as far as BOD and COD removal is concerned at inflow rates of about 235 lpm, which is equivalent to 675 lpm per m$^2$ of screen area and also 5 lpm/cm of screen width. The latter parameter is the usual design parameter for this type of screen. The results given in Table 1 indicate that for the screen with small size openings (0.10 cm) a low flow rate of 2 to 3 lpm per cm$^2$ gives best TS, BOD and COD removal efficiencies. For the screen with larger size openings flow rates in the range of 5 to 6 lpm per cm$^2$ of screen width are optimum.

The major problem with the stationary screen was the clogging up of the openings with a film of biomass between flushings. Flow onto the screen was intermittent. Between flushings a film would form

**TABLE 1. SOLIDS REMOVAL PERFORMANCE OF STATION-ARY SCREEN (PERCENT OF INDICATED PARAMETER RETAINED BY SCREEN).**

| | | | Size of openings, cm | |
|---|---|---|---|---|
| | Parameter | Units | 0.10 | 0.15 |
| 123 | lpm flow rate (352 lpm per $m^2$ loading rate) | | | |
| | Flow volume | Percent inflow | 2.1 | 0.4 |
| | TS volume | ,, ,, | 35.2 | 3.0 |
| | TS concentration | Percent wet basis | 9.1 | 10.9 |
| | TVS volume | Percent inflow | 21.5 | 17.7 |
| | BOD | ,, ,, | 62.2 | 13.9 |
| | COD | ,, ,, | 69.1 | 11.4 |
| 183 | lpm flow rate (526 lpm per $m^2$ loading rate) | | | |
| | Flow volume | Percent inflow | 2.9 | 0.3 |
| | TS volume | ,, ,, | 25.8 | 4.2 |
| | TS concentration | Percent wet basis | 7.2 | 6.4 |
| | TVS volume | Percent inflow | 30.5 | 0.9 |
| | BOD | ,, ,, | 38.4 | 4.1 |
| | COD | ,, ,, | 63.8 | 11.5 |
| 235 | lpm flow rate (675 lpm per $m^2$ loading rate) | | | |
| | Flow volume | Percent inflow | 2.5 | 0.3 |
| | TS volume | ,, ,, | 27.5 | 9.8 |
| | TS concentration | Percent wet basis | 7.6 | 6.0 |
| | TVS volume | Percent inflow | 33.7 | 5.3 |
| | BOD | ,, ,, | 51.7 | 34.0 |
| | COD | ,, ,, | 71.1 | 24.2 |
| 313 | lpm flow rate (897 lpm per $m^2$ loading rate) | | | |
| | Flow volume | Percent inflow | 0.6 | 0.4 |
| | TS volume | ,, ,, | 11.3 | 4.2 |
| | TS concentration | Percent wet basis | 6.9 | 6.0 |
| | TVS volume | Percent inflow | 13.7 | 5.6 |
| | BOD | ,, ,, | 13.7 | 5.6 |
| | COD | ,, ,, | 51.6 | 24.1 |

underneath the screen which would tend to plug the openings. It became necessary to brush the screen with a liquid disinfectant to mechanically break the film and also kill the organisms which helped form the film. A hot water heater was installed beside the screen. It was equipped with a high-pressure hose to spray hot water on the screen automatically. This did not operate satisfactorily, so daily brushing of the screen was continued.

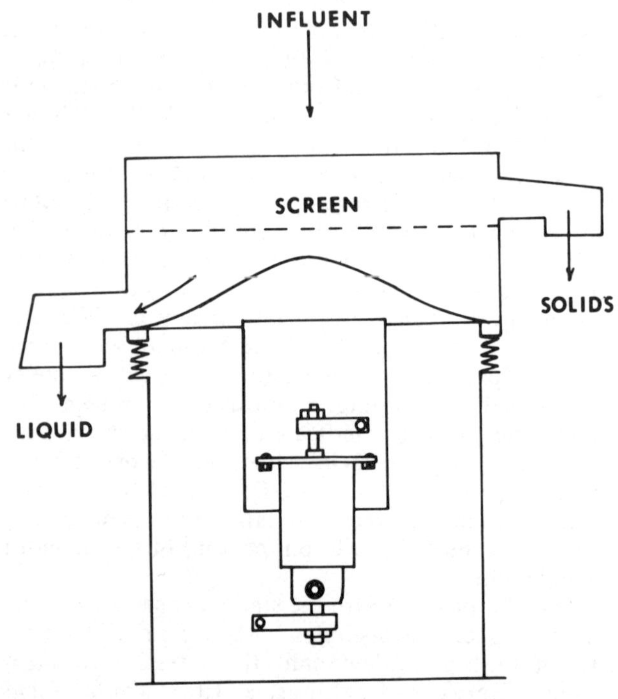


**FIG. 2. Vibrating screen.**

### Vibrating Screen

A vibrating screen of the type shown in Fig. 2 was tested using the same type of swine wastewaters used in the testing of the stationary screen. The screen tested was 46 cm in diameter and had a surface area of 1642 $cm^2$. Table 2 shows the results of the testing program where the screen was operated at 3 different flow rates and 4 different opening sizes. BOD tests were run only on one series of tests.

Analyses of the test data, as summarized in Table 2, indicates that each screen size has an optimum application rate, which in the screens tested was in the 67 lpm range. The screen with the largest openings (0.039 cm) produced the most desirable results. At an application rate of 67 lpm, solids were concentrated to 16.4 percent on wet basis with only 0.6 percent of the inflow being retained on the screen. The liquid effluent from the screen contained 78 percent of the applied TS, and 72 percent of the applied TVS, and 84 percent of the applied COD. The remaining TS, TVS and COD would be retained in the separated solids.

It is difficult to compare equitably the stationary and vibrating screen because of their differences in flow-through rates and different size of openings. However, on the basis of equivalent lpm level (110 to 123 lpm) the stationary screen (0.15 cm openings) and the vibrating screen (0.039 cm openings) produced similar removal efficiencies. On the basis of equivalent loading rate per unit of total screen area (around 670 lpm/$m^2$), the BOD and COD removal efficiencies by the stationary screen are higher than those of the vibrating screen.

### Liquid Cyclone

The liquid cyclone used for this study was a 7.6 cm diameter, 6 degree apex cone, constructed of polyvinyl chloride. The main features of the liquid cyclone are shown in Fig. 3. The testing program included 3 different underflow nozzle diameters (0.64, 0.47, and 0.32 cm), and 4 rates of pressure drops (1.4, 2.8, 4.2, and 5.6 Kg/$cm^2$). The swine manure slurry was pumped over a stationary run-down screen with 0.10 cm openings

**TABLE 2. SOLIDS REMOVAL PERFORMANCE OF VIBRAT-ING SCREEN (PERCENT OF INDICATED PARAMETER RETAINED IN SCREENED FRACTION).**

| | | | Size of openings, cm | | | |
|---|---|---|---|---|---|---|
| | Parameter | Units | 0.012 | 0.017 | 0.021 | 0.039 |
| 41 | lpm loading rate (248 lpm per $m^2$ loading rate) | | | | | |
| | Flow volume | Percent inflow | 0.2 | 0.5 | 2.8 | 0.4 |
| | TS volume | ,, ,, | 2.5 | 5.8 | 14.3 | 12.6 |
| | TS conc. (wb) | Percent wet basis | 5.7 | 3.9 | 8.7 | 12.2 |
| | TVS volume | Percent inflow | 3.3 | 8.4 | 12.4 | 4.3 |
| | BOD | ,, ,, | — | — | 4.5 | — |
| | COD | ,, ,, | — | 3.3 | 16.3 | 10.0 |
| 67 | lpm loading rate (411 lpm per $m^2$ loading rate) | | | | | |
| | Flow volume | Percent inflow | 1.2 | 0.7 | 0.7 | 0.6 |
| | TS volume | ,, ,, | 13.8 | 14.0 | 9.8 | 22.2 |
| | TS conc. (wb) | Percent wet basis | 8.5 | 10.9 | 10.8 | 16.4 |
| | TVS volume | Percent inflow | 18.5 | 17.0 | 12.9 | 28.1 |
| | BOD | ,, ,, | — | — | 2.4 | — |
| | COD | ,, ,, | — | 15.3 | 8.9 | 16.1 |
| 110 | lpm loading rate (670 lpm per $m^2$ loading rate) | | | | | |
| | Flow volume | Percent inflow | 2.1 | 0.8 | 0.9 | 1.6 |
| | TS volume | ,, ,, | 18.7 | 1.8 | 7.0 | 12.3 |
| | TS conc. (wb) | Percent wet basis | 4.8 | 1.9 | 4.8 | 4.9 |
| | TVS volume | Percent inflow | 42.8 | 2.4 | 9.0 | 17.2 |
| | BOD | ,, ,, | — | — | 3.9 | — |
| | COD | ,, ,, | — | 12.5 | 10.7 | 12.2 |

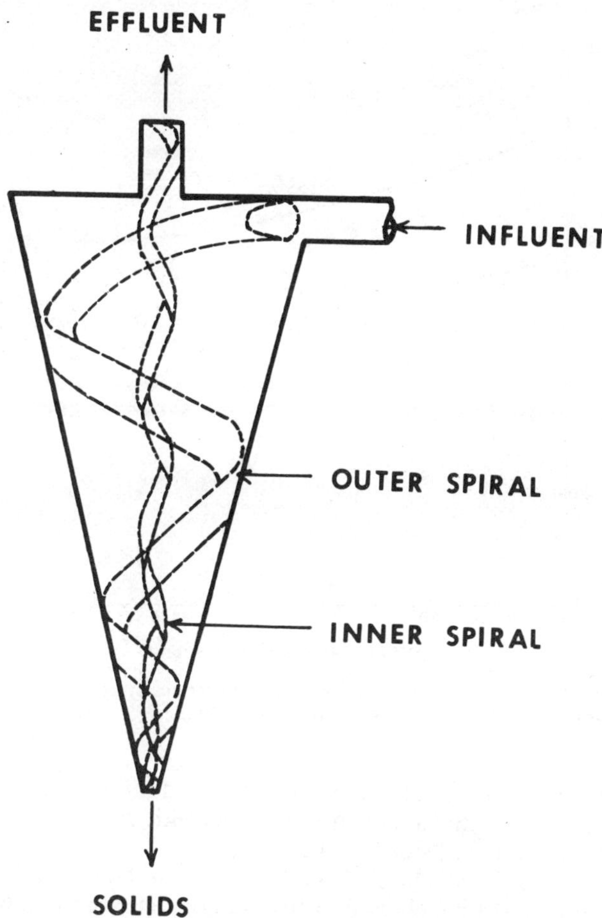


FIG. 3 Liquid cyclone.

before pumping through the liquid cyclone. Passing the slurry over the screen eliminated the coarse solids which would tend to plug the small diameter underflow nozzles.

The inflow rate was measured by a flow meter and the underflow rate was measured by collecting the discharge over a timed interval. The overflow rate was calculated by differences. A pressure gauge in the influent line measured the pressure drop across the cyclone. Samples of the influent, the underflow, and the overflow were collected for each run and analyzed in the laboratory for TS, TVS, and TSS (Shutt 1974).

The results of the liquid cyclone test are presented in Table 3. The largest concentrating effect for TSS was with the cyclone with a 0.32 cm diameter underflow nozzle. The best separation occurred at 2.8 $Kg/cm^2$ pressure drop. This pressure drop produces a flow rate of about 88 lpm. The cyclone was operated using manure slurries with TSS concentrations ranging from 0.10 percent to 0.50 percent. In this study, the cyclone's performance was superior when higher influent TSS concentrations were used.

The cyclone was able to achieve an 18 fold increase in TSS concentration, resulting in 9 percent TSS in the underflow discharge when operated with the 0.32 cm underflow nozzle diameter at 4.2 $Kg/cm^2$ pressure drop.

### Settling Chamber

The settling characteristics of swine waste slurries with TSS concentrations ranging from 0.5 percent to 4.0 percent were studied using a 20 cm (i.d.) by 183 cm deep cylinder. The cylinder was maintained at a constant temperature to preclude convection currents. (Thompson 1971).

Fig. 4 shows that with a detention time of 30 minutes in the sedimentation cylinder, about 75 percent of the original TSS and 55 percent of the original COD are removed from a slurry with an initial 5000 mg/l TSS concentration. Detention times longer than 30 minutes

TABLE 3. SOLIDS REMOVAL AND CONCENTRATION PERFORMANCE OF
LIQUID CYCLONE (PERCENT OF INDICATED PARAMETER RETAINED IN
UNDERFLOW DISCHARGE WITH INFLUENT CONCENTRATION
OF 0.5 PERCENT TSS).

| Parameter | Units | Underflow nozzle diameter, cm | | |
|---|---|---|---|---|
| | | 0.32 | 0.47 | 0.64 |
| 68 lpm flow rate and 1.4 $Kg/cm^2$ pressure drop | | | | |
| Flow volume | Percent inflow | 2.2 | 10.4 | 10.0 |
| TS volume | " " | 20.3 | 14.7 | 26.1 |
| TS concentration | Percent wet basis | 7.5 | 3.0 | 3.1 |
| TSS volume | Percent inflow | 11.3 | 4.5 | 30.2 |
| TSS concentration (wb) | Percent wet basis | 7.1 | 0.3 | 2.6 |
| TVS volume | Percent inflow | 5.5 | 2.3 | 2.3 |
| 88 lpm flow rate and 2.8 $Kg/cm^2$ pressure drop | | | | |
| Flow volume | Percent inflow | 2.1 | 4.8 | 11.2 |
| TS volume | " " | 26.5 | 14.1 | 24.1 |
| TS concentration | Percent wet basis | 8.4 | 2.4 | 1.6 |
| TSS volume | Percent inflow | 38.8 | 20.7 | 34.4 |
| TSS concentration | Percent wet basis | 8.2 | 2.2 | 1.3 |
| TVS volume | Percent inflow | 5.8 | 1.7 | 0.7 |
| 111 lpm flow rate and 4.2 $Kg/cm^2$ pressure drop | | | | |
| Flow volume | Percent inflow | 0.9 | 5.1 | 9.6 |
| TS volume | " " | 10.5 | 23.5 | 27.0 |
| TS concentration | Percent wet basis | 9.0 | 3.4 | 2.0 |
| TSS volume | Percent inflow | 15.7 | 34.0 | 40.7 |
| TSS concentration | Percent wet basis | 9.1 | 2.9 | 1.7 |
| TVS volume | Percent inflow | 6.6 | 2.6 | 1.4 |
| 127 lpm flow rate and 5.6 $Kg/cm^2$ pressure drop | | | | |
| Flow volume | Percent inflow | 1.8 | 5.0 | 10.4 |
| TS volume | " " | 15.8 | 17.1 | 24.2 |
| TS concentration | Percent wet basis | 5.8 | 3.7 | 1.6 |
| TSS volume | Percent inflow | 23.5 | 21.1 | 34.5 |
| TSS concentration | Percent wet basis | 5.5 | 3.3 | 1.3 |
| TVS volume | Percent inflow | 4.1 | 2.7 | 1.1 |

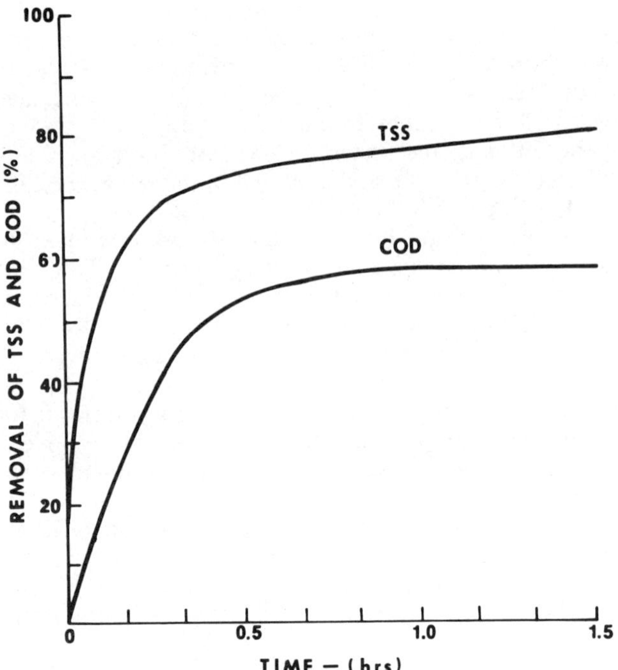

FIG. 4 TSS and COD removal efficiencies at various detention times for a slurry with initial TSS concentration of 5000 mg/l.

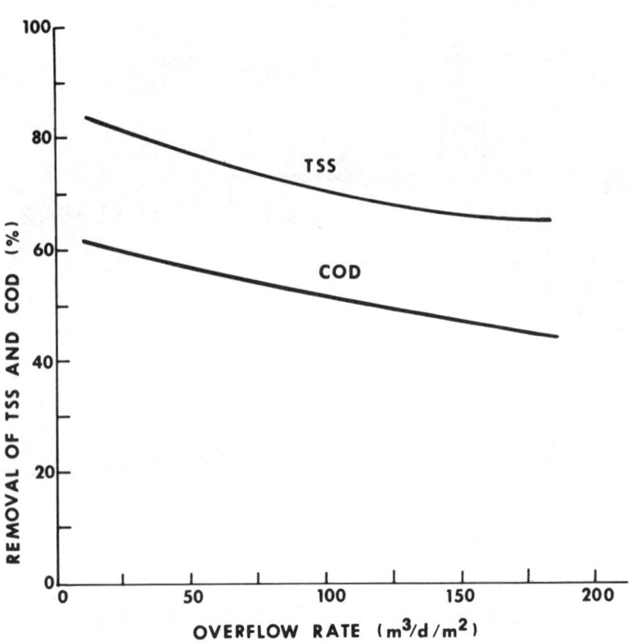

FIG. 5 TSS and COD removal efficiencies at various overflow rates for a slurry with initial TSS concentration of 5000 mg/l.

do not result in significantly greater removals of either TSS or COD. Fig. 5 shows that for the same slurry, TSS and COD removal efficiencies increase as the overflow rate decreases. The overflow rate is the design parameter used to determine the surface area of a clarifier, while detention time is a function of depth.

Solids settling in slurries with initial concentrations of TSS larger than 1 percent occurs as a zone settling process. Removal of solids from these slurries is governed by unit area relationships. Table 4 gives unit area requirements for slurries with initial TSS concentrations larger than 10,000 mg/l. Tests indicated that as the initial TSS concentration increases the unit area requirements increases. As higher underflow concentrations are sought, the unit area required increases. After 4 hours of detention time, TSS removal efficiencies were the same (90 percent) for all slurries tested. This was confirmed with field tests at the Botkins, Ohio, Swine Wastewater treatment facility (Taiganides and White 1971) where TSS removal efficiencies of 94 percent were achieved in the final clarifier irrespective of the TSS concentration of the incoming liquid (Thompson 1971).

### Laboratory Settling Tests

Settling tests were conducted with mixed liquor from an oxidation ditch treating swine wastes. One cylinder was filled with mixed liquor as it came from the ditch. Two others were filled with mixed liquor that had been diluted by 25 percent and 50 percent, respectively, with

distilled water. The purpose of the dilution was to provide settling data on mixed liquors with different TSS concentrations (Mote 1974).

A record was made of the height of the liquid-sludge interface as time elapsed. The data are plotted in Fig. 6. The curves show that as the TSS concentration of the mixed liquor increases, the rate of sludge settling

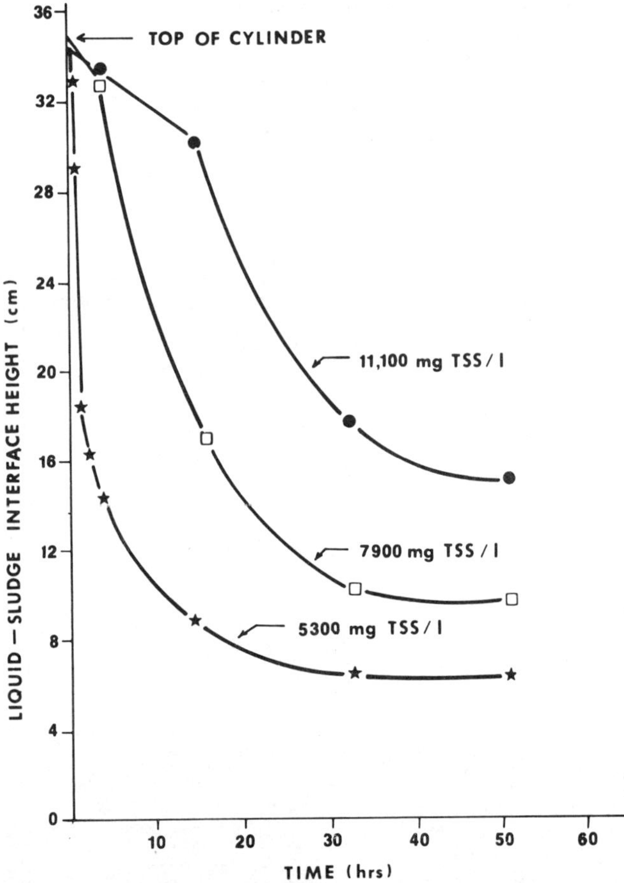


FIG. 6 Settling curves for oxidation ditch mixed liquor with different TSS concentrations.

TABLE 4. UNIT AREA AND OVERFLOW RATE REQUIREMENTS IN A SETTLING CHAMBER TO CONCENTRATE SOLIDS TO 8 PERCENT FOR SLURRIES WITH VARIOUS INITIAL SOLIDS CONCENTRATIONS.

| Initial TSS conc. | | Unit area required | Overflow rate | |
|---|---|---|---|---|
| Percent | mg/l | $m^2$/ton/d | $m^3$/d/$m^2$ | (gpd/ft$^2$) |
| 1 | 10,000 | 1.1 | 89 | 2180 |
| 2 | 20,000 | 3.0 | 17 | 415 |
| 4 | 40,000 | 6.6 | 4 | 92 |



MANAGING LIVESTOCK WASTES

declines. Fig. 6 shows that the sludge initially settles at a slower rate and that when the TSS concentration increases the length of the initial, slower settling period is longer.

Initial settling occurs at such a slow rate in mixed liquor with a TSS concentration of 7900 mg/l or above, that it is not practical to design settling systems to operate with TSS concentrations of 8000 mg/l or larger.

The prime consideration in the design of clarifiers for the purpose of providing a clarified effluent as opposed to concentrating the sludge, is the rate at which the sludge interface settles and leaves clear supernatant behind. The initial settling rate is four to nine times slower than the secondary settling rate. For design purposes it can be assumed that the settling velocity is zero for the time period equivalent to the initial settling phase. This means that the surface area of a clarifier should be designed on the basis of the second stage settling velocity. When a clarifier is built, provisions should be made for the clarifier to have a detention time equal to or greater than the time of the initial settling phase. Providing the required detention time should not be a problem since more than ample detention time is usually built into a clarifier to provide sufficient depth to accommodate sludge collection and removal.

## DISCUSSION AND EVALUATION

The four separation devices tested, a stationary screen, a vibrating screen, a liquid cyclone and a settling chamber, may be evaluated on the basis of separation efficiency, concentration of solids fraction and flow capacity. The better separation efficiency was obtained by the stationary screen, 35 percent of the TS volume as compared to 22 percent and 26 percent for the vibrating screen and cyclone, respectively. The best TS concentrations of the solids fraction obtained for the test conditions were 16, 9, and 8 percent for the vibrating, stationary screen and cyclone and smallest for the vibrating screen.

Based on the test conducted the stationary screen gave the better separation efficiency at the largest capacity. It must be recognized that flow capacity can be increased by making the equipment larger. The vibrating screen gave the highest concentration, i.e., the highest TS concentration of the solids fraction. From a management point of view the stationary screen will incur less operating costs because there is no power input to the screen. Both screens will require cleaning to avoid blanking due to a biological slime build-up. The cyclone will require a higher pump pressure than the screen for satisfactory operation.

On the basis of the removal of TSS and COD, the settling chamber performance is superior to the other devices. The final TS concentration, however, was less than the other devices. Therefore, the settling chamber is superior with respect to clarification of the supernatant. However, it is not as effective in concentrating solids. Smaller overflow rates will improve the clarification.

The settling tests conducted with oxidation ditch mixed liquor indicate that the design of settling chambers or clarifiers can be based on the more rapid, secondary phase settling; if the initial TSS concentraiton is low enough and if sufficient detention time is provided to allow for the initial settling phase. Sedimentation tanks are not effective when TSS concentration of the influent exceeds 8000 mg/l.


### References

1 Mote, C. R. 1974. A computer simulation of biological treatment, storage, and land disposal of swine wastes. Unpublished Ph.D. dissertation. The Ohio State University, Columbus, OH, 205 pp.

2 Shutt, J. W. 1974. Liquid cyclone for separation and concentration of swine manure solids. Unpublished M.S. thesis. The Ohio State University, Columbus, OH, 76 pp.

3 Taiganides, E. P. and R. K. White. 1971. Automated handling treatment and recycling of wastewater from an animal confinement production unit, pp. 146-148. IN: Livestock waste management and pollution abatement. ASAE Publication PROC-271, St. Joseph, MI 49085.

4 Taiganides, E. P. and R. K. White. 1975. Performance of an automated waste treatment and recycle system. IN: Proceedings International Symposium on Livestock Waste '75. ASAE, St. Joseph, MI 49085.

5 Thompson, A. V. 1971. Separation of solids from swine slurries. Unpublished M.S. Thesis, The Ohio State University, Columbus, OH, 91 pp.

# In-House Handling and Dehydration of Poultry Manure from a Caged Layer Operation: A Project Review

M. L. Esmay, C. J. Flegal, J. B. Gerrish, J. E. Dixon, C. C. Sheppard, H. C. Zindel, T. S. Chang

FELLOW
ASAE

ASSOC. MEMBER
ASAE

MEMBER
ASAE

EXEC. AFFILIATE
ASAE

ACCEPTABLE and economic handling and treatment of poultry wastes requires a combination of management and engineering solutions. This MSU/EPA demonstration project* was undertaken to obtain design and management information for the optimum handling of the wastes from a commercially sized cage-type layer house.

## OBJECTIVES

The overall goal of this project involved the management of poultry excreta in a closed ecological system. This concept included the production of a dehydrated product which has nutritive value as a feed ingredient. Odor abatement and pollution control were also major criteria that the excreta management system must satisfy.

Specific objectives of the project were:

1 To demonstrate and evaluate a complete poultry excreta handling system including in-house dehydration.

2 To determine the optimum dehydration conditions for the multiphase drying system.

3 To minimize the energy required to operate the system.

4 That the system must be adaptable to most existing commercial egg production units.

5 To determine the emissions from the system.

6 To determine the economics of the system.

This paper pertains mainly to the first three objectives listed above.

## REVIEW OF LITERATURE

Successful dehydration of poultry excreta has been described by several investigators. Ludington (1963) developed a type of poultry excreta dehydration and discussed the economics of that system. Surbrook (1969) and Surbrook, et al. (1971) dried poultry excreta by the use of a shaker-type heated-air dryer. Factors which were found to be important parameters of drying were temperature and particle diameter. The dried poultry waste product had a reduced odor (compared to undried manure), granular form and a bulk density that ranged from 12-20 pounds per cubic foot (192 to 321 gm/l).

A two-stage poultry excreta dehydration system has

been described by Bressler and Bergman (1971). In this system additional fans circulated air over the excreta which collected under a sloping wire floor (in-house drying). To accelerate the drying, a stirring device was also installed and operated periodically. By this technique, the excreta removed from the house varied in moisture content from 23.8 percent to 50.5 percent. Bressler and Bergman (1971) also described final dehydration of the excreta (down to about 10 percent moisture content wet basis) similar to that reported by Surbrook, et al. (1971).

Sobel (1972) reported on several systems of in-house drying of poultry waste in laying houses. The systems tested included mechanical devices (no forced air), forced air with a slot outlet and forced air in a high rise laying house. All the systems tested removed moisture to levels below 60 percent.

Previous progress report papers presented on this MSU/EPA project that pertain to the subject of this paper will provide additional background on the investigation (Flegal et al. 1974) (Sheppard et al. 1974) (Esmay et al. 1975) (Gerrish et al. 1973).

## MATERIALS AND METHODS

### House

A clear-span pole and truss building was remodeled to accommodate this project (Fig. 1). Four rows of wire cages (stacked, triple deck) 21.95 meters (72 feet) long with dropping boards below the top two decks confined the birds. The individual cages were 30.48 x 40.64 cm (12 inches x 16 inches). One-half of the cages contained three birds each and one-half of the cages contained four birds each. This cage system provided for a starting flock of 5,292 birds. The excreta from the upper two rows of cages were hand scraped daily. A commercial blade-type pit scraper was used daily to remove the excreta to a cross-conveyor (barn gutter cleaner) at the end of the house. The barn cleaner elevated and dropped the excreta onto the end of a continuous synthetic (PVC) belt located in the drying tunnel.

### Drying Tunnel

The drying tunnel was constructed at the side of the house opposite the ventilation air inlet slot. A 5.08 cm (2 inch) wide continuous slot was used for winter ventilation and a 15.2 cm (6 inch) wide continuous slot was used for summer ventilation. The drying tunnel was 1.88 meters (74 inches) wide and 2.44 m (8 feet) high and ran the entire length of the house. Four 283 m³/min (10,000 cfm) fans were located at one end of the drying tunnel (Fig. 1). One 60.96 cm, 28.3 to 141.5 m³/min (24 inch, 1,000 - 5,000 cfm) fan was located near the belt loading point of the cross conveyor. This fan was adjusted to run continuously at minimum output. All large fans were

---

*This project is cooperative with the United States Environmental Protection Agency.

Approved for publication as Journal Article No. 7202, Michigan Agricultural Experiment Station.

The authors are: M. L. ESMAY, Agricultural Engineering Dept., C. J. FLEGAL, Poultry Science Dept., J. B. GERRISH and J. E. DIXON, Agricultural Engineering Dept., C. C. SHEPPARD, H. C. ZINDEL and T. S. CHANG, Poultry Science Dept., Michigan State University, East Lansing.



MANAGING LIVESTOCK WASTES

single speed and thermostatically controlled; one fan to start operating at 15 C (59 F) and one more fan to start operating at each 2.8 C (5 F) increase in temperature.

The continuous PVC conveyor belt was 76.2 cm (30 inches) wide and was suspended in the drying tunnel from the ceiling. Ventilation air from the house was directed onto the full length of the belt by a series of adjustable baffles on the ventilation outlet slot. Normal daily excreta production was sufficient to provide a depth of about 5 cm (2 inches) to 7.5 cm (3 inches) on the belt. The conveyor belt was a sliding-tray type and geared to move about three meters per minute. The delivery of excreta into the dryer was controlled by a sensor in the dryer hopper.

### Heated Air Dryer

The dryer was of the same type described by Surbrook (1969) and was fitted with an afterburner. The dryer and afterburner exhaust gases were discharged down the drying tunnel. A heat exchanger (that consisted of a plenum chamber and five parallel 25 cm (10 inches) diameter stovepipes) was used to dissipate the heat from both the afterburner and dryer without causing a fire. The dried poultry waste was either bagged at the dryer or placed in a storage bin.

Appropriate monitoring equipment was installed to measure fuel consumption, electrical inputs, air movement, temperature and relative humidity.

## RESULTS AND DISCUSSION

The water was evaporated from the poultry excreta in four phases:
1 as deposited in the house by ventilation air.
2 as moved from pits and spread on the belt.
3 on the belt by ventilation air and waste dryer heat.

4 as run through the heated air dryer.

A daily excreta removal schedule was maintained. The excreta was first deposited in the house throughout a 24-hour period (approximately 7 a.m. to 7 a.m.) and then it was scraped from the dropping boards and pits into the cross-conveyor and from there immediately spread on the drying belt where it stayed for another 24 hours. After the belt drying period the excreta was conveyed directly into the dryer. A drying period of two or three hours of machine time was required depending on the moisture content of the excreta being dried. The handling system was thus operated on a daily basis even though it took approximately two days to move the excreta from its initial deposited location onto the belt, through the dryer and out of the house. For experimental and measurement purposes, only Tuesdays and Thursdays were specified as critical data taking days.

### Phase (1) In-House Drying

The moisture removal capability of a ventilation system depends on the initial temperature and humidity of the air as it enters the building, and the temperature and humidity maintained in the building. The condition of the incoming ventilation air is thus dependent on climatic conditions. The average monthly Michigan (East Lansing) temperatures are shown in Figs. 2, 3 and 4. The temperature of the experimental house was maintained at approximately 55 F to 60 F (13 to 16 C) and the relative humidity at less than 75 percent during the cold months. During the warmer months the house environmental temperature consistently ran a few degrees above outside temperatures. The optimum summer ventilation system is one that keeps inside temperatures a minimum margin above outside temperatures. To prevent undesirable hot weather, in-house

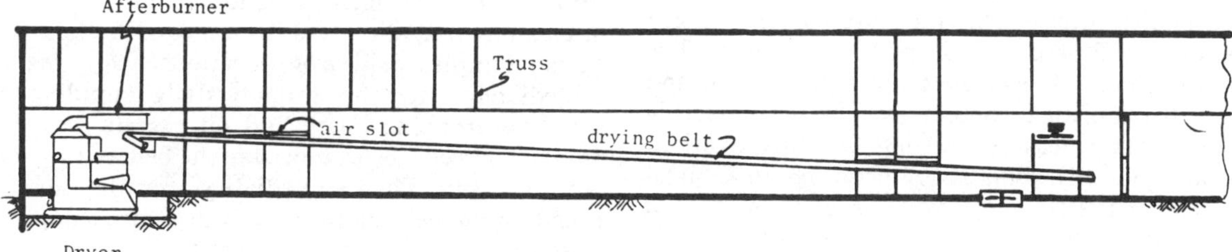


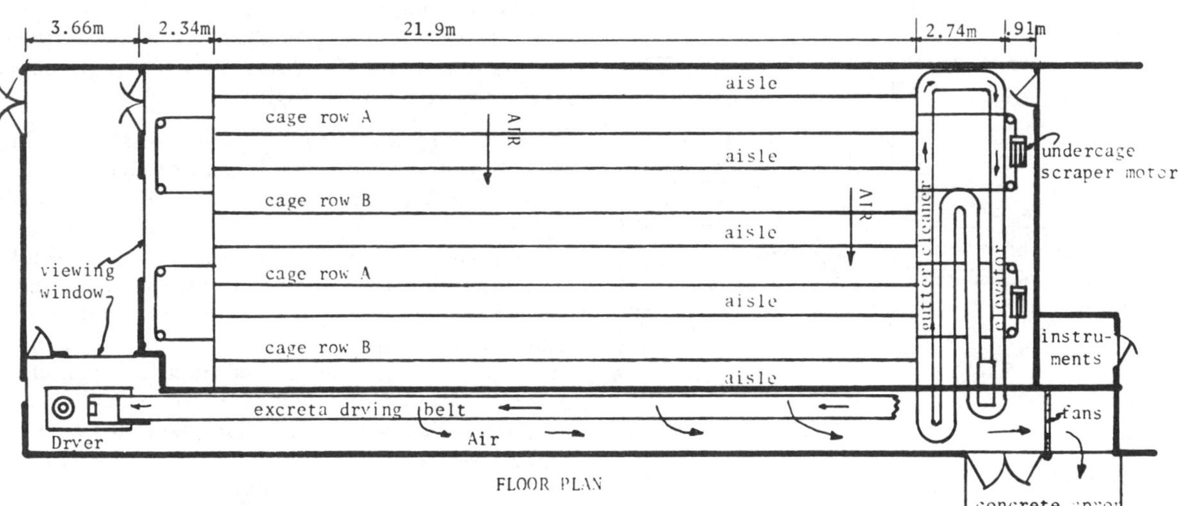


FIG. 1 Floor plan and section view of the modified poultry house. A heat exchanger consisting of five parallel runs of 10-in. stovepipe carries exhaust gases from the afterburner over the drying belt toward the fans. The heat exchanger is not shown in this figure.

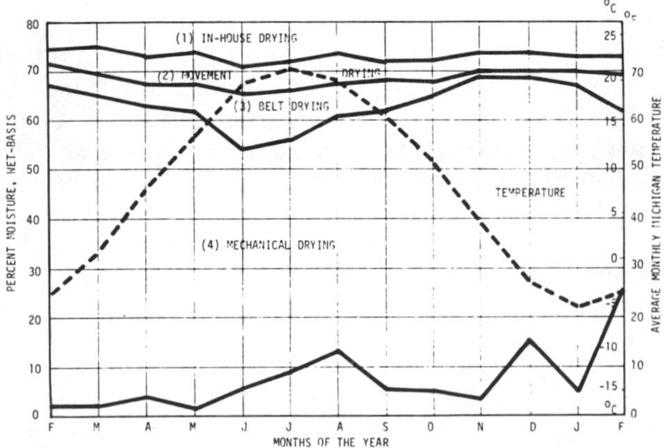

FIG. 2 Moisture content percentage [wet-basis] change in excreta during the dry phases.

temperature increases, the sensible heat from the birds must be used for evaporating moisture. The total bird heat is thus converted to latent heat so it does not increase the house environmental temperature. The ventilation system of this experimental house attempted to maximize evaporation during all seasons for; one, the maintenance of an optimum environment and two, to reduce the moisture content of the excreta.

The drying of excreta in the experimental house was summarized on a monthly basis over a period of one year and the results are shown in Figs. 2 and 3. These summarized results were obtained from hundreds of excreta moisture samples. Fig. 2 shows graphically in percent moisture content (wet basis) the amount of excreta drying (1) on the dropping boards and in the pits, (2) through movement of excreta, (3) on the belt in the drying tunnel, and (4) by the oil-fired mechanical dryer. The bottom curve indicates the final moisture content of the dried poultry waste. This moisture content ranged from extremes of 2 percent to 25 percent wet basis. The top moisture content line of the graph is 80 percent and represents the moisture content (wet basis) of freshly voided excreta as measured during this investigation. The average monthly climatic temperatures in Central Lower Michigan are also plotted on this graph in order to show seasonal trends.

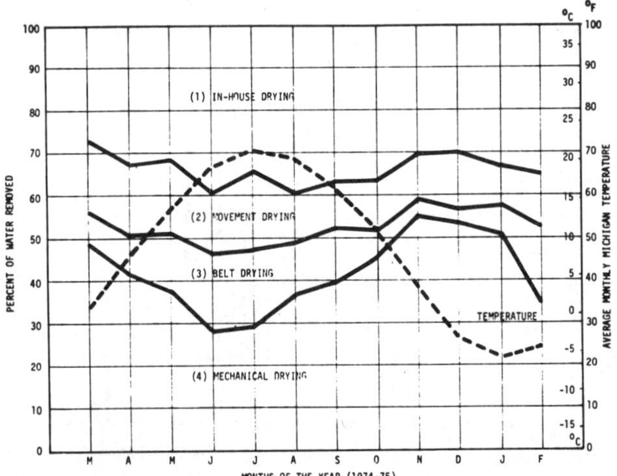

FIG. 3 Proportion of excreta water removed in the four drying phases of the experimental house.

Moisture percentages calculated on a wet basis misrepresent the true amount of water removal by various drying processes, so Fig. 3 was prepared. This graph presents the proportion of the total water removal accomplished by each of the four drying processes or phases. The proportion of the excreta drying that took place in the house is shown by the top part of the graph. It will be noted that the in-house drying reduced the excreta moisture to below 70 percent quite consistently throughout the year. In other words, about one-third of the excreta water removal was done by the ventilation air as it normally moved through the cage-type poultry house.

The somewhat surprising aspect of the in-house drying by ventilation air was (a) its quantity and (b) that it was not significantly season dependent. The consistent in-house evaporation prevailed in spite of the fact that the winter ventilation rates were much less (down to one-tenth) than for the summer months, and that the winter in-house air of from 55 to 60 F (13 to 16 C) could hold much less (approximately one-half) water vapor than the warmer air [80 F (27 C) or above] during hot weather. Also, the in-house ventilatin air must absorb all of the water vapor dissipated directly from the metabolic processes of the birds. The respired water vapor can amount to some three times the water evaporated from the excreta as shown by Fig. 3. A ventilation air psychrometric analysis of three typical months (Sept., Oct. and Nov., 1973) of the experimental house operation was made and presented in a previous paper by Flegal and Esmay et al. (1974).

### Phase (2) Movement Drying

The movement of the excreta by conventional cleaning methods of scrapers and flight-type cross conveyors proved to be a very effective way of reducing the moisture content. The movement moisture reduction during the pit cleaning operation was documented by taking eight excreta samples daily from locations below the cages. Immediately after the in-house daily sampling procedure, the dropping board and pit excreta was scraped into the cross conveyor, moved to the belt and spread to a uniform depth. Then immediately after the excreta was spread on the belt another set of moisture samples were regularly taken. There was consistently a three to four percent reduction in moisture content (wet basis) that could only be accounted for as water loss during movement. The percentage (wet basis) reduction during movement is plotted on Fig. 2 and shown as a proportion of total water removal on Fig. 3.

The loss of water by scraping, cross conveying and spreading shows up on Fig. 3 as a very significant portion of the total water removal. During the last six months (August, 1974 to February, 1975) of the experimental project when the entire operation was somewhat more stable than for the first six months, the movement water losses were consistently from 12 to 14 percent of the total excreta water removed. To further document the movement water losses, additional moisture samples were taken on two different days as shown in Table 1. Samples as indicated by row no. (3) were taken at the end of the pits where the scraper dumped the excreta into the cross conveyor. Likewise the row no. (4) samples were taken at the point where the cross conveyor dumped the excreta onto the belt. A little variation of a percent or so is in evidence between the two different sampling days which

| | | Percent moisture content, wet basis | |
|---|---|---|---|
| | | Jan. 20 | Feb. 8 |
| (1) | On the dropping boards, | 73.7 | 73.0 |
| | percent difference | 0.9 | 0.0 |
| (2) | In the dropping pits, | 72.8 | 73.0 |
| | percent difference | 1.3 | 0.7 |
| (3) | As scraped from pits, | 71.5 | 72.3 |
| | percent difference | 0.4 | 1.9 |
| (4) | As conveyed onto belt, | 71.1 | 70.4 |
| | percent difference | 0.4 | 1.4 |
| (5) | After spread on belt, | 70.7 | 69.0 |
| | overall percentage change | 3.0 | 4.0 |

The highly significant aspect of the experimental in-house manure handling and drying system was that during the summer months only from one-fourth to one-third of the total original excreta water remained to be removed by an oil-fired, heated air dryer. Even during the winter months up to nearly half of the excreta moisture was removed by the in-house drying systems prior to its introduction to the machine dryer. With the added turbulent air drying aid introduced in February, 1975, the remaining excreta water throughout the winter months could be reduced to the summer level of one-third of the total. No doubt with the turbulent air drying aid during the summer months, the remaining excreta moisture for machine drying could be further reduced to well within one-fourth of the total.

is within the sample accuracy of this investigation. The data do show however that each of the movement operations accounts for a portion of the water losses. The moisture samples taken January 20 showed the excreta on the dropping boards to have a 0.9 percent higher moisture content than the excreta in the pits. There was no explainable logic for this and the February 8 tests showed no difference. In general a percentage point of water is lost by each movement operation.

### Phase (3) Belt Drying

The reduction of excreta moisture content while on the belt throughout a 24-hour period in the drying tunnel is shown by the No. 3 phase on Figs. 2 and 3. The belt drying phase under conditions of this experiment was definitely season dependent. Belt drying accounted for up to 18 percent of the total excreta water removal during some of the summer months. During the latter winter months (Oct. to Jan.) of the investigation the belt drying accounted for only from four to six percent of the total excreta water removed. The seasonal difference evidently reflects the difference in the drying potential of the exhaust air from the poultry house.

Additional drying aids were investigated during the last month of the investigation (February, 1975). It was shown that the maximum drying capability of the house exhaust air had not been previously attained even during cold weather. The excreta drying while on the belt for February, 1975 (See Fig. 3) was about 18 percent of the total excreta moisture removal. This water removal equals the maximum attained during any summer month of the investigation. The increased excreta drying on the belt was accomplished with added air turbulence over the excreta and with periodic stirring with a disc-harrow type stirring device. Details of these drying aids are discussed by Dr. Sheppard in another paper at this Symposium entitled "Modifications on the Michigan State In-House Drying System".

### Phase (4) Heated Air Mechanical Drying

The fourth drying phase was that accomplished with an oil-fired dryer. In order to improve the efficiency of this mechanical dryer, its exhaust gases, as well as those from the afterburner, were directed down the tunnel for additional excreta drying. Moisture samples indicated that the waste dryer heat contributed to additional excreta drying of from one to two percentage points (wet basis) during the three or four hours of daily dryer operation.

### Heated Air Dryer Performance

Machine heated-air drying of poultry excreta has not yet been developed into a very efficient process. When an after burner is used, along with the oil-fired air heater for minimizing air pollution, as in this investigation, an equivalent fuel heat content of from 2208 to 2760 kcal/kg (4 to 5,000 Btu) was required to remove each pound of excreta water. Approximately 1,000 Btu's of heat are theoretically required to evaporate 552 kcal/kg (one pound) of water; thus, a machine drying efficiency of from 20 to 25 percent was accomplished. Elimination of the afterburner, if acceptable from a pollution standpoint, could double the fuel utilization efficiency.

The performance characteristics of the heated air dryer are shown by Fig. 4. The critical criterion of a dryer is the amount of water it will remove. A reliable mechanical dryer should be quite independent of the season of the year and the moisture content of the input excreta. This statement is made in view of the fact that some additional dryer heat losses could be expected during cold weather and that the last few percentage points of moisture are the most difficult to remove. For the last nine months of operation, as shown by Fig. 4, the machine dryer water removal rate was quite consistently about 59 kg (130 pounds) per hour for all except one month (September). This seasonal independent period extends from summer months through the coldest winter months. The much better water removal performance of the dryer during the first three months of the investigation (See Fig. 4) cannot be explained by other than possibly the better performance of the new dryer for a brief period and/or by the unreliability of the measurement data during the early months of the investigation.

If the water removal capability of a dryer is constant throughout all months of the year then the dry matter output must vary seasonally with the moisture content of the incoming excreta. As shown in Fig. 2, the moisture content (wet basis) of the excreta going into the dryer was down to the 55 percent range during the summer months and as high as 68 percent during the winter months. For comparison purposes a 50 percent moisture content (wet basis) sample contains only one-half as much water as a 67 percent sample. Much more dry matter was thus produced by the machine dryer during the summer months. This is shown to be the case quite consistently during the last nine months of stable operation of the experimental house (See Fig. 4).

MANAGING LIVESTOCK WASTES

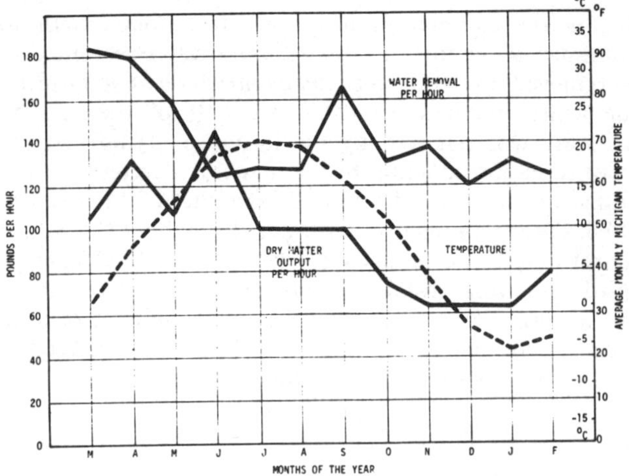



FIG. 4 Dryer performance on a monthly basis.

### Fuel Cost for Machine Drying

The fuel cost of the machine removal of excreta water was about 2.2¢/kg (1¢/pound) during the in-house drying investigation. This cost is based upon a fuel oil cost of 9¢/liter (34¢/gallon) and an assumed equivalent heat content of 9,620 kcal/liter (144,000 Btu/gallon). For 5,000 laying hens the daily fuel oil cost for machine drying of the excreta from 67 percent to five percent (wet basis) was $5.38. This cost was reduced by about one-half by pre-drying the excreta in the house to a 50 percent level before it goes into the dryer. Over the one year operational period for this experimental project the excreta dry-matter output per laying hen averaged 24.7 g/day/hen (0.0543 lb/day/hen). The calculated voided weight of excreta at 80 percent (wet basis) was then 123 g/day/hen (0.272 lb/day/hen) and the total water to be removed in drying the excreta to 5 percent was 97.6 gm/day/hen (0.2147 lb/day/hen). (See Table 2).

## SIGNIFICANT RESULTS

The following results were documented under the con-

### TABLE 2. EXCRETA PRODUCTION OF BIRDS.

| | gm/day/hen (lb/day/hen) | kg/day for 5000 hens (lb/day for 5000 hens) |
|---|---|---|
| Excreta dry matter | 24.7 (0.0543) | 123.5 (271.5) |
| Excreta wt. @ 5 percent m.c. w.b. | 26.1 (0.0573) | 130.0 (286.5) |
| Excreta wt. @ 80 percent m.c. w.b. | 123.0 (0.272) | 618.0 (1360.0) |
| Total water to be removed in reducing excreta m.c. from 80 percent to 5 percent | 97.6 (0.2147) | 488.0 (1073.5) |
| Total water to be removed in reducing excreta m.c. from 67 percent to 5 percent | 49.0 (0.1077) | 244.0 (538.5) |
| Total water to be removed in reducing excreta m.c. from 50 percent to 5 percent | 23.4 (0.0517) | 117.5 (258.5) |

ditions of this experimental project:

1. The excreta water removal by a conventional cross-flow ventilation system consistently (during all seasons) accounted for about one-third of the total excreta water removal in producing a five percent moisture content (wet basis) dried poultry waste product.

2. Excreta movement operations of scraping, cross-conveying and spreading on a belt consistently (during all seasons) accounted for a moisture reduction of from three to four percentage points (wet basis).

3. The poultry excreta was moved daily when it was in the moisture content range of 60 to 70 percent (wet basis) and the scraping, conveying and spreading accounted for from 12 to 14 percent of the total excreta water removed in producing a five percent dried poultry waste product.

4. Belt drying of excreta unlike the in-house drying was season dependent. The excreta drying in the tunnel accounted for up to 18 percent of the total excreta water removal during the summer months and down to four to six percent of the total during winter months.

5. Additional drying aids of air turbulence and excreta stirring while on the belt in the tunnel during the 24-hour periods increased belt drying to 18 percent of the total excreta water removal during February, 1975. This equalled the maximum for belt drying during the summer months.

6. In-house and belt drying reduced the excreta water during summer months to from one-fourth to one-third of the total excreta water to be removed. Even during the winter months one-half of the excreta water was removed by other than the heated air dryer. Additional air turbulence over the excreta on the belt was shown to reduce the excreta water during a winter month to one-third or less which had to be removed by a fuel oil fired, dryer.

7. The oil fired machine dryer with an afterburner consistently required 1260 kcal (4-5,000 Btu) of equivalent fuel oil heat to remove 0.454 kg (one pound) of water from the excreta. This is a thermal efficiency of from 20 to 25 percent.

8. For the last nine months of quite stable experimental project operation, the machine dryer removed quite consistently about 59 kg (130 pounds) of excreta water per hour of operation. The dry matter output of the machine dryer varied inversely with the moisture content of the excreta being dried from about 28.2 kg (62 pounds) per hour during the winter months to 45.4 kg (100 pounds) per hour during the summer months.

9. The fuel cost for machine drying of excreta was about one cent per pound of water removed with oil at 9¢/liter (34¢/gallon) and with an assumed fuel oil heat content of 9,620 Kcal/liter (144,000 Btu/gallon).

10. For a 5,000 layer hen house the fuel oil cost for machine drying of excreta to a five percent moisture level can be reduced by about one-half, from $5.38 per day to $2.58 per day by in-house drying of excreta to a 50 percent level as compared to 67 percent (wet basis).

### References


1 Bressler, G. O., and E. L. Bergman. 1971. Solving the poultry manure problem economically through dehydration. Proc. Int. Symp. on Livestock Wastes, Ohio State University. ASAE Publication Proc. 271:81-84.
2 Esmay, M. L., C. J. Flegal, C. C. Sheppard, J. B. Gerrish, J. E. Dixon, H. C. Zindel and T. S. Chang. 1975. Poultry manure


*(Continued on page 477)*

# Drying of Poultry Manure — An Economic and Technical Feasibility Study

J. B. Akers, B. T. Harrison, J. M. Mather

ONE facet of the disposal of poultry manure or litter is the potential for profit in its sale in the dried form as a source of crude protein in animal feeds or as a horticultural fertilizer in the domestic market.

Estimates for 1973 give an approximate quantity of 70,000 tonnes being dried in the United Kingdom. Of this, an estimated 50,000 tonnes go into animal feeds, and the remainder into miscellaneous horticultural products. For animal feeds use, drying to approximately 10 percent moisture is necessary. For horticultural products, 35 percent moisture seems acceptable.

The value of dried poultry manure in animal feeds will vary with the prices of the normal commodity ingredient; £1/unit of nitrogen was being paid in late 1973 and early 1974. In the horticultural market, dried poultry manure products average ca £90/tonne.

Altough not currently used in the agricultural market, the rising prices of phosphates, nitrogen and petroleum are beginning to improve the feasibility of this operation.

In the light of these possible markets, a critical appraisal of the different types of dryers available for the drying of animal manures has been made, from the point of view of economics and technical suitability. Consideration has been given to both small scale package units for on farm use, and large scale industrial dryers. Small scale dryers have the advantages of low overheads and applicability to the "small man". Large industrial dryers gain the benefits of the "economy of scale", and the control over product quality (e.g. analysis and sterility) which will be necessary for use in animal feeds.

## DESIRABLE DRYER CHARACTERISTICS

1 Sterility — for animal feeds, by maintaining a high enough temperature for sufficient time.

2 No odor problems — i.e. the drying process should not emit odors or should be amenable to easy treatment for odor removal.

3 Low labor requirement and simple to operate (especially for on-farm units).

4 Flexibility — the dryer should be capable of handling a variable composition feed and giving an adjustable product moisture content.

5 Dryer should be capable of handling "foreign bodies".

6 Materials of construction should be compatible with product specification and have good corrosion resistance.

7 Farm units should have simple, quick start-up/ shut-down procedures to maintain efficiency.

8 Economy — in terms of capital cost, installation and running costs.

The authors are: J. B. AKERS and B. T. HARRISON, Unilever Research Laboratory, Port Sunlight, Wirral, Merseyside, L62 4XN, United Kindom; and J. M. MATHER, Ciba Geigy, Grimsby, Lincolnshire, United Kingdom.

A detailed breakdown of the comparative economics of manure drying is contained in Tables 1-5. It must be appreciated that these figures contain no account of overheads (e.g. transportation, marketing). These will vary widely with circumstances and could greatly affect the final cost of the product.

## INDUSTRIAL DRYERS FOR MANURE APPLICATIONS

In the light of these factors, a survey was made of the available types of industrial dryers which might have application in manure drying on a large scale, e.g. at least 50,000 poultry.

*Tray dryers* are of relatively low capacity, being applicable to the drying of fine chemicals. They are quite inappropriate to the drying of poultry manure beacuse of costs, high labor and utilities and odor problems.

*Continuous band dryers* present odor control problems due to low gas velocity. Afterburners, stacks, filters and scrubbers require high gas velocities. Drying rates are low and labor requirements are high making this type of dryer unattractive.

*Batch agitated dryers* have a flexibility of operation which is advantageous to farm use, e.g. operates satisfactorily at intermittent rates, with variable feed moisture content, and with foreign matter other than stones. Other advantages are almost certain product sterility, low labor, running and capital costs. The chief disadvantage is that maximum capacity is limited to about 11 m³ (4 m³ charged) for vertical and 25 m³ (12 m³ charged) for horizontal dryers, thus losing economy of scale. Low thermal efficiency, and about a 15 percent depreciation rate are additional disadvantages.

*Direct heated rotary drum dryers* have advantages in product quality and handleability of variations in feed moisture. Labor requirements are low, foreign matter is readily handled and dryers can be constructed from various materials. Capital costs are high in comparison to pneumatic and batch agitated dryers. Thermal efficiency is poor during intermittent operation. Product cooling is also necessary for maintenance of thermal efficiency and packaging purposes, as material is discharged at 100 C.

*Pneumatic [flash] dryers* can approach 90 percent thermal efficiency when run continuously. Low gas flow rate (m³/s) and high velocity (m/s) help both with thermal efficiency and odor control equipment. Operation is co-current and space requirements are low. It is a gentle drying process and reduces loss of nutrients. Disadvantages to pneumatic dryers are possible problems of sterility because of short residence times, relatively low temperatures and the tendency to form "balls", and the necessity to macerate foreign matter in the feed.

We consider that pneumatic dryers represent the most

| Tonne dry product per annum | 250 | 500 | 750 | 125 | 250 | 375 |
|---|---|---|---|---|---|---|
| Installed cost | | £10,560 | | | £8,710 | |
| Running time hr/day 5 days/week | 8 | 16 | 24 | 8 | 16 | 24 |
| No of poultry | 15,600 | 31,200 | 46,500 | 7,800 | 15,600 | 23,250 |
| | All costs given below in £/tonne dried product | | | | | |
| Maintenance | 2.11 | 1.06 | 0.70 | 3.48 | 1.74 | 1.16 |
| Depreciation | 5.63 | 2.81 | 1.88 | 9.29 | 4.65 | 3.10 |
| Cost of capital | 7.43 | 3.72 | 2.48 | 11.99 | 5.99 | 4.00 |
| Labor | 1.13 | 1.13 | 5.27 | 2.26 | 2.26 | 10.54 |
| Fuel oil | 14.72 | 13.06 | 11.88 | 14.72 | 13.06 | 11.88 |
| Electricity | 2.26 | 2.26 | 2.26 | 3.39 | 3.39 | 3.39 |
| Total | 33.28 | 24.04 | 24.47 | 45.13 | 31.09 | 34.07 |

promising equipment for development into poultry manure dryers for large poultry holdings. We feel optimistic that problems of sterility can be overcome in development, if it is indeed a problem. Batch agitated dryers show economic advantages up to about 50,000 poultry. Economic analyses are presented in Tables 1-5 and are discussed in the following section.

## PACKAGED AGRICULTURAL POULTRY WASTE DRYERS

The main types of farm package available in the United Kingdom for the drying of manures are discussed here.

### Batch Agitated Pan Dryers

Vertical pan dryers to handle 22,500 and 45,000 poultry on a 16 hr/day, 7 day/week basis are available at costs of approximately £8,700 and £10,500. Vessels are of mild steel construction, clad with stainless. A one tonne batch of dry product takes 4 to 8 hr time and 1 hr labor. Final temperature is about 150 C. Heating is indirect and provided by oil or gas. A final direct heating section gives odorous off-gases. Odor control seems feasible however, with further development. Our calculated heating efficiencies are functions of operational continuity, varying from 50 to 75 percent. This dryer type seems well suited to small farm applications. Economic calculations are presented in Tables 1 and 6.

A horizontal pan dryer to handle 40,000 poultry is available. Operation is on a 16 hr/day, 5 day/week basis. Horizontal screw conveying is an advantage, both for manure recycling and bulk handling of product. Manure and combustion gases are fed co-currently. Construction is of mild steel, with stainless cladding. The dryer package includes sieves, hammer mills and bulk storage. The capital cost of £ 10,000 may include ancillary equipment but does not include any installation costs.

### Rotary Drum Dryers (Direct Heating)

Three main makes of dryer were located.

A double pass dryer is marketed to handle 40,000 poultry on a 16 hr/day, 5 day/week basis. The uninstalled cost is approximately £ 13,400, including ancilliaries. Construction is of mild steel, with stainless cladding. The reason for double pass construcion is not apparent, unless developed from a grass dryer.

A range of portable single pass dryers is available, to operate on an 8 hr/day, 5 Day/week basis. Capital costs range from £10,500 for a 20,000 poultry unit to £50,900 for a 150,000 poultry unit. Ancilliary equipment and installation costs are not included. LPG fired afterburner, cyclone dust collector, hopper and temperature controller are included in the costs. Insufficient information is available for a thorough technical assessment, or calculation of operating costs.

A single pass dryer is available, primarily for litter drying. Operation is on a 16 hr/day, 5 day/week basis. From available information, our calculations show the dryer could handle manure (not litter) from 15,000 poultry, assuming a 70 percent thermal efficiency. Construction is of mild steel. Ancilliary equipment includes a feed material pre-breaker, radial lifters in the dryer, and a product mill. No costs were available.

Operating costs for specimen rotary drum dryers are given in Tables 4 and 6.

### Pneumatic Dryer

Two continuous pneumatic dryers are available to handle manure from 20,000 and 40,000 poultry on a 16 hr/ day maximum and 5 day/week basis. Capital costs are approximately £9,300 and £20,700 uninstalled and £ 14,700 and £ 26,800 installed. We calculated the thermal efficiency at 63 percent from given data, and feel this is low. Thermal efficiency could be improved by redesigning solids feed and recycle without the presently required breaks in operation for cleaning. Better flue gas heat recovery also seems possible. Operating cost calculations are given for this dryer in Tables 2 and 6.

A combined rotary drum/pneumatic dryer is available to handle 40,000 poultry operating 16 hr/day. The installed cost is £18,100. Design is good, combining the advantages of both types of dryer. Mechanical attrition of product between dryers is provided. A cyclone, afterburner and feather sifter are also included.

TABLE 2. PNEUMATIC DRYERS. COSTS FOR FARM PACKAGES.

| Tonne dry product per annum | 320 | 640 | 960 |
|---|---|---|---|
| Running time hr/day 5 days/week | 8 | 16 | 24 |
| No of poultry | 20,000 | 40,000 | 60,000 |
| Installed cost | | £14,520 | |
| | All costs below given in £/tonne dried product | | |
| Maintenance | 2.27 | 1.13 | 0.76 |
| Depreciation | 4.54 | 2.27 | 1.51 |
| Cost of capital | 7.99 | 3.99 | 2.66 |
| Labor | 7.35 | 5.34 | 19.50 |
| Fuel oil | 14.12 | 11.88 | 9.90 |
| Electricity | 3.39 | 3.39 | 3.39 |
| Total | 39.66 | 28.00 | 37.72 |

TABLE 3. PNEUMATIC DRYERS. COSTS FOR INDUSTRIAL MODELS
(FROM 25 PERCENT SOLIDS TO 90 PERCENT SOLIDS)

| Tonne dry product per annum | 800 | 800 | 800 | 1,600 | 1,600 | 1,600 | 16,000 | 16,000 | 16,000 |
|---|---|---|---|---|---|---|---|---|---|
| Running time hr/day (5 days/week) | 8 | 16 | 24 | 8 | 16 | 24 | 8 | 16 | 24 |
| No of poultry | 50,000 | 50,000 | 50,000 | 100,000 | 100,000 | 100,000 | 1,000,000 | 1,000,000 | 1,000,000 |
| Installed cost £ | 27,400 | 20,000 | 16,800 | 56,800 | 42,200 | 35,100 | 228,400 | 169,000 | 113,500 |
| | | | | All costs below given in £/tonne dried product | | | | | |
| Maintenance | 1.71 | 1.25 | 1.05 | 1.78 | 1.32 | 1.10 | 0.71 | 0.53 | 0.35 |
| Depreciation | 3.43 | 2.50 | 2.10 | 3.55 | 2.64 | 2.19 | 1.43 | 1.06 | 0.71 |
| Cost of capital | 6.02 | 4.40 | 3.70 | 6.25 | 4.64 | 3.86 | 2.51 | 1.86 | 1.25 |
| Labor | 7.50 (2.94)* | 15.00 (2.26) | 22.50 | 3.75 | 7.00 | 11.25 | 0.70 | 1.40 | 2.10 |
| Fuel oil | 14.12 | 11.88 | 9.90 | 14.12 | 11.88 | 9.90 | 14.12 | 11.88 | 9.90 |
| Electricity | 3.39 | 3.39 | 3.39 | 2.26 | 2.26 | 2.26 | 2.26 | 2.26 | 2.26 |
| Direct farm Operating expenses (£/tonne) | 35.72 (31.16) | 40.68 (27.94) | 42.64 | 31.71 | 29.74 | 30.56 | 21.73 | 18.99 | 16.57 |

*Figures in brackets indicate the effect of taking agricultural labor costs.

## Continuous Band Dryers

Two makes of dryer are currently marketed.

From one manufacturer a series of both intermittent and continuous operation models are available. Capital costs range from £ 5,800 to £ 12,000 for continuous operation models to handle 18,000 to 120,000 poultry. Installation costs should not be expensive.

Combustion gases at 80-110 C pass up through the wet manure. Only poor odor control is possible, as this dryer uses a low gas velocity. An 88 percent dry product is possible, but the dryer requires feed manure to contain at least 25 percent solids. Thermal efficiency is implied at 88 percent.

While capital costs are low, labor and maintenance costs are high. One man is required continuously during drying, presumably to rake the drying manure, and for bulk handling. Shift working will inflate the already high labor costs.

The second make of dryer available is designed to handle manure from 22,000 poultry. Basic capital cost is approximately £ 3,750 and £ 4,000 with ancilliaries. Little reliable information is available about the operation and economics of this dryer. Additional features are storage of product in glass reinforced plastic hoppers, a stack is provided for an element of odor control, and the dryer is totally enclosed. Our comments on labor intensity are applicable to this dryer. Problems of variable product moisture is also noted.

## COST CALCULATIONS

Cost calculations are presented here for the types of dryer which in our opinion are best suited for manure drying on the different scales. Assumptions and calculations are shown in Tables 1-5 for batch agitated, pneumatic and rotary dryers to dry 25 percent solids feed to 90 percent and 65 percent solids.

### Batch Agitated Dryers. Costs for Farm Packages

The calculations assumed the use of a farm package dryer. It would be impractical and uneconomical to use a battery of dryers on a large industrial scale, due to the limited size of these dryers, and the considerable extra capital investment required to duplicate them and provide extra ancilliaries. No costing has been carried out on that scale. Assumptions were as follows:

1 Capital costs (installed £10,560 for the larger unit, £8,712 for the smaller (half size). Costs corrected from 1973 to January 1975.

2 Tax depreciation over 7 1/2 years (13.3 per annum).

3 Maintenance 5 percent per annum.

4 Cost of capital 17.6 per annum, based on 15 percent DCF, ROI in 10 years after deducting 52 percent corporation tax.

5 Labor. Large unit capacity 1 tonne/batch at 1 man-hour per batch and £1.13/man-hour smaller unit half size at same labor requirement. For 24 hr/day running 2,000 hr (night shift) at £1.70 per man-hour in addition to day-time labor.

6 Thermal efficiency 75 percent for 24 hr/day running, 67.5 percent for 16 hr/day running, 60 percent for 8 hr/day running.

7 Fuel oil cost 16 p/gallon. (Cost as of March 1974).

8 Electricity 30 kW large unit, 22.4 kW small unit, electricity at 1.855 p per kW hr.

### Pneumatic Dryers. Costs for Farm Packages

This has been taken as an example of the package unit capital cost for small farm units. Other capital costs were assessed for industrial dryers from Nonhebal and Moss, 1971, and a Lang factor of 2.0 taken for installation. Other assumptions were:

1 Capital cost (installed £ 14,520. Costs corrected from 1973 to January 1975.

2 Tax depreciation over 10 years (10 percent per annum).

3 Maintenance 5 percent per annum.

4 Cost of capital 17.6 percent per annum, based on 15 percent DCF, ROI in 10 years after deducting 52 percent corporation tax.

5 Labor. £1.13/man-hour farm labor (package unit

TABLE 4. ROTARY DRUM DRYERS.
COSTS FOR FARM PACKAGES.

| Tonne dry product per annum | 240 | 480 | 720 |
|---|---|---|---|
| Running time hr/day 5 days/week operation | 8 | 16 | 24 |
| No of poultry | 15,000 | 30,000 | 45,000 |
| Installed cost | | £17,200 | |
| | All costs given below in £/tonne dried product | | |
| Maintenance | 7.17 | 3.58 | 2.39 |
| Depreciation | 7.15 | 3.58 | 2.39 |
| Cost of capital | 12.61 | 6.31 | 4.20 |
| Labor | 9.41 | 7.06 | 28.25 |
| Fuel oil | 14.70 | 12.22 | 11.14 |
| Electricity | 3.39 | 3.39 | 3.39 |
| Total | 54.43 | 36.66 | 52.69 |

**TABLE 5. ROTARY DRUM DRYERS. INDUSTRIAL MODELS FROM 25 PERCENT TO 90 PERCENT SOLIDS.**

| | | | | | | | | | |
|---|---|---|---|---|---|---|---|---|---|
| Tonne dry product per annum | 800 | 800 | 800 | 1,600 | 1,600 | 1,600 | 16,000 | 16,000 | 16,000 |
| Running time hr/day (5 days/week) | 8 | 16 | 24 | 8 | 16 | 24 | 8 | 16 | 24 |
| Installed cost | £92,400 | 52,900 | 38,300 | 160,800 | 92,400 | 66,700 | 1,014,600 | 583,000 | 420,700 |
| No of poultry | 50,000 | 50,000 | 50,000 | 100,000 | 100,000 | 100,000 | 1,000,000 | 1,000,000 | 1,000,000 |
| | | | | | All costs given below in £/tonne dried product | | | | |
| Maintenance | 11.55 | 6.61 | 4.79 | 10.05 | 5.78 | 4.17 | 6.34 | 3.64 | 2.63 |
| Depreciation | 11.55 | 6.61 | 4.79 | 10.05 | 5.78 | 4.17 | 6.34 | 3.64 | 2.63 |
| Cost of capital | 20.33 | 11.64 | 8.43 | 17.69 | 10.16 | 7.34 | 11.16 | 6.41 | 4.63 |
| Labor | 7.50 (2.83)* | 15.00 (5.65) | 22.50 | 3.75 | 7.50 | 11.25 | 0.75 | 1.50 | 2.25 |
| Fuel oil | 14.70 | 12.72 | 11.14 | 14.70 | 12.72 | 11.14 | 14.70 | 12.72 | 11.14 |
| Electricity | 3.39 | 3.39 | 3.39 | 3.39 | 3.39 | 3.39 | 3.39 | 3.39 | 3.39 |
| Direct factory Operating Expenses | 69.02 (64.35) | 55.97 (46.62) | 55.04 | 59.63 | 45.33 | 41.46 | 42.68 | 31.30 | 26.67 |

*Figures in brackets indicate the effect of taking agricultural labor costs.

or small industrial dryer 8 or 16 hr/day operation), £3/man-hour plant labor and overhead (large units and 24 hr/day running). One man permanently small units, two men for large units.

6 Thermal efficiency 90 percent 24 hr/day running, 77 percent for 16 hr/day running, 63 percent for 8 hr/day running.

6 Thermal efficiency 90 percent for 24 hr/day running, 77 percent for 16 hr/day running, 63 percent for 8 hr/day running.

7 Fuel oil at 16 p/gallon.

8 Electricity £2.26/tonne dry product large unit, £3.39/tonne dry product package unit.

### Rotary Drum Dryers. Costs for Farm Packages

A farm package rotary drum dryer was taken as an example of a rotary drum package unit capital cost. Capital costs of industrial scale rotary drum dryers were obtained from Nonhebel and Moss, 1971, and from manufacturers. Other assumptions were:

1 Capital costs corrected to January 1975.

2 Tax depreciation over 10 years (10 percent per annum).

3 Maintenance 10 percent per annum.

4 Cost of capital 17.6 percent per annum, based on 15 percent DCF, ROI in 10 years after deducting 52 percent corporation tax.

5 Labor. £1.13/man-hour farm labor, £3/hr man-hour plant labor, one man permanently small units, two men large units.

6 Thermal efficiency 80 percent for 24 hr/day running, 70 percent for 16 hr/day running, 60 percent for 8 hr/day running.

7 Fuel oil 16 p/gallon.

8 Electricity £3.39/tonne dry product for all units.

## DISCUSSION AND CONCLUSIONS

As a result of our calculations, we believe that batch agitated pan dryers are the most attractive proposition for small farmers. Such units are best operated on a 16 hr/day basis.

Operating scales of $10^6$ poultry are required before industrial pneumatic dryers become attractive in comparison with batch agitated dryers on the small scale. Minimum direct farm operating costs for 90 percent dry manure would seem to be £16.57 and £24.04/tonne respectively for pneumatic and batch dried product, and £10.10 and £17.36/tonne for 65 percent dry manure.

Pneumatic dryers have scope for development for poultry manure drying on the large scale. Sterility may be a problem because of the relatively mild drying conditions and the occurrence of "balling". This should

**TABLE 6. COSTS OF DRYING POULTRY MANURE FROM 25 PERCENT TO 65 PERCENT SOLIDS.**

| Type of dryer | | Throughput (tonne product per annum) at each running time (hr/day, all 5 days/week) | | | Cost (£/tonne product) of drying at each running time (hr/day, all 5 days/week) | | |
|---|---|---|---|---|---|---|---|
| | | 8 hr/day | 16 hr/day | 24 hr/day | 8 hr/day | 16 hr/day | 24 hr/day |
| 1 Batch agitated | | | | | | | |
| - Package unit | (a) | 347 | 693 | 1,039 | 24.82 | 17.36 | 18.01 |
| - Package unit | (b) | 173 | 346 | 520 | 34.90 | 23.36 | 26.17 |
| 2 Pneumatic conveying | | | | | | | |
| - Package unit | | 520 | 1,040 | 1,560 | 24.39 | 17.22 | 23.20 |
| - Industrial dryers | (a) | <——— 1,300 ———> | | | 21.97 (19.16) | 25.02 (17.18) | 26.22 |
| - Industrial dryers | (b) | <——— 2,600 ———> | | | 19.50 | 18.29 | 18.79 |
| - Industrial dryers | (c) | <——— 26,000 ———> | | | 13.36 | 11.68 | 10.19 |
| 3 Rotary drum | | | | | | | |
| - Package unit | | 310 | 620 | 930 | 40.59 | 26.83 | 39.84 |
| - Industrial dryers | (a) | <——— 1,030 ———> | | | 52.17 (48.46) | 42.16 (34.74) | 41.71 |
| - Industrial dryers | (b) | <——— 2,060 ———> | | | 44.22 | 33.22 | 30.98 |
| - Industrial dryers | (c) | <——— 20,600 ———> | | | 31.26 | 22.58 | 19.18 |

*NB* Figures in brackets indicate the effect of taking farm labor costs.

be relatively easy to overcome.

From a technical point of view, rotary drum dryers are the most acceptable for poultry manure drying. Labor requirements on the small scale and capital costs on the large scale make rotary drum dryers economically non-competitive.

While our calculations show that all dryers, pneumatic dryers have the lowest direct farm operating expenses on the $10^6$ poultry scale, no indirect costs have been included in our calculations. Indirect costs are very variable, and are dependent on may circumstances. Inclusion of indirect costs could significantly influence a final choice of dryer.

### References


1 Heiering, A. G. and F. B. Hessa. 1973. Odourless drying of food production and processing wastes. 4th Joint Chem. Eng. Conf., Vancouver, 10 Sept.

2 Nonhebel, G. and A. A. H. Moss. 1971. Drying of solids in the chemical industry. Butterworths, London. 301 p.


---

## In-House Handling of Poultry Manure

*(Continued from page 472)*


dehydration by air drying and machine in a caged-layer house handling system. Poultry Pollution: Research Results, Farm Science M.S.U., Agricultural Experiment Station, East Lansing Research Report No. 269.

3 Flegal, C. J., M. L. Esmay, J. B. Gerrish, J. E. Dixon, C. C. Sheppard, H. C. Zindel and T. S. Chang. 1974. A complete system for collecting, handling, air-drying and machine dehydration of poultry manure in a caged layer production unit. Proceedings of the Sixth National Agricultural Waste Management Conference, Cornell University, March.

4 Gerrish, J. B., J. E. Dixon, M. L. Esmay, G. H. Quebe, C. J. Flegal, C. C. Sheppard and H. C. Zindel. 1973. Engineering aspects of handling and dehydrating poultry excreta: A progress report. ASAE Paper No. 73-4560, ASAE, St. Joseph, Mich. 49085.

5 Ludington, D. C. 1963. Dehydration and incineration of poultry manure. Proc. National Symposium on Poultry Waste Management.

6 Sheppard, C. C., C. J. Flegal, M. L. Esmay, J. B. Gerrish, J. E. Dixon, H. C. Zindel and T. S. Chang. 1974. A complete system for collecting, handling, air-drying and machine drying of poultry manure in a caged layer production unit. Proceedings of the XV World Poultry Congress, New Orleans, August.

7 Sobel, A. T. 1972. Undercage drying of laying hen manure. Proc. Cornell Ag. Waste Management Conf.: 187-200.

8 Surbrook, T. C. 1969. Evaluation of an animal excreta dryer. M.S. Thesis, Michigan State University, East Lansing, Mich. 48824.

9 Surbrook, T. C., C. C. Sheppard, J. S. Boyd, H. C. Zindel and C. J. Flegal. 1971. Drying poultry waste. Proc. Int. Symp. on Livestock Wastes, Ohio State University. ASAE Publication Proc. 271:192-194.

# Drying Dairy Wastes with Solar Energy

Brian Horsfield
ASSOC. MEMBER
ASAE

## INTRODUCTION

DEHYDRATION has always held some appeal as a method of treating and handling livestock wastes. The dried product, behaving like a solid, can be handled and conveyed with solid-handling equipment. If the moisture content is low enough, the product will be stable and relatively odor-free, storable for prolonged periods, and at much less cost than liquid manure. The loss of 80 to 90 percent of the bulk through evaporation greatly reduces problems of transport, and much less stream pollution results when this material is applied on damp or wet soils.

Partial drying of manure occurs naturally in the southwestern USA for the greater portion of the year. That is especially true in southern California, where, nearly the year around, sun-dried wastes are periodically scraped from feedlots and, if necessary, simply stored in piles ready for land application. Dairies in the midwestern and eastern USA are less fortunate because rainfall patterns there seldom allow open lots to remain dry for long. Liquid-manure handling has therefore been the usual waste-handling method under such conditions.

The drawbacks of liquid manure handling (expensive storage, odor potential, problems of year round land application) led to research on the use of solar energy for drying dairy wastes in the eastern USA.

Very little work has been done on solar drying of animal wastes. Hart (1964) investigated the drying of poultry wastes with solar energy. Moisture content was reduced to about 10 percent in thin layers of manure slurry exposed to the California sun. The author (1973) developed a computer simulation method which explored the effects of weather on the solar drying of hog manure from confined hog production in the Midwest. If sensible heat from the animals was used in combination with solar energy, it was concluded that a solar dryer of about the same area as the confinement unit could satisfactorily dry the wastes on a year-around basis. The concept underlying that computer simulation study was that, beginning sometime during the summer, the wastes would be periodically loaded into the solar dryer and that this uniform loading would continue until next summer, when the dried manure would be removed and the process repeated. This would permit once-a-year land application of the dried manure, allowing it to be spread over newly emerging crops without plant damage.

## CONCEPT FOR SOLAR DRYING
## OF DAIRY MANURE

The concept for solar drying of dairy manure is similar to the one used in that computer simulation. The solar dryer, similar in construction to a greenhouse (least expensive would be an air-supported double-walled greenhouse), should be large enough to hold a year's production of manure while capturing enough solar energy to reduce its moisture content to below 50 percent by June 15 of the following year. A device would mix the contents and spread the fresh manure uniformly within the solar dryer.

## EXPERIMENTAL PROCEDURE

Computer simulation indicated the approach to be feasible, but the concept required actual experimental confirmation not only to verify the drying rate but also to reveal any problems such as fly production, effects of possible aerobic decomposition, and effects of condensation on final weight loss.

Constructed for solar dryers were two plastic greenhouses similar to Plan No. 5946 of the USDA Plan Service. These were covered with 4-mil polyethylene plastic. Fig. 1 shows the two dryers in operation.

Each was to be loaded periodically (about three times per week), and the contents were to be stirred superficially at each loading and stirred thoroughly once a week. The computer simulation results indicated a loading rate of 0.334 lb of fresh manure per sq ft per day (1.63 kilos/sq meter/day).

Because the experiment was not started until November 7 (rather than in summer), some dry material was added to compensate for the delay. The computer simulation had indicated that loading beginning in early spring would yield a moisture content of 25-30 percent in November. Hence, on November 7, 1066 lb (484 kilos) of moldy chopped hay at 11.2 percent MC was mixed with 325 lb (148 kilos) of wet manure at 81.2 percent MC. This was equivalent to loading the dryer at the rate of 0.054 lb of dry material/sq ft/day (0.264 kilo/sq meter/day) since May 1.

Because of possible errors introduced through the addition of this dry material and because some dairy operations would find it more convenient to empty and begin reloading in the fall, considerably less dry material was added to the second dryer. The computer simulation had indicated that a starting date of August 15 would result in a moisture content of 45-50 percent by November 7. Therefore, 449 lb (204 kilos) of moldy chopped hay at 11.2 percent MC and 420 lb (191 kilos) of fresh manure were added, giving a

FIG. 1 Plastic covered greenhouses used for drying manure with solar energy near Lafayette, Indiana.

The author is BRIAN HORSFIELD, Assistant Professor, Department of Agricultural Engineering, University of California, Davis, California.

moisture content of 46 percent. At a loading rate of 0.054 lb of dry material/sq ft (0.264 kilo/sq meter), this was equivalent to starting the dryer on August 13.

## RESULTS AND DISCUSSION

After the initial loading, both dryers were loaded periodically. Fig. 2 illustrates the average daily loading rate of wet manure for each month of the experiment. Although inclement weather caused variations in loading rates, it was concluded that these variations would not seriously affect the outcome.

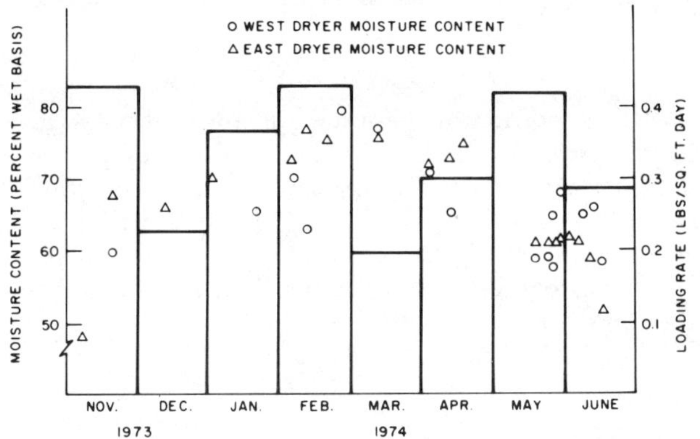


FIG. 2 Loading rate and moisture content (wet basis) for both dryers.

Moisture content was determined in each loading of wet manure, and was also determined periodically in the drying material. The temperature of the drying manure was determined periodically to indicate the level of aerobic decomposition using 6-in. dial thermometers. Fig. 3 shows the temperature range and average value for each dryer by month. In the early months the temperature exceeded 160 F (71 C) on several occasions. The average temperature dropped continually until March-April and then began to increase.

The range of temperature is probably less than indicated because of the shortcomings of the 6-in. dial thermometer. In addition, the wet and drier material often could not be mixed as thoroughly as desired, so there were pockets of material at higher moisture contents than others, which may have led to temperature variation within the drying material.

The moisture content of both dryers is plotted in Fig. 2 along with the loading rate. The difficulty in stirring the dryer contents thoroughly made it extremely hard to obtain accurate samples for moisture content. Only the first and last moisture-content readings are accurate. Even so, the general pattern is evident: an increasing and then decreasing moisture content. In addition the west dryer was in a very slight depression. Hence, despite a shallow ditch dug around both dryers to keep moisture out, a freezing-thawing pattern caused some water to leak into the west dryer and possibly into the east dryer.

Condensation was observed very frequently in both dryers. Hence, the doors were usually left open to promote moisture removal by air movement.

On days of low relative humidity and moderate winds, condensation completely disappeared. On humid still days, however, condensation dripped back onto the drying manure.

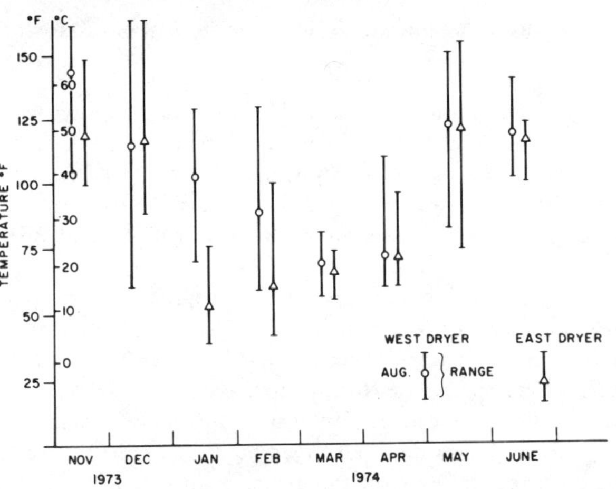


FIG. 3 Temperature of drying manure in both dryers.

Flies were observed in November but not in other months. Field mice occupied both dryers until late February. In February the polyethylene appeared to lose some of its transparency.

The contents of both dryers were removed and carefully weighed on June 14, 217 days after the experiment was started. Four moisture samples were taken from each dryer and the results averaged. Table 1 summarizes the total loading and the final contents of both dryers.

Even though water leaked into the west dryer and both dryers had condensation drip back onto the drying manure, both dryers lost over 79 percent of the moisture introduced. Further, the west dryer lost 56 percent of the dry material added, and the east dryer lost 31.4 percent, indicating that composting was active and probably accounts for some of the energy available for water evaporation.

TABLE 1. SUMMARY OF MANURE LOADED AND FINAL CONTENTS

| | | West dryer | | East dryer | |
|---|---|---|---|---|---|
| Material added | Dry weight | 2196 lb | ( 996 kg) | 1645 lb | ( 746 kg) |
| | Water fraction | 6593 lb | (2990 kg) | 6508 lb | (2952 kg) |
| | Total | 8889 lb | (4032 kg) | 8153 lb | (3698 kg) |
| | Average moisture content | 74 percent | | 80 percent | |
| Final material removed | Dry weight | 963 lb | ( 437 kg) | 1128 lb | ( 512 kg) |
| | Water fraction | 1345 lb | ( 611 kg) | 1217 lb | ( 552 kg) |
| | Total | 2310 lb | (1048 kg) | 2345 lb | (1064 kg) |
| | Final moisture content | 58 percent | | 52 percent | |
| Percent loss | Dry weight | 56.1 | | 31.4 | |
| | Water | 79.6 | | 81.3 | |
| | Total | 74.0 | | 71.2 | |

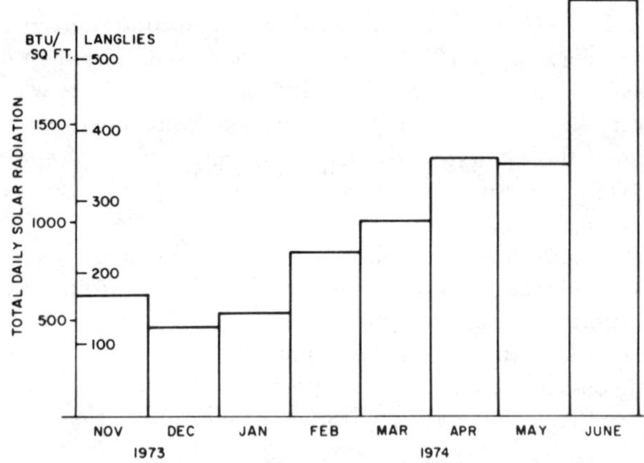



FIG. 4 Monthly average of total daily solar radiation.

Solar radiation incident on both dryers is summarized in Fig. 4. Total radiation between November 7, 1973, and June 14, 1974, was 216,796 Btu/sq ft (58,805 Langlies). At 1047 Btu/lb of water evaporated (582 calories/gram), solar efficiency was 24.8 percent for the west dryer and 25 percent for the east dryer.

Solar efficiency was considerably lower than anticipated. Several factors were responsible. When wet manure was placed in the driers during the first few months, it was very difficult to spread uniformly. As a result, only a portion of the surface was covered by wet manure, the remainder of the surface being material already dry. In addition, condensation dripping back on the drying manure and water leaks due to poor placement of the dryer added water that is not accounted for but nevertheless, required solar energy to be evaporated.

The loading rate of 0.054 lb of dry weight/sq ft/day (0.264 kg/ sq meter/day) is roughly equivalent to a dryer of 194 sq ft (18 sq meter) for a cow producing 65 lb (30 kilos) of manure per day. An 80-cow herd would thus require a solar dryer of about 16,000 sq ft (1486 sq meter). Estimated costs are as follows:

| EQUIPMENT | Initial Cost ($ USA) | Yearly Cost ($ USA) |
|---|---|---|
| Stirring device | 4000 | |
| Fans and controls | 900 | |
| Total | 4900 | 833* |

PLASTIC COVER

| | | |
|---|---|---|
| Ultraviolet resistent polyethylene (6 mil) | $0.02/sq ft | |
| inner layer polyethylene (2 mil) | $0.01/sq ft | |
| | $0.03 | |

For 16,000 sq ft

| Total | $480 | 240** |
|---|---|---|
| Total yearly cost | | $1073 |

* Assume yearly cost of 17 percent of initial investment.
**Assume material must be replaced every 2 years.

Or an annual cost of $13.41 USA per dairy cow.

For this annual cost the manure is reduced in total weight to about 71 to 74 percent of original and has been stored for an entire year.

## CONCLUSION

A large portion of the water content of dairy waste can be successfully removed by a solar dryer of proper size designed similarly to a greenhouse. For conditions in central Indiana it appears that a solar dryer with a surface area equivalent to 200 sq ft per dairy cow can remove approximately 79 percent of the total moisture added in the course of one year.

## SUGGESTIONS FOR IMPROVEMENTS

The experiment was conducted under less than ideal conditions. An adequate spreading and stirring device would have given better utilization of solar energy, especially during the initial phase of the experiment. In addition, proper stirring would undoubtedly have encouraged greater composting activity. A double-walled air-supported greenhouse type of solar collector would have helped prevent some condensation as well as provide positive constant air exchange. That would have removed more moisture. If this air-supported solar dryer were orientated on an east-west axis and the lower portion of the northern inner surface made of a reflective plastic, some of the solar energy which would normally pass through the dryer in winter, when the sun is low, would be reflected back onto the drying manure.


**References**

1   Hart, S. A. 1964. The spreading of slurried manures. Proc. Pacific N.W. Animal-Industry Waste Conf., Washington State, Pullman, Wash. Publ. 69, October.

2   Horsfield, B. C. 1973. Drying animal wastes with solar energy and exhaust ventilation air. ASAE Paper Number 73-411, ASAE, St. Joseph, Michigan 49085.

# High-Rate Mechanized Composting of Dairy Manure


**J. W. Hummel, G. B. Willson**

MEMBER     MEMBER
ASAE       ASAE


IN years past in the United States, animal manures have been low-value, odorous, undesirable materials; in other words, waste products of the animal production industry. Due to recent shortages and high costs of commercial fertilizers, coupled with increased demand for United States agricultural products in the world markets, animal manures are rapidly becoming a valuable resource, to be recovered and utilized. New environmental standards regulating animal feeding and housing facilities, the collection of animal manures and their storage and utilization are being established. These developments have resulted in increased interest in the composting process and predict an expanded role for composting in meeting our conservation and utilization objectives.

Laboratory investigations have established that composting has the potential to reduce the waste management problems associated with livestock production systems. Aerobic composting results in high decomposition rates and minimal odor production. Composting could be a method to treat and manage dairy cow manure to meet odor, fly and environmental limitations.

Research was initiated to investigate the concept of a mechanized aerobic composting system which could be sized for typical dairy operations in Northeastern United States and which would operated satisfactorily in that climate. Due to the significant rural/urban interface that exists throughout this region resulting in a market potential for the compost as a soil amendment, a uniform product of high quality was desired.

## CHANNEL DESIGN

The success of the composting trials with the laboratory system and the potential uses for the material produced culminated in a pilot scale mechanized composting system (Fig. 1) for the University's 80-cow dairy herd on the College Park Campus. The system concept provides oxygen to support the aerobic microorganisms, agitation to counteract moisture migration and settling, and moves the composting mass through a 18-meter long channel in approximately 15 days.

The channel itself was constructed on a concrete base. The channel sides of 2.5-cm marine plywood are 1.5-m high and are supported on 1.2-m centers by angle iron supports set in the concrete. Wood surfaces in contact with the composting materials were protected by pentachlorophenol. Rails of 7.6-cm square structural tubing attached at the top of the channel sides support the traversing elevator and its

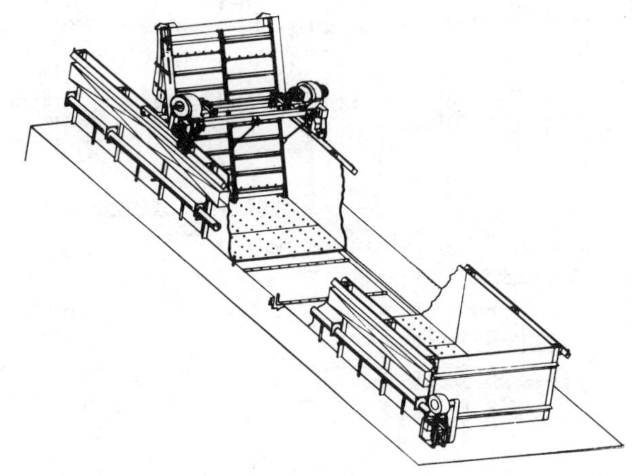

FIG. 1 Pilot mechanized composting system.

carriage.

Oxygen to support the aerobic microorganisms in the composting mass was supplied to the channel by a Model 22 Buffalo centrifugal fan operating continuously. The main air duct was divided into five sections, each supplying air to a 3.7-meter long section of the channel. Sharp-edge orifice plates located at the end of each section of the main air duct permitted flow measurement as well as some pressure regulation. The airflow rate to each section was controlled by the pressure in the laterals extending under the channel, and the number and size of the orifices in the laterals. A layer of 1.3-cm plywood with a series of 2.5-cm holes over the laterals provided an air diffusion chamber throughout the channel.

A carriage (Fig. 2), which was mounted on the channel rails, carried the elevator, an 11.2-kw electric motor, and all drive train components. The elevator (Fig. 3), which traversed the channel length each day at 0.5 cm/sec, stirred the composting mass and systematically advanced the material along the channel. The elevator was fitted with cam-controlled feathering toothbars (Fig. 4) to loosen the face of the composting mass prior to contact by the elevator flights. Teeth on succeeding bars were arranged to contact the compost at 5-cm intervals across the channel width.

The composting channel is housed in a 4.9-m x 24.4-m pole building with open sides. Insulated agricultural siding was used for roofing to guard against condensation.

## OPERATION

Manure placed in the composting system was received directly from the gutter cleaner at the University of Maryland dairy barns where the cows are housed in a stanchion system. This charge material was a mixture of feces, urine, sawdust bedding, and small amounts of waste feed removed from the research animals' mangers on a daily


The authors are: J. W. HUMMEL, Associate Professor, Agricultural Engineering Dept., University of Maryland, College Park, Md.; and G. B. WILLSON, Agricultural Engineer, ARS-USDA, University of Maryland, College Park, Md.

Scientific Article Number A2088 Contribution Number 5042 of the Maryland Agricultural Experiment Station (Department of Agricultural Engineering).

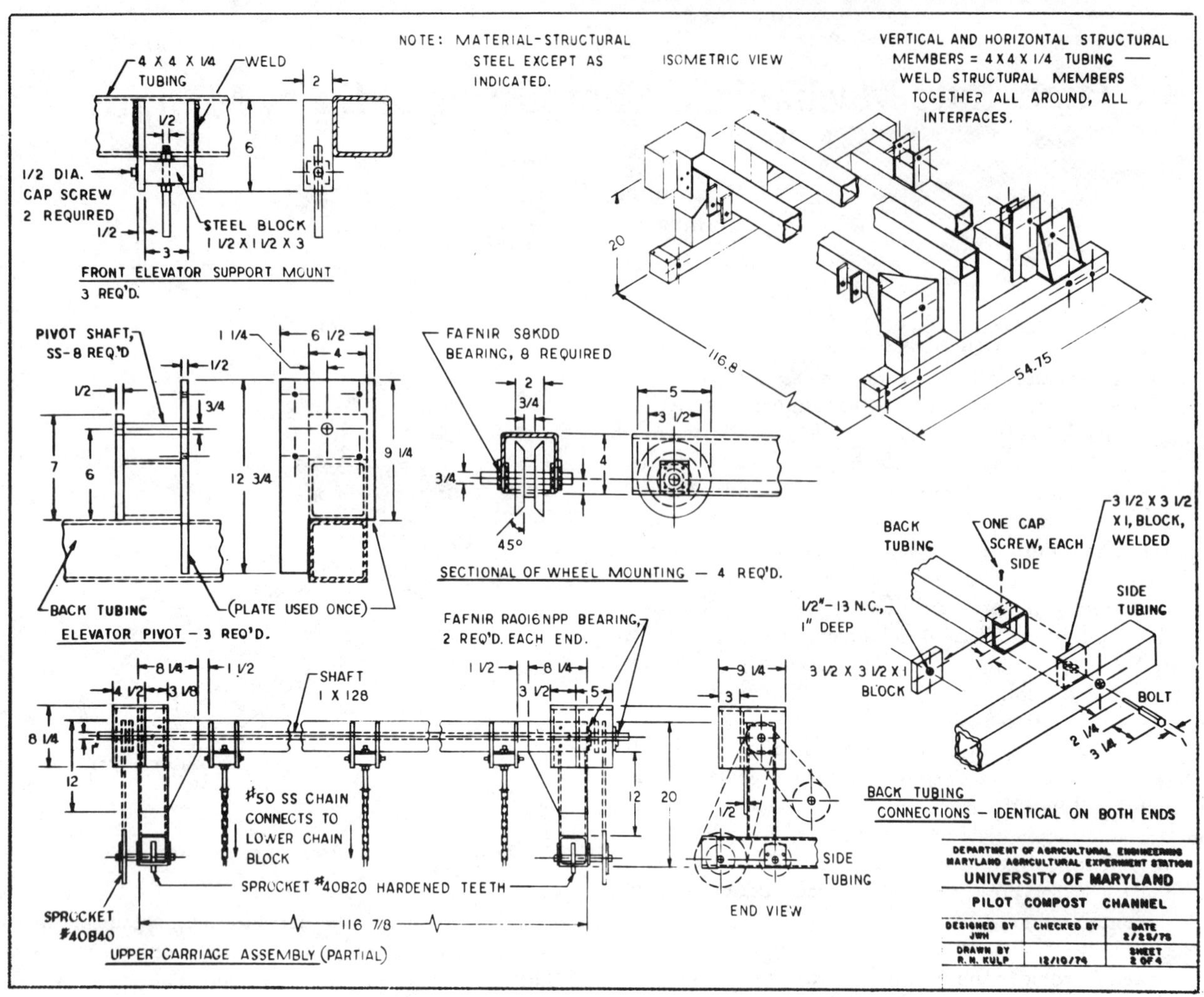

FIG. 2 Composting system carriage.

basis. In order to reach thermophilic composting temperatures (50 C. or higher) within 24 hours, the moisture content of this mixture of material must not exceed 70 percent W.B. (wet basis). The structure of the composting mass, which controls air diffusion, required little attention as long as the moisture content was controlled to the proper level. The moisture content was lowered, when necessary, by increasing the amount of bedding material used. Because of the high moisture content of the sawdust bedding (36 percent to 47 percent W.B.), the bedding rate was higher than normally expected. During November 1973, sawdust bedding was used at a rate of 8 kg per cow per day. The system, as operated, accepted the manure from approximately 70 cows.

Since increased bedding use results in increased operating costs, every effort was made to exclude excess moisture from the charge material. Rain and snow must be excluded, dumping of waste milk eliminated, and washing down of facilities minimized.

The manure from the stanchion barn system was moved directly to the compost channel on a daily basis. Each day of operation the elevating mechanism in the channel moved the composting waste approximately 1.2 meters farther along the channel, producing a semi-continuous flow system with raw manure introduced at one end of the channel, and expelled as partially stabilized compost at the opposite end of the channel about 15 days later. The aeration rates and operating conditions in each channel section are presented in Table 1. The volume values were estimated from data on overall volume reduction and observation of material depth in the channel. Data used to develop this table were collected during a two-month period of operation.

Moisture content of input and output materials was monitored regularly. Prior experience had shown that if the moisture content of the charge material was held to 70 percent W.B. or lower, and sufficient aeration supplied to keep the temperature of the composting mass below 70 C., moisture contents in the interior sections would be conducive to composting and therefore were not monitored. The average temperatures are based on measurements of the temperature in the interior of the composting material. All surface or boundary temperatures were excluded from the data presented.

The airflow rate to each section was controlled by the pressure in the laterals extending under the channel, and the number and size of the orifices in the laterals. For this system and the flowrates listed, the principal pressure drop is across the orifice. Since the pressure drop through the composting material is negligible, no adjustments were necessary during channel start-up. The airflow rate/m$^3$ of waste material was lower initially where the population of microorganisms was low (Section 1), increased as the population and heat production increased (Section 2), decreased as microorganism activity decreased (Section 3) and increased (Section 4 & 5) to achieve maximum moisture evaporation.

The output material of the compost channel, although



MANAGING LIVESTOCK WASTES

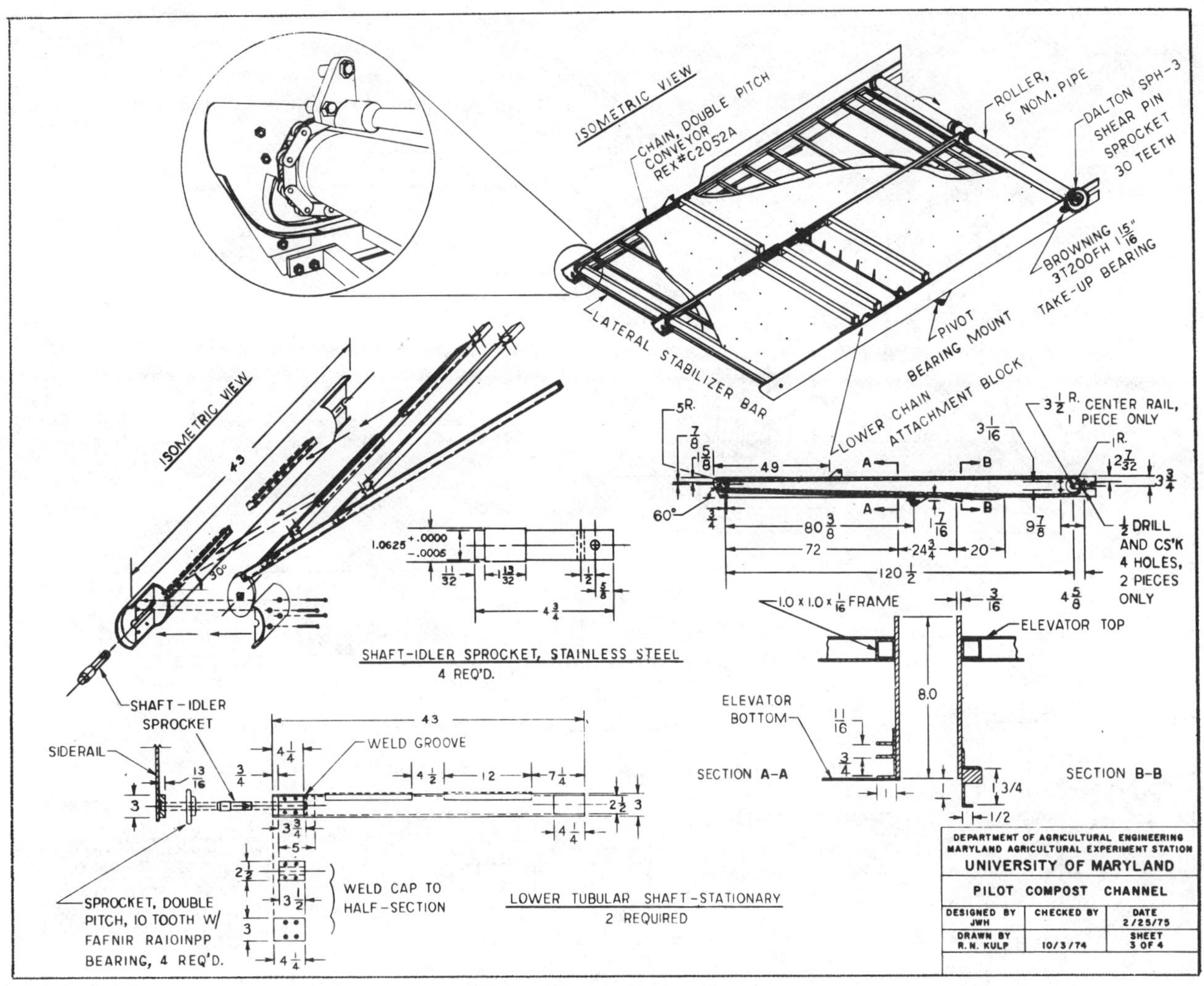

FIG. 3 Composting system elevator.

not completely decomposed, has lost its objectionable odor and can be utilized directly as a soil amendment or mulch. If further composting and/or storage are desired, the material is placed in a programmed windrow where convective currents aerate the compost. Some additional moisture reduction is realized in the windrow due to the elevated temperatures ($\simeq$ 70 C.) achieved and the ability of the windrow to shed precipitation.

The functional testing of this system has demonstrated its feasibility for accelerated composting of dairy manure. The material produced is attractive to urban homeowners and can be marketed. An economic analysis of the system was conducted to establish the cost of operation (Hummel and Lessley 1974). The analysis showed an investment cost of $17,340, and an annual cost of $84/cow. Presently planned modification should reduce the labor and repairs and maintenance costs and result in an annual cost of $60/cow, or $4.40/m$^3$ of compost produced. The system is not an economically feasible alternative unless some of the system's cost is offset by sale of the compost produced.

## CONCLUSIONS

The following conclusions can be made about this mechanized composting system for manure from stanchion barns:

1    The design can be sized to accommodate most of the range of individual dairy operations in Northeastern United States.

**TABLE 1. COMPOST CHANNEL OPERATING CONDITIONS.**

| Item | Section | | | | | | | | | |
|---|---|---|---|---|---|---|---|---|---|---|
| | 1 | | 2 | | 3 | | 4 | | 5 | |
| | $\overline{X}$ | S.D. | $\overline{X}$ | S.D. | $\overline{X}$ | S.D. | $\overline{X}$ | S.D. | $\overline{X}$ | S.D. |
| Volume, m$^3$ | 11.9 | | 11.9 | | 11.6 | | 10.9 | | 10.3 | |
| Moisture content, percent W.B. | 70.4 | 3.8 | — | | — | | — | | 66.1 | 3.1 |
| Avg. temp. C | 63.0 | 16.0 | 62.0 | 10.0 | 56.0 | 4.0 | 48.0 | 4.0 | 51.0 | 4.0 |
| Airflow rate, m$^3$/h | 98.4 | 3.7 | 105.0 | 4.8 | 69.2 | 2.7 | 82.1 | 1.0 | 80.2 | 1.2 |
| Specific airflow, m$^3$/h/m$^3$ | 8.3 | | 8.8 | | 6.0 | | 7.5 | | 7.8 | |
| Pressure, Pa | 2060 | 37.0 | 1920 | 35.0 | 1790 | 35.0 | 1660 | 32.0 | 1580 | 30.0 |

$\overline{X}$ — Mean
S.D. — Standard Deviation

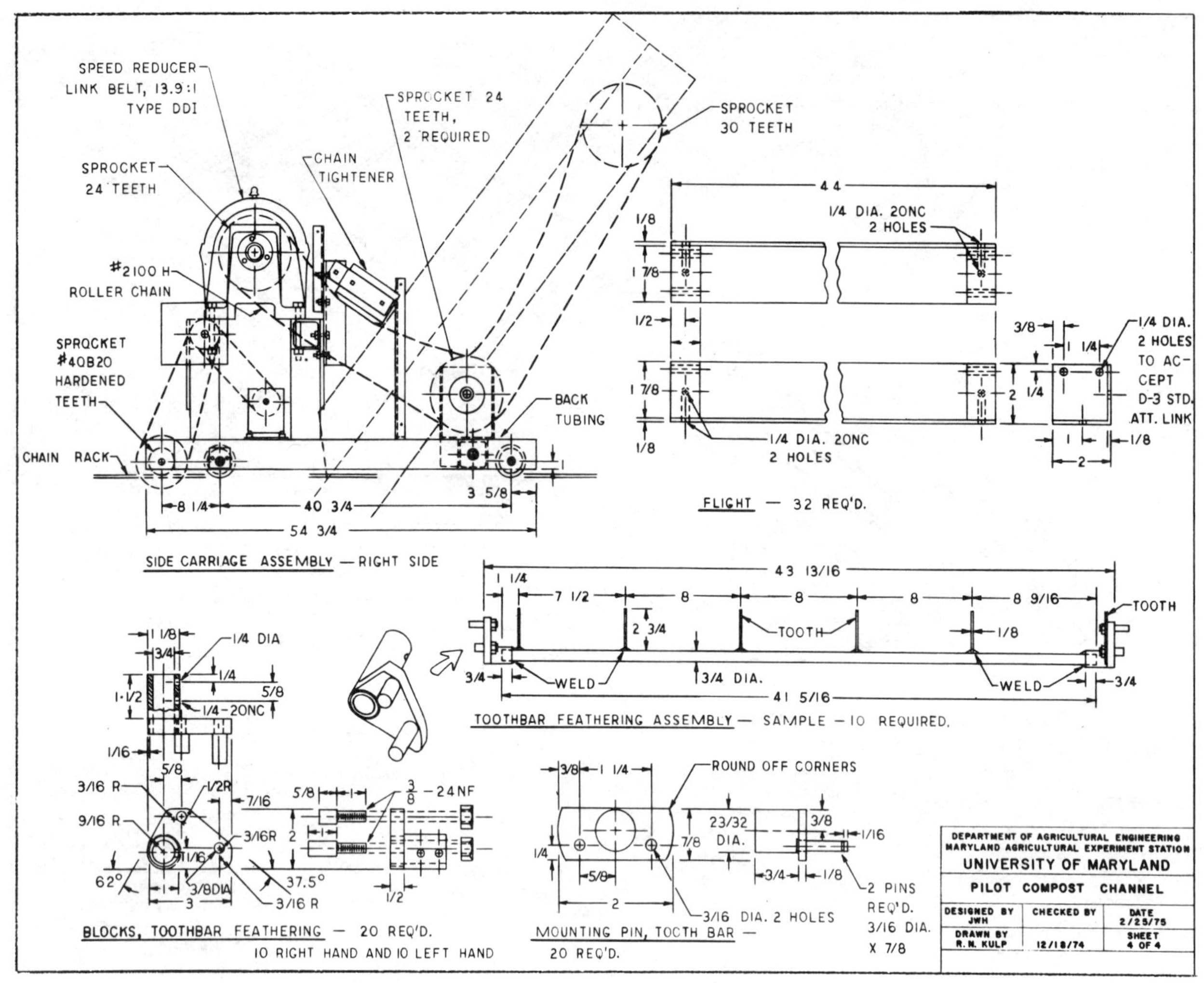


FIG. 4 Composting system toothbar and drive train configurations.

2 Charge material moisture content must be closely controlled.

3 A uniform product, appealing to home gardeners, is produced.

4 The product is cost competitive with similar or substitute materials already being marketed.

### References


1 Bell, R. G. 1973. The role of compost and composting in modern agriculture. Compost Science, 14(6):24-25.

2 Berge, O. J. 1971. Waste management . . . what does it cost? Hoards Dairyman, 116(7):420.

3 Grim, A. 1972. Dairy manure waste handling systems. Proc. Cornell Univ. Agricultural Waste Management Conference. Graphics Management Corp., Washington, pp 125-144.

4 Hummel, J. W. 1973. Equipment for composting of agricultural wastes. ASAE Paper No. NA73-109, ASAE, St. Joseph, Michigan 49085.

5 Hummel, J. W. and B. V. Lessley. 1974. Mechanized composting system evaluation. ASAE Paper No. 74-4513, ASAE, St. Joseph, Michigan.

6 Hummel, J. W., W. F. Schwiesow and G. B. Willson. 1974. Power requirements of a compost channel for animal wastes. TRANSACTIONS of the ASAE, 17(1):70-73, St. Joseph, Michigan.

7 Poincelot, R. P. 1972. The biochemistry and methodology of composting. Bulletin 727, Connecticut Agricultural Experiment Station.

8 Senn, C. L. 1974. Role of composting in waste utilization. Compost Science, 15(4):24-28.

9 Willson, G. B. 1971. Composting dairy cow wastes. Proc. International Symposium on Livestock Wastes, ASAE, St. Joseph, Michigan 49085, pp 163-165.

10 Willson, G. B. and J. W. Hummel. 1972. Aeration rates for rapid composting of dairy manure. Proc. Cornell Univ. Agricultural Waste Management Conference. Graphics Management Corp., Washington, pp 145-158.

# Aerobic Composting — New Built-Up Bed Technique

D. P. Stombaugh, R. K. White

ASSOC. MEMBER    ASSOC. MEMBER
ASAE           ASAE

AEROBIC composting is a process in which the organic solids in animal wastes are digested by bacterial action. This process generates few offensive odors. Once the process is completed, the organic solids are in a stable condition undergoing little or no further decomposition (Miner 1971). Composting reduces the weight and volume of the solids being treated, which can lead to significantly lower transport costs for material that must be disposed of on agricultural land.

The aerobic composting process produces heat that aids in the evaporation of large quantities of water. Under appropriate conditions, process temperatures may reach 50 to 75 C. These temperatures kill many pathogenic bacteria and eggs of parasites (Miner 1971).

Aerobic composting has generally been accomplished in open piles, by windrowing or by using mechanical mixing processes requiring a large initial investment. Climatic conditions, a tendency to become anaerobic and the possibility of runoff hinder the widespread adoption of aerobic composting in open piles. Windrowing the waste solids on the gound or on paving and the more sophisticated processes are prohibitively expensive for all but the largest producers because of the large initial investment required.

The objective of this study was to design, develop and evaluate in a pilot model installation a manure treatment and storage system for the small producer who must meet strict environmental standards due to increased urbanization. The aerobic composting process was selected because it was felt that the following criteria could be met.

1 Low odor levels.

2 Negligible leaching of soluble substances into groundwater supplies.

3 Minimal runoff into streams and lakes.

4 Elimination of pathogenic bacteria, parasites and the attractiveness to flies.

5 Storage could be provided to permit flexibility in land application of solids.

6 By utilizing existing facilities on many farms, the initial cost of the proposed system could be maintained at a reasonable level.

## BUILT-UP BED TECHNIQUE

As conceived, the built-up bed, aerobic composter provided both treatment and storage capability and consisted of two major components. The first major component was the structural facility for receiving and storing animal waste. This structure consisted of a large holding bin possibly constructed on a concrete base to prevent runoff and an open-sided, pole shelter to provide protection from high moisture loads due to precipitation. The second component was the mechanical aeration system. This system, which could be operated automatically and at various heights in the compost bin, provided an adequate oxygen supply for rapid aerobic composting by tilling or mixing the upper surface of the waste material. Aeration must occur throughout the upper layers which are undergoing the most rapid stabilization if the process is to proceed rapidly and without the evolution of obnoxious odors. With the daily addition of approximately 3 cm of waste material over the entire bin surface the waste undergoing the most rapid stabilization would be located within the upper 35 cm. Consequently, a mixing device compatible with the main storage structure and capable of aerating animal waste to a depth of at least 30 to 40 cm was developed. The stabilized compost stored and allowed to ripen beneath these upper layers was not aerated.

The composter was conceived to utilize existing manure handling and storage facilities presently available on many small dairy and beef farms. For example, for many 50 cow dairy herds manure is presently scraped from the alleyways or gutters, carried by a cross conveyor, elevated and then stacked in open piles. If a storage facility is needed, the aerobic composter as proposed could be installed for a slightly higher initial cost, but could, depending on circumstances, pay for itself in reduced hauling costs and in a smaller initial structure due to volume and moisutre reductions obtained during the composting process. For a 50 cow dairy herd, typical manure production would be 2.8 m³/day (99 ft³/day). If a compost bin 3.66 m (12 ft) wide by 25.5 m (70 ft) long by 3.66 m (12 ft) high were constructed, 3 cm of manure could be added on the surface daily and with some volume and moisture reduction, storage for approximately 180 days would be provided. Although the cost would be higher than for a typical elevator and approved solid manure storage, parasite problems and the reduced odor levels, especially during field application, could permit small producers to meet very strict environmental demands.

## PILOT MODEL DESIGN

To evaluate the proposed aerobic treatment system, a pilot model composter was designed, installed and evaluated at The Ohio Agricultural Research and Development Center's swine farm. The swine farm was

---

This paper was approved as Journal Article No. 34-75 of the Ohio Agricultural Research and Development Center, Wooster.

The authors are: D. P. STOMBAUGH and R. K. WHITE, Assistant Professors, Agricultural Engineering Dept., Ohio State University and Ohio Agricultural Research and Development Center, Columbus.

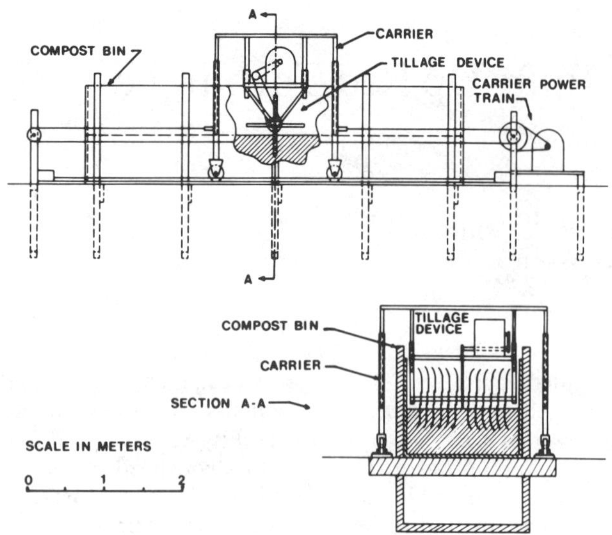

FIG. 1 Schematic diagram of pilot model, built-up bed, aerobic composter showing the compost bin, carrier, carrier power train and tillage device.



chosen because it utilized solid manure handling with bedding and because an existing, open sided pole structure and utilities were available.

Within this building a compost bin (one end removable for cleaning) 1.5 x 4.88 x 1.22 m was constructed from 2 in. x 6 in. x 16 ft penta-treated, tongue and groove lumber as shown in Fig. 1. The bottom and sides were supported by ten 4 in. x 4 in. x 8 ft posts which were braced at ground level and at a depth of 1 m. Additional posts were installed at each end to anchor the carrier drive mechanism.

The carrier and tillage device are shown in Fig. 2. The carrier constructed from tubular steel, rolled on runners (2 in. x 10 in. planks, with angle iron guides) at ground level (Figs. 1 and 3). A 2 hp gasoline engine with appropriate speed reduction and a truck transmission (to provide forward and reverse directions of travel) was mounted on a platform at one end of the compost bin and propelled the carrier. Two cables, shown in Figs. 1 and 3, were attached to the four corners of the carrier and connected the carrier to the gasoline engine. Using the system the composter traversed the length of the compost bin in 5 minutes (0.73 m/min).

FIG. 2 Carrier, tillage device and tiller power train for the aerobic composter.

FIG. 3 Pilot model aerobic composter installed at The Ohio Agricultural Research and Development Center swine farm.

The carrier provided a course height adjustment totaling 84 cm (33 in.) in 7.6 and 15.2 cm increments which was manually adjusted approximately once each week. The carrier supported a hanging platform which supported both the tillage mechanism and its power train. The platform carried a 1.5 hp gasoline engine and drive train for the tiller which included a clutch and a transmission used to reverse the direction of tiller rotation. The four corners of the platform were attached to the carrier by four gear and rack jacks (Hammer, Blow, Model 164). All four worm gears were attached to a reversible, 1 hp, electric motor also mounted on the tillage platform. Electrical controls mounted on the carrier frame permitted continuous vertical adjustment of the tillage platform through the 43 cm travel of the worm gears on the racks. Safety overrides, both top and bottom, were installed to limit vertical movement on the racks.

The tiller blades were mounted on a 1 1/2 in. diameter shaft and driven by a shielded roller chain drive at approximately 200 rpm (no load). Tiller blades were made from 3/8 x 2 in. cold rolled steel, by forming an elongated rounded point on one end and bending the tip with a smooth radius at an angle of approximately 30 degrees as seen in Fig. 2. Total blade length was 36 cm from the shaft center line. Six staggered sets of two blades each were mounted on the tiller shaft on each side of the centered chain drive as shown in Fig. 2.

## COMPOST TRIALS—PROCEDURES

Following two initial trials to establish operational guidelines and mechanical design parameters (carriage travel speed, tiller rpm, etc.), three composting trials to evaluate the built-up bed composting technique were completed. In each of the three tests reported approximately 12 cm of product from the previous test was left in the composter to facilitate the initial tillage operation.

### Manure Addition

Feces, urine, bedding and other waste products removed from pens at the OARDC swine farm were placed in wheelbarrows, weighed and dumped into the compost bin from a removable ramp. During tests 1 and 2 manure was added on Monday, Wednesday and Friday; during test 3 manure was added daily with the exception of Saturday and Sunday. Manure was normally dumped at one location and then spread over the surface of the composter by running the tiller across



MANAGING LIVESTOCK WASTES

TABLE 1. MOISTURE CONTENT, pH, pH  AFTER 48 HR ANAEROBIC FERMENTATION AND MAXIMUM TEMPERATURES FOR TRIAL NO. 1, 3/5/74 to 6/12/74.

| Week | Input | | | | Moisture content, percent | | | Average compost | | Maximum temperature, deg C |
| | Weight, kg | Moisture, percent | pH | 48 hr pH | 0-5 cm, | 15-25 cm, | bottom | pH | 48 hr, pH | |
|---|---|---|---|---|---|---|---|---|---|---|
| 1 | 349 | 78.2 | 8.14 | — | 75.3 | 74.3 | — | 9.14 | — | 15 |
| 2 | 557 | 73.1 | 8.20 | — | 70.8 | 77.5 | 75.3 | 8.97 | — | 27 |
| 3 | 602 | 81.6 | 8.16 | 6.31 | 77.8 | 78.3 | — | 8.58 | 6.92 | 33 |
| 4 | 620 | — | — | 6.69 | 79.7 | 79.0 | — | 8.30 | 6.60 | 14 |
| 5 | 392 | 79.1 | 7.34 | 6.41 | 79.1 | 79.4 | — | 7.58 | 6.60 | 22 |
| 6 | 358 | 83.2 | 6.93 | 6.30 | 76.5 | 77.6 | — | 7.34 | 6.89 | 31 |
| 7 | 320 | 79.4 | 8.46 | 6.65 | 76.1 | 76.8 | — | 8.85 | 6.82 | 49 |
| 8 | 415 | 78.9 | 8.08 | 6.80 | 70.1 | 74.9 | — | 8.79 | 7.30 | 59 |
| 9 | 432 | 72.1 | 8.14 | 6.12 | 71.9 | 71.5 | — | 9.10 | 7.04 | 59 |
| 10 | 570 | 68.7 | 8.14 | 6.56 | 57.7 | 61.1 | 57.7 | 8.95 | 6.81 | 59 |
| 11 | 314 | — | 8.22 | — | — | — | — | 8.60 | — | 60 |
| 12 | | | | | 59.0 | 64.9 | 64.1 | 8.63 | 6.70 | 54 |
| 13 | | | | | 64.9 | 61.9 | 50.6 | 8.48 | — | 50 |
| 14 (Final) | | | | | 44.0 | 49.4 | 73.8 | 8.66 | 6.52 | 41 |

the pen one or two times with the direction of rotation reversed and with the tiller blades at the desired final depth.

### Aeration

Manure in the composter was mechanically aerated (tilled) to a depth of 25 to 35 cm daily (with the exception of Sunday). This was performed during midmorning and immediately followed the addition of manure when it was added. This normally involved one pass in each direction of travel; however, on Mondays two passes in each direction were usually made because of the greater degree of settling which occurred on Sunday when no tillage occurred. Rotation of the tiller baldes at the top of the arc was in the direction of tiller travel. With the exception of the final week of test 3, aeration was continued following the cessation of manure input until the final samples were taken from the composter.

### Sampling

Grab samples were taken once each week from the input manure and from the material in the compost bin. Both input and output samples were taken immediately preceding manure addition. Output samples from the bin were taken, when total compost depth permitted, at three distinct levels (0-5 cm from top surface, 15-25 cm from the top surface, and midway between the bin's bottom surface and maximum tillage depth). Samples were taken from the center portion of the bin (i.e. no samples, except for comparison purposes, were taken within 30 cm of the bin wall).

Process temperatures (vertical gradient) were monitored at several locations within the bin once each week. These temperatures were taken before stirring and before taking grab samples each week. Normally temperatures were monitored again following mechanical aeration. Temperatures were measured with a thermister probe (calibrated to the nearest 0.5 C against an NBS thermometer) in a stainless steel tube taped to an aluminum shaft and pushed into the compost pile by hand.

Atmospheric oxygen samples were taken from the air spaces in the bin on two occasions from a depth of 20 cm. Each sample to be analyzed (approximately 1.5 l) was collected in a gas analysis bag by drawing approximately fifteen 100 ml samples. To collect each of the 15 samples, two 25 cm x 18 gauge stainless steel tubes were connected to the main sample bag by an airtight pump, a threeway stopcock and small bore (1.2 mm I.D.) tubing. Each small (100 ml) sample was taken at a point at least 30 cm from the previous sample.

### Analysis

Weekly input and compost samples were brought to the laboratory and analyzed for moisture, carbon and nitrogen contents, pH, pH after anaerobic incubation for 48 hours at 55 C and volatile solids (VS). From each grab sample approximately 100 g was removed and dried to

TABLE 2. MOISTURE CONTENT, pH, pH AFTER 48 HR ANAEROBIC FERMENTATION AND MAXIMUM TEMPERATURES FOR TRIAL NO. 2, 7/2/74 to 9/10/74.

| Week | Input | | | | Moisture content, percent | | | Average compost | | Maximum temperature, deg C |
| | Weight, kg | Moisture, percent | pH | 48 hr pH | 0-5 cm, | 15-25 cm, | bottom | pH | 48 hr, pH | |
|---|---|---|---|---|---|---|---|---|---|---|
| 1 | 456 | 74.9 | 8.05 | 6.79 | — | — | — | — | — | 34 |
| 2 | 768 | 69.7 | 8.17 | 6.86 | 69.0 | 64.1 | — | 8.89 | 7.21 | 33 |
| 3 | 804 | 77.1 | 8.61 | 6.74 | 69.3 | 69.4 | — | 9.04 | 6.94 | 57 |
| 4 | 736 | 71.6 | 8.40 | 6.74 | 66.6 | 63.8 | 48.9 | 8.75 | 6.74 | 59 |
| 5 | 720 | 70.8 | 7.94 | 6.98 | 47.1 | 78.9 | 36.9 | 8.55 | 7.24 | 68 |
| 6 | 837 | 78.5 | 7.97 | 6.88 | 61.8 | 58.8 | 55.6 | 8.55 | 6.90 | 66 |
| 7 | 261 | — | — | — | 69.2 | 65.9 | 52.4 | 8.62 | 7.34 | 61 |
| 8 | | | | | 28.0 | 37.6 | 36.2 | 8.48 | — | 53 |
| 9 | 227 | 100 | | | 28.5 | 26.6 | 27.0 | 8.63 | 7.34 | 29 |
| (after H₂O) | | | | | (56.1) | (61.4) | (65.4) | | | (57) |
| 10 | | | | | 53.2 | 44.8 | 51.5 | 8.76 | 7.40 | 57 |
| 11 | | | | | 48.6 | 36.8 | 33.5 | 8.76 | 7.55 | 35 |

TABLE 3. MOISTURE CONTENT, pH, pH AFTER 48 HR ANAEROBIC FERMENTATION AND MAXIMUM TEMPERATURES FOR TRIAL NO. 3, 10/21/74 TO 12/23/74.

| Week | Input Weight, kg | Input Moisture, percent | pH | 48 hr pH | Moisture content, percent 0-5 cm | 15-25 cm | bottom | Average compost pH | 48 hr, pH | Maximum temperature, deg C |
|---|---|---|---|---|---|---|---|---|---|---|
| 1 | 382 | 54.0 | 8.86 | 7.16 | 77.2 | — | — | 8.64 | 7.08 | 7 |
| 2 | 768 | 78.5 | 8.09 | 6.80 | 75.3 | 76.4 | — | 8.70 | 7.47 | 25 |
| 3 | 452 | 82.2 | 7.90 | 7.10 | 72.1 | 75.2 | — | 8.51 | 7.53 | 36 |
| 4 | 895 | — | — | — | 74.0 | 76.4 | — | 8.07 | 7.17 | 40 |
| 5 | 895 | 74.1 | 7.49 | 6.42 | 74.6 | 75.0 | 75.9 | 8.17 | 7.32 | 52 |
| 6 | 888 | 78.4 | 8.30 | 6.90 | 75.0 | 69.0 | 76.2 | 8.51 | 7.15 | 50 |
| 7 | | | | | 82.8 | 72.5 | 75.5 | 8.35 | 7.30 | 50 |
| 8 | | | | | 78.2 | 57.6 | 74.8 | 8.48 | 7.64 | 45 |
| 9 | | | | | 73.9 | 75.0 | 64.9 | 8.83 | 7.57 | 42 |
| 10 | | | | | 67.1 | 69.4 | 74.2 | 8.72 | 7.48 | 28 |

obtain moisture content. The remaining sample was placed in a blender and chopped (with the addition of water as necessary). After drying, approximately 5 g of this sample were stored at -18 C for later VS determination. Approximately 100 g of the diluted mixture was immediately sealed and stored at -18 C for total carbon and nitrogen determinations at the end of each trial. Results for VS and carbon and nitrogen contents were all expressed on a percent dry weight basis. The remaining, blended mixture was further diluted (10 g diluted mixture/500 ml water) and pH readings recorded (Carnes and Lossin 1970). As an indicator of the completion of composting a diluted mixture (5 g compost per 50 ml water) was anaerobically fermented at 55 C as described by Jann et al. (1960). Under these conditions, raw manure would undergo anaerobic decomposition and produce acidic products. If a neutral or alkaline reaction is maintained for 48 hours, it demonstrates that the organic material has stabilized (Jann et al. 1960).

Oxygen concentrations were measured with a calibrated Beckman F-3 oxygen analyzer.

## COMPOST TRAILS—RESULTS

Observations of the composter in operation indicated that the stirring device mixed, aerated and leveled in one or two passes (depending on the frequency of stirring) without clogging. The tiller blades reached the desired depth (35 cm) and mixed and aerated the waste by continuously slinging small pieces against the shield hanging from the rear of the tiller platform (Fig. 2). After hitting the shield, the compost fell into the void dug by the tiller blades leaving a level surface.

Data obtained from the weekly grab samples are shown in Tables 1-4. Once a manure depth of 20-30 cm was obtained in the compost bin, process temperatures between 40-68 C were rapidly developed and maintained (Tables 2 and 3). In all three trials, with average loading rates between 448 kg/week for trial 1 to 713 kg/week for trial 3 the manure was at least partially stabilized during the composting process. This stabilization was indicated by the reductions in VS shown in Table 4. Although VS reductions consistently occurred, 87 to 75 percent in trial 1, 89 to 86 percent in trial 2 and 83 to 78 percent in trial 3, the amount of reduction (12, 3 and 5 percent) was quite variable. C/N ratios (averages shown in Table 4) were quite variable with the only significant differences occurring in the carbon and nitrogen contents between trials. These differences were attributed to the percent of bedding contained in the samples. Large reductions in moisture content occurred during each trial (Tables 1, 2, and 3) and by itself would account for total weight reductions of 21 percent for trial 1, 46 percent for trial 2 (before the addition of water) and only 3 percent for trial 3 if the final moisture contents at each level and the input moisture contents are equally weighted. While the pH values obtained after 48 hours of anaerobic fermentation, Tables 1, 2, and 3, were not always neutral or alkaline (especially during trial 1) for the composter material, they were consistently more alkaline than the input samples.

During the first two weeks of the summer trial, the composter attracted large numbers of flies; however, as the depth of material began to increase and high process temperatures were obtained during the third week (Table 2) the fly problem disappeared. Attractiveness to flies was checked by placing samples of the input manure and samples from the composter in individual dishes outside the laboratory building. While the input material quickly

TABLE 4. AVERAGE OF WEEKLY C/N RATIOS, PERCENT C, PERCENT N AND VS FOR INPUT WASTE, COMPOST DURING MANURE ADDITION AND COMPOST FINAL VALUE.

| Trial no. | Input C/N, C, percent, N, percent | During C/N, C, percent, N, percent | Final C/N, C, percent, N, percent | Input VS, percent | During VS, percent 0-5 cm, 15-25 cm, bottom | Final VS, percent 0-5 cm, 15-25 cm, bottom |
|---|---|---|---|---|---|---|
| No. 1 | 15.1 / 52.2 / 3.48 | 14.1 / 45.6 / 3.25 | 11.9 / 42.2 / 3.59 | 86.9 | 80.0 / 80.0 / 76.0 | 75.0 / 73.1 / 77.9 |
| No. 2 | 8.4 / 22.6 / 2.70 | 9.6 / 17.2 / 1.79 | 15.1 / 32.2 / 2.13 | 89.3 | 88.9 / 86.1 / 86.5 | 88.5 / 88.1 / 81.1 |
| No. 3 | 5.4 / 18.3 / 3.40 | 6.4 / 21.1 / 3.30 | 5.6 / 21.6 / 3.84 | 83.4 | 82.6 / 78.4 / 77.5 | 79.4 / 73.7 / 79.7 |

attracted flies, the composter material attracted no more flies than empty control dishes.

The major operational problem encountered was the control of moisture content. During weeks 3 through 6 of trial 1 (Table 1) input moisture levels and consequently compost moisture levels increased and high process temperatures could not be obtained. Generally, whenever moisture levels in the composter exceeded 75 percent, the once per day tilling was not sufficient to maintain aerobic conditions. Following tillage the manure tended to settle and pack at these high moisture contents. When this settling occurred, the tillage procedure released odorous gases which were not encountered at lower moisture contents when small air spaces were present in the tilled layers.

Oxygen samples taken from the tilled layer approximately 3 and 22 hours after the last tilling operation indicated oxygen concentrations averaging 19.4 percent and 18.5 percent, respectively. These samples were taken during weeks 6 and 7 of trial 2 when process temperatures were high (55 to 66 C), moisture levels were between 52 and 69 percent and when the material had not settled appreciably.

During the summer trial (No. 2, Table 2) the material in the compost bin reached a moisture content of approximately 27 percent and process temperatures decreased (week 9). Following the addition of 227 kg of water, process temperatures again increased to 50 to 60 C within 2 days. Within two weeks the compost again began to dry out with decreasing process temperatures and the trial was terminated.

## DISCUSSION

Although the composting process was not complete in any of these trials, the high temperatures, the moisture reductions and the low odor levels when high moisture contents were avoided, provided the desired benefits. Although the product would not be suitable for sale as a commercial compost, the reduced weight and odor levels would be beneficial during field application.

If moisture content of input waste is controlled (below 75 percent) or if a suitable method for moisture removal is available, this built-up bed process could be utilized to meet strict environmental criteria and reduce fly attraction by the small producer. For a small producer faced with strict application of odor and pollution abatement criteria, an adaptation of this system could provide a reasonable alternative.

Development of field models should include automation of input, output, and especially tillage operations to reduce labor requirements. It is anticipated that either round or rectangular bins could be utilized with this system.

Another application of the concept which might be feasible for some operations would be to use an existing bunk silo as the holding bin if moisture levels are below 75 percent, and if a low cost roof can be installed. A small tractor equipped with mounted tiller blades could then be used to level, aerate and mix the waste material. With the increased power available from the tractor, the aeration process could be completed rapidly.

Further development at The Ohio Agricultural Research and Development Center is not anticipated because of changes in job responsibilities. It is felt that further development work is warranted.

## CONCLUSIONS

A pilot model, built-up bed, aerobic composter was designed, constructed and evaluated using swine waste. Based on the results, the following observations and conclusions were obtained:

1 The aerobic composter reached high process temperatures (50-68 C) and provided reductions in both moisture content and volatile solids, although commercial quality compost was not obtained.

2 Low odor levels and a reduction in the attractiveness to flies and other insects were evident.

3 The process should be applicable on small farms that must meet strict environmental criteria, that have some existing facilities that could be utilized and that have limited investment capital for waste treatment facilities.

References

1 Carnes, R. A. and R. D. Lossin. 1970. An investigation of the pH characteristics of compost. Compost Science 11(5):18-21.

2 Jann, G. J., D. H. Howard and A. J. Salle. 1960. Method for determining completion of composting. Compost Science 1(3):31-34.

3 Miner, J. Ronald (ed). 1971. Farm animal-waste management. North Central Regional Research Publication 206. Iowa Agr. Exp. Sta. Spec. Rep. 67.

# Conservation of Nitrogen in Dairy Manure During Composting


G. B. Willson, J. W. Hummel

MEMBER
ASAE

MEMBER
ASAE


THE current energy crisis and the resultant rapid inflation in commercial fertilizer cost has stimulated renewed interest in effectively utilizing the plant nutrients in animal manures. However, some methods of handling, processing, and storing fertilizers can result in substantial nutrients losses, especially nitrogen, before the manure is applied to the land. Ammonia may be lost by volatization, if the manure is not incorporated into the soil immediately after land application (Benne et al. 1961). Nitrates may be lost by leaching and become groundwater pollutants when heavy manure rates are applied to the soil (Bartlett and Marriott 1971). About one half of the nitrogen and potash are in the liquid portion of the manure and can be lost by drainage (MacLean and Hore 1974). Manure can also be a source of odors and weed seeds.

Composting manure eliminates undesirable odors, kills weed seeds, and improves the handling characteristics for dairy manure. Adding a bulking material to the dairy manure develops porosity for rapid aerobic thermophilic composting and eliminates nutrient losses due to drainage. However, nitrogen can be lost during the composting process (University of California 1953).

This research was conducted to determine the potential for conserving nitrogen through process control. Results are included on 48 tests in 10 l bench digestors and 19 tests in 0.85 $m^3$ bins. To verify laboratory results, a pilot composter at the University of Maryland dairy farm, composting manure from 70 dairy cows, was spot checked at 5-day intervals during the 15 day detention in the channel.

## PROCEDURE

Bench Composters tests — The six bench composters used were cylinders (60 cm high and 15 cm inside diameter) (Willson 1971) aerated at a rate to provide excess oxygen. For the 2 week composting period, wall temperatures were maintained at 60 C. Temperature of the manure was 60 C ± 5 C.

Carbon nitrogen ratio (C/N) and pH have been suggested as contributing factors for N losses (Gotaas 1956). Denitrification will also occur with N loss as gas under anaerobic conditions. To test the effect of available carbon on the N loss, sawdust, straw, perlite and composted manure were mixed with fresh manure at ratios of 3:1 and 6:1 by volume. Three bulking materials at these ratios were tested simultaneously to reduce the effect of probable variability in manure characteristics.

Dense lumps of manure that the air cannot penetrate decompose anaerobically. Denitrification would expectedly increase as moisture content and lumping tendency increase. To test this theory, one batch of manure was air dried, mixed with three parts sawdust, and then remoistened and placed in each of the six bench composters at moisture contents varying from 30 to 60 percent.

For the last set of bench composter tests straw and perlite were used as bulking materials. The ratio of straw to perlite varied from 0 to 100 percent at 20 percent increments.

Bin Composter Tests — Although the bench composters were satisfactory for evaluating some characteristics of the composting process, the larger bin tests were required to eliminate some of the scale factors, especially for the curing stage of the process when the rate of decomposition has slowed. Four open top plywood bins (1.22 by 0.76 by 0.91 m) were used to compost three batches of manure simultaneously (Willson 1971). About 400 kg dairy cow manure was placed in each bin and aerated at rates up to 3 $m^3$/hr. Perlite was added at the concentration of 1 to 2 percent by weight to prevent drainage. The bins were usually turned three times weekly into another bin. Most tests lasted for 4 wk. Data are reported for 19 tests.

The possible effects of partial anaerobic decomposition were tested by turning off the air supply for various lengths of time (30 min to 2 days) in 5 bins. During the first 10 days of composting, oxygen levels in the center of the bin dropped to zero in less than 1 hr after the air supply was turned off. One bin was not turned or aerated for the first 30 days. Then it was processed normally for 27 additional days.

Superphosphate is sometimes used in broiler houses for controlling odor and conserving nitrogen. Therefore, 5 percent (dry-weight basis) superphosphate was added to three bins and 2 percent to one bin.

One bin was inoculated with a commercially available bacteria that was advertised as able to eliminate odor and speed decomposition. When it was added to small quantities of manure stored at room temperature, it seemed to reduce odors. However, there was no apparent difference in temperature, oxygen comsumption, odor or or nitrogen transformation between inoculated and uninoculated bins tested simultaneously. This suggests that the cultured organisms could not adapt to the thermophilic temperatures.

## ANALYSIS

The manure was sampled for analysis before and after composting for all tests and weekly for the bin tests. All samples were analyzed for moisture content (MC), volatile solids (VS), chemical oxygen demand (COD), pH, ammonia, nitrate and Kjeldahl N.

Organic (Kjeldahl) nitrogen, ammonia, and COD were analyzed using APHA Standard Methods (1971) proce-


The authors are: G. B. WILLSON, Agricultural Engineer, ARS, USDA, and J. W. HUMMEL, Associate Professor, Agricultural Engineering Dept., University of Maryland, College Park.

TABLE 1. PERCENT REDUCTION OF TOTAL WEIGHT, VOLATILE SOLIDS, COD, AND TOTAL N DURING 2 WK IN BENCH COMPOSTERS FOR THREE BULKING MATERIALS.

|  | Sawdust | | Perlite | | Compost | |
|---|---|---|---|---|---|---|
|  | Avg | Range | Avg | Range | Avg | Range |
| Total weight | 36 | 6-55 | 13 | 12-14 | 26 | 16-45 |
| Volatile solids | 26 | 11-59 | 17 | 15-19 | 28 | 22-34 |
| COD | 40 | 19-62 | 33 | 19-47 | 22 | 8-40 |
| Total N | 26 | 2-60 | 23 | 11-32 | 20 | 5-35 |

TABLE 3. COMPARISON OF PERCENT REDUCTION FOR ALL BIN AND BENCH TESTS.

|  | Bin | | Bench | |
|---|---|---|---|---|
|  | Avg | Range | Avg | Range |
| Total weight | 51 | 30-72 | 26 | 6-55 |
| Volatile solids | 53 | 35-68 | 22 | 7-58 |
| COD | 51 | 24-80 | 31 | 6-69 |
| Total N | 36 | 10-65 | 28 | 2-59 |
| Ammonia | 86 | 54-99 |  |  |
| Nitrate | 57 | 28-84 |  |  |

dures. Nitrate was analyzed with a specific ion electrode using a resin to counteract the chloride ion interference. Ammonia and nitrates were reported as N concentration.

## RESULTS

Bench Tests — Table 1 presents the percent reductions in total weight, volatile solids, COD, and total N for the 2 wk bench composter tests with sawdust, perlite, and compost, as bulking materials. Data for straw are not included because of difficulties in obtaining representative samples for analysis. Carbon nitrogen ratios averaged 24, 13, and 30 for the sawdust, perlite and compost, respectively.

Since perlite is light weight and inert, its main effect was to make the manure porous, permitting air movement. The compost added relatively unavailable nutrients for the micro-organisms and the sawdust supplied some moderately available carbon.

Losses of total weight, VS and COD tended to be higher with the addition of nutrients in the sawdust and compost. Carbon addition (bulking material) did not seem to affect N loss.

Table 2 compares the percent losses of VS and total N for the two ratios of bulking material to manure. The differences were small indicating that the 3:1 ratio had adequately conditioned the manure for good air distribution so that the amount of anaerobic activity and the additional carbon source had little or no effect.

For the variable moisture content test, N losses did not show a clear trend; however, the highest N loss (59 percent) corresponded with the highest moisture content (60 percent). The initial N content for this set of tests was 0.54 percent or about one third the N concentration in most of the other test manures. The difference was attributed to several days of air drying required to lower the moisture content to 30 percent before rewetting to selected levels. Although the percentage N loss was higher for this set of tests than for most of the others, the absolute loss was about the same.

For the final set of bench tests a combined straw and perlite bulking material varying in 20 percent increments from all straw to all perlite was used. This test did not show any correlation between the added carbon in the straw and the N loss.

Bin Tests — Table 3 compares the percent reduction of total weight, VS, COD, total N, ammonia, and nitrates for all bench and bin tests. The process time for the bins was twice that for the bench composters. This is reflected in the approximately doubling of percent reductions for total weight, VS and COD. Bins lost about one third more N than bench composters. Usually, N concentration slightly increased since the manure was biologically oxidized faster that N was lost. Ammonia was substantially reduced, with the most reduction occuring near the end of the process. Nitrate concentrations remained essentially constant so that the total percent reduced was about the same percent as for VS.

Table 4 summarizes the N data for all tests. Total N concentrations increased slightly during the process and nitrates remained about the same. Ammonia concentrations decreased slightly for the 2 wk bench tests, but decreased substantially during the 4 wk of bin tests.

Ten of the 19 bin tests received little special treatment. The N losses for these tests ranged from 17 to 55 percent with an average of 34 percent. Only 10 percent N was lost for the first test using superphosphate (5 percent dry weight). For the 2 percent superphosphate added to one of two bins, N loss was 34 percent with superphosphate as compared to 40 percent without. For the next trial 5 percent superphosphate was added to two of three bins which were also subjected to a variation in the air supply. For the bin aerated conventionally with superphosphate, N loss was 18 percent. For the other bin with superphosphate aerated for 2 days and then allowed to become anaerobic for 2 days, N loss was 35 percent.

The third bin in this set used for the superphosphate test was not aerated (mostly anaerobic) for 30 days and then it was composted aerobically for an additional 27 days. During the anaerobic (30 day) test period, there was no heating and low concentrations of oxygen were detected at 15 cm below the surface on few occasions. For this trial N loss was 58 percent.

These tests suggest that the superphosphate addition may have reduced the N loss and that periods of anaerobic decomposition combined with periods of aerobic composting will contribute to N losses.

One other set of three bins was composted for variable periods without aeration. For one half of the 0.5 hr, 4 hr, and 8 hr cycles, aeration was used. After 2 day composting, the temperatures were 66, 55, and 42 C, respectively. Maximum temperatures observed were 67, 67, and 63 C for the 0.5 hr, 4 hr, and 8 hr, bins, respectively, and N losses were 17, 30, and 37 percent, increasing as cycle length increased.

No attempt was made to control the manure's pH. Its

*(Continued on page 496)*

TABLE 2. PERCENT REDUCTION IN VOLATILE SOLIDS AND TOTAL N DURING 2 WK IN BENCH COMPOSTERS BY BULKING RATIO.

|  | 3:1 Ratio | | 6:1 Ratio | |
|---|---|---|---|---|
|  | Avg | Range | Avg | Range |
| Volatile solids | 27 | 11-59 | 23 | 10-40 |
| Total N | 21 | 2-35 | 22 | 10-34 |

TABLE 4. SUMMARIZED NITROGEN DATA FOR ALL TESTS.

|  | Bin | | Bench | |
|---|---|---|---|---|
|  | Avg initial | Avg final | Avg initial | Avg final |
| Total N, percent | 2.0 | 2.1 | 1.5 | 1.7 |
| Nitrate, ppm, N | 470 | 410 | 600 | 690 |
| Ammonia, ppm, N | 4100 | 1100 | 2300 | 1700 |

# Composting of Swine Waste


**Mark E. Singley, Martin Decker, Stephen J. Toth**

MEMBER     MEMBER
ASAE      ASAE


A DEMONSTRATION project was conducted by an interdisciplinary group at the New Jersey Agricultural Experiment Station on the composting of swine waste. The work was initiated on July 1, 1969, under a USDA Cooperative Agreement and concluded on June 30, 1971. An extension of the project was initiated on January 1, 1972, and concluded on December 31, 1972.

The objectives of the original project were to investigate rapid composting as a method for handling and treating the organic waste from swine farms in New Jersey so that waste would be reduced in volume, be odor free and be suitable for agricultural land disposal or acceptable for landfill disposal. The swine wastes used were collected from an operation that used food wastes as feed and the waste was unsorted and contained items such as cans, bottles, bones, plastics, paper and soil as well as the feces and uneaten food waste. This swine waste was regarded as difficult to stabilize using the composting process. If the process proved feasible for this waste, it was anticipated that it would be acceptable for the less complex wastes of other swine production systems. Although the general feasibility of the method was demonstrated, a major problem was identified which had to be solved for the system to be practical (Decker et al. 1971). The initial waste was so dense that internal air movement and resulting composting rates were initially too slow. Also, the carbon content was deemed insufficient to satisfy the requirements for rapid composting. Therefore, additives were required which added bulk to the waste and provided nutrients for the microorganisms, thereby accelerating the process. Additions of straw or municipal refuse were shown to accelerate the process markedly (Savage 1972, Ali 1973).

Recognizing the need for the additives, the extension of the project was committed to the examination of the management system for composting animal wastes mixed with municipal refuse and similar wastes. This included an examination of methods for separating the nonbiodegradable materials from the completed compost so that it would be suitable for disposal on agricultural land. Throughout the project an important objective was to control objectionable odors.

For all tests, a windrow composting system was used. A Roto-Shredder*, a self-propelled over-the-windrow machine, was used to turn, aerate and shred the windrows which were formed and worked on a concrete platform 50 ft wide and 135 ft long (Fig. 1).

Street refuse was collected at curbside in nearby com-

FIG. 1 A front view of the ROTO-SHREDDER showing it over a windrow located on the composting site.

munities and delivered to the site by compactor truck. Heavy and bulky items such as automobile tires, washing machines, refrigerators, etc., were not included in the collections. They were normally picked up separately.

From March 1972 through November 1972 eleven windrows of different compositions were composted. The mixtures used are shown in Fig. 2. Each windrow was approximately 40 ft long and 12 ft wide. About 5 ft high at the beginning, the windrows became lower as composting progressed. All windrows except Windrow G were turned two times per day, weekdays; Windrow G was turned four times per day. Mixtures were composed on a volume basis. They are approximate because of the variability in moisture content and composition of the materials received over the period of time the project was conducted. No separation of nonbiodegradable material was made prior to composting. The materials were positioned by a front-end loader. The street refuse was placed first and the swine waste or sewage sludge added on top (Singley et al. 1973).

For the initial windrows the date for windrow termination was not so well recognized as for later windrows. As a result, composting time may have been extended for the early windrows. Windrow termination was later identified with: (a) disappearance of steam when turning the windrow; (b) granulation in particle form and reduction in size; (c) flowability of the compost from the loader bucket into the hopper of the shaker and over the shaker table; (d) temperature decline after reaching the thermophilic range.

The composting process is not complete at windrow termination. The separated composted material should be stored until temperatures have reached ambient conditions. There is a desirable condition for shaker separation of the nonbiodegradable materials from the final compost. This identified windrow termination. If the windrow is allowed to go "dead" before separation is effected, the materials will reconsolidate by absorbing moisture from rainfall, and


Paper of the Journal Series, New Jersey Agricultural Experiment Station, Rutgers University, Biological and Agricultural Engineering Dept., Cook College, New Brunswick, NJ.

The authors are: MARK E. SINGLEY, Professor, MARTIN DECKER, Associate Specialist, Biological and Agricultural Engineering Dept., and STEPHEN J. TOTH, Professor, Soils and Crops Dept., Cook College, Rutgers University, New Brunswick, NJ.

*Manufactured by Roto-Shredder Co., Division of Imco, Inc., Crestline, Ohio 44827. Reference to commercial products or trade names is made with the understanding that no discrimination or endorsement is intended or implied.

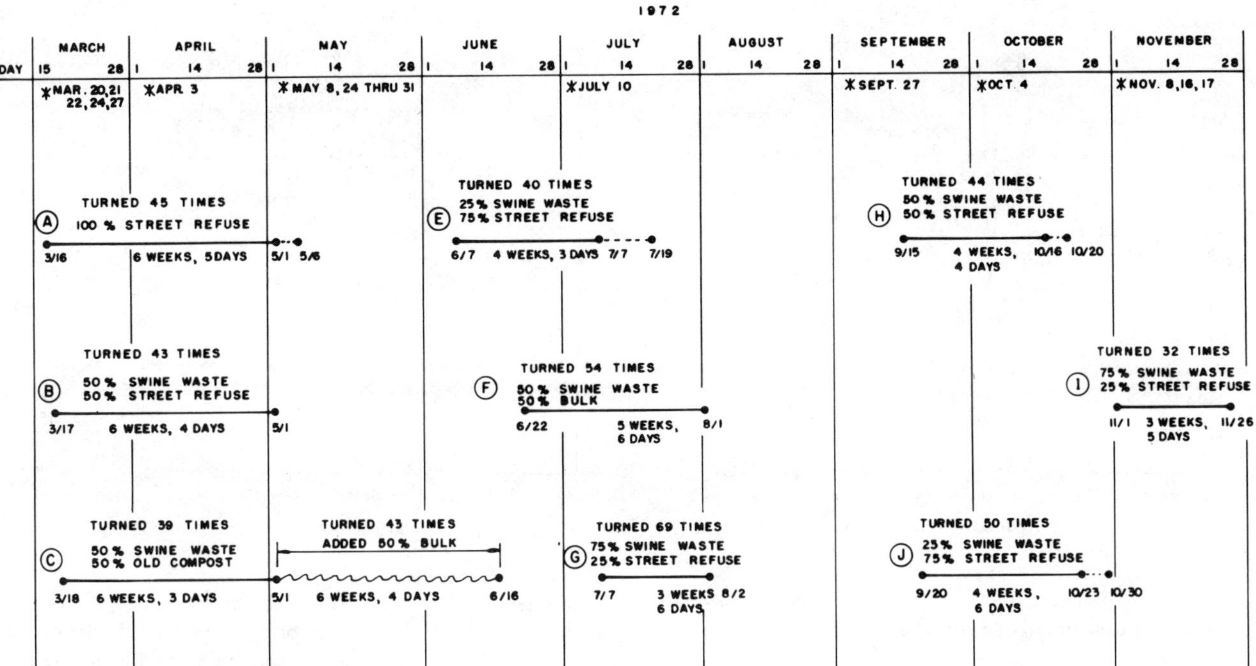


FIG. 2 Periods of composting, mixtures, times turned, and duration of turning.

separation will be more difficult.

## WINDROW TERMINATION

Fig. 2 shows a history of the project. While there are too few windrows of each mixture to be completely certain about the influence of mixtures on the exact time to reach windrow termination, the mixtures required less time than windrows of either 100 percent swine waste or 100 percent street refuse. The shortest period required was 3 weeks and 5 days for windrow I, composed of 75 percent swine waste and 25 percent street refuse, to compost.

Time limits of the experiment allowed only one windrow composed of 50 percent sewage sludge in cake form and 50 percent street refuse to be composted. This windrow's composting time, 4 weeks and 6 days, compared favorably with the 4 weeks and 4 days needed to compost Windrow H, composed of 50 percent swine waste and 50 percent street refuse. This is important because sewage sludge in the wet form has a consistency much like swine waste. Though mixed in the windrow as sewage cake, subsequent moisture accumulation in the windrow rewet the sludge.

Mixing street refuse with swine waste, regardless of the proportions tested, accelerated the composting as shown by the time to windrow termination. Temperature records and oxygen readings also confirmed this result.

## TEMPERATURE RECORDS

The hostile environment within the composting mass caused the failure of both steel and plastic temperature probes. As a result, the most complete records were obtained for the earliest windrows (Figs. 3, 4, 5, 6).

Windrow A did not sustain the high temperature for as long a period of time as Windrows B and D, and generally the temperatures were considerably lower. This may be a result of excessive bulking which made the windrow highly permeable to air, or of insufficient nitrogen to stimulate and sustain thermophilic bacteria.

Windrow C did not reach moderate temperature levels until a bulking material was added. The bulking material was that separated from previous windrows when passed over a 3/4 in. shaker screen.

The scatter of temperature readings both day to day and position to position within the windrow shows the temperature variability resulting from repositioning the temperature probes. When repositioned, required by windrow turning, the probes sensed new temperature locations. As a result,

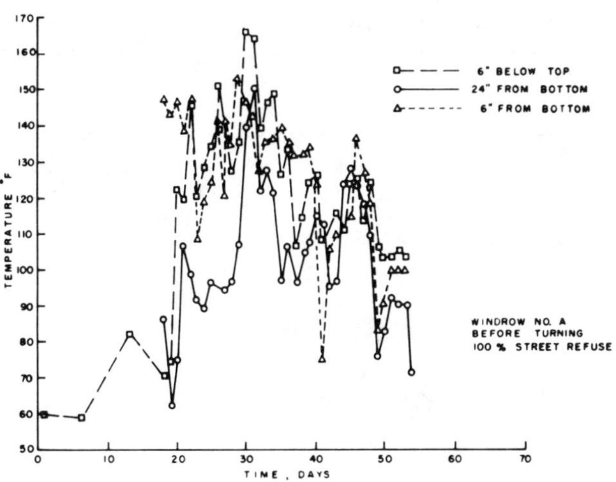


FIG. 3 Windrow temperatures.

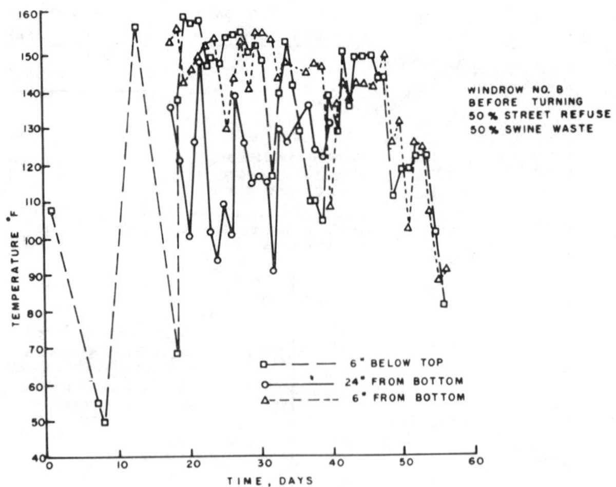


**FIG. 4 Windrow temperatures.**

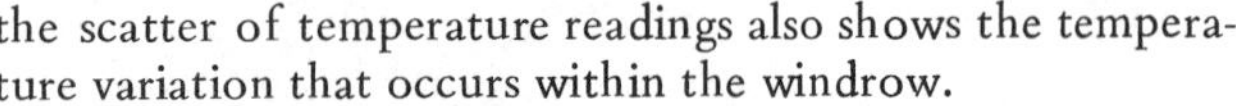

## OXYGEN RECORDS

The oxygen content readings for Windrow H are shown on Fig. 7. These records show the decrease in the presence of oxygen in the windrows as the composting process proceeds from beginning to termination. One reason for this is the increase in bulk density that occurs. This tends to exclude oxygen and diminish the amount present. In addition, as the composting process develops and intensifies, oxygen consumption increases and results in a decline in the measurable amount present.

The decline in oxygen present in the windrows indicates that there may be a more desirable turning schedule than that of turning a uniform number of times each day during the composting period, for example, two times each day. A better schedule might be to use a graduated turning schedule such as two times per day after the initial mixing period until the thermophilic temperature range is reached, then reduce to once per day and during the latter stages to once every other day. This should help to maintain the bulking of the windrow. A schedule of this kind was not tried in this demonstration, but it is recommended for any future demonstration.

## BULK DENSITY

Changes in bulk density are a measure of the consolidation that occurs during composting. Fig. 7 shows the bulk density determinations for Windrows H, J, and K. At the

the scatter of temperature readings also shows the temperature variation that occurs within the windrow.

termination of composting the bulk density is in the neighborhood of 40 to 45 pounds per cubic foot. The change that occurs during composting depends on the mixture used to compose the windrow. The record for Windrow J indicates the large change that occurs for a windrow containing a high percentage of street refuse. The bulk density for this windrow more than doubles. For Windrows H and K, which contain a large percentage of swine waste and sewage sludge, the bulk density increases by approximately 10 pounds per cubic foot, or about 1/3 of the original density of approximately 30 pounds per cubic foot.

It is difficult to measure bulk density when street refuse is one of the components of the mixture. Large solid objects, rags, plastics and broken containers require that it least a four cubic foot container be used to measure the

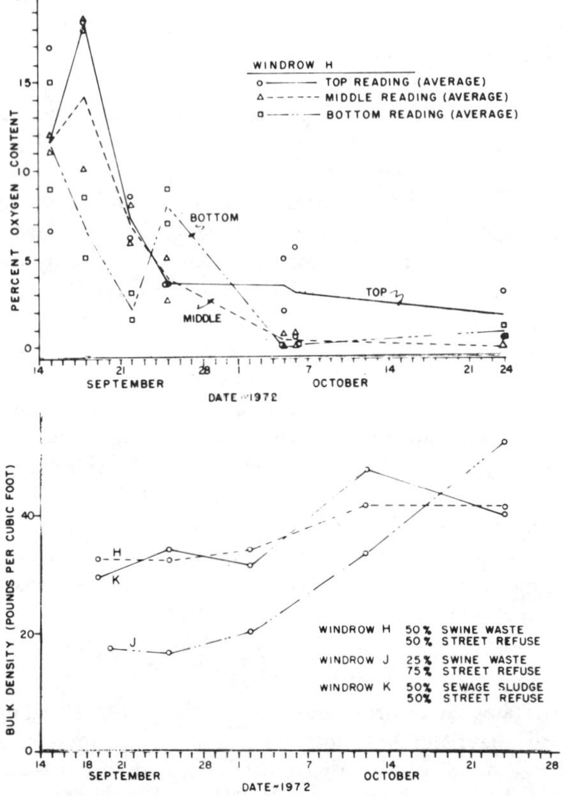


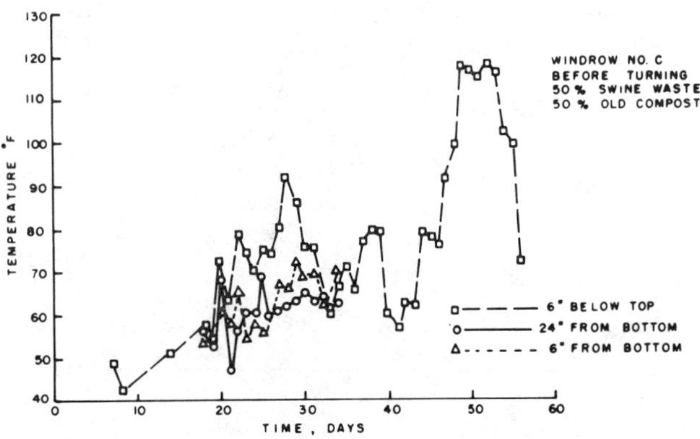


**FIG. 5 Windrow temperatures.**

**FIG. 7 Oxygen content and bulk density readings.**



MANAGING LIVESTOCK WASTES

| | Windrow numbers | | | | | | | |
| | A | | B | | C | | D | |
| | 4/13/72 | 4/27/72 | 4/13/72 | 4/27/72 | 4/13/72 | 4/27/72 | 4/13/72 | 4/27/72 |
|---|---|---|---|---|---|---|---|---|
| pH | 6.9 | 7.6 | 7.0 | 7.5 | 6.7 | 6.8 | 6.5 | 7.5 |
| Moisture percent | 47.7 | 63.9 | 60.1 | 65.1 | 58.4 | 63.7 | 59.0 | 78.7 |
| Inerts* percent | 70.2 | 90.0 | 74.5 | 75.6 | 46.6 | 51.6 | 60.9 | 71.7 |
| Ash percent | 70.5 | 79.3 | 72.2 | 75.8 | 77.7 | 76.6 | 70.1 | 71.8 |
| Insoluble ash percent | 52.6 | 70.9 | 61.6 | 64.8 | 66.9 | 65.1 | 60.1 | 58.9 |
| Organic matter percent | 29.5 | 20.7 | 27.8 | 24.2 | 22.3 | 23.4 | 29.9 | 28.2 |
| Total N percent | 1.01 | 0.78 | 1.01 | 0.99 | 1.17 | 1.23 | 1.25 | 1.06 |
| N content percent organic matter | 3.46 | 3.76 | 3.63 | 3.78 | 5.00 | 5.26 | 4.15 | 3.79 |
| $NH_4$ (N) percent | 0.06 | 0.06 | 0.07 | 0.05 | 0.00 | 0.05 | 0.00 | 0.09 |
| $NO_3$ (N) percent | 0.01 | 0.01 | 0.01 | 0.01 | 0.00 | 0.00 | 0.00 | 0.01 |
| C/N | 15.8 | 14.3 | 14.8 | 14.2 | 10.2 | 10.2 | 12.8 | 14.3 |
| Total $P_2O_5$ percent | 1.60 | 1.25 | 1.60 | 1.82 | 1.94 | 1.94 | 1.60 | 1.82 |
| Available $P_2O_5$ percent | 1.28 | 1.00 | 1.28 | 1.46 | 1.55 | 1.55 | 1.28 | 1.46 |
| Total $K_2O$ percent | 0.48 | 0.47 | 0.96 | 0.91 | 0.62 | 0.74 | 1.08 | 1.03 |
| Na percent | 0.10 | 0.10 | 0.18 | 0.18 | 0.09 | 0.10 | 0.18 | 0.18 |
| Ca percent | 1.63 | 1.55 | 1.93 | 2.13 | 2.13 | 2.25 | 2.05 | 2.13 |
| Mg percent | 0.12 | 0.10 | 0.16 | 0.16 | 0.13 | 0.13 | 0.15 | 0.13 |
| Cl percent | 0.14 | 0.14 | 0.25 | 0.25 | 0.14 | 0.19 | 0.25 | 0.23 |
| S percent | 0.20 | 0.15 | 0.17 | 0.18 | 0.16 | 0.16 | 0.16 | 0.16 |
| Fe percent | 0.75 | 1.00 | 0.95 | 1.07 | 0.73 | 0.66 | 0.62 | 0.81 |
| Mn ppm | 130 | 130 | 165 | 185 | 130 | 130 | 175 | 240 |
| Cu ppm | 118 | 10 | 8 | 25 | 10 | 10 | 8 | 8 |
| Zn ppm | 450 | 500 | 500 | 500 | 350 | 350 | 350 | 500 |
| Ni ppm | 30 | 45 | 30 | 45 | 45 | 45 | 30 | 45 |
| Cr ppm | 15 | 20 | 13 | 10 | 10 | 5 | 10 | 5 |
| Cd ppm | 3 | 3 | 0 | 0 | 0 | 3 | 0 | 0 |
| Pb ppm | 145 | 150 | 140 | 123 | 88 | 68 | 81 | 83 |

*Material not passing a 2 mm sieve.

sample. To transfer the sample from the windrow to the container and duplicate the way the sample occupied the windrow is most difficult. At best, all of the measurements are approximate.

## YIELD OF COMPOST

The yield of compost of mixtures is dependent on the proportions mixed and the composition of the street refuse. Several samples separated using a 1/4 in. screen, show the approximate yields expected. Using a street refuse, the yield was approximately 38 percent compost by weight; a 25 percent swine waste and 75 percent street refuse mixture would yield approximately 51 percent compost. A 50 percent sewage sludge and 50 percent street refuse mixture yielded approximately 60 percent compost. These were determined by weight and calculation. Volumetric determinations are extremely difficult to make. Since hollow volumes of nonbiodegradable materials may be filled when in place, essentially both materials occupy the same volume. Separated, they each contribute to the measured volume.

## COMPOSITION OF COMPOSTS

The data on the composts presented in Table 1, Windrows A, B, C and D, indicate that they are poor products due to the high content of ash and inert materials. On the average, the composts contain only about 25 percent organic matter and about 1 percent total N.

The nature of the organic matter present and its N content indicate that the composting process is satisfactory for producing high nitrogen residues at the completion of the process. This is shown by the N content of the organic fraction that ranges from 3.46 percent to 5.26 percent. The C/N ratios of the organic fraction 10.2-15.8 indicate that no problems will be encountered with N deficiency symptoms of plants when the composts are added to the soil.

As expected, the total $P_2O_5$ contents are high, but the P is relatively easily available for plant use. The $K_2O$ contents

TABLE 2. COMPOSITION OF COMPOSTS

| | Material* not passing 4/inch, percent | Material† not passing 2mm sieve, percent | Percent moisture‡ | Total N percent | Organic matter percent | C/N | N content O.M., percent | C/N O.M. | $K_2O$ percent | $P_2O_5$ percent | Ca percent | Fe | Mn | Cu | An | Ni | Cr | Cd | Pb |
|---|---|---|---|---|---|---|---|---|---|---|---|---|---|---|---|---|---|---|---|
| | | | | | | | | | | | | ppm | | | | | | | |
| **Windrow H** | | | | | | | | | | | | | | | | | | | |
| 9/19/72 | 57.4 | 43.6 | 29.0 | 1.10 | 16.3 | 8.0 | 6.74 | 1.3 | 0.30 | 2.17 | | | | | | | | | |
| 9/25/72 | 48.5 | 27.8 | 15.6 | 0.98 | 19.6 | 10.7 | 5.00 | 2.1 | 0.32 | 2.05 | | | | | | | | | |
| 10/ 2/72 | 64.5 | 33.8 | 19.8 | 0.87 | 16.8 | 10.3 | 5.18 | 1.7 | 0.25 | 1.94 | | | | | | | | | |
| 10/12/72 | 46.7 | 25.1 | 36.7 | 0.99 | 18.83 | 19.0 | 5.25 | 2.0 | 0.36 | 1.94 | 2.50 | 13,750 | 140 | 300 | 500 | 20 | 150 | 1.0 | 133 |
| 10/24/72 | 41.9 | 33.6 | 35.4 | 1.04 | 17.58 | 16.9 | 5.92 | 1.6 | 0.46 | 1.94 | 2.95 | 17,500 | 140 | 350 | 500 | 20 | 175 | 1.0 | 145 |
| **Windrow J** | | | | | | | | | | | | | | | | | | | |
| 9/19/72 | 30.3 | 44.6 | 17.3 | 1.06 | 18.4 | 9.3 | 5.76 | 1.7 | 0.30 | 1.94 | | | | | | | | | |
| 9/25/72 | 70.0 | 41.9 | 8.2 | 1.03 | 19.5 | 9.2 | 5.28 | 1.8 | 0.30 | 2.05 | | | | | | | | | |
| 10/ 2/72 | 75.5 | 35.3 | 12.5 | 0.88 | 16.6 | 10.1 | 5.00 | 1.8 | 0.24 | 1.94 | 2.05 | 12,500 | 130 | 300 | 550 | 23 | 160 | 1.0 | 210 |
| 10/12/72 | 60.9 | 34.1 | 35.9 | 0.90 | 16.60 | 18.4 | 5.42 | 1.7 | 0.36 | 1.60 | 2.50 | 15,500 | 175 | 450 | 600 | 30 | 90 | 1.0 | 210 |
| 10/24/72 | 37.8 | 36.0 | 28.5 | 1.14 | 19.84 | 17.4 | 5.73 | 1.9 | 0.14 | 1.82 | | | | | | | | | |
| **Windrow K** | | | | | | | | | | | | | | | | | | | |
| 9/19/72 | 53.0 | 30.3 | 25.5 | 1.06 | 17.6 | 9.0 | 6.02 | 1.5 | 0.12 | 1.03 | | | | | | | | | |
| 9/25/72 | 41.7 | 25.1 | 24.9 | 1.10 | 16.6 | 8.3 | 6.62 | 1.3 | 0.12 | 1.14 | | | | | | | | | |
| 10/ 2/72 | 39.1 | 30.0 | 27.4 | 0.88 | 17.8 | 10.9 | 4.94 | 1.9 | 0.12 | 1.03 | | | | | | | | | |
| 10/12/72 | 32.4 | 33.1 | 40.1 | 0.99 | 18.35 | 18.5 | 5.40 | 1.9 | 0.16 | 0.90 | 1.80 | 8,250 | 70 | 1,000 | 1,250 | 23 | 700 | 1.5 | 300 |
| 10/24/72 | 40.5 | 42.8 | 37.8 | 0.88 | 22.64 | 25.7 | 3.89 | 3.2 | 0.16 | 0.90 | 1.80 | 10,750 | 70 | 1,000 | 1,250 | 23 | 600 | 1.0 | 220 |

*Wet basis.
†Oven dried basis.
‡Material passing 4 opening to 1 inch.

are low and extra K must be added to the compost for best results.

The heavy metal contents are not excessively high so no difficulties can be expected upon their addition to soils. This fact has been verified by field tests using these composts and those made during the initial project.

Windrows H, J and K were examined more extensively as shown in Table 2. Compost from H and J seems to be satisfactory for the addition to soil. For Windrow K, where sewage sludge supplied by the City of Bridgeton, NJ, was substituted for swine waste, the suitability for addition to the soil is questioned. The sewage sludge increased the Cu and Zn contents of the compost to levels that might be toxic to plant growth if added to soils at rates in excess of 50 tons per acre.

A wide range of mixtures of swine waste and municipal refuse can be used to shorten the composting time on a concrete platform to less than six weeks. A shaker screen was used to separate the composted material from the large particle size nonbiodegradable materials. The composting process, especially during turning by machine, produces objectionable odors in the vicinity of the machine during the first two weeks or less, depending on the mixture. Odors then become less objectionable, acceptable and finally disappear.


### References
1   Ali, G. 1973. Thermal properties of composting material. M.S. thesis, Rutgers University.*
2   Decker, M. et al. 1971. Rapid controlled composting of organic wastes. Final report USDA Cooperative Agreement 12-14-100-10078 (42). Biological and Agricultural Engineering Department, Rutgers University.
3   Savage, J. 1972. Microbial aspects of composting hog wastes. M.S. thesis, Rutgers University.*
4   Singley, M. et al. 1973. Improving composting techniques of organic wastes. Final report, USDA Cooperative Agreement D.A.-11.160 (42). Biological and Agricultural Engineering Department, Rutgers University.

*Note: The two theses listed contain comprehensive bibliographies on composting.


## Nitrogen in Dairy Manure

*(Continued from page 491)*

initial values ranged from 7.0 to 8.7. Within 5 days all pH's were above 8.0. By the third wk of composting the pH usually began decreasing. By the end of the composting process the pH was usually between 7.0 and 8.0. Since ammonia volatization increases with pH, an attempt was made to relate N loss to pH. Within the range of pH's that occured no relationship was noted.

Pilot Test — For tests with the pilot channel composter Kjeldahl-N increased gradually from 0.95 to 1.04 percent during the 15 days. Since input and output weights of manure were not obtained, N loss could not be directly calculated. Assuming the solids were reduced 25 percent during the 15 days (as was observed in a smaller model of the channel), about 18 percent of the Kjeldahl-N was lost. The low N values obtained resulted from dilution with sawdust. Since only moist sawdust was available to condition the manure for composting a large amount of sawdust was required.

### CONCLUSIONS

These tests showed that nitrogen is lost from dairy manure during the composting process and that its loss is variable, depending on operating conditions. Within the ranges tested moisture content, pH, and bulking material had no effect on nitrogen loss. Periods of anaerobic activity during composting increased N loss. Adding superphosphates seemed to reduce N loss. Since nearly all N loss was less than the reduction in VS due to biooxidation, N concentration usually increased.


### References
1   Benne, E. J., C. R. Hoglund, E. D. Longnecker, R. L. Cook. 1961. Animal manures, what are they worth today: Circular Bulletin 231, Michigan State University, East Lansing. MI.
2   Bartlett, H. D., L. F. Marriott. 1971. Subsurface disposal of liquid manure. Proceedings of the International Symposium on Livestock Wastes. ASAE, St. Joseph, MI. 49085.
3   MacLean, A. J., F. R. Hore. 1974. Manures and compost publication 868. Information Division, Canada Department of Agriculture, Ottawa.
4   University of California. 1953. Reclamation of municipal refuse by composting. Technical Bulletin 9.
5   Willson, G. B. 1971. Composting dairy cow wastes. Proceedings of the International Symposium on Livestock Wastes. ASAE, St. Joseph, MI. 49085.
6   Gotaas, H. B. 1956. Composting, sanitary disposal and reclamation of organic wastes. World Health Organization. Geneva, Switzerland.
7   APHA. 1971. Standard methods for examination of water and wastewater. 13th ed. American Public Health Association, New York.
8   Hummel, J. W., G. B. Willson. 1975. High rate mechanized composting of dairy manure. Proceedings of the 3rd International Symposium on Livestock Wastes. ASAE, St. Joseph, MI. 49085.

# Liquid Composting of Dairy Manure

F. A. Grant

W ITH EPA and local support Chino Basin Municipal Water District, San Bernardino Co., California, conducted a study of liquid composting of dairy manure. Major objectives of this study were:

1 Establish design and operating parameters of a thermophilic and a tandem thermophilic-mesophilic aerobic stabilization process.

2 Determine if heat produced by microbial thermogenesis will maintain thermophilic temperatures without addition of external thermal energy.

3 Perform a preliminary economic evaluation of the processes based on extrapolations of the bench scale results.

Fig. 1 illustrates a conceptual treatment process. This process consists of thermophilic and mesophilic treatment followed by some method of solids separtion such as vacuum filtration. Solids concentration in the thermophilic step must be maintained at a high level to maximize biological heat generation per unit volume, and therefore the feed approaches the composition of fresh manure. Additional wastewater from dairy washing operations may be added to dilute the raw manure if necessary. Residence time in the thermophilic stage is adjusted to achieve a reasonable balance between degree of stabilization and rate of stabilization.

Material from the thermophilic reactor passes to the mesophilic reactor for further stabilization. This reactor receives the benefit of a hot influent, but operates with a lower solids concentration and a much lower temperature. This may be considered a finishing stage toward producing a well stablized material. The final desired degree of stabilization establishes the residence time in the reactor. Some recycle of solids after the filtration step may enhance the provess by maintaining a high concentration of microorganisms and solids in the mesophilic reactor.

The thermophilic treatment step provides the bulk of the waste stabilization. It is the most novel part of the overall process and was, therefore,the primary focus of the investigation.

The basic theoretical framework utilized in the investigation was that developed by Andrews and Kambhu (1971).

## EXPERIMENTAL EQUIPMENT AND PROCEDURES

The major study tool was the reactor shown in Fig. 2. A total of three were constructed and each was 72 cm (28.5 in.) high and 44.5 cm (17.5 in.) inside diameter. Insulation consisted of fiberglass and polyurethane. A variable speed mixer, 1/4 kW (1/3 hp), was supported

above the reactor with a 8.4 cm (3.3 in.) diameter stainless steel propeller to provide mixing. A bubble cutter at the surface of the liquid was also fitted to the shaft. Air was dispersed through a 30.5 cm by 5.1 cm (12 in. by 2 in.) diameter diffuser located on the bottom of the reactor. A heat exchanger was coiled inside the reactor. A pump circulated heated water through the heating coils.

Freshly dropped manure was collected from the feeding corral prior to experimental runs. According to design a specified amount of material was removed from the reactors and fresh manure added twice daily. During high feed rates especially, the manure was heated to minimize temperature shocks which would occur with addition of cold feed.

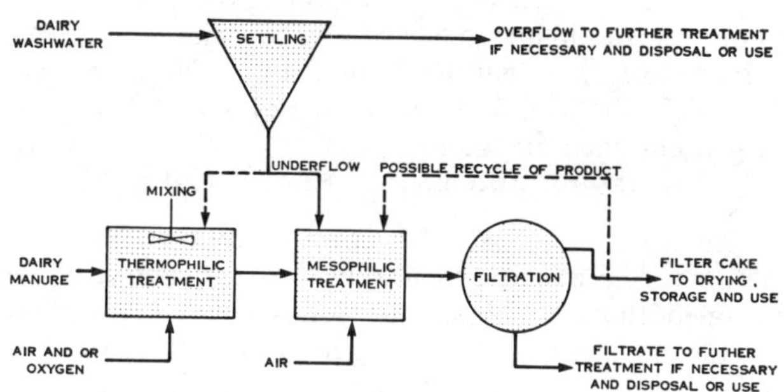



FIG. 1 Conceptual thermophilic-mesophilic treatment process.

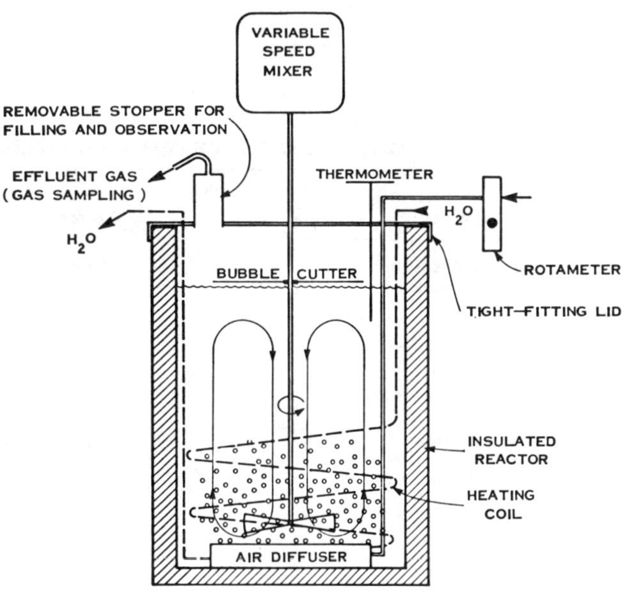



FIG. 2 Schematic diagram of batch reactor.


The author is: F. A. GRANT, James M. Montgomery, Consulting Engineers, Inc., 555 East Walnut Street, Pasadena, California 91101, Special Consultant to Chino Basin Municipal Water District, San Bernardino County, California.

## TABLE 1. EXPERIMENTAL DESIGN FOR CONTINUOUS THERMOPHILIC REACTIONS

| Week | 1 | | | 2 | | | 3 | | | 4 | | | 5 | | | 6 | | | 7 | | | 8 | | |
|---|---|---|---|---|---|---|---|---|---|---|---|---|---|---|---|---|---|---|---|---|---|---|---|---|
| Reactor | 1 | 2 | 3 | 1 | 2 | 3 | 1 | 2 | 3 | 1 | 2 | 3 | 1 | 2 | 3 | 1 | 2 | 3 | 1 | 2 | 3 | 1 | 2 | 3 |
| Temperature, deg C | ——45 - 50—— | | | | | | | | | ——55 - 60—— | | | | | | | | | ——45 - 50—— | | | | | |
| Feedrate level* | 1 | 1 | 2 | 1 | 1 | 2 | 3 | 1 | 2 | 3 | 1 | 2 | 3 | 2 | 1 | 3 | 2 | 1 | 3 | 2 | 3 | 3 | 2 | 3 |
| Aeration level† | 2 | 3 | 4 | 1 | 4 | 3 | 1 | 3 | 2 | 2 | 4 | 1 | 3 | 4 | 1 | 4 | 3 | 2 | 2 | 1 | 3 | 1 | 2 | 4 |

*Level 1 corresponds to   7.6 liters per day (2 gpd)
 Level 2 corresponds to 15.1 liters per day (4 gpd)
 Level 3 corresponds to 30.3 liters per day (8 gpd)

†Level 1 corresponds to 0.07 l/sec (0.15 cfm) of air and 0.05 l/sec (0.10 cfm) of oxygen
 Level 2 corresponds to 0.12 l/sec (0.25 cfm) of air
 Level 3 corresponds to 0.24 l/sec (0.50 cfm) of air
 Level 4 corresponds to 0.35 l/sec (0.75 cfm) of air

Liquid samples for 5-day BOD, COD, total solids, volatile solids and other constituents were collected in 0.12 l plastic jars for analyses. Effluent gas samples from the reactors were collected in plastic bags and analyzed at the study site.

A major portion of the testing involved a factorial type experiment wherein the following major variables were chosen for study:

1 feedrate - three levels correspondig to 2 1/2-, 5-, and 10-day hydraulic and solids residence time;

2 air/oxygen - four levels of 0.12, 0.24, 0.35 l/sec (0.25, 0.50, 0.75 cfm) of air, and 0.12 l/sec (0.25 cfm) of oxygen enriched air; and

3 temperature - two levels at 45-50 C and 55-60 C.

Two other variables, solids concentration and amount of mixing, were held at relatively constant values. It was reasoned that the higher the solids concentration the greater the microbial thermogenesis per unit volume, and therefore, the solids concentration was kept at a reasonable maximum of approximately 10-12 percent total solids. Each run spanned one week with three runs performed simultaneously. Prior to starting the runs there was a trial period of three weeks in which the reactors were brought slowly to temperature. After completion of the runs shown in Table 1, two weeks of additional runs were conducted to verify results.

## FINDINGS AND DISCUSSION

Fresh dairy cow manure is variable in character as evidenced by results of analyses presented in Table 2.

With manure fed to the reactors in accordance with the experimental design, the effects of aeration, feedrate, and temperature upon biological activity were analyzed according to the groupings presented in Table 3. Table 4 summarizes the analyses for effects of aeration, feedrate and temperature upon biological energy generation. The results show that aeration level $A_1$ produces far greater energy generation than does any other aeration level, suggesting that organism activity is limited by the oxygen that can be transferred to the solution. The lack of substantial effects among $A_2$, $A_3$, and $A_4$ suggests that the aeration system employed herein was not effective in transferring proportionately increased quantities of oxygen from air at airflow rates above the 0.12 l/sec level.

Feedrate produced a smaller effect upon biological activity, but the data does show an increase in activity between $F_1$, the lowest feedrate, and $F_2$, the intermediate feedrate. There is essentially no difference between $F_2$ and $F_3$, the highest feedrate. This result might indicate that available substrate begins to control biological activity somewhere between 5 and 10 days residence time (hydraulic and solids residence time were identical during the experiment).

## TABLE 2. EXPERIMENTAL CHARACTERIZATION OF DAIRY COW MANURE

| Component | Average | Range | Standard deviation | Coefficient of variation | Number of samples |
|---|---|---|---|---|---|
| Total solids, percent of manure | 15.4 | 12.9 - 19.8 | 2.17 | 0.14 | 21 |
| Volatile solids, percent of solids | 86.1 | 76.7 - 91.8 | 3.0 | 0.035 | 21 |
| Total COD, gms/l | 149 | 81  -284 | 57 | 0.38 | 17 |
| Soluble COD, gms/l | 33 | 19  - 53 | 10 | 0.30 | 16 |
| Total $BOD_5$, gms/l | 16.1 | 8.6 - 21.5 | 3.5 | 0.22 | 12 |
| Soluble $BOD_5$ gms/l | 9.3 | 4.6 - 14.4 | 3.0 | 0.32 | 11 |
| pH | 6.2 | 5.2 -  6.8 | 0.5 | 0.08 | 12 |
| Total COD, gm/gm TS | 0.96 | 0.56- 1.48 | 0.30 | 0.31 | 17 |
| Total COD, gm/gm VS | 1.10 | 0.66- 1.69 | 0.34 | 0.31 | 17 |
| Soluble COD, gm/gm TS | 0.22 | 0.11- 0.33 | 0.07 | 0.33 | 16 |
| Total $BOD_5$, gm/gm TS | 0.105 | 0.06- 0.13 | 0.025 | 0.24 | 12 |
| Total $BOD_5$, gm/gm VS | 0.123 | 0.06- 0.16 | 0.030 | 0.25 | 12 |
| Soluble $BOD_5$, gm/gm TS | 0.062 | 0.03- 0.10 | 0.023 | 0.36 | 11 |
| Total $BOD_5$/Total COD, gm/gm | 0.108 | 0.04- 0.15 | 0.038 | 0.35 | 10 |
| Soluble $BOD_5$/Soluble COD, gm/gm | 0.270 | 0.16- 0.41 | 0.075 | 0.27 | 9 |
| Total nitrogen (N) percent TS | 2.8 | 2.6 -  2.9 | | | 2 |
| Soluble phosphorous (P) | 0.25 | 0.17- 0.32 | | | 4 |
| Total potassium (K), percent TS | | 0.05- 5 | | | 7 |
| Heating value, kJ/kg | 19 000 | | | | 1 |



MANAGING LIVESTOCK WASTES

TABLE 3. REACTIONS INVOLVED AT LEVELS
OF INDEPENDENT VARIABLES

| Independent variable | Reaction numbers involved at level of variable |
|---|---|
| Aeration level | |
| $A_1$ (0.12 l/sec)* | 1, 5, 9, 13, 17, 21 |
| $A_2$ (0.12 l/sec) | 2, 6, 10, 14, 18, 22 |
| $A_3$ (0.24 l/sec) | 3, 7, 11, 15, 19, 23 |
| $A_4$ (0.35 l/sec) | 4, 8, 12, 16, 20, 24 |
| Feedrate† | |
| $F_1$ (7.6 l/day) | 1, 2, 3, 4, 13, 14, 15, 16 |
| $F_2$ (15.1 l/day) | 5, 6, 7, 8, 17, 18, 19, 20 |
| $F_3$ (30.3 l/day) | 9, 10, 11, 12, 21, 22, 23, 24 |
| Temperature | |
| $T_1$ (45-50 C) | 1 through 12 |
| $T_2$ (55-60 C) | 13 through 24 |

*0.07 l/sec of air and 0.05 l/sec of oxygen.
†Feedrates $F_1$, $F_2$, and $F_3$ correspond to 10.5, 5.25 and 2.63 days retention time, respectively.

TABLE 4. EFFECT OF VARIABLES ON BIOLOGICAL
ENERGY GENERATION

| Variable | Energy generation, kJ/hr |
|---|---|
| Aeration levle: | |
| $A_1$ | 251 |
| $A_2$ | 92 |
| $A_3$ | 106 |
| $A_4$ | 115 |
| Feedrate: | |
| $F_1$ | 103 |
| $F_2$ | 162 |
| $F_3$ | 157 |
| Temperature: | |
| $T_1$ | 116 |
| $T_2$ | 166 |
| Average for all reactions | 141 |

TABLE 5. EFFECT OF VARIABLE ON STABILIZATION
OF MANURE

| Variable | Soluble-COD/TS | Soluble-BOD/TS |
|---|---|---|
| Aeration level: | | |
| $A_1$ | 0.31 | 0.063 |
| $A_2$ | 0.37 | 0.098 |
| $A_3$ | 0.37 | 0.095 |
| $A_4$ | 0.37 | 0.093 |
| Feedrate: | | |
| $F_1$ | 0.42 | 0.108 |
| $F_2$ | 0.37 | 0.084 |
| $F_3$ | 0.29 | 0.071 |
| Temperature: | | |
| $T_1$ | 0.27 | 0.062 |
| $T_2$ | 0.44 | 0.113 |
| Average of all reactions during weeks 1 through 8 | 0.356 | 0.0875 |
| Raw manure as fed | | |
| Average | 0.22 | 0.062 |
| Standard deviation | 0.07 | 0.023 |

The effect of temperature is indicated by the difference in energy generation between $T_1$ and $T_2$. Apparently a 10 C rise from 45-50 C to 55-60 C produces approximately a 40-percent increase in energy generation. However, because of the oxygen-limited nature of many of the reactions, this value probably does not reflect the true increase in biological activity that would occur with unlimited oxygen.

In addition to the factorial group of reactions, six other reactions were conducted. These reactions indicated that lower solids concentration produced little net effect upon biological activity.

Considerable effort went towards modifying the aeration equipment and the basic mixing agitator to increase the biological activity of the thermophilic reactions. Several shapes of aerators, fabricated from plastic tubing, and several agitators, including some large, slow-turning types, were experimented with. Any substantial improvement over the basic diffuser and propeller could not be detected.

Significant observation for all of the reactions is that thermophilic treatment produced very minimal stabilization. The effect of variables upon stabilization, as measured by the ratios of soluble COD and soluble 5-day BOD to the total solids content, is shown in Table 5.

The F test was employed to determine whether differences measured as effects were significant or could be attributed to statistical error. For the biological energy generation analysis, the effects due to aeration, feedrate, and temperature were significant at approximately the 99-, 85-, and 90-percent levels, respectively. This means that the observed differences would occur by chance alone no more than 1, 15 and 10-percent of the time, respectively. The differences in stabilization as affected by aeration levels could be attributed to statistical error. For soluble-COD differences the effects due to feedrate and temperature were significant at approximately the 85-, and 99.5-percent levels respectively. For soluble-BOD differences the effects due to feedrate could be attributed to statistical error, but the effects due to temperature were significant at approximately the 97.5-percent level.

Other measurements and observations should be noted regarding the thermophilic reactions. The pH was higher in the reactors than it was in the raw manure. Values generally ranged above pH 7.0, and in no case was the pH below 6.6, so pH of the reactions was not considered a limitation to the biological activity. There was approximately a one-quarter reduction in the total nitrogen values from the raw manure. There were isolated incidents when an ammonia smell could be detected from the reactors. There was a short period of time of 2 or 3 days at the start of the experimentation when the three thermophilic reactors began to smell somewhat sour. An increase in mixing energy and a continual clean-up of the aerators throughout the experimentation prevented this from occurring again. Odor from the reactors generally smelled of fresh cow manure and was not objectionable to the operators. However, members of the dairy family at the study site did comment that they found the odors objectionable, which probably meant that the odors were somewhat different from the characteristic dairy odor. Foaming in reactors at times posed a problem which was solved by the use of bubble cutters fastened to the mixer shaft. At other times there was very little foam. The variation could not be identified with any operating procedure or parameter and remained an enigma. The bacterial population in the thermophilic reactors was examined under a microscope and found to be very undiversified when compared with the bacterial population in the raw manure. The thermophilic bacteria appeared to be practically all small, gram-positive cocci. The raw manure included many rods and many gram-negative and gram-positive cocci forms.

Mesophilic reaction remained unchanged throughout the experimentation. Product from the thermophilic reactors was fed to the mesophilic reactors. The only mixing was that produced by the aerators and that done manually to break up any solids accumulation on a once or twice a day basis. Airflow rate was adjusted to 0.24-0.35 l/sec for each 95 l reactor. However, it became necessary at times to reduce this flow because of foaming problems. the results again show minimal, overall stabilization of the manure. However, significantly reduced values of total-BOD, soluble-BOD were recorded. This would be expected because there was very little mechanical action occurring in the reactors to help dissolve additional organic material.

Experiments with a filter leaf testing apparatus were conducted on products from both the thermophilic and mesophilic reactors. Among 56 recorded tests no substantial differences were noted between the filter-ability of the two products. The addition of lime or alum alone as a filtering aid did not produce as high a filtration rate as did the addition of lime and alum together or the addition of lime, alum and ferric chloride together.

## ENGINEERING AND ECONOMIC CONSIDERATIONS

A large fraction of dairy cattle manure degrades rather slowly under aerobic conditions, and therefore, complete stabilization of the manure requires more time than does stabilization of normal domestic sludges. This has been attributed to the presence of greater quantities of lignins and cellulosic material in cattle manure.

Maximum rates of aerobic biological activity require an abundant supply of oxygen, and in a liquid system, a large excess of air is needed to obtain sufficient quantities of oxygen in solution. The transfer of oxygen is aided by an increase in diffusivity or diffusion coefficient with temperature. On the other hand, thermophilic temperatures and high solids content result in saturation oxygen concentrations reduced from that of pure water at ambient temperatures. As an example of the magnitude of these factors, the diffusivity of oxygen in an air-water system increases by approximately 18 percent for a temperature rise from 25 C to 60 C while the solubility of oxygen in pure water decreases by approximately 30 percent for the same temperature rise. The effect of high manure solids concentration is not precisely known but it substantially reduces both the diffusivity and the solubility of oxygen. It is apparent that the net effect of these factors is a substantial reduction in ability to transfer oxygen to solution under thermophilic digestion conditions when compared, for example, with an activiated sludge process. This is an unfortunate situation because less oxygen can be supplied precisely when greater biological activity that is potentially available at higher substrate concentrations and temperatures requires more oxygen.

A large excess supply of air could provide the necessary oxygen but this imposes an additional stress upon the process in terms of removing heat from the thermophilic reactor. This heat loss severely limits the maximum temperature that the reactor can reach without a separate external source of heat, which is a relatively costly operating item. The major heat loss with the airstream is latent heat of vaporization of water vapor that saturates the exit airstream. The quanity of water vapor, and therefore, the heat loss increase exponentially with temperature. (Thesensible heat required to raise the airstream from ambient temperature to reactor temperature is a relatively minor loss by comparison). The need to minimize heat losses precludes the use of large quantities of air at thermophilic temperature conditions. These conflicting requirements for quantity of air supply could not be adequately compromised within limitations of the experimental apparatus used in the current study. Improved conditions were recorded with an oxygen-enriched airstream, and this supports the suitability of a pure oxygen system. A pure oxygen system, particularly under pressure, would substantially increase transfer rates of oxygen and eliminate a major heat loss item.

Other pertinent observations affecting design of a large scale process were foaming and coating of equipment with solids. The foaming problem was not consistent during the experiments, and variations could not be identified with observed parameters. Bubble cutters kept the problem under control and appeared to be a practical solution. The severe clogging of air diffusers on the liquid side experienced in the study indicates unsuitability of small-opening air diffusers for service in this particular type operation. Mechanical aerators would appear to be more suitable toward overcoming the clogging problems.

Indications from the experiments are that a technically feasible process would include a pure oxygen system, mechanical aeration which insures plenty of mixing, and foam-control equipment such as mechanical bubble cutters. Elements of this visualized system exist on a commercial scale.

It was apparent that substantial, temperature control provisions would be necessary for thermophilic units if air were the oxygen source. With pure oxygen instead of air as the oxygen source, control is less critical because the rate of heat loss and, thereby, the potential for cooling in case of upset is less. Observations during experiments suggested that a separate heating system was unnecessary when heat losses were not excessive and that sufficient temerature control could be provided by varying energy input through mixing.

The economics of dairy operation and the cost of alternative treatment processes are vital to the economic evaluation of a liquid, aerobic composting process. At the time of this study (end of 1973), a dairy in the Chino-Corona area had capital investment estimated at approximately $1100 per cow. The estimated gross revenue was approximately $1100 to $1200 per cow per year. A profit of 10 percent on capital investment corresponds to a net revenue of approximately $110 per cow per year.

Cost estimates for a thermophilic-mesophilic treatment process, including vacuum filtration, for a 500 cow operation indicate a treatment cost of $230 per cow per year which is approximately 20 percent of gross revenue per cow per year and a factor of two greater than estimated profit per cow per year. This cost would be a substantial burden to absorb within current dairy expenses. On a much larger scale the costs would be reduced, but would remain prohibitive within the current economics of dairy operation.

An alternate treatment method which has been demonstrated successfully in the Chino-Corona area is conventional composting. The fact that evaporation substantially exceeds precipitation in this area makes this method particularly attractive because moisture content of manure can be easily reduced to optimum

*(Continued on page 505)*



MANAGING LIVESTOCK WASTES

# Liquid Composting Applied to Agricultural Wastes

A. Raymond Terwilleger, Lois S. Crauer
ASSOC. MEMBER
ASAE

AERATION of agricultural wastes at elevated temperatures is a relatively new treatment process. The De Laval Separator Company has developed systems which utilize this process, known as liquid composting or LICOM$^{TM}$* (Hoffman and Crauer 1973), to obtain a variety of treatment objectives. These range from simple deodorization to thorough treatment and separation. Three commercial-size installations are described to demonstrate the process, the equipment, and its capabilities.

## PROCESS CHARACTERISTICS

Liquid composting is characterized as an aerobic treatment process applied to a concentrated organic waste in such a manner as to obtain elevated temperatures from the conservation of excess bacterial metabolic heat.

One of the features most important to the success of this process is the reduction in viscosity of high total solid wastes which occurs at elevated temperatures. Fig. 1 demonstrates this property for both dairy and beef wastes. The drastically reduced viscosity in a thermophilic treatment system reduces the mixing requirements for maintenance of proper recirculation and also enhances oxygen transfer. The net effect is to make possible the successful aeration of more concentrated wastes (up to 9-10 percent TS), thus reducing dilution water requirements and total reactor volume.

Another characteristic of the process is intense development of foam to the extent that it must be mechanically "cut" or knocked down at all times. The foam contains material from the bio-mass and has a density 10-15 percent that of water meaning that in the normal operating configuration 6-7 percent of the total bio-mass is in the foam layer at any one time. The continuously regenerating foam layer is an ideal aerobic environment for bacteria, keeps injected air in contact with the liquid for a longer period of time, and affords good insulation for the heat developed in the bio-mass.

When the proper equipment and system design are used, the process is quite stable and is not susceptible to imbalance. There are no known pH, alkalinity, or ammonia toxicity problems with these systems. They are able to accept slug loads with rapid recovery. They require very little operator monitoring for correct operation other than normal control over the total solids content of the feed and the rate at which it is fed.

Warm Treatment identifies the De Laval process which operates in the mesophilic temperature range. This is a less intense treatment than the thermophilic LICOM process;

---

The authors are: A. RAYMOND TERWILLEGER, Process Engineer, and LOIS S. CRAUER, Chemical Research Dept., The De Laval Separator Company, Poughkeepsie, N.Y.
*LICOM is a registered trademark of the The De Laval Separator Company.

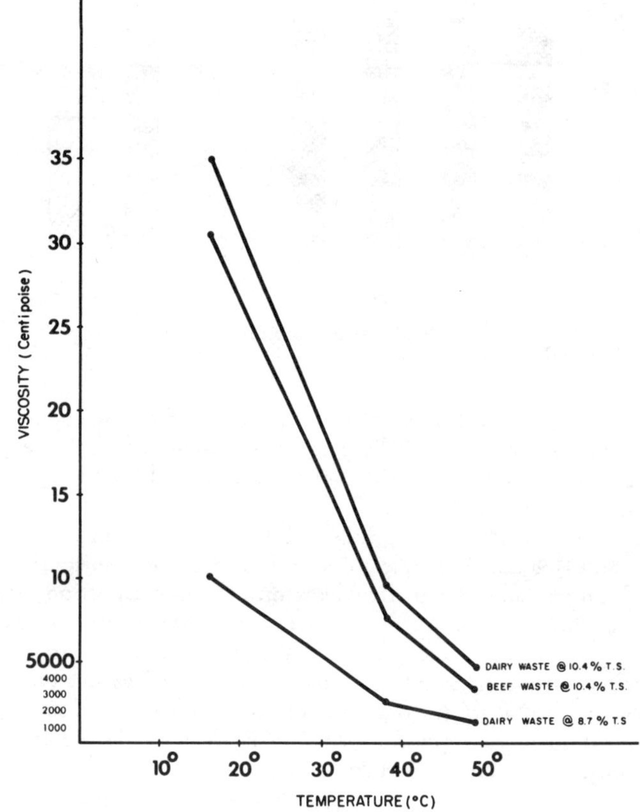



FIG. 1 Changes in viscosity of dairy and beef cattle wastes with temperature.

i.e. it operates with larger volumes per aerator and lower total solids (range 2-7 percent TS). Its intent is to deodorize the waste, raise its pH, and make it more suitable for pumping. TS, COD, and BOD reduction do take place in lightly loaded systems as would be expected but this is not necessary to success of the process. Systems can be of batch or continuous flow type in open or closed tanks.

LICOM identifies a thermophilic process in an enclosed insulated reactor. The normal total solids range is 5-10 percent for most agricultural wastes. LICOM II is a continuous flow process in which the waste is aerated for 7-14 days. LICOM II FS denotes an aeration system to which a vibrating screen has been added for solids separation.

## EQUIPMENT

The LICOM effect is achieved through the use of a specially designed aerator called the Centrirator. It is a down-draft type unit which mechanically mixes and aerates the wastes in circular treatment tanks. The Centrirator is powered by a 5 hp, 3 phase motor enclosed in a housing which allows incoming air to pass over the motor. The Centrirator makes effective use of the total energy

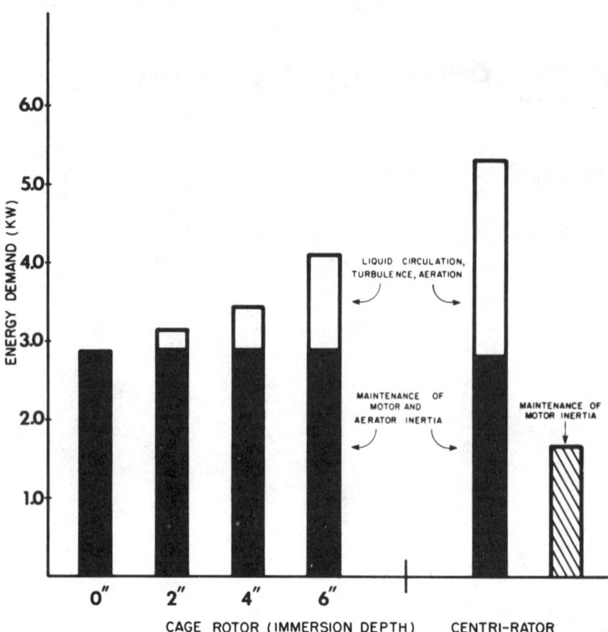

FIG. 2 Comparison of energy use for cage rotors (Wong Chong 1973) and the centrirator.

expended. Fig. 2 shows that 43 percent of the total energy used by the unit actually goes for circulation, turbulence, and aeration as compared to considerably small percentages for cage rotors.

The foam is mechanically cut by a 2 hp Foam Cutter unit. This unit is simply a disc with blades that liquifies the foam and allows it to flow back into the liquid portion of the reactor. The foam produced from cattle and swine wastes is dense and flows toward the area of the foam cutter from up to 50+ feet away. Agricultural wastes of 2 percent total solids concentration or greater produce sufficient foam to require a foam cutting device at least intermittently on enclosed tanks.

## RESULTS OF FIELD INSTALLATIONS

### Warm Treatment for Swine

This installation (Fig. 3) is located on a small hog growing operation in a suburban area. Approximately 250 hogs per year are fattened with 100-120 on hand at any one time. The hogs are fed a standard commercial growing ration. Wastes are moved out of the barn each day by mechanical scrapers and fall directly into the aerated tank. Dilution water comes from washdown and whatever is lost from the waterers. The tank is operated on a batch basis, taking 3-4 weeks to fill.

Table 1 lists major parameters for the system. The raw waste is the material as collected off the floor. Diluted waste is calculated from raw to bring it to the approximate concentration of the tank contents. Data from two different batches and different treatment periods is used to give an idea of treatment efficiency. Temperatures in the aeration tank averaged 29-32 C with occasional higher readings. The pH of the waste is raised from the acid range well into the alkaline range with obvious benefits for cropland. TS and TVS reduction becomes significant at longer detention times. COD and BOD reduction is quite significant for the detention times involved. Odor reduction as measured by the Threshold Dilution Number is quite dramatic. The farm manager and neighbors confirmed this finding when the waste was spread on the land. The aeration and mixing also make the waste easier to pump.

It is quite obvious from the data that as far as the deodorization objective is concerned the Centrirator is underloaded. It is estimated that this unit could handle 300-400+ hogs and still accomplish the Warm Treatment objective of deodorization. At its present loading it is a thorough biodegradation unit.

### LICOM II FS for Dairy

The De Laval Separator Company operates a Research Farm for its own use which includes a LICOM II FS system (Fig. 4) for treatment of the milking cow wastes. The free stall barn operates with 75-80 milkers. Mechanical scrapers move the waste to the mixing tank. Washwaters from the parlor and holding area are also delivered to the mixing tank. The De Laval Delta-Matic™† "chopper" pump is used manually to homogenize the mixing tank contents on a daily basis and is run on a time clock control to feed the first reactor at two hour intervals the rest of the time. The

---

†Delta-Matic is a trademark of the De Laval Separator Company.

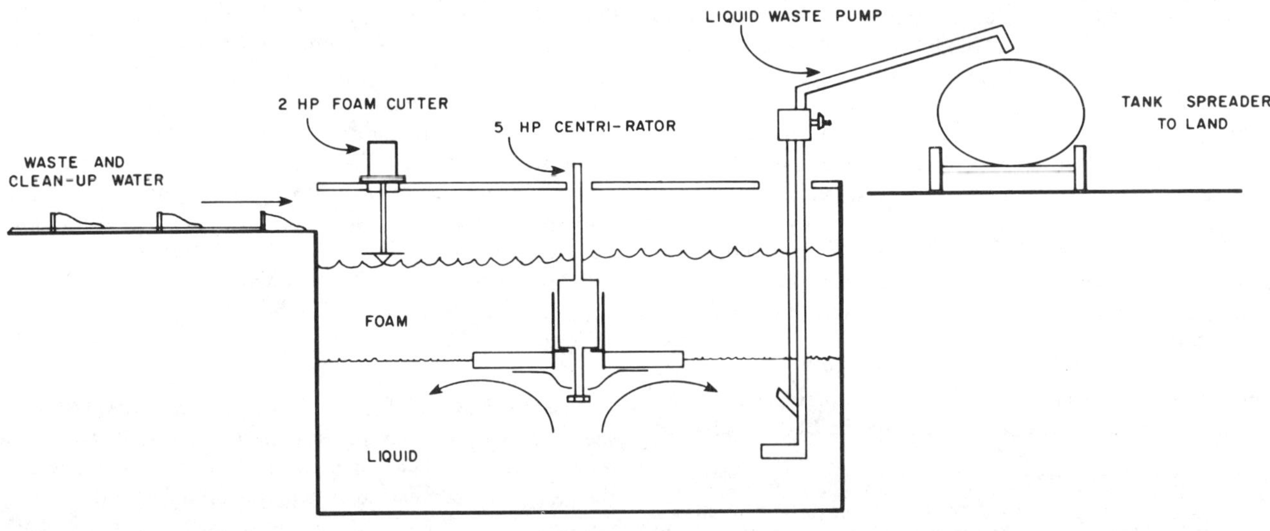


FIG. 3 Schematic of Warm Treatment System operating on swine wastes.

                    MANAGING LIVESTOCK WASTES

**TABLE 1. WARM TREATMENT — SWINE WASTE — AVERAGES OF DATA TAKEN.**

| Parameter§ | Raw waste | Diluted waste* | Aerated Waste | | | |
| --- | --- | --- | --- | --- | --- | --- |
| | | | Batch A† | | Batch B‡ | |
| | | | Concentration - | Reduction∥ | Concentration - | Reduction∥ |
| pH | 5.8 | 5.8 | 7.9 | — | 8.9 | — |
| TS, percent WB | 25.6 | ~2.0 | 1.9 | 7 | 2.34 | 65 |
| TVS, percent of TS | 87.0 | 87.0 | 86.0 | 8 | 63.0 | 74 |
| COD, mg/l | 54,500.0 | 4550.0 | 1215.0 | 73 | 1410.0 | 69 |
| $BOD_5$, mg/l | 35,300.0 | 2940.0 | 590.0 | 80 | 890.0 | 70 |
| Odor threshold dilution number | 32,800 | — | 64.0 | 500:1 | 64.0 | 500:1 |

*Calculated from raw waste diluted to range of tank contents.
†Data for 12 days total detention.
‡Data for 25 days total detention.
§ All parameters tested using procedures outlined in Standard Methods, 1971.
∥Percent reduction of TS and TVS based on comparison of organic and inorganic solid contents of diluted and aerated wastes to give percent reduction of quantity not percent of percent.

**TABLE 2. LICOM II FS — DAIRY CATTLE WASTE — AVERAGES OF DATA TAKEN.**

| Parameter | Raw cattle waste | Mixing tank | Aerated waste | | Screened liquid | | Screened solids |
| --- | --- | --- | --- | --- | --- | --- | --- |
| | | | Concentration - | Reduction† | Concentration - | Reduction† | |
| Quantity, liters/day | 3800* | 7300* | 6450-7300* | 2-10 | 5150-5850* | 22-32 | 1280-1460* |
| pH | 7.5-8.0 | 7.1-7.8 | 8.5-9.3 | — | 8.5-9.3 | — | 8.5-9.3 |
| TS, percent WB | 15.0 | 7.8 | 6.0 | 24 | 4.0 | 50 | 17.5 |
| TVS, percent of TS | 86.0 | 86 | 85 | 26 | 77 | 55 | 94.0 |
| COD,mg/l | 25,200 | 10,800 | 5750 | 47 | 3050 | 72 | — |
| $BOD_5$, mg/l | 16,800 | 6100 | 4000 | 35 | 1750 | 71 | — |
| Total coliform, #/cc | 100,000,000 | — | — | — | 0 | 99+ | — |
| TKN,mg/l | 3800 | 2540 | 1800 | 29 | 1820 | — | 1670 |
| $NO_3$mg/l | 370 | 560 | 250 | — | 242 | — | 260 |
| $NO_2$,mg/l | 21 | 26 | 22 | — | 25 | — | 21 |
| Odor threshold dilution number | 5770 | — | — | — | 191 | 32:1 | — |

*Represents best estimate from data available — not physically measured.
†Percent reduction from mixing tank contents — percent reduction of TS and TVS based on quantity, i.e. TVS reduction always greater than TS reduction.

waste passes through the three aeration tanks which have a detention time of approximately 12 days for the present loading rate. The treated waste is pumped from the last treatment tank to the 30 in. vibrating screen where separation takes place. Separation is enhanced by the low viscosity of the material as it comes out of the last treatment tank at 45+ C. The solids accumulate in a pile and the liquid effluent flows to a storage tank.

Table 2 lists average values for significant parameters of the system. The feed rate is calculated from the waste production of the cows adjusted to the solids level found in the mixing tank. Operating temperatures typically range from 40-52 C with the lower temperatures occurring in cold winter weather. The pH is raised well into the alkaline range which is an indication of a good process. TS and TVS reduction occur approximately half during aeration and

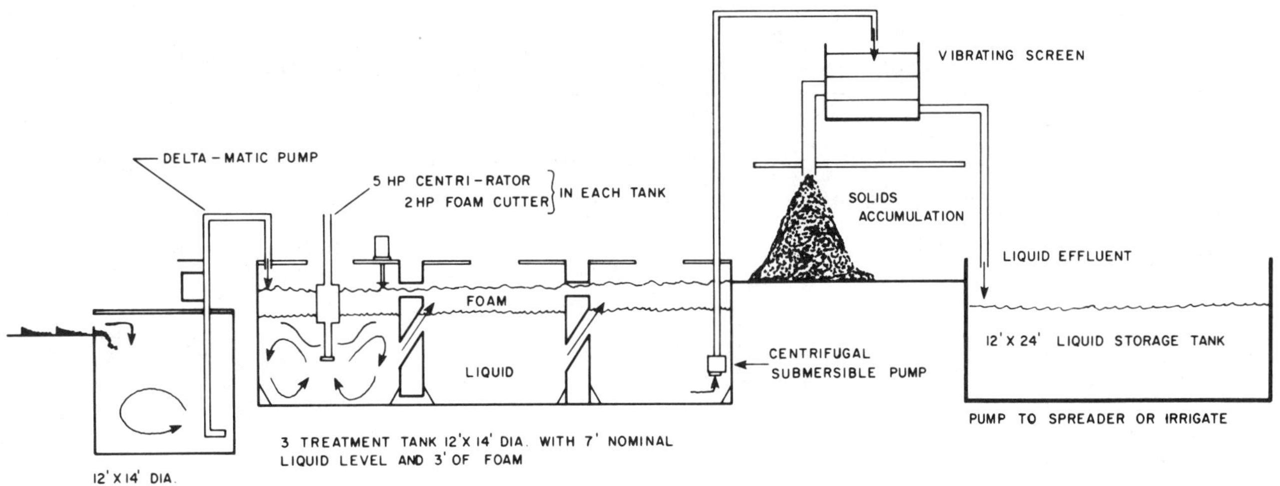


FIG. 4 Schematic of LICOM II FS System operating on dairy waste.

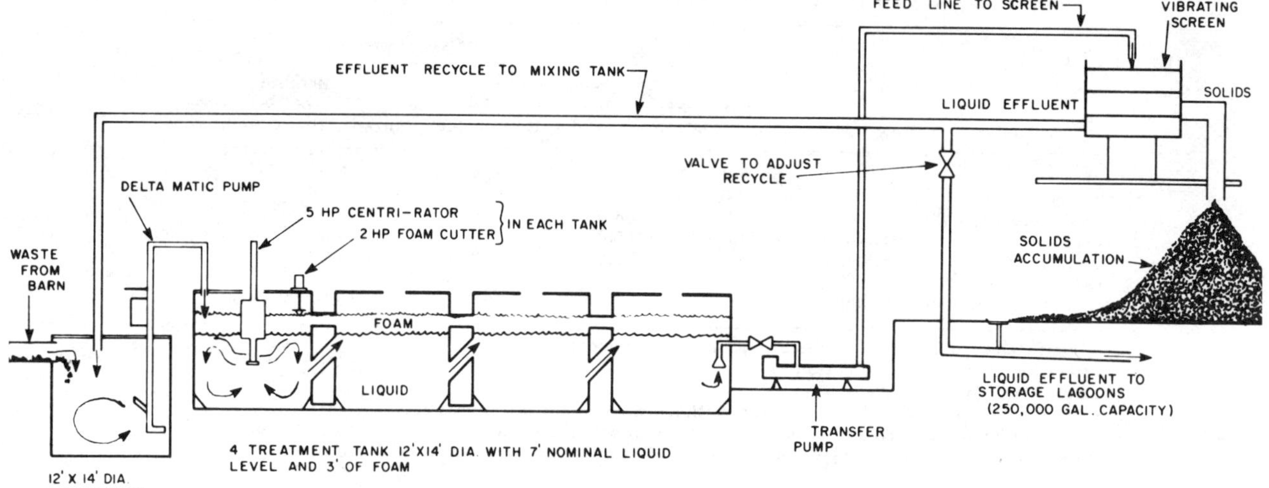



FIG. 5 Schematic of LICOM II FS System operating on beef cattle waste.

half during screen separation. Solids reduction during aeration is minimal with dairy cattle wastes because of the large percentage of solids resistant to biodegradation in the time period involved. This fact is also reflected in COD and BOD reduction figures. Odor is greatly reduced for this system also. Total coliform counts indicate complete eradication. Tests have shown that weed seed germination is reduced and eliminated in similar proportions and by the same conditions that reduce and eliminate coliforms. These conditions are elevated temperatures in combination with the biochemical environment of high pH and aggressive organisms. Seventy-one percent (71 percent) of the Kjeldahl N was retained through the aeration system. $NO_2$ and $NO_3$ concentrations changed very little through the system indicating that there was little if any nitrification taking place. Phosphorus and potassium carried through the system with little change. A majority of the N, P, K is found in the screen effluent. Research must confirm it, but it appears that the nutrients are in a more highly oxidized state than in raw waste and are thus more rapidly available to plants.

The liquid effluent, at 3-4 percent total solids, is easily irrigated or spread on the land. It can be applied directly to growing crops without any leaf burning. The solids dewater to 24-30 percent TS in a short time. They have good mulching properties and have appeal for greenhouses, etc. because they are pathogen free and weed seeds are destroyed. Also in drier climates the solids may be recycled for bedding.

## LICOM II FS Beef

A system was started fall 1974 to process the wastes from a confined beef feeding operation with a capacity of 320 steers (Fig. 5). The steers are in complete confinement over slats with mechanical scrapers for cleaning waste and spilled forage. The cross alley scraper empties into the mixing tank. This system is quite similar to the De Laval Research Farm with an additional aeration tank for greater waste treatment capacity. Also there is an extra line to recycle screened liquid back to the mixing tank for dilution purposes. The complete waste removal and treatment system including both pumps and the screen is set up to operate automatically on time clock control. The only manual functions are agitation of the mixing tank each day and observation of the entire system for proper operation and timing.

At the time the testing was performed there were 220-240 steers in the building which were estimated to produce 5,200 liters/day of waste including spilled forage (Table 3). Based on volume removed in biodegradation and screening, the liquid effluent is reduced to approximately 3,800 liters/day at 5-6 percent TS going to the lagoon. Aeration tank temperatures average 49 C with a range of 43-54 C. The only water added to the system is that which

**TABLE 3. LICOM II FS — BEEF CATTLE WASTE — AVERAGES OF DATA TAKEN.**

| Parameter | Raw cattle waste | Aerated waste | | Screened liquid | | Screened solids | Storage lagoons |
|---|---|---|---|---|---|---|---|
| | | Concentration - percent | Reduction | Concentration - percent | Reduction | | |
| Quantity, liters/day | 5200† | — | — | ≤3800† | 25-30 | ≤900† | — |
| pH | 6.7 | 8.5 | — | 8.5 | — | 8.5 | 7.8 |
| TS, percent WB | 13.8 | 6.7 | 55 | 5.3 | 61 | 13.5 | 0.85‡ |
| TVS, percent of TS | 86 | 79 | 64 | 75 | 71 | 89 | 64 |
| COD, mg/l | 19,200 | 6525 | 66 | 5350 | 72 | — | 570 |
| $BOD_5$, mg/l | 11,250 | 2850 | 75 | 1930 | 83 | — | 121 |
| Total coliform, #/cc | 3,150,000 | — | — | 0 | 99+ | — | — |
| TKN, percent WB | 0.415 | 0.23 | 43 | 0.23 | — | 0.22 | — |
| Odor threshold dilution number | 1024 | — | — | 32 to 64 | 16-32:1 | — | — |

*Percent reduction from raw waste — percent reduction of TS and TVS based on quantity.
†Represents best estimate from data available — not physically measured.
‡Liquid in lagoon diluted by ground water present at start.



MANAGING LIVESTOCK WASTES

leaks from waterers and the small amount used to lubricate the positive displacement pump. This water addition is easily offset by vaporization out of the aeration tanks. Thus the TS reduction from 13.8 going into the system to 5.3 going out is accomplished by degradation and separation. The TVS level through the system verifies the solids removal. COD and BOD reductions in the aeration step are substantially greater than for dairy wastes. COD and BOD removals through the screening step are also somewhat greater. The Threshold Dilution Number is greatly reduced as has been noted previously in other systems. Again the total coliform count is reduced to negligible value. TKN content is reduced more in this system than with the dairy waste, apparently as a result of the longer detention time in the aeration tanks. Phosphorus and potassium are carried through the system. Samples from the storage lagoons indicate that the contents are quite innocuous compared to the raw waste.

One of the most important factors for success of the process is the recycle feature. Thirty-forty percent (30-40 percent) of the screen effluent is recycled each day. Besides providing needed dilution (instead of adding water which would increase the volume of effluent) the recycle allows a longer detention time for suspended and dissolved solids which have a greater potential to be digested than the large particulates which are screened off. The ability of the LICOM process, in combination with the screen, to remove a significant percentage of the solids enables the recycle to work. It is this type of system which offers great potential where thorough treatment is required with a small system and at the same time minimizing the volume of treated waste to be handled.

## CONCLUSIONS

These LICOM installations are demonstrating the capabilities of waste treatment systems operating at elevated temperatures. A deodorization effect as well as some degradation can be achieved in a short time by mesophilic bacteria. Temperatures in the range of 27-32 C can be maintained during the aeration of the low TS wastes. Deodorization combined with a rise in pH and some degradation offer a solution to the producer who must adjust to the community around him and yet is not ready to provide thorough treatment.

Temperatures in the range of 40-54 C can be maintained for higher TS wastes simply from the effect of aeration and mixing combined with good heat economy. High temperature processes offer a substantial degree of degradation in a comparatively short period of time even with relatively stable wastes. When coupled with a solids separation device and a recycle feature these systems become very effective. LICOM systems offer distinct advantages wherever it is necessary to reduce the volume and pollution potential of any high strength organic waste.


**References**
1    Hoffman, B. and L. S. Crauer. 1973. Liquid composting of dairy cow waste. National Dairy Housing Conference Papers. ASAE SP-01-73:429-440.
2    Standard Methods for Examination of Water and Waste Water. 1971. American Public Health Association. 1740 Broadway, New York, N.Y.
3    Wong Chong, G. M., A. C. Anthonisen and R. C. Loehr. 1973. Comparison of the conventional cage rotor and Jet-Aero-Mix systems in oxidation ditch operations, 28th Purdue Industrial Waste Conference, West Lafayette, Indiana.


---

## Liquid Composting of Dairy Manure
*(Continued from page 500)*

levels of approximately 50 percent moisture. The cost of conventional composting is on the order of two dollars per cow per year. The obvious items of cost advantage over liquid composting can be identified with natural drying in place of filtration and predominantly natural aeration in place of mechanical aeration.

Transportation costs of getting compost to market are a substantial hurdle because the fertilizer value of compost is insufficient to cover transportation costs to potential market areas outside the Santa Ana River Basin. The cost of trucking compost to many of the hay fields is estimated to be $20 per cow per year. Dairymen have yet to implement a composting and trucking program which would cost them as much as $20 to $30 per cow per year. This fact substantiates the conclusion that the estimated cost of $230 per cow per year for liquid composting is much too high to be considered economically viable.

### CONCLUSIONS

1 Utilizing air as the oxygen source, insufficeint heat is generated biologically to compensate for energy losses and maintain thermophilic temperatures. A large energy input from mixing or some other source is needed to maintain thermophilic temperatures.

2 The necessary substrate-organisms-oxygen contact to produce rapid stabilization of organic material appears to be limited by the amount of oxygen that can be transferred from the air supply. Conventional equipment appears to be inadequate at thermophilic temperatures and high solids content to transfer the oxygen needed to maintain a high rate of biological activity.

3 Results with an oxygen-enriched air supply indicate that greater biological activity and reduced energy losses can be achieved with a pure oxygen supply.

4 The diversity of biological organisms in the treatment process appears to be diminished at elevated temperatures. This may tend to limit the range of organic material that can be stabilized at thermophilic temperatures.

5 Advantages, if any, of a liquid composting process over a conventional composting process presently appear to be more than offset by treatment limitations and costs.

6 Estimated costs of a complete treatment process appear to be too high for current dairy operations. Liquid composting is energy intensive and requires a substantial capital investment. Considering the value of energy and natural resources this type of treatment would probably become relatively more expensive in the future when compared with processes requiring less energy and equipment.


### References

1 Andrews, J. F., and K. Kambhu. 1971.thermophilic aerobic digestion of organic solid wastes. USDHEW, PHS, Office of Solid Wastes Research Grant U100550, Final Progress Report. Clemson University. 76 p.

# Investigations on the Procedure and the Turn-Over of Organic Matter by Hot Fermentation of Liquid Cattle Manure

K. Grabbe, R. Thaer, R. Ahlers

THE procedures which are normally used for aerobic treatment of liquid cattle manure in purification plants, e.g., oxidation ditches, surface aerators or aerators with compressed air, work at relatively low temperatures. The degradation of carbon and nitrogen compounds is relatively slow. Higher temperatures, which induce greater rates of consumption, make possible a rapid turn-over of organic material in a short time. Furthermore, the death of pathogenic organisms is stimulated. If the energy which is biochemically released is optimally used, the necessary heating of the substrate can be otained biologically without the addition of further energy sources (Poepel 1970b). The conditions fullfilling this requirement could be established by use of a down-draft turbine, a so-called Centrirator$^{(R)}$ (Poepel 1970a).

The practical experiments thus far made with this apparatus, in which liquid cattle-, hog- and hen-manure were used (e.g. Rueprich and Jaeger 1972, Hoffman and Crauer 1973), have given important data concerning the course and economics of the fermentation, but they have only to a small extent allowed general conclusions to be drawn. Therefore, the research reported here will be related to determination of the physical, chemical and biological factors which aid in optimizing the process. The hot fermentation, as used here, did not have the same goals as those of purification plants, which desire a separation of sludge and sewage (Riemann and Thaer 1972, Traulsen 1971). On the contrary, the alternative was of interest: namely the fermentation of undiluted material in such a manner that it could be stored sanitarily and odor free until it could be disposed of on agricultural land.

## METHODS AND MEASUREMENTS

Experiments were carried out in a thermo-insulated wooden tank which had a total volume of 25 m$^3$ (Fig. 1). The Centrirator mentioned above was used for aeration and stirring. The stirrer was formed as a centrifugal turbine which threw the fluid from the middle of a deflecting shell to its brim. Air was drawn into the system through a tube which surrounded the drive-shaft of the motor. In some experiments this tube was connected to a compressor to increase the air transmission. A second horizontally arranged motor, which had a two-bladed propeller at each end of the drive-shaft, served to control foam (Thaer et al. 1973). Beside the Centrirator, the suction aerator of Peters (Riemann and Jensen 1973) was used. The construction of this aerator was similar to that of the Waldhof fermentor, except that an additional aeration through a drive-shaft was not provided.

The authors are: DR. KLAUS GRABBE, Institut fuer Bodenbiologie, DR. RUDOLF THAER, project leader, and Ing. (grad.) ROLF AHLERS, Institut fuer Landmaschinenforschung. Forschungsanstalt fuer Landwirtschaft, D 33 Braunschweig, Federal Republic of Germany.

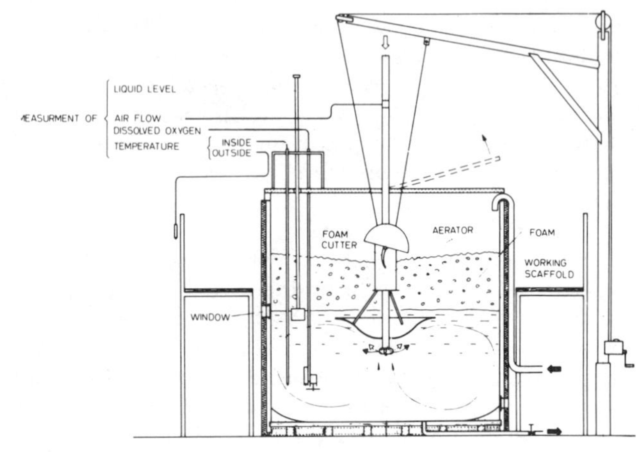


FIG. 1 Cross section of the fermentation tank. Inside diameter: 3.24 m; inside height: 3.01 m; total volume: 24.8 m$^3$; calculated heat transmission: 0.6 Kcal/m$^2$ h grd.

For the estimation of the dry matter content, and its percentage of organic substances, samples of the substrate were dried for 24 hours at 105 C and afterwards ashed for one hour at 600 C. Ammonia was steam distilled from liquid samples after addition of magnesium oxide (Bremner 1965); $N_{ox}$ = nitrite-N plus nitrate-N were evolved as ammonia after addition of Dewarda reagent. The content of nitrite-N was then determined as difference between $N_{ox}$ and the content of nitrate-N, which was separately measured in a second sample by a selective electrode. Dissolved oxygen was estimated by a conventional D.O. probe, pH-values by a so-called bridge electrode, and the COD by the dichromate method.

The liquid manure was collected from mobile cattle boxes. Its dry matter content varied from 10-13 percent. The total amount of organic matter, which depended on the kind of feed given to the cattle came to between 70 and 80 percent. The composition of the feed also influenced, above all, the content of ammonia in the manure; at the time of delivery, urea was scarcely present. The pH-values were 7 to 8.5. Initially batch fermentations took place without any removal of treated material. Later equal amounts of fresh manure and treated material were exchanged depending upon the processing and the condition of the substrate. The total volume of the fluid in the fermentation tank was between 7 and 13 m$^3$.

## RESULTS

Experiments dealing with the loading rate and the development of temperature showed (Fig. 2) that immediately after the addition of fresh liquid cattle manure, which caused a dry matter content between 6 and 8 percent in the fermentation tank, the temperatures rose to about 50 C.

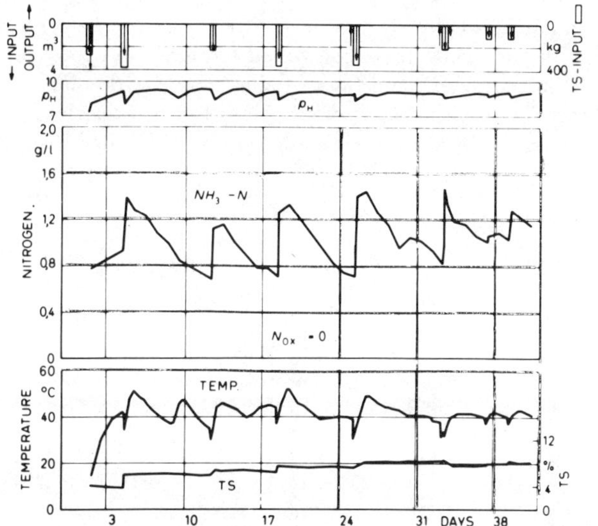

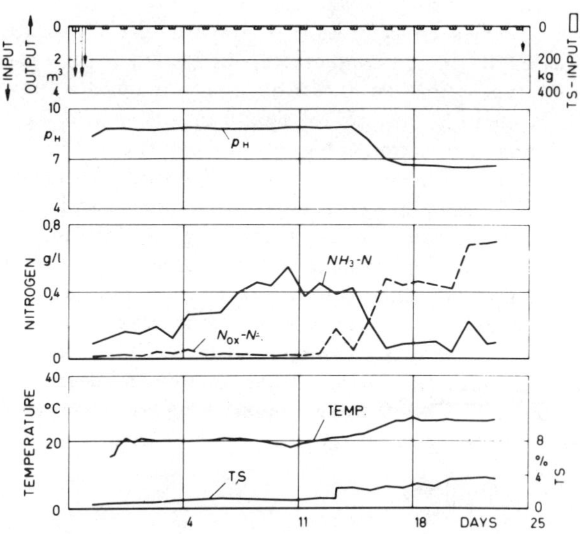

FIG. 2 Fermentation of liquid cattle manure at elevated temperature. Dependence of temperature, pH and ammonia liberation on
addition of fresh material to a fermentation tank containing material having a dry matter content between 6 and 8 percent.

FIG. 3 Fermentation of liquid cattle manure at elevated temperatures. Temperature development in response to addition of small
amounts of fresh manure and development of pH-values and turn-
over of nitrogen before and after nitrification ($N_{ox}$-N=$NO_2^-$
+$NO_3^-$-N).

Within two days the temperature dropped which meant
that the content of easily available substances was low.
Therefore, too small of supplements with fresh manure or
dilution with water which has to be heated up, as well as a
high heat transmission to the surroundings, should be
avoided. The influence of these factors becomes clear at the
end of the experiment, since the temperature only rose to
42 C. The rapid increase of the temperature corresponded
to the peak of liberation of ammonia. The easily available
organic material would therefore seem to consist of nitrogen containing metabolites. Between additions of fresh
manure, a large portion of the ammonia vaporized. The pH
rose to values between 8 and 9 and was scarcely affected by
the changes in ammonia concentrations. This phenomenon
can possibly be explained by the resin-like exchange properties of the lignin constituents in the manure (Grabbe, unpublished). In agreement with the literature (Loehr 1974),
no nitrification occurred at temperatures above 40 C. The
COD of the fresh manure decreased during the fermentation process from 1120 mg/g dry wt. to 850 mg/g dry wt.

The objective of the following experiments was to
combine nitrification with an efficient degradation of
organic material at temperatures between 30 and 40 C. The
fermentation process was initially started using 10 m³ of
municipal sewage, but later on it was found that this supplement was unnecessary. To use diluted manure ($\sim$ 1
percent dry wt.) at the beginning of fermentation, addition
of undiluted material, in small portions and at intervals of
some days, was quite sufficient. Using this procedure, the
temperature could be held below 30 C (Fig. 3). After two
weeks, the oxidation of nitrogen was in progress. The ammonia, which accumulated to 0.6 g/liter, disappeared rapidly. The $N_{ox}$-($NO_2^-$+$NO_3^-$-N) level rose to 0.7 g/liter, since
nitrogen was subsequently made available by the degradation processes. The pH decreased from 9.0 to 6.5, and a
rapid decrease in the odor was obvious. The starting phase
of the fermentation process was completed with the
initiation of nitrification. Attempts to increase the dry
matter content at this point caused difficulties in regulation
of the temperature. To hold the temperature below 40 C
during addition of larger amounts of liquid manure, the air
supply had to be doubled from 80 to 160 m³/h. In all
cases, the strong formation of ammonia progressed into a

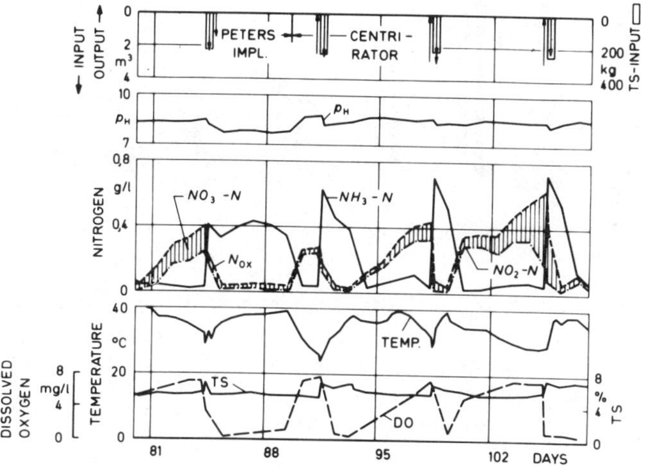

FIG. 4 Fermentation of liquid cattle manure. Course of temperature
and pH; turn-over of nitrogen in dependence of the temperature and
oxygen supply by different aerators; nitrate and nitrite content.

strong nitrification.

During the course of this experiment (Fig. 4), the Peters
aerator was tested. The oxygen supplied by this apparatus
was found to be insufficient and the oxidation of ammonia
was delayed. Therefore, the Centrirator and air compressor
were reinstalled. Thereafter, the nitrification returned to
original dimensions. Separate measurement of nitrite and
nitrate indicated that the nitrite content was high. This
result was in agreement with the literature (Loehr 1974)
which has described the inhibition of nitrate formation, in
the presence of more than 0.02 g/liter ammonia. The data
concerning the repression of the denitrification in presence
of oxygen are inconsistent. In all the cases reported here, it
was observed that a decrease in oxygen below 1 mg/liter,
which was caused by the activity of the micro-flora immediately after the addition of fresh manure, led to the
gradual decrease of nitrite- and nitrate-N. Assimilation reactions during biomass production, or development of anaerobic zones within either the substrate or the layer of heavy
material at the bottom of the tank, in which denitrifiers
would be active, could possibly explain these observations.

With a higher concentration of oxygen, the nitrification increased if the TS-content did not exceed 9 percent. It would appear that an excess of oxygen induced a conversion of nitrite to nitrate. No significant pH changes occurred during nitrification at higher temperatures.

Using the volume, dry weight and organic matter content of the samples which were both removed and added, mass balances could be calculated (Fig. 5). Fifteen to 42 percent of the fluid evaporated; the loss of the dry matter and organic substances through respiration was 18-33 percent and 26 to 42 percent, respectively. The organic portion of the dry matter was reduced from 70-80 percent to 60-70 percent. The average of evaporation and degradation amounted to 77-169 l/day and 8-17 kg/day, respectively. Two-thirds of the energy for heating was derived from

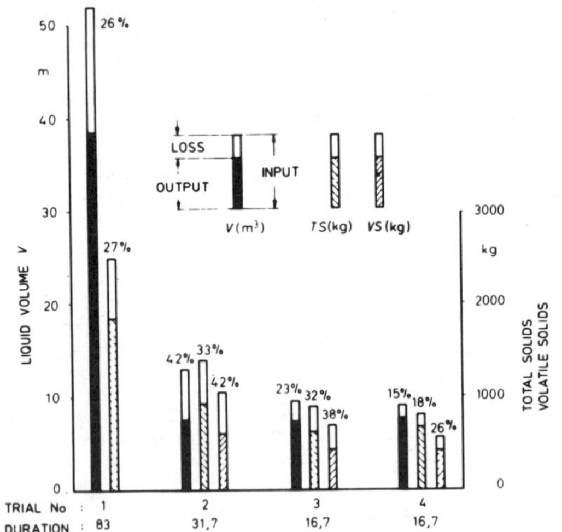

FIG. 5 Balances for changes of volumes and masses during four experiments.

metabolic processes, and the rest was derived from the engines. At the least, half of the available energy was consumed for the evaporation of water.

The results showed that temperatures between 45 and 55 C are obtainable if a correspondingly high content of dry matter is present and loss of heat is repressed. This is of great importance in case of liquid cattle manure which has only small amounts of easily degradable nitrogen and carbon. With this substrate the growth of microorganisms is soon limited.

The main source of odor during the hot fermentation was ammonia. Since the evaporation of ammonia is troublesome and takes a long period of time, a fully sufficient turn-over of this nitrogen source for the removal of odor can only be realized when efficient nitrification occurs.

Nitrification is essential to this process and required that the temperature during fermentation be carefully controlled. This was made possible by changing the aeration rate, but its influence on the fermentation process was both subtle and complex. Introduction of insufficient quantities of oxygen, which was caused by too slow a rate of aeration, resulted in reduction of both microbial acitivity and heat formation. High aeration rates resulted in a combination of maximum mineralization with a high rate of evaporation and an increased loss of heat. A special disadvantage of this was that the fermentation processes, conducted as described, needed a high degree of management control.

It seemed of practical value to try the efficiency of a two-tank procedure (Fig. 6). This meant the hot fermentation and the nitrification would be combined in separate but coupled systems. Fermentation was conducted in the

wooden tank after it had been fitted with the Peters aerator. In this tank an uncontrolled fermentation took place continuously. A metal tank of similar size with a high heat transmission was prepared for the nitrogen oxidation. This tank was aerated by a self-made aerator which had a hollow

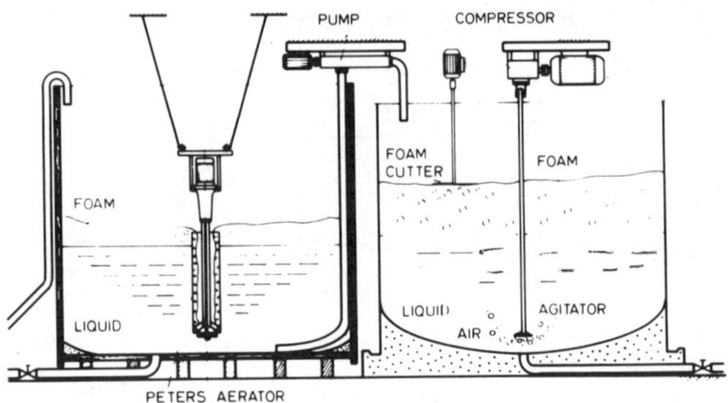


FIG. 6 Two tank system for the fermentation of liquid cattle manure.

drive-shaft through which air, under pressure, could be forced.

Currently, during manure processing either 3- or 4-day cycles are used. The fermented material removed from the metal tank is replaced with hot-fermented substrate from the wooden tank. The level of fluid in the wooden tank is reinstated by pumping in fresh manure. Eventually, up to 75 percent of the contents of the tank could be exchanged.

Temperatures reached 60 C when the dry matter content was 8 percent or more (Fig. 7). An additional effect was observed which was caused by the Peters aerator. In contrast to the Centrirator, the evaporation of ammonia was considerably restricted and a temporary accumulation of 1.6 g/liter ammonia was measured. No excessive odor accompanied these developments. Another point of interest is the sanitary effect of hot fermentation. Research in cooperation with Tierarztlichen Hochschule Hannover and the University of Hohenheim showed that temperatures of about 40 C are necessary to kill parasites and salmonella (Enigk et al. 1975, Strauch 1974).

By mixing the material from the nitrification tank with the substrate from the hot fermentation the concentration

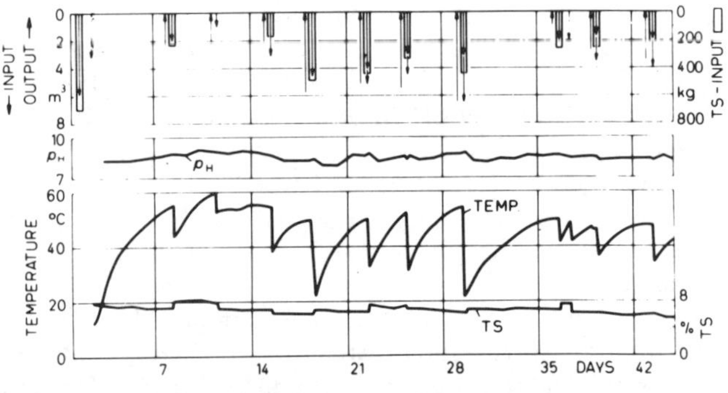


FIG. 7 Hot fermentation of liquid cattle manure. Course of pH and temperature; changes in ammonia and dry matter content.



MANAGING LIVESTOCK WASTES

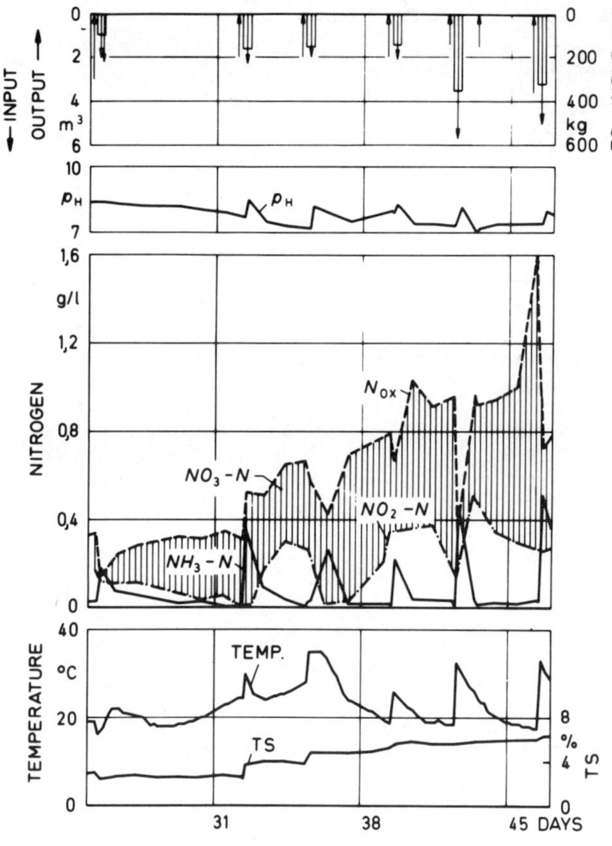



FIG. 8 Fermentation of liquid cattle manure. Course of temperature and pH. Turn-over of nitrogen to nitrite and nitrate.

of ammonia was lowered, and subsequently rapidly transformed to nitrite and partly nitrate (Fig. 8). Through addition of the hot fluid, the temperatures rose to 35 C and more. After one day they fell below 30 C and 20 C, respectively. The content of $N_{ox}$ ($NO_3^-+NO_2^-$-N) was temporarily higher than 2 g/liter. Analyses of the pH and nitrite were similar to those described before.

Intensive aeration made possible a stabilization of the concentration of the oxidized nitrogen. When aeration was slow, denitrifying organisms became active. The storage of fermented material under anaerobic or semi-aerobic conditions led to a consumption of $N_{ox}$.

## DISCUSSION

Processing of animal excrements from agricultural production is necessary to reduce their contribution to environment pollution, to conserve and improve their plant nutrition value or to reduce organic matter and nitrogen. Processes of the two-tank-system can be used to serve different goals. Hot fermentation can be of importance if the rapid reduction of volume and organic material is necessary because of transport or disposal on fields of limited area. Heat and alkalinity during the hot fermentation process kill pathogenic organisms and make the product sanitary. High aeration during nitrification makes retention of the nitrogen possible in the form of nitrite and nitrate, which are useful as fertilizers. By working with reduced air supply, the same equipment can be used to eliminate oxidized nitrogen through denitrification. This avoids pollution of groundwater when this material is disposed on fields in the autumn and winter.

The loss of mobile carbon and nitrogen leads to a partially stabilized liquid manure. The absence of odor depends on the condition used for storage. Without an air supply, and after the disappearance of nitrate and nitrite, respectively, the formation of $H_2S$, which is much more disturbing than ammonia or barn-yard smells, can not be excluded. Therefore, if the material can not be immediately disposed, an aerobic storage would be preferred.


## References

1 Bremner, J. M. 1965. Organic forms of nitrogen. p. 1238-1255. In. C. A. Black, (ed.) Methods of Soil Analysis. Amer. Soc. of Agron., Madison, Wisc.

2 Enigk, K., R. Thaer, A. Key-Hazra, R. Ahlers. 1975. Die Lebensfaehigkeit parasitaeron Dauerformen bei der Fermentation van Rindermist unter erhoehten Temperaturen. Zentralblatt fue Veterinaermedizin (in press).

3 Hoffmann, B. u. L. S. Crauer. 1973. Liquid composting of dairy cow waste. Nation. Dairy Housing Conf. Febr. 6-8, Mich. State Univ., East Lansing, Mich. ASAE Special Publication SP-01-73.

4 Loehr, R. C. 1974. Agricultural waste management. Academic Press. New York and London. 576 p.

5 Poepel, F. 1970a. Aufbau, Wirkungsweise und Foerderleistung von Umwalzbelueftern. Landt. Forschung 18, p. 138-140.

6 Poepel, F. 1970b. Selbsterwaermung bei der aeroben Reinigung hochkonzentrierter Substrate mit Hilfe von Umwaelzbelueftern. Landt. Forschung 18, p. 140-142.

7 Riemann, U. u. F. H. Jensen. 1973. Ein neues Geraet zur aeroben Dung-behandlung. Landtechnik 28, p. 319-320.

8 Riemann, U. u. R. Thaer. 1972. Behandlung tierischer Abfaelle in Nordamerika. Landtechnik 27, p. 345-348.

9 Rueprich, W. u. P. Jaeger. 1972. Probleme und bauliche Konsequenzen bei der Aufbereitung von Fluessigmist durch Umwaelzbelueftung. Beton-Landbau, Duess., p. 34-39.

10 Strauch, D. 1974. Bericht ueber die im Jahre 1973 durchgefuehrten Untersuchungen, erstattet vor der KTBL-Arbeitsgemeinschaft "Umweltschutz und Landwirtschaft", 5th Febr. 1974 in Braunschweig.

11 Thaer, R., R. Ahlers u. K. Grabbe. 1973. Untersuchungen zum Prozessverlauf und Stoffumsatz bei der Fermentation von Rinderfluessigmist bei erhoehten Temperaturen. Landbauforschung Voelkenrode 23, p. 117-126.

12 Traulsen, H. 1971. Verfahren zur Beseitigung tierischer Exkremente KTBL-Berichte ueber Landtechnik Nr. 147.

13 Wolfermann, H. F. 1970 Immissionen durch Tierhaltung und Moeglichkeiten ihrer Einschraenkungen. Bauen auf d. Lande 21, p. 179-182.

# Oxidation Ditches for Livestock Wastes


D. L. Day, D. D. Jones, A. C. Dale, D. Simons

MEMBER   ASSOC. MEMBER   FELLOW
ASAE          ASAE            ASAE


THIS is a status report and literature review on the use of oxidation ditches for the treatment of livestock wastes. It summarizes oxidation ditch development and present usage and includes an extensive list of references.

During the 1950's, the oxidation ditch waste treatment plant was developed at the Research Institute for Public Health Engineering (TNO) in the Netherlands as a low-cost method of purifying non-pretreated sewage emanating from small communities and industries (Pasveer 1963). The oxidation ditch is a modified form of the activated-sludge process and may be classed as extended aeration treatment. It is made up of two principle parts — a continuous open-channel ditch, usually shaped like a racetrack, and an aerator that continuously circulates and aerates the ditch contents.

In the mid 1960's, modified forms of the oxidation ditch treatment plant were developed for low-odor treatment of livestock wastes. One of the earliest was an external ditch for treating swine wastes in the Netherlands (Scheltinga 1966). Combination internal-external units were developed in Scotland (Baxter et al. 1966). An internal unit with the oxidation ditch beneath slotted floors was studied at the University of Illinois (Irgens and Day 1966) and a unit beneath the alleyway of a free stall dairy barn was tested at Purdue University (Schmisseur 1966). Large-scale use of the method was studied by McKinney and Bella (1967) on the Paul Smart swine farm near Lawrence, Kansas. These early studied were reviewed by Linn (1966), by Dale (1967), and by Newtson (1967).

Early development of livestock oxidation ditches was plagued with such problems as severe foaming, equipment malfunctions, and poor circulation and mixing (Day 1968). With increased usage, however, equipment reliability improved and the other major problems were solved.

Newtson (1967) reported about 150 oxidation ditch units in Colorado, Illinois, Indiana, Iowa, Kansas, New York, and Ohio. The early installations were in confinement swine buildings beneath partially or totally slotted floors, but installations have since been made in beef, dairy, and poultry buildings. Results of a recent survey of farm installations in the U.S.A. (similar states as above) are given in Table 1. Most oxidation ditches are located beneath slotted floors, such as in Fig. 1 and capitalize on the following features: (a) manure falls directly into the ditch so that uniform and continuous feeding of the biological treatment system is achieved with no extra waste collection and transportation equipment, (b) odor production is minimized due to the aerobic process, and (c) freezing problems are at a minimum. However, complete treatment for discharging an acceptable effluent into a stream is not normally achieved.

---


The authors are: D. L. DAY, Agricultural Engineering Dept., University of Illinois, Urbana; D. D. JONES and A. C. DALE, Agricultural Engineering Dept., Purdue University, West Lafayette, Ind.; and D. SIMONS, Agricultural Engineering Dept., University of Bonn, Federal Republic of Germany.


## TABLE 1. LIVESTOCK OXIDATION DITCHES, USA, 1975*

| Livestock species | Number of aerators | Number of animals (capacity at one time) |
|---|---|---|
| Swine | | |
|   Gestation | 9 | 1,260 sows |
|   Farrowing | 271 | 11,080 sows |
|   Nursery | 202 | 17,900 pigs |
|   Finishing | 404 | 79,990 hogs |
| Beef | 54 | 7,500 animals |
| Dairy | | |
|   Calves | 1 | 150 calves |
|   Heifers | 4 | 600 heifers |
| Poultry | 61 | 274,000 layers |
| | 1006 | |

*Results of a survey to manufacturers of aerators for livestock oxidation ditches as of Jan., 1975. Manufactueers responding to the survey were: Fairfield Engineering and Mfg. Co., Fairfield, Iowa; Honeybee Co., Inc., Everly, Iowa; Huskee Bilt Construction Co., Monmouth, Illinois; and Montair Mfg. Inc., Montgomery, New York.

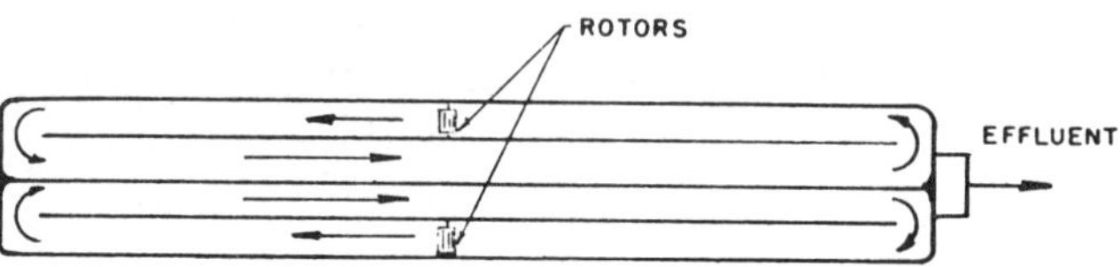


PLAN VIEW OF AN OXIDATION DITCH

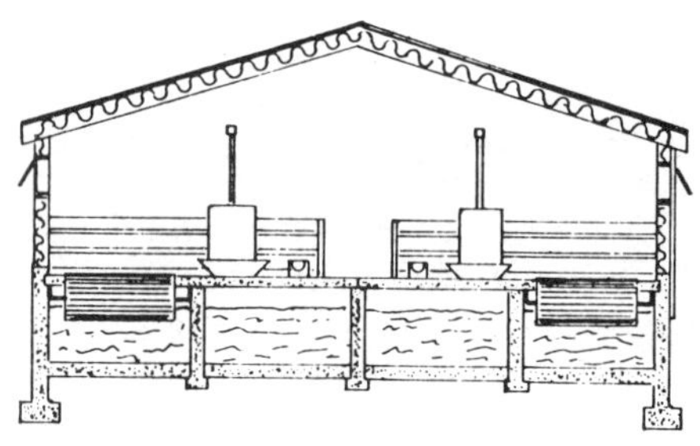

ELEVATION VIEW

FIG. 1 Totally slotted swine-confinement building with an oxidation ditch beneath slotted floors.

Fig. 2 depicts common methods of handling overflow from a livestock oxidation ditch.

Several years of research and development at the University of Illinois resulted in design and operational criteria (Jones et al. 1971) which are summarized in Table 2.

The ditch is first started in operation with water in the channel. Continuous aeration is required to maintain aerobic conditions (1 mg/l or more of dissolved oxygen) in the ditch. Methods of aerating involve the use of rotors,

TABLE 2. DESIGN RECOMMENDATIONS FOR IN-THE-BUILDING
OXIDATION DITCHES

| | Animal weight, | | Daily* $BOD_5$, | | Daily† req. oxygenation capacity, | | Ditch‡ vol., | | Daily § power reqmt., |
|---|---|---|---|---|---|---|---|---|---|
| | | | | | | | Ft³ | M³ | |
| Animal | Lb | Kg | Lb | Kg | Lb | Kg | | | kWh |
| SWINE | | | | | | | | | |
| Sow with litter | 375 | 170.1 | 0.79 | 0.36 | 1.58 | 0.716 | 23.7 | 0.671 | 0.83 |
| Growing pig | 65 | 29.5 | 0.14 | 0.064 | 0.28 | 0.127 | 4.2 | 0.119 | 0.15 |
| Finishing hog | 150 | 68.0 | 0.32 | 0.145 | 0.62 | 0.281 | 9.6 | 0.272 | 0.33 |
| DAIRY CATTLE | | | | | | | | | |
| Dairy cow | 1,300 | 589.7 | 2.21 | 1.002 | 4.42 | 2.00 | 66.0 | 1.88 | 2.33 |
| BEEF CATTLE | | | | | | | | | |
| Beef feeder | 900 | 408.2 | 1.35 | 0.612 | 2.70 | 1.22 | 40.0 | 1.13 | 1.42 |
| SHEEP | | | | | | | | | |
| Sheep feeder | 75 | 34.0 | 0.053 | 0.023 | 0.11 | 0.05 | 1.6 | 0.045 | 0.06 |
| POULTRY | | | | | | | | | |
| Laying hen | 4.5 | 2.04 | 0.0198 | 0.009 | 0.0396 | 0.018 | 0.6 | 0.017 | 0.021 |

*Use specific production data when known.
†Twice the daily $BOD_5$.
‡Based on 30 cubic feet per pound of daily $BOD_5$. (0.53 kg $BOD_5/m^3$)
§Based on 1.9 pounds of oxygen per KWH. (0.862 kg per kWh)

Source: Jones et al. (1971).

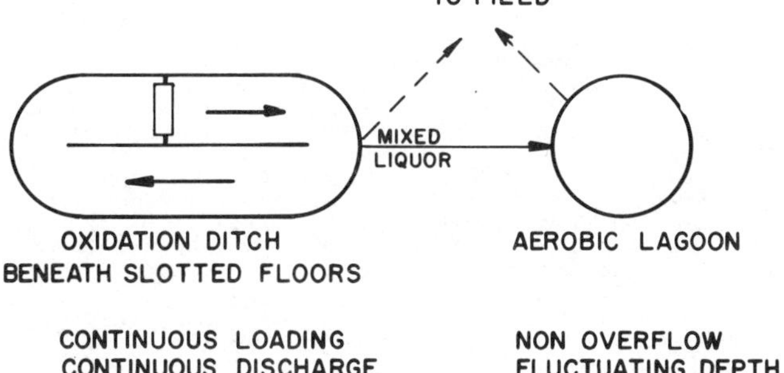



FIG. 2 Flow diagram for an oxidation ditch.

propellers, or pumps, but the rotor (a form of paddle wheel) has been most common. Manufacturer's recommendations should be followed for ditch size, liquid depth, and number and size of aerators for a particular situation.

Good circulation and mixing is also required. Poor circulation is usually caused by a buildup of total solids concentration above about 5 percent. Solids buildup is accelerated by the considerable evaporation that occurs with rotor aeration and may range up to 60 percent of the input volume of liquid manure (Hore and Spencer 1970).

Solids removal, necessary, on a regular basis is usually accomplished by adding water to maintain an overflow. Solids control by mechanical methods, however, has the advantages that the total volume of liquid that must be handled is not enlarged, and the biological stability of the system is not disturbed (Simons et al. 1974).

Oxidation ditch mixed liquor (ODML) contains a large amount of high quality protein (Holmes et al. 1971). Refeeding ODML in situ has recently given renewed interest to the use of livestock oxidation ditches (Harmon and Day 1975).

An extensive list of publications about oxidation ditches for livestock wastes is included at the end of this paper. Topical groupings of the publications follow.

Municipal wastes: Berk (undated), Pasveer (1954), Pasveer (1960), Pasveer (1963), Staff Report (1972), (Symposium—(1969).

Oxygenation and circulation: Anthonisen et al. (1974), Baker et al. (1974), Barrs and Muskat (1959), Baumann and Cleasby (1965), Gaudy and Yang (1973), Jones et al. (1969), Mitchell and Day (1973), Nelson et al. (1972A), Nelson et al. (1972), Ogilvie and Phillips (1972), Simons et al. (1974), Wong-Chong et al. (1973), Zwietering (1958).

Design recommendations: Jones et al. (1971), Kroeker and Loehr (1975), Simons et al. (1974).

General livestock wastes: Anthonisen and Loehr (1974), Collier (1968), Dale (1967), Dale and Morris (1966), Day (1968), Day (1970), Day and Harmon (1974), Day et al. (1970), Guilliame (1964), Hore and Spencer (1970), Jones et al. (1971), Kroeker and Loehr (1975), Larsson (1969), Linn (1966), Pos and Robinson (1973), Scheltinga and Poelma (1970), Simons (1974), Simons et al. (1974).

Swine: Baxter et al. (1966), Day and Harmon (1974A), Day et al. (1969), Foree and O'Dell (1969), Foree et al. (1973), Frank (1973), Grunloh (1972), Holmes et al. (1971), Institute Technique Du Porc (1973), Irgens and Day (1966), Jones et al. (1969), Jones and Patni (1972A), Jones and Patni (1972), Kim and Day (1975), Koch et al. (1975), Littlejohn (1974), KcKinney and Bella (1967), Newtson (1967), Robertson et al. (1974), Robinson (1974), Robinson et al. (1970), Robinson et al. (1971), Rueprich (1974), Scheltinga (1966), Smart et al. (1975), Smith et al. (1971), Taiganides and White (1971), Windt et al. (1971), Wolfermann et al. (1973).

Beef: Aasen et al. (1971), Bauling et al. (1973), Diesch et al. (1971), Hegg and Larson (1972), Hegg et al. (1974), Jones et al. (1971), Knepp (1970), Larson and Moore (1971), Moore et al. (1969), Moore et al. (1973), Self and Hoffman (1974), ten Have (1971), Vetter et al. (1972).

Dairy: Chang et al. (1971), Schmisseur (1966), Nye et al. (1974).

Poultry: Edwards and Robinson (1969), Hashimoto (1972), Loehr et al. (1971), Ludington et al. (1971), Martin and Loehr (1975), Prakasam et al. (1974), Pos et al. (1971), Stevenson and Roth (1972), Stewart and McIlwain (1971), Walker and Pos (1969).

External installations: Baxter et al. (1966), Institute Technique Du Porc (1973), Jones and Patni (1972), Scheltinga (1966), Smith et al. (1971), Taiganides and White (1971).

Microbiological: Diesch et al. (1971), Foree and O'Dell (1969), Grunloh (1972), Robinson et al. (1971).

Nutrients for feed supplements: Day and Harmon (1974A), Day and Harmon (1974), Harmon and Day (1975), Hegg et al. (1974), Holmes et al. (1971), Nye et al. (1974), Robinson (1974), Vetter et al. (1972).

## SUMMARY AND CONCLUSION

This paper gives a status report and literature review of the use of oxidation ditches for treating livestock wastes which began about ten years ago. Odor control is a main advantage of oxidation ditch treatment and high operational energy requirements and costs are a main disadvantage. The potential for utilizing ODML as a feed supplement in situ has given renewed emphasis to this method of waste management. The use of oxidation ditches is best suited for confinement buildings with liquid manure falling through slotted floors into the oxidation ditch.


## References

1 Aasen, A. K., J. B. McQuitty and P. H. Bouthilier. 1971. Storage of beef cattle wastes under aerobic and anaerobic conditions. Paper 71-202, CSAE Conf., July 6. 21 p. (Canada).

2 Anthonisen, A. C. and R. C. Loehr. 1974. The handling and treatment of mink wastes by liquid aeration. Proc., Agri. Waste Mgt. Conf., Cornell Univ., p. 295-308.

3 Baker, D. R., R. C. Loehr and A. C. Anthonisen. 1974. Oxygen transfer at high solids concentrations. ASCE, Environmental Engr. Div., (in press).

4 Barrs, J. K. and J. Muskat. 1959. Oxygenation of water by bladed rotors. Report 28, Research Institute for Public Health Engineering (TNO, Netherlands).

5 Bauling, D. B., W. D. Boston and D. L. Day. 1973. A beef confinement building with an oxidation ditch. ASAE Paper 73-4544, ASAE, St. Joseph, Mich. 49085.

6 Baumann, E. R. and J. L. Cleasby. 1965. Oxygenation efficiency of a bladed rotor. Engr. Exp. Sta. Progress Report, Iowa State Univ., 25 p.

7 Baxter, S. H., R. A. Pontin and J. S. Watson. 1966. Development of a prefabricated feeding piggery with waste treatment in Pasveer-type oxidation ditches. Farm Building Report 2, Scottish Farm Buildings Investigation Unit, Dec., 30 p. (Scotland).

8 Berk, W. L. (Undated). Theory, operation, and cost of the oxidation ditch process. RAD-90, Lakeside Engineering Corp.

9 Chang. A. C., A. C. Dale and J. M. Bell. 1971. Nitrogen transformation during aerobic digestion and denitrification of dairy cattle wastes. Proc., Int. Symp. on Livestock Wastes, ASAE PROC-271:272-274.

10 Collier, A. 1968. They seek a solution to slurry problem. Power Farming, 40(6):12-13. (Britain).

11 Dale, A. C. 1967. Aerobic treatment of animal wastes. ASAE Paper No. 67-927A, ASAE, St. Joseph, Mich. 49085.

12 Dale, A. C. and W. H. M. Morris. 1966. Tentative criteria for design, construction and operation of the Pasveer oxidation ditch system for the treatment of animal wastes. Unpublished report, Agri. Engr. Dept., Purdue Univ., 11 p.

13 Day, D. L. 1968. Oxidation ditches for waste disposal. Int. J. of Farm Bldg. Res. 3:2-7. (British).

14 Day, D. L. 1970. Reducing the pollution potential of livestock wastes with in-the-building oxidation ditches. Proc., Agri. Waste Mgt. Conf., Cornell Univ., p. 77-84.

15 Day, D. L. and B. G. Harmon. 1974A. A recycled feed source from aerobically processed swine wastes. TRANSACTIONS of the ASAE 17(1):82-84, 87.

16 Day. D. L. and B. G. Harmon. 1974. Nutritive value of aerobically treated livestock and municipal wastes. Proc., Wastewater Use in the Production of Food and Fiber, U.S. EPA-660/2-74-041, p. 240-255.

17 Day, D. L., D. D. Jones and J. C. Converse. 1970. Livestock waste management studies — termination report. Agri. Engr. Res. Report, Proj. 31-15-10-375, Univ. of Ill., 97 p.

18 Day, D. L., D. D. Jones, J. C. Converse, A. H. Jensen and E. L. Hansen. 1969. Oxidation ditch treatment of swine wastes-summary report. ASAE Paper 69-924, 16 p. (Condensation in AGRICULTURAL ENGINEERING 52(2):71-73, 1971).

19 Diesch, S. L., B. S. Pomroy and E. R. Allred. 1971. Survival and detection of leptospires in aerated beef cattle manure. Proc., Int. Symp. on Livestock Wastes, ASAE PROC-271:263-266.

20 Edwards, J. B. and J. B. Robinson. 1969. Changes in composition of continuously aerated poultry manure with special reference to nitrogen. Proc., Agri. Waste Mgt. Conf., Cornell Univ., p. 178-184.

21 Foree, G. R. and R. A. O'Dell. 1969. Farm waste disposal field studies utilizing a modified Pasveer oxidation ditch, settling tank, lagoon systems. Proc., Agri. Waste Mgt. Conf., Cornell Univ., p.185-192.

22 Foree, G. R. and O. Vanderslice and D. Pfost. 1973. Swine housing and waste disposal research and developments. ASAE Paper No. 73-4510, ASAE, St. Joseph, Mich. 49085.

23 Frank, J. F. 1973. The design and evaluation of an oxidation ditch system at SIU swine center. ASAE Paper No. 73-4520, ASAE, St. Joseph, Mich. 49085.

24 Gaudy, A. F. and P. Y. Yang. 1973. Control of biological solids concentration in the extended aeration process. 28th Industrial Waste Conf., Purdue Univ.

25 Grunloh, D. 1972. Identification of the predominant microorganisms in swine oxidation ditch mixed liquor. Unpublished report, Anim. Sci. Dept., Univ. of Ill., June.

26 Guilliame, F. 1964. Evaluation of the oxidation ditch as a means of treatment in Ontario. Res. Publ. No. 6, Ontario Water Resources Commission. (Canada).

27 Harmon, B. G. and D. L. Day. 1975. Nutrient availability from oxidation ditches. Proc., Int. Symp. on Livestock Wastes. ASAE PROC (this issue).

28 Hashimoto, A. G. 1972. Aeration under caged laying hens. TRANSACTIONS of the ASAE 15(6):1119-1123.

29 Hegg, R. O. and R. E. Larson. 1972. Solids balance on a beef cattle oxidation ditch. Proc., Agri. Waste Mgt. Conf., Cornell Univ., p. 555-562.

30 Hegg, R. O., J. C. Meiske, R. E. Larson and J. A. Moore. 1974. Beef oxidation ditch settled solids fed to steers. Proc., Agri. Waste Mgt. Conf., Cornell Univ., p. 382-386.

31 Holmes, L. W. J., D. L. Day and J. T. Pfeffer. 1971. Concentration of proteinaceous solids from oxidation ditch mixed liquor. Proc., Int. Symp. on Livestock Wastes. ASAE PROC-271:351-354.

32 Hore, F. R. and V. I. D. Spencer. 1970. Survey of operating farm oxidation ditches. ERDA 11, Canada Agri., Ottawa, p. 9-10. (Canada).

33 Institute Technique Du Porc. 1973. Annual Rapport, Station Experimentale, Region Sud., Villefranche. (France).

34 Irgens, R. L. and D. L. Day. 1966. Aerobic treatment of swine wastes. Proc., Nat. Symp. on Animal Waste Management. ASAE SP-0366:58-60.

35 Jones, D. D., D. L. Day and A. C. Dale. 1971. Aerobic treatement of livestock wastes. Ill. Agri. Exp. Sta. Bul. 737 in co-operation with Purdue Univ. (revised Apr.) 55 p.

36 Jones, D. D., D. L. Day and J. C. Converse. 1969. Oxygenation capacity of oxidation ditch rotors for confinement livestock buildings. Proc., 24th Industrial Waste Conf., Purdue Univ., p. 191-208.

37 Jones, D. D., D. L. Day and J. C. Converse. 1969. Field tests of oxidation ditches in confinement swine buildings. Proc., Agri. Waste Mgt. Conf., Cornell Univ., p. 160-171.

38 Jones, D. D., D. L. Day and U. S. Garrigus. 1971. Oxidation ditch in a confinement beef building. TRANSACTIONS of the ASAE 14(5):825-827.

39 Jones, P. H. and N. K. Patni. 1972A. A study of foaming problems in oxidation ditch treating swine waste. Proc., Agri. Waste Mgt. Conf., Cornell Univ., p. 503-515.

40 Jones, P. H. and N. K. Patni. 1972. An oxidation ditch-lagoon system in southern Ontario for handling swine wastes. Environmental Sciences and Engineering EG-8, Univ. of Toronto, 149 p. (Canada).

41 Kim, H. C. and D. L. Day. 1975. Energetics of alternative waste management systems. Proc., Int. Symp. on Livestock Wastes, ASAE PROC (this issue).

42 Knepp, Don. 1970. Experience in raising beef under three different management systems. ASAE Paper No. 70-903, ASAE, St. Joseph, Mich. 49085.

43 Koch, B. A., R. H. Hines, G. L. Allee and R. I. Lipper. 1975. K.S.U. aerobic swine waste handling system (six years of problems and progress). Proc., Int. Symp. on Livestock Wastes, ASAE PROC (this issue).

44 Kroeker, E. J. and R. C. Loehr. 1975. A design approach for the use of an oxidation ditch for livestock waste treatment. Proc., Int. Symp. on Livestock Wastes, ASAE PROC (this issue).

45 Larson, R. E. and J. A. Moore. 1971. Beef wastes and the oxidation ditch today and tomorrow. Proc., Int. Symp. on Livestock Wastes. ASAE PROC-271:217-219.

46 Larsson. S. 1969. Stabilisering av flytgödsel genom luftning, en litteraturöversikt (Stabilization of farm animal wastes by aeration, a literature review). 1024, Stenciltryck Fran Institutionen For Lantbrukets Byggnadsteknik Lantbrukshögskolan, Lund. (Sweden).

47 Linn. A. 1966. Whipping the manure problem. The Farm Quarterly, 21(6):56, 58-59, 115.

48 Littlejohn, L. (ed.) 1974. The treatment of piggery wastes. Scottish Farm Buildings Investigation Unit, June, 66 p. (Scotland).

49 Loehr, R. C., D. F. Anderson and A. C. Anthonisen. 1971. An oxidation ditch for the handling and treatment of poultry wastes. Proc., Int. Symp. on Livestock Wastes, ASAE PROC-271:209-212.

50 Ludington, D. C., A. T. Sobel, R. C. Loehr and A. G. Hashimoto. 1972. Pilot plant comparison of liquid and dry waste management systems for poultry manure. Proc., Agri. Waste Mgt. Conf., Cornell Univ., p. 569-580.

51 Martin, J. H., Jr. and R. C. Loehr. 1975. An evaluation of aeration systems for poultry wastes under commercial conditions. Proc., Int. Symp. on Livestock Wastes. ASAE PROC (this issue).

52 McKinney, R. E. and R. Bella. 1967. Water quality changes in confined hog waste treatment. Project Completion Report. Kansas Water Resources Res. Inst., Univ. of Kan., 88 p.

53 Mitchell, J. K. and D. L. Day. 1973. Oxygenation and flow characteristics of mechanical aerators. Proc., Ill. Livestock Waste Mgt. Conf., Univ. of Ill., Mar. 7-8, p. E1-E13.

54 Moore, J. A., R. E. Larson and E. R. Allred. 1969. Study of the use of the oxidation ditch to stabilize beef animal manure in

cold climates. Proc., Agri. Waste Mgt. Conf., Cornell Univ., p. 172-177.

55   Moore, J. A., R. E. Larson, R. O. Hegg and E. R. Allred. 1973. Beef confinement systems — oxidation ditch. TRANSACTIONS of the ASAE 16(1):168-171.

56   Nelson, G. L., J. J. Kolega and Q. B. Graves. 1972A. Basic performance parameters for oxygenation in rotor-aerated liquid waste systems. TRANSACTIONS of the ASAE 15(6):1138-1144.

57   Nelson, G. L., U. Agena and G. Hoffman. 1972. Performance parameters for rotor-driven liquid circulation patterns in ditches for waste treatment. TRANSACTIONS of the ASAE 15(6):1145-1149.

58   Newtson, K. 1967. Current status of oxidation ditch field application and results. Proc., 10th National Pork Industry Conference, Nov. 9-10 at Lincoln, Nebraska.

59   Nye, J. C., A. C. Dale, T. W. Perry, R. B. Harrington and E. J. Kirsch. 1974. Recovering protein from dairy cattle wastes. TRANSACTIONS of the ASAE 17(6):1155-1160.

60   Ogilvie, J. R. and P. Phillips. 1972. Modeling process variations in an oxidation ditch. Canadian Agricultural Engineering, 14(2):59-62. (Canada).

61   Pasveer, A. 1954. Research on activated sludge-IV. Purification with intense aeration. Sewage and Industrial Waste, 26(2):149-159.

62   Pasveer, A. 1960. New Developments in the application of Kessner brushes (aeration rotors) in the activated-sludge treatment of trade-waste water. p. 226-253. In: P. C. G. Isaac (ed.) Waste Treatment, Pergamon Press, New York.

63   Pasveer, A. 1963. Developments in activated sludge treatment in the Netherlands. p. 291-297. In: W. W. Eckenfelder and B. J. McGabe (ed's.). Advances in Biological Waste Treatment, Pergamon Press, New York.

64   Pos, J. and J. B. Robinson. 1973. Winter operation of aerated liquid animal waste storage systems. Canadian Agricultural Engineering, 15(1):43-48. (Canada).

65   Pos, J., R. G. Bell and J. B. Robinson. 1971. Aerobic treatment of liquid and solid poultry manure. Proc., Int. Symp. on Livestock Wastes, ASAE PROC-271:220-224.

66   Prakasam, T. B. S., E. C. Srinath, A. C. Anthonisen, J. H. Martin (Jr.) and R. C. Loehr. 1974. Approaches for the control of nitrogen with an oxidation ditch. Proc., Agri. Waste Mgt. Conf., Cornell Univ., p. 421-435.

67   Robertson, A. M., J. J. Clark and S. H. Baxter. 1974. The treatment of pig waste in a two stage anaerobic/aerobic system. Paper 74-II-113, Section II, Theme 1, VIIIth International Congress of Agricultural Engineering (CIGR) at Flevohof, Netherlands, Sept. 23-29. (Netherlands).

68   Robinson, K. 1974. Waste treatment with a protein bonus. Proc., Agri. Waste Mgt. Conf., Cornell Univ., p. 415-420.

69   Robinson, K., J. R. Saxon and S. H. Baxter. 1971. Microbiological aspects of aerobically treated swine wastes. Proc., Int. Symp. on Livestock Wastes, ASAE PROC-271:225-228.

70   Robinson, K., S. H. Baxter and J. R. Saxon. 1970. Aerobic treatment of farm wastes. Proc., Symp. Farm Wastes, Univ. of Newcastle upon Tyne, Paper No. 17, Jan., p. 122-131. (England).

71   Rueprich, W. 1974. Der mastschweinestall mit oxydationsgraben (fattening hog confinement units with oxidation ditches). Bauen auf dem Lande, 25(5):142-148. (W. Germany).

72   Scheltinga, H. M. J. 1966. Biological treatment of animal wastes. Proc., Nat. Symp. on Animal Waste Mgt., ASAE SP-0366:140-143.

73   Scheltinga, H. M. J. and H. R. Poelma. 1970. Treatment of farm wastes. Proc., Symposium Farm Wastes, Univ. of Newcastle upon Tyne, Paper No. 19, Jan., p. 138-146. (England).

74   Schmisseur, W. E. 1966. Oxidation of dairy manure via recirculating oxidation ditch. Progress Report, Dept. of Agri. Econ., Purdue Univ., April. 9 p.

75   Self, H. L. and M. P. Hoffman. 1974. Influence of environment on cattle feeding parameters. Proc., Int. Livestock Environment Symp., ASAE SP-0174:281-287.

76   Simons, D. 1974. Umweltfreundliche fluessigmistketten durch oxydationsgraben (environmental protective liquid manure systems by means of the oxidation ditch). Deutsche Gefluegelwirtschaft und Schweineproduktion, 26(48):1197-1202. (W. Germany).

77   Simons, D., D. D. Jones and A. C. Dale. 1974. Oxidation ditch system analysis and field evaluation of the aerob-a-jet. Proc., Agri. Waste Mgt. Conf., Cornell Univ., p. 436-454.

78   Smart, P., F. McCain, D. L. Day and B. G. Harmon. 1975. Oxidation ditch waste management system for a large confinement swine farm. Proc., Int. Symp. on Livestock Wastes. ASAE PROC (this issue).

79   Smith, R. J., T. E. Hazen and J. R. Miner. 1971. Manure management in a 700-head swine-finishing building; two approaches using renovated waste water. Proc., Int. Symp. on Livestock Wastes, ASAE PROC-271:149-153.

80   Stevenson, J. S. and L. J. Roth. 1972. Experiences with oxidation ditches in a pullet growing house. ASAE Paper No. 72-452, ASAE, St. Joseph, Mich. 49085.

81   Stewart, T. A. and R. McIlwain. 1971. Aerobic storage of poultry manure. Proc., Int. Symp. on Livestock Wastes, ASAE PROC-271:261-262, 266.

82   Symposium on low cost waste treatment. 1969. Proc., Tenth Symposium, Vol. 1, Central Public Health Engineering Research Institute, Nagpur. (India).

83   Taiganides, E. P. and R. K. White. 1971. Automated handling, treatment, and recycling of wastewater from an animal confinement production unit. Proc., Int. Symp. on Livestock Wastes, ASAE PROC-271:146-148.

84   ten Have, P. 1971. Aerobic biological breakdown of farm waste. Proc., Int. Symp. on Livestock Wastes, ASAE PROC-271:275-278.

85   U.S. Dept. of Transportation. 1972. Oxidation ditch sewage waste treatment process. Water Supply and Waste Disposal Series, 6. 52 p.

86   Vetter, R. L., R. D. Christensen and G. Frankl. 1972. Feeding value of animal waste nutrients from a cattle confinement oxidation ditch system. Leaflet R170, Animal Sci. Dept., Iowa State Univ., 7 p.

87   Walker, J. P. and J. Pos. 1969. Caged layer performance in pens with oxidation ditches and liquid manure tanks. Proc., Agri. Waste Mgt. Conf., Cornell Univ., p. 249-253.

88   Windt, T. A., N. R. Bulley and L. M. Staley. 1971. Design, installation and biological assessment of a Pasveer oxidation ditch on a large British Columbia swine farm. Proc., Int. Symp. on Livestock Wastes, ASAE PROC-271:213-216.

89   Wolfermann, H. F., J. Hornig and D. Klingenschmitt. 1973. Der oxydationsgraben im mastschweinestall mit ganzspaltenboden (the oxidation ditch in fattening hogs confinement units with totally slotted floors). Bauen auf dem Lande, 24(6):156-160. (W. Germany).

90   Wong-Chong, G. M., A. C. Anthonisen and R. C. Loehr. 1973. Comparison of the conventional cage rotor and jet-aero-mix systems in oxidation ditch operations. 28th Industrial Waste Conf., Purdue Univ.

91   Zwietering, Th. N. 1958. Suspending of solid particles in liquid by agitators. Chem. Engr. Sci. 8:244.

# Nitrogen Transformations in Aerated Beef Slurries

R. O. Hegg, E. R. Allred

ASSOC. MEMBER MEMBER
ASAE   ASAE

A FIELD and a laboratory study were conducted to determine the change in organic nitrogen for beef waste slurries under low temperatures and low aeration rates. Also determined were nitrifying and heterotrophic bacteria populations.

## DISCUSSION

Experimental pilot-scale studies on the aeration of beef animal wastes have shown buildups of nitrite and nitrate (Moree et al. 1973). An oxidation ditch located at the University of Minnesota, Agricultural Engineering Farmstead, Rosemount, Minnesota, receives waste from approximately 36 beef animals. The animals are housed on slatted floors directly above the race track shaped ditch (27.4 m long, 4.6 m wide, and 1.4 m deep). A continuously operating rotor moves the liquid around the ditch and introduces air into the liquid for aerobic decomposition of the wastes. This treatment facility operates as a continuously fed batch system, which means that some dilution water is added initially but thereafter only the wastes are added and no material is removed until the experiment is completed (normally about 3 to 6 months).

The laboratory experiment was designed based on observations of the nitrogen transformations in the oxidation ditch. Six containers with a capacity of about 20 l each were used as aerobic digesters. They were operated so that liquid temperatures and aeration rate could be controlled, with the contents being stirred continuously. Three trials (two 20 wk and one 15 wk) were run with two digesters operated at each of the three temperatures (1.7, 7.2, and 12.8 C). The operating conditions are summarized in Table 1.

The digesters were sampled weekly to determine pH, nitrate-N, nitrite-N, ammonium-N, organic-N, and dissolved oxygen level. Tests were also run to determine the *Nitrosomonas*, *Nitrobacter* and heterotrophic populations. Overall nitrogen balances were run on the digesters, including sampling for losses of ammonia gas.

### TABLE 1. SUMMARY OF OPERATING CONDITIONS

| | Length of acclimation period (all digesters at 12.8 C), weeks | Time to reduce temp. to normal operating temp., weeks | Total length of trial, weeks | Airflow rate, l/m |
|---|---|---|---|---|
| Trial I | 3 | 3 | 20 | 2 |
| Trial II | 1 | 1 | 15 | 1 |
| Trial III | 3 | 2 | 20 | 0.5 |

Approved for publication as Journal Article No. 8989 of the Minnesota Agricultural Experiment Station.

The authors are: R. O. HEGG, Agricultural Engineer, NCR, ARS, USDA, and E. R. ALLRED, Professor, Agricultural Engineering Dept., University of Minnesota, St. Paul.

The *Nitrosomonas* populations were determined by the dilution extinction technique (Most Probable Number) Standard Methods, 1971, and the *Nitrobacter* populations by the fluorscent antibody technique (Fliermans et al. 1974). The other analyses were done according to Standard Methods, 1971.

## NITRIFYING BACTERIA

Nitrification takes place in two basic steps: oxidation of ammonia to nitrite by *Nitrosomonas*, and oxidation of nitrite to nitrate by *Nitrobacter* bacteria. The aerobic, autotrophic bacteria *Nitrosomonas* (five genera) and *Nitrobacter* (two main genera) can live, grow, and reproduce in an environment free of organic matter, with carbon dioxide as the sole carbon source, but with other essential nutrients available. Ammonification converts the organic-N into ammonium-N. In the first step of the nitrification, coenzymes molecules are reduced, yielding considerable energy to the *Nitrosomonas* bacteria. In the last step, the conversion of nitrite to nitrate, only one coenzyme molecule is reduced, yielding less energy than in the first step. In the normal nitrification process, nitrite is rapidly converted to nitrate because *Nitrobacter* require three times as much energy as do *Nitrosomonas* to obtain the same amount of energy. Normally, nitrite does not buildup in an active nitrifying system (McKinney 1962).

Most bacteria are heterotrophic, meaning that they derive their energy from simple or complex organic compounds that have been synthesized by other heterotrophic or autotrophic organisms. Heterotrophic nitrification has been found in culture systems and in natural soils but their contribution is generally much smaller than autotrophic nitrification (Verstraete and Alexander 1973, McCoy 1972, Eylar and Schmidt 1959).

Nitrifiers require elements for growth, but high concentrations of these elements may be inhibitory. Although nitrifiers are sensitive to high concentrations of their own substrates, they are more sensitive to the substrate of the other species (Engel and Alexander 1958). The optimum pH range for nitrification is between 7.8 and 8.9 (Wild et al. 1971). Some research has shown that nitrate formation is inhibited more strongly than nitrite formation at the higher concentrations of both ammonium and total salts. The theory is that ammonia is toxic to *Nitrobacter*. So a combination of high pH and high ammonium concentrations create conditions conducive to high ammonia levels. With pH's of 6, 7, 8, 9, the respective percentages of ammonium existing as non-ionized ammonia are 0.1, 1.0, 10, and 50. Ammonia levels of above 10 mg/l were toxic. Ammonium is not toxic because, unlike ammonia, it will not readily cross cell walls. *Nitrobacter* can tolerate low ammonia concentrations because of their detoxifying mechanisms that form urea, glutamine, or asparagin (Smith 1964, Harada and Kai 1968, Chapman and Liebig 1952).

Nitrification can take place under a wide temperature range. Temperatures above 40 C are inhibitory to nitrifica-

TABLE 2. TRIAL I, AVERAGE AMMONIA CONCENTRATION IN A SOLUTION
OF ANIMAL WASTE, BASED ON IONIZATION CONSTANT OF AMMONIA,
K=1.4 x $10^{-5}$ at 1.7 C, K=1.5 x $10^{-5}$ at 7.2 C, and K=1.6 x $10^{-5}$ at 12.8 C.

| Week | 1.7 C | | | 7.2 C | | | 12.8 C | | |
|---|---|---|---|---|---|---|---|---|---|
| | $NH_4$—N, mg/l | pH | $NH_3$—N, mg/l | $NH_4$—N, mg/l | pH | $NH_3$—N, mg/l | $NH_4$—N, mg/l | pH | $NH_3$—N, mg/l |
| 1 | 9999 | *9999.0 | 9999.0 | 9999 | 9999.0 | 9999.0 | 9999 | 9999.0 | 9999.0 |
| 2 | 205 | 7.1 | 2.0 | 200 | 7.2 | 2.2 | 135 | 6.7 | 0.4 |
| 3 | 250 | 6.6 | 0.8 | 160 | 7.4 | 2.8 | 175 | 6.4 | 0.3 |
| 4 | 105 | 6.8 | 0.5 | 85 | 6.7 | 0.3 | 100 | 6.4 | 0.2 |
| 5 | 195 | 7.3 | 2.9 | 155 | 6.9 | 0.9 | 180 | 6.9 | 0.9 |
| 6 | 255 | 7.8 | 12.2 | 180 | 7.5 | 4.0 | 185 | 7.2 | 1.9 |
| 7 | 400 | 8.3 | 60.4 | 300 | 7.9 | 16.8 | 300 | 7.2 | 3.1 |
| 8 | 460 | 8.4 | 87.4 | 305 | 6.9 | 1.7 | 310 | 6.7 | 1.0 |
| 9 | 540 | 8.3 | 81.5 | 305 | 7.3 | 4.3 | 305 | 6.6 | 0.8 |
| 10 | 705 | 8.3 | 106.4 | 370 | 6.8 | 1.6 | 390 | 6.2 | 0.4 |
| 11 | 530 | 8.2 | 63.5 | 310 | 6.7 | 1.1 | 285 | 6.3 | 0.4 |
| 12 | 510 | 8.1 | 48.6 | 255 | 7.0 | 1.8 | 315 | 6.3 | 0.4 |
| 13 | 565 | 7.9 | 33.9 | 315 | 7.2 | 3.5 | 295 | 6.9 | 1.6 |
| 14 | 555 | 8.0 | 42.0 | 240 | 7.2 | 2.7 | 210 | 7.0 | 1.4 |
| 15 | 590 | 8.0 | 44.6 | 285 | 7.3 | 4.0 | 200 | 7.0 | 1.3 |
| 16 | 565 | 8.1 | 53.8 | 260 | 7.4 | 4.6 | 175 | 6.9 | 0.9 |
| 17 | 685 | 8.3 | 103.4 | 265 | 7.2 | 3.0 | 230 | 6.8 | 1.0 |
| 18 | 605 | 8.8 | 288.7 | 185 | 6.9 | 1.0 | 205 | 6.8 | 0.9 |
| 19 | 635 | 8.9 | 381.5 | 270 | 6.9 | 1.5 | 155 | 6.4 | 0.3 |
| 20 | 615 | 9999.0 | 9999.0 | 245 | 9999.0 | 9999.0 | 115 | 9999.0 | 9999.0 |

*9999, means insufficient data

tion, but from 40 C down to freezing it will proceed, although the rate decreases with temperatures.

## RESULTS

The results showed steady increases in the Kjeldahl-N concentrations for all three laboratory trials and in the pilot scale due to semicontinuous loading of the batch system. In the digesters, maximum Kjeldahl-N concentrations ranged from 1000 to 2000 mg/l.

The ammonium-N concentrations in the laboratory trials were normally above 100 mg/l and as high as 700 mg/l (Table 2). The ammonium-N level in each trial was highest at 1.7 C. In Trial III, with the low aeration rate, oxygen became limiting in all digesters, but at the warmest temperature first. After oxygen became limiting (dissolved oxygen less than 0.5 mg/l) the ammonium-N concentrations steadily increased.

Nitrite and nitrate built up in each of the three laboratory trials at all three temperatures and also in the pilot scale oxidation ditch. The Trial I (high aeration rate, 2 l/m) nitrite-N steadily built up at 7.2 and 12.8 C reaching approximately 500 mg/l (Fig. 1). The nitrate-N reached approximately 300 mg/l after 10 wk, at which level it stayed for the remaining 10 wk (Fig. 2).

In Trial II, there was a lag time before a buildup of nitrite at the two colder temperatures (4 wk for 7.2 C and 10 wk for 1.7 C), which likely was due to the shorter acclimation period at the beginning of the trial. The lag period was longer before nitrate built up in the corresponding digesters. Also, because of the lower aeration rate in Trial II (1 l/m), the digesters became oxygen deficient before the end of the trial due to the increase in suspended solids, which decreased the oxygen solubility. The warmer digesters (12.8 C) became oxygen deficient first, because the oxygen became less soluable as the temperature of the liquid increased.

Trial III showed a similar pattern of sharp drops in the nitrite and nitrate concentrations when oxygen became limiting (Figs. 3 and 4). Because the aeration rate was even lower (0.5 l/m) than in Trial II, the oxygen, became deficient sooner. At the beginning of Trial III, all digesters were

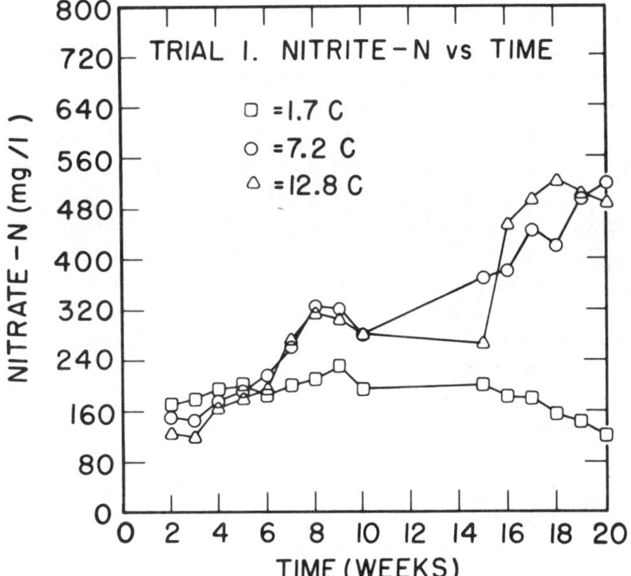


FIG. 1 Trial I, nitrite-N versus time at 1.7, 7.2, and 12.8 C.

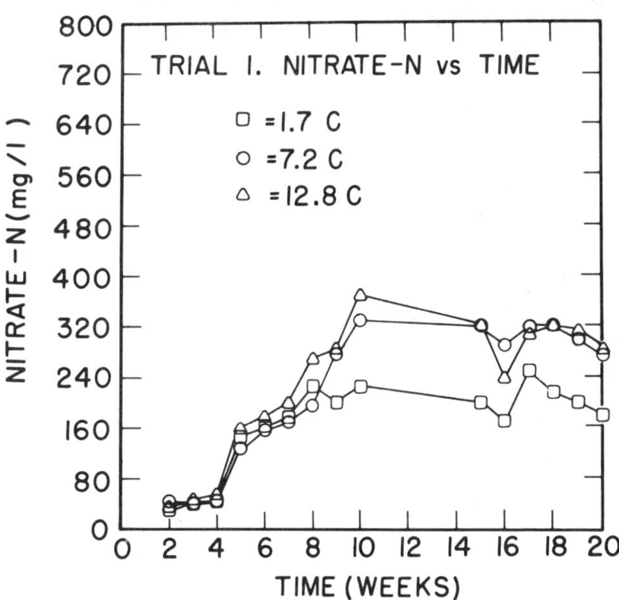


FIG. 2 Trial I, nitrate-N versus time at 1.7, 7.2, and 12.8 C.

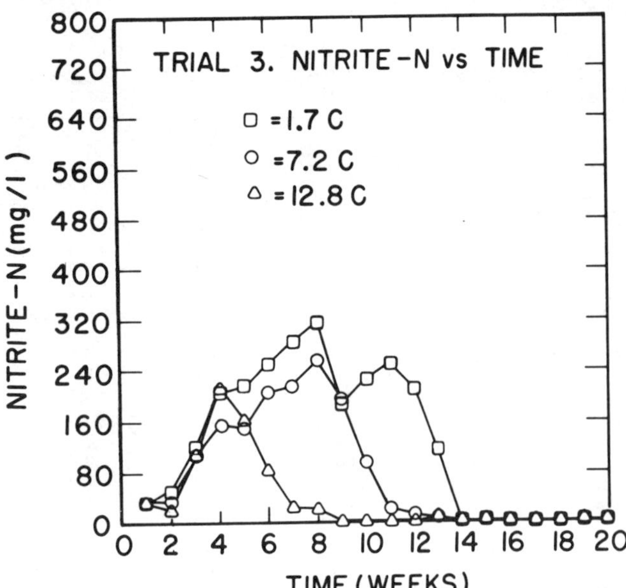


FIG. 3 Trial III, nitrite-N versus time at 1.7, 7.2, and 12.8 C.

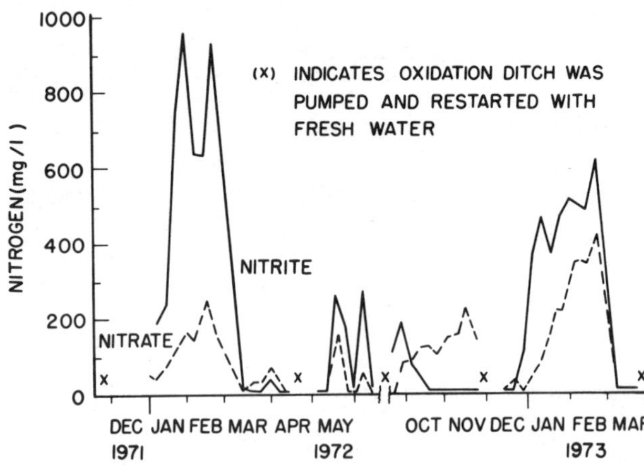


FIG. 5 Nitrite-N and nitrate-N versus time in a batch-operated beef oxidation ditch.

maintained at a temperature of 12.8 C for 3 wk, to enable nitrification to proceed at a rate so that no lag phase occurred, as in Trial II.

Nitrite and nitrate also built up in the oxidation ditch, where the nitrite-N level at one time reached 1000 mg/l (Fig. 5). The maximum nitrate-N level was over 400 mg/l (Fig. 5). The decreases in nitrite and nitrate in the oxidation ditch corresponded with low levels of dissolved oxygen. No data are shown for the summer period because no nitrite built up as a result of oxygen deficient conditions.

The reason for a nitrite buildup is an interesting question, because as was stated earlier, nitrite normally is rapidly converted to nitrate. Calculations were made to determine the amount of non-ionized ammonia at the different temperatures and pH's of the aerated slurry. The data shown in Table 2 are averages of the two digesters operating at each of the three temperatures. At all three temperatures the ammonia-N concentration remained below 3 mg/l for the first 5 wk. In the 1.7 C digesters, the pH increased which caused a corresponding increase in ammonia-N. These digesters had a relatively low value of nitrite-N (up to 150 mg/l) throughout the trial (Fig. 1). In contrast, nitrite-N at the 7.2 and 12.8 C temperatures increased almost

steadily, even though conditions for ammonia toxicity were not present (ammonia-N less than 4 mg/l). This seem's to indicate that something besides ammonia concentration was responsible for the nitrite buildup.

Concentrations of certain metallic ions can be inhibitory. Periodically throughout each trial, samples from the digesters were analyzed for Ca, Al, Na, Fe, Mg, Zn, Mo, Mn, B, and Cu. Concentrations of these elements, as compared with the studies of Meiklejohn (1954), were not inhibitory, except for Cu. Loveless and Painter (1968) stated that at 25 C 0.1 mg/l of Cu would completely inhibit oxygen uptake by *Nitrosomonas*. The maximum Cu concentrations in all digesters reached 2 to 3 mg Cu/l by the end of the trials.

Fig. 6 shows the *Nitrosomonas, Nitrobacter*, and heterotrophic populations from Trial III for all three temperatures. The heterotrophic population increased during the trial and was approximately 100 times the *Nitrosomonas* and *Nitrobacter* populations. The low aeration rate in Trial III, (0.5 l/m) resulted in anaerobic conditions in all digesters before the end of the trial. Oxygen became deficient (less than 0.5 mg/l dissolved oxygen) first in the 12.8 C digesters at 9 wk, and in the 7.2 C and 1.7 C digesters, at 12 and 16 wk, respectively. By the end of the trial, the total

(Continued on page 521)

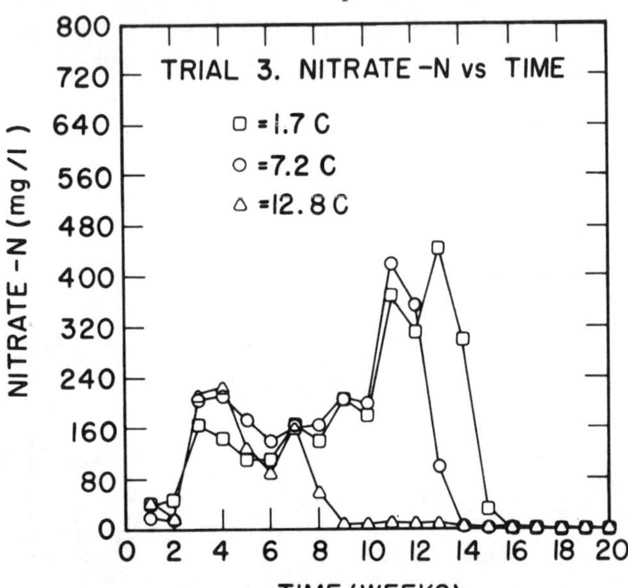


FIG. 4 Trial III, nitrate-N versus time at 1.7, 7.2, and 12.8 C.

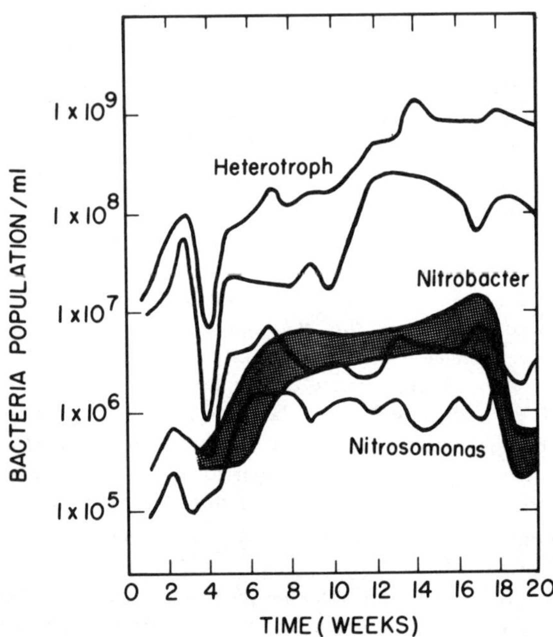


FIG. 6 The range in population for *Nitrosomonas, Nitrobacter,*, and heterotrophic bacteria for Trial III. The range includes the bacteria population at 1.7, 7.2, and 12.8 C.



MANAGING LIVESTOCK WASTES

# A Design Approach for the Use of an Oxidation Ditch for Livestock Waste Treatment

E. J. Kroeker, R. C. Loehr

MEMBER
ASAE

MEMBER
ASAE

THE oxidation ditch was developed in the Netherlands during the 1950's as a means of treating municipal wastes and has been adapted to the treatment of agricultural wastes. It is characterized as a completely mixed extended aeration process with long liquid detention times. The two principal components of the system are a continuous open channel and a rotor whose function is to transfer oxygen from the atmosphere to the ditch contents as well as to circulate the ditch contents at sufficient velocity to maintain the solids in suspension.

To date, design criteria for the oxidation ditch treatment of livestock wastes have been based on studies with specific livestock wastes. These results have been extrapolated to other situations and livestock wastes. Jones et al. (1971) developed an empirical approach using 5-day biochemical oxygen demand ($BOD_5$) loading rates to estimate ditch volume and aeration requirements. Design parameters and criteria based on the fundamentals of biological processes and the characteristics of municipal and industrial wastes have been developed by many individuals in an attempt to place the design of biological waste treatment systems on a more fundamental basis. Since livestock wastes differ in concentration and characteristics from the wastes used to establish these more rational design criteria, the criteria need to be used cautiously in the design of livestock waste treatment systems. The aerobic treatment of livestock wastes using the oxidation ditch takes advantage of relatively constant temperatures and loading rates and long detention times to produce a near ideal biological waste treatment system.

The object of the larger study from which this paper was taken (Kroeker 1974) was to develop an approach which could be used as a tool by practitioners for the rational design of in-house continuous flow oxidation ditch systems for the treatment of livestock wastes for three specific treatment objectives: (a) odor control, (b) nitrogen removal, and (c) nitrogen conservation. Fig. 1 enumerates the design decisions which must be made when an oxidation ditch is incorporated into a livestock production unit. Table 1 identifies the relevant design parameters and their related variables. A mathematical model was developed and verified by its application to the design of two existing large scale oxidation ditch systems used to treat caged layer poultry wastes.

This paper describes the development of the design criteria which relate oxygen requirements for waste treatment to aeration rotor size.

Acknowledgments: The authors express their appreciation to Dr. T. B. S. Prakasam, Dr. E. G. Srinath and J. H. Martin, Jr. for supplying the data used in the verification of the model.

The authors are: EDWIN J. KROEKER, Research Engineer, Agricultural Engineering Dept., University of Manitoba, Winnipeg, Man. Canada; and RAYMOND C. LOEHR, Professor, Civil and Agricultural Engineering Depts., and Director, Environmental Studies Program, Cornell University, Ithaca, N.Y.

**TABLE 1. DESIGN VARIABLES FOR AN OXIDATION DITCH**

| Design parameters | Related variables |
|---|---|
| Channel size | 1. Surface area required to collect wastes from the caged layers.<br>2. Quantity of wastes added.<br>3. Liquid retention time.<br>4. Mixed liquor solids concentration.<br>5. Rotor immersion depth. |
| Oxygen requirements | 1. Treatment objective(s).<br>2. Quantity of wastes added. |
| Rotor size for aeration | 1. Oxygen requirements.<br>2. Mixed liquor solids concentration.<br>3. Type of rotor.<br>4. Rotor immersion depth. |
| Rotor size for mixing | 1. Liquid depth and channel width.<br>2. Type of rotor. |

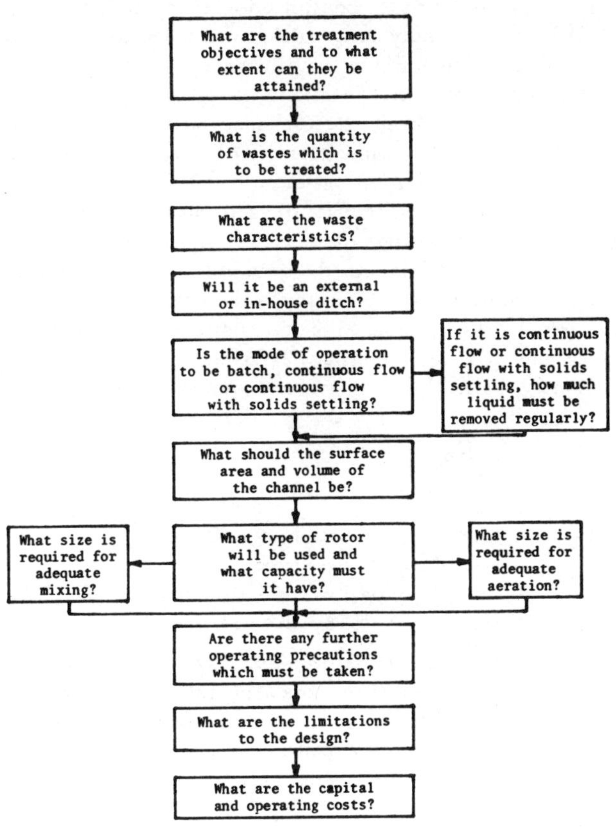



FIG. 1 Oxidation ditch design decisions.

## MATHEMATICAL RELATIONSHIPS

### Oxygen Demand

In an aerobic biological treatment system a reasonably

accurate estimate of the amount of oxygen-demanding material is necessary in order to properly size the aeration equipment. There are two important components of oxygen-demanding material in livestock wastes (Table 2). Oxygen is needed for the microorganisms to oxidize the organic carbon present in the waste. This carbonaceous oxygen demand can be estimated by the chemical oxygen demand (COD) if nitrite ($NO_2^-$) ions are not present. In addition, oxygen may also be needed to oxidize the ammonia which is present in the wastes and released from the organic matter.

### TABLE 2. BIOCHEMICAL OXYGEN DEMAND

A. *Carbonaceous demand*

$$\text{Organic matter} + O_2 \xrightarrow[\text{cells}]{\text{Microbial}} \text{New cells} + CO_2 + H_2O$$

B. *Nitrogenous demand*

$$NH_4^+ + 1.5\, O_2 \xrightarrow{\text{bacteria}} 2H^+ + NO_2^- + H_2O$$

$$NO_2^- + 0.5\, O_2 \xrightarrow{\text{bacteria}} NO_3^-$$

The total amount of oxygen which must be supplied to the mixed liquor will depend upon which of the three previously outlined objectives have been chosen. If the system will be operated only to control odor, then the rotor may be designed to supply only sufficient oxygen to meet the carbonaceous demand. On the other hand, if the system is used to minimize nitrogen losses, enough oxygen must be supplied to meet both the carbonaceous and the nitrogenous demands. This same quantity of oxygen also may be required for maximum nitrogen removal.

To control odors, adequate oxygen must be supplied to prevent the formation of reduced organic compounds which are responsible for the foul odors released under anoxic conditions. In this study, this oxygen requirement was assumed to be equal to the maximum quantity of COD which may be removed through aeration.

A mass balance for the organic matter in an oxidation ditch can be expressed by:

$$O_{acc} = O_i - O_o - O_d \quad . . . . . . . . . . . . . . . . . .[1]$$

in which

$O_{acc}$ = rate of accumulation, mass/time
$O_i$   = rate of input, mass/time
$O_o$   = rate of removal by oxidation, mass/time
$O_d$   = rate of removal by disposal, mass/time

In a continuous flow system at steady state, organic matter will not accumulate. Therefore, the rate of removal by oxidation is equal to the difference between the raw waste input rate and the rate of removal by disposal. This can be expressed by the following equation:

$$f_c \cdot O_i = O_i - O_d$$

in which $f_c$ = fraction of the influent organic carbon which is oxidized.

The oxidation of organic carbon can be expressed directly as an oxygen demand. Each unit by weight of COD which is oxidized is assumed to require one unit of oxygen. The following equation, derived from the mass balance, was used in the model to predict the rate at which oxygen must be supplied to the mixed liquor for odor control:

$$R = f_c \cdot S \cdot n \quad . . . . . . . . . . . . . . . . . . . . .[2]$$

in which

R = microbial oxygen demand, mass/time
S = rate of COD loading to the ditch, mass/animal-time
n = number of animals

In an aerobic biological treatment system, nitrogen may be removed by ammonia stripping or by the microbial formation of nitrite and nitrate (nitrification) followed by anaerobic denitrification. This latter is the more effective and controllable method for the removal of nitrogen in livestock wastes. The most efficient means of storing nitrogen in an aerobic biological system is to oxidize the ammonia to nitrate and to avoid subsequent denitrification Consequently, if the treatment objective is either to remove or conserve nitrogen, sufficient oxygen must be supplied to meet both the carbonaceous and the nitrogenous demands. The nitrogenous oxygen demand was assumed to be equal to the biodegradable fraction of the total Kjeldahl nitrogen (TKN).

According to the stoichiometric relationship in Table 2, 4.57 grams of oxygen are required for every gram of ammonium ion oxidized to nitrate. The total oxygen demand for an oxidation ditch system used to remove or conserve nitrogen will be the sum of the carbonaceous and nitrogenous demands; it was expressed by the following equation:

$$R = n \cdot (f_c \cdot S + 4.57 \cdot f_N \cdot S_N) \quad . . . . . . . . . .[3]$$

in which

$f_N$ = fraction of TKN which is biodegradable
$S_N$ = rate of TKN loading to the ditch, mass/animal-time

**Oxygen Transfer**

The fundamental relationship used to estimate the rate at which oxygen may be supplied by an aeration device to an aerobic treatment system is:

$$dC/dt = K_La \cdot (C_S - C_L) - R_r \quad . . . . . . . . . . .[4]$$

in which

$dC/dt$ = rate of change of dissolved oxygen concentration, mass/volume-time
$K_La$ = overall gas transfer coefficient, $\text{time}^{-1}$
$C_S$ = oxygen saturation concentration for given atmospheric composition and pressure, mass/volume
$C_L$ = actual oxygen concentration at time t, mass/volume
$R_r$ = oxygen utilization rate of microorganisms, mass/volume-time

At steady state conditions the rate of change of dissolved oxygen (DO) will be zero and the aeration rate should equal the oxygen utilization rate.

The oxygenation capacity of aeration equipment is usually determined in water at zero DO and is adjusted to 20 C. Under these conditions the oxygenation capacity of the rotor can be expressed as:

$$N_o = K_La_{20} \cdot C_S \cdot V \quad . . . . . . . . . . . . . . . .[5]$$

in which

$N_o$ = oxygenation capacity, mass/time



MANAGING LIVESTOCK WASTES

$K_L a_{20}$ = oxygen transfer coefficient in water at 20 C, time$^{-1}$

$V$ = volume of the liquid under aeration

The standard conditions which are assumed in Equation [5] rarely exist and modifications must be made to adjust for different $K_L a$, $C_S$, and atmospheric pressure. Temperature effects, however, are minimal, at least for the range 10 to 25 C (Baker 1973). The following equation can be used to describe oxygen transfer relationships under process conditions:

$$N_T = a \cdot K_L a_{20} \cdot (\beta \cdot C_S - C_L) \cdot \frac{P}{14.7} \cdot V \quad \ldots [6]$$

in which

$N_T$ = aeration capacity of the rotor under process conditions at temperature T, mass/time

$a$ = ratio of $K_L a$ in the mixed liquor at temperature T to $K_L a$ in water at temperature T

$C_S$ = oxygen saturation in water at temperature T, mass/volume

$C_L$ = dissolved oxygen concentration in the ditch under steady state conditions, mass/volume

$P$ = atmospheric pressure, mass/area

$\beta$ = ratio of DO concentration at saturation in wastewater to that in pure water.

$K_L a_{20}$ and V can be elimated from the oxygen transfer equation by dividing Equation [6] by Equation [5] to obtain the following:

$$N_T = N_o \cdot a \cdot \frac{(\beta \cdot C_S - C_L)}{C_S} \cdot \frac{P}{14.7} \quad \ldots \ldots \ldots [7]$$

The value of $N_o$ is a basic characteristic of the aerator (Jones et al. 1969). Equation [7] was used in the model to describe the oxygen transfer capacity of the rotor under process conditions.

Fig. 2 demonstrates the effect of total solids concentration on $a$. The relationships noted in the figure were used to estimate $a$ in this study.

In the application of Equation [7] to poultry waste treatment, the value of $\beta$ was assumed to be 1.0 (Baker 1973).

Equations [2] and [3] were used with Equation [7] in the following way to predict rotor length design:

$$L = R/N_T \quad \ldots \ldots \ldots \ldots \ldots \ldots \ldots [8]$$

in which

$L$ = design length of rotor, length

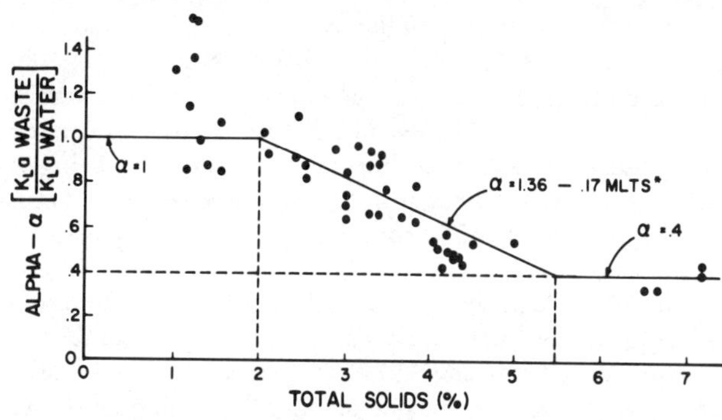


FIG. 2 Effect of total solids concentrations on $a$ in a poultry waste mixed liquor (after Calhoun, 1974). *Mixed Liquor Total Solids (percent)

**TABLE 3. COMPARISON OF MODEL PREDICTED ROTOR LENGTH REQUIREMENTS WITH EXPERIMENTAL DATA FOR OXIDATION DITCH I.**

| Period | Period length days | Mixed liquor total solids, mg/l | Oxidation Ditch I (7500 hens) Rotor length = 3.36 m Immersion depth = 15.25 cm Operating characteristics | Model predicted rotor length requirements (meters) | |
|---|---|---|---|---|---|
| | | | | Odor control | Nitrogen conservation |
| 1 | 59 | 39 000 | Average removal of TKN = 52 percent<br>Average removal of COD = 46 percent<br>Average NH$_3$-N = 1709 mg/l<br>No measurable nitrite and nitrate N<br>No residual DO for at least 70 percent of the ditch<br>Average pH = 8.1 | 3.14 | 7.56 |
| 2 | 62 | 25 700 | Average removal of TKN = 58 percent during first 40 days and 29 percent thereafter<br>Average removal of COD = 27 percent<br>Average NH$_3$-N = 660 mg/l<br>Average NO$_2$-N and NO$_3$-N = 9 mg/l<br>Measurable residual DO in 75 percent of ditch during first 27 days; no measurable residual for at least 70 percent of ditch thereafter<br>Average pH = 7.8 | 2.29 | 5.49 |
| 3 | 65 | 34 900 | Average removal of TKN = 50 percent<br>Average removal of COD = 44 percent<br>Average NH$_3$-N = 1400 mg/l<br>No measurable nitrite and nitrate N<br>No residual DO for at least 70 percent of the ditch<br>Average pH = 8.2 | 2.77 | 6.71 |

R  = microbial oxygen demand, described previously and expressed in the model in units of, mass/time

$N_T$ was expressed in units of, mass/time-unit length of rotor

## VERIFICATION AND DISCUSSION

Model predicted design parameters for two large scale oxidation ditch systems treating caged layer poultry wastes were compared with experimental data from the same systems in order to determine the validity of the model as a potential design tool. Raw waste characteristics had been previously determined for each system (Martin 1974, Nieswand 1974) and were used as inputs for the model. Poultry waste treatability studies at Cornell University have indicated that the fraction of COD which exerts an oxygen demand, $f_c$, is approximately 0.5 (Prakasam et al. 1974a) and that the maximum fraction of TKN which is biodegradable, $f_N$, is 0.9 (Prakasam et al. 1974b). These data together with experimentally determined rotor aeration characteristics (Jones et al. 1971, Martin 1974) were used to predict rotor design requirements for the two systems during experimental periods when the control parameters including mixed liquor total solids concentration remained relatively constant.

The design data from this study have been summarized and compared to previously determined experimental data (Martin 1974, Nieswand 1974) in Tables 3 and 4. Aeration requirements for odor control have been previously explained; aeration requirements for nitrogen conservation constitute complete conversion of ammonia to nitrate with a mixed liquor residual dissolved oxygen concentration, $C_L$, of 2 mg/l. Aeration requirements for nitrogen removal were predicted on the basis of cyclic rotor operation as reported by Prakasam et al. (1974b), but the unavailability of experimental data prevented a quantitative assessment of predicted values.

During the first three comparison periods (Table 3), rotor length (3.36 m) was adequate for effective odor control but the data indicate that aeration was insufficient for substantial nitrification in the mixed liquor. The high mixed liquor $NH_3$-N concentration, absence of measureable $NO_2$-N and $NO_3$-N, and alkaline pH during periods 1 and 3 typify conditions of minimum aeration for odor control. Predicted rotor length requirements for odor control for these periods appeared reasonable. During period 2, partial nitrification of the mixed liquor ammonia appeared to be occurring. This was indicated by the decrease in mixed liquor ammonia concentration, by the presence of measurable nitrites and nitrates, and by the slightly depressed pH. Model predicted rotor lengths for the period conformed to these conditions in which aerations requirements were approximately midway between those required to control odors and those required for maximum nitorgen conservation.

In the second oxidation ditch system used in the verification (Table 4), substantial nitrification was occurring. Mixed liquor $NH_3$-N concentrations were very low, there was measurable mixed liquor $NO_2$-N and $NO_3$-N and pH was depressed. Nitrification was not complete, however, since most of the oxidized nitrogen in the mixed liquor was measured as $NO_2$-N. The relatively low concentrations of all forms of nitrogen in the mixed liquor indicate that substantial denitrification was occurring. Model predicted rotor length requirements conformed favorably to these aeration conditions.

## SUMMARY

Comparison of predicted aerations requirements with actual operating conditions has indicated that the design approach made reasonably accurate predictions for rotor design requirements in two oxidation ditch treatment sys-

TABLE 3. COMPARISON OF MODEL PREDICTED ROTOR LENGTH REQUIREMENTS
WITH EXPERIMENTAL DATA FOR OXIDATION DITCH II.

| Period | Period length days | Mixed liquor total solids, mg/l | Operating characteristics | Model predicted rotor length requirements (meters) | |
|---|---|---|---|---|---|
| | | | Oxidation Ditch II (4000 hens) Rotor length = 1.68 m Immersion depth = 12.7 cm | Odor control | Nitrogen conservation |
| 4 | 23 | 19 900 | Average conversion of organic N to $NH_3$-N = 69 percent<br>Average $NH_3$-N = 200 mg/l<br>Average $NO_2$-N = 360 mg/l<br>Average $NO_3$-N = 12 mg/l<br>Average removal of COD = 38 percent<br>Residual DO greater than 3 mg/l at the 3/4 point most of the time<br>Average pH = 6.7 | 0.91 | 2.35 |
| 5 | 32 | 23 700 | Average conversion of organic N to $NH_3$-N = 67 percent<br>Average $NH_3$-N = 3 mg/l<br>Average $NO_2$-N = 10 mg/l<br>No measurable nitrate - N<br>Average removal of COD = 41 percent<br>Residual DO of 0.5 - 1.5 mg/l in at least 75 percent of the ditch<br>Average pH = 7.5 | 0.91 | 2.41 |



MANAGING LIVESTOCK WASTES

tems for poultry wastes. With a knowledge of raw waste characteristics and treatability, the approach should be applicable to the design of oxidation ditch systems for other livestock wastes. At the same time the approach allows a quick comparison of trade-offs of capital and operating costs for various treatment objectives and for various mixed liquor total solids concentrations.

## References


1 Baker, D. R. 1973. Oxygen transfer relationships in a poultry waste mixed liquor. Unpublished M.S. thesis, Cornell University, Ithaca, N.Y.

2 Calhoun, G. D. 1974. Design considerations for the oxidation ditch as a dairy manure management alternative. Unpublished M.Eng. report, Cornell University, Ithaca, N.Y.

3 Jones, D. D., D. L. Day and J. C. Converse. 1969. Oxygenation capacities of oxidation ditch rotors for confinement livestock buildings. Proc. 24th Industrial Waste Conference, Purdue University, West Lafayette, Indiana, pp. 191-208.

4 Jones, D. D., D. L. Day and A. C. Dale. 1971. Aerobic treatment of livestock wastes. Bulletin 737, University of Illinois, Urbana-Champaign.

5 Kroeker, E. J. 1974. A design and management model of the oxidation ditch for livestock waste treatment. Unpublished M.S. thesis, Cornell University, Ithaca, N.Y.

6 Martin, J. H., Jr. 1974. Unpublished research data.

7 Nieswand, S. P. 1974. An evaluation of a full-scale in-house oxidation ditch for poultry waste. Unpublished M.S. thesis, Cornell University, Ithaca, N.Y.

8 Prakasam, T. B. S., R. C. Loehr, P. Y. Yang, T. W. Scott and T. W. Bateman. 1974a. Design parameters for animal waste treatment systems. EPA-660/2-74-063. Office of Research and Development, Environmental Protection Agency, Washington, D.C., pp. 89-97.

9 Prakasam, T. B. S., Y. D. Joo, E. G. Srinath and R. C. Loehr. 1974b. Nitrogen removal from a concentrated waste by nitrification and denitrification. Proc. 29th Industrial Waste Conference, Purdue University, West Lafayette, Indiana, in press.


---

# Nitrogen Transformations

*(Continued from page 516)*

solids concentration of the aerated waste slurries reached 2.8 and 3.4 percent, with the lowest level in the warmest digesters and the highest in the coldest. *Nitrosomonas* bacteria are aerobic, but from Fig. 6 it can be seen that the populations remained relatively stable even though oxygen was limited. The *Nitrobacter* populations ranged from $2 \times 10^5$ to $2 \times 10^7$ per ml at all three temperatures for all digesters but decreased somewhat in the last few weeks of the trial.

Nitrogen balances were run on the digesters in each of the three trials. The amount of nitrogen entering each digester, in the form of feces and urine, was balanced against that in the aerated slurry and that lost as ammonia gas. The average weekly input of nitrogen was 223 mg/l.

Overall nitrogen loss was more than 50 percent or more in each trial at all three temperatures. In Trial I, the average losses were 54.9 percent at 1.7 C, 59.5 percent at 7.2 C and 64.3 percent at 12.8 C. The losses in Trial II were higher than in Trial I at each temperature (60.6 percent at 1.7 C, 69.4 percent at 7.2 C, and 55.4 percent at 12.8 C). A very small percentage of the nitrogen was being lost as ammonia. Therefore, it was likely that the nitrogen was being lost as nitrogen through denitrification. Studies have shown that nitrification and denitrification can occur concurrently (McCoy 1972).

## CONCLUSIONS

1 Laboratory, batch-operated, continuously fed, aerobic digesters produced nitrogen transformations similar to those produced by a batch-operated pilot-scale oxidation ditch.

2 Nitrite and nitrate concentrations reached several hundred mg/l under all temperatures from 2 to 13 C.

3 The heterotrophic population was 100 times greater than the nitrifying bacteria population. Temperatures from 2 to 13 C did not seem to affect the maximum bacteria population.

4 The nitrifying population appeared to be stable, even when oxygen was limited.

5 Overall nitrogen losses ranged from 50-75 percent.

6 Nitrite buildups were not due to ammonia toxicity of *Nitrobacter* bacteria.

## References


1 American Public Health Association, Inc. 1971. Standard Methods for the examination of water and wastewater. 13th ed.

2 Chapman, H. D. and F. G. Liebig. 1952. Field and laboratory studies of nitrite accumulation in soil, Soil Science Society of American Proceedings 16:276-282.

3 Engel, M. S. and M. Alexander. 1958. Growth and autotrophic metabolism of *Nitrosomonas* Europaea. Journal of Bacteriology 72:217-222.

4 Eylar, O. R. and E. L. Schmidt. 1959. A survey of heterotrophic microorganisms from soil for ability to form nitrite and nitrate. Journal General Microbiology 20:473-481.

5 Fliermans, C. B., B. B. Bohool and E. L. Schmidt. 1974. Autecological study of the chemautotroph *Nitrobacter* in immumofluorescence. Applied Microbiology. 27.

6 Harada, Togors and Kai Nedeaki. 1968. Studies on the environmental conditions controlling nitrification in soil. I. effect of ammonium and total salts in media on the rate of nitrification. Soil Science and Plant Nutrition 14(1):20-26.

7 Loveless, J. E. and H. A. Painter. 1968. The influence of metal ion concentration and pH value on the growth of a *Nitrosomonas* strain isolated from activated sludge. Journal of General Microbiology 52:1-14.

8 McKinney, R. E. 1962. Microbiology for Sanitary Engineers. McGraw-Hill Book Company, New York.

9 McCoy, E. F. March 1972. Role of bacteria in the nitrogen cycle in lakes. Water Pollution Control Research Series, 16010 EHR, Environmental Protection Agency.

10 Meiklejohn, J. 1954. Some aspects of the physiology of the nitrifying bacteria. Symposium on Autotrophic Microorganisms. Cambridge University Press, London, pp. 66-82.

11 Moore, J. A., R. E. Larson, R. O. Hegg and E. R. Allred. 1973. Beef confinement systems oxidation ditch. TRANSACTIONS of the ASAE 16(1):169-171.

12 Smith, J. H. 1969. Relations between soil cation-exchange capacity and toxicity of ammonia to the nitrification process. Soil Science Society of American Proceedings 33:640-643.

13 Verstraete, W. and M. Alexander. 1973. Heterotrophic nitrification in samples of natural ecosystem. Environmental Science and Technology 7(1):39-42.

14 Wild, H. E., C. N. Sawyer and T. C. McMahon. 1971. Factors affecting nitrification kinetics. Journal Water Pollution Control Federation 43:1845-1854.

# A Theoretical Description of Aerobic Treatment

J. L. Woods, J. R. O'Callaghan

A THEORETICAL model is used to provide a framework for the comparison of laboratory data and treatment plant results on the soluble C.O.D. output of the systems. On the basis of these results the soluble C.O.D. removal rate per unit aeration volume is optimized.

## INTRODUCTION

The wide variation in operation conditions for aerobic treatment systems places restrictions on the direct comparison of results. In order to increase the range of systems which can be compared, some framework of understanding giving rise to unifying parameters is necessary. With this objective in mind, Woods and O'Callaghan (1974) adapted a model, already successfully used in other continuously operated microbiological systems, to the animal waste treatment situation.

The theory proposed was tested against laboratory scale results from the Microbiology Department of the West of Scotland Agricultural College. Some of these results have been published by Owens et al. (1973). Constants in the theory were determined by graphical analysis of this laboratory data. The usefulness of the resulting model can only be assessed by comparison with field scale treatment plants, but before attempting this a basic outline of the theory is presented.

## THEORY

The analysis is based on a relationship between micro-organism growth and substrate concentration. The micro-organism growth rate is related to the micro-organism concentration, X by the equation,

$$dX/dt = \mu X \quad \dotfill [1]$$

where

$$\mu = \hat{\mu} S/(K_s + S) \quad \dotfill [2]$$

Equation [2] is the Monod expression for growth rate, relating it to the concentration of a substrate, S. This equation is plotted in Fig. 1

Applying Equations [1] and [2] to the treatment systems illustrated in Fig. 2 and assuming negligible micro-organism concentration in the feed, the balance of micro-organism growth and micro-organism washout gives

$$V X_1 \hat{\mu} S_1/(K_s + S_1) = \dot{x} \quad \dotfill [3]$$

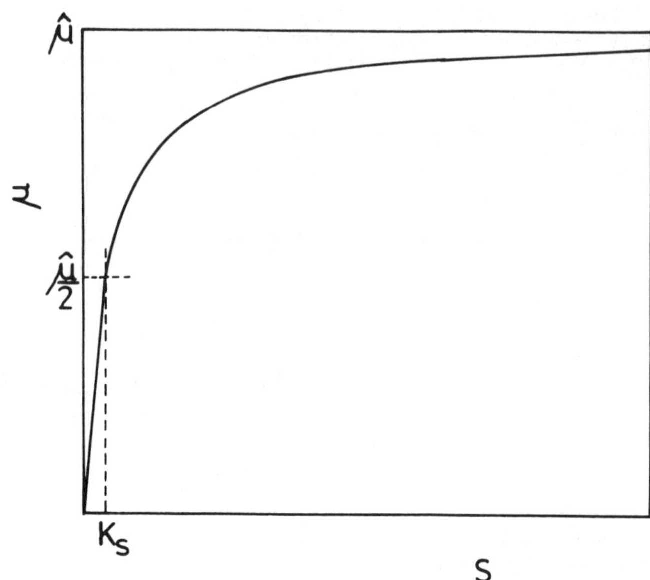



FIG. 1 Specific growth rate against substrate concentration.

Equation [3] can be rearranged to give

$$S_1 = K_s/(\hat{\mu}\Theta_m - 1) \quad \dotfill [4]$$

where

$$\Theta_m = V X_1/\dot{x} \quad \dotfill [5]$$

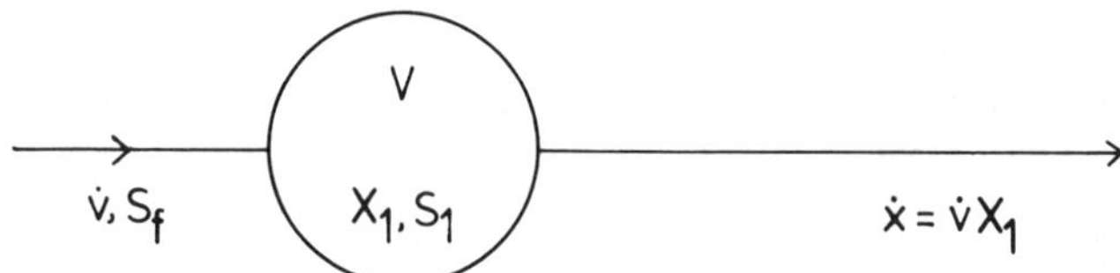



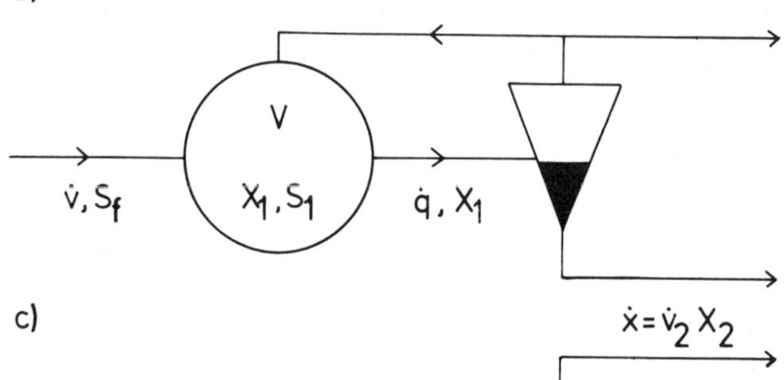



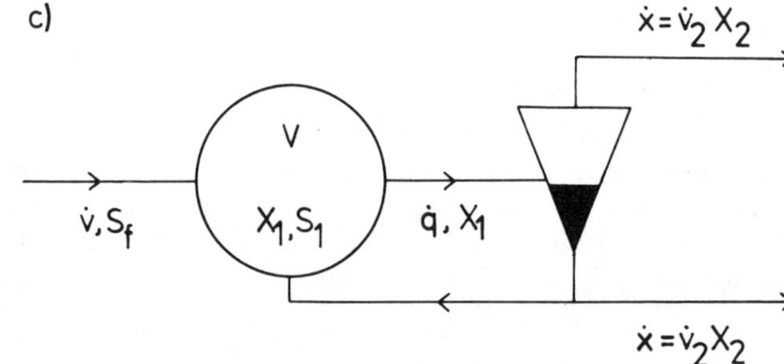



FIG. 2 Continuous treatment operating systems.

**Acknowledgments:** The authors are grateful to the Agricultural Research Council for provision of funds to the University for research work, to the Microbiology Dept., West of Scotland Agricultural College for use of their experimental data and to C. Neilsen, University of Reading and A. Robertson, Scottish Farm Buildings Investigation Unit for the communication of data on treatment systems.

The authors are: J. L. WOODS and J. R. O'CALLAGHAN, Agricultural Engineering Dept., The University, Newcastle upon Tyne, England.

522

For the straight through system illustrated in Fig. 2a, $\Theta_m$, the micro-organism retention time, can be written

$$\Theta_m = VX_1/\dot{v}X_1 = \Theta_h \quad \dots \dots \dots \dots \dots \dots [6]$$

This demonstrates that the micro-organism retention is equal to the hydraulic retention time, $\Theta_h$ for a straight through system.

For the systems with solids or supernatant recycle, as illustrated in Fig. 2b and c, the micro-organism retention time can be written

$$\Theta_m = VX_1/v_2X_2 = \Theta_h/r \dots \dots \dots \dots \dots [7]$$

where r is defined as

$$r = \dot{v}_2X_2/\dot{v}X_1 \quad \dots \dots \dots \dots \dots \dots \dots [8]$$

The value of r defines the ratio of micro-organism removal rate for a system with solids control to that of the equivalent straight through system. For supernatant recycle, r is greater than unity decreasing $\Theta_m$ relative to $\Theta_h$, while for solids recyle is less than unity and $\Theta_m$ is greater than $\Theta_h$.

Assuming a constant yield of micro-organism per unit substrate consumed and a soluble substrate, the output micro-organism concentration for all systems illustrated in Fig. 2 is given by

$$X_1 = Y (S_f - S_1) \quad \dots \dots \dots \dots \dots \dots \dots [9]$$

## Application to Animal Waste

In addition to the assumptions already made in the analysis, four further assumptions are necessary to relate the analysis to aerobic treatment.

1   The soluble or non-settleable C.O.D. can be considered as the sum of a biologically inert fraction and a biologically available fraction.

$$\text{C.O.D.}_s = \text{C.O.D.}_b + \text{C.O.D.}_i \dots \dots \dots \dots \dots [10]$$

2   The biologically available fraction of the soluble C.O.D. is a measure of substrate available to the micro-organisms and controls the growth according to Monod's Equation [2].

3   The micro-organism retention time is equal to the solids retention time.

4   The removal of solid material does not effect the breakdown of soluble biologically available C.O.D. except in its effect on micro-organism retention time.

The second assumption enables Equation [4] to be applied to the biologically available C.O.D. giving

$$\text{C.O.D.}_b = K_s/(\hat{\mu}\Theta_m - 1). \dots \dots \dots \dots \dots [11]$$

The inert fraction of the C.O.D. is unaffected by the treatment and can be related to the suspended solids level in the feed, $P_f$, to allow for dilution effects

$$\text{C.O.D.}_i = C\,P_f \quad \dots \dots \dots \dots \dots \dots \dots [12]$$

The comparison with the laboratory data indicated that

$$\hat{\mu}\Theta_m \gg 1$$

Combining this information with Equations [10], [11] and [12] and inputting the values of $K_s/\hat{\mu}$ and C obtained by graphical analysis, the soluble C.O.D. concentration in the output can be written

$$\text{C.O.D.}_s = 6.12/\Theta_m + 0.011\,P_f \quad \dots \dots \dots \dots [13]$$

Equation [11] is compared with the laboratory data in Fig. 3. The range of $P_f$ encountered was 10-80 g/l. The C.O.D.$_b$ values were calculated from the laboratory data using

$$\text{C.O.D.}_b = \text{C.O.D.}_s - 0.011\,P_f \quad \dots \dots \dots \dots [14]$$

The micro-organism retention time was calculated using the third assumption, as the ratio of solid material in the treatment vessel to the solids output rate.

## A COMPARISON WITH TREATMENT PLANTS

The treatment plants considered are described in Chapter 5 of the A.R.C. Bulletin (1975) with the exception of Craibstone, which is described as an internal report edited by Littlejohn (1974). The operating systems for the treatment plants are summarized in Table 1 and the results rele-

**TABLE 1. TREATMENT PLANT DESCRIPTION.**

| Treatment plant | Primary stage | Treatment vessel | Aeration device | Solids control |
|---|---|---|---|---|
| Craibstone | None | In house oxidation | Cage rotor | Solids removed by screen and hydrocyclone. |
| Culbae | None | Aeration tank | Surface aerator | None. |
| Trawscoed | Primary settling | Aeration tank | Surface aerator | Solids recycle by settling. |
| Great house I | Primary settling | Aeration tank | Surface aerator | Solids removal by settling. |
| Great house II | Primary settling | Aeration tank | Surface aerator | Solids recycle by settling. |

vant to this analysis presented in Table 2. The solids retention for three of the plants was calculated from the ratio of the total solid material in the aeration tank to that leaving. However, for Great House I and Great House II the solids output data was inadequate and an indirect calculation had to be made. This involved estimating the effect of treatment on the feed solids to predict the output solids.

For Great House I, the feed to the aeration tank is formed by the supernatant from an original waste of 86 g/l. This supernatant contains 2 g/l suspended solids and 14 g/l soluble solids. Recent work by the Microbiology Department at the West of Scotland Agricultural College suggests that the yield of cell material from this 14 g/l soluble fraction of pig waste would be of the order of 5 g/l. Allowing for a small solids reduction (5 percent), the solids retention time is given by

$$\Theta_m = V \times \text{M.L.S.S.}/(\dot{v} \times \text{S.S.}_{output})$$

$$= 34 \times 3 /(1.36 \times (2.0 + 5.0) \times 95) = 11.3 \text{ days}$$

For Great House II, the supernatant from the primary settling tank has a suspended solids level of 0.3 g/l. The soluble solids were not significant. Assuming a maximum solids removal efficiency of 45 percent for cow waste, the output suspended solids has a value 0.165 g/l. Since the M.L.S.S. is 2 g/l, the solids retention time is given by

$$\Theta_m = 12 \times 2 /(3.91 \times 0.165) = 37.2 \text{ days}$$

The results for these five treatment plants are presented in Fig. 3 for comparison with the theoretical interpretation of the laboratory data. The agreement appears to be good.

## OPTIMIZATION OF C.O.D. REMOVAL

The soluble C.O.D. for raw pig waste of 100 g/l total solids and 86 g/l suspended solids is 19.2 g/l (Chapter 2, A.R.C. Bulletin, 1975).

From these results the C.O.D. in the feed can be related to the suspended solids level to allow for dilution, thus

$$\text{C.O.D.}_{sf} = 0.2 \ P_f \dots \dots \dots \dots [15]$$

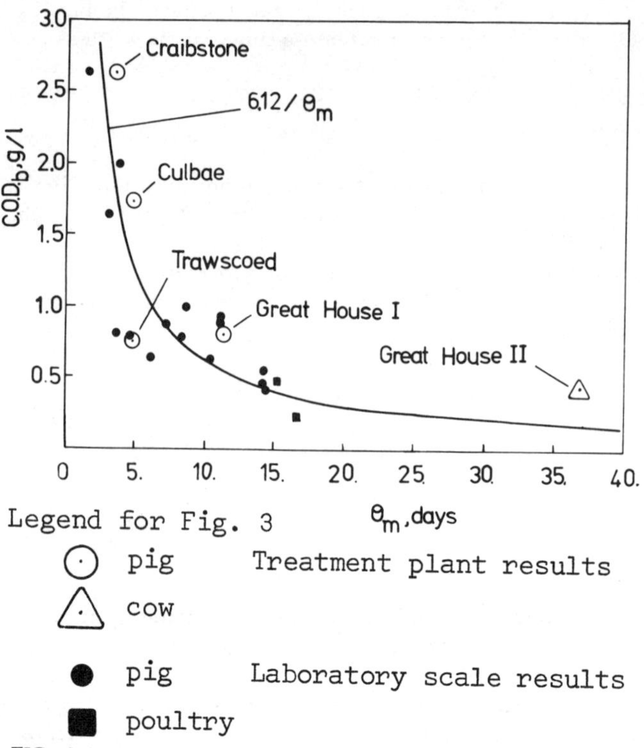



FIG. 3 A comparison between treatment plants, laboratory experiment and the theory.

Subtracting the expression for the output soluble C.O.D. in Equation [13], the change in soluble C.O.D. across the treatment system is given by

$$\Delta(\text{C.O.D.}) = 0.21 \ P_f - 6.12/\Theta_m \dots \dots \dots \dots [16]$$

A useful parameter to evaluate the utilization of a treatment plant is the specific substrate removal rate. In this case for a straight through system it is defined as

$$R = \dot{v} \Delta(\text{C.O.D.})/V = 0.21 \ P_f/\Theta_m - 6.12/\Theta_m^2 \dots \dots [17]$$

The maximum value of R occurs at

$$\Theta_{m \ opt} = 58.3/P_f \dots \dots \dots \dots \dots \dots [18]$$

TABLE 2. TREATMENT PLANT DATA.

| Treatment plant | SS, g/l (Slurry) | Treatment volume, m$^3$ | Aerator power, kW | Dissolved oxygen | Hydraulic retention | Solids retention | Soluble C.O.D., g/l | Soluble C.O.D._b, g/l | Soluble B.O.D._5, g/l |
|---|---|---|---|---|---|---|---|---|---|
| Craibstone (44 pigs) | 86.0 | 14.2 | 1.5 | 40-70 | 81.53 | 3.36 | 3.58 | 2.63 | 1.00 |
| Culbae (5000-6000 pigs) | 20.0 | 380.0 | 11.2 | Not detectable | 4.7 | 4.7 | 1.94 | 1.72 | 1.00 |
| Trawscoed (37 sows 260 followers) | 4.5 | 35.0 | 2.2 | 38 | 3.6 | 4.85 | 0.83 | 0.75 | 0.29 |
| Great house I (36 sows 206 followers) | 86.0 | 34.0 | 3.75 | - | 25.0 | 11.3* | 1.75 | 0.80 | 0.11 |
| Great house II (cows, pigs, silage) | 1.5 | 12.0 | 1.5 | High | 3.1 | 37.2* | 0.478 | 0.46 | 0.065 |

*Calculated indirectly as described in text.

At this value,

$$R_{opt} = 0.0018 \, P_f^2 \quad \dots\dots\dots\dots\dots\dots \quad [19]$$

$$C.O.D._b = 0.108 \, P_f \quad \dots\dots\dots\dots\dots\dots \quad [20]$$

This indicates how the optimum value of R with respect to $\Theta_m$ improves rapidly with increased feed concentration. However, the improvement is at the expense of an increase in the output concentration of the biologically available fraction of the C.O.D. The variation of R with $P_f$ and $\Theta_m$ as defined in Equation [17] is illustrated in Fig. 4.

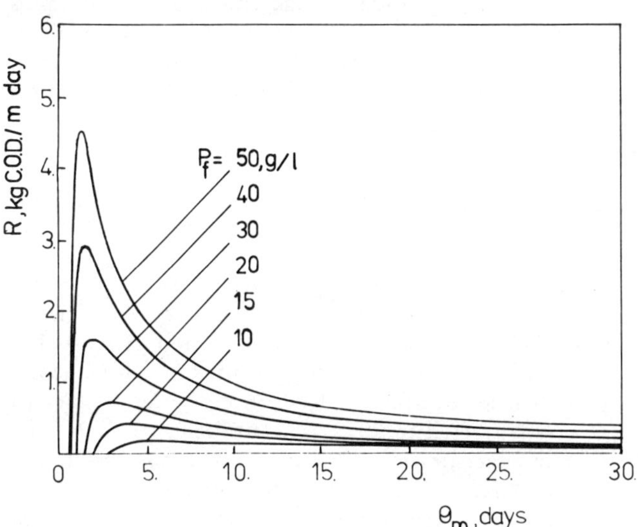


FIG. 4 Specific substrate removal rate against micro-organism retention time for various feed suspended solids levels.

## CONCLUSIONS

The results presented in Fig. 3 demonstrate that within the limits of experimental error the theory is successful in describing soluble C.O.D. removal by aerobic treatment. The parameters enabling this comparison are the solids retention time $(\Theta_m)$ and the feed suspended solids, $P_f$. Their independent effects suggest that the grouping of solids and volumetric loading into one loading rate parameter is undesirable. The independence of the biological component of output C.O.D. on the input concentration is due to the adjustment of the micro-organism population to consume the food available.

The $C.O.D._s$ removal rate per unit volume of the aeration tank, R would be a useful parameter for optimizing treatment. It is a measure of the pollutant removal relative to the capital cost. If the treated waste is to be spread on the land then the output $C.O.D._b$ is no longer a fixed constraint as it would be if the treated waste was discharged to a water course. This flexibility in output $C.O.D._b$ might allow an optimization of R along the lines described in this paper.

The numerical values of the coefficients in the theory could be improved with more data. The authors would therefore welcome co-operation with other researchers in the aerobic treatment field.

## NOTATION

| | |
|---|---|
| C | ratio of inert soluble C.O.D. to suspended solids in the waste. |
| $K_s$ | limiting substrate concentration |
| $P_f$ | suspended solids concentration in the feed |
| $\dot{q}$ | volume flow rate through separator |
| R | specific substrate removal rate |
| r | ratio of micro-organism removal rate with separation to that for straight through system |
| S | substrate concentration |
| t | time |
| V | aeration tank volume |
| $\dot{v}$ | volume flow rate through the treatment system |
| $\dot{v}_2$ | volume flow rate of settled material out of treatment system |
| X | micro-organism concentration |
| $X_2$ | micro-organism concentration in the settled material |
| $\dot{x}$ | micro-organism removal rate from treatment systems |
| Y | yield of micro-organisms per unit substrate |
| $\Delta(C.O.D._s)$ | change in soluble C.O.D. across the treatment system |
| $\Theta_h$ | hydraulic retention time |
| $\Theta_m$ | micro-organism retention time |
| $\mu$ | specific growth rate |
| $\hat{\mu}$ | maximum value of $\mu$ |

## SUBSCRIPTS

| | |
|---|---|
| b | biological soluble component |
| f | value in feed |
| i | inert soluble component |
| opt | optimum condition |
| s | soluble fraction |
| l | value in aeration vessel |


### References

1 Agricultural Research Council Bulletin. 1975. Studies on farm livestock wastes. To be published.

2 Littlejohn. L. 1974. The treatment of piggery wastes. Joint internal publication of the Scottish Farm Buildings Investigation Unit and the North of Scotland Agricultural College.

3 Owens, J. D., M. R. Evans, F. E. Thacker, R. E. Hissett and S. Baines. 1973. Aerobic treatment of piggery waste. Wat. Res. 7:1745-1767.

4 Woods, J. L. and J. R. O'Callaghan. 1974. Mathematical modeling of animal waste treatment, J. Agric. Eng. Res. 19:145-258.

# An Evaluation of Aeration Systems for Poultry Wastes under Commercial Conditions


**John H. Martin, Jr.,    Raymond C. Loehr**
ASSOC. MEMBER                MEMBER
ASAE                        ASAE


AEROBIC, biological stabilization of poultry and other animal manures has been shown to be an excellent method of odor control and waste stabilization by many investigators. Martin et al. (1974) have reported significant reductions in total and volatile solids and chemical oxygen demand (COD) resulting from aerobic stabilization of poultry wastes. Perhaps the most unique aspect of the aerobic stabilization process is its flexibility in the management of nitrogen. Prakasam et al. (1974) reported that nitrogen losses could be either maximized or minimized by modifying the method of operation of a pilot plant scale oxidation ditch. Losses of nitrogen ranged from 30 to 90 percent.

Although the fundamental aspects of aerobic stabilization of poultry wastes have been investigated, the practicality of this approach under commercial conditions remains unclear. To achieve acceptance by the poultry industry, aerobic systems should have a minimal economic impact and be reasonably free from operational problems.

Recognizing the limitations of laboratory and pilot plant scale studies, a full scale assessment of aerobic stabilization of poultry wastes by the Cornell Agricultural Waste Management Program over a period of two years is currently under way. The oxidation ditch is the system which was selected due to its ease of incorporation into confinement housing.

## OBJECTIVES

It is the purpose of this paper to report the following results from the first year of the study:

1   Odor control and waste stabilization efficiency as related to the level of oxygen input.

2   Operating costs in terms of total egg production costs.

## MATERIALS AND METHODS

### Oxidation Ditches

The oxidation ditches involved in this study are located at Manorcrest Farms, a commercial poultry and dairy operation near Camillus, New York. Each of two ditches receives waste from approximately 4000 caged, white Leghorn hens located directly above the ditch. Fig. 1 is a plan view of the system. The two ditches are identical in all respects with the exception of the type and oxygen transfer capacity of the aeration unit. A prototype brush aerator manufactured by the Montair Corporation is utilized in

Ditch I. This unit is driven by a two horsepower, gearhead motor. Ditch II is aerated with a five horsepower, Thrive Centers cage rotor.

Each ditch is connected to a series of four 9460 l (2500 gal) tanks which provide storage and a degree of solids separation through settling. Settled mixed liquor is periodically pumped from the final tank into the ditch causing an overflow into the first tank. This allows a degree of control of mixed liquor total solids concentrations and solids retention time, $\Theta_c$.

### Analytical Methods

Following construction of the ditches, the oxygen transfer characteristics of both aeration devices were determined. The nonequilibrium chemical method, Water Pollution Control Federation (1969), was used to determine the oxygen transfer coefficient, $K_L a$, in tap water. Values of $K_L a$ were corrected to 20 C using a $\Theta$ value of 1.024. The saturation concentration of tap water, $C_s$, was corrected to 760 mm of mercury. A Yellow Springs Instrument Company model 54 oxygen meter was used to measure dissolved oxygen concentrations during the oxygen transfer studies and later in the ditch mixed liquor.

In order to determine the waste stabilization efficiency, raw wastes and grab samples of mixed liquor were routinely analyzed for the appropriate parameters using the following analytical techniques. Total and fixed solids were determined as described in APHA (1971). Total Kjeldahl nitrogen (TKN) was determined by a micro-Kjeldahl method (McKenzie and Wallace 1954). Ammonia nitrogen determinations were made by the steam distillation procedure, and nitrate-nitrogen was determined by the salicylic acid method described by Prakasam et al. (1972). Nitrite-nitrogen was determined by the diazotization method presented by Montgomery and Dymock (1961). A modified chemical oxygen demand (COD) method (Jeris 1967) was used.


Acknowledgments: This study was supported by the Demonstration Grant No. S800863, U.S. Enviornmental Protection Agency. The guidance of Lee A. Mulkey, Environmental Protection Agency, Athens, Ga., who served as project officer is gratefully acknowledged. The authors are indebted to Earl Hudson, co-owner of Manorcrest Farms, for his interest and cooperation.

The authors are: JOHN H. MARTIN, Jr., Research Specialist, Agricultural Engineering Dept., and RAYMOND C. LOEHR, Director, Environmental Studies Progarm, College of Agriculture and Life Sciences, Cornell University, Ithaca, N.Y.


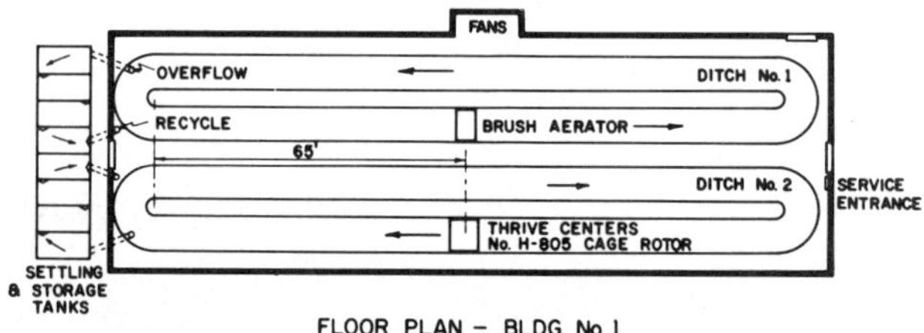


FIG. 1 Plan view of Manorcrest Farms oxidation ditches.

| Ditch | Level of oxygen transfer | Aerator speed, rpm | Immersion depth, cm | kwhr consumed | Oxygen transferred, kg/hr* | gm $O_2$ /kwhr |
|---|---|---|---|---|---|---|
| I | I | 252 | 15.2 (6 in.) | 1.403 | 1.40 (3.09 lb/hr) | 998 |
| I | II | 310 | 11.4 (4.5 in.) | 2.076 | 2.08 (4.57 lb/hr) | 1002 |
| II | III | 100 | 12.7 (5 in.) | 3.517 | 3.36 (7.40 lb/hr) | 955 |

**TABLE 1. DETAILS OF THE THREE LEVELS OF OXYGEN TRANSFER EVALUATED.**

*at 20 C, 760 mm Hg, in tap water

## RESULTS AND DISCUSSION

### Oxygen Transfer

Oxygen transfer studies in tap water were conducted to establish modes of aerator operation and to determine the rates of transfer for the three modes selected. The results of these studies for the three modes selected are noted in Table 1. Baker (1973) has reported that the oxygen transfer coefficient, $a$, decreases in aerated poultry wastewaters as mixed liquor total solids concentrations increase. Therefore, transfer under process conditions is less than the values reported in Table 1. The brush aerator in Ditch I supplied the two lowest levels of oxygen transfer. Input was increased by raising the aerator speed. The highest level of oxygen transfer was provided by the Thrive cage rotor in Ditch II. For purposes of simplicity, these three levels of oxygen transfer will be referred to as levels I, II, and III.

### Odor Control and Waste Stabilization

Odor control has been excellent over the range of oxygen transfer levels studied. At the lowest level, i.e., level I, the odor of ammonia due to ammonia stripping was detectable in the immediate vicinity of the aeration unit. No waste associated odors were present in other areas of the building or in the ventilation exhaust air streams at any level of oxygen transfer. In addition to odor control, the presence of house flies, *Musca domestica*, which breed in animal manures has been essentially eliminated. No other vermin problems have been observed.

The average characteristics of raw wastes entering the oxidation ditches are given in Table 2. The degree of waste stabilization was estimated during four different time periods and at three different levels of oxygen input using mass balances. Relationship between oxygen transferred and mass balance results are presented in Table 3. Oxygen transfer under process conditions were calculated using the relationship between mixed liquor total solids concentrations and the coefficient $a$ developed by Baker (1973). Average dissolved oxygen concentrations (DO) for each balance period are given in Table 4. The designations 1/4 and 3/4 point indicate the distance downstream from the aeration unit at which the DO measurements were made.

The results of the four mass balances show little difference in the percentages of total and volatile solids and COD that were biologically removed. The small increase in loss of solids and COD as the solids retention time, $\Theta_c$, increased indicates that the readily biodegradable fraction of the COD can be removed at a $\Theta_c$ of less than 15 days. This conclusion is supported by results of laboratory studies reported by Prakasam et al. (1974). They found that two or more removal rates occur in long term, aerobic treatment of

**TABLE 2. AVERAGE CHARACTERISTICS OF THE POULTRY WASTES ENTERING THE OXIDATION DITCHES.**

| Properties | Grams per bird day |
|---|---|
| Total solids | 29.6 |
| Volatile solids | 22.1 |
| Fixed solids | 7.5 |
| Total Kjeldahl nitrogen | 2.30 |
| COD | 19.24 |
| $BOD_5$ | 4.88 |
| Soluble COD | 4.71 |

**TABLE 4. DISSOLVED OXYGEN CONCENTRATIONS.**

| Mode of operation | Time period, days | Average dissolved oxygen Conc., mg/l 1/4 point | 3/4 point | Temp., deg C |
|---|---|---|---|---|
| I | 51 - 106 | 1.0 (0.7 - 1.55*) | 0.25 (0.1 - 0.4) | 13 |
| II | 110 - 165 | 1.0 (0.5 - 1.6) | 0.3 (0.1 - 0.5) | 12 / 12 |
| III | 91 - 124 | 4.45 (1.35 - 7.3) | 3.9 (0.35 - 6.9) | 13 / 13 |
| III | 189 - 224 | 2.6 (1.4 - 4.8) | 1.4 (0.4 - 32.) | 13 |

*Range of values.

**TABLE 3. RELATIONSHIP BETWEEN OXYGEN TRANSFERRED AND MASS BALANCE RESULTS.**

| Period of balance, days | Total solids concentration, mg/l | $\Theta_c$, days | $a$ Factor | Oxygen* transferred, grams/1000 bird-hr | Percent Loss Solids Total | Volatile | COD | Nitrogen Organic | Total |
|---|---|---|---|---|---|---|---|---|---|
| 56 | 9 960 | 15 | 1.0 | 350 | 38.9 | 50.8 | 32.9 | 55.0 | 48.8† |
| 56 | 13 550 | 21 | 1.0 | 520 | 37.9 | 49.1 | 31.2 | 60.4 | 60.4 |
| 34 | 19 800 | 27 | 0.97 | 813 | 35.4 | 47.7 | 31.4 | 64.7 | 47.7‡ |
| 36 | 23 830 | 36.5 | 0.94 | 788 | 40.8 | 54.3 | 35.0 | 63.3 | 63.3 |

*Calculated oxygen transfer under process conditions
†Nitrogen was accumulated as $NH_4$-N
‡Nitrogen was accumulated as $NH_4$-N, $NO_2$-N, and $NO_3$-N

poultry manure. Their studies indicated that the easily bio-degradable fraction of poultry waste is oxidized in a period of less than five days.

There was no significant increase in loss of solids or COD as the level of oxygen transfer increased. However, significant nitrification which was not observed at the lowest level of oxygen transfer, did occur as oxygen input was increased. It appears that the lowest level of oxygen transfer, 350 grams $O_2$ per 1000 bird-hours, was adequate to meet the exerted carbonaceous oxygen demand of the waste expressed as COD removed. Prakasam et al. (1974) concluded that the oxygen requirement for odor control in poultry wastes is about 400 grams $O_2$ per 1000 bird-hours. This indicates that the transfer of 350 to 400 grams $O_2$ per 1000 bird-hours is a realistic design value for odor control. It appears that a significant portion of the observed nitrogen loss at the lowest level of oxygen input was due to ammonia stripping. The undissociated mixed liquor ammonia-nitrogen concentration during the mass balance period was greater than 50 mg/l.

Increasing the amount of oxygen transferred to 2.08 kg $O_2$ per hour (520 grams $O_2$ per bird-hours) resulted in the decrease of the mixed liquor ammonia-nitrogen concentration from 1320 mg/l to 250 mg/l over the mass balance period. This was accompanied by a decrease in pH which reduced the potential for ammonia loss via stripping. The lack of nitrite or nitrate-nitrogen in the mixed liquor and the nitrogen balances indicate that simultaneous nitrification-denitrification was occurring. The losses of organic and total nitrogen are the same. Combining the quantity of ammonia released from degradation of organic nitrogen added during this period with the reduction of the ammonia residual, 150 percent of the nitrogeneous oxygen demand was satisfied. This indicates that the quantity of oxygen supplied was slightly greater than the demand in terms of waste added to the system during this period. Although the quantity of oxygen transferred was increased, average dissolved oxygen concentrations were unchanged. The low dissolved oxygen levels explain the simultaneous nitrification-denitrification which was observed.

Mass balances were made for two separate time periods at the highest level of oxygen transfer, 3.36 kg $O_2$ per hour in tap water (840 grams $O_2$ per 1000 bird-hours). During the first mass balance period, loss of total nitrogen was less than that of organic nitrogen with accumulation of nitrogen as ammonia, nitrite, and nitrate responsible for the difference. Mixed liquor concentrations of ammonia, nitrite, and nitrate nitrogen during this period are shown in Fig. 2.

During the second mass balance period, complete nitrification-denitrification occurred with only trace amounts of ammonia, nitrite, and nitrate nitrogen observed in the mixed liquor. With the relatively high concentrations of

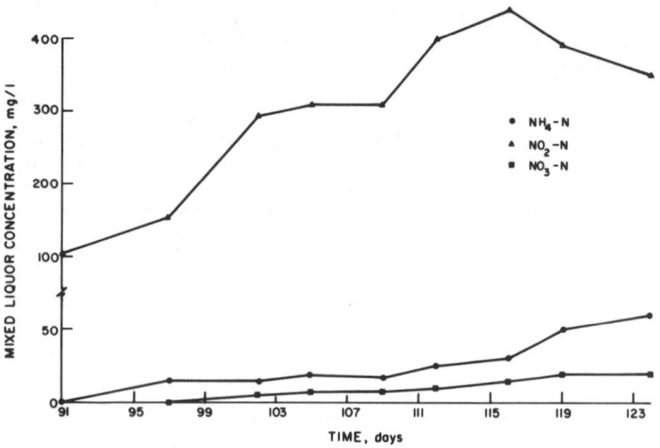


FIG. 2 Mixed liquor concentrations of $NH_4$, $NO_2$, and $NO_3$-N during the first mass balance at the oxygen transfer rate of 3.36 kg $O_2$ per hour in tap water.

dissolved oxygen in the ditch, it appears that most of the denitrification occurred in the storage tanks. If the mixed liquor was removed directly to the land, the nitrogen loss would be reduced. The decrease in dissolved oxygen levels from the first to the second balance period is attributed to an increase in COD loss and the decrease in the oxygen transfer coefficient, $a$.

### Economic Impact

The practicality of aeration as a poultry waste management tool will depend heavily on economic impact. The total economic assessment is not complete at this time. However, the power cost of aeration associated with the three levels of stabilization has been determined. These costs are presented in Table 5. Included is the operating cost of the cage rotor (Thrive Center) with worn bearings to illustrate the effect of maintenance on cost. Although two different aeration units are involved in this cost analysis, the tap water oxygen transfer studies indicated that the efficiency, grams $O_2$ per kWhr of both the brush aerator and the Thrive rotor are similar as shown in Table 1.

The costs determined in this study are significantly lower than those previously reported. Ludington et al. (1972) reported power costs between two and four cents per dozen eggs for a pilot plant oxidation ditch. Martin et al. (1974) estimated energy costs at one cent per dozen eggs for a full scale, 15 000 bird oxidation ditch system.

The economic impact of any waste management system on net income should be the criteria used in the economic assessment of that system. However, a 1973 survey of 40 New York State poultry farms, Bratton and Thacker (1974), showed that labor and management incomes varied

**TABLE 5. ENERGY COSTS ASSOCIATED WITH AERATION OF POULTRY WASTES.**

| Level of oxygen | Degree of stabilization | Aeration unit | Cost/* 1000 hens/yr | Cost† doz. eggs |
|---|---|---|---|---|
| I | Odor control | Brush aerator | $ 66.89 | $0.0033 |
| II | Odor control and nitrification-denitrification | Brush aerator | 91.62 | 0.0046 |
| III | Odor control and minimal in-ditch denitrification | Thrive rotor | 134.14 | 0.0067 |
| III | Level III with worn bearings | Thrive rotor | 161.97 | 0.0081 |

*Current energy cost of $0.0201 per kilowatt-hour
†Estimated production of 20 dozen eggs per hen-year



MANAGING LIVESTOCK WASTES

widely. Income ranged from minus values to over $30 000 per operator. They reported similar variations in 1971 and 1972. Variation in egg prices and management skills among producers appear to be the two major factors in this variability.

The use of production costs as a baseline to measure the economic impact of aeration as a waste management tool is the alternative approach which was selected. This procedure allows determination of economic impact in terms of minimum production costs resulting from skillful management. Production costs for 1971-1973 are presented in Table 6. The effect of management is reflected in the selected feed conversion efficiency and production values which are above average for the three years evaluated.

**TABLE 6. NEW YORK STATE EGG PRODUCTION COSTS.**

| Cost/hen-year | 1973 | 1972 | 1971 |
|---|---|---|---|
| Return to capital @ 7 percent | 0.52 | 0.43 | 0.48 |
| Labor* | 0.94 | 0.87 | 1.05 |
| Feed† | 4.87 | 3.08 | 3.35 |
| Hen‡ | 1.75 | 1.75 | 1.75 |
| Building repairs | 0.03 | 0.04 | 0.04 |
| Electricity | 0.11 | 0.10 | 0.09 |
| Taxes | 0.07 | 0.06 | 0.07 |
| Insurance | 0.11 | 0.09 | 0.09 |
| Total | 8.40 | 6.42 | 6.92 |
| Cost/doz eggs† | 0.420 | 0.321 | 0.346 |

*Includes operators labor
†Based on 4.0 lbs/doz. eggs and 20 doz. eggs/hen-year.
‡Estimated cost, $2.00/bird, less salvage value of $0.25/bird.

The economic impact of the power costs for the three levels of stabilization discussed are presented in Table 7. For the past three years, odor control would have increased production costs by a maximum of one percent. In considering nitrogen conservation, the trade-offs between increased energy cost and the value of conserved nitrogen should be considered.

**TABLE 7. EFFECT OF POWER COSTS FOR AEROBIC STABILIZATION ON EGG PRODUCTION COSTS.**

| Degree of stabilization | Percent 1973 | Production cost 1972 | Increase 1971 |
|---|---|---|---|
| Odor control | 0.8 | 1.0 | 0.9 |
| Nitrogen removal (nitrification-denitrification) | 1.1 | 1.4 | 1.3 |
| Nitrogen conservation (degree unknown) | 1.6 | 2.0 | 1.8 |

In evaluating these energy costs, it is important to realize that energy costs are a function of equipment oxygen transfer efficiency. Both aeration units in this study deliver approximately 1500 grams $O_2$ per kWhr (2.5 lb $O_2$ per hp-hr). Other types of surface aeration equipment with higher efficiencies, 2400 grams $O_2$ per kWhr (4 lb $O_2$ per hp-hr) are available, Metcalf and Eddy (1972). Use of more efficient aeration equipment would lower the operating cost.

No attempt will be made to rigorously estimate the total capital cost of the required aeration equipment at this time. Prices per unit of oxygen transferred vary depending on the manufacturer. Construction cost will vary in relation to the type of aeration system that is used and whether such construction is done in existing or new facilities. Also, the trade-offs due to replacement of other waste handling equipment such as mechanical scrapers and similar equipment exist. For the two aeration units used in this study, the capital investment per bird is $0.375 for the brush aerator and $0.625 for the Thrive rotor. This is in comparison to an average total capital investment of $7.43 per bird in New York State for 1973 as reported by Bratton and Thacker (1974). It should be recognized that the Thrive Center rotor has a higher total oxygen transfer capacity. The unit cost in terms of $/kg $O_2$ per hour transferred are similar.

## SUMMARY

Aeration of poultry wastes for odor and nutrient control using an oxidation ditch was evaluated under commerical conditions. The results of this study demonstrate the flexibility and feasibility of aeration as a poultry waste management technique. By controlling the level of oxygen input, the specific goal of odor control can be reached at a minimum cost. Increasing oxygen transfer will allow nitrogen management. The options of either nitrogen removal via simultaneous nitrification-denitrification or conservation of relatively high dissolved oxygen levels are available.

The power cost of odor control in this study was $0.0033 per dozen eggs. This represents an increase in production costs of about one percent. Conditions favoring conservation of nitrogen can be attained but energy costs may be doubled.

The degree of nitrogen conservation possible has not been clearly defined. In terms of odor control and energy cost, the results of this investigation indicate that aerobic stabilization of poultry wastes is a feasible waste management alternative for the poultry industry.

### References

1  American Public Health Association. 1971. Standard methods for the examination of water and wastewater. 13th ed., New York.

2  Baker, D. R. 1973. Oxygen transfer relationships in a poultry waste mixed liquor. Unpublished M.S. thesis, Cornell University, Ithaca, New York.

3  Bratton, C. A. and G. H. Thacker. 1971, 1972, 1973. Poultry farm business summary. Dept. of Agricultural Economics, Cornell University, Ithaca, New York.

4  Jeris, J. S. 1961. A rapid COD test. Water and Wastes Engr. 4:89-91.

5  Ludington, D. C., A. T. Sobel, R. C. Loehr and A. G. Hashimoto. 1972. Pilot plant comparison of liquid and dry waste management systems for poultry manure. Proc. of the 1972 Cornell University Agricultural Waste Management Conference, Syracuse, New York.

6  Martin, J. H., R. C. Loehr, A. C. Anthonisen and S. P. Nieswand. 1974. Aerobic treatment of poultry wastes. ASAE Paper No. 74-4029. ASAE, St. Joseph, Mich. 49085.

7  McKenzie, H. A. and H. S. Wallace. 1954. The Kjeldahl determination of nitrogen: A critical study of digestion conditions, temperature catalyst, and oxidizing agent. Aust. J. Chem. 7:55-71.

8  Metcalf and Eddy. 1972. Wastewater Engineering. McGraw-Hill, New York.

9  Montgomery, H. A. C. and J. F. Dymock. 1961. The determination of nitrite in water. Analyst 86:414-416.

10  Prakasam, T. B. S., et al. 1972. Evaluation of methods of analysis for the determination of physical, chemical and biochemical parameters of poultry wastewater. Paper presented at the pre-winter ASAE meeting, Chicago, Ill. pp. 72.

11  Prakasam, T. B. S., R. C. Loehr, P. Y. Yang, T. W. Scott and T. W. Bateman. 1974. Design parameters for animal waste treatment systems. Report of Project S800767, Office of Research and Development, U.S. Environmental Protection Agency, Washington, D.C.

12  Prakasam, T. B. S., E. G. Srinath, A. C. Anthonisen, J. H. Martin and R. C. Loehr. 1974. Approaches for the control of nitrogen with an oxidation ditch. Proceedings of the 1974 Cornell Agricultural Waste Management Conference, Rochester, New York.

13  Water Pollution Control Federation. 1969. Aeration in wastewater treatment — Manual of Practice No. 5, JWPCF 41:1863-1878.

# Turbine-Air Aeration System for Poultry Wastes

A. G. Hashimoto, Y. R. Chen

ASSOC. MEMBER
ASAE

A ERATION systems that successfully control odors and stabilize wastes from livestock operations have been documented. Aeration systems are currently being used commercially and actively investigated under experimental conditions. The relatively high operating cost is one of the most serious liabilities of available aeration systems. This study was undertaken to evaluate the Turbine-Air Aeration System (TAAS) in terms of: power requirements, oxygen transfer rates and efficiencies, and feasibility of treating poultry wastes. Studies were conducted using tap water, and batch and continuous feeding of poultry manure.

## EQUIPMENT AND METHODS

### Aeration System

The TAAS consisted of a 1.37 m diameter by 3.05 m deep tank, with four 0.102 m wide baffles along the circumference parallel to the tank center axis. Two Dravo "Leaf Spring" air diffusers were located at the tank bottom, and air was supplied by a 1.1 kW, three-lobe, positive displacement type blower. The turbine mixer was powered by a 1.5 kW, variable speed motor attached to a 0.457 m diameter, four-blade, radial, flat turbine located directly above the air diffusers. A schematic diagram of the TAAS is shown in Fig. 1.

Oxygen transfer and power consumption were studied at three liquid depths (volume): 1.22 m (1.80 m$^3$), 1.83 m (2.70 m$^3$), and 2.44 m (3.60 m$^3$). Fresh poultry manure, from 270 to 288 White Leghorn layers, was transported from the manure pit to the aeration tank by a manure flushing system and pump.

### Power Consumption

The net power consumption of the air blower and turbine mixer was determined for the TAAS at various airflow and mixing speeds for tap water and aerated poultry waste slurries (APWS). Net power (power actually delivered to the aerated medium) rather than gross power was used to eliminate the variability due to equipment efficiencies and to facilitate comparison with theoretical and semi-empirical relationships reported in the literature.

The net blower power ($P_b$) was calculated by the following equation:

$$P_b = 1.013 \times 10^5\, Q\,[k/(k\text{-}1)]\,[R_p^{\,(k\text{-}1)/k}\text{-}1] \qquad \dots \dots \dots [1]$$

Cooperative investigations of the Agricultural Research Service, USDA, and Cornell University Agricultural Experiment Station, Ithaca, N.Y. Mention of proprietary products does not imply endorsement by the USDA or CUAES.

Acknowledgment: The technical assistance of G. W. Hoffman, Y. C. Lee, and J. H. Martin, Jr. are greatly appreciated. Reveiw and comments by D. C. Ludington and J. H. Martin, Jr. are also appreciated.

The authors are: A. G. HASHIMOTO, Research Leader, and Y. R. CHEN, Agricultural Engineer, ARS, USDA, Agricultural Engineering Dept., Cornell University, Ithaca, N. Y.

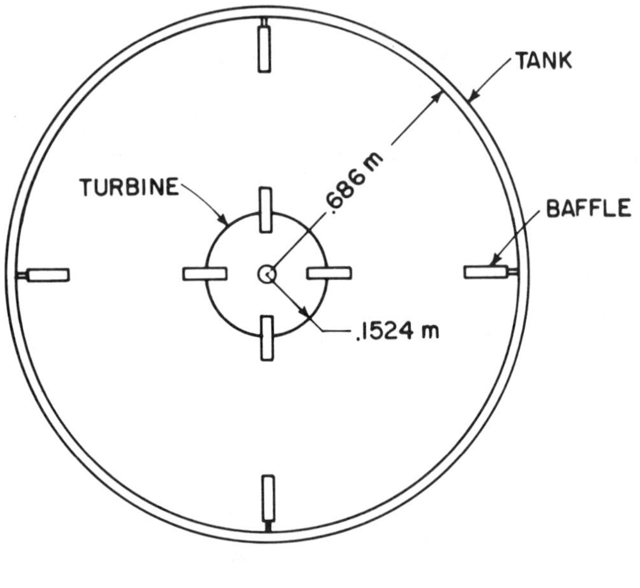


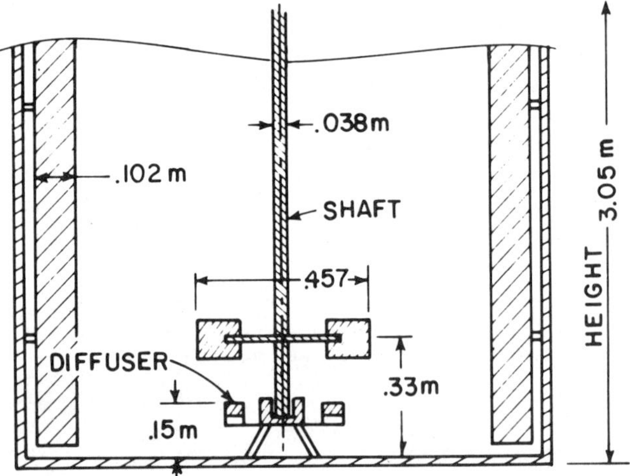


FIG. 1 Schematic diagram of turbine-air aeration system.

The air flow was measured by a rotometer equipped with a thermometer for measuring temperature, and a mercury manometer for measuring air pressure.

The net mixer power was obtained by multiplying the gross power (measured by a kilowatt hour meter attached to the mixer motor) by the mixer efficiency at the operating speed. The efficiency curves for different speeds and loads were determined by using a dynamometer.

### Dissolved Oxygen Measurement

Dissolved oxygen (DO) was measured by a Clark type, membrane covered, polarographic electrode and meter. Calibration was achieved by placing the electrode in moisture-saturated air, then adjusting the meter to correspond to the standard oxygen partial pressure (APHA, 1971) at the operating atmospheric pressure and temperature.

### Oxygen Transfer Coefficients

Oxygen transfer coefficients ($K_La$) for the TAAS were determined in tap water using the following procedure. The airflow rate and mixer speed were adjusted to the desired level, then the blower was turned off. The dissolved oxygen (DO) in the water was removed by the sodium sulfate — cobalt chloride procedure (WPCF, 1971). When the DO reached zero, the blower was restarted and the DO was measured at predetermined time intervals. $K_La$ was computed using subroutine BMDX-85 (Dixon 1969) to obtain nonlinear, least squares fit of the data to the following equation:

$$C_t = C_s [1 - exp(-K_La\,t)] \dots\dots\dots\dots\dots\dots [2]$$

Oxygen transfer coefficients in APWS ($K_La'$) were calculated using the following procedure. After stopping the feed pump and adjusting the airflow rate and mixer speed to the desired level, the blower was turned off and the oxygen uptake rate (rate of oxygen depletion) was measured. When the DO of the APWS reached zero, the blower was restarted and the DO measured periodically. Usually, several reaerations and uptakes were performed to obtain replicate data. The oxygen uptake rate (R) was determined by linear regression of the oxygen depletion data and $K_La'$ was determined by nonlinear least squares fit of the data to the following equation:

$$C_t = C_{eq} [1 - exp\,(-K_La'\,t)] \dots\dots\dots\dots\dots\dots [3]$$

$K_La$ and $K_La'$ were multiplied by $1.02^{(20-T)}$ (Eckenfleder 1966) to standardize the results to 20 C ($K_La_{20}$ and $K_La'_{20}$). The turbine mixer speeds used in this study (6.91 to 14.45 rad/s) did not produce sufficient turbulence to cause surface reaeration. This was confirmed by an experiment which showed no measurable DO in oxygen depleted tap water even after four hours of mixing.

### Effective Viscosity

APWS has been shown to be pseudo-plastic and can be described by the power law formula and the rheological behavior index (n) and consistency index (K) (Chen and Hashimoto 1974). The effective viscosity ($\mu_e$) of the APWS was calculated by the following equation (Calderbank and Moo-Young 1961):

$$\mu_e = K\,(11N/2\pi)^{n-1}\,[(3\,n + 1)/4\,n]^n \dots\dots\dots\dots [4]$$

RESULTS AND DISCUSSION

### Power Consumption

The turbine power requirement was evaluated in tap water and APWS under aerated ($P_g$) and nonaerated ($P_o$) conditions. Fig. 2 shows the nonaearted, turbine power characteristics (Power number vs generalized Reynolds number) provided by the manufacturer and by experimental data from the TAAS. The manufacturer's curve is for Newtonian liquids (water and oil) while the experimental data is for Newtonian (water) and non-Newtonian (pseudo-plastic APWS, 1 to 6 percent total solids) liquids. Fig. 2 shows that the TAAS power number ($N_p$) is constant in the turbulent region or:

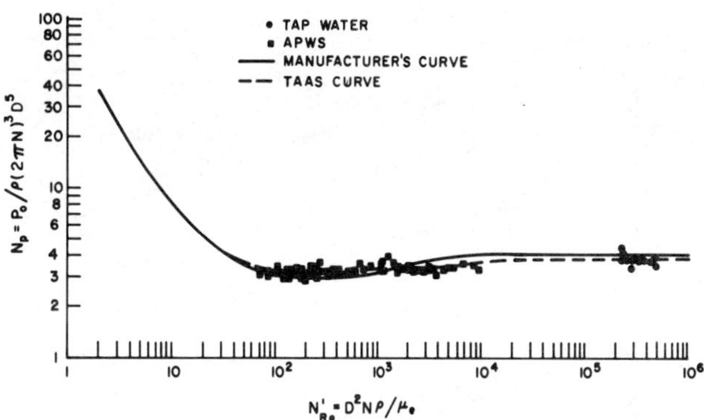



FIG. 2 Turbine power characteristics (power number versus generalized Reynolds number).

$$P_o = 0.285\,N^3,$$
$$\text{for tap water } (\# = 16, \hat{\sigma} = \pm 7\%) \dots\dots\dots\dots\dots [5]$$

Fig. 2 also shows that the TAAS operated in the transition region for APWS and that this region has a greater $N_p$ and encompasses a wider $N_{Re}'$ range than Newtonian liquids (manufacturer's curve). The maximum difference in $N_p$ between the two curves is about $\pm 15$ percent; therefore, the practical implication is that the manufacturer's curve can be used for initial design calculations. However, when the system is constructed and in operation, the actual power characteristics must be evaluated for a more realistic estimate of $P_o$. For the TAAS, $P_o$ was found to be:

$$P_o = 0.136\,N^{3.27},$$
$$\text{for APWS } (r = 0.997, \# = 91, \hat{\sigma} = \pm 7\%) \dots\dots\dots\dots [6]$$

Equation [6] indicates that $N_p$ for APWS is not constant in the transition region but increases with $N_{Re}'$ (Fig. 2). The relative standard deviation ($\hat{\sigma} = \pm 7$ percent) is significantly below the reported reliability ($\pm 20$ percent) for predicting power in non-Newtonian fluids (Bates et al. 1966).

The turbine power under aerated condition ($P_g$) was a function of N and the aeration number ($N_a$):

$$P_g = 0.0598\,N_a^{-0.34}\,N^3,$$
$$\text{for water } (r = 0.904, \# = 66, \hat{\sigma} = \pm 11\%) \dots\dots\dots\dots [7]$$

$$P_g = 0.158\,N_a^{-0.25}\,N^{2.70},$$
$$\text{for APWS } (r = 0.993, \# = 245, \hat{\sigma} = \pm 10\%) \dots\dots\dots\dots [8]$$

Equations [7] and [8] show that aeration decreases the turbine power requirement at a constant N and that the effect of $N_a$ on $P_g$ is more pronounced for water than for APWS. Fig. 3 illustrates the effect of $N_a$ on $P_g$ at various N values. At low $N_a$, $P_g$ for APWS is less than for water because of the lower power number for APWS (Fig. 2). However, $P_g$ for water decreases more rapidly than for APWS and falls below the APWS curve as $N_a$ increases.

### Oxygen Transfer Coefficients

$K_La_{20}$ was found to be a function of airflow rate (Q), turbine angular velocity (N) and liquid depth (H). However,

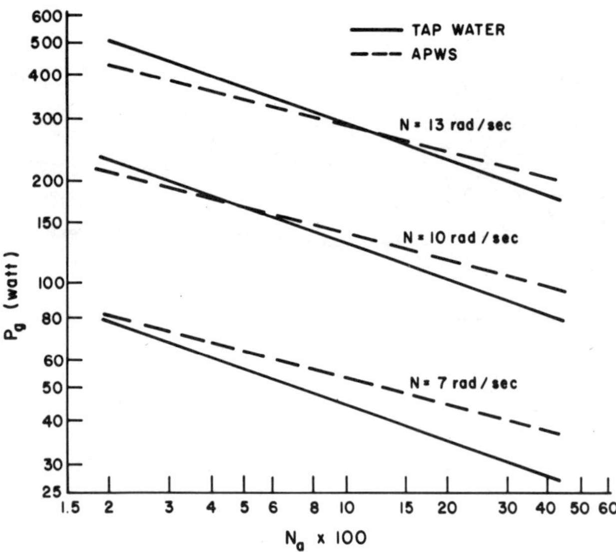



FIG. 3 Effect of aeration number ($N_a$) on turbine power ($P_g$).

a unified equation to predict $K_L a_{20}$ over the total operating range of the TAAS could not be obtained. Rather, two distinct aeration regions, corresponding to the degree of mixing, were observed as follows:

$$K_L a_{20} = 0.172\, N^{1.40}\, Q^{0.448}\, H^{-0.085},$$
$$(r = 0.983,\ \# = 31,\ \hat{\sigma} = \pm 7\%)$$
$$\text{for } N < 75.2\, Q^{0.434} \dots\dots\dots\dots\dots\dots [9]$$

$$K_L a_{20} = 1.756\, N^{0.908}\, Q^{0.613}\, H^{-0.558},$$
$$(r = 0.994,\ \# = 35,\ \hat{\sigma} = \pm 6\%)$$
$$\text{for } N > 75.2\, Q^{0.434} \dots\dots\dots\dots\dots [10]$$

Calderbank (1967) has attributed the two aeration regions to the air bubble size. He reports that as the agitator power increases, the bubble size decreases and the specific interfacial area (a) increases, while the film transfer coefficient ($K_L$) remains constant. $K_L$ decreases rapidly in the critical bubble size region and ultimately becomes constant

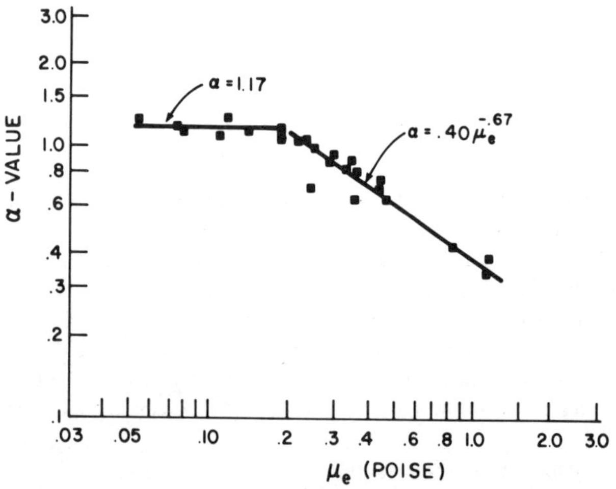



FIG. 4 Effect of effective viscosity ($\mu_e$) on relative oxygen transfer rate ($a$).

again when this region is passed. Concurrently, the interfacial area continues to increase at a decreasing rate as the agitator power is increased. Thus, $K_L a_{20}$ does not increase as rapidly with N in Equation [10] as in Equation [9].

$K_L a'_{20}$ was measured and compared to $K_L a_{20}$ using Equation [9] and [10] depending upon the relative magnitude on N and Q. The relative oxygen transfer coefficient ($a = K_L a'_{20}/K_L a_{20}$) was then calculated and found to be constant (1.17) until the effective viscosity ($\mu_e$) approached 0.2 poise, after which $a$ decreased in the following manner (Fig. 4):

$$a = 0.40\, \mu_e^{-0.67}$$
$$\text{for } \mu_e > 0.2 \text{ poise} \dots\dots\dots\dots\dots\dots\dots [11]$$

Thus, these results indicate that aeration systems for poultry wastes should be designed to operate at apparent viscosities below 0.2 poise for maximum oxygen transfer efficiencies. The effective viscosity of 0.2 poise corresponds to total solids concentrations of 2 to 3 percent depending upon the turbine angular velocity (Equation [2]) and the volume fraction index (Chen and Hashimoto 1974).

**Saturation DO in APWS**

Saturation DO concentrations in APWS ($C_s'$) were calculated using the following equation:

$$C_s' = C_{eq} + R/K_L a' \dots\dots\dots\dots\dots\dots\dots [12]$$

The relative DO saturation ($\beta = C_s'/C_s$) was calculated and found to be independent of total solids concentration (1 to 6 percent). The average value of $\beta$ for 50 observations was 0.97 with a standard deviation of $\pm$ 0.05. This confirms the assumption of $C_s = C_s'$ when using the Steady-State method for calculating $K_L a'$ (Hashimoto and Martin 1974).

**Oxygen Transfer Efficiency**

A plot of the oxygen transfer rate ($K_L a_{20} C_s' C_s = 9.2$ mg/l @ 20 C) vs total power input per unit volume (P/V) (Fig. 5) shows that the oxygen transfer efficiency ($L = K_L a C_s V/P$) increases to a maximum value ($L_m$) at a particular P/V then decreases as P/V continues to increase. Connecting the $L_m$ values yields the following relationship:

$$L_m = 3.91\, (P/V)^{-0.063} \dots\dots\dots\dots\dots\dots [13]$$

Equation [13] shows that $L_m$ decreases as P/V increases. This is logical since relatively more power is needed to maintain a higher oxygen transfer rate.

Oxygen transfer efficiency curves for various livestock wastes aeration systems are shown in Fig. 6. Included are the TAAS (Curve A) using net power input; cage rotor operating in a 1.52 m wide by 6.10 m long and 2.44 m deep tank using net power input (Curve B) (Baumann and Cleasby 1962); TAAS (Curve C) using gross power input (assuming 70 percent efficient blower and turbine); and oxidation ditches (Curve D) using gross power data for cage rotors from data of Jones et al. (1969 and 1970) and Cornell Agricultural Waste Management Pilot Plant (averages of data from Loehr et al. 1972, Baker 1973, Kroeker

532

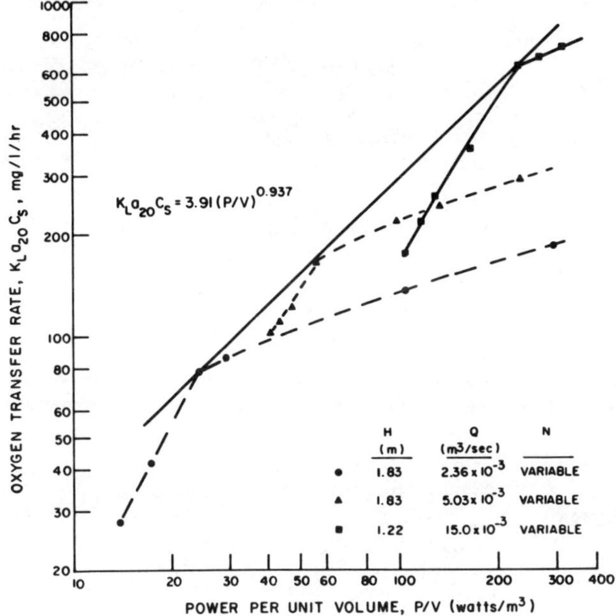


FIG. 5 Oxygen transfer efficiency curve for TAAS.

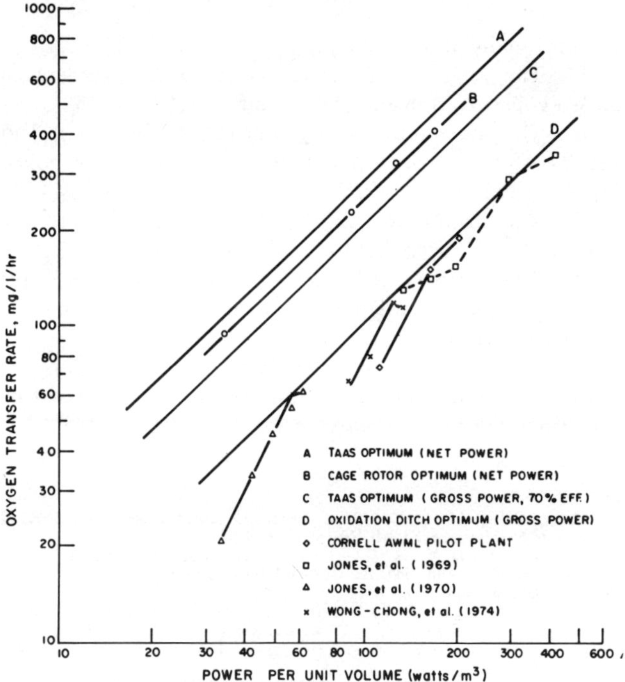


FIG. 6 Oxygen transfer efficiency curves for aeration systems.

1974) and for the jet-aero mix system (Wong-Chong et al. 1974). Curves B and D were drawn parallel to curve A and show good fit of the experimental data. Curves, B, C and D can be expressed as:

$$L_m = 3.40 (P/V)^{-0.063} \text{, Curve B} \dots \dots \dots \dots [14]$$

$$L_m = 2.80 (P/V)^{-0.063} \text{, Curve C} \dots \dots \dots \dots [15]$$

$$L_m = 1.37 (P/V)^{-0.063} \text{, Curve D} \dots \dots \dots \dots [16]$$

This consistent relationship between $L_m$ and P/V is useful in comparing relative oxygen transfer efficiencies of aeration systems and in optimizing the design of TAAS as well as the oxidation ditch.

Dividing Equation [15] by Equation [16] shows that the TAAS is 2.04 times more efficient in oxygen transfer than the oxidation ditch but only 1.15 times more efficient than the cage rotor (Equation [13] ÷ Equation [14]). Thus, it is apparent that the relative efficiency of the oxidation ditch is due to the geometric configuration of the ditch rather than inefficient aeration devices (rotors or jet pump). These findings indicate that for more efficient aeration of livestock wastes, serious consideration should be given to the TAAS as an alternative to the oxidation ditch.

### Operation and Treatment

A detailed discussion of the waste treatment efficiencies under various modes of operation of the TAAS is beyond the scope of this paper. The TAAS was capable of maintaining residual DO from 1 to 9 mg/l, depending upon the total power input and manure loading rate. Treatment efficiencies were similar to those reported for diffused aeration (Hashimoto and Martin 1974) and oxidation ditches

(Prakasam et al. 1974) depending upon the degree of aeration. Excellent odor control was attained by the system.

Foaming was considered a major operational problem of the TAAS. Foaming was generally acute during "start up" and when nitrification was inhibited by nitrite accumulation. Adding motor oil (about 0.1 to 0.2 mg/l of APWS) was found to effectively control foaming; however, oil addition reduced $a$ by about one-third (Table 1). This is a significant decrease which seriously questions the common practice of adding oil for foam control.

### CONCLUSIONS

The following conclusions were drawn from the results of this study:

1  The nonaerated, turbine power characteristics for APWS closely follows (within ± 15 percent) the manufacturer's curve; and therefore, the manufacturer's curve may be used for initial design calculations.

2  The turbine power consumption decreases with aeration rate, the decrease is more pronounced for water than APWS.

3  Oxygen transfer coefficient is a function of the turbine angular velocity, airflow rate, and liquid depth.

4  The relative oxygen transfer coefficient ($a$) is constant up to an effective viscosity of 0.2 poise, then de-

**TABLE 1. EFFECT OF OIL ADDITION ON RELATIVE OXYGEN TRANSFER RATE IN APWS.**

| $\mu_e$ (poise) | $a$ with oil | $a$ without oil |
|---|---|---|
| 0.034 | 0.89 | 1.17 |
| 0.054 | 0.94 | 1.17 |
| 0.069 | 0.43 | 1.17 |
| 0.091 | 0.79 | 1.17 |
| 0.148 | 0.62 | 1.17 |
| 0.189 | 0.80 | 1.17 |
| Ave. | 0.84 | 1.17 |

Reduction in $a$ due to oil addition = 100 (1.17 - .84)/1.17 = 36 percent

creases rapidly above 0.2 poise.

5   The relative DO saturation ($\beta$) is independent of APWS concentration and close to unity (0.97 ± 0.05).

6   The maximum oxygen transfer efficiency ($L_m$) of the TAAS and oxidation ditches decreases as the power input per unit volume (P/V) increases.

7   The maximum oxygen transfer efficiency of the TAA system is twice that for oxidation ditches.

8   The relative inefficiency of oxidation ditches is due to the geometric configuration of the ditch rather than an inefficient aeration device.

9   Adding motor oil to APWS (0.1 to 0.2 mg/l) is successful in suppressing foaming; however, $a$ is decreased by about one-third.

## References


1   APHS. 1971. Standard methods for the examination of water and wastewater. 13th Edition. Washington, D.C.

2   Baker, D. F. 1973. Oxygen transfer relationship in a poultry waste mixed liquor. Unpublished M.S. thesis. Cornell University, Ithaca, New York.

3   Bates, R. L., P. L. Fondy and J. G. Ferris. 1966. Impeller characteristics and power, p. 111-178. In: V. W. Uhl and J. B. Grey, (ed.). Mixing: Theory and Practice, Vol. 1. Academic Press, New York.

4   Baumann, E. R. and J. L. Cleasby. 1965. Oxygenation efficiency of a bladed rotor. Progress Report. Iowa State University Engineering Experiment Station, Ames, Iowa.

5   Calderbank, P. H. and M. B. Moo-Young. 1961. The power characteristics of agitators for the mixing of Newtonian and non-Newtonian fluids. Trans. Inst. Chem. Engrs. 39:338-347.

6   Calderbank, P. H. 1967. Mass transfer in fermentation equipment, p. 101-189. In: N. Blakebrough (ed.). Biochem. and Bio. Eng. Sci., Vol. 1. Adademic Press, New York.

7   Chen, Y. R. and A. G. Hashimoto. 1974. Rheological properties of aerated poultry waste slurries. ASAE Paper No. 4511, ASAE, St. Joseph, Mich. 49085.

8   Dixon, W. J. 1969. Biomedical computer programs-X-series supplement. p. 177-186. University of California Press. Berkeley, California.

9   Echkenfelder, W. W. 1966. Industrial water pollution control. p. 66. McGraw-Hill, Inc., New York.

10   Gill, T. T. 1941. Air and gas compression. p. 52. John Wiley and Sons, New York.

11   Hashimoto, A. G. and J. H. Martin, Jr. 1974. Design consideration for aeration under caged hens. TRANSACTIONS of the ASAE 9(3):500-504.

12   Jones, D. D., D. L. Day and J. C. Converse. 1969. Oxygenation capacities of oxidation ditch rotors for confinement livestock buildings. Proceedings 24th Purdue Industrial Wastes Conference. p. 191-208. Lafayette, Indiana.

13   Jones, D. D., D. L. Day and A. C. Dale. 1970. Aerobic treatment of livestock wastes. Bulletin 737, University of Illinois Agricultural Experiment Station, Urbana-Champaign, Illinois.

14   Kroeker, E. J. 1974. A design and management model of the oxidation ditch for livestock waste treatment. Unpublished M.S. thesis, Cornell University, Ithaca, New York.

15   Loehr, R. C., D. F. Anderson and A. C. Anthonisen. 1971. An oxidation ditch for the handling and treatment of poultry waste. p. 209-212. Proc. Int'l. Symp. on Livestock Wastes, Columbus, Ohio.

16   Prakasam, T. B. S., E. G. Srinath, A. C. Anthonisen, J. H. Martin, Jr. and R. C. Loehr. 1974. Approaches for the control of nitrogen with an oxidation ditch. Proceedings, 1974 Cornell Agricultural Waste Management Conference. p. 421-435. Cornell University, Ithaca, New York.

17   WPCF. 1971. Aeration in wastewater treatment. Manual of Practice No. 5. Washington, D. C.

18   Wong-Chong, G. M., A. C. Anthonisen and R. C. Loehr. 1974. Comparison of the conventional cage rotor and Jet-Aero-Mix system in oxidation ditch operations. Water Research. Vol. 8:761-768. Pergamon Press. Great Britain.


## LIST OF SYMBOLS

| | | |
|---|---|---|
| $a$ | = | specific interfacial area $cm^{-1}$ |
| APWS | = | Aerated Poultry Waste Slurry |
| $C_{eq}$ | = | equilibrium DO in APWS, mg $l^{-1}$ |
| $C_s$ | = | saturation DO in tap water, mg $l^{-1}$ |
| $C_s'$ | = | saturation DO in APWS, mg $l^{-1}$ |
| $C_t$ | = | DO at time t, mg $l^{-1}$ |
| $D$ | = | turbine diameter, m |
| DO | = | dissolved oxygen concentration, mg $l^{-1}$ |
| $H$ | = | liquid depth, m |
| $K$ | = | rheological consistency index, $s^n$ dyne $cm^{-2}$ |
| $K_L$ | = | film transfer coefficient, cm $min^{-1}$ |
| $K_L a$ | = | oxygen transfer coefficient in tap water, $min^{-1}$ |
| $K_L a_{20}$ | = | $K_L a$ at 20 C, $min^{-1}$ |
| $K_L a'$ | = | oxygen transfer coefficient in APWS, $min^{-1}$ |
| $K_L a_{20}'$ | = | $K_L a'$ at 20 C, $min^{-1}$ |
| $k$ | = | ratio of gas specific heats, dimensionless |
| $L$ | = | $K_L a_{20} C_s V/P$ = oxygen transfer efficiency, kg $O_2 kW^{-1} hr^{-1}$ |
| $L_m$ | = | maximum oxygen transfer efficiency, kg $O_2 kW^{-1} hr^{-1}$ |
| $N$ | = | turbine angular velocity, rad $s^{-1}$ |
| $N_a$ | = | $Q/(2\pi ND^3)$, aeration number, dimensionless |
| $N_p$ | = | $P_o/[\rho(2\pi N)^3 D^5]$, power number, dimensionless |
| $N_{Re}'$ | = | $D^2 N\rho/\mu_e$, generalized Reynolds number, dimensionless |
| $n$ | = | rheological behavior index, dimensionless |
| $P$ | = | $P_b + P_g$ = total power, watts |
| $P_b$ | = | blower power, watts |
| $P_g$ | = | turbine power with aeration, watts |
| $P_o$ | = | turbine power without aeration, watts |
| $Q$ | = | standard volumetric airflow rate, $m^3 s^{-1}$ |
| $R$ | = | oxygen uptake rate, mg $l^{-1} min^{-1}$ |
| $R_p$ | = | ratio of absolute discharge to standard atmospheric pressure, dimensionless |
| TAAS | = | Turbine-Air Aeration System |
| $T$ | = | temperature, C |
| $t$ | = | time, min |
| $V$ | = | liquid volume, $m^3$ |

Greek

| | | |
|---|---|---|
| $\alpha$ | = | $K_L a_{20}'/K_L a_{20}$ = relative oxygen transfer coefficient, dimensionless |
| $\beta$ | = | $C_s'/C_s$ = relative oxygen saturation concentration, dimensionless |
| $\mu_e$ | = | effective viscosity, poise |
| $\rho$ | = | liquid density, g $cm^{-3}$ |

Statistical

| | | |
|---|---|---|
| r | = | regression coefficient |
| # | = | number of observations |
| $\hat{\sigma}$ | = | relative standard deviation of calculated value to experimental value, percent |

# Sludge Management for Anaerobic Dairy Waste Lagoons

R. A. Nordstedt, L. B. Baldwin

MEMBER
ASAE

MEMBER
ASAE

ANAEROBIC lagoons are being used extensively for ruminant wastes. The characteristics of these wastes strongly suggest that the sludge accumulation rate will be more rapid than for nonruminant wastes, and recent research tends to confirm this hypothesis (Nordstedt and Baldwin 1975, Barth et al. 1973, and Barth and Bond 1974). Therefore, sludge management becomes an even more important aspect of the total waste management system.

Recently completed studies of the sludge accumulation rates in lagoons have made possible a more complete analysis of alternatives available in management of sludge. The loading rate on an anaerobic lagoon has been found to affect both the rate of sludge accumulation and the characteristics of the sludge. For a given rate of sludge and nutrient accumulation, the relative advantages can be determined for several alternative sludge removal methods.

## SLUDGE ACCUMULATION RATE

Studies of the sludge accumulation rate in an anaerobic lagoon on an 800 cow commercial dairy have shown that sludge accumulates at a rate of 0.0033 cu m per kg (0.053 cu ft per lb) of VS (volatile solids) added. This occurred at an average loading rate of 0.115 kg VS per cu m (0.007 lb VS per cu ft) per day (Nordstedt and Baldwin 1975).

Laboratory digestion studies have yielded similar results (Barth et al. 1973, and Barth and Bond 1974). These studies involved a wide range of loading rates and several temperatures. There was no apparent difference in the volume of sludge buildup at 25 and 10 C. However, since the solids reduction was two to three times greater at the higher temperature, the solids accumulated at a lower temperature had to be more concentrated.

Barth et al. (1973) also found that higher loading rates produced a more dense sludge. Hence, the most efficient storage of waste solids was obtained at the highest loading rates. The efficiency as treatment units was greater at lower loading rates.

## SLUDGE CHARACTERISTICS

Sludge in the above mentioned lagoon averaged 6.3 percent TS (total solids) after 56 months from initial loading of the lagoon. Sludge had accumulated to near the surface at that time. The sludge in the bottom 15 to 20 percent of the lagoon contained approximately 10 percent TS. The remainder of the sludge was a very light, fluid slurry which was formed by evolution of gas and the resultant mixing effect.

Total Kjeldahl nitrogen (TKN) content of the sludge after 56 months of loading was relatively high, averaging over 3000 mg per l. This concentration was substantially higher than the average of 182 mg per l of TKN after 40 months of loading. The percentage of nitrogen which was in the ammonium form decreased from 61 percent at 40 months to 34 percent at 56 months. This may be due to seasonal effects (Nordstedt 1975).

The physical properties of the sludge dictate the methods for economical removal from the anaerobic lagoon and for final disposition. Transportation of sludge which is over 90 percent water would be expensive, as would dehydration or drying. Sludge removal from the lagoon may be accomplished with normal excavating equipment if the lagoon can be taken out of use long enough for it to lose most of its supernatant through evaporation or a slow drawdown by discharging from the surface. If the lagoon is located where the groundwater level, or hillside seepage, will tend to keep it full or cause bank failure during drawdown, the sludge must be removed by pumping. Efforts to excavate sludge from a full lagoon with a dragline have not succeeded. Only a portion of the dense bottom sludge can be retained in a dragline bucket.

The ultimate disposal of anaerobic sludge will be on nearby land in most cases. Economics will also favor pumping from the lagoon to discharge on the land at acceptable rates of application.

Nitrogen is generally considered to be the limiting factor in determining the application rate. A representative sludge sample should be analyzed for nitrogen and the forms in which it occurs. The proper rate can then be selected, based upon the intended crop, soil characteristics and climate. The ammonium content of the sludge should be watched carefully with respect to inhibition of seed germination.

Sludge from dairy waste lagoons does not generate significant odor when small quantities are sampled from the lagoon. However, it was found that stored samples become very odorous after storage at 4 C for a few days. Hence, ponded wet sludge on the land could present serious odor problems. It will be necessary to disc or plow the treated fields soon after sludge application.

## SLUDGE MANAGEMENT ALTERNATIVES

Several sludge management schemes are possible, depending upon climate, farm management practices, lagoon dimensions, equipment availability, and other factors. This discussion will be limited to only three: (a) batch, (b) periodic, and (c) continuous.

Batch removal would involve complete or nearly complete sludge cleanout when the lagoon becomes full of

Florida Agricultural Experiment Station Journal Series No. 5835.

The authors are: R. A. NORDSTEDT and L. B. BALDWIN, Assistant Professors, Institute of Food and Agricultural Sciences, Agricultural Engineering Dept., University of Florida, Gainesville.

sludge or begins to discharge sludge in the effluent, perhaps at four to six year intervals. If the surrounding groundwater level is above the lagoon bottom, then dewatering of the lagoon would not be desirable. Since neither the light nor the dense sludge can be dug successfully if the lagoon is not dewatered, a dredging operation is indicated. Pumps are available which will handle these sludges. This procedure should be timed so that sludge is removed at a rate which will not lower the lagoon surface enough to induce bank failure. For lagoons with extremely long detention times, this operation will require an extended period of time for completion. Sludge removal should begin where sludge accumulation has become significant and proceed toward the lagoon outlet. The area near the inlet will probably contain only raw manure and very little sludge. The dredge pump intake should be moved commensurate with pumping rate along the remaining length of the lagoon.

If the groundwater level is lower than the lagoon bottom and dewatering is possible, then sludge removal could be accomplished in a much shorter period of time. However, very little is known about the dewatering characteristics of this type of sludge. In any case, a temporary waste handling system would have to be implemented to handle the continuing inflow of wastewater.

Every effort should be made to avoid disturbing the soil surface on the lagoon sides and bottom to minimize groundwater pollution potential. This would be less likely to occur during pumping or dredging.

Periodic sludge removal might be performed to reduce accumulation near the outlet structure and to reduce the amount of sludge in the effluent. Such an operation would remove the light sludge, particularly in elongated lagoons. Dense bottom sludge would continue to accumulate over most of the lagoon bottom, requiring eventual cleanout by the batch method. The advantages of periodic sludge removal include limited disruption of the lagoon and more efficient utilization of the nutrients in the sludge at yearly or six-month intervals.

Continuous sludge removal might be advantageous for lagoons which are destratified by mixing or recirculation. Effluent containing suspended solids could then be passed through a settling pond or clarifier where sludge would be removed as necessary. Continuous mixing of anaerobic lagoons may increase digestion efficiency (Mahan 1972), and sludge management would become a regular management operation. However, increased capital and operating costs would be necessary.

As previously discussed, sludge removal can best be accomplished by pumping in most cases. Complete sludge removal by the batch method would require specialized dredging equipment with long discharge pipes, as well as tractors and plows or discs. A labor force of two to four people would also be required for a one to three week period, depending on lagoon size and detention time. Such an operation indicates the use of hired equipment and labor. Specialized equipment has been developed to remove sludge from very large sewage and industrial lagoons. However, its use for most animal waste lagoons would not be practical. At the present time, there is little experience with batch cleanout of these lagoons, and most contractors are cautious about quoting prices for the job. Therefore, it is important to calculate the pump capacity needed, which

will be relatively small, and to make the prospective contractor aware of the nature of the material to be dredged. Costs will probably be lower if negotiated on an hourly or daily basis rather than for the total job.

Periodic or continuous cleanout of small quantities of sludge would involve less equipment and labor, indicating that the necessary equipment should be purchased and the cleanout operation performed as a part of routine farm duties. The fertilizer value of the sludge is such that cost return will be significant.

Utilization of the sludge for nonfarm purposes may be possible in isolated cases. Such uses might be for commercial composting, flower nurseries, high value cash crops, etc. In such cases, continuous removal of small quantities of sludge would provide a valuable fertilizer and soil conditioner.

The sludge should be spread on the soil at a rate that will not (a) result in surface runoff, (b) cause groundwater contamination, or (c) result in over-fertilization. Some contour diking may be necessary to prevent surface runoff. The sludge should be incorporated into the soil as soon as possible to prevent odor and insect problems. A grazed or harvested crop should be established to utilize the nutrients in the sludge.

## TOTAL ANAEROBIC LAGOON SYSTEM

The anaerobic animal waste lagoon has evolved into a complex system from its beginning as a single, nonoverflow pond. Increased use of water for waste collection resulted in a consideration of effluent management systems, including use of storage ponds for the effluent. The more recent use of the anaerobic lagoon for ruminant wastes has added another dimension to anaerobic lagoon design and management, that of sludge management. Swine and poultry lagoons have not presented this problem, since sludge accumulated at a much lower rate.

Sludge management should be considered from the initial planning stages of an anaerobic lagoon system. Water usage will affect lagoon volume in cases of extremely high water usage and will also affect the detention time and the design of the effluent dispersal system.

The organic loading rate on the lagoon will affect the sludge accumulation rate. This in turn will dictate the frequency of sludge removal. Consideration should then be given to the cost of sludge removal versus the cost of lagoon construction to decrease the frequency of sludge removal.

The type of sludge removal scheme—batch, periodic, or continuous—will depend on farm management preference, cropping practices, equipment availability, and off-farm markets for the sludge. The lagoon system design must be optimized for these conditions in each case.


**References**

1    Barth, C. L. and John H. Bond. 1974. Water Management in livestock waste handling systems. Completion Report, OWRR Project No. A-025-SC.
2    Barth, C. L., H. P. Lynn, and W. L. Northern. 1973. Progress report—aerobic and anaerobic lagooning of dairy and milking wastes. In: Proceedings of the National Dairy Housing Conference, ASAE Publication No. SP-01-73. pp. 371-380.
3    Mahan, Richard M. 1972. The stratification of an anaerobic dairy manure lagoon. M.S. Thesis, University of Florida, Gainsville.
4    Nordstedt, R. A. 1975. Unpublished data.
5    Nordstedt, R. A., and L. B. Baldwin. 1975. Sludge accumulation and stratification in anaerobic dairy waste lagoons. TRANSACTIONS of the ASAE, Vol. 18, No. 2, pp. 312-315.

# Trends and Variations in an Anaerobic Lagoon with Recycling


C. V. Booram,
ASSOC. MEMBER
ASAE

T. E. Hazen,
SENIOR MEMBER
ASAE

R. J. Smith
ASSOC. MEMBER
ASAE


LIVESTOCK waste lagoons came into widespread use during the early 1960's, but many of these lagoons were improperly designed because their expected performance was overestimated. Those properties of livestock manure needed for accurate design were largely unknown and performance predictions were drawn mostly from experience with lagoons used for treatment of human wastes. Typical of the thinking of that time was that an anaerobic lagoon was a method for completely disposing of livestock wastes (Ashmore 1966).

Later, more fundamental research (e.g., Willrich 1966, Lynn 1968 and Hill and Barth 1974) showed that a lagoon is no more than an effective treatment component usable in well-designed livestock-production systems. Also, specific design information now is becoming available from these findings and the accumulated experiences from operating lagoons.

The factors having significant bearing on design are: What are acceptable water quality parameters for the lagoon liquid? How rapidly does sludge accumulate? This paper examines these two questions by using data collected from an 11-year-old lagoon.

## RETROSPECT

The lagoon is part of the waste-manatement system for a 700-head swine-finishing building (Unit K) located at the Swine Nutrition Station of Iowa State University. The loading rate is about 0.08 kg-VS/m³-day and the detention time varies from 50 to 200 days. The waste-management system has undergone several modifications since its construction 14 years ago (1959). Initially, the system had a scraper in an open gutter to move manure into storage tanks. Three years later, the lagoon was added and scraping was replaced by intermittent flushing with fresh water. The final major revision, which followed shortly thereafter, was to recycle treated wastewater for flushing the gutters. Several treatment schemes have been tried, including an oxidation ditch used both singly and following treatment in an anaerobic lagoon (Smith 1971). These trials led to the present simplified system that recycles the anaerobic-lagoon liquid hourly as flushing water for the open duging gutters. Exposure of the pigs to the anaerobic-lagoon liquid does not seem a health hazard (Schmitt et al. 1974). Disposal of surplus liquids from the lagoon is by seasonal application to adjacent cropland via a solid-set irrigation system.

## TECHNIQUE

In July, 1973, the third sludge survey of the Unit K lagoon was made (the results are reported later in this paper). Willrich (1966) had made the initial survey after less than 3 years of operation and found that the deep cell contained an average depth of 24.4 cm of sludge and that the shallow cell contained an average of 33.5 cm. Five years later, Koelliker and Miner (1970) again surveyed the lagoon and found average sludge depths of 23.4 cm and 41.4 cm in the deep and shallow cells, respectively.

The latest survey in 1973 samples 27 locations in the lagoon, 15 in the deep cell and 12 in the shallow cell. The samples were collected by forcing a Plexiglas® tube through the sludge and into the bottom soil. Depth of sludge, depth of lagoon liquid, and some visual characteristics of the samples were recorded. The samples were then placed in plastic bags and frozen until each could be analyzed in the laboratory. Laboratory analyses were conducted for pH, Kjeldahl nitrogen, chemical oxygen demand (COD), and total phosphorus. Analytical procedures were similar to those described in Standard Methods (APHA 1971).

The lagoon liquid has also been analyzed at intervals over the past 12 years; e.g., Ashmore (1966), Willrich (1966, Vanderholm (1969), Koelliker (1972) and Booram (1975). These analyses have followed the methods for

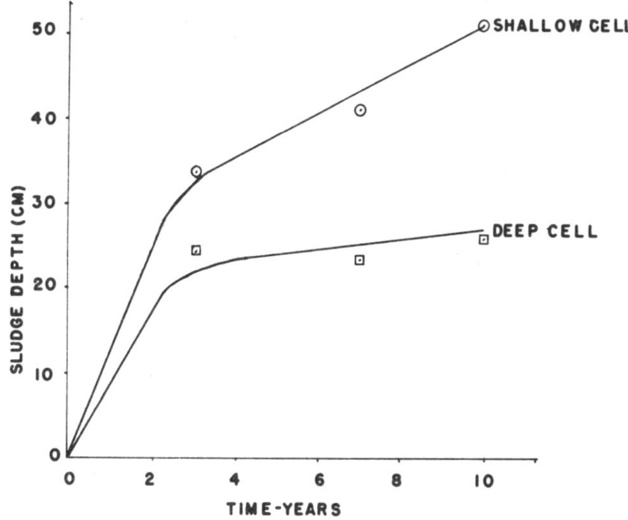



FIG. 1 Sludge accumulation in an anaerobic lagoon used to treat wastes flushed from a 700-head swine confinement building at Ames, Iowa.


Journal Paper No. J-8132 of the Iowa Agriculture and Home Economics Experiment Station, Ames, Iowa, Project No. 1842.

The authors are: C. V. BOORAM, Assistant Professor, Agricultural Engineering Dept., University of Georgia Coastal Plain Experiment Station, Tifton (formerly at Iowa State University); T. E. HAZEN, Assistant Director, Agriculture and Home Economics Experiment Station, and R. J. SMITH, Assistant Professor, Agricultural Engineering Dept., Iowa State University, Ames.

TABLE 1. AVERAGE VALUES OF pH, COD, KJELDAHL N AND TOTAL P FOR ANAEROBIC LAGOON (JULY, 1973) SLUDGE AT UNIT K, SWINE NUTRITION STATION, IOWA STATE UNIVERSITY, AMES, IOWA.

| Parameter | Value |
|---|---|
| pH | 7.8 |
| Chemical oxygen demand (mg/l) | 48,700 |
| Total Kjeldahl nitrogen (mg/l) | 3,580 |
| Total phosphorus (mg/l) | 1,490 |

TABLE 2. MASS OF CHEMICAL OXYGEN DEMAND AND NUTRIENTS IN UNIT K ANAEROBIC LAGOON LIQUID AND SLUDGE (JULY, 1973), SWINE NUTRITION STATION, IOWA STATE UNIVERSITY, AMES, IOWA.

| Parameter | Sludge mass, kg | Liquid mass, kg | Percent contained in sludge |
|---|---|---|---|
| Chemical oxygen demand | 58,900 | 4,950 | 92 |
| Nitrogen | 4,320 | 1,645 | 72 |
| Phosphorus | 1,810 | 215 | 89 |

water quality measurement as described in Standard Methods.

## COMPOSITION AND ACCUMULATION OF SLUDGE

The 1973 sludge survey shows that the Unit K lagoon contained an average of 26.7 cm of sludge in the deep cell and 50.8 cm of sludge in the shallow cell — both slightly higher accumulations than found in previous surveys (Fig. 1). The sludge in both cells was dark brown to black and contained hoghair and small pieces of undigested corn.

Average values of pH, the concentrations of COD, Kjeldahl nitrogen and total phosphorus for the sludge samples are given in Table 1. The total mass of COD, N, and P as well as the partition of each parameter between the sludge and the liquid are found in Table 2. The sludge contained most of the COD (92 percent), nitrogen (72 percent) and phosphorus (89 percent) in the system. The sludge depth in the deep cell has remained stable since 1966, and although the sludge in the shallow cell continues to increase, the rate of accumulation has been declining.

The accumulation of sludge in waste-treatment systems has been discussed in several reports; Al-Timini et al. (1965), Loehr (1968), Koon et al. (1970), and Hart (1970), to name a few. Hart predicted that the sludge in a lagoon would increase at a daily rate of from 0.02 to 1 percent of the lagoon volume. Accordingly, the Unit K lagoon should fill in less than 11 years; but these predictions obviously are in error for this particular lagoon. Two reasons seem to account for the difference. First, the solids degradation probably is more effective during the spring, summer and fall. Adequate capacity and nearly uniform daily loading of the lagoon undoubtedly have helped to create more favorable conditions for solids degradation. The second reason is related to the management of the excess liquid; i.e., we have observed that some suspended solids are removed by the irrigation system.

## WATER-QUALITY EVALUATION

The results of the analyses of liquid samples taken from the Unit K lagoon follow (t = 0, 1, 2 . . . 11 years, with 1962 being the base year).

### pH

From 1968 (t = 6) to 1972 (t = 10), the pH has decreased linearly (Fig. 2). A regression ananysis indicates that the following is the best fet of the data:

$$pH = 8.74 - 0.14t \quad (t = 6, 7, \ldots 10)$$

and the slope is significantly different from zero at the 1 percent level.

### Chlorides

The chloride-concentration data (1968 (t = 6)-1973 (t = 11) were analyzed, and a regression analysis was completed. The following equation resulted:

$$Cl(mg/l) = -159 + 42.2t \quad (t = 6, 7, \ldots 11)$$

and the slope is significantly different from zero at the 1 percent level. The chloride concentration has increased steadily since 1968 (t = 6) (Fig. 3). Although the 1973 (t = 11) concentration amounts are less than for 1972 (t = 10), the chloride concentration is increasing for the time period.

### Total Kjeldahl Nitrogen

The nitrogen concentration increased in the Unit K lagoon from 1968 (t = 6) - 1973 (t = 11). Fig. 4 indicates the general trend of the nitrogen concentration. The data yield the regression equation:

$$TKN(mg/l) = -13 + 48.4t \quad (t = 6, 7, 8, \ldots 11)$$

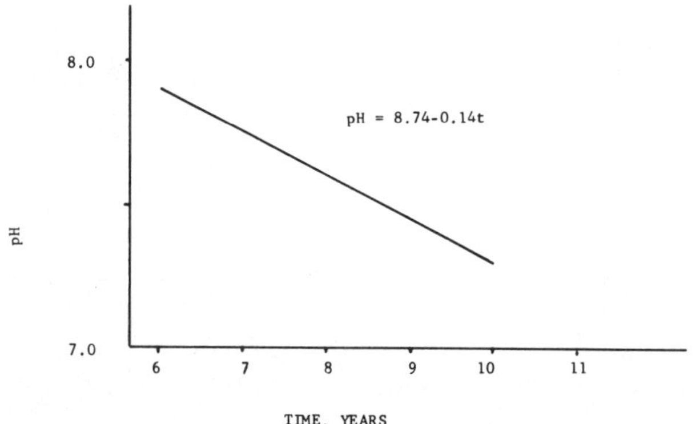


FIG. 2 pH of liquid in an anaerobic lagoon used to treat wastes flushed from a 700-head swine confinement building at Ames, Iowa.

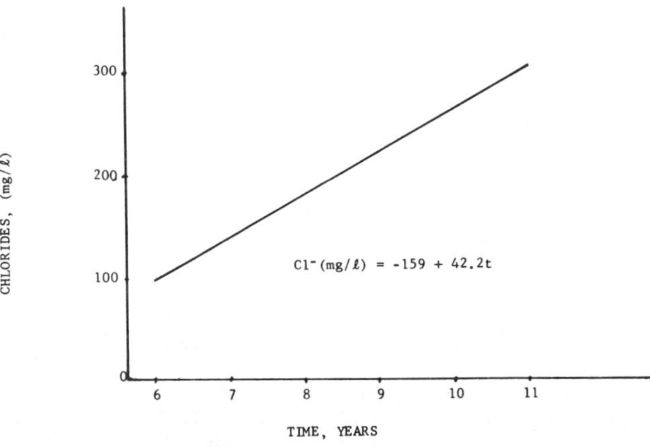


FIG. 3 Chloride concentration of liquid in an anaerobic lagoon used to treat wastes flushed from a 700-head swine confinement building at Ames, Iowa.

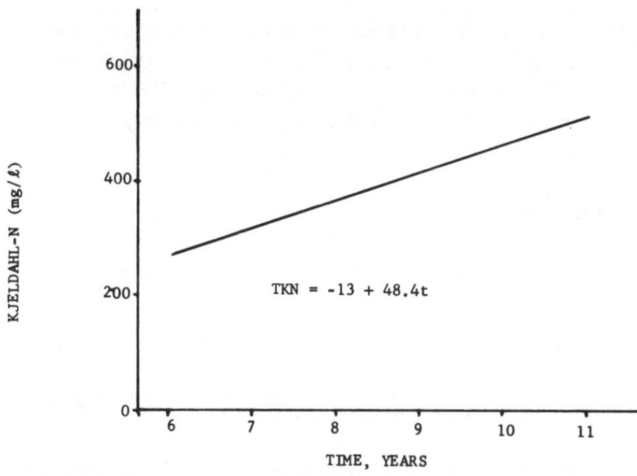



FIG. 4 Kjeldahl N concentration of liquid in an anaerobic lagoon used to treat wastes flushed from a 700-head swine confinement building at Ames, Iowa.

and the slope is significantly different from zero at the 1 percent level.

### Chemical Oxygen Demand

Analysis of the measurements for 1968 ($t = 6$) - 1973 ($t = 11$) shows an increase in COD with time since ($t = 6$) (Fig. 5). The regression equation is:

$$COD(mg/l) = -434.1 + 188.8t \quad (t = 6, 7, 8, \ldots 11)$$

and the slope is significantly different from zero at the 1 percent level. If we also consider COD data for 1964 ($t = 2$), 1965 ($t = 3$) and 1969 ($t = 6$) to 1973 ($t = 11$), the COD reaches a minimum between 1965 ($t = 3$) and 1968 ($t = 6$). The equation for all the measurements is:

$$COD(mg/l) = 1688.1 - 321.9t + 29.6t^2$$

### Phosphorus

From 1968 ($t = 6$) to 1973 ($t = 11$), no increasing or decreasing treand was observed in the phosphorus concentration. The regression equation is:

$$P(mg/l) = 60.8 + 1.2t \quad (t = 6, 7, 8, \ldots 11)$$

but the slope is not significantly different from zero at the 1 percent level.

## CONCLUSION

We can interpret these results by comparing the actual system to hypothetical system made up of steady flow through a continuously mixed reactor of constant volume. If a substance is introduced into the influent stream of the model at a steady rate, it will accumulate in the reactor, be removed by the effluent stream and possibly be removed by an alternate pathway (e.g., COD is converted to gas). Using the symbols $c_1$ for the concentration of substance in the influent, m for the rate of mass removal by the alternate pathway, V for the volume of the reactor, and q for the volume flow rate, we may show that the concentration c of the substance in the effluent after time t will be

$$c = (c_1 - m/q) [ 1 - exp(- qt/V) ]$$

We observe that V/q is the hydraulic detention time of the system. The Unit K lagoon has an annual flow-through rate of about 320,000 gal (1.2 gal daily per pig, 700 pigs) and a volume of 650,000 gal. The elapsed time after the start of recycling was 6 years by 1973.

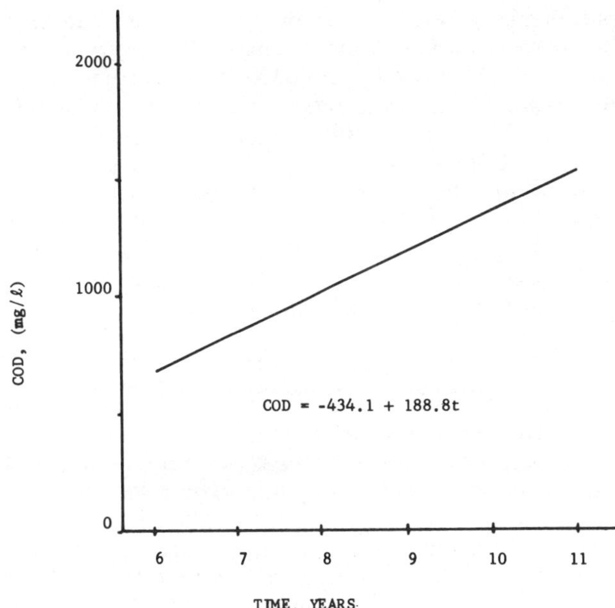



FIG. 5 Chemical oxygen demand concentration of liquid in an anaerobic lagoon used to treat wastes flushed from a 700-head swine confinement building at Ames, Iowa.

Substitution of these numbers into the equation shows that equilibrium should be better than 95 percent complete. The data presented for COD, Kjeldahl N, $Cl^-$, and pH seem to indicate otherwise.

The removal of COD from the lagoon (the "alternate" pathway) is by sedimentation and by biological conversion to gases. The continued upward trend in the COD concentration may indicate that biological activity is deteriorating or, perhaps more probably, that the settled solids are decomposing to release biologically refractory substances into the liquid. This trend also may account for the rise in Kjeldahl N and the decline in pH. It is well established that fatty acids are intermediate compounds in anaerobic fermentation; hence, a release of acids would lower pH and hold more $NH_4^+$ in solution to buffer the acidity. Another factor that may contribute to the increased concentrations of COD and Kjeldahl N may be an increase in suspended matter; this would be a physical consequence of the sludge decomposition. We are unable to validate this idea because we have not developed suitable techniques for suspended solids measurements.

The lack of equilibrium of the $Cl^-$ concentration is

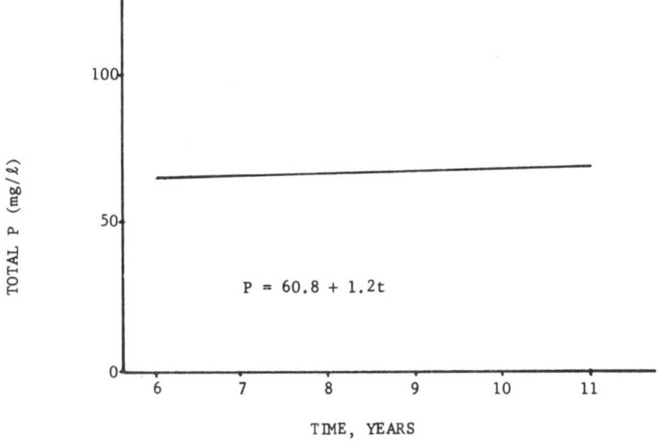



FIG. 6 Phosphorus concentration of liquid in an anaerobic lagoon used to treat wastes flushed from a 700-head swine confinement building at Ames, Iowa.

MANAGING LIVESTOCK WASTES

puzzling because Cl⁻ is required in only trace quantities by microorganisms. Approximately 3700 lb of Cl⁻ are added in the feed and about 2300 lb are passed through the animals into the anaerobic treatment system. Because this ion is so soluble, a pool of undissolved Cl⁻ in the bottom sludge seems unlikely. We believe the equilibrium shown by the phosphorus data can be explained by the soluble orthophosphate being in equilibrium with precipitated phosphate compounds in the sludge and with the $MgNH_4PO_4$ often found precipitated in the recycle plumbing.

## SUMMARY

In this paper we have discussed lagooning as only one treatment component of a totally integrated management system for livestock waste. Separation, transport, storage and disposal also play important roles. If a treatment system is to be useful for producers, it must be effective, and it must be simple to operate and maintain. Treatment-system longevity also is an important concern.

The management and operation of an anaerobic lagoon can be quite critical. In some areas of the United States, it is common to construct and operate anaerobic lagoons simply as settling basins. It seems obvious that, in these situations, the supernatant flows away, leaving solids that rapidly fill the structure.

The application of laboratory models or small field models for predicting solids accumulation in field-scale lagoons should be approached with caution. Many short-term studies have been conducted, but few, if any, have been observed for the extended period required for field-scale systems operating under ambient conditions to become stabilized.

The anaerobic-treatment system at Iowa State University has operated for more than 12 years without becoming filled with solids. Solids accumulation today is less rapid than during the initial startup period. Eventually, this system may fill with sludge, but the life will have extended far beyond that predicted by the literature.

The liquid in the Unit K anaerobic lagoon has been regularly monitored since 1968, at which time the lagoon had been in operation 6 years. Several of the water-quality parameters (COD, Kjeldahl N, and Cl⁻) have increased in concentration during this time. The pH is decreasing, but the phosphorus concentration is stable. The use of a hydraulic manure-transport system that recycles treated wastewater has reduced both the labor and the volume of liquid that must be discarded from the system. Impending failure of the anaerobic-treatment system does not seem imminent.


### References

1 Al-Timini, A. A., W. J. Owings and J. L. Adams. 1965. Effects of volume and surface area on the rate of accumulation of solids in indoor manure digestion tanks. Poultry Science 44:112-115.

2 Ammerican Public Health Association. 1971. Standard methods for the examination of water and wastewater. 13th ed. American Public Health Association. Washington, D.C.

3 Ashmore, L. E. 1966. Disposal of wastes from a confinement unit in anaerobic lagoons. M.S. thesis. Iowa State University Library, Ames.

4 Booram, C. V. 1975. Nutrient changes in plant tissue, soil and applied effluent for a system using an anaerobic lagoon with recycled swine wastes. 185 pp. Ph.D. thesis. Iowa State University. (Not yet catalogued by University Microfilms, Ann Arbor, Mich.(.

5 Hart, S. A. 1970. Animal manure lagoons, a questionable treatment system. pp. 320-325. (Proceedings, Second International Symposium for Waste Treatment Lagoons).

6 Hill D. T. and C. L. Barth. 1974. A fundamental approach to anaerobic lagoon analysis. pp. 387-404. In: Processing and Management of Agricultural Waste (Proceedings, Cornell Agricultural Waste Management Conference).

7 Koelliker, J. K. 1972. Sprinkler application of anaerobically treated swine wastes as limited by nitrogen concentration. 206 pp. Ph.D. thesis. Iowa State University. (Mic. 72-19,989, University Microfilms, Ann Arbor, Mich.).

8 Koelliker, J. K. and J. R. Miner. 1970. Pasture application of aerobic lagoon water: a practical experience. ASAE Paper No. 70-406. ASAE, St. Joseph, Mich. 49085.

9 Koon, J. L., R. E. Hermanson, T. A. McCaskey and A. E. Hiltbolb. 1970. Experiences with anaerobic swine lagoon. ASAE SE. ASAE, St. Joseph, Mich. 49085.

10 Loehr, R. C. 1968. Anaerobic lagoons; considerations in design and application. TRANSACTIONS of the ASAE 11(2):320-322, 330.

11 Lynn, H. P. 1968. The effects of loading rates on the design and operation of anaerobic swine lagoons. M.S. thesis. Clemson University Library, Clemson.

12 Schmitt, L. W., T. E. Hazen and R. J. Smith. 1974. Influence of ingestion of anaerobic lagoon effluent on growing swine. pp. 356-374. In: Processing and management of agricultural wastes (Proceedings, Cornell Agricultural Waste Management Conference).

13 Smith, R. J. 1971. A protype to renovate and recycle swine waste hydraulically. 188 pp. Ph.D. thesis. Iowa State University. (Mic. 72-5258, University Microfilms, Ann Arbor, Mich.).

14 Vanderholm, D. H. 1969. Field treatment and disposal of livestock lagoon effluent by soil percolation. M.S. thesis. Iowa State University library, Ames.

15 Willrich, T. L. 1966. Primary treatment of swine wastes by lagooning. pp. 70-74. In: Management of farm wastes (Proceedings, National Symposium). ASAE Paper No. SP-0366. ASAE, St. Joseph, Mich. 49085.

# A Lagoon-Grass Terrace System to Treat Swine Waste


D. M. Sievers, G. B. Garner, E. E. Pickett

ASSOC. MEMBER
ASAE


OFTEN, in planning a livestock waste management system, an engineer can take advantage of topography, location, existing facilities or other available resources to design a low-cost system. This is particularly true of small livestock enterprises. This paper describes the performance of a waste management system for a small swine finishing unit (200 hd).

The waste system consists of a slotted floor finishing house over an anaerobic lagoon (26 x 32 m) and a 259 m grassed terrace. All surface runoff is diverted around the lagoon such that no water other than wastewater and direct precipitation enters the lagoon. During precipitation events, the lagoon discharges to the terrace where renovation and dilution of the effluent occurs. Surface runoff from 2.8 ha of cropland is diverted by a second terrace to the upper end of the disposal terrace. This provides dilution of the lagoon effluent as it flows from the lagoon.

The terrace discharges to a grass waterway, flows approximately 113 m to a culvert under a county road and then 153 m through natural drainage to enter a small arm of a water supply reservoir. Fig. 1 is a sketch of the general layout.

The lagoon and grassed waste renovation terrace had been in service seven years as of January 1971. The lagoon was malfunctioning biologically and periodically caused odor problems. A depth profile was made using a boat and a long wooden dowel. The profile study revealed an accumulation of approximately two feet of sludge on the lagoon bottom. Lagoon volume was determined to be 588 cu m.

Three objectives of the study were (a) determine the effectiveness of the lagoon-grassed terrace system in providing treatment of the lagoon effluent; (b) assess the potential impact of the system on a nearby water supply reservoir during a precipitation event; and (c) attempt to improve the biological activity of the lagoon by removing the liquid fraction and diluting the sludge with fresh water.

## METHODS AND PROCEDURES

In March 1972, the liquid level of the lagoon was lowered 2 feet by pumping the liquids onto adjacent cropland. Fresh water from a nearby pond was then added to the lagoon, returning the liquids to the original level. The diluting of the lagoon solids was an attempt to restore good biological activity in the lagoon and was performed at least once a year. Since the lagoon was to be diluted routinely, it was felt that determining the renovative capacity of the grassed terrace after dilution would be more representative


Acknowledgments: Appreciation is expressed to Mr. Vernon Heins for his cooperation and for the use of his facilities. Mr. Bill Keffer, EPA Region VII, is recognized for his help in collection and analysis of water samples and data analysis.

The authors are: D. M. SIEVERS, Assistant Professor, Agricultural Engineering Dept., G. B. GARNER and E. E. PICKETT, Professors, Biochemistry Dept., University of Missouri, Columbia.


of lagoon management. Samples of the lagoon were taken before and after pumping. All samples were taken at a 0.61 m depth.

On April 19, 1972, a three and one-half hour precipitation event occurred, producing a total rainfall at the lagoon of 2.3 cm. Flow from the lagoon and at the end of the grass terrace was measured by type H flumes (Fig. 1).

Water samples were obtained from the flumes at 30-minute intervals. All samples were refrigerated until analyzed. All samples (lagoon, runoff, and reservoir) taken during the precipitation event study period were analyzed by the Environmental Protection Agency, Region VII Office at Kansas City, Missouri. Methods of analysis included Standard Methods (1971) or the Environmental Protection Agency (1971).

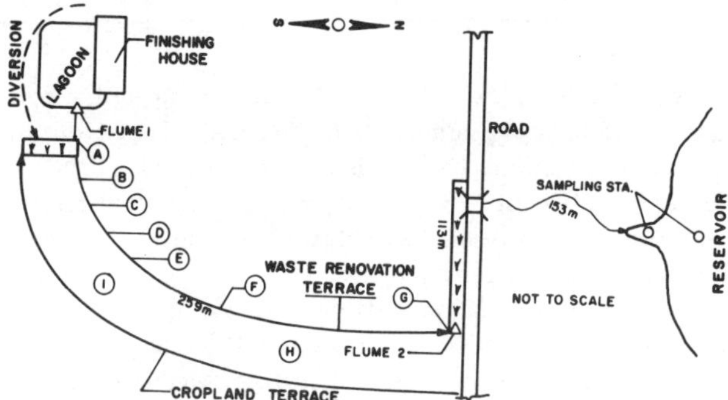



FIG. 1 General layout of the facilities. Circled letters are soil sampling station. Triangles are flumes.

In August 1974, soil samples at selected distances along the length of the terrace channel were obtained to check for chemical buildup. Samples were obtained with a stainless steel sampler, brought to the laboratory and air-dried for 48 hours. Samples representing a range of soil depths were thoroughly mixed before performing any analysis.

Nitrate was determined colormetrically on extracts using $1 \ N \ K_2SO_4$ (Bremner 1965). Sodium, phosphorus, copper, zinc and arsenic were determined from extracts prepared by boiling 10 grams of soil with 20 ml concentrated nitric acid. Copper and zinc were determined by atomic adsorption in air-acetylene flame. No compensation for matrix effects was attempted as they were expected to be small (Slavin 1968). Sodium was determined by flame emission in the air-acetylene flame without compensation for matrix effects (Dean 1960, Pickett and Koirtyohann 1969). Phosphorus was determined by the reduced phosphomolydbate colorimetric method (Assoc. Offic. Anal. Chem. 1965) and arsenic by the arsine-silver diethyldithiocarbamate colorimetric method (Liederman et al. 1959).

## RESULTS AND DISCUSSION

### Grass Terrace System

As measured during the April 1972 precipitation event,

substantial decreases in the concentrations of most parameters were noted in the effluent flowing from the terrace. Table 1 lists changes in the concentration of 14 parameters. Reductions in excess of 80 percent were achieved for total and volatile solids, Zn and Cu, while over 90 percent reductions occurred for COD, $BOD_5$, Org-N, $NH_3$-N, total P, Na, and K.

### TABLE 1. CHANGE IN EFFLUENT QUALITY AFTER FLOWING THROUGH TERRACE

| Parameter | Lagoon effluent, concentration | Grass terrace at flume 2, concentration | Percent change in concentration |
|---|---|---|---|
| $BOD_5$, mg/l | 931 | 25 | 97 |
| COD, mg/l | 2679 | 133 | 95 |
| TS, mg/l | 3645 | 700 | 81 |
| VS, mg/l | 1660 | 213 | 87 |
| Org-N, mg/l | 81 | 5 | 93 |
| $NH_3$-N, mg/l | 548 | 7 | 98 |
| Total P, mg/l | 131 | 8 | 94 |
| Na, mg/l | 220 | 20 | 91 |
| K, mg/l | 496 | 42 | 92 |
| Fe, mg/l | 13.04 | 12.45 | 4.5 |
| Zn, mg/l | 1.288 | 0.195 | 85 |
| Cu, mg/l | 0.206 | 0.025 | 88 |
| Spec. Cond., micromhos/cm | 5377 | 529 | 90 |
| Ca, mg/l | 106 | 42 | 60 |

Hydrographs obtained at each flume indicated a total flow of 5980 l at the end of the terrace and 1574 l from the lagoon. On a volume basis, this represents a 4:1 dilution of the lagoon effluent, 75 percent of the total flow coming from the cropland terrace. Dilution obviously helped reduce the concentrations.

Dilution of the lagoon effluent in the grassed terrace makes it somewhat difficult to assess the true removal of various materials by the grassed terrace. Materials actively removed by the grassed terrace are better determined through a mass balance. Mass balance removals were calculated for selected pollution parameters and are presented in Table 2. Mass balances indicate the terrace was removing a substantial amount of most parameters, particularly $BOD_5$, COD, N, and P. Lesser removals were obtained for dissolved material as indicated by the removals of K, Na, and total solids. Lagoons act as sedimentation basins, producing an effluent low in suspended solids but high in dissolved solids.

Water quality data were not available on the incoming flow from the 2.8 ha of cropland prior to mixing with the lagoon effluent. The mass balance removals of certain parameters, particularly dissolved materials, may be slightly low due to the addition of similar materials in the cropland runoff. Cropland runoff normally contains a large fraction of dissolved materials including small amounts of organic matter and soil (Loehr 1974). Concentrations of most parameters in the cropland runoff would be small in comparison to the lagoon effluent, but would tend to increase the mass balance removals if accounted for. Overall, it appears that the terrace is doing a reasonably good job of renovation.

### Reservoir Water Quality

Surface water samples were taken from the reservoir at two locations (Fig. 1). The near-shore station was located close to the entrance of the runoff. The offshore station

### TABLE 2. MASS BALANCE REMOVALS OF POLLUTANTS IN TERRACE CHANNEL

| Parameter | Mass reduction, percent |
|---|---|
| $BOD_5$ | 89.9 |
| COD | 81.1 |
| TS | 27 |
| VS | 51 |
| Org.-N | 75 |
| $NH_3$-N | 95 |
| Total P | 78 |
| Na | 65 |
| K | 68 |
| Zn | 42 |
| Cu | 54 |

provided a comparison to observed changes at the near-shore station. It was assumed that water quality conditions observed at the offshore station were representative of reservoir water quality unaffected by runoff from the swine farm. Quality of the runoff immediately prior to entering the reservoir was not available.

Table 3 compares the changes in water quality of the reservoir and flow from the terrace. Comparison of the two reservoir stations indicates no significant change in quality resulting from the inflowing runoff. Concentrations of all parameters were further reduced between flume 2 and the reservoir. This most likely is a result of adsorption by soil and plant materials, dilution, or both.

Data from the two reservoir sampling stations indicate a minimal influence of the lagoon on the reservoir water quality, but addition of non-point runoff makes it impossible to qualify that statement absolutely. Although the terraceway does reduce any potential pollution load from the lagoon to the reservoir, important factors would include the reduction of lagoon overflow to a minimum by exclusion of all surface runoff into the lagoon and providing dilution water via the second terrace.

### TABLE 3. CHANGES IN RESERVOIR WATER QUALITY

| Parameter | Grass terrace at flume 2, concentration | Near shore station, concentration | Off shore station, concentration |
|---|---|---|---|
| COD, mg/l | 133 | 23 | 16 |
| TS, mg/l | 700 | 350 | 335 |
| VS, mg/l | 213 | 93 | 104 |
| Total P, mg/l | 7.5 | 0.57 | 0.23 |
| Na, mg/l | 20 | 16 | 11 |
| Org-N, mg/l | 5 | 1.2 | 1.7 |

### Terrace Soil Analyses

Beginning at the point of lagoon outflow, soil samples were taken at 0 m (A), 15.2 m (B), 30.5 m (C), 45.7 m (D), 61 m (E), 112 m (F), and 243.8 m (G) along the terrace channel (Fig. 1). Samples were taken to a total depth of 106.7 cm. Zinc, copper, arsenic, and phosphorus concentrations were determined for the first 30.5 cm of soil only in depth increments of 0-7.6 cm, 7.6-15.2 cm, and 15.2-30.5 cm. Sodium and nitrate were determined for these three depth increments and at the 61.0 cm, 91.4 cm, and 106.7 cm depths. Two sites upslope from the terrace which had never received effluent were sampled as a control (Stations H and I, Fig. 1). The results of Stations H and I were averaged. The soils belong to the Sharpsburg-Grundy group.

Nitrate had accumulated in the top 30 cm of soil (Station A) but decreased rapidly with terrace length (Fig. 2). Little accumulation was observed below this depth. The summer had been droughty with little rain for leaching. The largest accumulation (48 ppm) remains below levels of concern, particularly when confined to such a small area. Much of the nitrogen deposited in the lagoon is undoubtedly lost as ammonia through desorption (Koelliker and Miner 1973). This is a benefit to the terrace system in terms of nitrogen accumulation.

The increase between 50 and 150 m can be attributed to a low area in the terrace. On numerous occasions, following precipitation, water was observed standing in this area for several days. This would allow increased leaching and contribute to increased concentrations. Increases in other parameters were noticed in this area.

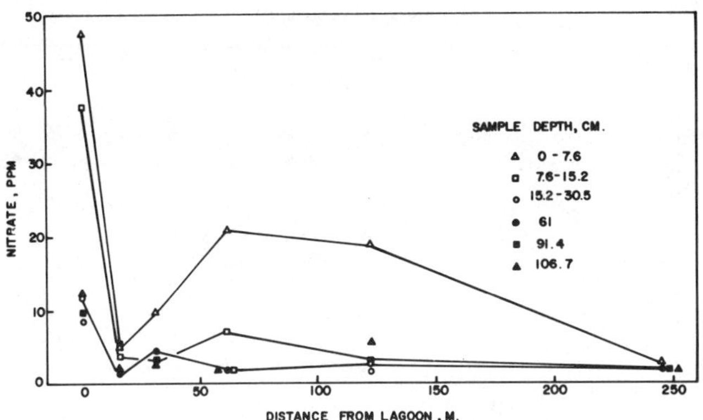

FIG. 2 Nitrate concentrations of terrace channel soil.

A definite accumulation of phosphorus exists within the top 15 cm of soil (Fig. 3). Accumulations exist throughout the terrace but are highest near the lagoon and in the low area previously mentioned. Maximum accumulations exceed 1200 ppm in these areas as compared to an average concentration of 459 ppm for the same soil depth in the control samples. Most soils possess a large capacity to fix phosphorus (Lindsay 1973) and there should be no concern for pollution problems as a result of the accumulations.

Sodium concentrations along the terrace are presented in Fig. 4. Sodium has accumulated in the channel when compared to the control values. Largest accumulations are found at Station A with gradual decreases along the channel.

Sodium is involved in soil exchange reactions and can be held at the exchange sites. Large accumulations can cause soil dispersion problems. Dispersion reportedly occurs when 10 to 20 percent of the soil cation exchange capacity (C.E.C.) is saturated with Na (Peterson et al. 1973). The

soils at the test site have a C.E.C. of 15 me/100 g (Christy and Fisher 1966). Averaging the Na concentrations for the first 60 cm of soil at Station A (point of highest concentrations) results in a value of 120 ppm Na. This represents 4.9 percent of the C.E.C. which is well below the 10 percent limit.

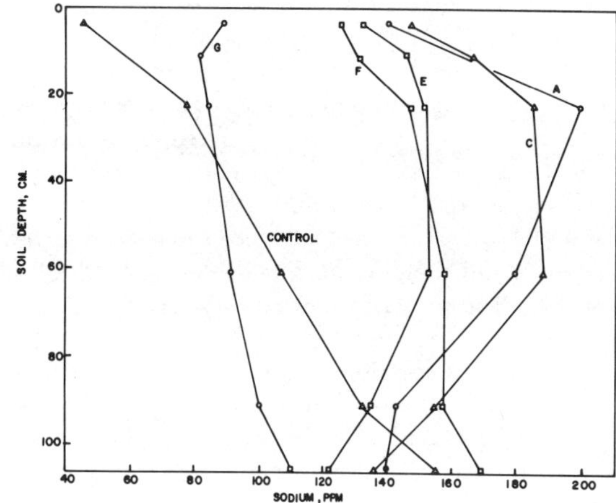

FIG. 4 Sodium concentrations of terrace channel soil.

Soil concentrations of three metals (copper, zinc, and arsenic) are presented in Fig. 5. Average concentrations over a 30 cm depth for the control samples were As-5.7 ppm; Cu-15.2 ppm; Zn-50.3 ppm. Only Zn appears to have accumulated to any degree. The accumulations are small and fall short of reported minimum toxic levels of 1 me Zn/100 g (Chapman 1966). It appears that the terrace is accumulating little of these metals and buildup should be no problem.

## SUMMARY

An anaerobic lagoon with its overflow passing over a grassed terrace was analyzed for its ability to renovate the wastes from a 200 hd swine finishing unit. The following conclusions are derived from the data.

1   The combined effects of passing the lagoon effluent over a 259 m grassed terrace and addition of dilution water

*(Continued on page 548)*

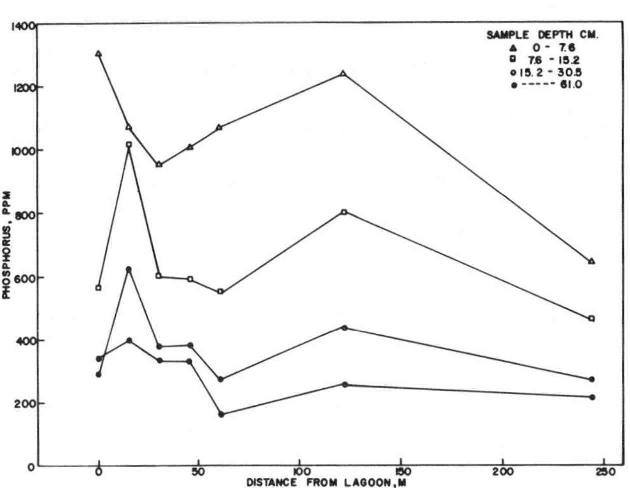

FIG. 3 Phosphorus concentrations of terrace channel soil.

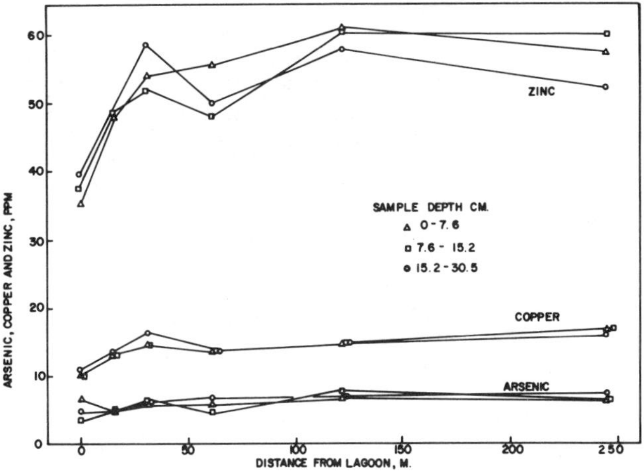

FIG. 5 Arsenic, copper, and zinc concentrations of terrace channel soil.

# Bioengineering Aspects of Anaerobic Digestion of Piggery Wastes

A. M. Robertson, G. A. Burnett, P. N. Hobson, Shiela Bousfield, R. Summers

THE management of agricultural wastes has achieved prominence as a problem area in animal production systems over the last decade. It is now widely recognized that at the planning and design stage of any enterprise provision must be made for re-use of livestock waste on the farm, while at the same time steps must be taken to minimize the impact on the environment of any handling operations. In the UK guidelines have been proposed governing the hydraulic and chemical loading of the land (O'Callaghan et al. 1971). The present situation however demands technological solutions to on-farm problems. Much of the work directed towards a solution of these problems has been on aerobic treatment systems which although capable of eliminating odors and achieving reductions in the pollutional loading of animal wastes, have the disadvantage that they cannot treat the solid particulate matter present in pig waste. Also, under aerobic conditions denitrification will take place leading to a reduction in nitrogen available as a plant nutrient in the effluent from the treatment system.

Anaerobic digestion on the other hand, has been shown to be efficient in reducing solids in waste while producing a final effluent containing plant nutrients equivalent to those in raw waste. Digestion also offers the possibility of a useful by-product, methane gas, which if produced surplus to the heating requirements of the digester could be utilized as a means of offsetting the cost of treatment.

## LABORATORY STUDIES

The work described is a continuation of the work carried out by Shaw (1971). Laboratory experiments were carried out first in 15 liter digesters and later in a 100 liter digester, all of which were loaded with waste from the below-slat channels of a growing-finishing piggery. Among factors

The authors are: A. M. ROBERTSON and G. A. BURNETT, The Scottish Farm Buildings Investigation Unit, Craibstone, Bucksburn, Aberdeen, Scotland; P. N. HOBSON, SHIELA BOUSFIELD and R. SUMMERS, Microbiology Dept., Rowett Research Institute, Bucksburn, Aberdeen, Scotland.

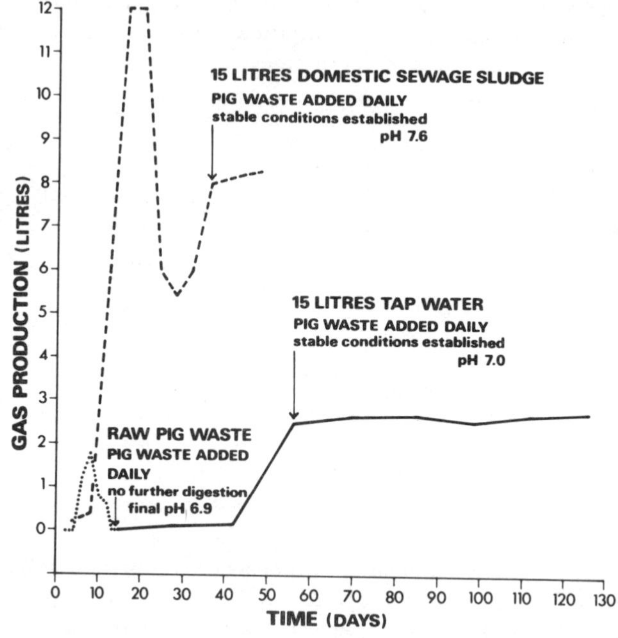



FIG. 1 Start-up procedure in small scale laboratory digesters.

investigated were start-up procedure, retention times, batch versus continuous loading and temperature.

### Start-up Procedure

Start-up procedure was investigated in relation to the establishment of stable digestion in a 15 liter digester. The three methods were (a) from raw pig waste, (b) from a digester filled with water to which was added raw pig waste and (c) from a digester filled with liquor from a domestic anaerobic digester to which was added raw pig waste. The benefits of an anaerobic inoculum is demonstrated in Fig. 1, in which the establishment of a stable digestion is shown in terms of the regular production of gas.

TABLE 1. PERFORMANCE OF 100 LITER DIGESTER AT DIFFERENT LOADING RATES.

| | 20 Day retention* | | | 10 Day retention | | | 7 Day retention | | |
|---|---|---|---|---|---|---|---|---|---|
| | Input | Output | Percent reduction | Input | Output | Percent reduction | Input | Output | Percent reduction |
| Total solids, percent | 3.2 | 1.2 | 62.5 | 3.3 | 2.1 | 36.0 | 3.66 | 2.6 | 29.0 |
| VFA, mg/liter | 2180 | 394 | 82.0 | 3697 | 274 | 92.6 | 5008 | 924 | 82.0 |
| $NH_3$, mg/liter | 890 | 1024 | Nil | 1237 | 1044 | 15.6 | 2156 | 2118 | 1.7 |
| $BOD_5$, mg/liter | 10750 | 1520 | 86.0 | 14672 | 2567 | 82.5 | 18917 | 2787 | 85.0 |
| COD, mg/liter | 30250 | 1676 | 94.0 | 46920 | 21956 | 53.2 | 79980 | 41910 | 47.6 |

*Diluted waste
VFA Volatile fatty acid
$BOD_5$ 5 Day Biochemical Oxygen Demand
COD Chemical Oxygen Demand

### TABLE 2. PERFORMANCE OF A 100 LITER DIGESTER OPERATING AT 30 DEG C AND 35 DEG C (10 DAY RETENTION TIME).

|  | 30 deg C | | 35 deg C | |
|  | Input | Output | Input | Output |
| --- | --- | --- | --- | --- |
| Total solids, percent | 3.35 | 2.19 | 3.3 | 2.1 |
| VFA, (mg/liter) | 4695 | 715 | 3697 | 274 |
| $NH_3$, mg/liter | 1734 | 1952 | 1237 | 1044 |
| $BOD_5$ mg/liter | 9550 | 2200 | 14672 | 2567 |
| COD, mg/liter | 41305 | 20971 | 46920 | 21956 |
| Gas production, liters/day | --- | 92.2 | --- | 98 |

### Retention Times

Table 1 shows the results obtained from the 100 liter digester operating at retention times of 20, 10 and 7 days. Each result is the mean of some months' stable digestion except for the 20 day retention which was over four weeks during start-up.

### Batch Versus Continuous Loading

The 15 liter digesters were loaded once daily. Maximum waste breakdown was obtained with an input of approximately 3 percent total solids at a retention time of 14 days (Hobson and Shaw 1973), but gas production and waste degradation were worse than those obtained in the 100 liter digester loaded every five minutes. As can be seen (Table 1), retention time can be reduced to ten or seven days without seriously increasing the levels of total solids, volatile fatty acids, ammonia, $BOD_5$ or COD in the effluent. Maximum permissible solids concentrations have not yet been identified for digester inputs but operation at a five day retention time, while giving a further loss in efficiency, has not resulted in digester failure. At a ten day retention time input, volatile fatty acids up to 8000 mg/liter and ammonia up to 3200 mg/liter did not impair stability. During continuous loading sudden changes in temperature caused large temporary decreases in gas production but stable digestion can be obtained at temperatures other than the accepted optimum, 35 C (Table 2).

### FIELD SCALE STUDIES

### Start-up Procedure

Fig. 2 gives details of the field scale anaerobic digester. After initial filling with liquor from a domestic digester, loading started with the same waste as used in the labora-tory digester; loading took place at 100 liters/day and was increased by 50 liters/day until 450 liters/day was reached.

### Waste Input and Effluent Output Characteristics

Table 3 gives mean values of total solids, volatile fatty acids, ammonia, $BOD_5$ and COD for the digester inputs and outputs, together with gas production over a period of several months' operation at retention times of 30, 20 and 15 days, the latter being the shortest retention time investi-gated so far. When the ten day retention time is investi-gated, the total solids input will be increased from the present value of approximately 5 percent.

### Gas Production

There is a basic gas production (per volume of liquid) from the volatile fatty acid in the input but the main gas production in both the 100 liter and field scale digesters (Tables 1 and 3) is influenced by, and proportional to, the concentration of organic solids in the input. The volume of gas produced corresponded closely to the calculated values predicted from analysis of the input, the degree of break-down achieved and theoretical conversions of the compo-nents to methane and carbon dioxide. Chemical analysis of the gas produced gave mean values in terms of quality of 65 percent methane and 35 percent carbon dioxide. Some-times the gas contained 1 percent nitrogen with traces of hydrogen sulphide and ammonia. The mean calorific value of the gas was 6.7 kW hr/m$^3$.

### Heating Requirements

The theoretical heating requirements of the digester ves-sel are made up of two components. Heat is required to maintain the digester contents at 35 C against heat loss under a wide range of climatic conditions and the tempera-ture of the input must be brought up to the digester operat-ing temperature. Fig. 3 shows the relationship between ambient temperature, the daily heat requirement of the digester and the mean daily energy available as gas pro-duced at retention times of 30 and 20 days. At 30 days the gross energy produced exceeds the heat requirement of the digester above an ambient temperature of 7.2 C, while at a 20 day retention time gas production provides sufficient energy for all the heat requirements of the digester. Allow-ing for a theoretical loss in energy through boiler and heat exchange unit of the order of 30 percent however, changes

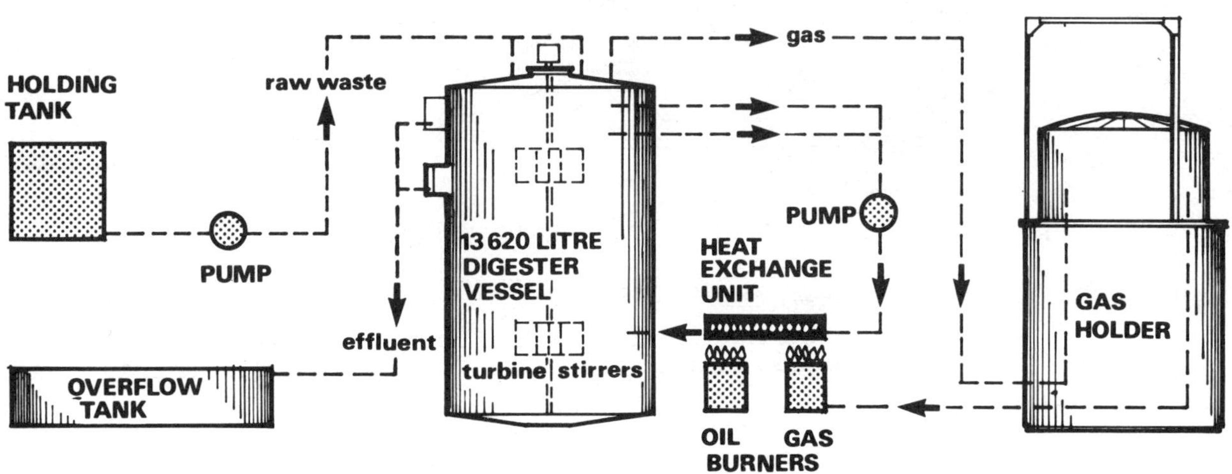


FIG. 2 Field scale anaerobic digester.

## TABLE 3. MEAN VALUES OF DIGESTER INPUT AND OUTPUT CHARACTERISTICS.

| | 30 Day retention | | | 20 Day retention | | | 15 Day retention | | |
|---|---|---|---|---|---|---|---|---|---|
| | Input | Output | Percent reduction | Input | Output | Percent reduction | Input | Output | Percent reduction |
| Total solids, percent | 4.3 | 3.1 | 28.0 | 5.3 | 3.3 | 37.7 | 4.9 | 2.7 | 44.8 |
| VFA, mg/liter | 4332 | 1354 | 68.7 | 5820 | 406 | 93.0 | 6560 | 1240 | 81.1 |
| $NH_3$, mg/liter | 2329 | 2402 | Nil | 2217 | 2263 | Nil | 2216 | 2173 | 1.9 |
| $BOD_5$, mg/liter | 17500 | 1800 | 89.7 | 28500 | 1700 | 94.0 | 22500 | 3666 | 83.7 |
| COD, mg/liter | 77000 | 36792 | 52.2 | 62933 | 43867 | 30.3 | 67526 | 31680 | 53.1 |
| Gas production, $m^3$/day | | 7.6 | | | 10.2 | | | 11.4 | |

the picture to such an extent that at 30 and 20 day retention times there is insufficient energy available to maintain the digester temperature without relying on supplementary heating as has been found in practice.

### Additional Energy Requirements

Additional power will be required for mixing the digester contents, circulating digester liquor through the heat exchange unit, loading on an hourly basis and mixing the input. This additional energy amounted to 75 kW hr/day of which 36 kW hr/day was accounted for by the pump circulating the digester contents through the heat exchanger.

### DISCUSSION

### Treatment

The performance of the field scale digester in terms of a waste treatment system is similar to that of the laboratory units (Tables 1 and 3). Laboratory results indicate that retention times can be reduced to seven to ten days without depressing the degree of treatment achieved. The effluent produced has good settling characteristics (one hour sludge volume of 20 percent) and although it is not completely odor-free, it is less objectionable on a subjective basis than untreated waste. The settled liquor from the digester is sim-

ilar in characteristics to lagoon supernatant and could be readily treated aerobically to give an effluent similar to that described for a treatment system consisting of a lagoon and oxidation ditch (Robertson et al. 1974). Laboratory studies have confirmed that further treatment of the liquid can be obtained by aeration. Alternatively digester effluent could be spread on land by conventional means.

### Gas Production

Gas production from the field scale digester is similar in volume and quality to that produced in laboratory experiments (0.40 $m^3$/kg organic matter added). No attempt has been made to remove impurities from the gas which burns readily in a normal domestic type gas boiler.

At the detention times (30, 20 and 15 days) and solids input rates (5 percent total solids) studied so far the calorific value of the gas produced from the pig waste is insufficient to maintain the required temperature in the digester under climatic conditions experienced in north-east Scotland. Although operation at 30 C would decrease the gas required to heat the digester, Table 2 shows that this might be offset by a small decrease in gas production at this temperature. Increasing input solids levels and reducing retention times appear to be better ways of increasing gas production.

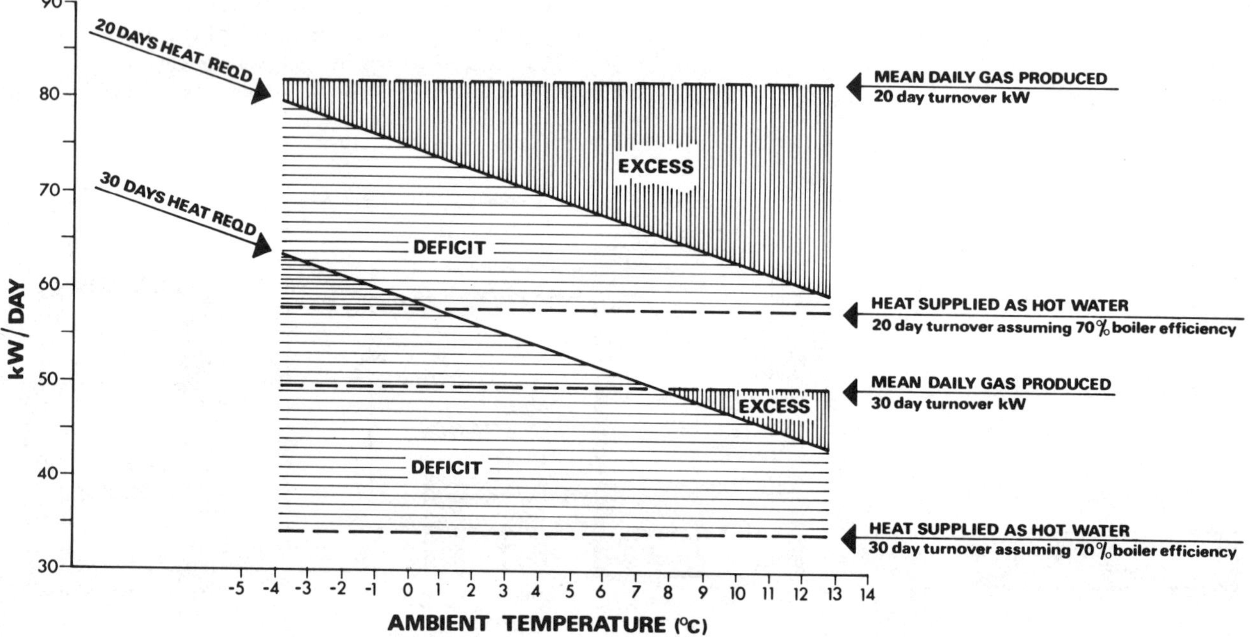


FIG. 3 Relationship between ambient temperature daily heat requirements and gas production.

## Energy Requirements

It is apparent that in order to produce surplus gas to offset the cost of treatment, economies in energy inputs must be made. At a 30 day retention time and ambient temperature of 7 C 32 percent of the heating requirements go towards heating the input up to working temperature and 68 percent goes towards balancing heat loss, while at a 20 day retention time the values are 48 percent and 52 percent, respectively. The effluent from the digester is discharged at 35 C; if this material could be used to pre-heat the input, economies could be made. As the loading rate increases this aspect becomes more important. At present the digester is insulated with 50 mm fiberglass (U-value 0.67 W/m$^2$ C). In the climatic conditions experienced it has been calculated that a minimum of the equivalent of 100 mm fiberglass (U-value 0.34 W/m$^2$ C) would be justified to further reduce heat loss. Apart from heating, the other single source of demand for energy is in the heat exchange circulation system. At present digester liquor is pumped through the heat exchanger regardless of whether the boilers are on demand. Initially this mode of operation was used as a method of agitating the digester contents but its only purpose now is to prevent possible caking of digester liquor in the heat exchanger tubes, thus reducing heat transfer efficiency. Further trials will show whether intermittent circulation is possible. In the meantime, however, this mode of operation is a source of heat loss in that the heat exchange unit works in reverse when the boilers are not in operation.

## Stability

Although some oscillations occur in any mixed-culture biological system, predictable performance of a digester depends upon an input of constant volume and solids concentration. To achieve this an intermittent stirrer system, coupled with the inlet pump was fitted to the feed tank. This arrangement also prevented pump blockages caused by settled solids. During experimental modifications the digester was left for up to six weeks without heating or loading but digestion started as soon as loading was resumed. Small leakages of air resulted in the production of incombustible gas containing nitrogen; in the long term oxygen can also contaminate the gas if leakage continues. Digestion will start again as soon as the leak is blocked. The digester contents have also been removed, stored in the open at ambient temperatures and returned to the digester without destroying the bacterial flora. The above situation would tend to indicate that the system is not unduly sensitive to management procedures.

## Economics

The present results show that anaerobic digestion of piggery waste does significantly decrease the pollutional loading of pig waste and is technically feasible on a large scale. The results also show that further studies are required to find ways of reducing energy inputs while at the same time increasing the gas produced as a means of offsetting treatment costs.

Although it is not possible to give capital costs for the existing digester because of the experimental features included in its design, extrapolation from the field of domestic waste treatment suggests that a digester of 45 m$^3$ in capacity, together with ancillary equipment would cost approximately $24 000. Writing off the equipment over ten

## TABLE 4. RELATIVE OPERATING COSTS OF TREATMENT SYSTEMS.

| Treatment system | Cost index | Order |
|---|---|---|
| Below house oxidation ditches | 243 | 4 |
| Surface aerator | 77 | 1 |
| Anaerobic lagoon/oxidation ditch | 100 | 2 |
| Anaerobic digester | 167 | 3 |

years and charging interest on half the capital would give an annual cost of $4.58/pig. At a ten day retention time running costs would amount to $3.25/pig place per year; based on the running cost of the existing digester, giving a total annual cost of $7.83. Apart from any treatment achieved as a result of anaerobic digestion, which is difficult to value, the process will (as has been shown) produce methane gas. In monetary terms the value of the surplus gas would be $4.75 (based on estimates made using information from the existing digester and valuing the gas in terms of current fuel oil prices of 13 cents/liter).

The digester effluent also has a value as a source of plant nutrients replacing inorganic fertilizer. At current fertilizer costs this is equivalent to $3.16/pig giving a total value for the digester outputs of $7.90. The validity of considering the value of plant nutrients in the effluent as a digester output could be questioned as the raw waste contains these nutrients before treatment. At current energy and fertilizer costs a digester for 1000 growing-finishing pigs would just recover the cost of treatment. If the effluent is to be returned to the land the spreading cost must also be taken into account; this could amount to a further $2.33/pig annum. This spreading charge applies to all systems where the aim is to return treated wastes to the land.

It is difficult to make direct comparisons between different methods of treatment because of differences in the degree of treatment achieved and also in the mode of operation adopted. Table 4 gives details of the relative operating costs of different treatment systems at the authors' unit in Scotland (anaerobic lagoon/oxidation ditch≡ 100).

The figures quoted do not include the value of recycling any of the by-products of the treatment system, a factor which may well change the ranking given to each method of treatment.

## Alternative Sources of Feed Stock

So far only pig waste has been used as feed stock for the field scale digester. In the laboratory, however, tests are being carried out to evaluate other materials, particularly those of high carbohydrate content to balance the high nitrogen content of fecal wastes. In particular straw, potato haulms, silage effluent and wastes from vegetable canning enterprises offer a ready source of feed stock.

## SUMMARY

Anaerobic digestion of pig waste is technically feasible on a large scale. Significant reductions in total solids (40 percent), volatile fatty acids (90 percent), BOD$_5$ (90 percent) and COD (40 percent) in the effluent and odor have been achieved. Further studies are required to improve the economic viability of anaerobic digestion. Several possible ways of achieving positive economic balance have been outlined. The most promising methods appear to be reducing retention times, increasing solids concentration per unit volume of input, reducing energy requirements through improved insulation, recovery of heat from the digester effluent and an examination of alternative feed stock.

## References

1   Hobson, P. N. and B. G. Shaw. 1973. The anaerobic digestion of waste from an intensive pig unit. Water Res. 7:437.
2   O'Callaghan, J. R., V. A. Dodd and K. A. Pollock. 1971. Land spreading of manure from animal production units. J. Agric. Engng. Res., 16:280-300.
3   Robertson, A. M., J. J. Clark and S. H. Baxter. 1974. The treatment of pig waste in a two stage anaerobic/aerobic system. C.I.G.R. 2nd Section 8th International Congress of Agricultural Engineering. Flevohof, The Netherlands.
4   Shaw, B. G. 1971. A practical and bacteriological study of the anaerobic digestion of waste from an intensive pig unit. PhD thesis. University of Aberdeen.

---

## Lagoon-Grass Terrace System

*(Continued from page 543)*

from cropland runoff resulted in reductions of over 80 percent in the concentrations of most pollution parameters. Reductions in COD, $BOD_5$, total P, Na, and K exceeded 90 percent.

2   Reductions of pollutants on a mass balance basis were lower than reductions in concentrations. A combined average mass reduction of 86 percent was achieved for COD, $BOD_5$, total P, and $NH_3$-N.

3   The performance of the grassed terrace was aided by the desorption of ammonia from the lagoon and exclusion of all surface runoff to the lagoon.

4   The terrace did contribute significantly to the protection of the reservoir.

5   No significant accumulation of chemicals in the terrace soil was observed except for phosphorus which had accumulated to a maximum of 1300 ppm in the surface 8 cm of soil. Minimum accumulations were observed for nitrate, sodium, and zinc.

## References

1   Assoc. Office of Analytical Chemists. 1965. Official methods of analysis, 10th ed.
2   Bremner, J. M. 1965. Inorganic forms of nitrogen, p. 1179-1237. In: C. A. Black (ed.), Methods of soil analysis. Part 1. American Society of Agronomy, Madison, Wis.
3   Chapman, H. D. 1966. Zinc, p. 484-499. In: H. D. Chapman (ed.), Diagnostic criteria for plants and soils. University of California, Division of Agricultural Sciences, Riverside, Calif.
4   Christy, C. M. and T. R. Fisher. 1966. Fertility levels of Missouri soils. University of Missouri Extension Publication C884. University of Missouri, Columbia, Mo. 37 p.
5   Dean, J. A. 1960. Flame photometry. McGraw-Hill. New York.
6   Environmental Protection Agency. 1971. Methods for chemical analysis of water and wastes. National Evnironmental Research Center, Cincinnati, Ohio.
7   Koelliker, J. K. and J. R. Miner. 1973. Desorption of ammonia from anaerobic lagoons. TRANSACTIONS of the ASAE 16(1):148-151.
8   Liederman, D., J. E. Bowen and O. I. Milner. 1959. Determination of arsenic in petroleum stocks and catalysts by evolution as arsine. Anal. Chem. 31:2052-2055.
9   Lindsay, W. L. 1973. Inorganic reactions of sewage wastes with soil. p. 91-96. In: Recycling municipal sludges and effluents on land. National Association of State Universities and Land Grant Colleges, Washington, D.C.
10   Loehr, R. C. 1974. Characteristic and comparative magnitude on non-point sources. Jour. WPCF 46(8):1849-1872.
11   Peterson, M., et al. 1973. A guide to planning and designing effluent irrigation disposal systems in Missouri. University of Missouri Extension Publication MP 337. University of Missouri, Columbia, Missouri. 90 p.
12   Pickett, E. E. and S. R. Koirtyohann. 1969. Emission flame photometry—a new look at an old method. Anal. Chem. 41(14):28A-42A.
13   Slavin, W. 1968. Atomic absorption spectroscopy. Wiley-Interscience Publications. New York.
14   Standard methods for the examination of water and waste-water. 1971. 13th ed. American Public Health Assoc., Washington, D.C.

# Simulation of Fundamental Anaerobic Lagoon Kinetics


D. T. Hill, C. L. Barth

ASSOC. MEMBER   MEMBER
ASAE            ASAE


THE work described in this report is an extension of previously published research. Approximately 2 years ago, studies were undertaken to aid in developing a mathematical model that could be used to simulate the anaerobic lagoon on the digital computer. The first report of this effort was made in March 1974 at the Cornell Agricultural Waste Management Conference. Since that time, additional laboratory work and modification of the mathematical model has been performed to give the best fit for the laboratory data.

The task of modeling the anaerobic lagoon has turned out to be considerably more challenging than had been anticipated. Once thought of as simple and technologically relaxed, the anaerobic lagoon upon closer examination is quite stubborn in yielding to any systematic analysis. Indeed, the methods and particularly the assumptions made in classic sanitary engineering to model continuous and rapid treatment processes such as activated sludge or trickling filters may not adequately describe the lightly loaded — long detention time biological processes such as anaerobic lagoons.

## BACKGROUND

The basic development of the mathematical model was reported by Hill and Barth (1974). The primary objective of the research is to develop a comprehensive computer model that will predict anaerobic lagoon performance based on volatile matter (VM) reduction and the concentration of volatile organic acids (VOA) in the lagoon. The VM reduction is important because it is the primary reason for installing anaerobic lagoons on farms. The VOA concentration is significant due to its inhibitory nature should it become high (greater than approximately 3000 to 4000 mg/l as acetic acid ($CH_3COOH$)) and because the methanogenic bacteria utilize VOA's as the substrate for methane formation.

The basic model was developed using conventional sanitary engineering techniques. These included chemical reactor theory and estimation of the chemical composition of the raw wastes, stoichiometry of the biological conversions, and dynamics and kinetics of microbial growth. Mathematical modeling interfaces all these fundamental characteristics and through computer simulation can provide basic understanding of the overall process.

The estimation of the kinetic parameters was obtained from the literature. These kinetic parameters were acknowledged by the authors in most cases to be within an order-of-magnitude. The development of the mathematical model was performed using a laboratory scale physical model at 10 C. In order to obtain the best fit, some of these parameters were modified slightly as reported by Hill and Barth (1974). The next stage of development came by utilizing the data of a 25 C laboratory scale physical simulation and comparing the results with the output of the mathematical model for the same temperature and organic loading rates. Table 1 summarizes the characteristics of each laboratory run. The theoretical detention time was 200 days. This varied slightly in both runs.

TABLE 1. PARAMETERS FOR LABORATORY SIMULATION RUNS

| Reactor | Detention Time, day | Temp, deg C | Feeding frequency | Feed rate Kg VS $m^3$ - day |
|---|---|---|---|---|
| | | | Run 1 | |
| 1 | 193 | 10 | once weekly | 0.0389 |
| 2 | 193 | 10 | once weekly | 0.0778 |
| 3 | 193 | 10 | once weekly | 0.1556 |
| 4 | 193 | 10 | once weekly | 0.3112 |
| 5 | 193 | 10 | once weekly | 0.6224 |
| 6 | 193 | 10 | once weekly | 1.2448 |
| 7 | 193 | 10 | twice weekly | 2.4896 |
| 8 | 193 | 10 | twice weekly | 4.9792 |
| | | | Run 2 | |
| 1 | 213 | 25 | once weekly | 0.0636 |
| 2 | 213 | 25 | once weekly | 0.0884 |
| 3 | 213 | 25 | once weekly | 0.1490 |
| 4 | 213 | 25 | once weekly | 0.2911 |
| 5 | 213 | 25 | once weekly | 0.5115 |
| 6 | 213 | 25 | once weekly | 1.0015 |
| 7 | 213 | 25 | twice weekly | 1.7680 |
| 8 | 213 | 25 | twice weekly | 4.8132 |

## UPDATING THE MODEL

The original model was used to simulate the data obtained for the laboratory run at 25 C. Fig. 1, repre-

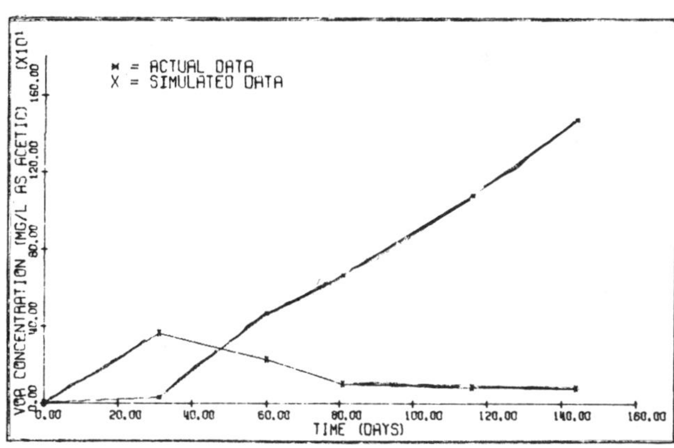


FIG. 1 Time variation of VOA concentration for Reactor 5, original model simulating Run 2.


This publication has been approved and designated as No. 1246 by the South Carolina Agricultural Experiment Station.

The authors are: D. T. HILL, Assistant Agricultural Engineer, C. L. BARTH, Associate Professor, Agricultural Engineering Dept., Clemson University, Clemson, S. C.

senting the time variation of the actual and simulated values of VOA for reactor 5, is typical of the response for the entire loading range. The trend in actual VOA concentration is not represented by the simulated data. Table 2 shows the maximum error in VOA concentration between the simulated and actual data. The largest error occurred at the end of the run for all reactors and the error increased as the loading rate increased. The agreement between VM in the reactors at the end of the run is within an order of magnitude.

With these results for the original model, an attempt was made to update the model to give the best possible fit for both Run 1 (10 C) and Run 2 (25 C). This necessitated the adjustment of three kinematic parameters. However, the values of these parameters that gave the best fit for both runs were not grossly changed from their previous values.

The yield coefficient for the acid formers was changed from 0.2 to 0.3 and the yield coefficient for the methane formers was adjusted to 0.01 from the previous value of 0.06. The final adjustment was made on the specific death coefficient for the methane formers. The original value of 0.02 was replaced with a slightly higher value of 0.04.

It is worth pointing out that the model can be forced to simulate either Run 1 or Run 2 extremely well, if the mathematical parameters are adjusted independently for each run. That is not the objective here. It is desired that the computer model simulate to within an order-of-magnitude for all conditions; not independently for one condition. With these objectives in mind, the updating procedure was completed with the adjustment of the three parameters mentioned.

### RESULTS OF THE UPDATING

The updated computer model provides the best fit for both the 10 C and 25 C runs with no changes in parameters except the organic loading rates and temperature. Table 3 provides a summary of the VOA data for all eight loading rates for Run 1 (10 C) using the updated computer model. The trend in VOA, that is, the steadily increasing concentration, is simulated quite well. The results of the four lightest loaded reactors (1, 2, 3, and 4) and the two heaviest loaded (7 and 8) are quite close in the quantitative prediction as well as the trend. In dealing with biological reactors, prediction of the general trends and quantitatively to within a order-of-magnitude is considered a satisfactory effort, particularly when dealing in uncharted areas of simulation. The results of reactors 5 and 6, the mid-range organic loading rates

$$(0.6224 \text{ and } 1.2448 \frac{\text{Kg VS}}{\text{m}^3\text{-day}}),$$

appear to indicate a transition range in that the actual VOA concentration jumped from a plateau of approximately 1600 mg/l in reactor 5 to a level of about 6000 mg/l in reactor 6. For a doubling in loading rate, the mathematical model did not predict a quadrupling of the VOA concentration. These kinds of results are quite common in biological processes.

The agreement between volatile matter in the reactor at the end of the run is much better than that simulated by the original model as reported by Hill and Barth (1974). Table 4 shows the volatile matter data for Run 1.

**TABLE 2. SUMMARY OF RESULTS OF ORIGINAL MODEL SIMULATING RUN 2**

| Reactor | Percent error in VOA concentration | Time, day | VM in reactor at end of test, g | |
|---|---|---|---|---|
| | | | Actual | Simulated |
| 1 | 36.3 | 144 | 17.2 | 27.4 |
| 2 | 56.0 | 144 | 22.5 | 35.4 |
| 3 | 81.9 | 144 | 28.5 | 70.9 |
| 4 | 82.9 | 144 | 49.7 | 111.6 |
| 5 | 94.5 | 144 | 95.5 | 195.1 |
| 6 | 96.4 | 144 | 130.0 | 384.0 |
| 7 | 97.8 | 144 | 265.0 | 643.4 |
| 8 | 99.3 | 144 | 988.0 | 1421.5 |

**TABLE 3. COMPARISON OF VOA CONCENTRATION FOR RUN 1 (10 C) USING UPDATED COMPUTER MODEL**

VOA concentration, mg/l
Reactor

| Day | 1 Actual | 1 Simulated | 2 Actual | 2 Simulated | 3 Actual | 3 Simulated | 4 Actual | 4 Simulated |
|---|---|---|---|---|---|---|---|---|
| 0 | 0 | 0 | 0 | 0 | 0 | 0 | 0 | 0 |
| 13 | 12 | 6 | 12 | 11 | 180 | 22 | 156 | 37 |
| 34 | 60 | 21 | 48 | 41 | 72 | 84 | 144 | 147 |
| 62 | 96 | 53 | 120 | 120 | 180 | 297 | 276 | 609 |
| 83 | 36 | 90 | 48 | 232 | 216 | 603 | 300 | 1173 |
| 104 | 84 | 142 | 120 | 379 | 240 | 885 | 480 | 1629 |
| 125 | 384 | 207 | 588 | 521 | 648 | 1150 | 804 | 2094 |
| 146 | 516 | 277 | 732 | 658 | 852 | 1424 | 1272 | 2567 |
| 167 | 588 | 346 | 936 | 795 | 1092 | 1699 | 1440 | 3046 |
| 188 | 672 | 413 | 972 | 932 | 1188 | 1974 | 1976 | 3525 |

| Day | 5 Actual | 5 Simulated | 6 Actual | 6 Simulated | 7 Actual | 7 Simulated | 8 Actual | 8 Simulated |
|---|---|---|---|---|---|---|---|---|
| 0 | 0 | 0 | 0 | 0 | 0 | 0 | 0 | 0 |
| 13 | 252 | 69 | 300 | 130 | 324 | 241 | 372 | 333 |
| 34 | 468 | 282 | 576 | 500 | 1704 | 826 | 2616 | 1061 |
| 62 | 1200 | 1322 | 1800 | 2287 | 3612 | 3106 | 5148 | 3447 |
| 83 | 960 | 2396 | 4224 | 4786 | 7644 | 6649 | 13068 | 7000 |
| 104 | 996 | 3340 | 5160 | 6500 | 9984 | 10953 | 14340 | 11200 |
| 125 | 1608 | 4269 | 4764 | 8490 | 9756 | 15085 | — | — |
| 146 | 1632 | 5220 | 5232 | 10400 | 10272 | 18721 | — | — |
| 167 | 1560 | 6183 | 5952 | 12360 | 11004 | 21845 | — | — |
| 188 | 1632 | 7153 | 6500 | 14300 | 12000 | 24413 | — | — |



MANAGING LIVESTOCK WASTES

The updated computer model provides a better fit for the overall range of loading rates than did the original model. Agreement in VM data is important because VM degradation is the primary objective of the anaerobic lagoon treatment of farm wastes. The data indicates the model would be valid for simulating solids destruction at low temperatures.

TABLE 4. VOLATILE MATTER AT END OF TEST IN EACH REACTOR FOR RUN 1 (10 C)

| Reactor | Input VM for test, g | Actual VM in reactor, g | Simulated VM in reactor, g |
|---|---|---|---|
| 1 | 36.7 | 29.5 | 21.5 |
| 2 | 73.5 | 45.9 | 41.5 |
| 3 | 147.0 | 66.4 | 81.1 |
| 4 | 257.0 | 103.9 | 140.5 |
| 5 | 514.0 | 201.6 | 280.0 |
| 6 | 1028.0 | 491.9 | 562.4 |
| 7 | 2056.0 | 1383.2 | 1175.8 |
| 8 | 2938.0 | — | 1800.1 |

The simulation of Run 2 did not provide the expected results. Instead of the characteristic rise in VOA followed by a sharp fall in concentration as the methanogenic bacteria population grew, a steady increase in VOA concentration in the actual data was noted. However, much was learned about some basic deficiencies in the model by comparing simulated and actual data for Run 2. These deficiencies could not be detected by the use of Run 1 alone. Table 5 shows VOA concentration for all loading rates for Run 2 (25 C) using the updated computer model. Fig. 2, a graph of the time variation of VOA concentration for Reactor 5, is typical of the response of the computer simulation and the actual VOA concentration in the first 7 reactors. The simulated VOA data for reactor 8 (see Table 5) shows that the concentration of VOA was sufficiently high to inhibit the methane formers and therefore no stable VOA concentration was attained in the simulated data.

The agreement between simulated and actual VM remaining in the reactor at the end of the test for Run 2 is generally good. The largest percentage error occurred for reactor 6, 54.6 percent. Table 6 gives the actual and simulated VM data for the end of the test for Run 2. Also included is the input volatile matter for purposes of illustration of the magnitude of the reduction.

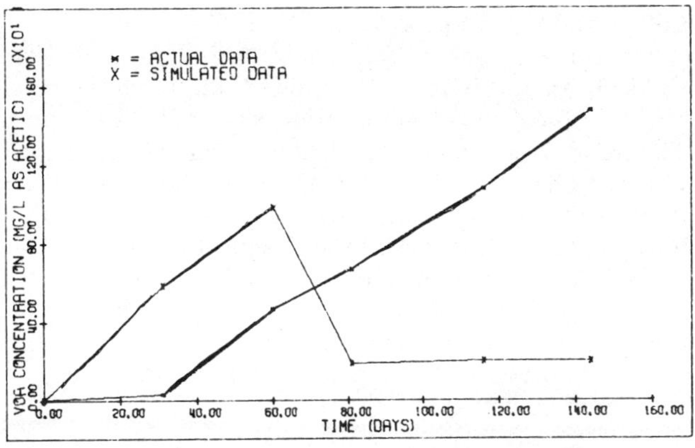


FIG. 2 Time variation of VOA concentration for Reactor 5, updated model simulating Run 2.

TABLE 6. VOLATILE MATTER AT END OF TEST IN EACH REACTOR FOR RUN 2 (25 C)

| Reactor | Input VM for test, g | Actual VM in reactor, g | Simulated VM in reactor, g |
|---|---|---|---|
| 1 | 35.8 | 17.22 | 15.5 |
| 2 | 50.4 | 22.5 | 19.9 |
| 3 | 84.6 | 28.5 | 38.8 |
| 4 | 144.7 | 49.7 | 59.9 |
| 5 | 254.3 | 95.5 | 95.4 |
| 6 | 497.9 | 130.0 | 201.4 |
| 7 | 833.6 | 265.0 | 336.0 |
| 8 | 2073.2 | 988.0 | 806.5 |

## DISCUSSION

The results of the simulation by the original model of Run 2 (25 C) data provides some insight into the deficiencies of the original model. After much experimentation with the model, changing kinetic parameters and basic equations, one must conclude that some serious deficiencies exist in the model due to assumptions made in its development. The basic fault lies in the assumption of a continuous flow reactor when in reality six of the reactors (1 through 6) were fed once weekly and the remaining two reactors were fed twice weekly. The objective here, however, is to obtain a working model upon which to base further refinements. The introduction of batch feeding into the mathematical model requires much additional programing of the computer and is not

TABLE 5. COMPARISON OF VOA CONCENTRATION FOR RUN 2 (25 C) USING UPDATED COMPUTER MODEL

VOA concentration, mg/l
Reactor

| Day | 1 Actual | 1 Simulated | 2 Actual | 2 Simulated | 3 Actual | 3 Simulated | 4 Actual | 4 Simulated |
|---|---|---|---|---|---|---|---|---|
| 0 | 0 | 0 | 0 | 0 | 0 | 0 | 0 | 0 |
| 31 | 24 | 52 | 24 | 69 | 36 | 160 | 48 | 287 |
| 60 | 60 | 197 | 48 | 260 | 84 | 513 | 120 | 729 |
| 81 | 156 | 325 | 60 | 400 | 84 | 519 | 252 | 426 |
| 116 | 140 | 470 | 216 | 434 | 372 | 172 | 360 | 186 |
| 144 | 168 | 276 | 216 | 179 | 432 | 191 | 444 | 200 |

| Day | 5 Actual | 5 Simulated | 6 Actual | 6 Simulated | 7 Actual | 7 Simulated | 8 Actual | 8 Simulated |
|---|---|---|---|---|---|---|---|---|
| 0 | 0 | 0 | 0 | 0 | 0 | 0 | 0 | 0 |
| 31 | 30 | 581 | 144 | 1245 | 300 | 2133 | 1452 | 4007 |
| 60 | 468 | 983 | 984 | 1684 | 2772 | 4340 | 5868 | 12333 |
| 81 | 668 | 185 | 1608 | 170 | 3528 | 4090 | 6360 | 19983 |
| 116 | 1080 | 200 | 2784 | 200 | 3984 | 170 | 12840 | 30334 |
| 144 | 1475 | 200 | 2268 | 200 | 3636 | 193 | 11460 | 42427 |

a "cut and dried" process.

The general trend of VOA concentration for Run 1 (10 C) has been approximated quite well, although the magnitude of the concentration varies between loading rates. The trend of ever increasing VOA concentration (Figure 1) is not characteristic of biological reactions except for the case of extremely depressed temperatures which slow biological growth as well as the chemical reactions of the enzymes of the biological floc. As evidenced by both the actual and simulated data, this is undoubtedly the case here. The rate limiting mechanism of the reactions is the severely depressed growth rate of the organisms and their associated chemical reactions.

The general trend in actual VOA concentration for Run 2 (25 C, Figure 2) does not exhibit the classic rise in VOA concentration followed by a reduction, (Andrews, in press) which is characteristic of biological reactions. The simulated data for reactors 1 through 7 shows this general trend which is associated with biological reactions. In this run, as in Run 1, a continuous rise in actual VOA concentration is noted (Table 5). This, however, can not be caused by depressed temperatures, suggesting that the rate limiting mechanism for anaerobic lagoons is not biological in nature but is probably physical or chemical in nature. This also suggests that another assumption made during development of the mathematical model is not valid. The assumption of complete mixing was used in development of the model and was thought to be justified by sufficient mixing when the reactors were loaded. After aproximately 100 days of testing, a considerable portion of sludge filled the contents of the heavier loaded reactors. This, coupled with the general trend of increasing VOA concentration throughout the entire test period suggests that the rate limiting mechanism might be diffusion. Most VM fed into the reactors was insoluable and remained in the sludge portion. Concentration of VOA in the sludge was high, thus inhibiting growth and the conversion kinetics of the bacterial floc. Since the removal of VOA from the sludge portion was dependent on diffusion out into the bulk of the reactor, the trend in VOA production in the liquid would follow the same pattern.

The suspicion of high VOA concentration in the sludge was verified by tests performed at the end of Run 2. The lowest level of 3000 mg/1 of VOA in the sludge occurred in reactor 1. All other reactors had levels between 3000 and 9000 mg/1 (Reactor 4). These levels are considerably above the levels considered by Buswell (1939), Lawrence (1971), and Alexander (1961) to be inhibitory to microbial cultures.

## CONCLUSION

Basic unknowns in the original mathematical model developed for data taken at 10 C have surfaced when using the original model to simulate data taken at 25 C. This does not suggest that the actual data taken for the reactors for either run is not characteristic of lagoons operating at these selected environmental conditions. The fact that the mathematical model depicts quite well the actual conditions in reactors at 10 C and then does not perform to the same degree for reactors at 25 C does not invalidate the computer simulation, for the model considers only what has been built into it, namely biological factors. For temperatures at 25 C, the microbial growth and enzyme kinetics are not depressed and should show exactly what they did in the simulation, that VOA concentration is a biological function. However, the actual case had phenomenon acting on it that was not considered in the modeling, probably diffusion, and the data of this study indicates the rate limiting mechanism perhaps is not biological in nature. Further study of this phenomenon should provide more definitive conclusions later.

Other possible deficiencies include: (a) the assumption of continuous flow in the model and (b) the assumption of complete mixing for mathematical simplicity. These are justifiable assumptions made in modeling high rate-short detention time biological processes. Their use in long detention time-lightly loaded and infrequently loaded biological processes does not appear to be justified.

**References**

1 Alexander, M. 1961. Soil microbiology. John Wiley and Sons, Inc. New York.

2 Andrews, J. F. (in press) Kinetics and mathematical modeling. To be published in: H. A. Hawkes and C. R. Curds, (ed.) Ecological aspects of wastewater treatment. Academic Press. London.

3 Buswell, A. M. 1939. Anaerobic fermentation. Illinois State Water Survey. Bull. 321.

4 Hill, D. T. and C. L. Barth. 1974. A fundamental approach to anaerobic lagoon analysis. IN: Proceedings of the 1974 Cornell agricultural waste management conference. Cornell University. Ithaca, N.Y. pp. 387-404.

5 Lawrence, A. W. 1971. Application of process kinetics to design of anaerobic processes. IN: Robert F. Gould, (ed.) Anaerobic biological treatment processes. Advances in Chemistry Series. Amer. Chem. Soc. Washington.

# Aerobic Treatment of Silage Effluent — A Field Study

T. A. Stewart

F RESH grass, particularly when cut at a young stage of growth or during showery weather can give rise to large volumes of effluent when directly ensiled. Various estimates of effluent output range from almost nil for grass with a dry matter or over 25 percent and up to 500 litres per ton from grass with a dry matter of around 15 percent. The wilting of grass prior to ensiling reduces this effluent output and may even eliminate it altogether. Unfortunately the unsettled showery weather encountered in Northern Ireland makes the field wilting of grass an unreliable process and it is not therefore widely practiced.

Silage effluent has a high $BOD_5$ value, ranging from 40 000 to 70 000 mg/l and can be a serious pollutant if allowed to enter watercourses. Fresh silage effluent has a pH of 4.0 to 5.5 and when disposed of by spreading on grassland may cause grass scorch and in severe cases, kill-out of plants. Consequently only limited quantities can be safely applied to grassland at any one time. Silage effluent can also have a corrosive effect on concrete structures.

Serious problems may therefore arise concerning the disposal of silage effluent on farms making large amounts of silage. Such a problem arose at the College following the construction of a new farmyard for animal feeding experiments involving rations based on silage.

## THE PROBLEM

Each year in this farmyard approximately 1200 tons of grass are ensiled without wilting in a number of small 50-70 ton capacity silos. From this silage an estimated 275 m³ effluent are produced. Most of the effluent is produced from mid-May until the end of September, the pattern of flow depending on the silage making program for the season. In addition there is a large volume of contaminated drainage water from the 0.25 hectare farmyard site. This has been estimated to amount to approximately 1400 m³ annually with the greater proportion produced during the winter months.

Both effluents could be a major source of pollution if not collected and handled properly. A concrete lined lagoon was therefore constructed with an effective capacity of 733 m³ into which all effluent from the site is drained. Because of the theoretical 6-fold dilution of the silage effluent obtainable by the addition of contaminated drainage water to the lagoon, aeration was considered as a possible method of treatment. A simple plant was constructed beside the lagoon to investigate the extent of purification possible by this means and to examine the practical problems involved in operating such a plant at farm level.

## TREATMENT PLANT

The plant operates on the extended aeration principle. It consists of an aeration tank which is approximately 6.4 x 6.4 m square and has a capacity of some 95 m³ when filled to a depth of 2.4 m, a 1.8 m x 1.8 m square settling tank with a cone shaped base (capacity 6.1 m³) and a third or final effluent holding tank (capacity 17.5 m³). Aeration is by means of a 2 HP 'Powell Duffryn' cavitair surface aerator which has an oxygenation capacity rating in clean water of approximately 125 kg per day. It is located centrally in a fixed position in the aeration tank. Aerator immersion depth is altered by means of an adjustable weir located in the outlet between the aeration and settling tanks.

Untreated effluent from the lagoon may be pumped continuously or intermittently into the aeration tank. The incoming effluent displaces a similar volume of mixed liquor into the settling tank from which the settled effluent passes by gravity to the third tank. From here the final effluent is irrigated over a 2 hectare area of grassland. The effluent retention period in the aeration tank is approximately 21 days when loaded at a daily rate of 4.5 m³.

Sludge from the settling tank may be either recycled to the aeration tank or pumped to drying beds. Mono pumps of various sizes are used because of the low pumping rates required to operate the plant.

## PLANT OPERATION

The treatment plant became operational in May 1972 and ran continuously for the next 12 months. Great difficulty was experienced during the winter months in keeping the pumps functioning properly. This led to erratic loading of the plant, variable removal of the sludge and a serious deterioration in the standad of treatment. Because of these difficulties only the results for the first 6 months of operation (Period 1) are reported.

The plant was closed down for the most of 1973 and was not started up again on a continuous basis until June 1974 when it operated for a 25 week period until December 1974 (Period 2).

During Period 1 untreated effluent was pumped to the treatment plant at a rate of 4.3 m³ per day with the loading pump operating for 30 minutes each hour. A similar loading rate of 4.6 m³ per day was maintained in 1974, but as a result of fitting a larger capacity pump, this rate was achieved by operating the pump for 2 to 3 minutes in each hour. Sampling of the untreated effluent entering the plant, the aeration tank mixed liquor and the settled treated effluent passing to the holding tank was carried out each Monday morning between 9 and 11 am.

The author is T. A. Stewart, Head, Experimental Department, Greenmount Agricultural and Horticultural College, Antrim, Northern Ireland.

**Acknowledgments:** The author wishes to acknowledge the assistance received from the field, laboratory and office staff of the Experimental Department at the College and in particular Miss S. McCloy with the chemical analysis of samples. He is also indebted to Dr. V. McAllister for his assistance with the preparation of the paper and to the Principal of the College and the Northern Ireland Department of Agriculture for providing the necessary research facilities.

Spot readings of the pH, temperature (T) and dissolved oxygen (DO) level of the mixed liquor were taken at the same time. The following determinations were carried out on all samples: pH, total solids (TS) volatile solids (VS) suspended (SS) and chemical oxygen demand (COD). The settling rate of the mixed liquor sludge was also measured each week to calculate the sludge volume index (SVI). In addition during Period 2, all samples were analyzed for total nitrogen (total N) and total phosphorus (total P) contents and occasional 5-day biochemical oxygen demand ($BOD_5$) determinations were also carried out.

An attempt was made to recycle sludge from the settling tank to the aeration tank. This was only partially successful during the first 12 month period because of pumping diffculties which led to problems with floating sludge in the settling tank from time to time.

During both operational periods the settling tank was completely cleaned out on several occasions using a vacuum tanker.

## RESULTS

### Period 1 (15 May to 13 November 1972)

The untreated effluent characteristics are given in Table 1. When treatment commenced the pH was 7.4 but with the addition of fresh silage effluent to the lagoon in May and June, it fell to its lowest level of 5.5 in July and then gradually returned during the autumn to its former level. The average pH during Period 1 was 6.8. The TS content of the untreated effluent increased during May and June from under 3000 mg/l to over 7000 mg/l and remained at about this level for the rest of Period 1. VS, approximately 57 percent of the total solids, followed a similar pattern as did the COD content. Fresh silage effluent should contain little or no suspended solids. The SS in this effluent were derived from small pieces of grass carried into the lagoon and from fungal growths which developed on the surface of the liquid during storage.

**TABLE 1. EFFLUENT CHARACTERISTICS DURING PERIOD 1 (MAY — NOVEMBER 1972)**

| Variable | Mean | Range | |
|---|---|---|---|
| Untreated effluent | | | |
| pH | 6.8 | 5.5 - | 8.4 |
| TS, mg/l | 6257 | 2193 - | 7405 |
| VS, mg/l | 3542 | 975 - | 4995 |
| SS, mg/l | 198 | 10 - | 874 |
| COD, mg/l | 5355 | 1840 - | 7920 |
| Settled treated effluent | | | |
| pH | 8.2 | 7.7 - | 8.4 |
| TS, mg/l | 3212 | 1811 - | 4873 |
| VS, mg/l | 1224 | 355 - | 2745 |
| SS, mg/l | 161 | 10 - | 888 |
| COD, mg/l | 620 | 257 - | 3010 |

The mean COD loading rate during Period 1 was 24 kg per day. It increased from 16 kg per day during the first week to a maximum of 34 kg per day at the end of July and then fell off to an average of 20 kg per day during October and November. The daily total and volatile solids entering the plant followed a similar pattern, the average volumes for the period being, TS - 28 kg per day VS - 16 kg per day.

The mixed liquor characteristics are shown in Table 2. There was little variation in pH despite the lower pH of the untreated effluent at certain times during Period 1.

**TABLE 2. MIXED LIQUOR CHARACTERISTICS DURING PERIOD 1 (MAY — NOVEMBER 1972)**

| Variable | Mean | Range | |
|---|---|---|---|
| pH | 8.4 | 7.6 - | 9.1 |
| TS, mg/l | 7331 | 2238 - | 12311 |
| VS, mg/l | 4487 | 1036 - | 9316 |
| SS, mg/l | 4485 | 187 - | 9072 |
| COD, mg/l | 5015 | 890 - | 8910 |
| SVI | 85 | 35 - | 159 |
| T, deg C | 11.4 | 4.5 - | 17.1 |
| DO, percent saturation | 56 | 50 - | 106 |

All sludge produced during the first 4 months was recycled to the aeration tank with the result that in the mixed liquor there was a steady buildup of TS, VS, SS and COD to levels of 12311, 9316, 9072 and 8910 mg/l respectively at the beginning of September. Problems were encountered towards the end of August and during September with floating sludge in the settling tank. This occurred because the sludge pump in use at the time (Mono CD 20) was not capable of recycling all the sludge resulting in a buildup of sludge in the bottom of the settling tank, some of which eventually floated to the surface. The settling tank was completely cleaned out on 3 occasions during September and again in October and the effluent removed replaced with clean water. This lowered the VS, SS and COD levels of the mixed liquor to below 5000 mg/l. They did not exceed this level for the remainder of Period 1.

The mixed liquor temperature increased from 7.8 C at the start of treatment to a maximum of 17.1 C in mid-July. Thereafter it declined gradually to about 10 C at which level it remained until mid-November when it dropped rapidly to 4.5 C.

Aeration was adequate at all times as indicated by the dissolved oxygen levels shown in Table 2. The quality and settling rate of the mixed liquor solids declined during July and August as the solids content increased but improved considerably after the solids level was reduced in September. A high quality floc with good settling characteristics (as indicated by a low SVI) was produced during October and November.

The settled effluent characteristics are shown in Table 1. The final effluent produced had an average COD of 620 mg/l and a SS of 161 mg/l. COD removal was high averaging 87 percent but weekly fluctuations ranging from 51 to 96 percent were recorded. VS removal averaged 65 percent with weekly variations ranging from 21 to 91 percent. Removal of TS was even lower averaging 46 percent with a range of 19 to 64 percent. This reflects the high non-volatile constituents in this type of effluent.

Some indication of the optimum conditions necessary to achieve maximum COD removal and produce a treated effluent approaching Royal Commission Standards (20 mg/l BOD: 30mg/l SS) may be obtained from the plant performance during a 6 week period from the end of September to the beginning of November. Details are given in Table 3. The average COD and VS loadings during this period were 20 and 15 kg per day respectively. Dissolved oxygen level of the mixed liquor averaged 79 percent saturation and the mixed liquor temperature averaged 11 C while the SVI averaged 54. Under these conditions 93 percent of the COD in the untreated effluent was removed by treatment producing a final effluent with a mean COD of 330 mg/l and SS of 32 mg/l.

The electrical power required to run the plant during Period 1 averaged 30.3 kW-hr per day.

TABLE 3. EFFLUENT CHARACTERISTICS WHEN PLANT OPERATING UNDER OPTIMUM CONDITIONS DURING PERIOD 1. (MEAN OF 6 WEEKS FROM 25 SEPTEMBER TO 6 NOVEMBER)

| Variable | Untreated effluent | Mixed liquor | Settled effluent |
|---|---|---|---|
| pH | 7.4 | 8.1 | 8.1 |
| VS, mg/l | 3582 | 3828 | 1426 |
| SS, mg/l | — | 4250 | 32 |
| COD, mg/l | 4887 | 4523 | 330 |

## Period 2 (24 June to 9 December 1974)

The untreated effluent characteristics are given in Table 4. Its composition on average was similar to that of the untreated effluent in Period 1. During this period however, it was at its strongest when the plant was started up in June and declined steadily to its lowest strength in December.

TABLE 4. EFFLUENT CHARACTERISTICS DURING PERIOD 2 (JUNE — DECEMBER 1974)

| Variable | Mean | Range | |
|---|---|---|---|
| Untreated effluent | | | |
| pH | 6.7 | 5.8 - | 7.9 |
| TS, mg/l | 6143 | 3821 - | 8559 |
| VS, mg/l | 3221 | 1707 - | 6406 |
| SS, mg/l | 117 | 10 - | 339 |
| COD, mg/l | 6142 | 1791 - | 14080 |
| Total N, mg/l | 537 | 350 - | 680 |
| Total P, mg/l | 124 | 32 - | 208 |
| Settled treated effluent | | | |
| pH | 7.8 | 6.6 - | 8.5 |
| TS, mg/l | 4382 | 3062 - | 8331 |
| VS, mg/l | 1882 | 1041 - | 5732 |
| SS, mg/l | 200 | 10 - | 554 |
| COD, mg/l | 1676 | 388 - | 5700 |
| Total N, mg/l | 228 | 30 - | 550 |
| Total P, mg/l | 48 | 4 - | 190 |

The mean COD loading rate during Period 2 was 28 kg per day. It was at a maximum of 40 kg per day when treatment commenced and declined steadily to its lowest level of 9 kg per day during December. The daily total and volatile solids entering the plant followed a similar pattern to the COD loading, average for the period being, TS - 28 kg per day and VS - 14 kg per day.

Details of the mixed liquor characteristics for Period 2 are shown in Table 5. Again the pH remained fairly constant with a mean value of 8.0. The level of solids was high when treatment started in June (TS - 8910 mg/l, VS - 6640 mg/l, SS - 4610 mg/l, COD - 6200 mg/l) and remained at approximately these levels until the beginning ofSeptember. During this period because of difficulties with the sludge pump it was not possible to recycle the sludge to the aeration tank. Consequently the settling tank was cleaned out by vacuum tanker at intervals of 1 to 2 weeks. This procedure while preventing a buildup of solids in the plant did not produce conditions which were conducive to good effluent clarification. A new sludge pump was fitted in September and the sludge was recycled to the aeration tank for the remainder of the period. During this period the SS level increased from about 3200 mg/l to 5100 mg/l.

The mixed liquor temperature remained between 14 and 15 C during July and August and then fell steadily during September and October to 6 C. It remained between 3 and 6 C during November and December.

TABLE 5. MIXED LIQUOR CHARACTERISTICS DURING PERIOD 2 (JUNE — DECEMBER 1974)

| Variable | Mean | Range | |
|---|---|---|---|
| pH | 8.0 | 7.3 - | 8.5 |
| TS, mg/l | 7955 | 5583 - | 11100 |
| VS, mg/l | 4499 | 2984 - | 6639 |
| SS, mg/l | 3836 | 2147 - | 6948 |
| COD, mg/l | 5316 | 2100 - | 11550 |
| Total N, mg/l | 488 | 250 - | 880 |
| Total P, mg/l | 200 | 122 - | 302 |
| SVI | 66 | 54 - | 142 |
| T, deg C | 10.4 | 3.0 - | 15.0 |
| DO, percent saturation | 64 | 26 - | 91 |

Aeration levels were again adequate throughout Period 2 as indicated in Table 5. The mixed liquor floc exhibited good settling characteristics throughout Period 2 as indicated by the satisfactory SVI figures.

From Table 4 it can be seen that the mean COD value of the treated settled effluent during Period 2 of 1676 mg/l was considerably higher than that recorded in Period 1. The higher level resulted from the much higher levels recorded during the first 13 weeks of operation when it averaged 3000 mg/l compared with 500 mg/l for the second 13 week period (mid-September to mid-December). This reflects the higher loading rate during the earlier part of Period 2.

$BOD_5$ determinations carried out over the last 11 weeks of Period 2 (October to December 1974) averaged 18 mg/l indicating that during this time the plant produced a treated effluent approaching the Royal Commission Standard for $BOD_5$. The SS content was reasonably consistent throughout Period 2, averaging 200 mg/l.

COD removal by treatment was lower than in Period 1 and averaged 72 percent with weekly fluctuations ranging from 40 to 87 percent. VS removal was also lower throughout this period compared with Period 1. The average VS removed was 38 percent with a weekly range from 11 to 69 percent. TS removal was again poor averaging 25 percent and ranging from 3 to 44 percent.

After treatment it was found that the total N and total P contents of the treated settled effluent were decreased by 56 and 62 percent respectively from the contents in the untreated effluent. The mean total N content of the settled treated effluent was 228 mg/l and the mean total P was 48 mg/l.

The electrical consumption during Period 2 averaged 26.5 kW-hr per day.

## DISCUSSION

The results reported clearly indicate that silage effluent may be aerobically treated to remove a high proportion of the bio-degradable matter present. It was possible to produce a treated effluent from the plant described which would meet the statutory $BOD_5$ requirements concerning direct discharge to a river, provided the COD loading rate did not exceed about 20 kg per day. At these low loading rates the existing plant could not cope with the quantity of effluent produced from the site. Furthermore the generally high suspended solids, non-volatile solids and phosphate levels of the best effluent produced would probably be unacceptable to local river authorities.

It is not known what standard of treatment could be achieved during the winter months but it is unlikely to be as high as during the warmer summer months. It is also unlikely that an effluent of a pre-determined standard

*(Continued on page 559)*

# Aerobic Treatment of Piggery Waste Prior to Land Treatment
## A Case Study

M. R. Evans, R. Hissett, D. F. Ellam, S. Baines

CULBAE Farm, Wigtownshire (Grid ref: NX 3848) is situated at the top of a water catchment area at 50 m and covers about 100 ha. Between 5000 and 6000 fattening pigs were housed on the farm and fed on a diet of whey supplemented with barley meal and other fibrous and carbohydrate material. Slurry drained from the under-slat channels into a 400 $m^3$ slurry tank and was disposed of by spraying through rain-guns onto the surrounding fields five days each week. The rate of slurry application to land was 150 $m^3$ per ha every 6 months. No attempt was made to keep grazing animals off the fields while the rain-guns were in operation.

All the fields drained into a series of open-ditches and these, in turn, discharged into a watercourse called The Canal. The soil of the area was dominated by a fine sandy loam (average analysis down to 60 cm depth: sand, 50 percent; silt, 20 percent; clay, 22 percent; pH value 4.7 to 5.7) which was freely drained. There were also areas of basin peat, some of which were drained by tile drains.

During spraying, odor was detectable for several kilometers down-wind. The drainage water leaving Culbae in The Canal, was heavily polluted. The water in the ditch nearest the spraying area was turbid and malodorous, but the water in the remaining ditches was clear, although a pale straw color similar to that of other waters draining from peat. There was, however, a thick slime growth on the bed of all the ditches.

Drainage water was examined to determine the concentrations of bacteria and chemical pollutants, to see if the factors affecting these concentrations were similar to those reported by Evans and Owens (1972 and 1973).

The effectiveness of partial aerobic treatment of the slurry prior to land treatment was evaluated in a preliminary experiment, using a surface aerator on the existing slurry tank. An aeration system based on the result of this experiment and design data obtained from laboratory studies (Owens et al. 1973) has been installed.

## MATERIALS AND METHODS

### Examination of Drainage Water

The quality of water draining from a 20 ha area of land, (Fig. 1) contained within the boundaries of Culbae, was monitored in the absence of recent slurry application. In January 1973, 160 $m^3$ of slurry were sprayed, over a period of 8 hr, onto the Longcastle field and the quality of water draining from the area was monitored. In January 1974 the

The authors M. R. EVANS, R. HISSETT, D. F. ELLAM and S. BAINES are with Microbiology Department, The West of Scotland Agricultural College, Auchincruive, Ayr, U.K.

Acknowledgements: The authors wish to thank Mr. Robinson of Culbae; members of staff at the West of Scotland Agricultural College, in particular Miss E. Thompson, Mrs. L. Innes, Mrs. L. Stokes and Mr. R. Wilson; the Solway River Purification Board for analyses of water samples. The work formed part of an investigation into the treatment of farm wastes supported by the AGRICULTURAL RESEARCH COUNCIL.

experiment was repeated, at a time of similar weather conditions and drainage rate, using a 110 $m^3$ of partially treated slurry.

The topography of the area and drainage properties of the soil were such that there was little chance of surface run-off. The Longcastle and Midden fields were undulating, whereas the Standing Stone field had a steep gradient (Fig. 1). An open ditch originated on the south east boundary of the Longcastle field, and continued through the peat along the boundary of the other two fields.

Water samples were collected in sterile evacuated Winchester bottles, with an attached length of silicone-rubber tubing (Zobell 1941), and analyzed bacteriologically within 6 hr of collection, by methods described by Evans and Owens (1972). Chemical analyses were carried out within 24 hr at the laboratories of the Solway River Purification Board.

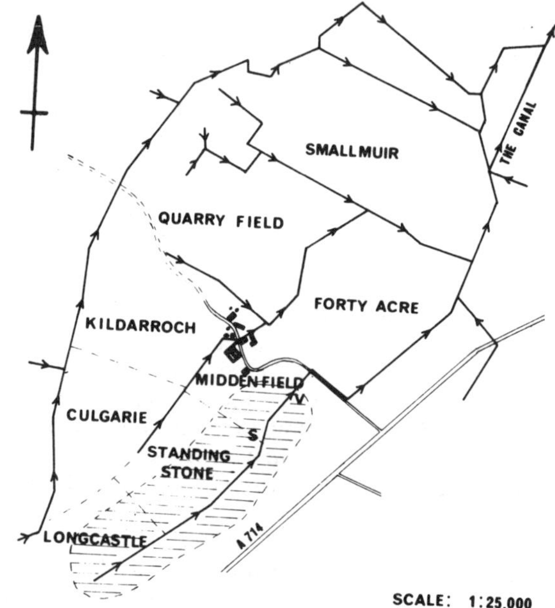


FIG. 1 Culbae farm, Wigtownshire (grid ref. NX 3848). ⟶ open ditches and direction of water flow; V, V-notch weir gauge; s, sampling point; shaded area, water catchment area.

### Aerobic Treatment

Initially a 11 kW floating surface aerator was installed on the slurry storage tank. The machine was designed to supply 23 kg $O_2$/h. Depending on the times of spraying, the mean residence time (SRT) of the slurry in the tank was about 3 to 5 days.

The second stage of development was the installation of another 400 $m^3$ tank equipped with three 5 kW surface aerators. Each aerator supplies 11 kg $O_2$/h, (Total 33 kg $O_2$/h). Slurry drains from the piggeries into the tank, which provides an SRT of 5 d. Mixed liquor overflows from this tank into the original tank which is used as a secondary treatment and aerobic storage unit. (Fig. 2).

Representative samples of slurry and mixed liquor were analyzed by the methods described by Owens et al. (1973). Continuous monitoring equipment for slurry flow into the main tank, and dissolved oxygen, pH value and temperature in both tanks was installed.

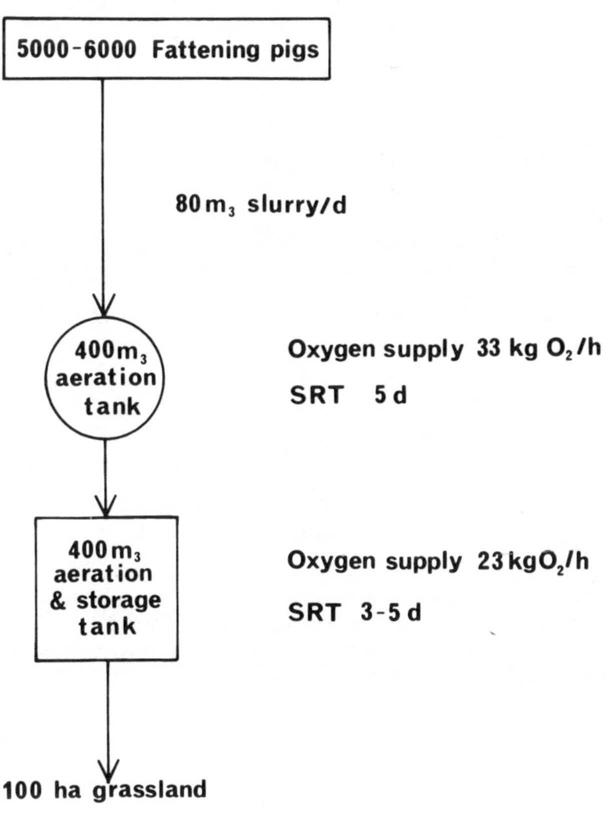


FIG. 2 Flow diagram of second stage aerobic treatment system.

## RESULTS AND DISCUSSION

### Pollution of Drainage Water

In the absence of recent slurry application to the land, the concentrations in the drainage water of *Escherichia coli* ($10^2$ to $10^5$ viable cells/l) and enterococci ($10^2$ to $10^4$ viable cells/l) were positively related to the drainage rate. This relationship was similar to that reported by Evans and Owens (1972). The concentrations of 5-day biological oxygen demand ($BOD_5$) and total suspended solids (TSS), in the drainage water were less than 8 mg/l and 14 mg/l, respectively.

All the parameters measured increased within 45 min of starting to spray untreated slurry (160 m$^3$) onto the Longcastle field (3 ha). The concentration of *E. coli* increased from $10^3$ to $10^6$ organisms/l (Fig. 3); $BOD_5$ from 1 to 360 mg/l (Fig. 4); and TSS from 11 to 117 mg/l. (Fig. 5).

The values of these parameters then declined during the next hour but increased again during the following 2 hr. This decline was probably due to changes in the position of the rain-guns on the field. The $BOD_5$ and TSS of the stream returned to their original concentrations within 12 hr and the *E. coli* returned to its original concentration within 36 h of starting to spray slurry on the field. These results are also similar to those reported by Evans and Owens (1972).

After the installation of an aerator on the slurry storage tank the slime layer on the stream beds was noticeably reduced. Water beetles, snails and aquatic plants also markedly increased during the summer. On repeating the experiment in January 1974, using partially treated slurry (110 m$^3$), similar changes in the concentrations of bacterial and chemical parameters occurred. (Figs. 3, 4 and 5). The maximum concentration *E. coli* and TSS were similar to those in the previous experiment but the maximum $BOD_5$ concentration was considerably less.

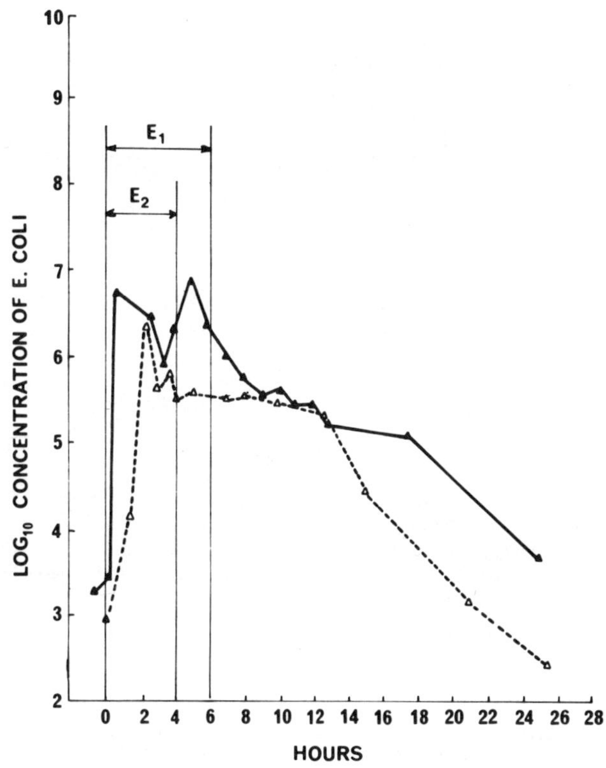


FIG. 3 Concentration of *E. coli* in drainage water following the application of untreated slurry (continuous line) and partially treated slurry (dashed line) onto the Longcastle field. E$_1$ and E$_2$ indicate periods during which the untreated slurry and partially treated slurry were applied respectively.

The quantities of $BOD_5$ discharged within 12 hr and *E. coli* discharged within 36 hr were also less than corresponding figures calculated from the first experiment (Table 1). The quantity of TSS discharged, however, was similar to that obtained in the earlier experiment.

### Aerobic Treatment

The quantity and composition of the slurry is shown in Table 2. From laboratory studies (Owens et al. 1973) it was calculated that the oxygen demand by a microbial population would be between 27 and 33 kg O$_2$/h, depending on the SRT. If sufficient oxygen was supplied, then reductions of up to 40 percent COD and 20 percent TSS could be expected. The total $BOD_5$ of the partially treated material would be 4 g/l and soluble $BOD_5$ less than 0.1 g/l. Odor should also be eliminated.

Since the 11 kW aerator only supplied 23 kg O$_2$/h, oxygen would be a limiting factor and would prevent the reduction of the chemical parameters to the predicted values.

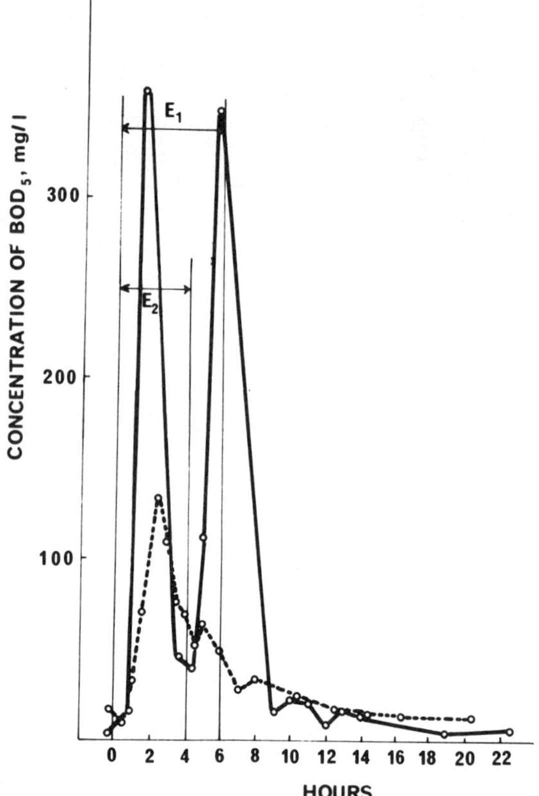

FIG. 4 Concentration of $BOD_5$ in drainage water following the application of untreated slurry (continuous line) and partially treated slurry (dashed line) onto the Longcastle field. $E_1$ and $E_2$ indicate periods during which the untreated slurry and partially treated slurry were applied respectively.

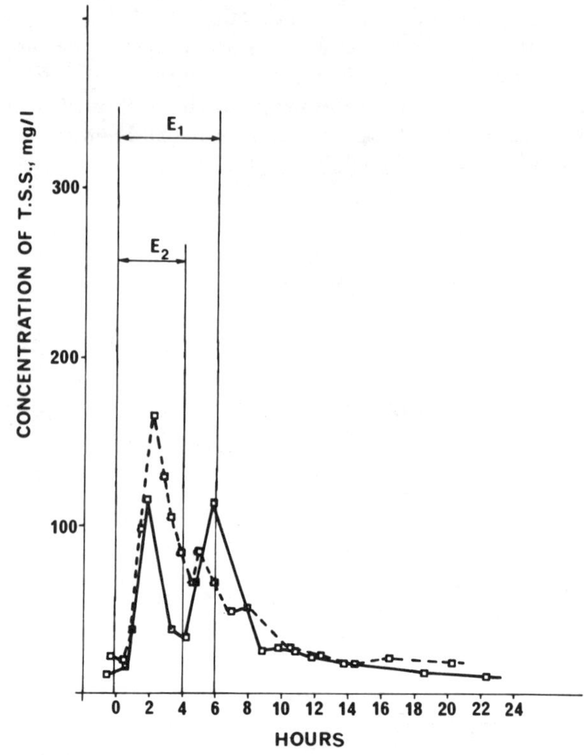

FIG. 5 Concentration of TSS in drainage water following the application of untreated slurry (continuous line) and partially treated slurry (dashed line) onto the Longcastle field. $E_2$ and $E_2$ indicate periods during which the untreated slurry and partially treated slurry were applied respectively.

TABLE 1. QUANTITIES OF BOD AND TSS, AND NUMBERS OF *E. Coli* DISCHARGED INTO DRAINAGE WATER AS A RESULT OF THE APPLICATION OF UNTREATED SLURRY (JANUARY 1973) AND PARTIALLY TREATED SLURRY (JANUARY 1974 ONTO THE LONGCASTLE FIELD.

|  | Untreated slurry | Partially treated slurry |
|---|---|---|
| $BOD_5$ discharged within 12 h of starting to spray slurry, $g/m^3$ slurry applied | 56 | 27 |
| TSS discharged within 12 h of starting to spray slurry, $g/m^3$ slurry applied | 25 | 36 |
| *E. coli* discharged within 36 h of starting to spray slurry, viable cells/$m^3$ slurry applied | $10^{12}$ | $10^{11}$ |

TABLE 2. ANALYSIS OF SLURRY DRAINING FROM PIGGERIES.

| | |
|---|---|
| Volume of slurry, l/pig.d | 14 |
| TSS, g/l | 14 |
| Total $BOD_5$, g/l | 12 |
| Soluble $BOD_5$, g/l | 7 |
| COD, g/l | 25 |
| Organic nitrogen, g/l | 0.9 |
| Ammoniacal nitrogen, g/l | 1.0 |

Steady state conditions developed in the aerated mixed liquor within the first month. About 30 cm of foam formed on the surface but only caused problems when the tank was filled to capacity on windy days. Odor from the tank and spraying areas was eliminated.

As expected, the dissolved oxygen tension was usually below 2 mm Hg. The pH value of the mixed liquor remained between 8.0 and 8.2. Microscopic examination of the mixed liquor demonstrated flocculation. No ciliated protozoa were seen, but high numbers of flagellated protozoa were present. Upon standing for about one hour, solids settled leaving a reasonably clear brown liquor which was considerably improved from the malodorous opaque grey liquor of the incoming slurry from the piggery. Results of chemical analysis of samples of mixed liquor are shown in Fig. 6.

Although insufficient oxygen was supplied, the quality of the mixed liquor was similar to that predicted. There was no significant reduction of suspended solids or organic nitrogen. Total COD was reduced by 24 percent and ammonia was reduced by 43 percent. The total $BOD_5$ of the mixed liquor was 5.3 g/l and the soluble $BOD_5$ was about 0.9 g/l.

Although the new aeration system has been installed, it has not yet been commissioned, due to economic problems and restrictions of animal movement within the U.K.

## CONCLUSIONS

The examination of samples of drainage water from the open ditches, before the installation of an aerobic treatment system, demonstrated that most of the pollutants

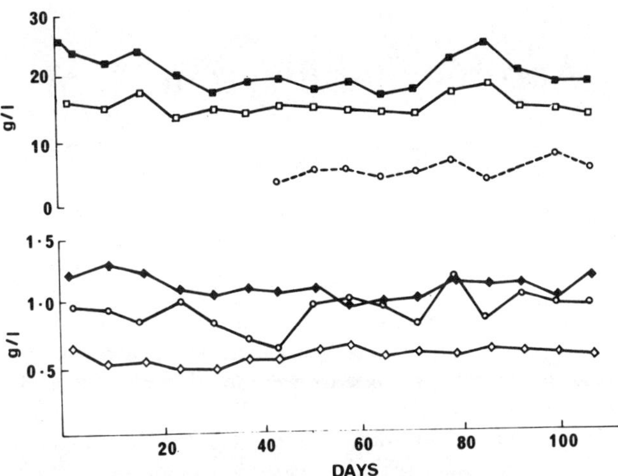

FIG. 6 Quality of mixed liquor in aeration tank: ■—■ COD; o---o total BOD$_5$; •—• soluble BOD$_5$; □—□ TSS; ◆—◆ organic nitrogen; ◇—◇ ammoniacal nitrogen.

gained access to the watercourse during, and immediately following, the time of spraying slurry onto the adjacent field. However, there was sufficient residual material remaining on the land, to be washed out following rainfall, and encourage slime growth on the bed of the ditches.

The installation of an aeration system, to partially treat the slurry prior to land treatment, reduced the organic load being applied to the land and eliminated odor during spraying.

The water quality in the ditches has improved. Aquatic plants and insects have markedly increased and microbial slime is no longer apparent. Transient pollution of ditches adjacent to spraying areas has been reduced but not eliminated. Following the commissioning of the second stage aeration system this problem should be overcome.


## SUMMARY

Between 5000 and 6000 fattening pigs, fed on a diet of whey supplemented by barley meal and other fibrous and carbohydrate material, were housed on a 100 ha farm situated at the top of a water catchment area. About 80 m$^3$/d of excreta drained from the piggeries into a 400 m$^3$ slurry tank before spraying onto land using rain-guns.

Odor and water pollution were causing concern. The pollutants gained access to drainage ditches during the time of spraying slurry onto adjacent fields and following rainfall.

A 11 kW floating surface aerator was installed on the slurry tank to evaluate the use of partial aerobic treatment. Odor was eliminated from the tank and during spraying. Pollution of the water in the ditches adjacent to sprayed fields was reduced.

A second 400 m$^3$ tank, equipped with three 5 kW surface aerators and equipment to continuously monitor effluent flow, dissolved oxygen, pH and temperature has been installed. The existing tank and aerator are retained as a secondary treatment and aerobic storage unit.



### References

1   Evans, M. R. and J. D. Owens. 1972. Factors affecting the concentration of faecal bacteria in land-drainage water. J. gen. Microbiol. 71:477-485.
2   Evans, M. R. and J. D. Owens. 1973. Soil bacteria in land-drainage water. Water Res. 7:1295-1300.
3   Owens, J. D., M. R. Evans, F. E. Thacker, R. Hissett, and S. Baines. 1973. Aerobic treatment of piggery waste. Water Res. 7:1745-1766.
4   Zobell, C. E. 1941. Apparatus for collecting water samples from different depths for bacteriological analysis. J. mar. Res. 4:173-188.


# Aerobic Treatment of Silage Effluent

*(Continued from page 555)*

could be consistently produced without the introduction of a two-stage treatment process. As the plant performed efficiently at much higher loading rates, (up to 40 kg COD per day) a more practical approach would appear to be to aim at partial treatment coupled with land disposal rather than complete treatment. Partially treated effluent of the type produced in either treatment period could be applied to grassland at rates of up to 500 m$^3$ per hectare per annum without leading to nutrient enrichment or other adverse effects.

The main problems encountered in operating the treatment plant were associatedc with the pumps used. These have now been largely overcome by fitting direct drive couplings rather than 'V' belts and using pumps with a minimum suction and delivery outlet of 30 mm.

No accurate estimates were possible of the sludge output from the plant although the indications are that the quantity should not be large. Foaming was not a problem at any time.


## SUMMARY

Experiences over two 6-month periods with the aerobic treatment of diluted silage effluent indicate that a high degree of purification is possible using a simple extended aeration type treatment plant. During the first treatment period (May-November 1972) an average of 87 percent of the untreated effluent COD was removed on treatment when the plant was loaded at an average rate of 24 kg COD per day. During Period 2 (June-December 1974) the average COD removal on treatment was 72 percent when the plant was loaded at an average rate of 28 kg COD per day. Removal of total solids was poor during both periods because of the high non-volatile solids content characteristic of silage effluent. Suspended solids content of the treated effluent averaged 161 mg/l in Period 1 and 200 mg/l in Period 2.

At COD loading rates of under 20 kg per day the plant produced an effluent with BOD$_5$ values approaching the Royal Commission Standard.

# Biologically-Controlled Loading of Aerobic Stabilization Plants

K. Robinson, D. Fenlon

URING a study of various aerobic methods for the stabilization of swine wastes (Report 1974) problems such as foaming and marked fluctuations in dissolved oxygen and pH of the mixed liquor occurred which were a disadvantage in successful day-to-day operation or might be expected to affect the efficiency of the stabilization process. It was considered that these problems were a reflection of the varying response of the microflora to differences in environmental conditions such as temperature, chemical load and the presence of end-products of the biological reactions.

The initial design data for a stabilization system is in effect a compromise relating the mean values for the quantity and BOD of the input waste to the size and aeration capacity of the system. In practice the actual values for volume and BOD may vary widely from the mean. The achievement of conditions approximating to a 'steady-state' in the mixed liquor is unlikely if the quantity and quality of the input varies and is even less likely if the waste is added at random time intervals, such as may occur with an in-house oxidation ditch, completely unrelated to state of biological activity within the aerobic stabilization system. In the absence of 'steady-state' conditions it is improbable that the system will operate with maximum efficiency and produce an end-product of predictable composition.

Nitrification is an important reaction occurring during the aerobic stabilization of an animal waste and the accumulation of nitrate is primarily responsible for the development of acid conditions. It was considered that either nitrification or oxygen utilization were two aspects of biological activity which could be used to control the frequency of loading in relation to activity. A comparison of pH-controlled and dissolved oxygen-controlled loading had shown that the former was preferable (Robinson and Fenlon 1975). The advantages of an automatically-controlled loading procedure are that it would eliminate the need for frequent analysis by skilled personnel and reduce the man-hours which must be allocated to maintenance and operation of a commercially operated stabilization process.

The purpose of the investigations reported here was to examine the feasibility of pH-controlled loading and to compare the efficiency of stabilization at different pH levels.

## MATERIALS AND METHODS

Aerobic stabilization was conducted using a continuous fermenter system (Chemap, Switzerland). The fermenter vessel had a working capacity of 12.6 l and the contents, maintained at 15 C, were mixed by stirring at 400 rpm. Air was supplied to the vessel at a rate of 400 cc/minute at which the oxygen transfer coefficient $K_T$ (Jones, Day and Converse 1969) in water at 15 C was 13.45 h$^{-1}$ ($K_{20}$ =

15.14 h$^{-1}$). Dissolved oxygen concentration was monitored using a Mackereth-type oxygen electrode. Changes in the pH of the fermenter contents were monitored using a pH meter/controller and chart recorder. The signal from the pH controller, produced when the pH fell below the preset value, was used to switch a peristaltic pump and supply unstabilized waste to the fermenter until the pH returned to its preset value. Stabilized waste was removed from the vessel through a constant level device which also served as an exit port for effluent air.

The unstabilized waste supply was the supernatant liquid from an anaerobic lagoon used for the storage of swine waste (Report 1974, Robinson 1974).

The period of investigation began using a batch of mixed liquor prepared by the aeration of unstabilized waste until it reached a pH of 6. During the course of the investigation of pH controller setting was changed after a suitable period of operation from pH 6.0 to 7.0 and finally 7.6. Each period of operation ran concurrently and the mixed liquor in the fermenter vessel was not changed but allowed to adapt to the new conditions.

The major parameters measured were (i) input volume, soluble and suspended COD and $NH_4^+$-N, $NO_2^-$-N and $NO_3^-$-N. The methods used for analysis have been described elsewhere (Murray, Parson and Robinson 1975, Robinson and Fenlon 1975).

## RESULTS

During fermenter operation a number of problems such as the blockage or splitting of influent air and feed lines and burnt out pump motors occurred. These were annoying because they produced a change from 'steady-state' conditions however, they were of some value in that it was possible to make some interpretation of the resultant effects. Similar problems occurring during the operation of field-scale plants are much more difficult to interpret because of the many other uncontrolled variables.

Reduction in the oxygen demand of raw wastes is an important function of the stabilization process. The efficiency of the process, as volumetric and COD load per day, for the pH values examined is shown in Table 1. The mean values for the COD load increase in the proportions 1:1.6:3.4 as the pH increases from 6 to 7.6 while the volumetric load increases in the proportion 1:3.6:5.1 and demonstrate the value of the pH loading control mechanism to relate loading to chemical and not volumetric load. It is

**TABLE 1. THE EFFECT OF pH ON VOLUMETRIC AND COD INPUT**

| pH | Influent volume mean, l/day | Total COD mean, mg/day | Influent COD mean, mg/l |
|---|---|---|---|
| 6.0 | 1.05 | 6539 | 6249 |
| 7.0 | 3.8 | 10484 | 2752 |
| 7.6 | 5.4 | 22517 | 4163 |

The authors are: K. ROBINSON and D. FENLON, Bacteriology Division, School of Agriculture, Aberdeen, Scotland.

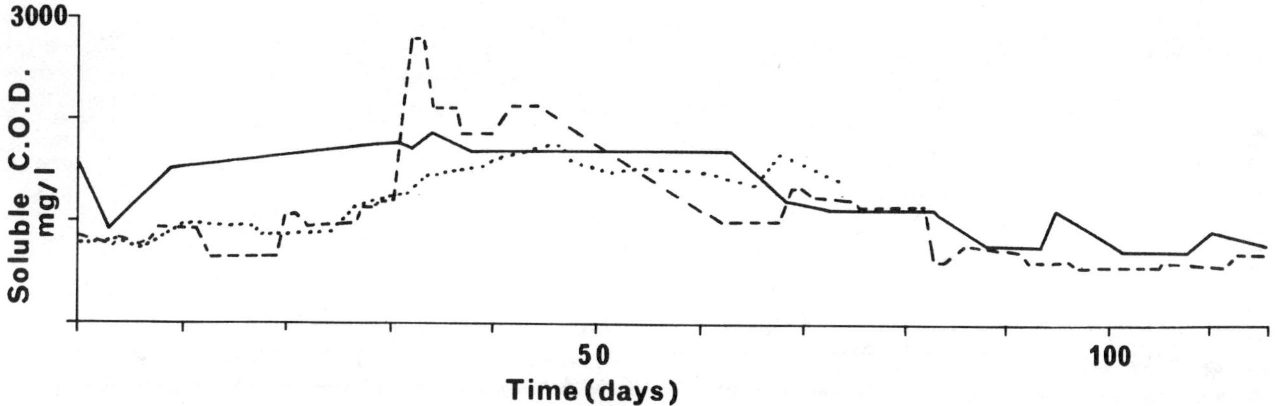


FIG. 1 Dissolved COD of the stabilized effluent produced at pH 6, 7 and 7.6.

important to note that the increased efficiency was achieved without changing the oxygen input and this is obviously important from the points of view of energy conservation and obtaining stabilization at minimum operating cost. The results also indicate that the use of input waste BOD as a design criterion is by itself of limited value.

As Fig. 1 shows the increase in loading rate was achieved without any significant change in the dissolved COD of the stabilized effluent from the fermenter. The relatively stable composition of the clarified effluent was also achieved despite very considerable fluctuations in the daily COD reduction rate. The plot for the removal of soluble COD il-lustrates that the daily removal rate was not constant (Fig. 2). For example, the curve at pH 6 is comparatively smooth until about Day 68 when there was a change to a new batch of unstabilized waste with a COD about twice that of the previous batch; other subsequent peaks can be associated with a failure of the pH meter and other mechanical problems. One effect of this peak load seems to be an initiation of a denitrifying sequence (Fig. 5). COD reduction was more erratic at pH 7 and even more so at pH 7.6 at which the retention times were often less than 2 days. At short retention times it is inevitable that 'wash-out' of micro-organisms must be taking place and if this were to occur in

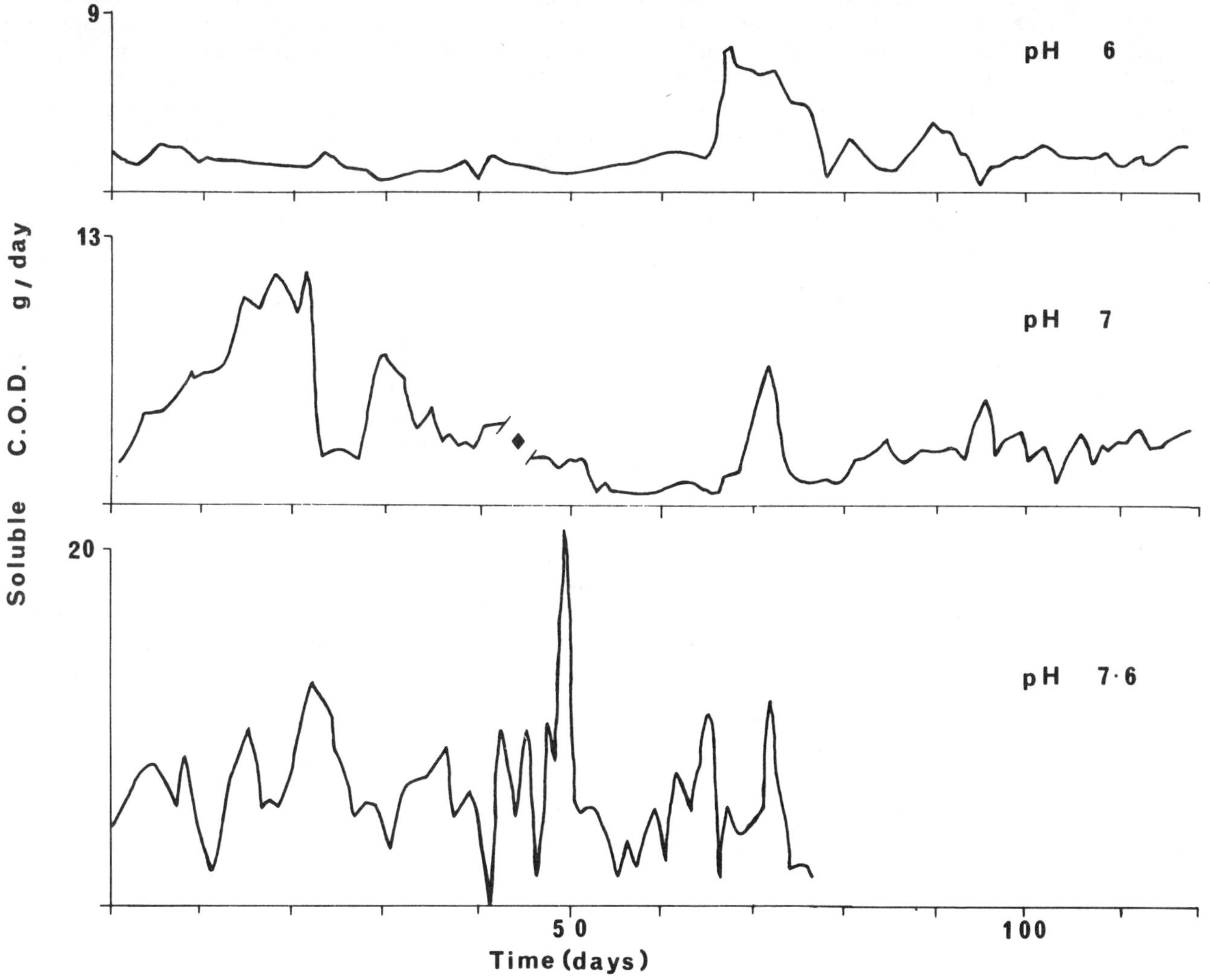

FIG. 2 Daily soluble COD reduction (COD in - COD out).

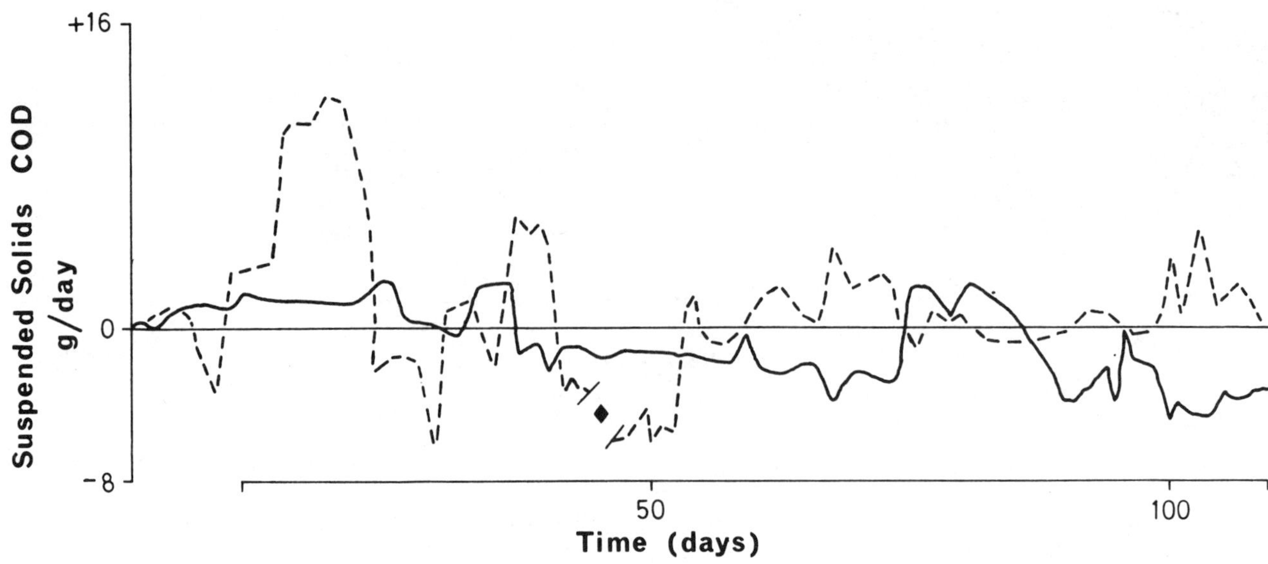

FIG. 3 Change in suspended solids (COD out - COD in) at pH 6 and 7.

a system receiving a standard volumetric load regardless of its biological substrate value it is inevitable that an overload condition would develop. The results are an excellent demonstration of the ability of pH-controlled loading to prevent the development of an overload condition and allow time for the system to recover before loading commences. The plots in Fig. 2 also show that the value, COD in - COD out, was always positive. In contrast the plot of COD out - COD in for the suspended solids (made on the assumption that suspended solids are a product of the stabilization process) had both negative and positive values (Fig. 3) and at pH 7.6 values from +30 000 mg to -30 000 mg COD were recorded. Possible explanations for this phenomenon are a change in the state of oxidation of the suspended solids (is it possible for example that oxygen can

be released from suspended solids in a manner equivalent to the denitrification of nitrate) or a digestion or solubilization of the suspended solids?

The effect of failures in operating procedure such as stoppages in the feed or air supply on the pH and dissolved oxygen concentration can be seen in Fig. 4. In the absence of faults a plot of dissolved oxygen showed a wave motion which had a amplitude of a few percent oxygen saturation at pH 6 and increasing slightly as the pH increased (Robinson 1974b). Presumably at the higher pH overshoot occurred as a result of larger load, slower mixing and probe response time. The failure of the air supply during Days 10-20 at pH 6 resulted in a rapid fall in dissolved oxygen concentration followed a little later by a rise in pH. A failure in the feed supply between Days 21-22 at pH 7

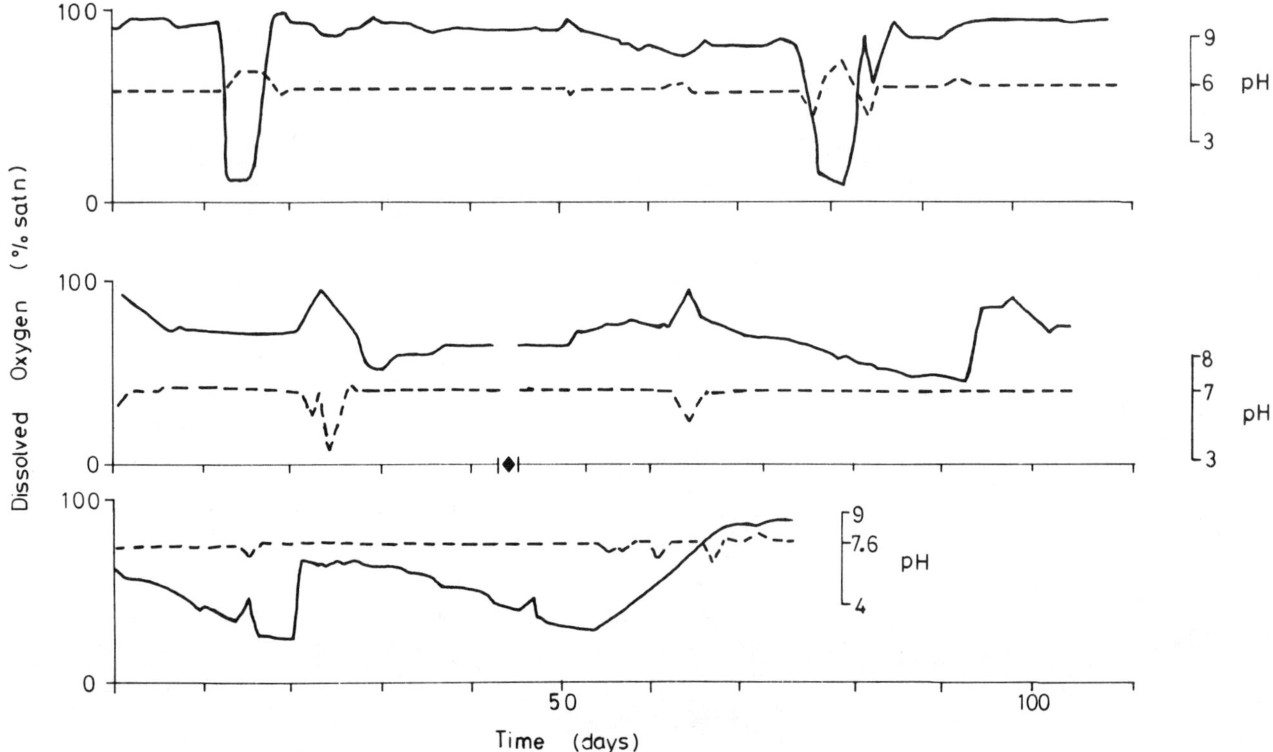

FIG. 4 pH and dissolved oxygen.



MANAGING LIVESTOCK WASTES

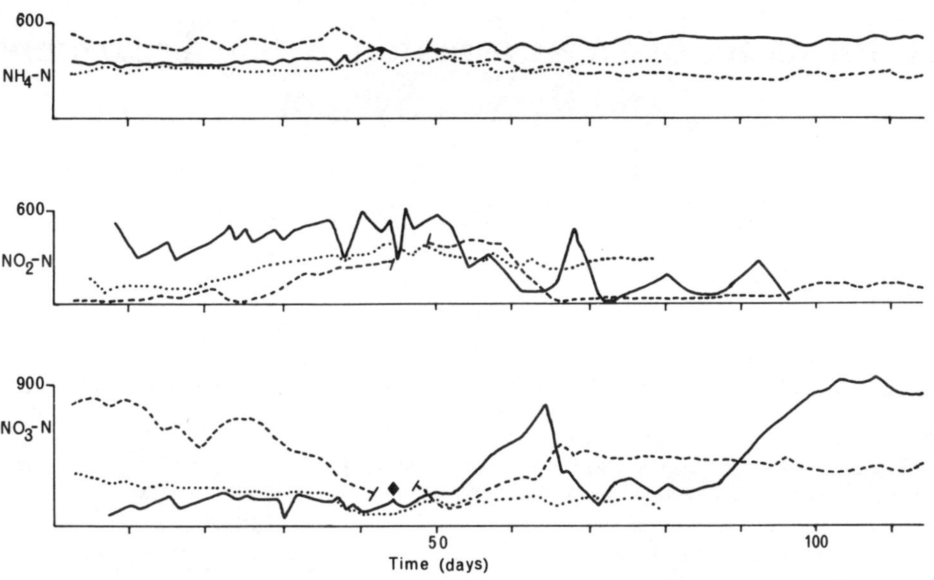

FIG. 5 Ammonium, nitrite and nitrate nitrogen (mg/l) in the mixed liquor at controlled pH.

caused a rise in dissolved oxygen, a fall in pH and an increase in $NO_3^-$-N concentration. It is, however, difficult to identify a particular pattern of events resulting from a failure because the onset of reactions such as nitrification or denitrification may influence the pattern.

Nitrification/denitrification and the relative concentrations of $NH_4^+$-N, $NO_2^-$-N and $NO_3^-$-N may be considered as major factors influencing the pH of the mixed liquor during aerobic stabilization. The changes which occurred in ammonium, nitrite and nitrate concentrations at pH 6, 7 and 7.6 are shown in Fig. 5. The ammonium concentration fell within the range 240-500 mg/l for all pH values, there were no major changes in concentration and there is no clear indication that a particular zone of concentration is associated with a particular pH value. The stability in ammonium level was maintained despite the considerable fluctuations in daily loading rate and the marked variations in oxidized nitrogen. There was no apparent reason for the fluctuations in nitrite concentration at pH 6 during the first half of the experiment but the fall beginning at Day 52 and accompanied by an increase in nitrate concentration, coincided with the feed reservoir running dry. The onset of nitrate loss (Day 62) and the almost immediate reappearance of nitrite coincided with an input having a COD (Fig. 2). From these results it appears that under or overfeeding has a very profound influence on nitrification/denitrification assuming that suitable species of micro-organisms are present. The plots of nitrite and nitrate concentration at pH 7 and 7.6 were much smoother that at pH 6 and where sudden changes did occur they could also be associated with unscheduled changes in the loading rate. The general trend appears to be that at low pH there is more nitrate than nitrite and more nitrite than nitrate at neutral and alkaline pH levels.

## SUMMARY AND CONCLUSIONS

The study of aerobic stabilization in a fermenter of the supernatant from an anaerobic lagoon storing swine waste in which the supply of supernatant was used to maintain the pH of the mixed liquor at a selected value has shown that:

1    The mean input COD increases in the proportion 1:3.4 as the pH is raised from 6 to 7.6.

2    The increase in COD input is unrelated to the mean volumetric input.

3    The retention times, particularly at pH 7.6, vary considerably on a day to day basis.

4    There are no significant differences in the dissolved COD of the effluents produced at pH 6, 7 and 7.6.

5    Increase in COD load was achieved without a change in agitation or aeration rate.

6    Overload conditions are prevented by operation of the pH controller in a 'fail-safe' mode.

7    Changes in the concentrations and relative proportions of nitrite and nitrate nitrogen as a result of nitrification/denitrification were usually the result of sudden and/or persistent changes in the input COD.

8    Increased efficiency is due to the ability of pH control to adapt the quantity of waste loaded into the system to the needs of the treatment microflora.

9    The application of small loads at frequent intervals prevents wide fluctuations in the pH and dissolved oxygen concentration in the mixed liquor.

Attempts have also been made to operate pH control at pH 8. Retention times of less than one day were obtained and the system behaved much more erratically than at the pH values reported in detail in this paper. One possible application of control at pH 8 might be for odor control.

The results reported are valuable in that they demonstrate that the efficiency of stabilization of a liquid waste can be improved using pH-controlled loading. Similar experiments are now in progress using swine waste from a slurry pit, after screening to remove coarse fiber, as the input. Preliminary results indicate that a pH-controlled slurry system operates differently.

### References


1    Jones, D. D., D. L. Day and J. C. Converse. 1969. Oxygenation capacities of oxidation ditch rotors for confinement livestock buildings. Proc. 24th Indust. Waste Conf., Purdue Univ., Indiana. p. 542.

2    Murray, I., J. Parsons and K. Robinson. 1975. Inter-relationships between nitrogen balance, pH and dissolved oxygen in an oxidation ditch treating farm animal waste. Water Res. 9, 25.

3    Report. 1974. Treatment of piggery wastes. North of Scotland College of Agriculture, Aberdeen, Scotland.

4    Robinson, K. 1974a. Waste treatment with a protein bonus. Symposium, 'Processing and management of agricultural waste'. Cornell Univ., Ithaca, N.Y. p. 415.

5    Robinson, K. 1974b. The use of aerobic processes for the stabilization of animal wastes. Crit. Review in Environ. Control, p. 193.

6    Robinson, K. and D. Fenlon. 1975. A comparison of pH and dissolved oxygen for the automatic control of loading to aerobic farm waste treatment systems. J. Appl. Bact. (in press).

# Performance of an Automated Waste Treatment and Recycle System

E. P. Taiganides, R. K. White

SENIOR MEMBER    ASSOC. MEMBER
ASAE         ASAE

A N automated treatment and recycle system for a nursery to finishing unit of 500 pigs was put into operation in April, 1971. The performance of the system unit was monitored for three years, from June 1971 to May 1974. The facility is located on the Research Farm of the Botkins Grain and Feed Company in Botkins, Ohio.

The purpose of this project was to demonstrate the technical and environmental feasibility of an automated system of waste removal, collection, treatment and recycle without creating pollution or public nuisance. The total system is shown in Fig. 1. The principal unit operations are the flushing system, stationary screen, aerobic solids stabilization, solids storage, the oxidation ditch, the clarifier and the effluent recycle component. Details of the design and operational parameters were given at the 1971 International Symposium on Livestock Wastes and were published (Taiganides and White 1971).

## ANIMAL PRODUCTION

The pigs in the animal unit came from a sow herd belonging to the farm. The number of animals in the production unit ranged from 184 to 593 as shown in Fig. 2. The total liveweight ranged from 7,120 kg (15,700 lb) to 24,700 kg (54,000 lb) and average weight of the pigs ranged from 22 to 66 kg (47 to 145 lb) per pig. Feed intake ranged from 5 percent of liveweight per day for pigs below 20 kg, to 4 percent of liveweight per day for pigs averaging between 20 and 70 kg, and less than 4 percent for heavier pigs; in other words, young pigs ate 1 kg of feed per day, growing pigs 2 kg/day, and finishing pigs 3 kg/day. Daily weight gain ranged from 0.5 to 1.0 kg/day. Feed efficiency was 3.3 kg of feed per kg of average weight gain. (Nelson 1974).

Based on a total wastewater and recycle water monitoring study, the volume of wastewater generated by the pigs is estimated to be 12 l/day for each 100 kg of liveweight (1.4 gpd for 100 lb-pig). This volume of wastewater is made up of feces, urine, wasted feed and drinking water spillage. On a weight basis, the wastewater generated daily from the building is estimated as 10-12 percent liveweight.

The total daily volume of wastewater flushed out of the building to the treatment system was 58 m³/day (15,400 gpd). On a per-animal basis this daily flow amounted to 330 l/day for each 100 kg pig (40 gpd for 100 lb-pig). The range of the daily flow was 38 to 106 m³/day. Table 1 summarizes the characteristic of the wastewater flushed from the building.

Approved for publication as Journal Article No. 46-75 of the Ohio Agricultural Research and Development Center (OARDC). The project was supported principally by Grant No. R-801125 from the US Environmental Protection Agency and was completed in cooperation with Botkins Grain and Feed Co., Botkins, OH.

The authors are: E. P. TAIGANIDES, Professor, and R. K. WHITE, Associate Professor, Agricultural Engineering Dept., OARDC and Ohio State University, Columbus.

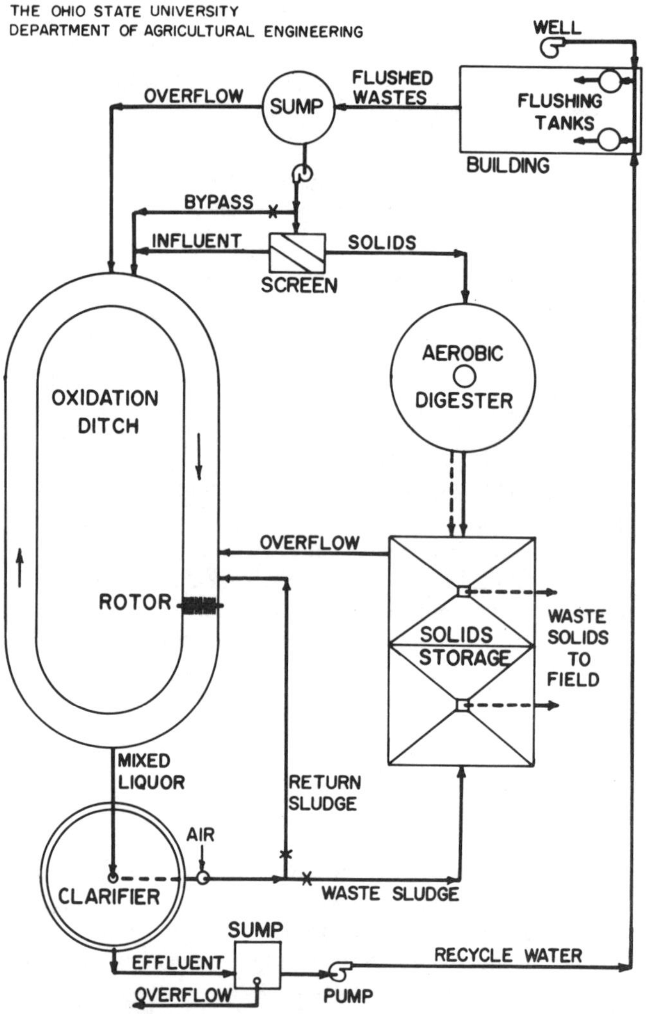

FIG. 1 A schematic of the flow lines of waste materials from the confinement building to the waste treatment units and back to the building for reflushing.

## PERFORMANCE OF EACH UNIT OPERATION

Unit operations were grouped as primary and secondary treatment. The objectives of the primary treatment were to separate settleable solids, stabilize and dispose of the solids. The stationary screen and aerobic stabilization of the solids were grouped as primary treatment. Secondary treatment included the oxidation ditch and final clarification. The initial flushing of wastes and the recycling of clarified effluent complete the system. Tertiary treatment was investigated with laboratory tests and are reported separately (Mehta and Taiganides 1975). A computer model for the total system was developed by Mote (1974).

### Flushing System

The automatic siphons used to periodically discharge a

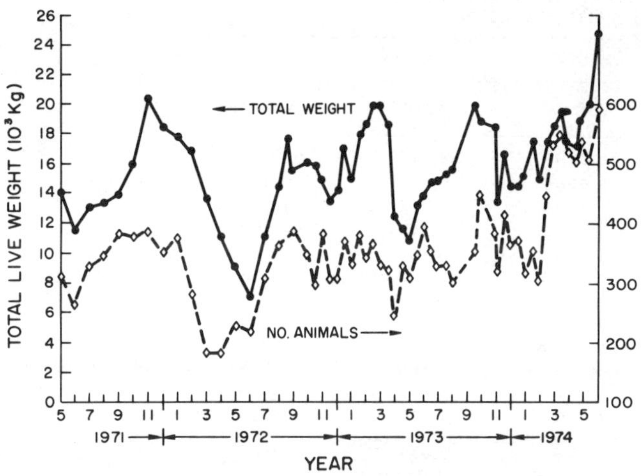

FIG. 2 Animal population and total live weight in swine production unit.

quantity of flushing water worked reliably throughout the study. With continuous recycling of clarified supernatant, the flushing interval was about one hour. This flushing interval was adequate to remove the feces, even though the slopes of the flushed areas were less than 1 percent, which is considered the minimum recommended slope for the water quantities used in flushing.

## Stationary Screen

The stationary, rundown type screen removed pig hair, grain hulls and other particles from the influent to the oxidation ditch. The screen was 46 cm (18 in.) wide and 76 cm (30 in.) long. The openings were 0.10 cm (40 mil) wide. Experimentation to determine the efficiency of solids separation on this screen was about 6 l/min/cm of screen width. The screened fraction had 9 percent TS on a wet basis and 35 percent of the influent TS by volume. Under most operating conditions the TS were about 7 percent, or less if blinding of the screen occurred.

The major problem with the screen was the clogging of the openings with a film of biomass. Because wastewater was being flushed about every hour, the screen remained moist and the film of biomass did not dry and flake off. It became necessary to brush the screen with a liquid disinfectant to mechanically break up the film.

## Aerobic Stabilization of Solids

Aeration was accomplished by using floating aerators with submerged air discharge. It was the initial hope that thermophilic temperatures could be reached in the aerobic digester. In a test operating the digester as a 22-day batch system, temperature of the tank contents was raised over the ambient air by 14 to 22 C. Fig. 3 shows that the highest temperature attained was 36 C, when the air temperature was 16 C. In this batch test the TS were reduced from 3.7

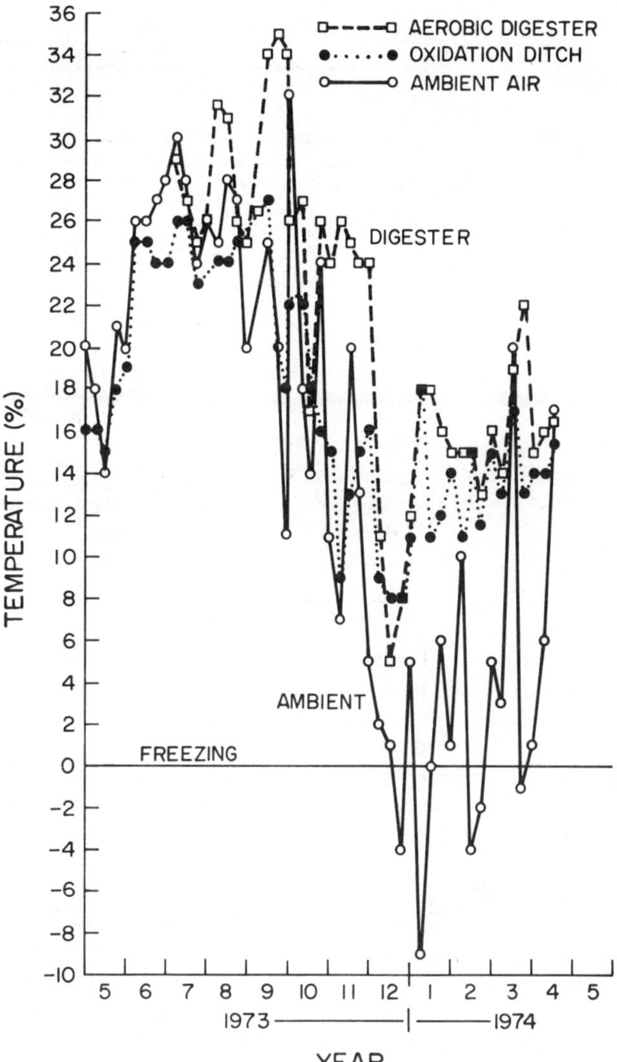



FIG. 3 Temperatures of aerobic digester, oxidation ditch and ambient air.

to 1.0 percent (wb), BOD from 5000 to 2100 mg/l and COD from 29,500 to 8,300 mg/l. Also, the pH rose from 7.0 to 8.0.

When operating on a continuous basis, the aerator maintained temperatures in the digester higher than that in the oxidation ditch as shown in Fig. 3. By using in-the-ditch aerators with submerged air outlets and covering the ditch with plywood, the oxidation ditch temperature was maintained above freezing even when ambient temperatures reached 9 C. (See Fig. 3).

## Stabilized Solids Storage

The effective capacity of the solids storage tank was over one month. The hoppered bottom of the tank proved effective in concentrating the solids and in allowing removal by a vacuum tank wagon. The capacity of the storage tank could

TABLE 1. CHARACTERISTIC OF THE WASTEWATER FLUSHED
OUT OF THE SWINE BUILDING

| Parameter | Symbol | Range mg/l | Average mg/l | kg/day |
|---|---|---|---|---|
| Total solids | TS | 2,100 - 34,400 | 8,500 | 500 |
| Total volatile solids | TVS | 1,040 - 27,900 | 7,200 | 420 |
| Total suspended solids | TSS | 540 - 17,200 | 4,300 | 250 |
| 5-day biochemical oxygen demand | BOD | 390 - 6,200 | 1,500 | 90 |
| Chemical oxygen demand | COD | 550 - 28,100 | 6,400 | 370 |

| Parameter | Unit | Average | | |
|---|---|---|---|---|
| | | Annual | Apr-Oct | Nov-Mar |
| Volume of flow* | l/day | 58,070 | 58,070 | 58,070 |
| Mixed liquor | mg/l | 8,555 | 7,600 | 10,026 |
| susp. sol. (MLSS) | kg | 915 | 813 | 1,073 |
| | (lb) | (2,013) | (1,789) | (2,360) |
| Total solids (TS) | mg/l | 5,826 | 5,196 | 6,949 |
| | $kg/m^3$ per day | 3.2 | 2.8 | 3.8 |
| | ($lb/1000\ ft^3$ per day) | (197) | (176) | (235) |
| Volatile solids (TVS) | mg/l | 3,367 | 2,967 | 4,093 |
| | $kg/m^3$ per day | 1.8 | 1.6 | 2.2 |
| | ($lb/1000\ ft^e$ per day) | (114) | (100) | (138) |
| | (lb/d per 100 lb MLSS) | (21) | (21) | (22) |
| BOD | mg/l | 1,400 | 1,215 | 1,733 |
| | $kg/m^3$ per day | 0.76 | 0.66 | 0.94 |
| | ($lb/1000\ ft^3$ per day) | (47) | (41) | (59) |
| | (lb/d per 100 lb MLSS) | (8.9) | (8.7) | (9.4) |
| COD | mg/l | 4,606 | 3,391 | 6,189 |
| | $kg/m^3$ per day | 2.5 | 1.8 | 3.4 |
| | ($lb/1000\ ft^3$ per day) | (156) | (115) | (209) |

*Assumed same for all periods

be extended by allowing supernatant to overflow back into the oxidation ditch. At no time was an offensive odor attributed to solids storage. Also, during field spreading on the surface of cropland, the odor was not so intense as to be a problem. A model for storage and land disposal of solids was developed and is reported separately. (Mote and Taiganides 1975).

## Oxidation Ditch

The average detention time in the oxidation ditch was approximately two days. Table 2 shows that the average BOD loading rate for the ditch was about three times the design loading rate of 0.28 $kg/m^3$/day. Also the loading rate for TVS was almost twice the design rate of 1.0 $kg/m^3$ (Taiganides and White 1971). The BOD loading per meter of rotor averaged 2.2 kg BOD/hr per meter of rotor (1.5 lb BOD/hr per ft of rotor), or twice the recommended loading rate (Loehr 1974). The ratio of influent BOD to the weight of MLSS averaged to be twice as high as that being recommended for dilute waste with one day detention time (Jones et al. 1971).

Fig. 4 shows the influent BOD data over the three year period of operation. There is considerable scatter in the data. However, a statistical analysis of all the data showed the median value for the influent BOD was smaller than the

mean for all the periods of the year. The same relationship between mean and median values were found to be true for the other parameters, such as TS, TVS, COD, etc.

Despite the higher than recommended loading rates, the oxidation ditch gave acceptable treatment. Effluent quality varied with the time of the year. Fig. 5 indicates that the monthly mean of the BOD in the effluent was much larger during the winter than the summer months. Fig. 6 shows that 50 percent of the time monthly effluent BOD was below 80 mg/l during April-October. The average for the same period was 157 mg/l or almost two times greater than the median. During the warm months of the year, April-October, BOD effluent was as low as 21 mg/l and was measured in the 20 to 50 mg/l range for several weeks of opera-

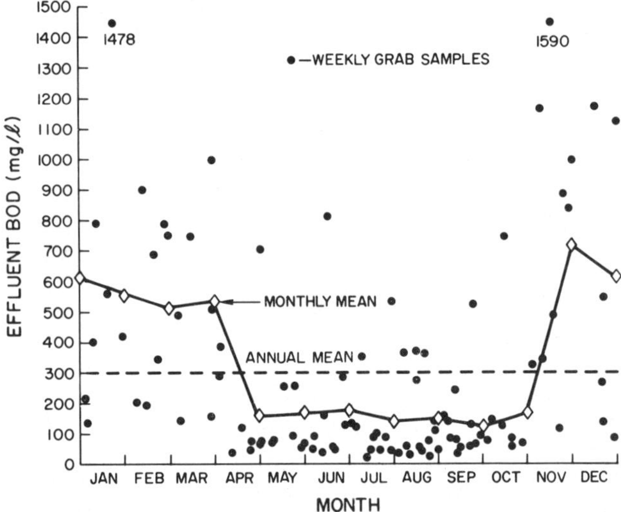


FIG. 5 Weekly average monthly and annual mean plant effluent BOD, June 1971 to May 1974.

tion during the summer.

The major problems in the operation of the oxidation ditch were foaming and freezing. Severe foaming occurred in the spring and the fall. Vegetable oil and commercial defoaming agents were tried. Both worked for a period of time and would have to be reapplied. Time and costs of materials made this method of foam control unsatisfactory. Control of foam was achieved by spraying the foam with liquids pumped from the wet well. Fig. 7 shows the spray-

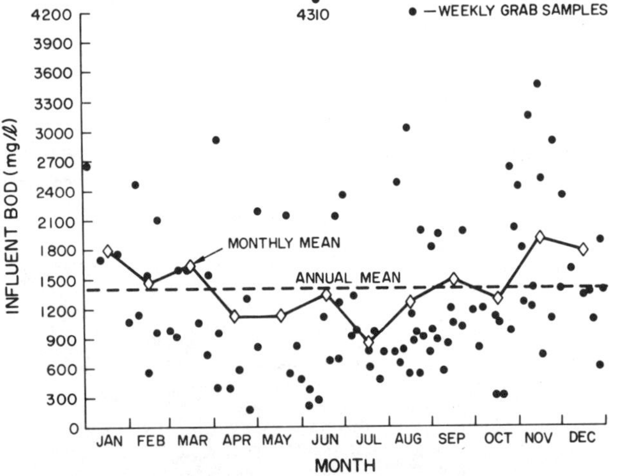


FIG. 4 Weekly, average monthly, and annual mean plant influent BOD, June 1971 to May 1974.



MANAGING LIVESTOCK WASTES

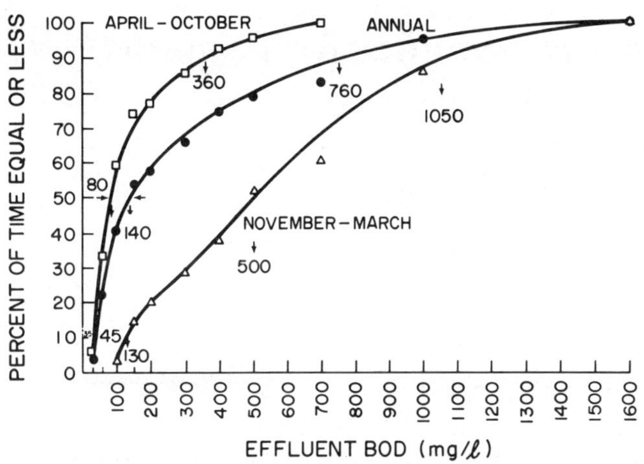


FIG. 6 Percent of time effluent BOD was equal or less than given value for the period April-October and for year-round operation.

FIG. 7 Nozzles of spraying equipment apparatus breaking down foam in oxidation ditch.

ing system in operation. Freezing was eliminated by partially covering the ditch with plywood sheets and with the use of aerators with submerged outlets. Such aerators did not cool the mixed liquor as much as the rotor would.

## Clarifier

The flow rate into the clarifier is determined by the pumping rate from the sump to the screen, which averaged 280 l/min giving an overflow rate of 28 m³/m² (690 gal/ft²) / day. This was only 11 m³/day per m² above the designed overflow rate. When MLSS were greater than 8000 mg/l poor settling occurred in the clarifier (Shutt et al. 1975).

## Recycle System

To the original 4.4 m³ (1000 gal) wet well another 4.4 m³ was added to provide surge capacity for the system. Sludge bulking in the clarifier caused clogging problems in the recycle lines. This was corrected by opening throttle valves, which would allow the pipes to be flushed out.

## OVERALL SYSTEM EVALUATION

The overall seasonal treatment efficiencies are given in Table 3. The average monthly BOD removal efficiency of the plant ranged from a minimum of 62 percent in the winter months to a maximum of 91 percent in the summer months, with the annual mean being 79 percent. Average annual removal efficiencies (monthly ranges) for other parameters were 67 percent for COD (51-76 percent), 43 percent for TS (31-52 percent), 57 percent for TVS (44-64 percent) and 82 percent for TSS (42-94 percent). Fig. 8 is a plot of the average monthly BOD and COD removal efficiencies. There is an expected seasonal effect upon removal efficiency. The BOD removal efficiency is improved during the summer months by about 15 to 20 percent.

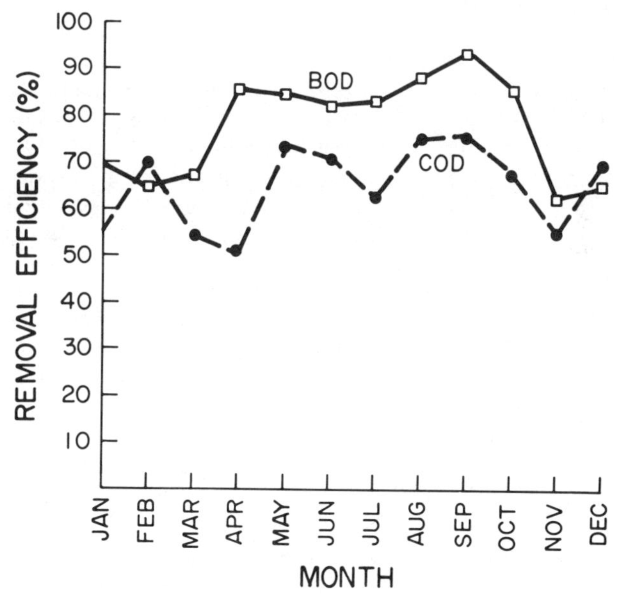


FIG. 8 Average monthly BOD and COD removal efficiencies.

## CONCLUSIONS

This demonstration project has shown that the handling and treatment of swine wastes may be automated with no resulting water pollution and with minimum odor nuisance. The average annual BOD removal was 79 percent for the

(Continued on page 575)

### TABLE 3. AVERAGE, ANNUAL AND SEASONAL TREATMENT SYSTEM EFFICIENCIES

| | Annual | | | Apr - Oct | | | Nov - Mar | | |
|---|---|---|---|---|---|---|---|---|---|
| | mg/l | | Percent removal | mg/l | | Percent removal | mg/l | | Percent removal |
| | Infl | Effl | | Infl | Effl | | Infl | Effl | |
| BOD | 1400 | 300 | 79 | 1200 | 160 | 87 | 1700 | 600 | 65 |
| COD | 4600 | 1500 | 67 | 3400 | 1140 | 66 | 6200 | 2300 | 63 |
| TS | 5800 | 3300 | 43 | 5200 | 2900 | 44 | 7000 | 4300 | 39 |
| TVS | 3400 | 1450 | 57 | 3000 | 1200 | 60 | 4100 | 2000 | 51 |
| TSS | 3900 | 700 | 82 | — | — | — | — | — | — |

MANAGING LIVESTOCK WASTES

# Surface Aeration: Design and Performance for Lagoons


F. J. Humenik, M. R. Overcash, T. Miller

MEMBER     ASSOC. MEMBER
ASAE       ASAE


DESIGN and operational criteria for aerated lagoons vary greatly depending upon equipment and intended treatment strategy. Aeration recommendations for odor control have varied from satisfying the five-day biochemical oxygen demand ($BOD_5$) to one-third that value (Muehling 1969). Aeration rates for complete treatment are reported to be from one to two times the daily $BOD_5$ input or about equal to the total COD load (Muehling 1969). Size requirements for aerated swine waste units vary from 6-8 $ft^3$/150-lb hog to 1 $ft^3$/lb of hog (Irgens and Day 1966, Jones et al. 1969, Jones et al. 1970, Muehling 1969).

### Model Field Units

Two pilot scale fabricated steel units 8 ft deep and 11.5 feet diameter were constructed so that a surface aerator could be fixed to a catwalk superstructure spanning the tank. A 1/4-hp variable speed mixer-aerator requiring submergence of the impeller just below the liquid surface was selected. The constant liquid level required for this aerator was maintained by a combination of an overflow pipe and water reservoir controlled by a mechanical float valve to counter-balance rainfall and evaporation, Fig. 1.

Impeller rotation for one unit was 65 rpm and 110 rpm for the other reactor. These rotor speeds require approximately 0.05 and 0.08 hp, respectively. Each reactor received weekly 224 gallons of raw swine waste from an underfloor storage pit diluted to a constant concentration of 40,000 mg of COD/l.

Calculations based upon the COD input and continuous aerator operation indicated that the 0.08 and 0.05 hp units received about 70 percent and 40 percent, respectively, of the oxygen required for complete COD stabilization based upon the manufacturer's estimated oxygen transfer rate of 3.5 lb $O_2$/hp/hr. After 15 months' operation, an overall mass balance indicated that the oxygen transfer was about 2.5 to 3.0 lb/hp/hr for both units, (Table 1). Thus the pilot-scale reactor with the higher horsepower input for oxygen transfer would be expected to have a lower COD supernatant concentration. However, steady state data summarized in Table 2 revealed that the 0.08 hp unit had higher COD, TOC, and orthophosphate supernatant concen-

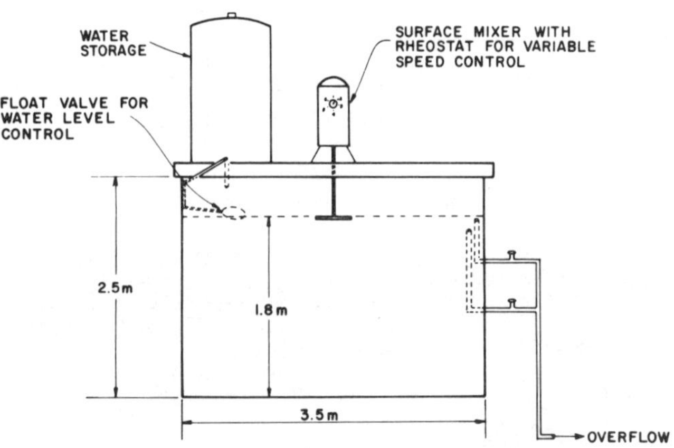


FIG. 1 Pilot scale surface aeration reactors.

trations but about the same nitrogen level as the aerated unit with the lower power input.

Concentration variations with depth in the reactor fluid were determined to evaluate the degree of mixing induced by the employed surface agitation. Orthophosphate concentrations with depth for the high and low aeration rates are shown in Fig. 2. The concentration-depth profiles for these units with similar management show a uniform upper zone and then an abrupt concentration increase indicative of a sludge blanket. The sludge depth was greater in the unit with the lower horsepower input (0.05 hp) and the phosphate supernatant concentrations were lower by a factor of about 3 below the high aeration unit (350 vs 125 mg P/l) indicating less sludge resuspension or greater initial raw waste settling with the low rate unit. The concentration-depth data also show that the larger horsepower input resulted in only a two-fold higher supernatant concentration of COD and TOC in comparison with the reactor with the lower energy input.

These data indicate that the higher concentration levels for the larger aeration input result from increased agitation and suspension of solids. Additionally, because the phosphate level is about three times higher at the greater aeration rate and the organic level is only twice as high (Table 2), the expected higher level of oxygen addition and thus organic waste stabilization for the higher horsepower input are corroborated. No dissolved oxygen has been recorded in either unit even in the surface layers.

The supernatant concentration of ammonia is 1250 to 1300 mg $NH_3$-N/l for the 0.05 hp input and 900 to 1000 mg $NH_3$-N/l for the 0.08 hp level. Ammonia concentrations in the two units are more similar than phosphate and COD supernatant levels indicating that the greatest impact of aeration is on ammonia volatilization. Surface aeration results in augmentation of gas phase transport due to increased surface renewal and additional surface area of


The authors are: F. J. HUMENIK, Associate Professor, M. R. OVERCASH, Assistant Professor, and T. MILLER, Graduate Student, Biological and Agricultural Engineering Dept., North Carolina State University, Raleigh.

The use of trade names in this publication does not imply endorsement by the North Carolina Agricultural Experiment Station of the products named, nor criticism of similar ones not mentioned.

Acknowledgement: This work was in part supported with funds from the North Carolina Agricultural Experiment Station and North Carolina Agricultural Extension Service in conjunction with EPA Grant No. R802203, "Design Criteria for Swine Waste Treatment Systems." The fullscale demonstration investigations were possible because of the excellent assistance of the two cooperating producers, Roger and Gails Phillips, Chick Sales Hatchery, Siler City, NC, and Dr. George D. Wetherill, Lexington Swine Breeders, Lexington, NC.

TABLE 1. OXYGEN MASS BALANCE FOR PILOT SCALE SURFACE AERATION REACTORS LOADED WITH SWINE WASTE.

| | Reactor X | Reactor Z |
|---|---|---|
| Operation period | a) 47 weeks, 7 days/week 0.08 hp<br>b) 20 weeks, 6 days/week 0.08 hp | a) 47 weeks, 7 days/week 0.054 hp<br>b) 20 weeks, 7 days/week 0.161 hp |
| Manufacturer's rating 3.5 lb $O_2$/hr/hp | 3030 lb oxygen | 3370 lb oxygen |
| Oxygen demand in input | 4015 lb oxygen | 4015 lb oxygen |
| Oxygen demand in effluent | 1070 lb oxygen | 725 lb oxygen |
| Oxygen demand in reactor | 490 lb oxygen | 605 lb oxygen |
| Oxygen demand unaccounted for | 2455 lb oxygen | 2685 lb oxygen |
| Oxygen transfer rate based on unaccounted COD | 2.9 lb $O_2$/hr/hp | 2.8 lb $O_2$/hr/hp |

TABLE 2. STEADY-STATE SUPERNATANT CONCENTRATIONS (MG/L) FOR PILOT SCALE SURFACE AERATION REACTORS LOADED WITH SWINE WASTE.

| Parameter | Aeration, watts | |
|---|---|---|
| | 37(0.05 hp) | 60(0.08 hp) |
| COD | 7750 | 15000 |
| TOC | 3000 | 5500 |
| TKN | 1500 | 1500 |
| $NH_3$-N | 1300 | 1000 |
| $O-PO_4$-P | 125 | 350 |

generated droplets. Thus the expected resuspension of nitrogen components from the sludge layer is counteracted partially by increased ammonia volatilization in the unit with the highest horsepower input.

A large number of minute bubbles were observed to be liberated when the aerators were stopped. These bubbles were much smaller than normally observed in lagoons as characteristic of gas released from bottom sludge. Concentrations of gas components for samples collected at the surface were highly variable as shown in Table 3 but generally products of anaerobic fermentation such as methane and carbon dioxide and a high nitrogen content were present. The explanation for these recorded gaseous quality and quantity data and corresponding mechanisms is not yet fully developed. Either surface molecular level nitrification followed by bulk phase liquid denitrification or entrained air bubbles which are stripped of oxygen present potential origins of these bubbles with a high nitrogen content. Nitrate levels were subject to analytical interferences because of the high color and organic content of the supernatant. Approximate tests yielded 10-15 mg $NO_3$N/l. This nitrate content is low in comparison with typical oxidation ditch levels but does not rule out significant nitrate generation or either premise for the bubble generation mechanism, thus the phenomenon investigation continues. Unfortunately a nitrogen mass balance will not indicate denitrification losses because of the confounding effect of ammonia volatilization.

After steady-state supernatant conditions were reached, the aerator was stopped for 24 hours after loading in the 0.08 hp unit and aeration to the lower horsepower unit (0.05 hp) was increased to 137 rpm or 0.161 hp. Allowing a quiescent period of 24 hours for settling did not appear to improve the supernatant quality as evidenced by concentrations recorded for this operational technique. The supernatant concentrations for the unit with the increased rotor speed immediately rose above those recorded earlier for the 0.08 hp input due to the greater scour of bottom sludge associated with the higher mixing intensity. Unexpectedly, TKN showed only a modest increase further verifying that the increased ammonia volatilization associated with a higher degree of aeration compensated for the greater suspension of nitrogenous solids.

## FULL SCALE LAGOONS

### Boar Testing Station

Wastes from two identical totally roofed swine facilities with concrete floors are scraped or flushed into collection troughs for daily water conveyance into two lagoons operated in series. The first lagoon, which was built in 1961 for the initial housing unit, overflows into the second lagoon, which was completed in 1971. Sludge accumulation in the initial lagoon has been about 18-26 inches. Average loading to the primary lagoon during the last year has resulted in about 65 ft$^3$/100-lb hog. Supernatant TKN concentrations show a summer and winter cycle of 250 mg/l to 350 mg/l respectively.

Aeration is provided to the second lagoon by a floating oxygenator specified as heavy duty, 5 hp, 4800 gpm, 13 ft/sec discharge and equipped with an anti-erosion shield to

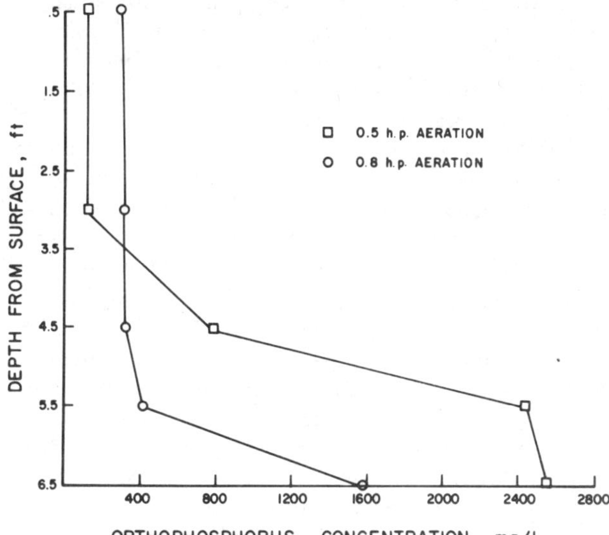


FIG. 2 Orthophosphorus concentrations with depth for pilot scale surface aeration reactors loaded with swine waste.

TABLE 3. TYPICAL COMPOSITION OF GAS BUBBLES COLLECTED AT SURFACE OF PILOT SCALE SURFACE AERATION REACTORS LOADED WITH SWINE WASTE.

| Date | Aeration horsepower | Parameter volume percentage | | | |
|---|---|---|---|---|---|
| | | $O_2$ | $N_2$ | $CH_4$ | $CO_2$ |
| 7/16/73 | 0.05 | 2.6 | 30.2 | 67.2 | trace |
| 7/19/73 | 0.05 | 2.0 | 35.6 | 61.0 | 1.4 |
| 8/ 1/73 | 0.08 | 4.0 | 91.0 | 4.0 | 1.0 |
| 7/ 2/74 | 0.08 | 2.0 | 96.0 | 0.0 | 2.0 |
| 7/11/74 | 0.08 | 4.0 | 76.0 | 18.0 | 2.0 |

promote surface pumpage.* The lagoon size was 90 ft x 75 ft resulting in an aeration rating of 1350 ft$^2$/hp.

The treatment strategy under investigation is to obtain maximum wastewater stabilization and to employ the minimum hp needed for complete surface agitation but not the energy levels required for complete reactor mixing. Surface scum and odor have been virtually non-existent in this aerated unit with an average surface dissolved oxygen content of about 10 mg/l, and supernatant concentrations of 150 mg COD/l, 30 mg NH$_3$-N/l, and 20 mg O-N/l. Thus this heavy duty model provided complete surface agitation at 1350 ft$^2$/hp for an influence area of about 90 ft by 70 ft.

### Lexington Swine Breeders

An aerated lagoon was designed to control odor and provide stabilization pretreatment prior to terminal land application for a feeder pig operation with a steady-state live weight population of 200,000 lb based on satisfying about one-half the input BOD$_5$ or about 1/3 of the input COD. At an oxygen transfer rate of 3 to 3.5 lb of oxygen per hp hour, about 6 hp were required. Two 3 hp oxygenators were purchased to obtain greater operational flexibility and to conform to the concept that two small units were better than a single large equivalent.

The lagoon was sized at about 80 ft by 100 ft to be between the complete mixed diameter criteria of 35-55 ft and total horizontal effect of 125-145 ft specified for the purchased oxygenators and not detention time. However, final surface measurements of the 8 ft deep lagoon were 90 ft by 140 ft which resulted in a surface rating of 2100 ft$^2$ per hp. Mixing was not sufficient to keep the total surface agitated, thus scum built up in the unaffected areas at lagoon corners resulting in unsightly and odorous conditions. Subsequently a 5 hp aerator was added to make the total input 11 hp and reduce the surface rating to 1145 ft$^2$/hp. This has resulted in almost complete surface mixing with only occasional dead spots existing about 8 to 10 ft from the shoreline where scum accumulates as a result of wind effects. The two adjacent 3 hp units keep the influent area of the lagoon well mixed, but the one 5 hp aerator at the effluent side does not always result in complete surface agitation.

Visible observation indicates that the surface agitation potential of a 5 hp oxygenator is somewhat less than two 3 hp units of the same manufacturer design series. Specifications for the 5 hp custom series oxygenator used at this site were 1800 rpm with a pumpage of 3500 gpm at 15.1 ft/sec discharge velocity. The surface area of influence for these aerators is specified to be about two-thirds that of the heavy duty model. Thus differences in liquid pumpage and surface impact were found to exist and to have a direct effect on satisfactory operation. This site has yielded data on the maximum surface area per horsepower input allowable to achieve complete surface mixing with this equipment which is defined as about 1000 ft$^2$ per hp.

The impact of surface aeration on improved ammonia loss was verified for these field units as with the pilot scale reactors. Depth samples for this aerated lagoon showed that the upper 5.5 ft was relatively homogeneous. The sludge depth was about 1 to 2 ft with virtually no sludge beneath the aerator despite the anti-erosion shield employed. The unaerated lagoon following this primary unit reflected the

---



**TABLE 4. LEXINGTON SWINE BREEDERS — SERIES AERATED-UNAERATED LAGOONS PERFORMANCE DATA.**

| Date | Aerator hp | Aerated unit Cod ------ mg/l ------ | TKN | Unaerated unit COD ------ mg/l ------ | TKN |
|---|---|---|---|---|---|
| 5/26/73 | 0 | --- | --- | | 470 |
| 8/24/73 | 0 | -- | --- | 4370 | 300 |
| 2/17/74 | 6 | 5580 | 710 | 1280 | 200 |
| 2/18/75 | 11 | 8920 | 920 | 800 | 175 |
| 2/26/75 | 11 | 9350 | 990 | 750 | 210 |

improved ammonia loss associated with the staged conversion from 2100 ft$^2$/hp to 1100 ft$^2$/hp. Original unaerated lagoon levels of about 4500 mg/l COD and 450 mg/l TKN dropped to about 1250 mg/l COD and 200 mg/l TKN for 6 hp primary input for three months, and to 800 mg/l COD and175 mg/l TKN after one month of 11 hp primary aeration, Table 4. Thus odor control and improved nitrogen removals were achieved with the custom series units at the current aeration level of 1100 ft$^2$/hp.

### Chick Sales Hatchery

The additional demonstration scale evaluation of lagoon aeration was conducted in conjunction with series aerated-unaerated lagoon pretreatment of hatchery waste prior to terminal land recycling. A schematic of the total system for domestic and process wastewater management at this hatchery is shown in Fig. 3. Aeration requirements were calculated on the basis of laboratory COD data for whole eggs and then extrapolated into the maximum waste load associated with the lowest expected hatchability rate of 80 percent. The projected double capacity load of about 150,000 no-hatches per week resulted in an oxygen demand of 44 lb/hour which required about 15 hp of aeration based upon a transfer of 3 lb of oxygen per hp hour. The domestic load oxygen demand was insignificant. Therefore three 5 hp aerators with anti-erosion shields were purchased to provide oxygenation for the total COD input associated with the poorest expected hatchability for double capacity load.

Criteria for complete mixing were influential in the design of this aerated lagoon because of problems experienced with achieving complete surface agitation at the swine facility. The 6 ft deep lagoon had water surface dimensions of 51 ft by 130 ft with a total capacity of 33,750 ft$^3$ for a volumetric rating of 2250 ft$^3$/hp or a surface rating of 445 ft$^2$/hp. Thus this lagoon was sized between the equipment manufacturer's recommendation of 1333 ft$^3$ or 167 ft$^2$/hp for complete mix and 4000 ft$^3$ or 766 ft$^2$/hp for complete surface agitation.

This aerated unit which provides a 63-day mean cell residence time has excellent total surface agitation with all three 5 hp units operating and good surface agitation with just the outside two aerators running. Characteristic aerated lagoon levels are about 1500 mg/l COD and 300 mg/l TKN with secondary unaerated lagoon values being 500 mg/l COD and 150 mg/l TKN. Odor control was achieved after initial start-up with just the center aerator operating until the total present production capacity of 285,000 eggs at 88 percent hatchability resulting in about 34,000 no-hatches per week was recently achieved. Thereafter the outside two 5 hp aerators have been required for odor control and in addition have provided total surface agitation. Foaming has been an intermittent problem, especially on overcast days. Generally the most effective management scheme is reduced aeration input by partial operation of oxygenator

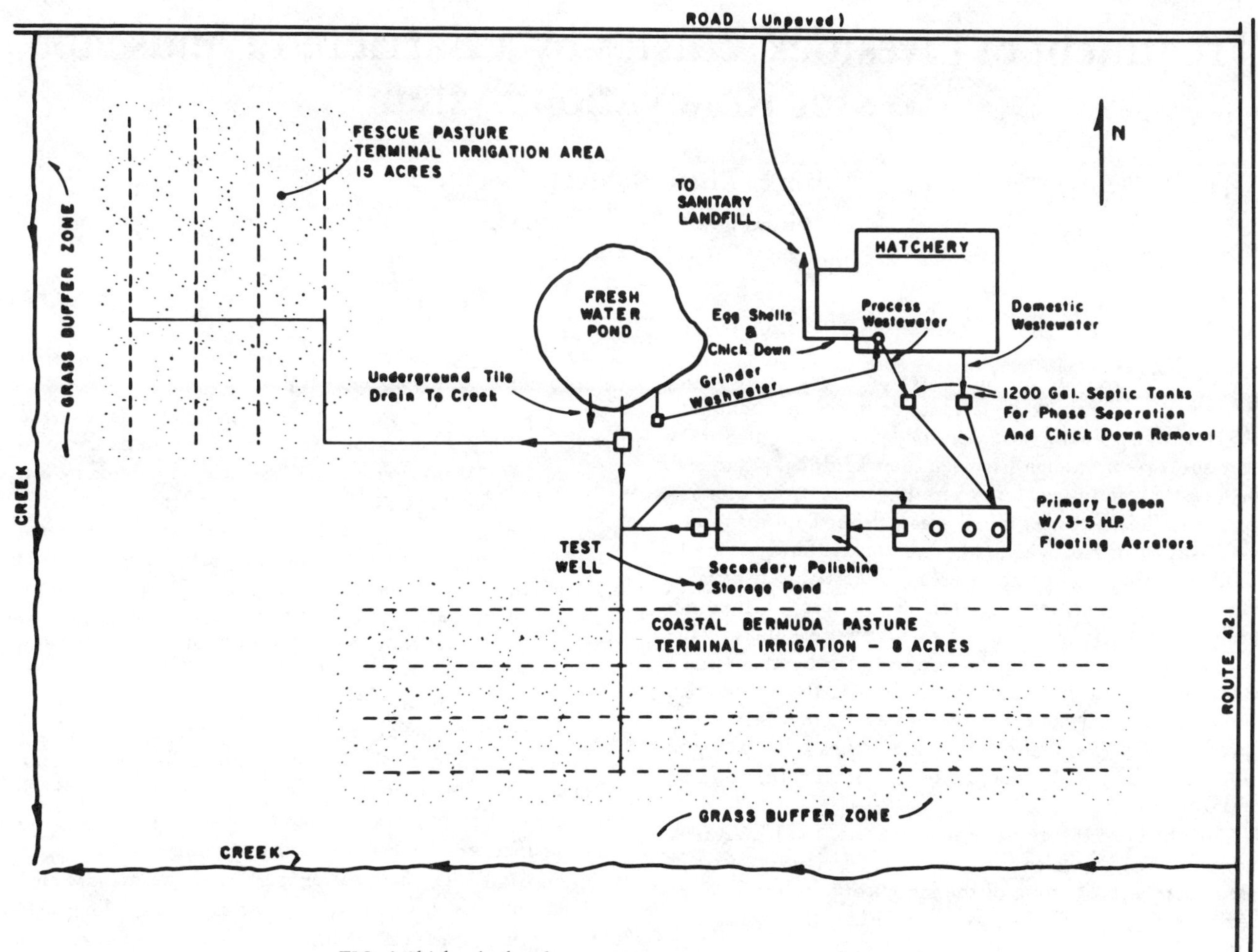



FIG. 3 Chick sales hatchery waste management system diagramatic

at the influent side. Thus this unit documents the minimum surface area required for complete surface mixing and helps set our recommendation range for achieving complete surface agitation at about 700 to 1000 ft$^2$/hp for these floating aerators.

## CONCLUSIONS

1   Ammonia concentrations for pilot scale field units with 0.05 to 0.16 hp aeration input are more similar than supernatant phosphate and COD levels indicating that the greatest impact of surface aeration is on nitrogen reduction by ammonia volatilization. Thus the expected resuspension of nitrogen components from bottom sludge with higher mixing intensities is counteracted partially by increased surface volatilization.

2   Aeration strategy employed for full-scale field lagoons is to accomplish odor and scum control through complete surface agitation by horizontal pumpage which requires a minimum of 1 hp/1000 ft$^2$ surface area for utilized equipment.


### References

1   Irgens, R. L. and D. L. Day. 1966. Laboratory studies of aerobic stabilization of swine waste. J. Agr. Engr. Res. 11(1):1-10.
2   Jones, D. D., J. C. Converse and D. L. Day. 1969. Aerobic digestion of swine waste. Paper presented at the International Congress of Agricultural Engineering. C.I.G.R. Baden-Baden, Germany.
3   Jones, D. D., D. L. Day and A. C. Dale. 1970. Aerobic treatment of livestock wastes. Univ. of Ill. Exp. Sta. Bull. 737. Urbana-Champaign.
4   Muehling, A. J. 1969. Swine housing and waste management. A research review. AEnd-873, Dept. of Ag. Engr., Univ. of Ill., Urbana-Champaign. 91 p. August.

# Treatment of Livestock Wastes by a Barriered Landscape Water Renovation System


**William F. Ritter, Robert P. Eastburn**

ASSOC.
MEMBER

AFFILIATE
MEMBER


OVER 50 percent of the wastes produced by livestock are from livestock in feedlots or in confinement rearing. Direct disposal of wastes on the land as fresh manure or after short-term storage has been the traditional practice of the livestock industry, but where large quantities of waste are generated in concentrated areas, direct land disposal is not always feasible. When large livestock enterprises are located near dense population areas, direct land disposal of manure often becomes a nuisance because of flies and odors, and sufficient land areas are often unavailable. Erickson et al. (1971) have developed a barriered landscape water renovation system (BLWRS) as a relatively inexpensive modified soil land disposal system. Their BLWRS reduced the nitrogen content in swine wastes from 340 ppm to less than 2 ppm and the phosphate content from 36 ppm to 0.07 ppm.

The object of this research is to evaluate the BLWRS as a means of treating dairy cattle wastes in Delaware. This paper is a report of some of the laboratory and preliminary field experiments associated with the BLWRS project.

## EXPERIMENTAL PROCEDURES

### Soil Columns

Prior to construction of a BLWRS, a laboratory experiment was conducted to evaluate soil type and the depth of soil through which livestock wastes would percolate. Soil columns constructed from acrylic plastic pipe with lengths of 91, 152, 183 and 244 cm and 10.2 cm in diameter were used in the experiment. The columns were packed with either Evesboro loamy sand, Sassafras loam or Matapeake silt loam soil. Table 1 shows some of the physical and chemical properties of the soils used in the experiment. Liquid dairy manure from a storage tank on the University of Delaware dairy farm was applied to the soil columns at loading rates of 1.25 and 2.5 cm/day. All of the effluent from each soil column was collected and analyzed weekly for COD, ortho phosphorus, total phosphorus, organic nitrogen and ammonia.

For all the experiments, COD analysis was done according to Standard Methods. Ortho phosphates and total phosphates were analyzed by the ascorbic acid and persulfate digestion methods (American Public Health Association 1971). Ammonia and nitrate-nitrite nitrogen analyses were done by the method of Bremner and Keeney (1965), and organic nitrogen by micro-Kjeldhal and nesslerization.

### Dilute Acid Soluble Phosphorus

Since phosphorus is one of the main pollutants removed by the BLWRS, a laboratory experiment was undertaken to measure the soil phosphorus changes when liquid dairy manure was added to soil. The objective of this part of the research was to investigate the changes in the dilute acid soluble phosphorus fraction of soil amended with liquid dairy manure. Soils used for the experiment were Evesboro loamy sand, Sassafras loam, and Matapeake silt loam. The physical and chemical properties of the soils used were similar to the properties presented in Table 1.

One hundred g air-dried soil samples that had passed through a 2 mm sieve were mixed with 3 percent and 7 percent total solids, dairy manure slurries in 500 ml containers. The samples were incubated for 21 days at 3, 20 and 30 C. New batches of manure slurries were used for each incubation temperature. Since manure is a heterogeneous material, the chemical properties of the manure slurries used for the different temperatures varied. Along with the dairy manure slurry samples a control sample for each soil type and temperature also was prepared. Control solutions were made by combining $NH_4Cl$, $NH_4NO_3$ and glucose in amounts to produce the same levels of $NH_4$, $NO_3$ and organic matter as found in the dairy manure slurries. Enough liquid dairy manure or control solution was added to the soil samples to bring them to field capacity. The samples were brought up to field capacity every second day during the 21-day incubation period by the addition of distilled water.

After incubation, the soil samples were air-dried and crushed to pass through a 2 mm sieve. Five subsamples of each treatment were analyzed for dilute acid soluble phosphorus by the Troug method (Jackson 1958).

### BLWRS

A BLWRS was constructed on the University of Delaware dairy farm in the spring of 1974. The dimensions of the BLWRS were 9.1 x 9.1 m with a depth of 2.4 m and 1:1 sideslopes. It was underlined with a 30 mil vinyl liner. The anaerobic zone of the BLWRS was constructed from a silt loam soil taken from the construction site while the


**Acknowledgments:** The research on which this paper is based was supported in part by funds from Hercules, Inc. and in part by funds from the U.S. Department of Interior under the Water Resources Act of 1964, Public Law 88-379.

Published with the approval of the Director of the Delaware Agricultural Experiment Station as Miscellaneous Paper No. 714.

The authors are: WILLIAM F. RITTER, Assistant Professor, and ROBERT P. EASTBURN, Research Assistant, Agricultural Engineering Dept., University of Delaware, Newark.


TABLE 1. PHYSICAL AND CHEMICAL PROPERTIES OF SOILS USED FOR SOIL COLUMNS AND DILUTE ACID SOLUBLE PHOSPHORUS STUDY.

| Soil Properties | Evesboro | Sassafras | Matapeake |
|---|---|---|---|
| Texture | Loamy sand | Loam | Silt loam |
| Percent moisture at 0.33 bars | 15.3 | 20.7 | 35.4 |
| Cation exchange capacity, me/100g | 3.84 | 6.65 | 9.46 |
| Organic matter content, percent | 0.8 | 3.1 | 4.2 |
| pH | 6.4 | 4.2 | 4.7 |



MANAGING LIVESTOCK WASTES

TABLE 2. AVERAGE CONCENTRATION OF COD, ORGANIC NITROGEN AND AMMONIA IN THE SOIL COLUMN EFFLUENT

| Soil type | Column length, cm | COD High | COD Low | Organic-N High | Organic-N Low | $NH_3$-N High | $NH_3$-N Low |
|---|---|---|---|---|---|---|---|
| | | | | mg/l | | | |
| Evesboro | 91 | 52 | 64 | 12.9 | 15.3 | 0.48 | 0.32 |
| | 152 | 63 | 69 | 13.2 | 14.9 | 0.48 | 0.41 |
| | 183 | 61 | 69 | 15.4 | 20.0 | 0.29 | 0.57 |
| | 244 | 78 | 82 | 9.0 | 12.4 | 0.33 | 1.10 |
| Sassafras | 91 | 35 | 68 | 11.2 | 14.5 | 17.1 | 10.1 |
| | 152 | 64 | 72 | 10.3 | 15.4 | 21.8 | 4.38 |
| | 183 | 485 | 690 | 14.1 | 17.6 | 2.02 | 2.43 |
| | 244 | 58 | 677 | 11.6 | 20.6 | 7.62 | 1.65 |
| Matapeake | 91 | 48 | 75 | 12.0 | 14.0 | 0.32 | 0.89 |
| | 152 | 49 | 97 | 12.8 | 15.5 | 1.27 | 1.17 |
| | 183 | 55 | 68 | 16.6 | 14.7 | 1.10 | 1.13 |
| | 244 | 88 | 89 | 14.4 | 13.3 | 5.35 | 0.91 |

*High loading rate of 2.5 cm/day and low loading rate of 1.25 cm/day.

TABLE 3. AVERAGE CONCENTRATION OF ORTHO PHOSPHORUS AND TOTAL PHOSPHORUS IN THE SOIL COLUMN EFFLUENT.

| Soil type | Column length, cm | Ortho-P High | Ortho-P Low | Total-P High | Total-P Low |
|---|---|---|---|---|---|
| | | | mg/l-P | | |
| Evesboro | 91 | 0.25 | 0.14 | 0.41 | 0.31 |
| | 152 | 0.25 | 0.10 | 0.31 | 0.16 |
| | 183 | 0.19 | 0.16 | 0.29 | 0.27 |
| | 244 | 0.16 | 0.14 | 0.24 | 0.25 |
| Sassafras | 91 | 0.04 | 0.07 | 0.14 | 0.18 |
| | 152 | 0.06 | 0.07 | 0.14 | 0.16 |
| | 183 | 0.09 | 0.06 | 0.36 | 0.24 |
| | 244 | 0.04 | 0.06 | 0.15 | 0.20 |
| Matapeake | 91 | 0.06 | 0.06 | 0.12 | 0.16 |
| | 152 | 0.05 | 0.07 | 0.12 | 0.25 |
| | 183 | 0.05 | 0.04 | 0.27 | 0.16 |
| | 144 | 0.03 | 0.05 | 0.10 | 0.18 |

*High loading rate of 2.5 cm/day and low loading rate of 1.25 cm/day.

aerobic zone was constructed from a loam soil that was being used as a cover for a nearby sanitary landfill.

Drainage for the BLWRS was provided by 10.2 cm diameter perforated plastic pipe laid at each end of the barrier. The pipes drained into two sumps constructed from steel drums. Semi-solid dairy manure was scraped from the holding area of the milking parlor and mixed with water in a 3.78m³ storage tank located near the BLWRS. Waste was applied to the BLWRS by an automated flow control system and it was spread on top of the BLWRS by a hose and slinger mechanism on a 3 x 3 m area.

Samples were collected three times a week for the first three months the BLWRS was operating and weekly during the later stages of operation from the sumps and liquid manure storage tank. All samples were analyzed for ortho phosphorus, COD, nitrate-nitrite nitrogen, organic nitrogen and ammonia.

## RESULTS AND DISCUSSION

### Soil Columns

Tables 2 and 3 show the average values of the COD, ammonia, organic nitrogen, ortho phosphorus and total phosphorus in the soil column effluent. The average values of the chemical properties of the liquid dairy manure applied to the soil columns are shown in Table 4. A total of 1752 mm of waste was applied to the high loading rate columns and 876 mm was applied to the low loading rate columns.

In all cases the COD was reduced by 95 percent or more except for three of the Sassafras columns. The high COD values of the effluent in these columns was probably caused by organic material being leached from the soil. All of the soil columns were leached with distilled water before the manure was applied and the effluent from these three Sassafras columns had high COD values with the distilled water. There was no definite trend in COD removal with soil type, column length or loading rate.

The amount of ammonia adsorbed by the soil varied with soil type. In general Evesboro loamy sand soil columns had the lowest concentrations of ammonia in the effluent, while the Sassafras loam had the highest levels of ammonia in the effluent. Ammonia removal was apparently not related to the cation exchange capacity of the soil.

In all cases over 99 percent of the ortho and total phosphorus was removed from the liquid dairy manure.

The results of the soil column study indicate that any of the three soils tested would give high COD and phosphorus reduction in a BLWRS. In general most of the COD, phosphorus, ammonia and organic nitrogen removal would take place in the top 91 cm of the soil so that most reduction would probably take place in the aerobic zone of the BLWRS. Of the three soils studied, Evesboro loamy sand would probably be the most satisfactory soil for a BLWRS since it has the highest infiltration rate and would probably remove 90 percent or more of the COD and phosphorus and give the greatest ammonia removal.

### Dilute Acid Soluble Phosphorus

Table 5 shows the average values of the Troug acid soluble phosphorus for the different soil types and incubation temperatures and the phosphorus content of the manure slurries used for each incubation temperature. An analysis of variance showed that the main effects of soil type, total solids level and temperature were significant at the 1 percent level together with all the interactions. This apparently indicates that the amount of dilute acid soluble phosphorus was the result of a soil biochemical equilibrium unique for each soil type. The data was normalized for variance caused by variation in the initial amounts of phosphorus in the dairy manure slurries before the analysis of variance was performed.

Table 6 shows the linear regression equations between dilute soluble acid phosphorus and the values of total soil phosphorus. The total soil phosphorus values are the sum of the pretreated soil total phosphorus and the dairy manure slurry total phosphorus. The linear regression analysis based on soil type indicates that there is a definite and characteristic rate of increase in Troug phosphorus with increasing amounts of dairy manure slurry. The linear regression equations on the basis of temperature indicate that the Troug phosphorus decreased only with the 30 C incubation temperature while on the basis of percent total solids, the

TABLE 4. AVERAGE CONCENTRATION OF CHEMICAL PARAMETERS OF THE MANURE APPLIED TO SOIL COLUMNS.

| Properties | Concentration, mg/l |
|---|---|
| COD | 2067 |
| Organic-N + $NH_3$-N | 61.4 |
| Ortho-P | 32.2 |
| Total-P | 43.5 |

**TABLE 5. MEAN DILUTE ACID SOLUBLE PHOSPHORUS VALUES OF INCUBATED SOILS AND TOTAL PHOSPHORUS VALUES OF MANURE SLURRIES.**

| Soil type | Incubation temperature, deg C | Dairy manure 3 percent | Slurry 7 percent | Control 3 percent | Solution 7 percent |
|---|---|---|---|---|---|
| | | | mg/l $PO_4$-P | | |
| Evesboro | 30 | 181.9 | 151.1 | 118.7 | 107.2 |
| | 20 | 151.3 | 223.0 | 133.2 | 109.2 |
| | 3 | 131.5 | 183.0 | 88.4 | 65.2 |
| Sassafras | 30 | 132.4 | 117.0 | 67.6 | 57.2 |
| | 20 | 96.9 | 208.8 | 59.6 | 66.4 |
| | 3 | 97.6 | 144.3 | 61.7 | 38.2 |
| Matapeake | 30 | 85.2 | 98.3 | 66.0 | 46.6 |
| | 20 | 102.9 | 203.9 | 50.9 | 39.4 |
| | 3 | 90.1 | 153.0 | 36.8 | 35.1 |
| | | | Total phosphorus | | |
| 7 percent solids | 30 | | 1750 | | |
| | 20 | | 5475 | | |
| | 3 | | 3233 | | |
| 3 percent solids | 30 | | 1275 | | |
| | 20 | | 1770 | | |
| | 3 | | 1343 | | |

Troug phosphorus increased only in the case of the 7 percent total solids slurry. This would suggest that application of a 7 percent slurry would result in a positive increase in the amount of dilute acid soluble phosphorus present. The resulting higher phosphorus released could leach through the soil profile until the phosphorus was adsorbed by the soil colloid matrix, used by soil microbes or absorbed by plant roots. It is possible with high leaching rates that some of the soluble phosphorus could enter the ground water. Application of a 3 percent total solids slurry would result in a decrease in the total dilute acid soluble phosphorus present, perhaps due to soil microbial utilization.

### BLWRS

The BLWRS began operation on May 30, 1974. Table 7 summarized the amount of waste applied to the BLWRS and rainfall from June 1 to November 31, while Table 8 shows the average concentration of COD, organic nitrogen, ammonia, nitrite-nitrate nitrogen and ortho phosphorus in the effluent for each month from June to November.

The ortho phosphorus in the dairy manure was reduced by over 99 percent each month and the COD was reduced by 90 percent or more each month. Even though the nitrate concentration was low in the effluent at all times, some organic nitrogen and ammonia appeared in the effluent. This suggests that the aerobic zone was not large enough for complete nitrification. At no time during the 6 months of operation was an energy source added.

The amount of waste applied was limited by the permeability of the soil. The soil was classified as a loam or silt loam with the clay content varying from 18 to 24 percent. High amounts of rainfall occurred during the latter part of August and the month of September. This rainfall limited the barrier operation to only 9 days in September. In July the barrier was operated for 30 days with an average application rate of 7.7 mm/day.

A major operating problem with the BLWRS was the clogging of the waste delivery system with pieces of straw and manure solids. The problem of the delivery pipeline clogging with solids is viewed as one of the major operational problems a dairy farmer would face with a BLWRS.

## CONCLUSIONS

The following conclusions can be drawn from the data presented in this paper:

1   Large amounts of COD, phosphorus and ammonia are removed in the top 91 cm of a soil column.

2   Ammonia removal varies with soil type.

3   The amounts of dilute acid soluble phosphorus resulting from the addition of dairy manure slurries are influenced by soil type, temperature and total solids content of the manure.

4   The BLWRS is capable of reducing COD and nitrogen by approximately 90 percent and phosphorus by 99 percent respectively in dairy cattle wastes.

5   The initial capacity of a BLWRS is limited by the permeability of the soil.

**TABLE 6. LINEAR REGRESSION BETWEEN DILUTE ACID SOLUBLE PHOSPHORUS VALUES(X) AND CALCULATED TOTAL SOIL PHOSPHORUS VALUES (Y).**

| Parameter | Equation | R | Significance level |
|---|---|---|---|
| **Soil type** | | | |
| Evesboro | Y = - 158.23 + 4.79X | 0.86 | 0.9995 |
| Sassafras | Y =  186.35 + 5.25X | 0.95 | 0.9995 |
| Matapeake | Y =  199.57 + 10.76X | 0.96 | 0.9995 |
| **Incubation temperature** | | | |
| 3 deg C | Y =   770.1 + 1.0X | 0.09 | 0.60 |
| 20 deg C | Y =   435.7 + 5.2X | 0.41 | 0.99 |
| 30 deg C | Y =  1944.6 - 8.3X | -0.88 | 0.9995 |
| **Total solids** | | | |
| 3 percent | Y = 1582.5 - 6.4X | -0.71 | 0.9995 |
| 7 percent | Y =  563.2 + 4.0X | 0.28 | 0.75 |

**TABLE 7. AMOUNT OF MANURE APPLIED TO THE BLWRS AND RAINFALL FOR JUNE TO NOVEMBER, 1974.**

| Month | Rainfall, mm | Manure applied, mm | Days operated per month |
|---|---|---|---|
| June | 98 | 213 | 15 |
| July | 65 | 230 | 30 |
| August | 172 | 168 | 24 |
| September | 136 | 63 | 9 |
| October | 56 | 158 | 21 |
| November | 32 | 177 | 23 |

| Month | COD | Organic-N | $NH_3$-N mg/l | $NO_3$-N [*] | $PO_4$-P |
|---|---|---|---|---|---|
| June | 122 | 6.57 | 2.79 | 0.31 | 0.05 |
| July | 85 | 2.99 | 2.86 | 0.21 | 0.07 |
| August | 105 | 5.20 | 3.20 | 0.27 | 0.16 |
| September | 58 | 2.68 | 3.66 | 0.73 | 0.03 |
| October | 69 | 1.87 | 5.02 | 0.66 | 0.04 |
| November | 65 | 1.94 | 4.14 | 0.77 | 0.02 |
| Manure | 1295 | 30.9 | 21.4 | 0.65 | 17.6 |

*Includes nitrate-nitrite nitrogen.

### References


1 American Public Health Association. 1971. Standard methods for the examination of water and wastewater. Washington, D.C.

2 Bremner, J. M. and D. R. Keeney. 1971. Stream distillation of ammonia, nitrate and nitrite. Anal. Chem. Acta 32:485-495.

3 Erickson, A. E., J. M. Tiedje, B. G. Ellis and C. M. Hanson. 1971. A barriered landscape water renovation system for removing phosphorus and nitrogen from liquid feedlot wastes. International Symposium on Livestock Wastes Proceedings, Columbus, Ohio. ASAE PROC 271:232-234, ASAE, St. Joseph, Mich., 49085.

4 Jackson, M. L. 1958. Soil chemical analysis. Prentice-Hall, Inc., Englewood, New Jersey.


---

# Automated Waste Treatment and Recycle System

*(Continued from page 567)*

aerobic treatment unit. The treated effluent had an average BOD for April-October of 160 mg/l and for November-March of 600 mg/l, which is not suitable for direct discharge into a stream. The use of the effluent as recycled flushing water is a key element in the automated system.

The oxidation ditch operated at a higher than recommended loading rate and still produced an acceptable performance. Tests indicated that the MLSS should be less than 8000 mg/l to provide suitable settling in the clarifier.

Based upon the experience in this demonstration project, the unit operations used may be applied to commercial production facilities.

### References


1 Jones, D. P., D. L. Day and A. C. Dale. 1971. Aerobic treatment of livestock wastes. Illinois Agricultural Experiment Station, Bulletin 737, Urbana, IL.

2 Loehr, R. 1974. Agricultural waste management. Academic Press, NY. 576. p.

3 Mehta, B. S. and E. P. Taiganides. 1975. Tertiary treatment of animal wastewaters by reverse osmosis membranes. In: Proceedings, International Symposium on Livestock Wastes - 1975. ASAE, St. Joseph, MI. 49085

4 Mote, C. R. 1974. A computer simulation of treatment, storage, and land disposal of swine waste. Unpublished PhD dissertation, The Ohio State University, Columbus, OH. 205 pp.

5 Mote, C. R. and E. P. Taiganides. 1975. A computer simulation of storage and land disposal of swine waste. In: Proceedings of International Symposium on Livestock Wastes - 1975. ASAE, St. Joseph, MI. 49085

6 Nelson, M. 1974. Personal communication. Research Division, Botkins Grain and Feed Company, Botkins, OH.

7 Shutt, J. W., R. K. White, E. P. Taiganides and C. R. Mote. 1975. Evaluation of solids separation devices. In: Proceedings, International Symposium on Livestock Wastes - 1975. ASAE, St. Joseph, MI. 49085

8 Taiganides, E. P. and R. K. White. 1971. Automated handling, treatment and recycling of waste water from an animal confinement production unit. P. 196-198. In: Livestock Waste Management and Pollution Abatement, ASAE Publication. PROC. 271. ASAE, St. Joseph, MI. 49085

# Tertiary Treatment of Animal Wastewaters
# by Reverse Osmosis Membranes


B. S. Mehta, E. Paul Taiganides

SENIOR MEMBER
ASAE


TO attain by 1984 the goal of "zero discharge" of pollutants from animal feedlots, as is stipulated by environmental laws, treatment methods must be used which improve the pollutant removal efficiencies of primary and secondary treatments (settling tanks, lagoons, aeration systems, etc.) and, at the same time produce effluents of reusable quality.

Primary treatment is capable of removing a majority of the settleable solids in a wastewater stream. Secondary treatment, consisting primarily of biochemical oxidation processes, removes 75 to 90 percent of the 5-day Biochemical Oxygen Demand (BOD) and of the total suspended solids (TSS). Tertiary treatment is designed to provide additional BOD and TSS removal, but, more importantly, to remove specific dissolved solids, such as nitrogen, salts, phosphorus, etc. Membrane systems have the ability to make one step separation of dissolved solids on a molecular or ionic level, and thus, are suited as tertiary treatment.

Reverse osmosis and other membrane systems have been used to produce high quality potable water by removing color, turbidity, bacteria and dissolved salts from a variety of water supplies. They also have been applied in food processing (Mehta 1973) and in industrial wastewaters (Lacey and Loeb 1972). However, no published information is available on the treatment of animal feedlot wastewaters by membrane separation techniques.

The purpose of this study was to test the efficiency of membrane separation systems in the clarification of biologically pre-treated animal wastewaters. In addition, tests were conducted to determine the effect of process variables on wastewater purification rate and power requirements.

## EXPERIMENTAL PROCEDURE

An experimental apparatus developed by Mehta (1973) was used in this study. Three types of swine wastewaters were tested: effluent from an anaerobic lagoon, effluent from the second stage aerated lagoon which received the effluent from the anaerobic lagoon, and effluent from the final clarifier of secondary biological treatment plant for pig wastes (Taiganides and White 1971)


Approved for publication as Journal Article No. 45-75 of the Ohio Agricultural Research and Development Center (OARDC); supported in part by Grant No. R-801125 from the US Environmental Protection Agency.

The authors are: B. S. MEHTA, Engineer, C. F. Braun & Co. (Former Post-Doctoral Research Associate, OARDC), and E. PAUL TAIGANIDES, Professor, Agricultural Engineering Dept., OARDC and Ohio State University, Columbus.


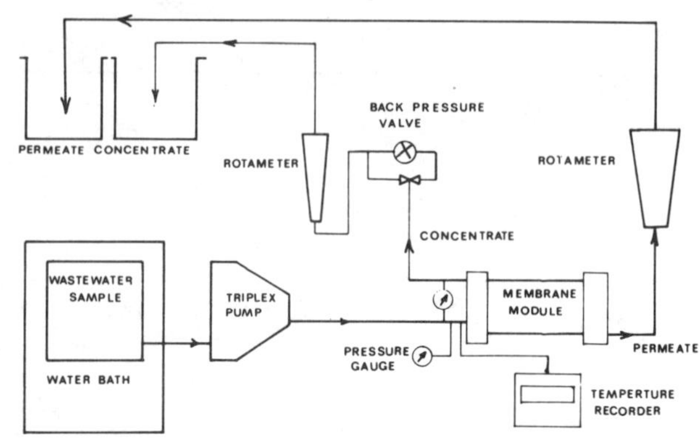


FIG. 1 Schematic diagram of laboratory apparatus for pressure driven membrane separation.

### Experimental Apparatus

The experimental apparatus used in this study is schematically presented in Fig. 1. A picture of the apparatus assembled is shown in Fig. 2. The main components of the apparatus are the membrane module, the high-pressure pump, and various recorders.

The membrane module used in this investigation was UOP No. 420 consisting of 18, 127 mm in diameter tubes. The 18 tubes in this module were connected to each other with "U" shape connectors in the module headers. The backing material was epoxy bonded fiberglass on which a cellulose acetate membrane was cast. The net membrane area was 930 cm$^2$. The pore size of the membrane varied from 10 to 30 Angstrom.

A high-pressure pump was equipped with a variable speed transmission unit to obtain flow rates from 1 to 6 liters per minute (l/m) at pressure levels ranging from 7

FIG. 2 Pressure driven membrane separation apparatus assembled for operation.

| Property | Symbol | Unit | Sample source | | |
|---|---|---|---|---|---|
| | | | Anaerobic lagoon | Aerobic lagoon | Clarifier |
| Total solids | TS | mg/l | 1265 | 740 | 5150 |
| Total suspended solids | TSS | mg/l | 125 | 370 | 2238 |
| Total dissolved solids | TDS | mg/l | 1140 | 370 | 2912 |
| Biochemical oxygen demand | BOD | mg/l | 770 | 466 | 473 |
| Chemical oxygen demand | COD | mg/l | 1162 | 532 | 2784 |
| Nitrogen | N | mg/l | 181 | 77 | 534 |
| Phosphorus | P | mg/l | 71 | 49 | 188 |
| Conductivity | C | $\mu$ mohs/cm | 1700 | 850 | 5900 |
| Color* | Co | * | 3.68 @ 230 NM | 2.6 @ 205 NM | 4.40 @ 290 NM |
| Turbidity | Tu | ppm $S_iO_2$ | 500 | 220 | 4500 |
| pH | pH | --- | 7.9 | 8.1 | 7.6 |

*Maximum absorbance from the spectrograph.

to 35 kg per cm$^2$. Rotameters were used to measure the flow rate of concentrate and permeate. A constant-temperature waterbath was used to maintain a constant wastewater temperature with an accuracy of 0.5 C. A continuous temperature recorder was used to record temperature of the feed solution at the module inlet.

### Procedure

The samples of biologically treated animal wastewaters were passed through the membrane module at five different operating pressure levels, five temperature levels for the sample, and five flow rates through the system. Only one process variable was changed at a time; all the others were held constant at predetermined levels. Permeation flux was recorded when the steady state was reached. At steady-state permeate samples of 100 to 120 ml were collected in plastic bottles and stored at 4 C for chemical analysis later. The process variable under study was then increased.

Pressure drop across the module was recorded for each test run. Power requirement of the unit was determined for some of the typical test runs by using a wattmeter.

Using the samples from the clarifier, the unit membrane separation process was operated for eight hours at 20 kg/cm$^2$ at 3 l/m and at 20 C. The permeation flux was recorded every 30 minutes to determine any membrane plugging problems.

After completion of each set of test runs, the module was flushed with tap water at 30 C for 30 minutes. Then, the module was sanitized with a dilute solution of a strong germicidal agent. The module's permeability was checked after every clean-up operation to determine any permanent damage to membrane.

### Analysis of Samples

Both wastewater and permeate samples were analyzed for total solids (TS), total suspended solids (TSS), total dissolved solids (TDS), biochemical oxygen demand (BOD), chemical oxygen demand (COD), and nitrogen (N), according to standard procedures (APHA 1971).

Colorimetric method was used to determine phosphorus (Harris and Poppat 1954). The conductivity of the samples was determined by using a Hach meter. The color of the samples was estimated by their absorbance characteristics in visual as well as in ultraviolet wavelength range on a Coleman Perkin Elmer 124 double beam spectrophotometer. Color of the samples was determined by double beam spectrophotometer. Turbidity of the samples was determined by Hellige turbidimeter.

The results of the analysis of animal wastewater samples from anaerobic lagoon, aerated lagoon and clarifier are given in Table 1. Wastewater from clarifier had the highest levels of TS, TDS, COD, N, conductivity, color and turbidity followed by anaerobic lagoon and then aerated lagoon. TSS in clarifier wastewater were higher than aerobic lagoon followed by anaerobic lagoon. BOD levels of clarifier and aerobic lagoon wastewaters were very close but about 40 percent lower than anaerobic lagoon. pH of wastewaters from aerobic lagoon, anaerobic lagoon and clarifier were 8.1, 7.9 and 7.6, respectively.

TABLE 2. POLLUTANT REMOVAL EFFICIENCIES ACHIEVED FOR
ANAEROBIC LAGOON WASTEWATER SAMPLE.

| Operating conditions | | | Percent removal of | | | | | | | |
|---|---|---|---|---|---|---|---|---|---|---|
| Pressure kg/cm$^2$ | Temperature deg C | Flow rate l/m | TS | BOD | COD | N | P | C | Co | Tu |
| 7 | 20 | 3.4 | 85 | | 84 | | | 59 | 98 | 98 |
| 14 | 20 | 3.4 | 90 | | 88 | | | 69 | 98 | 97 |
| 21 | 20 | 3.4 | 93 | | 88 | | | 73 | 99 | 98 |
| 28 | 20 | 3.4 | 96 | | 92 | | | 77 | 99 | 98 |
| 35 | 20 | 3.4 | 99 | 88 | 92 | 90 | 93 | 79 | 99 | 99 |
| 21 | 3 | 3.4 | 94 | | 96 | | | 78 | 99 | 99 |
| 21 | 10 | 3.4 | 92 | | 97 | | | 79 | 99 | 99 |
| 21 | 19 | 3.4 | 92 | | 96 | | | 80 | 98 | 99 |
| 21 | 29 | 3.4 | 93 | | 96 | | | 78 | 98 | 98 |
| 21 | 35 | 3.4 | 94 | 93 | 96 | 89 | 95 | 77 | 98 | 99 |
| 21 | 20 | 1.1 | 92 | | 91 | | | 78 | 99 | 99 |
| 21 | 20 | 2.3 | 94 | | 92 | | | 81 | 99 | 98 |
| 21 | 20 | 3.4 | 94 | | 92 | | | 81 | 98 | 98 |
| 21 | 20 | 4.5 | 97 | | 92 | | | 82 | 99 | 99 |
| 21 | 20 | 5.7 | 95 | 92 | 93 | 86 | 96 | 82 | 99 | 99 |

MANAGING LIVESTOCK WASTES

# RESULTS

Tables 2, 3, and 4 show removal efficiencies for the various pollutant parameters at the operating conditions investigated in this study. Table 5 shows that the minimum removal efficiencies obtained for each of the parameters for the three samples of wastewater treated were at least 83 percent except for conductivity. Conductivity was reduced by only 49 percent for the clarifier liquid when operating the apparatus at the lowest pressure of 7 kg/cm$^2$, but reached 86 percent when the operating pressure was increased to 35 kg/cm$^2$ (Table 4). Fig. 3 shows that the effluent was almost colorless, clear and of high enough quality to be reused on the farm as animal drinking water.

Power requirements increased from 0.17 to 0.85 hp when the operating pressure was increased from 7 to 35 kg/cm$^2$. The increase was almost linear for the intermediate pressure levels. Similarly, power requirements increased from 0.16 to 0.8 hp when the flow rate was increased from 1 to 5 l/m.

Rate of purification of the wastewater samples increased almost linearly with increase of applied pressure. Permeation flux rate increased from 7-10 ml/min-module at 7 kg/cm$^2$ operating pressure to 30-50 ml/min-module at 500 kg/cm$^2$. Only about 25 percent reduction in the permeation flux rate was observed

TABLE 3. POLLUTANT REMOVAL EFFICIENCIES ACHIEVED FOR AEROBIC LAGOON WASTEWATER SAMPLE.

| Operating conditions | | | Percent removal of | | | | | | | |
|---|---|---|---|---|---|---|---|---|---|---|
| Pressure kg/cm$^2$ | Temperature deg C | Flow rate l/m | TS | BOD | COD | N | P | C | Co | Tu |
| 7 | 20 | 3.4 | 83 | | 93 | | | 52 | 99 | 96 |
| 14 | 20 | 3.4 | 93 | | 97 | | | 67 | 98 | 95 |
| 21 | 20 | 3.4 | 93 | | 96 | | | 82 | 99 | 96 |
| 28 | 20 | 3.4 | 95 | | 96 | | | 87 | 98 | 94 |
| 35 | 20 | 3.4 | 99 | 84 | 99 | 90 | 90 | 89 | 99 | 94 |
| 21 | 3 | 3.4 | 90 | | 90 | | | 85 | 99 | 97 |
| 21 | 10 | 3.4 | 96 | | 95 | | | 85 | 98 | 97 |
| 21 | 19 | 3.4 | 95 | | 95 | | | 83 | 99 | 98 |
| 21 | 29 | 3,4 | 98 | | 95 | | | 83 | 99 | 97 |
| 21 | 35 | 3.4 | 98 | 89 | 96 | 84 | 94 | 79 | 99 | 97 |
| 21 | 20 | 1.1 | 89 | | 93 | | | 82 | 98 | 97 |
| 21 | 20 | 2.3 | 93 | | 98 | | | 85 | 98 | 97 |
| 21 | 20 | 3.4 | 93 | | 96 | | | 85 | 98 | 98 |
| 21 | 20 | 4.5 | 95 | | 98 | | | 86 | 99 | 97 |
| 21 | 20 | 5.7 | 95 | 91 | 96 | 88 | 92 | 86 | 99 | 97 |

TABLE 4. POLLUTANT REMOVAL EFFICIENCIES ACHIEVED FOR CLARIFIER WASTEWATER SAMPLE

| Operating Conditions | | | Percent removal of | | | | | | | |
|---|---|---|---|---|---|---|---|---|---|---|
| Pressure kg/cm$^2$ | Temperature deg C | Flow rate l/m | TS | BOD | COD | N | P | C | Co | Tu |
| 7 | 20 | 3.4 | 84 | | 97 | | | 49 | 98 | 99 |
| 14 | 20 | 3.4 | 90 | | 98 | | | 74 | 99 | 99 |
| 21 | 20 | 3.4 | 94 | | 99 | | | 81 | 99 | 99 |
| 28 | 20 | 3.4 | 95 | | 99 | | | 84 | 98 | 99 |
| 35 | 20 | 3.4 | 96 | 84 | 99 | 90 | 96 | 86 | 98 | 99 |
| 21 | 3 | 3.4 | 94 | | 99 | | | 82 | 98 | 98 |
| 21 | 10 | 3.4 | 94 | | 99 | | | 81 | 98 | 99 |
| 21 | 19 | 3.4 | 93 | | 98 | | | 82 | 97 | 99 |
| 21 | 29 | 3.4 | 94 | | 99 | | | 81 | 99 | 99 |
| 21 | 35 | 3.4 | 93 | 93 | 99 | 84 | 95 | 80 | 99 | 99 |
| 21 | 20 | 1.1 | 92 | | 99 | | | 79 | 99 | 99 |
| 21 | 20 | 2.3 | 94 | | 99 | | | 82 | 99 | 99 |
| 21 | 20 | 3.4 | 94 | | 98 | | | 83 | 98 | 99 |
| 21 | 20 | 4.5 | 94 | | 99 | | | 81 | 99 | 99 |
| 21 | 20 | 5.7 | 94 | 92 | 99 | 87 | 97 | 82 | 99 | 99 |

TABLE 5. MINIMUM AND MAXIMUM POLLUTANT REMOVAL EFFICIENCIES.

| Pollution parameter | | Anaerobic lagoon | | Aerobic lagoon | | Clarifier | |
|---|---|---|---|---|---|---|---|
| | | Min | Max | Min | Max | Min | Max |
| Total solids | TS | 85 | 99 | 83 | 99 | 84 | 96 |
| Total suspended solids | TSS | 100 | 100 | 100 | 100 | 100 | 100 |
| Total dissolved solids | TDS | 85 | 99 | 83 | 99 | 84 | 96 |
| Biochemical oxygen demand | BOD | 88 | 93 | 84 | 91 | 84 | 93 |
| Chemical oxygen demand | COD | 84 | 96 | 93 | 99 | 97 | 99 |
| Nitrogen | N | 86 | 90 | 84 | 90 | 84 | 90 |
| Phosphorus | P | 93 | 96 | 90 | 94 | 95 | 97 |
| Conductivity | C | 59 | 82 | 52 | 89 | 49 | 86 |
| Color | Co | 98 | 99 | 98 | 99 | 98 | 99 |
| Turbidity | Tu | 97 | 99 | 94 | 98 | 98 | 99 |



MANAGING LIVESTOCK WASTES

**FIG. 3 Typical clarity of permeate achieved by apparatus from the three samples fluxed.**

during 8 hours of continuous operation. The majority of this reduction occurs within the first 30 minutes of operation. The pressure drop across the module was around 0.4 kg/cm$^2$.

## DISCUSSION

Excellent reductions of the primary pollution parameters (TSS, BOD, N and P) were achieved over the whole range of operating conditions. All suspended solids were removed. The effluent contained no measurable suspended matter. From Tables 1 and 4, it can be shown that in the case of the clarifier liquid sample, the BOD could be reduced from 473 to 38 mg/l, nitrogen from 534 to 53 mg/l and phosphorus from 188 to 5.6 mg/l as is shown in Table 4. Color and turbidity removal was almost 100 percent.

Removal efficiency increased with increased operating pressure. Power requirements are directly proportional to the operating pressure. To achieve higher pollutant removal efficiencies, the energy input to the system must be increased by operating at higher pressure levels. Temperature of the influent apparently did not affect significantly the pollutant removal efficiencies; neither did the flow rates as may be concluded from a cursory study of Tables 2, 3 and 4.

Additional studies which may be of interest are tests with different size membrane separation modules on a pilot scale basis. Also, reuse of the concentrate for refeeding and of the permeate for water supply deserve additional studying.


### References

1 APHA. 1971. Standard methods for the examination of water and wastewater. 13th Ed. New York. American Public Health Association.

2 Harris, W. P. and D. Poppat. 1954. Determination of phosphorus contents of liquids. J. Am. Oil Chemists Soc. 31:124.

3 Lacey, R. E. and S. Loeb. 1972. Industrial processing with membranes. (eds.) New York, Wiley Interscience.

4 Mehta, B. S. Processing of model composition whey solutions with pressure driven membranes. Unpublished Ph. D. thesis. Columbus, OH. The Ohio State University Library. 299 pp.

5 Taiganides, E. P. and R. K. White. 1971. Automated handling, treatment and recycling of wastewater from an animal confinement production unit. In: Livestock Waste Management and Pollution Abatement. ASAE Publication PROC-271. pp. 146-148.

# Present Knowledge on the Effects of Land Application of Animal Waste

G. W. Wallingford, W. L. Powers, L. S. Murphy

## INTRODUCTION

THE fate of most animal waste has been and continues to be disposal onto land. With a trend towards the feeding of large numbers of livestock in confined areas and with the increased concern over the effect of animal waste on the environment, there has been an increased effort by scientists to determine the effects of animal waste on soil and plant conditions and to develop recommendations for safe disposal of animal waste on soils. There have been substantial amounts of information published in recent years on various land disposal systems. In order that this material could be evaluated and condensed, a "state-of-the-art" paper has been prepared in which pertinent land disposal literature has been reviewed with the purpose of con-

concentrations of the constituents in the waste are known, can one estimate such reactions as solute movement in the soil and plant nutrient availability or toxicity.

Waste composition data in the literature were compiled by climatic region, species and type of waste. No consistent trends attributable to climatic region were found. Extreme variations in the data were also found within most species and waste classifications. Table 1 contains the minimum and maximum total nitrogen, phosphorus and potassium in various waste type, animal species combinations. Mean values are not given because of the lack of accuracy obtained when data (some of which themselves are mean values) from divergent sources are averaged, and because the tremendous variability in waste composition makes land appli-

TABLE 1. THE MINIMUM AND MAXIMUM TOTAL NITROGEN, PHOSPHORUS AND POTASSIUM CONTENT OF ANIMAL WASTES FROM DATA FOUND IN THE LITERATURE COMBINED OVER ALL CLIMATIC RESIONS.

| Waste type | Constituent | Species | | | |
| --- | --- | --- | --- | --- | --- |
| | | Beef | Dairy | Swine | Poultry |
| | | percent | | | |
| Solid* | N | 0.60 —4.9 | 1.5 —3.9 | 2.0 — 7.5 | 1.1 —11 |
| | P | 0.11 —1.6 | 0.41—1.6 | 0.56— 2.5 | 0.38— 6.3 |
| | K | 0.053 —4.0 | 1.4 —3.4 | 1.5 — 4.9 | 0.73— 5.2 |
| Runoff† | N | 0.0011 —0.86 | ‡ | 0.40— 0.48 | ‡ |
| | P | 0.00032—0.52 | ‡ | 0.1 — 1.2 | ‡ |
| | K | 0.0097 —0.20 | 4.8 —6.9 | 0.2 — 3.6 | ‡ |
| Slurry-digested* | N | ‡ | ‡ | 8.8 —12 | 2.2 — 7.8 |
| | P | ‡ | ‡ | 3.1 — 6.4 | ‡ |
| | K | ‡ | ‡ | 2.9 — 8.3 | 2.5 — 4.3 |
| Slurry-undigested* | N | 1.9 —9.0 | 1.8 —5.1 | 3.4 —19 | ‡ |
| | P | 0.52 —2.3 | 0.43—1.2 | 1.2 — 3.7 | 1.5 — 2.3 |
| | K | 0.24 —3.4 | 1.9 —5.0 | 1.0 — 4.8 | ‡ |

*Dry-weight basis
†Wet-weight basis
‡Data not found

solidating land disposal information, recommending future research needs and establishing disposal guidelines. The discussion to follow is a summary of a publication prepared for the Environmental Protection Agency. If more specific information is desired, including a complete listing of references, then the original publication should be obtained from the Superintendent of Documents, publication EPA 66012-75-010.

## WASTE COMPOSITION

A prediction of the effects of animal waste application on soil properties and plant growth requires a foreknowledge of the composition of the waste material. Only if the

cation calculations based on mean values subject to large errors.

If the waste composition data were constant from one animal production site to the next, then the task of determining application rates would be less complicated. Unfortunately, factors such as climate, species, ration and management act together to make composition of animal waste extremely variable. Fig. 1 illustrates various factors which affect animal waste composition. If accurate disposal rate recommendations are to be made, this extreme variability necessitates the use of chemical analysis of the waste prior to disposal.

## EFFECT OF WASTE ON THE ENVIRONMENT

### Soil Properties

Most researchers who have measured the effects of animal waste on soil physical properties have found infiltration rates, hydraulic conductivity, bulk density, water-holding

Contribution No. 9040 Soil Science Dept., University of Minnesota, St. Paul, and No. 1494 Agronomy Dept., Kansas State University, Manhattan. This work was partially supported by the Environmental Protection Agency under Project 803021.

The authors are: G. W. WALLINGFORD, Assistant Professor, Northwest Experiment Station, Crookston, Minn.; W. L. POWERS, Associate Professor, and L. S. MURPHY, Professor, Agronomy Dept., Kansas State University, Manhattan.



MANAGING LIVESTOCK WASTES

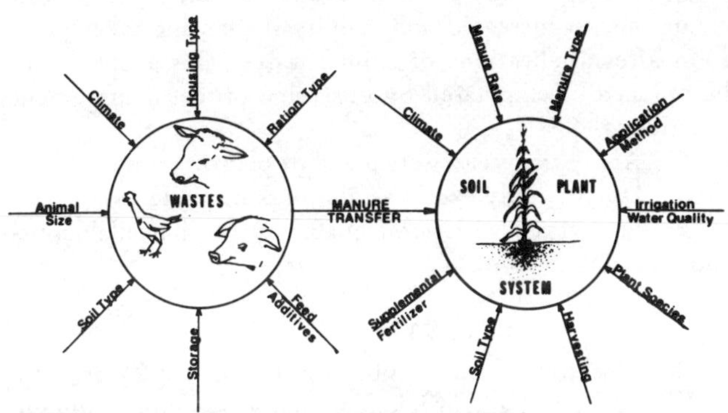


FIG. 1 Factors affecting animal waste composition and the fate of animal waste constituents after application on land.

capacity and aggregate stability to be improved after applications of animal waste. The organic matter contained in the wastes has been credited as being the waste constituent responsible for these improved soil physical properties. More specifically, Azevedo and Stout (1974) found the fiber content of animal manures to be the most important constituent controlling soil hydraulic conductivity after waste applications.

Deterioration of soil physical properties resulting from dispersion of soil colloids has been found less frequently and has been attributed to an imbalance of salts in the soil. Even though some animal wastes, particularly liquid waste, are relatively higher in monovalent cations, which can cause soil dispersion, the beneficial effects of the organic matter in the waste override the potential negative effects of the salts.

Soil chemical properties have primarily been investigated in relation to improved soil fertility due to nutrients added in the waste. When soil is used for waste applications, amounts of plant nutrients added are frequently in excess of those removed by the crop. Researchers recently began to study the possible deleterious effects of accumulation and movement of plant nutrients on the environment.

Soil nitrate nitrogen and soluble salts are the two soil chemical properties which have received the most attention in waste disposal research. The accumulation and downward leaching of nitrate nitrogen have been reported as being a potential threat to groundwater quality. Numerous researchers have measured nitrate leaching after waste applications and have found in some cases nitrate will leach if the nitrogen applied is greater than that used by plants; others have found no significant nitrate leaching even after nitrogen loadings far in excess of removal by crop uptake. These contrasting results arise from different leaching volumes and variation in the complicated factors that control nitrification and denitrification in soils. These data show that the depth of downward leaching, if any, is difficult to predict. The best way to avoid downward leaching of nitrate nitrogen is to apply rates of animal waste that do not provide nitrogen in excess of crop use.

Buildup of soluble salts in soils after waste application has been implicated in reducing the fertility of soil by increasing soil salinity. Most researchers who have measured soil salinity after waste applications have found high correlations between salinity and waste treatment and/or crop growth. Potassium is considered the specific soluble salt most responsible for increased soil salinity because of its high solubility and relatively high concentration in animal wastes.

## Rainfall and Snowmelt Runoff

Runoff loss of animal waste constituents is a function of the runoff quantity and quality. Most research has shown that animal waste applications reduce the quantity of runoff leaving a field by increasing the soil infiltration rate. Research has shown that applications of animal waste to frozen ground or to ground with vegetative cover will increase the likelihood of lowering the quality of surface runoff. Incorporating the waste after application into the soil is a management practice that reduces surface runoff contamination.

## Plant Growth

Improved soil productivity is the most beneficial result of applying animal waste to soils. Enhanced soil availability of plant macro- and micro-nutrients has been shown to be the major factor improving plant growth after waste applications. Positive effects on soil physical properties have also been shown to improve soil productivity.

Several researchers who applied large amounts of animal wastes have measured yields that were depressed relative to control plot yields or relative to plots obtaining maximum yields. Increased soil salinity was thought responsible in many studies in which yields were found to decrease (refs. 7, 8, 13, 16), and in only one case (Boorman, Loynachan, Koelliker 1974) was there found no accumulation of soluble salts in soils that had shown depressed plant growth. Several experiments have shown no negative effect on yields due to heavy rates of animal waste applications, but when soil salinity was measured, no buildup of soluble salts was found (refs. 4, 6, 14).

## APPLICATION RATES BASED ON WASTE CONSTITUENTS

There have been several criteria used to formulate application rates; most are based on providing ample nutrition for a crop without creating soil conditions that are toxic to plant growth. Numerous research studies have shown certain application rates to produce maximum crop yields or reduce crop growth, but many researchers have been reluctant to base application rate recommendations on research findings because of the tremendous variability that exists between local waste composition, management practices, soil types and weather. Published loading rate recommendations have been based both on experimental data and on criteria the researcher felt would be important in the future.

## Nitrogen

Because nitrogen is often the nutrient that limits plant growth and because nitrogen poses the greatest threat to groundwater contamination, it is often used as the basis for determining application rates.

The extension services of several states have developed animal waste application guidelines based partly on the nitrogen content of the waste. Indiana (Cooperative Extension Service 1972) recommends application rates which provide 225 pounds per acre of nitrogen, while North Carolina (Agricultural Extension Service 1973) determines application rates by calculating the quantity of the waste needed to satisfy a nitrogen soil test. Iowa (Vanderholm 1973) determines acres of land needed for application of waste based on the pounds of nitrogen to be added to the soil. Maine (Life Sciences and Agricultural Experiment Station

MANAGING LIVESTOCK WASTES

and the Cooperative Extension Service 1972) recommends application rates based on the soil type and on the nitrogen content of the waste.

An estimation of nitrogen mineralization using a decay series concept coupled with the nitrogen usage of the crop, with an awareness of potential salt toxicities, appears at this time to be the best criteria on which to base application rate calculations. As described by Pratt et al. (1973), the decay series concept is based on mineralization of organic nitrogen into inorganic or available nitrogen and provides a mathematical method to describe how much of the waste nitrogen becomes available for plant growth. The rate of mineralization is most rapid in the first year after application and decreases in subsequent years. Variables such as soil type, waste nitrogen content and climate must be accounted for when determining decay series constants. Determining application rates based on a nitrogen decay series concept could have widespread adaptability because the constants can be determined by experimental data. Some existing data on nitrogen availability might be used to calculate decay series constants, but most of these constants must come from experimental data yet to be gathered.

### Soluble Salts

It has been shown that large applications of animal waste can be toxic to plant growth by creating saline soil conditions. Except for wastes low in nitrogen, application rates based on salt content generally provide nitrogen in excess of crop use. Application guidelines based on a salt content of the waste will, therefore, not be used as a means to determine the most efficient use of the nitrogen but as a method to determine amounts that can be applied without decreasing crop growth. Maximum application rates permissible without causing excessive salt buildup in soil are based on the salt content of the waste, soil type and the amount and salt content of irrigation water. Powers et al. (1973, 1974) have developed that type of guideline for beef-feedlot manure and lagoon water.

PRACTICAL ASPECTS OF<br>APPLYING WASTES TO SOILS

The amount of waste to apply can be accurately determined only after a soil test has been taken to determine the needed nitrogen, the decay rate has been estimated, and the waste has been analyzed for nitrogen, phosphorus, potassium, sodium, calcium and magnesium. Some state extension agencies have published guidelines based on some or all of those factors. It is recommended that those state publications be referred to for specific application rate recommendations in your area.

Proper sampling of a waste stockpile is essential for accurate analysis results. Several samples from different locations in a stockpile should be taken and mixed for determination of the average composition. It is desirable to take samples several times during the year to further establish the average composition of the waste. Know if the analysis is reported on a dry-weight basis or on a wet-weight or "as is" basis.

Proper and timely spreading and incorporation of the waste into the soil is important because it can prevent reduced crop yields and decrease the potential for pollution. It is recommended that spreading be as uniform as possible on the field, that the waste be incorporated as soon as possible after application and that the waste not be applied on frozen soils.

Lowered germination and reduced seedling growth can occur due to increased soil salinity if planting takes place soon after applications of animal waste. This problem can be reduced or eliminated by observing proper management practices:

1    Apply the waste well ahead of planting time;
2    Do not apply excessive amounts of waste to soils;
3    Pre-irrigate with good quality water, if available, before planting.

RESEARCH NEEDS

The objective of land application research is to develop guidelines for safe and efficient animal waste application on land. The research needs mentioned here will help yield information needed to establish these guidelines. Guidelines should establish application rates which maintain maximum soil productivity and nutrient recycling while avoiding increased soil salinity and contamination of surface water and groundwater.

Standardization of data is needed so that research results from different locations can be compared. Waste analyses should always be expressed on a dry-weight basis except for liquids of low solids content (approximately 1 percent solids or lower). The location of the research should be included when reporting data so that the climatic region can be determined. Soil characteristics which might influence application rate calculations should be given: slope, soil texture, presence of impervious layers in the profile and depth to water table.

The variability of animal waste composition necessitates analysis of the waste before application. Waste sampling techniques and analytical procedures should be standardized so that errors in application rate calculations based on waste constituents can be minimized. Considerable data have been published which shows that animal wastes are extremely variable in composition; further data showing characterization variability will do no more than verify the extreme variabilities that do exist.

In most regions of the United States nitrogen appears to be the best waste constituent on which to base application rates. In some regions other factors such as salt buildup in soil, high copper and nitrate accumulation in plant tissues, or grass tetany may limit waste applications. Further research is needed in order that other specific limiting factors in region or management system may be determined.

More information is needed on organic nitrogen decay rates. Experimental data must be gathered which can be used to determine nitrogen decay constants for different soil types, for wastes of different nitrogen contents and for different climates. While it is recognized that denitrification can cause large losses of soil nitrogen to the atmosphere, little is known on how animal waste applications affect this process. Such variables as soil texture, climate and waste composition need to be examined as to how they affect denitrification. In some cases, denitrification may account for large errors in underestimating the application rate of a given agronomic system.

Most data have shown that if animal wastes are incorporated and if soil erosion is prevented, there will be little loss of waste constituents into surface runoff. Some data have shown that due to increased infiltration rates, surface runoff is reduced by waste applications. That positive aspect of land applications needs further investigation.

More long-term experiments are needed to study the effects of various application rates on nutrient availability

*(Continued on page 586)*

# Comparison of Lint Cotton Yields Following Applications of Beef Cattle Wastes and Commercial Nitrogen


W. I. Spurgeon, J. M. Anderson, J. W. Holloway


AGRONOMISTS are in general agreement that organic matter levels in the soils of the Mississippi Delta are low enough to affect yields of lint cotton in areas where intensive row crop production has been practiced. The need to introduce technologies capable of increasing soil organic matter, even on a temporary basis, is sufficient to warrant a research effort to increase the level of organic materials in the soil. More recently, the need to reduce the use of commercial fertilizers which require large expenditures of energy in their manufacture has been recognized (Fairbanks 1974, and Hoeft and Siemens 1975).

A new type of beef feedlot with a slatted floor and a manure collection pit has been developed (Holloway et al. 1974). This technology has resulted in a mechanized means for handling beef wastes, thus increasing practical usefulness of manure from a commercial farm feedlot operation. The objectives of the research included in this report were: (a) to evaluate cotton yield response to various rates of beef cattle wastes and commercial nitrogen on two soil types and (b) to compare injection of liquid manure with surface applications.

## MATERIALS AND METHODS

### Silt Loam Soil (Experiment 1)

Treatments were arranged in a split plot design with four replications on Dubbs silt loam soil to measure the effect of various manure and nitrogen rates on lint cotton yield. Main plots received rates of 0, 56, and 112 kg of nitrogen (urea-ammonium nitrate solution, 32 percent nitrogen) per hectare and subplots received injected applications of 40, 54, 67, and 81 metric tons of liquid beef wastes per hectare. Chemical analyses of beef wastes are included in Table 1. Both manure and nitrogen were injected in the spring just prior to planting in 1973 and 1974.

### Silt Loam Soil (Experiment 2)

A randomized complete block design with six treatments and four replications was used on Dubbs silt loam soil to measure the effect of manure placement on lint cotton yields in both 1973 and 1974. The six treatments were: (a) 54 metric tons per hectare of manure applied to the soil surface; (b) 54 metric tons per hectare of manure injected; (c) 81 metric tons per hectare of manure applied to the soil surface; (d) 81 metric tons per hectare of manure injected; (e) 134 kg of nitrogen per hectare; and (f) control which received neither manure nor nitrogen.

### Clay Soil (Experiment 3)

In 1973, an experiment with four treatments and four replications was arranged in a randomized complete block design on a Sharkey clay soil to measure the effect of manure rates on lint cotton yield. This soil is generally considered to be unproductive for cotton. The four treatments in this experiment included: 54, 81, and 108 metric tons per hectare of manure injected into the soil and no manure with 134 kg per hectare of nitrogen (liquid form). Inclem-

**TABLE 1. COMPOSITION OF CATTLE WASTES COLLECTED FROM A SLATTED FLOOR FEEDLOT, DELTA AREA OF MISSISSIPPI.**

| Item | Amount, as applied basis |
|---|---|
| Moisture, percent | 88.7 |
| Nitrogen | |
|     Ammonia, percent | 0.15 |
|     Nitrate, percent | 0.08 |
|     Urea, percent | 0.003 |
|     Total, percent | 0.24 |
| Calcium, percent | 0.25 |
| Phosphorus, percent | 0.08 |
| Potassium, percent | 0.28 |
| Iron, percent | 0.03 |
| Copper, ppm | 3.27 |
| Manganese, ppm | 1.13 |
| Zinc, ppm | 8.47 |

ent weather in 1974 prevented repetition of this experiment. However, residual effects were evaluated with all plots receiving 134 kg per hectare of nitrogen in 1974.

### General Production Practices

Production practices were those recommended for use on the particular soil types in the Delta area of Mississippi. These soils do not generally require application of phosphorus or potassium for cotton production. Stoneville variety 213 cottonseed was planted in all tests. Harvesting was done with a conventional spindle picker. Regression estimated were made from yields on injected plots on Experiments 1 and 2. Intercept differences for year and test were removed with dummy variables (Draper and Smith 1966).

## RESULTS AND DISCUSSION

### Experiment 1

The effects of various manure and nitrogen rates on lint cotton yields are shown in Table 2. For the 2-yr average, there were no differences between lint cotton yields for 0, 56, or 112 kg per hectare of nitrogen fertilizer. The 54-


The authors W. I. SPURGEON, J. M. ANDERSON, and J. W. HOLLOWAY are, Agronomist, Agricultural Economist, and former Assistant Animal Scientist, respectively, Delta Branch Mississippi Agricultural and Forestry Experiment Station, Stoneville, MS 38776.

TABLE 2. EFFECT OF NITROGEN AND LIQUID MANURE RATES ON THE YIELD OF LINT COTTON PRODUCED ON DUBBS SILT LOAM SOIL, AVERAGE OF 1973 AND 1974.

| Nitrogen per hectare, kg | Manure per hectare, metric tons | | | | Means* nitrogen rates |
|---|---|---|---|---|---|
| | 40 | 54 | 67 | 81 | |
| | ------- lint per hectare, kilograms ----------------- | | | | |
| 0 | 1089 | 1152 | 1141 | 1130 | 1128a |
| 56 | 1133 | 1216 | 1173 | 1110 | 1158a |
| 112 | 1128 | 1163 | 1142 | 1188 | 1155a |
| Means† manure rates | 1117b | 1177a | 1151ab | 1142ab | |

*Nitrogen rate means with the same letter are not different at the 5 percent level according to Duncan's New Multiple Range Test.

†Manure rate means with the same letter are not different at the 5 percent level according to Duncan's New Multiple Range Test.

TABLE 3. COMPARISON OF INJECTING AND SURFACE APPLICATION OF LIQUID MANURE WITH COMMERCIAL NITROGEN ON THE YIELDS OF LINT COTTON PRODUCED ON DUBBS SILT LOAM SOIL, 1973 AND 1974.

| Treatment | Lint yield* kilograms per hectare |
|---|---|
| No manure, no nitrogen | 576d |
| Commercial nitrogen, 134 kg per hectare | 1174a |
| Manure, 54 metric tons per hectare | |
| Surface | 904c |
| Injected | 1107a |
| Manure, 81 metric tons per hectare | |
| Surface | 1006b |
| Injected | 1169a |

*Means followed by the same letter are not different at the 5 percent level according to Duncan's New Multiple Range Test.

metric ton rate of manure resulted in lint cotton yields which averaged 1177 kg per hectare. This was similar to the 67-metric ton rate (1151 kg lint per hectare) and the 81-metric ton rate (1142 kg lint per hectare), but significantly more than the 40-metric ton rate (1117 kg lint per hectare).

There was a significant year X nitrogen rate interaction. In 1973 greater yields per hectare were obtained with 112 kg per hectare of nitrogen across all manure rates than for the 0 rates (1362 vs. 1236 kg lint per hectare). However, in 1974 the 0 nitrogen rate produced 72 kg per hectare more lint cotton than did the plots receiving 112 kg nitrogen per hectare. This interaction may be explained by the unusual rainfall distribution, the very low temperature and solar radiation levels experienced during September of 1974, or possibly by some differential residual effect of the manure applied in 1973.

There was no significant interaction between nitrogen rate and manure rate. However, with the exception of one comparison, 54 metric tons of manure per hectare resulted in greater lint cotton yields with each of the three nitrogen rates.

Based on these results and estimates by Fairbanks (1974) and Hoeft and Siemens (1975), liquid cattle wastes could be used to effectively substitute for commercially produced nitrogen fertilizer with a resulting savings of 120 to 133 cu m of natural gas per hectare of cotton produced. Based on current prices, substituting liquid beef wastes for anhydrous ammonia could eliminate about $25 per hectare of fertilizer costs associated with current cotton production technology on Dubbs silt loam soil. There would probably be a slightly higher cost involved with field application of beef wastes than is now incurred with use of anhydrous ammonia.

### Experiment 2

Injection of manure resulted in greater lint cotton yields per hectare than did surface applications of 54 metric tons or 81 metric tons of manure (1107 vs. 904 kg and 1169 vs. 1006 kg, respectively; Table 3). When the manure was applied to the surface, a significantly greater yield was obtained from 81 metric tons per hectare as compared with the 54-metric ton per hectare rate. This apparently resulted from a greater quantity of nitrogen being lost by volatilization when the manure was applied to the soil surface. Injection of manure at either rate resulted in lint cotton yields similar to commercial nitrogen applied at a rate of 134 kg per hectare. Equipment limitations prevented complete injection when rates of liquid wastes were 54 metric tons per hectare or greater. The size of the injection slot was not large enough to accommodate the higher rates without some overflow. The quantity of overflow increased as the manure rate increased.

### Experiment 3

Yields of lint cotton per hectare for Sharkey clay soil were the same (852 kg) for 134 kg of nitrogen per hectare and 108 metric tons of manure per hectare (Table 4). Although not significant, these yields were somewhat higher than those for 54 metric tons of manure per hectare (765 kg lint per hectare). The Sharkey clay soil responds to a higher level of nitrogen application than can be expected from the Dubbs silt loam soil; hence, the apparent response to higher per hectare rates of manure on Sharkey clay soil. No significant differences in lint cotton yields during the 1974 crop year indicated that practically no residual nitrogen was available for crop use from the manure applied in 1973 (Table 5).

TABLE 4. LINT COTTON YIELDS ON SHARKEY CLAY SOIL AS INFLUENCED BY VARIOUS RATES OF LIQUID MANURE, 1973.

| Treatment | Lint yield* kilograms per hectare |
|---|---|
| No manure, 134 kg per hectare of commercial nitrogen | 852a |
| 54 metric tons manure per hectare† | 765a |
| 81 metric tons manure per hectare† | 812a |
| 108 metric tons manure per hectare† | 852a |

*Means followed by the same letter are not different at the 5 percent level according to Duncan's New Multiple Range Test.
†No commercial nitrogen was applied to these plots.

TABLE 5. LINT COTTON YIELDS ON
SHARKEY CLAY SOIL AS INFLUENCED
BY VARIOUS RATES OF LIQUID MANURE
APPLIED THE PRECEDING YEAR, 1974.

| Treatment in 1973† | Lint yield* kilograms per hectare |
|---|---|
| No manure, 134 kg per hectare of commercial nitrogen | 526a |
| 54 metric tons manure hectare‡ | 492a |
| 81 metric tons manure hectare‡ | 517a |
| 108 metric tons manure per hectare‡ | 502a |

*Means followed by the same letter are not different at the 5 percent level according to Duncan's New Multiple Range Test.

†All plots received 134 kg per hectare of commercial nitrogen in 1974.

‡These plots received no commercial nitrogen in 1973.

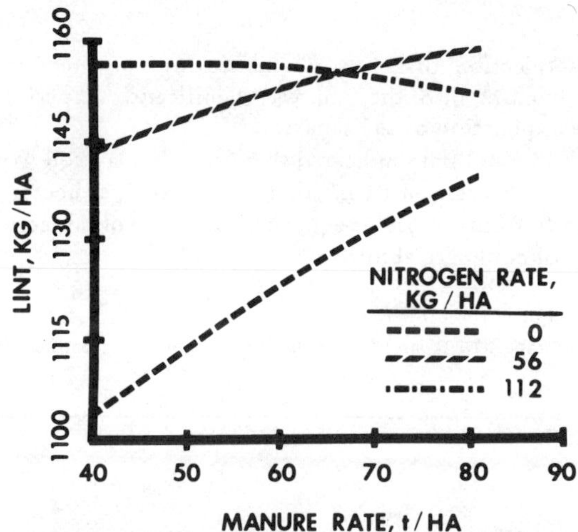



FIG. 1 Estimated effects on lint cotton yields for various rates of liquid beef cattle wastes given application of 0, 56, and 112 kg of commercial nitrogen per hectare, Dubbs silt loam soil.

## Regression Estimate

Data from experiments on Dubbs silt loam soil over the 2 yr were combined to estimate lint cotton yield response to various rates of soil injected beef wastes and commercial nitrogen. The following equation was derived:

$$\hat{Y} = 1063 + 1.3 X_1 - 0.004 X_1^2 + 1.4 X_2 - 0.005 X_2^2 - 0.009 X_1 X_2$$

$$S_{Y.X_1, X_2} = 92; S_{X_1} = 1.1; S_{X_1^2} = 0.006; \quad S_{X_2} = 3.1; S_{X_2^2} = 0.023; S_{X_1 X_2} = 0.0124$$

where:

$\hat{Y}$ = estimated lint cotton yields, kg per hectare

$X_1$ = commercial nitrogen rate, kg per hectare

$X_2$ = liquid manure rate, metric tons per hectare

$R^2$ = 73.6 percent

Expansion of this equation for different manure rates given the three levels of commercial nitrogen is illustrated in Fig. 1. A similar expansion for the different rates of commercial nitrogen given various levels of manure application is included in Fig. 2.

These figures suggest that no yield response resulted from adding commercial nitrogen above 56 kg per hectare when manure rates were 40 metric tons per hectare or greater. Compared to no nitrogen, the increase in lint cotton yield resulting from 56 kg of nitrogen per hectare declined as manure rates of application were increased from 40 to 81 metric tons per hectare. Applying up to 81 metric tons of manure per hectare along with 112 kg of nitrogen per hectare resulted in only slightly reduced lint cotton yields. However, this would be very inefficient use of the nutrients available in the manure.

Although not noted in this experiment, it is possible that continued use of beef cattle wastes could increase the weed control problems associated with cotton production.

## SUMMARY AND CONCLUSIONS

Various rates and methods of applying liquid cattle wastes and commercial nitrogen to two soil types over 2 yr were evaluated to determine their effects on lint cotton production. Conclusions which appear to be supported by these results are as follows:

1. Lint cotton production was not significantly increased by application of commercial nitrogen when liquid cattle wastes were applied to a Dubbs silt loam soil at rates of 54 to 81 metric tons per hectare.

2. Substitution of liquid cattle wastes for commercial nitrogen in cotton production could result in energy conservation of 120 to 133 cu m of natural gas per hectare of cotton. This results in a reduction of about $25 per hectare in fertilizer costs.

3. Greatest lint yields were obtained when liquid manure rates were 54 metric tons per hectare or greater on a Dubbs silt loam soil.

4. Liquid manure applied at a rate of 81 metric tons per hectare for two consecutive years did not adversely affect lint cotton yields on a Dubbs silt loam soil.

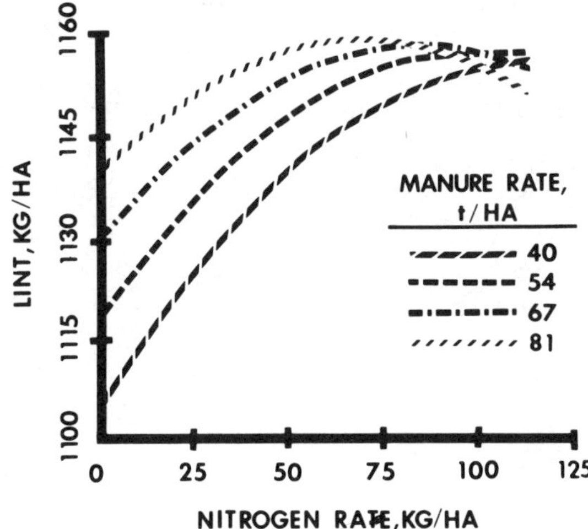



FIG. 2 Estimated effects on lint cotton yields for various rates of commercial nitrogen given injection of 40, 54, 67, and 81 metric tons of liquid beef wastes per hectare, Dubbs silt loam soil.

5. Injection of 54 or 81 metric tons per hectare of liquid manure into the soil was significantly superior to surface application of the same rates.

6. Limited data indicated that Sharkey clay soil would require application of 81 metric tons or more per hectare to achieve lint cotton yields equivalent to those obtained with 134 kg of commercial nitrogen.

### References


1 Draper, N. R. and H. Smith. 1966. Applied regression analysis. John Wiley and Sons, Inc., New York. 407 p.

2 Fairbanks, W. C. 1974. Fuel from livestock wastes: an economic analysis. AGRICULTURAL ENGINEERING 55(9):20-23, September.

3 Hoeft, R. G. and J. C. Siemens. 1975. Energy consumption and return from adding nitrogen to corn. Illinois Research 17(1):10-11.

4 Holloway, J. W., J. M. Anderson, W. A. Pund, W. D. Robbins, F. T. Cooke, Jr., and R. W. Rogers. 1974. Space allowances for cattle fed in a slatted-floor feedlot compared to those fed in a conventional slab-floor feedlot. Miss. Agri. and Forestry Experiment Station Tech. Bul. 69, 15 p.


---

## Land Application of Animal Waste

*(Continued from page 582)*

and plant toxicities. Only by continuing experiments on a long-term basis can the validity of application rate guidelines be evaluated.

### References


1 Azevedo, J. and P. R. Stout. 1974. Farm Animal manures: an overview of their role in the agricultural environment. Agricultural Experiment Station/Extension Service Manual 44. Agricultural Publications, University of California, Berkeley.

2 Boorman, C. V., T. E. Loynachan, J. K. Koelliker. 1974. Effects of sprinkler application of lagoon effluent on corn and grain sorghum. Processing and Management of Agricultural Wastes, Proc. 1974. Cornell Agricultural Waste Management Conf. (In press).

3 Cooperative Extension Service, Purdue University, Lafayette, Indiana. 1972. Waste handling and disposal guidelines for Indiana Beef Producers. ID-84. Prepared by the Purdue University Animal Waste Committee. 13 p.

4 Goodrich, P. R., E. C. Miller, J. J. Boedicker, S. D. Evans, G. W. Randall and A. E. Hanson. 1973. Effects of intensive applications of livestock manure on soil and crops. p. 99-115. 1973 Minnesota Cattle Feeders' Report. Dept. of Animal Sci., University of Minnesota.

5 The Life Sciences and Agriculture Experiment Station and the Cooperative Extension Service, University of Maine at Orono and The Main Soil and Water Conservation Commission. 1972. Maine guidelines for manure and manure sludge disposal on land. Miscellaneous Report 142. 22 p.

6 Linderman, C. L. and N. P. Swanson. 1972. Disposal of beef feedlot runoff on corn. Agron. Abstracts, p. 182.

7 Mathers, A. C. and B. A. Stewart. 1974. Corn silage yield and soil chemical properties as affected by cattle feedlot manure. J. Environ. Quality 3:143-147.

8 Murphy, L. S., G. W. Wallingford, W. L. Powers and H. L. Manges. 1972. Effects of solid beef feedlot wastes on soil conditions and plant growth. Waste Management Research, Proc. 1972 Cornell University Agricultural Waste Management Conf. p. 449-464. Graphics Management Corp., Washington, D.C.

9 North Carolina Agricultural Extension Service. 1973. Circular 568. Dairy Waste Management Alternatives. 25 p.

10 Powers, W. L., R. L. Herpich, L. S. Murphy, D. A. Whitney, H. L. Manges, G. W. Wallingford. 1973. Guidelines for land disposal of feedlot lagoon water. Kansas State University, Cooperative Extension Service, Manhattan, Kansas. 7 p.

11 Powers, W. L., G. W. Wallingford, L. S. Murphy, D. A. Whitney, H. L. Manges and H. E. Jones. 1974. Guidelines for applying beef feedlot manure to fields. Cooperative Extension Service, Kansas State University, Manhattan, Kansas. 11p.

12 Pratt, P. F., F. E. Broadbent and J. P. Martin. 1973. Using organic wastes as nitrogen fertilizers. California Agriculture. June. p. 10-13.

13 Shortall, J. G. and W. C. Liebhardt. 1974. Yield and growth of corn as affected by poultry manure. J. Environ. Quality (In press).

14 Swanson, N. P., C. L. Linderman and J. R. Ellis. 1973. Irrigation of perennial forage crops with feedlot runoff. ASAE Paper No. 73-241. ASAE, St. Joseph, Mich. 49085.

15 Vanderholm, D. H. 1973. Area needed for land disposal of beef and swine wastes. Iowa State University, Cooperative Extension Service, Ames, Iowa. Pm-552. (Rev.) 2 p.

16 Wallingford, G. W., L. S. Murphy, W. L. Powers and H. L. Manges. 1974. Effect of beef-feedlot-lagoon water on soil chemical properties and growth and composition of corn forage. J. Environ. Quality 3:74-78.

# On-The-Farm Determination of Animal Waste Disposal Rates for Crop Production

D. O. Turner

THERE is general agreement among research and extension workers in the Pacific Northwest that animal wastes can, and should, be used in the production of crops. The term "animal waste" in itself is misleading and inappropriate. Animal-origin fertilizer, or manure fertilizer, might be better terms. When manure is spread on farmland where it originated, there is relatively low energy cost involved in disposal. Because manure is a bulky material, costs increase rapidly with increasing transport distances. This is, of course, the main reason manure came to be called a waste material; value/unit of manure was simply too low to warrant movement to land areas which could receive it. The recent history of high cost and short supply of commercial fertilizers has stimulated a new interest in beneficial use of animal manures.

Livestock operators in the Pacific Northwest need a means whereby they can estimate manure loading capabilities of their soils. Average annual precipitation varies from less than 10 inches to nearly 200. Soils vary throughout the region. Seventy percent of the 300,000 dairy cows and most of the poultry flocks in Washington and Oregon are west of the Cascade mountains in areas having heavy winter rainfall.

There is a wide range of manure handling methods in use within the region. Beef cattle in dry feedlots (both large and small) are found mostly in arid sections. Dairy operators use anaerobic lagoons, liquid manure tanks and solid manure handling methods. Distribution of manure may be through manure pump and irrigation pipeline systems, manure tank wagons, or beater-spreaders.

The fate of the nitrogen excreted by the animal is determined by the manure handling system, climate, soil characteristics, and crop removal capabilities. Farmers are aware of the value of manure in crop production but there are questions as to how much manure to apply per acre, how many acres are needed, and to what extent can manure nitrogen be substituted for synthetic nitrogen for adequate crop production?

## NEED FOR MANURE DISPOSAL GUIDELINES

Inadequacies of existing information and the pressing need to provide farmers, county agents, representatives of regulatory agencies and others with assistance concerning manure management problems has prompted the preparation of Pacific Northwest Manure Management Guidelines.

Animal manure disposal problems can be divided into two general categories:

1 Operations where the land base associated with the enterprise is generally adequate for manure disposal and recycling through crops.

2 Operations with an inadequate land base which requires more sophisticated treatment of the manure or disposal through massive soil loading.

---

The author is: D. O. TURNER, Extension Soil Scientist and Agronomist, Washington State University, Pullman.

The guidelines as developed are designed to deal with category 1 only; they are intended to encourage the efficient use of manures in crop production. The specific purpose of these guidelines is to provide a method whereby animal manure can be used wholly, or in part, to meet the nitrogen requirements of crops. The decision of how much manure to apply should be made by the farm operator because he alone knows the character of the manure to be used, the crop to be grown, and the soil to which manure will be applied. These guidelines provide the framework by which the farm operator can determine the appropriate manure loading rate he should use to grow the desired crops without contributing to groundwater pollution with nitrates.

## USE OF MANURE TO MEET CROP NITROGEN NEEDS

Nitrogen is the major nutrient which more frequently limits crop yields and adversely affects water quality. The guidelines are intended to allow utilization of animal manure as a substitute for commercial fertilizers with the emphasis on nitrogen. Equations have been developed to accomplish this objective.

The animal waste guidelines for the Pacific Northwest are basically those suggested by Willrich et al. (1974) with minor modifications. The guidelines allow 3 basic determinations:

1 Estimation of total manure nitrogen available for crop recycling through soils after storage and application losses. This allows determination of total nitrogen load to be imposed on the crop-soil system.

$$SAMN = EMN \times NR$$

where

SAMN = soil-applied manure nitrogen

EMN = excreted manure nitrogen, based on excretion rates x animal days

NR = a fraction of nitrogen remaining after accounting for storage, treatment, and handling losses

2 Estimation of nitrogen mineralization rate, or nitrogen availability; estimation of denitrification, and a measure of crop nitrogen requirements. This provides a value as manure nitrogen/acre to be applied.

$$SIMN = \frac{FGN}{A \times D}$$

where

SIMN = soil incorporated manure nitrogen

FGN = fertilizer guide nitrogen, the rate of synthetic nitrogen suggested for the crop to be grown

A = availability coefficient, a fraction of the soil-incorporated total nitrogen of manure origin

D = denitrification coefficient, a fraction of the available, inorganic nitrogen

3   Using values obtained from 1 and 2, the area required for manure disposal can be calculated.

$$\frac{SAMN}{SIMN} = \text{number of acres required for disposal}$$

Tables 1, 2, 3, and 4 have been prepared to provide input values for use with the equations. These tables allow estimation of the amount of excreted nitrogen for various species; nitrogen remaining after accounting for storage, treatment and handling losses; manure nitrogen availability over a series of years; and an estimation of denitrification.

**TABLE 1. TOTAL NITROGEN EXCRETED IN FRESH MANURE.**

| Specie | Total N/1000 lb animal liveweight/day |
|---|---|
| | - - - - - - - -lb - - - - - - - - |
| Dairy cow | 0.41 |
| Beef feeder | 0.34 |
| Swine feeder | 0.45 |
| Sheep feeder | 0.45 |
| Layer | 0.72 |
| Broiler | 1.16 |

Source: American Society of Agricultural Engineers, data adapted SE-412 Committee report AW-D1 (Rev. 6/14/73).

**TABLE 2. NITROGEN REMAINING AFTER ACCOUNTING FOR STORAGE, TREATMENT, AND HANDLING LOSSES (NR).**

| Manure storage, treatment, and handling system | (NR) a fraction of N remaining |
|---|---|
| Oxidation ditch, anaerobic lagoon, irrigation or liquid spreading | 0.16 |
| Anaerobic lagoon, irrigation or liquid spreading | 0.22 |
| Deep pit storage, liquid spreading | 0.34 |
| Aerobic lagoon, irrigation or liquid spreading | 0.40 |
| Open lot surface storage, solid spreading | 0.40 |
| Roofed storage in manure pack, solid spreading | 0.65 |
| *Fresh manure, directly field spread* | |
| Time between application and incorporation: | |
| 1-4 days; warm, dry soil | 0.65 |
| 7 days or more; warm dry soil | 0.50 |
| 1-4 days; warm, wet soil | 0.85 |
| 7 days or more; warm wet soil | 0.70 |
| 7 days or more; cool, wet soil | 0.90 |

**TABLE 3. AVAILABILITY COEFFICIENTS FOR THE FIRST FIVE APPLICATION YEARS (A).**

| Type of manure | Availability coefficient for the year of application | | | | |
|---|---|---|---|---|---|
| | 1 | 2 | 3 | 4 | 5 |
| Poultry, fresh | 0.75 | 0.80 | 0.85 | 0.90 | 0.93 |
| Poultry, aged, covered | 0.60 | 0.75 | 0.80 | 0.84 | 0.87 |
| Dairy, fresh | 0.50 | 0.65 | 0.70 | 0.74 | 0.78 |
| Dairy, liquid manure tank | 0.42 | 0.54 | 0.60 | 0.64 | 0.68 |
| Dairy, anaerobic lagoon storage | 0.30 | 0.38 | 0.45 | 0.50 | 0.53 |
| Beef, feedlot stockpiled | 0.35 | 0.45 | 0.50 | 0.53 | 0.55 |

**TABLE 4. DENITRIFICATION COEFFICIENT (D)**

| Degree of soil drainage | Denitrification coefficient |
|---|---|
| Excessive or somewhat excessive | - - |
| Well to moderately well | 0.85 |
| Somewhat poorly to poorly | 0.70 |
| Very poorly | 0.60 |

To use these equations the farm operator must know:

1   The nature of the manure to be used: The nature of the manure has reference to that material as it is applied to the land. There is often a vast difference between manure as voided by the animal and that which is applied to the soil in terms of nitrogen content and potential nitrogen release rate.

2   The method of applying the manure: Manure may be applied through a conventional beater-spreader, liquid tank wagon, large bore irrigation nozzle, soil injection system, or by other means. Nitrogen use efficiency varies dependent upon the system used.

3   The crop to be grown: The Pacific Northwest states have used fertilizer guide sheets (FG's) for several years. These FG's are written for specific crops and each contains a recommendation for the amount of nitrogen that should be applied for each crop.

4   The soil type: Soil surveys of the Northwest are available. The Soil Conservation Service and Extension Service can provide information to enable a farmer to determine the soil types and hence, the degree of soil drainage on his farm. The degree of soil drainage has an influence on efficiency of manure nitrogen utilization.

## EXPERIMENTAL MANURE APPLICATIONS TO CROP AND SOIL SYSTEMS

Manure handling and transport systems in conjunction with field applications have been under study at the Washington State Department of Institutions' Monroe Honor Farm and the Western Washington Research and Extension Center at Puyallup, Wash.

The Monroe location has been in operation since 1969. This site had an average herd size of 350 mature Holsteins. Facilities as described by Turner and Proctor (1971) included total confinement barn, anaerobic lagoons, and pump-pipe distribution system. Total N applied as a manure slurry was measured and found to be about 6000 lb/acre over a 5 year period. The soil is a Puget silty clay loam located on an alluvial plain and was cropped annually with silage corn. Annual crop removal of nitrogen by the corn was determined and averaged 220 lb/acre for a 5 year total of 1100 lb of nitrogen.

Nitrogen availability coefficients for different manure handling systems, including anaerobic lagoons, are given in Table 3. Application of the availability coefficients to the 5-year loading schedule at this site indicates approximately 3325 lb of applied nitrogen should have remained in the soil as organic nitrogen not yet mineralized. This would leave 2675 lb of nitrogen that had either been removed by the crop, denitrified, or leached.

The Puget silty clay loam is poorly drained and denitrification may have been substantial (Turner and Proctor 1971). A denitrification coefficient of D = 0.60 (Table 4) was considered proper for this soil. Soil samples to a depth of 4 feet were taken periodically from April through October during the 5 year period. Nitrate-N concentrations normally ranged from 0.3 to 2.0 mg/l in the third and fourth foot. An occasional sample contained as high as 5.0 mg/l $NO_3^-$-N.

Koelliker et al. (1971) suggest 600 lb/acre/season as reasonable to allow crop removal of from 150 to 300 lb/acre of nitrogen and have denitrification remove the remaining mineralized portion on a poorly drained Iowa soil.

The 5 year total of 2675 lb of mineralized nitrogen, as estimated by the equations, amounted to 535 lb/acre/year that had been removed by the crop or disposed of through the soil. This quantity is close to the amount suggested in the Iowa study.

Although there appears to be no pollution threat to groundwater as a result of the loading imposed on this soil, the loss of more than 300 lb of nitrogen/year, presumably by denitrification, indicates the loading rate is too high to be consistent with the desire to achieve maximum utilization of manure nitrogen as a resource.

Work at the Puyallup location started in the spring of 1973. The dairy herd, including replacement heifers, totaled about 130 animals. An anaerobic lagoon was used for winter storage of the manure. This material was passed through a mixing tank and transported to the fields by means of a soil-injection equipped manure tank wagon. The soil injection equipment was a necessity because of the close proximity of non-farm residences. This means of application has been effective in providing odor control. It also prevented loss of nitrogen that might occur from volatization as manure was applied to the soil surface. Because treatments were made to land that had been tilled, it was necessary to equip the tractor drawing the tank wagon with dual wheels.

The experimental site consisted of four replications of five treatments and one treatment with synthetic nitrogen. A split-plot arrangement was used that allowed silage corn as a summer crop and a winter program with, and without, cereal rye which was removed as green-chop forage in the spring. The soil at this location is a well-drained Puyallup fine sandy loam.

The first treatments were made in the spring of 1973 with the soil injection tank wagon. Quantities of nitrogen contained in the manure slurry are shown in Table 5. Treatment 6 was given the fertilizer suggested in the Washington State University fertilizer guide for silage corn by Turner et al. (1970). All plots received fertilizer guide phosphorus rates in standard band placement in addition to the manure. After the corn was harvested, additional treatments were made to some plots (Table 5) and winter rye was seeded in the split-plot design. The nitrogen loading in 1974 was less than intended as unusually heavy winter rains had diluted the lagoon material.

Yields obtained from the two corn harvests and the one rye harvest are shown in Table 5. It appears that $T_1$ was inadequate, especially in 1973, in providing sufficient nitrogen for optimum growth of the corn.

The amount of nitrogen removed by the three crop harvests is also given in Table 5. Differences were small among $T_2$, $T_3$, $T_4$, and $T_5$ in either corn yield or nitrogen recycling. There is no advantage to be gained from the heavier loading of $T_3$; in fact, it represents a waste of nitro-

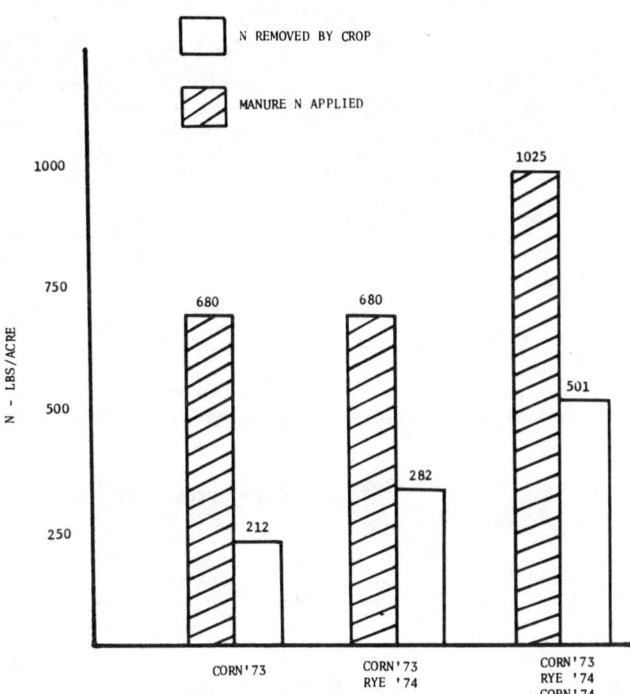


FIG. 1 Applied manure $N(T_2)$ compared to N removed by crop harvest in 1973 and 1974.

gen. The fall applications, $T_4$ and $T_5$, promoted more rye yield and nitrogen recovery. A comparison between applied manure nitrogen in $T_2$ and quantity of nitrogen removed by the crop system is shown in Fig. 1.

Nitrate-N concentrations were determined at four soil depths as the silage corn was growing and just after harvest (Table 6). These data show that the corn crop effectively utilized the nitrogen as it was mineralized and became available, with the exception of $T_3$. The $NO_3^-$-N concentrations for $T_3$ in the October sampling indicated that some nitrogen would be lost through the soil profile to ground water. Soils at the October sampling were extremely dry and there had been no opportunity for leaching to have occurred. Heavy rains occurred in November and subsequent samplings showed that $NO_3^-$-N loss did occur from $T_3$. Other treatments lost very little nitrogen.

These equations provide a simple procedure to relate livestock source, handling methods, crops and soils to efficient use of manure as a fertilizer source.

If the guidelines procedure had been followed at the Monroe location, it would have indicated a 5 year total application of 3500 lb manure nitrogen required with which to grow the crop, account for non-mineralized nitrogen, denitrification, and leaching. This is 2500 lb less man-

---

**TABLE 5. YIELD OF FORAGE AND AMOUNT OF NITROGEN REMOVED WITH HARVESTED CROP AS INFLUENCED BY TIME AND RATE OF APPLIED NITROGEN.**

| | | Nitrogen applied | | | Dry matter yield of forage | | | Nitrogen removed by crop | | |
|---|---|---|---|---|---|---|---|---|---|---|
| | | April | October | April | Silage corn | Winter rye | | Silage corn | | Winter rye |
| | | 1973 | 1973 | 1974 | 1973 | 1974 | Spring 1974 | 1973 | 1974 | Spring 1974 |
| | | lb/acre | | | tons/acre | | | lb/acre | | |
| $T_1$ | manure | 335 | - | 160 | 6.5 | 7.5 | 2.3 | 154 | 176 | 70 |
| $T_2$ | manure | 680 | - | 345 | 8.5 | 8.8 | 2.3 | 212 | 219 | 70 |
| $T_3$ | manure | 970 | - | 425 | 8.2 | 8.9 | 2.9 | 202 | 200 | 84 |
| $T_4$ | manure | 670 | 230 | 290 | 8.6 | 8.4 | 3.2 | 223 | 220 | 97 |
| $T_5$ | manure | 680 | 390 | 290 | 8.1 | 8.5 | 3.3 | 196 | 212 | 100 |
| $T_6$ | urea | 175 | 45 | 175 | 7.4 | 7.8 | 1.9 | 180 | 187 | 57 |

| Treat-ment | Nitrogen source and rate, 2 year total | $NO_3^-$-N concentrations at different soil depths, cm | | | |
|---|---|---|---|---|---|
| | | 0-30.5 | 30.5-61.0 | 61.0-91.5 | 91.5-122.0 |
| | - lb N/acre - - - - | - - - - - - - - - - - mg/l - - - - - - - - - - - - - | | | |
| | | August 5, 1974 sampling | | | |
| $T_1$ | manure 495 | 15.6 | 8.9 | 4.7 | 4.5 |
| $T_2$ | manure 1025 | 35.6 | 13.3 | 7.4 | 4.7 |
| $T_3$ | manure 1395 | 30.5 | 13.8 | 5.7 | 3.8 |
| $T_4$ | manure 1375 | 18.6 | 8.8 | 3.8 | 4.8 |
| $T_5$ | manure 1660 | 36.4 | 10.0 | 5.5 | 3.4 |
| $T_6$ | urea 395 | 41.0 | 10.7 | 6.3 | 5.1 |
| | | October 1, 1974 sampling | | | |
| $T_1$ | manure 495 | 12.4 | 5.7 | 4.0 | 2.8 |
| $T_2$ | manure 1025 | 27.0 | 10.5 | 4.1 | 3.9 |
| $T_3$ | manure 1395 | 59.8 | 21.1 | 6.7 | 5.0 |
| $T_4$ | manure 1375 | 17.1 | 6.1 | 5.1 | 4.9 |
| $T_5$ | manure 1660 | 15.4 | 7.0 | 3.3 | 5.2 |
| $T_6$ | urea 395 | 13.0 | 5.4 | 4.9 | 4.5 |

ure nitrogen than was actually applied. Manure nitrogen use efficiency should be far greater by following the outlined procedure than by the loading rates actually used which are similar to those manure applications made on nearby farms.

Manure nitrogen loading rates for the Puyallup location can be estimated by use of the equation system as:

$$SIMN = \frac{FGN}{A \times D}$$

where

FGN = 175, the recommended rate of N/acre for silage corn

A = 0.30, (Table 3, anaerobic lagoon, 1st year)

A = 0.38, (Table 3, anaerobic lagoon, 2nd year)

D = 0.85, (Table 4, well drained soil)

$$SIMN = \frac{175}{0.30 \times 0.85} = 685 \text{ lb N/acre, 1st year}$$

$$SIMN = \frac{175}{0.38 \times 0.85} = 540 \text{ lb N/acre, 2nd year}$$

The guideline procedure indicates a first year loading of 685 lb manure nitrogen and a second year application of 540 lb for a 2 year total of 1225 lb. Data in Table 5 show the first year estimate to be extremely close. The second year estimate may be high.

The guidelines concept offered here has been accepted by the Environmental Protection Agency, Region X; Washington State Department of Ecology; Soil Conservation Service; and the Cooperative Extension Service. The guidelines are now in use in Washington State.

There are weaknesses and causes for error in the guidelines procedure. These limitations should be recognized by those who use the system. Such limitations need to be balanced against the errors presently being made when manure is applied without consideration of any of the factors that influence the fate of nitrogen.

As a final precaution, it must be recognized that the manure loading rates are based on nitrogen only. In the event that nutrients other than nitrogen may be the limiting factor in obtaining optimum yields, the manure rates calculated by this method may prove inadequate as the sole source of fertilizer. Because the fertilizer guide is an integral part of this procedure, any weakness in it will appear in the animal waste guidelines.

### References

1   Koelliker, J. K., J. R. Miner, C. E. Beer and T. E. Hazen. 1971. Treatment of livestock-lagoon effluent by soil filtration, p. 329-333. In: ASAE Livestock waste management and pollution abatement. St. Joseph, Mich. 49085.

2   Turner, D. O., S. E. Brauen and A. R. Halvorson. 1970. Wash. State Univ. Agr. Ext. FG-18, Silage corn for western Washington. 4 p.

3   Turner, D. O. and D. E. Proctor. 1971. A farm scale dairy waste disposal system, p. 85-88. In: ASAE Livestock waste management and pollution abatement. St. Joseph, Mich. 49085.

4   Willrich, T. L., D. O. Turner and V. V. Volk. 1974. Manure application guidelines for the Pacific Northwest. ASAE Paper No. 74-4061. Oklahoma State Univ. Stillwater, Oklahoma.

# Disposal of Dairy Cattle Manure on Soil

Z. F. Lund, F. L. Long, B. D. Doss, F. E. Lowry

LAND spreading of dairy cattle manure is effective both in disposing of a waste product and in utilizing a valuable fertilizer resource. Caution must be exercised, however in selecting application rates that are optimum for forage production since they may produce forage detrimental to animal health. Forage grown at high soil nitrate levels may be toxic to ruminant animals (Crawford et al. 1966). Animals grazing pasture fertilized with manure have developed grass tetany (Grunes et al. 1970, Wilkinson et al. 1970); therefore, manure application rates must be carefully determined. We evaluated yield, forage quality, and runoff water quality in several disposal systems in these experiments.

## MATERIALS AND METHODS

### Runoff Water Quality

Our study was conducted on 0.04-ha (0.1-acre) plots on Norfolk sandy loam with <2 percent slope, and instrumented to measure and sample runoff water. Each spring for 3 yr two plots received manure incorporated into the surface 15 cm at the rate of 45 mt/ha (20 t/acre), on a dry-weight basis, which supplied 800, 175, and 400 kg/ha/yr of N, P, and K, respectively. The two check plots received mineral fertilizer at rates considered adequate to remove N, P, and K as growth-limiting factors in this soil. The plots were double-cropped with pearl millet (*Pennisetum americanum* [L.] K. Schum) var. 'Gahi-1', and Wren's 'Abruzzi' rye (*Secale cereale* L.), similar to the following incorporation studies.

We analyzed runoff water for nitrate N ($NO_3$-N), ammonium N ($NH_4$-N), pH, electrical conductivity (EC), chemical oxygen demand (COD), and biochemical oxygen demand (BOD).

### Incorporation Studies

Manure application rates were studied on two experiments: one initiated in April 1970 on a 2.6 by 2.3 m plot of Lucedale sandy loam (Rhodic Paleudults) at Thorsby, Alabama, and one in July 1970 on a 2.7 by 2.7 m plot of Dothan loamy sand (Plinthic Paleudults) at Auburn, Alabama. Fresh dairy manure was incorporated into the surface 15-cm annually at rates of 0, 22.5, 45, 90, 180, and 270 mt/ha on a dry weight basis. The check plots (0-manure) received broadcast applications of N, P, and K at the rate of 114 kg/ha/yr before planting. Supplemental N and K at rates of 57 kg/ha was surface-applied after each cutting on the check plots.

A randomized-block design with four replications was used in all tests. Millet and rye were used in a double-cropping system. The millet was harvested when it was 120 to 180 cm tall. Three cuttings each of millet and rye were made on the Lucedale soil each year. On the Dothan soil millet was harvested twice in 1970, four times in 1971 and 1972, and rye was harvested three times in 1970 and 1971, and twice in 1972. Forage was sampled at all harvest dates and analyzed for organic N, nitrate N ($NO_3$-N), K, Ca, Mg, and P.

### Surface-Application Studies

We evaluated surface application of manure on sod on Dothan and Lucedale soils. Established Coastal bermudagrass (*Cynodon dactylon* [L.] Pers.) was used as a test crop on both areas. Treatments consisted of a mineral fertilizer check and five manure treatments: solid at 45 and 90 mt/ha/yr, and liquid at 45, 90, and 135 mt/ha/yr on a dry-weight basis. Rates were divided into six applications at approximately 2-month intervals. Check plots received four applications of mineral fertilizer totaling 470, 225, and 470 kg/ha/yr of N, P, and K, respectively. Solid or liquid manure was spread over the surface. The number of harvests varied; four in 1971 and 1973, and five in 1972 on the Dothan soil; five for each of the 3 years on the Lucedale soil. Plot size, plant analyses, and statistical design were similar to those used in the incorporation studies.

## RESULTS AND DISCUSSION

### Runoff Water Quality

The incorporated manure had essentially no effect on the BOD of runoff water (Table 1). The 3-yr mean values for the check and manured plots were equal at 4.7 mg/liter. These values are low and are only about one-sixth of the generally accepted BOD values for such water. For all plots, the values were considerably lower than those reported for barnlot runoff (White and Edwards 1972). No buildup in the BOD of runoff water was indicated with repeated applications of manure during the 3-yr period. In fact, mean BOD values decreased from 8.0 mg/liter in 1970 to 1.8 mg/liter in 1973.

Manure did not appreciably affect the $NH_4$-N content of runoff water (data not shown). Mean 3-yr values were only 0.65 and 0.98 mg/liter for the check and manured plots, respectively. The $NO_3$-N content of the runoff water was essentially unaffected by the manure treatment. Mean

**TABLE 1. BIOCHEMICAL OXYGEN DEMAND (BOD) VALUES OF RUNOFF WATER FROM CHECK AND MANURED PLOTS**

| | BOD, mg/liter | | | | | |
| | Check | | | Manured | | |
| Year | Max. | Min. | Mean | Max. | Min. | Mean |
|---|---|---|---|---|---|---|
| 1970* | 11.9 | 2.5 | 8.0 | 10.7 | 3.6 | 8.0 |
| 1971 | 9.6 | 3.0 | 5.7 | 8.9 | 3.5 | 6.1 |
| 1972 | 9.9 | 3.0 | 5.0 | 6.3 | 1.5 | 4.3 |
| 1973† | 4.4 | 0.0 | 1.5 | 5.4 | 0.0 | 1.8 |
| 3-year mean‡ | | | 4.7 | | | 4.7 |

*July 23 through December
†January through April 23
‡Actual time is 2.75 yr

Contribution from Soil and Water Research, Southern Region, ARS, USDA, and Departments of Agronomy and Soils and Agricultural Engineering, Auburn University Agricultural Experiment Station, Auburn, Ala.

The authors are: Z. F. LUND, F. L. LONG, B. D. DOSS, Soil Scientists, ARS, USDA, and F. E. LOWRY, Research Associate, Auburn University, Auburn, Ala.

values for the 3-yr period were 1.48 and 1.59 mg/liter for the check and manured plots, respectively. These values are well within acceptable nitrate levels even for drinking water (USDHEW 1962).

Yearly rainfall runoff and N loss from check and manured plots are shown in Table 2. About 17 percent of rainfall ran off. Total runoff was highest during 1971, which was associated with the highest N loss. The N losses from check and manured plots, however, were less than 5 kg/ha each year. Over the 3-yr period the average N loss per year in runoff water was 4.22 and 3.15 kg/ha for the check and manured plots, respectively. These amounts of N are insignificant from the standpoint of nutrient loss or pollution.

## Incorporation Studies

High manure application rates (270 mt/ha) reduced millet yields on both Dothan and Lucedale soils the first year (Table 3). Ammonification of the first application in these sandy soils with low exchange capacities increased the soil pH to 9.0 (data not shown), resulting in poor emergence, stand loss, and reduced yield. Soil pH declined as the season progressed. Manure application in 1971 and 1972 resulted in very little increase in pH. The residue increased the cation exchange capacity (CEC) of the soil, thus increasing ammonium adsorption and reducing the effect on soil pH.

On Dothan soil for all 3 yr the 22.5 mt/ha manure rate produced as much millet forage as the check plot. Maximum yields were obtained with the 90 mt/ha application rate, which were 40 to 50 percent higher than those obtained with the check (Table 3). On Lucedale soil the 22.5 mt/ha rate did not yield as much as the check plot in 1971 and 1972, and a higher manure rate was required to obtain maximum yields.

Maximum rye forage yields varied among the high application rates of manure each year, but there was no statistical difference in yield between the 45 mt/ha/yr and higher rates of application the last 2 yr on the Dothan soil. There were differences in rye forage yield with application rates up to 180 mt/ha/yr on the Lucedale soil in 1970 and 1971, and in 1972 the 270 mt/ha treatment had the highest yields.

In addition, manure application has many effects on soil that stimulate plant growth. These include changes in bulk density, total water-holding capacity, color, and exchange capacity. Reduced bulk density decreases resistance to root penetration in the soil and enhances root growth. Increased total water-holding capacity allows more rapid movement of water and ions to the root surface, and darker color causes the soil to warm up earlier in the spring, promoting increased root growth on the early crops when the soil is cold. Increased exchange capacity increases the reserve of available ions and buffers change in pH.

Plant species and soils were important factors in considering application rates to produce forage with safe levels of nitrate (Fig. 1). Rye was lower in nitrate than millet at most levels of manure application. The tendency was greater for high nitrate in forage produced on Dothan than on Lucedale soil. All levels of application on Lucedale produced rye forage with less than 2 percent nitrate. Whether this is an effect of location, manure source, or soils is unknown.

The equivalent ratio of K to Ca plus Mg (K/(Ca+Mg)) exceeded 2.2 in forage produced in most manure treat-

### TABLE 3. YIELD OF MILLET FORAGE ON MANURE-TREATED PLOTS

| Treat-ment | Dothan soil | | | Lucedale soil | | |
|---|---|---|---|---|---|---|
| | 1970 | 1971 | 1972 | 1970 | 1971 | 1972 |
| | | | mt/ha | | | |
| Check | 11.6b* | 15.3c | 14.0c | 7.0c | 13.2bc | 16.3d |
| 22.5 | 11.9b | 16.1c | 13.8c | 8.0c | 9.0d | 13.0e |
| 45 | 14.0b | 19.2b | 17.9b | 10.1b | 11.3c | 17.2cd |
| 90 | 17.7a | 23.3a | 23.1a | 12.2a | 13.6b | 18.4bc |
| 180 | 12.0b | 24.1a | 24.1a | 10.0b | 14.1b | 19.9ab |
| 270 | 7.4c | 23.9a | 23.4a | 9.9b | 15.5a | 20.3a |

*Oven-dry yields within a column followed by the same letter are not significantly different at the 5 percent level using Duncan's Multiple Range Test.

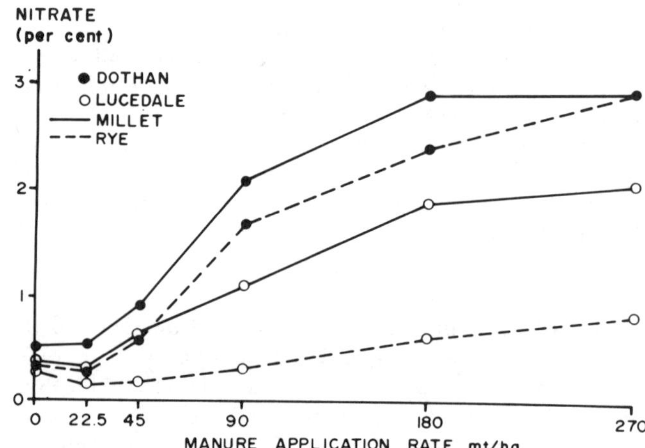


FIG. 1 Effect of annual manure application rate on 3-yr average nitrate level in forage.

### TABLE 2. RAINFALL, RUNOFF AND TOTAL N LOSS IN RUNOFF WATER FROM CHECK AND MANURED PLOTS

| Year | Rain-fall | Check | | | | Manured | | | |
|---|---|---|---|---|---|---|---|---|---|
| | | Run-off | NH$_4$-N | NO$_3$-N | Total N | Run-off | NH$_4$-N | NO$_3$-N | Total N |
| | cm | cm | kg/ha | kg/ha | kg/ha | cm | kg/ha | kg/ha | kg/ha |
| 1970* | 53.2 | 9.96 | 0.28 | 0.99 | 1.27 | 10.57 | 0.48 | 1.78 | 2.26 |
| 1971 | 155.5 | 33.68 | 1.61 | 3.12 | 4.73 | 20.78 | 1.46 | 2.77 | 4.23 |
| 1972 | 140.6 | 9.45 | 0.31 | 1.35 | 1.66 | 7.21 | 0.41 | 1.18 | 1.59 |
| 1973† | 50.4 | 10.97 | 0.26 | 2.10 | 2.36 | 3.86 | 0.10 | 0.41 | 0.51 |
| 3-yr mean‡ | 144.9 | 25.29 | 0.85 | 3.37 | 4.22 | 16.12 | 0.82 | 2.33 | 3.15 |

*July 23 through December
†January through April 23
‡Actual time is 2.75 yr



MANAGING LIVESTOCK WASTES

ments, indicating a tetany hazard (Fig. 2). Soils had an effect at the high rates of application; forage grown on Lucedale soil had higher K/(Ca+Mg) ratios than that on Dothan. Lucedale soil has a more clayey subsoil with a greater accumulation of K in the surface horizons than Dothan loamy sand, which could account for the high ratio. Only millet forage at the 22.5 mt/ha manure application rate produced safe forage. However, all treatments, including the check plots, had high rates of K application, and could cause high ratios.

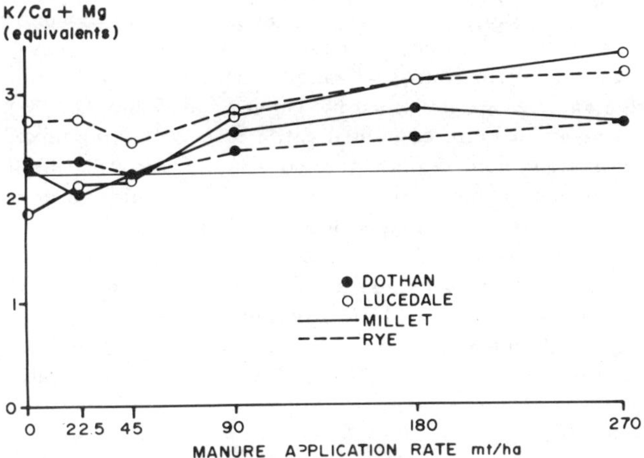


FIG. 2 Effect of annual manure application rate on 3-yr average equivalent ratio of K/(Ca+Mg).

### TABLE 4. YIELD OF RYE FORAGE ON MANURE-TREATED PLOTS

| Treat-ment | Dothan soil | | | Lucedale soil | | |
|---|---|---|---|---|---|---|
| | 1970 | 1971 | 1972 | 1970 | 1971 | 1972 |
| | | | —mt/ha— | | | |
| Check | 4.3c* | 10.2a | 8.1cd | 6.5b | 5.3bc | 3.3e |
| 22.5 | 3.8c | 7.6b | 7.8d | 3.8d | 2.4d | 2.2f |
| 45 | 5.3b | 10.1a | 9.9ab | 4.8c | 4.5c | 4.3d |
| 90 | 7.2a | 10.8a | 10.7a | 6.7b | 6.0b | 5.1c |
| 180 | 7.6a | 11.7a | 9.6ab | 8.1a | 8.0a | 5.5b |
| 270 | 6.1b | 11.2a | 8.9bc | 8.7a | 8.3a | 6.4a |

*Oven-dry yields within a column followed by the same letter are not significantly different at the 5 percent level using Duncan's Multiple Range Test.

### Surface Application

Surface manure applications were begun just before growth of Coastal bermudagrass in the spring. Only one-sixth of the yearly rate was used per application, and rates were insufficient to provide adequate nutrients for optimum growth the first part of the year (Table 5). Yields were low on both test areas and plants had N-deficiency symptoms. The $NO_3$-N and organic N were low, as shown by analysis of plant materials (data not shown). Nutrients applied as liquid more effectively increased forage yields than equivalent amounts of solids on both soils the first year.

On the Dothan soil, the 45 mt/ha manure rate produced as much forage as the check plots in 1972, but not on the Lucedale soil. This was surprising, since the more clayey Lucedale soil should have had much less leaching loss than the Dothan. The 90 and 135 mt/ha treatments had the

### TABLE 5. YIELD OF COASTAL BERMUDAGRASS ON MANURE-TREATED PLOTS

| Treat-ment | Dothan soil | | | Lucedale soil | | |
|---|---|---|---|---|---|---|
| | 1971 | 1972 | 1973 | 1971 | 1972 | 1973 |
| | | | —mt/ha— | | | |
| Check | 10.5b* | 19.3b | 14.9a | 17.8a | 18.1c | 17.8c |
| 45, solid | 4.5e | 19.1b | 14.3a | 7.5c | 13.3d | 18.6c |
| 90, solid | 6.3d | 21.5a | 16.4a | 11.5b | 21.3b | 22.0a |
| 45, liquid | 7.8c | 18.9b | 14.2a | 7.9c | 15.4d | 18.3c |
| 90, liquid | 11.4b | 22.4a | 17.3a | 11.0bc | 23.9a | 20.9b |
| 135, liquid | 13.3a | 22.6a | 16.0a | 13.2b | 25.8a | 21.8ab |

*Oven-dry yields within a column followed by the same letter are not significantly different at the 5 percent level using Duncan's Multiple Range Test.

highest yields on both soils in 1972. The lack of difference in yields of Coastal bermudagrass on Dothan soil for any treatment in 1973 was probably due to the infestation of weeds. Weed population seemed related to manure application rate. Obviously, some weeds were from seed in the manure, and others were from tubers infesting the soil. Both test areas showed the effect of accumulation of nutrients from manure from previous years since the 45 mt/ha treatment generally yielded as much as the commercial fertilizer (check) treatments. These tests indicated that during 1971 liquid manure applications may have been more effective in establishing sod than equivalent amounts applied in solid form. For subsequent years, however, with both forms of application nutrients moved into the root zone equally well.

Nitrate-N content of the Coastal forage produced on these tests was associated with organic-N content, as expected. The relationship was logarithmic (Fig. 3). Nitrate-N was converted to protein until organic-N content was about 2.5 percent, at which point $NO_3$-N accumulation proceeded very rapidly. Thus, $NO_3$-N accumulation in forage could exceed tolerance levels for ruminant animals if $NH_4$-N reached 3 percent or more.

*(Continued on page 601)*

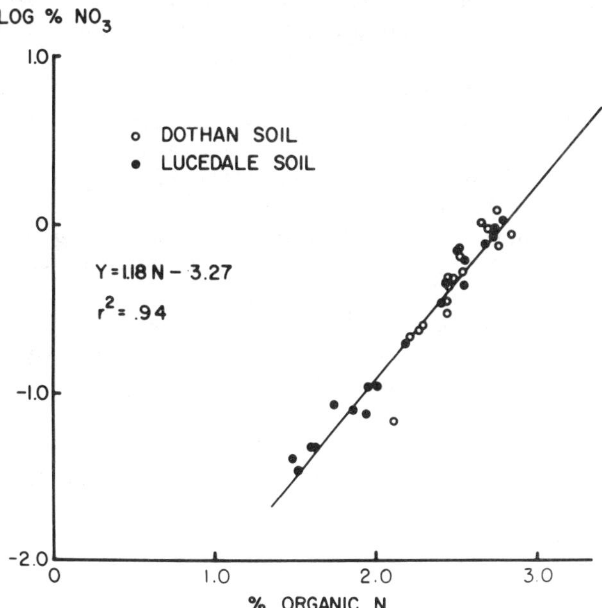


FIG. 3 Relationship between nitrate and organic N in Coastal bermudagrass.

# Fertilizer Value of Livestock Wastes

Hubert Tunney

THERE is wide variation in the plant nutrient composition of livestock wastes. Wastes should be applied to soil at rates that give adequate nutrients for crop needs. Normally it is necessary to supplement with chemical fertilizers in order to get maximum benefit from nutrients in wastes, and to ensure the correct balance of nutrients for the crop. Regular soil analyses is the most reliable way of ensuring that a satisfactory nutrient balance is maintained on land receiving wastes. This approach would facilitate the use of wastes in a fertilizer program, and minimize the risk of pollution.

## INTRODUCTION

In recent years the fertilizer value of livestock wastes has been receiving increased attention in many countries (Adams et al. 1975, Berryman 1970), Coppenet 1974, Lande Gremer 1972, Peterson et al. 1971, Tietjen 1972). However, the research effort devoted to this aspect is relatively small, considering that spreading on soil is the only accepted practical method for disposal of livestock wastes.

Over 90 percent of the 4.5 million ha of good agricultural land in the Republic of Ireland is devoted to grassland, with a population of 7 million cattle. These animals are out on pasture for most of the year and are normally housed for a 4 month winter feeding period. The winter feed is mainly grass silage or hay conserved from part of the farm during the summer. On most farms practically all the nutrients in the animal manure come from the conserved area of the farm, so it is possible to recycle the nutrients back to this area. With pig and poultry manure this simple recycling relationship does not exist as the feed is usually purchased, however, there is normally adequate land available locally where the manure can be spread.

Manure produced by cattle in the 4 month winter period and by pigs and poultry in 12 months totals about 30 million tons with 15 percent dry matter. Cattle manure accounts for over 90 percent of this total. On the basis of the nitrogen (N), phosphorus (P) and potassium (K) present this manure is worth about 100 million dollars, or about half the annual expenditure by Irish farmers on chemical fertilizers.

Much of our research efforts are devoted to integrating manures into a fertilizer program, and the results of three experiments are summarized in this paper.

## MATERIALS AND METHODS

All analyses were according to the methods (Byrne 1968) used in this laboratory. Soil was extracted with sodium acetate and acetic acid (Morgan's solution) at pH 4.8 for available P and K determinations.

The authors is: HUBERT TUNNEY, The Agricultural Institute, Johnstown Castle, Wexford, Ireland.

## Variations in Composition of Manures

In this study samples of each category of animal manure were collected from a number of representative intensive farms in 1973. Details of samples, methods of collection and analyses are described by Tunney and Molloy (1975a). Farmyard manure contained cattle feces and urine mixed with straw, dungstead manure consisted of feces and urine in an open compound with provision for liquids to drain off. Pig slurry consisted of feces plus urine stored in a tank with variable quantities of water. Cattle slurry samples were not included in this study. On the basis of results pig slurry samples were collected from 25 farms in 1974 to investigate the development of a simple field test for dry matter. Details of this study are described by Tunney and Molloy (1975b).

## Use of Cattle and Pig Slurry for Grass Silage

In this study there were three treatments that attempted to simulate three farming systems, employing fertilizer, pig slurry and cattle slurry and to gain practical experience of the problems involved, the trial began in spring 1973 and continued in 1974. There were 10 replicates for each of the three treatments, each plot was 0.4 ha in area, giving a total of 30 plots on 12 ha. The quantities of nutrients and slurry applied are summarized in Table 1.

It can be seen that there was a high and a low rate for the three treatments in 1974, this gave five replications for each subtreatment. Pig slurry was from a tank on a commercial pig farm, cattle slurry was from a slatted floor beef house.

TABLE 1. RATES OF N, P AND K APPLIED IN SLURRY AND FERTILIZER BEFORE SILAGE CUTS 1 AND 2 IN 1973 AND 1974.

| Treatment | | N | P | K | Slurry |
|---|---|---|---|---|---|
| | | | kg/ha | | m$^3$/ha |
| Cut 1, 1973 | | | | | |
| Fertilizer | | 112 | 0 | 0 | --- |
| Pig slurry | | 150 | 60 | 83 | 45 |
| Cattle slurry | | 166 | 34 | 181 | 45 |
| Cut 2, 1973 | | | | | |
| Fertilizer | | 64 | 0 | 0 | --- |
| Pig slurry | | 68 | 40 | 30 | 28 |
| Cattle slurry | | 58 | 11 | 77 | 28 |
| Cut 1, 1974 | | | | | |
| Fertilizer | --low | 108 | 15 | 59 | --- |
| | --high | 162 | 22 | 89 | --- |
| Pig slurry | --low | 171 | 121 | 52 | 34 |
| | --high | 270 | 180 | 87 | 56 |
| Cattle slurry | --low | 101 | 25 | 114 | 34 |
| | --high | 191 | 45 | 201 | 56 |
| Cut 2, 1974 | | | | | |
| Fertilizer | --low | 68 | 9 | 37 | --- |
| | --high | 108 | 15 | 59 | --- |
| Pig slurry | --low | 186 | 105 | 75 | 34 |
| | --high | 313 | 178 | 112 | 56 |
| Cattle slurry | --low | 103 | 24 | 140 | 34 |
| | --high | 165 | 39 | 224 | 56 |

TABLE 2. COMPOSITION OF ANIMAL MANURES FROM COMMERCIAL FARMS.

| | | N | P | K | Percent D.M. |
|---|---|---|---|---|---|
| | | | kg/10 m³, | | |
| Farmyard manure | Mean | 45 | 10 | 68 | 20 |
| (18 farms) | Range | 32 - 65 | 8 - 17 | 33 - 128 | 13 - 26 |
| Dungstead manure | Mean | 33 | 8 | 42 | 17 |
| (16 farms) | Range | 20 - 52 | 5 - 12 | 30 - 60 | 10 - 23 |
| Pig slurry | Mean | 43 | 18 | 20 | 8 |
| (20 farms) | Range | 12 - 70 | 1 - 45 | 6 - 34 | 1 - 21 |
| Poultry slurry | Mean | 142 | 51 | 57 | 24 |
| layers (8 farms) | Range | 100 - 170 | 40 - 62 | 40 - | 20 - 30 |
| Poultry deep litter | Mean | 256 | 82 | 122 | 47 |
| broilers (8 farms) | Range | 133 - 368 | 39 - 112 | 59 - 200 | 26 - 76 |

To ensure adequate fertility all plots received a basal fertilizer dressing of 39 kg P and 166 kg K/ha at the start of the experiment, in addition to treatments outlined in Table 1. In 1974 basal fertilizer treatments were 34 kg P and 140 kg K/ha on fertilizer plots, 123 kg K/ha on pig plots and 20 kg P on cattle plots. Slurry plots received no fertilizer nitrogen. The fertilizer dressings in Table 1 show that a straight N fertilizer (26 percent N) was used in 1973 and a compound (22 percent N, 3 percent P, 12 percent K) fertilizer was used in 1974.

Silage was cut at the end of May, July and September with a double chop forage harvester. Slurry was applied by vacuum tanker 7 to 8 weeks before cut 1 and cut 2. There was no application before cut 3 and this yield was taken as an indication of residual value of the treatments. The yield from each plot was recorded and subsampled for chemical analyses. Soil samples were taken to a depth of 10 cm.

The silage from the three treatments was fed to three groups of animals (avg. wt. 380 kg) over the winter period of 1973/74. The silage intake and live weight gain of the animals was monitored as a guide to silage palatability and quality. The 1974/75 feeding trial is in progress at time of writing.

### Response of Different Grass Species to Pig Slurry

There were 8 treatments with five replications for both slurry and fertilizer. The design was a randomized block, with 80 plots, each 15 m x 8 m. The treatments were:

1. Timothy (Canadian);
2. Tall fescue (S.170);
3. Cocksfoot (S.26);
4. Perennial (S.24);
5. Perennial (S.24) + Clover (Blanca);
6. Italian ryegrass (R.V.P.);
7. Meadow fescue (S.215);
8. Control (no seed sown).

On treatment 8 natural grasses and weeds were allowed to develop. The soil was tilled and seeds sown in Fall 1972.

All plots received a basal dressing of 35 kg P and 150 kg K/ha at the start of the experiment. Fertilizer plots received a basal dressing of 26 kg P and 112 kg K/ha in spring 1974. Slurry at 50 m³/ha and fertilizer was applied 6 to 8 weeks before each cut with the exception of the first cut in 1974. There were 3 cuts in 1973 and 5 cuts in 1974. In 1973 slurry application contained a total of 238 N, 128 P and 180 K in kg/ha. Totals in 1974 were 788 N, 361 P and 246 K in kg/ha. In 1973 fertilizer plots received a total of 236 kg N/ha and in 1974 received 314 kg N, 42 kg P and 171 kg K/ha in addition to the basal dressing. Pig slurry plots received no chemical fertilizer during the course of the experiment. Grass was cut by rotary motor scythe, yields were measured and subsampled for chemical analysis. Soil samples to a depth of 10 cm were taken for analysis.

## RESULTS AND DISCUSSION

### Variation in Composition of Animal Manure

The results of chemical analyses are summarized in Table 2. These results together with statistical data are discussed in detail by Tunney and Molloy (1975a). The cattle manure samples had approximately a two-fold difference between highest and lowest values. Pig slurry showed the greatest variation in composition between farms.

There was a significant correlation between dry matter and nitrogen composition of pig slurry, this is illustrated in Fig. 1. Phosphorus and magnesium were also significantly correlated with dry matter. A significant correlation between dry matter and nutrient content did not exist for the other manures.

These results emphasize the variability of animal manures and underlines the need for avoiding loss during storage and avoiding water entering storage tanks. Pig slurry normally contains about 9 percent dry matter (O'Callaghan et al. 1971). The lower values in Fig. 1 suggest considerable dilution with water. The higher values suggest poor agitation of the storage tanks and the liquid fraction being drawn off and spread. Composition of cattle slurry is being studied at present. An average sample contains about 12 percent dry matter with 45 kg N 8 kg P and 45 kg K per 10 m³.

Fig. 2 shows a straight line correlation between dry matter and specific gravity of pig slurry. There was not a significant difference (t = 1.71) between specific gravity measured volumetrically and by hydrometer. This relationship formed the basis of a hydrometer that was developed for

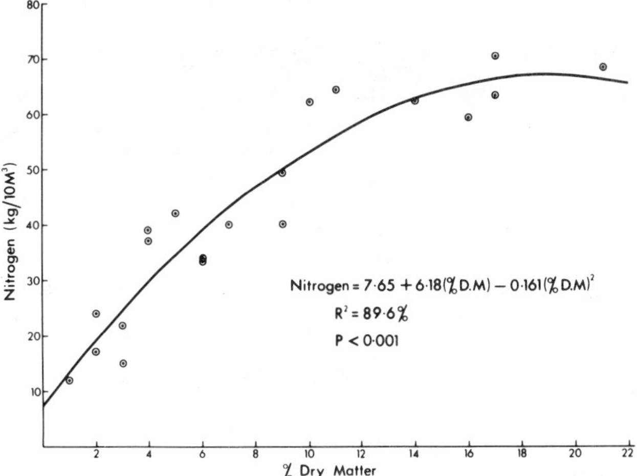


FIG. 1 Relationship between dry matter and nitrogen content of pig slurry.

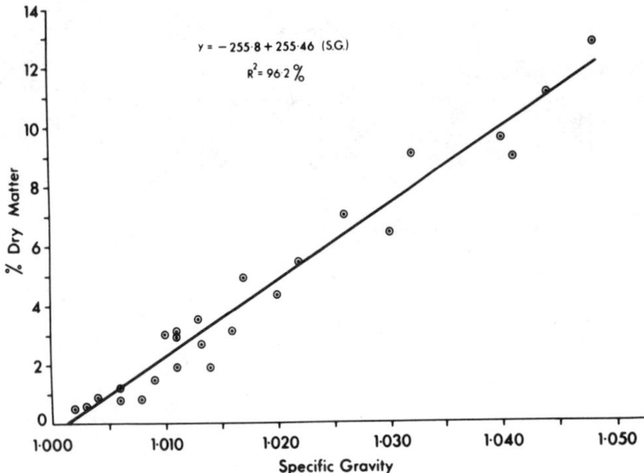



FIG. 2 Relationship between dry matter and specific gravity of pig slurry from 25 farms.

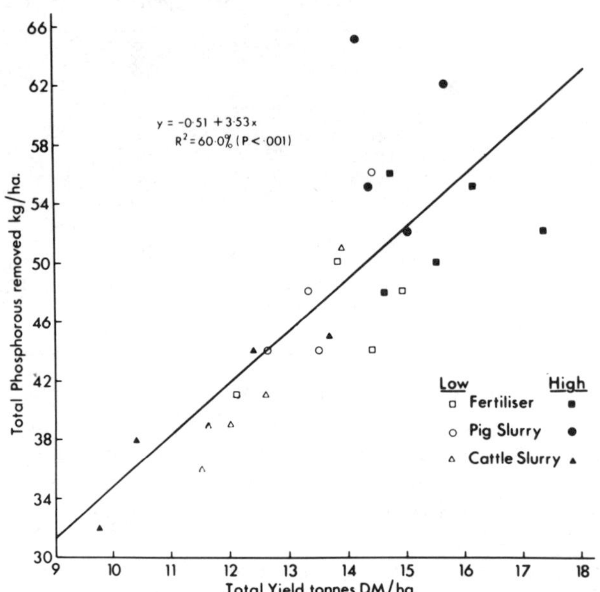



FIG. 3 Relationship between yield of grass and phosphorus removed in the crop.

estimating slurry composition, and calibrated in percent dry matter, and kg N and kg P per 10 m³. The relationship is excellent despite the narrow specific gravity range.

### Use of Cattle and Pig Slurry for Grass Silage

Grass yields for the different treatments are summarized in Table 3. In 1973 there was little difference between treatments. The marginal lower yield on cattle slurry plots may reflect lower nitrogen availability from cattle slurry due to high carbon nitrogen ratio (Tietjen 1966). In general 1974 yields were higher than 1973, and cattle slurry gave a significantly lower yield than the other treatments. The high rate of cattle slurry reduced yield about 0.5 ton/ha relative to the low cattle slurry. This suggests that some factor other than nitrogen was limiting yield, and that the high rate of cattle slurry had an adverse chemical or physical effect on grass growth. There is need for further investigations on this effect. There was a cover of slurry on grass on the cattle slurry plots for a considerable time after application. This was less noticable and less persistent with pig slurry. This cover of slurry may have reduced the photosynthetic rate on cattle slurry plots. Cattle slurry on the grass surface was particularly noticable in 1974 because of prolonged dry weather after application for both cut 1 and cut 2. Slurry fragments were noticable in the harvested grass.

The yield on cut 3 was higher on slurry plots indicating a significant residual nitrogen effect relative to the fertilizer plots. Grass growth was noticeably better on slurry plots in February 1975, indicating a residual effect from slurry applied in 1974.

There was not a significant difference in K content of grass between high and low cattle slurry plots despite the higher rate of K applied in the high rate of slurry. Fig. 3 shows a positive correlation between yield and phosphorus removed in the crop. There was also a correlation between yield and N and K removed by the crop. The distribution of points in Fig. 3 indicates a higher P removal from pig slurry plots, however, yield rather than treatments appeared to be the major factor influencing P removal from the soil.

High pig slurry plots had 27.6, 17.4 and 12.4 ppm copper for cut 1, cut 2 and cut 3 respectively, corresponding levels for high fertilizer plots were 8.0, 6.0 and 5.8 ppm. Pig slurry contained in the region of 100 ppm copper on a dry matter basis.

Results of soil analyses are summarized in Table 4. Pig slurry increased soil P considerably, reflecting the high P applied in the slurry. Cattle slurry maintained a high K status in the soil. Results of the animal feeding trial for 1973 silage showed that there was not a significant difference in live weight gain or silage intake between the three treatments, indicating that quality of the silage was not adversely affected by slurry. These results suggest that in order to get maximum benefit from nutrients in slurry for grass silage, pig slurry should be applied to give adequate P and supplemental N and K should come from chemical fertilizer. Cattle slurry should be applied to give adequate K,

### TABLE 3. SILAGE YIELDS FROM FERTILIZERS, PIG SLURRY AND CATTLE SLURRY TREATMENTS.

| | | Cut 1 | Cut 2 | Cut 3 | Total |
|---|---|---|---|---|---|
| | | | Yield tonnes d.m./ha, (means of 10 plots) 1973 | | |
| Fertilizer | | 5.43 | 4.35 | 1.86 | 11.64 |
| Pig slurry | | 5.27 | 4.38 | 2.31 | 11.96 |
| Cattle slurry | | 5.21 | 3.81 | 2.19 | 11.27 |
| Standard error | | .24 | .22 | .14 | 0.41 |
| | | | (means of 5 plots) 1974 | | |
| Fertilizer | low | 7.14 | 4.87 | 1.94 | 13.85 |
| | high | 7.04 | 6.20 | 2.29 | 15.53 |
| Pig slurry | low | 6.36 | 4.40 | 2.62 | 13.49 |
| | high | 5.94 | 5.46 | 3.30 | 14.63 |
| Cattle slurry | low | 6.00 | 3.74 | 2.56 | 12.30 |
| | high | 5.50 | 3.25 | 2.86 | 11.54 |
| Standard error | | .28 | .25 | .24 | 0.59 |

### TABLE 4. RESULTS OF AVAILABLE (MORGAN'S) SOIL P AND K.

| Date sampled | | Soil P, ppm | | | Soil K, ppm | | |
|---|---|---|---|---|---|---|---|
| | | Feb. '73 | Sept. '73 | Oct. '74 | Feb. '73 | Sept. '73 | Oct. '74 |
| Fertilizer | low | 5.5 | 7.0 | 8 | 135 | 104 | 103 |
| | high | | | 6.4 | | | 72 |
| Pig slurry | low | 5.8 | 16.2 | 18.4 | 155 | 127 | 106 |
| | high | | | 35.4 | | | 84 |
| Cattle slurry | low | 5.5 | 8.7 | 11.8 | 162 | 194 | 174 |
| | high | | | 9.8 | | | 176 |
| Standard error | | | 1.1 | 3.0 | 2.4 | 30 | 24 | 20 |



MANAGING LIVESTOCK WASTES

| Treatments* | Tonnes d.m./ha, | | | | | | | |
|---|---|---|---|---|---|---|---|---|
| | 1 | 2 | 3 | 4 | 5 | 6 | 7 | 8 |
| **1973 (3 cuts)** | | | | | | | | |
| Fertilizer | 11.7 | 12.1 | 12.4 | 11.0 | 10.4 | 13.6 | 12.2 | 10.4 |
| Pig slurry | 7.7 | 9.6 | 8.6 | 8.2 | 10.6 | 10.8 | 9.0 | 6.7 |
| | Standard error of the means = 0.55 | | | | | | | |
| **1974 (5 cuts)** | | | | | | | | |
| Fertilizer | 12.0 | 12.0 | 12.0 | 10.9 | 10.9 | 14.2 | 10.5 | 9.6 |
| Pig slurry | 9.9 | 12.2 | 10.4 | 9.8 | 11.0 | 12.5 | 9.7 | 8.9 |
| | Standard error of the means = 0.38 | | | | | | | |

*1 = Timothy; 2 = Tall fescue; 3 = Cocksfoot; 4 = Perennial ryegrass;
5 = Perennial ryegrass + clover; 6 = R.V.P. ryegrass; 7 = Meadow
fescue; 8 = Control

and N and P supplied as chemical fertilizer.

## Response of Different Grass Species to Pig Slurry

A dense sward established on all species, including indigenous grasses on the control. Early in 1975 bare areas were appearing in all R.V.P. plots. The total yields for the different treatments are summarized in Table 5. Slurry plots gave a noticeably lower yield in 1973, probably due to the low N applied in the relatively dilute pig manure, this was particularly true in cut 1 and 2. Visual differences were not apparent between slurry and fertilizer treatments in 1974. The R.V.P. gave the best yield on slurry and fertilizer plots in both years. The control treatment gave the lowest yields.

The clover ryegrass treatment was the only one that performed as well on slurry as fertilizer in both years. The clover growth was noticeably better on slurry plots particularly in 1973. There was a statistically significant yield increase on slurry plots due to the presence of clover with perennial ryegrass in both years. Tall fescue performed well on slurry plots in 1974. It is interesting to note that despite the high N applied in pig slurry in 1974, many of the grass species had a lower yield than the fertilizer plots, indicating that some factor other than N may have been limiting yield.

Pig slurry plots showed a significant buildup in soil P levels and this was reflected in higher P levels in the grasses. The average copper content for all slurry plots for cut 2 in 1974 was 9.9 ppm, the average for the fertilizer plots was 7.9 ppm. This is in contrast to experiment 2 and suggests that the high copper on slurry plots there, may have been due to uptake of slurry fragments by the vacuum action of the forage harvester.

In conclusion, none of the grass species appeared particularly unsuited to slurry relative to chemical fertilizer.

### References

1   Adams, S. N. and J. S. V. McAllister. 1975. P and K cycles on livestock farms. J. Agric. Sci. (in press)

2   Berryman, C. 1970. The problems of disposal of farm wastes with particular reference to maintaining soil fertility. Symposium on farm wastes, University of Newcastle upon Tyne.

3   Byrne, E. 1968. Methods of analysis. Agric. Institute, Johnstown Castle, Wexford.

4   Coppenet, M. 1974. L'espandage du lisier de porcherie. Annales agronomiques. 25:403-423.

5   Lande Cremer, L. C. N. de la. 1972. Gebruik de drijfmest, maar misbruik hem niet. Bedrijfsontwikkeling. 3:523-526.

6   O'Callaghan, J. R. et al. 1971. Characterization of waste treatment properties of pig manure. J. Agric. Engng. Res. 16:399-419.

7   Peterson, J. R. et al. 1971. Human and animal wastes as fertiliser. Fertiliser technology and use, 2nd Ed. p. 557-596. Soil Sci. Soc. Amer.

8   Tietjen, C. 1966. Plant response to manure nutrients and processing of organic wastes. Proceedings of ASAE symposium on management of farm animal wastes. p. 136-140.

9   Tietjen, C. 1972. Auswirkungen grosser abfallgaben auf den natzwert des bodens. Landbauforschung Volkenrode. 14:80-86.

10   Tunney, H. and S. Molloy. 1975a. Variations in N, P, K, Mg and dry matter of cattle, pig and poultry manures. Irish Jn. Agric. Res. 14:71-79.

11   Tunney, H. and S. Molloy. 1975b. Field test for estimating dry matter and fertiliser value of slurry — preliminary report. Irish Jn. Agric. Res. 14:84-86.

# Plant and Soil Effects of Swine Lagoon Effluent Applied to Coastal Bermudagrass


G. A. Cummings, J. C. Burns, R. E. Sneed, M. R. Overcash, F. J. Humenik

MEMBER ASAE     ASSOC. MEMBER ASAE     MEMBER ASAE


MANY swine producers utilize anaerobic lagoons for partial digestion of waste with ultimate land disposal of effluent. The ability of a soil-plant system to utilize applied waste is dependent primarily upon the crop and both the soil chemical and physical properties. In addition, optimum loading rate may vary with climate and nature of the waste.

The objectives of this study were to (a) determine maximum permissible loading rate of anaerobic swine lagoon effluent upon Coastal bermudagrass (Cynodon dactylon [L.] Pers) grown on a Norfolk sandy loam; and (b) determine fate of possible pollutants added to soils including crop utilization, soil absorption, and losses due to runoff, seepage and leaching. In this report the effects of three rates of effluent application upon yield and chemical content of Coastal bermudagrass and changes in soil levels of several nutrients to a depth of 30 inches are reported and discussed.

## PROCEDURE

In May 1972 Coastal bermudagrass was established by sprigging on a Norfolk sandy loam, 0 to 3 percent slope (Typic, Paleudult: fine-loamy, siliceous, thermic) after fumigation with methyl bromide to kill all existing vegetation. Drainage tile, backfilled with gravel, were installed in the B horizon on the up slope side of the experimental area. Metal flashing was installed on all but the down slope side of 30 by 30 feet plots to prevent entry of surface water into the plots. Metal flashing was also installed around a 7.5 by 15 feet area in the lowest corner of each plot and runoff was funneled into a submerged 55 gallon drum. Quarter circle rotary impact sprinklers were installed at each corner of the plot with application at one time limited to sprinklers on opposite corners to limit application to soil intake rate. Within, and 20 and 40 feet below each plot one bar ceramic cups were installed at the plow sole layer, at the surface of the B horizon and 10 inches below the surface of the B horizon.

Application of effluent from the Swine Evaluation Station, Clayton, North Carolina started in late April and was applied weekly until mid September in both 1973 and 1974. Rates were selected to supply 300, 600 and 1200 lb of N which required approximately 5, 10 and 20 inches of effluent respectively. These effluent rates were replicated three times for a total of nine plots. Forage was harvested and removed when growth reached a height of 12-14 inches. Yield estimates were obtained and subsamples taken for dry matter and mineral determinations and nutritive evaluation.

Soil samples were taken to a depth of 30 inches in 6 inch increments prior to initiation of the experiment. Starting in September, 1973, soil samples were taken every three months in 2 inch increments to a depth of 10 inches and in 4 inch increments from 10 to 30 inches.

Samples of effluent applied to the plots were collected by placing 400 ml beakers within the plot during effluent application. Volume of effluent in the beakers also served as a check of application rates. All soil and effluent samples were kept frozen until analyzed. Effluent, runoff, soils, forage, and solution samples from porous cups were analyzed for total N, P, K, Ca, Mg, Na, Cl and Cu. In addition effluent was analyzed for COD, TOC, pH, $NO_3$, $NH_3$ and in 1973 for Fe, Mn, and Zn; soils for pH, $NO_3$, $NH_3$, and initially for exchangeable Al, organic matter, cation exchange capacity, percent clay; forage for Mn, Zn, and Fe; runoff for pH, COD, TOC, and $NO_3$; and solution samples for pH, TOC, $NO_3$ and COD.

In the portion of the study reported herein, soil samples were extracted with 0.05 N HCl, 0.025 N $H_2SO_4$ for exchangeable K, Ca, Mg, Na, Cu and P; water for Cl and $NO_3$ and pH was determined on a 1:1 soil water ratio. Total N was determined on all samples by Kjeldahl; K and Na by flame photometer; Ca, Mg, Cu, Mn, Zn and Fe by atomic absorption; Cl by a chloridometer; $NO_3$ by direct scan on UV spectrophotometer; $NH_3$ by $NH_3$ electrode and P by a vanadate molybdate procedure. Chemical analysis of forage after dry ashing, and effluent samples employed the same procedures used for soil analysis.

## RESULTS AND DISCUSSION

### Crop Yield and Nutrient Removal

Concentrations of elements in the effluent did not vary


Paper Number 4649 of the Journal Series of the North Carolina Agricultural Experiment Station, Raleigh, N. C.

Acknowledgment: This work was supported by N. C. Agricultural Experiment Station and EPA Grant No. 13040 G.D.D. Design Criteria for Swine Waste Treatment Systems; Lynn Shuyler, Project Officer.

The authors are: G. A. CUMMINGS, Associate Professor, Soil Science Dept., J. D. BURNS, Associate Professor, Crop Science Dept., R. E. SNEED and M. R. OVERCASH, Assistant Professors, and F. J. HUMENIK, Associate Professor, Biological and Agricultural Engineering Dept., North Carolina State University, Raleigh.


TABLE 1. EFFLUENT RATES, ELEMENTAL CONTENT, AND POUNDS APPLIED PER ACRE 1973-1974.

| Element | Effluent, ppm | | Effluent, inches applied | | | | | |
| --- | --- | --- | --- | --- | --- | --- | --- | --- |
| | | | 1973 | | | 1974 | | |
| | 1973 | 1974 | 5.4 | 10.8 | 21.6 | 6 | 12 | 24 |
| | | | | | Pounds per acre | | | |
| N | 212 | 272 | 265 | 529 | 1058 | 355 | 710 | 1420 |
| P | 49 | 49 | 60 | 120 | 239 | 66 | 132 | 264 |
| K | 220 | 274 | 269 | 538 | 1076 | 364 | 728 | 1456 |
| Ca | 52 | 55 | 63 | 126 | 252 | 73 | 146 | 292 |
| Mg | 62 | 55 | 74 | 148 | 296 | 74 | 148 | 296 |
| Na | 86 | 113 | 105 | 210 | 420 | 152 | 304 | 607 |
| Cl | 139 | 187 | 170 | 341 | 681 | 245 | 489 | 978 |
| Cu | 0.5 | 0.4 | 0.6 | 1.2 | 2.4 | 0.6 | 1.2 | 2.4 |

greatly between years, although several elements were slightly higher in 1974 (Table 1). Variation in elemental content of the lagoon within years was also noted but could usually be accounted for by swine population liveweight or rainfall. In both years the amount of N, P, (Woodhouse 1969) and K (Woodhouse 1968) at the highest effluent application rate was more than double the amounts found to produce maximum yield in North Carolina. In addition, levels of Ca, Mg, Na, Cl and Cu, even at the lowest rate, were in excess of the amounts normally added as impurities or carrier ions in commercial fertilizer. Since the swine rations contained 250 ppm Cu, Cu content of the applied wastes was monitored each year. Concentration of Cu in the effluent, as well as Zn and Mn averaged about 0.5 ppm both years and Fe was less than 2 ppm. Lagoon pH was approximately 8 during the entire period and was below neutrality on only one sampling date.

Weekly application of swine lagoon effluent from late April through mid September, even at high rates, did not adversely influence Coastal bermudagrass over a two year period. The only visable symptom of the forage was a temporary chlorotic condition at the high rate of application which developed in the regrowth each September. Foliar analysis indicated no nutrient abnormalities and no permanent damage occurred.

High forage yields were obtained both years even at the lowest rate of application (Table 2). Crop recovery of N and K at the low rate was relatively high and crop removal of these elements accounted for about 70 percent of the amount applied. A slight increase in yield was obtained at the two highest rates compared to the low rate. Crop removal of elements increased with application rate but the ratio of the amount removed was much lower than the 1-2-4 application rate ratio. Crop removal accounted for less than half of almost all other elements determined and was extremely low for Na, Cl, and Cu at the high rate. Nutrient removal by crops at the low rate, of every element monitored except Cu, was higher in 1974 than in 1973 (Table 2). This indicates that the residual influence must be considered even at relatively low application rates.

Residual effects were also noted in regard to concentration of elements in the grass (Table 3). Wide differences in elemental content of the grass between the two higher rates were evident in 1973, but in 1974 elemental concentrations in the forage were almost the same at both rates. The data indicate that the ratio of elements supplied in the effluent has not resulted in an imbalance of nutrients in the grass. It also indicates that many elements, especially at the higher rates, were supplied in amounts much greater than needed for optimal yield.

### Soils

Bulk density of samples varied from about 1.50 at the surface and the 26-30 inch depth to 1.90 at 6-8 inches and these values were utilized at each depth sampled to convert elemental concentration to lb/acre (Table 4).

Analysis of soil samples taken after seasonal effluent application in September 1973 and September 1974 verify the build up of nutrients noted earlier with the crop removal data (Table 5). The values reported for acid extractable ions are only a portion of the total elements in the soil. After the first summer it appeared that the medium application rate supplied sufficient P for optimal yield and would maintain soil reserves. However, this soil fixes phosphorus (Woodruff and Kamprath 1965) and only a portion of the P added, until a specific soil level of P is reached, would be extracted by soil test or be available for crop growth. It appears that the P fixation capacity was satisfied the second year even at the low rate of P application. It is also apparent that most of the extractable P at all application rates has remained in the top 14 inches of soil after two years of effluent application (Table 6). There is also little consistent difference in total extractable P between the low and medium rates below a depth of 6 inches. However, since only a small portion of the supplied P is removed by crops it is apparent that soil reserves will be increased by all application rates.

TABLE 3. INFLUENCE OF APPLICATION OF SWINE LAGOON EFFLUENT UPON ELEMENTAL CONTENT OF COASTAL BERMUDAGRASS.

| Year | Effluent applied inches | Elemental content percent | | | | | | | PPM Cu |
|---|---|---|---|---|---|---|---|---|---|
| | | N | P | K | Ca | Mg | Na | Cl | |
| 1973 | 5.4 | 1.84 | 0.18 | 1.84 | 0.29 | 0.16 | 0.09 | 0.78 | 10.8 |
| | 10.8 | 2.31 | 0.22 | 2.32 | 0.36 | 0.22 | 0.16 | 0.88 | 14.6 |
| | 21.6 | 2.65 | 0.22 | 2.56 | 0.44 | 0.26 | 0.16 | 0.72 | 15.6 |
| 1974 | 6.0 | 2.43 | 0.23 | 2.14 | 0.46 | 0.26 | 0.16 | 1.06 | 9.3 |
| | 12.0 | 3.12 | 0.26 | 2.59 | 0.55 | 0.34 | 0.18 | 0.95 | 9.9 |
| | 24.0 | 3.12 | 0.28 | 2.61 | 0.57 | 0.34 | 0.18 | 0.81 | 9.5 |

TABLE 4. INITIAL SOIL TEST VALUES OF PLOTS UTILIZED FOR EFFLUENT APPLICATION.

| Depth, inches | Pounds per acre each soil layer | | | | |
|---|---|---|---|---|---|
| | N | P | K | Ca | Mg |
| 0-6 | 810 | 57 | 144 | 331 | 70 |
| 6-12 | 522 | 29 | 31 | 149 | 18 |
| 12-18 | 382 | 18 | 20 | 90 | 13 |
| 18-24 | 265 | 13 | 19 | 74 | 8 |
| 24-30 | 258 | 12 | 35 | 53 | 11 |
| Profile total | 2237 | 129 | 249 | 697 | 120 |

TABLE 2. INFLUENCE OF THREE EFFLUENT RATES UPON YIELD AND NUTRIENT REMOVAL OF COASTAL BERMUDAGRASS 1973-1974.

| Effluent applied, inches | Year | Yield, tons/acre | Crop removal of elements, lbs/acre | | | | | | | |
|---|---|---|---|---|---|---|---|---|---|---|
| | | | N | P | K | Ca | Mg | Na | Cl | Cu |
| 5.4 | 1973 | 5.0 | 183 | 18 | 183 | 30 | 13 | 10 | 89 | 0.11 |
| 10.8 | | 6.1 | 294 | 28 | 285 | 47 | 30 | 20 | 109 | 0.18 |
| 21.6 | | 6.3 | 383 | 32 | 323 | 64 | 37 | 20 | 78 | 0.20 |
| 6.0 | 1974 | 5.1 | 248 | 26 | 239 | 43 | 25 | 17 | 117 | 0.09 |
| 12.0 | | 6.3 | 407 | 35 | 347 | 67 | 42 | 23 | 128 | 0.13 |
| 24.0 | | 6.4 | 414 | 38 | 355 | 71 | 43 | 23 | 110 | 0.12 |

TABLE 5. EFFECT OF SWINE LAGOON EFFLUENT APPLICATION UPON SOIL NUTRIENTS TO A 30 INCH DEPTH (1973-1974).

| Nutrient | Initial soil levels | September 1973 | | | September 1974 | | |
|---|---|---|---|---|---|---|---|
| | | Effluent applied, inches | | | | | |
| | | 5.4 | 10.8 | 21.6 | 6 | 12 | 24 |
| | | --------Pounds per acre to depth of 30 inches-------- | | | | | |
| P | 129 | 97 | 125 | 155 | 158 | 210 | 336 |
| K | 249 | 196 | 273 | 389 | 374 | 441 | 661 |
| Ca | 697 | 582 | 711 | 686 | 681 | 639 | 790 |
| Mg | 120 | 125 | 160 | 198 | 167 | 187 | 210 |
| Na | 74 | 73 | 123 | 116 | 177 | 173 | 185 |
| Cl | | 439 | 455 | 538 | 319 | 346 | 350 |
| $NO_3$ | | 11 | 11 | 73 | 18 | 60 | 110 |
| Cu | | 12.8 | 22.8 | 15.8 | 9.9 | 8.3 | 9.6 |

**TABLE 6. EFFECT OF APPLICATION OF SWINE LAGOON EFFLUENT FOR TWO CONSECUTIVE YEARS UPON CONCENTRATION OF CERTAIN ELEMENTS TO SOIL DEPTH OF 30 INCHES.**

| Depth, inches | Inches effluent applied - 2 years | | | | | | | | |
|---|---|---|---|---|---|---|---|---|---|
| | 11.4 | 22.8 | 45.6 | 11.4 | 22.8 | 45.6 | 11.4 | 22.8 | 45.6 |
| | Ppm-P | | | Ppm-K | | | Ppm $NO_3$-N | | |
| 1-2 | 76 | 109 | 135 | 68 | 62 | 78 | 0. | 2.5 | 1.7 |
| 2-4 | 31 | 34 | 80 | 60 | 49 | 69 | 0. | 2.9 | 0. |
| 4-6 | 15 | 34 | 58 | 54 | 48 | 63 | 0.9 | 1.4 | 0. |
| 6-8 | 13 | 15 | 32 | 41 | 44 | 58 | 0.6 | 1.7 | 1.1 |
| 8-10 | 9 | 16 | 25 | 39 | 34 | 51 | 0.6 | 1.5 | 1.4 |
| 10-14 | 7 | 8 | 19 | 27 | 38 | 51 | 0.4 | 4.6 | 2.6 |
| 14-18 | 9 | 13 | 14 | 28 | 35 | 58 | 2.2 | 4.9 | 8.0 |
| 18-22 | 7 | 5 | 11 | 20 | 31 | 59 | 3.1 | 6.0 | 16.2 |
| 22-26 | 5 | 5 | 4 | 20 | 33 | 59 | 0.9 | 10.7 | 19.7 |
| 26-30 | 4 | 4 | 7 | 21 | 32 | 52 | 4.8 | 9.4 | 28.5 |

Soil K also increased with effluent application rate similar to the increase noted for P (Table 5). After the second year soil K levels at each rate were well above the levels determined when the study was initiated. In addition soil K levels increased throughout the profile to a depth of 30 inches as the application rate was increased (Table 6). The concentration of other exchangeable cations in the effluent were much lower than K. Increasing rates induced only a small change in soil Ca within rates or between years. The increase in soil Mg with loading rates was proportionally greater than Ca, but differences between years were similar. Levels of Na in the soil were much higher at the two higher rates compared to the low rate the first year. The second year soil Na levels were much higher than the first at all rates with little variation within rates. Levels of Cu in the soil were lower in 1974 than 1973 and the influence of rate varied between years.

Soil N was determined prior to effluent application in March 1973 and were determined every 3 months from September 1973 to December 1974. However, data are not reported because results have been variable and levels of N detected were not related to treatment. The only clear indication of N behavior were changes in soil $NO_3$-N related to date and rate (Table 5) and soil depth (Table 6). After termination of effluent application in 1973 nitrate levels were low for both lower rates but rather high at the heaviest rate. Levels were higher the second year at all rates with total soil content approaching the 1-2-4 ratio of rate of application. Regardless of the rate most of the nitrate was present in the lower portion of the soil profile (Table 6). Only minor differences among rates were evident to a 10 inch depth but large differences associated with rates were evident below that depth. The increase in $NO_3$ within and at the surface of the B horizon warrants further study in regard to depth of movement and amount of lateral flow on the surface of the B horizon. Check samples in an area fertilized with a maintenance rate of $NH_4NO_3$ contained only 5 lb of $NO_3$N per acre to a depth of 30 inches.

Only a small portion of the Cl applied at the higher rates could be accounted for by crop removal. A water extract of the soil revealed no consistent differences in soil content related to effluent rate. Soil samples taken in December and March revealed lower levels of Cl than samples taken in September after seasonal application of effluent. This indicates that Cl is lost rapidly from the soil.

Rainfall was probably the major uncontrolled variable influencing effluent concentration and soil nutrient movement each year. In 1973 rainfall was below normal in August and September, while in 1974 rainfall was adequate the entire growing season (Table 7). The original objective of the experiment was to apply 300, 600, or 1200 lb of N per year. Since the same effluent was applied to all plots the 1-2-4 ratio existed for liquid as well as nutrient load. However, even during periods of limited rainfall in 1973 soil samples taken in plots receiving the highest effluent rate were saturated in the lower portion of the profile. In June and September of 1974 the soil in all plots was at, or above, field capacity in the lower portion of the profile.

Surface runoff resulting from rainfall, during the period from late April to mid September, indicated only a small portion of the nutrients applied were lost through rainfall runoff during the first year. Amount of runoff occurring during effluent application was measured both years, but chemical analysis was determined only after mid summer the first year. From the limited data in 1973 amount of elements lost during effluent application probably exceeded the amount lost from rainfall runoff. However, during 1974, even though rainfall was greater, very little runoff resulted from effluent application. Runoff occurred only when effluent was applied when the soil was saturated from normal rainfall. When runoff did result from effluent application the surface flow was limited to only a few feet below the plots thus indicating, on this soil type and slope, only a few feet of buffer area would be required to prevent runoff from the application area.

**TABLE 7. RAINFALL ON EXPERIMENTAL PLOTS — CLAYTON, NORTH CAROLINA — 1973-1974.**

| | Year | | | |
|---|---|---|---|---|
| | 1973 | | 1974 | |
| Month | Inches | Accumulative total | Inches | Accumulative total |
| Jan. | 1.90 | | 3.11 | |
| Feb. | 4.99 | 6.89 | 3.74 | 6.85 |
| Mar. | 4.00 | 10.89 | 4.29 | 11.14 |
| Apr. | 5.14 | 16.03 | 1.80 | 12.94 |
| May | 4.48 | 20.51 | 6.22 | 19.16 |
| June | 5.14 | 25.65 | 3.45 | 22.61 |
| July | 3.57 | 29.22 | 3.50 | 26.11 |
| Aug. | 2.51 | 31.73 | 6.55 | 32.66 |
| Sept. | 1.71 | 33.44 | 6.51 | 39.17 |
| Oct. | 0.38 | 33.82 | 1.81 | 40.98 |
| Nov. | 0.81 | 34.63 | 1.85 | 42.83 |
| Dec. | 5.87 | 40.50 | 3.90 | 46.73 |

## CONCLUSIONS

Application of 24 inches of swine lagoon effluent that supplied over 1400 lb each of N and K per acre annually was not detrimental to yield or mineral content of Coastal bermudagrass in 1974. Slightly smaller amounts, 22 inches of effluent and approximately 1050 lb per acre each of N and K had been applied the previous year without apparent damage to the crop or soil. In both years yields of Coastal bermudagrass were approximately the same at the two higher application rates, 12 or 24 inches of effluent annually.

Total amounts of exchangeable P, K, and Mg increased in the soil profile as rate was increased. Even at the low rate of effluent application exchangeable P and K in the soil to a 30 inch depth increased the second year. Changes in total N in the soil induced by treatments were not detected over the two year period. However, nitrate levels were high, up to 28 ppm, in the lower portion of the profile the second year. Effluent rates have induced changes in soil levels of Mg and Na and both remain relatively low. Soil levels of Ca did not vary greatly among rates or between years. Chloride and Cu levels were lower in 1974 than in 1973. In both years Cl increased slightly as rate was increased, but changes in levels of Cu in the soil was not consistent with the increase in effluent application rate either year.

Although effluent rates have increased soil reserves of several elements no detrimental effect of effluent applications upon plants or soils has been noted. However, more years of application are needed to evaluate long term effects.


### References
1   Woodhouse, W. W., Jr. 1968. Long-term fertility requirements of Coastal bermudagrsss. I. Potassium. Agron. J. 60(5):508-512.
2   Woodhouse, W. W., Jr. 1969. Long-term fertility requirements of Coastal bermudagrass. II. Nitrogen, phosphorus and lime. Agron. J. 61(2):251-256.
3   Woodruff, J. R. and E. J. Kamprath. 1965. Phosphorus adsorption maximum as measured by the Langmuir isotherm and its relationship to phosphorus availability. Soil Sci. Soc. Amer. Proc. 29(2):148-150.


---

## Disposal of Cattle Manure on Soil

*(Continued from page 593)*

### CONCLUSIONS

Runoff water quality studies show that at least 45 mt/ha of dairy cattle manure can be incorporated into the surface of a Norfolk sandy loam soil with <2 percent slope for at least 3 consecutive years without appreciably affecting the quality of the runoff water. Dairy cattle manure at rates of 22.5 and 45 mt/ha incorporated into the soil produced millet and rye forage of good quality. Yields were highest with rates of 90 mt/ha or more, but this forage also had nitrate levels above 2 percent or K/(Ca+Mg) ratios above 2.2, either of which could be detrimental to animals fed this forage. Forage produced on the Dothan soil contained higher nitrate levels than that produced on the Lucedale soil. However, at an equal rate of manure application, forage produced on the Lucedale soil had higher K/Ca+Mg) ratios than that produced on the Dothan soil. Coastal bermudagrass sod required several manure applications before sufficient nutrients were available in the root zone for yields to equal those from the mineral fertilizer (check) treatment. These data indicated that nitrate could accumulate in Coastal bermudagrass in sufficient quantities to be potentially dangerous.


### References
1   Crawford, R. F., W. K. Kennedy, and M. J. Wright. 1966. Nitrate in forage crops and silage: Benefits, hazards, precautions. Cornell Misc. Bull. 37, Cornell Univ., Ithaca, N. Y.
2   Grunes, D. L., P. R. Stout, and J. R. Brownell. 1970. Grass tetany in ruminants. Pages 332-337 in N. C. Brady, ed. Advances in agronomy, Academic Press, New York.
3   U. S. Department of Health, Education and Welfare Public Health Service. 1962. Drinking water standards. Environmental control administration, Rockville, Md.
4   White, R. K. and W. M. Edwards. 1972. Beef barnlot runoff and stream water quality. Pages 225-235 in R. C. Loehr, chairman, Proc. 1972 Cornell Agr. Waste Manage. Conf., Cornell Univ., Ithaca, N. Y.
5   Wilkinson, S. R., W. A. Jackson, R. N. Dawson, and D. J. Williams. 1970. Progress report: Pasture fertilization using poultry litter. Proc. Poultry Waste Mange. Seminar, Univ. Georgia, Athens, Ga.

# Pollution Abatement of Poultry Manure
# by Maxi-Mixing Method

W. A. Aho, G. F. Griffin, Abu Kishk  Bakir

MAXI-MIXING is a term coined to describe a manure disposal system using a maximum amount of manure and a minimum amount of soil in a composting situation. It is a system of returning manure into the soil in massive quantities at low cost. The actual mixing is accomplished with a bulldozer or with a payloader. The manure and soil are folded together and windrowed. Odors are quelled almost immediately, and under ideal temperature and aeration composting occurs in several weeks. Several maxi-mixing operations have been carried on in Connecticut since 1969, mostly as emergency measures in lieu of disposal on available crop lands.

In 1969, an area in Lebanon, Connecticut, where 34 cubic meters (1200 cubic feet) of manure was maxi-mixed in the fall, was monitored for nitrate ($NO_3$) levels in corn leaf and stalk tissue the following summer. The composted material remained in the original disposal site. By late August leaf samples were within tolerable $NO_3$ levels for feeding. Stalk samples, however, were high in $NO_3$ at harvest time. Considering that the maxi-mix area was only 4.4 m x 26 m (14 ft x 85 ft) in a 5.3 hectare (13 acre) corn field, corn harvested from the maxi-mix site diluted by that harvested from the entire field would have made it safe for cattle feeding. Feeding either green chop in August or silage corn cut from only the disposal site would have been hazardous.

## METHODS AND PROCEDURE

### Anaerobic Maxi-Mixing

In November, 1970, 3333.9 metric tons (3675 tons) of poultry manure (53.2 metric tons nitrogen) were mixed in an area of eight tenths of a hectare in Colchester, Connecticut. The area was too small to provide enough soil for a windrowed compost. The mix remained in a wet anaerobic state with no mounding features. (The mix needs to be less than 50 percent moisture to form a mound). It lay fallow until 1974 when it became evident that the same area would be needed as a site to dispose of another 5715.3 metric tons (6300 tons) of poultry manure. This manure was mixed in the same site after the site was sampled. The remainder of this paper describes the monitoring and effect of these heavy manure applications on the levels of $NO_3$ and $NH_4$ in stream water, groundwater, well water and soil.

### Soil and Water Sampling

The area surrounding the maxi-mix area was sampled for nitrate and ammonia movement in 1971 and 1972. A brook adjacent to the mix was monitored at nine locations, 12 holes were drilled to groundwater depths, to obtain groundwater samples, and the farm well 76.22 m (250 ft drilled) was sampled (B and S series Fig. 1). Water samples were collected again from the farm well and brook ($C_1$, $C_2$, Fig.

The authors are: W. A. AHO, Animal Industries Dept., G. F. GRIFFIN, Plant Science Dept., University of Connecticut, Storrs; and ABU KISHK, BAKIR, Deputy Regional Officer Nazereth and Head of Regional Planning. Div., Ministry of Agriculture, State of Israel.

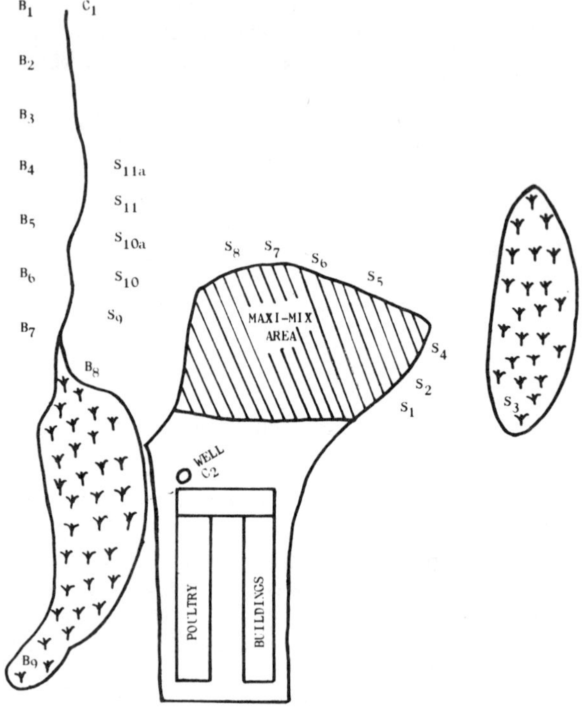



B   series are 1971 & 1972 water sample sites on adjacent brook and well

S   series are 1971, 1972 water sample sites from adjacent ground water holes

C   series are 1974 water sample sites on adjacent brook and well

Sample site $B_1$ and $C_1$, is the outcropping of the bedrock and lower than the Maxi-mix area

**FIG. 1** Map of maxi-mix area — sample sites and poultry buildings.

1) during May and June 1974, after the second maxi-mix in April, 1974. The brook alone was also sampled in October, 1974. The location of the 1974 water sampling site at the brook was determined through the cooperation of William Brown, a geologist from the U.S.D.A., Soil Conservation Service, Conn. The site was one half kilometer (0.3 mile) below the maxi-mixing area where the bedrock was exposed as an outcropping. A wooden weir was constructed at the outcropping ($C_1$ Fig. 1) to measure water discharge at sampling time. The 1974 sampling differed from the 1971-72 sampling in that only water samples from the outcropping and the well were analyzed.

Before the second application of manure in 1974, the maxi-mix site was examined by digging a short trench with a backhoe to a depth of 2.44 m (8 ft). Ledge (bedrock) was encountered at this point. This depth was at least 1.22 m (4 ft) lower than that of the original mix.

The vertical cut made by the backhoe revealed six distinct soil layers or zones from the soil surface to the bedrock. Fig. 2 shows the position and approximate thickness of these layers. Soil samples were taken by hand from each of the six layers, placed in plastic bags and sealed.

### Soil and Water Annalyses

Upon arrival at the University of Connecticut Soil Test-

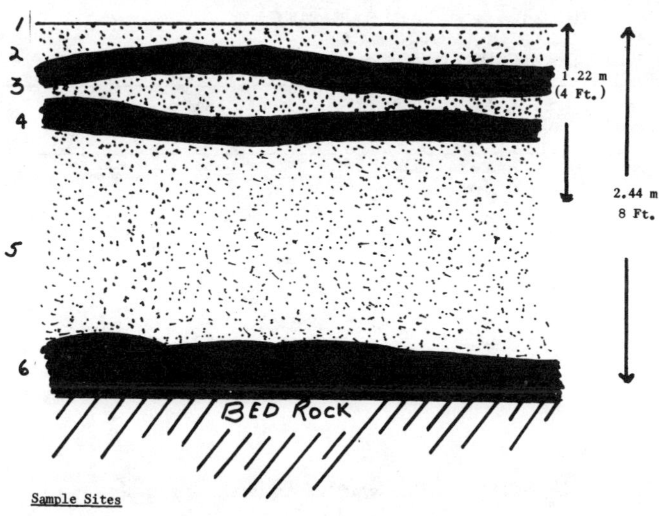



FIG. 2 Cross section of test trench of maxi-mix area, 1974.

ing Laboratory, the samples were immediately removed from the bags and weighed. A second weighing was made after the samples had dried at room temperature. The percentage of water in the samples was calculated by dividing each sample's water loss by its air-dry weight.

The air-dried samples were passed through a 2 mm sieve. Soil pH was measured on a 1:1, soil:water suspension with a pH meter. Organic matter content was estimated by a loss on ignition procedure in which the soil was ignited at 500 C for one hour after having been oven-dried overnight at 105 C.

Calcium, potassium and phosphorus were extracted from the soil by a modified Morgan procedure (McIntosh 1969): 4.00 g of air-dry soil were shaken for 15 minutes with 20 ml of $NH_4OAc$ solution (1.25 $\underline{N}$ $CH_3COOH$ + 0.625 $\underline{N}$ $NH_4OH$, pH 4.8). After the soil-extractant suspensions were filtered, calcium and potassium in the soil extracts were measured by a Technicon Auto-Analyzer flame photometer. Phosphorus in the soil extracts was measured colorimetrically using a Technicon Auto-Analyzer. Nitrate

and ammonium were extracted from the soil by the Spurway procedure: 4.00 g of air-dry soil were shaken for 20 sec with 20 ml of a weak solution of acetic acid (0.018 $\underline{N}$ $CH_3COOH$). Nitrate in the soil extracts was determined using a diphenylamine reagent. Ammonium was measured with Nessler's reagent. Soluble salt content was determined by mixing 50 cc of distilled $H_2O$ with 25 cc of air-dry soil. The electrical conductivity of the soil-water mixture was measured with a Soil Tester solu-bridge.

Nitrate in water samples was determined by an automated hydrazine reduction procedure initially developed by Kamphake (1966). Ammonium was measured using Nessler's reagent.

## RESULTS AND DISCUSSION

### Levels of Nitrate and Ammonium in Water Near Maxi-Mix Area

As shown in Tables 1, 2 and 3, levels of nitrate ($NO_3$) were well below 10 ppm N in samples from the well and brook in both 1971 and 1972. In 1971, $NO_3$ levels in groundwater samples were also consistently below 10 ppm N. In 1972, samples from five of the 12 groundwater sampling sites exceeded 10 ppm N. In two cases, external factors probably caused the high levels despite precautionary measures to prevent such occurrences: at site S7 (45 ppm N), two moles had fallen into the sample hole and had partially decayed; at site S10 (16.8 ppm N) there was evidence of surface runoff into the sampling hole. The nitrate levels in groundwater samples from sites S9, S19A and S11A were 11.9, 11.7 and 12.9 ppm N, respectively, values quite close to 10 ppm N.

Ammonium ($NH_4$) levels were consistently low in brook and well water samples in 1971 and 1972. Some groundwater samples, however, were very high in $NH_4$ in 1971. The levels were highest in sampling sites nearest the maxi-mix area (S1, S7, S8) and decreased with increasing distance from the area. By 1972, $NH_4$ levels in groundwater had dropped below 15 ppm N in samples from all sites.

We can only speculate as to the fate of the $NH_4$ found in 1971. Some of it may have been lost as $NH_3$. It is also possible that some of the ammonium moved beyond the

TABLE 1. NITROGEN LEVELS, 1971, IN STREAM WATER AND FARM WELL
WATER FROM SAMPLING SITES ADJACENT TO AN AREA WHERE
LARGE AMOUNTS OF POULTRY MANURE WERE MAXI—MIXED.

| DATE | | B1 | B2 | B3 | B4 | B5 | B6 | B7 | B8 | B9 | WELL |
|---|---|---|---|---|---|---|---|---|---|---|---|
| | | | | | | ppm N | | | | | |
| 3/12 | $NO_3$ | nd* | nd | nd | nd | | nd | nd | nd | nd | |
| | $NH_4$ | nd | nd | nd | nd | | nd | nd | nd | nd | |
| 3/22 | $NO_3$ | nd | nd | nd | nd | | 0.1 | 0.1 | nd | nd | nd |
| | $NH_4$ | 1 | 1 | nd | nd | | nd | 1 | 2 | 2 | nd |
| 3/31 | $NO_3$ | nd | nd | nd | ---† | | nd | 0.1 | nd | nd | --- |
| 4/30 | $NO_3$ | nd | nd | nd | nd | | nd | nd | nd | nd | nd |
| | $NH_4$ | nd | nd | 1 | 1 | | 1 | 1 | 1 | 1 | 1 |
| 5/20 | $NO_3$ | nd | nd | nd | --- | | nd | nd | nd | nd | 0.1 |
| | $NH_4$ | 1 | 1 | nd | --- | | nd | nd | nd | nd | 1 |
| 6/24 | $NO_3$ | nd | nd | --- | --- | | nd | --- | --- | nd | nd |
| 8/2 | $NO_3$ | nd | nd | --- | --- | | nd | --- | --- | --- | --- |
| | $NH_4$ | 1 | 1 | --- | --- | | 1 | --- | --- | --- | --- |

*None detectable

†No sample

| DATE | | S1 | S2 | S3 | S4 | S5 | S6 ppm N | S7 | S8 | S9 | S10 | S10a | S11 | S11a |
|---|---|---|---|---|---|---|---|---|---|---|---|---|---|---|
| 3/22 | $NO_3$ | 0.1 | --† | 2.3 | nd* | --- | nd | 1.0 | 0.7 | 0.1 | 0.1 | 0.1 | 0.3 | 0.2 |
|  | $NH_4$ | 130 | -- | --- | nd | --- | 7 | 80 | 160 | 40 | 40 | nd | nd | 2 |
| 3/31 | $NO_3$ | 0.1 | 0.4 | 0.5 | nd | --- | --- | 0.8 | -- | nd | 0.2 | 0.1 | --- | -- |
| 4/30 | $NO_3$ | 0.4 | 0.1 | 0.1 | 0.1 | --- | --- | 0.3 | 0.3 | nd | 0.2 | 0.2 | 0.2 | 0.6 |
|  | $NH_4$ | 125 | 50 | 12 | 1 | --- | --- | 250 | 110 | 35 | 30 | 6 | 2 | 1 |
| 5/20 | $NO_3$ | 0.4 | 0.1 | 0.1 | 0.1 | --- | --- | 0.2 | 0.1 | nd | 0.4 | 0.3 | 0.3 | 0.1 |
|  | $NH_4$ | 100 | 40 | 15 | 1 | --- | --- | 140 | 130 | 40 | 40 | 1 | 1 | nd |
| 6/24 | $NO_3$ | --- | --- | --- | 0.4 | --- | --- | --- | --- | --- | --- | --- | --- | --- |
| 8/2 | $NO_3$ | --- | -- | --- | --- | --- | --- | --- | --- | --- | --- | --- | --- | --- |
|  | $NH_4$ | --- | --- | --- | --- | --- | --- | --- | --- | --- | --- | --- | --- | --- |

*None detectable
†No sample

sampling sites. It is clear that some of the $NH_4$ was nitrified as evidenced by the uniformly higher levels of $NO_3$ in groundwater in 1972.

Table 4 shows that levels of $NO_3$ and $NH_4$ in brook and well water were very low in May and June, 1974. Brook water sampled in October, 1974 (Table 5) was also very low in $NO_3$ and $NH_4$.

### Soil Samples

Results of the analyses of the soil samples are shown in Table 6. They show several interesting trends especially with regard to soil pH and extractable nutrients.

The layer of soil identified in Table 6 as "1st black layer" probably represents a residual decomposed manure zone resulting from the folding action of the bulldozer. The samples from this layer had high pH values (8.1 and 8.2) relatively high soluble salts (83 to 90 x $10^{-5}$ mhos for the 1:2 soil-water mixture), very high $NH_4$ levels (400 ug N per g soil) and very high levels of extractable calcium (Ca), phosphorus (P), and potassium (K). Only a trace (less than 10 ug N/g soil) of nitrate was found in these samples.

The high pH values found in the black layer samples (as well as in most other samples) probably resulted from the formation of ammonium carbonate during manure decomposition in the soil. Similar pH changes were found by Hileman (1971). The high concentration of elements in this zone reflects the substantial contents of Ca, P, K and N in poultry manure.

The very low level of $NO_3$ in all soil layers except for one of the top layers was probably due to inhibition of nitrifying organisms by the combination of high $NH_4$ and high pH and/or denitrification of $NO_3$ produced. It is likely that nitrification did take place in the top layer of the mix as this layer dried out and became oxidized. This reaction would account for the 50 ug N as $NO_3$ per g of soil found in one of the top layer samples and the low level of $NH_4$ in the top layers. Downward movement of $NO_3$ effected by leaching could subject it to denitrification upon contact with the underlying waterlogged (reduced) layers, (Alexander 1965).

A substantial amount of the N in the manure which was maxi-mixed probably was lost to the atmosphere as ammonia ($NH_3$). Ludington and Sobel (1973) reported a 70 percent loss of N in poultry manure dried over a 30 day period. They found that N losses increased with drying time as did losses of total solids in the manure.

One puzzling aspect of the soil results is the combination of high pH and very low calcium found in samples below the first black layer and in the subsurface clear layer. because it is highly unlikely that a virgin or fallow Connecticut soil would have pH values above 6.0 to 6.5, some reaction resulting from the manure application must have

TABLE 3. NITROGEN LEVELS, MAY 3, 1972, IN GROUND WATER AND STREAM WATER FROM SAMPLING SITES ADJACENT TO AN AREA WHERE LARGE AMOUNTS OF POULTRY MANURE WERE MAXI-MIXED.

| STATION NO. | $NH_4$ ppm N | $NO_3$ ppm N |
|---|---|---|
| S1 | 12 | 0.5 |
| S2 | 6 | 2.2 |
| S3 | 6 | 8.2 |
| S4 | 4 | 8.5 |
| S5 | 2 | 0.3 |
| S7* | 10 | 45.0 |
| S8 | 6 | 5.7 |
| S9 | 4 | 11.9 |
| S10† | 2 | 16.8 |
| S10a | 2 | 11.7 |
| S11 | 2 | 5.3 |
| S11a | 2 | 12.9 |
| B1 | 2 | 2.1 |
| B2 | 2 | 1.0 |
| B3 | 2 | nd‡ |
| B4 | 2 | 0.2 |
| B5 | 2 | 0.2 |
| B6 | 2 | 0.2 |
| B7 | 2 | 0.2 |
| B8 | 2 | 0.2 |

*Two dead, partially decayed moles were found in the S7 hole. These may have contributed to the high $NO_3$ level here.
†Evidence of surface runoff into the sampling hole here.
‡None detectable.

TABLE 4. CHEMICAL ANALYSIS AND DISCHARGE OF STREAM WATER AT OUTCROPPING DOWN THE STREAM FROM MAXI-MIXING SITE AND FROM THE WELL — MAY, JUNE 1974.

| DATE | DIS-CHARGE G.P.M. | LOCATION | CHEMICAL COMPONENTS | | | |
|---|---|---|---|---|---|---|
|  |  |  | BOD | $NO_3$-N | $PO_4$ | pH |
|  |  |  |  | ppm | | |
| 5/ 8 | --† | WELL | -- | nd* | nd | 5.4 |
| 5/ 8 | 15 | WEIR | -- | nd | nd | 4.4 |
| 5/10 | 15 | WEIR | nd | nd | 0.05 | 4.4 |
| 5/14 | 30 | WEIR | nd | nd | 0.30 | 4.4 |
| 5/16 | 21 | WEIR | nd | nd | nd | 5.8 |
| 5/16 | -- | WELL | nd | nd | nd | 6.6 |
| 5/23 | 11 | WEIR | 36.0 | nd | nd | 4.7 |
| 5/30 | 12 | WEIR | 23.7 | nd | nd | 5.8 |
| 6/ 6 | 12 | WEIR | 18.4 | nd | nd | 6.5 |
| 6/18 | 11 | WEIR | --- | nd | nd | 6.4 |

*None detectable
†Analysis not made

TABLE 5. CHEMICAL ANALYSIS OF STREAM WATER AT OUTCROPPING DOWN THE STREAM FROM MAXI-MIXING SITE — OCTOBER 23, 1974.

| SAMPLE ID | pH | SOLUBLE SALTS mhos x $10^{-5}$ | Ca | K | P ppm | $NO_3$-N | $NH_4$-N | Mg |
|---|---|---|---|---|---|---|---|---|
| 1 | 5.2 | 10 | Trace* | Trace† | 0.2 | Trace‡ | 2 | 5 |
| 1A | 4.8 | 10 | Trace | Trace | 0.2 | Trace | 2 | 5 |
| 2 | 5.0 | 10 | Trace | Trace | 0.2 | Trace | 2 | 5 |
| 4 | 5.0 | 10 | Trace | Trace | 0.2 | Trace | 2 | 5 |

*Less than 2 ug Ca/ml
†Less than 1 ug K/ml
‡Less than 0.1 ug N/ml

effected the rise in pH. The formation of ammonium carbonate is one possibility.

## CONCLUSIONS

The advantage of maxi-mixing is its low cost per ton of manure disposed. Cost for manure disposal in the 1970 project was 62 cents per ton, while the 1974 cost was 60 cents per ton. By comparison, the U.S.D.A. Agricultural Economic Report No. 254, (1974) quotes costs of $2.15 per ton for land spreading of manure from 50,000 bird flocks with pit cleaners and cites a figure of $1.63 per ton with a two-stage drying system used by Bressler at Pennsylvania State University.

There are some disadvantages that limit the use of maxi-mixing: (a) there are odors from the mixing process for a short period necessitating allowances for residential area; (b) when land is expensive, the site cost may be a limiting factor; (c) some soil types (e.g. shallow to bedrock, excessively drained, steeply sloped) may not be compatible with maxi-mixing; (d) considering the existing scarcity and cost of nitrogen and phosphorus fertilizers, this approach does not make economic use of poultry manure as a source of nutrients for plants.

It would be a better conservation practive to add the manure to crop lands than to dispose of it in a maxi-mix system, but as an emergency action in a land-poor situation, it is difficult to find a cheaper method of manure disposal. The criteria for decision making from the farm's point of view should be based on equating marginal cost of maxi-mixing to marginal benefits of poultry manure application on agricultural crop land. Thus, rising prices of fertilizers will shift manure disposal to crop application, but when agricultural land is limited to a reasonable distance from the farm, or if the prices of fertilizers decline maxi-mixing may be the best means of manure disposal.

**References**

1   Alexander, M. 1965. Nitrification, pp. 307-343. In: W. V. Bartholomew and F. E. Clark, (ed.) Soil Nitrogen. American Soc. Agron., Madison, Wis.

2   Agricultural Economic Report No. 254. 1974. U.S.D.A. Recycling poultry waste as feed, will it pay?

3   Hileman, L. H. 1971. Effect of poultry manure application on selected soil chemical properties. Proceedings International Symposium on Livestock Wastes. pp 247-248, ASAE, St. Joseph, Mich. 49085.

4   Kamphake, L. J., S. A. Hannah and J. M. Cohen. 1966. Automated analysis for nitrate by hydrazine reduction. Federal Water Pollution Control Administration, U. S. Department of the Interior, Robert A. Taft Sanitary Engineering Center, Cincinnati, Ohio.

5   Ludington, D. C. and A. T. Sobel. 1973. Disposal and utilization of dairy and poultry manures by land application. Annual Progress Report: 1 January, 1971 to 31 December 1972, Hatch Project 406 contributing to NE63, New York State Experiment Station, Cornell University, Ithaca, New York.

6   McIntosh, J. L. 1969. Bray and Morgan soil extractants modified for testing acid soils from different parent materials. Agron. J. 61:259-265.

TABLE 6. VALUES FOR MOISTURE CONTENT, LOSS ON IGNITION, pH, SOLUBLE SALTS AND EXTRACTABLE $NH_4$, $NO_3$, Ca, P and K IN SAMPLES OF SOIL TAKEN FROM DIFFERENT HORIZONTAL LAYERS‡ IN AN AREA WHERE LARGE AMOUNTS OF POULTRY MANURE WERE MAXI-MIXED.

| LAYER AND DEPTH SAMPLED, CM | PERCENT $H_2O$* | PERCENT LOSS ON IGNITION | pH | SOLUBLE SALTS mhos x $10^{-5}$ | $NH_4$-N | $NO_3$-N | Ca | P | K |
|---|---|---|---|---|---|---|---|---|---|
| | | | | | | ug per g soil | | | |
| Top (0-6) | 26.9 | 4.46 | 7.4 | 11 | 20 | 50 | 2000 | 50+ | 300+ |
| Top (0-6) | | 4.32 | 7.5 | 11 | 20 | Trace† | 2000 | 16 | 300+ |
| Subsurface clear (6-30) | 12.2 | 1.35 | 7.9 | 38 | 150 | Trace | 100 | 4 | 300+ |
| 1st black (30-60) | 26.9 | 5.10 | 8.1 | 90 | 400 | Trace | 2000+ | 50+ | 300+ |
| 1st black (30-60) | 27.1 | 5.21 | 8.2 | 83 | 400 | Trace | 2000+ | 50+ | 300+ |
| 2nd black (70-90) | 16.8 | 0.77 | 7.1 | 10 | 50 | Trace | 100 | 5 | 235 |
| Below 2nd black (120-124) | 15.0 | 0.96 | 6.7 | 13 | 75 | Trace | 100 | 5 | 200 |
| Coal-like (220-240) | 19.6 | 1.03 | 7.5 | 24 | 75 | Trace | 150 | 5 | 300+ |
| Coal-like (220-240) | 20.2 | 1.07 | 7.5 | 34 | 75 | Trace | 200 | 6 | 300+ |
| Coal-like (220-240) | 15.5 | 4.39 | 7.2 | 21 | 60 | Trace | 150 | 4 | 245 |

*Air-dry soil basis
†Less than 10 ug N/g soil
‡See figure 2

# On Land Disposal of Liquid Organic Wastes Through Continuous Subsurface Injection


J. L. Smith, D. B. McWhorter, R. C. Ward

ASSOC. MEMBER  ASSOC. MEMBER   ASSOC. MEMBER
ASAE        ASAE         ASAE


LAND spreading has been used as an effective method of organic waste disposal for centuries. The presence of organic matter and certain plant nutrients makes the concept of recycling these wastes to the soil appealing. However, as new and more restrictive laws governing waste disposal are adopted, each proposed method and/or site must be carefully evaluated with respect to the following criteria: (a) potential for odors and flies or other insects, (b) runoff pollution, (c) aesthetics, (d) soil or groundwater pollution, and (e) economics. Subsurface injection generally eliminates problems associated with the first three criteria listed above. The last two will be discussed in this paper.

The research has involved the development and evaluation of an efficient, economical continuous subsurface injection machine. Disposal sites have been instrumented so the quantity and quality of water percolating beneath the injection zone can be measured. Wells located around the sites are used to monitor groundwater quality. Soil samples are taken periodically to determine nutrients, salts, heavy metal concentrations and bacteria movement and survival.

## DESIGN OF THE SUBSURFACE INJECTION EQUIPMENT

A series of laboratory tests (Gold 1973) was conducted to study various methods of injecting liquid organic wastes with up to 6 percent solids into the soil. The optimal procedure was determined to be one where the material was thoroughly mixed with the top 10 cm of soil. This procedure resulted in the most rapid drying of the liquid portion of the material and also facilitated repeated applications in the shortest time interval. Deep injections dried slowly and resulted in near mummification of the injected material.

The first subsurface injector system consisted of a 2 840 liter portable tank and the injector pulled behind the tank. The major accomplishment of this phase of the research was the development of the injector units. These consisted of commercially available chisel plow shanks with 40 cm wide—high lift sweeps. The liquid waste is discharged under the wings of the sweep through a tubular deflector. This provides thorough mixing of the liquid waste with soil and total coverage. Each deflector has two outlets with a maximum area of 19.4 cm$^2$ per outlet. Plugging is not a problem provided the material is pumped to the deflectors under a small pressure (34 kPa).

Two major problems were observed in testing this machine: (a) it was impossible to pull the portable tank through the field except under nearly ideal conditions, and

FIG. 1 Boulder injector, 3000 liter per minute capacity.

(b) the "turn around" time was too long and only 2 percent of the total operating time was spent injecting.

Major modifications of the above system were incorporated into subsurface injectors now operating to dispose municipal sewage sludge (Boulder machine) and liquid dairy manure (Fort Collins machine). The units are shown in Figs. 1 and 2.

Material is supplied continuously through an underground pipeline which is connected to a 200 m flexible hose which, in turn, is connected to the injector. A pump, located at the source or storage facility, delivers the material to the injector. This eliminates the need to pull heavy portable tanks through the field and provides continuous operation of the injector until the supply is exhausted.

The Fort Collins machine has five injector sweeps with a 7.6 cm supply hose. It is capable of injecting at rates up to 1 500 liters per minute over a 230 cm width. The 34 kW wheel-type tractor is capable of pulling the unit and hose in

FIG. 2 Fort Collins injector, 1500 liter per minute capacity.


This research was supported in part by the Environmental Protection Agency, Project S802940-01-0.

The authors are: J. L. SMITH, Associate Professor, D. B. McWHORTER and R. C. WARD, Assistant Professors, Agricultural Engineering Dept., Colorado State University, Fort Collins.




MANAGING LIVESTOCK WASTES

| | |
|---|---|
| Operating speed | 0.8 to 2.4 km/hr |
| Capacity per sweep | 230 to 450 liters/min |
| Operating depth | 7.6 to 20 cm |
| Field efficiency | 87 percent |
| Tractor power required | |
|    Wheel type | 34 to 90 kW |
|    Crawler type | 30 to 45 kW |
| Capacity | 187 000 to 748 000 liters/hectare |
| Maximum area covered per | |
|    hose attachment | 7.49 hectares |
| Pressure required at | |
|    hose attachment | 400 kPa |
| Solids content | 6 percent (nominal) |
| Frequency of injection | 2 to 7 days |

good weather at 2.4 km/hr. In moderately severe weather, the speed must be increased due to the torque characteristics of the tractor engine.

The Boulder machine consists of seven injector sweeps with a 11.4 cm supply hose. It is capable of injecting at rates up to 3 000 liters per minute over 305 cm width. A 31 kW crawler tractor has sufficient power to pull the unit at approximately 1.6 km/hr. Speeds as low as 0.8 km/hr can be used in good soil conditions. The tractor is equipped with wide tracks (84 cm) to provide flotation in adverse weather. This unit is used regularly in mud and snow; however, it will not operate when the frost depth is greater than 5 cm.

Specifications and approximate sizes for the equipment are given in Tables 1 and 2.

## DISPOSAL SITE DESCRIPTIONS

The manure injection site at Fort Collins consists of 1.05 hectares. Top soils at the site range from sandy loam to loam, approximately 45 cm thick, overlaying alluvial sands and gravels. The normal water table depth is approximately 2 m, but this decreases to about 1 m during the 2½ month irrigation season. The site is divided into plots which have received total dry weight loadings over a one year period of 20 000 kg/ha, 10 000 kg/ha and nothing. Manure from the collection pit ranges from 1 percent to 2 percent solids and a total quantity of 41 600 liters is disposed each week. Using the smaller (Fort Collins) injector, requires about 30 minutes operating time.

The sludge injection site at Boulder consists of 3.2 hectares. Soils at the site consist of 9 to 15 cm of silty or sandy loam overlaying from 0.6 to 2 m of sandy clay overlaying alluvial sand and gravel. The water table depth remains relatively stable at approximately 2 m. Approximately 265 000 liters of 4 to 6 percent solid anaerobically digested sludge is applied each week. Using the larger (Boulder) injector, this requires from 60 to 90 minutes. The maximum yearly dry solids loading rate achieved to date is 65 000 kg/ha.

## WATER QUALITY MONITORING

Subsurface injection of liquid wastes has a potential for contamination of surface and ground waters as a result of surface runoff and subsurface drainage from the disposal



| Hose diameter, cm | Flow, liter/min | Full hose weight, kg/m | Number of sweeps | Width, m |
|---|---|---|---|---|
| 7.6 | 1 500 | 4.6 | 5 | 2.27 |
| 10.1 | 3 000 | 8.0 | 7 | 3.20 |
| 11.4 | 3 800 | 10.3 | 9 or 11 | 5.03 |
| 12.7 | 5 300 | 12.8 | 13 | 5.94 |

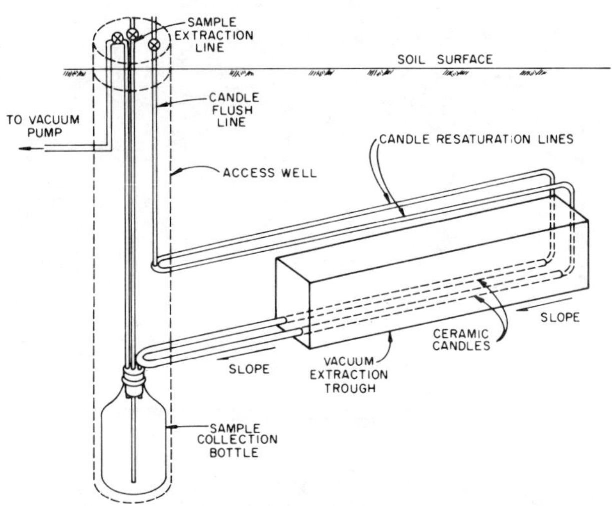


FIG. 3 Vacuum extractor system.

area. Unlike surface runoff, percolation toward the groundwater beneath a disposal area cannot be prevented by simple berms or dikes. Deep percolation, therefore, represents a more limiting constraint than does surface runoff, if the percolate is contaminated. For this reason, current water quality monitoring programs are oriented toward assessment of the nature and magnitude of groundwater quality degradation.

Rarely will the hydro-geologic boundaries of a groundwater body be coincident with the surface boundaries of an injection site. Thus, changes in the quantity and quality of groundwater induced by injection of liquid wastes are likely to be masked by occurrences totally unrelated to the disposal operation. To adequately evaluate the effects of disposal may require monitoring many other potential sources of pollution. Additional difficulties result from the large dilution effect as percolate from the disposal site enters the groundwater body and from the subsequent dispersion and movement with the groundwater flow.

The above problems can be overcome by measuring the quality and rate of percolation in the zone between the water table and the land surface. A vacuum extractor, suitable for measuring both the quality and quantity of percolate, has been developed by Duke and Haise (1973). The device is depicted schematically in Fig. 3. Components of the vacuum extractor are a stainless steel trough 15 cm wide, 250 cm long, and 20 cm deep; two strings of porous, hollow, ceramic candles; a collection jar; and a vacuum pump and control system. Water in the porous walls of the ceramic candles is maintained at negative gauge pressure by applying a vacuum to the collection vessel. The function of the ceramic candles is to allow water to be withdrawn from the soil at negative pressures without permitting air to enter the vacuum system. In most situations of practical interest, the water pressure in the soil at the top of the trough will adjust (approximately) to the ambient soil-water pressure (Corey 1974). Thus, the trough intercepts a quantity of water proportional to the top-area of the trough at a rate nearly equal to the ambient percolation rate. The volume of water in the collection vessel is measured periodically to determine the total quantity of percolate and the average percolation rate. Samples from the collection vessel are subjected to chemical analysis on a regular basis.

Installation of the vacuum extractor involves excavation of a narrow trench to a preselected depth (usually about 1 m) and placement of the trough in the trench on a slight grade so water will flow from the candles to the access well

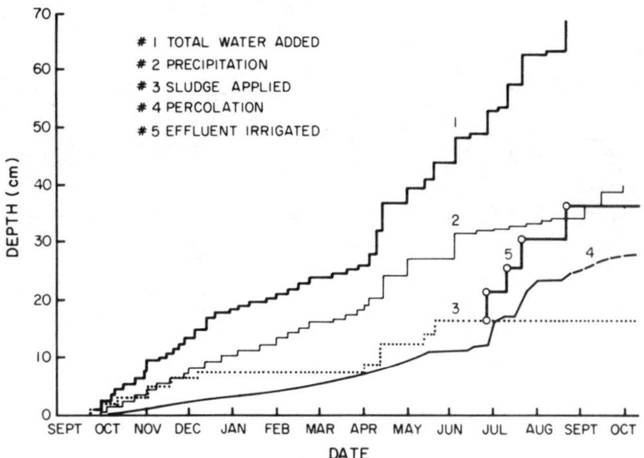

FIG. 4 Water balance for sludge plot.



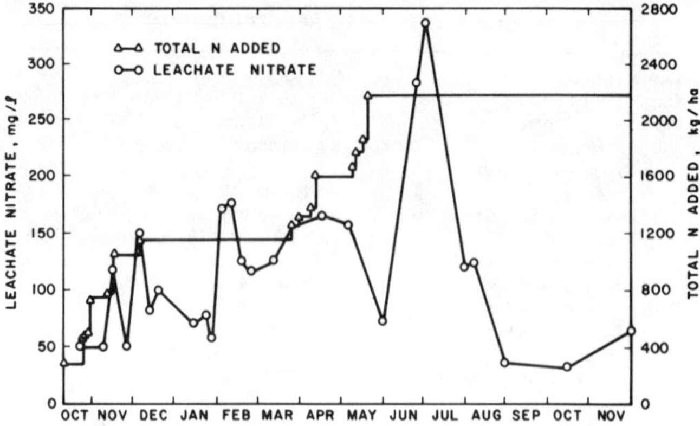

FIG. 5 Nitrogen added to sludge plot and leachate nitrate concentration.



and collection bottle. The trench is then carefully back-filled so the soil is replaced in approximately the same order as it was excavated. Procedures are also available for installing the trough "undisturbed." Errors resulting from disturbance should result in conservative estimates of the percolation rate. This and other potential errors are discussed in detail by Trout (1975).

Six vacuum extractors are installed at the Boulder sludge site. In addition, groundwater wells are located on a 45 m grid throughout the site (as required by state and local health officials). The manure site is instrumented with nine vacuum extractors, and a total of nine groundwater wells are located along the site boundary.

A water balance for one sludge plot is shown in Fig. 4. This plot received 65 000 dry kg/ha of approximately 4.5 percent solids sludge from September 1973 to May 1974. Sweet corn was planted on the plot during the middle of June and it was irrigated with treatment plant effluent during July and August.

Approximately 25 percent of the total water applied prior to July 1 was leached and approximately 75 percent of the first three irrigations was leached. On sandy plots, an average of 25 percent of the total water applied was leached, but the percentage varied considerably depending upon climatic conditions. No change in water table depth

could be associated with sludge or effluent applications.

Typical water quality data for the sludge, manure, leachate and groundwater are shown in Table 3 along with drinking water standards. Note that the relatively high levels of heavy metal in the sludge are reduced to levels well below drinking water standards as the sludge solution passes through 1 m of soil. These metals are believed to be precipitated as slightly soluble complexes as a result of contact with the soil in the alkaline pH range (7.2).

Nitrate in the leachate collected by the vacuum extractors significantly exceeded drinking water standards (45 ppm, Colorado Public Health), even though much of the inorganic nitrogen was retained in the upper 1 m of soil. Fig. 5 shows the variation in leachate nitrate concentration with time for the sludge plot treated with 65 000 dry kg/ha. Note that prior to sludge application, the leachate nitrate concentration approximately equalled the drinking water standard. It is also interesting to note that the corn crop, planted in June, removed sufficient available nitrate to reduce the leachate nitrate concentration to approximately the same level as prior to sludge application.

Using the data obtained in this research, and the principle of conservation of mass, it is possible to estimate the quantity of a given contaminant that goes to the groundwater or remains in the soil. In some cases, particularly with nitrogen, other estimates must be inferred from available

**TABLE 3. COMPOSITION OF WASTE MATERIALS, LEACHATE, AND GROUNDWATER**

| | Concentration, mg/l (ppm), (unless noted) | | | | | | |
| | Sludge plots | | | | Manure plots | | |
| Parameter | Sludge | Leachate | Groundwater | Drinking water standard | Manure | Leachate | Groundwater |
|---|---|---|---|---|---|---|---|
| Cr | 7.1 | <0.005 | <0.01 | 0.05* | | | |
| Cu | 17 | 0.06 | 0.01 | 1.0 * | | | |
| Ni | 1.0 | — | <0.04 | — | | | |
| Cd | 0.28 | 0.007 | 0.001 | 0.01* | | | |
| Zn | 27 | 0.29 | <0.01 | 5.0 * | | | |
| Fe | — | 0.06 | 0.15 | 0.3 * | | | |
| Pb | 53 | <0.005 | <0.002 | 0.05* | | | |
| Mn | — | <0.05 | <0.02 | 0.05* | | | |
| Ag | — | <0.0005 | <0.00025 | 0.05† | | | |
| Al | 1800 | — | <1.0 | — | | | |
| Hg | 15 | 0.0001 | <0.0005 | 0.002† | | | |
| Na | — | 55 | 36 | — | | | |
| Mg | — | 133 | 51 | 125 * | | | |
| Ca | — | 200 | 80 | — | | | |
| B | — | 0.35 | 0.36 | — | | | |
| Kjeldahl N | 1160 | 157 | 5 | 45 * | 290 | 7 | 10 |
| PO$_4$ | 4.7% | 1 | 0.8 | 4 * | 10 | — | 3.5 |
| K | 0.1% | — | — | — | — | 12 | 4.8 |
| Cl | 729 | — | 102 | — | — | 46 | 110 |
| NH$_3$ | — | — | — | — | — | 2.4 | — |
| TDS | — | — | — | — | 7000 | 6800 | 4800 |

* Colorado Public Health

† EPA proposed (1973)



MANAGING LIVESTOCK WASTES

**TABLE 4. CHEMICAL ANALYSIS OF SWEET CORN LEAF AND GRAIN TISSUE**

| Parameter | Concentration, mg/l (ppm) dry weight (unless noted) | | | Normal range[1] for midwest corn | |
|---|---|---|---|---|---|
| | Leaf sample 8/7/74 | Leaf sample 8/22/74 | Grain sample 8/22/74 | Leaf | Grain |
| Cd | 0.2 | 0.8 | 0.3 | 0-5.0 | 0.1-1.0 |
| Cr | 5 | 2 | 1 | 0.1-4.0 | 0.1-1.0 |
| Cu | 10 | 16 | 13 | 6-20 | 4-10 |
| Ni | 1 | 1 | 1 | 0-5.5 | 0-5 |
| Pb | — | 4 | 1 | 0-10 | 0.2-5.0 |
| Zn | 62 | 117 | 70 | 20-70 | 20-100 |
| N (total) | 3.6% | 2.56% | 2.25% | | |

[1]Bauer (1974)

literature. A major advantage of this method of analysis is that predictions can be made. If a given treatment adds a certain amount of an element to the soil or to the groundwater, the size of treatment can be predicted which would create a hazardous situation. Predictions can be adjusted for effects of weather. Details of this analysis are discussed by Trout (1975).

## SALINITY

Soil salinity, as measured by soil electroconductivity, has remained relatively constant. In fact, a slight lowering of the soil salinity at the Boulder site has been observed. This is reasonable since the electroconductivity of the untreated soil was high (1.7 mmhos/cm) due to a previously existing high groundwater table.

Salinity problems could be created with moderately heavy manure or sludge loadings in arid or semiarid climates. However, from the results of this research, the problems appear to be temporary. Any increase in salinity is removed by precipitation or water percolating from the injection zone. Assuming a salt balance is maintained, leaching all the salts from the 65 000 kg/ha sludge plot would result in a salt increase of 60 ppm in the groundwater passing under the plot. This degradation is small considering the effect of additional dilution and the 1962 United States Public Health Service Standard of 500 ppm.

## FECAL COLIFORMS

High fecal coliform counts have been observed in the soil on injected plots. However, below the first 0.3 m, counts decrease rapidly with depth. This indicates an effective filtering action by the soil. Fecal coliform counts were also found to decrease rapidly with time following completion of injection operations. For example, the fecal coliform count of the soil was greater than $2 \times 10^6$ per 100 ml of soil three days after injection ceased. Two months later the count had decreased to 1 700 per 100 ml.

Fecal coliform counts of the groundwater have remained relatively constant. At the Boulder sludge site, counts are normally less than 100 per 100 ml. Counts are much higher at the manure site, i.e., 3 000 per 100 ml, when the groundwater is high during irrigation. Since this occurred in all wells on the site, it was probably due to the location of various other waste disposal systems in the vicinity.

## SWEET CORN TISSUE ANALYSIS

Sweet corn was grown on plots treated with 65 000 kg/ha municipal sewage sludge. Results of chemical analysis of leaf and kernel samples are shown in Table 4 along with normal values from corn grown elsewhere. Only Ca in the grain sample and Zn and Cr in one leaf sample exceeded the normal values. No quantitative measure of production was made because the corn planted in untreated areas failed to grow; however, the corn grown in treated plots appeared to grow normally and the crop was consumed by humans with no ill effects. Additional details are given by Trout (1975).

## ECONOMICS

Economic analysis of the continuous subsurface injection system will be based on a comparison developed by the North Carolina Agricultural Extension Service in Circular 568 (Kriz et al. 1973). The circular uses a 100-cow herd of 635 kg dairy cows and includes a system which utilizes storage and spreading of the manure in liquid form. The dairy has a storage capacity which represents a 30-day cleanout interval and also has a pond to retain milkhouse wastewater and runoff from a 15 cm rainfall (a total runoff of 55 500 liters).

The estimated initial investment and annual costs for the liquid manure storage and spreading system are given in Table 5. Similar estimated costs are given in Table 6 for the

**TABLE 5. ESTIMATED INITIAL INVESTMENT AND ANNUAL COSTS OF DISPOSAL FOR A LIQUID MANURE STORAGE AND TANK SPREADING SYSTEM FOR A 100 COW FREE-STALL DAIRY WITH A 30-DAY STORAGE PERIOD (FROM KRIZ 1973)**

| Item | Estimated cost | | | | |
|---|---|---|---|---|---|
| | Initial investment | Ownership cost factor | Hours | Hourly rate | Annual |
| Scraper, rear mounted | $ 300 | 0.18* | | | $ 54 |
| Tractor use, 37-44 kW for scraping | | | 280 | 2.26 | 633 |
| Storage tank | 3,900 | 0.12† | | | 468 |
| Agitator pump | 1,900 | 0.27‡ | | | 513 |
| Tractor, 45-59 kW for agitating and pumping | | | 45 | 2.90 | 131 |
| Spreader, liquid manure | 2,100 | 0.23§ | | | 438 |
| Tractor, 67-74 kW for hauling and spreading | | | 105 | 3.52 | 370 |
| Retention pond | 150 | 0.12† | | | 18 |
| Pump, 1.5 kW, 76 liters/min | 500 | 0.27‡ | | | 135 |
| Irrigation equipment | 1,215 | 0.19‖ | | | 231 |
| Electricity | | | 188 | 0.06 | 11 |
| Labor | | | 440 | 2.50 | 1,100 |
| Total | $10,065 | | | | $4,147 |

* 8 years of life; 7 percent interest on ½ initial cost; repairs, 1 percent of initial cost; taxes and insurance, 1 percent of initial cost.
† 15 years of life; 7 percent interest on ½ initial cost; repairs, 0.5 percent of initial cost; taxes and insurance, 1 percent of initial cost.
‡ 6 years of life; 7 percent interest on ½ initial cost; repairs, 6 percent of initial cost; taxes and insurance, 1 percent of initial cost.
§ 6 years of life; 7 percent interest on ½ initial cost; repairs, 2 percent of initial cost; taxes and insurance, 1 percent of initial cost.
‖ 8 years of life; 7 percent interest on ½ initial cost; repairs, 2 percent of initial cost; taxes and insurance, 1 percent of initial cost.

TABLE 6. ESTIMATED INITIAL INVESTMENT AND ANNUAL COSTS OF DISPOSAL FOR A LIQUID MANURE STORAGE AND CONTINUOUS SUBSURFACE INJECTION SYSTEM FOR A 100 COW FREE-STALL DAIRY WITH A 30-DAY STORAGE PERIOD

| Item | Initial investment | Estimated cost | | | Annual |
| | | Ownership cost factor | Hours | Hourly rate | |
| --- | --- | --- | --- | --- | --- |
| Scraper, rear mounted | $ 300 | 0.18* | | | $ 54 |
| Tractor use, 37-44 kW for scraping | | | 280 | 2.26 | 633 |
| Storage tank | 3,900 | 0.12† | | | 468 |
| Agitator pump | 1,900 | 0.27‡ | | | 513 |
| Tractor, 45-59 kW for agitating, pumping, and injecting (pit and pond) | | | 44 | 2.90 | 128 |
| Installation of piping | 2,480 | 0.12† | | | 298 |
| Flexible hose | 1,300 | 0.27‡ | | | 351 |
| Subsurface injector | 1,000 | 0.23§ | | | 230 |
| Tractor, 67-74 kW for injecting | | | 32 | 3.52 | 113 |
| Retention pond | 150 | 0.12† | | | 18 |
| Labor | | | 324 | 2.50 | 810 |
| Total | $11,030 | | | | $3,616 |

* 8 years of life; 7 percent interest on ½ initial cost; repairs, 1 percent of initial cost; taxes and insurance, 1 percent of initial cost.

† 15 years of life; 7 percent interest on ½ initial cost; repairs, 0.5 percent of initial cost; taxes and insurance, 1 percent of initial cost.

‡ 6 years of life; 7 percent interest on ½ initial cost; repairs, 6 percent of initial cost; taxes and insurance, 1 percent of initial cost.

§ 6 years of life; 7 percent interest on ½ initial cost; repairs, 2 percent of initial cost; taxes and insurance, 1 percent of initial cost.

continuous subsurface injection system. Many of the costs for the two systems are identical; however, they do not necessarily utilize the same equipment. For example, the pump for the injection system must have a higher efficiency than the pump for the spreader system. Also, with subsurface injection, it is feasible to inject runoff water from the pond rather than irrigating it, and a single pump can be used for both manure and runoff.

In this particular comparison, 170 000 liters per month are being disposed, since the storage tank for the spreader system in Circular 568 contains this volume. A total time of only 22.5 hours is needed to continuously inject the 2 044 000 liters of waste produced during the year. The total annual hours of injection for both the pit and pond is approximately 32 hours compared to 150 hours for liquid spreading and 188 hours of pumping of the retention pond in the liquid spreading system. Twelve hours were permitted for agitating in Table 7.

The total initial investment for the continuous subsurface injection system is almost $1,000 higher. The major difference is reflected in the annual costs where there is $531 savings with the continuous subsurface injector over the liquid spreader system. Since the initial cost of the continuous system is higher, the major annual cost savings comes from the reduced time needed to dispose of the wastes.

Table 7 compares continuous subsurface injection with other methods of collection and disposal discussed in Circular 568. Rather than the liquid manure, 30-day storage being the highest annual cost, the continuous injector reduces the cost to a point where it is competitive with solid manure systems. Note that the cost of continuous subsurface injection will effectively decrease in comparison to most other methods as the quantity of material to be disposed increases.

The continuous subsurface injector may make the last three systems presented in Table 7 more economical if the irrigation component is replaced with the injector system. Since the irrigation system in continuous, the injector would not have as dramatic economic effects as it had with the liquid spreader.

Methods that utilize surface disposal may require additional field operation in order to "plow down" the manure. No attempt has been made to include this cost into the above economic comparison. This cost would not be necessary with the continuous injection system.

### References


1   Bauer Engineering. 1974. Wastewater Management Opportunities, Boulder, Colorado. Chicago, Ill.

2   Corey, P. R. 1974. Soil Water Monitoring. Unpublished Report to Department of Agricultural Engineering, Colorado State University, Ft. Collins, CO,

3   Duke, H. R. and H. R. Haise. 1973. Vacuum Extractors to Assess Deep Percolation Losses and Chemical Constituents of Soil Water. Soil Sci. Soc. of Am. Proc., Vol. 37, No. 6, p. 963-964.

4   Gold, R. C., J. L. Smith and R. D. Hall. 1973. Development of an Organic Waste Slurry Injector. ASAE Paper No. 73-4529. ASAE, St. Joseph, Mich. 49085.

5   Kriz, G. J., et al. 1973. Dairy Waste Management Alternatives. North Carolina Agricultural Extension Service Circular 568, North Carolina State University, Raleigh, N. C., September.

6   Trout, T. J., Smith, J. L. and McWhorter, D. B. 1975. Environmental Effects of Land Application of Digested Municipal Sewage Sludge, Report submitted to City of Boulder, CO. Colorado State University, Ft. Collins, CO.


TABLE 7. ESTIMATED INITIAL INVESTMENT, ANNUAL COSTS AND LABOR REQUIREMENTS FOR ALTERNATIVE WASTE MANAGEMENT SYSTEMS FOR A 100 COW FREE-STALL DAIRY*

| Waste management system | Initial investment | Annual† ownership costs | Annual labor requirement, hrs | Total annual cost |
| --- | --- | --- | --- | --- |
| Direct spreading in solid form | $ 4,415 | $1,873 | 504 | $3,206 |
| Storage and spreading solid form | 5,165 | 1,963 | 504 | 3,326 |
| Storage and spreading in liquid form (30-day storage and liquid spreader) | 10,065 | 1,902 | 440 | 4,147 |
| Storage and injection in liquid form (30-day storage and continuous injection) | 11,030 | 1,932 | 324 | 3,616 |
| Lagoon and irrigation | 7,680 | 1,391 | 325 | 2,924 |
| Flush, storage tank, and irrigation | 12,945 | 2,475 | 223 | 3,501 |
| Flush, lagoon, and irrigation | 17,095 | 2,898 | 213 | 3,982 |

* Data for all systems except the continuous injection system taken from Kriz (1973).

† Does not include tractor ownership cost.



MANAGING LIVESTOCK WASTES

# Soil Properties and Future Crop Production As Affected By Maximum Rates of Dairy Manure


G. W. Randall, R. H. Anderson, P. R. Goodrich

MEMBER
ASAE


ONDITIONS sometime exist in livestock operations where acreage, time and/or labor may not be sufficient to permit the application of manure to land just prior to planting or at conventional rates. In addition, the monetary value of the nutrients contained in the manure in relation to prices for inorganic fertilizers was relatively low in the early 1970's. As a result of these factors, heavy rates of manure were often applied or disposed of in localized areas; often close to the livestock facility.

With these conditions in mind an experiment was established to determine the maximum quantity of manure that can be applied and incorporated in a limited non-crop area. Primary objectives were to investigate: (a) the capacity of land to serve as a disposal medium for excessive rates of manure, (b) the accumulation and movement of nutrients in the soil profile and (c) the response of future crops to these high rates.

## EXPERIMENTAL PROCEDURES

During 1971, 1972 and 1973, beginning in mid-May and ending in mid-September, dairy cattle manure taken directly from the barn was applied to the surface of a Webster clay loam soil. This poorly drained soil containing 5-8 percent organic matter in the Ap horizon is common in south-central Minnesota. Manure was applied to the same 0.2-hectare area in both 1971 and 1972. In 1973, this area was split and manure was applied to one of the 0.10-ha areas. The manure was allowed to dry for 1 to 7 days before incorporating by disking, field cultivating or periodic plowing by either moldboard or chisel plow. Dry matter determined at 105 C and nutrient application rates were calculated by weighing each load of manure and by gathering random manure samples throughout the season for chemical anlaysis. Chemical properties of the manure applied is shown in Table 1.

In 1973, corn (*Zea mays L.*) was planted on the other 0.10-ha area and also on an adjacent border area that had received a relatively small amount of manure in 1970 (73 T/ha on a dry weight basis) and had been fallowed since. In 1974, corn was planted to both 0.10-ha areas as well as to another border area which had received 73 T manure/ha in 1970 and had been fallowed since. Corn was planted at 66,000 plants/ha in early May in both years. No commercial fertilizers or insecticides were used. Weeds were chemically controlled.

Soil samples in 30-cm increments were taken to a depth of 210 cm before initial treatment application in 1971 and again in November 1972, to 270 cm in April 1973 and 300 cm in May 1974. Samples were analyzed for ammonia, nitrite, nitrate, chloride, extractable phosphorus, exchangeable potassium and sodium.

Soil solution samples from the 180-cm depth were taken periodically from 1972 thru 1974 with the aid of suction lysimeters constructed from 5-cm PVC pipe fitted with porous ceramic cups.

**TABLE 1. CHEMICAL ANALYSIS OF DAIRY MANURE APPLIED**

| Year | Total solids, percent | Total N | Org-N | $NH_4$-N | P | K | Na | Cl |
|---|---|---|---|---|---|---|---|---|
| | | | | — percent (dry wt. basis) — | | | | |
| 1971 | 16.83 | 2.99 | 2.29 | 0.70 | 0.81 | 1.39 | 0.51 | 1.76 |
| 1972 | 20.68 | 2.95 | 2.23 | 0.72 | 0.81 | 1.75 | 0.37 | 2.20 |
| 1973 | 20.34 | 2.88 | 2.19 | 0.69 | 0.91 | 1.58 | 0.32 | 1.19 |

Early plant growth was determined approximately 6 weeks after planting by randomly selecting 10 plants, removing them by cutting at the soil surface and drying. Samples of the leaf opposite and below the ear leaf were taken at the tasseling stage. In addition, fodder samples (above ground portion minus the ear) were obtained at physiological maturity in 1974. All tissue samples were chemically analyzed by emission spectroscopy with N being determined by Kjeldahl analysis. Grain yields were obtained by hand harvesting measured areas.

## RESULTS AND DISCUSSION

### Manure Application

With some exceptions the entire production of manure by a 60-cow dairy herd was applied in weekly intervals and incorporated in this experimental area each summer. Manure application rates for 1971, 1972 and 1973 totaled 231, 215 and 324 T/ha, respectively, on a dry basis. These rates resulted in 446 and 770 T/ha applied during the 1971-72 and 1971-73 periods, respectively (Table 2).

Nutrient application rates of most elements were quite high (Table 2). Of the 22,590 kg N/ha applied over the 3-year period, 76 percent was in an organic form with only trace amounts of nitrate-N present. Ammonium-N constituted 24 percent of the total N; however, some may have been lost thru volatilization before incorporation. Relatively small amounts of the micronutrients Mn, Zn, Cu & B were applied.

Precipitation amounts varied each year (Table 3) and consequently did affect the application rates. Above-normal precipitation in June 1971 restricted manure application and incorporation somewhat. Dry conditions in August and September 1971 limited decomposition of the manure. This resulted in poor incorporation because of traction problems on the loose, manure-covered, dry soil. Hence, odors per-


The authors are: G. W. RANDALL, Soil Scientist, R. H. ANDERSON, Superintendent, Southern Experiment Station and P. R. GOODRICH, Assistant Professor, Agricultural Engineering, University of Minnesota. A contribution from the Minnesota Agricultural Experiment Station.

TABLE 2. MANURE RATES AND THE RESULTANT NUTRIENT
AMOUNTS APPLIED

|  | Period | |
|---|---|---|
|  | 1971-72 | 1971-73 |
| Manure rate, T/ha* | 446 | 770 |
| Nutrients, kg/ha |  |  |
| Total N | 13230 | 22590 |
| Org-N | 10070 | 17180 |
| $NH_4$-N | 3160 | 5400 |
| $NO_3$-N | 4 | 6 |
| P | 3610 | 6550 |
| K | 6960 | 12090 |
| Ca | 5210 | 8780 |
| Mg | 2190 | 3650 |
| Na | 1970 | 3010 |
| Cl | 8790 | 12650 |
| Fe | 880 | 1480 |
| Al | 680 | 1200 |
| Mn | 63 | 105 |
| Zn | 37 | 68 |
| Cu | 10 | 16 |
| B | 20 | 29 |

*dry weight basis.

TABLE 3. PRECIPITATION DURING THE MANURE
APPLICATIONS PERIODS

| Month | 1971 | 1972 | 1973 | 30-year normal |
|---|---|---|---|---|
|  | ————————— mm ————————— | | | |
| May | 80 | 109 | 159 | 98 |
| June | 135 | 43 | 86 | 121 |
| July | 70 | 197 | 118 | 102 |
| August | 12 | 39 | 147 | 91 |
| September | 41 | 135 | 108 | 88 |
| TOTAL | 338 | 523 | 618 | 500 |

sisted and fly problems became abundant. Consequently, manure applications were reduced. Wet soil conditions in July 1972 prevented manure application for 33 consecutive days in July and early August. In 1973, weather conditions were almost ideal. Manure applications were reduced only during a wet period in August.

Incorporation of manure into the soil is necessary to promote mineralization of the manure and to reduce volatilization, fly and odor problems. To obtain optimum incorporation, the areas receiving manure were tilled 15, 24 and 27 times during 1971, 1972 and 1973, respectively.

These results indicate that with optimum weather conditions approximately 350-400 T/ha of dairy manure (dry basis) could be surface-applied and incorporated into this soil each summer. If precipitation departed from normal excessively, we could expect application rates to be limited to less than 200 T/ha. Under those conditions incorporation methods would be restricted severely. Thus, a reduction in the number and/or rate of application would be necessary.

### Nutrients in Soil

Ammonium, phosphorus, potassium and sodium accumulated primarily in the upper 30-cm until 1974 when some appeared in the 30-60-cm layer (Table 4). This movement perhaps could have been a result of numerous tillage operations which may have physically incorporated some of the surface layer with soil at this depth. Accumulations of P and K were very high with lesser amounts of Na. Because $NH_4$ concentrations were generally low and $NO_2$ levels

were less than 1 ppm, toxicity to corn seedlings was never observed.

Largest accumulations of nitrate were also in the surface 30-cm layer. However, significant amounts had moved to 150 cm after 1 year (231 T/ha) and to 180 cm in 1973 and 1974 following the 446 and 770 T/ha rates. Slight but variable amounts were found at depths between 180 and 270 cm. Nitrate concentrations were unusually low in the 30-90-cm layer in 1973 and in the 30-120-cm layer in 1974 (446 T/ha). Even with 770 T/ha, nitrate levels at these depths were lower than after 231 T/ha (1972). In addition, total nitrate accumulation in the profile was less in 1973 and in 1974 at both rates than in 1972 after only 231 T/ha. This would indicate that denitrification losses of $NO_3$-N were occurring at these depths in the years of higher precipitation.

TABLE 4. NUTRIENT CONCENTRATIONS IN THE SOIL AS
INFLUENCED BY MANURE RATE AND TIME FOLLOWING
APPLICATION

|  | Time of sampling and manure applied, T/ha | | | | |
|---|---|---|---|---|---|
|  |  | 5/72 | 4/73 | 5/74 | |
| Depth, cm | Initial* | 231 | 446 | 446 | 770 |
| **$NH_4$-N, ppm** | | | | | |
| 0-30 | 6 | 21 | 17 | 8 | 8 |
| 30-60 | 4 | 4 | 10 | 10 | 10 |
| 60-90 | 4 | 4 | 6 |  |  |
| 90-120 | 4 | 5 | 5 |  |  |
| **$NO_3$-N, ppm** | | | | | |
| 0-30 | 16 | 53 | 34 | 44 | 44 |
| 30-60 | 12 | 40 | 8 | 7 | 31 |
| 60-90 | 8 | 32 | 11 | 4 | 20 |
| 90-120 | 6 | 23 | 14 | 8 | 17 |
| 120-150 | 4 | 15 | 12 | 13 | 12 |
| 150-180 | 4 | 8 | 10 | 10 | 11 |
| 180-210 | 3 | 6 | 6 | 7 | 9 |
| 210-240 |  |  | 5 | 5 | 7 |
| 240-270 |  |  | 5 | 3 | 6 |
| 270-300 |  |  | 4 | 3 |  |
| **Cl, ppm** | | | | | |
| 0-30 | 92 | 317 | 112 | 61 | 84 |
| 30-60 | 51 | 105 | 146 | 60 | 85 |
| 60-90 | 47 | 44 | 116 | 60 | 74 |
| 90-120 | 34 | 40 | 93 | 75 | 86 |
| 120-150 | 24 |  | 50 | 41 | 96 |
| 150-180 | 21 |  | 43 | 41 | 51 |
| 180-210 | 14 |  | 32 | 30 | 32 |
| 210-240 |  |  | 32 | 24 | 18 |
| 240-270 |  |  | 25 | 22 | 11 |
| 270-300 |  |  | 12 | 12 |  |
| **Exch. K, ppm** | | | | | |
| 0-30 | 345 | 755 | 1510 | 1595 | 2300 |
| 30-60 | 245 | 220 | 170 | 235 | 485 |
| 60-90 | 200 | 170 | 125 | 120 | 110 |
| 90-120 | 185 | 160 | 105 | 115 | 105 |
| **Exch. Na, ppm** | | | | | |
| 0-30 | 60 | 130 | 140 | 135 | 175 |
| 30-60 | 45 | 48 | 80 | 110 | 140 |
| 60-90 | 43 | 33 | 36 | 63 | 100 |
| 90-120 | 43 | 36 | 32 | 46 | 65 |
| **Bray no. 1 ext. P, ppm** | | | | | |
| 0-30 | 30 |  | 190 | 285 | 415 |
| 30-60 | 15 |  | 20 | 16 | 43 |
| 60-90 | 4 |  | 4 | 4 | 6 |
| 90-120 | 4 |  | 4 | 2 | 5 |

*Only the 73 T/ha which had been applied the previous year — 1970.

Chlorides showed somewhat similar rates of movement as did the nitrates; however, they were more evenly distributed throughout the profile. By 1973 chlorides had moved to 270 cm. Chloride concentrations in 1974 following 770 T/ha were not markedly higher than from the 446 T/ha rate. This probably reflected lower Cl concentrations in the



MANAGING LIVESTOCK WASTES

manure applied in 1973.

In considering the magnitude of the N application rates in relation to the nitrate concentrations found to date, the movement rates of the $NO_3$ and Cl ions and the surface accumulations of P, K and Na, we can foresee long-term effects of these manure rates on the soil. Nitrogen transformations such as mineralization and denitrification will have a pronounced effect on these results.

### Nutrients in Soil Solution

Nitrate concentrations in the soil solution at 180 cm have been consistently higher from the fallowed area than from the manured areas (Table 5). However, the concentrations in 1974 from the 446 T/ha treatment were approaching the levels from the fallow area. Lowest $NO_3$-N concentrations were found in 1974 following 770 T/ha. These low $NO_3$-N concentrations again suggest that denitrification reactions may be present during periods of high precipitation resulting in a loss of $NO_3$-N and/or that mineralization of the organic N to inorganic forms of N may be slow. This premise is substantiated by the chloride data.

Chlorides, which do not undergo transformations in the soil, were significantly higher from the 446 T/ha treatment than from the fallow area. In 1974, Cl levels were lower, probably due to the low Cl concentration in the manure

applied in 1973 and to below normal precipitation amounts resulting in slow downward movement. In addition, with 770 T/ha of manure applied the water holding capacity of the surface soil was increased markedly. As a result very little water moved down through the soil profile, which caused some difficulty in obtaining soil solution samples.

Although $NH_4$-N and $PO_4$-P concentrations were small, slightly higher levels were found from the manure areas.

### Early Growth and Yield of Corn

Early growth and grain yield of corn was depressed significantly by the manure treatment in 1973 (Table 6). This perhaps was due to the high salt content from the manure in the soil which increased the osmotic pressure of the soil solution and resulted in reduced availability of soil moisture. During periods of dryness (late June and again in mid-July), there was evidence of drought stress as shown by the rolling and curling of leaves from the manure-treated areas. This was not evident on the fallow area.

Neither early growth nor grain yield was affected significantly in 1974. A hot, dry period at tasseling and an extremely early frost (September 3) may have negated any yield differences. Silage yields were depressed slightly by the manure treatments.

Yields from the areas which received very high rates of manure were quite respectable each year (9542 kg/ha or 152 bu/A in 1973). Therefore, it appears that these areas can be brought back into production the year following manure application without seriously limiting corn yields.

### Nutrients in Corn Tissue

The macronutrients (N, P and K) in the corn leaf tissue were not affected by the manure treatments (Table 7). Even though very high soil tests did result from the manure rates, luxury consumption did not occur. Calcium in both years and magnesium in 1974 were reduced by the manure,

*(Continued on page 621)*

TABLE 5. NUTRIENT CONCENTRATIONS IN THE SOIL SOLUTION AT 180 cm AS INFLUENCED BY HIGH MANURE RATES

| Date | Fallow | Manure, T/ha | |
|---|---|---|---|
| | | 446 | 770 |
| | | $NO_3$-N, ppm | |
| 10/72 | 51 | 25 | |
| 4/73 | 34 | 21 | |
| 7/73 | 55 | 27 | |
| 8/73 | 49 | 38 | |
| 9/73 | 43 | 37 | |
| 6/74 | 52 | 46 | 25 |
| 8/74 | 42 | 35 | 11 |
| | | Cl, ppm | |
| 10/72 | 83 | 278 | |
| 4/73 | 78 | 186 | |
| 7/73 | 87 | 211 | |
| 8/73 | 87 | 266 | |
| 9/73 | 102 | 297 | |
| 6/74 | 105 | 238 | 111 |
| 8/74 | 114 | 226 | 73 |
| | | $NH_4$-N, ppm | |
| 10/73 | 0.33 | 0.67 | |
| 4/73 | 0.03 | 0.12 | |
| 7/73 | 0.14 | 1.34 | |
| 6/74 | 0.08 | 0.28 | 0.60 |
| | | $PO_4$-P, ppm | |
| 10/72 | 0.026 | 0.085 | |
| 4/73 | 0.004 | 0.009 | |
| 7/73 | 0.009 | 0.119 | |
| 6/74 | 0.061 | 0.140 | 0.285 |

TABLE 6. EARLY GROWTH AND YIELD OF CORN AS INFLUENCED BY MANURE

| Treatment | Early plant growth, g/10 plants | | Grain yield, kg/ha | | Silage yield, T/ha, DM |
|---|---|---|---|---|---|
| | 1973 | 1974 | 1973 | 1974 | 1974 |
| Fallow | 85 | 39 | 11965 | 8142 | 15.89 |
| Manure (446 T/ha) | 60 | 49 | 9542 | 7527 | 14.95 |
| Manure (770 T/ha) | | 46 | | 7351 | 13.44 |
| Significance* | † | ns | † | ns | + |

*Determined by t test in 1973 and ANOVA-nested analysis in 1974; † and + = significant at the 95 and 90 percent levels of probability, respectively.

TABLE 7. NUTRIENT CONCENTRATIONS IN THE CORN LEAVES AS INFLUENCED BY HIGH MANURE RATES

| Treatment | N | P | K | Ca | Mg | Fe | Mn | Zn | Cu | B |
|---|---|---|---|---|---|---|---|---|---|---|
| | —————percent————————— | | | | | —————————ppm————————— | | | | |
| **1973** | | | | | | | | | | |
| Fallow | 2.70a* | 0.31a | 2.94a | 0.61b | 0.24a | 142a | 47a | 26a | 4.5a | 18a |
| Manure (446 T/ha) | 2.79a | 0.31a | 2.96a | 0.44a | 0.20a | 139a | 106b | 42b | 4.8a | 19a |
| **1974** | | | | | | | | | | |
| Fallow | 3.03a | 0.31a | 3.37a | 0.77c | 0.28b | 248a | 63a | 29a | 7.1a | 13a |
| Manure (446 T/ha) | 3.22a | 0.33a | 3.31a | 0.65b | 0.28b | 237a | 79a | 28a | 7.4a | 18ab |
| Manure (770 T/ha) | 3.11a | 0.36a | 3.52a | 0.52a | 0.23a | 239a | 123b | 47b | 8.0a | 27b |

*Values accompanied by the same letter within a column for each year are not statistically different at the 0.10 level of probability.

MANAGING LIVESTOCK WASTES

# Composition of Poultry Manure and Effect of Heavy Application on Soil Chemical Properties and Plant Nutrition, British Columbia, Canada

A. A. Bomke, L. M. Lavkulich

THE Lower Fraser Valley of British Columbia, Canada, has a poultry population of approximately 8.5 million birds, mostly concentrated on small farms of usually less than five hectares. This is also the most densely populated area of the province with over 1 million people in Vancouver and other cities of the Lower Fraser Valley. In addition, this area supports the most intensive agriculture in the province with significant quantities of vegetables, fruits and forages. The climate in the region is inshore maritime with an abundance of rain in the winter followed by a deficiency in the summer. The mean annual temperature is about 10 C with maximum and minimum temperatures around 38 and -21 C (B.C.D.A. 1968).

Waste disposal has become an acute problem from the standpoint of environmental quality in the Lower Fraser Valley. The application of large amounts of animal wastes to land and the subsequent growth of crops also raises the question of crop quality. Thus, a program was initiated to assess the character of poultry manure under the practical existing management conditions and the effects of the existing methods of disposal. Objectives of the study were to determine:

1   The effect of heavy application of poultry manure on soil chemical properties and crop composition, and

2   Plant nutrient content of poultry manure deposited and stored in deep pits under laying cages.

## METHODS AND MATERIALS

The four soil series employed in this study were Abbotsford, Whatcom, Ross and Laxton (Runka and Kelley 1964). The Abbotsford soil is a podzol (Haplorthod) derived from loess overlying outwash gravels. The loess is generally less than 60 cm in thickness and usually of a loam to silt loam in texture. The Whatcom soil is a podzol developed from glacio-marine deposits. The texture is generally silt loam at the surface to silty clay loam in the subsoil. The Laxton soil is also a podzol developed from shallow loess overlying duned fine sand. Surface textures are loam to silt loam overlying sand at depths of less than 50 cm. The Ross soil is classified as a gleysol (Typic Humaquept) occurring on upland alluvial floodplains derived from glacio-marine materials. Surface textures range from silt loam to silty clay loam with gravels and sands occurring at depths greater than 60 cm.

During June, July and August, 1974 soil and plant samples were collected from manure disposal sites where no attempt was made to optimize plant growth, and grass hay and raspberry fields receiving poultry manure. Soil sampling was done on 15 cm depth increments, samples were divided

immediately after sampling and a part of the sample was oven dried at 65 C for $NH_4$-N, $NO_3$-N and total N analyses. The other portion of the sample was air dried for determination of exchangeable K, Ca, Mg and Na, available P, pH, and electrical conductivity. Samples were stored in plastic specimen containers. Determination of $NH_4$-N and $NO_3$-N was done by the semi-microkjeldahl technique following extraction in 2N KC1. Total N determinations were done by standard macrokjeldahl methods. Available phosphorus was determined colorimetrically following extraction with 0.03N $NH_4$F in 0.025N HC1. Exchangeable Ca, Mg, K and Na were extracted with 1N ammonium acetate (pH 7.0) and determined by atomic absorption spectrophotometry. Soil pH measurements were in a suspension of 1 part soil in 2 parts 0.01M $CaC1_2$. Electrical conductivity was determined on a saturated soil paste extract.

Grasses (whole plant samples) mostly orchardgrass (*Dactylis glomerata L.*) and Kentucky bluegrass (*Poa pratensis L.*) were collected during June, 1974 from disposal areas where possible, and hayfields receiving poultry manure. Raspberry (*Rubus daeus L.* "Williamette") leaves were collected during mid August, subsequent to fruit harvest. Each sample consisted of approximately 15 of the youngest mature leaves about 40 cm from the tips on lateral primocanes. Plant samples were dried at 80 C, ground and stored in tightly capped glass bottles. Total N was determined by semi-microkjeldahl, total P by colorimetry, and total K, Ca, Mg, Fe, Mn and Zn by atomic absorption spectrophotometry.

Manure samples were collected in regular depth increments from deep storage pits under laying cages in houses on two different farms. Manure had been stored up to 10 months in House I and 12 months in House II. A spade was used to expose a "profile" approximately one meter from the edge of the pit. Samples from the desired depths were collected from two "profiles" in House I and three profiles in House II. The samples were placed in plastic bags and frozen prior to laboratory analyses. Analyses of manure samples for total N, P, K, Ca, Mg, Na, Mn, and Zn were done employing similar procedures as for plant samples, except that total N analysis was performed on moist samples and a subsample dried for moisture determination.

Methods of soil, plant and manure analysis were adopted from Black (1964). In the field each manure disposal site was examined and sampled. Adjacent areas of the same soils which had not received manure applications were also examined and sampled to serve as a "control". Actual application rates of manure to the sites are unknown.

## RESULTS AND DISCUSSION

### Chemical Analysis of Heavily Manured Soils

Data are presented on pH, electrical conductivity,

---

The authors are: A. A. BOMKE and L. M. LAVKULICH, Soil Science Dept., University of British Columbia, Vancouver, BC, Canada.

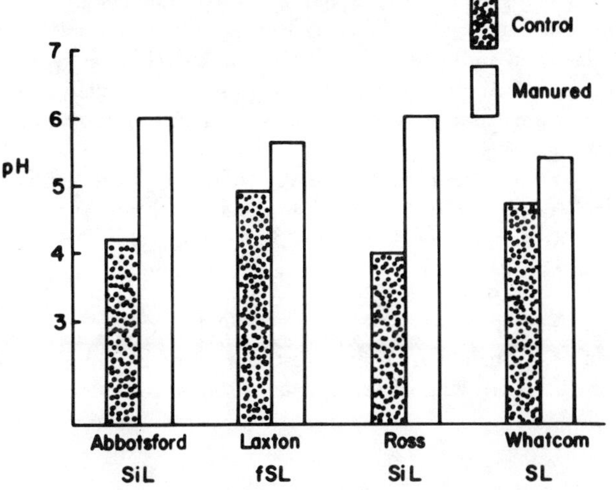



FIG. 1 Average pH of samples from control and manured areas on the four study soils.

exchangeable cations, available P (Bray $P_1$), $NH_4$-N and $NO_3$-N levels in soils from manure disposal sites and nearby control sites for each of four representative soils. Some difficulty was experienced in locating a "true" control in this heavily populated and intensively farmed area; hence there are some anomalous values for control soils. Chemical data for manure disposal sites tend to be quite variable, the result of extremely non-uniform manure application.

In agreement with Perkins et al. (1964), additions of poultry manure increased soil pH by as much as two pH units (4.0 to 6.0) in the Ross silt loam sites and by at least 0.7 pH units in the other sites (Fig. 1). This pH increase probably reflects release of $NH_3$ from the manure and should last until conditions become favorable for nitrification. High levels of $NH_4$-N and little $NO_3$-N in the manured Ross soil indicate that nitrification has not proceeded to any extent in that soil, and hence the large pH increase over the control site remains.

During rainless periods of the summer of 1974, white salt deposits were observed on the soil surface of some of the more heavily manured areas of the disposal sites. This observation was substantiated by electrical conductivity measurements shown in Table 1. Electrical conductivity was increased to a depth of 45 cm in all heavily manured

## TABLE 1. ELECTRICAL CONDUCTIVITY OF SOIL SATURATED-PASTE EXTRACTS OF SAMPLES FROM THE PROFILES OF THE FOUR STUDY SOILS

| Soil | Depth | Control | Manured | |
|---|---|---|---|---|
| | | | mean | range |
| | cm | | mmho/cm | |
| Abbotsford SiL | 0-15 | 0.14 | 2.40 | 0.72-4.93 |
| | 15-30 | 0.10 | 1.44 | 0.46-3.49 |
| | 30-45 | 0.08 | 1.76 | 0.57-4.10 |
| | 45-60 | 0.09 | 2.49 | 0.50-4.22 |
| Laxton fSL | 0-15 | 0.21 | 0.78 | 0.46-1.10 |
| | 15-30 | 0.22 | 0.66 | 0.18-1.13 |
| | 30-45 | 0.14 | 0.68 | 0.09-1.28 |
| | 45-60 | 0.05 | 0.13 | 0.07-0.19 |
| Ross SiL | 0-15 | 0.36 | 2.45 | 0.40-3.95 |
| | 15-30 | 0.12 | 1.72 | 0.22-3.38 |
| | 30-45 | 0.10 | 0.70 | 0.18-1.65 |
| | 45-60 | 0.15 | 0.27 | 0.16-0.41 |
| Whatcom SiL | 0-15 | 0.25 | 0.61 | 0.10-1.51 |
| | 15-30 | 0.20 | 0.43 | 0.17-0.88 |
| | 30-45 | 0.09 | 0.42 | 0.23-0.66 |
| | 45-60 | 0.11 | 0.31 | 0.26-0.36 |

## TABLE 2. AVAILABLE P IN CONTROL AND MANURED PROFILES FROM THE FOUR STUDY SOILS

| Soil | Depth | Control | Manured | |
|---|---|---|---|---|
| | | | mean | range |
| | cm | | ppm | |
| Abbotsford SiL | 0-15 | 68 | 1126 | 509-1484 |
| | 15-30 | 29 | 456 | 101-1216 |
| | 30-45 | 25 | 182 | 94- 359 |
| | 45-60 | 31 | 175 | 89- 327 |
| Laxton fSL | 0-15 | 250 | 585 | 430- 820 |
| | 15-30 | 118 | 382 | 188- 605 |
| | 30-45 | 36 | 313 | 24- 174 |
| | 45-60 | 9 | 104 | 6- 172 |
| Ross SiL | 0-15 | 17 | 379 | 284- 520 |
| | 15-30 | 11 | 277 | 228- 361 |
| | 30-45 | 18 | 106 | 86- 133 |
| | 45-60 | 11 | 81 | 56- 106 |
| Whatcom SiL | 0-15 | 46 | 27 | 19- 40 |
| | 15-30 | 37 | 9 | 6- 16 |
| | 30-45 | 51 | 5 | 2- 8 |
| | 45-60 | 69 | 16 | 4- 28 |

soils relative to the control. Little increase in salt content, as measured by electrical conductivity, was noted below 45 cm. With one or two exceptions salt content is not extremely high and even in those cases would limit plant growth only until sufficient precipitation occurs to cause leaching below the root zone. The movement of significant quantities of salts into groundwater may be a potential problem.

In general, an extremely high level of available P (Bray $P_1$) is a good indication of heavy manure applications (Table 2). Heavily manured areas on the Abbotsford, Laxton and Ross soils showed relatively high available P levels in the O - 15 cm depth of soil. The Whatcom site, which had received less extensive manure additions, showed less available P than the control for that soil. With the exception of the Whatcom soil, heavy manuring resulted in increases in available P at all depths sampled. This indicates some P leaching in these soils as mechanical incorporation would have only influenced the upper 0 - 30 cm zone.

All manure disposal sites included areas with higher total N concentrations than comparable controls (data not shown). The largest N increase with manuring occurred on the Abbotsford site where total N increased from 0.13 percent to an average value of 0.59 percent for the disposal area. Increases in inorganic N fractions such as $NH_4$-N or $NO_3$-N are usually quite marked with manure addition (Fig.

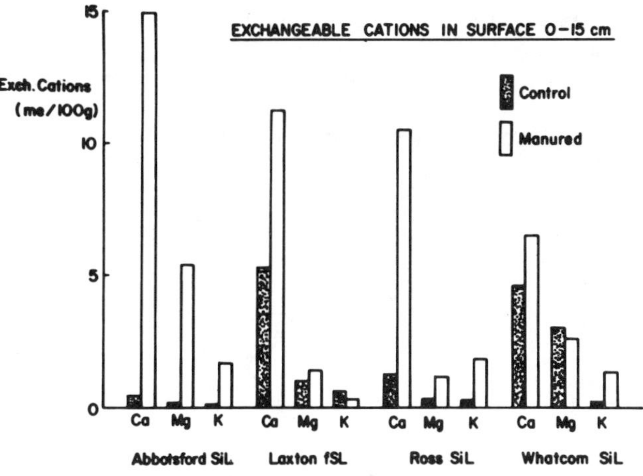



FIG. 2 Average concentration of exchangeable cations in samples from control and manured areas on the four study soils.

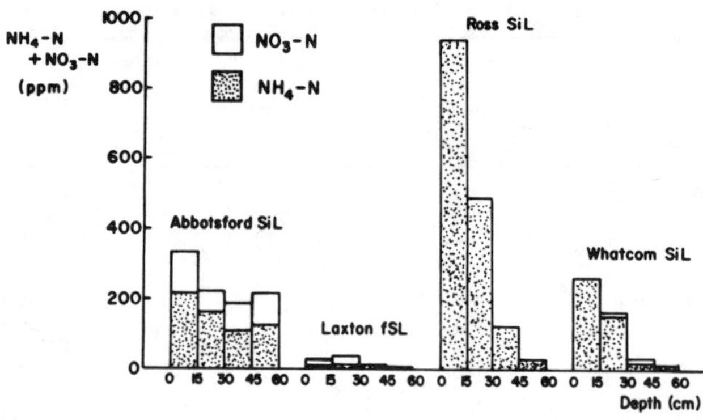



FIG. 3 Increase in $NH_4$-N and $NO_3$-N concentrations with manuring on the four study soils.

3). Levels of $NH_4$-N found in manured sites were usually higher than $NO_3$-N possibly because of the consumption of oxygen by the decomposing manure and resultant unfavorable conditions for nitrification. Also because the Abbotsford, Laxton and Ross soils are quite permeable, $NO_3$ leaching is also a possibility. This is indicated by $NO_3$-N concentration of 336.9 ppm at the 45 - 60 cm depth in the manured Abbotsford loam. Increases in $NH_4$-N below the surface 0 - 30 cm occurred in all four manured soils relative to control sites. Average $NH_4$+$NO_3$ concentrations in the surface 0 - 50 cm of 306.4, 58.9, 1013.0, and 290.3 ppm N in the Abbotsford, Laxton, Ross and Whatcom soils, respectively, indicate a potential threat to quality of drainage water, especially if conditions become favorable for nitrification. Leaching would not be expected to be a serious factor in the Whatcom silt loam because of its slowly permeable nature.

## PLANT COMPOSITION

The orchard grass and bluegrass samples represented in Table 3 were collected from the Laxton and Whatcom sites, respectively. Comparable samples were not obtained from the Abbotsford or Ross sites because there was either no vegetation or a grass-weed mixture.

Since most of the grasses in the area are cut for hay or grazed, quality of the plant material for animal nutrition is important. In this respect then, it appears that Mg content averaging 0.16 percent and 0.14 percent for the bluegrass and orchard grass samples, respectively, falls below the 0.20 percent level commonly referred to as critical (Grunes et al. 1970) for ruminants. The ratio of K to Ca + Mg could also be a factor in decreasing Mg availability as reported by Wilkinson et al. (1971) in Georgia. According to Grunes et al. (1970) the susceptibility of animals to grass tetany increases at K/Ca + Mg ratios greater than 2.2. Average K/Ca + Mg ratios of 3.5 and 2.4 were found in this study for the bluegrass and orchard grass samples, respectively. This could indicate potential Mg deficiencies for livestock fed from heavily manured forage grasses. Controlled experimentation has been initiated to substantiate this and other findings.

The most significant problem evident with the "Williamette" raspberry leaf samples is the high N content of 3.3 percent as compared to the 2.2 percent average found by John (1972) for this cultivar at a comparable sampling date. This was reflected in excessive cane growth and high disease incidence in plants from manured fields. Other raspberry composition data are similar to those found by John (1972) for plants growing under optimum soil fertility conditions.

## MANURE COMPOSITION

Although fans installed in the sides of each house dried the manure somewhat, moisture content remained quite high in both cases. Average moisture contents of 67.2 percent and 69.1 percent for House I and II, respectively were quite similar although moisture content varied by as much as 12 percent within House I and 10 percent within House II (Table 4 and 5). Manure moisture content in House I increased within the lower 40 cm of the pile probably due to leakage of the watering system early in the laying cycle.

N content was quite different between houses averaging 3.12 percent in House I and 5.34 percent in House II (Tables 4 and 5). Differences in N content of manure between the two houses can be related to ration and possibly the fact that a different strain of chicken was housed in each of the two houses studied. Percent protein in the ration fed to hens in House II was kept at a constant 16 percent during the laying cycle. This is reflected in the relatively similar N content throughout the pile. In contrast, hens in House I were fed a ration of 18 percent for the initial four months, 16 percent protein for the next two months and 14 percent protein for the remainder of the laying cycle. Manure in House I was higher in N in the lower 30 cm of the pile in manure which was produced from the highest protein ration. Higher percent N in fresh manure at the top of the pile is difficult to explain, although volatilization of $NH_3$ may still be lowering N content in the fresh material.

Other elements show a wide range in values within each manure pile and consistent trends are difficult to distinguish. Examples would be Ca in House I and Fe and Zn in both houses. K, Na, and Mn show relatively uniform distribution with depth in both houses. P concentration averaged

TABLE 3. COMPOSITION OF RASPBERRY LEAVES (SAMPLED AUGUST 14) AND WHOLE PLANT GRASS SAMPLES (SAMPLED JUNE) WITH HEAVY MANURE APPLICATION OF POULTRY MANURE

| Sample (no.) | N | P | K | Ca | Mg | K/Ca+Mg meq. basis | Fe | Mn | Zn |
|---|---|---|---|---|---|---|---|---|---|
| | | | percent | | | | | ppm | |
| Bluegrass (7) | | | | | | | | | |
| average | 3.2 | 0.46 | 3.6 | 0.22 | 0.16 | 3.5 | 491 | 129 | 41 |
| range | 1.9-3.5 | 0.25-0.50 | 1.9-4.0 | 0.16-0.27 | 0.11-0.20 | 3.1-4.1 | 90-1290 | 87-149 | 31-61 |
| Orchard grass (6) | | | | | | | | | |
| average | 2.2 | 0.31 | 2.2 | 0.25 | 0.14 | 2.4 | 343 | 45 | 29 |
| range | 1.0-3.7 | 0.21-0.42 | 1.4-3.2 | 0.12-0.37 | 0.11-0.18 | 1.5-2.8 | 80-1290 | 23-66 | 21-32 |
| Raspberries | | | | | | | | | |
| average | 3.3 | 0.18 | 1.6 | 0.72 | 0.37 | — | 216 | 194 | 22 |
| range | 2.8-3.5 | 0.14-0.22 | 1.2-1.8 | 0.62-0.95 | 0.34-0.40 | — | 100-330 | 80-348 | 18-31 |
| John (1972) | 2.2 | 0.19 | 1.6 | 0.90 | 0.37 | | 431 | 159 | 23 |



MANAGING LIVESTOCK WASTES

TABLE 4. COMPOSITION ON DRY WEIGHT BASIS OF CHICKEN (LAYER)
MANURE WITH DEPTH IN DEEP PIT HOUSE I

| Depth cm | Moisture | N | P | K | Ca | Mg | Na | Fe | Mn | Zn |
|---|---|---|---|---|---|---|---|---|---|---|
| | | | | percent | | | | | ppm | |
| 0.10 | 71.4 | 4.55 | 2.41 | 2.2 | 2.20 | 0.67 | 0.48 | 1340 | 265 | 230 |
| 10-20 | 66.4 | 3.18 | 2.96 | 2.1 | 0.71 | 0.72 | 0.44 | 2205 | 330 | 360 |
| 20-30 | 70.7 | 2.23 | 3.80 | 2.6 | 1.06 | 0.84 | 0.44 | 1905 | 335 | 418 |
| 30-40 | 67.4 | 2.14 | 3.85 | 2.4 | 0.96 | 0.78 | 0.48 | 1520 | 320 | 378 |
| 40-50 | 66.0 | 2.21 | 4.58 | 2.4 | 1.05 | 0.96 | 0.50 | 1900 | 408 | 453 |
| 50-60 | 60.8 | 2.42 | 3.50 | 2.0 | 1.30 | 0.86 | 0.48 | 1030 | 376 | 413 |
| 60-70 | 61.2 | 3.24 | 3.85 | 1.9 | 1.22 | 0.92 | 0.50 | 1070 | 344 | 348 |
| 70-80 | 67.7 | 3.79 | 3.18 | 1.9 | 0.88 | 0.97 | 0.45 | 1210 | 342 | 313 |
| 80-90 | 72.8 | 4.32 | 3.26 | 2.2 | 0.84 | 0.72 | 0.46 | 790 | 276 | 293 |
| Average | 67.2 | 3.12 | 3.49 | 2.2 | 1.14 | 0.83 | 0.47 | 1441 | 333 | 331 |

3.17 percent and was present in relatively uniform levels in House II although P analyses in House I showed a large variability.

If one considers the average elemental composition values presented in Tables 6 and 7, it appears that quantity of N would be the most critical factor in determining rate of usage for crops. Total N per dry ton applied would vary from approximately 31 kg for House I manure to 53 kg for House II. For raspberries, a crop sensitive to excessive N application, use of a highly variable fertilizer source such as poultry manure is more difficult than with grasses, which can assimilate higher quantities of N without detriment to yield and crop quantity.

## CONCLUSIONS

Heavy manure applications greatly altered various soil chemical properties. These include increases in pH, electrical conductivity, available P, exchangeable cations, total N, $NH_4$-N and, in most cases, $NO_3$-N. Increases in salts as measured by electrical conductivity and inorganic N represent potential threats to water quality.

Plant composition showed a potential problem of Mg unavailability to ruminants utilizing grass from heavily manured areas. Heavily manured raspberries were found to have excessively high N concentration in leaves.

Manure analyses showed considerable difference between the two Houses sampled. Differences in N content could be related in part to differences in percent protein in rations.


## References

1   British Columbia Department of Agriculture. 1968. Climate of British Columbia. Victoria, British Columbia.

2   Black, C. A. 1965. Methods of soil analysis. Am. Soc. Agron. Special Publ. No. 9. Madison, Wisconsin.

3   Grunes, D. L., P. R. Stout, and J. R. Brownell. 1970. Grass tetany of ruminants. Adv. Agron. 22:331-374.

4   John, M. K. and H. A. Daubeny. 1972. Influence of genotype, date of sampling, and age of plant on leaf chemical composition of red raspberry (*Rubus idaeus L.*). J. Amer. Soc. Hort. Sci. 97(6):740-742.

5   Perkins, H. F., M. B. Parker and M. L. Walker. 1964. Chicken manure—its production, composition and use as a fertilizer. Georgia Exp. Sta. Bul. N.S. 123.

6   Runka, G. G. and C. C. Kelley. 1964. Soil survey of Matsqui Municipality and Sumas Mountain. Preliminary Rept. No. 6. British Columbia Dept. of Agriculture, Kelowna, B.C.

7   Wilkinson, S. R., J. A. Stuedemann, D. J. Williams, J. B. Jones, R N. Dawson, and W. A. Jackson. 1971. Recycling broiler house litter on tall fescue pastures at disposal rates and evidence of beef cow health problems. Proc. Int. Symp. Livestock Wastes, pp 321-324.


TABLE 5. COMPOSITION ON DRY WEIGHT BASIS OF CHICKEN (LAYER)
MANURE WITH DEPTH IN DEEP PIT HOUSE II

| Depth cm | Moisture | N | P | K | Ca | Mg | Na | Fe | Mn | Zn |
|---|---|---|---|---|---|---|---|---|---|---|
| | | | | percent | | | | | ppm | |
| 0-15 | 61.9 | 4.96 | 3.04 | 1.7 | 0.63 | 0.61 | 0.61 | 830 | 319 | 375 |
| 15-30 | 70.7 | 5.59 | 2.86 | 1.4 | 0.59 | 0.63 | 0.62 | 1216 | 339 | 458 |
| 30-45 | 71.9 | 5.62 | 3.08 | 1.6 | 0.78 | 0.70 | 0.55 | 1050 | 379 | 421 |
| 45-60 | 71.5 | 5.50 | 3.57 | 1.4 | 0.64 | 0.64 | 0.56 | 1363 | 387 | 511 |
| 60-75 | 69.3 | 5.05 | 3.28 | 1.2 | 0.91 | 0.72 | 0.46 | 1820 | 398 | 635 |
| Average | 69.1 | 5.34 | 3.17 | 1.5 | 0.71 | 0.66 | 0.56 | 1256 | 364 | 480 |

# An Overland Flow — Lagoon Recycle System as a Pretreatment of Poultry Wastes


M. R. Overcash, J. W. Gilliam, F. J. Humenik

ASSOC. MEMBER
ASAE

MEMBER
ASAE


WASTEWATER application on soils at liquid rates which are designed to promote controlled surface runoff comprise the system of overland flow treatment. Treatment mechanisms are soil sorption, microbial decomposition and crop uptake which take place at the soil and microbial mat interface with the atmosphere. Conditions which promote or optimize this interfacial reaction zone can produce a highly treated effluent under a variety of climatological conditions. Early application of overland flow treatment was for cannery or agricultural processing wastes (Stevens and Dunn 1969, C. W. Thornwaite Assoc. 1969). As favorable experiences developed subsequent systems for feedlot runoff wastes (Thomas 1974), raw and treated domestic wastes (Thomas 1974a, Butler 1974), and animal wastes were begun.

The microbial relationships and initial design and operational criteria were developed from extensive cannery systems studies (C. W. Thornwaite Assoc. 1969). However several aspects of system behavior and performance remained unanswered especially for animal manure wastewaters which contain nearly ten times the nitrogen levels of cannery wastewater. System response to wastes of different C:N ratios is unclear. Also the stabilization rates or distance required for treatment is undetermined. Thus a poultry manure overland flow site was established to determine treatment feasibility and mechanisms for animal waste.

## SYSTEM DESCRIPTION AND RESEARCH OBJECTIVES

The field installation was designed to be a liquid recycling pretreatment facility for the manure from one alley (1400 birds) of a caged layer house. The flush water (recycled from the terminal lagoon) was stored in an outside tank. Once per day the manure was prewetted with 700 l to 1100 l and then the entire flush tank (8200 l) emptied into the alley. This method kept the floor clean requiring only occasional localized scraping.

The flushed wastes passed through a coarse feather trap (necessary only for this pilot scale research project) and into a mixing tank which upon filling activated a control system which turned on the slurry mixing pump. After agitation the slurry was pumped to the terrace overland flow system. The overland flow (OLF) site was three terrace systems, each containing three nearly identical terraces in series, Fig. 1. Differences between terrace systems were: T1 and T3 — 8 percent terrace slope and T2 — 5.5 percent terrace slope.

The waste slurry was pumped from the mix-tank to a distribution box which regulated the amount added to each terrace system. The raw waste slurry, diluted with water from the terminal lagoon, was applied to the first terraces by means of a gated, ground-level trough. This arrangement allowed flexibility of liquid and waste concentration loadings. The first operational arrangement pumped raw wastewater to the terrace systems for a total of 110 minutes at which time the slurry pump stopped and the recycled lagoon water was pumped to the flush tank.

After flowing across each terrace the wastewater was collected in a trough, passed through a V-shaped flume, and redistributed through a gated, level trough. Distribution was adjusted to provide nearly uniform flow onto the start of each terrace. Flow volume was measured by stage recorders at each terrace. Following the third terrace in each system the effluent entered a lagoon which overflowed to a larger polishing pond. From this pond water was pumped to flush the house or to dilute the manure slurry.

The terraces were seeded with a mixture of Reed Canary, Redtop, and fescue grass seed. There was a variable


The authors are: M. R. OVERCASH, Assistant Professor, Biological and Agricultural Engineering Dept., J. W. GILLIAM, Associate Professor, Soil Science Dept., and F. J. HUMENIK, Associate Professor, Biological and Agricultural Engineering Dept., North Carolina State University, Raleigh.

Acknowledgements: This work was supported by the North Carolina Agricultural Experiment Station with supplemental funds from the US Dept. of Interior Matching Grant program through the Water Resources Research Institute of the University of North Carolina.


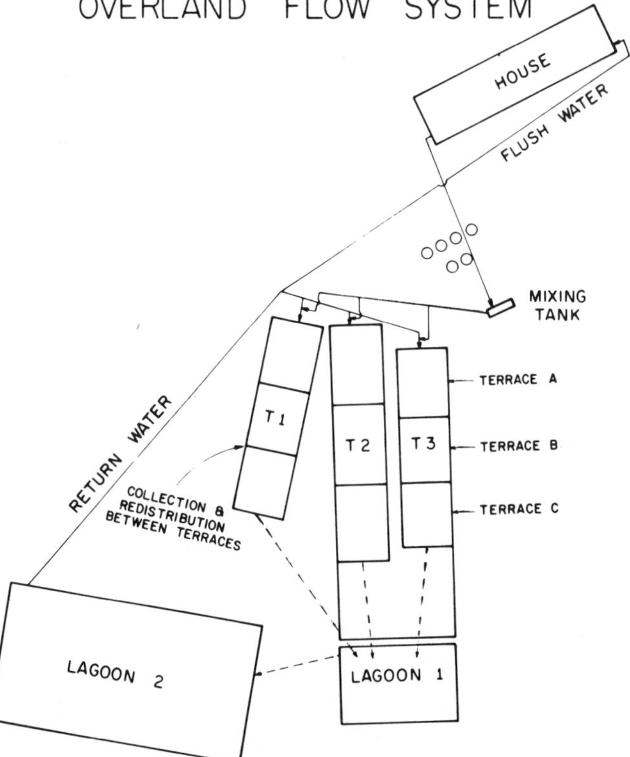


FIG. 1 Schematic of poultry manure overland flow—recycle system.

TABLE 1. POULTRY MANURE WASTEWATER INPUT TO
OVERLAND FLOW TERRACE SYSTEMS, BASED
ON 6.5 CM/WEEK OF WASTEWATER.

| | Concentration mg/l | Aeral loading Kg/ha/yr |
|---|---|---|
| Chemical oxygen demand | 500 | 17,000 |
| Total organic carbon | 95 | 3,200 |
| Total Kjeldahl nitrogen | 44 | 1,500 |
| Ammonia nitrogen | 15 | 500 |
| Orthophosphate | 10 | 340 |

mixture of these grasses from terrace to terrace but the most hardy grass under these experimental conditions should become dominant.

The total system was designed with the capability of varying the manure input, liquid loading, frequency and time of application, as well as obtaining samples at several system locations so that the following research objectives could be investigated:

1  Evaluate the overland flow distance required to achieve varying levels of treatment as measured by chemical oxygen demand (COD), total organic carbon (TOC), nitrogen removal or conversions, and phosphorus transformations.

2  Determine the pathways and amount of nitrogen removed per unit area in an OLF animal waste pretreatment system.

3  Evaluate the plant-soil system response to alternative loading strategies of animal waste.

## RESULTS AND DISCUSSION

A summary of the operating variables and wastewater characteristics used during the first phase of experimentation are given in Table 1. The terrace input concentration represented dilution water used to promote overland flow. As the wastewater moved down each terrace the potential for percolation or evapotranspiration losses existed. Water mass balances were used to determine moisture losses. During winter operation evapotranspiration losses were minimal so percolation was the principle water loss mechanism.

The hydraulic response of the system as measured at the end of each terrace is shown in Fig. 2. The time of initial flow at the 15.2, 30.4, and 45.6 m collection sites is about 20, 40, and 60 minutes, respectively, indicating fairly rapid flow conditions. Despite considerable caution and effort during construction each terrace had a degree of unevenness so that completely smooth sheet flow was not achieved. The micro-uniformity was however as good as could be obtained in any large scale installation. Despite unevenness, observations of soil wetness and microbial mat growth indicated that the entire soil surface was active in waste treat-

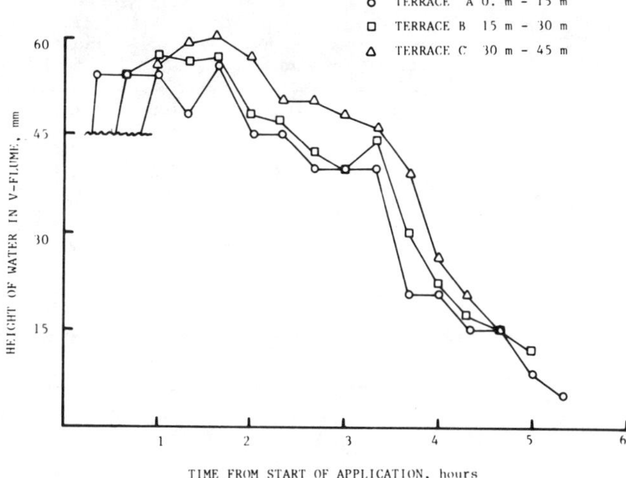



FIG. 2 Typical hydraulic response of one overland flow terrace system.

ment. The actual uniformity or sorptive capacity or biochemical conversions throughout a given terrace was not determined thus the need for ultrasmooth construction cannot be assessed. Observations to-date do indicate that to maintain smooth surface conditions for overland flow, care must be exercised in harvesting grass and using agricultural equipment.

The Cecil sandy clay loam soil present at the research site was not generally considered a highly impermeable soil. Estimated hydraulic conductivities have been reported (Lutz 1970) in the range of 1.5 to 10 cm per hour. Thus the capability of this soil type to produce large runoff volumes under the initial application rate (0.51 cm/hr) was investigated.

Water balances through the terrace system was performed by measuring flow at each terrace collection-distribution system over the total daily flow period. Typical results for two winter days on the 8 percent slope system were 6850 l, 6700 l, and 6500 l on 12/20/74 and 5700 l, 5010 l on 12/27/73 for total flow at distances of 15 m, 30 m and 45 m, respectively. Very little percolation occurred with 7 percent to 12 percent water loss through the system indicating effective soil sealing presumably by the organic loading and microbial growth. Based on these initial observations, the use of a wider range of soils with moderate hydraulic conductivities for animal waste application can be feasible for overland flow treatment.

Waste was added daily over a two hour period resulting in organics and ion concentration and mass reductions. The

TABLE 2. REPRESENTATIVE CONCENTRATION REDUCTIONS THROUGH
INCREMENTAL OVERLAND FLOW TREATMENT DISTANCES.

| Elapsed operation time | Flow distance, m | COD mg/l | COD percent reduction | O-PO$_4$-P mg/l | O-PO$_4$-P percent reduction | TKN mg/l | TKN percent reduction |
|---|---|---|---|---|---|---|---|
| initial liquid | 15 | 220 | 65 | 12 | 0 | 56 | 7 |
| | 30 | 180 | 71 | 10 | 0 | 45 | 25 |
| | 45 | 140 | 78 | 10 | 0 | 32 | 47 |
| 2 hrs | 15 | 103 | 83 | 5.4 | 55 | 24 | 60 |
| | 30 | 101 | 84 | 7.0 | 42 | 29 | 52 |
| | 45 | 109 | 82 | 9.2 | 24 | 34 | 44 |

**TABLE 3. AVERAGE SOIL WATER CONCENTRATIONS BENEATH TERRACES OF OVERLAND FLOW SYSTEM.**

| Terrace | Depth, m | Nitrate Initial mg/l | Nitrate After 5 months, mg/l | Chloride Initial mg/l | Chloride After 5 months, mg/l | Potassium Initial mg/l | Potassium After 5 months, mg/l |
|---|---|---|---|---|---|---|---|
| 0 - 15 m | 0.3 | 0.8 | 0 | 18.0 | 6.1 | 2.7 | 8.6 |
| | 0.9 | 4.6 | 4.2 | 12.2 | 11.8 | 2.0 | 6.9 |
| | 1.5 | 1.1 | 1.3 | 5.7 | 6.7 | 6.1 | 1.9 |
| | 2.4 | 3.3 | 2.0 | 4.2 | 8.9 | 2.7 | 3.5 |
| 15 - 30 m | 0.3 | 0 | 0 | 0.6 | 10.9 | 13.6 | 15.2 |
| | 0.9 | 2.0 | 1.4 | 8.6 | 10.5 | 4.2 | 2.3 |
| | 1.5 | 2.6 | 2.3 | 11.1 | 14.0 | 3.0 | 2.1 |
| | 2.4 | 5.7 | --- | 9.4 | 8.2 | 8.2 | 5.6 |
| 30 - 45 m | 0.3 | --- | 0.1 | --- | 22.0 | --- | 14.6 |
| | 0.9 | 8.3 | 0.6 | 22.8 | 24.0 | 8.5 | 3.4 |
| | 1.5 | 4.1 | 1.4 | 11.1 | 22.5 | 3.5 | 2.4 |
| | 2.4 | 4.4 | 2.7 | 6.0 | 6.5 | 4.2 | 3.4 |

treated effluent drained from the system and during the period between loadings (about 22 hours), drying and microbial reactions took place. As wastewater was again applied it could redissolve metabolic end products which would be carried as an initial pulse flushed from the terraces. Concentrations of the first liquid leaving the terraces indicated higher levels than samples taken after 2 hours of operation (Table 2). Under certain operating conditions the initial terrace effluent liquid may also contain any nitrate formed between loading events. The percent concentration reduction after each 15 m of overland flow, with the water balance showing little infiltration, indicated that successive terraces did not remove as much material as the first (Table 2). Organic removals of 90+ percent through the entire system have not been achieved for the operating conditions used. The cost of constructing successive terraces or flow distances must be balanced against the fact that even at 95 percent removals, the animal wastewater is still quite strong. Thus OLF can only be a pretreatment process and system length would be dictated by other costs associated with a total system.

Observation of fluid flow showed thick liquid layers with considerable physical transport of material. Thus system operation was changed to reduce the liquid application while maintaining the same waste input. Investigation of these conditions is currently underway.

There was a large response in grass cover crop yields with growth continuing through late November. Between December and February the grass remained green and thick but there was little growth. The nitrogen content of grass dry matter was 4.0 to 4.8 percent and with relatively uniform concentrations on all terraces. All terraces received a large amount of nitrogen for potential plant uptake based on wastewater concentrations (Table 2).

Percolation losses of nitrogen or other constituents was of primary environmental concern at the high waste nitrogen application employed. Thus four porous cups were located at depths of 0.3, 0.9, 1.5, and 2.4 meters at the center of each terrace to sample the soil solution.

Soil water samples were taken prior to system startup and monthly thereafter. Values for anions and cations likely to be most mobile after 5 months of operation and initial levels are shown in Table 3. Nitrate values were slightly reduced from the initial level as anticipated with the conditions of heavy waste loading. Chloride and potassium concentrations did not increase substantially from the initial levels. Considering the amount of poultry manure applied the lack of large increases in soil water concentra-

tions in the profile corroborated the soil-sealing hypothesis for this system.

Several oxidation-reduction measurements were taken in the soil profile to more fully elucidate the loss mechanisms, especially for nitrogen. These measurements were taken at depths between 2.5 cm and 75 cm and at three locations perpendicular to flow after about 14 m of overland flow (Table 4). Reduced conditions in comparison to background conditions were found at the surface layers. Deeper beneath the surface, the soil conditions were less reduced but oxidation potentials were below background levels. These data indicated that nitrification during overland flow at the organic and water application rates was improbable. Longer rest or less water application could improve the nitrification potential.

Conclusions from the system operated at conditions given in Table 1 were that:

1　Significant reductions in concentration of organics, nitrogen and phosphorus were achieved for the applied concentrated poultry manure wastewater. Greater removals of COD occurred than TKN with phosphorus removal the least efficient. These conclusions agree with data obtained with cannery and municipal systems.

2　Lower hydraulic rates may be needed for animal waste systems to achieve maximum treatment because large residual concentrations remained in terrace effluent for the high organic loadings and heavy liquid application (5-8 cm/wk) used in the first part of this study.

3　The total distance needed for the use of overland flow as a pretreatment based on economic considerations will probably be less than 45 m for animal waste systems because of the declining removal percentages found after 15 m of treatment. The exact distance needed will depend on

**TABLE 4. OXIDATION-REDUCTION POTENTIAL WITHIN OVERLAND FLOW TERRACE.**

| Depth, cm | Terrace location Left edge $E_{cal}$, mV | Terrace location center $E_{cal}$, mV | Terrace location Right edge $E_{cal}$, mV |
|---|---|---|---|
| 2.5 | - 90 | - 90 | - 90 |
| 5 | - 90 | - 90 | - 90 |
| 7.5 | - 90 | - 90 | - 90 |
| 15 | 0 | 0 | 0 |
| 30 | 20 | 70 | 10 |
| 45 | 175 | 175 | 175 |
| 61 | 175 | 180 | 175 |
| 76 | 175 | 180 | 175 |

treatment objectives.

4   Grass growth was very good under quite wet conditions resulting in uptake of about 40 percent of total nitrogen applied.

5   Effective soil sealing was achieved such that the majority of the applied water appeared as runoff and little downward movement of wastewater components was found.

**References**

1   Butler, R. M., E. A. Myers, J. N. Walters and J. V. Husted. 1974. Nutrient reduction in wastewater by grass filtration. ASAE Paper No. 74-4024. ASAE, St. Joseph, Mich. 49085. June.

2   Stevens, T. G. and G. G. Dunn. 1969. Grass filtration-pond stabilization of canning waste, a two-step process, Proc. 16th Ontario Industrial Waste Conf.:161-175.

3   Thomas, R. E. 1974. Feasibility of overland-flow treatment of feedlot runoff. EPA Publication, EPA-660/2-74-062.

4   Thomas, R. E. 1974a. Feasibility of overland flow for treatment of raw domestic wastewater. EPA-660/2-74-087.

5   Thornthwaite Assoc. 1969. An evaluation of cannery waste disposal by overland flow spray irrigation. Public in Climatology 22(2):1-73.

---

## Soil Affects from Dairy Manure

*(Continued from page 613)*

whereas manganese and zinc were increased significantly both years. Boron was also increased in 1974. Nutrient concentrations were not depressed below deficiency levels or increased to toxicity levels by the manure.

Fodder samples did contain relatively high concentra-

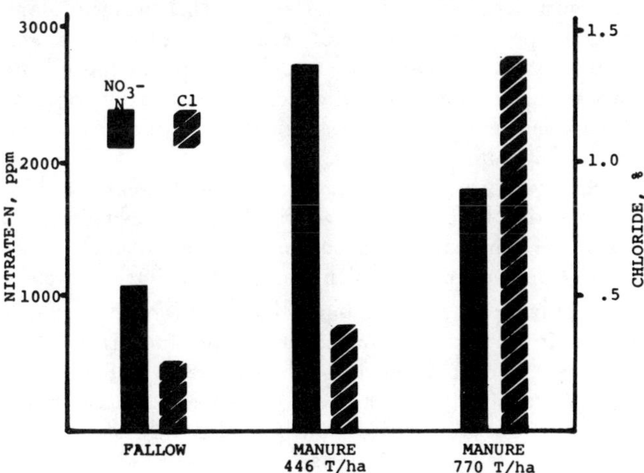


FIG. 1 Nitrate and chloride concentrations in the fodder in 1974 as influenced by manure treatment.

tions of $NO_3$-N, especially from the 446 T/ha treatment (Fig. 1). For silage purposes these $NO_3$-N values could be reduced approximately 50 percent because the ear, which would contain very low amounts of nitrate, would then be included. Nevertheless, a livestock farmer would not be advised to supply this forage as the sole source of feed for cattle. Chloride concentrations were substantially higher in corn grown on the 770 T/ha treatment than on the 447 T/ha or fallow treatments.

## CONCLUSIONS

1   Manure rates up to 400 T/ha (dry basis) could be applied and incorporated into the soil during the summer season under optimum weather conditions.

2   Because nutrient application rates greatly exceeded crop removal, high levels of immobile elements accumulated in the tilled layer with a downward movement of mobile elements. Manure mineralization rates along with denitrification will have pronounced, long-term effects on these soils.

3   Corn can be produced with little sacrifice in yield in the year following the manure application rates.

# Methods and Trends in Livestock Waste Management in Japan

Motoji Yamaguchi

HIGH density livestock production started in Japan after World War II by commercial companies. This trend towards concentration of animals in confined areas became prevalent among farmers and rural cooperatives as a result of policies of federal government encouraging the restructuring of Japanese agriculture. As animal agriculture in Japan became more specialized and larger in size, problems of odor and water pollution arose. Public nuisance from animal production units surfaced as a big problem in the 1960's.

In Japan, as in many other parts of the world, basic laws were enacted to prevent and abate public nuisance from animals. By 1975, 14 laws on environmental pollution were enacted by the Japanese Diet. Almost all of these laws pertain to agricultural pollution. Specifically, the water pollution control law, the law of prevention of odors and the law of treatment and purification of substances discharged from economic production units are directly related to animal production.

As a result of these developments, animal waste management in Japan is trying to develop ways to adequately and efficiently manage manures without creating any public nuisance and yet grow animals near urban areas. In an effort to meet the spirit of the law and avoid any public nuisance problems, many farmers built facilities and purchased expensive equipment without any knowledge of the effectiveness of these methods. As a consequence some tragic and at times comic events occurred in several instances. In most instances the machines broke down and never operated as designed either because the equipment was not suited for the operation or because it was overloaded.

The methods that are currently considered applicable under Japanese conditions are the following: (a) Disposal on land, (b) Hydraulic collection and transport, (c) Dehydration and incineration, (d) Refeeding, and (c) Miscellaneous.

The central federal government and state governments (Prefectures) have established rules and regulations by which animal wastes must be handled. At the same time there are subsidies to farmers and agricultural cooperatives to construct facilities and purchase equipment for disposal of livestock wastes. Educational programs have been developed for farmers; also, research funds have been allocated to study improved methods of animal production and manure disposal. Furthermore, subsidies have been voted for farmers and cooperatives to develop animal production units in areas away from urban centers. These subsidies will pay or defray the costs of moving existing facilities near open areas to rural-like districts.

Environmental pollution from livestock wastes is one of the most serious problems in Japan. This problem along with the perennial shortage of feed for animals would tend to depress the growth of livestock industry in Japan. The two problems are a major concern and a daily topic of discussion among Japanese agricultural officials, animal farmers and animal scientists.

However, what is encouraging is the enthusiasm and interest which have been shown by Japanese engineers and researchers and animal producers to the livestock waste management problem, which should result in the development of practical methods uniquely suited for the small-size production units in Japan. So far, no revolutionary device has been fabricated which would solve this problem completely. Perhaps there is no need to look for such a solution because it does not exist. What is needed is a system which will manage the problem, not solve it.

---

The author is with Department of Animal Husbandry, Faculty of Agriculture, Niigata University, 2-Igarashi, Niigata, Japan.

# Animal Waste Management in Finland

Martti Sipilä

## WASTE MANAGEMENT AND ECONOMY

**B**ECAUSE arable land is thought to be the best final place for manure, it is necessary to try to collect, dispose, store, load, transport, spread and plow under manure as economically as possible. Therefore, work is aimed at (a) examining methods that reduce and facilitate labor, (b) developing cheap construction and equipment, and (c) investigating the economic benefit that manure gives by means of crop farming.

### Labor

Labor requirements in cattle and pig raising have been examined at different sized cow farms and piggeries with different manure systems. Labor requirement has also been investigated to load, transport and spread manure with different crews of workmen and equipment. For example, in a stall cow barn with liquid manure, the cleaning, manure disposal and littering take only about half the time per cow compared to the time needed for the same jobs in a cow barn where solid manure is disposed by manual methods. Lightening of labor may be achieved by using mechanical manure disposal, but the saving of labor is smaller than with the liquid manure method.

The saving of labor is even greater in transporting and spreading liquid manure compared to solid manure. Loading, transporting, spreading and plowing the amount of liquid manure per animal takes less than half (even a quarter only) of the time that is needed for the same work per animal in the solid manure method. Manure transport distances are short in Finland, because the fields are usually situated in the immediate vicinity of the farm. The profitability of a mechanical manure clearing contrivance has been investigated, for example, by counting how much the contrivance would save labor and what would be the price of a saved working hour. The investment is profitable only when the saved time can be used elsewhere to raise income.

There is no reason to buy a manure clearing contrivance on small livestock farms without thorough consideration of economic factors.

### The Value of Manure in Crop Farming

Solid manure containing urine is accumulated at about six tons and liquid manure at about twelve tons per cow during the eight month indoor feeding period. The

nutrients of liquid manure are more suitable for plants than those in solid manure. This is the case in particular with nitrogen.

## LIQUID MANURE SYSTEM

The liquid manure system has come to Finland and will remain permanently; the system is in use in about 4000 piggeries and cow barns.

Most liquid manure systems in Finland work using the damming up methods. In stall cow barns and piggeries manure is gathered in a channel covered with a steel grille. The channel is emptied by opening a drain into a concrete collecting pit outside the building.

In stall cow barns you can also find the floating manure method, where liquid manure continually floats along a flat bottomed channel into the collecting pit. There are also loose housing barns with liquid manure systems.

Runoff is usually collected and handled with manure. The collecting pit is emptied in spring and in autumn. Industry has produced effective pumps, which make it possible to mix, break up litter, and also pump from the collecting pit during spreading. Spreading of liquid manure is done with slurry tankers on the surface of the field.

During the last few years, experiments have been conducted with placement slurry tankers. In comparison with tankers that spread on the surface, considerable yield increases, odorless spreading, prevention of flowing on the surface, and improved hygienic qualities of grass feed are obvious profits gained with placement.

Fertilizer placement machines came into rapid general use when yield increases due to placement were noticed. Yield increases are considerably greater when liquid manure is placed in the soil, so that the placement of slurry is even more important than in the case of fertilizers. It is obvious that the use of placement equipment for liquid manure will become more general. Because they are expensive, they are suitable when shared for common use on several farms in Finland.

When the number of domestic animals increases, livestock waste management turns out to be a problem. For small numbers of cattle, manure does not become difficult, but in large numbers, the amount of manure creates a real problem.

The advance of prices for energy and protein may teach us to appreciate manure in a new manner; the goal in treatment techniques for manure would be more exact than the earlier exploitation in crop production.

---

The author is: MARTTI SIPILÄ, Retired Director, Work Efficiency Assn., Vantaa, Finland.

# Studies on the Collection and Disposal of Slurry in Northern Ireland

J. S. V. McAllister

THE amount of slurry which can be safely spread on land is decided mainly by three interdependent factors — soil type, cropping system and climate. The most satisfactory conditions for disposal will be found in areas which are relatively dry with a considerable amount of arable cropping. Conditions deteriorate in humid climates with heavier rainfall, lower evaporation and a higher proportion of the land under grass. Northern Ireland is an example of such an area with 90 percent of the agricultural land under grass and intensive livestock production based largely on imported feeding stuffs. Excreta produced from housed farm livestock in Northern Ireland would provide, assuming a 1:1 dilution with water, an annual dressing of over 20 000 kg slurry for each hectare of arable land and grassland. Therefore, it is not surprising that considerable investigation has been undertaken into problems associated with the collection and disposal of slurry. Some of the main results of this work, outlined below, will be described and discussed in detail.

## Changes During Collection and Storage

Storage under anaerobic conditions will result in the breakdown of solids and the conversion of insoluble nitrogen compounds into more soluble forms. There will be a loss of nitrogen, mainly from ammonium compounds, when slurry is stored under open or exposed conditions. Other toxic gases are produced from stored slurry, the most dangerous being hydrogen sulfide.

The author is: J. S. V. McALLISTER, Reader in Agricultural and Food Chemistry Dept., Queen's University, Belfast, and Principal Scientific Officer, Dept. of Agriculture, Northern Ireland.

## Manurial Value

Slurry is a valuable manure and good responses to the nitrogen content and especially to the soluble nitrogen compounds have been obtained on grass and barley crops. Satisfactory responses to the phosphorus and potassium in slurry have also been obtained.

## Problems in Land Spreading

Pollution of the atmosphere and of water may occur when slurry is being spread or carelessly handled. Nutrient enrichment of drainage water can occur when slurry is applied under unsuitable soil or climatic conditions.

The use of excessive dressings can result in harmful physical and chemical conditions in soils and subsequently adversely affect plant growth and composition. The "nutrient balance" concept which is being tested may help to overcome these problems.

## Alternative Methods of Disposal

Aeration treatments on a farm scale are unlikely to provide adequate treatment to permit discharge of the effluent directly to a water course. If such treatment becomes necessary, it should be undertaken at a communal plant. Poultry slurry may be dried for use as ruminant food, and as cattle slurry is normally produced under conditions where there is adequate land for disposal, pig slurry may be considered to present the major problem. Removal of solids by centrifugation and subsequent incineration gives an ash with high contents of calcium and magnesium phosphates.

The use of effluents as a substrate for the production of single cell protein is being studied.

# Studies on the Use of Solid Substances of Pig Waste (Slurry) in the Feeding of Fattening Cattle

G. Flachowsky, H.-J. Löhnert, A. Hennig

FOUR years ago we were able to demonstrate that the crude protein of the feces of pigs being fattened can be digested at a rate of 57 percent by sheep. The energetic nutrition value of the feces resembles dried lucerne fodder.

As in large quantity production, the amount of liquid manure from pigs is high. We tested the solid substances of pig liquid manure in feeding tests with sheep and cattle. The solid substances of pigs being fattened contain, after the elimination of the liquid matter, 16 percent crude protein, 66 percent cell wall parts, 2 percent ether extracts and

The authors are: G. FLACHOWSKY, H.-J. LÖHNERT, A. HENNIG, Animal Feeding Dept., Animal Production and Veterinary Medicine Section, Karl-Marx University of Leipzig, Jena, German Democratic Republic.

16 percent ash. The digestibility of the organic substances gained in this way varies between 45 and 55 percent. One kg dry matter equals 90 g digestible protein and about 400 energetic feeding units for cattle.

Up to now we have tested the solid substances of pig liquid manure with 520 head of cattle being fattened. Having 30 percent of the solid substance mentioned worked up into ready-made pellet material, we noticed a weight growth up to 1000 or 1200 g over a long-time period.

A combination of these substances plus straw, grain and sugar beet has proved to be most efficient. If you offer a higher percentage of solid substances, the weight growth will be depressively influenced, the high amount of ash being the inherent reason. At the end of the fattening period, the bulls were slaughtered and various organs and body parts were biochemically, bacteriologically and gustatorially tested. The quality of meat of those animals had not been affected.

# Agricultural Conditions and Livestock Wastes in Norway


Olav Hjulstad

MEMBER
ASAE


THE total land area of Norway is 324,000 sq km, compared with Illinois' 147,000 sq km. The length from south to north is 1,750 km, but the total length of the shore line is twelve times that distance (21,000 km). If we also count the islands, the shore line would be 53,000 km. The latitude of the country is very much the same as Alaska, between the 58th and 72nd parallels. There are 4,000,000 inhabitants in Norway, plus 1,000,000 cattle, 800,000 pigs, and 1,600,000 poultry.

Only 2.8 percent of the total area is agricultural land, compared with 7.6 percent in Sweden, 8.6 percent in Finland, and 66 percent in Denmark. We are short of arable land due to many rocks, mountains and lakes. Twenty-seven percent of the area is occupied by forests and about 70 percent is non-productive area. However, this provides many opportunities for recreation and tourism.

In spite of our extreme northern location, we have a reasonably mild climate in most of the country. We can raise barley up to the arctic circle, potatoes and some vegetables can be raised almost any place where people are settled, and we have good grass crops in all parts of the country.

The country is self-supplied with livestock products such as milk, beef, pork, and eggs. We import concentrates, but on the other hand, we export fish and herring meal for animal feed. We have to import most of our small grains and fruits, and all of our sugar, tea, and coffee.

Livestock production is of rather high standard and provides most of the income for the farmers.

Most of the manure is stored in cellars beneath the animal pens. In some parts of the country, they have a system where the liquid part of the manure (the urine) is separated from the solid part and stored in separate tanks. The system works well for cattle when the feces are solid. With more intensive feeding, we will get more semi-liquid manure and separation will cause some problems.

In many parts of the country, we are short of bedding, and the manure is so liquid that we mostly have anaerobic fermentation. This has caused some gas and odor problems. In some cases, there may be greater risk for development of pathogenic agents like *salmonella*.

Larger herds, larger animals and increased feeding have resulted in larger quantities of manure. Extension of the manure cellar will, in many cases, be rather expensive and we will have to build additional manure tanks outside the buildings.

We are presently conducting some experiments with liquid composting manure in order to: (a) reduce poisonous gas and odor problems, (b) reduce development of pathogenic agents and parasites, and (c) bring the manure to a consistency which can be easily handled as a liquid (transport tank with sprinkler or pipelines with sprinklers).

Laboratory experiments with chemical additives have resulted in reduced gas production, but the results of full-scale tests on operating farms have not been so promising.


The author is: OLAV HJULSTAD, Professor, Agricultural Structures Dept., Agricultural University of Norway, N-1432 Aas-N.L.H.

# The Swedish Experience in Connection with Environment Protection at Animal Production Sites

Sven Berglund

SWEDISH agriculture is based on family farming. About four percent of the population is active in agriculture and seven percent of the land area is cultivated.

The Environment Protection Act of 1969 is both conprehensive and explicit. Some forms of pollution, such as the discharge of animal urine and silage effluent into streams and lakes, are specifically prohibited by law. Application for a pemit is compulsory when the number of animals exceeds a certain minimum (100 cattle or 1,000 swine). The Act is supplemented by the Environment Protection Ordinance and different branch regulations.

The 1969 Guidelines for Environment Protection at Animal Production Sites were further revised in late 1973. These regulations deal with location of new animal units, requirements of cultivated land area to number of animals, and manure storage and handling requirements. Enforcement and supervision are mainly implemented by the provincial administrations. This has been easy on farms big enough to be supported by state loans, but more difficult on smaller holdings. Special grants for water protection devices will help this situation.

Research regarding the efficiency of different methods of reducing malodors associated with spreading manure has been carried out. Emissions from animal stables are being investigated and water pollution problems connected with farming are being investigated.

After five years of legislation, the Swedish experience shows that cooperation on different levels among the authorities, local agricultural advisors, and farmers' unions has been successful. A variety of simple practices to improve the environment—such as the reuse of plant nutrients by saving animal manures at a time when prices of commercial fertilizers are on the rise—has proved attractive to farmers.

The author is: SVEN BERGLUND, Head of Section, National Swedish Protection Board, Fack, 171 20, SOLNA, Sweden.

AGRA-LABS, INC.
716 S. Randolph Street
Champaign, Illinois 61820

AGRA-TRONIX, INC.
7506 Hickman Road
Des Moines, Iowa 50322

AGRI-LINES CORP.
Box 248
Parma, Idaho 83660

AQUA AEROBIC SYSTEMS DIV.
(Solem Machine Co.)
6306 N. Alpine Road
Rockford, Illinois 61111

BADGER NORTHLAND INC.
215 W. Second St.
Kaukauna, Wisconsin 54130

THE BAUER BROTHERS CO.
P. O. Box 968
Springfield, Ohio 45501

BIG WHEELS, INC.
Box 278
Paxton, Illinois 60957

BITTERSWEET SYSTEMS INC.
1811 W. Adams
Kirkwood, Missouri 63122

BRIM CONCRETE, INC.
P. O. Box 481
Roanoke, Illinois 61561

THE CALUMET CO.
340 N. Water Street
Algoma, Wisconsin 54201

CERES ECOLOGY CORP.
1860 Lincoln Street
Denver, Colorado 80203

CHROMALLOY FARM SYSTEMS DIV.
Starline, Co.
300 E. Front
Harvard, Illinois 60033

CLARK EQUIPMENT CO.
Melroe Div.
Gwinner, North Dakota 58040

CLAY EQUIPMENT CORP.
101 Lincoln Street
Cedar Falls, Iowa 50613

CLEAN AIR SYSTEMS, INC.
201 S. Walnut Street
Rochester, Illinois 62563

COLT INDUSTRIES
Fairbanks Weighing Div.
711 E. Johnsbury Road
St. Johnsbury, Vermont 05819

DANFORTH AGRI-RESOURCES, INC.
Livestock Equipment Div.
755 New Ballas Road, South
St. Louis, Missouri 63141

THE DELAVAL SEPARATOR CO.
5724 N. Pulaskie Road
Chicago, Illinois 60643

DEMUTH STEEL PRODUCTS CO.
Lipp System USA
Schiller Park, Illinois 60176

DISTRIBUTORS PROCESSING INC.
17656 Avenue 168
Porterville, California 92357

DIXIE IRRIGATION CO.
P. O. Box F
Louisville, Kentucky 40219

ENPO-CORNELL PUMP CO.
420 E. Third St.
Piqua, Ohio 45356

ENVIRONETICS, INC.
8924 Industrial Drive
Bridgeview, Illinois 60455

ENVIRONMENT 2000
Denison, Iowa 51442

FAIRFIELD ENGINEERING & MFG. CO.
601 W. Kirkwood
Fairfield, Iowa 52556

FARMSTEAD INDUSTRIES, INC.
P. O. Box 279
Cedar Falls, Iowa 50613

JOHN FAYHEE & SONS, INC.
Prairie City, Illinois 61470

GCA/PRECISION SCIENTIFIC
3737 W. Cortland Street
Chicago, Illinois 60647

GAME SLATTED FLOOR CO.
West Plumb Street
Ransom, Illinois 60470

THE GORMAN-RUPP CO.
305 Bowman Street
Mansfield, Ohio 44902

GOULDS PUMPS, INC.
P. O. Box 330
Seneca Falls, New York 13148

GRAZON
Pharmaco Inc.
1617 Interstate Drive
Champaign, Illinois 61820

H-W SYSTEMS
600 W. Walnut
Fairbury, Illinois 61739

HACH CHEMICAL CO.
P. O. Box 907
Ames, Iowa 50010

HEDLUND MFG. CO., INC.
Boyceville, Wisconsin 54725

HONEYBEE CO., INC.
Box 278
Everly, Iowa 51338

HUSKEE-BILT STRUCTURES
P. O. Box 648
Monmouth, Illinois 61462

IDREX, INC.
1018 Lambrecht Drive
Frankfort, Illinois 60423

KRUSE IRRIGATION CO.
600 S. Schrader
Havana, Illinois 62644

MICHIGAN ORCHARD SUPPLY
P. O. Box 231
South Haven, Michigan 49090

NESSETH, INC.
Box 29, Route 1
Dafter, Michigan 49724

PCP MANUFACTURING, INC.
Box 53
Spencer, Iowa 51301

PEARSON BROTHERS CO., INC.
P. O. Box 192
Galva, Illinois 61434

PENBERTHY DIV.
Houdaille Industries
P. O. Box 112
Prophetstown, Illinois 61277

RANDOLPH CONCRETE PRODUCTS, INC.
321 Stark Street
Randolph, Wisconsin 53956

REN VAND, INC.
P. O. Box 268
Webster City, Iowa 50595

A. O. SMITH HARVESTORE PRODUCTS, INC.
P. O. Box 395
Arlington Heights, Illinois 60006
and
P. O. Box 584
Milwaukee, Wisconson 53201

THE SOIL MOVER MFG. CORP.
P. O. Box 667
Columbus, Nebraska 68601

SPRINKLER IRRIGATION ASSN.
13975 Connecticut Ave., Suite 310
Silver Spring, Maryland 20906

STARCRAFT — AGRI PRODUCTS DIV.
516 E. Madison
Goshen, Indiana 46526

SYDNOR HYDRODYNAMICS INC.
2111 Magnolia
Richmond, Virginia 23223

TASCO, INC.
1001 Cherry Street
Shell Rock, Iowa 50670

VALMONT INDUSTRIES, INC.
Valley, Nebraska 68064

WILDWOOD FARM SERVICES, INTL.
Route 1
Lodi, Wisconsin 53555

WILLIAMSTOWN IRRIGATION, INC.
Williamstown, New York 13493

Wm. T. LOWRY, Ph. D.

Wm. T. LOWRY, Ph. D.